Essentials of
PHYSICAL CHEMISTRY

28th Edition

Essentials of

PHYSICAL CHEMISTRY

(For BSc Students)

SI UNITS

ARUN BAHL

MS (Boston), PhD (Edinburgh), CChem FRSC (London), FAIC (USA)
Professor of Organic Chemistry, Panjab University, Chandigarh

B S Bahl

MSc, Formerly Principal and Head Postgraduate
Department of Chemistry, DAV College, Jalandhar

G D Tuli

MSc, PhD, Former Principal Shyam Lal College
Delhi University, Delhi

S. CHAND
PUBLISHING

S. Chand And Company Limited
(ISO 9001 Certified Company)

S. Chand And Company Limited

(ISO 9001 Certified Company)

Head Office: D-92, Sector–2, Noida – 201301, U.P. (India), Ph. 91-120-4682700

Registered Office: A-27, 2nd Floor, Mohan Co-operative Industrial Estate, New Delhi – 110 044, Phone: 011-49731800

www.**schandpublishing.com;** e-mail: **info@schandpublishing.com**

Marketing Offices:

Chennai	:	Ph: 23632120; chennai@schandpublishing.com
Guwahati	:	Ph: 2738811, 2735640; guwahati@schandpublishing.com
Hyderabad	:	Ph: 40186018; hyderabad@schandpublishing.com
Jalandhar	:	Ph: 4645630; jalandhar@schandpublishing.com
Kolkata	:	Ph: 23357458, 23353914; kolkata@schandpublishing.com
Lucknow	:	Ph: 4003633; lucknow@schandpublishing.com
Mumbai	:	Ph: 25000297; mumbai@schandpublishing.com
Patna	:	Ph: 2260011; patna@schandpublishing.com

Subsequent Editions and Reprints 1994 (Twice), 95, 96, 97, 99, 2000, 2001, 2002, 2003, 2004, 2005, 2006
First Multicolour Illustrative Revised Edition 2008
Subsequent Editions and Reprints 2009, 2010, 2011, 2012 (Twice), 2014, 2016 (Thrice), 2017, Reprint 2018
Revised Edition 2019; Reprint 2021

Low Priced Students' Paperback Edition (LPSPE), 2022; *Reprint 2022, 2023*

Reprint 2024

ISBN: 978-93-550-1060-5 **Product Code:** H6PCH68CHEM10ENAA21L

PRINTED IN INDIA

By Vikas Publishing House Private Limited, Plot 20/4, Site-IV, Industrial Area Sahibabad, Ghaziabad – 201 010 ,and Published by S. Chand And Company Limited, A-27, 2nd Floor, Mohan Co-operative Industrial Estate .New Delhi – 110 044

Preface

The *Essentials of Physical Chemistry* has been written for BSc students. It has been national best-seller for more than 75 years. It has been used by more than 3 million students. It is 28 editions old. It really has been that long. A lot of things have changed since then. We also changed with every edition so that you could get the best. In this new edition we have retained all those features that made it a classic. It is the most trusted Physical Chemistry textbook in India. Recent reviews from some teachers are reproduced. These sum up book's high-quality and study-approach :

Bible of Physical Chemistry !

This textbook is currently in use at hundreds of colleges and universities throughout the country and is a national best-seller. In this edition, the authors continue to do what they do best, focus on the important material of the course and explain it in a concise, clear way. I have found this book to be very easy to follow. There are hundreds of computer-generated colored diagrams, graphs, photos and tables which aid in understanding the text. The book goes step-by-step, so you don't get lost. No wonder it is a market-leader ! Blindly go for it.

Exactly what I needed. Gives me hope !

The acknowledged leader and standard in Physical Chemistry, this book maintains its effective and proven features – clear and friendly writing style, scientific accuracy, strong exercises, step-by-step solved problems, modern approach and design. The organization and presentation are done with marvelous clarity. The book is visually beautiful and the authors communicate their enthusiasm and enjoyment of the subject in every chapter.

Easy to read and understand. The coverage is very comprehensive !

The Essentials of Physical Chemistry is best summarized by "classic text, modern presentation". This simple phrase underlines its strong emphasis on fundamental skills and concepts. As in previous editions, clearly explained step-by-step problem-solving strategies continue to be the strength of this student-friendly text. This revision builds on its highly praised style that has earned this text a reputation as the *voice of authority* in Physical Chemistry. The authors have built four color art program that has yet to be seen in India !

Awesome Book ! Excellent for Self-Study !

This absolutely beautiful textbook of Physical Chemistry is a *must see, must do*. It includes very well organized discussions, quizzes, and visualizations of all the topics. This is an essential textbook for students and teachers. It feels like I am having a conversation with a personal tutor.

STUDENT FRIENDLY

Many BSc students do not have a good background in Physical Chemistry. This examination-oriented text is written with these students in mind. The language is simple, explanations clear, and presentation very systematic. Our commitment to simplicity is total !

Concept-density per page has been kept low. We feel that this is a big time saver and essential to quick-learning and retention of the subject matter.

STRESS IS ON UNDERSTANDING

This book will help you overcome the fear of Physical Chemistry. Stress is on understanding and not on memorization. Topics which usually confuse the students are explained in greater detail than commonly done. This text will help you learn Physical Chemistry faster and enjoy it more !

USEFUL FOR ENTRANCE TESTS

This is an important textbook for the Medical and Engineering College Entrance Exams. Your choice of a book can mean success or failure. Because today you need a book that can help you streak ahead of competition and succeed. No-one knows more about your needs than us. It is a tall claim, but it is true !

NEW IN THIS EDITION

The new edition of *Essentials of Physical Chemistry* contains numerous discussions, illustrations, and exercises aimed at overcoming common misconceptions. It has become increasingly clear from our own teaching experience that students often struggle with Physical Chemistry because they misunderstand many of the fundamental concepts. In this text, we have gone to great lengths to provide illustrations and explanations aimed at giving students more accurate pictures of the fundamental ideas of chemistry.

In this New Edition we have retained all that was judged good in the previous edition. However, a number of changes have been made in this new edition. Subject matter has been updated. This edition provides quick access to the important facts and concepts. It includes every important principle, equation, theorem, and concept. SI units have been used throughout.

VALUE ADDITION

1. **Problem-Solving.** To a great extent, a student's understanding of chemistry depends on his or her ability to solve and analyze problems. We have structured this book with the idea of weaving the techniques of problem-solving throughout the content, so that the student is systematically guided and challenged to view chemistry as a series of solvable problems.

   Question-style has changed over the years. Latest university questions are given at the end of each chapter to show these trends. Step-by-step answers are provided for the in-chapter problems. This book contains more than 1500 latest university questions. More than 115 new *Additional Solved Problems* grace the new edition that diligent students will find very useful. It also contains more than 1700 multiple-choice questions. By solving these problems you can precisely know your own success-level. This is the book which the examiners use !

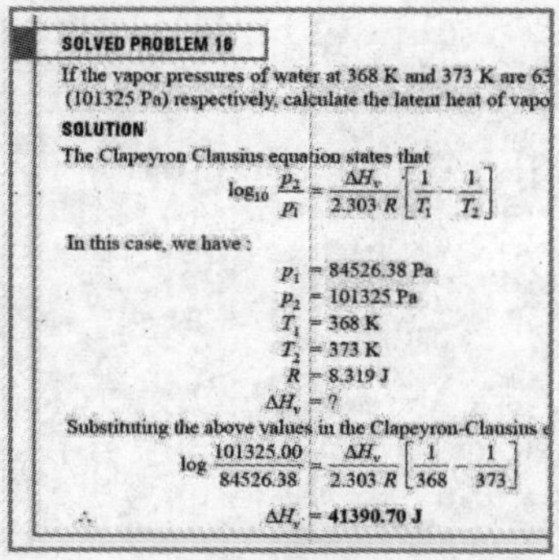

SOLVED PROBLEM 16

If the vapor pressures of water at 368 K and 373 K are 63
(101325 Pa) respectively, calculate the latent heat of vapo

SOLUTION

The Clapeyron Clausius equation states that

$$\log_{10}\frac{p_2}{p_1} = \frac{\Delta H_v}{2.303\,R}\left[\frac{1}{T_1}-\frac{1}{T_2}\right]$$

In this case, we have :

$$p_1 = 84526.38 \text{ Pa}$$
$$p_2 = 101325 \text{ Pa}$$
$$T_1 = 368 \text{ K}$$
$$T_2 = 373 \text{ K}$$
$$R = 8.319 \text{ J}$$
$$\Delta H_v = ?$$

Substituting the above values in the Clapeyron-Clausius e

$$\log\frac{101325.00}{84526.38} = \frac{\Delta H_v}{2.303\,R}\left[\frac{1}{368}-\frac{1}{373}\right]$$

$$\therefore \quad \Delta H_v = \mathbf{41390.70\ J}$$

1500	450	1700
Examination Questions	Solved Problems	Multiple Choice Questions

2. **Four-Color Art Program.** One of the distinctive features of the text is its visual impact. This is the first Indian Physical Chemistry textbook to be completely done in four-color and on computer. Color graphics, illustrations, and real pictures have been extensively used to highlight and reinforce the important points. Color has also been used to highlight change and concepts.

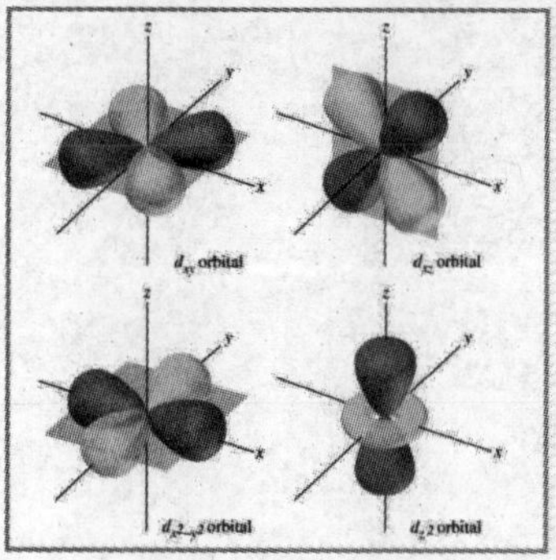

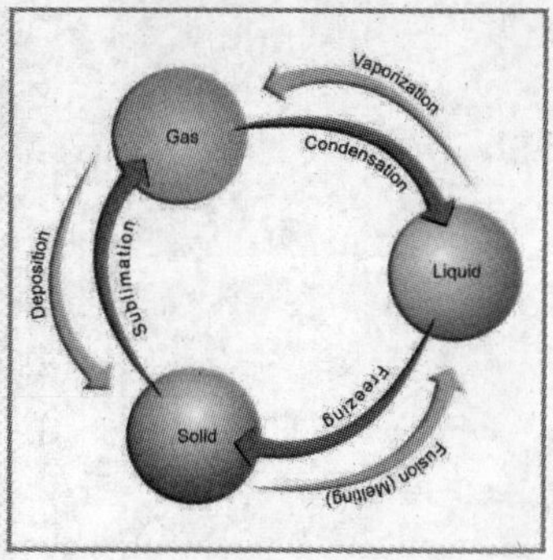

This is the first Indian Physical Chemistry textbook to be completely done in four-color

3. **Guidelines** are provided to help you understand concepts that are considered difficult and catch careless mistakes before exams.

4. **Scientific Accuracy** has been checked and rechecked. Subject matter is modern and error-free.

5. **Extensive** Index has been provided for quick cross-reference.

WE WISH YOU SUCCESS !

Yes, we think you will appreciate the thought and care that has gone into the making of this text. If you have the will, this book will show the way. We urge you to study regularly, and hope that this error-free book will make it easier for you to do so. You can depend on this book!

The book has everything you want to have in your Physical Chemistry text. In case you find something missing, please write at the following address:

Mail: 590, Sector 18-B, Chandigarh - 160018

e-mail: arunbahl2000@gmail.com

We would be glad to receive suggestions and comments for further improvements.

Authors

Brief Contents

Contents

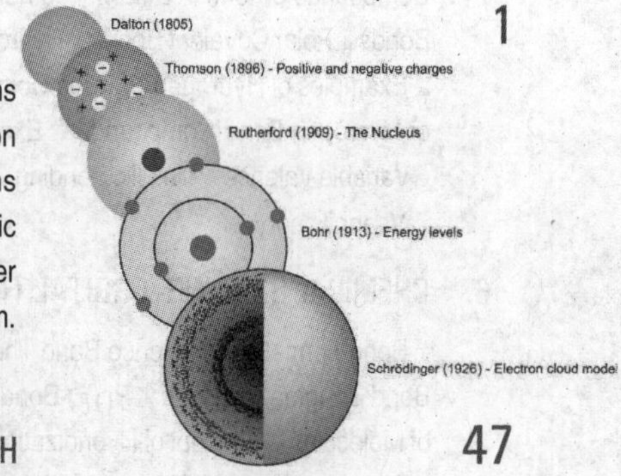

Dalton (1805)
Thomson (1896) - Positive and negative charges
Rutherford (1909) - The Nucleus
Bohr (1913) - Energy levels
Schrödinger (1926) - Electron cloud mode

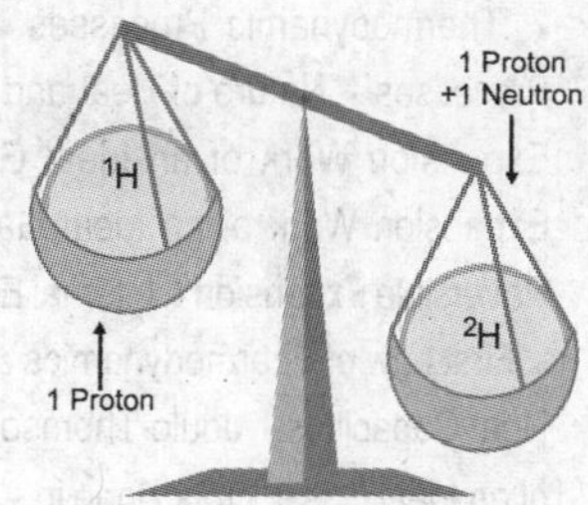

1 Proton
+1 Neutron
^1H
^2H
1 Proton

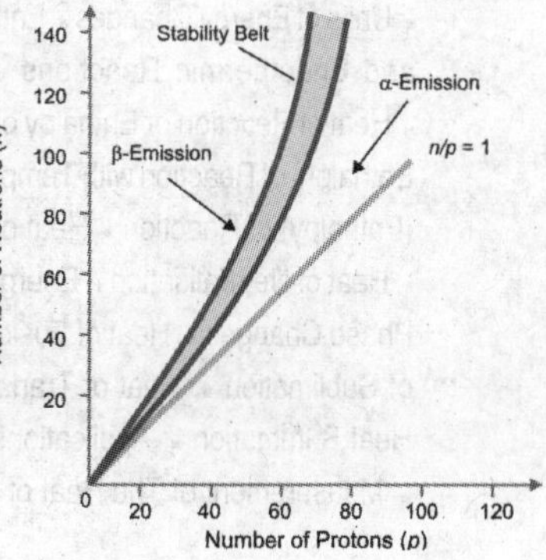

Stability Belt
α-Emission
β-Emission
n/p = 1
Number of Neutrons (n)
Number of Protons (p)
0
20
40
60
80
100
120
140

Na+
Cl

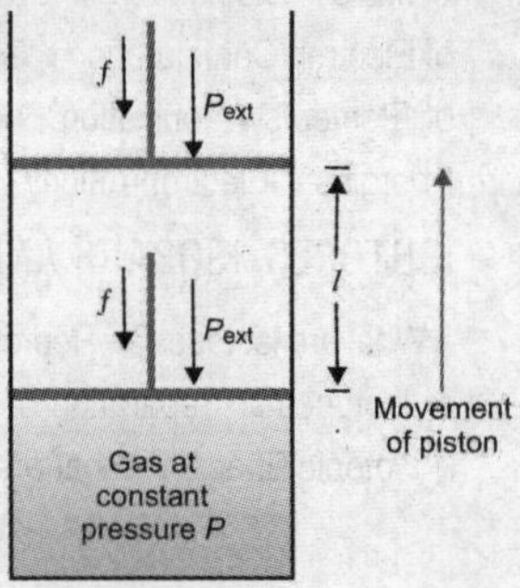
f
Pext
l
f
Pext
Movement
of piston
Gas at
constant
pressure P

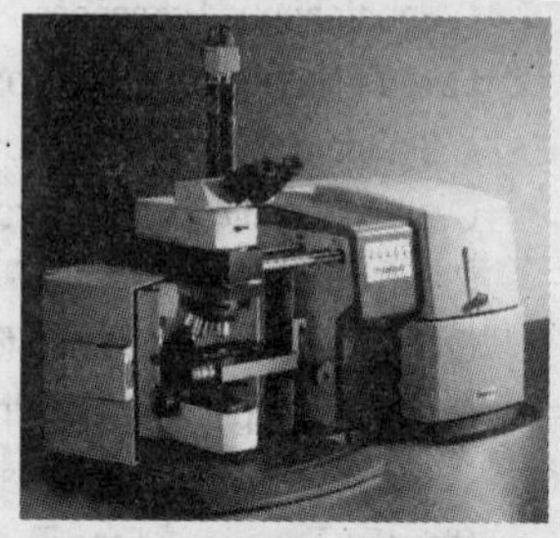

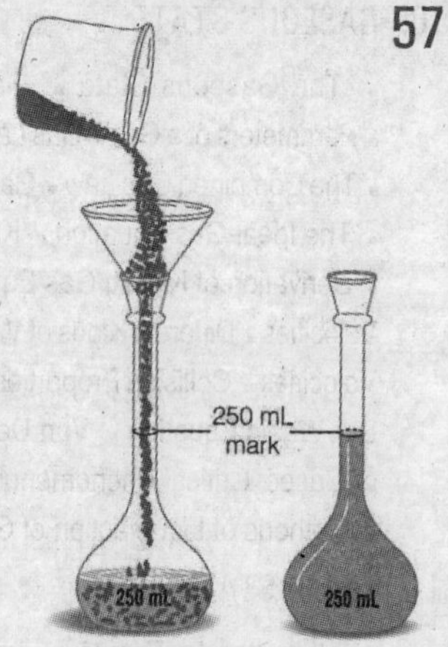

250 mL
mark
250 mL
250 mL

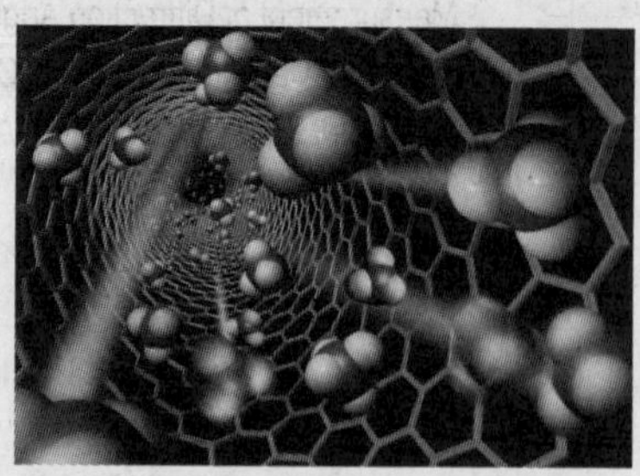

SPEED
LIMIT
65
X 10^{-4} mol/liter-sec

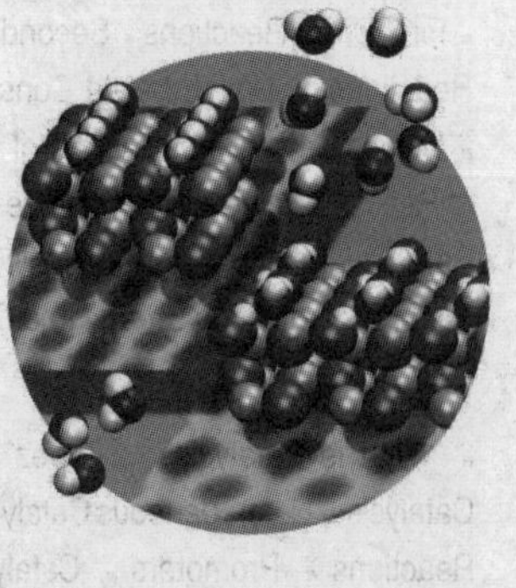

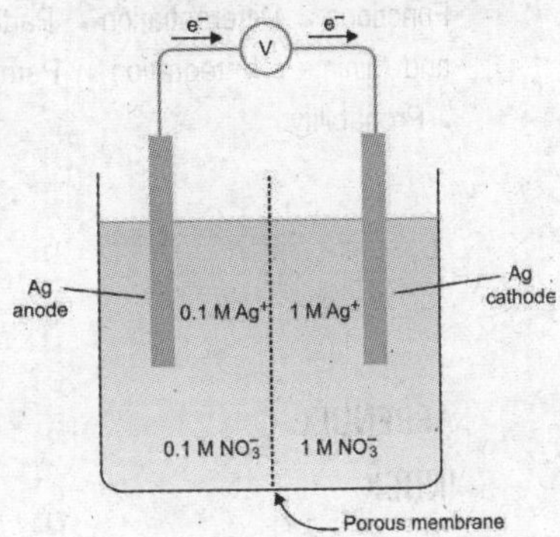

e⁻
V
e⁻
Ag anode
0.1 M Ag+
1 M Ag+
Ag cathode
0.1 M NO3−
1 M NO3−
Porous membrane

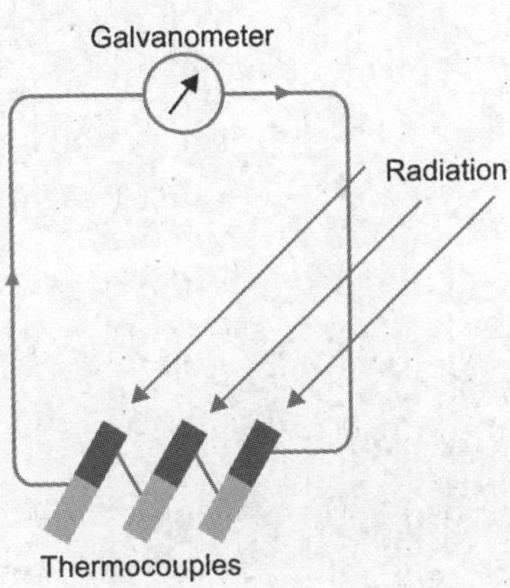

Galvanometer
Radiation
Thermocouples

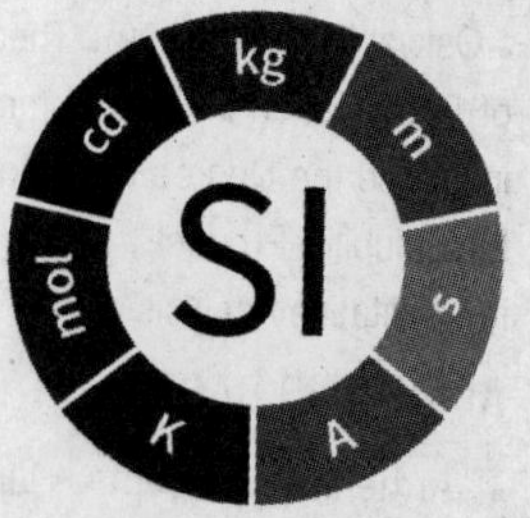
kg
m
s
A
K
mol
cd
SI

√144
√1
√4
√9
√16
√25
√36
√49
√64
√81
√100
√121

1 Structure of Atom – Classical Mechanics

CONTENTS

John Dalton (1805) considered that **all matter was composed of small particles called atoms.** He visualized the atom as a hard solid individual particle incapable of subdivision. At the end of the nineteenth century there accumulated enough experimental evidence to show that the atom is made of still smaller particles. These subatomic particles are called the **fundamental particles.** The number of subatomic particles now known is very large. For us, the three most important are the **proton, neutron and electron.** How these fundamental particles go to make the internal structure of the atom, is a fascinating story. The main landmarks in the evolution of atomic structure are :

1896 J.J. Thomson's discovery of the electron and the proton

1909	Rutherford's Nuclear Atom
1913	Mosley's determination of Atomic Number
1913	Bohr Atom
1921	Bohr-Bury Scheme of Electronic Arrangement
1932	Chadwick's discovery of the neutron.

CATHODE RAYS—THE DISCOVERY OF ELECTRON

The knowledge about the electron was derived as a result of the study of the electric discharge in the **discharge tube** (J.J. Thomson, 1896). The discharge tube consists of a glass tube with metal electrodes fused in the walls (Fig. 1.1). Through a glass side-arm air can be drawn with a pump. The electrodes are connected to a source of high voltage (10,000 Volts) and the air partially evacuated. The electric discharge passes between the electrodes and the residual gas in the tube begins to glow. If virtually all the gas is evacuated from within the tube, the glow is replaced by faintly luminous 'rays' which produce fluorescence on the glass at the end far from the cathode. **The rays which proceed from the cathode and move away from it at right angles in straight lines are called Cathode Rays.**

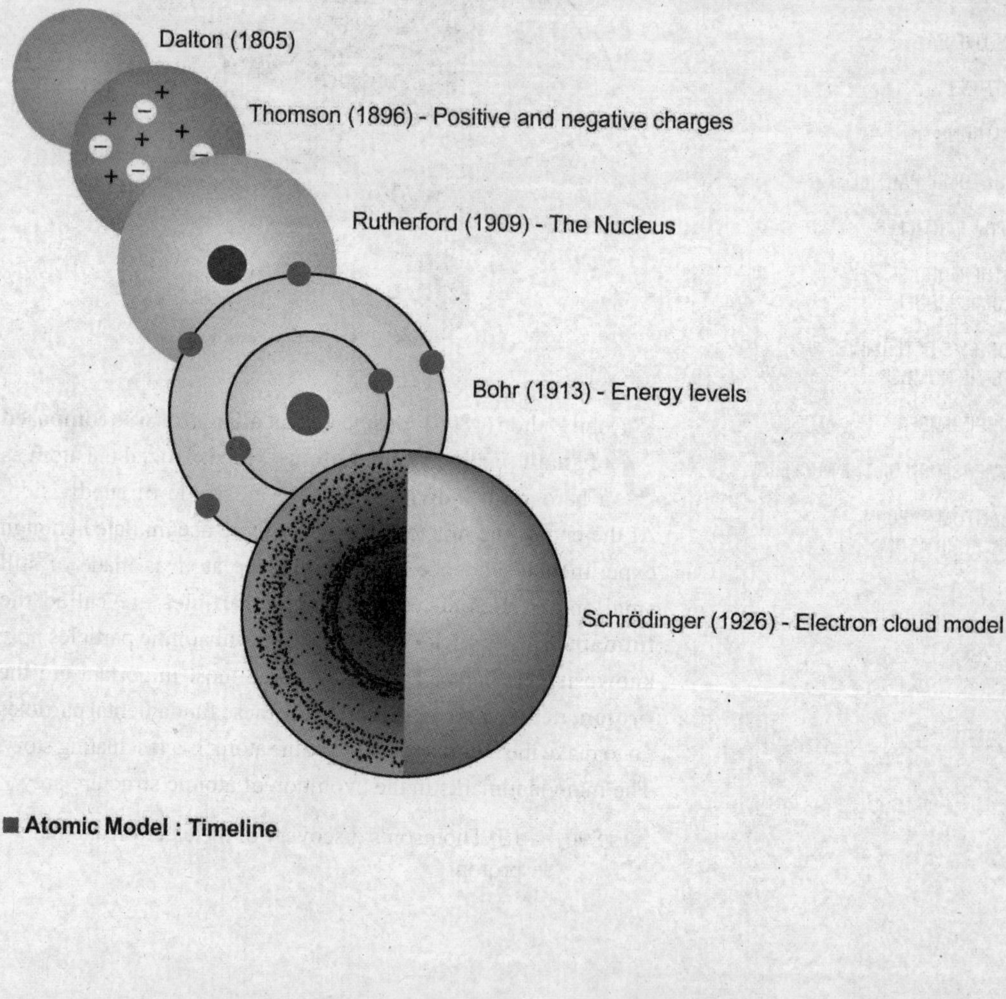

■ **Atomic Model : Timeline**

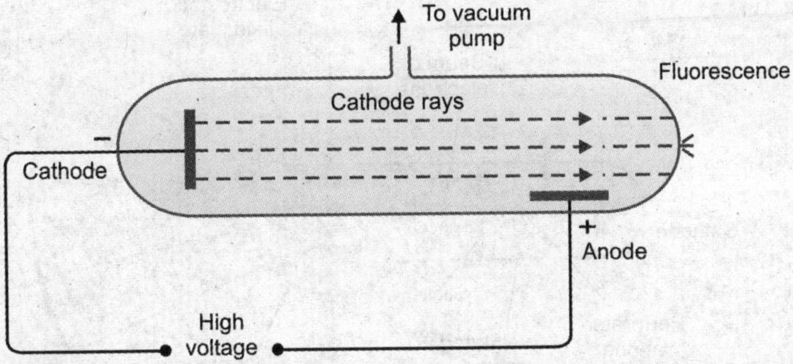

Figure 1.1
Production of cathode rays.

PROPERTIES OF CATHODE RAYS

1. They travel in straight lines away from the cathode and cast shadows of metallic objects placed in their path.
2. Cathode rays cause mechanical motion of a small pin-wheel placed in their path. Thus they possess kinetic energy and must be material particles.
3. They produce fluorescence (a glow) when they strike the glass wall of the discharge tube.
4. They heat up a metal foil to incandescence which they impinge upon.
5. Cathode rays produce X-rays when they strike a metallic target.
6. Cathode rays are deflected by the electric as well as the magnetic field in a way indicating that they are streams of minute particles carrying negative charge.

By counterbalancing the effect of magnetic and electric field on cathode rays. Thomson was able to work out the ratio of the charge and mass (*e/m*) of the cathode particle. The value of *e/m* of cathode particles is -1.76×10^8 coulomb per gram (C/g) or -1.76×10^{11} coulomb per kilogram (C/kg) in SI unit. As a result of several experiments, Thomson showed that the value of *e/m* of the cathode particle was the same regardless of both the gas and the metal of which the cathode was made. This proved that the particles making up the cathode rays were all identical and were constituent parts of the various atoms. Dutch Physicist H.A. Lorentz named them **Electrons.**

Electrons are also obtained by the action of X-rays or ultraviolet light on metals and from heated filaments. These are also emitted as β-particles by radioactive substances. Thus it is concluded that **electrons are a universal constituent of all atoms.**

MEASUREMENT OF e/m FOR ELECTRONS

The ratio of charge to mass (*e/m*) for an electron was measured by J.J. Thomson (1897) using the apparatus shown in Fig. 1.2. Electrons produce a bright luminous spot at X on the fluorescent screen. Magnetic field is applied first and causes the electrons to be deflected in a circular path while the spot is shifted to Y. The radius of the circular path can be obtained from the dimensions of the apparatus, the current and number of turns in the coil of the electromagnet and the angle of deflection of the spot. An electrostatic field of known strength is then applied so as to bring back the spot to its original position. Then from the strength of the electrostatic field and magnetic field, it is possible to calculate the velocity of the electrons.

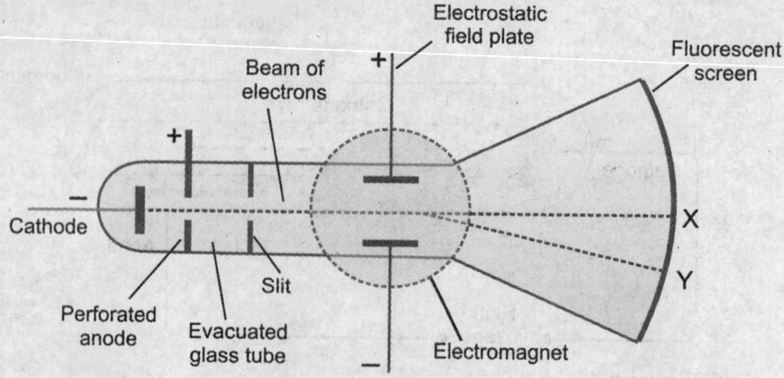

Figure 1.2
Measurement of *e/m* for electrons.

Equating magnetic force on the electron beam to centrifugal force.

$$Bev = \frac{mv^2}{r}$$

where B = magnetic field strength

v = velocity of electrons e = charge on the electron

m = mass of the electron

r = radius of the circular path of the electron in the magnetic field.

This means

$$\frac{e}{m} = \frac{v}{Br} \qquad ...(1)$$

The value of r is obtained from the dimensions of the tube and the displacement of the electron spot on the fluorescent screen.

When the electrostatic field strength and magnetic field strength are counterbalanced,

$$Bev = Ee$$

where E is the strength of the electrostatic field.

Thus $$v = \frac{E}{B} \qquad ...(2)$$

If E and B are known, v can be calculated and on substitution in equation (1), we get the value of e/m.

$$\frac{e}{m} = \frac{E}{B^2 r}$$

All the quantities on the right side of the equation can be determined experimentally. Using this procedure, the ratio e/m works out to be -1.76×10^8 coulomb per gram or -1.76×10^{11} coulomb per kilogram.

or e/m for the electron $= -1.76 \times 10^8$ C/g

$= -1.76 \times 10^{11}$ C/kg

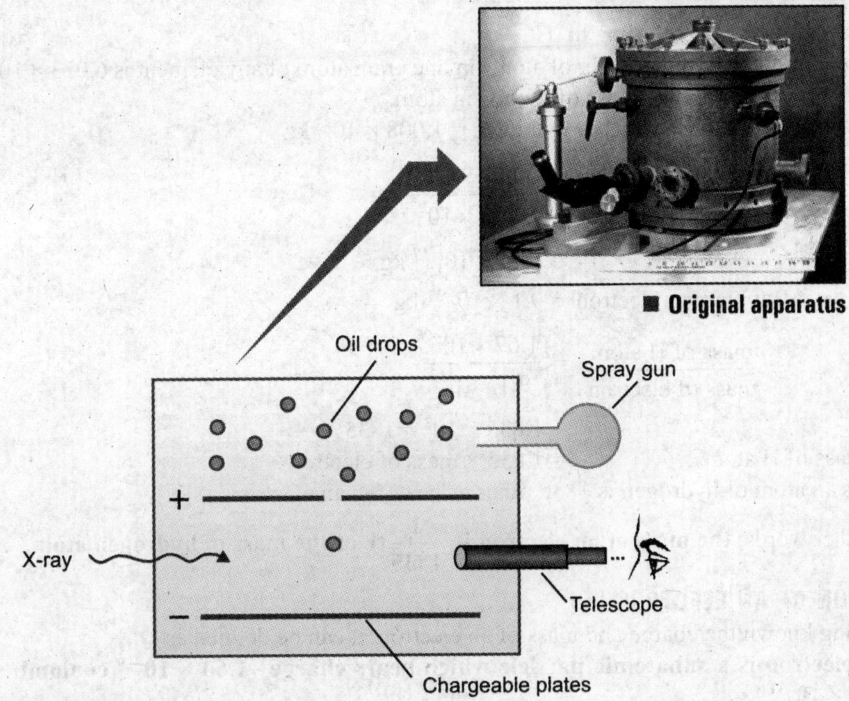

■ Figure 1.3
Milikan's apparatus for the Oil-drop experiment.

DETERMINATION OF THE CHARGE ON AN ELECTRON

The absolute value of the charge on an electron was measured by R.A. Milikan (1908) by what is known as the **Milikan's Oil-drop Experiment.** The apparatus used by him is shown in Fig. 1.3. He sprayed oil droplets from an *atomizer* into the apparatus. An oil droplet falls through a hole in the upper plate. The air between the plates is then exposed to X-rays which eject electrons from air molecules. Some of these electrons are captured by the oil droplet and it acquires a negative charge. When the plates are earthed, the droplet falls under the influence of gravity.

He adjusted the strength of the electric field between the two charged plates so that a particular oil drop remained suspended, neither rising nor falling. At this point, the upward force due to the negative charge on the drop, just equalled the weight of the drop. As the X-rays struck the air molecules, electrons are produced. The drop captures one or more electrons and gets a negative charge, Q. Thus,

$$Q = ne$$

where n = number of electrons and e = charge of the electron. From measurement with different drops, Milikan established that electron has the charge -1.60×10^{-19} coulomb.

Mass of Electron

By using the Thomson's value of e/m and the Milikan's value of e, the absolute mass of an electron can be found.

$$e/m = -1.76 \times 10^{8} \text{ coulomb/g (Thomson)}$$

$$e = -1.60 \times 10^{-19} \text{ coulomb (Milikan)}$$

$$\therefore \quad \frac{e}{e/m} = \frac{1.60 \times 10^{-19}}{1.76 \times 10^{8}}$$

hence $$m = 9.1 \times 10^{-28} \text{ g or } 9.1 \times 10^{-31} \text{ kg}$$

Mass of an Electron relative to H

Avogadro number, the number of atoms in one gram atom of any element is 6.023×10^{23}. From this we can find the absolute mass of hydrogen atom.

Mass of 6.023×10^{23} atoms of hydrogen = 1.008×10^{-3} kg

$$\therefore \quad \text{Mass of a hydrogen atom} = \frac{1.008 \times 10^{-3}}{6.023 \times 10^{23}} \text{ kg}$$

$$= 1.67 \times 10^{-27} \text{ kg}$$

$$\text{But mass of electron} = 9.1 \times 10^{-31} \text{ kg}$$

$$\therefore \quad \frac{\text{mass of H atom}}{\text{mass of electron}} = \frac{1.67 \times 10^{-27}}{9.1 \times 10^{-31}}$$

$$= 1.835 \times 10^{3} = 1835$$

$$\text{or mass of H atom} = 1.835 \times \text{mass of electron}$$

Thus an atom of hydrogen is 1835 times as heavy as an electron.

In other words, **the mass of an electron is $\frac{1}{1835}$th of the mass of hydrogen atom.**

DEFINITION OF AN ELECTRON

Having known the charge and mass of an electron, it can be defined as :

An electron is a subatomic particle which bears charge -1.60×10^{-19} coulomb and has mass 9.1×10^{-31} kg.

Alternatively, an electron may be defined as :

A particle which bears one unit negative charge and mass 1/1835th of a hydrogen atom.

Since an electron has the smallest charge known, it was designated as unit charge by Thomson.

POSITIVE RAYS

In 1886 Eugen Goldstein used a discharge tube with a hole in the cathode (Fig. 1.4). He observed that while cathode rays were streaming away from the cathode, there were colored rays produced simultaneously which passed through the perforated cathode and caused a glow on the wall opposite to the anode. Thomson studied these rays and showed that they consisted of particles carrying a positive charge. He called them **Positive rays.**

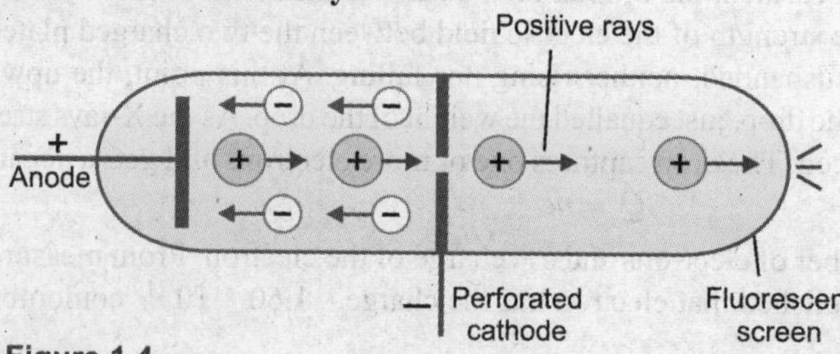

Figure 1.4
Production of Positive rays.

How are Positive rays produced ?

When high-speed electrons (cathode rays) strike molecule of a gas placed in the discharge tube, they knock out one or more electrons from it. Thus a positive ion results

$$M + e^- \longrightarrow M^+ + 2e^-$$

These positive ions pass through the perforated cathode and appear as positive rays. When electric discharge is passed through the gas under high electric pressure, its molecules are dissociated into atoms and the positive atoms (ions) constitute the positive rays.

PROPERTIES OF POSITIVE RAYS

(1) They travel in a straight line in a direction opposite to the cathode.
(2) They are deflected by electric as well as magnetic field in a way indicating that they are positively charged.
(3) The charge-to-mass ratio (e/m) of positive particles varies with the nature of the gas placed in the discharge tube.
(4) They possess mass many times the mass of an electron.
(5) They cause fluorescence in zinc sulfide.

Conclusions from the study of Positive rays

From a study of the properties of positive rays, Thomson and Aston (1913) concluded that atom consists of at least two parts :

(*a*) the electrons ; and
(*b*) a positive residue with which the mass of the atom is associated.

PROTONS

E. Goldstein (1886) discovered protons in the discharge tube containing hydrogen.

$$H \longrightarrow \underset{\text{proton}}{H^+} + e^-$$

It was J.J. Thomson who studied their nature. He showed that :

(1) The actual mass of proton is 1.672×10^{-27} kg. **On the relative scale, proton has mass 1 atomic mass unit (amu).**
(2) **The electrical charge of proton is equal in magnitude but opposite to that of the electron.** Thus proton carries a charge $+1.6 \times 10^{-19}$ coulomb or + 1 elementary charge unit.

Since proton was the lightest positive particle found in atomic beams in the discharge tube, it was thought to be a unit present in all other atoms. Protons were also obtained in a variety of nuclear reactions indicating further that **all atoms contain protons.**

Thus **a proton is defined as a subatomic particle which has a mass of 1 amu and charge + 1 elementary charge unit.**

A proton is a subatomic particle which has one unit mass and one unit positive charge.

NEUTRONS

In 1932 Sir James Chadwick discovered the third subatomic particle. He directed a stream of alpha particles $\left({}_{2}^{4}He\right)$ at a beryllium target. He found that a new particle was ejected. It has almost the same mass (1.674×10^{-27} kg) as that of a proton and has no charge.

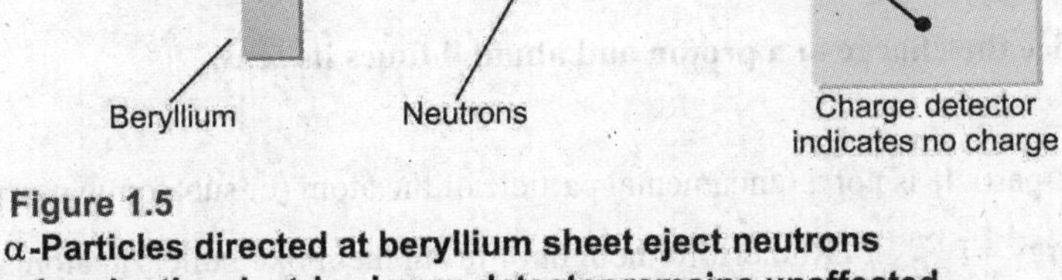

Figure 1.5
α-Particles directed at beryllium sheet eject neutrons whereby the electric charge detector remains unaffected.

He named it *neutron*. The assigned relative mass of a neutron is approximately one atomic mass unit (amu). Thus :

A neutron is a subatomic particle which has a mass almost equal to that of a proton and has no charge.

The reaction which occurred in Chadwick's experiment is an example of artificial transmutation where an atom of beryllium is converted to a carbon atom through the nuclear reaction.

$$^{4}_{2}\text{He} + {}^{9}_{4}\text{Be} \longrightarrow {}^{12}_{6}\text{C} + {}^{1}_{0}\text{n}$$

SUBATOMIC PARTICLES

We have so far studied the properties of the three principal fundamental particles of the atom, namely the *electron*, *proton*, and *neutron*. These are summarized in Table 1.1.

TABLE 1.1. CHARGE AND MASS OF ELECTRON, PROTON AND NEUTRON

Particle	Symbol	Mass		Charge	
		amu	Kilograms	Units	Coulomb (C)
Electron	e^-	$\frac{1}{1835}$	9.1×10^{-31}	-1	-1.60×10^{-19}
Proton	p^+	1	1.672×10^{-27}	$+1$	$+1.60 \times 10^{-19}$
Neutron	n or n^0	1	1.674×10^{-27}	0	0

Nearly all of the ordinary chemical properties of matter can be examined in terms of atoms consisting of electrons, protons and neutrons. Therefore for our discussion we will assume that atom contains only these three principal subatomic particles.

Other Subatomic Particles

Besides electrons, protons and neutrons, many other subatomic particles such as *mesons*, *positrons*, *neutrinos* and *antiprotons* have been discovered. A great deal of recent research is producing a long list of still other subatomic particles with names *quarks*, *pions* and *gluons*. With each discovery, the picture of atomic structure becomes increasingly complex. Fortunately, the three-particle (electron, proton, neutron) picture of the atom still meets the needs of the chemists.

ALPHA PARTICLES

Alpha particles are shot out from radioactive elements with very high speed. For example, they come from radium atoms at a speed of 1.5×10^7 m s^{-1}. Rutherford identified them to be **di-positive helium ions,** He^{2+} or $^{4}_{2}He$. Thus an alpha particle has 2+ charge and 4 amu mass.

α-Particles are also formed in the discharge tube that contains helium,

$$\text{He} \longrightarrow \text{He}^{2+} + 2e^-$$

It has twice the charge of a proton and about 4 times its mass.

Conclusion

Though α-particle is not a fundamental particle of the atom (or subatomic particle) but because of its high energy $\left(\frac{1}{2}\,mv^2\right)$, Rutherford thought of firing them like bullets at atoms and thus obtain information about the structure of the atom.

(1) **He bombarded nitrogen and other light elements by α-particles when H^+ ions or protons were produced. This showed the presence of protons in atoms other than hydrogen atom.**

(2) **He got a clue to the presence of a positive nucleus in the atom as a result of the bombardment of thin foils of metals.**

RUTHERFORD'S ATOMIC MODEL–THE NUCLEAR ATOM

Having known that atom contains electrons and a positive ion, Rutherford proceeded to perform experiments to know as to how and where these were located in the atom. In 1909 Rutherford and Marsden performed their historic **Alpha Particle-Scattering Experiment,** using the apparatus illustrated in Fig. 1.6. They directed a stream of very highly energetic α-particles from a radioactive source against a thin *gold foil* provided with a circular fluorescent zinc sulfide screen around it. Whenever an α-particle struck the screen, a tiny flash of light was produced at that point.

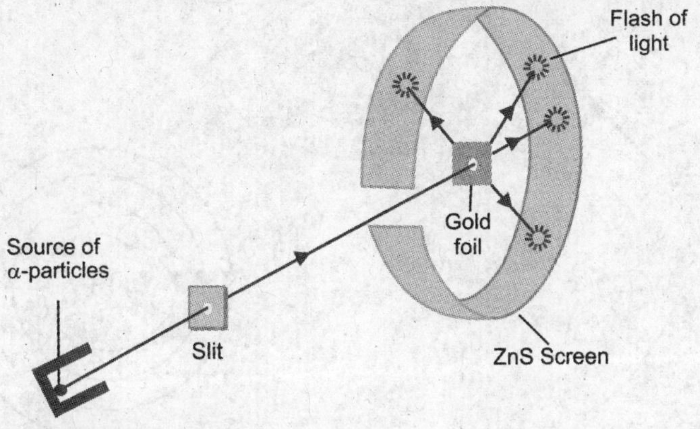

■ **Figure 1.6**
Rutherford and Marsden's α-particle scattering experiment.

Rutherford and Marsden noticed that most of the α-particles passed straight through the gold foil and thus produced a flash on the screen behind it. This indicated that gold atoms had a structure with plenty of empty space. To their great astonishment, tiny flashes were also seen on other portions of the screen, some time in front of the gold foil. This showed that gold atoms deflected or 'scattered' α-particles through large angles so much so that some of these bounced back to the source. Based on these observations, Rutherford proposed a model of the atom which is named after him. This is also called the **Nuclear Atom.** According to it :

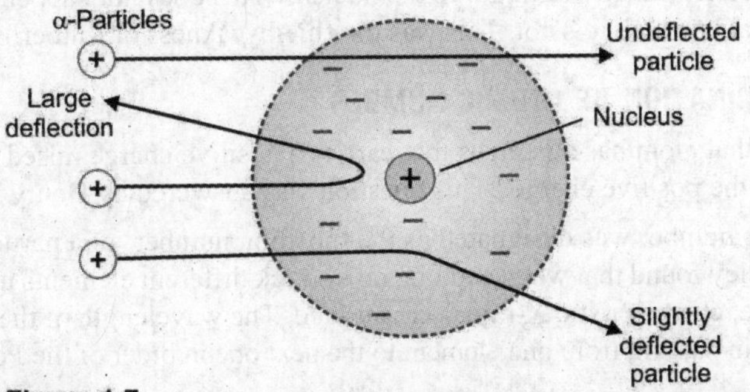

■ **Figure 1.7**
How nuclear atom causes scattering of α-particles.

(1) **Atom has a tiny dense central core or the nucleus which contains practically the entire mass of the atom, leaving the rest of the atom almost empty.** The diameter of the nucleus is about 10^{-15} m as compared to that of the atom 10^{-10} m. If the nucleus were the size of a football, the entire atom would have a diameter of about 8046.72 meters (5 miles). It was this empty space around the nucleus which allowed the α-particles to pass through undeflected.

(2) **The entire positive charge of the atom is located on the nucleus, while electrons were distributed in vacant space around it.** It was due to the presence of the positive charge on the nucleus that α-particle (He^{2+}) were repelled by it and scattered in all directions.

(3) **The electrons were moving in orbits or closed circular paths around the nucleus like planets around the sun.**

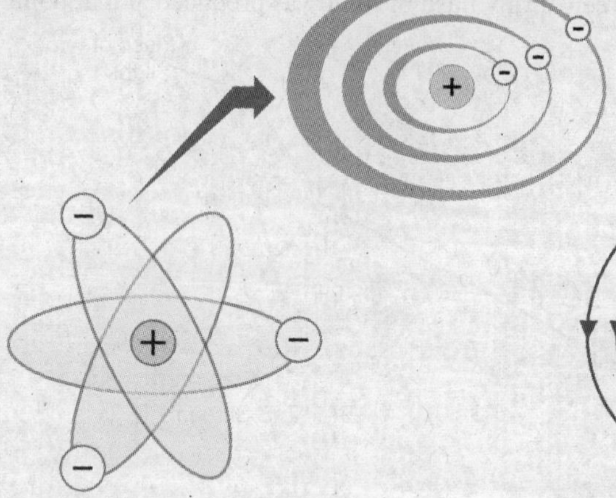

■ **Figure 1.8**
Rutherford's model of atom ; electrons orbiting around nucleus.

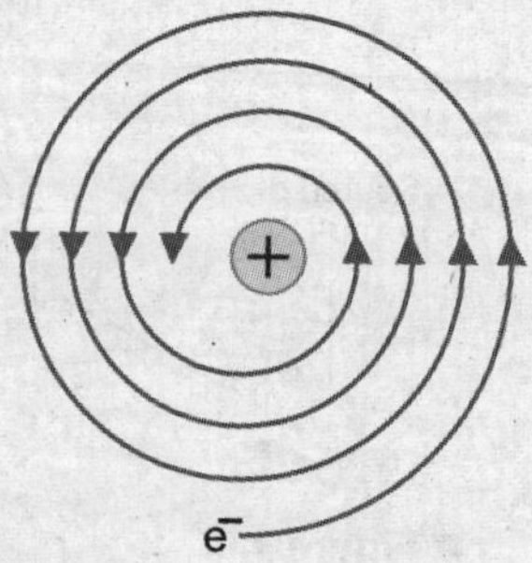

■ **Figure 1.9**
Orbiting electron would radiate energy and spiral into the nucleus.

Weakness of Rutherford Atomic Model

The assumption that electrons were orbiting around the nucleus was unfortunate. According to the classical electromagnetic theory if a charged particle accelerates around an oppositely charged particle, the former will radiate energy. If an electron radiates energy, its speed will decrease and it will go into spiral motion, finally falling into the nucleus. This does not happen actually as then the atom would be unstable which it is not. This was the chief weakness of Rutherford's Atomic Model.

MOSLEY'S DETERMINATION OF ATOMIC NUMBER

The discovery that atom has a nucleus that carries a positive charge raised the question : What is the magnitude of the positive charge? This question was answered by Henry Mosley in 1913.

Hitherto atomic number was designated as the 'position number' of a particular element in the Periodic Table. Mosley found that when cathode rays struck different elements used as anode targets in the discharge tube, characteristic X-rays were emitted. The wavelength of these X-rays decreases in a regular manner in passing from one element to the next one in order of the Periodic Table.

Mosley plotted the atomic number against the square root of the frequency of the X-rays emitted and obtained a straight line which indicated that atomic number was not a mere 'position number' but a fundamental property of the atom. He further made a remarkable suggestion that the wavelength (or frequency) of the emitted X-rays was related to the number of positive charges or protons in the nucleus. The wavelength changed regularly as the element that came next in the Periodic Table had one proton (one unit atomic mass) more than the previous one. Mosley calculated the number of units of positive charge on the nuclei of several atoms and established that :

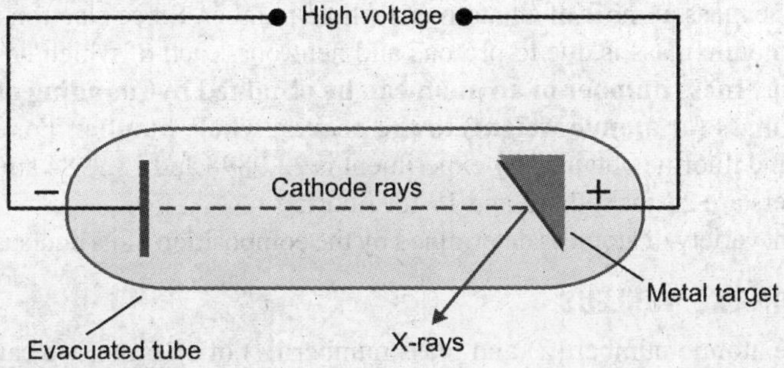

Figure 1.10
Production of X-rays.

Atomic Number of an element is equal to the number of protons in the nucleus of the atom of that element.

Since the atom as a whole is electrically neutral, the atomic number (Z) is also equal to the number of extranuclear electrons. Thus hydrogen (H) which occupies first position in the Periodic Table has atomic number 1. This implies that it has a nucleus containing one proton (+ 1) and one extranuclear electron (– 1).

Now the term Atomic Number is often referred to as the **Proton Number.**

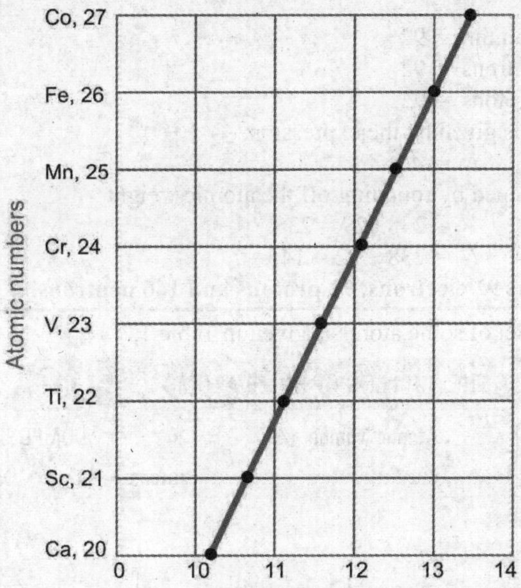

Figure 1.11
Mosley's plot of the square root of X-ray frequencies against atomic number for the elements calcium through cobalt.

WHAT IS MASS NUMBER ?

The total number of protons and neutrons in the nucleus of an atom is called the Mass Number, *A*, of the atom. In situations where it is unnecessary to differentiate between protons and neutrons, these elementary particles are collectively referred to as **nucleons. Thus mass number of an atom is equal to the total number of nucleons in the nucleus of an atom.**

Obviously, the mass number of an atom is a whole number. Since electrons have practically no mass, the entire atomic mass is due to protons and neutrons, each of which has a mass almost one unit. Therefore, **the mass number of an atom can be obtained by rounding off the experimental value of atomic mass (or atomic weight) to the nearest whole number.** For example, the atomic mass of sodium and fluorine obtained by experiment is 22.9898 and 18.9984 amu respectively. Thus their mass numbers are 23 for sodium and 19 for fluorine.

Each different variety of atom, as determined by the composition of its nucleus, is called a **nuclide.**

COMPOSITION OF THE NUCLEUS

Knowing the atomic number (*Z*) and mass number (*A*) of an atom, we can tell the number of protons and neutrons contained in the nucleus. By definition :

$$\text{Atomic Number, } Z = \text{Number of protons}$$

$$\text{Mass Number, } A = \text{Number of protons} + \text{Number of neutrons}$$

∴ The number of neutrons is given by the expression :

$$N = A - Z$$

SOLVED PROBLEM 1

Uranium has atomic number 92 and atomic weight 238.029. Give the number of electrons, protons and neutrons in its atom.

SOLUTION

Atomic Number of Uranium = 92

∴ Number of electrons = 92

and Number of protons = 92

Number of neutrons (*N*) is given by the expression

$$N = A - Z$$

Mass Number (*A*) is obtained by rounding off the atomic weight

$$= 238.029 = 238$$

$$\therefore \quad N = 238 - 92 = 146$$

Thus Uranium atom has 92 electrons, 92 protons and 146 neutrons.

The composition of nuclei of some atoms is given in Table 1.2.

TABLE 1.2. COMPOSITION OF THE NUCLEUS OF SOME ATOMS

Atom	Mass Number (A)	Atomic Number (Z)	COMPOSITION	
			Protons = Z	Neutrons = A – Z
Be	9	4	4	5
F	19	9	9	10
Na	23	11	11	12
Al	27	13	13	14
P	31	15	15	16
Sc	45	21	21	24
Au	197	79	79	118

QUANTUM THEORY AND BOHR ATOM

Rutherford model laid the foundation of the model picture of the atom. However it did not tell anything as to the position of the electrons and how they were arranged around the nucleus.

Rutherford recognized that electrons were orbiting around the nucleus. But according to the classical laws of Physics an electron moving in a field of force like that of nucleus, would give off radiations and gradually collapse into the nucleus. Thus Rutherford model failed to explain why electrons did not do so.

Neils Bohr, a brilliant Danish Physicist, pointed out that the old laws of physics just did not work in the submicroscopic world of the atom. He closely studied the behavior of electrons, radiations and atomic spectra. In 1913 Bohr proposed a new model of the atom based on the modern Quantum theory of energy. With his theoretical model he was able to explain as to why an orbiting electron did not collapse into the nucleus and how the atomic spectra were caused by the radiations emitted when electrons moved from one orbit to the other. Therefore to understand the Bohr theory of the atomic structure, it is first necessary to acquaint ourselves with the nature of electromagnetic radiations and the atomic spectra and also the Quantum theory of energy.

Electromagnetic Radiations

Energy can be transmitted through space by electromagnetic radiations. Some forms of *radiant energy* are radio waves, visible light, infrared light, ultraviolet light, X-rays and γ-radiations.

Electromagnetic radiations are so named because they consist of waves which have electrical and magnetic properties. An object sends out energy waves when its particles move up and down or *vibrate* continuously. Such a vibrating particle causes an intermittent disturbance which constitutes a wave. A wave conveys energy from the vibrating object to a distant place. The wave travels at right angle to the vibratory motion of the object.

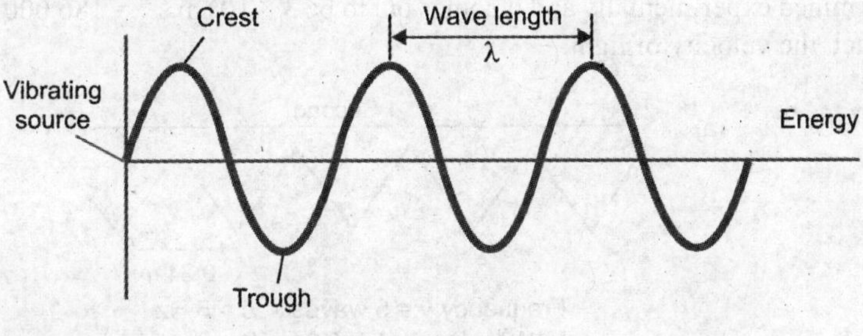

■ Figure 1.12
Illustration of wave motion caused by a vibrating source.

Waves similar to electromagnetic waves are caused when a stone is thrown in a pond of water. The stone makes the water molecules vibrate up and down and transmit its energy as waves on water surface. These waves are seen travelling to the bank of the pond.

A wave may be produced by the actual displacement of particles of the medium as in case of water or sound waves. However, electromagnetic waves are produced by a periodic motion of charged particles. Thus vibratory motion of electrons would cause a wave train of oscillating electric field and another of oscillating magnetic field. These electromagnetic waves travel through empty space with the speed or velocity of light.

Characteristics of Waves

A series of waves produced by a vibrating object can be represented by a wavy curve of the type shown in Fig. 1.12. The tops of the curve are called *crests* and the bottoms *troughs*. Waves are characterized by the following properties :

Wavelength

The wavelength is defined as the distance between two successive crests or troughs of a wave.

Wavelength is denoted by the Greek letter λ (lambda). It is expressed in centimeters or meters or in *angstrom* units. One angstrom, Å, is equal to 10^{-8} cm. It is also expressed in nanometers (1nm = 10^{-9} m). That is,

$$1 \text{ Å} = 10^{-8} \text{ cm} = 10^{-10} \text{ m} \quad \text{or} \quad 1 \text{ cm} = 10^{8} \text{ Å and } 1 \text{ m} = 10^{10} \text{ Å}$$

$$1 \text{ nm} = 10^{-9} \text{ m}$$

Frequency

The frequency is the number of waves which pass a given point in one second.

Frequency is denoted by the letter ν (nu) and is expressed in *hertz* (hz).

It is noteworthy that **a wave of high frequency (*b*) has a shorter wavelength, while a wave of low frequency (*a*) has a longer wavelength.**

Speed

The speed (or velocity) of a wave is the distance through which a particular wave travels in one second.

Speed is denoted by *c* and it is expressed in cm per second. If the speed of a wave is *c* cm/sec, it means that the distance travelled by the wave in one second is *c* cm. Speed is related to frequency and wavelength by the expression

$$c = \nu\lambda$$

or $$\text{Speed} = \text{Frequency} \times \text{Wavelength}$$

The various types of electromagnetic radiations have different wavelengths and frequencies. As evident from Fig. 1.13, all types of radiations travel with the same speed or velocity. This velocity has been determined experimentally and it comes out to be 3×10^8 m s^{-1} = 186,000 miles per second which is, in fact, the velocity of light.

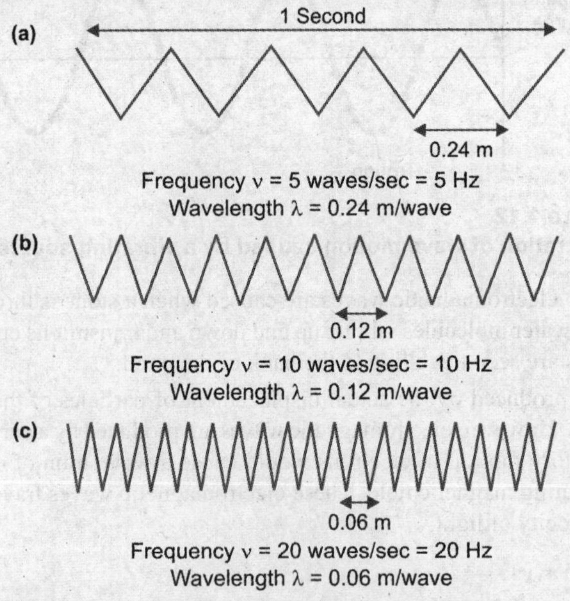

Figure 1.13
Waves of different wavelengths and frequencies.
In all three cases; velocity = λ x ν = 1.2 m s^{-1}.

Wave Number

Another quantity used to characterise radiation is the *wave number*. **This is reciprocal of the wavelength and is given the symbol $\bar{\nu}$ (nu bar).** That is,

$$\bar{\nu} = \frac{1}{\lambda}$$

The wave number is the number of wavelengths per unit of length covered. Its units are cm^{-1} or m^{-1}.

SOLVED PROBLEM 2

The wavelength of a violet light is 400 nm. Calculate its frequency and wave number.

SOLUTION. We know that

$$\text{frequency, } \nu = \frac{c}{\lambda}$$

Here $c = 3.0 \times 10^8 \text{ m s}^{-1}$; $\lambda = 400 \text{ nm} = 400 \times 10^{-9}$ m

$$\nu = \frac{c}{\lambda} = \frac{3.0 \times 10^8 \text{ m s}^{-1}}{400 \times 10^{-9} \text{ m}}$$

$$= \frac{3}{400} \times 10^{17} \text{ s}^{-1}$$

$$= \frac{3000}{400} \times 10^{14} \text{ s}^{-1}$$

$$= \mathbf{7.5 \times 10^{14}\ Hz}$$

Also, wave number $\bar{\nu} = \frac{1}{\lambda}$

$$\bar{\nu} = \frac{1}{400 \times 10^{-9} \text{ m}}$$

$$= \mathbf{25 \times 10^5\ m^{-1}}$$

SOLVED PROBLEM 3

The frequency of strong yellow line in the spectrum of sodium is $5.09 \times 10^{14} \text{ s}^{-1}$. Calculate the wavelength of the light in nanometers.

SOLUTION. We know that wavelength, $\lambda = \frac{c}{\nu}$

Here $c = 3.0 \times 10^8 \text{ m s}^{-1}$

$\nu = 5.09 \times 10^{14} \text{ s}^{-1}$ (given)

Wavelength $$\lambda = \frac{3.0 \times 10^8 \text{ m s}^{-1}}{5.09 \times 10^{14} \text{ s}^{-1}}$$

$$= \frac{3000}{5.09} \times 10^{-9} \text{ m}$$

$$= 589 \times 10^{-9} \text{ m}$$

$$= \mathbf{589\ nm} \qquad [\because 1 \text{ nm} = 10^{-9} \text{ m}]$$

SPECTRA

A spectrum is an array of waves or particles spread out according to the increasing or decreasing of some property. An increase in frequency or a decrease in wavelength represent an increase in energy.

THE ELECTROMAGNETIC SPECTRUM

Electromagnetic radiations include a range of wavelengths and this array of wavelengths is referred to as the *Electromagnetic radiation spectrum* or simply *Electromagnetic spectrum.* The electromagnetic spectrum with marked wavelengths is shown in Fig. 1.14.

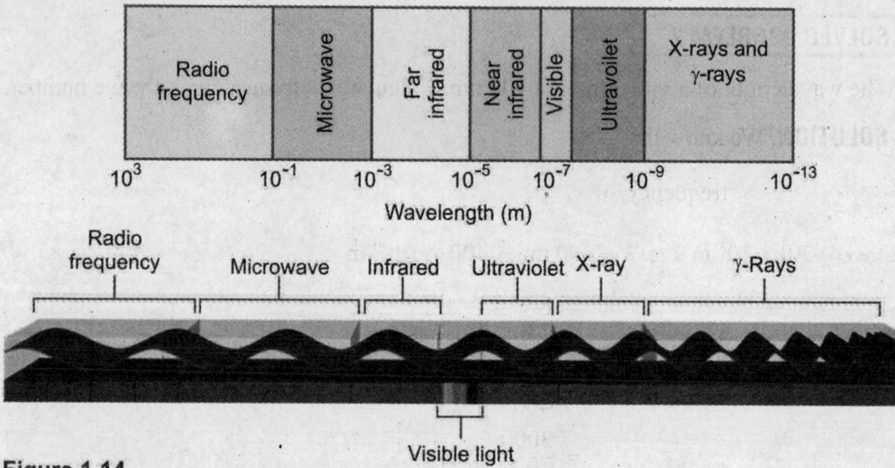

■ **Figure 1.14**
Electromagnetic spectrum. Wavelength boundaries of each region are approximate.

CONTINUOUS SPECTRUM

White light is radiant energy coming from the sun or from incandescent lamps. It is composed of light waves in the range 4000-8000 Å (4×10^{-7} m to 8×10^{-7} m). Each wave has a characteristic color. When a beam of white light is passed through a prism, different wavelengths are refracted (or bent) through different angles. When received on a screen, these form a continuous series of color bands : violet, indigo, blue, green, yellow, orange and red (VIBGYOR). **This series of bands that form a continuous rainbow of colors, is called a Continuous Spectrum.**

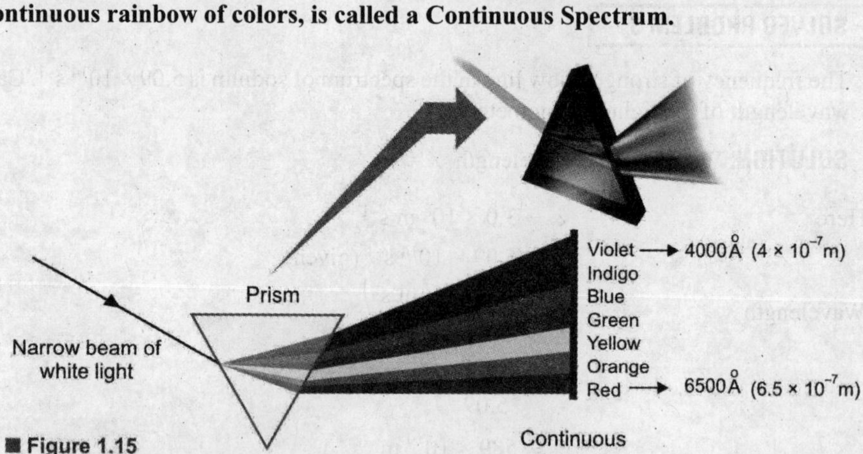

■ **Figure 1.15**
The continuous spectrum of white light.

The violet component of the spectrum has shorter wavelengths (4000 – 4250 Å or 4×10^{-7} to 4.25×10^{-7} meter) and higher frequencies. The red component has longer wavelengths (6500 – 7500Å or $6.5 \times 10^{-7} - 7.5 \times 10^{-7}$ meter) and lower frequencies. The invisible region beyond the violet is called **ultraviolet region** and the one below the red is called **infrared region.**

ATOMIC SPECTRA

When an element in the vapor or the gaseous state is heated in a flame or a discharge tube, the atoms are excited (energized) and emit light radiations of a characteristic color. The color of light produced indicates the wavelength of the radiation emitted.

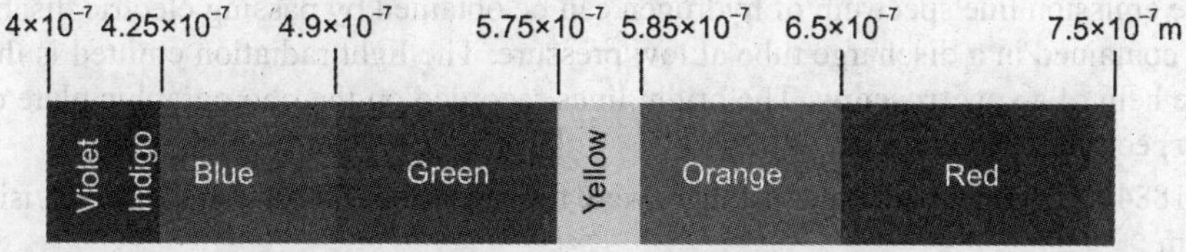

■ **Figure 1.16**
Wavelength range of color bands in Å of continuous spectrum.

For example, a Bunsen burner flame is colored yellow by sodium salts, red by strontium and violet by potassium. In a discharge tube, neon glows orange-red, helium-pink, and so on. If we examine the emitted light with a **Spectroscope** (a device in which a beam of light is passed through a prism and received on a photograph), the spectrum obtained on the photographic plate is found to consist of bright lines (Fig. 1.18). **Such a spectrum in which each line represents a specific wavelength of radiation emitted by the atoms is referred to as the Line spectrum or Atomic Emission spectrum** of the element. The emission spectra of some elements are shown in Fig. 1.17. An individual line of these spectra is called a **Spectral line.**

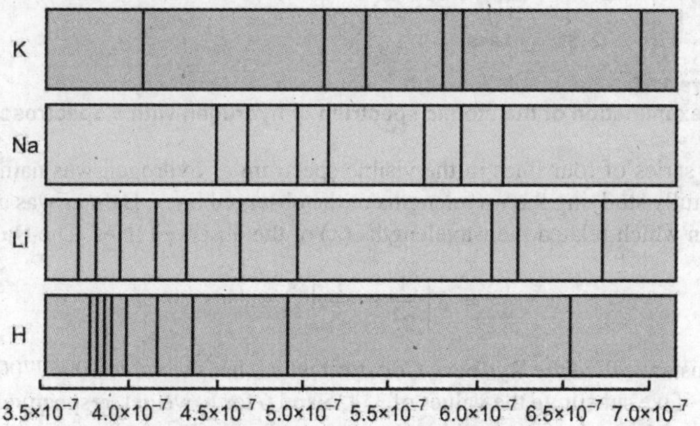

■ **Figure 1.17**
Emission spectra of K, Na, Li and H.

When white light composed of all visible wavelengths, is passed through the cool vapor of an element, certain wavelengths may be absorbed. These absorbed wavelengths are thus found missing in the transmitted light. The spectrum obtained in this way consists of a series of dark lines which is referred to as the **Atomic Absorption spectrum** or simply **Absorption spectrum.** The wavelengths of the dark lines are exactly the same as those of bright lines in the emission spectrum. The absorption spectrum of an element is the reverse of emission spectrum of the element.

Atomic spectral lines are emitted or absorbed not only in the visible region of the electromagnetic spectrum but also in the infrared region **(IR spectra)** or in the ultraviolet region **(UV spectra).**

Since the atomic spectra are produced by emission or absorption of energy depending on the internal structure of the atom, each element has its own characteristic spectrum. Today spectral analysis has become a powerful method for the detection of elements even though present in extremely small amounts. The most important consequence of the discovery of spectral lines of hydrogen and other elements was that it led to our present knowledge of atomic structure.

ATOMIC SPECTRUM OF HYDROGEN

The emission line spectrum of hydrogen can be obtained by passing electric discharge through the gas contained in a discharge tube at low pressure. The light radiation emitted is then examined with the help of a **spectroscope.** The bright lines recorded on the photographic plate constitute the atomic spectrum of hydrogen (Fig. 1.18).

In 1884 J.J. Balmer observed that there were four prominent colored lines in the visible hydrogen spectrum :

(1) a *red line* with a wavelength of 6.563×10^{-7} m
(2) a *blue-green line* with a wavelength 4.861×10^{-7} m
(3) a *blue line* with a wavelength 4.34×10^{-7} m
(4) a *violet line* with a wavelength 4.102×10^{-7} m

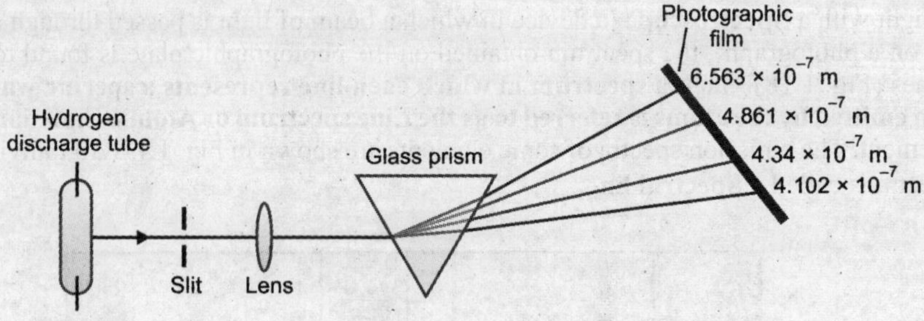

■ **Figure 1.18**
The examination of the atomic spectrum of hydrogen with a spectroscope.

The above series of four lines in the visible spectrum of hydrogen was named as the **Balmer Series.** By carefully studying the wavelengths of the observed lines, Balmer was able empirically to give an equation which related the wavelengths (λ) of the observed lines. The **Balmer Equation** is

$$\frac{1}{\lambda} = R\left(\frac{1}{2^2} - \frac{1}{n^2}\right)$$

where R is a constant called the **Rydberg Constant** which has the value 10967700 m^{-1} and $n = 3, 4, 5, 6$ etc. That is, if we substitute the values of 3, 4, 5 and 6 for n, we get, respectively, the wavelength of the four lines of the hydrogen spectrum.

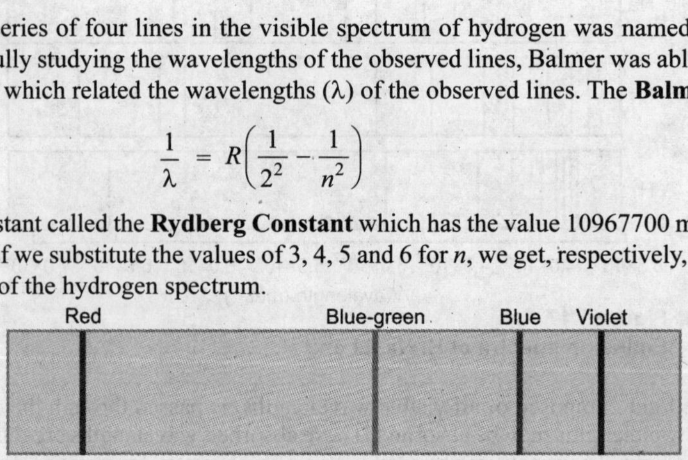

■ **Figure 1.19**
Balmer series in the Hydrogen spectrum.

In addition to Balmer Series, four other spectral series were discovered in the infrared and ultraviolet regions of the hydrogen spectrum. These bear the names of the discoverers. Thus in all we have **Five Spectral Series** in the atomic spectrum of hydrogen :

Name	Region where located
(1) Lyman Series	Ultraviolet
(2) Balmer Series	Visible
(3) Paschen Series	Infrared
(4) Brackett Series	Infrared
(5) Pfund Series	Infrared

Balmer equation had no theoretical basis at all. Nobody had any idea how it worked so accurately in finding the wavelengths of the spectral lines of hydrogen atom. However, in 1913 Bohr put forward his theory which immediately explained the observed hydrogen atom spectrum. Before we can understand Bohr theory of the atomic structure, it is necessary to acquaint ourselves with the quantum theory of energy.

QUANTUM THEORY OF RADIATION

The wave theory of transmission of radiant energy appeared to imply that energy was emitted (or absorbed) in continuous waves. In 1900 Max Planck studied the spectral lines obtained from hot-body radiations at different temperatures. According to him, light radiation was produced discontinuously by the molecules of the hot body, each of which was vibrating with a specific frequency which increased with temperature. Thus Planck proposed a new theory that a hot body radiates energy not in continuous waves but in small units of waves. The 'unit wave' or 'pulse of energy' is called **Quantum** (plural, *quanta*). In 1905 Albert Einstein showed that light radiations emitted by 'excited' atoms or molecules were also transmitted as particles or quanta of energy. These light quanta are called **photons.**

The general **Quantum Theory of Electromagnetic Radiation** in its present form may be stated as :

(1) **When atoms or molecules absorb or emit radiant energy, they do so in separate 'units of waves' called quanta or photons.** Thus light radiations obtained from energized or 'excited atoms' consist of a stream of photons and not continuous waves.

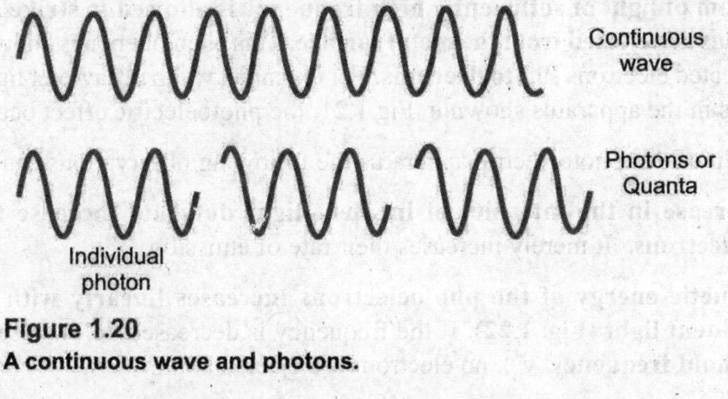

■ **Figure 1.20**
A continuous wave and photons.

(2) **The energy, *E*, of a quantum or photon is given by the relation**

$$E = h\nu \qquad ...(1)$$

where ν is the frequency of the emitted radiation, and h the **Planck's Constant.** The value of $h = 6.62 \times 10^{-27}$ erg s or 6.62×10^{-34} J s.

We know that c, the velocity of radiation, is given by the equation

$$c = \lambda\nu \qquad ...(2)$$

Substituting the value of ν from (2) in (1), we can write

$$E = \frac{hc}{\lambda}$$

Thus **the magnitude of a quantum or photon of energy is directly proportional to the frequency of the radiant energy, or is inversely proportional to its wavelength, λ.**

(3) **An atom or molecule can emit (or absorb) either one quantum of energy (*h*ν) or any whole number multiple of this unit.**

Thus radiant energy can be emitted as $h\nu$, $2h\nu$, $3h\nu$, and so on, but never as 1.5 $h\nu$, 3.27 $h\nu$, 5.9 $h\nu$, or any other fractional value of $h\nu$ *i.e.* $nh\nu$

Quantum theory provided admirably a basis for explaining the photoelectric effect, atomic spectra and also helped in understanding the modern concepts of atomic and molecular structure.

SOLVED PROBLEM 4

Calculate the magnitude of the energy of the photon (or quantum) associated with light of wavelength 6057.8 Å. (Å = 10^{-10} m)

SOLUTION

(*a*) Calculation of Frequency :

$$\nu = \frac{c}{\lambda} = \frac{3.0 \times 10^{8}\,\text{m s}^{-1}}{6057.8 \times 10^{-10}\,\text{m}}$$

$$= 4.952 \times 10^{14}\ \text{s}^{-1}$$

$$= 4.952 \times 10^{14}\ \text{Hz}$$

(*b*) Calculation of Energy :

$$E = h\nu = (6.625 \times 10^{-34}\ \text{J s})(4.952 \times 10^{14}\ \text{s}^{-1})$$

$$= \mathbf{3.281 \times 10^{-19}\ J}$$

PHOTOELECTRIC EFFECT

When a beam of light of sufficiently high frequency is allowed to strike a metal surface in vacuum, electrons are ejected from the metal surface. This phenomenon is known as **Photoelectric effect** and the ejected electrons **Photoelectrons.** For example, when ultraviolet light shines on Cs (or Li, Na, K, Rb) as in the apparatus shown in Fig 1.21, the photoelectric effect occurs.

With the help of this photoelectric apparatus the following observations can be made :

(1) **An increase in the intensity of incident light does not increase the energy of the photoelectrons.** It merely increases their rate of emission.

(2) **The kinetic energy of the photoelectrons increases linearly with the frequency of the incident light (Fig. 1.22).** If the frequency is decreased below a certain critical value **(Threshold frequency, ν_0)**, no electrons are ejected at all.

The Classical Physics predicts that the kinetic energy of the photoelectrons should depend on the intensity of light and not on the frequency. Thus it fails to explain the above observations.

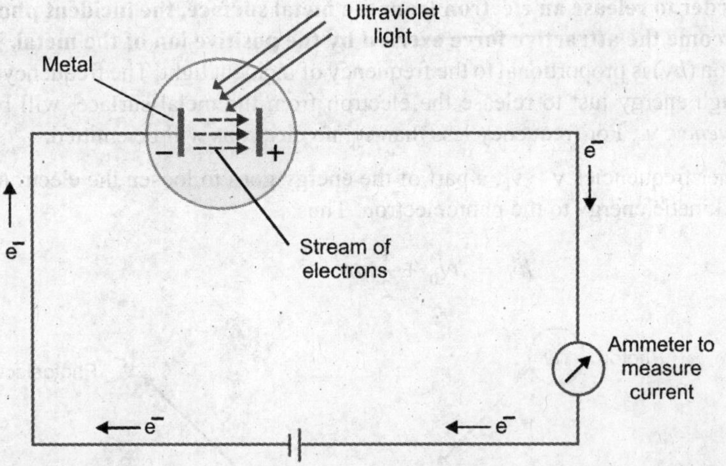

■ **Figure 1.21**
Apparatus for measuring the photoelectric effect. When ultraviolet light shines on the metal, the emitted electrons flow to the anode and the circuit is completed. This current can be measured with the help of an ammeter.

EINSTEIN'S EXPLANATION OF PHOTOELECTRIC EFFECT

In 1905 Albert Einstein, who was awarded Nobel Prize for his work on photons, interpreted the Photoelectric effect by application of the Quantum theory of light.

(1) A photon of incident light transmits its energy ($h\nu$) to an electron in the metal surface which escapes with kinetic energy $\frac{1}{2}mv^2$. **The greater intensity of incident light merely implies greater number of photons each of which releases one electron.** This increases the rate of emission of electrons, while the kinetic energy of individual photons remains unaffected.

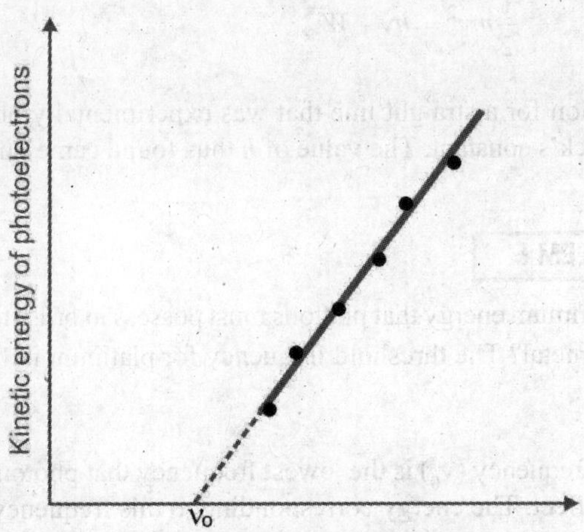

■ **Figure 1.22**
Kinetic energy of photoelectrons plotted against frequency of incident light.

(2) **In order to release an electron from the metal surface, the incident photon has first to overcome the attractive force exerted by the positive ion of the metal.** The energy of a photon ($h\nu$) is proportional to the frequency of incident light. The frequency which provides enough energy just to release the electron from the metal surface, will be the *threshold frequency*, ν_0. For frequency less than ν_0, no electrons will be emitted.

For higher frequencies $\nu > \nu_0$, a part of the energy goes to loosen the electron and remaining for imparting kinetic energy to the photoelectron. Thus,

$$h\nu = h\nu_0 + \frac{1}{2}mv^2 \qquad ...(1)$$

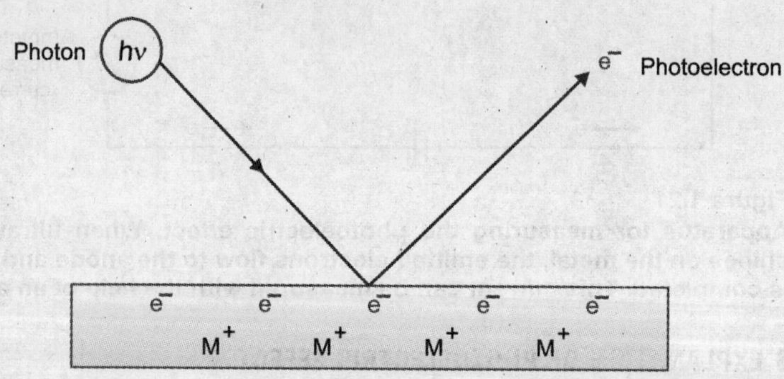

■ Figure 1.23
It needs a photon(*hv*) to eject an electron with energy ½ *mv*².

Where $h\nu$ is the energy of the incoming photon, $h\nu_0$ is the minimum energy for an electron to escape from the metal, and $\frac{1}{2}mv^2$ is the kinetic energy of the photoelectron. $h\nu_0$ is constant for a particular solid and is designated as W, the **work function.** Rearranging equation (1)

$$\frac{1}{2}mv^2 = h\nu - W \qquad ...(2)$$

This is the equation for a straight line that was experimentally obtained in Fig. 1.22. Its slope is equal to h, the Planck's constant. The value of h thus found came out to be the same as was given by Planck himself.

SOLVED PROBLEM 5

What is the minimum energy that photons must possess in order to produce photoelectric effect with platinum metal? The threshold frequency for platinum is $1.3 \times 10^{15}\ s^{-1}$.

SOLUTION

The threshold frequency (ν_0) is the lowest frequency that photons may possess to produce the photoelectric effect. The energy corresponding to this frequency is the minimum energy (E).

$$E = h\nu_0$$
$$= (6.625 \times 10^{-34}\ J\ s)\ (1.3 \times 10^{15}\ s^{-1})$$
$$= \mathbf{8.61 \times 10^{-19}\ J}$$

SOLVED PROBLEM 6

Calculate the kinetic energy of an electron emitted from a surface of potassium metal (work function = 3.62×10^{-19} Joule) by light of wavelength 5.5×10^{-10} m.

SOLUTION

$$\nu = \frac{c}{\lambda} \text{ where } c = \text{velocity of light } \left(3.0 \times 10^8 \text{m s}^{-1}\right)$$

For $\lambda = 5.5 \times 10^{-10}$ m

$$\nu = \frac{c}{\lambda} = \frac{3.0 \times 10^8 \text{ m s}^{-1}}{5.5 \times 10^{-10} \text{m}} = 5.5 \times 10^{17} \text{ s}^{-1}$$

$$\frac{1}{2}mv^2 = h\nu - W$$

$$= (6.6 \times 10^{-34} \text{ J s})\ (5.5 \times 10^{17} \text{ s}^{-1}) - 3.62 \times 10^{-19} \text{ J}$$

$$= 3.63 \times 10^{-16} \text{ J} - 3.62 \times 10^{-19} \text{ J}$$

$$= \mathbf{3.63 \times 10^{-16} \text{ J}}$$

Thus the electron will be emitted with kinetic energy of 3.63×10^{-16} J.

COMPTON EFFECT

In 1923 A.H. Compton provided one more proof to the quantum theory or the photon theory. He was awarded Nobel Prize in 1927 for his discovery of what is now called the **Compton Effect.** He demonstrated that : **When X-rays of wavelength λ' struck a sample of graphite, an electron was ejected and the X-rays scattered at an angle θ had longer wavelength λ.**

Explanation of Compton Effect

Compton said that it was like a ball hitting a stationary ball which is pushed away while the energy of the striking ball decreases. Thus he argued that light radiation (X-rays) consisted of particles (photons), as a continuous wave could not have knocked out the electron. **He visualized that a photon of incident light struck a stationary electron in graphite and hence lost some energy which resulted in the increase of wavelength.** This process could not have occurred unless light radiation consisted of particles or photons.

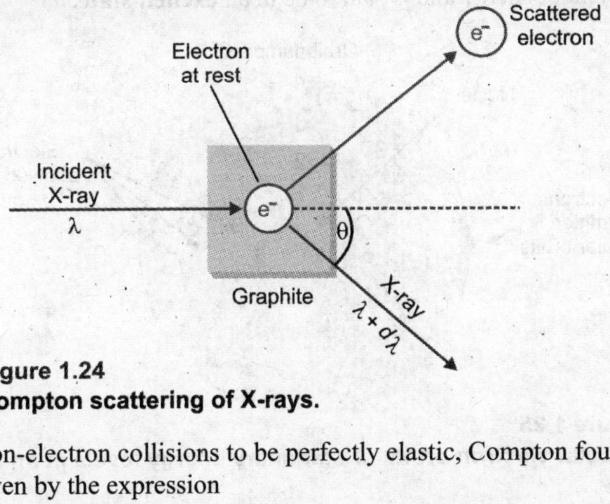

■ Figure 1.24
Compton scattering of X-rays.

By assuming photon-electron collisions to be perfectly elastic, Compton found that the shift in wavelength, $d\lambda$ was given by the expression

$$d\lambda = \frac{2h}{mc} \sin^2 \theta / 2$$

where h is Planck's constant, m the mass of an electron, c the velocity of light and θ the angle of scattering. The expression shows that $d\lambda$ is independent of the nature of the substance and wavelength of the incident radiation. Given the wavelength of a photon, one can calculate the momentum of the electron ejected.

BOHR MODEL OF THE ATOM

Rutherford's nuclear model simply stated that atom had a nucleus and the negative electrons were present outside the nucleus. It did not say anything as to how and where those electrons were arranged. It also could not explain why electrons did not fall into the nucleus due to electrostatic attraction. In 1913 Niels Bohr proposed a new model of atom which explained some of these things and also the emission spectrum of hydrogen. Bohr's theory was based on Planck's quantum theory and was built on the following postulates.

Postulates of Bohr's Theory

(1) **Electrons travel around the nucleus in specific permitted circular orbits and in no others.**

Electrons in each orbit have a definite energy and are at a fixed distance from the nucleus. The orbits are given the letter designation n and each is numbered 1, 2, 3, etc. (or K, L, M, etc.) as the distance from the nucleus increases.

(2) **While in these specific orbits, an electron does not radiate (or lose) energy.**

Therefore in each of these orbits the energy of an electron remains the same *i.e.* it neither loses nor gains energy. Hence the specific orbits available to the electron in an atom are referred to as **stationary energy levels** or simply **energy levels.**

(3) **An electron can move from one energy level to another by quantum or photon jumps only.**
When an electron resides in the orbit which is lowest in energy (which is also closest to the nucleus), the electron is said to be in the **ground state.** When an electron is supplied energy, it absorbs one quantum or photon of energy and jumps to a higher energy level. The electron then has potential energy and is said to be in an **excited state.**

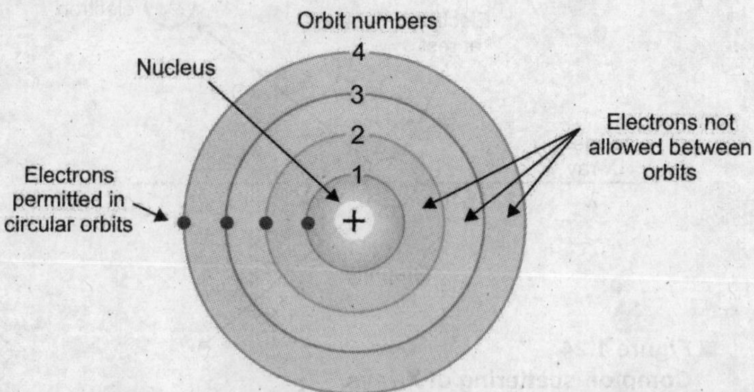

■ **Figure 1.25**
Circular electron orbits or stationary energy levels in an atom.

The quantum or photon of energy absorbed or emitted is the difference between the lower and higher energy levels of the atom

$$\Delta E = E_{high} - E_{low} = h\nu \qquad ...(1)$$

where h is Planck's constant and ν the frequency of a photon emitted or absorbed energy.

(4) **The angular momentum (*mvr*) of an electron orbiting around the nucleus is an integral multiple of Planck's constant divided by 2π.**

$$\text{Angular momentum} = mvr = n\frac{h}{2\pi} \qquad ...(2)$$

where m = mass of electron, v = velocity of the electron, r = radius of the orbit ; n = 1, 2, 3, etc., and h = Planck's constant.

By putting the values 1, 2, 3, etc., for n, the respective angular momentums are

$$\frac{h}{2\pi}, \frac{2h}{2\pi}, \frac{3h}{2\pi}, \text{ etc.}$$

There can be no fractional value of $h/2\pi$. Thus *the angular momentum is said to be quantized.* The integer n in equation (2) can be used to designate an orbit and a corresponding energy level n is called the atom's **Principal quantum number.**

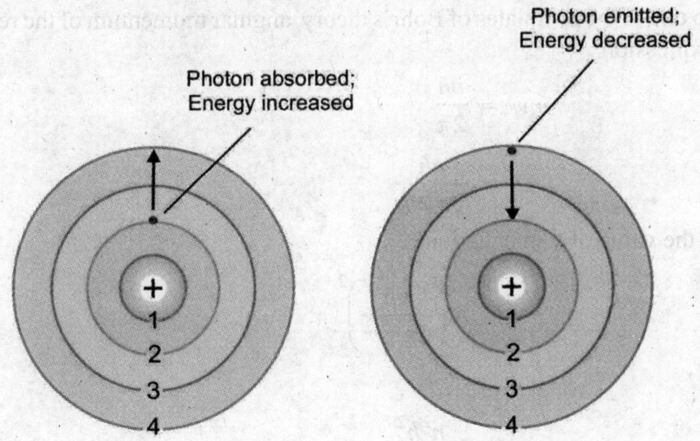

■ Figure 1.26
An electron absorbs a photon of light while it jumps from a lower to a higher energy orbit and a photon is emitted while it returns to the original lower energy level.

Using the above postulates and some classical laws of Physics, Bohr was able to calculate the radius of each orbit of the hydrogen atom, the energy associated with each orbit and the wavelength of the radiation emitted in transitions between orbits. The wavelengths calculated by this method were found to be in excellent agreement with those in the actual spectrum of hydrogen, which was a big success for the Bohr model.

Calculation of Radius of Orbits

Consider an electron of charge e revolving around a nucleus of charge Ze, where Z is the atomic number and e the charge on a proton. Let m be the mass of the electron, r the radius of the orbit and ν the tangential velocity of the revolving electron.

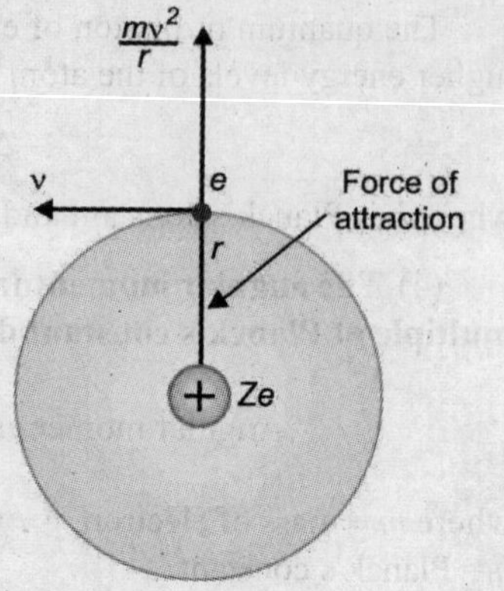

■ **Figure 1.27**
Forces keeping electron in orbit.

The electrostatic force of attraction between the nucleus and the electron (Coulomb's law),

$$= \frac{Ze \times e}{r^2}$$

The centrifugal force acting on the electron

$$= \frac{mv^2}{r}$$

Bohr assumed that these two opposing forces must be balancing each other exactly to keep the electron in orbit. Thus,

$$\frac{Ze^2}{r^2} = \frac{mv^2}{r}$$

For hydrogen $Z = 1$, therefore,

$$\frac{e^2}{r^2} = \frac{mv^2}{r} \quad \text{...(1)}$$

Multiplying both sides by r

$$\frac{e^2}{r} = mv^2 \quad \text{...(2)}$$

According to one of the postulates of Bohr's theory, angular momentum of the revolving electron is given by the expression

$$mvr = \frac{nh}{2\pi}$$

or

$$v = \frac{nh}{2\pi\, mr} \quad \text{...(3)}$$

Substituting the value of v in equation (2),

$$\frac{e^2}{r} = m\left(\frac{nh}{2\pi\, mr}\right)^2$$

Solving for r,

$$r = \frac{n^2h^2}{4\pi^2\, me^2} \quad \text{...(4)}$$

Since the value of h, m and e had been determined experimentally, substituting these values in (4), we have

$$r = n^2 \times 0.529 \times 10^{-10}\ \text{m} \quad \text{...(5)}$$

where n is the principal quantum number and hence the number of the orbit.

When $n = 1$, the equation (5) becomes

$$r = 0.529 \times 10^{-10}\ \text{m} = \alpha_0 \quad \text{...(6)}$$

This last quantity, α_0 called the first **Bohr radius** was taken by Bohr to be the radius of the hydrogen atom in the ground state. This value is reasonably consistent with other information on the size of atoms. When $n = 2, 3, 4$ etc., the value of the second and third orbits of hydrogen comprising the electron in the excited state can be calculated.

SOLVED PROBLEM 7

Calculate the first five Bohr radii.

SOLUTION

The equation (5) may be written as

$$r = n^2 \times 0.529 \times 10^{-10} \text{ m}$$

$$n = 1\ ;\ r = 1^2 \times 0.529 \times 10^{-10} = 0.529 \times 10^{-10} \text{ m}$$

$$n = 2\ ;\ r = 2^2 \times 0.529 \times 10^{-10} = 2.12 \times 10^{-10} \text{ m}$$

$$n = 3\ ;\ r = 3^2 \times 0.529 \times 10^{-10} = 4.76 \times 10^{-10} \text{ m}$$

$$n = 4\ ;\ r = 4^2 \times 0.529 \times 10^{-10} = 8.46 \times 10^{-10} \text{ m}$$

$$n = 5\ ;\ r = 5^2 \times 0.529 \times 10^{-10} = 13.2 \times 10^{-10} \text{ m}$$

Energy of Electron in each Orbit

For hydrogen atom, the energy of the revolving electron, E is the sum of its kinetic energy $\left(\frac{1}{2}mv^2\right)$ and potential energy $\left(-\frac{e^2}{r}\right)$.

$$E = \frac{1}{2}mv^2 - \frac{e^2}{r} \qquad \text{...(7)}$$

From equation (2)

$$mv^2 = \frac{e^2}{r}$$

Substituting the value of mv^2 in (7)

$$E = \frac{1}{2}\frac{e^2}{r} - \frac{e^2}{r}$$

or
$$E = -\frac{e^2}{2r} \qquad \text{...(8)}$$

Substituting the value of r from equation (4) in (8)

$$E = -\frac{e^2}{2} \times \frac{4\pi^2\ me^2}{n^2\ h^2}$$

$$= -\frac{2\pi^2\ me^4}{n^2\ h^2} \qquad \text{...(9)}$$

Substituting the values of m, e, and h in (9),

$$E = \frac{-2.179 \times 10^{-11}}{n^2} \text{ erg per atom} \qquad \text{...(10)}$$

or
$$E = \frac{-2.179 \times 10^{-18}}{n^2} \text{ J per atom}$$

or
$$E = \frac{-2.17 \times 10^{-18} \times 6.02 \times 10^{23}}{n^2} \text{ J per mol}$$

or $$E = \frac{-1311.8}{n^2} \text{ kJ/mol}$$

or $$E = \frac{-313.3}{n^2} \text{ kcal/mol}$$

By using proper integer for n (quantum or orbit number), we can get the energy for each orbit.

SOLVED PROBLEM 8

Calculate the five lowest energy levels of the hydrogen atom.

SOLUTION

From equation (10)

$$E = \frac{-2.179 \times 10^{-18} \text{ J/atom}}{n^2}$$

Therefore the energy associated with the first five energy levels (or orbits) is :

$$n = 1;\ E_1 = \frac{-2.179 \times 10^{-18}}{1^2} = -2.179 \times 10^{-18} \text{ J/atom or } -1311.8 \text{ kJ/mol}$$

$$n = 2;\ E_2 = \frac{-2.179 \times 10^{-18}}{2^2} = -0.5448 \times 10^{-18} \text{ J/atom or } -327.9 \text{ kJ/mol}$$

$$n = 3;\ E_3 = \frac{-2.179 \times 10^{-18}}{3^2} = -0.2421 \times 10^{-18} \text{ J/atom or } -147.5 \text{ kJ/mol}$$

$$n = 4;\ E_4 = \frac{-2.179 \times 10^{-18}}{4^2} = -0.1362 \times 10^{-18} \text{ J/atom or } -82.0 \text{ kJ/mol}$$

$$n = 5;\ E_5 = \frac{-2.179 \times 10^{-18}}{5^2} = -0.08716 \times 10^{-18} \text{ J/atom or } -52.44 \text{ kJ/mol}$$

Significance of Negative Value of Energy

The energy of an electron at infinity is arbitrarily assumed to be zero. This state is called *zero-energy* state. When an electron moves and comes under the influence of nucleus, it does some work and spends its energy in this process. Thus the energy of the electron decreases and it becomes less than zero *i.e.* it acquires a negative value.

Bohr's Explanation of Hydrogen Spectrum

The solitary electron in hydrogen atom at ordinary temperature resides in the first orbit ($n = 1$) and is in the lowest energy state (ground state). When energy is supplied to hydrogen gas in the *discharge tube*, the electron moves to higher energy levels *viz.*, 2, 3, 4, 5, 6, 7, etc., depending on the quantity of energy absorbed. From these high energy levels, the electron returns by jumps to one or other lower energy level. In doing so the electron emits the excess energy as a photon. This gives an excellent explanation of the various spectral series of hydrogen.

Lyman series is obtained when the electron returns to the ground state *i.e.*, $n = 1$ from higher energy levels ($n_2 = 2, 3, 4, 5$, etc.). Similarly, Balmer, Paschen, Brackett and Pfund series are produced when the electron returns to the second, third, fourth and fifth energy levels respectively as shown in Fig. 1.28.

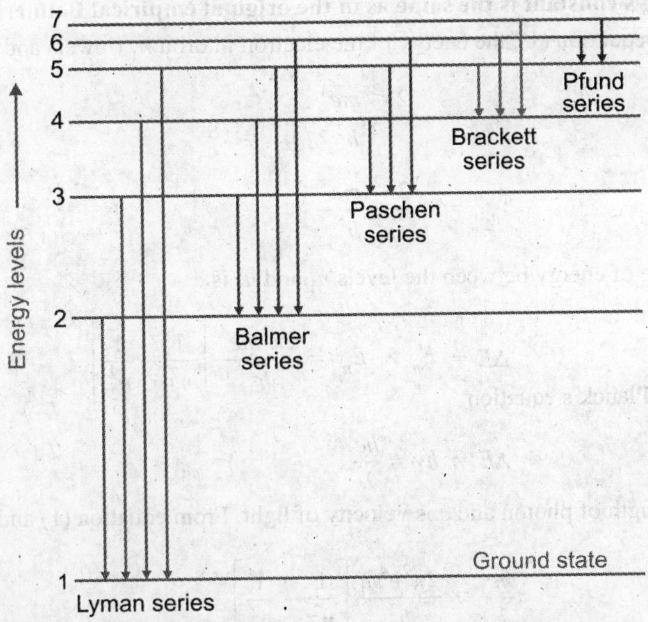

■ **Figure 1.28**
Hydrogen spectral series on a Bohr atom energy diagram.

TABLE 1.3. SPECTRAL SERIES OF HYDROGEN

Series	n_1	n_2	Region	Wavelength λ (m)
Lyman	1	2, 3, 4, 5, etc.	ultraviolet	$0.92 \times 10^{-7} - 1.2 \times 10^{-7}$m
Balmer	2	3, 4, 5, 6, etc.	visible	$4.0 \times 10^{-7} - 6.5 \times 10^{-7}$m
Paschen	3	4, 5, 6, 7, etc.	infrared	$9.5 \times 10^{-7} - 18.75 \times 10^{-7}$m
Brackett	4	5, 6, 7	infrared	$19.45 \times 10^{-7} - 40.50 \times 10^{-7}$m
Pfund	5	6, 7	infrared	$37.80 \times 10^{-7} - 75.00 \times 10^{-7}$m

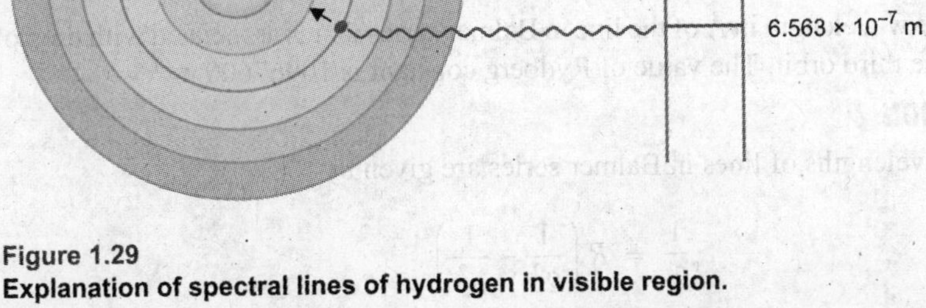

■ **Figure 1.29**
Explanation of spectral lines of hydrogen in visible region.

Value of Rydberg's constant is the same as in the original empirical Balmer's equation

According to equation (1), the energy of the electron in orbit n_1 (lower) and n_2 (higher) is

$$E_{n_1} = -\frac{2\pi^2 me^4}{n_1^2 h^2}$$

$$E_{n_2} = -\frac{2\pi^2 me^4}{n_2^2 h^2}$$

The difference of energy between the levels n_1 and n_2 is :

$$\Delta E = E_{n_2} - E_{n_1} = \frac{2\pi^2 me^4}{h^2}\left[\frac{1}{n_1^2} - \frac{1}{n_2^2}\right] \quad ...(1)$$

According to Planck's equation

$$\Delta E = h\nu = \frac{hc}{\lambda} \quad ...(2)$$

where λ is wavelength of photon and c is velocity of light. From equation (1) and (2), we can write

$$\frac{hc}{\lambda} = \frac{2\pi^2 e^4 m}{h^2}\left[\frac{1}{n_1^2} - \frac{1}{n_2^2}\right]$$

or

$$\frac{1}{\lambda} = \frac{2\pi^2 e^4 m}{h^3 c}\left[\frac{1}{n_1^2} - \frac{1}{n_2^2}\right]$$

$$= R\left[\frac{1}{n_1^2} - \frac{1}{n_2^2}\right] \quad ...(3)$$

where R is **Rydberg constant.** The value of R can be calculated as the value of e, m, h and c are known. It comes out to be 10967900 m^{-1} and agrees closely with the value of Rydberg constant in the original empirical Balmer's equation (10967700 m^{-1}).

Calculation of Wavelengths of the Spectral Lines of Hydrogen in the Visible Region

These lines constitute the Balmer series when $n_1 = 2$. Now the equation (3) above can be written as

$$\frac{1}{\lambda} = 10967900\left[\frac{1}{2^2} - \frac{1}{n_2^2}\right]$$

Thus the wavelengths of the photons emitted as the electron returns from energy levels 6, 5, 4 and 3 were calculated by Bohr. The calculated values corresponded exactly to the values of wavelengths of the spectral lines already known. This was, in fact, a great success of the Bohr atom.

SOLVED PROBLEM 9

Find the wavelength in Å of the line in Balmer series that is associated with drop of the electron from the third orbit. The value of Rydberg constant is 10967600 m^{-1}.

SOLUTION

The wavelengths of lines in Balmer series are given by

$$\frac{1}{\lambda} = R\left(\frac{1}{2^2} - \frac{1}{n^2}\right)$$

where λ = wavelength, R (Rydberg constant) = 10967600 m^{-1} ; $n = 3$.

$$\therefore \quad \frac{1}{\lambda} = 10967600 \left(\frac{1}{2^2} - \frac{1}{3^2}\right) = 10967600 \left(\frac{9-4}{36}\right)$$

$$= 10967600 \times \frac{5}{36}$$

$$\lambda = \frac{36}{10967600 \times 5} = 6.561 \times 10^{-7}\,\text{m}$$

$$\lambda \text{ in Å} = 6.561 \times 10^{-5} \times 10^{8} = 6561\ \text{Å}$$

$\therefore$ Wavelength of the spectral line is **6561 Å**

SHORTCOMINGS OF THE BOHR ATOM

(1) The great success of the Bohr theory was in its ability to predict lines in the hydrogen atom spectrum. But **it was spectacularly unsuccessful for every other atom containing more than one electron.**

(2) **We no longer believe in well-defined electron orbits as was assumed by Bohr.** In fact, in view of modern advances, like dual nature of matter, uncertainty principle, any mechanical model of the atom stands rejected.

(3) **Bohr's model of electronic structure could not account for the ability of atoms to form molecules through chemical bonds.** Today we only accept Bohr's views regarding quantization as nobody has explained atomic spectra without numerical quantization and no longer attempted description of atoms on classical mechanics.

(4) Bohr's theory could not explain the effect of magnetic field (Zeeman effect) and electric field (Stark effect) on the spectra of atoms.

SOMMERFELD'S MODIFICATION OF BOHR ATOM

When spectra were examined with spectrometers, each line was found to consist of several closely packed lines. The existence of these multiple spectral lines could not be explained on the basis of Bohr's theory. Sommerfeld modified Bohr's theory as follows. **Bohr considered electron orbits as circular but Sommerfeld postulated the presence of elliptic orbits also.** An ellipse has a major and minor axis. A circle is a special case of an ellipse with equal major and minor axis. The angular momentum of an electron moving in an elliptic orbit is also supposed to be quantized. Thus only a definite set of values is permissible. It is further assumed that the angular momentum can be an integral part of $h/2\pi$ units, where h is Planck's constant. Or that,

$$\text{angular momentum} = \frac{kh}{2\pi}$$

where k is called the **azimuthal quantum number,** whereas the quantum number used in Bohr's theory is called the **principal quantum number.** The two quantum numbers n and k are related by the expression :

$$\frac{n}{k} = \frac{\text{length of major axis}}{\text{length of minor axis}}$$

ELECTRONIC CONFIGURATION OF ELEMENTS
(ATOMIC NUMBERS ARE GIVEN AFTER THE SYMBOLS OF THE ELEMENTS)

Group 0	Group 1 A	Group 1 B	Group 2 A	Group 2 B	Group 3 A	Group 3 B	Group 4 A	Group 4 B	Group 5 A	Group 5 B	Group 6 A	Group 6 B	Group 7 A	Group 7 B	Group 8		
	H 1 1												H 1 1				
He 2 2	Li 3 2, 1		Be 4 2, 2			B 5 2, 3		C 6 2, 4		N 7 2, 5		O 8 2, 6		F 9 2, 7			
Ne 10 2, 8	Na 11 2, 8, 1		Mg 12 2, 8, 2			Al 13 2, 8, 3		Si 14 2, 8, 4		P 15 2, 8, 5		S 16 2, 8, 6		Cl 17 2, 8, 7			
Ar 18 2, 8, 8	K 19 2, 8, 8, 1		Ca 20 2, 8, 8, 2		Sc 21 2, 8, 9, 2		Ti 22 2, 8, 10, 2		V 23 2, 8, 11, 2		Cr 24 2, 8, 12, 2		Mn 25 2, 8, 13, 2		Fe 26 2,8,14,2	Co 27 2,8,15,2	Ni 28 2,8,16,2
		Cu 29 2, 8, 17, 2		Zn 30 2, 8, 18, 2		Ga 31 2, 8, 18, 3		Ge 32 2, 8, 18, 4		As 33 2, 8, 18, 5		Se 34 2, 8, 18, 6		Br 35 2, 8, 18, 7			
Kr 36 2, 8, 18, 8	Rb 37 2, 8, 18, 8, 1		Sr 38 2, 8, 18, 8, 2		Y 39 2, 8, 18, 9, 2		Zr 40 2,8,18,10,2		Nb 41 2,8,18,11,2		Mo 42 2, 8,18,12,2		Tc 43 2,8,18,13,2		Ru 44 2,8,18 14,2	Rh 45 2,8,18 15,2	Pd 46 2,8,18 16,2
		Ag 47 2, 8, 18, 17, 2		Cd 48 2, 8, 18, 18, 2		In 49 2, 8, 18, 18, 3		Sn 50 2,8,18,18,4		Sb 51 2,8,18,18,5		Te 52 2,8,18,18,6		I 53 2,8,18,18,7			
Xe 54 2,8,18,18,8	Cs 55 2,8,18,18,8,1		Ba 56 2,8,18,18,8,2		La and Rare Earths (57-71) 2, 8, 18, 18, 9, 2 to 2,8,18,32,9,2		Hf 72 2, 8, 18, 32, 10, 2		Ta 73 2, 8, 18, 32, 11, 2		W 74 2, 8, 18, 32, 12, 2		Re 75 2, 8, 18, 32, 13, 2		Os 76 2,8,18, 32,14,2	Ir 77 2,8,18, 32,15,2	Pt 78 2,8,18, 32,16,2
		Au 79 2, 8, 18, 32, 17, 2		Hg 80 2, 8, 18, 32, 18, 2		Tl 81 2, 8, 18, 32, 18, 3		Pb 82 2, 8, 18, 32, 18, 4		Bi 83 2, 8, 18, 32, 18, 5		Po 84 2, 8, 18, 32, 18, 6		At 85 2, 8, 18, 32, 18, 7			
Rn 86 2, 8, 18, 32, 18, 8	Fr 87 2, 8, 18, 32, 18, 8, 1		Ra 88 2, 8, 18, 32, 18, 8, 2		Ac 89 2, 8, 18, 32, 18, 9, 2		Th 90 2, 8, 18, 32, 19, 9, 2		Pa 91 2, 8, 18, 32, 20, 9, 2		U 92 2, 8, 18, 32 21, 9, 2						

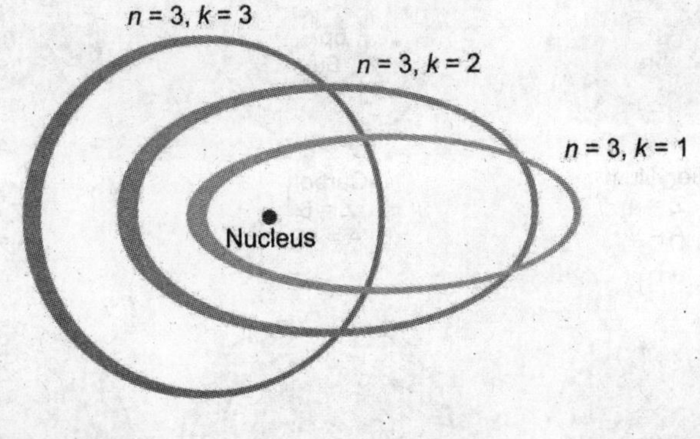

■ Figure 1.30
Sommerfeld orbits in hydrogen atom.

The values of k for a given value of n are $k = n - 1, n - 2, n - 3$ and so on. A series of elliptic orbits with different eccentricities result for the different values of k. When $n = k$, the orbit will be circular. In other words k will have n possible values (n to 1) for a given value of n. However, calculations based on wave mechanics have shown that this is incorrect and the Sommerfeld's modification of Bohr atom fell through.

ELECTRON ARRANGEMENT IN ORBITS

Having known that planetary electrons numerically equal to the atomic number are revolving about the atomic nucleus in closed orbits, the question arises as to how they are arranged in these orbits.

Langmuir Scheme

We are indebted to Langmuir for putting forward the first elaborate scheme of the arrangement of extranuclear electrons in 1919. His fundamental conception is that **the inert gases possess the most stable electron configuration and, therefore, contain complete electron orbits.** Since *helium* has two planetary electrons, the first orbit is considered fully saturated with 2 electrons. In the next inert gas *neon*, we have 10 planetary electrons and since 2 electrons would fully saturate the first orbit the remaining 8 will form the next stable orbit. Argon with atomic number 18 will similarly have the similar arrangement 2, 8, 8. Proceeding in this manner the successive orbits would contain 2, 8, 8, 18, 32 electrons. Langmuir's scheme although quite correct for the first few elements, failed to explain the behavior of higher elements.

Bohr-Bury Scheme

In 1921, Bury put forward a modification of Langmuir scheme which is in better agreement with the physical and chemical properties of certain elements. At about the same time as Bury developed his scheme on chemical grounds, Bohr (1921) published independently an almost identical scheme of the arrangement of extra-nuclear electrons. He based his conclusions on a study of the emission spectra of the elements. Bohr-Bury scheme as it may be called, can be summarized as follows :

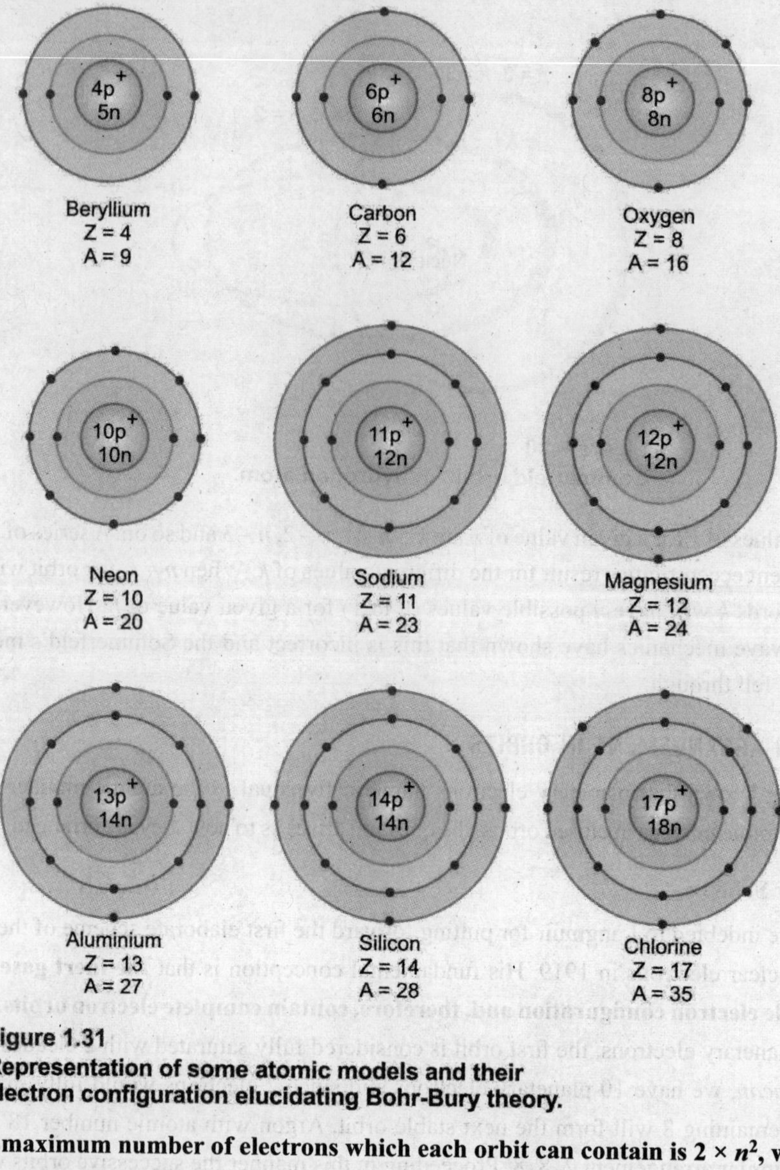

■ **Figure 1.31**
Representation of some atomic models and their electron configuration elucidating Bohr-Bury theory.

Rule 1. The maximum number of electrons which each orbit can contain is $2 \times n^2$, where *n* is the number of orbit.

The first orbit can contain $2 \times 1^2 = 2$; second $2 \times 2^2 = 8$; third $2 \times 3^2 = 18$; fourth $2 \times 4^2 = 32$, and so on.

Rule 2. The maximum number of electrons in the outermost orbit is 8 and in the next-to-the outermost 18.

Rule 3. It is not necessary for an orbit to be completed before another commences to be formed. In fact, a new orbit begins when the outermost orbit attains 8 electrons.

Rule 4. The outermost orbit cannot have more than 2 electrons and next-to-outermost cannot have more than eight so long as the next inner orbit, in each case, has not received the maximum electrons as required by rule (1).

According to Bohr-Bury scheme the configuration of the inert gases is given in the table below :

TABLE 1.4. ELECTRON CONFIGURATION OF INERT GASES

Inert Gas	Atomic Number	Electron Orbits					
		1st (K)	2nd (L)	3rd (M)	4th (N)	5th (O)	6th (P)
Helium (He)	2	2	–	–	–	–	–
Neon (Ne)	10	2	8	–	–	–	–
Argon (Ar)	18	2	8	8	–	–	–
Krypton (Kr)	36	2	8	18	8	–	–
Xenon (Xe)	54	2	8	18	18	8	–
Radon (Rn)	86	2	8	18	32	18	8

A complete statement of the electron configuration of elements elucidating the various postulates of Bohr-Bury scheme is given in the table on page 31 for ready reference.

ZEEMAN EFFECT

In 1896 Zeeman discovered that **spectral lines are split up into components when the source emitting lines is placed in a strong magnetic field.** It is called the Zeeman effect after the name of the discoverer. The apparatus used to observe Zeeman effect is shown in the Fig. 1.32.

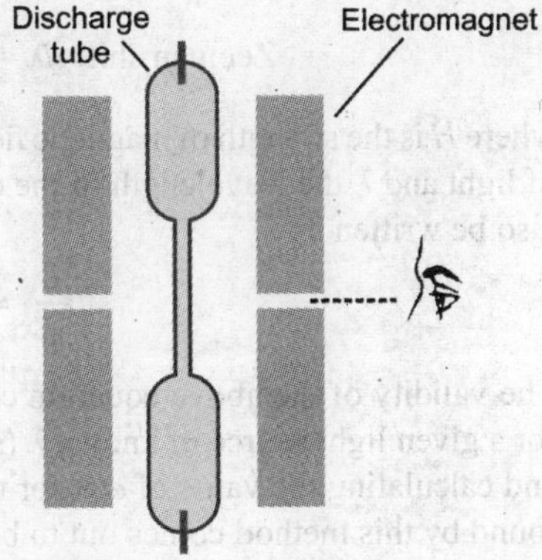

■ Figure 1.32 Zeeman effect.

It consists of electromagnets capable of producing strong magnetic field with pole pieces through which holes have been made lengthwise. Let a discharge tube or sodium vapor lamp emitting radiations be placed between the pole pieces. When the spectral lines are viewed axially through the hole in the pole pieces *i.e.*, parallel to the magnetic field, the line is found to split up into two components, one having shorter wavelength (higher frequency) and the other having higher wavelength (shorter frequency) than that of the original spectral line, which is no longer observable. The two lines are symmetrically situated around the position of the original line and the change in wavelength is termed the Zeeman shift (denoted as $d\lambda$). When viewed in a direction perpendicular to the applied field the lines split up into three, the central one having the same wavelength and frequency as that of the original line and the other two occupying the same position as observed earlier.

In order to explain Zeeman effect, let us consider motion of an electron in a particular orbit corresponding to its permitted angular momentum. The motion of the electron in an orbit is equivalent to a current in a loop of wire. If a current carrying loop of wire be placed in a magnetic field, it experiences a torque, and energy of the system depends upon the orientation of the loop with respect to magnetic field. The correct values of the energies are obtained if the components of the angular momentum of the electron along the direction of the magnetic field are restricted to the value

$$= m \times \frac{h}{2\pi}$$

where $m = 0, \pm 1, \pm 2, \ldots$ and so on. Corresponding to these values of m, a given line splits into as many lines. Hence for each frequency of a radiation emitted by the atom in the absence of magnetic field, there are several possible frequencies in the presence of it. This is, in fact, the cause of Zeeman Effect.

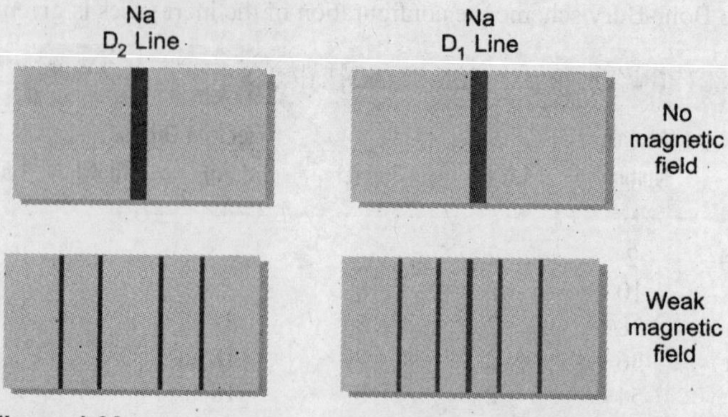

Figure 1.33
Splitting of the D_2 and D_1 lines in the sodium spectrum by a weak magnetic field (Illustration of Zeeman effect).

The shift in the frequency $d\lambda$ for each of the component lines is given by Lorentz's theoretically derived equation as

$$\text{Zeeman shift } d\lambda = \pm \frac{He\lambda^2}{4\pi mc}$$

where H is the strength of magnetic field, e the electronic charge, m the mass of electron, c the velocity of light and λ the wavelength of the original line in the absence of magnetic field. The equation can also be written as

$$\frac{e}{m} = \pm \frac{4\pi c d\lambda}{H\lambda^2}$$

The validity of the above equation can be tested experimentally by observing the Zeeman shift $d\lambda$ for a given light source of known λ (say D-line of sodium) for a magnetic field of known strength H and calculating the value of e/m for the above equation. Lorentz found that the e/m of the electrons found by this method comes out to be the same as by any other method.

ADDITIONAL SOLVED PROBLEMS

SOLVED PROBLEM 10

Calculate the radius of third orbit of hydrogen atom ($h = 6.625 \times 10^{-27}$ erg s; $m = 9.1091 \times 10^{-28}$ g; $e = 4.8 \times 10^{-10}$ esu).

SOLUTION

Formula used

$$r = \frac{n^2 h^2}{4\pi^2 m e^2}$$

Quantities given

$n = 3$ $\quad h = 6.625 \times 10^{-27}$ erg s

$m = 9.1091 \times 10^{-28}$ g $\quad e = 4.8 \times 10^{-10}$ esu

Substitution of values

We have

$$r = \frac{3^2 \times (6.625 \times 10^{-27} \text{ erg s})^2}{4 \times (3.14)^2 \times (9.1091 \times 10^{-28} \text{ g}) \times (4.8 \times 10^{-10} \text{ esu})^2}$$

$$= 0.04772 \times 10^{-6} \text{ cm}$$

$$= 4.772 \times 10^{-8} \text{ cm} = \mathbf{4.772 \times 10^{-10} \text{ m}}$$

SOLVED PROBLEM 11

Calculate the wavelength of first line in Lyman series of hydrogen spectrum. (R, Rydberg constant = 10967700 m^{-1}).

SOLUTION

Formula used $$\frac{1}{\lambda} = R Z^2 \left[\frac{1}{n_1^2} - \frac{1}{n_2^2}\right]$$

Quantities given

R = 10967700 m^{-1} Z = 1 (for hydrogen atom)

n_1 = 1 n_2 = 2 (for 1st line in Lyman series)

Substitution of values

$$\frac{1}{\lambda} = 10967700\ m^{-1} \times \left(\frac{1}{1^2} - \frac{1}{2^2}\right)$$

$$= 10967700\ m^{-1} \times \left(\frac{1}{1} - \frac{1}{4}\right)$$

$$= 10967700\ m^{-1} \times 0.75$$

$$= 8225775\ m^{-1}$$

or $$\lambda = \frac{1}{8225775\ m^{-1}}$$

$$= 1.22 \times 10^{-7}\ m = \mathbf{1220\ Å}$$

SOLVED PROBLEM 12

Calculate the energy difference, frequency and wavelength of light emitted when the electron in a hydrogen atom undergoes transition from 4th energy level to 2nd energy level (Rydberg constant = 1.09677 × 10^7 m^{-1}).

SOLUTION

Formula used

$$\Delta E = h\,c\,R\left[\frac{1}{n_1^2} - \frac{1}{n_2^2}\right]$$

Quantities given

n_1 = 2 n_2 = 4 R = 1.09677 × 10^7 m^{-1} c = 3 × 10^8 m s^{-1}

Substitution of values

$$\Delta E = (6.625 \times 10^{-34}\ J\ s^{-1}) \times (3 \times 10^8\ m\ s^{-1}) \times (1.09677 \times 10^7\ m^{-1})\left[\frac{1}{2^2} - \frac{1}{4^2}\right]$$

$$= 21.798 \times 10^{-34+8+7} \times 0.1875\ J$$

$$= \mathbf{4.087 \times 10^{-19}\ J}$$

Also frequency, $$\upsilon = \frac{\Delta E}{h} = \frac{4.087 \times 10^{-19}\ J}{6.625 \times 10^{-34}\ J\ s}$$

$$= 0.6169 \times 10^{15} \text{ s}^{-1}$$

$$= \mathbf{6.169 \times 10^{14} \text{ s}^{-1}}$$

$$\text{and wavelength, } \lambda = \frac{c}{\upsilon} = \frac{3 \times 10^8 \text{ m s}^{-1}}{6.169 \times 10^{14} \text{ m}}$$

$$= 0.4863 \times 10^{-6} \text{ m}$$

$$= \mathbf{4.863 \times 10^{-7} \text{ m}}$$

SOLVED PROBLEM 13

Iodine molecule dissociates after absorbing light of 4500Å. If one quantum radiation is absorbed by each molecule, calculate the kinetic energy of iodine atom. (Bond energy of I_2 = 240 × 10^3 J mol^{-1}).

SOLUTION

Formula used

$$\Delta E = \frac{hc}{\lambda}$$

Quantities given

$h = 6.625 \times 10^{-34}$ J s $\qquad c = 3 \times 10^8$ m s^{-1} $\qquad \lambda = 4500$ Å $= 4500 \times 10^{-10}$ m

Substitution of values

$$\Delta E = \frac{(6.625 \times 10^{-34} \text{ J s}) \times (3 \times 10^8 \text{ m s}^{-1})}{(4500 \times 10^{-10} \text{ m})}$$

$$= 4.417 \times 10^{-19} \text{ J}$$

$$\text{Now Bond energy of } I_2 \text{ per molecule} = \frac{240 \times 10^3}{\text{Avogadro's No.}} \text{ J/mol}$$

$$= \frac{240 \times 10^3}{6.023 \times 10^{23}} \text{ J/mol}$$

$$= 3.985 \times 10^{-19} \text{ J/mol}$$

Kinetic energy of iodine atoms = Energy absorbed – Bond energy

$$= 4.417 \times 10^{-19} \text{ J} - 3.985 \times 10^{-19} \text{ J}$$

$$= 0.432 \times 10^{-19} \text{ J}$$

$$\therefore \text{Kinetic energy per iodine atom} = \frac{0.432 \times 10^{-19}}{2} \text{ J}$$

$$= \mathbf{2.16 \times 10^{-20} \text{ J}}$$

SOLVED PROBLEM 14

Calculate the wavelength of radiation emitted, producing a line in Lyman series, when an electron falls from fourth stationary state in hydrogen atom. (R = 1.1 × 10^7 m^{-1})

SOLUTION

Formula used

$$\frac{1}{\lambda} = R\left[\frac{1}{n_1^2} - \frac{1}{n_2^2}\right]$$

Quantities given

$n_2 = 4 \qquad n_1 = 1$ (Lyman series) $\qquad R = 1.1 \times 10^7$ m^{-1}

Substitution of values

$$\frac{1}{\lambda} = (1.1 \times 10^7 \text{ m}^{-1}) \times \left[\frac{1}{1^2} - \frac{1}{4^2}\right]$$

$$= 1.1 \times 10^7 \text{ m}^{-1} \times \left[1 - \frac{1}{16}\right]$$

$$= 1.1 \times 10^7 \times \frac{15}{16} \text{ m}^{-1}$$

$$= 1.0313 \times 10^7 \text{ m}^{-1}$$

$$= 1.0313 \times 10^7 \text{ m}^{-1}$$

or

$$\lambda = \frac{1}{1.0313 \times 10^7 \text{ m}^{-1}}$$

$$= 0.9696 \times 10^{-7} \text{ m}$$

$$= \mathbf{9.696 \times 10^{-8} \text{ m}}$$

EXAMINATION QUESTIONS

1. Define or explain the following terms :

   (*a*) Neutrons (*b*) Nucleons

   (*c*) Atomic number (*d*) Mass Number

   (*e*) Photoelectric effect (*f*) Threshold energy

2. Give an account of the experiment which led Rutherford to conclude that every atom has a positively charged nucleus which occupies a very small volume. What were the drawbacks of Rutherford's nuclear model of the atom? How did Bohr rectify the drawbacks of Rutherford model?

3. (*a*) State the postulates of Bohr's theory of the hydrogen atom. Derive an expression for the *n*th orbit of a hydrogen atom. Derive an expression for the radius of any orbit in the atom.

   (*b*) Calculate the energy of transition involving $n_1 = 6$ to $n_2 = 3$ in a hydrogen atom, given that Rydberg constant $R = 10973732 \text{ m}^{-1}$ and $h = 6.63 \times 10^{-34}$ J s.

   Answer. (*b*) 1.818×10^{-19} J

4. (*a*) Discuss Bohr's model of an atom. Show how it successfully explains the spectra of hydrogen atom.

   (*b*) Calculate the velocity of the electron in the first Bohr's orbit. ($h = 6.625 \times 10^{-34}$ J s; $r = 0.529 \times 10^{-11}$m ; $m = 9.109 \times 10^{-31}$ kg)

   Answer. (*b*) $2.189 \times 10^{-16} \text{ m s}^{-1}$

5. (*a*) Explain Mosley's contribution towards the structure of the atom.

   (*b*) Give the defects of Rutherford's model of atom. What suggestions were given by Bohr to remove these defects?

6. Calculate the velocity of the third orbit of hydrogen atom. ($h = 6.625 \times 10^{-34}$ J s; $r = 5.29 \times 10^{-11}$ m; $m = 9.109 \times 10^{-31}$ kg; $e = 1.6 \times 10^{-19}$ C)

   Answer. (*b*) $6.5 \times 10^{-16} \text{ m s}^{-1}$

7. Calculate the wavelength of the first line in Balmer series of hydrogen spectrum. ($R = 10967700 \text{ m}^{-1}$)

   Answer. 1.215×10^{-7} m

8. (*a*) How does Bohr's theory explain the spectrum of hydrogen atom?
   (*b*) Calculate the wavelength associated with an electron moving with a velocity of 1×10^6 m/s^{-1} Mass of an electron = 9.1×10^{-31} kg

   Answer. (*b*) 7.28×10^{-10} m

9. A line at 434 nm in Balmer series of spectrum corresponds to a transition of an electron from the *n*th to 2nd Bohr orbit. What is the value of *n*?

   Answer. $n = 5$

10. (*a*) Explain Rutherford's atomic model. What are its limitations?
    (*b*) State and explain Ritz combination principle.
    (*c*) Calculate the radius of third orbit of hydrogen atom. (h=6.625 × 10^{-34} J s; m = 9.1091×10^{-31}kg; e = 4.8 × 10^{-10} esu)
    (*d*) Calculate the wavelength of first line in Balmer series of hydrogen spectrum. (R = Rydberg's constant = 10967700 m^{-1})

    Answer. (*c*) 4.772×10^{-19} m (*d*) 1.215×10^{-7} m

11. Describe various series in hydrogen spectrum and calculate energy levels of hydrogen atom.
12. Write Rutherford's experiment of scattering of α-particles and give the drawbacks of atomic model.
13. Write notes on :
    (*a*) Merits and demerits of Bohr's theory (*b*) Assumptions of Bohr's atomic model
14. Based on Bohr's calculations, establish the energy expression of the rotating electron in hydrogen like atomic species.
15. Give an account of Bohr's theory of atomic structure and show how it explains the occurrence of spectral lines in the atomic spectra of hydrogen.
16. The electron energy in hydrogen atom is given by $E = -21.7 \times 10^{-19}/n^2$ Joule. Calculate the energy required to remove an electron completely from the $n = 2$ orbit. What is the longest wavelength (in m) of light that can be used to cause this transition?

    Answer. 5.42×10^{-19} J; 3.67×10^{-7} m

17. In a hydrogen atom, an electron jumps from 3rd orbit to first orbit. Find out the frequency and wavelength of the spectral line.

    Answer. 1.0256×10^{-7} m *(Agra BSc)*

18. The energy of the electron in the second and third orbits of the hydrogen atom is 5.42×10^{-19} J and 2.41 × 10^{19} J respectively. Calculate the wavelength of the emitted radiation when the electron drops from third to second orbit.

    Answer. 6.6×10^{-7} m *(Kolkata BSc)*

19. Calculate the wavelength in Å of the photon that is emitted when an electron in Bohr orbit $n = 2$ returns to the orbit $n = 1$ in the hydrogen atom. The ionization potential in the ground state of hydrogen atom is 2.17×10^{-18} J per atom.

    Answer. 1.220×10^{-7} m *(Osmania BSc)*

20. What transition in the hydrogen spectrum would have the same wavelength as the Balmer transition $n = 4$ to $n = 2$ of He^+ transition?

    Answer. $n = 2$ to $n = 1$ *(Baroda BSc)*

21. (*a*) State postulates of Bohr's theory of an atom and derive an expression for radius of Bohr orbit of hydrogen atom.
    (*b*) Give any four limitations of Bohr's theory of an atom. *(Nagpur BSc)*
22. Describe Rutherford's model of the atom. How was it improved by Bohr? *(Arunachal BSc)*
23. Atomic hydrogen is excited to the 4th energy level from the ground state. Determine
    (*a*) the number of lines emitted and
    (*b*) the shortest wavelength present in the emission spectrum. (R_H = 1.09678 × 10^{-7} m^{-1})

    Answer. (*a*) 3; (*b*) 9.7255×10^{-8} m *(Vidyasagar BSc)*

24. Radius of the first Bohr orbit of H-atom is 5.29×10^{-11} m. Find the radii of the first and second Bohr orbit of Li^{2+} ion.

Answer. (*a*) 1.763×10^{-11} m; (*b*) 7.053×10^{-11} m *(Vidyasagar BSc)*

25. If the energy difference between the ground state of an atom and its excited state is 4.4×10^{-19} J, what is the wavelength of the photon required to produce this transition?

Answer. 4.517×10^{-7} m *(Madras BSc)*

26. Calculate the wavelength and energy of radiations emitted for the electronic transition from infinity (∞) to stationary state of the hydrogen atom. ($R = 1.09678 \times 10^7$ m^{-1}; $h = 6.625 \times 10^{-34}$ J s and $c = 2.9979 \times 10^8$ m s^{-1}) *(Gulbarga BSc)*

Answer. 9.11×10^{-6} m; 217.9×10^{-23} kJ

27. The energy transition in hydrogen atom occurs from $n = 3$ to $n = 2$ energy level. ($R = 1.097 \times 10^7$ m^{-1}).

(*a*) Calculate the wavelength of the emitted electron.

(*b*) Will this electron be visible?

(*c*) Which spectrum series does this photon belong to? *(Jadavpur BSc)*

28. Calculate the energy emitted when electrons of 1×10^{-3} kg of hydrogen undergo transition giving the spectral line of lowest energy in the visible region of its atomic spectrum ($R = 1.1 \times 10^7$ m^{-1}; $c = 3 \times 10^8$ m s^{-1}; $h = 6.62 \times 10^{-34}$ J s)

Answer. 2.1846×10^{-18} J *(Panjab BSc)*

29. In hydrogen atom the energy of the electron in first Bohr's orbit is -1312×10^5 J mol^{-1}. What is the energy required for the excitation of second Bohr's orbit ? *(Burdwan BSc)*

Answer. 9.84×10^5 J mol^{-1}

30. Calculate the wavelength in meter of the photon that is emitted when an electron in Bohr orbit $n = 2$ returns to the orbit $n = 1$ in the hydrogen atom. The ionization potential in the ground state of hydrogen atom is 2.17×10^{-18} J per atom. *(Kalayani BSc)*

Answer. 1.220×10^{-7} m

31. A line at 4.34×10^{-7} m in Balmer series of spectrum corresponds to a transition of an electron from the *n*th to 2nd Bohr orbit. What is the value of *n* ? *(Gulbarga BSc)*

Answer. $n = 5$

32. The energy transition in hydrogen atom occurs from $n = 3$ to $n = 2$ energy level. ($R = 1.097 \times 10^7$ m^{-1}). (*i*) Calculate the wavelength of the emitted electron (*ii*) Will this electron be visible ? (*iii*) Which spectrum series does this photon belong to ? *(Vikram BSc)*

Answer. 6.564×10^{-7} m ; Yes ; Balmer series

33. The energy of the electron in the second and third Bohr orbits of the hydrogen atom is -5.42×10^{-19} J and -2.41×10^{-19} J respectively. Calculate the wavelength of the emitted radiation when the electron drops from third to second orbit. *(Kalayani BSc)*

Answer. 6.6×10^{-7} m

34. Calculate the energy of transition involving $n_1 = 6$ to $n_2 = 3$ in a hydrogen atom, given that Rydberg constant $R = 10973732$ m^{-1} and $h = 6.63 \times 10^{-34}$ J s. *(Arunanchal BSc)*

Answer. -1.819×10^{-19} J

35. Calculate the wavelength associated with an electron moving with a velocity of 1×10^6 m s^{-1}. Mass of an electron = 9.1×10^{-31} kg. *(Kalyani BSc)*

Answer. 7.28×10^{-12} m

36. In the photoelectric effect experiment irradiation of a metal with light of frequency 5×10^{20} s^{-1} yields electrons with maximum kinetic energy 6.63×10^{-14} J. Calculate υ_0, threshold frequency of the metal.

Answer. 4×10^{20} s^{-1} *(Jamia Millia BSc)*

37. The kinetic energy of a subatomic particle is 4.55×10^{-25} J. Calculate the frequency of the particle wave. Planck's constant $h = 6.62 \times 10^{-34}$ kg m^2 s^{-1}. *(Delhi BSc)*

Answer. 1.37×10^{-9} s^{-1}

38. What is the frequency of emission spectrum when the electron in hydrogen atom falls from 5th to 2nd orbit ? *(Panjab BSc)*

Answer. 4.517×10^{-7} m

39. Calculate the wave number for the shortest wavelength transition in the Balmer series of hydrogen atom.

Answer. 2741925 m^{-1} *(Nagpur BSc)*

MULTIPLE CHOICE QUESTIONS

1. Cathode rays are deflected by
   (*a*) electric field only (*b*) magnetic field only
   (*c*) electric and magnetic field (*d*) none of these
   Answer. (*c*)
2. The *e*/*m* value for the particles constituting cathode rays is the same regardless of
   (*a*) the gas present in cathode rays tube (*b*) the metal of which cathode was made
   (*c*) both of these (*d*) none of these
   Answer. (*c*)
3. The charge to mass ratio (*e*/*m*) of positive particles
   (*a*) varies with the nature of gas in discharge tube
   (*b*) is independent of the gas in discharge tube
   (*c*) is constant
   (*d*) none of the above
   Answer. (*a*)
4. A subatomic particle which has one unit mass and one unit positive charge is known as
   (*a*) hydrogen atom (*b*) neutron
   (*c*) electron (*d*) proton
   Answer. (*d*)
5. Atomic number of an element is equal to the number of _______ in the nucleus of the atom.
   (*a*) neutrons (*b*) protons
   (*c*) both the neutrons and protons (*d*) electrons
   Answer. (*b*)
6. The mass number of an atom is equal to the number of _______ in the nucleus of an atom
   (*a*) protons (*b*) neutrons
   (*c*) electrons (*d*) nucleons
   Answer. (*d*)
7. If Z is the number of proton and A the number of nucleons, then the number of neutrons in an atom is given by
   (*a*) A + Z (*b*) A – Z
   (*c*) Z – A (*d*) none of these
   Answer. (*b*)

8. The mass number and atomic number of Phosphorus atom are 31 and 15 respectively. The number of neutrons in the nucleus is

(*a*) 15 (*b*) 16
(*c*) 31 (*d*) 46

Answer. (*b*)

9. In a sodium atom (atomic number = 11 and mass number = 23), the number of neutrons is

(*a*) equal to the number of protons (*b*) less than the number of protons
(*c*) greater than the number of protons (*d*) none of these

Answer. (*c*)

10. Which of the following is not correct for electromagnetic waves?

(*a*) the wavelength is the distance between two successive crests
(*b*) the frequency is the number of waves which pass a given point in one second
(*c*) the velocity of a wave is the distance covered by the particular wave in one second
(*d*) all electromagnetic waves have equal wavelengths

Answer. (*d*)

11. Which of the following relations is not correct?

(*a*) $c = \lambda \times \nu$ (*b*) $\bar{\nu} = \frac{1}{\nu}$
(*c*) $\bar{\nu} = \frac{1}{\lambda}$ (*d*) $\lambda = c \div \nu$

Answer. (*b*)

12. The unit in which wave number is measured

(*a*) hertz (*b*) sec^{-1}
(*c*) nanometer (*d*) cm^{-1}

Answer. (*d*)

13. In the spectrum of hydrogen atom, the series which falls in ultraviolet region is

(*a*) Lyman series (*b*) Balmer series
(*c*) Paschen series (*d*) Brackett series

Answer. (*a*)

14. The Balmer series in the spectrum of hydrogen atom falls in

(*a*) ultraviolet region (*b*) visible region
(*c*) infrared region (*d*) none of these

Answer. (*b*)

15. The energy of a photon is given by the relation

(*a*) $E = \frac{h\nu}{\lambda}$ (*b*) $E = \frac{hc}{\lambda}$
(*c*) $E = \frac{h \times c}{\nu}$ (*d*) $E = \frac{\lambda \times c}{h}$

Answer. (*b*)

16. When a beam of light of sufficiently high frequency is allowed to strike a metal surface in vacuum, electrons are ejected from the metal surface. This phenomenon is called

(*a*) Black body radiation (*b*) Photoelectric effect
(*c*) Zeeman effect (*d*) Stark effect

Answer. (*b*)

17. In photoelectric effect, the kinetic energy of the photoelectrons increases linearly with the

(*a*) wavelength of the incident light (*b*) frequency of the incident light
(*c*) velocity of the incident light (*d*) none of these

Answer. (*b*)

18. The kinetic energy of the photoelectrons emitted from the metal surface is given by the relation (v_o is the threshold frequency and v is the frequency of the incident light)

(a) $\frac{1}{2} m v^2 = hv - hv_o$ (b) $\frac{1}{2} mv^2 = hv + hv_o$

(c) $\frac{1}{2} mv^2 = hv$ (d) $\frac{1}{2} mv^2 = hv_o$

Answer. (a)

19. In Bohr's model of atom, the angular momentum of an electron orbiting around the nucleus is given by the relation.

(a) $m v r = \frac{h}{2\pi}$ (b) $m v r = \frac{nh}{2\pi}$

(c) $m v r = \frac{n^2 h^2}{4\pi}$ (d) $m v r = \frac{nh}{4\pi}$

Answer. (b)

20. The radius of first orbit in hydrogen atom according to Bohr's Model is given by the relation

(a) $r = \frac{h^2}{4\pi^2 m e^2}$ (b) $r = \frac{h}{4\pi^2 m e^2}$

(c) $r = \frac{h^2}{4\pi m e^2}$ (d) $r = \frac{h^2}{4\pi m e^4}$

Answer. (a)

21. The radius of first orbit in hydrogen atom is 5.29×10^{-11} m. The radius of second orbit is given by

(a) $\frac{1}{2} \times 5.29 \times 10^{-11}$ m (b) $2 \times 5.29 \times 10^{-11}$ m

(c) $4 \times 5.29 \times 10^{-11}$ m (d) $8 \times 5.29 \times 10^{-11}$ m

Answer. (c)

22. The energy of an electron in the first orbit in hydrogen atom is -1312.980 kJ mol^{-1}. The energy of the electron in 3rd orbit is given by the relation

(a) $E_3 = \frac{-1312.980}{3}$ kJ mol^{-1} (b) $E_3 = \frac{-1312.980}{2}$ kJ mol^{-1}

(c) $E_3 = \frac{-1312.980}{9}$ kJ mol^{-1} (d) $E_3 = 1312.980 \times 3$ kJ mol^{-1}

Answer. (c)

23. Lyman series is obtained when the electrons from higher energy levels return to

(a) 1st orbit (b) 2nd orbit

(c) 3rd orbit (d) 4th orbit

Answer. (a)

24. A line in Pfund series is obtained when an electron from higher energy levels returns to

(a) 1st orbit (b) 3rd orbit

(c) 5th orbit (d) 6th orbit

Answer. (c)

25. The energy of an electron in Bohr's atom _______ as we move away from the nucleus

(a) remains the same (b) decreases

(c) increases (d) sometimes increases, sometimes decreases

Answer. (c)

26. When an electron drops from a higher energy level to a lower energy level, then

(a) the energy is absorbed (b) the energy is released

(c) the nuclear charge increases (d) the nuclear charge decreases

Answer. (b)

27. The spectrum of hydrogen atom is similar to that of

(a) H^+ ion (b) He^+ ion

(c) Li^+ ion (d) Na^+ ion

Answer. (b)

28. If r is the radius of first orbit, the radius of nth orbit of hydrogen atom will be

(*a*) $n^2 r$ (*b*) $n r$

(*c*) n/r (*d*) r/n

Answer. (*a*)

29. The ratio of radii of second and first orbit of hydrogen atom according to Bohr's model is

(*a*) 2:1 (*b*) 1:2

(*c*) 4:1 (*d*) 1:4

Answer. (*c*)

30. The spectrum of helium is expected to be similar to that of

(*a*) H-atom (*b*) Li atom

(*c*) Li^+ ion (*d*) Na^+ ion

Answer. (*c*)

31. Electromagnetic radiations with minimum wavelength is

(*a*) ultraviolet (*b*) X-rays

(*c*) infrared (*d*) radiowaves

Answer. (*b*)

32. Which of the following statements is false?

(*a*) electrons travel around the nucleus in specific permitted circular orbits

(*b*) an electron does not lose energy as long as it moves in its specified orbits

(*c*) an electron can jump from one energy level to another by absorbing or losing energy

(*d*) the angular momentum of an electron is not quantised

Answer. (*d*)

33. The idea of stationary orbits was first given by

(*a*) Rutherford (*b*) J.J. Thomson

(*c*) Niels Bohr (*d*) Max Planck

Answer. (*c*)

34. The maximum number of electrons that can be accommodated in an orbit is

(*a*) $2n$ (*b*) n^2

(*c*) $2n^2$ (*d*) $2n + 1$

Answer. (*c*)

35. The maximum number of electrons in the outermost orbit is

(*a*) 2 (*b*) 8

(*c*) 18 (*d*) 32

Answer. (*b*)

36. When the source emitting lines is placed in a strong magnetic field the spectral lines are split into its components. This effect is called

(*a*) Compton effect (*b*) Zeeman effect

(*c*) Rydberg effect (*d*) Photoelectric effect

Answer. (*b*)

37. The number of electrons in the outermost shell of Potassium (at. no. 19) is

(*a*) 1 (*b*) 2

(*c*) 8 (*d*) 9

Answer. (*a*)

38. An atom of silicon with atomic number 14 has the following number of electrons in the outermost shell

(*a*) 1 (*b*) 2

(*c*) 4 (*d*) 8

Answer. (*c*)

39. Inert gases possess the most stable electronic configuration as they contain

(*a*) fully filled outermost shell
(*b*) half filled outermost shell
(*c*) two electrons in the outermost shell
(*d*) eight electrons in the outermost shell

Answer. (*d*)

40. The effect of electric field on the spectra of atoms is called

(*a*) Compton effect
(*b*) Photoelectric effect
(*c*) Stark effect
(*d*) Zeeman effect

Answer. (*c*)

41. Which one of the following species has the same number of electrons as an atom of Neon?

(*a*) O^{2-}
(*b*) Na
(*c*) Mg
(*d*) K^+

Answer. (*a*)

42. The energy of an electron in the first Bohr orbit for hydrogen is

(*a*) 2.17898×10^{-18} Joule
(*b*) -2.17898×10^{-18} Joule
(*c*) 2.17898×10^{-19} Joule
(*d*) -2.17898×10^{-19} Joule

Answer. (*b*)

43. The energy of hydrogen atom in its ground state is -2.17898×10^{-18} Joule. The energy of the level corresponding to $n = 3$ is

(*a*) -7.258×10^{-19} J
(*b*) -3.629×10^{-19} J
(*c*) -2.419×10^{-19} J
(*d*) none of these

Answer. (*c*)

44. $E_n = -1311.8$ kJ mol^{-1}. If the value of E is -52.44 kJ mol^{-1}, to which value 'n' corresponds?

(*a*) 2
(*b*) 3
(*c*) 4
(*d*) 5

Answer. (*d*)

45. The spectral line lies in the Lyman series. It corresponds to the radiation emitted by an electron jumping from higher energy states to

(*a*) first energy state
(*b*) second energy state
(*c*) third energy state
(*d*) fifth energy state

Answer. (*a*)

46. The ground state of an atom corresponds to a state of

(*a*) maximum energy
(*b*) minimum energy
(*c*) zero energy
(*d*) negative energy

Answer. (*b*)

47. Balmer series in the spectrum of hydrogen atom lies in

(*a*) ultraviolet region
(*b*) visible region
(*c*) infrared region
(*d*) none of these

Answer. (*b*)

48. The spectrum of H-atom is expected to be similar to that of

(*a*) Li^+
(*b*) Na^+
(*c*) He^+
(*d*) K^+

Answer. (*c*)

49. An atom of Calcium (at. no. 20) contains _______ electrons in the third energy level.

(*a*) 2
(*b*) 8
(*c*) 10
(*d*) 18

Answer. (*b*)

50. Out of the following pairs of elements which has the same number of electrons in the outer most energy level?

(*a*) helium and lithium
(*b*) boron and carbon
(*c*) carbon and nitrogen
(*d*) lithium and hydrogen

Answer. (*d*)

2 Structure of Atom —Wave Mechanical Approach

CONTENTS

WAVE MECHANICAL CONCEPT OF ATOM

Bohr, undoubtedly, gave the first quantitative successful model of the atom. But now it has been superseded completely by the modern *Wave Mechanical Theory.* The new theory rejects the view that electrons move in closed orbits, as was visualized by Bohr. The Wave mechanical theory gave a major breakthrough by suggesting that the electron motion is of a complex nature best described by its wave properties and probabilities.

While the classical 'mechanical theory' of matter considered matter to be made of discrete particles (atoms, electrons, protons etc.), another theory called the 'Wave theory' was necessary to interpret the nature of radiations like X-rays and light. According to the wave theory, radiations as X-rays and light, consisted of continuous collection of waves travelling in space.

The wave nature of light, however, failed completely to explain the photoelectric effect *i.e.* the emission of electron from metal surfaces by the action of light. In their attempt to find a plausible explanation of radiations from heated bodies as also the photoelectric effect, Planck and Einstein (1905) proposed that energy radiations, including those of heat and light, are emitted discontinuously as little 'bursts', quanta, or photons. This view is directly opposed to the wave theory of light and it gives particle-like properties to waves. **According to it, light exhibits both a wave and a particle nature, under suitable conditions.** This theory which applies to all radiations, is often referred to as the **'Wave Mechanical Theory'.**

With Planck's contention of light having wave and particle nature, the distinction between particles and waves became very hazy. In 1924 Louis de Broglie advanced a complimentary hypothesis for material particles. According to it, the dual character–the wave and particle–may not be confined to radiations alone but should be extended to matter as well. In other words, matter also possessed particle as well as wave character. This gave birth to the **'Wave mechanical theory of matter'.** This theory postulates that electrons, protons and even atoms, when in motion, possessed wave properties and could also be associated with other characteristics of waves such as wavelength, wave-amplitude and frequency. **The new quantum mechanics, which takes into account the particulate and wave nature of matter, is termed the Wave mechanics.**

de BROGLIE'S EQUATION

de Broglie had arrived at his hypothesis with the help of Planck's Quantum Theory and Einstein's Theory of Relativity. He derived a relationship between the magnitude of the wavelength associated with the mass '*m*' of a moving body and its velocity. According to Planck, the photon energy '*E*' is given by the equation

$$E = h\nu \qquad ...(i)$$

where h is Planck's constant and ν the frequency of radiation. By applying Einstein's mass-energy relationship, the energy associated with photon of mass '*m*' is given as

$$E = mc^2 \qquad ...(ii)$$

where c is the velocity of radiation

Comparing equations (*i*) and (*ii*)

$$mc^2 = h\nu = h\frac{c}{\lambda} \qquad \left(\because \nu = \frac{c}{\lambda}\right)$$

or $$mc = \frac{h}{\lambda} \qquad ...(iii)$$

or $$\text{mass} \times \text{velocity} = \frac{h}{\text{wavelength}}$$

or $$\text{momentum } (p) = \frac{h}{\text{wavelength}}$$

or $$\text{momentum} \propto \frac{1}{\text{wavelength}}$$

The equation (*iii*) is called **de Broglie's equation** and may be put in words as : **The momentum of a particle in motion is inversely proportional to wavelength, Planck's constant '*h*' being the constant of proportionality.**

The wavelength of waves associated with a moving material particle (matter waves) is called **de Broglie's wavelength.** The de Broglie's equation is true for all particles, but it is only with very small particles, such as electrons, that the wave-like aspect is of any significance. Large particles in motion though possess wavelength, but it is not measurable or observable. Let us, for instance consider de Broglie's wavelengths associated with two bodies and compare their values.

(*a*) For a large mass

Let us consider a stone of mass 0.1 kg moving with a velocity of 10 m/sec. The de Broglie's wavelength λ will be given as follows :

$$\lambda = \frac{6.6256 \times 10^{-34}}{0.1 \times 10} \qquad \left(\lambda = \frac{h}{\text{momentum}}\right)$$

$$= 6.63 \times 10^{-34} \text{ m}$$

This is too small to be measurable by any instrument and hence no significance.

(*b*) For a small mass

Let us now consider an electron in a hydrogen atom. It has a mass = 9.1091×10^{-31} kg and moves with a velocity 2.18×10^{6} m s^{-1}. The de Broglie's wavelength λ is given as

$$\lambda = \frac{6.63 \times 10^{-34}}{9.1091 \times 10^{-31} \times 2.18 \times 10^{6}}$$

$$= 3.32 \times 10^{-10} \text{ m}$$

This value is quite comparable to the wavelength of X-rays and hence detectable.

It is, therefore, reasonable to expect from the above discussion that **everything in nature possesses both the properties of particles (or discrete units) and also the properties of waves (or continuity).** The properties of large objects are best described by considering the particulate aspect while properties of waves are utilized in describing the essential characteristics of extremely small objects beyond the realm of our perception, such as electrons.

THE WAVE NATURE OF ELECTRON

de Broglie's revolutionary suggestion that moving electrons had waves of definite wavelength associated with them, was put to the acid test by Davison and Germer (1927). They demonstrated the physical reality of the wave nature of electrons by showing that a beam of electrons could also be *diffracted* by crystals just like light or X-rays. They observed that the diffraction patterns thus obtained were just similar to those in case of X-rays. It was possible that electrons by their passage through crystals may produce secondary X-rays, which would show diffraction effects on the screen. Thomson ruled out this possibility, showing that the electron beam as it emerged from the crystals, underwent deflection in the electric field towards the positively charged plate.

Davison and Germers Experiment

In their actual experiment, Davison and Germer studied the scattering of slow moving electrons by reflection from the surface of nickel crystal. They obtained electrons from a heated filament and passed the stream of electrons through charged plates kept at a potential difference of V *esu*. Due to the electric field of strength $V \times e$ acting on the electron of charge e, the electrons emerge out with a uniform velocity v units. The kinetic energy $\frac{1}{2}mv^2$ acquired by an electron due to the electric field shall be equal to the electrical force. Thus,

$$\tfrac{1}{2}mv^2 = Ve$$

or

$$v = \sqrt{\frac{2Ve}{m}}$$

Multiplying by m on both sides,

$$mv = m\sqrt{\frac{2Ve}{m}} = \sqrt{2mVe} \qquad ...(i)$$

But according to de Broglie's relationship

$$mv = \frac{h}{\lambda} \qquad ...(ii)$$

Comparing (*i*) and (*ii*)

$$\frac{h}{\lambda} = \sqrt{2mVe}$$

$$\therefore \quad \lambda = \frac{h}{\sqrt{2mVe}} = \sqrt{\frac{h^2}{2mVe}}$$

Substituting for $h = 6.6256 \times 10^{-34}$ J s, $m = 9.1 \times 10^{-31}$ kg, $e = 1.602 \times 10^{-19}$ coulomb and changing V *esu* to V volts by using the conversion factor $\frac{1}{3} \times 10^{-2}$, we have

$$\lambda = \sqrt{\frac{\left(6.6256 \times 10^{-34}\right)^2 \times 0.33 \times 10^{-2}}{2 \times 9.1 \times 10^{-31} \times 1.602 \times 10^{-19}\ V \text{ volts}}}$$

$$= \sqrt{\frac{14.4865299 \times 10^{-70}}{29.1564 \times 10^{-50}}}$$

$$= \sqrt{0.49685592 \times 10^{-20}}$$

$$= 0.705 \times 10^{-10} \text{ m}$$

If a potential difference of 150 volts be applied, the wavelength of electrons emerging out is $\lambda = 1$ Å (0.1×10^{-10} m). Similarly if a potential difference of 1500 volts be created, the electrons coming out shall have a wavelength 0.1 Å (0.1×10^{-10} m). It is clear, therefore, that electrons of different wavelengths can be obtained by changing the potential drop. These wavelengths are comparable with those of X-rays and can undergo diffraction.

$$\sqrt{\frac{150}{V \text{ volts}}} \times 10^{-8} \text{ cm} = \sqrt{\frac{150}{V \text{ volts}}} \text{Å} \qquad ...(iii)$$

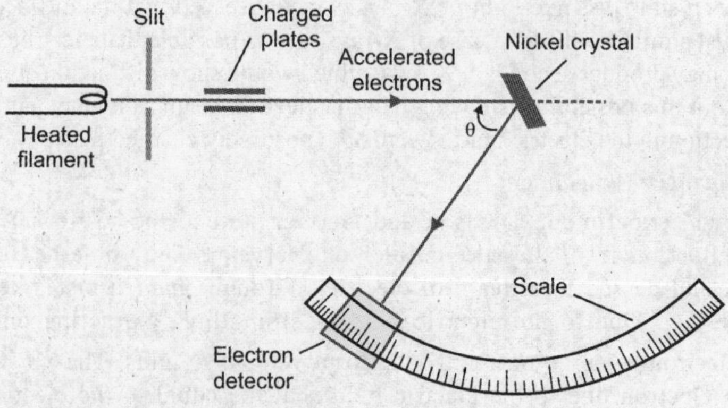

Figure 2.1
Schematic representation of the apparatus used by Davison and Germer.

The electrons when they fall upon the nickel crystal, get diffracted. Electrons of a definite wavelength get diffracted along definite directions. The electron detector measures the angle of diffraction (say θ) on the graduated circular scale. According to Bragg's diffraction equation, the wavelength λ of the diffracted radiation is given by $\lambda = d \sin \theta$, where d is a constant (= 2.15 for Ni crystal) and θ the angle of diffraction. By substituting the experimental value of θ in Bragg's equation ($\lambda = d \sin \theta$), the wavelength of electrons may be determined. This wavelength would be found to agree with the value of λ, as obtained from equation (*iii*).

Since diffraction is a property exclusively of wave motion, Davison and Germer's *electron diffraction* experiment established beyond doubt the wave nature of electrons. We have described earlier in this chapter that electrons behave like particles and cause mechanical motion in a paddle wheel placed in their path in the discharge tube. This proves, therefore, that electrons not only behave like 'particles' in motion but also have 'wave properties' associated with them. It is not easy at this stage to obtain a pictorial idea of this new conception of the motion of an electron. But the application of de Broglie's equation to Bohr's theory produces an important result. The quantum restriction of Bohr's theory for an electron in motion in the circular orbit is that the angular momentum (mvr) is an integral multiple (n) of $h/2\pi$. That is,

$$mvr = n \frac{h}{2\pi} \qquad \text{...Bohr Theory}$$

On rearranging, we get

$$2\pi r = n \frac{h}{mv}$$

Putting the value of $\frac{h}{mv}$ from equation (*i*), we have

$$2\pi r = n\lambda \qquad \left(\because \lambda = \frac{h}{mv}\right)$$

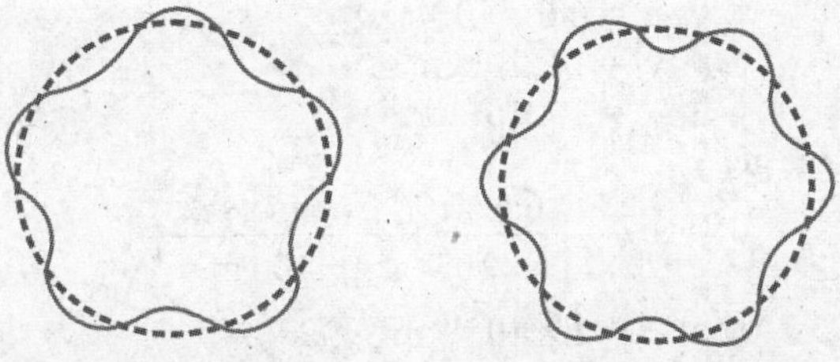

Figure 2.2
de Broglie's wave accommodated in Bohr's orbits. For these two wave trains the value of *n* is different.

Now the electron wave of wavelength λ can be accommodated in Bohr's orbit only if the circumference of the orbit, $2\pi r$, is an integral multiple of its wavelength. Thus de Broglie's idea of standing electron waves stands vindicated. However, if the circumference is bigger, or smaller than $n\lambda$, the wave train will go out of phase and the destructive interference of waves causes radiation of energy.

SOLVED PROBLEM 1

Calculate the wavelength of an electron having kinetic energy equal to 4.55×10^{-25} J. ($h = 6.6 \times 10^{-34}$ kg m^2 sec^{-1} and mass of electron = 9.1×10^{-31} kg).

SOLUTION

$$\text{Kinetic energy of an electron} = \frac{1}{2} mv^2$$
$$= 4.55 \times 10^{-25} \text{ J (given)}$$
$$= 4.55 \times 10^{-25} \text{ kg m}^2 \text{ s}^{-2}$$

or $$v^2 = \frac{2 \times 4.55 \times 10^{-25}}{m}$$

$$= \frac{2 \times 4.55 \times 10^{-25} \text{ kg m}^2 \text{ s}^{-2}}{9.1 \times 10^{-31} \text{ kg}}$$

or $$v^2 = 1 \times 10^6 \text{ m}^2 \text{ s}^{-2}$$

or $$v = 1 \times 10^3 \text{ m s}^{-1}$$

We know $$\lambda = \frac{h}{m \times v} \text{ (de Broglie equation)}$$

$$= \frac{6.6 \times 10^{-34} \text{ kg m}^2 \text{ s}^{-1}}{(9.1 \times 10^{-31} \text{ kg}) \times (1 \times 10^3 \text{ m s}^{-1})}$$

$$= 7.25 \times 10^{-7} \text{ m}$$

$$= 7.25 \times 10^{-7} \times 10^9 \text{ nm}$$

$$= \mathbf{725 \text{ nm}}$$

SOLVED PROBLEM 2

Calculate the wavelength of an α particle having mass 6.6×10^{-27} kg moving with a speed of 10^3 m s^{-1} ($h = 6.6 \times 10^{-34}$ kg m^2 s^{-1})

SOLUTION. We know $$\lambda = \frac{h}{mv} \text{ (de Broglie equation)}$$

Given
$$h = 6.6 \times 10^{-34} \text{ kg m}^2 \text{ s}^{-1}$$
$$m = 6.6 \times 10^{-27} \text{ kg}$$
$$v = 1 \times 10^5 \text{ cm s}^{-1}$$
$$= 1 \times 10^3 \text{ m s}^{-1}$$

On substitution, we get

$$\lambda = \frac{6.6 \times 10^{-34} \text{ kg m}^2 \text{ s}^{-1}}{6.6 \times 10^{-27} \text{ kg} \times 10^3 \text{ m s}^{-1}}$$

$$= \mathbf{1 \times 10^{-10} \text{ m}}$$

HEISENBERG'S UNCERTAINTY PRINCIPLE

One of the most important consequences of the dual nature of matter is the uncertainty principle developed by Werner Heisenberg in 1927. This principle is an important feature of wave mechanics and discusses the relationship between a pair of *conjugate properties* (those properties that are independent) of a substance. According to the uncertainty principle, it is impossible to know simultaneously both the conjugate properties accurately. For example, the position and momentum of a moving particle are interdependent and thus conjugate properties also. Both the position and the momentum of the particle at any instant cannot be determined with absolute exactness or certainty. If the momentum (or velocity) be measured very accurately, a measurement of the position of the particle correspondingly becomes less precise. On the other hand if position is determined with accuracy or precision, the momentum becomes less accurately known or uncertain. **Thus certainty of determination of one property introduces uncertainty of determination of the other.** The uncertainty in measurement of position, Δx, and the uncertainty of determination of momentum, Δp (or Δmv), are related by Heisenberg's relationship as

$$\Delta x \times \Delta p \geq \frac{h}{2\pi}$$

or $$\Delta x \times m\,\Delta v \geq \frac{h}{2\pi}$$

where h is Planck's constant.

It may be pointed out here that there exists a clear difference between the behavior of large objects like a stone and small particles such as electrons. **The uncertainty product is negligible in case of large objects.**

For a moving ball of iron weighing 0.5 kg, the uncertainty expression assumes the form

$$\Delta x \times m\,\Delta v \geq \frac{h}{2\pi}$$

or $$\Delta x \times \Delta v \geq \frac{h}{2\pi m}$$

$$\geq \frac{6.625 \times 10^{-34}}{2 \times 3.14 \times 0.5} \approx 5 \times 10^{-38} \text{ Js kg}^{-1}$$

which is very small and thus negligible. Therefore for large objects, the uncertainty of measurements is practically nil.

But for an electron of mass $m = 9.109 \times 10^{-31}$ kg, the product of the uncertainty of measurements is quite large as

$$\Delta x \times \Delta v \geq \frac{h}{2\pi m}$$

$$\geq \frac{6.625 \times 10^{-34}}{2 \times 3.14 \times 9.109 \times 10^{-31}} \approx 0.11581253 \times 10^{-3}$$

$$= 115.813 \times 10^{-6} \text{ J s kg}^{-1}$$

This value is large enough in comparison with the size of the electron and is thus in no way negligible. If position is known quite accurately *i.e.*, Δx is very small, the uncertainty regarding velocity Δv becomes immensely large and *vice versa*. It is therefore very clear that **the uncertainty principle is only important in considering measurements of small particles comprising an atomic system.**

Physical Concept of Uncertainty Principle

The physical concept of uncertainty principle becomes illustrated by considering an attempt to measure the position and momentum of an electron moving in Bohr's orbit. To locate the position of the electron, we should devise an instrument 'super microscope' to see the electron. A substance is said to be seen only if it could reflect light or any other radiation from its surface. Because the size of the electron is too small, its position at any instant may be determined by a super microscope employing light of very small wavelength (such as X-rays or γ-rays). A photon of such a radiation of small λ, has a great energy and therefore has quite large momentum. As one such photon strikes the electron and is reflected, it instantly changes the momentum of electron. Now the momentum gets changed and becomes more uncertain as the position of the electron is being determined (Fig. 2.3). Thus it is impossible to determine the exact position of an electron moving with a definite velocity (or possessing definite energy). It appears clear that the Bohr's picture of an electron as moving in an orbit with fixed velocity (or energy) is completely untenable.

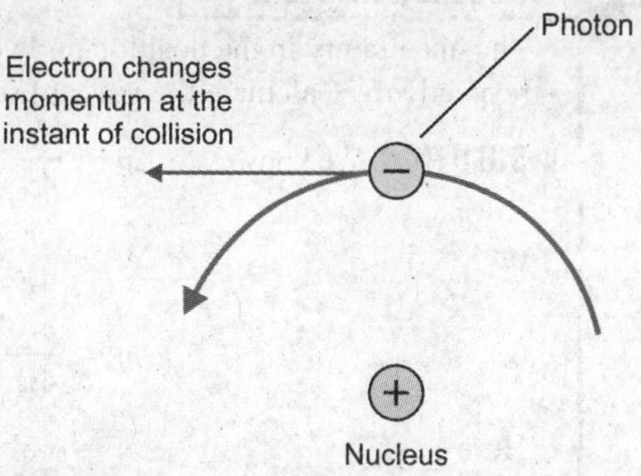

■ **Figure 2.3**
The momentum of the electron changes when a photon of light strikes it, so does its position.

As it is impossible to know the position and the velocity of any one electron on account of its small size, the best we can do is to speak of the probability or *relative chance* of finding an electron with a probable velocity. **The old classical concept of Bohr has now been discarded in favour of the probability approach.**

SOLVED PROBLEM 3

Calculate the uncertainty in position of an electron if the uncertainty in velocity is 5.7×10^5 m s^{-1}.

SOLUTION. According to Heisenberg's uncertainty principle

$$\Delta x \times \Delta p = \frac{h}{4\pi}$$

or
$$\Delta x \times m\,\Delta v = \frac{h}{4\pi}$$

or
$$\Delta x = \frac{h}{4\pi \mathrm{m} \times \Delta v}$$

Here
$$\Delta v = 5.7 \times 10^5 \text{ m s}^{-1}$$
$$h = 6.6 \times 10^{-34} \text{ kg } m^2 \text{ s}^{-1}$$
$$m = 9.1 \times 10^{-31} \text{ kg}$$

On substitution we get

$$\Delta x = \frac{6.6 \times 10^{-34} \text{ kg m}^2 \text{ s}^{-1}}{4 \times 3.14 \times (9.1 \times 10^{-31} \text{ kg})(5.7 \times 10^5 \text{ m s}^{-1})}$$

$$= \frac{6.6 \times 10^{-8}}{4 \times 3.14 \times 9.1 \times 5.7} m$$

$$= \mathbf{1 \times 10^{-10} \text{ m}}$$

SOLVED PROBLEM 4

The uncertainty in the position and velocity of a particle are 10^{-10} m and 5.27×10^{-24} m s^{-1} respectively. Calculate the mass of the particle.

SOLUTION. We know $\Delta x \times \Delta p = \dfrac{h}{4\pi}$

or
$$\Delta x \times m\,\Delta v = \frac{h}{4\pi}$$

or
$$m = \frac{h}{4\pi \times \Delta x \times \Delta v}$$

Here
$$h = 6.6 \times 10^{-34} \text{ kg m}^2 \text{ s}^{-1}$$
$$\Delta x = 1 \times 10^{-10} \text{ m}$$
$$\Delta v = 5.27 \times 10^{-24} \text{ m s}^{-1}$$

Substituting the values, we get

$$m = \frac{6.6 \times 10^{-34} \text{ kg m}^2 \text{ s}^{-1}}{4 \times 3.14 \times (1 \times 10^{-10} \text{ m})(5.27 \times 10^{-24} \text{ m s}^{-1})}$$

$$= 0.10 \text{ kg}$$

$$= \mathbf{100 \text{ g}}$$

SCHRÖDINGER'S WAVE EQUATION

In order to provide sense and meaning to the probability approach, Schrödinger derived an equation known after his name as **Schrödinger's Wave Equation.** Calculation of the probability of finding the electron at various points in an atom was the main problem before Schrödinger. **His equation is the keynote of wave mechanics and is based upon the idea of the electron as 'standing wave' around the nucleus.** The equation for the standing wave*, comparable with that of a stretched string is

$$\psi = A \sin 2\pi \frac{x}{\lambda} \qquad ...(a)$$

where ψ (pronounced as *sigh*) is a mathematical function representing the amplitude of wave (called *wave function*) x, the displacement in a given direction, and λ, the wavelength and A is a constant.

By differentiating equation (*a*) twice with respect to x, we get

$$\frac{d\psi}{dx} = A\frac{2\pi}{\lambda} \cos 2\pi \frac{x}{\lambda} \qquad ...(1)$$

and $$\frac{d^2\psi}{dx^2} = -A\frac{4\pi^2}{\lambda^2} \sin 2\pi \frac{x}{\lambda} \qquad ...(2)$$

But $$A \sin 2\pi \frac{x}{\lambda} = \psi$$

$\therefore$ $$\frac{d^2\psi}{dx^2} = -\frac{4\pi^2}{\lambda^2}\psi \qquad ...(3)$$

The K.E. of the particle of mass m and velocity v is given by the relation

$$\text{K.E.} = \frac{1}{2}mv^2 = \frac{1}{2}\frac{m^2v^2}{m} \qquad ...(4)$$

According to Broglie's equation

$$\lambda = \frac{h}{mv}$$

or $$\lambda^2 = \frac{h^2}{m^2v^2}$$

or $$m^2v^2 = \frac{h^2}{\lambda^2}$$

Substituting the value of m^2v^2, we have

$$\text{K.E.} = \frac{1}{2} \times \frac{h^2}{m\lambda^2} \qquad ...(5)$$

From equation (3), we have

$$\lambda^2 = \frac{-4\pi^2\psi}{\dfrac{d^2\psi}{dx^2}} \qquad ...(6)$$

Substituting the value of λ^2 in equation (5)

$$\text{K.E.} = -\frac{1}{2m}.\frac{h^2}{4\pi^2\psi}.\frac{d^2\psi}{dx^2}$$

$$= -\frac{h^2}{8\pi^2 m\psi}.\frac{d^2\psi}{dx^2}$$

* *For the derivation of equation for a 'standing wave' in a stretched string, the reader may refer to a book on Physics (Sound).*

The total energy E of a particle is the sum of kinetic energy and the potential energy

i.e., $$E = \text{K.E.} + \text{P.E.}$$

or $$\text{K.E.} = E - \text{P.E.}$$

$$= -\frac{h^2}{8\pi^2 m\psi}\cdot\frac{d^2\psi}{dx^2}$$

or $$\frac{d^2\psi}{dx^2} = -\frac{8\pi^2 m}{h^2}(E - \text{P.E.})\,\psi$$

$$\frac{d^2\psi}{dx^2} + \frac{8\pi^2 m}{h^2}(E - \text{P.E.})\,\psi = 0$$

This is Schrödinger's equation in one dimension. It need be generalized for a particle whose motion is described by three space coordinates x, y and z. Thus,

$$\frac{d^2\psi}{dx^2} + \frac{d^2\psi}{dy^2} + \frac{d^2\psi}{dz^2} + \frac{8\pi^2 m}{h^2}(E - \text{P.E.})\psi = 0$$

This equation is called the **Schrödinger's Wave Equation.** The first three terms on the left-hand side are represented by $\Delta^2\psi$ (pronounced as del-square sigh).

$$\Delta^2\psi + \frac{8\pi^2 m}{h^2}(E - \text{P.E.})\,\psi = 0$$

Δ^2 is known as **Laplacian Operator**.

The Schrödinger's wave equation is a second degree differential equation. It has several solutions. Some of these are imaginary and are not valid. If the potential energy term is known, the total energy E and the corresponding wave function ψ can be evaluated.

The wave function is always finite, single valued and continuous. It is zero at infinite distance. Solutions that meet these requirements are only possible if E is given certain characteristic values called **Eigen-values.** Corresponding to these values of E, we have several characteristic values of wave function ψ and are called **Eigen-functions.** As the eigen-values correspond very nearly to the energy values associated with different Bohr-orbits, the Bohr's model may be considered as a direct consequence of wave mechanical approach.

Significance of ψ and ψ^2

In Schrödinger's wave equation ψ represents the amplitude of the spherical wave. According to the theory of propagation of light and sound waves, the square of the amplitude of the wave is proportional to the intensity of the sound or light. A similar concept, modified to meet the requirement of uncertainty principle, has been developed for the physical interpretation of wave function ψ. This may be stated as the probability of finding an electron in an extremely small volume around a point. It is proportional to the square of the function ψ^2 at that point. If wave function ψ is imaginary, $\psi\psi^*$ becomes a real quantity where ψ^* is a complex conjugate of ψ. This quantity represents the probability ψ^2 as a function of x, y and z coordinates of the system, and it varies from one space region to another. Thus the probability of finding the electron in different regions is different. This is in agreement with the uncertainty principle and gave a death blow to Bohr's concept.

In Schrödinger's Wave Equation, **the symbol ψ represents the amplitude of the spherical wave.** For hydrogen atom, Schrödinger's Wave Equation gives the wave function of the electron (with energy $= -2.18 \times 10^{-18}$ Joules) situated at a distance 'r',

$$\psi = C_1 e - C_2 r$$

where C_1 and C_2 are constants. The square of the amplitude ψ^2 is proportional to the density of the wave. **The wave of energy or the cloud of negative charge is denser in some parts than in others.** Max Born interpreted the wave equations on the basis of probabilities. Even if an electron be considered as a particle in motion around the nucleus, the wave equation may be interpreted in terms of probability or relative chance of finding the electron at any given distance from the nucleus. The space characteristic of an electron is best described in terms of distribution function given by

$$D = 4\pi r^2 \psi^2$$

The numerical value of 'D' denotes the probability or chance of finding the electron in a shell of radius r and thickness dr, or of volume $4\pi r^2\, dr$. Substituting for ψ we have,

$$D = 4\pi r^2 (C_1 e - C_2 r)^2$$

The probability of finding the electron is clearly a function of 'r'. When $r = 0$ or $\propto$, the probability function D becomes equal to zero. In other words, there is no probability of finding the electron at the nucleus or at infinity. However, it is possible to choose a value of r such that there is 90-95 percent chance of finding the electron at this distance. For the hydrogen atom, this distance is equal to 0.53×10^{-10} m or 0.53 Å. If the probability distribution be plotted against the distance r from the nucleus, the curve obtained is shown in Fig. 2.4. The probability distribution is maximum at the distance 0.53×10^{-10} m and spherically symmetrical. This distance corresponds to Bohr's first radius a_0. The graph can be interpreted as representing a contour that encloses a high-percentage of charge.

When the electron gets excited and it is raised from n to higher energy levels (say $n = 2$ or $n = 3$), the solution of wave equation gives sets of value of ψ^2 which give different shapes to the space distribution of the electron.

CHARGE CLOUD CONCEPT AND ORBITALS

The Charge Cloud Concept finds its birth from wave mechanical theory of the atom. The wave equation for a given electron, on solving gives a three-dimensional arrangement of points where it can possibly lie. There are regions where the chances of finding the electron are relatively greater. Such regions are expressed in terms of **'cloud of negative charge'**. We need not know the specific location of the electrons in space but are concerned with the negative charge density regions. Electrons in atoms are assumed to be vibrating in space, moving haphazardly but at the same time are constrained to lie in regions of highest probability for most of the time. **The charge cloud concept simply describes the high probability region.**

The three-dimensional region within which there is higher probability that an electron having a certain energy will be found, is called an orbital.

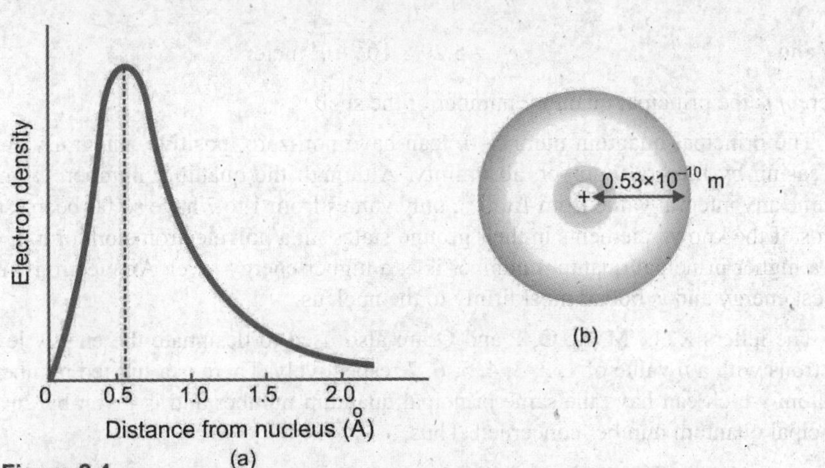

Figure 2.4
Shows the probability distribution of electron cloud :
(a) gives the graphical representation.
(b) depicts cross-section of the cloud.

An orbital is the most probable space in which the electron spends most of its time while in constant motion. In other words, it is the spatial description of the motion of an electron corresponding to a particular energy level. **The energy of electron in an atomic orbital is always the same.**

Each energy level corresponds to a three-dimensional electron wave which envelopes the nucleus. This wave possesses a definite 'size', 'shape' and 'orientation' and thus can be represented pictorially.

QUANTUM NUMBERS

Bohr's electronic energy shells or levels, designated as Principal Quantum Numbers '*n*', could hardly explain the hydrogen spectrum adequately. Spectra of other elements that are quite complex, also remained unexplained by this concept. Many single lines of the spectra are found to consist of a number of closely related lines when studied with the help of sophisticated instruments of high resolving power. Also the spectral lines split up when the source of radiation is placed in a magnetic field (**Zeeman Effect**) or in an electrical field (**Stark Effect**).

To explain these facts, it is necessary to increase the number of 'possible orbits' where an electron can be said to exist within an atom. In other words, it is necessary to allow more possible energy changes within an atom (or a larger number of energy states) to account for the existence of a larger number of such observed spectral lines. Wave mechanics makes a provision for three more states of an electron in addition to the one proposed by Bohr. Like the energy states of Bohr, designated by $n = 1, 2, 3...$, these states are also identified by numbers and specify the position and energy of the electron. Thus there are in all four such identification numbers called **quantum numbers** which fully describe an electron in an atom. Each one of these refers to a particular character.

Principal Quantum Number '*n*'

This quantum number denotes the principal shell to which the electron belongs. This is also referred to as **major energy level.** It represents the average size of the electron cloud *i.e.*, the average distance of the electron from the nucleus. This is, therefore, the main factor that determines the values of nucleus-electron attraction, or the energy of the electron. In our earlier discussion, we have found that the energy of the electron and its distance from the nucleus for hydrogen atom are given by

$$E_n = -\frac{-13.6}{n^2} \text{ eV}$$

and

$$r_n = 5.29 \times 10^{-11} n^2 \text{ meter}$$

where n is the principal quantum number of the shell.

The principal quantum number '*n*' can have non-zero, positive, integral values $n = 1, 2, 3...$ increasing by integral numbers to infinity. Although the quantum number '*n*' may theoretically assume any integral value from 1 to $\propto$, only values from 1 to 7 have so far been established for the atoms of the known elements in their ground states. In a polyelectron atom or ion, the electron that has a higher principal quantum number is at a higher energy level. An electron with $n = 1$ has the lowest energy and is bound most firmly to the nucleus.

The letters K, L, M, N, O, P and Q are also used to designate the energy levels or shells of electrons with a n value of 1, 2, 3, 4, 5, 6, 7 respectively. There is a limited number of electrons in an atom which can have the same principal quantum number and is given by $2n^2$, where n is the principal quantum number concerned. Thus,

Principal quantum number ($n =$)	1	2	3	4
Letter designation	K	L	M	N
Maximum number of electrons ($2n^2 =$)	2	8	18	32

Azimuthal Quantum number '*l*'

This is also called secondary or subsidiary quantum number. It defines the spatial distribution of the electron cloud about the nucleus and describes the angular momentum of the electron. In other words, **the quantum number *l* defines the shape of the orbital occupied by the electron and the angular momentum of the electron.** It is for this reason that '*l*' is sometimes referred to as *orbital* or *angular quantum* number. For any given value of the principal quantum number n, the azimuthal quantum number l may have all integral values from 0 to $n - 1$, each of which refers to an *Energy sublevel* or *Sub-shell.* **The total number of such possible sublevels in each principal level is numerically equal to the principal quantum number of the level under consideration.** These sublevels are also symbolized by letters s, p, d, f etc. For example, for principal quantum number $n = 1$, the only possible value for l is 0 *i.e.*, there is only one possible subshell *i.e.* s-subshell ($n = 1$, $l = 0$). For $n = 2$, there are two possible values of l, $l = 0$ and $l = 2 - 1 = 1$.

This means that there are two subshells in the second energy shell with $n = 2$. These subshells are designated as 2s and 2p. Similarly, when $n = 3$, l can have three values *i.e.* 0, 1 and 2. Thus there are three subshells in third energy shell with designations 3s, 3p and 3d respectively. For $n = 4$, there are four possible values of azimuthal quantum number l (= 0, 1, 2 and 3) each representing a different sublevel. In other words, the fourth energy level consists of four subshells which are designated as 4s, 4p, 4d and 4f. Thus for different values of principal quantum numbers we have

$n = 1$	$n = 2$	$n = 3$	$n = 4$	$n = 5$
$l = 0$ (1s)	$l = 0$ (2s)	$l = 0$ (3s)	$l = 0$ (4s)	$l = 0$ (5s)
	$l = 1$ (2p)	$l = 1$ (3p)	$l = 1$(4p)	$l = 1$ (5p)
		$l = 2$ (3d)	$l = 2$ (4d)	$l = 2$ (5d)
			$l = 3$ (4f)	$l = 3$ (5f)
				$l = 4$ (5g)

For a given value of principal quantum number the order of increasing energy for different subshells is

$$s < p < d < f \qquad \text{(except for H atom)}$$

Magnetic Quantum Number '*m*'

This quantum number has been proposed to account for the splitting up of spectral lines (Zeeman Effect). An application of a strong magnetic field to an atom reveals that electrons with the same values of principal quantum number '*n*' and of azimuthal quantum number '*l*', may still differ in their behavior. They must, therefore, be differentiated by introducing a new quantum number, the magnetic quantum number m. This is also called **Orientation Quantum Number** because it gives the

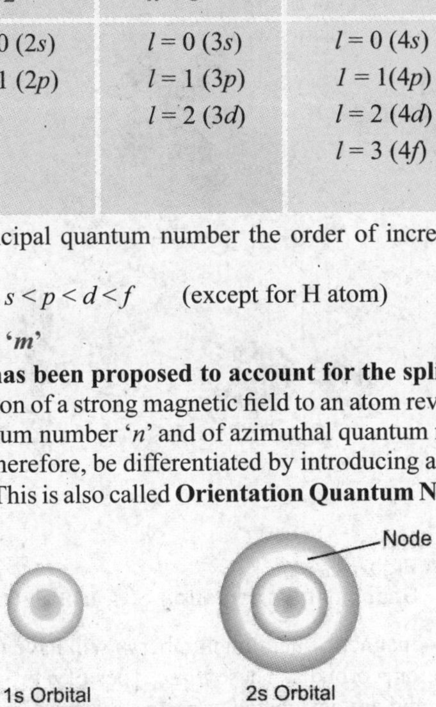

■ **Figure 2.5**
Spherical *s* orbitals, symmetrically disposed about the nucleus.

orientation or distribution of the electron cloud. For each value of the azimuthal quantum number '*l*', the magnetic quantum number m, may assume all the integral values between $+ l$ to $- l$ through zero *i.e.*, $+ l$, $(+ l - l)$,... 0..., $(- l + 1)$, $- l$. Therefore for each value of l there will be $(2l + 1)$ values of m_l.

Thus when $l = 0$, $m = 0$ and no other value. This means that **for each value of principal quantum number '*n*', there is only one orientation for $l = 0$ (*s* orbital) or there is only one *s* orbital.** For *s* orbital, there being only one orientation, it must be spherically symmetrical about the nucleus. There is only one spherically symmetrical orbital for each value of *n* whose radius depends upon the value of *n*.

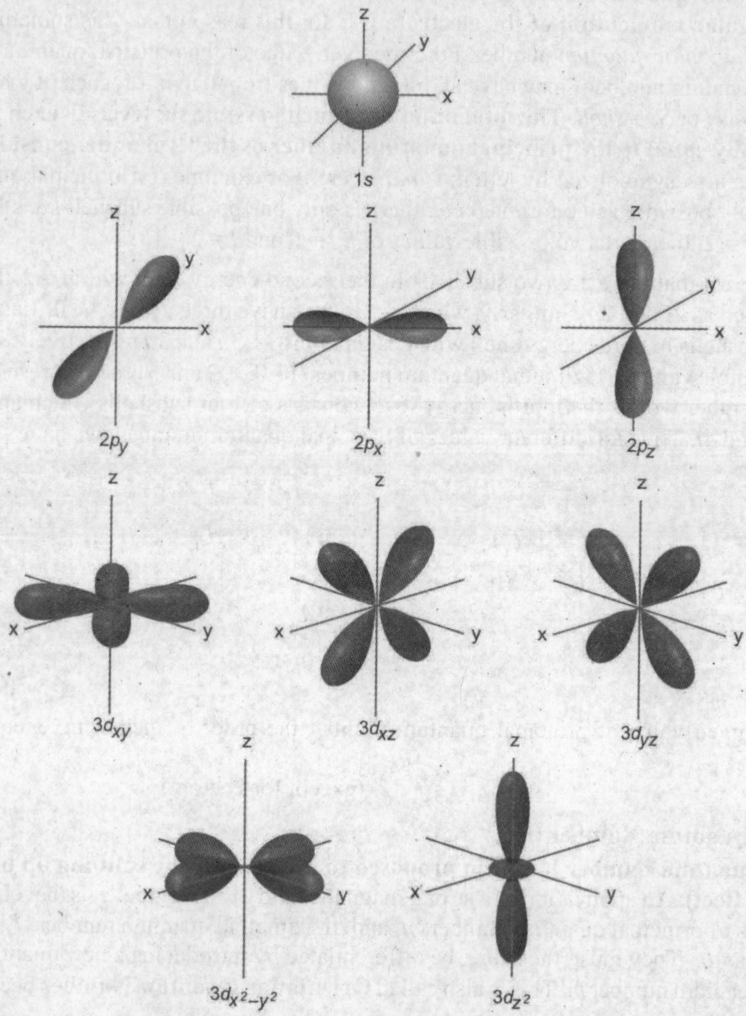

■ Figure 2.6
Shapes and orientation of *s*, *p* and *d* orbitals.

For $l = 1$ (*p* orbital), the magnetic quantum number *m* will have *three* values : + 1, 0 and – 1; so there are *three* orientations for *p* orbitals. These *three* types of *p* orbitals differ only in the value of magnetic quantum number and are designated as p_x, p_y, p_z depending upon the axis of orientation. The subscripts *x*, *y* and *z* refer to the coordinate axes. In the absence of a magnetic field, these three *p* orbitals are equivalent in energy and are said to be **three-fold degenerate** or **triply degenerate*.** In presence of an external magnetic field the relative energies of the three *p* orbitals vary depending upon their orientation or magnetic quantum number. This probably accounts for the existence of more

* *Different orbitals of equivalent energy are called degenerate orbitals and are grouped together.*

spectral lines under the influence of an external magnetic field. The *p* orbitals are of dumb-bell shape consisting of two lobes. The two lobes of a *p* orbital extend outwards and away from the nucleus along the axial line. Thus the two lobes of a *p* orbital may be separated by a plane that contains the nucleus and is perpendicular to the corresponding axis. Such plane is called a **nodal plane. There is no likelihood of finding the electron on this plane.** For a p_x orbital, the *yz* plane is the nodal plane. The shapes and orientations of the *p* orbitals are given in Fig. 2.6.

For $l = 2$ (*d* orbital), the magnetic quantum number are *five* $(2 \times 2 + 1)$; $+2, +1, 0, -1, -2$. Thus there are five possible orientations for *d* orbitals which are equivalent in energy so long as the atom is not under the influence of a magnetic field and are said to be **five-fold degenerate** (Different orbitals of equivalent energy are called degenerate orbitals and are grouped together). The *five d* orbitals are designated as d_{xy}, d_{yz}, d_{zx}, $d_{x^2-y^2}$, d_{z^2}. These orbitals have complex geometrical shapes as compared to *p* orbitals. The conventional boundary surfaces or shapes of five d_z^2 orbitals are shown in Fig. 2.6. The shape of the d_z orbitals is different from others.

When $l = 3$ (*f* orbital) the magnetic quantum number *m* can have *seven* $(2 \times 3 + 1)$ values as $+3, +2, +1, 0, -1, -2$ and -3. These *seven* orientations give rise to a set of **seven-fold degenerate orbitals.** These seven orbitals possess very complicated shapes and orientation in space. The shapes of s, *p* and *d* orbitals only are of interest to chemists.

Spin Quantum Number '*s*'

This quantum number has been introduced to account for the spin of electrons about their own axis. Since an electron can spin clockwise or anticlockwise (in two opposite directions), there are two possible values of *s* that are equal and opposite. As quantum numbers can differ only by unity from each other, there are two values given to *s* ; $+\frac{1}{2}$ and $-\frac{1}{2}$ depending upon whether the electron spins in one direction or the other. These spins are also designated by arrows pointing upwards and downward as ↓↑. **Two electrons with the same sign of the spin quantum numbers are said to have parallel spins** while those having opposite signs of the spin quantum numbers are said to have **opposite spin** or **antiparallel spin** or **paired-up spin.**

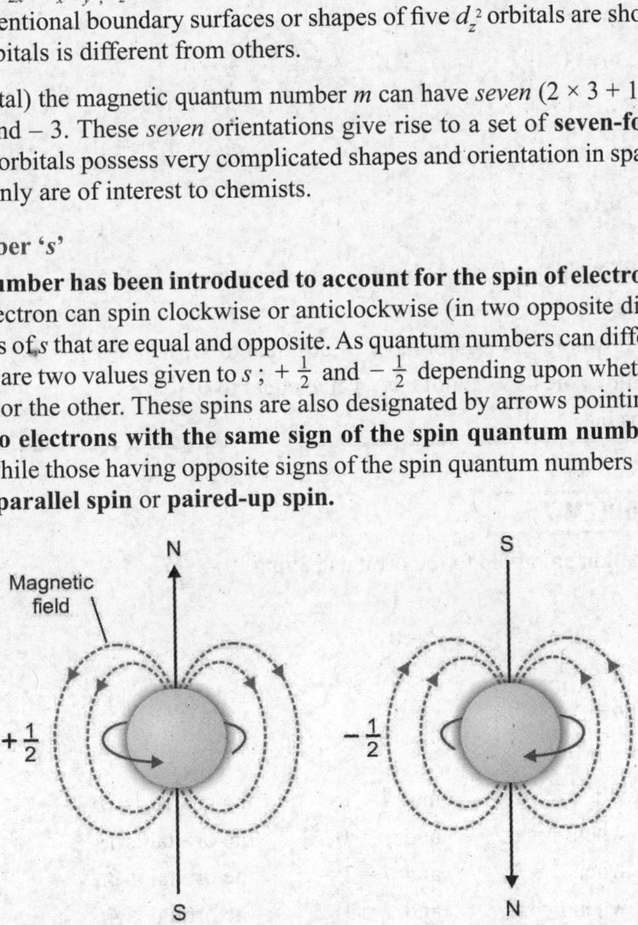

Figure 2.7
Clockwise and anticlockwise spins of electrons about their own axis produce opposite magnetic fields.

Since a spinning charge is associated with a magnetic field, an electron must have a magnetic moment associated with it.

The permitted values for each of these quantum numbers are given in the Table 2.1.

SOLVED PROBLEM 5

List all possible values of l and m for $n = 2$.

SOLUTION. Here, the principal quantum number $n = 2$. The azimuthal quantum number can have only two values. These are 0 and 1

When $l = 0$ $m = 0$

and $l = 1$ $m = +1, 0, -1$

SOLVED PROBLEM 6

Which of the following sets of quantum numbers are not allowable and why?

(*a*)	$n = 2$	$l = 2$	$m = 0$	$s = +\frac{1}{2}$
(*b*)	$n = 3$	$l = 1$	$m = 0$	$s = -\frac{1}{2}$
(*c*)	$n = 1$	$l = 1$	$m = +1$	$s = +\frac{1}{2}$
(*d*)	$n = 2$	$l = 0$	$m = -1$	$s = 0$
(*e*)	$n = 3$	$l = 2$	$m = +2$	$s = -\frac{1}{2}$

SOLUTION

(*a*) Not allowable as l cannot have value equal to 2 when $n = 2$.

(*b*) Allowable

(*c*) Not allowable as l cannot have value equal to 1 when $n = 1$

(*d*) Not allowable as s cannot have value equal to 0.

(*e*) Allowable

SOLVED PROBLEM 7

What designation are given to the orbitals having

(*a*)	$n = 2$	$l = 1$
(*b*)	$n = 1$	$l = 0$
(*c*)	$n = 3$	$l = 2$
(*d*)	$n = 4$	$l = 3$

SOLUTION

(*a*)	when $n = 2$	and $l = 1$	the orbital is 2*p*
(*b*)	when $n = 1$	and $p = 0$	the orbital is 1*s*
(*c*)	when $n = 3$	and $l = 2$	the orbital is 3*d*
(*d*)	when $n = 4$	and $l = 3$	the orbital is 4*f*

PAULI'S EXCLUSION PRINCIPLE

The nature of an electron, its position and energy, is fully implied only by mentioning the values of four quantum numbers ascribed to it. Each electron is, therefore, fully characterized by a set of four quantum numbers '*n*' – giving the size of electron orbital, l – its shape, and m – the orientation or disposition of the orbital and s the spin of the electron. Electrons having the same value of n, the principal quantum number, are said to belong to the same major energy level. However, the energies possessed by these electrons may yet be different owing to the different values of other quantum numbers assigned to them. In fact, the major energy levels are made of sublevels, given by the value of azimuthal quantum number 'l'. A particular energy sublevel may be designated by *s*, *p*, *d* and *f*. Within each energy level, the various sublevels have slightly different energies which increase in the same order as the value of the azimuthal quantum number l. Therefore, for the major energy level $n = 4$, which has an *s* orbital ($l = 0$), *p* orbitals ($l = 1$), *d* orbitals ($l = 2$) and *f* orbitals ($l = 3$), the energy increases in the order $s < p < d < f$. An electron with the principal quantum number n and azimuthal quantum number l has always lesser energy than that of an electron with principal quantum number $(n + 1)$ and the same azimuthal quantum number l *i.e.*, the energy of a 3*s* orbital is less than that of 4*s* orbital and energy of 4*p* orbitals is always more than the energy of 3*p* orbitals, and so on. The other two quantum numbers namely magnetic and spin quantum numbers determine the maximum number of electrons that can be accommodated in orbitals of a sublevel. It is, therefore, the assignment of the four quantum numbers to the electrons which ultimately count to determine its energy and location in space within an atom.

TABLE 2.1. QUANTUM NUMBERS AND ELECTRON ACCOMMODATION

Principal Q-number n		Azimuthal Q-number l		Magnetic Quantum Number m	Spin Quantum Number s	Number of Electrons accommodated	
1	K	0	*s*	0	$+\frac{1}{2}, -\frac{1}{2}$	2	2
2	L	0	*s*	0	$+\frac{1}{2}, -\frac{1}{2}$	2	8
		1	*p*	+ 1, 0, – 1	$+\frac{1}{2}, -\frac{1}{2}$	6	
3	M	0	*s*	0	$+\frac{1}{2}, -\frac{1}{2}$	2	18
		1	*p*	+ 1, 0, – 1	$+\frac{1}{2}, -\frac{1}{2}$	6	
		2	*d*	+ 2, + 1, 0, – 1, – 2,	$+\frac{1}{2}, -\frac{1}{2}$	10	
4	N	0	*s*	0	$+\frac{1}{2}, -\frac{1}{2}$	2	32
		1	*p*	+ 1, 0, – 1	$+\frac{1}{2}, -\frac{1}{2}$	6	
		2	*d*	+ 2, + 1, 0, – 1 – 2	$+\frac{1}{2}, -\frac{1}{2}$	10	
		3	*f*	+ 3, + 2, + 1, 0, – 1, – 2, – 3	$+\frac{1}{2}, -\frac{1}{2}$	14	

Wolfgang Pauli put forward an ingenious principle which controls the assignment of values of four quantum numbers of an electron. It applies certain restrictions on the values of electrons in an atom and hence the name **'exclusion principle'**. It is stated as : **No two electrons in an atom can have the same set of four identical quantum numbers.**

Even if two electrons have the same values for n, l and m, they must have different values of s. Thus every electron in an atom differs from every other electron in total energy and, therefore, there can be as many electrons in a shell as there are possible arrangements of different quantum numbers. The arrangements of electrons using permitted quantum numbers n, l, m and s are given in the Table 2.1. Let us find out the maximum number of electrons that can be accommodated in an orbital. We have seen that the first shell ($n = 1$) has only one orbital *i.e.*, 1*s*. The possible arrangements for the quantum numbers are only two in accordance with Pauli's exclusion principle.

n	*l*	*m*	*s*
1	0	0	$+\frac{1}{2}$ (1st electron)
1	0	0	$-\frac{1}{2}$ (2nd electron)

It follows, therefore, that a maximum of two electrons can be accommodated in an orbital and they must possess opposite spins.

Consider the second shell (n = 2), there being four orbitals, one s orbital (l = 0) and three p orbitals (l = 1), the possible number of electrons having different set of quantum numbers can be as follows :

n	*l*	*m*	*s*	
2	0	0	$+\frac{1}{2}$	*Two* electrons accommodated in 2*s* orbital (l = 0)
2	0	0	$-\frac{1}{2}$	
2	1	+ 1	$+\frac{1}{2}$	*Six* electrons in all the three 2*p* orbitals (l = 1), two electrons each in $2p_x$ (m = + 1), $2p_y$ (m = – 1) and $2p_z$ (m = 0) orbitals.
2	1	+ 1	$-\frac{1}{2}$	
2	1	– 1	$+\frac{1}{2}$	
2	1	– 1	$-\frac{1}{2}$	
2	1	0	$+\frac{1}{2}$	
2	1	0	$-\frac{1}{2}$	

The total number of electrons that can be accommodated in second shell is equal to 2 + 6 = 8. Similarly it can be shown that the maximum number of electrons in the third and fourth shells is equal to 18 and 32 respectively. On the basis of the above direction and the Table 2.1 it follows that *s* sublevel may contain upto *two* electrons, *p* sublevel upto *six, d* sublevel upto *ten* and *f* sublevel may have upto *fourteen* electrons. Each sublevel can accommodate at the most twice the number of available orbitals at that sublevel.

Pauli's exclusion principle is of immense value in telling the maximum number of electrons accommodated in any shell.

ENERGY DISTRIBUTION AND ORBITALS

In our earlier discussion we have seen that the energy of an electron is determined by the first two quantum numbers n and l, while the other two specify the orientation of the electron orbital in space and the spin. As we discuss the distribution of energy of the orbitals, the following two cases may arise :

(*a*) Hydrogen and Hydrogen-like atoms

Hydrogen is the simplest of all atoms since there is only one electron in it. This single electron is expected normally to be present in the lowest energy state n = 1. The values of l and m are both zero and spin quantum numbers can be either $+\frac{1}{2}$ or $-\frac{1}{2}$. When it absorbs energy, it may jump to higher energy levels given by n or a subshell thereon (represented by l values). The spectral study of hydrogen has revealed that its spectral lines correspond to the major energy levels only. Thus the energy value of an electron having a particular quantum is fixed, irrespective of the orbital to which it may belong. In other words, the energy associated with electrons in *s, p, d* and *f* orbitals of a particular principal quantum number is the same. For example, the energy level of 3*s*, 3*p* and 3*d* orbitals is equal (Fig. 2.8).

(*b*) · Polyelectron atoms

Let us first consider a two electron atom. The second electron which may differ from the first electron in spin only, is also accommodated in the 1*s* orbital, thereby completing the K shell. Unlike hydrogen, where there is no such completed shell, the energies of the subsequent electrons coming in various levels and sublevels will be affected. Thus it is this completed K shell that affects the energy of the electrons occupying subsequent energy levels. For atoms having more than two electrons, the nuclear charge is shielded from the outer electrons by the two K shell electrons. The effect of the completed K shell of electrons is to make the energy level of any orbital in a principal level *n* dependent upon the value of orbital quantum number *l*. The dependence of energy of orbitals of a shell on *l* ($l = 0$ or *s*, $l = 1$ or *p* etc.) is because of the fact that *s* orbital electrons ($l = 0$), for example, penetrate near the nucleus and are, therefore, less effectively shielded from the nuclear charge. The *s* electrons ($l = 0$) being less shielded are drawn inwards and possess lesser energy than *p* orbital ($l = 1$) electrons. The same argument can be extended for other values of *l*. Thus within each energy level '*n*', the various sublevels (different *l* values orbitals) exhibit slightly different energies. The orbitals at a principal level *n* get split up and come to possess different energies, which increase in the same order as the various values of *l*. Thus for a particular principal level, the energy of the sublevels is in the order $s < p < d < f$.

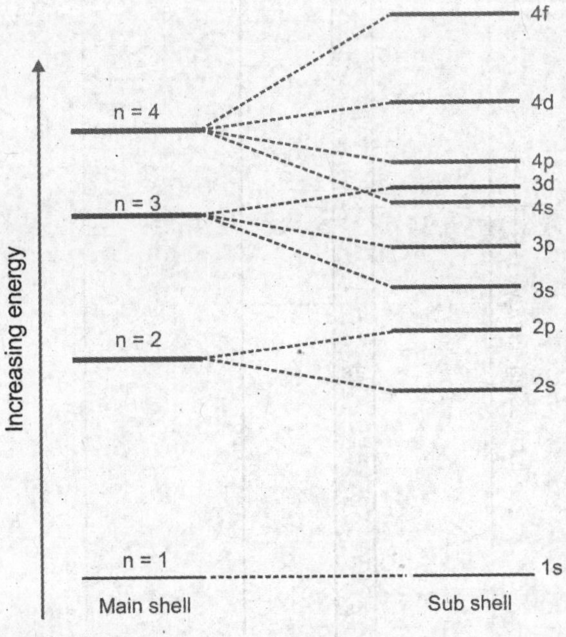

■ Figure 2.8
Energy level schemes of Hydrogen atom.

The energy levels of 3*s*, 3*p* and 3*d* orbitals are different even though they belong to the same shell $n = 3$. However, **it may be noted that the energy of electrons in the same orbital is the same.** Thus all 3*d* orbitals ($3d_{xy}$, $3d_{yz}$, $3d_{zx}$, $3d_{z^2}$, $3d_{x^2-y^2}$) or 4*p* orbitals ($4p_x$, $4p_y$, $4p_z$) are at the same level of energy, irrespective of their orientation. It is also noteworthy from the above diagram that the order of increase of energy values of various orbitals approximately follows the sequence given below :

$$1s < 2s < 2p < 3s < 3p < 4s < 3d < 4p < 5s < 4d < 5p < 6s < 4f < 5d$$

It is, therefore, clear that the 3*d* orbital electrons belonging to a lower shell ($n = 3$) possess more energy than 4*s* orbital electrons which belong to higher shell ($n = 4$). The 3*d* orbitals lie at a higher energy level than 4*s* orbital.

DISTRIBUTION OF ELECTRONS IN ORBITALS

In passing along the periodic table of the elements from one element to the other, we find that one electron is added every time to the next atom. Where should the incoming electron go? The answer is provided by the possible values of the quantum numbers that can be assigned to the electron in accordance with Pauli's exclusion principle–prohibiting an orbital to accommodate two electrons with the same set of quantum numbers.

On the basis of magnetic measurements, which also help to determine the electronic configuration of elements, Hund put forward another empirical rule, popularly known after his name as **Hund's Rule of Maximum Multiplicity.** It states that : **Electrons are distributed among the orbitals of a subshell in such a way as to give the maximum number of unpaired electrons and having the same direction of spin.**

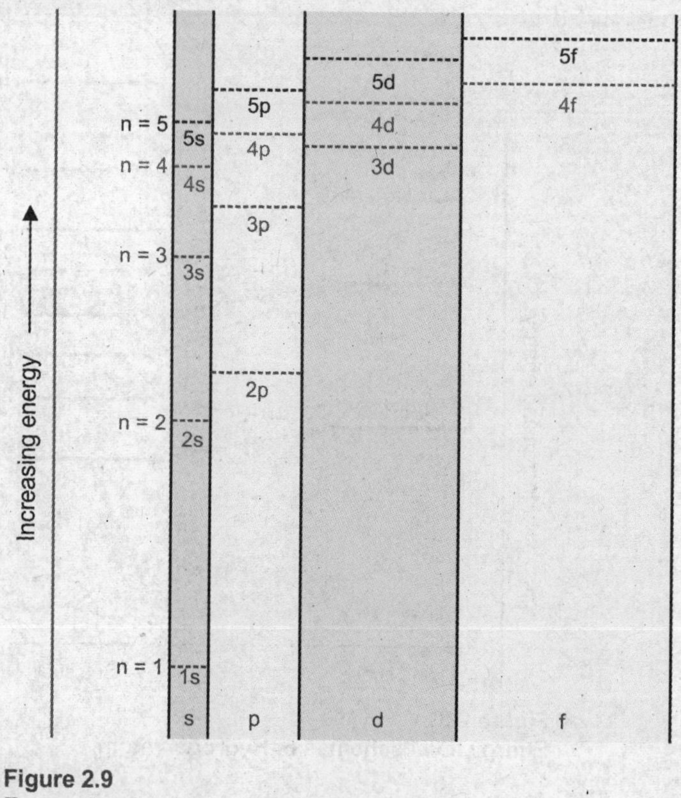

■ **Figure 2.9**
Energy level scheme of polyelectron atoms.

Thus the orbitals available at a subshell are first filled singly before they begin to pair. The following illustration shows the order of filling of electrons in the orbitals of n = 1 and n = 2 shells. The orbitals are shown by circles and the order of filling for the first ten electrons is indicated by the numbers entered in them.

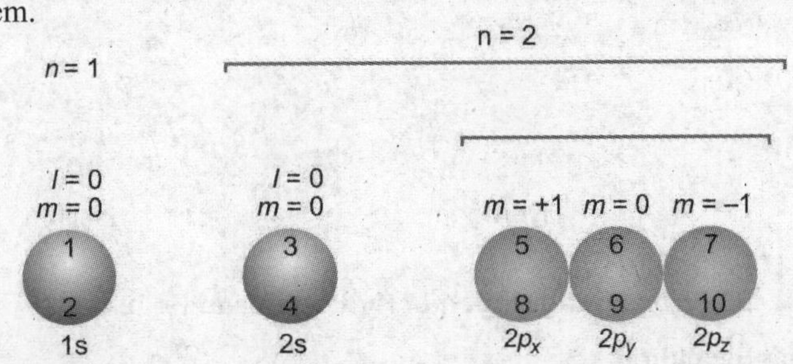

Figure 2.10
Order of filling of electrons in orbitals of $n = 1$ and $n = 2$ shells.

It is also clear from the illustration that no two electrons in an orbital have the same values of all four quantum numbers. In fact, three are identical while the fourth quantum number *i.e.*, the spin quantum number is invariably different. The electrons in the *p* orbitals are arranged and accommodated such that they have all obtained one electron first (5th in $2p_x$, 6th in $2p_y$, 7th in $2p_z$) and now they begin to pair up getting the 8th, 9th and 10th electrons respectively.

SCHEMATIC REPRESENTATION OF ELECTRON CONFIGURATION

We have seen above that to define completely the state of an atom it is obligatory to refer to all the four quantum numbers (n, l, m and s) of every electron in it. Since a simultaneous representation of all quantum numbers of each electron in a single symbolic notation seems quite difficult, it is customary to take into account the first two quantum numbers only while the other two can be inferred indirectly. The general symbolic notation employed for the purpose is nl^a where the numerical value of $n = 1$, 2, 3 etc., represents the principal quantum number, the letter designate of l (s for $l = 0$, p for $l = 1$ and so on) stands for the orbital and the superscript a gives the number of electrons in the orbital. Thus $3s^2$ indicates that two electrons are present in the first subshell s ($l = 0$) of the third shell ($n = 3$). For instance, the distribution of seven electrons (of N atom) may be schematically represented as $1s^2$; $2s^2, 2p^3$ or more elaborately as $1s^2$; $2s^2$, $2p_x^1$, $2p_y^1$, $2p_z^1$. By using the various designates of orbitals at a sublevel such as $2p_x$, $2p_y$ etc., the third quantum number m is also indicated (*e.g.*, $2p_x$, for $m = +1$, $2p_y$ for $m = 0$ and $2p_z$ for $m = -1$). Spin quantum numbers are indirectly inferred. Whenever there are two electrons in an orbital, one of these has $+\frac{1}{2}$ and the other $-\frac{1}{2}$ as their spin quantum number.

It is a common practice to denote an orbital by a horizontal line or a circle or square and an electron by an arrow over it. The direction of the arrow indicates the spin, an upward arrow representing a clockwise spin while the downward arrow stands for the anticlockwise direction of spin. When there are more than one orbitals in a subshell **(degenerate orbitals),** they are shown by an equivalent number of horizontal lines at the same energy level. Let us now describe the electron configuration of first ten elements.

(*a*) Hydrogen and Helium

These have one and two electrons respectively which are accommodated in 1*s* orbital while others remain vacant. The lone electron of hydrogen is filled in 1*s* orbital and for helium the second electron would also go in 1*s* orbital, since it could accommodate another electron with opposite spin.

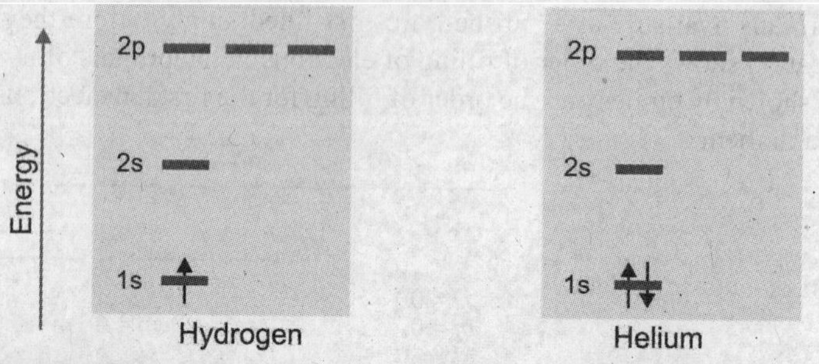

Figure 2.11
Electron configuration of Hydrogen and Helium.

(*b*) Lithium and Beryllium

These have three and four electrons respectively. The third electron of Li enters in the 2*s* orbital and the fourth electron of Be also enters in the same orbital, but has an opposite direction of spin.

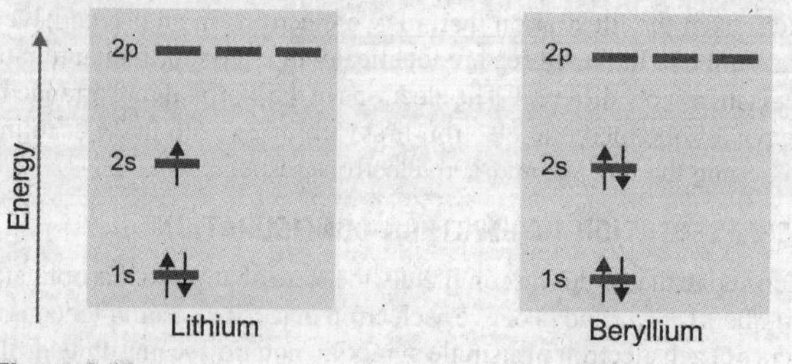

Figure 2.12
Electron configuration of Lithium and Beryllium.

(*c*) Boron and Carbon

These atoms have five and six electrons respectively. 1*s* and 2*s* orbitals being completely filled with four electrons, the fifth electron of boron would go in one of the 2*p* orbitals say $2p_x$. The sixth electron in carbon would prefer to be accommodated in another vacant 2*p* orbital say ($2p_y$) rather than going to $2p_z$ orbital (Hund's rule). The two unpaired electrons shall have similar spins as indicated.

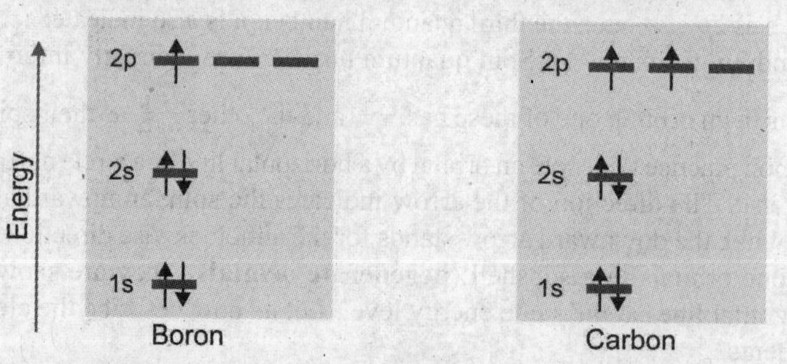

Figure 2.13
Electron configuration of Boron and Carbon.

(*d*) Nitrogen and Oxygen

These atoms have seven and eight electrons respectively. After six electrons have been accommodated as above, there is a vacant $2p_z$ orbital which will be the seat of the seventh electron possessing the same direction of spin. The eighth electron of the next element oxygen will go to pair up with the $2p_x$ electron and has an antiparallel spin as shown below.

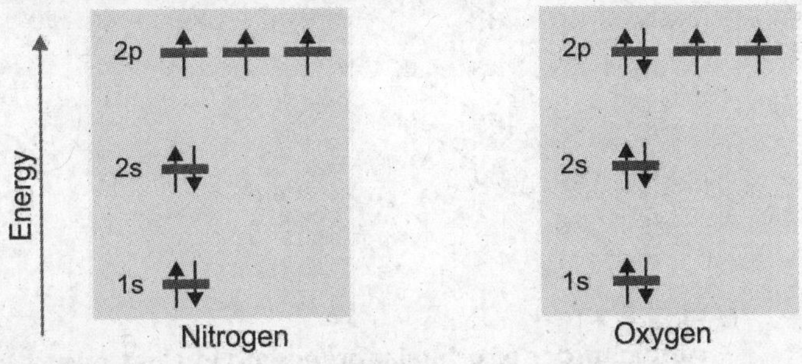

■ **Figure 2.14**
Electron configuration of Nitrogen and Oxygen.

(*e*) Fluorine and Neon

These atoms possess nine and ten electrons respectively which go to complete the other 2*p* orbitals as shown in Fig. 2.15.

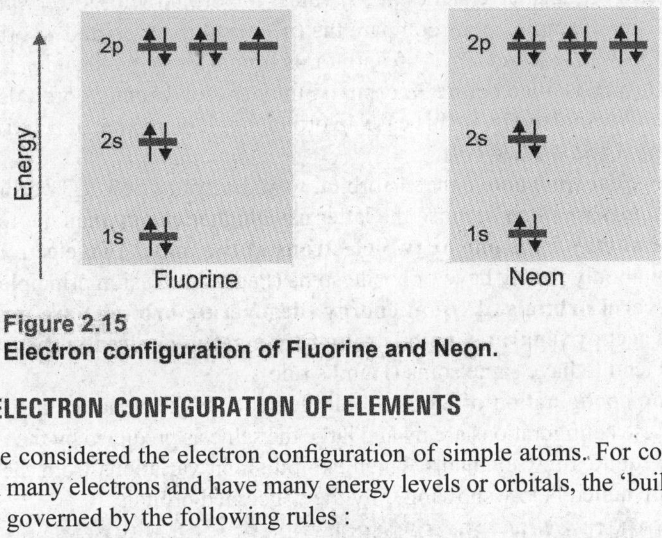

■ **Figure 2.15**
Electron configuration of Fluorine and Neon.

GROUND-STATE ELECTRON CONFIGURATION OF ELEMENTS

So far we have considered the electron configuration of simple atoms. For complicated atoms which may contain many electrons and have many energy levels or orbitals, the 'building up' process for the electrons is governed by the following rules :

Rule 1. Each electron shell can hold a maximum of $2n^2$ electrons where *n* is the shell number.

Rule 2. These electrons are accommodated in *s*, *p*, *d* and *f* orbitals, the maximum number of electrons in each type of orbitals being determined by its electron-holding capacity (for $s = 2, p = 6, d = 10$ and $f = 14$).

Rule 3. In the ground state of an atom, the electrons tend to occupy the available orbitals in the increasing order of energies, the orbitals of lower energy being filled first. This is called 'building up principle' or **Aufbau Principle** (*Aufbau* is a German expression meaning *building up or construction*). Lower energy orbitals are, therefore, better seats for electrons and better seats are occupied first. Fig 2.9 shows the energy level scheme of orbitals and this order can conveniently be remembered by the simple device given below.

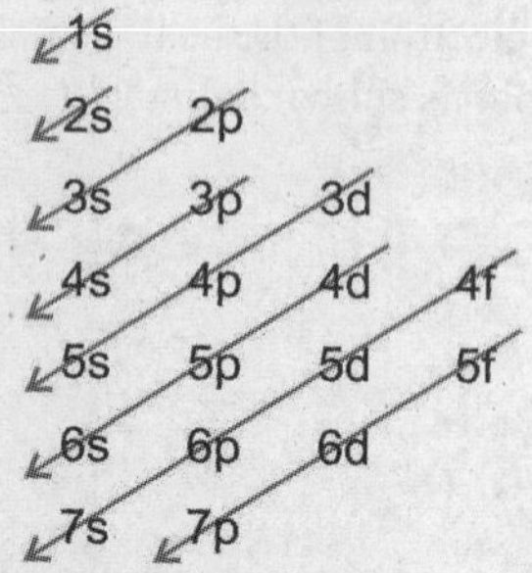

Figure 2.16
Aufbau order of orbitals for feeding in electrons.

The increasing order of energy of various orbitals is as follows :
$1s < 2s < 2p < 3s < 3p < 4s < 3d < 4p < 5s$ ......

The energy of an orbital is determined by the sum of principal quantum number (n) and the azimuthal quantum number (l). This rule is called **(n + *l*) rule.** There are two parts of this rule :

(*a*) The orbitals with the lower value of $(n + l)$ has lower energy than the orbitals of higher $(n + l)$ value.

(*b*) When two orbitals have same $(n + l)$ value, the orbital with lower value of n has lower energy. For example, let us compare the $(n + l)$ value for 3*d* and 4*s* orbitals.

For 3*d* orbital $n = 3$, $l = 2$ and $n + l = 5$ and for 4*s* orbital $n = 4$, $l = 0$ and $n + l = 4$.

Therefore, 4*s* orbital is filled before 3*d* orbital. Similarly, for 4*p* and 5*s* orbitals, the $(n + l)$ values are $(4 + 1)$ and $(5 + 0)$ respectively. In this case 4*p* orbital has lesser value of n and hence it has lower energy than 5*d* orbital and is filled first.

It is, therefore, clear from above that 4*s* orbital would be filled before 3*d* orbitals (belonging to a lower shell *i.e.*, third) are filled because the latter have higher energy than the former.

Rule 4. Any orbital may have one or two electrons at the most. Two electrons can occupy the same orbital only if they have opposite spins (Pauli's exclusion principle).

Rule 5. When several orbitals of equal energy (degenerate orbitals) are available, electrons prefer to occupy separate orbitals rather than getting paired in the same orbital. Such electrons tend to have same spins (Hund's rule).

Actual electron configuration of atoms of all elements of the periodic table is given in Table 2.3. We find that these configurations are by and large the same as predicted by the Aufbau procedure. However, there are quite a few elements which exhibit slight variations from the standard pattern. Some anomalies are tabled below showing only the concerned orbitals.

TABLE 2.2. ANOMALOUS ELECTRON CONFIGURATION OF SOME ELEMENTS

Element	At. No.	Expected Configuration	Actual Configuration
Cr	24	$3d^4\ 4s^2$	$3d^5\ 4s^1$
Cu	29	$3d^9\ 4s^2$	$3d^{10}\ 4s^1$
Mo	42	$4d^4\ 5s^2$	$4d^5\ 5s^1$
Pd	46	$4d^8\ 5s^2$	$4d^{10}\ 5s^0$
Ag	47	$4d^9\ 5s^2$	$4d^{10}\ 5s^1$
Pt	78	$5d^8\ 6s^2$	$5d^9\ 6s^1$
Au	79	$5d^9\ 6s^2$	$5d^{10}\ 6s^1$

TABLE 2.3. GROUND STATE ELECTRON CONFIGURATION OF ELEMENTS

Z	Element	Electron configuration	Z	Element	Electron configuration
1	H	$1s^1$	33	As	$[Ar]3d^{10}4s^24p^3$
2	He	$1s^2$	34	Se	$[Ar]3d^{10}4s^24p^4$
3	Li	$[He]2s^1$	35	Br	$[Ar]3d^{10}4s^24p^5$
4	Be	$[He]2s^2$	36	Kr	$[Ar]3d^{10}4s^24p^6$
5	B	$[He]2s^22p^1$	37	Rb	$[Kr]5s^1$
6	C	$[He]2s^22p^2$	38	Sr	$[Kr]5s^2$
7	N	$[He]2s^22p^3$	39	Y	$[Kr]4d^15s^2$
8	O	$[He]2s^22p^4$	40	Zr	$[Kr]4d^25s^2$
9	F	$[He]2s^22p^5$	41	Nb	$[Kr]4d^45s^1$
10	Ne	$[He]2s^22p^6$	42	Mo	$[Kr]4d^55s^1$
11	Na	$[Ne]3s^1$	43	Tc	$[Kr]4d^55s^2$
12	Mg	$[Ne]3s^2$	44	Ru	$[Kr]4d^75s^1$
13	Al	$[Ne]3s^23p^1$	45	Rh	$[Kr]4d^85s^1$
14	Si	$[Ne]3s^23p^2$	46	Pd	$[Kr]4d^{10}$
15	P	$[Ne]3s^23p^3$	47	Ag	$[Kr]4d^{10}5s^1$
16	S	$[Ne]3s^23p^4$	48	Cd	$[Kr]4d^{10}5s^2$
17	Cl	$[Ne]3s^23p^5$	49	In	$[Kr]4d^{10}5s^25p^1$
18	Ar	$[Ne]3s^23p^6$	50	Sn	$[Kr]4d^{10}5s^25p^2$
19	K	$[Ar]4s^1$	51	Sb	$[Kr]4d^{10}5s^25p^3$
20	Ca	$[Ar]4s^2$	52	Te	$[Kr]4d^{10}5s^25p^4$
21	Sc	$[Ar]3d^14s^2$	53	I	$[Kr]\ 4d^{10}5s^25p^5$
22	Ti	$[Ar]3d^24s^2$	54	Xe	$[Kr]\ 4d^{10}5s^25p^6$
23	V	$[Ar]3d^34s^2$	55	Cs	$[Xe]6s^1$
24	Cr	$[Ar]3d^54s^1$	56	Ba	$[Xe]6s^2$
25	Mn	$[Ar]3d^54s^2$	57	La	$[Xe]5d^16s^2$
26	Fe	$[Ar]3d^64s^2$	58	Ce	$[Xe]4f^15d^16s^2$
27	Co	$[Ar]3d^74s^2$	59	Pr	$[Xe]4f^36s^2$
28	Ni	$[Ar]3d^84s^2$	60	Nd	$[Xe]4f^46s^2$
29	Cu	$[Ar]3d^{10}4s^1$	61	Pm	$[Xe]4f^56s^2$
30	Zn	$[Ar]3d^{10}4s^2$	62	Sm	$[Xe]4f^66s^2$
31	Ga	$[Ar]3d^{10}4s^24p^1$	63	Eu	$[Xe]4f^76s^2$
32	Ge	$[Ar]3d^{10}4s^24p^2$	64	Gd	$[Xe]4f^75d^16s^2$

Z	Element	Electron configuration	Z	Element	Electron configuration
65	Tb	$[Xe]4f^{9}6s^{2}$	85	At	$[Xe]4f^{14}5d^{10}6s^{2}6p^{5}$
66	Dy	$[Xe]4f^{10}6s^{2}$	86	Rn	$[Xe]4f^{14}5d^{10}6s^{2}6p^{6}$
67	Ho	$[Xe]4f^{11}6s^{2}$	87	Fr	$[Rn]7s^{1}$
68	Er	$[Xe]4f^{12}6s^{2}$	88	Ra	$[Rn]7s^{2}$
69	Tm	$[Xe]4f^{13}6s^{2}$	89	Ac	$[Rn]6d^{1}7s^{2}$
70	Yb	$[Xe]4f^{14}6s^{2}$	90	Th	$[Rn]6d^{2}7s^{2}$
71	Lu	$[Xe]4f^{14}5d^{1}6s^{2}$	91	Pa	$[Rn]5f^{2}6d^{1}7s^{2}$
72	Hf	$[Xe]4f^{14}5d^{2}6s^{2}$	92	U	$[Rn]5f^{3}6d^{1}7s^{2}$
73	Ta	$[Xe]4f^{14}5d^{3}6s^{2}$	93	Np	$[Rn]5f^{4}6d^{1}7s^{2}$
74	W	$[Xe]4f^{14}5d^{4}6s^{2}$	94	Pu	$[Rn]5f^{6}7s^{2}$
75	Re	$[Xe]4f^{14}5d^{5}6s^{2}$	95	Am	$[Rn]5f^{7}7s^{2}$
76	Os	$[Xe]4f^{14}5d^{6}6s^{2}$	96	Cm	$[Rn]5f^{7}6d^{1}7s^{2}$
77	Ir	$[Xe]4f^{14}5d^{7}6s^{2}$	97	Bk	$[Rn]5f^{9}7s^{2}$
78	Pt	$[Xe]4f^{14}5d^{9}6s^{1}$	98	Cf	$[Rn]5f^{10}7s^{2}$
79	Au	$[Xe]4f^{14}5d^{10}6s^{1}$	99	Es	$[Rn]5f^{11}7s^{2}$
80	Hg	$[Xe]4f^{14}5d^{10}6s^{2}$	100	Fm	$[Rn]5f^{12}7s^{2}$
81	Tl	$[Xe]4f^{14}5d^{10}6s^{2}6p^{1}$	101	Md	$[Rn]5f^{13}7s^{2}$
82	Pb	$[Xe]4f^{14}5d^{10}6s^{2}6p^{2}$	102	No	$[Rn]5f^{14}7s^{2}$
83	Bi	$[Xe]4f^{14}5d^{10}6s^{2}6p^{3}$	103	Lr	$[Rn]5f^{14}6d^{1}7s^{2}$
84	Po	$[Xe]4f^{14}5d^{10}6s^{2}6p^{4}$	104	Rf	$[Rn]5f^{14}6d^{2}7s^{2}$

Note : The symbol in brackets indicates the electron core of the Noble gas.

We find from the table that irregularities involve the placing of one or two electrons from *ns* orbital in $(n-1)$ *d* orbitals. There is very little energy difference between such *s* and *d* orbitals so that there is very little to choose from energy point of view. The deviations occur when *d* level orbitals are either almost full (*e.g.*, Cu, Pd, Ag, Pt and Au) or half-full (Cr and Mo). The explanation for this deviation lies in the superior stability of completely filled or all half-filled orbitals than nearly filled or nearly half-filled orbitals. Thus d^5 and d^{10} configurations are much more stable than d^4 or d^8 or d^9. Spectroscopic data and magnetic properties of elements justify the statement that **half-filled and completely filled subshells contribute to the stability.**

IONIZATION ENERGY

The process of removing an electron from an isolated atom to form a positive ion is called ionization. Energy will be required to remove an electron from the atom against the force of attraction of the nucleus.

The ionization energy (IE) of an element is defined as the energy needed to remove a single electron from an atom of the element in the gaseous state. That is,

$$M(g) + IE \longrightarrow M^{+}(g) + e^{-}$$

PERIODIC TABLE OF ELEMENTS
(ELECTRON CONFIGURATIONS)

1 H $1s^1$																	2 He $1s^2$
3 Li $2s^1$	4 Be $2s^2$											5 B $2s^22p^1$	6 C $2s^22p^2$	7 N $2s^22p^3$	8 O $2s^22p^4$	9 F $2s^22p^5$	10 Ne $2s^22p^6$
11 Na $3s^1$	12 Mg $3s^2$											13 Al $3s^23p^1$	14 Si $3s^23p^2$	15 P $3s^23p^3$	16 S $3s^23p^4$	17 Cl $3s^23p^5$	18 Ar $3s^23p^6$
19 K $4s^1$	20 Ca $4s^2$	21 Sc $4s^23d^1$	22 Ti $4s^23d^2$	23 V $4s^23d^3$	24 Cr $4s^13d^5$	25 Mn $4s^23d^5$	26 Fe $4s^23d^6$	27 Co $4s^23d^7$	28 Ni $4s^23d^8$	29 Cu $4s^13d^{10}$	30 Zn $4s^23d^{10}$	31 Ga $4s^23d^{10}4p^1$	32 Ge $4s^23d^{10}4p^2$	33 As $4s^23d^{10}4p^3$	34 Se $4s^23d^{10}4p^4$	35 Br $4s^23d^{10}4p^5$	36 Kr $4s^23d^{10}4p^6$
37 Rb $5s^1$	38 Sr $5s^2$	39 Y $5s^24d^1$	40 Zr $5s^24d^2$	41 Nb $5s^14d^4$	42 Mo $5s^14d^5$	43 Tc $5s^24d^5$	44 Ru $5s^14d^7$	45 Rh $5s^14d^8$	46 Pd $4d^{10}$	47 Ag $5s^14d^{10}$	48 Cd $5s^24d^{10}$	49 In $5s^24d^{10}5p^1$	50 Sn $5s^24d^{10}5p^2$	51 Sb $5s^24d^{10}5p^3$	52 Te $5s^24d^{10}5p^4$	53 I $5s^24d^{10}5p^5$	54 Xe $5s^24d^{10}5p^6$
55 Cs $6s^1$	56 Ba $6s^2$	57 La $6s^25d^1$	72 Hf $6s^24f^{14}5d^2$	73 Ta $6s^24f^{14}5d^3$	74 W $6s^24f^{14}5d^4$	75 Re $6s^24f^{14}5d^5$	76 Os $6s^24f^{14}5d^6$	77 Ir $6s^24f^{14}5d^7$	78 Pt $6s^14f^{14}5d^9$	79 Au $6s^14f^{14}5d^{10}$	80 Hg $6s^24f^{14}5d^{10}$	81 Tl $6s^24f^{14}5d^{10}6p^1$	82 Pb $6s^24f^{14}5d^{10}6p^2$	83 Bi $6s^24f^{14}5d^{10}6p^3$	84 Po $6s^24f^{14}5d^{10}6p^4$	85 At $6s^24f^{14}5d^{10}6p^5$	86 Rn $6s^24f^{14}5d^{10}6p^6$
87 Fr $7s^1$	88 Ra $7s^2$	89 Ac $7s^26d^1$	104 Rf $7s^25f^{14}6d^2$	105 Db $7s^25f^{14}6d^3$	106 Sg $7s^25f^{14}6d^4$	107 Bh $7s^25f^{14}6d^5$	108 Hs $7s^25f^{14}6d^6$	109 Mt $7s^25f^{14}6d^7$	110 —	111 —	112 —		114 —		116 —		

58 Ce $6s^24f^15d^1$	59 Pr $6s^24f^3$	60 Nd $6s^24f^4$	61 Pm $6s^24f^5$	62 Sm $6s^24f^6$	63 Eu $6s^24f^7$	64 Gd $6s^24f^75d^1$	65 Tb $6s^24f^9$	66 Dy $6s^24f^{10}$	67 Ho $6s^24f^{11}$	68 Er $6s^24f^{12}$	69 Tm $6s^24f^{13}$	70 Yb $6s^24f^{14}$	71 Lu $6s^24f^{14}5d^1$
90 Th $7s^26d^2$	91 Pa $7s^25f^26d^1$	92 U $7s^25f^36d^1$	93 Np $7s^25f^46d^1$	94 Pu $7s^25f^6$	95 Am $7s^25f^7$	96 Cm $7s^25f^76d^1$	97 Bk $7s^25f^9$	98 Cf $7s^25f^{10}$	99 Es $7s^25f^{11}$	100 Fm $7s^25f^{12}$	101 Md $7s^25f^{13}$	102 No $7s^25f^{14}$	103 Lr $7s^25f^{14}6d^1$

Since one, two or more electrons may be removed from the same atom, one after the other, we have as many ionization energies of the element.

The **First ionization energy (IE_1),** is the energy needed to remove the first electron from the gaseous atom M to form M^+ ion.

The **Second ionization energy (IE_2),** is the energy needed to remove a second electron, from the gaseous M^+ ion to form M^{2+} ion.

Higher ionization energies can be defined in the same way. We can depict the first, second and third ionization energies in the form of equations as :

$$M(g) + IE_1 \longrightarrow M^+(g) + e^-$$

$$M^+(g) + IE_2 \longrightarrow M^{2+}(g) + e^-$$

$$M^{2+}(g) + IE_3 \longrightarrow M^{3+}(g) + e^-$$

Ionization energies are sometimes called **Ionization potentials.** Ionization energies are usually expressed in electron volts (eV) per atom, or in kilojoules per mole of atoms (kJ mol^{-1}). For conversion, 1eV $atom^{-1}$ = 96.48 kJ mol^{-1}.

MEASUREMENT OF IONIZATION ENERGIES

The amount of energy required to detach an electron from an atom can be measured by supplying the required energy as thermal energy, electrical energy, or radiant energy. Thus ionization energies can be determined from the spectrum of the element or by any of the two methods detailed below.

(1) The Electrical method

The apparatus used is shown in Fig. 2.17. The electrically heated tungsten wire emits electrons. The grid can be charged positively to different voltages which we read with a voltmeter. The plate opposite the grid has a small negative charge. When the potential to the grid is zero, no current flows between the grid and the plate. However if we give sufficient potential to the grid, the electrons emitted by the tungsten wire are accelerated towards the grid, pass through it and ionize the atoms between grid and plate. The electron ejected by each atom is attracted to grid and positive ion is attracted to plate. A current thus passes between grid and plate which is shown up by an ammeter. **The minimum grid voltage that just produces a current is called ionization potential.**

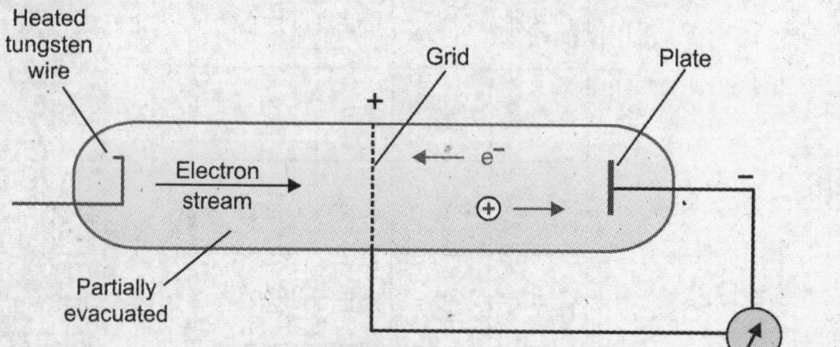

Figure 2.17
Measurement of Ionization energy.

If V be the ionization potential, the ionization energy (IE) is calculated as :

$$\begin{aligned} IE &= V \times \text{charge of electron} \times \text{Avogadro No} \\ &= V \times 1.60 \times 10^{-19} \times 6.02 \times 10^{23} \\ &= V \times 96.3 \text{ kJ mol}^{-1} \end{aligned}$$

(2) Photo-ionization Method

The gaseous atoms are introduced into a chamber containing two electrically charged plates (Fig. 2.18). As neutral atoms, they do not conduct electricity and no current flows between the plates. When radiant energy ($h\nu$) is supplied to the gaseous atoms, ionization will occur and electric current will flow. The frequency of the radiation used is gradually increased. The minimum frequency necessary to cause ionization of the gaseous atoms, as shown by the flow of an electric current is noted. From this frequency the ionization energy is calculated.

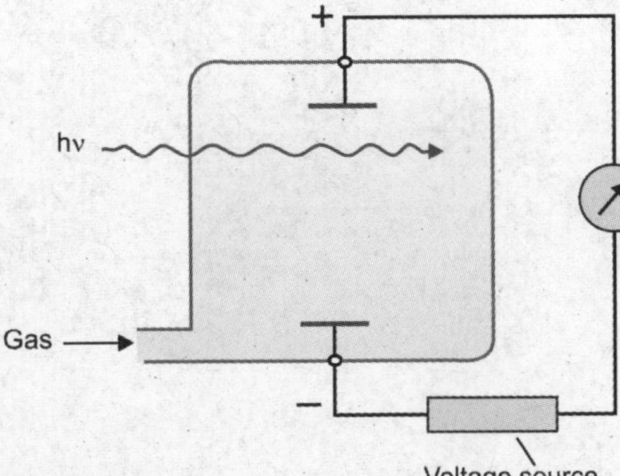

Figure 2.18
Measurement of ionization energy by photo-ionization method.

Order of Successive Ionization Energies

The second ionization energy (I E_2) is larger than the first ionization energy (I E_1) because it is more difficult to detach an electron from a + ve ion than a neutral atom. The third ionization energy (I E_3) is still larger as the third electron has to be detached from a 2 + ion. Thus in general successive ionization energies increase in magnitude. That is,

I E_1 < I E_2 < I E_3 < I E_4, and so on.

For illustration, the first four ionization energies for sodium and magnesium are listed below:

TABLE 2.4. IONIZATION ENERGIES IN KILO JOULE PER MOLE (kJ mol^{-1})

	I E_1	I E_2	I E_3	I E_4
Sodium	500	4600	6900	9500
Magnesium	740	1500	7700	10500

Principal Trends in Ionization Energies

A graph of the first ionization energies against atomic number (Z) for the first 18 elements of the Periodic Table is shown in Fig. 2.19.

The important trends as illustrated by the graph are:

(1) **Ionization energies increase across a period.** *e.g.*, Li to Ne.

(2) **Ionization energies decrease down a group** *e.g.*, Li, Na, K and so on.

(3) **There are regular discontinuities in the increase trend across a period** *e.g.*, Be to B, and N to O.

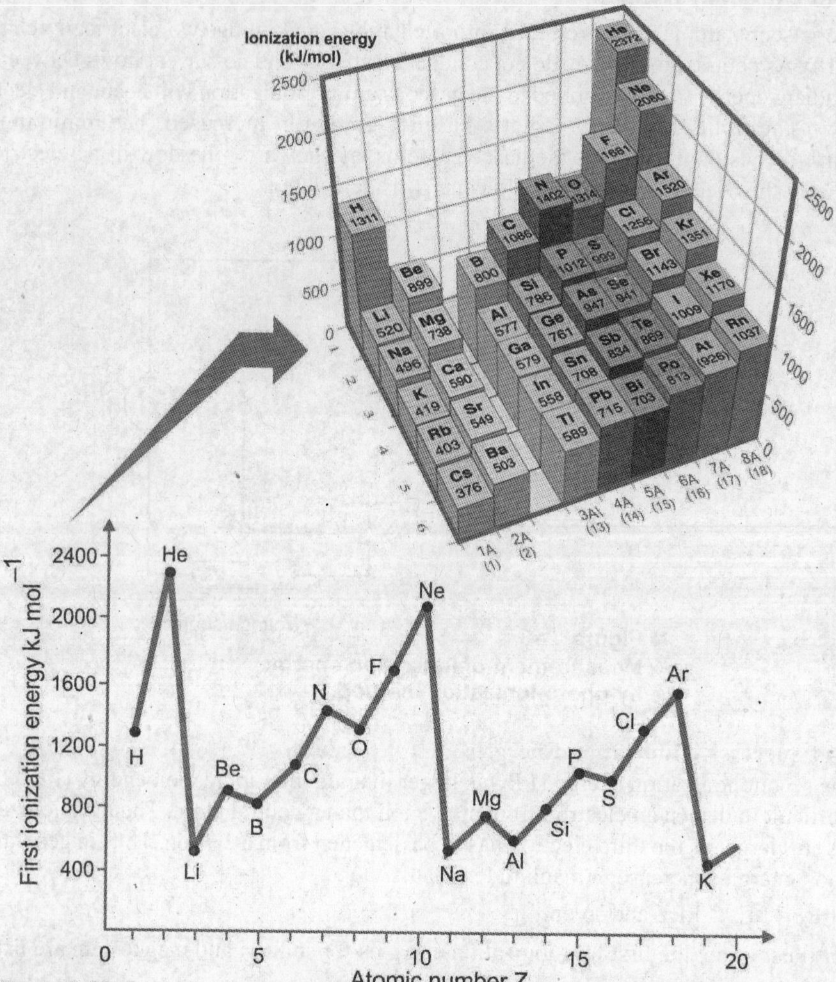

■ **Figure 2.19**
Graph showing the variation of first ionization energies with increase of atomic number.

Increase across a Period

As we pass from left to right in a period, the first ionization energy shows a steady increase. Thus in Period 2 from Li to N, we have

	Li	Be	B	C	N
IE_1 (k J mol^{-1})	525	906	805	1090	1400

Explanation

The outer-shell electrons in the elements of the same period are arranged in the same shell. For example, the build up of electrons in Period 2 from Li to B is shown in Fig. 2.20.

Moving from Li to B, the positive charge on the nucleus increases whereas the distance between the nucleus and valence electrons decreases. Therefore more energy is required to remove an electron as we go from left to right in the Period. Since the number of screening electrons remains the same, they do not upset the increase trend.

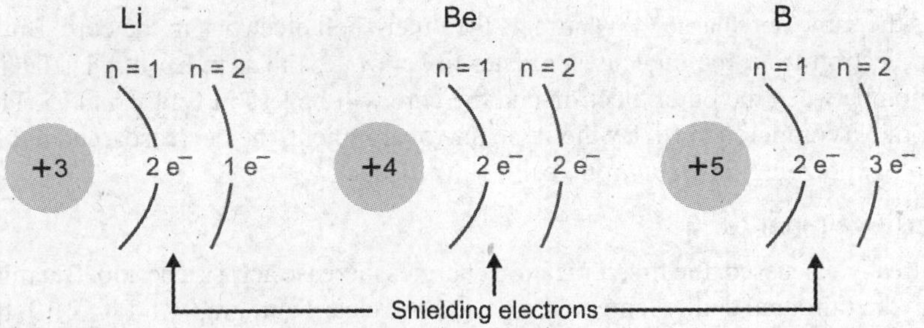

Figure 2.20
Build up of atoms of elements for Li, Be, B.

Decrease down a Group

In the elements of a vertical Group of the Periodic table, the number of outer shell electrons is the same. But the following changes are noted from top to bottom.

(1) **The principal quantum number *n* containing the valence electrons increases.**

(2) **The nuclear charge (At. No.) increases.**

(3) **The number of electrons in the inner shells (shielding electrons) increases.**

The net result of these changes is that the first ionization energies down a group record a progressive decrease. Thus for Group IA we have

	Li	Na	K	Rb	Cs
IE_1 ($kJ\ mol^{-1}$)	525	500	424	408	382

Let us explain the above decrease trend by taking example of lithium and sodium. They have the atomic structures.

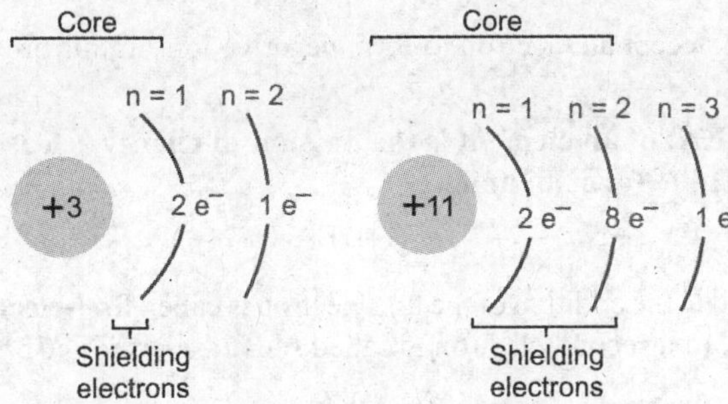

Lithium and sodium both have one outer-shell electron. The number of shielding electrons in sodium is 10 while in lithium it is 2. If we assume that the inner shell electrons provide hundred percent screening, the core charge attracting the outer-shell electron would be :

	Nuclear charge	Shielding electrons	Core charge attracting outer-shell electron
Li	3	2	3 − 2 = +1
Na	11	2, 8	11−10 = +1

Thus the same net charge (+ 1) attracts the outer-shell electrons to the core. But the distance of the outer electron from the nucleus is greater in Na ($n = 3$) than in Li ($n = 1$). Therefore the force of attraction between the outer electron and the core will be less in Na than in Li. That explains the lower I E of Na compared to Li. By the same line of argument, the decrease trend in I E from element to element while going down a Group can be justified.

Regular Discontinuities

As already discussed, the first ionization energies increase across a period. But this increase trend is upset at the third and sixth element in a period. As clear from graph in Fig. 2.19, there are breaks at B and O which occupy the third and fifth positions respectively in the 2nd period. The I E_1 of B is less than that of Be and the I E_1 of O is less than that of N.

Explanation

(*a*) The electronic configuration of Be and B are :

$$\text{Be} \quad 1s^2\,2s^2 \qquad\qquad \text{B} \quad 1s^2\,2s^2\,2p^1$$

The $2p$ orbital electron of B is already higher in energy than the $2s$ orbital electron. Therefore the removal of electron from B requires less energy and its I E_1 is lower.

(*b*) The electronic configuration of N and O is :

$$\text{N} \quad 1s^2\,2s^2\,2p^3 \qquad\qquad \text{O} \quad 1s^2\,2s^2\,2p^4$$

The $2p$ orbitals may be represented as

N ↑ ↑ ↑ O ↑↓ ↑ ↑

Whenever two electrons occupy a particular orbital, they repel each other. As a result it is easier to remove one of the paired $2p$ electrons from O than it is to remove an unpaired electron from N atom. Thus I E_1 of O is lower than that of N.

ELECTRON AFFINITY

A neutral atom can accept an electron to form negative ion. In this process, in general, energy is released.

Electron affinity (EA) of an element is the amount of energy released when an electron is added to a gaseous atom to form an anion.

$$X_{(g)} + e^- \longrightarrow X^-_{(g)} + \text{EA}$$

The energy involved in the addition of the first electron is called **first-electron affinity;** the energy involved in the addition of a second electron is called **second-electron affinity;** and so on. Thus,

$$X + e^- \longrightarrow X^- + \text{EA}_1$$

$$X^- + e^- \longrightarrow X^{2-} + \text{EA}_2$$

The electron affinity of an element measures the ease with which it forms an anion in the gas phase. Electron affinities are difficult to measure and accurate values are not known for all elements. They are expressed in kJ mol^{-1}.

Trends in Electron Affinities

The factors that determine the magnitude and sign of electron affinities are similar to those used to explain ionization energies of elements. In fact, the electron affinity of a neutral atom may be thought of simply as equivalent to the ionization energy of the singly charged negative ion of the atom.

$$A^- + IE \longrightarrow A + e^-$$
$$A + e^- \longrightarrow A^- + EA$$

The first-electron affinities of elements in the Periodic table are expected to show trends analogous to those of ionization energies.

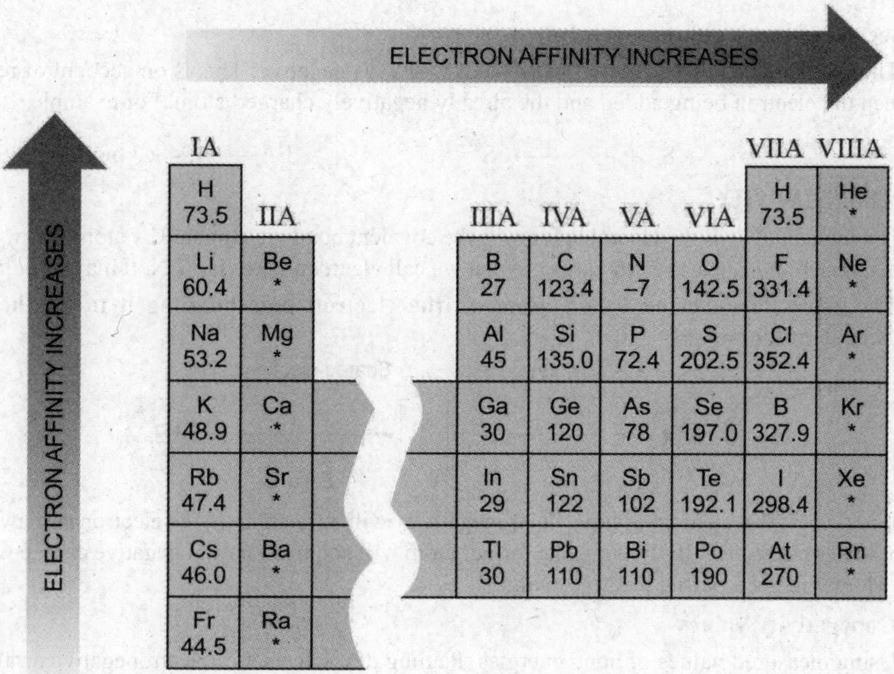

(*a*) Increase across a Period

The values of electron affinities for Period 2 are listed below.

Be	B	C	N	O	F	Ne
– 240	23	123	0.00	142	323	– 29 k J mol^{-1}

As we proceed from left to right, the general trend is the increase of electron affinities. Be, N and Ne are exceptions.

Explanation

Elements having relatively stable electronic configurations find it difficult to accept an electron readily. The atom of Be has the configuration $1s^2\ 2s^2$. The $1s$ subshell is completely filled and, therefore, the electron being added must go to a subshell of considerably higher energy. This gives rise to negative electron affinity for Be.

The atom of N $\left(1s^2\ 2s^2, 2p_x^1, 2p_y^1, 2p_z^1\right)$ has half-filled $2p$ subshells, a condition of extra stability. Therefore the electron affinity of N would be less than expected.

The electron affinity of Neon is low because it has a stable outer-shell octet. Its atom shows little tendency to start a new shell.

(*b*) Decrease down a Group

The values of electron affinities for halogens (Group VII) are given below.

F	Cl	Br	I
332.6	349	324.7	296 k J mol^{-1}

The electron affinities show a general decrease from top to bottom. This is so because the valence shell is progressively farther from the nucleus. The value for fluorine, however, is out of line as it has a smaller atomic size than that of chlorine.

(*c*) Second electron affinity negative

The second electron affinity of an element is always negative. This is on account of repulsion between the electron being added and the already negatively charged atom. For example,

$$S^- + e^- \longrightarrow S^{2-} \qquad EA = -194 \text{ k J mol}^{-1}$$

ELECTRONEGATIVITY

In a molecule A – B the electrons forming the covalent bond are attracted by atom A as well as by B. This attraction is measured in terms of what we call **electronegativity, EN.** It may be defined as :

The attraction exerted by an atom on the electron pair bonding it to another atom by a covalent bond.

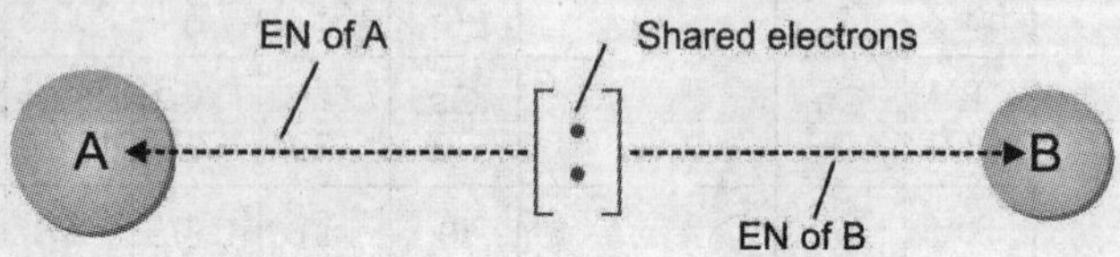

It is evident that an atom of high electronegativity will attract the shared electron pair away from one of lower electronegativity. Thus the former atom will acquire a partial negative charge while the other atom will get a partial positive charge.

Electronegativity Values

Using measured values of bond energies, Pauling devised a set of electronegativity values. He allotted a value of 4 to the most electronegative atom, namely fluorine, and assigned values to the atoms of other elements.

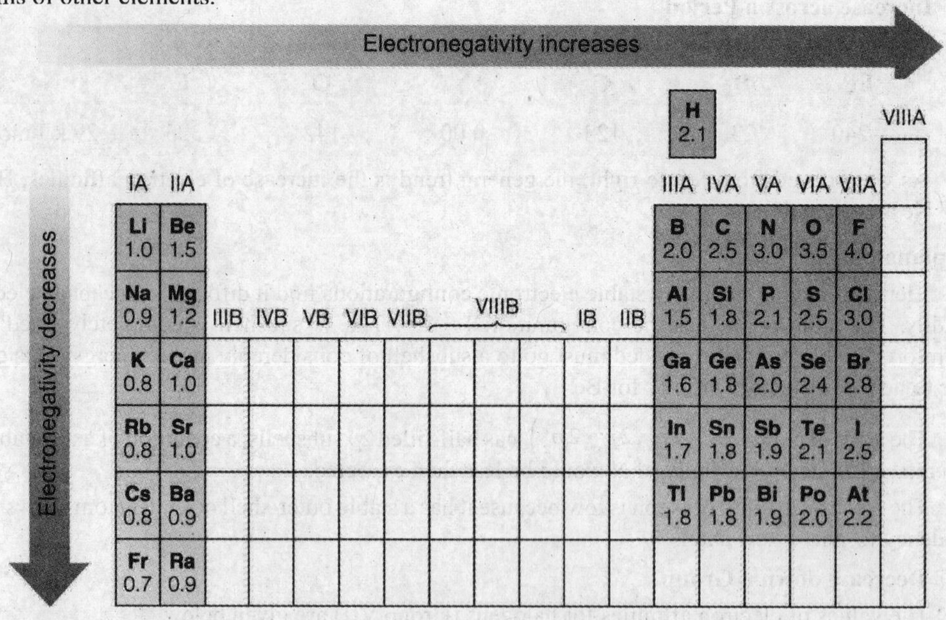

Trend in Electronegativities

The variations in electronegativities of elements in the Periodic table are similar to those of ionization energies and electron affinities.

(1) Increase across a Period

The values of electronegativities increase as we pass from left to right in a Period. Thus for Period 2 we have

Li	Be	B	C	N	O	F
1.0	1.5	2.0	2.5	3.0	3.5	4.0

This is so because the attraction of bonding electrons by an atom increases with increase of nuclear charge (At. No.) and decrease of atomic radius. Both these factors operate as we move to the right in a Period.

(2) Decrease down a Group

The electronegativities of elements decrease from top to bottom in a Group. Thus for Group VII we have

F	Cl	Br	I
4.0	3.0	2.8	2.5

The decrease trend is explained by more shielding electrons and larger atomic radius as we travel down a Group.

Importance of Electronegativity

The electronegativities of elements are widely used throughout the study of Chemistry. Their usefulness will be discussed at appropriate places. The important applications of electronegativities are listed below.

(1) **In predicting the polarity of a particular bond.** The polarity of a bond, in turn, shows the way how the bond would break when attacked by an organic reagent.

(2) **In predicting the degree of ionic character of a covalent bond.**

(3) **In predicting of inductive effects in organic chemistry.**

(4) **In understanding the shapes of molecules.**

ADDITIONAL SOLVED PROBLEMS

SOLVED PROBLEM 8

Calculate the wavelength of an electron of mass 9.1×10^{-31} kg moving with a velocity of 4.2×10^4 m s^{-1} ($h = 6.62 \times 10^{-34}$ J s).

SOLUTION

Formula used

$$\lambda = \frac{h}{m v}$$

Quantities given

$h = 6.62 \times 10^{-34}$ J s $\quad m = 9.1 \times 10^{-31}$ kg $\quad v = 4.2 \times 10^4$ m s^{-1}

Substitution of values

$$\lambda = \frac{6.62 \times 10^{-34} \text{ J s}}{(9.1 \times 10^{-31} \text{ kg}) \times (4.2 \times 10^4 \text{ m s}^{-1})}$$

$$= 0.1732 \times 10^{-34+31-4} \text{ m}$$
$$= 0.1732 \times 10^{-7} \text{ m}$$
$$= \mathbf{1.732 \times 10^{-8} \text{ m}}$$

SOLVED PROBLEM 9

An electron is moving with a kinetic energy of 4.55×10^{-25} J. Calculate de Broglie wavelength for it. (mass of the electron = 9.1×10^{-31} kg; $h = 6.6 \times 10^{-34}$ kg m^2 s^{-1}).

SOLUTION

(*a*) To calculate the velocity of the electron

Formula used

$$K.E. = \frac{1}{2} m v^2$$

or
$$v^2 = \frac{2 \times \text{K.E.}}{\text{m}}$$

Quantities given

$K.E. = 4.55 \times 10^{-25}$ J $\quad m = 9.1 \times 10^{-31}$ kg

Substitution of values

$$v^2 = \frac{2 \times 4.55 \times 10^{-25} \text{ kg m}^2 \text{ s}^{-2}}{9.1 \times 10^{-31} \text{ kg}}$$
$$= 1 \times 10^{6} \text{ m}^2 \text{ s}^{-2}$$

or
$$v = \sqrt{1 \times 10^{6} \text{ m}^2 \text{ s}^{-2}}$$
$$= 1 \times 10^{3} \text{ m s}^{-1}$$

(*b*) To calculate the wavelength of the electron

Formula used

$$\lambda = \frac{h}{m v}$$

Quantities given

$h = 6.6 \times 10^{-34}$ kg m^2 s^{-1} $\quad m = 9.1 \times 10^{-31}$ kg $\quad v = 1 \times 10^{3}$ m s^{-1}

Substitution of values

$$\lambda = \frac{6.6 \times 10^{-34} \text{ kg m}^2 \text{ s}^{-1}}{9.1 \times 10^{-31} \text{ kg} \times 1 \times 10^{3} \text{ m s}^{-1}}$$
$$= 0.7253 \times 10^{-34+31-3} \text{ m}$$
$$= 0.7253 \times 10^{-6} \text{ m}$$
$$= 7.253 \times 10^{-7} \text{ m}$$
$$= \mathbf{725.3 \text{ nm}}$$

SOLVED PROBLEM 10

Calculate the de Broglie's wavelength of an electron travelling with a speed of 20% of light.

SOLUTION

Formula used

$$\lambda = \frac{h}{m v}$$

Quantities given

$h = 6.625 \times 10^{-34}$ kg m^2 s^{-1} $\qquad m = 9.1 \times 10^{-31}$ kg

$$v = 20\% \text{ of } c = \frac{20 \times 3 \times 10^8}{100} \text{ m s}^{-1} = 0.6 \times 10^8 \text{ m s}^{-1}$$

Substitution of values

$$\lambda = \frac{6.625 \times 10^{-34} \text{ kg m}^2 \text{ s}^{-1}}{(9.1 \times 10^{-31} \text{ kg}) \times (0.6 \times 10^8 \text{ m s}^{-1})}$$

$$= 1.2134 \times 10^{-34+31-8} \text{ m}$$

$$= 1.2134 \times 10^{-11} \text{ m}$$

$$= 12.134 \times 10^{-12} \text{ m}$$

$$= \mathbf{12.134 \text{ pm}} \qquad [\therefore 1 \text{ pm} = 10^{-12} \text{ m}]$$

SOLVED PROBLEM 11

Calculate the certainty in position of an electron if uncertainty in velocity is 5.7×10^5 m s^{-1}. ($h = 6.6 \times 10^{-34}$ kg m^2 s^{-1} ; mass of electron = 9.1×10^{-31} kg).

SOLUTION

Formula used

$$\Delta x \times m = \frac{h}{4\pi}$$

or

$$\Delta x = \frac{h}{4 \pi \text{ m} \Delta v}$$

Quantities given

$h = 6.6 \times 10^{-34}$ kg m^2 s^{-1} $\qquad m = 9.1 \times 10^{-31}$ kg $\qquad \Delta v = 5.7 \times 10^5$ m s^{-1}

Substitution of values

$$\Delta x = \frac{6.6 \times 10^{-34} \text{ kg m}^2 \text{ s}^{-1}}{4 \times 3.14 \times 9.1 \times 10^{-31} \text{ kg} \times 5.7 \times 10^5 \text{ m s}^{-1}}$$

$$= 0.0103 \times 10^{-34+31-5} \text{ m}$$

$$= 0.0103 \times 10^{-8} \text{ m}$$

$$= \mathbf{1.03 \times 10^{-10} \text{ m}}$$

SOLVED PROBLEM 12

Calculate the wavelength of a 200 kg drum rolling at a speed of 30 km per hour or 8.33m/sec.

SOLUTION

Formula used

$$\lambda = \frac{h}{m v}$$

Quantities given

$h = 6.6 \times 10^{-34}$ kg m^2 s^{-1} $\qquad m = 200$ kg $\qquad v = 30 \text{ km hr}^{-1} = \dfrac{30 \times 10^3}{60 \times 60} \text{ m s}^{-1}$

$$= 8.33 \text{ m s}^{-1}$$

Substitution of values

$$\lambda = \frac{6.6 \times 10^{-34}\ \text{kg m}^2\ \text{s}^{-1}}{200\ \text{kg} \times 8.33\ \text{m s}^{-1}}$$

$$= \mathbf{3.96 \times 10^{-37}\ m}$$

SOLVED PROBLEM 13

Calculate the momentum of a particle which has a de Broglie wavelength of 1.1×10^{-8} m.

SOLUTION

Formula used

$$\lambda = \frac{h}{m\,v}$$

or

$$m\,v = \frac{h}{\lambda}$$

Quantities given

$h = 6.6 \times 10^{-34}$ kg m^2 s^{-1} $\quad\lambda = 110$ Å $= 110 \times 10^{-10}$ m

Substitution of values

$$m\,v = \frac{6.6 \times 10^{-34}\ \text{kg m}^2\ \text{s}^{-1}}{110 \times 10^{-10}\ \text{m}}$$

$$= \mathbf{6 \times 10^{-26}\ kg\ m\ s^{-1}}$$

SOLVED PROBLEM 14

By applying de Broglie's equation, calculate wavelength associated with the motion of the earth, a ball and an electron. Following data is given :

Mass of earth = 6.0×10^{24} kg.

Mass of ball = 0.1 kg

Mass of electron = 9.1×10^{-31} kg

Velocity of orbital motion of the earth = 3×10^4 m s^{-1}

Velocity of the ball = 100 m s^{-1}

Velocity of the electron = 1.0×10^6 m s^{-1}

$h = 6.6 \times 10^{-34}$ kg m^2 s^{-1}

In case of which of these three objects, will the wavelength be measurable ?

SOLUTION

Formula used

$$\lambda = \frac{h}{m\,v}$$

(*a*) For earth

Quantities given

$h = 6.6 \times 10^{-34}$ kg m^2 s^{-1} $\quad m = 6.0 \times 10^{24}$ kg $\quad v = 3 \times 10^4$ m s^{-1}

Substitution of values

$$\lambda = \frac{6.6 \times 10^{-34}\ \text{kg m}^2\ \text{s}^{-1}}{6.0 \times 10^{24}\ \text{kg} \times 3 \times 10^4\ \text{m s}^{-1}}$$

$$= \mathbf{3.66 \times 10^{-63}\ m}$$

(*b*) For ball

Quantities given

$h = 6.6 \times 10^{-34}$ kg m^2 s^{-1} $\quad m = 0.1$ kg $\quad v = 100$ m s^{-1}

Substitution of values

$$\lambda = \frac{6.6 \times 10^{-34}\ \text{kg m}^2\ \text{s}^{-1}}{0.1\ \text{kg} \times 100\ \text{m s}^{-1}}$$

$$= \mathbf{6.6 \times 10^{-35}\ m}$$

(*c*) For electron

Quantities given

$h = 6.6 \times 10^{-34}$ kg m^2 s^{-1} $\quad m = 9.1 \times 10^{-31}$ kg $\quad v = 1.0 \times 10^6$ m s^{-1}

Substitution of values

$$\lambda = \frac{6.6 \times 10^{-34}\ \text{kg m}^2\ \text{s}^{-1}}{(9.1 \times 10^{-31}\ \text{kg}) \times (1.0 \times 10^6\ \text{m s}^{-1})}$$

$$= \mathbf{7.25 \times 10^{-10}\ m}$$

Thus the wavelength is measurable in case of electron only.

EXAMINATION QUESTIONS

1. Define or explain the following terms :
   (*a*) Planck's constant
   (*b*) de Broglie equation
   (*c*) Heisenberg's uncertainty principle
   (*d*) Schrödinger's wave equation
   (*e*) Principal quantum number
   (*f*) Azimuthal quantum number
   (*g*) Magnetic quantum number
   (*h*) Zeeman effect
   (*i*) Spin quantum number
   (*j*) Pauli's exclusion principle
   (*k*) Aufbau principle
   (*l*) Hund's rule
2. **What do you mean by the 'ionization potential' of an element? Why the first ionization potential of an element is less than the second ionization potential? How does the ionization potential of an element vary with atomic volume?**
3. (*a*) What do you understand by the dual character of matter? Derive de Broglie's equation. How was it verified?

   (*b*) A particle having a wavelength 6.6×10^{-6} m is moving with a velocity of 10^4 m s^{-1}. Find the mass of the particle. Planck's constant = 6.62×10^{-34} J s.

   Answer. (*b*) 1.003×10^{-32} kg
4. (*a*) State and discuss Heisenberg's uncertainty principle.

   (*b*) Draw energy level diagram valid for hydrogen and multi-electron atoms.

   (*c*) Write Schrödinger wave equation.

5. Discuss the following :
   (a) Hund's Rule of Maximum multiplicity.
   (b) Pauli's Exclusion Principle.
6. What is the wavelength associated with a particle of mass 10^{-4} kg moving with a speed of 1×10^3 m s^{-1} ($h = 6.6 \times 10^{-34}$ J s)?
   Answer. 6.6×10^{-31} m
7. State Pauli's exclusion principle. Based on this principle show that the maximum number of electrons that can be accommodated in an orbit is 8 when $n = 20$.
8. The velocity of a ball being bowled by Kapil Dev is 25 m s^{-1}. Calculate the wavelength of the matter-wave associated with the ball. Derive the equation you used. Weight of cricket ball = 0.1585 kg and $h = 6.625 \times 10^{-34}$ J s.
   Answer. 1.67×10^{-34} m
9. Calculate the de Broglie wavelength for a ball of 0.2 kg mass moving with a velocity 3×10^8 m s^{-1} and an electron moving with the same velocity. What do these values indicate?
   Answer. 1.103×10^{-41} m; 2.4239×10^{-12} m
10. (a) Describe the Davison and Germer experiment for the verification of wave nature of electrons.
    (b) If an electron is accelerated by 100 volts, calculate the de Broglie wavelength associated with it. Also calculate the velocity acquired by the electron (Mass of electron = 9.1×10^{-31} kg; $h = 6.62 \times 10^{-34}$ J s).
    Answer. (b) 1.2267×10^{-10} m; 5.930×10^6 m s^{-1}
11. State Pauli's exclusion principle and show that the maximum number of electrons in a given shell is $2n^2$ where n is the principal quantum number of the shell.
12. Derive Schrödinger wave equation for the wave mechanical model of an atom and discuss its application to hydrogen atom. What is the significance of ψ and ψ^2 in it?
13. What are three quantum numbers used to describe an orbital? What property of an orbital is described by each quantum number? Specify the rule that governs the values of each quantum number.
14. What are quantum numbers? Mention all values of different quantum numbers when $n = 2$.
15. (a) Which other particles besides electron show particle-wave duality? Give two examples.
    (b) Calculate the wavelength of a particle of mass 0.0015 kg moving with a velocity of 250 m s^{-1}.
    Answer. (b) 1.766×10^{-33} m
16. (a) Write a note on Heisenberg's uncertainty principle. How this principle goes against Bohr's theory?
    (b) What is ionization energy? What are the factors which affect the ionization energy of an element? How ionization energy changes among the elements in a group and in a period?
17. Write short notes on :
    (a) Pauli's exclusion principle
    (b) Uncertainty principle
    (c) Photoelectric effect
    (d) Aufbau principle
18. Calculate the momentum of a particle which has a de Broglie's wavelength of 1×10^{-10} m.
    Answer. 6.6×10^{-24} kg m^2 s^{-1}
19. The kinetic energy of an electron is 4.55×10^{-25} J. Calculate its wavelength ($h = 6.6 \times 10^{-34}$ kg m^2 s^{-1}; mass of electron = 9.1×10^{-31} kg).
    Answer. 7.25×10^{-1} m
20. The kinetic energy of a subatomic particle is 5.60×10^{-25} J. Calculate the frequency of the particle wave (Planck's Constant $h = 6.6 \times 10^{-34}$ kg m^2 s^{-1})
    Answer. 1.696×10^9 s^{-1}

21. Calculate the wavelength associated with an electron moving with a velocity of 1×10^6 m s^{-1}. (mass of the electron = 9.1×10^{-31} kg).

Answer. 7.28×10^{-10} m

22. Calculate the uncertainty in the velocity of a bullet of mass 0.01 kg whose position at time *t* is known with uncertainty equal to 1.0×10^{-5} m.

Answer. 5.271×10^{-28} m s^{-1} *(Panjab BSc)*

23. Calculate the uncertainty in the velocity of an electron if the uncertainty in position is 1×10^{-10} m.

Answer. 5.76×10^5 m s^{-1} *(Madras BSc)*

24. Calculate the uncertainty in the position of a particle when the uncertainty in the momentum is

(*a*) 1×10^{-7} kg m s^{-1} and (*b*) zero

Answer. (*a*) 5.72×10^{-28} m; (*b*) ∞ *(Delhi BSc)*

25. Explain Heisenberg's uncertainty principle. How does it influence the concept of the electron? *(Lucknow BSc)*

26. What are postulates of Quantum Mechanics? *(MD Rohtak BSc)*

27. Write Schrödinger's wave equation for a single electron atom. *(Nagpur BSc)*

28. (*a*) Deduce de Broglie's relation for a matter wave and explain the terms involved in it.

(*b*) Draw and explain the angular probability distribution curve for 1*s* electron. *(Mizoram BSc)*

29. (*a*) What is the significance of the wave function, 'ψ'?

(*b*) State and explain Heisenberg's uncertainty principle. *(Mizoram BSc)*

30. Give wave mechanical interpretation of an atomic orbital. Discuss qualitatively the probability distribution curves of *s*, *p* and *d* orbitals and hence draw the contour diagrams of all the 2*p*, 3*p* and 3*d* orbitals on x, y and z axis, showing the sign of wave functions. *(Guru Nanak Dev BSc)*

31. Describe the shapes of *s* and *p* orbitals on the basis of quantum mechanical concept. *(Mizoram BSc)*

32. A moving ball weighing 0.2 kg is to be located within 2×10^{-11} m. What is the uncertainty in the velocity? Comment on your result. (Given $h = 6.626 \times 10^{-34}$ J s)

Answer. 1.3189×10^{-23} m s^{-1} *(Lucknow BSc)*

33. Calculate de Broglie's wavelength of a xenon atom moving with a velocity of 2.4×10^2 m s^{-1}. (Atomic weight of xenon is 2.2×10^{-25} kg)

Answer. 1.2538×10^{-11} m *(Arunachal BSc)*

34. (*a*) Give de Broglie's relation for a particle of mass, m, moving with a velocity, v.

(*b*) Calculate the uncertainty in velocity of a cricket ball of mass 0.15 kg if the uncertainty in position is 1×10^{-10} m. ($h = 6.63 \times 10^{-34}$ J s)

Answer. (*b*) 3.5138×10^{-24} m s^{-1} *(Arunachal BSc)*

35. An electron has a speed of 3.0×10^2 m s^{-1} accurate to 0.01%. Find out the uncertainty in the position of the electron. ($h = 6.625 \times 10^{-34}$ J s, $m = 9.11 \times 10^{-31}$ kg)

Answer. 17.582×10^{-32} m *(Vidyasagar BSc)*

36. (*a*) Discuss probability distribution curves for *s* and *p* orbitals.

(*b*) Discuss the following :

(*i*) Wave nature of an electron

(*ii*) Significance of wave function *(Jammu BSc)*

37. (*a*) Write the electronic configuration of the elements with atomic numbers : 10, 20, 29, 49 and 63

(*b*) Write Hund's rule of maximum multiplicity.

(*c*) What do you understand by wave function? (*Jammu BSc*)

38. (*a*) Give radial probability distribution curves for 2*p* and 3*p* orbitals and give their characteristics.

(*b*) Draw energy level diagram for a multi-electron atom.

(*c*) Write the ground state electronic configuration for Ce (Z = 58) and Sn (Z = 50).

(*Punjabi BSc*)

39. State Schrödinger's wave equation. How this equation led to quantisation of energy?

(*Nagpur BSc*)

40. Calculate uncertainty in momentum of electron, if uncertainty in position is 10^{-8} m ($h = 6.624 \times 10^{-34}$ J s).

Answer. 5.27×10^{-27} (*Nagpur BSc*)

41. (*a*) What is 'Effective Nuclear Charge'? What relation does it have with 'shielding' phenomenon?

(*b*) Given that the first I.E. of Al is less than that of Mg. What about the second I.E. of Al compared to that of Mg? Why? (*Delhi BSc*)

42. Calculate the energy of a photon of wavelength 4×10^{-7} m. ($h = 6.62 \times 10^{-34}$ J s.)

Answer. 4.965×10^{-19} J (*Sambalpur BSc*)

43. The wavelength of blue light is 4.8×10^{-7} m. Calculate the frequency and wavelength of the light. ($c = 3 \times 10^8$ m s^{-1})

Answer. 6.25×10^{14} s^{-1} and 1.6×10^{-15} m^{-1} (*Sambalpur BSc*)

44. The uncertainty in the momentum of a particle is found to be 2.5×10^{-17} kg m^{-1}. What is the uncertainty in its position? ($h = 6.626 \times 10^{-34}$ J s)

Answer. 2.108×10^{-18} m (*Sambalpur BSc*)

45. Determine the wavelength of a cricket ball having a mass 4.0×10^{-2} kg and velocity 30 m s^{-1}. ($h = 6.624 \times 10^{-34}$ J s)

Answer. 5.52×10^{-34} m (*Nagpur BSc*)

46. Calculate uncertainty in momentum of electron, if uncertainty in position is 10^{-6} m.

($h = 6.624 \times 10^{-34}$ J s)

Answer. 0.527×10^{-28} Ns (*Nagpur BSc*)

47. Calculate de Broglie wavelength associated with a ball weighing 0.15 kg thrown with a velocity of 30 m s^{-1}. ($h = 6.625 \times 10^{-34}$ s)

Answer. 0.0147×10^{-32} m (*Sambalpur BSc*)

48. Calculate the de Broglie wavelength of an electron moving with a velocity of 6×10^5 m s^{-1}.

Answer. 1.456×10^{-8} m (*Punjabi BSc*)

49. Calculate the uncertainty in position of an electron if uncertainty in velocity is 5.7×10^5 m s^{-1}. ($h = 6.6 \times 10^{-34}$ kg m^2 s^{-1}; mass of electron = 9×10^{-31} kg)

Answer. 10^{-10} m (*Agra BSc*)

50. A body moving with a speed of 100 m s^{-1} has a wavelength of 5×10^{-36} m. Calculate the mass of the body. ($h = 6.6 \times 10^{-34}$ kg m^2 s^{-1})

Answer. 1.32 kg (*Panjab BSc*)

51. Calculate the uncertainty in momentum of an electron if uncertainty in its position is approximately 100 pm (100×10^{-12} m) = 10^{-10} m.

Answer. 5.27×10^{-25} kg m s^{-1} (*Delhi BSc*)

52. Calculate the wavelength associated with an electron moving with a velocity of 1×10^6 m s^{-1}. (mass of the electron = 9.1×10^{-31} kg). *(Delhi BSc)*

Answer. 7.28×10^{-10} m

53. Calculate the de Broglie wavelength of an electron moving with a velocity of 6×10^5 m s^{-1}.

Answer. 1.456×10^{-8} m *(Tripura BSc)*

54. A body moving with a speed of 100 m s^{-1} has a wavelength of 5×10^{-36} m. Calculate the mass of the body. ($h = 6.6 \times 10^{-34}$ kg m^2 s^{-1}). *(Banaras BSc)*

Answer. 1.32 kg

55. Calculate the uncertainty in momentum of an electron if uncertainty in its position is approximately 100 pm. (1×10^{-12} m). *(Nagpur BSc)*

Answer. 5.271×10^{-23} kg m s^{-1}

56. Calculate the uncertainty in the velocity of a bullet weighing 0.01 kg whose position is known with an accuracy of $\pm 0.1 \times 10^{-9}$ m. *(Kanpur BSc)*

Answer. 0.527×10^{-27} m s^{-1}

57. Calculate the certainty in position of an electron if uncertainty in velocity is 5.7×10^5 m s^{-1}. ($h = 6.6 \times 10^{-34}$ kg m^2 s^{-1} ; mass of electron = 9.1×10^{-31} kg). *(Delhi BSc)*

Answer. 1.03×10^{-10} m

58. Calculate the momentum of a particle which has a de Broglie wavelength of 1.1×10^{-8} m.

Answer. 6×10^{-26} kg m s^{-1} *(Madhurai BSc)*

59. Calculate the wavelength of a particle of mass 1.5×10^{-3} kg moving with a velocity of 250 m s^{-1}.

Answer. 1.766×10^{-33} m *(Baroda BSc)*

60. Calculate the momentum of a particle which has de Broglie's wavelength of 1×10^{-10} m. ($h = 6.6 \times 10^{-34}$ kg m^2 s^{-1}). *(Banaras BSc)*

Answer. 6.6×10^{-24} kg m s^{-1}

MULTIPLE CHOICE QUESTIONS

1. According to de Broglie's equation, the momentum of a particle in motion is _______ proportional to wavelength.
(*a*) inversely (*b*) directly
(*c*) is not (*d*) none of these
Answer. (*a*)

2. The wavelength of large objects is of no significance as it is too _______ to be measurable.
(*a*) small (*b*) large
(*c*) heavy (*d*) none of these
Answer. (*a*)

3. de Broglie equation is
(*a*) $\lambda = h/m\,v$ (*b*) $\lambda = m\,v/h$
(*c*) $\lambda = h\,m\,v$ (*d*) $\lambda = \dfrac{h\,v}{m}$
Answer. (*a*)

4. "It is impossible to determine simultaneously the position and velocity with accuracy of a small particle like electron". This statement is
(*a*) Heisenberg's uncertainty principle (*b*) de Broglie principle
(*c*) Planck's law (*d*) Aufbau's principle
Answer. (*a*)

5. The relation $\Delta x \times \Delta p = \frac{h}{4\pi}$ represents

(*a*) de Broglie equation (*b*) Heisenberg's uncertainty principle
(*c*) Schrödinger's wave equation (*d*) Pauli's exclusion principle

Answer. (*b*)

6. In Schrödinger's wave equation, the symbol ψ represents the

(*a*) wavelength of the spherical wave (*b*) amplitude of the spherical wave
(*c*) frequency of the spherical wave (*d*) none of these

Answer. (*b*)

7. The energy of electron in an atomic orbital is always ________.

(*a*) different (*b*) zero
(*c*) infinite (*d*) same

Answer. (*d*)

8. An orbital is the space around the nucleus where the probability of finding electron is

(*a*) always zero (*b*) maximum
(*c*) minimum (*d*) always infinite

Answer. (*b*)

9. The Principal quantum number '*n*' represents

(*a*) average size of the electron cloud
(*b*) average energy of the electron
(*c*) average distance of the electron from the nucleus
(*d*) all of the above

Answer. (*d*)

10. The Principal quantum number is related to the

(*a*) orbital angular momentum (*b*) size and shape of the orbital
(*c*) orientation of the orbital (*d*) average size of the orbital

Answer. (*d*)

11. The quantum number that defines the shape of the orbital occupied by the electron is

(*a*) principal quantum number (*b*) azimuthal quantum number
(*c*) magnetic quantum number (*d*) spin quantum number

Answer. (*b*)

12. The angular momentum of the electron is defined by the quantum number that is denoted as

(*a*) n (*b*) ℓ
(*c*) m (*d*) s

Answer. (*b*)

13. The total number of sublevels in each principal level is equal to

(*a*) spin quantum number (*b*) magnetic quantum number
(*c*) azimuthal quantum number (*d*) principal quantum number

Answer. (*d*)

14. The quantum number which accounts for the splitting up of spectral lines (Zeeman effect) is

(*a*) principal quantum number (*b*) azimuthal quantum number
(*c*) magnetic quantum number (*d*) spin quantum number

Answer. (*c*)

15. For a given value of principal quantum number the order of increasing energy for different subshells is

(*a*) $s<p<d<f$ (*b*) $p<d<f<s$
(*c*) $d<f<p<s$ (*d*) $f<d<p<s$

Answer. (*a*)

16. The p_x, p_y and p_z orbitals are called degenerate orbitals as they have

(*a*) equal energy (*b*) same orientation in space

(*c*) same size (*d*) none of these

Answer. (*a*)

17. A nodal plane separates the two lobes of a *p*-orbital. There is _______ likelihood of finding the electron on this plane.

(*a*) no (*b*) every

(*c*) either of these (*d*) none of these

Answer. (*a*)

18. The total values of magnetic quantum number for a given value of azimuthal quantum number is

(*a*) 2ℓ (*b*) $2\ell + 1$

(*c*) $2\ell - 1$ (*d*) $2\ell - 2$

Answer. (*b*)

19. "No two electrons in an atom can have same set of four identical quantum numbers". It is the statement of

(*a*) Aufbau principle (*b*) Hund's rule

(*c*) Pauli's exclusion principle (*d*) none of these

Answer. (*c*)

20. The orbital with $n = 3$ and $\ell = 2$ is

(*a*) 3*s* (*b*) 3*p*

(*c*) 3*d* (*d*) 3*f*

Answer. (*c*)

21. 4*s* orbital has lesser energy than 3*d* orbital because it has

(*a*) greater value of *n* (*b*) lesser value of ℓ

(*c*) lesser value of $n + \ell$ (*d*) $\ell = 0$

Answer. (*c*)

22. The maximum number of electrons that can be accommodated in *f*-subshell is

(*a*) 5 (*b*) 7

(*c*) 10 (*d*) 14

Answer. (*d*)

23. The energy associated with electrons in *s*, *p*, *d* and *f* orbitals of a particular principal quantum number in hydrogen atom is in the order

(*a*) $s = p = d = f$ (*b*) $s < p < d < f$

(*c*) $p < d < f < s$ (*d*) $f < d < p < s$

Answer. (*a*)

24. For a multi-electron atom, the energy associated with electrons is *s*, *p*, *d* and *f* orbitals of a particular quantum number is in the order

(*a*) $s = p = d = f$ (*b*) $s < p < d < f$

(*c*) $p < d < f < s$ (*d*) $d < f < s < p$

Answer. (*b*)

25. The two electrons in the first shell will differ in the values for

(*a*) *n* (*b*) ℓ

(*c*) *m* (*d*) *s*

Answer. (*d*)

26. Which one of the following sets of quantum numbers is not allowed?

	n	ℓ	m	s
(*a*)	1	0	1	–S
(*b*)	2	1	0	+S
(*c*)	2	1	–2	+S
(*d*)	2	1	+2	0

Answer. (*d*)

27. Which of the following is incorrect for 3*d* orbital?

	n	ℓ	m	s
(*a*)	3	0	0	+S
(*b*)	3	1	0	+S
(*c*)	3	2	0	+S
(*d*)	3	1	2	+S

Answer. (*d*)

28. The value of azimuthal quantum number for last electron of N-atom is

(*a*) 0 (*b*) 1
(*c*) 2 (*d*) 3

Answer. (*b*)

29. The maximum number of electrons in a subshell is given by the equation

(*a*) n^2 (*b*) $2n^2$
(*c*) $2\ell-1$ (*d*) $2\ell+1$

Answer. (*d*)

30. Out of the following, which is the correct set of quantum numbers for the outermost electron of potassium atom (Z = 19)?

	n	ℓ	m	s
(*a*)	4	3	2	–S
(*b*)	4	2	0	–S
(*c*)	4	1	0	+S
(*d*)	4	0	0	–S

Answer. (*d*)

31. The number of unpaired electrons in oxygen atom is

(*a*) 1 (*b*) 2
(*c*) 3 (*d*) 4

Answer. (*b*)

32. The number of unpaired electrons in chromium atom (Z = 24) is

(*a*) 1 (*b*) 2
(*c*) 3 (*d*) 6

Answer. (*d*)

33. In nitrogen atom there are three unpaired electrons. These are having _______ direction of spin.

(*a*) same (*b*) different
(*c*) similar (*d*) none of these

Answer. (*a*)

34. The maximum number of electrons that can be accommodated in *s*, *p*, *d* and *f* orbitals is

(*a*) 1, 2, 3 and 4 respectively (*b*) 1, 2, 4 and 8 respectively
(*c*) 2, 4, 6 and 8 respectively (*d*) 2, 6, 10 and 14 respectively

Answer. (*d*)

35. The sum of all quantum numbers of the electron of hydrogen atom is

(*a*) –1/2 (*b*) 1
(*c*) 3/2 (*d*) +1/2

Answer. (*c*)

36. The sum of all quantum numbers of the last electron in lithium atom is

(*a*) 3/2 (*b*) 2

(*c*) 5/2 (*d*) 3

Answer. (*c*)

37. The value of azimuthal quantum number for the electrons present in 5*s*-orbital is

(*a*) 0 (*b*) 1

(*c*) 2 (*d*) 5

Answer. (*a*)

38. According to Pauli's exclusion principle two electrons can occupy the same orbital only if they have _______ direction of spin.

(*a*) different (*b*) same

(*c*) similar (*d*) none of these

Answer. (*a*)

39. In the ground state of an atom, the electrons tend to occupy the available orbitals in the _______ order of energies.

(*a*) increasing (*b*) decreasing

(*c*) any (*d*) none of these

Answer. (*a*)

40. Amongst 3*d*, 4*s* and 4*p* orbitals, the order of increasing energies is

(*a*) $3d < 4p < 4s$ (*b*) $4s < 3d < 4p$

(*c*) $4p < 4s < 3d$ (*d*) $3d < 4s < 4p$

Answer. (*d*)

41. While comparing the energies of two orbitals we compare their $(n + \ell)$ values the orbital with _______ $(n + \ell)$ value will have _______ energy.

(*a*) lower, lower (*b*) higher, lower

(*c*) lower, higher (*d*) none of these

Answer. (*a*)

42. When two orbitals have the same $(n + \ell)$ value, the orbital with lower value of _______ has lower energy.

(*a*) principal quantum number (*b*) azimuthal quantum number

(*c*) magnetic quantum number (*d*) spin quantum number

Answer. (*a*)

43. After filling the 4*p*-orbitals, an electron will enter in

(*a*) 4*d* (*b*) 4*f*

(*c*) 5*s* (*d*) 3*d*

Answer. (*c*)

44. If the electronic configuration of nitrogen (at no = 7) is written as $1s^2\ 2s^2\ 2p_x^2\ 2p_y^{\ 1}$, it would violate

(*a*) Aufbau principle (*b*) Pauli's exclusion principle

(*c*) Hund's rule of maximum multiplicity (*d*) none of these

Answer. (*c*)

45. The outermost electronic configuration of manganese (at. no. = 25) is

(*a*) $3d^5\ 4s^2$ (*b*) $3d^6\ 4s^1$

(*c*) $3d^7\ 4s^0$ (*d*) $3d^6\ 4s^2$

Answer. (*a*)

46. The subshell, which does not exist, has the quantum numbers

(*a*) $n = 2 \quad \ell = 0$ (*b*) $n = 2 \quad \ell = 1$

(*c*) $n = 2 \quad \ell = 2$ (*d*) $n = 3 \quad \ell = 0$

Answer. (*c*)

47. The ground state electronic configuration of carbon atom has _______ pairs and _______ unpaired electrons

(*a*) 2, 2 (*b*) 1, 2

(*c*) 2, 1 (*d*) 2, 3

Answer. (*a*)

48. Two electrons occupying the same orbital have different ______.
(*a*) principal quantum number (*b*) azimuthal quantum number
(*c*) magnetic quantum number (*d*) spin quantum number
Answer. (*d*)

49. If the value of azimuthal quantum number is 2, there will be _____ values for magnetic quantum number.
(*a*) 2 (*b*) 3
(*c*) 4 (*d*) 5
Answer. (*d*)

50. The energy needed to remove a single electron (most loosely bound) from an isolated gaseous atom is called
(*a*) ionization energy (*b*) electron affinity
(*c*) kinetic energy (*d*) electronegativity
Answer. (*a*)

51. Generally speaking, the ionization energies increase when we move
(*a*) from left to right in the periodic table (*b*) from top to bottom in a group
(*c*) from right to left in the periodic table (*d*) none of these
Answer. (*a*)

52. The ionization energy of Boron (Z = 5) is lesser than that of Beryllium (Z = 4). It is because
(*a*) Be has an incomplete 2*s* orbital
(*b*) Be has two pairs of electrons
(*c*) 2*p* orbital is already higher in energy than 2*s* orbital
(*d*) none of the above
Answer. (*c*)

53. A neutral atom can accept an electron to form an anion. This process involves
(*a*) loss of energy (*b*) gain of energy
(*c*) no change in energy (*d*) none of these
Answer. (*a*)

54. Electron affinity is expressed in
(*a*) g mol^{-1} (*b*) kJ mol^{-1}
(*c*) cal g^{-1} (*d*) kJ g^{-1}
Answer. (*b*)

55. When we move from left to right across a period, the electron affinity in general
(*a*) remains the same (*b*) decreases
(*c*) increases (*d*) becomes zero
Answer. (*c*)

56. The attraction exerted by an atom on the electron pair bonding it to another atom by covalent bond is called
(*a*) ionization energy (*b*) electron affinity
(*c*) electronegativity (*d*) none of these
Answer. (*c*)

57. The most electronegative element in the periodic table is
(*a*) cesium (*b*) chlorine
(*c*) fluorine (*d*) barium
Answer. (*c*)

58. The values of electronegativities ______ as we move from left to right in a period.
(*a*) increase (*b*) decrease
(*c*) remain the same (*d*) none of these
Answer. (*a*)

59. The electron affinities _______ from top to bottom in a group
(*a*) increase (*b*) decrease
(*c*) remain the same (*d*) none of these
Answer. (*b*)

60. The second electron affinity of an element is always
(*a*) zero (*b*) positive
(*c*) negative (*d*) infinity
Answer. (*c*)

3 Isotopes, Isobars and Isotones

CONTENTS

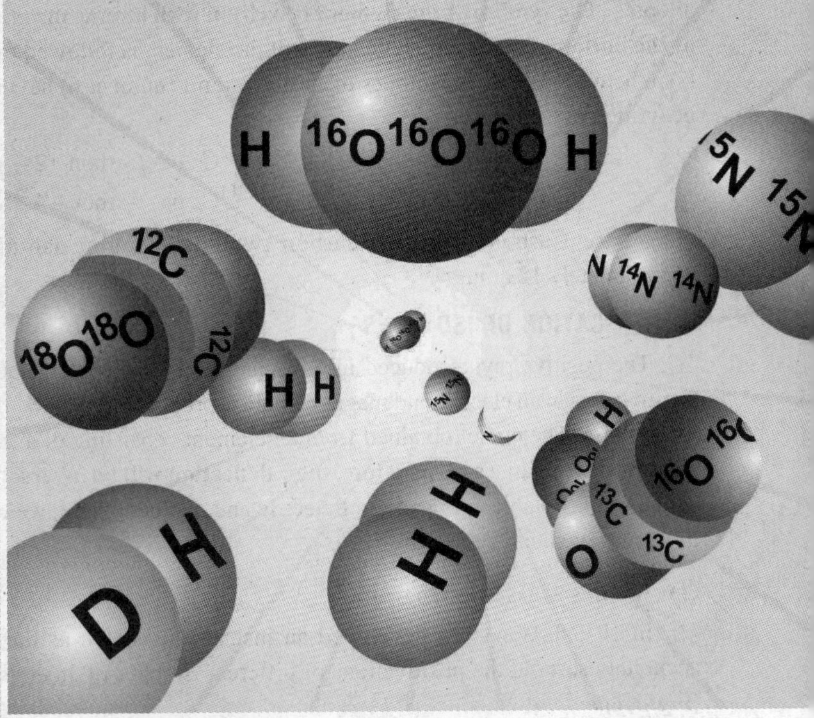

WHAT ARE ISOTOPES?

Contrary to Dalton's Atomic theory, all atoms of a given element are not necessarily identical. In fact, most elements have been shown to be composed of two or more types of atoms mixed in a fixed proportion.

(1) The different atoms of such an element contain equal number of protons and, therefore, have the same atomic number.

(2) The atoms which vary from one another have different number of neutrons in the nucleus. Thus they have different atomic masses.

The atoms of an element which have the same number of protons and different number of neutrons are called Isotopes.

Alternatively, isotopes may be defined as :

The atoms of an element which have the same atomic number but different atomic masses or mass numbers.

The name 'isotope' was assigned to them by Soddy because they have the same atomic number and hence occupied the same place in the periodic table (Greek, *isos* = same; *topos* = place). Isotopes have similar chemical properties as they have the same electronic configuration. However, they differ in respect of physical properties which depend on atomic mass.

SYMBOLIC REPRESENTATION OF ISOTOPES

In denoting particular isotopes of an element, the following notation has been internationally adopted. **The symbol of the element is written with atomic mass at the head and atomic number at the bottom.** Alternatively, the name of the element is followed by the atomic mass with a hyphen (-) in between. Thus the isotopes of carbon (atomic number 6) having atomic masses 12 and 14 may be written as

$$^{12}_{6}C \text{ or } ^{12}C \text{ or Carbon-12}$$
$$^{14}_{6}C \text{ or } ^{14}C \text{ or Carbon-14}$$

^{12}C or Carbon-12 reads 'carbon twelve', meaning isotope of carbon with a mass of approximately 12 amu.

IDENTIFICATION OF ISOTOPES

The positive rays produced in a discharge tube consist of nuclei of atoms. The deflection of positive rays in an electric and magnetic field is proportional to *e/m*, the charge on the particle divided by its mass. The nuclei obtained from an element consisting of a mixture of isotopes will have the same positive charge and, therefore, **their deflection will be inversely proportional to their masses.** Thus with a suitable application of electric and magnetic field, we can identify the isotopes present in a given element.

(1) Aston's Mass Spectrograph

In 1919 F.W. Aston developed an instrument known as the *Mass Spectrograph* which can accurately sort out the positive ions of different isotopes of an element and determine their masses (Fig. 3.1).

In this apparatus a beam of positive rays obtained from a gaseous element in the discharge tube, is rendered into a fine ribbon by passing through slits S_1 and S_2. The fine beam consisting of positive ions of the various isotopes of the element is then sent between the electrostatically charged plates P_1 and P_2. Here the beam of positive ions is deflected down toward the negative plate. The slow moving ions of the same isotope are deflected more than the faster ones which causes a broadening of the beam. Also, the heavier particles are deflected more (being slower) than the lighter ones and this brings about a separation of the various isotopes. The broadened beam of ions is then subjected to a magnetic field (shown by dashed circle) at right angles to the plane of the charged plates and is thus sent in a direction opposite to that caused by the electrostatic field, slower particles again being deflected most. By adjustment of the two fields all ions of the same mass come to focus on the same point on the photographic plate where a sharp line is obtained. Thus each line recorded on the photographic plate shows the existence of separate isotope. Further, the intensity of the line in comparison with the lines of other isotopes, gives the relative abundance of this particular isotope.

The mass of a particle corresponding to a line produced on the photographic plate is determined by comparing with a standard line produced by a particle of known mass (say, $O^+ = 16$). For example, the examination of a sample of neon and chlorine by Aston's Mass Spectrograph showed that they were made of Ne-20, Ne-22 and Cl-35, Cl-37 respectively. The intensities of their lines showed that the relative abundance was

$$^{20}Ne \ : \ ^{22}Ne \quad \text{as} \quad 9:1$$
$$^{35}Cl \ : \ ^{37}Cl \quad \text{as} \quad 3:1$$

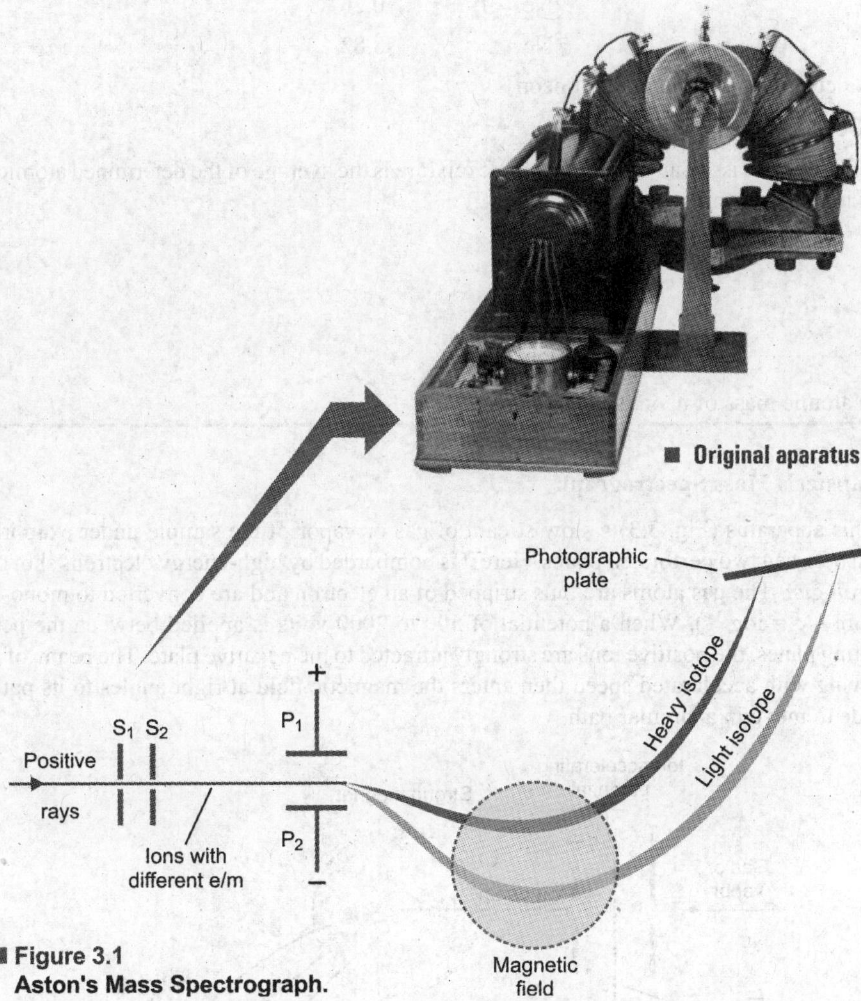

Figure 3.1
Aston's Mass Spectrograph.

Thus Aston's Mass spectrograph not only helped in identifying the isotopes present in an element but also helped in determining the average atomic mass of a given element.

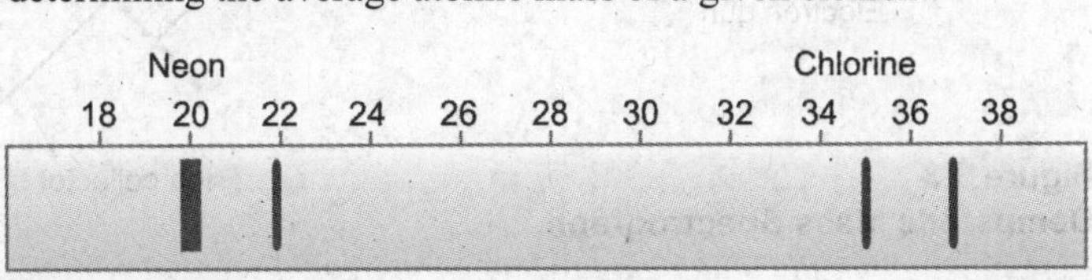

Figure 3.2
Mass spectrographs of Neon and Chlorine.

SOLVED PROBLEM 1

A Sample of neon is found to consist of ^{20}Ne, ^{21}Ne and ^{22}Ne in the following percentages :

^{20}Ne	90.92
^{21}Ne	0.26
^{22}Ne	8.82

Calculate the atomic mass of neon.

SOLUTION

The atomic mass of an ordinary isotopic mixture is the average of the determined atomic masses of individual isotopes. Thus :

$$20 \times 0.9092 = 18.18$$
$$21 \times 0.0026 = 0.055$$
$$22 \times 0.0882 = 1.94$$
$$\overline{\quad 20.18 \quad}$$

The atomic mass of neon is **20.18.**

(2) Dempster's Mass Spectrograph

In this apparatus (Fig. 3.3) a slow stream of gas or vapor of the sample under examination is passed in between two perforated plates. Here it is bombarded by high-energy electrons shot out from an *electron gun.* The gas atoms are thus stripped of an electron and are converted to mono-positive ions (atom – e = ion $^{1+}$). When a potential of 500 to 2000 volts is applied between the perforated accelerating plates, the positive ions are strongly attracted to the negative plate. The beam of positive ions moving with accelerated speed then enters the magnetic field at right angles to its path and is thus made to move in a circular path.

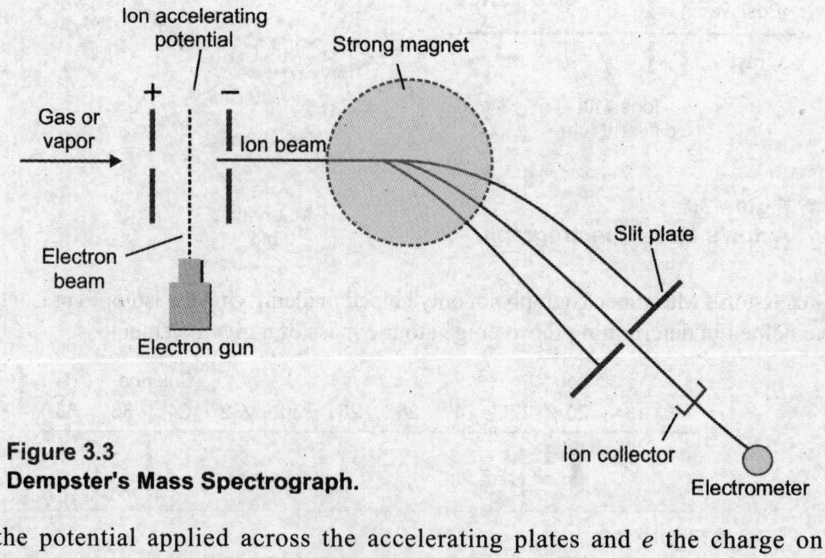

■ Figure 3.3
Dempster's Mass Spectrograph.

If V is the potential applied across the accelerating plates and e the charge on each positive particle, the electrical energy is Ve. This is imparted to the particles as kinetic energy, $1/2\ mv^2$. Thus,

$$Ve = \frac{1}{2} mv^2 \qquad ...(1)$$

In the magnetic field of strength H, the magnetic force on the particle Hev, exactly balances the centrifugal force, mv^2/r, r being the radius of the circular path. Thus,

$$\frac{mv^2}{r} = Hev \qquad ...(2)$$

Eliminating v between (1) and (2), we have

$$r = \sqrt{\frac{2Vm}{H^2 e}}$$

or

$$\frac{m}{e} = \frac{H^2 r^2}{2V} \qquad ...(3)$$

e, being the unit electrical charge, and r (depending on particular apparatus) are constant. If during an experiment magnetic field H is kept the same, from (3) it follows that

$$m \propto \frac{1}{V}$$

Thus by adjusting the accelerating potential (V), particles of mass m can be made to fall on the collector plate. Each ion sets up a minute electric current which passes to the electrometer. **The strength of the current thus measured, gives the relative abundance of the particles of mass *m*.** Similarly, the particles of the other isotopes having different masses are made to fall on the collector and current strength measured. The current strength in each case gives the relative abundance of the individual isotopes. By comparing the current strengths with an experiment performed with C-12 ion, the mass numbers of the various isotopes can be determined.

In the modern mass spectrographs, each ion strikes a detector, the ion current is amplified and fed to a recorder. The recorder makes a graph showing relative abundance plotted against mass number. A computer-plotted graph of neon isotopes is shown in Fig. 3.4.

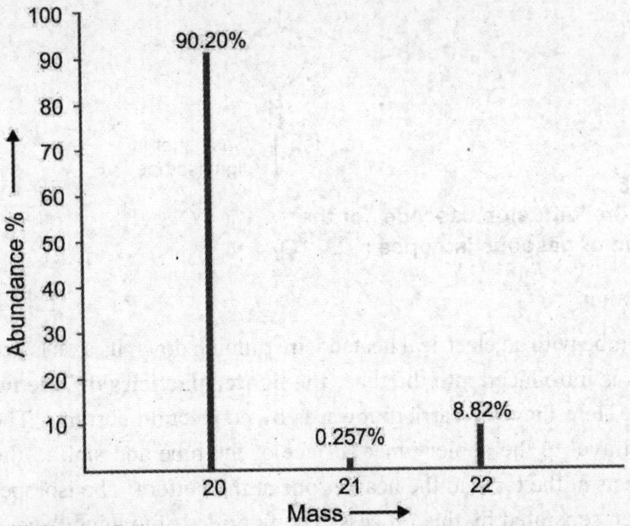

■ **Figure 3.4**
A computer-plotted graph showing relative abundances of the three isotopes of neon against mass number.

SEPARATION OF ISOTOPES

Since isotopes have exactly similar chemical properties, their separation by chemical means is not an easy task. **Their difference in those physical properties which depend on the mass of the atom has been utilized to effect their separation.** The methods commonly employed for the purpose are:

(1) Gaseous Diffusion.

The rate of diffusion of a gas is inversely proportional to the square root of the molecular weight (*Graham*'s *Law of Diffusion*).

$$\text{Rate of Diffusion} \propto \sqrt{\frac{1}{\text{Molecular weight}}}$$

Thus when a mixture of two gaseous isotopes is allowed to diffuse through a porous partition, **the lighter isotope passes through more rapidly than the heavier one.** The isotopes of neon (^{20}Ne, ^{22}Ne) and oxygen (^{16}O, ^{18}O) were separated by this method. The mixture of gaseous isotopes is passed through a porous tube sealed in an outer jacket (see Fig. 3.5). The lighter isotopes passes into the jacket, while the residual gas becomes richer in the heavier isotope. In actual practice a cascade of many 'Diffusion units' is used to achieve an appreciable separation. This process has been recently used for the separation of the gaseous fluorides $^{235}UF_6$ and $^{238}UF_6$. It provides a procedure for effective separation of the isotopes of uranium, namely, U-238 and U-235 (needed for atomic energy).

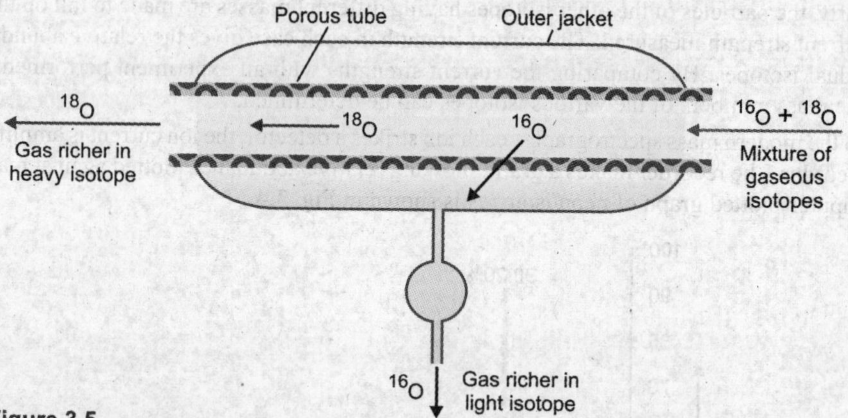

Figure 3.5
A unit of the 'diffusion cascade' for the separation of gaseous isotopes (^{16}O, ^{18}O).

(2) Thermal Diffusion

A long vertical tube with an electrical heated wire running down its axis is used. When a mixture of gaseous isotopes is introduced into the tube, **the lighter particles diffuse more rapidly to the central hot region.** Here they are carried upwards by convection currents. The heavier particles, on the other hand, travel to the cooler inner surface of the tube and sink to the bottom. Thus the lighter isotope collects at the top and the heavier one at the bottom. The isotopes of chlorine Cl-35 and Cl-37, have been separated by this process. The fluorides of uranium have also been separated by thermal diffusion.

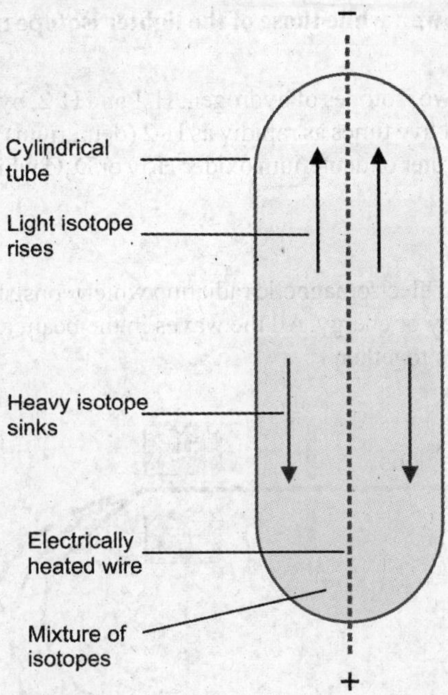

■ **Figure 3.6**
Separation of isotopes by thermal diffusion.

(3) Distillation

The lighter isotope will be distilled over first, leaving the heavier one behind. The isotopes of mercury were separated by this method. The frozen mercury from the cooled surface is removed, melted and evaporated under vacuum again. The whole process is repeated several times to separate the isotopes of mercury.

(4) Ultracentrifuge

The mixture of isotopes is rotated in a high speed centrifuge. The heavier isotope is concentrated at the periphery. The separation depends essentially on the molecular mass and not its square root, causing better separation. The gaseous fluorides of U-235 and U-238 have been separated by this method.

(5) Electromagnetic Separation

This method uses the principle of the Mass Spectrograph (Dempster). For example, the beam of ions of the isotopes of neon (Ne-20, Ne-21, Ne-22) as obtained in the mass spectrograph, is then passed between the poles of a magnet. **The different isotopes are deflected to different extents and are collected in cooled chambers placed in appropriate positions.** Although the quantities obtained by this method are very small indeed, the separation is complete.

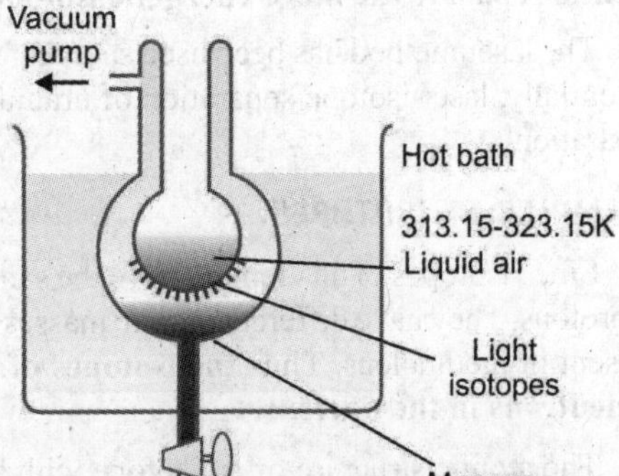

■ **Figure 3.7**
The separation of the isotopes of mercury by vacuum distillation.

(6) Fractional Electrolysis

Here the principle is that the rates of liberation of the isotopes of an element at an electrode during electrolysis are different. This is so because **the ions**

of the heavier isotope move slower, while those of the lighter isotope move faster to the opposite electrode.

Urey (1931) separated the two isotopes of hydrogen, H-1 and H-2, by the electrolysis of acidified water. H-1 (protium) is liberated five times as rapidly as H-2 (deuterium) at the cathode. The residual water becomes richer in heavy water or deuterium oxide ^2H_2O or D_2O which upon further electrolysis yields gas richer in deuterium.

(7) Laser Separation

A laser is a very fine beam of electromagnetic radiation which consists of photons corresponding to a single wavelength, frequency or energy. All the waves in the beam are in phase with all troughs and peaks moving through space together.

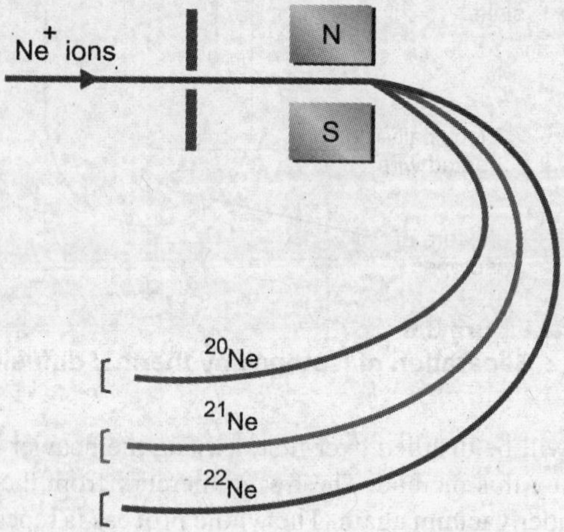

■ **Figure 3.8**
Separation of isotopes of Neon by using the principle of the Dempster's Mass Spectrograph.

In recent years, the development of lasers has been used for the separation of isotopes. If the laser light is of the appropriate wavelength, one isotope will absorb the energy, while another isotope will not. **The slight difference in absorption spectra of the two isotopes thus produced has been used to separate the more energetic isotope from the other.**

The laser method has been used successfully for the separation of isotopes of chlorine and sulfur. Potentially, laser isotope separation of uranium is 1000 times more efficient than gaseous diffusion separation.

EXAMPLES OF ISOTOPES

Since isotopes of an element have the same atomic number, each of these contains equal number of protons. They have different atomic masses which is accounted by the different number of neutrons present in the nucleus. Thus **the isotopes of an element are characterized by different number of neutrons in the nucleus.**

The atomic structure of an isotope with atomic number Z and mass number A (atomic mass in amu) can be given as follows :

(1) The number of extranuclear electrons = Z

(2) The number of protons in the nucleus = Z

(3) The mass number A is equal to the total number of protons (Z) and neutrons (N) in the nucleus. That is,

$$A = Z + N$$

$$\therefore \quad N = A - Z$$

Isotopes of Hydrogen

There are three isotopes of hydrogen : *protium* $\left({}^{1}_{1}H\right)$, *deuterium* $\left({}^{2}_{1}H \text{ or } D\right)$, and *tritium* $\left({}^{3}_{1}H \text{ or } T\right)$. Protium is by far the most abundant in natural hydrogen, deuterium about 0.015% and tritium only one out of 10,000,000 hydrogen atoms.

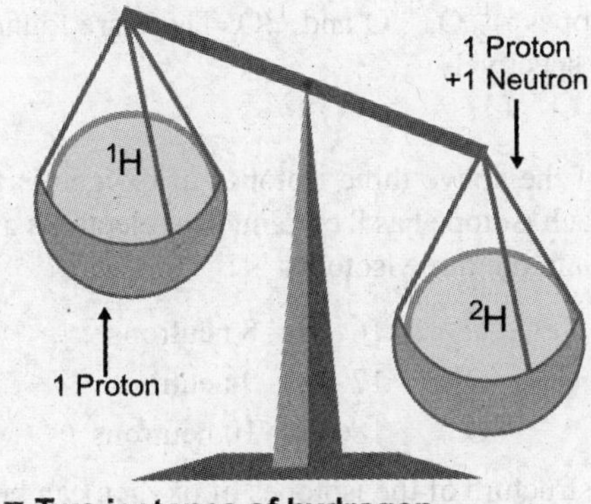

■ **Two isotopes of hydrogen.**

Structure

The atomic number of the three isotopes of hydrogen is 1, while their mass numbers are : protium 1, deuterium 2, and tritium 3. Therefore each of the three isotopes has one extranuclear electron and one proton in the nucleus. The nucleus of protium is made of one proton only, while the number of neutrons ($A - Z$) present in deuterium is 2 – 1 = 1, and in tritium 3 – 1 = 2. The structure of the three isotopes of hydrogen can be pictorially represented as :

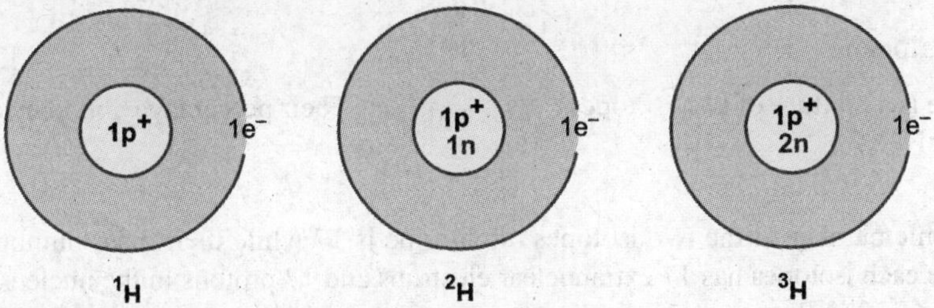

Isotopes of Neon

Neon has been found to consist of three isotopes : ${}^{20}_{10}Ne$, ${}^{21}_{10}Ne$ and ${}^{22}_{10}Ne$. Their percentage abundance is

^{20}Ne	^{21}Ne	^{22}Ne
90.92%	0.257%	8.82%

Structure

The atomic number of the three isotopes of neon is 10, while their mass numbers are 20, 21 and 22 respectively. Therefore each of these isotopes has ten extranuclear electrons and ten protons in the nucleus. The number of neutrons ($A - Z$) are : ^{20}Ne, 20 – 10 = 10 ; ^{21}Ne, 21 – 10 = 11 ; ^{22}Ne, 22 – 10 = 12. The atomic structure of the isotopes of neon can, therefore, be represented pictorially as:

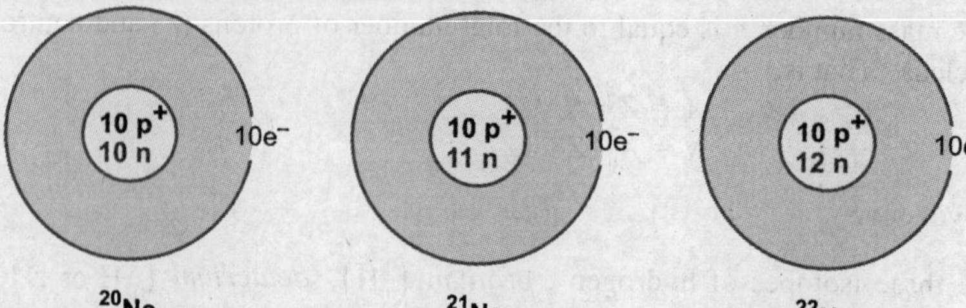

$^{20}_{10}Ne$ $^{21}_{10}Ne$ $^{22}_{10}Ne$

Isotopes of Oxygen

Oxygen has three isotopes : $^{16}_{8}O$, $^{17}_{8}O$ and $^{18}_{8}O$. These are found with the relative abundance of 99.759, 0.037 and 0.204 respectively.

Structure

The atomic number of the above three isotopes of oxygen is 8 while their mass numbers are 16, 17 and 18. Therefore each isotope has 8 extranuclear electrons and 8 protons in the nucleus. The number of neutrons (*A* – *Z*) in the three isotopes is :

^{16}O 16 – 8 = 8 neutrons
^{17}O 17 – 8 = 9 neutrons
^{18}O 18 – 8 = 10 neutrons

The complete atomic structure of the isotopes of oxygen can be pictorially represented as :

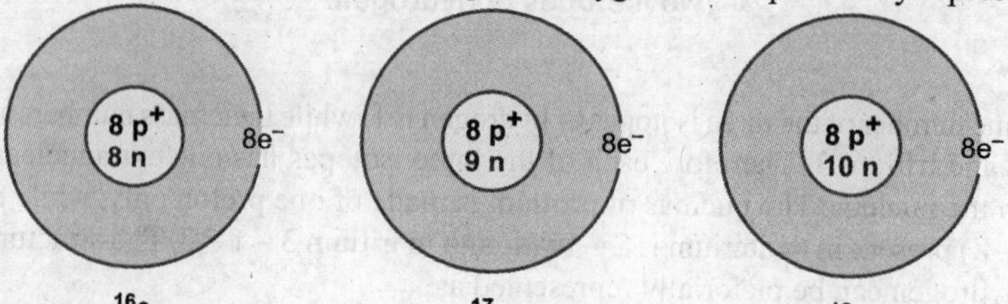

^{16}O ^{17}O ^{18}O

Isotopes of Chlorine

Chlorine is a mixture of two isotopes : $^{35}_{17}Cl$ and $^{37}_{17}Cl$. Their percentage abundance is 75.53 and 24.47 respectively.

Structure 3

The atomic number of the two isotopes of chlorine is 17 while their mass numbers are 35 and 37. Therefore each isotopes has 17 extranuclear electrons and 17 protons in the nucleus. The number of neutrons (*A* – *Z*) in these isotopes is :

^{35}Cl 35 – 17 = 18 neutrons
^{37}Cl 37 – 17 = 20 neutrons

The atomic structure of the isotopes of chlorine can be pictorially represented as :

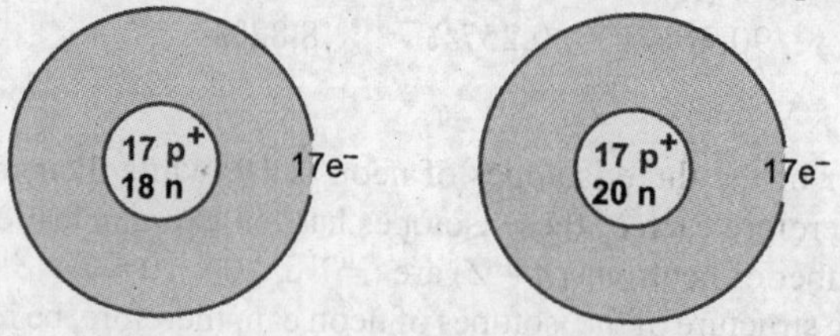

^{35}Cl ^{37}Cl

ISOTOPES OF CARBON

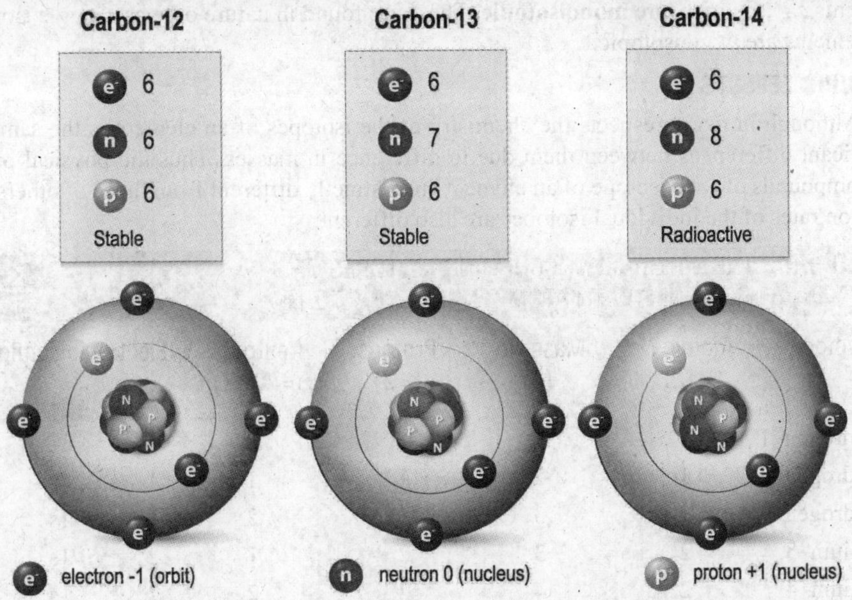

Isotopes of an element are atoms of the element that have different numbers of neutrons in their nuclei. Carbon has three naturally occurring isotopes. The isotopes of carbon are carbon-12, which constitutes 98.89 of all carbon atoms and serves as the standard for the atomic mass scale; carbon-13, which is the only magnetic isotope, making it very important for structural studies of compounds containing carbon; and carbon-14, which is produced by cosmic rays bombarding the atmosphere. Carbon-14 is radioactive, with a half-life of 5760 years. The amount of carbon-14 remaining in historical artifacts can be used to estimate their age.

Isotopes of Uranium

There are three isotopes of uranium :

$$^{238}_{92}U, \quad ^{235}_{92}U \text{ and } ^{234}_{92}U$$

Natural uranium consists almost entirely of ^{238}U, with about 0.72% of ^{235}U and 0.006% of ^{234}U. These isotopes are particularly important in atomic energy.

Structure

The atomic number of the three isotopes of uranium is 92 and their mass numbers are 238, 235 and 234. Thus each isotope has 92 extranuclear electrons and 92 protons. The number of neutrons $(A - Z)$ in these isotopes is :

^{238}U $238 - 92 = 146$ neutrons

^{235}U $235 - 92 = 143$ neutrons

^{234}U $234 - 92 = 142$ neutrons

The atomic structure of the three isotopes of uranium may be represented as :

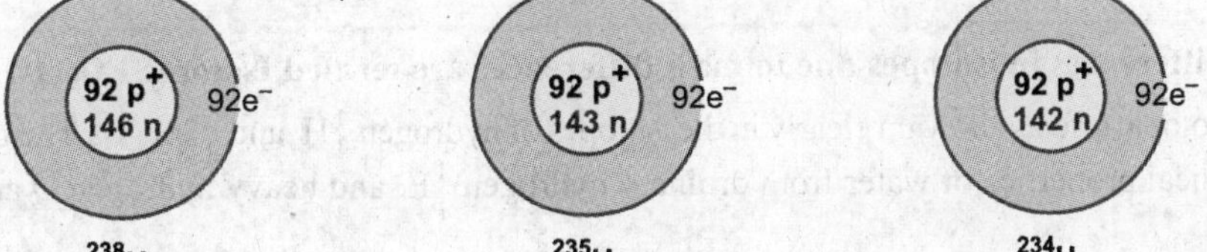

Almost every element in nature exists as a mixture of isotopes. The isotopes of the elements with atomic numbers 1 to 10, and their structure is listed in Table 3.1. It may be noted that some elements *e.g.*, fluorine, are **monoisotopic.** These are found in nature only as a single isotope. About 20 elements are monoisotopic.

ISOTOPIC EFFECTS

Although in many respects the chemistry of the isotopes of an element is the same, there are significant differences between them due to difference in masses. Thus the physical properties of the compounds of each isotope of an element are distinctly different from those of others. Similarly, reaction rates of the individual isotopes are also different.

TABLE 3.1. THE NATURALLY OCCURRING ISOTOPES OF ELEMENTS WITH ATOMIC NUMBERS 1 TO 10

Isotope	Atomic No. (Z)	Mass No. (A)	Protons (= Z)	Neutrons (= A – Z)	Electronic configuration			
Hydrogen–1	1	1	1	0	1	$1s^1$		
Hydrogen–2	1	2	1	1	1	$1s^1$		
Hydrogen–3	1	3	1	2	1	$1s^1$		
Helium–3	2	3	2	1	2	$1s^2$		
Helium–4	2	4	2	2	2	$1s^2$		
Lithium–6	3	6	3	3	2, 1	$1s^2$	$2s^1$	
Lithium–7	3	7	3	4	2, 1	$1s^2$	$2s^1$	
Beryllium–9	4	9	4	5	2, 2	$1s^2$	$2s^2$	
Boron–10	5	10	5	5	2, 3	$1s^2$	$2s^2$	$2p^1$
Boron–11	5	11	5	6	2, 3	$1s^2$	$2s^2$	$2p^1$
Carbon–12	6	12	6	6	2, 4	$1s^2$	$2s^2$	$2p^2$
Carbon–13	6	13	6	7	2, 4	$1s^2$	$2s^2$	$2p^2$
Carbon–14	6	14	6	8	2, 4	$1s^2$	$2s^2$	$2p^2$
Nitrogen–13	7	13	7	6	2, 5	$1s^2$	$2s^2$	$2p^3$
Nitrogen–14	7	14	7	7	2, 5	$1s^2$	$2s^2$	$2p^3$
Nitrogen–16	7	16	7	9	2, 5	$1s^2$	$2s^2$	$2p^3$
Oxygen–16	8	16	8	8	2, 6	$1s^2$	$2s^2$	$2p^4$
Oxygen–17	8	17	8	9	2, 6	$1s^2$	$2s^2$	$2p^4$
Oxygen–18	8	18	8	10	2, 6	$1s^2$	$2s^2$	$2p^4$
Fluorine–19	9	19	9	10	2, 7	$1s^2$	$2s^2$	$2p^5$
Neon–20	10	20	10	10	2, 8	$1s^2$	$2s^2$	$2p^6$
Neon–21	10	21	10	11	2, 8	$1s^2$	$2s^2$	$2p^6$
Neon–22	10	22	10	12	2, 8	$1s^2$	$2s^2$	$2p^6$

The differences in isotopes due to mass differences are termed Isotopic Effects.

The isotopic effects show up clearly in the isotopes of hydrogen ${}^{1}_{1}H$ and ${}^{2}_{1}H(D)$. The differences in the physical properties of water from ordinary hydrogen ${}^{1}_{1}H$ and heavy hydrogen D are listed in Table 3.2.

TABLE 3.2. PHYSICAL PROPERTIES OF H_2O AND D_2O

	H_2O	D_2O
Molecular weight	18.02	20.03
Density at 273K, kg/m ^{3}	1000	1105
Melting point, 273K	273.15 K	276.97 K
Boiling point, 273K	373.15 K	374.57 K

The isotopic effects are also exhibited by the difference in the reaction rates of the two isotopes of hydrogen. Under similar conditions, the reaction of the heavy isotopes with chlorine is about six times slower than that of light isotope.

$$\text{Cl}\,(g) + \text{D}_2\,(g) \longrightarrow \text{DCl}\,(g) + \text{D}\,(g) \quad \textit{SLOWER}$$

$$\text{Cl}\,(g) + \text{H}_2\,(g) \longrightarrow \text{HCl}\,(g) + \text{H}\,(g) \quad \textit{FASTER}$$

This difference in reaction rates is explained by the fact that the covalent bond formed by deuterium is slightly stronger than the corresponding bond with ordinary hydrogen. Hence a reaction that breaks a deuterium covalent bond is slower than the same reaction involving bond lighter hydrogen.

There are several useful applications of isotopic effects. One method of separation of isotopes of hydrogen depends on the fact that the electrolysis of heavy water (D_2O) is slower than the electrolysis of normal water. Pure heavy water is a valuable by-product of the electrolysis of water, since large quantities are required in nuclear reactors to moderate the rate of the uranium fission reaction.

WHAT ARE ISOBARS?

The atoms which have the same mass number but different atomic numbers are called isobars.

Isobars

The word isobar meaning 'equally heavy' is taken from the Greek *isos* = equal, and *barys* = heavy.

For example, $^{40}_{18}$Ar, $^{40}_{19}$K, and $^{40}_{20}$Ca are isobaric atoms. Similarly, $^{235}_{92}$U, $^{235}_{93}$Np, and $^{235}_{94}$Pu are isobars.

Structure

Since isobars have the same mass number, the number of protons plus neutrons in the nucleus in each of these is equal. The number of protons being given by atomic number (Z), the number of neutrons is, therefore, ($A - Z$) where A is the mass number. The number of extranuclear electrons is equal to (Z). Thus the atomic structure of the isobars $^{40}_{18}$Ar, $^{40}_{19}$K and $^{40}_{20}$Ca is shown below :

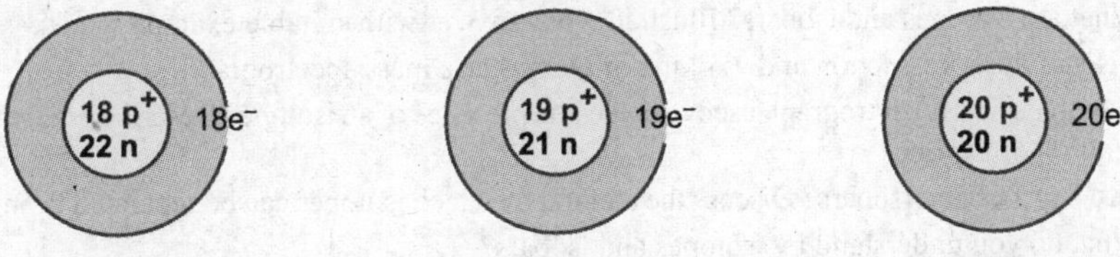

$^{40}_{18}$Ar $^{40}_{19}$K $^{40}_{20}$Ca

WHAT ARE ISOTONES?

Atoms which have different atomic number and different atomic masses but the same number of neutrons are called Isotones.

Examples of isotones

(1) $^{14}_{6}C$, $^{15}_{7}N$ and $^{16}_{8}O$ are isotones since each contains eight neutrons.

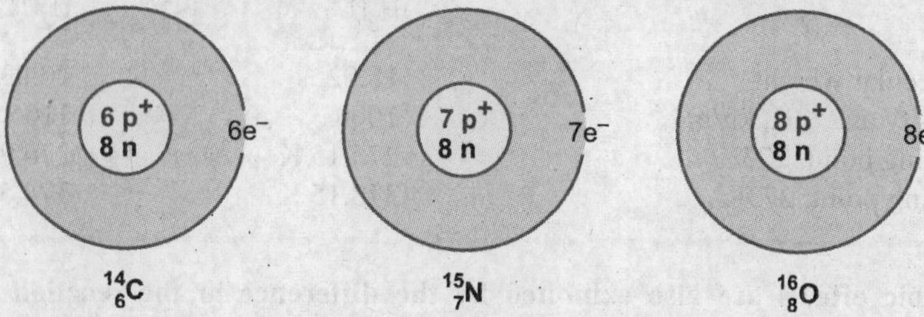

(2) $^{30}_{14}Si$, $^{31}_{15}P$ and $^{32}_{16}S$ are isotones. Each contains sixteen neutrons.

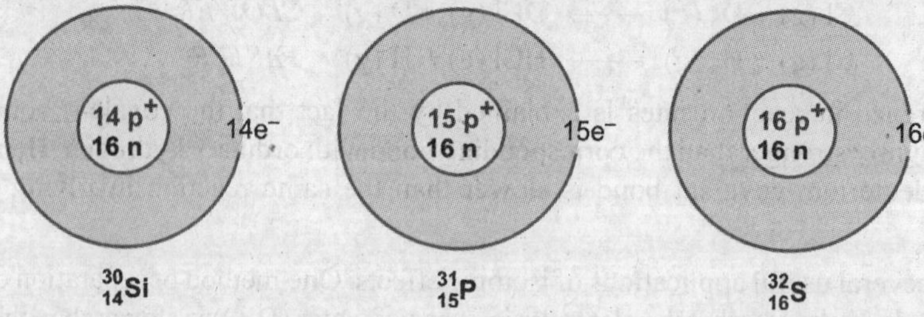

Isotones are different elements having entirely different atomic structure. They have different physical and chemical properties.

Some other examples of isotones are :

$$^{5}_{3}Li \text{ and } ^{6}_{4}Be$$

$$^{5}_{2}He \text{ and } ^{6}_{3}Li$$

$$^{6}_{2}He,\ ^{7}_{3}Li \text{ and } ^{8}_{4}Be$$

EXAMINATION QUESTIONS

1. Define or explain the following terms :
   (*a*) Isotopes (*b*) Rate of diffusion
   (*c*) Isobars (*d*) Isotones
2. (*a*) What are isotopes and isobars? Illustrate your answer with suitable examples.
   (*b*) Discuss the construction and working of Dempster's mass spectrograph.
3. How is Aston's mass spectrograph used to detect the presence of an isotope? Describe briefly two methods of separation of isotopes.
4. What are Isotopes and Isobars? Discuss the method by which isotopes can be separated from one another.
5. (*a*) What do you understand by isotopes and isobars?
   (*b*) Describe the working of Dempster's Mass Spectrograph.
6. Write three isotopes of hydrogen and draw their structure.
7. Discuss the construction, working and advantages of Aston's Mass Spectrograph.
8. Describe the construction, working and the use of Dempster's Mass Spectrograph.
9. Give the principle of the diffusion method for the separation of isotopes.

10. Magnesium has naturally occurring isotopes with the following masses and abundance.

Isotope	*Atomic mass (amu)*	*Fractional abundance*
$_{12}Mg^{24}$	23.985	0.7870
$_{12}Mg^{25}$	24.986	0.1013
$_{12}Mg^{26}$	25.983	0.1117

Calculate the atomic mass of magnesium.

Answer. 24.31 amu

11. Calculate the fractional abundance for the two naturally occurring isotopes of copper. The masses of isotopes are 62.9298 and 64.9278 amu. The atomic mass of copper is 63.546 amu.
Answer. 0.692; 0.308 *(Delhi BSc)*

12. Calculate the atomic mass of boron, B, from the following data :

Isotope	*Atomic mass (amu)*	*Fractional abundance*
$_5B^{10}$	10.013	0.1978
$_5B^{11}$	11.009	0.8022

Answer. 10.8119 amu *(Bundelkhand BSc)*

13. Naturally occurring boron consists of two isotopes whose atomic masses are 10.01 and 11.01. The atomic mass of naturally occurring boron is 10.81. Calculate the percentage of each isotope in natural boron.
Answer. Percentage of isotope with atomic mass 10.01 = 20 *(Dibrugarh BSc)*
Percentage of isotope with atomic mass 11.01 = 80

14. Describe in brief any two methods of separation of isotopes. *(Arunachal BSc)*
15. Discuss the detection and separation of isotopes. *(Madras BSc)*
16. What are isotones and nuclear isomerism? Give examples. *(Kerala BSc)*
17. Define the terms : Isotope and Isobar. Give one example of each. *(Kerala BSc)*
18. (*a*) How are nuclear masses determined with a mass spectrometer? Explain.
(*b*) Calculate the relative atomic mass of an element which consists of the following isotopes with the indicated relative abundance.

Isotope	*Isotopic mass*	*Natural abundance*
1	28.0	92.0
2	29.0	5.0
3	30.0	3.0

(Baroda BSc)

Answer. (*b*) 28.11

19. Silver has two naturally occurring isotopes with atomic masses 106.91 and 108.90 amu. The atomic mass of silver is 107.87 amu. Calculate the fractional abundance for these two isotopes.
Answer. 0.518; 0.482 *(North Eastern Hill BSc)*

20. Chromium, Cr, has following isotopic masses and fractional abundances :

Mass Number	*Mass (amu)*	*Fractional abundance*
50	49.9461	0.0435
52	51.9405	0.8379
53	52.9407	0.0950
54	53.9389	0.0236

Calculate the atomic mass of chromium.

Answer. 51.99 amu *(Madras BSc)*

21. Calculate the atomic mass of boron, B, from the following data :

Isotope	*Atomic mass (amu)*	*Fractional abundance*
$_5B^{10}$	10.013	0.1978
$_5B^{11}$	11.009	0.8022

Answer. 10.8120 amu *(Kerala BSc)*

MULTIPLE CHOICE QUESTIONS

1. The atoms of an element which have the same number of protons and different number of neutrons are called
(*a*) isotopes (*b*) isobars
(*c*) isotones (*d*) isomers
Answer. (*a*)

2. Isotopes of an element have _______ atomic number but _______ mass numbers.
(*a*) same, same (*b*) different, same
(*c*) same, different (*d*) different, different
Answer. (*c*)

3. _____occupy the same place in the periodic table.
(*a*) isobars (*b*) isotopes
(*c*) isotones (*d*) none of these
Answer. (*b*)

4. Isotopes have _______ chemical properties.
(*a*) same (*b*) similar
(*c*) different (*d*) none of these
Answer. (*b*)

5. The electronic configuration of isotopes of an element is _______.
(*a*) same (*b*) similar
(*c*) different (*d*) none of these
Answer. (*a*)

6. The physical properties of the isotopes which depend on the _______ of the atoms are used to separate them.
(*a*) electronic configuration (*b*) mass
(*c*) velocity (*d*) valence electrons
Answer. (*b*)

7. The mass number *A* is given by (*Z* is atomic number and *N* is number of neutrons)
(*a*) $A = Z - N$ (*b*) $A = Z + N$
(*c*) $A = N - Z$ (*d*) $A = 2\,(Z + N)$
Answer. (*b*)

8. An atom of _______ contains no neutrons.
(*a*) hydrogen (*b*) deuterium
(*c*) tritium (*d*) none of these
Answer. (*a*)

9. The difference between the number of neutrons and the protons is positive for
(*a*) hydrogen atom (*b*) deuterium atom
(*c*) tritium atom (*d*) none of these
Answer. (*c*)

10. An element with atomic number equal to one, exists in three isotopes namely ${}_1H^1$, ${}_1H^2$ and ${}_1H^3$. Which out of these has only one electron in its outermost shell.
(a) ${}_1H^1$ (b) ${}_1H^2$
(*c*) ${}_1H^3$ (*d*) all the three
Answer. (*d*)

11. In which isotope of oxygen out of O^{16}, O^{17} and O^{18} there are equal number of protons, electrons and neutrons?

(a) O^{16} (b) O^{17}
(c) O^{18} (d) none of these

Answer. (a)

12. Which isotope of chlorine out of $_{17}Cl^{35}$ and $_{17}Cl^{37}$ has greater number of neutrons than the protons?

(a) $_{17}Cl^{35}$ (b) $_{17}Cl^{37}$
(c) neither of the two (d) both

Answer. (d)

13. The reaction rates of the individual isotopes are

(a) the same (b) different
(c) sometimes the same, sometimes different (d) none of these

Answer. (b)

14. The atoms which have the same mass number but different atomic numbers are called

(a) isobars (b) isotopes
(c) isotones (d) isomers

Answer. (a)

15. Which is true about the isobars?

(a) they have same mass number and same atomic number
(b) they have same mass number and different atomic number
(c) they have different mass number and same atomic number
(d) they have different mass number and different atomic number

Answer. (b)

16. Which of the following statements holds good for $^{40}_{18}Ar$, $^{40}_{19}K$ and $^{40}_{20}Ca$

(a) they have equal number of protons and electrons
(b) they have equal number of protons and neutrons taken together
(c) they have equal number of neutrons in their respective nuclei
(d) none of the above

Answer. (b)

17. Atoms which have different atomic numbers, different mass numbers but the same number of neutrons are called

(a) isotopes (b) isobars
(c) isotones (d) isomers

Answer. (c)

18. Which of the following statement is true for $^{14}_{6}C$, $^{15}_{7}N$ and $^{16}_{8}O$?

(a) they have equal number of protons (b) they have equal number of electrons
(c) they have equal number of neutrons (d) they have equal mass number

Answer. (c)

19. Which of the following pairs represents isotones?

(a) $_1H^1$ and $_1H^2$ (b) $^{40}_{18}Ar$ and $^{40}_{19}K$
(c) $^{14}_{6}C$ and $^{16}_{8}O$ (d) $^{22}_{10}Ne$ and $^{23}_{11}Na$

Answer. (c)

20. Which of the following pairs represents isobars?

(a) $^{17}_{8}O$ and $^{18}_{8}O$ (b) $^{40}_{19}K$ and $^{40}_{20}Ca$
(c) $^{15}_{7}N$ and $^{16}_{8}O$ (d) $^{235}_{92}U$ and $^{238}_{92}U$

Answer. (b)

21. An isotone of $^{14}_{6}C$ is

(*a*) $^{16}_{8}O$ (*b*) $^{13}_{6}C$

(*c*) $^{17}_{8}O$ (*d*) $^{16}_{7}N$

Answer. (*a*)

22. Which of the following is isoelectronic with Cl^{-1}?

(*a*) S^{2-} (*b*) P^{3-}

(*c*) K^{+} (*d*) All

Answer. (*d*)

23. Two nuclei A and B are isotonic with mass numbers 15 and 16 respectively. If A contains 7 protons, then the number of protons in B would be

(*a*) 7 (*b*) 8

(*c*) 9 (*d*) 10

Answer. (*b*)

24. $^{76}_{32}Ge$ is isotonic with

(*a*) $^{77}_{32}Ge$ (*b*) $^{78}_{33}As$

(*c*) $^{78}_{32}Ge$ (*d*) $^{77}_{33}As$

Answer. (*d*)

25. Which of the following atoms contains the largest number of neutrons?

(*a*) $^{210}_{83}Bi$ (*b*) $^{208}_{83}Bi$

(*c*) $^{209}_{84}Pb$ (*d*) $^{208}_{84}Pb$

Answer. (*a*)

26. The number of neutrons present in $^{239}_{93}Np$ is

(*a*) 93 (*b*) 146

(*c*) 239 (*d*) 332

Answer. (*b*)

27. Which of the following properties of the element is a whole number?

(*a*) atomic mass (*b*) atomic volume

(*c*) atomic radius (*d*) atomic number

Answer. (*d*)

28. Which of the following is false about $^{16}_{8}O$ and $^{17}_{8}O$?

(*a*) both have eight protons (*b*) both have eight electrons

(*c*) both have eight neutrons (*d*) they have different rates of diffusion

Answer. (*c*)

29. What is the relationship between $^{27}_{13}Al^{3+}$ and $^{23}_{11}Na^{+}$?

(*a*) they are isotopes (*b*) they are isobars

(*c*) they are isotones (*d*) they are isoelectronic

Answer. (*d*)

30. $^{24}_{11}Na$ and $^{24}_{12}Mg$ are related to each other as

(*a*) isotopes (*b*) isobars

(*c*) isotones (*d*) isoelectronic

Answer. (*b*)

31. The *n*/*p* ratio for $^{16}_{8}O$ is the same as for

(*a*) $^{12}_{6}C$ (*b*) $^{14}_{7}N$

(*c*) both (*d*) none of these

Answer. (*c*)

32. In the nucleus of $^{40}_{18}Ar$, the difference between the number of neutrons and protons is

(*a*) 4 (*b*) 18

(*c*) 22 (*d*) 40

Answer. (*a*)

33. Amongst the three isotopes of Neon - $^{20}_{10}Ne$, $^{21}_{10}Ne$ and $^{22}_{10}Ne$ the nucleus with lowest *n*/*p* ratio is

(*a*) $^{22}_{10}Ne$ (*b*) $^{21}_{10}Ne$

(*c*) $^{20}_{10}Ne$ (*d*) all of these

Answer. (*c*)

34. The number of neutrons is greatest in

(*a*) $^{235}_{92}U$ (*b*) $^{236}_{92}U$

(*c*) $^{237}_{92}U$ (*d*) $^{238}_{92}U$

Answer. (*d*)

35. Calcium atom and Ca^{2+} ions have

(*a*) the same number of electrons (*b*) the same number of neutrons

(*c*) different number of protons (*d*) different number of neutrons

Answer. (*b*)

36. The set of isoelectronic species is

(*a*) Na^+ Ne Mg^+ (*b*) Na^+ Mg^{2+} Al^{3+}

(*c*) Na^+ K^+ Ne (*d*) Ne Cl^- Na

Answer. (*b*)

37. An atom of $^{94}_{36}Kr$ contains

(*a*) 36 protons, 36 electrons and 36 neutrons (*b*) 94 protons, 94 electrons and 94 neutrons

(*c*) 36 protons, 58 electrons and 36 neutrons (*d*) 36 protons, 36 electrons and 58 neutrons

Answer. (*d*)

38. In $^{10}_{5}B$, the *n*/*p* ratio is

(*a*) 1.0 (*b*) 1.25

(*c*) 1.50 (*d*) 2.0

Answer. (*a*)

39. The nuclei $_{52}Te^{130}$, $_{54}Xe^{130}$ and $_{56}Ba^{130}$ are isobars as they have

(*a*) different number of protons (*b*) different number of neutrons

(*c*) same mass number (*d*) different mass number

Answer. (*c*)

40. Isotones have different physical and chemical properties as they have

(*a*) same number of neutrons (*b*) different number of protons (or electrons)

(*c*) same mass numbers (*d*) different mass numbers

Answer. (*b*)

41. An element M has an atomic mass 19 and atomic number 9, its ion is represented by

(*a*) M^+ (*b*) M^{2+}

(*c*) M^- (*d*) M^{2-}

Answer. (*c*)

42. The nuclei, which are not identical but have the same number of nucleons, are called

(*a*) isotopes (*b*) isobars

(*c*) isotones (*d*) isoelectronic

Answer. (*b*)

43. Which of the following are isotopes?

(*a*) $17\,p + 18\,n$ and $17\,p + 20\,n$ (*b*) $19\,p + 20\,n$ and $20\,p + 21\,n$

(*c*) $18\,p + 20\,n$ and $19\,p + 21\,n$ (*d*) $20\,p + 20\,n$ and $19\,p + 22\,n$

Answer. (*a*)

44. In which of the species the number of neutrons is equal to the number of electrons?

(*a*) ${}_6C^{12}$ (*b*) ${}_7N^{15}$

(*c*) ${}_8O^{17}$ (*d*) ${}_9F^{19}$

Answer. (*a*)

45. The triad of nuclei that is isotonic is

(*a*) ${}_1H^1$ and ${}_1H^2$ and ${}_1H^3$ (*b*) ${}^{235}_{92}U$, ${}^{237}_{92}U$ and ${}^{238}_{92}U$

(*c*) ${}^{40}_{18}Ar$, ${}^{40}_{19}K$ and ${}^{40}_{20}Ca$ (*d*) ${}^{14}_{6}C$, ${}^{15}_{7}N$ and ${}^{16}_{8}O$

Answer. (*d*)

46. ${}^{24}_{12}Mg^{2+}$ is isoelectronic with

(*a*) ${}^{23}_{11}Na$ (*b*) ${}^{27}_{13}Al$

(*c*) ${}^{23}_{11}Na^+$ (*d*) ${}^{24}_{12}Mg$

Answer. (*c*)

47. In the nucleus of ${}_{42}Mo^{96}$ the number of nucleons is

(*a*) 42 (*b*) 54

(*c*) 96 (*d*) 138

Answer. (*c*)

48. The atomic mass of an element is not a whole number because

(*a*) nucleons are present in the nucleus

(*b*) the number of protons is different from the neutrons

(*c*) of the presence of isobars

(*d*) of the presence of isotopes

Answer. (*d*)

49. ${}_{88}Ra^{236}$ contains ______ neutrons in its nucleus.

(*a*) 88 (*b*) 236

(*c*) 148 (*d*) 324

Answer. (*c*)

50. An element has atomic number 20. Its cation has 2 units positive charge. The number of electrons and protons in its cation are

(*a*) 18 *e* and 20 *p* (*b*) 20 *e* and 18 *p*

(*c*) 20 *e* and 20 *p* (*d*) 18 *e* and 18 *p*

Answer. (*a*)

4

Nuclear Chemistry

CONTENTS

A nuclear reaction is different from a chemical reaction. In a chemical reaction, atoms of the reactants combine by a rearrangement of extranuclear electrons but the nuclei of the atoms remain unchanged. In a nuclear reaction, on the other hand, it is the nucleus of the atom which is involved. The number of protons or neutrons in the nucleus changes to form a new element. **A study of the nuclear changes in atoms is termed Nuclear Chemistry.**

RADIOACTIVITY

A number of elements such as uranium and radium are unstable. Their atomic nucleus breaks of its own accord to form a smaller atomic nucleus of another element. The protons and neutrons in the unstable nucleus regroup to give the new nucleus. This causes the release of excess particles and energy from the original nucleus, which we call **radiation.** The elements whose atomic nucleus emits radiation are said to be **radioactive.** The spontaneous breaking down of the unstable atoms is termed **radioactive disintegration or radioactive decay.**

The disintegration or decay of unstable atoms accompanied by emission of radiation is called Radioactivity.

TYPES OF RADIATIONS

The radioactive radiations are of three types. These were sorted out by Rutherford (1902) by passing them between two oppositely charged plates (Fig. 4.1). The one bending towards the negative plate carried positive charge and were named α **(alpha) rays.** Those bending towards the positive plate and carrying negative charge were called β **(beta) rays.** The third type of radiation, being uncharged, passed straight through the electric field and were named γ **(gamma) rays.** α, β and γ rays could be easily detected as they cause luminescence on the zinc sulfide screen placed in their path.

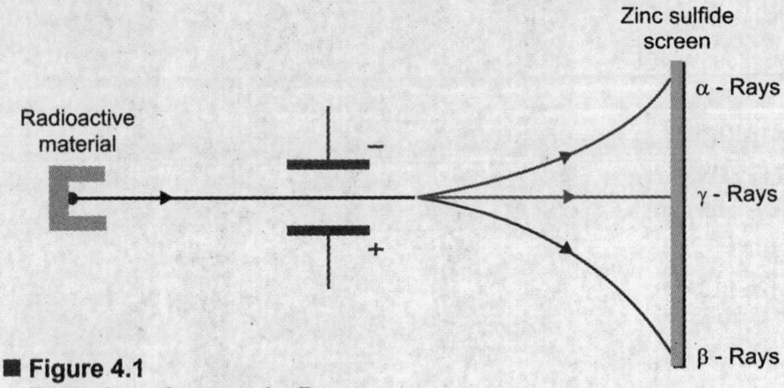

Figure 4.1
Detection of α, β and γ-Rays.

PROPERTIES OF RADIATIONS

Alpha (α), beta (β) and gamma (γ) rays differ from each other in nature and properties. There chief properties are : (*a*) Velocity; (*b*) Penetrating power; (*c*) Ionization.

ALPHA RAYS

(1) Nature. They consist of streams of α-particles. **By measurement of their e/m, Rutherford showed that they have a mass of 4 amu and charge of +2.** They are **helium nuclei** and may be represented as ${}^{4}_{2}\alpha$ or ${}^{4}_{2}He$.

(2) Velocity. α-particles are ejected from radioactive nuclei with very high velocity, about one-tenth that of light.

(3) Penetrating power. Because of their charge and relatively large size, α-particles have **very little power of penetration** through matter. They are stopped by a sheet of paper, 0.01 mm thick aluminium foil or a few centimeters of air.

(4) Ionization. They cause **intense ionization** of a gas through which they pass. On account of their high velocity and attraction for electrons, α-particles break away electrons from gas molecules and convert them to positive ions.

BETA RAYS

(1) Nature. They are streams of β-particles emitted by the nucleus. From their deflection in electric and magnetic fields, Becquerel showed that **β-particles are identical with electrons.** They have very small mass (1/1827 amu) and charge of – 1. A β-particle is symbolized as ${}^{0}_{-1}\beta$ or ${}^{0}_{-1}e$.

(2) Velocity. They travel about **10 times faster than α-particles.** Their velocity is about the same as of light.

(3) Penetrating power. β-Particles are 100 times more penetrating in comparison to α-particles. This is so because they have higher velocity and negligible mass. β-particles can be stopped by about 1 cm thick sheet of aluminium or 1 m of air.

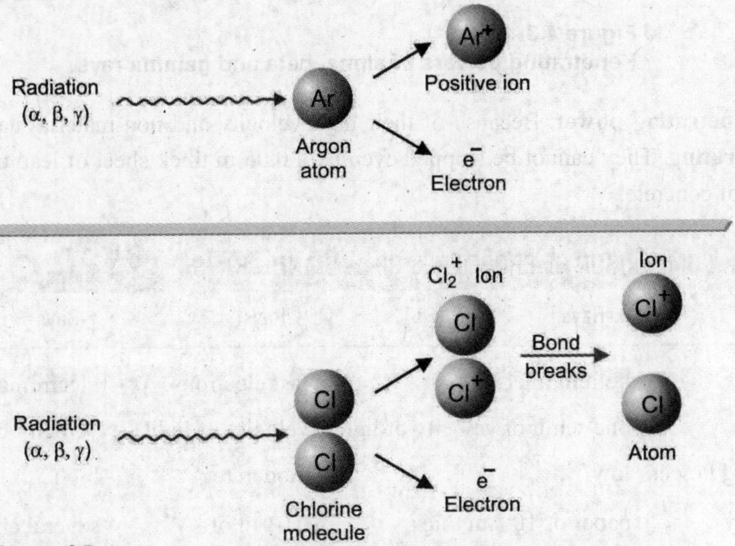

■ **Figure 4.2**
Radiation energy knocks electrons out of atoms or molecules. This produces positive ions. Bond breaking often occurs in unstable molecular ions.

(4) Ionization. The ionization produced by β-particles in a gas is about **one-hundredth of that of α-particles.** Though the velocity of β-particles is higher but the mass being smaller, their kinetic energy is much less than α-particles. Hence they are poor ionizers.

GAMMA RAYS

(1) Nature. Unlike α and β-rays, they do not consist of particles of matter. **γ-Rays are a form of electromagnetic radiation** of shorter wavelength than X-rays. They could be thought of as high-energy photons released by the nucleus during α- or β-emissions. They have no mass or charge and may be symbolized as ${}^{0}_{0}\gamma$.

(2) Velocity. Like all forms of electromagnetic radiation, **γ-rays travel with the velocity of light.**

(3) Ionizing power. Their ionizing power is very weak in comparison to α- and β-particles. A γ-photon displaces an electron of the gas molecule to yield a positive ion. Since the chances of photon-electron collisions are small, γ-rays are weak ionizers.

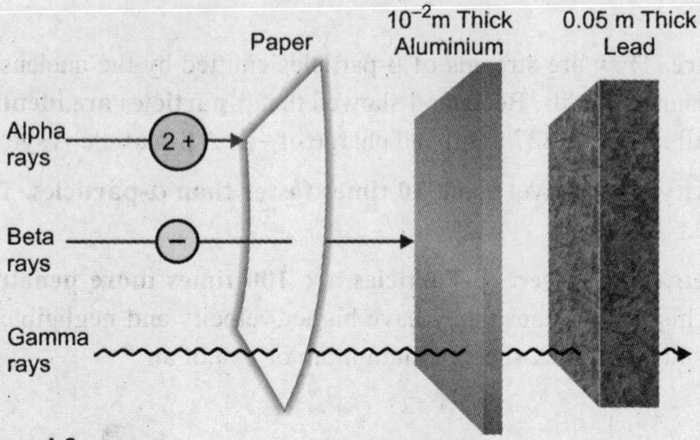

■ **Figure 4.3**
Penetrating powers of alpha, beta and gamma rays.

(4) **Penetrating power.** Because of their high velocity and non-material nature, **γ-rays are most penetrating.** They cannot be stopped even by a 0.05 m thick sheet of lead or several metres thick layer of concrete.

TABLE 4.1. COMPARISON OF PROPERTIES OF α, β AND γ-RAYS

Property	α-Rays	β-Rays	γ-Rays
Nature	helium nuclei ^4_2He	fast electrons $^{0}_{-1}e$	electromagnetic radiation
Velocity	one-tenth of velocity of light	velocity of light	velocity of light
Penetrating Power	low	moderate	high
Stopped by	paper or 10^{-5} m thick aluminium sheet	0.01 m of aluminium sheet	several cm thick lead/ concrete layer

DETECTION AND MEASUREMENT OF RADIOACTIVITY

The radioactive radiation can be detected and measured by a number of methods. The important ones used in modern practice are listed below.

(1) Cloud Chamber

This technique (Fig. 4.4) is used for detecting radioactivity. The chamber contains air saturated with water vapor. When the piston is lowered suddenly, the gas expands and is supercooled. When α- or β-particle is passed through this gas, ions are created along its path. These ions provide nuclei upon which droplets of water condense. The trail or cloud thus produced marks the track of the particle. The track can be seen through the window above and immediately photographed.

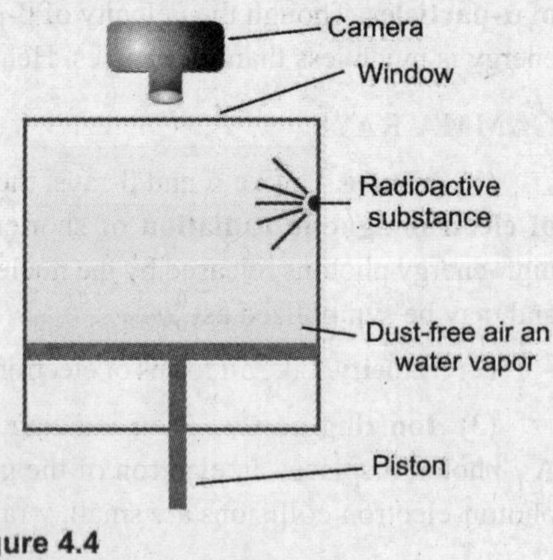

■ **Figure 4.4**
Principle of Cloud Chamber.

Similarly, α- or β-particles form a trail of bubbles as they pass through liquid hydrogen. The **bubble chamber method** gives better photographs of the particle tracks.

(2) Ionization Chamber

This is the simplest device used to measure the strength of radiation. An ionization chamber is fitted with two metal plates separated by air. When radiation passes through this chamber, it knocks electrons from gas molecules and positive ions are formed. The electrons migrate to the anode and positive ions to the cathode.

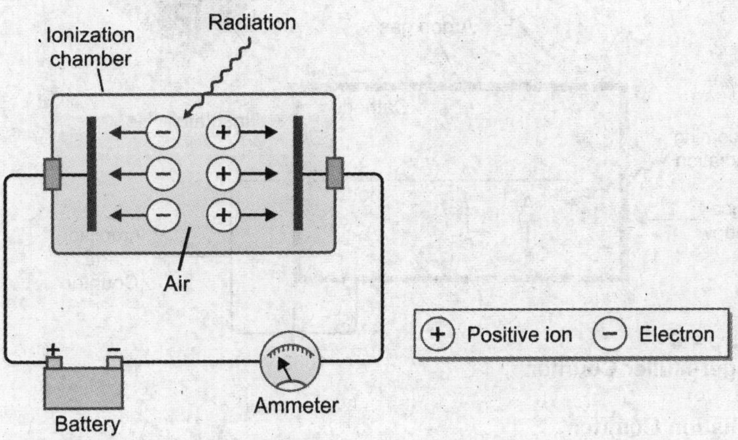

Figure 4.5
An Ionization chamber used to measure the strength of radiation.

Thus a small current passes between the plates. This current can be measured with an ammeter, and gives the strength of radiation that passes through the ionization chamber. In an ionization chamber called **Dosimeter,** the total amount of electric charge passing between the plates in a given time is measured. This is proportional to the total amount of radiation that has gone through the chamber.

(3) Geiger-Muller Counter

This device (Fig. 4.6) is used for detecting and measuring the rate of emission of α- or β-particles. It consists of a cylindrical metal tube (cathode) and a central wire (anode). The tube is filled with argon gas at reduced pressure (10132.5 Pascal). A potential difference of about 1000 volts is applied across the electrodes. When an α- or β-particle enters the tube through the mica window, it ionizes the argon atoms along its path.

$$Ar \longrightarrow Ar^+ + e^-$$

The argon ions (Ar^+) are drawn to the cathode and electrons to anode. Thus for a fraction of a second, a pulse of electrical current flows between the electrodes and completes the circuit around. **Each electrical pulse marks the entry of one α- or β-particle into the tube and is recorded in an automatic counter.** The number of such pulses registered by a radioactive material per minute, gives the intensity of its radioactivity.

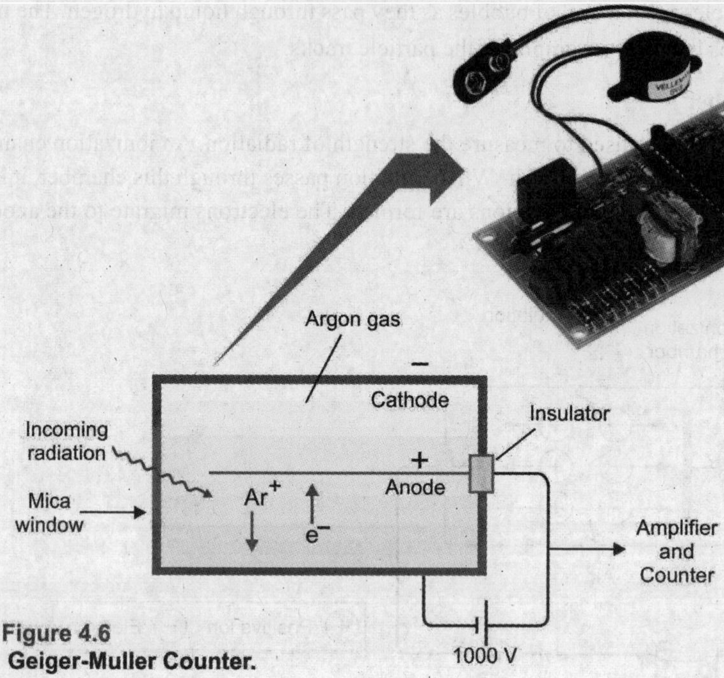

■ **Figure 4.6**
Geiger-Muller Counter.

(4) Scintillation Counter

Rutherford used a spinthariscope (Fig. 4.7) for the detection and counting of α-particles. The radioactive substance mounted on the tip of the wire emitted α-particles. Each particle on striking the zinc sulfide screen produced a flash of light. These flashes of light (scintillations) could be seen through the eye-piece. With this device it was possible to count α-particles from 50 to 200 per second.

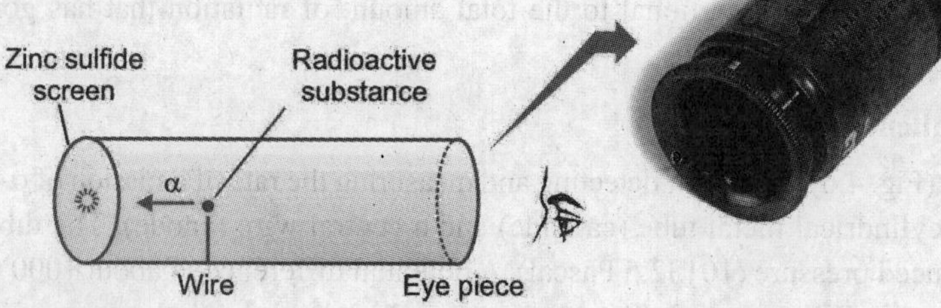

■ **Figure 4.7**
Spinthariscope.

A modern scintillation counter also works on the above principle and is widely used for the measurement of α- or β-particles. Instead of the zinc sulfide screen, a crystal of sodium iodide with a little thallium iodide is employed. The sample of the radioactive substance contained in a small vial, is placed in a 'well' cut into the crystal. The radiation from the sample hit the crystal wall and produce scintillations. **These fall on a photoelectric cell which produces a pulse of electric current for each flash of light.** This is recorded in a mechanical counter. Such a scintillation counter can measure radiation upto a million per second.

(5) Film Badges

A film badge consists of a photographic film encased in a plastic holder. When exposed to radiation, they darken the grains of silver in photographic film. The film is developed and viewed under a powerful microscope.

As α- or β-particles pass through the film, they leave a track of black particles. These particles can be counted. In this way the type of radiation and its intensity can be known. However, γ-radiation darken the photographic film uniformly. The amount of darkening tells the quantity of radiation.

A film badge is an important device to monitor the extent of exposure of persons working in the vicinity of radiation. The badge-film is developed periodically to see if any significant dose of radiation has been absorbed by the wearer.

TYPES OF RADIOACTIVE DECAY

According to the theory put forward by Rutherford and Soddy (1903), radioactivity is a nuclear property. The nucleus of a radioactive atom is unstable. It undergoes decay or disintegration by spontaneous emission of an α- or β-particle. This results in the change of proton-neutron composition of the nucleus to form a more stable nucleus. **The original nucleus is called the parent nucleus and the product is called the daughter nucleus.**

As evident from above, there are two chief types of decay:

(1) α-decay (2) β-decay

α-Decay

When a radioactive nucleus decays by the emission of an α-particle (**α-emission**) from the nucleus, the process is termed α-decay. An alpha particle has four units of atomic mass and two units of positive charge. If Z be the atomic number and M the atomic mass of the parent nucleus, the daughter nucleus will have

$$\text{atomic mass} = M - 4$$

$$\text{atomic number} = Z - 2$$

Thus an α-emission reduces the atomic mass by 4 and atomic number by 2.

For example, Radium decays by α-emission to form a new element Radon,

$$\underset{\text{(parent)}}{{}^{226}_{88}\text{Ra}} - {}^{4}_{2}\alpha = \underset{\text{(daughter)}}{{}^{222}_{86}\text{Rn}}$$

β-Decay

When a radioactive nucleus decays by β-particle emission (**β-emission**), it is called β-decay. A free β-particle or electron does not exist as such in the nucleus. It is produced by the conversion of a neutron to a proton at the moment of emission.

$$\text{Neutron} = \text{p} + \text{e} \nearrow$$

This results in the increase of one positive charge on the nucleus. The loss of a β-particle from the nucleus does not alter its atomic mass. For a parent nucleus with atomic mass M and atomic number Z, the daughter nucleus will have

$$\text{atomic mass} = M$$

$$\text{atomic number} = Z + 1$$

Thus a β-emission increases the atomic number by 1 with no change in atomic mass.

An example of β-decay is the conversion of lead-214 to bismuth-214,

$$\underset{\text{(parent)}}{{}^{214}_{82}\text{Pb}} - {}^{0}_{-1}\beta = \underset{\text{(daughter)}}{{}^{214}_{83}\text{Bi}}$$

It is noteworthy that a β-emission results in the production of an isobar. Thus, ${}^{214}_{82}\text{Pb}$ and ${}^{214}_{83}\text{Bi}$ are isobaric as they have the same mass number 214 but different atomic numbers (82 and 83).

One α-emission and two β-emissions yield an isotope. Let us consider the following series of changes.

$$\underset{\text{(parent)}}{{}^{218}_{84}\text{Po}} \xrightarrow{\alpha} {}^{214}_{82}\text{Pb} \xrightarrow{\beta} {}^{214}_{83}\text{Bi} \xrightarrow{\beta} \underset{\text{(isotope)}}{{}^{214}_{84}\text{Po}}$$

The parent element ${}^{218}_{84}\text{Po}$ emits an α-particle and subsequently two β-particles, resulting in the formation of ${}^{214}_{84}\text{Po}$ which is an isotope of the parent. Both the parent and the end-product have the same atomic number 84 but different mass numbers (218 and 214).

SOLVED PROBLEM 1

How many α and β particles are emitted in passing down from ${}^{232}_{90}\text{Th}$ to ${}^{208}_{82}\text{Pb}$.

SOLUTION. Let *a* be the number of α particles and *b* be the number of β particles emitted during the radioactive transformation. It can be represented as

$${}^{232}_{90}\text{Th} \longrightarrow {}^{208}_{82}\text{Pb} + a\,{}^{4}_{2}\text{He} + b\,e^{0}_{-1}$$

Comparing the mass numbers, we get

$$232 = 208 + 4 \times a + b \times 0$$

or $4a = 232 - 208 = 24$

or $a = 6$

Comparing the atomic numbers, we get

$$90 = 82 + 2 \times a + b\,(-1)$$

Substituting the value of a, we get

$$90 = 82 + 2 \times 6 - b$$

or $b = 94 - 90$

or $b = \mathbf{4}$

Thus the number of α particles emitted = 6 and the number of β particles emitted = 4

SOLVED PROBLEM 2

${}^{210}_{82}\text{Pb}$ is a β-emitter and ${}^{226}_{88}\text{Ra}$ is an α-emitter. What will be the atomic masses and atomic numbers of daughter elements of these radioactive elements ? Predict the position of daughter elements in the periodic table.

SOLUTION. (*a*) ${}^{210}_{82}\text{Pb}$ undergoes β-decay *i.e.*

$${}^{210}_{82}\text{Pb} \longrightarrow {}^{0}_{-1}\text{e} + {}^{b}_{a}\text{X}$$

Comparing the atomic masses, we have

$$210 = 0 + b$$

or $$b = \mathbf{210}$$

and comparing the atomic numbers, we get

$$82 = -1 + a \quad \text{or} \quad a = \mathbf{83}$$

Thus the daughter element will have the same atomic mass **210** and its atomic number will be **83.** It will occupy **one position right** to the parent element.

(*b*) $^{236}_{88}\text{Ra}$ undergoes α decay *i.e.*

$$^{236}_{88}\text{Ra} \longrightarrow {}^{4}_{2}\text{He} + {}^{b}_{a}\text{X}$$

Comparing the atomic masses, we get

$$236 = 4 + b \quad \text{or} \quad b = \mathbf{232}$$

and comparing the atomic number, we get

$$88 = 2 + a \quad \text{or} \quad a = \mathbf{86}$$

Thus the daughter element will have atomic mass **232** and its atomic number will be **86.** It will occupy **two positions to the left** of the parent element.

THE GROUP DISPLACEMENT LAW

The position number of an element in a Group of the Periodic Table corresponds to its atomic number. If the atomic number of a given element is changed, its Group also changes accordingly. We know that an α-emission decreases the atomic number of the parent element by 2 and a β-emission increases the atomic number by 1. Thus : **in an α-emission, the parent element will be displaced to a Group two places to the left and in a β-emission, it will be displaced to a Group one place to the right.**

This is called the **Group Displacement Law.** It was first stated by Fajans and Soddy (1913) and is often named after them as '*Fajans-Soddy Group Displacement Law*'.

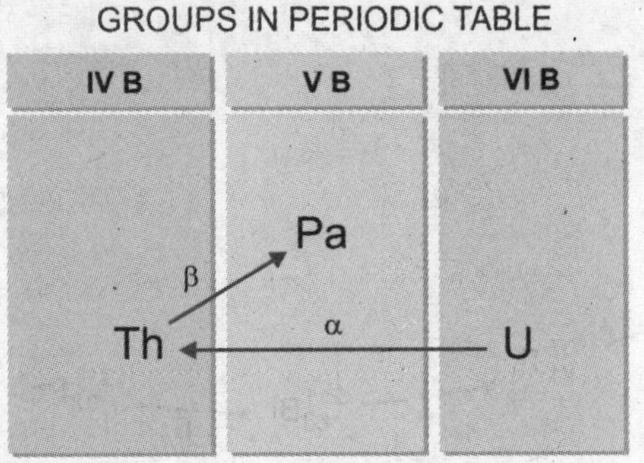

Figure 4.8
Illustration of Group Displacement Law.

RADIOACTIVE DISINTEGRATION SERIES

A radioactive element disintegrates by the emission of an α- or β- particle from the nucleus to form a new '*daughter element*'. This again disintegrates to give another '*daughter element*'. The process of disintegration and formation of a new element continues till a non-radioactive stable element is the product.

The whole series of elements starting with the parent radioactive element to the stable end-product is called a Radioactive Disintegration Series.

Sometime, it is referred to as a **Radioactive Decay Series** or simply **Radioactive Series.** All the natural radioactive elements belong to one of the three series :

(1) The Uranium Series

(2) The Thorium Series

(3) The Actinium Series

The Uranium Series

It commences with the parent element uranium-238 and terminates with the stable element lead-206. It derives its name from uranium-238 which is the prominent member of the series and has the longest half-life. The Uranium series is illustrated in Fig. 4.9.

The Thorium Series

It begins with the parent element thorium-232 and ends with lead-208 which is stable. This series gets its name from the prominent member thorium-232.

The Actinium Series

It starts with the radioactive element uranium-235. The end-product is the stable element lead-207. This series derives its name from the prominent member actinium-227.

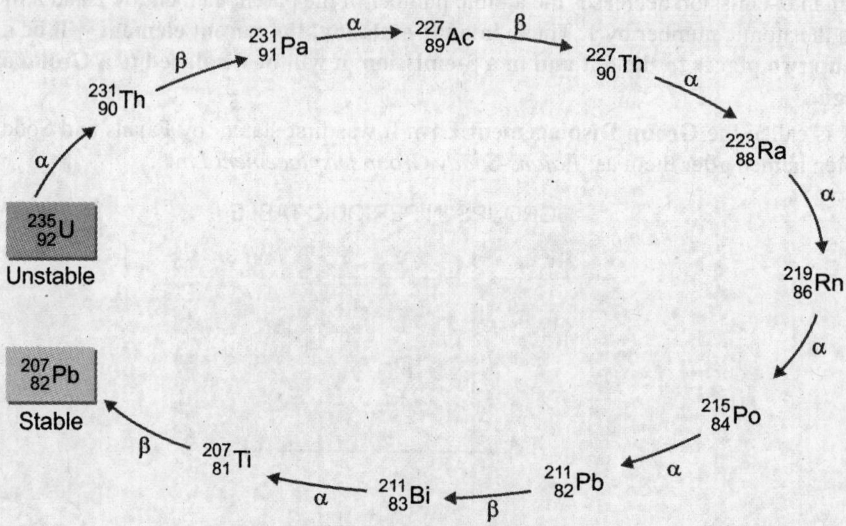

THE ACTINIUM SERIES

The Neptunium Series

This series consists of elements which do not occur naturally. It commences with neptunium-237 and terminates at bismuth-200. It derives its name from the prominent member neptunium-237.

THE URANIUM DISINTEGRATION SERIES

NUCLIDE	PARTICLE PRODUCED	HALF-LIFE
Uranium-238 ($^{238}_{92}U$)	α	4.51×10^9 years
↓		
Thorium - 234 ($^{234}_{90}Th$)	β	24.1 days
↓		
Protactinium-234 ($^{234}_{91}Pa$)	β	6.75 hours
↓		
Uranium-234 ($^{234}_{92}U$)	α	2.48×10^5 years
↓		
Thorium-230 ($^{230}_{90}Th$)	α	8.0×10^4 years
↓		
Radium-226 ($^{226}_{88}Ra$)	α	1.62×10^3 years
↓		
Radon-222 ($^{222}_{86}Rn$)	α	3.82 days
↓		
Polonium-218 ($^{218}_{84}Po$)	α	3.1 minutes
↓		
Lead-214 ($^{214}_{82}Pb$)	β	26.8 minutes
↓		
Bismuth-214 ($^{214}_{83}Bi$)	β	19.7 minutes
↓		
Polonium-214 ($^{214}_{84}Po$)	α	1.6×10^{-4} second
↓		
Lead-210 ($^{210}_{82}Pb$)	β	20.4 years
↓		
Bismuth-210 ($^{210}_{83}Bi$)	β	5.0 days
↓		
Polonium-210 ($^{210}_{84}Po$)	α	138.4 days
↓		
Lead-206 ($^{206}_{82}Pb$)	—	Stable

Figure 4.9
The Uranium Series.

RATE OF RADIOACTIVE DECAY

The decay of a radioactive isotope takes place by disintegration of the atomic nucleus. It is not influenced by any external conditions. Therefore the rate of decay is characteristic of an isotope and depends only on the number of atoms present. If N be the number of undecayed atoms of an isotope present in a sample of the isotope, at time t,

$$-\frac{dN}{dt} \propto N$$

or
$$-\frac{dN}{dt} = \lambda N \qquad ...(1)$$

where $-\frac{dN}{dt}$ means the rate of decrease in the number of radioactive atoms in the sample; and λ is the proportionality factor. This is known as the **decay constant** or **disintegration constant.** Putting $dt = 1$ in equation (1) we have

$$-\frac{dN}{N} = \lambda \qquad ...(2)$$

Thus decay constant may be defined as the proportion of atoms of an isotope decaying per second.

UNITS OF RADIOACTIVITY

The standard unit of radioactivity (*i.e.* rate of disintegration) is *Curie* (c). A curie is a quantity of radioactive material decaying at the same rate as 0.001 kg of Radium (3.7×10^{10} Bq.). Rutherford is a more recent unit.

$$1 \text{ Rutherford} = 10^6 \text{ dps}$$

The S.I. unit is Becquerel

$$1 \text{ Bq} = 1 \text{ dps}$$

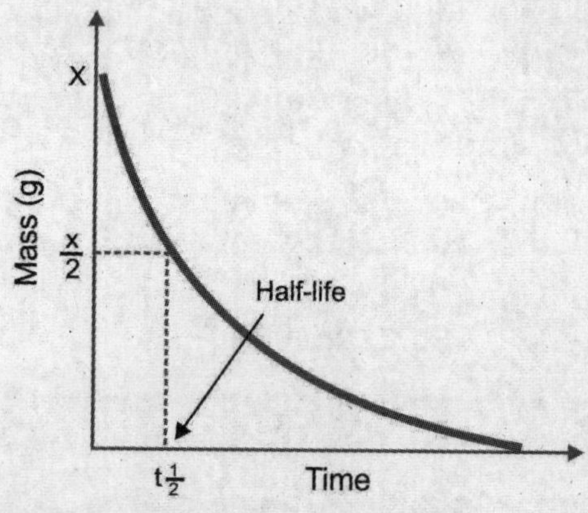

Figure 4.10
Decay curve of a radioactive isotope of mass x.

HALF-LIFE

The half-life or half-life period of a radioactive isotope is the time required for one-half of the isotope to decay. Or, it may be defined as the time for the radioactivity of an isotope to be reduced to half of its original value. Half-life period is characteristic of a radioactive element. For example, the half-life of radium is 1620 years (51088320000 seconds). This means that 0.001 kg. of radium will be reduced to 0.00005 kg in 51088320000 seconds; and so on. Some other radioactive elements may have half-life of a fraction of a second and for others it may be millions of years. The unit of half-life period is time^{-1}.

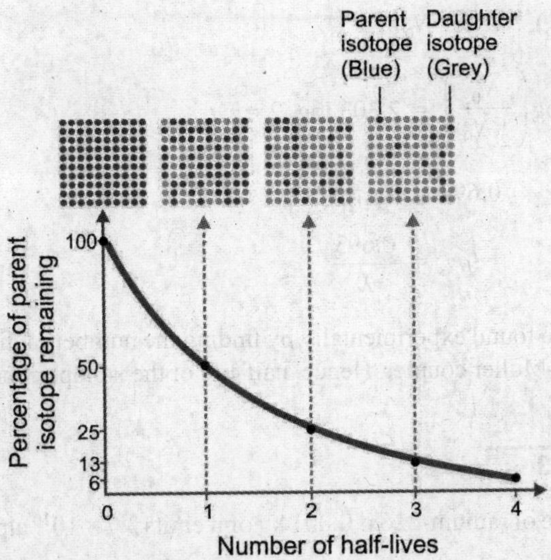

THE ACTIVITY OF A RADIOACTIVE SUBSTANCE

It is defined as the **rate of decay or the number of disintegrations per unit time.** The activity of a sample is denoted by A. It is given by the expression :

$$A = \frac{dN}{dt} = \lambda N$$

The unit of activity is the **curie (Ci)** which is the rate of decay of 3.7×10^{10} Bq. (disintegrations per second). The SI unit of activity is **becquerel (Bq)** which is defined as one disintegration per second.

The activity of a radioactive sample is usually determined experimentally with the help of a Geiger-Muller counter.

CALCULATION OF HALF-LIFE

From equation (1) we can write

$$-\frac{dN}{N} = \lambda\, dt$$

On integration,

$$-\int\frac{dN}{N} = \lambda \int dt$$

or $$-\ln N = \lambda t + X \text{ (constant)} \qquad ...(3)$$

If N_0 is the number of atoms at time $t = 0$, $X = -\ln N_0$

Substituting the value of X in (3)

$$-\ln N = \lambda t - \ln N_0$$

$$\ln\left(\frac{N_0}{N}\right) = \lambda t$$

or

Using ordinary logs,

$$2.303 \log\left(\frac{N_0}{N}\right) = \lambda t \qquad ...(4)$$

At half-life time ($t_{1/2}$), $N = 1/2\ N_0$

$$\therefore \quad 2.303 \log\left(\frac{N_0}{\frac{1}{2}N_0}\right) = 2.303 \log\ 2 = \lambda\, t_{1/2}$$

or
$$0.693 = \lambda\, t_{1/2}$$

or
$$t_{1/2} = \frac{0.693}{\lambda} \quad \text{... (5)}$$

The value of λ can be found experimentally by finding the number of disintegrations per second with the help of a Geiger-Muller counter. Hence, half-life of the isotope concerned can be calculated by using the relation (5).

SOLVED PROBLEM 3

Calculate the half-life of radium-226 if 0.001 kg of it emits 3.7×10^{10} alpha particles per second.

SOLUTION

Rate of decay = Rate of emission of α-particles

We know that

$$\frac{dN}{dt} = \lambda N = 3.7 \times 10^{10} \text{ per second} \quad \text{...(1)}$$

The number of atoms of radium present (N) in 1g of sample, $= \dfrac{6.023 \times 10^{23}}{226}$

From equation (5) stated earlier $\lambda = \dfrac{0.693}{t_{1/2}}$

Substituting the value of λ and N in equation (1) above

$$\frac{dN}{dt} = \frac{0.693}{t_{1/2}} \times \frac{6.023 \times 10^{23}}{226} = 3.7 \times 10^{10}$$

Hence
$$t_{1/2} = \frac{0.693 \times 6.023 \times 10^{23}}{3.7 \times 10^{10} \times 226}$$

$$= \mathbf{4.9915 \times 10^{10}\ s}$$

SOLVED PROBLEM 4

Calculate the disintegration constant of cobalt 60 if its half-life to produce nickel –60 is 1.6×10^8 seconds.

SOLUTION

From equation (5) stated earlier $t_{1/2} = \dfrac{0.693}{\lambda}$

or
$$\lambda = \frac{0.693}{t_{1/2}}$$

Substituting the value of $t_{1/2}$, we have $\lambda = \dfrac{0.693}{1.6 \times 10^8\ s^{-1}} = 0.433 \times 10^8\ s^{-1} = \mathbf{4.33 \times 10^{-9}\ s^{-1}}$

SOLVED PROBLEM 5

The half-life period of radon is 3.825 days. Calculate the activity of radon. (atomic weight of radon = 222)

SOLUTION

We know that

$$dN = \lambda N \qquad \text{... (2)}$$

where dN is the number of atoms disintegrating per second, λ is the decay constant and N is the number of atoms in the sample of radon.

Calculation of λ :

$$\lambda = \frac{0.693}{t_{1/2}} = \frac{0.693}{3.825 \times 24 \times 60 \times 60} = 2.096 \times 10^{-6}\ \text{s}^{-1}$$

Calculation of N :

From expression (1) above

$$N = \frac{dN}{\lambda} = \frac{3.7 \times 10^{10}}{2.096 \times 10^{-6}} = 1.7653 \times 10^{16}\ \text{atoms}$$

Mass of 6.02×10^{23} atoms of radon = 0.222 kg

Mass of 1.7653×10^{16} atoms of radon $= \frac{0.222}{6.02 \times 10^{23}} \times 1.7653 \times 10^{16}$

$$= 6.51 \times 10^{-9}\ \text{kg}$$

By definition, the activity of radon is its mass in grams which gives 3.7×10^{10} disintegrations per second. Therefore activity of radon $= 6.51 \times 10^{-9}\ \text{kg} \times 3.7 \times 10^{10}$ dps

$= 240.87$ kg dps $=$ **240.87 kg Bq**

CALCULATION OF SAMPLE LEFT AFTER TIME T

It follows from equation (4) stated earlier that

$$\log\left(\frac{N_0}{N}\right) = \frac{\lambda t}{2.303}$$

Knowing the value of λ, the ratio of N_0/N can be calculated. If the amount of the sample present to start with is given, the amount left after lapse of time t can be calculated.

SOLVED PROBLEM 6

Cobalt-60 disintegrates to give nickel-60. Calculate the fraction and the percentage of the sample that remains after 15 years (473040000 seconds). The disintegration constant of cobalt-60 is 0.13 yr^{-1} (4099680 s^{-1}).

SOLUTION

$$\log\frac{N_0}{N} = \frac{\lambda t}{2.303} = \frac{(0.13\text{yr}^{-1})(15\text{yr})}{2.303} = 0.847$$

$$\frac{N_0}{N} = \text{antilog } 0.847 = 7.031$$

The fraction remaining is the amount at time t divided by the initial amount.

$$\frac{N}{N_0} = \frac{1}{7.031} = \mathbf{0.14}$$

Hence the fraction remaining after 15 years is 0.14 or 14 per cent of that present originally.

SOLVED PROBLEM 7

How much time would it take for a sample of cobalt-60 to disintegrate to the extent that only 2.0 per cent remains ? The disintegration constant λ is 0.13 yr^{-1}(4099680 s^{-1}).

SOLUTION

$$\frac{N}{N_0} = \frac{2}{100} = 0.02$$

or $$\frac{N_0}{N} = \frac{1}{0.02} = 50$$

From equation (4) stated earlier

$$\log\left(\frac{N_0}{N}\right) = \frac{\lambda t}{2.303}$$

$$\log 50 = \frac{(0.13\, yr^{-1})t}{2.303}$$

or $$t = \frac{2.303 \log 50}{0.13\, yr^{-1}} = \mathbf{30\ yr} = \mathbf{946080000\ s}$$

SOLVED PROBLEM 8

A sample of radioactive^{133}I gave with a Geiger counter 3150 counts per minute at a certain time and 3055 counts per unit exactly after one hour later. Calculate the half life period of ^{133}I.

SOLUTION. Here $N_0 = 3150$; $N = 3055$; $\frac{t_1}{2} = 1\,\text{hour}$.

We know $$\lambda = \frac{0.693}{\frac{t_1}{2}} \text{ and } \lambda = \frac{2.303}{t} \log \frac{N_0}{N}$$

or $$\frac{0.693}{t_{1/2}} = \frac{2.303}{t} \log \frac{N_0}{N} = \frac{2.303}{1} \log \frac{3150}{3055}$$

or $$t_{1/2} = \mathbf{22.63\ yr} = \mathbf{713659680\ s}$$

AVERAGE LIFE

In a radioactive substance, some atoms decay earlier and others survive longer. **The statistical average of the lives of all atoms present at any time is called the Average life.** It is denoted by the symbol τ and has been shown to be reciprocal of decay constant, λ.

$$\tau = \frac{1}{\lambda}$$

The average life of a radioactive element is related to its half-life by the expression :

Average life = 1.44 × Half-life

or $$\tau = 1.44 \times t_{1/2}$$

The average life is often used to express the rate of disintegration of a radioactive element. The average life of radium is 7.56864 × 10^{10} s.

RADIOACTIVE EQUILIBRIUM

Let a radioactive substance A decay to give another radioactive substance B which decays to form substance C. If λ_A and λ_B are the decay constants for the two changes, we can write

$$A \xrightarrow{\lambda_A} B \xrightarrow{\lambda_B} C$$

The rate of disintegration of A is also the rate of formation of B. When the rate of disintegration of A (or formation of B) is equal to the rate of disintegration of B, the amount of B does not change with lapse of time. Then the radioactive equilibrium is said to be established between the substance A and the substance B. At this stage

$$\frac{dN_A}{dt} = \frac{dN_B}{dt}$$

where N_A and N_B are atoms of A and B present at the equilibrium.

Since $\quad \lambda_A N_A = \lambda_B N_B, \quad \frac{N_A}{N_B} = \frac{\lambda_B}{\lambda_A}$

But $\quad \lambda \propto t_{1/2}$

$\therefore \quad \frac{N_A}{N_B} = \frac{t_{1/2} \text{ of } B}{t_{1/2} \text{ of } A}$

Thus the atoms of A and B are present in the ratio of their half-lives.

The radioactive equilibrium differs from a chemical equilibrium in that it is irreversible.

RADIOACTIVE DATING

The age of an old piece of wood can be determined by radioactive dating technique. The atmosphere contains radioactive carbon dioxide, $^{14}CO_2$, and ordinary carbon dioxide, $^{12}CO_2$, in a fixed ratio. A plant while alive takes up both types of carbon dioxide and converts them to carbon-14 and carbon-12 photosynthesis. Thus a living plant contains radioactive carbon-14 and stable carbon-12 in a fixed ratio. When the plant dies, the uptake of carbon from the atmosphere stops. Now onward, carbon-12 remains unchanged but carbon-14 decays by beta-emission.

$$^{14}_{6}C \longrightarrow {}^{14}_{7}N + {}^{0}_{-1}\beta \qquad \text{(Half-life = 5730 yr)}$$

(radioactive)

$$^{12}_{6}C \longrightarrow \textit{No change}$$

(stable)

As a result, $^{14}C/^{12}C$ decreases with lapse of time.

Therefore the concentration of carbon-14 declines with time. The concentration of carbon-14 can be measured by counting radioactivity. Knowing the concentration of carbon-14 in a given sample of old wood and that in a living plant, the age of the sample can be calculated.

SOLVED PROBLEM 9

The amount of carbon-14 in a piece of wood is found to be one-sixth of its amount in a fresh piece of wood. Calculate the age of old piece of wood.

SOLUTION

Calculation of Decay constant :

The half-life of carbon-14 $= 5730$ yr

$= 1.8070128 \times 10^{11}$ s

$$\therefore \quad \lambda = \frac{0.693}{t_{1/2}} = \frac{0.693}{1.8070128 \times 10^{11}\,\text{s}}$$

Calculation of Age of Wood : From expression (4) on page 127 $t = \log \frac{^{14}C_0}{^{14}C_t} \times \frac{2.303}{\lambda}$

or
$$t = \log \frac{1}{1/6} \times \frac{2.303 \times 1.8070128 \times 10^{11}}{0.693}$$
$$= \log 6 \times \frac{2.303 \times 1.8070128 \times 10^{11}}{0.693}$$
$$= \frac{0.7782 \times 2.303 \times 1.8070128 \times 10^{11}}{0.693}$$
$$= 467316216000 \text{ s} = \mathbf{4.6731 \times 10^{11}\ s}$$

In a new method for determining the age of old wood (or fossil) the measurement of radioactivity is avoided. The ratio $^{14}C/^{12}C$ is found with the help of a mass spectrometer in the old wood and fresh wood from a living plant. It is assumed that the ratio $^{14}C/^{12}C$ in the fresh wood today is the same as it was at the time of death of the plant.

Let the ratio in the fresh plant at $t = 0$ be
$$\frac{^{14}C_0}{^{12}C_0} = x \quad ...(1)$$
Let the ratio in the old piece of wood at time t be
$$\frac{^{14}C_t}{^{12}C_t} = y \quad ...(2)$$
Dividing (1) by (2) $\frac{^{14}C_0}{^{12}C_0} \times \frac{^{12}C_t}{^{14}C_t} = \frac{x}{y}$

The concentration of carbon-12 does not change with time and $^{12}C_0 = {^{12}C_t}$

Therefore $\frac{^{14}C_0}{^{14}C_t} = \frac{x}{y} = \frac{1}{y/x}$

where the ratio in the old wood is y/x times the ratio in the living plant.

Knowing the value of y/x, the value of t can be found from the expression
$$t = \log \frac{^{14}C_0}{^{14}C_t} \times \frac{2.303}{\lambda}$$

SOLVED PROBLEM 10

A bone taken from a garbage pile buried under a hill-side had $^{14}C/^{12}C$ ratio 0.477 times the ratio in a living plant or animal. What was the date when the animal was buried?

SOLUTION

Half-life of carbon-14 is 5730 years (1.807×10^{11} seconds)
$$\therefore \quad \lambda = \frac{0.693}{t_{1/2}} = \frac{0.693}{1.807 \times 10^{11}\ \text{s}}$$
Substituting values in the expression
$$t = \log \frac{^{14}C_0}{^{14}C_t} \times \frac{2.303}{\lambda}$$

$$= \log \frac{1.0}{0.477} \times \frac{2.303 \times 1.807 \times 10^{11}}{0.693} = 6.1 \times 10^3$$

$$= \mathbf{1.9237 \times 10^{11}\ s}$$

The animal was buried 1.9237×10^{11} seconds ago or 6100 years ago.

NUCLEAR REACTIONS

In a chemical reaction there is merely a rearrangement of extranuclear electrons. The atomic nucleus remains intact. A nuclear reaction involves a change in the composition of the nucleus. The number of protons and neutrons in the nucleus is altered. The product is a new nucleus of another atom with a different atomic number and/or mass number. Thus,

A nuclear reaction is one which proceeds with a change in the composition of the nucleus so as to produce an atom of a new element.

The conversion of one element to another by a nuclear change is called **transmutation.**

We have already considered the nuclear reactions of radioactive nuclei, producing new isotopes. Here we will consider such reactions caused by artificial means.

DIFFERENCES BETWEEN NUCLEAR REACTIONS AND CHEMICAL REACTIONS

Nuclear Reactions	Chemical Reactions
1. Proceed by redistribution of nuclear particles.	1. Proceed by the rearrangement of extra nuclear electrons.
2. One element may be converted into another.	2. No new element can be produced.
3. Often accompanied by release or absorption of enormous amount of energy.	3. Accompanied by release or absorption of relatively small amount of energy.
4. Rate of reaction is unaffected by external factors such as concentration, temperature, pressure and catalyst.	4. Rate of reaction is influenced by external factors.

NUCLEAR FISSION REACTIONS

In these reactions an atomic nucleus is broken or fissioned into two or more fragments. This is accomplished by bombarding an atom by alpha particles ($^4_2$He), neutrons (^1_0n), protons ($^1_1$H), deutrons ($^2_1$H), etc. All the positively charged particles are accelerated to high kinetic energies by a device such as a *cyclotron.* This does not apply to neutrons which are electrically neutral. The projectile enters the nucleus and produces an unstable '**compound nucleus**'. It decomposes instantaneously to give the products. For example, $^{14}_7$N when struck by an α-particle first forms an intermediate unstable compound nucleus, $^{18}_9$F, which at once cleaves to form stable $^{17}_8$O.

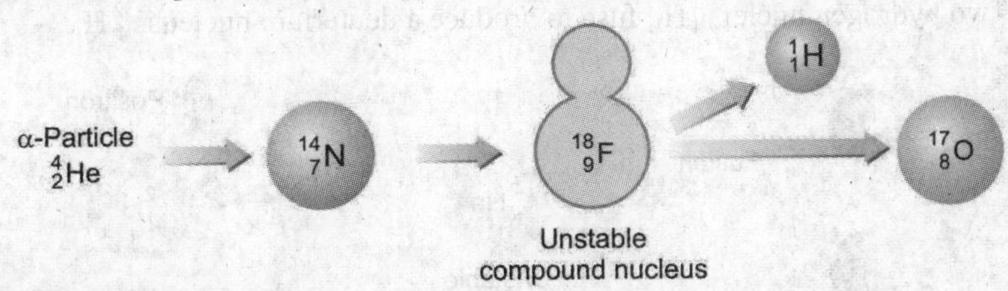

■ **Figure 4.11**
Mechanism of a Nuclear fission reaction.

Other examples are,

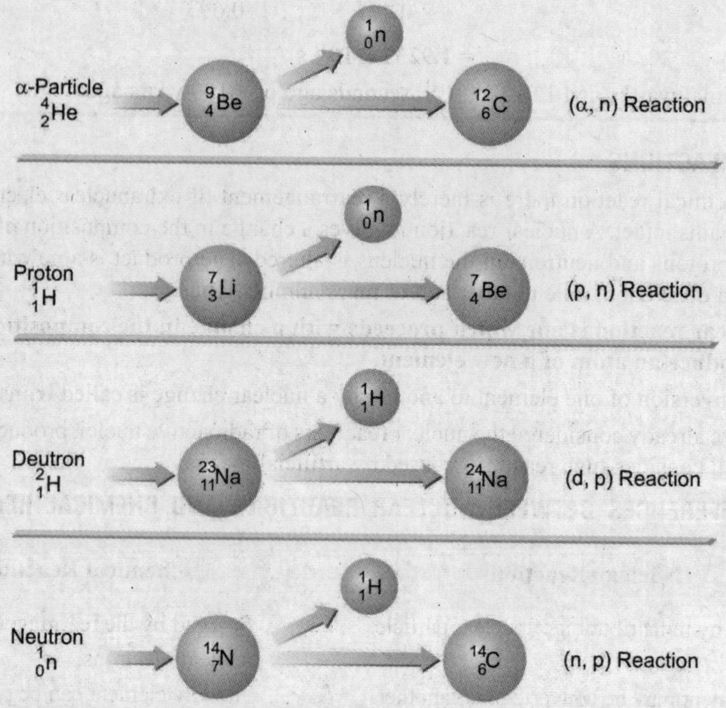

■ **Figure 4.12**
Representation of some Nuclear Fission reactions. Unstable compound nucleus is not shown.

Nuclear fission reactions are classified according to the projectile used and the particle that is emitted. In Fig. 4.12 the type of the reaction has been stated.

It is noteworthy that neutrons are particularly useful as projectile. Sir James Chadwick obtained these by bombarding beryllium-9 with α-particles. Being electrically neutral, neutrons pierce the positive nucleus easily.

NUCLEAR FUSION REACTIONS

These reactions take place by combination or fusion of two small nuclei into a larger nucleus. At extremely high temperatures the kinetic energy of these nuclei overweighs the electrical repulsions between them. Thus they coalesce to give an unstable mass which decomposes to give a stable large nucleus and a small particle as proton, neutron, positron, etc. For example :

(1) Two hydrogen nuclei, $^{1}_{1}H$, fuse to produce a deuterium nucleus, $^{2}_{1}H$.

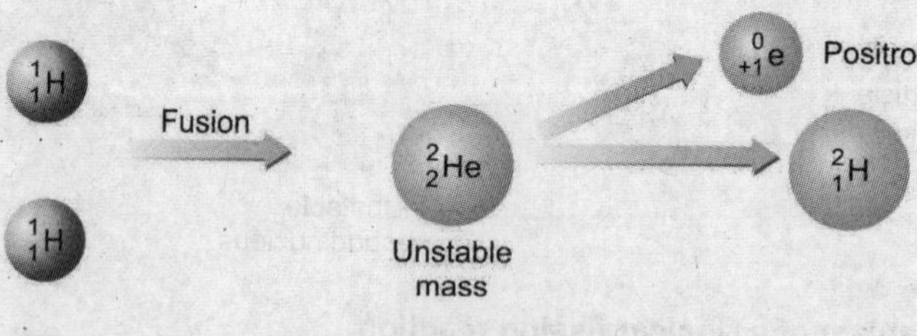

(2) Deuterium nucleus, ${}^{2}_{1}H$, and tritium nucleus, ${}^{3}_{1}H$, combine to give helium nucleus, ${}^{5}_{2}He$ with the expulsion of a neutron.

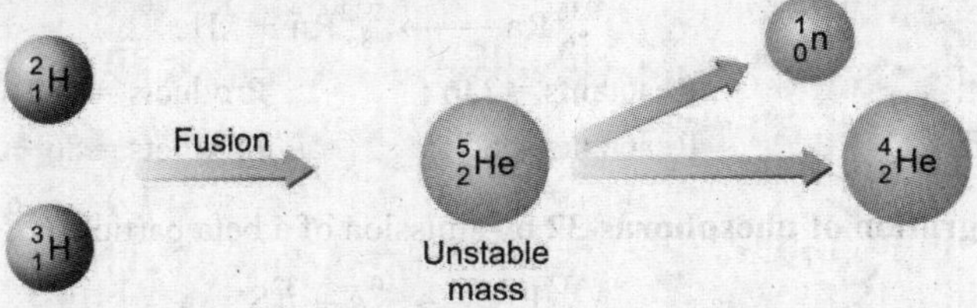

DIFFERENCES BETWEEN NUCLEAR FISSION AND NUCLEAR FUSION

Nuclear Fission	Nuclear Fusion
1. A bigger (heavier nucleus splits into smaller lighter) nuclei.	1. Lighter nuclei fuse together to form the heavier nucleus.
2. It does not require high temperature.	2. Extremely high temperature is required for fusion to take place.
3. A chain reaction sets in.	3. It is not a chain reaction.
4. It can be controlled and energy released can be used for peaceful purposes.	4. It cannot be controlled and energy released cannot be used properly.
5. The products of the reaction are radioactive in nature.	5. The products of a fusion reaction are non-radioactive in nature.
6. At the end of the reaction nuclear waste is left behind.	6. No nuclear waste is left at the end of fusion reaction.

NUCLEAR EQUATIONS

Similar to a chemical reaction, nuclear reactions can be represented by equations. **These equations involving the nuclei of the reactants and products are called nuclear equations.** The nuclear reactions occur by redistribution of protons and neutrons present in the reactants so as to form the products. Thus the total number of protons and neutrons in the reactants and products is the same. Obviously, **the sum of the mass numbers and atomic numbers on the two sides of the equation must be equal.**

If the mass numbers and atomic numbers of all but one of the atoms or particles in a nuclear reaction are known, the unknown particle can be identified.

How to write a Nuclear equation?

(1) Write the symbols of the nuclei and particles including the mass numbers (superscripts) and atomic numbers (subscripts) on the left (reactants) and right (products) of the arrow.

(2) Balance the equation so that the sum of the mass numbers and atomic numbers of the particles (including the unknown) on the two sides of the equation are equal. Thus find the atomic number and mass number of the unknown atom, if any.

(3) Then look at the periodic table and identify the unknown atom whose atomic number is disclosed by the balanced equation.

Examples of Nuclear equations

(*a*) **Disintegration of radium-236** by emission of an alpha particle ($^{4}_{2}He$),

$$^{236}_{88}Ra \longrightarrow ^{232}_{86}Rn + ^{4}_{2}He$$

Mass Nos :	Reactants = 236 ;	Products = 232 + 4 = 236
Atomic Nos :	Reactants = 88 ;	Products = 86 + 2 = 88

(*b*) **Disintegration of phosphorus-32** by emission of a beta particle ($^{0}_{-1}e$),

$$^{32}_{15}P \longrightarrow ^{0}_{-1}e + ^{32}_{16}S$$

Mass Nos :	Reactants = 32 ;	Products : 0 + 32 = 32
Atomic Nos :	Reactants = 15 ;	Products : 16 – 1 = 15

(*c*) **Fission of argon-40** by bombardment with a proton $^{1}_{1}H$,

$$^{40}_{18}Ar + ^{1}_{1}H \longrightarrow ^{40}_{19}K + ^{1}_{0}n$$

Mass Nos :	Reactants = 40 + 1 = 41 ;	Products = 40 + 1 = 41
Atomic Nos :	Reactants = 18 + 1 = 19 ;	Products = 19 + 0 = 19

(*d*) **Fission of uranium-235** by absorption of a neutron ($^{1}_{0}n$),

$$^{235}_{92}U + ^{1}_{0}n \longrightarrow ^{141}_{56}Ba + ^{92}_{36}Kr + 3\,^{1}_{0}n$$

Mass No :	Reactants = 235 + 1 = 236;	Products = 141 + 92 + 3 = 236
Atomic Nos :	Reactants = 92 + 0 = 92;	Products = 56 + 36 + 0 = 92

(*e*) **Fusion of lithium-7** and proton, $^{1}_{1}H$,

$$^{7}_{3}Li + ^{1}_{1}H \longrightarrow 2\,^{4}_{2}He$$

Mass Nos :	Reactants = 7 + 1 = 8 ;	Products = 2 × 4 = 8
Atomic Nos :	Reactants = 3 + 1 = 4 ;	Products = 2 × 2 = 4

SOLVED PROBLEM 11

Write the nuclear equation for the change that occurs in radium-226 when it emits an alpha particle.

SOLUTION

Step 1 Write the symbol of the parent atom with its mass number and atomic number (from periodic table) on the left-hand side of the equation.

$$^{226}_{88}Ra \longrightarrow$$

Step 2 Write the symbol for the alpha particle on the right-hand side of the equation.

$$^{226}_{88}Ra \longrightarrow ^{4}_{2}He + ?$$

Step 3 Complete the equation by writing the symbol of an isotope that has an atomic number 88 – 2 = 86 and mass number 226 – 4 = 222. As shown by the periodic table, the isotope with atomic number 86 is radon, Rn. Thus,

$$^{226}_{88}Ra \longrightarrow ^{4}_{2}He + ^{222}_{86}Rn$$

Step 4 Check that the mass numbers and atomic numbers on the two sides of the equation are balanced.

Mass Nos : 226 = 4 + 222

Atomic Nos : 88 = 2 + 86

SOLVED PROBLEM 12

Cobalt–60 decays by emission of a beta particle. Predict the atomic number, mass number, and name of the isotope formed.

SOLUTION

Step 1 Write the symbol of the cobalt with mass number and atomic number (from the periodic table) on the left-hand side of the equation.

$$^{60}_{27}\text{Co} \longrightarrow$$

Step 2 If the unknown product isotope is X with mass number M and atomic number A, the nuclear equation may be written as

$$^{60}_{27}\text{Co} \longrightarrow {}^{\text{M}}_{\text{A}}X + {}^{0}_{-1}e$$

Step 3 Since the sum of mass numbers and atomic numbers are equal on the two sides of the equation,

$$M = 60$$
$$A = 27 + 1 = 28$$

Step 4 Consult the periodic table and find the element whose atomic number is 28. This is nickel, Ni. Therefore, the complete equation for the decay of cobalt-60 is

$$^{60}_{27}\text{Co} \longrightarrow {}^{60}_{28}\text{Ni} + {}^{0}_{-1}e$$

SOLVED PROBLEM 13

Complete the nuclear equation

$$^{238}_{92}\text{U} + {}^{4}_{2}\text{He} \longrightarrow ? + {}^{1}_{0}n$$

SOLUTION

Let the unknown atom be X with mass number M and atomic number A.

We can write the above nuclear equation as

$$^{238}_{92}\text{U} + {}^{4}_{2}\text{He} \longrightarrow {}^{\text{M}}_{\text{A}}\text{X} + {}^{1}_{0}n$$

But the sum of mass numbers on the two sides of the equation is equal. Thus,

$$238 + 4 = M + 1$$

or, $$M = 242 - 1 = 241$$

Also $$92 + 2 = A + 0$$

or $$A = 94$$

Find the isotope of atomic number 94 from the periodic table. It is plutonium, Pu. Thus the completed nuclear equation is

$$^{238}_{92}\text{U} + {}^{4}_{2}\text{He} \longrightarrow {}^{241}_{94}\text{Pu} + {}^{1}_{0}n$$

ARTIFICIAL RADIOACTIVITY

Many stable nuclei when bombarded with high speed particles produce unstable nuclei that are radioactive. The radioactivity produced in this manner by artificial means is known as **artificial radioactivity** or **induced radioactivity.** The artificial isotopes disintegrate in a definite fashion and have specific half-life. For example, aluminium-27 when bombarded with a neutron emits an alpha particle and forms sodium-24 which is radioactive. It disintegrates spontaneously by emission of a beta particle (${}^{0}_{-1}e$) and the product is magnesium-24. Sodium-24 has half-life of 7.56864×10^8 sec.

$$^{1}_{0}n + ^{27}_{13}Al \longrightarrow ^{24}_{11}Na + ^{4}_{2}He$$

radioactive

$$^{24}_{11}Na \longrightarrow ^{24}_{12}Mg + ^{0}_{-1}e$$

NUCLEAR ISOMERISM

Sometimes α and β-decays may produce a pair of nuclei that have the same number of protons and neutrons but different radioactive properties.

A pair of nuclei having same number of protons and neutrons but different half-lives are called nuclear isomers. The phenomenon is called **nuclear isomerism.**

Example of Nuclear isomerism

Uranium-Z and Uranium-X_2 constitute a pair of nuclear isomers. Both nuclei contain 91 protons and 143 neutrons, and are isotopes. They exhibit β-ray activity with half-lives 6.7 hr and 1.14 min respectively.

Explanation

The α- or β-decay of a radionuclide first leaves it in an excited state. This is then converted into the ground state nucleus. The excited and the ground state nuclei thus produced are called nuclear isomers. The nuclear isomers may be isotopic or isobaric.

ENERGY RELEASED IN NUCLEAR REACTIONS

Einstein's Equation Relating Mass and Energy

According to Albert Einstein, mass can be converted into energy and *vice versa.* His famous equation relating mass and energy is

$$E = mc^2 \qquad ...(1)$$

where E = energy ; m = mass and c = velocity of light. In nuclear reactions, a change in mass, Δm, is accompanied by release of energy, ΔE. Thus equation (1) may be written as

$$\Delta E = \Delta mc^2 \qquad ...(2)$$

If we substitute the value 3.00×10^8 m/sec for the velocity of light, the equation (2) directly gives the relation between the energy change in joules and the mass change in kilograms.

$$\Delta E \text{ (in Joules)} = 9.00 \times 10^{16} \times \Delta m \text{ (in kilograms)} \qquad ...(3)$$

Making use of the conversion factor 1 Joule = 2.39×10^{-4} kcal, we can express the energy change in k cals.

$$\Delta E = 9.00 \times 10^{16} \text{ erg} \times \frac{2.39 \times 10^{-4}\, kcal}{1\text{J}} \times \Delta m$$

or

$$\Delta E \text{ (in kcal)} = 2.15 \times 10^{13} \times \Delta m \text{ (in kilograms)} \qquad ...(4)$$

Very often, in a nuclear reaction, the mass of the products is less than that of the reactants. The mass difference is converted into energy. Therefore by using equation (4), we can calculate the amount of energy released in a particular reaction. For example, in the equation

$^{7}_{3}Li$	+	$^{1}_{1}H$	$\longrightarrow$	$^{4}_{2}He$	+	$^{4}_{2}He$	+	energy
7.160×10^{-3} kilograms		1.0078×10^{-3} kilograms		4.0026×10^{-3} kilograms		4.0026×10^{-3} kilograms		

The atomic mass difference between the reactants and products is 0.0186×10^{-3} kilogram. Using equation (4)

$$\begin{aligned} \Delta E &= 2.15 \times 10^{13} \times \Delta m \\ &= 2.15 \times 10^{13} \times 0.0186 \times 10^{-3} \\ &= 4.0 \times 10^{8} \text{ kcal} \\ &= 1.675 \times 10^{9} \text{ kJ} \end{aligned}$$

MASS DEFECT

We know that atomic nucleus consists of protons and neutrons; collectively known as nucleons. It is found that the measured mass of nucleus is always less than the sum of the masses of the individual protons and neutrons which make it up. Let us take the example of helium, ${}^{4}_{2}He$. It consists of two protons and two neutrons. Its mass may be calculated as :

$$\text{mass of the protons} = 2 \times 1.00815$$
$$\text{mass of the neutrons} = 2 \times 1.00899$$
$$4.03428$$

However, the experimental mass of the helium nucleus is only 4.00388. This is less by 0.03040 amu than that calculated above. This is called the mass defect of helium nucleus.

The difference between the experimental and calculated masses of the nucleus is called the Mass defect or Mass deficit.

(experimental mass of nucleus) – (mass of protons + mass of neutrons) = mass defect

NUCLEAR BINDING ENERGY

Atomic nucleus is made of protons and neutrons closely packed in a small volume. Although there exist intensive repulsive forces between the component protons, the nucleus is not split apart. This is so because the nucleons are bound to one another by very powerful forces. **The energy that binds the nucleons together in the nucleus is called the Nuclear binding energy.**

When a nucleus is formed from individual protons and neutrons, there occurs a loss of mass (mass defect). According to Einstein's theory, it is this mass defect which is converted into binding energy. **Hence binding energy is the energy equivalent of the mass defect.** The various nuclei have different binding energies.

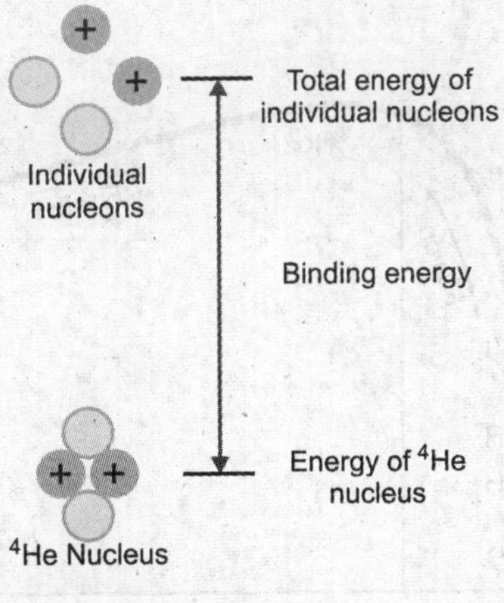

■ **Figure 4.13**
A nucleus has a lower energy and a smaller mass than free nucleons.

Binding energy is a measure of the force that holds the nucleons together. Hence an energy equivalent to the binding energy is required to disrupt a nucleus into its constituent protons and neutrons. Since the nuclear energy is of an extremely high order, it is not easy to fission a nucleus.

Calculation of Binding Energy

The binding energy of a nucleus can be calculated from its mass defect by using Einstein's equation, $\Delta E = \Delta m \times c^2$.

SOLVED PROBLEM 14

What is the binding energy for $^{11}_{5}B$ nucleus if its mass defect is 0.08181 amu?

SOLUTION

$$\Delta E = \Delta m \times c^2 \qquad \text{...Einstein's equation}$$

Here,

$$\Delta m = 0.08181 \times 10^{-3} \text{ kg/mol}$$

$$c = 3 \times 10^8 \text{ m/s}$$

Substituting values in Einstein's equation,

$$\Delta E = (0.08181 \times 10^{-3} \text{ kg/mol})(3 \times 10^8 \text{ m/s})^2$$

$$= 7.4 \times 10^{12} \text{ J/mol}$$

No. of nuclei in one mole is 6.02×10^{23} (Avogadro's Law).

$\therefore$ Binding energy for $^{11}_{5}B$ nucleus, ΔE, may be expressed as

$$\Delta E = \frac{7.4 \times 10^{12}}{6.02 \times 10^{23}} = \mathbf{1.2 \times 10^{-11}\ J / nucleus}$$

Binding energy per nucleon

It can be calculated by dividing the total binding energy by the sum of the number of protons and neutrons present in the nucleus.

$$\text{Binding energy per nucleon} = \frac{\text{Binding energy of nucleus}}{\text{No. of protons + No. of neutrons}}$$

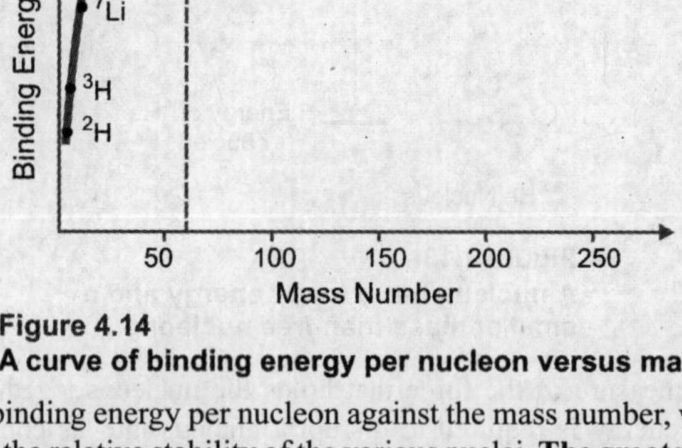

Figure 4.14
A curve of binding energy per nucleon versus mass number.

By plotting the binding energy per nucleon against the mass number, we get the graph shown in Fig. 4.14 This shows the relative stability of the various nuclei. **The greater the binding energy per nucleon the more stable is the nucleus.** Thus the nuclei of about 60 atomic mass having maximum energy per nucleon are most stable *e.g.*, ^{56}Fe. The nuclei that are heavier or lighter than this have lower binding energies per nucleon and are less stable. Thus ^{235}U undergoes fission into lighter and more stable isotopes as ^{139}Ba and ^{94}Kr with the release of energy. Similarly two or more lighter nuclei (^2H, ^3H) with lower binding energy per nucleon combine or fuse together into a heavier and more stable nucleus. This is also accompanied by release of energy.

Equivalence of amu and Energy

Since 1 amu is exactly equal to $\frac{1}{12}$th of the mass of C^{12} atom, therefore

$$1 \text{ amu} = \frac{1}{12} \times \text{Mass of } C^{12} \text{ atom}$$

$$= \frac{\frac{1}{12} \times \text{gram atomic mass of } C^{12}}{\text{Avogadro's Number}}$$

$$= \frac{1}{12} \times \frac{12}{6.02 \times 10^{23}} g$$

$$= 1.66 \times 10^{-27} \text{ kg}$$

Also $E = mc^2$

$\therefore$ $E = 1.66 \times 10^{-27} \text{ kg} \times (3 \times 10^8)^2$

$$= 1.494 \times 10^{-10} \text{ J}$$

$$= \frac{1.494 \times 10^{-3}}{10^7} \text{J} \quad [\because 1 \text{ eV} = 1.602 \times 10^{-19} \text{ J}]$$

$$= 931.5 \times 10^6 \text{ eV}$$

$$= 931.5 \text{ Mev}$$

Thus **1 amu = 931.5 Mev**

SOLVED PROBLEM 15

Calculate the binding energy per nucleon (in Joules) in He atom ^4_2He which has a mass of 4.00260 amu. Mass of neutron = 1.008655 amu and mass of 1 hydrogen atom = 1.007825 mass.

SOLUTION. In Helium atom there are 2 electrons, 2 protons and 2 neutrons.

$\therefore$ Mass of 2p + 2e = 2 × mass of 1 H atom

= 2 × 1.007825 amu

= 2.01565 amu

and Mass of 2 neutrons = 2 × 1.008665 amu

Mass of He atom = 2.01565 + 2.01733

= 4.03298 amu

Actual mass of the atom = 4.0026 amu (given)

∴ Mass defect = 4.0398 – 4.00260

= **0.03038 amu**

Binding Energy = 0.03038 × 931.5 Mev

= 28.298 Mev [∵ 1 MeV = 1.6×10^{-13} J]

= 45.2768×10^{-13} J

= 4.52768×10^{-12} J

Binding energy per nucleon

$$= \frac{\text{Binding Energy}}{\text{No. of Nucleons}}$$

$$= \frac{4.52768 \times 10^{-12}}{4}$$

$$= \mathbf{1.13192 \times 10^{-12}\ J}$$

NEUTRON-PROTON RATIO AND NUCLEAR STABILITY

Nuclei are composed of protons and neutrons. The protons would tend to fly apart due to repulsive forces between them. But the neutrons in some way hold the protons together within the nucleus. **The stability of a nucleus seems to depend on the neutron-to-proton ratio (*n/p*) in the nucleus.** Fig. 4.15 shows the neutron-to-proton ratios for all known stable elements. Each point on the graph indicates the number of protons and neutrons in a particular stable nucleus. It is clear from the graph that :

(1) **The lower elements (up to Z = 20), the stable nuclei have about equal number of protons and neutrons i.e.,** n/p = 1.

(2) **For higher elements to be stable, there must be more neutrons than protons i.e., n/p > 1.**

(3) **The shaded portion in Fig. 4.15 represents the region or belt of stability.** The element whose n/p ratios lie inside the belt are stable.

(4) **A nucleus whose n/p lies above or below the stability belt is radioactive or unstable on account** of unfavorable n/p ratio. It emits α- or β- particles so as to move into the stability range.

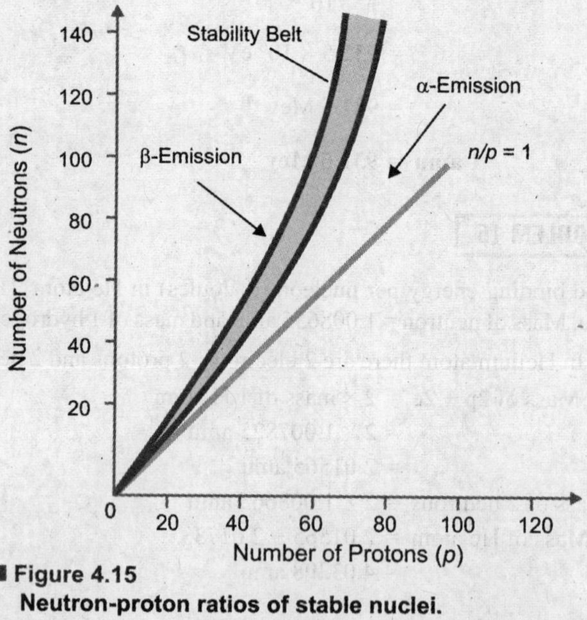

■ **Figure 4.15**
Neutron-proton ratios of stable nuclei.

(*a*) A nucleus that is above the stability belt emits a β-particle whereby a neutron is converted to proton. Thus *n*/*p* decreases and the nucleus becomes more stable or enters the stability belt. For example,

$$^{14}_{6}C \longrightarrow {}^{14}_{7}N + {}^{0}_{-1}\beta$$

n/*p* 1.33 1.0

(*b*) A nucleus whose *n*/*p* lies below the stability belt emits an α-particle and loses 2 protons and 2 neutrons. This results in a net increase of *n*/*p* and the new nucleus may enter the stability belt. For example,

$$^{238}_{92}U \longrightarrow {}^{234}_{90}Th + {}^{4}_{2}\alpha$$

n/*p* 1.565 1.60

The radioactive nuclei continue to emit α- or β-particles, one after the other, till a stable nucleus is the end-product.

NUCLEAR FISSION PROCESS

In 1939, Hahn and Stassmann discovered that a heavy atomic nucleus as of uranium-235 upon bombardment by a neutron splits apart into two or more nuclei. U-235 first absorbs a neutron to form an unstable 'compound nucleus'. The excited 'compound nucleus' then divides into two daughter nuclei with the release of neutrons and large amount of energy.

The splitting of a heavy nucleus into two or more smaller nuclei is termed nuclear fission.

The smaller nuclei formed as a result of fission are called **fission products.** The process of fission is always accompanied by the ejection of two or more neutrons and liberation of vast energy.

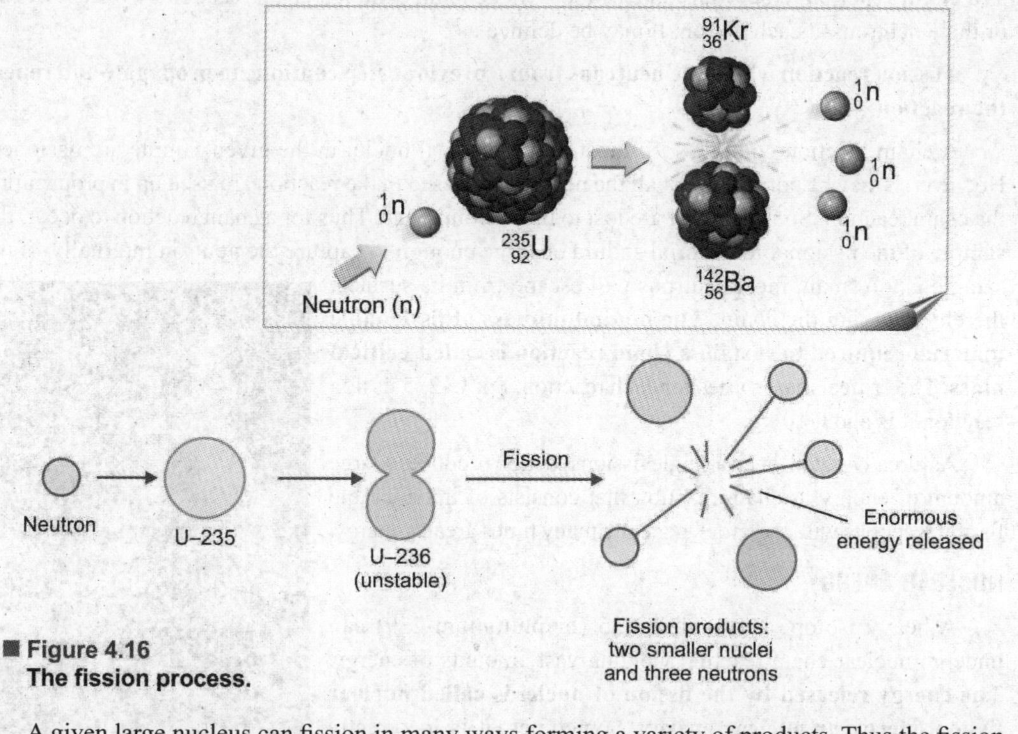

■ Figure 4.16
The fission process.

A given large nucleus can fission in many ways forming a variety of products. Thus the fission of U-235 occurs in about 35 ways. Two of these are given below in the form of equations.

$$^{235}_{92}U + ^{1}_{0}n \longrightarrow ^{139}_{56}Ba + ^{94}_{36}Kr + 3\,^{1}_{0}n + energy$$

$$^{235}_{92}U + ^{1}_{0}n \longrightarrow ^{106}_{42}Mo + ^{128}_{50}Sn + 2\,^{1}_{0}n + energy$$

In these fission reactions, the mass of the products is less than the mass of the reactant. A loss of mass of about 0.2 amu per uranium atom occurs. This mass is converted into a fantastic quantity of energy which is 2.5 million times of that produced by equivalent amount of coal.

CHARACTERISTICS OF NUCLEAR FISSION

(1) Upon capturing a neutron, a heavy nucleus cleaves into two or more nuclei.
(2) Two or more neutrons are produced by fission of each nucleus.
(3) Vast quantities of energy are produced as a result of conversion of small mass into energy.
(4) All the fission products are radioactive, giving off beta and gamma radiations.

NUCLEAR CHAIN REACTION

We know that U-235 nucleus when hit by a neutron undergoes the reaction,

$$^{235}_{92}U + ^{1}_{0}n \longrightarrow ^{139}_{56}Ba + ^{94}_{36}Kr + 3\,^{1}_{0}n$$

Each of the three neutrons produced in the reaction strikes another U-235 nucleus, thus causing nine subsequent reactions. These nine reactions, in turn, further give rise to twenty seven reactions. This process of propagation of the reaction by multiplication in threes at each fission, is referred to as a **chain reaction**. Heavy unstable isotopes, in general, exhibit a chain reaction by release of two or three neutrons at each fission. It may be defined as :

a fission reaction where the neutrons from a previous step continue to propagate and repeat the reaction.

A chain reaction continues till most of the original nuclei in the given sample are fissioned. However, it may be noted that not all the neutrons released in the reaction are used up in propagating the chain reaction. Some of these are lost to the surroundings. Thus for a chain reaction to occur, the sample of the fissionable material should be large enough to capture the neutron internally. If the sample is too small, most neutrons will escape from its surface, thereby breaking the chain. **The minimum mass of fissionable material required to sustain a chain reaction is called critical mass.** The critical mass varies for each reaction. For U-235 fission reaction it is about 10 kg.

As already stated, even a single fission reaction produces a large amount of energy. A chain reaction that consists of innumerable fission reactions will, therefore, generate many times greater energy.

NUCLEAR ENERGY

A heavy isotope as uranium-235 (or plutonium-239) can undergo nuclear chain reaction yielding vast amounts of energy. **The energy released by the fission of nuclei is called nuclear fission energy or nuclear energy.** Sometimes, it is incorrectly referred to as **atomic energy.**

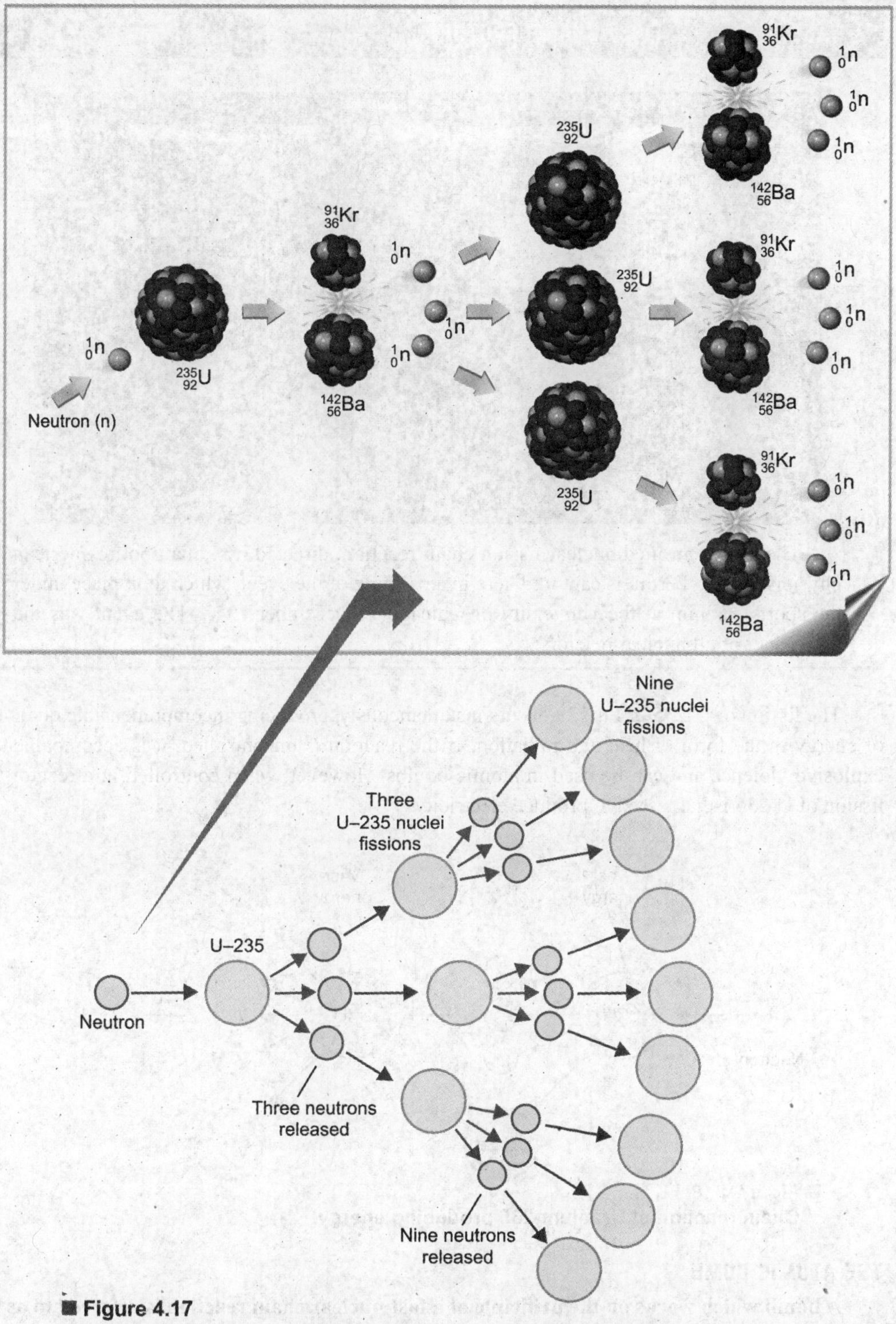

■ **Figure 4.17**

Illustration showing how U-235 fission chain reaction is propagated and multiplied. The products $^{139}_{56}$Ba and $^{94}_{36}$Kr are not shown.

FIRST CHAIN REACTION

The first controlled nuclear fission chain reaction, directed by Italian-born American physicist Enrico Fermi, is captured here in a painting of the event, which took place under the sports stadium at the University of Chicago in December 1942. The event was the forerunner of all nuclear reactors.

The fission of U-235 or Pu-239 occurs instantaneously, producing incomprehensible quantities of energy in the form of heat and radiation. If the reaction is uncontrolled, it is accompanied by explosive violence and can be used in atomic bombs. However, when controlled in a reactor, the fission of U-235 is harnessed to produce electricity.

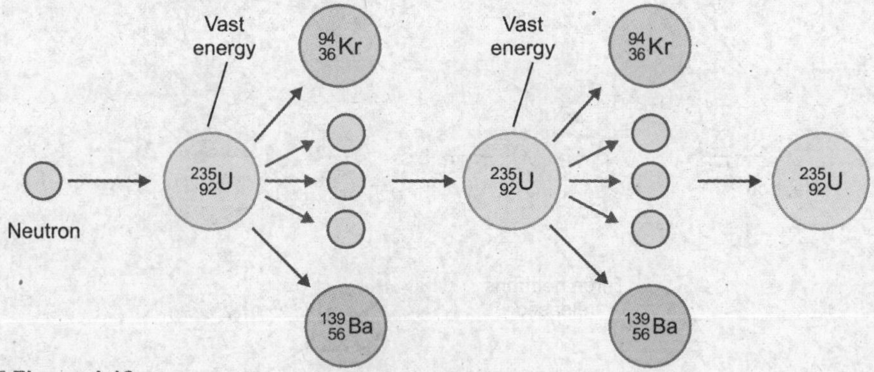

Figure 4.18
Chain reaction of Uranium-235 producing energy.

THE ATOMIC BOMB

A bomb which works on the principle of a fast nuclear chain reaction is referred to as the atomic bomb. A design of such a bomb is shown in Fig. 4.19. It contains two subcritical masses of fissionable material, ^{235}U or ^{239}Pu. It has a mass of trinitrotoluene in a separate pocket. When TNT is detonated, it drives one mass of ^{235}U into the other. A supercritical mass of the fissionable material is obtained. As a result of the instantaneous chain reaction, the bomb explodes with the release of tremendous heat energy. **Temperature developed in an atomic bomb is believed to be 10 million °C (temperature of the sun).** Besides many radionuclei and heat, deadly gamma rays are released.

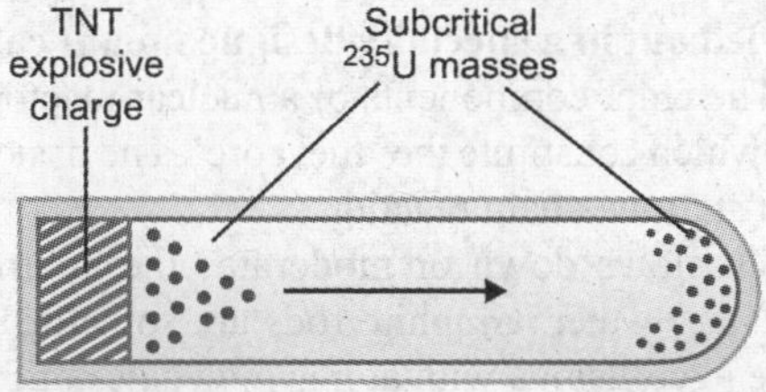

■ Figure 4.19
A design used in atomic bombs to bring together two subcritical masses of ^{235}U.

LITTLE BOY AND FAT MAN

■ The First Atomic Bombs

Little Boy was the first nuclear weapon used in warfare. It exploded approximately 1,800 feet or 548.64 meter over Hiroshima, Japan, on the morning of August 6, 1945, with a force equal to 13,000 tons or 1.3×10^7 kg of TNT. Immediate deaths were between 70,000 to 130,000.

Fat Man was the second nuclear weapon used in warfare. Dropped on Nagasaki, Japan, on August 9, 1945, *Fat Man* devastated more than two square miles of the city and caused approximately 45,000 immediate deaths.

While *Little Boy* was a uranium gun-type device, *Fat Man* was a more complicated and powerful plutonium implosion weapon that exploded with a force equal to 20 kilotons of TNT.

These play havoc with life and environment. If the bomb explodes near the ground, it raises tons of dust into the air. The radioactive material adhering to dust known as **fall out.** It spreads over wide areas and is a lingering source of radioactive hazard for long periods.

NUCLEAR REACTOR

It has been possible to control fission of U-235 so that energy is released slowly at a usable rate. **Controlled fission is carried out in a specially designed plant called a nuclear power reactor or simply nuclear reactor.** The chief components of a nuclear reactor are :

(1) **U-235 fuel rods** which constitute the 'fuel core'. The fission of U-235 produces heat energy and neutrons that start the chain reaction.

(2) **Moderator** which slows down or moderates the neutrons. The most commonly used moderator is ordinary water. Graphite rods are sometimes used. Neutrons slow down by losing energy due to collisions with atoms/molecules of the moderator.

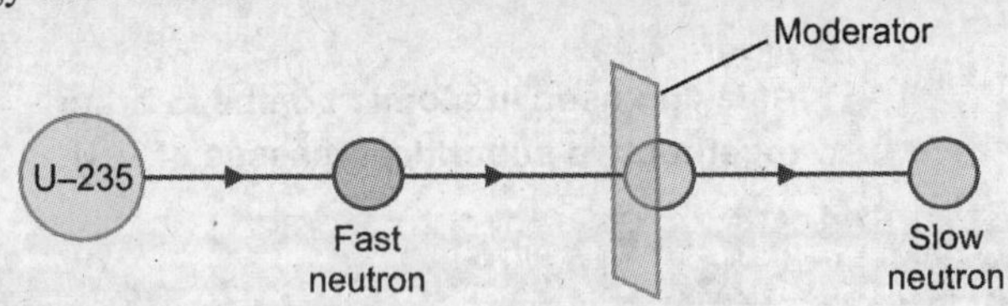

■ **Figure 4.20**
Moderator slows down a fast neutron.

(3) **Control rods** which control the rate of fission of U-235. These are made of boron-10 or cadmium, that absorbs some of the slowed neutrons.

$$^{10}_{5}B + ^{1}_{0}n \longrightarrow ^{7}_{3}Li + ^{4}_{2}He$$

Thus the chain reaction is prevented from going too fast.

(4) **Coolant** which cools the fuel core by removing heat produced by fission. Water used in the reactor serves both as moderator and coolant. Heavy water (D_2O) is even more efficient than light water.

(5) **Concrete shield** which protects the operating personnel and environments from destruction in case of leakage of radiation.

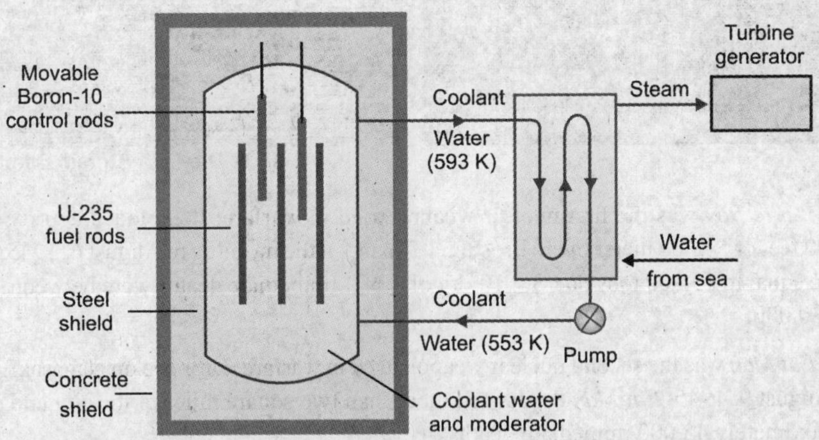

■ **Figure 4.21**
A light-water reactor producing electricity.

Light-water Nuclear power plant

Most commercial power plants today are 'light-water reactors'. In this type of reactor, U-235 fuel rods are submerged in water. Here, water acts as coolant and moderator. The control rods of boron-10 are inserted or removed automatically from spaces in between the fuel rods.

The heat emitted by fission of U-235 in the fuel core is absorbed by the coolant. The heated coolant (water at 573 K) then goes to the *exchanger*. Here the coolant transfers heat to sea water which is converted into steam. The steam then turns the turbines, generating electricity. **A reactor once started can continue to function and supply power for generations.**

About 15 per cent of consumable electricity in U.S.A. today is provided by light water reactors. India's first nuclear plant went into operation in 1960 at Tarapur near Mumbai. Another plant has been set up at Narora in Uttar Pradesh. While such nuclear power plants will be a boon for our country, they could pose a serious danger to environments. In May 1986, the leakage of radioactive material from the Chernobyl nuclear plant in USSR played havoc with life and property around.

Disposal of reactor waste poses another hazard. The products of fission *e.g.,* Ba-139 and Kr-92, are themselves radioactive. They emit dangerous radiation for several hundred years. The waste is packed in concrete barrels which are buried deep in the earth or dumped in the sea. But the fear is that any leakage and corrosion of the storage vessels may eventually contaminate the water supplies.

Breeder Reactor

We have seen that uranium-235 is used as a reactor fuel for producing electricity. But our limited supplies of uranium-235 are predicted to last only for another fifty years. However, nonfissionable uranium-238 is about 100 times more plentiful in nature. This is used as a source of energy in the so-called breeder reactors which can supply energy to the world for 5,000 years or more.

Here the uranium-235 core is covered with a layer or 'blanket' of uranium-238. The neutrons released by the core are absorbed by the blanket of uranium-238. This is then converted to fissionable plutonium-239. It undergoes a chain reaction, producing more neutrons and energy.

$$^{238}_{92}\text{U} + ^{1}_{0}n \longrightarrow ^{239}_{94}\text{Pu} + 2\,^{0}_{-1}e$$

$$^{239}_{94}\text{Pu} + ^{1}_{0}n \longrightarrow ^{90}_{38}\text{Sr} + ^{147}_{56}\text{Ba} + 3\,^{1}_{0}n$$

The above reaction sequence produces three neutrons and consumes only two. The excess neutron goes to convert more uranium to plutonium-239. **Thus the reactor produces or 'breeds' its own fuel and hence its name.** Several breeder reactors are now functioning in Europe. However, there is opposition to these reactors because the plutonium so obtained can be used in the dreaded H-bomb.

NUCLEAR FUSION PROCESS

This process is opposite of nuclear fission. Nuclear fusion may be defined as : **the process in which two light-weight nuclei combine or fuse to form a single heavier nucleus.**

For example, two nuclei of deuterium ($^{2}_{1}\text{H}$) undergo nuclear fusion to yield a heavier nucleus of helium-3. This fusion reaction takes place at a temperature of about 100 million °C. The above fusion reaction may be stated in the form of an equation as :

$$2\,^{2}_{1}\text{H} \longrightarrow ^{3}_{2}\text{He} + ^{1}_{0}n + \textit{energy}$$

Since fusion reactions occur at extremely high temperatures, these are also called **thermonuclear reactions.**

In a fusion reaction, the mass of the reacting nuclei is greater than that of the nucleus formed. The differential mass is manifested in the great amount of energy released in the reaction. For example,

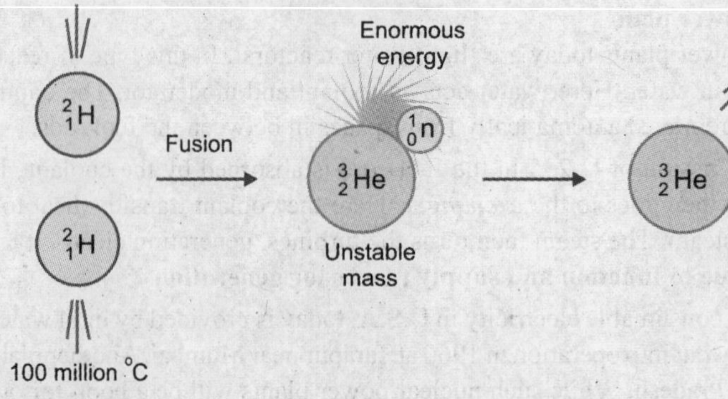

■ **Figure 4.22**
Illustration of fusion of two deuterium ($^{2}_{1}H$) nuclei to form a single nucleus of helium ($^{3}_{2}He$) with the release of a neutron and enormous energy.

$$\underset{3.01495\ \text{amu}}{^{3}_{1}H} + \underset{1.00782\ \text{amu}}{^{1}_{1}H} \longrightarrow \underset{4.00260\ \text{amu}}{^{4}_{2}He} + \textit{energy}$$

The total mass of the reactants is 4.02277 amu which is 0.02017 amu greater than the mass of the product. The mass that lost is converted into lot of energy. A pair of reacting nuclei induces fusion of another pair of nuclei. In this way fantastic amounts of energy are generated. This is the basis of the H-bomb or **Hydrogen bomb.**

How Fusion occurs?

Let us explain the mechanism of fusion by taking the example of the fusion of deuterium ($^{2}_{1}H$) cited above. At extremely high temperatures, 373 million kelvin or more, atoms do not exist as such. At these temperatures, deuterium atoms are completely stripped of orbital electrons. Thus results a system containing positive nuclei and electrons, called **plasma.** In this state, the high kinetic energy of the nuclei can overcome electrostatic repulsions between them. The nuclei collide with such great force that they merge or fuse to form larger nuclei. (Fig. 4.23)

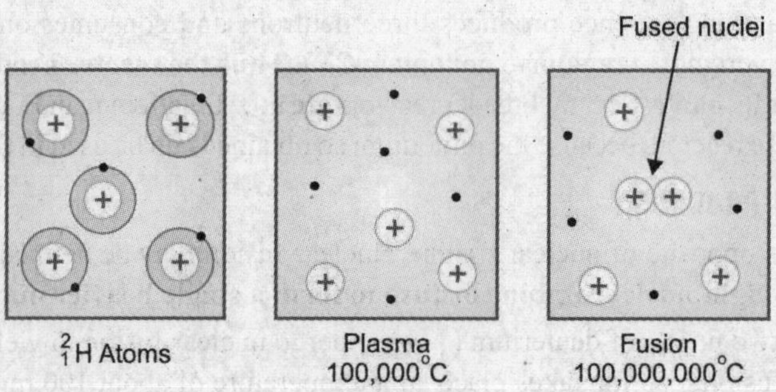

■ **Figure 4.23**
Mechanism of fusion of $^{2}_{1}H$.

SOLAR ENERGY

The energy released by the sun results from a series of nuclear fusion reactions. The overall reaction consists of the fusion of four hydrogen nuclei (protons) to form helium nucleus. One mechanism suggested for the process is:

$$^1_1H + ^1_1H \longrightarrow ^2_1H + ^0_1e$$

$$^2_1H + ^1_1H \longrightarrow ^3_2He$$

$$^3_2He + ^1_1H \longrightarrow ^4_2He + ^0_1e$$

Overall $$4\,^1_1H \longrightarrow ^4_2He + 2\,^0_1e \text{ (positrons)}$$

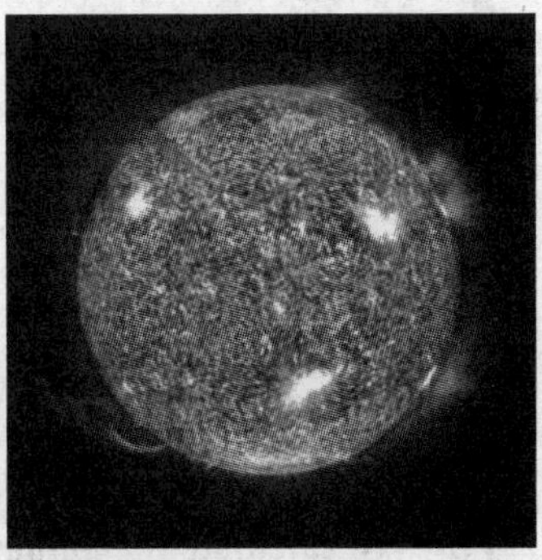

The fusion reactions in the sun take place at exceedingly high temperatures-greater than 313 million kelvin. Every second the sun loses 4.3×10^9 kg (420,000 tons) of mass by the fusion reactions. This mass is converted to energy. But the total mass of the sun is so great that its loss of mass is imperceptible. It is hoped that the sun will continue to pour energy on the earth for billions of years.

HYDROGEN BOMB OR H-BOMB

This deadly device makes use of the nuclear fusion of the isotopes of hydrogen. It consists of a small plutonium fission bomb with a container of isotopes of hydrogen. While the exact reaction used is a strictly guarded military secret, a fusion reaction between H-2 and H-3 may be the possible source of the tremendous energy released.

$$^2_1H + ^3_1H \longrightarrow ^4_2He + ^1_0n + \begin{bmatrix} \textit{tremendous} \\ \textit{energy} \end{bmatrix}$$

The 'fusion bomb' produces the high temperature required for nuclear fusion and triggers the H-bomb. The explosion of such a bomb is much more powerful than that of a fission bomb or atomic bomb. Fortunately, the H-bombs have been tested and not used in actual warfare. If they are ever used, it may mean the end of civilisation on earth.

FUSION AS A SOURCE OF ENERGY IN 21st CENTURY

Almost likely that the world's energy source in the twenty-first century will be a fusion reactor. As indicated by the trends of research, it will be based on the reactions as:

$$^2_1H + ^2_1H \longrightarrow ^3_2He + ^1_0n + \textit{energy}$$

$$^2_1H + ^3_1H \longrightarrow ^4_2He + ^1_0n + \textit{energy}$$

A fusion reactor thus developed will be any time superior to a fission reactor for generating electricity.

1. The fusion fuel, deuterium (^2_1H), can be obtained in abundance from heavy water present in sea water. The supplies of U-235 needed for a fission reactor are limited.
2. A fusion reaction produces considerably greater energy per gram of fuel than a fission reaction.
3. The products of fusion (^3_2He, ^4_2He) are not radioactive. Thus, there will be no problem of waste disposal.

So far, it has not been possible to set up a fusion reactor. The chief difficulty is that the reactant nuclei must be heated to very high temperatures. A mixture of deuterium and tritium nuclei, for example, requires 303 million kelvin before they can fuse. So far no substance is known which can make a container that could with stand such high temperatures. However, scientists are making efforts to effect fusion at a lower temperature with the help of laser beams.

SOLVED PROBLEM 16

It is found that a sample of radioactive matter is half integrated in 18 hours (64800 seconds). How much of it will remain after 42.5 hours (153000 seconds)?

SOLUTION

(*a*) To calculate disintegration constant

Formula used $\lambda = \dfrac{0.693}{t_{½}}$

Quantities given $t_{1/2} = 64800 \text{ s}$

Substitution of values $\lambda = \dfrac{0.693}{64800 \text{ s}} = 1.069 \times 10^{-5} \text{ s}^{-1}$

(*b*) To calculate the amount of sample left after 15300 seconds

Formula used $\lambda = \dfrac{1}{t} \log \dfrac{a}{a-x}$

Quantities given

$\lambda = 1.069 \times 10^{-5}$ $t = 153000$ seconds $a = 100$ $a - x = 100 - x$

Substitution of values

$$1.069 \times 10^{-5} \text{ s}^{-1} = \frac{1}{153000 \text{sec}} \log \frac{100}{100-\text{x}}$$

$$\text{or} \quad \log \frac{100}{100-x} = 1.63557$$

$$\text{or} \quad \frac{100}{100-x} = \text{Antilog } 1.63557$$

$$= 5.13$$

$$\text{or} \quad 100 = 5.13 \times (100 - x)$$

$$100 = 513 - 5.13\, x$$

$$\text{or} \quad x = \frac{413}{5.13}$$

$$= 80.51$$

$$\therefore \quad \text{Amount left} = 100 - 80.51$$

$$= \mathbf{19.49\%}$$

SOLVED PROBLEM 17

The activity of a radioactive isotope reduces by 25% after 6000 seconds . Calculate the decay constant and half life period.

SOLUTION

(*a*) To calculate decay constant

Formula used $\lambda = \dfrac{2.303}{t} \log \dfrac{a}{a-x}$

Quantities given

$t = 6000$ seconds $\qquad a = 100 \qquad a - x = 25$

Substitution of values

$$\lambda = \frac{2.303}{6000} \log \frac{100}{25}$$

$$= 3.8 \times 10^{-4} \times \log 4$$

$$= 0.00038 \times 0.6021$$

$$= 0.000228798 \text{ s}^{-1}$$

$$= \mathbf{2.28798 \times 10^{-4} \text{ s}^{-1}}$$

(b) To calculate Half Life period

Formula used $$t_{1/2} = \frac{0.693}{2.28798 \times 10^{-4} \text{ s}^{-1}}$$

$$= \mathbf{3028.873 \text{ s}}$$

SOLVED PROBLEM 18

How many α and β particles will be emitted by an element $_{84}A^{218}$ is changing to a stable isotope of $_{82}B^{206}$?

SOLUTION

Let x and y be the number of α and β particles respectively emitted during the change

$$_{84}A^{218} \longrightarrow {}_{82}B^{206} + x\,{}_{2}He^{4} + y\,{}_{-1}e^{0}$$

Comparing the mass number

$$218 = 206 + 4x + 0y$$

or $$4x = 12$$

or $$x = \mathbf{3}$$

Now comparing the atomic numbers

$$84 = 82 + 2x - y$$

$$84 = 82 + 2 \times 3 - y \qquad [\because x = 3]$$

or $$y = \mathbf{4}$$

SOLVED PROBLEM 19

0.002 kg of a radioactive element degraded to 0.0005 kg in 2.16×10^5 s. In what time will it be reduced 10% of its original amount ?

SOLUTION

(a) To calculate the disintegration constant

Formula used

$$\lambda = \frac{2.303}{t} \log \frac{a}{a - x}$$

Quantities given

$t = 2.16 \times 10^5$ s $\qquad a = 0.002$ kg $\qquad a - x = 0.0005$ kg

Substitution of values

$$\lambda = \frac{2.303}{2.16 \times 10^5 \text{ s}} \times \log \frac{0.002}{0.0005}$$

$$= \frac{2.303}{2.16 \times 10^5 \text{ s}} \times 0.6021$$

$$= \mathbf{6.4 \times 10^{-6} \ s^{-1}}$$

(*b*) To calculate the time t

Formula used

$$\lambda = \frac{2.303}{t} \log \frac{a}{a-x}$$

or

$$t = \frac{2.303}{\lambda} \log \frac{a}{a-x}$$

Quantities given

$\lambda = 6.4 \times 10^{-6} \text{ s}^{-1}$ $\quad a = 0.002$ kg $\quad a - x = 90\%$ of $a = 0.0018$ kg

Substitution of values

$$t = \frac{2.303}{6.4 \times 10^{-6} \text{ s}^{-1}} \times \log \frac{0.002}{0.0018}$$

$$= \frac{2.303}{6.4 \times 10^{-6} \text{ s}^{-1}} \times \log 1.11$$

$$= \frac{2.303}{6.4 \times 10^{-6} \text{ s}^{-1}} \times 0.0453$$

$$= \mathbf{1.63 \times 10^4 \ s}$$

EXAMINATION QUESTIONS

1. Define or explain the following terms :
   (*a*) Radioactivity
   (*b*) Radioactive substances
   (*c*) Radioactive Decay
   (*d*) Half life period
   (*e*) Radioactive equilibrium
   (*f*) Group displacement law
   (*g*) Nuclear reactions
   (*h*) Nuclear fission reactions
   (*i*) Nuclear fusion reactions
   (*j*) Mass defect
   (*k*) Binding energy
2. (*a*) What is group displacement law of radioactivity? How does it throw light on the idea of radioactive isotopes?
   (*b*) Radium has atomic number 226 and a half-life of 1600 years. Calculate the number of disintegrations produced per second from one gram of radium.
   Answer. (*b*) 3.660×10^{10} atoms
3. (*a*) Derive an expression for the decay constant for disintegration of a radioactive substance.
   (*b*) The activity of a radioactive isotope reduces by 25% after 6000 seconds. Calculate the decay constant and half-life period.
   Answer. (*b*) 2.301×10^{-4} s^{-1}; 2999 s
4. Explain what you understand by the term Radioactive Dating. How the age of earth was determined ?
5. How many do α and β particles will be emitted by an element $^{218}_{84}$A in changing to a stable isotope of $^{206}_{82}$B ?
   Answer. 3 α; 4 β

6. (*a*) State and explain the group displacement law in radioactivity.
   (*b*) Calculate the decay constant for ^{108}Ag, if its half-life is 138.60 seconds.
   Answer. (*b*) 0.005 s^{-1}
7. (*a*) Write the significance of half-life period of a radioactive substance.
   (*b*) The half-life of ^{232}Th is 4.415 × 10^{17} seconds. Calculate its disintegration constant.
   Answer. (*b*) 1.5696 × 10^{-18} s^{-1}
8. Complete the following nuclear reactions :
   (*i*) $^{15}_{8}O \rightarrow {}^{0}_{-1}e + ?$ (*ii*) $^{238}_{92}U \rightarrow {}^{234}_{90}Th + ?$
   (*iii*) $^{236}_{92}U \rightarrow {}^{143}_{56}Ba + ? + {}^{1}_{0}n$ (*iv*) $^{238}_{92}U + ? \rightarrow {}^{239}_{92}U$
   Answer. (*i*) $^{15}_{7}N$; (*ii*) $^{4}_{2}He$; (*iii*) $^{92}_{36}Kr$; (*iv*) $^{1}_{0}n$
9. (*a*) Define radioactivity. Discuss the liquid drop model for nuclear structure.
   (*b*) Complete the following nuclear reactions :
   (*i*) $^{12}_{6}C + {}^{2}_{1}M \rightarrow ? + {}^{4}_{2}He$ (*ii*) $^{27}_{13}Al + {}^{1}_{0}n \rightarrow {}^{24}_{11}Na + ?$
   (*iii*) $^{14}_{7}N + ? \rightarrow {}^{11}_{6}C + {}^{4}_{2}He$
   Answer. (*b*) (*i*) $^{10}_{5}B$; (*ii*) $^{4}_{2}He$; (*iii*) $^{1}_{1}H$
10. (*a*) Explain what is meant by radioactive equilibrium. How does it differ from chemical equilibrium?
    (*b*) A radioactive substance having half-life of 3.8 days, emitted initially 7 × 10^4 alpha particles per second. In what time will its rate of emission reduces to 2 × 10^4 alpha particles per second?
    Answer. (*b*) 24732 s
11. Show how C–14 is used for radio-carbon dating. A freshly cut piece of plant gives 20.4 counts per minute per gram. A piece of wood antique gives 12.18 counts per minute per gram. What is the age in years of antique? It is assumed that the radioactivity is entirely due to C–14. The half life period of C–14 is 5760 years.
    Answer. 1.35 × 10^{11} s
12. Calculate the rate of disintegration of one gram of ^{232}Th if its decay constant is 1.58 × 10^{-18} s^{-1}.
    Answer. 4.0998 × 10^3 dps
13. (*a*) Calculate the mass defect, binding energy and the binding energy per nucleon of $^{4}_{2}He$ which has an isotopic mass of 4.0026 amu ($^{1}_{1}H$ = 1.0081 amu; $^{1}_{0}H$ = 1.0089 amu).
    (*b*) With the help of a diagram discuss the salient features of a nuclear power reactor.
    (*c*) Explain the principle, construction and working of cyclotron.
    Answer. (*a*) 0.0314 amu; 29.249 Mev; 7.312275 Mev
14. Write short notes on
    (*a*) Radioactive series (*b*) Nuclear Fission
    (*c*) Tracer Technique (*d*) Geiger-Muller counter
15. (*a*) What do you mean by Tracer Technique? Write different applications of Tracer-Techniques.
    (*b*) The mass number and atomic number of a radioactive element Actinium are 227 and 89 respectively. Calculate the number of α and β particles emitted, if the mass number and atomic number of the new element Lead are 207 and 82 respectively.
    Answer. (*b*) 5 α and 3 β
16. $_{84}Po^{210}$ decays with alpha to $_{82}Pb^{206}$ with a half life of 498240 seconds. If 1 × 10^{-3} kg of Po^{210} is placed in a sealed tube, how much helium will accumulate in 249120 seconds? Express the answer in m^3 at STP.
    Answer. 3.567 m^3
17. (*a*) What is group displacement law in radioactivity? What is binding energy and how it is related to Mass number of the nucleus?
    (*b*) A radioactive isotope has half-life of 72000 seconds. What is the amount of isotope left over after 144000 seconds if the initial amount is 5 × 10^{-3} kg?
    Answer. (*b*) 1.34 kg

18. (*a*) Define radioactive constant and derive the relation between decay constant and number of particles left at time t.

(*b*) Half-life period of thorium is 24.5 minutes. How much thorium would be left after 30 minutes if the initial amount of thorium is one gram?

Answer. (*b*) 0.429 g

19. Calculate the number of λ and β particles emitted in the conversion of Thorium, $_{90}Th^{232}$, to Lead, $_{82}Pb^{206}$.

Answer. 6 α and 4 β (*Delhi BSc*)

20. A sample of U^{238} (half life = 4.5×10^9 years) ore is found to contain 23.8 g of U^{238} and 20.6 g of Pb^{206}. Calculate the age of the ore.

Answer. 4.715×10^{17} s (*Bundelkhand BSc*)

21. The half life of Cobalt-60 is 5.26 years. Calculate the percentage activity after eight years.

Answer. 22.36% (*Nagpur BSc*)

22. Calculate the time required for a radioactive sample to lose one-third of the atoms of its parent Isotope. The half life is 33 min.

Answer. 1019.255 s (*Bhopal BSc*)

23. An old wooden article shows 2.0 counts per minute per gram. A fresh sample of wood shows 15.2 counts per minute per gram. Calculate the age of the wooden article. ($t_{½}$ of C^{14} = 5760 years)

Answer. 5.317×10^{11} s (*Gulbarga BSc*)

24. The mass number and atomic number of a radioactive element Actinium are 227 and 89 respectively. Calculate the number of α and β particles emitted if the mass number and atomic number of the new element Lead are 207 and 82 respectively.

Answer. 5 α and 3 β (*Calicut BSc*)

25. Calculate the age of the tooth in which C^{14} activity is 20% of the activity found at the present time (t½ for C^{14} = 5580 years)

Answer. 4.0877×10^{11} s (*Delhi BSc*)

26. A piece of wood recovered in excavation has 30% as much $_6C^{14}$ as a fresh wood today. Calculate the age of excavated piece assuming half life period of $_6C^{14}$ as 5700 years.

Answer. 3.1245×10^{11} s (*Madurai BSc*)

27. The activity of a radioactive sample falls to 85% of the initial value in four years. What is the half life of the sample? Calculate the time by which activity will fall by 85%.

Answer. 5.3763×10^8 s; 1.4720×10^9 s (*Mysore BSc*)

28. Write short notes on

(*a*) Carbon dating (*b*) Nuclear reaction (*Lucknow BSc*)

29. Derive an expression for the disintegration constant of a radioactive element and show that decay of the element is exponential. (*Lucknow BSc*)

30. (*a*) Write a note on Nuclear Fission and Nuclear Fusion.

(*b*) What are nuclear reactions? How are they classified? Explain with examples.

(*HS Gaur BSc*)

31. (*a*) What is Group displacement law? Explain with examples.

(*b*) How does discovery of isotopes help in explaining fractional atomic weight of elements? Mention some applications of radioactivity. (*Arunachal BSc*)

32. Write notes on

(*a*) Nuclear Fusion (*b*) Nuclear Binding Energy

(*c*) Tracers and their applications (*Mizoram BSc*)

33. Define binding energy of a nucleus. Explain with the help of binding energy curve, the stability of the nuclei.

(*Nagpur BSc*)

34. (*a*) Give the details of the application of C^{14} isotope in carbon dating. How the age of an old wooden sample is determined?

(*b*) The rate of disintegration of an old wooden sample was found 25% as compared to a fresh identical sample. Find out the age of the old sample. (Half life of C^{14} = 5760 years)

Answer. (*b*) 3.635×10^{11} s *(Bundelkhand BSc)*

35. (*a*) How many α and β-particles are emitted for the transformation

$$^{232}_{90}Th \rightarrow ^{208}_{82}Pb$$

(*b*) What is a nuclide? Mention different kinds of nuclides with suitable examples.

(*c*) What is a nuclear reactor? Describe its main parts briefly. *(Sambalpur BSc)*

36. (*a*) Calculate the (*i*) Mass Defect (*ii*) Binding energy per nucleon of $_6C^{12}$. (Given C^{12} = 12.0038 amu; H^1 = 1.0081 amu; n^1 = 1.0089 amu)

(*b*) What is artificial radioactivity? Give two examples. *(Goa BSc)*

37. (*a*) Discuss the stability of nucleus under the following heads:

(*i*) binding energy per nucleon

(*ii*) neutron-proton ratio and odd-even rule

(*b*) What is the difference between Fission and Fusion reactions?

(*c*) How do two isotopes of elements differ in the number of nucleons. *(Aligarh BSc)*

38. (*a*) What is group displacement law of radioactivity? How does it throw light on the idea of radioactive isotopes?

(*b*) The half-life of radioactive isotope is 47.2 sec. Calculate N/N_o left after one hour.

Answer. (*b*) 0.3912 *(Arunachal BSc)*

39. Give the postulates of theory of radioactive emissions. *(Arunachal BSc)*

40. Write short notes on :

(*a*) Mass defect

(*b*) Packing fraction

(*c*) Nuclear Fission

(*d*) Half life period *(Arunachal BSc)*

41. (*a*) What is group displacement law of radioactivity? How does it throw light on the idea of radioactive isotopes?

(*b*) Explain :

(*i*) Radioactive dating

(*ii*) Average life

(*iii*) Radioactive equilibria *(Arunachal BSc)*

42. Explain :

(*a*) Why $_{13}Al^{27}$ is stable while $_{13}Al^{24}$ is radioactive in nature?

(*b*) The atomic weight of lithium is variable while in other members of group I it is fixed. *(Delhi BSc)*

43. Calculate the half life and average life period of a radioactive element if its decay constant is 7.37×10^{-3} hr^{-1}.

Answer. 0.0261 s; 0.0376 s *(Sambalpur BSc)*

44. $_{92}U^{238}$ undergoes a series of changes emitting α and β particles and finally $_{82}Pb^{206}$ is formed. Calculate the number of α and β particles which must have been ejected during the series.

Answer. 8 α and 6 β *(Panjab BSc)*

45. Radioisotope $_{15}P^{32}$ has a half life of 15 days. Calculate the time in which the radioactivity of its 1 mg quantity will fall to 10% of the initial value.

Answer. 4.307×10^6 s *(Osmania BSc)*

46. An old wooden article shows 2.0 counts per minute per gram. A fresh sample of wood shows 15.2 counts per minute per gram. Calculate the age of the wooden article ($t_{½}$ of C^{14} = 5460 years)

Answer. 5.0397×10^{11} s *(Banaras BSc)*

47. Explain the difference between Nuclear Fission and Nuclear Fusion. *(Agra BSc)*

48. (*a*) Write briefly on nuclear fission and nuclear fusion.

(*b*) A radioactive isotope has half-life period of 20 days. What is the amount of the isotope left over after 40 days if the initial concentration is 5 g?

Answer. (*b*) 1.78×10^{-3} kg *(Osmania BSc)*

49. (*a*) Discuss the stability of nucleus in terms of neutron-proton ratio and binding energy.

(*b*) 2 g of radioactive element degraded to 0.5 g in 60 hours. In what time will it be reduced to 10% of its original amount?

Answer. (*b*) 3.587×10^5 s *(Budelkhand BSc)*

50. The nucleidic ratio $_1H^3$ to $_1H^1$ in a sample of water is . Tritium undergoes decay with half life period of 12.3 years. How many tritium atoms would 10 g of such a sample contain 40 years after the original sample is collected.

Answer. 5.35×10^6 atoms *(Jiwaji BSc)*

51. An experiment requires minimum Beta activity produced at the rate of 346 Beta particles per minute. The half life of $_{42}Mo^{99}$, which is a Beta emitter is 66.6 hours. Find the minimum amount of $_{42}Mo^{99}$ required to carry out the experiment 6.909 hours.

Answer. 135.63×10^{-15}g *(Sri Venkateshwara BSc)*

52. A sample of U^{238} (half life = 4.5×10^9 years) ore is found to contain 23.8 g of U^{238} and 20.6 g of Pb^{206}. Calculate the age of the ore.

Answer. 1.4177×10^8 s *(Mysore BSc)*

53. Calculate the mass defect, binding energy and the binding energy per nucleon of $_2He^4$ which has an isotopic mass of 4.0026 amu ($_1H^1$ = 1.0081 amu; $_0H^1$ = 1.0089 amu).

Answer. 11.70×10^{-18} J *(Delhi BSc)*

54. A freshly cut piece of plant gives 20.4 counts per minute per gram. A piece of wood antique gives 12.18 counts per minute per gram. What is the age in years of antique? It is assumed that the radioactivity is entirely due to C^{14}. The half life period of C^{14} is 5760 years.

Answer. 1.35186×10^{11} s *(Panjab BSc)*

MULTIPLE CHOICE QUESTIONS

1. Which is the correct symbol for an alpha particle?

(*a*) ^4_2He (*b*) ^1_0n

(*c*) $^0_{-1}e$ (*d*) ^1_1p

Answer. (*a*)

2. Of the following, which is the most damaging when ingested?

(*a*) beta emitters (*b*) alpha emitters

(*c*) gamma emitters (*d*) all of these

Answer. (*b*)

3. An alpha particle is
(*a*) an electron
(*b*) one neutron and one proton
(*c*) two protons and two neutrons
(*d*) an X-ray emission
Answer. (*c*)

4. By what type of decay process might ^{214}Pb convert to ^{214}Bi?
(*a*) beta decay
(*b*) alpha decay
(*c*) gamma decay
(*d*) electron capture
Answer. (*a*)

5. The curie is a measure of the
(*a*) lethal threshold for radiation exposure
(*b*) number of alpha particles emitted by exactly 1 g of a radioactive substance
(*c*) number of disintegrations per second of a radioactive substance
(*d*) total energy absorbed by an object exposed to a radioactive source
Answer. (*c*)

6. Of the following processes, which one does not change the atomic number?
(*a*) gamma emission
(*b*) alpha emission
(*c*) positron emission
(*d*) beta emission
Answer. (*a*)

7. Which of these nuclides is most likely to be radioactive?
(*a*) $^{39}_{19}K$
(*b*) $^{27}_{13}Al$
(*c*) $^{127}_{53}I$
(*d*) $^{243}_{95}Am$
Answer. (*d*)

8. The beta particle consists of
(*a*) high-energy rays
(*b*) 1 neutron
(*c*) 2 neutrons and 2 protons
(*d*) 1 electron
Answer. (*d*)

9. Lab coats and gloves provide shielding from
(*a*) alpha radiation
(*b*) alpha and beta radiation
(*c*) alpha, beta, and gamma radiation
(*d*) gamma radiation
Answer. (*b*)

10. When an alpha particle is released in nuclear decay, the mass number of the nucleus undergoing decay?
(*a*) stays the same
(*b*) increases by 4
(*c*) decreases by 4
(*d*) decreases by 2
Answer. (*c*)

11. Which one of the following processes could not result in the conversion of strontium to rubidium?
(*a*) gamma emission
(*b*) proton emission
(*c*) electron capture
(*d*) positron emission
Answer. (*a*)

12. Alpha decay of ^{226}Ra will yield which of the following nuclides?
(*a*) ^{222}Rn
(*b*) ^{230}Th
(*c*) ^{222}Fr
(*d*) ^{222}Th
Answer. (*a*)

13. In a Geiger-Müller counter, one "count" is directly due to

(*a*) a secondary electron (*b*) a primary electron

(*c*) many electrons and ions (*d*) a beta particle

Answer. (*c*)

14. $^{8}_{3}Li$ decays to $^{8}_{4}Be$. What type of decay is this?

(*a*) positron emission (*b*) beta

(*c*) γ-ray (*d*) alpha

Answer. (*b*)

15. _______ rays consist of He nuclei, while _______ rays are electromagnetic radiation.

(*a*) beta, alpha (*b*) alpha, beta

(*c*) alpha, gamma (*d*) gamma, beta

Answer. (*c*)

16. Which type of radioactive decay results in an increase in atomic number?

(*a*) positron emission (*b*) alpha emission

(*c*) gamma emission (*d*) beta emission

Answer. (*d*)

17. When ^{249}Cf is bombarded with ^{10}B, ^{257}Lr is formed. What other particle(s) is/are produced?

(*a*) $^{0}_{-1}e$ (*b*) $^{1}_{0}n$

(*c*) $^{0}_{1}e$ (*d*) $^{1}_{1}p$

Answer. (*b*)

18. Which combination of the number of protons and the number of neutrons is most common among the naturally occurring non-radioactive nuclides?

(*a*) even protons; even neutrons (*b*) odd protons; even neutrons

(*c*) even protons; odd neutrons (*d*) odd protons; odd neutrons

Answer. (*a*)

19. Which form of radioactivity is most penetrating?

(*a*) alpha particles (*b*) beta particles

(*c*) neutrons (*d*) gamma rays

Answer. (*d*)

20. In electron capture,

(*a*) gamma rays are emitted (*b*) a neutron is formed

(*c*) a positron is formed (*d*) an alpha particle is emitted

Answer. (*b*)

21. The spontaneous transformation of one nuclide into others occurs only if

(*a*) the process is endothermic

(*b*) the process results in a neutron/proton ratio of 1:0 in the products

(*c*) sufficient energy can be absorbed from the surroundings to drive the process

(*d*) the combined mass of the products is less than the mass of the original nuclide

Answer. (*d*)

22. What particle is released when Ga-75 decays to Ge-75?

(*a*) neutron (*b*) beta

(*c*) gamma (*d*) alpha

Answer. (*b*)

23. What particle is missing in the following bombardment reaction? $Al^{27} + ? = {}^{1}n + P^{30}$

(*a*) neutron (*b*) beta

(*c*) proton (*d*) alpha

Answer. (*d*)

24. What would be the immediate product of neutron absorption by ^{107}Ag?

(*a*) ^{107}Pd (*b*) ^{109}In

(*c*) ^{108}Cd (*d*) ^{108}Ag

Answer. (*d*)

25. When a nuclide undergoes beta decay,

(*a*) the atomic number remains unchanged and the mass number increases by one

(*b*) the mass number remains unchanged and the atomic number decreases by one

(*c*) the mass number remains unchanged and the atomic number increases by one

(*d*) the atomic number remains unchanged and the mass number decreases by one

Answer. (*c*)

26. The half-life of ^{32}P, which is used as a label on red blood cells to determine blood volume, is 1.2×10^6 s. How many days are required for the activity of a sample of ^{32}P to drop to 5.00% of its initial level?

(*a*) 2.31×10^6 s (*b*) 3.69×10^6 s

(*c*) 5.34×10^6 s (*d*) 1.8×10^4 s

Answer. (*c*)

27. The half-life of Sulfur-35 is 88 days. If 8.0 g of Sulfur-35 exists on day one, what fraction will remain after 264 days?

(*a*) 0.5 g (*b*) 4.0 g

(*c*) 0 g (*d*) 1.0 g

Answer. (*d*)

28. When Xenon-123 emits a gamma ray, what is the product?

(*a*) $^{123}_{54}Xe$ (*b*) $^{123}_{53}I$

(*c*) $^{119}_{52}Te$ (*d*) $^{123}_{55}Cs$

Answer. (*a*)

29. The cloth shroud from around a mummy is found to have a ^{14}C activity of 8.9 disintegrations per minute per gram of carbon as compared with living organisms that undergo 15.2 disintegrations per minute per gram of carbon. From the half-life for ^{14}C decay, 1.807×10^{11} s, calculate the age of the shroud.

(*a*) 2.996×10^3 s (*b*) 1.39×10^{11} s

(*c*) 2.0498×10^3 s (*d*) 6.05×10^{10} s

Answer. (*b*)

30. ^{198}Au has a half-life of 2.33×10^{-5} s. Assuming you start with a 10^{-5} kg sample of ^{198}Au, how much will remain after 8.64×10^{-5} s?

(*a*) 2.46×10^{-5} kg (*b*) 1.32×10^{-5} kg

(*c*) 9.58×10^{-5} kg (*d*) 2.39×10^{-5} kg

Answer. (*c*)

31. The half-life for the beta decay of ^{233}Pa is 27.4 days. How many days must pass to reduce a 5.00 g sample of ^{233}Pa to 0.625 g?

(*a*) 9.47×10^6 s (*b*) 4.73×10^6 s

(*c*) 7.105×10^6 s (*d*) 2.37×10^6 s

Answer. (*c*)

32. The half-life of tritium (Hydrogen-3) is 3.8789×10^8 s. If 48×10^{-6} kg of tritium is released from a nuclear power plant during the course of an accident, what mass of this nuclide will remain after 1.55×10^9 s?

(*a*) 6×10^{-6} kg (*b*) 3×10^{-6} kg

(*c*) 24×10^{-6} kg (*d*) 12.0×10^{-6} kg

Answer. (*b*)

33. How old is a fossil bone whose ^{14}C content is 15.0 percent that of living bone?

(*a*) 8.010×10^{11} s (*b*) 4.92×10^{11} s

(*c*) 1.198×10^{10} s (*d*) 2.148×10^{11} s

Answer. (*b*)

34. Iodine-131 has a half-life of 6.912×10^5 s. How many grams of I-131 in a 4×10^{-3} kg sample remain after 2.074×10^6 s?

(*a*) 2×10^{-3} kg (*b*) 1.0×10^{-3} kg

(*c*) 39×10^{-3} kg (*d*) 0.0×10^{-3} kg

Answer. (*c*)

35. The half-life of ^{45}Ca is 1.4256×10^7 s. After 3.1536×10^7 s., what percentage of the original sample of ^{45}Ca remained?

(*a*) 10.9 percent (*b*) 99.6 percent

(*c*) 2.16 percent (*d*) 21.6 percent

Answer. (*d*)

36. What particle is produced when Phosphorus-29 decays to silicon-29?

(*a*) positron (*b*) beta

(*c*) gamma (*d*) alpha

Answer. (*a*)

37. The bombardment of which isotope by a neutron produces ^{198}Au and proton?

(*a*) $^{199}_{81}Tl$ (*b*) $^{197}_{80}Hg$

(*c*) $^{198}_{80}Hg$ (*d*) $^{197}_{78}Pt$

Answer. (*c*)

38. Complete and balance the following nuclear equation by selecting the missing particle.

$$^{24}_{12}Mg + ^{2}_{1}H \rightarrow ^{4}_{2}He\ ?$$

(*a*) $^{22}_{13}Al$ (*b*) $^{22}_{11}Na$

(*c*) $^{26}_{13}Al$ (*d*) $^{20}_{10}Ne$

Answer. (*b*)

39. What is the product of beta decay of $^{159}_{64}Gd$?

(*a*) $^{159}_{65}Tb$ (*b*) $^{159}_{63}Eu$

(*c*) $^{159}_{64}Gd$ (*d*) $^{155}_{62}Sm$

Answer. (*a*)

40. Which of the following is the nuclear equation for bismuth-214 undergoing beta decay?

(*a*) $^{214}_{83}\text{Bi} \rightarrow {}^{0}_{1}\text{e} + {}^{214}_{84}\text{Po}$ (*b*) $^{214}_{83}\text{Bi} \rightarrow {}^{0}_{1}\text{e} + {}^{214}_{82}\text{Pb}$

(*c*) $^{214}_{83}\text{Bi} \rightarrow {}^{0}_{1}\text{e} + {}^{214}_{84}\text{Po}$ (*d*) $^{214}_{83}\text{Bi} \rightarrow {}^{0}_{1}\text{e} + {}^{214}_{82}\text{Pb}$

Answer. (*a*)

41. When Zinc-73 decays by beta emission, the product of decay is

(*a*) $^{74}_{29}\text{Cu}$ (*b*) $^{73}_{29}\text{Cu}$

(*c*) $^{73}_{31}\text{Ga}$ (*d*) $^{72}_{30}\text{Zn}$

Answer. (*c*)

42. When $^{27}_{13}\text{Al}$ is bombarded by an alpha particle, the products are a proton and another nuclide. What is the other nuclide?

(*a*) $^{31}_{16}\text{S}$ (*b*) $^{27}_{12}\text{Mg}$

(*c*) $^{30}_{14}\text{Si}$ (*d*) $^{31}_{13}\text{Al}$

Answer. (*c*)

43. The nucleus of an atom of radioactive I-131 used in thyroid imaging contains

(*a*) 53 protons and 75 neutrons (*b*) 53 protons and 78 neutrons

(*c*) 53 electrons and 78 neutrons (*d*) 53 neutrons and 78 protons

Answer. (*b*)

44. What is the product of the alpha decay of Rn-220?

(*a*) Po-216 (*b*) Rn-220

(*c*) Rn-216 (*d*) Ra-224

Answer. (*a*)

45. A short-time exposure to a radiation dose of 100 to 200 rems will result in what health effect?

(*a*) death of half the exposed population within 30 days after exposure

(*b*) there are no detectable clinical effects

(*c*) nausea and marked decrease in white blood cells

(*d*) a slight temporary decrease in white blood cell counts

Answer. (*c*)

46. When a nuclide undergoes electron capture,

(*a*) the mass number remains unchanged and the atomic number decreases by one

(*b*) the atomic number remains unchanged and the mass number increases by one

(*c*) the mass number remains unchanged the atomic number increases by one

(*d*) the atomic number remains unchanged and the mass number decreases by one

Answer. (*a*)

47. What nuclide forms as a result of the positron emission of $^{12}_{7}\text{N}$?

(*a*) $^{13}_{6}\text{C}$ (*b*) $^{11}_{7}\text{N}$

(*c*) $^{12}_{6}\text{C}$ (*d*) $^{12}_{8}\text{O}$

Answer. (*c*)

48. $^{232}_{90}\text{Th}$ undergoes a series of alpha and beta emissions resulting in the final stable product, $^{208}_{82}\text{Pb}$. How many alpha and beta emissions occur in this series?

(*a*) 4 alpha, 6 beta (*b*) 8 alpha, 6 beta

(*c*) 12 alpha, 4 beta (*d*) 6 alpha, 4 beta

Answer. (*d*)

49. What is the product formed from ^{207}Po by positron emission?

(*a*) ^{203}Pb (*b*) ^{207}Bi

(*c*) ^{208}At (*d*) ^{206}Bi

Answer. (*b*)

50. Which of the following statements about the biological effects of radiation is false?

(*a*) Radiation can cause leukemia

(*b*) Ionizing radiation is more dangerous than non-ionizing radiation

(*c*) Radon is absorbed through the skin

(*d*) Radon is harmful because it decays to polonium

Answer. (*c*)

51. When a ^{235}U nucleus is struck by a thermal neutron, fission occurs with the release of neutrons. If the fission fragments are ^{90}Sr and ^{144}Xe, how many neutrons are released?

(*a*) 3 (*b*) 4

(*c*) 1 (*d*) 2

Answer. (*d*)

52. The nuclide Iron-56 has a mass defect of 0.52840 amu. What is the binding energy per nucleon in Mev?

(*a*) 8.81 Mev (*b*) 9.79×10^{-17} Mev

(*c*) 8.81×10^{3} Mev (*d*) 494 Mev

Answer. (*a*)

53. The mass defect for ^{12}C is 0.0990, meaning that six protons and six neutrons have a mass that is 0.0990 amu more than the mass of the ^{12}C nucleus. What is the binding energy of the ^{12}C nucleus per nucleon in kJ?

(*a*) 1.23×10^{-15} kJ (*b*) 1.23×120^{-12} kJ

(*c*) 4.11×10^{-24} kJ (*d*) 1.48×10^{-14} kJ

Answer. (*a*)

54. The combustion of one mol of graphite releases 393.5 kJ of energy. What is the mass change that accompanies the loss of this energy?

(*a*) 4.372×10^{-12} kg (*b*) 3.542×10^{22} kg

(*c*) 1.312×10^{-3} kg (*d*) none of these

Answer. (*a*)

55. Plutonium-239 is produced from ^{238}U in breeder reactor. This might involve

(*a*) neutron absorption followed by emission of two positrons

(*b*) alpha particle absorption followed by emission of three protons

(*c*) proton absorption followed by positron emission

(*d*) neutron absorption followed by emission of two beta particles

Answer. (*d*)

56. What particle is produced when Plutonium-242 decays to Uranium-238?

(*a*) gamma (*b*) alpha

(*c*) positron (*d*) beta

Answer. (*b*)

57. In a living organism, the ^{14}C concentration

(a) remains approximately constant

(b) continually increases

(c) continually decreases

(d) varies unpredictably during the lifetime of the organism

Answer. (a)

58. Which equation represents alpha emission from ^{196}Po?

(a) $^{196}_{84}Po \rightarrow ^{192}_{82}Pb + ^{4}_{2}He$

(b) $^{196}_{84}Po + ^{4}_{2}He \rightarrow ^{192}_{82}Pb$

(c) $^{196}_{84}Po \rightarrow ^{200}_{86}Rn + ^{4}_{2}He$

(d) $^{196}_{84}Po + ^{4}_{2}He \rightarrow ^{200}_{86}Rn$

Answer. (a)

59. Hydrogen burning

(a) is the reaction sequence that is proposed to take place in normal stars

(b) requires very large kinetic energy in the colliding nuclei in order to overcome huge activation energies associated with electrostatic repulsion

(c) results in the conversion of hydrogen nuclei to helium nuclei

(d) all of above

Answer. (d)

60. The energy produced by the Sun involves which of the following nuclei?

(a) Pu (b) Li

(c) U (d) H

Answer. (d)

61. Radiation is used in cancer treatment to

(a) destroy cancer causing substances (b) relieve pain

(c) obtain images of the diseased region (d) destroy cancer cells

Answer. (d)

62. The amount of fissionable material large enough to maintain the chain reaction in nuclear fission is called the

(a) moderator (b) critical mass

(c) mass defect (d) nuclear binding energy

Answer. (b)

63. The radioisotopes used for diagnosis in nuclear medicine

(a) have short half lives (b) travel rapidly through tissue

(c) are usually gamma emitters (d) all of these

Answer. (d)

64. The purpose of a moderator in a nuclear reactor is

(a) to prevent corrosion of the core components

(b) to slow the fission neutrons so that they can be captured to sustain the chain reaction

(c) to cool the core to prevent melt down

(d) to dissolve the fission products for disposal

Answer. (b)

65. Fission reactions can be run continuously to generate electric power commercially because

(*a*) the reactors generate more readily fissionable fuel than they consume

(*b*) more neutrons are produced in the fission reactions than are consumed

(*c*) supercritical neutrons split into protons and electrons

(*d*) the different isotopes of uranium interconvert under reaction conditions to form the necessary Uranium-235

Answer. (*b*)

66. A cyclotron is used to get

(*a*) energetic ions (*b*) positrons

(*c*) magnetic fields (*d*) neutrons

Answer. (*a*)

67. The reaction of H^3 and H^2 to form He^4 and a neutron is an example of

(*a*) a fission reaction (*b*) a fusion reaction

(*c*) both fission and fusion reactions (*d*) neither a fission nor fusion reaction

Answer. (*b*)

68. Nuclear fusion

(*a*) takes place in the sun (*b*) uses large nuclides

(*c*) is used in atomic bombs (*d*) takes place in a fusion reactor

Answer. (*a*)

69. A moderator in a nuclear reactor serves to

(*a*) accelerate neutrons (*b*) diminish the nuclear binding energy

(*c*) provide fissionable isotopes (*d*) slow neutrons

Answer. (*d*)

5 Chemical Bonding – Lewis Theory

CONTENTS

TERMS AND DEFINITIONS

Chemical Bond

Molecules of chemical substances are made of two or more atoms joined together by some force, acting between them. This force which results from the interaction between the various atoms that go to form a stable molecule, is referred to as a *Chemical Bond.*

A chemical bond is defined as a force that acts between two or more atoms to hold them together as a stable molecule.

As we will study later, there are three different types of bonds recognized by chemists :

(1) Ionic or Electrovalent bond
(2) Covalent bond
(3) Coordinate covalent bond

There is a fourth type of bond, namely, the metallic bond which we will consider later in this chapter.

Definition of Valence

The term valence (or valency) is often used to state the potential or capacity of an element to combine with other elements. At one time, it was useful to define valence of an element as : the number of hydrogen atoms or twice the number of oxygen atoms with which that element could combine in a binary compound (containing two different elements only).

In hydrogen chloride (HCl), one atom of chlorine is combined with one atom of hydrogen and the valence of chlorine is 1. In magnesium oxide (MgO), since one atom of magnesium holds one atom of oxygen, the valence of magnesium is 2.

By the above definition, we would assign a valence of 2 to sulfur in H_2S, but 4 to sulfur in SO_2. Some elements have fractional valence in certain compounds, while there are elements that have variable valencies. The concept of valence as a mere number could not explain these facts. This concept, in fact, was very confusing and has lost all value.

As already stated, there are three different type of bonds that are known to join atoms in molecules. Although no precise definition of valence is possible, we can say that : **Valence is the number of bonds formed by an atom in a molecule.**

Valence Electrons

The electrons in the outer energy level of an atom are the ones that can take part in chemical bonding. These electrons are, therefore, referred to as the valence electrons.

The electronic configuration of Na is 2, 8, 1 and that of Cl is 2, 8, 7. Thus sodium has one valence electron and chlorine 7. It is important to remember that **for a group element of the periodic table (H, O, K, F, Al etc.) the group number is equal to the number of valence electrons.**

Bonding and Non-bonding Electrons

The valence electrons actually involved in bond formation are called bonding electrons. The remaining valence electrons still available for bond formation are referred to as non-bonding electrons. Thus :

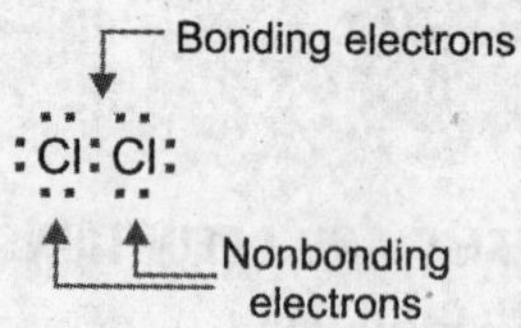

TERMS AND DEFINITIONS

Lewis Symbols of Elements

A Lewis symbol of an element consists of an element's symbol and surrounding dots to represent the number of valence electrons. In this notation, **the symbol of an element represents the nucleus plus the inner normally filled levels (or shells) of the atom.** For illustration, the symbol Na stands for the nucleus of sodium atom plus 2, 8 electrons in the inner two levels.

To represent a Lewis symbol for an element, write down the symbol of the element and surround the symbol with a number of dots (or crosses) equal to the number of valence electrons. The position of dots around the symbol is not really of any significance. The bonding electrons are shown at appropriate positions, while the rest of the electrons are generally given in pairs. The Lewis symbols for hydrogen, chlorine, oxygen and sulfur may be written as :

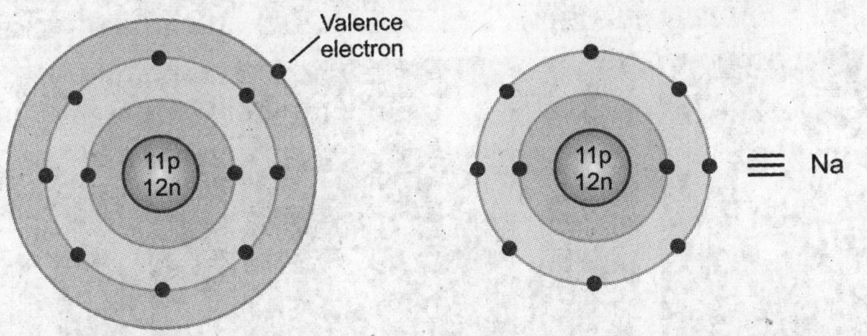

■ **Figure 5.1**
The Lewis symbol Na represents the nucleus and the electrons arranged in the inner two levels as 2, 8, minus the valence electrons.

The structural formulae of compounds built by union of Lewis symbols for the component atoms, are referred to as Electron-dot formulas, or Electron-dot structures or Lewis structures. For this purpose, the valence electrons actually involved in bond formation may be shown by crosses (x) or dots (·) for the sake of distinction.

Now we will proceed to discuss the common types of chemical bonds in the light of the electronic theory of valence.

ELECTRONIC THEORY OF VALENCE

As Bohr put forward his model of the atom so electronic configuration of elements was known. G.N. Lewis and W. Kossel, working independently, used this knowledge to explain 'why atoms joined to form molecules'. They visualized that noble gas atoms had a stable electronic configuration, while atoms of all other elements has unstable or incomplete electronic configuration. In 1916, they gave **the electronic theory of valence.** It states that : **In chemical bond formation, atoms interact by losing, gaining, or sharing of electrons so as to acquire a stable noble gas configuration.** Each noble gas, except helium, has a valence shell of eight electrons (Table 5.1).

TABLE 5.1. ELECTRONIC CONFIGURATION OF NOBLE GASES

Noble gas	At. No.	Electrons in principal shells
He	2	2
Ne	10	2, 8
Ar	18	2, 8, 8
Kr	36	2, 8, 18, 8
Xe	54	2, 8, 18, 18, 8
Rn	86	2, 8, 18, 32, 18, 8

While atoms of noble gases possess a stable outer shell of eight electrons or **octet,** atoms of most other elements have incomplete octets. They may have less than 8 electrons or in excess. Thus, the electronic theory or valence could well be named as the **Octet theory of Valence.** It may be stated as : **Atoms interact by electron-transfer or electron-sharing, so as to achieve the stable outer shell of eight electrons.**

The tendency for atoms to have eight electrons in the outer shell is also known as the **Octet Rule** or the **Rule of Eight.** Since helium has two electrons in the outer shell, for hydrogen and lithium, having one and three (2, 1) electrons respectively, it is the **Rule of two** which will apply. We will see later in this chapter that there are quite a few exceptions to the rule of eight in covalent compounds.

LINUS CARL PAULING

Linus Pauling won the Nobel Prize in Chemistry in 1954 for his work on chemical bonding. He also received the Nobel Peace Prize in 1962 for his campaign against nuclear testing.

Linus Carl Pauling (February 28, 1901 – August 19, 1994) was an American quantum chemist and biochemist. Pauling is widely regarded as the premier chemist of the twentieth century. He pioneered the application of quantum mechanics to chemistry and in 1954 was awarded the Nobel Prize in chemistry for his work describing the nature of chemical bonds. He also made important contributions to crystal and protein structure determination, and was one of the founders of molecular biology. He came near to discovering the "double helix," the ultrastructure of DNA, when Watson and Crick made the discovery in 1953.

Pauling received the Nobel Peace Prize in 1962 for his campaign against above-ground nuclear testing, and is the only person to win two unshared Nobel prizes. Later in life, he became an advocate for greatly increased consumption of vitamin C and other nutrients. He generalized his ideas to define orthomolecular medicine, which is still regarded as unorthodox by conventional medicine. He popularized his concepts, analyses, research and insights in several successful but controversial books centered around vitamin C and orthomolecular medicine.

IONIC BOND

This type of bond is established by transfer of an electron from one atom to another. Let us consider a general case when an atom *A* has one electron in the valence shell and another atom *B* has seven electrons. *A* has one electron in excess and *B* has one electron short than the stable octet. Therefore, *A* transfers an electron to *B* and in this transaction both the atoms acquire a stable electron-octet. The resulting positive ion (cation) and negative ion (anion) are held together by electrostatic attraction.

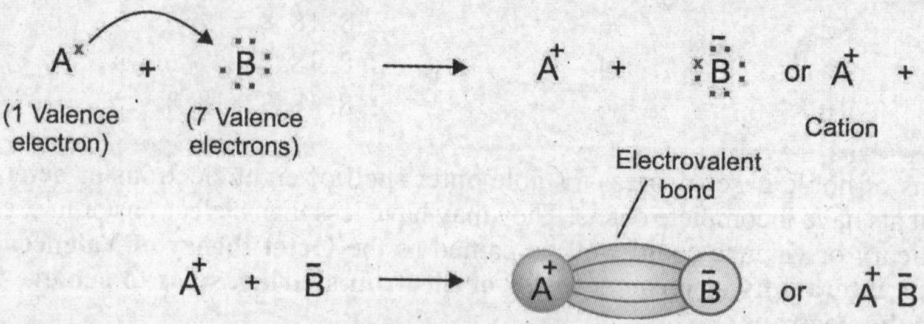

The electrostatic attraction between the cation (+) and anion (–) produced by electron-transfer constitutes an Ionic or Electrovalent bond.

The compounds containing such a bond are referred to as **Ionic** or **Electrovalent Compounds.**

CONDITIONS FOR FORMATION OF IONIC BOND

The conditions favorable for the formation of an ionic bond are :

(1) Number of valence electrons

The atom *A* should possess 1, 2 or 3 valence electrons, while the atom *B* should have 5, 6 or 7 valence electrons. The elements of group IA, IIA and IIIA satisfy this condition for atom *A* and those of groups VA, VIA, and VIIA satisfy this condition for atom *B*.

(2) Net lowering of Energy

To form a stable ionic compound, there must be a net lowering of the energy. In other words energy must be released as a result of the electron transfer and formation of ionic compound by the following steps :

(*a*) The removal of electron from atom *A* ($A - e^- \rightarrow A^+$) requires input of energy, which is the ionization energy (IE). It should be low.

(*b*) The addition of an electron to *B* ($B + e^- \rightarrow B^-$) releases energy, which is the electron affinity of *B* (EA). It should be high.

(*c*) The electrostatic attraction between A^+ and B^- in the solid compound releases energy, which is the electrical energy. It should also be high.

If the energy released in steps (*b*) and (*c*) is greater than the energy consumed in step (*a*), the overall process of electron transfer and formation of ionic compound results in a net release of energy. Therefore, ionization of *A* will occur and the ionic bond will be formed. For example, in case of formation of sodium chloride (NaCl), we have

$$Na - e^- \longrightarrow Na^+ \quad - 498.23\,kJ$$
$$Cl + e^- \longrightarrow Cl^- \quad + 355.88\,kJ$$
$$Na^+ + Cl^- \longrightarrow Na^+Cl^- \quad + 782.93\,kJ$$

The net energy released is 782.93 + 355.88 – 498.23 = 640.58 kJ. Since the overall process results in a lowering of energy, the ionic bond between Na and Cl will be formed.

(3) Electronegativity difference of A and B

From the line of argument used in (2), we can say that atoms *A* and *B* if they have greatly different electronegativities, only then they will form an ionic bond. In fact, a difference of 2 or more is necessary for the formation of an ionic bond between atoms *A* and *B*. Thus Na has electronegativity 0.9, while Cl has 3.0. Since the difference is (3.0 – 0.9) = 2.1, Na and Cl will form an ionic bond.

FACTORS GOVERNING THE FORMATION OF IONIC BOND

(1) Ionization Energy

The ionization energy of the metal atom which looses electron(s) should be low so that the formation of +vely charged ion is easier. *Lower the ionization energy greater will be the tendency of the metal atom to change into cation and hence greater will be the ease of formation of ionic bond.* That is why *alkali metals and alkaline earth metals form ionic bonds easily.* Out of these two, alkali metals form ionic bonds easily as compared to alkaline earth metals. *In a group the ionization energy decreases as we move down the group and therefore, the tendency to form ionic bond increases in a group downward.* Due to this reason Cs is the most electropositive atom among the alkali metals.

(2) Electron Affinity

The atom which accepts the electron and changes into anion should have high electron affinity. *Higher the electron affinity more is the energy released and stable will be the anion formed.* The elements of group VI A and VII A have, in general, higher electron affinity and have high tendency to form ionic bonds. Out of these two, *the elements of group* VII A (*halogens*) *are more prone to the formation of ionic bond than the elements of group* VI A. In moving down a group the electron affinity decreases and, therefore, the tendency to form ionic bond also decreases.

(3) **Lattice Energy**

After the formation of cations and anions separately, they combine to form ionic compound.

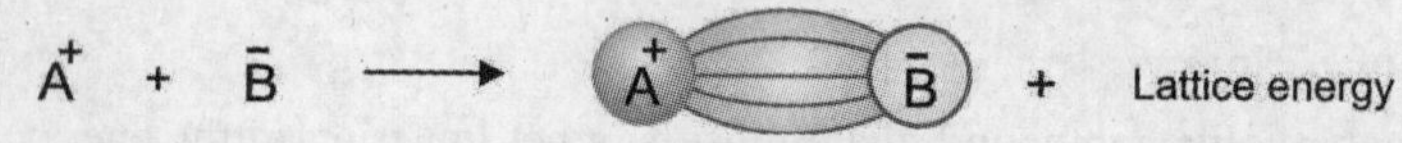

In this process, energy is released. It is called Lattice Energy. It may be defined as "*the amount of energy released when one mole of an ionic compound is formed from its cations and anions.*"

Greater the lattice energy, greater the strength of ionic bond. The value of lattice energy depends upon the following two factors :

(*a*) **Size of the ions**

In order to have the greater force of attraction between the cations and anions their size should be small as the force of attraction is inversely proportional to the square of the distance between them.

(*b*) **Charge on Ions**

Greater the charge on ions greater will be the force of attraction between them and, therefore, greater will be the strength of the ionic bond.

Necessary for the formation of an ionic bond between atoms *A* and *B*. Thus Na has electronegativity 0.9, while Cl has 3.0. Since the difference is (3.0 – 0.9) = 2.1, Na and Cl will form an ionic bond.

SOME EXAMPLES OF IONIC COMPOUNDS

Here we will discuss the formation of **Lewis formula** or **Electron dot formula** of some binary ionic compounds, for illustration.

Sodium Chloride, NaCl

A simple sodium chloride molecule is formed from an atom of sodium (Na) and one atom of chlorine (Cl). Na (2, 8, 1) has one valence electron, while Cl (2, 8, 7) has seven. Na transfers its valence electron to Cl, and both achieve stable electron octet. Thus Na gives Na^+ and Cl gives Cl^- ion, and the two are joined by an ionic bond.

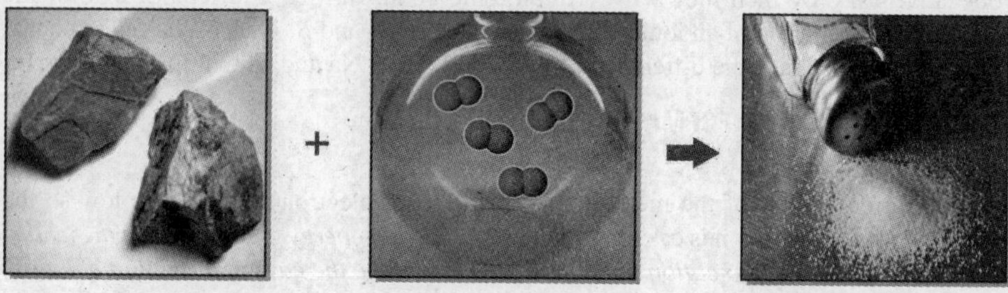

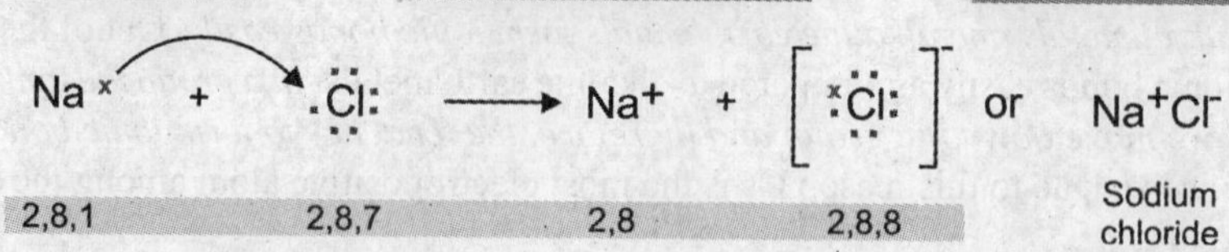

Ionic Compounds Exist as Crystals. The (+) and (–) ions attract each other with electrostatic force that extends in all directions. This means that ions will be bonded to a number of oppositely charged ions around them. Therefore in solid state, single ionic molecules do not exist as such. Rather many (+) and (–) ions are arranged systematically in an alternating cation-anion pattern called the crystal lattice. The crystal lattice of NaCl is shown in Fig. 5.2. It will be noticed that here a large number of Na^+ and Cl^- ions are arranged in an orderly fashion so as to form a cubic crystal. Each Na^+ ion is surrounded by 6 Cl^- ions and each Cl^- ion is surrounded by 6 Na^+ ions. This makes a network of Na^+ and Cl^- ions which are tightly held together by electrostatic forces between them.

Although discrete molecules Na^+Cl^- do not exist in the solid form of ionic compounds, independent molecules do exist in the vapor form of such compounds.

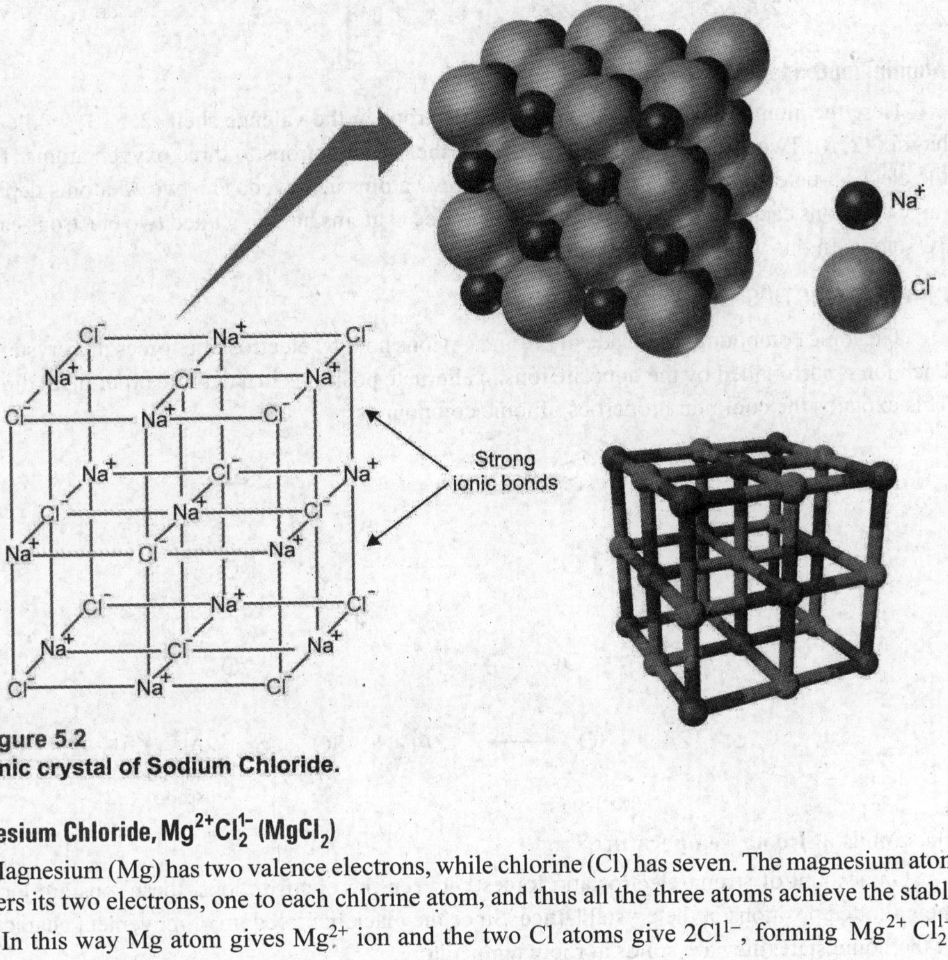

■ **Figure 5.2**
Ionic crystal of Sodium Chloride.

Magnesium Chloride, $Mg^{2+}Cl_2^{1-}$ ($MgCl_2$)

Magnesium (Mg) has two valence electrons, while chlorine (Cl) has seven. The magnesium atom transfers its two electrons, one to each chlorine atom, and thus all the three atoms achieve the stable octet. In this way Mg atom gives Mg^{2+} ion and the two Cl atoms give $2Cl^{1-}$, forming $Mg^{2+}Cl_2^{1-}$ (or $MgCl_2$).

$$Mg^{\times}_{\times} + \cdot\ddot{\underset{..}{Cl}}: + \cdot\ddot{\underset{..}{Cl}}: \longrightarrow Mg^{2+} + Cl^- + Cl^-$$

2, 8, 2	2, 8, 7	2, 8, 7	2, 8	2, 8, 8	2, 8, 8

$$\text{or} \quad Mg + 2Cl \longrightarrow Mg^{2+} + 2Cl^- \text{ or } Mg^{2+}Cl_2^{1-}$$

Calcium Oxide, $Ca^{2+}O^{2-}$ (CaO)

Calcium (Ca) has two valence electrons, while oxygen (O) has six. Calcium atom transfers its two valence electrons to the same oxygen atom. Thus both Ca and O achieve the stable electron-octet, forming Ca^{2+} and O^{2-} ions. Thus is obtained the molecule of calcium oxide, $Ca^{2+}O^{2-}$.

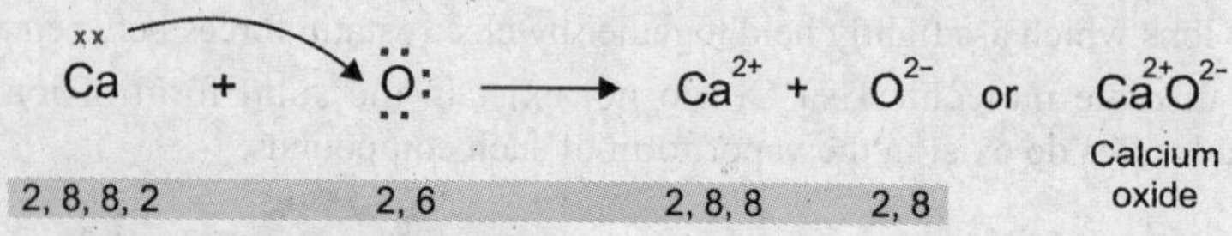

Aluminium Oxide, $Al_2^{3+}O_3^{2-}$ (Al_2O_3)

Here the aluminium atom (Al) has three electrons in the valence shell (2, 8, 3), while oxygen has six (2, 6). Two atoms of aluminium transfer their six electrons to three oxygen atoms. Thus are the electron-octets of the two Al atoms and three O atoms achieved. The two Al atoms deprived of three electrons each, give $2Al^{3+}$ ions, while the three O atoms having gained two electrons each give $3O^{2-}$ ions. In this way, we get $Al_2^{3+}O_3^{2-}$ or Al_2O_3.

CHARACTERISTICS OF IONIC COMPOUNDS

The ionic compounds are made of (+) and (–) ions held by electrostatic forces in a crystal lattice. Each ion is surrounded by the opposite ions in alternate positions in a definite order in all directions. This explains the common properties of ionic compounds.

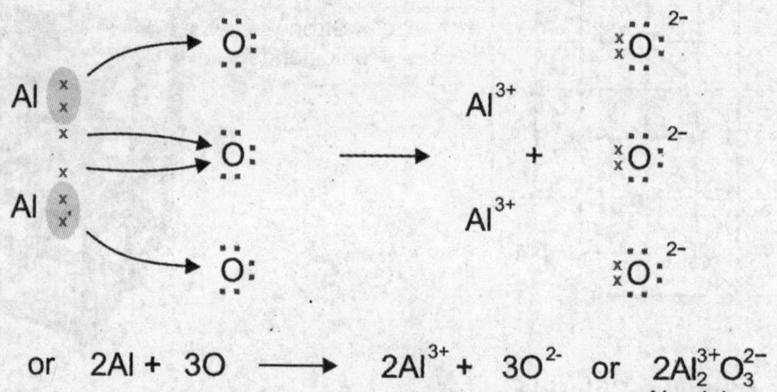

$$\text{or} \quad 2Al + 3O \longrightarrow 2Al^{3+} + 3O^{2-} \quad \text{or} \quad 2Al_2^{3+}O_3^{2-}$$

Aluminium oxide

(1) Solids at Room Temperature

On account of strong electrostatic forces between the opposite ions, these ions are locked in their allotted positions in the crystal lattice. Since they lack the freedom of movement characteristic of the liquid state, they are solids at room temperature.

(2) High Melting Points

Ionic compounds have high melting points (or boiling points). Since the (+) and (–) ions are tightly held in their positions in the lattice, only at high temperature do the ions acquire sufficient kinetic energy to overcome their attractive forces and attain the freedom of movement as in a liquid. Thus ionic compounds need heating to high temperatures before melting.

(3) Hard and brittle

The crystals of ionic substances are hard and brittle. Their hardness is due to the strong electrostatic forces which hold each ion in its allotted position.

These crystals are made of layers of (+) and (–) ions in alternate positions so that the opposite ions in the various parallel layers lie over each other. When external force is applied to a layer of ions (Fig. 5.3), with respect to the next, even a slight shift brings the like ions in front of each other. The (+) and (–) ions in the two layers thus repel each other and fall apart. The crystal cleaves here.

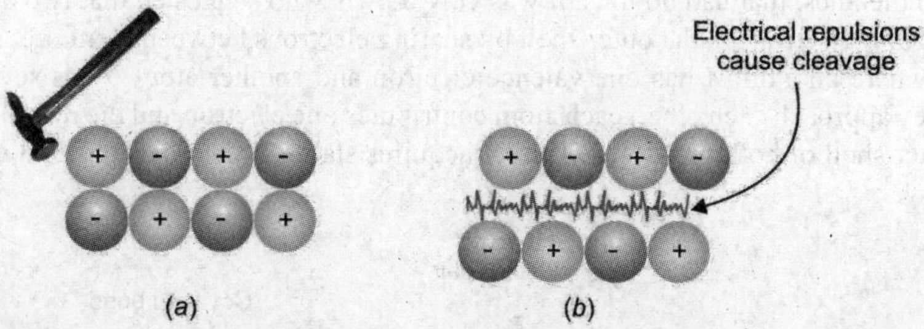

Figure 5.3
(*a*) Two layers of (+) and (–) ions in a crystal. (*b*) When force is applied to one layer it slips over the other so that similar ions come above one another and electrical repulsions between them cause cleavage of the crystal.

(4) Soluble in water

When a crystal of an ionic substance is placed in water, the polar water molecules detach the (+) and (–) ions from the crystal lattice by their electrostatic pull. These ions then get surrounded by water molecules and can lead an independent existence and are thus dissolved in water. By the same reason, non-polar solvents like benzene (C_6H_6) and hexane (C_6H_{14}) will not dissolve ionic compounds.

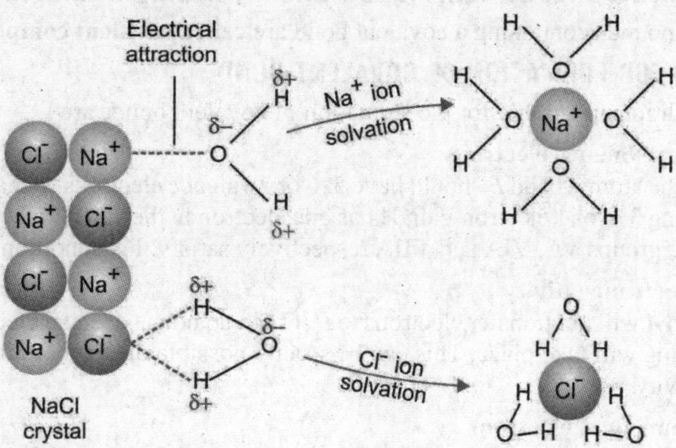

Figure 5.4
Solvation of NaCl in water.

(5) Conductors of electricity

Solid ionic compounds are poor conductors of electricity because the ions are fixed rigidly in their positions. In the molten state and in water solutions, ions are rendered free to move about. Thus molten ionic compounds or their aqueous solutions conduct a current when placed in an electrolytic cell.

(6) Do not exhibit isomerism

The ionic bond involving electrostatic lines of force between opposite ions, is non-rigid and non-directional. The ionic compounds, therefore, are incapable of exhibiting stereoisomerism.

(7) **Ionic reactions are fast**

Ionic compounds give reactions between ions and these are very fast.

COVALENT BOND

The electron transfer theory could not explain the bonding in molecules such as H_2, O_2, Cl_2 etc., and in organic molecules, that had no ions. It was G.N. Lewis who suggested that two atoms could achieve stable 2 or 8 electrons in the outer shell by sharing electrons between them. Let us consider a general case where an atom *A* has one valence electron and another atom *B* has seven valence electrons. As they approach each other, each atom contributes one electron and the resulting electron pair fills the outer shell of both the atoms. Thus *A* acquires stable 2 electrons and *B*, 8 electrons in the outer shell.

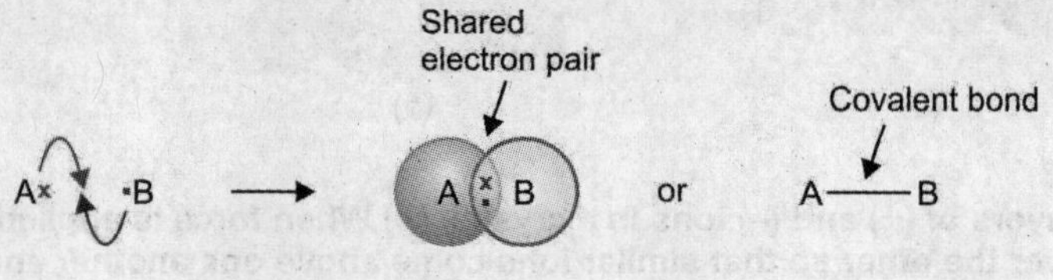

The shared pair is indicated by a dash (–) between the two bonded atoms. A shared pair of electrons constitutes a **Covalent bond** or **Electron-pair bond.**

In fact, the positive nuclei of atoms *A* and *B* are pulled towards each other by the attraction of the shared electron pair. At the same time, the nuclei of two atoms also repel each other as do the two electrons. It is the net attractive force between the shared electrons and the nuclei that holds the atoms together. Thus an alternative definition of a covalent bond would be :

The attractive force between atoms created by sharing of an electron-pair.

The compounds containing a covalent bond are called **covalent compounds.**

CONDITIONS FOR FORMATION OF COVALENT BOND

The conditions favorable for the formation of covalent bonds are :

(1) **Number of valence electrons**

Each of the atoms *A* and *B* should have 5, 6 or 7 valence electrons so that both achieve the stable octet by sharing 3, 2 or 1 electron-pair. H has one electron in the valence shell and attains duplet. The non-metals of groups VA, VIA and VIIA respectively satisfy this condition.

(2) **Equal electronegativity**

The atom *A* will not transfer electrons to *B* if both do not have equal electronegativity, and hence electron sharing will take place. This can be strictly possible only if both the atoms are of the same electronegativity.

(3) **Equal sharing of electrons**

The atoms *A* and *B* should have equal (or nearly equal) electron affinity so that they attract the bonding electron pair equally. Thus equal sharing of electrons will form a nonpolar covalent bond. Of course, precisely equal sharing of electrons will not ordinarily occur except when atoms *A* and *B* are atoms of the same element, for no two elements have exactly the same electron affinity.

SOME EXAMPLES OF COVALENT COMPOUNDS

The construction of Lewis structures of simple covalent compounds will be discussed.

Hydrogen, H_2

Hydrogen molecule is made of two H atoms, each having one valence electron. Each contributes an electron to the shared pair and both atoms acquire stable helium configuration. Thus stable H_2 molecule results.

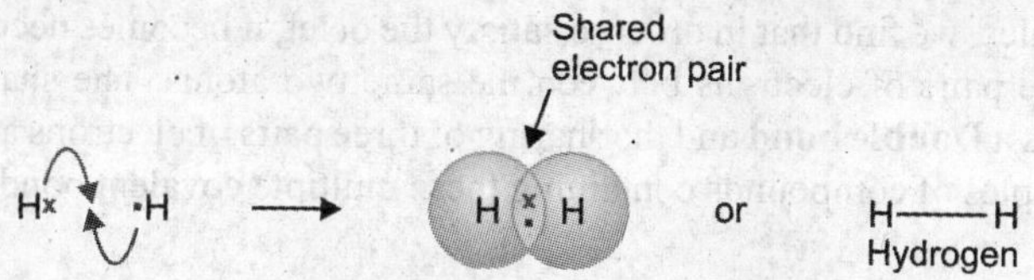

Chlorine, Cl_2

Each Cl atom (2, 8, 7) has seven valence electrons. The two Cl atoms achieve a stable electron octet by sharing a pair of electrons.

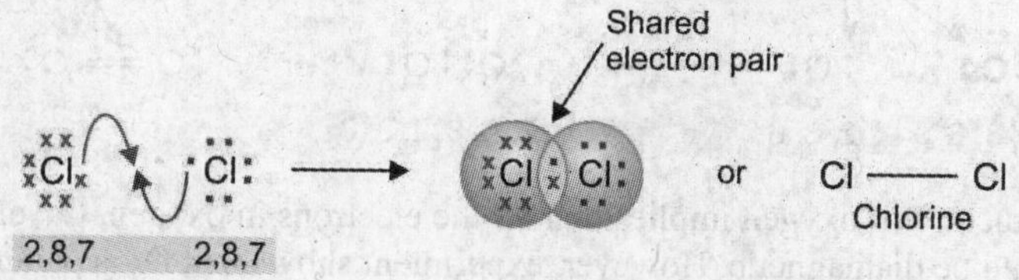

Water, H_2O

Oxygen atom (2, 6) has six valence electrons and can achieve the stable octet by sharing two electrons, one with each H atom. Thus Lewis structure of water can be written as :

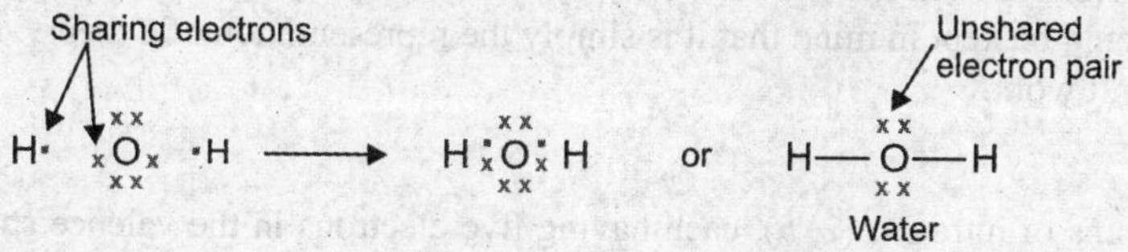

Ammonia, NH_3

Nitrogen atom (2, 5) has five valence electrons and can achieve the octet by sharing three electrons, one each with three H atoms. This gives the following Lewis structure for ammonia :

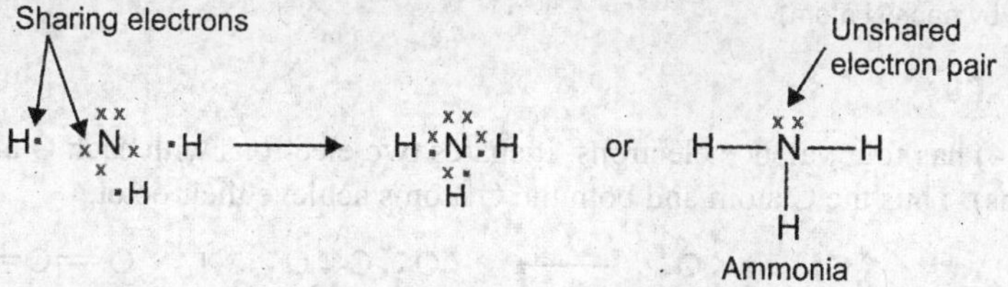

Methane, CH_4

Carbon atom (2, 4) has four electrons in the valence shell. It can achieve the stable octet by sharing these electrons with four H atoms, one with each H atom. Thus the Lewis structure of methane can be written as :

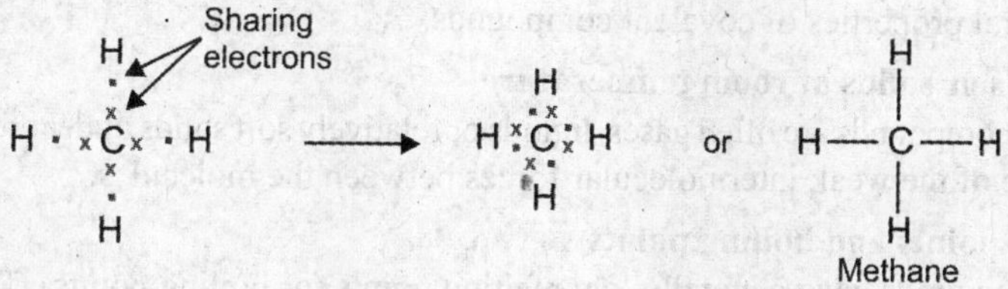

EXAMPLES OF MULTIPLE COVALENT COMPOUNDS

In many molecules, we find that in order to satisfy the octet, it becomes necessary for two atoms to share two or three pairs of electrons between the same two atoms. The sharing of two pairs of electrons is known as a **Double bond** and the sharing of three pairs of electrons a **Triple bond.** Let us consider some examples of compounds containing these multiple covalent bonds in their molecules.

Oxygen, O_2

The conventional Lewis structure of oxygen is written by sharing of two pairs of electrons between two O atoms (2, 6). In this way both the O atoms achieve the octet.

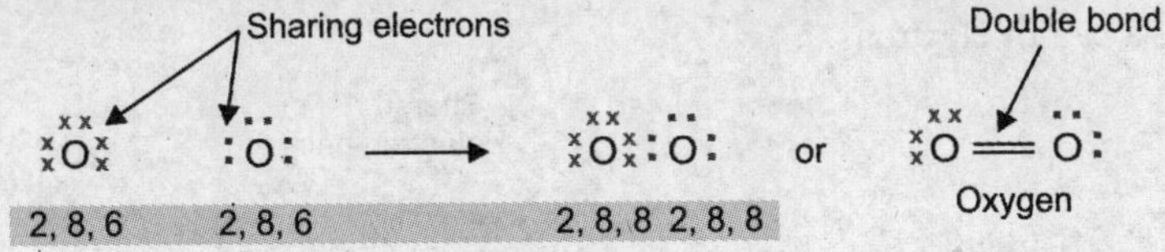

The above structure of oxygen implies that all the electrons in oxygen, O_2, are paired whereby the molecule should be diamagnetic. However, experiment shows that O_2 is paramagnetic with two unpaired electrons. This could be explained by the structure.

Although writing Lewis structures work very well in explaining the bonding in most simple molecules, it should be kept in mind that it is simply the representation of a theory. In this case, the theory just doesn't work.

Nitrogen, N_2

The two atoms of nitrogen (2, 5), each having five electrons in the valence shell, achieve the octet by sharing three electron pairs between them.

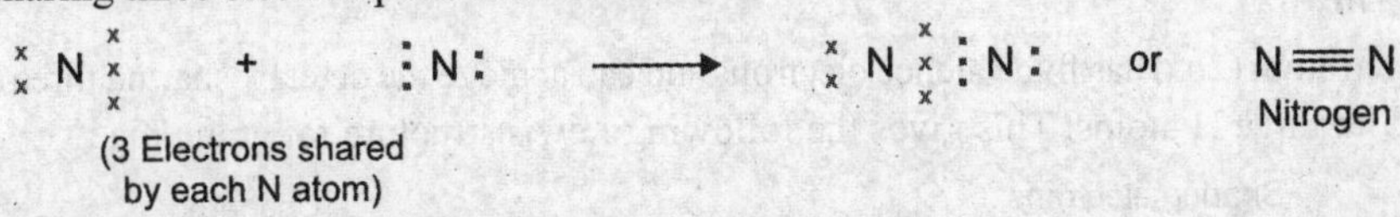

Carbon Dioxide, CO_2

Carbon (2, 4) has four valence electrons. It shares two electrons with each O atom (having six valence electrons). Thus the C atom and both the O atoms achieve their octet.

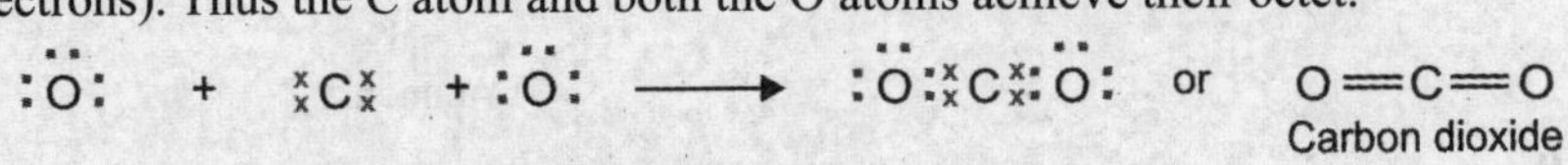

CHARACTERISTICS OF COVALENT COMPOUNDS

While the atoms in a covalent molecule are firmly held by the shared electron pair, the individual molecules are attracted to each other by weak van der Waals forces. Thus the molecules can be separated easily as not much energy is required to overcome the intermolecular attractions. This explains the general properties of covalent compounds.

(1) Gases, liquids or solids at room temperature

The covalent compounds are often gases, liquids or relatively soft solids under ordinary conditions. This is so because of the weak intermolecular forces between the molecules.

(2) Low melting points and boiling points

Covalent compounds have generally low melting points (or boiling points). The molecules are held together in the solid crystal lattice by weak forces. On application of heat, the molecules are readily pulled out and these then acquire kinetic energy for free movement as in a liquid. For the same reason, the liquid molecules are easily obtained in the gaseous form which explains low boiling points of covalent liquids.

(3) **Neither hard nor brittle**

While the ionic compounds are hard and brittle, covalent compounds are neither hard nor brittle. There are weak forces holding the molecules in the solid crystal lattice. A molecular layer in the crystal easily slips relative to other adjacent layers and there are no 'forces of repulsion' like those in ionic compounds. Thus the crystals are easily broken and there is no sharp cleavage between the layers on application of external force.

(4) **Soluble in organic solvents**

In general, covalent compounds dissolve readily in nonpolar organic solvents (benzene, ether). The kinetic energy of the solvent molecules easily overcomes the weak intermolecular forces.

Covalent compounds are insoluble in water. Some of them (alcohols, amines) dissolve in water due to hydrogen-bonding.

(5) **Non-conductors of electricity**

Since there are no (+) or (–) ions in covalent molecules, the covalent compounds in the molten or solution form are incapable of conducting electricity.

(6) **Exhibit Isomerism**

Covalent bonds are rigid and directional, the atoms being held together by shared electron pair and not by electrical lines of force. This affords opportunity for various spatial arrangements and covalent compounds exhibit stereoisomerism.

(7) **Molecular reactions**

The covalent compounds give reactions where the molecule as a whole undergoes a change. Since there are no strong electrical forces to speed up the reaction between molecules, these reactions are slow.

CO-ORDINATE COVALENT BOND

In a normal covalent bond, each of the two bonded atoms contributes one electron to make the shared pair. In some cases, a covalent bond is formed when both the electrons are supplied entirely by one atom. Such a bond is called **coordinate covalent** or **dative bond.** It may be defined as : **a covalent bond in which both electrons of the shared pair come from one of the two atoms (or ions).** The compounds containing a coordinate bond are called **coordinate compounds.**

If an atom *A* has an unshared pair of electrons (**lone pair**) and another atom *B* is short of two electrons than the stable number, coordinate bond is formed. *A* donates the lone pair to *B* which accepts it. Thus both *A* and *B* achieve the stable 2 or 8 electrons, the lone pair being held in common.

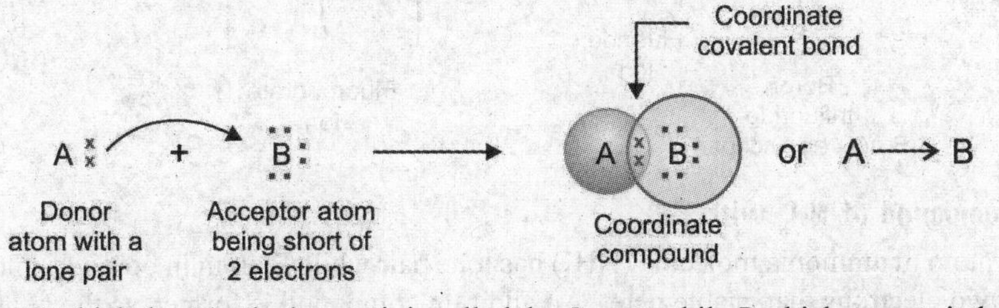

The atom *A* which donates the lone pair is called the **donor,** while *B* which accepts it the **acceptor.** The bond thus established is indicated by an arrow pointing from *A* to *B*. Although the arrow head indicates the origin of the electrons, once the coordinate bond is formed it is in no way different from an ordinary covalent bond.

The molecule or ion that contains the donor atom is called the ligand.

SOME EXAMPLES OF COORDINATE COMPOUNDS OR IONS

Lewis structures of some common molecules or ions containing a coordinate covalent bond are listed below.

Ammonium ion, NH_4^+

In ammonia molecule, the central N atom is linked to three H atoms and yet N has an unshared pair of electrons. The H^+ ion furnished by an acid has no electron to contribute and can accept a pair of electrons loaned by N atom. Thus, NH_3 donates its unshared electrons to H^+ forming ammonium ion.

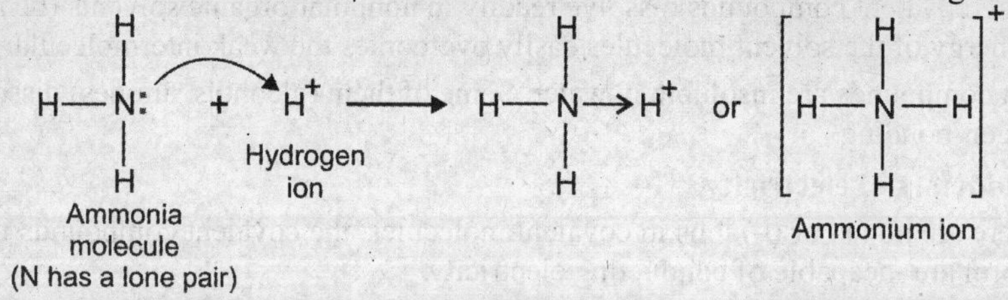

All the N–H bonds in NH_4^+ are identical, once the coordinate bond $N \rightarrow H^+$ is established.

Hydronium ion, H_3O^+

The oxygen atom in water molecule is attached to two H atoms by two covalent bonds. There are still two unshared pairs of electrons with the O atom. The O atom donates one of these pairs of electrons to H^+ ion and the hydronium ion is thus formed.

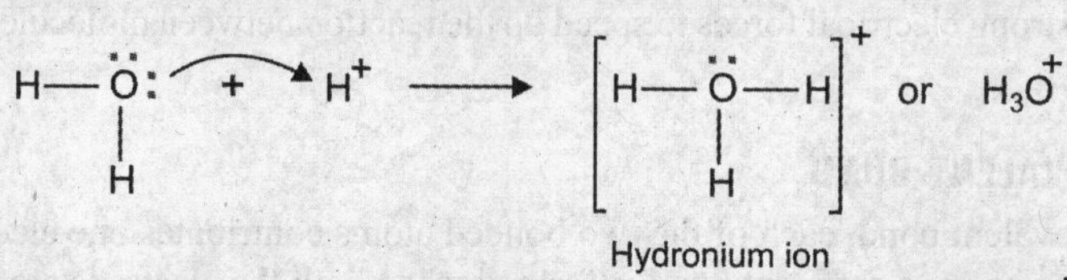

Fluoroborate ion, BF_4^-

It is formed when a boron trifluoride molecule (BF_3) shares a pair of electrons supplied by fluoride ion (F^-).

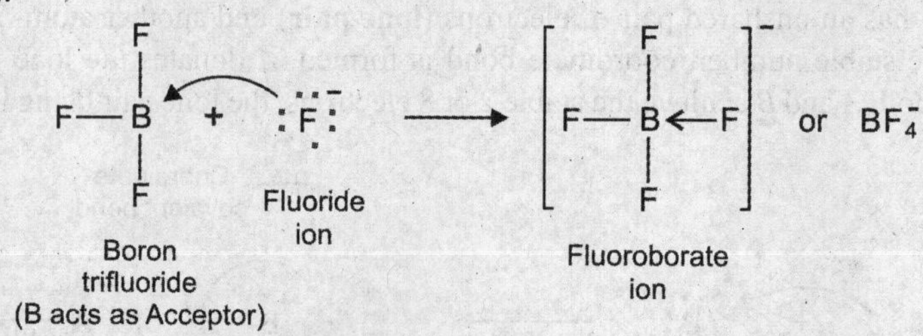

Addition compound of NH_3 with BCl_3

The N atom of ammonia molecule (NH_3) has lone pair while B atom in boron trichloride (BCl_3) is short of two electrons than stable octet. An addition compound is formed as the N atom donates its lone pair to B atom of BCl_3.

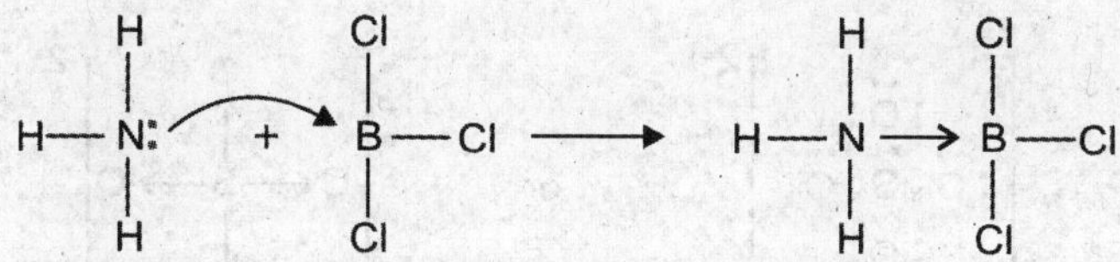

(N is Donor) (B is Acceptor) Addition compound

Nitromethane, CH_3NO_2

The Lewis structure of nitromethane is shown below. Here the N atom has five valence electrons, three of which are used in forming a covalent bond with C atom and two covalent bonds with O atom. The N atom is still left with two unshared electrons which are donated to another O atom.

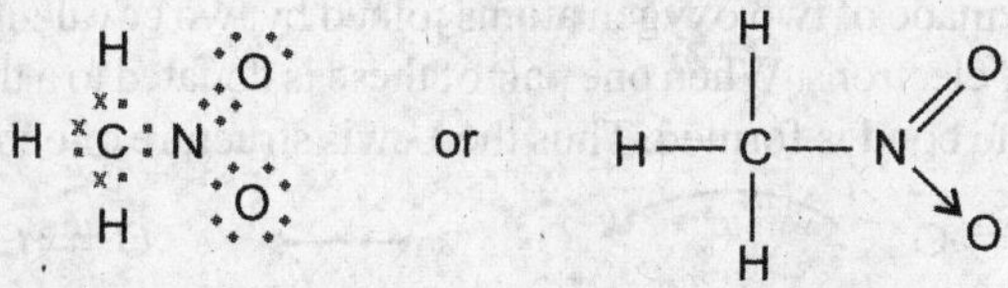

Nitromethane

Sulfur dioxide, SO_2, and Sulfur trioxide, SO_3

Sulfur achieves its octet by forming two covalent bonds with one O atom, giving SO. The S atom in SO has two lone pairs, one of which is shared with a second O atom to form sulfur dioxide, SO_2. The S atom in SO_2 still has one lone pair which it donates to a third O atom forming the sulfur trioxide (SO_3) molecule.

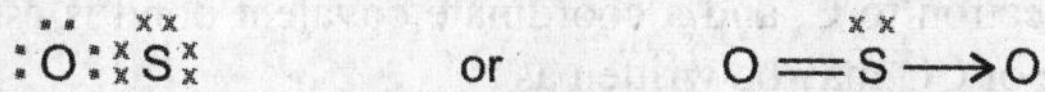

Sulfur dioxide

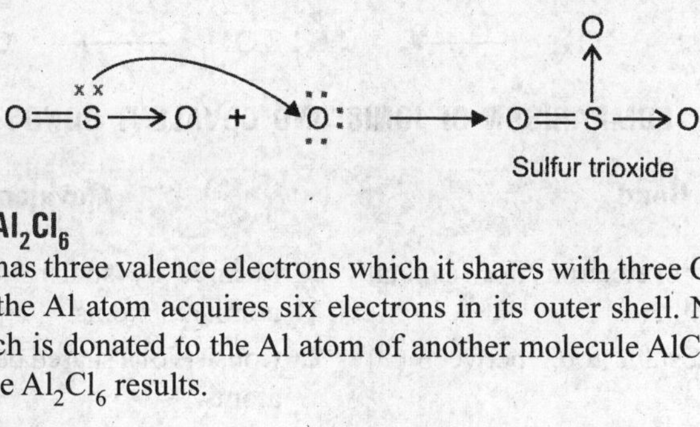

Sulfur trioxide

Aluminium Chloride, Al_2Cl_6

Aluminium atom has three valence electrons which it shares with three Cl atoms, forming three covalent bonds. Thus the Al atom acquires six electrons in its outer shell. Now Cl atom has three lone pairs, one of which is donated to the Al atom of another molecule $AlCl_3$. Thus both Al atoms achieve octet and stable Al_2Cl_6 results.

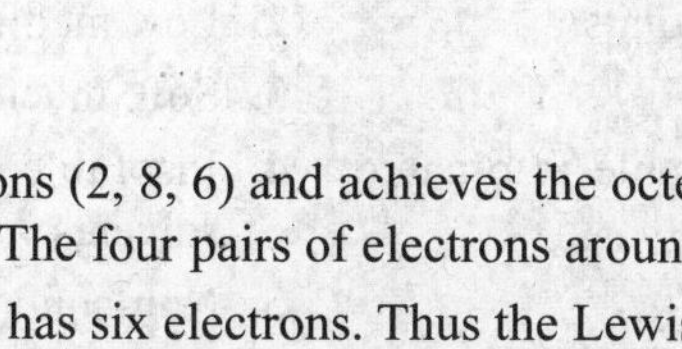

Sulfate ion, SO_4^{2-}

Sulfur has six valence electrons (2, 8, 6) and achieves the octet by gaining two electrons from metal atoms (say two Na atoms). The four pairs of electrons around the S atom are then donated to four oxygen atoms each of which has six electrons. Thus the Lewis structure for SO_4^{2-} ion may be written as :

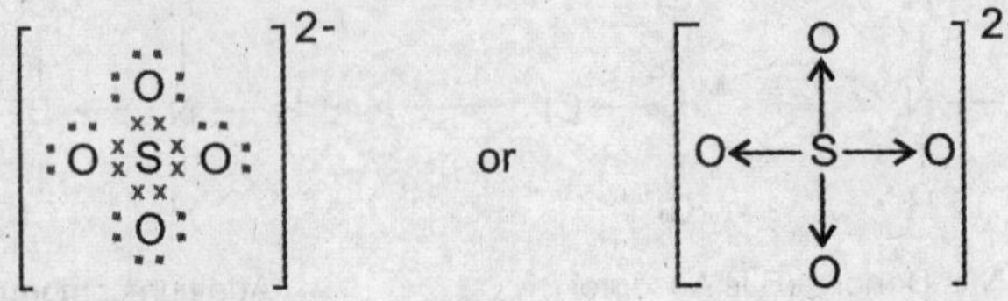

S gains two electrons from metal atoms and completes octet

Sulfate ion

Ozone, O_3

Oxygen molecule is made of two oxygen atoms joined by two covalent bonds. Each O atom in O_2 has two unshared pairs of electrons. When one pair of these is donated to a third O atom which has only six electrons, a coordinate bond is formed. Thus the Lewis structure of ozone may be represented as :

Oxygen (2 Atoms of O)

Ozone

Carbon Monoxide, CO

Carbon atom has four valence electrons while oxygen atom has six. By forming two covalent bonds between them, O atom achieves octet but C atom has only six electrons. Therefore O donates an unshared pair of electron to C, and a coordinate covalent bond is established between the two atoms. Lewis structure of CO may be written as :

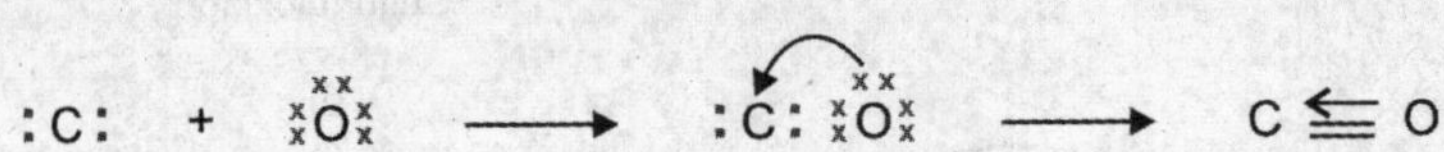

COMPARISON OF IONIC AND COVALENT BONDS

Ionic Bond	Covalent Bond
1. Formed by transfer of electrons from a metal to a non-metal atom.	**1.** Formed by sharing of electrons between nonmetal atoms.
2. Consists of electrostatic force between (+) and (–) ions.	**2.** Consists of a shared pair of electrons between atoms.
3. Non-rigid and non-directional; cannot cause isomerism.	**3.** Rigid and directional : causes stereoisomerism.
Properties of Compounds	**Properties of Compounds**
1. Solids at room temperature.	**1.** Gases, liquids or soft solids.
2. High melting and boiling points.	**2.** Low melting and boiling points.
3. Hard and brittle.	**3.** Soft, much readily broken
4. Soluble in water but insoluble in organic solvents.	**4.** Insoluble in water but soluble in organic solvents.
5. Conductors of electricity	**5.** Non-conductors of electricity.
6. Undergo ionic reactions which are fast.	**6.** Undergo molecular reactions which are slow.

POLAR COVALENT BONDS

In the H_2 or Cl_2 molecule, the two electrons constituting the covalent bond are equally shared by the two identical nuclei. Due to even distribution of (+) and (–) charge, the two bonded atoms remain electrically neutral. Such a bond is called **nonpolar covalent bond.** However, when two different atoms are joined by a covalent bond as in HCl, the electron pair is not shared equally.

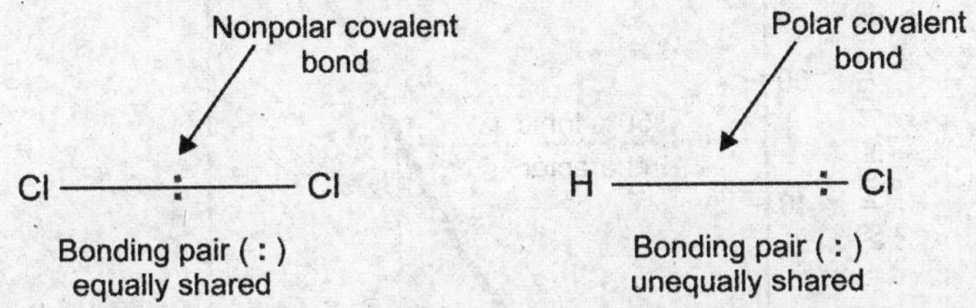

Due to a greater attraction of one nucleus (Cl) for the electrons, the shared pair is displaced towards it. This makes one end of the bond partially positive (δ+) and the other partially negative (δ–).

$$\overset{\delta+}{H}\text{———}\overset{\delta-}{Cl} \quad \text{or} \quad H\overset{+\!\!\longrightarrow}{\text{———}}Cl$$

A covalent bond in which electrons are shared unequally and the bonded atoms acquire a partial positive and negative charge, is called a polar covalent bond.

A molecule having partial positive and negative charge separated by a distance is commonly referred to as a **Dipole** (*two poles*). The dipole of a bond is indicated by an arrow from positive to negative end with a crossed tail as shown above in HCl molecule.

Since two atoms of different elements do not have exactly the same attraction for electrons in a bond, all bonds between unlike atoms are polar to some extent. The amount of polarity of a bond is determined by the difference of electronegativity (or tendency to attract electrons) of the two bonded atoms. **The greater the difference of electronegativity between two atoms, greater the polarity.** A graph showing the % age ionic character and difference in electronegativity between the two atoms is shown in Fig. 5.5.

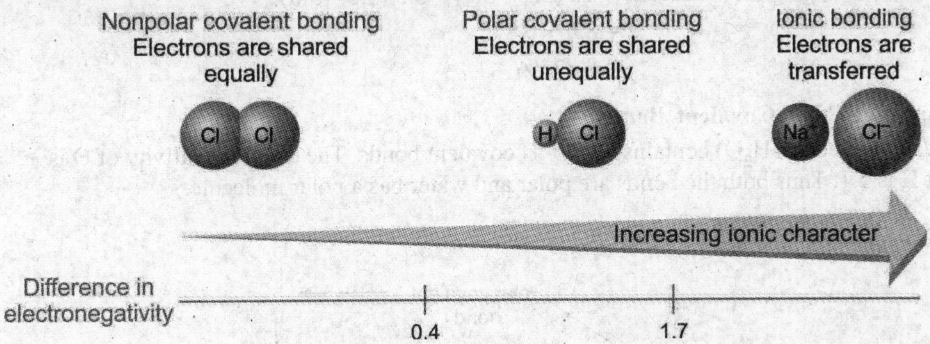

As a matter of fact, if this difference is around 1.9 and 2.9, the bond is generally ionic, meaning that one atom has gained complete control of the electron pair in the bond.

The percentage ionic character of a bond can be calculated by using the equation

$$\%\text{age ionic character} = 16\,[X_A - X_B] + 3.5\,[X_A - X_B]^2$$

This equation was given by Hannay and Smith.

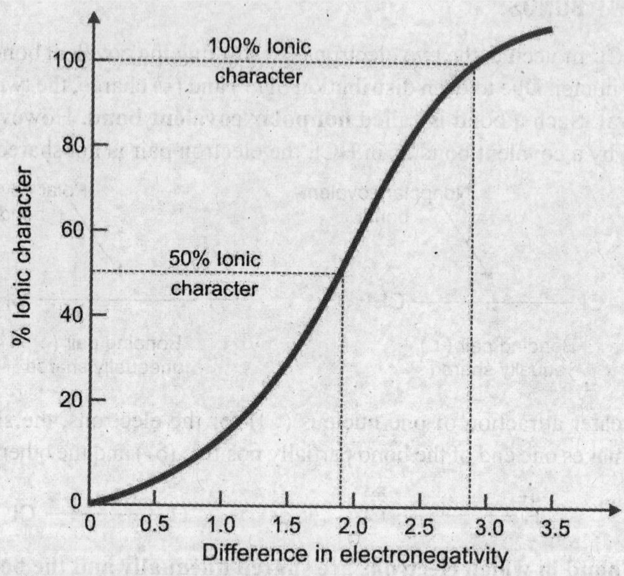

■ **Figure 5.5**
Graph between % ionic character and difference in electronegativity.

SOLVED PROBLEM 1

Calculate the percentage ionic character of C–Cl bond in CCl_4 if the electronegativities of C and Cl are 3.5 and 3.0 respectively.

SOLUTION

$$\%\text{age ionic character} = 16\,[X_A - X_B] + 3.5\,[X_A - X_B]^2$$

Given $X_A = 3.5$ and $X_B = 3.0$

$$\therefore \quad \%\text{age ionic character} = 16\,(3.5 - 3.0) + 3.5\,(3.5 - 3.0)^2$$
$$= 8.0 + 0.875$$
$$= \mathbf{8.875\%}$$

Examples of Polar Covalent Bonds

Water molecule (H_2O) contains two O–H covalent bonds. The electronegativity of O is 3.5 and that of H is 2.1. Thus both the bonds are polar and water has a polar molecule.

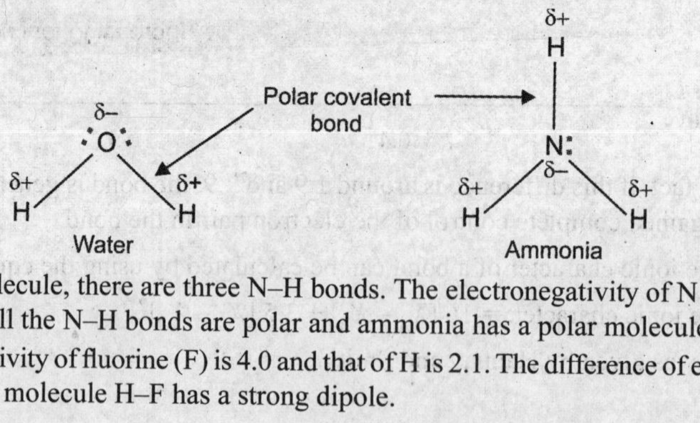

In ammonia molecule, there are three N–H bonds. The electronegativity of N is 3.0 and that of H is 2.1. Therefore all the N–H bonds are polar and ammonia has a polar molecule.

The electronegativity of fluorine (F) is 4.0 and that of H is 2.1. The difference of electronegativities being very great, the molecule H–F has a strong dipole.

HYDROGEN BONDING (H-Bonding)

When hydrogen (H) is covalently bonded to a highly electronegative atom X (O, N, F), the shared electron pair is pulled so close to X that a strong dipole results.

$$\overset{\delta-}{X} \text{———} \overset{\delta+}{H} \quad \text{or} \quad X \overset{\longleftarrow\!\!+}{\text{———}} H$$

Dipole

Since the shared pair is removed farthest from H atom, its nucleus (the proton) is practically exposed. The H atom at the positive end of a polar bond nearly stripped of its surrounding electrons, exerts a strong electrostatic attraction on the lone pair of electrons around X in a nearby molecule. Thus :

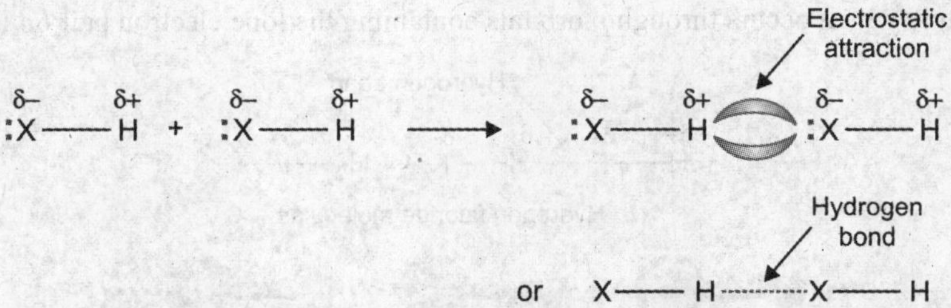

The electrostatic attraction between an H atom covalently bonded to a highly electronegative atom X and a lone pair of electrons of X in another molecule, is called Hydrogen Bonding. Hydrogen bond is represented by a dashed or dotted line.

POINTS TO REMEMBER

(1) **Only O, N and F which have very high electronegativity and small atomic size, are capable of forming hydrogen bonds.**

(2) **Hydrogen bond is longer and much weaker than a normal covalent bond.** Hydrogen bond energy is less than 10 kcal/mol, while that of covalent bond is about 120 kcal/mol.

(3) Hydrogen bonding results in long chains or clusters of a large number of 'associated' molecules like many tiny magnets.

(4) **Like a covalent bond, hydrogen bond has a preferred bonding direction.** This is attributed to the fact that hydrogen bonding occurs through *p* orbitals which contain the lone pair of electrons on X atom. This implies that the atoms X–H···X will be in a straight line.

CONDITIONS FOR HYDROGEN BONDING

The necessary conditions for the formation of hydrogen bonding are

(1) **High electronegativity of atom bonded to hydrogen**

The molecule must contain an atom of high electronegativity such as F, O or N bonded to hydrogen atom by a covalent bond. The examples are HF, H_2O and NH_3.

(2) **Small size of Electronegative atom**

The electronegative atom attached to H-atom by a covalent bond should be quite small. Smaller the size of the atom, greater will be the attraction for the bonded electron pair. In other words, the polarity of the bond between H atom and electronegative atom should be high. This results in the formation of stronger hydrogen bonding. For example, N and Cl both have 3.0 electronegativity. But hydrogen bonding is effective in NH_3 in comparison to that in HCl. It is due to smaller size of N atom than Cl atom.

EXAMPLES OF HYDROGEN-BONDED COMPOUNDS

When hydrogen bonding occurs between different molecules of the same compound as in HF, H_2O and NH_3, it is called **Intermolecular hydrogen bonding.** If the hydrogen bonding takes place within single molecule as in 2-nitrophenol, it is referred to as **Intramolecular hydrogen bonding.** We will consider examples of both types.

Hydrogen Fluoride, HF

The molecule of HF contains the strongest polar bond, the electronegativity of F being the highest of all elements. Therefore, hydrogen fluoride crystals contain infinitely long chains of H–F molecules in which H is covalently bonded to one F and hydrogen bonded to another F. The chains possess a zig-zag structure which occurs through *p* orbitals containing the lone electron pair on F atom.

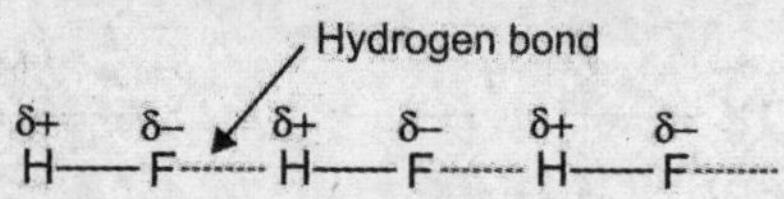

Hydrogen fluoride molecules

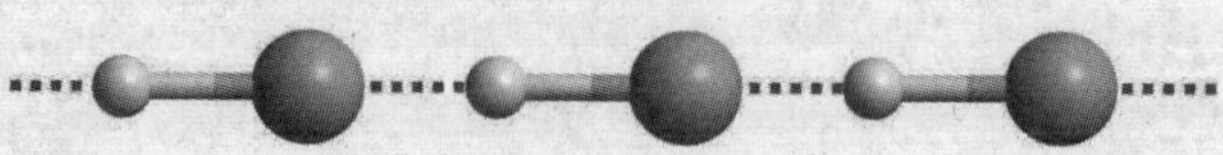

Water, H_2O

In H_2O molecule, two hydrogen atoms are covalently bonded to the highly electronegative O atom. Here each H atom can hydrogen bond to the O atom of another molecule, thus forming large chains or clusters of water molecules.

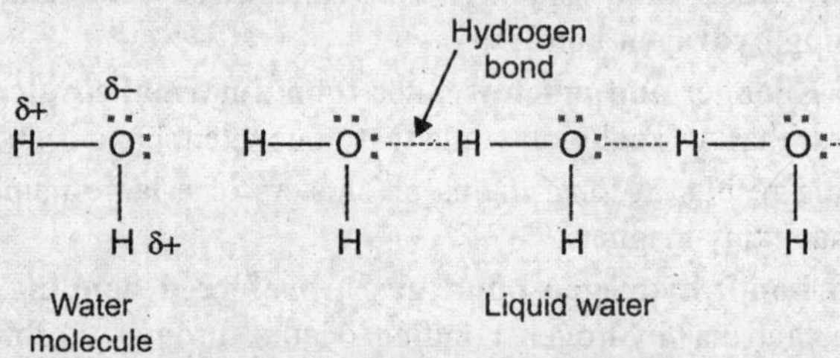

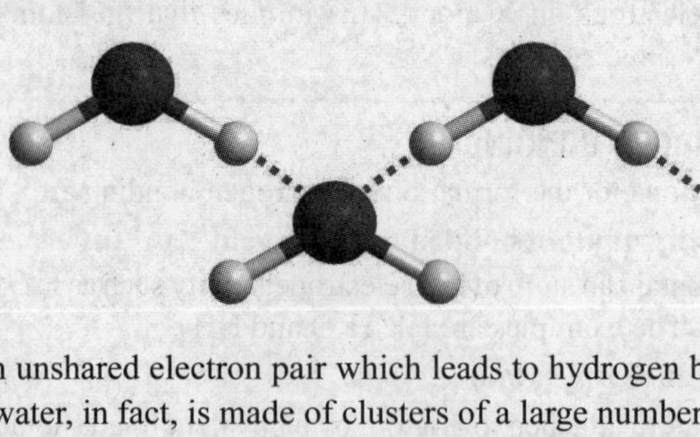

Each O atom still has an unshared electron pair which leads to hydrogen bonding with other water molecules. Thus liquid water, in fact, is made of clusters of a large number of molecules.

Ammonia, NH_3

In NH_3 molecules, there are three H atoms covalently bonded to the highly electronegative N atom. Each H atom can hydrogen bond to N atom of other molecules.

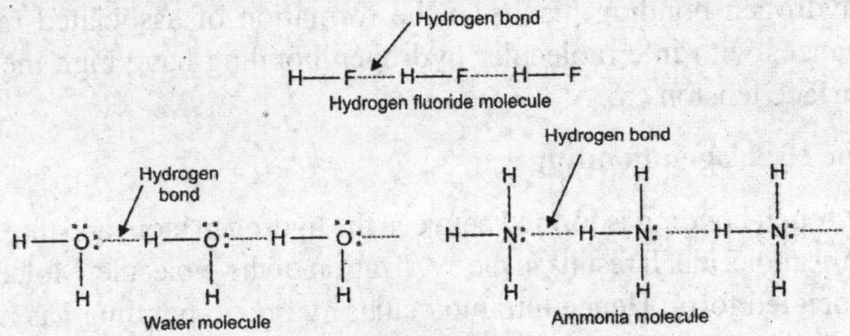

2-Nitrophenol

Here hydrogen bonding takes place within the molecule itself as O–H and N–H bonds are a part of the same one molecule.

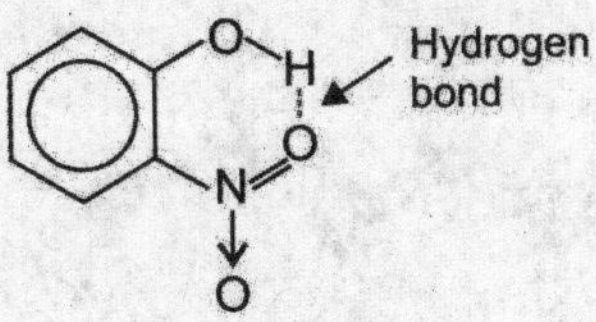

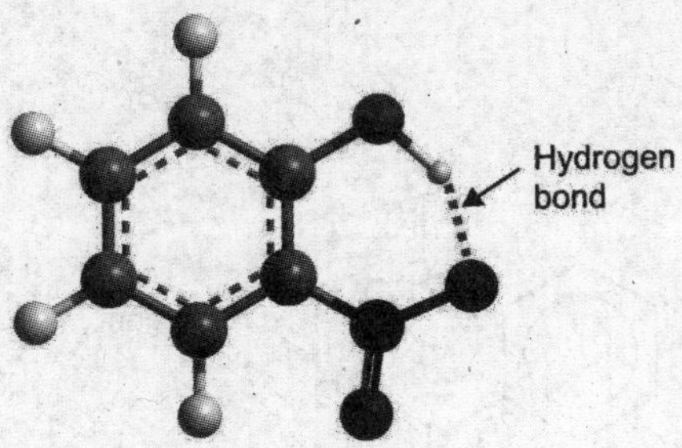

TYPES OF HYDROGEN-BONDING

Hydrogen bonding is of two types :

(1) **Intermolecular Hydrogen bonding**

This type of hydrogen bonding is formed between two different molecules of the same or different substances *e.g.* hydrogen bonding in HF, H_2O, NH_3 etc. It is shown in the following diagram (Fig. 5.6).

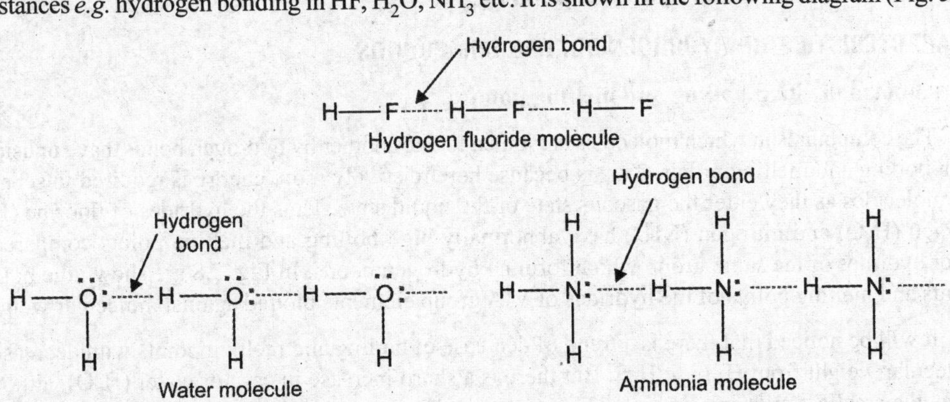

■ Figure 5.6
Intermolecular hydrogen bonding in HF, H_2O and NH_3.

This type of hydrogen bonding results in the formation of associated molecules. Generally speaking, the substances with intermolecular hydrogen bonding have high melting points, boiling points, viscosity, surface tension etc.

(2) Intramolecular Hydrogen bonding

This type of hydrogen bonding is formed between the hydrogen atom and the electronegative atom present within the same molecule. It results in the cyclization of the molecule. Molecules exist as discrete units and not in associated form. Hence intramolecular hydrogen bonding has no effect on physical properties like melting point, boiling point, viscosity, surface tension, solubility etc. For example intramolecular hydrogen bonding exists in 2-nitrophenol, 2-nitrobenzoic acid etc. as shown below :

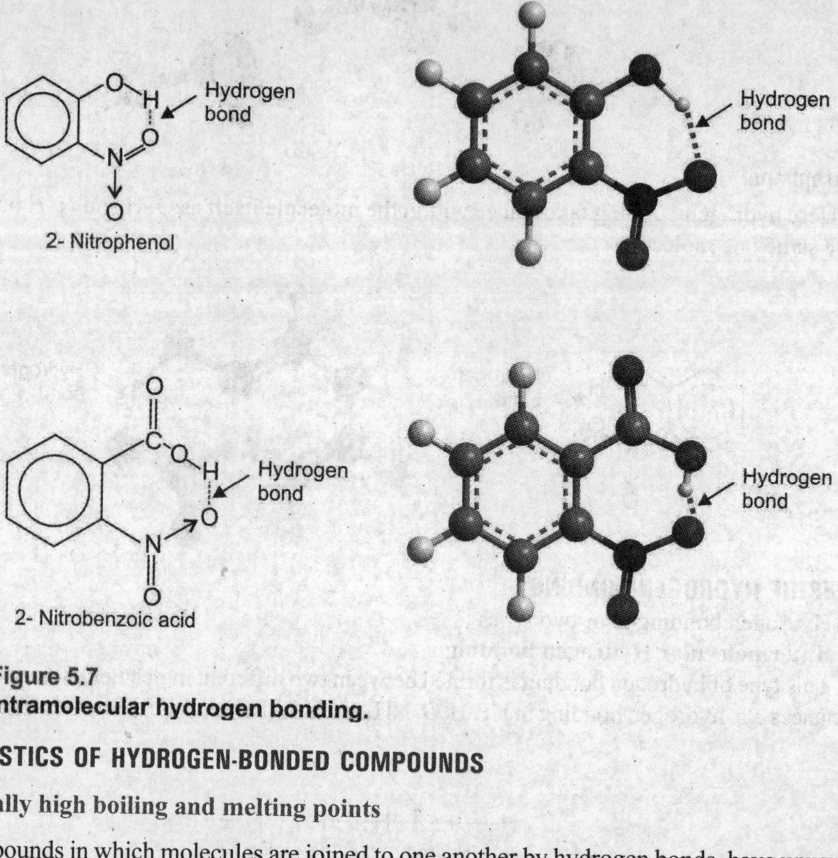

Figure 5.7
Intramolecular hydrogen bonding.

CHARACTERISTICS OF HYDROGEN-BONDED COMPOUNDS

(1) Abnormally high boiling and melting points

The compounds in which molecules are joined to one another by hydrogen bonds, have unusually high boiling and melting points. This is because here relatively more energy is required to separate the molecules as they enter the gaseous state or the liquid state. Thus the hydrides of fluorine (HF), oxygen (H_2O) and nitrogen (NH_3) have abnormally high boiling and melting points compared to other hydrides of the same group which form no hydrogen bonds. In Fig. 5.8 are shown the boiling points and melting points of the hydrides of VIA group elements plotted against molecular weights.

It will be noticed that there is a trend of decrease of boiling and melting points with decrease of molecular weight from H_2Te to H_2S. But there is a sharp increase in case of water (H_2O), although it has the smallest molecular weight. The reason is that the molecules of water are 'associated' by hydrogen bonds between them, while H_2Te, H_2Se and H_2S exist as single molecules since they are incapable of forming hydrogen bonds.

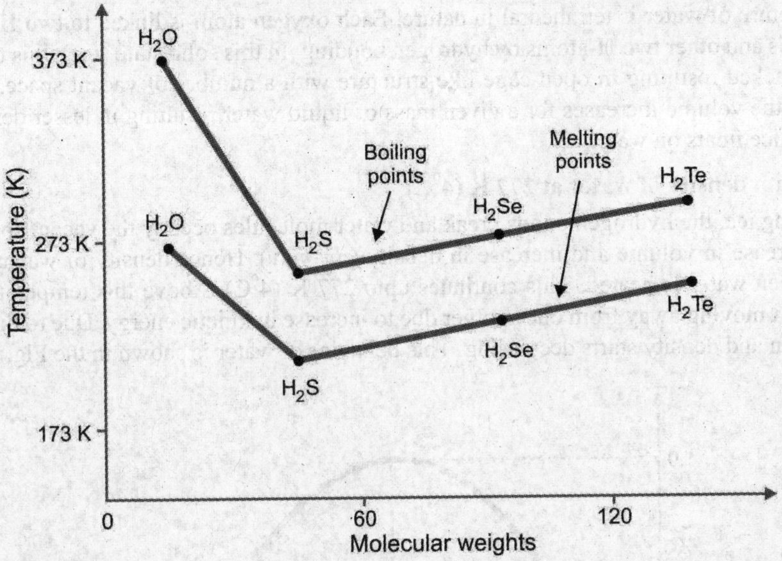

■ **Figure 5.8**
Boiling and melting point curves of the hydrides of VIA group showing abrupt increase for water (H_2O) although it has the lowest molecular weight.

(2) **High solubilities of some covalent compounds**

The unexpectedly high solubilities of some compounds containing O, N and F, such as NH_3 and CH_3OH in certain hydrogen containing solvents are due to hydrogen bonding. For example, ammonia (NH_3) and methanol (CH_3OH) are highly soluble in water as they form hydrogen bonds.

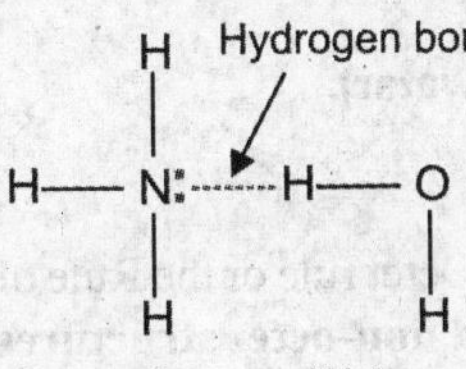

Ammonia   Water

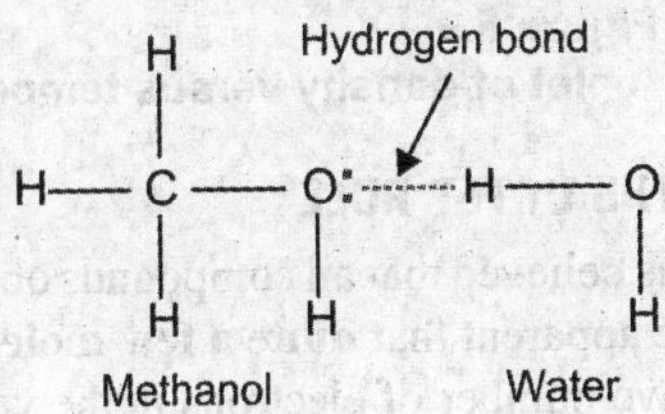

Methanol   Water

(3) **Three dimensional crystal lattice**

As already stated, **hydrogen bonds are directional and pretty strong to form three dimensional crystal lattice.** For example, in an ice crystal the water molecules (H_2O) are held together in a tetrahedral network and have the same crystal lattice as of diamond. This is so because the O atom in water has two covalent bonds and can form two hydrogen bonds. These are distributed in space like the four covalent bonds of carbon. The tetrahedral structural units are linked to other units through hydrogen bonds as shown in Fig. 5.6.

Since there is enough empty space in its open lattice structure ice is lighter than water, while most other solids are heavier than the liquid form.

Water as an Interesting Liquid

Water is very interesting solvent with unusual properties. It dissolves many ionic compounds and polar organic compounds. It has high heat of vaporisation, high heat of fusion, high specific heat with melting point 273 K and boiling point 373 K. Its structure as shown above is very interesting as it explains many properties :

(1) **Ice (Solid) is lighter than water (Liquid)**

The structure of water is tetrahedral in nature. Each oxygen atom is linked to two H-atoms by covalent bonds and other two H-atoms by hydrogen bonding. In this solid state (Ice), this tetrahedral structure is packed resulting in open cage like structure with a number of vacant space. Hence in this structure the volume increases for a given mass of liquid water resulting in lesser density. Due to this reason ice floats on water.

(2) **Maximum density of water at 277 K (4°C)**

On melting ice, the hydrogen bonds break and water molecules occupy the vacant spaces. This results in decrease in volume and increase in density ($d = m/v$). Hence density of water keeps on increasing when water is heated. This continues upto 277 K (4°C). Above this temperature water molecules start moving away from one another due to increase in kinetic energy. Due to this volume increases again and density starts decreasing. This behavior of water is shown in the Fig. 5.9.

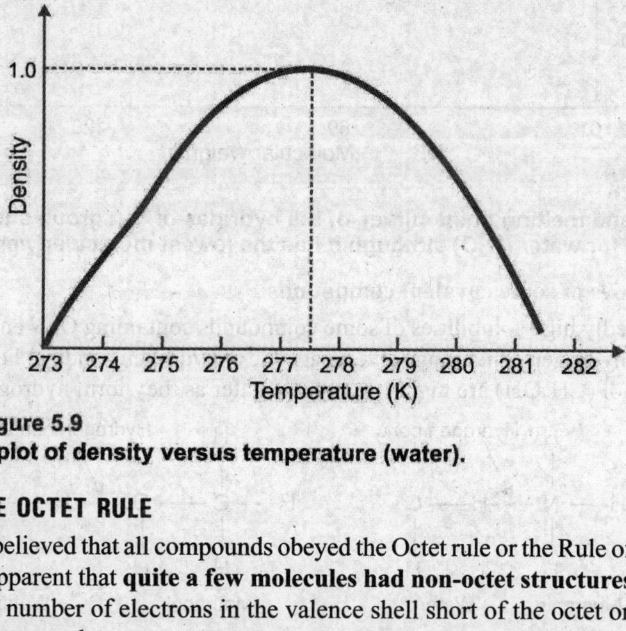

Figure 5.9
A plot of density versus temperature (water).

EXCEPTIONS TO THE OCTET RULE

For a time it was believed that all compounds obeyed the Octet rule or the Rule of eight. However, it gradually became apparent that **quite a few molecules had non-octet structures.** Atoms in these molecules could have number of electrons in the valence shell short of the octet or in excess of the octet. Some important examples are :

(1) **Four or six electrons around the central atom**

A stable molecule as of beryllium chloride, $BeCl_2$, contains an atom with four electrons in its outer shell.

(4 Electrons around Be)

The compound boron trifluoride, BF_3, has the Lewis structure :

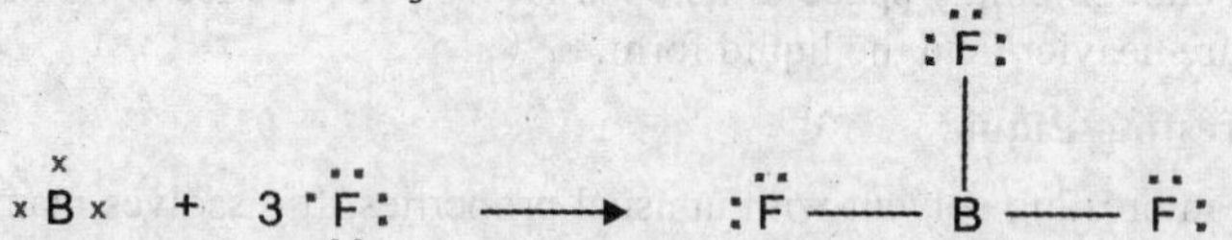

(6 Electrons around B)

The boron atom has three valence electrons.

Beryllium chloride and boron trifluoride are referred to as **electron-deficient compounds.**

(2) Seven electrons around the central atom

There are a number of relatively stable compounds in which the central atom has seven electrons in the valence shell. A simple example is chlorine dioxide, ClO_2.

Chlorine dioxide

The chlorine atom in ClO_2 has seven electrons in its outer shell.

Methyl radical (CH_3) has an odd electron and is very short lived. When two methyl free radicals collide, they form an ethane molecule (C_2H_6) to satisfy the octet of each carbon atom. **Any species with an unpaired electron is called a free radical.**

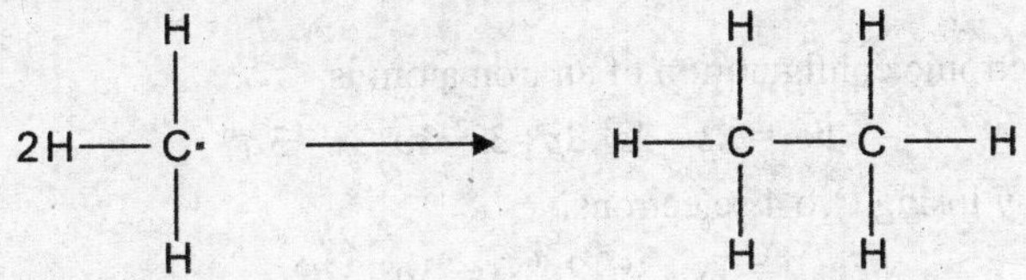

(C has 7 electrons) Ethane

(3) Ten or more electrons around the central atom

Non-metallic elements of the third and higher periods can react with electronegative elements to form structures in which the central atom has 10, 12 or even more electrons. The typical examples are PCl_5 and SF_6.

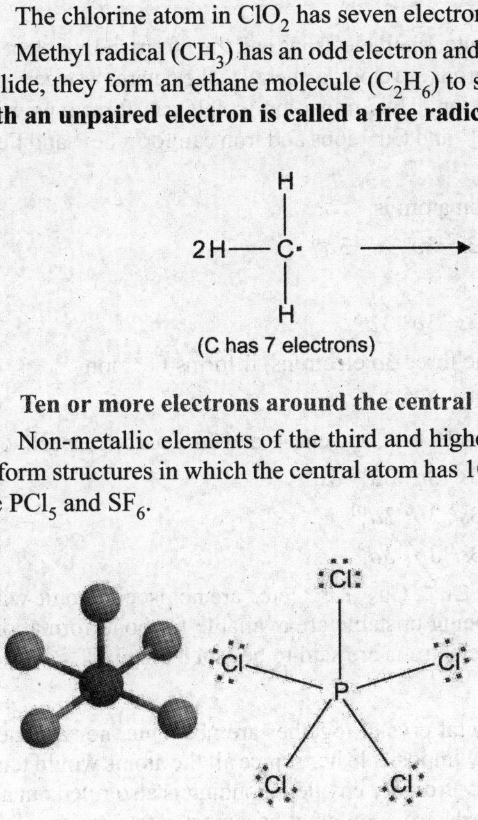

(10 Electrons about P)

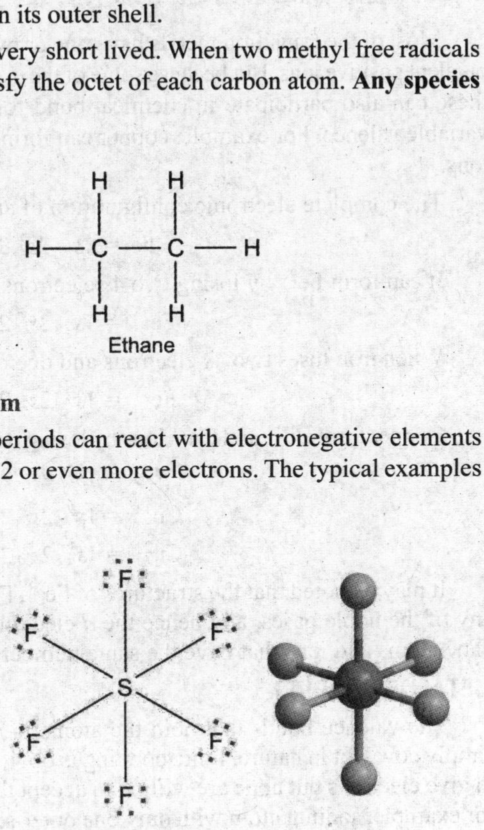

(12 Electrons about S)

The molecules with more than an octet of electrons are called **superoctet structures.**

In elements C, N, O and F the octet rule is strictly obeyed because only four orbitals are available (one 2*s* and three 2*p*) for bonding. In the elements P and S, however, 3*s*, 3*p*, and 3*d* orbitals of their atoms may be involved in the covalent bonds they form. Whenever an atom in a molecule has more than eight electrons in its valence shell, it is said to have an **expanded octet.**

VARIABLE VALENCE

Some elements can display two or more valences in their compounds. The transition metals belong to this class of elements. The Electronic Structure of some of these metals is given below :

TABLE 5.2. ELECTRONIC STRUCTURE OF THE TWO OUTER-MOST SHELLS OF SOME TRANSITION METALS

Sc	$3d^1 4s^2$	Ag	$4d^{10} 5s^1$
Cr	$3d^5 4s^1$	La	$5d^1 6s^2$
Mn	$3d^5 4s^2$	Os	$5d^6 6s^2$
Fe	$3d^6 4s^2$	Ir	$5d^7 6s^2$
Co	$3d^7 4s^2$	Pt	$5d^9 6s^2$
Cu	$3d^{10} 4s^1$	Au	$5d^{10} 6s^1$

Most of the transition metals have one or two outer-shell electrons and they form monovalent or bivalent positive ions. But because some of the *d* electrons are close in energy to the outermost electrons, these can also participate in chemical bond formation. Thus transition metals can form ions with variable valence. For example, copper can form Cu^{1+} and Cu^{2+} ions and iron can form Fe^{2+} and Fe^{3+} ions.

The complete electronic configuration of an iron atom is

$$Fe = 1s^2\ 2s^2\ 2p^6\ 3s^2\ 3p^6\ 4s^2\ 3d^6$$

It can form Fe^{2+} by losing two 4*s* electrons,

$$Fe^{2+} = 1s^2\ 2s^2\ 2p^6\ 3s^2\ 3p^6\ 3d^6$$

When iron loses two 4*s* electrons and one of the three 3*d* electrons, if forms Fe^{3+} ion

$$Fe^{3+} = 1s^2\ 2s^2\ 2p^6\ 3s^2\ 3p^6\ 3d^5$$

Copper form Cu^{1+} and Cu^{2+} ions by losing one 4*s* electron, and one 4*s* and 3*d* electron respectively

$$Cu = 1s^2\ 2s^2\ 2p^6\ 3s^2\ 3p^6\ 3d^{10}\ 4s^1$$

$$Cu^{1+} = 1s^2\ 2s^2\ 2p^6\ 3s^2\ 3p^6\ 3d^{10}$$

$$Cu^{2+} = 1s^2\ 2s^2\ 2p^6\ 3s^2\ 3p^6\ 3d^9$$

It may be noted that the structures of Fe^{2+}, Fe^{3+}, Cu^{1+}, Cu^{2+}, Cr^{3+}, etc., are not isoelectronic with any of the noble gases, and hence the *d* electrons being unstable are available for bond formation. (The atoms and ions that have the same number of electrons are said to be **Isoelectronic**).

METALLIC BONDING

The valence bonds that hold the atoms in a metal crystal together are not ionic, nor are they simply covalent in nature. Ionic bonding is obviously impossible here since all the atoms would tend to give electrons but none are willing to accept them. Ordinary covalent bonding is also ruled out as, for example, sodium atom with only one outer-shell electron could not be expected to form covalent bonds with 8 nearest neighboring atoms in its crystal. **The peculiar type of bonding which holds the atoms together in metal crystal is called the Metallic Bonding.**

Many theories have been proposed to explain the metallic bonding. Here we will discuss the simplest of these : The Electron Sea Model.

THE ELECTRON SEA MODEL

Metal atoms are characterized by :

(1) **Low ionization energies** which imply that the valence electrons in metal atoms can easily be separated.

(2) **A number of vacant electron orbitals in their outermost shell.** For example, the magnesium atom with the electron configuration $1s^2\ 2s^2\ 2p^6\ 3s^2\ 3p^0$ has three vacant 3*p* orbitals in its outer electron shell.

There is considerable overlapping of vacant orbitals on one atom with similar orbitals of adjacent atoms, throughout the metal crystal. Thus it is possible for an electron to be delocalized and move freely in the vacant molecular orbital encompassing the entire metal crystal. The delocalized electrons no longer belong to individual metal atoms but rather to the crystal as a whole.

As a result of the delocalization of valence electrons, the positive metal ions that are produced, remain fixed in the crystal lattice while the delocalized electrons are free to move about in the vacant space in between. The metal is thus pictured as a network or lattice of positive ions of the metal immersed in a 'sea of electrons' or 'gas of electrons'. This relatively simple model of metallic bonding is referred to as the **Electron Sea model** or the **Electron Gas model** (Fig. 5.10.)

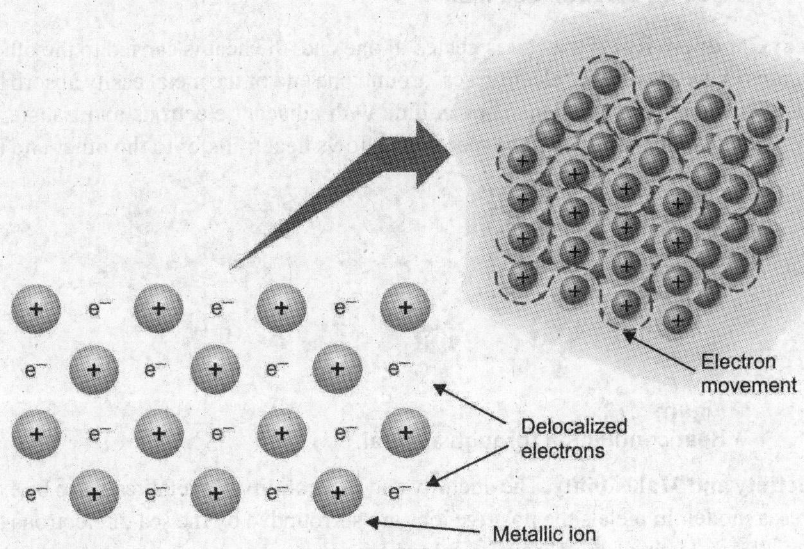

■ Figure 5.10
The Electron Sea model of metallic bonding.

A metallic bond is the electrostatic force of attraction that the neighbor positive metallic ions have for the delocalized electrons.

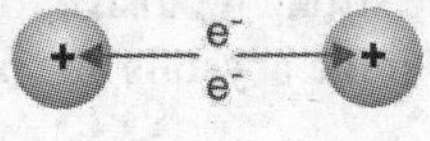

The electron sea model of metallic bonding explains fairly well the most characteristic physical properties of metals.

(1) **Luster or Reflectivity.** The delocalized mobile electrons of the 'electron sea' account for this property. Light energy is absorbed by these electrons which jump into higher energy levels and return immediately to the ground level. In doing so, the electrons emit electromagnetic radiation (light) of the same frequency. Since the radiated energy is of same frequency as the incident light, we see it as a reflection of the original light.

(2) **Electrical Conductivity.** Another characteristic of metals is that they are good conductors of electricity. According to the electron sea model, the mobile electrons are free to move through the vacant space between metal ions. When electric voltage is applied at the two ends of a metal wire, it causes the electrons to be displaced in a given direction. The best conductors are the metals which attract their outer electrons the least (low ionization energy) and thus allow them the greatest freedom of movement.

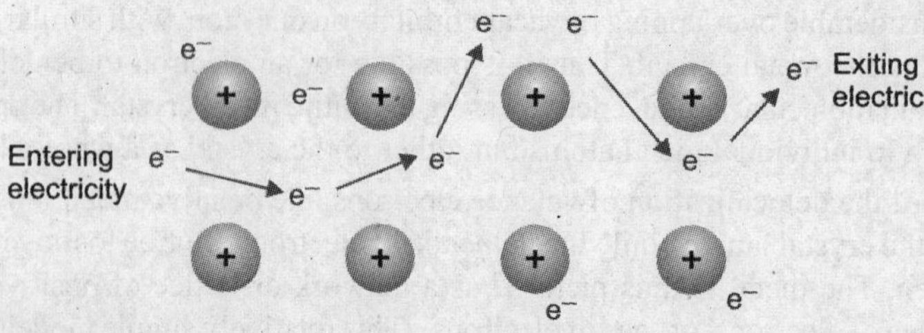

■ **Figure 5.11**
Electrical conductivity by flow of electrons based on Electron Sea model.

(3) **Heat Conductivity.** If a metal is heated at one end, the heat is carried to the other end. The mobile electrons in the area of the 'electron sea' around one end of the metal easily absorb heat energy and increase their vibrational motion. They collide with adjacent electrons and transfer the added energy to them. Thus the mobility of the electrons allows heat transfer to the other end (Fig. 5.12).

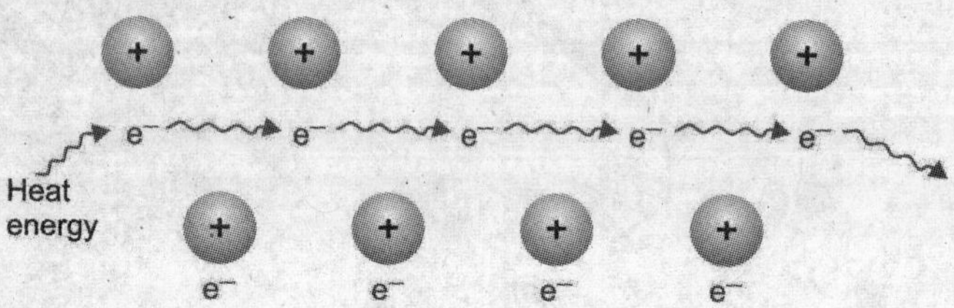

■ **Figure 5.12**
Heat conduction through a metal.

(4) **Ductility and Malleability.** The ductility and malleability of metals can also be explained by the electron sea model. In metals the positive ions are surrounded by the sea of electrons that 'flows' around them. If one layer of metal ions is forced across another, say by hammering, the internal structure remains essentially unchanged (Fig. 5.13). The sea of electrons adjusts positions rapidly and the crystal lattice is restored. This allows metals to be ductile and malleable. However, in ionic crystals of salts *e.g.*, sodium chloride, displacement of one layer of ions with respect to another brings like charged ions near to each other. The strong repulsive forces set up between them can cause the ionic crystals to cleave or shatter. **Thus ionic crystals are brittle.**

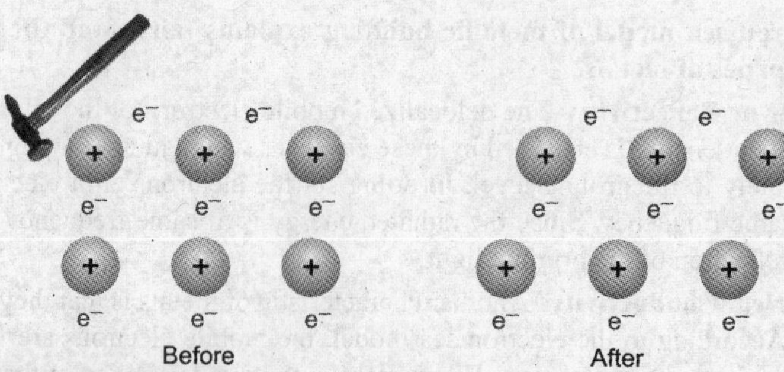

■ **Figure 5.13**
When force is applied to the upper layer of cations it slips to the right without changing the environments. Illustration of malleability and ductility.

(5) **Electron Emission.** When enough heat energy is applied to a metal to overcome the attraction between the positive metal ions and an outer electron, the electron is emitted from the metallic atom. When the frequency and, therefore, the energy of the light that strikes the metal is great enough to overcome the attractive forces, the electron escapes from the metal with a resultant decrease in the energy of the incident photon **(Photoelectric effect)**.

GEOMETRIES OF MOLECULES

So far we have depicted molecules by Lewis structures in the flat plane of paper. But **all molecules containing three or more atoms are three-dimensional.** The shape of a particular molecule is determined by the specific arrangement of atoms in it and the bond angles. Molecular shapes may be linear, bent (or angular), trigonal planar, pyramidal or tetrahedral.

The shapes of molecules can be determined in the laboratory by modern methods such as X-ray and electron diffraction techniques. Molecular shapes are important because they are helpful in the investigation of molecular polarity, molecular symmetry or asymmetry. Physical and chemical properties of compounds depend on these factors. VSEPR theory throws light on the three dimensional shapes of molecules.

VSEPR THEORY

The Lewis structure of a molecule tells us the number of pairs of electrons in the valence shell of the central atom. These electron pairs are subject to electrostatic attractions between them. On this basis, R.G.Gillespie (1970) proposed a theory called the **Valence-Shell Electron Pair Repulsion or VSEPR (pronounced as 'Vesper') theory.** It states that : **The electron pairs (both lone pairs and shared pairs, surrounding the central atom will be arranged in space as far apart as possible to minimise the electrostatic repulsion between them).**

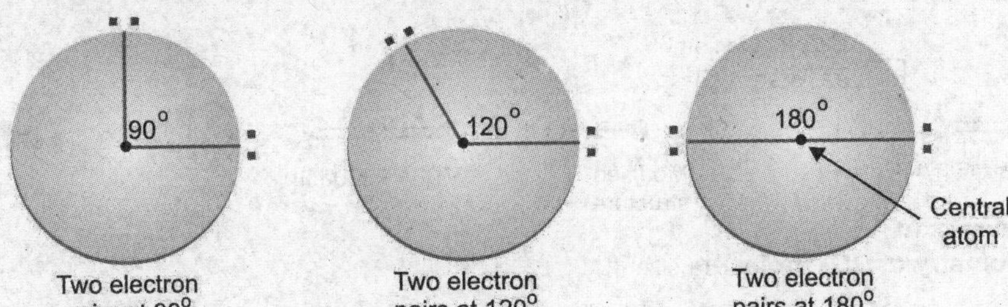

Figure 5.14
Arrangement of two electron pairs on circle at 90°, 120° and at 180°. Placement of electron pairs at 180° puts them the farthest apart, thereby minimising the electrostatic repulsion.

Let us consider the simplest case of an atom with two electron pairs. We wish to place the electron pairs on the surface of a sphere such that they will be as far apart as possible so as to minimise repulsion between them. Fig. 5.14 illustrates it by showing some possible placements of the two electron pairs. The arrangement in which the electron pair-central atom-electron pair angles is 180°, makes the electron pairs farthest apart. **This arrangement is called linear because the electron pairs and the central atom are in a straight line.**

VSEPR theory is simple but remarkably powerful model for predicting molecular geometries and bond angles. While working out the shapes of molecules from this theory, it must be remembered :

(1) **Multiple bonds behave as a single electron-pair bond** for the purpose of VSEPR. They represent a single group of electrons.

(2) **Order of repulsions** between lone pair and lone pair (*lp-lp*), lone pair and bonding pair (*lp-bp*), and bonding pair and bonding pair (*bp-bp*) is

$$lp - lp >> lp - bp > bp - bp$$

When a molecule has lone pairs of electrons, the bonding electron pairs are pushed closer and thus the bond angle is decreased.

Now we proceed to work out the shapes of some common molecules with the help of VSEPR theory.

(1) **Linear Molecules**

(*a*) **Beryllium chloride, $BeCl_2$.** It has the Lewis structure

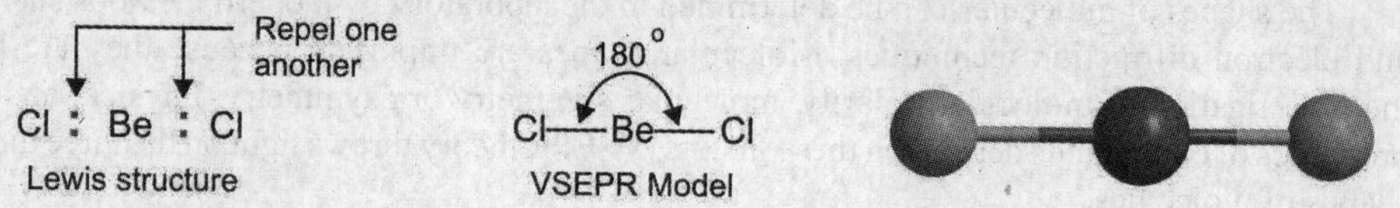

Figure 5.15
Geometry of $BeCl_2$ molecule.

The central atom Be has two bonding electron pairs and no unshared electron. According to VSEPR theory, the bonding pairs will occupy positions on opposite sides of Be forming an angle of 180°. An angle of 180° gives a straight line. Therefore, $BeCl_2$ molecule is linear. In general, **all molecules as A–B–A which have only two bonds and no unshared electrons are linear.**

(*b*) **Carbon dioxide, CO_2.** It has the structure

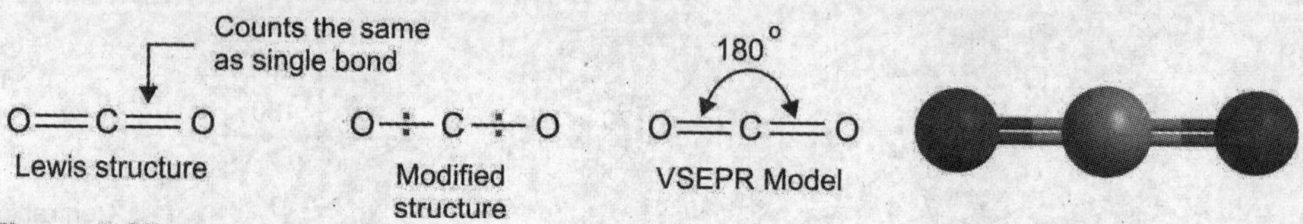

Figure 5.16
Geometry of CO_2 molecule.

The central C atom has no unshared electron. We know that a double bond counts the same as a single bond in VSEPR model. Thus CO_2 is a linear molecule.

Similarly, it can be shown that hydrogen cyanide (H – C ≡ N) and acetylene (H – C ≡ C – H) are linear molecules.

(2) **Trigonal Planar Molecules**

(*a*) **Boron trifluoride, BF_3.** Its Lewis structure shows that the central atom B has three bonding electron pairs and no unshared electrons. VSEPR theory says that the three bonding electron pairs will be as far apart as possible. This can be so if these electron pairs are directed to the corners of an equilateral triangle. Thus VSEPR model of BF_3 molecule has three F atoms at the corners of the triangle with B atom at its center. All the four atoms (three *F* and one B) lie in the same plane. Therefore, the shape of such a molecule is called **trigonal planar.** The bond angle is 120°.

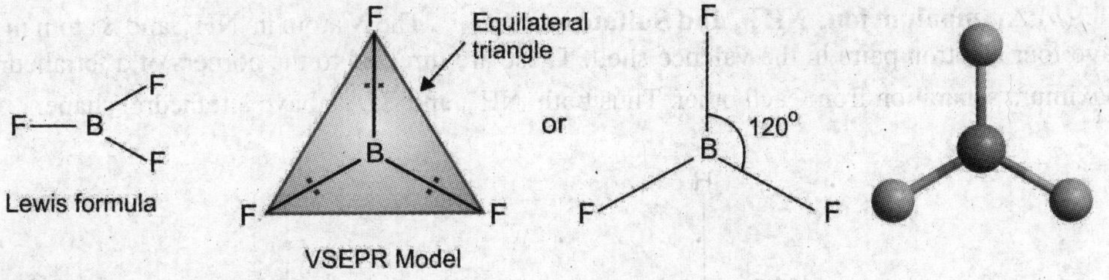

■ **Figure 5.17**
Geometry of BF_3 molecule.

(*b*) **Sulfur trioxide, SO_3.** In the Lewis structure of SO_3, the central S atom is joined with two O atoms by covalent bonds. The third O atom is joined with S by a double bond. But a double bond is counted as a single electron pair for the purpose of VSEPR model. Therefore, in effect, S has three electron pairs around it. Thus like BF_3, SO_3 has trigonal planar geometry.

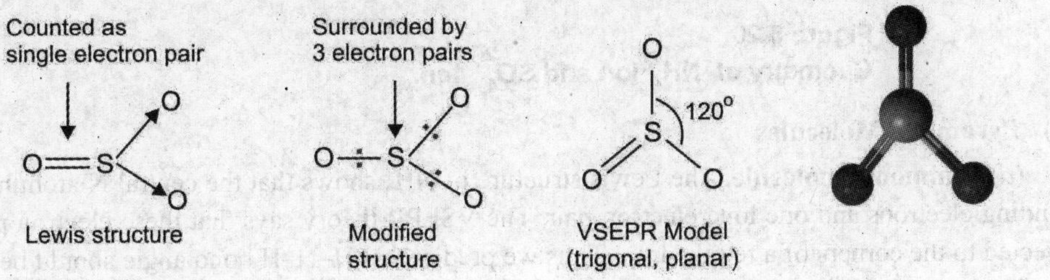

■ **Figure 5.18**
Geometry of SO_3 molecule.

(3) **Tetrahedral Molecules**

(*a*) **Methane, CH_4.** Lewis structure of methane shows that the central C atom has four bonding electron pairs. These electron pairs repel each other and are thus directed to the four corners of a regular tetrahedron. A regular tetrahedron is a solid figure with four faces which are equilateral triangles. All bond angles are 109.5°.

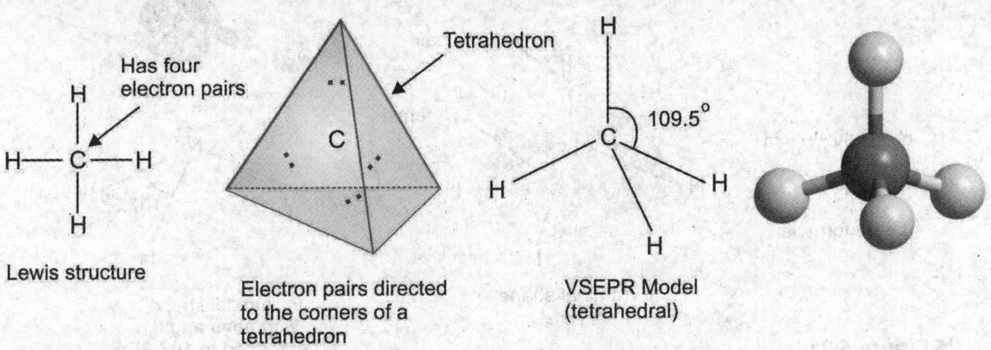

■ **Figure 5.19**
Geometry of CH_4 molecule.

Similarly, CCl_4 in which the central C atom is bonded to four other atoms by covalent bonds has tetrahedral shape.

(*b*) **Ammonium ion, NH_4^+, and Sulfate ion, SO_4^{2-}.** The N atom in NH_4^+ and S atom in SO_4^{2-} have four electron pairs in the valence shell. These are directed to the corners of a tetrahedron for maximum separation from each other. Thus both NH_4^+ and SO_4^{2-} have tetrahedral shape.

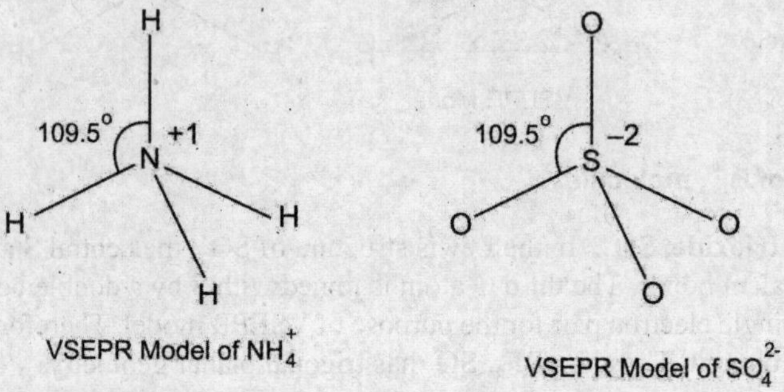

■ Figure 5.20
Geometry of NH_4^+ ion and SO_4^{2-} ion.

(4) **Pyramidal Molecules**

(*a*) **Ammonia molecule.** The Lewis structure of NH_3 shows that the central N atom has three bonding electrons and one lone electron pair. The VSEPR theory says that these electron pairs are directed to the corners of a tetrahedron. Thus we predict that H–N–H bond angle should be 109.5°. But the shape of a molecule is determined by the arrangement of atoms and not the unshared electrons. Thus, if we see only at the atoms, we can visualize NH_3 molecule as a pyramid with the N atom located at the apex and H atoms at the three corners of the triangular base.

According to VSEPR theory, **a lone pair exerts greater repulsion on the bonding electron pairs than the bonding pairs do on each other.** As a result, the bonds of NH_3 molecule are pushed slightly closer. This explains why the observed bond angle H–N–H is found to be 107.3° instead of 109.5° predicted from tetrahedral geometry.

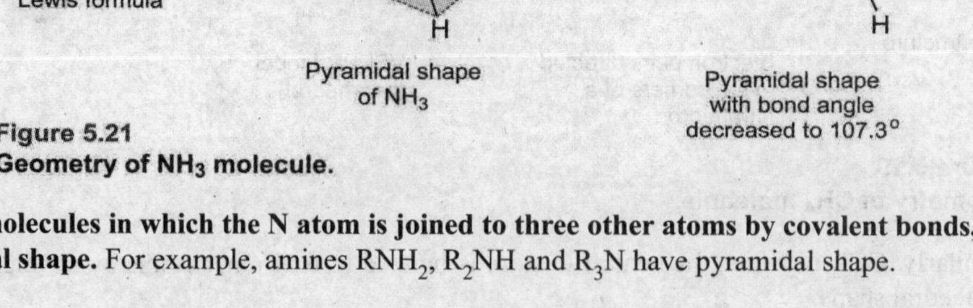

■ Figure 5.21
Geometry of NH_3 molecule.

All molecules in which the N atom is joined to three other atoms by covalent bonds, have pyramidal shape. For example, amines RNH_2, R_2NH and R_3N have pyramidal shape.

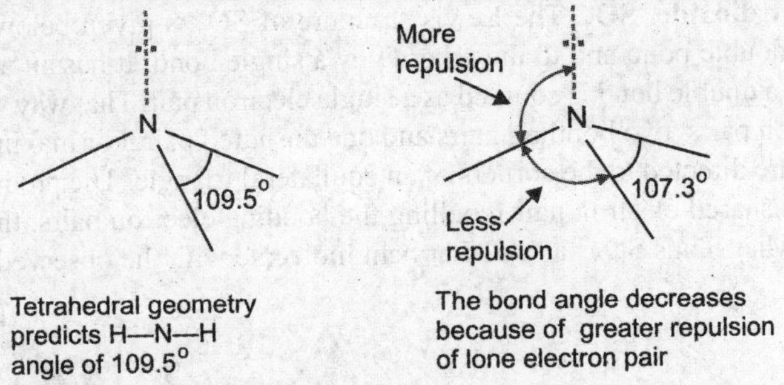

■ **Figure 5.22**
Why the angle H—N—H in NH_3 molecule is 107.3° while the tetrahedral angle is 109.5°?

(*b*) **Phosphorus trichloride, PCl_3.** The structural formula indicates that the central phosphorus atom has three bonding electron pairs and one lone electron pair. Thus, like NH_3 it has pyramidal shape and the observed bond angle Cl–P–Cl is 100°.

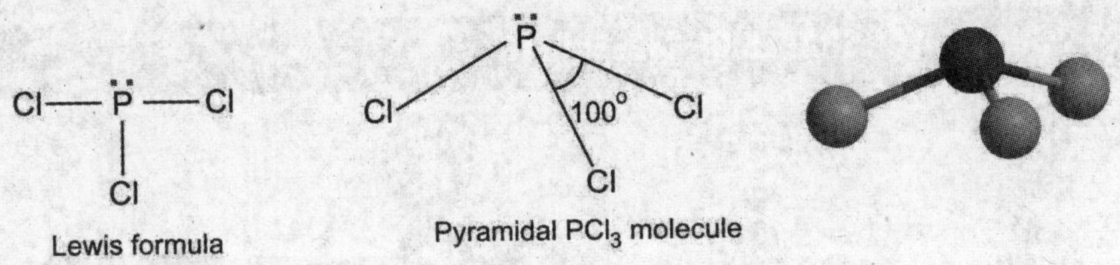

■ **Figure 5.23**
Geometry of PCl_3 molecule.

(5) **Bent or Angular Molecules**

(*a*) **Water, H_2O.** In the structural formula of H_2O, the O atom is bonded to two H atoms by covalent bonds and has two lone pairs. Thus O is surrounded by two bonding electron pairs

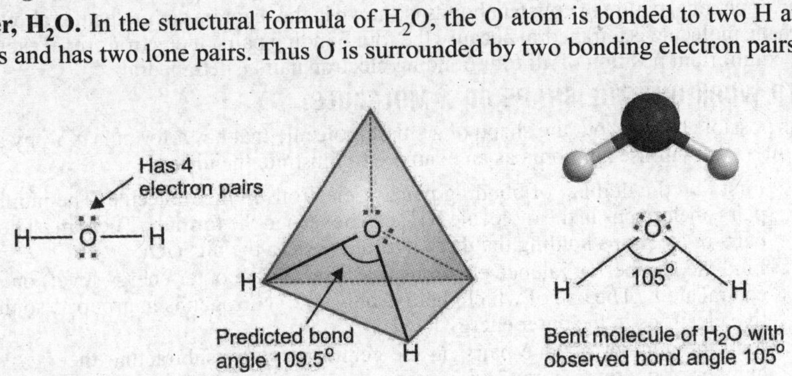

■ **Figure 5.24**
Geometry of H_2O molecule.

and two unshared electron pairs. VSEPR theory says that in order to secure maximum separation between them, the four electron pairs are directed to the corners of a tetrahedron. If we look at the three atoms (and ignore the unshared pairs), the atoms HOH lie in the same plane and the predicted bond angle is 109.5°. But with two unshared pairs repelling the bonding pairs, the bond angle is compressed to 105°, the experimental value. Thus the H_2O molecule is flat and bent at an angle at the O atom. Such a molecule is called a **bent molecule** or **angular molecule**.

(*b*) **Sulfur dioxide, SO_2.** The Lewis structure of SO_2 is given below. The S atom is bonded to one O by a double bond and to the other O by a single bond. It has an unshared electron pair. In VSEPR model a double bond is counted as a single electron pair. That way, the S atom is surrounded by three electron pairs, two bonding pairs and one unshared pair. For maximum separation the three electron pairs are directed to the corners of an equilateral triangle. The predicted bond angle is 120°. But with the unshared electron pair repelling the bonding electron pairs, the bond angle is actually reduced somewhat. Thus SO_2 has a planar bent molecule with the observed bond angle 119.5°.

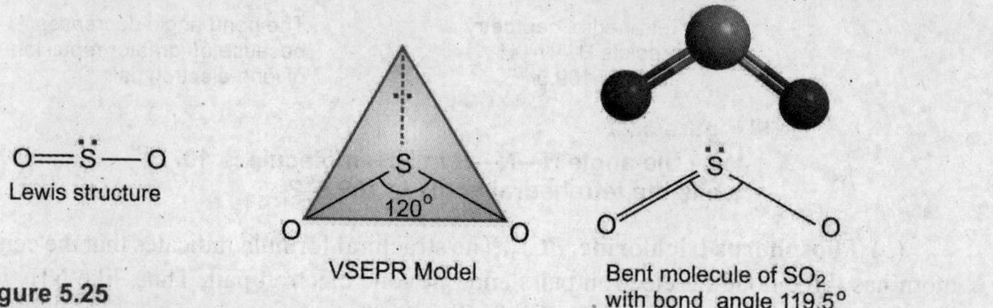

■ Figure 5.25
Geometry of SO_2 molecule.

SUMMARY : SHAPES OF MOLECULES

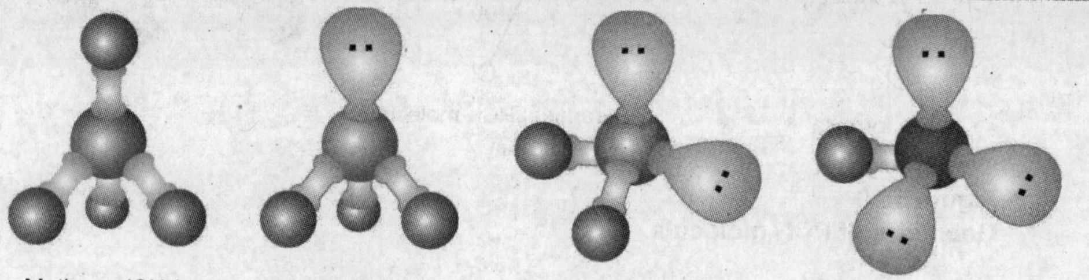

The directional nature of covalent bonds is shown in the diagrams of molecules above. The shape of the methane molecule is tetrahedral because the four bonding pairs of electrons repel each other equally, and the equilibrium position of all four bonding electron pairs is tetrahedral.

HOW TO WORK OUT THE SHAPE OF A MOLECULE?

It is possible to work out the shape of a small molecule that has a formula XY_n by applying a few simple rules. We will use ammonia as an example to illustrate the idea.

Rule 1 First find the number of bonding pairs of electrons in the molecule. The number of bonding pairs of electrons in the molecule NH_3 can be seen in the formula. There must be three bonding pairs of electrons holding the three hydrogens onto the nitrogen.

Rule 2 Find the number of valence electrons (electrons in the outer energy level) on an atom of the central atom (The one of which there is only one.) Nitrogen is in group V, so the nitrogen has five electrons in the outer energy level.

Rule 3 Find the number of lone pairs on the central atom by subtracting the number of bonding pairs (3) from the valence electrons (5) to find the number of electrons (2) that will make up lone pairs of electrons. Divide this number by 2 to find the number of lone pairs, 2/2 = 1.

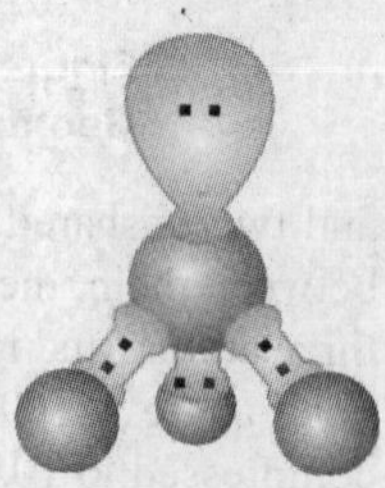

Rule 4 Distribute all the electron pairs around the central atom and learn the angles they will make from molecules with no lone pairs.

Rule 5 Learn that the repulsion between lone pairs of electrons is greater than the repulsion between bonding pairs, and subtract 2° from the bond angles for every lone pair.

Rule 6 Learn the names of the shapes. The shapes are named from the position of the atoms and not the position of the orbitals.

TABLE OF SHAPES

Formula	$BeCl_2$ Beryllium chloride	BCl_3 Boron trichloride	CH_4 Methane	NH_3 Ammonia	H_2O Water
Bonding Pairs	2	3	4	3	2
Valence Electrons	2	3	4	5	6
Lone Pairs	0	0	0	1	2
Angles between bonding pairs	180°	120°	109.5°	107°	105°
Name of shape	Linear	Trigonal Planar	Tetrahedral	Trigonal Pyramid	Bent
		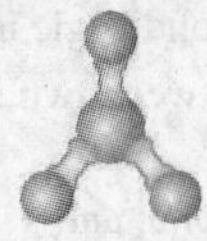		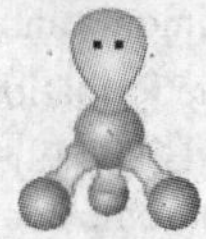	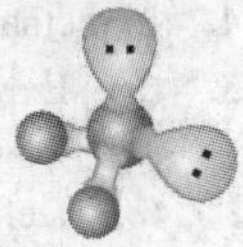

There is one more rule to learn, and it concerns the shape of polyatomic ions.

Rule 2(a) If the molecule is an ion, e.g. ammonium $(NH_4)^+$, subtract 1 from the number of valence electrons for every + charge on the ion and add 1 to the valence number for every - charge, then proceed as before.

SOME MORE EXAMPLES

Formula	NH_4^+	PCl_5	SF_6	XeF_4	ICl_3
Bonding Pairs	4	5	6	4	3
Valence Electrons	5	5	6	8	7
Rule 2(a)	5 – 1 = 4				
Lone Pairs	0	0	0	2	2
Angles between bonding pairs	109.5°	90° and 120°	90°	90°	90°
Name of shape	Tetrahedral	Trigonal Bipyramid	Octahedron	Square	T shape
		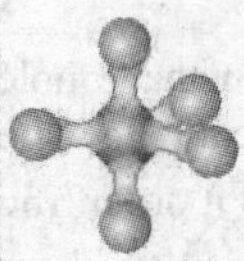	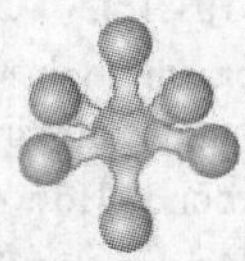	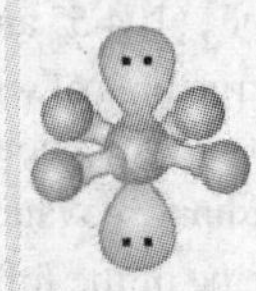	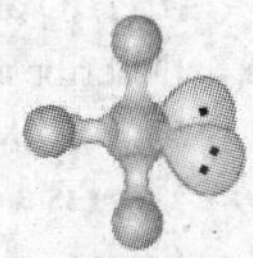

EXAMINATION QUESTIONS

1. Define or explain the following terms :
   (*a*) Octet rule (*b*) Ionic bond
   (*c*) Covalent bond (*d*) Coordinate covalent bond
   (*e*) Polar covalent bond (*f*) Hydrogen bonding
   (*g*) Intermolecular H-bonding (*h*) Intramolecular H-bonding
   (*i*) VSEPR theory
2. (*a*) Compare the properties of ionic and covalent compounds.
   (*b*) State whether the following compounds are ionic or covalent?
   (*i*) $AlCl_3$ (*ii*) HgF_2
3. (*a*) Draw the structure of NaCl crystal and give the coordination number of Na^+.
   (*b*) Which of the two is more covalent and why in the following pairs?
   (*i*) AgCl and AgI (*ii*) LiCl and KCl
4. In methane, ammonia and water molecule the bond angle is decreasing. Explain giving reasons.
5. (*a*) Explain the formation of covalent bond between two atoms of chlorine in a chlorine molecule on the basis of octet rule.
   (*b*) Define (*i*) Ionic bond; (*ii*) Co-ordinate bond; and (*iii*) Metallic bond
6. (*a*) What do you understand by 'Stable configuration'? What are the ways by which an atom can attain stable configuration?
   (*b*) Write the electronic configuration of any two of the following compounds :
   (*i*) Phosphorus pentachloride
   (*ii*) Sulfuric acid
   (*iii*) Lithium fluoride
7. What type of bonds do you expect in the following cases? Give reasons :
   (*i*) between a very small cation and a large anion.
   (*ii*) between atoms having a very large difference in electronegativities,
   (*iii*) between atoms of the same element.
8. Explain qualitatively the valence bond theory with reference to Hydrogen molecule.
9. Compare the properties of ionic and covalent compounds. Give two examples of each type of compounds.
10. Indicate the type of bonding that exists in the following solids :
    (*i*) Ice (*ii*) Naphthalene
    (*iii*) Diamond (*iv*) Potassium chloride
11. Write Lewis dot formulae of : (*a*) HOCl (*b*) BF_3 (*c*) NH_4^+.
12. Show the formation of a coordinate bond in ozone molecule and discuss briefly the electron gas model of the metallic bond and how it explains the electrical conductivity of metals.
13. What is electronegativity? How is the concept of electronegativity used to predict the bond types between hetero atoms?
14. Account for the variation of bond angles between the pairs (*i*) H_2O and H_2S (104.5° and 92°) and (*ii*) H_2O and OF_2 (104.5° and 101.1°).
15. Describe the structures of water, ammonia and methane molecules in terms of the electron pair repulsion theory. Explain why the bond angles are different in the three molecules.
16. What is a coordinate covalent bond? How does it differ from a normal covalent bond?
17. Discuss the shape of the following molecules on the basis of VSEPR theory :

$$NH_3,\ CH_4,\ PCl_3$$

18. Explain the formation of NH_3 molecule if no hybridization of *s* and *p*-orbitals of nitrogen is assumed. Give diagrammatic representation also.

19. Explain :
(*a*) The structure of H_2 molecule according to V.B. theory.
(*b*) Ionic bond and Metallic bond.

20. Two elements *X* and *Y* occur in the same period and their atoms have two and seven valence electrons respectively. Write down the electronic structure of the most probable compound between *X* and *Y*. Will the bond between *X* and *Y* be predominantly ionic or covalent?
Answer. XY_2; Ionic

21. Using VSEPR theory, identify the type of hybridization and draw the structure of OF_2. What are the oxidation states of O and F?

22. Account for : The experimentally found N——F bond length in NF_3 is greater than the sum of single covalent radii of N and F.

23. Which of the following compounds contain bonds that are predominantly ionic in character : MgO, Ca_3P_2, $AlCl_3$, Mg_2Si and CsF.
Answer. CsF, Mg_2Si and MgO

24. Classify the bonds in the following as ionic, polar covalent or covalent : (*a*) HCl (*b*) NaCl and (*c*) NCl_3.
Answer. HCl - Polar covalent, NaCl - Ionic and NCl_3 - Covalent

25. Predict the geometry of the following molecules using VSEPR theory.
(*a*) CCl_4 (*b*) $AlCl_3$
(*c*) H_2Se
Answer. (*a*)Tetrahedral (*b*) Trigonal planar (*c*) Bent

26. Predict the geometry of the following ions having VSEPR model.
(*a*) H_3O^+ (*b*) NO_2^-
(*c*) ClO_2^-
Answer. (*a*) Pyramidal (*b*) Bent (*c*) Bent

27. Calculate the percentage ionic character of C—Cl bond in CCl_4, if the electronegativities of C and Cl are 3.5 and 3.0 respectively.
Answer. 8.875%

28. The experimentally determined dipole moment, *m*, of KF is 2.87×10^{-29} coulomb meter. The distance, *d*, separating the centers of charge in a KF dipole is 2.66×10^{-3} m. Calculate the percent ionic character of KF.
Answer. 67.4%

29. The dipole moment of KCl is 3.336×10^{-29} coulomb meter which indicates that it is highly polar molecule. The interionic distance between K^+ and Cl^- in this molecule is 2.6×10^{-10} m. Calculate the dipole moment of KCl molecule if there were opposite charges of one fundamental unit localized at each nucleus. Calculate the percentage ionic character of KCl.
Answer. 80%

30. What is meant by an ionic bond? What are the conditions necessary for the formation of an ionic bond? (*Agra BSc*)

31. Describe the basic ideas of the VSEPR theory. Explain the application of the theory for predicting the shapes of the molecules, BCl_3, NH_3, H_2O and SF_6. (*Delhi BSc*)

32. (*a*) What are electrovalent compounds? Discuss various factors which affect the formation of these compounds.
(*b*) What do you understand by hydrogen bonds? Classify them with examples. Explain why water has abnormally high boiling point. (*Baroda BSc*)

33. Why bond angles of H_2O and NH_3 are 104.5° and 107° respectively although central atoms are sp^3 hybridized? (*Aligarh BSc*)

34. Define Lattice energy. Discuss the factors on which it depends.

35. (*a*) Why melting and boiling points of ionic compounds are usually higher than covalent compounds?

(*b*) Discuss the geometry and shape of PF_5 molecule.

(*c*) Write a short note on hydrogen bonding. (*Arunachal BSc*)

36. Each of the concepts of covalency and electrovalency relates to an idealized state of chemical bonding which often does not exist in real compounds. Discuss how far this statement is valid and give two examples with suitable explanation of cases where such non-ideality infact arises. (*DelhiBSc*)

37. Strength of hydrogen bond in H–F is more than in H_2O but still HF is a gas and H_2O is a liquid at room temperature. Explain. (*Delhi BSc*)

38. (*a*) The bond angle ∠HNH in ammonia is 107° while bond angle∠HOH in water is about 104°. Why?

(*b*) A covalent bond is stronger than a metallic bond. Why? (*Agra BSc*)

39. Explain intermolecular and intramolecular hydrogen bonding with one example for each. (*Baroda BSc*)

40. Based on metallic bond, explain why metals are :

(*a*) good conductors of electricity (*b*) malleable and ductile

(*c*) having characteristic lustre (*Sambalpur BSc*)

41. The experimentally determined dipole moment, μ, of KF is 2.87×10^{-29} coulomb meter. The distance, *d*, separating the centers of charge in a KF dipole is 2.66×10^{-3} m. Calculate the percent ionic character of KF. (*Madurai BSc*)

Answer. 67.4%

42. Predict the geometry of the following molecules using VSEPR theory.

(a) CCl_4 (b) $AlCl_3$ (c) H_2Se. (*Baroda BSc*)

Answer. (a) Tetrahedral (b) Trigonal planar (c) Bent

43. What geometry is expected for the following molecules according to the VSEPR theory?

(a) PF_5 (b) SCl_4 (c) BrF_3 (*Baroda BSc*)

Answer. (a) Trigonal bipyramidal (b) See-saw (c) T-Shaped

MULTIPLE CHOICE QUESTIONS

1. The valency of an element is

(*a*) the combining capacity of one atom of it

(*b*) the number of bonds formed by its one atom

(*c*) the number of hydrogen atoms that combine with one atom of it

(*d*) all the above

Answer. (*d*)

2. The octet rule is

(*a*) the tendency of atoms to have eight electrons in the outermost shell

(*b*) the tendency of atoms to have eight pairs of electrons in the valence shell

(*c*) the tendency of the molecule to have a total of eight electrons

(*d*) the tendency of atoms to have eight non-bonding electrons

Answer. (*a*)

3. An ionic bond is formed between

(*a*) two metal atoms (*b*) two non-metal atoms

(*c*) one metal atom and one non-metal atom (*d*) one metal atom and one metalloid atom

Answer. (*c*)

4. Factors governing the formation of an ionic bond are
   (*a*) low ionization energy of metal and high electron affinity of non-metal atom
   (*b*) high ionization energy of metal and high electron affinity of non-metal atom
   (*c*) low ionization energy of metal atom and low electron affinity of non-metal atom
   (*d*) high ionization energy of metal and low electron affinity of non-metal atom
   Answer. (*a*)
5. The lattice energy is the amount of energy that
   (*a*) is released when one cation combines with one anion
   (*b*) is released when one mole of cations combine with one mole of anions
   (*c*) is released when one mole of an ionic compound is formed from its cations and anions
   (*d*) is absorbed when one mole of an ionic compound is formed from its cation and anions
   Answer. (*c*)
6. The most favorable conditions for the formation of an ionic compound is
   (*a*) low charge on ions, small cation and small anion
   (*b*) high charge on ions, large cation and large anion
   (*c*) high charge on ions, small cation and large anion
   (*d*) low charge on ions, large cation and small anion
   Answer. (*c*)
7. Ionic compounds are generally
   (*a*) solids having large melting points and good conductors of electricity
   (*b*) gases having low melting points and poor conductors of electricity
   (*c*) solids having low melting points and good conductors of electricity
   (*d*) solids having high melting points and bad conductors of electricity
   Answer. (*a*)
8. A covalent bond involves
   (*a*) sharing of electrons between a metal and a non-metal atom
   (*b*) sharing of electrons between two metal atoms
   (*c*) sharing of electrons between two atoms having similar electronegativity
   (*d*) sharing of electrons between two atoms having a large difference in electronegativity
   Answer. (*c*)
9. The total number of electron pairs in a nitrogen molecule is
   (*a*) 2 (*b*) 3 (*c*) 5 (*d*) 7
   Answer. (*d*)
10. The covalent compounds are soluble in
   (*a*) all acids (*b*) all bases (*c*) all solvents (*d*) non-polar solvents
   Answer. (*d*)
11. The compounds which contain both ionic and covalent bonds are
   (*a*) $CHCl_3$ and CCl_4 (*b*) KCl and $AlCl_3$ (*c*) KCN and NaOH (*d*) H_2 and CH_4
   Answer. (*c*)
12. A coordinate bond is formed by
   (*a*) complete transfer of electrons
   (*b*) sharing of electrons contributed by both the atoms
   (*c*) sharing of electrons contributed by one atom only
   (*d*) none of these
   Answer. (*c*)
13. The types of bonds present in sulfuric acid molecules are
   (*a*) only covalent (*b*) ionic and covalent
   (*c*) coordinate and covalent (*d*) coordinate, covalent and ionic
   Answer. (*d*)
14. The common feature among the species O_3, SO_4^{2-}, H_3O^+ and $AlCl_3$ is that
   (*a*) they contain only ionic bonds (*b*) they contain only covalent bonds
   (*c*) they contain coordinate bond (*d*) they contain covalent and ionic bonds
   Answer. (*c*)

15. The species CO, CN^- and N_2 are
(*a*) isoelectronic (*b*) having coordinate bond
(*c*) having low bond energies (*d*) having polar bonds
Answer. (*a*)

16. The polarity of a covalent bond is due to
(*a*) lesser electronegativity difference between two atoms
(*b*) greater electronegativity difference between two atoms
(*c*) lesser bond energy
(*d*) greater bond energy
Answer. (*b*)

17. A CO_2 molecule contains two polar bonds but the net dipole moment is zero. It is because
(*a*) the molecule has symmetrical linear geometry
(*b*) the molecule is non-linear
(*c*) the electronegativity difference between the two atoms is too large
(*d*) the electronegativity difference between the two atoms is too small
Answer. (*a*)

18. Among BeF_2, BF_3, NH_3 and CCl_4, the molecule with net dipole moment is
(*a*) BeF_2 (*b*) BF_3 (*c*) NH_3 (*d*) CCl_4
Answer. (*c*)

19. The common feature among the molecules HF, H_2O, HCl and NH_3 is
(*a*) intramolecular H-bonding (*b*) intermolecular H-bonding
(*c*) that they contain no polar bonds (*d*) that their dipole moment is zero
Answer. (*b*)

20. Methanol is soluble in water due to
(*a*) covalent bond nature (*b*) ionic bond nature
(*c*) hydrogen bonding (*d*) its poisonous nature
Answer. (*c*)

21. Among H_2O, H_2S, H_2Se and H_2Te, the substance with highest boiling point is
(*a*) H_2O; due to hydrogen bonding
(*b*) H_2S; due to large size of S atom
(*c*) H_2Se; due to large electronegativity difference
(*d*) H_2Te; due to largest size of Te atom
Answer. (*a*)

22. In ice crystal, the H_2O molecules are held together in a
(*a*) planar structure (*b*) linear structure
(*c*) tetrahedral three dimensional structure (*d*) none of these
Answer. (*c*)

23. The density of ice (solid) is lesser than that of water (liquid) because it has
(*a*) open cage like structure with no empty spaces
(*b*) open cage like structure with large empty spaces
(*c*) intermolecular H-bonding
(*d*) intramolecular H-bonding
Answer. (*b*)

24. The density of water is maximum at
(*a*) 273 K (*b*) 277 K (*c*) 281 K (*d*) 285 K
Answer. (*b*)

25. Among $BeCl_2$, $CHCl_3$, CCl_4 and PCl_5, the octet rule is not observed in
(*a*) $BeCl_2$ only (*b*) PCl_5 only (*c*) $BeCl_2$ and PCl_5 (*d*) $CHCl_3$ and CCl_4
Answer. (*c*)

26. An example of electron deficient compound among BF_3, CF_4, PF_5 and SF_6 is
(*a*) BF_3 (*b*) CF_4 (*c*) PF_5 (*d*) SF_6
Answer. (*a*)

27. The transition metals show variable valency because of
(*a*) the availability of vacant *d*-orbitals
(*b*) their tendency to form complex ions
(*c*) their ability to form colored ions
(*d*) none of these
Answer. (*a*)

28. The electrical conductivity of metals is due to
(*a*) mobile protons is the nucleus (*b*) mobile nucleus in the nucleus
(*c*) mobile electrons in outer vacant spaces (*d*) none of these
Answer. (*c*)

29. According to VSEPR theory,
(*a*) the lone pairs only decide the structure of the molecule
(*b*) the bond pairs only decide the structure of the molecule
(*c*) the lone pairs and bond pairs both decide the structure of the molecule
(*d*) none of these
Answer. (*c*)

30. In which of the following, the central atom is surrounded by four electron pairs
(*a*) H_2O (*b*) NH_3 (*c*) CH_4 (*d*) All
Answer. (*d*)

31. The molecule among CCl_4, PCl_3, SF_4 and NH_3 that does not contain lone pairs of electrons around the central atom is
(*a*) CCl_4 (*b*) PCl_3 (*c*) SF_4 (*d*) NH_3
Answer. (*a*)

32. Which of the following are isostructural
(*a*) SO_2 and CO_2 (*b*) SO_2 and H_2O (*c*) BCl_3 and $CHCl_3$ (*d*) NH_3 and CH_4
Answer. (*b*)

33. The molecular shapes of H_2O, NH_3 and CH_4 are
(*a*) similar with 2, 1 and 0 lone pairs of electrons respectively
(*b*) similar with 0, 1 and 2 lone pairs of electrons respectively
(*c*) different with 0, 1 and 2 lone pairs of electrons respectively
(*d*) different with 2, 1 and 0 lone pairs of electrons respectively
Answer. (*d*)

34. The molecule of NH_3 is
(*a*) tetrahedral with bond angle 109° 28′
(*b*) pyramidal with bond angle 107° 20′
(*c*) trigonal with bond angle 120°
(*d*) linear with bond angle 180°
Answer. (*b*)

35. The NH_4^+ and SO_4^{2-} ions have
(*a*) tetrahedral geometry (*b*) triangular geometry
(*c*) pyramidal geometry (*d*) square planar geometry
Answer. (*a*)

36. Which is incorrect?
(*a*) all molecules with polar bonds have dipole moment
(*b*) all molecules with polar bonds may or may not have dipole moment
(*c*) the greater the difference in electronegativity between two atoms, greater is the polarity
(*d*) if the electronegativity difference between two atoms is greater than 1.7, the bond will be ionic
Answer. (*a*)

37. The favorable conditions for the formation of H–bonding are

(*a*) high electronegativity and small size of the atom bonded to H–atom
(*b*) low electronegativity and large size of the atom bonded to H–atom
(*c*) high electronegativity and large size of the atom bonded to H–atom
(*d*) low electronegativity and small size of the atom bonded to H–atom

Answer. (*a*)

38. The strength of hydrogen bonding lies in between

(*a*) covalent and ionic bond (*b*) metallic and covalent bond
(*c*) van der Waal's and covalent bond (*d*) metallic and ionic bond

Answer. (*d*)

39. The bond angles in a trigonal bipyramid molecules are

(*a*) 90° (*b*) 120° (*c*) 109.5° (*d*) 120°, 90°

Answer. (*d*)

40. CO_2 has zero dipole moment whereas H_2O has a dipole moment. It is because

(*a*) H_2O is linear while CO_2 is a bent molecule
(*b*) of intermolecular H–bonding in H_2O molecules
(*c*) CO_2 is linear while H_2O is a bent molecule
(*d*) CO_2 is a gas while H_2O is a liquid at room temperature

Answer. (*c*)

41. Which of the following does not obey the octet rule?

(*a*) PCl_5 (*b*) H_2O (*c*) NH_3 (*d*) CCl_4

Answer. (*a*)

42. The total number of electrons that take part in forming bonds in O_2 is

(*a*) 2 (*b*) 4 (*c*) 6 (*d*) 8

Answer. (*d*)

43. CO is isoelectronic with

(*a*) C_2H_2 (*b*) CN^- (*c*) O_2^+ (*d*) O_2^-

Answer. (*b*)

44. CO_2 is isostructural with

(*a*) H_2O (*b*) NO_2 (*c*) H_2S (*d*) C_2H_2

Answer. (*d*)

45. In a bond between two atoms X and Y, the shared electron pair does not lie in the center. The bond is

(*a*) single bond (*b*) non-polar bond (*c*) polar bond (*d*) coordinate bond

Answer. (*c*)

46. The maximum number of Hydrogen bonds formed by a water molecule is

(*a*) 1 (*b*) 2 (*c*) 3 (*d*) 4

Answer. (*b*)

47. Out of the following, intramolecular Hydrogen bonding exists in

(*a*) water (*b*) H_2S (*c*) 2-nitrophenol (*d*) 4-nitrophenol

Answer. (*c*)

48. Which one of the following is the most polar?

(*a*) H—F (*b*) H—Cl (*c*) H—Br (*d*) H—I

Answer. (*a*)

6 Chemical Bonding – Orbital Theory

CONTENTS

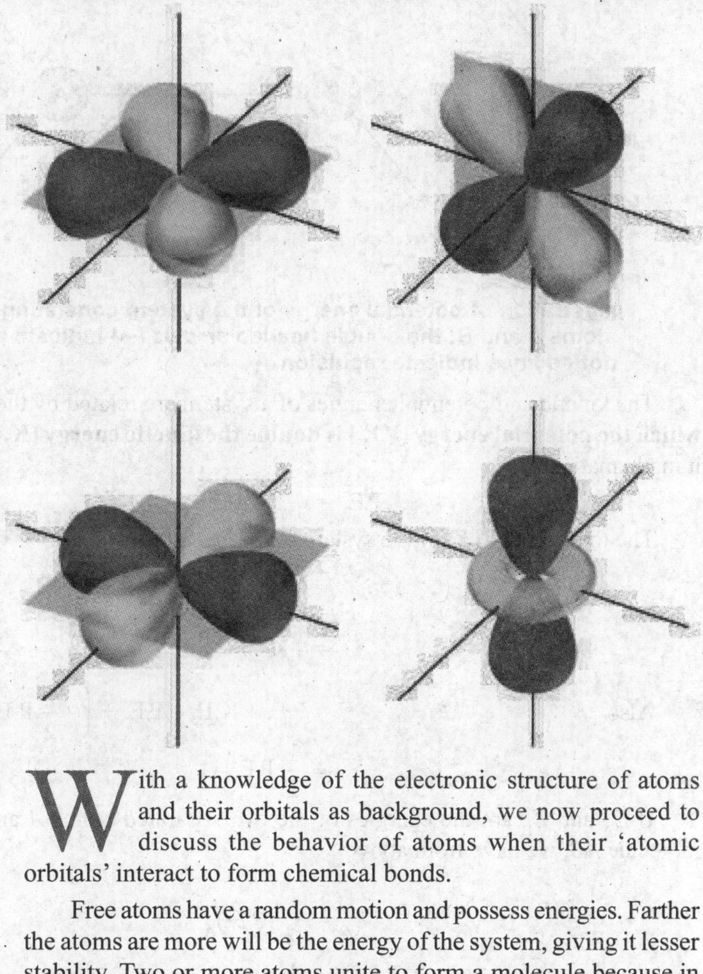

With a knowledge of the electronic structure of atoms and their orbitals as background, we now proceed to discuss the behavior of atoms when their 'atomic orbitals' interact to form chemical bonds.

Free atoms have a random motion and possess energies. Farther the atoms are more will be the energy of the system, giving it lesser stability. Two or more atoms unite to form a molecule because in doing so the energy of the system is lowered and thus the 'molecule' becomes stable in comparison to separate atoms. **In other words, a stable chemical union or bond between two (or more) atoms comes into existence only if the energy is lowered when the atoms come in close proximity.**

By the term 'energy' here, we mean the sum of the kinetic and potential energies. The kinetic energy of atoms finds its origin in the unabated motion of the electrons. The potential energy of a system in which atoms A and B approach each other can be due to

(*a*) Electron-electron repulsions

(*b*) Nucleus-nucleus repulsions

(*c*) Nucleus-electron attractions.

This is beautifully illustrated by the following diagram (Fig. 6.1).

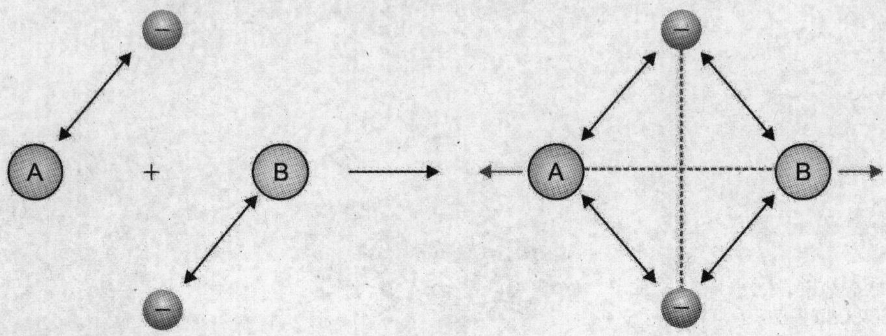

■ **Figure 6.1**
Illustration of potential energy of the system consisting of approaching atoms A and B; the double headed arrows (↔) indicate attraction and dotted lines indicate repulsion.

The kinetic and potential energies of a system are related by the **Virial Theorem according to which the potential energy (P.E.) is double the kinetic energy (K.E.)** and opposite in sign. Putting it mathematically,

$$\text{P.E.} = -2\,\text{K.E.} \quad \text{...(1)}$$

The total energy (E) of the system is given as

$$E = \text{P.E.} + \text{K.E.}$$

or

$$E = -2\,\text{K.E.} + \text{K.E.}$$

$$= -\text{K.E.} \quad \text{...(2)}$$

Also

$$E = \text{P.E.} + \text{K.E.} = \text{P.E.} + \left(-\frac{1}{2}\,\text{P.E.}\right)$$

$$= \frac{1}{2}\,\text{P.E.} \quad \text{...(3)}$$

If E_A and E_B are the energies of the two separated atoms A and B and E_{AB} the energy of the molecule AB, we have from above

$$E_A = \frac{1}{2}\,(\text{P.E.})_A$$

$$E_B = \frac{1}{2}\,(\text{P.E.})_B$$

$$E_{AB} = \frac{1}{2}\,(\text{P.E.})_{AB}$$

Let us now examine the change in energy (ΔE) as the atoms A and B come together :

$$\Delta E = E_{AB} - (E_A + E_B)$$

$$= \frac{1}{2}\,(\text{P.E.})_{AB} - \left[\frac{1}{2}\,(\text{P.E.})_A + \frac{1}{2}\,(\text{P.E.})_B\right]$$

or

$$\Delta E = \frac{1}{2}\,\Delta\text{P.E.} \quad \text{...(4)}$$

where ΔE is the change in energy of the system. As stated earlier, the change in the energy of the system should be negative in order a stable bond be formed between A and B. In other words, **the net potential energy of the system must decrease as a result of the formation of a chemical bond by electronic rearrangements.** Out of the three factors determining the potential energy, the first two are 'repulsive' in nature and, therefore, cannot lower the potential energy in any manner. As the atoms approach each other, the electron-nucleus attractions contribute to the lowering of the potential energy, since each electron is now attracted towards the two positive nuclei at the same time, as compared to one nucleus in the isolated atoms.

BOND FORMATION (VALENCE BOND THEORY)

Bond formation between atoms to give chemical compounds can be interpreted admirably in terms of the orbital theory of atomic structure. Heitler and London believed that electron cloud of the valence orbital on one atom 'overlaps' the electron cloud of the other bonding atom to form a covalent linkage. On the contrary, the electrovalent bond formation involves a physical transfer of the electron and the orbital concept is not very useful for their explanation. The theory of '*maximum overlap*', affords an excellent interpretation of covalent bond formation. The essential conditions for the overlap of electron waves of orbitals are :

(1) **The orbitals entering into combination must have only one electron.**

The orbitals containing a pair of electrons are not capable of combination. In fact, half-filled orbitals on one atom have a tendency to combine with half-filled orbitals on other atom, and the resulting orbital acquires a pair of electrons of opposite spins.

(2) **The atoms with valence or bonding orbital (half-filled) should approach sufficiently close to one another with the axis of their orbitals in proper alignment.**

The strength of a covalent bond depends upon the extent of overlapping. Greater the overlapping between the atomic orbitals stronger is the bond formed between the two atoms.

The new arrangement has lesser energy than the isolated atoms and, is therefore, more stable. The amount of energy given off or released per mole at the time of overlapping of atomic orbitals to form a bond is termed as **Bond Energy** or **Stabilization Energy**. It also implies that this much energy must be put in to separate the atoms in a molecule.

Since the overlapping of atomic orbitals involves a release of energy, it must produce a stabilizing effect on the system. The merger or overlapping of the atomic orbitals halts at a stage when the atomic nuclei have come close enough to exercise a repulsive force on one another, which exactly balances the force of merger. **This equilibrium distance at which the two atomic nuclei are now held is called the Bond length.**

	Covalent Bond	
Bond	**Energy (kJ/mole)**	**Length r_c (m)**
H – H	432.078	0.74×10^{-10} m
C – H	413.237	1.10×10^{-10} m
C – C	347.504	1.54×10^{-10} m
C = C	611.273	1.34×10^{-10} m
C ≡ C	837.360	1.20×10^{-10} m
C – N	294.332	1.52×10^{-10} m
C – N	619.646	1.29×10^{-10} m
C ≡ N	819.788	1.15×10^{-10} m
C – O	351.691	1.43×10^{-10} m
C = O	741.064	1.24×10^{-10} m

In the bond formation, greater the overlap greater is the energy set free, *i.e.*, the higher will be the bond strength. This implies that for a bond to be stronger, greater should be the overlap which in turn shortens the distance between the nuclei. **A stronger bond has, therefore, a shorter bond length.**

Formation of H_2 molecule

Let us consider the formation of hydrogen molecule (H_2) from two isolated H atoms each having an electron in its 1*s* orbital. When two H atoms approach each other, their 1*s* orbitals overlap, resulting in the formation of a bigger electron cloud known as *molecular orbital*. This new molecular orbital contains both the electrons. As a result, the two H atoms are held together in the form of Hydrogen molecule (H–H). This overlapping of atomic orbitals of hydrogen atoms is shown in Fig. 6.2.

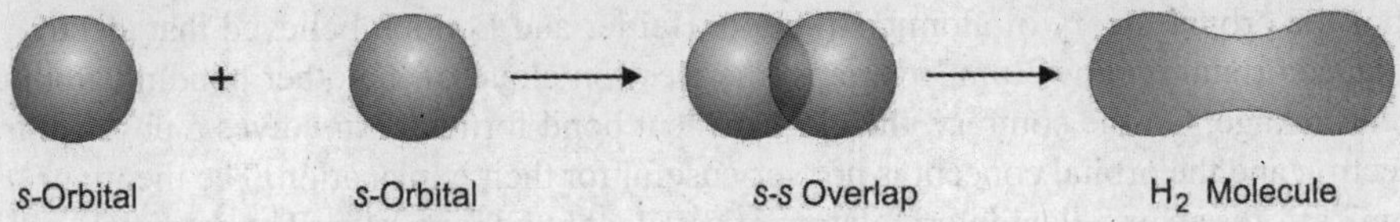

Figure 6.2
Formation of H_2 molecules (*s*-*s* overlap).

As the two atoms approach closer and closer for the overlap of half-filled 1*s* orbitals on them, the energy of the system goes on decreasing till it registers a minimum. At this point the overlapping halts, since the positive cores (nuclei) of the two atoms are apart by a certain equilibrium distance, the stabilizing effect of overlapping is completely balanced by the repulsion between the positive nuclei. Beyond this point the energy of repulsion predominates and there is a steep rise in the energy curve with a further decrease of internuclear distance. The point *A* in the diagram below represents a large value of internuclear distance corresponding to the two isolated atoms. By convention, the energy corresponding to this point is taken to be zero and for stabler systems, the energy of the system is negative. The energy values go on falling as the internuclear distance decreases. At point *X* on the curve, the internuclear distance decreases and so does the energy value. The minima 'B' represents the point of maximum overlap. The energy value and the internuclear distance corresponding to this point are respectively the bond energy and bond length. For the H_2 molecule, the bond length is 0.74×10^{-10} m and bond energy is 432.078 kJ/mole.

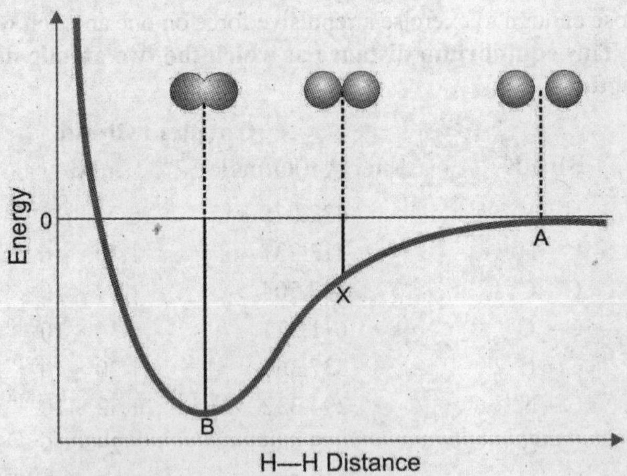

Figure 6.3
Energy diagram of *s*-*s* overlap.

Types of overlapping and Nature of Covalent Bond

A covalent bond is of two types depending on the type of overlapping between the two atoms

(1) Sigma (σ) bond

(2) Pi (π) Bond

(1) Sigma (σ) Bond

When there is end to end overlapping of atomic orbitals along the internuclear axis, the bond resulted is called sigma (σ) bond. This type of overlapping between the atomic orbitals is also called "head-on" overlapping or "axial" overlapping. It results when one of the following types of overlapping takes place :

(*a*) ***s–s* overlapping**

Here *s*-orbital of one atom overlaps with the *s*-orbital of other atom. An example of this type of overlapping is the formation of hydrogen molecule from two H-atoms.

(*b*) ***s – p* overlapping**

In this type of overlap *s*-orbital of one atom overlaps with the half filled *p*-orbital of the other atom as shown below :

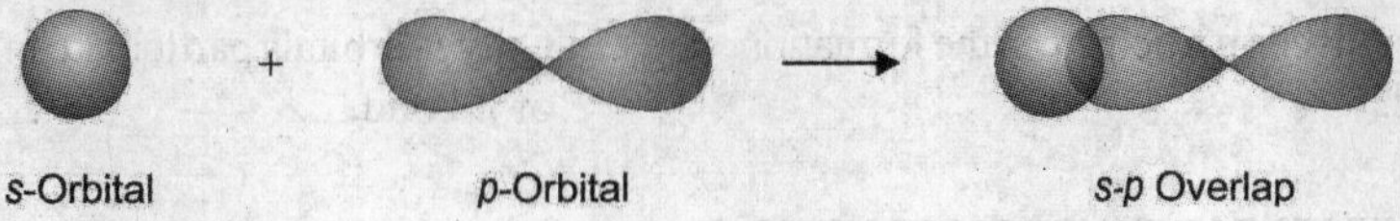

Examples : HF, HCl etc.

(c) ***p – p* overlapping**

Here *p*-orbital of one atom overlaps with the *p*-orbital of the other atom on internuclear axis. It is shown below :

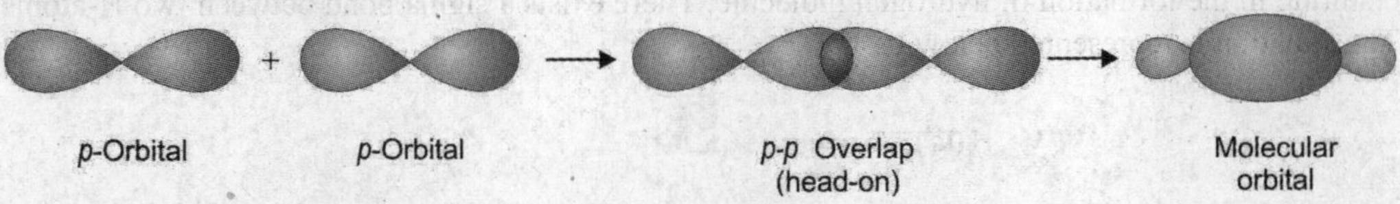

Examples : F_2, Cl_2, Br_2 etc.

(2) Pi (π) Bond

This type of covalent bond is formed by the sidewise overlap of the half filled atomic orbitals. It is also called **lateral** or **sidewise** overlap. This type of overlapping takes place perpendicular to the internuclear axis as shown below :

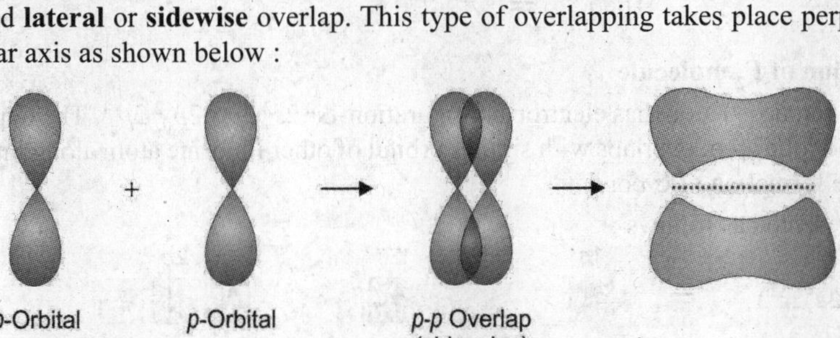

Strength of σ and π bonds

As mentioned earlier, the strength of a covalent bond depends upon the extent of overlapping between the atomic orbitals of participating atoms. During the formation of σ bond the extent of overlapping is more and hence *a Sigma bond is stronger than Pi bond.*

Differences Between Sigma and Pi bonds

Sigma (σ) bond	Pi (π) Bond
1. It is formed by *end to end* overlapping of half filled atomic orbitals.	1. It is formed by the sidewise overlapping of half filled *p*-orbitals only.
2. Overlapping takes place along internuclear axis.	2. Overlapping takes place perpendicular to internuclear axis.
3. The extent of overlapping is large and bond formed is *stronger*.	3. The extent of overlapping is small and bond formed is *weaker*.
4. The molecular orbital formed as a result of overlapping is symmetrical about the internuclear axis.	4. The molecular orbital formed as a result of overlapping consists of two lobes above and below the internuclear axis.
5. There is free rotation about σ bond and no geometrical isomers are possible.	5. There is no free rotation about π bond and geometrical isomers are possible.
6. The bond can be present alone.	6. The bond is always formed in addition to sigma (σ) bond.
7. *s* and *p* orbitals can participate in the formation of σ bond.	7. Only *p*-orbitals participate in the formation of π bond.

ORBITAL REPRESENTATION OF MOLECULES

(*a*) Formation of H_2 molecule

Each hydrogen atom has one electron in 1*s*-orbital. Two such atoms join to form a molecule of hydrogen. In this case *s–s* overlapping between two 1*s*-orbitals of hydrogen atoms takes place resulting in the formation of hydrogen molecule. There exists a sigma bond between two H-atoms. The situation is represented below :

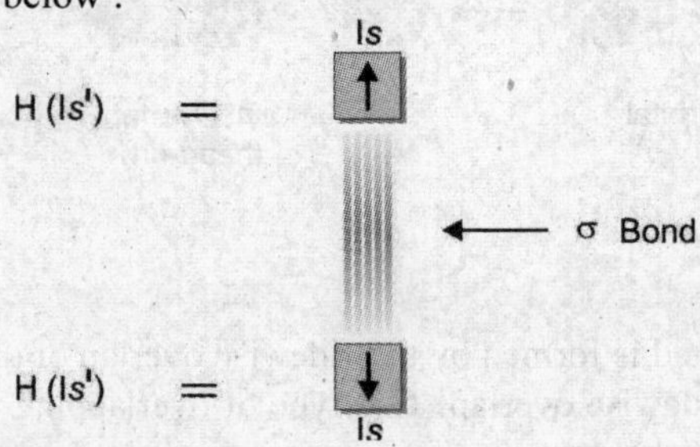

(*b*) Formation of F_2 molecule

Fluorine (at. no. 9) atom has electron configuration $1s^2\ 2s^2\ 2p_x^{\ 2}\ 2p_y^{\ 2}\ 2p_z^{\ 1}$. The unpaired electron in $2p_z$ orbital of one atom overlaps with similar orbital of other Fluorine atom along internuclear axis leading to the formation of σ bond.

It is represented as follows :

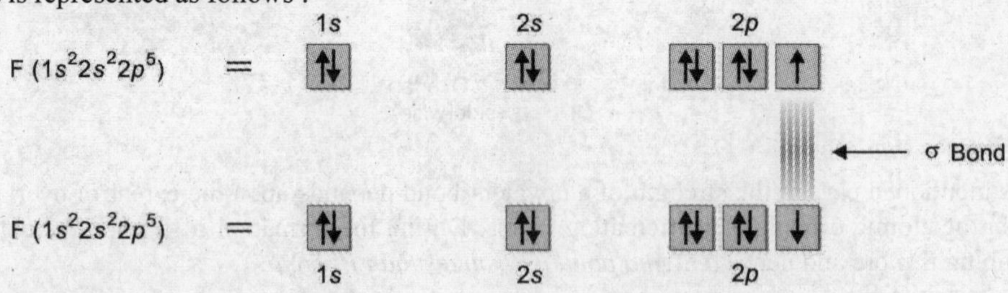

(c) Formation of HCl molecule

During the formation of HCl molecule, $1s$ orbital of Hydrogen atom having an unpaired electron overlaps with $3p_z$ orbital of chlorine atom having electronic configuration $1s^2\ 2s^2\ 2p^6\ 3s^2\ 3p_x^{\ 2}\ 3p_y^{\ 2}\ 3p_z^{\ 1}$. This overlapping takes place along internuclear axis leading to the formation of σ bond as shown below :

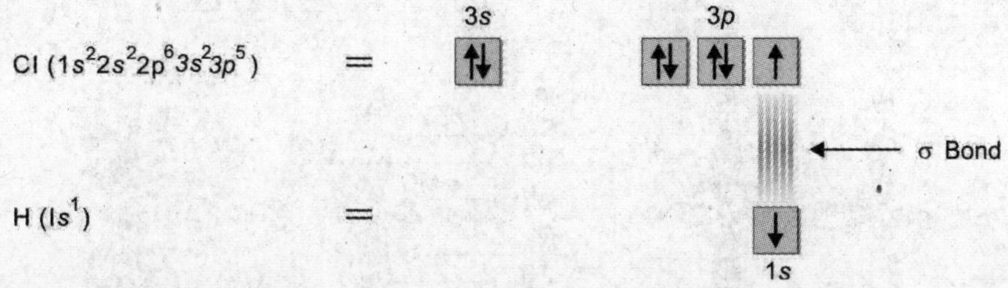

(d) Formation of Oxygen molecule

Oxygen (at. no. 8) atom has electronic configuration $1s^2\ 2s^2\ 2p_x^{\ 2}\ 2p_y^{\ 1}\ 2p_z^{\ 1}$. In the formation of O_2 molecule $2p_z$ orbital of one oxygen atom overlaps with the similar orbital of the other atom along internuclear axis leading to the formation of a sigma (σ) bond. The other $2p_y$ orbital of one oxygen atom overlaps with the similar orbital of the another oxygen atom in sidewise manner forming a pi (π) bond. Thus an oxygen molecule contains a double bond consisting of a sigma (σ) bond and a pi (π) bond. The orbital overlap is represented below :

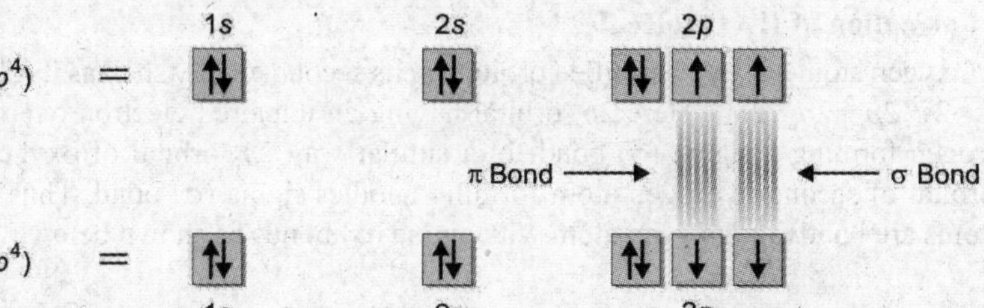

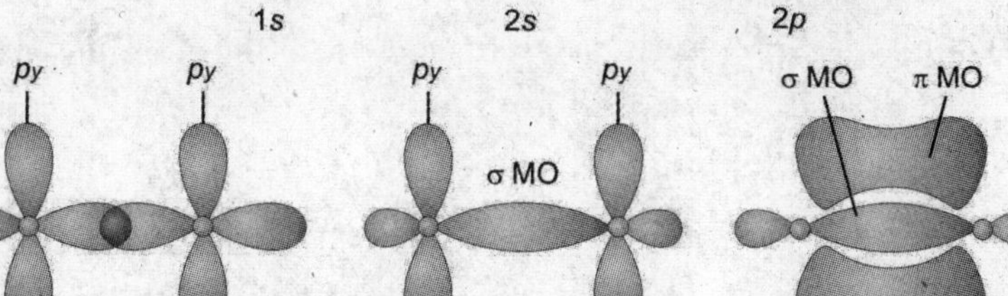

■ Figure 6.4
Schematic overlaps of atomic orbitals of the two oxygen atoms in the formation of O_2 molecule (inner filled orbitals not shown). (*a*) represents a head on overlap forming; (*b*) the σ bond; and (*c*) represents a π MO enveloping a σ MO.

(e) Formation of Nitrogen molecule

Nitrogen atom has three bonding electrons $2p_x^{\ 1}\ 2p_y^{\ 1}\ 2p_z^{\ 1}$ in its valence shell. These electrons are present in three p-orbitals which are directed along three axes at right angles. When two N atoms approach each other, $2p_z$ orbitals overlap in head-on manner to form a sigma (σ) bond, leaving two sets of parallel $2p_y$ and $2p_x$ orbitals. Now the electron clouds in these orbitals interact to produce two pi (π) bonds by the sidewise overlaps. Thus in N_2 molecule two N atoms are bonded by a triple bond consisting of one sigma (σ) and two pi (π) bonds. The orbital overlap in N_2 molecule is shown below:

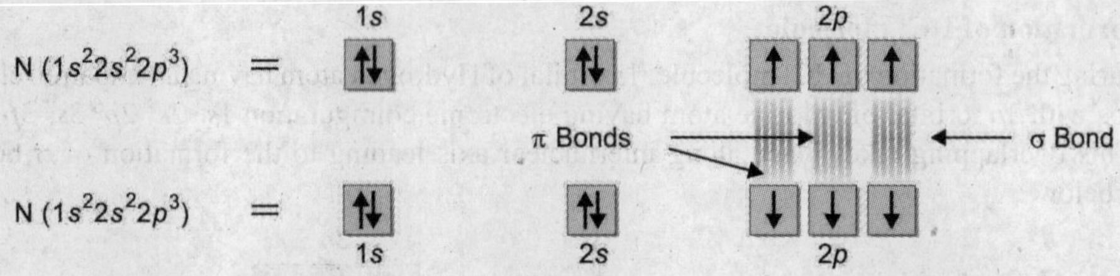

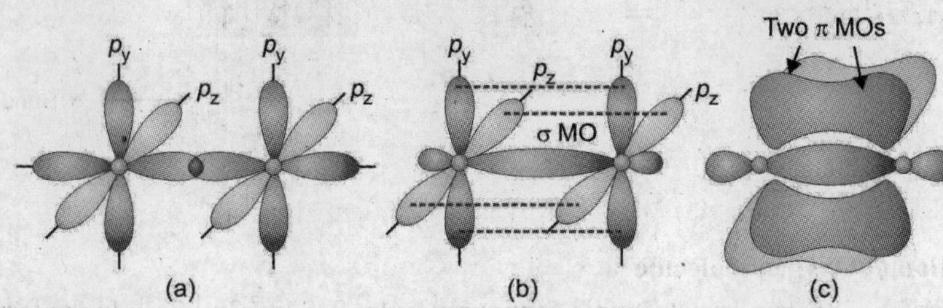

■ **Figure 6.5**
Orbital representation of N_2 molecule formation; (*a*) indicates the approach of p_x orbitals for head on overlap; (*b*) the formation of σ bond; and (*c*) the formation of two π MOs by sidewise overlaps.

Thus there is one sigma and two pi bonds in a molecule of nitrogen. It is equivalent to a triple bond in the classical theory.

(*f*) **Formation of H_2O molecule**

Oxygen atom has two half filled orbitals in its second energy shell as its electronic configuration is $1s^2\ 2s^2\ 2p_x^{\ 2}\ 2p_y^{\ 1}\ 2p_z^{\ 1}$. Here $2p_z$ orbital having an unpaired electron overlaps with $1s$ orbital of hydrogen forming a sigma (σ) bond. In a similar way $2p_y$ orbital of oxygen atom overlaps with $1s$ orbital of second hydrogen atom forming another sigma (σ) bond. Thus in H_2O molecule two H atoms are bonded to oxygen atom with sigma (σ) bonds as shown below :

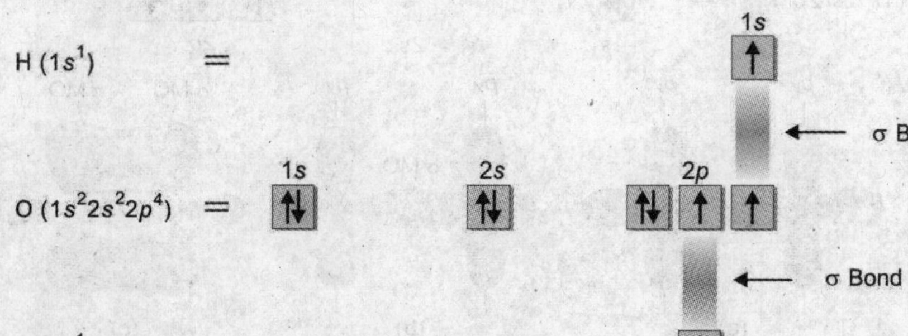

CONCEPT OF HYBRIDIZATION

While formation of simple molecules could be explained adequately by overlap of atomic orbitals, the formation of molecules of Be, B and C present problems of greater magnitude having no solution with the previous theory. To explain fully the tendency of these atoms to form bonds and the shape or geometry of their molecules, a new concept called **Hybridization** is introduced. **According to this concept, we may mix any number of atomic orbitals of an atom, which differ in energy only** slightly, to form new orbitals called Hybrid orbitals. The mixing orbitals generally belong to the same energy level (say $2s$ and $2p$ orbitals may hybridize). The total number of hybrid orbitals formed after mixing, is invariably equal to the number of atomic orbitals mixed or hybridized.

An important characteristic of hybrid orbitals is that they are all identical in respect of energy and directional character. They, however, differ from the original atomic orbitals in these respects. They may also differ from one another in respect of their arrangement in space. *i.e.*, orientation. Like pure atomic orbitals, the hybrid orbitals of an atom shall have a maximum of two electrons with opposite spin. Hybrid orbitals of an atom may overlap with other bonding orbitals (pure atomic or hybrid) on other atoms or form molecular orbitals and hence new bonds.

Thus hybridization may precisely be defined as the phenomenon of mixing up (or merging) of orbitals of an atom of nearly equal energy, giving rise to entirely new orbitals equal in number to the mixing orbitals and having same energy contents and identical shapes.

RULES OF HYBRIDIZATION

For hybridization to occur, it is necessary for the atom to satisfy the following conditions :

(1) **Orbitals on a single atom only would undergo hybridization.**

(2) **There should be very little difference of energy level between the orbitals mixing to form hybrid orbitals.**

(3) **Number of hybrid orbitals generated is equal to the number of hybridizing orbitals.**

(4) **The hybrid orbitals assume the direction of the dominating orbitals.** For example, if *s* and *p* orbitals are to hybridize, the *s* orbital having no directional character, does not contribute towards the direction whereas *p* orbitals determine the directional character of the hybrid orbitals.

(5) **It is the orbitals that undergo hybridization and not the electrons.** For example, four orbitals of an oxygen atom $(2s^2, 2p_x^2, 2p_y^1, 2p_z^1)$ belonging to second level (*i.e.*, $2s$, $2p_x$, $2p_y$, $2p_z$) can hybridize to give four hybrid orbitals, two of which have two electrons each (as before) and the other two have one electron each.

(6) **The electron waves in hybrid orbitals repel each other and thus tend to be farthest apart.**

TYPES OF HYBRIDIZATION

Since hybridization lends an entirely new shape and orientation to the valence orbitals of an atom, it holds a significant importance in determining the shape and geometry of the molecules formed from such orbitals. Depending upon the number and nature of the orbitals undergoing by hybridization, we have various types of hybrid orbitals. For instance *s*, *p*, and *d* orbitals of simple atoms may hybridize in the following manner.

(*a*) sp Hybridization

The mixing of an *s* and a *p* orbital only leads to two hybrid orbitals known as **sp hybrid orbitals** after the name of an *s* and a *p* orbital involved in the process of hybridization. The process is called **sp hybridization.** Each *sp* orbital has 50%, *s*-character and 50% *p*-character. Orbitals thus generated are the seat of electrons which have a tendency to repel and be farther apart. In order to do so the new orbitals arrange themselves along a line and are, therefore, often referred to as **Linear hybrid orbitals.** This gives an angle of 180° between the axes of the two orbitals. It is clear from Fig. 6.6 that an *sp* orbital has two lobes (a character of *p* orbital) one of which is farther than the corresponding *s* or *p* orbitals and also protrudes farther along the axis. It is this bigger lobe that involves itself in the process of an overlap with orbitals of other atoms to form bonds. It will be seen later on that the smaller lobes of hybrid orbitals are neglected while considering bond formation.

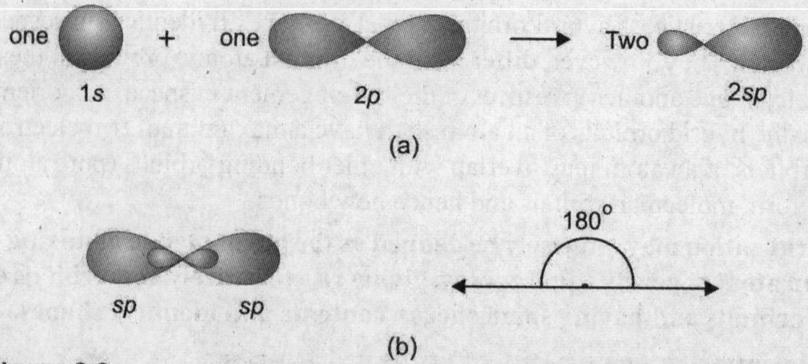

■ **Figure 6.6**
(*a*) *sp* Hydridization. (*b*) Disposition of hybrid orbitals along a line.

Examples : BeF_2, $BeCl_2$, etc.

(*b*) **sp² Hybridization**

When an *s* and two *p* orbitals mix up to hybridize, there result three new orbitals called **sp² hybrid orbitals** (spoken as '*ess pee* two'). Each sp^2 hybrid orbital has 33% *s*-character and 67% *p*-character. As the three orbitals undergoing hybridization lie in a plane, so do the new orbitals. They have to lie farthest apart in a plane which can happen if they are directed at an angle 120° to one another as shown in Fig. 6.7 (*b*). It is for this reason that sp^2 hybrid orbitals are also called **Trigonal hybrids,** the process being referred to as **Trigonal hybridization.** The sp^2 hybrid orbitals resemble in shape with that of *sp* hybrid orbitals but are slightly fatter.

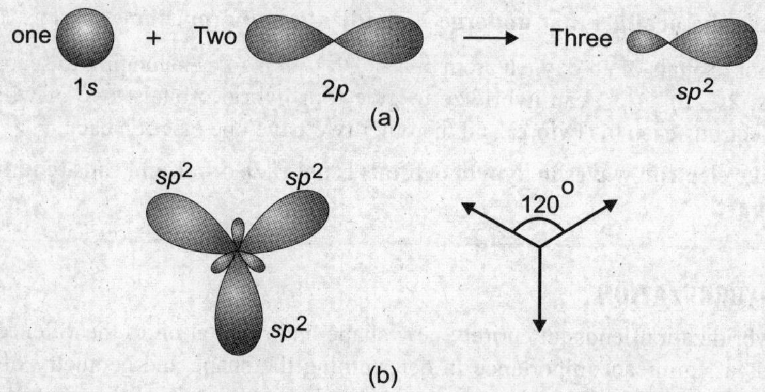

■ **Figure 6.7**
(*a*) sp² Hybridization; and (*b*) Disposition of hybrid orbitals in the same plane at an angle 120°to one another.

Examples : BF_3, NO_3^-, etc.

(*c*) **sp³ Hybridization**

The four new orbitals formed by mixing an *s* and three *p* orbitals of an atom are known as **sp³ hybrid orbitals.** Each sp^3 hybrid orbital has 25% *s*-character and 75% *p*-character. Since mixing of orbitals takes place in space, the four hybrid orbital would also lie in space. An arrangement in space that keeps them farthest apart is that of a tetrahedron. Thus each of the four hybrid orbitals are directed towards the four corners of a regular tetrahedron as shown in Fig. 6.8 (*b*). Because of their tertrahedral disposition, this type of hybridization is also called **Tetrahedral hybridization.** They are of the same shape as that of the previous two types but bigger in size. They are disposed in manner such that the angle between them is 109.5° as shown in Fig. 6.8.

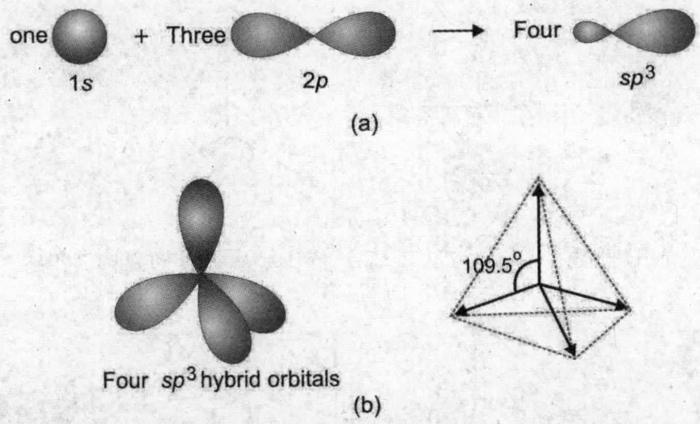

Figure 6.8
(a) sp^3 Hybridization (b) The four hybrid orbitals are directed towards the four corners of a tetrahedron and are at an angle 109.5° to one another.

Examples : CH_4, SO_4^{2-}, ClO_4^-, etc.

(d) Hybridization involving *d* orbitals

There are several types of hybridization involving *d* orbitals. Since the *d* orbitals have a relatively complex shape, we will consider here only some of the common types. The most important of these are sp^3d hybridization, sp^3d^2 hybridization and sp^2d hybridization.

In **sp^3d hybridization** the orbitals involved are one of *s* type, three of *p* type and one of *d* type. The five new orbitals will be farthest apart by arranging three of them in a plane at an angle of 120° to one another and the other two in a direction perpendicular to the plane. The figure obtained by joining the ends assumes the shape of a trigonal bipyramid (Fig. 6.9). This type of hybridization is, therefore, called **Trigonal bipyramidal hybridization.**

When two *d* type of orbitals take part in hybridization with one *s* type and three *p* type **orbitals, six hybrid orbitals called sp^3d^2 hybrid orbitals** are created. To be away from one another four of them are dispersed in a plane at an angle of 90° each and the rest two are directed up and below this plane in a direction perpendicular to it. On joining their corners, an octahedron results and this type of hybridization also gets the name **Octahedral hybridization** (Fig. 6.9).

So far we have been considering the hybridization of orbitals belonging to the same energy level (say 3*s*, 3*p*, and 3*d* orbitals) of an atom. But this may not necessarily be so always. In fact, there is very little energy difference between 3*d*, 4*s* and 4*p* orbitals which may undergo **sp^2d hybridization.** The *d* orbital involved in this type of hybridization has the same planar character as the two *p* orbitals have and the hybrid orbitals will also be planar, dispersed in such a way so as to be farthest apart *i.e.*, subtending an angle of 90° between them. This gives a square planar arrangement for them and the hybridization is, therefore, called **Square planar hybridization.**

The directional characters of the types of hybridization discussed above are summarized in Fig. 6.9.

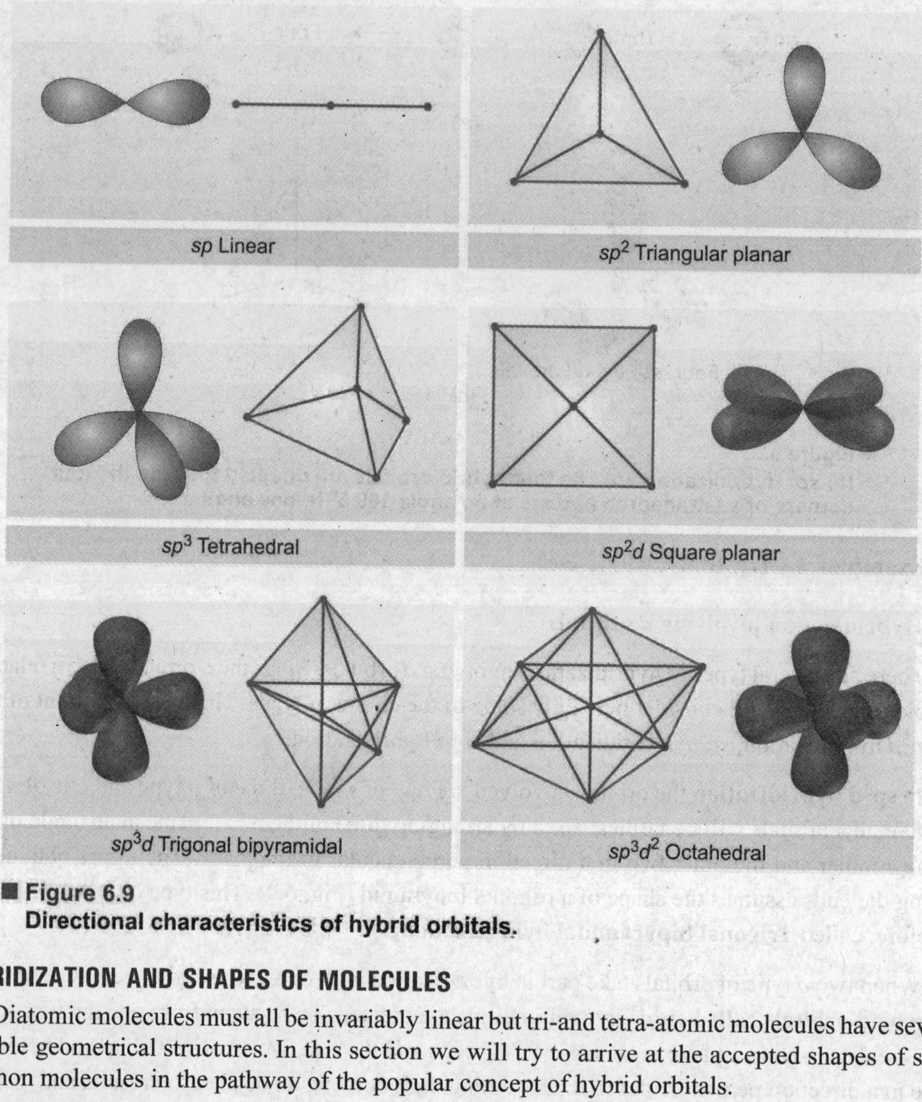

Figure 6.9
Directional characteristics of hybrid orbitals.

HYBRIDIZATION AND SHAPES OF MOLECULES

Diatomic molecules must all be invariably linear but tri-and tetra-atomic molecules have several possible geometrical structures. In this section we will try to arrive at the accepted shapes of some common molecules in the pathway of the popular concept of hybrid orbitals.

Shape of BeF_2 molecule

An isolated Be atom in its ground state has the electronic configuration $1s^2$, $2s^2$. At the first thought, one would expect Be to be chemically inert like He since it has all its orbitals completely filled (no bonding orbital). But Be behaves differently because its $2s$ orbital though complete, possesses another empty $2p$ level lying in the same shell. This is not so for He ($1s^2$). Since orbitals on the same atom lying very close to one another in energy possess an unusual quality of mixing with one another additively forming new hybrid orbitals, there is ample scope for hybridization here. The Be atom, therefore, gets excited so that one of its $2s^2$ electrons is 'promoted' to the next available $2p_x$ orbital before the atom participates in chemical bonding. But this process requires energy which will be available from the heat of reaction released at the time of bond formation. Now the excited atom acquires the structure $1s^2$, $2s^1$, $2p_x^{\ 1}$. If at this stage the atom forms two bonds by suitable overlaps with two fluorine atoms, these bonds will not be identical, one involving the s and the other p orbitals. Thus the s and p orbitals first hybridize to form two new and completely equivalent sp hybrid orbitals.

These hybrid orbitals of Be are now capable of forming bonds. The two *sp* hybrid orbitals overlap two 2*p* orbitals from two fluorine atoms in the 'head on' manner to form two σ bonds. The two *sp* orbitals being linear, lend a linear shape to BeF_2 molecule with Be atom lying in the center and two F atoms on its either side so that F—Be—F angle is equal to 180° as shown in Fig. 6.10.

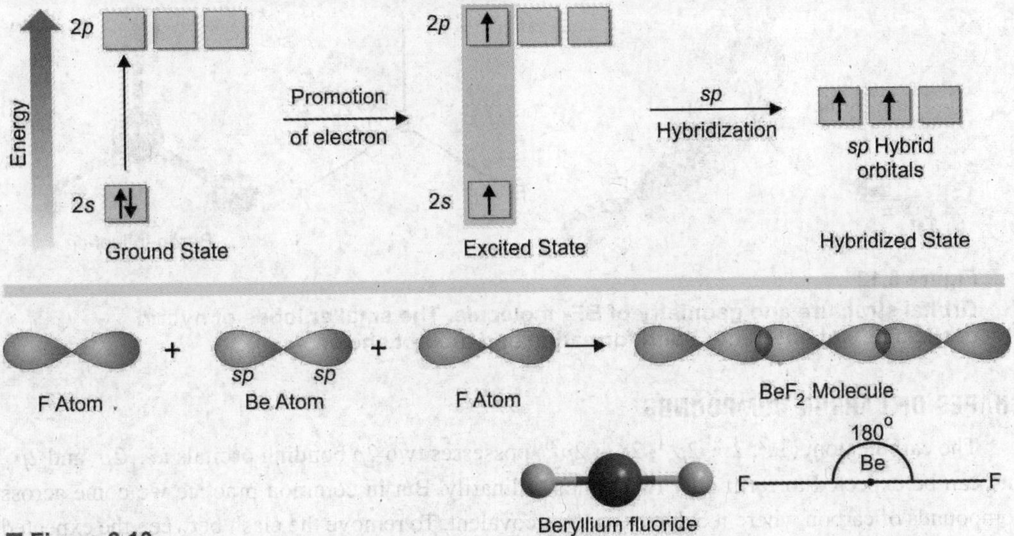

■ **Figure 6.10**
Various states of Be atom. Its orbitals undergo *sp* hybridization in the excited state. Overlaps and geometry of BeF_2 are also shown.

Shape of BF_3 molecule

The orbital electronic configuration of Boron (B) is $1s^2, 2s^2, 2p_x^{\ 1}, 2p_y^{\ 0}, 2p_z^{\ 0}$. As there is only one bonding orbital $2p_x$, B is expected to form only one bond. Boron, in fact, is known to form compounds such as BCl_3, BF_3, BH_3 etc., indicating its capacity to form three bonds. What actually happens is that the two electrons of 2*s* orbital get unpaired when it is excited just like Be. One of these unpaired electrons thus gets promoted to the vacant $2p_y$ orbital lying close to it. Thus in the excited state of Boron there are three half-filled orbitals available for bonding. If as such it were to form bonds by overlap, the nature of these bonds would be different owing to their different types. One 2*s* and two 2*p* orbitals undergo sp^2 hybridization giving three sp^2 hybrid orbitals lying in one plane *i.e.*, *xy*, and subtending an angle of 120°. The equivalent hybrid orbitals can now enter into bond formation by overlapping with three 2*p* orbitals of three fluorine atoms as illustrated in Fig. 6.12. Thus BF_3 molecule is planar and each F—B—F angle is equal to 120°.

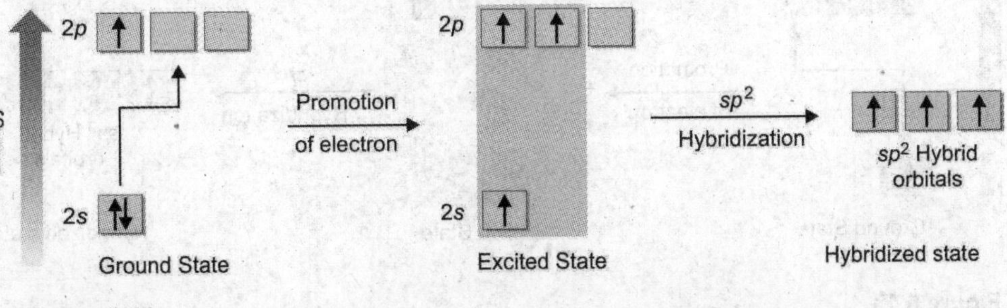

■ **Figure 6.11**
Various states of B atom. The orbitals in the valence shell undergo sp^2 hybridization after excitation of electron.

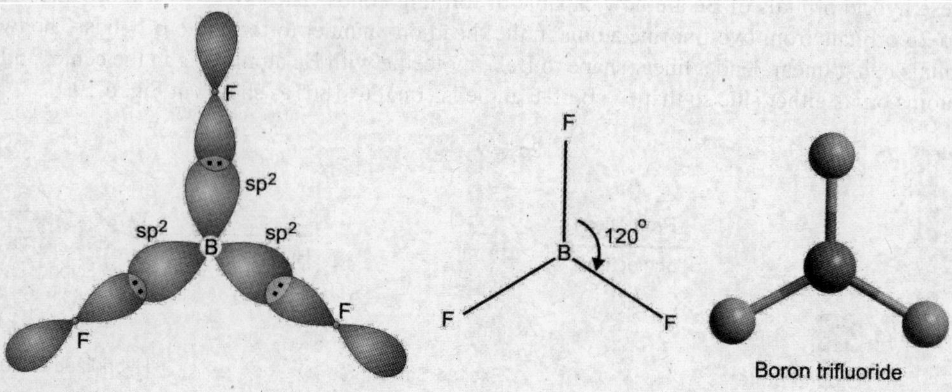

Figure 6.12
Orbital structure and geometry of BF_3 molecule. The smaller lobes of hybrid orbitals do not enter into bond formation and are not shown.

SHAPES OF CARBON COMPOUNDS

The carbon atom ($1s^2$, $2s^2$, $2p_x^{\ 1}$, $2p_y^{\ 1}$, $2p_z^{\ 0}$) possesses two $2p$ bonding orbitals *i.e.*, $2p_x$ and $2p_y$, and can be expected to form only two bonds ordinarily. But in common practice we come across compounds of carbon where it behaves as tetra-covalent. To remove the clash between the expected and the actual, the concept of hybridization comes to our rescue. Carbon can undergo three types of hybridization.

(*a*) sp³ Hybridization of Carbon

It is proposed that from $2s$ orbital, being quite near in energy to $2p$ orbitals, one electron may be promoted to the vacant $2p_z$ orbital thus obtaining the excited atom. At this stage the carbon atom undoubtedly has four half-filled orbitals and can form four bonds. But by the strength of the argument extended in case of Be and B, it is assumed that the orbitals of carbon atom first undergo hybridization before forming bonds. In the excited atom all the four valence shell orbitals may mix up to give identical sp^3 hybrid orbitals also four in number. Each of these four sp^3 orbitals possesses one electron and overlaps with $1s$ orbitals of four H-atoms thus forming four equivalent bonds in methane molecule. Due to the tetrahedral disposition of sp^3 hybrid orbitals, the orbitals are inclined at an angle of 109.5°. Thus all the HCH angles are equal to 109.5°.

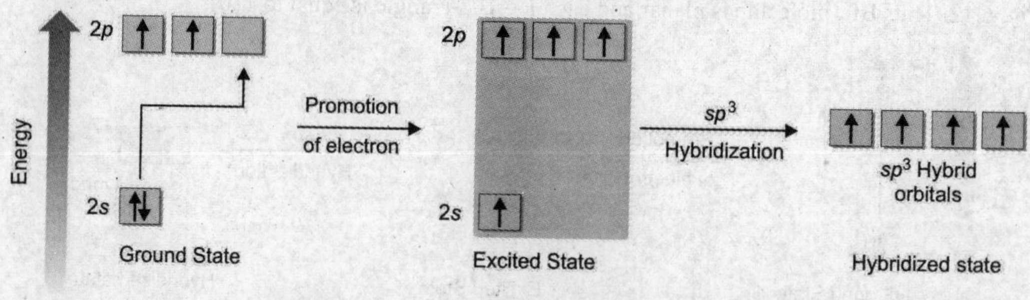

Figure 6.13
Various states of carbon atom; the orbitals here are undergoing sp^3 hybridization.

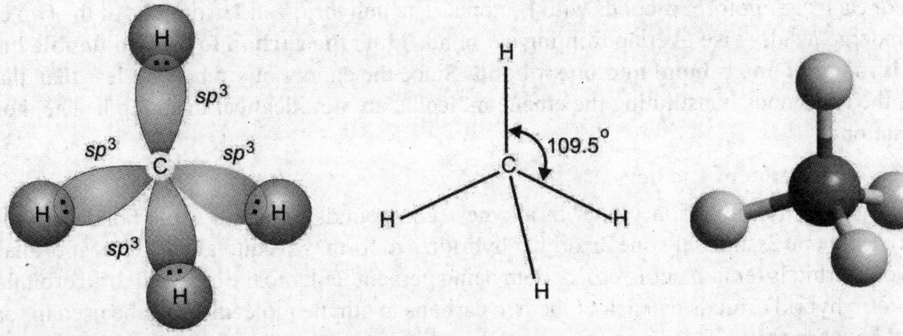

■ **Figure 6.14**
Shape and formation of methane molecule.

(*b*) **sp² Hybridization of Carbon**

When three out of the four valence orbitals hybridize, we have three sp^2 hybrid orbitals lying in a plane and inclined at an angle of 120°. If $2s$, $2p_x$ and $2p_y$ orbitals of the excited carbon atom are hybridized, the new orbitals lie in the *xy* plane while the fourth pure $2p_z$ orbital lies at right angles to the hybridized orbitals with its two lobes disposed above and below the plane of hybrid orbitals. Two such carbon atoms are involved in the formation of alkenes (compounds having double bonds). In the formation of ethene two carbon atoms (in sp^2 hybridization state) form one sigma bond by 'head-on' overlap of two sp^2 orbitals contributed one each by the two atoms. The remaining two sp^2

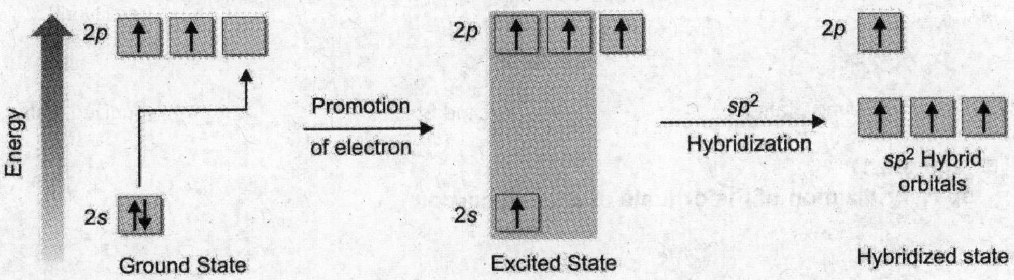

■ **Figure 6.15**
Excited carbon atom undergoing *sp²* hybridization leaves a pure 2*p*z orbital.

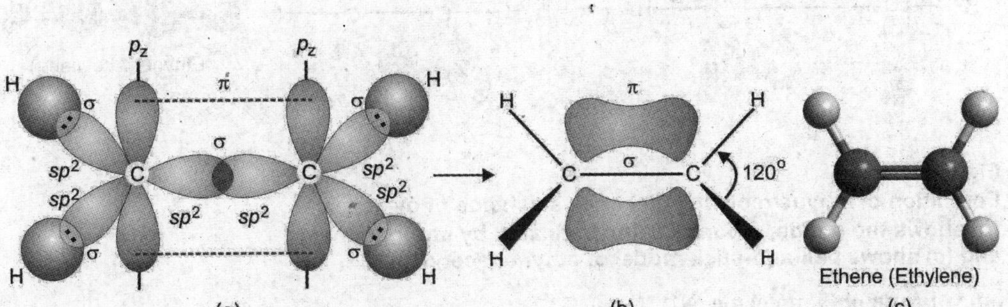

■ **Figure 6.16**
Orbital model of ethene molecule. (*a*) shows scheme of overlaps; (*b*) shows the bonds, sigma bonds indicated by straight lines; and (*c*) shows ball-and-stick model of ethene (ethylene).

orbitals of each carbon form σ bonds with H atoms. The unhybridized $2p_z$ orbitals of the two carbon atoms undergo a side-wise overlap forming a π bond. **Thus the carbon to carbon double bond in ethene is made of one σ bond and one π bond.** Since the energy of a π bond is less than that of a σ bond, the two bonds constituting the ethene molecule are not identical in strength. The molecule is a planar one.

(*c*) **sp Hybridization of Carbon**

This type of hybridization is met in alkynes (compounds having a triple bond between two carbons). Here one 2*s* and only one 2*p* orbital hybridize to form two equivalent collinear orbitals; the other two 2*p* orbitals remain undisturbed, both being perpendicular to the axis of hybrid orbitals. One of the two *sp* hybrid orbitals on each of the two carbons in ethyne molecule, may be used in forming a σ bond between them. This leaves two pure 2*p* orbitals ($2p_y$ and $2p_z$) on each carbon atom. Both these are mutually perpendicular to H–C–C–H nuclear axis, the C–H bonds being formed by overlap of the remaining *sp* orbital with 1*s* orbitals of hydrogens (see Fig. 6.17). These pure 2*p* orbitals are capable of forming two π bonds by side-wise overlaps.

Thus ethyne molecule contains one σ and a two π bonds between the two carbons and each carbon is linked with one H-atom through σ bonds.

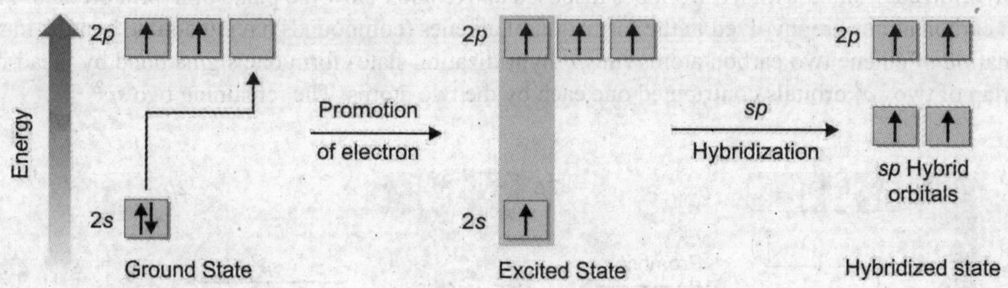

■ **Figure 6.17**
sp Hybridization of the orbitals of excited carbon.

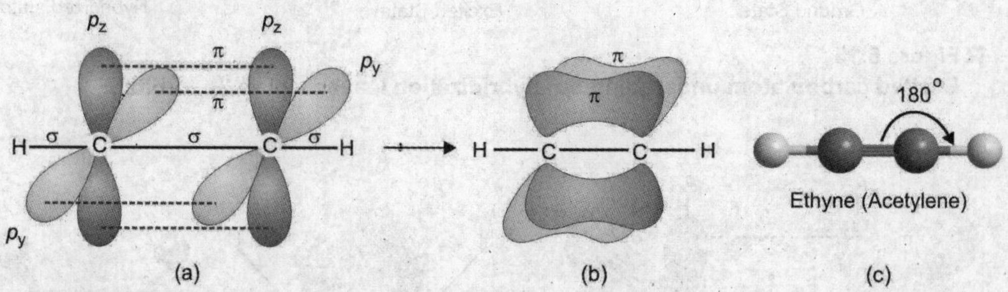

■ **Figure 6.18**
Formation of ethyne molecule. (*a*) shows scheme of overlaps; (*b*) shows the bonds, σ bonds being indicated by straight lines; and (*c*) shows ball-and-stick model of ethyne (acetylene).

Shape of Ammonia molecule, NH_3

There are three 2*p* bonding orbitals on the nitrogen atom $(2p_x^1, 2p_y^1, 2p_z^1)$. It will form three σ bonds by overlap with three 1*s* orbitals of the three H atoms. The orbital overlaps are shown in Fig. 6.19. The orbital structure of NH_3 has three orbital overlap axes inclined at an angle of 90° with N-atom as the origin. The four atoms in the molecule of NH_3 do not lie in the same plane but form a *pyramid* at whose base are the three H-atoms and the N-atom is at the apex of the pyramid. What we can predict about the H–N–H bond angles is that they are 90°, the angle between the axes. But this is erroneous and does not agree with the experimental value of 107°. The anomaly can be explained satisfactorily employing (*a*) the concept of hybrid orbitals and (*b*) the electron pair interactions.

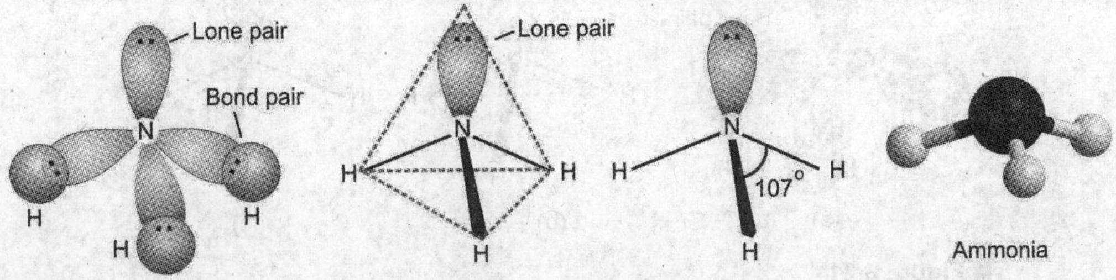

■ **Figure 6.19**
Orbital overlaps in the formation of Ammonia molecule.

(*a*) **sp³ Hybridization in ammonia molecule**

It is assumed that the valence orbitals of the central N-atom undergo hybridization before affecting overlaps with 1*s* orbitals of hydrogen. When the orbitals of the second energy level of N-atom ($2s^2$, $2p_x^1$, $2p_y^1$, $2p_z^1$) undergo sp^3 hybridization, four new hybrid orbitals result. One of these will have two electrons (like the original atom) and is non-bonding while the other three have one electron each and can form bonds by overlap. Now these hybrid orbitals are tetrahedrally dispersed with an angle of 109.5° between them. After hybridization, let the 1*s* orbitals of three H-atoms overlap to form three σ bonds (Fig. 6.21*a*). The tetrahedral angle 109.5° is quite near the experimental value 107°, and a difference of 2.5° can be explained by taking into consideration the electron pair interactions.

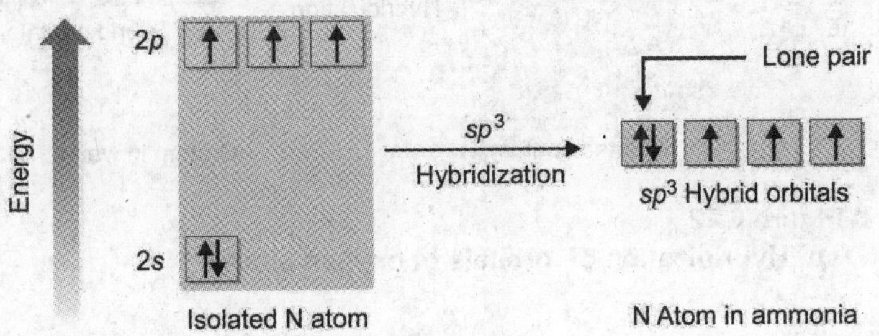

■ **Figure 6.20**
sp³ Hybridization of valence orbitals of N-atom.

(*b*) **Electron pair repulsion**

We have seen that the symmetrical central N-atom has in its valence shell, three bond pairs (*bp*), each shared with one of the three H-atoms, and also a lone pair (*lp*) of electrons. The three bond pairs and one lone pair may get arranged tetrahedrally about the central atom. The central atom exercises different pulls on them. The lone pair is attracted more towards the N-atom than the bond pairs which belongs to the H-atoms and N-atom jointly. This is because of the fact that the lone pair belongs *only* to the N-atom and hence its electron cloud is more concentrated near the N-atom. The lone pair is, therefore, capable of exerting a greater repulsion on a bond pair than a bond pair can repel another bond pair. As a result three bonds of ammonia molecule are forced slightly closer than in the normal tetrahedral arrangement. Therefore each of the HNH bond angles is 107° rather than the anticipated tetrahedral angle of 109.5°. It is also clear from the above reasoning that **more the number of lone pairs greater will be their influence in decreasing the bond angles.**

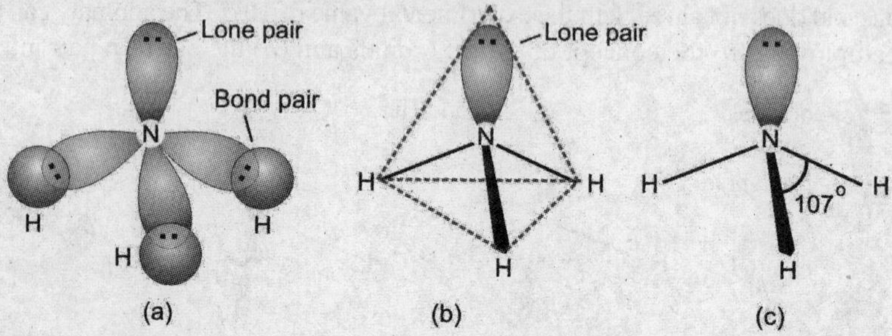

Figure 6.21
(*a*) Tetrahedral ammonia molecule (NH_3) with a lone-pair orbital in the vicinity of N-atom. (*b*) Tetrahedral geometry of NH_3.(*c*) Electron pair repulsions : *lp-bp* repulsion are greater than *bp-bp* repulsion.

Shape of Water molecule, H_2O

In the central oxygen atom of the molecule, there are two bonding orbitals ($2p_y^1$ and $2p_z^1$). These may overlap with 1*s* orbitals of two atoms of opposite spins. Each of these two overlaps results in the formation of a σ MO, giving two σ bonds in the molecule as a whole. The pictorial representation of the orbital overlaps is shown in Fig. 6.23. Since the molecule involves two 2*p* orbitals at right angles and the bond established by an orbital retains the directional character of the

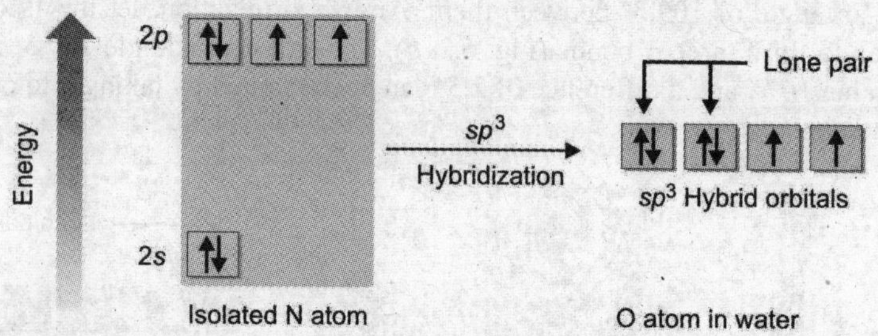

Figure 6.22
***sp*³ Hybridization of orbitals of oxygen atom.**

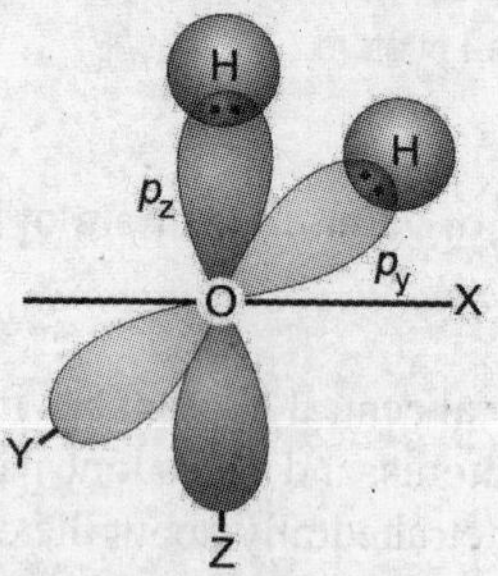

Figure 6.23
Overlap of $2p_y$ and $2p_z$ orbitals of O-atom with ls orbitals of two H-atoms.

bonding orbital, it is reasonable to expect the bond angle to be equal to 90°. But careful experiments reveal the HOH angle to be 104.3° rather than the predicted 90°. The discrepancy between the expected and the experimental values of the bond angle is best explained with the help of hybridization concept. The valence orbitals *i.e.*, of the second energy shell of oxygen atom all hybridize giving four tetrahedrally dispersed sp^3 hybrid orbitals. Two of these four hybrid orbitals have pairs of electrons with opposite spins and are non-bonding, while the rest two having one electron each are capable of bonding. They do so by overlaps with orbitals of two H atoms. An adequate guess of the HOH angle would be 109.5°, tetrahedral angle (Fig. 6.24 *a* & *b*). This is certainly in better agreement with the experimental value of 104.3° than our earlier contention of 90° on the basis of pure $2p$ orbital overlaps. But this is not all. The lone pair-bond pair repulsions have also to play their role. In water molecule there are two lone pairs in the vicinity of the central O-atom which has two bond pairs also. The repulsive forces operating are shown in Fig. 6.24 (*c*) above. Here we would expect the two lone pairs to repel each other more strongly than do a lone pair and a bond pair, and of course, even more strongly than two bond pairs. As a result, the two lone pairs of water force the two (O–H) bond pairs closer together than the one lone pair in case of ammonia forces together the three (N–H) bond pair. Thus the HOH angle is smaller (104.3°) than the HNH bond angles of 107°. In the light of the above discussions we can explain the molecular geometry of PH_3, PCl_3, NF_3, H_2S, etc.

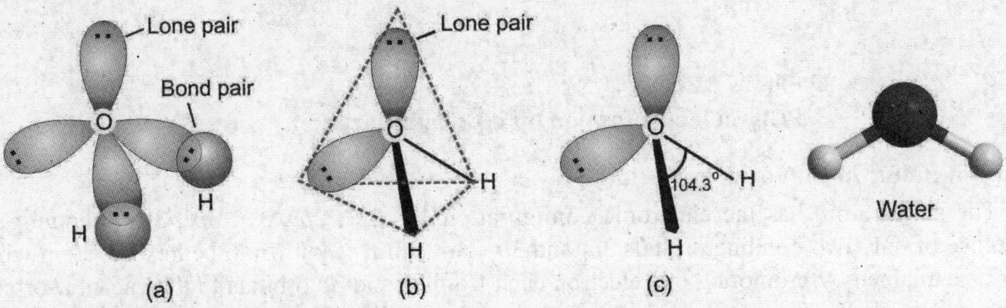

Figure 6.24
Tetrahedral model of water molecule showing two lone pairs on O-atom; (c) shows *bp-bp, lp-bp* and *lp-lp* repulsions.

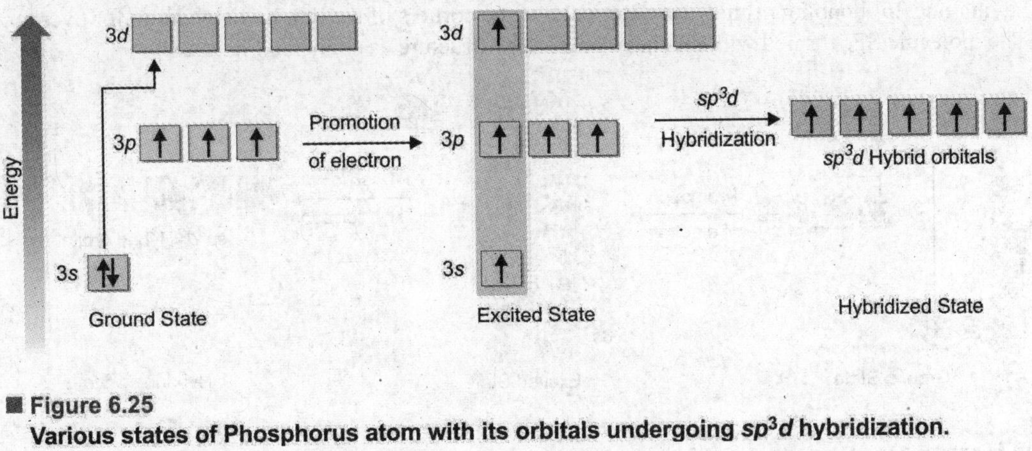

Figure 6.25
Various states of Phosphorus atom with its orbitals undergoing *sp³d* hybridization.

Shape of Phosphorus pentachloride molecule, PCl_5

In PCl_5 molecule, phosphorus is the central atom which has the electronic configuration $1s^2$, $2s^2$, $2p^6$, $3s^2$, $3p^3$, $3d^0$. In the ground state, it has only three bonding orbitals in the valence shell. One of the two $3s$ electrons uncouples itself and is promoted to the $3d$ orbital. The orbitals now hybridize in accordance with sp^3d type as follows. Three of these five bonding orbitals lie in a plane inclined at an angle of 120°, while the other two are directed above and below the plane in a direction perpendicular to the plane, taking the shape of a trigonal bipyramid. These orbitals of phosphorus atom can overlap with those of five chlorine atoms forming the PCl_5 molecule which will therefore have trigonal bipyramidal shape as shown in Fig. 6.26. Here, some of the bond angles are 90° while other bonds have an angle of 120° between them. This geometry of the molecule explains high reactivity of two of the five Cl atoms in PCl_5 molecule.

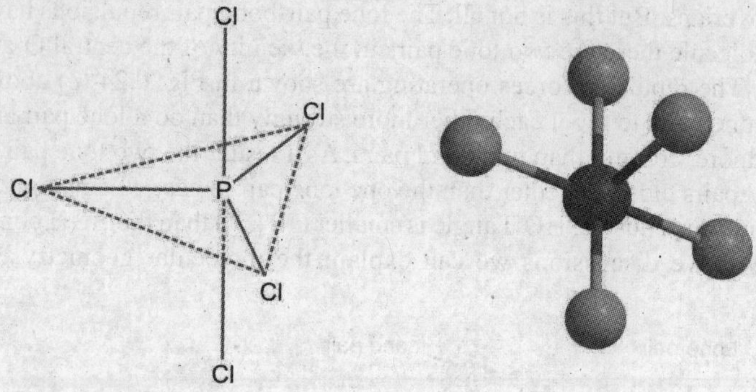

Figure 6.26.
PCl_5 molecule formed by sp^3d hybridization.

Shape of Sulfur hexafluoride molecule, SF_6

The sulfur atom has the electronic configuration $1s^2$, $2s^2$, $2p^6$, $3s^2$, $3p^4$, $3d^0$, showing the existence of only two bonding orbitals $3p_y$ and $3p_z$. But sulfur is known to be hexacovalent which may be explained by promoting one electron each from 3*s* and 3*p* orbitals to the vacant *d*-orbitals of the valence shell. The orbitals of the excited atom then undergo sp^3d^2 hybridization to produce six equivalent hybrid orbitals each having one electron. These hybrid orbitals are now available for the overlap after getting octahedrally dispersed (four of them lying in one plane inclined at an angle of 90° while the other two directed above and below the plane perpendicularly). Six fluorine atoms (each having one $2p^1$ bonding orbital) may approach at the corners of the regular octahedron for overlap. The molecule SF_6 formed will thus have octahedral structure as shown in Fig. 6.28.

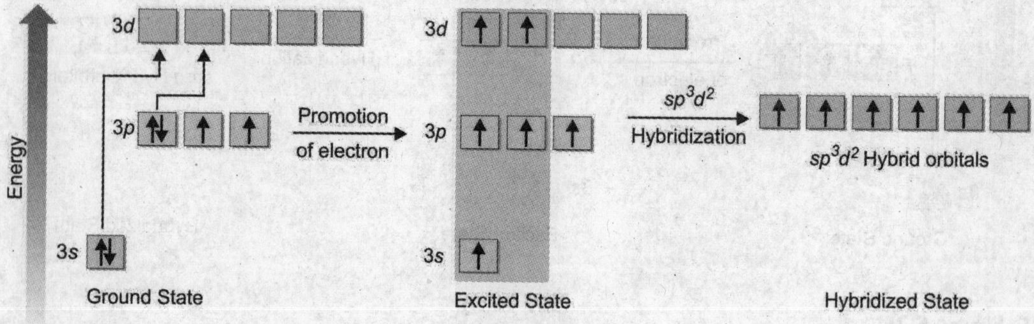

Figure 6.27
Various states of sulfur atom, the orbitals of the excited atom undergo sp^3d^2 hybridization.

Limitations of Valence Bond Theory

Valence Bond theory proposed by Heitler and London in 1927 explains the bonding in simple molecules. According to this theory a covalent bond is formed by the sharing of electron between the participating atoms. This results if the overlapping between the half filled atomic orbitals takes place and the strength of the bond formed depends upon the extent of overlapping between the atomic orbitals of two atoms. Though the valence bond theory explains the geometry of many molecules and ions yet it has following limitations :

(1) The presence of other nuclei should affect the electronic arrangement of all the atoms in the molecule

(2) Sometimes a single electronic structure does not explain all known properties of that molecule or ion and we therefore assume many electronic structures called resonating structures.

(3) Valence Bond theory fails to explain the bonding in electron deficient compounds.

(4) It fails to explain the paramagnetic character of oxygen molecule.

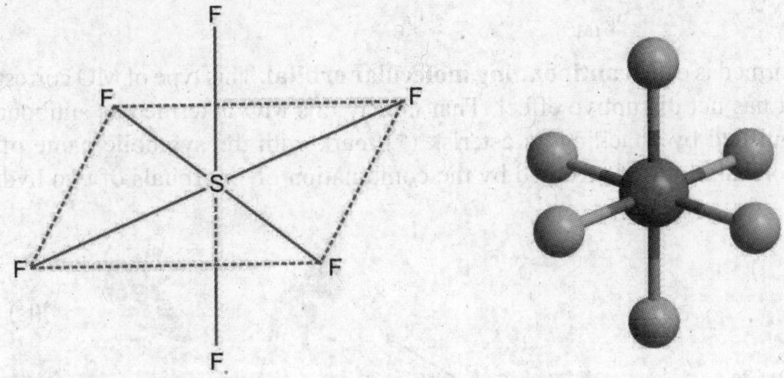

Figure 6.28
Formation of SF_6 molecule by sp^3d^2 hybridization.

MOLECULAR ORBITAL THEORY

Molecular orbital theory proposed by Hund and Mulliken in 1932 explains the formation of a covalent bond in a better way. According to molecular orbital theory all atomic orbitals of the atoms participating in molecule formation get disturbed when the concerned nuclei approach nearer. They all get mixed up to give rise to an equivalent number of new orbitals that belong to the molecule now. These are called *molecular orbitals*. Thus every molecule is supposed to have molecular orbitals associated with it in much the same way as a single isolated atom has. The main features of molecular orbital theory can be summed up as follows :

(1) A molecule is quite different from its constituent atoms. All the electrons belongs to the constituent atom and are considered to be moving under the influence of all nuclei.

(2) Atomic orbitals of individual atoms combine to form molecular orbitals and these MOs are filled up in the same way as atomic orbitals are formed. In other words, Pauli's exclusion principle, Aufbau principle and Hund's rule of maximum multiplicity are followed.

(3) The molecular orbitals have definite energy levels.

(4) The shapes of MOs formed depend upon the shape of combining atomic orbitals.

Linear Combination of Atomic orbitals (LCAO method)

According to wave mechanics the atomic orbitals can be expressed as wave functions (ψ_s) which represent the amplitude of electron waves. Their values can be calculated from the solutions of Schrödinger's wave equation. In a similar way, Schrödinger's wave equation can be applied to molecules but with slight difficulty. To cope with the difficulty an approximate method known as Linear Combination of atomic orbitals is applied.

According to this method, the molecular orbitals are formed by the linear combination (addition or subtraction) of atomic orbitals of the constituent atoms of the molecule. Let us consider the simplest case of H_2 molecule consisting of two hydrogen atoms represented by H_A and H_B. The atomic orbitals of these atoms are represented by the wave functions ψ_A and ψ_B. When these atoms approach each other there come two possibilities.

(1) Molecular orbital is formed by the **addition** of wave functions of atomic orbitals. It can be represented by

$$\psi_{(MO)} = \psi_A + \psi_B \qquad ...(i)$$

The M.O. formed is called **bonding molecular orbital.** It lowers the energy and brings about the stability in the system.

(2) Molecular orbital is formed by the **subtraction** of wave functions of atomic orbitals. It can be represented by

$$\psi^*_{(MO)} = \psi_A - \psi_B \qquad ...(ii)$$

The MO formed is called **antibonding molecular orbital.** This type of MO corresponds to higher energy state. It has net disruptive effect. That is why this MO is termed as antibonding molecular orbital, distinguished by attaching an asterisk (*) mark with the symbolic name of the molecular orbital. The molecular orbitals formed by the combination of 1*s* orbitals of two hydrogen atoms is shown below :

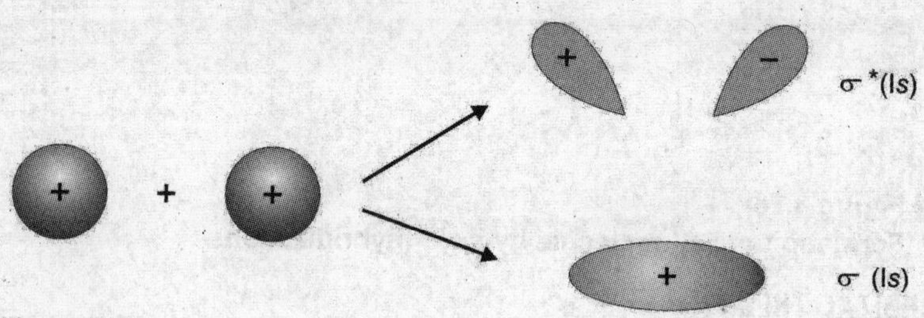

Figure 6.29
Formation of molecular orbitals from two 1s orbitals of hydrogen atoms.

The wave functions either have (+) or (–) sign. The positive sign shown on 1*s* orbitals of hydrogen atoms shows that these orbitals are spherically symmetrical and their wave functions are associated with +ve sign in all directions. In case of *p*-orbitals one lobe has +ve sign while other has –ve sign. The overlapping of atomic orbitals takes place if they have similar signs on their lobes.

The probability density in bonding and antibonding molecular orbitals is given by the square of the wave functions shown in the equations (*i*) and (*ii*)

$$\psi^2_{(MO)} = (\psi_A + \psi_B)^2$$

$$= \psi_A^2 + \psi_B^2 + 2\psi_A\psi_B \qquad ... (iii)$$

and

$$\psi^{*2}_{(MO)} = (\psi_A - \psi_B)^2$$

$$= \psi_A^2 + \psi_B^2 - 2\psi_A\psi_B \qquad ... (iv)$$

From equation (*iii*) it is clear that probability density of bonding molecular orbital is greater than the sum of the probability densities of isolated atoms by the factor $2\psi_A\,\psi_B$. Thus the probability density of electrons in bonding MO is greater than that in either of atomic orbitals ($\psi_A{}^2$ or $\psi_B{}^2$).

From equation (*iv*) it is clear that probability density of antibonding MO is lesser than the sum of probability densities of isolated atoms by a factor $2\psi_A\,\psi_B$. Thus the probability density of electrons in antibonding MO is less than that in either of atomic orbitals ($\psi_A{}^2$ or $\psi_B{}^2$). In a similar way, the combination of two 2*s* orbitals produces a bonding σ (2*s*) and antibonding σ* (2*s*) molecular orbitals.

Relative Energies of bonding and antibonding MOs

We have seen above that when atomic orbitals combine, an equivalent number of new orbitals is formed. For example, when two 1*s* orbitals combine two new molecular orbitals are formed. One of these pertains to the bonding molecular orbital with lower energy while the other corresponds to higher energy as compared to both the atomic orbitals concerned in the process. Thus we find that the number of molecular orbitals formed from atomic orbitals is equal to the number of atomic orbitals responsible for their formation and are of two types as shown below. (Fig. 6.30)

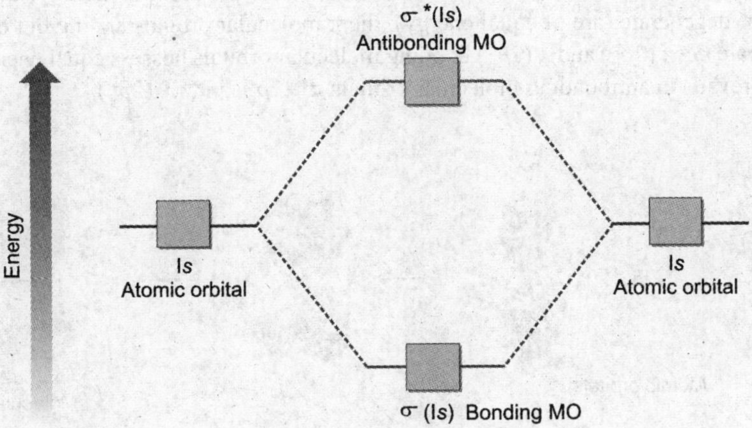

■ Figure 6.30
Energy level diagram for bonding and antibonding molecular orbitals arising from 1*s* atomic orbital.

Molecular Orbitals from p-Atomic Orbitals

Combination of *p*-orbitals is slightly complicated. There being three types of 2*p* orbitals namely $2p_x$, $2p_y$ and $2p_z$ directed along the cartesian coordinates x, y and z respectively. By convention we assume that z-axis is the intermolecular axis.

(1) **Combination of $2p_z$ atomic orbitals**

When two $2p_z$ orbitals on two atoms approach each other along the internuclear axis they combine by the addition of electron wave. This results in the formation of bonding σ $(2p_z)$ orbital while subtraction of electron waves give rise to antibonding molecular orbital σ* $(2p_z)$. It is represented as follows :

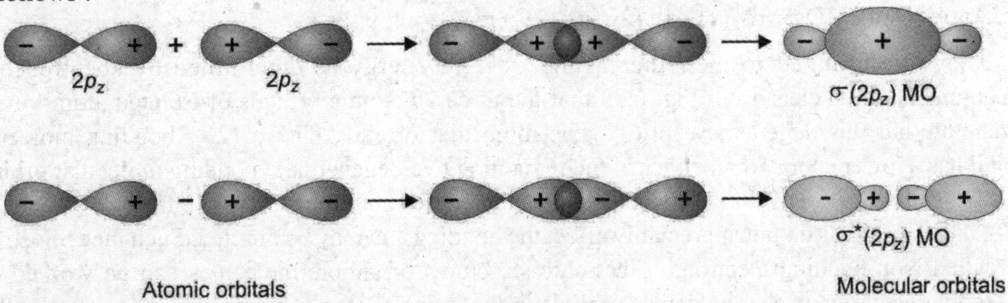

■ Figure 6.31
Combination of $2p_z$ orbitals to form bonding σ $(2p_z)$ and antibonding σ*$(2p_z)$ molecular orbitals.

(2) Combination of $2p_y$ and $2p_x$ atomic orbitals

When two $2p_y$ or $2p_x$ orbitals on two atoms are approaching in a manner that their axes are mutually parallel, they interact to give rise to the formation of molecular orbitals that are not symmetrical about the internuclear axis and are called molecular orbitals represented by π ($2p_y$) or π ($2p_x$) bonding orbitals. These π ($2p_y$) and π ($2p_x$) bonding MOs have zero electron density on the plane that contains the nuclear axis (nodal plane), while the electron density is concentrated in two regions above and below the nodal plane. The π^* ($2p_y$) and π^* ($2p_x$) antibonding MOs have higher energy than their corresponding molecular and atomic orbitals. Since $2p_y$ and $2p_x$ atomic orbitals on an atom are degenerate (are of equal energy), the π molecular orbitals arising out of them will also be degenerate *i.e.* π ($2p_y$) and π ($2p_x$) bonding molecular orbitals possess equal energy and the same situation prevails in antibonding molecular orbitals π^* ($2p_y$) and π^* ($2p_x$).

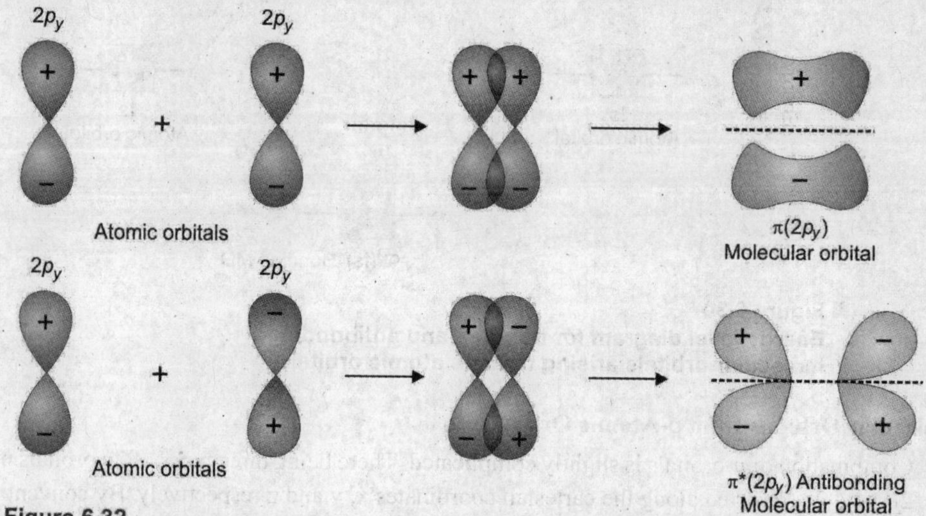

■ **Figure 6.32**
Formation of bonding and antibonding molecular orbitals by the side-wise overlapping of $2p_y$ atomic orbitals. $2p_x$ orbitals also combine in the similar way.

Since we are concerned here with few simple molecules, the only molecular orbitals which need to be considered are

Bonding MOs = σ (1s) σ (2s) σ $2p_z$ π ($2p_y$) π $2p_x$

Antibonding MOs = σ* (1s) σ* (2s) σ* ($2p_z$) π* ($2p_y$) π* ($2p_x$)

They are arranged in their increasing order of energy as determined by spectroscopic measurements. It is clear from Fig. 6.33 that the three 2p atomic orbitals of isolated atoms are all degenerate but this degeneracy splits in the molecular orbitals. The σ ($2p_z$) bonding molecular orbital has a lower energy than either of the remaining two degenerate π bonding molecular orbitals –π ($2p_y$) and π ($2p_x$). For diatomic molecules obtained from atom having more than one electronic shell, the inner shells do not appreciably affect the bonding and may be omitted. Such inner electrons are called Non-bonding electrons. Electronic structures of simple molecules can be worked out by feeding electrons in the molecular orbitals in the order of increasing energy. Here also Aufbau principle–feeding electrons in the increasing order of energy of orbitals repeats itself. This order shown in Fig. 6.33 may be written as :

$$\sigma (1s)\ \sigma^* (1s)\ \sigma (2s)\ \sigma^* (2s)\ \sigma (2p_z)\ \pi (2p_y) = \pi (2p_x)\ \pi^* (2p_y) = \pi^* (2p_x)\ \sigma^* (2p_z)$$

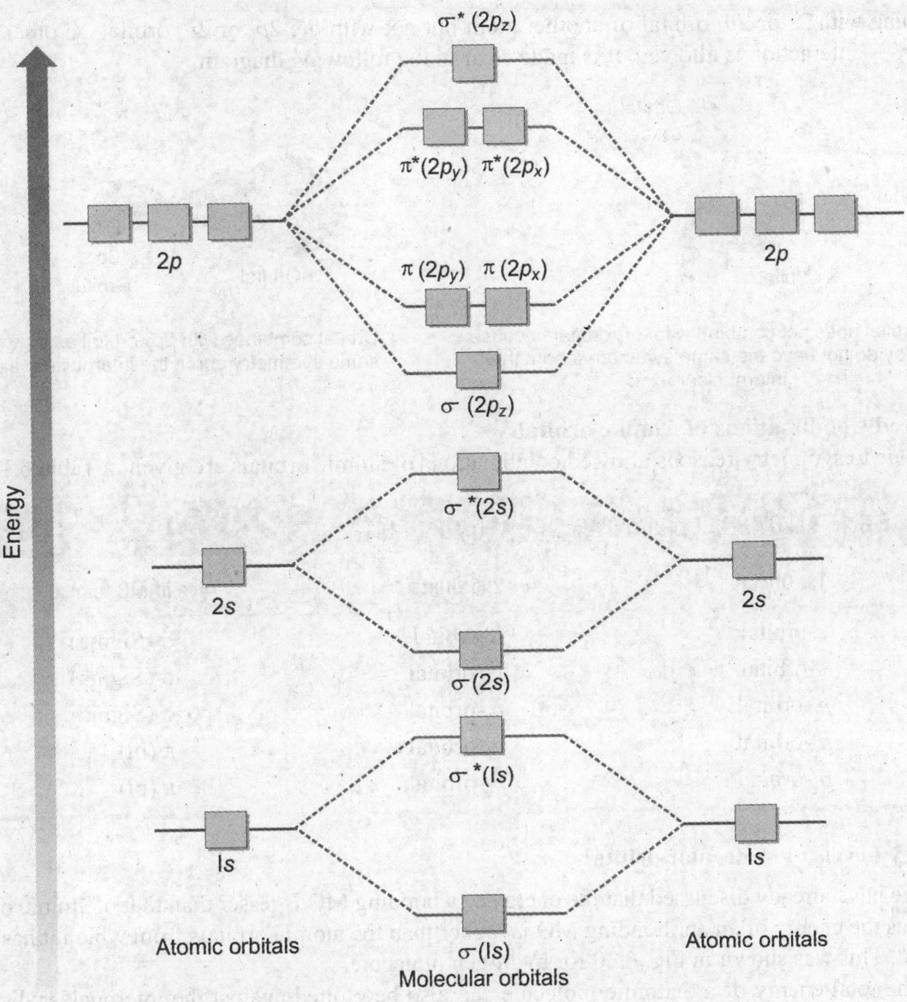

■ **Figure 6.33**
Schematic energy level representation of molecular orbitals for simple molecules of elements.

Conditions for the combination of Atomic Orbitals

The atomic orbitals may combine when certain conditions are met with. For effective combinations of atomic orbitals following are the conditions.

(1) The energies of atomic orbitals should be comparable. For homonuclear diatomic molecules of the type X_2 similar atomic orbitals combine *i.e.* 1*s* orbital of one atom combines with 1*s* orbital of another and similarly 2*s* orbital combines with 2*s* orbital of another atom and so on. This means that 1*s* orbital cannot combine with 2*s* orbital because of large energy difference between the two orbitals. In case homonuclear molecules this condition may not be valid.

(2) The extent of overlap between the atomic orbitals should be to a considerable extent. Greater the overlap between the atomic orbitals, greater is charge density between the nuclei.

(3) The symmetry of the combining atomic orbitals should be the same. The atomic orbitals should have the same symmetry about the internuclear axis. For example, 2*s* orbital of one atom can combine with $2s$ or $2p_z$ orbital of another atom but not with the $2p_y$ or $2p_x$ orbital. In other words ++ or – – interaction is allowed. It is made clear in the following diagram :

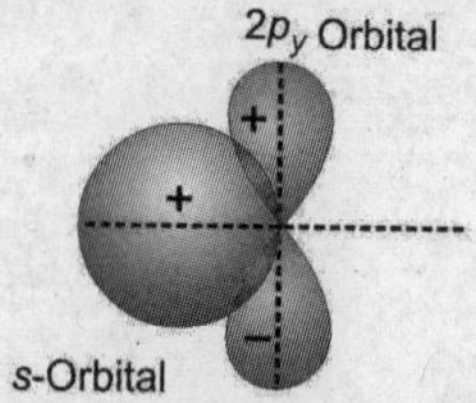

s-Orbital does not combine with $2p_y$ orbital because they do not have the same symmetry about the internuclear axis.

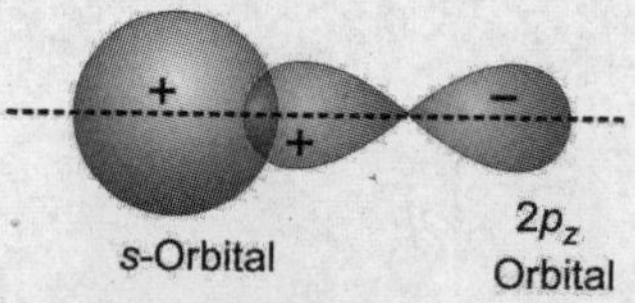

s-Orbital combines with $2p_z$ orbital as they have same symmetry about the internuclear axis.

Allowed Combinations of Atomic orbitals

Due to symmetry reasons allowed combinations of atomic orbitals are given in Table 6.1.

TABLE 6.1. ALLOWED COMBINATIONS OF ATOMIC ORBITALS

1st Orbital	2nd Orbital	Type of MO formed
s-orbital	*s*-orbital	σ (Sigma)
s-orbital	p_z-orbital	σ (Sigma)
p_z-orbital	p_z-orbital	σ (Sigma)
p_y-orbital	p_y-orbital	π (pi)
p_x-orbital	p_x-orbital	π (pi)

Energy Levels of Molecular orbitals

We have already discussed that the energy of a bonding MO is lesser than that of atomic orbital whereas the energy of an antibonding MO is higher than the atomic orbitals from which it has been formed. This was shown in Fig. 6.30 for hydrogen molecule.

The total energy of a diatomic molecule can also be plotted against the internuclear distance between its atoms. When these two atoms are brought closer the energy changes are shown in Fig. 6.34.

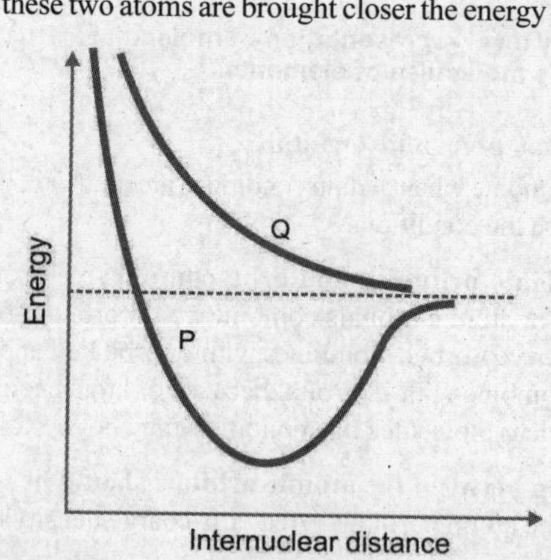

Figure 6.34
Plot of energy of diatomic molecules versus internuclear distance.

The curve P represents the formation of a bonding molecular orbital and the curve Q represents the formation of an antibonding molecular orbital. It is clear from the curve Q that repulsive forces dominate during the formation of an antibonding molecular orbital. During the formation of a bonding molecular orbital the energy starts decreasing and reaches to a minimum value which corresponds to the bond length between the two atoms (r_0). At a distance less than r_0 the repulsive forces start dominating over the attractive forces and energy starts increasing.

Difference Between Bonding and Antibonding Molecular Orbitals

The difference between bonding and antibonding orbitals is summarized below :

TABLE 6.2. DIFFERENCE BETWEEN BONDING AND ANTIBONDING MOLECULAR ORBITALS

Bonding MO	Antibonding MO
1. It is formed by **addition** overlap of atomic orbitals.	1. It is formed by **subtraction** overlap of atomic orbitals.
2. The wave function of a bonding MO is given by $\psi_{(MO)} = \psi_A + \psi_B$	2. The wave function of an antibonding MO is given by $\psi^*_{(MO)} = \psi_A - \psi_B$
3. Its formation takes place when the lobes of atomic orbitals have **same** signs.	3. Its formation takes place when the lobes of atomic orbitals have **different** signs.
4. The energy of bonding MO is **lower** than that of atomic orbitals from which it is formed.	4. The energy of antibonding MO is **higher** than that of atomic orbitals from which it is formed.
5. The electron density is **high** in the region between the nuclei of bonded atoms.	5. The electron density is **low** in the region between the nuclei of bonded atoms.
6. Every electron in bonding MO contributes towards the **attractive force.**	6. Every electron in antibonding contributes towards **repulsive force.**

Combination of s and p_z-orbitals

s and p_z-orbitals can also combine in a similar way along internuclear axis to form bonding and antibonding molecular orbitals as shown in Fig. 6.35.

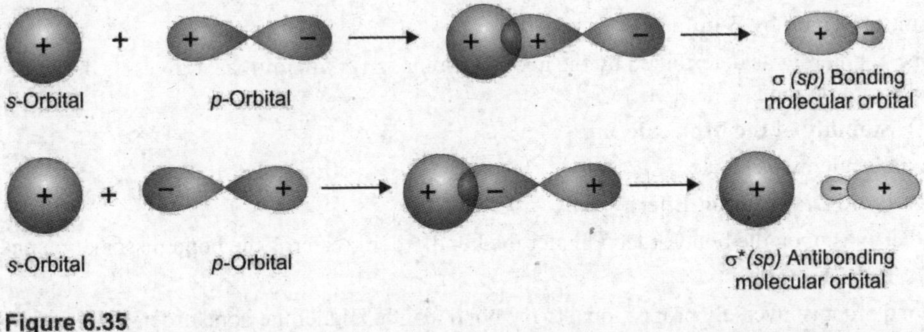

Figure 6.35
Formation of bonding and antibonding orbitals by the combination of s and p_z orbitals.

Difference between Sigma and Pi molecular orbitals

The main points of difference between sigma and Pi molecular orbitals can be summed up in the Table 6.3.

TABLE 6.3. DIFFERENCE BETWEEN SIGMA AND PI MOLECULAR ORBITALS

Sigma (σ) MO	Pi (π) MO
1. It is formed by **head to head** overlapping of atomic orbitals.	1. It is formed by the **sidewise** overlapping of p_y or p_x atomic orbitals.
2. The overlap is **along internuclear axis.** As a result of this, **strong** bond is formed.	2. The overlap is **perpendicular** to **internuclear** axis. As a result of this, **weak** bond is formed.
3. **Rotation** is possible around σ MO.	3. **Rotation** is **not** possible around π MO.
4. The region of overlap is **greater.**	4. The region of overlap is **lesser.**

BOND ORDER

How many are the bonds in a molecule or **Bond order** is correctly predicted by the Molecular Orbital theory. The number of bonds in a molecule is one-half of the difference of the number of electrons in the bonding molecular orbitals and the number of electrons in the antibonding molecular orbitals. Mathematically,

$$\text{Bond order} = \frac{\begin{pmatrix}\text{No. of electrons in}\\ \text{bonding orbitals}\end{pmatrix} - \begin{pmatrix}\text{No. of electrons in}\\ \text{antibonding orbitals}\end{pmatrix}}{2} = \frac{N_b - N_a}{2}$$

In common practice, only MOs formed from valence orbitals are considered for determining bond order.

The predicted bond orders are found to be in conformity with other methods of determining it. Thus Molecular Orbital theory helps in establishing whether the existence of a bond is feasible or not. For instance, the formation of He_2 molecule, discussed in the following section, is completely ruled out as the bond order predicted is zero.

In general,

(*a*) A molecule is stable if $N_b > N_a$

(*b*) A molecule is unstable if $N_b < N_a$

Information given by Bond order

The bond order as determined by the above formula is very helpful as it gives us the following valuable information :

(1) Stability of the Molecule/Ion

A molecule/ion is stable if $N_b > N_a$

(2) Bond Dissociation Energy

It depends upon the bond order. Greater the bond order greater is the bond dissociation energy.

(3) Bond Length

Bond order is inversely proportional to the bond length. Higher the bond order, smaller the bond length.

(4) Magnetic Properties

The presence of unpaired electrons in Molecular orbitals of a species makes it paramagnetic in nature. Greater the number of unpaired electrons, the more will be its paramagnetic character. On the other hand, if there are no unpaired electrons in MOs the species will be diamagnetic in nature.

HOMONUCLEAR DIATOMIC MOLECULES

After having discussed the basic principles of Molecular Orbital theory, we are now able to take up the electronic structures and bonding properties of some of the homonuclear diatomic molecules.

(1) **Hydrogen, H_2**

Hydrogen molecule is formed from $1s^1$ atomic orbitals of two atoms. They give rise to two molecular orbitals $\sigma(1s)$ and $\sigma^*(1s)$. The molecule has two electrons to be accommodated in these two molecular orbitals and are filled as desired by Aufbau order of filling. Both these electrons go to the lower energy bonding molecular orbital and the antibonding orbital remains vacant. The electronic configuration of the molecule is represented by the equation :

$$2H\,(1s^1) \longrightarrow H_2[\sigma\,(1s^2)\,\sigma^*\,(1s^0)]$$

The number of bonds in the molecule, accordingly is $= \frac{1}{2}\,(2-0) = 1$. Thus two hydrogen atoms are bonded through only one bond in the molecule.

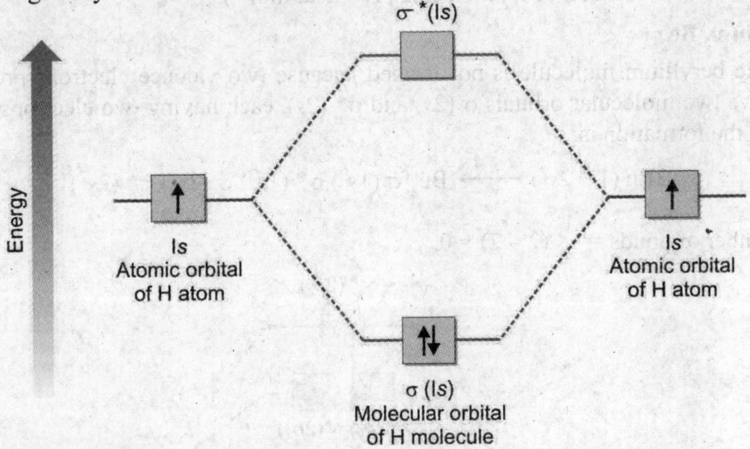

■ **Figure 6.36**
Molecular orbital diagram of H_2 molecule.

(2) **Helium, He_2**

Helium atoms have the configuration $1s^2$. For the formation of He_2, the molecular orbitals required are $\sigma\,(1s)$ and $\sigma^*\,(1s)$ which have to accommodate four electrons (two each contributed by the atoms). Each one of these molecular orbitals would accommodate two electrons as shown below.

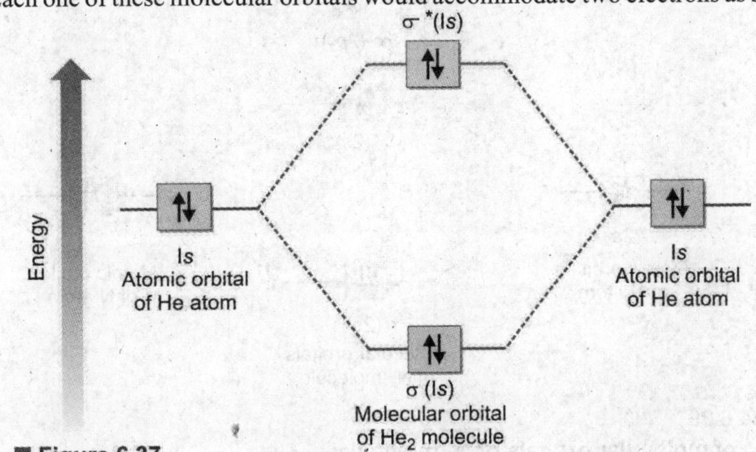

■ **Figure 6.37**
Molecular orbital diagram of He_2 molecule.

The molecule of helium is represented by the equation,

$$2He\ (1s^2) \longrightarrow He_2[\sigma\ (1s^2)\ \sigma^*\ (1s^2)]$$

There are two electrons in each of the bonding and antibonding orbitals giving zero as the number of bonds, $\frac{1}{2}\ (2-2) = 0$ *i.e.*, this molecule does not exist.

(3) Lithium, Li_2

Lithium atom has one electron in the valence shell $2s^1$, the other electrons are non-bonding. Two such atoms give a molecule having two molecular orbitals $\sigma\ (2s)$ and $\sigma^*\ (2s)$ and only two electrons are to be filled in them. These are accommodated in the molecular bonding orbital while the antibonding molecular orbital remains vacant. The number of bonds = $\frac{1}{2}\ [(2-0)] = 1$ *i.e.*, two lithium atoms form a molecule involving a single covalent bond. The bond energy is about 25 kcal mol^{-1} which is much less than of H_2 molecule (103 kcal mol^{-1}).

(4) Beryllium, Be_2

Diatomic beryllium molecule is not formed because two valence electrons present as $2s^2$ in Be atoms give two molecular orbitals $\sigma\ (2s)$ and $\sigma^*\ (2s)$, each having two electrons. The equation representing the formation is,

$$2Be\ (1s^2\ 2s^2) \longrightarrow Be_2[\sigma\ (1s^2)\ \sigma^*\ (1s^2)\ \sigma\ (2s^2)\ \sigma^*\ (2s^2)]$$

The number of bonds = $\frac{1}{2}\ (2-2) = 0.$

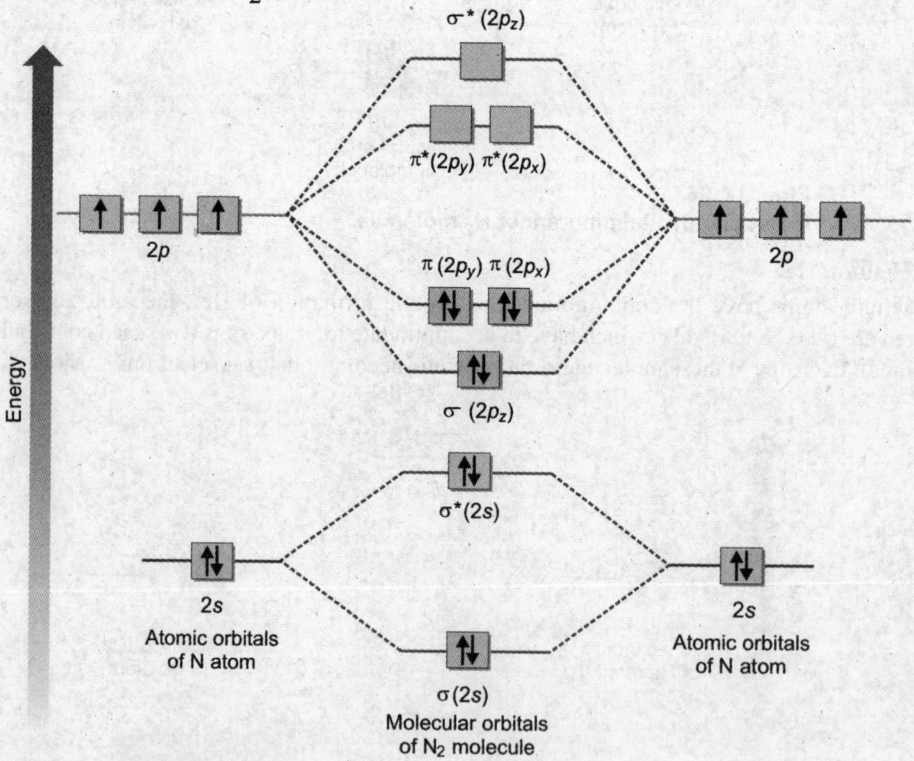

Figure 6.38
Filling of molecular orbitals of N_2 molecule; three anitbonding orbitals are vacant.

(5) Nitrogen, N_2

Each of the two N-atoms $(1s^2, 2s^2, 2p_x^1, 2p_y^1, 2p_z^1)$ contribute, (2 + 1 + 1 + 1) = 5 valence electrons and four atomic orbitals towards the molecule. The molecule would have eight molecular orbitals and ten electrons to be accommodated in them. The electronic configuration in the molecule is represented as given in Fig. 6.38.

There are eight electrons in the bonding molecular orbitals and only two in antibonding orbital. This gives the number of bonds $\frac{1}{2}$ (8 – 2) = 3. The two N atoms are bonded with three covalent bonds *i.e.*, a triple bond. N_2 molecule is diamagnetic as it has no unpaired electron.

(6) Oxygen, O_2

The two participating oxygen atoms $(1s^2, 2s^2, 2p_x^2, 2p_y^1, 2p_z^1)$ contribute a total of 2 (2 + 2 + 1 + 1) = 12 valence electrons and 2 × 4 = 8 atomic orbitals towards the oxygen molecule. There are eight molecular orbitals and 12 electrons are accommodated as shown in Fig. 6.39.

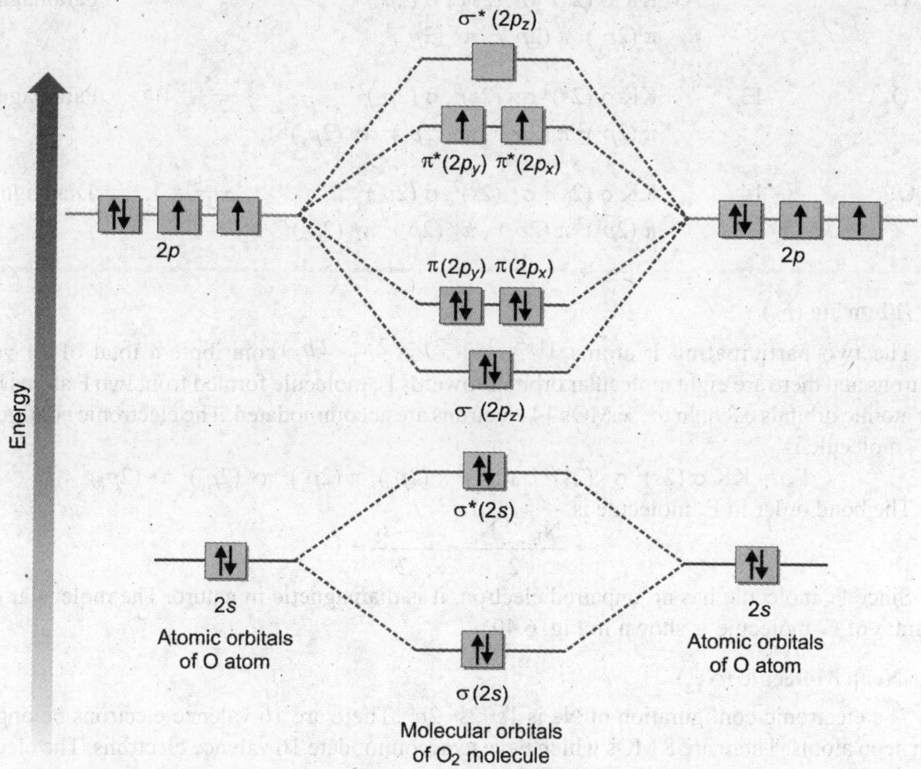

■ **Figure 6.39**
Molecular orbitals of O_2 molecule, having one full, two half-filled and one vacant antibonding orbitals; and all the bonding molecular orbitals are full.

We find from Fig. 6.39 that Aufbau order of filling of molecular orbital leaves two unpaired electrons in each of degenerate $\pi_y^*(2p_y)$ and $\pi_z^*(2p_x)$ orbitals. The electronic configuration of the molecular orbitals accounts admirably for the paramagnetic behavior of oxygen molecule. It is the lone electron that creates a magnetic field around it due to its spin. But if there are two electrons spinning near each other in opposite directions *i.e.*, with their spins paired, the magnetic fields created by these electrons cancel each other. However, if the electrons spin in the same direction, the total magnetic field would be the sum of the fields of each of the two electrons. This situation prevails in oxygen molecule and the experimentally established extent or magnitude of paramagnetism is consistent with two unpaired electrons per molecule. This is the greatest success of the Molecular Orbital theory. The number of bonds in O_2 molecules $= \frac{1}{2}(8-4) = 2$.

Proceeding on these lines, the electronic configurations of other simple species like O_2^-, O_2^+ can be worked out. The electronic configuration, bond order and magnetic character of these species is summed in the Table 6.4.

TABLE 6.4. ELECTRONIC CONFIGURATION, BOND ORDER AND MAGNETIC CHARACTER OF O_2^+, O_2^- AND O_2^{2-}

Species	No. of valence Electrons	Electronic Configuration	Bond Order	Magnetic Character
O_2^+	11	KK $\sigma(2s)^2\ \sigma^*(2s)^2,\ \sigma(2p_z)^2$ $\pi(2p_y)^2\ \pi(2p_x)^2,\ \pi^*(2p_y)^1$	$\frac{8-3}{2} = 2.5$	Paramagnetic
O_2^-	13	KK $\sigma(2s)^2\ \sigma^*(2s)^2,\ \sigma(2p_z)^2$ $\pi(2p_y)^2\ \pi(2p_x)^2,\ \pi^*(2p_y)^2\ \pi^*(2p_x)^1$	$\frac{8-5}{2} = 1.5$	Paramagnetic
O_2^{2-}	14	KK $\sigma(2s)^2\ \sigma^*(2s)^2,\ \sigma(2p_z)^2$ $\pi(2p_y)^2\ \pi(2p_x)^2,\ \pi^*(2p_z)^2\ \pi^*(2p_x)^2$	$\frac{8-6}{2} = 1$	Diamagnetic

(7) **Fluorine (F_2)**

The two participating F atoms $(1s^2, 2s^2, 2p_x^2, 2p_y^2, 2p_z^1)$ contribute a total of 14 valence electrons and there are eight molecular orbitals towards F_2 molecule formed from two F atoms having four atomic orbitals each. In these MOs 14 electrons are accommodated. The electronic configuration of F_2 molecule is

$$F_2 = KK\ \sigma(2s)^2\ \sigma^*(2s)^2\ \sigma(2p_z)^2\ \pi(2p_y)^2\ \pi(2p_x)^2\ \pi^*(2p_y)^2\ \pi^*(2p_x)^2$$

The bond order in F_2 molecule is

$$= \frac{N_b - N_a}{2} = \frac{8-6}{2} = 1$$

Since F_2 molecule has no unpaired electron, it is diamagnetic in nature. The molecular orbital diagram of F_2 molecule is shown in Fig. 6.40.

(8) **Neon Molecule (Ne_2)**

The electronic configuration of Ne is $1s^2\ 2s^2\ 2p^6$. There are 16 valence electrons belonging to both neon atoms. There are 8 MOs which have to accommodate 16 valence electrons. The electronic configuration of Ne_2 molecule is given below :

$$Ne_2 = KK\ \sigma(2s)^2\ \sigma^*(2s)^2\ \sigma(2p_z)^2\ \pi(2p_y)^2\ \pi(2p_x)^2\ \pi^*(2p_y)^2\ \pi^*(2p_x)^2\ \pi^*(2p_z)^2$$

$$\text{The bond order} = N_b - N_a$$

$$= \frac{8-8}{2} = 0$$

Since no bond is formed between two Ne atoms, this molecule does not exist.

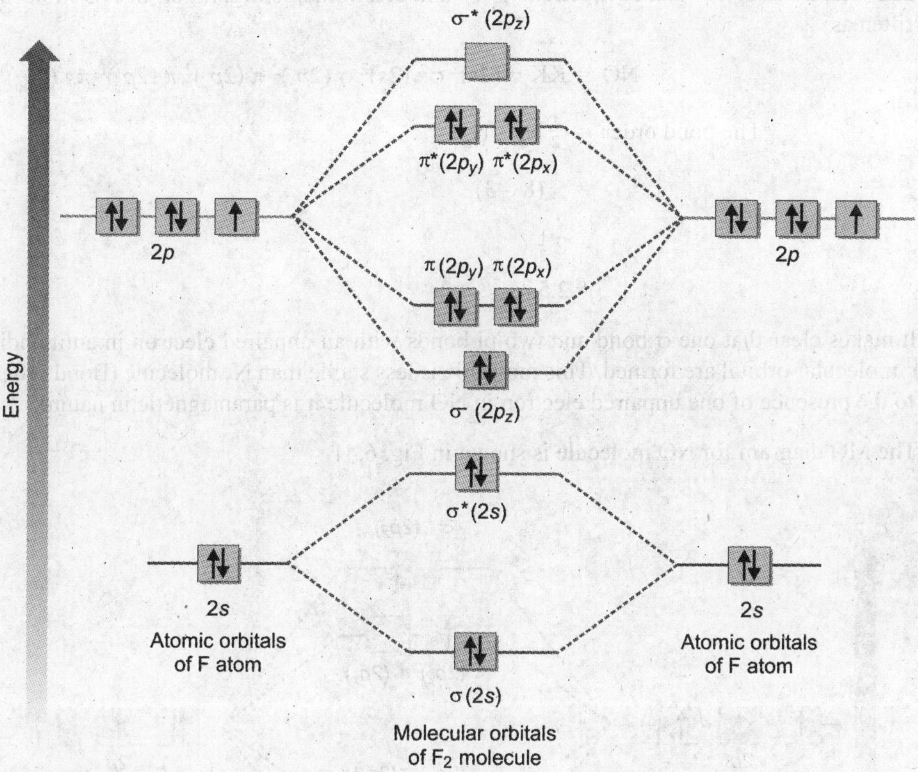

■ **Figure 6.40**
Molecular orbital diagram of F_2 molecule.

HETERONUCLEAR DIATOMIC MOLECULES

When two different atoms are bonded together, their molecule is called heteronuclear molecule. The same general principle applies to heteronuclear molecules but the main difference is that in heteronuclear molecule different atoms contribute unequally to the wave function *i.e.*

$$\psi_{MO} = C_1 \psi_1 \text{AO} + C_2 \psi_2 \text{AO}$$

where C_1 and C_2 are two constants having different values for different atoms. Also the molecular orbitals formed are unsymmetrical due to difference in electronegativities. In the construction of Molecular Orbital diagrams for heteronuclear molecules the bonding MOs are shown closed to electronegative atoms while antibonding MOs are shown closer to lesser electronegative atom.

It may be noted that if two different molecules have the same total number of electrons their molecular energy diagram will be similar.

Let us now construct the molecular energy diagram for some heteronuclear molecules.

(1) **Nitric Oxide (NO)**

The electronic configuration of participating N and O atoms are

$$N = 1s^2\ 2s^2\ 2p_x^{\ 1}\ 2p_y^{\ 1}\ 2p_z^{\ 1} \text{ (5 valence electrons)}$$

$$O = 1s^2\ 2s^2\ 2p_x^{\ 2}\ 2p_y^{\ 1}\ 2p_z^{\ 1} \text{ (6 valence electrons)}$$

The total number of valence electrons is 11 and electronic configuration of NO molecule can be written as

$$NO = KK\ \sigma (2s)^2\ \sigma^* (2s)^2\ \sigma (2p_z)^2\ \pi (2p_y)^2\ \pi (2p_x)^2\ \pi^* (2p_y)^1$$

$$\begin{aligned} \text{The bond order} &= \tfrac{1}{2} N_b - N_a \\ &= \tfrac{1}{2}(8 - 3) \\ &= 2\tfrac{1}{2} \\ &= 2.5 \end{aligned}$$

It makes clear that one σ bond and two pi bonds with an unpaired electron in antibonding $\pi^* (2p_y)^1$ molecular orbital are formed. This molecule is less stable than N_2 molecule (Bond order = 3). Due to the presence of one unpaired electron in NO molecule it is paramagnetic in nature.

The MO diagram for NO molecule is shown in Fig. 6.41.

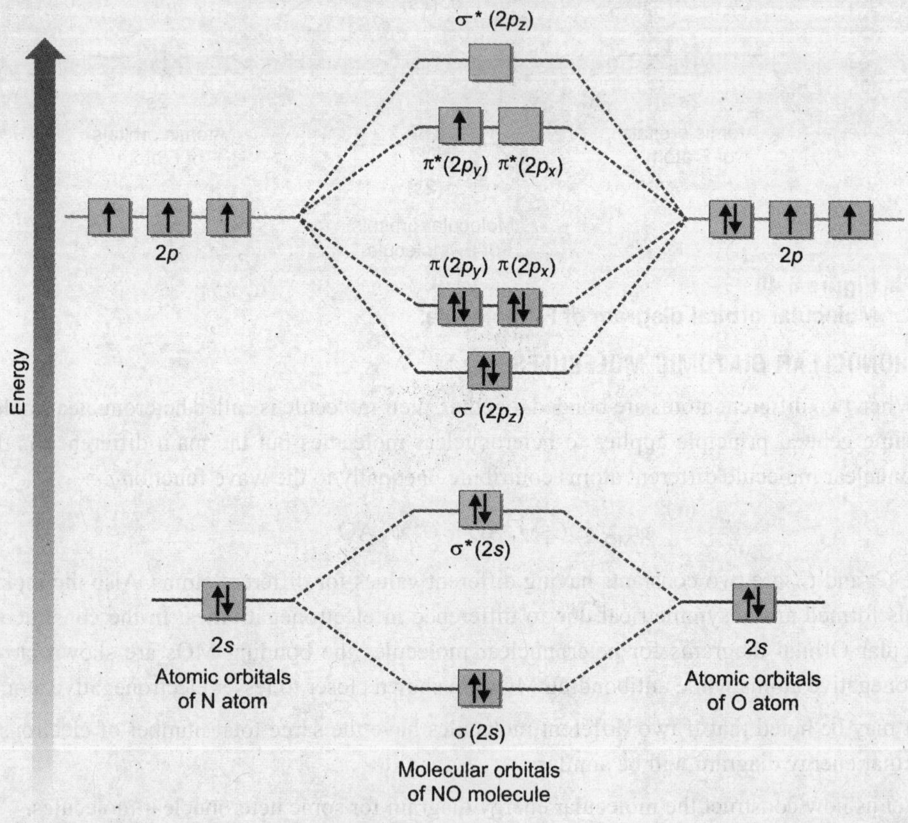

■ Figure 6.41
Molecular orbital diagram of NO molecule.

(2) **NO^+ and NO^- Ions**

On similar lines we can write the electronic configuration, bond order and magnetic character of these ions. These are summarized in Table 6.5.

TABLE 6.5. ELECTRONIC CONFIGURATION, BOND ORDER AND MAGNETIC CHARACTER OF NO^+ AND NO^- IONS

Species	No. of Electrons	Electronic Configuration	Bond Order	Magnetic Character
NO^+	14	KK $\sigma(2s)^2\ \sigma^*(2s)^2\ \sigma(2p_z)^2$ $\pi(2p_y)^2\ \pi(2p_x)^2$	$\frac{8-2}{2} = 3$	Diamagnetic
NO^-	16	KK $\sigma(2s)^2\ \sigma^*(2s)^2\ \sigma(2p_z)^2$ $\pi(2p_y)^2\ \pi(2p_x)^2$ $\pi^*(2p_y)^1\ \pi^*(2p_x)^1$	$\frac{8-4}{2} = 2$	Paramagnetic

(3) **Carbon Monoxide (CO)**

The electronic configurations of participating C and O atoms are

$$C = 1s^2\ 2s^2\ 2p_x^{\ 1}\ 2p_y^{\ 1}\ 2p_z^{\ 0} \text{ (4 valence electrons)}$$
$$O = 1s^2\ 2s^2\ 2p_x^{\ 2}\ 2p_y^{\ 1}\ 2p_z^{\ 1} \text{ (6 valence electrons)}$$

The total number of valence electrons is 10 and the electronic configuration of CO molecule can be written as

$$CO = KK\ \sigma(2s)^2\ \sigma^*(2s)^2\ \sigma(2p_z)^2\ \pi(2p_y)^2\ \pi(2p_x)^2$$

$$\text{The bond order} = \frac{N_b - N_a}{2} = \frac{1}{2}(8-2) = 3$$

This shows that carbon and oxygen atom in CO are bonded to each other by a triple bond (one σ bond and two π bonds). There is no unpaired electron in CO molecule and hence it is diamagnetic in nature. The molecular orbital diagram for CO molecule is shown in Fig. 6.42.

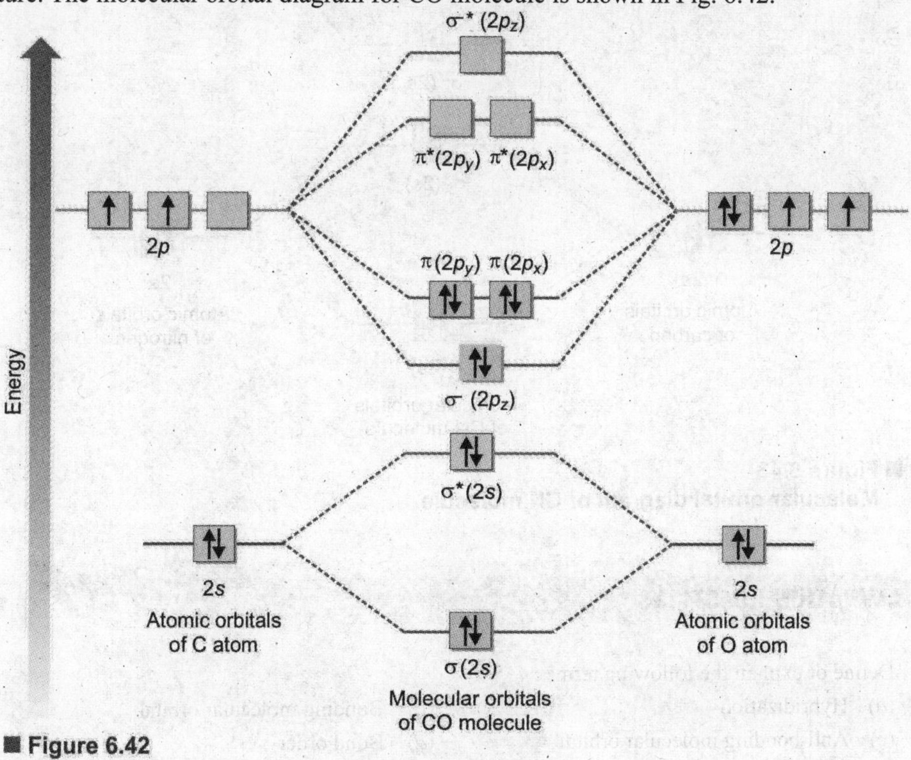

■ **Figure 6.42**
Molecular orbital diagram of CO molecule.

(4) CN Molecule

The electronic configuration of participating C and N atoms are

$$C = 1s^2\ 2s^2\ 2p_x^{\ 1}\ 2p_y^{\ 1}\ 2p_z^{\ 0} \text{ (4 valence electrons)}$$
$$N = 1s^2\ 2s^2\ 2p_x^{\ 1}\ 2p_y^{\ 1}\ 2p_z^{\ 1} \text{ (5 valence electrons)}$$

The total number of valence electrons is 9 and the electronic configuration of CN molecule can be written as

$$CN = KK\ \sigma\ (2s)^2\ \sigma*\ (2s)^2\ \sigma\ (2p_z)^2\ \pi\ (2p_y)^2\ \pi\ (2p_x)^1$$

$$\text{The bond order} = \frac{1}{2}\ (N_b - N_a) = \frac{1}{2}\ (7 - 2) = 2\ \frac{1}{2}$$

Since the bond order in CN molecule is lesser than CO molecule, the former is less stable than the latter. Also, there is one unpaired electron in CN molecule. It is paramagnetic in nature. The molecular orbital diagram of CN molecule is shown in Fig. 6.43.

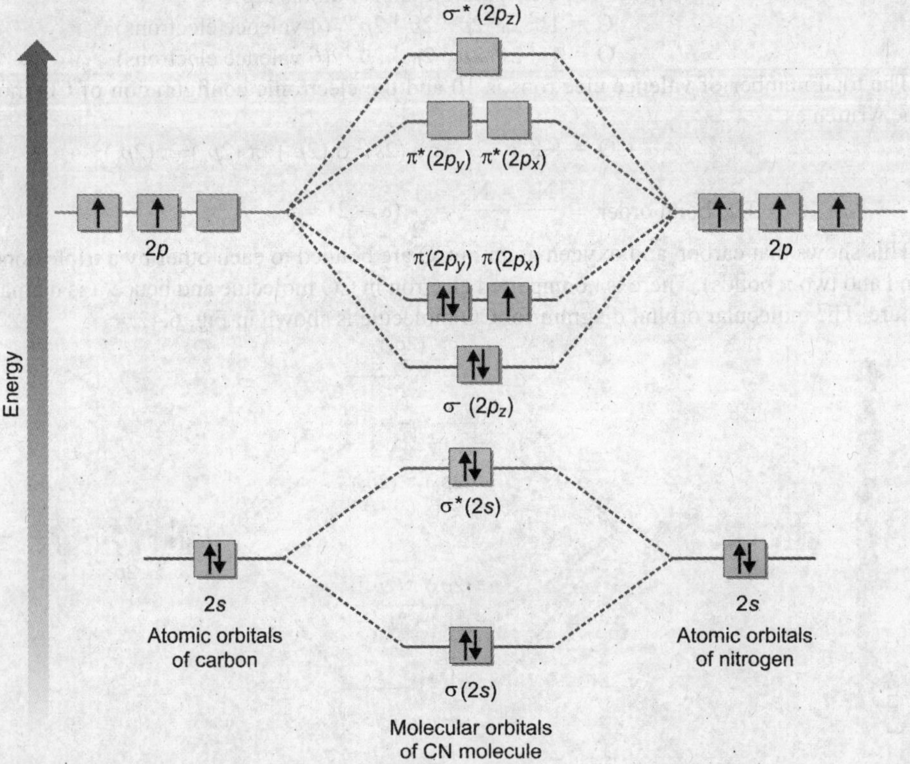

■ **Figure 6.43**
Molecular orbital diagram of CN molecule.

EXAMINATION QUESTIONS

1. Define or explain the following terms :

(*a*) Hybridization (*b*) Bonding molecular orbital

(*c*) Anti-bonding molecular orbital (*d*) Bond order

2. What is meant by hybridization? Write the hybridization involved in (*i*) Beryllium dichloride and (*ii*) Boron trifluoride.
3. The molecules of CH_4 (methane), NH_3 (Ammonia) and H_2O (water) all involve sp^3 hybridization of the central atom but the bond angles in these molecules are 109° 28′, 107° 18′ and 104° 30′ respectively. Explain.
4. Explain the following on the basis of molecular orbital theory :
   (*a*) Br_2 molecule is not stable
   (*b*) N_2 has a stronger bond than B_2
   (*c*) F_2 is diamagnetic while O_2 is paramagnetic
5. Explain clearly the concept of hybrid bond orbitals and show how it gives a better description of the geometry of molecules.
6. (*a*) What is a molecular orbital? In what ways is a molecular orbital different from a valence bond orbital?
   (*b*) Sketch and describe the molecular orbital resulting from the following overlaps of atomic orbitals:
   (*i*) Two *s*-orbitals
   (*ii*) Head-on overlap of two *p*-orbitals
7. (*a*) What is hybridization? Write down the rules for hybridization of orbitals.
   (*b*) CH_4, NH_3 and H_2O have tetrahedral geometry yet their bond angles are different. Why?
8. Describe sp^3, sp^2 and *sp* hybridization of carbon in organic molecules. Draw the model of acetylene and label the bonds as sigma (σ) and pi (π) bonds.
9. (*a*) Indicate the type of hybridization of C atoms in $CH_2{=}C{=}CH_2$
   (*b*) Draw the molecular orbital structure of oxygen molecule and answer the following :
   (*i*) How many bonding and anti-bonding orbitals are there in the molecule?
   (*ii*) How many unpaired electrons are there?
   (*iii*) How many σ and π bonds are there?
   (*c*) Draw the formation of
   (*i*) three sp^2 hybrid orbitals and
   (*ii*) orbital structure of acetylene
   (*iii*) Why is the shape of methane molecule tetrahedral?
10. (*a*) Explain the terms bonding and molecular orbitals with examples.
    (*b*) Explain with the help of MO theory, the paramagnetic character of oxygen.
11. (*a*) What are σ and π molecular orbitals and give their characteristics?
    (*b*) Explain the formation of HF molecule with MO diagram.
12. Explain the state of hybridization of oxygen atom in water and indicate its shape.
13. What is *pi* bond? Explain the formation of *pi* bond in ethylene and acetylene.
14. What is meant by hybridization?
15. Explain *sp*, sp^2 and sp^3 hybridization with the help of simple organic molecules. Indicate the shape of the molecule in each case.
16. Discuss with the help of M.O. theory, the formation of N_2, O_2 and F_2 molecules? How does this theory explain the difference in reactivity of N_2, O_2 and F_2 molecules?
17. (*a*) Fill up the blanks :
    (*i*) HF molecules are bonded through ______ bond.
    (*ii*) The geometrical shape of CH_4 molecule is ______.
    (*b*) Explain the type of hybridization and geometrical shape of PCl_5 molecule with the help of orbital diagram.
    (*c*) Explain the shape of NH_3 and H_2O with the help of VSEPR theory.

Answer. (*a*).(*i*) Hydrogen; (*ii*) Tetrahedral

18. (*a*) What are the characteristics of hybrid orbitals?
(*b*) What is tetrahedral hybridization? Explain the formation of C–H bonds in CH_4.
(*c*) What are sigma (σ) and *pi* (π) bonds? Distinguish between them.

19. (*a*) Discuss the wave mechanical concept of covalent bond.
(*b*) What is meant by π bond?
(*c*) Give the electronic configuration of HCl molecule. Draw its energy level diagram.

20. (*a*) What is meant by hybrid orbital? On the basis of concept of hybridization how do you account for the observed H–O–H bond angle in water molecule?
(*b*) Compare the polarity character of the following pairs :
(*i*) NH_3 and NF_3 (*ii*) CO_2 and SO_2

21. (*a*) Compare V.B.T. and M.O.T.
(*b*) Draw the molecular orbital picture of NO.
(*c*) What are the differences between bonding and antibonding molecular orbitals?
(*d*) Explain the structure of SF_4 on the basis of hybridization.

22. Write the molecular orbital configuration of C_2^{2-} and calculate the bond order of the acetylide ion C_2^{2-}.
Answer. Three

23. Write the molecular orbital configuration of O_2^- ion.
Answer. KK $\sigma(2s^2)\ \sigma^*(2s^2)\ \sigma(2p^2_z)\ \pi(2p^2_y)\ \pi(2p^2_x)\ \pi^*(2p^2_y)\ \pi^*(2p^1_x)$

24. Write MO configuration for O_2 and O_2^{2-} and which of them is paramagnetic?
Answer. O_2 is paramagnetic

25. On the basis of bond order, predict which of the following species is the most stable?

$$O_2^-,\ O_2 \text{ and } Ne_2^+$$

Answer. O_2 is most stable

26. Write the ground state electronic configuration of O_2^+ on the basis of MO theory.
Answer. KK $\sigma(2s^2)\ \sigma^*(2s^2)\ \sigma(2p^2_z)\ \pi(2p^2_y)\ \pi(2p^2_y)\ \pi^*(2p^1_x)$

27. Which has greater bond dissociation energy N_2 or N_2^{+}?
Answer. N_2, as it has bond order of three

28. Write the electronic configuration of Be_2 molecule and calculate its bond order.
Answer. KK $\sigma(2s^2)\ \sigma^*(2s^2)$; zero

29. Write the ground state electronic configuration of N_2^- on the basis of MO theory and calculate its bond order.
Answer. KK $\sigma(2s^2)\ \sigma^*(2s^2)\ \sigma(2p^2_z)\ \pi(2p^2_y)\ \pi(2p^2_x)\ \pi^*(2p^1_y)$; 2.5

30. Calculate the number of antibonding electrons in O_2^{2-} molecule on the basis of M.O. theory
Answer. Four

31. (*a*) Why the bond angle in H_2S is less than H_2O?
(*b*) Explain the structure and hybridization of SO_4^{2-} ion.
(*c*) Explain the structure and various bond angles in IF_7.
(*d*) Why is the bond angle of OF_2 smaller than that of Cl_2O?

32. Explain the structure of O_2 and N_2 molecules by M.O. theory. Show that O_2 is paramagnetic according to MO theory.

33. Account for the following with reasons :
(*a*) Melting point of $MgCl_2$ is much higher than that of $AlCl_3$.
(*b*) KI is soluble in alcohol but KCl is not.

34. Write short notes on :
(*a*) Lattice Energy (*b*) Polarity of bonds (*Agra BSc*)

35. Explain why σ bond is stronger than π bond. (*Madras BSc*)

36. (*a*) What are the main differences between valence bond and molecular orbital theories?
(*b*) Discuss the main features of the molecular orbital approach in context of bond formation in diatomic molecules. (*Purvanchal BSc*)

37. Draw figures for *p-p* axial and *p-p* lateral overlap. (*Nagpur BSc*)

38. Differentiate between bonding and non-bonding molecular orbitals. Draw the molecular structure of N_2. (*Kanpur BSc*)

39. (*a*) What are the factors which influence the formation of an ionic bond?
(*b*) Draw M.O. diagram for CO and compare its stability with CO^+.
(*c*) Distinguish between bonding and anti-bonding molecular orbitals. (*Panjab BSc*)

40. Explain the structures of following pairs of molecules in terms of valence bond theory.
(*i*) BF_3 and NH_3 (*ii*) H_2O and H_2S (*Arunachal BSc*)

41. What do you understand by LCAO in molecular orbital theory? Differentiate bonding, anti-bonding and non-bonding molecular orbital. (*Arunachal BSc*)

42. Describe the valence bond theory proposed by Heitler-London for the formation of H_2 molecule. (*Arunachal BSc*)

43. Write the electronic configuration of Be_2 molecule and calculate its bond order.
Answer. KK $\sigma(2s^2)\ \sigma^*(2s^2)$; zero (*Baroda BSc*)

44. Write the ground state electronic configuration of N_2^- on the basis of MO theory and calculate its bond order.
Answer. KK $\sigma(2s^2)\ \sigma^*(2s^2)\ \sigma(2p^2_z)\ \pi(2p^2_y)\ \pi(2p^2_x)\ \pi^*(2p^1_y)$; 2.5 (*Jabalpur BSc*)

45. Calculate the number of antibonding electrons in O_2^{2-} molecule on the basis of M.O. theory.
Answer. Four (*Delhi BSc; Madurai BSc*)

46. Calculate the bond order in He^+ on the basis of M.O. theory.
Answer. 0.5 (*Delhi BSc*)

47. (*a*) Draw M.O. diagram of CO and show its bond order.
(*b*) Nitrogen molecule is diamagnetic while oxygen molecule is paramagnetic. Explain.
(*c*) Discuss the conditions for the combination of atomic orbitals to form M.O. (*Punjabi BSc*)

MULTIPLE CHOICE QUESTIONS ???

1. When a chemical bond between two atoms is formed, the potential energy of the system
(*a*) decreases (*b*) increases
(*c*) remains the same (*d*) cannot be predicted
Answer. (*a*)

2. According to valence bond theory, a bond between two atoms is formed when
(*a*) half filled atomic orbitals overlap (*b*) fully filled atomic orbitals overlap
(*c*) non-bonding atomic orbitals overlap (*d*) electrons of the two atoms overlap
Answer. (*a*)

3. The axial overlap between the two orbitals leads to the formation of a
(*a*) sigma bond (*b*) pi bond
(*c*) multiple bond (*d*) none of these
Answer. (*a*)

4. The free rotation about a bond exists when the bond is a
(*a*) sigma bond (*b*) pi bond
(*c*) double bond (*d*) hydrogen bond
Answer. (*a*)

5. The strength of a bond depends upon
(*a*) free rotation about σ bond
(*b*) extent of overlapping between the orbitals
(*c*) resonance in the molecule
(*d*) whether the overlap is axial or sidewise
Answer. (*b*)

6. In a N_2 molecule there are
(*a*) one sigma and one pi bond
(*b*) two sigma and one pi bond
(*c*) one sigma and two pi bonds
(*d*) three sigma bonds
Answer. (*c*)

7. Out of the following, which statement is incorrect
(*a*) half filled orbitals only take part in hydridization
(*b*) the orbitals taking part in hydridization should have very little energy difference
(*c*) the number of hybrid orbitals formed is equal to the number of orbitals intermixed
(*d*) the energy of hybrid orbitals is less than that of atomic orbitals
Answer. (*a*)

8. When one *s* and two *p* orbitals hybridize we get
(*a*) three new orbitals at 90° to each other
(*b*) three new orbitals at 120° to each other
(*c*) two new orbitals at 180° to each other
(*d*) two new orbitals at 90° to each other
Answer. (*b*)

9. In sp^3, sp^2 and *sp* hybridized carbon atom the *p* character is maximum in
(*a*) sp^3
(*b*) sp^2
(*c*) *sp*
(*d*) all have equal *p*-character
Answer. (*a*)

10. In H_2O, NH_3 and CH_4 molecules the oxygen, nitrogen and carbon atom are
(*a*) sp^3 hybridized
(*b*) sp^3, sp^2 and *sp* respectively
(*c*) *sp*, sp^2 and sp^3 respectively hybridized
(*d*) sp^2 hybridized
Answer. (*a*)

11. sp^3 hybridization leads to
(*a*) trigonal geometry with bond angles 120° each
(*b*) tetrahedral geometry with bond angles 109.5° each
(*c*) tetrahedral geometry with bond angles 90° each
(*d*) square planar geometry with bond angles 90° each
Answer. (*b*)

12. In dsp^2 hybridization, the new orbitals have the following geometry:
(*a*) square planar
(*b*) tetrahedral
(*c*) trigonal
(*d*) trigonal bipyramid
Answer. (*d*)

13. The shape of BF_3 molecule is planar with bond angles equal to 120° each. It is due to
(*a*) sp^3 hybridized B atom
(*b*) sp^2 hybridized B atom
(*c*) *sp* hybridized B atom
(*d*) dsp^2 hybridized B atom
Answer. (*b*)

14. The total number of orbitals taking part in *sp* hybridization in carbon atom is
(*a*) one
(*b*) two
(*c*) three
(*d*) four
Answer. (*b*)

15. A molecule of acetylene contains
(*a*) one σ and two π bonds
(*b*) two σ and two π bonds
(*c*) three σ and two π bonds
(*d*) two σ and three π bonds
Answer. (*c*)

16. In PCl_5 molecule the phosphorus atom is

(*a*) sp^3 hybridized (*b*) sp^3d hybridized
(*c*) sp^3d^2 hybridized (*d*) sp^2 hybridized

Answer. (*b*)

17. Valence bond theory was proposed by

(*a*) Rutherford (*b*) Neils Bohr
(*c*) Heitler and London (*d*) Hund and Mulliken

Answer. (*c*)

18. In the compound $HC{\equiv}C{-}C^*H{=}CH_2$, C^* is

(*a*) *sp* hybridized (*b*) sp^2 hybridized
(*c*) sp^3 hybridized (*d*) none of these

Answer. (*b*)

19. The carbon-carbon bond length is maximum in

(*a*) ethane (*b*) ethene
(*c*) ethyne (*d*) equal in all

Answer. (*a*)

20. The carbon-hydrogen bond length is shortest in

(*a*) ethane (*b*) ethene
(*c*) ethyne (*d*) CH_4

Answer. (*c*)

21. In SO_2 molecule, S atom is

(*a*) sp^3 hybridized (*b*) sp^2 hybridized
(*c*) *sp* hybridized (*d*) dsp^2 hybridized

Answer. (*b*)

22. The molecular orbitals formed as a result of LCAO method obey

(*a*) Pauli's exclusion principle (*b*) Hund's rule of maximum multiplicity
(*c*) Aufbau principle (*d*) all of these

Answer. (*d*)

23. The energy of atomic orbitals taking part in molecular orbital formation

(*a*) is equal to the molecular orbitals formed
(*b*) is less than the molecular orbitals formed
(*c*) is greater than the molecular orbitals formed
(*d*) cannot be predicted

Answer. (*c*)

24. A He_2 molecule is not formed because

(*a*) $N_a > N_b$ (*b*) $N_b = N_a$
(*c*) $N_b > N_a$ (*d*) $N_b - N_a = +ve$

Answer. (*b*)

25. A molecule or ion is stable if

(*a*) $N_b = N_a$ (*b*) $N_b < N_a$
(*c*) $N_a < N_b$ (*d*) $N_a - N_b = +ve$

Answer. (*c*)

26. The bond order in H_2 molecule as compared to H_2^+ ion is

(*a*) double (*b*) half
(*c*) equal (*d*) cannot be predicted

Answer. (*a*)

27. The bond order in He_2^+ ion is

(*a*) 0.5
(*b*) 1.0
(*c*) 1.5
(*d*) 2.0

Answer. (*a*)

28. The O_2 molecule is paramagnetic. It can be explained on the basis of

(*a*) hybridization
(*b*) valence bond theory
(*c*) molecular orbital theory
(*d*) none of these

Answer. (*c*)

29. The bond order in O_2^+, O_2^- and O_2^{2-} respectively is

(*a*) 1, 1.5, 2.0
(*b*) 1.5, 2.0, 2.5
(*c*) 2.5, 1.5, 1
(*d*) 2.5, 1.0, 1.5

Answer. (*c*)

30. Which among the species O_2^+, O_2^-, O_2 and O_2^{2-} is diamagnetic

(*a*) O_2^+
(*b*) O_2^-
(*c*) O_2
(*d*) O_2^{2-}

Answer. (*d*)

31. The diamagnetic species among NO^+, NO and NO^- is

(*a*) NO^+
(*b*) NO
(*c*) NO^-
(*d*) none of these

Answer. (*a*)

32. The number of bonding and antibonding electrons respectively in CO molecule is

(*a*) 8, 2
(*b*) 2, 8
(*c*) 4, 2
(*d*) 2, 4

Answer. (*a*)

33. Among CO, NO and CN molecules, the one which contains no unpaired electrons and hence is diamagnetic

(*a*) CO
(*b*) NO
(*c*) CN
(*d*) none of these

Answer. (*a*)

34. The species with the highest bond order among NO, CO, CN and O_2 is

(*a*) NO
(*b*) CO
(*c*) CN
(*d*) O_2

Answer. (*b*)

35. Which one of the following has a zero bond order?

(*a*) He_2
(*b*) F_2
(*c*) N_2
(*d*) H–F

Answer. (*a*)

36. Which among the following pairs are paramagnetic?

(*a*) O_2 and N_2
(*b*) O_2 and CO
(*c*) O_2 and NO
(*d*) CO and NO

Answer. (*c*)

37. The bond order of a molecule is given by

(*a*) $\dfrac{N_a - N_b}{2}$
(*b*) $\dfrac{N_b - N_a}{2}$
(*c*) $\dfrac{N_b \div N_a}{2}$
(*d*) $\dfrac{N_a \div N_b}{2}$

Answer. (*b*)

38. Oxygen molecule is paramagnetic because it has

(*a*) less N_b than N_a (*b*) more N_b than N_a

(*c*) all electrons are paired (*d*) unpaired electrons

Answer. (*d*)

39. The increasing bond order in the species O_2^-, O_2^+ and O_2^{2-} is

(*a*) $O_2^{2-} < O_2^+ < O_2^-$ (*b*) $O_2^- < O_2^{2-} < O_2^+$

(*c*) $O_2^{2-} < O_2^- < O_2^+$ (*d*) $O_2^+ < O_2^- < O_2^{2-}$

Answer. (*c*)

40. The molecules which are iso-electronic among CO, N_2, O_2 and NO are

(*a*) O_2 and N_2 (*b*) O_2 and NO

(*c*) N_2 and NO (*d*) CO and N_2

Answer. (*d*)

41. The molecule with the highest bond order among CO, CN, NO and O_2 is

(*a*) CO (*b*) CN

(*c*) NO (*d*) O_2

Answer. (*a*)

42. Which of the following is not true regarding LCAO method?

(*a*) the energies of atomic orbitals should be comparable

(*b*) the atomic orbitals should overlap to a considerable extent

(*c*) the symmetry of the combining orbitals should be the same

(*d*) the energy of resulting antibonding orbital is less than that of bonding orbital

Answer. (*d*)

43. Which is not true about bonding MO?

(*a*) it is formed by the addition overlap of atomic orbitals

(*b*) the wave function of a bonding MO is given by $\psi_{MO} = \psi_A - \psi_B$

(*c*) the lobes of atomic orbitals should have the same signs

(*d*) every electron in bonding MO contributes towards the attractive force

Answer. (*b*)

7 First Law of Thermodynamics — Basic Concepts

CONTENTS

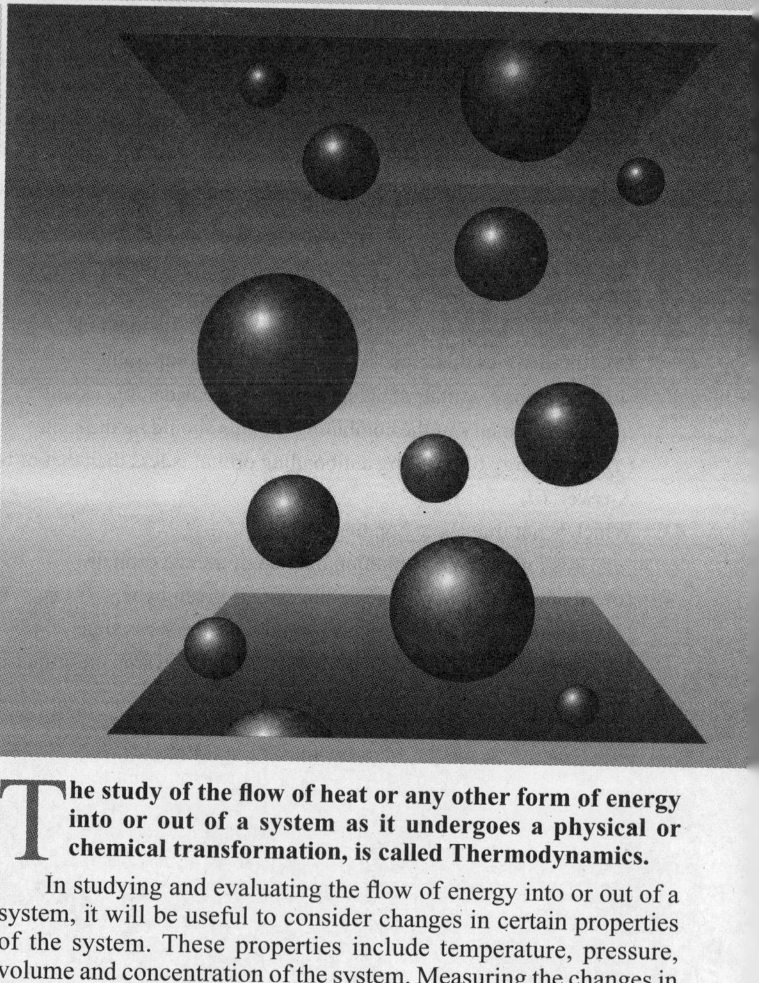

The study of the flow of heat or any other form of energy into or out of a system as it undergoes a physical or chemical transformation, is called Thermodynamics.

In studying and evaluating the flow of energy into or out of a system, it will be useful to consider changes in certain properties of the system. These properties include temperature, pressure, volume and concentration of the system. Measuring the changes in these properties from the initial state to the final state, can provide information concerning changes in energy and related quantities such as heat and work.

THE THREE EMPIRICAL LAWS

The study of thermodynamics is based on three broad generalizations derived from well established experimental results. These generalizations are known as **the First, Second and Third law of thermodynamics.** These laws have stood the test of time and are independent of any theory of the atomic or molecular structure. The discussion of these laws will be the subject of our study in the two chapters on thermodynamics.

Scope of Thermodynamics

(1) Most of the important laws of Physical Chemistry, including the van't Hoff law of lowering of vapor pressure, Phase Rule and the Distribution Law, can be derived from the laws of thermodynamics.

(2) It tells whether a particular physical or chemical change can occur under a given set of conditions of temperature, pressure and concentration.

(3) It also helps in predicting how far a physical or chemical change can proceed, until the equilibrium conditions are established.

Limitations of Thermodynamics

(1) **Thermodynamics is applicable to macroscopic systems** consisting of matter in bulk and not to microscopic systems of individual atoms or molecules. **It ignores the internal structure of atoms and molecules.**

(2) Thermodynamics does not bother about the time factor. That is, it does not tell anything regarding the rate of a physical change or a chemical reaction. **It is concerned only with the initial and the final states of the system.**

THERMODYNAMIC TERMS AND BASIC CONCEPTS

An important part of the study of thermodynamics is few terms and definitions which must be understood clearly.

SYSTEM, BOUNDARY, SURROUNDINGS

A system is that part of the universe which is under thermodynamic study and the rest of the universe is surroundings.

The real or imaginary surface separating the system from the surroundings is called the boundary.

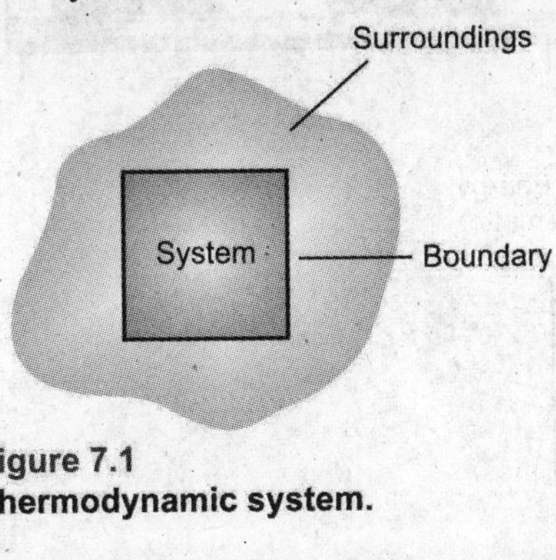

Figure 7.1
Thermodynamic system.

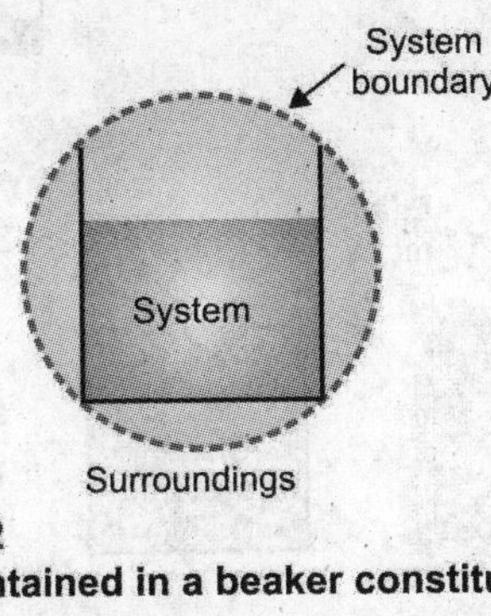

Figure 7.2
Water contained in a beaker constitutes a system.

In experimental work, a specific amount of one or more substances constitutes the system. Thus 0.200 kg of water contained in a beaker constitutes a thermodynamic system. The beaker and the air in contact, are the surroundings.

Similarly 1 mole of oxygen confined in a cylinder fitted with a piston, is a thermodynamic system. The cylinder and the piston and all other objects outside the cylinder, form the surroundings. Here the boundary between the system (oxygen) and the surroundings (cylinder and piston) is clearly defined.

HOMOGENEOUS AND HETEROGENEOUS SYSTEMS

When a system is uniform throughout, it is called a Homogeneous System. Examples are : a pure single solid, liquid or gas, mixtures of gases, and true solution of a solid in a liquid. A homogeneous system is made of one phase only. A **phase** is defined as a homogeneous, physically distinct and mechanically separable portion of a system.

A heterogeneous system is one which consists of two or more phases. In other words it is not uniform throughout. Examples of heterogeneous systems are : ice in contact with water, ice in contact with vapor etc. Here ice, water and vapor constitute separate phases.

TYPES OF THERMODYNAMIC SYSTEMS

There are three types of thermodynamic systems depending on the nature of the boundary. If the boundary is closed or sealed, no *matter* can pass through it. If the boundary is insulated, no *energy* (say heat) can pass through it.

(1) Isolated System

When the boundary is both sealed and insulated, no interaction is possible with the surroundings. Therefore, **an isolated system is one that can transfer neither matter nor energy to and from its surroundings.**

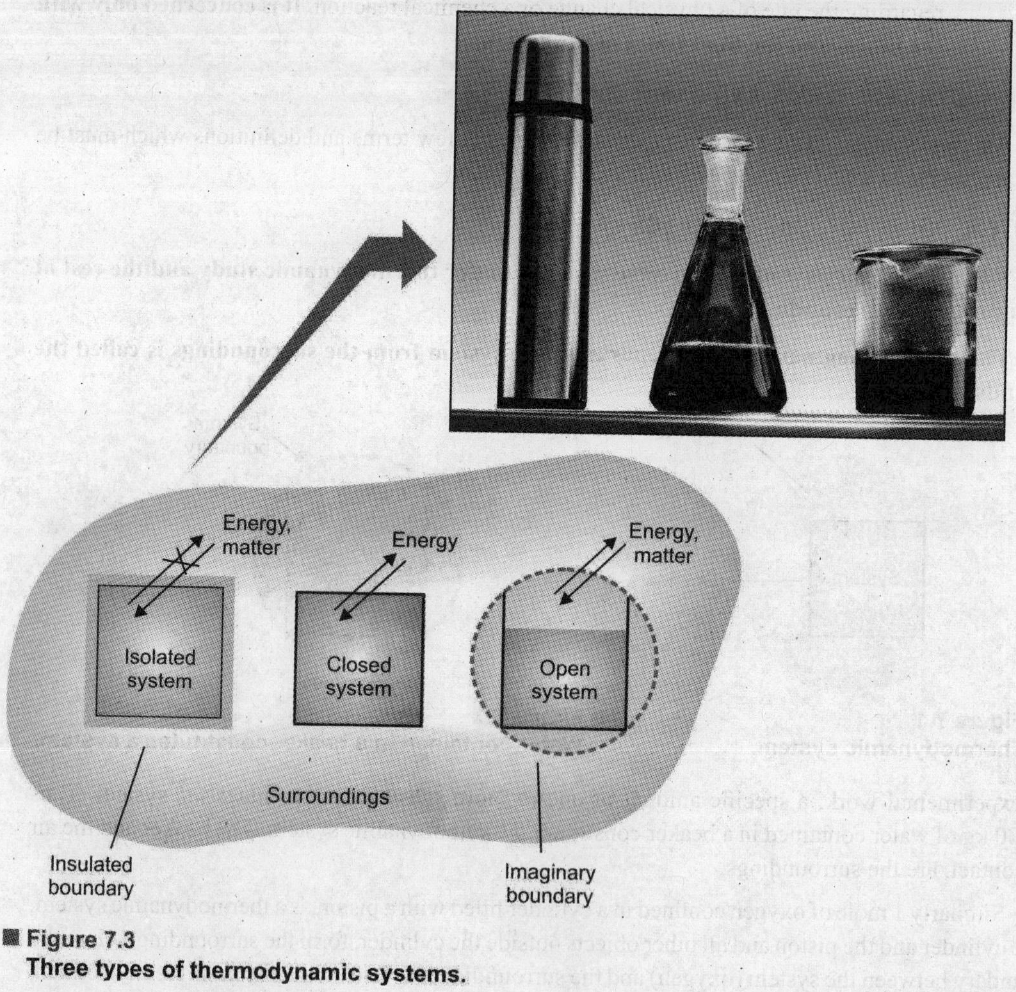

■ Figure 7.3
Three types of thermodynamic systems.

Let us consider a system 100 ml (1×10^{-6} m^3) of water in contact with its vapor in a closed vessel which is insulated. Since the vessel is sealed, no water vapor (matter) can escape from it. Also, because the vessel is insulated, no heat (energy) can be exchanged with the surroundings.

A substance, say boiling water, contained in a *thermos flask,* is another example of an isolated system.

(2) Closed System

Here the boundary is sealed but not insulated. Therefore, **a closed system is one which cannot transfer matter but can transfer energy in the form of heat, work and radiation to and from its surroundings.**

A specific quantity of hot water contained in a sealed tube, is an example of a closed system. While no water vapor can escape from this system, it can transfer heat through the walls of the tube to the surroundings.

A gas contained in a cylinder fitted with a piston constitutes a closed system. As the piston is raised, the gas expands and transfers heat (energy) in the form of work to the surroundings.

(3) Open System

In such a system the boundary is open and un-insulated. Therefore, **an open system is one which can transfer both energy and matter to and from its surroundings.**

Hot water contained in a beaker placed on laboratory table is an open system. The water vapor (matter) and also heat (energy) is transferred to the surroundings through the imaginary boundary.

Zinc granules reacting with dilute hydrochloric acid to produce hydrogen gas in a beaker, is another example of open system. Hydrogen gas escapes and the heat of the reaction is transferred to the surroundings.

What are Adiabatic Systems?

Those systems in which no thermal energy passes into or out of the system, are said to be adiabatic systems.

INTENSIVE AND EXTENSIVE PROPERTIES

The macroscopic or bulk properties of a system (volume, pressure, mass, etc.) can be divided into two classes :

(*a*) Intensive properties
(*b*) Extensive properties

Intensive Properties

A property which does not depend on the quantity of matter present in the system, is known as Intensive Property.

Some examples of intensive properties are *pressure, temperature, density*, and *concentration.* If the overall temperature of a glass of water (our system) is 293K, then any drop of water in that glass has a temperature of 293K. Similarly if the concentration of salt, NaCl, in the glass of water is 0.1 mole/liter (0.0001 m^3), then any drop of water from the glass also has a salt concentration of 0.0001 m^3.

Extensive Properties

A property that does depend on the quantity of matter present in the system, is called an Extensive Property.

Some examples of extensive properties are *volume, number of moles, enthalpy, entropy, and Gibbs' free energy*. Some of these properties are unfamiliar to you but these will be defined and illustrated later.

By definition, **the extensive properties are additive while intensive properties are not.** Let us consider the system 'a glass of water'. If we double the mass of water, the volume is doubled and so is the number of moles and the internal energy of the system.

TABLE 7.1. COMMON PROPERTIES OF A SYSTEM

Intensive properties		Extensive properties
Temperature	Surface tension	Mass
Pressure	Refractive index	Volume
Density	Viscosity	Internal energy
Boiling point	Freezing point	Enthalpy, Entropy

STATE OF A SYSTEM

A thermodynamic system is said to be in a certain state when all its properties are fixed.

The fundamental properties which determine the state of a system are pressure (P), temperature (T), volume (V), mass and composition. Since a change in the magnitude of such properties alters the state of the system, these are referred to as **State variables** or **State functions** or **Thermodynamic parameters.** It also stands to reason that a change of system from the *initial state* to the *final state* (2nd state) will be accompanied by change in the state variables.

It is not necessary to state all the properties (state variables) to define a system completely. For a pure gas, the composition is fixed automatically, as it is cent per cent. The remaining state variables P, V, T are interrelated in the form of an algebraic relationship called the **Equation of State.** Thus for one mole of a pure gas, the equation of state is :

$$PV = RT$$

where R is gas constant. If of the three state variables (P, V, T), P and T are specified, the value of third (V) is fixed automatically and can be calculated from the equation of state. The variables (P and T) which must be necessarily specified to define the state of a system, are designated as **Independent state variables.** The remaining state variable (V) which depends on the value of P and T, is called **Dependent state variable.**

An important characteristic of a state variable (or state function) is that when the state of a system is altered, the change in the variable depends on the initial and final states of the system. For example, if we heat a sample of water from 273.15K to 298.15K, the change in temperature is equal to difference between the initial and final temperatures.

$$\Delta T = T_{\text{final}} - T_{\text{initial}} = 298.15 \text{ K}$$

The way in which the temperature change is brought about has no effect on the result.

EQUILIBRIUM AND NON-EQUILIBRIUM STATES

A system in which the state variables have constant values throughout the system is said to be in a state of thermodynamic equilibrium.

Suppose we have a gas confined in a cylinder that has a frictionless piston. If the piston is stationary, the state of the gas can be specified by giving the values of pressure and volume. The system is then in a *state of equilibrium.*

A system in which the state variables have different values in different parts of the system is said to be in a non-equilibrium state.

If the gas contained in a cylinder, as stated above, is compressed very rapidly by moving down the piston, it passes through states in which pressure and temperature cannot be specified, since these properties vary throughout the gas. The gas near the piston is compressed and heated and that at the far end of the cylinder is not. The gas then would be said to be in non-equilibrium state.

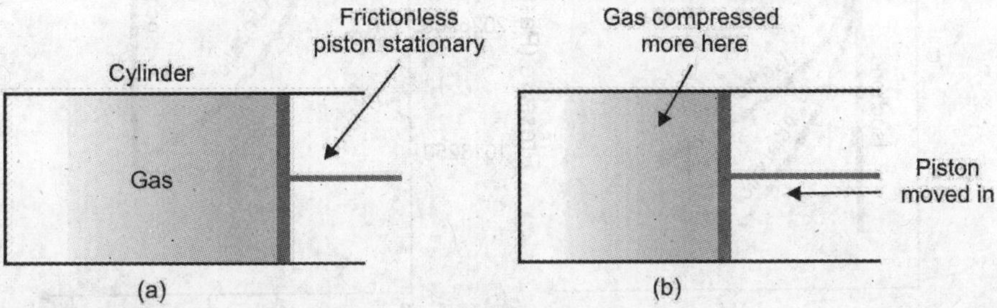

■ **Figure 7.4**
***(a)* Gas in equilibrium state; *(b)* Rapidly compressed gas in non-equilibrium state.**

Thermodynamics is concerned only with equilibrium states.

The Criteria for Equilibrium

(1) **The temperature of the system must be uniform and must be the same as the temperature of the surroundings (thermal equilibrium).**

(2) **The mechanical properties must be uniform throughout the system (mechanical equilibrium).** That is, no mechanical work is done by one part of the system on any other part of the system.

(3) **The chemical composition of the system must be uniform with no net chemical change (chemical equilibrium).**

If the system is heterogeneous, the state variables of each phase remain constant in each phase.

THERMODYNAMIC PROCESSES

When a thermodynamic system changes from one state to another, the operation is called a Process. These processes involve the change of conditions (temperature, pressure and volume). The various types of thermodynamic processes are :

(1) **Isothermal Processes**

Those processes in which the temperature remains fixed, are termed isothermal processes. This is often achieved by placing the system in a thermostat (a constant temperature bath).

For an isothermal process $dT = 0$

(2) **Adiabatic Processes**

Those processes in which no heat can flow into or out of the system, are called adiabatic processes. Adiabatic conditions can be approached by carrying the process in an insulated container such as 'thermos' bottle. High vacuum and highly polished surfaces help to achieve thermal insulation.

For an adiabatic process $dq = 0$

(3) **Isobaric Processes**

Those processes which take place at constant pressure are called isobaric processes. For example, heating of water to its boiling point and its vaporisation take place at the same atmospheric pressure. These changes are, therefore, designated as isobaric processes and are said to take place isobarically. For an isobaric process $dp = 0$

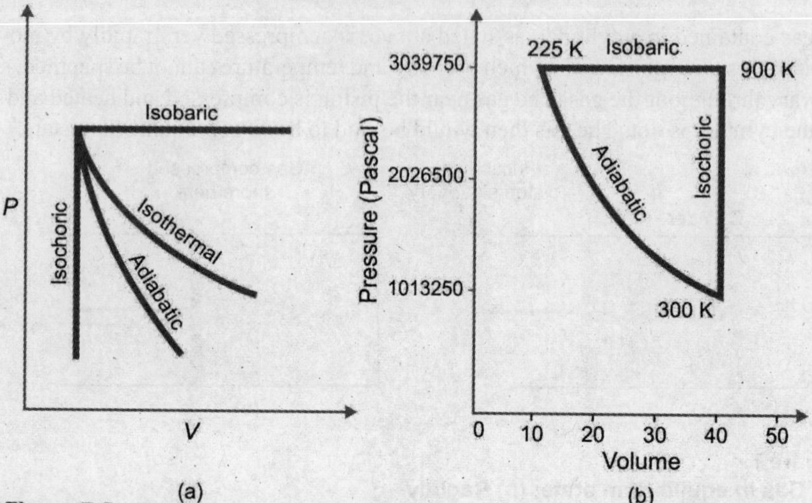

■ **Figure 7.5**
(*a*) Four basic thermodynamic processes; (*b*) A cyclic process.

(4) **Isochoric Processes**

Those processes in which the volume remains constant are known as isochoric processes. The heating of a substance in a non-expanding chamber is an example of isochoric process.

For isochoric processes $dV = 0$.

(5) **Cyclic Process**

When a system in a given state goes through a number of different processes and finally returns to its initial state, the overall process is called a cycle or cyclic process.

For a cyclic process $dE = 0$, $dH = 0$.

REVERSIBLE AND IRREVERSIBLE PROCESSES

A thermodynamic reverse process is one that takes place infinitesimally slowly and its direction at any point can be reversed by an infinitesimal change in the state of the system.

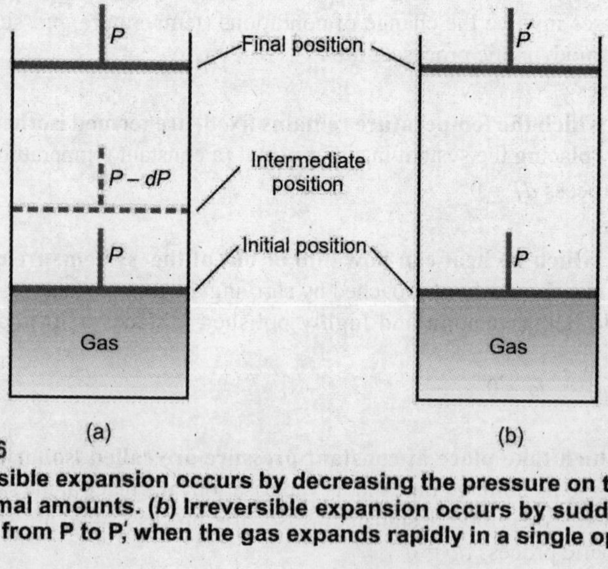

■ **Figure 7.6**
(*a*) Reversible expansion occurs by decreasing the pressure on the piston by infinitesimal amounts. (*b*) Irreversible expansion occurs by sudden decrease of pressure from P to P′, when the gas expands rapidly in a single operation.

In fact, a reversible process is considered to proceed from the initial state to the final state through an infinite series of infinitesimally small stages. At the initial, final and all intermediate stages, the system is in equilibrium state. This is so because an infinitesimal change in the state of the system at each intermediate step is negligible.

When a process goes from the initial to the final state in a single step and cannot be carried in the reverse order, it is said to be an irreversible process. Here the system is in equilibrium state in the beginning and at the end, but not at points in between.

Consider a certain quantity of a gas contained in a cylinder having a weightless and frictionless piston. The expansion of the gas can be carried by two methods illustrated in Fig. 7.6.

Let the pressure applied to the piston be P and this is equal to the internal pressure of the gas. Since the external and internal pressures are exactly counterbalanced, the piston remains stationary and there is no change in volume of the gas. Now suppose the pressure on the piston is decreased by an infinitesimal amount dP. Thus the external pressure on the piston being $P - dP$, the piston moves up and the gas will expand by an infinitesimal small amount. The gas will, therefore, be expanded infinitely slowly *i.e.*, by a thermodynamically *reversible process*. At all stages in the expansion of the gas, dP being negligibly small the gas is maintained in a state of equilibrium throughout. If at any point of the process the pressure is increased by dP, the gas would contract reversibly.

On the other hand, the expansion is irreversible (Fig. 7.6 *b*) if the pressure on the piston is decreased suddenly. It moves upward rapidly in a single operation. The gas is in equilibrium state in the initial and final stages only. The expansion of the gas, in this case, takes place in an irreversible manner.

DIFFERENCES BETWEEN REVERSIBLE AND IRREVERSIBLE PROCESSES

Reversible Process	Irreversible Process
1. It takes place in infinite number of infinitesimally small steps and it would take *infinite time* to occur.	1. It takes place *infinite time.*
2. It is *imaginary* as it assumes the presence of frictionless and weightless piston.	2. It is *real* and can be performed actually.
3. It is in equilibrium state at *all stages* of the operation.	3. It is in equilibrium state only at the *initial and final stages* of the operation.
4. All changes are *reversed* when the process is carried out in reversible direction.	4. After this type of process has occurred all changes *do not return* to the initial state by themselves.
5. It is extremely slow.	5. It proceeds at *measureable speed.*
6. Work done by a reversible process is *greater* than the corresponding irreversible process.	6. Work done by a irreversible process is *smaller* than the corresponding reversible process.

NATURE OF HEAT AND WORK

When a change in the *state* of a system occurs, energy is transferred to or from the surroundings. This energy may be transferred as heat or mechanical work.

We shall refer the term 'work' for mechanical work which is defined as force × distance.

Units of Work

In CGS system the unit of work is erg which is defined as the work done when a resistance of 1 dyne is moved through a distance of 1 centimeter. Since the erg is so small, a bigger unit, the **joule (J)** is now used.

$$1 \text{ joule} = 10^7 \text{ ergs}$$

or

$$1 \text{ erg} = 10^{-7} \text{ J}$$

We often use **kilojoule** (kJ) for large quantities of work

$$1 \text{ kJ} = 1000 \text{ J}$$

Units of Heat

The unit of heat, which was used for many years, is **calorie** (cal). A calorie is defined as the quantity of heat required to raise the temperature of 1 gram of water by 1°C in the vicinity of 15°C.

Since heat and work are interrelated, **SI unit of heat is the joule (J)**:

$$1 \text{ joule} = 0.2390 \text{ calories}$$

$$1 \text{ calorie} = 4.184 \text{ J}$$

or

$$1 \text{ kcal} = 4.184 \text{ kJ}$$

Sign Convention of Heat

The symbol of heat is q. If the heat flows from the surroundings into the system to raise the energy of the system, it is taken to be positive, $+q$. If heat flows from the system into the surroundings, lowering the energy of the system, it is taken to be negative, $-q$.

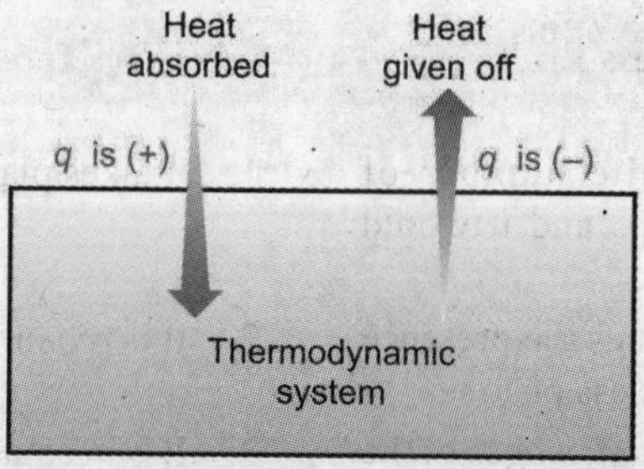

■ **Figure 7.7**
Sign convention for heat flow in a thermodynamic system.

Sign Conventions of Work

The symbol of work is w. If work is done on a system by the surroundings and the energy of the system is thus increased, it is taken to be positive, $+w$. If work is done by the system on the surroundings and energy of the system is decreased, it is taken to be negative, $-w$.

SUMMARY OF SIGN CONVENTIONS

Heat flows into the system, q is +ve	Heat flows out of the system, q is –ve
Work is done on the system, w is +ve	Work is done by the system, w is –ve

PRESSURE–VOLUME WORK

In physics, mechanical work is defined as force multiplied by the distance through which the force acts. In elementary thermodynamics the only type of work generally considered is the work done in expansion (or compression) of a gas. This is known as **pressure-volume work** or **PV work** or **expansion work.**

Consider a gas contained in a cylinder fitted with a frictionless piston. The pressure (force per unit area) of the gas, P, exerts a force on the piston. This can be balanced by applying an equal but opposite pressure from outside on the piston. Let it be designated as P_{ext}. **It is important to remember that it is the external pressure, P_{ext}, and not the internal pressure of the gas itself which is used in evaluating work.** This is true whether it be expansion or contraction.

If the gas expands at constant pressure, the piston would move, say through a distance l. We know that

$$work = force \times distance \text{ (by definition)}$$

or $$w = f \times l \qquad ...(1)$$

Since pressure is force per unit area,

$$f = P_{ext} \times A \qquad ...(2)$$

where A is the cross-section area of the piston.

From (1) and (2), we have

$$w = P_{ext} \times A \times l$$
$$= P_{ext} \times \Delta V$$

where ΔV is the increase in volume of the gas.

Since the system (gas) is doing work on the surroundings (piston), it bears negative sign. Thus,

$$w = -P_{ext} \times \Delta V$$

Proceeding as above the work done in compression of a gas can also be calculated. In that case the piston will move down and sign of the work will be positive.

$$w = P_{ext} \times \Delta V$$

As already stated, work may be expressed in dynes-centimetres, ergs, or joules. PV work can as well be expressed as the product of pressure and volume units *e.g.*, in liter or atmospheres.

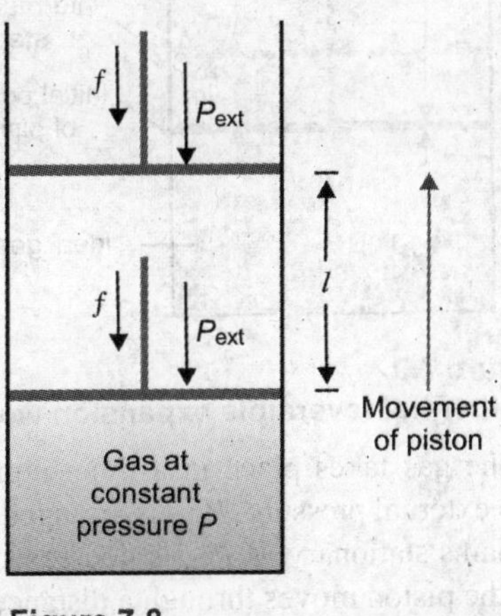

Figure 7.8
Pressure volume work of a system (gas) confined by a frictionless piston.

It may be noted that **the work done by a system is not a state function.** This is true of the mechanical work of expansion. We shall show presently that the work is related to the process carried out rather than to the internal and final states. This will be evident from a consideration of the *reversible expansion* and an *irreversible process*.

SOLVED PROBLEM 1

Calculate the pressure-volume work done when a system containing a gas expands from $0.001 m^3$ to $0.002 m^3$ against a constant external pressure of 1013250.00 Pa. Express the answer in calories and joules.

SOLUTION

$$w = -P_{ext}(V_2 - V_1)$$
$$= -1013250.00 \text{ Pa } (0.002 \text{ m}^3 - 0.001 \text{ m}^3)$$
$$= -1013250.00 \text{ Pa} \times 0.001 \text{ m}^3$$
$$= -1013.250 \text{ Pa m}^3$$
$$= 1013.250 \times 8.3191/8.314$$
$$w = \mathbf{1013.8715 \ J}$$

ISOTHERMAL REVERSIBLE EXPANSION WORK OF AN IDEAL GAS

Consider an ideal gas confined in a cylinder with a frictionless piston. Suppose it expands in a reversible manner from volume V_1 to V_2 at a constant temperature. The pressure of the gas is successively reduced from P_1 to P_2.

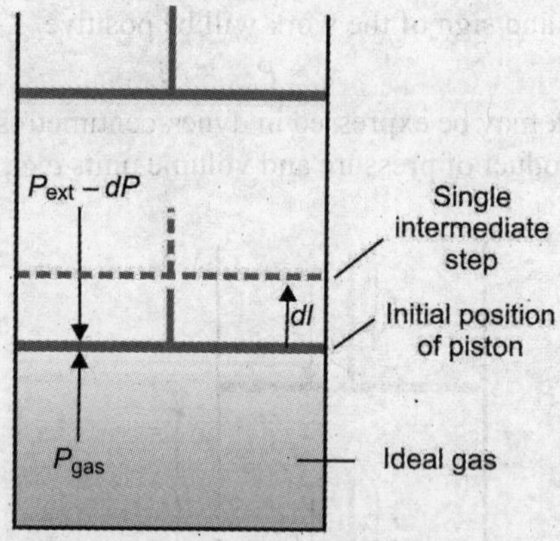

■ **Figure 7.9**
Isothermal reversible expansion work.

The reversible expansion of the gas takes place in a finite number of infinitesimally small intermediate steps. To start with the external pressure, P_{ext}, is arranged equal to the internal pressure of the gas, P_{gas}, and the piston remains stationary. If P_{ext} is decreased by an infinitesimal amount dP the gas expands reversibly and the piston moves through a distance dl.

Since dP is so small, for all practical purposes,

$$P_{ext} = P_{gas} = P$$

The work done by the gas in one infinitesimal step dw, can be expressed as

$$dw = P \times A \times dl \qquad (A = \text{cross-sectional area of piston})$$
$$= P \times dV$$

where dV is the increase in volume

The total amount of work done by the isothermal reversible expansion of the ideal gas from V_1 to V_2 is, therefore,

$$w = -\int_{V_1}^{V_2} PdV \qquad ...(1)$$

By the ideal gas equation

$$P = \frac{nRT}{V}$$

$$\therefore \quad w = -\int_{V_1}^{V_2} \frac{nRT}{V}\, dV$$

$$= -\, nRT \int_{V_1}^{V_2} \frac{dV}{V}$$

which integrates to give

$$w = -\, nRT \ \ln \frac{V_2}{V_1}$$

Since

$$P_1V_1 = P_2V_2$$
$$V_2/V_1 = P_1/P_2$$

$$\therefore \quad w = -\, nRT \ \ln \frac{P_1}{P_2} = -2.303\ nRT \ \log \frac{P_1}{P_2}$$

Isothermal compression work of an ideal gas may be derived similarly and it has exactly the same value with the sign changed. Here the pressure on the piston, P_{ext}, is increased by dP which reduces the volume of the gas.

ISOTHERMAL IRREVERSIBLE EXPANSION WORK OF AN IDEAL GAS

Suppose we have an ideal gas contained in a cylinder with a piston. This time the process of expansion of the gas is performed irreversibly *i.e.*, by instantaneously dropping the external pressure, P_{ext}, to the final pressure P_2. The work done by the system is now against the pressure P_2 throughout the whole expansion and is given by the following expression :

$$w = -\, P_{ext} \int_{V_1}^{V_2} dV$$

$$= P_2\,(V_2 - V_1)$$
$$= P_2\, dV$$

MAXIMUM WORK DONE IN REVERSIBLE EXPANSION

The isothermal expansion of an ideal gas may be carried either by the *reversible process* or *irreversible process* as stated above.

The reversible expansion is shown in Fig. 7.10 in which the pressure is falling as the volume increases. The reversible work done by the gas is given by the expression

$$-\, w_{irr} = \int_{V_1}^{V_2} PdV$$

which is represented by the shaded area.

If the expansion is performed irreversibly by suddenly reducing the external pressure to the final pressure P_2, the irreversible work is given by

$$-\, w_{irr} = P_2\,(V_2 - V_1)$$

which is shown by the shaded area in Fig. 7.10 (*b*).

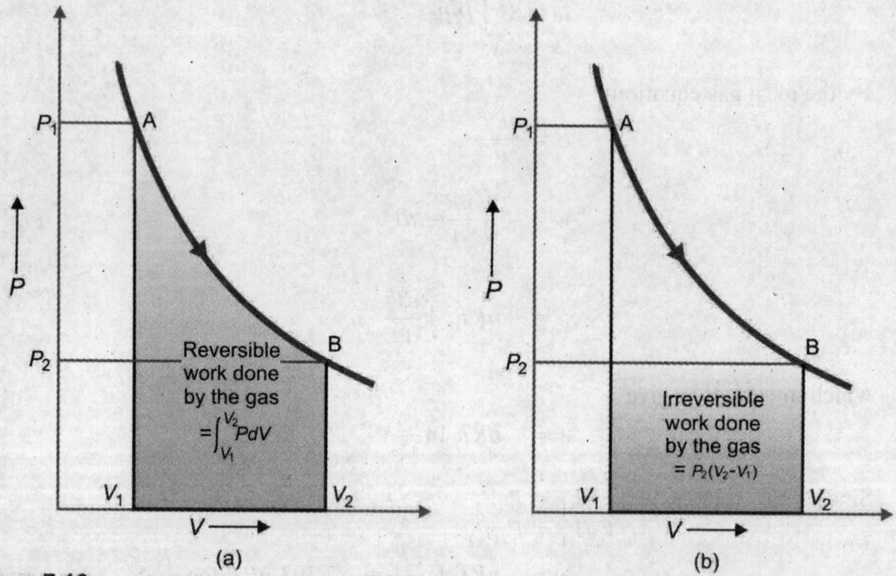

Figure 7.10
***(a)* The reversible work of expansion; *(b)* The irreversible work done by the gas when the external pressure is at once dropped to the final value P_2.**

In both the processes, the state of the system has changed from *A* to *B* but the work done is much less in the irreversible expansion than in the reversible expansion. **Thus mechanical work is not a state function as it depends on the path by which the process is performed rather than on the initial and final states. It is a path function.**

It is also important to note that **the work done in the reversible expansion of a gas is the maximum work that can be done by a system (gas) in expansion between the same initial (A) and final state (B).** This is proved as follows :

We know that the work always depends on the external pressure, P_{ext}; the larger the P_{ext} the more work is done by the gas. But the P_{ext} on the gas cannot be more than the pressure of the gas, P_{gas} or a compression will take place. Thus the largest value P_{ext} can have without a compression taking place is equal to P_{gas}. But an expansion that occurs under these conditions is the reversible expansion. Thus, maximum work is done in the reversible expansion of a gas.

SOLVED PROBLEM 2

One mole of an ideal gas at 298 K is allowed to expand reversibly at constant temperature from a volume of 0.01 m³ to 0.02 m³. Calculate the work done by the gas in joules and calories.

SOLUTION

$$w = -\,nRT\ \ln\frac{V_2}{V_1} = -2.303\ nRT\ \log\frac{V_2}{V_1}$$

$$= -2.303 \times 8.314 \times 298 \log\frac{0.02}{0.01} = -1717.46\ \text{J}$$

$$= -\,\mathbf{410.46\ cal}$$

Notes. (1) $\ln x = 2.303 \log x$

(2) The units of R will determine the units of w in this expression.

(3) The temperature must be expressed in degrees Kelvin.

SOLVED PROBLEM 3

Find the work done when one mole of the gas is expanded reversibly and isothermally from 5 atm (506625.00 Pa) to 1 atm (101325.00 Pa) at 298 K.

SOLUTION

$$w = -nRT \ln \frac{P_1}{P_2}$$

$$= -1 \times 8.314 \times 298 \times 2.303 \log 5$$

$$= -3988 \text{ J}$$

$$= \mathbf{-3.988 \text{ kJ}}$$

INTERNAL ENERGY

A thermodynamic system containing some quantity of matter has within itself a definite quantity of energy. This energy includes not only the translation kinetic energy of the molecules but also other molecular energies such as rotational, vibrational energies. The kinetic and potential energy of the nuclei and electrons within the individual molecules also contribute to the energy of the system.

The total of all the possible kinds of energy of a system, is called its Internal Energy.

The word 'internal' is often omitted and the energy of a system always implies internal energy.

The internal energy of a system, like temperature, pressure, volume, etc., is determined by the state of a system and is independent of the path by which it is obtained. **Hence internal energy of a system is a state function.**

For example, we consider the heating of one mole of liquid water from 273 K to 373 K. The change in energy is 7.5362 kJ and is the same regardless of the form in which this energy is transferred to the water by heating, by performing work, by electrical means or in any other way.

Since the value of internal energy of a system depends on the mass of the matter contained in a system, it is classed as an **extensive property.**

Symbol Representation of Internal Energy and Sign Conventions

The internal energy of a system is represented by the symbol E (Some books use the symbol U). It is neither possible nor necessary to calculate the absolute value of internal energy of a system. In thermodynamics we are concerned only with the energy changes when a system changes from one state to another. If ΔE be the difference of energy of the initial state (E_{in}) and the final state (E_f), we can write

$$\Delta E = E_f - E_{in}$$

ΔE is +ve if E_f is greater than E_{in} and –ve if E_f is less than E_{in}.

A system may transfer energy to or from the surroundings as heat or as work, or both.

UNITS OF INTERNAL ENERGY

The SI unit for internal energy of a system is the joule (J). Another unit of energy which is not an SI unit is the calorie, 1 cal = 4.184 J.

FIRST LAW OF THERMODYNAMICS

The first law of thermodynamics is, in fact, an application of the broad principle known as the Law of Conservation of Energy to the thermodynamic system. It states that : **the total energy of an isolated system remains constant though it may change from one form to another.**

When a system is changed from state *A* to state *B*, it undergoes a change in the internal energy from E_A to E_B. Thus, we can write

$$\Delta E = E_B - E_A$$

This energy change is brought about by the evolution or absorption of heat and/or by work being done by the system. Because the total energy of the system must remain constant, we can write the **mathematical statement of the First Law** as :

$$\Delta E = q - w \qquad ...(1)$$

where q = the amount of heat supplied to the system

w = work done by the system

Thus First Law may also be stated as : **the net energy change of a closed system is equal to the heat transferred to the system minus the work done by the system.**

To illustrate the mathematical statement of the First Law, let us consider the system 'expanding hot gas' (Fig. 7.11).

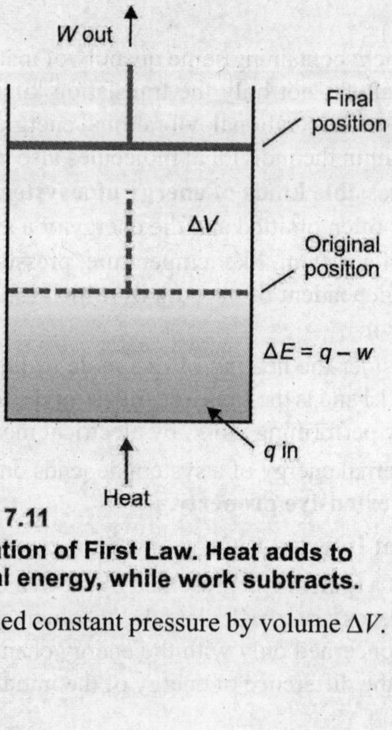

■ **Figure 7.11**
Illustration of First Law. Heat adds to internal energy, while work subtracts.

The gas expands against an applied constant pressure by volume ΔV. The total mechanical work done is given by the relation

$$w = P \times \Delta V \qquad ...(2)$$

From (1) and (2), we can restate

$$\Delta E = q - P \times \Delta V$$

Other Definitions of First Law of Thermodynamics

(1) Whenever energy of a particular type disappears equivalent amount of another type must be produced.

(2) Total energy of a system and surroundings remains constant (or conserved)

(3) It is impossible to construct a perpetual motion machine that can produce work without spending energy on it.

Some Special Forms of First Law of Thermodynamics

Mathematical statement of the First law of Thermodynamics is

$$\Delta E = q - w$$

Case 1 : For a cyclic process involving isothermal expansion of an ideal gas

$$\Delta E = o$$

$$\therefore \quad q = w$$

Case 2 : For an isochoric process (no change in volume) there is no work of expansion *i.e.* $w = 0$. Hence

$$\Delta E = q_v$$

Case 3 : For an adiabatic process there is no change in heat gained or lost *i.e.* $q = 0$. Hence

$$\Delta E = -w$$

In other words, the decrease in internal energy is exactly equal to the work done on the system by surroundings.

Case 4 : For an isobaric process there is no change in pressure, *i.e.* P remains constant. Hence

$$\Delta E = q - w$$

or

$$\Delta E = q - P\Delta V$$

SOLVED PROBLEM 4

Find ΔE, q and w if 2 moles of hydrogen at 303975.00 Pa pressure expand isothermally at 323 K and reversibly to a pressure of 101325.00.

SOLUTION

Since the operation is isothermal and the gas is ideal

$$\Delta E = 0$$

From the First Law

$$\Delta E = q - w$$

$$\therefore \quad q - w = 0$$

when $\Delta E = 0$

or $q = w$

For a reversible process

$$w = -nRT \ln (P_1/P_2) \text{ or } -2.303\, nRT \log \frac{P_1}{P_2}$$

$$= -2 \times 8.314 \text{ JK}^{-1} \text{ mol}^{-1} \times 323 \times 2.303 \times \log \frac{303975.00}{101325.00}$$

$$= -2 \times 8.314 \text{ JK}^{-1} \text{ mol}^{-1} \times 323 \times 2.303 \times \log 3$$

$$= -5903.387 \text{ J}$$

Since $q = w$

$$q = \mathbf{-5903.387\ J}$$

SOLVED PROBLEM 5

0.001 kg of water at 373 K is converted into steam at the same temperature. The volume of water becomes 0.001671 m^3 on boiling. Calculate the change in the internal energy of the system if the heat of vaporisation is 2260980 J/kg.

SOLUTION

As the vaporisation takes place against a constant pressure of 1 atmosphere, work done for an irreversible process, w, is

$$w = P(V_2 - V_1)$$

$$= nRT$$

$$= \frac{1}{18} \times 8.314 \text{ JK}^{-1} \text{ mol}^{-1} \times 373 \text{ K}$$

$$= 171667 \text{ J/kg}$$

Now $\quad q = 2260980$ J/kg

Since $\quad \Delta E = q - w$ (First Law)

$$= 2260980 - 171667$$

$\therefore \quad \Delta E =$ **2089313** J/kg

SOLVED PROBLEM 6

A gas contained in a cylinder fitted with a frictionless piston expands against a constant external pressure of 101325.00 Pa from a volume of 0.005 m^3 to a volume of 0.01 m^3. In doing so it absorbs 400 J thermal energy from its surroundings. Determine ΔE for the process.

SOLUTION

$$\Delta E = q - w \quad ...(1)$$

Here $\quad q = 400$ J

$$w = -P(V_2 - V_1) = -101325\,(0.01 - 0.005)$$
$$= -101325 \times 0.005$$
$$= -506.625 \text{ J}$$

Substituting values in (1)

$$\Delta E = 400 \text{ J} - (-506.625 \text{ J})$$
$$= 400 \text{ J} + 506.625 \text{ J}$$
$$= \mathbf{906.625\ J}$$

SOLVED PROBLEM 7

Calculate the maximum work done when pressure on 0.01 kg of hydrogen is reduced from 2026500.00 Pa to 101325.00 Pa at a constant temperature of 273 K. The gas behaves ideally. Will there be any change in internal energy ? Also calculate 'q'.

SOLUTION: We know

$$w = -2.303\, nRT \log\left(\frac{P_1}{P_2}\right)$$

Here

$$n = \frac{10}{2} = 5 \text{ moles; } T = 273 \text{ K}$$

$$P_1 = 2026500.00 \text{ Pa; } P_2 = 101325.00 \text{ Pa}$$

Substituting the values we get

$$w = -2.303 \times 5 \times 8.314 \times 273 \log \frac{2026500}{101325}$$

$$= -2.303 \times 5 \times 8.314 \times 273 \log \frac{20}{1}$$

$$= \mathbf{-34024.658\ J}$$

Since there is no change in temperature

$$\Delta E = 0$$

Hence $\quad q = \Delta E + w$

$$= 0 + (-34024.658)$$
$$= \mathbf{-34024.658}$$

ENTHALPY OF A SYSTEM

In a process carried at constant volume (say in a sealed tube), the heat content of a system is the same as internal energy (E), as no PV work is done. But in a constant-pressure process, the system (a gas) also expands energy in doing PV work. Therefore, **the total heat content of a system at constant pressure is equivalent to the internal energy E plus the PV energy.** This is called the **Enthalpy** (Greek *en* = in; *thalpos* = heat) of the system and is represented by the symbol H. Thus enthalpy is defined by the equation :

$$H = E + PV \qquad ...(1)$$

Enthalpy – A Function of State

In the equation (1) above, E, P, V are all state functions. Thus H, the value of which depends on the values of E, P, V must also be a function of state. Hence its value is independent of the path by which the state of the system is changed.

Change in Enthalpy

If ΔH be the difference of enthalpy of a system in the final state (H_2) and that in the initial state (H_1),

$$\Delta H = H_2 - H_1 \qquad ...(2)$$

Substituting the values of H_2 and H_1, as from (1) and (2), we have

$$\begin{aligned}\Delta H &= (E_2 + P_2V_2) - (E_1 + P_1V_1)\\ &= (E_2 - E_1) + (P_2V_2 - P_1V_1)\\ &= \Delta E + \Delta PV\end{aligned}$$

If P is constant while the gas is expanding, we can write

$$\Delta H = \Delta E + P\Delta V$$

or

$$\Delta H = \Delta E + w \quad (w = \text{work}) \qquad ...(3)$$

According to the First Law,

$$\Delta E = q - w \qquad ...(4)$$

where $\quad q$ = heat transferred

From equations (3) and (4)

$$\Delta H = q \text{ when change in state occurs at constant pressure}$$

This relationship is usually written as

$$\Delta H = q_p$$

where subscript p means constant pressure.

Thus ΔH can be measured by measuring the heat of a process occurring at constant pressure.

Units and Sign Conventions of Enthalpy

Since

$$\Delta H = H_2 - H_1$$

ΔH is positive if $H_2 > H_1$ and the process or reaction will be endothermic. ΔH is negative if $H_1 > H_2$ and the reaction will be exothermic.

In case of a chemical reaction carried in the laboratory in an open vessel,

$$\Delta H = H\text{ products} - H\text{ reactants} = q_p$$

The heat of reaction at one atmosphere pressure is usually shown along with the equation. Thus,

$$H_2(g) + \frac{1}{2}O_2(g) \longrightarrow H_2O(l) + 286.042\,\text{kJ}$$

The quantity of heat 286.042 kJ on the right hand represents – ΔH of the reaction.

The units of ΔH are kilocalories (kcal) and its SI unit is kilojoules (kJ).

Relation Between ΔH and ΔE

Calorific values of many gaseous fuels are determined in constant volume calorimeters. These values are, therefore, given by the expression

$$q_v = \Delta E$$

When any fuel is burnt in the open atmosphere, additional energy of expansion, positive or negative, against the atmosphere is also involved. The value of q thus actually released, *i.e.*, $q_p = \Delta H$, may be different from the equation

$$\Delta H = \Delta E + P\Delta V \qquad ...(1)$$

If gases are involved in a reaction, they account for most of the volume change as the volumes of solids and liquids are negligibly small in comparison.

Suppose we have n_1 moles of gases before reaction, and n_2 moles of gases after it. Assuming ideal gas behavior, we have

$$P\,V_2 = n_2\,RT$$

$$P\,V_1 = n_1\,RT$$

$$\therefore \quad P\,(V_2 - V_1) = (n_2 - n_1)\,RT$$

or

$$P\Delta V = \Delta n\,RT$$

Substituting in equation (1) we have,

$$\Delta H = \Delta E + \Delta n\,RT$$

SOLVED PROBLEM 8

For the reaction

$$H_2F_2\,(g) \longrightarrow H_2(g) + F_2(g)$$

$$\Delta E = -\,59.4526\ \text{kJ/mol at 298 K}$$

Calculate ΔH for the reaction.

SOLUTION

$$\Delta H = \Delta E + \Delta\,n\,RT$$

$$\Delta n = n_2 - n_1$$

Now

$$n_2 = 1 + 1 = 2$$

$$n_1 = 1$$

$$n_2 - n_1 = 2 - 1 = 1$$

$$\therefore \quad \Delta H = \Delta E + 1 \times 8.314 \times 298$$

$$= -\,59.4526 + 2477.572$$

$$= \mathbf{-\,2418.1194\ kJ/mol}$$

MOLAR HEAT CAPACITIES

By heat capacity of a system we mean the capacity to absorb heat and store energy. As the system absorbs heat, it goes into the kinetic motion of the atoms and molecules contained in the system. This increased kinetic energy raises the temperature of the system.

If q calories is the heat absorbed by mass m and the temperature rises from T_1 to T_2, the heat capacity (c) is given by the expression

$$c = \frac{q}{m \times (T_2 - T_1)} \qquad ...(1)$$

Thus heat capacity of a system is the heat absorbed by unit mass in raising the temperature by one degree (K or °C) at a specified temperature.

When mass considered is 1 mole, the expression (1) can be written as

$$C = \frac{q}{T_2 - T_1} = \frac{q}{\Delta T} \qquad ...(2)$$

where C *is denoted as Molar heat capacity.*

The molar heat capacity of a system is defined as the amount of heat required to raise the temperature of one mole of the substance (system) by 1 K.

Since the heat capacity (C) varies with temperature; its true value will be given as

$$C = \frac{dq}{dT}$$

where dq is a small quantity of heat absorbed by the system, producing a small temperature rise dT.

Thus the molar heat capacity may be defined as the ratio of the amount of heat absorbed to the rise in temperature.

Units of Heat Capacity

The usual units of the molar heat capacity are calories per degree per mole (cal K^{-1} mol^{-1}), or joules per degree per mole (J K^{-1} mol^{-1}), the latter being the SI unit.

Heat is not a state function, neither is heat capacity. It is, therefore, necessary to specify the process by which the temperature is raised by one degree. The two important types of molar heat capacities are those : (1) at constant volume; and (2) at constant pressure.

Molar Heat Capacity at Constant Volume

According to the first law of thermodynamics

$$dq = dE + PdV \qquad ...(i)$$

Dividing both sides by dT, we have

$$\frac{dq}{dT} = \frac{dE + PdV}{dT} \qquad ...(ii)$$

At constant volume $dV = 0$, the equation reduces to

$$C_V = \left(\frac{dE}{dT}\right)_V$$

Thus the heat capacity at constant volume is defined as the rate of change of internal energy with temperature at constant volume.

Molar Heat Capacity at Constant Pressure

Equation (ii) above may be written as

$$C = \frac{dE}{dT} + \frac{PdV}{dT} \qquad ...(iii)$$

We know $\qquad H = E + PV$

Differentiating this equation w.r.t. T at constant pressure, we get

$$\left(\frac{dH}{dT}\right)_P = \left(\frac{dE}{dT}\right)_P + P\left(\frac{dV}{dT}\right)_P \qquad ...(iv)$$

comparing it with equation (*iii*) we have

$$C_p = \left(\frac{dH}{dT}\right)_p$$

Thus heat capacity at constant pressure is defined as the rate of change of enthalpy with temperature at constant pressure.

Relation Between C_p and C_v

From the definitions, it is clear that two heat capacities are not equal and C_p is greater than C_v by a factor which is related to the work done. At a constant pressure part of heat absorbed by the system is used up in increasing the internal energy of the system and the other for doing work by the system. While at constant volume the whole of heat absorbed is utilized in increasing the temperature of the system as there is no work done by the system. Thus increase in temperature of the system would be lesser at constant pressure than at constant volume. Thus **C_p is greater than C_v.**

We know

$$C_p = \frac{dH}{dT} \qquad ...(i)$$

and

$$C_v = \frac{dE}{dT} \qquad ...(ii)$$

By definition $H = E + PV$ for 1 mole of an ideal gas

or $$H = E + RT \qquad (\because PV = RT)$$

Differentiating w.r.t. temperature, T, we get

$$\frac{dH}{dT} = \frac{dE}{dT} + R$$

or $$C_p = C_v + R \qquad \text{[By using equations (i) and (ii)]}$$

or $$C_p - C_v = R$$

Thus C_p is greater than C_v by a gas constant whose value is 1.987 cal K^{-1} mol^{-1} or 8.314 J K^{-1} mol^{-1} in S.I. units.

Calculation of ΔE and ΔH

(A) ΔE : For one mole of an ideal gas, we have

$$C_v = \frac{dE}{dT}$$

or $$dE = C_v \times dT$$

For a finite change, we have

$$\Delta E = E_2 - E_1 = C_v (T_2 - T_1)$$

and for *n* moles of an ideal gas we get

$$\Delta E = E_2 - E_1 = n \times C_v \times (T_2 - T_1)$$

(B) ΔH : We know

$$\begin{aligned}
\Delta H &= \Delta (E + PV) \\
&= \Delta E + \Delta (PV) \\
&= \Delta E + \Delta RT \qquad [\because PV = RT] \\
&= \Delta E + R\Delta T \\
&= C_v (T_2 - T_1) + R (T_2 - T_1) \\
&= (C_v + R) (T_2 - T_1) \\
&= C_p (T_2 - T_1) \qquad [\because C_p - C_v = R]
\end{aligned}$$

and for *n* moles of an ideal gas we get

$$\Delta H = n \times C_p \times (T_2 - T_1)$$

SOLVED PROBLEM 9

Calculate the value of ΔE and ΔH on heating 0.064 kg of oxygen from 273 K to 373 K. C_v and C_p on an average are 20.93 J mol^{-1} K^{-1} and 29.31 J mol^{-1} K^{-1}.

SOLUTION. We know

$$\Delta E = n \times C_v \times (T_2 - T_1) \quad ...(i)$$

and

$$\Delta H = n \times C_p \times (T_2 - T_1) \quad ...(ii)$$

Here

$$n = \frac{0.064}{0.016} = 4 \text{ moles}; \; T_2 = 100^\circ C = 373 \text{ K}$$

$$T_1 = 0^\circ C = 273 \text{ K}$$

On substitution we get

$$\Delta E = 4 \times 20.93 \times (373 - 273)$$

$$= 4 \times 20.93 \times 100 = \mathbf{8372\ J}$$

and

$$\Delta H = 4 \times 29.31 \times (373 - 273)$$

$$= 4 \times 29.31 \times 100$$

$$= \mathbf{11724\ J}$$

SOLVED PROBLEM 10

Calculate the amount of heat necessary to raise 0.2135 kg of water from 298 K to 373 K. Molar heat capacity of water is 75.36 J mol^{-1} K^{-1}.

SOLUTION

By definition

$$C = \frac{q}{T_2 - T_1}$$

or

$$q = C\,(T_2 - T_1) \qquad \text{– for 1 mole}$$

$$q = n\,C\,(T_2 - T_1) \qquad \text{– for } n \text{ moles} \quad ...(1)$$

In the present case

$$n = 0.2135/75.36$$

$$C = 75.36 \text{ J mol}^{-1} \text{ K}^{-1}$$

$$T_2 - T_1 = (373 - 298)\text{K}$$

Substituting the value in (1)

$$q = \frac{0.2135}{75.36} \times 75.36 \times (373 - 298)$$

$$= \mathbf{16.012\ kJ}$$

SOLVED PROBLEM 11

Three moles of an ideal gas (C_v = 20.93 J K^{-1} mol^{-1}) at 1013250 Pa and 273 K are converted to 202650.00 Pa at 323 K. Find ΔE and ΔH for the change.

$$R = 8.314 \text{ J mol}^{-1} \text{ K}^{-1}$$

SOLUTION.

(*a*)

$$\Delta E = n\,C_v dT$$

$$= 3 \times 20.93 \times (323 - 273)$$

$$= \mathbf{3139.5\ J}$$

(*b*)
$$\Delta H = n\, C_p dT = n\,(C_v + R)\, dT$$
$$= 3 \times (20.93 + 8.314) \times (323 - 273)$$
$$= 3 \times 29.244 \times 50 = \mathbf{4386.6\ J}$$

JOULE-THOMSON EFFECT

Joule and Thomson (later Lord Kelvin) showed that when a compressed gas is forced through a porous plug into a region of low pressure, there is appreciable cooling.

The phenomenon of producing lowering of temperature when a gas is made to expand adiabatically from a region of high pressure into a region of low pressure, is known as Joule-Thomson Effect or Joule-Kelvin Effect.

Joule-Thomson Experiment

The apparatus used by Joule and Thomson to measure the temperature change on expansion of a given volume of gas is illustrated in Fig. 7.12. An insulated tube is fitted with a porous plug in the middle and two frictionless pistons *A* and *B* on the sides. Let a volume V_1 of a gas at pressure P_1 be forced through the porous plug by a slow movement of piston *A*. The gas in the right-hand chamber is allowed to expand to volume V_2 and pressure P_2 by moving the piston *B* outward. The change in temperature is found by taking readings on the two thermometers.

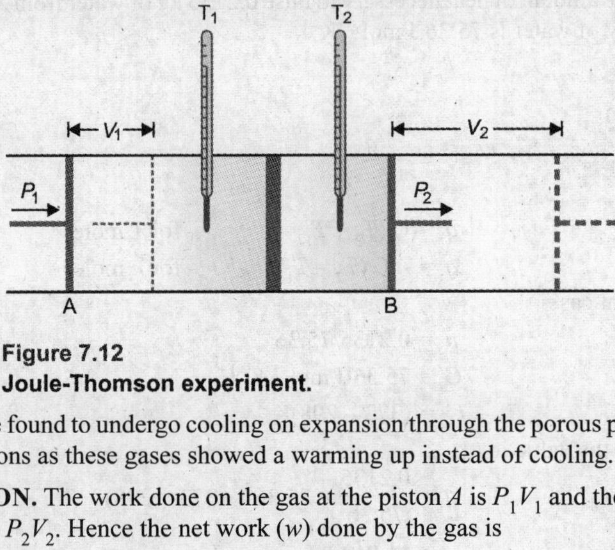

Figure 7.12
Joule-Thomson experiment.

Most gases were found to undergo cooling on expansion through the porous plug. Hydrogen and helium were exceptions as these gases showed a warming up instead of cooling.

EXPLANATION. The work done on the gas at the piston *A* is P_1V_1 and the work done by the gas at the piston *B* is P_2V_2. Hence the net work (*w*) done by the gas is

$$w = P_2V_2 - P_1V_1$$
$$\Delta E = q - w \quad \text{(First Law)}$$

But the process is adiabatic and, therefore, $q = 0$

$$\therefore \quad \Delta E = E_2 - E_1 = -w = -(P_2V_2 - P_1V_1)$$

or
$$E_2 - E_1 = -(P_2V_2 - P_1V_1)$$

Rearranging,

$$E_2 + P_2V_2 = E_1 + P_1V_1$$
$$H_2 = H_1 \text{ or } \Delta H = 0$$

Thus the process in Joule-Thomson experiment takes place at constant enthalpy.

Joule-Thomson Coefficient

The number of degrees temperature change produced per atmosphere drop in pressure under constant enthalpy conditions on passing a gas through the porous plug, is called Joule-Thomson coefficient. It is represented by the symbol μ. Thus,

$$\mu = \frac{dT}{dP}$$

If μ is positive, the gas cools on expansion; if μ is negative, the gas warms on expansion. The temperature at which the sign changes is called the **Inversion temperature.** Most gases have positive Joule-Thomson coefficients and hence they cool on expansion at room temperature. Thus liquefaction of gases is accomplished by a succession of Joule-Thomson expansion.

The inversion temperature for H_2 is –193.15 K. Above the inversion temperature, μ is negative. Thus at room temperature hydrogen warms on expansion. Hydrogen must first be cooled below –193.15 K (with liquid nitrogen) so that it can be liquefied by further Joule-Thomson expansion. So is the case with helium.

Explanation of Joule-Thomson Effect

We have shown above that Joule-Thomson expansion of a gas is carried at constant enthalpy. But

$$H = E + PV$$

Since H remains constant, any increase in PV during the process must be compensated by decrease of E, the internal energy. This leads to a fall in temperature *i.e.*, $T_2 < T_1$. For hydrogen and helium PV decreases with lowering of pressure, resulting in increase of E and $T_2 > T_1$. Below the inversion temperature, PV increases with lowering of pressure and cooling is produced.

ADIABATIC EXPANSION OF AN IDEAL GAS

A process carried in a vessel whose walls are perfectly insulated so that no heat can pass through them, is said to be **adiabatic.** In such a process there is no heat exchange between a system and surroundings, and $q = 0$.

According to the First law

$$\Delta E = q - w = 0 - w$$

or $$\Delta E = -w \qquad \text{...(1)}$$

Since the work is done at the expense of internal energy, the internal energy decreases and the temperature falls.

Consider 1 mole of an ideal gas at pressure P and a volume V. For an infinitesimal increase in volume dV at pressure P, the work done by the gas is $-PdV$. The internal energy decreases by dE.

According to equation (1)

$$dE = -PdV \qquad \text{...(2)}$$

By definition of molar heat capacity at constant volume

$$dE = C_v dT \qquad \text{...(3)}$$

From (2) and (3)

$$C_v dT = -PdV$$

For an ideal gas

$$P = RT/V$$

and hence

$$C_v dT = -RT\frac{dV}{V}$$

or
$$C_v \frac{dT}{T} = -R\frac{dV}{V}$$

Integrating between T_1, T_2 and V_1, V_2 and considering C_v to be constant,
$$C_v \int_{T_1}^{T_2} \frac{dT}{T} = -R.\int_{V_1}^{V_2} \frac{dV}{V}$$

Thus
$$C_v \ln\frac{T_2}{T_1} = -R\ln\frac{V_2}{V_1}$$

Since $R = C_p - C_v$, this equation may be written as
$$\ln\frac{T_2}{T_1} = -\frac{(C_p - C_v)}{C_v}\ln\frac{V_2}{V_1} \qquad ...(4)$$

The ratio of C_p to C_v is often written as γ,
$$\gamma = \frac{C_p}{C_v}$$

and equation (4) thus becomes
$$\ln\frac{T_2}{T_1} = -(\gamma - 1)\ln\frac{V_2}{V_1}$$

Replacing – ve sign by inverting V_2/V_1 to V_1/V_2 and taking antilogarithms
$$\frac{T_2}{T_1} = \left(\frac{V_1}{V_2}\right)^{\gamma-1} \qquad ...(5)$$
$$T_2V_2^{\gamma-1} = T_1V_1^{\gamma-1}$$
or
$$TV^{\gamma-1} = \text{a constant}$$

We can also eliminate the temperature by making use of the ideal gas relationship
$$\frac{T_2}{T_1} = \frac{P_2V_2}{P_1V_1} = \left[\frac{V_1}{V_2}\right]^{\gamma-1} \qquad ...(6)$$

Equating the right-hand sides of equations (5) and (6)
$$\frac{P_2}{P_1} = \left(\frac{V_1}{V_2}\right)^{\gamma}$$
$$P_1V_1^{\gamma} = P_2V_2^{\gamma} \quad \text{or} \quad PV^{\gamma} = k$$

Comparison between Isothermal and Adiabatic Expansions

Boyle's law describes pressure-volume relations of an ideal gas under isothermal conditions (T, constant). This is similar to the relation derived for adiabatic expansion.

$PV = constant$ (*Boyle's law*)

$PV^{\gamma} = constant$ (*Adiabatic expansion*)

γ for an ideal monoatomic gas = 1.67. The difference between the two processes is : **in an isothermal process, temperature of a system remains constant while in an adiabatic process, temperature must change.**

EXPLANATION. In an isothermal process heat is absorbed to make up for the work done by the gas in expansion and the temperature remains unchanged. On the other hand, adiabatic expansion takes place at the expense of internal energy which decreases and the temperature falls. For the same reason, the curve for the adiabatic process (Fig. 7.13) is steeper than that for the isothermal process.

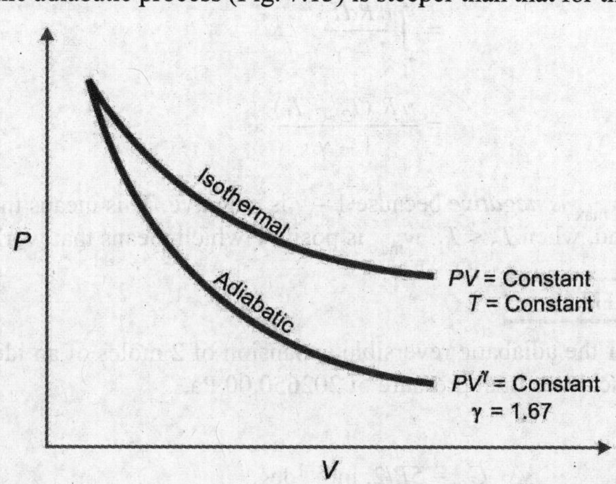

■ **Figure 7.13**
Curves for isothermal and adiabatic expansions for a monoatomic ideal gas.

WORK DONE IN ADIABATIC REVERSIBLE EXPANSION

Step 1. Value of V*d*P from adiabatic equation

For an adiabatic process

$$PV^{\gamma} = \text{constant}$$

Differentiating it, we have

$$\gamma PV^{\gamma-1}\, dV + V^{\gamma} dp = 0$$

Dividing by $V^{\gamma-1}$, we get

$$\gamma PdV + VdP = 0$$

or

$$VdP = -\gamma PdV \quad \text{...(1)}$$

Step 2. Value of *VdP* from ideal gas equation

For 1 mole of an ideal gas

$$PV = RT$$

Complete differentiation gives

$$PdV + VdP = RdT$$

$$VdP = RdT - PdV \quad \text{...(2)}$$

Step 3. Substitution

Substituting the value of *VdP* from (2) in (1) we get

$$RdT - PdV = -\gamma PdV$$

or

$$RdT = P(1-\gamma)\, dV$$

or

$$PdV = \frac{RdT}{1-\gamma}$$

If there are *n* moles of a gas

$$PdV = \frac{nRdT}{1-\gamma}$$

Step 4. Integration

Integrating from T_1, V_1 to T_2, V_2 with γ constant

$$w_{max} = P(V_2 - V_1)$$
$$= \int_{T_1}^{T_2} \frac{n\,RdT}{1-\gamma}$$
$$= \frac{nR\,(T_2 - T_1)}{1-\gamma}$$

When $T_2 > T_1$, w_{max} *is negative* because $1 - \gamma$ is negative. This means that work is done on the gas. On the other hand, when $T_2 < T_1$, w_{max} is positive which means that work is done by the gas.

SOLVED PROBLEM 12

Calculate w for the adiabatic reversible expansion of 2 moles of an ideal gas at 273.2 K and 2026500.00 Pacal to a final pressure of 202650.00 Pa.

SOLUTION

Given
$$C_p = 5R/2,\ \text{mol}^{-1}\text{deg}^{-1}$$
$$C_v = 3R/2,\ \text{mol}^{-1}\text{deg}^{-1}$$
$$R = 8.314\text{J /mol}$$

Step 1. To calculate the value of T_2, the final temperature, using the equation

$$(T_2 / T_1)^{\gamma} = (P_2 / P_1)^{\gamma - 1}$$
$$\gamma = \frac{C_p}{C_v} = \frac{5}{3}$$

Substituting the value of γ in (1)

$$(T_2 / 273.2)^{5/3} = (202650.00 / 2026500.00) = (2/20)$$

Solving it, we get

$$T_2 = \mathbf{108.8\ K}$$

Step 2. To calculate maximum work under adiabatic conditions

$$w_{max} = \frac{nR\,(T_2 - T_1)}{1-\gamma}$$
$$= \frac{2 \times 8.314\,(108.8 - 273.2)}{1 - 5/3}$$
$$= 4100\text{ J} = \mathbf{4.1\ kJ}$$

ALTERNATIVE SOLUTION

The work done under adiabatic conditions may be obtained by calculating decrease in internal energy.

$$w = -\Delta E = -nC_v\,(T_2 - T_1)$$
$$= -2 \times 3/2 \times 8.314\,(108.8 - 273.2)$$
$$= 4100\text{ J} = 4.1\text{ kJ}$$

ADDITIONAL SOLVED PROBLEMS

SOLVED PROBLEM 13

At 298 K for the combustion of 1 mole of liquid benzene the heat of reaction at constant pressure is given by

$$C_6H_6\,(l) + 7\tfrac{1}{2}\,O_2(g) \longrightarrow 6CO_2(g) + 3H_2O(1)\ \Delta H = -\,3269807.017 \text{ Joule}$$

What would be the heat of reaction at constant volume?

SOLUTION. We have

$$\Delta H = \Delta E + \Delta n\,R\,T$$

Here $\Delta n = n_p - n_R = 6 - 7.5 = -\,1.5$

Thus

$$\begin{aligned}\Delta E &= \Delta H - \Delta n\,(RT)\\ &= -\,3269807 - (-1.5) \times (8.314 \times 298)\\ &= -\,3269807 + 3716.358\\ &= -\,3266090.642 \text{ J}\\ &= \mathbf{-\,3266.090\ kJ}\end{aligned}$$

SOLVED PROBLEM 14

Calculate the maximum work done when 0.044 kg of Neon gas expands reversibly and isothermally from 1013250.00 Pa to 202650.00 Pa pressure at constant temperature 300 K. (Neon at. mass = 20; R = 8.3736 J K^{-1} mol^{-1}).

SOLUTION

Formula used

$$w = 2.303 \times n\,R\,T \quad \log \frac{P_1}{P_2}$$

Quantities given

$n = \frac{44}{20} = 2.2$ moles $\qquad R = 8.3736$ J K^{-1} mol^{-1}

$T = 27 + 273 = 300$ K $\qquad P_1 = 1013250.00$ Pa $\qquad P_2 = 202650.00$

Substitution of values

$$\begin{aligned}w &= 2.303 \times 2.2 \text{ mol} \times 8.3736 \text{ J K}^{-1}\text{ mol}^{-1} \times 300 \text{ K} \times \log \frac{1013250.00}{202650.00}\\ &= 12727.705 \text{ J} \times 0.69897\\ &= \mathbf{8896.28\ J}\end{aligned}$$

SOLVED PROBLEM 15

How many calories of heat are required to heat 1 mole of Argon from 313 K to 373 K at (i) Constant Volume and (ii) at Constant pressure ?

Molar heat capacity of Ar at constant volume = 12.56 J K^{-1}

Molar heat capacity of Ar at constant pressure = 20.93 J K^{-1}

SOLUTION : (i) At constant volume

Formula used

$$\Delta E = n \times C_v \times (T_2 - T_1)$$

Quantities given

$n = 1$ mole $\qquad C_v = 12.56\ J\ K^{-1}$

$T_2 = 373$ K $\qquad T_1 = 313$ K

Substitution of values

$$\Delta E = 1 \times 12.56 \times (373 - 313)\ K$$
$$= \mathbf{753.62\ J}$$

(ii) At constant pressure

Formula used

$$\Delta H = n \times C_p \times (T_2 - T_1)$$

Quantities given

$n = 1$ mole $\qquad C_p = 20.93\ J\ K^{-1} T_2 = 373$ K $\qquad T_1 = 313$ K

Substitution of values

$$\Delta H = 1 \times 20.93 \times (373 - 313)$$
$$= \mathbf{1256.04\ J}$$

SOLVED PROBLEM 16

Calculate the value of ΔE and ΔH on heating 0.064 kg of oxygen from 273 K to 373 K. C_v and C_p on an average are 20.93 J mol^{-1} K^{-1} and 29.308 J mol^{-1} K^{-1} respectively.

SOLUTION : (i) To calculate ΔE

Formula used

$$\Delta E = n \times C_v \times (T_2 - T_1)$$

Quantities given

$n = \frac{0.064\ kg}{0.032\ kg} = 2$ mol $\qquad C_v = 20.93\ J\ K\ mol^{-1}$

$T_2 = 373$ K $\qquad T_1 = 273$ K

Substitution of values

$$\Delta E = 2\ mol \times 20.93\ J\ K^{-1}\ mol^{-1} \times (373\ K - 273\ K)$$
$$= \mathbf{4186\ J}$$

(i) To calculate ΔH

Formula used

$$\Delta H = n \times C_p \times (T_2 - T_1)$$

Quantities given

$n = 2$ mol $\qquad C_p = 29.308\ J\ K^{-1}\ mol^{-1} \qquad T_2 = 373$ K $\qquad T_1 = 273$ K

Substitution of values

$$\Delta H = 2\ mol \times 29.308\ J\ K^{-1}\ mol^{-1} \times (373\ K - 273\ K)$$
$$= \mathbf{5861.6\ J}$$

SOLVED PROBLEM 17

Calculate the maximum work obtained when 2 moles of nitrogen were expanded isothermally and reversibly from 0.01 m³ to 0.02 m³ at 298 K.

SOLUTION

Formula used

$$w = 2.303\ n\ RT\ \log \frac{V_2}{V_1}$$

Quantities given

$n = 2$ mol $T = 298$ K $R = 8.314$ J K^{-1} mol^{-1} $V_1 = 0.01$ m^3 $V_2 = 0.02$ m^3

Substitution of values

$$\begin{aligned} w &= 2.303 \times 2 \text{ mol } \times 8.314 \text{ J K}^{-1} \text{ mol} \times 298 \text{ K} \times \log \frac{0.02 \text{ m}^3}{0.01 \text{ m}^3} \\ &= 11411.69 \text{ J} \times \log 2 \\ &= 11411.69 \text{ J} \times 0.3010 \\ &= \mathbf{3434.92\ J} \end{aligned}$$

SOLVED PROBLEM 18

Five moles of a perfect gas with $C_v = 20.93$ J K^{-1} mol^{-1} are compressed adiabatically and reversibly from a volume of 0.005 m³ at 101325.00 Pa to a pressure of 10132500.00 Pa. Calculate (i) the final volume of the gas (ii) the final temperature of the gas.

SOLUTION : (*a*) To calculate the final volume of the gas

Formula used

$$P_1 V_1^{\gamma} = P_2 V_2^{\gamma}$$

or

$$\frac{P_1}{P_2} = \left(\frac{V_2}{V_1}\right)^{\gamma}$$

Quantities given

$P_1 = 101325.00$ Pa $P_2 = 10132500.00$ Pa $V_1 = 0.005$ m^3

$$\gamma = \frac{C_p}{C_v} = \frac{C_v + R}{C_v} = \frac{20.93 + 8.314}{20.93} = \frac{29.244}{20.93} = 1.4$$

Substitution of values

$$\frac{101325}{10132500} = \left(\frac{V_2}{0.005}\right)^{1.4}$$

Taking logarithms

$$\begin{aligned} \log \frac{1}{100} &= 1.4 (\log V_2 - \log 0.005) \\ -2 &= 1.4 (\log V_2 - (-3.6990)) \\ \frac{-2}{1.4} &= \log V_2 + 3.6990 \\ &= -3.6990 - 1.4286 = \log V_2 \end{aligned}$$

or

$$\begin{aligned} V_2 &= \text{Antilog } 5.1276 \\ &= \mathbf{0.00186\ m^3} \end{aligned}$$

(b) Calculate the final temperature of the gas

Formula used

$$P_2 V_2 = n R T_2$$

or
$$T_2 = \frac{P_2 V_2}{n R}$$

Quantities given

$n = 5$ moles $\quad P_2 = 10132500.00$ Pa $\quad R = 8.314$ J K^{-1} mol^{-1} $\quad V_2 = 0.00186$ m^3

Substitution of values

$$T_2 = \frac{10132500 \times 0.00186}{5 \times 8.314 \text{ J K}^{-1} \text{ mol}^{-1}}$$

$$= \mathbf{453 \text{ K}}$$

EXAMINATION QUESTIONS

1. Define or explain the following terms :
   (a) First law of thermodynamics
   (b) An isothermal reversible expansion
   (c) Irreversible expansion
   (d) Internal energy
   (e) Enthalpy
   (f) Molar heat capacities
   (g) Molar heat capacity
   (h) Adiabatic expansion
   (i) Ideal gas
2. Explain the following terms :
   (a) State of a System
   (b) Extensive Properties
   (c) Closed System
   (d) Isothermal Process
3. What do you understand by C_P and C_V of gases? Why is the value of C_P always greater than that of C_V? How are they related?
4. State the first law of thermodynamics in as many ways as possible. Obtain the mathematical expression for the law with sign conventions.
5. Give a concise statement of the first law of thermodynamics. Deduce its mathematical form and explain the terms involved.
6. (a) What do you understand by thermodynamic system and surroundings?
   (b) Which of the following are intensive properties :
   (i) Density
   (ii) Surface Tension
   (iii) Volume
   (iv) Entropy
7. (a) Show thermodynamically that for an ideal gas $C_P - C_V = R$
   (b) A dry gas at NTP is expanded adiabatically from 0.001 m^3 to 0.005 m^3. Calculate the final temperature and pressure assuming ideal behavior ($C_P/C_V = 1.4$)

   Answer. (b) $T_2 = 143.31$ K; $P_2 = 10639.125$ Pa
8. (a) State and explain First law of Thermodynamics
   (b) Prove that $Q_V = \Delta E$ and $Q_P = \Delta H$ and discuss the relation amongst heat, internal energy and work.
9. Derive the relationship $\Delta H = \Delta E + \Delta n\, RT$
10. Explain heat capacity at constant volume thermodynamically.
11. Distinguish between isothermal and adiabatic process.
12. How would the energy of an ideal gas change if it is made to expand into vacuum at constant temperature?

13. (*a*) Derive the expression for maximum work done when *n* moles of an ideal gas are expanded isothermally and reversibly from V_1 to V_2 volume.

(*b*) Describe different types of thermodynamic processes.

14. (*a*) Explain the term enthalpy.

(*b*) State the first law of thermodynamics.

(*c*) Define C_P and C_V. State their relation.

15. (*a*) What are state functions? How do these differ from path functions?

(*b*) State the first law of thermodynamics. Give its mathematical statement and explain each term involved.

16. (*a*) Describe open, closed and isolated systems.

(*b*) What do you understand by the terms : Extensive properties and Intensive properties. Give two examples of each category.

(*c*) Calculate the work of expansion of one mole of an ideal gas at 298 K under isothermal conditions, the pressure being changed from 101325.00 to 506625.00 Pa

Answer. (*c*) –3988.2 J

17. (*a*) Give two definitions of First law of thermodynamics.

(*b*) State "Kirchoff's Law". Derive it with the help of first law of thermodynamics.

(*c*) Calculate the value of ΔE and ΔH on heating 0.064 kg of oxygen from 273 K to 1273 K. C_V and C_P on an average are 20.93 and 29.31 J mol^{-1} K^{-1} respectively.

Answer. (*c*) 83720 J; 17240 J

18. (*a*) Under what conditions $\Delta E = \Delta H$ for a chemical reaction?

(*b*) One mole of an ideal gas expands isothermally and reversibly from 0.001 m^3 to 0.1 m^3 at 300 K. Calculate *w*, *q*, ΔE, ΔH and ΔS for the process.

Answer. (*b*) 5744.1426 J; 5744.1426 J; zero; 2494.2 J, 19.471 J K^{-1}

19. State the first law of thermodynamics. With the help of this law show that:

(*i*) Heat absorbed by a system at constant volume is equal to increase in internal energy of the system.

(*ii*) Heat absorbed by a system at constant pressure is equal to the increase in the enthalpy.

20. Explain the difference between isothermal and adiabatic processes.

21. (*a*) Distinguish between open, closed and isolated system. Give examples.

(*b*) Differentiate between reversible and irreversible processes.

22. (*a*) Prove that the value of Joule Thomson coefficient is zero for an ideal gas.

(*b*) Explain First law of Thermodynamics. Calculate the work done in an isothermal and reversible expansion process of an ideal gas.

23. (*a*) Derive an expression for the work done by a gas in isothermal reversible expansion of an ideal gas.

(*b*) One mole of an ideal gas at 298 K is allowed to expand reversibly at constant temperature from volume 0.01 m^3 to 0.02 m^3. Calculate the work done by the gas in Joules and calories.

Answer. (*b*) 1717.46 J *(Guru Nanak Dev BSc)*

24. Distinguish between

(*a*) Isothermal and adiabatic process

(*b*) State function and path function

(*c*) Reversible and irreversible process

(*d*) Gibbs and Helmholtz free energy

(Andhra BSc)

25. Calculate *q*, *w*, ΔH, ΔS_{system}, ΔS_{surr} and $\Delta S_{universe}$ for one mole of an ideal gas which expands from V_1 to $10V_1$ at 300 K isothermally under reversible conditions (Use $R = 8.314$ J K^{-1} mol^{-1})

Answer. 5744.14 J, 5744.14 J, 0, 0, 19.147 J K^{-1}, –19.147 J K^{-1}, 0 *(Delhi BSc)*

26. Two moles of an ideal gas expanded isothermally and reversibly at 300 K to twice the original volume. Calculate q, w, ΔE and ΔH (Given log 2 = 0.3010, $R = 8.314 \text{ J K}^{-1} \text{ mol}^{-1}$)

Answer. 3457.97 J, 3457.97 J, 0, 4988.4 *(Sambalpur BSc)*

27. What is the basic principle of Joule-Thomson effect? *(Nagpur BSc)*

28. Which of the following parameters are state functions : q, H, E and w. The terms have their usual meanings. Show that the work done in an isothermal expansion of an ideal gas is greater than that of a van der Waal's gas. *(Kalyani BSc)*

29. Distinguish between :
 (*a*) Isothermal and adiabatic process
 (*b*) Reversible and Irreversible process *(Panjab BSc)*

30. Define heat capacity at constant pressure and heat capacity at constant volume. *(Arunachal BSc)*

31. How many joules of heat are required to heat 0.018 kg of Argon from 313 K to 373 K at
 (*i*) constant volume and (*ii*) constant pressure
 Molar heat capacity of Argon at constant volume is 12.56 J/K and molar heat capacity of Argon at constant pressure is 20.93 J/K.

Answer. 753.62 J; 1256.04 J *(Meerut BSc)*

32. (*a*) Calculate the maximum work done when 0.044 kg of Neon gas expands reversibly and isothermally from 1013250.00 Pa to 202650.00 Pa pressure at constant temperature 300 K. (Neon at wt. = 20; $R = 8.37 \text{ J K}^{-1} \text{ mol}^{-1}$)
 (*b*) Derive Kirchoff's equation.

Answer. (*a*) 8896.28 J *(Dibrugarh BSc)*

33. (*a*) Explain the following :
 Heat capacity, Thermodynamic state, Isolated state, Intensive properties.
 (*b*) Calculate the maximum work obtained when 2 moles of nitrogen were expanded isothermally and reversibly from 0.01 to 0.02 m^3 at 298 K.

Answer. (*b*) 3434.9 J *(Sri Venkateswara BSc)*

34. (*a*) Write notes on objectives and limitations of Thermodynamics.
 (*b*) Calculate ΔE and ΔH when the temperature of one mole of water is increased from 283K to 343K. The density of water is 977.8 kg/m^3 and 999.7 kg/m^3 at 343 K and 383 K respectively.

Answer. (*b*) 4521.74 K mol^{-1}; 4521.70 K mol^{-1} *(Gulbarga BSc)*

35. Calculate w and ΔE for the conversion of 1 mole of water into 1 mole of steam at a temperature of 373K and 101325.00 Pa. Latent heat of vaporisation of water 2.26098 J kg^{-1}.

Answer. 37.59 Kilojoule *(Delhi BSc)*

36. Calculate the work done when 5 moles of hydrogen gas expand isothermally and reversibly at 298 K from 0.01 m^3 to 0.05 m^3.

Answer. 19.95 Kilojoule *(Bangalore BSc)*

37. When one mole of liquid Br_2 is converted to Br_2 vapor at 298 K and 101325 Pa, 30.56 kJ of heat is absorbed and 2.47 kJ of expansion work is done by the system. Calculate ΔE for this process.

Answer. + 28.09 kJ *(Nagpur BSc)*

38. Find the work done when 2 moles of an ideal gas expand isothermally from 0.002 m^3 to 0.005 m^3 against a constant pressure of 101325 Pa at 298 K.

Answer. 4540.35 J *(Madras BSc)*

39. Calculate ΔE for the combustion of one mole of Magnesium in an open container at 298 K and 101325 Pa pressure, if $\Delta H_{combustion} = -602.48$ kJ ?

Answer. – 600.002 kJ *(Meerut BSc)*

40. Calculate the work done in expanding 2 moles of an ideal gas from 0.002 to 0.005 m^3 at 546 K.

(*Delhi BSc*)

Answer. 8218.91 J

41. (*a*) Write short notes on following :

(*i*) Enthalpy (*ii*) Internal Energy

(*iii*) Free Energy

(*b*) Heat supplied to a Carnot engine is 1899 kJ. How much useful work can be done by the engine between 273 K and 373 K?

(*Sambalpur BSc*)

Answer. (*b*) 509114.87 kJ

42. What is the efficiency of Carnot engine working between 500 K and 850 K? If the engine absorbs 5024 J of work, how much heat is given to sink?

(*Andhra BSc*)

Answer. 0.4117; 12203.0605 J

43. Calculate maximum amount of heat withdrawn from a hot reservoir at 410 K to obtain work equivalent to 15 kJ per cycle. The temperature of the sink is 290 K.

(*Kalyani BSc*)

Answer. 512.50 kJ

44. A gas at 1013250.00 Pa pressure occupies a volume of 0.01 m^3 at 300 K. It is allowed to expand at the constant temperature of 300 K under a constant external pressure till the volume equilibrates at 0.1 m^3. Calculate the work done.

(*Panjab BSc*)

Answer. 91192.5 J

MULTIPLE CHOICE QUESTIONS

1. The study of the flow of heat or any other form of energy into or out of a system undergoing physical or chemical change is called

(*a*) thermochemistry (*b*) thermokinetics

(*c*) thermodynamics (*d*) thermochemical studies

Answer. (*c*)

2. Thermodynamics is applicable to

(*a*) microscopic systems only (*b*) macroscopic systems only

(*c*) homogeneous systems only (*d*) heterogeneous systems only

Answer. (*b*)

3. Which is not true about thermodynamics?

(*a*) it ignores the internal structure of atoms and molecules

(*b*) it involves the matter in bulk

(*c*) it is concerned only with the initial and final states of the system

(*d*) it is not applicable to macroscopic systems

Answer. (*d*)

4. A system that can transfer neither matter nor energy to and from its surroundings is called

(*a*) a closed system (*b*) an isolated system

(*c*) an open system (*d*) a homogeneous system

Answer. (*b*)

5. A thermos flask is an example of

(*a*) isolated system (*b*) closed system

(*c*) open system (*d*) heterogeneous system

Answer. (*a*)

6. A closed system is one which cannot transfer matter but transfer ________ to and from its surrounding
   (*a*) heat
   (*b*) work
   (*c*) radiations
   (*d*) all of these
   Answer. (*d*)
7. A gas contained in a cylinder fitted with a piston constitutes
   (*a*) an open system
   (*b*) a heterogeneous system
   (*c*) a closed system
   (*d*) an isolated system
   Answer. (*c*)
8. A system that can transfer both energy and matter to and from its surroundings is called
   (*a*) an isolated system
   (*b*) a closed system
   (*c*) an open system
   (*d*) a heterogeneous system
   Answer. (*c*)
9. Zinc granules reacting with dilute hydrochloric acid in an open beaker constitutes
   (*a*) an isolated system
   (*b*) an open system
   (*c*) a closed system
   (*d*) a heterogeneous system
   Answer. (*b*)
10. A system in which no thermal energy passes into or out of the system is called
    (*a*) adiabatic system
    (*b*) an open system
    (*c*) a reversible system
    (*d*) a closed system
    Answer. (*a*)
11. An intensive property does not depend upon
    (*a*) nature of the substance
    (*b*) quantity of matter
    (*c*) external temperature
    (*d*) atmospheric pressure
    Answer. (*b*)
12. Which out of the following is not an intensive property?
    (*a*) pressure
    (*b*) concentration
    (*c*) density
    (*d*) volume
    Answer. (*d*)
13. A property that depends upon the quantity of matter is called an extensive property. Which of the following is not an extensive property?
    (*a*) mass
    (*b*) volume
    (*c*) density
    (*d*) internal energy
    Answer. (*c*)
14. Which of the following sets of properties constitute intensive properties?
    (*a*) temperature, pressure and volume
    (*b*) mass, density and volume
    (*c*) density, pressure and temperature
    (*d*) internal energy, density and pressure
    Answer. (*c*)
15. A system in which state variables have constant values throughout the system is called in a state of
    (*a*) equilibrium
    (*b*) non-equilibrium
    (*c*) isothermal equilibrium
    (*d*) none of these
    Answer. (*a*)
16. In an adiabatic process ________ can flow into or out of the system.
    (*a*) no heat
    (*b*) heat
    (*c*) matter
    (*d*) no matter
    Answer. (*a*)

17. Which of the following conditions holds good for an adiabatic process?

(*a*) $dq < 0$ (*b*) $dq > 0$
(*c*) $dq = 0$ (*d*) $dq = \alpha$

Answer. (*c*)

18. An isobaric process takes place at constant ______.

(*a*) temperature (*b*) pressure
(*c*) volume (*d*) concentration

Answer. (*b*)

19. Which is true for an isobaric process?

(*a*) $dp > 0$ (*b*) $dp < 0$
(*c*) $dp = \alpha$ (*d*) $dp = 0$

Answer. (*d*)

20. An isochoric process takes place at constant ______.

(*a*) volume (*b*) temperature
(*c*) pressure (*d*) concentration

Answer. (*a*)

21. For a cyclic process, the change in internal energy of the system is

(*a*) always positive (*b*) always negative
(*c*) equal to zero (*d*) equal to infinity

Answer. (*c*)

22. Which out of the following is incorrect?

(*a*) heat flow into the system is +ve (*b*) heat flow out of the system is –ve
(*c*) work done on the system is –ve (*d*) none of these

Answer. (*c*)

23. The units erg, joule and calorie are interconvertible, which of the following is incorrect?

(*a*) 10^7 ergs = 1 J (*b*) 4.184 J = 1 cal
(*c*) 1 J = 0.2390 cal (*d*) 1 erg = 4.184 cal

Answer. (*d*)

24. A gas expands from 0.01 m^3 liters to 0.02 m^3 against a constant external pressure of 1013250.00 Pa. The pressure-volume work done by the system is

(*a*) 10120 J (*b*) –10120 J
(*c*) 1012 J (*d*) –1012 J

Answer. (*b*)

25. Which out of the following is incorrect, for an ideal gas?

(*a*) $PV = nRT$ (*b*) $V = \dfrac{nRT}{P}$
(*c*) $P = \dfrac{nRT}{V}$ (*d*) all are correct

Answer. (*d*)

26. The work done in the reversible expansion of a gas from the initial state A to final state B is

(*a*) maximum (*b*) minimum
(*c*) equal to zero (*d*) equal to infinity

Answer. (*a*)

27. The first law of thermodynamics is

(*a*) the total energy of an isolated system remains constant though it may change from one form to another

(*b*) total energy of a system and surroundings remains constant
(*c*) whenever energy of one type disappears, equivalent amount of another type is produced
(*d*) all of the above

Answer. (*d*)

28. The mathematical relation for the first law of thermodynamics is
(*a*) $\Delta E = q - w$ (*b*) $\Delta E = 0$ for a cyclic process
(*c*) $\Delta E = q$ for an isochoric process (*d*) all of these

Answer. (*d*)

29. For an adiabatic process, according to first law of thermodynamics,
(*a*) $\Delta E = -w$ (*b*) $\Delta E = w$
(*c*) $\Delta E = q - w$ (*d*) none of these

Answer. (*a*)

30. The change in internal energy for an isobaric process is given by
(*a*) $\Delta E = q + p\,\Delta v$ (*b*) $\Delta E = q - p\,\Delta v$
(*c*) $\Delta E = q$ (*d*) $\Delta E = p\,\Delta v$

Answer. (*b*)

31. Which of the following properties is not a function of state?
(*a*) concentration (*b*) internal energy
(*c*) enthalpy (*d*) entropy

Answer. (*a*)

32. The change in enthalpy of a system is measured by measuring
(*a*) heat of the process at constant volume (*b*) heat of the process at constant temperature
(*c*) heat of the process at constant pressure (*d*) none of these

Answer. (*c*)

33. The enthalpy change, ΔH of a process is given by the relation
(*a*) $\Delta H = \Delta E + p\,\Delta v$ (*b*) $\Delta H = \Delta E + \Delta n\,R\,T$
(*c*) $\Delta H = \Delta E + w$ (*d*) all of these

Answer. (*b*)

34. The amount of heat required to raise the temperature of one mole of the substance by 1 K is called
(*a*) heat capacity (*b*) molar heat capacity
(*c*) molar heat (*d*) molar capacity

Answer. (*b*)

35. Which of the following statements is not correct?
(*a*) heat is not a state function (*b*) heat capacity is not a state function
(*c*) neither of these (*d*) both

Answer. (*c*)

36. Heat capacity at constant pressure is the change in
(*a*) internal energy with temperature at constant volume
(*b*) internal energy with temperature at constant pressure
(*c*) enthalpy with temperature at constant volume
(*d*) enthalpy with temperature at constant pressure

Answer. (*d*)

37. Heat capacity at constant volume is the change in

(*a*) internal energy with temperature at constant volume

(*b*) internal energy with temperature at constant pressure

(*c*) enthalpy with temperature at constant volume

(*d*) enthalpy with temperature at constant pressure

Answer. (*a*)

38. Which of the following relations is true ?

(*a*) $C_p > C_v$ (*b*) $C_v > C_p$

(*c*) $C_p = C_v$ (*d*) $C_p = C_v = 0$

Answer. (*a*)

39. The heat capacity at constant pressure is related to heat capacity at constant volume by the relation

(*a*) $C_p - R = C_v$ (*b*) $C_v - R = C_p$

(*c*) $C_p - C_v = R$ (*d*) $R - C_p = C_v$

Answer. (*c*)

40. The phenomenon of lowering of temperature when a gas is made to expand adiabatically from a region of high pressure into a region of low pressure is known as

(*a*) First law of thermodynamics (*b*) Second law of thermodynamics

(*c*) Le Chatlier's principle (*d*) Joule Thomson effect

Answer. (*d*)

41. Which of the following relations is applicable to adiabatic expansion of an ideal gas?

(*a*) $T_1 V_1^{\gamma-1} = T_2 V_2^{\gamma-1}$ (*b*) $P_1 V_1^{\gamma} = P_2 V_2^{\gamma}$

(*c*) both (*d*) none of these

Answer. (*c*)

42. In an adiabatic process _______ must change

(*a*) pressure (*b*) volume

(*c*) concentration (*d*) temperature

Answer. (*d*)

43. The enthalpy change of a reaction is independent of

(*a*) state of the reactants and products

(*b*) nature of the reactants and products

(*c*) initial and final enthalpy change of the reaction

(*d*) different intermediate reaction

Answer. (*d*)

44. Which of the following is not correct?

(*a*) $H = E + P V$ (*b*) $H - E = P V$

(*c*) $H - E - P V = 0$ (*d*) $H = E - P V$

Answer. (*d*)

45. When the total energy change in an isothermal cycle is zero, it represents

(*a*) a reversible cycle (*b*) an adiabatic change

(*c*) a thermodynamic equilibrium (*d*) an irreversible cycle

Answer. (*a*)

46. One mole of an ideal gas at 300 K is expanded isothermally from 0.001 m^3 volume to 0.01 m^3 volume. ΔE for this process is (R = 8.314 J K^{-1} mol^{-1})

(*a*) 1256 J (*b*) 2512 J

(*c*) 5024 J (*d*) 0 J

Answer. (*d*)

47. A system absorbs 100 kJ heat and performs 50 kJ work on the surroundings. The increase in internal energy of the system is

(*a*) 50 kJ (*b*) 100 kJ

(*c*) 150 kJ (*d*) 5000 kJ

Answer. (*a*)

48. For the reaction $H_2 + I_2 \rightleftharpoons 2HI$, ΔH is equal to

(*a*) $\Delta E + 2RT$ (*b*) $\Delta E - 2RT$

(*c*) ΔE (*d*) $\Delta E + RT$

Answer. (*c*)

49. The work done when 1 mole of a gas expands reversibly and isothermally from 506625 Pa to 101325 Pa at 300 K is

(*a*) – 4015 J (*b*) +4015 J

(*c*) zero (*d*) 150 J

Answer. (*a*)

50. Three moles of an ideal gas (C_v = 20.93 J K^{-1} mol^{-1}) at 1013250 Pa and 273 K are converted to 202650.00 Pa at 323 K. The ΔE for the process is

(*a*) 628 J (*b*) 1256 J

(*c*) 3140 J (*d*) 6280 J

Answer. (*c*)

8 Thermochemistry

CONTENTS

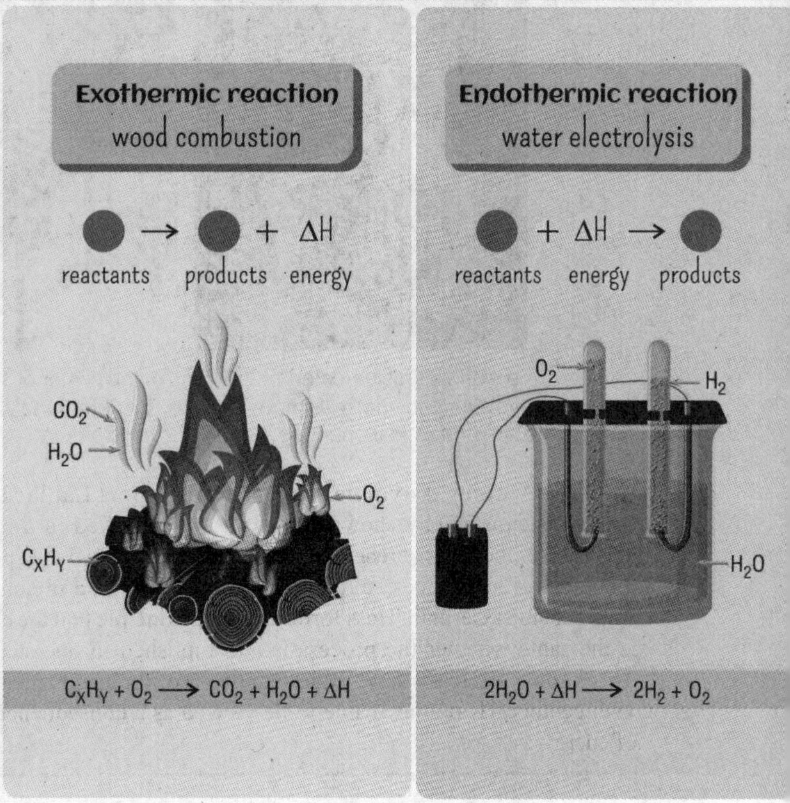

It is noticed that energy in the form of heat (thermal energy) is generally evolved or absorbed as a result of a chemical change.

Thermochemistry is the branch of physical chemistry which deals with the thermal or heat changes caused by chemical reactions.

We have studied in the previous chapter that every substance has a definite amount of energy known as the **intrinsic energy or internal energy, E.** Its exact value cannot be determined but the change in internal energy, ΔE, can be accurately measured experimentally.

When the internal energy of reactants (E_r) is greater than the internal energy of the products (E_p), the difference of internal energy, ΔE, is released as heat energy.

$$\Delta E = E_{\text{products}} - E_{\text{reactants}}$$

or

$$\Delta E = E_p - E_r$$

HISTORY OF THERMOCHEMISTRY

The world's first ice-calorimeter, used in the winter of 1782-83, by Antoine Lavoisier and Pierre-Simon Laplace, to determine the heat evolved in various chemical changes.

In 1782 Antoine Lavoisier and Pierre-Simon Laplace laid the foundations of "thermochemistry" by showing that the heat evolved in a reaction is equal to the heat absorbed in the reverse reaction. They also investigated the specific heat and latent heat of a number of substances, and amounts of heat evolved in combustion. Similarly, in 1840 Swiss chemist Germain Hess formulated the principle that the evolution of heat in a reaction is the same whether the process is accomplished in one-step or in a number of stages. This known as Hess's law. With the advent of the mechanical theory of heat in the early 19th century, Hess's law came to be viewed as a consequence of the law of conservation of energy.

Such a reaction is called **exothermic reaction.** If the internal energy of the products (E_p) is greater than that of the reactants (E_r), heat is absorbed from the surroundings. Such a reaction is called **endothermic reaction.** The amount of heat released or absorbed in a chemical reaction is termed the **heat of reaction.**

The energy changes in chemical reactions are largely due to the breaking of existing bonds between the atoms and the formation of new bonds. **Thus thermochemistry provides useful information regarding the bond energies.**

UNITS OF ENERGY CHANGES

The energy changes are usually expressed as the calorie (cal.), kilocalorie (1 kcal = 1000 cal), Joule (J) and kilojoule (kJ). It may be noted that 1 cal = 4.18 J and 1 kcal = 4.18 kJ.

ENTHALPY OF A REACTION

Thermochemical measurements are made either at (*a*) constant volume or (*b*) constant pressure. The magnitudes of changes observed under the two conditions are different.

The change in internal energy (ΔE) is the heat change accompanying a chemical reaction at constant volume because no external work is performed.

However at constant pressure not only does the change in internal energy take place but work is also involved because of expansion or contraction. In the laboratory most of the chemical reactions are carried out at constant pressure (atmospheric pressure) rather than at constant volume. In order to study the heat changes for reactions taking place at constant pressure and constant temperature, chemists have introduced a new term called **enthalpy.**

The enthalpy of a system is defined as the sum of the internal energy and the product of its pressure and volume. That is,

$$H = E + PV$$

where E is the internal energy, P is the pressure and V is the volume of the system. It is also called **Heat content.**

Just like internal energy, enthalpy is also a function of the state and it is not possible to measure its absolute value. However a change in enthalpy (ΔH) accompanying a process can be measured accurately and is given by the expression

$$\begin{aligned}\Delta H &= H_{\text{products}} - H_{\text{reactants}} \\ &= H_{\text{p}} - H_{\text{r}}\end{aligned}$$

Thus if ΔV be the change in volume in case of a reaction at constant temperature and pressure, the thermal effect observed will be the sum of the change in internal energy (ΔE) and the work done in expansion or contraction. That is,

$$\Delta H = \Delta E + P \times \Delta V$$

Therefore, while the heat change in a process is equal to its change in internal energy ΔE at constant volume, it gives at constant pressure the enthalpy change ΔH. That is,

$$\begin{aligned}\Delta E &= \text{Heat change in a reaction at constant volume} \\ \Delta H &= \text{Heat change in a reaction at constant pressure}\end{aligned}$$

For reactions involving solids and liquids only the change in volume (ΔV) is very small and the term $P \times \Delta V$ is negligible. For such reactions ΔH is equal to ΔE. In case of gases, however, we must specify whether the reaction has taken place at constant volume or at constant pressure because the value of $P \times \Delta V$ is appreciable. Most of such reactions are, however, studied at constant pressure and change in enthalpy (ΔH) is involved.

EXOTHERMIC AND ENDOTHERMIC REACTIONS

Let us consider a general reaction at constant pressure,

$$\text{A} + \text{B} \longrightarrow \text{C} + \text{D}$$

If H_A, H_B, H_C and H_D be the enthalpies of A, B, C and D respectively, the heat of reaction at constant pressure *viz.*, ΔH is equal to the difference in enthalpies of the products and the reactants *i.e.*,

$$\begin{aligned}\Delta H &= H_{\text{products}} - H_{\text{reactants}} \\ &= (H_C + H_D) - (H_A + H_B)\end{aligned}$$

The value of ΔH may be either zero, negative or positive. Where ΔH is zero, the enthalpies of the products and reactants being the same, the heat is evolved or absorbed. In case ΔH is negative, the sum of enthalpies of the products is less than that of the reactants and the difference in enthalpy is given out in the form of heat.

Such reactions which are accompanied by the evolution of heat energy are called Exothermic reactions.

When ΔH is positive, the enthalpy or heat content of the reactants and an equivalent of heat is absorbed by the system from the surroundings.

EXAMPLES OF EXOTHERMIC AND ENDOTHERMIC PROCESSES

When trying to classify a process as exothermic or endothermic, watch how the temperature of the surroundings changes. An exothermic process releases heat, and causes the temperature of the immediate surroundings to rise. An endothermic process absorbs heat and cools the surroundings.

Exothermic processes	Endothermic processes
Making ice cubes	Melting ice cubes
Formation of snow in clouds	Conversion of frost to water vapor
Condensation of rain from water vapor	Evaporation of water
Mixing sodium sulfite and bleach	Baking bread
Rusting iron	Cooking an egg
Burning sugar	Producing sugar by photosynthesis
Forming ion pairs	Separating ion pairs
Mixing water and strong acids	Mixing water and ammonium nitrate
Mixing water with an anhydrous salt	Making an anhydrous salt from a hydrate
Crystallizing liquid salts (as in sodium acetate in chemical handwarmers)	Melting solid salts
Nuclear fission	Reaction of barium hydroxide octahydrate crystals with dry ammonium chloride
Mixing water with calcium chloride	Reaction of thionyl chloride ($SOCl_2$) with cobalt(II) sulfate heptahydrate

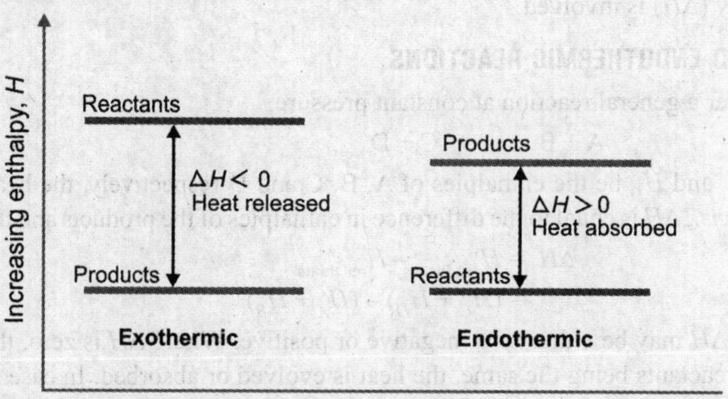

■ **Figure 8.1**
Enthalpy diagram for an exothermic and endothermic reaction.

Such reactions which are accompanied by absorption of heat are called Endothermic reactions.

Thus for an exothermic reaction $H_p < H_r$ and $\Delta H = -$ ve, for an endothermic reaction $H_p > H_r$ and $\Delta H = +$ve.

Sign of ΔH and ΔE

A negative sign of ΔH or ΔE shows that heat is evolved and the reaction is exothermic. A positive sign of ΔH or ΔE indicates that heat energy is absorbed and the reaction is endothermic.

TABLE 8.1. SIGN CONVENTIONS FOR ENERGY

Energy	Terms used	Sign
Released	Exothermic	–
Absorbed	Endothermic	+

Calculation of ΔH from ΔE and *vice versa*

The enthalpy change of a reaction at constant pressure (ΔH) and internal energy change (ΔE) are related to each other as

$$\Delta H = \Delta E + P \times \Delta V \qquad \text{... }(i)$$

where ΔV is the change in volume due to expansion or contraction when measurement is done at constant pressure, P. Though heat changes of reactions are usually measured at constant pressure, it is sometimes necessary to carry out the reaction at constant volume as, for example, in the measurement of heat of combustion in a bomb calorimeter. The above relationship can be used, if desired, for the conversion of ΔH into ΔE and *vice versa*.

Let us consider a reaction

$$\text{aA} + \text{bB} \longrightarrow \text{cC} + \text{dD}$$

Change in number of moles

$$= \text{No. of moles of products} - \text{No. of moles of reactants}$$

$$= (\text{c} + \text{d}) - (\text{a} + \text{b})$$

$$= \Delta n$$

Let the volume occupied by one mole of the gas be V. Then, change in volume, ΔV = change in number of moles × volume occupied by one mole of the gas.

$$\Delta V = \Delta n \times V$$

or

$$P \times \Delta V = P(\Delta n \times V)$$

$$P \times \Delta V = PV \times \Delta n \qquad \text{...}(ii)$$

But

$$PV = RT \qquad \text{(for one mole of gas)}$$

Putting RT in place of PV in equation (ii) we get

$$P\Delta V = RT\,\Delta n$$

Substituting the value of $P\Delta V$ in equation (i) we get

$$\Delta H = \Delta E + \Delta n\,RT$$

It may be pointed out that while determining the value of ΔH, only the number of moles of gaseous reactants and products are taken into consideration. The value of gas constant R is taken either in calories or joules per degree per mol and is 1.987 cal (approximately 2 calories) or 8.314 joules.

SOLVED PROBLEM 1

The heat of combustion of ethylene at 290 K and at constant volume is – 1390.81 kJ. Calculate the heat of combustion at constant pressure considering water to be in liquid state. ($R = 8.37$ J K^{-1} mol^{-1})

SOLUTION

The chemical equation for the combustion of ethylene is

$$\underset{1 \text{ mole}}{C_2H_4(g)} + \underset{3 \text{ moles}}{3O_2(g)} \longrightarrow \underset{2 \text{ moles}}{2CO_2(g)} + \underset{\text{negligible volume.}}{2H_2O(l)}$$

No. of moles of the products = 2

No. of moles of the reactants = 4

$\therefore$ $\Delta n = (2 - 4) = -2$

We know that $\Delta H = \Delta E + \Delta n\, RT$

Given that $\Delta E = -1390.81$ kJ

$T = 290$ K

$R = 8.37$ J K^{-1} mol^{-1}

$\therefore$ $\Delta H = -1390.8 + 8.37 \times -2 \times 290$

$= $ **– 1395.46 kJ**

SOLVED PROBLEM 2

The heat of combustion of carbon monoxide at constant volume and at 290 K is – 283.3 kJ. Calculate its heat of combustion at constant pressure ($R = 8.314$ J K^{-1} mol^{-1}).

SOLUTION

$$\underset{1 \text{ mole}}{CO(g)} + \underset{\frac{1}{2} \text{ mole}}{\tfrac{1}{2}\, O_2(g)} \longrightarrow \underset{1 \text{ mole}}{CO_2(g)}$$

No. of moles of products = 1

No. of moles of reactants = $1\frac{1}{2}$

Δn = No. of moles of products – No. of moles of reactants

$$= 1 - 1\frac{1}{2} = -\frac{1}{2}$$

Given that $\Delta E = -283.3$ kJ

$T = 290$ K

and $R = 8.314$ J or 8.314×10^{-3} kJ

Substituting these values in the equation

$$\Delta H = \Delta E + \Delta n \times RT$$

we get

$$\Delta H = -283.3 + \left[-\frac{1}{2} \times (8.314 \times 10^{-3}) \times 290\right]$$

$$= -283.3 - 1.20$$

$$= -284.5 \text{ kJ}$$

$\therefore$ Heat of combustion of CO at constant pressure is – **284.5 kJ.**

HOT PACKS / COLD PACKS

Heat therapy has become a standard treatment for ailing muscles among athletes, the disabled and elderly people. Heat packs provide relief by dilating the blood vessels of nearby muscles and allowing the soft tissue to stretch, and cold packs reduce the swelling and inflammation of injured body parts.Chemical advances have led to the development of very convenient types of heat packs and cold packs, which can now be found in most emergency first aid kits.

Hot Packs

There are a number of types of chemical heat packs used. Some packs consist of two plastic bags, the inner bag contains water, and the area between the inner bag and the outer bag is filled with a dry salt. When the inside bag is broken, the solid and the water react in an exothermic reaction, releasing heat. However, the most frequently used heat packs today involve one sealed plastic pouch containing a metal disk and a salt solution; commonly calcium chloride, magnesium sulfate, or sodium acetate.

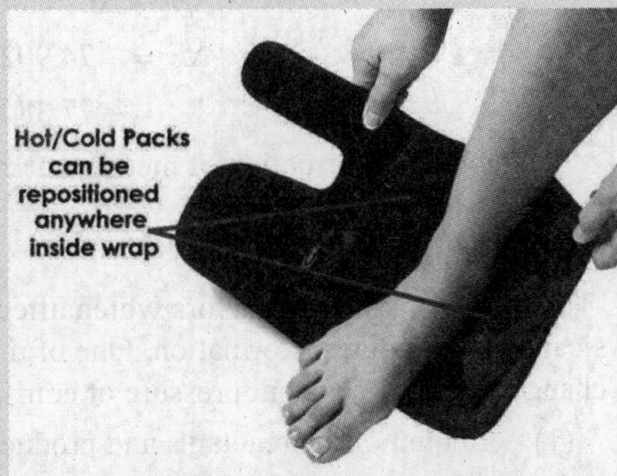

Cold Packs

A cold pack comes in a plastic bag made of tough white plastic. This bag is filled with a smaller bag and ammonium nitrate crystals. The smaller bag contains water, and is made of a thin weak plastic, so it is easy to break. When a cold pack is used, it must be "broken" by rupturing the inner bag. Breaking the bag releases the water, which dissolves the ammonium nitrate. The water and ammonium nitrate react completely, and within fifteen to twenty the cold pack will no longer feel cold.

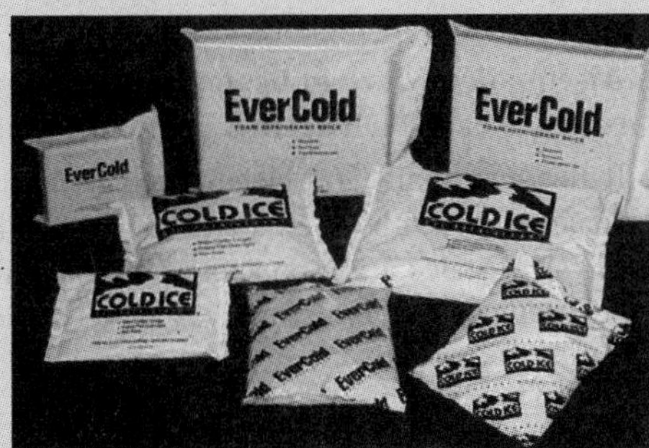

Cold packs make use of the heat transfer that occurs during chemical reactions, but in contrast to heat packs, utilize endothermic reactions. In the endothermic reaction between ammonium nitrate crystals and water, the heat required for the reaction to proceed from reactants to products is absorbed from the surrounding environment, resulting in a decrease in temperature of the pack noticeable to the touch.

SOLVED PROBLEM 3

The heat of formation of methane at 298 K at constant pressure is – 74.902 kJ. Calculate its heat of formation at constant volume. ($R = 8.314 \text{ J K}^{-1} \text{ mol}^{-1}$)

SOLUTION

The thermochemical equation for the heat of formation of methane at 298 K at constant pressure is :

$$C(s) + 2H_2(g) \longrightarrow CH_4(g) \quad \Delta H = -74.90 \text{ kJ}$$

No. of moles of gaseous products = 1

No. of moles of the gaseous reactants = 2

Change in No. of moles, $\Delta n = 1 - 2 = -1$

Given that $\Delta H = -74.902$ kJ; T = 298 K

and R = 8.314 J K^{-1} mol^{-1}

Substituting these values in the equation

$$\Delta H = \Delta E + \Delta n \times RT$$

$$\Delta E = -74.902 + [-1 \times (8.314) \times 298]$$

$$= -77.38 \text{ kJ}$$

∴ The heat of formation of methane at constant volume is – 77.38 kJ.

THERMOCHEMICAL EQUATIONS

There are a number of factors which affect the quantity of heat evolved or absorbed during a physical or chemical transformation. One of these factors has already been discussed *viz.*, whether the change occurs at constant pressure or constant volume. The other factors are :

(1) Amount of the reactants and products
(2) Physical state of the reactants and products
(3) Temperature
(4) Pressure

An equation which indicates the amount of heat change (evolved or absorbed) in the reaction or process is called a Thermochemical equation.

It must essentially : (*a*) be balanced; (*b*) give the value of ΔE or ΔH corresponding to the quantities of substances given by the equation; (*c*) mention the physical states of the reactants and products. The physical states are represented by the symbols (*s*), (1), (*g*) and (aq) for solid, liquid, gas and gaseous states respectively.

Example of Thermochemical Equation

The equation :

$$H_2 + \tfrac{1}{2}O_2 \longrightarrow H_2O \qquad \Delta H = -286.042 \text{ kJ}$$

indicates that when 1 mole of hydrogen reacts with 0.5 mole of oxygen, one mole of water is formed and 286.042 kJ of heat is evolved at constant pressure. If two moles of hydrogen are burnt, the heat evolved would be (2 × 286.042) kJ. This equation, however, is not a complete thermochemical equation because it does not specify whether water is in the form of steam or liquid. There is difference in the value of ΔH if water is in the liquid or gaseous state as shown below :

$$H_2(g) + \tfrac{1}{2}O_2(g) \longrightarrow H_2O(l) \qquad \Delta H = -286.042 \text{ kJ}$$

$$H_2(g) + \tfrac{1}{2}O_2(g) \longrightarrow H_2O(g) \qquad \Delta H = -241.997 \text{ kJ}$$

HEAT OF REACTION OR ENTHALPY OF REACTION

The heat of a reaction is simply the amount of heat absorbed or evolved in the reaction. We also know that the amount of heat absorbed or evolved at constant temperature and pressure is called enthalpy. Therefore **the amount of heat change during a reaction at constant temperature and pressure may also be called enthalpy change.** Its value depends upon the number of moles of the reactants which have reacted in the given chemical reaction. Thus,

Heat of reaction may be defined as the amount of heat absorbed or evolved in a reaction when the number of moles of reactants as represented by the balanced chemical equation change completely into the products. For example, the heat change for the reaction of one mole of carbon monoxide with 0.5 mole of oxygen to form one mole of carbon dioxide is – 284.5 kJ. This means that 284.5 kJ of heat is evolved during the reaction and is the heat of reaction. It can be represented as

$$CO(g) + \tfrac{1}{2}O_2(g) \longrightarrow CO_2(g) \quad \Delta H = -284.5 \text{ kJ}$$

It is very important to note that heat of reaction varies with the change in temperature. Therefore, we must mention the temperature at which the reaction is taking place. It is also convenient for comparison to fix up some temperature as standard or reference. According to the conventions prevalent in thermodynamics, the temperature of 298 K under a pressure of one atmosphere has been fixed as the *standard state.* **The heat change accompanying a reaction taking place at 298 K and one atmospheric pressure is called the standard heat change or standard enthalpy change.** It is denoted by ΔH^o.

VARIATION OF HEAT (OR ENTHALPY) OF REACTION WITH TEMPERATURE

The heat of reaction changes with change in temperature of a gas due to variation in its specific heat. The equations representing the variation of heat change of reaction with temperature are known as **Kirchoff's equations.**

At constant volume, the heat of reaction, ΔE, is given by the relation

$$\Delta E = E_2 - E_1$$

where E_1 and E_2 are the internal energies of the reactants and products.

Differentiating this equation with respect to temperature at constant volume, we get

$$\left[\frac{d(\Delta E)}{dT}\right]_v = \left(\frac{dE_2}{dT}\right)_v - \left(\frac{dE_1}{dT}\right)_v \quad \text{...(1)}$$

But we have already seen that

$$\left(\frac{dE}{dT}\right)_v = C_v$$

$$\therefore \quad \frac{d(\Delta E)}{dT} = (C_v)_2 - (C_v)_1 = \Delta C_v \quad \text{...(2)}$$

where $(C_v)_2$ and $(C_v)_1$ are heat capacities of the products and reactants respectively. Or, **change in heat of reaction at constant volume per degree change in temperature is equal to the difference in heat capacities at constant volume of products and reactants.**

Integrating the above equation between temperatures T_1 and T_2, we have

$$\Delta E_2 - \Delta E_1 = \int_{T_1}^{T_2} \Delta C_v \, dT$$

or
$$\Delta E_2 - \Delta E_1 = \Delta C_v \, [T_2 - T_1] \quad \text{...(3)}$$

where ΔE_2 and ΔE_1 are heats of reaction at temperatures T_2 and T_1 respectively.

Similarly, at **constant pressure** the heat of reaction ΔH is given by the reaction

$$\Delta H = H_2 - H_1$$

where H_2 is the heat content (enthalpy) of the products and H_1 being that of the reactants.

Differentiating with respect to temperature at constant pressure, we have

$$\left(\frac{d(\Delta H)}{dT}\right)_P = \left(\frac{dH_2}{dT}\right)_P - \left(\frac{dH_1}{dT}\right)_P \quad ...(4)$$

According to the equation, Chapter 7, we have

$$\left(\frac{dH}{dT}\right)_P = C_P$$

$$\therefore \quad \left(\frac{d(\Delta H)}{dT}\right)_P = (C_P)_2 - (C_P)_1 = \Delta C_P \quad ...(5)$$

where $(C_P)_2$ and $(C_P)_1$ are the heat capacities of products and reactants respectively.

or $$d(\Delta H) = \Delta C_P \times dT$$

Change in heat of reaction at constant pressure per degree change of temperature is equal to difference in heat capacities of products and reactants at constant pressure.

Integrating the equation between temperature T_1 and T_2, we have

$$\Delta H_2 - \Delta H_1 = \int_{T_1}^{T_2} \Delta C_p \, dT$$

or $$\Delta H_2 - \Delta H_1 = \Delta C_p [T_2 - T_1] \quad ...(6)$$

The relations (2), (3), (5) and (6) were first derived by Kirchoff and are called **Kirchoff's equations.** These equations may be used for calculating heat of reaction at a given temperature when it is known at some other temperature and when the heat capacities of products and reactants are known.

SOLVED PROBLEM 4

The heat of reaction $\frac{1}{2}H_2 + \frac{1}{2}Cl_2 \rightarrow HCl$ at 300 K is – 92.53 kJ. Calculate the heat of reaction at 350 K. The molar heat capacities at constant pressure at 300 K for hydrogen, chlorine and HCl are 28.55, 32.24 and 28.47 J mol^{-1} respectively.

SOLUTION

Here,

$$\frac{1}{2}H_2 + \frac{1}{2}Cl_2 \rightarrow HCl \qquad \Delta H = -92.53 \text{ kJ}$$

ΔC_P = Heat capacities of products – Heat capacities of reactants

$$= 28.47 - \left[\frac{1}{2}(28.55) + \frac{1}{2}(32.24)\right]$$
$$= 28.47 - [14.275 + 16.12]$$
$$= 28.47 - 30.395$$
$$= -1.925 \text{J}$$

$$T_2 = 273 + 77 \text{ K} = 350 \text{ K}; \; T_1 = 273 + 27 = 300 \text{ K}$$
$$T_2 - T_1 = (350 - 300) \text{ K} = 50 \text{ K}$$

Substituting these values in Kirchoff's equation, we have

$$\Delta H_2 - \Delta H_1 = \Delta C_p (T_2 - T_1)$$
$$= -1.925 \times 10^{-3} \times 50$$
$$= -96.25 \times 10^{-3} \text{ kJ}$$
$$= -96.25 \text{ kJ}$$

$\therefore$ Heat of reaction at 350 K is – **96.25 kJ**

SOLVED PROBLEM 5

The heat of reaction $N_2 + 3H_2 \rightarrow 2NH_3$ at 300 K was found to be –92.00911 kJ. What will be the heat of reaction at 323 K ?

The molar heat capacities at constant pressure and at 300 K for nitrogen, hydrogen and ammonia are 28.47, 28.34 and 37.095 J mol^{-1} K^{-1}.

SOLUTION

Here, $\Delta H = -92.00911$ kJ

$$T_2 = 273 + 50 = 323 \text{ K}$$
$$T_1 = 273 + 27 = 300 \text{ K}$$
$$[T_2 - T_1] = (323 - 300) \text{ K} = 23 \text{ K}$$

ΔC_p = Heat capacities of products – Heat capacities of reactants

$$= (2 \times 37.095) - [28.47 + (3 \times 28.34)]$$
$$= 74.19 - (28.47 + 85.02) = -39.3 \text{ J}$$
$$= -3.931 \times 10^{-2} \text{ kJ}$$

Substituting these values in Kirchoff's equation, we have

$$\Delta H_2 = \Delta H_1 + (T_2 - T_1)\,\Delta C_p$$
$$= -92.00911 + [23 \times (-3.931 \times 10^{-2})]$$
$$= -92.00911 - 0.90413 = -92.91323 \text{ kJ}$$

$\therefore$ Heat of reaction at 323 K is = **92.91 kJ**

DIFFERENT TYPES OF HEAT (ENTHALPY) OF REACTION

The heat or enthalpy changes accompanying chemical reactions are expressed in different ways, depending on the nature of the reaction. These are discussed below.

HEAT OF FORMATION

The heat of formation of a compound is defined as :

The change in enthalpy that takes place when one mole of the compound is formed from its elements.

It is denoted by ΔH_f. For example, the heat of formation of ferrous sulfide and acetylene may be expressed as :

$$Fe(s) + S(s) \longrightarrow FeS(s) \qquad \Delta H_f = -100.483 \text{ kJ}$$
$$2C(s) + H_2(g) \longrightarrow C_2H_2(g) \qquad \Delta H_f = +222.487 \text{ kJ}$$

Similarly, the reaction between gaseous hydrogen and gaseous chlorine to form gaseous hydrogen chloride is represented by the equation

$$H_2(g) + Cl_2(g) \longrightarrow 2HCl(g) \qquad \Delta H = -184.22 \text{ kJ}$$

It may be noted in this case that – 184.22 kJ is not the heat of formation of hydrogen chloride because this amount of heat is evolved when two moles of hydrogen chloride are formed. The heat of formation of hydrogen chloride, therefore, would be – 184.22/2 = – 92.11 kJ and the equation can be written as

$$\tfrac{1}{2}H_2(g) + \tfrac{1}{2}Cl_2(g) \longrightarrow HCl(g) \qquad \Delta H_f = -92.11 \text{ kJ}$$

STANDARD HEAT OF FORMATION

The standard heat of formation of a compound is defined as :

The change in enthalpy that takes place when one mole of a compound is formed from its elements, all substances being in their standard states (298 K and 101325.00 Pascal).

The standard heat of formation of some compounds are given in Table 8.2.

TABLE 8.2. STANDARD HEAT OF FORMATION OF SOME COMPOUNDS

Substance	ΔH_f^o kcal mol^{-1}	ΔH_f^o kJ mol^{-1}	Substance	ΔH_f^o kcal mol^{-1}	ΔH_f^o kJ mol^{-1}
$H_2O(g)$	– 57.84	– 242.16	CH_4 (g)	– 17.89	– 74.90
$H_2O(l)$	– 68.38	– 286.20	C_2H_6 (g)	– 20.23	– 84.70
$HCl(g)$	– 22.08	– 92.45	C_6H_6 (g)	+ 11.86	+ 49.66
$HBr(g)$	– 8.70	– 36.43	C_2H_2 (g)	+ 53.14	+ 222.47
$NH_3(g)$	– 11.02	– 46.14	CH_3OH (l)	– 57.17	– 239.36
$CO_2(g)$	– 94.13	– 394.10	C_2H_5OH (l)	– 66.26	– 277.42
$SO_2(g)$	– 71.00	– 297.26	CH_3COOH (l)	– 115.83	– 484.96

By convention the standard heat of formation of all elements is assumed to be zero.

Standard Heat of Reaction (ΔH^o) from Standard Heat of Formation (ΔH_f^o)

We can calculate the heat of reaction under standard conditions from the values of standard heat of formation of various reactants and products. **The standard heat of reaction is equal to the standard heat of formation of products minus the standard heat of formation of reactants.**

That is,

$$\Delta H^o = \begin{bmatrix}\text{Total standard heat} \\ \text{of formation of products}\end{bmatrix} - \begin{bmatrix}\text{Total standard heat} \\ \text{of formation of reactants}\end{bmatrix}$$

$$\Delta H^o = \Delta H_f^o \text{ (products)} - \Delta H_f^o \text{ (reactants)}$$

Let us consider a general reaction

$$\text{aA} + \text{bB} \longrightarrow \text{cC} + \text{dD}$$

The standard heat of reaction is given by

$$\Delta H^o = \Delta H_f^o \text{ (products)} - \Delta H_f^o \text{ (reactants)}$$

$$= [\text{c} \times \Delta H_f^o (\text{C}) + \text{d} \times \Delta H_f^o (\text{D})] - [\text{a} \times \Delta H_f^o (\text{A}) + \text{b} \times \Delta H_f^o (\text{B})]$$

SOLVED PROBLEM 6

Calculate ΔH^o for the reaction

$$CO_2(g) + H_2(g) \longrightarrow CO(g) + H_2O(g)$$

given that ΔH_f^o for $CO_2(g)$, $CO(g)$ and $H_2O(g)$ are – 393.5, – 111.31 and – 241.80 kJ mol^{-1} respectively.

SOLUTION

Here we have

$$CO_2(g) + H_2(g) \longrightarrow CO(g) + H_2O(g)$$

$$\Delta H^o = \Delta H_f^o\,(\text{products}) - \Delta H_f^o\,(\text{reactants})$$
$$= [\Delta H_f^o\,[CO\,(g)] + \Delta H_f^o\,[H_2O(g)]] - [\Delta H_f^o\,[CO_2\,(g)] + \Delta H_f^o\,[H_2(g)]]$$
$$= [-111.3 + (-241.8)] - [-393.5 + 0]$$
$$= -353.1 + 393.5$$
$$= \mathbf{40.4\ kJ}$$

SOLVED PROBLEM 7

The standard heats of formation of $C_2H_5OH(l)$, $CO_2(g)$ and $H_2O(l)$ are – 277.0, –393.5 and –285.5 kJ mol^{-1} respectively. Calculate the standard heat change for the reaction

$$C_2H_5OH(l) + 3O_2(l) \longrightarrow 2CO_2(g) + 3H_2O(l)$$

SOLUTION

We know that :

$$\Delta H^o = \Delta H_f^o\,(\text{products}) - \Delta H_f^o\,(\text{reactants})$$

In the present case

$$\Delta H^o = [2 \times \Delta H_f^o\,[CO_2(g)] + 3 \times \Delta H_f^o\,[H_2O(l)]]$$
$$= -\Delta H_f^o\,[C_2H_5OH(l)] + 3 \times \Delta H_f^o\,[O_2(g)]]$$
$$= [2 \times (-393.5) + 3 \times -285.5] - [-277.0 - 0]$$
$$= -1643.5 - (-277)$$
$$= \mathbf{-1366.5\ kJ}$$

HEAT OF COMBUSTION

The heat of combustion of a substance is defined as : **the change in enthalpy of a system when one mole of the substance is completely burnt in excess of air or oxygen.**

It is denoted by ΔH_c. As for example, heat of combustion of methane is –87.92 kJ as shown by the equation

$$CH_4(g) + 2O_2(g) \longrightarrow CO_2(g) + 2H_2O(l) \qquad \Delta H_c = -87.92$$

Now consider the chemical equations

$$C(s) + O_2(g) \longrightarrow CO_2(g) \qquad \Delta H = -394.82\ \text{kJ}$$

$$C(s) + \tfrac{1}{2}O_2(g) \longrightarrow CO(g) \qquad \Delta H = -108.86\ \text{kJ}$$

It may be noted that – 394.82 and not – 87.92 is the heat of combustion of carbon as the combustion is complete only in the first reaction. In the second case, oxidation has converted carbon to carbon monoxide and is by no means complete as carbon monoxide can be further oxidized to carbon dioxide.

It should be noted clearly that the **heat of combustion of a substance (ΔH_c) is always negative.** Heat energy is evolved during the process of combustion *i.e.*, ΔH_c = – ve.

APPLICATIONS OF THE HEAT OF COMBUSTION

(1) **Calculation of heat of formation.** Since the heats of combustion of organic compounds can be determined with considerable ease, these are employed to calculate their heats of formation. The direct determination of these is often impossible.

(2) **Calorific value of foods and fuels.** The calorific value is defined as : **the amount of heat produced in calories (or joules) when one gram of a substance is completely burnt.**

It is expressed in cal g^{-1} or kcal g^{-1} or kJ/ kg^{-1}. Let us compare the calorific values of methane and ethane. Their heats of combustion are – 890.3 kJ and – 1559.7 kJ. These combustion reactions are expressed as

$$CH_4(g) + 2O_2(g) \longrightarrow CO_2(g) + 2H_2O(l) \qquad \Delta H_c = -890.3 \text{ kJ}$$

$$C_2H_6(g) + 3\tfrac{1}{2}O_2(g) \longrightarrow 2CO_2(g) + 3H_2O(l) \qquad \Delta H_c = -1559.7 \text{ kJ}$$

In case of methane heat produced per gram is $890.3/16 = 55.64 \times 10^{-3}$ kJ/*kg* while for ethane it is $-1559.7/30 = 51.90 \times 10^{-3}$ kJ/*kg*. Thus methane has better fuel efficiency than ethane as it produces more heat per gram.

(3) **Deciding constitution.** Heat of combustion of organic compounds is to a large extent an additive property, as shown by the fact that in a homologous series the difference between the heats of combustion of successive members is nearly constant and is equal to 661.072 Joules. Constants corresponding to the heats of combustion of various atoms and linkages have been worked out. The heat of combustion of an organic substance can be calculated from its probable structural formula by adding up the values of the constants corresponding to the atoms and linkages involved therein. If the value so obtained comes out to be the same as the experimental value of the heat of combustion of the compound, the assumed formula must be correct. In this way Kekule's formula for benzene with alternate double and single linkages has been supported as the calculated value of the heat of combustion of benzene according to this formula agrees with the actual heat of combustion. Heat of combustion of organic compounds has thus proved very valuable in deciding their chemical constitution.

SOLVED PROBLEM 8

Calculate the standard heat of formation of propane (C_3H_8) if its heat of combustion is – 2220.2 kJ mol^{-1}. The heats of formation of $CO_2(g)$ and $H_2O(l)$ are – 393.5 and – 285.8 kJ mol^{-1} respectively.

SOLUTION

We are given

(*i*) $C_3H_8(g) + 5O_2(g) \longrightarrow 3CO_2(g) + 4H_2O(l) \qquad \Delta H_c = -2220.2 \text{ kJ}$

(*ii*) $C(s) + O_2(g) \longrightarrow CO_2(g) \qquad \Delta H = -393.5 \text{ kJ}$

(*iii*) $H_2(g) + \tfrac{1}{2}O_2(g) \longrightarrow H_2O(l) \qquad \Delta H = -285.8 \text{ kJ}$

We should manipulate these equations in a way so as to get the required equation

$$3C(s) + 4H_2(g) \longrightarrow C_3H_8(g) \qquad \Delta H = ?$$

Multiplying equation (*ii*) by 3 and equation (*iii*) by 4 and adding up we get

$$3C(s) + 3O_2(g) \longrightarrow 3CO_2(g) \qquad \Delta H = -1180.5 \text{ kJ}$$

$$4H_2(g) + 2O_2(g) \longrightarrow 4H_2O(l) \qquad \Delta H = -1143.2 \text{ kJ}$$

(*iv*) $3C(s) + 4H_2(g) + 5O_2(g) \longrightarrow 3CO_2(g) + 4H_2O(l) \qquad \Delta H = -2323.7 \text{ kJ}$

Subtracting equation (*i*) from equation (*iv*), we have

$$3C(s) + 4H_2(g) + 5O_2(g) - 5O_2(g) \longrightarrow C_3H_8(g) \qquad \Delta H = -103.5 \text{ kJ}$$

$\therefore$ The heat of formation of propane is – **103.5 kJ mol^{-1}.**

HEAT OF SOLUTION

Heat changes are usually observed when a substance is dissolved in a solvent. When a reaction takes place in solution, the heat of solution of reactants and products must be taken into consideration. The heat of solution is defined as : **the change in enthalpy when one mole of a substance is dissolved in a specified quantity of solvent at a given temperature.**

For example, when one mole of copper sulfate is dissolved in water so that we get one molar solution, the heat absorbed is 78.5 kJ. If the solution so obtained is further diluted, there will again be a change in enthalpy. If we go on diluting the solution, a stage will come when further dilution produces no thermal effect. This state is called the state of infinite dilution. To avoid the quantity of the solvent, we have to incorporate the idea of infinite dilution in our definition which may be stated as: **the heat of solution is the change in enthalpy when one mole of a substance is dissolved in a solvent so that further dilution does not give any change in enthalpy.**

The heat of solution can also be expressed as :

$$KCl(s) + H_2O(l) \longrightarrow KCl(aq) \qquad \Delta H = -18.42 \text{ kJ}$$

$$MgSO_4(s) + H_2O(l) \longrightarrow MgSO_4(aq) \qquad \Delta H = -84.91 \text{ kJ}$$

The heat of solution of an electrolyte may be due to energy change involved during ionization or some hydrate formation as in case of sulfuric acid. Usually heat is absorbed when ions are torn apart from each other in the process of solution and heat is evolved during hydrate formation. With a salt as sodium chloride the heat of separation of ions just equals the heat of hydration and there is very little heat effect.

HEAT OF NEUTRALIZATION

The heat of neutralization is defined as : **the change in heat content (enthalpy) of the system when one gram equivalent of an acid is neutralized by one gram equivalent of a base or *vice versa* in dilute solution.**

The following may be considered as typical examples of the heat of neutralization.

$$HNO_3(aq) + NaOH(aq) \longrightarrow NaNO_3(aq) + H_2O(l) \qquad \Delta H = -57.32 \text{ kJ}$$

$$HNO_3(aq) + KOH(aq) \longrightarrow KNO_3(aq) + H_2O(l) \qquad \Delta H = -58.07 \text{ kJ}$$

$$HCl(aq) + NaOH(aq) \longrightarrow NaCl(aq) + H_2O(l) \qquad \Delta H = -57.28 \text{ kJ}$$

$$HCl(aq) + LiOH(aq) \longrightarrow LiCl(aq) + H_2O(l) \qquad \Delta H = -57.36 \text{ kJ}$$

It may be concluded from the above data that **the heat of neutralization of a strong acid and strong base is –57.36 kJ, no matter which acid or base is employed.** This regularity has been explained satisfactorily with the help of the theory of ionization. If HA and BOH represent any strong acid and any strong base respectively and equivalent amounts of these in dilute solution be mixed, we have

$$H^+(aq) + \bar{A}(aq) + \overset{+}{B}(aq) + OH^-(aq) \longrightarrow \bar{A}(aq) + \overset{+}{B}(aq) + H_2O(l) \qquad \Delta H = -57.36 \text{ kJ}$$

Disregarding the ions which are present on both sides of the equation, we get

$$H^+(aq) + OH^-(aq) \longrightarrow H_2O(l) \qquad \Delta H = -57.36 \text{ kJ}$$

Thus the heat of neutralization of an acid and a base is merely the heat of formation of water from hydrogen and hydroxyl ions.

When weak acids or weak bases are neutralized by strong bases or strong acids respectively, the heat of neutralization differs widely from – 57.36 kJ. This is shown by the following examples :

$$HCl(aq) + NH_4OH(aq) \longrightarrow NH_4Cl(aq) + H_2O(l) \qquad \Delta H = -51.50 \text{ kJ}$$
$$HCN(aq) + NaOH(aq) \longrightarrow NaCN(aq) + H_2O(l) \qquad \Delta H = -51.50 \text{ kJ}$$
$$HNO_3(aq) + NH_4OH(aq) \longrightarrow NH_4NO_3(aq) + H_2O(l) \qquad \Delta H = -51.50 \text{ kJ}$$
$$HCOOH(aq) + NH_4OH(aq) \longrightarrow HCOONH_4(aq) + H_2O(l) \qquad \Delta H = -49.82 \text{ kJ}$$

In such cases the neutralization process involves not only the union of hydrogen and hydroxyl ions but also the dissociation of the weak acid or base. The measured heat of neutralization is, therefore, equal to the heat given out in the union of H^+ *(aq)* and OH^- *(aq)* ions plus the heat accompanying the dissociation of weak acid or weak base. The neutralization of NH_4OH with HCl, for example, can be represented as :

$$NH_4OH(aq) \longrightarrow NH_4^+(aq) + OH^-(aq) \qquad \Delta H = Q \text{ kcal}$$
$$\text{and } H^+(aq) + OH^-(aq) + Cl^-(aq) \longrightarrow H_2O(l) + Cl^-(aq) \qquad \Delta H = -57.36 \text{ kJ}$$
$$NH_4OH(aq) + H^+(aq) + Cl^-(aq) \longrightarrow NH_4^+(aq) + Cl^-(aq) + H_2O(l) \quad \Delta H = Q - 57.36 \text{ kJ}$$

But the measured heat of neutralization is – 51.50 kJ. Therefore,

$$Q - 57.36 = -51.50$$
$$Q = 57.36 - 51.50 = 5.86 \text{ kJ}$$

Hence the heat of dissociation of NH_4OH is 5.86 kJ *i.e.*, 5.86 kJ of heat is absorbed when one mole of ammonium hydroxide is dissociated into ions. In general, **the heat of dissociation of a weak acid or weak base may be defined as the change in enthalpy of the system when one mole of it is dissociated into ions.**

ENERGY CHANGES DURING TRANSITIONS OR PHASE CHANGES

The three states of matter – solid, liquid and gas differ from one another in the arrangement of their constituent particles. The magnitudes of intermolecular forces acting between the particles in these states are also different. It is a common observation that when a solid is converted into the liquid state, energy is to be supplied. This energy is spent in breaking the intermolecular forces in the solid which are of high magnitude. Whenever there is a change in the state of matter (solid → liquid or liquid → gas), the process is called phase change or transition. It is also accompanied by the change in enthalpy or heat content of the system.

HEAT OF FUSION

It is defined as : **the heat change (or enthalpy change) when one mole of a solid substance is converted into the liquid state at its melting point.**

As an example, we can take the melting of one mole of ice at its melting point, 0°C or 273 K. The process can be represented as

$$\underset{\text{ice}}{H_2O(s)} \longrightarrow \underset{\text{water}}{H_2O(l)} \qquad \Delta H = +5.98 \text{ kJ}$$

and is accompanied by the absorption of 5.98 kJ of heat. From the values of fusion of various substances we can compare their magnitudes of intermolecular forces. **Greater the heat of fusion of a substance higher the magnitude of intermolecular forces.**

HEAT OF VAPORIZATION

The heat of vaporization is defined as : **the heat change (or enthalpy change) when one mole of liquid is converted into vapor or gaseous state at its boiling point.**

For example, when one mole of water is converted into steam at 100°C or 373 K, the heat absorbed is 40.65 kJ which is the heat of vaporisation of water. The change can be represented as :

$$\underset{\text{water}}{H_2O(l)} \longrightarrow \underset{\text{steam}}{H_2O(g)} \qquad \Delta H = +40.65 \text{ kJ}$$

The heats of vaporisation of ethyl alcohol (C_2H_5OH) and benzene (C_6H_6) are 30.52 kJ mol^{-1} and 30.81 kJ mol^{-1} respectively. The values of heats of vaporisation can also be used for the comparison of the magnitude of intermolecular forces of attraction in liquids.

HEAT OF SUBLIMATION

Sublimation is a process when a solid changes directly into gaseous state without changing into liquid state. It occurs at a temperature below the melting point of the solid. Heat of sublimation is defined as : **the heat change (or enthalpy change) when one mole of a solid is directly converted into the gaseous state at a temperature below its melting point.**

For example, the heat of sublimation of iodine is 62.47 kJ mol^{-1}. It can be represented as

$$I_2(s) \longrightarrow I_2(g) \qquad \Delta H = +\,62.47 \text{ kJ mol}^{-1}$$

HEAT OF TRANSITION

The heat of transition is defined as : **the change in enthalpy which occurs when one mole of an element changes from one allotropic form to another.**

For example, the transition of diamond into amorphous carbon may be represented as

$$C_{\text{diamond}} \longrightarrow C_{\text{amorphous}} \qquad \Delta H = +\,13.82 \text{ kJ}$$

Similarly,

$$S_{\text{monoclinic}} \longrightarrow S_{\text{rhombic}} \qquad \Delta H = -\,6.70 \times 10^{-2} \text{ kJ}$$

and

$$P_{\text{white}} \longrightarrow P_{\text{red}} \qquad \Delta H = -\,4.30 \text{ kJ}$$

where – 0.067 kJ and – 4.30 kJ are heats of transition of monoclinic sulfur to rhombic sulfur and white phosphorus to red phosphorus respectively.

HESS'S LAW OF CONSTANT HEAT SUMMATION

We have already seen that heat changes in chemical reactions are equal to the difference in internal energy (ΔE) or heat content (ΔH) of the products and reactants, depending upon whether the reaction is studied at constant volume or constant pressure. Since ΔE and ΔH are functions of the state of the system, the heat evolved or absorbed in a given reaction must be independent of the manner in which the reaction is brought about. Thus it depends only on the initial state and final states of the system and not the manner or the steps in which the change takes place. This generalization is known as **Hess's Law** and may be stated as : **If a chemical change can be made to take place in two or more different ways whether in one step or two or more steps, the amount of total heat change is same no matter by which method the change is brought about.**

The law also follows as a mere consequence of the first law of thermodynamics. Let us suppose that a substance A can be changed to Z directly.

$$A \longrightarrow Z + Q_1 \qquad \Delta H_1 = -\,Q_1$$

where Q_1 is the heat evolved in the direct change. When the same change is brought about in stages :

$$A \longrightarrow B + q_1 \qquad \Delta H_2 = -\,q_1$$

$$B \longrightarrow C + q_2 \qquad \Delta H_2 = -\,q_2$$

$$C \longrightarrow Z + q_3 \qquad \Delta H_2 = -\,q_3$$

the total evolution of heat $= q_1 + q_2 + q_3 = Q_2$

According to Hess's law $Q_1 = Q_2$. If it be not so, let $Q_2 > Q_1$. Then by transforming A to Z through stages and retransforming directly back to A, there would be gain of heat energy $= Q_2 - Q_1$. By repeating the process again and again an unlimited heat energy will be developed in an isolated system. This goes against the first law of thermodynamics. Hence Q_1 must be equal to Q_2.

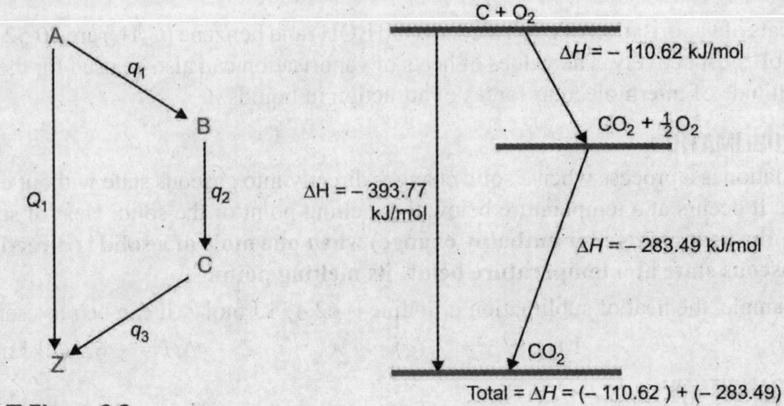

Figure 8.2
Illustration of Hess's Law.

Hess's law has been tested experimentally and shown to be true.

Illustrations of Hess's Law

(1) **Burning of carbon to CO_2.** Carbon can be burnt to carbon dioxide directly or it may first be changed to carbon monoxide which may then be oxidized to carbon dioxide.

1st way :

$$C(s) + O_2(g) \longrightarrow CO_2(g) \qquad \Delta H = -393.77 \text{ kJ}$$

2nd way :

$$C(s) + \tfrac{1}{2}O(g) \longrightarrow CO(g) \qquad \Delta H = -110.62 \text{ kJ}$$
$$CO(g) + \tfrac{1}{2}O_2(g) \longrightarrow CO_2(g) \qquad \Delta H = -283.49 \text{ kJ}$$

$$\textit{overall change} \quad C(s) + O_2(g) \longrightarrow CO_2(g) \qquad \Delta H = -394.10 \text{ kJ}$$

It is evident from above that the total heat energy evolved is the same in the two cases (Fig. 8.2)

(2) **Formation of Sodium hydroxide from Na.** The formation of sodium hydroxide from metallic sodium presents another example of Hess's law. The process can be carried out in two ways.

1st way :

$$2Na(s) + \tfrac{1}{2}O_2(g) \longrightarrow Na_2O(s) \qquad \Delta H = -418.68 \text{ kJ}$$
$$Na_2O(s) + H_2O(l) \longrightarrow 2NaOH(aq) \qquad \Delta H = -234.46 \text{ kJ}$$

$$2Na(s) + H_2O(l) + \tfrac{1}{2}O_2(g) \longrightarrow 2NaOH(aq) \qquad \Delta H = -653.14 \text{ kJ}$$

2nd way :

$$2Na(s) + 2H_2O(l) \longrightarrow 2NaOH(aq) + H_2(g) \qquad \Delta H = -368.44 \text{ kJ}$$
$$H_2(g) + \tfrac{1}{2}O_2(g) \longrightarrow H_2O(g) \qquad \Delta H = -286.80 \text{ kJ}$$

$$2Na(s) + H_2O(l) + \tfrac{1}{2}O_2(g) \longrightarrow 2NaOH(aq) \qquad \Delta H = -655.23 \text{ kJ}$$

It may be observed that the total heat evolved in carrying the reaction in two different ways is the same. The difference of 2.09 kJ is within the experimental error.

It is obvious from the above examples that by the addition of a series of chemical equations we can obtain not only the resultant products of this series of reactions but also the net heat effect. It is, therefore, clear that : **thermochemical equations may be multiplied, added or subtracted like ordinary algebraic equations.** For this reason, Hess's law has been of great service in the indirect determination of heats of formation.

APPLICATIONS OF HESS'S LAW

(1) **Determination of heat of formation of substances which otherwise cannot be measured experimentally.** The substances like methane, carbon monoxide, benzene, etc., cannot be prepared by uniting their elements. Therefore it is not possible to measure the heats of formation of such compounds directly. These can be determined indirectly by using Hess's law. This will be illustrated by the following examples.

SOLVED PROBLEM 9

Calculate the heat of formation of potassium hydroxide from the following data.

(*i*) $K(s) + H_2O\,(aq) \longrightarrow KOH + \frac{1}{2}H_2$ $\quad \Delta H = -200.97$ kJ

(*ii*) $H_2(g) + \frac{1}{2}O_2(g) \longrightarrow H_2O(l)$ $\quad \Delta H = -286.80$ kJ

(*iii*) $KOH(s) \longrightarrow KOH(aq)$ $\quad \Delta H = -58.62$ kJ

SOLUTION

We should aim at finding the value of ΔH for the equation, which is the heat of formation.

$$K(s) + \tfrac{1}{2}O_2(g) + \tfrac{1}{2}H_2(g) \longrightarrow KOH(s) \qquad \Delta H = ?$$

Adding equations (*i*) and (*ii*) and subtracting equation (*iii*) will give

$$K(s) + H_2O(l) + H_2(g) + \tfrac{1}{2}O_2(g) - KOH(s) \longrightarrow KOH(aq) + \tfrac{1}{2}H_2(g) + H_2O(l) - KOH(aq)$$

$$\Delta H = -200.97 + (-286.80) - (-58.62)$$

$$= -429.15 \text{ kJ}$$

or $\quad K(s) + \frac{1}{2}H_2(g) + \frac{1}{2}O_2(g) \longrightarrow KOH(s) \qquad \Delta H = -429.15$ kJ

Thus heat of formation of KOH is – **429.15 kJ**

SOLVED PROBLEM 10

The heat of combustion of ethyl alcohol is –1381.64 kJ. If the heat of formation of $CO_2(g)$ and $H_2O(l)$ be –394.82 kJ and –286.80 kJ respectively, calculate the heat of formation of ethyl alcohol.

SOLUTION

We are given

(*a*) $C_2H_5OH(l) + 3O_2(g) \longrightarrow 2CO_2(g) + 3H_2O(l)$ $\quad \Delta H = -1381.64$ kJ

(*b*) $C(s) + O_2(g) \longrightarrow CO_2(g)$ $\quad \Delta H = -394.82$ kJ

(*c*) $H_2(g) + \frac{1}{2}O_2(g) \longrightarrow H_2O(l)$ $\quad \Delta H = -286.80$ kJ

We have to manipulate these equations so as to get the required equation :

$$2C(s) + 3H_2(g) + \tfrac{1}{2}O_2(g) \longrightarrow C_2H_5OH(l) \qquad \Delta H = ?$$

where ΔH is the heat of formation of ethyl alcohol.

Multiplying equation (*b*) by 2 and equation (*c*) by 3 and adding up these, we get

$$2C(s) + 3H_2(g) + \tfrac{3}{2}O_2(g) \longrightarrow 2CO_2(g) + 3H_2O(l)$$

$$\Delta H = -1650.01 \text{ kJ}$$

Subtracting equation (*c*) from the above equation, we have

$$2C(s) + 3H_2(g) + \tfrac{1}{2}O_2(g) \longrightarrow C_2H_5OH(l)$$
$$\Delta H = -1650.01 - (-1381.64)$$
$$= -268.37 \text{ kJ}$$

Thus the heat of formation of ethyl alcohol is – **268.37 kJ**

SOLVED PROBLEM 11

Determine ΔH of the reaction

$$C(s) + 2H_2(g) \longrightarrow CH_4(g)$$

from the following data :

(*i*) $C(s) + O_2(g) \longrightarrow CO_2(g)$ $\quad \Delta H = -393.7$ kJ

(*ii*) $H_2(g) + \tfrac{1}{2}O_2(g) \longrightarrow H_2O(l)$ $\quad \Delta H = -285.7$ kJ

(*iii*) $CH_4(g) + 2O_2(g) \longrightarrow CO_2(g) + 2H_2O(l)$ $\quad \Delta H = -890.3$ kJ

SOLUTION

Multiplying equation (*ii*) by 2 and adding to equation (*i*), we get

(*iv*) $C(s) + O_2(g) + 2H_2(g) + O_2(g) \longrightarrow CO_2(g) + 2H_2O(l)$

$$\Delta H = (2 \times -285.7) + (-393.7)$$
$$= -965.1 \text{ kJ}$$

On subtracting equation (*iii*) from equation (*iv*), we get

$$C(s) + 2H_2(g) = CH_4(g) \qquad \Delta H = -965.1 - (-890.3)$$
$$\Delta H = -74.8 \text{ kJ}$$

Thus the heat of formation of methane is – **74.8 kJ.**

(2) **Determination of Heat of Transition**

The heat of transition of one allotropic form to another can also be calculated with the help of Hess's law. For example, the enthalpy of transition from monoclinic sulfur to rhombic sulfur can be calculated from their heats of combustion which are :

(*i*) $S_{rhombic} + O_2(g) \longrightarrow SO_2(g)$ $\quad \Delta H = -291.4$ kJ

(*ii*) $S_{monoclinic} + O_2(g) \longrightarrow SO_2(g)$ $\quad \Delta H = -295.4$ kJ

Subtracting equation (*ii*) from (*i*) we get

$$S_{rhombic} - S_{monoclinic} + O_2(g) - O_2(g) \longrightarrow SO_2(g) - SO_2(g)$$
$$\Delta H = -291.4 - (-295.4)$$

or

$$S_{rhombic} = S_{monoclinic} \qquad \Delta H = 4.0 \text{ kJ}$$

Thus heat of transition of rhombic sulfur to monoclinic sulfur is **4.0 kJ.**

(3) **Determination of heats of various reactions**

By using Hess's law we can calculate the heats or enthalpies of many reactions which otherwise cannot be measured directly. For example, from the following equations the enthalpy of dimerization of NO_2 can be calculated.

(*i*) $N_2(g) + 2O_2(g) \longrightarrow 2NO_2(g)$ $\quad \Delta H = 67.9$ kJ

(*ii*) $N_2(g) + 2O_2(g) \longrightarrow N_2O_4(g)$ $\quad \Delta H = 9.3$ kJ

Subtracting equation (*i*) from equation (*ii*) we have

$$2NO_2(g) \longrightarrow N_2O_4(g) \qquad \Delta H = (9.3 - 67.9) \text{ kJ}$$
$$= -58.6 \text{ kJ}$$

Thus the heat of dimerization of NO_2 is – **58.6 kJ.**

BOND ENERGY

When a bond between two atoms is formed, there is a release of energy. The same amount of energy is absorbed when the bond is broken.

The bond energy is defined as the average amount of energy required to break all bonds of a particular type in one mole of the substance.

Thus the bond energy of H – H bond is the energy required to break all the bonds in one mole of the gas. It is expressed in kcal mol^{-1} or kJ mol^{-1}. For example, the bond energy of H – H bond is 433 kJ mol^{-1} or 103.58 kcal mol^{-1}. The bond energies of some common bonds are listed below :

Bond	Bond Energy
Cl–Cl	243 kJ mol^{-1}
O=O	499.0 kJ mol^{-1}
C–H	414.0 kJ mol^{-1}
O–H	460.0 kJ mol^{-1}

Bond Energy is a Measure of Strength of the Bond

In other words, bond energy is the force with which the atoms are bonded together. It depends upon :

(*i*) Size of the atom

(*ii*) Electronegativity

(*iii*) Bond length

A knowledge of bond enthalpy is useful for calculating heats of reaction for gaseous reactions for which no thermal data is available and which involve substances having covalent bonds.

Suppose we desire to determine the bond energy of C–H bond in methane. For this purpose we need to know the enthalpy change for the reaction

$$C(g) + 4H(g) \longrightarrow CH_4(g)$$

This is obtained by combining the heat of formation of methane from $C(s) + H_2(g)$ with the heat of sublimation of carbon *i.e.*, $C(s) \rightarrow C(g)$ and the heat of dissociation of hydrogen into atoms *i.e.*, $H_2(g) \rightarrow 2H(g)$, which have been determined by spectroscopic methods. The value so obtained is 398 kcal mol^{-1} (or 1663.64 kJ mol^{-1}). This represents the bond energy of four C–H bonds. Since all the bonds in methane are identical, the bond energy of C–H bond is 1663.64/4 = 415.91 kJ mol^{-1}.

In a similar manner the bond energies of other types of bonds have been calculated. **When a bond is broken, the bond energy is positive because heat is absorbed.** It is written with a minus sign when a bond is formed and heat is evolved. The calculation of heat of reaction with the help of bond energies is illustrated in the following examples.

SOLVED PROBLEM 12

Given that energies for H–H, O=O and O–H bonds are 435.43, 494.04 and 464.73 kJ mol^{-1} respectively, calculate the heat of the reaction

$$H_2(g) + \tfrac{1}{2}O_2(g) \longrightarrow H_2O(g)$$

SOLUTION

In this reaction, two O–H bonds are formed and one H–H bond is broken. Therefore we can write for ΔH

$$\Delta H = -2\,\Delta H\,(\text{O}-\text{H}) + \Delta H\,(H-H) + \tfrac{1}{2}\Delta H\,(\text{O}=\text{O})$$

$$= (-2 \times 464.73) + 435.43 + \left(\tfrac{1}{2} \times 494.04\right)$$

$$= -929.46 + 435.43 + 247.02$$

$$= -247.02 \text{ kJ mol}^{-1}$$

∴ The heat of the given reaction is – **247.02 kJ mol^{-1}**

SOLVED PROBLEM 13

Calculate the bond energy of HCl, given that H–H bond energy is 433 kJ mol^{-1}, Cl – Cl bond energy is 242 kJ mol^{-1} and ΔH_f for HCl is – 91 kJ mol^{-1}.

SOLUTION

Given :

(*i*) $\frac{1}{2}H_2(g) + Cl_2(g) \longrightarrow HCl(g)$ $\quad \Delta H = -91$ kJ

(*ii*) $H_2(g) \longrightarrow 2H(g)$ $\quad \Delta H = +433$ kJ

(*iii*) $Cl_2(g) \longrightarrow 2Cl(g)$ $\quad \Delta H = +242$ kJ

Multiplying equation (*i*) by 2, we get

(*iv*) $H_2(g) + 2Cl_2(g) \longrightarrow 2HCl(g)$ $\quad \Delta H = -182$ kJ

Adding equation (*ii*) and (*iii*), we have

(*v*) $H_2(g) + Cl_2(g) \longrightarrow 2H(g) + 2Cl(g)$ $\quad \Delta H = +675$ kJ

Subtracting equation (*iv*) from (*v*), one has

$2HCl(g) \longrightarrow 2H(g) + 2Cl(g)$ $\quad \Delta H = +857$ kJ

Dividing this equation by 2 we get

$HCl(g) \longrightarrow H(g) + Cl(g)$ $\quad \Delta H = 428.5$ kJ

Thus the bond energy of H–Cl is 428.5 kJ.

MEASUREMENT OF THE HEAT OF REACTION

The experimental measurement of the heat of reaction or enthalpy change is known as **calorimetry.** The name evidently finds its origin in the unit of heat–the calorie. The heat given out or absorbed in a chemical reaction is measured in a suitable apparatus called a **calorimeter.** These calorimeters vary considerably in their construction and designs. They are adapted to suit the requirements of a particular reaction under study. For instance, to measure the heats of reactions involving (*i*) solutions only, (*ii*) gases, (*iii*) very reactive chemicals etc., different types of calorimeters are employed. We discuss below two of the common type of calorimeters.

Water Calorimeter

This is a convenient apparatus for finding the heat changes accompanying chemical reactions taking place in solutions.

The apparatus consists essentially of a water-bath with thermally insulated walls. A reaction chamber consisting of two limbs is suspended in the water-bath. Through the lid of the water-bath pass (*a*) *thermometer* that records the temperature variations and (*b*) a *stirrer* that stirs water in the water-bath. A known quantity of water (say *W* gms) is taken in the water-bath and its temperature is noted. The reacting substances are filled in the two limbs as shown in Fig. 8.3. The reacting chamber is now turned upside down (position II) to allow the solutions to mix. They react and the heat produced during the reaction is taken up by water, raising its temperature. If the rise in temperature (Final reading – Initial reading) is t°C, the heat absorbed by water 'Q' is given by

$$Q = W \times 1 \times t \text{ calories}$$

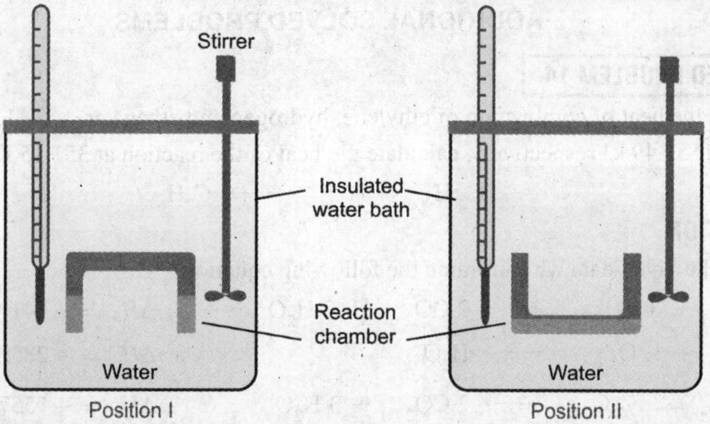

■ **Figure 8.3**
Water Calorimeter.

But heat produced in the reaction is equal to that absorbed by water, hence heat of the reaction can be calculated.

Bomb Calorimeter

This apparatus was devised by Berthelot (1881) to measure the heat of combustion of organic compounds. A modified form of the apparatus shown in Fig. 8.4 consists of a sealed combustion chamber, called a *bomb*, containing a weighed quantity of the substance in a dish along with oxygen under about 2026500.00 pascal pressure. The *bomb* is lowered in water contained in an insulated copper vessel. This vessel is provided with a stirrer and a thermometer reading up to 1/100th of a degree. It is also surrounded by an outer jacket to ensure complete insulation from the atmosphere. The temperature of water is noted before the substance is ignited by an electric current. After combustion, the rise in temperature of the system is noted on the thermometer and heat of combustion can be calculated from the heat gained by water and the calorimeter.

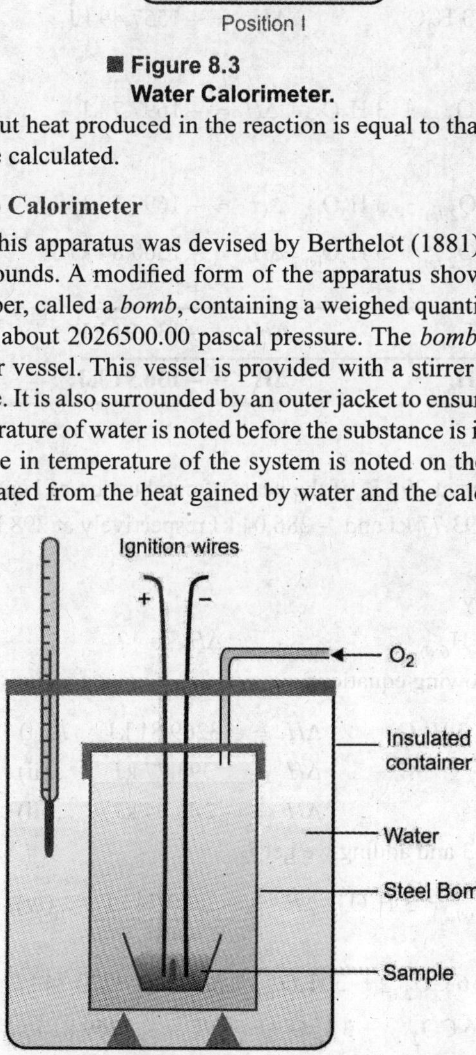

■ **Figure 8.4**
Bomb Calorimeter.

ADDITIONAL SOLVED PROBLEMS

SOLVED PROBLEM 14

Given the heat of combustion of ethylene, hydrogen and ethane are –1411.79 kJ, –285.96 kJ and –1557.49 kJ respectively, calculate the heat of the reaction at 351.15 K for

$$C_2H_{4(g)} + H_{2(g)} \longrightarrow C_2H_{6(g)}$$

SOLUTION

From the given data we can frame the following equations

$$C_2H_{4(g)} + 3O_{2(g)} \longrightarrow 2CO_{2(g)} + 2H_2O \qquad \Delta H = -1411.79 \text{ kJ} \qquad ...(i)$$

$$H_{2(g)} + \tfrac{1}{2}O_{2(g)} \longrightarrow H_2O_{(l)} \qquad \Delta H = -285.96 \text{ kJ} \qquad ...(ii)$$

$$C_2H_{6(g)} + \tfrac{7}{2}O_{2(g)} \longrightarrow 2CO_{2(g)} + 3H_2O_{(g)} \qquad \Delta H = -1557.49 \text{ kJ} \qquad ...(iii)$$

Adding (i) and (ii) we get

$$C_2H_{4(g)} + \tfrac{7}{2}O_{2(g)} + H_{2(g)} \longrightarrow 2CO_{2(g)} + 3H_2O_{(g)} \quad \Delta H = -1697.7 \text{ kJ} \qquad ...(iv)$$

Subtracting (iii) from (iv) we have

$$C_2H_{4(g)} + \tfrac{7}{2}O_{2(g)} + H_{2(g)} \longrightarrow 2CO_{2(g)} + 3H_2O_{(g)} \quad \Delta H = -1697.7 \text{ kJ}$$

$$C_2H_{6(g)} + \tfrac{7}{2}O_{2(g)} \longrightarrow 2CO_{2(g)} + 3H_2O_{(g)} \quad \Delta H = -1560.84 \text{ kJ}$$

$$- \qquad - \qquad\qquad - \qquad - \qquad\qquad +$$

$$C_2H_{4(g)} + H_{2(g)} \longrightarrow C_2H_{6(g)} \quad \Delta H = -136.91 \text{ kJ}$$

or $\mathbf{C_2H_{4(g)} + H_{2(g)} \longrightarrow C_2H_{6(g)} \quad \Delta H = -136.91 \text{ kJ}}$

SOLVED PROBLEM 15

Calculate the heat of formation of benzene at 298 K, if the heats of combustion of benzene, carbon and hydrogen are – 3269.81 kJ, 393.77 kJ and – 286.04 kJ respectively at 298 K.

SOLUTION

The formation of benzene is represented by

$$6C_{(s)} + 3H_{2(g)} \longrightarrow C_6H_{6(l)} \qquad \Delta H = ?$$

From the given data we can frame the following equations

$$C_6H_{6(l)} + \tfrac{15}{2}O_{2(g)} \longrightarrow 6CO_{2(g)} + 3H_2O_{(g)} \qquad \Delta H = -3269.81 \text{ kJ} \qquad ...(i)$$

$$C_{(s)} + O_{2(g)} \longrightarrow CO_{2(g)} \qquad \Delta H = 393.77 \text{ kJ} \qquad ...(ii)$$

$$H_{2(g)} + \tfrac{1}{2}O_{2(g)} \longrightarrow H_2O_{(g)} \qquad \Delta H = -286.04 \text{ kJ} \qquad ...(iii)$$

Multiplying equation (ii) by 6 and (iii) by 3 and adding we get

$$6C_{(s)} + \tfrac{15}{2}O_{2(g)} + 3H_{2(g)} \longrightarrow 6CO_{2(g)} + 3H_2O_{(g)} \quad \Delta H = -3220.74 \text{ kJ} \qquad ...(iv)$$

Subtracting (i) from (iv), we have

$$6C_{(s)} + \tfrac{15}{2}O_{2(g)} + 3H_{2(g)} \longrightarrow 6CO_{2(g)} + 3H_2O_{(g)} \quad \Delta H = -3220.74 \text{ kJ}$$

$$C_6H_{6(l)} + \tfrac{15}{2}O_{2(g)} \longrightarrow 6CO_{2(g)} + 3H_2O_{(g)} \quad \Delta H = -3269.81 \text{ kJ}$$

$$- \qquad - \qquad\qquad - \qquad - \qquad\qquad +$$

$$6C_{(s)} + 3H_{2(g)} \longrightarrow C_6H_{6(l)} \quad \Delta H = +49.07 \text{ kJ}$$

Thus the heat of formation of benzene = + 49.07 kJ

SOLVED PROBLEM 16

The heat of formation of methane at 300 K is – 80.81 kJ when the measurements are made at constant pressure. What will be the heat of formation at constant volume?

SOLUTION

Given that

$$C_{(s)} + 2H_{2\,(g)} \longrightarrow CH_{4\,(g)} \qquad \Delta H = -80.81 \text{ kJ}$$

Formula used

$$\Delta H = \Delta E + \Delta n RT \qquad \text{or} \qquad \Delta E = \Delta H - \Delta n RT$$

Quantities given

$$\Delta H = n_P - n_R = 1 - 2 = -1 \qquad R = 8.31 \times 10^{-3} \text{ kJ}$$

Substitution of values

$$\Delta E = -80.81 \text{ kJ} + [(-1) \times 8.31 \times 10^{-3} \text{ kJ K}^{-1} \text{ mol}^{-1} \times 300 \text{ K}]$$

$$= -80.81 \text{ kJ} - 2.50 \text{ kJ}$$

$$= \mathbf{-83.30 \text{ kJ}}$$

SOLVED PROBLEM 17

Bond energies of F_2 and Cl_2 are 153.24 and 2428.34 kJ mol^{-1} respectively. Heat liberated in the reaction $F_2 + Cl_2 \longrightarrow 2\,FCl$ is –111.37 kJ. Find the bond energy of F — Cl bond.

SOLUTION

The given reaction is

$$F—F + Cl—Cl \longrightarrow 2\,F—Cl \qquad \Delta H = -111.37 \text{ kJ}$$

The heat of the reaction is given by

ΔH = [2 × Bond energy of F — Cl] – [Bond energy of F — F + Bond energy of Cl — Cl]

Quantities given

$\Delta H = -111.37$ kJ *B.E.* of F — F = 153.24 kJ mol^{-1} *B.E.* of Cl — Cl = 2428.34 kJ mol^{-1}

Substitution of values

– 111.37 kJ = [2 × Bond energy of F — Cl] – [153.24 kJ + 2428.34 kJ]

– 111.37 kJ = 2 × Bond energy of F — Cl – 2581.58 kJ

∴ 2 × Bond energy of F — Cl = 2581.58 kJ – 153.24 kJ

Bond energy of F — Cl = 2428.34/2 kJ

= 1214.17 kJ

Thus bond energy of F — Cl bond = 1214.17 kJ

SOLVED PROBLEM 18

0.16 g of methane was subjected to combustion at 27°K in a bomb calorimeter. The temperature of calorimeter system including water was found to rise by 0.5°K. Calculate the heat of combustion of methane at (i) constant volume and (ii) at constant pressure. The thermal capacity of the calorimeter system is 17.7 kJ K^{-1}. ($R = 8.314$ J K^{-1} mol^{-1})

SOLUTION

(i) To calculate the heat of combustion at constant volume

Quantity of heat liberated = Thermal capacity of calorimeter × Rise in temperature for 0.16 g of methane

$$= 17.7 \text{ kJ K}^{-1} \times 0.5 \text{ K}$$

$$= -8.85 \text{ kJ}$$

$$\Delta E = \text{Quantity of heat liberated for one mole of methane}$$

$$= \frac{8.85 \text{ kJ} \times 16 \text{ g mol}^{-1}}{0.16 \text{ g}}$$

$$= -885 \text{ kJ}$$

(ii) To calculate the heat of combustion at constant pressure

The equation representing the combustion of methane is

$$CH_{4\,(g)} + 2\,O_{2\,(g)} \longrightarrow CO_{2\,(g)} + 2\,H_2O_{\,(l)}$$

Formula used

$$\Delta H = \Delta E + \Delta n\, R\, T$$

Quantities given

$\Delta E = -885 \text{ kJ mol}^{-1}$ $\quad R = 8.314 \text{ J K}^{-1} \text{ mol}^{-1} = 8.314 \times 10^{-3} \text{ kJ K}^{-1} \text{ mol}^{-1}$

$T = 27 + 273 = 300 \text{ K}$ $\quad \Delta n = n_P - n_R = 1 - 3 = -2$

Substitution of values

$$\Delta H = -885 \text{ kJ mol}^{-1} + (-2) \times 8.314 \times 10^{-3} \text{ kJ K}^{-1} \text{ mol} \times 300 \text{ K}$$

$$= -885 - 4.9884 \text{ kJ mol}^{-1}$$

$$= \mathbf{-889.9884 \text{ kJ mol}^{-1}}$$

EXAMINATION QUESTIONS

1. Define or explain the following terms :
   (*a*) Thermochemistry (*b*) Internal energy (*c*) Enthalpy of a reaction
   (*d*) Thermochemical equations (*e*) Heat of reaction (*f*) Standard heat of reaction
   (*g*) Heat of formation (*h*) Standard heat of formation (*i*) Heat of combustion
   (*j*) Heat of solution (*k*) Heat of neutralization (*l*) Heat of vaporisation
   (*m*) Heat of sublimation (*n*) Heat of transition (*o*) Hess's law
   (*p*) Bond energy
2. (*a*) Explain the terms : heat of reaction at constant pressure and heat of reaction at constant volume. How are they related?
   (*b*) If the heat of formation of methane at constant pressure is –74.94 kJ/mol at 298 K, what is its value at constant volume?
   Answer. (*b*) –72.465 kJ
3. ΔH values for the combustion reaction of ethyl alcohol, carbon and hydrogen are –1381.64, –394.81 and –286.80 kJ respectively. Calculate the heat of formation of ethyl alcohol.
   Answer. – 268.37 kJ
4. State Hess' Law of constant heat summation and explain some of its important applications.
5. Graphite and diamond are two forms of carbon. The enthalpy of combustion of graphite at 298 K is –393.51 kJ mol^{-1} and that of diamond is –395.41 kJ mol^{-1}. What is the enthalpy change in the process Graphite $\rightarrow$ Diamond at the same temperature?
   Answer. 1.90 kJ mol^{-1}
6. (*a*) State and explain heat of formation, heat of combustion with examples.
   (*b*) The standard heats of formation of CO(g) and H_2O(g) at 298 K are 100.5 kJ and –241.8 kJ/mol respectively. Calculate the heat change of the following reaction at this temperature.
   $H_2O(g) + C(s) \rightarrow H_2(g) + CO(g)$
   Answer. (*b*) 342.3 kJ

7. Explain the term enthalpy of ionization. The enthalpy of neutralization of HCl with NaOH is -57.3 kJ mol^{-1} and that of neutralization of CH_3COOH with NaOH is -55.2 kJ mol^{-1}. Calculate the enthalpy of ionization of CH_3COOH.

Answer. +2.1 kJ

8. Calculate the standard enthalpy of formation of naphthalene ($C_{10}H_8$) when the standard enthalpy of combustion of naphthalene is -5153 kJ and the standard enthalpies of formation of CO_2 and H_2O are -393.5 kJ mol^{-1} and -285.8 kJ mol^{-1} respectively.

Answer. +74.8 kJ

9. (*a*) Discuss the variation of heat of reaction with temperature.

(*b*) Calculate the heat of formation of carbon disulfide. The heat of combustion of CS_2, Sulfur and Carbon are -110.95 kJ, -394.82 kJ and -297.60 kJ respectively at 298 K.

Answer. (*b*) -976.28 kJ

10. 2.0 g of C_6H_6 was burnt in excess of O_2 in a bomb calorimeter :

$$C_6H_6(\ell) + 7½O_2(g) \rightarrow 6CO_2(g) + 3H_2O(\ell)$$

If the temperature rise is 313.5 K and the heat capacity of the system is 2.0 kJ K^{-1}, calculate the enthalpy of combustion of C_6H_6 at constant volume and constant pressure.

Answer. – 2.4414 kJ mol^{-1}; –24410.283 kJ mol^{-1} (*Guru Nanak Dev BSc*)

11. The enthalpies of combustion of benzene, Carbon and hydrogen are -3267.7 kJ, -393.5 kJ and -286.2 kJ respectively. Calculate the enthalpy of formation of benzene.

Answer. – 48.1 kJ (*Nagpur BSc*)

12. Calculate heat of fusion of benzene at its freezing point, 278 K, if the rate of change of melting point with pressure is 0.0129 K atm^{-1} (1.273×10^{-7} K pascal) and ΔV is 0.059 cm^{-3} g^{-1} (59 kg/m^3). (Mol. wt. of benzene is 78)

Answer. 99.174 kJ mol^{-1} (*Aligarh BSc*)

13. Define enthalpy of Neutralization. When 100 ml of 1N HCl is neutralized by equivalent amount of NaOH, 5.273 kJ of heat is evolved. Calculate the heat of neutralization of HCl.

Answer. 52.73 kJ (*Nagpur BSc*)

14. Determine ΔH of the reaction :

$$C(s) + 2H_2(g) \rightarrow CH_4(g)$$

from the following data :

(*i*) $C(s) + O_2(g) \rightarrow CO_2(g)$ $\Delta H = -393.7$ kJ

(*ii*) $H_2(g) + ½O_2(g) \rightarrow H_2O(\ell)$ $\Delta H = -285.7$ kJ

(*iii*) $CH_4(g) + 2O_2(g) \rightarrow CO_2(g) + 2H_2O$ $\Delta H = -890$ kJ

Answer. -75.1 kJ (*Guru Nanak Dev BSc*)

15. (*a*) State Hess's law of constant heat summation and describe its applications.

(*b*) Calculate ΔH for the reaction :

$$C_{(graphite)} + 2H_2(g) \rightarrow CH_4(g)$$

from the following data.

(*i*) $CH_4(g) + 2O_2(g) \rightarrow CO_2(g) + 2H_2O(l)$ $\Delta H = -890.35$ kJ

(*ii*) $H_2(g) + ½O_2(g) \rightarrow H_2O(\ell)$ $\Delta H = -286.2$ kJ

(*iii*) $C_{(graphite)} + O_2(g) \rightarrow CO(g)$ $\Delta H = -393.51$ kJ

Answer. -75.56 kJ (*Allahabad BSc*)

16. (*a*) Show that

$$\Delta H = \Delta E + \Delta n RT$$

(*b*) If the heat of formation of methane at constant pressure is -74.94 kJ/mol at 298 K, what is its value at constant volume?

Answer. (*b*) -72.465 kJ mol^{-1} (*Guru Nanak Dev BSc*)

17. Define bond energy. Calculate ΔH for the reaction :

$$C_2H_4(g) + 3O_2(g) \rightarrow 2CO_2(g) + 2H_2O(g)$$

from the following values of bond energies :

Bond	C–H	O=O	C=O	O–H	C=C
Bond Energy (kJ)	414	499	724	460	619

Answer. – 960 kJ (*Nagpur BSc*)

18. Calculate the standard enthalpy of formation of *n*-butane, given that the standard enthalpies of combustion of *n*-butane, $c_{(graphite)}$ and $H_2(g)$ are –2878.5 kJ mol^{-1}, –393.5 kJ mol^{-1} and –285.8 kJ mol^{-1} respectively.

Answer. –124.5 kJ mol^{-1} (*Delhi BSc*)

19. The enthalpy of formation of methane at constant volume at 298 K is –72.3 kJ. What will be the heat of formation at constant pressure? (R = 8.314 J K^{-1} mol^{-1})

Answer. –74.777 kJ mol^{-1} (*Arunachal BSc*)

20. $\Delta G°$ for the formation of CO, H_2O, CO_2 and H_2 gases are –137.37, –228.77, –394.65 and zero kJ mol^{-1} respectively. Calculate $\Delta G°$ and K_p for the forward reaction in the following equilibrium :

$$CO + H_2O \rightleftharpoons H_2 + CO_2$$

Answer. 28.51 kJ mol^{-1} (*Delhi BSc*)

21. The heat of combustion of carbon, hydrogen and sucrose are –393.5, –286.2 and –5644.2 kJ mol^{-1} respectively. Calculate the heat of formation of sucrose.

Answer. 1566.0 kJ mol^{-1} (*Jamia Millia BSc*)

22. Calculate enthalpy of formation of ethane from the following data :

(*i*) $C(s) + O_2(g) \rightarrow CO_2(g)$ $\Delta H = -393.4$ kJ

(*ii*) $H_2(g) + ½O_2(g) \rightarrow H_2O(\ell)$ $\Delta H = -284.61$ kJ

(*iii*) $C_2H_6(g) + 7/2\ O_2(g) \rightarrow 2CO_2(g) + 3H_2O(\ell)$ $\Delta H = -1561.0$ kJ

Answer. –79.63 kJ (*Sambalpur BSc*)

23. Define bond energy. Explain how it can be used to calculate the heat of a reaction. Calculate the heat of reaction between propene and chlorine to give 1, 2-dichloropropane from the values of bond energies :

$$E_{C-C} = 347.3 \text{ kJ mol}^{-1}$$
$$E_{C=C} = 615 \text{ kJ mol}^{-1}$$
$$E_{C-H} = 413 \text{ kJ mol}^{-1}$$
$$E_{C-Cl} = 330.5 \text{ kJ mol}^{-1}$$
$$E_{Cl-Cl} = 242.7 \text{ kJ mol}^{-1}$$

Answer. –150.6 kJ mol^{-1} (*Nagpur BSc*)

24. (*a*) Derive Kirchoff's equation when C_P values are

(*i*) independent on temperature; and

(*ii*) dependent on temperature

(*b*) Calculate the enthalpy of vaporization of SO_2 at –248 K if the same at its boiling point (i.e. –260K) be 24911.46 J mol^{-1}. Given for SO_2, molar heat capacities in liquid and vapor phase are 862.48 J K^{-1} mol^{-1} and 38.94 J K^{-1} mol^{-1} respectively.

Answer. (*b*) 25621.12 J mol^{-1} (*Delhi BSc*)

25. The standard enthalpy of formation (ΔH_f°) of $NH_3(g)$ as indicated by the chemical equation is – 46.19 kJ.

$$1/2\ N_2(g) + 3/2\ H_2(g) \rightarrow NH_3(g)$$

Standard entropies of $N_2(g)$, $H_2(g)$ and $NH_3(g)$ are 191.62, 130.60 and 192.46 JK^{-1} mol^{-1} respectively. Calculate the standard free energy of formation of $NH_3(g)$.

Answer. –16.613 kJ mol^{-1} (*Kalyani BSc*)

26. The heat of combustion of benzoic acid at constant pressure is –3229697.47 J at 298 K. Calculate the heat of combustion at constant volume.

Answer. – 3228458.68 J (*Madurai BSc*)

27. Heats of combustion of hydrated copper sulfate and anhydrous copper sulfate are –11.72 and 66.53 kJ respectively. Calculate the heat of hydration of copper sulfate.

Answer. –78.25 kJ (*Delhi BSc*)

28. (*a*) Derive thermodynamically Kirchoff's equation.

(*b*) Calculate the heat of formation of acetic acid its heat of combustion is –869.0 kJ mol^{-1}. The heats of formation of $CO_2(g)$ and $H_2O(\ell)$ are –390.0 kJ mol^{-1} and –285.0 kJ mol^{-1} respectively.

Answer. (*b*) – 481 kJ mol^{-1} (*Agra BSc*)

29. The heat of combustion of $C_2H_4(g)$, $C_2H_6(g)$ and $H_2(g)$ are –1409 kJ, –1558.3 kJ and – 285.645 respectively. Calculate the heat of hydrogenation of ethylene.

Answer. 136.8 kJ (*Assam BSc*)

30. When 2 moles of ethane are completely burnt 3129 kJ of heat is liberated. Calculate the heat of formation, ΔH_f for C_2H_6. ΔH_f for $CO_2(g)$ and $H_2O(l)$ are – 395 and – 286 kJ respectively.

Answer. – 83.5 kJ (*Purvanchal BSc*)

31. The bond dissociation energies of gaseous H_2, Cl_2 and HCl are 435.43, 242.83 and 431.24 kJ mol^{-1} respectively. Calculate the enthalpy of formation of HCl.

Answer. – 92.11 kJ (*Madras BSc*)

32. In an experiment 0.00548 kg of formic acid, HCOOH, was burnt and the amount of heat evolved was 30.3 kJ. Calculate ΔH per mole of formic acid.

Answer. 42.79 kJ (*Nehu BSc*)

33. Calculate the heat of formation of ammonia from the heats of combustion of ammonia and hydrogen which are 37.93 and 288.47 kJ respectively.

Answer. – 394.77 kJ (*Madurai BSc*)

34. Calculate the standard heat of formation of carbon disulfide (*l*). Given that the standard heat of combustion of carbon(*s*), sulfur(*s*) and carbon disulfide(*l*) are –393.3, – 293.72 and – 1108.76 kJ mol^{-1} respectively.

Answer. + 128.02 kJ (*Kerala BSc*)

35. The standard heat of formation of $CH_4(g)$, $CO_2(g)$ and $H_2O(g)$ are – 76.2, – 394.8 and –241.6 kJ mol^{-1} respectively. Calculate the amount of heat evolved by burning 1 m^3 of methane measured under normal condition.

Answer. 3.58 × 10^4 kJ (*Agra BSc*)

36. The heat of combustion of methane is – 890.65 kJ mol^{-1} and heats of formation of CO_2 and H_2O are – 393.5 and –286.0 kJ mol^{-1} respectively. Calculate the heat of formation of methane.

Answer. –74.85 kJ mol^{-1} (*Madurai BSc*)

37. (*a*) Derive Kirchoff's equation.

(*b*) Bond energies of F_2 and Cl_2 are 153.24 and 2428.34 kJ/mol respectively. Heat liberated in the reaction

$$F_2 + Cl_2 \rightarrow 2FCl$$

is 111.37 kJ. Find the bond energy of F—Cl bond.

Answer. (*b*) 1214.17 kJ (*Bhavnagar BSc*)

38. (*a*) Define the heat of reaction. Derive the relationship between heat of reaction at constant volume and at constant pressure.

(*b*) The heat of combustion of glucose at 290 K and at constant pressure is –2725606.76 J. Calculate heat of combustion at constant volume at 290 K.

Answer. (*b*) – 2725606.76 J (*Gujarat BSc, 2010*)

39. The heat of combustion of liquid benzene is –326.7 kJ. The heat of formation of CO_2 and H_2O are –393.5 kJ and –286.2 kJ respectively. Calculate the heat of formation of liquid benzene.

Answer. –2892.9 kJ *(Baroda BSc)*

40. The heat liberated on complete combustion of 0.0078 kg^{-1} benzene is 327 kJ. This heat has been measured at constant volume and at 300 K. Calculate the heat of combustion of benzene at constant pressure. ($R = 8.314$ J mol^{-1} K^{-1})

Answer. 3273.741 kJ mol^{-1} *(Osmania BSc)*

MULTIPLE CHOICE QUESTIONS

1. The branch of chemistry which deals with the heat changes caused by chemical reactions is called

(*a*) thermodynamics (*b*) thermal chemistry
(*c*) thermochemistry (*d*) none of these

Answer. (*c*)

2. In an exothermic reaction, the internal energy of the products is _______ internal energy of the reactants

(*a*) greater than (*b*) lesser than
(*c*) equal to (*d*) none of these

Answer. (*a*)

3. In an endothermic reaction

(*a*) $E_R > E_P$ (*b*) $E_R < E_P$
(*c*) $E_R = E_P$ (*d*) None of these

Answer. (*a*)

4. The enthalpy of a system is defined by the relation

(*a*) $H = E + P V$ (*b*) $H = E - P V$
(*c*) $E = H + P V$ (*d*) $P V + E - H$

Answer. (*a*)

5. For exothermic reactions, ΔH is _______ while for endothermic reactions it is _______.

(*a*) positive, negative (*b*) positive, positive
(*c*) negative, negative (*d*) negative, positive

Answer. (*d*)

6. In an exothermic reaction the heat energy is _______ while in endothermic reaction it is _______.

(*a*) released, released (*b*) released, absorbed
(*c*) absorbed, released (*d*) absorbed, absorbed

Answer. (*b*)

7. Which one of the following is correct for a reaction aA + bB → cC + dD where $\Delta n = (c + d) - (a + b)$

(*a*) $P\Delta V = \Delta H - \Delta E$ (*b*) $\Delta H = \Delta E + \Delta n R T$
(*c*) $\Delta n R T = \Delta H - \Delta E$ (*d*) all of these

Answer. (*d*)

8. For the reaction $N_2 + 3H_2 \rightleftharpoons 2NH_3$, the change in enthalpy is given by

(*a*) $\Delta H = \Delta E - 2 R T$ (*b*) $\Delta H = \Delta E + 2 R T$
(*c*) $\Delta H = \Delta E + 3 R T$ (*d*) $\Delta H = \Delta E + R T$

Answer. (*a*)

9. $\Delta H = \Delta E$ is true for the reaction

(*a*) $H_2 + Cl_2 \rightleftharpoons 2HCl$ (*b*) $2NO \rightleftharpoons N_2 + O_2$
(*c*) $2HI \rightleftharpoons H_2 + I_2$ (*d*) all of these

Answer. (*d*)

10. Which of the following relations holds good for the reaction?

$CO(g) + ½O_2(g) \rightarrow CO_2(g)$

(*a*) $\Delta H = \Delta E$ (*b*) $\Delta H = \Delta E + RT$
(*c*) $\Delta H = \Delta E + ½ RT$ (*d*) $\Delta H = \Delta E - ½ RT$

Answer. (*d*)

11. The change in internal energy with temperature at a constant volume is given by the relation

(*a*) $\Delta E = \Delta C_v (T_1 - T_2)$ (*b*) $\Delta E = \Delta C_p (T_2 - T_1)$
(*c*) $\Delta E = \Delta C_v (T_2 - T_1)$ (*d*) $\Delta E = \Delta C_p (T_1 - T_2)$

Answer. (*c*)

12. The Kirchoff's equation is

(*a*) $\Delta H_2 - \Delta H_1 = \Delta C_p (T_2 - T_1)$ (*b*) $\Delta E_2 - \Delta E_1 = \Delta C_v (T_2 - T_1)$
(*c*) both (*d*) neither (*a*) nor (*b*)

Answer. (*c*)

13. The change in enthalpy that takes place when one mole of the compound is formed from its elements is called

(*a*) heat of formation of compound (*b*) heat of synthesis
(*c*) heat of combustion (*d*) standard heat of formation

Answer. (*a*)

14. By convention, the standard heat of formation of all elements is assumed to be

(*a*) zero (*b*) negative
(*c*) positive (*d*) infinity

Answer. (*a*)

15. The change in enthalpy of a system when one mole of the substance is completely burnt in excess of air or oxygen is called

(*a*) heat of reaction (*b*) heat of formation
(*c*) heat of oxidation (*d*) heat of combustion

Answer. (*d*)

16. The calorific value is defined as the amount of heat produced in calories when _______ of a substance is completely burnt.

(*a*) one gram (*b*) one mole
(*c*) 100 grams (*d*) 1 kg

Answer. (*a*)

17. The fuel efficiency of methane (mol mass = 16) is _______ that of ethane (mol mass = 30).

(*a*) lesser than (*b*) greater than
(*c*) equal to (*d*) none of these

Answer. (*b*)

18. Which of the following always has a negative value?

(*a*) heat of formation (*b*) heat of reaction
(*c*) heat of combustion (*d*) heat of solution

Answer. (*c*)

19. The change in enthalpy when one mole of a substance is dissolved in a specified quantity of solvent at a given temperature is called

(*a*) heat of reaction (*b*) heat of combustion
(*c*) heat of solvation (*d*) heat of solution

Answer. (*d*)

20. The change in enthalpy of the system when one _______ of an acid is neutralized by one _______ of a base or *vice versa* in dilute solution is called heat of neutralization.

(*a*) g-equivalent, mole (*b*) mole, g-equivalent
(*c*) g-equivalent, g-equivalent (*d*) mole, mole

Answer. (*c*)

21. The heat of neutralization of a strong acid and a strong base is always

(*a*) zero (*b*) constant

(*c*) positive (*d*) changing

Answer. (*b*)

22. The heat of neutralization of a weak acid by a strong base is _______ that of a strong acid by a strong base.

(*a*) lesser than (*b*) greater than

(*c*) equal to (*d*) none of these

Answer. (*a*)

23. The heat change in equations $H^+(aq) + OH^-(aq) \rightarrow H_2O(\ell)\ \Delta H = -57.36$ kJ represents

(*a*) heat of neutralization (*b*) heat of formation of water

(*c*) heat of dissociation of water (*d*) heat of solution

Answer. (*a*)

24. The heat change (or enthalpy change) when one mole of a solid substance is converted into the liquid state at its _______ is called heat of fusion.

(*a*) boiling point (*b*) melting point

(*c*) freezing point (*d*) evaporation point

Answer. (*b*)

25. The process depicted by the equation

$H_2O(s) \rightarrow H_2O(\ell)\quad \Delta H = +5.99$ kJ represents

(*a*) melting (*b*) boiling

(*c*) evaporation (*d*) fusion

Answer. (*d*)

26. Greater the heat of fusion of a substance ______ is the magnitude of intermolecular forces

(*a*) higher (*b*) lower

(*c*) zero (*d*) infinite

Answer. (*a*)

27. The heat change when one mole of a liquid is converted into vapor or gaseous state at its boiling point is called

(*a*) heat of liquefaction (*b*) heat of fusion

(*c*) heat of vaporization (*d*) heat of sublimation

Answer. (*c*)

28. The process of sublimation for a solid occurs

(*a*) at the melting point (*b*) at a temperature below its melting point

(*c*) at a temperature above its melting point (*d*) none of these

Answer. (*b*)

29. The change in enthalpy which occurs when one mole of an element changes from one allotropic form to another is called.

(*a*) heat of allotropy (*b*) heat of transition

(*c*) heat of transformation (*d*) heat of sublimation

Answer. (*b*)

30. The thermochemical equations may be

(*a*) multiplied (*b*) added

(*c*) subtracted (*d*) all of these

Answer. (*d*)

31. In a chemical process, the amount of total heat change is same no matter by which method the change is brought about. This is statement of

(*a*) law of thermochemistry (*b*) Hess's law

(*c*) thermodynamic law (*d*) law of heat of reaction

Answer. (*b*)

32. The average amount of energy required to break all bonds of a particular type in one mole of the substance is called

(*a*) heat of reaction (*b*) bond energy
(*c*) heat of transition (*d*) heat of bond formation

Answer. (*b*)

33. The bond energy depends upon

(*a*) size of the atom (*b*) electronegativity
(*c*) bond length (*d*) all of these

Answer. (*d*)

34. When a bond is broken, the bond energy is positive because heat ________

(*a*) is evolved (*b*) is absorbed
(*c*) remains unchanged (*d*) none of these

Answer. (*b*)

35. For an endothermic reaction

(*a*) ΔH is –ve (*b*) ΔH is +ve
(*c*) ΔE is –ve (*d*) ΔH is zero

Answer. (*b*)

36. The enthalpy change in the reaction

$$2\,CO + O_2 \rightarrow 2CO_2$$

is termed as

(*a*) enthalpy of reaction (*b*) enthalpy of fusion
(*c*) enthalpy of formation (*d*) enthalpy of combustion

Answer. (*a*)

37. For a reaction

$$A_2(g) + B_2(g) \rightarrow 2AB(g) \qquad \Delta H = x \text{ kcal}$$

According to this reaction, the heat of formation of AB will be

(*a*) x kcal (*b*) –x kcal
(*c*) x/2 kcal (*d*) –x/2 kcal

Answer. (*c*)

38. For a gaseous reaction

$$X(g) + Y(g) \rightarrow 4Z(g)$$

ΔE at 300 K is 83.74 kJ. Taking $R = 8.314$ J K^{-1} mol^{-1}, ΔH for the above reaction will be

(*a*) 78.71 kJ (*b*) 81.22 kJ
(*c*) 83.74 kJ (*d*) 88.76 kJ

Answer. (*d*)

39. The heat of neutralization of an acid A with a base B is 57.36 kJ. Which of the following is true

(*a*) A is weak and B is also weak (*b*) A is strong and B is weak
(*c*) A is weak and B is strong (*d*) A is strong and B is also strong

Answer. (*d*)

40. For which of the following reactions, will ΔH be equal to ΔE?

(*a*) $N_2(g) + 3H_2(g) \rightarrow 2NH_3$ (*b*) $2HI(g) \rightarrow H_2(g) + I_2(g)$
(*c*) $2NO_2 \rightarrow N_2O_4(g)$ (*d*) $2SO_2(g) + O_2(g) \rightarrow 2SO_3(g)$

Answer. (*b*)

41. The neutralization of an acid with a base is represented by

$$H^+(aq) + OH^-(aq) \rightarrow H_2O(\ell) \qquad \Delta H = -57.36 \text{ kJ}$$

The heat of neutralization of one mole of H_2SO_4 with sodium hydroxide will be

(*a*) –57.36 kJ (*b*) –114.72 kJ
(*c*) +114.72 kJ (*d*) –28.68 kJ

Answer. (*b*)

42. Given that heat of neutralization of strong acid and strong base is –57.36 kJ. The heat produced when one mole of HCl is mixed with 0.5 mole of NaOH will be

(*a*) 28.68 kJ (*b*) – 28.68 kJ
(*c*) –114.72 kJ (*d*) – 57.36 kJ

Answer. (*b*)

43. The apparatus used to measure heat changes during chemical reactions is called

(*a*) polarimeter (*b*) colorimeter
(*c*) calorimeter (*d*) none of these

Answer. (*c*)

44. Hess's law is used to determine

(*a*) heat of formation of substances which are otherwise difficult to measure
(*b*) heat of transition
(*c*) heats of various other reactions like dimerization
(*d*) all of the above

Answer. (*d*)

45. ΔH° represent the enthalpy change

(*a*) at 273 K and 101325.00 Pa (1 atm) pressure (*b*) at 0 K and 101325.00 Pa (1 atm) pressure
(*c*) at 25 K and 101325.00 Pa (1 atm) pressure (*d*) none of these

Answer. (*c*)

46. The heats of formation of CO(g) and CO_2(g) are –110.53 kJ and –393.56 kJ respectively. The heat of combustion of carbon monoxide according to Hess's law will be

(*a*) +110.53 kJ (*b*) + 393.56 kJ
(*c*) – 283.03 kJ (*d*) –504.09 kJ

Answer. (*c*)

47. The heat of combustion of ethane (C_2H_6) is –1410.95 kJ at 298 K. The heat of the reaction when 0.003 kg of ethane is burnt completely is

(*a*) –14.1 kJ (*b*) +14.11 kJ
(*c*) –141.10 kJ (*d*) +141.10 kJ

Answer. (*c*)

48. For the reaction $CaCO_3(s) \rightarrow CaO(s) + CO_2(g)$ which one of the following is true?

(*a*) $\Delta H = \Delta E$ (*b*) $\Delta H > \Delta E$
(*c*) $\Delta H < \Delta E$ (*d*) $\Delta H = ½\,\Delta E$

Answer. (*b*)

49. Consider the reaction

$$CO(g) + ½O_2(g) \rightarrow CO_2(g)$$

at constant temperature and pressure. Which one of the following is correct?

(*a*) $\Delta H < \Delta E$ (*b*) $\Delta H > \Delta E$
(*c*) $\Delta H = \Delta E$ (*d*) none of these

Answer. (*a*)

50. The energy required to dissociate 0.006 kg of gaseous hydrogen into free gaseous atoms is 13.06 kJ at 298 K. The bond energy of H–H bond will be

(*a*) 13.06 kJ (*b*) 6.53 kJ
(*c*) 4.35 kJ (*d*) 42.04 kJ

Answer. (*c*)

9 Second Law of Thermodynamics

CONTENTS

The first law of thermodynamics states that one form of energy can change into another form but the total amount of energy remains the same. Once we specify a particular process or change, this law helps us to balance the internal energy, heat released and work done in the process. But the first law says nothing about whether the process of change we specify can, in fact, occur and if so in which direction. This question concerns the second law of thermodynamics. Before we can state the second law in a usable form, we must define some terms.

SPONTANEOUS PROCESSES

A process which proceeds of its own accord, without any outside assistance, is termed a spontaneous or natural process.

The reverse process which does not proceed on its own, is referred to as a nonspontaneous or unnatural process.

In general, the tendency of a process to occur naturally is called the spontaneity.

Examples of Spontaneous Processes

(1) **Cliff Jumping and Mountain Climbing**

Cliff jumping is a spontaneous process, whereas mountain climbing requires work.

(2) **Heat Flow**

When two balls of metal, one hot and one cold, are connected, heat flows spontaneously from the hot ball to the cold one, never from cold to hot. It requires work to transfer heat from one place to the other, say, by means of refrigerator pump.

(3) **Gas Flow**

When a vessel containing a gas is connected to another evacuated vessel, the gas spreads throughout spontaneously unless the pressure is the same in both the vessels. The reverse process of compressing the gas into the original vessel cannot occur unless work is done on it.

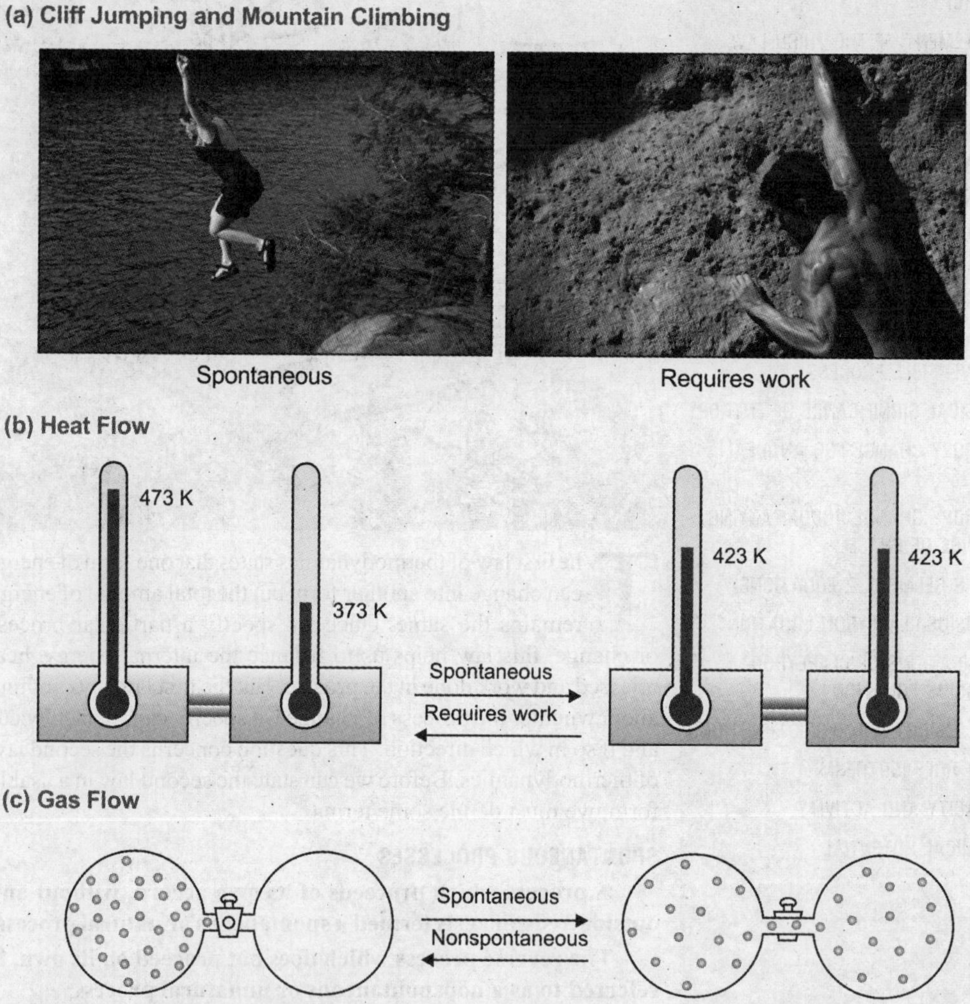

■ **Figure 9.1**
Spontaneous processes (Illustration).

CRITERIA OF SPONTANEITY

Some important criteria of spontaneous physical and chemical changes are listed below.

(1) **A spontaneous change is one-way or unidirectional.** For reverse change to occur, work has to be done.

(2) **For a spontaneous change to occur, time is no factor.** A spontaneous reaction may take place rapidly or very slowly.

(3) **If the system is not in equilibrium state (unstable), a spontaneous change is inevitable.** The change will continue till the system attains the state of equilibrium.

(4) **Once a system is in equilibrium state, it does not undergo any further spontaneous change in state if left undisturbed.** To take the system away from equilibrium, some external work must be done on the system.

(5) **A spontaneous change is accompanied by decrease of internal energy or enthalpy (ΔH).**

It implies that only such reactions will occur which are exothermic. But the melting of ice and evaporation of rain water are endothermic processes which proceed spontaneously. Clearly, there is some other factor in addition to ΔH which governs spontaneity. It is the second law of thermodynamics which introduces this new factor that is called *entropy*.

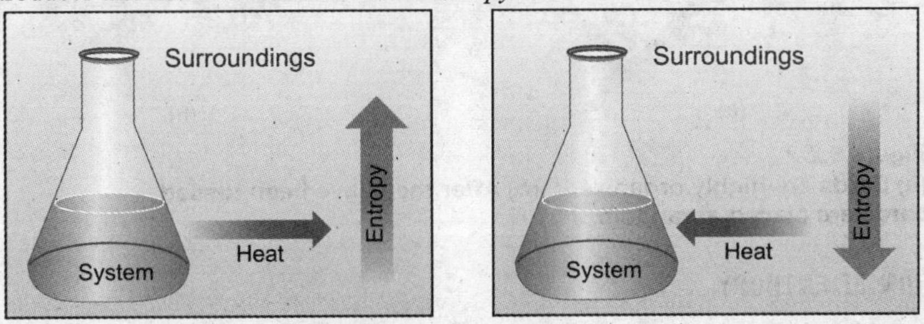

ENTROPY

For many years scientists believed that only exothermic changes resulting in a lowering of internal energy or enthalpy could occur spontaneously. But melting of ice is an endothermic process and yet occurs spontaneously. On a warm day, ice melts by itself. The evaporation of water is another example of a spontaneous endothermic process. Thus arose the need of inventing another driving force that affects the spontaneity. This was known as the entropy change, ΔS.

Spontaneity and Randomness

Careful examination shows that in each of the processes *viz.*, melting of ice and evaporation of water, there is an increase in randomness or disorder of the system. The water molecules in ice are arranged in a highly organized crystal pattern which permits little movement. As the ice melts, the water molecules become disorganized and can move more freely. The movement of molecules becomes freer still when the water evaporates into space as now they can roam about throughout the entire atmosphere. In other words, we can say that the randomness of the water molecules increases as ice melts into water or water evaporates into space.

Melting of Ice is a classic example of 'Increasing Entropy'.

Increase in Randomness Favors a Spontaneous Change

A change that brings about randomness is more likely to occur than one that brings about order. Let us suppose we have a suit of playing cards arranged numerically (Fig. 9.2 *a*).

We can see that the sequence of cards is certainly highly organized. Now, if we throw the cards into the air, collect them and re-stack them, we will almost surely find that they have been placed at random. This is expected because when the cards are tossed, there are many ways for them to be disordered, while there is only one way for them to come together again in their original sequence. Thus on the basis of pure chance a disordered sequence is far more probable than the ordered one with which we started. The same law of chance applies to any physical or chemical process.

(a) (b)

Figure 9.2
(*a*) Cards are highly organized; (*b*) After they have been tossed, cards are placed at random.

DEFINITION OF ENTROPY

Entropy is a thermodynamic state quantity that is a measure of the randomness or disorder of the molecules of the system.

The symbol of entropy is *S*, while the change in disorder accompanying a process from start to completion is represented by ΔS. The entropy of a system is a state function and depends only on the initial and final states of the system. The change in entropy, ΔS, for any process is given by the equation,

$$\Delta S = S_{\text{final}} - S_{\text{initial}}$$

When $S_{\text{final}} > S_{\text{initial}}$, ΔS is positive.

A process accompanied by an increase in entropy tends to be spontaneous. Let us consider a molecular system in states *A* and *B* (Fig. 9.3). In state *A* all the molecules are arranged and highly ordered, while in state *B* the molecules are present at random and it is highly disordered.

(1) By definition, the entropy of *A* is low and that of *B* high. Thus an increase of entropy occurs in the change from *A* to *B*.

(2) According to the law of chance (probability), *A* is less probable and *B* is more probable. Therefore, the change from *A* to *B* is spontaneous.

(3) From (1) and (2), it follows that the change from *A* to *B* which is accompanied by increase of entropy will tend to be spontaneous.

Hence we can say, in general, that a change in a system which is accompanied by an increase in entropy, tends to be spontaneous.

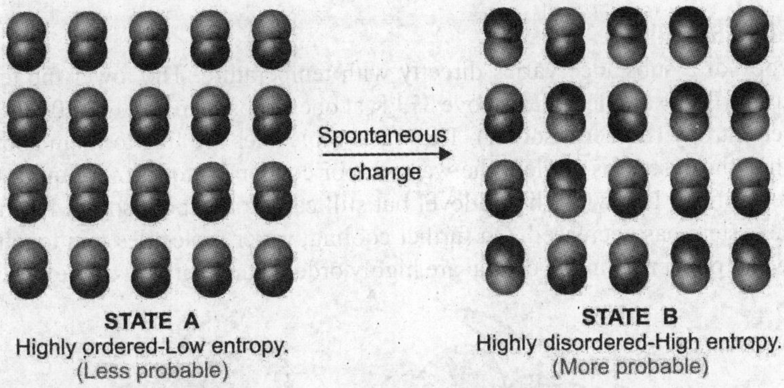

■ **Figure 9.3**
(a) State A is highly ordered, low entropy and less probable;
(b) State B is highly disordered high entropy and more probable.

STATEMENT OF THE SECOND LAW

The second law of thermodynamics states that : **whenever a spontaneous process takes place, it is accompanied by an increase in the total energy of the universe.**

More specifically, we take the term 'universe' to mean the system and the surroundings. Thus,

$$\Delta S_{univ} = \Delta S_{syst} + \Delta S_{surr}$$

The second law, as stated above, tells us that when an irreversible spontaneous process occurs, the entropy of the system and the surroundings increases. In other words $\Delta S_{univ} > 0$. When a reversible process occurs, the entropy of the system remains constant. $\Delta S_{univ} = 0$. Since the entire universe is undergoing spontaneous change, the second law can be most generally and concisely stated as : **the entropy of the system is constantly increasing.**

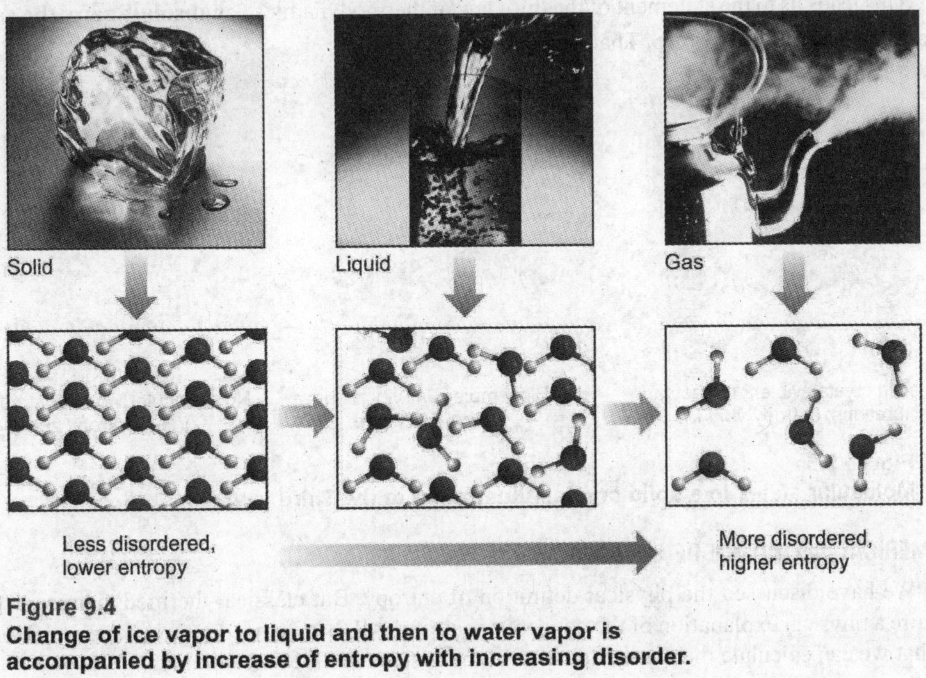

■ **Figure 9.4**
Change of ice vapor to liquid and then to water vapor is accompanied by increase of entropy with increasing disorder.

STATEMENT OF THE THIRD LAW

The entropy of a substance varies directly with temperature. The lower the temperature, the lower the entropy. For example, water above 373 K at one atmosphere (101325.00 Pa) exists as a gas and has higher entropy (higher disorder). The water molecules are free to roam about in the entire container. When the system is cooled, the water vapor condenses to form a liquid. Now the water molecules are confined below the liquid level but still can move about somewhat freely. Thus the entropy of the system has decreased. On further cooling, water molecules join together to form ice crystal. The water molecules in the crystal are highly ordered and entropy of the system is very low.

"If we have everlasting life, what about entropy?"

If we cool the solid crystal still further, the vibration of molecules held in the crystal lattice gets slower and they have very little freedom of movement (very little disorder) and hence very small entropy. Finally, at absolute zero all molecular vibration ceases and water molecules are in perfect order. Now the entropy of the system will be zero.

This leads us to the statement of the third law of thermodynamics : **at absolute zero, the entropy of a pure crystal is also zero.** That is, $S = 0$ at $T = 0$ K.

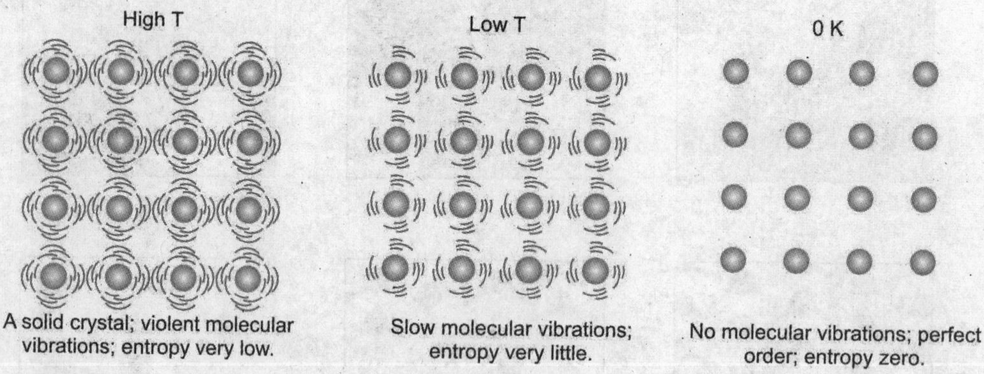

■ Figure 9.5
Molecular states in a solid crystal (Illustration of the Third law).

NUMERICAL DEFINITION OF ENTROPY

We have discussed the physical definition of entropy. But classical thermodynamics does not require a physical explanation of the concept of entropy. All that we need is an operational definition so that we can calculate the entropy change of the system and the surroundings.

In 1850 Clausius introduced a numerical definition of entropy. According to him entropy of a system (not undergoing chemical or physical changes), is a constant quantity when there is no communication of heat. When heat (q) flows into a system, the entropy increases by q/T. Heat flowing out of a system produces a corresponding decrease. Thus entropy could be precisely defined as : **for a reversible change taking place at a fixed temperature (T), the change in entropy (ΔS) is equal to heat energy absorbed or evolved divided by the temperature (T).** That is,

$$\Delta S = \frac{q}{T}$$

If heat is absorbed, then ΔS is positive and there will be increase in entropy. If heat is evolved, ΔS is negative and there is a decrease in entropy.

UNITS OF ENTROPY

As stated above, entropy is equal to heat energy divided by absolute temperature. Therefore, it is measured in entropy units (*'eu'*) which are calories per degree per mole *i.e.*, cal mol^{-1} K^{-1}.

In the SI system, the units are joules per mole per degree *i.e.*, J mol^{-1} K^{-1}. These are represented by *eu*.

$$1 eu = 4.184$$

STANDARD ENTROPY

From the third law, we know that the entropy of a pure crystal is zero at absolute zero (K) Therefore, it is possible by measurement and calculation to find the actual amount of entropy that a substance possesses at any temperature above 0 K. It is often referred to as **absolute entropy.**

TABLE 9.1. STANDARD ENTROPIES OF SOME SUBSTANCES (298K, 101325.00 PASCAL OR 25°C, 1 ATM)

Substance	Entropy, S° cal mol^{-1} K^{-1}	Entropy, S° J mol^{-1} K^{-1}	Substance	Entropy, S° cal mol^{-1} K^{-1}	Entropy, S° J mol^{-1} K^{-1}
Ag (*s*)	41.32	172.9	$H_2(g)$	31.21	130.6
AgCl (*s*)	58.5	24.5	$H_2O(g)$	45.11	188.7
Al (*s*)	6.77	28.3	$H_2O(l)$	16.72	69.96
Al_2O_3 (*s*)	12.19	51.0	HCl (*g*)	44.62	186.7
C (*s*, graphite)	0.58	2.4	HNO_3 (*l*)	37.19	155.6
CO (*g*)	47.30	197.9	H_2SO_4 (*l*)	37.5	157.0
CO_2 (*g*)	51.06	213.6	Hg (*l*)	18.2	76.1
CH_4 (*g*)	44.50	186.2	K (*s*)	38.30	160.2
CH_3OH (*l*)	30.3	126.8	KCl (*s*)	57.24	239.5
$CO(NH_2)_2$ (*s*)	25.0	104.6	K_2SO_4 (*s*)	42.0	176.0
C_2H_2 (*g*)	48.0	200.8	N_2 (*g*)	45.77	191.5
C_2H_6 (*g*)	54.85	229.5	NH_3 (*g*)	46.01	192.5
Ca (*s*)	36.99	154.8	Na (*s*)	36.72	153.6
$CaCO_3$ (*s*)	22.2	92.9	NaCl (*s*)	17.30	72.88
Cl_2 (*g*)	53.29	223.0	O_2 (*g*)	49.0	205.0
Fe (*s*)	6.5	27.0	S (*s*)	7.62	31.9
Fe_2O_3 (*s*)	21.5	90.0	SO_2 (*g*)	59.40	248.5

The absolute entropy of a substance at 25°C (298 K) and one atmosphere (101325.00 Pa) pressure is called the standard entropy; S^o. The absolute entropy of elements is zero only at 0 K in a perfect crystal, and standard entropies of all substances at any temperature above 0 K always have positive values. The standard entropies of a number of elements and compounds are given in Table 9.1.

Once we know the entropies of a variety of substances, we can calculate the standard entropy change, ΔS^o, for chemical reactions.

$$\Delta S^o = \Sigma S^o \text{ (products)} - \Sigma S^o \text{ (reactants)}$$

STANDARD ENTROPY OF FORMATION

It is the entropy of formation of 1 mole of a compound from the elements under standard conditions. It is denoted ΔS^o_f. We can calculate the value of entropy of a given compound from the values of S^o of elements.

$$\Delta S^o_f = S^o \text{ (compound)} - \Sigma S^o \text{ (elements)}$$

SOLVED PROBLEM 1

Urea hydrolyses in the presence of water to produce ammonia and carbon dioxide.

$$CO(NH_2)_2\,(aq) + H_2O\,(l) \longrightarrow CO_2(g) + 2NH_3(g)$$

What is the standard entropy change for this reaction when 1 mole of urea reacts with water? The standard entropies of reactants and products are listed below :

Substance	S°(J/mol K)
$CO(NH_2)_2$	173.96
$H_2O\,(l)$	70.00
$CO_2\,(g)$	213.78
$NH_3\,(g)$	192.63

SOLUTION

We know that

$$\Delta S^o = \Sigma S^o \text{ (products)} - \Sigma S^o \text{ (reactants)}$$

or

$$\Delta S^o = \left(S^o{}_{CO_2} + 2S^o{}_{NH_3}\right) - \left(S^o{}_{CO(NH_2)_2} + S^o{}_{H_2O}\right)$$

$$= [213.78 + 2 \times 192.63] - [173.96 + 70.00]\ JK^{-1}$$

$$= \mathbf{355.08\ JK^{-1}}$$

SOLVED PROBLEM 2

Calculate the standard entropy of formation, ΔS^o_f, of $CO_2(g)$. Given the standard entropies of $CO_2(g)$, $C(s)$, $O_2(g)$, which are 213.6, 5.740, and 205.0 JK^{-1} respectively.

SOLUTION

We know that

$$\Delta S^o_f = S^o{}_{\text{compound}} - \Sigma S^o{}_{\text{elements}}$$

or

$$S^o_f = S^o{}_{CO_2(g)} - \left[S^o{}_{C(s)} + S^o{}_{O_2(g)}\right]$$

Substituting the values

$$S^o_f = 213.6 - [5.740 + 205.0]\ JK^{-1}$$

$$= (213.6 - 210.74)\ JK^{-1} = \mathbf{2.86\ JK^{-1}}$$

SOME USEFUL DEFINITIONS

(1) Cyclic Process

When a system undergoes a series of changes and in the end returns to its original state, it is said to have completed a cycle. The whole process comprising the various changes is termed as **cyclic process.**

Since the internal energy of a system depends upon its state, it stands to reason that in cyclic process the net change of energy is zero. Or, we can say that the work done by the system during all these changes should be equal to the heat absorbed by the system.

$$\Delta E = 0 = q - w \text{ or } q = w$$

(2) Heat Engines

The flow of heat from a hotter body to a colder body is spontaneous process. The heat that flows out spontaneously can be used to do work with the help of a suitable device.

A machine which can do work by using heat that flows out spontaneously from a high-temperature source to a low-temperature sink, is called a heat engine.

A heat engine takes heat energy from a high-temperature reservoir and converts some of it into work, returning the unconverted heat to a low-temperature sink. A basic heat engine is illustrated in Fig. 9.6. A steam engine is a typical heat engine. It takes heat from the boiler (high-temperature source), converts some heat to work and returns the unused heat to the surroundings (low-temperature sink).

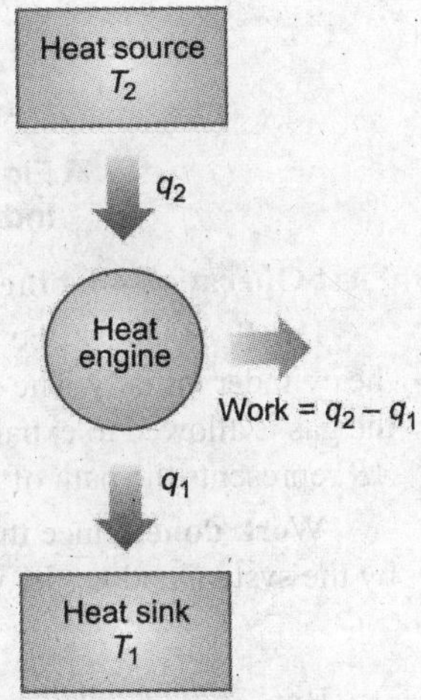

Figure 9.6 Principle of heat engine (illustration).

A heat engine running on a periodic cyclic process can yield work continuously.

(3) Efficiency of a Heat Engine

The ratio of the work obtained in a cyclic process (w) to the heat taken from the high-temperature reservoir (q) is referred to as the efficiency of a heat engine.

No heat engine, no matter how well constructed, can convert all the heat from the high-temperature reservoir into work. Such an engine would be 100% efficient. Sadi Carnot was the first scientist to realize this and deduce an expression showing the limitations of heat engines.

THE CARNOT CYCLE

In 1824 Sadi Carnot proposed a theoretical heat engine to show that the efficiency was based upon the temperatures between which it operated. Carnot's imaginary engine could perform a series of operations between temperatures T_1 and T_2, so that at the end of these operations the system was restored to the original state. **This cycle of processes which occurred under reversible conditions is referred to as the Carnot cycle.** The medium employed in operating Carnot's engine was one mole of an ideal gas which could be imagined to be contained in a cylinder fitted with a frictionless piston.

The Carnot cycle comprises four operations or processes.

(1) Isothermal reversible expansion

(2) Adiabatic reversible expansion

(3) Isothermal reversible compression

(4) Adiabatic reversible compression

The above four processes are shown in the indicator diagram of Carnot cycle (Fig. 9.7).

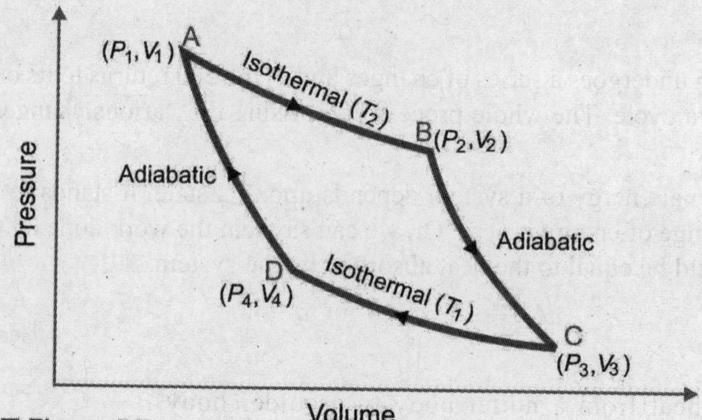

■ **Figure 9.7**
Indicator diagram of the Carnot cycle.

First Operation – Isothermal Reversible Expansion

Let T_2, P_1 and V_1 be the temperature, pressure and volume respectively of the gas enclosed in the cylinder initially. The cylinder is placed in the heat reservoir at the higher temperature (T_2). Now the gas is allowed to expand isothermally and reversibly so that the volume increases from V_1 to V_2. AB represents the path of the process in the diagram.

Work done. Since the process in operation 1 is isothermal, $\Delta E = 0$. If q_2 be the heat absorbed by the system and w_1 the work done by it, according to the first law equation ($\Delta E = q - w$),

$$q_2 = w_1$$

But $$w_1 = RT_2 \ln \frac{V_2}{V_1}$$

Therefore, $$q_2 = RT_2 \ln \frac{V_2}{V_1} \qquad ...(1)$$

Second Operation – Adiabatic Reversible Expansion

The gas at B is at a temperature T_2 and has volume V_2 under the new pressure P_2. The gas is now allowed to expand reversibly from volume V_2 to V_3 when the temperature drops from T_2 to T_3 (along BC).

Work done. Since this step is adiabatic, $q = 0$. If w_2 be the work done, according to the first law equation ($\Delta E = q - w$),

$$\Delta E = -w_2$$

or $$w_2 = -\Delta E$$

But $$\Delta E = C_v (T_1 - T_2)$$

Therefore, $$w_2 = C_v (T_2 - T_1) \qquad ...(2)$$

Third Operation – Isothermal Reversible Compression

Now the cylinder is placed in contact with a heat reservoir at a lower temperature, T_1. The volume of the gas is then compressed isothermally and reversibly from V_3 to V_4 (represented by CD in diagram).

Work done. During compression, the gas produces heat which is transferred to the low temperature reservoir. Since the process takes place isothermally, $\Delta E = 0$. If q_1 is the heat given to the reservoir and w_3 the work done on the gas, using proper signs for q and w, we have

$$-q_1 = -w_3 = RT_1 \ln \frac{V_4}{V_3} \qquad ...(3)$$

Fourth Operation – Adiabatic Reversible Compression

The gas with volume V_4 and temperature T_1 at D is compressed adiabatically (along *DA*) until it regains the original state. That is, the volume of the system becomes V_1 and its temperature T_2.

Work done. In this step work is done on the system and, therefore, bears the negative (–) sign. If it is denoted by w_4, we can write

$$-w_4 = -C_v (T_2 - T_1) \qquad ...(4)$$

Net Work Done in One Cycle

Adding up the work done (w) in all the four operations of the cycle as shown in equations (1), (2), (3) and (4), we have

$$w = w_1 + w_2 + (-w_3) + (-w_4)$$

$$= RT_2 \ln\frac{V_2}{V_1} + C_v (T_2 - T_1) + RT_1 \ln\frac{V_4}{V_3} - C_v (T_2 - T_1)$$

$$= RT_2 \ln\frac{V_2}{V_1} + RT_1 \ln\frac{V_4}{V_3}$$

Net Heat Absorbed in One Cycle

If q is the net heat absorbed in the whole cycle.

$$q = q_2 - q_1$$

where q_2 is heat absorbed by the system in operation 1 and q_1 is the heat transferred to the sink reservoir.

From (1) and (3)

$$q = q_2 - q_1 = RT_2 \ln\frac{V_2}{V_1} + RT_1 \ln\frac{V_4}{V_3}$$

or $$q = RT_2 \ln\frac{V_2}{V_1} - RT_1 \ln\frac{V_3}{V_4} \qquad ...(5)$$

According to the expression governing adiabatic changes,

$$\frac{T_2}{T_1} = \left(\frac{V_3}{V_2}\right)^{\gamma - 1} \qquad ...\textit{for adiabatic expansion}$$

$$\frac{T_1}{T_2} = \left(\frac{V_1}{V_2}\right)^{\gamma - 1} \qquad ...\textit{for adiabatic compression}$$

or $$\frac{V_3}{V_2} = \frac{V_4}{V_1}$$

or $$\frac{V_3}{V_4} = \frac{V_2}{V_1}$$

Therefore, substituting the value of V_3/V_4 in equation (5), the value of net heat may be given as

$$q = RT_2 \ln\frac{V_2}{V_1} - RT_1 \ln\frac{V_2}{V_1}$$

$$= R (T_2 - T_1) \ln \frac{V_2}{V_1} \qquad ...(6)$$

Calculation of Thermodynamic Efficiency

Since the total work done in a cycle is equal to net heat absorbed, from (6) we can write

$$w = R(T_2 - T_1) \ln \frac{V_2}{V_1} \quad ...(7)$$

The heat absorbed, q_2, at higher temperature T_2 is given by equation (1),

$$q_2 = RT_2 \ln \frac{V_2}{V_1} \quad ...(8)$$

Dividing (7) by (8)

$$\frac{w}{q_2} = \frac{R(T_2 - T_1) \ln V_2/V_1}{RT_2 \ln V_2/V_1} = \frac{T_2 - T_1}{T_2}$$

or

$$\frac{w}{q_2} = \frac{T_2 - T_1}{T_2} \quad ...(9)$$

The factor w/q_2 is called **thermodynamical efficiency.** It is denoted by η and gives the fraction of the heat taken from the high-temperature reservoir which is possible to convert into work by a heat engine. Therefore, the efficiency of a Carnot engine, the most ideal of all engines, is limited by the operating temperatures of the engine. The larger the temperature difference $(T_2 - T_1)$ between the high and the low temperature reservoirs, the more the heat converted to work by the heat engine. For a given temperature of the high-temperature reservoir, the lower the temperature of the sink, the greater will be the efficiency of the machine. Similarly, for a given temperature of the sink, the efficiency will be increased by using a high temperature of the source reservoir.

Carnot Theorem

We have shown above that

$$\frac{w}{q_2} = \frac{T_2 - T_1}{T_2}$$

This result deduced for a perfect gas depends upon the temperature limits between which the cycle operates. It is independent of all other factors. Thus Carnot stated an important relation known as the **Carnot theorem.** It states that : **every perfect engine working reversibly between the same temperature limits has the same efficiency, whatever be the working substance.**

MORE STATEMENTS OF THE SECOND LAW

From equation (9)

$$\frac{w}{q_2} = \frac{T_2 - T_1}{T_2} = 1 - \frac{T_1}{T_2}$$

Evidently w/q_2 is less than 1, or q_2 is greater than w. This means that heat transferred by a spontaneous process is never completely converted into work (If so, w/q_2 would be 1). This leads to another statement of the Second law (Lord Kelvin).

It is impossible to take heat from a hotter reservoir and convert it completely into work by a cyclic process without transferring a part of heat to a cooler reservoir.

This statement recognizes the fact that heat engines could never be 100% efficient, since some heat must be returned to a low-temperature reservoir. Another statement of the Second law was given by Clausius.

It is impossible for a cyclic process to transfer heat from a body at a lower temperature to one at higher temperature without at the same time converting some work to heat.

WONDERFUL ACHIEVEMENTS IN SCIENCE AND ENGINEERING

The Second Law of Thermodynamics

"A spontaneous change is accompanied by an increase in the total entropy of the system and its surroundings."

$$\Delta S_{Sys} + \Delta S_{Surr} > 0$$

ΔS_{Sys} is the entropy change of the system.
ΔS_{Surr} is the entropy change of the surroundings.

Scientists responsible for the formulation and development of the Second Law include Rudolph Clausius (1822-1888), Lord Kelvin (1824-1907), Josiah Willard Gibbs (1839-1903) and Ludwig Boltzmann (1844-1906). The Second Law explains why the rusting of Iron is inevitable, why car engines can never be 100% efficient and why batteries eventually run down. It governs the direction of all biochemical reactions including those responsible for glucose catabolism, DNA replication and protein synthesis—even life obeys the Second Law of Thermodynamics.

This statement recognizes that heat flows spontaneously from hot objects to cold objects and to get it flow in the opposite direction, we have to do some work on it.

SOLVED PROBLEM 3

An engine operating between 423K and 298K takes 500 J heat from a high temperature reservoir. Assuming that there are no frictional losses, calculate the work that can be done by this engine.

SOLUTION

From equation (9)

$$\text{Efficiency} = \frac{w}{q_2} = \frac{T_2 - T_1}{T_2}$$

Here $T_2 = 423 \text{ K};\ T_1 = 298 \text{ K}$

Substituting the values $\dfrac{w}{500} = \dfrac{423 - 298}{423}$

$$\therefore \quad w = \mathbf{147.75\ J}$$

SOLVED PROBLEM 4

The boiling point of water at a pressure of 5066250.00 Pa is 538 K. Compare the theoretical efficiencies of a steam engine operating between the boiling point of water at (*i*) 1 atmosphere (101325.00 Pa) (*ii*) 50 atmosphere (5066250.00 Pa) assuming the temperature of the sink to be 308 K in each case.

SOLUTION

(*i*) *At 1 atmosphere (101325.00 Pa) :*

T_2 (boiling point of water) = 373 K
T_1 (temperature of sink) = 308 K

$$\text{Efficiency} = \frac{T_2 - T_1}{T_2} = \frac{373 - 308}{373} = 0.174$$

(ii) At 50 atmosphere (5066250.00 Pa) :

T_2 (boiling point of water) = 265°C or 538 **K**

T_1 (temperature of sink) = 35°C or 308 **K**

$$\text{Efficiency} = \frac{T_2 - T_1}{T_2} = \frac{538 - 308}{538} = \mathbf{0.428}$$

The possible increase of efficiency is very marked.

SOLVED PROBLEM 5

If a Carnot engine operating between two heat reservoirs at 500 K and 300 K absorbs 4186.80 J from the 500 K reservoir per cycle, how much heat is discharged into the 300 K reservoir and how much work is done per cycle ? What is the efficiency of the cycle ?

SOLUTION

(a) We know that :

$$\text{Efficiency} = \frac{w}{q_2} = \frac{T_2 - T_1}{T_2}$$

or

$$w = \frac{q_2\,(T_2 - T_1)}{T_2}$$

$$= \frac{4186.80\,(500 - 300)}{500}$$

$$= 1674.72 \text{ J}$$

∴ The work done per cycle is **1674.72 J**

(b) The heat from the high-temperature reservoir (q_2) minus the heat discharged into the low-temperature reservoir (q_1) is converted into work (w). Thus,

$$q_2 - q_1 = w$$

∴

$$4186.80 - q_1 = 1674.72 \text{ J}$$

or

$$q_1 = 2512.08 \text{ J}$$

(c) Efficiency :

$$\frac{w}{q_2} = \frac{T_2 - T_1}{T_2}$$

$$= \frac{500 - 300}{500} = \frac{200}{500}$$

$$= 0.4$$

Therefore, efficiency of the engine is **0.4**

DERIVATION OF ENTROPY FROM CARNOT CYCLE

Because processes cannot be 100% efficient, a term to describe the energy available for doing useful work becomes necessary. Although we have discussed the concept of entropy already, its classical derivation deserves attention.

In a Carnot cycle, q_2 has a positive value and q_1 has a negative value, since former is *taken up* at a higher temperature and the latter is *given out* at the lower temperature. Thus thermodynamic efficiency may be expressed as

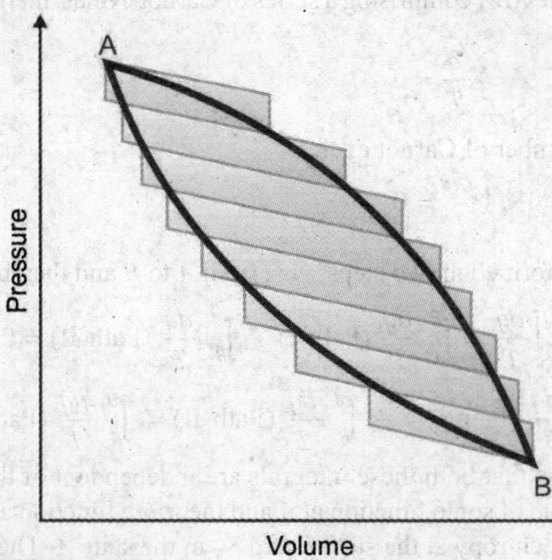

Figure 9.8
Carnot cycle.

$$\eta = \frac{q_2 - q_1}{q_2} = \frac{T_2 - T_1}{T_2}$$

or $$1 - \frac{q_1}{q_2} = 1 - \frac{T_1}{T_2}$$

or $$\frac{q_1}{q_2} = \frac{T_1}{T_2}$$

or $$\frac{q_1}{T_1} = \frac{q_2}{T_2} \qquad ...(i)$$

using sign convention, heat absorbed (*i.e.*, q_2) is given the +ve sign and heat lost (*i.e.* q_1) is given the – ve sign. Equation (*i*) becomes

$$+\frac{q_2}{T_2} = -\frac{q_1}{T_1}$$

or $$\frac{q_2}{T_2} + \frac{q_1}{T_1} = 0$$

or $$\Sigma\frac{q}{T} = 0$$

Any reversible cycle may be regarded as made up of a number of Carnot cycles. Consider, for example, the cycle represented in Fig. 9.8 by the closed curve *ABA*. Imagine a series of isothermal and adiabatic curves drawn across the diagram so that a number of Carnot cycles are indicated. Starting at *A* and going through all the cycles successively from *A* to *B*, it can be shown that all paths inside the diagram cancel each other leaving only zigzag outer path. The larger the number of cycles taken in this manner, the closer will the resultant path correspond to *ABA* which represents the reversible cycle under consideration. The reversible cycle can, therefore, be regarded as being made up of an infinite number of Carnot cycles, for each of which the sum of the two *q*/*T* terms involved is zero *i.e.*,

$$\frac{q_1}{T_1} + \frac{q_2}{T_2} = 0$$

For the reversible cycle *ABA* comprising a series of Carnot cycles, therefore, the above expression takes the form

$$\Sigma \frac{q}{T} = 0$$

and for an infinite number of Carnot cycles

$$\int \frac{dq}{T} = 0$$

Since the cycle is performed in two steps, *viz.*, from *A* to *B* and then back from *B* to *A*, we have:

$$\int \frac{dq}{T} = \int_A^B \frac{dq}{T} \text{ (Path I)} + \int_B^A \frac{dq}{T} \text{ (Path II)} = 0$$

$$\int_A^B \frac{dq}{T} \text{ (Path I)} = -\int_B^A \frac{dq}{T} \text{ (Path II)} = \int_A^B \frac{dq}{T} \text{ (Path II)}$$

It is evident, therefore, that both these integrals are independent of the path taken from *A* to *B*. Both depend upon the value of some function at *A* and the same function at *B*. This function is called **entropy** (*S*). Let S_B be the entropy at the state *B* and S_A in the state *A*. Then, the increase in entropy, ΔS, is given by the expression

$$\Delta S = S_B - S_A = \int_A^B \frac{dq}{T}$$

and for each infinitesimally small change

$$dS = \frac{dq}{T}$$

Like ΔE and ΔH, ΔS is dependent only on the state of the system and can be calculated if the substance can be brought reversibly from one state to the other. It is independent of the path taken.

SOLVED PROBLEM 6

Calculate the entropy change in the evaporation of 1 mole of water at 373 K. Latent heat of evaporation of water is 40402.62 J per mole.

SOLUTION

Entropy change on the evaporation of 1 mole of water is obtained by dividing the latent heat of evaporation of 1 mole of water by the absolute temperature

$$\therefore \quad \Delta S = \frac{40402.62}{373} = 108.31 \text{ JK}^{-1} \text{ mol}^{-1}$$

SOLVED PROBLEM 7

Calculate the increase in entropy when one gram (1×10^{-3} kg) molecular weight of ice at 273K melts to form water. Latent heat of fusion of ice = 334.94 J.

SOLUTION

$$\Delta S = \frac{q}{T}$$

$$q \text{ for one mole of ice} = 334.94 \times 18 \text{ J}$$

$$\text{and} \quad T = 273 \text{ K}$$

$$\therefore \quad \Delta S = \frac{334.94 \times 18}{273} = 22.09 \text{ JK}^{-1} \text{ mol}^{-1}$$

ENTROPY CHANGE IN AN IRREVERSIBLE PROCESS

The efficiency of an irreversible Carnot cycle is always less than that of a reversible one operating between the same two temperatures.

$$\frac{q_2 - q_1}{q_2} < \frac{T_2 - T_1}{T_2}$$

where q_2 is the heat absorbed at temperature T_2 and q_1 is the heat returned at temperature T_1.

$$\frac{q_2}{T_2} - \frac{q_1}{T_1} < 0$$

In other words $\int \frac{dq}{T}$ for an irreversible cycle is always less than zero and the entropy of the final state is always greater than that of the initial state. **As most of the processes going on in nature are spontaneous and irreversible, it has been said that "the entropy of the universe always tends towards a maximum".**

PHYSICAL SIGNIFICANCE OF ENTROPY

Entropy is a measure of disorder or randomness of a system. The entropy of the system increases if it goes from less orderly state to more orderly state and vice–versa. This concept of entropy has led to the conclusion that *all substances in their normal crystalline state at absolute zero would be the most ordered state with zero entropy*. At this state, all motion ceases. In case of a perfect crystal the entropy is zero. This is **third law of Thermodynamics.**

ENTROPY CHANGE FOR AN IDEAL GAS

Entropy is a state function and its value depends on two of three variables *T*, *P* and *V*.

(*a*) *T* and *V* as Variables

Let us consider *n* moles of an ideal gas occupying volume *V* at pressure *P* and temperature *T*. If the system absorbs dq_{rev} heat reversibly then increase in entropy is given by

$$dS = \frac{dq_{rev}}{T} \quad \text{...}(i)$$

According to the first law of thermodynamics

$$dq_{rev} = dE + PdV$$

Putting in equation (*i*), we get

$$dS = \frac{dE + PdV}{T} \quad \text{...}(ii)$$

we know $\quad PV = nRT$

or $\quad P = \frac{n}{V}RT$

and $\quad dE = n\,C_v\,dT \qquad \left[\because C_v = \frac{dE}{dT} \text{ for 1 mole}\right]$

where C_v is the molar heat at constant volume.

Substituting this value in equation (*ii*), we have

$$dS = \frac{n\,C_v\,dT + \left(\frac{n}{V}RT\right)dV}{T}$$

or

$$dS = n\,C_v\,\frac{dT}{T} + n\,R\,\frac{dV}{V}$$

Integrating between the limits S_1, S_2 ; T_1, T_2 ; and V_1, V_2 we get

$$\int_{S_1}^{S_2} dS = \int_{T_1}^{T_2} nC_v \frac{dT}{T} + \int_{V_1}^{V_2} nR\frac{dV}{V}$$

or $$\Delta S = S_2 - S_1 = n\,C_v \ln \frac{T_2}{T_1} + nR \ln \frac{V_2}{V_1} \quad ...(iii)$$

or $$\Delta S = 2.303\, n\, C_v \log \frac{T_2}{T_1} + 2.303\, R \log \frac{V_2}{V_1} \quad ...(iv)$$

For 1 mole of an ideal gas

$$\Delta S = 2.303 C_v \log \frac{T_2}{T_1} + 2.303\, R \log \frac{V_2}{V_1}$$

(*b*) *P* and *T* as Variables

Let P_1 be the pressure in the initial state and P_2 in the final state then

$$\frac{P_1 V_1}{T_1} = \frac{P_2 V_2}{T_2}$$

or $$\frac{V_2}{V_1} = \frac{P_1 T_2}{P_2 T_1}$$

Substituting in equation (*iii*) we get

$$\Delta S = n\,C_v \ln \frac{T_2}{T_1} + nR \ln \frac{P_1 T_2}{P_2 T_1}$$

$$= n\,C_v \ln \frac{T_2}{T_1} + nR \ln \frac{T_2}{T_1} + n\,R \ln \frac{P_1}{P_2}$$

$$= n\,(C_v + R) \ln \frac{T_2}{T_1} + nR \ln \frac{P_1}{P_2}$$

$$= nC_P \ln \frac{T_2}{T_1} + nR \ln \frac{P_1}{P_2} \quad [\because C_P = C_v + R]$$

or $$\Delta S = 2.303 \times n \times C_p \log \frac{T_2}{T_1} + 2.303 \times n \times R \log P_1/P_2 \quad ...(v)$$

For 1 mole of an ideal gas

$$\Delta S = 2.303\, C_p \log \frac{T_2}{T_1} + 2.303\, R \log \frac{P_1}{P_2}$$

From this equation entropy change can be calculated.

Case 1. At constant temperature for an isothermal process

In this case $T_1 = T_2$, the equation (*iv*) and (*v*) reduce to

$$\Delta S_T = 2.303 \times n \times R \log \frac{V_2}{V_1}$$

and $$\Delta S_T = 2.303 \times n \times R \log \frac{P_1}{P_2}$$

In an isothermal expansion $V_2 > V_1$ or $P_1 > P_2$ hence ΔS_T is positive whereas in isothermal contraction $V_2 < V_1$ or $P_1 < P_2$, ΔS_T is negative

Case 2 : At constant pressure (Isobaric process)

In this case $P_1 = P_2$

The equation (*iv*) reduces to

$$\Delta S_P = 2.303\ n\ C_P \log \frac{T_2}{T_1}$$

Case 3 : At constant volume for an isobaric process
In this case $V_1 = V_2$
The equation (*iv*) reduces to

$$\Delta S_v = 2.303 \times n \times C_v \log \frac{T_2}{T_1}$$

SOLVED PROBLEM 8

Calculate the entropy change involved in thermodynamic expansion of 2 moles of a gas from a volume of 5×10^{-3} m³ to a volume of 50×10^{-3} m³ at 303 K.

SOLUTION. Here $n = 2$; $V_1 = 5 \times 10^{-3}$ m³; $V_2 = 50 \times 10^{-3}$ m³

using the relation $\Delta S_T = 2.303 \times n \times R \log \frac{V_2}{V_1}$

on substituting the values we get

$$\Delta S_T = 2.303 \times 2 \times 8.314 \times \log \frac{50 \times 10^{-3}}{5 \times 10^{-3}}$$

$$= \mathbf{38.29\ JK^{-1}}$$

SOLVED PROBLEM 9

Calculate the entropy change when 2 moles of an ideal gas are allowed to expand isothermally at 293 K from a pressure of 10 atmosphere (1013250.00 Pa) to a pressure of 2 atmosphere (202650.00 Pa).

SOLUTION. We know

$$\Delta S_T = 2.303 \times n \times R \times \log \frac{P_1}{P_2}$$

Here $n = 2$; $R = 8.314$ J
$P_1 = 1013250.00$ Pa (10 atm); $P_2 = 202650.00$ Pa (2 atm).

Substituting the values we get

$$\Delta S_T = 2.303 \times 2 \times 8.314 \times \log \frac{1013250.00}{202650.00}$$

$$= 2.303 \times 2 \times 8.314 \times 0.6990$$

$$= \mathbf{26.77\ JK^{-1}}$$

ENTROPY CHANGE ACCOMPANYING CHANGE OF PHASE

When there is a change of state from solid to liquid or liquid to vapors or solid to vapors (melting, evaporation and sublimation respectively), there is a change in entropy. This change may be carried out at constant temperature reversibly as two phases are in equilibrium during the change.

Let us consider the process of melting of 1 mole of the substance being carried out reversibly. It would absorb molar heat of fusion at temperature equal to its melting point. The entropy change is given by

$$\Delta S_f = \frac{\Delta H_f}{T_f}$$

where ΔH_f is the Molar heat of fusion at its melting point, T_f at constant pressure.

Similarly, when one mole of liquid is boiled reversibly it would absorb molar heat of vaporisation at a temperature equal to its boiling point. In this case entropy change is given by

$$\Delta S_v = \frac{\Delta H_v}{T_b}$$

where ΔH_v is Molar heat of vaporisation at its boiling point at constant pressure.

On similar lines we can calculate the change in entropy when one mole of a solid changes reversibly from one allotropic form to another at its transition temperature. We can write

$$\Delta S_t = \frac{\Delta H_t}{T_t}$$

Where ΔH_t is the Molar heat of transition at its transition temperature T_t.

SOLVED PROBLEM 10

Calculate the entropy change when 1 mole of ethanol is evaporated at 351 K. The molar heat of vaporisation of ethanol is 39.84 kJ mol^{-1}.

SOLUTION

We know

$$\Delta S_v = \frac{\Delta H_v}{T_b}$$

Here

$$\Delta H_v = 39.84 \text{ kJ mol}^{-1}$$
$$= 39840 \text{ J mol}^{-1}$$
$$T_b = 351 \text{ K}$$

$$\therefore \quad \Delta S_v = \frac{39840}{351}$$
$$= \mathbf{113.5 \ JK^{-1} \ mol^{-1}}$$

SOLVED PROBLEM 11

30.4 kJ is required to melt one mole of sodium chloride. The entropy change during melting is 28.4 J mol^{-1} K^{-1}. Calculate the melting point of sodium chloride.

SOLUTION

We know

$$\Delta S_f = \frac{\Delta H_f}{T_f}$$

Here $\quad \Delta S_f = 28.4 \text{ JK}^{-1} \text{ mol}^{-1}$

and $\quad \Delta H_f = 30.4 \text{ kJ K}^{-1} \text{ mol}^{-1}$

$$= 30400 \text{ JK}^{-1} \text{ mol}^{-1}$$

On substitution, we get

$$T_f = \frac{\Delta H_f}{\Delta S_f} = \frac{30400}{28.4}$$
$$= \mathbf{1070.4 \ K}$$

Free Energy Function (*G*) and Work Function (*A*)

The free energy function (*G*) is defined as

$$G = H - TS$$

where *H* is the heat content or enthalpy of the system, T is its temperature and *S* its entropy. It is a single valued function of thermodynamic state of the system and is an extensive property.

Let us consider a system which undergoes a change of state from (1) to (2) at constant temperature.

We have

$$G_2 - G_1 = (H_2 - H_1) - T(S_2 - S_1)$$

or $$\Delta G = \Delta H - T\,\Delta S$$

We know

$$\Delta H = \Delta E + P\Delta V$$

$\therefore$ $$\Delta G = \Delta E + P\Delta V - T\Delta S$$

Also $$\Delta A = \Delta E - T\Delta S$$

$\therefore$ $$\Delta G = \Delta A + P\Delta V \qquad \text{(At constant } P \text{ \& } T\text{)}$$

But $P\Delta V$ represents the work done due to expansion against a constant external pressure P. Therefore, it is clear that the decrease in free energy ($-\Delta G$) accompanying a process taking place at constant temperature and pressure is equal to the maximum work obtainable from the system other than the work of expansion. This quantity is referred to as the net work, so that

$$\text{Net work} = w - P\Delta V = -\Delta G$$

This quantity is of great importance in thermodynamics because the change in free energy is a measure of net work which may be electrical, chemical or surface work.

The work function (A) is defined as

$$A = E - TS$$

Where E is the energy content of the system, T its absolute temperature and S its entropy. Since E, T and S depend upon the thermodynamic state of the system only and not on its previous history, it is evident that the function A is also a single valued function of the state of the system.

Consider an isothermal change at temperature T from the initial state indicated by subscript (1) to the final state indicated by subscript (2) so that

$$A_1 = E_1 - TS_1 \qquad ...(i)$$

and $$A_2 = E_2 - TS_2 \qquad ...(ii)$$

Subtracting (*ii*) from (*i*), we have

$$A_2 - A_1 = (E_2 - E_1) - T(S_2 - S_1)$$

$$\Delta A = \Delta E - T\Delta S \qquad ...(iii)$$

Where ΔA is the change in work function A, ΔE is the corresponding change in internal energy and ΔS as change in the entropy of the system.

Since $\Delta S = \frac{q_{rev}}{T}$ where q_{rev} is the heat taken up when the change is carried out in a reversible manner and constant temperature, we have

$$\Delta A = \Delta E - q_{rev} \qquad ...(iv)$$

According to first law of thermodynamics

$$dE = q_{rev} - w_{rev}$$

or $$-w = \Delta E - q_{rev}$$

substituting the value in equation (*iv*), we get

$$-\Delta A = w_{rev}$$

i.e., **decrease in the work function A in any process at constant temperature gives the maximum work that can be obtained from the system during any change.**

Variation of Free Energy with Temperature and Pressure

By definition

$$G = H - TS$$

and $$H = E + PV$$

$\therefore$ $$G = E + PV - TS$$

Differentiating, we get

$$dG = dE + PdV + VdP - TdS - SdT \quad ...(i)$$

For an infinitesimal stage of a reversible process

$$dq = dE + dw$$

and $$dS = \frac{dq}{T}$$

If the work done is restricted to work of expansion, then

$$dq = dE + PdV$$

and $$dS = dE + PdV \quad ...(ii)$$

comparing equation (*i*) and (*ii*) we have

$$dG = dE + PdV + VdP - TdE - TPdV - SdT$$
$$= VdP - SdT$$

If the pressure remains constant

$$dG_p = -SdT_p$$

or $$\left(\frac{dG}{dT}\right)_p = -S \quad ...(iii)$$

But at constant temperature $dT = 0$ and we have

$$(dG)_T = VdP$$

or $$\left(\frac{dG}{dP}\right)_T = V \quad ...(iv)$$

Equations (*iii*) and (*iv*) give the variation of free energy with temperature and pressure respectively.

Isothermal change in Free Energy

By definition

$$G = H - TS$$

and $$H = E + PV$$

$\therefore$ $$G = E + PV - TS$$

on differentiating we get

$$dG = dE + PdV + VdP - TdS - SdT \quad ... (i)$$

and from first law of thermodynamics we have

$$dq = dE + PdV \quad ...(ii)$$

From equation (*i*) and (*ii*) we get

$$dG = dq + VdP - TdS - SdT \quad ...(iii)$$

For a reversible process

$$dS = \frac{dq}{T} \quad \text{or} \quad TdS = dq$$

Substituting this in equation (*iii*) we get

$$dG = TdS + VdP - TdS - SdT$$
$$= VdP - SdT \quad ...(iv)$$

Since the process is isothermal (there is no change in temperature) $dT = 0$, the equation (*iv*) reduces to

$$dG = VdP$$

Integrating with in the limits G_1 and G_2; P_1 and P_2, we get

$$\Delta G = G_2 - G_1 = \int_{P_1}^{P_2} VdP$$

For 1 mole of the gas $PV = RT$

or

$$V = \frac{RT}{P}$$

On substitution we get

$$dG = RT \int_{P_1}^{P_2} \frac{dP}{P}$$

$$= RT \ln \frac{P_2}{P_1} \quad ...(v)$$

For *n* moles of the gas we have

$$\Delta G = nRT \ln \frac{P_2}{P_1} \quad ...(vi)$$

We know that $P_1 V_1 = P_2 V_2$

or

$$\frac{P_2}{P_1} = \frac{V_1}{V_2}$$

From equation (*v*) we get

$$\Delta G = RT \ln \frac{V_1}{V_2}$$

and from equation (*vi*) we get

$$\Delta G = n\ RT \ln \frac{V_1}{V_2}$$

Changing to natural logarithms, we get

$$\Delta G = 2.303 \times nRT \log \frac{P_2}{P_1}$$

or

$$\Delta G = 2.303 \times nRT \log \frac{V_1}{V_2}$$

With the help of these equations we can calculate the change in free energy in isothermal process having an ideal gas.

SOLVED PROBLEM 12

Four moles of an ideal gas expand isothermally from $1 \times 10^{-3} m^3$ (1 liter) to $10 \times 10^{-3} m^3$ (10 liters) at 300 K. Calculate the change in free energy of the gas. ($R = 8.314$ JK^{-1} mol^{-1})

SOLUTION

For an isothermal process

$$\Delta G = 2.303 \times nRT \log \frac{V_1}{V_2}$$

Here $V_1 = 1 \times 10^{-3}$ m^3 ; $V_2 = 10 \times 10^{-3}$ m^3

$T = 300$ K; $R = 8.314$ JK^{-1} mol^{-1}

$n = 4$

Substituting the values in the formula we get

$$\Delta G = 2.303 \times 8.314 \times 300 \log \frac{1\times10^{-3}}{10\times10^{-3}}$$

$$= 22976.5 \times [-1.0]$$

$$= -22976.5 \text{ J}$$

$$= \mathbf{-22.9765\ kJ}$$

SOLVED PROBLEM 13

Two moles of an ideal gas are allowed to expand reversibly and isothermally at 300 K from a pressure of 1 atm (101325.00 Pa) to a pressure of 0.1 atm (10132.50.00 Pa). What is the change in Gibbs free energy?

SOLUTION. We know for an isothermal process

$$\Delta G = 2.303 \times nRT \log \frac{P_2}{P_1}$$

Here $P_1 = 101325.00$ Pa; $P_2 = 10132.50.00$ Pa.

$n = 2$ moles $R = 8.314$ JK^{-1} mol^{-1}

$T = 300$ K

On substitution we get

$$\Delta G = 2.303 \times 2 \times 8.314 \times 300 \times \log \frac{10132.50}{101325.00}$$

$$= 11488.3 \times [-1.0]$$

$$= -11488.3 \text{ J}$$

$$= \mathbf{-11.4883\ kJ}$$

GIBB'S HELMHOLTZ EQUATIONS

These are two equations derived by J.W. Gibbs and H.Von Helmholtz and are known as Gibbs Helmholtz equations. One equation can be expressed in terms of changes in free energy (ΔG) and enthalpy (ΔH), while the other can be expressed in terms of changes in internal energy (ΔE) and work function (ΔA). The former is generally employed and is applicable to all processes, chemical or physical, but in a closed system.

(a) In Terms of Free Energy and Enthalpy

We know that

$$dG = VdP - SdT$$

At constant pressure $dP = 0$, then

$$dG = -SdT$$

Let G_1 be the initial free energy at temperature T, $G_1 + dG_1$, be the free energy at temperature $T + dT$. Then

$$dG_1 = -S_1 dT \quad ...(i)$$

Where S_1 is the entropy of the system in the initial state. Now suppose the free energy of the system in the final state is G_2 at temperature T, and $G_2 + dG_2$ is the free energy at temperature $T + dT$ in the final state.

Then

$$dG_2 = -S_2\, dT \quad ...(ii)$$

where S_2 is the entropy of the system in the final state.

Subtracting (*ii*) from (*i*) we get

$$dG_2 - dG_1 = -(S_2 - S_1)\, dT$$

or

$$d(\Delta G) = -\Delta S\, dT$$

At constant pressure

$$d\left(\frac{\Delta G}{dT}\right)_p = -\Delta S \quad ...(iii)$$

We know

$$\Delta G = \Delta H - T\Delta S$$

or

$$-\Delta S = \frac{\Delta G - \Delta H}{T} \quad ...(iv)$$

Comparing equations (*iii*) and (*iv*)

$$\frac{\Delta G - \Delta H}{T} = d\left(\frac{\Delta G}{dT}\right)_p$$

or

$$\Delta G = \Delta H + Td\left(\frac{\Delta G}{dT}\right)_p$$

This equation is called Gibb's Helmholtz equation in terms of free energy and enthalpy change at constant pressure.

(*b*) In terms of Internal Energy and Work Function

By definition the work function

$$A = E - TS \quad ...(i)$$

or

$$\Delta A = \Delta E - T\Delta S$$

or

$$-\Delta S = \frac{\Delta A - \Delta E}{T} \quad ...(ii)$$

Differentiating equation (*i*) we get

$$dA = dE - TdS - SdT \quad ...(iii)$$

From the first law of thermodynamics

$$dq = dE + PdV$$

and at constant volume $dV = 0$

$$dq = dE$$

For a reversible change

$$dS = \frac{dq}{T}$$

or $$dq = TdS - dE \qquad ...(iv)$$

Comparing equations (*iii*) and (*iv*) we get

$$dA = -SdT$$

Let A_1 be the work function in its initial state at temperature T and $A_1 + dA_1$ be the work function in its initial state at $T + dT$. And A_2 be the work function in its final state at temperature T and $A_2 + dA_2$ be the work function in its final state at $T + dT$. Then

$$dA_1 = -S_1\,dT \qquad ...(v)$$

and $$dA_2 = -S_2\,dT \qquad ...(vi)$$

where S_1 and S_2 are the entropies of the system in initial and final states of the system respectively.

Subtracting equation (*v*) from equation (*vi*) we get

$$dA_2 - dA_1 = -(S_2 - S_1)\,dT$$

or $$d(\Delta A) = -\Delta S\,dT$$

At constant volume

$$d\left(\frac{\Delta A}{dT}\right)_v = -\Delta S$$

From equation (*ii*) we have

$$\frac{\Delta A - \Delta E}{T} = -\Delta S$$

On comparison we have

$$\frac{\Delta A - \Delta E}{T} = d\left(\frac{\Delta A}{dT}\right)_v$$

or $$\Delta A = \Delta E + T\,d\left(\frac{\Delta A}{dT}\right)_v$$

This equation is called Gibbs Helmholtz equation in terms of internal energy and work function at constant volume.

Importance of Gibb's Helmholtz Equations

Gibb's Helmholtz equations are used to calculate the heats of reaction (ΔH or ΔE) when ΔG or ΔA at two temperatures are given. This point is made clear in the following examples.

SOLVED PROBLEM 14

For the reaction

$$H_2(g) + \frac{1}{2}O_2(g) \longrightarrow H_2O(l)$$

The values of enthalpy change and free energy change are – 286.04 and – 237.35 kJ respectively. Calculate the value of free energy change at 298 K.

SOLUTION. Using the Gibb's Helmholtz equation

$$\Delta G = \Delta H + Td\left(\frac{\Delta G}{dT}\right)_p$$

Here $$\Delta G = -237.35 \text{ kJ}$$

$$\Delta H = -286.04 \text{ kJ}$$

and $\qquad T = 298 \text{ K}$

On substitution we get

$$-237.35 = -286.04 + 298\left[d\left(\frac{\Delta G}{dT}\right)_p\right]$$

or
$$d\left(\frac{\Delta G}{dT}\right)_p = \frac{-237.35 + 286.04}{298}$$

$$= 0.1632 \text{ kJ}$$

Assuming that $d\left(\frac{\Delta G}{dT}\right)_p$ remains constant over this temperature range. At 303 K we can write

$$\Delta G = -286.04 + 303 \times 0.1632$$
$$= -286.04 + 49.45$$
$$= \mathbf{-236.59\ kJ}$$

SOLVED PROBLEM 15

For the following reaction

$$N_2(g) + 3H_2(g) = 2NH_3(g)$$

The free energy changes at 298 K and 308 K are – 33.089 and – 28.018 kJ respectively. Calculate the heat of reaction.

SOLUTION. We know

$$\Delta G = \Delta H + Td\left(\frac{\Delta G}{dT}\right)_p$$

Here $\qquad G_1 = -33.089 \text{ kJ} \quad T_1 = 298 \text{ K}$

$\qquad G_2 = -28.018 \text{ kJ} \quad T_2 = 308 \text{ K}$

$\therefore$
$$d\left(\frac{\Delta G}{dT}\right)_p = \frac{-28.018 - (-33.089)}{308 - 298}$$

$$= 0.507 \text{ kJ}$$

At 308 K $\qquad \Delta G = -28.018 \text{ kJ}$

$\qquad T = 308 \text{ K}$

$\therefore$
$$\Delta G = \Delta H + Td\left(\frac{\Delta G}{dT}\right)_p$$

$$-28.018 = \Delta H + 308 \times 0.507$$

or
$$\Delta H = 28.018 + 156.156$$
$$= \mathbf{184.174\ kJ}$$

Conditions of Equilibrium and Criterion for a Spontaneous Process

(*a*) In Terms of Entropy Change

The entropy of a system remains unchanged in a reversible change while it increases in an irreversible change *i.e.*

$\sum dS = 0$ For Reversible change

and $\sum dS > 0$ For Irreversible change

For a system with its surroundings we can write

$$dS_{\text{system}} + dS_{\text{surroundings}} = 0$$

and $$dS_{\text{system}} + dS_{\text{surroundings}} > 0$$

combining the two relations, we have

$$dS_{\text{system}} + dS_{\text{surroundings}} \geq 0 \quad ...(i)$$

If we assume the change in surroundings as reversible and surroundings evolve dq heat reversibly, then

$$dS_{\text{surrounding}} = -\frac{dq}{T}$$

From the first law of Thermodynamics

we know

$$dS = -\frac{dq}{T} = \frac{dE + dw}{T}$$

where dE is the increase in internal energy of the system and dw be the work done by the system.

From equation (i) we get

$$dS_{\text{system}} - \frac{dE + dw}{T} \geq 0$$

or $$TdS_{\text{system}} - dE - dw \geq 0$$

or $$TdS_{\text{system}} - dE - PdV \geq 0 \quad ...(ii)$$

$[\because dw = PdV]$

Case I. When E and V are constant

In this case $dE = 0$ and $dV = 0$ the equation (i) reduces to

$$dS_{E.V} \geq 0$$

or $dS_{E.V} > 0$ for an irreversible change (spontaneous)

and $dS_{E.V} = 0$ for a reversible change (equilibrium)

Case II. When S and V are constant

Here $dS = 0$ and $dV = 0$

The equation (i) reduces to $-dE \geq 0$

or $-dE > 0$ for an irreversible change (spontaneous)

and $-dE = 0$ for a reversible change (equilibrium)

(b) In Terms of Enthalpy Change

We know

$$H = E + PV$$

on differentiating

$$dH = dE + PdV + VdP$$

or $$-dE - PdV = -dH + VdP$$

Putting in equation (ii)

$$TdS - dH + VdP \geq 0$$

or $$dH - VdP - TdS \leq 0$$

At constant S and P

Here $dS = 0$ and $dP = 0$

$\therefore$ we have $dH \le 0$

or $\qquad dH < 0$ for an irreversible change (spontaneous)

and $\qquad dH = 0$ for a reversible change (equilibrium)

(c) In Terms of Free Energy Change

We know

$$G = H + TS$$

or $$G = E + PV + TS \qquad [\because H = E + PV]$$

on differentiating we get

$$dG = dE + PdV + VdP + TdS + SdT$$

$$-TdS - dE - PdV = -dG + VdP + SdT$$

Substituting in equations (*ii*) we get

$$-dG + VdP - SdT \ge 0$$

or $$dG - VdP + SdT \le 0$$

At constant pressure in an isothermal process (*T* is also constant) this equation reduces to

$$dG \le 0$$

or $\qquad dG < 0$ for an irreversible change (spontaneous)

and $\qquad dG = 0$ for a reversible change (equilibrium)

Thus the conditions for spontaneity and equilibrium may be summed up in the Table 9.2.

TABLE 9.2. CONDITIONS FOR SPONTANEITY AND EQUILIBRIUM

Conditions	Irreversible Process (Spontaneous)	Reversible Process (Equilibrium)
At Constant *E*, *V*	$dS > 0$	$dS = 0$
At Constant *S*, *V*	$-dE > 0$	$-dE = 0$
At Constant *S*, *P*	$dH < 0$	$dH = 0$
At Constant *P*, *T*	$dG < 0$	$dG = 0$

THE CLAPEYRON EQUATION

A useful thermodynamic relation which gives us important information about a system consisting of any two phases of a single substance in chemical equilibrium is the Clapeyron equation. It is derived from the Gibbs-Helmholtz equation mentioned above.

Let the system studied be

Liquid $\quad$ *Vapour*

Consider one gram mole of a liquid confined in a cylinder by a frictionless piston. Let the volume of the liquid be V_1 and its vapor pressure equal to p. Now allow the liquid to evaporate reversibly at a constant temperature T and when the whole of it has vaporised, let the volume of the vapor be V_2.

$\therefore$ Work done during evaporation $w = p\,(V_2 - V_1)$ ...(*i*)

Differentiating equation (*i*) with respect to temperature at constant $(V_2 - V_1)$, we get,

$$\frac{dw}{dT} = (V_2 - V_1)\frac{dp}{dT} \qquad ...(ii)$$

Heat absorbed from the surroundings is the latent heat of vaporisation L which on substitution in the first law equation gives us

$$\Delta E = w - L \qquad ...(iii)$$

On substitution of expressions (*ii*) and (*iii*) in the Gibbs-Helmholtz equation, we have

$$w + \Delta E = T \frac{dw}{dT}$$

$$w + (L - w) = T(V_2 - V_1) \frac{dp}{dT}$$

or

$$L = T \frac{dp}{dT}(V_2 - V_1)$$

or

$$\frac{dp}{dT} = \frac{L}{T(V_2 - V_1)}$$

This is the Clapeyron equation which in its general form may be written as

$$\frac{dp}{dT} = \frac{\Delta H}{T(V_2 - V_1)} \qquad ...(1)$$

where ΔH is the heat of transition when a volume V_1 of a definite weight of one form changes to a volume V_2 of the same weight of other form at the temperature T.

CLAUSIUS-CLAPEYRON EQUATION

The above equation can be simplified by neglecting the small volume of the liquid in comparison with the volume of the vapor. Equation (1) given above in such a case becomes

$$\frac{dp}{dT} = \frac{\Delta H}{TV}$$

Supposing the vapor obeys the ideal gas laws

$$V_2 = \frac{RT}{p}$$

$$\frac{dp}{dT} = \frac{\Delta H\, p}{RT^2}$$

or

$$\frac{1}{p} \times \frac{dp}{dT} = \frac{\Delta H}{RT^2}$$

But

$$\frac{1}{p} \times \frac{dp}{dT} = \frac{d \log_e p}{dT}$$

$\therefore$

$$d \log_e \frac{p}{dT} = \frac{\Delta H}{RT^2} \qquad ...(2)$$

Equation (2) above is known as the **Clausius-Clapeyron equation** and though approximate is of very great value.

If ΔH is regarded as constant, we may integrate the above equation

$$d \log_e P = \Delta H \frac{dT}{RT^2}$$

$$\int d \log_e p = \frac{\Delta H}{R} \int \frac{dT}{T^2}$$

or
$$\log_e p = \frac{-\Delta H}{RT} + \text{Constant}$$

$$\log_{10} p = \frac{-\Delta H}{2.303\ RT} + C$$

If p_1 is the vapor pressure at T_1 and p_2 the vapor pressure at T_2 we have

$$\log_{10} p_1 = \frac{-\Delta H}{2.303\ RT_1} + C \qquad ...(i)$$

and
$$\log_{10} p_2 = \frac{-\Delta H}{2.303\ RT_2} + C \qquad ...(ii)$$

Subtracting (*i*) from (*ii*)

$$\log_{10} \frac{p_2}{p_1} = \frac{\Delta H}{2.303\ R}\left(\frac{1}{T_1} - \frac{1}{T_2}\right) \qquad ...(3)$$

APPLICATIONS OF CLAPEYRON-CLAUSIUS EQUATION

(1) Calculation of Latent Heat of Vaporisation

If the vapor pressure of a liquid at two temperatures T_1 and T_2 be p_1 and p_2 respectively, the molar heat of vaporisation ΔH_v can be calculated by substituting these values in Clapeyron-Clausius equation.

SOLVED PROBLEM 16

If the vapor pressures of water at 368 K and 373 K are 634 mm (84526.38 Pa) and 760 mm (101325 Pa) respectively, calculate the latent heat of vaporisation per mole.

SOLUTION

The Clapeyron Clausius equation states that

$$\log_{10} \frac{p_2}{p_1} = \frac{\Delta H_v}{2.303\ R}\left[\frac{1}{T_1} - \frac{1}{T_2}\right]$$

In this case, we have :

$$p_1 = 84526.38 \text{ Pa}$$
$$p_2 = 101325 \text{ Pa}$$
$$T_1 = 368 \text{ K}$$
$$T_2 = 373 \text{ K}$$
$$R = 8.319 \text{ J}$$
$$\Delta H_v = ?$$

Substituting the above values in the Clapeyron-Clausius equation, we have :

$$\log \frac{101325.00}{84526.38} = \frac{\Delta H_v}{2.303\ R}\left[\frac{1}{368} - \frac{1}{373}\right]$$

$$\therefore \qquad \Delta H_v = \mathbf{41390.70\ J}$$

In a similar manner, if the vapor pressures at two different temperatures of a solid in equilibrium with its liquid phase are known, the **latent heat of fusion** can be calculated.

(2) Calculation of Boiling Point or Freezing Point

If the freezing point or the boiling point of a liquid at one pressure is known, it is possible to calculate it at another pressure by the use of the Clapeyron-Clausius equation.

SOLVED PROBLEM 17

At what temperature will water boil under a pressure of 787 mm (104924.71 Pa)? The latent heat of vaporisation is 2244.13×10^3 per kg.

SOLUTION

The data is as follows

$$p_1 = 760 \text{ mm (101325.00 Pa) } T_1 = 373 \text{ K}$$
$$p_2 = 787 \text{ mm (104924.71 Pa) } T_2 = ?$$
$$\Delta H_v = 2244.13 \times 10^3 \text{ J/kg}$$

According to Clapeyron-Clausius equation, we have :

$$\log_{10} \frac{p_2}{p_1} = \frac{\Delta H_v}{4.576}\left[\frac{1}{T_1} - \frac{1}{T_2}\right]$$

Substituting the above values, we have :

$$\log_{10} \frac{104924.71}{101325.00} = \frac{2244.13 \times 10^3 \times 18 \times 10^{-3}}{4.576}\left[\frac{T_2 - 373}{373\ T_2}\right]$$

$$T_2 = \mathbf{374\ K}$$

$\therefore$ Water will boil at 374 K under a pressure of 787 mm (104924.71 Pa).

(3) Calculation of Vapor Pressure at Another Temperature

If the mean heat of vaporisation is available, it is possible to calculate the vapor pressure of a liquid at given temperature if the vapor pressure at another temperature is known.

SOLVED PROBLEM 18

Calculate the vapor pressure of water at 363 K if its value at 373 K is 76.0cm (1049247.02 Pa). The mean heat of vaporisation of water in the temperature range 363 K and 373 K is 2269.25×10^3 J/kg.

SOLUTION

In this problem, we have :

$$\Delta H_v = 2269250 \times 18 \text{ J/mol}$$
$$p_2 = ?$$
$$p_1 = 1049247.02 \text{ Pa}$$
$$T_2 = 363 \text{ K}$$
$$T_1 = 373 \text{ K}$$

Using the Clapeyron-Clausius equation we have :

$$\log_{10} \frac{p_2}{p_1} = \frac{\Delta H_v}{4.576}\left[\frac{T_2 - T_1}{T_1 T_2}\right]$$

or

$$\log_{10} \frac{p_2}{76} = \frac{2269250 \times 18}{4.576}\left[\frac{363 - 373}{363 \times 373}\right]$$

$\therefore$

$$p_2 = \text{Antilog } 659.2549 \times 76 = \mathbf{70500.87\ Pa}$$

(4) Calculation of Molal Elevation Constant

The molal elevation constant (K_b) of a solvent is defined **as the elevation in boiling point which may theoretically be produced when one mole of any solute is dissolved in 1 kg of the solvent.** Accordingly, if w g of a solute of molecular weight M is dissolved in 1 kg of the solvent and ΔT is the elevation produced, the molal elevation K_b is given by the equation

$$K_b = \frac{M\,\Delta T}{w}$$

Let the boiling point of the pure solvent be T and that of the solvent $(T + \Delta T)$ when the atmospheric pressure is p_1. While p_1 is the vapor pressure of the solution at $(T + \Delta T)$ and is also the vapor pressure of the solvent at T, p_2 the vapor pressure of the solvent at $(T + \Delta T)$ can be calculated by the application of Clapeyron-Clausius equation.

$$\log_e \frac{p_2}{p_1} = \frac{L}{R}\left[\frac{1}{T} - \frac{1}{(T + \Delta T)}\right]$$

$$= \frac{L}{R}\left[\frac{\Delta T}{T\,(T + \Delta T)}\right]$$

$$= \frac{L}{R} \times \frac{\Delta T}{T^2} \text{ when } \Delta T \text{ is very small.}$$

Now $\log_e \frac{p_2}{p_1} = \log_e \left(1 + \frac{p_2 - p_1}{p_1}\right) = \frac{p_2 - p_1}{p_1}$ since $(p_2 - p_1)$ is very small and the remainder of the terms can be neglected.

According to Raoult's law, the relative lowering of vapor pressure in a dilute solute is equal to the molar fraction of the solute in solution

$$\therefore \quad \frac{p_2 - p_1}{p_2} = \frac{n}{N}$$

When the difference between p_2 and p_1 is very small as has just been supposed, $\frac{p_2 - p_1}{p_2}$ may be taken equal to $\frac{p_2 - p_1}{p_1}$.

$$\therefore \quad \frac{p_2 - p_1}{p_2} = \frac{n}{N} = \frac{L\,\Delta T}{RT^2}$$

$$\text{or} \quad \Delta T = \frac{RT^2}{L} \times \frac{n}{N}$$

$$\text{But} \quad n = \frac{w}{m} \text{ and } N = \frac{W}{M}$$

$$\therefore \quad \Delta T = \frac{RT^2}{L} \times \frac{wM}{mW}$$

If 1 mole of the solute is dissolved in 1000 g of the solvent, the above equation becomes :

$$\Delta T = \frac{RT^2}{L \times 1000}$$

$$\text{But} \quad K_b = \frac{M\,\Delta T}{w}$$

$$\therefore \quad K_b = \frac{M\,RT^2}{L \times 1000}$$

But $\frac{\text{L}}{\text{M}} = l$, the latent heat of vaporisation per gram of the solvent.

$$\therefore \quad K_b = \frac{RT^2}{1000l}$$

Putting $R = 0.002 \text{ kg} \times 4.18 \text{ J} = 0.00836 \text{ J}$

$$K_b = \frac{0.00836\ T^2}{l}$$

SOLVED PROBLEM 19

The heat of vaporisation of 0.001 kg of carbon disulfide is 363.08 J and the boiling point is 319 K. Calculate the molal elevation constant.

SOLUTION

The molal elevation constant of a solvent is given by the expression

$$K_b = \frac{0.00836\ T^2}{l}$$

Here $T = 319$ K

$l = 363.08$ J

$$\therefore \quad K_b = \frac{0.00836 \times 319 \times 319}{363.08}$$

$= \mathbf{2.34^\circ}$ **per 1 kg of carbon disulfide**

(5) Calculation of Molal Depression Constant

By an exactly similar reasoning, we can calculate the molal depression constant of a solvent as

$$K_f = \frac{RT^2}{1000l_f}$$

where l_f is the latent heat of fusion per gram of the solvent,

or
$$K_f = \frac{0.00836\ T^2}{l_f}$$

SOLVED PROBLEM 20

The latent heat of fusion of ice per kilogram is 334.94 J and the freezing point of water is 273 K. Calculate the molal depression constant of water.

SOLUTION

The molal depression constant of a solvent is given by the expression

$$K_f = \frac{0.00836\ T^2}{l_f}$$

In this problem :

$T = 273$ K

$l_f = 334.94$ J

$$\therefore \quad K_f = \frac{0.00836 \times 273 \times 273}{334.94 \text{ Joules}}$$

$= \mathbf{1.86^\circ}$

FREE ENERGY AND WORK FUNCTIONS

Besides heat content (*H*), internal energy (*E*) and entropy (*S*), there are two other thermodynamic functions depending upon the state of the system which utilize *E*, *H* or *S* in their derivation and are more convenient for use. These are **Work** and **Free energy functions** represented by *A* and *G* respectively.

The Work Function

The work function (*A*) is defined by

$$A = E - TS$$

where *E* is the energy content of the system, *T* is its absolute temperature and *S* its entropy. Since *E*, *T* and *S* depend upon the thermodynamic state of the system only and not on its previous history, it is evident that the function *A* is also a single valued function of the state of the system.

Consider an isothermal change at temperature *T* from the initial state indicated by subscript 1 to the final state indicated by subscript 2, so that

$$A_1 = E_1 - TS_1 \quad ...(1)$$

and

$$A_2 = E_2 - TS_2 \quad ...(2)$$

Subtracting (1) from (2), we have :

$$A_2 - A_1 = (E_2 - E_1) - T(S_2 - S_1)$$

or

$$\Delta A = \Delta E - T\Delta S \quad ...(3)$$

where ΔA is the increase in function *A*, ΔE is the corresponding increase in internal energy and ΔS is the increase in the entropy of the system.

Since $\Delta S = q_{rev}/T$ where q_{rev} is the heat taken up when the change is carried out in a reversible manner at a constant temperature, we have :

$$\Delta A = \Delta E - q_{rev} \quad ...(4)$$

According to the first law of thermodynamics, $\Delta E = (q_{rev} - w_{rev})$

or

$$-w_{rev} = \Delta E - q_{rev} \quad ...(5)$$

Substituting this value in equation (4), we get

$$-\Delta A = w_{rev}$$

i.e., **decrease in the work function *A* in any process at constant temperature gives the maximum work that can be obtained from the system during any change.**

VAN'T HOFF ISOTHERM

The van't Hoff isotherm gives the net work that can be obtained from a gaseous reactant at constant temperature when both the reactants and the products are at suitable arbitrary pressures. It can be derived by using the "equilibrium box" which is a theoretical device with the supposition that among its four walls, one is permeable to *A*, the second to *B*, the third to *C* and the fourth to *D* when the gaseous reaction to be considered is

$$A + B \rightleftharpoons C + D$$

Let the initial pressures of *A* and *B* be p_a and p_b and the final pressure of *C* and *D* be p_c and p_d respectively and let the equilibrium pressure of the four be P_A, P_B, P_C and P_D respectively.

The following theoretical operations may be performed :

(*i*) Change the pressure on *A* from the initial pressure p_a to the equilibrium pressure P_A.

$\therefore$ Work done by the gas $= RT \ln \dfrac{p_a}{P_A}$

(*ii*) Change the pressure on B from p_b to P_B.

$\therefore$ Work done by the gas $= RT \ln \dfrac{p_b}{P_B}$

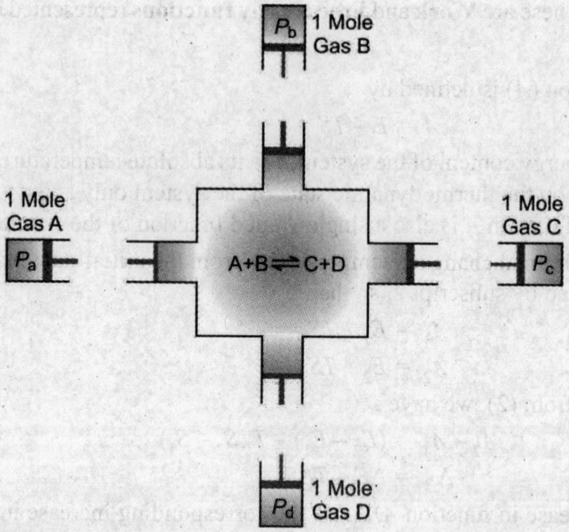

■ **Figure 9.9**
van't Hoff equilibrium box.

(*iii*) Introduce 1 gm mole of *A* and 1 gm mole of the gas *B* through their respective semipermeable membranes into the equilibrium box which contains the reactants and the products at the equilibrium pressure. This will not involve any work as the partial pressures of *A* and *B* inside the box are equal to the pressure of the gases coming in. *A* and *B* react to form 1 gm mole of *C* and 1 gm mole of *D*.

(*iv*) Withdraw 1 gm mole of *C* and 1 gm mole of *D* from the equilibrium box through their respective semipermeable walls. No work is done in this process as the gases come out at the equilibrium pressure P_C and P_D.

(*v*) Now alter the pressure on the gas from the equilibrium pressure P_C and P_D to the final pressure p_c and p_d.

Work done by the gas $C = RT \ln \dfrac{P_C}{p_c}$

Work done by the gas $D = RT \ln \dfrac{P_D}{p_d}$

As the change in free energy is equal to the total work done by the gases :

$$-\Delta G = RT \ln \frac{P_C \times P_D}{P_A \times P_B} - RT \ln \frac{p_c \times p_d}{p_a \times p_b}$$

$$= RT \ln K_p - RT \ln \frac{p_c \times p_d}{p_a \times p_b}$$

If the reaction is started with reactants at a partial pressure of 1 atmosphere and the resulting products are also at 1 atmosphere pressure we have,

$$-\Delta G = RT \ln K_p - RT \ln 1$$

or

$$-\Delta G = RT \ln K_p$$

i.e., the net work of the reaction is equal to the decrease in free energy of the system and is given by the expression, **RT 1n K_p** or **2.303 RT log K_p.**

It will be observed that ΔG is + ve when K_p is less than unity. It is – ve when K_p is greater than one and is zero when $K_p = 1$.

VAN'T HOFF ISOCHORE

The van't Hoff Isochore is obtained by combining the van't Hoff Isotherm with the Gibbs Helmholtz equation.

$$\Delta G = \Delta H + T\left[\frac{d(\Delta G)}{dT}\right]_p$$

or

$$-\Delta H = T\left[\frac{d(\Delta G)}{dT}\right]_p - \Delta G$$

Dividing both sides by T^2 gives

$$-\frac{\Delta H}{T^2} = \frac{T\left[d\left(\dfrac{\Delta G}{dT}\right)\right]_p - \Delta G}{T^2}$$

The right-hand side of the above expression is obtained by differentiating $\Delta G/T$ w.r.t. T at constant pressure

i.e.,

$$\left[\frac{d(\Delta G/T)}{dT}\right]_p = -\frac{T\left[\dfrac{d(\Delta G)}{dT}\right]_p - \Delta G}{T^2}$$

$$\therefore \quad -\frac{\Delta H}{T^2} = \left[\frac{d(\Delta G/T)}{dT}\right]_p \qquad ...(i)$$

According to van't Hoff isotherm,

$$-\Delta G = RT \ln K_p \qquad ...(ii)$$

combining this with equation (*i*), we have

$$\frac{\Delta H}{T^2} = \frac{Rd(\ln K_p)}{dT}$$

$$\frac{\Delta H}{RT^2} = \frac{d(\ln K_p)}{dT}$$

The above equation is known as van't Hoff Isochore.

For applying the Isochore to any particular reaction, it is essential to integrate it. If ΔH remains constant over a range of temperature, we have on integration

$$\ln K_p = \int \frac{\Delta H}{RT^2}\, dT$$

$$= \frac{-\Delta H}{RT} + Constant$$

Applying the limits T_1 and T_2 at which the equilibrium constants are K_{p_1} and K_{p_2} respectively, we have

$$\ln K_{p_2} - \ln K_{p_1} = -\frac{\Delta H}{R}\left[\frac{1}{T_2} - \frac{1}{T_1}\right]$$

$$\ln K_{p_2} - \ln K_{p_1} = \frac{\Delta H}{R}\left[\frac{1}{T_1} - \frac{1}{T_2}\right]$$

$$\ln \frac{K_{p_2}}{K_{p_1}} = \frac{\Delta H}{R}\left[\frac{T_2 - T_1}{T_1 T_2}\right]$$

or
$$\log \frac{K_{p_2}}{K_{p_1}} = \frac{\Delta H}{2.303 \times R}\left[\frac{T_2 - T_1}{T_1 T_2}\right]$$

Knowing the equilibrium constant at two different temperatures it is possible, therefore, to calculate the change in heat content.

SOLVED PROBLEM 21

The equilibrium constant K_p for a reaction

$A + B \rightleftharpoons C + D$ is 10^{-12} at 600 K and 10^{-7} at 700 K. Calculate the enthalpy of the reaction.

$(R = 8.314 \text{ JK}^{-1} \text{ mol}^{-1})$

SOLUTION

Applying van't Hoff's equation

$$\log \frac{K_{p_2}}{K_{p_1}} = \frac{\Delta H}{2.303 R}\left[\frac{T_2 - T_1}{T_1 T_2}\right]$$

Here
$$K_{p_2} = 10^{-7}; \qquad T_2 = 700 \text{ K}$$
$$K_{p_1} = 10^{-12}; \qquad T_1 = 600 \text{ K}$$
$$R = 8.314 \text{ JK}^{-1} \text{ mol}^{-1}$$

On substitution we get

$$\log \frac{10^{-7}}{10^{-12}} = \frac{\Delta H}{2.303 \times 8.314}\left[\frac{700 - 600}{700 \times 600}\right]$$

$$\Delta H = \log 10^5 \times \frac{2.303 \times 8.314 \times 700 \times 600}{100}$$

$$= 402089.98 \text{ J}$$
$$= \mathbf{402.08998 \text{ kJ}}$$

FUGACITY AND ACTIVITY

It has already been pointed out that equation $\Delta G = RT \log_e P_2/P_1$ is applicable only to ideal gases. When applied to real gases, particularly at higher pressures, it is found that this expression does not reproduce the change in free energy, the reason being that under these conditions V is not equal to nRT/P. In order to apply this equation to non-ideal systems, Lewis introduced two new thermodynamic quantities, **fugacity** and **activity.**

Consider a system composed of liquid water and its vapor. Liquid water has a tendency to escape into the vapor phase while the vapor tends to escape the gaseous state and come into the liquid phase by condensation. When the system is in equilibrium, these two escaping tendencies become equal and we observe a constant vapor pressure at a constant temperature. In general, **it may be stated that each substance in a given state has a tendency to escape from that state and this escaping tendency denoted by f is called fugacity.** It is related to the free energy content (G) by the expression

$$G = RT \ln f + B$$

where B is a constant depending upon the temperature and the nature of the substance. It is not possible to evaluate B since the absolute values of the free energy are not known. To circumvent this difficulty, all free energy measurements for any given substance are referred to as standard reference point. If we represent by $G°$ the free energy per mole and $f°$ the fugacity in this standard state, then $G°$ is given by

$$G^{\circ} = RT \ln f^{\circ} + B$$

If G is the free energy of the substance in any state, then the free energy difference between this state and the standard state is given by

$$G - G^{\circ} = RT \ln f/f^{\circ}$$

or

$$G = G^{\circ} + RT \ln f/f^{\circ} \qquad ...(i)$$

The ratio f/f° is called **activity** and is denoted by the symbol a. The activity of any substance may therefore, be defined as the *ratio of fugacity of the substance in the given state to the fugacity of the same substance in the standard state.*

$$G = G^{\circ} + RT \ln a$$

In the standard state, $G = G^{\circ}$ $\therefore$ $RT \ln a = 0$ or $a = 1$

i.e., in the standard state the activity of a substance is equal to unity. In any other state the value of activity will depend upon the difference $(G - G^{\circ})$. The difference in free energy per mole caused on passing from one state in which the free energy is G_1 and the activity a_1 to another state in which these are G_2 and a_2 respectively, is given by the expression

$$\Delta G = G_2 - G_1 = (G^{\circ} + RT \ln a_2) - (G^{\circ} + RT \ln a_1)$$

or

$$\Delta G = RT \ln \frac{a_2}{a_1}$$

The similarity between the above equation and the equation

$$\Delta G = RT \ln P_2/P_1$$

suggests that **activity is the thermodynamic counterpart of the gas pressure.**

For the standard state of any gas at the given temperature, the fugacity is taken as equal to unity, *viz*, $f^{\circ} = 1$ and on the basis of this definition the activity of any gas becomes equal to fugacity

$$a = \frac{f}{f^{\circ}} = \frac{f}{1} = f$$

The equation (*i*) can, therefore, be written as

$$G = G^{\circ} + RT \log_e f$$

For an *ideal gas* the fugacity is equal to pressure and $f/\mathrm{P} = 1$. For a *real gas*, the fugacity is not equal to P and the ratio f/P varies. It is observed, however, that on decreasing the pressure, the behavior of the gas approaches that of an ideal gas. It may be stated, therefore, that f approaches P as P approaches zero

i.e.,

$$\lim_{p \to 0} \frac{f}{P} = 1$$

or

$$\frac{f}{P} \to 1 \text{ as } P \to 0$$

The ratio f/P is called **activity coefficient** of a gas and is represented by the symbol γ. It gives a direct measure of the extent to which any gas deviates from ideal behavior at any given pressure and temperature for the farther this ratio is from unity, the greater is the non-ideality of the gas.

CHEMICAL POTENTIAL

Partial Molar Properties

We have so far studied the thermodynamic systems in which there was a change in thermodynamic properties with the variation of one or more state variables. In such systems there was no transfer of mass taking place (closed systems). For studying the systems containing two or more phases or components G.N. Lewis introduced the concept of **partial molar properties** as in these cases both mass and composition vary (open systems). Consider any extensive thermodynamic property X of such a system, the value of which is determined by the temperature, pressure and the amounts of various constituents present. Let the system consist of J constituents and let n_1, n_2, n_3 .... n_j be the number of moles of the various constituents present. Evidently X must be a function of P, T and the number of moles of various constituents present, *i.e.*

$$X = f(T, P, n_1, n_2, n_3 \ldots n_j) \qquad \ldots(1)$$

If there is a small change in the temperature and pressure of the system as well as the amounts of its constituents, the change in the property X is given by

$$dX = \left(\frac{\delta X}{\delta T}\right)_{P \; n_1, n_2 \ldots n_j} dT + \left(\frac{\delta X}{\delta p}\right)_{T \; n_1, n_2 \ldots n_j} dp + \left(\frac{\delta X}{\delta n_1}\right)_{T,P \; n_2 \ldots n_j} dn_1$$

$$+ \left(\frac{\delta X}{\delta n_2}\right)_{T,P \; n_1, n_3 \ldots n_j} dn_2 + \left(\frac{\delta X}{\delta n_3}\right)_{T,P \; n_1, n_2 \ldots n_j} dn_3 + \ldots + $$

$$\left(\frac{\delta X}{\delta n_j}\right)_{T,P \; n_1, n_2 \ldots} dn_j \qquad \ldots(ii)$$

The quantity $\left(\frac{\delta X}{\delta n_1}\right)_{T,P,n_2 \ldots n_j}$ is called partial molar property for the constituent 1. It is represented by writing a bar over its symbol for the particular property i.e. $\bar{X}$ so that

$$\bar{X}_1 = \left(\frac{\delta X}{\delta n_1}\right)_{T,P,n_2 \ldots n_j} ; \; \bar{X}_2 = \left(\frac{\delta X}{\delta n_2}\right)_{T,P,n_1,n_3 \ldots n_j}$$

The equation (*ii*) may be written as :

$$dX = \left(\frac{\delta X}{\delta T}\right)_{P, \; n_1, n_2 \ldots n_j} dT + \bar{X} = \left(\frac{\delta X}{\delta P}\right)_{T, \; n_1, n_2 \ldots n_j} dP$$

$$+ \bar{X}_1 \, dn_1 + \bar{X}_2 \, dn_2 + \bar{X}_3 \, dn_3 + \ldots \bar{X}_j \, dn_j$$

If the temperature and the pressure of the system are kept constant dT and dP are zero so that

$$dX = \bar{X}_1 \, dn_1 + \bar{X}_2 \, dn_2 + \ldots \bar{X}_j \, dn_j \qquad \ldots(iii)$$

and this on integration for a system of definite composition represented by the number of moles n_1, $n_2 \ldots n_j$ gives

$$X = n_1 \bar{X}_1 + n_2 \bar{X}_2 + n_3 \bar{X}_3 + \ldots n_j \bar{X}_j \qquad \ldots(iv)$$

i.e., the partial molal property $\bar{X}$ of any constituent may be regarded as the contribution of 1 mole of that constituent to the total value of the property of the system under specified conditions.

Partial Molar Free Energy : Chemical Potential

If the extensive property under study is free energy (G), $\bar{G}$ will represent the partial molar free energy so that

$$\bar{G}_1 = \left(\frac{\delta G}{\delta n_1}\right)_{T,P,n_2 n_3 \ldots n_j} \quad \text{and } \bar{G}_j = \left(\frac{\delta G}{\delta n_j}\right)_{T,P,n_1 n_2 \ldots n_{j-1}}$$

This quantity is, for most purposes, identical with the function known as **chemical potential** represented by the symbol μ. Accordingly we have

$$\mu_1 = \bar{G}_1 = \left(\frac{\delta G}{\delta n_1}\right)_{T,P,n_2 \ldots n_j}$$

Thus, it is the partial derivative of the free energy with n_i when all other variables are kept constant.

Physical Significance of Chemical Potential

By definition the chemical potential of a given substance is the change in free energy of the system produced on addition of one mole of the substance at constant temperature and pressure to a large bulk of the mixture so that its composition does not undergo any change. It is an intensive property and it may be regarded as the force which drives the chemical system to equilibrium. At equilibrium the chemical potential of the substance in the system must have the same value through the system. In other words, the matter flows spontaneously from a region of high chemical potential to low chemical potential. The chemical potential may also be regarded as the escaping tendency of that system. Greater the chemical potential of a system greater will be its escaping tendency.

Gibbs Duhem Equation

It has already been discussed that free energy G is an intensive thermodynamic property. It can be determined by fixing the variables T, P and number of moles of various constituents (composition of the mixture under study). Mathematically, we can write.

$$G = f\,(T, P, n_1, n_2 \,.....\, n_j) \qquad ...(i)$$

where $n_1, n_2 \,.......\, n_j$ are the number of moles of various constituents.

Differentiating equation (i), we get

$$dG = \left(\frac{\delta G}{\delta T}\right)_{P,\, n_1, n_2 .....n_j} dT + \left(\frac{\delta G}{\delta P}\right)_{T,\, n_1, n_2 .....n_j} dP +$$

$$\left(\frac{\delta G}{\delta n_1}\right)_{T, P,\, n_2 .....n_j} dn_1 + \left(\frac{\delta G}{\delta n_2}\right)_{T, P,\, n_1 .....n_j} dn_2 + \left(\frac{\delta G}{\delta n_j}\right)_{T_1, P,\, n_1 .....n_{j-1}} dn_j$$

We know the chemical potential is given by

$$\mu_i = \left(\frac{\delta G}{\delta n_i}\right)_{T, P, n_1, n_2 .....} = \bar{G}_1$$

Substituting in equation (*ii*) we get

$$dG = \left(\frac{\delta G}{\delta T}\right)_{P,\, n_1, n_2 .....n_j} dT + \left(\frac{\delta G}{\delta P}\right)_{T,\, n_1, n_2 .....n_j} dP + \mu_1\, dn_1, + \mu_2\, dn_2, + ... + \mu_j\, dn_j \qquad ...(iii)$$

For a closed system there is no change in the composition and equation (*iii*) reduces to

$$dG = \left(\frac{\delta G}{\delta T}\right)_{P,\, n_1, n_2 .....n_j} dT + \left(\frac{\delta G}{\delta P}\right)_{T,\, n_1, n_2 .....n_j} dP \qquad ...(iv)$$

But we know

$$dG = -\,SdT + VdP \qquad ...(v)$$

$$[\because\ G = H - TS$$

$$H = E + PV$$

and

$$G = dE + PdV + VdP - TdS - SdT]$$

comparing equation (*iv*) and (*v*)

$$\left(\frac{\delta G}{\delta T}\right)_{P, n_1 n_2 ....n_j} = -\,S$$

$$\left(\frac{\delta G}{\delta T}\right)_{T, n_1 n_2 ....n_j} = -\,V$$

and

Putting these values in equation (*iii*) we get

$$dG = -SdT + VdP + \mu_1\, dn_1 + \mu_2\, dn_2 + ... \mu_j\, dn_j \quad ...(vi)$$

At constant temperature and pressure equation (*vi*) reduces to

$$(dG)_{T.P} = \mu_1\, dn_1 + \mu_2\, dn_2 + ... \mu_j\, dn_j \quad ...(vii)$$

Integrating equation (*vii*) we get the following for a system of definite composition

$$(G)_{T.P} = \mu_1\, n_1 + \mu_2\, n_2 + \mu_3\, n_3 + ... \mu_j\, n_j \quad ...(viii)$$

Differentiating equation (*viii*) we get

$$(dG)_{T.P} = \mu_1\, dn_1 + n_1\, d\mu_1 + \mu_2\, dn_2 + n_2\, d\mu_2 + ... \mu_j\, dn_j + n_j\, d\mu_j \quad ...(ix)$$

Comparing equation (*vii*) and (*ix*) we get

$$n_1 d\,\mu_1 + n_2 d\,\mu_2 + ... n_j d\,\mu_j = 0$$

or

$$\Sigma\, n_j d\,\mu_j = 0 \quad ...(x)$$

Equation (*x*) is called Gibbs Duhem equation. It is applicable to a system at constant temperature and pressure.

Variation of Chemical Potential with Temperature and Pressure

(*a*) With Temperature

We know that chemical potential of a constituent *i* in a system is given by

$$\mu_i = \left(\frac{\delta G}{\delta n_i}\right)_{T,P,n_1....n_j} = \bar{G}_i \quad ...(i)$$

Differentiating equation (*i*) w.r.t. *T* at constant pressure *P*, we get

$$\left(\frac{\delta \mu_i}{\delta T}\right)_{P,n_1 n_2....n_j} = \frac{\delta^2 G}{\delta n_i \delta T} \quad ...(ii)$$

We also know that

$$\left(\frac{\delta G}{\delta T}\right)_{P,n_1 n_2....n_j} = S \quad ...(iii)$$

Differentiating equation (*iii*) w.r.t. n_j at constant temperature and pressure.

$$-\left(\frac{\delta S}{\delta n_i}\right)_{T,P,n_1....n_j} = \frac{\delta^2 G}{\delta T \delta n_i} \quad ...(iv)$$

Comparing equation (*ii*) and (*iv*), we have

$$\left(\frac{\delta \mu_i}{\delta T}\right)_{P,n_1 n_2....n_j} = \left(\frac{\delta S}{\delta n_i}\right)_{T,P,n_1....n_j} = \bar{S}_i$$

[By definition of Partial Molal Property]

Thus

$$\left(\frac{\delta \mu_i}{\delta T}\right)_{P,n_1 n_2....n_j} = \bar{S}_i = \text{Partial Molal Entropy}$$

This equation gives the variation of chemical potential of any constituent *i* with temperature.

(*b*) With Pressure

By definition, chemical potential is given by

$$\mu_i = \left(\frac{\delta G}{\delta n_i}\right)_{T,P,n_1....n_j} = \bar{G}_i \quad ...(i)$$

THERMODYNAMIC EQUILIBRIUM (ZEROTH LAW)

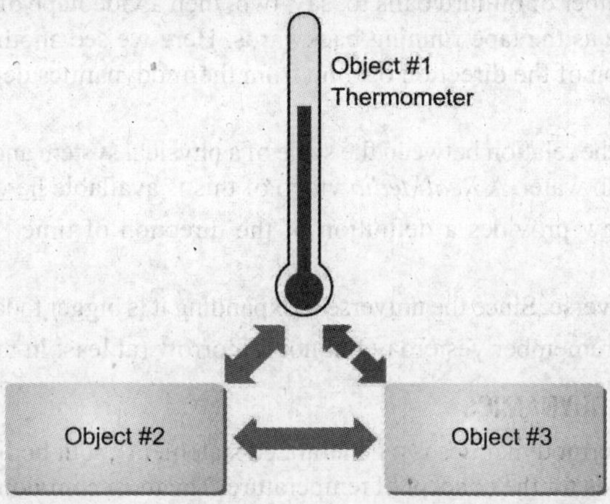

When two objects are separately in thermodynamic equilibrium with a third object, they are in equilibrium with each other.

Objects in thermodynamic equilibrium have the same temperature.

Differentiating equation (*i*) w.r.t. pressure at constant temperature

$$\left(\frac{\delta \mu_i}{\delta P}\right)_{T, n_1 n_2 \ldots n_j} = \frac{\delta^2 G}{\delta n_i \delta P} \qquad \ldots(ii)$$

But we know that

$$\left(\frac{\delta G}{\delta P}\right)_{T, n_1 n_2 \ldots n_j} = V \qquad \ldots(iii)$$

Differentiating equation (*iii*) w.r.t. n_i at constant T, P and n_j

$$\left(\frac{\delta V}{\delta n_i}\right)_{T, P, n_1 n_2 \ldots n_j} = \frac{\delta^2 G}{\delta P \, \delta n_i} = \overline{V}_i \qquad \ldots(iv)$$

(By definition of Partial Molal Property)

Comparing equations (*ii*) and (*iv*)

$$\left(\frac{\delta \mu_i}{\delta P}\right)_{T, n_1 n_2 \ldots n_j} = \overline{V}_i = \text{Partial Molal Volume} \qquad \ldots(v)$$

Equation (*v*) gives the variation of Chemical potential of any constituent *i* with pressure. From this equation we can define the partial molar volume of a constituent *i* as the rate of change of chemical potential of a constituent *i* with pressure at constant temperature.

TIME'S ARROW

The Second Law of Thermodynamics provides a definition of time's arrow. The law states that the entropy (usually) increases. And the increase occurs in time.

Imagine that we make a videotape of a "break" in a game of pool. We see 15 colored balls in a triangular configuration and a white ball coming in and breaking them up; the balls scatter in all directions.

If we run the videotape backwards we know immediately that this is (probably) not the correct direction for time.

If we reduce the number of billiard balls to, say, two, then a videotape of one ball colliding with another is just a possible as the tape running backwards. Here we see another manifestation of the fact that the determination of the direction of time from thermodynamics depends on having lots of objects in the system.

Another example is the relation between the scale of a physical system and the "correct" direction of time is milk swirling in water. A *RealMedia* video of this is available here. File size is 530k.

Thus the Second Law provides a definition of the direction of time. We know of two other definitions:

Expansion of the universe. Since the universe is expanding it is bigger today than it was yesterday.

Consciousness. We remember yesterday but not tomorrow (at least in the usual case).

ZEROTH LAW OF THERMODYNAMICS

The zeroth law of thermodynamics is a generalized statement about bodies in contact at thermal equilibrium and is the basis for the concept of temperature. The most common definition of the zeroth law of thermodynamics is: **If two thermodynamic systems are in thermal equilibrium with a third,** they are also in thermal equilibrium with each other.

The term zeroth law was coined by Ralph H. Fowler. In many ways, the law is more fundamental than any of the others. However, the need to state it explicitly as a law was not perceived until the first third of the 20th century, long after the first three laws were already widely in use and named as such, hence the zero numbering. There is still some discussion about its status in relation to the other three laws.

A system in thermal equilibrium is a system whose macroscopic properties (like pressure, temperature, volume, etc.) are not changing in time. A hot cup of coffee sitting on a kitchen table is not at equilibrium with its surroundings because it is cooling off and decreasing in temperature. Once its temperature stops decreasing, it will be at room temperature, and it will be in thermal equilibrium with its surroundings.

Two systems are said to be in thermal equilibrium when (*a*) both of the systems are in a state of equilibrium, and (*b*) they remain so when they are brought into contact, where 'contact' is meant to imply the possibility of exchanging heat, but not work or particles. And more generally, two systems can be in thermal equilibrium without thermal contact if one can be certain that *if they were* thermally connected, their properties would not change in time. Thus, thermal equilibrium is a relation between thermodynamical systems. Mathematically, the zeroth law expresses that this relation is an equivalence relation.

ADDITIONAL SOLVED PROBLEMS

SOLVED PROBLEM 22

Calculate the work done in the expansion of 3 moles of hydrogen reversibly and isothermally at 300 K from 0.021 m^3 to 0.0703 m^3.

SOLUTION

Formula used

$$w = 2.303 \times n R T \log \frac{V_2}{V_1}$$

Quantities given

$n = 3$ moles $\qquad R = 8.314$ J

$T = 300$ K $\quad V_1 = 0.021$ m^3 $\quad V_2 = 0.0703$ m^3

Substitution of values

$$w = 2.303 \times 3 \text{ moles} \times 8.314 \text{ J} \times 300 \text{ K} \times \log \frac{0.0703 \text{ m}^3}{0.021 \text{ m}^3}$$
$$= 17232.43 \times \log \frac{70.3}{21}$$
$$= 17232.43 \times \log 3.3490$$
$$= 17232.43 \times 0.5249$$
$$= \mathbf{9045.30 \text{ J}}$$

SOLVED PROBLEM 23

A Carnot's engine works between the temperatures 300 K and 400 K. Calculate the efficiency of the engine.

SOLUTION

Formula used :

$$\textit{Efficiency}, \eta = \frac{T_2 - T_1}{T_2}$$

Quantities given

$T_1 = 300$ K $\quad T_2 = 400$ K

Substitution of values

$$\text{Efficiency}, \eta = \frac{400 \text{ K} - 300 \text{ K}}{400 \text{ K}}$$
$$= \frac{100}{400}$$
$$= \mathbf{0.25 \text{ or } 25\%}$$

SOLVED PROBLEM 24

Two moles of an ideal gas undergo isothermal reversible expansion from 0.015 m^3 to 0.03 m^3 at 300 K. Calculate the work done and change in entropy.

SOLUTION

(i) To calculate the work done

Formula used

$$w = 2.303 \times n\,R\,T \log \frac{V_2}{V_1}$$

Quantities given

$n_2 = 2$ moles $\quad R = 8.314$ J K^{-1} mol^{-1} $\quad T = 300$ K

$V_1 = 0.015$ m^3 $\quad V_2 = 0.03$ m^3

Substitution of values

$$w = 2.303 \times 2 \text{ mole} \times 8.314 \text{ JK}^{-1} \text{ mol}^{-1} \times 300 \times \log \frac{0.03}{0.015}$$
$$= 11495.43 \text{ J} \times 0.3010$$
$$= \mathbf{3459.97 \text{ J}}$$

(ii) To calculate the change in entropy

Formula used

$$\Delta S = \frac{q}{T}$$

Quantities given

$q = 3459.97$ J $\qquad T = 300$ K

Substitution of values

$$\Delta S = \frac{3459.97 \text{ Joule}}{300 \text{ K}}$$

$$= \mathbf{11.53 \ J \ K^{-1}}$$

SOLVED PROBLEM 25

An engine operates between 373 K and 273 K and another engine operates between 373 K and 0 K (absolute zero). Find the efficiency in two cases.

SOLUTION

(i) For engine A operating between 373 K and 273 K

Formula used

$$\eta = \frac{T_2 - T_1}{T_2}$$

Quantities given

$T_2 = 373$ K $\qquad T_1 = 273$ K

Substitutions of values

$$\eta = \frac{373 \text{ K} - 273\text{K}}{373 \text{ K}}$$

$$= \frac{100}{373}$$

$$= 0.268 \quad \text{or} \quad 26.8\%$$

(ii) For engine B operating between 373 K and 0 K

Formula used

$$\eta = \frac{T_2 - T_1}{T_1}$$

Quantities given

$T_2 = 373$ K $\qquad T_1 = 0$ K

Substitution of values

$$\eta = \frac{373 \text{ K} - 0 \text{ K}}{373 \text{ K}}$$

$$= \frac{373}{373}$$

$$= 1 \quad \text{or} \quad 100\%$$

SOLVED PROBLEM 26

Calculate the change in free energy for the expansion of one mole of an ideal gas from 202650.00 Pa (2.0 atm) to 20265.00 Pa (0.2 atm) at 298 K (R = 8.314 JK^{-1} mol^{-1}).

SOLUTION

Formula used

$$\Delta G = 2.303 \times n\,R\,T \log \frac{P_2}{P_1}$$

Quantities given

$n = 1$ mole $\quad T = 298$ K $\quad R = 8.314$ J K^{-1} mol^{-1}

$P_1 = 202650.00$ Pa $\quad P_2 = 20265.00$ Pa

Substitution of values

$$\Delta G = 2.303 \times 1 \text{ mol} \times 8.314 \text{ JK}^{-1} \text{ mol}^{-1} \times 298 \text{ K} \times \log \frac{20265.00 \text{ Pa}}{202650.00 \text{ Pa}}$$

$$\Delta G = 5705.848 \text{ J} \times \log \frac{1}{10}$$

$$= 5705.8483 \text{ J} \times (-1.0)$$

$$= \mathbf{-5705.8483 \text{ J}}$$

SOLVED PROBLEM 27

Calculate the entropy change in transforming 0.024 kg of Ice into water at 273 K. Molar heat of fusion = 6009 J mol^{-1}.

SOLUTION

Formula used

$$\Delta S_f = \frac{\Delta H_f}{T_f}$$

Quantities given

$\Delta H_f = 6009$ J mol^{-1} $\quad T_f = 273$ K

$$\Delta H_f \text{ for 24 g of water} = \frac{24}{18} \times 6009 \text{ J} = 8012 \text{ J}$$

Substitution of values

$$\Delta S_f = \frac{8012 \text{ J}}{273 \text{ K}} = \mathbf{29.3480 \text{ JK}^{-1}}$$

SOLVED PROBLEM 28

For the reaction

$$M_2O_{(s)} \longrightarrow 2\,M_{(s)} + \tfrac{1}{2}\,O_{2\,(g)}$$

$\Delta H = 30$ kJ mol^{-1} and $\Delta S = 0.07$ kJ K^{-1} mol^{-1} at 101325.00 Pa pressure. Calculate upto which temperature the reaction would not be spontaneous.

SOLUTION

Formula used

$$\Delta G = \Delta H - T\Delta S$$

or

$$T = \frac{\Delta H}{\Delta S} \qquad \left[\because \Delta G = 0 \text{ at Equilibrium} \right]$$

Quantities given

$\Delta H = 30$ kJ mol^{-1} $\quad \Delta S = 0.07$ kJ K^{-1} mol^{-1}

Substitution of values

$$T = \frac{30 \text{ kJ mol}^{-1}}{0.07 \text{ kJ K}^{-1} \text{ mol}^{-1}}$$

$$= 428.57 \text{ K}$$

Thus the reaction would not be spontaneous up to 428.57 K

EXAMINATION QUESTIONS

1. Define or explain the following terms :
   (*a*) Entropy
   (*b*) Second law of thermodynamics
   (*c*) Carnot cycle
   (*d*) Third law of thermodynamics
   (*e*) Gibbs free energy
   (*f*) Gibbs Helmholtz equation
   (*g*) Spontaneous reactions
   (*h*) Clausius clapeyron equation
   (*i*) van't Hoff isotherm
2. (*a*) Derive the concept of entropy from the Second law of Thermodynamics. Show that entropy is a state function.
   (*b*) What is the change in entropy when 1 mole of helium gas is heated from 200 K to 400 K at constant pressure? (Given : C_P for helium = 20.93 J/K/mol)
   Answer. 14.51 J/K
3. (*a*) Derive an expression for entropy change for ideal gas associated with temperature and pressure changes.
   (*b*) Calculate the total entropy change when 0.005 kg of ice at 273 K is converted into steam at 373K. (Latent heat of evaporation of water = 2260 × 10^3 J/kg; C_p for water = 75.36 J/mol; Latent heat of water = 334.94 Joule)
   Answer. (*b*) 42.98 J/K
4. Calculate the work performed when 0.002 kg of hydrogen gas is expanded isothermally and reversibly at 300 K from 0.01 to 0.1 m^3. What is the amount of heat absorbed? What is the change in the internal energy?
   Answer. 5743.1077 J
5. Water boils at 373 K at 101325.00 Pa pressure. At what temperature will it boil at a hill station where the atmospheric pressure is 66661.18 Pa?
   (Latent heat of vaporisation of water is 2300 kJ/kg and $R = 8.314\ JK^{-1}\ mol^{-1}$)
   Answer. 361.65 K
6. A Carnot's engine works between the temperature 300 K and 400 K. Calculate the efficiency of the engine.
   Answer. 25%
7. (*a*) Explain the term Thermodynamic efficiency.
   (*b*) Calculate the work done on the system if one mole of an ideal gas at 300 K is compressed isothermally and reversibly to one-fifth of its original volume.
   Answer. (*b*) 4014.26 J
8. What do you understand by the term enthalpy?
   Calculate the work done in the expansion of 3 moles of hydrogen reversibly and isothermally at 300 K from 0.021 m^3 to 0.0703 m^3.
   Answer. 9045.30 J
9. (*a*) "It is not profitable to carry out a process reversibly although maximum work can be obtained by doing so". Comment.
   (*b*) Two moles of hydrogen are compressed adiabatically from NTP conditions to occupy a volume of 0.00448 m^3. Calculate the final pressure and temperature ($\gamma = 1.41$)
   (*c*) Derive a relation between pressure and volume for an adiabatic reversible expansion of an ideal gas.
   Answer. (*b*) 2604052.50 Pa; 702.25 K
10. (*a*) Explain giving reason "The net entropy of the universe tends to increase".
    (*b*) For the reaction
    $H_2(g) + \frac{1}{2}O_2(g) \rightarrow H_2O(\ell)$

the values of enthalpy change and free energy change are –286.04 and –237.35 kJ respectively at 298K. Calculate the value of free energy change at 303 K.

(*c*) Write down the applications of Gibb's Helmholtz equation.

Answer. (*b*) 236.53 kJ

11. (*a*) Bring about clearly the criteria for reversibility and irreversibility in terms of *S*, *E*, *H* and *G*.

(*b*) 1.0 mole of steam is condensed at 373 K and water is cooled to 273 K and frozen to ice. Calculate the entropy change for the process. Latent heat of fusion of ice and evaporation of water are 334.94×10^3 and 2260.87×10^3 J/kg respectively.

Answer. (*c*) 107.67 J K^{-1} mol^{-1}

12. (*a*) Write a note on "Carnot's Cycle".

(*b*) How are work function and free energy related? Discuss the criteria of spontaneity of a chemical reaction.

(*c*) Calculate the entropy increase in the evaporation of a mole of water at 373 K (Heat of vaporization = 2260.87×10^3J/kg)

13. (*a*) Give the expression for Gibb's free energy change (ΔG) for the reaction

$$nA + mB \rightarrow pC + qD$$

(*b*) How is ΔG of a chemical reactions is related to ΔS and ΔH?

(*c*) The enthalpy change for the transition of liquid water to steam is 40.8 kJ mol^{-1} at 373 K. Calculate ΔG for the process.

Answer. (*b*) 109.38 3 JK^{-1} mol^{-1}

14. Calculate ΔS when 0.028 kg of N_2 gas expands reversibly from 0.002 m^3 to 0.02 m^3 at 300 K.

Answer. 38.294 JK^{-1}

15. (*a*) Prove that Enthalpy remains constant when a real gas passes through a porous plug in adiabatic expansion.

(*b*) Explain the relationship between entropy and probability.

(*c*) Derive an expression for the efficiency of a Carnot's engine working between the two temperatures T_1 and T_2.

16. (*a*) Define standard heat of formation and standard entropy change of a reaction.

(*b*) Explain why in case of non-polar solvents, the ΔS is nearly equal to 88 JK^{-1} mol^{-1}.

(*c*) Calculate the entropy change for the reaction :

$$N_2(g) + 3H_2(g) \rightarrow 2NH_3(g)$$

Given : $S°$ of N_2, H_2 and NH_3 as 191.5, 130.6 and 192.2 JK^{-1} mol^{-1} respectively.

Answer. –198.9 J K^{-1} mol^{-1} (*Delhi BSc*)

17. (*a*) Determine the entropy change for an ideal gas when temperature and volume are varied.

(*b*) Calculate the entropy change involved in isothermal expansion of 2 moles of the gas from a volume of 0.005 m^3 to a volume of 0.05 m^3 at 303 K.

Answer. 38.29 JK^{-1} (*Guru Nanak Dev BSc*)

18. Calculate entropy change if 2 moles of water at 373 K are evaporated to vapors at 373 K. Give its units also. (Given molar heat of vaporisation of water is 40402.62 Joule)

Answer. 216.49 JK^{-1} (*Andhra BSc*)

19. 0.0005 kg of nitrogen is enclosed in a cylinder fitted with a piston at 298 K. Calculate the change in entropy, if the gas is expanded adiabatically to double its volume.

Answer. zero (*Guru Nanak Dev BSc*)

20. (*a*) Starting from appropriate definition of chemical potential of a component in an ideal gas solution, derive an expression for ΔG_{mixing} for the formation of ideal binary solution and show further that ΔH_{mixing}=0 for the solution.

(*b*) A solution is prepared by mixing 2 moles of CS_2 and 3 moles of CCl_4 at 298 K and 101325.00 (1atm) pressure Assuming ideal behavior, calculate ΔG_{mixing} for the solution.

Answer. –3620.22 J (*Guru Nanak Dev BSc*)

21. Calculate entropy change for the fusion of one mole of a solid which melts at 300 K. The latent heat of fusion is 2.51 k J mol^{-1}.

Answer. 8.366 J *(Arunachal BSc)*

22. Calculate the change in entropy when 3 moles of an ideal gas is heated from 323 K to 423 K at a constant volume. (C_v = 32.94 JK^{-1} mol^{-1})

Answer. 26.658 JK^{-1} *(Nagpur BSc)*

23. ΔG and ΔH values for a reaction at 300 K are : –66.944 kJ and –41.84 kJ respectively. Calculate the free energy change at 330 K, assuming that ΔH and ΔS remain constant over this temperature range.

Answer. –69.454 kJ *(Nagpur BSc)*

24. (*a*) Derive the integral Clausius-Clapeyron equation in the form

$$\log \frac{P_2}{P_1} = \frac{\Delta H}{2.303\,R}\left[\frac{T_2 - T_1}{T_1\,T_2}\right]$$

for an ideal gas.

(*b*) At 373.6 K and 372.6 K the vapor pressure of $H_2O(\ell)$ are 103148.85 and 99501.15 Pa respectively. What is the heat of vaporization of water? (R = 8.314 Joules)

Answer. (*b*) 41675.02 J *(Jamia Millia BSc)*

25. Explain the term fugacity. How is fugacity of a gas determined? *(Panjab BSc)*

26. (*a*) State Carnot's theorem and second law of thermodynamics.

(*b*) Define Chemical potential. Derive effect of temperature and pressure on chemical potential. *(Indore BSc)*

27. Explain the following :

(*a*) Under what conditions can an isothermal expansion of a gas become a free expansion process.

(*b*) Increase in volume of a gas for a given decrease in pressure is less in an adiabatic expansion than in isothermal expansion.

(*c*) All spontaneous processes lead to increase the entropy of the universe.

(*d*) Free Energy of formation of an element at 101325.00 Pa (1 atm) and 298 K is assumed to be zero but entropy is not zero under the same conditions. *(Gulbarga BSc)*

28. (*a*) Describe Carnot's cycle for establishing the maximum convertibility of heat into work. How does it lead to the definition of Second Law of Thermodynamics?

(*b*) Heat supplied to a Carnot engine is 1897.86 kJ. How much useful work can be done by the engine which works between 273 K and 373 K?

Answer. (*b*) 508.80 kJ *(Madurai BSc)*

29. The heat of vaporisation, ΔH_{vap} of carbon tetrachloride, CCl_4, at 298 K is 43 kJ mol^{-1}. If 1 mole of liquid carbon tetrachloride at 298 K has an entropy of 214 JK^{-1}, what is the entropy of 1 mole of the vapor in equilibrium with the liquid at this temperature ?

Answer. 358 JK^{-1} mol^{-1} *(Kalyani BSc)*

30. Calculate the amount of the heat supplied to Carnot cycle working between 378 K and 293 K if the maximum work obtained is 837.36 Joule ?

Answer. 3723.9 Joule *(Panjab BSc)*

31. The enthalpy change involved in oxidation of glucose is – 2880 kJ mol^{-1}. Twenty five percent of this energy is available for muscular work. If 100 kJ of muscular work is needed to walk one kilometer, what is the maximum distance a person will be able to walk after eating 0.12 kg of glucose ?

Answer. 4800 m *(Delhi BSc)*

32. What is the entropy change for conversion of one mole of ice to water at 273 K and 101325.00 Pa (1 atm) pressure (Given ΔH of ice = 6.025 kJ mol^{-1}).

Answer. 22.069 J mol^{-1} *(Bundelkhand BSc)*

33. Calculate the efficiency of steam engine operating between 373 K and 298 K. What would be the efficiency of the engine if the boiler temperature is raised to 423 K, the temperature of the sink remaining same ?

Answer. 20.1% ; 29.55% *(Agra BSc)*

34. At 373 K the entropy change for the transition of liquid water to steam is 109 JK^{-1} mol^{-1}. Calculate the enthalpy change ΔH_{vap} for the process.

Answer. 40.657 kJ mol^{-1} *(Madurai BSc)*

35. Ethanol boils at 351.55 K and standard enthalpy of vaporisation of ethanol is 42 kJ mol^{-1}. Calculate the entropy of vaporisation of ethanol.

Answer. 120.66 JK^{-1} mol^{-1} *(Kanpur BSc)*

36. Two moles of an ideal gas undergo isothermal reversible expansion from 0.015 m^3 to 0.03 m^3 at 300 K. Calculate the work done and the change in entropy.

Answer. 3460.39 Joule; 11.53 JK^{-1} *(Annamalai BSc)*

37. (*a*) Derive the Clapeyron-Clausius equation giving the temperature dependence of water pressure indicating clearly the assumption involved.

(*b*) An engine operates between 373 K and 273 K and another engine operates between 373 K and 0 K (absolute zero). Find the efficiencies in two cases.

Answer. (*b*) 26.8%; 100 % *(Calicut BSc)*

38. Calculate the value of dT/dP for water $\rightleftharpoons$ ice system at 273 K. ΔH_f for ice is 6007.8 J mol^{-1}; Molar Volume of water and ice are 0.018×10^{-3} m^3 mol^{-1} and 0.1963×10^{-3} m^3 mol^{-1} respectively.

(Given 1 J = 9.87 m^3 atm or 9.87 × 101325 m^3 Pa)

Answer. 0.0100654 K atm^{-1} *(Gujarat BSc)*

39. Calculate standard Gibb's Free energy change for the combustion of methane :

$$CH_4(g) + 2O_2(g) \rightarrow CO(g) + 2H_2O(g)$$

at 298 K. $\Delta H° = -803.03$ kJ and $\Delta S° = 5.02$ J K^{-1}

Answer. 2300.23 Joule *(Kakatiya BSc)*

MULTIPLE CHOICE QUESTIONS

1. A process which proceeds of its own accord, without any outside assistance, is called

(*a*) non-spontaneous process (*b*) spontaneous process
(*c*) reversible process (*d*) irreversible process

Answer. (*b*)

2. The tendency of a process to occur naturally is called

(*a*) momentum of the reaction (*b*) spontaneity of the reaction
(*c*) equilibrium of the reaction (*d*) equilibrium of the reaction

Answer. (*b*)

3. Which of the following is true about the criteria of spontaneity?

(*a*) a spontaneous change is unidirectional
(*b*) a spontaneous change to occur, time is no factor
(*c*) once a system is in equilibrium, a spontaneous change is inevitable
(*d*) all of the above

Answer. (*d*)

4. A spontaneous change is accompanied by _______ of internal energy or enthalpy.
(*a*) increase
(*b*) decrease
(*c*) neither increase nor decrease
(*d*) none of these
Answer. (*b*)

5. Mixing of two or more gases is a
(*a*) spontaneous process
(*b*) non-spontaneous process
(*c*) reversible process
(*d*) none of these
Answer. (*a*)

6. Entropy is a measure of _______ of the molecules of the system.
(*a*) concentration
(*b*) velocity
(*c*) zig-zag motion
(*d*) randomness or disorder
Answer. (*d*)

7. The second law of thermodynamics states that
(*a*) whenever a spontaneous process occurs, it is accompanied by an increase in the total energy of the universe
(*b*) the entropy of the system is constantly increasing
(*c*) neither of the above
(*d*) both (*a*) and (*b*)
Answer. (*d*)

8. The entropy of a pure crystal is zero at absolute zero. This is statement of
(*a*) first law of thermodynamics
(*b*) second law of thermodynamics
(*c*) third law of thermodynamics
(*d*) none of these
Answer. (*c*)

9. The entropy is measured in
(*a*) cal K^{-1} mol^{-1}
(*b*) JK^{-1} mol^{-1}
(*c*) entropy unit
(*d*) all of these
Answer. (*d*)

10. The standard entropy, $S°$, of a substance is
(*a*) its entropy at 273 K and 101325 Pa (1 atm) pressure
(*b*) its entropy at 0 K and 101325 Pa (1 atm) pressure
(*c*) its entropy at 298 K and 101325 Pa (1 atm) pressure
(*d*) its entropy at 25 K and 101325 Pa (1 atm) pressure
Answer. (*c*)

11. The change in entropy of a reaction is given by
(*a*) $\Delta S = \Sigma S_{\text{Reactants}} + \Sigma S_{\text{Products}}$
(*b*) $\Delta S = \Sigma S_{\text{Products}} - \Sigma S_{\text{Reactants}}$
(*c*) $\Delta S = \Sigma S_{\text{Reactants}} - \Sigma S_{\text{Products}}$
(*d*) none of these
Answer. (*b*)

12. Which of the following is true for a cyclic process?
(*a*) $\Delta E = 0$
(*b*) $\Delta E = q - w$
(*c*) $q = w$
(*d*) all of these
Answer. (*d*)

13. A machine that can do work by using heat which flows out spontaneously from a high-temperature source to a low-temperature sink is called
(*a*) Carnot machine
(*b*) cyclic machine
(*c*) heat machine
(*d*) heat engine
Answer. (*d*)

14. The efficiency of a heat engine is the ratio of
(*a*) work obtained in a cyclic process (w) to the heat taken from the high temperature reservoir (q)
(*b*) heat taken from the high temperature reservoir (q) to the work obtained in a cyclic process
(*c*) work obtained in a cyclic process (w) to the heat taken from the low temperature sink (q)
(*d*) none of the above
Answer. (*a*)

15. The cycle of processes which occurs under reversible conditions is referred to as
(*a*) cyclic process (*b*) closed process
(*c*) Carnot cycle (*d*) reversible cycle
Answer. (*c*)

16. The efficiency of a heat engine is given by
(*a*) $\frac{w}{q_2} = \frac{T_2 - T_1}{T_2}$ (*b*) $\frac{w}{q_2} = \frac{T_1 - T_2}{T_2}$
(*c*) $\frac{w}{q_2} = \frac{T_2 - T_1}{T_1}$ (*d*) $\frac{w}{q_2} = \frac{T_1 - T_2}{T_1}$
Answer. (*a*)

17. The second law of thermodynamics may be stated as
(*a*) it is impossible to take heat from a hotter reservoir and convert it completely into work by a cyclic process without transferring a part of heat to a cooler reservoir.
(*b*) it is impossible to transfer heat from a body at a lower temperature to one at higher temperature
(*c*) the efficiency of heat engine is always less than 1
(*d*) all of the above
Answer. (*d*)

18. The efficiency of a heat operating between 400 K and 300 K is
(*a*) 1.0 (*b*) 0.75
(*c*) 0.50 (*d*) 0.25
Answer. (*d*)

19. The efficiency of heat engine operating between 1000 K and 300 K is _______ the engine operating between 1000 K and 500 K.
(*a*) greater than (*b*) lesser than
(*c*) is equal to (*d*) none of these
Answer. (*a*)

20. The entropy of the system increases in the order
(*a*) gas < liquid < solid (*b*) solid < liquid < gas
(*c*) gas < solid < liquid (*d*) none of these
Answer. (*b*)

21. The efficiency of an irreversible Carnot cycle is always _______ that of a reversible one operating between the same two temperatures
(*a*) less than (*b*) greater than
(*c*) equal to (*d*) none of these
Answer. (*a*)

22. The free energy function (G) is defined as
(*a*) $G = H + TS$ (*b*) $G = H - TS$
(*c*) $G = TS - H$ (*d*) none of these
Answer. (*b*)

23. The change in free energy is a measure of :
(*a*) net work done
(*b*) net change is entropy
(*c*) net change in enthalpy
(*d*) net change in internal energy
Answer. (*a*)

24. The work function (A) is defined as
(*a*) $A = E - TS$
(*b*) $A = E + TS$
(*c*) $A = TS - E$
(*d*) none of these
Answer. (*a*)

25. The change in free energy of a system is given by
(*a*) $\Delta G = \Delta A + P\,\Delta V$
(*b*) $\Delta G = \Delta H - T\,\Delta S$
(*c*) $\Delta G = \Delta E + P\,\Delta V - T\,\Delta S$
(*d*) all of these
Answer. (*d*)

26. Which out of the following is not a state function?
(*a*) free energy
(*b*) work function
(*c*) entropy
(*d*) work done
Answer. (*d*)

27. The variation of free energy with temperatures at constant pressure is given by the relation:
(*a*) $dG_P = -SdT_P$
(*b*) $\left(\dfrac{dG}{dT}\right)_P = -S$
(*c*) neither of these
(*d*) both (*a*) and (*b*)
Answer. (*d*)

28. The variation of free energy with pressure at constant temperature is given by
(*a*) $(dG)_T = V\,d\,P_T$
(*b*) $dG_P = -SdT_P$
(*c*) $\left(\dfrac{dG}{dT}\right)_P = -S$
(*d*) none of these
Answer. (*a*)

29. The change in free energy in an isothermal process for *n* moles of the gas is given by
(*a*) $\Delta G = 2.303 \times n\,RT \log \dfrac{P_2}{P_1}$
(*b*) $\Delta G = 2.303 \times RT \log \dfrac{V_1}{V_2}$
(*c*) $\Delta G = 2.303 \times RT \log \dfrac{P_2}{V_1}$
(*d*) none of these
Answer. (*a*)

30. The Gibb's Helmholtz equation is applicable to
(*a*) all processes, chemical or physical
(*b*) all process, chemical or physical but in a closed system
(*c*) all chemical processes in a closed system
(*d*) all physical processes in a closed system
Answer. (*b*)

31. For a spontaneous process
(*a*) $\Delta G > 0$
(*b*) $\Delta G < 0$
(*c*) $\Delta G = 0$
(*d*) none of these
Answer. (*b*)

32. A process is in the equilibrium state when
(*a*) $\Delta G > 0$
(*b*) $\Delta G < 0$
(*c*) $\Delta G = 0$
(*d*) none of these
Answer. (*c*)

33. Which of the following equation is used to calculate the heats of reaction when ΔG at two temperature are given?

(*a*) Gibbs Helmholtz equation
(*b*) Clapeyron equation
(*c*) Kirchoff's equation
(*d*) none of these

Answer. (*a*)

34. The equation $\frac{dP}{dT} = \frac{\Delta H}{T(V_2 - V_1)}$ is called

(*a*) Gibb's Helmholtz equation
(*b*) Kirchoff's equation
(*c*) Clapeyron equation
(*d*) Clausius Clapeyron equation

Answer. (*c*)

35. The Clausius Clapeyron equation helps to calculate

(*a*) latent heat of vaporization
(*b*) boiling point or freezing point
(*c*) vapor pressure at one temperature, if at another temperature is given
(*d*) all of the above

Answer. (*d*)

36. The decrease in the work function A in any process at constant temperature gives the ______ that can be obtained from the system during any change.

(*a*) minimum work
(*b*) maximum work
(*c*) useful work
(*d*) net work

Answer. (*b*)

37. The equation for van't Hoff isotherm is

(*a*) $-\Delta G = 2.303\ RT \log K_P$
(*b*) $\Delta G = 2.303\ RT \log K_P$
(*c*) $-\Delta G = 2.303\ RT^2 \log K_P$
(*d*) $\Delta G = 2.303\ RT^2 \log K_P$

Answer. (*a*)

38. The equation $\frac{\Delta H}{RT^2} = \frac{d\,(\ln K_P)}{dT}$ is known as

(*a*) van't Hoff equation
(*b*) van't Hoff isochore
(*c*) Gibbs equation
(*d*) Gibbs Duhem equation

Answer. (*b*)

39. Each substance in a given state has a tendency to escape from that state and this escaping tendency is called

(*a*) spontaneity
(*b*) Gibbs free energy
(*c*) fugacity
(*d*) activity

Answer. (*c*)

40. When water is cooled to ice, its entropy

(*a*) increases
(*b*) decreases
(*c*) remains the same
(*d*) becomes zero

Answer. (*b*)

41. Which of the following sets of conditions makes a process spontaneous at all temperatures?

(*a*) $\Delta H = 0; \Delta S > 0$
(*b*) $\Delta H = 0; \Delta S < 0$
(*c*) $\Delta H > 0; \Delta S > 0$
(*d*) $\Delta H < 0; \Delta S < 0$

Answer. (*c*)

42. The increase in entropy is maximum in

(*a*) $CaCO_3(s) \rightarrow CaO(s) + CO_2(g)$
(*b*) $CO(g) + ½O_2(g) \rightarrow CO_2(g)$
(*c*) $N_2(g) + 3H_2(g) \rightarrow 2NH_3(g)$
(*d*) $H_2(g) + I_2(g) \rightarrow 2HI(g)$

Answer. (*a*)

43. A spontaneous reaction proceeds with a decrease in

(*a*) entropy (*b*) enthalpy

(*c*) free energy (*d*) internal energy

Answer. (*c*)

44. In a process $\Delta H = 100$ kJ and $\Delta S = 100$ JK^{-1} at 400 K. The value of ΔG will be

(*a*) zero (*b*) 100 kJ

(*c*) 50 kJ (*d*) 60 kJ

Answer. (*d*)

45. A chemical reaction proceeds with decrease in both the enthalpy and entropy. This reaction will be spontaneous if

(*a*) $\Delta H = T\,\Delta S$ (*b*) $\Delta H < T\,\Delta S$

(*c*) $\Delta H > T\,\Delta S$ (*d*) none of these

Answer. (*b*)

46. Which is the correct unit for entropy?

(*a*) kJ mol (*b*) JK^{-1} mol

(*c*) JK^{-1} mol^{-1} (*d*) kJ mol

Answer. (*c*)

47. The efficiency of heat engine is maximum when

(*a*) temperatures of source and sink are maximum

(*b*) temperatures of source and sink are minimum

(*c*) temperature of source is minimum and that of sink is maximum

(*d*) temperature of source is maximum and that of sink is minimum

Answer. (*d*)

48. A reaction proceeds with increase in both the enthalpy and entropy. It will be spontaneous if

(*a*) $\Delta H = T\,\Delta S$ (*b*) $\Delta H > T\,\Delta S$

(*c*) $\Delta H < T\,\Delta S$ (*d*) none of these

Answer. (*c*)

49. A spontaneous reaction is not possible if

(*a*) ΔH and $T\,\Delta S$ are both negative (*b*) ΔH and $T\,\Delta S$ are both positive

(*c*) ΔH is +ve and $T\,\Delta S$ is –ve (*d*) ΔH is –ve and $T\,\Delta\,S$ is +ve

Answer. (*c*).

10 Gaseous State

CONTENTS

All matter exists in three states: *gas*, *liquid* and *solid*. A molecular level representation of gaseous, liquid and solid states is shown in Fig. 10.1

A **gas** consists of molecules separated wide apart in empty space. The molecules are free to move about throughout the container.

A **liquid** has molecules touching each other. However, the intermolecular space, permit the movement of molecules throughout the liquid.

A **solid** has molecules, atoms or ions arranged in a certain order in fixed positions in the crystal lattice. The particles in a solid are not free to move about but vibrate in their fixed positions.

Of the three states of matter, the gaseous state is the one most studied and best understood. We shall consider it first.

GENERAL CHARACTERISTICS OF GASES

1. Expansibility

Gases have limitless expansibility. They expand to fill the entire vessel they are placed in.

2. Compressibility

Gases are easily compressed by application of pressure to a movable piston fitted in the container.

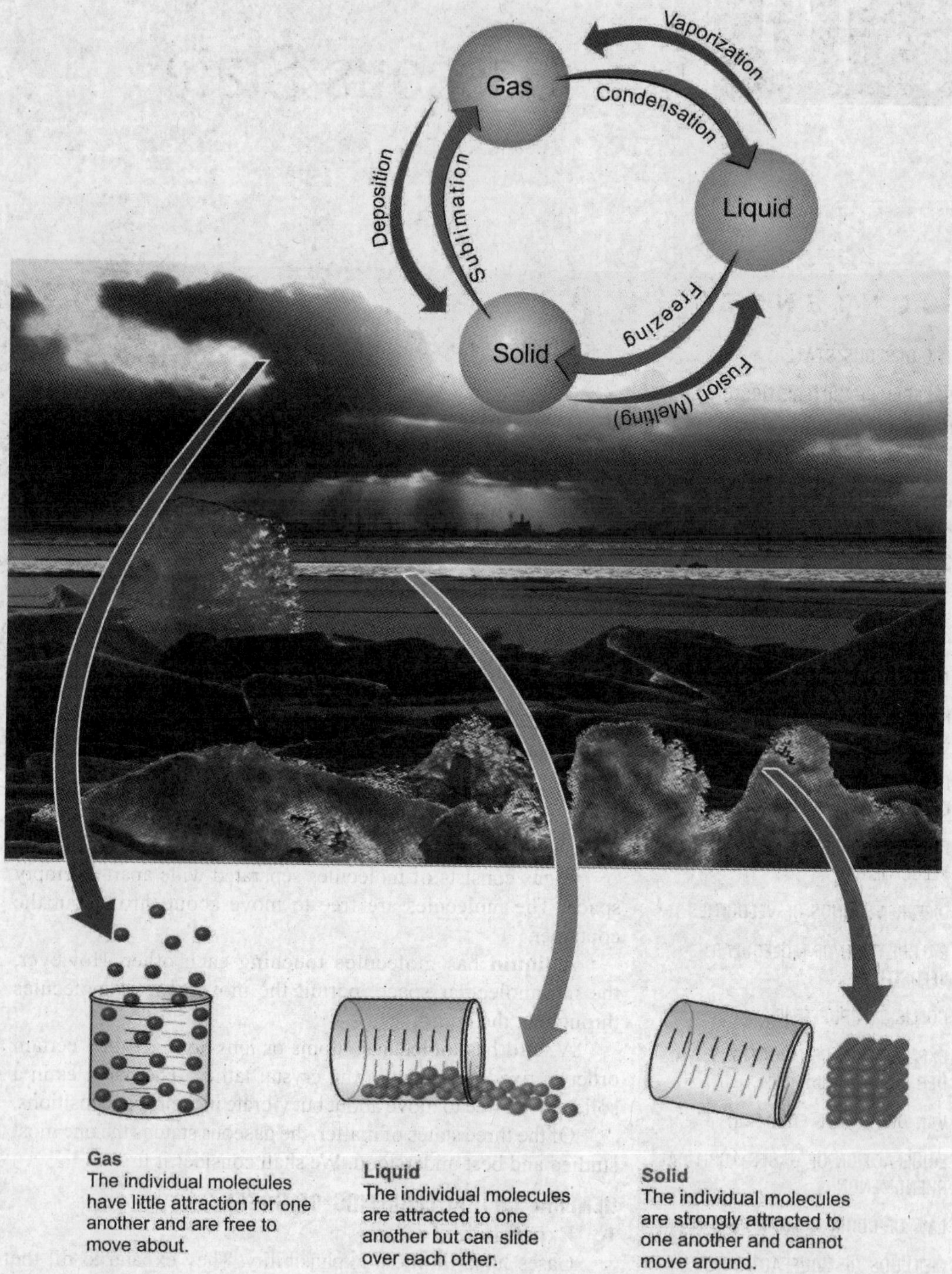

■ **Figure 10.1**
Molecular representation of the gaseous, liquid and solid states.

3. **Diffusibility**
   Gases can diffuse rapidly through each other to form a homogeneous mixture.
4. **Pressure**
   Gases exert pressure on the walls of the container in all directions.
5. **Effect of Heat**
   When a gas, confined in a vessel is heated, its pressure increases. Upon heating in a vessel fitted with a piston, volume of the gas increases.

The above properties of gases can be easily explained by the Kinetic Molecular Theory which will be considered later in the chapter.

PARAMETERS OF A GAS

A gas sample can be described in terms of four parameters (measurable properties):

(1) the volume, V of the gas
(2) its pressure, P
(3) its temperature, T
(4) the number of moles, n, of gas in the container

The Volume, V

The volume of the container is the volume of the gas sample. It is usually given in liter (l or L) or milliliters (ml or mL).

$$1 \text{ liter(l)} = 1000 \text{ ml and } 1 \text{ ml} = 10^{-3} \text{ l}$$

One milliliter is practically equal to one cubic centimeter (cc). Actually

$$1 \text{ liter(l)} = 1000.028 \text{ cc}$$

The SI unit for volume is cubic meter (m^3) and the smaller unit is decimeter^3 (dm^3).

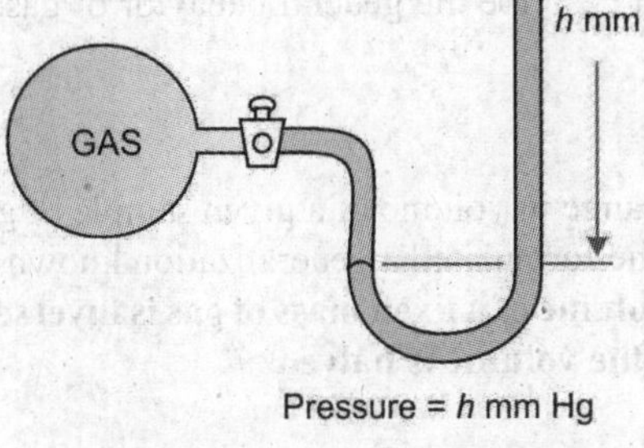

■ **Figure 10.2**
A mercury manometer.
The gas container is connected with a U-tube containing Hg having vacuum in the closed end. The difference in Hg height in two limbs gives the gas pressure in mm Hg.

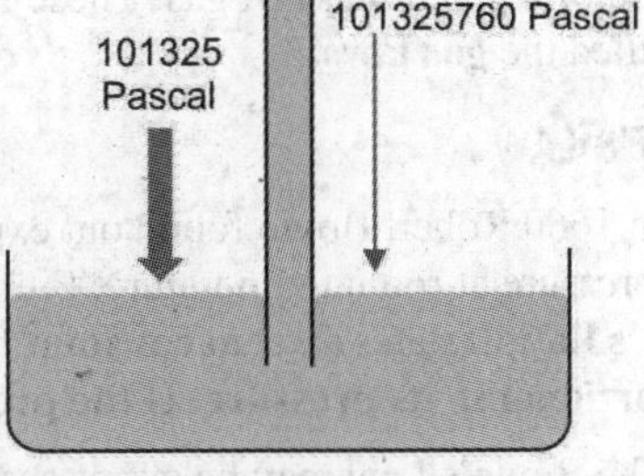

■ **Figure 10.3**
A mercury barometer.
A long tube (80 × 1 cm) filled with Hg inverted into dish containing Hg. The atmospheric pressure is equal to 760 mm Hg column supported by it at sea level.

The pressure of a gas is defined as the force exerted by the impacts of its molecules per unit surface area in contact. The pressure of a gas sample can be measured with the help of a mercury manometer (Fig. 10.2) Similarly, the atmospheric pressure can be determined with a mercury barometer (Fig. 10.3).

The pressure of air that can support 760 mm Hg column at sea level, is called one atmosphere (1 atm). The unit of pressure, millimeter of mercury, is also called **torr.**

Thus,

$$1 \text{ atm} = 760 \text{ mm Hg} = 760 \text{ torr}$$

The SI unit of pressure is the **Pascal** (Pa). The relation between atmosphere, torr and pascal is :

$$1 \text{ atm} = 760 \text{ torr} = 1.013 \times 10^5 \text{ Pa}$$

The unit of pressure 'Pascal" is not in common use.

Temperature, *T*

The temperature of a gas may be measured in **Centigrade degrees** (°C) or **Celsius degrees.** The SI unit of temperature is **Kelvin (K)** or **Absolute degree.** The centigrade degrees can be converted to **kelvins** by using the equation.

$$\text{K} = {}^\circ\text{C} + 273$$

The Kelvin temperature (or absolute temperature) is always used in calculations of other parameters of gases. **Remember that the degree sign (°) is not used with K.**

The Moles of a Gas Sample, *n*

The number of moles, n, of a sample of a gas in a container can be found by dividing the mass, m, of the sample by the molar mass, M (molecular mass).

$$\text{moles of gas } (n) = \frac{\text{mass of gas sample } (m)}{\text{molecular mass of gas } (M)}$$

THE GAS LAWS

The volume of a given sample of gas depends on the temperature and pressure applied to it. Any change in temperature or pressure will affect the volume of the gas. As results of experimental studies from 17th to 19th century, scientists derived the relationships among the pressure, temperature and volume of a given mass of gas. These relationships, which describe the general behavior of gases, are called the **gas laws.**

BOYLE'S LAW

In 1660 Robert Boyle found out experimentally the change in volume of a given sample of gas with pressure at room temperature. From his observations he formulated a generalization known as **Boyle's Law.** It states that : **at constant temperature, the volume of a fixed mass of gas is inversely proportional to its pressure. If the pressure is doubled, the volume is halved.**

The Boyle's Law may be expressed mathematically as

$$V \propto 1/P \qquad (T, n \text{ are constant})$$

or

$$V = k \times 1/P$$

where k is a proportionality constant.

$$PV = k$$

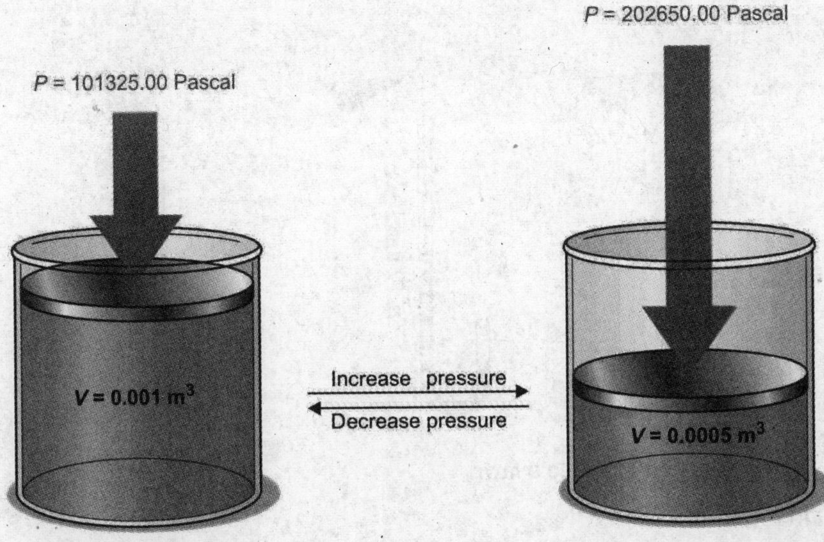

■ Figure 10.4
Boyle's law states that at constant temperature, the volume of a fixed mass of gas is inversely proportional to its pressure. If the pressure is doubled, the volume is halved.

If P_1, V_1 are the initial pressure and volume of a given sample of gas and P_2, V_2 the changed pressure and volume, we can write

$$P_1 V_1 = k = P_2 V_2$$

or

$$P_1 V_1 = P_2 V_2$$

This relationship is useful for the determination of the volume of a gas at any pressure, if its volume at any other pressure is known.

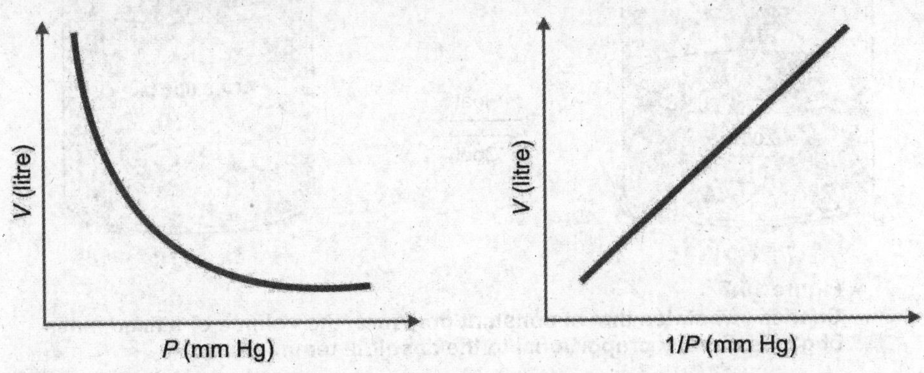

■ Figure 10.5
Graphical representation of Boyle's law. (*a*) a plot of *V* versus *P* for a gas sample is hyperbola; (*b*) a plot of *V* versus 1/*P* is a straight line.

The Boyle's law can be demonstrated by adding liquid mercury to the open end of a J-tube. As the pressure is increased by addition of mercury, the volume of the sample of trapped gas decreases. Gas pressure and volume are inversely related; one increases when the other decreases.

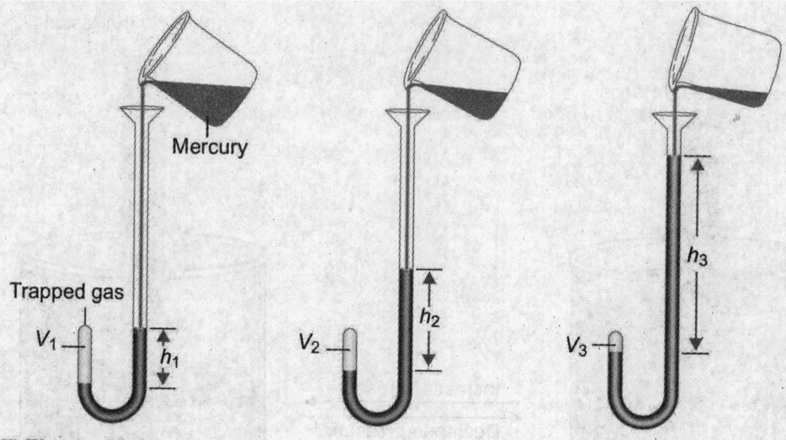

Figure 10.6
Demonstration of Boyle's law.

CHARLES'S LAW

In 1787 Jacques Charles investigated the effect of change of temperature on the volume of a fixed amount of gas at constant pressure. He established a generalization which is called the **Charles' Law.** It states that : **at constant pressure, the volume of a fixed mass of gas is directly proportional to the Kelvin temperature of absolute temperature. If the absolute temperature is doubled, the volume is doubled.**

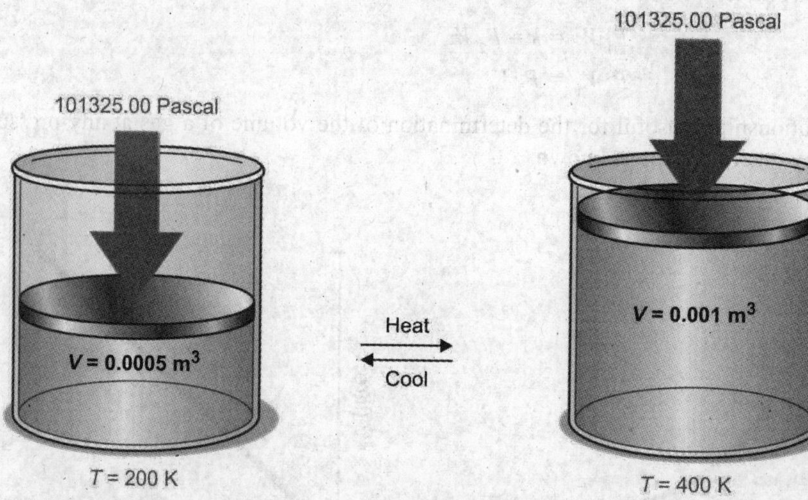

Figure 10.7
Charles law states that at constant pressure, the volume of a fixed mass of gas is directly proportional to the absolute temperature.

Charles' Law may be expressed mathematically as

$$V \propto T \qquad (P, n \text{ are constant})$$

or

$$V = k\,T$$

where k is a constant.

$$\frac{V}{T} = k$$

or

If V_1, T_1 are the initial volume and temperature of a given mass of gas at constant pressure and V_2, T_2 be the new values, we can write

$$\frac{V_1}{T_1} = k = \frac{V_2}{T_2}$$

$$\frac{V_1}{T_1} = \frac{V_2}{T_2}$$

or

Using this expression, the new volume V_2, can be found from the experimental values of V_1, T_1 and T_2.

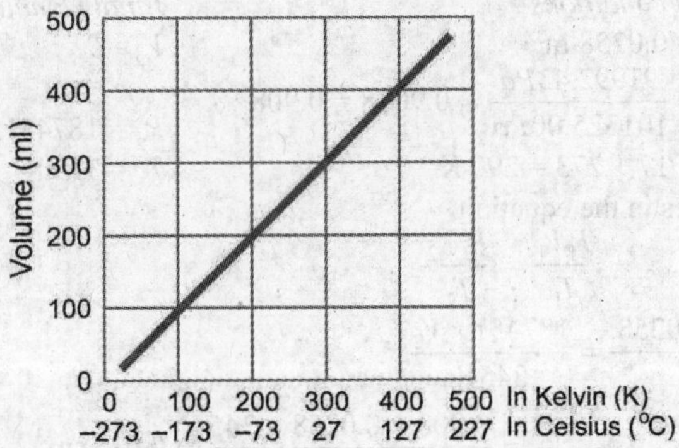

■ **Figure 10.8**
Graph showing that at constant pressure, volume of a given mass of gas is directly proportional to the Kelvin temperature.

THE COMBINED GAS LAW

Boyle's Law and Charles' Law can be combined into a single relationship called the **Combined Gas Law.**

Boyle's Law	$V \propto \frac{1}{P}$	(T, n constant)
Charles' Law	$V \propto T$	(P, n constant)
Therefore,	$V \propto \frac{T}{P}$	(n constant)

The combined law can be stated as : **for a fixed mass of gas, the volume is directly proportional to kelvin temperature and inversely proportional to the pressure.**

If k be the proportionality constant,

$$V = \frac{kT}{P} \quad (n \text{ constant})$$

$$\frac{PV}{T} = k \quad (n \text{ constant})$$

or

If the pressure, volume and temperature of a gas be changed from P_1, V_1 and T_1 to P_2, T_2 and V_2, then

$$\frac{P_1V_1}{T_1} = k \qquad \frac{P_2V_2}{T_2} = k$$

or
$$\frac{P_1V_1}{T_1} = \frac{P_2V_2}{T_2}$$

This is the form of combined law for two sets of conditions. It can be used to solve problems involving a change in the three variables *P*, *V* and *T* for a fixed mass of gas.

SOLVED PROBLEM 1

25.8×10^{-3} m³ liter of a gas has a pressure of 690 torr (91992.43 Pa) and a temperature of 290 K. What will be the volume if the pressure is changed to 1.85 atm (187451.00 Pa) and the temperature to 345 K.

SOLUTION

Initial conditions :

$V_1 = 0.0258$ m³

$$P_1 = \frac{91992.43 Pa}{101325.00 Pa} = 0.9078 = 0.908$$

$T_1 = 17 + 273 = 290$ K

Final Conditions :

$V_2 = ?$

$P_2 = 187451.00$ Pa

$T_2 = 345$ K

Substituting values in the equation

$$\frac{P_1V_1}{T_1} = \frac{P_2V_2}{T_2}$$

$$\frac{0.908 \times 0.0258}{290\text{K}} = \frac{187451 \times V_2}{345\text{K}}$$

Hence,
$$V_2 = \frac{0.908 \times 0.0258 \times 345}{290 \times 187451} = \mathbf{0.0151\,m^3}$$

GAY LUSSAC'S LAW

In 1802 Joseph Gay Lussac as a result of his experiments established a general relation between the pressure and temperature of a gas. This is known as **Gay Lussac's Law or Pressure-Temperature Law.** It states that : **at constant volume, the pressure of a fixed mass of gas is directly proportional to the Kelvin temperature or absolute temperature.**

The law may be expressed mathematically as

$$P \propto T \qquad \text{(Volume, } n \text{ are constant)}$$

or
$$P = kT$$

or
$$\frac{P}{T} = k$$

For different conditions of pressure and temperature

$$\frac{P_1}{T_1} = k = \frac{P_2}{T_2}$$

or
$$\frac{P_1}{T_1} = \frac{P_2}{T_2}$$

Knowing P_1, T_1, and T_2, P_2 can be calculated.

AVOGADRO'S LAW

Let us take a balloon containing a certain mass of gas. If we add to it more mass of gas, holding the temperature (*T*) and pressure (*P*) constant, the volume of gas (*V*) will increase. It was found experimentally that the amount of gas in moles is proportional to the volume. That is,

$$V \propto n \qquad (T \text{ and } P \text{ constant})$$

or
$$V = A\,n$$

where A is constant of proportionality.

or
$$\frac{V}{n} = A$$

For any two gases with volumes V_1, V_2 and moles n_1, n_2 at constant T and P,

$$\frac{V_1}{n_1} = A = \frac{V_2}{n_2}$$

If
$$V_1 = V_2, \qquad n_1 = n_2$$

Thus for equal volumes of the two gases at fixed T and P, number of moles is also equal. This is the basis of **Avogadro's Law** which may be stated as : **equal volumes of gases at the same temperature and pressure contain equal number of moles or molecules. If the molar amount is doubled, the volume is doubled.**

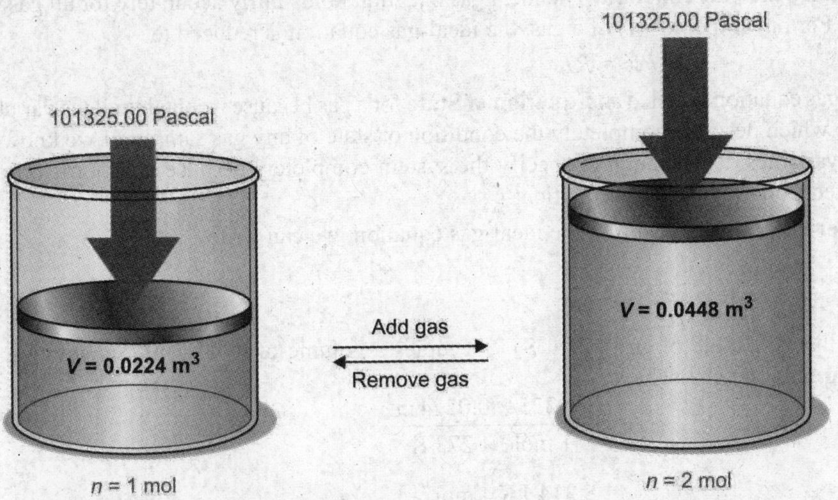

■ Figure 10.9
Avogadro's law states that under equal conditions of temperature and pressure, equal volumes of gases contain an equal number of molecules.

The Molar Gas Volume. It follows as a corollary of Avogadro's Law that one mole of any gas at a given temperature (T) and pressure (P) has the same fixed volume. It is called the molar gas volume or **molar volume.** In order to compare the molar volumes of gases, chemists use a fixed reference temperature and pressure. This is called standard temperature and pressure (abbreviated, **STP**). The standard temperature used is 273 K (0°C) and the standard pressure is 1 atm (101325.00 Pa). At STP we find experimentally that one mole of any gas occupies a volume of 0.0224 m³. To put it in the form of an equation, we have

$$1 \text{ mole of a gas at STP} = 0.0224 \text{ m}^3$$

THE IDEAL GAS EQUATION

We have studied three simple gas laws :

Boyle's Law	$V \propto \frac{1}{P}$	(T, n constant)
Charles' Law	$V \propto T$	(n, P constant)
Avogadro's Law	$V \propto n$	(P, T constant)

These three laws can be combined into a single more general gas law :

$$V \propto \frac{nT}{P} \qquad ...(1)$$

This is called the **Universal Gas Law.** It is also called **Ideal Gas Law** as it applies to all gases which exhibit ideal behavior *i.e.*, obey the gas laws perfectly. The ideal gas law may be stated as : **the volume of a given amount of gas is directly proportional to the number of moles of gas, directly proportional to the temperature, and inversely proportional to the pressure.**

Introducing the proportionality constant R in the expression (1) we can write

$$V = R\frac{nT}{P}$$

or

$$PV = nRT \qquad ...(2)$$

The equation (2) is called the **Ideal-gas Equation** or simply the general **Gas Equation.** The constant R is called the **Gas constant.** The ideal gas equation holds fairly accurately for all gases at low pressures. For one mole ($n = 1$) of a gas, the ideal-gas equation is reduced to

$$PV = RT \qquad ...(3)$$

The ideal-gas equation is called an **Equation of State** for a gas because it contains all the variables (T, P, V and n) which describe completely the condition or state of any gas sample. If we know the three of these variables, it is enough to specify the system completely because the fourth variable can be calculated from the ideal-gas equation.

The Numerical Value of *R*. From the ideal-gas equation, we can write

$$R = \frac{PV}{nT} \qquad ...(1)$$

We know that one mole of any gas at STP occupies a volume of 0.0224 m³. Substituting the values in the expression (1), we have

$$R = \frac{101325 \times 0.0224\,\text{m}^3}{1\text{ mole} \times 273\text{ K}}$$

$$= 8.314\text{ J K}^{-1}\text{ mol}^{-1}$$

It may be noted that the unit for R is complex; it is a composite of all the units used in calculating the constant.

If the pressure is written as force per unit area and volume as area times length, from (1)

$$R = \frac{(\text{force/area}) \times \text{area} \times \text{length}}{n \times T} = \frac{\text{force} \times \text{length}}{n \times T}$$

$$= \frac{\text{work}}{n\,T}$$

Hence R can be expressed in units of work or energy per degree per mole. The actual value of R depends on the units of P and V used in calculating it. The more important values of R are listed in Table 10.1.

TABLE 10.1. VALUE OF R IN DIFFERENT UNITS

0.0821	liter-atm K^{-1} mol^{-1}	8.314×10^7	erg K^{-1} mol^{-1}
82.1	ml-atm K^{-1} mol^{-1}	8.314	J K^{-1} mol^{-1}
62.3	liter-mm Hg K^{-1} mol^{-1}	1.987	cal K^{-1} mol^{-1}

DALTON'S LAW OF PARTIAL PRESSURES

John Dalton visualized that in a mixture of gases, each component gas exerted a pressure as if it were alone in the container. The individual pressure of each gas in the mixture is defined as its **Partial Pressure.** Based on experimental evidence, in 1807, Dalton enunciated what is commonly known as the **Dalton's Law of Partial Pressures.** It states that : **the total pressure of a mixture of gases is equal to the sum of the partial pressures of all the gases present (Fig. 10.10).**

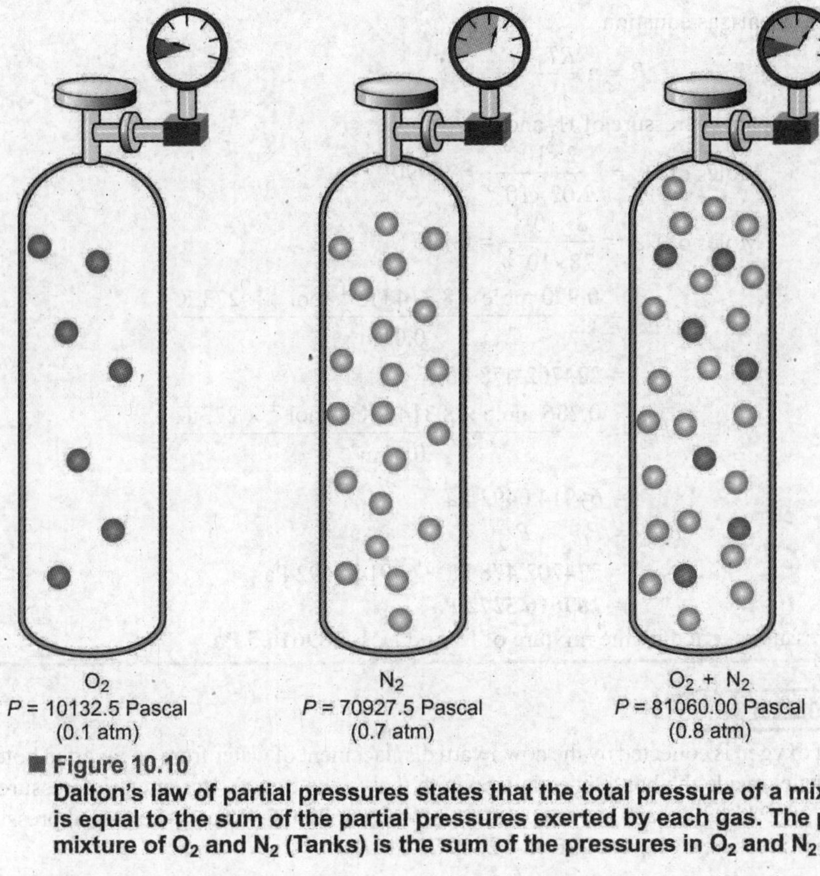

■ Figure 10.10
Dalton's law of partial pressures states that the total pressure of a mixture of gases is equal to the sum of the partial pressures exerted by each gas. The pressure of the mixture of O_2 and N_2 (Tanks) is the sum of the pressures in O_2 and N_2 tanks.

Mathematically the law can be expressed as $P_{total} = P_1 + P_2 + P_3 ...$ (V and T are constant) where P_1, P_2 and P_3 are partial pressures of the three gases 1, 2 and 3; and so on.

Dalton's Law of Partial Pressures follows by application of the ideal-gas equation $PV = n\,RT$ separately to each gas of the mixture. Thus we can write the partial pressures P_1, P_2 and P_3 of the three gases

$$P_1 = n_1\left(\frac{RT}{V}\right) \qquad P_2 = n_2\left(\frac{RT}{V}\right) \qquad P_3 = n_3\left(\frac{RT}{V}\right)$$

where n_1, n_2 and n_3 are moles of gases 1, 2 and 3. The total pressure, P_t, of the mixture is

$$P_t = (n_1 + n_2 + n_3)\frac{RT}{V}$$

or

$$P_t = n_t\,\frac{RT}{V}$$

In the words, the total pressure of the mixture is determined by the total number of moles present whether of just one gas or a mixture of gases.

SOLVED PROBLEM 2

What pressure is exerted by a mixture of 0.002 kg of H_2 and 0.008 kg of N_2 at 273 K in a 0.01 m³ (10 liter) vessel ?

SOLUTION

Applying the ideal-gas equation

$$P = n\frac{RT}{V}$$

we can find the partial pressure of H_2 and N_2

$$\text{Moles of } H_2 = \frac{2\times10^{-3}}{2.02\times10^{-3}} = 0.990$$

$$\text{Moles of } N_2 = \frac{8\times10^{-3}}{28\times10^{-3}} = 0.286$$

$$\therefore \quad P_{H_2} = \frac{0.990 \text{ mole} \times 8.314\,\text{J K}^{-1}\text{ mol}^{-1} \times 273 \text{ K}}{0.01\text{m}^3}$$

$$= 224702.478 \text{ Pa}$$

and

$$P_{N_2} = \frac{0.286 \text{ mole} \times 8.314\,\text{J K}^{-1}\text{ mol}^{-1} \times 273 \text{ K}}{0.01\text{m}^3}$$

$$= 64914.0492 \text{ Pa}$$

Thus

$$P_{\text{total}} = P_{H_2} + P_{N_2}$$

$$= 224702.478 \text{ Pa} + 64914.0492 \text{ Pa}$$

$$= \mathbf{289616.5272 \text{ Pa}}$$

Thus the pressure exerted by the mixture of H_2 and N_2 is **289616.5 Pa.**

SOLVED PROBLEM 3

A sample of oxygen is collected by the downward displacement of water from an inverted bottle. The water level inside the bottle is equalized with that in the trough. Barometeric pressure is found to be 100925.03 Pa, and the temperature of water is 296 K. What is the partial pressure of O_2 ? Vapor pressure of H_2O at 296 K = 2639.78 Pa.

SOLUTION

The total pressure inside the bottle is

$$P_{\text{total}} = P_{O_2} + P_{H_2O}$$

Since the water levels inside and outside the bottle were equalized, the total gas pressure inside the bottles must be equal to P_{atm}.

$$\therefore \quad P_{\text{total}} = P_{\text{atm}} = P_{O_2} + P_{H_2O}$$

But P_{atm} is given as (100925.03 Pa)

$$\therefore \quad P_{O_2} = 100925.03 - P_{H_2O}$$

$$= 100925.03 \text{ Pa} - 2639.78 \text{ Pa}$$

$$= \mathbf{98285.25 \text{ Pa}}$$

Thus the partial pressure of O_2 is **98285.25 Pa.**

GRAHAM'S LAW OF DIFFUSION

When two gases are placed in contact, they mix spontaneously. This is due to the movement of molecules of one gas into the other gas. This process of mixing of gases by random motion of the molecules is called **Diffusion.** Thomas Graham observed that molecules with smaller masses diffused faster than heavy molecules.

■ **Figure 10.11**
A light molecule diffuses quicker than a heavy molecule.

In 1829 Graham formulated what is now known as Graham's Law of Diffusion. It states that : **under the same conditions of temperature and pressure, the rates of diffusion of different gases are inversely proportional to the square roots of their molecular masses.**

Mathematically the law can be expressed as

$$\frac{r_1}{r_2} = \sqrt{\frac{M_2}{M_1}}$$

where r_1 and r_2 are the rates of diffusion of gases 1 and 2, while M_1 and M_2 are their molecular masses.

When a gas escapes through a pin-hole into a region of low pressure of vacuum, the process is called **Effusion.** The rate of effusion of a gas also depends, on the molecular mass of the gas.

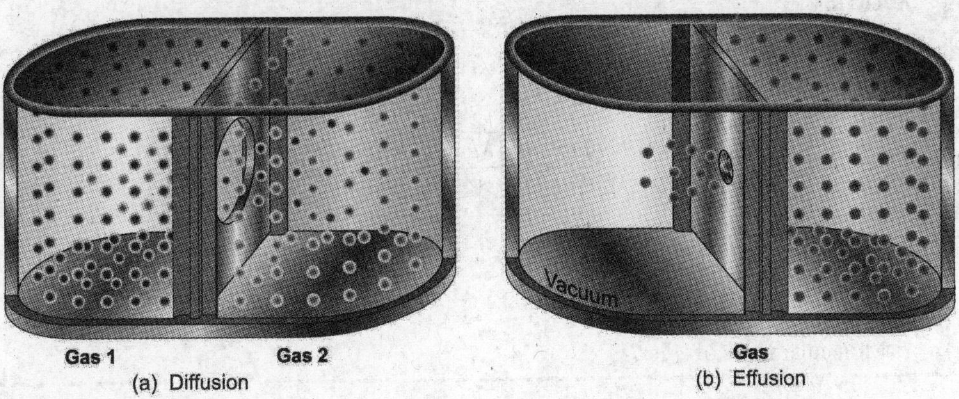

■ **Figure 10.12**
(*a*) Diffusion is mixing of gas molecules by random motion under conditions where molecular collisions occur. (*b*) Effusion is escape of a gas through a pinhole without molecular collisions.

Dalton's law when applied to effusion of a gas is called the Dalton's Law of Effusion. It may be expressed mathematically as

$$\frac{\text{Effusion rate of Gas 1}}{\text{Effusion rate of Gas 2}} = \sqrt{\frac{M_2}{M_1}} \qquad (P, T \text{ constant})$$

The determination of rate of effusion is much easier compared to the rate of diffusion. Therefore, Dalton's law of effusion is often used to find the molecular mass of a given gas.

SOLVED PROBLEM 4

If a gas diffuses at a rate of one-half as fast as O_2, find the molecular mass of the gas.

SOLUTION

Applying Graham's Law of Diffusion,

$$\frac{r_1}{r_2} = \sqrt{\frac{M_2}{M_1}}$$

$$\frac{\frac{1}{2}}{1} = \sqrt{\frac{32}{M_1}}$$

Squaring both sides of the equation.

$$\left(\frac{1}{2}\right)^2 = \frac{32}{M_1} \text{ or } \frac{1}{4} = \frac{32}{M_1}$$

Hence, $M_1 = 128$

Thus the molecular mass of the unknown gas is **128.**

SOLVED PROBLEM 5

0.00005 m^3 of gas A effuse through a pin-hole in 146 seconds. The same volume of CO_2 under identical conditions effuses in 115 seconds. Calculate the molecular mass of A.

SOLUTION

$$\frac{\text{Effusion rate of } CO_2}{\text{Effusion rate of } A} = \sqrt{\frac{M_A}{M_{CO_2}}}$$

$$\frac{5\times10^{-5}/115}{50\times10^{-5}/146} = \sqrt{\frac{M_A}{44}}$$

or $$(1.27)^2 = \frac{M_A}{44}$$

Hence $M_A = 71$

∴ Molecular mass of A is **71.**

KINETIC MOLECULAR THEORY OF GASES

Maxwell and Boltzmann (1859) developed a mathematical theory to explain the behavior of gases and the gas laws. It is based on the fundamental concept that **a gas is made of a large number of molecules in perpetual motion.** Hence the theory is called the **kinetic molecular theory** or simply the **kinetic theory of gases** (The word kinetic implies motion). The kinetic theory makes the following assumptions.

Assumptions of the Kinetic Molecular Theory

(1) **A gas consists of extremely small discrete particles called molecules dispersed throughout the container.** The actual volume of the molecules is negligible compared to the total volume of the gas. The molecules of a given gas are identical and have the same mass (m).

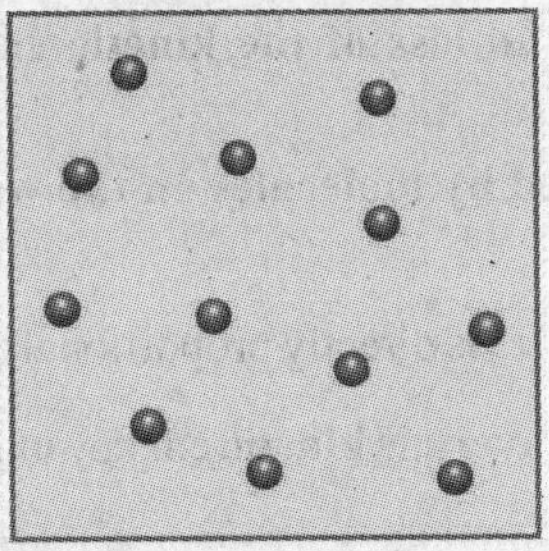

■ **Figure 10.13**
A gas is made of molecules dispersed in space in the container.

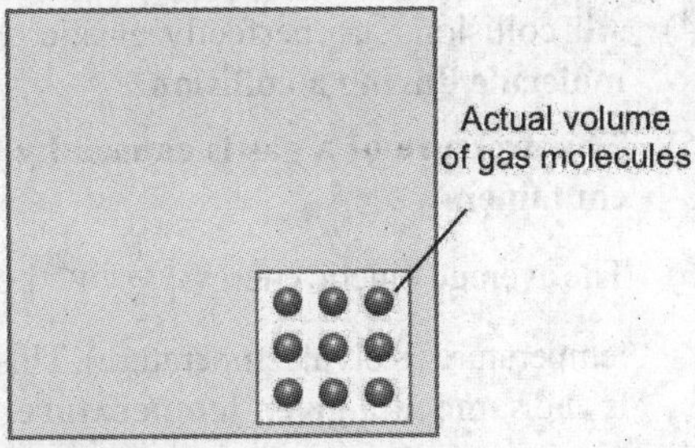

■ **Figure 10.14**
Actual volume of the gas molecules is negligible.

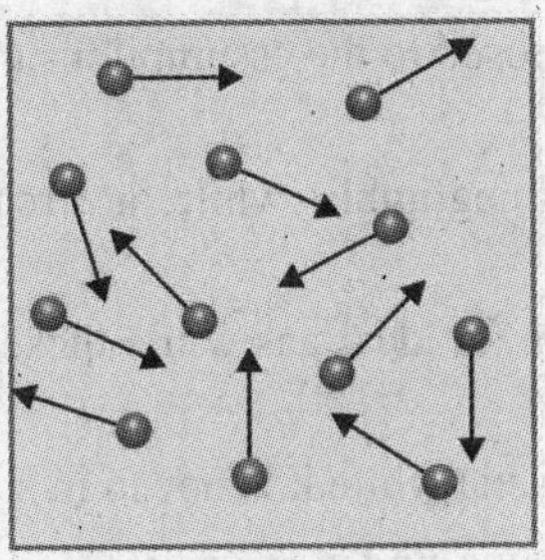

■ **Figure 10.15**
Gas molecules are in constant motion in all possible directions.

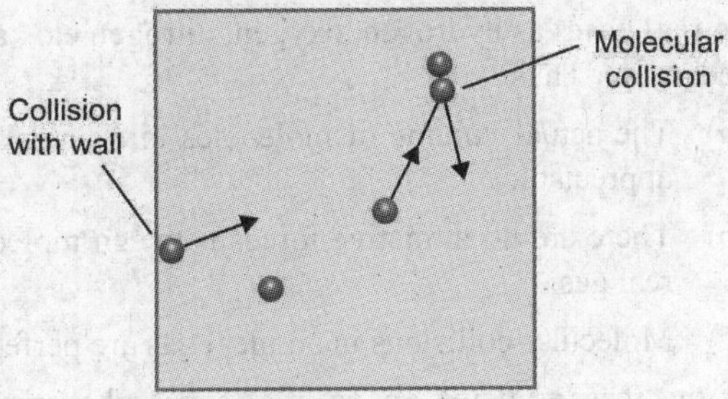

■ **Figure 10.16**
Molecules move in straight line and change direction on collision with another molecule or wall of container.

(2) **Gas molecules are in constant random motion with high velocities.** They move in straight lines with uniform velocity and change direction on collision with other molecules or the walls of the container. Pool table analogy is shown in Fig.10.17.

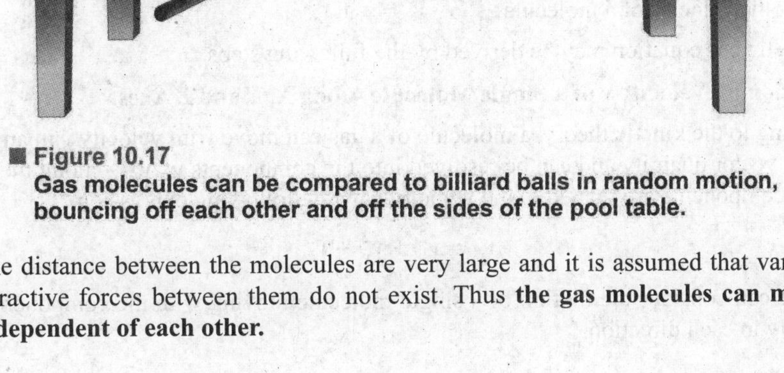

■ **Figure 10.17**
Gas molecules can be compared to billiard balls in random motion, bouncing off each other and off the sides of the pool table.

(3) The distance between the molecules are very large and it is assumed that van der Waals attractive forces between them do not exist. Thus **the gas molecules can move freely, independent of each other.**

(4) All collisions are perfectly elastic. Hence, **there is no loss of the kinetic energy of a molecule during a collision.**

(5) **The pressure of a gas is caused by the hits recorded by molecules on the walls of the container.**

(6) The average kinetic energy $\left(\frac{1}{2}mv^2\right)$ of the gas molecules is directly proportional to absolute temperature (Kelvin temperature). This implies that **the average kinetic energy of molecules is the same at a given temperature.**

How Does an Ideal Gas Differ from Real Gases?

A gas that confirms to the assumptions of the kinetic theory of gases is called an ideal gas. It obeys the basic laws strictly under all conditions of temperature and pressure.

The **real gases** as hydrogen, oxygen, nitrogen etc., are opposed to the assumptions (1), (2) and (3) stated above. Thus :

(*a*) The actual volume of molecules in an ideal gas is negligible, while in a real gas it is appreciable.

(*b*) There are no attractive forces between molecules in an ideal gas while these exist in a real gas.

(*c*) Molecular collisions in an ideal gas are perfectly elastic while it is not so in a real gas.

For the reasons listed above, real gases obey the gas laws under moderate conditions of temperature and pressure. At very low temperature and very high pressure, the clauses (1), (2) and (3) of kinetic theory do not hold. Therefore, under these conditions the real gases show considerable deviations from the ideal gas behavior.

DERIVATION OF KINETIC GAS EQUATION

Starting from the postulates of the kinetic molecular theory of gases we can develop an important equation. This equation expresses PV of a gas in terms of the number of molecules, molecular mass and molecular velocity. This equation which we shall name as the *Kinetic Gas Equation* may be derived by the following clauses.

Let us consider a certain mass of gas enclosed in a cubic box (Fig. 10.18) at a fixed temperature. Suppose that :

the length of each side of the box $= l$ cm

the total number of gas molecules $= n$

the mass of one molecule $= m$

the velocity of a molecule $= v$

The kinetic gas equation may be derived by the following steps :

(1) Resolution of Velocity v of a Single Molecule Along X, Y and Z Axes

According to the kinetic theory, a molecule of a gas can move with velocity v in any direction. Velocity is a vector quantity and can be resolved into the components v_x, v_y, v_z along the X, Y and Z axes. These components are related to the velocity v by the following expression.

$$v^2 = v_x^2 + v_y^2 + v_z^2$$

Now we can consider the motion of a single molecule moving with the component velocities independently in each direction.

(2) The Number of Collisions Per Second on Face *A* Due to One Molecule

Consider a molecule moving in *OX* direction between opposite faces *A* and *B*. It will strike the face *A* with velocity v_x and rebound with velocity $-v_x$. To hit the same face again, the molecule must travel *l* cm to collide with the opposite face *B* and then again *l* cm to return to face *A*. Therefore,

$$\text{the time between two collisions of face } A_v = \frac{2l}{v_x} \text{ seconds}$$

$$\text{the number of collisions per second on face } A = \frac{v_x}{2l}$$

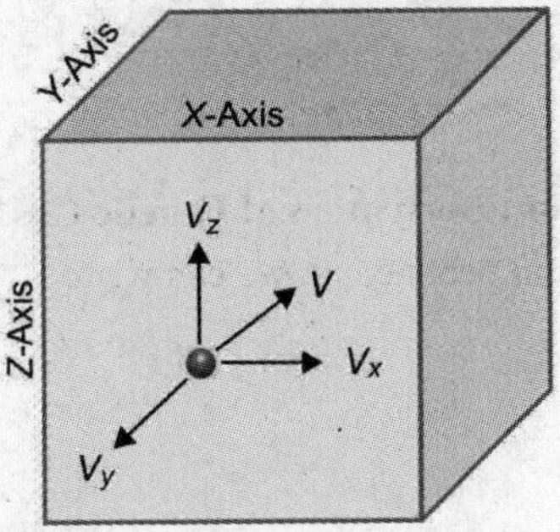

■ Figure 10.18
Resolution of velocity v into components V_x, V_y and V_z.

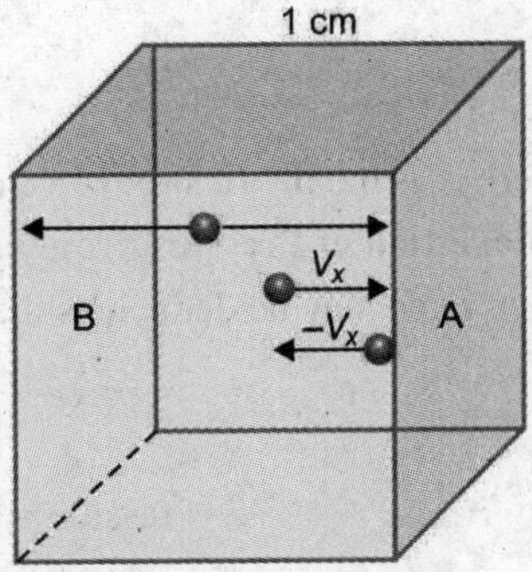

■ Figure 10.19
Cubic box showing molecular collisions along *X* axis.

(3) The Total Change of Momentum on All Faces of the Box Due to One Molecule Only

Each impact of the molecule on the face *A* causes a change of momentum (mass × velocity) :

$$\begin{aligned} &\text{the momentum before the impact} &&= mv_x \\ &\text{the momentum after the impact} &&= m(-v_x) \\ \therefore\ &\text{the change of momentum} &&= mv_x - (-mv_x) \\ & &&= 2\,mv_x \end{aligned}$$

But the number of collisions per second on face *A* due to one molecule $= \dfrac{v_x}{2l}$

Therefore, the total change of momentum per second on face *A* caused by one molecule

$$= 2m\,v_x \times \left(\frac{v_x}{2l}\right) = \frac{m\,v_x^2}{l}$$

The change of momentum on both the opposite faces *A* and *B* along *X*-axis would be double *i.e.*, $2mv_x^2/l$ similarly, the change of momentum along *Y*-axis and *Z*-axis will be $2mv_y^2/l$ and $2mv_z^2/l$ respectively. Hence, the overall change of momentum per second on all faces of the box will be

$$= \frac{2mv_x^2}{l} + \frac{2mv_y^2}{l} + \frac{2mv_z^2}{l}$$

$$= \frac{2m}{l}(v_x^2 + v_y^2 + v_z^2)$$

$$= \frac{2m\,v^2}{l} \qquad (\because\ v^2 = v_x^2 + v_y^2 + v_z^2)$$

(4) Total Change of Momentum Due to Impacts of All the Molecules on All Faces of the Box

Suppose there are N molecules in the box each of which is moving with a different velocity v_1, v_2, v_3, etc. The total change of momentum due to impacts of all the molecules on all faces of the box

$$= \frac{2m}{l}(v_1^2 + v_2^2 + v_3^2 + ...)$$

Multiplying and dividing by n, we have

$$= \frac{2mN}{l}\left(\frac{v_1^2 + v_2^2 + v_3^2 + ...}{n}\right)$$

$$= \frac{2mN\,u^2}{l}$$

where u^2 is the **mean square velocity.**

(5) Calculation of Pressure from Change of Momentum; Derivation of Kinetic Gas Equation

Since force may be defined as the change in momentum per second, we can write

$$\text{Force} = \frac{2mN\,u^2}{l}$$

But
$$\text{Pressure} = \frac{\text{Total Force}}{\text{Total Area}}$$

$$P = \frac{2mNu^2}{l} \times \frac{1}{6l^2} = \frac{1}{3}\frac{mNu^2}{l^3}$$

Since l^3 is the volume of the cube, V, we have

$$P = \frac{1}{3}\frac{mNu^2}{V}$$

or
$$P\,V = \frac{1}{3}mNu^2$$

This is the fundamental equation of the kinetic molecular theory of gases. It is called the **Kinetic Gas equation.** This equation although derived for a cubical vessel, is equally valid for a vessel of any shape. The available volume in the vessel could well be considered as made up of a large number of infinitesimally small cubes for each of which the equation holds.

Significance of the term *u*. As stated in clause (4) u^2 is the mean of the squares of the individual velocities of all the N molecules of the gas. But $u = \sqrt{u^2}$. Therefore u is called the **Root Mean Square (or RMS) Velocity.**

KINETIC GAS EQUATION IN TERMS OF KINETIC ENERGY

If N be the number of molecules in a given mass of gas,

$$P\,V = \frac{1}{3}mNu^2 \quad \text{(Kinetic Gas equation)}$$

$$= \frac{2}{3}N \times \frac{1}{2}mu^2$$

$$= \frac{2}{3}N \times e$$

where e is the average kinetic energy of a single molecule.

$$\therefore \quad PV = \frac{2}{3}Ne = \frac{2}{3}E$$

or $$PV = \frac{2}{3}E \qquad ...(1)$$

where E is the total kinetic energy of all the N molecules. The expression (1) may be called the kinetic gas equation in terms of kinetic energy.

We know that the General ideal gas equation is

$$PV = nRT \qquad ...(2)$$

From (1) and (2)

$$\frac{2}{3}E = nRT \qquad ...(3)$$

For one mole of gas, the kinetic energy of N molecules is,

$$E = \frac{3RT}{2} \qquad ...(4)$$

Since the number of gas molecules in one mole of gas in N_0 (Avogadro number),

$$e = \frac{E}{N_0} = \frac{3RT}{2N_0}$$

or $$e = \frac{3RT}{2N_0} \qquad ...(5)$$

substituting the values of R, T, N_0, in the equation (5), the average kinetic energy of a gas molecule can be calculated.

SOLVED PROBLEM 6

Calculate the average kinetic energy of a hydrogen molecule at 273 K.

SOLUTION

$$e = \frac{3RT}{2N_0}$$

Here $R = 8.314 \text{ J K}^{-1} \text{ mol}^{-1}$

$T = 273 \text{ K}$; $N_0 = 6.02 \times 10^{23}$

$$e = \frac{3}{2} \times \frac{8.314 \times 273}{6.02 \times 10^{23}} = \mathbf{5.66 \times 10^{-21}} \textbf{ Joule}$$

$\therefore$

Thus the average kinetic energy of H_2 at 273 is **5.66×10^{-21} J**

SOLVED PROBLEM 7

Calculate the kinetic energy of two moles of N_2 at 300 K. ($R = 8.314 \text{ JK}^{-1} \text{ mol}^{-1}$)

SOLUTION

We know $$E = \frac{3}{2}nRT$$

Here, $T = 300 \text{ K}$; $n = 2$; $R = 8.314 \text{ JK}^{-1} \text{ mol}^{-1}$

Substituting these values, we have

$$E = \frac{3}{2} \times 2 \times 8.314 \times 300 = \mathbf{7482.6 \text{ J}}$$

Therefore the kinetic energy of two moles of N_2 is **7482.6 J**

DEDUCTION OF GAS LAWS FROM THE KINETIC GAS EQUATION

(*a*) Boyle's Law

According to the Kinetic Theory, there is a direct proportionality between absolute temperature and average kinetic energy of the molecules *i.e.*,

$$\frac{1}{2}mNu^2 \propto T$$

or $$\frac{1}{2}mNu^2 = kT$$

or $$\frac{3}{2}\times\frac{1}{3}mNu^2 = kT$$

or $$\frac{1}{3}mNu^2 = \frac{2}{3}kT$$

Substituting the above value in the kinetic gas equation $PV = \frac{1}{3}mNu^2$, we have

$$PV = \frac{2}{3}kT$$

The product *PV*, therefore, will have a constant value at a constant temperature. This is Boyle's Law.

(*b*) Charles' Law

As derived above,

$$PV = \frac{2}{3}kT$$

or $$V = \frac{2}{3}\times\frac{k}{P}T$$

At constant pressure,

$$V = k'T \qquad \text{where } \left(k' = \frac{2}{3}\times\frac{k}{P}\right)$$

or $$V \propto T$$

That is, at constant pressure, volume of a gas is proportional to Kelvin temperature and this is Charles' Law.

(*c*) Avogadro's Law

If equal volume of two gases be considered at the same pressure,

$$PV = \frac{1}{3}m_1N_1u_1^2 \qquad \text{...Kinetic equation as applied to one gas}$$

$$PV = \frac{1}{3}m_2N_2u_2^2 \qquad \text{...Kinetic equation as applied to 2nd gas}$$

$$\therefore \quad \frac{1}{3}m_1N_1u_1^2 = \frac{1}{3}m_2N_2u_2^2 \qquad ...(1)$$

When the temperature (*T*) of both the gases is the same, their mean kinetic energy per molecule will also be the same.

i.e., $$\frac{1}{3}m_1u_1^2 = \frac{1}{3}m_2u_2^2 \qquad ...(2)$$

Dividing (1) by (2), we have

$$N_1 = N_2$$

Or, under the same conditions of temperature and pressure, equal volumes of the two gases contain the same number of molecules. This is Avogadro's Law.

(*d*) Graham's Law of Diffusion

If m_1 and m_2 are the masses and u_1 and u_2 the velocities of the molecules of gases 1 and 2, then at the same pressure and volume

$$\frac{1}{3}m_1N_1u_1^2 = \frac{1}{3}m_2N_2u_2^2$$

By Avogadro's Law $\quad N_1 = N_2$

$\therefore \quad m_1u_1^2 = m_2u_2^2$

or

$$\left(\frac{u_1}{u_2}\right)^2 = \frac{m_2}{m_1}$$

If M_1 and M_2 represent the molecular masses of gases 1 and 2,

$$\left(\frac{u_1}{u_2}\right)^2 = \frac{M_2}{M_1}$$

$$\frac{u_1}{u_2} = \sqrt{\frac{M_2}{M_1}}$$

The rate of diffusion (r) is proportional to the velocity of molecules (u), Therefore,

$$\frac{\text{Rate of diffusion of gas 1}}{\text{Rate of diffusion of gas 2}} = \frac{r_1}{r_2} = \sqrt{\frac{M_2}{M_1}}$$

This is Graham's Law of Diffusion.

DISTRIBUTION OF MOLECULAR VELOCITIES

While deriving Kinetic Gas Equation, it was assumed that all molecules in a gas have the same velocity. But it is not so. When any two molecules collide, one molecule transfers kinetic energy ($\frac{1}{2}mv^2$) to the other molecule. The velocity of the molecule which gains energy increases and that of the other decreases. Millions of such molecular collisions are taking place per second. Therefore, the velocities of molecules are changing constantly. Since the number of molecules is very large, a fraction of molecules will have the same particular velocity. In this way there is a broad distribution of velocities over different fractions of molecules. In 1860 James Clark Maxwell calculated the distribution of velocities from the laws of probability. He derived the following equation for the distribution of molecular velocities.

$$\frac{dN_c}{N} = 4\pi\left(\frac{M}{2\pi RT}\right)^{3/2} e^{\frac{-MC^2}{2RT}} C^2\, dc$$

where $\quad dN_c$ = number of molecules having velocities between C and $(C + dc)$

N = total number of molecules

M = molecular mass

T = temperature on absolute scale (K)

The relation stated above is called **Maxwell's law of distribution of velocities.** The ratio dn_c/n gives the fraction of the total number of molecules having velocities between C and $(C + dc)$. Maxwell plotted such fractions against velocity possessed by the molecules. The curves so obtained illustrate the salient features of Maxwell distribution of velocities.

Fig. 10.20. Shows the distribution of velocities in nitrogen gas, N_2, at 300 K and 600 K. It will be noticed that :

(1) A very small fraction of molecules has either very low (close to zero) or very high velocities.

(2) Most intermediate fractions of molecules have velocities close to an average velocity represented by the peak of the curve. This velocity is called the **most probable velocity.** It may be defined as the velocity possessed by the largest fraction of molecules corresponding to the highest point on the Maxvellian curve.

(3) At higher temperature, the whole curve shifts to the right (dotted curve at 600 K). This shows that **at higher temperature more molecules have higher velocities and fewer molecules have lower velocities.**

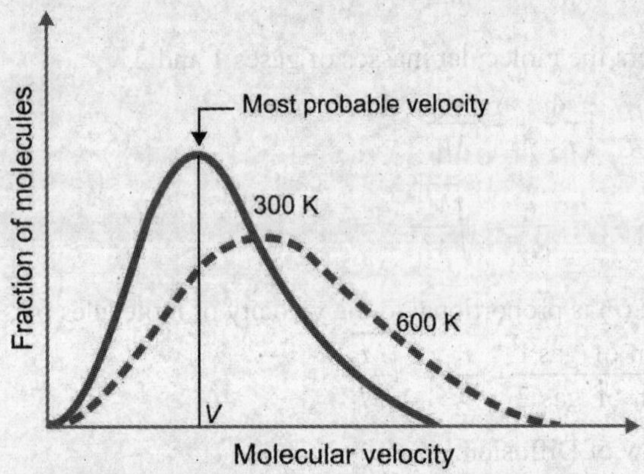

Figure 10.20
Distribution of molecular velocities in nitrogen gas, N_2, at 300 K and 600 K.

DIFFERENT KINDS OF VELOCITIES

In our study of kinetic theory we come across three different kinds of molecular velocities :

(1) the Average velocity ($\overline{V}$)

(2) the Root Mean Square velocity (μ)

(3) the Most Probable velocity (v_{mp})

Average Velocity

Let there be n molecules of a gas having individual velocities v_1, v_2, v_3 ..... v_n. The ordinary average velocity is the arithmetic mean of the various velocities of the molecules.

$$\overline{v} = \frac{v_1 + v_2 + v_3 \; ..... + v_n}{n}$$

From Maxwell equation it has been established that the average velocity v_{av} is given by the expression

$$\overline{v} = \sqrt{\frac{8RT}{\pi M}}$$

Substituting the values of R, T, π and M in this expression, the average value can be calculated.

Root Mean Square Velocity

If $v_1, v_2, v_3 \ldots\ldots v_n$ are the velocities of n molecules in a gas, μ^2, the mean of the squares of all the velocities is

$$\mu^2 = \frac{v_1^2 + v_2^2 + v_3^2 \ldots\ldots + v_n^2}{n}$$

Taking the root

$$\mu = \sqrt{\frac{v_1^2 + v_2^2 + v_3^2 \ldots\ldots + v_n^2}{n}}$$

μ is thus the **Root Mean Square velocity or RMS velocity.** It is denoted by u.

The value of the RMS of velocity u, at a given temperature can be calculated from the Kinetic Gas Equation.

$$PV = \frac{1}{3} mNu^2 \quad \text{...Kinetic Equation}$$

$$u^2 = \frac{3PV}{mN}$$

For one mole of gas

$$PV = RT$$

Therefore,

$$u^2 = \frac{3RT}{M} \quad \text{...}M \text{ is molar mass}$$

$$u = \sqrt{\frac{3RT}{M}}$$

By substituting the values of R, T and M, the value of u (RMS velocity) can be determined.

RMS velocity is superior to the average velocity considered earlier. With the help of u, the total Kinetic energy of a gas sample can be calculated.

Most Probable Velocity

As already stated the most probable velocity is possessed by the largest number of molecules in a gas. According to the calculations made by Maxwell, the most probably velocity, v_{mp}, is given by the expression.

$$v_{mps} = \sqrt{\frac{2RT}{M}}$$

Substituting the values of R, T and M in this expression, the most probably velocity can be calculated.

Relation between Average Velocity, RMS Velocity and Most Probable Velocity

We know that the average velocity, $\bar{v}$, is given by the expression

$$\bar{v} = \sqrt{\frac{8RT}{\pi M}}$$

and

$$\mu = \sqrt{\frac{3RT}{M}}$$

$$\therefore \quad \frac{\bar{v}}{\mu} = \sqrt{\frac{8RT}{\pi M}} \times \sqrt{\frac{M}{3RT}} = \sqrt{\frac{8}{3\pi}}$$

$$= 0.9213$$

or $$\bar{v} = \mu \times 0.9213 \qquad ...(1)$$

That is, **Average Velocity = 0.9213 × RMS Velocity**

The expression for the most probably velocity, v_{mp}, is

$$v_{mp} = \sqrt{\frac{2RT}{M}}$$

and $$\mu = \sqrt{\frac{3RT}{M}}$$

$$\therefore \quad \frac{v_{mp}}{\mu} = \sqrt{\frac{2RT}{M}} \times \sqrt{\frac{M}{3RT}} = \sqrt{\frac{2}{3}} = 0.8165$$

or $$v_{mp} = \mu \times 0.8165 \qquad ...(2)$$

That is, **Most Probable Velocity = 0.8165 × RMS Velocity**

RMS can be easily calculated by the application of Kinetic Gas equation. Knowing the value of RMS, we can find the average velocity and the most probable velocity from expressions (1) and (2).

CALCULATION OF MOLECULAR VELOCITIES

The velocities of gas molecules are exceptionally high. Thus velocity of hydrogen molecule is 1,838 meters s^{-1}. While it may appear impossible to measure so high velocities, these can be easily calculated from the Kinetic Gas equation. Several cases may arise according to the available data.

While calculating different types of velocities, we can also make use of the following expressions stated already.

RMS velocity, $$\mu = \sqrt{\frac{3RT}{M}}$$

Average velocity, $$\bar{v} = \sqrt{\frac{8RT}{\pi M}}$$

Most Probable velocity, $$v_{mp} = \sqrt{\frac{2RT}{M}}$$

Case 1. Calculation of Molecular Velocity when temperature alone is given

$$PV = \frac{1}{3} mNu^2 \qquad \text{(Kinetic Gas equation)}$$

where $$N = N_0 \qquad \text{(Avogadro's number)}$$

Thus we have,

$$M = m \times N_0 = \text{molecular mass of the gas}$$

$$\therefore \quad u = \sqrt{\frac{3PV}{M}} = \sqrt{\frac{3RT}{M}} \qquad (\because PV = RT \text{ for 1 mole})$$

But $$R = 8.314 \text{ J K}^{-1} \text{ mol}^{-1}$$

$$\therefore \quad u = \sqrt{\frac{3 \times 0.8314 \times T}{M}}$$

$$= \mathbf{1.58 \times 10^2 \times \sqrt{\frac{T}{M}} \textbf{ m sec}^{-1}}$$

where T is Kelvin temperature and M the molar mass.

SOLVED PROBLEM 8

Calculate the root mean square velocity of CO_2 molecule at 1273 K

SOLUTION

$$T = 1273 \text{ K}; M = 44$$

Applying the equation

$$u = 1.58 \times 10^4 \times \sqrt{\frac{T}{M}}$$

we have
$$u = 1.58 \times 10^4 \times \sqrt{\frac{1273}{44}}$$

$$\mathbf{u = 84985 \text{ cm s}^{-1} \quad \text{or} \quad 849.85 \text{ m s}^{-1}}$$

Case 2. Calculation of Molecular Velocity when temperature and pressure both are given.

In such cases we make use of the following relation based on Kinetic Gas equation.

$$u = \sqrt{\frac{3PV}{M}}$$

We know that 1 mole of a gas at STP occupies a volume of 0.0224 m³ (22400 ml) (known as molar volume). But before applying this relation the molar volume is reduced to the given conditions of temperature and pressure.

SOLVED PROBLEM 9

Calculate the RMS velocity of chlorine molecules at 285 K and 103991.44 Pa pressure.

SOLUTION

At STP :	At given conditions :
$V_1 = 0.0224 \text{ m}^3$	$V_2 = ?$
$T_1 = 273 \text{ K}$	$T_2 = 285 \text{ K}$
$P_1 = 101325.00 \text{ Pa}$	$P_2 = 103991.44 \text{ Pa}$

Applying
$$\frac{P_1 V_1}{T_1} = \frac{P_2 V_2}{T_2}$$

we have
$$V_2 = \frac{P_1 V_1 T_2}{T_1 P_2} = \frac{101325 \times 0.224 \times 285}{273 \times 103991.44} = 0.022785 \text{ m}^3$$

we know that
$$u = \sqrt{\frac{3PV}{M}}$$

$$P = hdg = 103991.44 \text{ Pa}$$

$$V = 0.022785 \text{ m}^3; \; M = 71$$

$$\therefore \quad u = \sqrt{\frac{3 \times 103991.44 \times 0.022785}{71 \times 10^{-3}}}$$

$$\mathbf{u = 31652 \text{ cm s}^{-1} \text{ or } 316.52 \text{ m s}^{-1}}$$

Case 3. Calculation of Molecular Velocity at STP

Here we use the relation

$$u = \sqrt{\frac{3PV}{M}}$$

where $P = 1 \text{ atm} = 101325 \text{ Pa}$

$V = 0.0224 \text{ m}^3$

$M = \text{Molar mass of the gas}$

SOLVED PROBLEM 10

Calculate the average velocity of nitrogen molecule at STP.

SOLUTION

In this example we have,

$P = 101325 \text{ Pa (1 atm)}$

$V = 22{,}400 \text{ ml } (0.0224 \text{ m}^3)$

$M = 28\text{g} = 28 \times 10^{-3} \text{ kg}.$

Substituting these values in the equation

$$u = \sqrt{\frac{3PV}{M}}$$

we have

$$= \sqrt{\frac{3 \times 101325 \times 0.0224}{28 \times 10^{-3}}}$$

$$= 493.13 \text{ s}^{-1}$$

$$\therefore \text{ Average velocity} = 0.9213 \times 493.13 \text{ s}^{-1}$$

$$= \mathbf{454.32 \text{ m s}^{-1}}$$

Case 4. Calculation of Molecular Velocity when pressure and density are given

In this case we have

$$u = \sqrt{\frac{3PV}{M}} \quad \text{or} \quad u = \sqrt{\frac{3P}{D}} \qquad \left[\because \frac{M}{V} = D\right]$$

where P is expressed in dynes cm^{-2} and D in gm ml^{-1}.

SOLVED PROBLEM 11

Oxygen at 1 atm (101325.00 Pa) pressure and 0°C (273 K) has a density of 1.429 kg/m^3. Find the RMS velocity of oxygen molecules.

SOLUTION

We have $P = 1 \text{ atm} = 101325.00 \text{ Pa}$

$D = 1.429 \text{ kg/m}^3$

Applying

$$u = \sqrt{\frac{3P}{D}}$$

we get

$$u = \sqrt{\frac{3 \times 101325.00}{1.429}} = \mathbf{461.21 \text{ m / sec}}$$

Case 5. Calculation of most probable velocity

In this case we have

$$v_{mp} = 1.29 \times 10^4 \sqrt{\frac{T}{M}}$$

where T expressed in Kelvin and M to mass.

SOLVED PROBLEM 12

Calculate the most probable velocity of nitrogen molecules, N_2, at 288 K.

SOLUTION

$$T = 273 + 15 = 288 \text{ K}$$

We know that

$$v_{mp} = 1.29 \times 10^4 \sqrt{\frac{T}{M}}$$

$$= 1.29 \times 10^4 \sqrt{\frac{288}{28\times 10^{-3}}}$$

$$= \mathbf{4.137 \times 10^2 \ m\ s^{-1}}$$

COLLISION PROPERTIES

In the derivation of Kinetic gas equation we did not take into account collisions between molecules. The molecules in a gas are constantly colliding with one another. The transport properties of gases such as diffusion, viscosity and mean free path depend on molecular collisions. We will now discuss some properties of gases which determine the frequency of collisions.

The Mean Free Path

At a given temperature, a molecule travels in a straight line before collision with another molecule. The distance travelled by the molecule before collision is termed *free path*. The free path for a molecule varies from time to time. **The mean distance travelled by a molecule between two successive collisions is called the Mean Free Path.** It is denoted by λ. If l_1, l_2, l_3 are the free paths for a molecule of a gas, its free path

$$\lambda = \frac{l_1 + l_2 + l_3 + ... + l_n}{n}$$

where n is the number of molecules with which the molecule collides. Evidently, the number of molecular collisions will be less at a lower pressure or lower density and longer will be the mean free path. The mean free path is also related with the viscosity of the gas.

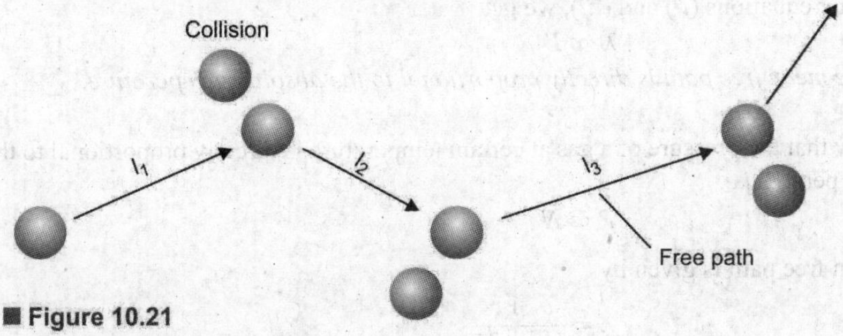

■ **Figure 10.21**
The mean free path illustrated.

The mean free path, λ, is given by the expression

$$\lambda = \eta\sqrt{\frac{3}{Pd}}$$

where P = pressure of the gas
d = density of the gas
η = coefficient of viscosity of the gas

By a determination of the viscosity of the gas, the mean free path can be readily calculated. At STP, the mean free path for hydrogen is 1.78×10^{-5} cm and for oxygen it is 1.0×10^{-5} cm.

Effect of Temperature and Pressure on Mean Free Path

(*a*) Temperature

The ideal gas equation for n moles of a gas is

$$PV = nRT \qquad ...(i)$$

where n is the number of moles given by

$$n = \frac{\text{Number of molecules}}{\text{Avogadro's Number}} = \frac{N}{N_0}$$

Substituting this in equation (*i*) we get

$$PV = \frac{N}{N_o}RT$$

or

$$\frac{N}{V} = \frac{PN_0}{RT}$$

At constant pressure

$$N \propto \frac{1}{T} \qquad ...(ii)$$

The mean free path is given by

$$\lambda = \frac{\text{Distance travelled by the molecule per second}}{\text{Number of collisions per c.c.}}$$

$$= \frac{\bar{v}}{\sqrt{2}\,\pi\sigma^2\bar{v}\bar{N}}$$

$$= \frac{1}{\sqrt{2}\,\pi\sigma^2\bar{N}} \qquad ...(iii)$$

combining equations (*ii*) and (*iii*), we get

$$\lambda \propto T$$

Thus, *the mean free path is directly proportional to the absolute temperature.*

(*b*) Pressure

We know that the pressure of a gas at certain temperature is directly proportional to the number of molecules per c.c. *i.e.*

$$P \propto \bar{N}$$

and mean free path is given by

$$\lambda = \frac{1}{\sqrt{2}\pi\sigma^2\bar{N}}$$

Combining these two equations, we get

$$\lambda \propto \frac{1}{P}$$

Thus, **the mean free path of a gas is directly proportional to the pressure of a gas at constant temperature.**

SOLVED PROBLEM 13

At 273 K and 101325.00 Pa(1 atm) pressure the molecular diameter of a gas is 4×10^{-10} m. Calculate the mean free path of its molecule.

SOLUTION. The mean free path is given by

$$\lambda = \frac{1}{\sqrt{2}\,\pi\sigma^2 \bar{N}}$$

where σ is the molecular diameter

and $\bar{N}$ is the no. of molecules per c.c.

Here $\sigma = 4\text{Å} = 4 \times 10^{-10}$ m.

We know 0.0224 m^3 of a gas at 273 K and 1 atm. pressure (101325.00 Pa) contains 6.02×10^{23} molecules.

$$\therefore \quad \text{No. of molecules per m}^3, \bar{N} = \frac{6.02 \times 10^{23}}{0.0224}$$

$$= 2.689 \times 10^{13} \text{ molecules}$$

Substituting the values, we get

$$\sigma = \frac{1}{1.414 \times 3.14 \times (4 \times 10^{-10})^2 \times 2.689 \times 10^{13}}$$

$$= \frac{1}{1.414 \times 3.14 \times 16 \times 2.689 \times 10^{-7}}$$

$$= \mathbf{0.524 \times 10^{-7}\ m}$$

SOLVED PROBLEM 14

The root mean square velocity of hydrogen at STP is 1.83×10^3 m s^{-1} and its mean free path is 1.78×10^{-7} m. Calculate the collision number at STP.

SOLUTION. Here root mean square velocity

$$\mu = 1.83 \times 10^3 \text{ m s}^{-1}$$

We know average velocity $\bar{v} = 0.9213 \times$ RMS velocity

$$= 0.9213 \times 1.83 \times 10^3 \text{ m s}^{-1}$$

$$= 1.6869 \times 10^3 \text{ m s}^{-1}$$

$$\text{The mean free path} = \frac{\text{Average velocity}}{\text{Collision Number}}$$

$$\therefore \quad \text{Collision Number} = \frac{\text{Average velocity}}{\text{Mean free path}}$$

$$= \frac{1.6869 \times 10^3 \text{ m sec}^{-1}}{1.78 \times 10^{-7} \text{ m}}$$

$$= \mathbf{9.4769 \times 10^9\ s^{-1}}$$

The Collision Diameter

When two gas molecules approach one another, they cannot come closer beyond a certain distance. **The closest distance between the centers of the two molecules taking part in a collision is called the Collision Diameter.** It is denoted by σ. Whenever the distance between the centers of two molecules is σ, a collisoin occurs.

The collision diameter is obviously related to the mean free path of molecules. **The smaller the collision or molecular diameter, the larger is the mean free path.**

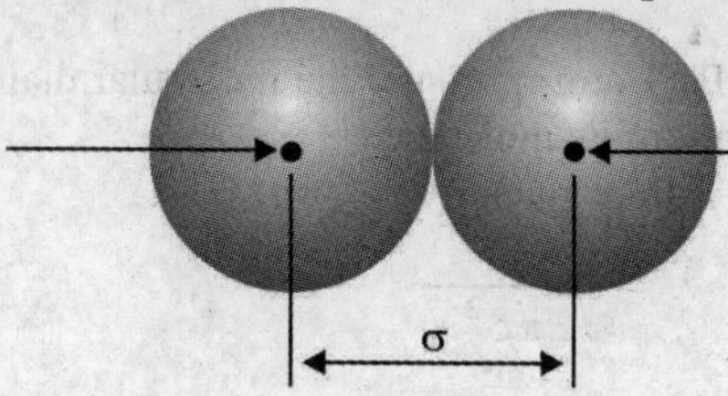

Figure 10.22
Collision diameter of molecules.

The collision diameter can be determined from viscosity measurements. The collision diameter of hydrogen is 2.74×10^{-10} m and that of oxygen is 3.61×10^{-10} m.

The Collision Frequency

The collision frequency of a gas is defined as :

the number of molecular collisions taking place per second per unit volume (c.c.) of the gas.

Let a gas contain $\bar{N}$ molecules per c.c. From kinetic consideration it has been established that the number of molecules, n, with which a single molecule will collide per second, is given by the relation

$$n = \sqrt{2}\,\pi\,\bar{v}\,\sigma^2 N$$

where $\bar{v}$ = average velocity; σ = collision diameter.

If the total number of collisions taking place per second is denoted by Z, we have

$$Z = \sqrt{2}\,\pi\,\bar{v}\,\sigma^2 N \times N$$

$$= \sqrt{2}\,\pi\,\bar{v}\,\sigma^2 N^2$$

Since each collision involves two molecules, the number of collision per second per c.c. of the gas will be $Z/2$.

$$\text{Hence the collision frequency} = \frac{\sqrt{2}\,\pi\bar{v}\sigma^2 N^2}{2}$$

$$= \frac{\pi\bar{v}\sigma^2 N^2}{\sqrt{2}}$$

Evidently, **the collision frequency of a gas increases with increase in temperature, molecular size and the number of molecules per c.c.**

Effect of Temperature and Pressure on Collision Frequency

(i) Effect of Temperature

We know collision frequency is given by

$$Z = \frac{\pi\bar{v}\sigma^2 \bar{N}^2}{\sqrt{2}} \qquad ...(i)$$

From this equation it is clear that

$$Z \propto \overline{v}$$

But $$\mu \propto \sqrt{T}$$

or $$Z \propto \sqrt{T}$$

Hence collision frequency is directly proportional to the square root of absolute temperature.

***(ii)* Effect of Pressure**

From equation (*i*), we have

$$Z \propto \overline{N}^2 \qquad ...(ii)$$

where $\overline{N}$ is the number of molecules per c.c. But we know that the pressure of the gas at a certain temperature *i.e.*

$$P \propto \overline{N} \qquad ...(iii)$$

combining equation (*ii*) and (*iii*) we get

$$Z = P^2$$

Thus the collision frequency is directly proportional to the square of the pressure of the gas.

SPECIFIC HEAT RATIO OF GASES

The Specific heat is defined as the amount of heat required to raise the temperature of one gram of a substance through 273 K. It may be measured at constant volume or at a constant pressure and though the difference in the two values is negligible in case of solids and liquids, it is appreciable in case of gases and a ratio of the two values gives us valuable information about the atomicity of a gas molecule.

Specific Heat at Constant Volume

It is the amount of heat required to raise the temperature of one gas through 273 K while the volume is kept constant and the pressure allowed to increase. It is denoted by the symbol C_v. In Physical Chemistry it is more common, however, to deal with one gram mole of the gas and the heat required in such case is called **Molecular Heat** and is represented at constant volume by C_v.

It is possible to calculate its value by making use of the Kinetic theory.

Consider one mole of a gas at the temperature *T*. Its kinetic energy is $\frac{1}{2}mnu^2$. From the kinetic gas equation

$$PV = \frac{1}{2}mnu^2$$

$$= \frac{2}{3} \times \frac{1}{2}mnu^2 = RT$$

or $$\frac{1}{2}mnu^2 \text{ (= KE)} = \frac{3}{2}RT$$

If the temperature is raised by 273 K to $(T + 1)K$ kinetic energy becomes $\frac{3}{2}R\ (T + 1)$.

$$\therefore \quad \text{Increase in kinetic energy} = \frac{3}{2}R\ (T + 1) - \frac{3}{2}RT$$

$$= \frac{3}{2}R$$

If, therefore, it be assumed that the heat supplied to a gas at constant volume is used up entirely in increasing the kinetic energy of the moving molecules, and consequently increasing the temperature, the value of C_v should be equal $\frac{3}{2}R$. It is actually so for monoatomic gases and vapors because such molecules can execute only translatory motion along the three co-ordinate axes. Motion of monoatomic gas molecules is the simplest and can be resolved into three perpendicular components along the co-ordinate axes. Thus the energy of such a molecule can be considered to be composed of three parts as

$$\frac{1}{2}mv^2 = \frac{1}{2}mv_x^2 + \frac{1}{2}mv_y^2 + \frac{1}{2}mv_z^2$$

The number of square terms involved in determining the total kinetic energy of a molecule is often referred to as the **Degrees of freedom of motion.** Such molecules have three degrees of freedom of motion. According to the principle of equipartition of energy, total energy of the molecule is equally distributed among all its degrees of freedom. But in the case of diatomic and polyatomic molecules, the heat supplied may not only increase this **kinetic energy of translation** of the molecules as a whole but also cause an increase in the energy in the inside of the molecules which we may call as **intramolecular energy.** This intramolecular energy may be the **vibration energy** *i.e.*, energy of the atoms executing vibrations with respect to each other along their line of centers or **rotational energy** which manifests itself in the rotation of the molecules about axes perpendicular to the line of centers. There will be other degrees of freedom for rotational and vibrational modes of motion also. For such cases the heat needs will be complex and are denoted by 'x' – a factor which depends upon vibrational and rotational degrees of freedom. Vibrational degrees of freedom rapidly increase with the increase in the total number of atoms in a molecule but the degrees of freedom are two for linear diatomic and three for non-linear diatomic molecules in case of rotational motion.

Consequently in such cases the molecular heat will be greater than $\frac{3}{2}R$ by the factor x.

or $$C_v = \frac{3}{2}R + x$$

The value of x varies from gas to gas and is zero for monoatomic molecules.

Specific Heat at Constant Pressure

It may be defined as the amount of heat required to raise the temperature of one gram of gas through 273 K (1°C), the pressure remaining constant while the volume is allowed to increase. It is written as c_p and the Molecular heat in this gas is represented as C_p.

Now, whenever a gas expands it has to do work against external pressure. It means that when a gas is heated under constant pressure, the heat supplied is utilized in two ways :

(1) **in increasing the kinetic energy of the moving molecules** and this has already been shown to be equal to 3/2 $R + x$ cal.

(2) **in performing external work done by the expanding gas.** The work done by the gas is equivalent to the product of the pressure and the change in volume. Let this change in volume be ΔV when the constant pressure is P and the initial volume is V.

For 1 g mole of the gas at temperature T,

$$PV = RT \qquad ...(i)$$

At temperature $(T + 1)$ K

$$P(V + \Delta V) = R(T + 1) \qquad ...(ii)$$

Subtracting (*i*) from (*ii*)

$$P \times \Delta V = R$$

Hence R cal must be added to the value of $\frac{3}{2}R$ cal in order to get the thermal equivalent of the energy supplied to one gram mole of the gas in the form of heat when its temperature is raised by 273 K.

$$\therefore \quad C_p = \frac{3}{2}R + R = \frac{5}{2}R \quad \text{(for monoatomic molecules)}$$

For di– and polyatomic molecules, it will be $\frac{3}{2}R + x$.

Specific Heat Ratio

The ratio of the molecular heats will be the same as the ratio of the specific heats. It is represented by the symbol γ.

$$\gamma = \frac{C_p}{C_v} = \frac{\frac{5}{2}R + x}{\frac{3}{2}R + x}$$

For monoatomic molecules, $x = 0$

$$\gamma = \frac{C_p}{C_v} = \frac{\frac{5}{2}R}{\frac{3}{2}R} = \frac{5}{3} = \mathbf{1.667}$$

For diatomic molecules in most cases, $S = R$

$$\gamma = \frac{C_p}{C_v} = \frac{\frac{7}{2}R}{\frac{5}{2}R} = \frac{7}{5} = \mathbf{1.40}$$

For polyatomic molecules, very often $x = \frac{3}{2}R$

$$\gamma = \frac{C_p}{C_v} = \frac{\frac{5}{2}R + \frac{3}{2}R}{\frac{3}{2}R + \frac{3}{2}R} = \frac{8}{6} = \mathbf{1.33}$$

These results are found to be in accord with experimental observations at 288 K given in the Table that follows and thus **specific heat ratio helps us to determine the atomicity of gas molecules.** The theoretical difference between C_p and C_v as calculated above is R and its observed value also shown in the table below comes out to about 2 calories.

Gas	C_p	C_v	$C_p - C_v = R$	$g = C_p/C_v$	Atomicity
Helium	5.00	3.01	1.99	1.661	1
Argon	4.97	2.98	1.99	1.667	1
Mercury vapor	6.93	4.94	1.99	1.40	2
Nitrogen	6.95	4.96	1.99	1.40	2
Oxygen	6.82	4.83	1.99	1.41	2
Carbon dioxide	8.75	6.71	2.04	1.30	3
Hydrogen sulfide	8.62	6.53	2.09	1.32	3

DEVIATIONS FROM IDEAL BEHAVIOR

An ideal gas is one which obeys the gas laws or the gas equation $PV = RT$ at all pressures and temperatures. However no gas is ideal. Almost all gases show significant deviations from the ideal behavior. Thus the gases H_2, N_2 and CO_2 which fail to obey the ideal-gas equation are termed **nonideal** or **real gases.**

Compressibility Factor

The extent to which a real gas departs from the ideal behavior may be depicted in terms of a new function called the **Compressibility factor,** denoted by Z. It is defined as

$$Z = \frac{PV}{RT}$$

The deviations from ideality may be shown by a plot of the compressibility factor, Z, against P.

For an ideal gas, $Z = 1$ and it is independent of temperature and pressure. The deviations from ideal behavior of a real gas will be determined by the value of Z being greater or less than 1. **The difference between unity and the value of the compressibility factor of a gas is a measure of the degree of nonideality of the gas.**

For a real gas, the deviations from ideal behavior depend on (*i*) pressure; and temperature. This will be illustrated by examining the compressibility curves of some gases discussed below with the variation of pressure and temperature.

Effect of Pressure Variation on Deviations

Fig. 10.23 shows the compressibility factor, Z, plotted against pressure for H_2, N_2 and CO_2 at a constant temperature.

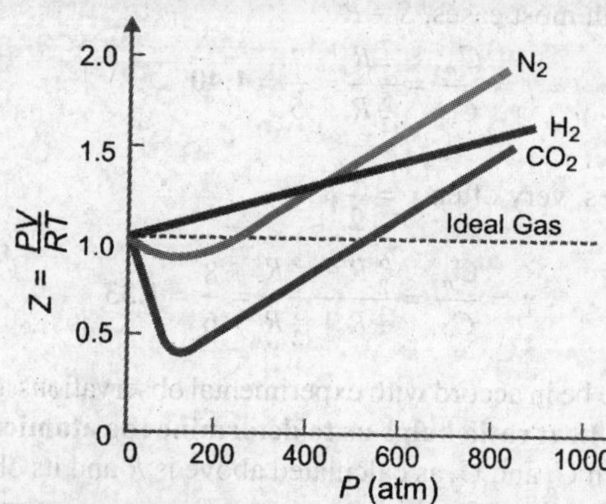

Figure 10.23
Z versus P plots for H_2, N_2 and CO_2 at 300 K.

At very low pressure, for all these gases Z is approximately equal to one. This indicates that at low pressures (upto 10 atm or 1013250.00 Pa), real gases exhibit nearly ideal behavior. As the pressure is increased, H_2 shows a continuous increase in Z (from $Z = 1$). Thus the H_2 curve lies above the ideal gas curve at all pressures.

For N_2 and CO_2, Z first decreases ($Z < 1$). It passes through a minimum and then increases continuously with pressure ($Z > 1$). For a gas like CO_2 the dip in the curve is greatest as it is most easily liquefied.

Effect of Temperature on Deviations

Fig 10.24 shows plots of Z or PV/RT against P for N_2 at different temperatures. It is clear from the shape of the curves that **the deviations from the ideal gas behavior become less and less with increase of temperature.** At lower temperature, the dip in the curve is large and the slope of the curve is negative. That is, $Z < 1$. As the temperature is raised, the dip in the curve decreases. At a certain temperature, the minimum in the curve vanishes and the curve remains horizontal for an appreciable range of pressures. At this temperature, PV/RT is almost unity and the Boyle's law is obeyed. Hence this temperature for the gas is called **Boyle's temperature.** The Boyle temperature of each gas is characteristic *e.g.*, for N_2 it is 332 K.

Conclusions

From the above discussions we conclude that :

(1) **At low pressures and fairly high temperatures, real gases show nearly ideal behavior and the ideal-gas equation is obeyed.**

(2) **At low temperatures and sufficiently high pressures, a real gas deviates significantly from ideality and the ideal-gas equation is no longer valid.**

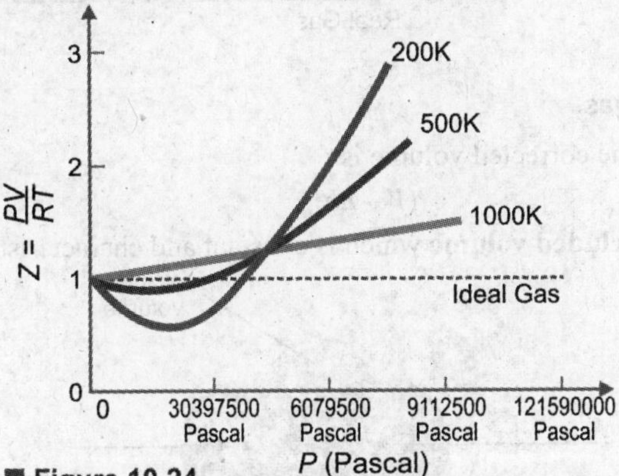

Figure 10.24
***Z* versus *P* plots for N_2 at different temperatures.**

(3) **The closer the gas is to the liquefaction point, the larger will be the deviation from the ideal behavior.**

EXPLANATION OF DEVIATIONS – VAN DER WAALS EQUATION

van der Waals (1873) attributed the deviations of real gases from ideal behavior to two erroneous postulates of the kinetic theory. These are :

(1) **the molecules in a gas are point masses and possesses no volume.**

(2) **there are no intermolecular attractions in a gas.**

Therefore, the ideal gas equation $PV = nRT$ derived from kinetic theory could not hold for real gases. van der Waals pointed out that both the pressure (P) and volume (V) factors in the ideal gas equation needed correction in order to make it applicable to real gases.

Volume Correction

The volume of a gas is the free space in the container in which molecules move about. **Volume V of an ideal gas is the same as the volume of the container.** The dot molecules of ideal gas have zero-volume and the entire space in the container is available for their movement. However, van der Waals assumed that molecules of a real gas are rigid spherical particles which possess a definite volume.

The volume of a real gas is, therefore, ideal volume minus the volume occupied by gas molecules (Fig. 10.25). If b is the effective volume of molecules per mole of the gas, the volume in the ideal gas equation is corrected as :

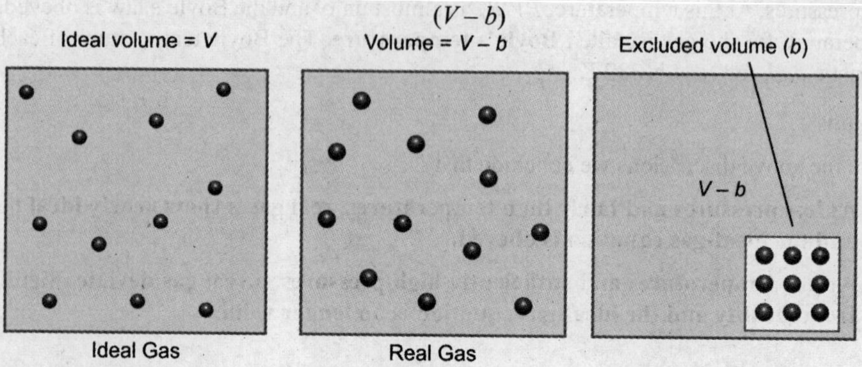

Figure 10.25
Volume of a Real gas.

For n moles of the gas, the corrected volume is :

$$(V - nb)$$

where b is termed the **excluded volume** which is constant and characteristic for each gas.

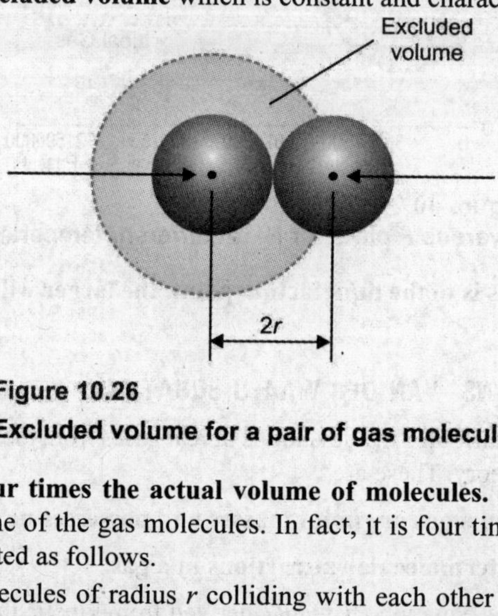

Figure 10.26
Excluded volume for a pair of gas molecules.

Excluded volume is four times the actual volume of molecules. The excluded volume is not equal to the actual volume of the gas molecules. In fact, it is four times the actual volume of molecules and can be calculated as follows.

Let us consider two molecules of radius r colliding with each other (Fig. 10.26). Obviously, they cannot approach each other closer than a distance ($2r$) apart. Therefore, the space indicated by the dotted sphere having radius ($2r$) will not be available to all other molecules of the gas. In other words the dotted spherical space is excluded volume per pair of molecules. Thus,

$$\text{excluded volume for two molecules} = \frac{4}{3}\pi(2r)^3$$

$$= 8\left(\frac{4}{3}\pi r^3\right)$$

$$\text{excluded volume per molecule } (V_e) = \frac{1}{2} \times 8\left(\frac{4}{3}\pi r^3\right)$$

$$= 4\,V_m$$

where V_m is the actual volume of a single molecule.

Therefore, in general, excluded volume of the gas molecules is four times the actual volume of molecules.

Pressure Correction

A molecule in the interior of a gas is attracted by other molecules on all sides. These attractive forces cancel out. But a molecule about to strike the wall of the vessel is attracted by molecules on one side only. Hence it experiences an inward pull (Fig. 10.27).

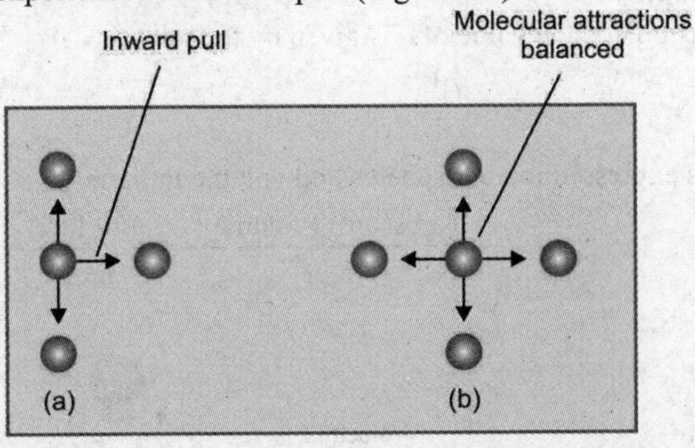

■ Figure 10.27
***(a)* A molecule about to strike the wall has a net inward pull;**
***(b)* A molecule in the interior of gas has balanced attractions.**

Therefore, it strikes the wall with reduced velocity and the actual pressure of the gas, P, will be less than the ideal pressure. If the actual pressure P, is less than P_{ideal} by a quantity p, we have

$$P = P_{\text{ideal}} - p$$

or $$P_{\text{ideal}} = P + p$$

p is determined by the force of attraction between molecules (A) striking the wall of container and the molecules (B) pulling them inward. The net force of attraction is, therefore, proportional to the concentration of (A) type molecules and also of (B) type of molecules. That is,

$$p \propto C_A \times C_B$$

or $$p \propto \frac{n}{V} \times \frac{n}{V}$$

or $$p = \frac{an^2}{V^2}$$

where n is total number of gas molecules in volume V and a is proportionality constant characteristic of the gas. Thus the pressure P in the ideal gas equation is corrected as:

$$\left(P + \frac{an^2}{V^2}\right)$$

for n moles of gas.

VAN DER WAALS EQUATION

Substituting the values of corrected pressure and volume in the ideal gas equation, $PV = nRT$, we have

$$\left(p + \frac{an^2}{V^2}\right)(V - nb) = nRT$$

This is known as **van der Waals equation** for n moles of a gas. For 1 mole of a gas ($n = 1$), van der Waals equation becomes

$$\left(p + \frac{a}{V^2}\right)(V - b) = RT$$

Constant a and b in van der Waals equation are called **van der Waals constants.** These constants are characteristic of each gas.

Determination of *a* and *b*

From the expression (1), the value of a is given by the relation

$$a = \frac{pV^2}{n^2}$$

If the pressure is expressed in atmospheres and volume in liters,

$$a = \frac{(\text{pressure})\ (\text{volume})^2}{\text{mol}^2} = \frac{\text{atm litre}^2}{\text{mol}^2}$$

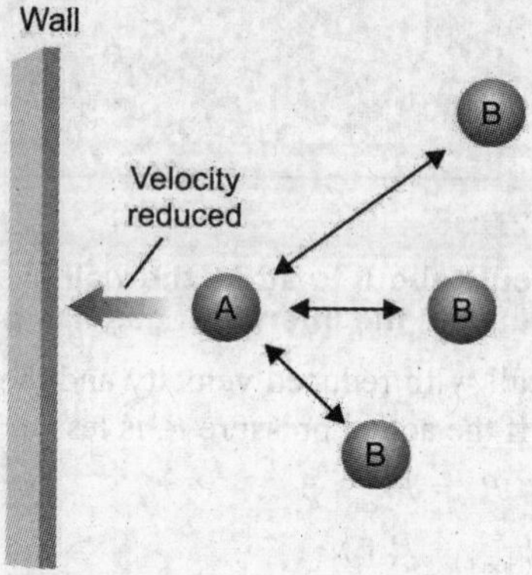

■ **Figure 10.28**
The striking molecule A is pulled inward by molecules B which reduces the velocity of A and causes the decrease of pressure.

Thus a is expressed in **atm liter^2 mol^{-2}** units.

Since nb is excluded volume for n moles of gas,

$$b = \frac{\text{volume}}{n} = \frac{\text{litre}}{\text{mole}}$$

If volume is expressed in liters, **b is expressed in liter mol^{-1}** units.

SI units of *a* and *b*. If pressure and volume are taken in SI units, we have

$$a = \frac{(\text{pressure})\ (\text{volume})}{(\text{mol})^2} = \frac{(\text{Nm}^{-2})\ (\text{m}^3)^2}{(\text{mol})^2}$$

$$= N\,\text{m}^4\ \text{mol}^{-2}$$

and $$b = \text{Volume mol}^{-1}$$

$$= \mathbf{m^3\ mol^{-1}}$$

The values of (a) and (b) can be determined by knowing the P, V and T of a gaseous system under two different conditions. Table 10.2 gives values of a and b for some common gases.

TABLE 10.2. VAN DER WAALS CONSTANTS FOR SOME COMMON GASES

Gas	a		b	
	atm liter^2 mol^{-2}	N m^4 mol^{-2}	liter mol^{-1}	10^3 m^3 mol^{-1}
Hydrogen	0.245	0.0247	0.0266	0.0266
Oxygen	1.360	0.1378	0.0318	0.0318
Nitrogen	1.390	0.1408	0.0391	0.0391
Chlorine	6.493	0.6577	0.0562	0.0562
Carbon dioxide	3.590	0.3637	0.0428	0.0428
Ammonia	4.170	0.4210	0.0371	0.0371
Sulfur dioxide	6.710	0.6780	0.0564	0.0564

SOLVED PROBLEM 15

Calculate the pressure exerted by 1.00 mole of methane (CH_4) in a 0.00025 m^3 container at 300 K using van der Waals equation. What pressure will be predicted by ideal gas equation ?

a = 0.2283 m^3 Pa mol^{-2}, b = 0.0428 × 10^{-3} m^3 mol^{-1}; R = 8.319 m^3 Pa K^{-1}.

SOLUTION

$$\left(P + \frac{n^2 a}{V^2}\right)(V - nb) = nRT \qquad \text{...van der Waals equation}$$

Dividing by ($V - nb$) and solving for P

$$P = \frac{nRT}{V - nb} - \frac{n^2 a}{V^2}$$

Substituting n = 1, R = 8.319 m^3 Pa K^{-1}, V = 0.00025 m^3, T = 300 K and the values of a and b, we have

$$P = \frac{1 \times 8.314 \times 300}{0.00025 - (1 \times 0.0000428)} - \frac{1^2 \times 0.2283}{(0.00025)^2}$$

$$= 8384844.79 \text{ Pa}$$

The ideal gas equation predicts that

$$P = \frac{nRT}{V} = \frac{1 \times 8.319 \times 300}{0.00025}$$

$$= \mathbf{9982800 \text{ Pa}}$$

Limitations of van der Waals Equation

van der Waals equation explains satisfactorily the general behavior of real gases. It is valid over a wide range of pressures and temperatures. However, **it fails to give exact agreement with experimental data at very high pressures and low temperatures.** Dieterici (1899) proposed a modified van der Waals equation. This is known as **Dieterici equation.** For one mole of gas, it may be stated as

$$P(V - b) = RTe^{-a/VRT}$$

Here the terms (a) and (b) have the same significance as in van der Waals equation.

Interpretation of Deviations from van der Waals equation

For one mole of gas, van der Waals equation is

$$\left(P + \frac{a}{V^2}\right)(V - b) = RT \quad ...(1)$$

or

$$PV = RT + Pb - \frac{a}{V} + \frac{ab}{V^2} \quad ...(2)$$

Now we proceed to interpret the deviations of real gases from ideal behavior as depicted in Figs. 10.21 and 10.22.

***(a)* At low pressure.** When P is small, V will be large. Thus both the terms Pb and ab/V^2 in equation (2) are negligible compared to a/V. Ignoring these,

$$PV = RT - \frac{a}{V}$$

or

$$\frac{PV}{RT} = 1 - \frac{a}{VRT}$$

or

$$Z = 1 - \frac{a}{VRT}$$

Thus **at low pressure, the compressibility factor is less than 1.** This explains the initial portions of Z/P curves of N_2 and CO_2 which lie below the ideal curve. As the pressure is increased, V decreases and the value of Z increases. Hence the curves show upward trend.

***(b)* At high pressures.** When P is large, V will be small. Therefore the terms a/V and ab/V^2 are negligible in comparison with Pb. Hence equation (2) is reduced to

$$PV = RT + Pb$$

or

$$\frac{PV}{RT} = 1 + \frac{Pb}{RT}$$

or

$$Z = 1 + \frac{Pb}{RT}$$

Thus at high pressures, Z is greater than 1 and Z/P lies above the ideal gas curve. With the increase of pressure, the value of Z will be still higher. This accounts for the rising parts of the curves in Fig. 10.18.

***(c)* At extremely low pressures.** At extremely low pressures, V becomes very large. Hence all the terms Pb, a/V and ab/V^2 in equation (2) are negligibly small. These could be ignored compared to RT. Thus equation (2) reduces to

$$PV = RT$$

Hence, **at low pressures real gases behave ideally.**

***(d)* At high temperatures.** At high temperatures, volume will be large ($V \propto T$). Hence P will be small. Then in the equation (2) the term RT predominates the other terms and the equation is reduced to

$$PV = RT$$

Thus **at extremely high temperatures real gases tend to show ideal behavior.**

However, at low temperatures, both P and V will be small and the net result of Pb, $-a/V$, and ab/V^2 will be appreciable. Therefore the deviations would be quite prominent.

(e) Exceptional behavior of hydrogen. Because of the small mass of H_2 molecule, the attractions between the molecules are negligible. Hence the term '*a*' is extremely small and the terms a/V and ab/V^2 in equation (2) can be ignored. The equation now becomes

$$PV = RT + Pb$$

or $$\frac{PV}{RT} = 1 + \frac{Pb}{RT}$$

or $$Z = 1 + \frac{Pb}{RT}$$

Since Z is always greater than 1, **the Z/P curve throughout lies above the ideal curve.**

SOLVED PROBLEM 16

One mole of water vapor is confined to a 0.02 m³ flask at 300 K. Calculate its pressure using

(*a*) van der Waal's equation

(*b*) Ideal gas equation

Given that $a = 0.553639\ m^6\ Pa\ mol^{-1}$

$b = 0.0305 \times 10^{-3}\ m^3\ mol^{-1}$

$R = 8.319\ m^3\ Pa\ mol^{-1}$

SOLUTION. (*a*) using van der Waal's equation

$$\left(P + a\frac{n^2}{V^2}\right)(V - nb) = nRT$$

Here $n = 1$ mole ; $T = 300$ K

$R = 8.319\ m^3\ Pa\ mol^{-1}$; $a = 0.553639\ m^6\ Pa\ mol^{-1}$ and $b = 0.0305 \times 10^{-3}\ m^3\ mol^{-1}$; $P = ?$

$V = 0.02\ m^3$. Substituting the values, we get

$$\left[P + \frac{0.553639 \times 1}{(0.02)^2}\right][0.02 - 1 \times 0.0000305] = 1 \times 8.319 \times 300$$

$$[P + 1384.0975]\,[0.0019695] = 2495.7$$

or $$P + 1384.0975 = \frac{2495.7}{0.0019695}$$

$$= 1267174.41 - 1384.0975 = \mathbf{1265790.3125\ Pa}$$

(*b*) using Ideal gas equation

$$PV = n\,RT$$

or $$P = \frac{n}{V}RT$$

Substituting the values, we get

$$P = \frac{1}{0.002} \times 8.319 \times 300 = \mathbf{124785.00\ Pascal}$$

LIQUEFACTION OF GASES–CRITICAL PHENOMENON

A gas can be liquefied by lowering the temperature and increasing the pressure. At lower temperature, the gas molecules lose kinetic energy. The slow moving molecules then aggregate due to attractions between them and are converted into liquid. The same effect is produced by the increase of pressure. The gas molecules come closer by compression and coalesce to form the liquid.

Andres (1869) studied the $P-T$ conditions of liquefaction of several gases. He established that for every gas there is a temperature below which the gas can be liquefied but above it the gas defies liquefaction. This temperature is called the critical temperature of the gas.

The **critical temperature,** T_c, of a gas may be defined as that temperature above which it cannot be liquefied no matter how great the pressure applied.

The **critical pressure,** P_c, is the minimum pressure required to liquefy the gas at its critical temperature.

The **critical volume,** V_c, is the volume occupied by a mole of the gas at the critical temperature and critical pressure.

Tc, Pc and Vc are collectively called the **critical constants** of the gas. All real gases have characteristic critical constants.

TABLE 10.4. THE CRITICAL CONSTANTS OF SOME COMMON GASES

Gas	Critical temperature (K)	Critical pressure (atm / Pa)	Critical volume (ml/mol / m³/mol)
Helium	5.3	2.26 / 228994.5	57.8 / 57.8 × 10^{-6}
Hydrogen	33.2	12.8 / 1296960.0	65.0 / 65.0 × 10^{-6}
Nitrogen	126.0	33.5 / 3394387.5	90.1 / 90.1 × 10^{-6}
Oxygen	154.3	50.1 / 5076382.5	74.4 / 74.4 × 10^{-6}
Carbon dioxide	304.0	72.9 / 7386592.5	94.0 / 94.0 × 10^{-6}
Ammonia	405.5	111.5 / 11297737.5	72.1 / 72.1 × 10^{-6}
Chlorine	407.1	76.1 / 7751362.5	123.8 / 123.8 × 10^{-6}
Sulfur dioxide	430.3	77.7 / 7852687.5	122.3 / 122.3 × 10^{-6}

At critical temperature and critical pressure, the gas becomes identical with its liquid and is said to be in **critical state.** The smooth merging of the gas with its liquid is referred to as the **critical phenomenon.** Andrews demonstrated the critical phenomenon in gases by taking example of carbon dioxide.

Andrews Isotherms of Carbon Dioxide

The P-V curves of a gas at constant temperature are called **isotherms** or **isothermals.** For an ideal gas $PV = nRT$ and the product PV is constant if T is fixed. Hence the isotherms would be rectangular parabolas.

For an ideal gas $PV = nRT$ and the product PV is constant if π is fixed. Hence the isotherms would be rectangular parabolas.

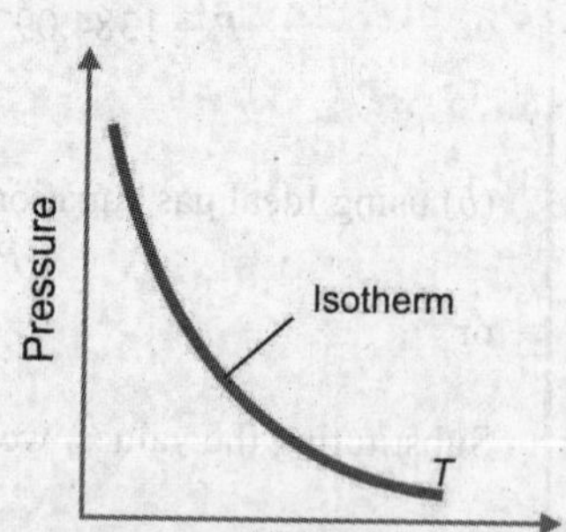

■ **Figure 10.29 Isotherm of an ideal gas at temperature *T*.**

Andrews plotted the isotherms of carbon dioxide for a series of temperatures. From Fig. 10.30 can be seen that there are three types of isotherms *viz*., those above 304 K, those below 304 K; and the one at 304 K.

(*a*) **Isotherms above 304 K.** The isotherm at 298 K is a rectangular hyperbola and approximates to the isotherm of ideal gas. So are all other isotherms above 304 K. Thus in the region above the isotherm at 304 K, carbon dioxide always exists in the gaseous state.

(*b*) **Isotherms below 304 K.** The isotherms below 304 K are discontinuous. For example, the isotherm of 294 K consists of three parts.

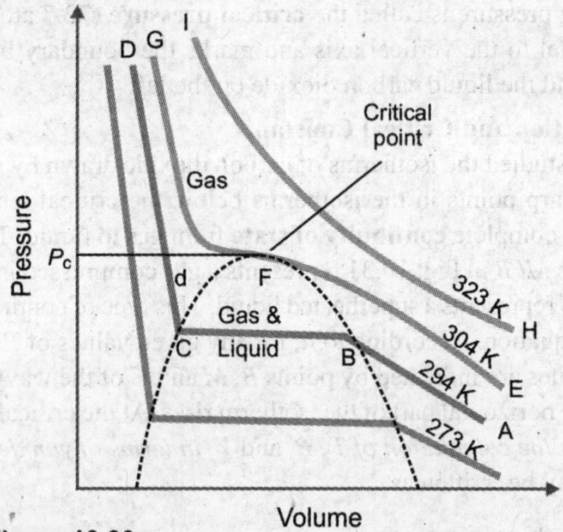

■ **Figure 10.30**
Andrews isotherms of CO_2 at different temperatures.

(*i*) *The curve AB*. It is a *PV* curve for gaseous carbon dioxide. Along *AB*, the volume decreases gradually with the increase of pressure. At *B* the volume decreases suddenly due to the formation of liquid carbon dioxide having higher density.

(*ii*) *The horizontal portion BC*. Along the horizontal part *BC* of the isotherm, the liquefaction continues while the pressure is held constant. At *C* all the gas is converted to liquid.

(*iii*) *The vertical curve CD*. This part of the isotherm is, in fact, the *P-V* curve of liquid carbon dioxide. This is almost vertical since the liquid is not very compressible.

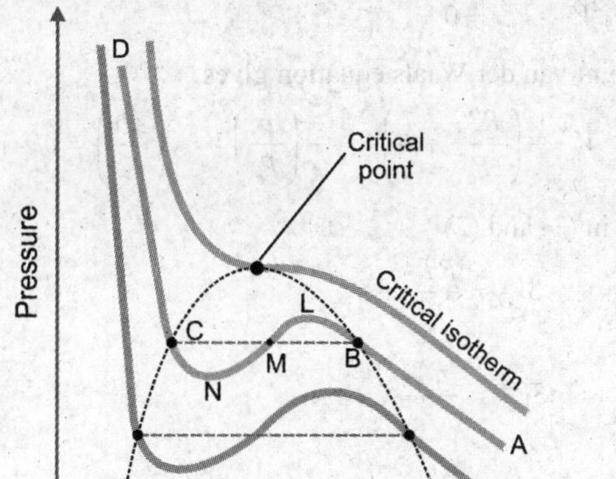

■ **Figure 10.31**
Isotherms of carbon dioxide according to van der Waals equation.

(*c*) Isotherm at 304 K. Andrews noted that above 304 K there was no possibility of liquefaction of carbon dioxide however great the pressure applied. The **critical temperature** of carbon dioxide is, therefore, 304 K. The isotherm *EFG* at this temperature is called the **critical isotherm.** The *EF* portion of the critical isotherm represents the *P-V* curve of carbon dioxide gas. At the point *F*, the curve records a twist which is coincident with the appearance of liquid carbon dioxide. Here the gas and the liquid have the same density and are indistinguishable. The point is called the **critical point**

and the corresponding pressure is called the **critical pressure** (72.7 atom). Beyond F the isotherm becomes nearly parallel to the vertical axis and marks the boundary between the gaseous carbon dioxide on the right and the liquid carbon dioxide on the left.

van der Waals' Equation and Critical Constants

Thomson (1871) studied the isotherms of carbon dioxide drawn by Andrews. He suggested that there should be no sharp points in the isotherms below the critical temperature. These isotherms should really exhibit a complete **continuity of state** from gas to liquid. This he showed a theoretical wavy curve. The curve *MLB* in Fig. 10.31 represents a gas compressed in a way that it would remain stable. The curve *MNC* represents a superheated liquid. This type of continuity of state is predicated by van der Waals cubic equation. According to it, for any given values of P and T there should be three values of V. These values are indicated by points B, M and C of the wavy curve. The three values of V become closer as the horizontal part of the isotherm rises. At the critical point, these values become identical. *This enables the calculation of* T_c, P_c and V_c *in terms of van der Waals constants.* The van der Waals equation may be written as

$$\left(P + \frac{a}{V^2}\right)(V - b) = RT$$

$$PV - Pb + \frac{a}{V} - \frac{ab}{V^2} - RT = 0$$

Multiplying the whole equation by V^2

or $$PV^3 - (RT + Pb)\,V^2 + aV - ab = 0$$

At the critical point

$$V^3 = V_c$$

$$(V - V_c)^3 = 0$$

or $$V^3 - 3V_cV^2 + 3V_c^2V - V_c^3 = 0 \qquad \text{...(1)}$$

Thus at the critical point van der Waals equation gives

$$V^3 - \left(\frac{RT_c}{P_c} + b\right)V^2 + \left(\frac{a}{P_c}\right)V - \left(\frac{ab}{P_c}\right) = 0 \qquad \text{...(2)}$$

Equating coefficients in (1) and (2)

$$3V_c = \frac{RT_c}{P_c} + b \qquad \text{...(3)}$$

$$3V_c^2 = \frac{a}{P_c} \qquad \text{...(4)}$$

$$V_c^3 = \frac{ab}{P_c} \qquad \text{...(5)}$$

From (4) and (5)

$$V_c = 3b \qquad \text{...(6)}$$

Substituting the value in (4)

$$P_c = \frac{a}{27b^2} \qquad \text{...(7)}$$

Substituting the values of V_c and P_c in (3)

$$T_c = \frac{8a}{27Rb} \qquad \text{...(8)}$$

Knowing a and b which can be deduced from deviations of ideal gas equation, the critical constants can easily be calculated. Conversely, since P_c and T_c can often be determined experimentally with comparative ease, these values may be employed to calculate the constants a and b.

$$a = 3\,V_c^2 P_c \qquad b = \frac{V_c}{3}$$

SOLVED PROBLEM 17

van der Waals' constants for hydrogen chloride gas are $a = 371862.75 \times 10^{-6}$ Pa m^{-6} and $b = 0.0000408$ m^3. Find the critical temperature and critical pressure of the gas.

SOLUTION

$$T_c = \frac{8a}{27Rb}$$

$$= \frac{8 \times 371862.75 \times 10^{-6}}{27 \times 8.34 \times 0.0000408\,\text{m}^3}$$

$$= 324.82\ \text{K}$$

$$P_c = \frac{a}{27b^2} = \frac{371862.75 \times 10^{-6}}{27 \times (0.0000408\,\text{m}^3)^2}$$

$$= \mathbf{8278252.5\ Pa}$$

SOLVED PROBLEM 18

The critical constants for water are $T_c = 647$ K, $P_c = 2208850.00$ Pa, $V_c = 0.000057$ m^3/mol. Calculate van der Waals constants.

SOLUTION

$$a = 3\,P_c V_c^2 = 3 \times 2208850 \times (0.000057)^2$$

$$= 0.0212\ \text{m}^3\ \text{Pa mol}^{-2}$$

$$b = \frac{V_c}{3} = \frac{0.000057}{3}$$

$$= \mathbf{0.019 \times 10^{-3}\ m^3\ mol^{-1}}$$

Experimental Determination of Critical Constants

The actual determination of critical constants of a substance is often a task of considerable difficulty. Of these the critical temperature and critical pressure can be measured relatively easily with the help of Cagniard de la Tour's apparatus. It consists of a stout glass U-tube blown into a bulb at the lower end. The liquid under examination is contained in the bulb and the rest of the apparatus is filled with mercury. The upper end of the U-tube is sealed leaving a little air in it so that this can be used as a manometer.

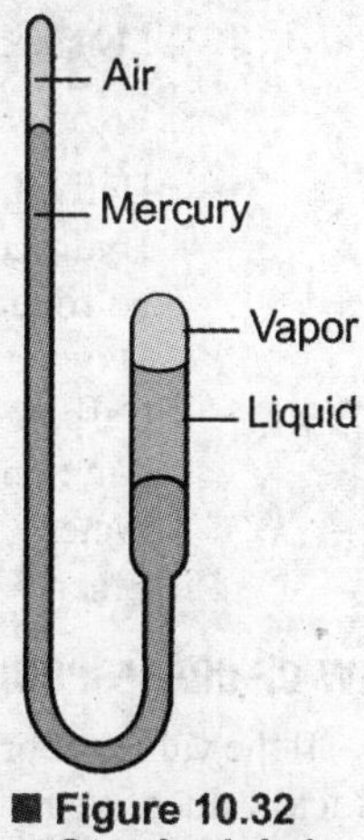

■ **Figure 10.32 Cagniard de la Tour's apparatus.**

The temperature of the bulb containing the liquid and its vapor is raised gradually by means of the heating jacket. A point is reached when the meniscus of the liquid becomes faint and then disappears leaving the contents of the bulb perfectly homogeneous. On allowing the bulb to cool again, a mist first forms in the gas which quickly settles with the reappearance of the meniscus. **The mean of the temperatures of disappearance and reappearance of the meniscus in the bulb, is the critical temperature.** The pressure read on the manometer at the critical temperature, gives the critical pressure.

The critical volume is the volume at critical temperature and critical pressure. It is much more difficult to measure since even a slight change in temperature of pressure at the critical point produces a large change in volume.

The most accurate method of determining the critical volume was given by Amagat. It consists of measuring the densities of a liquid and its vapor at a number of temperatures near the critical point, and plotting these two densities against the temperature. When the two curves representing the densities of the liquid and vapor thus drawn are extended, they naturally meet at the critical temperature because here the density of the liquid and vapor becomes identical.

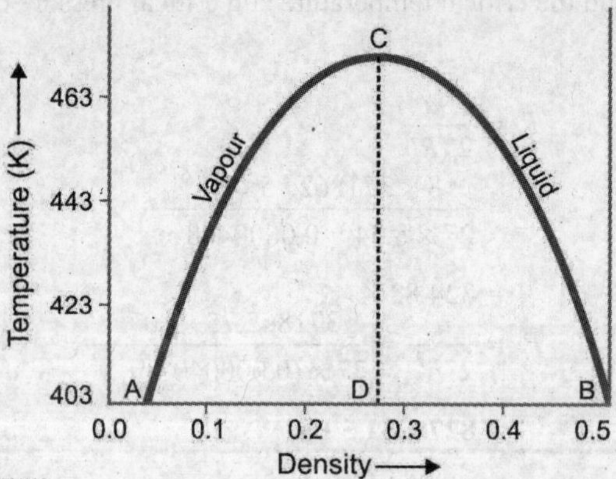

Figure 10.33

Illustration of the determination of critical volume (drawn for *n*-pentane).

The mean values of the densities are then plotted against the various temperatures where straight line *DC* is obtained which will obviously pass through the critical temperature which will be given by the point where this line will cut the curve *AB* (point *C* in Fig. 10.33). The density corresponding to the point *C* in the diagram is the **critical density.**

The **critical volume** is obtained by dividing the molecular weight of the liquid by **critical density.**

The following Table gives the critical temperature and critical pressure of a few substances.

Substance	Critical temp ()		Critical pressure (Atm. / Pa)	
	(°C)	(K)	(Atm)	(Pa)
Helium	– 269	– 542	2.3	233047.50
Hydrogen	– 249	– 522	11	1114575.00
Nitrogen	– 146	– 419	35	3546375.00
Oxygen	– 119	– 392	51	5167575.00
Carbon dioxide	+ 31	+ 304	73	7396725.00
Ammonia	+ 131	+ 404	113	11449725.00
Water	+ 374	+ 647	217	21987525.00
Ether	+ 194	+ 467	35	3546375.00

LAW OF CORRESPONDING STATES

If the values of pressure, volume and temperature be expressed as fractions of the corresponding critical values, we have

$$\frac{P}{P_c} = \pi, \quad \frac{V}{V_c} = \phi, \quad \frac{T}{T_c} = \theta$$

where π, ϕ and θ are termed the **reduced pressure,** the **reduced volume,** and the **reduced temperature** respectively.

If now we replace P, V and T by πP_c, ϕV_c and θT_c respectively in van der Waals equation

$$\left(P + \frac{a}{V^2}\right)(V - b) = RT$$

we have,

$$\left(\pi P_c + \frac{a}{\phi^2 V_c^2}\right)(\phi V_c - b) = R\theta T_c$$

Substituting the values of P_c, V_c and T_c in terms of a, b and R as given in equation (6), (7) and (8), we get

$$\left(\frac{\pi a}{27b^2} + \frac{a}{9\phi^2 b^2}\right)(3\phi b - b) = R\theta \frac{8a}{27\,Rb}$$

Dividing this equation throughout by $\frac{a}{27b^2}$, we get

$$\left(\pi + \frac{3}{\phi^2}\right)(3\phi - 1) = 8\theta \qquad ...(9)$$

This is known as **van der Waals reduced equation of state.** In this equation the quantities a, b, P_c, T_c, V_c which are characteristics of a given gas have cancelled out, thus making it applicable to all substances in the liquid or gaseous state irrespective of their specific nature. From equation (9) it is clear that **when two substances have the same reduced temperature and pressure, they will have the same reduced volume.** This is known as the **Law of Corresponding States** and when two or more substances are at the same reduced temperature and pressure, they are said to be in the **Corresponding states.** In practice this means that the properties of liquids should be determined at the same reduced temperature because pressure has very slight effect on them. Since it has been found that boiling points of liquids are approximately 2/3rd of the critical temperature, it follows that liquids are at their boiling points (in degrees absolute) approximately in corresponding states. Therefore in studying the relation between the physical properties of liquids and the chemical constitution, the physical properties may be conveniently determined at the boiling points of liquids.

SOLVED PROBLEM 19

The reduced volume (ϕ) and reduced temperature (θ) of a gas are 10.2 and 0.7. What will be its pressure if its critical pressure (P_c) is 430631.25 Pa ?

SOLUTION. It is given that

Reduced volume, $\phi = 10.2$; Reduced temperature, $\theta = 0.7$

Critical pressure, $P_c = 430631.25$ Pa

Applying reduced equations of state

$$\left[\pi + \frac{3}{\phi^2}\right][3\phi - 1] = 8\theta$$

Substituting the values we get

$$\left[\pi + \frac{3}{(10.2)^2}\right][3 \times 10.2 - 1] = 8 \times 0.7$$

or $$\left[\pi + \frac{3}{104.04}\right][30.6 - 1] = 5.6$$

or $$(\pi + 0.0288)(29.6) = 5.6$$

or $$\pi = \frac{5.6}{29.6} - 0.0288$$

$$= 0.18918 - 0.0288$$

$$= \mathbf{0.160389}$$

We know $$\frac{P}{P_c} = \pi$$

or $$P = \pi \times P_c = 0.160389 \times 430631.25$$

$$= \mathbf{69068.5156\ Pa}$$

SOLVED PROBLEM 20

The critical temperature of hydrogen gas is 306.35 K and its critical pressure is 1256430.00. Find out the values of 'a' and 'b' for the gas

SOLUTION. We know

$$T_c = \frac{8a}{27Rb} \quad ...(i); \qquad P_c = \frac{a}{27b^2} \quad ...(ii)$$

Dividing (*i*) by (*ii*) we get

$$\frac{T_c}{P_c} = \frac{8a}{27\,Rb} \times \frac{27\,b^2}{a} = \frac{8b}{R} \quad ...(iii)$$

Given $T_c = 306.35$ K

and $P_c = 1256430.00$ Pa ; $R = 8.314$ J K^{-1} mol^{-1}

Substituting the values in equation (*iii*), we get

$$\frac{306.35}{12.4} = \frac{8 \times b}{0.082}$$

or $$b = \frac{306.35 \times 0.082}{12.4 \times 8}$$

$$= \mathbf{0.253 \times 10^{-3}\ m^3\ mol^{-1}}$$

Now, substituting the value of 'b' in equation (*i*) we have

$$T_c = \frac{8a}{27Rb} \quad \text{or} \quad 306.2 = \frac{8 \times a}{27 \times 8.314 \times 10^{-3}}$$

or $$a = \mathbf{2.1748\ Pa\ m^6\ mol^{-1}}$$

METHODS OF LIQUEFACTION OF GASES

The general behavior of gases with the decrease of temperature and increase of pressure is shown by the Andrews isotherms of CO_2 (Fig. 10.31). **If a gas is cooled below its critical temperature and then subjected to adequate pressure, it liquefies.** The various methods employed for the liquefaction of gases depend on the technique used to attain low temperature. The three important methods are :

(1) *Faraday's method* in which cooling is done with a freezing mixture
(2) *Linde's method* in which a compressed gas is released at a narrow jet (*Joule-Thomson effect*)
(3) *Claude's method* in which a gas is allowed to do mechanical work

FARADAY'S METHOD

Faraday (1823) used freezing mixtures of ice with various salts for external cooling of gases. The melting of ice and dissolution of salts both are endothermic processes. The temperature of the mixture is lowered up to a temperature when the solution becomes saturated.

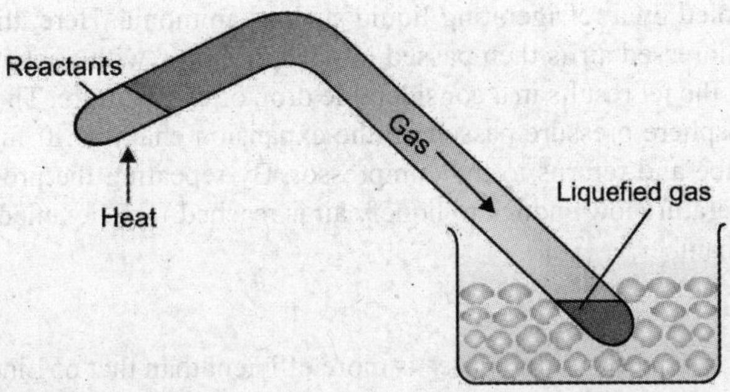

■ **Figure 10.34**
Faraday's method for the liquefaction of gases.

Faraday succeeded in liquefying a number of gases such as SO_2, CO_2, NO and Cl_2 by this method. He employed a V-shaped tube in one arm of which the gas was prepared. In the other arm, the gas was liquefied under its own pressure.

The gases liquefied by this method had their critical temperature above or just below the ordinary atmospheric temperature. The other gases including H_2, N_2 and O_2 having low critical points could not be liquefied by Faraday's method.

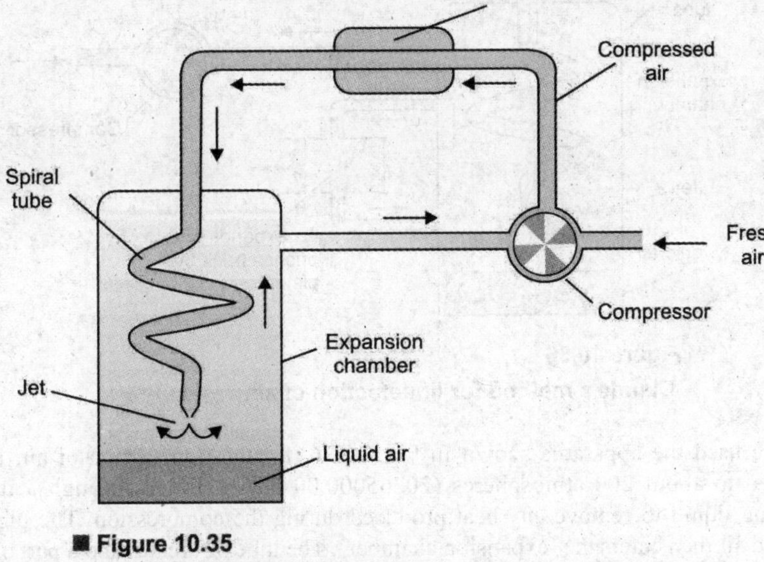

■ **Figure 10.35**
Linde's method for liquefaction of air.

LINDE'S METHOD

Linde (1895) used Joule Thomson effect as the basis for the liquefaction of gases. **When a compressed gas is allowed to expand into vacuum or a region of low pressure, it produces intense cooling.** In a compressed gas the molecules are very close and the attractions between them are appreciable. As the gas expands, the molecules move apart. In doing so, the intermolecular attraction must be overcome. The energy for it is taken from the gas itself which is thereby cooled.

Linde used an apparatus worked on the above principle for the liquefaction of air (see Fig. 10.35). Pure dry air is compressed to about 200 atmospheres (20265000.00 Pa). It is passed through a pipe cooled by a refrigerating liquid such as ammonia. Here, the heat of compression is removed. The compressed air is then passed into a spiral pipe with a jet at the lower end. The free expansion of air at the jet results in a considerable drop of temperature. The cooled air which is now at about one atmosphere pressure passed up the expansion chamber. It further cools the incoming air of the spiral tube and returns to the compressor. By repeating the process of compression and expansion, a temperature low enough to liquefy air is reached. The liquefied air collects at the bottom of the expansion chamber.

CLAUDE'S METHOD

This method for liquefaction of gases is more efficient than that of Linde. Here also the cooling is produced by free expansion of compressed gas. But in addition, the gas is made to do work by driving an engine. The energy for it comes from the gas itself which cools. **Thus in Claude's method the gas is cooled not only by overcoming the intermolecular forces but also by performance of work.** That is why the cooling produced is greater than in Linde's method.

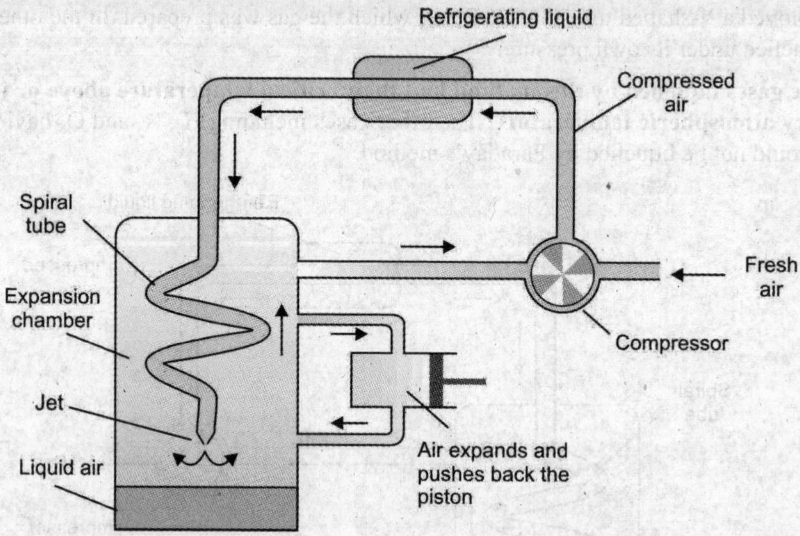

■ **Figure 10.36**
Claude's method for liquefaction of air.

Claude used the apparatus shown in Fig. 10.36. for the liquefaction of air. Pure dry air is compressed to about 200 atmospheres (20265000.00 Pa). It is led through a tube cooled by refrigerating liquid to remove any heat produced during the compression. The tube carrying the compressed air then enters the 'expansion chamber'. The tube bifurcates and a part of the air passes through the side-tube into the cylinder of an engine. Here it expands and pushes back the piston. Thus the air does mechanical work whereby it cools. The air then enters the expansion chamber and cools the incoming compressed air through the spiral tube. The air undergoes further cooling by expansion at the jet and liquefies. The gas escaping liquefaction goes back to the compressor and the whole process is repeated over and over again.

ADDITIONAL SOLVED PROBLEMS

SOLVED PROBLEM 21

One mole of diethyl ether occupies 0.015 m³ liters at 500 K. Calculate the pressure if van der Waals constants for diethyl ether are $a = 1.76$ Pa m^{-6} mol^{-2} and $b = 0.134 \times 10^{-3}$ m mol^{-1}

SOLUTION

Formula used

$$\left(P + \frac{n^2 a}{V^2}\right)(V - n\,b) = n\,R\,T$$

or

$$P = \frac{n\,R\,T}{V - n\,b} - \frac{n^2\,a}{V^2}$$

Quantities given

$n = 1$ mol $\quad V = 15$ lit (0.015 m³) $\quad R = 8.314$ J K^{-1} mol^{-1}

$a = 1.76$ Pa m^{-6} mol^{-2} $\quad T = 500$ K $\quad b = 0.134 \times 10^{-3}$ m mol^{-1}

Substitution of values

$$P = \frac{1\text{ mol} \times 8.314\text{ J K}^{-1}\text{ mol}^{-1} \times 500\text{ K}}{0.015\text{ m}^3 - 0.134 \times 10^{-3}\text{ m}^3} - \frac{(1\text{ mol})^2 \times 1.76\text{ Pa m}^{-6}\text{ mol}^{-2}}{0.015\text{ m}^3}$$

$$= 279631.3736 - 117.33$$

$$= \mathbf{279514.0436}$$

SOLVED PROBLEM 22

A vessel of volume 1×10^{-3} liter contains 10^{25} gas molecules of mass 10^{-27} kg each. If r. m. s. velocity is 10^3 m / s, calculate the total kinetic energy and temperature.

SOLUTION

(*a*) To calculate the total Kinetic energy

Formula used

$$K.E. = \tfrac{1}{2}\, m\, v^2$$

Quantities given

Mass of the gas = no. of molecules × mass of each molecule

$= 10^{25} \times 10^{-27}$ kg = 0.01 kg

$v = 10^3$ m / s

Substitution of values

K.E. of the gas = $\tfrac{1}{2} \times 0.01$ kg $\times (10^3 \text{ m / s})^2$

= **5000 J**

(*b*) To calculate the Temperature

Formula used

$$K.E. = \frac{3}{2}\,R\,T \quad\text{or}\quad T = \frac{2\,K.E.}{3\,R}$$

Quantities given

$K.E. = 5 \times 10^3$ J $\qquad R = 8.314$ J K^{-1} mol^{-1}

Substitution of values

$$T = \frac{2 \times 5 \times 10^3 \text{ J}}{3 \times 8.314 \text{ J K}^{-1} \text{ mol}^{-1}}$$

$$= \mathbf{400.93 \text{ K}}$$

SOLVED PROBLEM 23

Calculate the pressure developed in a 0.005 m^3 vessel containing 0.088 kg of CO_2 at 300 K. Assume CO_2 as van der Waals gas (C = 12, O = 6) ; a = 0.363 Pa m^6 mol^{-2} ; b = 0.043 × 10^{-3} m^3 mol^{-1}.

SOLUTION

Formula used

$$\left(P + \frac{n^2 a}{V^2}\right)(V - nb) = nRT$$

or

$$P = \frac{nRT}{V - nb} - \frac{n^2 a}{V^2}$$

Quantities given

$n = \dfrac{0.088 \text{ kg}}{0.044 \text{ mol}^{-1}} = 2$ mol R = 8.314 J K^{-1} mol^{-1} V = 0.005 m^3

T = 300 K a = 0.363 Pa m^6 mol^{-2} b = 0.043 × 10^{-3} m^3 mol^{-1}

Substitution of values

$$P = \frac{2 \text{ mol} \times 8.314 \text{ J K}^{-1} \times 300 \text{ K}}{0.005 \text{ m}^3 - 2 \times 0.043 \times 10^{-3} \text{ m}^3} - \frac{(2 \text{ mol})^2 \times 0.363 \text{ Pa m}^6 \text{ mol}^{-2}}{(0.005)^2}$$

$$= \frac{4988.4}{0.004914} - \frac{1.452}{(0.005)^2}$$

$$= 1015140.4151 - 58080 = \mathbf{957060.4151 \text{ Pa}}$$

SOLVED PROBLEM 24

Calculate the root mean square velocity, average velocity and the most probable velocity of a gas at 300 K, where the mass of gas molecule is 4.4688 × 10^{-26} kg.

SOLUTION

(*a*) To calculate the molecular mass of the gas

Mass of a molecule of the gas = 4.4688 × 10^{-26} kg

Molecular mass = Avogadro's number × Mass of one molecule

= 6.023 × 10^{23} mol^{-1} × 4.4688 × 10^{-26} kg

= 26.92 × 10^{-3} kg mol^{-1}

(*b*) To calculate the root mean square velocity

Formula used

$$\mu = \sqrt{\frac{3RT}{M}}$$

Quantities given

R = 8.314 J K^{-1} mol^{-1} T = 300 K M = 26.92 × 10^{-3} kg mol^{-1}

Subsitution of values

$$\mu = \sqrt{\frac{3 \times 8.314 \times 300 \text{ K}}{26.92 \times 10^{-3}}}$$

$$= \sqrt{27.796 \times 10^4}$$

$$= \mathbf{527.2 \text{ m s}^{-1}}$$

(*c*) To calculate the average velocity

$$v = 0.9213 \times \mu$$

$$= 0.9213 \times 527.2 \text{ m s}^{-1}$$

$$= \mathbf{485.71 \text{ m s}^{-1}}$$

(*d*) To calculate the most probable velocity

$$\mu_{mp} = \sqrt{\frac{2}{3}} \times \mu$$

$$= \sqrt{\frac{2}{3}} \times 527.2 \text{ m s}^{-1}$$

$$= 0.816 \times 527.2 = \mathbf{430.1952 \text{ m s}^{-1}}$$

SOLVED PROBLEM 25

Calculate the root mean square velocity of nitrogen at 300 K and 93325.66 Pa (700 mm) pressure.

SOLUTION

(*a*) To convert Gram molar volume to given conditions

At NTP $P_1 = 101325.00$ Pa $V_1 = 0.0224 \text{ m}^3$ $T_1 = 273$ K

At given conditions $P_2 = 93325.66$ Pa $T_2 = 300$ K

Formula used

$$\frac{P_1 V_1}{T_1} = \frac{P_2 V_2}{T_2} \quad \text{or} \quad V_2 = \frac{P_1 V_1 T_2}{T_1 P_2}$$

Substitution of values

$$V_2 = \frac{101325.00 \times 0.0224 \times 300}{273 \times 93325.66}$$

$$= \mathbf{0.026725 \text{ m}^3}$$

(*b*) To calculate root mean square velocity, μ

Formula used

$$\mu = \sqrt{\frac{3PV}{M}}$$

Quantities given

$V = 0.026725 \text{ m}^3$ $P = 93325.66$ Pa $M = 0.028 \text{ kg mol}^{-1}$

Substitution of values

$$\mu = \sqrt{\frac{3 \times 93325.66 \times 0.026725}{0.028}}$$

$$= 517.12 \text{ m s}^{-1}$$

$$= \mathbf{5.1712 \times 10^2 \text{ m s}^{-1}}$$

SOLVED PROBLEM 26

Calculate the volume occupied by 0.005 kg of acetylene gas at 323 K and 98658.55 Pa.

SOLUTION

Formula used

$$PV = nRT \quad \text{or} \quad V = \frac{nRT}{P}$$

Quantities given

$$n = \frac{5 \times 10^{-3}\ \text{kg}}{26 \times 10^{-3}\ \text{kg mol}^{-1}} = 0.1923\ \text{mol}^{-1} \qquad R = 8.314\ \text{J K}^{-1}\ \text{mol}^{-1}$$

$$T = 323\ \text{K} \qquad P = 98658.55\ \text{Pa}$$

Substitution of values

$$V = \frac{0.1923 \times 8.314 \times 323}{98658.55}$$

$$= \mathbf{0.005239\ m^3}$$

SOLVED PROBLEM 27

At 300 K hydrogen is leaked through a tiny hole into a vessel for 1200 seconds. Another unknown gas at the same temperature and pressure, as that of H_2, is leaked through the same hole for 1200 seconds. After the effusion of gases the mixture exerts a pressure of 607950.00 Pa. The hydrogen content of the mixture is 0.7 mole. If the volume of the container is 0.003 m^3, what is the molecular mass of the unknown gas.

SOLUTION

(*a*) To calculate the total number of moles in the mixture

Formula used

$$PV = nRT \quad \text{or} \quad n = \frac{PV}{RT}$$

Quantities given

$P = 607950.00$ Pa $\qquad V = 0.003$ m^3 $\qquad T = 300$ K

$R = 8.314$ J K^{-1} mol^{-1}

Substitution of values

$$n = \frac{607950.00\ \text{Pascal} \times 0.003\ \text{m}^3}{8.314\ \text{J K}^{-1}\ \text{mol}^{-1} \times 300\ \text{K}}$$

$$= 0.7312\ \text{mol}$$

Number of moles of unknown gas = Total number of moles – moles of hydrogen

= 0.7312 – 0.7 mol

= 0.0312 mol

(*b*) To calculate the molecular mass of the unknown gas

Formula used

$$\frac{r_{H2}}{r_{gas}} = \sqrt{\frac{M_{gas}}{M_{H2}}}$$

Quantities given

$$\frac{r_{H2}}{r_{gas}} = \frac{\text{amt of } H_2}{\text{amt of gas}} = \frac{0.7 \text{ mol}}{0.0312 \text{ mol}}$$

$$M_{H2} = 0.002 \text{ kg mol}^{-1}$$

Substitution of values

$$\frac{0.7}{0.0312} = \sqrt{\frac{M_{gas}}{0.002}}$$

Squaring both sides

$$\frac{0.7 \times 0.7}{0.0312 \times 0.0312} = \frac{M_{gas}}{0.002}$$

or $M_{gas} = \mathbf{1.0067 \text{ kg mol}^{-1}}$

SOLVED PROBLEM 28

A balloon of diameter 20 m. weighs 100 kg. Calculate its pay-load if it is filled with Helium at 101325.00 Pa and 300 K. Density of air 1.2 kg m^{-3} (R = 8.314 J K^{-1} mol^{-1})

SOLUTION

(*a*) To calculate the mass of Helium gas present in the balloon

Formula used

$$PV = nRT \quad \text{or} \quad n = \frac{PV}{RT}$$

Quantities given

P = 101325.00 Pa $\quad V = \frac{4}{3}\pi \quad r^3 = \frac{4}{3} \times 3.14 \times (10\text{m})^3 = 4186.66 \text{ m}^3$

R = 8.314 J K^{-1} mol^{-1} $\quad T$ = 300 K

Substitution of values

$$n = \frac{101325 \times 4186.66}{8.314 \times 300 \text{ K}}$$

$$= 170.079 \times 10^3 \text{ mol}$$

Mass of Helium present $= 4 \times 10^{-3}$ kg mol^{-1} $\times$ 170.079 mol

= 680.319 kg

∴ Mass of the filled balloon = Mass of the balloon + Mass of Helium present

= 100 kg + 680.319 kg

= 780.319 kg

(*b*) To calculate the pay load

Mass of the displaced air = Volume × density

= 4186.66 m^3 × 1.2 kg m^{-3}

= 5023.99 kg

∴ Pay load = Mass of displaced air – Total mass of balloon

= 5023.99 kg – 780.319 kg

= **4243.671 kg**

SOLVED PROBLEM 29

An LPG (liquified petroleum gas) cylinder weighs 14.8 kg when empty. When full it weighs 29.0 kg and shows a pressure of 253312.5 Pa (2.5 atm). In the course of use at 300 K the weight of the full cylinder reduces to 23.2 kg. Find out the volume of the gas in cubic meters used up at the normal usage conditions and the final pressure inside the gas. Assume LPG to be *n*-butane with normal boiling of 273 K.

SOLUTION

(*a*) To calculate the volume of the gas initially present

Mol mass of LPG (n-C_4H_{10}) = $12 \times 4 + 10 \times 1 = 58.0 \text{ g mol}^{-1} = 58 \times 10^{-3} \text{ kg mol}^{-1}$

Weight of the full cylinder = 29.0 kg

Weight of the empty cylinder = 14.8 kg

Weight of gas initially present = 29.0 kg – 14.8 kg = 14.2 kg

$$\text{No. of moles of } n\text{-butane initially present} = \frac{14.2 \text{ kg}}{58 \times 10^{-3} \text{ kg mol}^{-1}}$$

$$= 244.827 \text{ mol}$$

Pressure of the gas before use = 253312.5 Pa

$$\text{Initial volume of the gas, } V = \frac{nRT}{P} = \frac{244.827 \times 8.314 \times 300}{253312.5}$$

$$= \mathbf{2.410648 \text{ m}^3}$$

(*b*) To calculate the pressure of the gas after use

Weight of cylinder after = 23.2 kg

Weight of empty cylinder = 14.8 kg

∴ Weight of the gas unused gas = 23.2 kg – 14.8 kg

= 8.4 kg

$$\text{Pressure of gas after use } P = \frac{nRT}{V}$$

$$= \frac{8.4 \text{ kg}}{58 \times 10^{-3} \text{ kg mol}} \times \frac{8.314 \text{ J K}^{-1} \text{ mol}^{-1} \times 300 \text{ K}}{2.410648 \text{ m}^3}$$

$$= \mathbf{149847.247 \text{ Pa}}$$

(*c*) To calculate the volume of the gas after used

Weight of the gas initially present = 14.2 kg

Weight of gas after use = 8.4 kg

Weight of the gas used up = 14.2 kg – 8.4 kg

= 5.8 kg

Pressure under normal conditions of usage = 101325 Pa

$$\therefore \text{ Volume of the gas used up, } V = \frac{nRT}{P}$$

$$= \frac{5.8 \text{ kg} \times 8.314 \text{ J K}^{-1} \text{ mol}^{-1} \times 300 \text{ K}}{58 \times 10^{-3} \text{ kg mol}^{-1} \times 101325 \text{ Pascal}}$$

$$= \mathbf{2.462 \text{ m}^3}$$

SOLVED PROBLEM 30

0.020 m³ of SO_2 diffuse through a porous partition in 60 sec. What volume of O_2 will diffuse under similar conditions in 30 sec ?

SOLUTION

Formula used

$$\frac{r_{SO_2}}{r_{O_2}} = \sqrt{\frac{M_{O_2}}{M_{SO_2}}}$$

Quantities given

$$r_{SO_2} = \frac{\text{Vol. of } SO_2}{\text{Time taken}} = \frac{0.020 \text{ m}^3}{60 \text{ sec}} \qquad M_{SO_2} = 0.064 \text{ kg mol}^{-1}$$

$$r_{O_2} = \frac{\text{Vol. of } O_2}{\text{Time taken}} = \frac{0.020 \text{ m}^3}{30 \text{ sec}} \qquad M_{O_2} = 0.032 \text{ kg mol}^{-1}$$

Substitution of values

$$\frac{\dfrac{0.020 \text{ m}^3}{60 \text{ sec}}}{\dfrac{V_{O2} \text{ m}^3}{30 \text{ sec}}} = \sqrt{\frac{32 \times 10^{-3} \text{ kg mol}^{-1}}{64 \times 10^{-3} \text{ kg mol}^{-1}}}$$

$$\frac{0.010}{V_{O_2}} = \sqrt{\frac{1}{2}}$$

or

$$V_{O_2} = 0.010 \times \sqrt{2} \text{ m}^3$$

$$= 0.010 \times 1.414 \text{ m}^3$$

$$= \mathbf{0.01414 \text{ m}^3}$$

EXAMINATION QUESTIONS

1. Define or explain the following terms :

(*a*) Boyle's law (*b*) Charle's law (*c*) Absolute zero
(*d*) Avogadro's law (*e*) Combined gas laws (*f*) Ideal gas equation
(*g*) Gas constant (*h*) Dalton's law (*i*) Graham's law
(*j*) Mean free path (*k*) Collision frequency (*l*) Specific heat
(*m*) Molar heat capacity (*n*) Real gases (*o*) van der Waals equation
(*p*) Critical temperature (*q*) Critical pressure (*r*) Critical volume

2. At what temperature would ethane molecules have the same r.m.s. velocity as methane molecules at 300 K.

Answer. 562.50 K

3. (*a*) What is the law of corresponding states? How it is derived from the van der Waal's equation?

(*b*) Calculate the root mean square velocity of oxygen molecules at 300 K.

Answer. (*b*) 4.835×10^2 m s^{-1}

4. Calculate the critical constants (V_c, P_c and T_c) for C_2H_2 using van der Waal's constants $a = 4.390$ atm litre^2 mol^{-2}; $b = 0.05136 \times 10^{-3}$ m^3 mol^{-1} ($R = 8.314$ J K^{-1} mol^{-1})
Answer. $V_c = 0.1540$ liter; $P_c = 61.63$ atm; $T_c = 15.86$ K

5. (*a*) Derive an expression for the pressure of an ideal gas by means of the kinetic theory and show that the total kinetic energy of the molecules in one mole of an ideal gas is equal to $3/2RT$.
(*b*) Calculate the root mean square velocity of oxygen molecules at 298 K.
Answer. 4.819×10^2 m s^{-1}

6. (*a*) Derive the kinetic gas equation for an ideal gas.
(*b*) Calculate the root mean square velocity for oxygen molecules at 300 K, given that gas constant is 8.314 J mol^{-1} deg^{-1}.
Answer. 4.8365×10^4 m s^{-1}

7. (*a*) Derive van der Waal's reduced equation of state. Explain the significance of the law.
(*b*) Critical density of a substance having molecular weight is 555 kg/m^3 and P_c = 4863600.00 Pa. Calculate van der Waal's constants '*a*' and '*b*'.
Answer. (*b*) $a = 0.5720$ Pa m^6 mol^{-2}; $b = 0.066$ m^3 mol^{-1}

8. (*a*) Indicate what do you mean by 'mean' and r.m.s. velocity of gas molecules.
(*b*) Define critical constants of a gas. Obtain the relation from van der Waal's gas.

$$\frac{RT_c}{P_cV_c} = \frac{8}{3}$$

where symbols have their usual meanings.
Answer. (*b*) 4.1238×10^2 m s^{-1}

9. For ammonia gas van der Waals constants *a* and *b* are 0.405 Pa m^6 mol^{-2} and 0.036×10^{-3} m^3 mol^{-1} respectively. Calculate critical volume. ($R = 8.314$ J K^{-1} mol^{-1})
Answer. 0.108×10^{-3} m^3

10. (*a*) What do you understand by the distribution of molecular velocities of a gas? Illustrate your answer with the help of a diagram.
Answer. 4.124×10^2 m s^{-1}

11. (*a*) Discuss the causes of deviation of real gases from ideal gas behavior. How are they accounted for in the van der Waal's equation?
(*b*) What is the 'mean free path' of a gas? Give its relationship with the collision diameter as well as with the viscosity of a gas. (*Delhi BSc*)

12. Write short note on 'critical constants and their experimental determination'. (*Lucknow BSc*)

13. (*a*) Prove that for gases $C_p - C_v = R$. Define the terms involved. How does it help in determining the aromaticity of the gases?
(*b*) Explain the law of corresponding state. (*Lucknow BSc*)

14. Calculate critical constants V_c, P_c and T_c for C_2H_2 using van der Waal's constants.
$a = 0.448$ Pa m^3 mol^{-1}, $b = 0.5316 \times 10^{-3}$ m^3 mol^{-1}. ($R = 8.314$ J K^{-1} mol^{-1})
Answer. 1.5948×10^{-3} m^3 ; 6523.815 Pa and 30.03 K (*Delhi BSc; Nagpur BSc*)

15. (*a*) What are the limitations of the equation $PV = RT$? What improvements have been suggested by van der Waal?
(*b*) How van der Waal's equation can be applied for the calculation of Boyle's temperature? Also define Boyle's temperature.
(*c*) Show that van der Waal's equation reduces to ideal gas equation at Boyle's temperature. (*MD Rohtak BSc*)

16. (*a*) How does the van der Waal's equation explain the behavior of gases at
(*i*) High pressure and (*ii*) Low pressure
(*b*) Discuss the critical constants of a gas. (*Jamia Millia BSc*)

17. Calculate the average internal energy of a diatomic molecule at 300 K using law of equipartition of energy.
Answer. 6235.5 J (*Panjab BSc*)

18. Explain the deviation of real gases from ideal gas behavior and derive the van der Waal's equation for one mole of a gas

$$\left(P + \frac{a}{V^2}\right)(V - b) = RT$$

(Allahabad BSc)

19. (*a*) What is Boyle's temperature? Give its significance.

(*b*) Define specific heat and heat capacity. What is the difference between the two?

(*c*) Describe the temperature dependence of molecular distribution of energies.

(Guru Nanak Dev BSc)

20. (*a*) Explain the term degree of freedom.

(*b*) State law of corresponding state. *(Nagpur BSc)*

21. Derive the van der Waal's equation for *n* moles of a gas and write down the units in which van der Waal's constants are expressed. *(Arunachal BSc)*

22. Calculate the root mean square velocity of nitrogen molecule at 300 K.

Answer. 1.63×10^3 m s^{-1} *(Vidyasagar BSc)*

23. (*a*) If the equation of state for one mole of a gas is $P(v - b) = RT$, find the value of

$$\left(\frac{\partial P}{\partial V}\right)_T \left(\frac{\partial V}{\partial T}\right)_P \left(\frac{\partial T}{\partial P}\right)_V$$

(*b*) Show that $C_p - C_v = T\left(\frac{\partial P}{\partial T}\right)_V \left(\frac{\partial V}{\partial T}\right)_P$ and hence for an ideal gas (1 mole) $C_p - C_v = R$

(*c*) Show that at equilibrium : $(\partial G)_{T,P} = 0,\ (\partial A)_{T,V} = 0$ *(Jamia Millia BSc)*

24. (*a*) What are ideal and non-ideal gases? Explain why the real gases deviate from ideal behavior?

(*b*) The critical volume of a gas is 0.105×10^{-3} m^3 mol^{-1}. Assuming the molecules of the gas to be spherical, calculate the diameter of the gas molecule.

Answer. (*b*) 5.856×10^{-2} m *(Guru Nanak Dev BSc)*

25. (*a*) What is kinetic gas equation? Derive Boyle's law and Charle's law from it.

(*b*) Define mean free path and collision diameter.

(*c*) On the basis of kinetic theory of gases, show that for monoatomic gases $C_p - C_v = R$.

(Punjabi BSc)

26. (*a*) Explain the significance of van der Waal's constant.

(*b*) Why do gases fail to obey ideal gas equation at high pressure and low temperature?

(Punjabi BSc)

27. (*a*) What are the real gases? In which conditions real gases deviate from ideal gas behavior? Describe these deviations and derive van der Waal's equation.

(*b*) What do you understand by the liquefaction of gases? Describe two methods briefly.

(HS Gaur BSc)

28. (*a*) Deduce Avogadro's law from the kinetic gas equation.

(*b*) What will be the real gas equation if the pressure is too high?

(*c*) Why is volume correction necessary for a real gas? What is the unit of '*b*'. *(Arunachal BSc)*

29. State the principle of equipartition of energy and explain the high temperature limiting value of molar heat capacity of a diatomic molecule at constant volume can be evaluated with the help of this principle.

(Vidyasagar BSc)

30. What is compressibility factor? How van der Waal's equation of state accounts for the non-ideal behavior of gases? *(Aligarh BSc)*

31. (*a*) What do you understand by C_p and C_v of gases?
(*b*) State Kirchoff's law. *(Madurai BSc)*

32. Write down the expression for the Maxwell's distribution of molecular speed and discuss its characteristics. *(Vidyasagar BSc)*

33. (*a*) Explain how real gases deviate from ideal behavior. What are the causes of deviations.
(*b*) What is the law of corresponding states? Deduce the equation for law of corresponding states. *(Arunachal BSc)*

34. (*a*) What is meant by degree of freedom of a molecule? How is it classified into different forms?
(*b*) Explain giving reasons why the heat capacity of a diatomic gas is greater than that of a monoatomic gas. Show that the ratio C_p / C_v for monoatomic gases is 1.66. *(Mizoram BSc)*

35. (*a*) Write the wrong assumptions of kinetic theory of gases and derive the van der Waal's equation.
(*b*) Explain deviation of gases from ideal gas behavior with the help of a graph. *(Jamia Millia BSc)*

36. (*a*) What are the van der Waal's forces? How do they originate in non-polar molecules? Explain giving examples.
(*b*) Calculate molecular diameter of helium. Given $b = 2.4 \times 10^{-3}$ m^{-1} mol^{-1}

Answer. (*b*) 0.2396 m

37. Derive the reduced equation of state for a gas obeying van der Waal's equation (Assume the necessary expression for P_c, V_c, and T_c. Explain significance. *(Vidyasagar BSc)*

38. Derive the relation between C_p and C_v for an ideal gas. *(Nagpur BSc, 2002)*

39. Starting from kinetic gas equation derive
(*i*) Avogadro's law; and (*ii*) Graham's law of diffusion *(Nagpur BSc)*

40. Calculate the root mean square velocity of hydrogen gas at 273 K. (R = 8.314 JK^{-1} mol^{-1})

Answer. 1845.1 m s^{-1} *(Arunachal BSc [H])*

41. What are ideal and non-ideal gases? What are the chief causes of deviation of real gases from ideal behavior? Derive van der Waal's equation and show how these are accounted for in this equation. *(Guru Nanak Dev BSc)*

42. Show that the total pressure (P) exerted by an ideal gas containing 'N' molecules and occupying volume 'V' is given by :

$$P = \frac{m N \mu^2}{3V}$$ *(Guru Nanak Dev BSc)*

43. What is critical phenomenon? Derive expression for the critical constants of a gas using van der Waal's equation of state. How do you find out the van der Waal's constants from the critical values of pressure, temperature and volume? *(Sambalpur BSc)*

44. (*a*) Explain why beyond a certain temperature gases can not be liquefied whatever the pressure may be.
(*b*) Describe the effect of temperature and pressure on mean free path.
(*c*) Describe law of corresponding state. *(Guru Nanak Dev BSc)*

45. (*a*) Discuss the effect of temperature on molecular velocity.
(*b*) Derive a relationship between C_p and C_v for an ideal gas. *(Nagpur BSc)*

46. At what temperature would ethane molecule have the same r.m.s. velocity as methane molecules at 300 K?

Answer. 562.4 K *(Delhi BSc)*

47. (*a*) Derive the reduced equation of state for a gas.
(*b*) What is the law of corresponding states? *(Delhi BSc)*

48. Find the relationship between kinetic energy and temperature. *(Arunachal BSc)*

49. A 0.01 m^3 flask contains 0.064 kg of oxygen at 300 K. Calculate its pressure using

(*i*) van der Waal's equation and (*ii*) ideal gas equation

Given that $a = 0.42$ Pa m^6 mol^{-2} and $b = 0.037 \times 10^{-3}$ m^3 mol^{-1}

Answer. 485758.94 Pa; 498840 Pa *(Arunachal BSc [H])*

50. (*a*) What is kinetic gas equation? Explain the concept of absolute zero.

(*b*) What do you understand by the degree of freedom of motion? Briefly explain the different type of degrees of freedom possessed by linear and non-linear molecules. *(Panjab BSc)*

51. (*a*) Deduce the equation for mean free path of a gas molecule and hence show that how it varies with temperature and pressure.

(*b*) Explain using different curves that real gases can behave ideally at very low pressure and at fairly high temperature. *(Kalyani BSc)*

52. Starting from van der Waal's equation, derive the values of critical volume and critical pressure in terms of van der Waal's constants '*a*' and '*b*'. van der Waal's constants for carbon dioxide are $a = 0.365$ Pa m^6 mol^{-2} and $b = 4.28 \times 10^{-5}$ m^3 mol^{-1}. Calculate critical volume and critical pressure of the gas.

Answer. $V_c = 1.284 \times 10^{-4}$ m^3 mol^{-1}; $P_c = 7.3798 \times 10^6$ Pa m^3 mol^{-1} *(Nagpur BSc)*

53. Calculate RMS velocity of methane molecule at 370 K ($R = 8.314$ JK^{-1} mol^{-1}, $M = 16 \times 10^{-3}$ kg)

Answer. 7.5946×10^2 m s^{-1} *(Nagpur BSc)*

54. Oxygen has a density of 1.429×10^{-6} kg per m^3 at NTP. Calculate the r.m.s. and average velocity of its molecules.

Answer. 13.487 m s^{-1}; 12.426 m s^{-1} *(Delhi BSc)*

55. (*a*) What is meant by root mean square velocity of gaseous molecules? How is it different from average velocity?

(*b*) Calculate the average kinetic energy per molecule of CO_2 gas at 300 K. ($R = 8.314$ J K^{-1} mol^{-1})

Answer. (*b*) 3743.63 J *(Sambalpur BSc)*

56. What do you mean by fugacity? Explain clearly that the fugacity of a gas can be less than as well as more than the pressure. *(Arunachal BSc)*

57. Oxygen has a density of 1.429×10^{-6} kg m^3 at NTP. Calculate the r.m.s. and average velocity of its molecules.

Answer. 0.4612 m s^{-1}; 0.4249 m s^{-1} *(Delhi BSc)*

58. (*a*) What were the reasons which led van der Waals to modify the ideal gas equation? Write down the modified equation.

(*b*) One mole of diethyl ether occupies 0.015 m^3 at 500 K. Calculate the pressure if van der Waal's constants for diethyl ether are $a = 1.761$ Pa m^6 mol^{-2} and $b = 0.134 \times 10^{-3}$ m^3 mol^{-1}.

Answer. 265309.38 Pa *(Anna BSc)*

59. (*a*) Distinguish between root mean square velocity and most probable velocity.

(*b*) Calculate molar volume of an ideal gas at 400 K and 101325 Pa pressure.

(*c*) Why van der Waal's equation is applicable to real gases? Define compressibility factor and Boyle's temperature.

Answer. (*b*) 32.84×10^{-3} m^3 *(Burdwan BSc)*

60. Two gases P and Q having molecular masses 44 and 64 respectively are enclosed in a vessel. Their masses are 0.5×10^{-3} kg and 0.3×10^{-3} kg respectively and the total pressure of the mixture is 98658.55 Pa. Calculate the partial pressures of the two gases.

Answer. 69860.91 Pa ; 28797.63 Pa *(Burdwan BSc)*

61. Calculate the total pressure in a 0.01 m^3 cylinder which contains 0.4×10^{-3} kg of helium, 1.6×10^{-3} kg of oxygen and 1.4×10^{-3} kg of nitrogen at 300 K. Also calculate the partial pressure of helium gas in the cylinder. Assume ideal behavior for gases.

Answer. 37413 Pa ; 12471 Pa *(Arunachal BSc)*

62. Calculate the average kinetic energy in Joules of the molecules in 8.0×10^{-3} kg of methane at 300 K.

Answer. 1870.65 J *(Mysore BSc)*

63. Calculate the root mean square velocity of Ozone kept in a closed vessel at 293 K and 109324.34 Pa mercury pressure.

Answer. 3.9032×10^2 m s^{-1} *(Purvanchal BSc)*

64. Calculate the volume occupied by 7×10^{-3} kg of nitrogen gas at 300 K and 99991.77 Pa pressure.

Answer. 6.23×10^{-3} m^3 *(Kalyani BSc)*

65. Calculate the total pressure in a mixture of 4×10^{-3} kg of oxygen and 3×10^{-3} kg of hydrogen confined in a total volume of 10^{-3} m^3 at 273 K.

Answer. 3688298.25 Pa *(Madras BSc)*

66. (*a*) How and why real gases deviate from ideal behavior? Derive van der Waal's equation for '*n*' moles.

(*b*) What do you understand by the mean free path and collision diameter of a gas molecule? Give the effect of temperature and pressure on the free path. *(Barathidasan BSc)*

67. Define critical constants. Explain the experimental determination of critical constants. *(Delhi BSc)*

68. (*a*) Write the expression for the distribution of molecular velocities. What is root mean square velocity?

(*b*) Explain the principle of continuity of state. *(Gujarat BSc)*

69. (*a*) Expected values of heat capacities for gases are observed only at high temperatures. Explain.

(*b*) The compressibility factor is unity at Boyle temperature of gas. Why?

(*c*) Calculate '*r*' for diatomic molecule using equipartition principle. *(Kakatiya BSc)*

70. (*a*) Write van der Waal's equation. What are the limitations of this equation? Give the units of van der Waal's constants.

(*b*) Define mean free path and collision frequency. Explain how mean free path depends upon collision frequency and molecular size.

(*c*) Calculate the critical temperature of a van der Waal's gas for which P_c is 10132500 Pa. and b is 50×10^{-6} m^3 mol^{-1}.

Answer. (*c*) 487.35 K *(Mysore BSc)*

MULTIPLE CHOICE QUESTIONS

1. According to Boyle's law the volume of a fixed mass of a gas, at constant temperature, is

(*a*) directly proportional to its pressure
(*b*) inversely proportional to its pressure
(*c*) the square root of its pressure
(*d*) none of these

Answer. (*b*)

2. Mathematically, Boyle's law can be represented as

(*a*) $V \propto 1/P$
(*b*) $V = k/P$
(*c*) $VP = k$
(*d*) all of these

Answer. (*d*)

3. At constant temperature, the pressure of the gas is reduced to one third, the volume

(*a*) reduces to one third
(*b*) increases by three times
(*c*) remains the same
(*d*) cannot be predicted

Answer. (*b*)

4. At constant pressure, the volume of a fixed mass of a gas is

(*a*) directly proportional to its temperature
(*b*) directly proportional to its absolute temperature
(*c*) inversely proportional to its temperature
(*d*) inversely proportional to its absolute temperature

Answer. (*b*)

5. Which of the following is the correct mathematical relation for Charle's law at constant pressure?
(*a*) $V \propto T$ (*b*) $V \propto t$
(*c*) $V = kt$ (*d*) none of these
Answer. (*a*)

6. According to Gay Lussac's law for a fixed volume of a given gas
(*a*) P/T = a constant (*b*) $P \propto 1/T$
(*c*) $P = k/T$ (*d*) $PT = k$
Answer. (*a*)

7. "Equal volume of all gases at the same temperature and pressure contain equal number of molecules" is the statement of
(*a*) Combined gas law (*b*) Charle's law
(*c*) Avogadro's law (*d*) Boyle's law
Answer. (*c*)

8. For one mole of a gas, the ideal gas equation is
(*a*) $PV = RT$ (*b*) $PV = 1/2\ RT$
(*c*) $PV = 3/2\ RT$ (*d*) $PV = 5/2\ RT$
Answer. (*a*)

9. The units of R, the gas constant are
(*a*) erg K^{-1} mol^{-1} (*b*) cal K^{-1} mol^{-1}
(*c*) J K^{-1} mol^{-1} (*d*) all of these
Answer. (*d*)

10. In lit atm K^{-1} mol^{-1} the numerical value of R, the gas constant, is
(*a*) 0.821 (*b*) 0.0821
(*c*) 0.00821 (*d*) 0.000821
Answer. (*b*)

11. "The total pressure of a mixture of gases (non–reacting) is equal to the sum of the partial pressures of all the gases present" is the statement of
(*a*) Graham's law of diffusion (*b*) Dalton's law of partial pressures
(*c*) Avogadro's law of partial pressures (*d*) none of these
Answer. (*b*)

12. Under same conditions of temperature and pressure, the rates of diffusion of different gases are
(*a*) directly proportional to the square roots of the molecular masses
(*b*) directly proportional to the square roots of their vapor densities
(*c*) inversely proportional to the square roots of their molecular masses
(*d*) inversely proportional to the square roots of their molar volumes
Answer. (*c*)

13. The average kinetic energy of the gas molecules is
(*a*) inversely proportional to its absolute temperature
(*b*) directly proportional to its absolute temperature
(*c*) equal to the square of its absolute temperature
(*d*) directly proportional to the square root of its absolute temperature
Answer. (*b*)

14. For one mole of a gas the kinetic energy is given by
(*a*) $E = 1/2\ RT$ (*b*) $E = 3/2\ RT$
(*c*) $E = 5/2\ RT$ (*d*) $E = 7/2\ RT$
Answer. (*b*)

15. The kinetic gas equation is given by the relation

(*a*) $PV = 1/3\, m N \mu^2$ (*b*) $PV = 1/2\, m N \mu^2$

(*c*) $PV = 3/2\, m N \mu^2$ (*d*) $PV = 2/3\, m N \mu^2$

Answer. (*a*)

16. The average velocity of a gas is given by

(*a*) $\bar{v} = \sqrt{\frac{8RT}{n\,M}}$ (*b*) $\bar{v} = \sqrt{\frac{3RT}{n\,M}}$

(*c*) $\bar{v} = \sqrt{\frac{2RT}{n\,M}}$ (*d*) $\bar{v} = \sqrt{\frac{RT}{n\,M}}$

Answer. (*a*)

17. The root mean square velocity of gas molecules is given by the relation

(*a*) $\mu = \sqrt{\frac{RT}{M}}$ (*b*) $\mu = \sqrt{\frac{2RT}{M}}$

(*c*) $\mu = \sqrt{\frac{3RT}{M}}$ (*d*) $\mu = \sqrt{\frac{8RT}{M}}$

Answer. (*c*)

18. The root mean square velocity gas molecules is given by the relation

(*a*) $\mu = \sqrt{\frac{3PV}{M}}$ (*b*) $\mu = \sqrt{\frac{3RT}{M}}$

(*c*) $\mu = \sqrt{\frac{3P}{D}}$ (*d*) all of these

Answer. (*d*)

19. The free path is the distance travelled by the molecule

(*a*) before collision (*b*) in one second

(*c*) after collision (*d*) in one minute

Answer. (*a*)

20. The mean free path is

(*a*) directly proportional to the pressure of the gas

(*b*) directly proportional to the root mean square velocity of gas

(*c*) directly proportional to the temperature of the gas

(*d*) directly proportional to the absolute temperature of the gas

Answer. (*d*)

21. The collision frequency of a gas is

(*a*) directly proportional to the square root of absolute temperature

(*b*) directly proportional to the absolute temperature

(*c*) inversely proportional to the pressure of the gas

(*d*) inversely proportional to the absolute temperature

Answer. (*b*)

22. The value of γ, the specific heat ratio, for a monoatomic gas is

(*a*) 0 (*b*) 1.40

(*c*) 1.667 (*d*) 1.33

Answer. (*a*)

23. The compressibility factor, *z i.e.* the extent to which a real gas deviates from ideal behavior is given by

(*a*) $z = \dfrac{PV}{RT^2}$ (*b*) $z = \dfrac{PV}{2RT}$

(*c*) $z = \dfrac{PV}{RT}$ (*d*) $z = \dfrac{2PV}{RT}$

Answer. (*c*)

24. The real gases show nearly ideal behavior at

(*a*) low pressures and low temperatures (*b*) high pressures and low temperatures

(*c*) high pressures and high temperatures (*d*) low pressures and high temperatures

Answer. (*d*)

25. Excluded volume is _____ times the actual volume of molecules.

(*a*) ½ (*b*) two

(*c*) three (*d*) four

Answer. (*d*)

26. The pressure *P* in the ideal gas equation is replaced by

(*a*) $\left(P + \dfrac{a\,n^2}{V^2}\right)$ (*b*) $\left(P - \dfrac{a\,n^2}{V^2}\right)$

(*c*) $\left(P + \dfrac{2\,n^2}{V^2}\right)$ (*d*) $\left(P + \dfrac{n^2}{2\,V^2}\right)$

Answer. (*a*)

27. The units of '*a*' the van der Waal's constant are

(*a*) atm lit mol^{-1} (*b*) atm lit^{-1} mol^{-1}

(*c*) atm lit^{-2} mol^{-2} (*d*) atm lit^{-1} mol^{-2}

Answer. (*c*)

28. Which one of the following is incorrect?

(*a*) the critical temperature, T_c, of a gas is that temperature above which it can be liquefied no matter how high pressure is applied

(*b*) the critical pressure, P_c , is the minimum pressure required to liquefy the gas at its critical temperature

(*c*) the critical volume, P_c , is the volume occupied by one mole of the gas at critical temperature and critical volume

(*d*) none of these

Answer. (*d*)

29. Which one of the following relations is correct?

(*a*) $V_c = 3b$ (*b*) $P_c = \dfrac{a}{27\,R\,b^2}$

(*c*) $T_c = \dfrac{8a}{27\,R\,b}$ (*d*) none of these

Answer. (*d*)

30. The van der Waal's reduced equation of state is

(*a*) $\left[\pi + \dfrac{3}{\phi^2}\right](3\,\phi - 1) = 8\theta$ (*b*) $\left[\pi - \dfrac{3}{\phi^2}\right](3\,\phi + 1) = 8\theta$

(*c*) $\left[\pi + \dfrac{3}{\phi}\right](3\,\phi + 1) = 8\theta$ (*d*) $\left[\pi + \dfrac{3}{\phi^2}\right]\left(\dfrac{\phi}{3} + 1\right) = 8\theta$

Answer. (*a*)

31. How many molecules are present in 0.0002 kg of hydrogen?

(*a*) 6.023×10^{23} (*b*) 6.023×10^{22}

(*c*) 3.0125×10^{23} (*d*) 3.0125×10^{22}

Answer. (*b*)

32. Gas A diffuses twice as fast as another gas B. If the vapor density of the gas A is 2, the molecular mass of gas B is

(*a*) 2 (*b*) 4

(*c*) 8 (*d*) 16

Answer. (*d*)

33. A container contains a gas at 1 atm (101325.00 Pa) pressure. To compress it to 1/3rd of its initial volume, pressure to be applied is

(*a*) 101325 Pa (*b*) 202650 Pa

(*c*) 303975 Pa (*d*) 607950 Pa

Answer. (*c*)

34. The ratio of most probable velocity, average velocity and root mean square velocity of molecules of a gas is

(*a*) 1 : 1.128 : 1.224 (*b*) 1.128 : 1 : 1.224

(*c*) 1.128 : 1.224 : 1 (*d*) 1.224 : 1 : 1.128

Answer. (*a*)

35. Which of the following gases will have the highest rate of diffusion?

(*a*) CH_4 (*b*) NH_3

(*c*) N_2 (*d*) CO_2

Answer. (*a*)

36. 0.00029 m^3 of a gas at 290 K is cooled to –260 K at constant pressure. The new volume of the gas will be

(*a*) 0.00026 m^3 (*b*) 0.00027 m^3

(*c*) 0.00028 m^3 (*d*) 0.00029 m^3

Answer. (*a*)

37. The volume of a gas at 273 K is 0.000273 m^3. Its volume at 285 K and the same pressure will be

(*a*) $0.000273 + \dfrac{285}{0.000273}$ ml (*b*) $0.000273 + \dfrac{0.000273}{285}$ ml

(*c*) 0.000273 + 285 ml (*d*) 0.000273 – 285 ml

Answer. (*c*)

38. If the pressure and absolute temperature of 0.003 m^3 of a gas are doubled, its volume would be

(*a*) 0.002 m^3 (*b*) 0.003 m^3

(*c*) 0.006 m^3 (*d*) 0.012 m^3

Answer. (*b*)

39. The mass of 0.00224 m^3 of CO_2 at NTP will be

(*a*) 0.004 kg (*b*) 0.0044 kg

(*c*) 0.0088 kg (*d*) 0.008 kg

Answer. (*b*)

40. The mass of 0.000224 m^3 of N_2 on liquefaction will be

(*a*) 0.028 kg (*b*) 0.014 kg

(*c*) 0.0014 kg (*d*) 0.0028 kg

Answer. (*d*)

41. The root mean square velocity of a certain gas at 300 K is y m s^{-1}. The temperature at which its velocity will be $2y$ is

(*a*) 327 K (*b*) 381 K

(*c*) 600 K (*d*) 1200 K

Answer. (*d*)

42. Out of the following pairs of gases, which will diffuse through a porous plug with the same rate of diffusion?

(*a*) NO, CO (*b*) CH_4, O_2

(*c*) NO_2, CO_2 (*d*) NO, C_2H_6

Answer. (*d*)

43. 0.008 kg of CH_4 and 0.002 kg of hydrogen are mixed and kept at 101325 Pa pressure at 273 K. The total volume occupied by the mixture will be

(*a*) 0.0112 m^3 (*b*) 0.0224 m^3

(*c*) 0.0336 m^3 (*d*) 0.0448 m^3

Answer. (*c*)

44. The root mean square velocity of a certain gas at 300 K is a m s^{-1}. Its root mean square velocity at 1200 K is

(*a*) $a/2$ m s^{-1} (*b*) $2a$ m s^{-1}

(*c*) $3a$ m s^{-1} (*d*) $6a$ m s^{-1}

Answer. (*b*)

45. In a closed flask of 0.001 m^3, 0.002 kg of hydrogen gas is heated from 300 K to 600 K. Which of the following is incorrect?

(*a*) the pressure of the gas increases

(*b*) the kinetic energy of gaseous molecules increases

(*c*) the rate of collision increases

(*d*) the number of moles of the gas increases

Answer. (*d*)

46. Which of the following gases will have the lowest rate of diffusion?

(*a*) H_2 (*b*) N_2

(*c*) F_2 (*d*) O_2

Answer. (*c*)

47. A gas is heated at constant temperature. Then

(*a*) the no. of molecules of the gas increases

(*b*) the kinetic energy of the gas molecules decreases

(*c*) the kinetic energy of the gas molecules remains unaltered

(*d*) the kinetic energy of the gas molecules increases

Answer. (*c*)

48. Equal volumes of methane and ethane are mixed in an empty container at 298 K. The fraction of total pressure exerted by ethane is

(*a*) 1/2 (*b*) 2/3

(*c*) 8/15 (*d*) 3/2

Answer. (*a*)

49. In van der Waal's equation of state for a non-ideal gas the net force of attraction among the molecules is given by

(*a*) $\frac{a\,n^2}{V^2}$ (*b*) $P + \frac{a\,n^2}{V^2}$

(*c*) $P - \frac{a\,n^2}{V^2}$ (*d*) $-\frac{a\,n^2}{V^2}$

Answer. (*a*)

50. The compressibility factor, *z*, for an ideal gas is

(*a*) zero (*b*) less than one

(*c*) greater than one (*d*) equal to one

Answer. (*d*)

11 Liquid State

CONTENTS

We have seen that the molecules in a gas are in constant random motion. The spaces between them are large and the intermolecular attractions negligible. However, in a liquid the molecules are in contact with each other. The forces of attraction between the molecules are strong enough to hold them together. All the same, the molecules are able to move past one another through available intermolecular spaces. The molecules in a liquid move in a random fashion. At any instant the molecules may form clusters, leaving vacant space or 'hole' here and there. A liquid may be defined as : **a collection of molecules held close to each other and executing random motion through intervening spaces.**

Most of the physical properties of liquids are actually controlled by the strengths of intermolecular attractive forces. Therefore, before discussing the properties of liquids, we must have a look at the nature of intermolecular forces.

INTERMOLECULAR FORCES IN LIQUIDS

Intermolecular forces in liquids are collectively called **van der Waals forces**. These forces are essentially electrical in nature and result from the attraction of charges of opposite sign.

■ **Figure 11.1**
Molecular model of a liquid with holes indicated.

The principal kinds of intermolecular attractions are :

(1) Dipole-dipole attractions
(2) London forces
(3) Hydrogen bonding.

The relative size of these interactions is important so the relative effects are understood.

TABLE 11.1 : RELATIVE STRENGTHS FOR THE DIFFERENT INTERACTIONS

Covalent Bonds	Hydrogen Bonding	Dipole-Dipole Attractions	London Forces
1674.72 kJ	50.24-66.99 kJ	8.37-2.09 kJ	Less than 4.186 kJ

Clearly normal covalent bonds are almost 40 times the strength of hydrogen bonds. Covalent bonds are almost 200 times the strength of dipole-dipole forces, and more than 400 times the size of London dispersion forces.

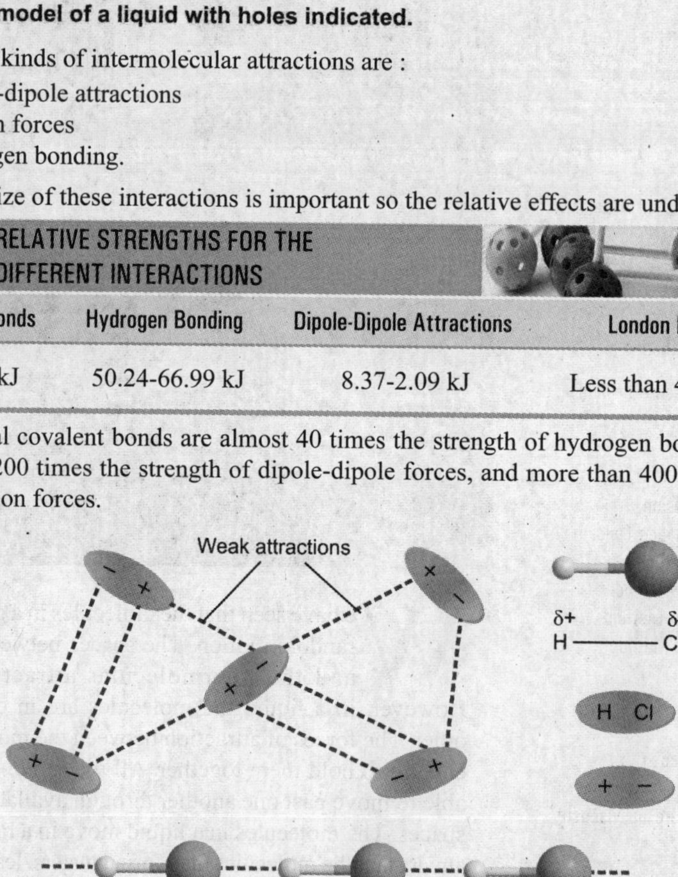

■ **Figure 11.2**
Weak attractions between polar HCl molecules.

DIPOLE–DIPOLE ATTRACTIONS

Dipole-dipole attractions exist between molecules that are polar. This requires the presence of polar bonds and an unsymmetrical molecule. These molecules have a permanent separation of positive and negative charge. In the illustration the H and of HCl is permanently slightly positive charge. The Cl end of HCl has a permanent slight negative charge. The H atom in one molecule is attracted to the Cl in a neighbor. The intermolecular force is weak compared to a covalent bond, but this dipole-dipole interaction is one of the stronger intermolecular attractions.

LONDON DISPERSION FORCES

London dispersion forces exist in nonpolar molecules. These forces result from temporary charge imbalances. The temporary charges exist because the electrons in a molecule or ion move randomly in the structure. The nucleus of one atom attracts electrons from the neighboring atom. At the same time, the electrons in one particle repel the electrons in the neighbor and create a short lived charge imbalance.

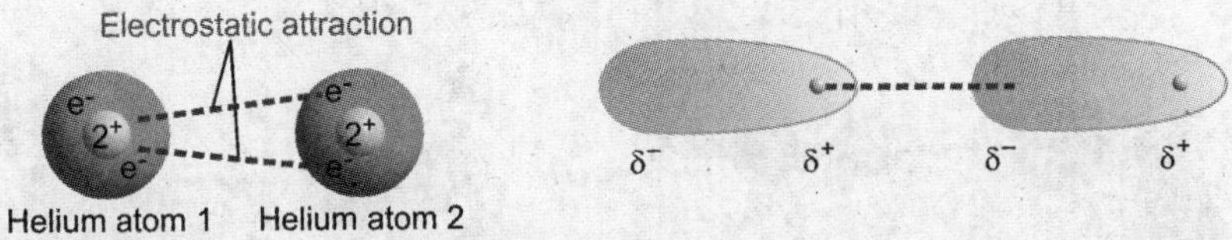

■ **Figure 11.3**
Explanation of London Forces.

These temporary charges in one molecule or atom attract opposite charges in nearby molecules or atoms. A local slight positive charge δ^+ in one molecule will be attracted to a temporary slight negative charge δ^- in a neighboring molecule.

LONDON FORCES IN HYDROCARBONS AND ORGANIC MOLECULES

The temporary separations of charge that lead to the London force attractions are what attract one nonpolar organic molecule to its neighbors. The possibilities for these interactions go up with increasing molecular size and surface. The larger surface increases the chances for the "induced" charge separation. If the molecules are linear they have more surface area than if they are folded into a sphere. The linear molecules have higher melting and boiling points because of the increased attractions.

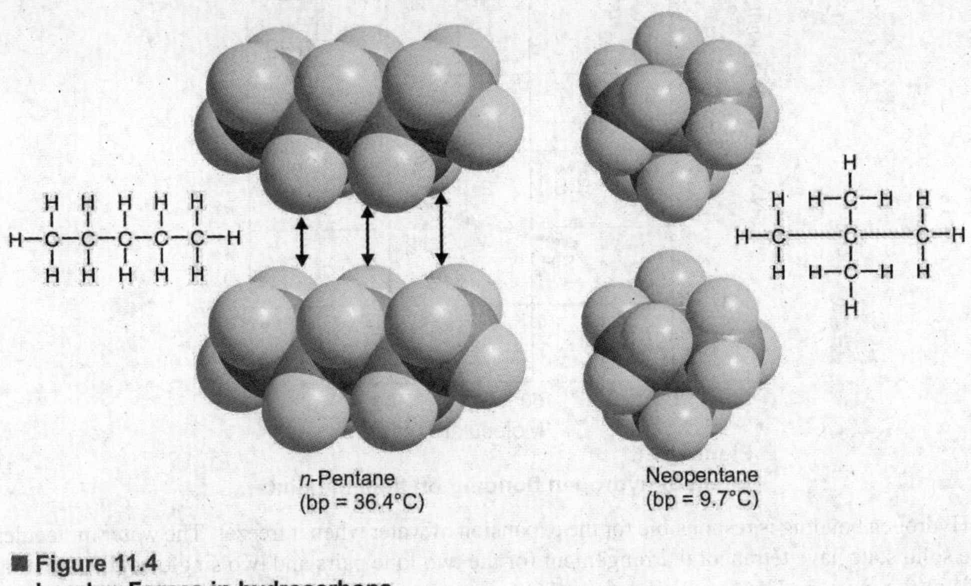

■ **Figure 11.4**
London Forces in hydrocarbons.

HYDROGEN BONDING

Hydrogen bonding is a unique type of intermolecular attraction. There are two requirements:

(1) Covalent bond between a H atom and either F, O, or N. These are the three most electronegative elements.

(2) Interaction of the H atom in this kind of polar bond with a lone pair of electrons on a nearby atom like F, O or N.

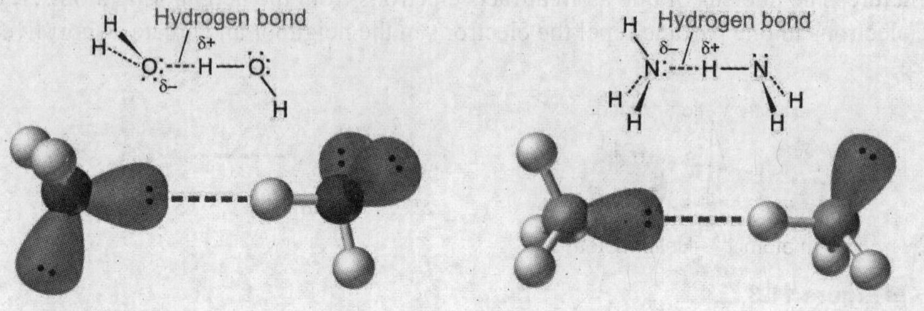

■ Figure 11.5
Hydrogen bonding in water and ammonia.

The normal boiling point for water is 373 K (100°C). The observed boiling point is high compared to the expected value. The predicted boiling point from the trend of boiling points for H_2Te, H_2Se, H_2S and H_2O is very low. If the trend continued the predicted boiling point would be below 211 K. The "anomalous" boiling point for water is the result of hydrogen bonding between water molecules.

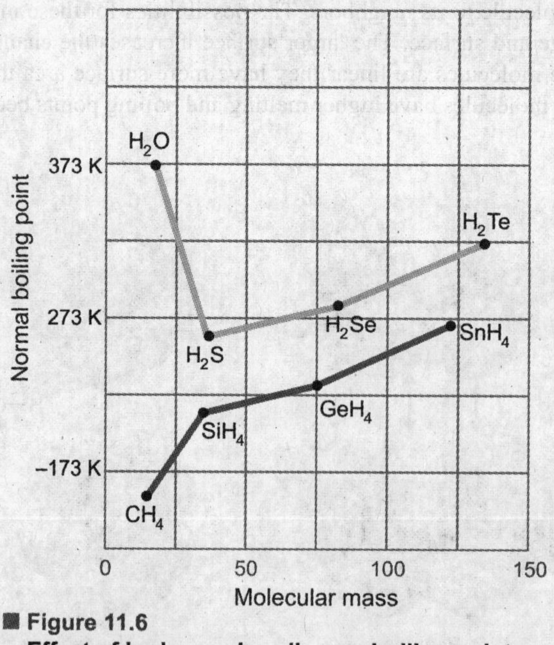

■ Figure 11.6
Effect of hydrogen bonding on boiling points.

Hydrogen bonding is responsible for the expansion of water when it freezes. The water molecules in the solid state have tetrahedral arrangement for the two lone pairs and two single bonds radiating out from the oxygen. The lone pairs on the "O" atom can be attracted to nearby water molecules through hydrogen bonds. A cage like structure results.

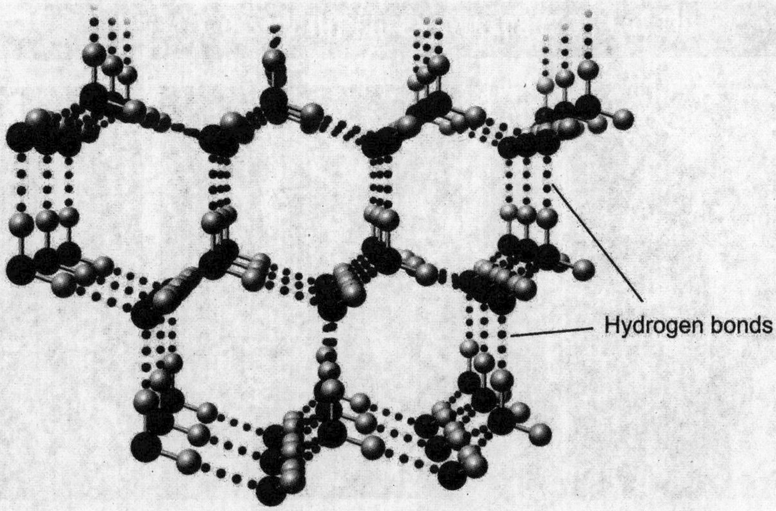

■ **Figure 11.7**
Hydrogen bonding in ice.

SUMMARY OF TYPES OF INTERMOLECULAR FORCES

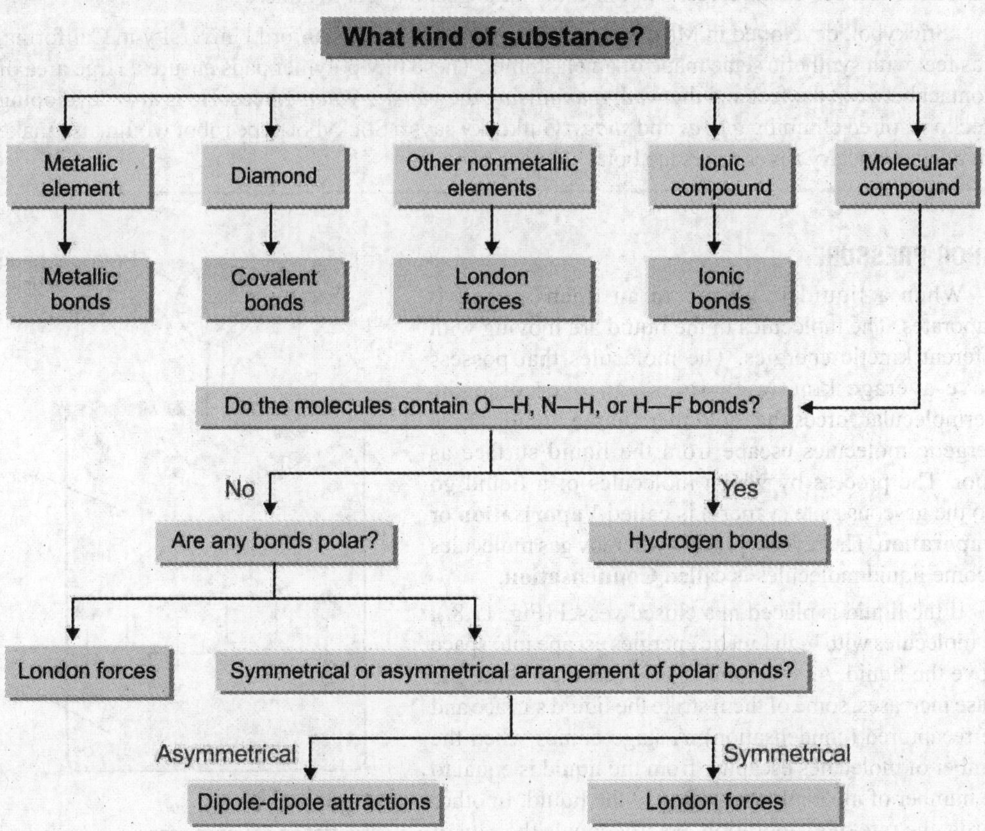

GECKO-LIKE ROBOT WALKS–UP THE WALLS

Gecko-like robot developed in May, 2006 by Mark Cutkosky, Stanford University, USA

A Gecko-like robot with sticky feet could soon be scampering up a wall near you.

Geckos can climb up walls and across ceilings thanks to the millions of tiny hairs, or setae, on the surface of their feet. Each of these hairs is attracted to the wall by an intermolecular force called the van der Waals force, and this allows the gecko's feet to adhere.

Stickybot, developed in May, 2006, by Mark Cutkosky at Stanford University in California, has feet with synthetic setae made of an elastomer. These tiny polymer pads ensure a large area of contact between the feet and the wall, maximising the van der Waals forces. He is also developing gecko inspired climbing *gloves* and *shoes*. Gutkosky says a Stickybot-type robot would also make an adept planetary rover or rescue bot.

VAPOR PRESSURE

When a liquid is placed in an open vessel, it evaporates. The molecules in the liquid are moving with different kinetic energies. The molecules that possess above-average kinetic energies can overcome the intermolecular forces that hold them in the liquid. These energetic molecules escape from the liquid surface as vapor. The process by which molecules of a liquid go into the gaseous state (vapors) is called **Vaporisation** or **Evaporation.** The reverse process whereby gas molecules become liquid molecules is called **Condensation.**

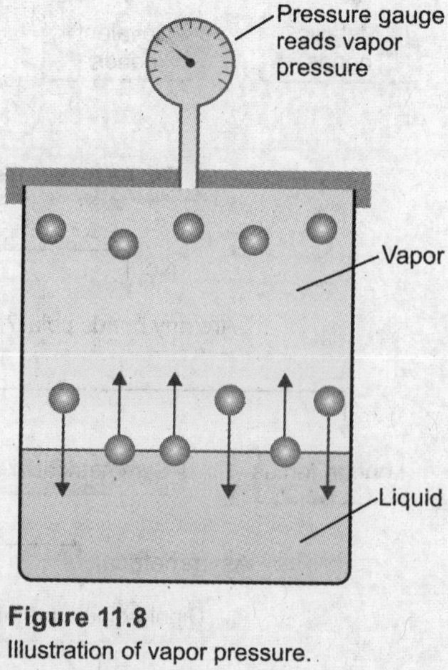

Figure 11.8
Illustration of vapor pressure.

If the liquid is placed in a closed vessel (Fig. 11.8.), the molecules with high kinetic energies escape into space above the liquid. As the number of molecules in the gas phase increases, some of them strike the liquid surface and are recaptured (condensation). A stage comes when the number of molecules escaping from the liquid is equal to the number of molecules returning to the liquid. In other words, the rate of evaporation exactly equals the rate of condensation. Thus a dynamic equilibrium is established between the liquid and the vapor at the given temperature.

$$\text{Liquid} \rightleftharpoons \text{Vapor}$$

Now the concentration of the vapor in the space above the liquid will remain unchanged with lapse of time. Hence the vapor will exert a definite pressure at the equilibrium. The vapor pressure of a liquid is defined as : **the pressure exerted by the vapor in equilibrium with the liquid at a fixed temperature.**

The vapor pressures of various liquids differ considerably, depending upon the identity of the liquid with its particular intermolecular forces. Thus ethanol having weaker hydrogen bonding than water, evaporates faster than water. Hence we expect that ethanol will have higher vapor pressure than water at a given temperature. As shown by the actual plot vapor pressure versus temperature, the vapor pressures of ethanol and water at 313 K are about 46662.82 Pa (350 torr) and 19998.35 Pa (150 torr) respectively.

Effect of Temperature on Vapor Pressure

If the temperature of the liquid is increased, the vapor pressure will increase. This is so because at higher temperature more molecules in the liquid will have larger kinetic energy and will break away from the liquid surface. Therefore the concentration of vapor molecules will increase before the equilibrium is re-established. Also, at higher temperature, the average kinetic energy of the vapor molecules will increase. Both vapor concentration and kinetic energy are proportional to temperature. Therefore, any increase of temperature will result in the increase of vapor pressure. From the experimental curves shown in Fig. 11.9, it is clear that for both ethyl alcohol and water, the vapor pressure rises with increase of temperature.

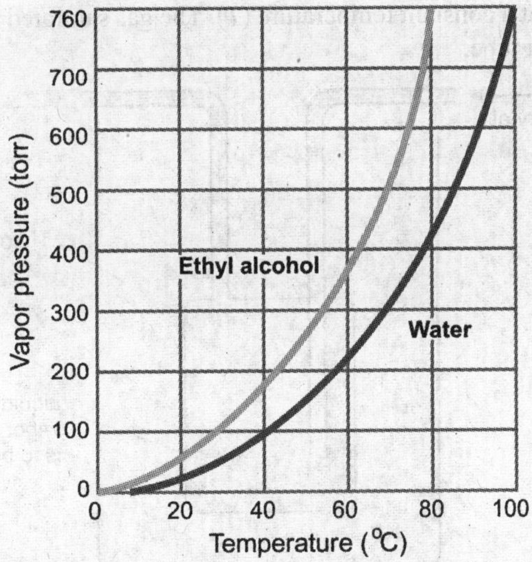

Figure 11.9
Vapor pressure increases with temperature for ethanol and water.

Determination of Vapor Pressure

The vapor pressure of a given liquid can be measured by *Static method* or *Dynamic method.*

(1) The Static Method

A simplified apparatus used for the static method is shown in Fig. 11.10. A sufficient amount of the liquid whose vapor pressure is to be determined is placed in the bulb connected to a mercury manometer and a vacuum pump. All the air from the bulb is removed by working the vacuum pump and the stopcock closed. A part of the liquid evaporates. The system is then maintained at a fixed temperature for enough time so that the equilibrium is established. **The difference in the levels of mercury in the manometer is equal to the vapor pressure of the liquid.** By adjusting the thermostat at a different temperature, the vapor pressure of the liquid at another temperature can be determined. This method is used for liquids having vapor pressures up to one atmosphere.

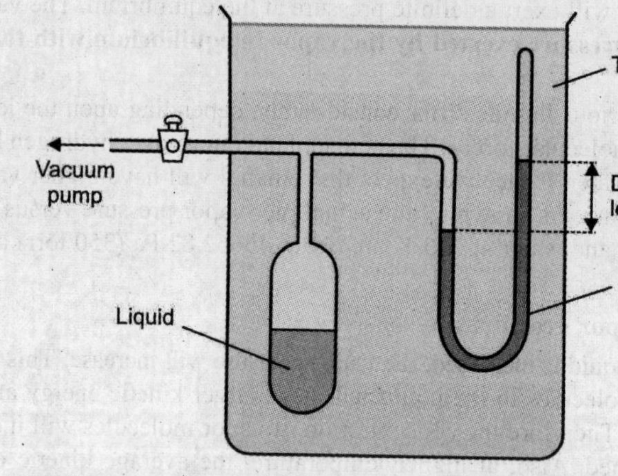

■ Figure 11.10
Determination of vapor pressure by Static method.

(2) The Dynamic Method

The apparatus used for the dynamic method is illustrated in Fig. 11.11. An inert gas is passed through the given liquid at a constant temperature (T). The gas saturated with the vapor of the liquid leaves the flask at the exit tube.

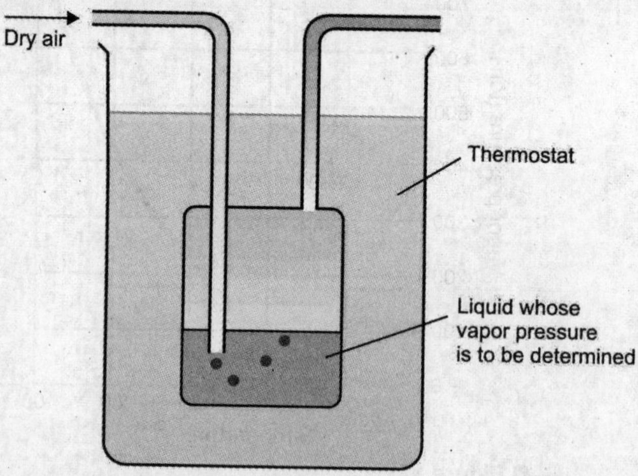

■ Figure 11.11
Determination of vapor pressure by Dynamic method.

If V be the volume of the gas passed and m the loss in weight of the liquid, the vapor pressure is given by the expression

$$\text{Vapour pressure} = \frac{m}{MV} \times RT$$

where M = molecular weight of the liquid and R = gas constant. This method is particularly suited for liquids of very low vapor pressure.

Effect of Vapor Pressure on Boiling Points

When a liquid is heated, tiny bubbles are formed in it. These rise to the liquid surface and burst. The temperature at which it happens is the boiling point of the liquid. Let us consider an individual bubble. The liquid vaporises into it and the vapor pressure in the bubble keeps it in form. However, the pressure of the atmosphere exerted on the liquid top tends to collapse the bubble. As the bubble goes to the surface, the vapor pressure in the bubble equals the atmospheric pressure. Thus the bubble collapses. The **boiling point** of the liquid may, therefore, be defined as **the temperature at which the vapor pressure of the liquid is equal to the atmospheric pressure.**

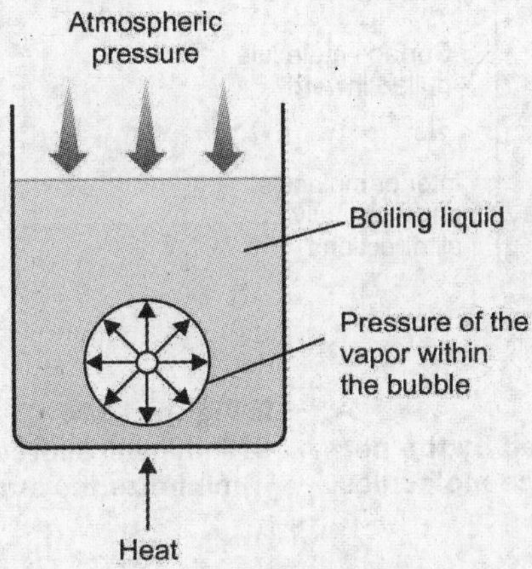

Figure 11.12
A liquid boils when the pressure of the vapor within the bubble equals the atmospheric pressure exerted on the bubble at the liquid surface.

Because the atmospheric pressure varies with altitude and other conditions, the boiling points are reported at 760 torr (1 atm or 101325.00 Pascal). Therefore the **normal boiling point** of a liquid is the temperature at which the vapor pressure of the liquid is 101325.00 Pascal. As evident from Fig. 11.9, the boiling point of ethanol is 351 K and of water, 373 K.

The boiling point of a liquid can be lowered by reducing the external pressure by vacuum pump. Then the vapor pressure of the liquid is equal to the external pressure at a lower temperature. The boiling point of a liquid can be increased by raising the external pressure. Thus the vapor pressure of the liquid is equal to the external pressure at a higher temperature. A domestic pressure cooker works on this principle. The pressure inside the cooker is maintained above 101325 Pascal and the liquid contained in it would boil at a higher temperature than 373 K. Thus the food is cooked in a shorter time.

SURFACE TENSION

This property of liquids arises from the intermolecular forces of attraction. A molecule in the interior of a liquid is attracted equally in all directions by the molecules around it. A molecule in the surface of a liquid is attracted only sideways and toward the interior. The forces on the sides being counterbalanced the surface molecule is pulled only inward the liquid. Thus there is a tendency on the part of the surface molecules to go into the bulk of the liquid. The liquid surface is, therefore, under tension and tends to contract to the smallest possible area in order to have the minimum number of molecules at the surface. It is for this reason that in air, drops of a liquid assume spherical shapes because for a given volume a sphere has the minimum surface area.

The surface tension (γ) is defined as : **the force in dynes acting along the surface of a liquid at right angle to any line 0.01 m in length.**

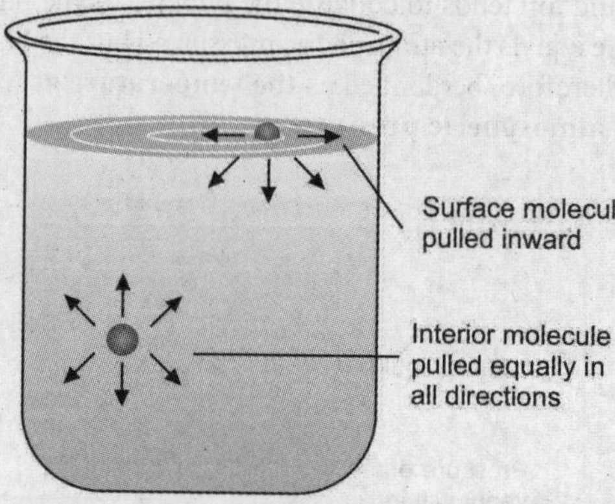

Figure 11.13
Surface tension is caused by the net inward pull on the surface molecules.

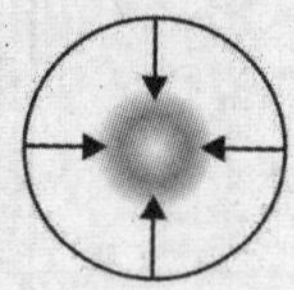

Figure 11.14
The inward forces on the surface molecules minimize the surface area and form a drop.

Units of Surface Tension

As included in the above definition the unit of surface tension in CGS system is dynes per centimeter (dyne cm^{-1}). In SI system, the unit is Newton per meter (Nm^{-1}). Both these units are related as : 1 dyne cm^{-1} = 1 m Nm^{-1}

Effect of Temperature on Surface Tension

A change in temperature causes a change in surface tension of a liquid. When temperature increases, there is an increase in kinetic energy of liquid molecules (KE $\propto$ T), thereby decreasing intermolecular forces. It results in decrease in the inward pull functioning on the surface of the liquid. In other words, **surface tension decreases with increase in temperature.** W. Ramsay and J. Shields gave the following relationship between surface tension of a liquid and its temperature

$$\gamma\,(M/\rho)^{2/3} = k\,(t_c - t - 6) \qquad ...(i)$$

where k is a constant (temperature coefficient), t_c is critical temperature and t any other temperature, $(M/\rho)^{2/3}$ represents molar surface energy of the liquid.

TABLE 11.2. SURFACE TENSION OF SOME LIQUIDS AT VARIOUS TEMPERATURES (N/m)

Liquid	293 K	313 K	333 K	353 K
Water	72.75×10^{-3}	69.56×10^{-3}	66.18×10^{-3}	62.61×10^{-3}
Ethyl alcohol	22.27×10^{-3}	20.60×10^{-3}	19.01×10^{-3}	–
Methyl alcohol	22.6×10^{-3}	20.9×10^{-3}	–	–
Acetone	23.7×10^{-3}	21.2×10^{-3}	18.6×10^{-3}	16.2×10^{-3}
Toluene	28.43×10^{-3}	26.13×10^{-3}	23.81×10^{-3}	21.53×10^{-3}
Benzene	28.9×10^{-3}	26.3×10^{-3}	23.7×10^{-3}	21.3×10^{-3}

SURFACE TENSION

Surface tension helps dew drops stick to the grass and prevents them from spreading.

Surface tension prevents this flower from sinking.

Surface tension prevents a paper clip from sinking.

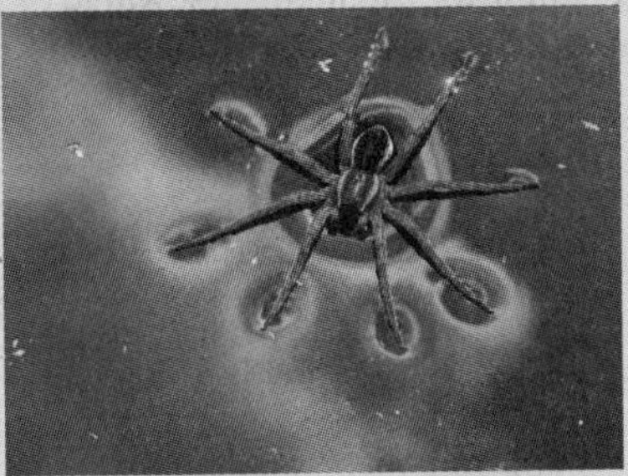

Surface tension helps insects to walk on water.

Surface tension is an effect within the surface layer of a liquid that causes the layer to behave as an elastic sheet. It is the effect that allows insects (such as the water strider) to walk on water, and causes capillary action.

Surface tension is caused by the attraction between the molecules of the liquid, due to various intermolecular forces. In the bulk of the liquid each molecule is pulled equally in all directions by neighboring liquid molecules, resulting in a net force of zero. At the surface of the liquid, the molecules are pulled inwards by other molecules deeper inside the liquid, but there are no liquid molecules on the outside to balance these forces.

All of the molecules at the surface are therefore subject to an inward force of molecular attraction which can be balanced only by the resistance of the liquid to compression. Thus the liquid squeezes itself together until it has the lowest surface area possible.

Determination of Surface Tension

The methods commonly employed for the determination of surface tension are :

(1) Capillary-rise Method

A capillary tube of radius r is vertically inserted into a liquid. The liquid rises to a height h and forms a concave meniscus. The surface tension (γ) acting along the inner circumference of the tube exactly supports the weight of the liquid column.

By definition, surface tension is force per 0.01 m acting at a tangent to the meniscus surface. If the angle between the tangent and the tube wall is θ, the vertical component of surface tension is

$\gamma \cos \theta$. The total surface tension along the circular contact line of meniscus is $2\pi r$ times. Therefore,

$$\text{Upward force} = 2\pi r\gamma \cos \theta$$

where r is the radius of the capillary. For most liquids, θ is essentially zero, and $\cos \theta = 1$. Then the upward force reduces to $2\pi r\gamma$.

The downward force on the liquid column is due to its weight which is *mass × gravity*. Thus,

$$\text{Downward force} = h\pi r^2 dg$$

where d is the density of the liquid.

But

$$\text{Upward force} = \text{Downward force.}$$

or

$$2\pi r\gamma = h\pi r^2 dg$$

$$\gamma = \frac{hr\ dg}{2} \text{ dynes/cm}$$

$$\gamma = \frac{hrdg \times 10^{-3}}{2} \text{ N/m} \quad ...(1)$$

In order to know the value of γ, the value of h is found with the help of a travelling microscope and density (d) with a pyknometer.

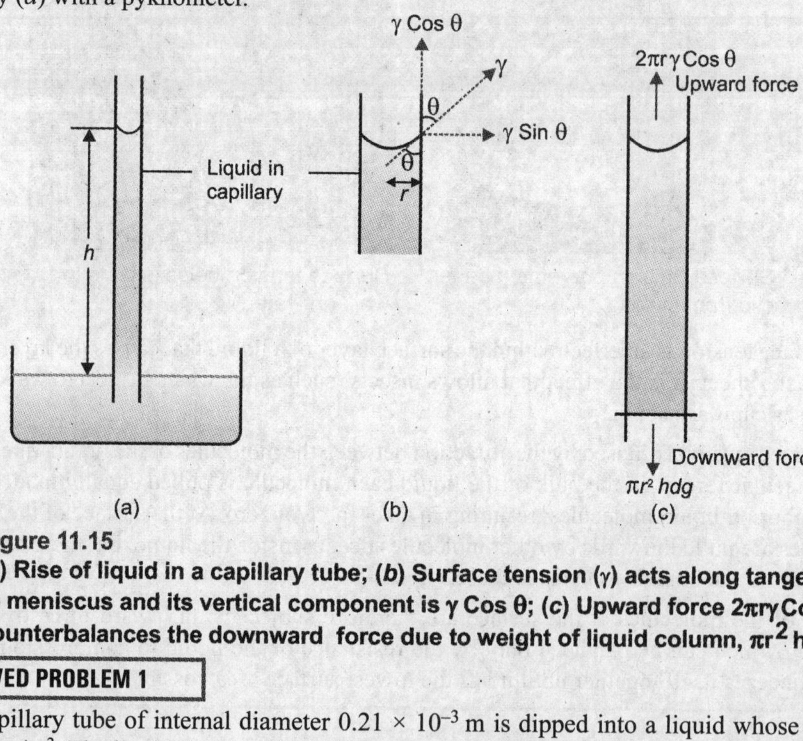

■ **Figure 11.15**
(*a*) Rise of liquid in a capillary tube; (*b*) Surface tension (γ) acts along tangent to meniscus and its vertical component is γ Cos θ; (*c*) Upward force 2πrγ Cos θ counterbalances the downward force due to weight of liquid column, πr² hgd.

SOLVED PROBLEM 1

A capillary tube of internal diameter 0.21×10^{-3} m is dipped into a liquid whose density is 790 kg/m^3. The liquid rises in this capillary to a height of 6.30×10^{-2} m. Calculate the surface tension of the liquid. ($g = 9.8$ m s^{-2})

SOLUTION

We know :

$$\gamma = \frac{hrdg \times 10^{-3}}{2} \text{ N/m} \quad ...(1)$$

where

h = height of liquid in capillary
r = radius of capillary
d = density of liquid
g = acceleration due to gravity

Substituting the values from the above example,

$$h = 6.30 \times 10^{-2}\ \text{m}$$

$$r = \frac{0.21 \times 10^{-3}}{2} = 0.105 \times 10^{-3}\ \text{m}$$

$$d = 790\ \text{kg/m}^3$$

$$g = 9.8\ \text{m/s}^2$$

In relation (1), we have

$$\gamma = \frac{6.30 \times 10^{-2} \times 0.105 \times 10^{-3} \times 790 \times 9.8}{2}$$

$$= \mathbf{25.6 \times 10^{-3}\ kg/s^2}$$

Thus the surface tension of the given liquid is **25.6×10^{-3} kg/s^2.**

SOLVED PROBLEM 2

How high will sap rise in a plant if the capillaries are 0.00001 meter diameter, the density of the fluid is 1300 kg/m^3 and its surface tension 0.065 Nm^{-1}. ($g = 9.8$ m s^{-2})

SOLUTION

We know that

$$h = \frac{2\gamma}{rdg}\ \text{metres} \qquad \text{...(1)}$$

where in SI units :

h = height of liquid
γ = surface tension
r = radius of capillary
d = density
g = acceleration due to gravity

In this example,

$$\gamma = 0.065\ \text{Nm}^{-1}$$

$$r = \frac{1}{2}(0.00001) = \frac{10^{-5}}{2}\ \text{metres}$$

$$d = 1.3 \times 10^3\ \text{kg m}^{-3}$$

$$g = 9.81\ \text{ms}^{-2}$$

Substituting these values in equation (1)

$$h = \frac{2 \times 2 \times 0.065}{10^{-5} \times 1.3 \times 10^3 \times 9.81} = \mathbf{2.04\ m}$$

Therefore in a plant with capillaries of 1.3×10^3 kg/m^3 diameter, the sap will rise to a height of **2.04 meters.**

(2) Drop Formation Method

A drop of liquid is allowed to form at the lower end of a capillary tube (Fig. 11.17). The drop is supported by the upward force of surface tension acting at the outer circumference of the tube. The weight of the drop (mg) pulls it downward. When the two forces are balanced, the drop breaks. Thus at the point of breaking,

$$m\,g = 2\,\pi\,r\,\gamma \qquad \text{...(1)}$$

where
m = mass of the drop
g = acceleration due to gravity
r = outer radius of the tube

The apparatus employed is a glass pipette with a capillary at the lower part. This is called a **Stalagmometer** or **Drop pipette (Fig. 11.16).** It is cleaned, dried and filled with the experimental liquid, say upto mark A. Then the surface tension is determined by one of the two methods given below.

(*a*) **Drop-weight Method.** About 20 drops of the given liquid are received from the drop-pipette in a weighing bottle and weighed. Thus weight of one drop is found. The drop-pipette is again cleaned and dried. It is filled with a second reference liquid (say water) and weight of one drop determined as before.

Then from equation (1)

$$m_1 g = 2 \pi r \gamma_1 \quad ...(2)$$

$$m_2 g = 2 \pi r \gamma_2 \quad ...(3)$$

Dividing (2) by (3)

$$\frac{\gamma_1}{\gamma_2} = \frac{m_1}{m_2} \quad ...(4)$$

Knowing the surface tension of reference liquid from Tables, that of the liquid under study can be found.

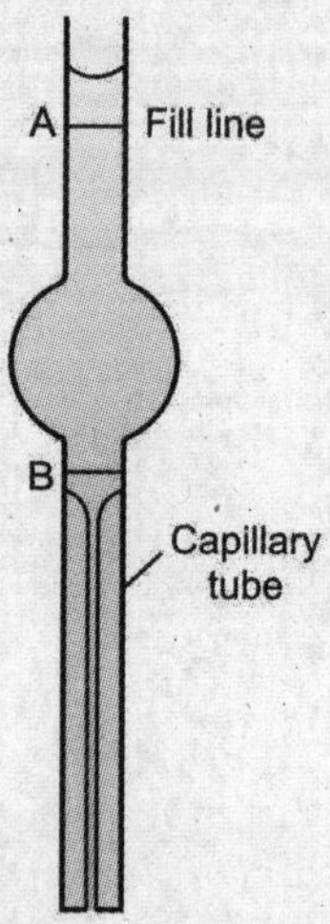

Figure 11.16
A stalagmometer.

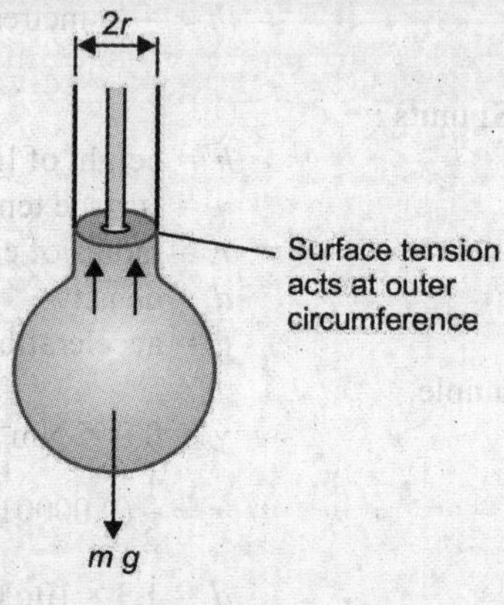

Figure 11.17
A drop forming from a tube of radius *r*.

(*b*) **Drop-number Method.** The drop-pipette is filled upto the mark A with the experimental liquid (No. 1). The number of drops is counted as the meniscus travels from A to B. Similarly, the pipette is filled with the reference liquid (No. 2) as the meniscus passes from A to B. Let n_1 and n_2 be the number of drops produced by the same volume V of the two liquids. Thus,

The volume of one drop of liquid 1 $= V/n_1$

The mass of one drop of liquid 1 $= (V/n_1)d_1$

where d_1 is the density of liquid 1.

Similarly,

The mass of one drop of liquid 2 $= (V/n_2)d_2$

Then from equation (4)

$$\frac{\gamma_1}{\gamma_2} = \frac{(V/n_1)\,d_1}{(V/n_2)\,d_2} = \frac{n_2 d_1}{n_1 d_2}$$

The value of d_1 is determined with a pyknometer. Knowing d_2 and γ_2 from reference tables, γ_1 can be calculated.

SOLVED PROBLEM 3

In the determination of surface tension of a liquid by the drop-number method, it gives 55 drops while water gave 25 drops for the same volume. The densities of the liquid and water are 996 and 800 kg/m^3 respectively. Find the surface tension of the liquid if that of water is 72.0×10^{-3} N/m.

SOLUTION

We know that

$$\frac{\gamma_1}{\gamma_2} = \frac{n_2 d_1}{n_1 d_2} \quad \text{...(1)}$$

where γ_1 = surface tension of liquid; γ_2 = surface tension of water; n_1 = number of drops of liquid; n_2 = number of drops of water.

In the present case we have :

$$\gamma_2 = 72 \times 10^{-3} \text{ N/m.}$$

$$n_1 = 55 \qquad n_2 = 25$$

$$d_1 = 996 \text{ kg/m}^3 \qquad d_2 = 800 \text{ kg/m}^3$$

From equation (1) we have

$$\gamma_1 = \gamma_2 \times \frac{n_2 d_1}{n_1 d_2} \quad \text{...(2)}$$

Substituting values in (2)

$$\gamma_1 = 72 \times \frac{996 \times 25}{800 \times 55} = 40.7 \times 10^{-3} \text{ N/m}$$

Therefore, the surface tension of the given liquid is **40.7×10^{-3} N/m.**

(3) Ring-detachment Method

In this method the force required to detach a platinum ring (du Nouy ring) from the liquid surface is measured. **This force (*F*) is exactly equal to the downward pull due to surface tension γ acting along the circumference of the ring.** Twice the length of the circumference ($2 \times 2\pi r$) is taken since the liquid is in contact with both the inside and the outside of ring. Thus,

$$F = 4 \pi r \gamma$$

or

$$\gamma = \frac{F}{4 \pi r} \quad \text{...(1)}$$

where r is the radius of the ring.

The apparatus employed is called the **du Nouy Tensiometer.** Its essential parts are shown in Fig. 11.19. One end of the torsion wire is fixed while the other is attached to a knob carrying a pointer. The pointer moves on a fixed scale. The scale is previously calibrated by taking different weights on the beam and noting the scale reading when it is lifted from the horizontal position. The liquid whose surface tension is to be determined is placed in a watch glass so that the Pt-ring just touches its surface. The knob of the torsion wire is then slowly turned till the ring is just detached from the surface. The reading shown by the pointer on the scale gives the force F. The surface tension is then calculated from equation (1).

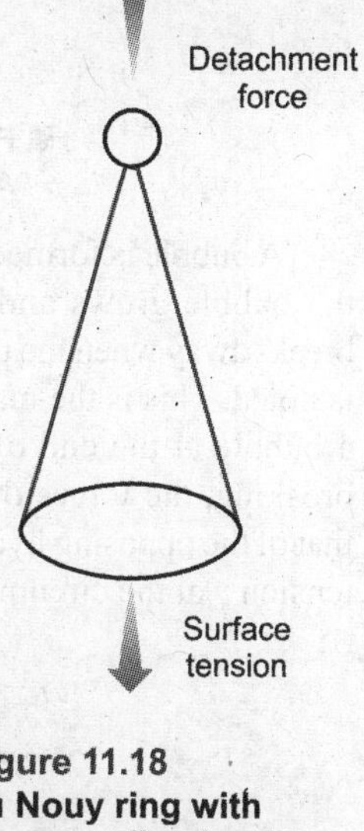

■ **Figure 11.18 du Nouy ring with a suspending hook.**

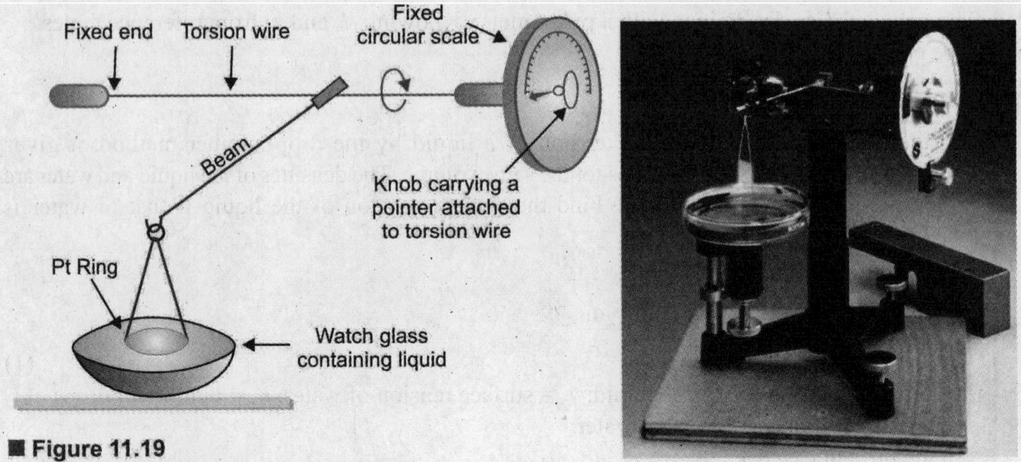

■ **Figure 11.19**
du Nouy Tensiometer.

(4) Maximum Bubble Pressure Method

In this method air-pressure is applied slowly through a capillary tube dipping in the experimental liquid (See Fig. 11.20).

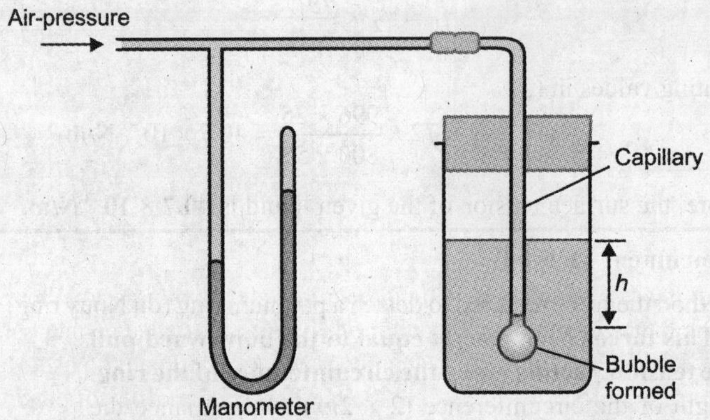

■ **Figure 11.20**
A simple apparatus for maximum bubble pressure method.

A bubble is formed at the end of the capillary. Slowly the bubble grows and becomes hemispherical. Then it breaks away when the pressure recorded by the manometer is noted. This is the maximum pressure required to make a bubble at the end of the capillary. At the moment of breaking, the forces due to maximum pressure P equals that of the opposing hydrostatic pressure P_h and the surface tension γ at the circumference of the capillary. Thus,

$$P \pi r^2 = P_h \pi r^2 + 2 \pi r \gamma$$

$$P = P_h + \frac{2\gamma}{r}$$

or

$$P = h d g + \frac{2\gamma}{r}$$

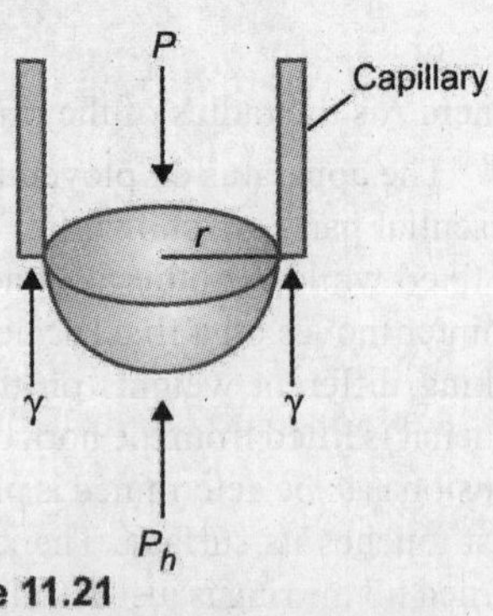

■ **Figure 11.21**
Applied pressure on bubble is opposed by hydrostatic pressure and surface tension.

where r = radius of capillary; d = density of the liquid; h = depth of liquid.

Knowing the value of P, h, d and r, γ can be found.

VISCOSITY

A liquid may be considered to be consisting of molecular layers arranged one over the other. When a shearing force is applied to a liquid, it flows. However, the forces of friction between the layers offer resistance to this flow. **Viscosity of a liquid is a measure of its frictional resistance.**

Let us examine a liquid flowing on a glass surface (Fig. 11.22). The molecular layer in contact with the stationary surface has zero velocity. The successive layers above it move with increasingly higher velocities in the direction of the flow.

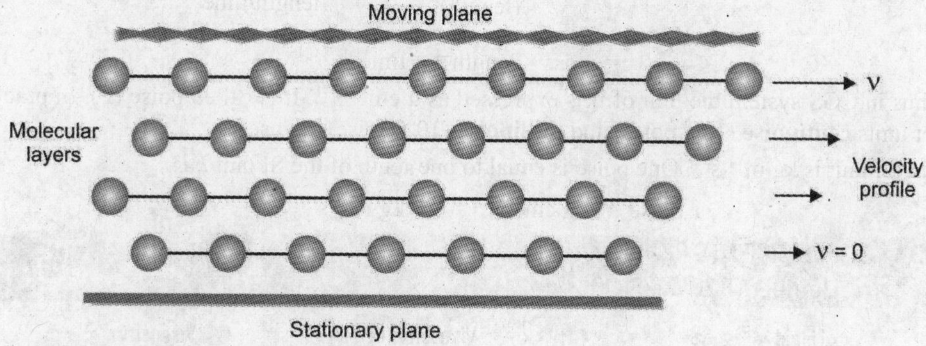

Figure 11.22
Flow of a liquid on a glass surface.

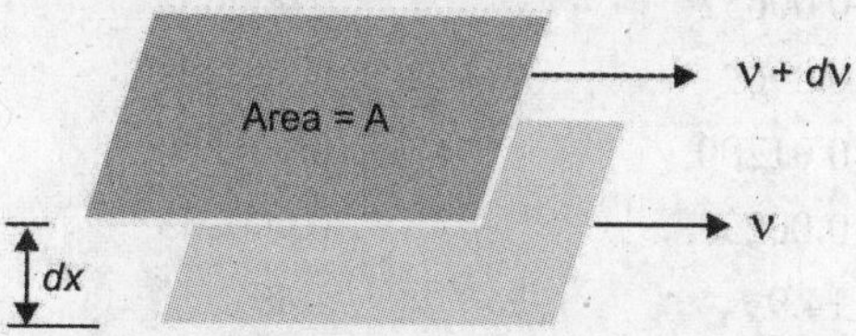

Figure 11.23
Two parallel layers moving in a liquid.

Now consider two adjacent moving layers of a liquid (Fig. 11.23). Let these be separated by a distance dx and have a velocity difference dv. The force of friction (F) resisting the relative motion of the two layers is directly proportional to the area A and the velocity difference dv, while it is inversely proportional to the distance between the layers.

That is,

$$F \propto A \frac{dv}{dx}$$

or

$$F = \eta A \frac{dv}{dx} \quad ...(1)$$

or

$$\eta = \frac{F}{A} \times \frac{dx}{dv} \quad ...(2)$$

where η (Greek letter *eta*) is the proportionality constant. It is known as the **Coefficient of Viscosity** or simply viscosity of a liquid. η has a specific value for a given liquid at the same temperature. It may be defined from equation (2) as : **the force of resistance per unit area which will maintain unit velocity difference between two layers of a liquid at a unit distance from each other.**

The reciprocal of viscosity is called Fluidity and is denoted by ϕ.

$$\phi = \frac{1}{\eta}$$

Units of Viscosity

The dimensions of the coefficient of viscosity (η) may be derived from equation (2).

$$\eta = \frac{F}{A} \times \frac{dx}{dv} = \frac{\text{force}}{\text{area}} \times \frac{\text{distance}}{\text{velocity}}$$

or
$$\eta = \frac{\text{mass} \times \text{length} \times \text{time}^{-2}}{(\text{length})^2} \times \frac{\text{length}}{\text{length/time}}$$

$$= \text{mass} \times \text{length}^{-1} \times \text{time}^{-1}$$

Thus in CGS system the unit of η is expressed as g cm^{-1} s^{-1}. It is called poise (P). In practice smaller units **centipoise** (10^{-2} poise) and **millipoise** (10^{-3} poise) are used.

The SI unit is kg m^{-1} s^{-1}. One poise is equal to one-tenth of the SI unit *i.e.*

$$1 \text{ poise} = 1 \text{ g cm}^{-1} \text{ s}^{-1} = 0.1 \text{ kg m}^{-1} \text{ s}^{-1}$$

TABLE 11.3. COEFFICIENTS OF VISCOSITY OF SOME LIQUIDS AT 20°C (293 K)

Liquid	Viscosity (P) CGS units	Viscosity SI units
Benzene	0.00652	0.000652
Carbon tetrachloride	0.0969	0.000969
Ethanol	0.01200	0.001200
Ether	0.00233	0.000233
Glycerine	14.9	1.49
Water	0.0101	0.00104

Measurement of Viscosity—The Ostwald Method

Viscosity of a liquid can be determined with the help of **Pioseulle's equation.** This expression which governs the flow of a liquid through a capillary may be written as :

$$\eta = \frac{\pi P r^4 t}{8 l V} \qquad ...(1)$$

where V is the volume of the liquid flowing through capillary in time t, P the pressure-head, r the radius of the tube and l its length. The experimental measurement of P, r, l and V offers considerable difficulty. Therefore, it is not possible to find the absolute coefficient of viscosity (η) straight away from Poiseulle's equation.

Ordinarily, the viscosity of a liquid is determined with respect to that of water. This is called **Relative Viscosity.** Let t_1 and t_2 be the times of flow of a fixed volume (V) of the two liquids through the same capillary. The expression for relative viscosity (η_1/η_2) can be derived from equation (1).

$$\frac{\eta_1}{\eta_2} = \frac{\pi P_1 r^4 t_1}{8 l V} \cdot \frac{8 l V}{\pi P_2 r^4 t_2} = \frac{P_1 t_1}{P_2 t_2} \qquad ...(2)$$

Since the pressure-head is proportional to density (d) of the liquid, from (2) we have :

$$\frac{\eta_1}{\eta_2} = \frac{d_1\, t_1}{d_2\, t_2} \qquad \text{...(3)}$$

Substituting the value of the viscosity coefficient of water (η_2) in (3), we can find the **absolute viscosity** of the given liquid (η_1).

Ostwald Viscometer

The apparatus commonly used for the determination of relative viscosity of a liquid is known as **Ostwald viscosimeter** *or* **viscometer.** A simple form of it is shown in Fig 11.24. The left-hand limb is essentially a pipette with two calibration marks A and B. A length of capillary tube joins the pipette to a bulb C in the right-hand limb.

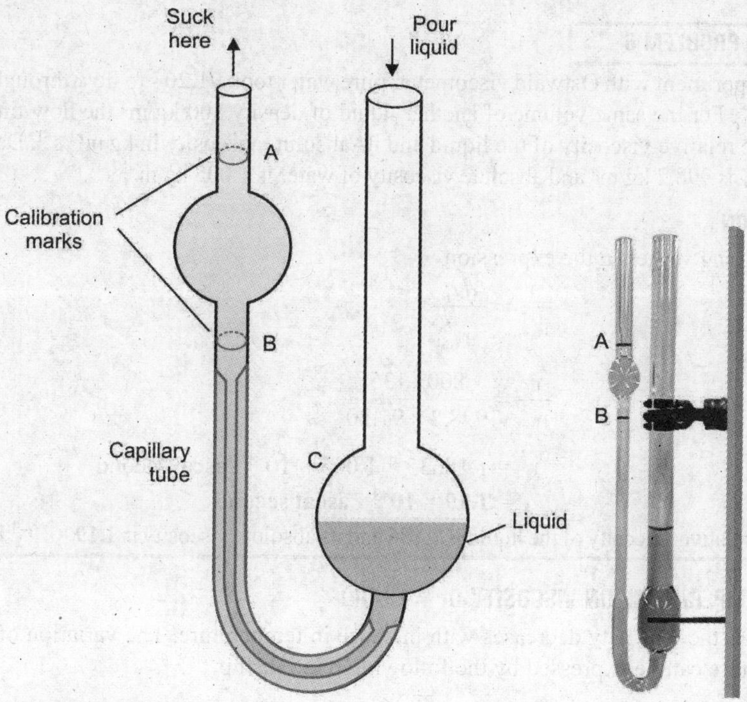

■ **Figure 11.24**
Ostwald Viscometer.

A definite volume of liquid (say 25 ml) is poured into the bulb C with a pipette. The liquid is sucked up near to the top of the left-limb with the help of a rubber tubing attached to it. The liquid is then released to flow back into the bulb C. The time (t_1) from A to B is noted with a stopwatch. Then the apparatus is cleaned and the experiment repeated with water, taking about the same volume. The time of flow of water (t_2) from A to B is recorded. The density of the liquid (d) and water (d_w) are determined with the help of a *pyknometer*. The relative viscosity coefficient is calculated from the expression

$$\frac{\eta}{\eta_w} = \frac{d\, t_1}{d_w\, t_2}$$

Knowing the value of the viscosity coefficient of water (η_w) at the temperature of the experiment, the absolute viscosity coefficient (η) of the given liquid can be found.

SOLVED PROBLEM 4

In an experiment with Ostwald viscometer, the times of flow of water and ethanol are 80 s and 175 s at 293 K. The density of water = 998 kg/m^3 and that of ethanol = 790 kg/m^3. The viscosity of water at 293 K is 0.001008 kg m^{-1}s^{-1}. Calculate the viscosity of ethanol.

SOLUTION

Substituting values in the expression

$$\eta = \eta_w \frac{d\ t_1}{d_w\ t_2}$$

or

$$\eta = 0.001008 \times \frac{790 \times 175}{998 \times 80}$$

$$= 0.001747 \text{ kg m}^{-1}\text{s}^{-1}$$

SOLVED PROBLEM 5

In an experiment with Ostwald viscometer, pure water took 91.20 s to flow through the capillary at 293 K. For the same volume of another liquid of density 800 kg/m^3 the flow time was 135 s. Find the relative viscosity of the liquid and its absolute viscosity in kg m^{-1} s^{-1}. Density of water at 293 K is 998.2 kg/m^3 and absolute viscosity of water is 1.005 kg m^{-1} s^{-1}.

SOLUTION

Substituting values in the expression

$$\frac{\eta}{\eta_w} = \frac{d t_1}{d_w t_2}$$

we have

$$\frac{\eta}{\eta_w} = \frac{800 \times 135}{998.2 \times 91.20}$$

$$\therefore \quad \eta = 1.18635 \times 1.005 \times 10^{-3} \text{ Pascal second}$$

$$= \mathbf{1.19 \times 10^{-3} \text{ Pascal second}}$$

Thus the relative viscosity of the liquid is **1.186** and its absolute viscosity is **1.19×10^{-3} Pascal second.**

EFFECT OF TEMPERATURE ON VISCOSITY OF A LIQUID

In general, the viscosity decreases with increase in temperature. The variation of viscosity (η) with temperature can be expressed by the following relationship

$$\eta = Ae^{-E/RT} \qquad ...(i)$$

where *A* and *E* are constants.

Taking logarithms on both sides, we get

$$\text{In } \eta = A + E/RT$$

or

$$\text{In } \eta = \frac{E}{R} \times \frac{1}{T} + A \qquad ...(ii)$$

comparing equation (*ii*) with

$$y = m\text{x} + c \qquad \text{(equation of straight line)}$$

A plot of log η versus 1/T should be a straight line (Fig. 11.25). It has been verified for a variety of liquids.

It has also been found that there is 2% decrease in viscosity for every increase in one degree of temperature of the liquid.

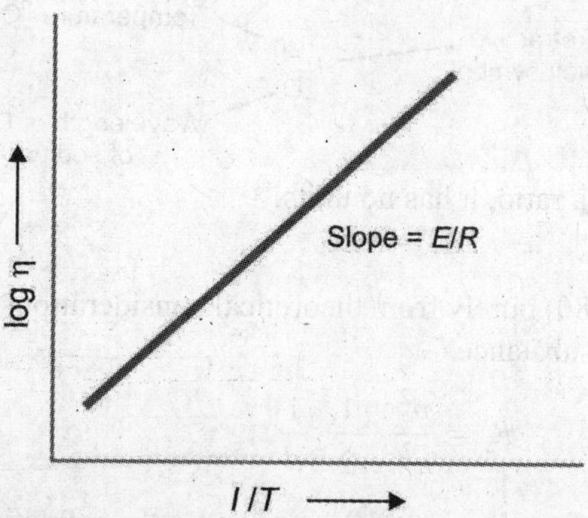

■ **Figure 11.25**
Plot of log η versus I/T

REFRACTIVE INDEX

The refractive index (n) of a substance is defined as the ratio of the velocity of light in vacuum or air, to that in the substance :

$$n = \frac{\text{Velocity of light in substance}}{\text{Velocity of light in air}}$$

When a ray of light passes from air into a liquid, its direction is changed. This change of direction is called *refraction*. The refractive index of the liquid with respect to air is given by **Snelle's Law.** According to it,

$$n = \frac{\text{Sin } i}{\text{Sin } r}$$

where i is the angle of incidence and r the angle of refraction.

The refractive index of a liquid can be easily determined to a high degree of accuracy. It is a characteristic property of a liquid. It varies with temperature and wavelength of light used.

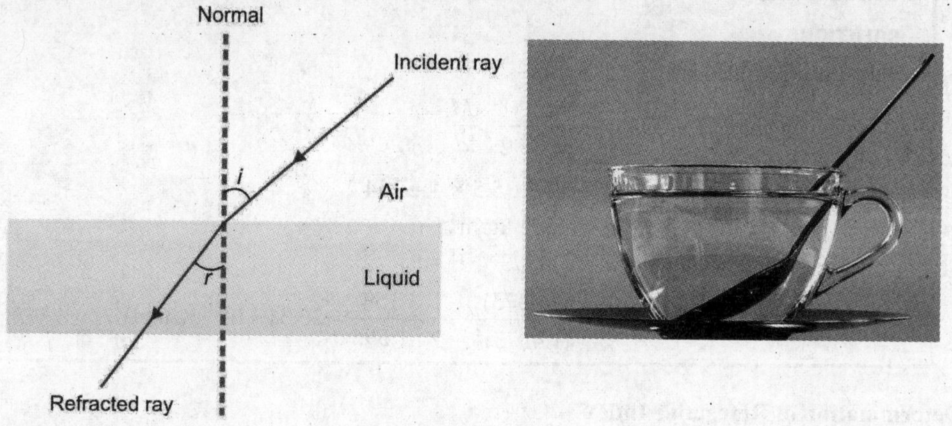

■ **Figure 11.26**
Refraction of light through a denser liquid medium.

The wavelength of D-line of the sodium spectrum is generally used for the purpose. If the refractive index of a liquid is measured at 293 K and using D-line of sodium, it is represented by the following symbol.

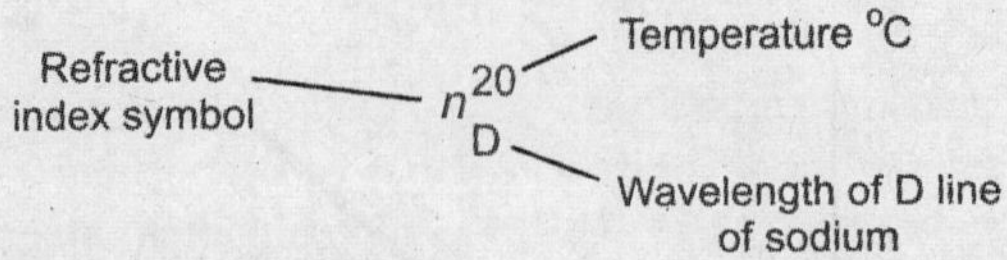

Because refractive index is a ratio, it has no units.

SPECIFIC REFRACTION

Lorenz and Lorenz (1880) purely from theoretical considerations derived the following relation for the refractive power of substance.

$$R_s = \frac{n^2 - 1}{n^2 + 2} \cdot \frac{1}{d} \qquad ...(1)$$

where R_s is the **Specific Refraction,** d the density and n the refractive index. The value of R_s was constant at all temperatures.

MOLAR REFRACTION

It is defined as **the product of specific refraction and molecular mass.** Thus molar refraction of a liquid (R_M) is obtained by multiplying equation (1) by molecular mass (M).

$$R_M = \frac{n^2 - 1}{n^2 + 2} \cdot \frac{M}{d} \qquad ...(2)$$

The value of molar refraction is characteristic of a substance and is temperature-independent. It can be determined by substituting the values of n, M and d in the equation (2). Since it depends on the wavelength of light, the values of molar refraction are generally reported for D-line of sodium. Since the value of refractive index (n) is dimensionless, from equation (2) it is evident that R_M has the units of the molar volume *i.e.*, cm^3 mol^{-1}.

SOLVED PROBLEM 6

The refractive index of carbon tetrachloride for D-line of sodium has been found to be 1.4573. Calculate its molar refraction if the density is 1595 kg/m^3.

SOLUTION

Substituting values in the expression

$$R_M = \frac{n^2 - 1}{n^2 + 2} \cdot \frac{M}{d}$$

Here $M = 12 + 35.5 \times 4 = 154$

$d = 1595 \text{ kg/m}^3$

$n = 1.4573$

$$\therefore \quad R_M = \frac{(1.4573)^2 - 1}{(1.4573)^2 + 2} \times \frac{154}{1.595} = 26.31 \times 10^{-6} \text{m}^3 \text{mol}^{-1}$$

Determination of Refractive Index

The refractive index of a liquid can be determined with the help of an instrument called **Abbe Refractometer** (Fig. 11.27). A thin film of the liquid is placed between the two prisms. Light from a sodium lamp is made to fall on lower side of the lower prism with the help of a mirror. The hypotenuse surface of the lower prism is ground and, therefore, light enters the liquid at all angles of incidence. However, no ray can enter the upper prism with greater angle of refraction than the grazing incidence (*i.e.*, at an angle) slightly less than 363 K. Thus the view in the telescope appears to be divided into two bands, one bright and one dark. The prism assembly is rotated with the help of a side knob till the cross wire of the telescope coincides with the edge of the bright band. A pointer attached to the prism assembly indicates the refractive index on the scale calibrated to read refractive indices directly.

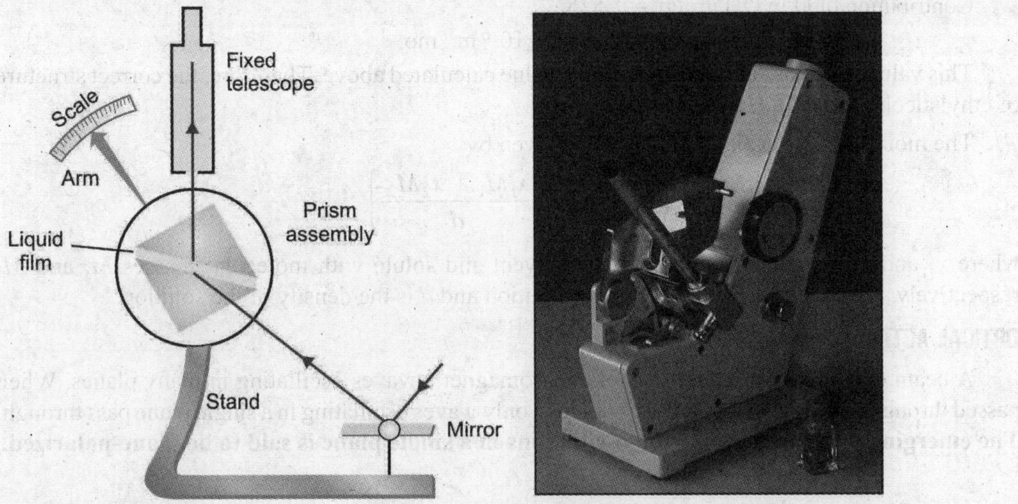

■ Figure 11.27
The Abbe refractometer.

Molar Refraction and Chemical Constitution

The molar refraction (R_M) is an additive property *i.e.* molar refraction of a molecule is the sum of the molar refraction of its constituent atoms. It is helpful in determining the constitution of a compound. The molar refraction values are calculated for various possible structures and the formula which is in accordance with the calculated molar refraction is the correct formula (or structure) of that compound. The molar refraction values for some atoms and bonds are given in Table 11.4.

TABLE 11.4. MOLAR REFRACTIONS (m^3 mol^{-1})

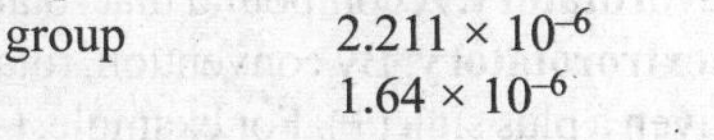

Carbon C	2.418×10^{-6}	3-membered ring	0.710×10^{-6}
Hydrogen H	1.100×10^{-6}	4-membered ring	0.480×10^{-6}
Chlorine Cl	5.967×10^{-6}	6-membered ring	0.15×10^{-6}
Bromine Br	8.861×10^{-6}	O in OH group	1.525×10^{-6}
Iodine I	13.900×10^{-6}	O is C = O group	2.211×10^{-6}
Double bond	1.733×10^{-6}	O in ethers	1.64×10^{-6}
Triple bond	2.398×10^{-6}		

Let us take the example of ethyl alcohol (CH_3CH_2OH). Its refractive index at 295.9 K for D-line is 1.3611 and its density is 788.5 kg/m^3. The molar mass of ethyl alcohol is 46 (2 × 12 + 1 × 6 + 1 × 16). The molar refraction can be calculated using the relation

$$R_M = \frac{n^2 - 1}{n^2 + 2} \cdot \frac{M}{d}$$

$$= \frac{(1.3611)^2 - 1}{(1.3611)^2 + 2} \times \frac{46}{788.5}$$

$$= 12.9105 \times 10^{-6} \text{ m}^3 \text{ mol}^{-1}$$

Let us now compute the value of molar refraction using the values in Table 11.4.

Contribution of 6 Hydrogens = 6 × 1.100 = 6.600
Contribution of 2 Carbons = 2 × 2.418 = 4.836
Contribution of O in OH group = 1.525
Total contribution = $12.961 \times 10^{-6}\ m^3\ mol^{-1}$

This value is in close agreement with the value calculated above. Therefore, the correct structure of ethyl alcohol is CH_3CH_2OH.

The molar refraction R_M of a solution is given by

$$R_M = \frac{n^2 - 1}{n^2 + 2}\left[\frac{x_1 M_1 + x_2 M_2}{d}\right]$$

where x_1 and x_2 are mole fractions of the solvent and solute with molecular masses M_1 and M_2 respectively, n is the refractive index of the solution and d is the density of the solution.

OPTICAL ACTIVITY

A beam of ordinary light consists of electromagnetic waves oscillating in many planes. When passed through a polarizer (*e.g.*, a Polaroid lens), only waves oscillating in a single plane pass through. **The emerging beam of light having oscillations in a single plane is said to be plane-polarized.**

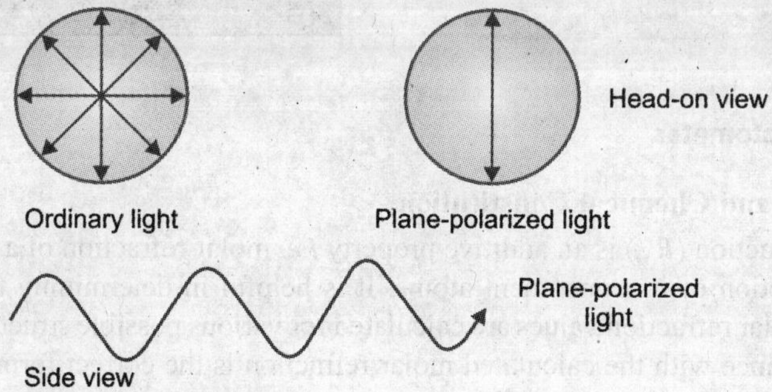

Figure 11.28
Plane-polarized light.

When plane-polarized light is passed through certain organic compounds, the plane of polarized light is rotated. **A compound that can rotate the plane of polarized light is called optically active.** This property of a compound is called **optical activity.**

A compound which rotates the plane-polarized light to the left (anticlockwise), is said to be **levorotatory.** A compound that rotates the plane-polarized light to the right (clockwise), is said to be **dextrorotatory**. By convention, rotation to the left is given a minus sign (–) and rotation to the right is given a plus sign (+). For example, (–)-lactic acid is levorotatory and (+)-lactic acid is dextrorotatory.

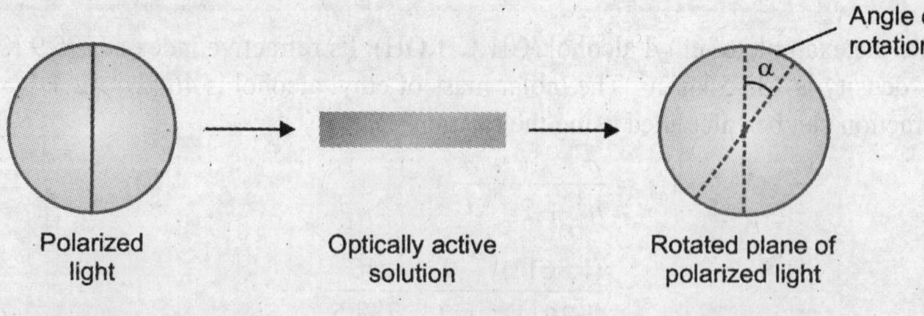

Figure 11.29
Optical Activity illustrated.

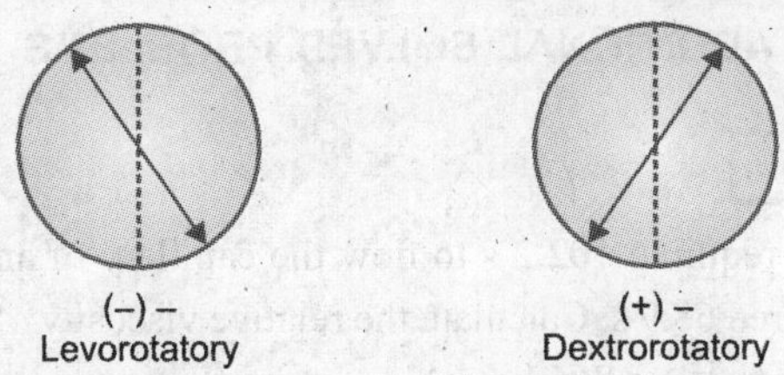

SPECIFIC ROTATION

The rotation of plane-polarized light is an intrinsic property of optically active molecules. When a polarized beam of light is passed through the solution of an optically active compound, its plane is rotated through an angle α (**angle of rotation**). This rotation depends on the number of optically active molecules encountered. Therefore, α is proportional to both the concentration and the length of the sample solution.

The **specific rotation** which is characteristic of an optically active substance, is expressed as

$$[\alpha] = \frac{\alpha}{l \times c} \quad ...(1)$$

where $[\alpha]$ = specific rotation in degrees
α = observed angle of rotation in degrees
l = length of the sample solution in decimeters (dm; 10 cm)
c = concentration of the sample solution

Thus from equation (1), the specific rotation can be defined as **the observed angle of rotation at a concentration of 10^{-3} kg/m³ and path length of 0.1 m.** Conventionally, a specific rotation is reported as $[\alpha]_D^t$, where *t* stands for temperature and D for D-line of sodium used for determination.

Measurement of Optical Activity

Optical activity is measured with the help of an instrument known as **polarimeter** (Fig. 11.30). This is basically a system of *polarizers* with a sample tube placed in between. First, an optically inactive medium (air or solvent) fills the sample tubes and polarized sodium light emerging from the *polarizer* passes through it. The *analyzer* is then turned to establish a dark field. This gives a zero reading on the circular scale around the analyzer.

Then the solution of the given optically active compound is placed in the sample tube. The plane of polarized light passing through it is rotated. The analyzer is turned to re-establish the dark field. The angle of rotation (α) is then noted in degrees on the circular scale. The specific rotation is calculated using the expression (1) above.

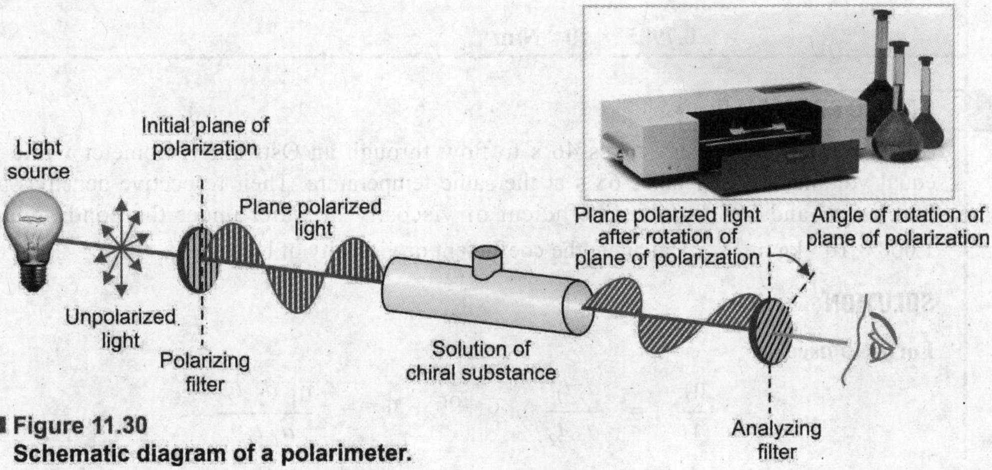

■ **Figure 11.30**
Schematic diagram of a polarimeter.

ADDITIONAL SOLVED PROBLEMS

SOLVED PROBLEM 7

At 293 K, pure water required 102.2 s to flow the capillary of an Ostwald viscometer while toluene at 293 K required 68.9 s. Calculate the relative viscosity of toluene. Densities of water and toluene are 998 kg/m^3 and 866 kg/m^3.

SOLUTION

Formula used

$$\frac{\eta_1}{\eta_2} = \frac{d_1\, t_1}{d_2\, t_2} \quad \text{or} \quad \eta_2 = \frac{\eta_1\, d_2\, t_2}{d_1\, t_1}$$

Quantities given

$\eta_1 = 1 \times 10^{-1}$ kg m^{-1} s^{-1} $\quad d_1 = 998$ kg/m^3 $\quad t_1 = 102.2$ s

$d_2 = 866$ kg/m^3 $\quad t_2 = 68.9$ s

Substitution of values

$$\eta_2 = \frac{10^{-1}\text{ kg m}^{-1}\text{ s}^{-1} \times 866\text{ kg/m}^3 \times 68.9\text{ s}}{998\text{ kg/m}^3 \times 102.2\text{ sec}}$$

$$= \mathbf{0.058\ kg\ m^{-1}\ s^{-1}}$$

SOLVED PROBLEM 8

The rise of water level in a capillary of radius 0.2×10^{-3} m at 293 K is 7.4×10^{-2} m. Calculate the surface tension of water taking its density as 1000 kg/m^3 at 293 K.

SOLUTION

Formula used

$$\gamma = \frac{(h + \frac{1}{3}\, r)\, r \rho g}{2}$$

Quantities given

$h = 7.4 \times 10^{-2}$ m $\quad r = 0.2 \times 10^{-3}$ m $\quad \rho = 1000$ kg/m^{-3} $\quad g = 9.8$ m s^{-2}

Substitution of values

$$\gamma = \frac{(7.4 + \frac{1}{3} \times 0.2 \times 10^{-3})\text{ m} \times 0.2 \times 10^{-3} \times 1000\text{ kg/m}^3 \times 9.8\text{ m s}^{-2}}{2}$$

$$= \mathbf{0.7905 \times 10^{-3}\ N/m}$$

SOLVED PROBLEM 9

Some volume of benzene takes 46 s to flow through an Ostwald viscometer while an equal volume of water takes 68 s at the same temperature. Their respective densities are 800 kg/m^3 and 998 kg/m^3. Coefficient of viscosity of water under the conditions is 1.008×10^{-3} kg m^{-1} s^{-1}. Calculate the coefficient of viscosity of benzene.

SOLUTION

Formula used

$$\frac{\eta_1}{\eta_2} = \frac{d_1\, t_1}{d_2\, t_2} \quad \text{or} \quad \eta_2 = \frac{\eta_1\, d_2\, t_2}{d_1\, t_1}$$

Quantities given

$\eta_1 = 1.008 \times 10^{-3}$ kg m^{-1} s^{-1} $\quad d_1 = 998$ kg/m^3 $\quad t_1 = 68$ s

$d_2 = 800$ kg/m^3 $\quad t_2 = 46$ s

Substitution of values

$$\eta_2 = \frac{1.008 \text{ kg m}^{-1}\text{ s}^{-1} \times 800 \text{ kg/m}^3 \times 46 \text{ s}}{998 \text{ kg/m}^3 \times 68 \text{ s}}$$

$$= \mathbf{546.599 \text{ kg m}^{-1}\text{ s}^{-1}}$$

SOLVED PROBLEM 10

In an experiment at 293 K, water rose in a capillary of diameter 0.078 m × 10^{-2} m to a height of 3.96 × 10^{-2} m. Calculate the surface tension of water if the density of water at this temperature is 998 kg/m^3 and g = 9.8 m/s^2.

SOLUTION

Formula used

$$\gamma = \frac{h\,r\,d\,g}{2}$$

Quantities given

$h = 3.96 \times 10^{-2}$ m $\quad r = \frac{0.078}{2} = 0.039 \times 10^{-2}$ m

$d = 998$ kg/m^3 $\quad g = 9.8$ m/s^2

Substitution of values

$$\gamma = \frac{0.039 \times 10^{-2} \text{ m} \times 3.96 \times 10^{-2} \text{ m} \times 998 \text{ kg/m}^3 \times 9.8 \text{ m/s}^2}{2}$$

$$= \frac{151.048}{2} \text{ dynes cm}^{-1} \qquad [\because 1 \text{ dyne} = 1 \text{ g cm s}^{-2}]$$

$$= \mathbf{75.524 \text{ dynes cm}^{-1} = 0.075524 \text{ N/m}}$$

SOLVED PROBLEM 11

The number of drops of water counted in a stalgamometer at 298 K is 300 whereas the number of drops for ethyl alcohol is 320. Calculate the surface tension of ethyl alcohol given that the density of water = 998 kg/m^3, density of ethyl alcohol = 961.4 kg/m^3 surface tension of water = 72.75 × 10^{-3} N/m.

SOLUTION :

Formula used

$$\frac{\gamma_1}{\gamma_2} = \frac{n_1}{n_2} \times \frac{d_1}{d_2}$$

Quantities given

$\gamma_1 = 72.75 \times 10^{-3}$ N/m $\quad n_1 = 300$ $\quad d_1 = 998$ kg/m^3

$n_2 = 320$ $\quad d_2 = 961.4$ kg/m^3

Substitution of values

$$\frac{72.75 \times 10^{-3} \text{ N/m}}{\gamma_2} = \frac{300 \times 998 \text{ kg/m}^3}{320 \times 961.4 \text{ kg/m}^3}$$

or $$\gamma_2 = \frac{72.75 \times 10^{-3} \times 320}{300} \times \frac{961.4}{998}$$

$$= \mathbf{0.0747541\ N/meter}$$

SOLVED PROBLEM 12

In a measurement of viscosity with Ostwald's viscometer, water takes 580 s to flow through a given volume, while an organic liquid takes 395 s. Calculate the viscosity of organic liquid. Density of organic liquid and water are786.7 kg/m^3 and 998.4 kg/m^3 respectively. Viscosity of water is 1.01×10^{-3} kg m^{-1}s.

SOLUTION

Formula used

$$\frac{\eta_1}{\eta_2} = \frac{d_1\, t_1}{d_2\, t_2}$$

Quantities given

$\eta_1 = 1.01 \times 10^{-3}$ kg m^{-1}s $d_1 = 998.4$ kg/m^3 $t_1 = 580$ s

$d_2 = 786.7$ kg/m^3 $t_2 = 395$ s

Substitution of values

$$\frac{1.01 \times 10^{-3}\ \text{kg m}^{-1}\text{s}}{\eta_2} = \frac{998.4\ \text{kg/m}^3 \times 580\ \text{s}}{786.7\ \text{kg/m}^3 \times 395\ \text{s}}$$

or $$\eta_2 = \frac{1.01 \times 10^{-3} \times 786.7 \times 395}{998.4 \times 580}$$

$$= \mathbf{0.542\ kg\ m^{-1}\ s^{-1}}$$

SOLVED PROBLEM 13

The refractive index of carbon tetrachloride is 1.4573 and its density is 1595.2 kg/m^3. Calculate its molar refraction.

SOLUTION

Formula used

$$R_m = \frac{n^2 - 1}{n^2 + 2} \times \frac{M}{d}$$

Quantities given

$n = 1.4573$ $d = 1595.2$ kg/m^3 $M = 12 + 4 \times 35.5 = 154 \times 10^{-3}$ kg mol^{-1}

Substitution of values

$$R_m = \frac{(1.4573)^2 - 1}{(1.4573)^2 + 2} \times \frac{154}{1595.2} \times 10^{-3}$$

$$= \frac{1.1237}{4.1237} \times \frac{154}{1595.2} \times 10^{-3}$$

$$= \mathbf{26.30 \times 10^{-6}\ m^3\ mol^{-1}}$$

EXAMINATION QUESTIONS

1. Define or explain the following terms :
   (*a*) van der Waal's forces (*b*) Hydrogen bond
   (*c*) Vapor pressure (*d*) Boiling point
   (*e*) Surface tension (*f*) Viscosity
   (*g*) Ostwald's method (*h*) Refractive index
   (*i*) Specific Refraction (*j*) Molar Refraction
   (*k*) Optical activity (*l*) Specific rotation
2. (*a*) Define the term coefficient of viscosity. Describe Ostwald's viscometer method for determining it.
   (*b*) In a measurement of viscosity by Ostwald's viscometer at 293.35 K, water takes 30 s to flow between the upper and lower marks while the flow of another liquid of density 1500 kg/m^3, takes 35 s. Taking density of water at 293.35 K to be 998.2 kg/m^3, calculate the viscosity of the other liquid if the viscosity of water at this temperature is 0.01 kg m^{-1} s^{-1}.
   Answer. (*b*) 0.01753 kg m^{-1} s^{-1}
3. The density of a liquid at 298 K is 1.203 × 10^3 kg/m^3. Its refractive index n^{25}_D is 1.552. Calculate its molar refractivity. (Molecular weight of liquid = 123)
   Answer. 32.66 × 10^{-3} m mol^{-1}
4. In an Ostwald viscometer, the flow times for two liquids A and B are in the ratio 4:5. If the density of liquid B is twice as high as that of A, calculate the ratio of their viscosities.
   Answer. 2:5
5. (*a*) Define viscosity and coefficient of viscosity. Describe one method of determining the viscosity of liquids.
   (*b*) It takes 100 s for water to flow through a capillary tube at fixed pressure. How long will it take toluene to do so under similar conditions. Given that: (*i*) density of water = 998 kg/m^3; absolute viscosity of water = 0.001 Pascal second; density of toluene = 866 kg m^{-3}; viscosity = 0.0006 Pascal second.
   Answer. (*b*) 69.14 s
6. (*a*) Define surface tension. What is the influence of temperature on surface tension?
   (*b*) Describe drop number method of determining surface tension of a liquid.
7. Define coefficient of viscosity. How does it vary with temperature? (*Punjabi BSc*)
8. Explain the following terms :
   (*a*) Fluidity (*b*) Surface Tension (*Kurukshetra BSc*)
9. What is the effect of temperature on surface tension of a liquid? (*MD Rohtak BSc*)
10. Why do you use the same viscometer for the liquid and water during the experimental determination of viscosity by the Ostwald viscometer? Describe the experiment. (*Delhi BSc*)
11. (*a*) How is the surface tension of a liquid affected by temperature?
    (*b*) Describe the relationship between the viscosity of a liquid and its chemical constitution.
    (*Madurai BSc*)
12. What is meant by viscosity of a liquid? How it is experimentally determined? Discuss the effect of temperature on it. (*Lucknow BSc*)
13. (*a*) Define the term viscosity, coefficient of viscosity and fluidity. What are the factors affecting the viscosity of liquid?
    (*b*) Describe a method for measuring surface tension of a liquid.
    (*Arunachal BSc*)

14. Describe the terms : Viscosity and coefficient of viscosity. Give their units. *(Guru Nanak Dev BSc)*

15. What is the effect of temperature on viscosity of a liquid? *(Nagpur BSc)*

16. How is relative surface tension of a liquid with reference to that of water measured by drop weight method? Write the dimensions of surface tension. *(Jammu BSc)*

17. Explain : Viscosity of liquids decreases with increase in temperature. *(Delhi BSc)*

18. Define surface tension. What are its units? Describe drop weight method for the determination of the surface tension. Why do liquids have surface tension? Show diagrammatically the direction in which surface tension acts. *(Guru Nanak Dev BSc)*

19. What do you mean by laminar flow of a liquid? Define coefficient of viscosity. Describe a method for its determination. *(Kalyani BSc)*

20. (*a*) When a liquid is warmed, initially many small bubbles begin to form. Why?
    (*b*) Surface tension of a liquid vanishes at its critical temperature. Explain.
    (*c*) What is coefficient of viscosity of a liquid? What are its units in CGS and SI systems? How do you determine it for a liquid by capillary method? How does it vary with temperature? *(Guru Nanak Dev BSc)*

21. (*a*) What is surface tension? Why rain drops are spherical? Write SI units of surface tension.
    (*b*) Write short note on parachor. *(Arunachal BSc)*

22. Define the terms : relative viscosity, specific viscosity, reduced viscosity and intrinsic viscosity as applied to the solutions of high polymers. How is the intrinsic viscosity related to molecular mass of a polymer? *(Burdwan BSc)*

23. Describe stalagmometer and what for it is used? *(HS Gaur BSc)*

24. What are the factors on which surface tension of a liquid depends? What are its units? Describe drop number method to measure surface tension of liquid. *(Delhi BSc)*

25. (*a*) What is stalagmometer and how is it used in the determination of surface tension of a liquid?
    (*b*) When 0.000012 m^3 of water is allowed to flow through a stalagmometer, 50.0 drops of water were obtained and when the same volume of liquid is allowed to flow, the number of drops formed was 35. If the density of the liquid is 1200 kg/m^3 and the surface tension of water is 0.072 N/m, calculate the surface tension of the liquid.

    Answer. (*b*) 0.082 N/m *(Jammu BSc)*

26. (*a*) Find the dimension of viscosity coefficient η and translate poise into SI units.
    (*b*) At 293 K flow times of water and another liquid in a viscometer are 144 s and 60 s respectively. Take density of water = 1000 kg/m^3, density of liquid = 800 kg/m^3, viscosity coefficient of water = 0.001 kg m^{-1} s^{-1} (all at 293 K). Calculate the coefficient of viscosity for the liquid at 293 K.

    Answer. (*b*) 0.33×10^{-3} kg m^{-1} s^{-1} *(Sambalpur BSc)*

27. The refractive index of a liquid of molar mass 72 is 1.34 and its density is 980 kg/m^3. Find its molar refraction.

    Answer. 0.02091 *(Jiwaji BSc)*

28. The surface tension of a liquid at 300 K is 0.0271 N/m and its density at this temperature is 988.00 kg/m^3. What is the radius of the largest capillary that will allow the liquid to rise 2×10^{-2} m ? (Assume the angle of contact to be zero and $g = 9.81$ m/s^2)

    Answer. 0.28 or 0.279 *(Panjab BSc)*

29. Calculate the optical rotation when sodium D light passes through a solution of a compound in chloroform, containing 0.025 kg per 10^{-2} m^3, placed in a cell of 0.2 m. [α] for the compound in chloroform is 37.5°C.

    Answer. 18.75° *(Delhi BSc)*

30. A glass capillary of diameter 0.001 m is dipped into water. Calculate the level of the water that rises in the capillary if surface tension of water = 0.07275 N/m density = 998.4 kg/m^3 and contact angle for water = 10°.

Answer. 0.0249 m *(Madurai BSc)*

31. The refractive index of a liquid at 298 K is 1.6 and its density is 870 kg/m^3. Find the molar refraction. (Molar mass of the liquid is 78)

Answer. 0.03067 *(Agra BSc)*

32. The molar refraction of a liquid is 12.95 × 10^{-6} m^3 mol^{-1} and its density is 1046 kg/m^3. Calculate the refractive index (molar mass of liquid is 0.060 kg mol^{-1})

Answer. 1.369 *(Panjab BSc)*

33. Define viscosity of fluids. Explain determination of coefficient of viscosity for a liquid using Ostwald's viscometer. How does the change in temperature affect the viscosity of a liquid. *(Goa BSc)*

34. Water passes through a viscometer in 30 s. The same volume of oil required 2263.7 s. Calculate the viscosity of oil if its density is 1.1 × 10^3 kg m^{-3}. Density of water is 0.998 × 10^3 kg m^{-3}, viscosity of water is 0.00101 kg m^{-1} s^{-1}.

Answer. 0.8400 × 10^{-1} kg m^{-1} s^{-1} *(Kerala BSc)*

35. At 293 K, toluene rises 1.95 × 10^{-2} m in a capillary tube of radius 0.3412 × 10^{-3} m. Calculate the surface tension of toluene. The density of toluene at 293 K is 866 kg/m^3.

Answer. 0.02823 N/m *(Osmania BSc)*

MULTIPLE CHOICE QUESTIONS

1. The molecules which have partial positive charge at one end and a partial negative charge at the other are called

(*a*) ion-pairs (*b*) charged molecules

(*c*) dipoles (*d*) electric molecules

Answer. (*c*)

2. The momentary attractions between the molecules of a liquid caused by instantaneous dipole and induced-dipole attractions are called

(*a*) van der Waals forces (*b*) polar forces

(*c*) French forces (*d*) London forces

Answer. (*d*)

3. The strength of H-bonding is about _______ as that of a covalent bond

(*a*) 1/2 (*b*) 1/3

(*c*) 1/10 (*d*) 1/100

Answer. (*c*)

4. Which of the following statements is incorrect?

(*a*) the pressure exerted by the vapor in equilibrium with the liquid at a fixed temperature is called vapor pressure

(*b*) the vapor pressure increases with rise in temperature

(*c*) at equilibrium, liquid $\rightleftharpoons$ vapor

(*d*) at equilibrium, the rate of evaporation is not equal to the rate of condensation

Answer. (*d*)

5. The boiling point of a liquid is that temperature at which
   (*a*) the vapor pressure of the liquid is equal to the atmospheric pressure
   (*b*) the vapor pressure of the liquid is less than the atmospheric pressure
   (*c*) the vapor pressure of the liquid is greater than the atmospheric pressure
   (*d*) the vapor pressure of the liquid is equal to the square root of the atmospheric pressure
   Answer. (*a*)
6. With rise in temperature, the surface tension of a liquid
   (*a*) increases (*b*) decreases
   (*c*) remains the same (*d*) none of the above
   Answer. (*b*)
7. The unit in which surface tension is measured is
   (*a*) dyne cm (*b*) dyne cm^{-1}
   (*c*) $dyne^{-1}$ cm (*d*) $dyne^{-1}$ cm^{-1}
   Answer. (*b*)
8. The formula used for the determination of surface tension by capillary rise method is
   (*a*) $2\gamma = hrdg$ (*b*) $2\gamma = hr^2dg$
   (*c*) $2\gamma = \pi\, r \cos\theta$ (*d*) $2\gamma = \pi\, hr^2dg$
   Answer. (*a*)
9. In drop weight method for determination of surface tension of liquids, the formula used is
   (*a*) $\frac{\gamma_1}{\gamma_2} = \frac{m_1}{m_2}$ (*b*) $\frac{\gamma_1}{\gamma_2} = \frac{m_2}{m_1}$
   (*c*) $\frac{\gamma_1}{\gamma_2} = \frac{m_2^2}{m_1^2}$ (*d*) $\frac{\gamma_1}{\gamma_2} = \frac{m_1^2}{m_2^2}$
   Answer. (*a*)
10. The formula used in drop-number method for determination of surface tension of liquids is
   (*a*) $\frac{\gamma_1}{\gamma_2} = \frac{n_2\, d_1}{n_1\, d_2}$ (*b*) $\frac{\gamma_1}{\gamma_2} = \frac{n_1\, d_1}{n_2\, d_2}$
   (*c*) $\frac{\gamma_1}{\gamma_2} = \frac{n_1\, d_2}{n_2\, d_1}$ (*d*) $\frac{\gamma_1}{\gamma_2} = \frac{n_2\, d_2}{n_1\, d_1}$
   Answer. (*a*)
11. Viscosity of a liquid is a measure of
   (*a*) repulsive forces between the liquid molecules
   (*b*) frictional resistance
   (*c*) intermolecular forces between the molecules
   (*d*) none of the above
   Answer. (*b*)
12. The reciprocal of viscosity is called
   (*a*) surface tension (*b*) frictional resistance
   (*c*) fluidity (*d*) none of these
   Answer. (*c*)
13. The dimensions in which coefficient of viscosity is expressed are
   (*a*) mass × length × $time^{-1}$ (*b*) mass × $length^{-1}$ × $time^{-1}$
   (*c*) mass × $length^{-1}$ × time (*c*) $mass^{-1}$ × $length^{-1}$ × $time^{-1}$
   Answer. (*b*)

14. The SI unit of coefficient of viscosity is

(*a*) kg m^2 s (*b*) kg m s^{-1}

(*c*) kg m^{-1} s (*d*) kg m^{-1} s^{-1}

Answer. (*d*)

15. The formula used for the determination of relative viscosity by Ostwald's method is

(*a*) $\frac{\eta_1}{\eta_2} = \frac{d_1\, t_1}{d_2\, t_2}$ (*b*) $\frac{\eta_1}{\eta_2} = \frac{d_1\, t_2}{d_2\, t_1}$

(*c*) $\frac{\eta_1}{\eta_2} = \frac{d_2\, t_1}{d_1\, t_2}$ (*d*) $\frac{\eta_1}{\eta_2} = \frac{d_2\, t_2}{d_1\, t_1}$

Answer. (*a*)

16. In general, the viscosity _______ with temperature

(*a*) decreases (*b*) increases

(*c*) remains the same (*d*) none of these

Answer. (*a*)

17. The specific refraction is given by the formula

(*a*) $R_s = \frac{n^2 - 1}{n^2 + 2} \times \frac{1}{d}$ (*b*) $R_s = \frac{n^2 + 1}{n^2 - 2} \times \frac{1}{d}$

(*c*) $R_s = \frac{n^2 - 1}{n^2 - 2} \times \frac{1}{d}$ (*d*) $R_s = \frac{n^2 + 1}{n^2 + 2} \times \frac{1}{d}$

Answer. (*a*)

18. The molar refraction is given by the formula

(*a*) $R_M = \frac{n^2 + 1}{n^2 + 2} \cdot \frac{M}{d}$ (*b*) $R_M = \frac{n^2 - 1}{n^2 + 2} \cdot \frac{M}{d}$

(*c*) $R_M = \frac{n^2 + 1}{n^2 - 2} \cdot \frac{M}{d}$ (*d*) $R_M = \frac{n^2 - 1}{n^2 - 2} \cdot \frac{M}{d}$

Answer. (*b*)

19. The units in which molar refraction is expressed are

(*a*) cm^3 mol^1 (*b*) cm^{-3} mol

(*c*) cm^{-3} mol^{-1} (*d*) cm^3 mol^{-1}

Answer. (*d*)

20. The specific rotation, a characteristic of an optically active substance, is expressed as

(*a*) $[\alpha] = \frac{\alpha}{\ell \times c}$ (*b*) $[\alpha] = \frac{\alpha}{\ell^2 \times c}$

(*c*) $[\alpha] = \frac{\alpha}{\ell \times c}$ (*d*) $[\alpha] = \frac{\alpha^2}{\ell \times c}$

Answer. (*c*)

21. Association of molecules in water is due to

(*a*) surface tension (*b*) viscosity

(*c*) hydrogen bonding (*d*) optical activity

Answer. (*c*)

22. With the increasing molecular mass of a liquid, the viscosity

(*a*) increases (*b*) decreases

(*c*) no effect (*d*) none of these

Answer. (*a*)

23. Which of the following liquids has the maximum viscosity?
(*a*) water (*b*) acetone
(*c*) ethyl alcohol (*d*) glycerine
Answer. (*d*)

24. The fluidity of liquids _______ with increase in temperature
(*a*) decreases (*b*) remains the same
(*c*) increases (*d*) none of these
Answer. (*c*)

25. A liquid rises in a capillary tube. It is due to
(*a*) surface tension (*b*) viscosity
(*c*) osmosis (*d*) effusion
Answer. (*a*)

26. A liquid is in equilibrium with its vapors at its boiling point. On the average, the molecules in the two phases have equal
(*a*) potential energy (*b*) kinetic energy
(*c*) van der Waal's forces (*d*) total energy
Answer. (*b*)

27. The rate of evaporation of a liquid depends upon
(*a*) surface area (*b*) temperature
(*c*) nature of liquid (*d*) all of these
Answer. (*d*)

28. The highest temperature at which vapor pressure of a liquid can be measured is
(*a*) boiling point of the liquid (*b*) freezing point of the liquid
(*c*) inversion temperature (*d*) critical temperature
Answer. (*d*)

29. Small droplets are spherical in shape. It is due to
(*a*) high viscosity
(*b*) their tendency to acquire minimum surface area
(*c*) their tendency to acquire maximum surface area
(*d*) less viscosity
Answer. (*b*)

30. Insects can walk on the surface of water due to
(*a*) viscosity (*b*) surface tension
(*c*) refractivity (*d*) optical activity
Answer. (*b*)

31. The cleansing action of soaps and detergents is due to
(*a*) internal friction (*b*) high hydrogen bonding
(*c*) surface tension (*d*) viscosity
Answer. (*c*)

32. The work in ergs required to be done to increase the surface area by 1 sq. cm is called
(*a*) surface tension (*b*) internal friction
(*c*) fluidity (*d*) surface energy
Answer. (*d*)

33. While determining viscosity of a liquid with the help of a viscometer, the liquid was sucked with great difficulty. The liquid may be
(*a*) water (*b*) acetone
(*c*) ether (*d*) glycerine
Answer. (*d*)

34. The vapor pressure of a liquid at a given temperature is given by

(*a*) $\frac{n}{V}RT$ (*b*) $\frac{n}{V}RT^2$

(*c*) $\frac{n}{2V}RT$ (*d*) $\frac{n}{V}RT^3$

Answer. (*a*)

35. Which of the following is incorrect?

(*a*) the rate of condensation is equal to the rate of evaporation at equilibrium in a closed container

(*b*) the concentration of the molecules in vapor phase decreases continuously

(*c*) the concentration of the molecules in vapor phase remains unchanged at equilibrium

(*d*) none of the above

Answer. (*b*)

36. The vapor pressure of non-polar liquids is

(*a*) fairly high as compared to polar liquids (*b*) fairly low as compared to polar liquids

(*c*) the same as that of polar liquids (*d*) none of these

Answer. (*a*)

37. The boiling point of a liquid _______ with external pressure

(*a*) changes (*b*) always increases

(*c*) always decreases (*d*) remains the same

Answer. (*a*)

38. The units of coefficient of viscosity are

(*a*) dyne cm^{-2} s (*b*) poise

(*c*) centipoise (*d*) all of these

Answer. (*d*)

39. Liquids with high molecular masses have _______ viscosity

(*a*) greater (*b*) lesser

(*c*) zero (*d*) none of these

Answer. (*a*)

40. The liquids with high intermolecular attractive forces have _______ viscosity

(*a*) low (*b*) intermediate

(*c*) high (*d*) none of these

Answer. (*c*)

12

Solid State

CONTENTS

Gases and liquids can flow and take up the shape of their container. Solids, on the other hand, have a definite volume and shape. They are rigid and lack the ability to flow.

In both gases and liquids, atoms, ions and molecules continually move. They translate randomly as well as rotate and vibrate. This determines the ability of gases and liquids to flow. In solids, atoms, ions and molecules are held together by relatively strong chemical forces-ionic bond, covalent bond, or by intermolecular van der Waals' forces. They do not translate although they vibrate to some extent in their fixed positions. This explains why solids are rigid and have definite shape.

TYPES OF SOLIDS

Broadly speaking, solids are of two types :

(*a*) Crystalline solids; also called true solids
(*b*) Amorphous solids

A **crystalline solid** exists as small crystals, each crystal having a characteristic geometrical shape. In a crystal, the atoms, molecules or ions are arranged in a regular, repeating three-dimensional pattern called the **crystal lattice.** Sugar and salt are crystalline solids.

An **amorphous solid** (Gr *amorphous* = no form) has atoms, molecules or ions arranged at random and lacks the ordered crystalline lattice. Examples are rubber, plastics and glass. In their disordered structure, amorphous solids resemble liquids. Thus glasses are to be regarded as super-cooled or highly viscous liquids. The liquid nature of glass is sometimes apparent in very old window panes that have become slightly thicker at the bottom due to gradual downward flow.

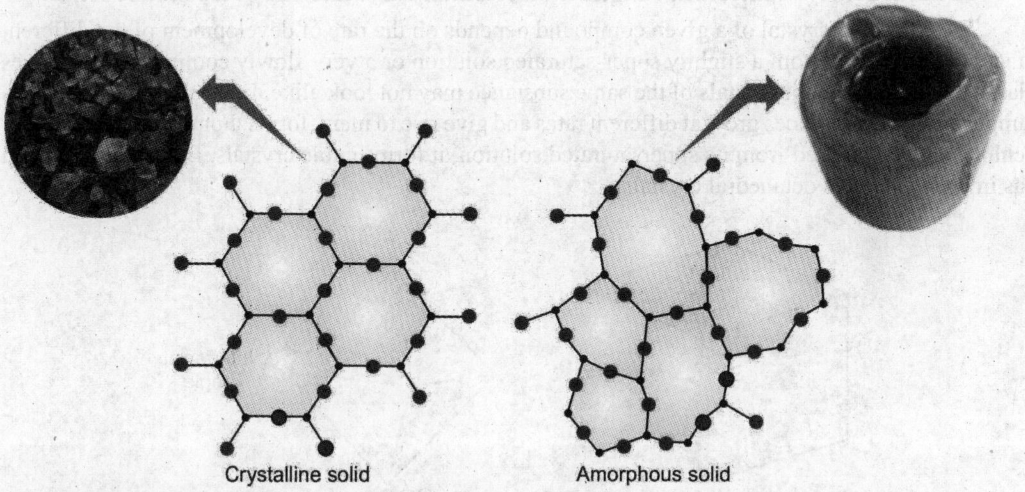

■ **Figure 12.1**
Two-dimensional representation of crystalline solid and an amorphous solid. Each substance has the formula A_2O_3. *A* is shown by • and *O* is shown by ●.

ISOTROPY AND ANISOTROPY

Amorphous substances are said to be isotropic because they exhibit the same value of any property in all directions. Thus refractive index, thermal and electrical conductivities, coefficient of thermal expansion in amorphous solids are independent of the direction along which they are measured.

Crystalline substances, on the other hand, are anisotropic and the magnitude of a physical property varies with directions. For example, in a crystal of silver iodide, the coefficient of thermal expansion is positive in one direction and negative in the other. Similarly, velocity of light in a crystal may vary with direction in which it is measured. Thus a ray of light passing through a Nicol prism splits up into two components, each travelling with different velocity (double refraction).

Explanation of Isotropy and Anisotropy

In amorphous substances, as in liquids, the arrangement of particles is random and disordered. Therefore all directions are equivalent and properties are independent of direction.

On the other hand, the particles in a crystal are arranged and well ordered. Thus the arrangement of particles may be different in different directions. This is illustrated in Fig. 12.2 in which a simple two-dimensional arrangement of two different types of particles has been shown. When a

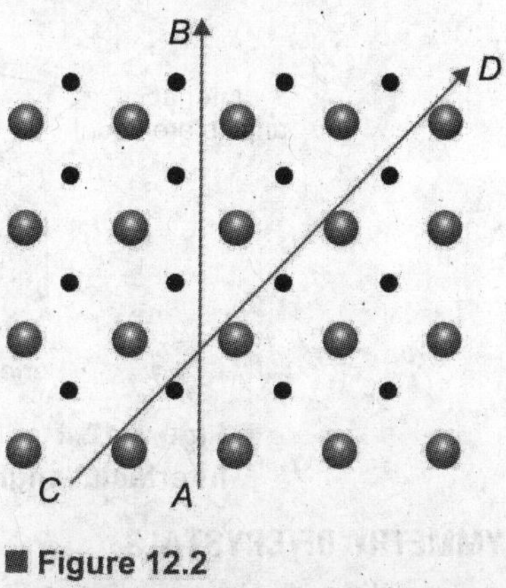

■ **Figure 12.2**
Anisotropy in crystals is due to different arrangements of particles in different directions.

property is measured along *AB* and *CD*, the value along *CD* will be different from that along *AB*. This is so because there is only one type of particles along *AB* while there are two types of particles in alternate positions along *CD*. This explains why crystalline substances are anisotropic.

THE HABIT OF A CRYSTAL

The external shape is called the **habit** of the crystal. The plane surfaces of the crystal are called **faces.** The angles between the faces are referred to as the **interfacial angles.** The interfacial angles for a given crystalline substance are always the same.

The consistency of interfacial angles is an essential characteristic of crystalline solids.

The habit of a crystal of a given compound depends on the rate of development of the different faces. Slow growth from a slightly super-saturated solution or a very slowly cooling solution gives large crystals. Different crystals of the same substance may not look alike. In the presence of certain impurities, different faces grow at different rates and give rise to many forms. For example, if sodium chloride is crystallized from its supersaturated solution, it forms cubic crystals. But if urea is added as impurity, it gives octahedral crystals.

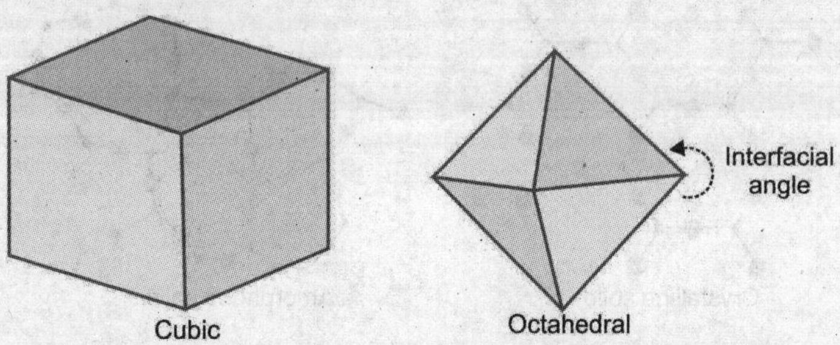

■ **Figure 12.3**
Shapes of sodium chloride crystals obtained under different conditions.

As already stated, the external appearance or size of crystals of a substance may be different. It depends on the relative development of the different faces under appropriate conditions. **But the interfacial angles are always the same.** This is illustrated diagrammatically in Fig. 12.4.

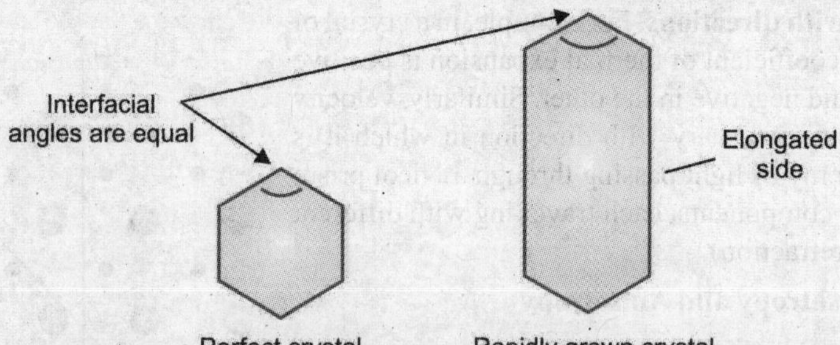

■ **Figure 12.4**
Interfacial angles are always the same.

SYMMETRY OF CRYSTALS

In addition to the angles, another important property of crystals is their symmetry. Symmetry in crystals may be due to a plane, a line or a point. Accordingly there are three types of symmetry elements associated with a crystal. These are called the **Elements of Symmetry.**

(1) Plane of Symmetry

A crystal is said to have a plane of symmetry if it can be divided by an imaginary plane into two equal parts, each of which is the exact mirror image of the other.

(2) Axis of Symmetry

An axis of symmetry is an imaginary line drawn through the crystal such that during rotation of the crystal through 360°, the crystal presents exactly the same appearance more than once. If similar view appears twice, it is called an axis of two fold symmetry or **diad axis.** If it appears thrice, it is an axis of three fold symmetry or **triad axis,** and so on.

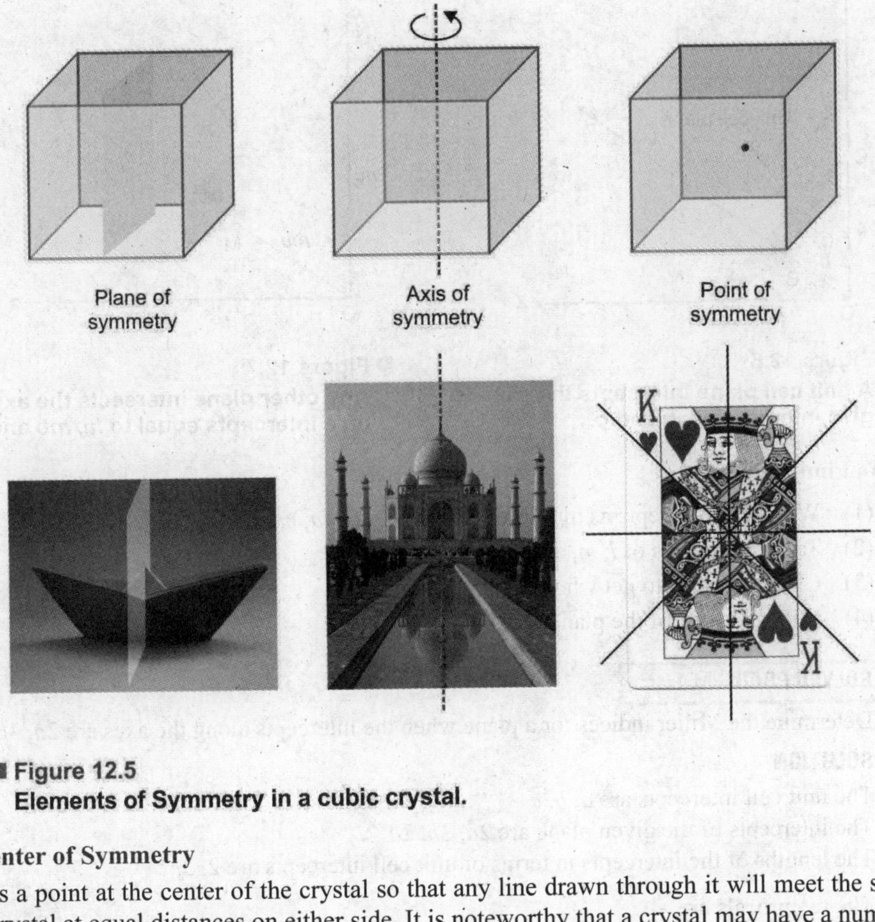

■ **Figure 12.5**
Elements of Symmetry in a cubic crystal.

(3) Center of Symmetry

It is a point at the center of the crystal so that any line drawn through it will meet the surface of the crystal at equal distances on either side. It is noteworthy that a crystal may have a number of planes of symmetry or axis of symmetry but it can have only one center of symmetry.

MILLER INDICES

We know that crystal lattice consists of unit cells arranged in parallel planes. Thus each crystal plane lies parallel to the crystal face as also to the unit cell face. These planes cut the three axes along the three crystallographic axes (OX, OY, OZ). Hauy proposed that a given crystal plane could be described in terms of intercepts along the axes (Law of Rational intercepts). The reciprocals of these intercepts are small whole numbers. These numbers, *h*, *k* and *l* are called **Miller indices** after the name of the British scientist W.H. Miller. Thus Miller indices of a plane may be defined as **the reciprocals of the intercepts which the plane makes with the axes.**

For illustration, let us consider a crystal system with the axes OX, OY and OZ. In Fig. 12.6 ABC represents a unit cell surface while LMN in Fig. 12.7 depicts another crystal plane under study. The intercepts of the unit cell plane are OA, OB and OC which have the lengths a, b and c respectively. The intercepts of the plane under study are OL, OM and ON. These can be expressed as multiples of the intercepts a, b, c, *i.e.*, la, mb and nc. Here l, m, n are either integral whole numbers or fractions of whole numbers. The reciprocals of these numbers are written together in brackets (h, k, l) to give the Miller indices of the plane under study.

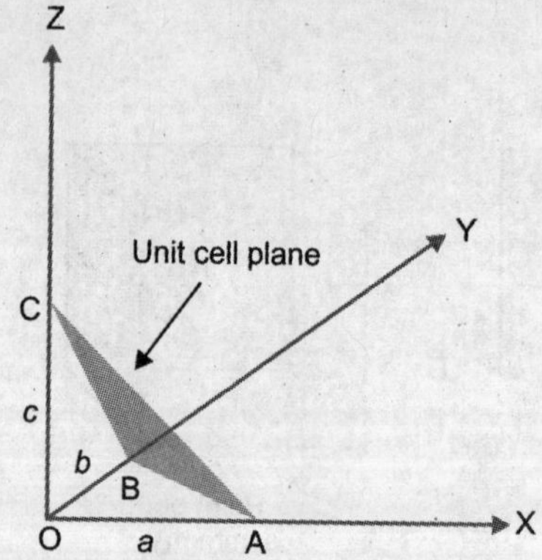

Figure 12.6
A unit cell plane intercepts the axes to give intercepts *a*, *b* and *c*.

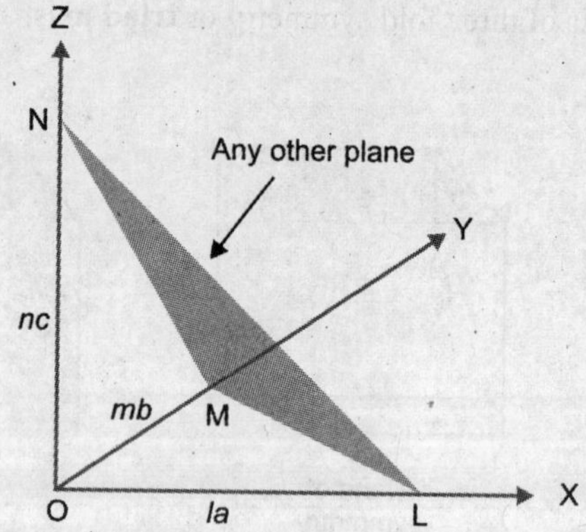

Figure 12.7
Any other plane intersects the axes to give intercepts equal to *la*, *mb* and *nc*.

How to Find Miller Indices?

(1) Write the intercepts as multiples of a, b, c, say, la, mb and nc.
(2) Take reciprocals of l, m, and n,
(3) Clear fractions to get whole numbers h, k, l,
(4) Miller indices of the plane are (h, k, l).

SOLVED PROBLEM 1

Determine the Miller indices for a plane when the intercepts along the axes are $2a$, $3b$ and $2c$.

SOLUTION

The unit cell intercepts are a, b, c
The intercepts of the given plane are $2a$, $3b$, $2c$
The lengths of the intercepts in terms of unit cell intercepts are 2, 3, 2
The reciprocals are $\frac{1}{2}, \frac{1}{3}, \frac{1}{2}$
Clear fractions by multiplying with 6 which gives the whole numbers 3, 2, 3
Thus the Miller indices of the given plane are (3, 2, 3).

CRYSTAL STRUCTURE

The particles (atoms, ions or molecules) in crystals are highly ordered as is evident from their geometrical shapes, smooth surfaces and specific interfacial angles.

The particles in crystals are arranged in regular patterns that extend in all directions. The overall arrangement of particles in a crystal is called the **Crystal lattice, Space lattice or Simply lattice.**

To describe the structure of a crystal it is convenient to view it as being made of a large number of basic units. The simple basic unit or the building block of the crystal lattice is called the **Unit cell.** This is illustrated in Fig. 12.8.

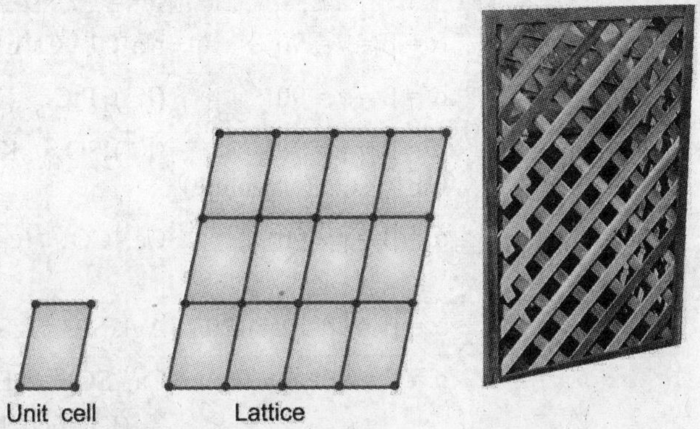

■ **Figure 12.8**
Two-dimensional illustration of unit cell and crystal lattice.

How to Represent Crystal Lattice and Units Cells?

The crystal lattice of a substance is depicted by showing the position of particles (**structural units**) in space. These positions are represented by bold dots (or circles) and are referred to as **lattice points** or **lattice sites.** The overall shape and structure of a crystal system is governed by that of the unit cell of which it is composed.

A unit cell has one atom or ion at each corner of the lattice. Also, there may be atoms or ions in faces and interior of the cell. A cell with an interior point is called the **body centered cell.** A cell which does not contain any interior points is known as the **primitive cell.** That is, **a primitive cell is a regular three-dimensional unit cell with atoms or ions located at its corners only.**

Parameters of the Unit Cells

In 1850, August Bravais, a French mathematician observed that the crystal lattice of substances may be categorized into seven types. These are called **Bravais lattices** and the corresponding unit cells are referred to as **Bravais unit cells.** The unit cells may be characterized by the following parameters :

(*a*) relative lengths of the edges along the three axes (a, b, c).

(*b*) the three angles between the edges (α, β, γ).

The parameters of a unit cell can be illustrated as in Fig. 12.9. Parameters of the seven Bravais unit cells are listed in Table 12.1.

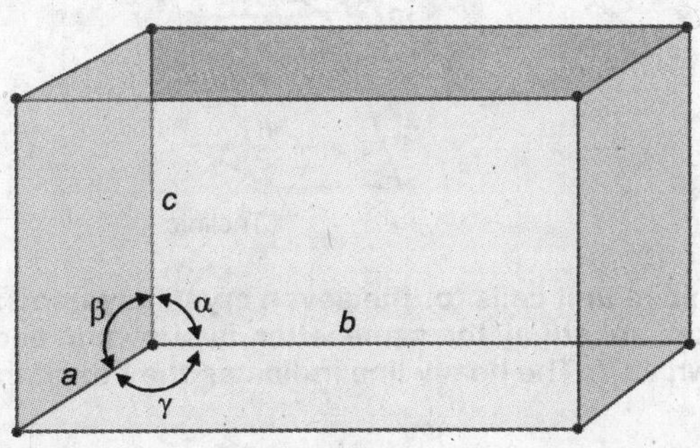

■ **Figure 12.9**
Illustration of parameters of a unit cell.

TABLE 12.1. THE SEVEN UNIT CELLS

Crystal system	Relative axial length	Angles	Examples
Cubic (isometric)	$a = b = c$	$\alpha = \beta = \gamma = 90°$	Na^+C1^-, Cs^+Cl^-, $Ca^{2+}(F^-)_2$, $Ca^{2+}O^{2-}$
Tetragonal	$a = b \neq c$	$\alpha = \beta = \gamma = 90°$	$(K^+)_2 PtCl_6^{2-}$, $Pb^{2+}WO_4^{2-}$, $NH_4^+Br^-$
Orthorhombic	$a \neq b \neq c$	$\alpha = \beta = \gamma = 90°$	$(K^+)_2SO_4^{2-}$, $K^+NO_3^-$, $Ba^{2+}SO_4^{2-}$, $Ca^{2+}CO_3^{2-}$ (aragonite)
Rhombohedral (trigonal)	$a = b = c$	$\alpha = \beta = \gamma \neq 90°$	$Ca^{2+}CO_3^{2-}$ (calcite), $Na^+NO_3^-$
Hexagonal	$a = b \neq c$	$\alpha = \beta = 90°, \gamma = 120$	Agl, SiC, HgS
Monoclinic	$a \neq b \neq c$	$\alpha = \beta = 90°, \gamma \neq 90°$	$Ca^{2+}SO_4^{2-}$, $2H_2O$, $K^+ClO_3^-$, $(K^+)_4Fe(CN)_6^{4-}$
Triclinic	$a \neq b \neq c$	$\alpha \neq \beta \neq \gamma \neq 90°$	$Cu^{2+}\ SO_4^{2-} . 5H_2O$, $(K^+)_2Cr_2O_7^{2-}$

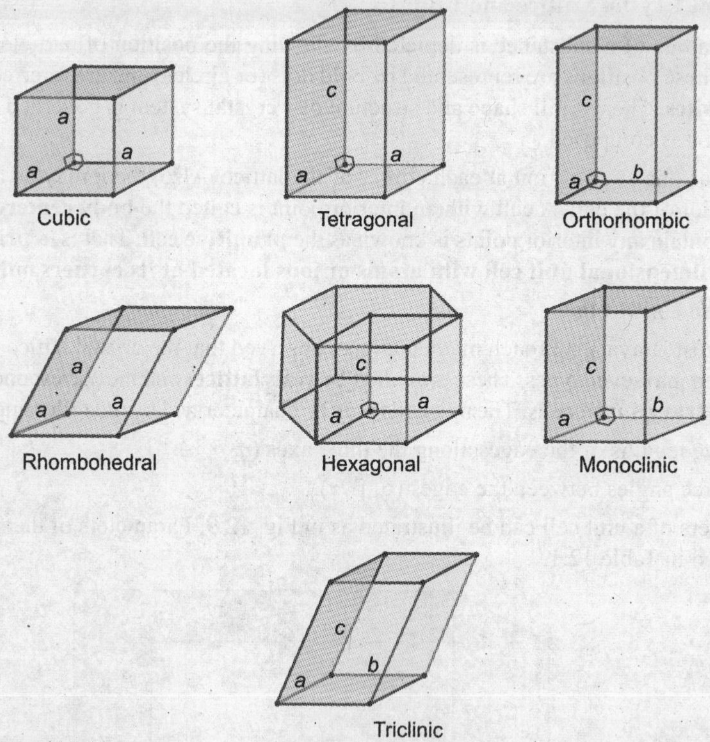

■ **Figure 12.10**
The primitive unit cells for the seven crystal systems. Where two or more of the axes are equal, the same letter is shown in each. Right angles (90°) are shown as 7. The heavy line indicates the hexagonal unit cell.

CUBIC UNIT CELLS

These are the simplest unit cells. These unit cells are particularly important for two reasons. First, a number of ionic solids and metals have crystal lattices comprising cubic unit cells. Second, it is relatively easy to make calculations with these cells because in them all the sides are equal and the cell angles are all 90°.

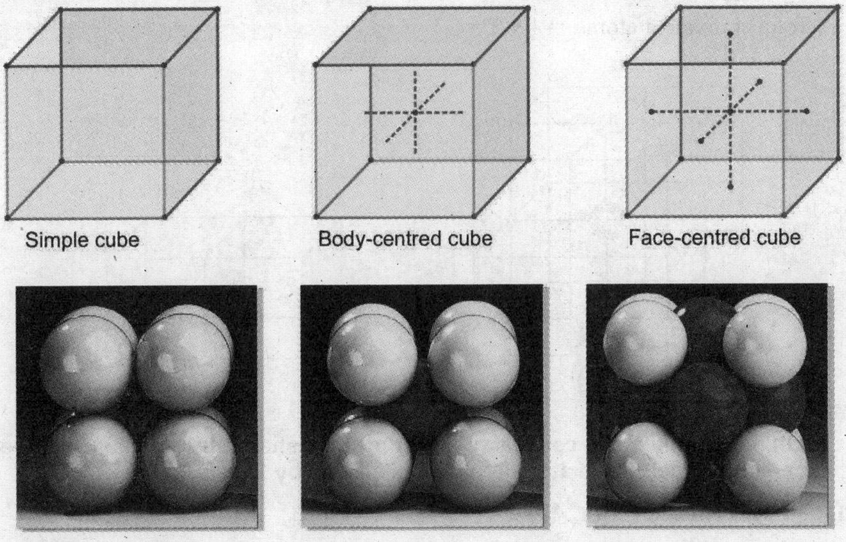

■ **Figure 12.11**
Three cubic unit cells.

Three types of Cubic Unit Cells

There are three types of cubic unit cells :

(1) Simple cubic unit cell

(2) Body-centered cubic unit cell

(3) Face-centered cubic unit cell

A **simple cubic unit cell** is one in which the atoms or ions are occupying only the corners of the cube.

A **body-centered cubic unit cell** has one particle at the center of the cube in addition to the particles at the corners.

A **face-centered cubic unit cell** has one particle at each of the six faces of the cube apart from the particles at the corners.

How to Compute Atoms or Points in a Unit Cell?

In counting the number of atoms (or points) per unit cell, we must keep in mind that atoms on corners of faces are shared with adjoining cells. Therefore the number of atoms in a cubic unit cell may be computed as follows.

(1) Eight unit cells share each corner atom. Therefore **the simple cubic unit cell contains the equivalent of one atom.**

At each corner we have = 1/8 atom
At 8 corners we have = 1/8 × 8 = 1 atom

(2) Each face-centered atom is shared by two unit cells. Therefore **the face-centered unit cell contains the equivalent of four atoms.**

At 8 corners, 1/8 each = 1 atom
6 face-centered sites, 1/2 each = 3 atoms
Total equivalent atoms = 1 + 3 = 4

(3) A body centered unit cell contains the equivalent of two atoms.

At 8 corners at 1/8 each = 1 atom

central unshared = 1 atom

Total equivalent atoms = 1 + 1 = 2

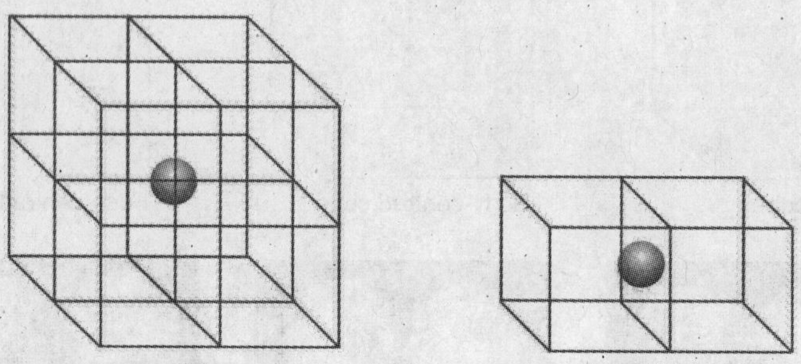

(a) (b)

■ Figure 12.12
***(a)* In the simple cubic cell, each corner atom is shared by 8 cells; *(b)* In the face-centred cell, each atom in a face is shared by 2 cells.**

Calculation of Mass of the Unit Cell

Mass of the unit cell can be calculated from the number of atoms in the unit cells. For this multiply the mass of one atom by the number of atoms in the unit cell.

$$\text{Thus the mass of one atom} = \frac{\text{molar mass of the substance}}{\text{Avogadro's number}}$$

SOLVED PROBLEM 2

The unit cell of metallic gold is face-centered cubic.

(*a*) How many atoms occupy the gold unit cell ?

(*b*) What is the mass of a gold unit cell ?

SOLUTION

eight corners at 1/8 atom each = 1 atom

six faces at 1/2 atom each = 3 atoms

∴ 4 atoms occupy the gold unit cell.

$$\text{Mass of gold atom} = \frac{\text{molar mass}}{\text{Avogadro's number}}$$

$$\frac{197 \times 10^{-5}\ \text{kg/mol}}{6.022 \times 10^{23}\ \text{Au atoms/mol}} \times \frac{1}{4} = \mathbf{8.178 \times 10^{-26}\ kg}$$

SOLVED PROBLEM 3

By X-ray diffraction it is found that nickel crystals are face-centered cubic. The edge of the unit cell is 3.52×10^{-10} m. The atomic mass of nickel is 58.7×10^{-3} and its density is 8940 kg/m^3. Calculate Avogadro's number from the data.

SOLUTION

Number of atoms in nickel unit cell = 4

Let Avogadro's number = N

$$\text{Thus the mass of one Ni atom} = \frac{58.7\times10^{-3}}{N}$$

$$\text{the mass of one unit cell} = 4\left(\frac{58.7\times10^{-3}}{N}\right)kg$$

$$\text{the volume of one unit cell} = (3.52 \times 10^{-10})^3 \text{ m}^3$$

$$\text{the density of nickel cell} = \frac{4(58.7\times10^{-3}/N)kg}{(3.52\times10^{-10})^3 \text{ m}^3} = 8940\,kg/\text{m}^3 \text{ (given)}$$

$$\therefore \quad N = \mathbf{6.02 \times 10^{23}}$$

What is Coordination Number of a Crystal Lattice?

The coordination number of a crystal structure is the number of particles (ions, atoms or molecules) immediately adjacent to each particle in the lattice. It is characteristic of a given space lattice and is determined by inspection of the model. In a simple cubic lattice *e.g.*, NaCl, each particle is surrounded by six other particles and so the coordination number is six. It will be observed that the coordination numbers for body-centered and face-centered cubic lattice are 8 and 12 respectively.

X-RAY CRYSTALLOGRAPHY

A crystal lattice is considered to be made up of regular layers or planes of atoms equal distance apart. Since the wavelength of X-rays is comparable to the interatomic distances, Laue (1912) suggested that crystal can act as grating to X-rays. Thus when a beam of X-rays is allowed to fall on a crystal, a large number of images of different intensities are formed. If the diffracted waves are in the same phase, they reinforce each other and a series of bright spots are produced on a photographic plate placed in their path. On the other hand, if the diffracted waves are out of phase, dark spots are caused on the photographic plate. From the overall diffraction patterns produced by a crystal, we can arrive at the detailed information regarding the position of particles in the crystal. **The study of crystal structure with the help of X-rays is called X-ray crystallography.**

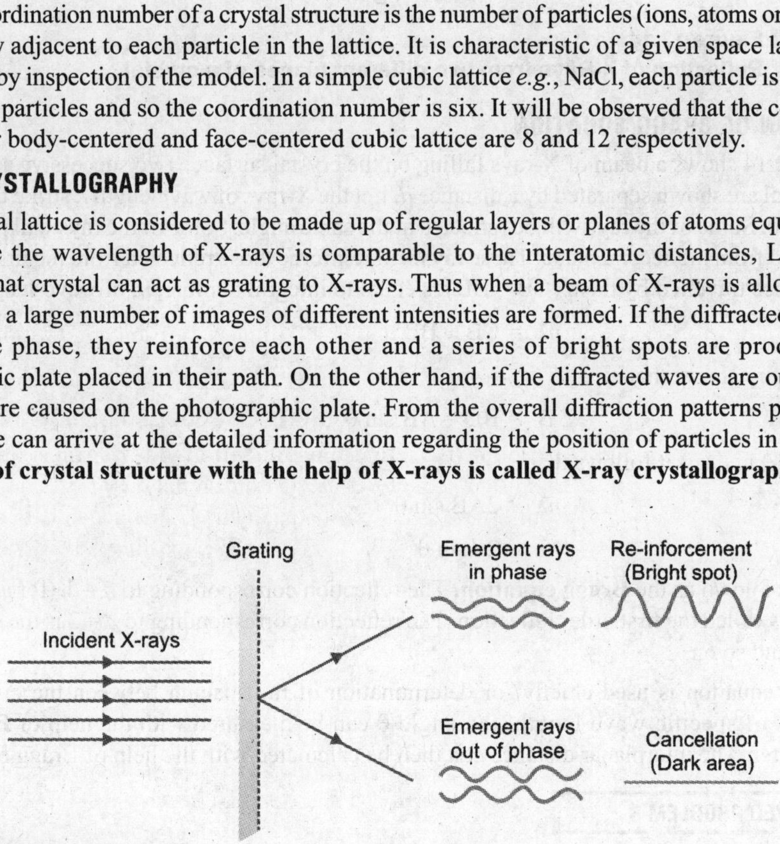

■ Figure 12.13
Diffraction patterns produced by crystals.

BRAGG'S EQUATION

In 1913 the father-and-son, W.L. Bragg and W.H. Bragg worked out a mathematical relation to determine interatomic distances from X-ray diffraction patterns. This relation is called the **Bragg equation.** They showed that :

(1) **the X-ray diffracted from atoms in crystal planes obey the laws of reflection.**

(2) **the two rays reflected by successive planes will be in phase if the extra distance travelled by the second ray is an integral number of wavelengths.**

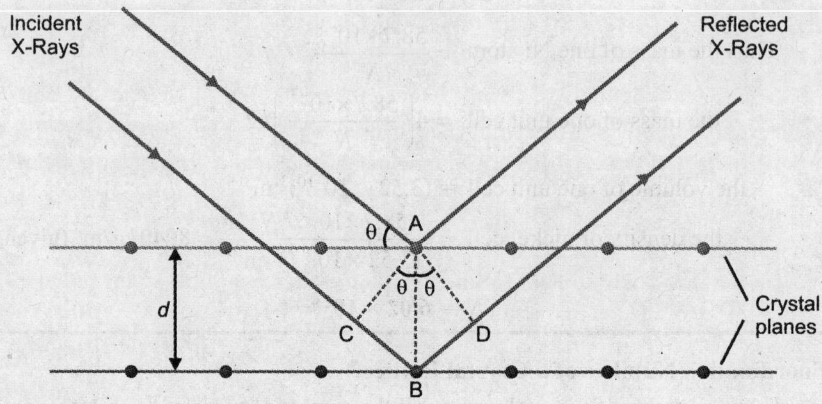

Figure 12.14
Reflection of X-Rays from two different planes of a crystal.

DERIVATION OF BRAGG EQUATION

Fig. 12.14 shows a beam of X-rays falling on the crystal surface. Two successive atomic planes of the crystal are shown separated by a distance *d*. Let the X-rays of wavelength λ strike the first plane at an angle θ. Some of the rays will be reflected at the same angle. Some of the rays will penetrate and get reflected from the second plane. These rays will reinforce those reflected from the first plane if the extra distance travelled by them (CB + BD) is equal to integral number, *n*, of wave-lengths. That is,

$$n\lambda = \text{CB} + \text{BD} \qquad ...(i)$$

Geometry shows that

$$\text{CB} = \text{BD} = \text{AB} \sin\theta \qquad ...(ii)$$

From (*i*) and (*ii*) it follows that

$$n\lambda = 2\text{AB}\sin\theta$$

or

$$n\lambda = 2d\sin\theta$$

This is known as the **Bragg equation.** The reflection corresponding to $n = 1$ (for a given series of planes) is called the first order reflection. The reflection corresponding to $n = 2$ is the second order reflection and so on.

Bragg equation is used chiefly for determination of the spacing between the crystal planes. For X-rays of specific wave length, the angle θ can be measured with the help of **Bragg X-ray spectrometer.** The interplanar distance can then be calculated with the help of Bragg equation.

SOLVED PROBLEM 4

Find the interplanar distance in a crystal in which a series of planes produce a first order reflection from a copper X-ray tube ($\lambda = 1.539 \times 10^{-10}$ m) at an angle of 295.5 K.

SOLUTION

From Bragg equation

$$n\lambda = 2d\sin\theta$$

We have $d = \dfrac{\lambda}{2\sin\theta}$ when $n = 1$

Hence $d = \dfrac{1.539\ \text{Å}}{2\ \sin(22.5)}$

$$= \frac{1.539\times10^{-10}}{2(0.383)} = \mathbf{2.01\times10^{-10}\ m}$$

SOLVED PROBLEM 5

Diffraction angle 2 θ equal to 16.8° for a crystal having inter planar distance in the crystal is 0.4×10^{-9} m when second order diffraction was observed. Calculate the wavelength of X-rays used.

SOLUTION. Here

$$n = 2\ ;\ d = 0.400\ \text{nm} = 0.4 \times 10^{-9}\text{m}$$

$$2\,\theta = 16.8° \text{ or } \theta = 8.4°$$

Using Bragg's equation

$$n\lambda = 2d \sin\theta$$

Substituting the values, we get

$$2 \times \lambda = 2 \times 0.4 \times 10^{-9} \times \sin 8.4$$

or $$\lambda = \frac{2 \times 0.4 \times 10^{-9} \times 0.146}{2}$$

or $$\lambda = 0.0584 \times 10^{-9}\ \text{m}$$

or $$\lambda = 0.584 \times 10^{-10}\text{m}$$

MEASUREMENT OF DIFFRACTION ANGLE

The measurement of diffraction angle, θ, required for Bragg equation can be done in two ways.

(1) The rotating crystal method (Braggs, 1913)

(2) The powder method (Debye and Scherrer, 1916)

(1) The Rotating Crystal Method

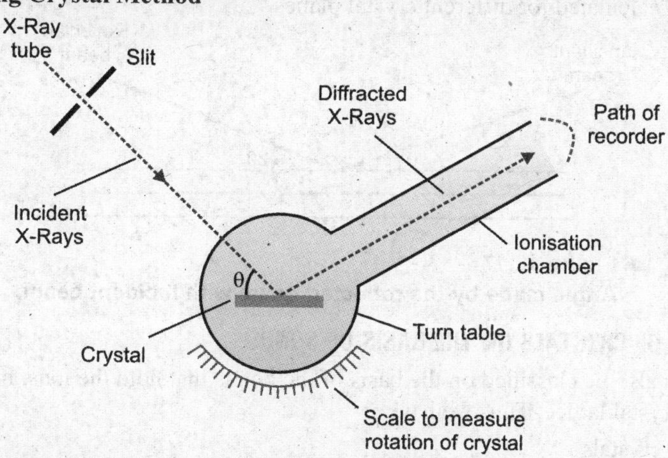

■ **Figure 12.15**
Rotating crystal method to determine the angle of diffraction θ.

The apparatus used by Bragg is shown in Fig. 12.15. A beam of X-rays of known wave length falls on a face of the crystal mounted on a graduated **turn table.** The diffracted rays pass into the ionization chamber of the **recorder.** Here they ionize the air and a current flows between the chamber wall and an electrode inserted in it which is connected to an electrometer. The electrometer reading is proportional to the intensity of X-rays. As the recorder along with the crystal is rotated, the angles of maximum intensity are noted on the scale. Thus values of θ for $n = 1, 2, 3$, etc. are used to calculate the distance d between the lattice planes parallel to the face of the crystal.

(2) The Powder Method

The rotating crystal method could only be used if a single undistorted crystal is available .To overcome this limitation, the powder method was devised. In this method the crystalline material contained in a capillary tube is placed in the camera containing a film strip (Fig. 12.16). The sample is rotated by means of a motor. The X-rays pass through the gap between the ends of the film.

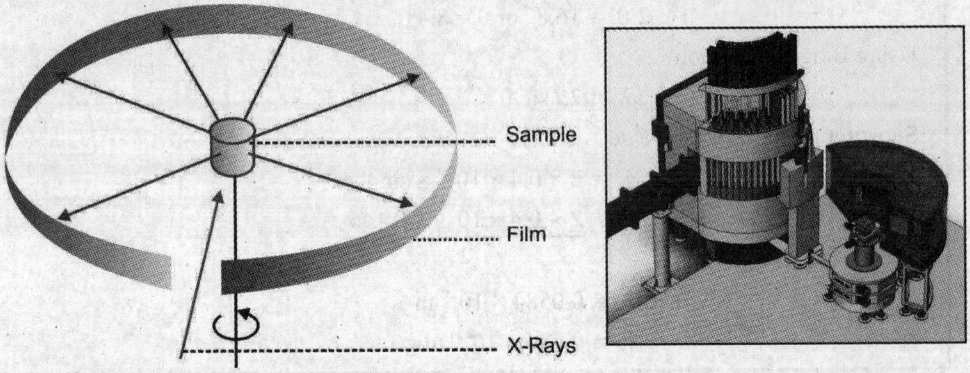

■ **Figure 12.16**
The Powder method.

The powdered sample contains small crystals arranged in all orientations. Some of these will reflect X-rays from each lattice plane at the same time. The reflected X-rays will make an angle 2θ with the original direction. Hence on the photo are obtained lines of constant θ. From the geometry of the camera, θ can be calculated for different crystal planes.

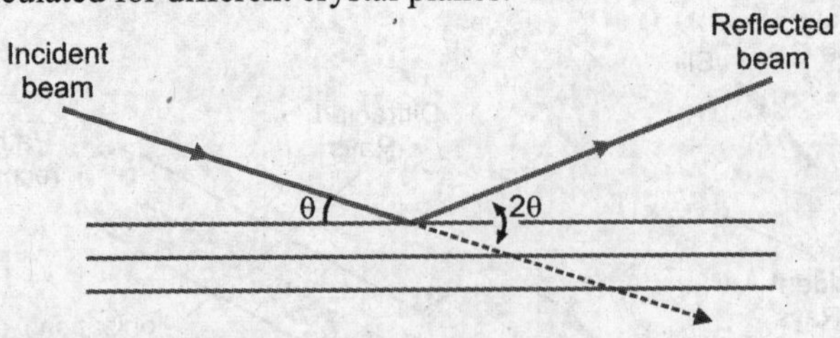

■ **Figure 12.17**
Angle made by the reflected beam with incident beam.

CLASSIFICATION OF CRYSTALS ON THE BASIS OF BONDS

Crystals can also be classified on the basis of the bonds that hold the ions, molecules or atoms together in the crystal lattice. Thus we have :

(*a*) Ionic crystals
(*b*) Molecular crystals
(*c*) Network covalent crystals
(*d*) Metallic crystals

IONIC CRYSTALS

In an ionic crystal the lattice is made of positive and negative ions. These are held together by ionic bonds – the strong electrostatic attractions between oppositely charged ions. Consequently, the cations and anions attract one another and pack together in an arrangement so that the attractive forces maximize. The sodium chloride lattice shown in Fig. 12.18 is an example. Each ion is surrounded by neighbors of opposite charge and there are no separate molecules. Since the ions are fixed in their lattice sites, typical ionic solids are hard and rigid with high melting points. In spite of their hardness, ionic solids are brittle. They shatter easily by hammering (Fig. 12.19). By hammering, a layer of ions slips away from their oppositely charged neighbors and brings them closer to ions of like charge. The increase of electrostatic repulsions along the displaced plane causes the crystal to break.

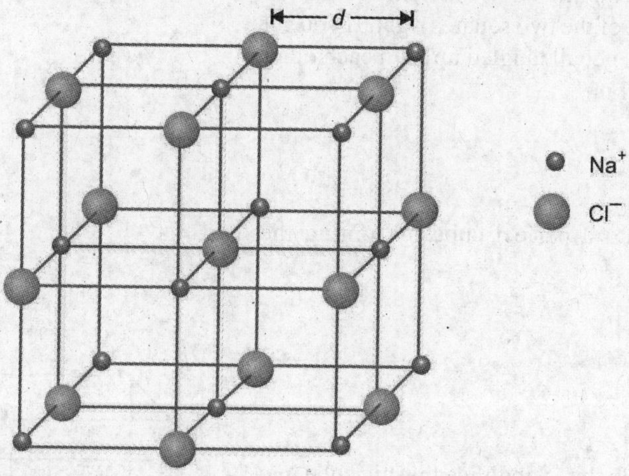

■ Figure 12.18
Sodium chloride crystal lattice. $d = 2.814 \times 10^{-10}$ m

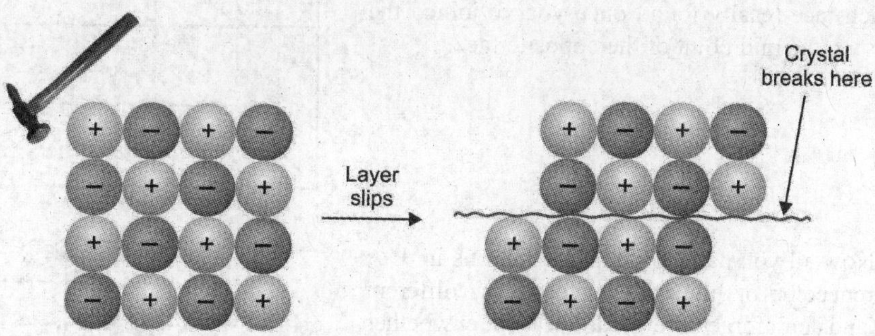

■ Figure 12.19
An ionic crystal is brittle and breaks by hammering.

Ionic solids are non-conducting because the ions are in fixed positions. However, in the fused state the ions are allowed freedom of movement so that it becomes possible for them to conduct electricity.

Sodium Chloride Crystal

The ionic lattice of NaCl is shown in Fig. 12.18. Each sodium ion is surrounded by six chloride ions and each chloride ion is clustered by six sodium ions. The co-ordination number for this crystal lattice is six as required by simple cubic type. In this cubic system, the planes can be passed through the atoms having Miller indices (100), (110) or (111) and the relative spacings for the unit cell of

HOW TO DRAW STRUCTURE OF NaCl ?

Draw a perfect square.

Now draw an identical square behind this one and offset a bit. You might have to practice a bit to get the placement of the two squares right. If you get it wrong, the ions get all tangled up with each other in your final diagram.

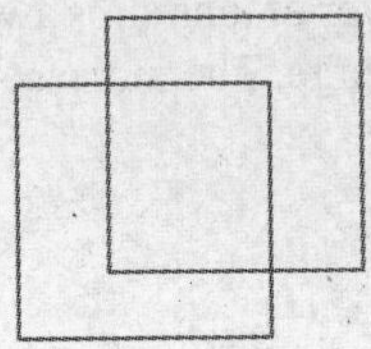

Turn this into a perfect cube by joining the squares together.

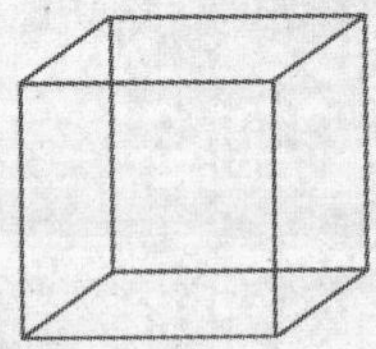

Now the tricky bit! Subdivide this big cube into 8 small cubes by joining the mid point of each edge to the mid point of the edge opposite it. To complete the process you will also have to join the mid point of each face (easily found once you've joined the edges) to the mid point of the opposite face.

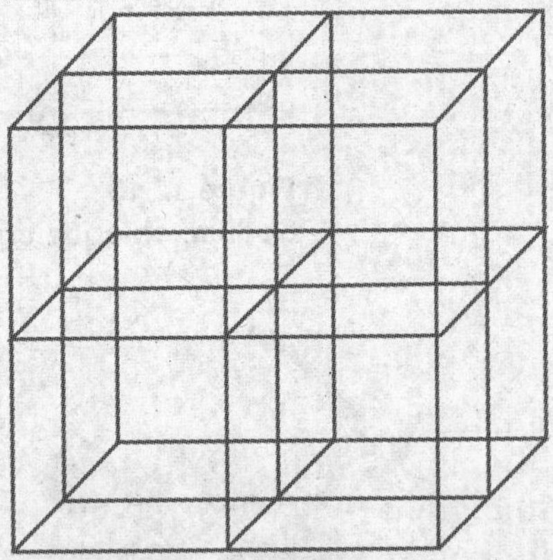

Now all you have to do is put the ions in. Use different colors or different sizes for the two different ions, and don't forget a key. It doesn't matter whether you end up with a sodium ion or a chloride ion in the center of the cube – all that matters is that they alternate in all three dimensions.

You should be able to draw a perfectly adequate free-hand sketch of this in under two minutes – less than one minute if you're not too fussy!

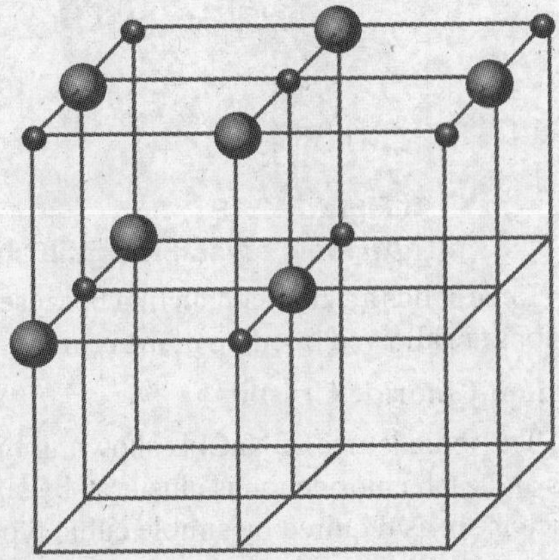

a face-centered cubic lattice are $\frac{a}{2}, \frac{a}{2\sqrt{2}}$ and $\frac{a}{\sqrt{3}}$ while it is $a : \frac{a}{\sqrt{2}} : \frac{a}{\sqrt{3}}$ for simple cubic and $\frac{a}{2} : \frac{a}{\sqrt{2}} : \frac{a}{2\sqrt{3}}$ for body-centered cubic lattice. For face-centered cubic lattice,

$$d_{100} : d_{110} : d_{111} = \frac{a}{2} : \frac{a}{2\sqrt{2}} : \frac{a}{\sqrt{3}}$$

$$= 1 : 0.707 : 1.154 \qquad ...(i)$$

In the case of sodium chloride the *first order* reflections from (100), (110) and (111) faces using *K* line from palladium anti-cathode are 5.9°, 8.4° and 5.2° respectively. From Bragg equation $n\lambda = 2d \sin \theta$, we have

$$d = \frac{n\lambda}{2 \sin \theta}$$

Since $n = 1$ and λ is the same in each case, the ratio of the spacings parallel to the three principal planes are

$$d_{100} : d_{110} : d_{111} = \frac{1}{\sin 5.9^\circ} : \frac{1}{\sin 8.4^\circ} : \frac{1}{\sin 5.2^\circ}$$

$$= 9.731 : 6.844 : 11.04$$

$$= 1 : 0.704 : 1.136 \qquad ...(ii)$$

This ratio is almost identical with the ratio required in the case of a face-centered cubic lattice as shown in (*i*).

X-ray diffraction studies reveal that unlike sodium chloride, potassium chloride has a simple cubic lattice. The edge length of the unit cell is 3.1465×10^{-10} m.

Cesium Chloride Crystal

Cesium chloride, CsCl, has a body-centered cubic structure. In its crystal lattice, each Cs^+ ion is surrounded by 8 Cl^- ions and its coordination number is eight. The value of *d* (Fig. 12.20) as determined by Bragg spectrometer is 3.510 E.

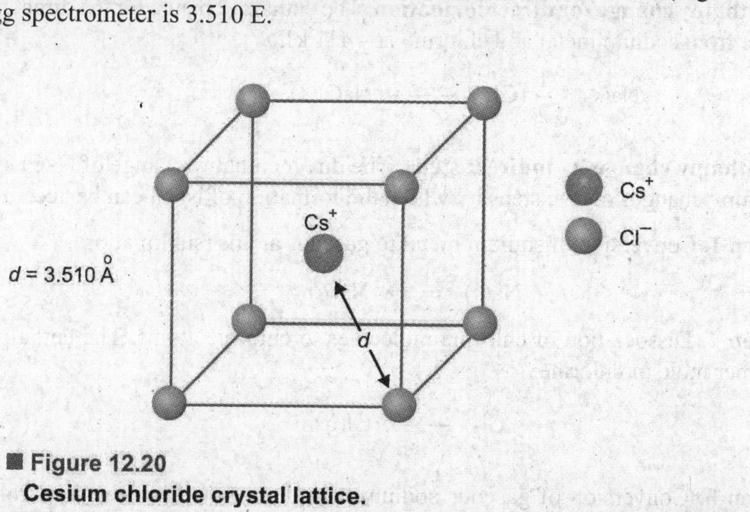

■ **Figure 12.20**
Cesium chloride crystal lattice.

Let us assume that the ions are hard spheres. Further, the anions are in contact with cations *i.e.*, ions are as closely packed as possible. Cs^+Cl^- is, therefore, $r_+ + r_-$, where r_+ and r_- represent radius of cations and the anions respectively. The radius ratio r_1/r_2 has been determined to be 1.37. The radius ratio for sodium chloride structure is 2.44. Knowing the radius ratio of a particular crystal, we can say whether it has CsCl or NaCl structure.

Lattice Energy of an Ionic Crystal (Born-Haber Cycle)

The positive and negative ion in an ionic crystal are held together by electrostatic forces. The bond energy is expressed in terms of the **lattice energy** which may be defined as :

the change in enthalpy (heat change) that occurs when 1 mole of a solid crystalline substance is formed from its gaseous ions.

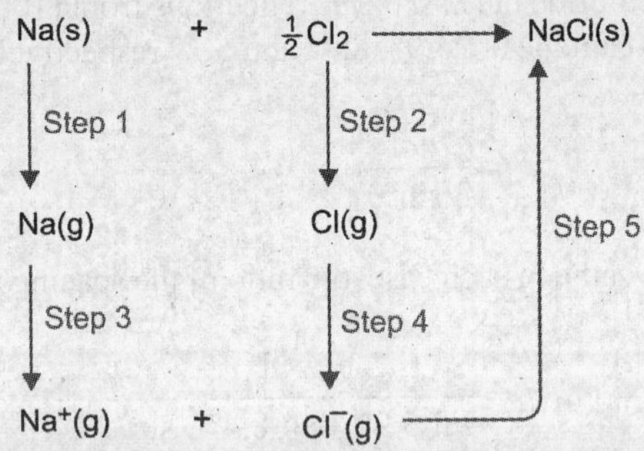

Figure 12.21
A Born-Haber cycle for the formation of NaCl crystal from its elements.

The lattice energy of NaCl, for example, is the change in enthalpy, ΔH°, when Na^+ and Cl^- ions in the gas phase come together to form 1 mole of NaCl crystal.

Determination of Lattice Energy

The lattice energy of an ionic crystal can be found by applying Hess's law. The formation of an ionic solid from its elements can be thought of as occurring directly in one step or indirectly as the sum of the series of steps. Let us illustrate by taking example of NaCl.

Enthalpy change for direct formation. The enthalpy change for the direct formation of sodium chloride from sodium metal and chlorine is – 411 kJ.

$$Na(s) + \frac{1}{2}(Cl_2) \longrightarrow NaCl(s) \qquad \Delta H° = -411 \text{ kJ}$$

Enthalpy change by indirect steps. The direct enthalpy change of NaCl stated above is equal to enthalpy changes of five steps by which the formation of NaCl can be accomplished.

Step 1. Conversion of sodium metal to gaseous atoms (sublimation).

$$Na(s) \longrightarrow Na(g) \qquad \Delta H_1° = +108 \text{ kJ}$$

Step 2. Dissociation of chlorine molecules to chlorine atoms. The enthalpy of dissociation is 121 kJ per mole of chlorine.

$$\frac{1}{2}Cl_2 \longrightarrow Cl(g) \qquad \Delta H_2° = +121 \text{ kJ}$$

Step 3. Conversion of gaseous sodium to sodium ions by losing electrons. The enthalpy of ionization is the ionization energy, 495 kJ/per mole of sodium atoms.

$$Na(g) \longrightarrow Na^+(g) + e^- \qquad \Delta H_3° = +495 \text{ kJ}$$

Step 4. Chlorine atoms gain an electrons to form chloride ions. The energy released is the electron affinity of chlorine –348 kJ.

$$Cl + e^- \longrightarrow Cl^-(g) \qquad \Delta H_4° = -348 \text{ kJ}$$

Step 5. Sodium and chloride ions get together and form the crystal lattice. The energy released in this process is the negative of lattice energy.

$$Na^+(g) + Cl^-(g) \longrightarrow NaCl(s) \qquad \Delta H_5° = -\text{ (lattice energy)}$$

The lattice energy can be calculated by equating the enthalpy of formation of NaCl, – 411 kJ, to the sum of the enthalpy changes for the five steps.

$$\Delta H°_1 + \Delta H°_2 + \Delta H°_3 + \Delta H°_4 + \Delta H°_5 = -411 \text{ kJ}$$

$$108 \text{ kJ} + 121 \text{ kJ} + 495 \text{ kJ} - 348 \text{ kJ} - \text{lattice energy} = -411 \text{ kJ}$$

By solving this equation we get

$$\text{Lattice energy} = +787 \text{ kJ mol}^{-1}$$

The cycle of changes shown in Fig. 12.21, is called the **Born-Haber Cycle** after Max Born and Fritz Haber, who devised this method of calculating lattice energies.

MOLECULAR CRYSTALS

In molecular crystals, molecules are the structural units. These are held together by van der Waals' forces. As in case of ionic crystals, the molecules are packed together in a tightly packed pattern because the forces of attraction are non-directional. When this type of crystal melts, it is only the weak van der Waals forces that must be overcome. Therefore molecular solids have low melting points. Most organic substances are molecular solids.

Crystal lattice of dry CO_2. Dry ice, or frozen carbon dioxide, is the best example of a molecular solid. The van der Waals' forces holding the CO_2 molecules together are weak enough so that dry ice passes from solid state to gaseous state at 195 K.

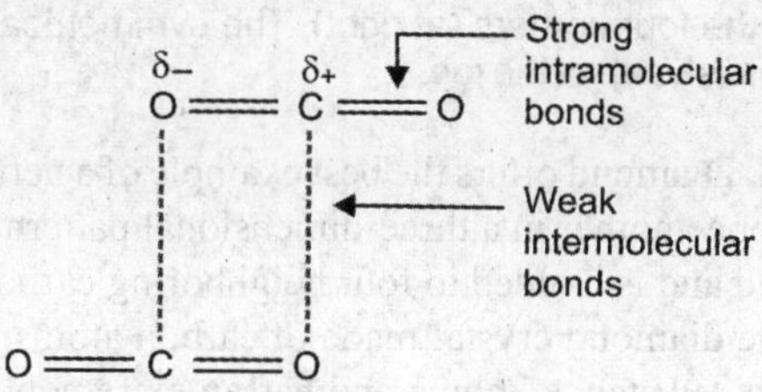

The pattern in which CO_2 molecules are arranged in the crystal lattice is shown in Fig. 12.22.

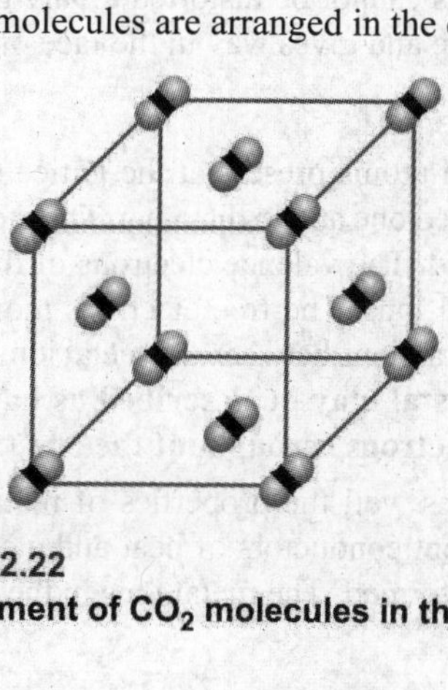

Figure 12.22
Arrangement of CO_2 molecules in the crystal lattice.

NETWORK COVALENT CRYSTALS

In this type of crystals atoms occupy the lattice sites. These atoms are bonded to one another by covalent bonds. The atoms interlocked by a network of covalent bonds produce a crystal which is considered to be a single giant molecule. Such a solid is called a **network covalent solid** or simply **covalent solid.** Since the atoms are bound by strong covalent bonds, these crystals are very hard and have very high melting points.

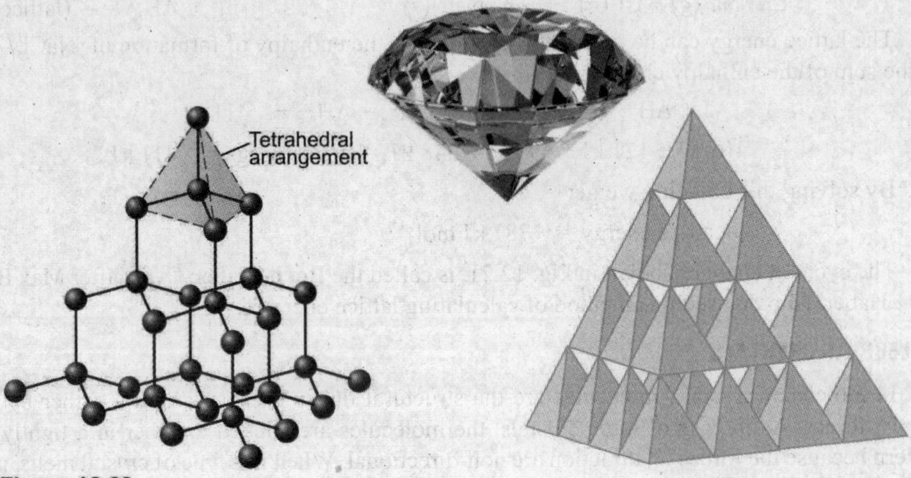

■ **Figure 12.23**
Crystal structure of diamond. Diamond owes its unique physical properties (like unmatched hardness and thermal conductivity) to strong covalent bonds connecting the carbon atoms (blue spokes in the diagram at left). These bonds form where adjacent carbon atoms share a pair or electrons. Each atom has 4 electrons available for sharing, which results in 4 bonds leading away from each of the atoms. Alternatively, the diamond structure may be visualized as a framework of tetrahedra with carbon atoms both in the center and at each of the four corners (at right). The symmetrical perfection of diamond is beautifully captured and shown at the top.

Structure of Diamond. Diamond offers the best example of a network covalent solid. It consists of carbon atoms bonded to one another in a three-dimensional pattern to form the crystal. However, each carbon is sp^3 hybridized and is bonded to four neighboring carbon atoms arranged towards the corners of a tetrahedron. The diamond crystal made of carbon atoms joined by a network of strong covalent bonds is the hardest substance known and has an extremely high melting point (3823 K). Since the bonds between two carbons cannot be distorted to any great extent, the diamond crystal cannot be distorted but is rather brittle and gives way in the face of a shearing force.

METALLIC CRYSTALS

The crystals of metals consist of atoms present at the lattice sites. The atoms are arranged in different patterns, often in layers placed one above the other. The atoms in a metal crystal are viewed to be held together by a metallic bond. The valence electrons of the metal atoms are considered to be delocalized leaving positive metal ions. The free electrons move throughout the vacant spaces between the ions. The electrostatic attractions between the metal ions and the electron cloud constitute the metallic bond. **Thus a metal crystal may be described as having positive ions at the lattice positions surrounded by mobile electrons throughout the crystal.**

The **electron sea model** explains well the properties of metals. The mobile electrons in the crystal structure make metals excellent conductors of heat and electricity. On application of force, say with a hammer, metals can be deformed. The metal ions in the crystal change positions without

BUCKMINSTERFULLERENE

Buckminsterfullerene molecules consist of 60 carbon atoms linked together to form an almost spherical ball with the chemical formula C_{60}. The bonds between atoms form a pattern of joined hexagons and pentagons that is similar to the panels on a soccer ball. The allotrope was given its name because its structure resembles the elaborate geometrical structures invented by American architect Buckminster Fuller. The individual molecules have become known as **buckyballs**.

For many years it was believed that the element carbon occurred as only three allotropes: diamond, graphite, and amorphous carbon. In each of these allotropes, the carbon atoms are linked together in a different arrangement, giving the form of the element different properties. In 1985, however, a new family of allotropes was discovered. Of these allotropes, which are called fullerenes, buckminsterfullerene has become the most famous. Other fullerenes have more carbon atoms, and their shapes resemble elongated versions of the original, football-shaped buckminsterfullerene. Once buckminsterfullerene could be produced in large amounts, a solid form, fullerite, was also produced. In this transparent yellow solid, the molecules are stacked together in a close-packed arrangement like a pile of cannon balls. Tubular versions of fullerenes are also available in solid form.

The original method of preparation of buckminsterfullerene was to produce it in a molecular beam, and only very small quantities could be made. However, it was soon found that the molecules were produced in large numbers in an electric arc between two carbon electrodes in a helium atmosphere. Scientists now believe that buckminsterfullerene is likely to be formed in sooty flames, and there is a possibility that it is abundant in the universe, particularly near red-giant stars.

The versatility of fullerene molecules has led to a great deal of research exploring their properties. One potentially useful property is that atoms of different elements can be placed inside the molecular cage formed by the carbon atoms, producing a "shrink wrapped" version of these elements. When metal atoms are introduced into fullerene tubes, the resulting material is like a one-dimensional insulated wire. Another important property is that certain compounds of buckminsterfullerene (notably K_3C_{60}) are superconducting at low temperatures. Compounds made by adding thallium and rubidium ions (electrically charged atoms) to fullerenes become superconducting at 45 K (–378° F). This temperature is relatively high compared to the cooling required by other superconducting materials. Derivatives of buckminsterfullerene have been found to be biologically active and have been used to attack cancer. It is believed that the molecules can enter the active sites of enzymes and block their action.

making material difference in the environments. The attractive force between ions and the electron cloud remains the same. The crystal, therefore, does not break.

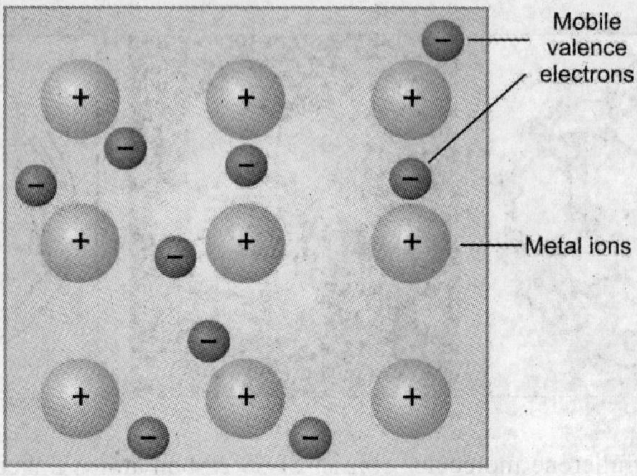

■ Figure 12.24
A representation of a metallic crystal structure.

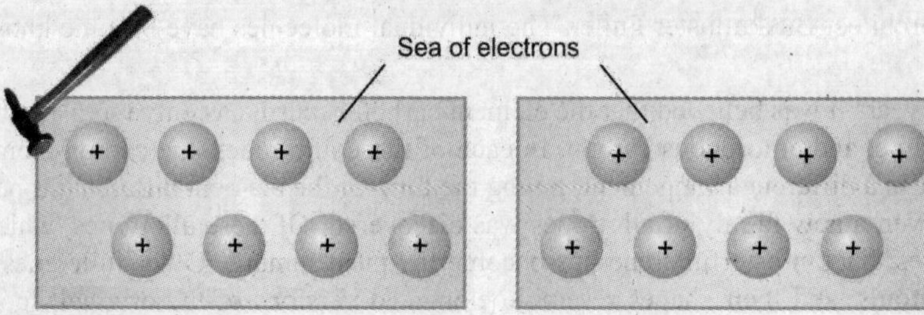

■ Figure 12.25
On application of force positive kernels of metallic atoms can be moved within the sea of electrons without materially changing the environments. That is why metals can be worked into sheets or foils without cracking.

STRUCTURE OF METAL CRYSTALS

The individual atoms in a metallic crystal lattice can be thought of as hard spheres. The spherical atoms are packed together in the lattice very efficiently in geometrical arrangements so as to leave minimum interspaces. A layer of uniform spheres can be arranged either as in Fig. 12.26(*a*) or (*b*). Clearly the second of the patterns uses space more efficiently. Here the spheres fit into the hollows between the adjacent spheres. Thus the vacant spaces (voids) between the spheres are smaller than in the first pattern. The metallic crystals are of the second type *i.e.*, close packing.

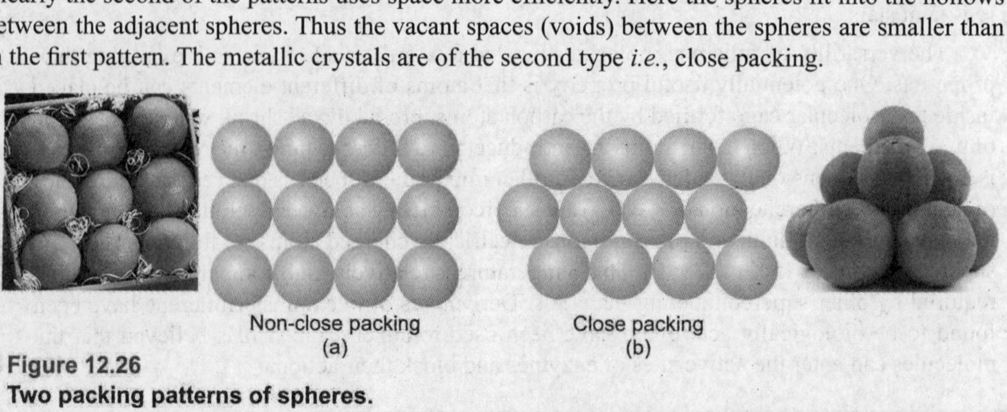

■ Figure 12.26
Two packing patterns of spheres.

As clear from Fig. 12.26(*b*), each sphere in a closely packed layer is in contact with four others. Thus each ball touches six other at the corners of a hexagon. Three dimensional metallic crystals consist of closely packed layers stacked one over the other. The spheres forming the second layer fill the holes or voids in the first layer and the spheres of the third layer fill the voids in the second layer. Depending upon the geometrical arrangements of spheres in the three layers, the close-packed metallic crystals are of two types :

(*a*) Hexagonal close-packed (hcp)
(*b*) Cubic close-packed (ccp)

Hexagonal Close-Packed Structure

The hexagonal close-packed structure of metallic crystals is shown in Fig. 12.27. It consists of three layers of spherical atoms packed one over the another. The bottom layer (*A*) and the top layer (*A*) have three spheres in similar orientation. The middle layer (*B*) consists of six spherical atoms. The three spheres in the top and the bottom layer fit into the same voids on either side of the middle layer. It is noteworthy that each sphere in the structure is in contact with 12 neighboring spheres, six in its own layer, three in the layer above and three in the layer below. Thus the coordination number of the close-packed structure is 12. In the overall close-packed structure, the layers repeat in the manner ABABAB.

The examples of metals having hexagonal close-packed structures are Ba, Co, Mg and Zn.

Cubic Close-Packed Structure

The cubic close-packed (ccp) pattern of a metallic crystal is illustrated in Fig. 12.27. Its coordination number is also 12. Like the hcp structure, it consists of three layers of spherical atoms lying over one another. There are three spheres in the top layer (C), six in the middle layer (B), and three in the bottom layer (A). However, the overall ccp differs in structure from the hcp structure in respect of the orientation of the three spheres in the top layer. In hcp structure both the top and the bottom layers have the same orientation. But in ccp structure, they are oriented in opposite directions. Therefore, the three spheres in the top layer do not lie exactly on the spheres in the bottom layers. In ccp structure, the layers are repeated in the order ABCABCABC. By turning the whole crystal you can see that the ccp structure is just the face-centered cubic structure.

Many metals including Ag, Au, Ca, Co, Cu, Ni, crystallize in ccp structures.

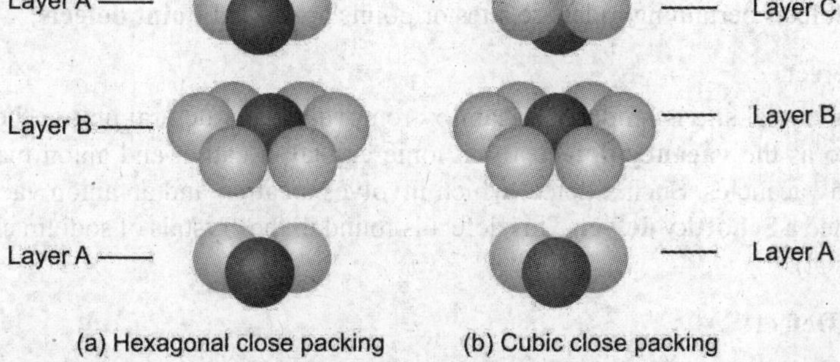

■ Figure 12.27
Hexagonal and cubic close packed structures.

Body-Centered Cubic Structure

About one-third of the metals pack in a body-centered cubic structure in which the coordination number is only 8. Each atom touches four atoms in the layer above and four atoms in the layer below.

When a square-packed layer (non-close packed) is packed on another layer (Fig. 12.28), a simple cubic pattern of spherical atoms results. The large holes remaining in the middle of each cube on slight expansion can accommodate another sphere to form a body-centered cube.

Li, Na and K crystallize in body-centered cubic structures.

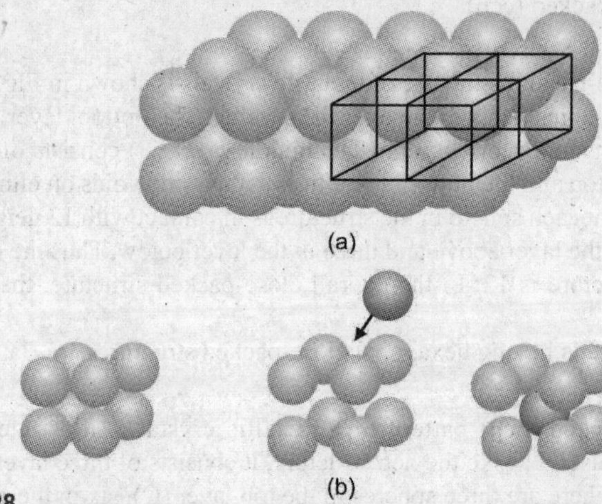

■ Figure 12.28
(*a*) Layers of non-close spheres stacked one on the other. (b) Fifth atom when inserted in a simple cubic structure forms a body-centred cubic pattern.

CRYSTAL DEFECTS

So far in our discussion of crystalline substances, we have assumed them to be **perfect crystals.** A perfect crystal is one in which all the atoms or ions are lined up in a precise geometric pattern. But crystals are never actually perfect. The **real crystals** that we find in nature or prepare in the laboratory always contain imperfections in the formation of the crystal lattice. These crystal defects can profoundly affect the physical and chemical properties of a solid.

The common crystal defects are :

(*a*) Vacancy defect
(*b*) Interstitial defect
(*c*) Impurity defects

These defects pertaining to lattice sites or points are called **Point defects.**

Vacancy Defect

When a crystal site is rendered vacant by removal of a structural unit in the lattice, the defect is referred to as the **vacancy defect.** In an ionic crystal, a cation and anion may leave the lattice to cause two vacancies. Such a defect which involves a cation and an anion vacancy in the crystal lattice is called a **Schottky defect.** This defect is found in the crystals of sodium chloride and cesium chloride (CsCl).

Interstitial Defect

Here, an ion leaves its regular site to occupy a position in the space between the lattice sites (interstitial position). This causes a defect known as **Interstitial defect or Frenkel defect.** As shown in Fig. 12.29, ordinarily the cation moves as it is smaller than the anion and can easily fit into the vacant spaces in the lattice. Thus in AgCl crystal, Ag^+ ion occupies an interstitial position leaving a vacancy (or hole) at the original site.

Impurity Defect

These defects arise due to the corporation of foreign atoms or ions in regular lattice sites or interstitial sites.

When foreign particles are substituted for normal lattice particles, it is called **substitution impurity.**

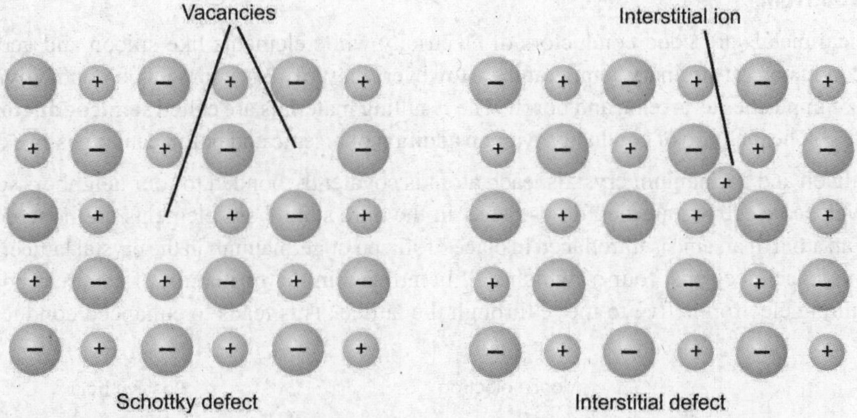

■ **Figure 12.29**
Two common types of defects in ionic crystals.

When foreign particles are trapped in vacant interstitial spaces, it is called **interstitial impurity.** Both types of impurities can have drastic effect on the properties of solids.

METAL ALLOYS

Other elements can be introduced into a metallic crystal to produce substances called **alloys.** Alloys are of two types.

(1) **a substitutional alloy** in which the host metal atoms are replaced by other metal atoms of similar size. For example, in brass (an alloy of copper and zinc) about one-third of the copper atoms have been substituted by zinc atoms.

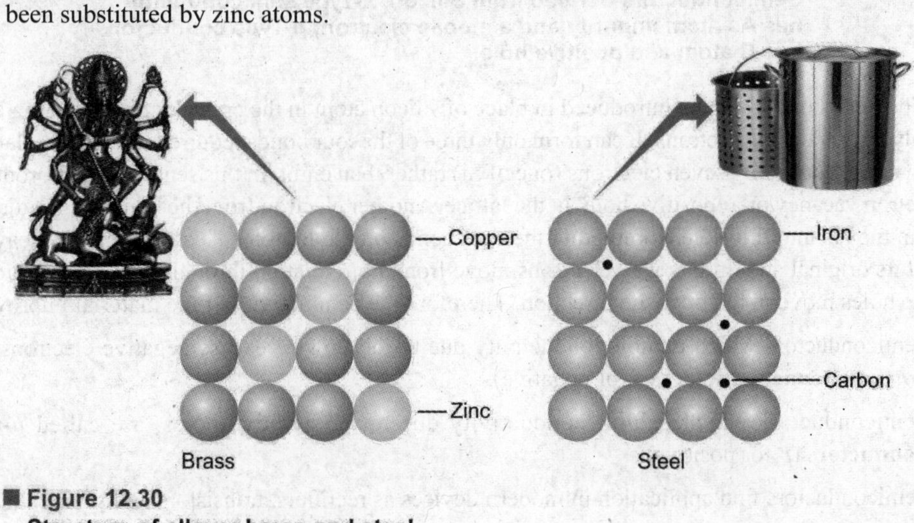

■ **Figure 12.30**
Structure of alloys; brass and steel.

(2) **an interstitial alloy** in which some of the interstices (holes) in close-packed metal structure are occupied by small atoms. For example, steel is an alloy of iron and carbon. It contains carbon atoms in the holes of an iron crystal. The presence of interstitial atoms changes the properties of the host metal. Thus pure iron is relatively soft, malleable and ductile. The introduction of the carbon atoms forms the directional carbon-iron bonds (C–Fe). This makes the relative movement of spherical iron atoms rather difficult and the resulting steel is harder, stronger and less ductile compared to pure iron.

SEMICONDUCTORS

Typical metals are good conductors of electricity while elements like silicon and germanium are nonconductors at ordinary temperature. However, they exhibit appreciable conductivity upon addition of impurities as arsenic and boron. The resulting materials are called **semiconductors** (poor conductors). The increase of conductivity upon addition of arsenic and boron can be easily explained.

In silicon and germanium crystals, each atom is covalently bonded to four neighbors so that all its four valence electrons are tied down. Thus in the pure state these elements are nonconductors. Suppose an atom of arsenic is introduced in place of silicon or germanium in the crystal lattice. Arsenic has five valence electrons, four of which will be utilized in the formation of covalent bonds and the remaining electron is free to move through the lattice. This leads to enhanced conductivity.

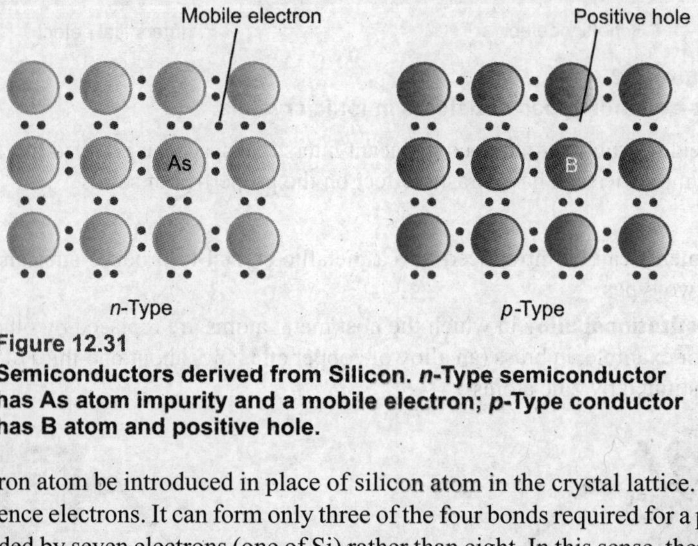

Figure 12.31
Semiconductors derived from Silicon. *n*-Type semiconductor has As atom impurity and a mobile electron; *p*-Type conductor has B atom and positive hole.

Now let a boron atom be introduced in place of silicon atom in the crystal lattice. A boron atom has only three valence electrons. It can form only three of the four bonds required for a perfect lattice. Thus it is surrounded by seven electrons (one of Si) rather than eight. In this sense, there is produced an electron vacancy or a 'positive hole' in the lattice. Another electron from the bond of the adjacent Si atom moves into this hole, completing the four bonds on the *B* atom. This electron also leaves a hole at its original site. In this way electrons move from atom to atom through the crystal structure and the holes move in the opposite direction. Therefore the conductivity of the material improves.

Semiconductors which exhibit conductivity due to the flow of excess negative electrons, are called ***n*-type semiconductors** (*n* for negative).

Semiconductors which exhibit conductivity due to the positive holes, are called ***p*-type semiconductors** (*p* for positive).

Semiconductors find application in modern devices as rectifiers, transistors and solar cells.

SOLAR CELL

It is a device to convert light energy into electrical energy. It is made of a thin wafer of silicon containing a tiny amount of arsenic (*n*-type semiconductor). A thin layer of silicon containing a trace of boron is placed on the surface of the water. Thus a junction, called the *p-n* junction, exists between the *p*-type silicon and *n*-type silicon. When the cell is exposed to sunlight, energy from sunlight excites electrons from *n*-type silicon to the holes of the *p*-type silicon. From the *p*-type silicon, the electrons flow through the external circuit as an electric current.

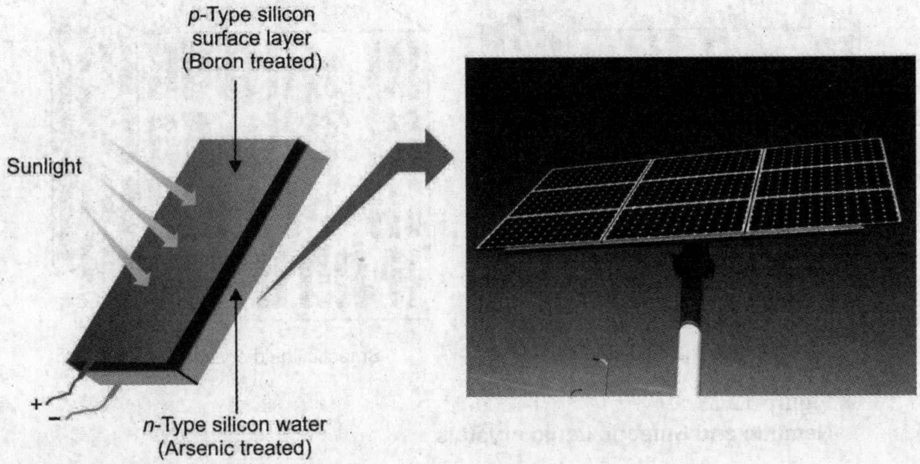

■ **Figure 12.32**
A solar cell.

The voltage of the solar cell is not large. Therefore a series of such cells are used to power communication devices in satellites and space crafts which have to remain in space for long periods.

WHAT ARE LIQUID CRYSTALS?

Some organic solids having long rod-like molecules which do not melt to give the liquid substance directly. They, instead, pass through an intermediate state called the **liquid crystal state,** often referred to as the **liquid crystal.** Thus the liquid crystal state is intermediate between the liquid state and the solid state.

One such substance that forms liquid crystal is

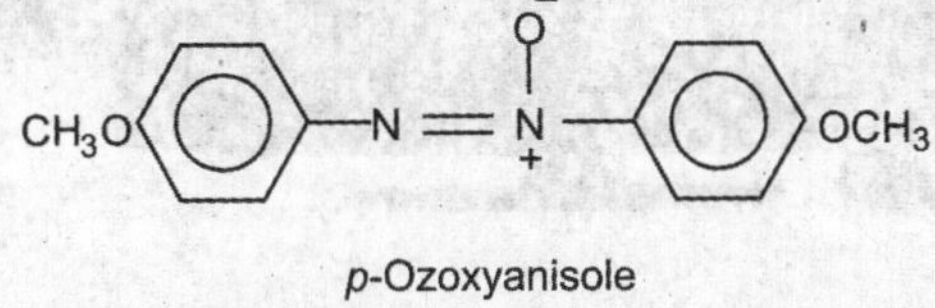

p-Ozoxyanisole

The liquid crystals have a structure between that of a liquid and that of a crystalline solid. In a liquid the molecules have a random arrangement and they are able to move past each other. In a solid crystal the molecules have an ordered arrangement and are in fixed positions. In a liquid crystals, however, molecules are arranged parallel to each other and can flow like a liquid. **Thus the liquid crystals have the fluidity of a liquid and optical properties of solid crystals.**

Types of Liquid Crystals

According to their molecular arrangement, the liquid crystals are classified into three types.

Nematic liquid crystals. They have molecules parallel to each other like soda straws but they are free to slide or roll individually.

Smectic liquid crystals. The molecules in this type of crystal are also parallel but these are arranged in layers. The layers can slide past each other.

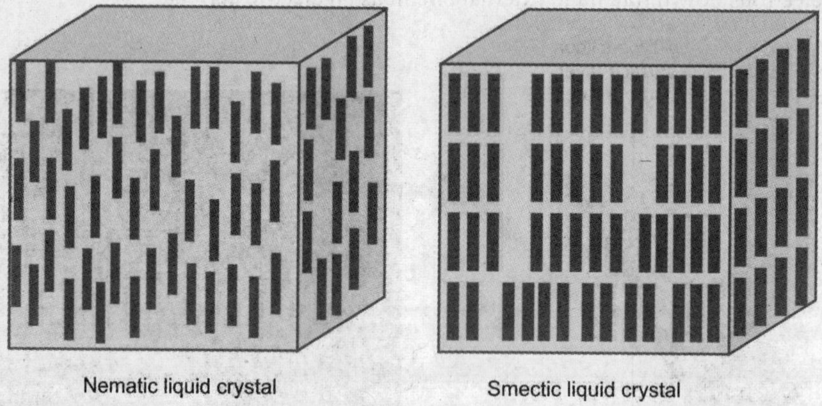

■ **Figure 12.33**
Nematic and Smectic liquid crystals.

Chloesteric liquid crystals. As in nematic crystals, in this type of crystal the molecules are parallel but arranged in layers. The molecules in successive layers are slightly rotated with respect to the layers above and below so as to form a spiral structure.

APPLICATIONS OF LIQUID CRYSTALS

On account of their remarkable optical and electrical properties, liquid crystals find several practical applications. Some of these are shown in Fig.12.34.

■ **Figure 12.34**
The technical application of liquid crystals in flat panel displays for desktop and notebook-computers or in the displays of cellular phones has become an indispensable

Number Displays

When a thin layer of nematic liquid crystal is placed between two electrodes and an electric field is applied, the polar molecules are pulled out of alignment. This causes the crystal to be opaque.

Transparency returns when electrical signal is removed. This property is used in the number displays of digital watches, electronic calculators, and other instruments.

Monitoring Body Temperature

Like the solid crystals, liquid crystals can diffract light. Only one of the wavelengths of white light is reflected by the crystal which appears colored. As the temperature changes, the distance between the layers of molecules also changes. Therefore the color of the reflected light changes correspondingly. These cholesteric liquid crystal undergoes a series of color changes with temperature. These crystals are used in indicator tapes to monitor body temperature or to spot areas of overheating in mechanical systems.

ADDITIONAL SOLVED PROBLEMS

SOLVED PROBLEM 6

Gold has face-centered cubic structure with a unit length 4.07×10^{-10} m, a density of 19300 kg/m^3. Calculate the Avogadro's number from the data (Atomic mass of Au = 197)

SOLUTION

Number of atoms in Gold unit cell $= 4$ [∴ f c c structure]

Let Avogadro's number be $= N_0 \text{ mol}^{-1}$

$$\text{Mass of gold atom} = \frac{\text{Atomic mass}}{\text{Avogadro's number}} = \frac{197 \times 10^{-3} \text{ kg mol}^{-1}}{N_o}$$

and $$\text{Mass of one gold unit cell} = \frac{4 \times 197 \times 10^{-3} \text{ kg mol}^{-1}}{N_o \text{ mol}^{-1}} = \frac{0.788}{N_o} \text{kg}$$

Edge length of gold unit cell, $a = 4.07 \times 10^{-10}$ m (given)

Volume of one gold unit cell, $a^3 = (4.07 \times 10^{-10} \text{ m})^3$

$$\therefore \quad \text{Density} = \frac{\text{Mass}}{\text{Volume}} = \frac{\frac{0.788}{No}}{(4.07 \times 10^{-10} \text{ m})^3} = 19300 \text{ kg/m}^3 \quad \text{(given)}$$

Substitution of values

$$\frac{0.788 \text{ kg}}{N_o \times (4.07 \times 10^{-10})^3 \text{ m}^3} = 19300 \text{ kg/m}^3$$

or $$N_o = \frac{0.788 \text{ kg}}{(4.07 \times 10^{-10})^3 \times 19300}$$

$$= 0.6056 \times 10^{24}$$

$$= \mathbf{6.056 \times 10^{23}}$$

SOLVED PROBLEM 7

Polonium crystallizes in a simple cubic unit cell. It has an atomic mass = 209 and density = 91.5 kg m^{-3}. What is the edge length of its unit cell ?

SOLUTION

Number of atoms of Po in unit cell $= 1$ [∴ Simple cube]

$$\text{Mass of one Po atom} = \frac{\text{Atomic mass}}{\text{Avogadro's number}}$$

$$= \frac{209 \times 10^{-3}\ \text{kg mol}^{-1}}{6.02 \times 10^{23}\ \text{mol}^{-1}}$$

$$= 34.7176 \times 10^{-26}\ \text{kg}$$

$$\text{Mass of one Po unit cell} = 1 \times 34.7176 \times 10^{-23}\ \text{g} \quad [\because \text{Simple Cube} = 1 \text{ atom}]$$

$$\text{Let the length of the unit cell be} = a$$

$$\text{Volume of the unit cell} = a^3$$

$$\text{And} \quad \text{Density} = \frac{\text{Mass}}{\text{Volume}} = 91.5\ \text{kg m}^{-3} \quad \text{(given)}$$

Substitution of values

$$0.0915\ \text{g cm}^{-3} = \frac{34.7176 \times 10^{-23}\ \text{g}}{a^3}$$

$$\text{or} \quad a^3 = \frac{34.7176 \times 10^{-26}\ \text{kg}}{91.5\ \text{kg/m}^3}$$

$$= 379.427 \times 10^{-29}\ \text{m}^3$$

$$= 3794.27 \times 10^{-30}\ \text{m}^3$$

$$\text{or} \quad a^3 = \sqrt[3]{3794.27 \times 10^{-30}\ \text{m}^3}$$

$$= \mathbf{15.597 \times 10^{-10}\ m}$$

SOLVED PROBLEM 8

Calculate the angle at which first order diffraction will occur in an X-ray diffractometer when X-rays of wavelength 1.54×10^{-10} m are diffracted by the atoms of a crystal, given that the interplanar distance is 4.04×10^{-10} m.

SOLUTION

Formula used

$n\lambda = 2d \text{ Sin } \theta$

Quantities given

$n = 1$ $\qquad \lambda = 1.54 \times 10^{-10}$ m $\qquad d = 4.04 \times 10^{-10}$ m

Substitution of values

$$1 \times 1.54 \times 10^{-10}\ \text{m} = 2 \times 4.04 \times 10^{-10}\ \text{m} \times \text{Sin } \theta$$

$$\text{or} \quad \text{Sin } \theta = \frac{1.54 \times 10^{-10}\ \text{m}}{2 \times 4.04 \times 10^{-10}\ \text{m}}$$

$$= 0.19059$$

$$\text{or} \quad \theta = \sin^{-1} 0.19059$$

$$= \mathbf{10.987° \text{ (This is in Radian)}}$$

SOLVED PROBLEM 9

Platinum crystallizes in a face-centered cubic lattice with all atoms at the lattice points. It has a density of 21450 kg/m^3 and an atomic mass of 195.08 amu. From these data, Calculate the length of a unit-cell edge.

SOLUTION

Molecular mass of Pt $= 195.08$ amu (Given)

$\therefore$ Mass of one atom of Pt $= \dfrac{\text{Molecular mass}}{\text{Avogadro number}}$

$= \dfrac{195.08}{6.02 \times 10^{23}}$ g

$= 3.240 \times 10^{-27}$ kg

Number of Pt atom in unit cell $= 4$ [$\because$ f c c structure]

$\therefore$ Mass of unit cell of Pt $= 4 \times 3.240 \times 10^{-22}$ g

$= 1.296 \times 10^{-24}$ kg

and volume of the unit cell, $V = \dfrac{M}{D} = \dfrac{1.296 \times 10^{-24} \text{ kg}}{21450 \text{ kg/m}^3}$

$= 6.042 \times 10^{-29}$ m^3

Edge of the unit cell, $a = \sqrt[3]{6.042 \times 10^{-29} \text{ m}^3}$

$=$ **3.924×10^{-10} m**

SOLVED PROBLEM 10

A body centered cubic element of density 10300 kg/m^3 has a cell edge of 314×10^{-12} m. Calculate the atomic mass of the element. (Avogadro's number $= 6.023 \times 10^{23}$)

SOLUTION

Number of atoms in bcc structure $= \dfrac{1}{8} \times 8 + 1$ atoms

$= 2$ atoms

Density of element $= 10300$ kg/m^3

Edge of the unit cell, $l = 314 \times 10^{-12}$ m

$\therefore$ Volume of the unit cell, $V = l^3 = (314 \times 10^{-12})^3$ m^3

Mass of the unit cell $=$ Density $\times$ Volume

$= 10300 \text{ kg/m}^3 \times (314 \times 10^{-12})^3 \text{ m}^3$

$= 3.1188 \times 10^{-27}$ kg

Mass of one atom $= \dfrac{3.1188 \times 10^{-27}}{2}$ kg [$\because$ unit cell = 2 atoms]

$= 1.5594 \times 10^{-27}$ kg

Atomic mass $=$ Avogadro's number $\times$ mass of 1 atom

$= 6.023 \times 10^{23} \times 1.5594 \times 10^{-22} \times 10^{-3}$ kg

$=$ **94×10^{-3} kg mol^{-1}**

EXAMINATION QUESTIONS

1. Define or explain the following terms :
   (*a*) Molecular solid (*b*) Ionic solid
   (*c*) Covalent network solid (*d*) Metallic solid
   (*e*) Crystalline solid (*f*) Amorphous solid

(*g*) Isotropy
(*h*) Anisotropy
(*i*) Crystal lattice
(*j*) Unit cell
(*k*) Cubic unit cells
(*l*) Body centered cubic unit cell
(*m*) Face centered cubic unit cell
(*n*) Bragg's equation

2. Aluminium forms face-centered cubic crystals. The density of Al is 2700 kg/m^3. Calculate the length of the side of the unit cell of Al. (At. wt. of Al = 27)
   Answer. 4.053×10^{-10} m
3. (*a*) Describe the theory of Bragg's method of crystal analysis.
   (*b*) Differentiate between the cubic close packing and hexagonal close packing of spheres.
4. How is Avogadro's number determined from X-ray diffraction of crystals? Explain.
5. (*a*) Derive the relation $n\lambda = 2d \sin \theta$ in crystallography.
   (*b*) Draw diagrams to represent (*i*) F.C.C. lattice, (*ii*) B.C.C. lattice.
6. (*a*) Write a short note on Bravis Lattices.
   (*b*) Calculate the angle at which first order diffraction will occur in an X-ray diffractometer when X-rays of wavelength 1.54×10^{-10} m are diffracted by the atoms of a crystal, given that the interplanar distance is 4.04×10^{-10} m.
   Answer. 10.987°
7. (*a*) What do you understand by the packing efficiency of a crystal?
   (*b*) Explain the terms point groups and space groups.
   (*c*) The ionic radii of Cs is 169×10^{-12} m and Br is 195×10^{-12} m. What kind of unit cell would be expected for CsBr crystal? Calculate the unit cell dimensions and the density of CsBr crystal (At. wt. of Cs = 133; Br = 80)
   Answer. (*c*) 390×10^{-12} m; 5962 kg/m^3
8. Derive a relationship between the interplanar spacing of a crystal and the wavelength of X-ray diffracted by it. (*Delhi BSc*)
9. Differentiate between isomorphism and polymorphism. (*Delhi BSc*)
10. (*a*) Discuss Bragg's equation. Describe briefly the experimental set up used to record rotating crystal X-ray diffraction photograph. What are the limitations of this method?
   (*b*) What is the minimum number of molecules per unit cell in a crystal having body centered cubic crystal lattice?
   (*c*) Why radiation of wavelength about 1.0×10^{-10} m is used to determine crystal structure by X-ray diffraction method? (*Jamia Millia BSc*)
11. (*a*) Enumerate various elements of symmetry of a cubic type of unit cell.
   (*b*) What is the law of rational indices? (*Delhi BSc*)
12. (*a*) Identify the crystal system to which some solids having the following dimensions for their unit cell belong. Give examples of the solid.
   (*i*) $a \neq b \neq c \quad \alpha \neq \beta \neq \gamma \neq 90°$ (*ii*) $a \neq b \neq c \quad \alpha = \beta = \gamma = 90°$
   (*b*) What are various types of crystals? (*MD Rohtak BSc*)
13. Calculate Miller indices of a crystal plane which is cut through the crystal axes 2a, –3b, –c.
   Answer. 3, –2, –6 (*Guru Nanak Dev BSc*)
14. (*a*) Explain the following with examples.
   (*i*) Primitive unit cell and non-primitive unit cell.
   (*ii*) Plane of symmetry and axis of symmetry.
   (*b*) Define and explain the law of constancy of interfacial angles and the law of rational indices. (*MD Rohtak BSc*)
15. Calculate the angle at which second order diffraction will appear in a X-ray spectrophotometer when X-rays of wavelength 1.5×10^{-10} m are used and interplanar distance is 4.04×10^{-10} m.
   Answer. 21.795° (*Guru Nanak Dev BSc*)

16. Explain what is meant by ionic crystal, molecular crystal and covalent crystal. Give examples. *(Jammu BSc)*

17. Tabulate all possible crystal systems alongwith geometrical characteristics of their lattices. *(Guru Nanak Dev BSc)*

18. What is meant by unit cell of crystal? Sketch the unit cell of simple body centered and face centered cubic space lattice and calculate the number of atoms per unit cell in these systems. *(Nagpur BSc)*

19. (*a*) Discuss various elements of symmetry of a cubic crystal lattice.
(*b*) What are different Bravias lattice types of a cubic crystal? *(Vidyasagar BSc)*

20. What do you understand by
(*i*) axis of four fold symmetry and
(*ii*) axis of three fold symmetry.
How many such axes are present in cubic crystals? *(Nagpur BSc)*

21. Calculate the angle at which first order diffraction will occur in X-ray diffractometer when X-rays of wavelength 1.54×10^{-10} m are diffracted by the atoms of crystal, given the interplanar distance is 4.04×10^{-10} m.
Answer. 10.9872 *(Delhi BSc)*

22. A crystal plane has intercepts of 3, 4 and 2 units with x, y and z axes respectively. Calculate its Miller Indices.
Answer. 4, 3, 6 *(Vidyasagar BSc)*

23. Lithium borohydride crystallizes as an orthorhombic system with four molecules per unit cell. The unit cell dimensions are $a = 6.81 \times 10^{-10}$ m, $b = 4.43 \times 10^{-10}$ m and $c = 7.2 \times 10^{-10}$ m. If the molar mass of $LiBH_4$ is 21.76×10^{-3} kg/mol, Calculate the density of crystal.
Answer. 665.30 kg/m^3 *(Guru Nanak Dev BSc)*

24. A body centered cubic element of density 10300 kg/m^3 has a cell edge of 314×10^{-12} m. Calculate the atomic mass of the element (Avogadro's number = 6.023×10^{23})
Answer. 0.09603 kg *(Delhi BSc)*

25. (*a*) Discuss powder method of crystal analysis?
(*b*) Potassium crystallizes with a body-centered cubic lattice and has a density of 856 kg/m^3. Calculate the length of side of the unit cell '*a*' and the distance between 200, 110 and 222 planes. *(Jamia Millia BSc)*
Answer. 5.28×10^{-10} m; 2.64×10^{-10} m; 3.734×10^{-10} m; 1.52×10^{-10} m

26. Derive the law of constancy of interfacial angles and law of symmetry. *(Agra BSc)*

27. Differentiate between crystalline and amorphous solid. What are Miller indices? Draw (110) plane in a face centered cubic lattice. Write a note on point defects in ionic crystal. *(Patna BSc)*

28. Discuss X-ray diffraction in elucidating structures of crystals and powders. *(Jiwaji BSc)*

29. Derive Bragg's equation for the diffraction of X-rays by crystal lattice. *(Madras BSc)*

30. Define unit cell and crystal lattice. *(Kerala BSc)*

31. Calculate the co-ordinate number of an atom in
(*a*) A body centered cubic
(*b*) A face centered cubic unit cell *(Burdwan BSc)*

32. Calculate the value of Avogadro's number from the data : Density of NaCl = 2165 kg/m^3 ; Distance between Na^+ and Cl^- in NaCl structure = 281×10^{-12} m.
Answer. 6.089×10^{23} *(Sambalpur BSc)*

33. Copper crystal has fcc cubic lattice structure. Its density is 8930 kg/m^3. What is the length of the unit cell? ($N_o = 6.023 \times 10^{23}$; Atomic mass of Cu = 63.5)
Answer. 3.614×10^{-10} m^3 *(Jiwaji BSc)*

34. A unit cell of sodium chloride has four formula unit. The edge length of the unit cell is 0.564×10^{-9} m. What is the density of sodium chloride?

Answer. 2165.6 kg/m^3

(Kanpur BSc)

35. A body centered cubic element of density 10300 kg/m^3 has a cell edge of 314×10^{-12} m. Calculate the atomic mass of element. (Avogadro's constant = 6.023×10^{23})

Answer. 94×10^{-3} kg

(Madras BSc)

36. Sodium chloride crystal has fcc structure. Its density is 2163 kg/m^3. Calculate the edge of the unit cell cube. (M_{NaCl} = 58.45×10^{-3} kg mol^{-1} ; N_o = 6.023×10^{23} mol^{-1})

Answer. 5.640×10^{-10} m

(Jiwaji BSc)

37. The density of Lithium metal is 530 kg/m^3 and the separation of 100 planes of metal is 350×10^{-12} m. Determine whether the lattice is fcc or bcc. (molar mass of Lithium = 6.941×10^{-3} kg/mol^{-1})

Answer. bcc

(Baroda BSc)

38. A certain solid X having atomic mass 30, crystallizes in the fcc arrangement. Its density is 3000 kg/m^3. What is the unit cell length? (N = 6.023×10^{23})

Answer. 4.049×10^{-10} m

(Kakatiya BSc)

39. Sodium chloride crystallizes in face-centered cubic (fcc) structure. Its density is 2165 kg/m^3. If the distance between Na^+ and its nearest Cl^- is 281×10^{-12} m, find out the Avogadro's number (Na = 23×10^{-3} kg/mol; Cl = 35.44×10^{-3} kg/mol)

Answer. 6.08×10^{-23} mol^{-1}

(Delhi BSc)

40. Calculate the wavelength of X-ray which shows a second order Bragg refraction angle of 14° from the plane of KCl. The density of KCl is 1984.9 kg/m^3 and there are four atoms in the unit cell.

Answer. 1966.26 m

(Madurai BSc)

41. Gold has a face-centered cubic structure with a unit length 4.07×10^{-10} m, a density of 19300 kg/m^3. Calculate the Avogadro's number from the data (At. wt. of Au = 197).

Answer. 6.056×10^{23}

(Panjab BSc)

MULTIPLE CHOICE QUESTIONS

1. Which is not true about the solid state?
   (*a*) they have definite shape and volume
   (*b*) they have high density and low compressibility
   (*c*) they have high attractive forces among molecules
   (*d*) they have high vapor pressure
   Answer. (*d*)
2. The melting point is that temperature at which
   (*a*) solid and liquid forms of the substance do not co-exist at equilibrium
   (*b*) solid and liquid forms of the substance have same vapor pressure
   (*c*) vapor pressure is equal to one atmospheric pressure
   (*d*) none of the above
   Answer. (*b*)
3. Amorphous solids do not have
   (*a*) sharp melting point
   (*b*) characteristic geometrical shapes
   (*c*) regularity of the structure
   (*d*) all of these
   Answer. (*d*)

4. A crystalline solid has
(*a*) definite geometrical shape
(*b*) flat faces
(*c*) sharp edges
(*d*) all of these
Answer. (*d*)

5. Amorphous substances are isotropic because
(*a*) they have same value of any property in all directions
(*b*) they have different values of physical properties in different directions
(*c*) they have definite geometrical shape
(*d*) none of the above
Answer. (*a*)

6. The elements of symmetry are
(*a*) plane of symmetry
(*b*) axis of symmetry
(*c*) center of symmetry
(*d*) all of these
Answer. (*d*)

7. The amorphous solid among the following is
(*a*) table salt
(*b*) diamond
(*c*) plastic
(*d*) graphite
Answer. (*c*)

8. A crystalline solid does not have one of the following properties. It is
(*a*) anisotropy
(*b*) sharp melting points
(*c*) isotropy
(*d*) definite and regular geometry
Answer. (*c*)

9. For tetragonal crystal system, which of the following is not true
(*a*) $a = b \neq c$
(*b*) $\alpha = \beta = \gamma = 90°$
(*c*) $a \neq b \neq c$
(*d*) none of these
Answer. (*c*)

10. For a orthorhombic crystal system, which is incorrect?
(*a*) $a \neq b \neq c$
(*b*) $\alpha = \beta = \gamma = 90°$
(*c*) $a = b \neq c$
(*d*) none of these
Answer. (*a*)

11. Na^+Cl^-, Cs^+Cl^- are the example of
(*a*) cubic crystal system
(*b*) tetragonal crystal system
(*c*) orthorhombic crystal system
(*d*) rhombohedral crystal system
Answer. (*a*)

12. The total number of atoms in a body centered cubic unit cell is
(*a*) 1
(*b*) 2
(*c*) 3
(*d*) 4
Answer. (*b*)

13. If there are 4 atoms in unit cell in a cubic system, it is an example of
(*a*) simple cubic unit cell
(*b*) body centered cubic unit cell
(*c*) face centered cubic unit cell
(*d*) none of these
Answer. (*c*)

14. Which is incorrect for a hexagonal crystal system?
(*a*) $a = b = c$
(*b*) $a = b \neq c$
(*c*) $\alpha = \beta = 90°, \gamma = 120°$
(*d*) none of these
Answer. (*a*)

15. The co-ordination number of Na^+ in Na^+Cl^- crystal is

(*a*) 2 (*b*) 4
(*c*) 6 (*d*) 8

Answer. (*c*)

16. The co-ordination number of body centered cubic lattice is

(*a*) 2 (*b*) 4
(*c*) 6 (*d*) 8

Answer. (*d*)

17. The Bragg's equation for diffraction of X-rays is

(*a*) $n\lambda = 2d^2 \sin\theta$ (*b*) $n\lambda = 2d\sin\theta$
(*c*) $n\lambda = 2d\sin^2\theta$ (*d*) $n\lambda = d\sin\theta$

Answer. (*b*)

18. In Bragg's equation $n\lambda = 2d\sin\theta$, '*n*' represents

(*a*) the number of moles (*b*) the principal quantum number
(*c*) the Avogadro's number (*d*) the order of reflection

Answer. (*d*)

19. The change in enthalpy that occurs when one mole of a solid crystalline substance is formed from the gaseous ions

(*a*) lattice energy (*b*) ionic energy
(*c*) Born-Haber cycle (*d*) crystalline energy

Answer. (*a*)

20. In cubic close packed (ccp) pattern of a metallic crystal, the co-ordination number is

(*a*) 12 (*b*) 8
(*c*) 6 (*d*) 4

Answer. (*a*)

21. In an ionic crystal, a cation and an anion leave the lattice to cause two vacancies. This defect is called

(*a*) Schottky defect (*b*) Frenkel defect
(*c*) interstitial defect (*d*) none of these

Answer. (*a*)

22. Which of the following defects is generally found in sodium chloride and cesium chloride?

(*a*) Frenkel defect (*b*) interstitial defect
(*c*) Schottky defect (*d*) none of these

Answer. (*c*)

23. In a solid lattice, a cation has left a lattice site and is present in interstitial position, the lattice defect is

(*a*) Schottky defect (*b*) Frenkel defect
(*c*) vacancy defect (*d*) interstitial defect

Answer. (*b*)

24. In Frenkel defect,

(*a*) some of the lattice sites are vacant
(*b*) an ion occupies interstitial position
(*c*) some of the cations are replaced by foreign ions
(*d*) none of the above

Answer. (*b*)

25. A device used to convert light energy into electrical energy is called

(*a*) a semiconductor (*b*) a solar cell
(*c*) an irreversible cell (*d*) an electrochemical cell

Answer. (*b*)

26. For an ionic crystal of formula AX, the radius ratio lies between 0.732 and 0.414. Its co-ordination number is

(*a*) 4 (*b*) 6

(*c*) 8 (*d*) 12

Answer. (*b*)

27. The radius ratio in an ionic crystal lies between 0.732–1.000, the co-ordination number is

(*a*) 3 (*b*) 4

(*c*) 6 (*d*) 8

Answer. (*d*)

28. The permitted co-ordination number in an ionic crystal is 6, the arrangement of anions around the cation will be

(*a*) plane triangular (*b*) tetrahedral

(*c*) octahedral (*d*) body centered cubic

Answer. (*c*)

29. A solid AB has the NaCl structure. If radius of the cation is 120×10^{-12} m, the minimum value of radius of the anion B^- will be

(*a*) $\dfrac{0.414}{12\times10^{-12}}$ (*b*) $\dfrac{120\times10^{-12}}{0.414}$

(*c*) $\dfrac{0.732}{120\times10^{-12}}$ (*d*) $\dfrac{120\times10^{-12}}{0.732}$

Answer. (*d*)

30. The number of atoms per unit cell in a simple cubic, *fcc* and *bcc* are

(*a*) 1, 2, 4 (*b*) 1, 4, 2

(*c*) 4, 2, 1 (*d*) 2, 4, 1

Answer. (*b*)

31. In a crystal, the atoms are located at the positions where potential energy is

(*a*) maximum (*b*) zero

(*c*) minimum (*d*) infinite

Answer. (*c*)

32. Potassium crystallizes in a *bcc* structure. The co-ordination number of potassium in potassium metal is

(*a*) 2 (*b*) 4

(*c*) 6 (*d*) 8

Answer. (*d*)

33. In an ionic crystal of general formula AX, the co-ordination number is six. The value of radius ratio is in the range

(*a*) 0.155 – 0.215 (*b*) 0.215 – 0.414

(*c*) 0.414 – 0.732 (*d*) 0.732 – 1

Answer. (*c*)

34. The number of atoms in a unit cell of a cube is

(*a*) 0 (*b*) 1

(*c*) 2 (*d*) 3

Answer. (*c*)

35. The number of atoms in a unit cell of a face centered cube is

(*a*) 2 (*b*) 4

(*c*) 6 (*d*) 8

Answer. (*b*)

36. NaCl is an example of

(*a*) covalent solid (*b*) metallic solid

(*c*) ionic solid (*d*) molecular solid

Answer. (*c*)

37. Which of the following defects results in the decrease of density of crystal?

(*a*) Schottky defect (*b*) Frenkel defect

(*c*) interstitial defect (*d*) impurity defect

Answer. (*a*)

38. Which of the following is a non-crystalline solid?

(*a*) rubber (*b*) ZnS

(*c*) HgS (*d*) PbI

Answer. (*a*)

39. Particles of quartz are packed by

(*a*) ionic bonds (*b*) van der Waal's forces

(*c*) hydrogen bonds (*d*) covalent bonds

Answer. (*d*)

40. LiF is an example of

(*a*) molecular crystal (*b*) covalent crystal

(*c*) ionic crystal (*d*) metallic crystal

Answer. (*c*)

41. In silicon crystal each atom is covalently bonded to _______ neighbors.

(*a*) 2 (*b*) 4

(*c*) 6 (*d*) 8

Answer. (*b*)

42. Silicon is an example of

(*a*) non-conductor (*b*) good conductor

(*c*) semi conductor (*d*) metallic conductor

Answer. (*c*)

43. Semiconductors which exhibit conductivity due to the flow of excess negative electrons are called

(*a*) *n*-type conductor (*b*) *p*-type conductor

(*c*) good conductors (*d*) none of these

Answer. (*a*)

44. In *p*-type semiconductors, the conductivity is due to

(*a*) negative holes (*b*) positive holes

(*c*) mobile electrons (*d*) valence electrons

Answer. (*b*)

45. Super conductors are substances which

(*a*) conduct electricity in liquid crystal state (*b*) conduct electricity at low temperatures

(*c*) conduct electricity at high temperatures (*d*) offer no resistance to the flow of current

Answer. (*d*)

46. The liquid crystals have

(*a*) properties of super cooled liquid

(*b*) properties of amorphous solids

(*c*) the fluidity of a liquid and optical properties of a solid

(*d*) none of these

Answer. (*c*)

47. Silicon and Germanium in the pure state are

(*a*) non-conductors
(*b*) good conductors
(*c*) metallic conductors
(*d*) metal complexes

Answer. (*a*)

48. *p*-Ozoxyanisole is an example of

(*a*) semi conductor
(*b*) super conductor
(*c*) liquid crystal
(*d*) none of these

Answer. (*c*)

49. When an arsenic atom is introduced in place of silicon in a crystal lattice, the conductivity

(*a*) increases
(*b*) decreases
(*c*) remains the same
(*d*) sometimes increases and sometimes decreases

Answer. (*a*)

50. The voltage of a solar cell is

(*a*) very high
(*b*) high
(*c*) not very large
(*d*) none of these

Answer. (*c*)

13 Physical Properties and Chemical Constitution

CONTENTS

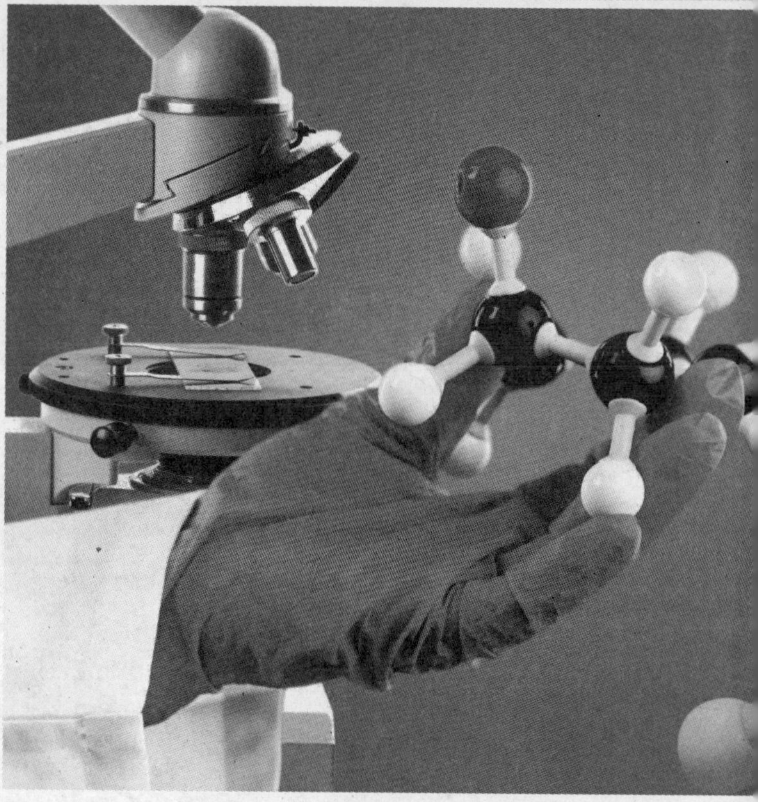

Physical properties of a substance depend on the intermolecular forces which originate in the internal structure or the constitution of the molecule. Thus the determination of properties such as surface tension, viscosity, refractive index etc., can give valuable information about the structure of molecules. In the modern times the molecular spectra of substances recorded by spectroscopic techniques have proved extremely helpful in elucidating the structure of organic molecules.

Physical properties may be classified into the following types :

(1) Additive Property

When a property of a substance is equal to the sum of the corresponding properties of the constituent atoms, it is

called an additive property. For example, molecular mass of a compound is given by the sum of the atomic masses of the constituent atoms.

(2) Constitutive Property

A property that depends on the arrangement of atoms and bond structure, in a molecule, is referred to as a constitutive property. Surface tension and viscosity and optical activity are examples of constitutive property.

(3) Additive and Constitutive Property

An additive property which also depends on the intramolecular structure, is called additive and constitutive property. Surface tension and viscosity are such properties.

In this chapter we will discuss the application of some important physical properties for elucidating the constitution of molecules.

SURFACE TENSION AND CHEMICAL CONSTITUTION

What are Parachors ?

From a study of a large number of liquids, Macleod (1923) showed that

$$\frac{\gamma^{1/4}}{D - d} = C \qquad ...(1)$$

where γ is the surface tension, D its density and d the density of vapor at the same temperature, C is a constant. Sugden (1924) modified this equation by multiplying both sides by M, the molecular weight of the liquid,

$$\frac{M\gamma^{1/4}}{D - d} = MC = [P] \qquad ...(2)$$

The quantity $[P]$, which was constant for a liquid, was given the name **Parachor.** As d is negligible compared to D the equation (2) reduces to

$$\frac{M}{D}\gamma^{1/4} = [P]$$

or

$$V_m\gamma^{1/4} = [P] \qquad ...(3)$$

where V_m is the molar volume of the liquid. If surface tension (γ) is unity, from equation (3), we may write

$$[P] = V_m$$

Thus, **the parachor [*P*] may be defined as the molar volume of a liquid at a temperature so that its surface tension is unity.**

Use of Parachor in Elucidating Structure

Sugden examined the experimental parachor values of several organic compounds of known molecular structure. He showed that **the parachor is both an additive and constitutive property.** That is, the parachor of an individual compound can be expressed as a sum of :

(1) **Atomic Parachors** which are the contributions of each of the atoms present in the molecule.

(2) **Structural Parachors** which are the contributions of the various bonds and rings present in the molecule.

By correlating the experimental values of parachor with molecular structure, Sugden (1924) calculated the atomic and structural parachors listed in Table 13.1. These values were further revised by Vogel (1948) on the basis of more accurate measurements of surface tension.

TABLE 13.1. SOME ATOMIC AND STRUCTURAL PARACHORS

Atom	Parachor		Bond or Ring	Parachor	
	Sugden	Vogel		Sugden	Vogel
C	4.8	8.6	Single bond	0	0
H	17.1	15.7	Double bond	23.2	19.9
O	20.0	19.8	Coordinate bond	–1.6	0
N	12.5	—	3- membered ring	17.0	12.3
Cl	54.3	55.2	6- membered ring	6.1	1.4

We can now illustrate the usefulness of parachor studies in the elucidation of molecular structure.

(1) Structure of Benzene (Vogel)

If the Kekule formula for benzene be accepted, the value of its parachor can be calculated by using Vogel's data.

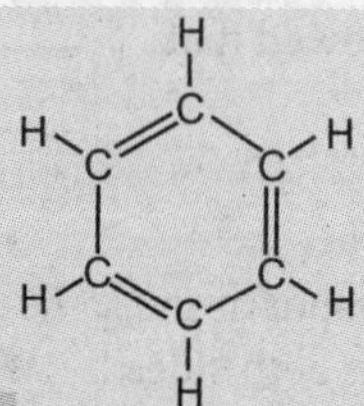

Kekule formula

6C	6 × 8.6 =	51.6
6H	6 × 15.7 =	94.2
3 (=)	3 × 19.9 =	59.7
	6-membered ring =	1.4
	Parachor of benzene =	**206.9**

The experimental value of the parachor of benzene is 206.2. Since the calculated parachor tallies with that determined by experiment, the Kekule structure for benzene is supported.

(2) Structure of Quinone (Sugden)

The two possible structural formulas proposed for quinone are :

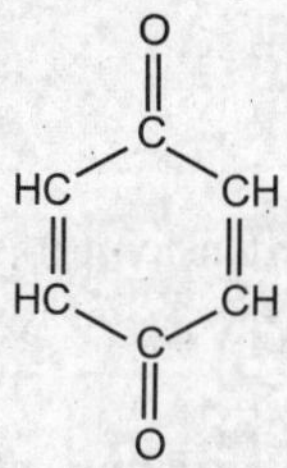

A

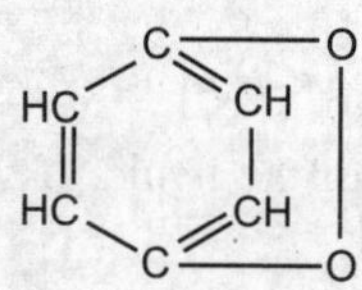

B

The parachors calculated for the two structures are :

Structure A		Structure B	
6 C	6 × 4.8 = 28.8	6 C	6 × 4.8 = 28.8
4 H	4 × 17.1 = 68.4	4 H	4 × 17.1 = 68.4
2 O	2 × 20.0 = 40.0	2 O	2 × 20.0 = 40.0
4 (=)	4 × 23.2 = 92.8	3 (=)	3 × 23.2 = 69.6
1 six-membered ring	1 × 6.1 = 6.1	2 six-membered rings	2 × 6.1 = 12.2
	Total = 236.1		**Total = 219.0**

The experimental value of parachor for quinone is 236.8. This corresponds to the parachor calculated from structure *A*. Therefore, the structure *A* represents quinone correctly.

(3) Structure of Nitro group (Sugden)

The parachor has also been found useful in providing information regarding the nature of bonds present in certain groups. The nitro group ($-NO_2$), for example, may be represented in three ways :

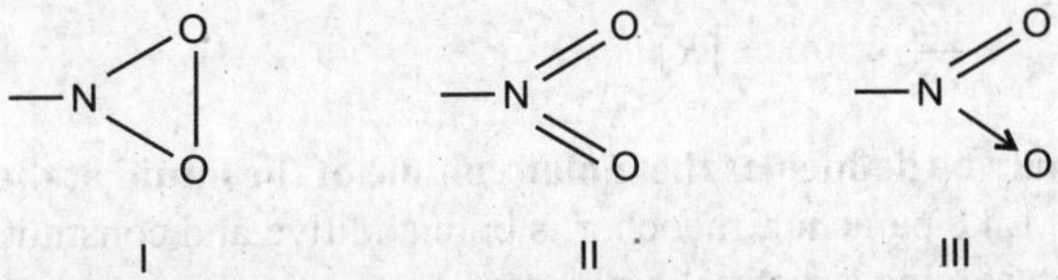

The calculated parachors are :

Structure I		Structure II		Structure III	
1 N	1 × 12.5 = 12.5	1 N	1 × 12.5 = 12.5	1 N	1 × 12.5 = 12.5
2 O	2 × 20.0 = 40.0	2 O	2 × 20.0 = 40.0	2 O	2 × 20.0 = 40.0
3-membered ring	1 × 17.0 = 17.0	2 (=)	2 × 23.2 = 46.4	1 (=)	1 × 23.2 = 23.2
				1 (→)	1 × (– 1.6) = – 1.6
Total = 69.5		Total = 98.9		Total = 74.1	

The experimental value of parachor for $-NO_2$ group has been found to be 73.0. This approximates to the calculated parachor for structure III which is, therefore, the appropriate structure of $-NO_2$ group.

VISCOSITY AND CHEMICAL CONSTITUTION

Viscosity is largely due to the intermolecular attractions which resist the flow of a liquid. Therefore, some sort of relationship between viscosity and molecular structure is to be expected. Viscosity is also dependent on the shape, size and mass of the liquid molecules. The following general rules have been discovered.

(1) Dunstan Rule

Dunstan (1909) showed that viscosity coefficient (η) and molecular volume (*d*/*M*) were related as:

$$\frac{d}{M} \times \eta \times 10^6 = 40 \textit{ to } 60$$

This expression holds only for normal (unassociated) liquids. For associated liquids this number is much higher than 60. For example, the number for benzene (C_6H_6) is 73, while for ethanol (C_2H_5OH) it is 189. This shows that benzene is a normal liquid, while ethanol is an associated one. Thus Dunstan rule can be employed to know whether a given liquid is normal or associated.

(2) Molar Viscosity

The molar surface of a liquid is $(M/d)^{2/3}$. **The product of molar surface and viscosity is termed molar viscosity.** That is,

$$\text{Molar Viscosity} = \text{Molar surface} \times \text{Viscosity}$$

$$= \left(\frac{M}{d}\right)^{2/3} \times \eta$$

Thorpe and Rodger (1894) found that molar viscosity is an additive property at the boiling point. They worked out the molar viscosity contributions of several atoms (C, H, O, S, etc) and groups.

From these, they calculate the molar viscosity of a liquid from its proposed structure. By comparing this value with the experimental one, they were able to ascertain the structure.

(3) Rheochor

Newton Friend (1943) showed that if molecular volume (*M/d*) be multiplied by the eighth root of the coefficient of viscosity, it gives a constant value [*R*]. The quantity [*R*] is termed *Rheochor*.

$$\frac{M}{d} \times \eta^{1/8} = [R]$$

The Rheochor may be defined as the molar volume of the liquid at the temperature at which its viscosity is unity. Like parachor, rheochor is both additive and constitutive. However it has not proved of much use in solving structural problems.

DIPOLE MOMENT

In a molecule such as HCl, the bonding electron pair is not shared equally between the hydrogen atom and the chlorine atom. The chlorine atom with its greater electronegativity, pulls the electron pair closer to it. This gives a slight positive charge (+*q*) to the hydrogen atom and a slight negative charge (– *q*) to the chlorine atom.

$$H—:Cl \quad \text{or} \quad \overset{+q}{H}—\overset{-q}{Cl}$$

Such a molecule with a positive charge at one end and a negative charge at the other end is referred to as an **electric dipole** or simply **dipole.** The degree of polarity of a polar molecule is measured by its dipole moment, μ (Greek *mu*).

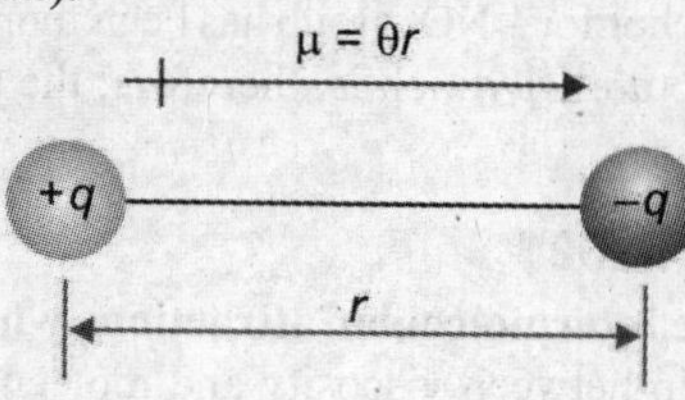

Figure 13.1
An electric dipole of the magnitude μ = θr.

The dipole moment of a polar molecule is given by the product of the charge at one end and the distance between the opposite charges. Thus,

$$\mu = q \times r$$

The dipole moment (μ) is a vector quantity. **It is represented by an arrow with a crossed tail.** The arrow points to the negative charge and its length indicates the magnitude of the dipole moment. Thus a molecule of HCl may be represented as

Unit of Dipole Moment

The CGS unit for dipole moment is the **debye,** symbolized by **D,** named after the physical chemist Peter Debye (1884-1966). A debye is the magnitude of the dipole moment (μ) when the charge (*q*) is 1×10^{-10} esu (electrostatic units) and distance (*r*) is 1E (10^{-8} cm).

$$\mu = q \times r = 1 \times 10^{-10} \times 10^{-8} = 1 \times 10^{-18} \text{ esu cm}$$

Thus $\qquad 1 \text{ D} = 1 \times 10^{-18}$ esu cm

In SI system, the charge is stated in Coulombs (*C*) and distance in meters (*m*). Thus dipole moment is expressed in Coulomb meters (*Cm*). The relation of debye to SI units is given by the expression coulomb:

$$1 \text{ D} = 3.336 \times 10^{-30} \text{ Cm}$$

Determination of Dipole Moment

Electric condenser. The dipole moment of a substance can be experimentally determined with the help of an electric condenser (Fig. 13.2). The parallel plates of the condenser can be charged by connecting them to a storage battery. When the condenser is charged, an electric field is set up with field strength equal to the applied voltage (*V*) divided by the distance (*d*) between the plates.

Polar molecules are electric dipoles. The net charge of a dipole is zero. When placed between the charged plates, it will neither move toward the positive plate nor the negative plate. On the other hand, it will rotate and align with its negative end toward the positive plate and positive end toward the negative plate. Thus all the polar molecules align themselves in the electric field. This orientation of dipoles affects the electric field between the two plates as the field due to the dipoles is opposed to that due to the charge on the plates.

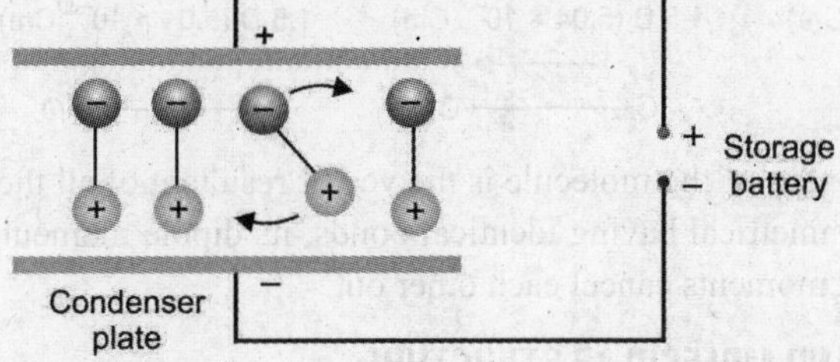

■ **Figure 13.2**
Polar molecules rotate and align in electric field.

The plates are charged to a voltage, say *V*, prior to the introduction of the polar substance. These are then disconnected from the battery. On introducing the polar substance between the plates, the voltage will change to a lower value, *V'*. Just how much the voltage changes depends on the nature of the substance. The ratio $\varepsilon = V/V'$ is a characteristic property of a substance called the **dielectric constant.** The experimentally determined value of dielectric constant is used to calculate the dipole moment.

Use of Rotational Spectra

The rotational spectrum of a polar molecule is examined in the gas phase. It is found that the spectral lines shift when the sample is exposed to a strong electric field. From the magnitude of this effect (**Stark effect**), the dipole moment can be determined very accurately. The dipole moments of some simple molecules are listed in Table 13.2.

TABLE 13.2. DIPOLE MOMENTS OF SOME SIMPLE MOLECULES IN THE VAPOR PHASE

Formula	μ(D)	μ(Cm)	Formula	μ(D)	μ(Cm)
H_2	0	0	CO_2	0	0
Cl_2	0	0	CH_4	0	0
HF	1.91	6.4176×10^{-30}	CH_3Cl	1.87	6.2832×10^{-30}
HCl	1.08	3.6288×10^{-30}	CH_2Cl_2	1.55	5.208×10^{-30}
HBr	1.80	6.048×10^{-30}	CCl_4	0	0
HI	0.42	1.4112×10^{-30}	NH_3	1.47	4.9392×10^{-30}
BF_3	0	0	H_2O	1.85	6.216×10^{-30}

BOND MOMENT

Any bond which has a degree of polarity has a dipole moment. This is called **Bond moment.** The dipole moment of H—H bond is zero because it is nonpolar. The dipole moment of the H—Cl bond is 3.6288×10^{-30} Cm because it is polar.

In a diatomic molecule, the bond moment corresponds to the dipole moment of the molecule. The dipole moments of the halogen halides shown below also indicate their bond moments.

1.91 D (6.4176×10^{-30} Cm) 1.08 D (3.6288×10^{-30} Cm) 1.80 D (6.048×10^{-30} Cm) 0.42 D (1.4112×10^{-30} Cm)

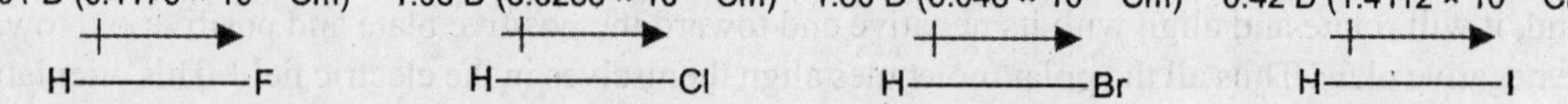

The bond moment decreases with decreasing electronegativity of the halogen atom.

When a molecule contains three or more atoms, each bond has a dipole moment. For example,

0.4 D (1.344×10^{-30} Cm) 1.5 D (5.04×10^{-30} Cm) 1.5 D (5.04×10^{-30} Cm) 1.3 D (4.368×10^{-30} Cm)

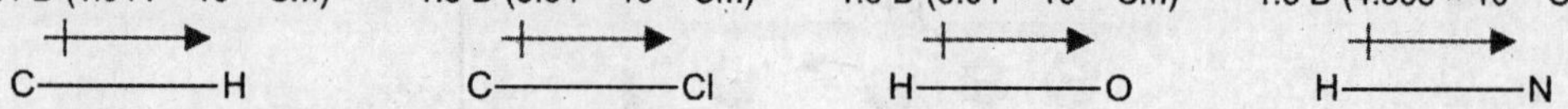

The net dipole moment of the molecule is the vector resultant of all the individual bond moments. If a molecule is symmetrical having identical bonds, its dipole moment is zero. That is so because the individual bond moments cancel each other out.

DIPOLE MOMENT AND MOLECULAR STRUCTURE

Dipole moment can provide important information about the geometry of molecular structure. If there are two or more possible structures for a molecule, the correct one can be identified from a study of its dipole moment.

(1) H_2O has a Bent Structure

Water molecule (H_2O) can have a linear or bent structure.

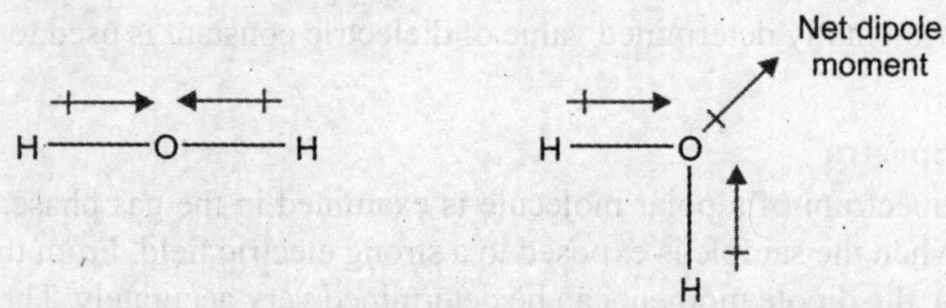

(a) (b)

The dipole moments of the two O—H bonds in structure (*a*) being equal in magnitude and opposite in direction will cancel out. The net dipole moment (μ) would be zero. In structure (*b*) the bond moment will add vectorially to give a definite net dipole moment. Since water actually has a dipole moment 1.85 D (6.216×10^{-30} Cm) ; its linear structure is ruled out. Thus water has a bent structure as shown in (*b*).

(2) CO_2 has a Linear Structure and SO_2 a Bent Structure

Carbon dioxide has no dipole moment (μ = 0). This is possible only if the molecule has a linear structure and the bond moments of the two C = O units cancel each other.

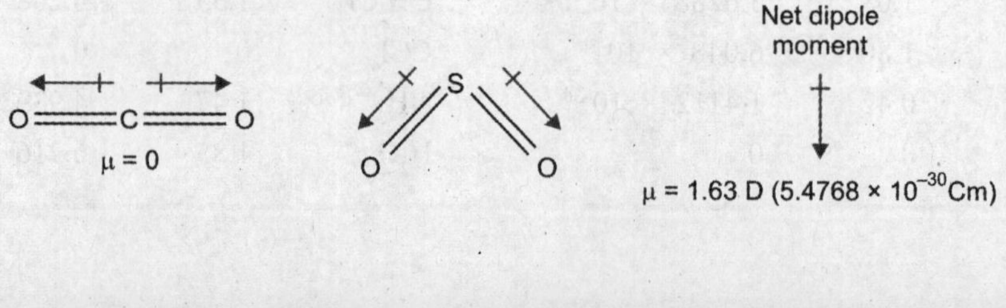

On the other hand, SO_2 has a dipole moment (μ = 1.63 D) 5.4768×10^{-30} Cm. Evidently, here the individual dipole moments of the two S = 0 bonds are not canceled. Thus the molecule has a bent structure. The vector addition of the bond moments of the two S = 0 units gives the net dipole moment 1.63 D (5.4768×10^{-30} Cm).

(3) BF_3 has a Planar and NH_3 a Pyramidal Structure

The dipole moment of boron trifluoride molecule is zero. This is possible if the three B—F bonds are arranged symmetrically around the boron atom in the same plane. The bond moments of the three B—F bonds cancel each others effect and the net μ = 0.

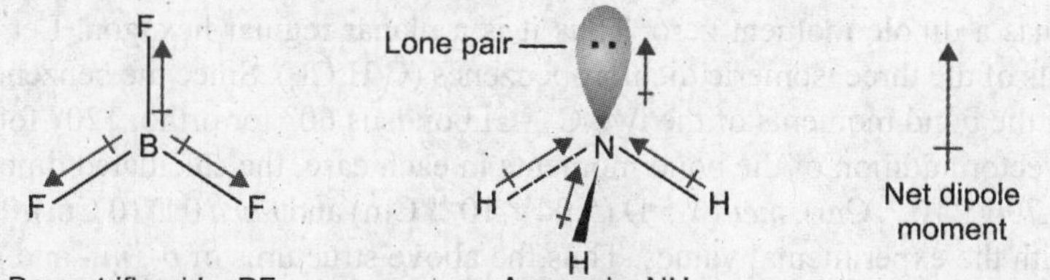

Boron trifluoride, BF_3 (μ = 0) | Ammonia, NH_3 (μ = 1.47 D or 4.9392×10^{-30} Cm)

Ammonia molecule (NH_3) has a dipole moment (μ = 1.47 D) 4.9392×10^{-30} Cm. This is explained by its pyramidal structure. The three H atoms lie in one plane symmetrically with N atom at the apex of the regular pyramid. The dipole moments of the three N—H bonds on vector addition contribute to the net dipole moment. In addition, there is a lone pair of electrons on the N atom. Since it has no atom attached to it to neutralize its negative charge, **the lone pair makes a large contribution to the net dipole moment.** Thus the overall dipole moment of ammonia molecule is the resultant of the bond moments of three N—H bonds and that due to lone-pair.

It may be recalled that the high dipole moment of water (H_2O) can also be explained by the presence of two lone-pairs of electrons on the oxygen atom.

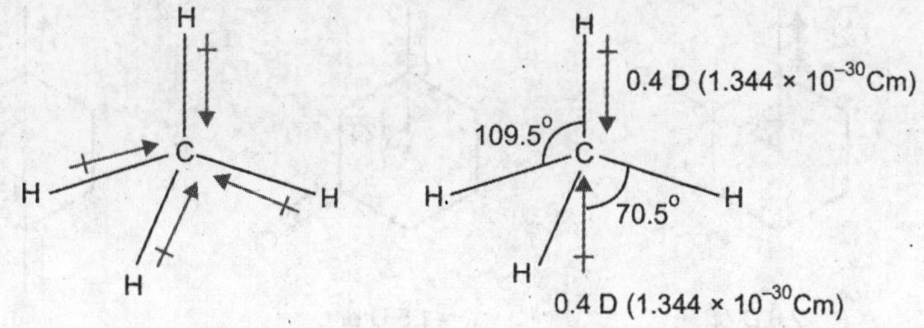

(a) (b)

■ **Figure 13.3**
Dipole moment of methane molecule.

(4) CH_4 has Tetrahedral Structure

Methane (CH_4) has zero dipole moment, despite the fact that each C—H bond possesses a dipole moment of 0.4 D (1.344×10^{-30} Cm). This can be explained if the molecule has a symmetrical tetrahedral structure (Fig. 13.3).

Each C—H bond in the pyramidal CH_3 group contributes $\frac{1}{3}$ μ (μ cos 70.5) to the resultant dipole moment. Thus the net dipole moment of CH_3 group is equal to μ. This acts in a direction opposite to that of the fourth C—H bond moment, thereby canceling each other.

(5) Identification of *cis* and *trans* Isomers

The dipole moment can be used to distinguish between the *cis* and *trans* isomers. The *cis* isomer has a definite dipole moment, while the *trans* isomer has no dipole moment (μ = 0). For example,

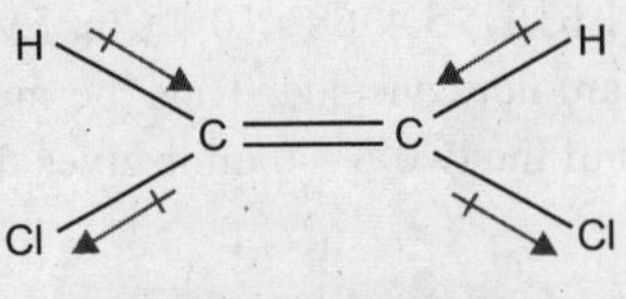

cis-Dichloroethylene
(μ = 2.96 D or 9.9456 × 10^{-30} Cm)

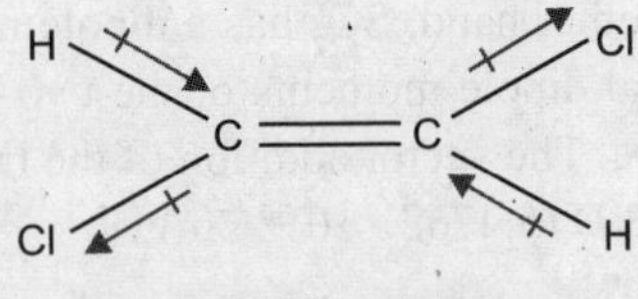

trans-Dichloroethylene
(μ = 0)

In the *cis* isomer, the bond moments add vectorially to give a net dipole moment. The *trans* isomer is symmetrical and the effects of opposite bond moments cancel so that $\mu = 0$.

(6) Identification of *ortho*, *meta* and *para* Isomers

Benzene has a dipole moment zero. Thus it is a planar regular hexagon. Let us examine the dipole moments of the three isomeric dichlorobenzenes ($C_6H_4Cl_2$). Since the benzene ring is flat, the angle between the bond moments of the two C—Cl bonds is 60° for ortho, 120° for *meta* and 180° for *para*. On vector addition of the bond moments in each case, the calculated dipole moments are *ortho* 2.6 D (8.736 × 10^{-30} Cm), *meta* 1.5 D (5.04 × 10^{-30} Cm) and *para* 0 D (0 Cm). These calculated values tally with the experimental values. Thus the above structures of *o*-, *m* - and *p*-isomers stand confirmed. In general, **a para disubstituted benzene has zero dipole moment, while that of the ortho isomer is higher than of meta isomer.** This provides a method for distinguishing between the isomeric *ortho*, *meta* and *para* disubstituted benzene derivatives.

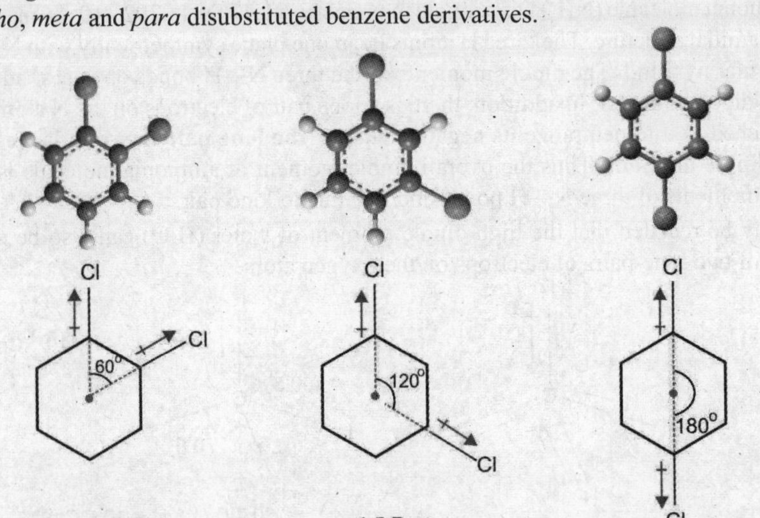

μ = 2.6 D or 8.736 × 10^{-30} Cm

μ = 1.5 D or 5.04 × 10^{-30} Cm)

μ = 0 or 0 Cm

Figure 13.4
Dipole moments of *ortho*, *meta* and *para* dichlorobenzenes.

Dipole Moment and Ionic Character

The magnitude of the dipole moment of a diatomic molecule determines its ionic character. Let us consider an HBr molecule whose measured dipole moment (μ_{exp}) is 0.79 D (2.654 × 10^{-30} cm) and bond distance (r) = 1.41Å (1.41 × 10^{-10} m).

If the molecule were completely ionic (H^+Br^-), each of the ions will bear a unit electronic charge, e (4.8 × 10^{-10} esu or 1.6013 × 10^{-19} Coulomb). Thus the dipole moment of the ionic molecule (μ_{ionic}) can be calculated.

$$\mu_{ionic} = e \times r = (1.6013 \times 10^{-19}\ \text{C})\ (1.41 \times 10^{-10}\ \text{m})$$
$$= 2.2578 \times 10^{-29}\ \text{Cm}\ (6.77\ \text{D})$$

But the experimental dipole moment (μ_{exp}) of $H^{\delta+}$—$Br^{\delta-}$, which determines its actual fractional ionic character, is 0.79 D (2.654 × 10^{-30} Cm). Therefore,

$$\% \text{ ionic character of HBr} = \frac{\mu_{expt}}{\mu_{ionic}} \times 100$$

$$= \frac{2.6544 \times 10^{-30}}{22.7472 \times 10^{-30}} \times 100 = 11.6$$

Hence HBr is 12% ionic in character.

MOLAR REFRACTION AND CONSTITUTION

The molar refraction (R_M) is an additive and constitutive property. **The molar refraction of a molecule is thus a sum of the contributions of the atoms (atomic refractions) and bonds (bond refractions) present.** From the observed values of R_M of appropriate known compounds, the atomic refractions of different elements and bonds have been worked out. Some of these are listed in Table 13.3.

TABLE 13.3. SOME ATOMIC AND BOND REFRACTIONS IN m³ mol⁻¹ FOR D LINE (VOGEL 1948)

Atom	R_{atomic}	Bond	R_{bond}
C	2.591	C – H	1.676×10^{-6}
H	1.028	C = C	1.575×10^{-6}
O (in C = O)	2.010	C ≡ C	1.977×10^{-6}
O (in –O–)	1.643	6C ring	– 0.15
O (in –O–H)	1.518	5C ring	– 0.10
Cl	5.844	4C ring	0.317
Br	8.741		

The molar refraction of the proposed structure of a molecule can be computed from the known atomic and bond refractions. If this value comes out to be the same as the experimental value, the structure stands confirmed. Some examples are given below for illustrating the use of molar refractions in elucidating molecular structure.

(1) **Acetic acid.** The accepted structural formula of acetic acid is

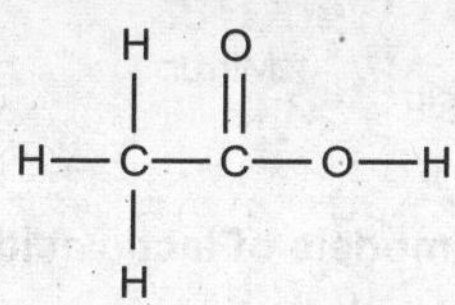

The molar refraction (R_M) may be computed from the atomic refractions of the constituent atoms as

2 C	$2 \times 2.591 \times 10^{-6}$	=	5.182×10^{-6}
4 H	$4 \times 1.028 \times 10^{-6}$	=	4.112×10^{-6}
1 O (in OH)	$1 \times 1.518 \times 10^{-6}$	=	1.518×10^{-6}
1 O (in C = O)	$1 \times 2.010 \times 10^{-6}$	=	2.010×10^{-6}
	Total R_M	=	**12.822×10^{-6} m³ mol⁻¹**

The value of molar refraction of acetic acid found by determination of its refractive index is 13.3 cm³ mol⁻¹. There is a fairly good agreement between the calculated and experimental values. It confirms the accepted formula of acetic acid.

(2) **Benzene.** The molar refraction of benzene (C_6H_6) on the basis of the much disputed Kekule formula may be calculated as :

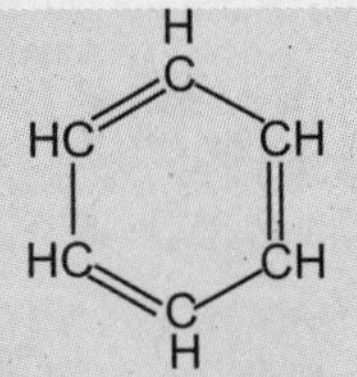

Kekule formula for benzene

6 C	$6 \times 2.591 \times 10^{-6}$	=	15.546×10^{-6}
6 H	$6 \times 1.028 \times 10^{-6}$	=	6.168×10^{-6}
3 C=C	$3 \times 1.575 \times 10^{-6}$	=	4.725×10^{-6}
1 6C ring		=	-0.150×10^{-6}
	Total R_M	=	**26.289×10^{-6} m³ mol⁻¹**

The observed values of R_M for benzene is 25.93×10^{-6}. This is in good agreement with the calculated value. Hence the Kekule formula for benzene is supported.

(3) **Optical Exaltation.** A compound containing conjugated double bonds (C=C—C=C) has a higher observed R_M than that calculated from atomic and bond refractions. The molar refraction is thus said to be exalted (raised) by the presence of a conjugated double bond and the phenomenon is called **optical exaltation.** For example, for **hexatriene,**

$$CH_2{=}CH{-}CH{=}CH{-}CH{=}CH_2$$

the observed value of R_M is 30.58×10^{-6} m³ mol⁻¹ as against the calculated value 28.28×10^{-6} m³ mol⁻¹.

If present in a closed structure as benzene, the conjugated double bonds do not cause exaltation.

OPTICAL ACTIVITY AND CHEMICAL CONSTITUTION

Optical activity is a purely constitutive property. This is shown by a molecule which is **dissymmetric or chiral** (pronounced ky-ral). A chiral molecule has no plane of symmetry and cannot be superimposed on its mirror image. One such molecule is lactic acid, $CH_3CHOHCOOH$. Assuming that carbon has a tetrahedral structure, lactic acid can be represented by two models I and II.

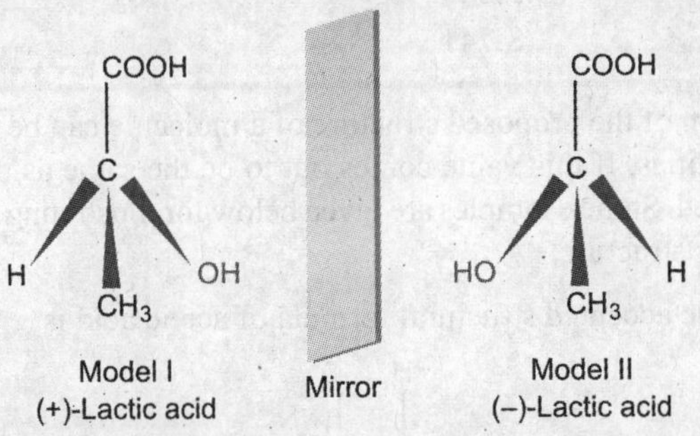

Model I (+)-Lactic acid

Model II (–)-Lactic acid

■ Figure 13.5
Two chiral models of lactic acid.

If you try to place model II on model I so that any two similar groups coincide, the remaining two will clash. Suppose you try to coincide COOH and H by rotating model II, the groups OH and CH_3 will go in opposite positions. The fact that model II cannot be superimposed on model I shows that they represent different molecules. **The molecules that are nonsuperimposable mirror images, are called enantiomers.**

Lactic acid is actually known to exist in two enantiomeric forms (optical isomers). They have the same specific rotation but with sign changed.

(+)-Lactic acid $[\alpha]_D^{25} = +3.8°$

(–)-Lactic acid $[\alpha]_D^{25} = -3.8°$

A third variety of lactic acid is obtained by laboratory synthesis. It is, in fact, a mixture of equimolar amounts of (+)- and (–)-lactic acid. It is represented as (±)-lactic acid and is termed **racemate** or **racemic mixture.** Evidently, a racemic mixture has zero optical rotation.

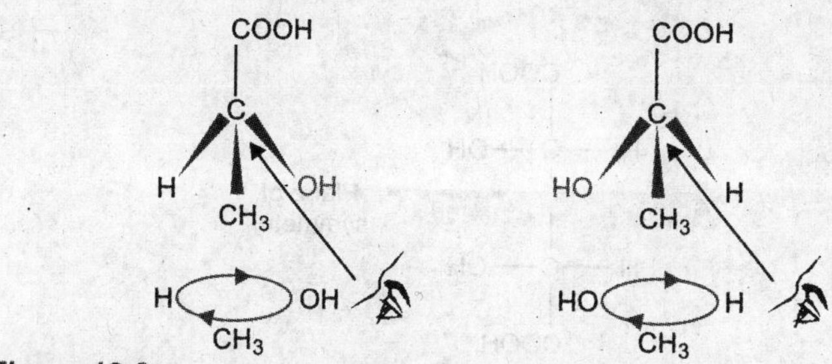

■ **Figure 13.6**
The interaction of polarized light with the opposite orientation of groups produces opposite rotatory powers in two enantiomers of lactic acid.

What Makes a Molecule Optically Active?

Lactic acid molecule is chiral because it contains a carbon joined to four different groups H, OH, CH_3, COOH. A carbon bearing four different groups is called an **asymmetric** or **chiral carbon.** It may be noticed that the arrangement of groups about the asymmetric carbon is either clockwise or anticlockwise. It is the interaction of polarized light with these opposite arrangements of groups which is responsible for the opposite rotatory powers of (+)- and (–)-lactic acids.

Presence of Chiral Carbons, Not a Necessary Condition for Optical Activity

A molecule containing two or more chiral (or asymmetric) carbons may not be optically active. **It is, in fact, the chirality of a molecule that makes it optically active.**

Let us consider the example of tartaric acid which contains two chiral carbon atoms. It exists in three stereoisomeric forms.

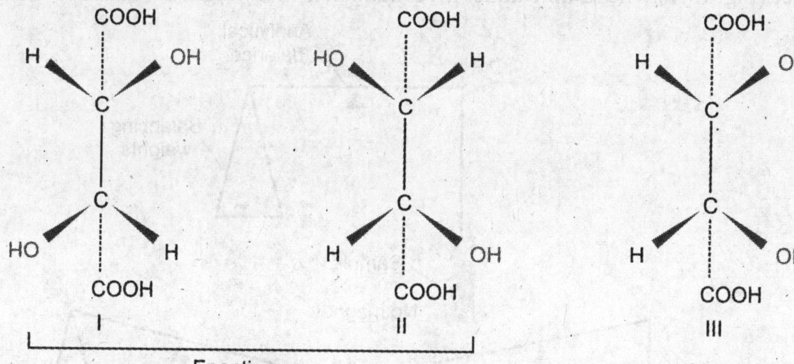

The forms I and II are nonsuperimposable mirror images. They represent an enantiomeric pair. The form III has a plane of symmetry and divides the molecule into two identical halves. The arrangement of groups about the two chiral carbons is opposite to each other. The optical rotation due to the upper half of the molecule is canceled by the rotation of the lower half. **Thus the form III although containing two chiral carbons, is optically inactive.** This is called the **meso form.**

Tartaric acid is actually known to exist in three stereoisomeric forms :

(+)-Tartaric acid	$[\alpha]_D^{20} = +12.7°$
(–)-Tartaric acid	$[\alpha]_D^{20} = -12.7°$
meso-Tartaric acid	$[\alpha]_D^{20} = 0°$

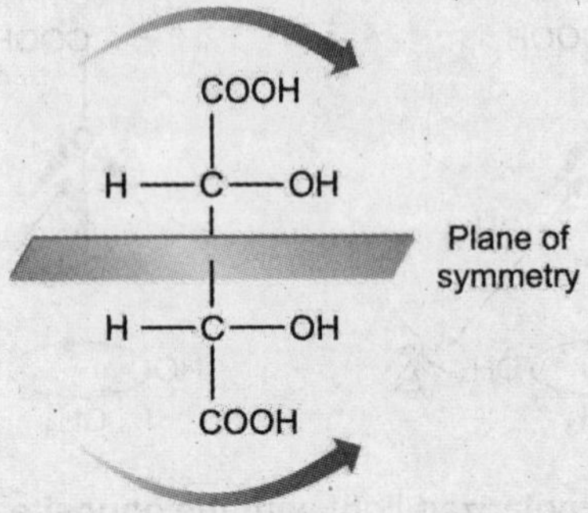

An equimolar mixture of (+)-tartaric acid and (–)-tartaric acid is referred to as **racemic tartaric** acid or (±)-tartaric acid.

MAGNETIC PROPERTIES

On the basis of their behavior in a magnetic field, substances can be divided into two classes.

(1) **Diamagnetic substances** which are slightly repelled or pushed out of the magnetic field. Most substances belong to this class.

(2) **Paramagnetic substances** which are slightly attracted or pulled into the magnetic field. Some substances belong to this class.

A few substances are intensely paramagnetic and retain their magnetic property when removed from the magnetic field. These are known as **ferromagnetic substances.** Examples are iron, cobalt and nickel.

Measurement of Magnetic Properties

The magnetic properties of substances can be measured with the help of a **magnetic balance** or **Gouy balance** (Fig. 13.7). The sample under investigation is first weighed without the magnetic field.

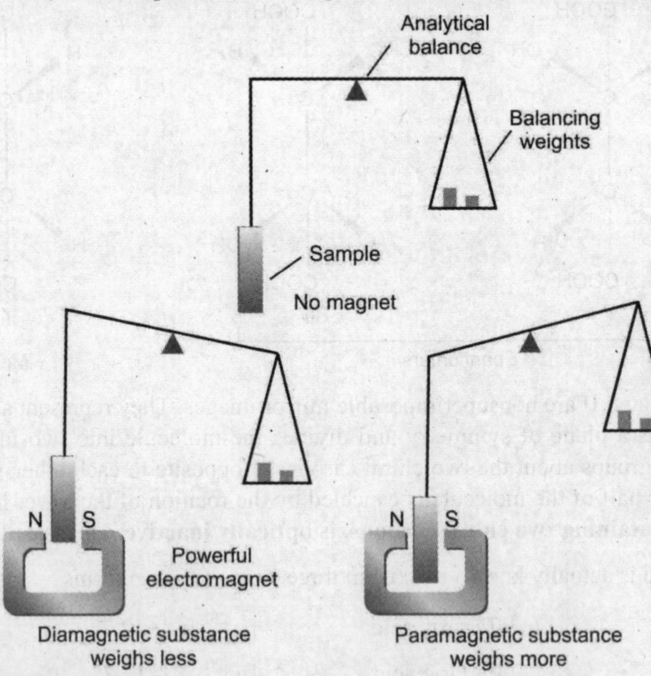

Figure 13.7
Whether a substance is diamagnetic or paramagnetic can be determined by a Gouy balance.

Then it is suspended between the poles of a strong electromagnet. A diamagnetic substance is pushed out of the field and weighs less. On the other hand, a paramagnetic substance is pulled into the field and weighs more. **The difference in the sample weight when there is no magnetic field and the weight on the application of the field, determines the magnetic susceptibility of the substance.**

Why a Substance is Paramagnetic or Diamagnetic?

A single electron spinning on its own axis generates a magnetic field and behaves like a small magnet. Therefore a substance with an orbital containing an unpaired electron will be attracted into the poles of an electromagnet. It follows that **any atom, ion or molecule that contains one or more unpaired electrons will be paramagnetic.** When an orbital contains two electrons (↑↓), their spins are opposed so that their magnetic fields cancel each other. Thus an atom, ion or molecule in which all electrons are paired will not be paramagnetic.

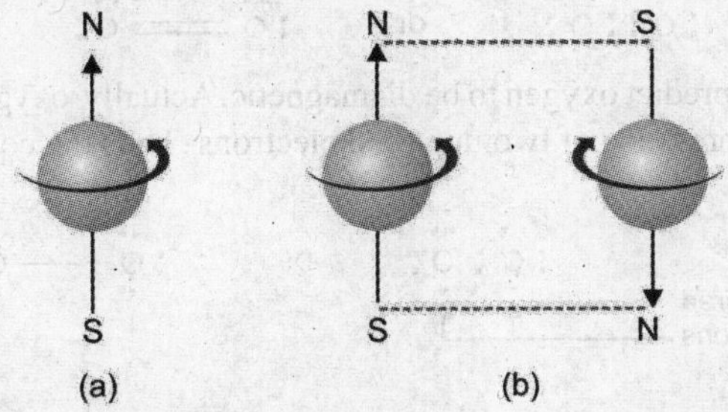

■ **Figure 13.8**
***(a)* An electron spinning on its axis behaves like a tiny magnet; *(b)* Two electrons with opposite spins cancel the magnetic field of each other.**

Most atoms or molecules have all their electrons paired (↑↓). These are repelled by a magnetic field and are said to be **diamagnetic.** This can be explained in a simple way. When an external magnetic field is applied to such a substance, it sets up tiny currents in individual atoms as if in a wire. These electric currents create new electric field in the opposite direction to the applied one. This causes the repulsion of the substance when placed in a magnetic field.

The fact that an atom, ion or molecule is paramagnetic if it contains one or more unpaired electrons and diamagnetic if it contains all paired electrons, can be illustrated by taking simple examples. The configuration of sodium shows that it has one unpaired electron and it is paramagnetic. Magnesium having all paired electrons is diamagnetic.

Na (Ne) $3s^1$ [↑]

Mg (Ne) $3s^2$ [↑↓]

Iron has four incompleted *d* orbitals having single electrons with spins aligned. The magnetic field due to these four unpaired electrons adds up to make iron strongly paramagnetic or **ferromagnetic.**

[↑↓] [↑] [↑] [↑] [↑]

3d orbitals of Fe

Magnetic Properties and Molecular Structure

A paramagnetic molecule or ion contains one or more unpaired electrons. **Each spinning electrons behaves like a magnet and has a magnetic moment.** The magnetic moment (μ) of a molecule or ion due to the spin of the unpaired electrons is expressed by the formula

$$\mu = \sqrt{n(n+2)}$$

where μ is the magnetic moment in **magnetons** and *n* the number of unpaired electrons. The number of unpaired electrons can be determined by measuring magnetic moment with the help of **Gouy balance.** This has proved of great use in establishing the structure of certain molecules and complex ions.

(1) Oxygen Molecule

The Lewis structure of oxygen molecule postulates the presence of a double bond between the two oxygen atoms.

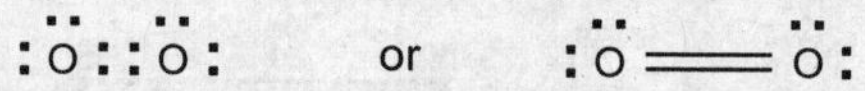

Such a structure would predict oxygen to be diamagnetic. Actually, oxygen is found to be slightly, paramagnetic showing the presence of two unpaired electrons. Thus the correct structure of oxygen molecule is written as

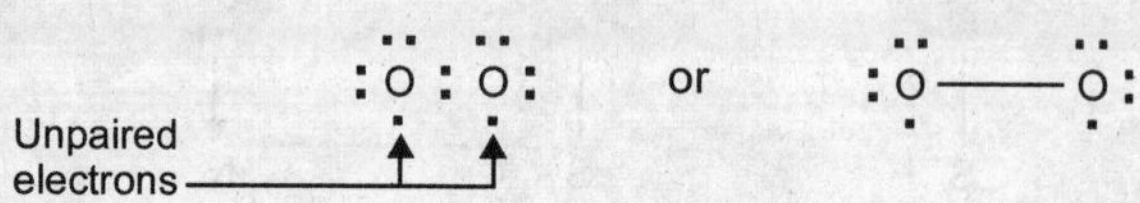

(2) Complex Ions

The geometrical structures of complex ions can be inferred from a study of their magnetic moments. For example, the complex K_2NiCl_4 is paramagnetic with a magnetic moment corresponding to two unpaired electrons. Thus the complex ion $NiCl_4^{2-}$ must have the same configuration of the 3*d* electrons as in nickel.

The 4*s* and three 4*p* orbitals could be involved in sp^3 hybridization, giving a tetrahedral structure for $NiCl_4^{2-}$ ion. This has been confirmed experimentally.

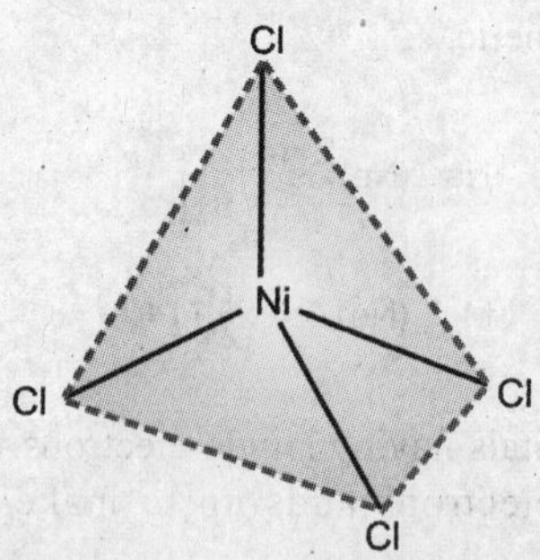

Figure 13.9
Tetrahedral structure of $NiCl_4^{2-}$ as predicted by a study of its dipole moment.

MOLECULAR SPECTRA

We have studied previously (Chapter 1) how hydrogen atom exhibits an **atomic spectrum** by interaction with electromagnetic radiation. This was explained in terms of energy levels in the atom. An electron can pass from one energy level to another by emission or absorption of energy from the incident radiation. When an electron drops from a higher energy level E_2 to another of lower energy E_1, the surplus energy is emitted as radiation of frequency v. That is

$$E_2 - E_1 = hv$$

where h is Planck's constant. Each frequency of emitted radiation records a bright line in the spectrum. On the other hand, if the transition of an electron occurs from a lower energy level E_1, to another of higher energy E_2, energy (hv) is absorbed. This records a dark line in the spectrum.

A spectrum which consists of lines of different frequencies is characteristic of atoms and is termed **Line Spectrum or Atomic Spectrum.**

A spectrum which consists of bright lines produced by emission of electromagnetic radiation is termed **Emission Spectrum.**

A spectrum which consists of dark lines produced by absorption of incident radiation, is termed **Absorption Spectrum.**

Similar to atoms, molecules also have energy levels. Thus molecules exhibit **Molecular spectra** by interaction with electromagnetic radiations. These spectra may consist of bright or dark bands composed of groups of lines packed together. Such spectra are termed **Band spectra.** Molecules generally exhibit spectra made of dark bands caused by absorption of incident radiation. These are referred to as **Absorption Band spectra** and provide valuable information about the chemical structure of molecules. However, molecular spectra are relatively complex. Before taking up their systematic study, it is necessary to acquaint the student with the Electromagnetic spectrum as also the Molecular energy levels.

ELECTROMAGNETIC SPECTRUM

Visible light, X-rays, microwaves, radio waves, etc., are all electromagnetic radiations. Collectively, they make up the **Electromagnetic spectrum** (Fig. 13.10).

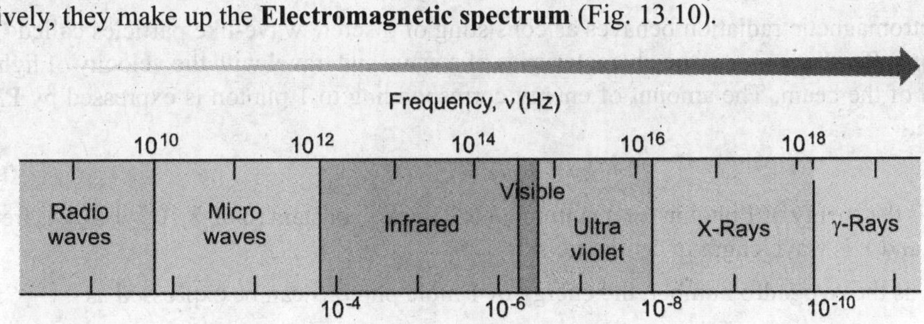

■ Figure 13.10
The Electromagnetic Spectrum.

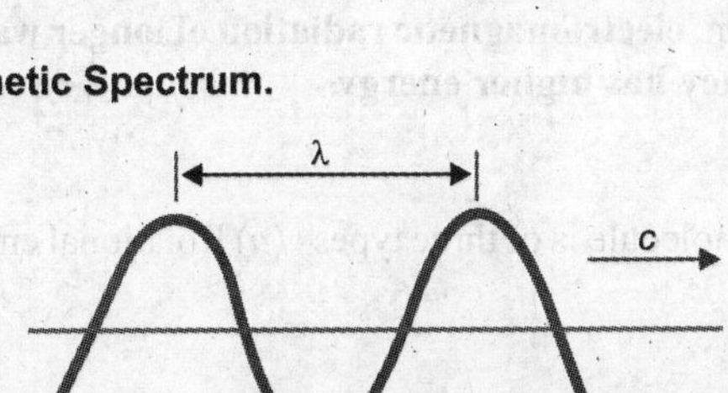

■ Figure 13.11
A single electromagnetic wave.

Electromagnetic radiations consist of electrical and magnetic waves oscillating at right angles to each other. They travel away from the source with the velocity of light (C) in vacuum. Electromagnetic waves are characterized by :

(1) **Frequency (ν) :** It is the number of successive crests (or troughs) which pass a stationary point in one second. The unit is **hertz;** $1\text{Hz} = 1\text{s}^{-1}$.

(2) **Wavelength (λ) :** It is the distance between successive crests (or troughs). λ is expressed in centimeters (cm), meters (m), or nanometers ($1\text{nm} = 10^{-9}$ m).

(3) **Wave number $\bar{\nu}$:** It is the reciprocal of wavelength, Its unit is cm^{-1}.

$$\bar{\nu} = \frac{1}{\lambda}$$

Relation Between Frequency, Wavelength and Wave Number

Frequency and wavelength of an electromagnetic radiation are related by the equation

$$\nu\lambda = c \qquad ...(1)$$

or

$$\nu = \frac{c}{\lambda} \qquad ...(2)$$

where c is the velocity of light. It may be noted that wavelength and frequency are inversely proportional. That is, **higher the wavelength lower is the frequency;** lower the wavelength higher is the frequency.

Further

$$\bar{\nu} = \frac{1}{\lambda}\ \text{cm}^{-1} \qquad ...(3)$$

From (2) and (3)

$$\nu = c\bar{\nu}$$

Energy of Electromagnetic Radiation

Electromagnetic radiation behaves as consisting of discrete wave-like particles called **Quanta** or **Photons.** Photons possess the characteristics of a wave and travel with the velocity of light in the direction of the beam. The amount of energy corresponding to 1 photon is expressed by **Planck's equation.**

$$E = h\nu = h\,c/\lambda$$

where E is the energy of 1 photon (or quantum), h is Planck's constant (6.62×10^{-34} J-s); ν is frequency in hertz; and λ is wavelength in centimeters.

If N is the Avogadro number, the energy of 1 mole photons can be expressed as

$$E = \frac{Nhc}{\lambda} = \frac{1.1932 \times 10^{-2}}{\lambda}\ \text{kJ/mol}$$

This shows that per photon, **electromagnetic radiation of longer wavelength has lower energy. Radiation of higher frequency has higher energy.**

MOLECULAR ENERGY LEVELS

The internal energy of a molecule is of three types : (*a*) Rotational energy; (*b*) Vibrational energy; and (*c*) Electronic energy.

Rotational Energy

It involves the rotation of molecules about the center of gravity or of parts of molecules.

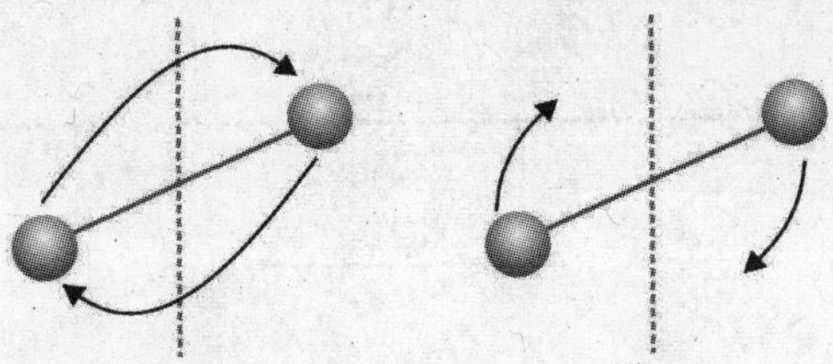

■ **Figure 13.12**
Molecular rotations in a linear molecule.

Vibrational Energy

It is associated with stretching, contracting or bending of covalent bonds in molecules. The bonds behave as spirals made of wire.

Electronic Energy

It involves changes in the distribution of electrons by splitting of bonds or the promotion of electrons into higher energy levels.

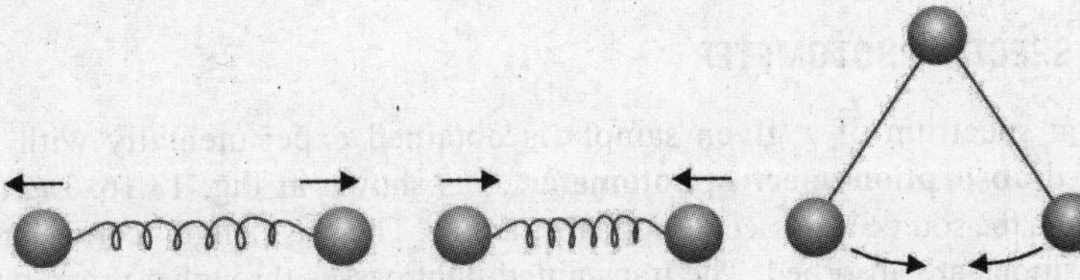

■ **Figure 13.13**
Vibrations within molecules.

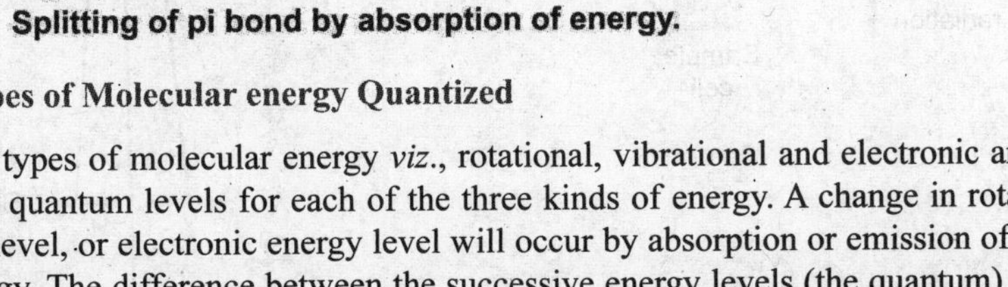

■ **Figure 13.14**
Splitting of pi bond by absorption of energy.

The Three types of Molecular energy Quantized

The three types of molecular energy *viz.*, rotational, vibrational and electronic are quantized. Thus there are quantum levels for each of the three kinds of energy. A change in rotational level, in vibrational level, or electronic energy level will occur by absorption or emission of one or more quanta of energy. The difference between the successive energy levels (the quantum) of electronic changes is 450 kJ mol^{-1}; for vibrational changes 5 to 40 kJ mol^{-1}; and for rotational changes, about 0.02 kJ mol^{-1}.The three types of energy levels are illustrated in Fig. 13.15. It may be noted that electronic energy levels are relatively far apart. For each electronic level there exist a number of closely spaced vibrational energy levels. Likewise each vibrational level contains a large number of rotational energy levels.

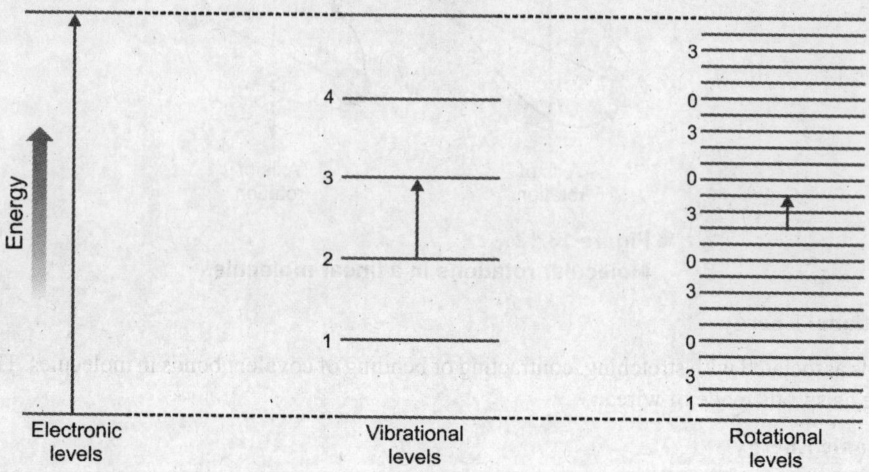

■ **Figure 13.15**
An illustration of the electronic, vibrational and rotational energy levels.

ABSORPTION SPECTROPHOTOMETER

Absorption spectrum of a given sample is obtained experimentally with the help of an apparatus called **absorption spectrophotometer.** It is shown in Fig. 13.16. Light of a range of wavelengths from the source is passed through the sample. The wavelengths corresponding to allowed molecular transitions are absorbed. The transmitted light passes through a prism which resolves it into various wavelengths. It is then reflected from the mirror onto a detector. The prism is rotated so that light of each given wavelength is focused on the detector. The response of the detector is recorded on a chart by means of a motor-driven pen synchronized with the prism movement. The pen recorder records the intensity of radiation as a function of frequency and gives the absorption spectrum of the sample.

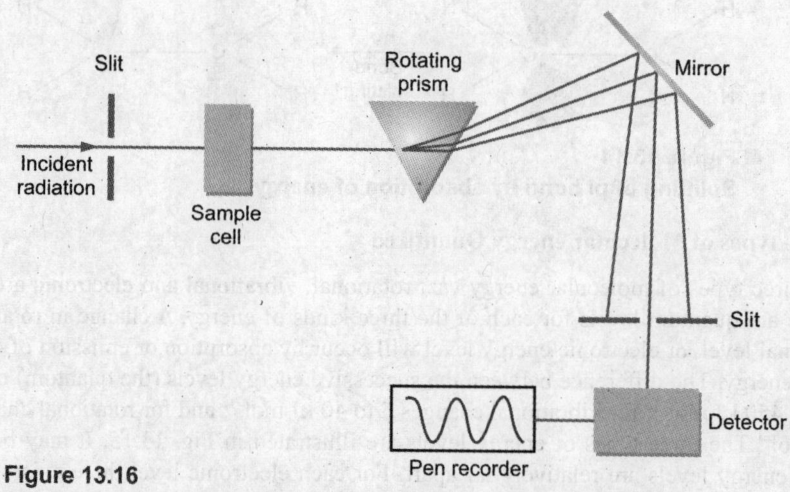

■ **Figure 13.16**
Schematic diagram of an Absorption Spectrophotometer.

TYPES OF MOLECULAR SPECTRA

To obtain a molecular spectrum, the substance under examination is exposed to electromagnetic radiations of a series of wavelengths. Molecules of the substance absorb certain wavelengths in order to be excited to higher electronic, vibrational or rotational energy levels. **The series of wavelengths absorbed in each case gives a distinct molecular spectrum.** Thus there are three main types of molecular spectra.

(1) Electronic Spectra

These are caused by absorption of high energy photons which can send electrons to higher energy levels. The *electronic spectra* are within the visible or ultraviolet regions of the electromagnetic spectrum.

(2) Vibrational Spectra

Lower energy photons cause changes in vibrational energy levels. The spectra thus obtained are referred to as *vibrational spectra*. These are in the near infrared region (near to the visible region).

(3) Rotational Spectra

Still lower energy photons cause changes in the rotational levels in the molecules. These spectra are called the *rotational spectra*. These spectra are in the far infrared or microwave region.

In order to use numbers of reasonable size, different units are used for different types of spectra. In ultraviolet and visible spectra the radiation is stated as a wavelength in nm or as a wave-number in cm^{-1}. In infrared spectra, wave-numbers in cm^{-1} are often used. In microwave or NMR spectra, frequencies are expressed in Hz (or MHz).

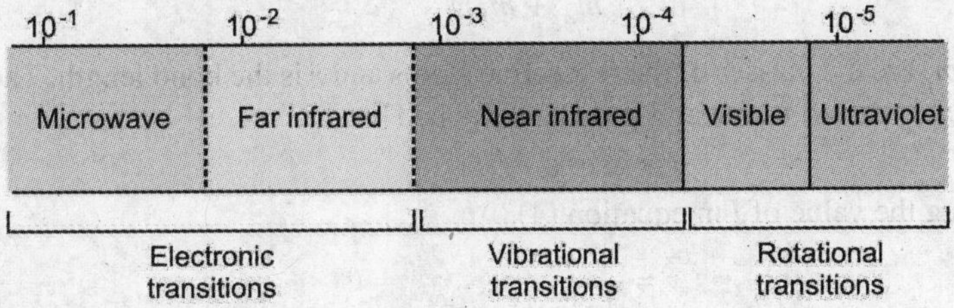

Figure 13.17
The regions of the electromagnetic spectrum and associated energy transitions in molecules.

When energy of the incident photon is large enough to produce changes in electronic levels, it can also bring, about changes in vibrational and rotational levels. Therefore some spectra are **electronic-vibrational-rotational spectra.** Similarly, the photon that is capable of bringing about changes in vibrational levels can also cause rotational energy changes. Thus vibrational -rotational spectra result. Pure **rotational spectra** involve only small energy changes which cannot bring about vibrational or electronic changes.

ROTATIONAL SPECTRA

Rotational spectra are used to determine the bond lengths in heteronuclear molecules, A–B. Such molecules have a permanent dipole moment and while rotating produce electric field. Thus electromagnetic radiations interact with these molecules and produce rotational spectra. The rotational spectra are observed in the far infrared and microwave regions.

We have shown in Fig. 13.15 that the rotational energy levels are spaced equally. For a molecule *A–B*, the energies of the rotational levels are given by the formula

$$E_J = \frac{h^2}{8\pi^2 I} J(J+1) \quad ...(1)$$

where J is the rotational quantum number and I the moment of inertia.

As a rule, a **molecule can be excited to only the next higher rotational level by absorption of energy.** If the molecule is raised from quantum number $J = 0$ to $J = 1$, the energy difference ΔE is given by

$$\Delta E = E_{J=1} - E_{J=0} \quad ...(2)$$

Using equation (1), we can write

$$\Delta E = \frac{h^2}{8\pi^2 I} 1(1+1) - 0 = \frac{h^2}{4\pi^2 I} \quad ...(3)$$

From (2) it follows

$$h\nu = \Delta E = \frac{h^2}{4\pi^2 I}$$

or

$$\nu = \frac{h}{4\pi^2 I}$$

In terms of the wave numbers,

$$\bar{\nu} = \frac{\nu}{c} = \frac{h}{4\pi^2 Ic} \quad ...(4)$$

The **moment of inertia** of the molecule $A - B$ is defined as

$$I = \left(\frac{m_A m_B}{m_A + m_B}\right) r^2$$

where m_A and m_B are the masses of the respective atoms and r is the bond length. The term inside the bracket is called **reduced mass** and is denoted by μ. Thus,

$$I = \mu r^2$$

Substituting the value of I in equation (4),

$$\bar{\nu} = \frac{h}{4\pi^2 c\mu r^2}$$

Hence

$$r = \sqrt{\frac{h}{4\pi^2 c\mu\bar{\nu}}}$$

$\bar{\nu}$ is measured from the experimental spectrum and all other quantities under the square root are known for a given molecule. Hence r can be calculated.

SOLVED PROBLEM 1

The spacing between the lines in the rotational spectrum of HCl is 20.68×10^2 m^{-1}. Calculate the bond length.

SOLUTION

Applying the equation

$$\bar{\nu} = \frac{h}{4\pi^2 Ic}$$

$$20.68 \times 10 = \frac{h}{4\pi^2 Ic} = \frac{55.96 \times 10^{-4}\text{ kg m}}{I}$$

$$I = 2.71 \times 10^{-47}\text{ kg m}^2$$

$$\mu = \frac{m_A m_B}{m_A + m_B}$$

Since m_A and m_B are atomic masses in grams divided by the Avogadro number, we have

$$\mu = \frac{(1.008 \times 10^{-3})\,(35.457 \times 10^{-3})\,/\,(6.02 \times 10^{23})}{(1.008 + 35.457) \times 10^{-3}\,/\,(6.02 \times 10^{23})} = 0.980 \times 10^{-3}\ \text{kg}$$

Hence $r = \sqrt{\frac{I}{\mu}} = \sqrt{\frac{2.71 \times 10^{-47}\ \text{kg m}^2}{0.980 \times 10^{-3}}} = 1.663 \times 10^{-22}\ \text{m}$

$= \mathbf{1.663 \times 10^{-22}\ m}$

VIBRATIONAL SPECTRA

Atoms in a molecule are in constant vibrational motion about mean or equilibrium positions. The modes of vibration may be bond stretching, bond bending, rocking motions and the like. The vibratory motion in a diatomic molecule is illustrated in Fig. 13.18.

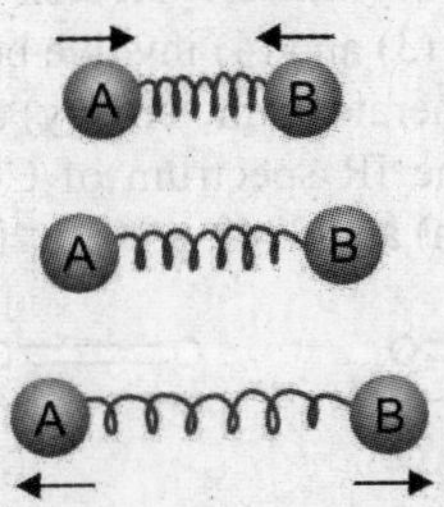

Figure 13.18
Vibratory motion in a diatomic molecule, A–B.

The energy involved in each mode of vibration is quantized and any change in the energy levels produces absorption at particular wavelengths in the infrared spectrum. Thus we can state that : **the vibrational spectra are those caused in the infrared region by the transitions in the vibrational levels in different modes of vibrations.**

The vibrational energy levels of a molecule are given by the relation

$$E_v = \left(v + \frac{1}{2}\right) h\nu_0 \qquad \text{...(1)}$$

where v is the vibrating quantum number and ν_0 is the frequency of the vibration. Thus the energy absorbed in promoting a molecule from its lowest energy level to the next highest, that is from v_0 to v_1 is given by the relation

$$\Delta E = \frac{3}{2} h\nu_0 - \frac{1}{2} h\nu_0 = h\nu_0 \qquad \text{...(2)}$$

Also for a diatomic system as shown in Fig. 13.18. executing a harmonic motion, we have

$$\nu_0 = \frac{1}{2\pi}\sqrt{\frac{k}{\mu}} \qquad \text{...(3)}$$

where μ is reduced mass and k is stretching **force constant.** The force constant is a measure of the strength of the bond between two atoms. Knowing the value of ν_0 from the absorption spectrum, the value of k can be calculated.

Combining equations (3) and (1), the energy difference, ΔE, between two adjacent vibrational levels for a diatomic molecule is given by the equation

$$\Delta E = \frac{h}{2\pi}\sqrt{\frac{k}{\mu}}$$

A molecule whose vibration can cause a change in its dipole moment, absorbs energy in the infrared region. Such a molecule is called **infrared active.** Thus H – Cl molecule possesses a dipole moment (atomic charge × bond distance). Its stretching vibration alter the bond distance and its dipole moment alters. Therefore H – Cl molecule is infrared active. On the other hand, a molecule of hydrogen, H – H, has no dipole moment, nor produces one on vibration. It is infrared inactive.

Most molecules containing three or more atoms have several modes of vibration. Some of these modes are infrared active, while others are not. The infrared active modes only show prominent peaks in the infrared spectrum. For example, CO_2 molecule has four modes of vibration which can be represented as in Fig. 13.19. The mode (1) is called a **symmetrical stretch** which does not involve a dipole moment change. It is IR inactive. The **asymmetric stretch** in mode (2) does involve a dipole moment change and is IR active. The modes (3) and (4) involve bending in the same plane and out of plane respectively. These modes are degenerate (same energy) and cause only one characteristic absorption in the IR spectrum. Actually, the IR spectrum of CO_2 shows prominent absorption frequencies of $667 \times 10^2\ m^{-1}$ (bending motion) and $2349 \times 10^2\ m^{-1}$(asymmetric stretch).

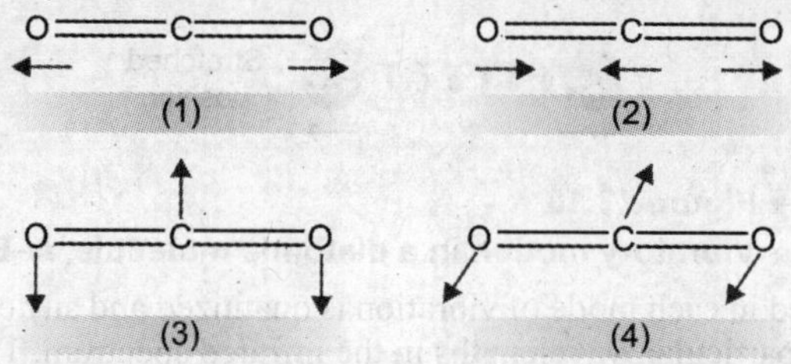

Figure 13.19
The vibrational modes of carbon dioxide molecule.

INFRARED SPECTROSCOPY

An infrared (IR) spectrometer subjects a compound to infrared radiation in the 5000-667 cm^{-1} (2μm) range. Although this radiation is weak, it does supply sufficient energy for bonds in the molecule to vibrate by **Stretching** or **Bending** (Fig. 13.20). The atoms of a molecule can be considered as linked by springs that are set in motion by the application of energy. As the molecule is subjected to the individual wavelengths in the $5000 \times 10^2 - 667 \times 10^2\ m^{-1}$ range, it absorbs only those possessing exactly the energy required to cause a particular vibration. Energy absorptions are recorded as bands (peaks) on chart paper.

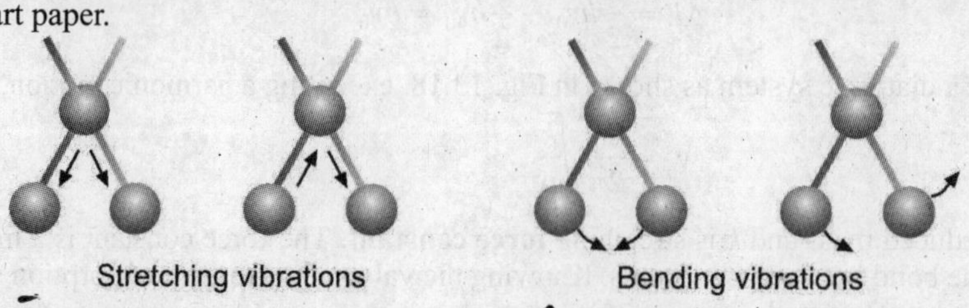

Figure 13.20
Molecular vibrations caused by infrared radiation.

Since different bonds and functional groups absorb at different wavelengths, an infrared spectrum is used to determine the structure of organic molecules. For example, carbon-carbon triple-bond is stronger than a carbon-carbon double bond and requires a shorter wavelength (greater energy) to stretch. The same considerations apply to carbon-oxygen and carbon-nitrogen bonds.

C≡C	C=O	C≡N
210000–220000 m^{-1}	169000–175000 m^{-1}	221000–226000 m^{-1}
C=C	C—O	C—N
162000–168000 m^{-1}	105000–140000 m^{-1}	105000–140000 m^{-1}

Thus, from the position of an absorption peak, one can identify the group that caused it. Fig. 13.21 shows the general areas in which various bonds absorb in the infrared.

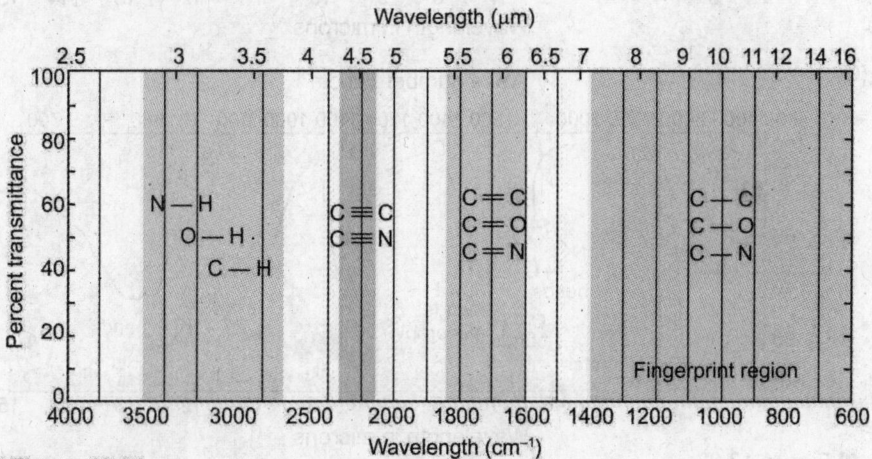

■ **Figure 13.21**
Area of absorption for various bonds in the infrared.

An infrared spectrum is usually studied in two sections :

(1) **Functional Group Region.** The area from 5000 × 10^2 m to 1300 × 10^2 m is called the functional group region. The bands in this region are particularly useful in determining the type of functional groups present in the molecule.

(2) **Fingerprint Region.** The area from 1300 × 10^2 m to 667 × 10^2 m is called the fingerprint region. A peak-by-peak match of an unknown spectrum with the spectrum of the suspected compound in this region can be used, much like a fingerprint, to confirm its identity. Table 9.1 shows some characteristic infrared absorption bands. Fig. 13.22 shows some examples of infrared spectra.

(A)

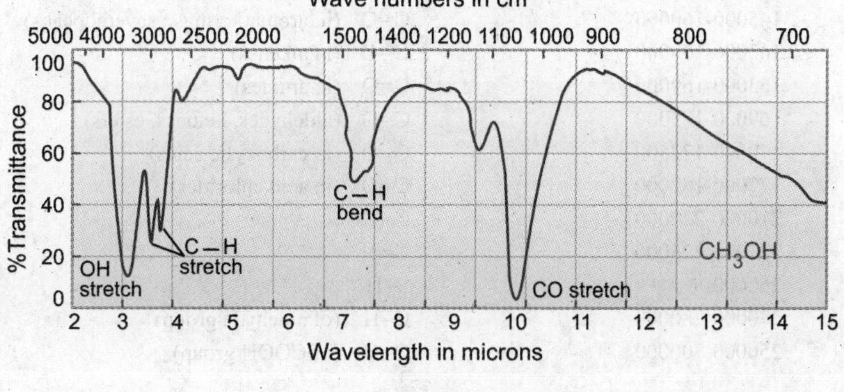

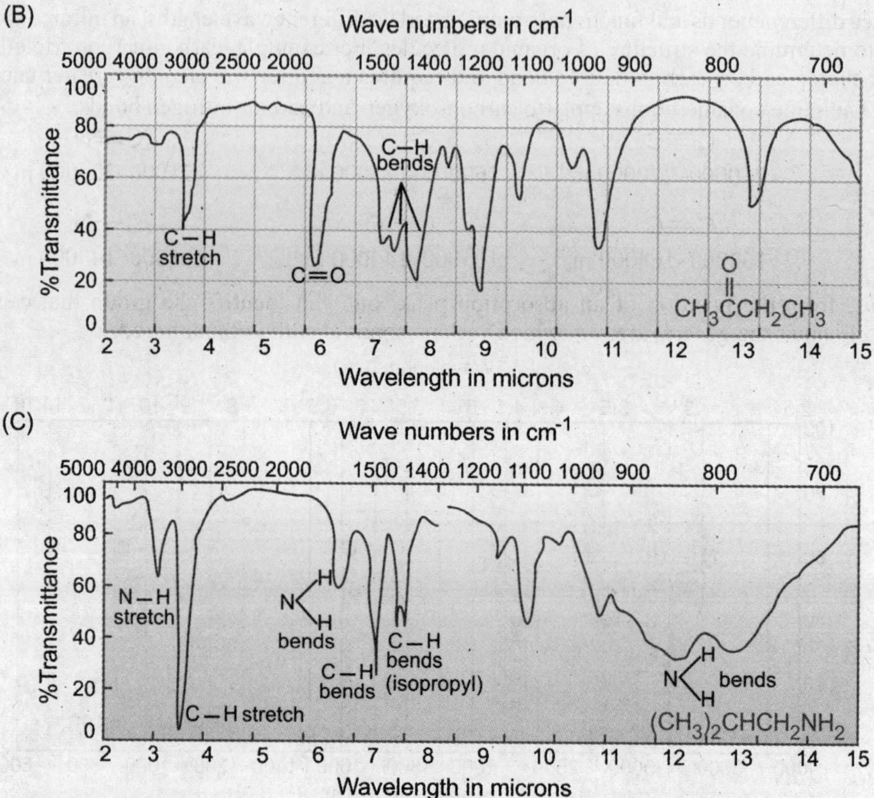

■ **Figure 13.22**

Some examples of IR spectra. (A) is Methanol. (B) is Butanone. (C) is Isobutylamine.

TABLE 13.4 : SOME CHARACTERISTIC IR ABSORPTION BANDS

Range in m^{-1}	Bond (Remarks)
105000-140000	C–O (in ethers, alcohols, esters)
105000-140000	C–N (in amines)
131500-147500	C–H (in alkanes)
134000-150000	NO_2 (two peaks)
145000-160000	C=C (in aromatic rings ; several peaks)
162000-168000	C=C (in alkenes)
163000-169000	C=O (in amides)
169000-175000	C=O (aldehydes, ketones, esters)
170000-172500	C=O (in carboxylic acids)
177000-182000	C=O (in acid chlorides)
210000-220000	C≡C
221000-226000	C≡N
250000	S–H
270000-280000	C–H (of aldehyde group)
250000-300000	O–H (of COOH group)

300000-310000	C–H	(C is part of aromatic ring)
333000	C–H	(C is part of C≡C)
302000-308000	C–H	(C is part of C=C)
280000-300000	C–H	(in alkanes)
330000-350000	N–H	(in amines, amides)
320000-360000	O–H	(in H-bonded ROH)
360000-365000	O–H	

SUMMARY OF IR SPECTROSCOPY

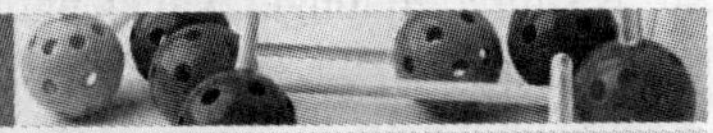

(1) Absorption of infrared radiation causes covalent bonds within the molecule to be promoted from one vibrational energy level to a higher vibrational energy level.

(2) Stronger bonds require greater energy to vibrate (stretch or bend). Therefore, such bonds absorb infrared radiation of shorter wavelengths.

(3) Different functional groups absorb infrared radiation at different wavelengths, and their presence or absence in a molecule can be determined by examination of an IR spectrum.

(4) No two compounds have exactly identical infrared spectra.

ULTRAVIOLET–VISIBLE SPECTROSCOPY

In ultraviolet-visible (UV-Vis) spectroscopy, the $200 \times 10^{-9} - 750 \times 10^{-9}$ m region of the ultraviolet spectrum is used. This includes both the visible region ($400 \times 10^{-9} - 750 \times 10^{-9}$ m) and near ultraviolet region ($200 \times 10^{-9} - 400 \times 10^{-9}$ m). Radiation of these wavelengths is sufficiently energetic to cause the promotion of loosely held electrons, such as nonbonding electrons or electrons involved in a π-bond to higher energy levels. For absorption in this particular region of ultraviolet spectrum, the molecule must contain conjugated double bonds. If the conjugation is extensive, the molecule will absorb in the visible region.

The ultraviolet-visible spectrum is composed of only a few broad bands of absorption (Fig. 13.23). The wavelength of maximum absorbance is referred to as λ_{max}. The following points should be kept in mind while interpreting a UV-Vis spectrum.

(1) Nonconjugated alkenes show an intense absorption below 200×10^{-9} m and is therefore inaccessible to most commonly used UV spectrometers. For example, ethylene has $\lambda_{max} = 171 \times 10^{-9}$ m. This absorption comes from the light-induced promotion of a π-electron to the next higher energy level.

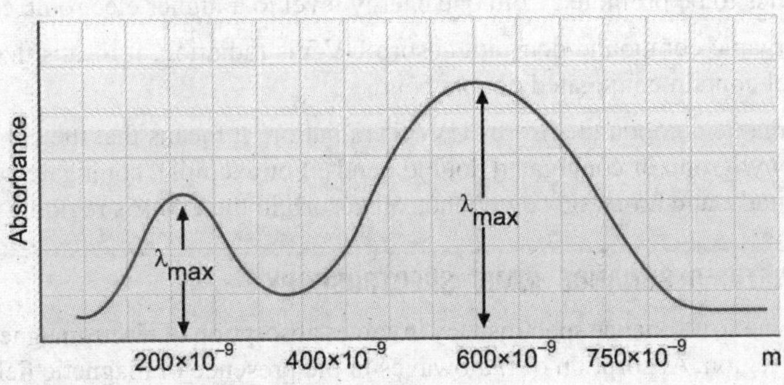

Figure 13.23
A sample UV-Vis spectrum.

(2) Nonconjugated carbonyl compounds have a very weak absorption band in the $200 \times 10^{-9} - 300 \times 10^{-9}$ m region. This band arises from excitation of one of the nonbonding electrons (from an unshared pair) to the next higher energy level. For example,

$CH_3-\overset{\overset{\displaystyle O}{\|}}{C}-CH_3$

Acetone
279 nm

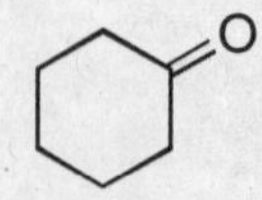

Cyclohexanone
291 nm

(3) When a molecule contains two or more nonconjugated carbon-carbon double bonds, the UV spectrum is that expected of a simple alkene. However, when the double bonds are conjugated, λ_{max} is shifted to longer wavelengths. For example,

$CH_2{=}CH-CH_2-CH_2-CH{=}CH_2$
1,5-Hexadiene
178×10^{-9} m

$CH_3CH{=}CH-CH{=}CHCH_3$
2,4-Hexadiene
227×10^{-9} m

(4) Conjugation of a carbon-carbon double bond and a carbonyl group shifts the λ_{max} of both groups to longer wavelengths. For example,

$(CH_3)_2C{=}CH-CH_2CH_3$ 180×10^{-9} m

$(CH_3)_2CH-CH_2-\overset{\overset{\displaystyle O}{\|}}{C}-CH_3$ 283×10^{-9} m

$(CH_3)_2C{=}CH-\overset{\overset{\displaystyle O}{\|}}{C}-CH_3$ 230×10^{-9} m (for C=C)
327×10^{-9} m (for C=O)

(5) As the number of double bonds in conjugation increases, λ_{max} also increases. For example,

$CH_2{=}CH-CH{=}CH_2$
1,3-Butadiene
215×10^{-9} m

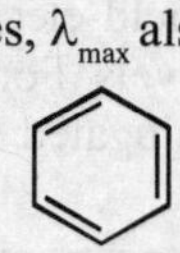

Benzene
257×10^{-9} m

SUMMARY OF UV-VIS SPECTROSCOPY

(1) Absorption of ultraviolet-visible radiation (200×10^{-9}-700×10^{-9} m) causes electrons within molecules to be promoted from one energy level to a higher electronic energy level.

(2) If an organic compound does not absorb UV-Vis radiation, it means that the compound does not contain conjugated double bonds.

(3) If an organic compound absorbs UV-Vis radiation, it means that the compound contains a carbonyl group or conjugated double bonds. For example, conjugated dienes, carbonyl compounds, and aromatic compounds all absorb in the UV-Vis region.

NUCLEAR MAGNETIC RESONANCE (NMR) SPECTROSCOPY

Nuclear magnetic resonance spectroscopy involves absorption of electromagnetic radiation in the radio frequency region. Absorption of radiowaves in the presence of magnetic field is accompanied by a special type of nuclear transition, and for this reason we call this type of spectroscopy nuclear magnetic resonance (NMR) spectroscopy.

The nuclei of certain atoms behave as if they are spinning charges. Any spinning charge creates a magnetic field and behaves as if it were a tiny bar magnet. Of the three nuclei most common in organic compounds (^{1}H, ^{12}C, and ^{16}O)only the hydrogen nucleus proton behaves in this manner.

When a proton in an organic molecule is placed in a strong magnetic field, it can align with the field or against it. In the more stable low-energy state, it is aligned *with* the magnetic field (Fig. 13.24). If energy is supplied in the form of radiowaves of exactly the right frequency, radiation will be absorbed and the nucleus will "flip" and align *against* the applied magnetic field in the higher energy state. For example, a proton in an applied field of 14000 gauss or 1.4 Tesla requires a frequency of 60 million hertz (cycles per second) for this particular transition. When this exact frequency is applied, it is absorbed, and the absorption recorded on a chart paper. In practice, either the magnetic field can be held constant and the radio frequency varied, or more commonly, the radio frequency can be held constant and the magnetic field varied. Fig. 13.25 shows the various components of a NMR spectrometer.

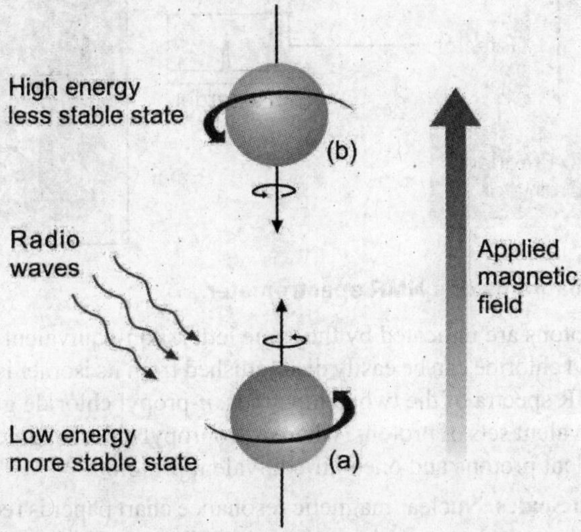

■ Figure 13.24
Orientation of a spinning hydrogen nucleus in an applied magnetic field. (*a*) shows the nucleus in more stable low-energy state. (*b*) shows the nucleus in less stable high-energy state. Applying radiowaves *flips* the nucleus from the more stable 'aligned' state (*a*) to less stable 'nonaligned' state (*b*).

Position of Signals ; Chemical Shifts. The main purpose of NMR is not to detect the presence of protons in a molecule. It can distinguish between protons in different chemical environments within the molecule. Protons on benzene ring, or on a carbon bearing a chlorine, or on a carbon adjacent to a carbonyl group absorb radio frequency energies at different applied magnetic fields, and appear at different locations (chemical shifts) on the recording paper. Also, the position of absorption is relatively constant for protons in a particular chemical or structural environment. Hence, *the number of signals recorded on the NMR chart paper indicates the number of different types of protons in a molecule.* The position of the peak can give information about the molecular structure in the vicinity of the proton.

To understand the application of NMR, we should know what are **Equivalent** and **Non-equivalent protons**. Equivalent protons are protons in the same molecule that are chemically indistinguishable. In other words, equivalent protons are positioned in structurally and chemically equivalent areas in the molecule. As an example, let us consider the following compounds.

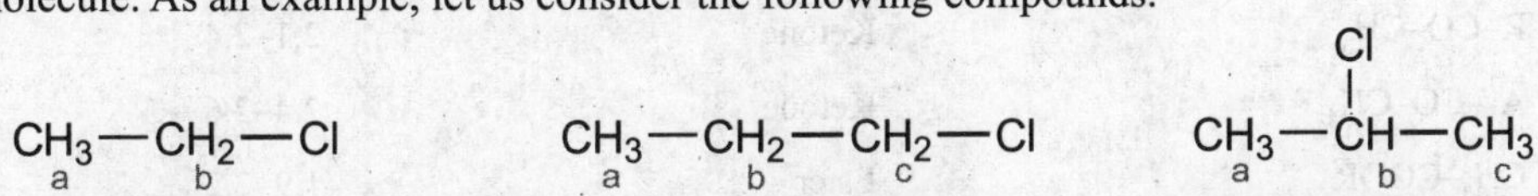

Ethyl chloride (2 Signals) | *n*-Propyl chloride (3 Signals) | Isopropyl chloride (2 Signals)

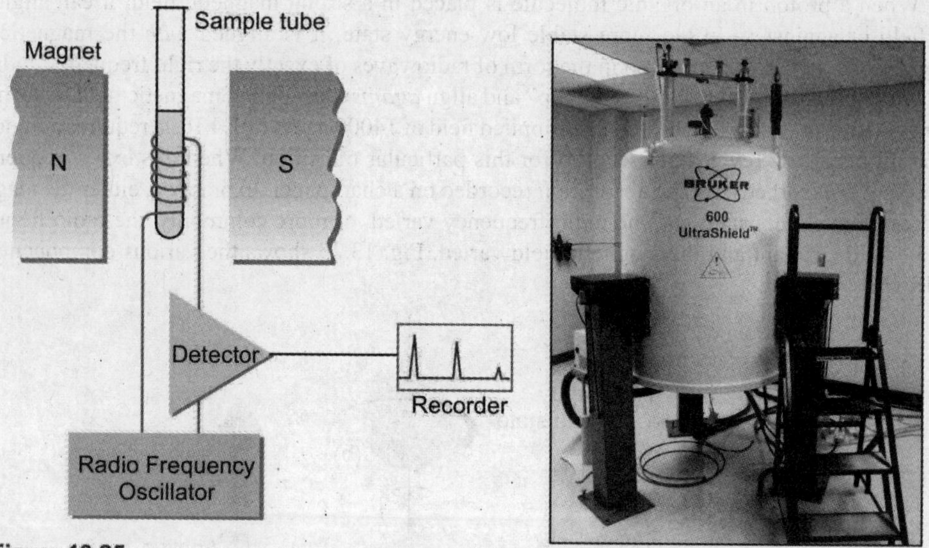

Figure 13.25
Schematic diagrams of a NMR spectrometer.

Equivalent protons are indicated by the same letter, non-equivalent protons by different letters. Notice that *n*-propyl chloride can be easily distinguished from its isomer isopropyl chloride, by simply examining the NMR spectra of the two compounds. *n*-propyl chloride gives three NMR signals and has three non-equivalent sets of protons, whereas isopropyl chloride gives two NMR signals and has two sets of equivalent protons and one non-equivalent proton.

Delta and Tau Scales. Nuclear magnetic resonance chart paper is rectangular with a linear scale of δ (*delta*) units across the bottom. Most chart papers have scales from $\delta = 0$ to $\delta = 8$ or 9 ppm (parts per million). To every sample a small amount of **Tetramethylsilane** (TMS), $(CH_3)_4Si$, is added as a reference (the TMS signal is at $\delta = 0$). All other signals are relative to TMS. Sometimes τ (*tau*) scale is used where the TMS signal is at 10.0 ppm. Hence $\tau = 10 - \delta$.

Table 13.5 lists some characteristic chemical shifts (from TMS) of different types of protons.

Table 13.5 : SOME TYPICAL NMR CHEMICAL SHIFTS OF PROTONS

Type of Proton	Type of Compound	δ Value
$R{-}CH_3$	Alkane	0.8–1.2
$Ar{-}CH_3$	Arene	2.2–2.5
$C{=}C{-}CH_3$	Alkene	1.6–1.9
$R{-}O{-}CH_3$	Ether	3.2–3.3
$Ar{-}O{-}CH_3$	Ether	3.7–4.0
$R{-}CO{-}CH_3$	Ketone	2.1–2.4
$Ar{-}CO{-}CH_3$	Ketone	2.4–2.6
$CH_3{-}COOR$	Ester	1.9–2.2
$R{-}COOCH_3$	Ester	3.0–3.9

$CH_3–N<$	Amine	2.2–2.6
$R–CH_2–R$	Alkane	1.1–1.5
$R–CH_2–Ar$	Arene	2.5–2.9
$R–CH_2–X$	Alkyl halide	2.7–4.5
$R–CH_2–NO_2$	Nitroalkane	4.4–4.6
$R–CH_2–OH$	Alcohol	3.4–4.0
$Ar–CH_2–OH$	Alcohol	4.0–5.0
$R–CH_2–COOH$	Acid	2.0–2.6
$R_3C–H$	Alkane	1.4–1.6
Ar–H	Arene	6.0–7.5
C=C–H	Alkene	4.5–6.0
C≡C–H	Alkyne	1.7–1.9
R–CHO	Aldehyde	9.0–10.0
R–COOH	Acid	10–12
R–OH	Alcohol	2–6
Ar–OH	Phenol	6–8
$R–NH_2$	Amine	1–5

As an example, let us consider the NMR spectrum of benzyl alcohol (Fig. 13.26A). Notice that there are four signals in this spectrum.

(1) The signal on the far right at δ = 0 is that of the TMS reference.

(2) The signal at δ = 7.3 is a single sharp peak due to the five chemically equivalent phenyl protons.

(3) The signal at δ = 4.6 is a single sharp peak due to the two chemically equivalent methylene ($–CH_2–$) protons.

(4) The signal at δ = 2.4 is a single sharp peak due to the hydroxy (–OH) proton.

(A)

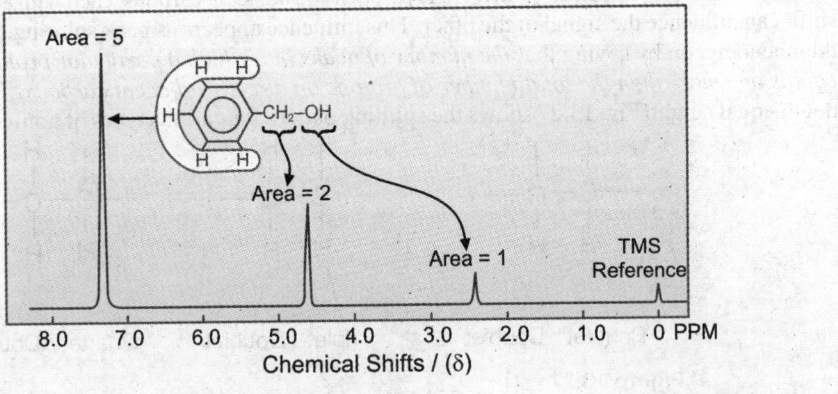

(B)

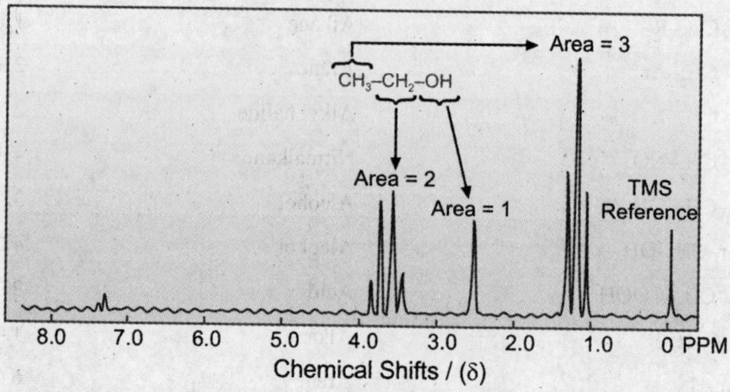

(C)

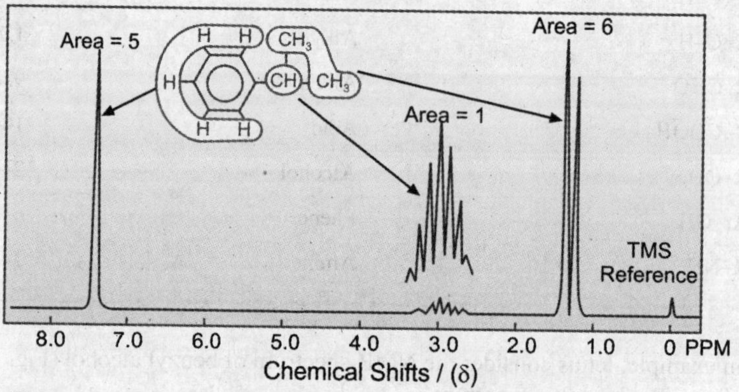

■ Figure 13.26

Some examples of NMR spectra. (A) is Benzyl alcohol. (B) is Ethyl alcohol. (C) is Isopropylbenzene.

Peak Area : Integration. The relative areas under the various peaks of a NMR spectrum are in proportion to the number of protons contributing to each signal. These areas can be electronically integrated by a NMR spectrometer. Comparison of the areas provides the ratio among the various kinds of protons in the molecule. For example, consider the NMR spectrum of benzyl alcohol (Fig. 13.26A). The hydrogens in the molecule are in a 1 : 2 : 5 ratio, like the corresponding areas in the spectrum.

Peak Splitting : Splitting Pattern. Hydrogens on adjacent carbons, each with a different chemical shift, can influence the signal of the other. This influence appears as peak splitting. We can generalize the phenomenon by saying that *the number of peaks into which a particular proton's signal is split, equals one more than the total number of protons on directly adjacent carbons*. The split peaks are not of equal height. Fig. 13.27 shows the splitting pattern for various types of non-equivalent protons.

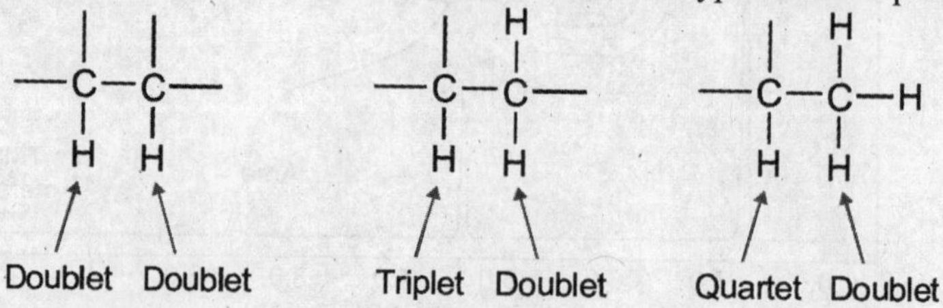

■ Figure 13.27

Splitting pattern for various types of non-equivalent protons.

MAGNETIC RESONANCE IMAGING (MRI)

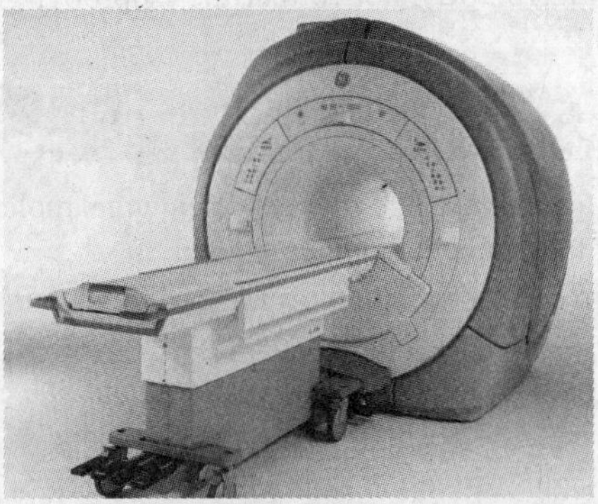

MRI is a diagnostic technique that generates images of the internal body. It specializes in creating a thin-section image of any part of the body – especially the heart, veins, arteries, brain and central nervous system – from any angle or direction.

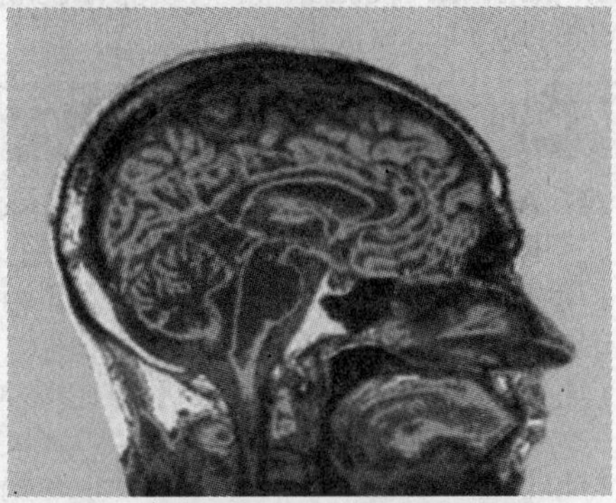

This MRI scan of a normal adult head shows the brain, airways, and soft tissues of the face. The large cerebral cortex, appearing in yellow and green, forms the bulk of the brain tissue; the circular cerebellum, center left, in red, and the elongated brainstem, center, in red, are also prominently shown.

Magnetic Resonance Imaging (MRI), medical diagnostic technique that combines strong magnetic fields, radio waves, and computer technology to create images of the body using the principles of *nuclear magnetic resonance*. A versatile, powerful, and sensitive tool, MRI can generate thin-section computerized images of any part of the body–including the heart, arteries, and veins–from any angle and direction, without surgical invasion and in a relatively short period of time. MRI also creates "maps" of biochemical compounds within any cross section of the human body. These maps give basic biomedical and anatomical information that provides new knowledge and may allow early diagnosis of many diseases. In 2003 Paul Lauterbur of the United States and Sir Peter Mansfield of the United Kingdom shared the Nobel Prize in physiology or medicine for their contributions to MRI technology.

MRI is possible in the human body because the body is filled with small biological "magnets," the most abundant and responsive of which is the proton, the nucleus of the hydrogen atom. The principles of MRI take advantage of the random distribution of protons, which possess fundamental magnetic properties. Once the patient is placed in the cylindrical magnet, the diagnostic process follows three basic steps. First, MRI creates a steady state within the body by placing the body in a steady magnetic field that is 30,000 times stronger than Earth's magnetic field. Then MRI stimulates the body with radio waves to change the steady-state orientation of protons. It then stops the radio waves and "listens" to the body's electromagnetic transmissions at a selected frequency. The transmitted signal is used to construct internal images of the body using principles similar to those developed for computerized axial tomography, or CAT scanners.

In current medical practice, MRI is preferred for diagnosing most diseases of the brain and central nervous system. MRI scanners provide equivalent resolution and superior contrast resolution to that of X-Ray CAT scanners. MRI scanners also provide imaging complementary to X-Ray images because MRI can distinguish soft tissue in both normal and diseased states.

In the NMR spectrum of ethyl alcohol (Fig. 13.26 B), notice that the ethyl group is indicated by a quartet and a triplet.

In the NMR spectrum of isopropyl benzene (Fig.13.26C), notice that the isopropyl group is indicated by a heptet and a doublet.

SUMMARY OF NMR SPECTROSCOPY

(1) Absorption of radiowaves in the presence of a magnetic field causes nuclei within molecules to be promoted from one spin energy level to a higher spin energy level.

(2) The number of signals in the NMR spectrum corresponds to the number of different types of protons in the molecule.

(3) The position (chemical shift) of each signal gives information about the structural environment of the protons.

(4) The relative areas (Integration) under the signals give the ratio of the numbers of each type of proton in the molecule. If the molecular formula is known, the actual number of each type of proton can be determined.

(5) The splitting pattern of each signal gives us the number of protons on neighboring carbons. The number of peaks into which a signal is split is one more than the total number of protons on directly adjacent carbons.

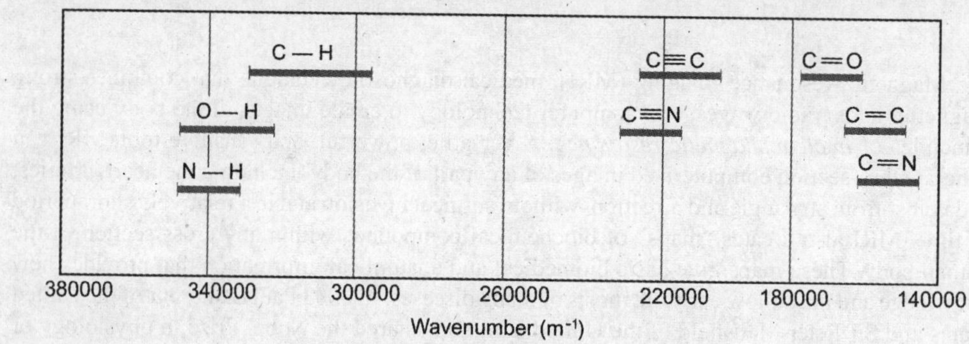

Group Assignments in Infrared Spectra

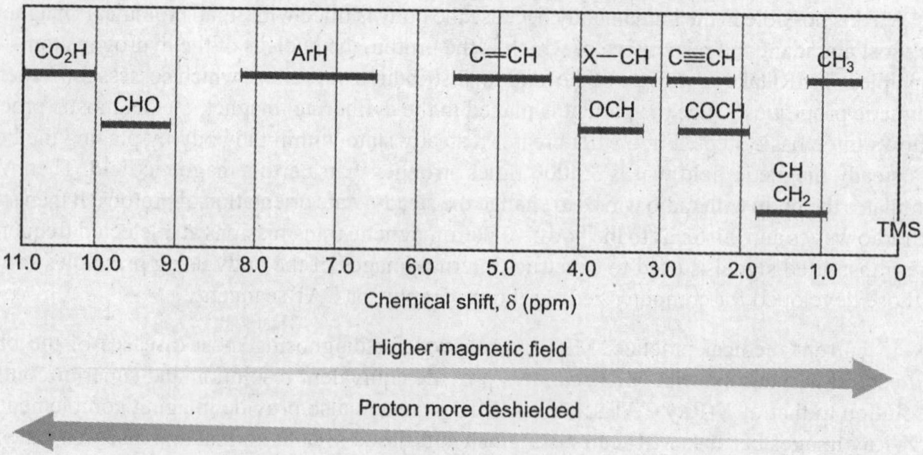

Group Assignments in NMR Spectra

IMPORTANT TERMS USED IN NMR SPECTROSCOPY

Chemical shift : The separation between the peak of the reference standard (TMS) and any other peak in an nmr spectrum.

δ scale : An nmr scale in which the TMS peak is set at zero and peaks to the left of it (downfield) have increasingly positive values, in parts per million (ppm).

TMS : Abbreviation for tetramethylsilane, $(CH_3)_4Si$, the reference standard in nmr spectrometry.

Downfield shift : The protons of most organic compounds are less shielded than those of TMS and therefore require less energy for resonance to take place. The peaks appear therefore at lower fields, or *downfield* of the TMS reference peak.

Spin-spin coupling : The splitting of a signal into two or more peaks because of magnetic interactions between neighboring protons.

$(n + 1)$ rule : A rule that allows us to predict into how many smaller peaks a given signal is split by neighboring protons. The rule states that a signal is split into $(n + 1)$ peaks, where n represents the number of equivalent *next-door protons*.

MASS SPECTROSCOPY

In a mass spectrometer, organic molecules are vaporized and bombarded with a beam of very high-energy electrons (Fig. 13.28). The resulting collisions impart considerable energy to the molecule, which in turn emit electrons to produce positively charged ions. These ions possess so much energy that they often fragment through various bond cleavages to produce new positively charged ions.

$$\mathrm{ABC} \xrightarrow[\text{beam}]{\text{electron}} \begin{array}{l} \mathrm{ABC^{+}} \quad \text{(Molecular ion)} \\ \mathrm{AB^{+}} \quad \mathrm{C^{+}} \\ \mathrm{A^{+}} \quad \mathrm{BC^{+}} \end{array}$$

The positive ions are accelerated toward a negatively charged plate by passing them through a magnetic field, which deflects the ions. The ions of lighter mass are deflected more than heavier ions. Each kind of ion has a particular mass to charge (m/e) ratio. For most ions produced in the fragmentation of the molecule, the charge is +1 so, that m/e usually represents the mass of the ion.

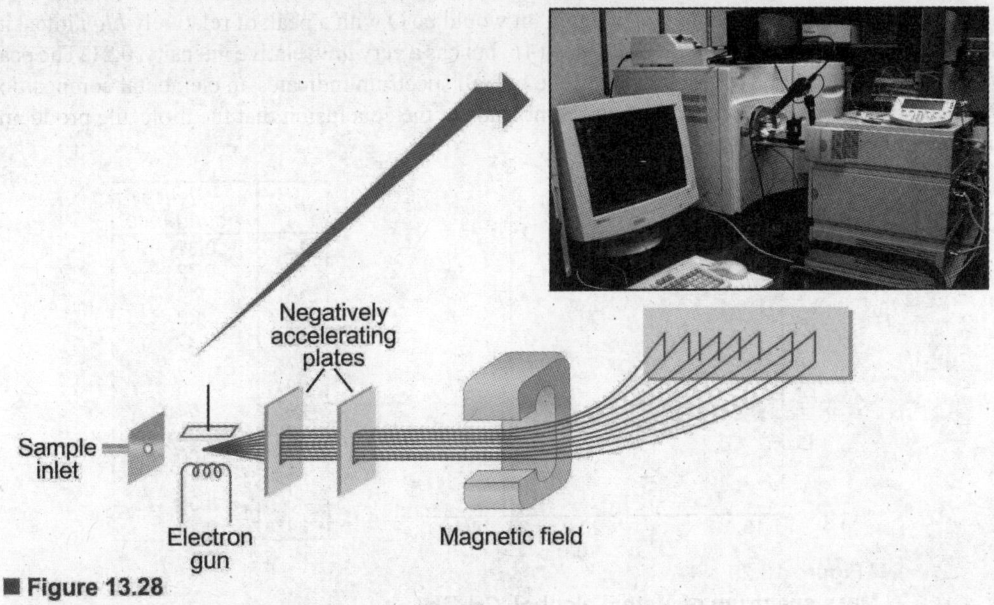

Figure 13.28
Schematic diagram of a Mass spectrometer.

Base Peak. The set of ions produced from a molecule can be analyzed since each has its own *m/e* ratio, and produces a signal whose intensity is due to the relative abundance of that ion. *The largest peak found in a mass spectrum (that of highest intensity) is called the base peak,* and is given the numerical value of 100. The intensities of all other peaks are expressed relative to the height of the base peak. A mass spectrum is highly characteristic of a particular compound.

Molecular Ion. *The ion formed by removing one electron from the parent molecule is called molecular ion or parent ion.* The molecular ion peak is usually represented as M^+. It may or may not be the peak of highest intensity. The molecular ion is the most important ion, since its mass is the **Molecular Weight** of the parent molecule.

The use of mass spectroscopy is not limited to the determination of the molecular weight of compounds. It also has great utility in the elucidation of the structure of a molecule. The fragmentation processes of organic molecules follow certain patterns in that the fragments broken off are related to bond strengths and the stability of the species formed. Since certain structural features in molecule produce definite characteristic fragmentation patterns, the identification of these fragments and their relative intensities in a mass spectrum can be used to determine the structure of organic compounds. The following examples illustrate how a mass spectrum can be used to determine the structure of organic molecules :

Example 1. Let us consider the following isomers.

$$CH_3CH_2CH_2CH_2CH_2CH_2CH_2CH_3 \quad n\text{-Octane}$$

$$CH_3CH_2CH_2\underset{\displaystyle |}{\overset{\displaystyle CH_3}{C}}HCH_2CH_2CH_3 \quad \text{4-Methylheptane}$$

n-Octane can be easily distinguished from 4-methylheptane, since the methyl group branched at C-4 in the latter compound, is easily broken off as CH_3^+ fragment and will give a peak of large intensity at *m/e* = 15

Example 2. Consider the mass spectrum shown in Fig. 13.29. The fragmentation pattern and relative abundance of peaks is as follows. A molecular weight of 32, which corresponds to O_2, is ruled out, since the only other possible fragment would be O with a peak of relatively *high* intensity at *m/e* = 16. An *m/e* peak is present at a value of 16, but has a very *low* relative intensity, 0.21. The peak at *m/e* = 15 is due to a CH_3^+ fragment, and the overall spectrum indicates an elemental composition of CH_4O. Consideration of other peaks present leads to the conclusion that the molecule producing this spectrum is methyl alcohol (CH_3OH).

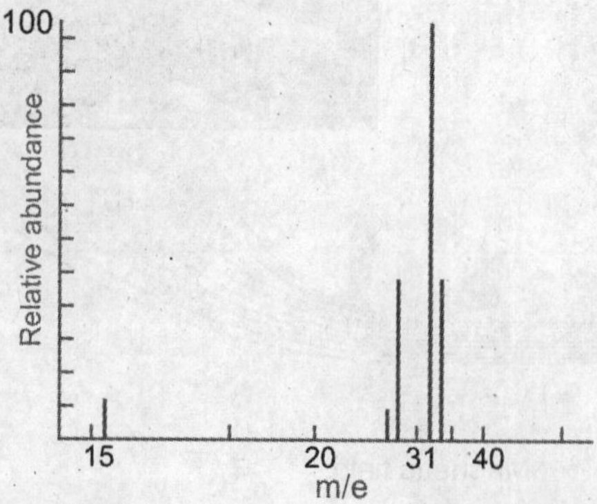

m/e	Relative abundance
12	0.32
13	0.72
14	2.4
15	1.3
16	0.21
17	1.0
28	6.3
29	64
30	3.8
31	100
32	66
33	0.98
34	0.14

■ **Figure 13.29**
Mass spectrum of Methyl alcohol, CH_3OH

SUMMARY OF MASS SPECTROSCOPY

(1) Mass spectroscopy involves organic molecules being bombarded by a very high-energy electron beam.

(2) The peak of highest intensity in a mass spectrum is referred to as the base peak.

(3) Fragmentation processes can produce numerous fragments, from which the structures of organic molecules can be deduced.

(4) When one electron is removed from a molecule, a molecular ion is produced. The *m/e* value of the molecular ion peak is the molecular weight of the compound being investigated. The molecular weights obtained by mass spectroscopy are extremely accurate.

RAMAN SPECTRA

When an intense beam of light is passed through a liquid or gas, some of the light is scattered by the molecules of the medium. Most of the scattered light has the same frequency as the incident light. This type of scattering which occurs without a change in the incident frequency is termed **Rayleigh scattering.** In 1928, Indian physicist Sir C.V. Raman discovered that a very small fraction of the scattered light has frequencies higher or lower than the incident frequency. The scattering which occurs with a change in the frequency of the incident radiation is called **Raman scattering.** The phenomenon is called **Raman effect.** The spectrum obtained from scattered radiations of a sample can be observed with the help of the apparatus shown in Fig. 13.30. Laser beam is now used as incident radiation. Raman originally used ultraviolet light which gave weak spectral lines.

The general pattern of spectrum recorded by scattered light is illustrated in Fig. 13.31. The sharp line in the middle corresponds to Rayleigh scattered light and is referred to as **Rayleigh line.** Raman scattered light records a series of lines on either side of the Rayleigh line. These are called **Raman lines** and constitute **Raman spectrum.**

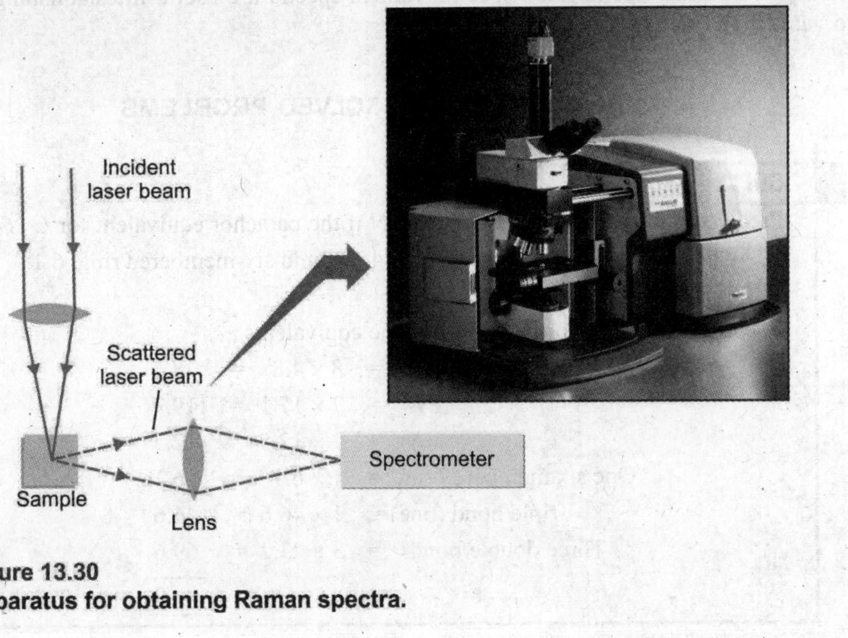

■ **Figure 13.30**
Apparatus for obtaining Raman spectra.

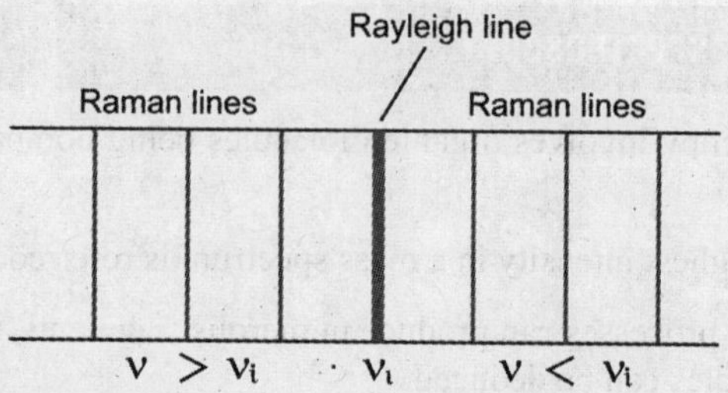

■ Figure 13.31
Raman spectrum (an illustration).

Explanation of Raman Spectrum

The change in frequency by scattering occurs due to the exchange of energy between the incident photon and the scattering molecule. If the photon collides with an 'excited' molecule, it acquires energy from the molecule. Thus the scattered light emerges with a higher frequency. If the photon collides with a molecule in the 'ground state', it loses energy to the molecule. This lowers the frequency of the scattered light. It explains why Raman spectrum of a substance shows lines of higher frequency and others of lower frequency than the incident line. The frequency difference (ΔV) between each Raman line and the incident line, is called the **Raman frequency** (v_{Raman}). The Raman frequencies shown by a substance are independent of the incident frequency and are characteristic of the substance.

Raman Spectra and Molecular Structure

Raman spectra provide valuable information about the shape and structure of molecules. When studied together, Raman and *IR* spectra can tell us whether : (*i*) a molecule is linear or non-linear; (*ii*) a molecule is symmetrical or asymmetrical.

The most important application of Raman spectra is in the determination of structure of organic molecules. Definite frequencies are associated with different types of bonds. The Raman frequency for C–H bond is about 280000 m^{-1}, for C ≡ C bond about 210000 m^{-1} and for C = C bond about 160000 m^{-1}. Thus like infrared spectra; Raman spectra are useful in establishing the structure of organic molecules.

ADDITIONAL SOLVED PROBLEMS

SOLVED PROBLEM 2

Predict the parachor for $CH_3–C_6H_4–CN$ if the parachor equivalent for *C* is 4.8 ; *H*, 17.1 ; *N*, 12.5 ; double bond, 23.2 ; triple Bond, 46.6 ; and six-membered ring, 6.1.

SOLUTION

The predicted parachor is the sum of the equivalents *i.e.*

8 *C*	=	8 × 4.8	=	38.4
7 *H*	=	7 × 17.1	=	119.7
1 *N*	=	1 × 12.5	=	12.5
One six membered ring	=	1 × 6.1	=	6.1
Triple bond (one)	=	1 × 46.6	=	46.6
Three double bonds	=	3 × 23.2	=	69.6
			Total =	**292.9**

SOLVED PROBLEM 3

Parachor of decane is found to be 424.2. Each $-CH_2-$ group has a parachor value of 39. Find out the parachor of hydrogen and carbon.

SOLUTION

Formula used

Parachor of decane = 2 × Parachor of CH_3- + 8 × Parachor of CH_2-

Quantities given

Parachor of decane = 424.2   *Parachor of* $-CH_2-$ = 39

Substitution of values

$$424.2 = 2 \times \text{Parachor of } CH_3- + 8 \times 39$$

or $$2 \times \text{Parachor of } CH_3- = 424.2 - 8 \times 39$$

or $$\text{Parachor of } CH_3- = \frac{112.2}{2} = 56.1$$

$$\text{Parachor of H} = \text{Parachor of } CH_3- \quad \text{Parachor of } -CH_2-$$
$$= 56.1 - 39 = \mathbf{17.1}$$

$$\text{Parachor of C} = \text{Parachor of } CH_3- - 3 \times \text{Parachor of C}$$
$$= 56.1 - 3 \times 17.1$$
$$= \mathbf{4.8}$$

SOLVED PROBLEM 4

The density of hexane at 298 K is 687.4 kg/m^3. Calculate the surface tension of hexane if the parachors of ethane and propane are 110.5 and 150.8 respectively.

SOLUTION

(i) To calculate the parachor of C_6H_{14}

$$[P]_{C_2H_6} = 2 \times [P]_{CH_3} = 110.5$$

$$[P]_{C_3H_8} = 2 \times [P]_{CH_3} + [P]_{CH_2} = 150.8$$

$$\therefore \quad [P]_{CH_2} = [P]_{C_3H_8} - [P]_{C_2H_6}$$
$$= 150.8 - 110.5 = 40.3$$

$$[P]_{C_6H_{14}} = [P]_{C_3H_8} + 3 \times [P]_{CH_2}$$
$$= 150.8 + 3 \times 40.3$$
$$= 271.7$$

Molar mass of Hexane = 86

(ii) To calculate the surface tension of C_6H_{14}

Formula used

$$\frac{M}{D}\gamma^{1/4} = [P]$$

Quantities given

$[P]$ = 271.7 M = 86 D = 687.4 kg/m^3

Substitution of values

$$\frac{86 \times 10^{-3}}{0.6874} \times \gamma^{1/4} = 271.7$$

or $\gamma^{1/4} = \dfrac{271.7 \times 687.4}{86 \times 10^{-3}}$ dyne cm^{-1}

$\gamma^{1/4} = 2.172$ dynes cm^{-1}

and $\gamma = (2.172)^4$ dynes cm^{-1}

$= \mathbf{22.25}$ **dyne cm^{-1} = 0.02225 N/m**

SOLVED PROBLEM 5

The density of paraldehyde is 994.3 kg/m^3. Calculate the parachor if the surface tension is 0.0259 N/m. The molar mass of paraldehyde is 132.16 × 10^{-3} kg mol^{-1}.

SOLUTION

Formula used

$$\frac{M}{D}\,\gamma^{1/4} = [P]$$

Quantities given

$M = 132.16 \times 10^{-3}$ kg/mol $D = 994.3$ kg/m^3 $\gamma = 0.0259$ dyne N/m

Substitution of values

$$[P] = \frac{132.16 \times 10^{-3}}{994.3} \times (25.9)^{1/4}$$

$$= 132.92 \times 2.256$$

$$= \mathbf{299.87}$$

EXAMINATION QUESTIONS

1. Define or explain the following terms :
   (*a*) Surface tension (*b*) Parachor
   (*c*) Viscosity (*d*) Rheochor
   (*e*) Dipole moment (*f*) Molar Refraction
   (*g*) Diamagnetic substances (*h*) Paramagnetic substances
   (*i*) Electromagnetic spectrum (*j*) Fingerprint region
   (*k*) Chemical shifts (*l*) TMS
2. (*a*) Define dipole moment. How can it be used in elucidation of molecular structure?
   (*b*) Discuss a method for determining magnetic susceptibility and discuss the magnetic behavior of $K_3[Fe(CN)_6]$ (*Jamia Millia BSc*)
3. (*a*) Calculate the normal modes of vibration in CO_2 and CH_4 molecules.
   (*b*) Differentiate between permanent and induced electric dipole moments. (*Panjab BSc*)
4. Calculate the percentage ionic character of HBr from the following data :
   Dipole moment of HBr = 2.654 × 10^{-30} cm
   Bond length of HBr = 1.41 × 10^{-10} m
   Theoretically expected dipole moment of HBr = 2.6544 × 10^{-30} cm
   Answer. 11.6% (*Mizoram BSc*)
5. Define optical density and molar extinction coefficient. On which the magnitude of these quantities depend? (*Guru Nanak Dev BSc*)

6. (*a*) Explain the terms (*i*) Specific refractivity and (*ii*) Molar refractivity.
   (*b*) Discuss the use of dipole moment measurements in the study of structure of compounds. (*Mizoram BSc*)
7. Write the expression for the vibrational energy of a diatomic molecule. What are the selection rules for the vibrational transition in a diatomic molecule and what type of vibrational spectrum is expected for a diatomic molecule taking it as simple harmonic oscillator? (*Panjab BSc*)
8. How many fundamental vibrational frequencies would you expect for CO_2? How do you represent these modes? (*Sambalpur BSc*)
9. Calculate the molar refractivity of allyl chloride at a temperature at which its density is 938 kg/m³. The experimentally observed value of refractive index at this temperature is 1.3715.
   Answer. 0.01867 (*Delhi BSc*)
10. Define specific and molar refraction. (*Kerala BSc*)
11. Define molecular refraction. Briefly describe the principle, construction and working of the instrument used in determining the molecular refraction of a liquid. (*Sri Venkateswara BSc*)
12. Discuss the principle and applications of NMR spectroscopy in structure determination of organic compounds. (*HS Gaur BSc*)
13. What is meant by the term Chemical Shift of a particular proton in NMR spectroscopy ? Draw the splitting patterns of protons in NMR spectrum of ethyl alcohol. (*Kalyani BSc*)
14. Give an account of pure rotational spectra. How force constant and bond length can be calculated for simple diatomic molecule? (*Jadavpur BSc*)
15. (*a*) Describe various normal modes of vibration pictorially for water molecules.
    (*b*) Define stretching force constant. How does it change from fluorine to iodine molecule?
    (*c*) Write briefly applications of NMR spectroscopy. (*Punjabi BSc, 2004*)
16. How will you distinguish between benzene and anthracene by UV spectroscopy? (*Bharathidasan BSc*)
    Answer. Anthracene will have higher λ_{max} than benzene. This is because anthracene is more conjugated than benzene.
17. Define dipole moment and discuss its ability in determining the molecular structure. (*Allahabad BSc*)
18. Two isomeric dienes (A) and (B) having the molecular formula C_5H_8 absorb at λ_{max} 223 × 10^{-9} m and λ_{max} 178 × 10^{-9} m respectively. Write the structures of the two isomers. (*Madras BSc*)
    Answer. (A) is $CH_2{=}CH{-}CH{=}CH{-}CH_3$; 1,3-Pentadiene
    (B) is $CH_2{=}CH{-}CH_2{-}CH{=}CH_2$; 1,4-Pentadiene
19. Arrange the following compounds in the increasing order of their UV absorption maxima :
    (a) Ethylene (b) Naphthalene
    (c) Anthracene (d) 1,3-Butadiene
    Answer. (a) < (d) < (c) < (b)
20. Discuss the principle of IR spectroscopy. How will you distinguish between the following pairs of compounds on the basis of IR spectroscopy :
    (a) Ethyl alcohol and diethyl ether
    (b) Acetic acid and ethyl acetate (*Jiwaji BSc*)
    Answer. (a) Ethyl alcohol shows a strong/broad absorption at 320000 – 350000 m^{-1} (due to O–H bond) while diethyl ether does not ; and (b) Acetic acid shows a strong/broad absorption at 250000 – 350000 m^{-1} (due to O–H bond) while ethyl acetate does not.
21. How many NMR signals do you expect from each of the following compounds. Indicate also the splitting pattern of the various signals. (*Kalyani BSc*)
    (a) CH_3OCH_3 (b) $CH_3OCH_2CH_3$ (c) CH_3CH_2OH
    Answer. (a) One signal (singlet) ; (b) Three signals (singlet, quartet, triplet) ; and (c) Two signals (singlet, quartet).

22. How many signals (ignoring the splitting patterns) would you see in the NMR spectra of the following compounds? *(Mysore BSc)*

(a) Butanone (b) Cyclobutane (c) p-xylene (d) 2-Propanol

Answer. (a)Three signals ; (b) One signal ; (c) Two signals ; and (d) Three signals.

23. What type of information one can obtain from : (i) IR spectrum ; (ii) UV spectrum ; and (iii) NMR spectrum. *(Kanpur BSc)*

24. What type of molecular transitions are brought about by absorption of :

(a) Ultraviolet-visible radiation

(b) Infrared radiation *(Madurai BSc)*

25. Write notes on :

(a) UV spectroscopy *(Banaras BSc)*

(b) IR spectroscopy *(Agra BSc)*

(c) NMR spectroscopy *(Jadavpur BSc)*

26. (*a*) Describe NMR spectroscopy.

(*b*) How is NMR method used to determine structure of ethyl alcohol? *(Panjab BSc)*

27. The NMR spectrum of a compound C_4H_9Br consists of a single sharp line. What is the structure of the compound? *(Delhi BSc)*

Answer. The compound is *tert*-Butyl bromide, $(CH_3)_3CBr$.

28. Suggest the structure of a compound each with the following NMR spectral features :

(a) An alcohol with two NMR peaks

(b) A compound C_5H_{10} with a single NMR peak

(c) A compound C_4H_6 with a single NMR peak. *(Madras BSc)*

Answer. (a) Methyl alcohol ; (b) Cyclopentane ; (c) 2-Butyne

29. Calculate the percentage ionic character of H – Cl bond if the distance between the two atoms is 1.275×10^{-10} m and its dipole moment is 3.4608×10^{-30} Cm. *(Sambalpur BSc)*

Answer. 16.885%

30. The parachor equivalent for C is 4.8; for H, 17.1 ; and for 0, 20.0 Calculate the parachor for methanol. *(Kalyani BSc)*

Answer. 93.2

31. The parachor equivalents for C is 4.8 ; for H, 17.1 ; and for O, 20.0. Calculate the parachor for ethyl alcohol. *(Mysore BSc)*

Answer. 132.2

32. Two compounds, (A) and (B), have the same molecular formula C_2H_6O. They have different IR spectra. Compound (A) shows a strong/broad absorption at 3400×10^2 m^{-1}, while compound (B) does not. Suggest formulas for (A) and (B) which account for the difference. *(Delhi BSc)*

Answer. (A) is ethyl alcohol (CH_3CH_2OH) ; (B) is dimethyl ether (CH_3OCH_3). The strong absorption at 3400×10^{-2} m^{-1} in (A) is due to the O–H bond.

33. What is meant by Chemical shift? On what factors does it depend? *(Osmania BSc)*

34. How will you distinguish between 1,3-pentadiene and 1,4-pentadiene by UV spectroscopy ? *(Bharthidasan BSc)*

Answer. 1,3-Pentadiene will have higher l_{max} than 1,4-pentadiene. This is because 1,3-Pentadiene is conjugated ; whereas 1,4-pentadiene is not.

35. (*a*) What is meant by molar volume, molar refraction, polarization and dipole moment?

(*b*) Calculate the percent ionic character in HCl molecule, if the observed dipole moment is 3.6288×10^{-30} Cm and bond length is 1.2746×10^{-10} m.

Answer. 17.78% *(Devi Ahilya BSc)*

MULTIPLE CHOICE QUESTIONS

1. The molecular mass of a compound is an example of
   (*a*) additive property (*b*) constitutive property
   (*c*) additive as well as constitutive property (*d*) none of these
   Answer. (*a*)
2. The molar volume of a liquid at a temperature where its surface tension is unity is called
   (*a*) molar volume (*b*) molar surface tension
   (*c*) molar viscosity (*d*) parachor
   Answer. (*d*)
3. The parachor is
   (*a*) an additive property (*b*) a constitutive property
   (*c*) both an additive and constitutive property (*d*) none of these
   Answer. (*c*)
4. The molar viscosity is the
   (*a*) product of molar surface and viscosity (*b*) sum of molar surface and viscosity
   (*c*) difference of molar surface and viscosity (*d*) none of these
   Answer. (*a*)
5. The Rhechor is the
   (*a*) molar volume of the liquid at the temperature at which viscosity is zero
   (*b*) molar volume of the liquid at the temperature at which viscosity is unity
   (*c*) molar volume of the liquid at the temperature at which viscosity is infinite
   (*d*) none of the above
   Answer. (*b*)
6. The quantity Rheochor [R] is given by the formula where *M* is molar surface and η is the viscosity
   (*a*) $[R]=\dfrac{M}{d}\times\eta^{1/4}$ (*b*) $[R]=\dfrac{M}{d}\times\eta^{1/6}$
   (*c*) $[R]=\dfrac{M}{d}\times\eta^{1/8}$ (*d*) $[R]=\dfrac{M}{d}\times\eta^{1/16}$
   Answer. (*c*)
7. The Rheochor is
   (*a*) an additive property (*b*) a constitutive property
   (*c*) both an additive and constitutive property (*d*) none of these
   Answer. (*c*)
8. The dipole moment of a polar substance is given by the formula, where *q* is charge at one end and *r* is the distance between the opposite charges
   (*a*) $\mu = q + r$ (*b*) $\mu = q - r$
   (*c*) $\mu = q \times r$ (*d*) $\mu = q \times r$
   Answer. (*c*)
9. The net dipole moment of the molecule is
   (*a*) sum of all individual bond moments
   (*b*) product of all individual bond moments
   (*c*) vector resultant of all the individual bond moments
   (*d*) none of the above
   Answer. (*c*)

10. Dichlorobenzene exists in three isomers-*ortho*, *meta* and *para* isomers. Out of these three isomers, one with highest dipole moment will be

(*a*) *ortho*-isomer (*b*) *meta*-isomer
(*c*) *para*-isomer (*d*) all will have the same dipole moment

Answer. (*a*)

11. The molar refraction of a molecule is _______ of the contribution of atoms (atomic refraction) and bonds (bond refractions) present

(*a*) difference (*b*) sum
(*c*) product (*d*) none of these

Answer. (*b*)

12. Optical activity is

(*a*) an additive property (*b*) a constitutive property
(*c*) both an additive and constitutive property (*d*) none of these

Answer. (*b*)

13. A Chiral molecule has

(*a*) no plane of symmetry (*b*) one plane of symmetry
(*c*) infinite planes of symmetry (*d*) none of these

Answer. (*b*)

14. The molecules that are non-superimposable mirror images are called

(*a*) optical isomers (*b*) racemic isomers
(*c*) enantiomers (*d*) none of these

Answer. (*c*)

15. A racemic mixture has

(*a*) zero optical rotation (*b*) positive optical rotation
(*c*) negative optical rotation (*d*) infinite optical rotation

Answer. (*a*)

16. The necessary condition for a compound to be optically active is

(*a*) the presence of no chiral atoms (*b*) the presence of at least one asymmetric atoms
(*c*) the presence of chirality in the molecule (*d*) none of these

Answer. (*c*)

17. An equimolar mixture of (+)-tartaric acid and (–)-tartaric acid is called

(*a*) enantiomers (*b*) optically active mixture
(*c*) racemic mixture (*d*) asymmetric mixture

Answer. (*c*)

18. A diamagnetic substance is _______ by/in the magnetic field

(*a*) attracted (*b*) repelled
(*c*) rotated (*d*) revolved

Answer. (*b*)

19. The substances which retain their magnetic field when removed from the magnetic field are called

(*a*) paramagnetic (*b*) diamagnetic
(*c*) ferrimagnetic (*d*) ferromagnetic

Answer. (*d*)

20. The paramagnetism is due to the presence of

(*a*) paired electrons (*b*) unpaired electrons
(*c*) both paired as well as unpaired electrons (*d*) none of these

Answer. (*b*)

21. Iron, cobalt and nickel are examples of
(*a*) diamagnetic substances
(*b*) paramagnetic substances
(*c*) ferromagnetic substances
(*d*) sometimes diamagnetic & sometimes paramagnetic substances
Answer. (*c*)

22. The magnetic moment, μ, is given by the formula in which *n* is the number of unpaired electrons
(*a*) $\mu = n \times (n+2)$ (*b*) $\mu = \sqrt{n \times (n+1)}$
(*c*) $\mu = \sqrt{n \times (n+2)}$ (*d*) $\mu = n \times (n+1)$
Answer. (*c*)

23. The frequency and wave length are related to each other by the formula
(*a*) $\nu = c\,\bar{\nu}$ (*b*) $\bar{\nu} = c\,\nu$
(*c*) $\nu = \hbar\,\bar{\nu}$ (*d*) $\bar{\nu} = \hbar\,\nu$
Answer. (*a*)

24. The energy of one mole of photon is given by the relation
(*a*) $E = \dfrac{N\,h\,c}{\lambda}$ (*b*) $E = \dfrac{N\,h\,c}{\lambda^2}$
(*c*) $E = \dfrac{N\,h\,c^2}{\lambda}$ (*d*) $E = \dfrac{N\,h^2\,c}{\lambda}$
Answer. (*a*)

25. The internal energy of a molecule is its
(*a*) rotational energy (*b*) vibrational energy
(*c*) translational energy (*d*) all of these
Answer. (*d*)

26. The electronic spectra are caused by
(*a*) the absorption of light (*b*) the emission of light
(*c*) the absorption of high energy photons (*d*) none of these
Answer. (*c*)

27. The rotational spectra involve
(*a*) a very high energy changes (*b*) small energy changes
(*c*) no energy change (*d*) none of these
Answer. (*b*)

28. A molecule can be excited to only the next higher rotational level by
(*a*) absorption of energy (*b*) release of energy
(*c*) the electric current (*d*) applying magnetic field
Answer. (*a*)

29. The spectra caused in the infrared region by the transition in vibrational levels in different modes of vibrations are called
(*a*) rotational spectra (*b*) electronic spectra
(*c*) vibrational spectra (*d*) none of these
Answer. (*c*)

30. The IR spectra of a compound helps in
(*a*) proving the identity of compounds
(*b*) showing the presence of certain functional groups in the molecule
(*c*) neither of the above
(*d*) both of the above
Answer. (*d*)

31. The electronic spectra consists of

(*a*) a large number of absorption bands (*b*) a large number of closely packed lines

(*c*) a large number of peaks (*d*) none of these

Answer. (*b*)

32. In the Raman spectrum, the middle line is called

(*a*) Raman line (*b*) Rayleigh line

(*c*) functional group line (*d*) none of these

Answer. (*b*)

33. The change in frequency by scattering (Raman Effect) occurs due to ______ of energy between the incident photon and the scattering molecule.

(*a*) release (*b*) absorption

(*c*) exchange (*d*) none of these

Answer. (*c*)

34. NMR spectroscopy is useful for determining the type and number of ______ atoms in organic molecules.

(*a*) hydrogen (*b*) carbon

(*c*) halogen (*d*) hetero atoms

Answer. (*a*)

35. The magnetic moment of a molecule is expressed in

(*a*) magnetic tons (*b*) magnetons

(*c*) debye units (*d*) g cm s^{-1}

Answer. (*b*)

36. The electronic spectra lies within

(*a*) infrared region (*b*) visible or ultraviolet regions

(*c*) radio wave region (*d*) micro waves region

Answer. (*b*)

37. The wave numbers are expressed in

(*a*) s^{-1} (*b*) cm s^{-1}

(*c*) cm^{-1} (*d*) cm^2 s^{-1}

Answer. (*c*)

38. The magnetic moment of oxygen molecule is 2.8×10^{-3} J T^{-1}. The number of unpaired electrons in its molecule is

(*a*) 0 (*b*) 1

(*c*) 2 (*d*) 3

Answer. (*c*)

39. The magnetic moment of a molecule is 1.732×10^{-3} J T^{-1} magnetons. The number of unpaired electrons is

(*a*) 0 (*b*) 1

(*c*) 2 (*d*) 3

Answer. (*b*)

40. The electromagnetic radiations of higher wavelengths has _______ energy.

(*a*) higher (*b*) lower

(*c*) intermediate (*d*) zero

Answer. (*b*)

41. The Raman and IR spectra can tell us whether

(*a*) a molecule is linear or non-linear (*b*) a molecule is symmetrical or asymmetrical

(*c*) neither of the above (*d*) both of the above

Answer. (*d*)

42. Define "wavenumber."
(*a*) 1/wavelength (*b*) it is the same as wavelength
(*c*) wavelength x speed of light
Answer. (*a*)

43. What does IR spectroscopy allow us to determine?
(*a*) the molecular formula of a compound (*b*) the kinds of bonds in a compound
(*c*) the molecular weight of a compound (*d*) the number of carbons in a compound
Answer. (*b*)

44. Which of the following will show an absorption band at the greatest wavenumber?
(*a*) C=C (*b*) C≡C
(*c*) C=O (*d*) C–N
Answer. (*b*)

45. What is a molecular ion?
(*a*) it is a compound that has gained an electron
(*b*) it is a compound that has lost an electron.
(*c*) it is a compound that has gained a pair of electrons.
(*d*) it is a compound that has lost a pair of electrons.
Answer. (*b*)

46. What is plotted on the *x* and *y* axes in a mass spectrum?
(*a*) *y* axis = abundance of ion; *x* axis = charge/mass
(*b*) *y* axis = abundance of ion; *x* axis = mass/charge
(*c*) *y* axis = abundance of ion; *x* axis = charge
(*d*) *y* axis = abundance of ion; *x* axis = mass
Answer. (*b*)

47. What is the "base peak" in a mass spectrum?
(*a*) the peak with the lowest intensity
(*b*) the peak with the highest *m/z* value
(*c*) the peak with the lowest *m/z* value
(*d*) the peak with the greatest intensity
Answer. (*d*)

48. Where in an IR spectrum would you find an absorption band due to a carbon-oxygen double bond?
(*a*) 178000 – 165000 m^{-1} (*b*) 125000 – 105000 m^{-1}
(*c*) 168000 – 160000 m^{-1} (*d*) 160000 – 150000 m^{-1}
Answer. (*a*)

49. Which of the following is not true about electromagnetic radiation?
(*a*) frequency is directly proportional to wavelength
(*b*) frequency is directly proportional to wavenumber
(*c*) energy is inversely proportional to wavelength
(*d*) energy is inversely proportional to wavenumber
Answer. (*a*)

50. Which of the following bonds will show an absorption band at the greatest wavenumber?
(*a*) C=O (*b*) C=C
(*c*) O–H (*d*) C–H
Answer. (*c*)

51. Which of the following compounds will show an absorption band at 2200 × 10^2 m^{-1}?
(*a*) $CH_3CH_2CH_2CH{=}O$ (*b*) $CH_3CH_2CH_2C{\equiv}N$
(*c*) $CH_3CH_2CH_2OH$ (*d*) $CH_3CH_2CH_2COOH$
Answer. (*b*)

52. An absorption band at 1600×10^2 m^{-1} indicates that a compound contains which of the following groups?

(*a*) C=O (*b*) C=C

(*c*) O–H (*d*) C–O

Answer. (*b*)

53. What type of radiation is used in nuclear magnetic resonance (NMR) spectroscopy?

(*a*) ultraviolet light (*b*) microwaves

(*c*) radio waves (*d*) infrared radiation

Answer. (*c*)

54. Which of the following absorption bands could be used to distinguish between dimethylamine and trimethylamine?

(*a*) 175000 – 170000 m^{-1} (*b*) 210000 m^{-1}

(*c*) 350000 – 330000 m^{-1} (*d*) 295000 m^{-1}

Answer. (*c*)

55. Which part of the IR spectrum is called the "fingerprint region"?

(*a*) 300000-200000 m^{-1} (*b*) 300000-100000 m^{-1}

(*c*) 200000-100000 m^{-1} (*d*) 100000-60000 m^{-1}

Answer. (*d*)

56. Which of the following bonds will show an absorption band at the greatest wavenumber?

(*a*) N–H (*b*) C=O

(*c*) C–O (*d*) C=C

Answer. (*a*)

57. What part of an IR spectrum is called the functional group region?

(*a*) 400000-300000 m^{-1} (*b*) 400000-200000 m^{-1}

(*c*) 400000-100000 m^{-1} (*d*) 300000-100000 m^{-1}

Answer. (*c*)

58. In what region of an IR spectrum would you find a C–H stretch?

(*a*) 210000 m^{-1} (*b*) 160000-150000 m^{-1}

(*c*) 360000-340000 m^{-1} (*d*) 330000-270000 m^{-1}

Answer. (*d*)

59. What is the effect of hydrogen-bonding on the position and shape of an O–H stretch?

(*a*) it is shifted to lower wavenumbers and remains sharp.

(*b*) it stays at the same wavenumbers and broadens.

(*c*) it is shifted to lower wavenumbers and broadens.

(*d*) it is shifted to higher wavenumbers and broadens.

Answer. (*c*)

60. Which of the following is required in order to obtain a mass spectrum?

(*a*) the compound must lose an electron.

(*b*) the compound must lose a pair of electrons.

(*c*) the compound must be water soluble.

(*d*) the compound must gain an electron

Answer. (*a*)

61. Where in an IR spectrum does a carbon-carbon triple bond show a stretching vibration?

(*a*) 310000 m^{-1} (*b*) 270000 m^{-1}

(*c*) 210000 m^{-1} (*d*) 330000 m^{-1}

Answer. (*c*)

62. How many kinds of chemically non-equivalent hydrogen nuclei are there in ethylbenzene?
(*a*) 3 (*b*) 4
(*c*) 5 (*d*) 6
Answer. (*c*)

63. What is the multiplicity of the methylene hydrogens of a carbon that is attached to a methyl group and to a methoxy group?
(*a*) doublet (*b*) singlet
(*c*) quartet (*d*) triplet
Answer. (*c*)

64. Each of the following compounds has a singlet in its ^{1}H NMR spectrum. Which compound would have its singlet at the highest frequency?
(*a*) 1-bromo-2,2-dimethylpropane (*b*) ethyl methyl ether
(*c*) ethyl-*tert*-butyl ether (*d*) 1,2-dibromoethane
Answer. (*b*)

65. Which of the following compounds would have the signal for its methylene hydrogens farthest downfield?
(*a*) ethyl chloride (*b*) ethyl iodide
(*c*) ethyl bromide (*d*) ethyl fluoride
Answer. (*d*)

66. How many signals would you expect to see in the ^{1}H NMR spectrum of chlorocyclobutane?
(*a*) 1 (*b*) 3
(*c*) 4 (*d*) 5
Answer. (*d*)

67. What is the multiplicity of the OH group of an alcohol in a solution that contains a trace amount of acid?
(*a*) doublet (*b*) quintet
(*c*) quartet (*d*) singlet
Answer. (*d*)

68. How many signals would you expect to see in the ^{1}H NMR spectrum of butane?
(*a*) 1 (*b*) 2
(*c*) 3 (*d*) 4
Answer. (*b*)

69. Which of the following will occur farthest downfield?
(*a*) the hydrogens of benzene
(*b*) the hydrogens of ethyne
(*c*) the hydrogens of dimethyl ether
(*d*) the hydrogens of $H-\overset{\overset{O}{\|}}{C}-H$
Answer. (*d*)

70. How many signals will vinyl chloride have in ^{1}H NMR spectrum?
(*a*) 1 (*b*) 2
(*c*) 3 (*d*) 4
Answer. (*c*)

71. How many signals does 2,2,4-trimethylpentane have in its ^{1}H NMR spectrum?
(*a*) 2 (*b*) 3
(*c*) 4 (*d*) 5
Answer. (*c*)

72. How many signals would you expect to see in the ^{1}H NMR spectrum of 2-methylpropene?

(*a*) 2 (*b*) 3

(*c*) 4 (*d*) 5

Answer. (*a*)

73. Which of the following compounds is frequently used as an internal reference in ^{1}H NMR spectroscopy?

(*a*) TNS (*b*) DMF

(*c*) TMS (*d*) DMSO

Answer. (*c*)

74. Which of the following describes the ^{1}H NMR spectrum of 1,2-dichloroethane?

(*a*) one doublet (*b*) two triplets

(*c*) two doublets (*d*) one singlet

Answer. (*d*)

75. What is the multiplicity of the CH_2, hydrogens in $CH_3OCH_2CH_2OCH_3$?

(*a*) a singlet (*b*) a triplet

(*c*) a quartet (*d*) a doublet

Answer. (*a*)

76. What would be the splitting of the protons on the CH_2 groups of butane?

(*a*) doublet (*b*) sextet

(*c*) triplet (*d*) singlet

Answer. (*c*)

14 Solutions

CONTENTS

A solution is a homogeneous mixture of two or more substances on molecular level. The constituent of the mixture present in a smaller amount is called the **Solute** and the one present in a larger amount is called the **Solvent.** For example, when a smaller amount of sugar (solute) is mixed with water (solvent), a homogeneous solution in water is obtained. In this solution, sugar molecules are uniformly dispersed in molecules of water. Similarly, a solution of salt ($Na^+ Cl^-$) in water consists of ions of salt (Na^+,Cl^-) dispersed in water.

CONCENTRATION OF A SOLUTION

The concentration of a solution is defined as : **the amount of solute present in a given amount of solution.**

Concentration is generally expressed as the quantity of solute in a unit volume of solution.

$$\text{Concentration} = \frac{\text{Quantity of solute}}{\text{Volume of solution}}$$

A solution containing a relatively low concentration of solute is called **Dilute solution.** A solution of high concentration is called **Concentrated solution.**

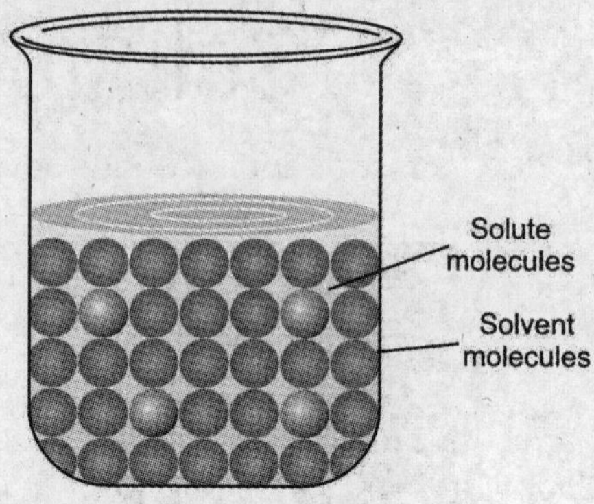

■ **Figure 14.1**
Molecular model of a solution.

TYPES OF SOLUTIONS

The common solutions that we come across are those where the solute is a solid and the solvent is a liquid. In fact, substance in any three states of matter (solid, liquid, gas) can act as solute or solvent. Thus there are **seven types of solutions** whose examples are listed in Table 14.1.

TABLE 14.1. TYPES AND EXAMPLES OF SOLUTIONS

State of Solute	State of Solvent	Example
1. Gas	Gas	Air
2. Gas	Liquid (Carbonated drinks)	Oxygen in water, CO_2 in water
3. Gas	Solid	Adsorption of H_2 by palladium
4. Liquid	Liquid	Alcohol in water
5. Liquid	Solid	Mercury in silver
6. Solid	Liquid	Sugar, Salt
7. Solid	Solid (Steel)	Metal alloys : Carbon in iron

Of the seven types of solutions mentioned in Table 14.1, we will discuss only the important ones in detail.

WAYS OF EXPRESSING CONCENTRATION

There are several ways of expressing concentration of a solution :

(*a*) Per cent by weight
(*b*) Mole fraction
(*c*) Molarity
(*d*) Molality
(*e*) Normality

Per cent by Weight

It is the **weight of the solute as a per cent of the total weight of the solution.** That is,

$$\% \text{ by weight of solute} = \frac{\text{Wt. of solute}}{\text{Wt. of solution}} \times 100$$

For example, if a solution of HCl contains 36 per cent HCl by weight, it has 36×10^{-3} kg of HCl for 100×10^{-3} kg of solution.

SOLVED PROBLEM 1

What is the per cent by weight of NaCl if 1.75×10^{-3} kg of NaCl is dissolved in 5.85×10^{-3} kg of water.

SOLUTION

$$\text{Wt. of solute (NaCl)} = 1.75 \times 10^{-3} \text{ kg}$$

$$\text{Wt. of solvent } (H_2O) = 5.85 \times 10^{-3} \text{ kg}$$

$$\therefore \quad \text{Wt. of solution} = 1.75 + 5.85 = 7.60 \times 10^{-3} \text{ kg}$$

Hence concentration of NaCl % by weight

$$= \frac{1.75 \times 10^{-3}}{7.60 \times 10^{-3}} \times 100$$

$$= \mathbf{23.0}$$

MOLE FRACTION

A simple solution is made of two substances : one is the solute and the other solvent. Mole fraction, X, of solute is defined as **the ratio of the number of moles of solute and the total number of moles of solute and solvent.** Thus,

$$X_{\text{solute}} = \frac{\text{Moles of solute}}{\text{Moles of solute} + \text{Moles of solvent}}$$

If n represents moles of solute and N number of moles of solvent,

$$X_{\text{solute}} = \frac{n}{n + N}$$

Notice that mole fraction of solvent would be

$$X_{\text{solvent}} = \frac{N}{n + N}$$

Mole fraction is unitless and

$$X_{\text{solute}} + X_{\text{solvent}} = 1$$

SOLVED PROBLEM 2

Calculate the mole fraction of HCl in a solution of hydrochloric acid in water, containing 36 per cent HCl by weight

SOLUTION

The solution contains 36×10^{-3} kg of HCl and 64×10^{-3} kg of H_2O

$$\text{Number of Moles of HCl} = (36 \times 10^{-3} \text{ kg HCl})\left(\frac{1 \text{ mol HCl}}{36.5 \times 10^{-3} \text{ kg HCl}}\right)$$

$$= 0.99$$

$$\text{Number of Moles of } H_2O = (64 \times 10^{-3} \text{ kg } H_2O)\left(\frac{1 \text{ mol } H_2O}{18 \times 10^{-3} \text{ kg } H_2O}\right)$$

$$= 3.55$$

$$X_{\text{HCl}} = \frac{\text{moles of HCl}}{\text{moles of HCl} + \text{moles of } H_2O}$$

$$= \frac{0.99}{3.55 + 0.99} = \mathbf{0.218}$$

MOLARITY

In current practice, concentration is most often expressed as molarity. Molarity (symbol *M*) is defined as **the number of moles of solute per liter of solution.** If *n* is the number of moles of solute and *V* liters the volume of solution,

$$\text{Molarity} = \frac{\text{Moles of solute}}{\text{Volume in litres}}$$

or

$$M = \frac{n}{V\text{ (in litres)}} \qquad ...(1)$$

For one mole of solute dissolved in one liter of solution $M = 1$ *i.e.*, molarity is one. Such a solution is called $1M$ (read "1 molar"). A solution containing two moles of solute in one liter is $2M$ ("two molar"); and so on. As evident from expression (1), **unit of molarity is mol/m³ or mol m⁻³.**

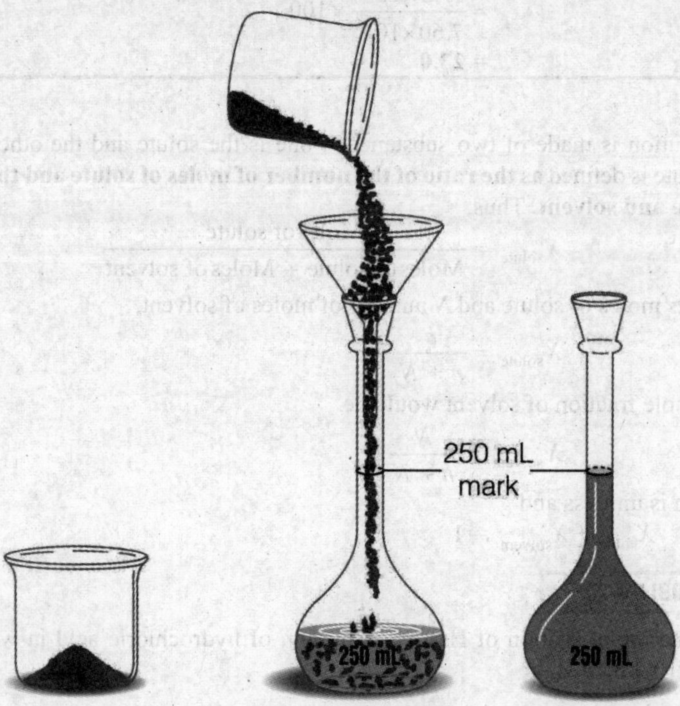

$$\text{Molarity} = \frac{\text{Moles of solute}}{\text{Volume in litres}}$$

Calculation of Molarity

Molarity of a solution can be calculated with the help of the expression (1) if moles of solute (*n*) and volume *V* (in liters or m³) are known. When the amount of solute is given in grams or kilograms and its molecular weight is *MW*, it can be converted to moles :

$$n = x \text{ kg} \times \frac{1 \text{ mol}}{MW \text{ kg}} \text{ mol}$$

Substituting in expression (1)

$$M = x \text{ kg} \times \frac{1 \text{ mol}}{MW \text{ kg}} \times \frac{1}{V} \qquad ...(2)$$

From expression (1) can also be found the amount of solute in kilograms if molarity is given.

SOLVED PROBLEM 3

What is the molarity of a solution prepared by dissolving 75.5×10^{-3} kg of pure KOH in $0.00054\ m^3$ of solution.

SOLUTION

$$\text{Molecular mass of KOH} = 39.1 + 16.0 + 1.0$$
$$= 56.1$$

Volumes in $m^3 = 0.00054\ m^3$

Calculation of volume in liters :

$$540\,\text{ml} \times \frac{1\,\text{litre}}{1000\,\text{ml}} = 0.540\,\text{litres} = 0.00054\,\text{m}^3$$

Calculation of Molarity :

$$M = \frac{n}{V}$$

$$= \frac{1.35\,\text{mol}}{0.00054\,\text{m}^3} = 2.50 \times 10^3\,\text{M}$$

Thus the solution is **2.50×10^3 M**

SOLVED PROBLEM 4

What weight of HCl is present in $0.155 \times 10^{-3}\ m^3$ of a 0.540×10^3 M solution ?

SOLUTION

$$M = 0.540 \times 10^3\ \text{M}$$

$$V\,\text{in litres} = \frac{155}{1000} = 0.155 \times 10^{-3}\ \text{m}^3$$

$$n = M \times V$$
$$= 0.540 \times 10^3\ \text{mol/m}^3 \times 0.155 \times 10^{-3}\ \text{m}^3$$
$$= 0.0837\ \text{mol of HCl}$$

$$0.0837\,\text{mol of HCl} \times \frac{36.5 \times 10^{-3}\,\text{kg}}{\text{mol HCl}} = 3.06 \times 10^{-3}\,\text{kg HCl}$$

Thus **3.06×10^{-3} kg of HCl** is present in $0.155 \times 10^{-3}\ m^3$ of the given solution.

MOLALITY

Molality of a solution (symbol *m*) is defined as **the number of moles of solute per kilogram of solvent :**

$$\text{Molality } (m) = \frac{\text{Moles of solute}}{\text{Mass of solvent in kilograms}}$$

A solution obtained by dissolving one mole of the solute in 1000 g or 1 kg of solvent is called **one molal** or **1*m* solution.**

Notice the difference between molality and molarity. **Molality is defined in terms of mass of solvent while molarity is defined in terms of volume of solution.**

SOLVED PROBLEM 5

What is the molality of a solution prepared by dissolving 0.005 kg of toluene (C_7H_8) in 225×10^{-3} kg of benzene (C_6H_6) ?

SOLUTION

Calculation of number of moles of solute :

$$\text{Molecular mass of toluene} = 12 \times 7 + 1 \times 8 = 0.092 \text{ kg}$$

$$\text{No. of moles 0.005kg of toluene} = \frac{5 \times 10^{-3}}{92 \times 10^{-3}} = 0.0543$$

$$\text{Mass of solvent in kg} = \frac{5 \times 10^{-3}}{92 \times 10^{-3}} = 0.0543$$

$$\text{Molality} = \frac{\text{No. of moles of solute}}{\text{Mass of solvent in kg}} = \frac{0.0543}{0.225}$$

$$= \mathbf{0.24\ m}$$

NORMALITY

Normality of a solution (symbol *N*) is defined *as number of equivalents of solute per liter of the solution :*

$$\text{Normality } (N) = \frac{\text{Equivalents of solute}}{\text{Volume of solution in m}^3}$$

Thus, if 0.040 kg of NaOH (eq. wt. = 40) be dissolved in one liter (10^{-3} m^3) of solution, normality of the solution is one and the solution is called 1*N* (one-normal). A solution containing 0.0040 kg of NaOH is 1/10 *N* or 0.1 *N* or decinormal.

SOLVED PROBLEM 6

0.005 kg of NaCl is dissolved in 1 kg of water. If the density of the resulting solution is 997 kg/m^3, calculate the molality, molarity, normality and mole fraction of the solute, assuming volume of the solution is equal to that of solvent.

SOLUTION

$$\text{Number of moles of NaCl} = \frac{\text{Mass of NaCl}}{\text{Molecular mass NaCl}}$$

$$= \frac{5 \times 10^{-3}}{58.5 \times 10^{-3}} = 0.0854$$

By definition,

$$\text{Molality} = \frac{\text{No. of moles of solute}}{\text{Mass of solvent in kg}}$$

$$= \frac{0.0854}{1\text{kg}} = \mathbf{0.0854\,m}$$

$$\text{Volume of the solution} = \frac{\text{Mass of solution in kg}}{\text{Density in kg per m}^3}$$

$$= \frac{1+0.005}{997} = 0.001008\,m^3$$

$$= \mathbf{0.001008\ m^3}$$

Now Molarity $= \dfrac{\text{Number of moles of solute}}{\text{Volume of solution in } m^3}$

$$= \frac{0.0854}{0.001008} = \mathbf{84.72\,M}$$

and Normality $= \dfrac{\text{Number of kg equivalents of solute}}{\text{volume of solution in } m^3}$

$$= \frac{0.0854}{0.001008} \qquad [\because \text{Mol mass of NaCl} = \text{Eq. wt. of NaCl}]$$

$$= \mathbf{84.7\ N}$$

To Calculate Mole Fraction of the Solute

No. of moles of water in 0.001 m^3 $= \dfrac{0.001}{18 \times 10^{-3}} = 0.0556$

$[\because$ 1 ml of Water = 1 g of water]

Total number of moles = No. of moles of solute + No. of moles of solvent

$= 0.0854 + 0.0556$

$= 0.141$

Mole fraction of NaCl $= \dfrac{\text{No. of moles of NaCl}}{\text{Total moles}}$

$$= \frac{0.0854}{0.141}$$

$$= \mathbf{0.6057}$$

SOLUTIONS OF GASES IN GASES

When a gas is mixed with another gas a completely homogeneous solution results provided of course they do not react chemically. It is obvious that such gaseous solutions will have the following characteristic properties :

(1) **Complete miscibility.** According to the *Kinetic Theory*, a gas consists of tiny molecules moving about in vacant space and thus when one gas is dissolved in another gas they form a homogeneous solution quite readily. In such a gaseous mixture, the components can be present to an unlimited extent.

(2) **Dalton's law of Partial pressures.** Since in a gaseous mixture the constituent molecules exist separately, it is obvious that the properties of the mixture will be the sum of properties of the components. Thus Dalton (1842) was the first to show that **the total pressure exerted by a gaseous mixture is the sum of the individual or partial pressures of the component gases.** If $p_1, p_2, p_3...$ be the partial pressures of the constituents, the total pressure P of the mixture is given the expression

$$P = p_1 + p_2 + p_3 + \ldots$$

Like other gas laws, Dalton's law holds strictly only when the partial pressures are not too high. This law can be experimentally tested by comparing the total pressure of a gaseous mixture with the sum of the individual pressure of each gas before mixing.

SOLVED PROBLEM 7

At constant temperature, 0.0025 m^3 of nitrogen under 95992.10 Pa and 380 × 10^{-6} oxygen under 86659.54 Pa were put into a 10^{-3} m^3 flask. What will be the final pressure of the mixture?

SOLUTION

Since $PV = P'V'$ at constant temperature *(Boyle's Law)*

$$P_{N_2} \times 10^{-3} = 95992.10 \times 250 \times 10^{-6} \quad \text{or} \quad P_{N_2} = 23998.03 \text{ Pa}$$

$$P_{O_2} \times 10^{-3} = 380 \times 10^{-6} \times 86659.54 \quad \text{or} \quad P_{N_2} = 32930.62 \text{ Pa}$$

$$\text{Total pressure } P = P_{N_2} + P_{O_2} \quad (\textit{Dalton's Law})$$

$$= 23998.08 + 32930.62 = \mathbf{56928.65\ Pa}$$

HENRY'S LAW

The solubility of a gas in a solvent depends on the pressure and the temperature. When a gas is enclosed over its saturated solution, the following equilibrium exists.

$$\text{gas} \rightleftharpoons \text{gas in solution}$$

If pressure is increased on the system, the equilibrium will move in the direction which will reduce the pressure (Le Chatelier Principle). The pressure can be reduced by more gas dissolving in solvent. Thus solubility or concentration of a gas in a given solvent is increased with increase of pressure. A kinetic molecular explanation of the effect of pressure on gas-solution system is illustrated in Fig. 14.2.

The relationship between pressure and solubility of a gas in a particular solvent was investigated by William Henry (1803). He gave a generalization which is known as **Henry's Law.** It may be stated as : **for a gas in contact with a solvent at constant temperature, concentration of the gas that dissolves in the solvent is directly proportional to the pressure of the gas.**

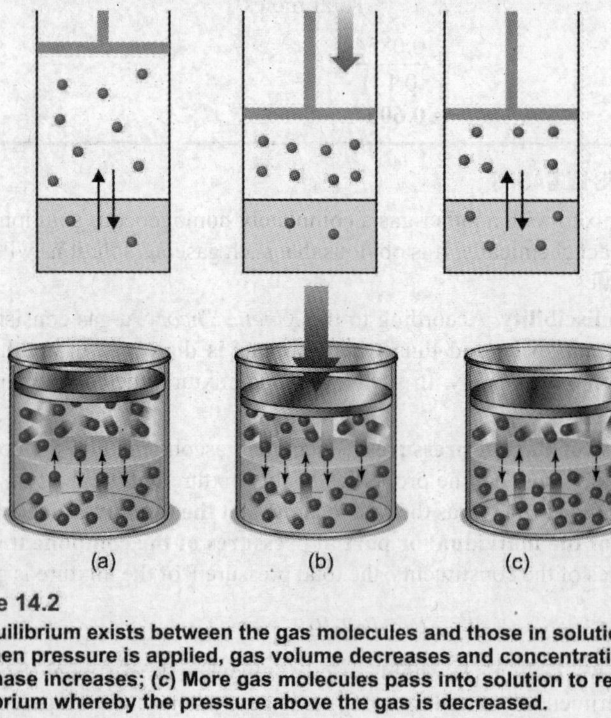

Figure 14.2

(*a*) Equilibrium exists between the gas molecules and those in solution; (*b*) When pressure is applied, gas volume decreases and concentration in gas phase increases; (*c*) More gas molecules pass into solution to reestablish equilibrium whereby the pressure above the gas is decreased.

Mathematically, Henry's Law may be expressed as

$$C \propto P$$

or

$$C = kP$$

where P = pressure of the gas; C = concentration of the gas in solution; and k = proportionality constant known as **Henry's Law Constant.** The value of k depends on the nature of the gas and solvent, and the units of P and C used.

Limitations of Henry's Law

It applies closely to gases with nearly ideal behavior.

(1) at moderate temperature and pressure.

(2) if the solubility of the gas in the solvent is low.

(3) the gas does not react with the solvent to form a new species. Thus ammonia (or HCl) which react with water do not obey Henry's Law.

$$NH_3 + H_2O \rightleftharpoons NH_4^+ + OH^-$$

(4) the gas does not associate or dissociate on dissolving in the solvent.

SOLVED PROBLEM 8

The solubility of pure oxygen in water at 293 K and 101325 Pa pressure is 1.38 mol/m^3. Calculate the concentration of oxygen at 293 K and partial pressure of 21278.25 Pa.

SOLUTION

Calculation of k :

$$C = kP$$

or

$$k = \frac{C}{P}$$

Substituting values for pure oxygen,

$$k = \frac{1.38 \text{ mol/m}^3}{101325 \text{ Pa}} = \frac{1.38 \text{ mol/m}^3}{101325 \text{ Pa}}$$

Calculation of Concentration of O_2 at given pressure

$$C = kP$$

$$\text{Concentration of } O_2 = \frac{1.38 \text{ mol/m}^3}{101325 \text{ Pa}} \times 21278.25 \text{ mol/m}^3$$

$$= \mathbf{2.9 \times 10^{-1} = 0.29 \text{ mol/m}^3}$$

SOLVED PROBLEM 9

At 293 K the solubility of nitrogen gas in water is 0.0150 kg/m^3 when the partial pressure of N_2 is 77326.97 Pa. Find the solubility of N_2 in H_2O at 293 K when its partial pressure is 106657.89 Pa

SOLUTION

We know that Henry's law for a gas at two different concentration and pressures can be written as

$$\frac{C_1}{P_1} = k \qquad \frac{C_2}{P_2} = k$$

Since k has the same value in two cases,

$$\frac{C_1}{P_1} = \frac{C_2}{P_2} \qquad \text{...(1)}$$

In this example,

$$C_1 = 0.0150 \text{ kg/m}^3 \qquad C_2 = ?$$
$$P_1 = 77326.97 \text{ Pa} \qquad P_2 = 106657.89 \text{ Pa}$$

Substituting values in equation (1), we have

$$\frac{0.0150 \text{kg}/\text{m}^3}{77326.97 \text{Pascal}} = \frac{C_2}{106657.89 \text{Pascal}}$$

Thus

$$C_2 = \frac{(106657.89 \text{Pascal})(0.0150 \text{kg}/\text{m}^3)}{(77326.97 \text{Pascal})}$$

$$= \mathbf{0.0207 \text{ kg/m}^3}$$

Thus solubility of N_2 in H_2O at 293 K at pressure of 106657.89 Pa is **0.0207 g / liter.**

Note. The above solution eliminates the calculation of *k* as was done in example 1.

SOLUTIONS OF LIQUIDS IN LIQUIDS

The solutions of liquids in liquids may be divided into three classes as follows :

Liquid-Liquid Systems

- (1) Liquids that are completely miscible e.g., alcohol and water
- (2) Liquids that are partially miscible e.g., ether and water
- (3) Liquids that are practically immiscible e.g., benzene and water

We will now proceed to consider the more important properties of these three classes of solutions.

SOLUBILITY OF COMPLETELY MISCIBLE LIQUIDS

Liquids like alcohol and ether mix in all proportions and in this respect they could be compared to gases. The properties of such solutions, however, are not strictly additive, and therefore their study has not proved of much interest. Generally the volume decreases on mixing but in some cases it increases. Sometimes heat is evolved when they are mixed while in others it is absorbed. The separation of this type of solutions can be effected by fractional distillation.

SOLUBILITY OF PARTIALLY MISCIBLE LIQUIDS

A large number of liquids are known which dissolve in one another only to a limited extent *e.g.*, ether and water. Ether dissolves about 1.2% water; and water also dissolves about 6.5% ether. Since their mutual solubilities are limited, they are only partially miscible. When equal volumes of ether and water are shaken together, two layers are formed, one of a saturated solution of ether in water and the other of a saturated solution of water in ether. These two solutions are referred to as **conjugate** solutions. The effect of temperature on the mutual solubility of these mixtures of conjugate solutions is of special interest. We will study the effect of temperature on the composition of such mixtures with reference to three typical systems :

(1) Phenol-Water system
(2) Triethylamine-Water system
(3) Nicotine-Water system

PHENOL-WATER SYSTEM

The curve in Fig. 14.3 (*a*) represents the miscibility of phenol and water. The left hand side of the parabolic curve represents one of the two conjugate solutions which depicts the percentage of phenol dissolved in water at various temperatures. The solubility of phenol increases with temperature. The right hand side of the curve represents the other conjugate solution layer that gives the percentage of water in phenol. The solubility of water in phenol also increase with increase of temperature. The two solution curves meet at the maxima on the temperature-composition curve of the system. This point here corresponds to temperature 338 K and composition of phenol as 33%. Thus at a certain maximum temperature the two conjugate solutions merge, become identical and only one layer results. The temperature at which the two conjugate solutions (or layers) merge into one another to form one layer, is called the **Critical Solution Temperature (CST)** or Upper Consolute Temperature. This is characteristic of a particular system and is influenced very much by the presence of impurities. The determination of critical solution temperature may, therefore, be used for testing the purity of phenol and other such substances.

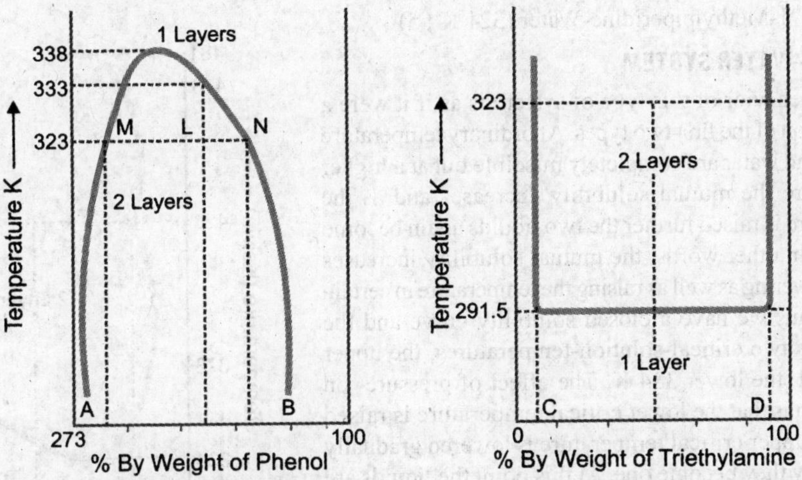

(a) Phenol-Water system

(b) Triethylamine-Water system

■ **Figure 14.3**
Temperature-composition curves.

At any temperature above the critical solution temperature, phenol and water are miscible in all proportions. Outside the curve there is complete homogeneity of the system, *i.e.*, one layer only exists; and under the curve there may be complete miscibility but it depends upon the composition of the mixture. It is clear from Fig. 14.3 (*a*) that at a temperature below 323 K a mixture of 90% phenol and 10% water or 5% phenol and 95% water, will be completely miscible since the corresponding points do not lie under the curve. Two layers will always separate out below the curve and the curve gives compositions of the conjugate solutions constituting the two layers. At 323 K a mixture of equal proportion of phenol and water (50% each) will form two layers whose compositions are given by *A* and *B*. The line joining the points (*M* and *N*) corresponding to the compositions *A* and *B* is called the **tie line.** This line helps in calculating the relative amounts of the two layers, which is here given by the ratio *MN/ML*.

Some other liquid pairs behaving like phenol-water system are given below with their CST values and the percentage of the first component being given in bracket.

(*a*) Methanol-Cyclohexane (322 K ; 29)

(*b*) Hexane-Aniline (332.75 K ; 52)

(*c*) Carbon disulfide-Methanol (322.65 K, 80)

TRIETHYLAMINE–WATER SYSTEM

The temperature-composition curve of mutual solubilities of triethylamine and water is given in Fig. 14.3 (*b*). The left hand side of the curve indicates the solubility curve of triethylamine in water and the right hand side of that of water in triethylamine. Unlike phenol-water system, the solubilities decrease with the increase in temperature in this system. The two conjugate solutions mix up completely at or below 291.5 K. This temperature is also called the critical solution temperature or the lower consolute temperature. As in the above case, any point above the horizontal line corresponds to heterogeneity of the system (two layers) while below it is complete homogeneity (one layer). Thus an equi-component mixture (50–50) will be completely miscible at 283 K but at 323 K there will be separating out of two layers having compositions corresponding to the points *C* and *D*.

Common examples of this system with their lower critical solution temperatures and percentage of the first component are given below.

(*a*) Diethylamine-Water (316 K ; 13)

(*b*) 1-Methylpiperidine-Water (321 K ; 5)

NICOTINE–WATER SYSTEM

The behavior of this type of system is as if it were a combination of the first two types. At ordinary temperature nicotine and water are completely miscible but at a higher temperature the mutual solubility decreases and as the temperature is raised further the two liquids again become miscible. In other words, the mutual solubility increases both on lowering as well as raising the temperature in certain ranges. Thus we have a closed solubility curve and the system has two critical-solution-temperatures, the upper 481 K and the lower 334 K. The effect of pressure on this system is that the lower critical temperature is raised while the upper critical temperature is lowered gradually until finally they become one. At this point the liquids are miscible at all the temperatures (Fig. 14.4).

Glycerine – *m*-Toluidine ; and Water – β-Picoline are other examples of this type.

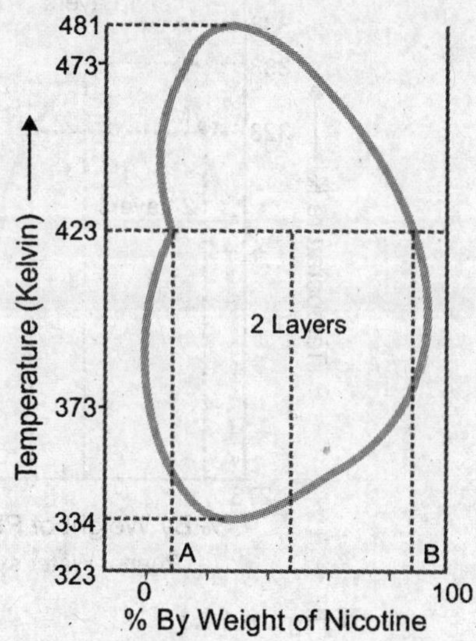

Figure 14.4
Miscibility of nicotine and water.

VAPOR PRESSURES OF LIQUID–LIQUID SOLUTIONS

The study of the vapor pressures of mixtures of completely miscible liquids has proved of great help in the separation of the liquids by fractional distillation. The vapor pressures of two liquids with varying composition have been determined at constant temperature. By plotting the vapor pressure against composition it has been revealed that, in general, mixtures of the miscible liquids are of three types.

First Type of Mixtures of Miscible Liquids

For this type of solutions the vapor pressure curve exhibits a minimum. If we take a mixture which has an excess of *X* (more volatile component), we are somewhere at *C* on the curve. When this is distilled the vapor will contain excess of *X* and thus the remaining mixture will get richer in *Y*. Finally we reach the point *D* *where vapor pressure is minimum and thus boiling point is maximum*. Here the mixture will distill unchanged in composition. Exactly similarly if we take a mixture having

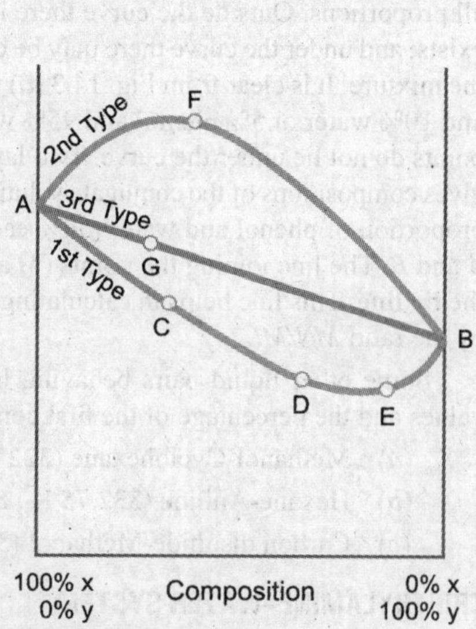

Figure 14.5
Vapor pressure of liquids.

a greater proportion of Y (point E), on distillation Y will pass over leaving the residue richer in X till in this way the minimum point D is again reached when the mixture will distill unchanged. It is obvious that **complete separation of this type of solutions into components is impossible.**

At best it can be resolved into one pure component and the constant boiling mixture. Solutions of this type which distill unchanged at a constant temperature and show a maximum boiling point are called **maximum boiling point azeotropic solutions.** The best known example of this type is shown by hydrochloric acid which forms a constant boiling mixture at 383 K and containing 20.24% of the acid. If a mixture of any composition is distilled, either hydrochloric acid or water will pass over, the composition will move to the point of minimum vapor pressure when it distills without any change in composition.

Second Type of Mixtures of Miscible Liquids (*Minimum boiling point azeotropic solutions*)

Here the vapor pressure curve records a maximum at F. At this point the mixture has the highest vapor pressure and, therefore, the lowest boiling point. Thus in this type of solutions the first fraction will consist of a constant boiling mixture with a fixed composition corresponding to the maximum point until whole of one component has been exhausted. After this the temperature will rise and the other component will pass over. **In this kind of solutions also it is not possible to effect a complete separation by fractional distillation.** At best we can resolve it into a constant-boiling mixture and one component in the pure state. Ethanol and water mixtures offer a good example of this type. Ethanol-water mixture containing 95.6 per cent ethanol boils at the minimum temperature 351 K. Thus it is very difficult to obtain pure absolute alcohol by distillation. This difficulty has, however, been overcome by adding benzene which form a low boiling mixture with water and on distillation it comes over leaving pure ethanol behind.

Third Type of Mixtures of Miscible Liquids

In this case the vapor pressures of mixtures always lie between the vapor pressures of pure components and thus the vapor-pressure composition curve is a straight line. Suppose we have a mixture containing excess of Y which is represented by point G on the curve. On distillation X component being more volatile will be obtained in greater proportion in the distillate and we gradually travel along the curve AB. The latter fractions will, of course, be poorer in X and richer in Y till we reach the 100 per cent Y-axis, when all the X will have passed over. By repeating the process of distillation with the fresh distillate which is now richer in X, we can get almost pure component X. Only in this type of solutions we can completely separate the components by fractional distillation. Thus methyl alcohol-water mixtures can be resolved into pure components by distillation. Liquid mixtures which distill with a change in composition are called **zeotropic mixtures.**

Azeotropes are Mixtures and not Pure Compounds

Although the azeotropes boil at a constant temperature and distill over without change in composition at any given pressure like a pure chemical compound, these cannot be regarded as chemical compounds. The reason is that when the total pressure is changed, both the boiling point and the composition of the azeotrope change whereas for a chemical compound the composition must remain constant over a range of temperature and pressures.

THEORY OF FRACTIONAL DISTILLATION

We have discussed above the vapor-pressure composition curves for the three types of solutions from a study of which we conclude that it is only in the case of the third type that a complete separation by distillation is possible. However to understand the process of fractional distillation we must have an idea of the composition of the vapor phase and that of the liquid mixtures at different boiling temperatures. Thus for this purpose it is not the vapor-pressure composition curve but rather the temperature-composition curve that is important. If we plot the boiling point of liquid mixture against its composition and the composition of the vapor in contact with it, we get two separate curves for each type of solutions. The curves obtained for the third type are shown diagrammatically in Fig. 14.6.

The curves *AEB* and *ADB* are the temperature composition curves for the vapor and liquid respectively. At any boiling temperature *C* the composition of liquid mixture is represented by *J* and that of the vapor in equilibrium by *K*. Obviously, the more volatile component *Y* is present in greater proportion in the vapor than the liquid mixture. Thus the condensed vapor or the distillate will be richer in *X*. If the distillate so obtained be now subjected to distillation, it will boil at *F* and the fresh distillate will have the composition *L* corresponding to *I*. Thus the proportion of *Y* in the second distillate is greater than in the first one. In this way by repeating the process of fractional distillation it is obvious that we can get almost pure *Y*.

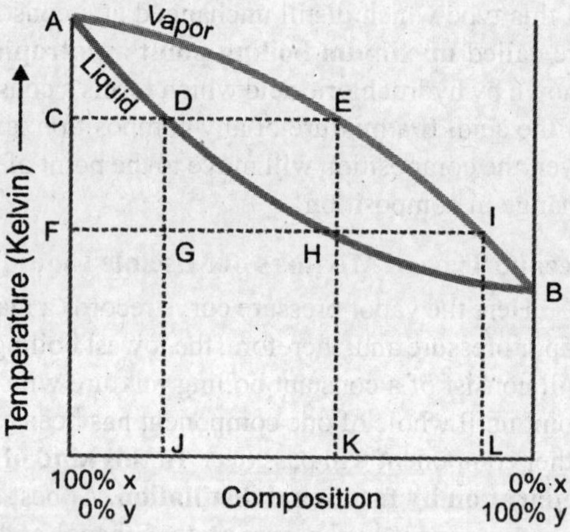

Figure 14.6
Curves showing the composition of vapor and liquid at various boiling temperatures.

In first type of solutions (Fig. 14.7 *a*) if we have a boiling mixture represented by *Y* its vapor will be poor in *Y* than the liquid mixture and the boiling point would gradually rise till we reach the maximum point *C* where the composition of liquid and vapor is the same. Here the distillation proceeds without change of composition. Similarly in the second type (Fig. 14.7 *b*), if we have a boiling mixture represented by the point *X'*, the amount of *Y* in vapor is higher and gradually the boiling point falls to the minimum *C'* where the vapor and the liquid mixtures have the same composition. At this temperature the mixture boils without any change in composition. Thus it is proved that **the second and first type of solutions are not capable of being separated by fractional distillation.**

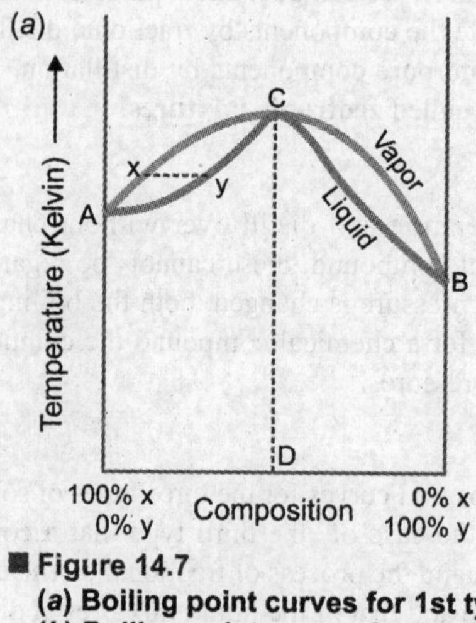

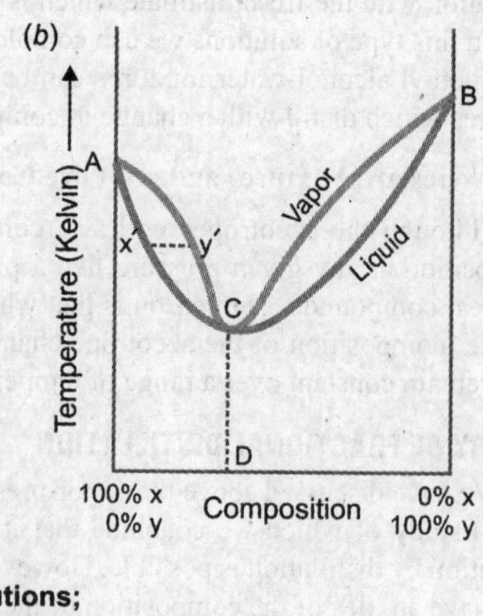

Figure 14.7
(*a*) Boiling point curves for 1st type of solutions;
(*b*) Boiling point curves for 2nd type of solutions.

The efficiency of the process of fractional distillation is considerably enhanced by the use of the so-called **Fractionating columns.** These are of different designs. An effective and simple fractionating column usually employed for laboratory use consists of a long glass tube packed with glass beads (Fig. 14.8 *a*) or specially made porcelain rings. The glass tube blown into bulbs at intervals may also constitute a fractionating column. For industrial purposes a fractionating tower (Fig. 14.8 *b*) is employed. A fractionating tower is divided into several compartments by means of tray that are set one above the other. There is a hole in the center of every tray which is covered by **bubble cap.** Each tray has an overflow pipe that joins it with the tray below by allowing the condensed liquid to flow down.

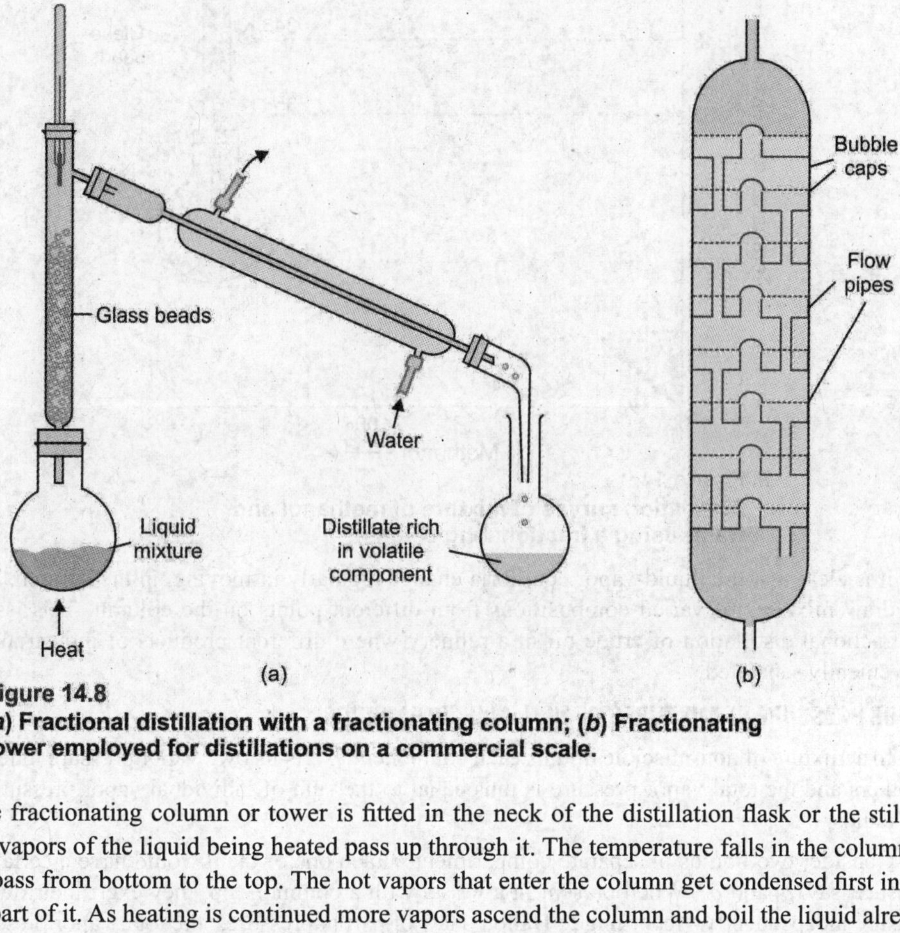

■ **Figure 14.8**
***(a)* Fractional distillation with a fractionating column; *(b)* Fractionating tower employed for distillations on a commercial scale.**

The fractionating column or tower is fitted in the neck of the distillation flask or the still so that the vapors of the liquid being heated pass up through it. The temperature falls in the column as vapors pass from bottom to the top. The hot vapors that enter the column get condensed first in the lowest part of it. As heating is continued more vapors ascend the column and boil the liquid already condensed, giving a vapor which condenses higher up in the column. This liquid is heated in turn by more vapors ascending the column. Thus the liquid condensed in the lowest part is distilled on to the upper part. In this manner a sort of distillation and condensation goes on along the height of the column which results in the increase of the proportion of the volatile component in the outgoing vapors. At every point in the column there exists an equilibrium between liquid and vapor. This is established quickly by and upward flow of vapors and the downward flow of liquid, a large surface area and a slow rate of distillation. A simple distillation of a mixture of methanol and water and the liquid vapor equilibrium states are depicted in Fig. 14.9.

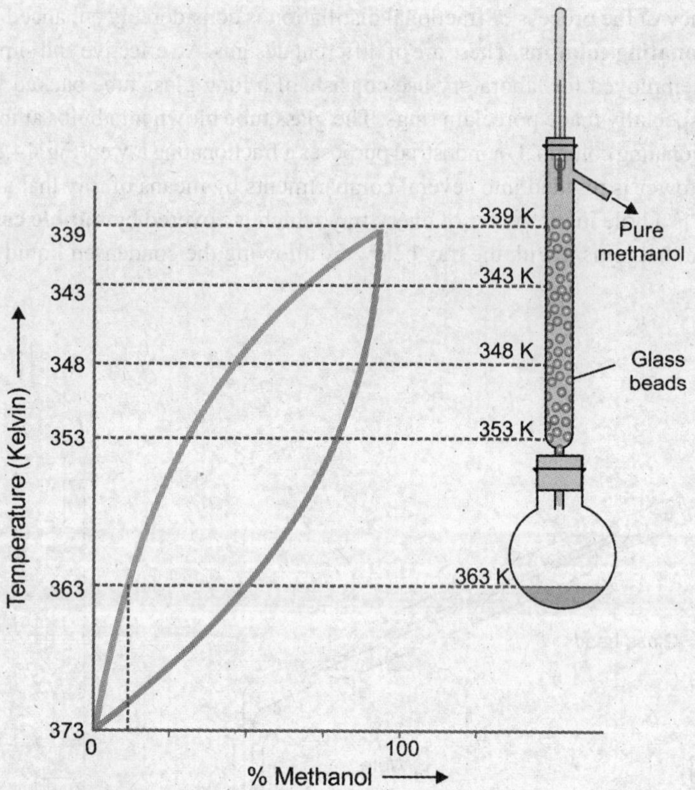

Figure 14.9
Distillation curves of mixture of methanol and water using a fractionating column.

It is clear that the liquid-vapor equilibria change regularly in moving up the column. We may withdraw mixtures of varied compositions from different points on the column. This is done in the fractional distillation of crude oil in a refinery where different products of industrial use are conveniently separated.

VAPOR PRESSURE OF MIXTURES OF NON-MISCIBLE LIQUIDS

In a mixture of non-miscible liquids each component exerts its own vapor pressure independent of others and the total vapor pressure is thus equal to the sum of individual vapor pressures of all the liquids.

Consider two liquids in separate compartments. Each one of them would have a certain vapor pressure, say, p_1 and p_2. When present in a vessel with a common top, they exert their own vapor pressure independently. According to Dalton's law of partial pressures, the total vapor pressure will be equal to the sum of their individual pressures. This generalization which is the basic principle of steam distillation, has been tested experimentally in several cases. Some of the results obtained by Regnault are given below:

Temperature	Vapor Pressure of Water	Vapor Pressure of Carbon disulfide	Sum	Vapor Pressure of Mixture (Observed)
285.22	10.5 mm/1399.88 Pa	216.7 mm/28890.95 Pa	272.2 mm/36290.35 Pa	225.9 mm/30117.52 Pa
300.02	26.2 mm /3493.05 Pa	388.7 mm/51822.40 Pa	415.0 mm/55328.78 Pa	412.3 mm/54968.81 Pa

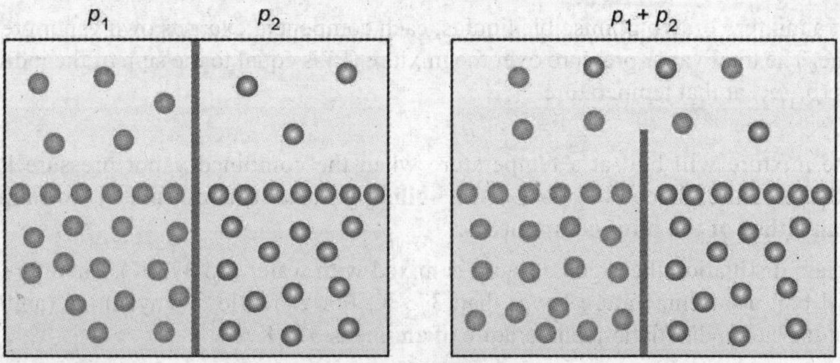

■ **Figure 14.10**
Vapor pressure of mixtures of non-miscible liquids (illustration).

The observed vapor pressure of the mixture is a little less than the sum of the individual vapor pressures of water and carbon disulfide and that is to be expected since each liquid is slightly soluble in the other.

STEAM DISTILLATION

Distillation carried in a current of steam is called steam distillation. This technique is widely used for purification of organic liquids which are steam volatile and immiscible with water (*e.g.*, aniline). The impure organic liquid admixed with water containing nonvolatile impurities is heated and steam passed into it (Fig. 14.11). The vapor of the organic liquid and steam rising from the boiling mixture pass into the condenser. The distillate collected in the receiver consists of two layers, one of the pure organic liquid and the other of water. The pure liquid layer is removed by means of a separatory funnel and further purified.

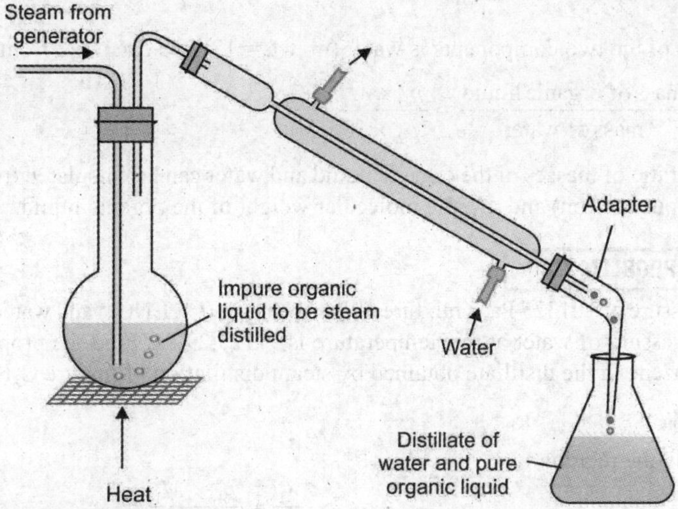

■ **Figure 14.11**
Steam Distillation.

Theory of Steam Distillation

The vapor pressure of a liquid rises with increase of temperature. When the vapor pressure equals the atmospheric pressure, the temperature recorded is the boiling point of the given liquid.

In case of a mixture of two immiscible liquids, each component exert its own vapor pressure as if it were alone. The total vapor pressure over the mixture (P) is equal to the sum of the individual vapor pressures (p_1, p_2) at that temperature.

$$P = p_1 + p_2$$

Hence the mixture will boil at a temperature when the combined vapor pressure P, equals the atmospheric pressure. Since $P > p_1$ or p_2, **the boiling point of the mixture of two liquids will be lower than either of the pure components.**

In steam distillation the organic liquid is mixed with water (bp 373 K). Therefore the organic liquid will boil at a temperature lower than 373 K. For example, phenylamine (aniline) boils at 457 K but the steam distillation temperature of aniline is 372 K.

Steam distillation is particularly used for the purification of an organic liquid (such as phenylamine) which decomposes at the boiling point and ordinary distillation is not possible.

Relative Amounts of Organic Liquid and Water Distilling Over

The number of molecules of each component in the vapor will be proportional to its vapor pressure *i.e.*, to the vapor pressure of the pure liquid at that temperature.

Hence,
$$\frac{n_1}{n_2} = \frac{p_1}{p_2} \quad \text{...(1)}$$

where n_1 and n_2 are the number of moles of the two components in the vapor.

Now, from (1), we have

$$\frac{w_1 / M_1}{w_2 / M_2} = \frac{p_1}{p_2} \quad \text{...(2)}$$

where w_1 and w_2 are the masses of the two liquids distilling over, and M_1 and M_2 their molecular weights.

Since one of the two components is water (m .wt. = 18), we can write from (2)

$$\frac{\text{mass of organic liquid}}{\text{mass of water}} = \frac{p_1 \times M_1}{p_2 \times 18}$$

Thus, the ratio of masses of the organic liquid and water can be calculated from the given values of p_1, p_2 (aqueous tension) and M_1, the molecular weight of the organic liquid.

SOLVED PROBLEM 10

At a pressure of 101325 Pa, a mixture of nitrobenzene ($C_6H_5NO_2$) and water boils at 372 K. The vapor pressure of water at this temperature is 97725.29 Pa. Find the proportion of water and nitrobenzene in the distillate obtained by steam distillation of impure $C_6H_5NO_2$.

SOLUTION

We apply the relation

$$\frac{w_1}{w_2} = \frac{p_1 \times M_1}{p_2 \times 18} \quad \text{...(1)}$$

Here, p_1, vapor pressure of $C_6H_5NO_2$ $= 101325 - 97725.29$

$= 3599.7$ Pa

p_2, vapor pressure of water $= 97725.29$ Pa

M_1, molecular mass of nitrobenzene, $C_6H_5NO_2$

$$= 12 \times 6 + 5 \times 1 + 14 \times 1 + 16 \times 2 = 123$$

Substituting the values in (1)

$$\frac{101325 \times 123 \times 10^{-3}}{97725.29 \times 18 \times 10^{-3}} = \frac{\mathbf{1}}{\mathbf{0.1411}}$$

∴ Proportion of water to $C_6H_5NO_2$ in the distillate is approximately 0.14 : 1

SOLVED PROBLEM 11

A mixture of water and bromobenzene (C_6H_5Br) distills at 368 K, and the distillate contains 1.6 times as much C_6H_5Br as water by mass. At 368 K the vapor pressure of water and C_6H_5Br are 85326.31 Pa Hg and 15998.68 Pa Hg respectively. Calculate the molecular weight of bromobenzene.

SOLUTION

We use the relation

$$\frac{w_1}{w_2} = \frac{p_1 \times M_1}{p_2 \times 18} \qquad ...(1)$$

Here, $w_1/w_2 = 1.6$

p_1, vapor pressure of bromobenzene = 15998.68 Pa

p_2, vapor pressure of water = 85326.31 Pa

Substituting values in (1)

$$1.6 = \frac{15998.68 \times M_1}{85326.31 \times 18}$$

$$\therefore \quad M_1 = \frac{1.6 \times 85326.31 \times 18}{15998.68} = \mathbf{153.6}$$

Thus the molecular mass of bromobenzene is **153.6**

Note : This method of determination of molecular weight of organic liquids is not accurate and gives only approximate value.

SOLUTIONS OF SOLIDS IN LIQUIDS

Solutions of this type are most commonly met with. The process of solution of a solid substance in a solvent is explained by the electrical forces operating between the molecules or ions of the solute and the molecules of the solvent. It is a common observation that polar solutes dissolve easily in polar solvents while they remain insoluble in non-polar solvents. For example, sodium chloride (an electrolyte) is fairly soluble in water which is highly polar solvent, while it is insoluble in a non-polar solvent like chloroform. On the other hand, a non-polar solute does not dissolve in a polar solvent *e.g.*, benzene which is non-polar is insoluble in water. The electrical attraction between the oppositely charged ends of the solute and the solvent molecules results to form a solution.

Water being highly polar is one of the best solvents for ionized solutes. An ionic substance, when placed in water furnishes cations (+) and anions (–). These ions are surrounded by solvent molecules with their oppositely charged ends directed towards the ion. The ion enveloped by a layer of the solvent molecules in this manner, is called a **Solvated ion** or **Hydrated ion** in case water is the solvent. Thus sodium chloride dissolves in water to give Na^+ and Cl^- ions. The Na^+ ion is hydrated to have around a layer of water molecules so that their negative ends are directed towards it. The Cl^- ion, on the other hand, attracts positive ends of water molecules which envelop it (Fig. 14.12).

It is customary to represent the hydrated sodium and chloride ions as $Na^+ (aq)$ and $Cl^- (aq)$. This representation indicates that the ions are in the aqueous phase.

The mechanism of solution of sodium crystal in water could be explained as follows. The polar water molecules try to pull out the Na^+ and Cl^- ions from the crystal by hydration. This becomes possible since the forces operating between the ion (Na^+ or Cl^-) and water molecules are strong enough to overcome the force binding the ion in the crystal. The ions detached from the crystal are

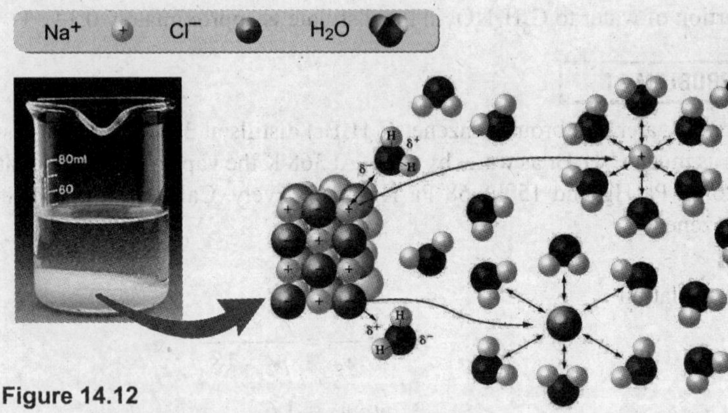

■ **Figure 14.12**
How sodium chloride dissolves in water?

now surrounded by cluster of water molecules. The layer of water molecules enveloping the ions effectively shields them and prevents them from coming in contact with each other. Thus they hardly aggregate into a crystal and remain in solution.

Dissolution Due to Hydrogen Bonding

Many non-ionic substance such as sugar also dissolve in water. Here the dissolution is due to hydrogen bonding that occurs between water and sugar molecules. The hydrogen bonding takes place through the hydroxyl group of the sugar molecules.

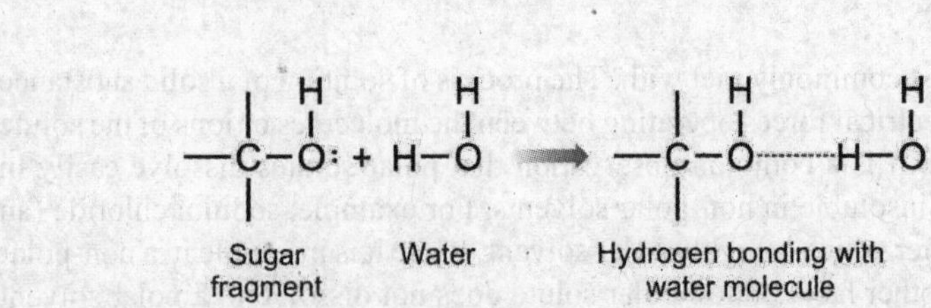

The water molecules are thus able to pull away the molecules of sugar from the crystal, which dissolve. In fact, every sugar molecule is surrounded by a number of water molecules, and these aggregates are free to migrate throughout the solution.

SOLUBILITY–ITS EQUILIBRIUM CONCEPT

When a solid is placed in a solvent, molecules or ions, as the case may be, break away from the surface and pass into the solvent. The particles of the solid thus detached are free to diffuse throughout the solvent to give a uniform solution. The solute and the solvent molecules are constantly moving about in the solution phase because of kinetic energy possessed by them. Some of the particles are deflected back towards the solid on account of collisions with other molecules. These then strike the solid surface and may get entangled in its crystal lattice and thus get deposited on it. This process by which the solute particles from solution are 'redeposited' or 'recrystallized' is often spoken of as **recrystallization** or **precipitation.**

In a solution in contract with solid solute, therefore, two opposing processes are operating simultaneously :

(*a*) **Dissolution** – the particles of the solute leaving the solid and passing into solution.

(*b*) **Recrystallization** – the particles of the solute returning from the solution and depositing (or precipitating) on the solid.

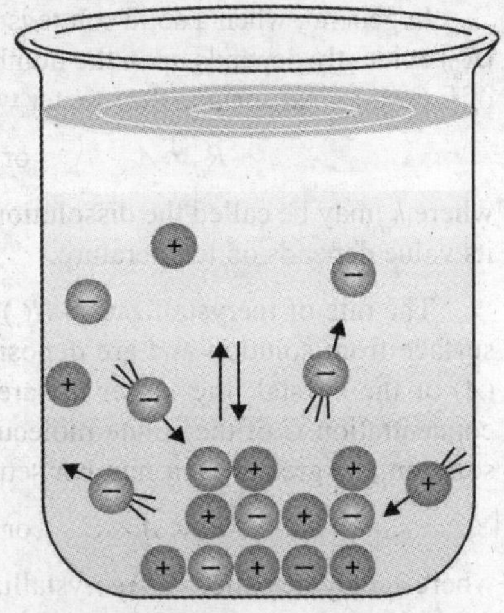

■ **Figure 14.13**
Dissolution and recrystallization.

To start with the rate at which the particles leave the solid is much greater than the rate at which they return to it. As the number of particles of the solute in solution increases, the rate at which they are returned to the solid also increases. Eventually, if there is excess of solid present the rate of dissolution and the rate of recrystallization become equal. At this stage, a state of equilibrium between the molecules of the solute in solution and the solid solute is said to have been reached. Thus,

$$\text{Solute solid} \rightleftharpoons \text{Solute dissolved}$$

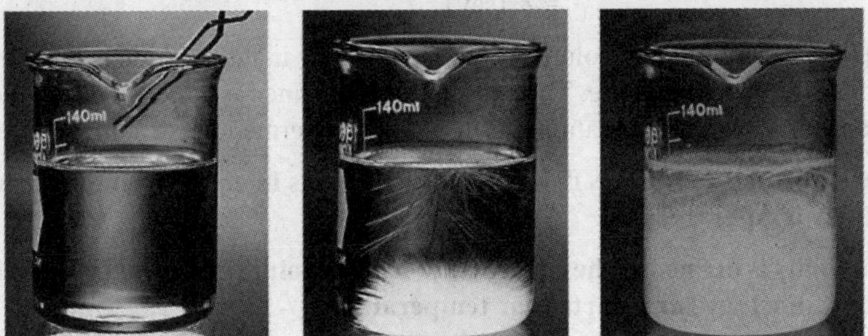

Henceforth neither the amount of the solute in solution nor the solid phase present in contact with it, will change with lapse of time. This equilibrium state will remain so, provided the 'kinetic energy' of the molecules is not changed by a change in temperature. The situation is illustrated in Fig. 14.13. The competition between the two processes and the eventual equality of the rates of the two processes, points to an important phenomenon in chemistry called the **dynamic equilibrium.** The term dynamic refers to the fact that both the processes are occurring continuously but due to the equality of the two rates (equilibrium) no net change in the amount of the solute in solution phase occurs with the passage of time.

The dynamic nature of solubility equilibrium can be demonstrated by putting a crystal of sugar having a hole in the surface, in a solution of sugar which has already attained the state of equilibrium. It will be seen that after some time the hole is filled with solid sugar and the concentration of the solution remains unchanged. This is explained by saying that the molecules from the solution get settled in the hole (process of crystallization) while some molecules from other parts of the crystal go into the solution (process of dissolution), thereby changing the shape of the crystal.

In general, when a solid solute is in dynamic equilibrium with its solution, the rate of dissolution (R_d) evidently depends upon the number of molecules leaving the crystal surface. The larger the area of liquid-crystal surface the greater will be the rate of dissolution. That is,

$$R_d \propto A \qquad \text{or} \qquad R_d = k_d \times A \qquad ...(1)$$

where k_d may be called the dissolution constant. Its value is characteristic of a particular system and its value depends on temperature.

The rate of recrystallization (R_r) is the rate at which the solute molecules return to the crystal surface from solution and are deposited on it. This is determined by two factors : (*a*) the surface (*A*) or the crystal; the larger the area, the greater the number of molecules settling on it; (*b*) the concentration *C* of the solute molecules in solution; the higher the number of solute molecules in solution, the greater their number settling down. Thus,

$$R_r \propto A \times C \qquad \text{or} \qquad R_r = k_r \times A \times C \qquad ...(2)$$

where k_r may be called the recrystallization constant. Its value is also characteristic of a system and depends upon temperature.

At equilibrium the rate of dissolution and the rate of recrystallization are equal.

$$R_d = R_r$$

or, from (1) and (2) $\quad k_d \times A = k_r \times A \times C$

or $\quad C = \dfrac{k_d}{k_r} = \text{constant}$

$$= K \text{ (say)}$$

Hence the concentration of solute at equilibrium state in the solution is constant for a particular solvent and at a fixed temperature. The solution thus obtained is called a 'Saturated solution' of the solid substance and the concentration of this solution is termed its 'Solubility'.

Thus a saturated solution is defined as one which is in equilibrium with the excess of solid at a particular temperature.

The solubility is defined as the concentration of the solute in solution when it is in equilibrium with the solid substance at a particular temperature.

Each substance has a characteristic solubility in a given solvent. The solubility of a substance is often expressed in terms of number of grams of it that can be dissolved in 0.01 kg of the solvent. For example, a saturated solution of sodium chloride in water at 273 K contains 35.7×10^{-3} kg of NaCl in 0.01 kg of H_2O. That is, the solubility of NaCl in water at 273 K is 35.7×10^{-3} kg/0.1 kg.

An increase in the temperature generally causes a rise in the solubility. Thus the solubility of copper sulfate in water at 273 K is 14.3×10^{-3} kg/0.1 kg, while at 273 K it is 75.4×10^{-3} kg/0.1 kg.

When a saturated solution prepared at a higher temperature is cooled, it gives a solution which would contain more solute than the saturated solution at that temperature. Such a solution is called a **supersaturated solution.** Supersaturated solutions are quite unstable and change to the saturated solution when excess of solute precipitates out. This fact is utilized in the purification of chemical substances.

DETERMINATION OF SOLUBILITY

The solubility of a substance is determined by preparing its saturated solution and then finding the concentration by evaporation or a suitable chemical method.

Saturated solution of a solid substance may be prepared by shaking excess of it with the solvent in a vessel placed in a constant temperature bath and filtering the clear solution. A known volume of this saturated solution is evaporated in a china dish and from the weight of the residue the solubility can easily be calculated. This method though simple, does not yield accurate results. During filtration, cooling would take place and thus some solid may be deposited on the filter paper or in the stem of the funnel. However, this method is quite good for the determination of solubility at room temperature. The evaporation of a liquid is a highly undesirable operation as it is not possible to avoid loss of the liquid caused by spurting. This difficulty can, however, be overcome whenever a chemical method of analysis is available. Another defect in this method is that it takes a long time to establish the equilibrium between the solid and the solution so that the preparation of saturated solution by simple agitation with the solvent is delayed. This difficulty may be overcome by first preparing the saturated solution at a higher temperature and then cooling it to the desired temperature at which solubility is to be determined.

SOLUBILITY CURVES

A curve drawn between solubility and temperature is termed Solubility Curve. It shows the effect of temperature on the solubility of a substance. The solubility curves of substances like calcium acetate and calcium chromate show decrease in solubility with increase of temperature while there are others like those of sodium nitrate and lead nitrate which show a considerable increase of solubility with temperature. The solubility curve of sodium chloride shows very little rise with increase of temperature. In general, the solubility curves are of two types :

(1) Continuous solubility curves
(2) Discontinuous solubility curves

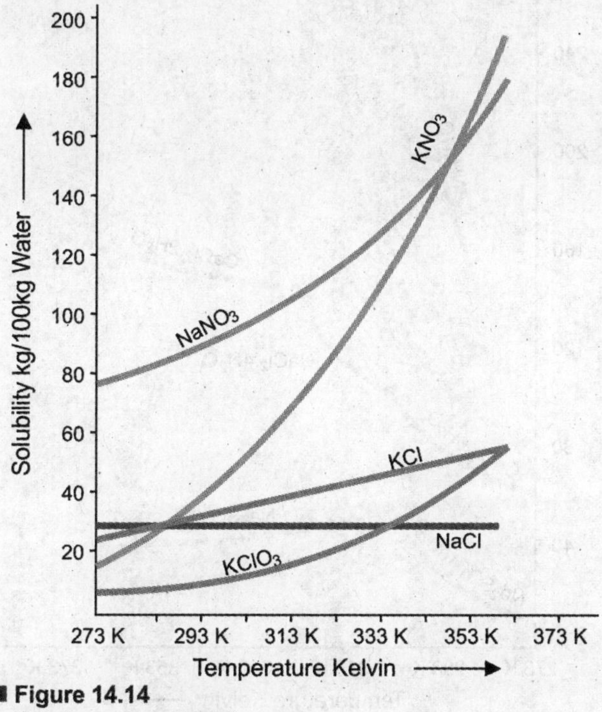

■ **Figure 14.14**
Continuous solubility curves.

Solubility curves of calcium salts of fatty acids, potassium chlorate, lead nitrate and sodium chloride are **Continuous solubility curves** as they show no sharp breaks anywhere. In case of $CaSO_4.2H_2O$, no doubt the curve first shows, a rise and then a fall but it remains continuous at the maximum point. Sometimes the solubility curves exhibit sudden changes of direction and these curves are, therefore, called **Discontinuous solubility curves.** The popular examples of substances which show discontinuous solubility curves are sodium sulfate, calcium chloride, ammonium nitrate etc. In fact, at the break a new solid phase appears and another solubility curve of that new phase begins. The break in a solubility curve thus shows a point where the two different curves meet each other. The solubility curves of sodium sulfate and ferric chloride will be discussed in detail in the chapter on 'Phase Rule'.

SOLUBILITY OF SOLIDS IN SOLIDS

Solution of a solid in another solid can be prepared by melting them together and subsequent cooling of the mixture. For example, gold and silver when mixed together, melted and cooled, yield solid solutions which are perfectly homogeneous. Sometimes solid solutions may be obtained by simply pressing together the two metals and thus establishing better contact when one metal would diffuse into the other. Solutions of gold and lead have been obtained by this method. The study of solutions of solids in solids is of great practical importance in metallurgy.

The formation of solid solutions is not limited to metals only. Organic substances like naphthalene and β-naphthol when melted together form the so-called mixed crystals on cooling which are a solid solution of one of them in the other.

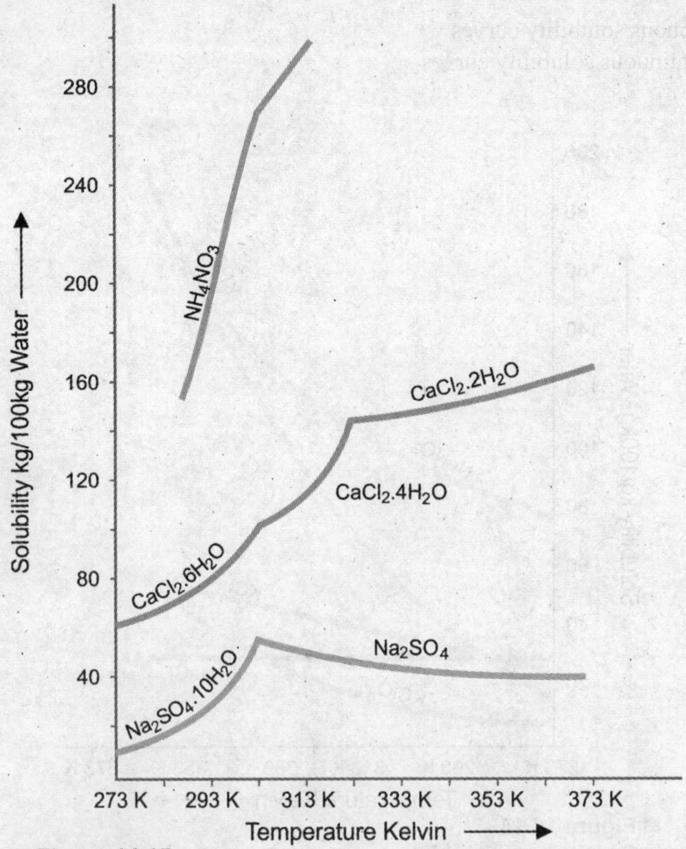

■ Figure 14.15
Discontinuous solubility curves.

ADDITIONAL SOLVED PROBLEMS

SOLVED PROBLEM 12

Calculate the normality of a solution containing 6.3×10^{-3} kg of oxalic acid crystals (molecular mass = 126×10^{-3} kg) dissolved in 0.5 m^3 of the solution.

SOLUTION

Formula used

$$\textit{Normality, N} = \frac{\textit{No. of gram equivalents}}{\textit{volume of solution in liters}}$$

Quantities given

Amount of oxalic acid in 500 m^3 solution = 6.3×10^{-3} kg

$$\text{No. of gram equivalents of oxalic acid} = \frac{\textit{Amt. of solute}}{\textit{gram equivalent weight}}$$

$$= \frac{6.3 \times 10^{-3}\ \text{kg}}{63 \times 10^{-3}\ \text{kg equiv}^{-1}}$$

$$= 0.1\ \text{equiv}$$

volume of solution = 0.5 m^3

Substitution of values

$$\text{Normality, } N = \frac{0.1}{0.5}$$

$$= \mathbf{0.2\ N}$$

SOLVED PROBLEM 13

Calculate the molality of a solution of sodium hydroxide which contains 0.2×10^{-3} kg of sodium hydroxide in 50×10^{-3} kg of the solvent.

SOLUTION

Formula used

$$\textit{Molality, m} = \frac{\textit{No. of moles of solute}}{\textit{mass of solvent in kg}}$$

Quantities given

amount of NaOH = 0.2×10^{-3} kg and no. of moles of NaOH = $\frac{0.2 \times 10^{-3}\ \text{kg}}{40 \times 10^{-3}\ \text{kg mol}^{-1}}$

= 0.005 mol

mass of solvent = 50×10^{-3} kg = 0.050 kg

Substitution of values

$$\text{Molality, } m = \frac{0.005}{0.050}$$

$$= \mathbf{0.1\ m}$$

SOLVED PROBLEM 14

Determine the molality of a solution containing 0.08653 kg of sodium carbonate (molar mass = 105.99×10^{-3} kg) per m^3 in water at 293 K. The density of the solution at this temperature is 1081.6 kg/m^3.

SOLUTION

(*a*) To calculate the amount of Na_2CO_3 in 1 kg of water

Density of solution = $1081.6\ kg/m^3$

$\therefore$ Mass of 1 liter solution = $1081.6\ kg/m^3 \times 10^{-3}\ m^3$ [$\because$ M = density × vol.]

= 1.0816 kg

Amt. of sodium carbonate in $1081.6\ kg/m^3$ of = 0.08653 kg (given)

$$\therefore \text{Amt. of } Na_2CO_3 \text{ in 1 kg of water} = \frac{0.08653 \times 1}{1.0816} = 0.080 \text{ kg}$$

(*b*) To calculate the molality

Formula used

$$\textit{Molality, m} = \frac{\textit{Amount of } Na_2CO_3 \textit{ in 1 kg of water}}{\textit{molar mass}}$$

Quantities given

Amount of Na_2CO_3 *in 1 kg of water* = 0.080 kg

Molar mass of Na_2CO_3 = 105.99×10^{-3} kg/mol

Substitution of values

$$\textit{Molality, m} = \frac{0.080}{105.99 \times 10^{-3}} = \mathbf{0.7548\ m}$$

SOLVED PROBLEM 15

A solution contains 25% water, 25% ethanol and 50% ethanoic acid. Calculate the mole fraction of each component.

SOLUTION

(*a*) To calculate the total number of moles

Let the mass of solution be = 100×10^{-3} kg

Amount of H_2O = 25 % = 25×10^{-3} kg

Amount of C_2H_5OH = 25 % = 25×10^{-3} kg

Amount of CH_3COOH = 50 % = 50×10^{-3} kg

$$\text{Number of moles of water} = \frac{25 \times 10^{-3}\text{ kg}}{18 \times 10^{-3}\text{ kg mol}^{-1}} = 1.3888 \text{ mol}$$

$$\text{Number of moles of } C_2H_5OH = \frac{25 \times 10^{-3}\text{ kg}}{46 \times 10^{-3}\text{ kg mol}^{-1}} = 0.5435 \text{ mol}$$

$$\text{Number of moles of } CH_3COOH = \frac{50 \times 10^{-3}\text{ kg}}{60 \times 10^{-3}\text{ kg mol}^{-1}} = 0.8333 \text{ mol}$$

Total number of moles = 1.3888 + 0.5434 + 0.8333 mol

= 2.7655 mol

(*b*) To calculate the mole fraction of each component

Formula used

$$X_{Component} = \frac{\text{No. of moles of component}}{\text{Total no. of moles}}$$

Substitution of values

Mole fraction of H_2O, X_{H2O} $= \frac{1.3888 \text{ moles}}{2.7655 \text{ moles}} =$ **0.502**

Mole Fraction of C_2H_5OH, X_{C2H5OH} $= \frac{0.5435 \text{ mole}}{2.7655 \text{ moles}} =$ **0.1965**

Mole Fraction of CH_3COOH, $X_{CH3COOH}$ $= \frac{0.8333 \text{ mole}}{2.7655 \text{ mole}} =$ **0.3013**

SOLVED PROBLEM 16

What is molarity and molality of a 13% solution (by weight) of sulfuric acid ? Its density is 1.09 kg m^{-3}.

SOLUTION

Let the mass of solution be $= 100 \times 10^{-3}$ kg

Amount of H_2SO_4 in solution $= 13 \times 10^{-3}$ kg

Amount of H_2O in solution $= 87 \times 10^{-3}$ kg

Number of moles of H_2SO_4 $= \frac{13 \times 10^{-3} \text{ kg}}{98 \times 10^{-3} \text{ kg mol}^{-1}} = 0.1326$ mol

(*a*) To calculate the molality

Formula used

$$\textit{Molality, m} = \frac{\textit{No. of moles of } H_2SO_4}{\textit{Mass of solvent in kg}}$$

Quantities given

No. of moles H_2SO_4 = 0.1326

Mass of solvent = 87×10^{-3} kg

Substitution of value

$$\text{Molality, } m = \frac{0.1326}{87 \times 10^{-3} \text{ kg}}$$

= **1.524 m**

(*b*) To calculate the molarity

Volume of 100×10^{-3} kg of solution $= \frac{\text{Mass}}{\text{Density}} = \frac{100 \times 10^{-3} \text{ kg}}{1.09 \text{ kg ml}^{-1}}$

$= 91.7 \times 10^{-3} \text{ m}^3$

$= 0.09174 \text{ m}^3$

$$\text{Molarity, } M = \frac{\text{No. of moles of } H_2SO_4}{\text{Volume of solution in liter}}$$

$= \frac{0.1326}{0.09174}$

= **1.4454 M**

SOLVED PROBLEM 17

45×10^{-3} kg of glucose, $C_6H_{12}O_6$, are dissolved in 500×10^{-3} kg of water. Calculate the molality of the solution.

SOLUTION

Amount of glucose in 500×10^{-3} kg water $= 45 \times 10^{-3}$ kg

No. of moles of glucose in 500 g water $= \frac{45 \times 10^{-3} \text{ kg}}{180 \times 10^{-3} \text{ kg mol}^{-1}}$

$$= 0.25 \text{ mol}$$

$$\text{Mass of water} = 500 \times 10^{-3} \text{ kg}$$

$$\textit{Molality, m} = \frac{\text{No. of moles of glucose}}{\text{Mass of water in kg}}$$

$$= \frac{0.25}{500 \times 10^{-3}}$$

$$= \mathbf{0.5\ m}$$

SOLVED PROBLEM 18

2.82×10^{-3} kg of glucose (mol mass = 180×10^{-3} kg) are dissolved in 30×10^{-3} kg of water. Calculate the (i) molality of the solution (ii) mole fraction of glucose and water.

SOLUTION

(*a*) To calculate the molality of the solution

$$\text{Number of moles of glucose} = \frac{\text{mass of glucose}}{\text{molar mass}}$$

$$= \frac{2.82 \times 10^{-3} \text{ kg}}{180 \times 10^{-3} \text{ kg mol}^{-1}} = 0.01567$$

$$\text{Amount of solvent} = 30 \text{ g} = 30 \times 10^{-3} \text{ kg}$$

$$\text{Molality of glucose, } m = \frac{\text{No. of moles of glucose}}{\text{amt. of water in kg}}$$

$$= \frac{0.01567}{30 \times 10^{-3}}$$

$$= \mathbf{0.5223\ m}$$

(*b*) To calculate the mole fraction of glucose and water

$$\text{No. of moles of } H_2O = \frac{30 \times 10^{-3} \text{ kg}}{18 \times 10^{-3} \text{ kg mol}^{-1}} = 1.6667 \text{ mol}$$

$$\text{Number of moles of glucose} = 0.01567 \text{ mol}$$

$$\text{Total number of glucose} = 0.01567 + 1.6667 \text{ mol}$$

$$= 1.68237$$

Formula used

$$X_{component} = \frac{\text{No. of moles of component}}{\text{Total number of moles}}$$

$$X_{glucose} = \frac{0.01567 \text{ mole}}{1.68237 \text{ mole}} = \mathbf{0.0093}$$

$$X_{H2O} = \frac{1.6667 \text{ mole}}{1.68237 \text{ mole}} = \mathbf{0.9907}$$

SOLVED PROBLEM 19

The density of a 3 M sodium thiosulfate solution is 1.25 kg/m^3. Calculate (i) the % by weight of sodium thiosulfate, (ii) the mole fraction of sodium thiosulfate, and (iii) the molality of Na^+ and $S_2O_3^{2-}$ ions.

SOLUTION

(a) To calculate the % by weight of sodium thiosulfate

Density of the solution = 1.25 kg/m^3

Mass of 1000 ml of solution = Density × volume

$= 1.25 \text{ kg/m}^3 \times 1000 \text{ ml} = 1.25 \text{ kg}$

Amt of $Na_2S_2O_3$ in 3 M solution = 3 × molar mass of $Na_2S_2O_3$

$= 3 \times 158 \times 10^{-3}$ kg

$= 474 \times 10^{-3}$ kg

1250 g of solution contains $Na_2S_2O_3$ = 474×10^{-3} kg

$$\% \text{ by weight of } Na_2S_2O_3 = \frac{474 \times 10^{-3} \text{ kg}}{1.25 \text{ kg}} \times 100 = \frac{47.4}{1.25}$$

$= \mathbf{37.92\ \%}$

(b) To calculate the mole fraction of $Na_2S_2O_3$

Amount of $Na_2S_2O_3$ in 100g solution = 37.92

Amount of H_2O in 100 g solution = 100 – 37.92 = 62.08×10^{-3} kg

$$\text{Number of moles of } Na_2S_2O_3 = \frac{37.92 \times 10^{-3} \text{ kg}}{158 \times 10^{-3} \text{ kg mole}} = 0.24 \text{ mol}$$

$$\text{Number of moles of water} = \frac{62.08 \times 10^{-3} \text{ kg}}{18 \times 10^{-3} \text{ kg mol}^{-1}} = 3.449 \text{ mol}$$

Total no. of moles = 0.24 mol + 3.449 mol = 3.689 mol

$$\text{Mole fraction of } Na_2S_2O_3,\ X_{Na2S2O3} = \frac{0.24 \text{ mole}}{3.689 \text{ mole}} = \mathbf{0.065}$$

(c) To calculate molality of Na^+ and $S_2O_3^{2-}$ ions

No. of moles of Na_2 in 68.02×10^{-3} kg water = 0.24 mol

No. of moles of $Na_2S_2O_3$ in 1000×10^{-3} kg water = 3.528 mol

Molality of Na^+ ions = 2 × 3.528 mol [∵ $Na_2S_2O_3 \equiv 2\ Na^+$]

= **7.056 m**

and Molality of $S_2O_3^{2-}$ ions = 1 × 3.528 mol [∵ $Na_2S_2O_3 \equiv S_2O_3^{2-}$]

= **3.528 m**

EXAMINATION QUESTIONS

1. Define or explain the following terms :

(*a*) Mole fraction (*b*) Molarity

(*c*) Molality (*d*) Normality

(*e*) Formality (*f*) Henry's law

2. Write notes on the following

(*a*) Freezing mixtures

(*b*) Azeotropic distillation

(*c*) Theory of fractional distillation

3. Write briefly on Raoult's law and azeotropes.

4. (*a*) State and explain Henry's Law. Show that the volume of a gas dissolved in a given volume of a solvent is independent of the pressure of the gas.
   (*b*) Explain the following :
   (*i*) on opening a carbonated cold drink bottle, bubbles of gas come out.
   (*ii*) on adding common salt to a bottle of carbonated cold drink froth comes out.
5. Mixture of alcohol and water cannot be separated by fractional distillation. Explain.
6. Nitrobenzene is completely miscible with water. A mixture of the two liquids boils at 372 K and 100391.74 Pa pressure. The vapor pressure of water is 97725.29 Pa at this temperature. Find out the weight composition of liquid mixture.
   Answer. 0.186 : 1
7. Explain giving reasons :
   (*a*) Solubility of a gas in a given liquid decreases with rise in temperature.
   (*b*) A salt solution fails to quench thirst.
8. Which out of two, a molal solution or a molar solution, is more concentrated and why? Give reasons. (*Nehu BSc*)
9. Define an ideal solution. (*Kerala BSc*)
10. When two components 'A' and 'B' are mixed and the resulting solution formed is almost an ideal solution. Derive an expression for free energy change of mixing. (*Madurai BSc*)
11. Draw only, well labelled vapor pressure-composition diagram of binary miscible real solutions at constant temperature. (*Panjab BSc*)
12. (*a*) Define the following terms :
   (*i*) Mole fraction (*ii*) Molal solution
   (*b*) Obtain a relationship for molal elevation constant from thermodynamic considerations. (*Purvanchal BSc*)
13. Define the following terms :
   (*i*) Molality (*ii*) Molarity
   (*iii*) Mole fraction (*Kathmandu BSc*)
14. Write short notes on :
   (*i*) Steam distillation (*ii*) Azeotropic mixture (*Lucknow BSc*)
15. (*a*) Explain the process of fractional distillation.
   (*b*) Give a short note on 'azeotropes'. (*MDU Rohtak BSc*)
16. Write how molarity, molality and mole fraction are used to express the concentration of a solution. (*Guru Nanak Dev BSc*)
17. (*a*) Give brief accounts of
   (*i*) Raoult's law (*ii*) Henry's law
   (*b*) Discuss the principle of fractional distillation of miscible liquid pairs and describe the use of fractionating column in this connection. (*Guru Nanak Dev BSc*)
18. Calculate the ionic strength of a solution containing 0.2×10^3 M NaCl and 0.1×10^3 M Na_2SO_4
   Answer. $Na^+ = 9.2$ kg m^3; $Cl^- = 7.1$ kg m^3; $SO_4^{2-} = 9.6$ kg m^3 (*Madras BSc*)
19. Calculate the amount of Na^+ and Cl^- ions in kilograms present in 500×10^{-3} of 1.5×10^{-3} M NaCl solution.
   Answer. 17.3×10^{-3} kg Na^+ and 26.6×10^{-3} kg Cl (*Arunachal BSc*)
20. Calculate the molarity and normality of a solution containing 5.3×10^{-3} kg of Na_2CO_3 dissolved in 1m^3 solution.
   Answer. 0.05 M, 0.10 *N* (*Gulbarga BSc*)
21. Calculate the molarity of a solution containing 0.331kg of HCl dissolved in sufficient water to make 2 dm^3 of solution.
   Answer. 4.534×10^3 M (*Vidyasagar BSc*)
22. What is molarity of Fe^{2+} ions in a solution containing 0.200 kg of $FeCl_2$ per 10^{-3} m^3 of solution ?
   Answer. 1.574×10^3 M (*Mizoram BSc*)

23. Calculate the number of molecules of sugar present in 1 m^3 of 10% sugar solution having density = 1.20 kg/m^3.
Answer. 2.1×10^{20} *(Kalyani BSc)*

24. What is the normality of a solution containing 28.0×10^{-3} kg of KOH dissolved in sufficient water to make 400×10^{-3} m^3 of solution ?
Answer. 1.27 *N* *(Jamia Millia BSc)*

25. 0.005 kg of NaCl is dissolved in 1 kg of water. If the density of the solution is 997 kg/m^3, calculate the molarity, molality and mole fraction of the solute.
Answer. 0.085×10^3 M, 0.0847 *m*, 0.00153 *(Kerala BSc)*

26. (*a*) 49×10^{-3} kg of H_2SO_4 are dissolved in 250×10^{-3} m^3 of solution. Calculate the molarity of the solution.
(*b*) 45×10^{-3} kg of glucose, $C_6H_{12}O_6$, are dissolved in 500×10^{-3} kg of water. Calculate molality of the solution.
Answer. (*a*) 2 M; (*b*) 0.5 m *(Madurai BSc)*

27. (*a*) Give illustration of maximum and minimum azeotropic mixtures.
(*b*) A sample of spirit contains 92% of ethanol by weight, the rest being water. What is the mole fraction of its constituents?
Answer. (*b*) 0.1803; 0.8197 *(Sambalpur BSc)*

MULTIPLE CHOICE QUESTIONS

1. The concentration of a solution is defined as
(*a*) the amount of solvent present in a given amount of solution
(*b*) the amount of solute present in a given amount of solution
(*c*) the amount of solute present in a given amount of solvent
(*d*) the amount of solvent present in a given amount of solute
Answer. (*b*)

2. If *n* represents the number of moles of a solute and *N* represents the number of moles of a solvent, the mole fraction of the solvent is given by
(*a*) $\frac{n}{n+N}$ (*b*) $\frac{N}{n+N}$
(*c*) $\frac{n+N}{n}$ (*d*) $\frac{n+N}{N}$
Answer. (*b*)

3. The molarity is defined as the number of moles of solute present in
(*a*) one m^3 of the solvent (*b*) one m^3 of the solution
(*c*) one kilogram of the solvent (*d*) one kilogram of the solution
Answer. (*b*)

4. The molality of a solution is defined as the number of moles of solute present in
(*a*) one liter of the solvent (*b*) one liter of the solution
(*c*) one kilogram of the solvent (*d*) one kilogram of the solution
Answer. (*c*)

5. Normality of a solution is the number of ______ of solute per m^3 of the solution.
(*a*) moles (*b*) equivalents
(*c*) formula weight (*d*) mole fraction
Answer. (*b*)

6. For a gas in contact with a solvent at a constant temperature, the concentration of the gas that dissolves in the solvent is directly proportional to the pressure of the gas. This law is

(*a*) Henry's law (*b*) Raoult's law
(*c*) Dalton's law (*d*) van't Hoff's law

Answer. (*a*)

7. The weight percent of a solute in a solution is given by

(*a*) $\frac{\text{wt of the solvent}}{\text{wt of the solute}} \times 100$ (*b*) $\frac{\text{wt of the solute}}{\text{wt of the solvent}} \times 100$

(*c*) $\frac{\text{wt of the solute}}{\text{wt of the solution}} \times 100$ (*d*) $\frac{\text{wt of the solution}}{\text{wt of the solute}} \times 100$

Answer. (*c*)

8. The Henry's law is applicable if

(*a*) the temperature and pressure are moderate
(*b*) the solubility of the gas in the solvent is low
(*c*) the gas does not react with the solvent to form a new species
(*d*) all of the above

Answer. (*d*)

9. The temperature at which two conjugate solutions (or layers) merge into one another to form one layer is called the

(*a*) critical temperature (*b*) critical solution temperature
(*c*) distillation temperature (*d*) Dalton's temperature

Answer. (*b*)

10. A saturated solution is defined as one which is ________ with the excess of solid at a particular temperature

(*a*) in contact (*b*) in equilibrium
(*c*) contains impurities (*d*) none of these

Answer. (*b*)

11. In a super saturated solution, the amount of solute is ________ in comparison to saturated solution.

(*a*) more (*b*) less
(*c*) equal to (*d*) none of these

Answer. (*a*)

12. The solubility generally rises with

(*a*) increase in temperature (*b*) decrease in temperature
(*c*) increases in volume of the solvent (*d*) none of these

Answer. (*a*)

13. A saturated solution of KCl on heating becomes

(*a*) unsaturated (*b*) supersaturated
(*c*) hydrated (*d*) none of these

Answer. (*b*)

14. Mathematically, Henry's law can be expressed as (C is the concentration of the gas in solution and P is the pressure of the gas)

(*a*) $C \propto P$ (*b*) $C = kP$
(*c*) $k = C/P$ (*d*) all of these

Answer. (*d*)

15. The Henry's law gives the relationship between

(*a*) the pressure and solubility of a gas in a particular solvent
(*b*) the temperature and solubility of a gas in a particular solvent
(*c*) the composition of the mixture and solubility of a gas in a particular solvent
(*d*) none of these

Answer. (*a*)

16. When two non-reacting gases are mixed, a ________ is obtained
(*a*) homogeneous mixture (*b*) heterogeneous mixture
(*c*) equilibrium mixture (*d*) none of these
Answer. (*a*)

17. The completely miscible solution can be separated by
(*a*) a separating funnel (*b*) evaporation
(*c*) fractional distillation (*d*) none of these
Answer. (*c*)

18. The liquid mixtures which distill with a change in composition are called
(*a*) azeotropic mixtures (*b*) equilibrium mixtures
(*c*) zeotropic mixtures (*d*) nonequilibrium mixtures
Answer. (*c*)

19. The organic liquids which are steam volatile and immiscible with water can be separated by
(*a*) distillation (*b*) fractional distillation
(*c*) steam distillation (*d*) evaporation
Answer. (*c*)

20. Sugar dissolves in water due to the formation of
(*a*) covalent bonds (*b*) ionic bonds
(*c*) co-ordinate bonds (*d*) hydrogen bonding
Answer. (*d*)

21. The rate of crystallization is the rate at which the
(*a*) solid molecules go into the solvent
(*b*) solute molecules return to the crystal surface and are deposited on it
(*c*) solute molecules leave the surface of the solution
(*d*) solute molecules condense on the surface of the solution
Answer. (*b*)

22. When a saturated solution prepared at a higher temperature is cooled, we get
(*a*) super cooled solution (*b*) super saturated solution
(*c*) an equilibrium mixture (*d*) one molar solution
Answer. (*b*)

23. An alloy is a homogeneous mixture of
(*a*) two solids (*b*) two liquids
(*c*) two metals (*d*) two non-metals
Answer. (*c*)

24. The organic liquids which decompose at the boiling point are purified / separated by
(*a*) distillation (*b*) steam distillation
(*c*) fractional distillation (*d*) none of these
Answer. (*b*)

25. In a saturated solution there exists an equilibrium between
(*a*) solvent and excess of solid (*b*) solution and excess of solid
(*c*) solid and excess of solvent (*d*) solid and excess of solution
Answer. (*b*)

26. In one molal solution that contains 0.5 mole of a solute there is
(*a*) 1 kg of solvent (*b*) 1 m^3 of solvent
(*c*) 0.5 m3 of solvent (*d*) 0.5 kg of solvent
Answer. (*d*)

27. Which of the following does not depend upon the temperature?
(*a*) molarity (*b*) molality
(*c*) formality (*d*) normality
Answer. (*b*)

28. 36×10^{-3} kg of glucose (molecular mass 180×10^{-3} kg) is present in 500×10^{-3} kg of water, the molality of the solution is
(*a*) 0.2 (*b*) 0.4
(*c*) 0.8 (*d*) 1.0
Answer. (*b*)

29. The number of moles of a solute per kilogram of the solvent is called
(*a*) formality (*b*) normality
(*c*) molarity (*d*) molality
Answer. (*d*)

30. The mole fraction of ethyl alcohol in a solution containing 36×10^{-3} kg of H_2O and 46×10^{-3} kg of ethyl alcohol is
(*a*) 1/3 (*b*) 1/2
(*c*) 2/3 (*d*) 3/4
Answer. (*a*)

31. The molarities of 0.1 N solution of HCl and 0.1 N solution of H_2SO_4 are respectively
(*a*) 0.1 M and 0.1 M (*b*) 0.1 M and 0.05 M
(*c*) 0.05 M and 0.1 M (*d*) 0.1 M and 0.2 M
Answer. (*b*)

32. For the preparation of 0.1 m^3 of 0.1 N solution of sulfuric acid, the amount of H_2SO_4 needed is
(*a*) 4.9×10^{-3} kg (*b*) 9.8×10^{-3} kg
(*c*) 14.7×10^{-3} kg (*d*) 19.6×10^{-3} kg
Answer. (*a*)

33. The normalities of 1 M solution of HCl, 1 M H_2SO_4 and 1 M H_3PO_4 are respectively
(*a*) 1 N, 2 N and 3 N (*b*) 2 N, 3 N and 1 N
(*c*) 0.1 N, 0.2 N and 0.3 N (*d*) 3 N, 2 N and 1 N
Answer. (*a*)

34. How many grams of glucose are present in 100 ml of 0.1 M solution
(*a*) 180×10^{-3} kg (*b*) 18.0×10^{-3} kg
(*c*) 1.8×10^{-3} kg (*d*) 3.6×10^{-3} kg
Answer. (*c*)

35. In 1.0 M solution of sulfuric acid, the moles of hydrogen ions present will be
(*a*) 1.0 M (*b*) 2.0 M
(*c*) 3.0 M (*d*) 4.0 M
Answer. (*b*)

36. The normality of a solution of sulfuric acid is $\frac{N}{10}$. Its molarity will be
(*a*) $\frac{M}{5}$ (*b*) $\frac{M}{10}$
(*c*) $\frac{M}{20}$ (*d*) $\frac{M}{40}$
Answer. (*c*)

37. A solution of KCl contains 74.5×10^{-3} kg of it in 90×10^{-3} kg of water. The mole fraction of KCl will be
(*a*) 1/6 (*b*) 1/5
(*c*) 1/4 (*d*) $\frac{74.5}{90}$
Answer. (*a*)

38. 49×10^{-3} kg of H_2SO_4 are dissolved in 250×10^{-3} m^3 of the solution. Its molarity will be

(*a*) 0.5×10^3 M (*b*) 1.0×10^3 M

(*c*) 1.5×10^3 M (*d*) 2.0×10^3 M

Answer. (*d*)

39. The amount of Na^+ in 0.1×10^3 M Na_2SO_4 is

(*a*) 2.3×10^{-3} kg (*b*) 4.6×10^{-3} kg

(*c*) 23×10^{-3} kg (*d*) 46×10^{-3} kg

Answer. (*b*)

40. 45 g of glucose is dissolved in 500×10^{-3} kg of water. The molality of the solution is

(*a*) 0.25 (*b*) 0.5

(*c*) 0.75 (*d*) 1.0

Answer. (*b*)

41. How many grams of water are present in 0.1 m^3 of 0.1×10^{-3} m solution of sulfuric acid?

(*a*) 0.010 kg (*b*) 0.020 kg

(*c*) 0.050 kg (*d*) 0.100 kg

Answer. (*d*)

42. The total weight of 100 ml of 2 M solution of HCl will be

(*a*) 107.3 kg (*b*) 0.1073 kg

(*c*) 0.01073 kg (*d*) 0.001073 kg

Answer. (*b*)

43. 1 kg of a solution of $CaCO_3$ contains 1×10^{-3} kg of calcium carbonate. Concentration of the solution will be

(*a*) 1 ppm (*b*) 10 ppm

(*c*) 100 ppm (*d*) 1000 ppm

Answer. (*d*)

44. A solution contains 180 g of glucose in 180 g of water. The total number of moles of both solute and solvent in the solution will be

(*a*) 1 (*b*) 10

(*c*) 11 (*d*) 21

Answer. (*c*)

45. A sample of H_2O_2 used for bleaching of hair contains 5.1×10^{-3} kg of H_2O_2 in 90×10^{-3} kg of water. The mole fraction of H_2O_2 is

(*a*) $\frac{0.15}{5}$ (*b*) $\frac{0.15}{5.15}$

(*c*) $\frac{5}{5.15}$ (*d*) $\frac{5}{0.15}$

Answer. (*b*)

46. Which one of the following has molality equal to one?

(*a*) 0.0365 kg of HCl in 0.5 m^3 of water (*b*) 0.0365 kg of HCl in 0.1 m^3 of water

(*c*) 0.0365 kg of HCl in 0.1 m^3 of water (*d*) 0.0365 kg of HCl in 0.5 kg of water

Answer. (*c*)

47. Which weighs the most?

(*a*) 1×10^{-3} M solution of HCl (*b*) 25×10^{-3} kg of H_2SO_4

(*c*) 2 moles of water (*d*) 1 kg of $CaCO_3$

Answer. (*a*)

48. The weight of urea required to prepare 0.2 m^3 of 2×10^3 M solution will be

(*a*) 0.012 kg (*b*) 0.024 kg

(*c*) 0.020 kg (*d*) 0.060 kg

Answer. (*b*)

49. What is the molality of a solution prepared by dissolving 9.2×10^{-3} kg toluene (C_7H_8) in 500×10^{-3} kg of benzene

(*a*) $1/2 \times 10^3$ (*b*) $1/4 \times 10^3$

(*c*) $1/5 \times 10^3$ (*d*) $1/10 \times 10^3$

Answer. (*c*)

50. A solution of urea is prepared by dissolving 0.180 kg of it in one kg of water. The mole fraction of urea in the solution

(*a*) $\frac{1}{55.5}$ (*b*) $\frac{1}{56.5}$

(*c*) $\frac{1}{100}$ (*d*) 1.8

Answer. (*b*)

15 Theory of Dilute Solutions

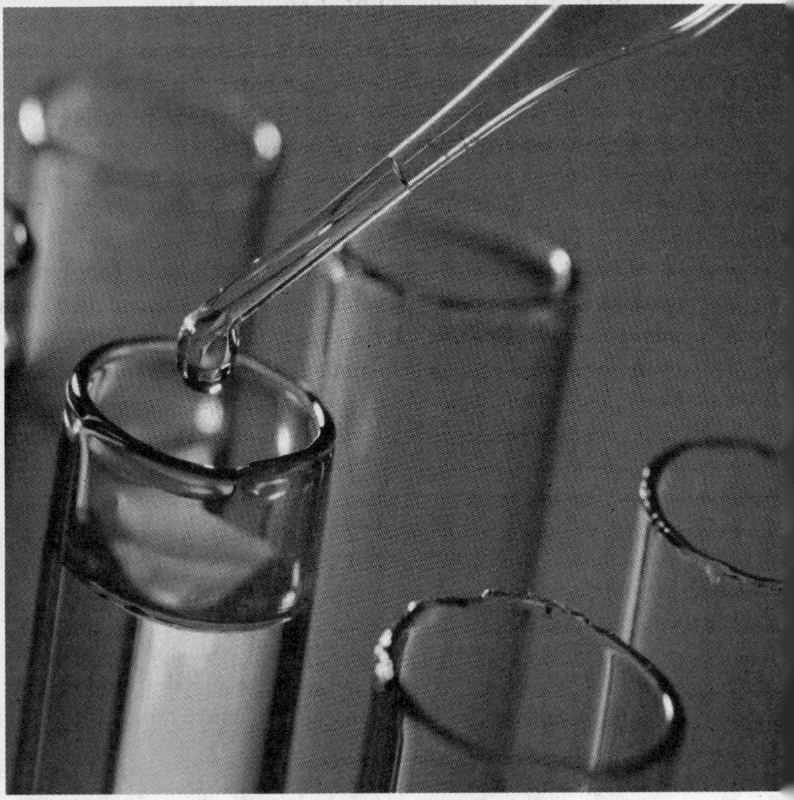

CONTENTS

In this chapter we will restrict our discussion to solutions in which a solid is dissolved in a liquid. The solid is referred to as the **solute** and the liquid as the **solvent.**

COLLIGATIVE PROPERTIES

Dilute solutions containing non-volatile solute exhibit the following properties :

(1) Lowering of the Vapor Pressure
(2) Elevation of the Boiling Point
(3) Depression of the Freezing Point
(4) Osmotic Pressure

The essential feature of these properties is that they depend *only* on the number of solute particles present in solution. Being closely related to each other through a common explanation, these have been grouped together under the class name **Colligative Properties** (Greek *colligatus* = Collected together).

A colligative property may be defined as one which depends on the number of particles in solution and not in any way on the size or chemical nature of the particles.

Consequent to the above definition, each colligative property is exactly related to any other. Thus if one property is measured, the other can be calculated. The colligative properties of dilute solutions are particularly important as these provide valuable methods for finding the molecular weights of the dissolved substances.

LOWERING OF VAPOR PRESSURE : RAOULT'S LAW

The vapor pressure of a pure solvent is decreased when a non-volatile solute is dissolved in it. If p is the vapor pressure of the solvent and p_s that of the solution, the lowering of vapor pressure is $(p - p_s)$. This lowering of vapor pressure relative to the vapor pressure of the pure solvent is termed the **Relative lowering of Vapor pressure.** Thus,

$$\text{Relative Lowering of Vapor Pressure} = \frac{p - p_s}{p}$$

As a result of extensive experimentation, Raoult (1886) gave an empirical relation connecting the relative lowering of vapor pressure and the concentration of the solute in solution. This is now referred to as the **Raoult's Law.** It states that : **the relative lowering of the vapor pressure of a dilute solution is equal to the mole fraction of the solute present in dilute solution.**

Raoult's Law can be expressed mathematically in the form :

$$\frac{p - p_s}{p} = \frac{n}{n + N}$$

where n = number of moles or molecules of solute

N = number of moles or molecules of solvent.

Derivation of Raoult's Law

The vapor pressure of the pure solvent is caused by the number of molecules evaporating from its surface. When a nonvolatile solute is dissolved in solution, the presence of solute molecules in the surface blocks a fraction of the surface where no evaporation can take place.

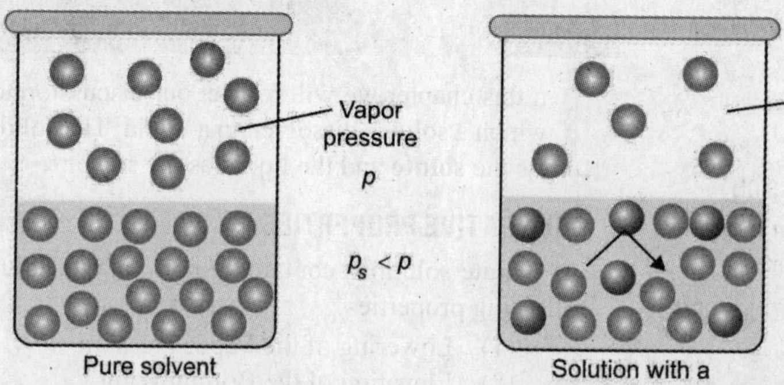

■ **Figure 15.1**
Lowering of vapor pressure by a nonvolatile solute. The particles of the solute block the escape of solvent molecules from the surface of the solution.

This causes the lowering of the vapor pressure. The vapor pressure of the solution is, therefore, determined by the number of molecules of the solvent present at any time in the surface which is proportional to the mole fraction. That is,

$$p_s \propto \frac{N}{n+N}$$

where N = moles of solvent and n = moles of solute.

or
$$p_s = k\,\frac{N}{n+N} \qquad ...(1)$$

k being proportionality factor.

In case of pure solvent $n = 0$ and hence

$$\text{Mole fraction of solvent} = \frac{N}{n+N} = \frac{N}{0+N} = 1$$

Now from equation (1), the vapor pressure $p = k$

Therefore the equation (1) assumes the form

$$p_s = p\frac{N}{n+N}$$

or
$$\frac{p_s}{p} = \frac{N}{n+N}$$

$$1 - \frac{p_s}{p} = 1 - \frac{N}{n+N}$$

$$\frac{p - p_s}{p} = \frac{n}{n+N}$$

This is Raoult's Law.

SOLVED PROBLEM 1

Calculate the vapor pressure lowering caused by the addition of 0.1 kg of sucrose (mol mass = 342g or 0.342 kg) to 1 kg of water if the vapor pressure of pure water at 298 K is 3.173×10^3 Pa.

SOLUTION

Using Raoult's Law Equation

$$\frac{p - p_s}{p} = \frac{\Delta p}{p} = \frac{n}{n+N} \qquad ...(1)$$

where
Δp = lowering of vapor pressure

p = vapor pressure of water = 3.173×10^3 Pa

n = moles of sucrose $= \frac{0.1}{0.342} = 0.292$ mol

N = moles of water $= \frac{1}{0.018} = 55.5$ mol

Substituting values in equation (1)

$$\frac{\Delta p}{3.173 \times 10^3} = \frac{0.292}{0.292 + 55.5}$$

$$\Delta p = 3.173 \times 10^3 \times \frac{0.292}{55.792} = \mathbf{0.0167 \times 10^3\ Pa}$$

Thus the lowering of vapor pressure = **0.0167×10^3 Pa**

Ideal Solutions and Deviations from Raoult's Law

A solution which obeys Raoult's law strictly is called an **Ideal solution.** A solution which shows deviations from Raoult's law is called a **Nonideal** or **Real solution.**

Suppose the molecules of the solvent and solute are represented by A and B respectively. Now let γ_{AB} be the attractive force between A and B, and γ_{AA} between A and A. If

$$\gamma_{AB} = \gamma_{AA}$$

the solution will show the same vapor pressure as predicted by Raoult's law and it is an ideal solution. However, if

$$\gamma_{AB} > \gamma_{AA}$$

molecule A will escape less readily and the vapor pressure will be less than that predicted by Raoult's law (**Negative deviation**). On the other hand, if

$$\gamma_{AB} < \gamma_{AA}$$

A molecule will escape from the solution surface more readily and the vapor pressure of the solution will be higher than predicted by Raoult's law **(Positive deviation).**

In very dilute solutions of nonelectrolytes, the solvent and solute molecules are very much alike in both molecular size and molecular attractions. Thus such solutions tend to approach the ideal behavior and obey Raoult's law fairly accurately.

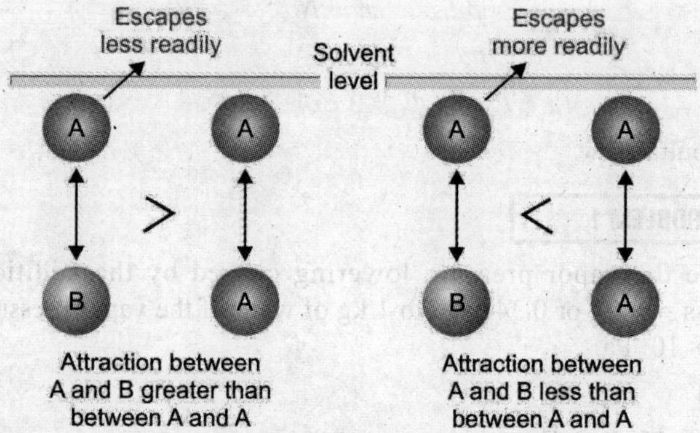

Figure 15.2
Negative deviation and Positive deviation.

Determination of Molecular Mass from Vapor Pressure Lowering

The molecular mass of a nonvolatile solute can be determined by measuring the lowering of vapor pressure $(p - p_s)$ produced by dissolving a known weight of it in a known weight of the solvent.

If in a determination w grams of solute is dissolved in W grams of the solvent, m and M are molecular masses of the solute and solvent respectively, we have :

$$\text{No. of Moles of solute } (n) = \frac{w}{m}$$

$$\text{No. of Moles of solvent } (N) = \frac{W}{M}$$

Substituting these values in the Raoult's law Equation,

$$\frac{p - p_s}{P} = \frac{n}{n + N} \qquad \text{...(Raoult's Law Eq.)}$$

$$\frac{p - p_s}{P} = \frac{w/m}{w/m + W/M} \quad ...(1)$$

Since for very dilute solution, the number of moles (molecules) of solute (w/m), is very small, it can be neglected in the denominator. The equation (1) can now be written as

$$\frac{p - p_s}{P} = \frac{wM}{mW} \quad ...(2)$$

Knowing the experimental value of $p - p_s/p$, and the molecular mass of the solvent (M), the molecular weight of solute (m) can be calculated from (1) or (2).

SOLVED PROBLEM 2

The vapor pressure of ether (mol mass = 74 or 0.074) is 5.8928×10^4 Pa at 293 K. If 0.003 kg of a compound A are dissolved in 0.05 kg of ether at this temperature, the vapor pressure falls to 5.6795×10^4 Pa. Calculate the molecular mass of A. Assume that the solution of A in ether is very dilute.

SOLUTION

Here the approximate form of the Raoult's law Equation will be used.

$$\frac{p - p_s}{p} = \frac{\frac{w}{m}}{W/M} = \frac{wM}{mW} \quad ...(1)$$

In this case :

w, the mass of solute (A) = 0.003 kg
W, the mass of solvent (ether) = 0.05 kg
m, the mol mass of solute A = ?
M, the mol mass of solvent (ether) = 0.074 kg
p, the vapor pressure of solvent (ether) = 5.8928×10^4 Pa
p_s, the vapor pressure of solution = 5.6795×10^4 Pa

Substituting the values in equation (1),

$$\frac{5.8928\times10^4 - 5.6795\times10^4}{5.8928\times10^4} = \frac{0.003 \times 0.074}{m \times 0.05}$$

$$\frac{0.2133\times10^4}{5.8928\times10^4} = \frac{0.003 \times 0.074}{m \times 0.05}$$

$$m = \frac{0.003 \times 0.074\times5.8928\times10^4}{0.05 \times 0.2133}$$

Hence $m = 0.12267 = \mathbf{0.123\ kg}$

Thus **the molecular weight of A is 0.123 kg or 123 g**

SOLVED PROBLEM 3

0.0182 kg of urea is dissolved in 0.1 kg of water at 323 K. The lowering of vapor pressure produced is 0.0666×10^4 Pa. Calculate the molecular mass of urea. The vapor pressure of water at 323 K is 1.2265×10^4 Pa.

SOLUTION

Since the solution is not very dilute, the complete Raoult's Law Equation applied is

$$\frac{p - p_s}{p} = \frac{w/m}{W/M + w/m} \quad ...(1)$$

In this case :

w, the mass of solute (urea)	= 0.0182 kg
W, *the* mass of solvent (water)	= 0.1 kg
m, the mol mass of solute (urea)	= ?
M, the mol mass of solvent (water)	= 0.018 kg
$p - p_s$, the lowering of vapor pressure	= 0.0666×10^4 Pa
p, the vapor pressure of solvent (water)	= 1.2265×10^4 Pa

Substituting these values in equation (1),

$$\frac{0.0666\times10^4}{1.2265\times10^4} = \frac{0.0182/m}{0.0182/m + 0.1/0.018}$$

Hence m, the molecular mass of urea = **0.05705 kg or 57.05 grams**

MEASUREMENT OF LOWERING OF VAPOR PRESSURE

(1) Barometric Method

Raoult measured the individual vapor pressure of a liquid and then the solution by this method. He introduced the liquid or the solution into Toricellian vacuum of a barometer tube and measured the depression of the mercury level. This method is neither practicable nor accurate as the lowering of vapor pressure is too small.

(2) Manometric Method

The vapor pressure of a liquid or solution can be conveniently measured with the help of a manometer (see Fig. 15.3). The bulb B is charged with the liquid or solution. The air in the connecting tube in then removed with a vacuum pump. When the stopcock is closed, the pressure inside is due only to the vapor evaporating from the solution or liquid. This method is generally used for aqueous solutions. The manometric liquid can be mercury or n-butyl phthalate which has low density and low volatility.

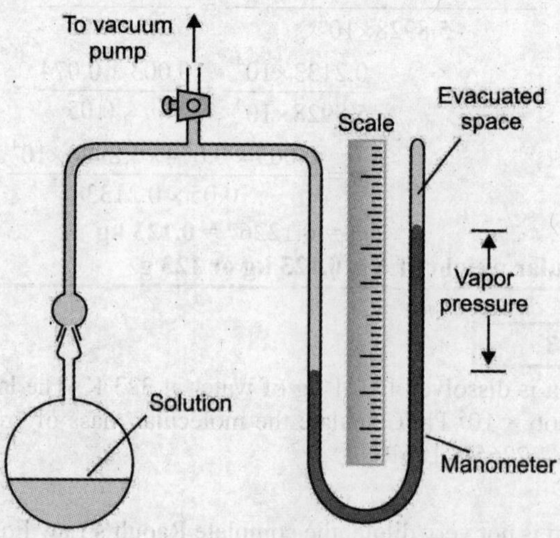

Figure 15.3
Measurement of vapor pressure of aqueous solutions with a manometer.

(3) **Ostwald and Walker's Dynamic Method** (*Gas Saturation Method*)

In this method the relative lowering of vapor pressure can be determined straightway. The measurement of the individual vapor pressures of a solution and solvent is thus eliminated.

Procedure. The apparatus used by Ostwald and Walker is shown in Fig. 15.4. It consists of two sets of bulbs :

(*a*) Set *A* containing the solution

(*b*) Set *B* containing the solvent

Each set is weighed separately. A slow stream of dry air is then drawn by suction pump through the two sets of bulbs. At the end of the operation, these sets are reweighed. From the loss of weight in each of the two sets, the lowering of vapor pressure is calculated. The temperature of the air, the solution and the solvent must be kept constant throughout.

Calculations. As the air bubbles through set *A* it is saturated up to the vapor pressure p_s of solution and then up to vapor pressure p of solvent in set *B*. Thus the amount of solvent taken up in set *A* is proportional to p_s and the amount taken up in set *B* is proportional to $(p - p_s)$.

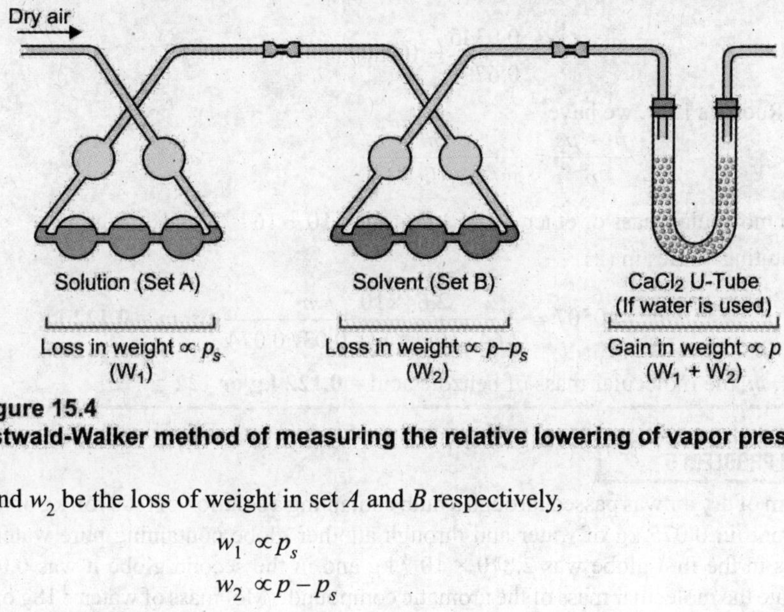

Figure 15.4
Ostwald-Walker method of measuring the relative lowering of vapor pressure.

If w_1 and w_2 be the loss of weight in set *A* and *B* respectively,

$$w_1 \propto p_s \quad ...(1)$$

$$w_2 \propto p - p_s \quad ...(2)$$

Adding (1) and (2),we have

$$w_1 + w_2 \propto p_s + p - p_s$$

$$\propto p \quad ...(3)$$

Dividing (2) by (3), we can write

$$\frac{p - p_s}{p} = \frac{w_2}{w_1 + w_2} \quad ...(4)$$

Knowing the loss of mass in set *B* (w_2) and the total loss of mass in the two sets ($w_1 + w_2$), we can find the relative lowering of vapor pressure from equation (4).

If water is the solvent used, a set of calcium chloride tubes (or a set of bulbs containing conc. H_2SO_4) is attached to the end of the apparatus to catch the escaping water vapor. Thus the gain in mass of the $CaCl_2$-tubes is equal to $(w_1 + w_2)$, the total loss of mass in sets *A* and *B*.

SOLVED PROBLEM 4

A current of dry air was passed through a solution of 0.00264 kg of benzoic acid in 0.03 kg of ether ($C_2H_5OC_2H_5$) and then through pure ether. The loss in weight of the solution was 0.000645 kg and the ether 0.0000345 kg. What is the molecular mass of benzoic acid?

SOLUTION

According to the theory of Ostwald-Walker method,

$$\frac{p - p_s}{p} = \frac{w_2}{w_1 + w_2} \qquad ...(1)$$

In this case,

w_1, loss of mass of solution $= 0.645 \times 10^{-3}$ kg

w_2, loss of mass of solvent $= 0.0345 \times 10^{-3}$ kg

Substituting values in equation (1)

$$\frac{p - p_s}{p} = \frac{0.0345}{0.645 + 0.0345}$$

$$= \frac{0.0345}{0.6795} = 0.0507$$

From Raoult's Law, we have

$$\frac{p - p_s}{p} = \frac{w/m}{w/m + W/M} \qquad ...(2)$$

M the molecular mass of ether, $(C_2H_5)_2O = 48 + 10 + 16 = 74 \times 10^{-3}$ kg

Substituting values in (2)

$$0.0507 = \frac{2.64\times10^{-3}/m}{2.64\times10^{-3}/m + 0.03/0.074} \quad \text{or} \quad \mathbf{m = 0.122\ kg}$$

Hence, m, the molecular mass of benzoic acid = **0.122 kg or 122 g**

SOLVED PROBLEM 5

A stream of dry air was passed through a bulb containing a solution of 0.00750 kg of an aromatic compound in 0.075 kg of water and through another globe containing pure water. The loss in mass in the first globe was 2.810×10^{-3} kg and in the second globe it was 0.000054 kg. Calculate the molecular mass of the aromatic compound. (Mol mass of water = 18g or 0.018 kg)

SOLUTION

According to the theory of Ostwald-Walker method,

$$\frac{p - p_s}{p} = \frac{w_2}{w_1 + w_2} \qquad ...(1)$$

In the present case,

w_1, loss of mass of solution $= 2.810 \times 10^{-3}$ kg

w_2, loss mass of solvent (water) $= 0.054 \times 10^{-3}$ kg

Substituting values in (1)

$$\frac{p - p_s}{p} = \frac{0.054\times10^{-3}}{2.810\times10^{-3} + 0.054\times10^{-3}} = \frac{0.054}{2.864} = 0.0188 \qquad ...(2)$$

From Raoult's Law,

$$\frac{p - p_s}{p} = \frac{w/m}{w/m + W/M}$$

Substituting values

$$0.0188 = \frac{7.50\times10^{-3}/m}{7.50\times10^{-3}/m + 75.0\times10^{-3}/18\times10^{-3}} \quad \text{or } \mathbf{m = 0.0936\ kg}$$

Hence m, mol mass of solute = **0.0936 kg** or **93.6 g**

SOLVED PROBLEM 6

In an experiment air was drawn successively through a solution of sugar (0.03889 kg per 0.1 kg water) and distilled water, and then through anhydrous calcium chloride. It was found that the water lost was 0.921×10^{-4} kg and calcium chloride tubes gained 0.5163×10^{-2} kg. Find the molecular mass of the sugar. (Mol mass of H_2O = 0.018 kg)

SOLUTION

According to the theory of Ostwald-Walker method,

$$\frac{p - p_s}{p} = \frac{w_2}{w_1 + w_2} \quad ...(1)$$

In this case,

w_2, the loss of mass of water = 0.921×10^{-4} kg

$(w_1 + w_2)$, the total loss of solution and solvent = 0.5163×10^{-2} kg

Substituting values in equation (1)

$$\frac{p - p_s}{p} = \frac{0.0921\times10^{-3}}{5.163\times10^{-3}} = 0.01784$$

From Raoult's Law,

$$\frac{p - p_s}{p} = \frac{w/m}{w/m + W/M}$$

Substituting values

$$0.0178 = \frac{38.89\times10^{-3}/m}{38.89\times10^{-3}/m + 100\times10^{-3}/18}$$

Hence m, mol mass of sugar = **0.386 kg or 386.27 g**

ELEVATION OF BOILING POINT

Relation between Elevation of Boiling Point and Lowering of Vapor-pressure

When a liquid is heated, its vapor pressure rises and when it equals the atmospheric pressure, the liquid boils. The addition of a non volatile solute lowers the vapor pressure and consequently elevates the boiling point as the solution has to be heated to a higher temperature to make its vapor pressure become equal to atmospheric pressure. If T_b is the boiling point of the solvent and T is the boiling point of the solution, the difference in the boiling points (ΔT) is called the **elevation of boiling point.**

$$T - T_b = \Delta T$$

The vapor pressure curves of the pure solvent, and solutions (1) and (2) with different concentrations of solute are shown in Fig.15.5.

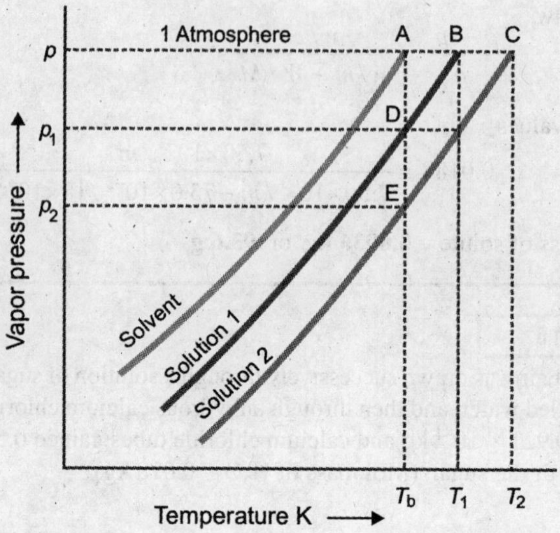

Figure 15.5
Ostwald-Walker method of measuring the relative lowering of vapor pressure.

For dilute solutions, the curves *BD* and *CE* are parallel and straight lines approximately. Therefore for similar triangles *ACE* and *ABD*, we have

$$\frac{AB}{AC} = \frac{AD}{AE}$$

or
$$\frac{T_1 - T_b}{T_2 - T_b} = \frac{p - p_1}{p - p_2}$$

where $p - p_1$ and $p - p_2$ are lowering of vapor pressure for solution 1 and solution 2 respectively. Hence **the elevation of boiling point is directly proportional to the lowering of vapor pressure.**

or
$$\Delta T \propto p - p_s \qquad ...(1)$$

Determination of Molecular Mass from Elevation of Boiling Point

Since p is constant for the same solvent at a fixed temperature, from (1) we can write

$$\Delta T \propto \frac{p - p_s}{p} \qquad ...(2)$$

But from Raoult's Law for dilute solutions,

$$\frac{p - p_s}{p} \propto \frac{wM}{Wm} \qquad ...(3)$$

Since M (mol mass of solvent) is constant, from (3)

$$\frac{p - p_s}{p} \propto \frac{w}{Wm} \qquad ...(4)$$

From (2) and (4)

$$\Delta T \propto \frac{w}{m} \times \frac{1}{W}$$

or
$$\Delta T = K_b \times \frac{w}{m} \times \frac{1}{W} \qquad ...(5)$$

where K_b is a constant called **Boiling point constant or Ebulioscopic constant of molal elevation constant.** If $w/m = 1$, $W = 1$, $K_b = \Delta T$. Thus,

Molal elevation constant may be defined as the boiling-point elevation produced when 1 mole of solute is dissolved in one kg (1000 g) of the solvent.

If the mass of the solvent (W) is given in grams, it has to be converted into kilograms. Thus the expression (5) assumes the form

$$\Delta T = K_b \times \frac{w}{m} \times \frac{1}{W/1000} \quad ...(6)$$

$$m = \frac{1000 \times K_b \times w}{\Delta T \times W} \quad ...(7)$$

Hence

where ΔT = elevation of boiling point; K_b = molal elevation constant; w = mass of solute in grams; m = mol mass of solute; and W = mass of solvent in grams.

Sometimes the value of K_b is given in K per 0.1 kg (100 g). In that case, the expression (6) becomes

$$m = \frac{100 \times K_b \times w}{\Delta T \times W}$$

The value of K_b. The value of K_b can be determined by measurement of ΔT by taking a solute of known molecular mass (m) and substituting the values in expression (7).

Units of K_b. From equation (6), we have

$$K_b = \frac{\Delta T \times W/1000}{w/m} = \frac{\Delta T \times \text{kg-solvent}}{\text{mol-solute}}$$

Thus the units of K_b are

$$\frac{\text{k kg-solvent}}{\text{mol-solute}}$$

The constant K_b, which is characteristic of a particular solvent used, can also be calculated from thermodynamically derived relationship

$$K_b = \frac{RT_b^2}{1000 \times L_v}$$

where R = gas constant; T_b = boiling point of solvent; L_v = molar latent heat of vaporization. Thus for water $R = 8.314$ J mol^{-1}; $T = 373$ K : $L_v = 2260$ J g^{-1}

Therefore,

$$K_b = \frac{8.314 \times 373 \times 373}{1000 \times 2260} = 0.52 \text{ K Kg}^{-1}$$

The molal boiling point constant for some common solvents are listed in Table 15.1

TABLE 15.1. MOLAL BOILING-POINT CONSTANTS

Solvent	K_b per kg (1000 g)	K_b per 0.1 kg (100 g)
Water	0.52	5.2
Propanone (acetone)	1.70	17.0
Ethoxyethane (ether)	2.16	21.6
Ethanoic acid (acetic acid)	3.07	30.7
Ethanol	1.75	11.5
Benzene	2.70	27.0
Trichloromethane (chloroform)	3.67	36.7

SOLVED PROBLEM 7

The boiling point of a solution containing 0.00020 kg of a substance X in 0.020 kg of ether is 0.17 K higher than that of pure ether. Calculate the molecular mass of X. Boiling point constant of ether per 1 Kg is 2.16 K.

SOLUTION

Applying the expression $$m = \frac{1000 \times K_b \times w}{\Delta T \times W}$$

In this case, we have

$$\Delta T = 0.17 \text{ K}$$
$$K_b = 2.16$$
$$w = 0.20 \times 10^{-3} \text{ kg}$$
$$W = 20 \times 10^{-3} \text{ kg}$$

Substituting values

$$m = \frac{1000 \times 2.16 \times 0.20 \times 10^{-3}}{0.17 \times 20.00 \times 10^{-3}}$$

$$m = \mathbf{0.000127\ kg\ \ or\ \ 127.06\ g}$$

SOLVED PROBLEM 8

Acetone boils at 329.53 K and a solution of 1.41×10^{-3} kg of an organic solid in 0.020 kg of acetone boils at 330.03 K. If K for acetone per 0.1 kg is 1.67, calculate the mass of one mole of the organic solid.

SOLUTION

Applying the expression

$$m = \frac{1000 \times K_b \times w}{\Delta T \times W}$$

In this case, we have

$$\Delta T = 330.03 \text{ K} - 329.53 \text{ K} = 0.50 \text{ K}$$
$$K = 16.7$$
$$w = 1.41 \times 10^{-3} \text{ kg}$$
$$W = 20 \times 10^{-3} \text{ kg}$$

Substituting the values

$$m = \frac{1000 \times 1.67 \times 1.41 \times 10^{-3}}{0.50 \times 20 \times 10^{-3}}$$

Hence $m =$ **0.235 kg** or molar mass = **235 *g***

SOLVED PROBLEM 9

In a Cottrell determination, 0.022 kg of benzene was used as solvent. The readings on the differential thermometer before and after adding 0.000586 kg of naphthalene (mol mass = 128g = 128×10^{-3} kg), were 1.262 and 1.799 respectively. In a separate experiment, using the same amount of benzene but this time adding 0.000627 kg of an organic compound X, the temperature readings were 1.269 and 1.963. Calculate the molecular mass of X.

SOLUTION

Calculation of K_b

The value of K_b is found from the results of the first experiment.

$$\Delta T = 1000\, K_b \frac{w}{m \times W}$$

$$\Delta T = 1.799 - 1.262 = 0.537$$

$$w = 0.586 \times 10^{-3} \text{ kg}$$

$$m = 128 \times 10^{-3} \text{ kg}$$

$$W = 22.0 \times 10^{-3} \text{ kg}$$

Hence,

$$0.537 = 1000 \times K_b \times \frac{0.586 \times 10^{-3}}{128 \times 10^{-3} \times 22 \times 10^{-3}}$$

$$\therefore \quad K_b = \frac{0.537 \times 128 \times 10^{-3} \times 22.0 \times 10^{-3}}{1000 \times 0.586 \times 10^{-3}} = \mathbf{0.00258053}$$

$$= \mathbf{2.58 \times 10^{-3}\ K_b \text{ per kg}}$$

Molecular weight of X

Knowing the value of K_b, m, the molecular mass can be found from the expression

$$m = \frac{1000 \times K_b \times w}{\Delta T \times W}$$

In this case,

$$\Delta T = 1.963 - 1.269 = 0.694$$

$$w = 0.627 \times 10^{-3} \text{ kg}$$

$$W = 22 \times 10^{-3} \text{ kg}$$

Substituting these values,

$$m = \frac{1000 \times 0.00258 \times 0.627 \times 10^{-3}}{0.694 \times 22 \times 10^{-3}}$$

Hence $\quad m = \mathbf{0.10595\ kg = 106\ g}$

MEASUREMENT OF BOILING–POINT ELEVATION

There are several methods available for the measurement of the elevation of boiling point. Some of these are outlined below :

(1) Landsberger-Walker Method

This method was introduced by Landsberger and modified by Walker.

Apparatus. The apparatus used in this method is shown in Fig. 15.6 and consists of : (*i*) An *inner tube* with a hole in its side and graduated in ml; (*ii*) A *boiling flask* which passes solvent vapor in to the graduated tube through a 'rosehead' (a bulb with several holes') (*iii*) An *outer tube* which receives hot solvent vapor issuing from the side-hole of the inner tube; (*iv*) A *thermometer* reading to 0.01 K, dipping in solvent or solution in the inner tube.

Procedure. Pure solvent is placed in the graduated tube and vapor of the same solvent boiling in a separate flask is passed into it. The vapor causes the solvent in the tube to boil by its latent heat of condensation. When the solvent starts boiling and temperature becomes constant, its boiling point is recorded.

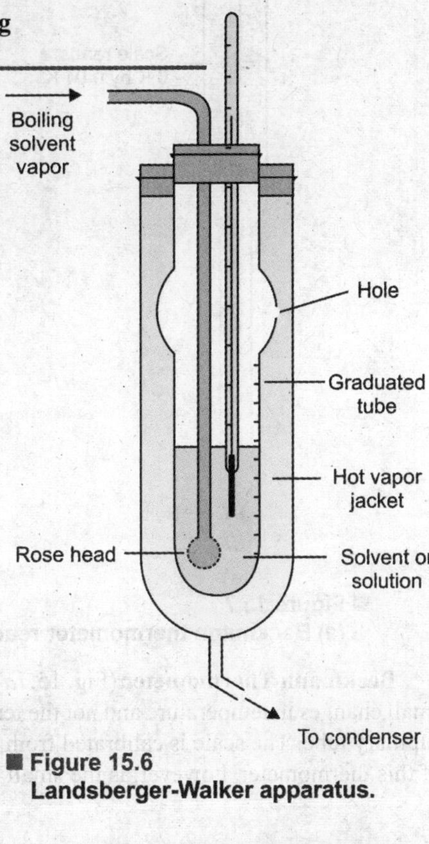

■ **Figure 15.6**
Landsberger-Walker apparatus.

Now the supply of vapor is temporarily cut off and a weighed pellet of the solute is dropped into the solvent in the inner tube. The solvent vapor is again passed through until the boiling point of the solution is reached and this is recorded. The solvent vapor is then cut off, thermometer and rosehead raised out of the solution, and the volume of the solution read.

From a difference in the boiling points of solvent and solution, we can find the molecular weight of the solute by using the expression

$$m = \frac{1000 \times K_b \times w}{\Delta T \times W}$$

where w = weight of solute taken, W = weight of solvent which is given by the volume of solvent (or solution) measured in m^3 multiplied by the density of the solvent at its boiling point.

(2) **Cottrell's Method**

A method better than Landsberger-Walker method was devised by Cottrell (1910).

Apparatus. It consists of : (*i*) a graduated *boiling tube* containing solvent or solution; (ii) a reflux condenser which returns the vaporized solvent to the boiling tube; (*iii*) a thermometer reading to 0.01 K, enclosed in a glass hood; (*iv*) A small inverted funnel with a narrow stem which branches into three jets projecting at the thermometer bulb.

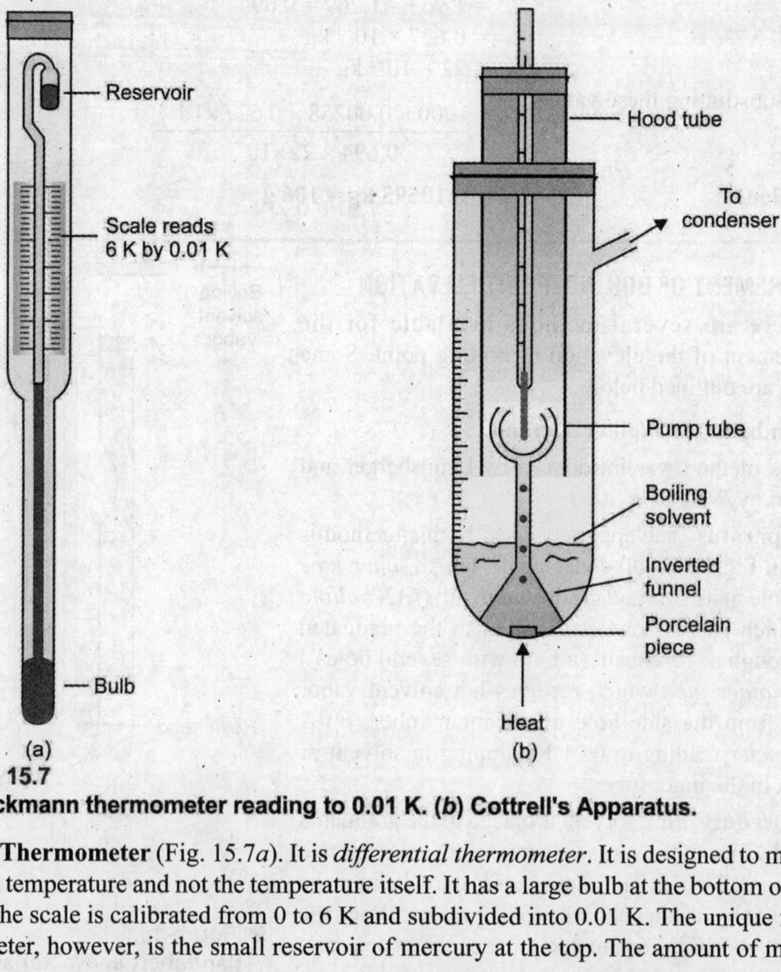

■ **Figure 15.7**
***(a)* Beckmann thermometer reading to 0.01 K. *(b)* Cottrell's Apparatus.**

Beckmann Thermometer (Fig. 15.7*a*). It is *differential thermometer*. It is designed to measure small changes in temperature and not the temperature itself. It has a large bulb at the bottom of a fine capillary tube. The scale is calibrated from 0 to 6 K and subdivided into 0.01 K. The unique feature of this thermometer, however, is the small reservoir of mercury at the top. The amount of mercury in this reservoir can be decreased or increased by tapping the thermometer gently. In this way the thermometer is adjusted so that the level of mercury thread will rest at any desired point on the scale when the instrument is placed in the boiling (or freezing) solvent.

Procedure. The apparatus is fitted up as shown in Fig. 15.7 (*b*). Solvent is placed in the boiling tube with a porcelain piece lying in it. It is heated on a small flame (micro burner). As the solution starts boiling, solvent vapor arising from the porcelain piece pump the boiling liquid into the narrow stem. Thus a mixture of solvent vapor and boiling liquid is continuously sprayed around the thermometer bulb. The temperature soon becomes constant and the boiling point of the pure solvent is recorded.

Now a weighed pellet of the solute is added to the solvent and the boiling point of the solution noted as the temperature becomes steady. Also, the volume of the solution in the boiling tube is noted. The difference of the boiling temperatures of the solvent and solute gives the elevation of boiling point. While calculating the molecular weight of solute the volume of solution is converted into mass by multiplying with density of solvent at its boiling point.

FREEZING–POINT DEPRESSION

Relation between Depression of Freezing-point and Lowering of Vapor-pressure

The vapor pressure of a pure liquid changes with temperature as shown by the curve *ABC*, in Fig. 15.8. There is a sharp break at B where, in fact, the freezing-point curve commences. Thus the point *B* corresponds to the freezing point of pure solvent, T_f. The vapor pressure curve of a solution (*solution* 1) of a nonvolatile solute in the same solvent is also shown in Fig. 15.8. It is similar to the vapor pressure curve of the pure solvent and meets the freezing point curve at *F*, indicating that T_1 is the freezing point of the solution. The difference of the freezing point of the pure solvent and the solution is referred to as the **Depression of freezing point.** It is represented by the symbol ΔT or ΔT_f.

$$T_f - T_1 = \Delta T$$

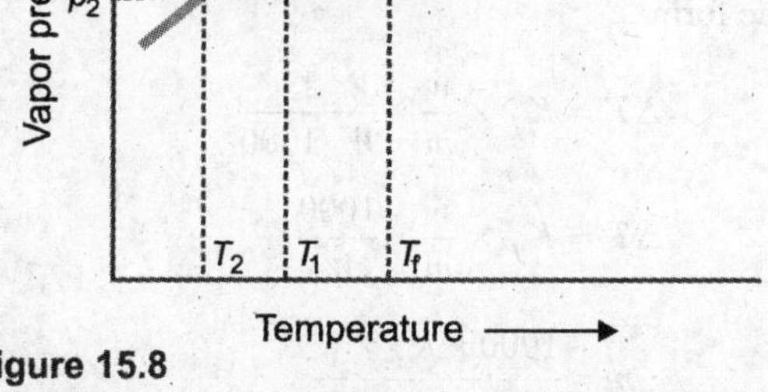

Figure 15.8
Relation between lowering of vapor-pressure and depression of freezing point.

When more of the solute is added to the solution 1, we get a more concentrated solution (*solution* 2.) The vapor pressure of *solution* 2 meets the freezing-point at *C*, indicating a further lowering of freezing point to T_2.

For dilute solutions *FD* and *CE* are approximately parallel straight lines and *BC* is also a straight line. Since the triangles *BDF* and *BEC* are similar,

$$\frac{DF}{EC} = \frac{BD}{BE}$$

or $$\frac{T_f - T_1}{T_f - T_2} = \frac{p - p_1}{p - p_2}$$

where p_1 and p_2 are vapor pressure of *solution* 1 and *solution* 2 respectively. **Hence depression of freezing point is directly proportional to the lowering of vapor pressure.**

or $$\Delta T \propto p - p_s \qquad ...(1)$$

Determination of Molecular Weight from Depression of Freezing point

Since p is constant for the same solvent at a fixed temperature, from (1) we can write

$$\Delta T \propto \frac{p - p_s}{p} \qquad ...(2)$$

But from Raoult's Law for dilute solutions,

$$\frac{p - p_s}{p} = \frac{wM}{Wm} \qquad ...(3)$$

Since M (mol wt) of solvent is constant, from (3)

$$\frac{p - p_s}{p} = \frac{w}{Wm} \qquad ...(4)$$

from (2) and (4)

$$\Delta T \propto K_f \times \frac{w}{m} \times \frac{1}{W}$$

or $$\Delta T = K_f \times \frac{w}{m} \times \frac{1}{W} \qquad ...(5)$$

where K_f is a constant called **Freezing-point constant or Cryoscopic constant or Molal depression constant.** If $w/m = 1$ and $W = 1$, $K_f = \Delta T$. Thus,

Molal depression constant may be defined as the freezing-point depression produced when 1 mole of solute is dissolved in one kg (1000 g) of the solvent.

If the mass of solvent (W) is given in grams, it has to be converted into kilograms. Thus the expression (5) assumes the form

$$\Delta T = K_f \times \frac{w}{m} \times \frac{1}{W/1000}$$

or $$\Delta T = K_f \times \frac{w}{m} \times \frac{1000}{W}$$

Hence $$m = \frac{1000 \times K_f \times w}{\Delta T \times W} \qquad ...(6)$$

where m = molecular mass of solute ; K_f = molal depression constant ; w = mass of solute ; ΔT = depression of freezing point ; W = mass of solvent.

Given the value of K_f, the molecular mass of solute can be calculated.

Sometimes the value of K_f is given in K per 0.1 kg. (100 g.) In that case, the expression (6) becomes

$$m = \frac{100 \times K_f \times w}{\Delta T \times W}$$

The value of K_f. The value of K_f can be determined by measurement of ΔT by taking a solute of known molecular mass (m) and substituting the values in expression (6). The constant K_f, which is characteristic of a particular solvent, can also be calculated from the relation

$$K_f = \frac{RT_f^2}{1000\ L_f}$$

where T_f = freezing point of solvent in K; L_f = molar latent heat of fusion; R = gas constant. Hence for water, T_f = 273 K and L_f = 336 × 10³ J kg. Therefore,

$$K_f = \frac{8.32 \times 273 \times 273}{1000 \times 336} = 1.85 \text{ K kg}^{-1}$$

The molal depression constant for some common solvents are given in Table 15.2

TABLE 15.2. MOLAL FREEZING-POINT CONSTANTS

Solvent	K_f per kg (1000 g)	K_f per 0.1 kg (100 g)
Water	1.86	18.6
Ethanoic acid (acetic acid)	3.90	39.0
Benzene	5.10	51.0
Camphor	40.0	400.0

SOLVED PROBLEM 10

0.000440 kg of a substance dissolved in 0.0222 kg of benzene lowered the freezing point of benzene by 0.567 K. Calculate the molecular mass of the substance. (K_f = 5.12 K per kg)

SOLUTION

We can find the molecular mass by applying the expression

$$m = \frac{1000 \times K_f \times w}{\Delta T \times W}$$

In this case :

$$w = 0.440 \times 10^{-3} \text{ kg}$$
$$\Delta T = 0.567 \text{ K}$$
$$W = 22.2 \times 10^{-3} \text{ kg}$$
$$K_f = 5.12 \text{ K kg}^{-1}$$

Substituting the values, $m = \dfrac{1000 \times 5.12 \times 0.440 \times 10^{-3}}{0.567 \times 22.2 \times 10^{-3}} = 178.9$

∴ Molecular mass of substance = **178.9**

SOLVED PROBLEM 11

0.001250 kg of naphthalene was dissolved in 60 × 10⁻⁶ m³ of benzene and freezing point of the solution was found to be 277.515 K, while that of benzene 278.495 K. Density of benzene = 880 kg/m³, K_f = 5.1 K per 1kg benzene. Calculate the molecular mass of naphthalene.

SOLUTION

Let us apply the expression

$$m = \frac{1000 \times K_f \times w}{\Delta T \times W}$$

In this case

$$K_f = 5.1 \text{ K}$$
$$w = 1.250 \times 10^{-3} \text{ kg}$$
$$W = 60 \times 10^{-6} \times 880 \text{ kg}$$
$$\Delta T = 278.495 - 277.515 = 0.980 \text{ K}$$

Substituting the values,

$$m = \frac{1000 \times 5.1 \times 1.250 \times 10^{-3}}{0.980 \times 60 \times 10^{-6} \times 0.880} = 123$$

Thus the molecular weight of naphthalene is **123**.

SOLVED PROBLEM 12

A solution of 0.000124 kg of a substance, X, in 0.025 m^3 of ethanoic acid (acetic acid) has a freezing point 0.324 K below that of the pure acid 16.6°C. Calculate the molecular mass (relative molecular mass) of X, given that the specific latent heat of fusion of ethanoic acid is 180.75×10^3 J kg^{-1}.

SOLUTION

Calculation of Molal depression Constant

We know that

$$K_f = \frac{RT_f^2}{L_f \times 1000} \qquad ...(1)$$

Here, freezing point of benzene, $T_f = 273.2 + 16.6 = 289.8$ K

Specific latent heat of fusion $L_f = 180.75 \times 10^3$ J kg^{-1}

Substituting in the equation (1)

$$K_f = \frac{8.314 \times (289.8)^2}{180.75 \times 1000} = 3.86° \text{ K}$$

Calculation of Molecular weight

Applying the expression

$$m = \frac{1000 \times K_f \times w}{\Delta T \times W}$$

We have

$$m = \frac{1000 \times 3.86 \times 0.124 \times 10^{-3}}{0.324 \times 0.025} = 59.09$$

Thus the molecular weight (or relative molecular) mass of X is **59.09**.

MEASUREMENT OF FREEZING-POINT DEPRESSION

The depression of freezing point can be measured more correctly and with less difficulty. Two simple methods commonly used are outlined below.

(1) Beckmann's Method (1903)

Apparatus. It consists of (*i*) A *freezing tube* with a side-arm to contain the solvent or solution, while the solute can be introduced through the side-arm; (*ii*) An outer larger tube into which is fixed the freezing tube, the space in between providing an air jacket which ensures a slower and more uniform rate of cooling; (*iii*) A *large jar* containing a freezing mixture *e.g.*, ice and salt, and having a stirrer.

Procedure. 0.015 kg to 0.020 kg of the solvent is taken in the freezing point of the solvent by directly coding the freezing point tube and the apparatus set up as shown in Fig. 15.9 so that the bulb of the thermometer is completely immersed in the solvent. First determine the approximate freezing point of the solvent by directly cooling the freezing point tube in the cooling bath. When this has been done, melt the solvent and place the freezing-point tube again in the freezing bath and allow the temperature to fall.

When it has come down to within about a degree of the approximate freezing point determined above, dry the tube and place it cautiously in the air jacket. Let the temperature fall slowly and when it has come down again to about 273.5 K below the freezing point, stir vigorously. This will cause the solid to separate and the temperature will rise owing to the latent heat set free. Note the highest temperature reached and repeat the process to get concordant value of freezing point.

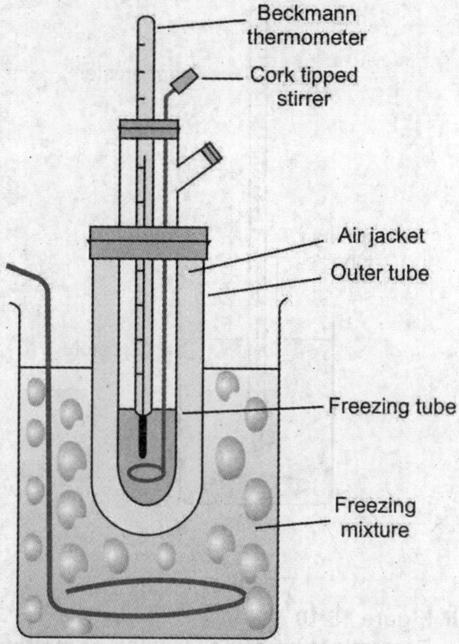

Figure 15.9
Relation between lowering of vapor-pressure and depression of freezing point.

The freezing point of the solvent having been accurately determined, the solvent is remelted by removing the tube from the bath, and a weighed amount (0.0001–0.0002 kg) of the solute is introduced through the side tube. Now the freezing point of the solution is determined in the same way as that of the solvent. A further quantity of solute may then be added and another reading taken. Knowing the depression of the freezing point, the molecular weight of the solute can be determined by using the expression

$$m = \frac{1000 \times K_f \times w}{\Delta T \times W}$$

This method gives accurate results, if the following precautions are observed :

(*a*) The supercooling should not exceed 273.5 K or 0.5°C.

(*b*) The stirring should be uniform at the rate of about one movement per second.

(*c*) The temperature of the cooling bath should not be 277 K to 278 K below the freezing point of the liquid.

(2) **Rast's Camphor Method**

This method due to Rast (1922) is used for determination of molecular weights of solutes which are soluble in molten camphor. The freezing point depressions are so large that an ordinary thermometer can be used.

Pure camphor is powdered and introduced into a capillary tube which is sealed at the upper end. This is tied along a thermometer and heated in a glycerol bath (see Fig. 15.10). The melting point of camphor is recorded. Then a weighed amount of solute and camphor (about 10 times as much) are melted in test-tube with the open end sealed. The solution of solute in camphor is cooled in air. After solidification, the mixture is powdered and introduced into a capillary tube which is sealed. Its melting point is recorded as before. The difference of the melting point of pure camphor and the mixture, gives the depression of freezing point. In modern practice, electrical heating apparatus is used for a quick determination of melting points of camphor as also the mixture.

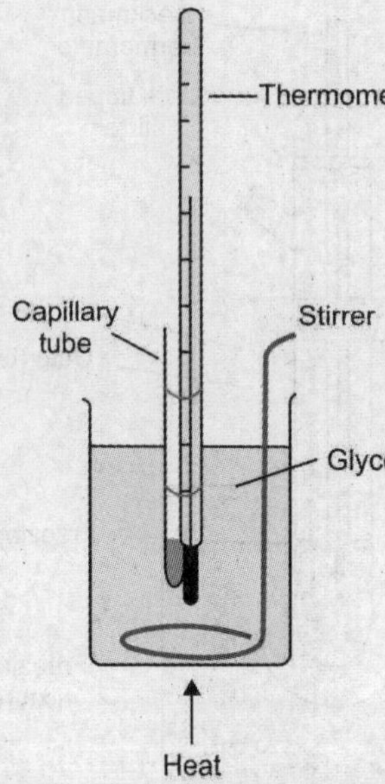

■ **Figure 15.10**
Determination of depression of melting point by capillary method.

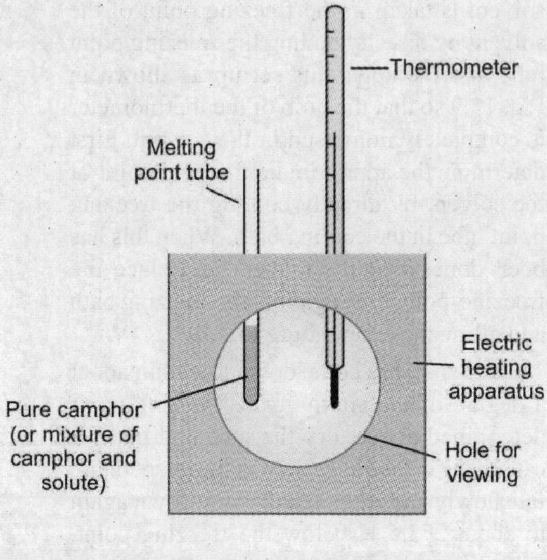

■ **Figure 15.11**
Determination of depression of melting point by electrical apparatus.

The molal depression constant of pure camphor is 313 K. But since the laboratory camphor may not be very pure, it is necessary to find the depression constant for the particular sample of camphor used by a preliminary experiment with a solute of known molecular weight.

SOLVED PROBLEM 13

A sample of camphor used in the Rast method of determining molecular masses had a melting point of 449.5 K. The melting point of a solution containing 0.522 × 10^{-3} kg camphor and 0.0386 × 10^{-3} kg of an unknown substance was 431.8 K. Find the molecular mass of the substance. K_f of camphor per kg is 37.7.

SOLUTION

Applying the expression

$$m = \frac{1000 \times K_f \times w}{\Delta T \times W}$$

to the present case, we have

$$\Delta T = 449.5 - 431.8 = 17.7$$

$$K_f = 37.7$$

$$w = 0.0386 \times 10^{-3} \text{ kg}$$

$$W = 0.522 \times 10^{-3} \text{ kg}$$

Substituting these values

$$m = \frac{1000 \times 37.7 \times 0.0386 \times 10^{-3}}{17.7 \times 0.522 \times 10^{-3}} = 157$$

COLLIGATIVE PROPERTIES OF ELECTROLYTES

The colligative properties of solutions *viz*, vapor-pressure lowering, boiling-point elevation, freezing-point depression and osmotic pressure, all depend solely on the total number of solute particles present in solution. The various electrolytes ionize and yield more than one particle per formula unit in solution. Therefore, **the colligative effect of an electrolyte solution is always greater than that of a nonelectrolyte of the same molal concentration.**

To represent the colligative properties of electrolytes by means of the same relations as for nonelectrolytes, van't Hoff (1880) suggested the use of a factor *i*. This is now known as **van't Hoff factor** and is defined as : **the ratio of the colligative effect produced by an electrolyte solution to the corresponding effect for the same concentration of a nonelectrolyte solution.**

Applying this definition of *i* to the freezing-point depression, we have

$$i = \frac{\Delta T_f}{[\Delta T_f]_0} \quad ...(1)$$

where ΔT_f is the freezing-point depression for the electrolyte measured experimentally and $[\Delta T_f]_0$ is the corresponding value calculated for a nonelectrolyte of the equal concentration. In other words, $[\Delta T_f]_0$ is the value of depression of freezing-point of the electrolyte solution assuming no ionization.

From (1)

$$\Delta T_f = i\,[\Delta T_f]_0$$

We know that

$$[\Delta T_f]_0 = \frac{1000\,K_f\,w}{mW} \quad ...(2)$$

$$\therefore \quad \Delta T_f = i \times \frac{1000\,K_f\,w}{mW} \quad ...(3)$$

A similar relation will hold for the observed elevation of boiling point of an electrolyte solution *i.e.*,

$$\Delta T_b = i \times \frac{1000\,K_b\,w}{mW} \quad ...(4)$$

The value of *i* can, therefore, be calculated from the experimental data as mentioned in (4).

It has been found that once *i* is known for a particular concentration of an electrolyte for one colligative property, the same value of *i* is essentially valid for other properties at the same concentration. That is,

$$i = \frac{\Delta T_f}{[\Delta T_f]_0} = \frac{\Delta T_b}{[\Delta T_b]_0} = \frac{\Delta p}{[\Delta p]_0}$$

$$= \frac{\pi}{[\pi]_0}$$

where the subscript zero refers in each case to the effect produced by a solute that is a nonelectrolyte.

ABNORMAL MOLECULAR MASSES OF ELECTROLYTES

Dividing (2) by (3), it is evident that

$$\frac{\Delta T_f}{[\Delta T_f]_0} = \frac{m_0}{m} = i \quad \text{or} \quad m = i \times m_0$$

where m_0 = molecular mass calculated from formula

m = molecular mass from observed data

Since the value of i is always greater than 1 **the experimental molecular weight will always be less than the theoretical value calculated from the formula.**

Relation Between van't Hoff Factor and Degree of Dissociation

Since colligative properties depend on the number of particles in solution, the van't Hoff factor may be interpreted as the ratio of the **number of particles present in solution to the number obtained assuming no ionization.** That is,

$$i = \frac{\text{Actual number of particles}}{\text{Number of particles for no ionization}}$$

The **degree of dissociation** is the fraction of an electrolyte which is dissociated into ions in aqueous solution. If an electrolyte is 50 per cent dissociated, its degree of dissociation is 0.5. The degree of dissociations is usually represented by α.

Suppose 1 mole of an electrolyte is capable of forming v ('nu') ions on complete dissociation. If the degree of dissociation be α, the total number of particles in solution are :

Number of undissociated molecules $= 1 - \alpha$

Number of ions produced $= \alpha v$

$\therefore$ Total number of particles $= 1 - \alpha + \alpha v$

Hence $$i = \frac{1 - \alpha + \alpha v}{1}$$

or $$i - 1 = \alpha (v - 1)$$

$$\alpha = \frac{i - 1}{v - 1}$$

This expression states the relationship between van't Hoff factor and the degree of dissociation. Knowing the value of i from colligative property measurements, α can be calculated.

SOLVED PROBLEM 14

A solution containing 1.5×10^{-3} kg of barium nitrate in 0.1 kg of water freezes at 272.720 K. Calculate the apparent degree of dissociation of the salt.

$$K_f = 1.86$$

Molecular mass of $Ba(NO_3)_2 = 261$

SOLUTION

Calculation of Depression of Freezing point $[\Delta T]_0$ for no ionization :

$$\Delta T = \frac{1000\, K_f \times w}{mW} = \frac{1000 \times 1.5 \times 10^{-3} \times 1.86}{261 \times 10^{-3} \times 0.1}$$

$$= 106.89 \text{ K}$$

Calculation of van't Hoff Factor

$$i = \frac{\Delta T}{[\Delta T]_0} = \frac{272.720}{106.89} = \frac{272.20}{106.89} = 2.55$$

Calculation of Degree of dissociation :

$$\alpha = \frac{i - 1}{v - 1}$$

Here $$Ba(NO_3)_2 \longrightarrow Ba^{2+} + 2NO_3^-$$

and 1 formula unit yields 3 particles on dissociation.

$\therefore \quad v = 3$

and $\quad \alpha = \dfrac{2.55 - 1}{3 - 1} = \dfrac{1.55}{2} = 0.775$ or **77.5%**

SOLVED PROBLEM 15

Find the degree of ionization for HF (hydrofluoric acid) in 0.100 m aqueous solution if the freezing point of the solution is – 0.197 K. (K_f for water = 1.86 K per Kg)

SOLUTION

Calculation of Freezing-point depression, $[\Delta T_f]_0$ for no ionization

$$[\Delta T_f]_0 = K_f \times m$$

where m = molal concentration (moles of solute/kg solvent)

$$\therefore \quad (\Delta T_f) = 1.86 \times 0.100$$

Calculation of i

$$i = \frac{\Delta T}{[\Delta T]_0}$$

Substituting values

$$i = \frac{0.197}{1.86 \times 0.100} = 1.06$$

Calculation of Degree of dissociation

$$\alpha = \frac{i - 1}{v - 1}$$

Since HF dissociates as

$$HF \rightleftharpoons H^+ + F^-$$

$$v = 2$$

Substituting values

$$\alpha = \frac{1.06 - 1}{2 - 1} = 0.06$$

Therefore, the degree of dissociation of 0.100 m HF is 0.06, or it is 6% ionized.

SOLVED PROBLEM 16

Phenol (C_6H_5OH) associates in water to double molecules. When 0.6677×10^{-3} kg of phenol is dissolved in 35.5×10^{-3} kg of water; it depresses the freezing point by 0.215 K. Calculate the van't Hoff factor and the degree of association of phenol. Molal depression constant of water = 1.85 K mol^{-1}.

SOLUTION

Calculation of van't Hoff Factor

Normal molecules weight calculated from formula = 6 × 12 + 5 + 16 + 1 = 94

Observed molecular mass $= \dfrac{1000 K_f\, w}{\Delta T\, W} = \dfrac{1000 \times 1.85 \times 0.6677 \times 10^{-3}}{0.215 \times 35.5 \times 10^{-3}}$

$= 161.84$

van't Hoff factor (i) $= \dfrac{\text{normal mol. mass}}{\text{observed mol. mass}}$

$$= \frac{94 \times 10^{-3}}{161.84 \times 10^{-3}}$$

Calculation of Degree of association

$$i = \frac{\text{Number of particles after association}}{\text{Number of particles for no association}}$$

If x be the degree of association, we have

$$\underset{1-x}{\overset{1}{2C_6H_5OH}} \rightleftharpoons \underset{x/2}{(C_6H_5OH)_2}$$

$\therefore$ Number of particles on association $= 1 - x + x/2 = 1 - x/2$

Number of particles for no association = 1

$$\therefore \quad i = \frac{94 \times 10^{-3}}{161.84 \times 10^{-3}} = \frac{1 - x/2}{1}$$

Hence x =0.838 or **Phenol is 83.8% associated**

CONCEPT OF ACTIVITY AND ACTIVITY COEFFICIENT

From the studies of solutions, G.N. Lewis discovered that **the experimentally determined value of concentration whether of molecules or ions in solution is less than the actual concentration.**

The apparent value of concentration is termed 'activity'. It may be defined as : **the effective concentration of a molecule or ion in a solution.**

The **activity coefficient** 'γ'is defined as : **the ratio between the activity, denoted by *a*, or effective concentration and actual concentration of the molecule or ion in solution.** That is,

$$\gamma = \frac{\text{effective concentration}}{\text{actual concentration}} = \frac{a}{c}$$

or

$$a = \gamma c$$

γ can be determined experimentally and is given in Tables. Thus the value of activity can be calculated by applying the above relation.

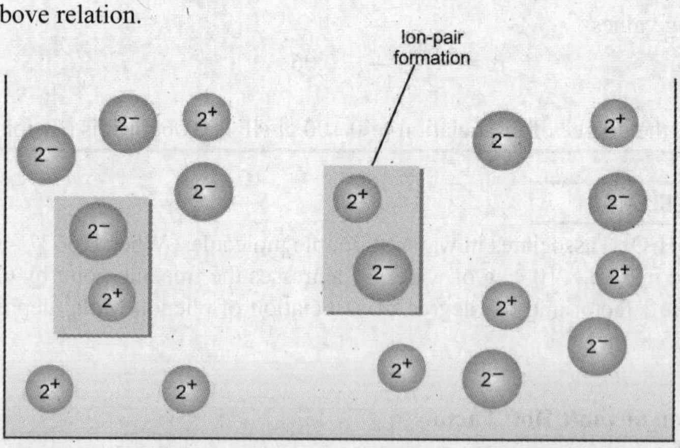

Figure 15.12
Ion-pair formation of a 2⁺ cation and a 2⁻ anion in a solution reduces the effective concentration of ions compared to actual concentration.

Explanation

The effective number of cations and anions in solution becomes less as these tend to form ion-pairs (Fig. 15.12) because of strong electrostatic attractions. Thus the effective concentration of the ions becomes less than the actual concentration.

SOLVED PROBLEM 17

Calculate the effective concentration of a 0.0992×10^3 M (in metre^3) solution of NaCl at 298 K for which activity coefficient is 0.782.

SOLUTION

We know that : $a = \gamma c$

Substituting the values, we have

$$a = 0.782 \times 0.0992 \times 10^3 \text{ M (in metre}^3)$$
$$= 0.0774 \times 10^3 \text{ M (in metre}^3) = 0.0744 \text{ M}$$

CONCLUSIONS

(1) The activity coefficient, γ of strong electrolytes is always less than 1.

(2) The value of γ decreases with dilution at the same temperature, and at infinite dilution it approaches 1.

(3) In the mathematical expressions of the various laws of physical chemistry such as Raoult's law, Henry's law, law of Mass action and Ostwald's law the actual concentration (C) has to be replaced with effective concentration (γ C) for accurate experimental work.

ADDITIONAL SOLVED PROBLEMS

SOLVED PROBLEM 18

0.010 kg of a non-volatile solute was dissolved in 10^{-1} kg of acetone (molecular mass = 58×10^{-3} kg) at 298 K. The vapor pressure of the solution was found to be 25664.56 Pa. Calculate the molecular mass of the solute. The vapor pressure of pure acetone at 298 K is 25997.86 Pa.

SOLUTION

Formula used

$$\frac{p - p_s}{p} = \frac{w \times M}{m \times W}$$

Quantities given

$p = 25997.86$ Pa $\quad p_s = 25664.56$ Pa $\quad w = 10 \times 10^{-3}$ kg

$W = 100 \times 10^{-3}$ kg $\quad M = 58 \times 10^{-3}$ kg

Substitution of values

$$\frac{25997.86 \text{ Pa} - 25664.56 \text{ Pa}}{25997.86 \text{ Pa}} = \frac{10 \times 10^{-3} \times 58 \times 10^{-3}}{100 \times 10^{-3} \times m}$$

or
$$\frac{333.3}{25997.86 \text{ Pa}} = \frac{10 \times 10^{-3} \times 58 \times 10^{-3}}{100 \times 10^{-3} \times m}$$

or
$$m = \frac{10 \times 10^{-3} \times 58 \times 10^{-3} \times 25997.86}{100 \times 10^{-3} \times 333.3}$$

$$= \mathbf{0.4524 \text{ kg} = 452.4 \text{ g}}$$

SOLVED PROBLEM 19

A solution of urea in water has a freezing point of – 273.55 K. Calculate the boiling point of the solution. (K_f = 275.01 K ; K_b = 273.66 K)

SOLUTION

Formula used

$$\text{molecular mass, } m = \frac{1000 \times K_f \times w}{\Delta T_f \times W} \quad \text{... (i)}$$

and

$$\text{molecular mass, } m = \frac{1000 \times K_b \times w}{\Delta T_b \times W} \quad \text{... (ii)}$$

Quantities given

ΔT_f = 273.55 K $\quad$ K_f = 275.01 K $\quad$ K_b = 273.66 K

Substitution in (i) and (ii) we have

$$m = \frac{1000 \times 275.01 \times w}{273.55 \times W} \quad \text{and} \quad m = \frac{1000 \times 273.66 \times w}{\Delta T_b \times W}$$

On equating we get

$$\frac{1000 \times 275.01 \times w}{273.55 \times W} = \frac{1000 \times 273.66 \times w}{\Delta T_b \times W}$$

or

$$1005.337 = \frac{273660}{\Delta T_b}$$

or

$$\Delta T_b = 272.20 \text{ K}$$

∴ Boiling point of the solution = Boiling point of water + Elevation in boiling point

= 373.15 + 272.20

= **645.35 K**

SOLVED PROBLEM 20

A 0.1 molar solution of urea at the room temperature freezes at 272.9 K at normal pressure. What would be the approximate freezing point of 0.1 molal aqueous solution of aluminium chloride at room temperature assuming complete ionization ?

SOLUTION

Formula used

$$\Delta T_f = K_f \times m \quad \text{or} \quad K_f = \frac{\Delta T_f}{m}$$

Quantities given

ΔT_f = 272.9 K $\quad$ m = 0.1

Substitution of values

$$K_f = \frac{272.9 \text{ K}}{0.1} = 27.29 \text{ K kg}^{-1}$$

Now, for $AlCl_3$

$$\Delta T_f = K_f \times m$$

$$= 27.29 \text{ K kg}^{-1} \times 0.1$$

$$= \mathbf{272.9\ K}$$

SOLVED PROBLEM 21

A 0.5 percent solution of potassium chloride was found to freeze at – 0.24 K. Calculate the degree of dissociation of potassium chloride. (Molal depression constant for water = 1.80 K)

SOLUTION

(a) To calculate the molecular mass of KCl

Formula used

$$m = \frac{1000 \times K_f \times w}{\Delta T_f \times W}$$

Quantities given

$K_f = 1.80$ K $\quad \Delta T_f = 0.24$ K $\quad w = 5 \times 10^{-3}$ kg $\quad W = 99.5 \times 10^{-3}$ kg

Substitution of values

$$m = \frac{1000 \times 1.80 \times 0.5 \times 10^{-3} \text{ kg}}{0.24 \times 99.5 \times 10^{-3} \text{ kg}}$$

$$= \mathbf{37.69}$$

(b) To calculate the degree of dissociation of KCl

The observed molecular mass = 37.69

and the normal molecular mass = 38 + 35.5 = 73.5

$$\text{van't Hoff factor, } i = \frac{\text{Normal molecular mass}}{\text{observed molecular mass}}$$

$$= \frac{73.5}{37.69} = 1.950$$

The degree of dissociation is related to van't Hoff factor by the relation

$$i = 1 + \alpha$$

or $\quad 1.950 = 1 + \alpha$

or $\quad \alpha = \mathbf{0.95}$ or **95%**

SOLVED PROBLEM 22

An immiscible liquid system composed of water and an organic liquid boils at 368.15 K when the barometer reads 98658.55 Pa. The vapor pressure of pure liquid at this temperature is 8399.31 Pa. The distillate contains 55% of the weight of the organic compound. Calculate the molecular mass of the organic compound.

SOLUTION

Formula used

$$\frac{w_1}{w_2} = \frac{p_1}{p_2} \times \frac{M_1}{M_2}$$

Quantities given

$w_1 = 45 \quad w_2 = 55 \quad M_1 = 18 \times 10^{-3}$

p_1 (water) = Vap. pressure of mixture – Vap. pressure of liquid

= 98658.55 Pa – 8399.31 Pa = 90259 Pa

Substitution of values

$$\frac{45}{55} = \frac{90259.24}{8399.31} \times \frac{18 \times 10^{-3}}{M_2}$$

or $$M_2 = \frac{89356.6476}{377968.95} = \mathbf{0.2364\ kg = 236.4\ g}$$

SOLVED PROBLEM 23

The relative molar mass of an ionic compound is 58.5×10^{-3} kg. If the experimentally observed molar mass is 30×10^{-3}, calculate van't Hoff factor.

SOLUTION

Formula used

$$\text{van't Hoff factor, } i = \frac{\text{Normal molecular mass}}{\text{Observed molecular mass}}$$

Quantities given

Normal molecular mass = 58.5×10^{-3} Observed molecular mass = 30×10^{-3}

Substitution of values

$$\text{van't Hoff factor, } i = \frac{58.5 \times 10^{-3}}{30 \times 10^{-3}}$$

$$= \mathbf{1.95}$$

SOLVED PROBLEM 24

2×10^{-3} kg of benzoic acid dissolved in 25×10^{-3} kg of benzene shows a depression in freezing point equal to 1.62 K. Molal depression constant K_f of benzene is 4.9 K kg mol^{-1}. What is the percentage association of the acid ?

SOLUTION

(*a*) To calculate the molecular mass of benzoic acid

Formula used

$$m = \frac{1000 \times K_f \times w}{\Delta T_f \times W}$$

Quantities given

K_f = 4.9 K kg mol^{-1} $w = 2 \times 10^{-3}$ kg ΔT_f = 1.62 K $W = 25 \times 10^{-3}$ kg

Substitution of values

$$m = \frac{1000 \times 4.9 \text{ K kg mol}^{-1} \times 2 \times 10^{-3} \text{ kg}}{1.62 \text{ K} \times 25 \times 10^{-3} \text{ kg}}$$

$$= \mathbf{241.975}$$

(*b*) To calculate the degree of association of benzoic acid

observed mol. mass of benzoic acid = 241.975×10^{-3} kg

normal mol. mass of benzoic acid = 122×10^{-3} kg

$$\text{van't Hoff factor, } i = \frac{\text{Normal molecular mass}}{\text{Observed molecular mass}}$$

$$= \frac{122 \times 10^{-3}}{241.975 \times 10^{-3}} = 0.504$$

$$2\,CH_3COOH \longrightarrow (C_6H_5COOH)_2$$

Degree of association x is related to van't Hoff factor by the relation

$$i = 1 - \frac{x}{2}$$

or $$0.504 = 1 - \frac{x}{2}$$

or $$1.008 = 2 - x$$

or $$x = 2 - 1.008$$

$$= \mathbf{0.992 \text{ or } 99.2\%}$$

SOLVED PROBLEM 25

In a cold climate water gets frozen causing damage to the radiator of a car. Ethylene glycol is used as an antifreezing agent. Calculate the amount of ethylene glycol to be added to 4 kg of water to prevent it from freezing at – 6 K. (K_f for water = 1.85 K kg mol^{-1})

SOLUTION

Formula used

$$m = \frac{1000 \times K_f \times w}{\Delta T_f \times W}$$

Quantities given

K_f = 1.85 K kg mol^{-1} $\qquad \Delta T_f$ = 6 K

m of ethylene glycol [$OHCH_2 - CH_2\,OH$] = 62 $\qquad W$ = 4 kg

Substitution of values

$$62 = \frac{1000 \times 1.85 \text{ K kg mol}^{-1} \times w}{6 \text{ K} \times 4}$$

or $$w = \frac{62 \times 6 \times 4}{1000 \times 1.85} = \mathbf{0.80432\ kg\ or\ 804.324\ g}$$

SOLVED PROBLEM 26

What mass of the non-volatile solute, urea, ($NH_2\,CO\,NH_2$) needs to be dissolved in 0.1 kg of water in order to decrease the vapor pressure of water by 25%? What will be the molality of the solution ?

SOLUTION

Formula used

$$\frac{p - p_s}{p} = \frac{\frac{w}{m}}{\frac{w}{m} + \frac{W}{M}}$$

Quantities given

$p = p$ Pa $\qquad p_s$ = 75% of p Pa = $\frac{3p}{4}$ Pa

m = 60 × 10^{-3} kg $\qquad W$ = 0.1 kg $\qquad M$ = 18 × 10^{-3} kg

Substitution of values

$$\frac{p - \frac{3}{4}p}{p} = \frac{\frac{w}{60}}{\frac{w}{60 \times 10^{-3}} + \frac{100}{18 \times 10^{-3}}}$$

$$4 \times \frac{w}{60 \times 10^{-3}} = \frac{w}{60 \times 10^{-3}} + 5555.6$$

or $$\frac{4w}{60} - \frac{w}{60 \times 10^{-3}} = 5555.6$$

or $$\frac{3w}{60 \times 10^{-3}} = 5555.6$$

or $$w = \frac{60 \times 10^{-3} \times 5555.6}{3} = \mathbf{111\ kg}$$

or $$\text{Molality} = \frac{\text{moles of solute}}{\text{mass of solvent in kg}}$$

$$= \frac{111 \text{ kg}}{60 \times 10^{-3} \times 0.1} = \mathbf{18.50 \times 10^3\ m}$$

EXAMINATION QUESTIONS

1. Define or explain the following terms :
   (*a*) Colligative properties
   (*b*) Raoult's law
   (*c*) Molecular mass
   (*d*) Boiling point elevation
   (*e*) Boiling point constant
   (*f*) Molal Elevation constant
   (*g*) Freezing point depression
   (*h*) Electrolytes
2. (*a*) State Raoult's law.
   (*b*) A solution of 8.585×10^{-3} kg of sodium nitrate in 0.1 kg of water freezes at –270.11 K. Calculate the molecular mass of sodium nitrate and account for the abnormal value. (K_f for water = 1.86 K mol^{-1})

   Answer. (*b*) 52.52×10^{-3} kg
3. (*a*) Discuss van't Hoff theory of dilute solutions. What is van't Hoff factor?
   (*b*) The values of molal elevation constant and molal depression constant for water are 0.52 and 1.86 K kg mol^{-1} respectively. If the elevation in boiling point by dissolving a solute is –0.2 K, what will be the depression in freezing point of this solution?

   Answer. (*b*) 0.71 K
4. When a certain amount of solute is added to 0.1 kg of water at 298 K, the vapor pressure reduces to one-half of that for pure water. The vapor pressure of water is 3167.74 Pa. Find the amount of salt added.

   Answer. 2.78 mol
5. 0.3×10^{-3} kg of camphor (molar mass 154.4×10^{-3} kg mol^{-1}) when added to 25.2×10^{-3} kg of chloroform raised the boiling point of the solvent by 0.299 K. Calculate the molar elevation constant of chloroform.

   Answer. 3.88×10^{-3} K
6. (*a*) How is the molecular mass of a solute determined from elevation of boiling point?
   (*b*) Calculate the value of K_b for water, given that pure water boils at 373 K and the latent heat of its vaporization is 2260871.968 J/kg.

   Answer. (*b*) 0.512 K
7. (*a*) Derive a relationship between the elevation in boiling point of a solution and the mole fraction of the solute from thermodynamic consideration.
   (*b*) The molal elevation constant (K_b) and the boiling point for carbon tetra chloride are 5.02 deg/molal and 76.8 K respectively. Calculate the boiling point of 1.0 molal solution of naphthalene in carbon tetrachloride.

   Answer. (*b*) 81.82 K
8. When 0.000946 kg of a sugar is dissolved in 0.150 kg of water, the resulting solution is observed to have a freezing point of – 0.0651 K. What is the molecular mass of the sugar? K_f for water is 1.86 K.

   Answer. 180
9. A freezing point depression of 1/200 molal solution of sodium sulfate in water was found to be 0.0265 K. Calculate the degree of dissociation of the salt at this concentration (K_f for water is 1.86 K mol^{-1})

   Answer. 0.404
10. Why the boiling point 0.1 m $BaCl_2$ solution is more than 0.1 m NaCl solution?
11. Show that Raoult's law is a special case of Henry's law. (*Delhi BSc*)
12. Explain the following :
    (*a*) Vapor pressure of a liquid does not depend upon the size of the container.
    (*b*) Boiling point of a liquid increases on adding non-volatile solute in it.
    (*c*) Vapor pressure of a liquid varies with temperature. (*Kathmandu BSc*)

13. (*a*) What are isotonic solutions? Explain

(*b*) Write a note on van't Hoff factor. (*Punjabi BSc*)

14. (*a*) Explain the term lowering of vapor pressure and relative lowering of vapor pressure.

(*b*) What are the colligative properties? Explain. Why electrolytes have abnormally high values of colligative properties? (*Arunachal BSc*)

15. State and explain Raoult's law and Henry's law. Show that in any solution if the solvent obeys Raoult's law, the solute obeys Henry's law. (*Lucknow BSc*)

16. 0.05394 kg of a substance of molecular mass 0.182 kg is dissolved in 1 kg of water at 293 K. At this temperature the vapor pressure of water is 2333.14 Pa. Calculate the vapor pressure of this dilute solution.

Answer. 2320.69 Pa (*Andhra BSc*)

17. (*a*) Describe the determination of molecular weight of a solute from relative lowering of vapor pressure.

(*b*) Calculate the molecular weight of a non-volatile solute, if 1.2×10^{-3} kg of it is dissolved in 0.1 kg of acetone at 292 K. Vapor pressure of the solution = 24331.33 Pa and vapor pressure of acetone = 24664.64 Pa.

Answer. 0.05150 kg (*Nagpur BSc*)

18. Describe a method for determining the molar mass of a non-volatile solute by the relative lowering in vapor pressure of the solvent. (*Arunachal BSc*)

19. What is molal depression constant? How is it related to the latent heat of fusion?

(*Punjabi BSc*)

20. Draw a vapor pressure composition diagram showing positive deviations from Raoult's law. Why do positive deviations occur? (*Guru Nanak Dev BSc*)

21. (*a*) Derive a relation for the depression of freezing point of a solution with its molality.

(*b*) Derive a relation for the elevation in boiling point of a solution with its molality.

(*Sambalpur BSc*)

22. State Raoult's law for ideal solutions. Draw a vapor pressure composition diagram for an ideal binary solution obeying this law and compare this with similar diagram showing deviation from Raoult's law. Why do these deviations occur? (*Panjab BSc*)

23. Write a short note on 'Elevation in boiling point'. (*Arunachal BSc*)

24. What do you understand by depression of freezing point? Derive an expression relating the freezing point depression of a solution with the mole fraction of the dissolved solute. (*Arunachal BSc*)

25. Explain giving reasons :

(*a*) Addition of non-volatile solute lowers the freezing point and elevates the boiling point of a solvent.

(*b*) Equimolar solutions of sucrose and sodium chloride in water are not isotonic. (*Arunachal BSc*)

26. What do you mean by van't Hoff factor? Explain it when solute is associated in solution.

(*Arunachal BSc*)

27. Acetic acid associates in benzene to form double molecule. 1.65×10^{-3} kg of acetic acid when dissolved in 100×10^{-3} kg of benzene raised the boiling point by 0.36 K. Calculate the van't Hoff's factor and degree of association of acetic acid in benzene (K for benzene = 2.57 K kg^{-1})

Answer. 0.982; 0.509 (*Delhi BSc*)

28. A solution containing 2.44×10^{-3} kg of a solute dissolved in 75×10^{-3} kg of water boiled at 373.413 K. Calculate the molar mass of the solute. (K_b for water = 0.52 K kg mol^{-1})

Answer. 40.96×10^{-3} kg mol^{-1} (*Nagpur BSc*)

29. (*a*) Describe a method for determining the molar mass of a non-volatile solute by the relative lowering of vapor pressure of solvent.

(*b*) The vapor pressure of water at 293 K is 2.337×10^3 Nm^{-2} or Pascal and the vapor pressure of a solution of 0.108 kg of a non-volatile solute in 1 kg of water at the same temperature is 2.313×10^3 Nm^2. Calculate the molar mass of the solute.

Answer. 0.187 kg mol^{-1} (*Arunachal BSc (H)*)

30. (*a*) State and derive Raoult's law for lowering of vapor pressure. How is this law used for the determination of molecular mass of a non-volatile solute?

(*b*) Calculate the freezing point of the solution that contains 0.0072 kg of glucose in 0.200 kg of water. Normal freezing point of water is 0 K and cryoscopic constant for water is 1.86 K kg^{-1}.

Answer. 0.372 K (*Sambalpur BSc*)

31. (*a*) Give the thermodynamic derivation of the expression for boiling point elevation ΔT_b in the form

$$\Delta T_b = K_b m$$

where m is the solute molality and K_b the ebullioscopic constant.

(*b*) Calculate the boiling point elevation resulting from dissolution of 0.002 kg of C_6Cl_6 (mol. mass = 0.285 kg) in 0.050 kg of CCl_4. (Ebullioscopic constant for solution in CCl_4 is 5.22 K kg mol^{-1})

Answer. 73.2 K (*Guru Nanak Dev BSc*)

32. (*a*) Define the term colligative properties.

(*b*) A solution of 0.0125 kg of an unknown solute in 0.170 kg of water gave a boiling point elevation of 0.63 K. Calculate the molar mass of the solute. (K_b = 0.52 K m^{-1})

Answer. (*b*) 60.69 (*Nagpur BSc*)

33. The molal depression constant for benzene is 5.12 K mol^{-1}. A solution containing 0.001 kg solute per 0.1 kg benzene freezes at 5.1 K. What is the molar mass of the solute. (Freezing point of pure benzene is 5.5 K)

Answer. 128 (*Gulbarga BSc*)

34. (*a*) Derive an expression for the elevation of boiling point of a liquid when a solute is dissolved in it and explain as to how the molecular mass of solute can be evaluated by the boiling point method.

(*b*) The molecular mass of a water soluble non-electrolyte is 0.058 kg. Compute the boiling point of a solution containing 0.024 kg of the solute and 0.6 kg of water at atmospheric pressure (K_{1000} for water = 0.573)

Answer. (*b*) 373.395 K (*Osmania BSc*)

35. The boiling point of a solution containing 0.00256 kg of substance A per 0.1 kg of water is higher by 0.052 K than the boiling point of pure water. Calculate the molar mass of the substance if molar elevation constant of water is 0.512 K kg m^{-1}.

Answer. 252.06 kg/mol (*Agra BSc*)

36. The vapor pressure of methyl alcohol at 298 K is 12798.95 Pa. Its mole fraction in a solution with ethyl alcohol is 0.305, what is its vapor pressure if it obeys Raoult's law.

Answer. 3903.68 Pa (*Baroda BSc*)

37. Calculate the amount of CS_2 present as an impurity in 0.050 kg of Chloroform which raises the boiling point of Chloroform by 0.3 K. (K_b = 2.40 K m^{-1})

Answer. 0.00397 kg (*Aligarh BSc*)

38. (*a*) What is molal elevation constant? How is it related to mole fraction of a non-volatile solute?

(*b*) A solution containing 0.005 kg of KCl per $metre^3$ of water boils at 100.065 K at 101325 Pa pressure. Determine the degree of dissociation of KCl (K_b for water is 0.54 K)

Answer. (*b*) 0.7935 (*Delhi BSc*)

39. An aqueous sucrose solution freezes at – 0.31 K. Calculate the molality of the solution and also its boiling point. Given that K_f = 1.86 and K_b = 0.51 for water.

Answer. 0.1666 molal; 373.085 K (*Gulbarga BSc*)

40. What will be the boiling point of solution containing 0.6×10^{-3} kg of urea (molecular mass = 0.060 kg or 60 g) in 0.0025 kg of water? (Boiling point of water = 373 K; K_b = 0.52 K mol^{-1} kg^{-1})

Answer. 2.080 K (*Panjab BSc*)

MULTIPLE CHOICE QUESTIONS

1. A colligative property depends upon

(*a*) chemical nature of the particles (*b*) size of the particles

(*c*) number of particles (*d*) temperature of the solution

Answer. (*c*)

2. When a non-volatile solute is dissolved in a pure solvent, the vapor pressure of the pure solvent

(*a*) increases (*b*) decreases

(*c*) remains the same (*d*) none of these

Answer. (*b*)

3. The relative lowering of vapor pressure is represented by

(*a*) $\dfrac{p-p_s}{p}$ (*b*) $\dfrac{p_s-p}{p}$

(*c*) $\dfrac{p_s+p}{p}$ (*d*) $\dfrac{p}{p_s+p}$

Answer. (*a*)

4. Mathematically, Raoult's law can be expressed as

(*a*) $\dfrac{p_s-p}{p}=\dfrac{n}{n+N}$ (*b*) $\dfrac{p_s-p}{p}=\dfrac{N}{n+N}$

(*c*) $\dfrac{p_s+p}{p}=\dfrac{n}{n+N}$ (*d*) $\dfrac{p-p_s}{p}=\dfrac{n}{n+N}$

Answer. (*d*)

5. A real solution is that which

(*a*) obeys Raoult's law (*b*) does not obey Raoult's law

(*c*) obeys Henry's law (*d*) does not obey Henry's law

Answer. (*b*)

6. The molecular mass of the solute (*m*) can be calculated by measuring the lowering of vapor pressure using the formula

(*a*) $\dfrac{wM}{mW}=\dfrac{p-p_s}{p}$ (*b*) $\dfrac{wM}{mW}=\dfrac{p_s-p}{p}$

(*c*) $\dfrac{wM}{mW}=\dfrac{p+p_s}{p}$ (*d*) $\dfrac{wM}{mW}=\dfrac{p_s}{p-p_s}$

Answer. (*a*)

7. A liquid boils when its vapor pressure becomes equal to

(*a*) one atmospheric pressure (*b*) zero

(*c*) very high (*d*) very low

Answer. (*a*)

8. The addition of a non-volatile solute _______ the vapor pressure

(*a*) enhances (*b*) lowers

(*c*) diminishes (*d*) none of these

Answer. (*b*)

9. The elevation in boiling point is given by the formula

$$\Delta T = K_b \times \frac{w}{m} \times \frac{1}{W}$$

where K_b is called

(*a*) boiling point constant (*b*) ebulioscopic constant

(*c*) molal elevation constant (*d*) all of these

Answer. (*d*)

10. Molal elevation constant is the boiling point elevation when _______ of the solute is dissolved in one kg of the solvent

(*a*) one gram (*b*) one kg

(*c*) one mole (*d*) none of these

Answer. (*c*)

11. When a non-volatile solute is added to a solvent, the freezing point of the solvent _______.

(*a*) increases (*b*) remains the same

(*c*) decreases (*d*) none of these

Answer. (*c*)

12. The depression in freezing point is measured by using the formula

$$\Delta T = K_f \times \frac{w}{m} \times \frac{1}{W}$$

where K_f is called

(*a*) molal depression constant (*b*) freezing point depression constant

(*c*) cryoscopic constant (*d*) all of these

Answer. (*d*)

13. Freezing point depression is measured by

(*a*) Beckmann's method (*b*) Rast's camphor method

(*c*) both (*d*) none of these

Answer. (*c*)

14. The colligative effect of an electrolyte is always _______ that of a non-electrolyte of the same molal concentration.

(*a*) greater (*b*) smaller

(*c*) equal to (*d*) none of these

Answer. (*a*)

15. The ratio of the colligative effect produced by an electrolyte solution to the corresponding effect for the same concentration of a non-electrolyte solution is known as

(*a*) degree of dissociation (*b*) degree of association

(*c*) activity coefficient (*d*) van't Hoff factor

Answer. (*d*)

16. Abnormal molecular masses are obtained when there exists

(*a*) dissociation of molecules (*b*) association of molecules

(*c*) either of the two (*d*) none of these

Answer. (*c*)

17. The degree of dissociation is the fraction of an electrolyte which

(*a*) is associated to from bigger molecules (*b*) is dissociated into ions

(*c*) does not pass into solution (*d*) none of these

Answer. (*b*)

18. The law of the relative lowering of vapor pressure was given by

(*a*) van't Hoff (*b*) Ostwald

(*c*) Raoult (*d*) Henry

Answer. (*c*)

19. The vapor pressure of a dilute aqueous solution of glucose is 98658.55 Pa of mercury at 373 K. The mole fraction of the solute is

(*a*) $\frac{1}{20}$ (*b*) $\frac{1}{38}$

(*c*) $\frac{1}{76}$ (*d*) $\frac{1}{740}$

Answer. (*b*)

20. Which of the following is a colligative property?

(*a*) molar refractivity (*b*) optical rotation

(*c*) depression in freezing point (*d*) viscosity

Answer. (*c*)

21. Which of the following is not a colligative property?

(*a*) relative lowering of vapor pressure (*b*) surface tension

(*c*) elevation in boiling point (*d*) depression in freezing point

Answer. (*b*)

22. The mole fraction of the solvent in a solution of non-volatile solute is 0.950. The relative lowering of vapor pressure is

(*a*) 0.050 (*b*) 0.020

(*c*) 0.010 (*d*) 0.095

Answer. (*a*)

23. An aqueous solution of Sodium chloride in water has vapor pressure

(*a*) equal to that of water (*b*) more than that of water

(*c*) less than that of water (*d*) none of these

Answer. (*c*)

24. Which of the following is a colligative property?

(*a*) K_b (*b*) K_f

(*c*) ΔT_f (*d*) degree of ionization

Answer. (*c*)

25. The ratio of the value of colligative property for K_2SO_4 solution to that of KCl solution at the same concentration is nearly

(*a*) 1.25 (*b*) 1.5

(*c*) 2.0 (*d*) 2.5

Answer. (*b*)

26. The freezing point of 1 molal K_2SO_4 solution assuming K_2SO_4 to be 100% ionised in water is

(*a*) –1.86 K (*b*) –3.72 K

(*c*) –5.58 K (*d*) +5.58 K

Answer. (*c*)

27. Which of the following 0.05 M aqueous solution will have the highest boiling point?

(*a*) glucose (*b*) NaCl

(*c*) K_2SO_4 (*d*) $Al(NO_3)_3$

Answer. (*d*)

28. The depression in freezing point of an unknown solution is equal to the depression in freezing point of 0.1 molal aqueous solution of urea. The concentration of the unknown solution is

(*a*) 0.1 M (*b*) 0.2 M
(*c*) 0.5 M (*d*) 1.0 M

Answer. (*a*)

29. The study of depression in freezing point of a solution is called

(*a*) osmotic pressure (*b*) ebullioscopy
(*c*) cryoscopy (*d*) none of these

Answer. (*c*)

30. The freezing point of 2% aqueous solution of Potassium nitrate is

(*a*) equal to 0°C (*b*) less than 0°C
(*c*) higher than 0°C (*d*) none of these

Answer. (*b*)

31. The molal depression constant is given by the expression

(*a*) $\Delta T_f \times M$ (*b*) $\Delta T_f \times m$
(*c*) $\Delta T_f \times m$ (*d*) $\Delta T_f \times M$

Answer. (*c*)

32. Ebullioscopy deals with study of

(*a*) osmotic pressure and membranes (*b*) depression in freezing point
(*c*) relative lowering of vapor pressure (*d*) elevation in boiling point

Answer. (*d*)

33. The depression in freezing point for 1 M urea, 1 M glucose and 1 M NaCl are in the ratio

(*a*) 1:2:2 (*b*) 1:2:1
(*c*) 1:1:1 (*d*) 1:1:2

Answer. (*d*)

34. One mole of sugar is dissolved in three moles of water. The relative lowering of vapor pressure is

(*a*) 1/2 (*b*) 1/3
(*c*) 1/4 (*d*) 1/5

Answer. (*c*)

35. Which one of the following salts would have the same value of the van't Hoff factor as that of $Al(NO_3)_3$?

(*a*) KCl (*b*) K_2SO_4
(*c*) $Al_2(SO_4)_3$ (*d*) $K_3[Fe(CN)_6]$

Answer. (*d*)

36. Which one of the following is true for the van't Hoff factor, *i* ?

(*a*) $i = \Delta T_f / [\Delta T_f]_o$ (*b*) $i = \Delta T_b / [\Delta T_b]_o$
(*c*) $i = \Delta p / [\Delta p]_o$ (*d*) all of these

Answer. (*d*)

37. The colligative effect of an electrolyte solution is always _______ that of a non-electrolyte of the same molal concentration.

(*a*) equal to (*b*) lesser than
(*c*) greater than (*d*) none of these

Answer. (*c*)

38. The degree of dissociation is related to the van't Hoff factor by the expression (v is the no. of ions on complete dissociation of the electrolyte)

(*a*) $\alpha = \dfrac{i+1}{\nu+1}$ (*b*) $\alpha = \dfrac{i-1}{\nu+1}$

(*c*) $\alpha = \dfrac{i+1}{\nu-1}$ (*d*) $\alpha = \dfrac{i-1}{\nu-1}$

Answer. (*d*)

39. Which one of the following solutions would produce maximum elevation in boiling point?

(*a*) 0.1 M glucose (*b*) 0.2 M sucrose

(*c*) 0.1 M $Ba(NO_3)_2$ (*d*) 0.1 M $MgSO_4$

Answer. (*c*)

40. Which one of the following solutions will boil at highest temperature?

(*a*) 1% glucose (*b*) 1% urea

(*c*) 1% KCl (*d*) 1% K_2SO_4

Answer. (*d*)

41. At high altitudes, water boils at a temperature which is

(*a*) higher than the normal boiling point (*b*) lower than the normal boiling point

(*c*) is equal to normal boiling point (*d*) none of these

Answer. (*b*)

42. The value of van't Hoff factor is _______ for an electrolyte

(*a*) greater than one (*b*) less than one

(*c*) equal to one (*d*) equal to zero

Answer. (*a*)

43. The ratio of the value of elevation in boiling point for NaCl solution to that of urea solution is nearly

(*a*) 0.5 (*b*) 1.0

(*c*) 1.5 (*d*) 2.0

Answer. (*d*)

44. On dissolving common salt in water it is observed that

(*a*) boiling point of the solution increases

(*b*) boiling point of the solution decreases

(*c*) vapor pressure is increased

(*d*) freezing point is raised

Answer. (*a*)

45. The highest boiling point will be observed in 1% solution of which of the following substances?

(*a*) $NH_2–CO–NH_2$ (*b*) $C_{12}H_{22}O_{11}$

(*c*) NaCl (*d*) Na_2SO_4

Answer. (*d*)

46. Benzoic acid when dissolved in benzene undergoes

(*a*) dissociation (*b*) association

(*c*) no change (*d*) none of these

Answer. (*b*)

47. The value of van't Hoff factor '*i*' in case of dimerization of benzoic acid in benzene is given by the expression

(*a*) $i = 1 - \alpha$ (*b*) $i = 1 + \alpha$

(*c*) $i = 1 - \alpha/2$ (*d*) $i = 1 + \alpha/2$

Answer. (*c*)

48. The ratio between the effective concentration and the actual concentration of the molecule or ions in a solution is called

(*a*) van't Hoff factor (*b*) colligative property

(*c*) active concentration (*d*) activity coefficient

Answer. (*d*)

49. The activity coefficient of strong electrolytes is

(*a*) always equal to zero (*b*) always equal to one

(*c*) always less than one (*d*) always greater than one

Answer. (*c*)

16 Osmosis and Osmotic Pressure

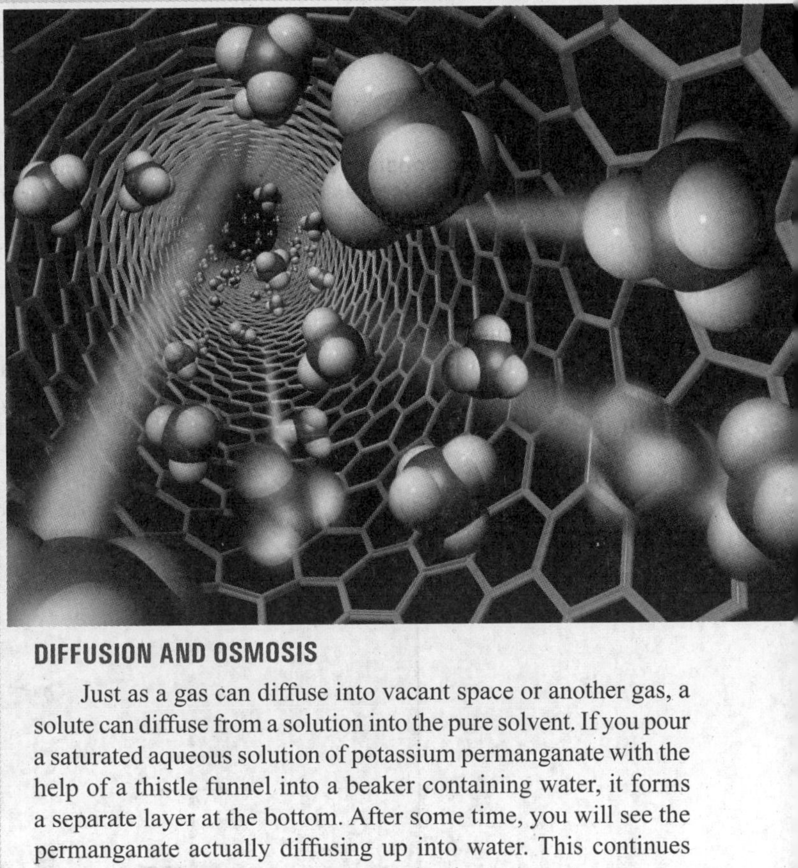

CONTENTS

DIFFUSION AND OSMOSIS

Just as a gas can diffuse into vacant space or another gas, a solute can diffuse from a solution into the pure solvent. If you pour a saturated aqueous solution of potassium permanganate with the help of a thistle funnel into a beaker containing water, it forms a separate layer at the bottom. After some time, you will see the permanganate actually diffusing up into water. This continues until a homogeneous solution is obtained.

The diffusion of solute into solvent is, in fact, a bilateral process. It consists of: (1) the solute molecules moving up into solvent; and (2) the solvent molecules moving down into solution. This intermingling of solute and solvent molecules goes on, so that ultimately a solution of uniform concentration results.

It is the tendency to equalize concentration in all parts of the solution which is responsible for the diffusion of the solute.

Thus diffusion of solute will also take place when two solutions of unequal concentrations are in contact. Solvent molecules will pass from the dilute to the concentrated solution and solute molecules will pass from the concentrated to the dilute solution until equality of concentration is achieved.

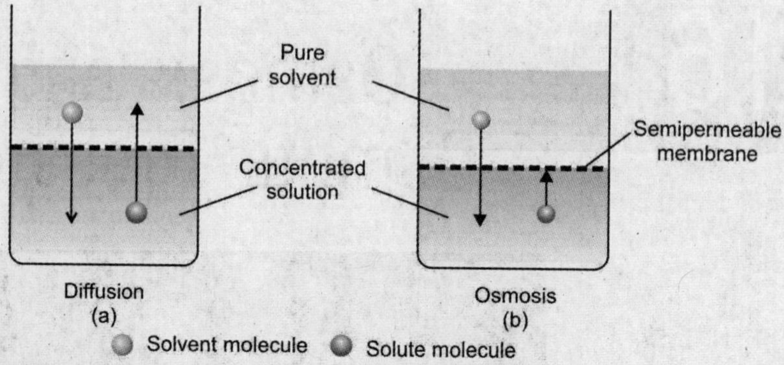

■ **Figure 16.1**
(*a*) Diffusion consists of solute molecules moving upward into the solvent and solvent molecules moving down into concentrated solution;
(*b*) Osmosis consists of solvent molecules passing through the semipermeable membrane, while large solute particles are held up.

WHAT IS OSMOSIS?

Let us consider a pure solvent and solution separated by a membrane which permits the passage to solvent molecules but not to solute molecules. Only the solvent will diffuse through the membrane into solution. **A membrane which is permeable to solvent and not to solute, is called a semipermeable membrane.**

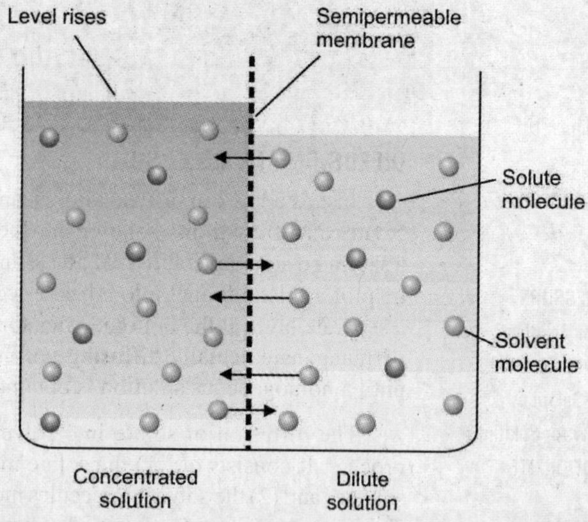

■ **Figure 16.2**
Illustration of the process of osmosis. The solvent molecules can pass through the membrane from either side of the semipermeable membrane. There is a net flow of solvent from the dilute solution to the concentrated solution. It shows up in the rise of level in the concentrated solution compartment.

The flow of the solvent through a semipermeable membrane from pure solvent to solution, or from a dilute solution to concentrated solution, is termed Osmosis (Greek *Osmos* = to push).

It must be clearly understood that the diffusion of solvent molecules through a semipermeable membrane is taking place in both directions. That is, solvent molecules are passing from solvent to solution, and also from solution to solvent. But since the diffusion from solvent to solution or from dilute to concentrated solution, is more rapid, **the net flow of the solvent is from low to high concentration.**

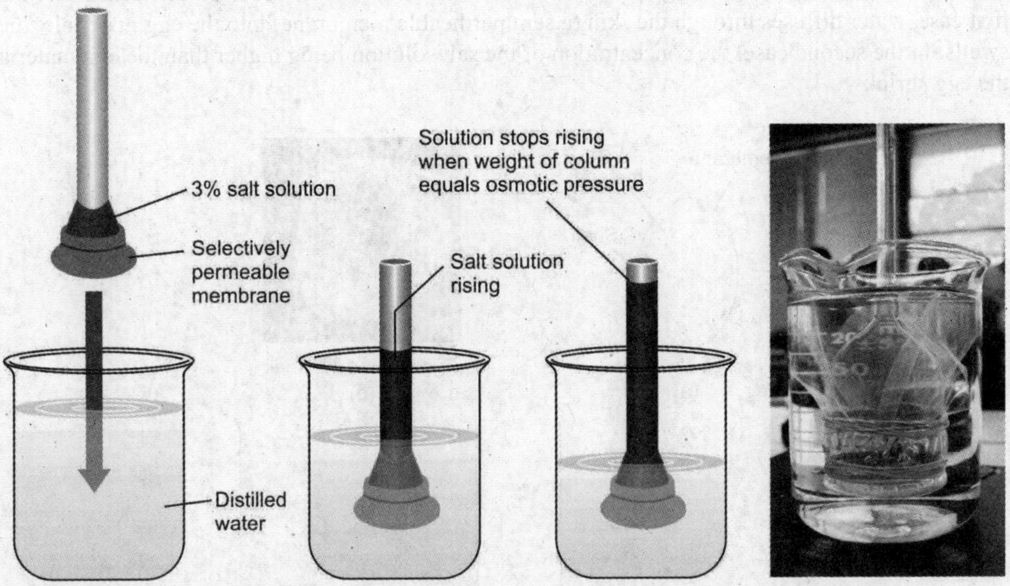

Figure 16.3
Demonstration of Osmosis.

The phenomenon of osmosis can be demonstrated by fastening a piece of animal bladder or cellophane over a funnel as shown in Fig. 16.3. An aqueous salt solution is placed inside the funnel which is then immersed in water. The osmosis takes place through the semipermeable membrane from water to the salt solution. The flow of water into the funnel shows up as the solution is seen rising in the tube remarkably.

SOME INTERESTING EXPERIMENTS DEMONSTRATING OSMOSIS

Many natural membranes are semipermeable *e.g.*, pig's bladder, skin around white of an egg, membrane around the red blood corpuscle, and the membrane in the cell of the plant. Some interesting experiments demonstrating osmosis are listed below.

(1) Silica Garden

Crystals of many salts *e.g.*, ferrous sulfate, nickel chloride, cobalt nitrate and ferric chloride are placed in a solution of glass material (sodium silicate). The layers of metallic silicates formed on the surface of crystals by double decomposition are semipermeable. The water from outside enters through these membranes which burst and form what we call a *Silica Garden*.

Figure 16.4
Silica garden.

(2) The Egg Experiment

The outer hard shell of two eggs of the same size is removed by dissolving in dilute hydrochloric acid. One of these is placed in distilled water and the other in saturated salt solution. After a few hours it will be noticed that the egg placed in water swells and the one in salt solution shrinks. In the first case, water diffuses through the skin (a semipermeable membrane) into the egg material which swells. In the second case, the concentration of the salt solution being higher than the egg material, the egg shrinks.

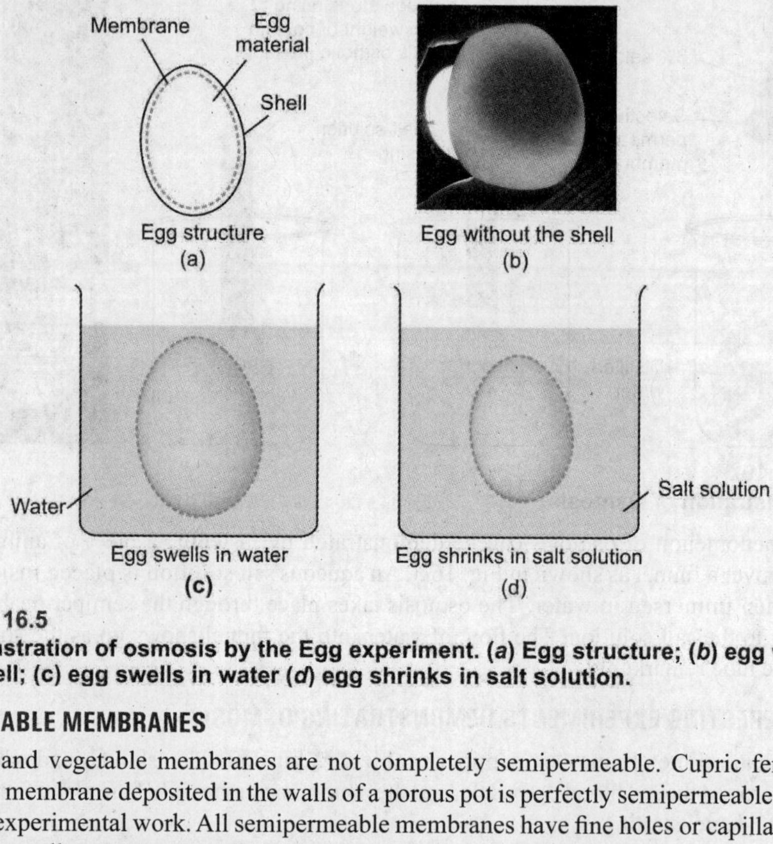

Figure 16.5
Demonstration of osmosis by the Egg experiment. (*a*) Egg structure; (*b*) egg without the shell; (c) egg swells in water (*d*) egg shrinks in salt solution.

SEMIPERMEABLE MEMBRANES

Animal and vegetable membranes are not completely semipermeable. Cupric ferrocyanide, $Cu_2Fe(CN)_6$, membrane deposited in the walls of a porous pot is perfectly semipermeable and is used for accurate experimental work. All semipermeable membranes have fine holes or capillaries in their structure. These allow passage to solvent molecules but not to larger solute molecules.

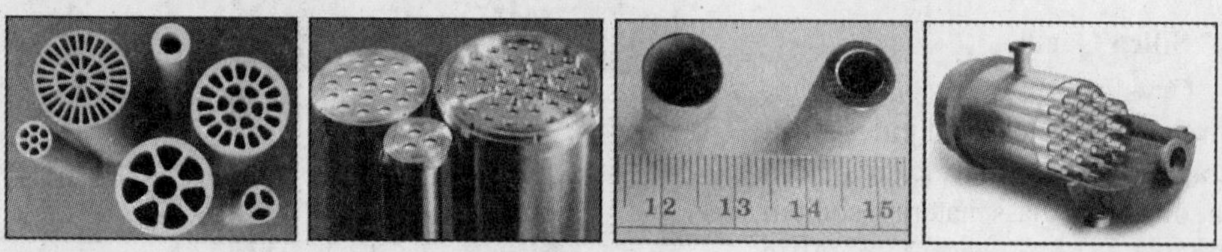

Figure 16.6
Some examples of semipermeable membranes.

Preparation of Cupric Ferrocyanide Membrane

An unglazed porcelain battery pot filled with copper sulfate solution is placed in a solution of potassium ferrocyanide, $K_4Fe(CN)_6$. The two solutions permeate into porous walls of the pot from opposite sides. A gelatinous precipitate of cupric ferrocyanide is formed in the middle of the pores of the walls.

$$2\ Cu^{2+} + Fe(CN)_6^{4-} \longrightarrow Cu_2Fe(CN)_6 \downarrow$$

The electrical method which gives a more compact and stout membrane in a short time is shown in Fig. 16.8. The resistance of the cell rises as the membrane is completely formed and the bell rings no more.

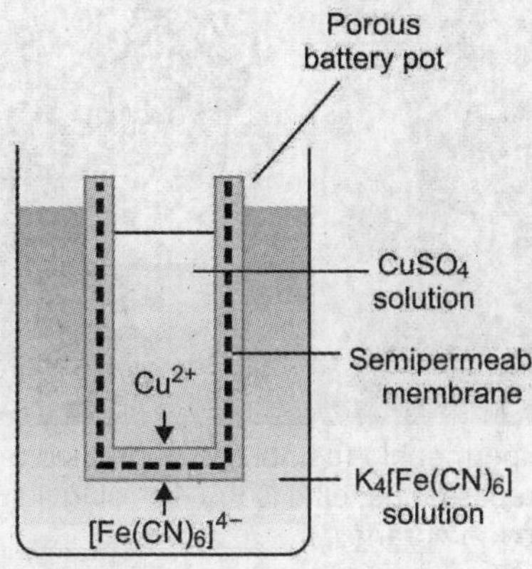

■ **Figure 16.7**
Preparation of semipermeable membrane.

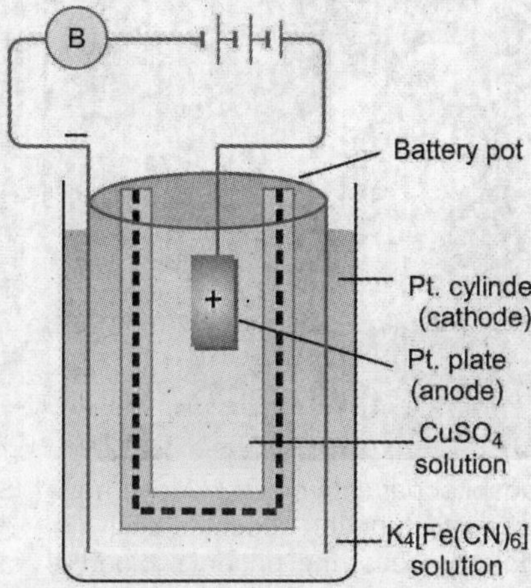

■ **Figure 16.8**
Electrical deposition of semipermeable membrane.

WHAT IS OSMOTIC PRESSURE?

A porous pot with cupric ferrocyanide membrane deposited in its walls is fitted with a rubber stopper having a long glass tube (Fig. 16.9). It is filled with concentrated aqueous sugar solution and immersed in distilled water. The osmosis of water through the membrane from water to the sugar solution takes place. As a result, the solution level in the long tube rises over a period of time. After a few days the level attains a definite maximum value. This marks the stage when the hydrostatic pressure set up due to the column of sugar solution counterbalances the flow of pure water (or osmosis) into the solution.

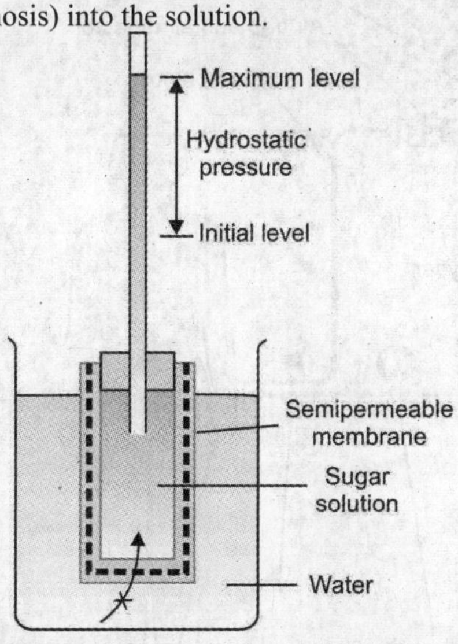

■ **Figure 16.9**
Hydrostatic pressure necessary to stop osmosis is equal to osmotic pressure.

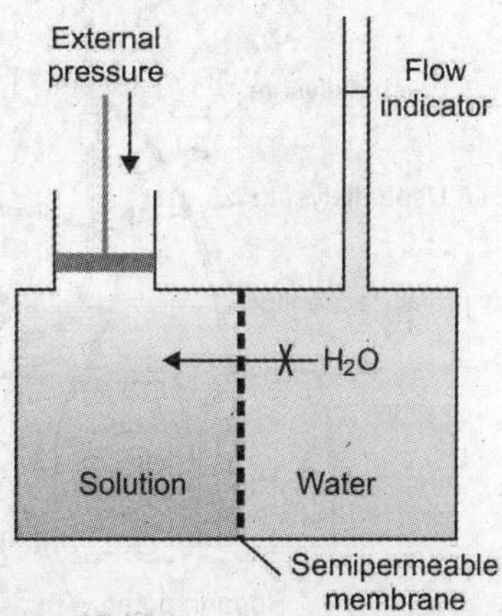

■ **Figure 16.10**
The external pressure applied on solution to stop osmosis is equal to osmotic pressure.

KIDNEY DIALYSIS

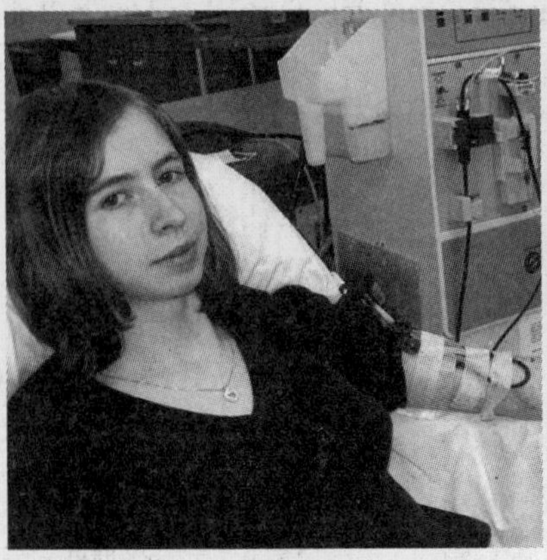

A hospital patient whose kidneys have ceased to function receives dialysis. In this process, the patient's blood is pumped through a dialysis machine, where it is filtered to remove waste products, then returned to the patient's body through a vein.

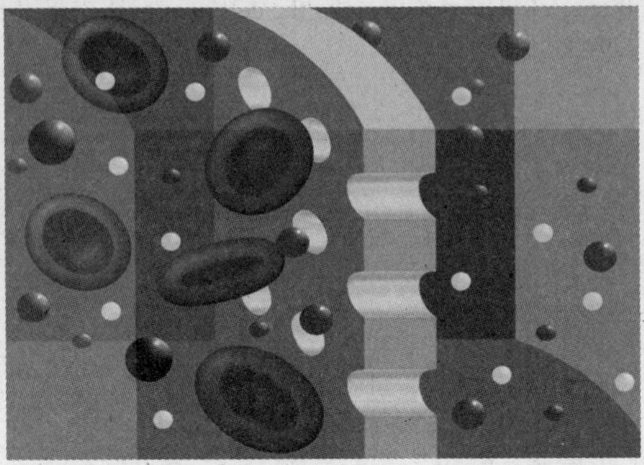

Scheme of semipermeable membrane during kidney dialysis, where red is blood, blue is the dialysing fluid, and yellow is the membrane.

Kidney Dialysis, also known as *hemodialysis*, is a medical treatment used to remove waste materials from the blood of patients lacking renal function. Blood from an artery is pumped through a dialyzer, or artificial kidney, where it flows past a semipermeable membrane. Dialysis fluid passing on the other side of the membrane removes unwanted elements in the blood by diffusion. The blood is then returned to the body through a vein.

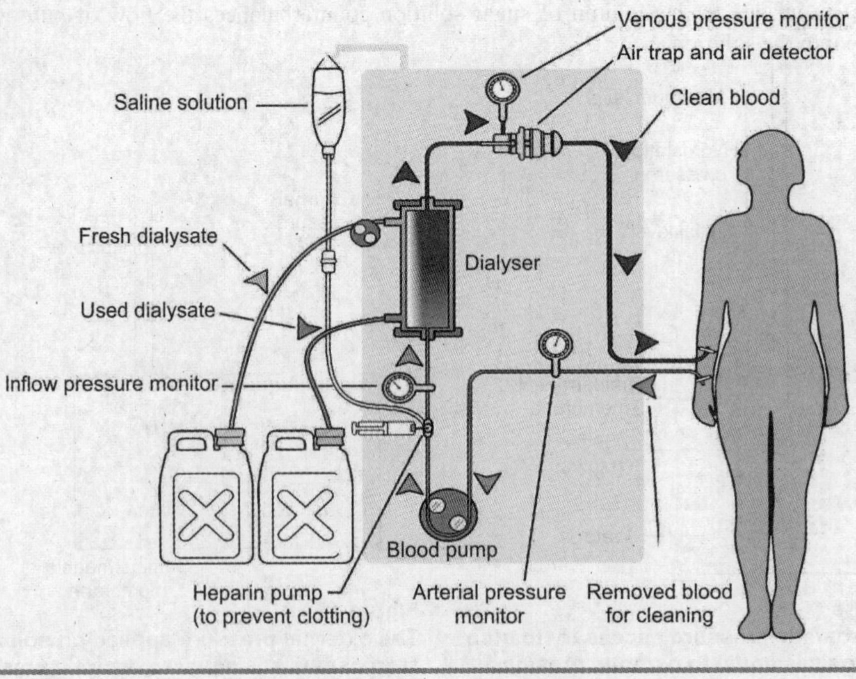

The hydrostatic pressure built up on the solution which just stops the osmosis of pure solvent into the solution through a semipermeable membrane, is called Osmotic Pressure.

Osmosis can be counteracted not only by hydrostatic pressure but also by application of external pressure on the solution. The external pressure may be adjusted so as to prevent the osmosis of pure water into solution. This provides another definition of osmotic pressure.

Osmotic pressure may be defined as the external pressure applied to the solution in order to stop the osmosis of solvent into solution separated by a semipermeable membrane.

This definition can be illustrated by means of the apparatus shown in Fig. 16.10. The external pressure on the solution is applied with the help of the piston and the progress of osmosis is shown by the movement of the liquid in the **flow indicator.** The pressure required just to arrest the movement of the liquid level in the **flow indicator,** is equal to the osmotic pressure of the solution.

DETERMINATION OF OSMOTIC PRESSURE

The osmotic pressure of a given solution can be determined experimentally by the methods detailed below. The apparatus used for the purpose is often referred to as **osmometer.**

(1) Pfeffer's Method

The apparatus used by Pfeffer (1877) for determination of osmotic pressure is shown in Fig. 16.11. It consists of a porous pot with copper ferrocyanide membrane deposited in its walls and having a glass top cemented to it. The glass top has an open tube and a closed mercury manometer in the side.

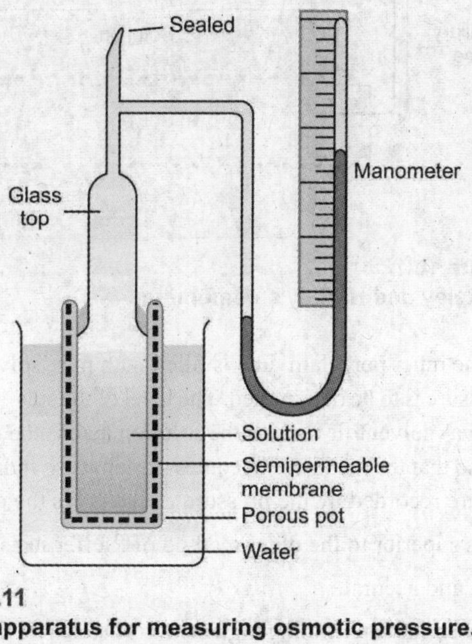

■ Figure 16.11
Pfeffer's apparatus for measuring osmotic pressure.

The apparatus is filled with a solution under examination through the tube which is then sealed off. The pot is now placed in pure water maintained at constant temperature. The water passes across the membrane into the solution and develops a pressure on the manometer. The highest pressure registered by the manometer gives the osmotic pressure of the solution. The Pfeffer's method suffers from two disadvantages :

(*a*) It is slow and it takes a few days before the highest pressure is reached.

(*b*) It cannot be used for measuring high osmotic pressures as the ferrocyanide membrane being weak ruptures.

Morse and Frazer (1910) used essentially the same method but deposited the copper ferrocyanide membrane electrolytically. The membrane so produced being more stout, they were able to measure osmotic pressures up to about 303975.00 Pa (300 atmospheres).

(2) **Berkeley and Hartley's Method**

Berkeley and Hartley (1904-1909) employed the technique of applying external pressure on the solution just enough to prevent osmosis. The osmometer used by them is illustrated in Fig. 16.12. A porcelain tube with copper ferrocyanide membrane deposited in its walls is enclosed in a metallic jacket. The tube is fitted with a reservoir of pure solvent (water) at one end and a capillary tube at the other. Mechanical pressure can be applied on the solution with a piston connected to a pressure gauge.

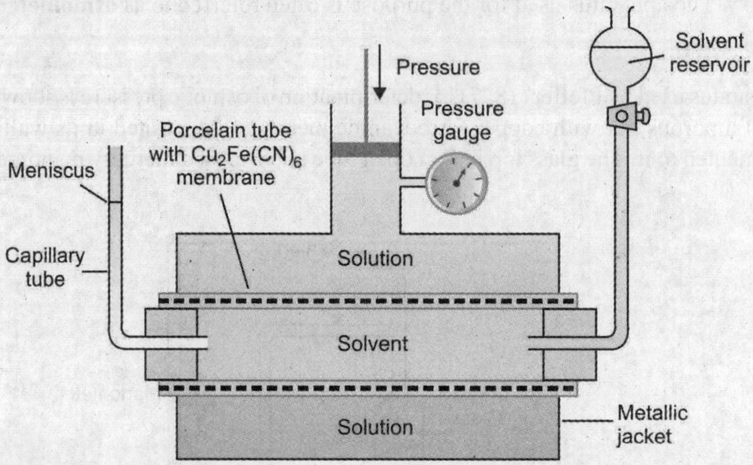

■ **Figure 16.12**
Berkeley and Hartley's osmometer.

Procedure. The inner porcelain tube is filled with pure solvent and the jacket with the solution whose osmotic pressure is to be determined. The level of the solvent meniscus in the capillary tube will tend to move down as solvent flows into the solution across the membrane. Pressure is then applied through the piston so that the meniscus becomes stationary. It indicates that osmosis has been stopped and now the pressure recorded by the pressure gauge gives the osmotic pressure of the solution.

This method is superior to the older method of Pfeffer in as much as :

(*a*) It is quick and accurate.

(*b*) It can be used for determining high osmotic pressures. The osmotic pressure being balanced by the external pressure, there is no strain left on the membrane and the danger of its bursting is eliminated.

(3) **A Modern Osmometer**

A modern apparatus for the determination of osmotic pressure is shown in Fig. 16.13.

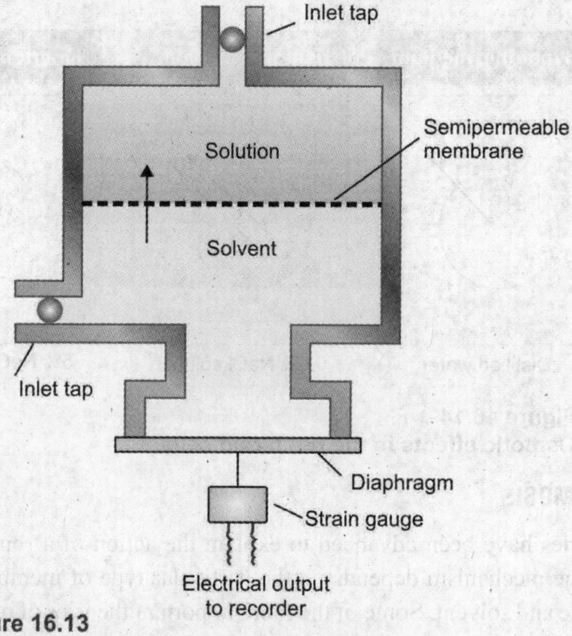

■ **Figure 16.13**
Stainless steel cell of a modern osmometer.

It consists of a stainless steel cell with a rigid fixed semipermeable membrane. The membrane divides the cell into solute and solvent compartments. A flexible diaphragm is fixed in the solvent compartment. When the solution and the solvent compartments are filled and the taps closed, osmosis occurs. The solvent flows into the solution across the semipermeable membrane. This reduces the pressure in the solvent compartment, causing the diaphragm to distort. Eventually the pressure becomes low enough to stop the occurrence of osmosis. The degree of diaphragm distortion is related to the osmotic pressure of the solution. The diaphragm distortion is measured by a device called **strain gauge.** The strain gauge provides an electric current that is proportional to the extent of distortion. The gauge is calibrated to give osmotic pressure directly.

ISOTONIC SOLUTIONS

When two solutions are separated by a semipermeable membrane and there is no flow of water across the membrane, the solutions are said to be Isotonic. If the membrane is completely semipermeable, isotonic solutions are also **iso-osmotic** *i.e.*, have the same osmotic pressure since osmotic pressure depends on the number of molecules. **The isotonic solutions have equimolar concentrations.**

Of the two solutions separated by a semipermeable membrane if one is of lower osmotic pressure, it is said to be **hypotonic** relative to the second solution. If it has a higher osmotic pressure than the second solution, it is said to be **hypertonic** relative to the second solution. Thus when red blood cells are placed in distilled water (*hypotonic medium*), water flows into the cells and they swell or burst. If, on the other hand, they are placed in 5% NaCl solution (*hypertonic medium*), water comes out of the cells and they shrink. A 0.16 *M* sodium chloride solution (0.95%) is isotonic with blood cells and they neither swell nor shrink as no osmosis takes place. However, since the membrane of the cells is not completely semipermeable and is leaky, the medium outside the cell and the inner fluid, although isotonic, is not necessarily iso-osmotic.

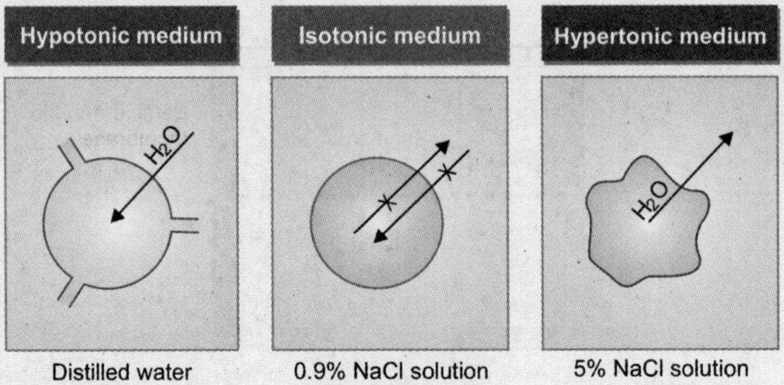

Figure 16.14
Osmotic effects in the red blood cells.

THEORIES OF OSMOSIS

Several theories have been advanced to explain the action of a semipermeable membrane. It is probable that the mechanism depends on the particular type of membrane used and also on the nature of the solute and solvent. Some of the more important theories of osmosis across a membrane are outlined below :

(1) The Molecular Sieve Theory

According to this theory, the membrane contains lots of fine pores and acts as a sort of 'molecular sieve'. **Smaller solvent molecules can pass through the pores but the larger molecules cannot.** Solvent molecules flow from a region of higher solute concentration to one of lower concentration across such a membrane. But since some membranes can act as sieves even though the solute molecules are smaller than the solvent molecules, this theory remains in doubt.

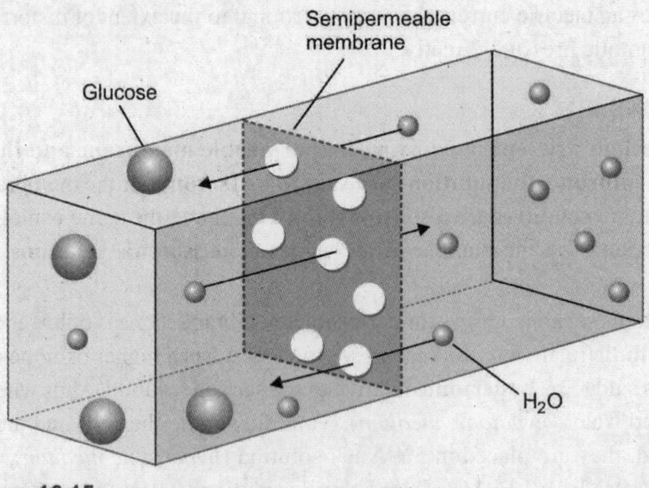

Figure 16.15
A semipermeable membrane can separate particles on the basis of size. It allows the passage of small H_2O molecules in both directions. But it prevents the passage of glucose molecules which are larger than water molecules. *(Illustration of Sieve theory)*

Recently it has been shown that the pores or capillaries between the protein molecules constituting an animal membrane are lined with polar groups ($-COO^-$, $-NH_3^+$, $-S^{2-}$, etc.). Therefore, the membrane acts not simply as a sieve but also regulates the passage of solute molecules by electrostatic or 'chemical interactions'. In this way even solute molecules smaller than solvent molecules can be held back by the membrane.

(2) **Membrane Solution Theory**

Membrane proteins bearing functional groups such as $-COOH$, $-OH$, $-NH_2$, etc., dissolve water molecules by hydrogen bonding or chemical interaction. Thus membrane dissolves water from the pure water (solvent) forming what may be called 'membrane solution'. The dissolved water flows into the solution across the membrane in a bid to equalize concentrations. In this way water molecules pass through the membrane while solute molecules being insoluble in the membrane do not.

(3) **Vapor Pressure Theory**

It suggests that a semipermeable membrane has many fine holes or capillaries. The walls of these capillaries are not wetted by water (solvent) or solution.

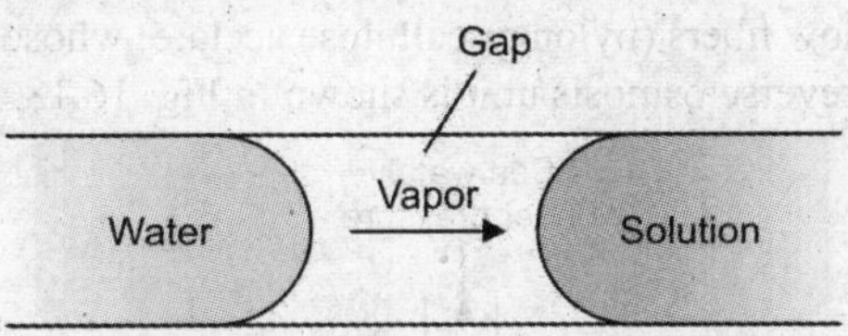

■ **Figure 16.16**
Water vapors diffuse into solution across the gas in a capillary of the membrane.

Thus neither solution nor water can enter the capillaries. Therefore each capillary will have in it solution at one end and water at the other, separated by a small gap. **Since the vapor pressure of a solution is lower than that of the pure solvent, the diffusion of vapor will occur across the gap from water side to solution side.** This will result in the transfer of water into the solution. The vapor theory offers a satisfactory explanation of the mechanism of osmosis in most cases.

(4) **Membrane Bombardment Theory**

This theory suggests that osmosis results from an unequal bombardment pressure caused by solvent molecules on the two sides of the semipermeable membrane. On one side we have only solvent molecules while on the other side there are solute molecules occupying some of the surface area. Thus there are fewer bombardments per unit area of surface on the solution side than on the solvent side. Hence the solvent molecules will diffuse more slowly through the membrane on the solution side than on the solvent side. The net result causes a flow of the solvent from the pure solvent to the solution across the membrane.

REVERSE OSMOSIS

When a solution is separated from pure water by a semipermeable membrane, osmosis of water occurs from water to solution. This osmosis can be stopped by applying pressure equal to osmotic pressure, on the solution. If pressure greater than osmotic pressure is applied, osmosis is made to proceed in the reverse direction to ordinary osmosis *i.e.*, from solution to water.

The osmosis taking place from solution to pure water by application of pressure greater than osmotic pressure, on the solution, is termed Reverse Osmosis.

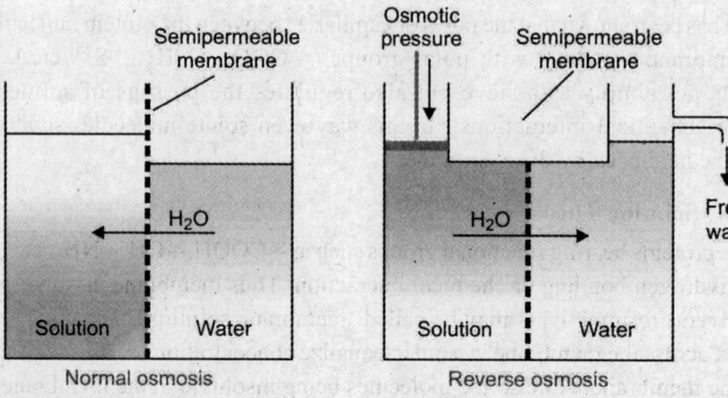

Figure 16.17
Reverse osmosis versus ordinary osmosis.

Desalination of Sea Water by Hollow-fiber Reverse Osmosis

Reverse osmosis is used for the desalination of sea water for getting fresh drinking water. This is done with the help of hollow fibers (nylon or cellulose acetate) whose wall acts as semipermeable membrane. A hollow-fiber reverse osmosis unit is shown in Fig. 16.18.

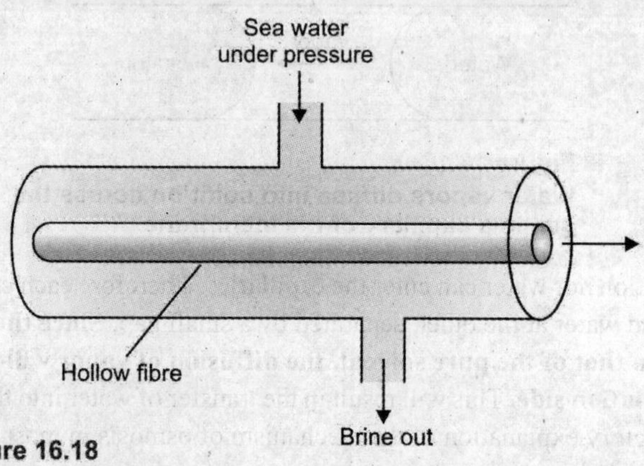

Figure 16.18
Desalination of sea water by reverse osmosis in a hollow fibre unit.

Water is introduced under pressure around the hollow fibers. The fresh water is obtained from the inside of the fiber. In actual practice, each unit contains more than three million fibers bundled together, each fiber is of about the diameter of a human hair.

LAWS OF OSMOTIC PRESSURE

From a study of the experimental results obtained by Pfeffer, van't Hoff showed that for dilute solutions :

(*a*) The osmotic pressure of a solution at a given temperature is directly proportional to its concentration.

(*b*) The osmotic pressure of a solution of a given concentration is directly proportional to the absolute temperature.

From the above findings, van't Hoff (1877) established the laws of osmotic pressure and pointed out that these were closely related to the gas laws.

(1) Boyle-van't Hoff Law for Solutions

If π is the osmotic pressure and C its concentration, from (*a*) we can write $\pi \propto C$, if temperature is constant.

NANOTUBE MEMBRANES

Nanotube Membranes Offer Possibility of Cheaper Desalination

A nanotube membrane on a silicon chip the size of a rupee coin may offer a cheaper way to remove salt from water.

Researchers at Lawrence Livermore National Laboratory (USA) have created a membrane made of carbon nanotubes and silicon that may offer, among many possible applications, a less expensive desalinization.

The nanotubes, special molecules made of carbon atoms in a unique arrangement, are hollow and more than 50,000 times thinner than a human hair. Billions of these tubes act as the pores in the membrane.

Molecules flowing through a carbon nanotube less than two nanometers in diameter.

The super smooth inside of the nanotubes allow liquids and gases to rapidly flow through, while the tiny pore size can block larger molecules. This previously unobserved phenomenon opens a vast array of possible applications.

Membranes that have carbon nanotubes as pores could be used in desalination. Salt removal from water, commonly performed through reverse osmosis, uses less permeable membranes, requires large amounts of pressure and is quite expensive. However, these more permeable nanotube membranes could reduce the energy costs of desalination by up to 75 percent compared to conventional membranes used in reverse osmosis.

If the concentration of the solute is expressed in moles per liter and V is the volume of the solution that contains 1 mol of solute,

$$C = \frac{1}{V}$$

Thus

$$\pi \propto \frac{1}{V} \quad \text{at constant temperature}$$

This relationship is similar to the *Boyle's law* for gases and is known as the *Boyle-van't Hoff law for solutions*.

(2) **Charles-van't Hoff Law for Solutions**

If T is the absolute temperature, from the statement (*b*), we can write

$$\pi \propto T, \text{ if temperature is constant}$$

This relationship is similar to the *Charles' Law for gases* and is known as *Charles-van't Hoff law for solutions*.

(3) **van't Hoff Equation for Solutions**

As shown above the osmotic pressure (π) of a dilute solution is inversely proportional to the volume (V) containing 1 mole of the solute and is directly proportional to the absolute temperature (T). This is,

$$\pi \propto \frac{1}{V} \qquad ...(1)$$

$$\pi \propto T \qquad ...(2)$$

Combining (1) and (2) van't Hoff gave the general relationship

$$\pi V = R'T \qquad ...(3)$$

where R' is a constant. He showed that this equation was parallel to the general *Gas Equation* ($PV = RT$), as the value of R' calculated from the experimental values of π, V, and T came out to be almost the same as of the *Gas constant, R*.

It is noteworthy that the *van't Hoff Equation* (3) was derived for 1 mole of solute dissolved in V liters. If n moles of solute are dissolved in V liters of solution, this equation assumes the form

$$\pi V = nRT$$

SOLVED PROBLEM 1

A solution of cane-sugar (mol mass = 342×10^{-3}) containing 34.2 kg/m³ has an osmotic pressure of 243180.00 Pa (2.4 atm) at 293 K. Calculate the value of R in m³ pascal.

SOLUTION

From van't Hoff equation

$$R = \frac{\pi V}{T} \qquad ...(1)$$

where π = osmotic pressure, V = volume of solution in m³ containing 1 mole of solute, T = absolute temperature.

In the present case,

$$\pi = 243180.00 \text{ Pa}$$

$$V = \frac{1}{34.2} \times 342 \times 10^{-3} = 10^{-2} \text{ m}^3$$

$$T = 20 + 273 = 293 \text{ K}$$

Substituting the values in the expression (1),

$$R = \frac{243180 \times 10^{-2}}{293} = \mathbf{8.30\,m^3 pascal\,K^{-1}\,mol^{-1}}$$

SOLVED PROBLEM 2

Experiment shows that 0.010 kg of cane sugar (mol mass = 342×10^{-3} kg) in 1×10^{-3} m³ of solution produces an osmotic pressure of 6.68×10^4 N m^{-2} at 273 K. Calculate the value of R in *SI* units.

SOLUTION

$$R = \frac{\pi V}{T} \qquad \text{...van't Hoff equation}$$

In this case,

$$\pi = 6.68 \times 10^4 \text{ N m}^{-2}$$

$$V = \frac{1 \times 10^{-3} \times 342 \times 10^{-3}}{10 \times 10^{-3}} = 3.42 \times 10^{-2} \text{ m}^3$$

Substituting the values in van't Hofff equation,

$$R = \frac{6.68 \times 10^4 \times 3.42 \times 10^{-2}}{273}$$

$$= \mathbf{8.36\ J\ mol^{-1}\ K^{-1}}$$

(4) **Avogadro-van't Hoff Law for Solutions**

Writing Van't Hoff equation for two different dilute solutions,

$$\pi_1 V_1 = n_1 RT_1 \qquad ...(1)$$

$$\pi_2 V_2 = n_2 RT_2 \qquad ...(2)$$

where n_1 and n_2 are the number of moles (molecules) of the solute in V_1 and V_2 liters of the two solutions.

If $\pi_1 = \pi_2$ and $T_1 = T_2$, it follows from (1) and (2) that

$$\frac{n_1}{V_1} = \frac{n_2}{V_2}$$

Hence **when osmotic pressure and temperature are the same, equal volumes of solutions would contain equal number of moles (molecules) of the solute.** This relationship is analogous to Avogadro's law of gases and is referred to as **Avogadro-van't Hoff law for solutions.** Alternatively, this law may be stated as:

Isotonic solutions at the same temperature have equimolar concentrations.

VAN'T HOFF THEORY OF DILUTE SOLUTIONS

Van't Hoff noted the striking resemblance between the behavior of dilute solutions and gases. Dilute solutions obeyed laws analogous to the gas laws. To explain it van't Hoff visualized that gases consist of molecules moving in vacant space (or vacuum), while in solutions the solute particles are moving in the solvent. The exact parallelism between solutions and gases is illustrated by experimental demonstration shown in Fig. 16.19.

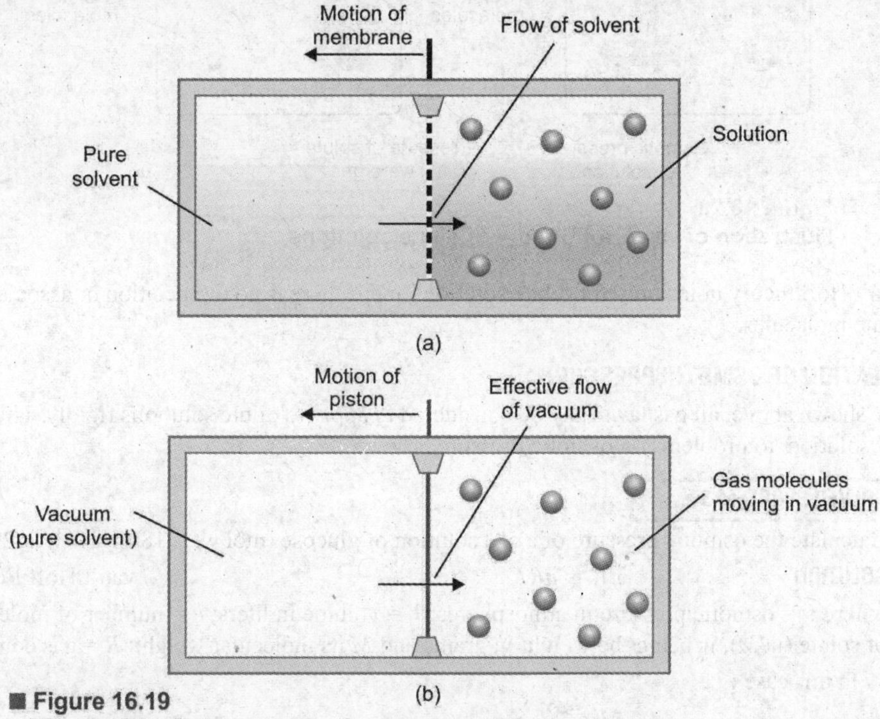

■ **Figure 16.19**
The analogy between osmotic pressure and gas pressure.

As shown in Fig. 16.19(*a*), the pure solvent flows into the solution by osmosis across the semipermeable membrane. The solute molecules striking the membrane cause osmotic pressure and the sliding membrane is moved towards the solvent chamber. In case of a gas (Fig. 16.19*b*), the gas molecules strike the piston and produce pressure that pushes it towards the empty chamber. Here it is the vacuum which moves into the gas. This demonstrates clearly that there is close similarity between a gas and a dilute solution.

Thinking on these lines, van't Hoff propounded his theory of dilute solution. The van't Hoff theory of dilute solutions states that : **a substance in solution behaves exactly like a gas and the osmotic pressure of a dilute solution is equal to the pressure which the solute would exert if it were a gas at the same temperature occupying the same volume as the solution.**

Suppose a dilute sugar solution to be contained in an exactly full, covered beaker. Then suppose the water to vanish, leaving the sugar molecules suspended in otherwise empty space. The sugar would then function as a gas, exerting pressure equal to its former osmotic pressure.

According to the van't Hoff theory of dilute solutions, all laws or relationship obeyed by gases would be applicable to dilute solutions.

From van't Hoff theory it follows that just as 1 mole of a gas occupying 0.0224 m^3 at 273 K exerts 1 atmosphere pressure, so, **1 mole of any solute dissolved in 0.0224 m^3 would exert 101325.00 Pa osmotic pressure.**

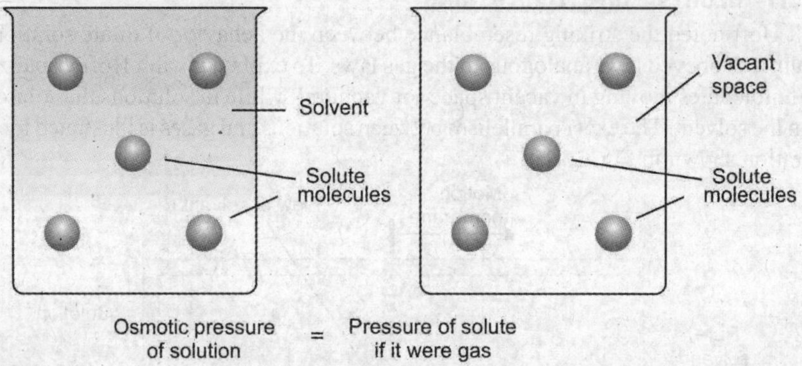

Figure 16.20
Illustration of van't Hoff theory of dilute solutions.

van't Hoff theory holds only for dilute solutions and if there is no dissociation or association of the solute molecules.

CALCULATION OF OSMOTIC PRESSURE

As shown above, all gas laws may be considered to apply to dilute solutions rigidly. This gives an easy solution to problems on osmotic pressure.

SOLVED PROBLEM 3

Calculate the osmotic pressure of a 5% solution of glucose (mol wt = 180 × 10^{-3}) at 291 K.

SOLUTION $\pi V = nRT$ ...van't Hoff Equation

where π = osmotic pressure in atmospheres; V = volume in liters; n = number of moles of solute (w/M), w being the weight in grams and M its molecular weight; R = gas constant.

In this case :

$$\pi = ?$$

$$V = 0.0001 \text{ m}^3 \text{ or } 10^{-4} \text{ m}^3$$

$$n = \frac{5}{M} = \frac{5\times10^{-3}}{180\times10^{-3}}$$

$$R = 8.314 \text{ J k}^{-1} \text{ mol}^{-1}$$

$$T = 291 \text{ K}$$

Substituting the values in van't Hoff Equation

$$\pi \times 10^{-4} = \frac{5}{180} \times 8.314 \times 291$$

$$\pi = \frac{5}{180} \times \frac{8.314}{10^{-4}} \times 291 = \mathbf{672048.33}$$

SOLVED PROBLEM 4

Find the osmotic pressure in pascal at 15°C of a solution of naphthalene ($C_{10}H_8$) in benzene containing 14×10^{-3} kg of naphthalene per liter of solution.

SOLUTION

$$\pi V = nRT \qquad \text{...vant Hoff equation}$$

Here,

$$\pi = ? \text{ atmosphere}$$

$$V = 10^{-3} \text{ m}^3$$

$$n = \frac{14}{128} \text{ (mol mass of naphthalene = 128)}$$

$$R = 8.314 \text{ J K}^{-1} \text{ mol}^{-1}$$

$$T = 15 + 273 = 288 \text{ K}$$

Substituting values in van't Hoff equation,

$$\pi \times 10^{-3} = \frac{14\times 10^{-3}}{128\times 10^{-3}} \times 8.314\times 288$$

$$\pi = 261891 \text{ Pa}$$

SOLVED PROBLEM 5

Calculate the osmotic pressure of solution obtained by mixing (*a*) 0.0001 m^3 of 3.4 per cent solution of urea (mol mass = 60) and (*b*) 0.0001 m^3 of 1.6 per cent solution of cane sugar (mol mass = 342) at 293 K.

SOLUTION

Osmotic pressure of urea

$$\pi V = nRT$$

$$\pi = ? \text{ atmosphere}$$

$$V = 0.0002 \text{ m}^3$$

$$n = \frac{3.4}{60}$$

$$R = 8.314 \text{ J K}^{-1} \text{ mol}^{-1}$$

$$T = 293 \text{ K}$$

Substituting values in van't Hoff equation

$$\pi \times 0.0002 = \frac{3.4}{60} \times 8.314 \times 293$$

where $\pi = \mathbf{690200.51\ Pa}$

Osmotic pressure of cane sugar

$$\pi V = nRT$$

$$V = 0.0002\ m^3$$

$$n = \frac{1.6}{342}$$

$$R = 8.314\ J\ K^{-1}\ mol^{-1}$$

$$T = 293\ K$$

Substituting values

$$\pi \times 0.0002 = \frac{1.6}{342} \times 8.314 \times 293$$

Hence $\pi = 56982$ Pa

According to the van't Hoff theory of dilute solutions, dissolved substances behave like gases. Thus they obey Dalton's Law of Partial Pressures. Hence the total osmotic pressure of the mixture

$$= 690200.57 + 56982 = \mathbf{747182.57\ Pa}$$

DETERMINATION OF MOLECULAR WEIGHT FROM OSMOTIC PRESSURE

Osmotic pressure measurements are less accurate and difficult to carry out. Therefore, this method is not widely used in the determination of molecular weights. The chief use of this method is in finding molecular weights of polymers *e.g.*, plastics, proteins, and starch, which have very high molecular weights.

Knowing the osmotic pressure of a given solution, the molecular weight of the solute can be calculated as follows from van't Hoff equation.

$$\pi V = nRT$$

$$= \frac{w}{M} RT$$

or

$$M = \frac{wRT}{\pi V}$$

where M = molecular mass of the solute

w = amount of solute in grams

$R = 8.314\ J\ K^{-1}\ mol^{-1}$

$T = (t°C + 273)\ K$

π = osmotic pressure in pascal

V = volume of solution in m^3

SOLVED PROBLEM 6

Morse and Frazer observed that a solution of glucose containing 18 kg/m^3 had an osmotic pressure of 242166.75 Pa (2.39 atmospheres) at 296 K. Find the molecular weight of glucose.

SOLUTION

We know that:

$$M = \frac{wRT}{\pi V} \qquad ...(1)$$

Here, $w = 18\ kg/m^3$

$R = 8.314\ J\ K^{-1}\ mol^{-1}$

$$T = 296 \text{ K}$$
$$\pi = 242166.75 \text{ Pa}$$
$$V = 10^{-3} \text{ m}^3$$

Substituting these values in relation (1)

$$M = \frac{18 \times 8.314 \times 296}{242166.75 \times 10^{-3}}$$
$$= 182.92 = \mathbf{183.0}$$

SOLVED PROBLEM 7

A solution of glycol containing 1.821 kg/m^3 has an osmotic pressure of 69060.98 Pa of mercury at 283 K. What is the molecular mass of glycol?

SOLUTION

We know that from van't Hoff equation

$$M = \frac{wRT}{\pi V} \qquad ...(1)$$

In this case,

$$w = 1.821 \text{ kg}$$
$$R = 8.314 \text{ J K}^{-1} \text{ mol}^{-1}$$
$$T = 283 \text{ K}$$
$$\pi = 69060.98 \text{ Pa}$$
$$V = 10^{-3} \text{ m}^3$$

Substituting these values in expression (1),

$$M = \frac{1.821 \times 8.314 \times 283}{69060.98 \times 10^{-3}}$$
$$= \mathbf{62.04}$$

RELATION BETWEEN VAPOR PRESSURE AND OSMOTIC PRESSURE

Consider an aqueous solution contained in a tube closed at the lower end by a semipermeable membrane and immersed in pure water (Fig. 16.21). The whole apparatus is enclosed under a bell-jar which is evacuated. When the osmotic equilibrium is reached, the bell jar is saturated with water vapor. Let the height of solution in the tube be h cm. If the solution is sufficiently dilute, the density of the solution will be approximately the same as that of the pure solvent, d.

Since at equilibrium, the osmotic pressure of solution (π) is exactly balanced by hydrostatic pressure due to h cm column of solution of density d, we can write,

$$\pi = h \times d \qquad ...(1)$$

If p is the vapor pressure of the solvent at the given temperature and p_s the vapor pressure of the solution, the pressure at all points on a level with A will be p and at all points on a level with B will be p_s. Now, the pressure at A is greater than at B due to the weight of the column of solvent vapor of height h. Thus,

$$p - p_s = hD \qquad ...(2)$$

where D = density of solvent vapor at pressure p.

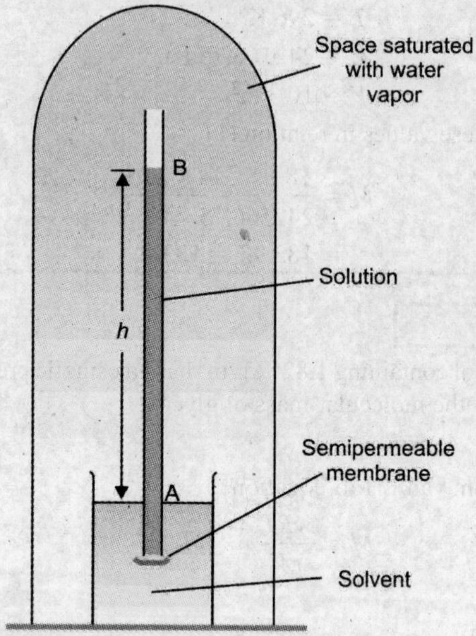

Figure 16.21
Vapor pressure and osmotic pressure.

Dividing (1) by (2), we have

$$\frac{\pi}{p - p_s} = \frac{hd}{hD} = \frac{d}{D} \quad \text{...(3)}$$

At a given temperature, d and D are constant. That is

$$\frac{\pi}{p - p_s} = \text{constant}$$

or $\quad \pi = \text{constant} \times (p - p_s)$

or $\quad \pi \propto (p - p_s)$

This shows that **osmotic pressure is directly proportional to the lowering of vapor pressure.**

Relative Lowering of Vapor Pressure and Osmotic Pressure

When the osmotic equilibrium is established in apparatus shown in Fig. 16.21, we have

$$\pi = h \times d, \text{ for dilute solution} \quad \text{...(1)}$$

$$p - p_s = h \times D, \text{ for solvent vapor} \quad \text{...(2)}$$

Dividing (2) by (1) and simplifying

$$p - p_s = \pi \times \frac{D}{d} \quad \text{...(3)}$$

where D is the density of the solvent vapor and d is the density of the solution.

From the gas equation for one mole of the solvent vapor ($pV = RT$),

$$p = \frac{RT}{V} \quad \text{...(4)}$$

If M is the molecular mass of the solvent vapor, its density

$$D = \frac{M}{V}$$

or $$V = \frac{M}{D}$$

Substituting the value of V in equation (4)

$$p = \frac{RTD}{M} \qquad ...(5)$$

Dividing (3) by (5)

$$\frac{p - p_s}{p} = \pi \times \frac{D}{d} \times \frac{M}{RTD}$$

or $$\frac{p - p_s}{p} = \pi \times \frac{M}{dRT} \qquad ...(6)$$

At a fixed temperature d is constant and hence the factor $\frac{M}{dRT}$ has a constant value. Therefore

we can write $\frac{p - p_s}{p} \propto \pi$

That is, **relative lowering of vapor pressure is directly proportional to osmotic pressure.**

DERIVATION OF RAOULT'S LAW

According to van't Hoff equation for solutions,

$$\pi V = nRT$$

where V = volume of solution and n = number of moles of solute.

Hence, $$\pi = \frac{nRT}{V}$$

Substituting the values of π in (6)

$$\frac{p - p_s}{p} = \frac{nRT}{V} \times \frac{M}{dRT} = \frac{n}{Vd/M}$$

or $$\frac{p - p_s}{p} = \frac{n}{N}$$

where N = number of moles of the solvent.

This is the **Raoult's law equation** in which n is the denominator is ignored when the solution is very dilute.

OSMOTIC PRESSURE OF ELECTROLYTES

Like other colligative properties, observed osmotic pressure of electrolytes (π) in aqueous solutions is higher than the value calculated using van't Hoff equation, $\pi_0 V = nRT$. Expressing in terms of **van't Hoff factor *i*,**

$$\frac{\pi}{\pi_0} = i$$

Consequently, the **van't Hoff equation for solutions** of electrolytes assumes the form

$$\pi V = inRT$$

Since electrolytes are ionized in solution, total number of particles increases as compared to the number of particles if no ionization had taken place. The osmotic pressure of solution being proportional to the number of particles in solution, we can write

$$\frac{\pi}{\pi_0} = \frac{\text{actual number of particles on ionization}}{\text{number of particles if no ionization occurred}}$$

For example, if NaCl is ionized completely as at infinite dilution, one mole of NaCl will give two moles of particles (Na^+ and Cl^-, one mole each). Therefore, the observed osmotic pressure will be twice the calculated value for no ionization. That is,

$$\frac{\pi}{\pi_0} = 2$$

Similarly, the value of i for $CaCl_2$ at infinite dilution will be 3; for $FeCl_3$ it will be 4.

Ordinarily, electrolytes are practically ionized in aqueous solution. If α be the degree of ionization for a salt AB,

$$\underset{1-\alpha}{AB} \rightleftharpoons \underset{\alpha}{A^+} + \underset{\alpha}{B^-}$$

The total number of moles of particles in aqueous solution will be $1 - \alpha + 2\alpha = 1 + \alpha$ as compared to 1 mole of salt for no ionization.

Therefore, $$\frac{\pi}{\pi_0} = \frac{1+\alpha}{1}$$

Thus the value of i will be more than 1 and less than 2.

SOLVED PROBLEM 8

The osmotic pressure of a 0.01 m solution of $CaCl_2$ and sucrose at 298.15 K are 61301.625 Pa and 22696.8 Pa respectively. Calculate the van't Hoff factor and degree of dissociation of $CaCl_2$.

SOLUTION

$$i = \frac{\pi}{\pi_0} = \frac{61301.625\,\text{Pa}}{22696.8\,\text{Pa}} = 2.70$$

If α is the degree of dissociation,

$$\underset{1-\alpha}{\overset{1}{CaCl_2}} \rightleftharpoons \underset{\alpha}{Ca^{2+}} + \underset{2\alpha}{2Cl^-}$$

$$\therefore \quad \frac{1+2\alpha}{1} = 2.70$$

and $\alpha = \mathbf{0.85}$

As shown above the van't Hoff factor $i = 2.70$ and the degree of dissociation (α) = 0.85

SOLVED PROBLEM 9

Find the osmotic pressure of an aqueous solution of $BaCl_2$ at 288 K containing 0.39×10^{-3} kg per 0.00006 m^3. The salt is 60% dissociated. (Ba = 137; Cl = 35.5)

SOLUTION

For an electrolyte solution, van't Hoff equation assumes the form

$$\pi V = inRT, \text{ where } i \text{ is van't Hoff factor} \qquad ...(1)$$

In this case,

$$V = 6 \times 10^{-5}\ m^3$$

$$n = \frac{0.39 \times 10^{-3}\ \text{kg}}{208 \times 10^{-3}}$$

$$T = 288\ K$$

Calculation of i

$$\underset{1-\alpha}{\overset{1}{BaCl_2}} \rightleftharpoons \underset{\alpha}{Ba^{2+}} + \underset{2\alpha}{2Cl^-}$$

$$\therefore \qquad i = \frac{1 - \alpha + \alpha + 2\alpha}{1} = \frac{1 + 2\alpha}{1}$$

But $\qquad \alpha = 0.60$

$$\therefore \qquad i = \frac{1 + 2 \times 0.60}{1} = 2.2$$

Substituting values in equation (1)

$$\pi = 2.2 \times \frac{0.39 \times 10^{-3}}{208 \times 10^{-3}} \times \frac{8.314 \times 288}{6 \times 10^{-5}}$$

$$= \mathbf{164617.2\ Pa}$$

ADDITIONAL SOLVED PROBLEMS

SOLVED PROBLEM 10

An aqueous solution contains 0.020 kg of glucose 10^{-3} m^3. Calculate the pressure of the solution at 298 K (molecular mass of glucose = 0.180 kg).

SOLUTION

Formula used

$$\pi = \frac{n}{V} R T$$

Quantities given

$$n = \frac{20 \times 10^{-3} \text{ kg}}{180 \times 10^{-3} \text{ kg mol}^{-1}} = 0.111 \text{ mol} \qquad V = 10^{-3} \text{ m}^3$$

$$T = 298\text{K} \qquad R = 8.314 \text{ J K}^{-1} \text{ mol}^{-1}$$

Substitution of values

$$\pi = \frac{0.111 \text{ mol} \times 8.314 \text{ J K}^{-1} \text{ mol}^{-1} \times 298\text{K}}{10^{-3} \text{ m}^3}$$

$$= \mathbf{275010.49\ Pa}$$

SOLVED PROBLEM 11

A 1% solution of potassium chloride is 80% dissociated. What is its osmotic pressure at 300 K? ($R = 8.314$ J K^{-1} mol^{-1} ; K = 39 ; Cl = 35.5)

SOLUTION : (i) To calculate the van't Hoff factor, *i*

$$\text{No. of moles of KCl in 1\% solution} = \frac{1 \times 10^{-3} \text{ kg}}{74.5 \times 10^{-3} \text{ kg mol}^{-1}} = 0.01342 \text{ mol}^{-1}$$

	KCl	⟶	K	+	Cl^{-1}
Initial Concentration	0.01342		0		0
Final Concentration	(0.01342 – α)		0.01342α		0.01342α

$$\text{Total number of particles} = 0.01342\,(1-\alpha) + 0.01342\alpha + 0.01342\alpha$$

$$= 0.01342\,(1+\alpha)$$

Since α = 80% or 0.8, we have

Total number of particles = 0.01342 (1 + 0.8)

= 0.024156

$$\text{van't Hoff factor, } i = \frac{0.02416}{0.01342} = 1.80$$

(ii) To calculate the osmotic pressure

Formula used

$$\pi = \frac{i\,n\,R\,T}{V}$$

Quantities given

i = 1.8 $\quad$ n = 0.01342 mol $\quad$ V = 10^{-4} m^3

R = 8.314 J K^{-1} mol^{-1} $\quad$ T = 300 K

Substitution of values

$$\pi = \frac{1.8 \times 0.01342 \times 8.314 \text{ J K}^{-1} \text{ mol}^{-1} \times 300 \text{ K}}{10^{-4} \text{ m}^3}$$

$$= \mathbf{602498.95 \text{ Pa}}$$

SOLVED PROBLEM 12

Calculate the molality and osmotic pressure of 0.5% solution of urea (molecular mass = 60) in water at 273 K. (R = 8.314 J K^{-1} mol^{-1}).

SOLUTION : (*a*) To calculate the molality

Formula used

$$\text{molality, } m = \frac{\text{mass of urea} \times 1000}{\text{mol. mass of urea} \times \text{mass of solvent}}$$

Quantities given

mass of urea = 0.5 × 10^{-3} kg $\quad$ mass of solvent = 99.5 × 10^{-3} kg

mol mass of urea = 60 × 10^{-3} kg mol^{-1}

Substitution of values

$$\text{molality, } m = \frac{0.5 \times 10^{-3} \times 1000 \times 10^{-3}}{60 \times 10^{-3} \times 99.5 \times 10^{-3}}$$

$$= \mathbf{0.0838 \text{ m}}$$

(*b*) To calculate the osmotic pressure

Formula used

$$\pi = \frac{n}{V} R T$$

Quantities given

$$n = \frac{0.5 \times 10^{-3} \text{ kg}}{60 \times 10^{-3} \text{ kg mol}^{-1}} = 0.00833 \text{ mol} \qquad R = 8.314 \text{ J K}^{-1} \text{ mol}^{-1}$$

V = 10^{-4} m^3 $\quad$ T = 273 K

$$\pi = \frac{0.00833 \text{ mol} \times 8.314 \text{ J K}^{-1} \text{ mol}^{-1} \times 273 \text{ K}}{10^{-4} \text{ m}^3}$$

$$= \mathbf{189067.843 \text{ Pa}}$$

SOLVED PROBLEM 13

Calculate the osmotic pressure of 0.745% solution of KCl at 300 K assuming to be 86% dissociated.

SOLUTION : ***(a)* To calculate the van't Hoff factor *i***

$$\text{Number of moles of KCl in 0.745\% solution} = \frac{0.745 \times 10^{-3}\text{ kg}}{74.5\text{ kg mol}^{-1}} = 0.01\text{ mol}$$

	KCl ⟶	K^+ +	Cl^{-1}
Initial Concentration	0.01	0	0
Final Concentration	0.01(1–α)	0.01α	0.01α

$$\text{Total number of particles} = 0.01(1-\alpha) + 0.01\alpha + 0.01\alpha$$
$$= 0.01(1+\alpha)$$

Since $\alpha = 86\%$ or 0.86, we have

$$\text{Total number of particles} = 0.01(1 + 0.86)$$
$$= 0.0186$$
$$\text{van't Hoff factor, } i = \frac{0.0186}{0.01} = 1.86$$

***(b)* To calculate the osmotic pressure**

Formula used

$$\pi = i\frac{n}{V}RT$$

Quantities given

$i = 1.86$ $n = 0.01$ m $T = 300$ K

$R = 8.314\text{ J K}^{-1}\text{ mol}^{-1}$ $V = 0.01 \times 10^{-3}\text{ m}^3$

Substitution of values

$$\pi = \frac{1.86 \times 0.01\text{ m} \times 8.314\text{ J K}^{-1}\text{ mol}^{-1} \times 300\text{ K}}{0.01 \times 10^{-3}\text{ m}^3}$$
$$= \mathbf{463663.20\ Pa}$$

SOLVED PROBLEM 14

0.010 kg of urea of molecular mass 60×10^{-3} is dissolved in $1 \times 10^{-3}\text{ m}^3$ of water. Calculate the osmotic pressure of the solution at 293 K.

SOLUTION

Formula used

$$\pi = \frac{n}{V}RT$$

Quantities given

$n = \dfrac{0.010}{60 \times 10^{-3}} = 0.1667$ mol $R = 8.314\text{ J K}^{-1}\text{ mol}^{-1}$

$T = 293$ K $V = 1 \times 10^{-3}\text{ m}^3$

Substitution of values

$$\pi = \frac{0.1667\text{ mol} \times 8.314\text{ J K}^{-1}\text{ mol}^{-1} \times 293\text{ K}}{1 \times 10^{-3}\text{ m}^3}$$
$$= \mathbf{406081.533\ Pa}$$

SOLVED PROBLEM 15

A solution contains 0.005 kg of urea (mol mass 60×10^{-3}) per 0.1 kg of water. If the density of water at 298 K is 998 kg m^{-3}, calculate the osmotic pressure of the solution at the same temperature ($R = 8.314 \text{ J K}^{-1} \text{ mol}^{-1}$).

SOLUTION : **(*a*) To calculate the volume of 0.01 kg of water**

$$\text{Density of water} = 998 \text{ kg m}^{-3}$$

$$\text{Volume of 0.1 kg of water} = 0.1 \text{ kg} \div 998 \text{ kg m}^{-3} \qquad [\because V = \frac{M}{D}]$$

$$= 0.1002 \times 10^{-3} \text{ m}^3$$

(ii) To calculate the osmotic pressure

Formula used

$$\pi = \frac{n}{V} R T$$

Quantities given

$$n = \frac{5 \times 10^{-3} \text{ kg}}{60 \times 10^{-3} \text{ kg mol}^{-1}} = 0.0833 \text{ mol} \qquad R = 8.314 \text{ J K}^{-1} \text{ mol}^{-1}$$

$$T = 298 \text{ K} \qquad V = 0.1002 \times 10^{-3} \text{ m}^3$$

Substitution of values

$$\pi = \frac{0.0833 \text{ mol} \times 8.314 \text{ J K}^{-1} \text{ mol}^{-1} \times 298 \text{ K}}{0.1002 \times 10^{-3} \text{ m}^3}$$

$$= \mathbf{2059698.08 \text{ Pa}}$$

EXAMINATION QUESTIONS

1. Define or explain the following terms :

(*a*) Osmosis
(*b*) Osmotic pressure
(*c*) van't Hoff equation
(*d*) Raoult's law
(*e*) Isotonic solutions

2. (*a*) Explain the term 'osmotic pressure'. How is the osmotic pressure of a solution is affected by concentration of a solute and by temperature?

(*b*) Calculate the osmotic pressure of a 5% solution of glucose ($C_6H_{12}O_6$) at 291 K. ($R = 8.314 \text{ J K}^{-1} \text{ mol}^{-1}$)

Answer. (*b*) 672102.097 Pa

3. (*a*) State the laws of osmotic pressure. Explain how osmotic pressure is analogous to gas pressure. How can the molecular mass of a substance in solution be determined from its osmotic pressure?

(*b*) An aqueous solution contains 0.020 kg of glucose per 10^{-3} m^3. Calculate the osmotic pressure of the solution at 298 K (mol. mass of glucose = 180×10^{-3} kg).

Answer. 275010.492 Pa

4. Explain : Equimolar aqueous solutions of glucose and potassium nitrate do not have equal osmotic pressures at room temperature.

5. (*a*) What do you mean by osmosis? State van't Hoff's law of osmotic pressure and deduce osmotic pressure equation $p = CRT$. The symbols have their usual significance.

   (*b*) A solution contains 0.005 kg of urea (mol mass = 60×10^{-3}) per 0.100 kg in water. If the density of water at 298 K is 998 kg/m^3, calculate the osmotic pressure of the solution at the same temperature ($R = 8.314$ J K^{-1} mol^{-1})

   Answer. (*b*) 2059698.08 Pa

6. Calculate the concentration (kg/m^3) of aqueous glucose (molecular mass = 180×10^{-3}) which is isotonic with blood (osmotic pressure for blood is 775136.25 Pa at 310 K. and $R = 8.314$ J K^{-1} mol^{-1})

   Answer. 54.135 kg/m^3

7. State and explain the laws of osmosis. Derive the equation relating osmotic pressure and molecular mass of the solute.

8. Given the following solutions at 300 K (*a*) 0.01 M sucrose; (*b*) 0.1 M NaCl; and (*c*) 0.1 M $BaCl_2$. Which one of these will have highest osmotic pressure? Why?

   Answer. None

9. (*a*) What is osmosis? Derive a relationship between osmotic pressure and concentration of a dissolved solute.

   (*b*) Calculate osmotic pressure of solutions at 298 K containing (*i*) 5.85×10^{-3} kg of NaCl, and (*ii*) 5.0×10^{-3} kg of urea per 10^{-3} m^3 of solutions.

   Answer. (*b*) (*i*) 247757.20 Pa and (*ii*) 206381.748 Pa

10. (*a*) Making use of the concept of chemical potential, derive the van't Hoff equation $p = CRT$, where p is the osmotic pressure of the solution and C is its molar concentration. Also derive a relationship between osmotic pressure and lowering of vapor pressure of an ideal solution.

    (*b*) What are isotonic solutions? Show that for isotonic solutions at the same temperature, the molar concentrations are the same. *(Madurai BSc)*

11. What is the difference between osmosis and diffusion? *(Punjabi BSc)*

12. Derive the relationship between lowering of vapor pressure and osmotic pressure of a solvent. *(Lucknow BSc)*

13. Define osmosis. How molecular weight of a substance can be determined by osmotic pressure method? *(Arunachal BSc)*

14. From thermodynamic considerations, obtain the expression; $p = CRT$ for the osmotic pressure. *(Aligarh BSc)*

15. Calculate osmotic pressure of human blood at body temperature (310 K) which shows a freezing depression of 273 K (assumed to contain no associating or dissociating substances) Molal depression constant of water is equal to 275.01 K kg mol^{-1}.

    Answer. 2558.53 Pa *(Andhra BSc)*

16. The osmotic pressure π of blood is 775136.25 Pa at 310 K. To inject glucose to a patient intravenously it has to be isotonic with blood. What would be the concentration of glucose solution in kg per m^3? ($M_{glucose} = 180 \times 10^{-3}$ kg mol^{-1}, $R = 8.314$ J K^{-1} mol^{-1})

    Answer. 54.135 kg/m^3 *(Guru Nanak Dev BSc)*

17. A 0.1 M aqueous solution of KNO_3 shows an osmotic pressure of 455962.5 Pa at 300 K while the calculated one is 253312.5 Pa. What is the van't Hoff factor '*i*' for the solution?

    Answer. 0.55 *(Guru Nanak Dev BSc)*

18. Discuss the laws of osmotic pressure and van't Hoff's theory. Prove that osmotic pressure is a colligative property. *(Arunachal BSc)*

19. (*a*) From thermodynamic considerations, derive the expression for osmotic pressure of a solution.

(*b*) Obtain a relationship between osmotic pressure and elevation in boiling point of a solution.

(*Kalyani BSc*)

20. Establish the relationship between osmotic pressure and vapor pressure lowering of an ideal solution.

(*Arunachal BSc*)

21. Describe one method for measurement of osmotic pressure. (*Purvanchal BSc*)

22. (*a*) Define Osmosis and Osmotic pressure. How molecular mass of solute is determined from osmotic pressure measurement?

(*b*) What is the concentration of cane sugar which is isotonic with solution containing 6×10^{-3} kg of urea per m^3. (Molecular masses of urea and cane sugar are 60×10^{-3} and 342×10^{-3} respectively).

Answer. 342×10^{-3} kg/m^3 (*Mysore BSc*)

23. (*a*) How can you determine Osmotic pressure by Berkeley and Hartley's method?

(*b*) Calculate the osmotic pressure of a 0.01 N solution of cane sugar at 303 K.

Answer. 25191.42 Pa

(*Dibrugarh BSc*)

24. Calculate the osmotic pressure of a solution at 298 K obtained by mixing 0.0001 m^3 of 4.6 % of glucose and 0.0002 m^3 of 4.5% solution of glucose.

Answer. 844239.9 Pa (*Delhi BSc*)

25. Calculate the osmotic pressure of a glucose solution containing 0.015 kg in 0.001 m^3 solution at 300 K.

Answer. 2077162.5 Pa (*Gulbarga BSc*)

26. The osmotic pressure of blood at 310 K is 770070.00 Pa. A solution that is given intravenously must have the same osmotic pressure as blood. What should be the molarity of a glucose solution to give an osmotic pressure of 770070.00 Pa at 310 K ?

Answer. 298.785 (*Panjab BSc*)

27. The osmotic pressure of a solution containing 0.002 kg of protein in 0.0002 m^3 of aqueous solution is 20265 Pa at 300K. Calculate the molar mass of the protein. ($R = 8.314$ J K^{-1} mol^{-1})

Answer. 1.230 kg mol^{-1} (*Banglore BSc*)

28. A 1.8% solution of an unknown solute is isotonic with 10.26% solution of sugar cane (molecular mass = 342×10^{-3} kg). Calculate the molar mass of the solute.

Answer. 0.060×10^{-2} kg mol^{-1}

(*Baroda BSc*)

29. Calculate the osmotic pressure of a sugar solution in which 0.00342 kg of sugar has been dissolved in 0.0001 m^3 of solution at 300 K.

Answer. 249420 Pa

(*Kalyani BSc*)

30. How is relative lowering of vapor pressure of solutions of non-volatile solute and volatile solvent is related to osmotic pressure? Deduce the relationship. (*Madras BSc*)

31. (*a*) State Raoult's law and show that relative lowering of vapor pressure of a solution is directly proportional to its osmotic pressure.

(*b*) 0.00358 kg NaCl was dissolved in 0.120 kg of water at 350 K. If the osmotic pressure of solution is 2665860.75 Pa, then calculate the degree of dissociation of NaCl.

Answer. (*b*) 79.6%

(*Goa BSc*)

32. Write van't Hoff equation for osmotic pressure of a solution. How can it be modified to determine molecular weight of polymers?

(*Panjab BSc*)

33. Phenol associates in water to form double molecules. When 0.00075 kg of phenol is dissolved in 0.045 kg of water, the freezing point is depressed by 273.375 K. Calculate the degree of association of phenol. (K_f for water is 1.88 per kg)

Answer. 0.647

(*Arunachal BSc*)

MULTIPLE CHOICE QUESTIONS

1. A semipermeable membrane allows the passage of _______ through it.
   (*a*) solvent only (*b*) solute only
   (*c*) solvent and solute both (*d*) either solvent or solute
   Answer. (*a*)
2. The flow of solvent through a semipermeable membrane from pure solvent to solution is called
   (*a*) diffusion (*b*) osmosis
   (*c*) effusion (*d*) none of these
   Answer. (*b*)
3. The flow of solvent molecules can take place from
   (*a*) dilute to concentrated solution (*b*) concentrated to dilute solution
   (*c*) anyone solution to the other (*d*) none of these
   Answer. (*a*)
4. The external pressure applied to the solution in order to stop the passage of solvent molecules into solution by a semipermeable membrane is called
   (*a*) osmosis (*b*) osmotic pressure
   (*c*) optical pressure (*d*) none of these
   Answer. (*b*)
5. Osmotic pressure can be measured by an instrument called
   (*a*) manometer (*b*) barometer
   (*c*) osmometer (*d*) nanometer
   Answer. (*c*)
6. Isotonic solutions have the same
   (*a*) vapor pressure (*b*) osmotic pressure
   (*c*) atmospheric pressure (*d*) internal pressure
   Answer. (*b*)
7. Which of the following is a colligative property?
   (*a*) atmospheric pressure (*b*) critical pressure
   (*c*) osmotic pressure (*d*) none of these
   Answer. (*c*)
8. The isotonic solutions have
   (*a*) equimolar concentrations (*b*) different osmotic pressure
   (*c*) different colligative properties (*d*) same chemical properties
   Answer. (*a*)
9. Two solutions are separated by a semipermeable membrane. The one which is of lower osmotic pressure is called
   (*a*) hypotonic (*b*) hypertonic
   (*c*) iso-osmotic (*d*) isomorphic
   Answer. (*a*)
10. A hypertonic solution is the one which has _______ osmotic pressure than the other.
    (*a*) lower (*b*) equal
    (*c*) higher (*d*) none of these
    Answer. (*c*)

11. The osmosis taking place from solution to pure water by application of pressure greater than osmotic pressure on the solution is called

(*a*) equilibrium osmosis (*b*) irreversible osmosis
(*c*) reversible osmosis (*d*) reverse osmosis

Answer. (*d*)

12. The osmotic pressure of a solution at a given temperature is directly proportional to

(*a*) atmospheric pressure (*b*) concentration
(*c*) internal energy (*d*) volume of the solution

Answer. (*b*)

13. The osmotic pressure of a solution of a given concentration is directly proportional to

(*a*) room temperature (*b*) atmospheric pressure
(*c*) absolute temperature (*d*) critical temperature

Answer. (*c*)

14. The van't Hoff equation for *n* moles of solute dissolved in *v* liters of solution is

(*a*) $\pi = nRT$ (*b*) $\pi V = nRT$
(*c*) $\pi P = nRT$ (*d*) $\pi P = nRT / V$

Answer. (*b*)

15. When osmotic pressure and temperature are the same, equal volumes of solution would contain equal number of moles (molecules) of the solute. This relationship is called

(*a*) van't Hoff law for solutions (*b*) Avogadro's law for solutions
(*c*) Boyle's van't Hoff law for solutions (*d*) Avogadro-van't Hoff law for solutions

Answer. (*d*)

16. The osmotic pressure of a solution containing 1 mole of solute in 0.0224 m^3 would be equal to

(*a*) 101325 Pa (*b*) 50662.5 Pa
(*c*) 202650 Pa (*d*) 2269680.00 Pa

Answer. (*a*)

17. Osmotic pressure is directly proportional to

(*a*) vapor pressure of the pure solvent (*b*) lowering of vapor pressure
(*c*) relative lowering of vapor pressure (*d*) none of these

Answer. (*c*)

18. The molecular mass 'M' of the solute can be calculated by using the formula

(*a*) $M = \dfrac{wRT}{\pi V}$ (*b*) $M = \dfrac{wRT}{\pi V^2}$

(*c*) $M = \dfrac{wRT^2}{\pi V}$ (*d*) $M = \dfrac{wRT}{\pi^2 V^2}$

Answer. (*a*)

19. The van't Hoff factor, *i*, is related to the osmotic pressure by the formula

(*a*) $i = \pi / \pi°$ (*b*) $i = \pi° / \pi$
(*c*) $i = \pi - \pi°/ \pi°$ (*d*) $i = \pi° - \pi / \pi$

Answer. (*a*)

20. The value of van't Hoff factor, *i*, will be equal to 2 for

(*a*) urea (*b*) glucose
(*c*) sucrose (*d*) NaCl

Answer. (*d*)

21. The value of van't Hoff factor *i* for the electrolytes which are partially ionized in aqueous solution will be

(*a*) equal to 1 (*b*) less than 1
(*c*) between 1 and 2 (*d*) greater than 2

Answer. (*c*)

22. Two solutions with equal osmotic pressure are called
(*a*) molar solutions (*b*) molal solutions
(*c*) isotonic solution (*d*) isomorphic solutions
Answer. (*c*)

23. The osmotic pressure of a solution increases if
(*a*) volume of the solutions is increased (*b*) the no. of the solute particles is increased
(*c*) the no. of solute particles is decreased (*d*) the temperature is decreased
Answer. (*b*)

24. The highest osmotic pressure will be observed in
(*a*) 0.1M urea (*b*) 0.1M glucose
(*c*) 0.1M NaCl (*d*) 0.1M $Al_2(SO_4)_3$
Answer. (*d*)

25. The ratio of osmotic pressures of $BaCl_2$, NaCl and urea will be
(*a*) 3:2:1 (*b*) 1:2:3
(*c*) 2:1:3 (*d*) 3:1:2
Answer. (*a*)

26. The osmotic pressure of equimolar solutions of $CaCl_2$, NaCl and urea will be in the order
(*a*) $CaCl_2 > NaCl > urea$ (*b*) $CaCl_2 < NaCl < urea$
(*c*) $urea > CaCl_2 > NaCl$ (*d*) $urea < CaCl_2 < NaCl$
Answer. (*a*)

27. The osmotic pressure of a solution containing 0.02 mole of solute at 300 K will be
(*a*) $0.02 \times 8.314 \times 300$ Pa (*b*) $\frac{0.02 \times 8.314}{300}$ Pa
(*c*) $\frac{0.02 \times 300}{8.314}$ Pa (*d*) $0.02 \times 8.314 \times 300$ Pa
Answer. (*a*)

28. The outer hard shell of two eggs of the same size is removed by dissolving in dil HCl. One is placed in water and other in saturated NaCl solution. It will be observed after few hours that
(*a*) the first one shrinks while the other swells (*b*) the first one swells while the other shrinks
(*c*) both eggs swell (*d*) both eggs shrink
Answer. (*b*)

29. Blood is isotonic with
(*a*) 0.16 M sodium chloride solution (*b*) 0.31 M sodium chloride solution
(*c*) 0.40 M sodium chloride solution (*d*) 0.80 M sodium chloride solution
Answer. (*a*)

30. The osmotic pressure of a solution of urea containing 8.6 kg/m^3 is equal to that of 5% solution of a non-volatile solute. The molecular mass of the solute will be
(*a*) 349 (*b*) 34.9
(*c*) 430 (*d*) 860
Answer. (*a*)

31. The osmotic pressure of 2% solution of urea (mol mass = 60) at 300 K would be (α is the degree of dissociation)
(*a*) 8314.00 Pa (*b*) 83140.00 Pa
(*c*) 8314000.00 Pa (*d*) none of these
Answer. (*c*)

32. The osmotic pressure of the solution of urea is
(*a*) $5\% < 3\% < 2\% < 1\%$ (*b*) $1\% < 2\% > 3\% > 5\%$
(*c*) $1\% < 2\% < 3\% < 5\%$ (*d*) $3\% > 2\% > 1\% > 5\%$
Answer. (*c*)

33. Which of the following inorganic precipitates acts as a semipermeable membrane?

(*a*) $Al(OH)_3$ (*b*) $BaCO_3$

(*c*) $CuSO_4$ (*d*) $Cu_2[Fe(CN)_6]$

Answer. (*d*)

34. The osmotic pressure of a solution increases if

(*a*) temperature is lowered (*b*) volume is increased

(*c*) concentration of solute particles is increased (*d*) solution constant is increased

Answer. (*c*)

35. The concentration in kg/m^3 of a solution of cane sugar (mol mass = 342 × 10^{-3}) is isotonic with a solution containing 3 kg of urea per m^3. The concentration of cane sugar is

(*a*) 17.1 kg/m^3 (*b*) 34.2 kg/m^3

(*c*) 3.42 kg/m^3 (*d*) 1.71 kg/m^3

Answer. (*a*)

36. The osmotic pressure of 10% solution of cane sugar at 342 K is

(*a*) 8318.78 Pa (*b*) 83187.825 Pa

(*c*) 831878.25 Pa (*d*) 1663756.50 Pa

Answer. (*c*)

37. Osmotic pressure is not determined by

(*a*) Pfeffer's method (*b*) Berkley and Hertley's method

(*c*) Abbe's refractometer (*d*) all of these

Answer. (*c*)

38. Which of the following is not correct?

(*a*) in osmosis solvent molecules flow in one direction

(*b*) in diffusion solute and solvent molecule flow in opposite direction

(*c*) for osmosis a semipermeable membrane is required

(*d*) for diffusion, a semipermeable membrane is required

Answer. (*d*)

39. The osmotic pressure of a solution of cane sugar (mol mass = 342 × 10^{-3} kg) at 342 K is 831878.25 Pa. The concentration of the solution is

(*a*) 1% (*b*) 2%

(*c*) 5% (*d*) 10%

Answer. (*d*)

40. The volume of the solution containing one mole of solute that shows an osmotic pressure of 101325 Pa is

(*a*) 22.4 × 10^{-3} m^3 (*b*) 2.24 × 10^{-3} m^3

(*c*) 0.224 × 10^{-3} m^3 (*d*) 1.0 × 10^{-3} m^3

Answer. (*a*)

41. Which of the following types of compounds will have van't Hoff factor = 1?

(*a*) strong electrolytes (*b*) weak electrolytes

(*c*) non-electrolytes (*d*) none of these

Answer. (*c*)

42. For binary electrolytes of the type AB, the abnormal mol. mass is given by (α is the degree of dissociation)

(*a*) $\dfrac{M_{\text{normal}}}{1+\alpha}$ (*b*) $\dfrac{M_{\text{normal}}}{1-\alpha}$

(*c*) $\dfrac{M_{\text{normal}}}{1+2\alpha}$ (*d*) $\dfrac{M_{\text{normal}}}{1+3\alpha}$

Answer. (*a*)

43. Which one of the following is not correct?

(*a*) the equimolar solutions have same osmotic pressure at different temperatures

(*b*) the equimolar solutions have same osmotic pressure at same temperatures

(*c*) the osmotic pressure is directly proportional to the number of moles of solutions

(*d*) the osmotic pressure is directly proportional to the relative lowering of vapor pressure

Answer. (*a*)

44. Which of the following statements is true?

(*a*) Osmosis can be reversed by application of pressure on the solution

(*b*) Beckmann thermometer can be used to measure elevation in boiling point

(*c*) Osmosis and diffusion are same in solutions

(*d*) Barometer can be used to measure osmotic pressure

Answer. (*a*)

45. The van't Hoff factor for dilute aqueous solutions of urea and NaCl respectively are

(*a*) 1, greater than 1 (*b*) 1, 0

(*c*) 0, 1 (*d*) greater than 1, 1

Answer. (*a*)

46. The observed molecular mass in case of electrolytes of the type AB_2 is

(*a*) equal to the normal molecular mass (*b*) less than the normal molecular mass

(*c*) greater than the normal molecular mass (*d*) none of these

Answer. (*b*)

47. The compounds which undergo association in solutions have observed molecular mass always

(*a*) equal to their normal molecular mass (*b*) less than their normal molecular mass

(*c*) greater than their normal molecular mass (*d*) none of these

Answer. (*c*)

48. The van't Hoff factors for urea, sucrose and glucose are in the ratio

(*a*) 1:2:1 (*b*) 1:1:2

(*c*) 2:1:1 (*d*) 1:1:1

Answer. (*d*)

49. The value of van't Hoff factor is more than one in dissociation and less than one in association. This statement is

(*a*) false (*b*) true

(*c*) sometimes false, sometimes true (*d*) none of these

Answer. (*c*)

50. For electrolytes of the type A_2B or AB_2 the abnormal molecular mass is given by

(*a*) $\dfrac{M_{\text{normal}}}{1+\alpha}$ (*b*) $\dfrac{M_{\text{normal}}}{1+2\alpha}$

(*c*) $\dfrac{M_{\text{normal}}}{1+3\alpha}$ (*d*) $\dfrac{M_{\text{normal}}}{1-\alpha}$

Answer. (*b*).

17 Chemical Equilibrium

CONTENTS

REVERSIBLE REACTIONS

Not all chemical reactions proceed to completion. In most reactions two or more substances react to form products which themselves react to give back the original substances. Thus A and B may react to form C and D which react together to reform A and B.

$$A + B \longrightarrow C + D \quad \text{(Forward reaction)}$$

$$A + B \longrightarrow C + D \quad \text{(Reverse reaction)}$$

A reaction which can go in the forward and backward direction simultaneously is called a Reversible reaction. Such a reaction is represented by writing a pair of arrows between the reactants and products.

$$A + B \rightleftharpoons C + D$$

AN EQUILIBRIUM ANALOGY

Nuts and Bolts Representation of Chemical Equilibrium

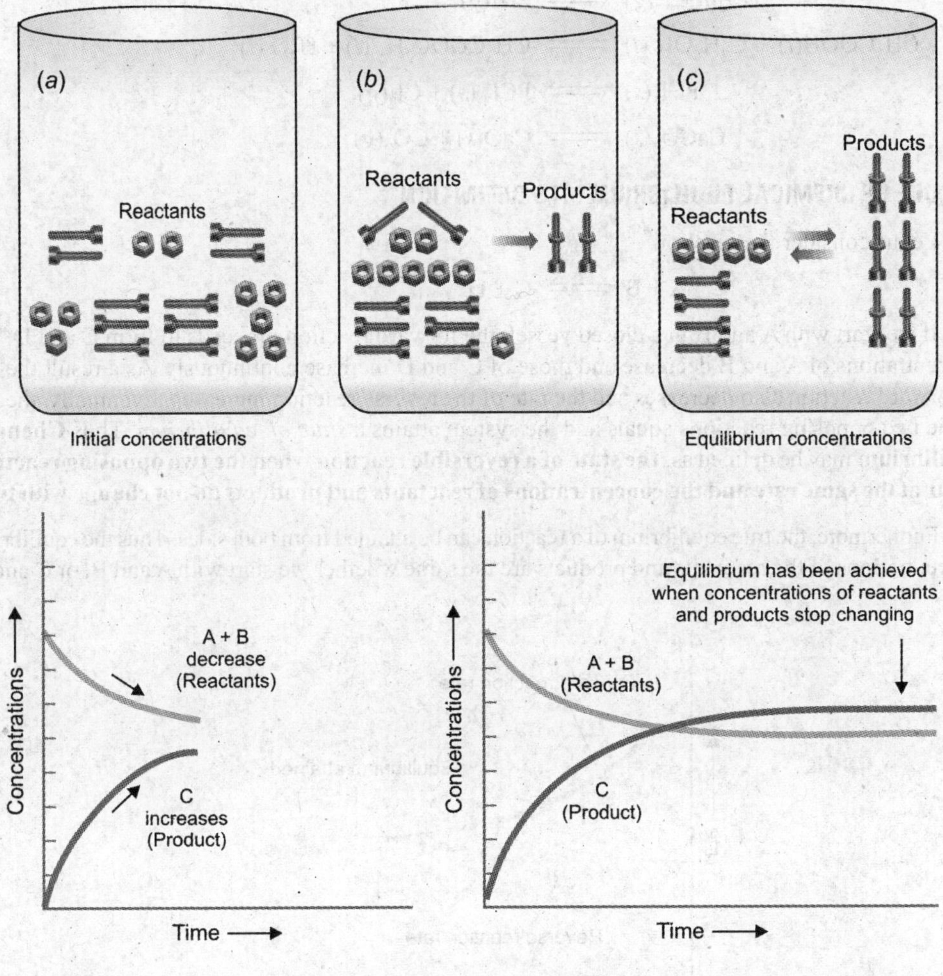

A "nuts and bolts" representation of chemical equilibrium for a reversible reaction. Initially (*a*) there are 100 nuts and 100 bolts in the beakers represented in the figure by 10 of each. The concentration is 100 bolts per beaker for both nuts and bolts. As time passes (*b*) concentrations of reactants decrease and concentrations of products—assembled units—increase. Equilibrium is established (*c*) when concentrations stop changing. For this equilibrium, there are 40 nuts per beaker, 40 bolts per beaker, and 60 assembled units per beaker. The rate of the forward reaction (assembling units) is equal to the rate of the reverse reaction (taking the units apart).

The arrow pointing right indicates the forward reaction, while that pointing left shows the reverse reaction.

Some Examples of Reversible Reactions

A few common examples of reversible reactions are listed below :

$$2NO_2(g) \rightleftharpoons N_2O_4(g)$$

$$H_2(g) + I_2(g) \rightleftharpoons 2HI(g)$$

$$CH_3COOH(l) + C_2H_5OH(l) \rightleftharpoons CH_3COOC_2H_5(l) + H_2O(l)$$

$$PCl_5(s) \rightleftharpoons PCl_3(s) + Cl_2(g)$$

$$CaCO_3(s) \rightleftharpoons CaO(s) + CO_2(g)$$

NATURE OF CHEMICAL EQUILIBRIUM : ITS DEFINITION

Let us consider the reaction

$$A + B \rightleftharpoons C + D$$

If we start with A and B in a closed vessel, the forward reaction proceeds to form C and D. The concentrations of A and B decrease and those of C and D increase continuously. As a result the rate of forward reaction also decreases and the rate of the reverse reaction increases. Eventually, the rate of the two opposing reactions equals and the system attains a *state of equilibrium.* Thus **Chemical equilibrium** may be defined as: **the state of a reversible reaction when the two opposing reactions occur at the same rate and the concentrations of reactants and products do not change with time.**

Furthermore, the true equilibrium of a reaction can be attained from both sides. Thus the equilibrium concentrations of the reactants and products are the same whether we start with A and B, or C and D.

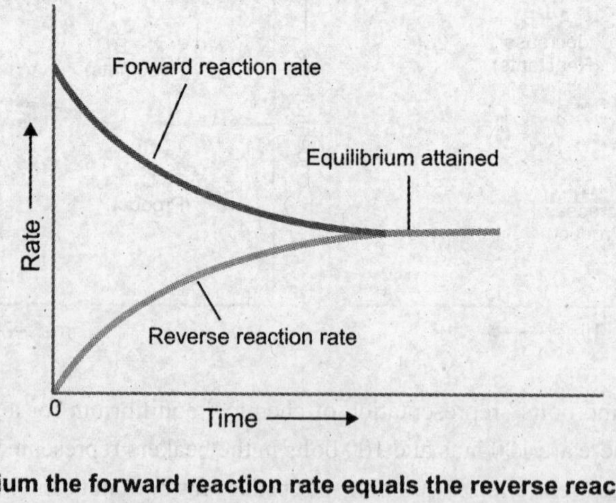

Figure 17.1
At equilibrium the forward reaction rate equals the reverse reaction rate.

Chemical Equilibrium is Dynamic Equilibrium

We have shown above that as the reaction $A + B \rightleftharpoons C + D$ attains equilibrium, the concentrations of A and B, as also of C and D remain constant with time. Apparently it appears that the equilibrium is dead. But it is not so. The equilibrium is dynamic. Actually, the forward and the reverse reactions are taking place at equilibrium but the concentrations remain unchanged.

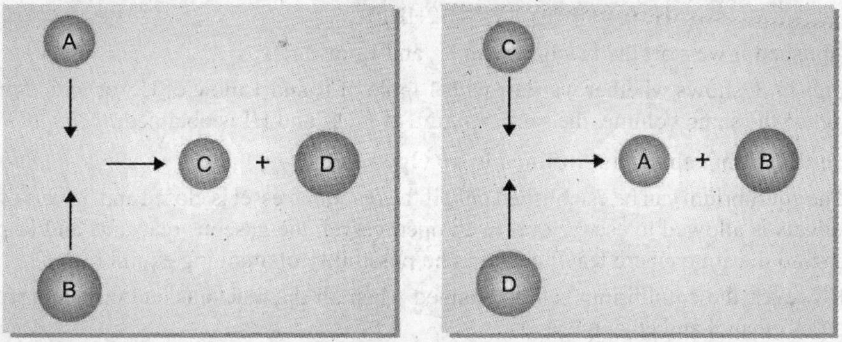

■ **Figure 17.2**
Molecules of A and B colliding to give C and D, and those of C and D colliding to give A and B.

The dynamic nature of chemical equilibrium can be easily understood on the basis of the kinetic molecular model. The molecules of A and B in the equilibrium mixture collide with each other to form C and D. Likewise C and D collide to give back A and B. The collisions of molecules in a closed system is a ceaseless phenomenon. Therefore collisions of A and B giving C and D (Forward reaction) and collisions of C and D giving back A and B (reverse reaction) continue to occur even at equilibrium, while concentrations remain unchanged.

CHARACTERISTICS OF CHEMICAL EQUILIBRIUM

Before we take up the mathematical study of chemical equilibria, let us understand the chemical characteristics of chemical equilibrium.

(1) **Constancy of concentrations**

When a chemical equilibrium is established in a closed vessel at constant temperature, concentrations of the various species in the reaction mixture become constant.

The reaction mixture at equilibrium is called **Equilibrium mixture.**

The concentrations at equilibrium are called **Equilibrium concentrations.** The equilibrium concentrations are represented by square brackets with subscript *eq* or *eqm,* $[\,]_{eq}$. Thus $[A]_{eq}$ denotes the equilibrium concentration of substance A in moles per liter. In modern practice the subscript *eq* is not used.

(2) **Equilibrium can be initiated from either side**

The state of equilibrium of a reversible reaction can be approached whether we start with reactants or products, for example, the equilibrium

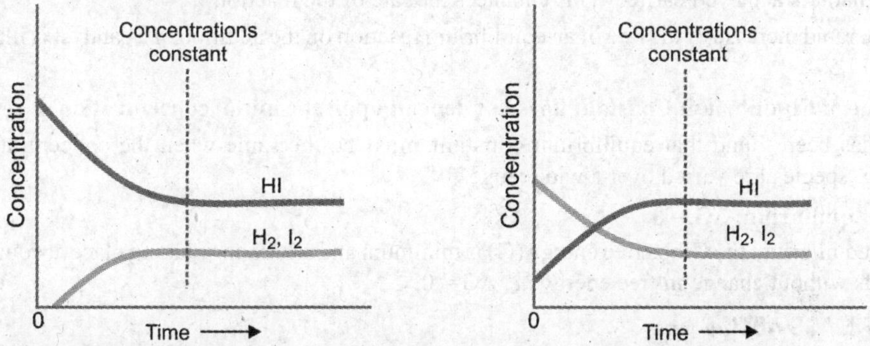

■ **Figure 17.3**
The left graph shows how equilibrium is attained for start with H_2 and I_2. The right graph depicts the attainment of equilibrium for start with 2HI. Equilibrium concentrations in both cases are the same.

$$H_2(g) + I_2(g) \rightleftharpoons 2HI(g)$$

is established if we start the reaction with H_2 and I_2, or 2 HI.

Fig. 17.3. shows whether we start with 1 mole of I_2 and 1 mole of H_2, or with 2 moles of HI in a vessel of the same volume, the same mixture of H_2, I_2 and HI is obtained.

(3) Equilibrium cannot be Attained in an Open Vessel

The equilibrium can be established only if the reaction vessel is closed and no part of the reactants or products is allowed to escape out. In an open vessel, the gaseous reactants and/or products may escape into the atmosphere leaving behind no possibility of attaining equilibrium.

However, the equilibrium can be attained when all the reactants and products are in the same phase *i.e.,* ethanol and ethanoic acid.

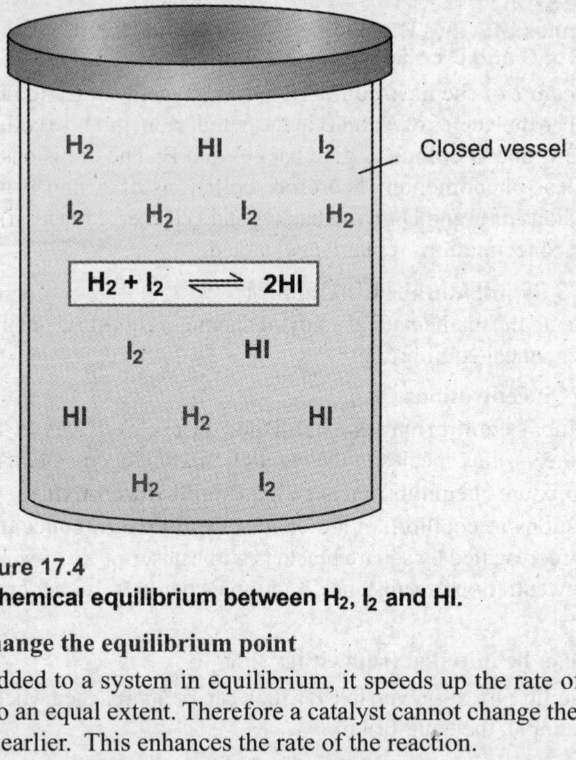

Figure 17.4
A chemical equilibrium between H_2, I_2 and HI.

(4) A catalyst cannot change the equilibrium point

When a catalyst is added to a system in equilibrium, it speeds up the rate of both the forward and the reverse reaction to an equal extent. Therefore a catalyst cannot change the equilibrium point except that it is achieved earlier. This enhances the rate of the reaction.

The rapid increase of the rate of an equilibrium reaction on the addition of a catalyst is illustrated in Fig. 17.5.

(5) Value of Equilibrium Constant does not depend upon the initial concentration of reactants

It has been found that equilibrium constant must be the same when the concentrations of reacting species are varied over a wide range.

(6) At Equilibrium $\Delta G = 0$

At equilibrium the Gibbs free energy (G) is minimum and any change taking place at equilibrium proceeds without change in free energy *i.e.* $\Delta G = 0$.

LAW OF MASS ACTION

Two Norwegian chemists, Guldberg and Waage, studied experimentally a large number of equilibrium reactions. In 1864, they postulated a generalization called the **Law of Mass action.** It states that : **the rate of a chemical reaction is proportional to the active masses of the reactants.**

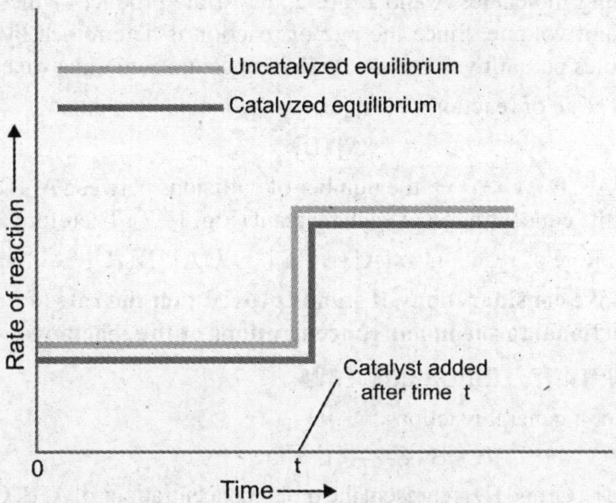

■ **Figure 17.5**
The vertical part of the curve indicates rapid increase of rate of reaction on addition of catalyst.

By the term 'active mass' is meant the molar concentration *i.e.,* number of moles per liter. It is expressed by enclosing the formula of the substance in square brackets. Thus a gas mixture containing 0.001 kg of hydrogen (H_2) and 0.127 kg of iodine (I_2) per meter^3 has the concentrations

$$[H_2] = 0.5 \text{ and } [I_2] = 0.5$$

Explanation of the Law of Mass Action Based on the Molecular Collision Theory

We assume that a chemical reaction occurs as a result of the collisions between the reacting molecules. Although some of these collisions are ineffective, the chemical change produced is proportional to the number of collisions actually taking place. Thus at a fixed temperature the rate of a reaction is determined by the number of collisions between the reactant molecules present in unit volume.

Let us consider four boxes of one cubic centimeter volume; containing different number of reacting molecules A and B (Fig 17.6). They undergo collisions to form the products C and D, the rate of reaction being governed by the number of possible collisions between them.

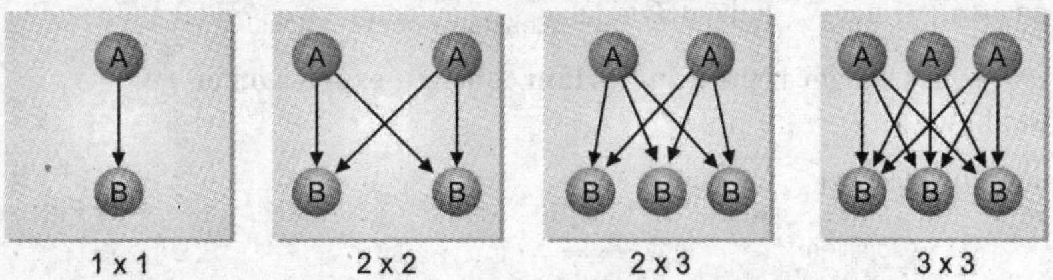

■ **Figure 17.6**
Possible collisions between molecules of the reactants A and B.

In the first box there is one molecule of A and one molecule of B, and the possibilities of collision at any instant is $1 \times 1 = 1$. In the second box, the number of molecules of A and B are two each and the possibilities of collisions are $2 \times 2 = 4$. In the third box, there are two molecules of A and three molecules of B, and the number of possibilities of collisions between A and B are increased to $3 \times 2 = 6$. In the fourth box, we have three molecules of each of A and B and the total number of collisions between them is $3 \times 3 = 9$. Thus in general, we can say that the possibilities of collisions between the reacting molecules A and B are equal to the product of the number of molecules of each species per unit volume. Since the rate of reaction is determined by molecular impacts, it is proportional to moles per unit volume *i.e.*, molar concentration. Thus we can write

$$\text{Rate of reaction} \propto [A][B]$$
$$= k[A][B]$$

In a reaction, $A + B + C \longrightarrow$ the number of collisions between A, B and C depends on the concentration of each reactant (Fig. 17.7). Therefore,

$$\text{Rate of reaction} \propto [A][B][C] = k[A][B][C]$$

From the above considerations, it stands proved that the rate of a reaction is proportional to the molar concentrations of the reactants.

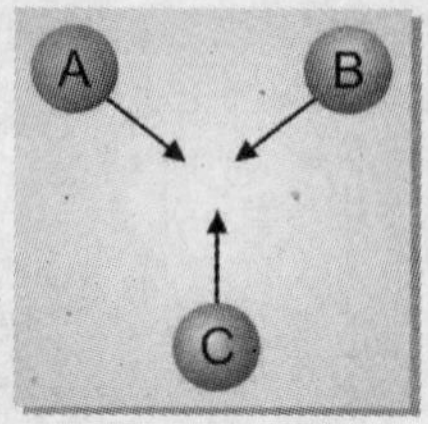

Figure 17.7

EQUILIBRIUM CONSTANT : EQUILIBRIUM LAW

Let us consider a general reaction

$$A + B \rightleftharpoons C + D$$

and let [A], [B], [C] and [D] represent the molar concentrations of A, B, C and D at the equilibrium point. According to the Law of Mass action.

$$\text{Rate of forward reaction} \propto [A][B] = k_1[A][B]$$
$$\text{Rate of reverse reaction} \propto [C][D] = k_2[C][D]$$

where k_1 and k_2 are rate constants for the forward and reverse reactions.

At equilibrium, rate of forward reaction = rate of reverse reaction.

Therefore,

$$k_1[A][B] = k_2[C][D]$$

or

$$\frac{k_1}{k_2} = \frac{[C][D]}{[A][B]} \qquad \text{... (1)}$$

At any specific temperature k_1/k_2 is constant since both k_1 and k_2 are constants. The ratio k_1/k_2 is called **Equilibrium constant** and is represented by the symbol K_c, or simply k. The subscript '*c*' indicates that the value is in terms of concentrations of reactants and products. The equation (1) may be written as

$$\underset{\text{Equilibrium constant}}{k_c} = \frac{[C][D] \;\text{— Product concentrations}}{[A][B] \;\text{— Reactant concentrations}}$$

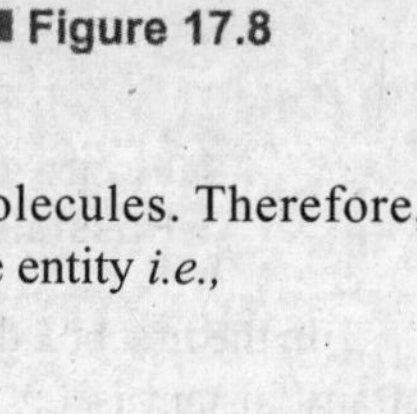

Figure 17.8

This equation is known as the **Equilibrium constant expression or Equilibrium law.**

Consider the reaction

$$2A \rightleftharpoons C + D$$

(Coefficient of A: 2)

Here, the forward reaction is dependent on the collisions of each of two A molecules. Therefore, for writing the equilibrium expression, each molecule is regarded as a separate entity *i.e.*,

$$A + A \rightleftharpoons C + D$$

Then the equilibrium constant expression is

$$k_c = \frac{[C][D]}{[A][A]} = \frac{[C][D]}{[A]^2}$$

(Power equal to coefficent of A)

As a general rule, if there are two or more molecules of the same substance in the chemical equation, **its concentration is raised to the power equal to the numerical coefficient of the substance in the equation.**

Equilibrium Constant Expression for a Reaction in General Terms

The general reaction may be written as

$$aA + bB \rightleftharpoons cC + dD$$

where *a, b, c* and *d* are numerical quotients of the substance, A, B, C and D respectively. The equilibrium constant expression is

$$K_c = \frac{[C]^c \ [D]^d}{[A]^a \ [B]^b}$$

where K_c is the **Equilibrium constant.** The general definition of the equilibrium constant may thus be stated as : **the product of the equilibrium concentrations of the products divided by the product of the equilibrium concentrations of the reactants, with each concentration term raised to a power equal to the coefficient of the substance in the balanced equation.**

How to Write the Equilibrium Constant Expression?

(1) Write the balanced chemical equation for the equilibrium reaction. By convention, the substances on the left of the equation are called the **reactants** and those on the right, the **products.**

(2) Write the product of concentrations of the 'products' and raise the concentrations of each substance to the power of its numerical quotient in the balanced equation.

(3) Write the product of concentrations of 'reactants' and raise the concentration of each substance to the power of its numerical quotient in the balanced equation.

(4) Write the equilibrium expression by placing the **product concentrations in the numerator and reactant concentrations in the denominator.** That is,

$$K_c = \frac{\text{Product of concentrations of 'products' form Step (2)}}{\text{Product of concentrations of 'reactants' form Step (3)}}$$

SOLVED PROBLEM 1

Give the equilibrium constant expression for the reaction

$$N_2(g) + 3H_2\ (g) \rightleftharpoons 2NH_3(g)$$

SOLUTION

(1) The equation is already balanced. The numerical quotient of H_2 is 3 and of NH_3 is 2.
(2) The concentration of the 'product' NH_3 is $[NH_3]^2$
(3) The product of concentrations of the reactants is $[N_2]\ [H_2]^3$
(4) Therefore, the equilibrium constant expression is

$$K_c = \frac{[NH_3]^2}{[N_2][H_2]^3}$$

SOLVED PROBLEM 2

Write the equilibrium constant expression for the reaction

$$N_2O_5(g) \rightleftharpoons NO_2(g) + O_2\ (g)$$

SOLUTION

(1) The equation as written is not balanced. Balancing yields

$$2N_2O_5 \rightleftharpoons 4NO_2(g) + O_2$$

(2) The quotient of the product NO_2 is 4 and of the reactant N_2O_5 it is 2.

(3) The product of the concentrations of products is

$$[NO_2]^4 [O_2]$$

(4) The concentration of the reactant is

$$[N_2O_5]^2$$

(5) The equilibrium constant expression can be written as

$$K_c = \frac{[NO_2]^4 [O_2]}{[N_2O_5]^2}$$

SOLVED PROBLEM 3

What is the equilibrium constant expression of the reaction

$$CH_4 (g) + H_2O(g) \rightleftharpoons CO(g) + 3H_2(g)$$

SOLUTION

Write the product of concentrations of 'products' divided by the product of concentrations of 'reactants'. The concentration of H_2 is to be raised by its quotient in the equation. Thus, the equilibrium constant expression is:

$$K_c = \frac{[CO]\,[H_2]^3}{[CH_4][H_2O]}$$

EQUILIBRIUM CONSTANT EXPRESSION IN TERMS OF PARTIAL PRESSURES

When all the reactants and products are gases, we can also formulate the equilibrium constant expression in terms of partial pressure. The relationship between the partial pressure (p) of any one gas in the equilibrium mixture and the molar concentration follows from the general ideal gas equation

$$pV = nRT \quad \text{or} \quad p = \left(\frac{n}{V}\right) RT$$

The quantity $\frac{n}{V}$ is the number of moles of the gas per unit volume and is simply the molar concentration. Thus,

$$p = (\text{Molar concentration}) \times RT$$

i.e., **the partial pressure of a gas in the equilibrium mixture is directly proportional to its molar concentration at a given temperature.** Therefore, we can write the equilibrium constant expression in terms of partial pressure instead of molar concentrations. For a general reaction

$$l\,L(g) + mM(g) \rightleftharpoons yY(g) + zZ(g)$$

the equilibrium law or the equilibrium constant may be written as

$$K_p = \frac{(p_Y)^y (p_Z)^z}{(p_L)^l (p_M)^m}$$

Here K_p is the equilibrium constant, the subscript p referring to partial pressure. Partial pressures are expressed in atmospheres.

SOLVED PROBLEM 4

Write the equilibrium constant expression for the synthesis of ammonia,

$$N_2(g) + 3H_2(g) \rightleftharpoons 2NH_3(g)$$

SOLUTION

$$K_p = \frac{(p_{NH_3})^2}{(p_{N_2})(p_{H_2})^3}$$

SOLVED PROBLEM 5

Using partial pressures, write the equilibrium law for the reaction

$$H_2(g) + I_2(g) \rightleftharpoons 2HI(g)$$

SOLUTION

$$K_p = \frac{(p_{HI})^2}{(p_{H_2})(p_{I_2})}$$

SOLVED PROBLEM 6

What is the expression K_p for the reaction

$$N_2O_4(g) \rightleftharpoons 2NO_2(g)$$

SOLUTION

For K_p we use the partial pressure in the equilibrium constant expression. Therefore,

$$K_p = \frac{(p_{NO_2})^2}{(p_{N_2O_4})}$$

HOW K_c AND K_p ARE RELATED?

Let us consider a general reaction

$$jA + kB \rightleftharpoons lC + mD$$

where all reactants and products are gases. We can write the equilibrium constant expression in terms of partial pressures as

$$K_p = \frac{(p_C)^l (p_D)^m}{(p_A)^j (p_B)^k} \quad ...(1)$$

Assuming that all these gases constituting the equilibrium mixture obey the ideal gas equation, the partial pressure (p) of a gas is

$$p = \left(\frac{n}{V}\right) RT$$

Where $\frac{n}{V}$ is the molar concentration. Thus the partial pressures of individual gases, A, B, C and D are:

$$P_A = [A]\,RT;\ p_B = [B]\,RT;\ p_C = [C]\,RT;\ p_D = [D]\,RT$$

Substituting these values in equation (1), we have

$$K_p = \frac{[C]^l (RT)^l [D]^m (RT)^m}{[A]^j (RT)^j [B]^k (RT)^k}$$

or

$$K_p = \frac{[C]^l [D]^m}{[A]^j [B]^k} \times \frac{(RT)^{l+m}}{(RT)^{j+k}}$$

$$K_p = K_c \times (RT)^{(l+m)-(j+k)}$$

$$\therefore \quad K_p = K_c \times (RT)^{\Delta n} \qquad ...(2)$$

where $\Delta n = (l + m) - (j + k)$, the difference in the sums of the coefficients for the gaseous products and reactants.

From the expression (2) it is clear that when $\Delta n = 0$, $K_p = K_c$.

SOLVED PROBLEM 7

At 773 K, the reaction between N_2 and H_2 to form ammonia has $K_c = 6.0 \times 10^{-2}$. What is the numerical value of K_p for the reaction?

SOLUTION

Here, we will use the general expression

$$K_p = K_c (RT)^{\Delta n}$$

For the reaction $N_2 + 3H_2 \rightleftharpoons 2NH_3$

we have Δn = (sum of quotients of products) – (sum of quotients of reactants)

$$= 2 - 4 = -2$$

$$K_c = 6.0 \times 10^{-2}$$

$$T = 773 \text{ K}$$

$$R = 8.3145 \text{ J/ mol. k}$$

Substituting the value of R, T, K_c and Δn in the general expression, we have

$$K_p = (6.0 \times 10^{-2}) [(8.3145 \times 773)]^{-2}$$

$$= \mathbf{1.5 \times 10^{-9}}$$

SOLVED PROBLEM 8

The value of K_p at 298 K for the reaction

$$2NO(g) + Cl_2(g) \rightleftharpoons 2NOCl(g)$$

is 1.9×10^8 Pa (192517500.00). Calculate the value of K_c at the same temperature.

SOLUTION

We can write the general expression as

Here, $$K_p = K_c (RT)^{\Delta n} \text{ or } K_c = \frac{K_p}{(RT)^{\Delta n}}$$

$$T = 298 \text{ K}$$

$$R = 8.3145$$

$$\Delta n = 2 - (2 + 1) = -1$$

$$K_p = 1.9 \times 10^8 \text{ Pa} (192517500.00)$$

Substituting these values in the general expression

$$K_c = \frac{1.9 \times 10^8}{(8.3145 \times 298)^{-1}}$$

$$= \mathbf{4.7 \times 10^{11}}$$

Calculation of K_c from Experimental Information

To determine the value of K_c of a reaction, write the balanced equation. Then write the equilibrium constant expression. Substitute in it the equilibrium concentrations of the reactants and products found experimentally. Thus calculate the value of K_c.

SOLVED PROBLEM 9

At equilibrium for the reaction

$$2SO_2(g) + O_2(g) \rightleftharpoons 2SO_3(g)$$

the concentrations of reactants and products at 1000 K were found to be : SO_2 = 270 mol/m³; O_2 = 400 mol/m³; and SO_3 = 330 mol/m³. What is the value of the equilibrium constant, K_c, at this temperature?

SOLUTION

Write the equilibrium constant expression for the reaction

$$K_c = \frac{[SO_3]^2}{[SO_2]^2[O_2]}$$

We know that:

$$[SO_3] = 330 \text{ mol/m}^3; [SO_2] = 270 \text{ mol/m}^3; [O_2] = 400 \text{ mol/m}^3$$

Substituting these values in the equilibrium constant expression, we have

$$K_c = \frac{(330)^2}{(270)^2 \times 400} = \mathbf{0.003734568 / m^3}$$

SOLVED PROBLEM 10

Some nitrogen and hydrogen gases are pumped into an empty five-liter glass bulb at 773 K. When equilibrium is established, 3.00 mol of N_2, 2.10 mol of H_2 and 0.298 mol of NH_3 are found to be present. Find the value of K_c for the reaction

$$N_2(g) + 3H_2(g) \rightleftharpoons 2NH_3(g) \text{ at 773 K.}$$

SOLUTION

The equilibrium concentrations are obtained by dividing the number of moles of each reactant and product by the volume, 5.00×10^{-3} m³. Thus,

$$[N_2] = 3.00 \text{ mol}/5 \times 10^{-3} \text{ m}^3 = 600 \text{ M}$$

$$[H_2] = 2.10 \text{ mol}/5 \times 10^{-3} \text{ m}^3 = 420 \text{ M}$$

$$[NH_3] = 0.298 \text{ mol}/5 \times 10^{-3} \text{ m}^3 = 59.6 \text{ M}$$

Substituting these concentrations (not number of moles) in the equilibrium constant expression, we get the value of K_c.

$$K_c = \frac{[NH_3]^2}{[N_2][H_2]^3} = \frac{(59.6)^2}{(600)(420)^3} = 0.80 \times 10^{-9}$$

Thus, for the reaction of H_2 and N_2 to form NH_3 at 773 K, we can write

$$K_c = \frac{[NH_3]^2}{[N_2][H_2]^3} = \mathbf{0.80 \times 10^{-9}}$$

SOLVED PROBLEM 11

Calculate the equilibrium constant at 298 K for the reaction

$$2NOCl(g) \rightleftharpoons 2NO(g) + Cl_2(g)$$

In an experiment, 2.00 mol of NOCl were placed in a 0.001 m^3 flask and the concentration of NO after equilibrium achieved was 660 mol/m^3.

SOLUTION

Let us write the equilibrium constant expression for the balanced chemical equation.

$$K_c = \frac{[NO]^2[Cl_2]}{[NOCl]^2}$$

Next we proceed to find the equilibrium concentrations of NOCl, NO and Cl_2. Since the volume of the reaction vessel is one m^3, the moles of various reagents also represent their molar concentrations.

The initial amount of NOCl is 2.00 mol. Let us assume that x moles of it have reacted to achieve the equilibrium. The moles of different reagents at equilibrium are indicated in the equation

$$\underset{(2-x)}{2NOCl} \rightleftharpoons \underset{x}{2NO} + \underset{x/2}{Cl_2}$$

According to this equation, one mole of NOCl gives one mole of NO and half a mole of Cl_2. The experimental value of equilibrium concentration of NO is 660 mol/m^3. Therefore, the equilibrium concentration Cl_2 is half of it *i.e.* 330 mol/m^3. Since x = 660 mol/m^3, the equilibrium concentration of NOCl = (2000 – 660 = 1340). To find K_c, these values of equilibrium concentrations are substituted in the equilibrium constant expression.

$$K_c = \frac{[NO]^2[Cl_2]}{[NOCl]^2} = \frac{(660)^2 \times 330}{(1340)^2} = \mathbf{80.056}$$

SOLVED PROBLEM 12

At a certain temperature, 100 mol of H_2 and 100 mol of I_2 were placed in a one-liter (10^{-3} m^3) flask. The purple color of iodine vapor was used to monitor the reaction. After a time, the equilibrium

$$H_2 + I_2 \rightleftharpoons 2HI$$

was established and it was found that the concentration of I_2 decreased to 20 mol/m^3. Calculate the value of K_c for the reaction at the given temperature.

SOLUTION

The equilibrium constant expression for the reaction

$$H_2 + I_2 \rightleftharpoons 2HI$$

is

$$K_c = \frac{[HI]^2}{[H_2][I_2]}$$

Let us proceed to find the equilibrium concentrations of the various species.

The initial moles of H_2, I_2 and HI present at equilibrium are

$$H_2 \quad + \quad I_2 \quad \rightleftharpoons \quad 2HI$$

moles at equilibrium: (100–x) (100–x) 2x

Since the volume of the reaction vessel is 10^{-3} m^3, the moles at equilibrium also represent the equilibrium concentrations. Thus,

$$[I_2] = 100 - x = 20 \text{ (given)}$$

or $$x = 100 - 20$$

$$= 80$$

Knowing the value of x, we can give the equilibrium concentrations of H_2 , I_2 and HI.

$$[H_2] = 100 - x = 100 - 80 = 20$$

$$[I_2] = 20 \text{ (given)}$$

$$[HI] = 2x = 160$$

Substituting these values in the equilibrium constant expression, we get the value of K_c.

$$K_c = \frac{[HI]^2}{[H_2][I_2]} = \frac{(160)^2}{20 \times 20} = \mathbf{64}$$

SOLVED PROBLEM 13

0.0000135 m^3 of HI are produced by the interaction of 0.0000081 m^3 of hydrogen and 0.0000093 m^3 of iodine vapor at 717 K. Calculate the equilibrium constant at this temperature of the reaction.

$$H_2(g) + I_2(g) \rightleftharpoons 2HI(g)$$

SOLUTION

The equilibrium constant expression for the reaction is

$$K_c = \frac{[HI]^2}{[H_2][I_2]}$$

Let us proceed to find the equilibrium concentrations.

Since the number of moles in a gas under the same conditions of temperature and pressure are proportional to volumes, m^3 of a gas may be used instead of molar concentrations. Thus,

	H_2	+	I_2	$\rightleftharpoons$	2HI
Equilibrium volumes :	$(8.1-x)$		$(9.3-x)$		$2x$

The experimental volume of HI = 0.0000135 m^3 (given)

$\therefore$ $$2x = 0.0000135 \text{ m}^3$$

or $$x = 0.00000675 \text{ m}^3$$

Thus, the equilibrium concentrations are:

$$[H_2] = 8.1 \times 10^{-6} - 6.75 \times 10^{-6} = 1.35 \times 10^{-6} = 1.35 \times 10^{-6} \text{ m}^3$$

$$[I_2] = 9.3 \times 10^{-6} - x = 9.3 \times 10^{-6} - 6.75 \times 10^{-6} = 2.55 \times 10^{-6} \text{ m}^3$$

$$[HI] = 1.35 \times 10^{-5} \text{ m}^3$$

Substituting these values in the equilibrium constant expression, we have

$$K_c = \frac{[HI]^2}{[H_2][I_2]} = \frac{(1.35 \times 10^{-5})^2}{1.35 \times 10^{-6} \times 2.55 \times 10^{-6}} = \mathbf{52.94}$$

Calculation of Equilibrium Concentrations

SOLVED PROBLEM 14

One mole of H_2 and one mole of I_2 were heated in a 0.001 m^3 sealed glass box at 663 K till the equilibrium was reached. Assuming that the equilibrium constant is 45.9, find the final concentrations of H_2, I_2 and HI.

SOLUTION

The equation for the reaction is

$$H_2 + I_2 \rightleftharpoons 2HI$$

If x mole of H_2 has reacted on attainment of equilibrium, the moles of components are

$$\begin{array}{ccccc} H_2 & + & I_2 & \rightleftharpoons & 2HI \\ (1-x) & & (1-x) & & 2x \end{array}$$

Since the reaction vessel is a 1 liter box, the concentrations at equilibrium are the same as the moles. Thus,

$$[H_2] = 1 - x$$
$$[I_2] = 1 - x$$
$$[HI] = 2x$$

Substituting the values in the equilibrium constant expression

$$K = \frac{[HI]^2}{[H_2][I_2]} = \frac{(2x)^2}{(1-x)(1-x)} = 45.9$$

or

$$\frac{4x^2}{1 - 2x + x^2} = 45.9$$

or

$$41.9\,x^2 - 91.8x + 45.9 = 0$$

Using the quadratic formula $x = 0.79$. Hence the equilibrium concentrations are

$$[H_2] = 1 - 0.79 = 0.21 \text{ mol per liter } (210 \text{ mol/m}^3)$$
$$[I_2] = 1 - 0.79 = 0.21 \text{ mol per liter } (210 \text{ mol/m}^3)$$
$$[HI] = 2x = 2(0.79) = \mathbf{1.58 \text{ mol per liter } (1580 \text{ mol/m}^3)}$$

SOLVED PROBLEM 15

At a certain temperature, K for the reaction $3C_2H_2\,(g) \rightleftharpoons C_6H_6(g)$ is 4. If the equilibrium concentration of C_2H_2 is 0.5 mol/liter (500 mol/m^3), what is the concentration of C_6H_6?

SOLUTION

The reaction can be written as

$$3C_2H_2 \rightleftharpoons C_6H_6$$

The equilibrium constant expression is

$$K = \frac{[C_6H_6]}{[C_2H_2]^3}$$

It is given that:

$$K = 4$$
$$[C_2H_2] = 500 \text{ mol/m}^3$$

Substituting these values,

$$4 = \frac{[C_6H_6]}{[0.5]^3}$$

$$\therefore \quad [C_6H_6] = 4 \times (0.5)^3 = \mathbf{0.5 \text{ mols/liter } (500 \text{ mol/m}^3)}$$

SOLVED PROBLEM 16

For the reaction

$$CO_2(g) + H_2(g) \rightleftharpoons CO(g) + H_2O(g)$$

the value of K at 825 K is 0.137. If 5 mol of CO_2, 5.0 mol of H_2, 1.0 mol of CO and 1.0 mol of H_2O are initially present, what are the actual concentrations of CO_2, H_2, CO and H_2O at equilibrium?

SOLUTION

If x moles of CO_2 has reacted at equilibrium, x mole of CO and x mole of H_2O will be produced. The moles at equilibrium due to the reaction will be

$$CO_2 + H_2 \rightleftharpoons CO + H_2O$$

Moles at equilibrium : $5–x$ $\quad 5–x \quad x \quad x$

But 1 mole each of CO and H_2O were present initially. Therefore the actual moles at equilibrium would be

$$CO_2 = 5 - x$$
$$H_2 = 5 - x$$
$$CO = 1 + x$$
$$H_2O = 1 + x$$

If V metre^3 be the volume of the reaction vessel, the concentration of the various species are:

$$[CO_2] = (5 - x)/V$$
$$[H_2] = (5 - x)/V$$
$$[CO] = (1 + x)/V$$
$$[H_2O] = (1 + x)/V$$

Substituting the values in the equilibrium constant expression, we have

$$K = \frac{[CO]\,[H_2O]}{[CO_2]\,[H_2]}$$

$$0.137 = \frac{\left[(1+x)/V\right]\left[(1+x)/V\right]}{\left[(5-x)/V\right]\left[(5-x)/V\right]} = \frac{(1+x)^2\,V^2}{(5-x)^2\,V^2}$$

$$0.137 = \frac{(1+x)^2}{(5-x)^2} = \frac{1 + 2x + x^2}{25 - 10x + x^2}$$

or $\quad 0.137\,(25) - 0.137\,(10)x + 0.137\,x^2 = 1 + 2x + x^2$

$$0.863\,x^2 + 3.370\,x - 3.425 = 0$$

Solving for x using the quadratic formula

$$x = 0.62$$

Thus the final equilibrium concentrations are:

$$[CO] = 1 + x = 1 + 0.62 = 1.62 \text{ mol l}^{-1} \text{ (1620 mol/m}^3\text{)}$$
$$[H_2O] = 1 + x = 1 + 0.62 = 1.62 \text{ mol l}^{-1} \text{ (1620 mol/m}^3\text{)}$$
$$[CO_2] = 5 - x = 5 - 0.62 = 4.38 \text{ mol l}^{-1} \text{ (4380 mol/m}^3\text{)}$$
$$[H_2] = 5 - x = 5 - 0.62 = \mathbf{4.38\ mol\ l^{-1}} \text{ (4380 mol/m}^3\text{)}$$

SOLVED PROBLEM 17

For the reaction $I_2(g) \rightleftharpoons 2I(g)$, $K = 3.76 \times 10^{-5}$ at 1000 K. Let 1 mole of I_2 be injected into a 0.002 m^3 glass box at 1000 K. What will be the concentrations of I_2 and I when the equilibrium is attained?

SOLUTION

The balanced equation for the reaction is

$$I_2 \rightleftharpoons 2I$$

1 mole of I_2 is injected in a 0.002 m^3 box. Thus the initial concentration of I_2 is

$$\frac{1 \text{ mole}}{0.002 \text{ m}^3} = 0.5 \text{ M or } 500 \text{ mole/m}^3$$

Let x moles of I_2 decompose at equilibrium. Then the moles of the two species are:

$$\begin{array}{cc} I_2 & \rightleftharpoons 2I \\ 0.5-x & 2x \end{array}$$

Therefore the equilibrium concentrations are:

$$[I_2] = 0.5 - x$$
$$[I] = 2x$$

and $K = 3.76 \times 10^{-5}$

Substituting the values in the equilibrium constant expression, we have

$$K = \frac{[I]^2}{[I_2]} = 3.76 \times 10^{-5}$$

or $$\frac{(2x)^2}{(0.5 - x)} = 3.76 \times 10^{-5}$$

Solving for x, using the quadratic equation formula,

$$x = 2.17 \times 10^{-3}$$

Thus the concentrations of I and I_2 at equilibrium are:

$[I] = 2x =$ **4.34×10^{-3} mol l^{-1} or 4.34 mol/m^3**

$[I_2] = 0.5 - x =$ **0. 498 mol l^{-1} or 498 mol/m^3**

CALCULATIONS INVOLVING K_p

It may be noted that the value of K_p depends on the units used to express pressures, which are usually atmospheres.

Calculation of K_p

SOLVED PROBLEM 18

The reaction

$$2NO(g) + Cl_2(g) \rightleftharpoons 2NOCl(g)$$

was studied at 298 K. The partial pressures at equilibrium were found to be

$$p_{NOCl} = 1.2 \text{ atm } (121590.00 \text{ Pa})$$
$$p_{NO} = 5.0 \times 10^{-2} \text{ atm } (5066.25 \text{ Pa})$$
$$p_{Cl_2} = 3.0 \times 10^{-1} \text{ atm } (30397.5 \text{ Pa})$$

Calculate K_p for the reaction at 298 K.

SOLUTION

The equilibrium constant expression in terms of partial pressures is

$$K_p = \frac{(p_{NOCl})^2}{(p_{NO})^2\,(p_{Cl_2})}$$

Substituting the values of partial pressures

$$K_p = \frac{(121590)^2}{5066.25 \times 30397.5} = \mathbf{1.92 \times 10^3}$$

SOLVED PROBLEM 19

At 313 K and a total pressure of 101325.00 Pa dinitrogen tetroxide, N_2O_4, is 50% dissociated into nitrogen dioxide, NO_2.

$$N_2O_4(g) \rightleftharpoons 2NO_2(g)$$

Calculate the value of K_p at this temperature.

SOLUTION

Let us have 1 mole of N_2O_4 to start with of which x mole dissociates at equilibrium. Then the moles of N_2O_4 and NO_2 at equilibrium are:

$$N_2O_4 \rightleftharpoons 2NO_2$$

moles at equilibrium : $(1 - x)$ $\quad 2x$

Since N_2O_4 is 50% dissociated, $x = 0.5$ mol and the equilibrium mixture contains

$$NO_2 = 2x = 2 \times 0.5 = 1 \text{ mol}$$

$$N_2O_4 = (1 - x) = (1 - 0.5) = 0.5 \text{ mol}$$

That is, the moles of N_2O_4 and NO_2 are present in the ratio 1 : 2

$\therefore$ The partial pressure of $N_2O_4 = \frac{1}{3}$ atm (33437.25 Pa)

The partial pressure of $NO_2 = \frac{2}{3}$ atm (67583.775 Pa)

Substituting these values in the equilibrium constant expression, we have

$$K_p = \frac{(p_{NO_2})^2}{p_{N_2O_4}} = \frac{(67583.775 \text{ Pa})^2}{633437.25} = \mathbf{1.33}$$

Calculation of Equilibrium Partial Pressures

SOLVED PROBLEM 20

At 1000 K the equilibrium constant for the reaction

$$2SO_2(g) + O_2(g) \rightleftharpoons 2SO_3(g)$$

is $K_p = 354637.5$ Pa. If the total pressure in the reaction flask is 101325.00 Pa, and the partial pressure of O_2 at equilibrium is 10132.5 Pa, calculate the partial pressures of SO_2 and SO_3.

SOLUTION

Let the partial pressure of SO_2 = x pascal

Then the partial pressure of SO_3 $= 101325 - p_{O_2} - x$

$= 101325 - 1013.25 - x$

since the partial pressure of O_2 $= 91192.5 - x$ Pa

Substituting these values in the equilibrium constant expression, we have

$$K_p = \frac{(p_{SO_3})^2}{(p_{SO_2})^2 (p_{O_2})} = \frac{(91192.5 - x)^2}{x^2 \times 10132.5} = 354637.5\,\text{Pascal}$$

Taking the square root, we have

$$\frac{(91192.5 - x)^2}{x} = (10132.5)^{\frac{1}{2}} (354637.5)^{\frac{1}{2}}$$

or $x = 0.48$

$p_{SO_2} = \mathbf{0.48\ Pascal}$

$p_{SO_3} = \mathbf{0.52\ Pascal}$

Calculation of Degree of Dissociation

Case 1. By determining equilibrium concentrations. The equilibrium constant (*K*) for the combination reaction is first calculated using equilibrium concentrations. The equilibrium law is then applied to the reverse (dissociation) reaction when the equilibrium constant is 1/*K*.

SOLVED PROBLEM 21

0.000025 m³ of H_2 and 0.000018 m³ of I_2 vapor were heated in a sealed glass bulb at 738 K when at equilibrium 0.0000308 m³ of HI was formed. Calculate the degree of dissociation of pure HI at 738 K.

SOLUTION

Step 1. Calculation of *K*

The combination of H_2 and I_2 to form HI can be written as

$$H_2 + I_2 \rightleftharpoons 2HI$$

m³ at equilibrium : $(a - x)$ $(b - x)$ $2x$

At equilibrium, HI formed is $2x$ m³

or $2x = 30.8 \times 10^{-6}$ m³

$x = 15.4 \times 10^{-6}$ m³

∴ Concentrations at equilibrium are:

$[H_2] = (a - x) = 25 \times 10^{-6} - 15.4 \times 10^{-6} = 9.6 \times 10^{-6} = 0.96 \times 10^{-6}$ mol/m³

$[I_2] = (b - x) = 18 \times 10^{-6} - 15.4 \times 10^{-6} = 2.6 \times 10^{-6} = 0.26 \times 10^{-6}$ mol/m³

$[HI] = 3.08 \times 10^{-5}$ mol/m³

Substituting the values in the equilibrium constant expression

$$K = \frac{[HI]^2}{[H_2][I_2]} = \frac{3.8 \times 10^{-5} \times 3.8 \times 10^{-5}}{0.96 \times 10^{-5}\,\text{m}^3 \times 0.26 \times 10^{-5}\,\text{m}^3} = \mathbf{38}$$

Step 2. Calculation of Degree of dissociation, *x*

If we inject 10^{-6} m³ of HI in a sealed glass bulb at 738 K, $x \times 10^{-6}$ m³ of it will decompose at equilibrium, where *x* represents the degree of dissociation.

$$2HI \rightleftharpoons H_2 + I_2$$

m³ at equilibrium : $1 - x$ $x/2$ $x/2$

Substituting the values in the equilibrium constant expression

$$K' = \frac{[x/2]^2}{(1-x)^2} = \frac{x^2}{4(1-x)^2}$$

Since $K' = 1/K$, we have

$$\frac{x^2}{4(1-x)^2} = \frac{1}{38}$$

Hence $x = 0.245$

Thus the degree of dissociation of pure HI at 738 K is **0.245**.

SOLVED PROBLEM 22

For the reaction $PCl_5 \rightleftharpoons PCl_3 + Cl_2$, $K = 33.3$ at 760 K. 0.001 kg of PCl_5 is injected into a 0.0005 m^3 evacuated flask and allowed to come to equilibrium. What percent of the PCl_5 will dissociate at equilibrium? Molecular mass of $PCl_5 = 208.235$.

SOLUTION

$$\text{Initial moles of } PCl_5 = \frac{\text{grams (mass)}}{\text{mol. mass}} = \frac{1}{208.235} = 0.0048 \text{ mole}$$

$$\text{Initial concentration of } PCl_5 = \frac{0.0048 \text{ mole}}{0.5 \times 10^{-3} \text{ m}^3} = 9.6 \text{ mol/m}^3$$

At equilibrium if x mole of PCl_5 is dissociated, the equilibrium concentrations are:

$$\begin{array}{ccccc} PCl_5 & \rightleftharpoons & PCl_3 & + & Cl_2 \\ 0.0096 - x & & x & & x \end{array}$$

Substituting values in the equilibrium constant expression,

$$\frac{[PCl_3][Cl_2]}{[PCl_5]} = 33.3$$

$$\frac{(x)(x)}{(0.0096 - x)} = 33.3$$

Solving for x using the quadratic equation obtained by you, we have

$$x = 9.597 \times 10^{-3}$$

$\therefore$ The initial concentration of $PCl_5 = 9.6$ mol/m^3

The final concentration of $PCl_5 = 9.6 - x$

$$= 9.6 - 9.597 \times 10^{-3}$$

$$= 0.03 \text{ mol/m}^3$$

Thus the amount of PCl_5 dissociated at equilibrium is

$$100 \times \frac{9.6 - 0.03}{9.6} = \mathbf{99.68\%}$$

Case 2. By density measurements. This method is used for the determination of the degree of dissociation of gases in which one molecule produces two or more molecules. For example, PCl_5 dissociates to give two molecules PCl_3 and Cl_2.

$$PCl_5 \rightleftharpoons PCl_3 + Cl_2$$

Thus at constant temperature and pressure the volume increases. The density at constant pressure then decreases. The degree of dissociation can be calculated from the difference between the density of the undissociated gas and that of the partially dissociated gas at equilibrium.

If we start with one mole of gas (PCl_5) and the degree of dissociation is x, we have

$$\begin{array}{lccc} & PCl_5 & \rightleftharpoons & PCl_3 + Cl_2 \\ \text{moles at equilibrium :} & (1-x) & & x \quad\quad x \end{array}$$

Thus the total number of moles at equilibrium

$$= (1-x) + x + x = 1 + x$$

The density of an ideal gas at constant temperature and pressure is inversely proportional to the number of moles for a given weight.

Hence the ratio of density ρ_1 of the undissociated gas to the density ρ_2 of the dissociated gas at equilibrium is given as

$$\frac{\rho_1}{\rho_2} = \frac{1+x}{1}$$

whence $\rho_1 = \rho_2 + x\rho_2$

or $x = \dfrac{\rho_1 - \rho_2}{\rho_2}$

Substituting the values of ρ_1 and ρ_2 the value of x, the degree of dissociation, can be calculated.

SOLVED PROBLEM 23

When PCl_5 is heated it gasifies and dissociates into PCl_3 and Cl_2. The density of the gas mixture at 473 K is 70.2. Find the degree of dissociation of PCl_5 at 473 K.

SOLUTION

We will use the expression $x = \dfrac{\rho_1 - \rho_2}{\rho_2}$

where x = degree of dissociation

ρ_1 = density of undissociated PCl_5

ρ_2 = density of gas mixture at equilibrium (partially dissociated)

The density (ρ_1) of undissociated PCl_5 would be

$$VD = \frac{\text{mol mass}}{2} = \frac{31 + 177.5}{2} = 104.25$$

Observed density $\rho_2 = 70.2$

$$\therefore \quad x = \frac{\rho_1 - \rho_2}{\rho_2} = \frac{104.25 - 70.2}{70.2} = \mathbf{0.485}$$

SOLVED PROBLEM 24

At 363 K the vapor density of nitrogen tetroxide (N_2O_4) is 24.8. Calculate the percentage dissociation into NO_2 molecules at this temperature.

SOLUTION

The dissociation of N_2O_4 can be represented by the equation

$$N_2O_4 \rightleftharpoons NO_2 + NO_2$$

Since the number of molecules on the two sides of the equation is different, we can calculate the degree of dissociation by using the expression

$$x = \frac{\rho_1 - \rho_2}{\rho_2}$$

The density of undissociated N_2O_4 is given by

$$\rho_1 = \frac{\text{mol mass}}{2} = \frac{28 + 64}{2} = \frac{92}{2} = 46$$

$$\rho_2 = 24.8 \text{ (given)}$$

$$\therefore \quad x = \frac{46 - 24.8}{24.8} = 0.8548$$

Hence, percentage dissociation = $100 \times x =$ **85.48**

Case 3. By molecular mass determination. Molecular masses are proportional at constant temperature and pressure to the gas densities. Therefore, we can substitute molecular masses for the densities in the expression

$$x = \frac{\rho_1 - \rho_2}{\rho_2}$$

which gives
$$x = \frac{M_1 - M_2}{M_2}$$

where M_1 is the molecular mass of the undissociated gas and M_2 is the average molecular mass of the gases at the equilibrium.

SOLVED PROBLEM 25

0.001588 kg of nitrogen tetroxide gives a total pressure of 1 atm (101325 Pa) when partially dissociated at equilibrium in a 0.0005 m^3 glass vessel at 298 K. What is the degree of dissociation at this temperature.

SOLUTION

M_1, molecular mass of undissociated N_2O_4 gas

$$= 14 \times 2 + 16 \times 4 = 92.0$$

M_2, molecular mass of partially dissociated gas at equilibrium can be determined by using the general gas equation.

$$PV = nRT$$

where
$$n = \frac{\text{mass of substance}}{M_2} = \frac{m}{M_2}$$

or
$$PV = \frac{m}{M_2} RT \quad \text{... (1)}$$

Rearranging the expression (1), we have

$$M_2 = \frac{RT}{P} \times \frac{m}{V} \quad \text{... (2)}$$

Here, $R = 8.3145 \text{ J mol}^{-1} \text{ K}^{-1}$; $T = 298$ K

$P = 1$ atm (101325 Pa); $V = 0.0005 \text{ m}^3$; $m = 0.001588$ kg

Substituting the values in the expression (2)

$$M_2 = \frac{8.3145 \times 298 \times 0.001588}{(101325.00)\ (0.00005 \text{ m}^3)} = 0.017663$$

Therefore,
$$x = \frac{M_1 - M_2}{M_2} = \frac{0.092 - 0.07766}{0.07766}$$

$$= \mathbf{0.1846}$$

UNITS OF EQUILIBRIUM CONSTANT

In the equilibrium expression for a particular reaction, the concentrations are given in units of moles/liter or mol/L, and the partial pressure are given in atmospheres (atm). The units of K_c and K_p, depend on the specific reaction.

(1) When the total number of moles of reactants and products are equal

In the equilibrium expression of these reactions, the concentration or pressure terms in the numerator and denominator exactly cancel out. Thus K_c or K_p for such a reaction is without units.

Taking example of the reaction

$$H_2(g) + I_2(g) \rightleftharpoons 2HI(g)$$

$$K_c = \frac{[HI]^2}{[H_2][I_2]} = \frac{(mol/L)^2}{(mol/L)(mol/L)} \quad \text{(No units)}$$

$$K_p = \frac{(P_{HI})^2}{(P_{H_2})(P_{I_2})} = \frac{(atm)^2}{(atm)(atm)} \quad \text{(No units)}$$

(2) When the total number of moles of the reactants and products are unequal

In such reactions K_c will have units (mol/liter)^n and K_p will have units (atm)^n, where n is equal to the total number of moles of products minus the total number of moles of reactants. Thus for the reaction

$$N_2O_4(g) \rightleftharpoons 2NO_2(g)$$

$$K_c = \frac{[NO_2]^2}{[N_2O_4]} = \frac{(mol/L)^2}{(mol/L)} = mol/L$$

$$K_p = \frac{(p_{NO_2})^2}{(P_{N_2O_4})} = \frac{(atm)^2}{(atm)} = atm$$

Thus units for K_c are mol/L and for K_p units are atm. For the reaction

$$N_2(g) + 3H_2(g) \rightleftharpoons 2NH_3(g)$$

the units for K_c and K_p may be found as follows

$$K_c = \frac{[NH_3]^2}{[N_2][H_2]^3} = \frac{(mol/L)^2}{(mol/L)(mol/L)^3} = mol^{-2}\,L^2$$

$$K_p = \frac{(p_{NH_3})^2}{(p_{N_2})(p_{H_2})^3} = \frac{(atm)^2}{(atm)(atm)^3} = atm^{-2}$$

Thus units of K_c are mol^{-2} L^2 and units of K_p are atm^{-2}.

It may be noted, however, that the units are often omitted when equilibrium constants are listed in tables.

Thermodynamic Derivation of Law of Chemical Equilibrium

Let us consider a general reaction

$$aA + bB + ... \rightleftharpoons cC + dD + ...$$

The chemical potential of a substance in a mixture is related to its activity by the expression

$$\mu = \mu^\circ + RT \ln a \quad ...(i)$$

where μ° is the chemical potential of the pure substance in standard state of unit activity, R is gas constant and T the absolute temperature.

For *a* mole of the substance A we can write using the equation (*i*)

$$a\mu_A = a(\mu° + RT \ln a_A)$$

and similarly

$$b\mu_B = b(\mu° + RT \ln a_B)$$
$$c\,\mu_c = c(\mu° + RT \ln a_c)$$
$$d\,\mu_D = d(\mu° + RT \ln a_D)$$

The change in free energy for the reaction is given by

$$\Delta G = G_{products} - G_{reactants}$$

On substitution we get

$$\Delta G = (c\mu_c + d\mu_D + ...) - (a\mu_A + b\mu_B + ...)$$
$$= [c\{\mu_c° + RT \ln a_c\} + d\{\mu°_D + RT \ln a_D\}]$$
$$- [a\{\mu°_a + RT \ln a_A\} + b\{\mu°_B + RT \ln a_B\}]$$

$$= [\{c\,\mu_c° + d\,\mu°_D + ...\} - \{a\,\mu°_A + b\,\mu°_B + ...\}] + RT \ln \frac{a_C^c \times a_D^d \times ...}{a_A^a \times a_B^b \times ...}$$

$$\Delta G = \Delta G° + RT \ln \frac{a_C^c \times a_D^d \times ...}{a_A^a \times a_B^b \times ...} \quad ...\ (ii)$$

where $\Delta G°$ is the difference in free energy of the reaction when all reactants and products are in their standard state. It is given by

$$\Delta G° = \{c\mu_c° + d\,\mu_D° + ...\} - \{a\mu_A° + b\,\mu_B° + ...\}$$

In equation (*ii*) the factor A given by

$$\frac{a_C{}^c \times a_D{}^d \times ...}{a_A{}^a \times a_B{}^b \times ...}$$

stands for the reaction quotient of activities of the product and reactants. It may be denoted by *J*. The equation (*ii*) becomes

$$\Delta G = \Delta G° + RT \ln J \quad ...\ (iii)$$

The equation (*iii*) is called **van't Hoff reaction isotherm**

For the reaction at equilibrium

$$\Delta G = 0$$

therefore,

$$\Delta G° = - RT \ln J$$

or

$$\Delta G = - RT \ln \frac{a_C^c \times a_D^d \times ...}{a_A^a \times a_B^b \times ...}$$

As $\Delta G°$ is the free energy of the reaction in the standard state and is constant at a given temperature.

Also, the gas constant *R* and *T* are constant, the factor

$$\frac{a_C^c \times a_D^d \times ...}{a_A^a \times a_B^b \times ...}$$

is a constant i.e.

$$\frac{a_C^c \times a_D^d \times ...}{a_A^a \times a_B^b \times ...} = a \text{ constant} = K$$

It is nothing but the law of chemical equilibrium. Therefore, from equation (*iii*) we have

$$\Delta G° = -RT \ln K \qquad ...(iv)$$

or

$$\Delta G° = -2.303\ RT \log K \qquad ...(v)$$

The equation (*iv*) is also called van't Hoff Isotherm. It may be used to calculate the change in free energy of a reaction in the standard ($\Delta G°$) from the equilibrium constant and vice-versa.

The sign of $\Delta G°$ indicates whether the forward or reverse reaction is spontaneous. Considering the equation (*iv*), we can have three possibilities depending on the sign of $\Delta G°$ for the reaction.

(1) If **$\Delta G°$ is negative,** $\log K$ must be positive and the reaction proceeds spontaneously in the forward reaction.

(2) If **$\Delta G°$ is positive,** $\log K$ must be negative and K is less than one. The reverse reaction is then spontaneous.

(3) If **$\Delta G° = 0$**, $\log K = 0$ and $K = 1$. The reaction is at equilibrium.

Meaning of the Magnitude of the Free Energy Change

If $\Delta G°$ is a large negative number, K will be much greater than one and the forward reaction will proceed nearly to completion. On the other hand if $\Delta G°$ is a large positive number, K will be a small fraction. Then the reverse reaction will proceed almost to completion.

SOLVED PROBLEM 26

Calculate K for reaction which has $\Delta G°$ value – 83.7 kJ at 298 K.

SOLUTION

We know that

$$\Delta G° = -2.303\ RT \log K \qquad ...(a)$$

where $\Delta G°$ is standard free energy; R is gas constant; T is absolute temperature; and K is equilibrium constant. It must be remembered that the same units of $\Delta G°$, R and T are to be used in a given problem.

If $\Delta G°$ is given in Joules

$$R = 8.3145 \text{ J mol}^{-1} \text{ K}^{-1}$$

$$T = 25 + 273 = 298 \text{ K}$$

The value of K from expression (*a*) may be calculated as

$$-\log K = \frac{-\Delta G°}{(2.303)(8.3145)(298)}$$

$$= \frac{-83700}{5706.53} = 14.7$$

Taking antilogs, $\log K = 0.7 + 14.0$

Thus $K = 5 \times 10^{14}$

SOLVED PROBLEM 27

The standard free energy change for the reaction

$$N_2(g) + O_2(g) \rightleftharpoons 2NO(g)$$

is +173.1 kJ. Calculate K_p for the reaction at 298 K.

SOLUTION

Here, we will use the expression

$$\Delta G° = -2.303\ RT \log K_p$$

or

$$\log K_p = -\frac{\Delta G°}{2.303\ RT}$$

Substituting the values

$$\log K_p = \frac{1.73 \times 10^5}{(2.303)(8.314)(298)}$$

$$= -30.34$$

Taking antilogs

$$K_p = 10^{0.66} \times 10^{-31}$$

$$= \mathbf{4.6 \times 10^{-31}}$$

SOLVED PROBLEM 28

The equilibrium constant, K_p for the reaction

$$C_2H_4(g) + H_2(g) \rightleftharpoons C_2H_6\,(g)$$

is 4.97×10^{12} Pa^{-1} at 298 K. Calculate $\Delta G°$.

SOLUTION

$$\Delta G° = -2.303\, RT \log K_p \qquad ...(a)$$

Here, $R = 8.314$; $T = 298$ K; $K_p = 4.97 \times 10^{12}$ Pa^{-1}

Substituting the values in the expression (*a*)

$$\Delta G° = -2.303\ (8.314)\ (298) \log (4.97 \times 10^{12})$$

$$= -5706 \times \log (4.97 \times 10^{12})$$

$$= -5706 \times (12.6964)$$

$$= -72445.6584 \text{ J}$$

Thus $\Delta G° = -7.245 \times 10^4$ J = **−72.45 kJ**

Temperature Dependence of Equilibrium Constant (van't Hoff's Equation)

There is appreciable change in the value of equilibrium constant with change in temperature. It can be determined thermodynamically using the relation

$$\Delta G° = -RT \ln K_p \qquad ...(i)$$

where $\Delta G°$ is the change in standard free energy of the reaction.

Differentiating equation (*i*) w.r.t. at constant pressure *P*, we get

$$\left(\frac{\delta(\Delta G°)}{\delta T}\right)_p = -R \ln K_p - RT\left(\frac{\delta \ln K_p}{\delta T}\right)_p$$

Multiplying both sides by *T*, we get

$$T\left(\frac{\delta(\Delta G°)}{\delta T}\right)_p = -RT \ln K_p - RT^2\left(\frac{\delta \ln K_p}{\delta T}\right)_p$$

From equation (*i*) we get

$$T\left(\frac{\delta(\Delta G°)}{\delta T}\right)_p = \Delta G° - RT^2\left(\frac{\delta(\ln K_p)}{\delta T}\right)_p \qquad ...(ii)$$

We know that Gibb's Helmholtz equation for a reaction in the standard state can be written as

$$\Delta G° = \Delta H° + T\left(\frac{\delta(\Delta G°)}{\delta T}\right)_p$$

or

$$T\left(\frac{\delta(\Delta G)}{\delta T}\right)_p = \Delta G° - \Delta H° \qquad ...(iii)$$

Comparing (*ii*) and (*iii*) we get

$$\Delta G^\circ = RT^2\left(\frac{\delta(\ln K_p)}{\delta T}\right)_p$$

or $$\frac{\Delta H^\circ}{RT^2} = \frac{\delta(\ln K_p)}{\delta T} \quad \text{... }(iv)$$

This equation is known as **vàn't Hoff equation,** where ΔH° is the enthalpy change of the reaction at constant pressure when all reactants and products are in their standard state. It has been found that the enthalpy change does not vary much with change in partial pressures of the reactants or products.

Therefore, ΔH° may be replaced by ΔH. The equation (*iv*) becomes

$$\frac{\Delta H}{RT^2} = \frac{d(\ln K_p)}{dT} \quad \text{... }(v)$$

Integrating equation (*v*) we get

$$ln\ K_p = -\frac{\Delta H}{RT} + C$$

or $$\log K_p = -\frac{\Delta H}{2.303\,RT} + C \quad \text{... }(vi)$$

where C is a constant of integration. When a graph of log K_p against $1/T$ is plotted we get straight line having slope equal to $-\dfrac{\Delta H}{2.303\,R}$ (Fig 17.9).

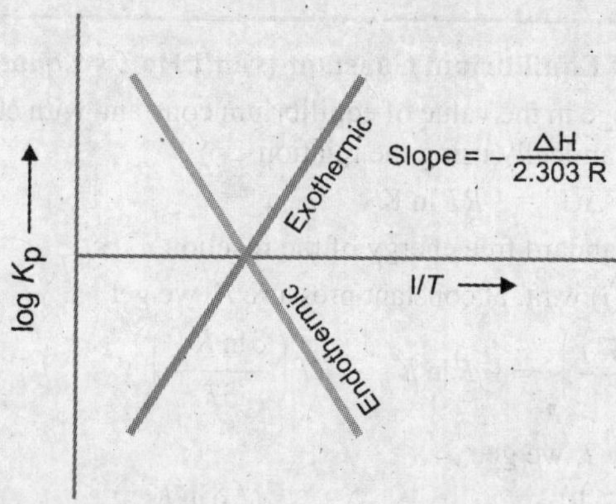

■ **Figure 17.9**
Graph between log *Kp* and *I/T*.

From the graph it is clear that

(*a*) **For exothermic reaction ΔH is negative and K_p decreases with increase in temperature**

(*b*) **For endothermic reaction ΔH is positive and K_p increases with increase in temperature.**

Integrating equation (*v*) within the limits K_{p2} at temperature T_2 and K_{p1} at temperature T_1

$$\int_{K_{p1}}^{K_{p2}} d\ \ln K_p = \int_{T_1}^{T_2} \frac{\Delta H}{RT^2} dT$$

or $$\ln\frac{K_{p2}}{K_{p1}} = -\frac{\Delta H}{R}\left[\frac{1}{T_2} - \frac{1}{T_1}\right]$$

$$= \frac{\Delta H}{R}\left[\frac{1}{T_1} - \frac{1}{T_2}\right]$$

$$\ln \frac{K_{p2}}{K_{p1}} = \frac{\Delta H}{R}\left[\frac{T_2 - T_1}{T_1 T_2}\right]$$

or
$$\log \frac{K_{p2}}{K_{p1}} = \frac{\Delta H}{2.303\,R}\left[\frac{T_2 - T_1}{T_1 T_2}\right]$$

From this equation the heat of reaction can be determined if the values of equilibrium constant K_{p2} and K_{p1} at temperature T_2 and T_1 are known and vice-versa.

van't Hoff Equation in Terms of *K*

We know that equilibrium constant in terms of partial pressure (K_p) and in terms of concentration (K_c) are related to each other by the relation

$$K_p = K_c\,(\mathrm{RT})^{\Delta n}$$

Taking logarithms we get

$$\ln K_p = \ln K_c + \Delta n \ln RT$$

Differentiating w.r.t. temperature, we get

$$\frac{d(\ln K_p)}{dT} = \frac{d(\ln K_c)}{dT} + \frac{\Delta n}{T}$$

or
$$\frac{d(\ln K_c)}{dT} = \frac{d(\ln K_P)}{dT} - \frac{\Delta n}{T}$$

From equation (v) we have

$$\frac{d(\ln K_c)}{dT} = \frac{\Delta H}{RT^2} - \frac{\Delta n}{T}$$

$$= \frac{\Delta H}{RT^2} - \frac{\Delta n\, RT}{TRT}$$

or
$$\frac{d(\ln K_c)}{dT} = \frac{\Delta H - \Delta n\, RT}{RT^2}$$

But we know

$$\Delta H = \Delta E + \Delta n\, RT$$

$\therefore$
$$\frac{d(\ln K_c)}{dT} = \frac{\Delta E}{RT^2}$$

where ΔE is the heat of the reaction at constant volume.

SOLVED PROBLEM 29

The partial pressure of CO_2 in the reaction

$$CaCO_3(s) \rightleftharpoons CaO(s) + CO_2(g)$$

is 103.05 Pa at 773 K. Calculate K_p at 873 K for the above reaction. ΔH for the reaction is 180.87 kJ per mol and does not change in the given range of temperatures.

SOLUTION

$$CaCO_3(s) \rightleftharpoons CaO(s) + CO_2(g)$$

$$K_p = P_{CO2}$$

Given
$$K_{p1} = 103.05 \text{ Pa at } 773 \text{ K}$$

$$\Delta H = 180.87 \text{ kJ per mol}$$
$$= 180870 \text{ J per mol}$$
$$T_2 = 873 \text{ K}; K_{p2} = ?$$

using the relation

$$\log \frac{K_{p2}}{K_{p1}} = \frac{\Delta H}{2.303\,R}\left[\frac{T_2 - T_1}{T_1 T_2}\right]$$

Substituting the values we get

$$\log \frac{K_{p2}}{K_{p1}} = \frac{180870}{2.303 \times 8.314}\left[\frac{873 - 773}{873 \times 773}\right]$$

$$K_{p2} = 1607.868 \text{ Pa}$$

∴ Equilibrium constant for the given reaction at 873 K is **1607.868 Pa**

SOLVED PROBLEM 30

The value of K_p at 298 K for the reaction

$$\tfrac{1}{2}N_2 + \tfrac{3}{2}H_2 \rightleftharpoons NH_3$$

is found to be 826.0, partial pressures being measured in pascal units. Calculate $\Delta G°$ at 298 K.

SOLUTION. We know

$$\Delta G° = -2.303\,RT \log K_p$$

Here $R = 8.314 \text{ J mol}^{-1} \text{K}^{-1}; T = 298 \text{ K}$

$$K_p = 826; \Delta G° = ?$$

Substituting the values we get

$$\Delta G° = -2.303 \times 8.314 \times 298 \times \log 826$$
$$= \mathbf{-16654.1690 \text{ Joule}}$$

SOLVED PROBLEM 31

The equilibrium constant K_p for the reaction

$$2NH_3(g) \rightleftharpoons 3H_2(g) + N_2(g)$$

is 1.22×10^{-3} at 298 K and 2.16 at 498 K. Calculate $\Delta H°$ for the reaction.

SOLUTION. We know

$$\log \frac{K_{p2}}{K_{p1}} = \frac{\Delta H°}{2.303\,R}\left[\frac{T_2 - T_1}{T_1\,T_2}\right]$$

Here $K_{p2} = 2.16; K_{p1} = 1.22 \times 10^{-3}$

$$T_2 = 498 \text{ K}; T_1 = 298 \text{ K}$$
$$R = 8.314 \text{ J mol}^{-1} \text{K}^{-1}$$

Substituting the values we get

$$\log \frac{2.16}{1.22 \times 10^{-3}} = \frac{\Delta H°}{2.303 \times 8.314}\left[\frac{498 - 298}{498 \times 298}\right]$$

$$\log \frac{2.16}{1.22 \times 10^{-3}} = \frac{\Delta H°}{2.303 \times 8.314} \times \frac{200}{498 \times 298}$$

or $\Delta H° = 46175.80 \text{ Joule}$

$$= \mathbf{46.176 \text{ kJ}}$$

LIQUID SYSTEMS

The chemical equilibrium in which all the reactants and products are in the liquid phase, are referred to as the **liquid equilibria.** Like the gas- phase equilibria, the liquid equilibria are also called **homogeneous equilibria.** For example, alcohols and acids react to form esters and water.

Reaction between Acetic acid and Ethyl alcohol to form Ethyl acetate

$$CH_3COOH(l) + C_2H_2OH(l) \rightleftharpoons CH_3COOC_2H_5(l) + H_2O(l)$$

Let us start with *a* moles of acetic acid and *b* moles of alcohol. If *x* moles of acetic acid react with *x* moles of ethyl alcohol, *x* moles of ester and *x* moles of water are produced when the equilibrium is established. Now the moles present at equilibrium are:

$$CH_3COOH = (a - x) \text{ moles}$$
$$C_2H_5OH = (b - x) \text{ moles}$$
$$CH_3COOC_2H_5 = x \text{ moles}$$
$$H_2O = x \text{ moles}$$

If *V* liter be the total volume of the equilibrium mixture, the concentrations of the various species are:

$$[CH_3COOH] = \frac{a - x}{V}$$

$$[C_2H_5OH] = \frac{b - x}{V}$$

$$[CH_3COOC_2H_5] = \frac{x}{V}$$

$$[H_2O] = \frac{x}{V}$$

The equilibrium constant expression may be written as

$$K = \frac{\frac{x}{V} \times \frac{x}{V}}{\frac{a - x}{V} \times \frac{b - x}{V}} \quad \text{or} \quad \frac{x^2}{(a - x)(b - x)}$$

It may be noted that the volume terms *V* in the numerator and denominator cancel out. **In liquid systems when there is a change in the number of moles as a result of the reaction, it is necessary to consider the volume *V* while calculating the equilibrium constant *K*.**

SOLVED PROBLEM 32

At 298 K one mole of acetic acid was allowed to react with one mole of ethyl alcohol until equilibrium was established. The equilibrium mixture was found to contain 0.333 mole of unused acid. Calculate the equilibrium constant of the reaction at the same temperature.

SOLUTION

The equation for the reaction is

$$CH_3COOH(l) + C_2H_5OH\,(l) \rightleftharpoons CH_3COOC_2H_5(l) + H_2O(l)$$

The moles of unused acid = 0.333

∴ moles reacted at equilibrium = 1 – 0.333 = 0.666

One mole of acid reacts with one mole of alcohol to form one mole of ester and one mole of water. Therefore, moles of the various species present at equilibrium are:

$$CH_3COOH = 0.333 \qquad CH_3COOC_2H_5 = 0.666$$
$$C_2H_5OH = 0.333 \qquad H_2O = 0.666$$

Since volume term V cancels out, moles may be taken instead of concentrations and the equilibrium constant expression may be written as

$$K = \frac{[CH_3COOC_2H_5][H_2O]}{[CH_3COOH][C_2H_5OH]}$$

$$= \frac{0.666 \times 0.666}{0.333 \times 0.333} = 4$$

Thus K for the reaction between acetic acid and ethyl alcohol at 298 K is **4**.

SOLVED PROBLEM 33

2 mol of acetic acid and 3 mol of ethyl alcohol are allowed to come to equilibrium at 298 K. If the equilibrium constant of the reaction at this temperature is 4, what is the equilibrium concentration of ethyl acetate?

SOLUTION

Let x mole of acetic acid react with x mole of ethyl alcohol to form x moles of ethyl acetate and x mole of water at equilibrium. Thus moles at equilibrium are:

$$\underset{2-x}{CH_3COOH} + \underset{3-x}{C_2H_5OH} \rightleftharpoons \underset{x}{CH_3COOC_2H_5} + \underset{x}{H_2O}$$

Since the volume terms cancel out in the equilibrium constant expression, we have

$$K = \frac{[CH_3COOC_2H_5][H_2O]}{[CH_3COOH][C_2H_5OH]} = \frac{x \times x}{(2-x)(3-x)} = 4$$

or

$$\frac{x^2}{(2-x)(3-x)} = 4$$

Rearranging gives

$$3x^2 - 20x + 24 = 0$$

Solving this quadratic equation, we get

$$x = 1.57 \text{ or } 5.09$$

Both roots are positive of which only 1.57 is acceptable. Since 2 mol of acetic acid can yield only 2 mol of the ester, the value 5.09 is rejected. The equilibrium conc. of ethyl acetate is **1.57 mol.**

HETEROGENEOUS EQUILIBRIA

So far we have discussed chemical equilibria in which all reactants and products are either gases or liquids. Such equilibria in which the reactants and products are not all in the same phase, are called **heterogeneous equilibria.** The decomposition of calcium carbonate upon heating to form calcium oxide and carbon dioxide is an example of heterogeneous equilibrium. If the reaction is carried in a closed vessel, the following equilibrium is established.

$$CaCO_3(s) \rightleftharpoons CaO(s) + CO_2(g)$$

Equilibrium Constant Expression

Straight forward application of Equilibrium law to the decomposition reaction of calcium carbonate leads to the equilibrium constant expression.

$$K = \frac{[CO_2][CaO]}{[CaCO_3]}$$

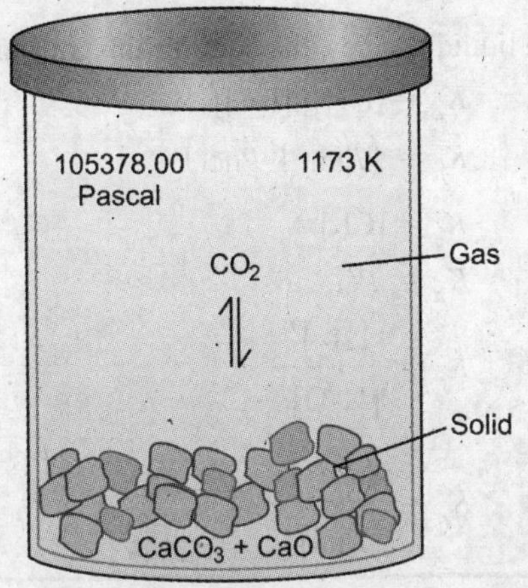

■ **Figure 17.10**
When solid $CaCO_3$ is heated in a closed vessel at 1173 K, equilibrium is established when the pressure of CO_2 reaches 1.04 atm (105378.00 Pascal).

But $CaCO_3$ and CaO are pure solids. The concentration (moles per unit volume) of a pure solid (or liquid) is fixed and cannot vary. Thus **the concentrations of pure solids or liquids are not included in the equilibrium constant expression.**

Ignoring the concentrations of $CaCO_3$ and CaO, the equilibrium constant expression for the decomposition of $CaCO_3$ may be written as

$$K_c = [CO_2]$$

In terms of partial pressures

$$K_p = p_{CO_2}$$

Similarly, the equilibrium constant expression for the decomposition of liquid water

$$2H_2O(l) \rightleftharpoons 2H_2(g) + O_2(g)$$

would be

$$K = \frac{[H_2]^2 [O_2]}{[H_2O]}$$

Ignoring the concentration of liquid water, we have

$$K_c = [H_2]^2 [O_2]$$

$$K_p = \left(p_{H_2}\right)^2 \left(p_{O_2}\right)$$

SOLVED PROBLEM 34

Write expressions for the equilibrium constants K_c and K_p for the equilibria:

(*a*) $NH_4Cl(s) \rightleftharpoons NH_3(g) + HCl(g)$

(*b*) $PCl_5(s) \rightleftharpoons PCl_3(l) + Cl_2(g)$

(*c*) $3Fe(s) + 4H_2O(g) \rightleftharpoons Fe_3O_4(s) + 4H_2(g)$

SOLUTION

Ignoring the solid and liquid species, the equilibrium constant expressions may be written as:

(*a*)
$$K_c = [NH_3]\,[HCl]$$
$$K_p = (p_{NH_3})(p_{HCl})$$

(*b*)
$$K_c = [Cl_2]$$
$$K_p = p_{Cl_2}$$

(*c*)
$$K_c = \frac{[H_2]^4}{[H_2O]^4}$$
$$K_p = \frac{(p_{H_2})^4}{(p_{H_2O})^4}$$

SOLVED PROBLEM 35

Carbon dioxide upon heating with carbon at high temperature is reduced to carbon monoxide:

$$CO_2(g) + C(s) \rightleftharpoons 2CO(g)$$

K_p for the reaction is 192517.50 Pa (1.90 atm).

In a particular experiment the total pressure at equilibrium was found to be 202650.00 Pa (2.00 atm). What were the partial pressures of CO and CO_2?

SOLUTION

Ignoring solid carbon, the equilibrium constant expression can be written as

$$K_p = \frac{(p_{CO})^2}{p_{CO_2}}$$

If the partial pressure of CO is x pascal, partial pressure of CO_2 is $(202650.00 - x)$

Substituting the values,

$$K_p = \frac{x^2}{202650.00 - x} = 192517.50\,\text{Pascal}$$

$$\therefore \qquad x^2 + 192517.50\,x - 3901361375 = 0$$

Solving the quadratic equation, $x = 123616.50$ Pa

Hence
$$p_{CO} = x \text{ pascal} = 123616.50 \text{ Pa}$$
$$p_{CO_2} = \mathbf{79035.50\ Pa}$$

LE CHATELIER'S PRINCIPLE

In 1884, the French Chemist Henry Le Chatelier proposed a general principle which applies to all systems in equilibrium. This important principle called the **Le Chatelier's principle** may be stated as : **when a stress is applied on a system in equilibrium, the system tends to adjust itself so as to reduce the stress.**

There are three ways in which the stress can be caused on a chemical equilibrium:

(1) Changing the concentration of a reactant or product.

(2) Changing the pressure (or volume) of the system.

(3) Changing the temperature.

Thus when applied to a chemical reaction in equilibrium, Le Chatelier's principle can be stated as: **if a change in concentration, pressure or temperature is caused to a chemical reaction in equilibrium, the equilibrium will shift to the right or the left so as to minimize the change.**

Now we proceed to illustrate the above statement by taking examples of each type of stress.

EFFECT OF A CHANGE IN CONCENTRATION

We can restate Le Chatelier's principle for the special case of concentration changes: **when concentration of any of the reactants or products is changed, the equilibrium shifts in a direction so as to reduce the change in concentration that was made.**

Addition of an inert gas

Let us add an *inert gas* to an equilibrium mixture while the volume of the reaction vessel remains the same. The addition of the inert gas increases the total pressure but the partial pressures of the reactants and products are not changed. Thus **the addition of an inert gas has no effect on the position of the equilibrium.**

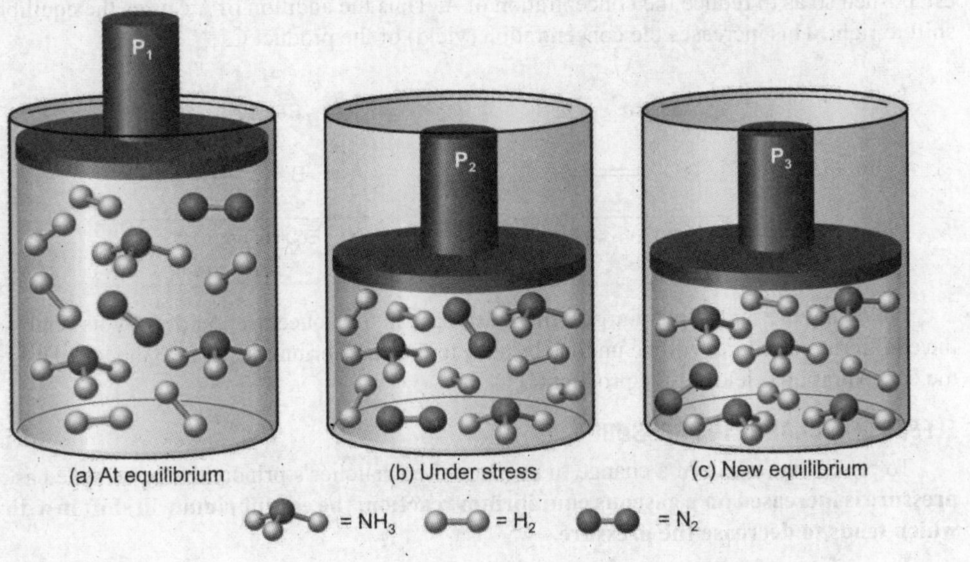

Figure 17.11
Illustration of Le Chatelier's principle. (*a*) System at equilibrium with $10H_2$, $5N_2$, and $3NH_3$, for a total of 18 molecules. (*b*) The same molecules are forced into a smaller volume, creating a stress on the system. (*c*) Six H_2 and $2N_2$ have been converted to $4NH_3$. A new equilibrium has been established with $4H_2$, $3N_2$, and $7NH_3$, a total of 14 molecules. The stress is partially relieved by the reduction in the total number of molecules.

Effect of change of concentration on Ammonia Synthesis reaction

Let us illustrate the effect of change of concentration on a system at equilibrium by taking example of the ammonia synthesis reaction:

$$N_2(g) + 3H_2(g) \rightleftharpoons 2NH_3(g)$$

When N_2 (or H_2) is added to the equilibrium already in existence (equilibrium I), the equilibrium will shift to the right so as to reduce the concentration of N_2 (Le Chatelier's principle). The concentration of NH_3 at the equilibrium II is more than at equilibrium I. The results in a particular case after the addition of one mol/liter are given below.

Equilibrium I		Equilibrium II
$[N_2] = 0.399$ M		$N_2 = 1.348$ M
$[H_2] = 1.197$ M	$\xrightarrow[\text{of } N_2 \text{ added}]{\text{1 mole/L}}$	$[H_2] = 1.044$ M
$[NH_3] = 0.202$ M		$[NH_3] = 0.304$ M

Obviously, the addition of N_2 (a reactant) increases the concentration of NH_3, while the concentration of H_2 decreases. Thus to have a better yield of NH_3, one of the reactants should be added in excess.

A change in the concentration of a reactant or product can be effected by the addition or removal of that species. Let us consider a general reaction

$$A + B \rightleftharpoons C$$

When a reactant, say, A is added at equilibrium, its concentration is increased. The forward reaction alone occurs momentarily. According to Le Chatelier's principle, a new equilibrium will be established so as to reduce the concentration of A. Thus the addition of A causes the equilibrium to shift to right. This increases the concentration (yield) of the product C.

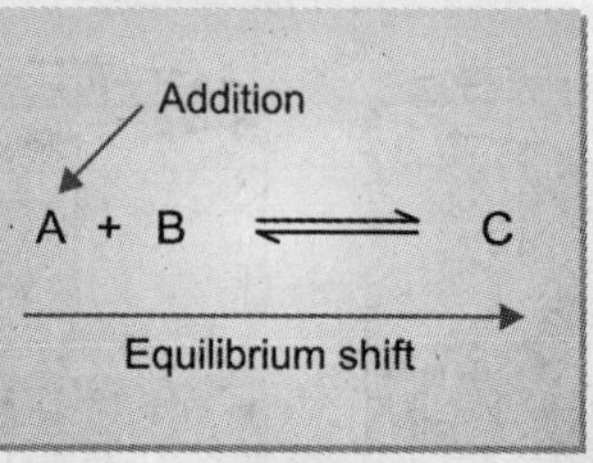

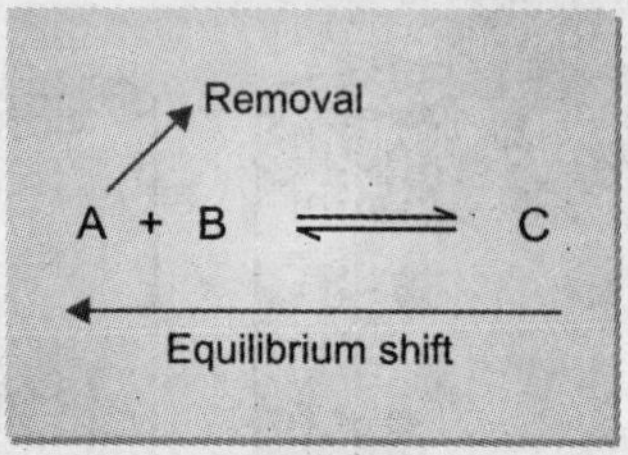

Following the same line of argument, a decrease in the concentration of A by its removal from the equilibrium mixture, will be undone by shift to the equilibrium position to the left. This reduces the concentration (yield) of the product C.

EFFECT OF A CHANGE IN PRESSURE

To predict the effect of a change of pressure, Le Chatelier's principle may be stated as : **when pressure is increased on a gaseous equilibrium reaction, the equilibrium will shift in a direction which tends to decrease the pressure.**

The pressure of a gaseous reaction at equilibrium is determined by the total number of molecules it contains. If the forward reaction proceeds by the reduction of molecules, it will be accompanied by a decrease of pressure of the system and *vice versa*.

Let us consider a reaction,

$$A + B \rightleftharpoons C$$

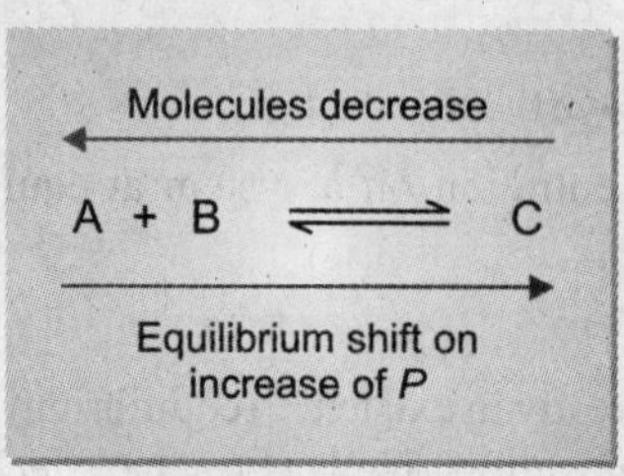

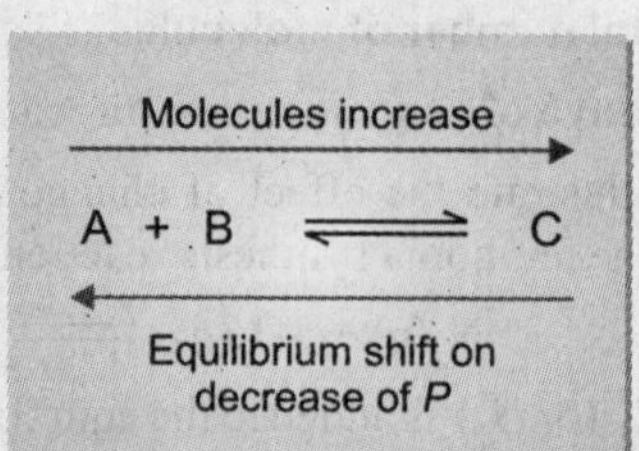

The combination of A and B produces a decrease of number of molecules while the decomposition of C into A and B results in the increase of molecules. Therefore, by the increase of pressure on the equilibrium it will shift to right and give more C. A decrease in pressure will cause the opposite effect. The equilibrium will shift to the left when C will decompose to form more of A and B.

The reactions in which the number of product molecules is equal to the number of reactant molecules,

e.g., $$H_2(g) + I_2(g) \rightleftharpoons 2HI$$

are unaffected by pressure changes. In such a case the system is unable to undo the increase or decrease of pressure.

In light of the above discussion, we can state a general rule to predict the effect of pressure changes on chemical equilibria.

The increase of pressure on a chemical equilibrium shifts it in that direction in which the number of molecules decreases and vice-versa. This rule is illustrated by the examples listed in Table 17.1.

TABLE 17.1. EFFECT OF PRESSURE ON VARIOUS GASEOUS EQUILIBRIA

System	Pressure increased	Pressure decreased
(1) $N_2O_4 \rightleftharpoons 2NO_2$	$\longleftarrow$	$\longrightarrow$
(2) $PCl_5 \rightleftharpoons PCl_3 + Cl_2$	$\longleftarrow$	$\longrightarrow$
(3) $2SO_2 + O_2 \rightleftharpoons 2SO_3$	$\longrightarrow$	$\longleftarrow$
(4) $N_2 + 3H_2 \rightleftharpoons 2NH_3$	$\longrightarrow$	$\longleftarrow$
(5) $H_2 + I_2 \rightleftharpoons 2HI$	No effect	No effect
(6) $N_2 + O_2 \rightleftharpoons 2NO$	No effect	No effect

EFFECT OF CHANGE OF TEMPERATURE

Chemical reactions consist of two opposing reactions. If the forward reaction proceeds by the evolution of heat (exothermic), the reverse reaction occurs by the absorption of heat (endothermic). Both these reactions take place at the same time and equilibrium exists between the two. If temperature of a reaction is raised, heat is added to the system. The equilibrium shifts in a direction in which heat is absorbed in an attempt to lower the temperature. Thus the effect of temperature on an equilibrium reaction can be easily predicted by the following version of the Le Chatelier's principle.

When temperature of a reaction is increased, the equilibrium shifts in a direction in which heat is absorbed.

Let us consider an exothermic reaction

$$A + B \rightleftharpoons C + \text{heat}$$

When the temperature of the system is increased, heat is supplied to it from outside. According to Le Chatelier's principle, the equilibrium will shift to the left which involves the absorption of heat. This would result in the increase of the concentration of the reactants A and B.

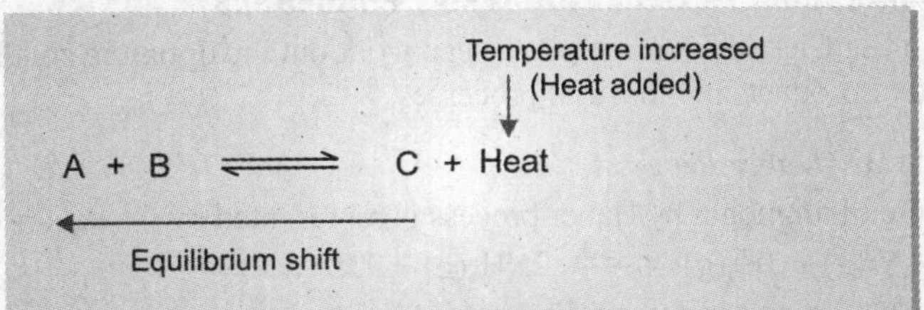

In an endothermic reaction

$$X + Y + \text{heat} \rightleftharpoons Z$$

the increase of temperature will shift the equilibrium to the right as it involves the absorption of heat. This increases the concentration of the product Z.

In general, we can say that **the increase of temperature favors the reverse change in an exothermic reaction and the forward change in an endothermic reaction.**

Temperature increased
(Heat added)
↓

$$X + Y + \text{Heat} \rightleftharpoons Z$$

⟶
Equilibrium Shift

Formation of Ammonia from N_2 and H_2

The synthesis of ammonia from nitrogen and hydrogen is an exothermic reaction.

$$N_2 + 3H_2 \rightleftharpoons 2NH_3 + 22.2 \text{ kcal}$$

When the temperature of the system is raised, the equilibrium will shift from right-to-left which absorbs heat (*Le Chatelier's principle)* This results in the lower yield of ammonia. On the other hand, by lowering the temperature of the system, the equilibrium will shift to the right which evolves heat in an attempt to raise the temperature. This would increase the yield of ammonia. But with decreasing temperature, the rate of reaction is slowed down considerably and the equilibrium is reached slowly. Thus in the commercial production of ammonia, it is not feasible to use temperature much lower than 773 K. At lower temperature, even in the presence of a catalyst, the reaction proceeds too slowly to be practical.

TABLE 17.2. THE DIRECTION OF EQUILIBRIUM SHIFT IN SOME REACTIONS ON INCREASE OF TEMPERATURE

Reaction	Equilibrium Shift
$4HCl + O_2 \rightleftharpoons 2H_2O + 2Cl_2 + 1189.05 \text{ kJ}$	⟵
$H_2 + Cl_2 \rightleftharpoons 2HCl + 185.06 \text{ kJ}$	⟵
$CH_4 + 2O_2 \rightleftharpoons CO_2 + 2H_2O + 890.95 \text{ kJ}$	⟵
$2CO_2 + 566.06 \text{ kJ} \rightleftharpoons 2CO + O_2$	⟶
$N_2O_4 + 58.62 \text{ kJ} \rightleftharpoons 2NO_2$	⟶
$H_2 + I_2 + 51.91 \text{ kJ} \rightleftharpoons 2HI$	⟶

CONDITIONS FOR MAXIMUM YIELD IN INDUSTRIAL PROCESSES

With the help of Le Chatelier's principle we can work out the optimum conditions for securing the maximum yield of products in industrial processes.

Synthesis of Ammonia (*Haber Process)*

The manufacture of ammonia by Haber process is represented by the equation

$$N_2(g) + 3H_2(g) \rightleftharpoons 2NH_3(g) + 92.11 \text{ kJ}$$

A look at the equation provides the following information:

(*a*) the reaction is exothermic

(*b*) the reaction proceeds with a decrease in the number of moles.

(1) **Low temperature.** By applying Le Chatelier's principle, low temperature will shift the equilibrium to the right. This gives greater yield of ammonia. In actual practice a temperature of about 718 K is used when the percentage of ammonia in the equilibrium mixture is 15.

(2) **High pressure.** High pressure on the reaction at equilibrium favors the shift of the equilibrium to the right. This is so because the forward reaction proceeds with a decrease in the number of moles. A pressure of about 20265000.00 Pa is applied in practice.

(3) **Catalyst.** As already stated, low temperature is necessary for higher yield of ammonia. But at relatively low temperatures, the rate of reaction is slow and the equilibrium is attained in a long time. To increase the rate of reaction and thus quicken the attainment of equilibrium, a catalyst is used. Finely divided iron containing molybdenum is employed in actual practice. Molybdenum acts as a promoter that increases the life and efficiency of the catalyst.

Pure N_2 and H_2 gases are used in the process. Any impurities in the gases would poison the catalyst and decrease its efficiency.

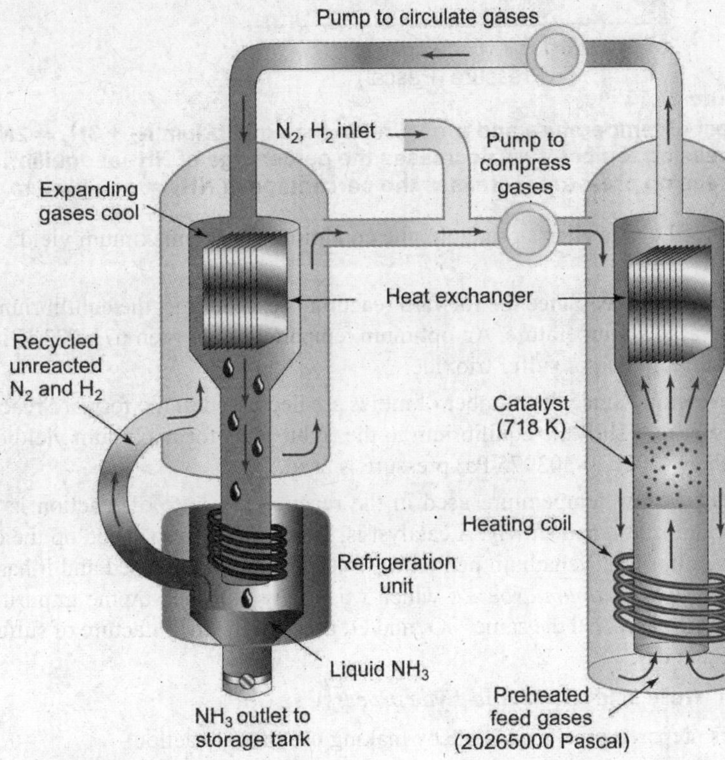

■ **Figure 17.12**
Synthesis of ammonia by Haber process.

Manufacture of Sulfuric acid (*Contact Process*)

The chief reaction used in the process is

$$2SO_2(g) + O_2(g) \rightleftharpoons 2SO_3(g) + 175.728 \text{ kJ}$$

Following information is revealed by the above equation:

(*a*) the reaction is exothermic.

(*b*) the reaction proceeds with a decrease in number of moles.

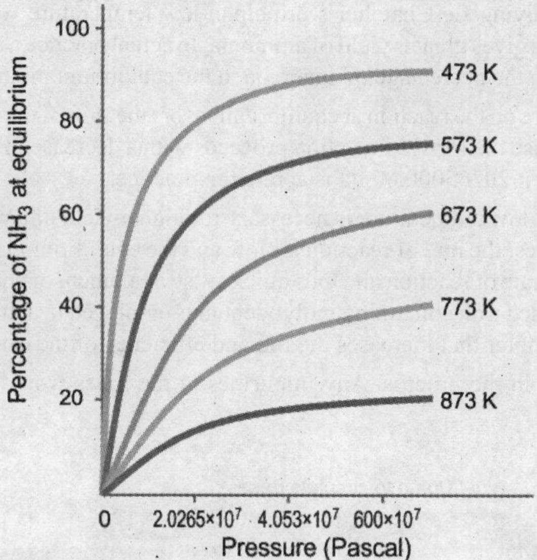

Figure 17.13

Effect of temperature and pressure on the equilibrium $N_2 + 3H_2 \rightleftharpoons 2NH_3$. Increasing temperature decreases the percentage of NH_3 at equilibrium. Increasing pressure increases the percentage of NH_3 at equilibrium.

On the basis of Le Chatelier's principle, the conditions for the maximum yield can be worked out as below:

(1) **Low temperature.** Since the forward reaction is exothermic, the equilibrium will shift on the right at low temperature. An optimum temperature between 673-723 K is required for the maximum yield of sulfur trioxide.

(2) **High pressure.** Since the number of moles are decreased in the forward reaction, increase of pressure will shift the equilibrium to the right. Thus for maximum yield of SO_3, 2 to 3 atmosphere (202650-303975 Pa) pressure is used.

(3) **Catalyst.** At low temperature used in the reaction, the rate of reaction is slow and the equilibrium is attained slowly. A catalyst is, therefore, used to speed up the establishment of the equilibrium. Vanadium pentoxide, V_2O_5, is commonly used and it has replaced the earlier catalyst *platinum asbestos* which was easily poisoned by the impurities present in the reacting gases. All the same, SO_2 and O_2 used for the manufacture of sulfuric acid must be pure and dry.

Manufacture of Nitric acid (*Birkeland-Eyde process*)

Nitric acid is prepared on a large scale by making use of the reaction

$$N_2(g) + O_2(g) \rightleftharpoons 2NO(g) - 180.87 \text{ kJ}$$

The equation tells us that:

(*a*) the reaction proceeds with no change in the number of moles.

(*b*) the reaction is endothermic and proceeds by absorption of heat.

The favorable conditions for the maximum yield of NO are:

(1) **High temperature.** Since the forward reaction is endothermic, increase of temperature will favor it (Le Chatelier's principle). Thus a high temperature of the order of 3273 K is employed to get high yield of nitric acid.

(2) **No effect of pressure.** Since the forward reaction involves no change in the number of moles, a change in pressure has no effect on the equilibrium.

(3) **High concentration.** The formation of nitric oxide is favored by using high concentrations of the reactants *i.e.* N_2 and O_2.

ADDITIONAL SOLVED PROBLEMS

SOLVED PROBLEM 36

One mole of PCl_5 is heated in closed two-liter vessel. At equilibrium 40% of PCl_5 is dissociated. Calculate the equilibrium constant of the reaction

SOLUTION

The dissociation of PCl_5 is represented by

	PCl_5 ⇌	PCl_3 +	Cl_2
Initial concentration	1.0 mol	0	0
Equilibrium concentration	0.6 mol	0.4 mol	0.4 mol

Formula used

$$K = \frac{[PCl_3]\,[Cl_2]}{[PCl_5]}$$

Quantities given

$$[PCl_5] = \frac{0.6 \text{ mole}}{2 \times 10^{-3} \text{ m}^3} = 300 \text{ mol/m}^3 \qquad [PCl_3] = \frac{0.4 \text{ mole}}{2 \times 10^{-3} \text{ m}^3} = 200 \text{ mol/m}^3$$

$$[Cl_2] = \frac{0.4 \text{ mole}}{2 \times 10^{-3} \text{ m}^3} = 200 \text{ mol/m}^3$$

Substitution of value

$$K = \frac{200 \text{ m}^3 \times 200 \text{ mol/m}^3}{300 \text{ mol/m}^3}$$

$$= \mathbf{133.33 \text{ mol/m}^3}$$

SOLVED PROBLEM 37

Calculate the equilibrium constant of the reaction from the data given below:

$$A + B \rightleftharpoons 2\,C$$

The reaction was started with 2000 mol/m^{3}of A and 2000 mol/m^3 of B and the equilibrium constant of C was found to be 320 mol/m^3.

SOLUTION : The equilibrium reaction is

	A +	B ⇌	2 C
Initial conc.	2000 mol	2000 mol	0
Eq. conc.	1240 mol	1840 mol	320 mol

Formula used

$$K \quad \frac{[C]^2}{[A]\;\;[B]}$$

Quantities given

$[C] = 320 \text{ mol/m}^3$ $[A] = 1840 \text{ mol/m}^3$ $[B] = 1840 \text{ mol/m}^3$

Substitution of values

$$K = \frac{(320 \text{ mol/m}^3)^2}{(1840 \text{ mol/m}^3) \times (1840 \text{ mol/m}^3)}$$

$$= \mathbf{30.2 \text{ mol/m}^3}$$

SOLVED PROBLEM 38

On heating in a closed vessel PCl_5 dissociates into PCl_3 & Cl_2. At 473 K the vapor density of the gaseous mixture is 75.5 Calculate the degree of dissociation of PCl_5 (P = 31; Cl = 35.5).

SOLUTION

Formula used

$$\frac{\rho_1}{\rho_2} = \frac{1+x}{1}$$

Quantities given

ρ_1, the density of undissociated gas $= \frac{\text{mol. mass of } PCl_5}{2} = \frac{31 + 5 \times 35.5}{2} = 104.25$

ρ_2, the density of dissociated gas = 75.5

Substitution of values

$$\frac{104.25}{75.5} = \frac{1+x}{1}$$

or $1.381 = 1 + x$

or $x = 1.381 - 1$

$= \mathbf{0.381}$

SOLVED PROBLEM 39

For the reaction

$$CO_{(g)} + 2H_{2(g)} \rightleftharpoons CH_3OH_{(g)}$$

hydrogen gas introduced into a 5×10^{-3} m^3 flask at 600 K containing 0.2 mol of $CO_{(g)}$ and a catalyst, until the pressure is 498519.00 Pa. At this point 0.1 mol of $CH_3OH_{(g)}$ is formed. Calculate the equilibrium constant K_p and K_c.

SOLUTION

(*a*) To calculate the total number of moles of all gas at equilibrium

Formula used

$$PV = nRT \quad \text{or} \quad n = \frac{PV}{RT}$$

Quantities given

$P = 498519.00$ Pa $\quad V = 0.005$ m^3 $\quad R = 8.314$ J/mol. K

$T = 327 + 273 = 600$K.

Substitution of values

$$n = \frac{498519 \times 0.005}{8.314 \times 600}$$

$= \mathbf{0.5 \text{ mol}}$

(*b*) To calculate the value of K_c

Formula used

$$K_c = \frac{[CH_3\,OH]}{[CO]\,[H_2]^2}$$

Quantities given

No. of moles of CH_3 OH = 0.1 $\quad \therefore [CH_3\,OH] = \frac{0.1}{5\times10^{-3}} = 20 \text{ mol/m}^3$

No. of moles of CO = 0.1 $\quad \therefore [CO] = \frac{0.1}{5\times10^{-3}} = 20 \text{ mol/m}^3$

No. of moles of H_2 = 0.5 – (0.1 + 0.1)

= 0.3 $\quad \therefore [H_2] = \frac{0.3}{5\times10^{-3}} = 60 \text{ mol/m}^3$

Substitution of values

$$K_c = \frac{20 \text{ mol/m}^3}{20 \text{ mol/m}^3 \times 60 \text{ mol/m}^3 \times 60 \text{ mol/m}^3}$$

$$= \mathbf{277.78 \times 10^{-6}\ mol^{-2}\ m^{-6}}$$

(*c*) To calculate the value of K_p

Formula used

$$K_p = K_c \times (R\,T)^{\Delta n}$$

Quantities given

K_c = 277.78 × 10^{-6} mol^{-2} m^{-6} $\qquad R$ = 8.314 J mol^{-1} K^{-1}

T = 600K $\qquad \Delta n = n_P - n_R = 1 - 3 = -2$

Substitution of values

$$K = 277.78 \times 10^{-6} \times (8.314 \text{ J mol}^{-1} \text{ K}^{-1} \times 600 \text{ K})^{-2}$$

$$= 277.78 \times 10^{-6} (4988.4 \text{ J mol}^{-1})^{-2}$$

$$= \mathbf{0.0000111629 \times 10^{-6}}$$

SOLVED PROBLEM 40

An equilibrium mixture at 300 K contains N_2O_4 and NO_2 at 0.28 at 111457.50 Pa (1.1 atm) pressure respectively. If the volume of the container is doubled, calculate the new equilibrium pressure of the two gases.

SOLUTION

(*a*) To calculate the value K_p

$$\underset{28371 \text{ Pa}}{N_2O_{4\,(g)}} \rightleftharpoons \underset{111457.50 \text{ Pa}}{2\,NO_{2\,(g)}}$$

Formula used

$$K_p = \frac{p^2_{NO_2}}{p_{N_2O_4}}$$

Quantities given

p_{NO_2} = 1.1 atm (111457.50 Pa) $\qquad p_{N_2O_4}$ = 0.28 atm (28371 Pa)

Substitution of values

$$K_p = \frac{(111457.50)^2}{28371} = 437868.74 \text{ Pa}$$

(b) To calculate the new equilibrium pressures when volume is doubled

When volume is doubled the pressure reduces to one half we have

	$N_2O_{4\,(g)}$ $\rightleftharpoons$	$2NO_{2\,(g)}$
Initial pressure	0.28/2 atm (14185 Pa)	1.1/2 atm (55728.75 Pa)
Equilibrium pressure	(14185 – x) Pa	(55728.75 + 2x) Pa

Formula used

$$K_p = \frac{p^2_{NO_2}}{p_{N_2O_4}}$$

Quantities given

$p_{NO_2} = (55728.75 + 2x)$ Pa $\qquad p_{N_2O_4} = (14185 - x)$ Pa

$K_p = 437868.74$ Pa

Substitution of values

$$437868.74 = \frac{(55728.75 + 2x)^2\ Pa^2}{(14185 - x)\ Pa}$$

or $4x^2 + 222915\,x + 3105693576.6 = 6209114940 - 437724\,x$

or $4x^2 + 660639\,x - 3103421364 = 0$

or $x^2 + 165159.75\,x - 775855341 = 0$

or $x = \dfrac{-6.52 \pm \sqrt{42.510 + 4.837}}{8}$ $\left[x = \dfrac{-b \pm \sqrt{b^2 - 4ac}}{2a}\right]$

$= \dfrac{-6.52 \pm 6.88}{8}$

$= 0.045$ or -1.675

$\therefore$ $p_{NO2} = 0.55 + 2 \times 0.045 =$ **0.64 atm** (Rejecting – 1.675)

and $p_{N2O4} = 0.14 - x =$ **0.095 atm or 9625.875 Pa**

SOLVED PROBLEM 41

The equilibrium constant for the reaction

$$CO_{(g)} + H_2O_{(g)} \rightleftharpoons CO_{2\,(g)} + H_{2\,(g)}$$

is 0.63. A mixture of 1 mol of water vapor and 3 mol of CO is allowed to come to equilibrium at a total pressure of 2 atmospheres (202650.00 Pa). Calculate

(i) the no. of moles of hydrogen present at equilibrium

(ii) the partial pressure of each gas in the equilibrium mixture.

SOLUTION

(a) To calculate the no. of moles of hydrogen at equilibrium

The reaction equilibrium is

	$CO_{(g)}$	+ $H_2O_{(g)}$ $\rightleftharpoons$	$CO_{2\,(g)}$	+ $H_{2\,(g)}$
Initial concentration	3	1	0	0
Equilibrium concentration	(3 – x)	(1 – x)	x	x

Total number of moles at equilibrium $= (3 - x) + (1 - x) + x + x$

$= (3 + 1)$

Formula used

$$K_p = \frac{[CO_2]\,[H_2]}{[CO]\,[H_2O]}$$

Quantities given

$K_p = 0.63$ $[CO] = 3 - x$ $[H_2O] = 1 - x$

Substitution of values

$$0.63 = \frac{x \times x}{3 - x \times 1 - x}$$

or $0.63 \times (x^2 - 4x + 3) = x^2$

or $0.373 x^2 + 2.52 x - 1.89 = 0$

or $x = 0.68$

Thus the no. of moles of hydrogen at equilibrium = **0.68**

(*b*) To calculate the partial pressure of each gas

Formula used

$$\text{Partial pressure} = \frac{\text{No. of moles of the gas}}{\text{Total no. of moles}} \times \text{Total pressure}$$

Quantities given

Total pressure = 2 atm (202650.00 Pa) *Total no. of moles* = 4

Substitution of values

$$p_{CO2} = \frac{x \times 202650 \text{ Pa}}{4} = \frac{0.68 \times 202650 \text{ Pa}}{4} = \mathbf{0.34 \text{ atm } (34450.50 \text{ Pa})}$$

$$p_{H2} = \frac{x \times 202650 \text{ Pa}}{4} = \frac{0.68 \times 202650 \text{ Pa}}{4} = \mathbf{0.34 \text{ atm } (34450.50 \text{ Pa})}$$

$$p_{CO} = \frac{(3 - x) \times 202650 \text{ Pa}}{4} = \frac{2.32 \times 202650 \text{ Pa}}{4} = \mathbf{1.16 \text{ atm } (117537.00 \text{ Pa})}$$

$$p_{H2O} = \frac{(1 - x) \times 202650 \text{ Pa}}{4} = \frac{0.32 \times 202650 \text{ Pa}}{4} = \mathbf{0.16 \text{ atm } (16212.00 \text{ Pa})}$$

SOLVED PROBLEM 42

Calculate the percent dissociation of $H_2S_{(g)}$ if 0.1 mol of H_2S is kept in 0.0004 m^3 vessel at 1000 K. For the reaction.

$$2H_2S_{(g)} \rightleftharpoons 2H_{2\,(g)} + S_{2\,(g)}$$

the value of K_c is 1.0×10^{-6}.

SOLUTION

The chemical equation representing the equilibrium is

	$2H_2S_{(g)}$	$\rightleftharpoons$	$2H_{2\,(g)}$	+	$S_{2\,(g)}$
Initial concentration	0.1		0		0
Equilibrium concentration	$0.1 - x$		x		$x/2$
Molar Concentrations	$\frac{0.1 - x}{V}$		$\frac{x}{V}$		$\frac{x}{2V}$

Formula used

$$K_c = \frac{[H_2]^2 [S_2]}{[H_2S]^2}$$

Quantities given

$K_c = 1.0 \times 10^{-6}$ $[H_2] = \frac{x}{V}$

$[S_2] = \frac{x}{2V}$ $[H_2S] = \frac{0.1 - x}{V}$ where $V = 0.4$ liter (0.0004 m^3)

Substitution of values

$$1.0 \times 10^{-6} = \frac{\left(\frac{x}{V}\right)^2 \left(\frac{x}{2V}\right)}{\left(\frac{0.1 - x}{V}\right)^2}$$

or $$1.0 \times 10^{-6} = \frac{x^3}{2V(0.1)^2} \qquad [\because 0.1 - x \cong 0.1]$$

or $$x^3 = 2 \times 0.4 \times 0.01 \times 1.0 \times 10^{-6} \quad [\because V = 0.0004 \text{ m}^3]$$

or $$x^3 = 8 \times 10^{-12}$$

$$x = \sqrt{8 \times 10^{-12}}$$

$$= 2 \times 10^{-4}$$

$$= 0.0002$$

$$\% \text{ Dissociation} = \frac{\text{Amt. dissociated}}{\text{Initial concentration}} \times 100$$

$$= \frac{0.002}{0.1} \times 100$$

$$= \mathbf{2\%}$$

EXAMINATION QUESTIONS

1. Define or explain the following terms :
   (*a*) Chemical equilibrium
   (*b*) Law of mass action
   (*c*) Equilibrium constant
   (*d*) Heterogeneous equilibria
2. One mole of PCl_5 is heated in a closed two-liter (0.002 m³) vessel. At equilibrium 40% of the PCl_5 is dissociated. Calculate the equilibrium constant of the reaction.
   Answer. 133.33
3. (*a*) Derive the law of mass action for the expression :
   $$n\,\text{A} + m\,\text{B} \rightleftharpoons p\,\text{C} + q\,\text{D}$$
   (*b*) The concentration equilibrium constant, K_c for the reaction $N_2 + 3H_2 \rightleftharpoons 2NH_3$ at 673 K is 0.5. Find the value of K_p. ($R = 8.3145$ J mol^{-1} K^{-1})
   Answer. (*b*) 1.5968×10^{-8}
4. For the reaction $CO_2(g) + H_2(g) \rightleftharpoons CO(g) + H_2O(g)$
   the equilibrium constant at 1000 K is 0.53.
   (*a*) If a mixture at equilibrium in a 1×10^{-3} m³ vessel contains 0.25 mol of CO 0.5 mol of CO_2 and 0.6 mol of H_2, how many moles of H_2O are there in the vessel?
   (*b*) 5 mol of inert gas are added to the equilibrium mixture containing 1 mol of H_2 and 1 mol of CO_2 in 1×10^{-3} m³ vessel. Predict equilibrium concentration of CO_2 and H_2O.
   Answer. (*a*) 0.636 mol; (*b*) 0.4 mol; 0.4 mol
5. (*a*) What is standard free energy change? Derive a relationship between standard free energy change and equilibrium constant of a reaction at a given temperature.
   (*b*) The equilibrium constant K_p for the reaction :
   $$N_2(g) + 3H_2(g) \rightleftharpoons 2NH_3(g)$$
   is 1.64×10^{-4} at 673 K and 0.144×10^{-4} at 773 K. Calculate the mean heat of formation of ammonia from its elements in this temperature range.
   Answer. (*b*) –52.6174 kJ

6. (*a*) Derive thermodynamically the law of chemical equilibrium.
(*b*) For the reaction $N_2(g) + 3H_2(g) \rightleftharpoons 2NH_3(g)$
K_p is 1.64×10^{-4} at 673 K. Calculate
(*i*) ΔG; (*ii*) ΔG when the partial pressure of N_2, H_2 and NH_3 are 1013250.00 Pa, 3039750.00 Pa and 303975.00 Pa respectively; (*iii*)Is the reaction spontaneous?
Answer. (*b*) (*i*) 48.775 kJ; (*ii*) 186.6943 kJ; (*iii*) No

7. Calculate $\Delta G°$ K_p for the following reaction at 298 K

$$CO(g) + H_2O(g) \rightarrow CO_2(g) + H_2(g)$$

Given that $\Delta G°$ for CO(g), CO_2(g) and H_2O(g) are –32.807, –97.26 and –54.64 J mol^{-1} respectively.
Answer. –9.813 J

8. (*a*) Why chemical equilibrium is called a dynamic equilibrium?
(*b*) In what direction the following equilibrium will be shifted if some chlorine gas is introduced into the system at equilibrium?

$$COCl_2(g) \rightleftharpoons CO(g) + Cl_2(g)$$

(*c*) Calculate the ratio of K_p to K_c at 298 K for the equilibrium reaction:

$$C_2H_6(g) \rightleftharpoons C_2H_4(g) + H_2(g)$$

Answer. (*c*) 2477.572

9. Derive a mathematical relation to calculate the temperature dependence of equilibrium constant of a chemical reaction.

10. Derive van't Hoff reaction isotherm for the reaction :

$$aA + bB + ... \rightleftharpoons lL + mM + ...$$

11. Derive the relation between K_c and K_p

12. Prove that $\Delta G° = -$ RT ln K_p

13. For a reaction $K_p = K_c$: What do you infer from this?

14. (*a*) Describe homogenous and heterogeneous equilibria.
(*b*) For the dissociation of water :

$$H_2O(g) \rightleftharpoons H_2(g) + ½O_2(g)$$

at 1773 K, the value of K_p is 1.87×10^{-6} atm (0.189478 Pa). Assuming ideal behavior of gases, calculate the value of K_c.
Answer. (*b*) 1.56×10^{-3}

15. What are the limitations of equation for chemical equilibrium? (*Himachal BSc*)

16. What is the vapor density of PCl_5 at 523 K, if it has dissociated to the extent of 80%?
Answer. 57.9 (*Bundelkhand BSc*)

17. Write a short note on "Le Chatelier's Principle". (*Purvanchal BSc*)

18. (*a*) Define equilibrium constant and show that it can have two different values depending on how you express concentration. Derive relationship between these two values.
(*b*) Equilibrium constant of the reaction $H_2 + I_2 \rightleftharpoons 2HI$ is 64 at a certain temperature. If 0.012 kg of hydrogen and 0.762 kg of iodine be kept in a closed vessel at this temperature to attain equilibrium what weight of HI will be present in the vessel?
Answer. 0.765 kg (*Purvanchal BSc*)

19. Differentiate between the rate constant from equilibrium constant. (*Kathmandu BSc*)

20. Deduce the law of mass action thermodynamically. (*Allahabad BSc*)

21. (*a*) Discuss the law of chemical equilibrium for the synthesis of ammonia.
(*b*) What is Le Chatelier's principle? Discuss its applications. (*Purvanchal BSc*)

22. Give the thermodynamical derivation of law of chemical equilibrium for a general reaction :
$aA + bB + ... \rightleftharpoons lL + mM + ...$ (*Mizoram BSc*)

23. (*a*) State law of Mass Action. Does equilibrium reaction stop at equilibrium?
(*b*) Establish a relationship between K_p and K_c. (*Punjabi BSc*)

24. The equilibrium constants K_p for a reaction at 700 K and 720 K are 1×10^{-12} and 5×10^{-12} respectively. Considering $\Delta H°$ to be constant in above temperature range, calculate $\Delta H°$ for the reaction.

Answer. 12.78×10^4 J *(Allahabad BSc)*

25. Derive the relationship between K_c and K_p. Calculate K_c for the following reaction at 673 K :

$$2H_2O(g) + 2Cl_2(g) \rightleftharpoons 4HCl(g) + O_2(g)$$

(Given $K_p = 0.035$; $R = 8.314$ J K^{-1} mol^{-1})

Answer. 6.25×10^{-6} *(Nagpur BSc)*

26. Given $\Delta G°$ for ionization of acetic acid in aqueous solution is 27.18 kJ mol^{-1}. Estimate the ionization constant of acetic acid at 298 K.

Answer. 1.72×10^{-5} *(Guru Nanak Dev BSc)*

27. What is the value of $K_p : K_c$ for the equilibrium

$$N_2(g) + O_2(g) \rightleftharpoons 2NO(g) \text{ at 300 K?}$$

Answer. 1 *(Vidyasagar BSc)*

28. (*a*) Derive thermodynamical expression for chemical equilibrium.

(*b*) How are K_p, K_c and K_a related? *(Arunachal BSc)*

29. Consider the reaction :

$$PCl_5(g) \rightleftharpoons PCl_3(g) + Cl_2(g)$$

How would the equilibrium be affected by

(*i*) the addition of Cl_2

(*ii*) decrease in the volume of the container? *(Jamia Millia BSc)*

30. (*a*) "Equilibrium constant is a measure of free energy change". Explain and establish a relation between them.

(*b*) Describe an experiment to establish that equilibrium constant of a reaction is constant at constant temperature. *(Kalyani BSc)*

31. (*a*) Derive van't Hoff equations

(*i*) $\dfrac{d}{dt}(\log K_p) = \dfrac{\Delta H°}{RT^2}$ (*ii*) $\dfrac{d}{dt}(\log K_c) = \dfrac{\Delta H°}{RT^2}$

(*b*) Write expression for K_p and K_c for the following equilibrium :

$$N_2O_4(g) \rightleftharpoons 2NO_2(g)$$

(Panjab BSc)

32. (*a*) Give van't Hoff's reaction isotherm for the reaction :

$$2A + 3B \rightleftharpoons 5C + 4B$$

(*b*) Give the relationship between K_p and K_c for the reaction :

$$N_2(g) + 3H_2(g) \rightleftharpoons 2NH_3(g)$$

(Nagpur BSc)

33. (*a*) The dissociation of N_2O_4 is given by the equation :

$$N_2O_4(g) \rightleftharpoons 2NO_2(g)$$

If α is the degree of dissociation of N_2O_4 and P is the total pressure of the system, show that

$$K_p = \frac{4\alpha^2 P}{1 - \alpha^2}$$

(*b*) Calculate K_c for the reaction :

$$N_2O_4(g) \rightleftharpoons 2NO_2(g)$$

for which $K_p = 0.157$ atm (15908.025 Pa) at 300 K and 1 atm (101325 Pa) pressure.

Answer. 6.38 *(Lucknow BSc)*

34. The value of K_p for the water gas reaction is

$$CO + H_2O \rightleftharpoons CO_2 + H_2$$

is 1.06×10^5 at 298 K. Calculate the standard state free energy change $\Delta G°$ of the reaction at 298 K. ($R = 8.314$J K^{-1} mol^{-1})

Answer. 28.673 kJ *(Sambalpur BSc)*

35. (*a*) Derive the following relationship.

$$\frac{d \ln K_p}{dT} = \frac{\Delta H^\circ}{RT^2}$$

(*b*) The equilibrium constant K_p for the reaction

$$N_2 + 3H_2 \rightleftharpoons 2NH_3$$

at 673 K is 1.64 × 10^{-4}. What will be the equilibrium constant at 773 K if the heat of the reaction in this temperature range is 10.18 kJ?

Answer. 2.07 × 10^{-4} (*Delhi BSc*)

36. (*a*) Deduce the equation of van't Hoff isochore for equilibrium constants K_{p1} and K_{p2}.

(*b*) The equilibrium constant for K_p for the reaction

$$H_2(g) + S(g) \rightleftharpoons H_2S(g)$$

is 2046765.00 Pa at 1218 K and 933203.25 Pa at 1338 K. Calculate the heat of reaction.

Answer. 88.67 kJ (*Arunachal BSc (H)*)

37. (*a*) Derive the relation between the energy change and equilibrium constant.

(*b*) Calculate the free energy change (ΔG°) associated with chemical reaction for which the equilibrium constant at 298 K is 0.01. (R = 8.314 J mol^{-1} K^{-1}).

Answer. (*b*) 11411.68 J (*Kalyani BSc*)

38. (*a*) Derive the relation between K_p and K_c.

(*b*) On heating in a closed vessel PCl_5 dissociates into PCl_3 and Cl_2. At 473 K the vapor density of the gaseous mixture is 75.5. Calculate the degree of dissociation of PCl_5. (P = 31, Cl = 35.5)

Answer. (*b*) 0.3807 (*Gulbarga BSc*)

39. The value of K_c for the following reaction at 1173K is 0.28.

$$CS_2(g) + 4H_2(g) \rightleftharpoons CH_4(g) + 2H_2S(g)$$

Calculate the value of K_p at this temperature.

Answer. 2.94 × 10^{-9} (*Nagpur BSc*)

40. Alcohol and acetic acid were mixed in equimolar proportions in aqueous medium at room temperature. At equilibrium 50% alcohol is converted into ester. Calculate how much ester will be formed if 2 mol of acetic acid and 1 mol of alcohol were mixed.

Answer. 0.67 mol (*Sri Venkateswara BSc*)

41. At 298 K and 101325 Pa (1 atm) pressure the partial pressure in an equilibrium mixture of N_2O_4 and NO_2 are 70927.50 (0.7 atm) and 30397.50 (0.3 atm) pascal respectively. Calculate the partial pressures of these gases when they are in equilibrium at 298 K and a total pressure of 506625.00 Pa (5 atm).

Answer. 452922.75 Pa and 53702.25 Pa (4.47 atm and 0.53 atm) (*Madras BSc*)

42. At 300 K and 101325 Pa (1 atm) pressure N_2O_4 is 20% dissociated in the NO_2. Calculate the value of K_p under these conditions.

Answer. 16887.5 Pa (*Bundelkhand BSc*)

43. Calculate the value of K_c for the following reaction

$$N_2O_4(g) \rightleftharpoons 2NO_2(g)$$

when K_p is 11449.725 Pa (0.113 atm) at 298 K.

Answer. 4.624 mol/m^3 (*Madurai BSc*)

44. (*a*) State and explain the law of mass action. Derive a relationship between K_p and K_c for the reaction:

$$2SO_2(g) + O_2(g) \rightleftharpoons 2SO_3(g)$$

(*b*) Explain how equilibrium constant changes with temperature for exothermic and endothermic reaction. (*Madras BSc*)

45. (*a*) What is the thermodynamical equilibrium constant? Derive the expression showing the effect of temperature on chemical equilibrium.

(*b*) The equilibrium constant K_p for a reaction

A + B $\rightleftharpoons$ C + D is 10^{-12} at 600 K and 10^{-7} at 700 K. Calculate the enthalpy of the reaction. (R = 8.314 J K^{-1} mol^{-1})

Answer. (*b*) 402.08998 kJ (*Panjab BSc*)

46. For the reaction

$$N_2(g) + 3H_2(g) \rightleftharpoons 2NH_3(g)$$

$K_p = 1.60 \times 10^{-4}$ at 673 K. Calculate $\Delta G°$ for the reaction.

Answer. 48.913 kJ *(Agra BSc)*

47. Derive van't Hoff equation showing the variation of equilibrium constant *K* with temperature. Also obtain its integrated form. *(Guru Nanak Dev BSc)*

48. The standard free energy change for the reaction :

$$N_2(g) + 3H_2(g) \rightleftharpoons 2NH_3(g)$$

is –33.2 kJ mol^{-1} at 298 K. Calculate the equilibrium constant. (Given $R = 8.314$ J K^{-1} mol^{-1})

Answer. 6.60×10^5 *(Kalyani BSc)*

MULTIPLE CHOICE QUESTIONS

1. A chemical system is at equilibrium

(*a*) when the rate of the forward reaction becomes zero

(*b*) when the rates of the forward reaction and the reverse reaction are equal

(*c*) when all of the reactants have been used up

(*d*) when the rates of the forward reaction and the reverse reaction are both zero

Answer. (*b*)

2. Equilibrium reactions are characterized by

(*a*) going to completion

(*b*) being non-spontaneous

(*c*) the presence of both reactants and products in a definite proportion

(*d*) (*a*) and (*b*)

Answer. (*c*)

3. A dynamic equilibrium

(*a*) is when the rate of the forward reaction is equal to the rate of the reverse reaction

(*b*) is a form of static equilibrium

(*c*) only occurs in chemical equilibrium

(*d*) involves radioactivity

Answer. (*a*)

4. Which of the following represent equilibrium constants?

(*a*) weak acid or weak base dissociation constant

(*b*) K_c for a reaction

(*c*) concentration of a strong acid in water

(*d*) (*a*) and (*b*) represent equilibrium constants

Answer. (*d*)

5. If the equilibrium constant for a reaction is large, what can be said about the reaction?

(*a*) very little product is formed

(*b*) very little reactant remains at equilibrium

(*c*) the reaction goes to completion

(*d*) large quantities of reactants will remain at equilibrium

Answer. (*b*)

6. Which of the following will change the equilibrium constant for a reaction mixture?
(*a*) changing temperature (*b*) adding an inert gas
(*c*) increasing pressure by decreasing volume (*d*) all of these
Answer. (*a*)

7. A reaction is at equilibrium. What happens to the value of the equilibrium constant if an additional quantity of reactant is added to the reaction mixture?
(*a*) the equilibrium constant is shifted to favor production of more reactant
(*b*) the equilibrium constant is shifted to favor production of more product
(*c*) the equilibrium constant is increased
(*d*) the equilibrium constant stays the same
Answer. (*d*)

8. What effect does a catalyst have on the equilibrium position of a reaction?
(*a*) a catalyst favors the formation of products
(*b*) a catalyst favors the formation of reactants
(*c*) a catalyst does not change the equilibrium position of a reaction
(*d*) a catalyst may favor reactants or product formation, depending upon the direction in which the reaction is written
Answer. (*c*)

9. Which of the following can change the value of the equilibrium constant for a reaction?
(*a*) changing the concentration of the reactants (*b*) adding a catalyst
(*c*) changing the solvent (*d*) removing the products as they are formed
Answer. (*c*)

10. Which of the following changes the value of *K*?
(*a*) adding reactant (*b*) adding product
(*c*) changing temperature (*d*) adding a catalyst
Answer. (*c*)

11. What is the expression for K_{eq} for the reaction $2N_2O(g) + O_2(g) \rightleftharpoons 4NO(g)$?
(*a*) $\frac{[N_2][O_2]}{[NO]}$ (*b*) $\frac{[NO]^4}{[N_2O]^2}$
(*c*) $\frac{[N_2O]^2\,[O_2]}{[NO]^4}$ (*d*) $\frac{[NO]^4}{[N_2O]^2\,[O_2]}$
Answer. (*d*)

12. In writing equilibrium constant expressions, which of the following quantities can be used to represent the amount of reactants and products?
(*a*) concentrations (*b*) partial pressures
(*c*) mole fractions (*d*) (*a*) and (*b*) only
Answer. (*d*)

13. Which one of the following statements is incorrect?
(*a*) adding products shifts the equilibrium to the left
(*b*) adding reactants shifts the equilibrium to the right
(*c*) exothermic reactions shift the equilibrium to the left with increasing temperature
(*d*) adding a catalyst shifts the equilibrium to the right
Answer. (*d*)

14. For which of the following reactions is the value of K_{eq} dependent on only one substance in the reaction?
(*a*) $C(s) + CO_2(g) \rightleftharpoons 2CO(g)$ (*b*) $H_2(g) + Cl_2(g) \rightleftharpoons 2HCl(g)$
(*c*) $CaCO_3(s) \rightleftharpoons CaO(s) + CO_2(g)$ (*d*) $N_2O_4(g) \rightleftharpoons 2NO_2(g)$
Answer. (*c*)

15. Which one of the following is incorrect?
(*a*) equilibrium is dynamic, as some molecules are always reacting
(*b*) the equilibrium constant is just the ratio of forward to reverse rate constants
(*c*) at equilibrium the concentrations no longer change with time
(*d*) the equilibrium constant is not affected by temperature changes
Answer. (*d*)

16. Which of the following reactions goes to completion because a gas is evolved?
(*a*) $2H_2 + O_2 \rightarrow 2H_2O$ (*b*) $N_2 + 3H_2 \rightarrow 2NH_3$
(*c*) $2CO + O_2 \rightarrow 2CO_2$ (*d*) $Zn + 2HCl \rightarrow ZnCl_2 + H_2$
Answer. (*d*)

17. A catalyst will increase the rate of a chemical reaction by
(*a*) shifting the equilibrium to the right (*b*) lowering the activation energy
(*c*) shifting the equilibrium to the left (*d*) increasing the activation energy
Answer. (*b*)

18. The yield of AB(g)

$$A(g) + B(g) \rightarrow AB(g) + \text{heat}$$

would be increased by
(*a*) decreasing the pressure
(*b*) adding additional AB to the reaction mixture
(*c*) decreasing the temperature
(*d*) adding a non-reactive liquid to the reaction mixture
Answer. (*c*)

19. In which of the following reactions is K_{eq} independent of the pressure?
(*a*) $CaCO_3(s) \rightleftharpoons CaO(s) + CO_2(g)$ (*b*) $2CO(g) + O_2(g) \rightleftharpoons 2CO_2(g)$
(*c*) $N_2(g) + 3H_2(g) \rightleftharpoons 2NH_3(g)$ (*d*) none of these
Answer. (*d*)

20. The equilibrium constant for the reaction
$CO(g) + O_2(g) \rightarrow CO_2(g)$ (not balanced) may be expressed as
(*a*) $K = K = \dfrac{[CO_2]}{[CO][O_2]}$ (*b*) $K = K = \dfrac{[CO][O_2]}{[CO_2]}$
(*c*) $K = [CO]^2[O_2][CO_2]^2$ (*d*) $K = K = \dfrac{[CO_2]^2}{[CO]^2[O_2]}$
Answer. (*d*)

21. $N_2 + 3H_2 \rightleftharpoons 2NH_3\uparrow + \text{heat}$
In this reversible reaction, the equilibrium shifts to the right because of all the following factors except:
(*a*) adding heat (*b*) adding reactant amounts
(*c*) formation of ammonia gas (*d*) increasing pressure on reactants
Answer. (*a*)

22. Which statement is true for a liquid/gas mixture in equilibrium?

(*a*) the equilibrium constant is dependent on temperature

(*b*) the amount of the gas present at equilibrium is independent of pressure

(*c*) all interchange between the liquid and gas phases has ceased

(*d*) all of the above

Answer. (*a*)

23. Consider the following reversible reaction :

$$N_2 + 3H_2 \rightleftharpoons 2NH_3$$

Its equilibrium constant "*K*" is expressed as :

(*a*) $\dfrac{[NH_3]}{[N_2][H_2]^3}$ (*b*) $\dfrac{[NH_3]^2}{[N_2]^3[H_2]}$

(*c*) $\dfrac{[NH_3]}{[N_2][H_2]}$ (*d*) $[N_2]^2[H_2]^3$

Answer. (*b*)

24. Why doesn't catalyst shift the equilibrium position?

(*a*) it speeds up both the forward and reverse reaction

(*b*) it is recoverable unchanged at the end of a reaction

(*c*) it increases the concentration of both the reactants and products equally

(*d*) it provides a surface site at which the reaction can occur

Answer. (*a*)

25. $H_2 + S \rightleftharpoons H_2S$ + energy

In this reversible reaction, select the factor that will shift the equilibrium to the right.

(*a*) adding heat (*b*) adding H_2S

(*c*) blocking hydrogen gas reaction (*d*) removing hydrogen sulfide gas

Answer. (*a*)

26. The result of adding a small crystal of sodium chloride to a saturated solution of NaCl would be

(*a*) the same crystal would precipitate (*b*) a larger amount of NaCl would precipitate

(*c*) the crystal would dissolve in solution (*d*) the same amount of NaCl would precipitate

Answer. (*d*)

27. Consider the reaction below :

$2SO_3(g) \rightleftharpoons 2SO_2(g) + O_2(g)$ $\Delta H° = + 198$ kJ

All of the following changes would shift the equilibrium to the left except one. Which one would not cause the equilibrium to shift to the left?

(*a*) removing some SO_3

(*b*) adding some SO_2

(*c*) decreasing the temperature

(*d*) adding a catalyst that speeds up the decomposition of SO_3

Answer. (*d*)

28. Consider the reaction below :

$$CO(g) + H_2O(g) \rightleftharpoons CO_2(g) + H_2(g) \quad \Delta H° = -41 \text{ kJ}$$

All of the following changes would shift the equilibrium to the right except one. Which one would not cause the equilibrium to shift to the right?

(*a*) decreasing the container volume (*b*) adding some CO

(*c*) removing some CO_2 (*d*) decreasing the temperature

Answer. (*a*)

29. To an equilibrium mixture of $2SO_2(g) + O_2(g) \rightleftharpoons 2SO_3(g)$, some helium, an inert gas, is added at constant volume. The addition of helium causes the total pressure to double. Which of the following is true?

(*a*) The concentrations of all three gases are unchanged

(*b*) $[SO_3]$ increases

(*c*) The number of moles of SO_3 increases

(*d*) $[SO_2]$ increases

Answer. (*a*)

30. For the endothermic reaction $CaCO_3(s) \rightarrow CaO(s) + CO_2(g)$, which of the following actions would favor shifting the equilibrium position to form more CO_2 gas?

(*a*) decreasing the system temperature

(*b*) both decreasing the system temperature and increasing the system pressure

(*c*) increasing the system pressure

(*d*) increasing the system temperature

Answer. (*d*)

31. Consider the following exothermic reaction :

$$N_2(g) + 3H_2(g) \rightleftharpoons 2NH_3(g)$$

Which of the following changes would not increase the amount of NH_3 produced from given quantities of N_2 and H_2?

(*a*) decrease in V
(*b*) increase in P
(*c*) increase in T
(*d*) remove some NH_3 and re-establish equilibrium

Answer. (*c*)

32. The equilibrium expression, $k = [Ag^+]\,[Cl^-]$ describes the reaction :

(*a*) $AgCl \rightarrow Ag^+ + Cl^-$
(*b*) $Ag^+ + Cl^- \rightarrow AgCl$
(*c*) $Ag^+ + Cl^- \rightarrow Ag + Cl$
(*d*) $Ag + Cl \rightarrow Ag^+ + Cl^-$

Answer. (*a*)

33. For which one of the following equilibrium equations will K_p equal K_c?

(*a*) $PCl_5 \rightleftharpoons PCl_3 + Cl_2$
(*b*) $COCl_2 \rightleftharpoons CO + Cl_2$
(*c*) $H_2 + I_2 \rightleftharpoons 2HI$
(*d*) $3H_2 + N_2 \rightleftharpoons 2NH_3$

Answer. (*c*)

34. Which of the following shifts the equilibrium of the following reaction to the right?

$$A(g) + B(g) + C(g) \rightleftharpoons A(g) + BC(g)$$

(*a*) addition of more A
(*b*) removal of B
(*c*) increasing the pressure
(*d*) decreasing the temperature

Answer. (*c*)

35. In the reaction below, how could we cause the equilibrium to shift to the left?

$CH_4(g) + 2\,H_2S(g) \rightleftharpoons CS_2(g) + 4\,H_2(g)\quad K = 2.5 \times 10^{-3}$

(*a*) increasing the pressure by adding an inert gas at constant volume

(*b*) increasing the pressure by decreasing the volume

(*c*) increasing the volume by adding an inert gas at constant pressure

(*d*) increasing the pressure and the volume by adding an inert gas

Answer. (*b*)

36. The equilibrium expression, $K_c = [CO_2]$ represents the reaction :

(*a*) $C(s) + O_2(g) \rightleftharpoons CO_2(g)$
(*b*) $CO(g) + ½O_2(g) \rightleftharpoons CO_2(g)$
(*c*) $CaCO_3(s) \rightleftharpoons CaO(s) + CO_2(g)$
(*d*) $CaO(s) + CO_2(g) \rightleftharpoons CaCO_3(s)$

Answer. (*c*)

37. Which of the following statements dealing with equilibria is incorrect?

(*a*) catalysts have no effect on the value of the equilibrium constant (*b*) the vapor pressure of liquid can be expressed as an equilibrium constant.

(*c*) at equilibrium in a reversible reaction, the rate of the forward reaction is equal to the rate of the reverse reaction

(*d*) K_p and K_c are numerically equal if a reversible reaction involves only gases

Answer. (*d*)

38. Hydrogen molecules (H_2) can be dissociated into hydrogen atoms (H). Which one of the following changes will not increase the number of atoms present at equilibrium?

(*a*) increasing the total pressure | (*b*) increasing the temperature

(*c*) adding H atoms | (*d*) increasing the volume of the container

Answer. (*a*)

39. Which of the following will increase the amount of Fe_2O_3 present at equilibrium in the reaction below?

$$Fe(s) + 3\,H_2O(g) \rightleftharpoons Fe_2O_3(s) + 3\,H_2(g)$$

(*a*) removing $H_2O(g)$ | (*b*) adding Fe(s)

(*c*) removing $H_2(g)$ | (*d*) cannot increase, since Fe_2O_3 is a solid

Answer. (*c*)

40. If a mixture where Q=K is combined, what happens?

(*a*) nothing appears to happen, but forward and reverse reactions are continuing at the same rate

(*b*) the reaction shifts toward products

(*c*) the reaction shifts toward reactants

(*d*) nothing happens

Answer. (*a*)

41. For the reaction of hydrogen cyanide with water to form hydrogen ions and cyanide ions, the equilibrium constant at 298 K is 5×10^{-10}. What does this tell you about the position of the equilibrium and the rate of reaching equilibrium?

(*a*) the equilibrium lies toward reactants and the rate is slow

(*b*) the equilibrium lies toward products and you can tell nothing about the rate

(*c*) the equilibrium lies toward products and the rate is fast

(*d*) the equilibrium lies toward reactants and you can tell nothing about the rate

Answer. (*d*)

42. Equilibrium is established in the reversible reaction

$4\,HCl(g) + O_2(g) \rightleftharpoons 2\,H_2O(g) + 2\,Cl_2(g) \quad \Delta H^\circ = -114.4$ kJ

Which one of the following changes will not increase the amount of Cl_2 present at equilibrium?

(*a*) removing $H_2O(g)$ | (*b*) increasing the total gas pressure in the system

(*c*) decreasing the volume of the container | (*d*) raising the temperature

Answer. (*d*)

43. When solid NH_4HS is placed in a closed flask at 301 K, the solid dissociates according to the equation:

$$NH_4HS(s) \rightleftharpoons NH_3(g) + H_2S(g)$$

The total pressure of the equilibrium mixture is 0.766 Pa. Determine K_{eq} at this temperature.

(*a*) 0.147 | (*b*) 0.766

(*c*) 0.587 | (*d*) 0.383

Answer. (*a*)

44. Consider the reaction C(graphite) + $CO_2(g) \rightleftharpoons 2CO(g)$, which has the equilibrium constant $K_{eq} = 3.7 \times 10^{-23}$ at 298 K. What is the concentration of CO at equilibrium if we start with 2.2×10^{-4} kg of CO_2 per liter?

(*a*) 2.1×10^{-12} M | (*b*) 3.7×10^{-23} atm

(*c*) 1.1×10^{-12} M | (*d*) 2.9×10^{-12} M

Answer. (*a*)

45. Which of the following reactions goes essentially to completion at equilibrium?

(*a*) $Br_2(g) + Cl_2(g) = 2BrCl(g)$ $K_c = 7.0$

(*b*) $C(graphite) + O_2(g) = CO_2(g)$ $K_c = 1.3 \times 10^{69}$

(*c*) $2CH_4(g) = C_2H_2(g) + 3H_2(g)$ $K_c = 0.154$

(*d*) $Br_2(g) = 2Br(g)$ $K_c = 4 \times 10^{-18}$

Answer. (*b*)

46. K_c for the equilibrium $2CO_2(g) = 2CO(g) + O_2(g)$ is 6.4×10^{-7}. What can be said about a mixture that contains $[CO_2] = 1.03 \times 10^{-1}$, $[CO] = 2.4 \times 10^{-2}$ and $[O_2] = 1.18 \times 10^{-5}$?

(*a*) the mixture is at equilibrium

(*b*) the mixture will shift to generate more products

(*c*) the mixture will shift to generate more reactants

(*d*) because these are all gases, we need to know K_p in order to predict how this mixture will change

Answer. (*a*)

47. For the equilibrium $N_2(g) + 3H_2(g) = 2NH_3(g)$, the value of K_c at 500 K is 152. What is the value of K_p at 500 K? Assume that pressures are measured in pascal.

(*a*) 8.80×10^{-6} (*b*) 12.5

(*c*) 347 (*d*) 2.79×10^{-4}

Answer. (*a*)

48. Of the following equilibria, which one will shift to the left in response to a decrease in volume?

(*a*) $CO_2(g) + H_2O(l) \rightleftharpoons H_2CO_3(aq)$ (*b*) $H_2(g) + Cl_2(g) \rightleftharpoons 2HCl(g)$

(*c*) $N_2(g) + 3H_2(g) \rightleftharpoons 2NH_3(g)$ (*d*) $2SO_3(g) \rightleftharpoons 2SO_2(g) + O_2(g)$

Answer. (*d*)

49. Dinitrogen tetroxide undergoes the decomposition $N_2O_4(g) \rightleftharpoons 2NO_2(g)$ with $K_{eq} = 5.85 \times 10^{-3}$ at 298 K. If we begin with 10132.50 Pa (0.100 atm) dinitrogen tetroxide, what is the partial pressure of nitrogen dioxide at equilibrium?

(*a*) 20265 Pa (0.200 atm) (*b*) 506.625 Pa (0.0050 atm)

(*c*) 2310.21 Pa (0.0228 atm) (*d*) 7599.375 Pa (0.0750 atm)

Answer. (*c*)

50. Calculate the equilibrium partial pressure of PCl_3 in a 0.003 m^3 vessel that was charged with 12462.98 Pa (0.123 atm) of PCl_5.

$PCl_5(g) \rightleftharpoons PCl_3(g) + Cl_2(g)$ $(K_{eq} = 0.0121)$

(*a*) 12462.98 Pa (0.123 atm) (*b*) 1114.56 Pa (0.0110 atm)

(*c*) 3343.73 Pa (0.0330 atm) (*d*) 9119.25 Pa (0.0900 atm)

Answer. *c*)

51. What is the solubility of AgCl in water if $K_{sp} = 1.6 \times 10^{-10}$?

(*a*) 1.6×10^{-10} (*b*) 3.2×10^{-10}

(*c*) 1.3×10^{-5} (*d*) 1.6×10^{-5}

Answer. (*c*)

52. For the following reaction at 500 K

$C(s) + CO_2(g) \rightleftharpoons 2CO(g)$

the equilibrium mixture contained CO_2 and CO at partial pressures of 7.6 Pa and 3.2 Pa respectively. The value of the K_p is

(*a*) 2.4 Pa (*b*) 18.1 Pa

(*c*) 0.6 Pa (*d*) 1.0 Pa

Answer. (*b*)

53. If the equilibrium concentrations found in the reaction $2A(g) + B(g) \rightleftharpoons C(g)$ are $[A] = 2.4 \times 10^{-2}$ M, $[B] = 4.6 \times 10^{-3}$ M, and $[C] = 6.2 \times 10^{-3}$ M, calculate the value of K_c.

(*a*) 2.3×10^3 (*b*) 1.8×10^{-2}
(*c*) 4.3×10^{-4} (*d*) 5.8×10^2

Answer. (*a*)

54. Given the reaction $A + B \rightleftharpoons C + D$, find the equilibrium constant for this reaction if 0.7 mol of C are formed when 1 mol of A and 1 mol of B are initially present.

(*a*) 5.44 (*b*) 6.22
(*c*) 9.67 (*d*) 1.23

Answer. (*a*)

55. The K_{sp} for PbI_2 is 8.7×10^{-9}. What is the molar solubility of PbI_2?

(*a*) 1.3×10^{-3} (*b*) 8.7×10^{-3}
(*c*) 9.3×10^{-5} (*d*) 1×10^{-9}

Answer. (*a*)

56. At a certain temperature, K_{eq} for the reaction $3C_2H_2 \rightleftharpoons C_6H_6$ is 4. If the equilibrium concentration of C_2H_2 is 500 mol/m^3, what is the concentration of C_6H_6?

(*a*) 1.2×10^9 M (*b*) 0.1×10^9 M
(*c*) 2.3×10^9 M (*d*) 0.5×10^9 M

Answer. (*d*)

57. For the reaction

$2NO(g) + Br_2(g) \rightleftharpoons 2NOBr(g)$ $K_c = 1.32 \times 10^{-2}$ at 1000 K.

Determine the value of K_p that corresponds to this value of K_c at that temperature.

(*a*) 1.08 (*b*) 0.161
(*c*) 1.61×10^{-4} (*d*) 2.39×10^{-8}

Answer. (*c*)

58. A chemist dissolves an excess of $BaSO_4$ in pure water at 298 K. If its $K_{sp} = 1 \times 10^{-10}$, what is the concentration of the barium in the water?

(*a*) 10^{-4} M (*b*) 10^{-5} M
(*c*) 10^{-6} M (*d*) 10^{-15} M

Answer. (*d*)

59. A reaction proceeds five times as fast at 333 K as it does at 303 K. Estimate its energy of activation in kcal/mol.

(*a*) 1.61 (*b*) 13.4
(*c*) 10.8 (*d*) 11.6

Answer. (*c*)

60. The K_{sp} of silver chromate (Ag_2CrO_4) is 1.4×10^{-12}. What is the solubility in moles per m^3 of silver chromate?

(*a*) 7.0×10^{-2} mol/m^3 (*b*) 3.5×10^{-10} mol/m^3
(*c*) 6.2×10^{-12} mol/m^3 (*d*) 1.0×10^{-15} mol/m^3

Answer. (*a*)

61. At 298 K, for $3O_2(g) \rightleftharpoons 2O_3(g)$ $K_c = 2.9 \times 10^{-59}$.

If at equilibrium, the concentration of O_2 is 0.040 M, what is the concentration of O_3?

(*a*) 1.9×10^{-63} M (*b*) 4.3×10^{-32} M
(*c*) 2.3×10^{31} M (*d*) 1.2×10^{21} M

Answer. (*b*)

62. Equilibrium is established in a sealed 0.00175 m^3 vessel at 523 K in the reaction

$PCl_5(g) \rightleftharpoons PCl_3(g) + Cl_2(g)$

The quantities found at equilibrium are 0.562 × 10^{-3} kg PCl_5, 1.950 × 10^{-3} kg PCl_3, and 1.007 × 10^{-3} kg Cl_2.

(*a*) $K_c = 3.49$ (*b*) $K_c = 0.0139$

(*c*) $K_c = 2.00$ (*d*) $K_c = 0.0427$

Answer. (*d*)

63. The equilibrium constant for the isomerization of butane at 298 K is $K_c = 7.94$.

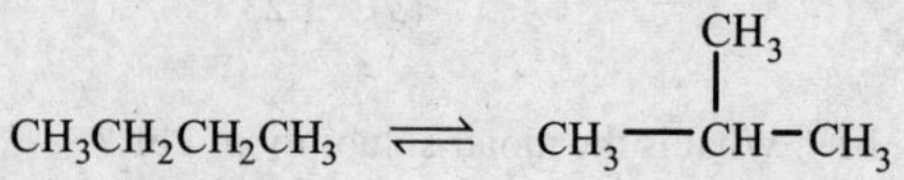

Butane Isobutane

If 5 kg butane is introduced into a 12.5 m^3 flask at 298 K, what mass of isobutane will be present when equilibrium is reached?

(*a*) 4.44 kg (*b*) 0.00613 kg

(*c*) 0.684 kg (*d*) 39.7 kg

Answer. (*a*)

64. For the synthesis of phosgene at 668 K

$CO(g) + Cl_2(g) \rightleftharpoons COCl_2(g)$ $K_c = 1.2 \times 10^3$

If 0.700 mol CO and 0.500 mol Cl_2 are placed in a 8.05 m^3 reaction vessel at 668 K and equilibrium is established, how many moles of $COCl_2$ will be present?

(*a*) 0.090 mol $COCl_2$ (*b*) 0.060 mol $COCl_2$

(*c*) 0.72 mol $COCl_2$ (*d*) 0.48 mol $COCl_2$

Answer. (*d*)

65. For the reaction $N_2(g) + O_2(g) \rightleftharpoons 2\,NO(g)$, $K_c = 4.08 \times 10^{-4}$ at 2000 K. What is the value of K_c at 2000 *K* for the reaction $NO(g) \rightleftharpoons \frac{1}{2} N_2(g) + \frac{1}{2} O_2(g)$?

(*a*) 6.01 × 10^6 (*b*) 2.45 × 10^3

(*c*) 2.02 × 10^{-2} (*d*) 49.5

Answer. (*d*)

66. For the reaction

$CO(g) + H_2O(g) \rightleftharpoons CO_2(g) + H_2(g)$ $K_c = 23.2$ at 600 K

If 0.250 mol each of CO and H_2O are introduced into a reaction vessel and equilibrium is established, how many moles of CO will be present?

(*a*) 1.20 mol CO (*b*) 0.043 mol CO

(*c*) 0.315 mol CO (*d*) 0.207 mol CO

Answer. (*b*)

67. For the reaction

$C(s) + 2H_2(g) \rightleftharpoons CH_4(g)$ $K_p = 0.0263$ at 1273 K.

Calculate the pressure of H_2 when 0.100 mol CH_4 and an excess of C(s) are brought to equilibrium at 1273 K in a 4.16 m^3 reaction vessel.

(*a*) 3.29 Pa (*b*) 0.183 Pa

(*c*) 2.28 Pa (*d*) 1.14 Pa

Answer. (*c*)

68. For the reaction below, $K_p = 110$ at 800 K.

$CO_2(g) + C(s) \rightleftharpoons 2\,CO(g)$

Suppose we introduce 10 Pa CO_2 into a 10 m^3 flask at 800 K. What is the minimum number of kilograms of C we must add to allow the system to achieve equilibrium?

(*a*) 79 (*b*) 1.8

(*c*) 18.3 (*d*) 14.2

Answer. (*d*)

69. For the reaction below, $K_p = 2.0$ at 1200 K. If we introduce 0.400 mol COF_2 into a 0.003 m^3 flask at 1200 K, what is the partial pressure of COF_2 at equilibrium?

$2COF_2(g) \rightleftharpoons 2CO(g) + 2F_2(g)$

(*a*) 6.1 Pa (*b*) 8.3

(*c*) 5.4 (*d*) 35 Pa

Answer. (*d*).

18 Distribution Law

CONTENTS

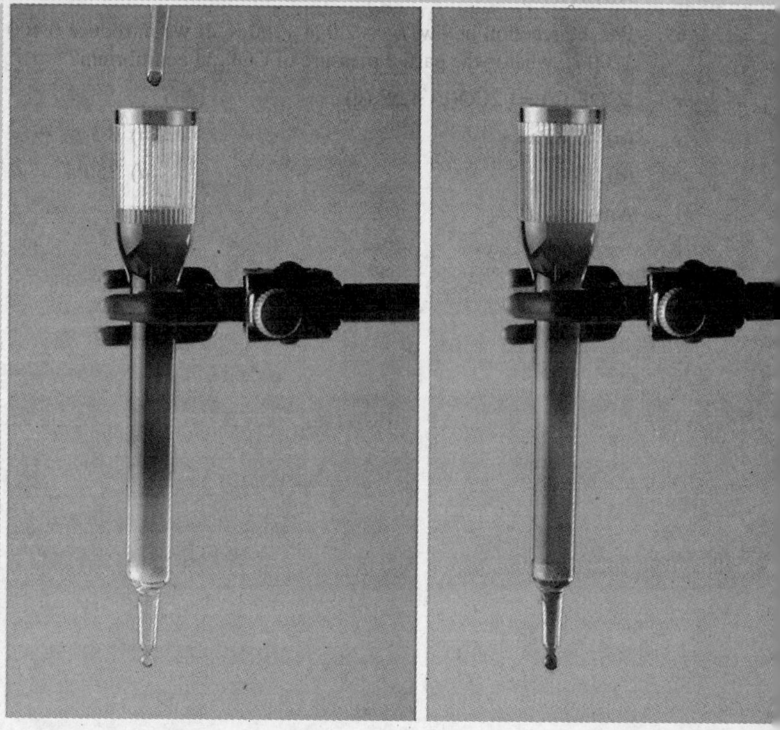

If we take two immiscible solvents *A* and *B* in a beaker, they form separate layers. When a solute *X* which is soluble in both solvents is added, it gets *distributed* or *partitioned* between them. Molecules of *X* pass from solvent *A* to *B* and from solvent *B* to *A*. Finally a dynamic equilibrium is set up. At equilibrium, the rate, at which molecules of *X* pass from one solvent to the other is balanced (18.1).

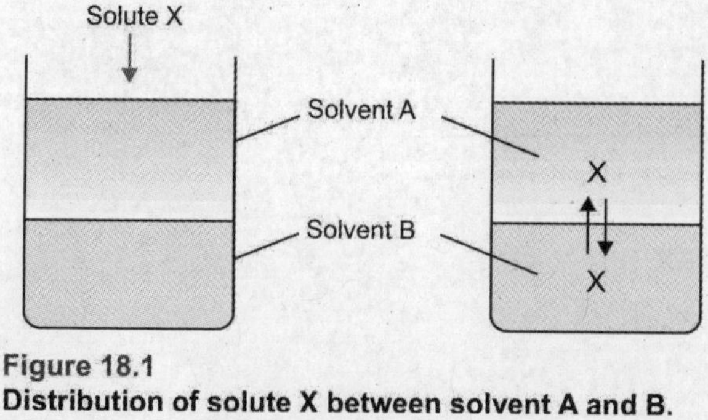

■ **Figure 18.1**
Distribution of solute X between solvent A and B.

$$\frac{\text{Concentration of } X \text{ in } A}{\text{Concentration of } X \text{ in } B} = a \text{ Constant}$$

For example, the following results from the distribution of succinic acid between ether and water illustrate the constancy of the ratio of the concentrations in each layer.

TABLE 18.1. ILLUSTRATION OF CONSTANCY OF CONCENTRATION RATIO OF A SOLUTE BETWEEN LAYERS OF TWO IMMISCIBLE SOLVENTS

Concentration of acid in ether (C_1)	Concentration of acid in water (C_2)	$\frac{C_1}{C_2}$
0.0046	0.024	0.191
0.013	0.069	0.188
0.022	0.119	0.185
0.031	0.164	0.189

STATEMENT OF NERNST'S DISTRIBUTION LAW

Nernst (1891) studied the distribution of several solutes between different appropriate pairs of solvents. He gave a generalization which governs the distribution of a solute between two non-miscible solvents. This is called **Nernst's Distribution law** (or *Nernst's Partition law*) or simply *Distribution law* or *Partition law.*

If a solute X distributes itself between two immiscible solvents A and B at constant temperature and X is in the same molecular condition in both solvents,

$$\frac{\text{Concentration of } X \text{ in } A}{\text{Concentration of } X \text{ in } B} = K_D$$

If C_1 denotes the concentration of the solute in solvent A and C_2 the concentration in solvent B, Nernst's Distribution law can be expressed as

$$\frac{C_1}{C_2} = K_D$$

The constant K_D (or simply K) is called the **Distribution coefficient** or **Partition coefficient** or **Distribution ratio.**

SOLVED PROBLEM 1

A solid X is added to a mixture of benzene and water. After shaking well and allowing to stand, 0.00001 m^3 of the benzene layer was found to contain 0.13×10^{-3} kg of X and 0.0001 m^3 of water layer contained 0.22×10^{-3} kg of X. Calculate the value of distribution coefficient.

SOLUTION

$$\text{Concentration of } X \text{ in benzene } (C_B) = \frac{0.13 \times 10^{-3} \text{ kg}}{10^{-5} \text{ m}^3} = 13 \text{ kg m}^3$$

$$\text{Concentration of } X \text{ in water } (C_W) = \frac{0.22 \times 10^{-3} \text{ kg}}{10^{-4} \text{ m}^3} = 2.2 \text{ kg m}^3$$

According to Distribution law :

$$\frac{C_B}{C_W} = \frac{13}{2.2} = \mathbf{5.9}$$

SOLVED PROBLEM 2

In the distribution of succinic acid between ether and water at 288 K, 0.00002 m³ of the ethereal layer contains 0.092×10^{-3} kg of the acid. Find out the weight of the acid present in 0.00005 m³ of the aqueous solution in equilibrium with it if the distribution coefficient for succinic acid between water and ether is 5.2.

SOLUTION

Let the weight of succinic acid in aqueous layer be x g

$$\text{Concentration in aqueous layer} = \frac{x}{0.00005}\ \text{kg/m}^3$$

$$\text{Concentration in ethereal layer} = \frac{0.092\times10^{-3}}{0.00002}\ \text{kg/m}^3$$

$$\frac{C_{\text{water}}}{C_{\text{ether}}} = \frac{x/0.00005}{0.092\times10^{-3}/0.00002}$$

or $$5.2 = \frac{x}{5\times10^{-5}} \times \frac{2\times10^{-5}}{0.0092\times10^{-3}}$$

Hence $$x = \mathbf{1.196 \times 10^{-3}\ kg}$$

SOLVED PROBLEM 3

An aqueous solution of succinic acid at 288 K, containing 0.07×10^{-3} kg in 0.00001 m³ is in equilibrium with an ethereal solution which has 0.013×10^{-3} kg in 0.00001 m³. The acid has its normal molecular weight in both the solvents. What is the concentration of the ethereal solution which is in equilibrium with an aqueous solution containing 0.024×10^{-3} kg in 0.00001 m³?

SOLUTION

$$\text{Concentration in aqueous solution } C_W = \frac{0.07\times10^{-3}}{10^{-5}}\ \text{kg/m}^3$$

$$\text{Concentration in ethereal solution } C_E = \frac{0.013\times10^{-3}}{10^{-5}}\ \text{kg/m}^3$$

$$\text{Partition coefficient } \frac{C_W}{C_E} = \frac{0.07\times10^{-3}}{10^{-5}} \times \frac{10^{-5}}{0.013\times10^{-3}} = \frac{0.07}{0.013} = \mathbf{5.38}$$

Concentration in aqueous solution in the second case

$$= \frac{0.024\times10^{-3}\text{kg}}{10^{-5}\text{m}^3}$$

Let concentration in ethereal solution in the second case be = x kg/m³

$$\frac{0.024\times10^{-3}}{10^{-5}\times x} = \frac{C_W}{C_E} = \frac{0.07}{0.013}$$

or $$0.024 \times 10^{-3} \times 0.013 = 0.07 \times 10^{-5} x$$

Hence $$x = \frac{0.024\times10^{-3}\times 0.013}{0.07\times10^{-5}} = 0.044\ \mathbf{kg/m^3}$$

SOLUBILITIES AND DISTRIBUTION LAW

When a solute is shaken with two non-miscible solvents, at equilibrium both the solvents are saturated with the solute. Since the solubility also represents concentration, we can write the distribution law as

$$\frac{C_1}{C_2} = \frac{S_1}{S_2} = K_D$$

where S_1 and S_2 are the solubilities of the solute in the two solvents.

Hence knowing the value of the Distribution coefficient (K_D) and the solubility of solute in one of the solvents, the solubility of solute in the second solvent can be calculated.

SOLVED PROBLEM 4

At 298 K an aqueous solution of iodine containing 0.0516 kg/m^3 is in equilibrium with a carbon tetrachloride (CCl_4) solution containing 4.412 kg/m^3. The solubility of iodine in water at 298 K is 0.34 kg/m^3. Find the solubility of iodine in carbon tetrachloride.

SOLUTION

(*a*) **To find the value of K_D**

Concentration of I_2 in water (C_{H_2O}) = 0.0516 kg/m^3

Concentration of I_2 in CCl_4 (C_{CCl_4}) = 4.412 kg/m^3

∴ The value of distribution coefficient is

$$\frac{C_{CCl_4}}{C_{H_2O}} = \frac{4.412}{0.0516} = 85.5$$

(*b*) **Calculation of solubility**

Applying Distribution law,

$$\frac{\text{Solubility of iodine in } CCl_4}{\text{Solubility of iodine in water}} = 85.5$$

$$\therefore \quad \frac{\text{Solubility of iodine in } CCl_4}{0.34} = 85.5$$

or $\quad$ Solubility of iodine in CCl_4 = 85.5 × 0.34

= **29.07 kg/m^3**

EXPLANATION OF DISTRIBUTION LAW

This is an equilibrium law. When the distribution of the solute *X* has reached dynamic equilibrium, the rate (R_1) at which molecules of *X* pass from solvent *A* to *B* is proportional to its concentration (C_1) in *A*. The rate (R_2) at which molecules of *X* pass from solvent *B* to *A* is proportional to its concentration (C_2) in *B*. Also, at equilibrium, the rate of migration of solute from one solvent to the other is equal. Thus we have,

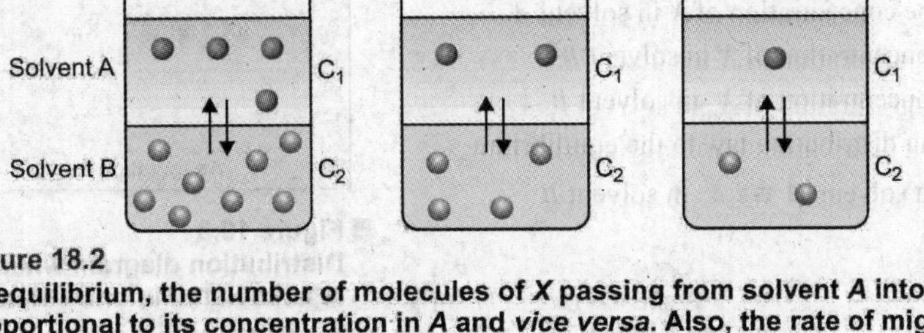

■ **Figure 18.2**
At equilibrium, the number of molecules of *X* passing from solvent *A* into *B* is proportional to its concentration in *A* and *vice versa*. Also, the rate of migration of solute molecules from *A* to *B* and *B* to *A* is equal (Illustration).

$$R_1 \propto C_1$$

or $\quad R_1 = k_1 \times C_1 \quad$ where k_1 is a constant

Again $\quad R_2 \propto C_2$

or $\quad R_2 = k_2 \times C_2 \quad$ where k_2 is a constant

Since at equilibrium $R_1 = R_2$

$$k_1 \times C_1 = k_2 \times C_2$$

or $$\frac{C_1}{C_2} = \frac{k_2}{k_1} = K_D$$

or $$\frac{C_1}{C_2} = K_D$$

This is the Nernst's Distribution law equation. Since k_1 and k_2 are constants at the same temperature, the distribution coefficient K_D is also constant if temperature is fixed.

LIMITATIONS OF DISTRIBUTION LAW

The conditions to be satisfied for the application of the Nernst's Distribution law are :

1. **Constant temperature.** The temperature is kept constant throughout the experiment.
2. **Same molecular state.** The molecular state of the solute is the same in the two solvents. The law does not hold if there is *association* or *dissociation* of the solute in one of the solvents.
3. **Equilibrium concentrations.** The concentrations of the solute are noted after the equilibrium has been established.
4. **Dilute solutions.** The concentration of the solute in the two solvents is low. The law does not hold when the concentrations are high.
5. **Non-miscibility of solvents.** The two solvents are non-miscible or only slightly soluble in each other. The extent of mutual solubility of the solvents remains unaltered by the addition of solute to them.

HOW IS DISTRIBUTION LAW MODIFIED BY CHANGE IN MOLECULAR STATE?

It was pointed out by Nernst that C_1/C_2 is constant only if the solute exists as simple molecules in the two solvents. If the solute undergoes *association* or *dissociation* in one of the solvents, it is found that C_1/C_2 is not constant. In these cases, distribution law applies only to that part of the solute which is present as simple molecules.

(1) When Solute undergoes Association

Suppose the solute is present as simple molecules X in solvent A. In solvent B, n molecules of X associate to form X_n molecules. Assuming that a few single molecules X are also present in solvent B the equilibria that exist in the two solvents are shown in Fig. 18.3.

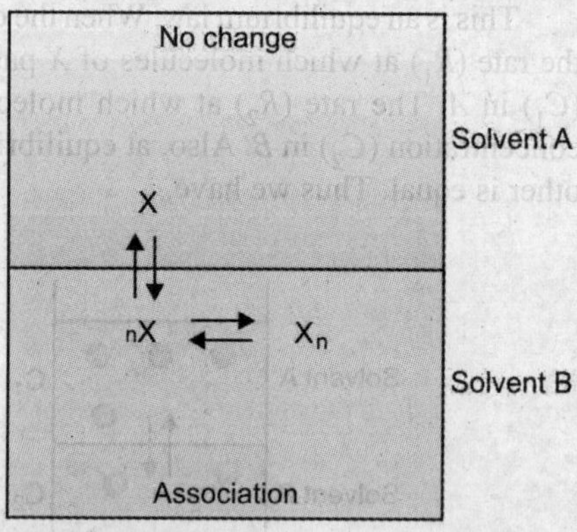

■ Figure 18.3
Distribution diagram when the solute is associated in one solvent.

Let C_1 be concentration of X in solvent A.

C_3 be concentration of X in solvent B

C_2 be concentration of X_n in solvent B.

Applying distribution law to the equilibrium

$$X \text{ in solvent } A \rightleftharpoons X \text{ in solvent } B$$

we have

$$\frac{C_1}{C_3} = K_D \quad ...(i)$$

Applying Mass Law to the chemical equilibrium

$$nX \rightleftharpoons X_n$$

$$\frac{[X]^n}{[X_n]} = K_C \quad \text{or} \quad \frac{C_3^n}{C_2} = K_C \quad ...(ii)$$

Taking nth root on both sides in equation (*ii*)

$$\frac{C_3}{\sqrt[n]{C_2}} = \sqrt[n]{K_C} \qquad ...(iii)$$

Dividing equation (*i*) by (*iii*)

$$\frac{C_1}{\sqrt[n]{C_2}} = \frac{K_D}{\sqrt[n]{K_C}} = K \text{ (a constant)}$$

Thus when association occurs in one solvent, the distribution equation is modified as

$$\frac{C_1}{\sqrt[n]{C_2}} = K_D$$

Since the solute exists largely as associated molecules, the total concentration of X determined experimentally in solvent B is taken as the concentration of the associated molecules X_n.

SOLVED PROBLEM 5

When benzoic acid was shaken with mixtures of benzene and water at constant temperature, the following results were obtained :

Concentration of acid in benzene (C_1)	0.24	0.55	0.93
Concentration of acid in water (C_2)	0.015	0.022	0.029

Comment on the results.

SOLUTION. Calculating the ratio C_1/C_2 for each case;

$$\frac{0.24}{0.015} = 16 \qquad \frac{0.55}{0.022} = 25 \qquad \frac{0.93}{0.029} = 32$$

Thus the distribution coefficient is not constant. Therefore, benzoic acid does not exist as single molecules in both solvents.

On calculating $\sqrt{C_1/C_2}$, we have

$$\sqrt{\frac{0.24}{0.015}} = 4 \qquad \sqrt{\frac{0.55}{0.022}} = 5 \qquad \sqrt{\frac{0.93}{0.029}} = \mathbf{5.66}$$

The constant value of partition coefficient now arrived at, suggests that benzoic acid is associated into double molecules in the benzene layer.

(2) **When Solute undergoes Dissociation**

Suppose the solute is present as normal molecules X in solvent A and it dissociates into $A + B$ in solvent B. The equilibria set up in the two solvents are shown in Fig. 18.4.

Let C_1 be the concentration of X in solvent A and C_2 the total concentration of X (dissociated and undissociated) in solvent B.

If the degree of dissociation in solvent B is x,

$$\underset{1-x}{X} \rightleftharpoons \underset{x}{A} + \underset{x}{B}$$

Hence the concentration of the *undissociated* (or normal) molecules in solvent B is $C_2(1 - x)$.

Applying distribution law to normal molecules

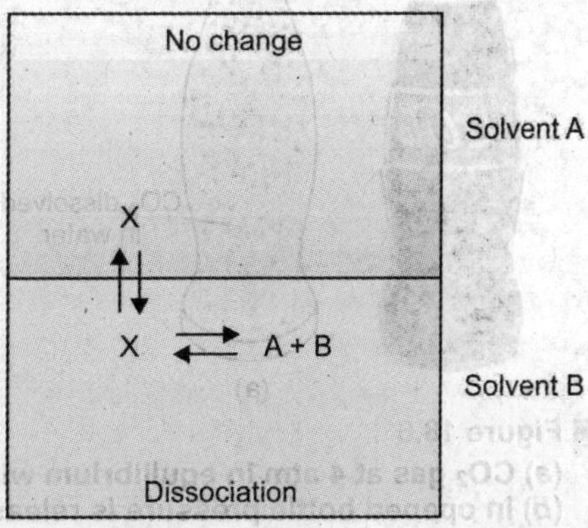

■ Figure 18.4
Distribution diagram when dissociation occurs in one solvent.

in the two solvents

$$\frac{C_1}{C_2(1-x)} = K_D$$

This is the modified distribution law equation when there is dissociation in one of the solvents.

A case of this type arises in the distribution of a weak acid (*e.g.,* succinic acid or oxalic acid) between ether and water. C_1 and C_2 can be determined by direct titration of the two layers against standard alkali solution. The value of x can be found by measuring electrical conductance of solution in solvent B.

HENRY'S LAW–A FORM OF DISTRIBUTION LAW

Henry's law states that **at a constant temperature the solubility of a gas in a liquid is proportional to the pressure of the gas above it.** Henry's law may be mathematically expressed as

$C = kP$

where C is the solubility (or concentration), k is a constant, and P is the pressure of the gas, k is called **Henry's constant.**

Explanation

Henry's law is, in fact, a form of Distribution law. If a vessel containing a liquid and a gas is shaken, at equilibrium the gas can be regarded as distributed between the liquid (Phase A) and the space above (Phase B).

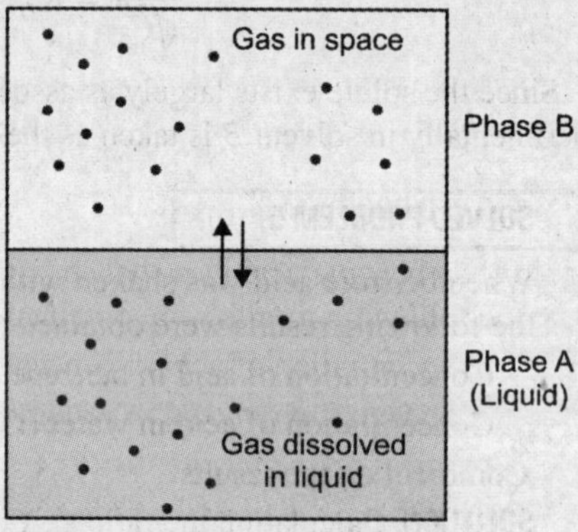

Figure 18.5
Distribution diagram illustrating Henry's law.

The influence of partial pressure on solubility is utilized in making carbonated beverages like beer, champagne, and many soft drinks. So called '*soda water*' is bottled under a carbon dioxide pressure of about 405300.00 Pa (4 atm). When the bottle is opened to the air, the partial pressure of CO_2 above the solution is decreased (about 101.325 Pa), and CO_2 bubbles out.

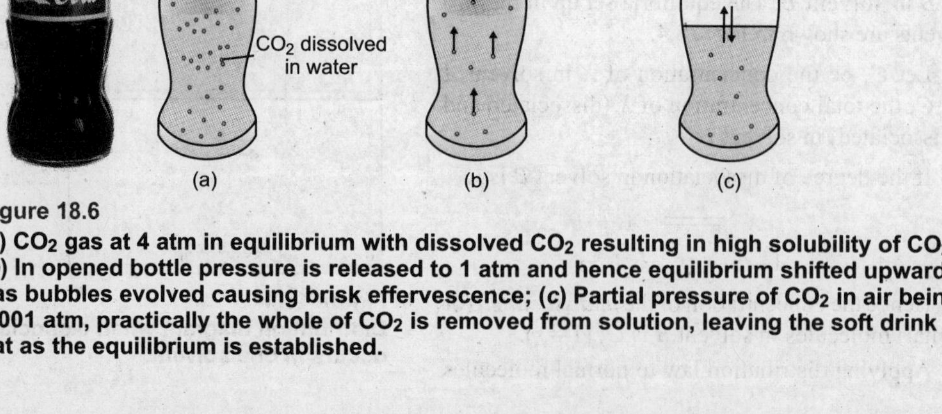

Figure 18.6
(*a*) CO_2 gas at 4 atm in equilibrium with dissolved CO_2 resulting in high solubility of CO_2; (*b*) In opened bottle pressure is released to 1 atm and hence equilibrium shifted upward, gas bubbles evolved causing brisk effervescence; (*c*) Partial pressure of CO_2 in air being 0.001 atm, practically the whole of CO_2 is removed from solution, leaving the soft drink flat as the equilibrium is established.

Let C_1 be the concentration of the gas in phase B

C be the concentration of the gas in phase A

Applying the Distribution law

$$\frac{C}{C_1} = K_D \quad \text{(a constant)}$$

We know that molar concentration of gas is proportional to its pressure, P.

Hence $$\frac{C}{P} = k \quad \text{(a constant)}$$

or $$C = k \times P$$

This is **Henry's Law equation.**

Like distribution law, Henry's law holds good for dilute solutions of gases which do not react with the solvent.

If a mixture of gases is in contact with a liquid, the partial pressure of an individual gas, not their total pressure, determines the mass of each gas dissolving. In such a case, **the solubility of each gas is proportional to its partial pressure.**

SOLVED PROBLEM 6

The solubility of pure oxygen in water at 293 K and 101325 Pa pressure is 1.38 mol/m^3. Calculate the concentration of O_2 (mol/liter) at 293 K and a partial pressure of 21278.25 Pa.

SOLUTION

Calculation of Henry's Constant

$$k = \frac{\text{conc } O_2}{\text{pressure } O_2}$$

$$= \frac{1.38 \text{ mol/m}^3}{101325 \text{ Pa}}$$

$$= 1.38 \frac{\text{mol/m}^3}{\text{Pascal}}$$

Calculation of concentration of O_2

$$C = k \times P \quad \text{(Henry's law)}$$

$$\text{Conc. of } O_2 = \frac{1.38 \text{ mol/m}^3}{101325 \text{ Pa}} \times 21278.25 \text{ Pa}$$

$$= \mathbf{0.2898 \text{ mol/m}^3}$$

SOLVED PROBLEM 7

A soda-water bottle was opened and the soft drink allowed to come to equilibrium at 298 K with air containing carbon dioxide at a pressure of 40.53 Pa. Find the concentration of CO_2 in the soda after it had stood open and come to equilibrium. The Henry's constant for aqueous solution of CO_2 at 298 K is

$$k = 3.2 \frac{\text{mole/litre}}{\text{pascal}}$$

SOLUTION

$$C = kP \qquad \text{(Henry's law)}$$

Substituting into Henry's law equation, we get for the opened soft drink at equilibrium with atmospheric CO_2,

$$C = \left(\frac{3.2 \text{ mol/m}^3}{\text{pascal}}\right)(40.53 \text{ Pa})$$

$$= \mathbf{129.696 \text{ mol/m}^3}$$

DETERMINATION OF EQUILIBRIUM CONSTANT FROM DISTRIBUTION COEFFICIENT

Distribution law helps in determining the equilibrium constant of a reaction when one of the reactants is soluble in two non-miscible solvents. Let us illustrate by taking example of the reaction when potassium iodide reacts with iodine to form potassium triiodide.

$$KI + I_2 \rightleftharpoons KI_3$$

This reaction can be carried in water, while iodine is soluble in both water and benzene.

Procedure

(1) **To find the distribution coefficient of I_2.** Iodine is shaken with water and benzene in a bottle. The concentration of iodine in the two layers is then determined by titration against standard thiosulfate solution. Thus,

$$\frac{\text{Concentration of } I_2 \text{ in water}}{\text{Concentration of } I_2 \text{ in benzene}} = K \text{ (Distribution coefficient)}$$

(2) **To find the equilibrium constant using the value of K.** A solution of KI of concentration a is shaken with iodine in a bottle. To this is added some benzene and shaken. On standing, the mixture separates into two layers. The equilibria that are set up are shown in Fig. 18.7.

The concentration of iodine is determined in the two layers by titration against standard thiosulfate solution.

Let b be the concentration of I_2 in benzene layer, c be the concentration of I_2 in water layer which is really the total of the concentration of free I_2 and KI_3.

K is the value of distribution coefficient of iodine between water and iodine as determined earlier. Thus,

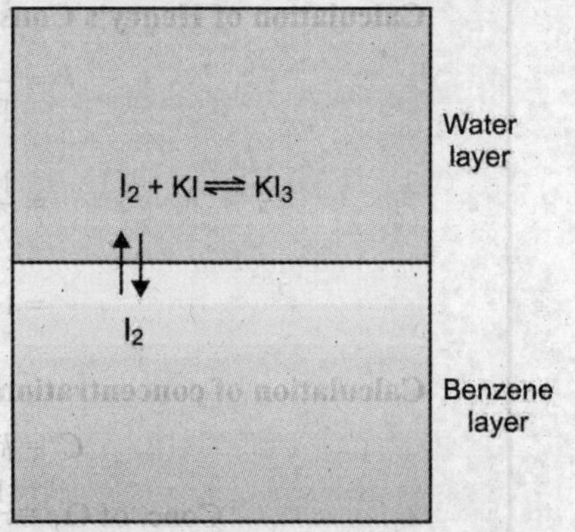

Figure 18.7
Determination of equilibrium constant.

$$\frac{\text{Conc of } I_2 \text{ in water}}{\text{Conc of } I_2 \text{ in benzene}} = K$$

$$\therefore \quad \text{Conc of } I_2 \text{ (free) in water layer} = K \times b$$

$$\text{Hence conc of } KI_3 \text{ in water layer} = c - Kb$$

$$\therefore \quad \text{Conc of KI in water layer} = a - (c - Kb)$$

$$= a - c + Kb$$

We can now say that the equilibrium constant of the reaction

$$I_2 + KI \rightleftharpoons KI_3$$

$$K_c = \frac{[KI_3]}{[KI][I_2]} = \frac{c - Kb}{(a - c + Kb)\,Kb}$$

The square brackets denote concentrations in water layer.

Since all the quantities on the right-hand side are known, the equilibrium constant (K_c) can be calculated.

STUDY OF COMPLEX IONS

If the value of K_c comes out to be constant by taking different concentrations of KI and I_2 to start with, it proves the existence of KI_3 or the complex ion I_3^-.

$$I_2 + I^- \rightleftharpoons I_3^-$$

The distribution law is, therefore, helpful in the study of complex ions.

Other examples of complex formation are :

$$CuSO_4 + 4NH_3 \rightleftharpoons CuSO_4.4NH_3 \text{ or } [Cu(NH_3)_4]^{2+} SO_4^{2-}$$

$$\underset{\text{Aniline}}{C_6H_5NH_2} + HCl \rightleftharpoons \underset{\text{Anilinium chloride}}{C_6H_5NH_3^+Cl^-}$$

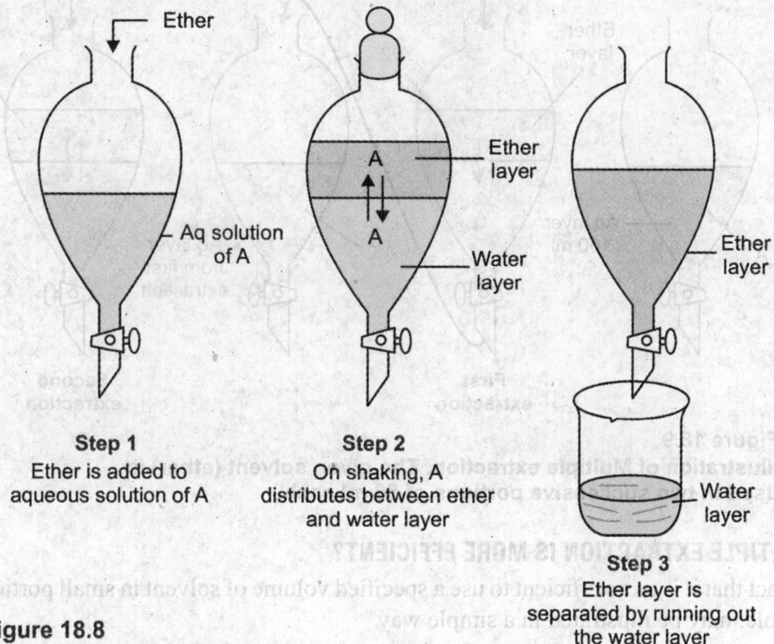

■ **Figure 18.8**
Extraction in a separatory funnel.

EXTRACTION WITH A SOLVENT

The extraction (removal by a solvent) of an organic substance from an aqueous solution is probably the most important application of Distribution law. The process is carried by shaking the aqueous solution with a immiscible organic solvent, say ether (ethoxyethane), in a separatory funnel. The distribution ratio being in favor of ether, most of the organic substance passes into ether layer. On standing, the aqueous and ether layers separate in the funnel. The lower aqueous layer is run out, leaving the ether layer behind. This is then transferred to a distillation flask. Ether is distilled over while the organic substance is left as residue in the flask.

If desired, the process may be repeated with aqueous layer left after the first extraction with a fresh quantity of the solvent.

The other common solvents used for extraction are hexane, benzene, chloroform, acetone, carbon disulfide, etc.

The greater the distribution ratio is in favor of the organic solvent, the greater will be the amount extracted in any one operation.

MULTIPLE EXTRACTION

The process of extraction when carried with the total amount of the given solvent in a single operation, is referred to as **simple extraction. To recover the maximum amount of the substance from aqueous solution, the extraction is made in two or more successive operations using small portions of the solvent provided.** This is called **multiple extraction** or **multi-step extraction.**

In such a process the aqueous solution is first extracted with a portion of the solvent in a separatory funnel. The aqueous layer from which some substance has been removed is then transferred to another funnel. This is shaken with a second portion of the solvent. Similarly, the aqueous layer from the second extraction is treated with a third portion of solvent, and so on.

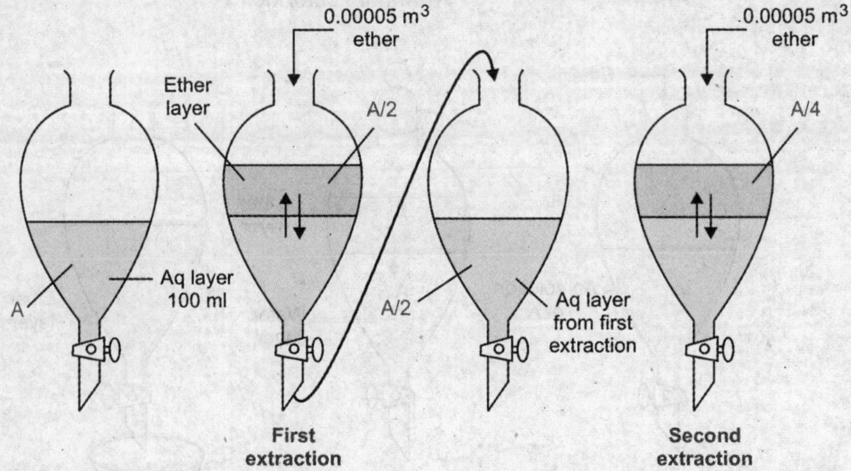

Figure 18.9
Illustration of Multiple extraction. The given solvent (ether) is used in two successive portions of 50 ml each.

WHY MULTIPLE EXTRACTION IS MORE EFFICIENT?

The fact that it is more efficient to use a specified volume of solvent in small portions rather than in one whole, may be illustrated in a simple way.

Suppose we have 0.0001 m^3 of an aqueous solution containing A kilograms of an organic substance. We can extracts the substance with ether (ethoxyethane), its distribution ratio being twice in favor of ether. We are provided with 0.0001 m^3 of ether which may be used in one lot or in two portions of 50×10^{-3} m^3 each.

(1) **Using all the ether in one lot.** Let x kilograms be the weight of the substance extracted in the solvent layer. Then the amount of substance left in the water layer = $A - x$ kilograms. Therefore,

Concentration in ether layer = $x/100$

Concentration in water layer = $(A - x)/100$

$$\frac{C_{\text{ether}}}{C_{\text{water}}} = K \qquad \text{(Distribution law)}$$

$$\therefore \qquad \frac{x/100}{(A-x)/100} = 2$$

Hence $\qquad x = 2/3\,A$

Thus 66% of substance is extracted.

(2) **Using two 50 × 10^{-3} m^3 portions of ether.** Let x_1 kilograms of substance be extracted in the first operation with 50 × 10^{-3} m^3 ether. Thus,

Concentration in ether layer = $x_1/50 \times 10^{-3}$

Concentration in water layer = $(A - x_1)/100 \times 10^{-3}$

$$\frac{C_{\text{ether}}}{C_{\text{water}}} = 2$$

or
$$\frac{x_1/50\times10^{-3}}{(A-x_1)/100\times10^{-3}} = 2$$

Hence
$$\frac{x_1}{50\times10^{-3}} \times \frac{100\times10^{-3}}{A-x_1} = 2$$

$$x = 1/2\,A$$

$\therefore$ 50% of substance is extracted.

The substance left in water layer is $\frac{1}{2}A$. Let x_2 kilograms be the substance removed from the water layer when it is extracted with another 50 × 10^{-3} m^3 portion of ether. Hence,

Concentration in ether layer = $x_2/50 \times 10^{-3}$

Concentration in water layer = $1/2\,A - x_2$

$$\frac{x_2/50\times10^{-3}}{(1/2A-x_2)/100\times10^{-3}} = 2$$

Hence $\qquad x_2 = 1/4\,A$

$\therefore$ 25% of substance is extracted.

Thus 75 per cent of substance is extracted by using two 50 × 10^{-3} m^3 portions of the solvent as against 66 per cent when 100 × 10^{-3} m^3 solvent is used in one lot.

Similarly, it can be shown that by using four 250 × 10^{-3} m^3 portions of ether it is possible to extract 80.2% of substance. Five 200 × 10^{-3} m^3 portions of ether would likewise remove 83.8%.

It is clear from above that it is more advantageous to use a given quantity of the solvent in as many portions as conveniently possible. It must be understood, however, that **it is not possible to remove the whole of the dissolved substance, however large the number of extractions may be.** Every time, the substance distributes itself between the two solvents and a quantity, howsoever small, must always be left behind.

SOLVED PROBLEM 8

The distribution coefficient of *X* for benzene and water is 10. Find the amount of *X* extracted if 1 × 10^{-3} kg of it dissolved in 0.00001 m^3 of water is equilibrated in a separatory funnel with 0.00001 m^3 of benzene.

SOLUTION

Let amount of *X* extracted with 0.00001 m^3 of benzene = x kg

The amount of *X* left in 0.00001 m^3 water = $1 - x$ kg

$$\frac{C_{\text{benzene}}}{C_{\text{water}}} = 10$$

$$\therefore \qquad \frac{x/10^{-4}}{(1-x)/10^{-4}} = 10$$

or

$$x = 10(1 - x)$$
$$11x = 10$$
$$x = \frac{10}{11} = 0.909$$

Hence the amount of X extracted with 10^{-4} m³ benzene is **0.909 × 10⁻³ kg**.

SOLVED PROBLEM 9

An aqueous solution contains 0.010 kg of solute per m³. When 0.001 m³ of the solution is treated with 0.00001 m³ of ether, 0.006 kg of the solute are extracted. How much more of the solute would be extracted from the aqueous solution by a further 0.0001 m³ ether? Assume that the molecular state of the solute is the same in ether and water.

SOLUTION

(*a*) **To find distribution coefficient**

In the first extraction :

$$\text{Concentration in ether } (C_{\text{ether}}) = \frac{6\times10^{-3}}{100\times10^{-3}} = \frac{6}{100}$$

$$\text{Concentration in water } (C_{\text{water}}) = \frac{10\times10^{-3} - 6\times10^{-3}}{1000\times10^{-3}} = \frac{4}{1000}$$

Applying distribution law, $$\frac{C_{\text{ether}}}{C_{\text{water}}} = \frac{6}{100} \times \frac{1000}{4} = 15$$

(*b*) **Substance recovered in second extraction**

Let x be the substance extracted with a further 100 ml of ether.

$$\text{The concentration in ether } (C_{\text{ether}}) = \frac{x}{10^{-4}}$$

$$\text{The concentration in water } (C_{\text{water}}) = \frac{4 - x}{10^{-3}}$$

Applying distribution law,

$$\frac{C_{\text{ether}}}{C_{\text{water}}} = \frac{x}{10^{-4}} \times \frac{10^{-3}}{4 - x} = 15$$

$$10x = 15(4 - x)$$
$$= 60 - 15x$$
$$25x = 60$$
$$\therefore \quad x = \mathbf{2.4}$$

Hence a further amount of **0.0024 kg** will be extracted.

SOLVED PROBLEM 10

The distribution coefficient of isobutyric acid between ether and water is 3 at 298 K. What will be the amount of isobutyric acid removed if 0.004 kg of isobutyric acid in 10^{-4} m³ of water is extracted with 10^{-4} m³ of ethoxyethane (ether) at 298 K. What would the effect have been if two successive 5×10^{-5} m³ portions of ether had been used to extract the aqueous layer?

SOLUTION

(*a*) **Extraction with 10^{-4} m³ ether**

$$\frac{C_{\text{ether}}}{C_{\text{water}}} = 3 \text{ (at 298 K)}$$

Let x g be the amount of the acid extracted with 10^{-4} m^3 of ether.

$$\therefore \qquad \frac{x/10^{-4}}{(0.004 - x)/10^{-4}} = 3$$

Hence $\qquad x = 0.003$ kg

Thus we can separate 3 kg (75%) of isobutyric acid from the aqueous layer with 10^{-4} m^3 ether.

(*b*) **Two extractions with 5 × 10^{-5} m^3 ether each :**

Let x_1 be the acid removed in the first extraction,

$$\frac{x_1/5\times10^{-5}}{(0.004 - x_1)/10^{-4}}$$

Whence $\qquad x_1 = 0.0024$ kg

The acid left in water layer is (4 – 2.4) = 1.6 kg

The 1.6 kg of the acid is then extracted with a second 5 × 10^{-5} m^3 portion of ether. If x_2 is the amount removed by ether

$$\frac{x_2/5\times10^{-5}}{(0.0016 - x_2)/10^{-4}} = 3$$

Hence $\qquad x_2 = 0.00096$ kg

That is, two successive extractions with 0.00005 m^3 portion of ether each would remove (0.0024 kg + 0.00096 kg) = **0.00336 kg (84%)** of isobutyric acid from aqueous layer.

SOLVED PROBLEM 11

A solution of 6 × 10^{-3} kg of substance *X* in 5 × 10^{-5} m^3 of aqueous solution is in equilibrium, at room temperature, with a solution of *X* in diethyl ether (ethoxyethane) containing 108 × 10^{-3} kg of *X* in 10^{-4} m^3. Calculate the mass of *X* extracted by shaking 10^{-4} m^3 of an aqueous solution containing 10 × 10^{-3} kg of *X* with (*a*) 10^{-4} m^3 of ether; (*b*) 5 × 10^{-5} m^3 of ether twice at room temperature.

SOLUTION

(*a*) **Calculation of Distribution coefficient**

$$\frac{C_{\text{ether}}}{C_{\text{water}}} = \frac{108\times10^{-3}/10^{-4}}{6\times10^{-3}/5\times10^{-5}}$$

$$= \frac{108\times10^{-3}}{10^{-4}} \times \frac{5\times10^{-5}}{6\times10^{-3}} = \frac{9}{1}$$

(*b*) **Mass extracted with 10^{-4} m^3 of ether**

Let x kg of *X* be extracted from aqueous solution. Then

$$\frac{x/10^{-4}}{(10\times10^{-3} - x)/10^{-4}} = 9$$

$$\frac{x}{10^{-4}} \times \frac{10^{-4}}{10^{-2} - x} = 0$$

$$x = 0.09 - 9x$$

$$9x + x = 0.09$$

$$x = 0.009$$

i.e., 9 × 10^{-3} kg of *X* pass into ether.

(*c*) **Two extractions with 5 × 10^{-5} m^3 each time**

Let x_1 g of substance be extracted with 5 × 10^{-5} m^3 ether.

Then,

$$\frac{x_1/5\times10^{-5}}{(10\times10^{-3}-x_1)/10^{-4}}=\frac{9}{1}$$

$$\frac{x_1}{5\times10^{-5}}\times\frac{10^{-4}}{10^{-2}-x_1}=\frac{9}{1}$$

$$x_1 = 4.5\,(0.01-x_1)$$

$$x_1 = 0.045-4.5\,x_1$$

$$5.5\,x_1 = 0.045$$

$$x_1 = 0.0082 \text{ kg (approx.)}$$

The amount of X left in aqueous solution after extraction with 5×10^{-5} m³ ether is

$$10\times10^{-3}-0.0082 = 0.0018 \text{ kg}$$

When shaken with fresh 5×10^{-5} m³ ether, let the amount extracted be x_2. Thus,

$$\frac{x_2/5\times10^{-5}}{(0.0018-x_2)/10^{-4}}=9$$

$$\frac{x_2}{5\times10^{-5}}\times\frac{10^{-4}}{0.0018-x_2}=9$$

$$x_2 = 4.5(0.0018-x_2)$$

$$5.5\,x_2 = 0.0018$$

$$x_2 = 0.0018/5.5$$

$$x_2 = 0.00147 \text{ kg}$$

Hence the total X extracted by two 5×10^{-5} m³ portions is

$$(0.0082+0.00147) = \mathbf{0.00967\ kg}$$

General Formula for Substance Left Unextracted

Suppose V ml of an aqueous solution contain A grams of an organic substance. This solution is extracted with v ml of the given organic solvent at a time

(1) **First extraction.** Let x_1 be the substance left unextracted in aqueous solution in the first operation. Now,

$$\text{Concentration in aqueous layer} = \frac{x_1}{V}$$

$$\text{Concentration in solvent} = \frac{A-x_1}{v}$$

$$\frac{C_{\text{water}}}{C_{\text{solvent}}} = K \qquad \text{(Distribution law)}$$

Thus,

$$\frac{x_1/V}{(A-x_1)/v} = K$$

Hence

$$x_1 = A\,\frac{KV}{KV+v} \qquad ...(1)$$

(2) **Second extraction.** Let x_2 be the substance left unextracted in the aqueous layer in the second extraction made with fresh v ml solvent. Now,

$$\text{Concentration in aqueous layer} = \frac{x_2}{V}$$

$$\text{Concentration in solvent} = \frac{x_1-x_2}{v}$$

Substituting the values of concentration in the Distribution law equation

$$\frac{x_2/V}{(x_1-x_2)/v} = K$$

Hence
$$x_2 = x_1 \frac{KV}{KV + v}$$

Substituting the value of x_1 from equation (1)

$$x_2 = A \frac{KV}{KV + v} \times \frac{KV}{KV + v}$$

$$= A \left(\frac{KV}{KV + v} \right)^2 \quad ...(2)$$

(3) **n-th extraction.** Following the above procedure for first and second extraction, the amount of substance left unextracted (x_n) in the n-th extraction would be

$$x_n = A \left(\frac{KV}{KV + v} \right)^n \quad ...(3)$$

This general formula enables the calculation of the amount of substance left unextracted after a specified number of extractions with v ml portion of the solvent each time.

(4) **Extraction with whole lot of solvent.** If the entire quantity of the given solvent is used for a single extraction, the amount of substance unextracted (x') would be

$$x' = A \left(\frac{KV}{KV + nv} \right) \quad ...(4)$$

Since the quantity within the bracket is less than unity, x_n is smaller than x'. Also, x_n will be smaller the greater the value of n. Hence it is more economical to use the solvent in small portions in a number of extractions.

SOLVED PROBLEM 12

The distribution coefficient of an alkaloid between chloroform and water is 20 in favor of chloroform. Compare the weights of the alkaloid remaining in aqueous solution when 10^{-4} m³ containing 10^{-3} kg has been shaken with : (*a*) 10^{-4} m³ chloroform; and (*b*) two successive 5×10^{-5} m³ portions.

SOLUTION

$$K = \frac{C_{\text{water}}}{C_{\text{chlorofom}}} = \frac{1}{20}$$

(1) When 10^{-4} m³ is used in one lot, the amount unextracted is

$$x' = 1 \times \left(\frac{KV}{KV + v} \right)$$

$$= 1 \times 10^{-3} \left(\frac{(1/20) \times 10^{-4}}{(1/20) \times 10^{-4} + 10^{-4}} \right) = \frac{1}{21} = \mathbf{0.0476 \times 10^{-3}\ kg}$$

(2) When 5×10^{-5} m³ chloroform is used in each of two stages, the amount unextracted is

$$x_2 = 1 \times \left(\frac{KV}{KV + v} \right)^2$$

$$= 1 \times 10^{-3} \left(\frac{(1/20) \times 10^{-4}}{(1/20) \times 10^{-4} + 5 \times 10^{-5}} \right)^2 = \left(\frac{1}{11} \right)^2 = \mathbf{0.0083 \times 10^{-3}\ kg}$$

LIQUID–LIQUID CHROMATOGRAPHY (Partition Chromatography)

This is a versatile and wide-spread technique for separation of a mixture of small amounts of organic substances. The operation is carried in a glass tube packed with an inert solid material (silica) soaked in a solvent 1 (water). The mixture paste is applied to the column top. Then a layer of another immiscible solvent 2 (*hexane*) is allowed to flow down the column (Fig. 18.10). A film of solvent 1 held by the silica forms the **stationary liquid phase,** while the running solvent 2 the **mobile liquid phase.** Each component of the mixture distributes itself between the stationary liquid phase and the mobile liquid phase. Thus,

$$\frac{C_s}{C_m} = K_D$$

where C_s = concentration in stationary liquid phase and C_m = concentration in the mobile liquid phase.

Different components in the mixture have different distribution coefficient. They pass into the mobile phase in order of their distribution coefficients. **The component (say S_2) with higher distribution coefficient is extracted (eluted) first and is the first to move down the column. The location of the components is easy if these are colored.** Otherwise they are visualized with the help of a locating agent. The portion of the solvent (or band) containing each component is received in a separate vessel and the solute recovered from it by distillation.

APPLICATIONS OF DISTRIBUTION LAW

There are numerous applications of distribution law in the laboratory as well as in industry. Here we will discuss some more important ones by way of recapitulation.

(1) Solvent Extraction

This is the process used for the separation of organic substances from aqueous solutions. The aqueous solution is shaken with an immiscible organic solvent such as ether (or benzene) in a separatory funnel. The distribution ratio being in favor of ether, most of the organic substance passes into the ethereal layer. The ethereal layer is separated and ether distilled off. Organic substance is left behind.

The process of extraction is more efficient if the solvent is used in a number of small portions than in one whole lot. Thus multiple extraction is preferred to simple extraction.

The multiple extraction is also the basis of the industrial **counterflow solvent extraction.** A dilute solution of the substance enters at the top of a column or tower. At the same time the extracting solvent enters at the bottom. As the aqueous solution and the solvent pass each other, the solute distributes itself in favor of the solvent. Since the process of extraction is repeated many times at different levels in the tower, maximum quantity of substance is removed in the solvent.

(2) Partition Chromatography

This is a modern technique of separating a mixture of small amounts of organic materials. A paste of the mixture is applied at the top of a column of silica soaked in water. Another immiscible solvent, say hexane, is allowed to flow down the column. Each component of the mixture is partitioned between the stationary liquid phase (*water*) and the mobile liquid phase (*hexane*). The various components of the mixture are extracted by *hexane* in order of their distribution coefficients. Thus the component with the highest distribution coefficient is first to move down in the flowing *hexane* which is collected separately. Similarly, a component with a lower distribution ratio comes down later and is received in another vessel.

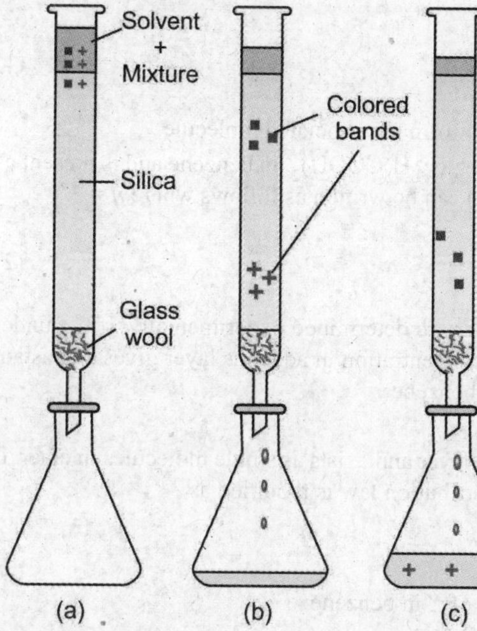

■ Figure 18.10
Details of a chromatography column.

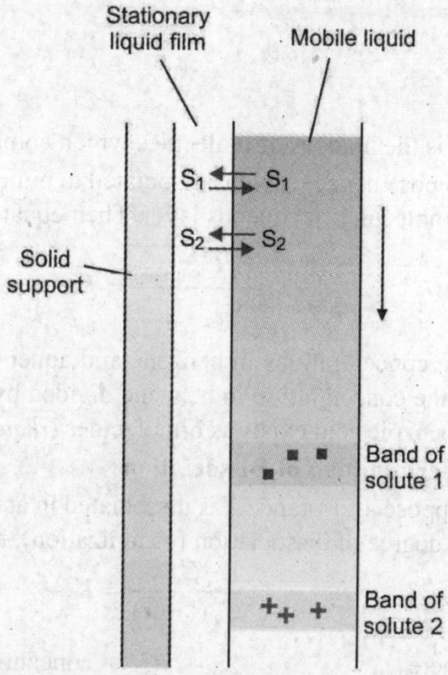

■ Figure 18.11
In partition chromatography, the liquid film on the solid support is the stationary phase. The second phase is an immiscible solvent flowing down. Each solute (S_1 and S_2) is distributed between the stationary liquid film and the mobile liquid phase. The solute (say S_2) with higher distribution ratio comes down first.

(3) Desilverization of Lead (*Parke's Process*)

When molten zinc is added to molten lead containing silver (*argentiferous lead*), zinc and lead form immiscible layers and silver is distributed between them. Since the distribution ratio is about 300 in favor of zinc at 1073 K, most of silver passes into the zinc layer. On cooling the zinc layer, an alloy of silver and zinc separates. The Ag-Zn alloy is distilled in a retort when zinc passes over leaving silver behind.

The lead layer still contains unextracted silver. This is treated with fresh quantities of molten zinc to recover most of the silver.

(4) Confirmatory Test for Bromide and Iodide

The salt solution is treated with chlorine water. Small quantity of bromine or iodine is thus liberated. The solution is then shaken with trichloromethane (chloroform). On standing chloroform forms the lower layer. The free bromine or iodine being more soluble in chloroform concentrates into the lower layer, making it red for bromine and violet for iodine.

(5) Determination of Association

When a substance is associated (or polymerized) in solvent *A* and exists as simple molecules in solvent *B*, the Distribution law is modified as

$$\sqrt[n]{\frac{C_A}{C_B}} = K \qquad ...(1)$$

when n is the number of molecules which combine to form an associated molecule.

Suppose benzoic acid is associated to bimolecules, $(C_6H_5COOH)_2$, in benzene and is present as simple molecules in aqueous layer. Then equation (1) can be written as follows where $n = 2$.

$$\frac{-\sqrt[2]{C_{\text{benzene}}}}{C_{aqueous}} = K \qquad ...(2)$$

The concentrations in benzene and aqueous layer are determined experimentally. If the under root of the concentration in benzene divided by the concentration in aqueous layer gives a constant value benzoic acid exists as bimolecules (*dimer*) in benzene.

(6) Determination of Dissociation

Suppose a substance X is dissociated in aqueous layer and exists as single molecules in ether. If x is the degree of dissociation (or ionization), the distribution law is modified as

$$\frac{C_1}{C_2(1-x)} = K$$

where C_1 = concentration of X in benzene

C_2 = concentration of X in aqueous layer

The value of x can be determined from conductivity measurements, while C_1 and C_2 are found experimentally. Thus the value of K can be calculated. Using this value of K, the value of x for any other concentrations of X can be determined.

(7) Determination of Solubility

Suppose the solubility of iodine in benzene is to be determined. Iodine is shaken with water and benzene. At equilibrium concentrations of iodine in benzene (C_b) and water (C_w) are found experimentally and the value of distribution coefficient calculated.

$$\frac{C_b}{C_w} = K_D$$

But
$$\frac{S_b}{S_w} = K_D$$

where S_b = solubility in benzene; and S_w = solubility in water.

If the solubility of iodine in water (S_w) is known, the solubility in benzene can be calculated.

(8) Deducing the Formula of a Complex Ion (I_3^-)

Some iodine is added to a solution of KI and the reaction mixture shaken with benzene.

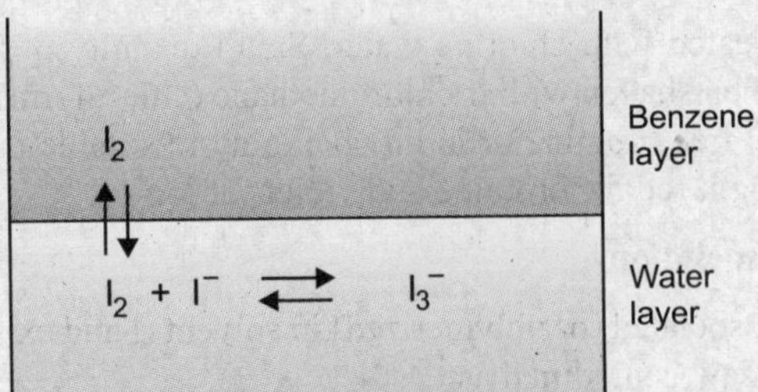

(*a*) The $[I_2]$ in water layer can be found knowing the value of K_D (determined separately) and concentration of iodine in benzene (determined by titration against thiosulfate).

(*b*) The total concentration of iodine, $[I_2] + [I_3^-]$ in water layer is found by titration against thiosulfate. Knowing $[I_2]$ from (a), $[I_3^-]$ can be calculated.

(*c*) The initial concentration of KI taken is represented by the equilibrium concentrations $[I^-] + [I_3^-]$. Knowing $[I_3^-]$ from (*b*), $[I^-]$ can be found.

Substituting the above values of concentrations in the law of Mass Action equation of the reaction in water layer,

$$\frac{[I_3^-]}{[I_2][I^-]} = K$$

the value of equilibrium constant *K* can be calculated. If it comes out to be constant for different concentrations of iodine, it stands confirmed that the formula of the complex I_3^-, which we assumed is correct.

(9) **Distribution Indicators**

In iodine titrations, the end point is indicated by adding starch suspension which turns blue. A greater sensitivity can be obtained by using what we may call 'Distribution Indicator'. A few drops of an immiscible organic solvent such as chloroform (or carbon tetrachloride) is added to the solution. The bulk of any iodine present passes into the organic layer and imparts intense violet color to it.

ADDITIONAL SOLVED PROBLEMS

SOLVED PROBLEM 13

If the distribution coefficient of benzoic acid between water and benzene is 3.304 at 293 K, calculate the number of moles of benzoic acid which may be extracted from 1×10^{-4} m^3 of 0.2 molar aqueous solution by 1×10^{-5} m^3 of benzene.

SOLUTION

Formula used

$$K = \frac{C_{benzene}}{C_{water}} = \frac{x/V}{(A-x)\,v}$$

Quantities given

$K = 3.304$ $\quad V = 1 \times 10^{-5}$ m^3 $\quad v = 1 \times 10^{-4}$ m^3

$A - x = 0.2$ mol

Substitution of values

$$3.304 = \frac{\dfrac{x}{10^{-5}}}{\dfrac{0.2}{10^{-4}}}$$

or $$\frac{x}{10^{-5}} \times \frac{10^{-4}}{0.2} = 3.304$$

or $$\frac{10\,x}{0.2} = 3.304$$

or $$x = \frac{0.6608}{10} = \mathbf{0.0660\ mol}$$

SOLVED PROBLEM 14

Benzoic acid was shaken between benzene and water. From the following concentrations of the acid in kilograms in 1×10^{-5} m^3 of the aqueous and benzene layers, show that acid exists in benzene as $(C_6H_5COOH)_2$

Acid in H_2O layer	0.150×10^{-3}	0.195×10^{-3}	0.289×10^{-3}
Acid in Benzene layer	0.242×10^{-3}	0.412×10^{-3}	0.970×10^{-3}

SOLUTION

Formula used

$$K = \frac{C_1}{\sqrt[n]{C_2}}$$

Quantities given

$n = 2$ (dimerization)

$C_1 = 0.150 \times 10^{-3}$ $\quad 0.195 \times 10^{-3}$ $\quad 0.289 \times 10^{-3}$

$C_2 = 0.242 \times 10^{-3}$ $\quad 0.412 \times 10^{-3}$ $\quad 0.970 \times 10^{-3}$

Substitution of values

$$K = \frac{0.150\times10^{-3}}{\sqrt[2]{0.242 \times 10^{-3}}} = 0.00964$$

$$K = \frac{0.195\times 10^{-3}}{\sqrt[2]{0.412 \times 10^{-3}}} = 0.00961$$

$$K = \frac{0.289\times 10^{-3}}{\sqrt[2]{0.970 \times 10^{-3}}} = 0.00928$$

Since the value of K is almost constant, benzoic acid exists as dimer in benzene.

SOLVED PROBLEM 15

At 298 K the iodine solution in water which contains 0.0516 kg/m^3 is in equilibrium with CCl_4 solution containing 4.1280 kg/m^3 of iodine. If at 298 K iodine solutions contains 0.410 kg/m^3 of iodine find out the solubility of iodine in CCl_4 solution.

SOLUTION

(*a*) **To calculate the distribution constant**

Formula used

$$K = \frac{C_{water}}{C_{CCl4}}$$

Quantities given

$C_{water} = 0.0516$ kg/m^3 $\qquad C_{CCl4} = 4.1280$ kg/m^3

Substitution of values

$$K = \frac{0.0516 \text{ kg/m}^3}{4.1280 \text{ kg/m}^3} = 0.0125$$

(*b*) **To calculate the solubility of I_2 in CCl_4**

Formula used

$$\frac{Conc.\ of\ I_2\ in\ water}{Conc.\ of\ I_2\ in\ CCl_4} = K$$

Quantities given

$K = 0.0125$ $\quad$ *Conc. of* I_2 *in water* = 0.410 kg/m³

Substitution of values

$$0.0125 = \frac{0.410 \text{ kg/m}^3}{\text{Conc. of } I_2 \text{ in } CCl_4}$$

or $$\text{conc. of } I_2 \text{ in } CCl_4 = \frac{0.410}{0.0125} \text{ kg/m}^3 = \mathbf{32.8 \text{ kg/m}^3}$$

SOLVED PROBLEM 16

The following data have been obtained on the distribution of phenol between water and Chloroform

C_1	0.094	0.164	0.254	0.436
C_2	0.254	0.761	1.850	5.430

where C_1 is concentration in the aqueous layer and C_2 is the concentration in the Chloroform layer. What conclusions do you draw from these results regarding molecular condition of phenol in the chloroform solution ?

SOLUTION

Formula used

$$K = \frac{C_{Water}}{\sqrt{C_{CHCl3}}} \quad \text{(for association)}$$

Quantities given

C_{water}	0.094	0.164	0.254	0.436
C_{CHCl3}	0.254	0.761	1.850	5.430

Substitution of values

$$K = \frac{0.094}{\sqrt{0.254}} = 0.1865 \qquad K = \frac{0.164}{\sqrt{0.761}} = 0.1880$$

$$K = \frac{0.254}{\sqrt{1.850}} = 0.1868 \qquad K = \frac{0.436}{\sqrt{5.430}} = 0.1871$$

Since the values of K are constant, phenol exists as ***dimer*** in Chloroform.

SOLVED PROBLEM 17

An organic acid was dissolved in two immiscible solvents (A) and (B). At equilibrium the concentration of the acid in (A) and (B) was found to be 400 mol/m³ and 640 mol/m³ respectively. Calculate the distribution coefficient of the organic acid, if it forms dimer in the solvent B.

SOLUTION

Formula used

$$K = \frac{C_A}{\sqrt{C_B}}$$

Quantities given

$C_A = 0.40$ $\quad$ $C_B = 640 \text{ mol/m}^3$

Substitution of values

$$K = \frac{400}{\sqrt{640}} = \frac{0.40}{0.8} = \mathbf{0.5}$$

SOLVED PROBLEM 18

Calculate how much succinic acid would be extracted from 1×10^{-4} m^3 of water containing 0.005 kg of acid if extracted with 50×10^{-5} m^3 of ether. Partition coefficient of succinic acid between water and ether is 5.5.

SOLUTION

Formula used

$$K = \frac{C_{water}}{C_{ether}}$$

Quantities given

$K = 5.5$ $\quad C_{water} = \frac{5 \times 10^{-3}}{1 \times 10^{-4}}$ $\quad C_{ether} = \frac{(5-x)\times 10^{-3}}{5 \times 10^{-5}}$

Substitution of values

$$5.5 = \frac{\frac{5 \times 10^{-3}}{1 \times 10^{-4}}}{\frac{(5-x)\times 10^{-3}}{5 \times 10^{-5}}}$$

$$5.5 = \frac{5 \times 10^{-3}}{1 \times 10^{-4}} \times \frac{5 \times 10^{-5}}{(5-x)\times 10^{-3}}$$

$$5.5\,(5 - x) = 25 \times 10^{-1}$$

$$27.5 - 5.5x = 2.5$$

$$25 = 5.5x$$

$$x = \mathbf{4.545\ kg}$$

SOLVED PROBLEM 19

Experiments in the study of rate of distribution of phenol between H_2O and chloroform gave the following results :

Concentration in aqueous solution	C_1	0.094	0.103	0.254
Concentration in Chloroform	C_2	0.254	0.761	1.850

Find whether phenol exists as monomer or dimer.

SOLUTION

In case phenol exists as monomer the value of K obtained from the following formula should be constant

Formula used

$$K = \frac{C_{water}}{C_{CHCl3}}$$

Quantities given

C_{water} =	0.094	0.103	0.254
C_{CHCl3} =	0.254	0.761	1.850

Substitution of values

$K = \frac{0.094}{0.254} = 0.3701$ $\quad K = \frac{0.103}{0.761} = 0.135$ $\quad K = \frac{0.254}{1.850} = 0.1373$

Since the value of K is not constant, phenol does not exists as monomer.

Now

Formula used

$$K = \frac{C_{water}}{\sqrt{C_{CHCl3}}}$$

Substitution of values

$$K = \frac{0.094}{\sqrt{0.254}} = 0.1865 \quad K = \frac{0.103}{\sqrt{0.761}} = 0.1868 \quad K = \frac{0.254}{\sqrt{1.850}} = 0.1867$$

Since the value of K is constant, phenol exists as **dimer** in chloroform.

SOLVED PROBLEM 20

The distribution constant of a certain solid X between two immiscible solvents A and B is 10 at 298 K. If the solubility of the solid in solvent A is 4800 mol/m^3, what is the solubility in solvent B?

SOLUTION

Formula used

$$K = \frac{\textit{Solubility in solvent A}}{\textit{Solubility in solvent B}}$$

Quantities given

$K = 10$ *Solubility in solvent A* = 4800 mol/m^3

Substitution of values

$$10 = \frac{4800 \text{ mole/m}^3}{\text{Solubility in solvent B}}$$

or *Solubility in solvent B* $= \frac{4800}{10}$

$=$ **480 mol/m^3**

SOLVED PROBLEM 21

In the distribution of benzoic acid between water and benzene the following values have been obtained in equal volumes of solvents :

$C_{water\ (g)}$	1.50	1.95	2.97
$C_{Benzene\ (g)}$	24.2	41.2	97.0

What is the molecular condition of benzoic acid in benzene ?

SOLUTION

Formula used

$$K = \frac{C_{water}}{C_{benzene}}$$

Quantities given

C_{water}	= 1.5	1.95	2.97
$C_{benzene}$	= 24.2	41.2	97.0

Substitution of values

$$K = \frac{1.5}{24.2} = 0.061983 \quad K = \frac{1.95}{41.2} = 0.04733 \quad K = \frac{2.97}{97.0} = 0.030619$$

Since the value of K is not constant, benzoic acid does not exist as monomer in benzene. Now

Formula used

$$K = \frac{C_{water}}{\sqrt{C_{benzene}}}$$

Substitution of values

$$K = \frac{1.5}{\sqrt{24.2}} = \frac{1.5}{4.919} = 0.3049$$

$$K = \frac{1.95}{\sqrt{41.2}} = \frac{1.95}{6.4187} = 0.3038$$

$$K = \frac{2.97}{\sqrt{97.0}} = \frac{2.97}{9.8488} = 0.30156$$

Now the value of distribution coefficient is constant. Hence benzoic acid exists as dimer in benzoic acid.

EXAMINATION QUESTIONS

1. Define or explain the following terms :
   (*a*) Nernst's distribution law (*b*) Henry's law
2. State Nernst distribution law. What are its important applications?
3. State and explain Nernst Distribution law. The solubility of a substance is twice as greater in ether as in water. Compare the quantities extracted from 10^{-4} m^3 of aqueous solution by
   (*a*) 10^{-4} m^3 of ether in a single operation
   (*b*) two successive operations of 5×10^{-5} m^3 ether in each.
4. Give thermodynamic basis for the Nernst Distribution law. Indicate its applications.
5. Deduce an expression for the Nernst Distribution law. State conditions under which the law is strictly valid.
6. Deduce the formulas for distribution law in the case of :
   (*a*) Association of solute in one of the phases
   (*b*) Dissociation of solute in one of the phases
7. The solubility of a substance is twice as great in ether as in water. Compare the quantities extracted from 10^{-4} m^3 of aqueous solution by:
   (*a*) 10^{-4} m^3 ether in a single operation
   (*b*) 5×10^{-5} m^3 ether in two successive operations
8. (*a*) Apply Distribution law to the case when the solute forms a dimer in one of the solvents.
   (*b*) Show that the process of extraction by a given volume of the solvent is more efficient if it is carried out in parts.
9. (*a*) What is Nernst Distribution law?
   (*b*) Nernst reported the following results for the distribution of acetic acid between water and CCl_4 :

Conc of acetic acid in CCl_4 (mol/m^3)	292	725	1410
Conc of acetic acid in H_2O (mol/m^3)	4870	7980	10700

   Assuming that acetic acid has its normal molecular weight in water, calculate its molecular weight in CCl_4.

   Answer. 120

10. If Henry's law constant for oxygen at 298 K is 4.40×10^{-5} Pa, find the solubility of oxygen in 1 kg of water at 298 K and a partial pressure of 25331.25 Pa.

Answer. 0.825×10^{-7} kg /m^3

11. Succinic acid was shaken with a mixture of water and ether. After distribution, upon analysis the concentrations of the acid in two layers are found as follows :

In aqueous layer (mol/m^3)	25.2	71	121
In ethereal layer (mol/m^{31})	4.6	13	22

If succinic acid has similar molecular weights in ether and water, determine its partition coefficient.

Answer. 5.4798

12. (*a*) State Distribution law and give the conditions under which the law is applicable.

(*b*) Succinic acid was shaken with a mixture of water and ether. The concentrations of acid in two layers per 10^{-4} m^3 of the solution are as follows:

Concentration in water layer (C_W)	25.4	33.2	42.6
Concentration in ether layer (C_E)	4.2	5.5	7.1

Find out the partition coefficient. If succinic acid has normal molecular weight of 118 in water, find its molecular weight in ether.

Answer. (*b*) 6.02; 118

13. The distribution coefficient of iodine between water and carbon disulfide is 0.0017. 10^{-3} m^3 of aqueous solution containing 10^{-3} kg of iodine is shaken with 10^{-4} m^3 of carbon disulfide till the equilibrium is reached. Find the amount of iodine extracted by carbon disulfide.

Answer. 0.983×10^{-3} kg

14. At 298 K, an aqueous solution of iodine containing 0.0387×10^{-3} kg per dm^{-3} of solution is in equilibrium with 3.31×10^{-3} kg/dm^3 of iodine in carbon tetrachloride. If solubility of iodine in CCl_4 is 0.0291 kg dm^{-3}, what is the solubility of iodine in water?

Answer. 0.00034 kg/dm^{-3}

15. (*a*) Explain how is distribution law modified when there is a change in molecular state.

(*b*) State Nernst's Distribution law explaining the significance of the term distribution coefficient.

(*c*) A solid X is added to a mixture of benzene and water. After shaking well and allowing it to stand. 10^{-4} m^3 of benzene layer was found to contain 13×10^{-4} kg of X, and 10^{-4} m^3 of water layer contained 2.2×10^{-4} kg of X. Calculate the value of distribution coefficient.

Answer. (*b*) 5.9090

16. (*a*) Define a solution and discuss factors affecting solubility. Explain solution of solids and liquids with examples.

(*b*) When a substance is distributed between two immiscible liquids and its molecule associates in one of them, the distribution constant is given by

(*i*) $\frac{C_1}{C_2}$ (*ii*) $\frac{C_1}{\sqrt[n]{C_2}}$ (*iii*) $\sqrt{\frac{C_1}{C_2}}$ (*Jiwaji BSc*)

17. (*a*) Explain Nernst distribution law. How is it modified when a solute undergoes (*i*) association in a solvent and (*ii*) dissociation in the solvent?

(*b*) Briefly describe the various applications of distribution law. (*MD Rohtak BSc*)

18. What are the conditions of validity for Nernst distribution law? (*Arunachal BSc*)

19. How is Nernst distribution law used in the study of complex ions? (*MD Rohtak BSc*)

20. Discuss the applications of Nernst distribution law in studying the molecular state of benzoic acid in benzene. (*Panjab MSc*)

21. Describe Nernst distribution law from thermodynamic considerations (*Arunachal BSc*)

22. (*a*) State Nernst distribution law. How is it used in the process of extraction of solutes?

(*b*) 0.012 kg of an organic substance A is present in 0.1 kg of its aqueous solution. How much of it would be left behind after extracting the solution with two successive applications of 5×10^{-5} m^3 each of ether? The distribution coefficient of A between water and ether is 2 in favor of ether.

Answer. (*b*) 0.01198 kg *(Bangalore BSc)*

23. (*a*) Derive distribution law from thermodynamic consideration. How is it modified when the solute undergoes association in one of the solvent?

(*b*) An organic compound is twice more soluble in ether than in water. Calculate the amount of the compound extracted when 10^{-4} m^3 of aqueous solution containing 0.001 kg of compound is shaken with

(*i*) 10^{-4} m^3 of ether at one time (*ii*) two successive quantities of 5×10^{-5} m^3 each

Answer. *(b)* (*i*) 0.67 kg; (*ii*) 1 kg *(Baroda BSc)*

24. The partition coefficient of a substance between benzene and water is 10.1. How much substance will be extracted from its 0.00025 m^3 aqueous solution containing 2.5×10^{-3} kg of it by using 0.0002 m^3 benzene in

(*i*) two equal portions and (*ii*) a single lot? *(Madurai BSc)*

Answer. (*i*) 0.8633×10^{-4} kg (*ii*) 0.0117×10^{-4} kg

25. The distribution coefficient of an organic acid between benzene and water is 80. An aqueous solution of the acid containing 2.25×10^{-3} kg in 10^{-4} m^3 was extracted with (i) 10^{-5} m^3 of benzene in one lot and (ii) twice with 5×10^{-6} m^3 benzene each time. Calculate the amount of acid extracted in each experiment.

Answer. (i) 2×10^{-3} kg (ii) 2.7×10^{-3} kg *(Madras BSc)*

26. In an experiment on distribution of succinic acid between water and ether at 298 K, 1×10^{-5} m^3 of ethereal layer contains 0.046×10^{-3} kg of acid. Calculate the amount of succinic acid present in 2.5×10^{-5} m^3 of the aqueous solution in equilibrium with it if the distribution coefficient of succinic acid between water and ether is 5.2.

Answer. 0.598×10^{-3} kg *(Allahabad BSc)*

27. The solubility of iodine in water is 0.35 kg/m^3. Calculate its solubility in carbon tetrachloride, if the distribution constant of iodine in CCl_4 and water is 88.

Answer. 30.8 kg/m^3 *(Bhopal BSc)*

28. 0.1 kg of an acid was dissolved in 10^{-3} m^3 of water. The distribution coefficient of acid between ether and water is 3. 10^{-3} m^3 of ether in one lot was used to extract acid. Calculate the amount of acid extracted.

Answer. 37.5 *(Jiwaji BSc)*

29. A solid *Y* was added to mixture of benzene and water. After shaking well and allowing it to stand, 2×10^{-5} m^3 of the benzene was found to contain 0.26×10^{-3} kg of *Y* and 2×10^{-4} m^3 of water layer contained 0.44×10^{-3} kg of *Y*. Calculate the value of distribution constant.

Answer. 5.9 *(Arunachal BSc)*

30. (*a*) How is distribution law modified when the solute undergoes dissociation in one of the solvents? What would happen if the solute is completely dissociated in one of the solvents?

(*b*) An organic acid is distributed between 5×10^{-4} m^3 each of a solvent A and water. In water it is dissociated. The amount of the acid in aqueous layer was 6×10^{-3} kg and in solvent A, it was 0.72×10^{-3} kg. If the distribution coefficient of the acid between the solvent A and water is 0.16, calculate the degree of dissociation, assuming that the acid has normal molecular weight in A.

Answer. (*b*) 0.25 *(Agra BSc)*

31. If the distribution coefficient of benzoic acid between water and benzene is 3.304 at 293 K, calculate the number of molecules of benzoic acid which may be extracted from 1×10^{-4} m^3 of 0.2 molar aqueous solution by 1×10^{-5} m^3 of benzene.

Answer. 0.06608 mol *(Panjab BSc)*

MULTIPLE CHOICE QUESTIONS

1. The expression $K_D = \frac{C_1}{C_2}$ where C_1 and C_2 denote the concentration of a solute in two solvents A & B, and K_D is the constant, is called

   (*a*) equilibrium law (*b*) rate law
   (*c*) Nernst distribution law (*d*) none of these

   Answer. (*c*)

2. For the study of distribution law the two solvents should be

   (*a*) miscible (*b*) non-miscible
   (*c*) volatile (*d*) reacting with each other

   Answer. (*b*)

3. While studying the distribution law

   (*a*) the temperature should be constant throughout
   (*b*) there should be no association or dissociation of the solute
   (*c*) the concentration of the solute in solvents
   (*d*) all of the above

   Answer. (*d*)

4. When a solute undergoes association in one of the solvent, the following expression of distribution law is applicable

   (*a*) $\frac{C_1}{C_2} = K_D$ (*b*) $\frac{C_1}{C_2^2} = K_D$
   (*c*) $\frac{C_1}{\sqrt[n]{C_2}} = K_D$ (*d*) $\frac{C_1}{\sqrt{C_2}} = K_D$

   Answer. (*c*)

5. The modified distribution law for the solute undergoing dissociation in one of the solvents is

   (*a*) $K_D = \frac{C_1}{\sqrt{C_2}}$ (*b*) $K_D = \frac{C_1}{C_2(1-\alpha)}$
   (*c*) $K_D = \frac{C_1}{C_2(\alpha-1)}$ (*d*) $K_D = \frac{C_1}{C_2^2}$

   Answer. (*b*)

6. At a constant temperature, the solubility of a gas in a liquid is proportional to the pressure of the gas above it. This law is

   (*a*) Nernst's distribution law (*b*) Henry's law
   (*c*) Le Chatelier's principle (*d*) Raoult's law

   Answer. (*b*)

7. If a mixture of gases is in contact with a liquid, the partial pressure of the individual gas determines the mass of each gas dissolving *i.e.*, the solubility of each gas is proportional to its ______

   (*a*) total pressure (*b*) concentration of the liquid
   (*c*) partial pressure (*d*) temperature

   Answer. (*c*)

8. The Henry's law equation is

   (*a*) $C = k \times P$ (*b*) $k = C \times P$
   (*c*) $P = C \times k$ (*d*) $C = k \times \sqrt{P}$

   Answer. (*a*)

9. The Nernst distribution law equation $C_1/C_2 = K_D$ applies when

(*a*) the molecular state of the solute is the same in both the solvents

(*b*) the molecular state of the solute is different in both the solute

(*c*) the molecular state of the solute may be the same or different in both the solute

(*d*) none of the above

Answer. (*a*)

10. Benzoic acid when shaken with mixtures of benzene and water undergoes dimerization in benzene. The distribution law applicable here is

(*a*) $\frac{C_1}{C_2} = K_D$ (*b*) $\frac{C_1}{\sqrt[3]{C_2}} = K_D$

(*c*) $\frac{C_1}{\sqrt[2]{C_2}} = K_D$ (*d*) none of these

Answer. (*c*)

11. The greater the distribution ratio in favor of the organic solvent, the _______ will be the amount extracted in any one operation

(*a*) greater (*b*) lesser

(*c*) equal (*d*) none of these

Answer. (*a*)

12. Multiple extraction is _______ efficient as compared to single extraction

(*a*) more (*b*) less

(*c*) the same (*d*) none of these

Answer. (*a*)

13. With the help of distribution law we can

(*a*) determine the association or dissociation of the solute in one of the solvents

(*b*) determine the solubility of solute in various solvents

(*c*) reduce the formula of complex ions

(*d*) all of the above

Answer. (*d*)

14. The _______ extraction is more efficient

(*a*) single (*b*) double

(*c*) multiple (*d*) none of these

Answer. (*c*)

15. The distribution law is applied in the

(*a*) Haber's process for the manufacture of NH_3

(*b*) Parke's process for the extraction of Ag

(*c*) Contact process for the manufacture of H_2SO_4

(*d*) none of the above

Answer. (*b*)

16. In the first extraction, the amount of the substance left unextracted is given by the formula (*K* is distribution coefficient, *V* ml of the aqueous solution contain A gram of an organic substance)

(*a*) $x_1 = A\frac{KV}{KV + v}$ (*b*) $x_1 = A\frac{KV}{KV + v^2}$

(*c*) $x_1 = A^2\frac{KV}{KV + v}$ (*d*) $x_1 = A\frac{K^2V}{KV + v}$

Answer. (*a*)

17. The Nernst distribution law $K_D = C_1/C_2$ is not applicable if the solute undergoes
(*a*) association in one of the solvents
(*b*) dissociation is one of the solvents
(*c*) association or dissociation in one of the solvents
(*d*) none of the above
Answer. (*c*)

18. When succinic acid or oxalic acid is shaken with ether and water it _______ in water.
(*a*) associates to form dimer (*b*) dissociates into ions
(*c*) remains the same (*d*) none of these
Answer. (*b*)

19. When a bottle of soda-water is opened, the partial pressure of CO_2 _______.
(*a*) decreases (*b*) increases
(*c*) remains the same (*d*) none of these
Answer. (*a*)

20. Is it possible to extract the whole of the dissolved substance by using large number of extractions?
(*a*) yes (*b*) no
(*c*) sometimes (*d*) none of these
Answer. (*b*)

21. In liquid-liquid chromatography technique, the component with _______ distribution coefficient is extracted first
(*a*) lower (*b*) higher
(*c*) intermediate (*d*) none of these
Answer. (*b*)

22. The Nernst's distribution law does not hold good if
(*a*) the temperature throughout the experiment is not constant
(*b*) concentration of the solute in two solvents are high
(*c*) there is association or dissociation of the solute in one of the solvents
(*d*) all of the above
Answer. (*d*)

23. The Nernst's distribution law is also known as
(*a*) partition law (*b*) Henry's law
(*c*) Raoult's law (*d*) equilibrium law
Answer. (*a*)

24. In the process of extraction of a substance from aqueous solution, the use of a _______ is made
(*a*) conical flask (*b*) measuring flask
(*c*) burette (*d*) separating funnel
Answer. (*d*)

25. When the solute undergoes association in one of the solvent, the Nernst distribution law as modified as
(*a*) $C_1 = \sqrt{C_2} \times K_D$ (*b*) $\sqrt{C_2} \times C_1 = K_D$
(*c*) $C_1 = K_D \times C_2^2$ (*d*) $C_1 = K_D \times C_2^3$
Answer. (*a*).

19

The Phase Rule

CONTENTS

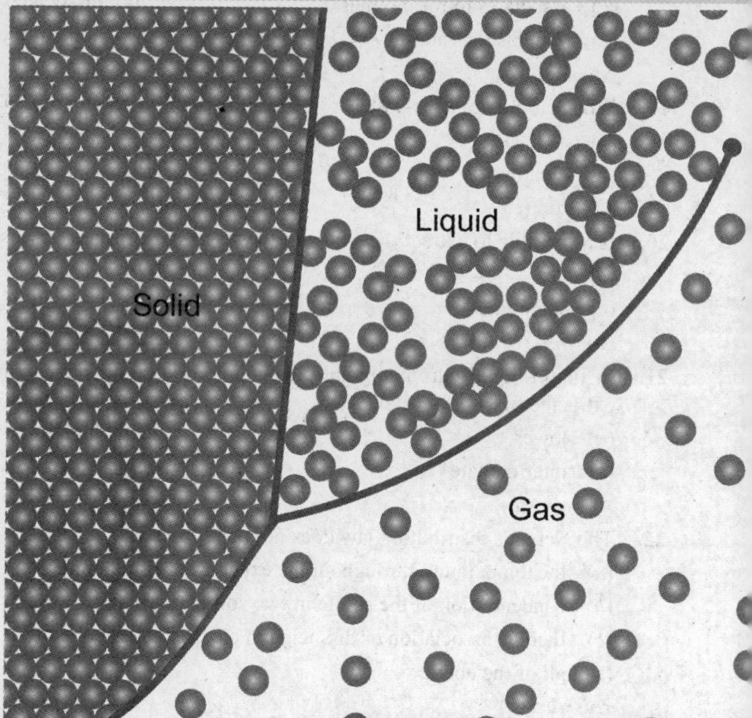

THE STATEMENT

The phase Rule is an important generalization dealing with the behavior of heterogeneous systems. In general it may be said that with the application of *phase rule* it is possible to predict qualitatively by means of a diagram the effect of changing pressure, temperature and concentration on a heterogeneous system in equilibrium. This relationship governing all heterogeneous equilibria was first discovered as early as 1874 by an American physicist Willard Gibbs. **Gibb's Phase Rule** is free from flaws and limitations which are a common feature of all other generalizations of Physical Chemistry based on hypothetical assumptions as to the nature of the constitution of matter. It may be stated mathematically as follows :

$$F = C - P + 2$$

where F is the number of *degrees of freedom*, C is the number of *components* and P is the number of *phases* of the system. The terms 'phase', 'component', and 'degree of freedom' involved in the statement of the Phase Rule have a special significance and a clear understanding of these terms is essential before we proceed further with the subject.

WHAT IS MEANT BY A 'PHASE'?

A phase may be defined as : **any homogeneous part of a system having all physical and chemical properties the same throughout.** A system may consist of one phase or more than one phases.

(1) A system containing only liquid water is **one-phase** or **1-phase system** ($P = 1$)

(2) A system containing liquid water and water vapor (a gas) is a **two-phase** or **2-phase system** ($P = 2$).

(3) A system containing liquid water, water vapor and solid ice is a **three-phase** or **3-phase system.**

A system consisting of one phase only is called a **homogeneous system.**

A system consisting of two or more phases is called a **heterogeneous system.**

Explanation of the Term 'Phase'

Ordinarily three states of matter-gas, liquid, and solid are known as phases. However in *phase rule,* **a uniform part of a system in equilibrium is termed a 'phase'.** Thus a liquid or a solid mixture could have two or more phases.

Let us consider a few examples to understand the meaning of the term phase as encountered in phase rule.

(1) **Pure substances.** A pure substance (solid, liquid, or gas) made of one chemical species only, is considered as one phase. Thus oxygen (O_2), benzene (C_6H_6), and ice (H_2O) are all 1-phase systems. *It must be remembered that a phase may or may not be continuous.* Thus, whether ice is present in one block or many pieces, it is considered one phase.

(2) **Mixtures of gases.** All gases mix freely to form homogeneous mixtures. Therefore any mixture of gases, say O_2 and N_2, is a 1-phase system.

(3) **Miscible liquids.** Two completely miscible liquids yield a uniform solution. Thus a solution of ethanol and water is a 1-phase system.

(4) **Non-miscible liquids.** A mixture of two non-miscible liquids on standing forms two separate layers. Hence a mixture of chloroform ($CHCl_3$) and water constitutes a 2-phase system.

(5) **Aqueous solutions.** An aqueous solution of a solid substance such as sodium chloride (or sugar) is uniform throughout. Therefore it is a 1-phase system.

However, a saturated solution of sodium chloride in contact with excess solid sodium chloride is a 2-phase system.

(6) **Mixtures of solids.** (*i*) By definition, **a phase must have throughout the same physical and chemical properties.** Ordinary sulfur as it occurs in nature is a mixture of monoclinic and rhombic sulfur. These allotropes of sulfur consist of the same chemical species but differ in physical properties. Thus mixture of two allotropes is a 2-phase system.

(*ii*) **A mixture of two or more chemical substances contains as many phases.** Each of these substances having different physical and chemical properties makes a separate phase. Thus a mixture of calcium carbonate ($CaCO_3$) and calcium oxide (CaO) constitutes two phases.

Let us consider the equilibrium system : the **Decomposition of Calcium carbonate.** When calcium carbonate is heated in a closed vessel, we have

$$\underset{\text{(solid)}}{CaCO_3} \rightleftharpoons \underset{\text{(solid)}}{CaO} + \underset{\text{(gas)}}{CO_2}$$

There are two solid phases and one gas phase. Hence it is a 3-phase system.

WHAT IS MEANT BY 'COMPONENTS'?

A system 'C' in the *Phase Rule equation* stands for the number of components of a system in equilibrium. The term component may be defined as : **the least number of independent chemical constituents in terms of which the composition of every phase can be expressed by means of a chemical equation.**

Explanation of the Term 'Component'

To understand the above definition and to use it for finding the number of components of a system, remember that :

(*a*) The chemical formula representing the composition of a phase is written on LHS.

(*b*) The rest of the chemical constituents existing independently in the system as represented by chemical formulas are placed on RHS.

(*c*) The quantities of constituents on RHS can be made minus (–) or zero (0) to get at the composition of the phase on LHS.

Now let us proceed to discuss the number of components of some systems.

(1) **Water and sulfur systems are 1-component systems.** Water system has three phases : *ice, water,* and *water vapor.* The composition of all the three phases is expressed in terms of one chemical individual H_2O. Thus water system has one component only.

Sulfur system has four phases : rhombic sulfur, monoclinic sulfur, liquid sulfur and sulfur vapor. The composition of all these phases can be expressed by one chemical individual sulfur (S). Hence it is a 1-component system.

As is clear from above, when all the phases of a system can be expressed in terms of one chemical individual, it is designated as a one-component or 1-component system.

(2) **Mixture of gases.** A mixture of gases, say O_2 and N_2, constitutes one phase only. Its composition can be expressed by two chemical substances O_2 and N_2.

Phase		**Components**
Gaseous Mixture	=	$x\,O_2 + y\,N_2$

Hence a mixture of O_2 and N_2 has two components. In general the number of components of a gaseous mixture is given by the number of individual gases present.

(3) **Sodium chloride solution.** A solution of sodium chloride in water is a 1-phase system. Its composition (xNaCl.$y$$H_2O$) can be expressed in terms of two chemical individuals, sodium chloride and water.

Phase		**Components**
Aq. solution of sodium chloride	=	$NaCl + y\,H_2O$

Therefore an aqueous solution of sodium chloride or any other solute is a two-component or 2-component system.

A saturated solution of sodium chloride, in contact with excess solid sodium chloride has two phase, namely aqueous solution and solid sodium chloride. The composition of both phase can be expressed in terms of two chemical individuals NaCl and H_2O.

Phase		Components
Aq. solution of sodium chloride	=	x NaCl + y H_2O
Solid sodium chloride	=	NaCl + 0 H_2O

Hence a saturated solution of sodium chloride or any other solute in contact with solid solute, is 2-component system.

(4) **Decomposition of Calcium carbonate.** When calcium carbonate is heated in a closed vessel, the following equilibrium system results.

$$\underset{\text{(solid)}}{CaCO_3} \rightleftharpoons \underset{\text{(solid)}}{CaO} + \underset{\text{(gas)}}{CO_2}$$

It has three phase : calcium carbonate, calcium oxide, and carbon dioxide. The composition of all the phase can be expressed in terms of any two of the three chemical substances in equilibrium.

Let us select calcium oxide (CaO) and carbon dioxide (CO_2) as the components. Then we can write,

Phase		Components
$CaCO_3$	=	CaO + CO_2
CaO	=	$CaCO_3$ – CO_2
CO_2	=	$CaCO_3$ – CaO

Again, selecting calcium carbonate ($CaCO_3$) and calcium oxide (CaO) as the components, we have

Phase		Components
$CaCO_3$	=	$CaCO_3$ + 0 CaO
CaO	=	0 $CaCO_3$ + CaO
CO_2	=	$CaCO_3$ – CaO

Thus decomposition of calcium carbonate is a 2-components system.

Furthermore, as is clear from the above examples, **by the components of a system is meant the number of chemical individuals and not any particular chemical substances by name.**

(5) **Dissociation of Ammonium chloride.** Ammonium chloride when heated in a closed vessel exists in equilibrium with the products of dissociation, ammonia (NH_3) and hydrogen chloride gas (HCl).

$$\underset{\text{(solid)}}{NH_4Cl} \rightleftharpoons \underset{\text{(gas)}}{NH_3} + \underset{\text{(gas)}}{HCl}$$

The system consists of two phase, namely, solid NH_4Cl and the gaseous mixture containing NH_3 and HCl. The constituents of the mixture are present in the same proportion in which they are combined in solid NH_4Cl. The composition of both the phase can, therefore, be expressed in terms of the same chemical individual NH_4Cl.

Phase		Components
Solid	=	NH_4Cl
Gaseous	=	x NH_3 + x HCl or x NH_4Cl

Thus dissociation of ammonium chloride is a one-component system.

DEGREES OF FREEDOM

The term **Degree of Freedom** represented by F in the phase Rule equation ($F = C - P + 2$) is defined as follows : **the least number of variable factors (concentration, pressure and temperature) which must be specified so that the remaining variables are fixed automatically and the system is completely defined.**

A system with $F = 0$ is known as **nonvariant** or having no degree of freedom.

A system with $F = 1$ is known as **univariant** or having one degree of freedom.

A system with $F = 2$ is known as **bivariant** or having two degrees of freedom.

Explanation of the Term Degree of Freedom

A system is defined completely when it retains the same state of equilibrium (or can be reproduced exactly) with the specified variables. Let us consider some examples.

(1) **For a pure gas, $F = 2$.** For a given sample of any pure gas $PV = RT$. If the values of pressure (P) and temperature (T) be specified, volume (V) can have only one definite value, or that the volume (the third variable) is fixed automatically. Any other sample of the gas under the same pressure and temperature as specified above, will be identical with the first one. Hence **a system containing a pure gas has two degrees of freedom ($F = 2$).**

(2) **For a mixture of gases, $F = 3$.** A system containing a mixture of two or more gases is completely defined when its composition, temperature and pressure are specified. If pressure and temperature only are specified, the third variable *i.e.*, composition could be varied. Since it is necessary to specify three variables to define the system completely, **a mixture of gases has three degrees of freedom ($F = 3$).**

(3) **For water $\rightleftharpoons$ water vapor, $F = 1$.** The system water in equilibrium with water vapor, has two variables *temperature* and *pressure.* At a definite temperature the vapor pressure of water can have only one fixed value. Thus if one variable (temperature or pressure) is specified, the other is fixed automatically. Hence **the system water has one degree of freedom ($F = 1$).**

(4) **For saturated NaCl solution, $F = 1$.** The saturated solution of sodium chloride in equilibrium with solid sodium chloride and water vapor.

$$\underset{\text{(solid)}}{\text{NaCl}} \rightleftharpoons \text{NaCl-solution} \rightleftharpoons \text{water vapor}$$

Thus the system is completely defined if we specify temperature only. The other two variables *i.e*,. *the* composition of NaCl-solution (solubility) and vapor pressure have a definite value at a fixed temperature. Hence **the system has one degree of freedom.**

(5) **For ice-water-vapor system, $F = 0$.** In the system ice $\rightleftharpoons$ water $\rightleftharpoons$ vapor, the three phases coexist at the freezing point of water. Since the freezing temperature of water has a fixed value, the vapor pressure of water has also a definite value. The system has two variables (temperature and pressure) and both these are already fixed. Thus the system is completely defined automatically, there being no need to specify any variable. Hence **it has no degree of freedom ($F = 0$).**

DERIVATION OF THE PHASE RULE

Consider a heterogeneous system in equilibrium of C components in which P phases are present. We have to determine the degrees of freedom of this system *i.e.*, the number of variables which must be arbitrarily fixed in order to define the system completely. Since the state of the system will depend upon the temperature and the pressure, these *two variables* are always there. The concentration variables, however, depend upon the number of phases. In order to define the composition of each phase it is necessary to specify the concentration of $(C - 1)$ constituents of each phase, the concentration of the remaining component being determined by difference. For P phases, therefore, the total number of concentration variables will be $P(C - 1)$ and these along with the two variables mentioned above *viz.*, temperature and pressure, make the total number of the variables of the system equal to $[P(C - 1) + 2]$.

On **thermodynamic consideration** when a system is in equilibrium, the partial molal free energy of each constituent of a phase is equal to the partial molal free energy of the same constituent in every other phase. Since the partial molal free energy of the constituents of a phase is a function of the temperature, pressure and $(C-1)$ concentration variables, it follows that if there is one component in two phases, it is possible to write one equation amongst the variables and if there is one component in three phases, this fact may be written with the help of *two equations*. In general, therefore, when P phases are present, $(P-1)$ equations are available for each component and for C components, the total number of equations or variables are $C(P-1)$.

Since the number of equations is equal to the number of variables, the number of *unknown variables* or degrees of freedom (F) will be :

$$
\begin{aligned}
F &= \text{No. of variables} - \text{Number of equations} \\
&= [P(C-1)+2] - [C(P-1)] \\
&= PC - P + 2 - PC + C \\
&= C - P + 2 \\
F &= C - P + 2
\end{aligned}
$$

ONE-COMPONENT SYSTEM

For a one-component system we can write the phase rule equation as :

$$F = C - P + 2 = 1 - P + 2 = 3 - P$$

Three cases may arise :

Case 1. When only **one phase** is present,

$$\therefore \quad F = 3 - 1 = 2$$

Thus the system is **bivariant.** It can be completely defined by specifying the two variables, temperature and pressure. Or that, both the temperature and pressure can be varied independently. Therefore **a single phase is represented by an area on P, T-graph.**

Case 2. When **two phases** are in equilibrium,

$$F = 3 - 2 = 1$$

The system then has one degree of freedom and is termed **monovariant.** This means that the pressure cannot be changed independently if we change the temperature. The pressure is fixed automatically for a given temperature. **A two-phase system is depicted by a line on a P, T-graph.**

Case 3. When **three phases** are in equilibrium,

$$F = 3 - P = 3 - 3 = 0$$

$$\therefore \quad F = 0$$

The system has zero degree of freedom and is termed **nonvariant or invariant.** This special condition can be attained at a definite temperature and pressure. The system is, therefore, defined completely and no further statement of external conditions is necessary. **A three-phase system is depicted by a point on the P, T-graph.** At this point the three phases (solid, liquid, vapor) are in equilibrium and, therefore, it is referred to as the **Triple point.**

PHASE DIAGRAMS

A phase diagram is a **plot showing the conditions of pressure and temperature under which two or more physical states can exist together in a state of dynamic equilibrium.** Fig. 19.1 is a typical phase diagram for a one-component system. The diagram consists of : (*a*) the Regions or Areas; (*b*) the Lines or Curves; and (*c*) the Triple point.

The significance of the characteristic features (Areas, Curves, Triple point) of a phase diagram are discussed below.

(1) Regions or Areas

The diagram is divided into three regions or areas which are labelled as 'solid', 'liquid', and 'vapor'. These areas in Fig. 19.1 are *COB*, *COA* and *AOB*. **Each of the three areas shows the conditions of temperature and pressure under which the respective phase can exist.** Applying the phase rule to the system when only one phase is present, we have

$$F = 1 - 1 + 2 = 2$$

i.e., each single phase has two degrees of freedom.

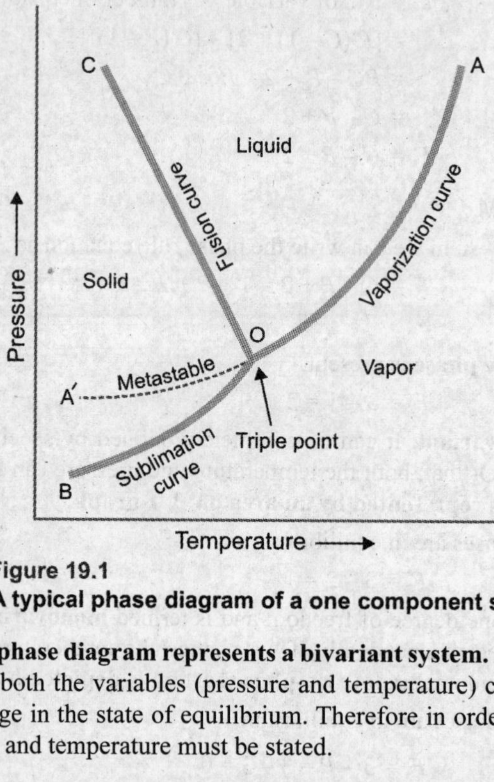

Figure 19.1
A typical phase diagram of a one component system.

Thus **each area of phase diagram represents a bivariant system.** At conditions wholly within one of the three areas, both the variables (pressure and temperature) can be varied independently without causing a change in the state of equilibrium. Therefore in order to define the condition of the phase both pressure and temperature must be stated.

(2) Lines or Curves

There are three lines or curves separating the regions or areas. These **curves show the conditions of equilibrium between any two of the three phases** *i.e.*, solid/liquid, liquid/vapor, solid/vapor.

(*a*) Solid/liquid line (*OC*) which represents the equilibrium *Solid* $\rightleftharpoons$ *Liquid,* is referred to as the **Melting curve** or **Fusion curve.**

(*b*) Liquid/vapor line (*OA*) which represents the equilibrium *Liquid* $\rightleftharpoons$ *Vapor,* is referred to as the **Vapor Pressure curve** or **Vaporisation curve** for the liquid.

(*c*) Solid/vapor line (*OB*), which represents the equilibrium *Solid* $\rightleftharpoons$ *Vapor,* is referred to as the **Sublimation curve.**

Applying phase rule to a one-component two-phase system.

$$F = C - P + 2 = 1 - 2 + 2 = 1$$

Thus phase rule predicts that the two phase equilibria stated above will have one degree of freedom. Along any of three lines on the phase diagram when one variable (pressure or temperature) is specified, the other is fixed automatically.

(3) Triple Point

The *three* boundary lines enclosing the *three areas* on the phase diagram intersect at a common point called the *Triple point.* **A triple point shows the conditions under which all the three phases (solid, liquid, vapor) can coexist in equilibrium.** Thus the system at the triple point may be represented as :

$$\text{Solid} \rightleftharpoons \text{Liquid} \rightleftharpoons \text{Vapor}$$

Applying the phase rule equation, we have

$$F = C - P + 2 = 1 - 3 + 2 = 0$$

which predicts that the system has no degree of freedom.

At the triple point both pressure and temperature on the diagram are fixed and, therefore, the system is nonvariant. This implies that if we try to change temperature or pressure, the equilibrium will be disturbed. For example, if we lower the pressure on the system, all the liquid will vaporise, leaving only two phases.

In case of water system, the temperature and pressure for the triple point are 0.0076° and 4.58 mm Hg respectively.

Metastable Equilibrium

The vapor pressure curve *AO* of the liquid phase terminates at *O*, when the liquid freezes (or solidifies). However by careful cooling of the liquid under conditions that crystals do not form, the curve *AO* can be extended to *A'*. This means that the liquid can be cooled far below the freezing point or 'supercooled' without separation of the crystals. The supercooled liquid is in an unstable condition. On the slightest disturbance as introduction of a seed crystal, the entire liquid solidifies rapidly. Thus the dashed curve *OA'* represents a **metastable equilibrium,**

$$\text{Supercooled Liquid} \rightleftharpoons \text{Vapor}$$

This system at once reverts to the true stable system

$$\text{Solid} \rightleftharpoons \text{Vapor}$$

under suitable conditions. It is noteworthy that the dashed curve of the metastable liquid lies above the normal sublimation curve (*BO*). This implies that **the vapor pressure of the metastable phase is always higher than that of the stable phase at the same temperature.**

POLYMORPHISM

The occurrence of the same substance in more than one crystalline forms is known as Polymorphism. This phenomenon is shown by both elements and compounds. In the case of elements the term **allotropy** is often used. The individual crystalline forms of an element are referred to as **polymorphs** or **allotropes.** Rhombic and monoclinic sulfur are two polymorphs or allotropes of sulfur. The polymorphic or allotropic forms of an element have distinct physical properties and constitute separate phases.

Allotropy can be divided into three types : Enantiotropy, Monotropy and Dynamic allotropy.

(1) Enantiotropy

In some cases one polymorphic form (or allotrope) can change into another at a definite temperature when the two forms have a common vapor pressure. This temperature is known as the **transition temperature.** One form is stable above this temperature and the other form below it. **When the change of one form to the other at the transition temperature is reversible, the phenomenon is called enantiotropy and the polymorphic forms enantiotropes.** For example, rhombic sulfur (α-Sulfur) on heating changes to monoclinic sulfur (β-Sulfur) at 368.6 K (transition temperature).

Also, monoclinic sulfur, on cooling, again changes to rhombic sulfur at 368.6 K. That is,

$$\alpha\text{-Sulfur} \underset{}{\overset{368.6\text{ K}}{\rightleftharpoons}} \beta\text{-Sulfur}$$

Thus α-Sulfur and β-Sulfur are enantiotropic.

(2) Monotropy

It occurs when one form is stable and the other metastable. **The metastable changes to the stable form at all temperatures and the change is not reversible.** Thus there is no transition temperature as the vapor pressures are never equal. This type of polymorphism is exhibited by phosphorus,

$$\text{White phosphorus} \longrightarrow \text{Red phosphorus}$$

Another example is graphite and diamond, graphite being stable and diamond metastable, although the change is infinitely slow.

(3) Dynamic allotropy

Some substances have several forms which can coexist in equilibrium over a range of temperature. The amount of each is determined by the temperature. The separate forms usually have different molecular formulas but the same empirical formula. **This form of allotropy, known as dynamic allotropy, resembles enantiotropy in that it is reversible but there is no fixed transition point.**

An example of dynamic allotropy is provided by liquid sulfur which consists of three allotropes S_μ, S_π and S_λ.

$$S_\mu \rightleftharpoons S_\pi \rightleftharpoons S_\lambda$$

These three forms of sulfur differ in molecular structure. S_λ is S_8, S_π is S_4 while formula of S_μ is not known. The composition of the equilibrium mixture at 393 K and 717.8 K (b.p. of sulfur) is :

393 K	S_μ 0%	S_π 3.7%	S_λ 96.3%
717.8 K	S_μ 37%	S_π 4%	S_λ 59%

EXPERIMENTAL DETERMINATION OF TRANSITION POINT

The temperature at which a polymorphic substance changes from one form to another, is known as the transition temperature or transition point. For example, rhombic variety of sulfur is converted to the monoclinic form of sulfur at 368.6 K at atmospheric pressure. The transition temperature in a particular case can be determined by measuring a change in physical properties such as color, density, solubility, etc.

(1) Color change

If a little mercury (II) iodide is placed in a melting point tube attached to a thermometer and heated in some form of apparatus (*e.g.*, electrical heater), it is possible to record temperature at which the red mercury (II) iodide changes to the yellow form.

(2) Density change

As rhombic sulfur changes to monoclinic sulfur, there is a decrease in density and, therefore, an increase in volume. The change in volume is employed to measure the transition temperature by using an apparatus known as **Dilatometer** shown in Fig. 19.2.

Some powdered rhombic sulfur is placed in the glass bulb and liquid paraffin (an inert liquid) is introduced above the sulfur. The apparatus is then immersed in a heating water-bath, the temperature of which is raised. The scale reading and the temperature is recorded every minute. A plot of liquid level in the capillary against temperature gives a curve as in Fig. 19.2 (*b*). On cooling of the dilatometer, reverse changes take place but due to thermal lag the curve assumes the form shown in the figure. The transition temperature is taken as the mean of the respective temperatures where expansion starts

(T_1) and contraction begins (T_2).

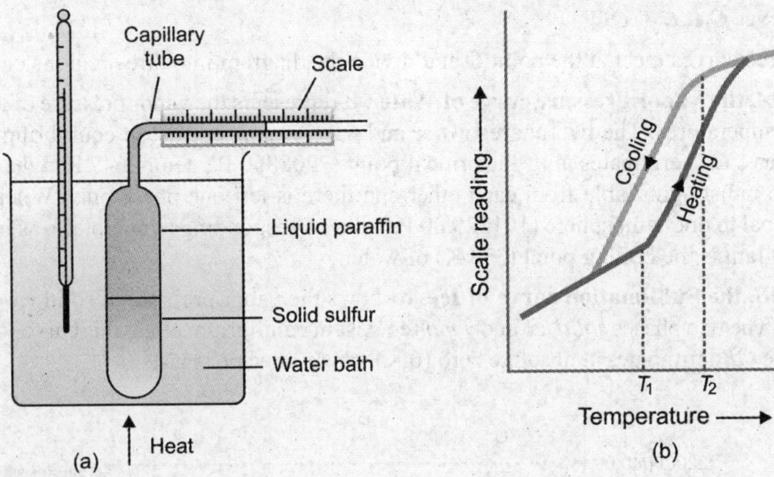

■ **Figure 19.2**
(*a*) A dilatometer; (*b*) Change in volume plotted against temperature recorded with a dilatometer.

(3) Solubility change

Two forms of the same substance have different solubilities but at the transition point they have identical solubility. Thus if solubility-temperature graph is plotted for the two forms, it is found to consist of two parts with a sharp break. While one part represents the solubility curve for one form, the second part represents that for the other. **At the meeting point of the two curves, the solubility of the two forms is the same and it indicates the transition temperature.** For example, in the diagram for the system sodium sulfate-water, the solubility curves of Na_2SO_4 (rhombic) and $Na_2SO_4.10H_2O$ meet at 305.2 K. Thus 305.2 K is the transition temperature where $Na_2SO_4.10H_2O$ changes to Na_2SO_4.

(4) Cooling curve method

There is often an evolution or absorption of heat when one form passes to the other. Suppose that form *A* is converted into form *B* on heating. Now let *B* be allowed to cool and a curve obtained by plotting the temperature against the time. **The otherwise steady curve has a distinct break at a temperature corresponding to the transition point because here heat is evolved from *B*.**

This method is suitable for determining the transition temperature between different hydrates of a salt or between a hydrate and an anhydrous salt (*e.g.*, $Na_2SO_4.10H_2O$ to Na_2SO_4), or for different forms of a metal.

THE WATER SYSTEM

Under normal conditions the system 'water' is a **three-phase, one-component system.** The three phases involved are *liquid water, ice, water vapor.* All these phases can be represented by one chemical entity H_2O and hence one component of the system. The number of phases which can exist in equilibrium any time depends on the conditions of temperature and pressure. The phase diagram or *PT*-graph of the system/ water/ice/vapor is shown in Fig. 19.3. The **salient features of the phase diagram** are listed below.

(1) The Curves *OA*, *OB*, *OC*

(2) The Triple Point *O*

(3) The Areas *AOC*, *AOB*, *BOC*

Let us proceed to discuss the significance of each of these features.

(1) The Curves *OA*, *OB*, *OC*

These three curves meet at the point *O* and divide the diagram into three regions or areas.

Curve *OA*, the Vapor Pressure curve of Water. It represents the vapor pressure of liquid water at different temperatures. The two phases *water* and *water vapor* coexist in equilibrium along this curve. The curve *OA* terminates at *A*, the critical point (2208850 Pa, temp. 647 K) when the liquid and vapor are indistinguishable from each other and there is left one phase only. When the vapor pressure is equal to one-atmosphere (101325.00 Pa), the corresponding temperature, as indicated on the phase diagram is the boiling point (373 K) of water.

Curve *OB*, the Sublimation curve of Ice. It shows the vapor pressure of solid ice at different temperatures. The two phases *solid ice* and *vapor* coexist in equilibrium along this curve. At the lower limit the curve *OB* terminates at absolute zero (0 K) where no vapor exists.

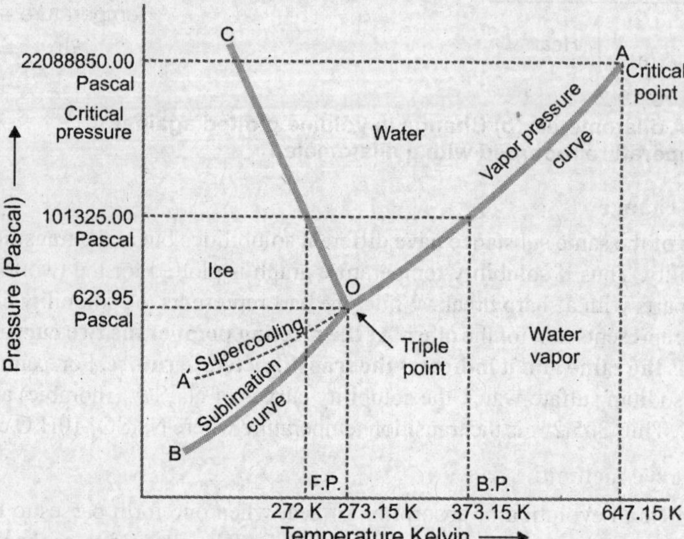

Figure 19.3
The phase diagram of the system 'Water'.

Curve *OC*, the Fusion curve of Ice. It depicts the effect of pressure on the melting point of ice. Here *ice* and *water* coexist in equilibrium. The fact that *OC* slopes to the left indicates that the melting point of ice decreases with increase of pressure. Since ice melts with decrease in volume by Le Chatelier's principle the melting point is lowered by an increase of pressure. It may be noted that the 1.0 atmosphere (101325 Pa) line meets the fusion curve at 273 K which is the normal melting point of ice.

Along the curves *OA*, *OB*, *OC* there are two phases in equilibrium and one component. Therefore,

$$F = C - P + 2 = 1 - 2 + 2 = 1$$

Hence each two-phase system :

water/water vapor represented by *OA*

ice/water vapor represented by *OB*

ice/water represented by *OC*

has one degree of freedom *i.e.*, is **monovariant.**

(2) The Triple point '*O*'

The curves *OA*, *OB* and *OC* meet at the triple point '*O*' where all the three phases *liquid water/ice/vapor* are in equilibrium. This occurs at 273.1576 K and vapor pressure 4.58 mm Hg (610.62 Pa). Since there are three phases and one component, we have

$$F = C - P + 2 = 1 - 3 + 2 = 0$$

i.e., the system at the triple point is **nonvariant.** Thus if either pressure or temperature is changed, the three phases would not exist and one of the phases would disappear.

(3) Area *AOC, AOB, BOC*

The areas or regions between the curves show the conditions of temperature and pressure under which a single phase–**ice, water or vapor** is capable of stable existence. Thus,

Area *AOC* represents conditions for the one-phase system **water.**

Area *AOB* represents conditions for the one-phase system **water vapor.**

Area *BOC* represents conditions for the one-phase system **ice.**

In all the three areas there being one-phase and one-component, we have

$$F = C - P + 2 = 1 - 1 + 2 = 2$$

Thus each system *water, water vapor,* or *ice* has 2 degrees of freedom *i.e.,* the system is **bivariant.**

(4) Metastable system : Supercooled Water/Vapor System

The vapor pressure curve of water *AO* can be continued past the triple point as shown by the dashed line *OA'*. That is, water can be *supercooled* by carefully eliminating solid particles. The supercooled water/vapor system is metastable (unstable). It at once reverts to the stable system ice/vapor on the slightest disturbance or introducing a crystal of ice.

THE Sulfur SYSTEM

It is a **one-component, four-phase system.** The four phases are :

(*a*) Two solid polymorphic forms :

(*i*) Rhombic Sulfur (S_R)

(*ii*) Monoclinic Sulfur (S_M)

(*b*) Sulfur Liquid (S_L)

(*c*) Sulfur Vapor (S_V)

All the four phase can be represented by the only chemical individual 'sulfur' itself and hence one component of the system.

The two crystalline forms of sulfur S_R and S_M exhibit enantiotropy with a transition point at 95.6°C. Below this temperature S_R is stable, while above it S_M is the stable variety. At 95.6°C each form can be gradually transformed to the other and the two are in equilibrium. At 120°C, S_M melts. Thus,

$$S_R \xrightleftharpoons{95.61^o} S_M \xrightleftharpoons{120^o} S_L$$

The phase diagram for the sulfur system is shown in Fig. 19.4. The **salient features of the phase diagram** are described below.

(*i*) The six curves *AB, BC, CD, BE, CE, EG*

(*ii*) The three Triple points *B, C, E*

(*iii*) The four areas :

ABG marked 'solid Rhombic'

BEC marked 'solid Monoclinic'

GECD marked 'liquid Sulfur'

ABCD marked 'Sulfur vapor'

Let us now proceed to discuss the significance of these features :

(1) The curves *AB, BC, CD, BE, CE, EG*

These *six curves* divide the diagram into *four* areas.

Curve *AB*, the Vapor Pressure curve of S_R. It shows the vapor pressure of solid rhombic sulfur (S_R) at different temperatures. Along this curve the two phases S_R and sulfur vapor (S_V) are in equilibrium. The system S_R/S_V has one degree of freedom,

$$F = C - P + 2 = 1 - 2 + 2 = 1$$

i.e., it is *monovariant.*

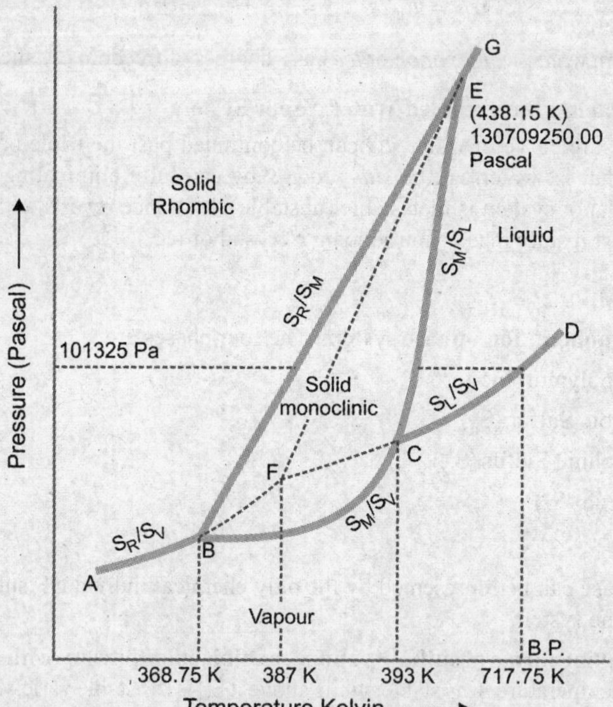

■ **Figure 19.4**
The phase diagram of the system 'Sulfur'.

Curve *BC*, the Vapor Pressure curve of S_M. It shows variation of the vapor pressure of monoclinic sulfur (S_M) with temperature. S_M and S_V coexist in equilibrium along this curve. The system S_M/S_V is *monovariant.*

Curve *CD*, the Vapor Pressure curve of S_L. It depicts the variation of the vapor pressure of liquid sulfur (S_L) with temperature. S_L and S_V are in equilibrium along *CD*. The two phase system S_L/S_V is monovariant. One atmosphere line meets this curve at a temperature (717.75 K) which is the boiling point of sulfur.

Curve *BE*, the Transition curve. It shows the effect of pressure on the transition temperature for S_R and S_M. As two solid phases are in equilibrium along the curve, the system S_R/S_M is *monovariant.* The transformation of S_R and S_M is accompanied by increase of volume (density of S_R = 2.04; S_M = 1.9) and absorption of heat *i.e.,*

$$S_R + Q \text{ (heat energy)} \rightleftharpoons S_M$$

Thus the increase of pressure will shift the equilibrium to the left (*Le Chatelier's Principle*) and the transition temperature will, therefore, be raised. This is why the line *BE* slopes away from the pressure axis showing thereby that the transition temperature is raised with increase of pressure.

Curve *CE*, the Fusion curve of S_M. It represents the effect of pressure on the melting point of S_M. The two phases in equilibrium along this curve are S_M and S_L. The system S_M/S_L is *monovariant.* As the melting or fusion of S_M is accompanied by a slight increase of volume, the melting point will rise by increase of pressure (*Le Chatelier's principle*). Thus the curve *CE* slopes slightly away from the pressure axis. The curve ends at *E* because S_M ceases to exist beyond this point.

Curve *EG*, the Fusion curve for S_R. Here the two phases in equilibrium are S_R and S_L. The number of phases being two, the system S_R/S_L is *monovariant.*

(2) The Triple points *B*, *C*, *E*

Triple point B. This is the meeting point of the three curves *AB*, *BC* and *BE*. Three phases, solid S_R, solid S_M and S_V are in equilibrium at the point B. There being three phases and one component, the system $S_R/S_M/S_L$ is *nonvariant.*

$$F = C - P + 2 = 1 - 3 + 2 = 0$$

At *B*, S_R is changed to S_M and the process is reversible. Thus the temperature corresponding to *B* is the *transition temperature* (95.6°C or 368.75 K).

Triple point *C*. The curves *BC*, CD, CE meet at this point. The three phases in equilibrium are S_M, S_L and S_V. There being three phases and one component, the system $S_M/S_L/S_V$ is *nonvariant.* The temperature corresponding to *C* as indicated on the phase diagram is 120°C (393 K). This is the melting point of S_M.

Triple point *E*. The two lines *CE* and *BE*, having different inclinations away from the pressure axis, meet at E where a third line *EG* also joins. The three phases S_R, S_M and S_L are in equilibrium and the system at the point *E* is *nonvariant*. This point gives the conditions of existence of the system $S_R/S_M/S_L$ at 155°C (428 K) and 1290 atmospheres (130709250.00 Pa) pressure.

(3) The Areas

The phase diagram of the sulfur system has **four areas** or **regions.** These are labelled as rhombic sulfur, monoclinic sulfur, liquid sulfur and vapor. These represent single phase systems which have two degrees of freedom,

$$F = C - P + 2 = 1 - 1 + 2 = 2$$

That is, each of the systems S_R, S_M, S_L, and S_V are *bivariant.*

(4) Metastable Equilibria

The change of S_R to S_M takes place very slowly. If enough time for the change is not allowed and S_R is heated rapidly, it is possible to pass well above the transition point without getting S_M. In that case, there being three phases (S_R, S_L, S_V) only and one component, the phase diagram, like that of water system, will consist of three curves, one triple point and three areas.

The dashed curve *BF*, the Vapor Pressure curve of metastable S_R. This is a continuation of the vapor pressure curve *AB* of stable S_R. The metastable phases S_R and S_V are in equilibrium along this curve. It is a *monovariant system.*

The dashed curve CF, the Vapor Pressure curve of supercooled S_L. On supercooling liquid sulfur, the dashed curve *CF* is obtained. It is, in fact, the back prolongation of *DC*. The curve *CF* represents the metastable equilibrium between supercooled S_L and S_V. Thus it may be designated as the vapor pressure curve of supercooled S_L. It meets the dashed curve *BF* at *F*.

The dashed curve *FE*, the Fusion curve of metastable S_R. The two metastable phases S_R and S_L are in equilibrium along this curve and the system is *monovariant.* This shows that the melting point of metastable S_R is increased with pressure. Beyond *E*, this curve depicts the conditions for the stable equilibrium S_R/S_L as the metastable S_R disappears.

The metastable Triple point *F*. At this point, three metastable phases S_R, S_L and S_V are in equilibrium. The system is a metastable triple point with no degree of freedom. The corresponding temperature is the melting point of metastable S_R (114°C or 387 K).

TWO-COMPONENT SYSTEMS

When a single phase is present in a two-component system, the degree of freedom is three,

$$F = 2 - 1 + 2 = 3$$

This means that three variables must be specified in order to describe the condition of the phase. Thus in such a system, in addition to pressure and temperature the concentration of one of the components has also to be given. For graphic representation of these variables, three coordinate axes at right angles to each other would be required. Therefore the phase diagram obtained would be a solid model.

For the sake of having simple plane diagrams we generally consider only two variables, the third one being a constant. For example, for a solid/liquid equilibrium, the gas phase is usually absent and the effect of pressure on the equilibrium is very small. Thus when a two-component system consists of solid and liquid phases only, the effect of pressure may be disregarded. Then it is necessary to take into account the remaining variables *viz.*, temperature and concentration. **Such a solid/liquid system with the gas phase absent is called a condensed system.**

The experimental measurements of temperature and concentration in condensed systems are usually carried out under atmospheric pressure. Since the degree of freedom in such a case is reduced by one, we may write the **Reduced Phase rule** as

$$F' = C - P + 1$$

where *F'* gives the remaining degrees of freedom of the system. The reduced phase rule is more convenient to apply to solid/liquid two-component condensed system.

Since the only variables for two-component solid/liquid systems are temperature and composition, the phase diagrams for such systems consist of Temperature-Concentration graphs (*TC graphs*).

Simple Eutectic Systems

The general form of the phase diagram of such a 2-component condensed system is shown in Fig. 19.5. Here the two components *A* and *B* are completely miscible in the liquid state, and these solutions on cooling yield only pure *A* or pure *B* as solid phases.

The diagram consists of :

Curve *AC*; the Freezing point curve of *A*. The point *A* represents the freezing point of *A*. The curve *AC* shows that the freezing point of *A* falls by the addition of *B* to *A*. Thus along this curve, the solid *A* is in equilibrium with the liquid solution of *B* in *A*.

Curve *BC*; the Freezing point curve of *B*. The point *B* shows the freezing point of *B*. The curve *BC* exhibits the fall of freezing point by the addition of *A* to *B*. Along this curve, the solid *B* is in equilibrium with the liquid solution of *A* in *B*.

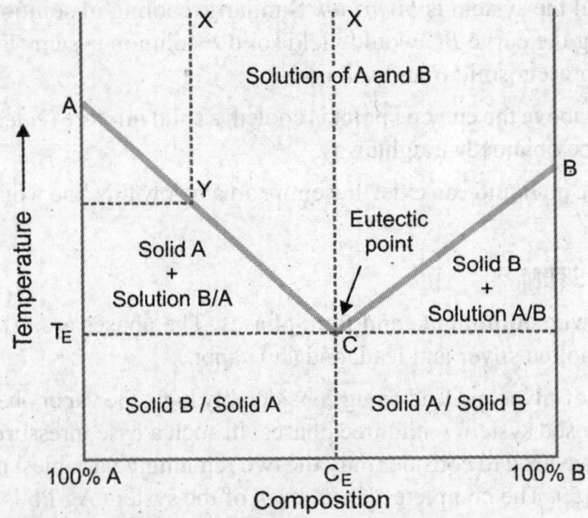

Figure 19.5
Simple Eutectic diagram.

Applying the reduced phase rule equation to the equilibria represented by the curve *AC* and *CB* *i.e.*, solid *A*/solution and solid *B*/solution respectively, we have

$$F = C - P + 1 = 2 - 2 + 1 = 1$$

The degree of freedom is one *i.e.*, both equilibria are *monovariant.*

The Eutectic point *C*. The two curves *AC* and *BC* meet at the point *C*. Here both the solids *A* and *B* must be in equilibrium with the solution phase (solution of *A* and *B*). The number of phases is 3. By applying the reduced phase rule equation, we have

$$F' = C - P + 1 = 2 - 3 + 1 = 0$$

Thus the system represented by the point *C* is *nonvariant.* In other words, both the temperature and composition of the system solid *A*/solid *B*/solution are fixed.

The mixture of components *A* and *B* at point *C* melts at the lowest temperature T_E indicated on the graph. The *point C* is therefore, called **the Eutectic point** (Greek *eutectos* = easy melting). The corresponding composition (C_E) and temperature (T_E) are known as the **eutectic composition** and the **eutectic temperature** respectively of the system.

The **eutectic mixture, although it has a definite melting point, is not to be regarded as a compound.** The reasons are : (*i*) the components are not in stoichiometric proportions; and (*ii*) on examination under a microscope these reveal the existence of separate crystals of the components.

The Area above the curves *AC* and *BC*. Here the two components *A* and *B* are present as *liquid solutions* of varying compositions. As a homogeneous solution of *A* and *B* constitutes one phase only, this system is *bivariant.*

$$F = C - P + 1 = 2 - 1 + 1 = 2$$

Therefore to define the system at any point in this area, both temperature and composition have to be specified.

Effect of Cooling

When the *A*/*B* solution at any point in the area above *ACB* is cooled, the cooling dashed line meets the curve *AC*, say at *Y*. Here solid *A* separates and the equilibrium shifts down along the curve *AC*. The change of composition and temperature continues till the eutectic point *C* is reached when solid *B* also separates. Thus in the area below *AC* and above T_E line, there exist two phases *viz.*, solid

A and solution *A/B*, and the system is *bivariant.* Similarly, cooling of solution *B/A* on the other side of eutectic, on reaching the curve *BC* would yield solid *B*/solution system. Thus the area below *BC* up to T_E line would represent solid *B* and solution.

If the solution just above the eutectic point is cooled, a solid mixture (*eutectic mixture*) of eutectic composition *CE*, will be obtained straightaway.

Since below T_E line no liquid can exist, the entire area below this line would represent the system solid *A*/solid *B*.

THE SILVER–LEAD SYSTEM

This system has **two components** and **four phases.** The phases are : (*i*) solid silver; (*ii*) solid lead; (*iii*) solution of molten silver and lead; and (*iv*) vapor.

The boiling points of silver and lead being considerably high, the vapor phase is practically absent. Thus Ag/Pb is a condensed system with three phases. In such a case, pressure can have no effect on the system. Therefore we need to consider only the two remaining variables, namely the temperature (*T*) and concentration (*C*). The complete *TC diagram* of the system Ag/Pb is shown in Fig. 19.6.

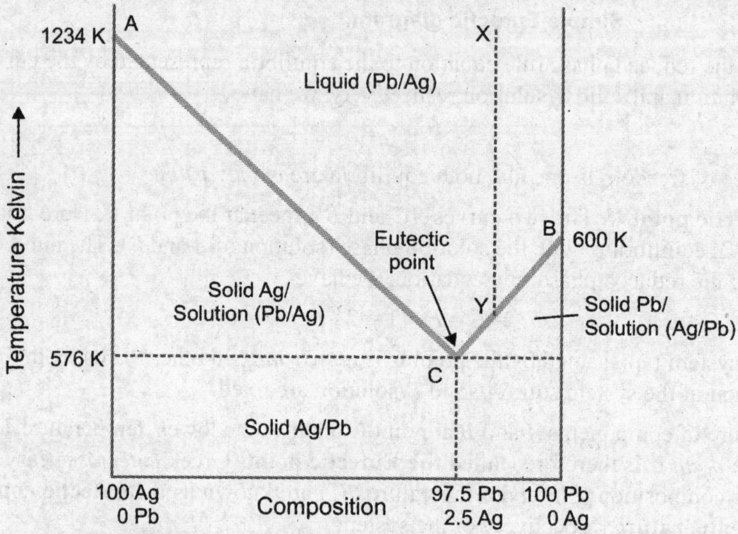

Figure 19.6
The phase diagram of Ag/Pb system.

The salient features of the diagram are :

(*a*) Two curves, *AC* and *BC*

(*b*) Eutectic point, *C*

(*c*) Three areas : (*i*) *above ACB*; (*ii*) *below AC;* (*iii*) *below BC*

Curve AC; the Freezing point curve of Ag. *A* represents the freezing point or melting point of solid silver (961°C or 1234 K) and the curve *AC* shows that the addition of lead lowers the melting point along it. The phases in equilibrium along *AC* are solid silver and solution of silver and lead. Applying the reduced phase rule equation

$$F' = C - P + 1 = 2 - 2 + 1 = 1$$

Thus the system Ag/solution is *monovariant.*

Curve *B*; the Freezing point curve of Pb. *B* represents the melting point of solid lead (327°C or 600 K) and the curve *BC* shows that the melting point is lowered by addition of silver. The phases in equilibrium along *BC* are solid lead and solution. The system is *monovariant.*

The Eutectic point *C*. The curves *AC* and *BC* intersect at *C*, which is called the *eutectic point.* Here three phases solid Ag, solid Pb, and solution are in equilibrium. Applying the reduced phase rule equation

$$F' = C - P + 1 = 2 - 3 + 1 = 0$$

Thus the system Ag/Pb/solution at *C* is *nonvariant.* Both the variables, temperature (303°C or 576 K) and composition (97.5% Pb, 2.5% Ag) are fixed. If you raise the temperature above the eutectic temperature, the solid phases Ag and Pb disappear and if you cool below it, you will land in the solid Ag/Pb area where solution phase is nonexistent.

The Area above *AOC*. This region represents the single phase system, the solution of molten Ag and Pb. Applying the reduced phase rule equation, we have

$$F' = C - P + 1 = 2 - 1 + 1 = 2$$

Thus the system solution Ag/Pb is *bivariant.*

The area below *AC* represents the phases Ag + solution, while that below *BC* the phases Pb + solution. The area below the temperature 576 K, represents solid Ag + solid Pb. All these areas have two phases and one degree of freedom,

$$F = C - P + 1 = 2 - 2 + 1 = 1$$

Pattinson's Process for the Desilverization of Argentiferous Lead

This process of recovery of silver from argentiferous lead is based on the facts contained in the diagram (Fig. 19.6). The argentiferous lead containing small amount of silver (less than 0.1%) is melted well above the melting temperature of pure lead (327°C or 600 K). Let the point *X* represent the system 'molten lead' on the diagram. It is then allowed to cool when the temperature of the melt falls along the dashed line *XY*. As the temperature corresponding to *Y* on the curve *BC* is reached solid lead begins to separate and the solution would contain relatively larger amount of silver. On further cooling, more of lead separates and we travel along the curve *BC* until the eutectic point *C* is reached. Lead is continuously removed by means of ladles and the percentage of silver in the melt goes on increasing. At *C*, an alloy containing 2.5% Ag and 97.5% *Pb* is obtained. This is treated for the recovery of silver profitably.

THE ZINC–CADMIUM SYSTEM

This is another example of a metal/metal system with a eutectic. The phase diagram is shown in Fig 19.7 which is similar to the one for silver/lead system. It consists of :

Curve *AO*; the Freezing point curve of Zinc. *A* represents the freezing point (or melting point) of zinc (419°C or 692 K). The curve *AO* shows that the melting point of zinc is lowered by the addition of cadmium. The phases in equilibrium along *AO* are solid zinc and liquid solution of zinc and cadmium. Applying the reduced phase rule equation.

$$F' = C - P + 1 = 2 - 2 + 1 = 1$$

Thus the system solid zinc/solution is *monovariant.*

Curve *BO*; the Freezing point curve of Cadmium. The point *B* represents the melting point of cadmium (321°C or 594 K). The curve *BO* shows the fall of melting point of cadmium on the addition of zinc. Along this curve, solid cadmium is in equilibrium with the liquid solution of zinc and cadmium.

Applying the reduced phase rule equation to the equilibrium cadmium/solution,

$$F' = C - P + 1 = 2 - 2 + 1 = 1$$

Thus the degree of freedom is one and the equilibrium is *monovariant.*

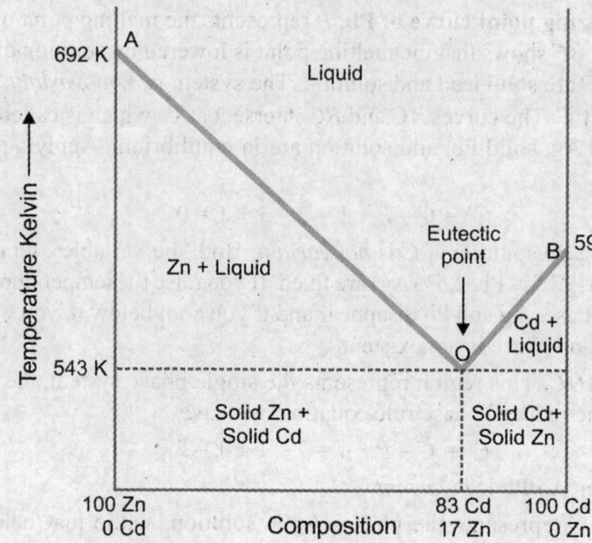

Figure 19.7
Phase diagram of Zn-Cd system.

The Eutectic point *O*. The curves *AO* and *BO* meet at *O*. Here, solid zinc, solid cadmium and solution are in equilibrium. The number of phases is, therefore, three.

$$F = C - P + 1 = 2 - 3 + 1 = 0$$

Thus the system at point *O* is *nonvariant.* The point *O* is called the **eutectic point.** Both the temperature and composition being fixed, the system has no degree of freedom. The eutectic temperature is 543 K and the eutectic composition is 83% cadmium 17% zinc.

The Area above *AO*. In this area, both the components zinc and cadmium are present as liquid solution of varying composition. The solution of the two metals being homogeneous, constitutes one phase only. Thus the system represented by the area above *AO* is *bivariant.*

$$F = C - P + 1 = 2 - 1 + 1 = 2$$

The Area above *BO*. In this area, zinc and cadmium exist as liquid solution. The composition of the solution is indicated on the composition axis. The liquid solution, regardless of its composition, represents one phase only. Thus the system in the area above *BO* is *bivariant.*

$$F = C - P + 1 = 2 - 1 + 1 = 2$$

The effect of cooling. If the solution in the area above *AO* is cooled, zinc separates as the curve itself is reached. This continues till the point *O* when the solid eutectic mixture (83% Cd + 17% Zn) separates. Similarly, the solution above *BO* on cooling allows the separation of cadmium. This continues till the point *O* is reached at 543 K and eutectic mixture separates. As clear from the diagram, the area below the curve *AO* represents zinc and solution; the area below *BO* shows cadmium and solution. The area below the eutectic temperature represents the system solid zinc/solid cadmium.

POTASSIUM IODIDE–WATER SYSTEM

It has **four phases** : (*i*) Solid KI ; (*ii*) Solution of KI in water ; (*iii*) Ice; and (*iv*) Vapor. Only two chemical constituents KI and H_2O being necessary to depict the composition of all the four phases, it is a **two-component system.**

Since the conditions for the existence of the various phases are studied at atmospheric pressure the vapor phase is ignored and the system KI-H_2O is regarded as a *condensed system.* Pressure being constant, the two variables, temperature and concentration will be considered. The *TC* diagram of the system is shown in Fig. 19.8. It consists of :

(*a*) The Curves *AO* and *OB*

(*b*) The Eutectic Point *O*

(*c*) The area above *AOB* and the areas below the curves *OA* and *BO*

The Curve *AO*; the Freezing point curve of Water. The point *A* represents the freezing point of water or the melting point of ice (0°C or 273 K) under normal conditions. The curve *AO* shows that the melting point of ice falls by the addition of solid KI. As more and more of KI is added, the concentration of solution and the melting temperature changes along the curve *AO*. The phases in equilibrium along the curve *AO* are ice and solution. Applying the reduced phase rule equation to the condensed system ice/solution, we have

$$F' = 2 - 2 + 1 = 1$$

Thus the system is *monovariant.*

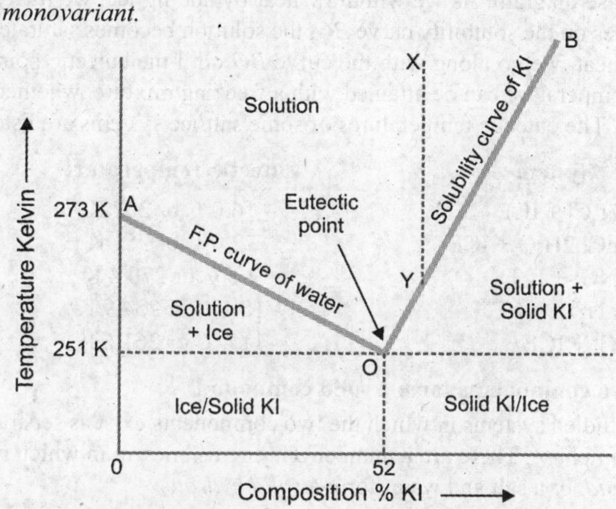

Figure 19.8
The phase diagram of the system KI-H_2O.

The Eutectic point. The lowest point attainable by the addition of KI along the curve *OA* is *O*. Here the solution becomes saturated with KI and the solid KI appears as the third phase. This point is termed the *Eutectic Point* or *Cryohydric Point* as one of the components in the system is water. Applying the reduced phase rule equation to the system ice/solid KI/solution at point *O*.

$$F' = 2 - 3 + 1 = 0$$

Hence the system is *nonvariant.* That is, both the temperature (– 22°C) and composition (52% KI + 48% ice) are fixed.

The Curve *BO*; the Solubility curve of KI. At *O*, the solution is saturated with *KI.* Thus the curve *BO* depicts the effect of temperature on the concentration of saturated solution or the solubility of KI. The phases in equilibrium along the curve are solid KI and solution. Applying the reduced phase rule equation, we have

$$F = C - P + 1 = 2 - 2 + 1 = 1$$

That is, the condensed system solid KI/solution is *monovariant.*

The Area above *AOB*. It represents the single phase system 'solution'. Applying the phase rule equation,

$$F = C - P + 1 = 2 - 1 + 1 = 2$$

Therefore the system is *bivariant.*

As labelled in the diagram, the area below *AO* shows the existence of ice and solution, while the area below *BO* depicts the presence of solid KI and solution. Below the eutectic temperature line, there can exist ice and solid KI only.

Cooling Produced by Freezing Mixtures

KI-H_2O is a typical eutectic system with a salt and water as components. Many other salts as sodium chloride, sodium nitrate, ammonium chloride and ammonium nitrate constitute eutectic systems with water and form similar phase diagrams as shown in Fig. 19.8. The facts contained in this diagram explain the **theory of freezing mixtures** which are obtained by mixing salt and ice.

When we add a salt, common salt (NaCl) to melting ice, we follow along the curve *AO*. Thus the addition of salt to the system ice/water produces a continued lowering of temperature until the eutectic point O is reached. Thus the lowest temperature attained in this way will be the eutectic temperature (251 K). Alternatively, we may start with a concentrated solution of the salt represented by point *X* on the phase diagram. As we withdraw heat by adding ice, we travel along the dashed line *XY*. At *Y* which lies on the solubility curve *BO*, the solution becomes saturated with the salt. On further withdrawing heat, we go along with the curve *BO* until the eutectic point is reached. Thus the same minimum temperature can be attained with a freezing mixture, whether we add salt to ice or ice to salt solution. The eutectic temperatures of some salt/ice systems are listed below.

System	Eutectic Temperature
NH_4Cl + ICE	– 16.0°C or 267 K
NaCl.$2H_2O$ + ICE	– 22.0°C or 251 K
KNO_2 + ICE	– 2.6°C or 270.4 K
$NaNO_2$ + ICE	– 18.1°C or 254.9 K
KCl + ICE	– 11.4°C or 261.6 K

Systems in which two components form a solid compound

So far we have studied systems in which the two components exist as separate entities whether in the solid or solution phase. There are a number of systems known, in which two metals form an *intermetallic compound,* or a salt and water form a *solid hydrate.*

Let us consider a general case with two components *A* and *B* which form a stable compound *AB*. The phase diagram for such a system will be of the type shown in Fig. 19.9.

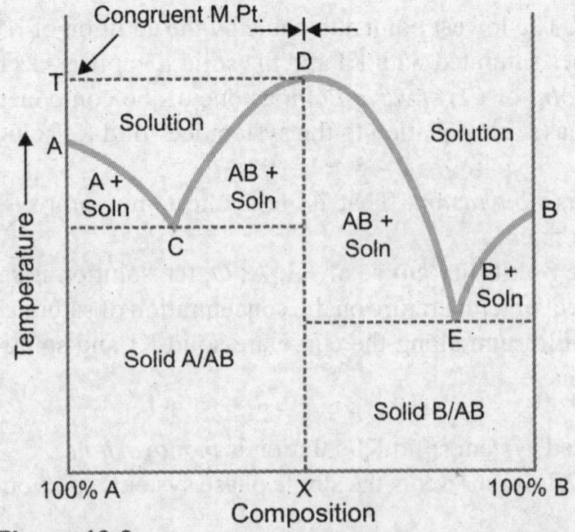

■ **Figure 19.9**
Compound formation with congruent melting point.

The system will involve three solid phases, *A*, *B* and *AB*. The fourth phase will be a liquid (or solution) containing varying concentration of *A* and *B*. There are three freezing point or melting point curves *viz.*, *AC*, *CDE* and *EB*. The various phases in equilibrium along these curves and in different regions are labelled in the phase diagram.

What is Congruent Melting Point?

Along the freezing point curve *CDE* of the compound *AB*, solid *AB* is in equilibrium with the solution phase. At the maximum point *D* of this curve, the composition of the solid compound and the liquid phase in contact is the same, which is indicated by *X* on the composition axis. The corresponding temperature *T*, therefore, coincides with the melting point of the compound *AB*.

When the solid compound *AB* and the liquid phase have identical composition at the maximum point on the freezing point curve, the corresponding temperature is said to be the Congruent Melting Point of the compound.

At point *D* since both the phases have the same composition, the system has one component only. Hence it is *nonvariant.*

$$F = C - P + 1 = 1 - 2 + 1 = 0$$

The diagram in Fig. 19.9 has the appearance of two simple eutectic diagrams joined together at the position of arrows. The diagram to the left represents the eutectic system *A*/*AB*, while that on the right the system *AB*/*B*. The phase diagram of the system Zinc-Magnesium is exactly of the type shown in Fig. 19.9.

The salt-water systems often involve the formation of several compounds (hydrates). For example, the system Ferric chloride-Water forms as many as four hydrates. Thus the composite phase diagram could be regarded as made of four simple eutectic diagrams.

Some common 2-component systems involving the formation of one or more compounds with congruent melting points are discussed below.

THE MAGNESIUM–ZINC SYSTEM

It is a typical 2-component system which involves the formation of an intermetallic compound $MgZn_2$. It has four phases : solid magnesium (Mg), solid zinc (Zn), solid $MgZn_2$ and the liquid solution of Mg and Zn.

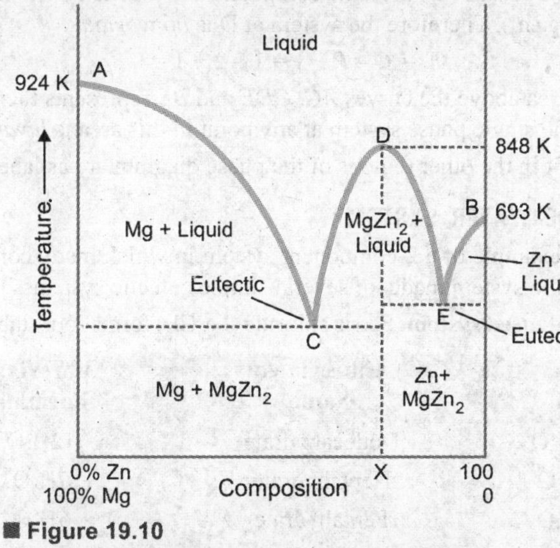

■ **Figure 19.10**
The phase diagram of the system Mg-Zn.

The complete phase diagram of the system magnesium-zinc is shown in Fig. 19.10. It appears to be made of two simple eutectic diagrams. The one toward the left represents the eutectic system Mg-$MgZn_2$, while the one to the right the system Zn-$MgZn_2$.

The curves *AC*, *CDE* and *BE*. *AC* is the freezing point curve of magnesium; *BE* is the freezing point curve of zinc; and *CDE* is that of the compound $MgZn_2$.

The curve *AC* shows that the melting point of magnesium (651°C or 924 K) is lowered on the addition of zinc. This continues until the point *C* is reached. Here a new phase, solid $MgZn_2$ appears.

The curve *CD* shows the increase of concentration of zinc in the melt with the rise of temperature. At the maximum point *D*, the composition of the melt and the solid compound becomes the same *i.e.*, $MgZn_2$. The point D, therefore, represents the melting point of $MgZn_2$ (575°C or 848 K). The curve DE now shows the lowering of the melting point with the addition of zinc until the lowest point is attained. Here solid zinc appears.

The curve *BE* exhibits that the melting point of zinc (420°C or 693 K) falls with the addition of magnesium until the point *E* is reached.

Along the freezing point curves *AC*, *CDE* and *BE*, there are two phases in equilibrium *viz*., one solid phase (Mg, Zn, or $MgZn_2$) and the other liquid phase. Applying the reduced phase rule equation, we have

$$F = C - P + 1 = 2 - 2 + 1 = 1$$

This shows that the system Mg/liquid, Zn/liquid and $MgZn_2$/liquid are all *monovariant.*

Eutectic points *C* and *E*. There are two eutectic points in the phase diagram. The systems at the points *C* and *E* have two components and three phases in equilibrium.

	Phases Present
C	Solid Mg, Solid $MgZn_2$, Liquid
E	Solid Zn, Solid $MgZn_2$, Liquid

These systems are, therefore, *nonvariant.*

$$F = C - P + 1 = 2 - 3 + 1 = 0$$

Congruent Melting Point. As already stated, the composition of the compound $MgZn_2$ and the melt at *D* is identical. The corresponding temperature is the *congruent melting point* of the compound. Here the system has two phases *viz.*, the solid compound and the melt. Both these can be represented by one component ($MgZn_2$). Therefore the system at D is *nonvariant*,

$$F = C - P + 1 = 1 - 2 + 1 = 0$$

The Areas. The area above the curves *AC*, *CDE* and *BE* represents the solution of magnesium and zinc (the melt). The single phase system at any point in this area is *bivariant.*

The phases present in the other regions of the phase diagram are as labelled.

THE FERRIC CHLORIDE–WATER SYSTEM

This provides an example of a 2-component system in which many compounds (hydrates) are formed. It is a composite system made of several simple eutectic systems.

Seven-phase condensed system. Ferric chloride (Fe_2Cl_6) forms four stable crystalline hydrates :

Formula	Abbreviated Name	Abbreviated Formula
$Fe_2Cl_6.12H_2O$	Dodecahydrate	$12H_2O$
$Fe_2Cl_6.7H_2O$	Heptahydrate	$7H_2O$
$Fe_2Cl_6.5H_2O$	Pentahydrate	$5H_2O$
$Fe_2Cl_6.4H_2O$	Tetrahydrate	$4H_2O$

The other three phases involved are ice, anhydrous ferric chloride and solution. Since all measurements of temperature and concentration are made at atmospheric pressure, the vapor phase is ignored and the system Fe_2Cl_6/H_2O is considered a **condensed system.**

Two components. As the composition of all the seven phases can be represented by the constituents Fe_2Cl_6 and H_2O, it is a two-component system.

Phase Diagram of Fe_2Cl_6/H_2O System

The phase diagram of the ferric chloride-water system is shown in Fig. 19.11.

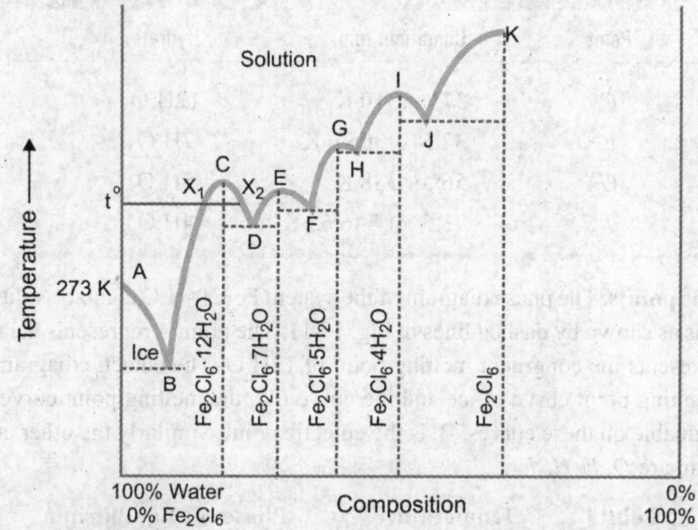

Figure 19.11
The Ferric chloride-Water system.

The Curves. The point *A* represents the normal freezing point of water. The curve *AB* shows the lowering of freezing point on the addition of ferric chloride and hence is named as the **freezing point curve of water.** The freezing point falls till the point *B* is reached where a new phase $Fe_2Cl_6.12H_2O$ separates. Now as the temperature is raised and more of Fe_2Cl_6 is added, we travel along the curve *BCD*. $Fe_2Cl_6.12H_2O$ and saturated solution are in equilibrium along this curve which is, therefore, called the **solubility curve of dodecahydrate.** It will be noted that both the increase as also the decrease in the proportion of Fe_2Cl_6 in the solution at *C* will cause a lowering of solubility. Thus at a temperature such as $t°$, the $12H_2O$ has two distinct solubilities in water, X_1 and X_2. **This phenomenon of having two solubilities at the same temperature, characteristic of systems producing solid compounds, is called retroflex solubility.**

The solubility curve *BCD* of $12H_2O$ terminates at *D* where another hydrate $7H_2O$ separates. The curve *DEF* is the solubility curve of the heptahydrate. At *F*, the $5H_2O$ appears and *FGH* is its solubility curve. AT *H*, $4H_2O$ separates and *HIJ* is its solubility curve. At *J*, solid ferric chloride (Fe_2Cl_6) appears and *JK* is the solubility curve of the anhydrous salt.

Along the curves *AB*, *BCD*, *DEF*, *FGH*, *HIJ* and *JK*, one solid phase and one liquid (solution) phase are in equilibrium. Therefore applying the reduced phase rule equation we have

$$F = C - P + 1 = 2 - 2 + 1 = 1$$

Hence the systems ice/solution, $12H_2O$/solution, $7H_2O$/solution, $5H_2O$/solution, $4H_2O$/solution, and Fe_2Cl_6/solution represented by the various curves are *monovariant*.

The Congruent Melting Points. At points *C*, *E*, *G*, *I* the composition of the solution in equilibrium with the respective hydrate is identical. These points, therefore, represent the congruent melting points of the different hydrates (Table 19.1).

Since a solid phase (hydrate) and liquid phase (solution) are identical at the congruent melting point, the system at points *C*, *E*, *G*, *I* has two phases and one component. It is, therefore, *nonvariant*,

$$F = C - P + 1 = 1 - 2 + 1 = 0$$

TABLE 19.1. CONGRUENT MELTING POINTS OF HYDRATES OF Fe_2Cl_6

Point	Congruent m.p.	Hydrate
C	37° or 310 K	$12H_2O$
E	32.5° or305.6 K	$7H_2O$
G	56° or 338 K	$5H_2O$
I	73.5° or 346.65 K	$4H_2O$

The Eutectic points. The phase diagram of the system Fe_2Cl_6-H_2O could be split into five simple eutectic diagrams as shown by dashed lines in Fig. 19.11. The point A represents the melting point of ice, while *C* represents the congruent melting point of $12H_2O$. The eutectic diagram is made of the curve *AB*, the melting point curve of ice and the curve *CB*, the melting point curve of $12H_2O$. The lowest point attainable on these curves, B, is the eutectic point. Similarly the other eutectic points in the phase diagram are *D*, *F*, *H*, *J*.

Eutectic Point	Temperature	Phases in Equilibrium
B	– 55° or 218 K	ice, $12H_2O$, solution
D	27.4° or 300 K	$12H_2O$, $7H_2O$, solution
F	30° or 303 K	$7H_2O$, $5H_2O$, solution
H	55° or 328 K	$5H_2O$, $4H_2O$, solution
J	66° or 339 K	$4H_2O$, Fe_2Cl_6, solution

At each of these eutectic points, the system has three phases and two components. Applying the reduced phase rule equation, we have

$$F = C - P + 1 = 2 - 3 + 1 = 0$$

That is, the system is *nonvariant.*

THE SODIUM SULFATE–WATER SYSTEM

Sodium sulfate forms two hydrates $Na_2SO_4.10H_2O$ (decahydrate) and $Na_2SO_4.7H_2O$ (heptahydrate). Furthermore, the anhydrous salt can exist in two enantiotropic crystalline forms, *viz.*, **rhombic** and **monoclinic.** The remaining phases are solid ice, solution, and vapor. Since all determinations are made at atmospheric pressure, the vapor phase is ignored. Thus the system Na_2SO_4-H_2O is a **six phase condensed system.**

The composition of all the six phases can be represented by the constituents Na_2SO_4 and H_2O. Hence the sodium sulfate-water system has **two components.**

The phase diagram (*or TC graph*) is shown in Fig. 19.12.

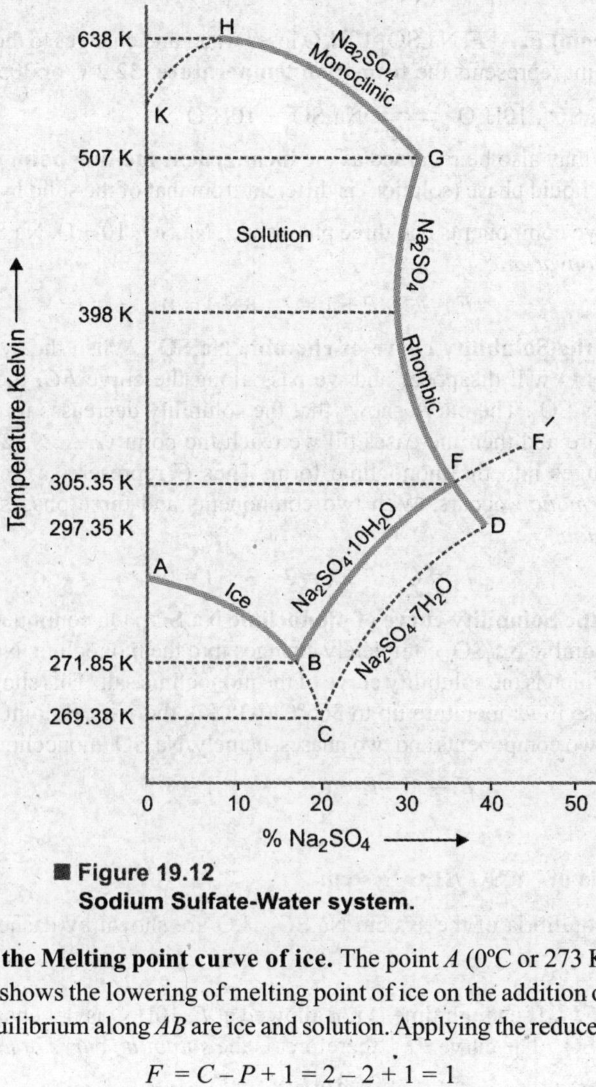

Figure 19.12
Sodium Sulfate-Water system.

The Curve *AB*, the Melting point curve of ice. The point *A* (0ºC or 273 K) is the melting point of ice. The curve *AB* shows the lowering of melting point of ice on the addition of anhydrous Na_2SO_4. The two phases in equilibrium along *AB* are ice and solution. Applying the reduced phase rule equation,

$$F = C - P + 1 = 2 - 2 + 1 = 1$$

That is the system ice/solution is *monovariant.*

The Eutectic point B. At *B*, a new solid phase $Na_2SO_4.10H_2O$ appears. This is the lowest temperature (– 1.3ºC or 271.85 K) attainable or the **Eutectic Point** where three phases, namely, ice, $Na_2SO_4.10H_2O$, and solution can coexist in equilibrium. Applying reduced phase rule equation,

$$F = C - P + 1 = 2 - 3 + 1 = 0$$

and the system is *nonvariant.*

The Curve *BF*, the Solubility curve of $10H_2O$. If the system at *B* is heated and more of anhydrous Na_2SO_4 added, we travel along the curve *BF*. Along this curve, saturated solution of sodium sulfate and $Na_2SO_4.10H_2O$ is in equilibrium. Thus the curve *BF* shows that the solubility of $10H_2O$ increases with temperature until the point *F* is reached. The system $Na_2SO_4.10H_2O$/solution is *monovariant.*

$$F = C - P + 1 = 2 - 2 + 1 = 1$$

The Transition point *F*. At *F*, $Na_2SO_4.10H_2O$ loses water and changes to the anhydrous rhombic Na_2SO_4. Thus, this point represents the **transition temperature** (32.2°C or 305.35 K).

$$Na_2SO_4.\ 10H_2O \rightleftharpoons Na_2SO_4 + 10H_2O$$

This temperature may also be regarded as the **incongruent melting point** of $Na_2SO_4.10H_2O$ as the composition of the liquid phase (solution) is different from that of the solid hydrate in equilibrium.

AT *F*, there are two components and three phases *viz.*, $Na_2SO_4.10H_2O$, Na_2SO_4 and the solution. Hence the system is *nonvariant.*

$$F = C - P + 1 = 2 - 3 + 1 = 0$$

The Curve *FG*, the Solubility curve of rhombic Na_2SO_4. When the system at *F* is heated, ultimately all the $10H_2O$ will disappear and we pass along the curve *FG*, the solubility curve of anhydrous rhombic Na_2SO_4. The curve shows that the solubility decreases up to 125°C(398.15 K) with rise of temperature and then increases till we reach the point *G*. At *G* (234°C or 507 K), the rhombic Na_2SO_4 changes into the monoclinic form. Thus *G* represents a transition point where *enantiotropic transformation* occurs. With two components and three phases in equilibrium, the system at *G* is *nonvariant.*

$$F = C - P + 1 = 2 - 3 + 1 = 0$$

The Curve *GH*, the Solubility curve of monoclinic Na_2SO_4. On continued heating the system at *G*, ultimately the rhombic Na_2SO_4 completely changes into the monoclinic Na_2SO_4. Then we pass along the curve *GH*, which is the solubility curve of the monoclinic salt. This shows that the solubility falls gradually with rise in temperature up to 365°C (638 K), the critical point of the solution. The system along *GH* has two components and two phases, namely, Na_2SO_4 monoclinic and solution. Thus,

$$F = C - P + 1 = 2 - 2 + 1 = 1$$

the system is *monovariant.*

Metastable Equilibria of Na_2SO_4/H_2O System

The metastable equilibria of the system Na_2SO_4-H_2O are shown by dashed lines in the phase diagram.

1. **The Curve *FF'*.** If enough time is not allowed at *F*, $10H_2O$ is not changed into anhydrous rhombic Na_2SO_4. The curve *FF'*, therefore, is the *solubility curve of metastable* $10H_2O$.
2. **The Curve *BC*.** If by careful cooling, $10H_2O$ does not separate at *B*, we pass along the dashed curve *BC*. This is the *freezing point curve of metastable supercooled solution saturated with respect to* $10H_2O$.
3. **The Eutectic *C*.** At *C* (– 3.77° or 269.38 K), metastable $7H_2O$ appears. Therefore, the point *C* is the eutectic point where three metastable phases *viz.*, supercooled solution, ice, and $7H_2O$ coexist in equilibrium.
4. **The Curve *CD*.** By raising the temperature of the system at *C*, we pass along the dashed curve *CD*, the solubility curve of metastable $7H_2O$.
5. **The Curve FD.** When a saturated solution with respect to rhombic Na_2SO_4 is cooled quickly $10H_2O$ does not appear at F (32.2° or 305.35 K) and we pass on to D. The dashed curve FD is, therefore, *the solubility curve of metastable* $10H_2O$.
6. **The Transition point *D*.** The point *D* represents the transition point where the metastable $7H_2O$ is changed into metastable rhombic Na_2SO_4.

EXAMINATION QUESTIONS

1. Define or explain the following terms :
   (*a*) Phase rule (*b*) Degree of freedom
   (*c*) Phase (*d*) Component
2. Explain the term component. How many components are present in the following systems?
   (*a*) Water $\rightleftharpoons$ Water - vapor (*b*) KCl + Water $\rightleftharpoons$ KCl hydrate
   Answer. (*a*) 1; (*b*) 2
3. How many phases are present in each of the following systems :
   (*i*) a drop of water placed in a stoppered bottle.
   (*ii*) a piece of molten ice placed in a beaker covered with a watch glass.
   (*iii*) mixture of N_2, H_2 and O_2.
   Answer. (*i*) 2; (*ii*) 2; (*iii*) 1
4. Draw a phase diagram for such a component system which contains more than one solid phase. Explain the following terms with the help of the diagrams :
   (*a*) Triple point
   (*b*) Invariant system
   (*c*) Metastable equilibrium
5. The number of components present in the following system :
   (*i*) a solution of common salt
   (*ii*) $MgCO_3(s) \rightleftharpoons MgO(s) + CO_2(g)$ (in a closed vessel)
   (*iii*) $NH_4Cl(s) \rightleftharpoons NH_3(g) + HCl(g)$
   Answer. (*i*) 2; (*ii*) 2; (*iii*) 1
6. For one component system, the triple point is an invariant point. Discuss.
7. A saturated solution of sodium sulfate with excess of the salt is at equilibrium with its vapor in a closed vessel. How many phases and components are present? What are the degrees of freedom in this system?
8. (*a*) Derive Gibbs phase rule from thermodynamic considerations.
   (*b*) Explain why the fusion curve of ice has a negative slope whereas the sublimation curve has positive slope in the phase diagram.
9. Explain the following briefly :
   (*i*) NH_4Cl in equilibrium with its dissociation product is a one component system.
   (*ii*) Sulfur system at any of its triple point is a non-variant system.
10. (*a*) State and explain the phase rule. Explain various terms involved.
    (*b*) Draw and explain the phase diagram of one component three phase system.
11. Explain the following terms :
    (*a*) Eutectic point (*b*) Eutectic mixture
    (*c*) Condensed system (*d*) Triple point
12. Draw the complete phase diagram for water system and prove that the conclusions in regard to the degree of freedom as derived from the diagram are the same as the deduction from the phase rule. (*Bhopal BSc*)
13. (*a*) Draw a well labelled phase diagram of KI-water system and discuss its salient features.
    (*b*) Calculate the number of components, number of phases and degrees of freedom of the following systems:
    (*i*) A liquid at critical temperature
    (*ii*) A binary azeotrope (*Mysore BSc*)

14. The melting point curve of ice in the water system has a negative slope. Explain it with the help of the phase diagram of water system. *(Agra BSc)*

15. (*a*) Describe KI-H_2O system on the basis of phase rule.
(*b*) Explain critical solution temperature. *(Jiwaji BSc)*

16. Distinguish between Eutectic point and Peritectic point. *(Delhi BSc)*

17. What is the difference between Transition temperature and melting temperature? *(Delhi BSc)*

18. Determine the number of phases, components and number of degrees of freedom for the following systems :
(*i*) $N_2(g) + O_2(g) \rightleftharpoons 2NO(g)$
(*ii*) $Na_2SO_4 . 10H_2O \rightleftharpoons Na_2SO_4 + 10H_2O(g)$ *(Allahabad BSc)*

19. Calculate the number of phases, components and degree of freedom in the following systems :
(*i*) $N_2O_4(g) \rightleftharpoons 2NO_2(g)$
(*ii*) Solid carbon in equilibrium with gaseous CO, CO_2 and O_2 at 100°C or 373 K. *(Jamia Millia BSc)*

20. Briefly explain the following giving one example of each type :
(*i*) System with upper critical solution temperature.
(*ii*) System with lower critical solution temperature.
(*iii*) System with both lower and upper critical solution temperatures. *(MD Rohtak BSc)*

21. (*a*) Discuss the phase equilibria of a system involving two solids and liquids.
(*b*) Explain triangular plots. *(Jamia Millia BSc)*

22. Draw graph of water system and explain it in detail. *(Jiwaji BSc)*

23. What is the criterion for multiphase system to be in equilibrium? How are degrees of freedom calculated? *(Panjab BSc, 2002)*

24. (*a*) In LiI, the equilibrium interionic distance, r_o is much greater than the sum of r_{Li^+} and r_{I^-}. How can you account for this?
(*b*) $MgCO_3$ is thermally less stable than $CaCO_3$. Explain. *(Guru Nanak Dev BSc)*

25. In the phase diagram for water
(*i*) What is upper limit of liquid-vapor equilibrium line?
(*ii*) Why is solid-liquid equilibrium line almost vertical and slightly tilted to the left? *(Panjab BSc)*

26. Draw a vapor pressure composition diagram showing positive deviations from Raoult's law. *(Guru Nanak Dev BSc)*

27. Find out the number of degrees of freedom in the following systems :
(*i*) Sulfur(ℓ) $\rightleftharpoons$ Sulfur(vap)
(*ii*) Saturated solution of NaCl
(*iii*) $CaCO_3(s) \rightleftharpoons CaO(s) + CO_2(g)$
(*iv*) A mixture of water and water vapor at the triple point of water
(*v*) $NH_4Cl(s) \rightleftharpoons NH_3(g) + HCl(g)$
(*vi*) $Na_2SO_4 . 10H_2O \rightleftharpoons Na_2SO_4(s) + 10H_2O$ *(Allahabad BSc)*

28. Determine the number of components, number of phases and degrees of freedom for the following systems :
(*i*) $H_2O(s) \rightleftharpoons H_2O(\ell) \rightleftharpoons H_2O(g)$
(*ii*) $CaCO_3(s) \rightleftharpoons CaO(s) + CO_2(g)$ *(Arunachal BSc)*

29. (*a*) Discuss the main features of phase diagram of a two component system (solids) that is completely miscible is the liquid phase.

(*b*) How many number of phases and components are present in the following systems?

(*i*) A mixture of molten lead, silver and bismuth.

(*ii*) $CaCO_3(s) \rightleftharpoons CaO(s) + CO_2(g)$ (*MD Rohtak BSc*)

30. Apply phase rule to water system. (*HS Gaur BSc*)

31. Deduce the phase rule equation on the basis of thermodynamics. (*Nagpur BSc*)

32. Is it possible to obtain pure ethanol by distillation a mixture of ethanol and water? Explain. (*Kolkata BSc*)

33. (*a*) What is a condensed system? Write the reduced phase rule equation.

(*b*) Discuss salient features of lead-silver system. (*Arunachal BSc*)

34. (*a*) In phase equilibria, what is understood by the number of degrees of freedom and reduced phase rule.

(*b*) Deduce Gibbs phase rule equation from thermodynamic considerations, making it clear what are these considerations. (*Guru Nanak Dev BSc*)

35. (*a*) Explain the terms : component, degree of freedom, eutectic and triple point. Calculate the number of components, number of phases and degrees of freedom of a liquid at its critical temperature.

(*b*) How do the phase diagram of $KI.H_2O$ and Pb-Ag systems differ from each other? (*Banaras BSc*)

36. Write short notes on the following :

(*a*) Phase rule for condensed systems

(*b*) Functioning of freezing mixtures (*Agra BSc*)

37. (*a*) Derive Gibbs Phase rule thermodynamically.

(*b*) Describe an experimental method for determining a phase diagram of a two-component system. (*Madras BSc*)

38. Calculate the degrees of freedom and number of components for a system of sodium chloride solution in water containing undissolved salt, in equilibrium with water vapor.

Answer. 1 and 2 (*Baroda BSc*)

39. How many degrees of freedom will be present in a solution of sodium sulfate in equilibrium with water vapor ?

Answer. 2 (*Madurai BSc*)

40. The vapor pressure of 2,2,4-trimethylpentane at 293.85 K and 802.25 K are 40 and 60 torr respectively. Calculate the enthalpy of vaporization of this compound.

Answer. 1563.33 J mol^{-1} (*Delhi BSc*)

41. Determine the number of phases, components and degrees of freedom in the system : Ice, water and vapor in equilibrium.

Answer. 3,1 and 0 (*Bundelkhand BSc*)

42. Describe phenol-water system with respect to critical solution temperature. (*Panjab BSc*)

43. (*a*) Explain Triple point and Eutectic point with examples.

(*b*) Explain why the melting point curve is inclined towards pressure axis in the phase diagram of water. (*Utkal BSc*)

44. State the phase rule. Explain the various terms used in it. Discuss the derivation of the phase rule from thermodynamic considerations. (*Mysore BSc*)

45. Discuss the salient features of phase diagram of sulfur system. Why can four phases of heterogeneous system not exist at equilibrium? (*Madras BSc*)

46. Discuss main features of the phase diagram of water system, explaining especially why the slope of solid-liquid line is negative for water. (*Delhi BSc*)

MULTIPLE CHOICE QUESTIONS ????

1. The phase rule was first discovered by

(*a*) Nernst (*b*) Gibbs

(*c*) Arrhenius (*d*) Le Chatelier

Answer. (*b*)

2. Mathematically, the phase rule can be expressed by (F is the number of degrees of freedom, C the number of components and P is the number of phases of the system)

(*a*) $F + P = C + 2$ (*b*) $F = C - P + 2$

(*c*) $P = C - F + 2$ (*d*) all of these

Answer. (*d*)

3. A system containing liquid water and water vapor has the number of phases equal to

(*a*) 0 (*b*) 1

(*c*) 2 (*d*) 3

Answer. (*c*)

4. A mixture of three gases O_2, N_2 and CO_2 is

(*a*) 1-phase system (*b*) 2-phase system

(*c*) 3-phase system (*d*) 4-phase system

Answer. (*a*)

5. A mixture of two miscible liquids (ethanol and water) has the number of phases equal to

(*a*) zero (*b*) one

(*c*) two (*d*) three

Answer. (*b*)

6. A mixture of two immiscible liquids (chloroform and water) constitutes a system having the number of phases equal to

(*a*) zero (*b*) one

(*c*) two (*d*) three

Answer. (*c*)

7. Decomposition of calcium carbonate represented by the equation $CaCO_3(s) \rightleftharpoons CaO(s) + CO_2(g)$ in a closed vessel constitutes a system with number of phases equal to

(*a*) 0 (*b*) 1

(*c*) 2 (*d*) 3

Answer. (*d*)

8. A saturated solution of sodium chloride is a

(*a*) one phase system (*b*) two phase system

(*c*) three phase system (*d*) none of these

Answer. (*b*)

9. Water system has three phases - ice, water and vapors. The number of components in the system is

(*a*) one (*b*) two

(*c*) three (*d*) four

Answer. (*a*)

10. A mixture of gases O_2 and N_2 constitutes one phase only. The number of components in the system is

(*a*) zero (*b*) one

(*c*) two (*d*) three

Answer. (*c*)

11. The decomposition of $CaCO_3$ in a closed vessel is represented by the equation

$CaCO_3(s) \rightleftharpoons CaO(s) + CO_2(g)$

The number of phases and components respectively are

(*a*) 2 and 3 (*b*) 3 and 2
(*c*) 2 and 2 (*d*) 3 and 3

Answer. (*b*)

12. A saturated solution of sodium chloride in contact with solid solute has the phases and components equal to

(*a*) 2 and 2 (*b*) 2 and 3
(*c*) 3 and 2 (*d*) 3 and 3

Answer. (*a*)

13. The total number of variable factors which must be specified so that the remaining variables are fixed automatically and the system is completely defined. It is known as

(*a*) a phase (*b*) a component
(*c*) degrees of freedom (*d*) none of these

Answer. (*c*)

14. For a pure gas and mixture of gases, the degrees of freedom are

(*a*) 2 and 2 (*b*) 2 and 3
(*c*) 3 and 2 (*d*) 3 and 3

Answer. (*b*)

15. A saturated solution of NaCl in water has degrees of freedom equal to

(*a*) one (*b*) two
(*c*) three (*d*) four

Answer. (*a*)

16. For one component system the phase rule is

(*a*) $F = 3 - P$ (*b*) $F = 2 - P$
(*c*) $F = 1 - P$ (*d*) none of these

Answer. (*a*)

17. The vapor pressure of the metastable phase is always _______ than that of the stable phase at the same temperature

(*a*) lower (*b*) higher
(*c*) constant (*d*) none of these

Answer. (*b*)

18. At a triple point

(*a*) both the temperature and pressure are fixed
(*b*) only the temperature is fixed
(*c*) only the pressure is fixed
(*d*) sometimes pressure and sometime temperature is fixed

Answer. (*a*)

19. The occurrence of the same substance in more than one crystalline forms is known as

(*a*) isomerism (*b*) racemization
(*c*) polymorphism (*d*) none of these

Answer. (*c*)

20. The temperature at which a polymorphic substance changes from one form to another is called

(*a*) equilibrium temperature (*b*) triple point
(*c*) Boyle's temperature (*d*) transition temperature

Answer. (*d*)

21. For one phase and one component system, the degrees of freedom are equal to
(*a*) 1 (*b*) 2
(*c*) 3 (*d*) 4
Answer. (*b*)

22. When a single phase is present in a two component system, the degree of freedom is
(*a*) zero (*b*) one
(*c*) two (*d*) three
Answer. (*d*)

23. The number of components present in the following systems
(*i*) Water $\rightleftharpoons$ Water vapor
(*ii*) KCl + Water $\rightleftharpoons$ KCl hydrate
(*a*) 1 and 1 (*b*) 1 and 2
(*c*) 2 and 1 (*d*) 2 and 3
Answer. (*b*)

24. The decomposition of NH_4Cl is represented by the equation
$NH_4Cl(s) \rightleftharpoons NH_3(g) + HCl(g)$
The number of components present in the system is
(*a*) 0 (*b*) 1
(*c*) 2 (*d*) 3
Answer. (*d*)

25. The number of components in a solution of common salt is
(*a*) 0 (*b*) 1
(*c*) 2 (*d*) 3
Answer. (*c*)

26. A drop of water is placed in a stoppered bottle. How many phases are present in the system?
(*a*) 0 (*b*) 1
(*c*) 2 (*d*) 3
Answer. (*c*)

27. A piece of molten ice is placed in a beaker covered with a water glass. How many phases are present in the system?
(*a*) 0 (*b*) 1
(*c*) 2 (*d*) 3
Answer. (*c*)

28. A system with zero-degree of freedom is known as
(*a*) monovariant (*b*) bivariant
(*c*) invariant (*d*) none of these
Answer. (*c*)

29. For a three phase system with one component, the degree of freedom according to phase rule is
(*a*) 0 (*b*) 1
(*c*) 2 (*d*) 3
Answer. (*a*)

30. The transition temperature of a substance is that temperature at which
(*a*) one enantiomer changes into another enantiomer
(*b*) one allotropic form changes to another
(*c*) all the three phases (solid, liquid and gas) can co-exist in equilibrium
(*d*) none of the above
Answer. (*b*)

31. A dilatometer is an apparatus used to measure
(*a*) transition temperature (*b*) triple point
(*c*) eutectic point (*d*) all of these
Answer. (*a*)

32. The reduced phase rule for a condensed system is

(*a*) $F = C - P + 2$ (*b*) $F' = C - P + 1$

(*c*) $F' = C - P$ (*d*) $F = C - P + 3$

Answer. (*b*)

33. The phase rule is applicable to

(*a*) homogenous systems

(*b*) reversible systems

(*c*) irreversible systems

(*d*) heterogeneous system whether physical or chemical

Answer. (*d*)

34. At a triple point

(*a*) three phases co-exist in equilibrium

(*b*) the vapor pressure is equal to the atmospheric pressure

(*c*) there are three components in equilibrium

(*d*) there are three degrees of freedom

Answer. (*a*)

35. For one component system, at triple point the number of degrees of freedom is

(*a*) zero (*b*) one

(*c*) two (*d*) three

Answer. (*a*)

36. For one component system, there does not exist a quadruple point as the number of degrees of freedom cannot be

(*a*) zero (*b*) –1

(*c*) 1 (*d*) 2

Answer. (*b*)

37. The sulfur system has four phases: rhombic, monoclinic, liquid and vapor sulfur. It is

(*a*) one-component system (*b*) two-component system

(*c*) three-component system (*d*) four-component system

Answer. (*a*)

38. Which one of the following is not a heterogenous system?

(*a*) $CaCO_3(s) \rightleftharpoons CaO(s) + CO_2(g)$ (*b*) Water vapor $\rightleftharpoons$ Water

(*c*) $NH_4Cl(s) \rightleftharpoons NH_3(g) + HCl(g)$ (*d*) $H_2 + I_2 \rightleftharpoons 2HI$

Answer. (*d*)

39. For a bivariant system, the degrees of freedom are

(*a*) one (*b*) two

(*c*) three (*d*) four

Answer. (*c*)

40. A saturated solution of NaCl has one degree of freedom. It is

(*a*) monovariant (*b*) non-variant

(*c*) bi-variant (*d*) none of these

Answer. (*a*)

41. For a two component system in a single phase, the degree of freedom is

(*a*) zero (*b*) one

(*c*) two (*d*) three

Answer. (*d*)

42. A one component system has four phases. Can the four phases co-exist in equilibrium?

(*a*) no (*b*) yes

(*c*) sometimes (*d*) none of these

Answer. (*a*).

20 Chemical Kinetics

CONTENTS

So far we have studied equilibrium reactions. In these reactions, the rates of the two opposing reactions are equal and the concentrations of reactants or products do not change with lapse of time. But most chemical reactions are **spontaneous reactions.** These reactions occur from left to right till all the reactants are converted to products. A spontaneous reaction may be slow or it may be fast. For example, the reactions between aqueous sodium chloride and silver nitrate is a fast reaction. The precipitate of AgCl is formed as fast as $AgNO_3$ solution is added to NaCl solution. On the contrary, the rusting of iron is a slow reaction that occurs over the years.

The branch of Physical chemistry which deals with the rate of reactions is called Chemical Kinetics. The study of Chemical Kinetics includes :

(1) The rate of the reactions and rate laws.

(2) The factors as temperature, pressure, concentration and catalyst, that influence the rate of a reaction.

(3) The mechanism or the sequence of steps by which a reaction occurs.

The knowledge of the rate of reactions is very valuable to understand the chemical of reactions. It is also of great importance in selecting optimum conditions for an industrial process so that it proceeds at a rate to give maximum yield.

REACTION RATE

The rate of a reaction tells as to what speed the reaction occurs. Let us consider a simple reaction

$$A \longrightarrow B$$

The concentration of the reactant A decreases and that of B increases as time passes. The **rate of reactions is defined as the change in concentration of any of reactant or products per unit time**. For the given reaction the rate of reaction may be equal to the rate of disappearance of A which is equal to the rate of appearance of B.

Thus

$$\text{rate of reaction} = \text{rate of disappearance of } A$$
$$= \text{rate of appearance of } B$$

or

$$\text{rate} = -\frac{d[A]}{dt}$$
$$= +\frac{d[B]}{dt}$$

where [ ] represents the concentration in moles per liter whereas 'd' represents infinitesimally small change in concentration. Negative sign shows the concentration of the reactant A decreases whereas the positive sign indicates the increase in concentration of the product B.

UNITS OF RATE

Reactions rate has the units of concentration divided by time. We express concentrations in moles per liter (mol/liter or mol/1 or mol 1^{-1}) but time may be given in any convenient unit second (s), minutes (min), hours (h), days (d) or possible years. The SI unit is mol/m^3 s. The units of reaction rates may be

mol/liter s	or	mol l^{-1} s
mol/liter min	or	mol l^{-1} min^{-1}
mol/liter hour	or	mol l^{-1} h^{-1} and, so on

Average Rate of Reaction is a Function of Time

Let us consider the reaction between carbon monoxide (CO) and nitrogen dioxide.

$$CO(g) + NO_2(g) \longrightarrow CO_2(g) + NO(g)$$

The average rate of reaction may be expressed as

$$\text{rate} = \frac{-\Delta[CO]}{\Delta t} = -\frac{d[CO]}{dt}$$

The concentration of CO was found experimentally every 10 seconds. The results of such an experiment are listed below.

Conc. of CO	100	67	50	40	33
Time (s)	0	10	20	30	40

The results are also shown graphically in Fig. 20.1.

As the reaction proceeds the concentration of CO decreases rapidly in the initial stages of the reaction. Then the concentration of CO decreases more and more slowly. Obviously the rate of reaction is a function of time.

Over the first 10 seconds, the average rate is

$$\frac{-d[\text{CO}]}{dt} = \frac{-(0.067 - 0.100)}{(10 - 0)} = \frac{0.033}{10} = 0.0033\,\text{mol l}^{-1}\,\text{s}^{-1} = 3.3\,\text{mol/m}^3\,\text{s}$$

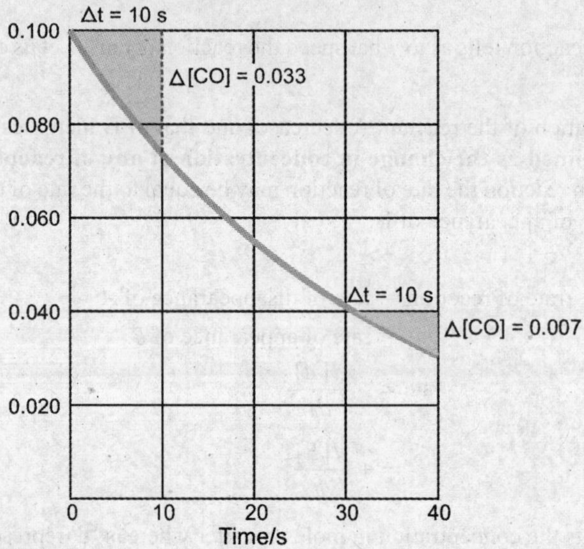

■ **Figure 20.1**
Graph of CO concentrations *versus time*. The average rate is equal to the slope of the curve. The average rates for the first 10 and the last 10 seconds are shown.

In the time interval between 30 and 40 seconds, the average rate is much smaller.

$$\frac{-d[\text{CO}]}{dt} = \frac{-(0.033 - 0.040)}{(40 - 30)} = \frac{0.007}{10} = 0.0007\,\text{mol l}^{-1}\,\text{s}^{-1} = 0.7\,\text{mol m}^{-3}\,\text{s}^{-1}$$

The reaction in indeed slowing down with time.

We shall see that average rates are not always useful. They cover a large time interval during which the rate of reaction changes significantly. So, a better way to estimate the rate of reaction, we need to make the time interval as small as possible.

Instantaneous Rate of Reaction

The average rates obtained by finding the slope of the curve are not always useful. They cover a large time interval during which the rate of reaction changes significantly. So, a better way to estimate the rate of a reaction is to make the time interval as small as possible. If the interval be infinitesimally small (that is as Δt approaches zero), the rate is referred to as the **instantaneous rate** and is written in calculus as

$$\text{rate} = \frac{d[\ ]_t}{dt}$$

where $[\ ]_t$ is the concentration at time t. In the present case,

$$\text{the instantaneous rate} = \frac{-d[\text{CO}]_t}{dt}$$

Thus at any time the instantaneous rate is equal to the slope of a straight line drawn tangent to the curve at that time. For example, in Fig. 20.2 the instantaneous rate at 10 seconds is found to be 2.2 mol^{-3} s^{-1}.

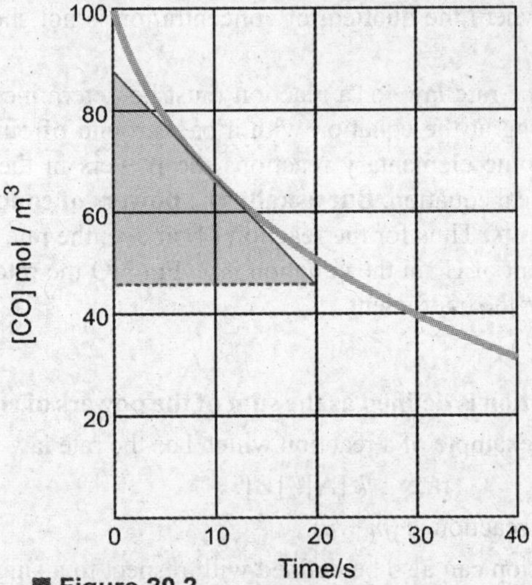

■ **Figure 20.2**
The slope of tangent at 10 seconds is equal to the instantaneous rate.

RATE LAWS

At a fixed temperature the rate of a given reaction depends on concentration of reactants. The exact relation between concentration and rate is determined by measuring the reaction rate with different initial reactant concentrations. By a study of numerous reactions it is shown that : **the rate of a reaction is directly proportional to the reactant concentrations, each concentration being raised to some power.**

Thus for a substance A undergoing reaction,

rate $\quad \text{rate} \propto [A]^n$

or $\quad \text{rate} = k\,[A]^n \quad$ ...(1)

For a reaction

$$2A + B \longrightarrow \text{products}$$

the reaction rate with respect to A or B is determined by varying the concentration of one reactant, keeping that of the other constant. Thus the rate of reaction may be expressed as

$$\text{rate} = k\,[A]^m\,[B]^n \quad \text{...(2)}$$

Expressions such as (1) and (2) tell the relation between the rate of a reaction and reactant concentrations.

An expression which shows how the reaction rate is related to concentrations is called the rate law or rate equation.

The power (exponent) of concentration n or m in the rate law is usually a small whole number integer (1, 2, 3) or fractional. The proportionality constant k is called the **rate constant** for the reaction.

Examples of rate law :

	REACTIONS	RATE LAW
(1)	$2N_2O_5 \longrightarrow 4\,NO_2 + O_2$	rate = $k\,[N_2O_5]$
(2)	$H_2 + I_2 \longrightarrow 2HI$	rate = $k\,[H_2]\,[I_2]$
(3)	$2NO_2 \longrightarrow 2NO + O_2$	rate = $k\,[NO_2]^2$
(4)	$2NO + 2H_2 \longrightarrow N_2 + 2H_2O$	rate = $k\,[H_2]\,[NO]^2$

In these rate laws where the quotient or concentration is not shown, it is understood to be 1. That is $[H_2]^1 = [H_2]$.

It is apparent that the rate law for a reaction must be determined by experiment. It cannot be written by merely looking at the equation with a background of our knowledge of Law of Mass Action. However, for some elementary reactions the powers in the rate law may correspond to coefficients in the chemical equation. **But usually the powers of concentration in the rate law are different from coefficients.** Thus for the reaction (4) above, the rate is found to be proportional to $[H_2]$ although the quotient of H_2 in the equation is 2. For NO the rate is proportional to $[NO]^2$ and power '2' corresponds to the coefficient.

ORDER OF A REACTION

The order of a reaction is defined as the sum of the powers of concentrations in the rate law.

Let us consider the example of a reaction which has the rate law

$$\text{rate} = k\,[A]^m\,[B]^n \qquad ...(1)$$

The order of such a reaction is $(m + n)$.

The order of a reaction can also be defined with respect to a single reactant. Thus the reaction order with respect to A is m and with respect to B it is n. The **overall order of reaction** $(m + n)$ may range from 1 to 3 and can be fractional.

Examples of reaction order :

RATE LAW	REACTION ORDER
rate = $k\,[N_2O_5]$	1
rate = $k\,[H_2]\,[I_2]$	$1 + 1 = 2$
rate = $k\,[NO_2]^2$	2
rate = $k\,[[H_2]\,[NO]^2$	$1 + 2 = 3$
rate = $k\,[CHCl_3]\,[Cl_2]^{1/2}$	$1 + \frac{1}{2} = 1\frac{1}{2}$

Reactions may be classified according to the order. If in the rate law (1) above

$m + n = 1$, it is **first order reaction**

$m + n = 2$, it is **second order reaction**

$m + n = 3$, it is **third order reaction**

ZERO ORDER REACTION

A reactant whose concentration does not affect the reaction rate is not included in the rate law. In effect, the concentration of such a reactant has the power 0. Thus $[A]^0 = 1$.

A zero order reaction is one whose rate is independent of concentration. For example, the rate law for the reaction

$$NO_2 + CO \longrightarrow NO + CO_2$$

at 473 K is

$$\text{rate} = k\,[NO_2]^2$$

Here the rate does not depend on [CO], so this is not included in the rate law and the power of [CO] is understood to be zero. The reaction is **zeroth order** with respect to CO. The reaction is second order with respect to $[NO_2]$. The overall reaction order is $2 + 0 = 2$.

MOLECULARITY OF A REACTION

Chemical reactions may be classed into two types :

(*a*) Elementary reactions

(*b*) Complex reactions

An **elementary reaction** is a simple reaction which occurs in a single step.

A **complex reaction** is that which occurs in two or more steps.

Molecularity of an Elementary Reaction

The molecularity of an elementary reaction is defined as : **the number of reactant molecules involved in a reaction.**

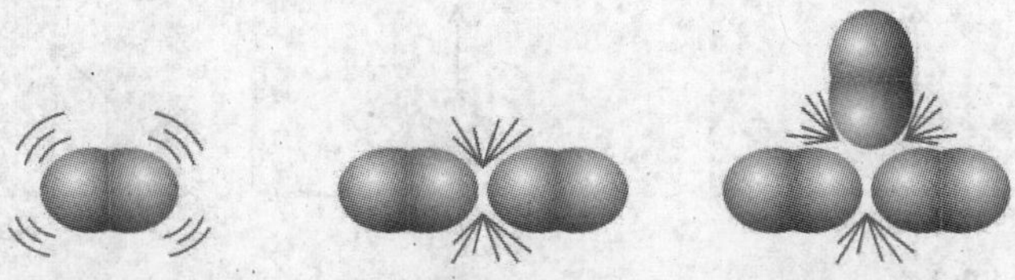

Unimolecular Bimolecular Termolecular

Thus the molecularity of an elementary reaction is 1, 2, 3, etc., according as one, two or three reactant molecules are participating in the reaction. The elementary reactions having molecularity 1, 2 and 3 are called **unimolecular, bimolecular and termolecular** respectively. Thus we have :

(*a*) **Unimolecular reactions :** (molecularity = 1)

$$A \longrightarrow \text{product}$$

Examples are : (*i*) $Br_2 \longrightarrow 2Br$

(*ii*)
```
H—C—COOH                 H—C—COOH
  ||          ──→           ||
H—C—COOH            COOH—C—H
maleic acid              fumaric acid
```

(*b*) **Bimolecular reactions :** (molecularity = 2)

$$A + B \longrightarrow \text{products}$$
$$A + A \longrightarrow \text{products}$$

Examples are :

(*i*) $CH_3COOC_2H_5 + H_2O \longrightarrow CH_3COOH + C_2H_5OH$
Ethyl acetate → acetic acid + ethyl alcohol

(*ii*) $2HI \longrightarrow H_2 + I_2$

(*c*) **Termolecular reactions** : (molecularity = 3)

$$A + B + C \longrightarrow \text{products}$$

Examples are :

$$2NO + O_2 \longrightarrow 2NO_2$$
$$2NO + Cl_2 \longrightarrow 2NOCl$$

Why High Molecularity Reactions are Rare?

Most of the reactions involve one, two or at the most three molecules. The reactions involving four or more molecules are very rare. The rarity of reactions with high molecularity can be explained on the basis of the kinetic molecular theory. According to this theory, the rate of a chemical reaction is proportional to the number of collisions taking place between the reacting molecules. The chances of simultaneous collision of reacting molecules will go on decreasing with increase in number of molecules. Thus the possibility of three molecules colliding together is much less than in case of bimolecular collision. For a reaction of molecularity 4, the four molecules must come closed and collide with one another at the same time. The possibility of their doing so is much less than even in the case of termolecular reaction. Hence the reactions involving many molecules proceed through a series of steps, each involving two or three or less number of molecules. Such a reaction is called a complex reaction and the slowest step determines the overall rate of the reactions.

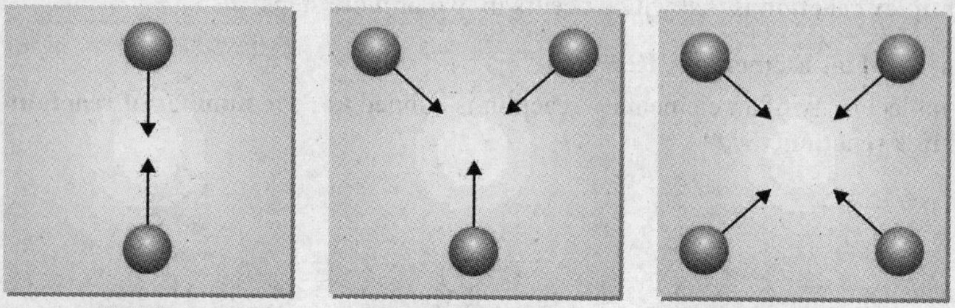

■ **Figure 20.3**
Chances of simultaneous collision between reacting molecules decrease as the molecularity increases.

Molecularity of a Complex Reaction

Most chemical reactions are complex reactions. These occur in a series of steps. Each step is an elementary reaction. The stepwise sequence of elementary reactions that convert reactions to products is called the **mechanism of the reaction.** In any mechanism, some of the steps will be fast, others will be slow. A reaction can proceed no faster than its slowest step. Thus the slowest step is the **rate-determining step** of the reaction.

The decomposition of N_2O_5,

$$2N_2O_5 \longrightarrow 4NO_2 + O_2$$

is an example of a complex reaction. It occurs by the following steps :

Step 1	$2N_2O_5$	$\longrightarrow$	$2NO_2 + 2NO_3$	(*slow*)
Step 2	$NO_2 + NO_3$	$\longrightarrow$	$NO + NO_2 + O_2$	(*slow*)
Step 3	$NO + NO_3$	$\longrightarrow$	$2NO_2$	(*fast*)
Overall reaction	$2N_2O_5$	$\longrightarrow$	$4NO_2 + O_2$	

Each elementary reaction has its own molecularity equal to the number of molecules or atoms participating in it. It is meaningless to give the molecularity of the overall reaction because it is made of several elementary reactions, each, perhaps with a different molecularity. At best could be thought of as : **the number of molecules or atoms taking part in the rate-determining step.**

Thus step 2 in the above mechanism is rate-determining and has molecularity '2' which could be considered as the molecularity of the decomposition reaction of N_2O_5.

MOLECULARITY VERSUS ORDER OF REACTION

The term molecularity is often confused with order of a reaction.

The total number of molecules or atoms which take part in a reaction as represented by the chemical equation, is known as the **molecularity of reaction.**

The sum of the powers to which the concentrations are raised in the rate law is known as the **order of reaction.**

Molecularity and Order are Identical for Elementary Reactions or Steps

The rate of an elementary reaction is proportional to the number of collisions between molecules (or atoms) of reactions. The number of collisions in turn is proportional to the concentration of each reactant molecule (or atom). Thus for a reaction.

$$2A + B \longrightarrow \text{products}$$

$$\text{rate} \propto [A]\,[A]\,[B]$$

or $$\text{rate} = k\,[A]^2\,[B] \qquad \textit{(rate law)}$$

Two molecules of A and one molecule of B are participating in the reaction and, therefore, molecularity of the reaction is 2 + 1 = 3. The sum of powers in the rate law is 2 + 1 and hence the reaction order is also 3. Thus **the molecularity and order for an elementary reaction are equal.**

TABLE 20.1. MOLECULARITY AND ORDER FOR ELEMENTARY REACTIONS

Reactions	Molecularity	Rate law	Order
$A \longrightarrow$ products	1	rate = $k\,[A]$	1
$A + A \longrightarrow$ products	2	rate = $k\,[A]^2$	2
$A + B \longrightarrow$ products	2	rate = $k\,[A]\,[B]$	2
$A + 2B \longrightarrow$ products	3	rate = $k[A]\,[B]^2$	3
$A + B + C \longrightarrow$ products	3	rate = $k\,[A]\,[B]\,[C]$	3

Differences Between Order and Molecularity

Order of a Reaction	Molecularity of a Reaction
1. It is the **sum of powers** of the concentration terms in the rate law expression.	1. It is number of **reacting species** undergoing simultaneous collision in the elementary or simple reaction.
2. It is an **experimentally** determined value.	2. It is a **theoretical** concept.
3. It can have **fractional** value.	3. It is always a **whole number.**
4. It can assume **zero value.**	4. It **can not have zero value.**
5. Order of a reaction **can change** with the conditions such as pressure, temperature, concentration.	5. Molecularity is **invariant** for a chemical equation.

PSEUDO-ORDER REACTIONS

A reaction in which one of the reactants is present in a large excess shows an order different from the actual order. **The experimental order which is not the actual one is referred to as the pseudo order.** Since for elementary reactions molecularity and order are identical, pseudo-order reactions may also be called **pseudo molecular reactions.**

Let us consider a reaction

$$A + B \longrightarrow \text{products}$$

in which the reactant B is present in a large excess. Since it is an elementary reaction, its rate law can be written as

$$\text{rate} = k\,[A]\,[B]$$

As B is present in large excess, its concentration remains practically constant in the course of reaction. Thus the rate law can be written as

$$\text{rate} = k'\,[A]$$

where the new rate constant $k' = k\,[B]$. Thus the actual order of the reaction is second-order but in practice it will be first-order. Therefore, the reaction is said to have a **pseudo-first order.**

Examples of Pseudo-order Reactions

(1) **Hydrolysis of an ester.** For example, ethyl acetate upon hydrolysis in aqueous solution using a mineral acid as catalyst forms acetic acid and ethyl alcohol.

$$\underset{\text{ethyl acetate}}{CH_3COOC_2H_5} + \underset{\text{(excess)}}{H_2O} \longrightarrow \underset{\text{acetic acid}}{CH_3COOH} + \underset{\text{ethyl alcohol}}{C_2H_5OH}$$

Here a large excess of water is used and the rate law can be written as

$$\begin{aligned} \text{rate} &= k\,[CH_3COOH]\,[H_2O] \\ &= k'\,[CH_3COOH] \end{aligned}$$

The reaction is actually second-order but in practice it is found to be first-order. Thus it is a pseudo-first order reaction.

(2) **Hydrolysis of sucrose.** Sucrose upon hydrolysis in the presence of a dilute mineral acid gives glucose and fructose.

$$\underset{\text{sucrose}}{C_{12}H_{22}O_{11}} + \underset{\text{(excess)}}{H_2O} \longrightarrow \underset{\text{glucose}}{C_6H_{12}O_6} + \underset{\text{fructose}}{C_6H_{12}O_6}$$

If a large excess of water is present, $[H_2O]$ is practically constant and the rate law may be written as

$$\begin{aligned} \text{rate} &= k\,[C_{12}H_{22}O_{11}]\,[H_2O] \\ &= k\,[C_{12}H_{22}O_{11}] \end{aligned}$$

The reaction though of second-order is experimentally found to be first-order. Thus it is a pseudo-first-order reaction.

ZERO ORDER REACTIONS

In a zero order reaction, rate is independent of the concentration of the reactions. Let us consider a zero-order reaction of the type

	A	$\longrightarrow$	Products
Initial conc.	a		0
Final conc.	$a - x$		x

$$\text{Rate of reaction} = \frac{-d[A]}{dt} = k_0\,[A]^\circ$$

or
$$\frac{dx}{dt} = \frac{-d\,(a-x)}{dt} = k_0(a-x)^\circ = k_0$$

On integrating we get

$$k_0 = \frac{x}{t} \quad \text{or} \quad x = k_0 t$$

where k_0 is the rate constant of a zero-order reaction, the unit of which is concentration per unit time. In zero order reaction, the **rate constant is equal to the rate of reaction at all concentrations.**

FIRST ORDER REACTIONS

Let us consider a first order reaction

$$A \longrightarrow \text{products}$$

Suppose that at the beginning of the reaction ($t = 0$), the concentration of A is a mol liter^{-1}. If after time t, x moles of A have changed, the concentration of A is $a - x$. We know that for a first order reaction, the rate of reaction, dx/dt, is directly proportional to the concentration of the reactant. Thus,

$$\frac{dx}{dt} = k(a - x)$$

or
$$\frac{dx}{a - x} = k\,dt \quad ...(1)$$

Integration of the expression (1) gives

$$\int \frac{dx}{a - x} = \int k\,dt$$

or
$$-\ln(a - x) = kt + \text{I} \quad ...(2)$$

where I is the constant of integration. The constant k may be evaluated by putting $t = 0$ and $x = 0$.

Thus,

$$\text{I} = -\ln a$$

Substituting for I in equation (2)

$$\ln\frac{a}{a - x} = kt \quad ...(3)$$

or
$$k = \frac{1}{t}\ln\frac{a}{a - x}$$

Changing into common logarithms

$$k = \frac{2.303}{t}\log\frac{a}{a - x} \quad ...(4)$$

The value of k can be found by substituting the values of a and $(a - x)$ determined experimentally at time interval t during the course of the reaction.

Sometimes the integrated rate law in the following form is also used :

$$k = \frac{2.303}{t_2 - t_1}\log\frac{(a - x_1)}{(a - x_2)}$$

where x_1 and x_2 are the amounts decomposed at time intervals t_1 and t_2 respectively from the start.

Examples of First order Reactions

Some common reactions which follow first order kinetics are listed below :

(1) **Decomposition of N_2O_5 in CCl_4 solution.** Nitrogen pentoxide in carbon tetrachloride solution decomposes to form oxygen gas,

$$N_2O_5 \longrightarrow 2NO_2 + \tfrac{1}{2}O_2$$

The reaction is carried in an apparatus shown in Fig. 20.4. The progress of the reaction is monitored by measuring the volume of oxygen evolved from time to time.

If V_t be the volume of O_2 at any time t and V_∞ the final volume of oxygen when the reaction is completed, the V_∞ is a measure of the initial concentration of N_2O_5 and $(V_\infty - V_t)$ is a measure of undecomposed N_2O_5 $(a - x)$ remaining at time t. Thus,

$$k = \frac{2.303}{t}\log_{10}\frac{V_\infty}{V_\infty - V_t}$$

On substituting values of V_∞, $(V_\infty - V_t)$ at different time intervals, t, the value of k is found to be constant. Thus it is a reaction of the first order.

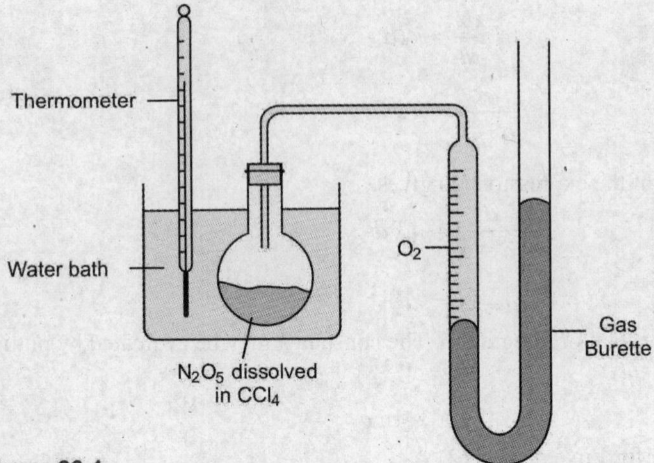

Figure 20.4
An apparatus for monitoring the volume of O_2 evolved in the decomposition of N_2O_5 dissolved in carbon tetrachloride.

SOLVED PROBLEM 1

From the following data for the decomposition of N_2O_5 in CCl_4 solution at 321 K, show that the reaction is of the first order

t (secs)	600	900	1200	∞
Vol of O_2 evolved	6.30	8.95	11.40	34.75

SOLUTION

For a first order reaction the integrated rate equation is

$$\frac{1}{t}\log\frac{V_\infty}{V_\infty - V_t} = k$$

In this example, $V_\infty = 34.75$

t	$V_\infty - Vt$	$\frac{1}{t}\log\frac{V_\infty}{V_\infty - V_t}$	=	k
600	28.45	$\frac{1}{600}\log\frac{34.75}{28.45}$	=	0.000145
900	25.80	$\frac{1}{900}\log\frac{34.75}{25.80}$	=	0.000144
1200	23.35	$\frac{1}{1200}\log\frac{34.75}{23.35}$	=	0.000144

Since the value of k is fairly constant, it is a **first order reaction.**

(2) **Decomposition of H_2O_2 in aqueous solution.** The decomposition of H_2O_2 in the presence of Pt as catalyst is a first order reaction.

$$H_2O_2 \xrightarrow{Pt} H_2O + O$$

The progress of the reaction is followed by titrating equal volumes of the reaction mixture against standard $KMnO_4$ solution at different time intervals.

SOLVED PROBLEM 2

A solution of H_2O_2 when titrated against $KMnO_4$ solution at different time intervals gave the following results :

t (secs)	0	600	1200
Vol $KMnO_4$ used for 10^{-5} m^3 H_2SO_4	23.8×10^{-5} m^3	14.7×10^{-5} m^3	9.1×10^{-5} m^3

Show that the decomposition of H_2O_2 is a first order reaction.

SOLUTION

The integrated rate equation for first order reaction is

$$k = \frac{2.303}{t} \log \frac{a}{a - x}$$

Since volume of $KMnO_4$ used in the titration is measure of concentration of H_2O_2 in solution,

$$a = 23.8 \times 10^{-5} \text{ m}^3$$
$$(a - x) = 14.7 \times 10^{-5} \text{ m}^3 \quad \text{when } t = 600 \text{ s}$$
$$(a - x) = 9.1 \times 10^{-5} \text{ m}^3 \quad \text{when } t = 1200 \text{ s}$$

Substituting these values in the rate equation above, we have

$$k = \frac{2.303}{600} \log \frac{23.8 \times 10^{-5}}{14.7 \times 10^{-5}}$$
$$= 0.00384 (\log 23.8 - \log 14.7)$$
$$= 0.00384 (1.3766 - 1.1673)$$
$$= 0.0008$$

and

$$k = \frac{2.303}{1200} \log \frac{23.8 \times 10^{-5}}{9.1 \times 10^{-5}}$$
$$= 0.00192 (\log 23.8 - \log 9.1)$$
$$= 0.00192 (1.3766 - 0.9595) = 0.0008$$

Since the value of k is almost constant, the decomposition of H_2O_2 is a **first order reaction.**

(3) **Hydrolysis of an Ester.** The hydrolysis of ethyl acetate or methyl acetate in the presence of a mineral acid as catalyst, is a first order reaction.

$$\underset{\text{ethyl acetate}}{CH_3COOC_2H_5} + H_2O \xrightarrow{H^+} \underset{\text{acetic acid}}{CH_3COOH} + C_2H_5OH$$

For studying the kinetics of the reaction, a known volume of ethyl acetate is mixed with a relatively large quantity of acid solution, say N/2 HCl. At various intervals of time, a known volume of the reaction mixture is titrated against a standard alkali solution. Hydrolysis of the ester produces acetic acid. Therefore as the reaction proceeds, the volume of alkali required for titration goes on increasing.

SOLVED PROBLEM 3

The following data was obtained on hydrolysis of methyl acetate at 298 K in 0.35N hydrochloric acid. Establish that it is a first order reaction.

t (secs)	0	4500	7140	∞
m^3 alkali used	24.36×10^{-5} m^3	29.32×10^{-5} m^3	31.72×10^{-5} m^3	47.15×10^{-5} m^3

SOLUTION

For a first order reaction,

$$k = \frac{2.303}{t} \log \frac{a}{a - x}$$

At any time, the volume of alkali used is needed for the acid present as catalyst and the acid produced by hydrolysis.

The volume of alkali used for total change from t_0 to t_∞ gives the initial concentration of ester. Thus,

a = 47.15 – 24.36 = 2.279 × 10^{-5} m^3

$(a - x)$ after 4500 s = 47.15 – 29.32 = 1.783 × 10^{-5} m^3

$(a - x)$ after 7140 s = 47.15 – 31.72 = 1.543 × 10^{-5} m^3

Substituting values in the rate equation above, we have

$$k = \frac{2.303}{4500} \log \frac{2.279 \times 10^{-5}}{1.783 \times 10^{-5}} = 0.00005455$$

$$k = \frac{2.303}{7140} \log \frac{2.279 \times 10^{-5}}{1.543 \times 10^{-5}} = 0.0000546$$

Since the values of k in the two experiments are fairly constant, the reaction is of the first order.

(4) **Inversion of Cane sugar (sucrose).** The inversion of cane sugar or sucrose catalyzed with dil HCl,

$$C_{12}H_{22}O_{11} + H_2O \xrightarrow{H^+} \underset{\text{D-glucose}}{C_6H_{12}O_6} + \underset{\text{D-fructose}}{C_6H_{12}O_6}$$

follows the first order kinetics. The progress of the reaction is followed by noting the optical rotation of the reaction mixture with the help of a polarimeter at different time intervals. The optical rotation goes on changing since D-glucose rotates the plane of polarized light to the right and D-fructose to the left. **The change in rotation is proportional to the amount of sugar decomposed.**

Let the final rotation be r_∞, the initial rotation r_0 while the rotation at any time t is r_t

The initial concentration, a is ∞ $(r_0 - r_\infty)$.

The concentration at time t, $(a - x)$ is ∝ $(r_t - r_\infty)$

Substituting in the first order rate equation,

$$k = \frac{2.303}{t} \log_{10} \frac{a}{a - x}$$

we have

$$k = \frac{2.303}{t} \log_{10} \frac{(r_0 - r_\infty)}{(r_t - r_\infty)}$$

If the experimental values of t $(r_0 - r_\infty)$ and $(r_t - r_\infty)$ are substituted in the above equation, a constant value of k is obtained.

SOLVED PROBLEM 4

The optical rotation of sucrose in 0.9N HCl at various time intervals is given in the table below.

time (secs)	0	430.8	1080	1626	∞
rotation (degree)	+24.09	+21.4	+17.7	+15	–10.74

Show that inversion of sucrose is a first order reaction.

SOLUTION

The available data is substituted in the first order rate equation for different time intervals.

$$k = \frac{2.303}{t} \log_{10} \frac{r_0 - r_\infty}{r_t - r_\infty}$$

$r_0 - r_\infty = 24.09 - (-10.74) = 34.83$ for all time intervals. Thus, the value of rate constant can be found.

time (t)	$r_t = r_\infty$	$k = \frac{1}{t} \log \frac{(r_0 - r_\infty)}{(r_t - r_\infty)}$
430.8	32.14	$k = \frac{1}{430.8} \log \frac{34.83}{32.14} = 8.103 \times 10^{-5}$
1080	28.44	$k = \frac{1}{1080} \log \frac{34.83}{28.44} = 8.150 \times 10^{-5}$
1626	25.74	$k = \frac{1}{1626} \log \frac{34.83}{25.74} = 8.078 \times 10^{-5}$

Since the value of k comes out to be constant, the inversion of sucrose is a **first order reaction**.

SECOND ORDER REACTIONS

Let us take a second order reaction of the type

$$2A \longrightarrow \text{products}$$

Suppose the initial concentration of A is $a \times 10^3$ mol/m^3. If after time t, x mol of A have reacted, the concentration of A is $(a - x)$. We know that for such a second order reaction, rate of reaction is proportional to the square of the concentration of the reactant. Thus,

$$\frac{dx}{dt} = k(a - x)^2 \qquad ...(1)$$

where k is the rate constant, Rearranging equation (1), we have

$$\frac{dx}{(a-x)^2} = k\,dt \qquad ...(2)$$

On integration, it gives

$$\frac{1}{a-x} = kt + I \qquad ...(3)$$

where I is integration constant. I can be evaluated by putting $x = 0$ and $t = 0$. Thus,

$$I = \frac{1}{a} \qquad ...(4)$$

Substituting for I in equation (3)

$$\frac{1}{a-x} = kt + \frac{1}{a}$$

$$kt = \frac{1}{a-x} - \frac{1}{a}$$

Thus $$k = \frac{1}{t} \cdot \frac{x}{a(a-x)}$$

This is the integrated rate equation for a second order reaction.

Examples of Second order Reaction

Hydrolysis of an Ester by NaOH. This is typical second order reaction.

$$\underset{\text{ethyl acetate}}{CH_3COOC_2H_5} + NaOH \longrightarrow CH_3COONa + \underset{\text{ethyl alcohol}}{C_2H_5OH}$$

The reaction is carried in a vessel at a constant temperature by taking. equimolar amounts of ethyl acetate and NaOH. Measured volumes of the reaction mixture (say, 2.5×10^{-5} m^3) are withdrawn at various times and titrated against a standard acid. The volume of the acid used is a measure of the concentration of NaOH or ester. Thus the volume of the acid used when $t = 0$, gives the initial concentration (a) of the reactants. The volume of acid consumed at any other time t gives $(a - x)$. The value of x can be calculated. The rate constant k can be determined by substituting values in the second order integrated rate equation.

SOLVED PROBLEM 5

Hydrolysis of ethyl acetate by NaOH using equal concentration of the reactants, was studied by titrating 2.5×10^{-5} m^3 of the reaction mixture at different time intervals against standard acid. From the data given below, establish that this is a second order reaction.

t (secs)	0	300	900	1500
m^3 acid used	1.6×10^{-5}	1.024×10^{-5}	6.13×10^{-6}	4.32×10^{-6}

SOLUTION

The second order integrated rate equation is

$$k = \frac{1}{at} \cdot \frac{x}{a(a - x)} \quad \text{...(1)}$$

The volume of acid used at any time is a measure of concentration of the unreacted substances at that time.

Therefore,

a, initial concentration $= 1.6 \times 10^{-5}$

after 300 seconds $(a - x) = 1.024 \times 10^{-5}$ and $x = 5.76 \times 10^{-6}$ m^3

after 900 seconds $(a - x) = 6.13 \times 10^{-6}$ and $x = 9.85 \times 10^{-6}$ m^3

after 1500 seconds $(a - x) = 4.32 \times 10^{-6}$ and $x = 11.68 \times 10^{-5}$ m^3

Substituting values in the rate equation (1), we have

$$k = \frac{1}{1.6 \times 10^{-5} \times 300} \cdot \frac{5.76 \times 10^{-6}}{1.024 \times 10^{-5}} = 0.012 \times 10^{4}$$

$$k = \frac{1}{1.6 \times 10^{-5} \times 900} \cdot \frac{79.85 \times 10^{-6}}{6.13 \times 10^{-6}} = 0.010 \times 10^{4}$$

$$k = \frac{1}{1.6 \times 10^{-5} \times 1500} \cdot \frac{1.168 \times 10^{-5}}{4.32 \times 10^{-6}} = 0.016 \times 10^{4}$$

The values of k being fairly constant, this reaction is of the second order.

THIRD ORDER REACTIONS

Let us consider a simple third order reaction of the type

$$3A \longrightarrow \text{products}$$

Let the initial concentration of A be a mol m^{-3} and after time t, x, mol have reacted. Therefore, the concentration of A becomes $(a - x)$. The rate law may be written as :

$$\frac{dx}{dt} = k(a - x)^3 \quad \text{...(1)}$$

Rearranging equation (1), we have

$$\frac{dx}{(a - x)^3} = k\,dt \quad \text{...(2)}$$

On integration, it gives

$$\frac{1}{2(a-x)^2} = kt + I \qquad ...(3)$$

where I is the integration constant. I can be evaluated by putting $x = 0$ and $t = 0$. Thus,

$$I = \frac{1}{2a^2}$$

By substituting the value of I in (3), we can write

$$kt = \frac{1}{2(a-x)^2} - \frac{1}{2a^2}$$

Therefore,

$$k = \frac{1}{t} \cdot \frac{x(2a-x)}{2a^2(a-x)^2}$$

This is the integrated rate equation for a third order reaction.

Examples of Third order Reactions

There are not many reactions showing third order kinetics. A few of the known examples are :

(*i*) $2FeCl_3(aq) + SnCl_2(aq) \longrightarrow 2FeCl_2 + SnCl_4$

(*ii*) $2NO(g) + O_2(g) \longrightarrow 2NO_2(g)$

(*iii*) $2NO(g) + Cl_2(g) \longrightarrow 2NOCl(g)$

UNITS OF RATE CONSTANT

The units of rate constant for different orders of reactions are different.

Units of Zero order Rate constant

For a zero order reaction, the rate constant k is given by the expression

$$k = \frac{d[\mathrm{A}]}{dt} = \frac{\mathrm{mol}}{\mathrm{litre}} \times \frac{1}{\mathrm{time}}$$

Thus the units of k are

mol m^{-3} s^{-1}

Time may be given in seconds, minutes, days or years.

Units of First order Rate constant

The rate constant of a first order reaction is given by

$$k = \frac{2.303}{t} \log \frac{[\mathrm{A}]_0}{[\mathrm{A}]_t}$$

Thus the rate constant for the first order reaction is independent of the concentration. It has the unit

time^{-1}

Units of Second order Rate constant

The rate constant for a second order reaction is expressed as

$$k = \frac{1}{t} \times \frac{x}{[\mathrm{A}]_0([\mathrm{A}]_0 - x)}$$

or

$$= \frac{\mathrm{concentration}}{\mathrm{concentration} \times \mathrm{concentration}} \times \frac{1}{\mathrm{time}}$$

$$= \frac{1}{\text{concentration}} \times \frac{1}{\text{time}}$$

$$= \frac{1}{\text{mole/litre}} \times \frac{1}{\text{time}}$$

$$= \text{mol}^{-1}\ \text{l time}^{-1}$$

Thus the units of k for a second order reactions are

$mol^{-1}\ m^3\ s^{-1}$

Units of Third order Rate constant

The rate constant for a third order reaction is

$$k = \frac{1}{t} \cdot \frac{x(2a - x)}{2a^2 (a - x)^2}$$

or
$$k = \frac{\text{concentration} \times \text{concentration}}{(\text{concentration})^2 \times (\text{concentration})^2} \times \frac{1}{\text{time}}$$

$$= \frac{1}{(\text{concentration})^2} \times \frac{1}{\text{time}}$$

$$= \frac{1}{(\text{mol/litre})^2} \times \frac{1}{\text{time}}$$

Thus the units of k for third order reaction are

$mol^{-2}\ m^6\ s^{-1}$

HALF–LIFE OF A REACTION

Reaction rates can also be expressed in terms of **half-life** or **half-life period.** It is defined as : **the time required for the concentration of a reactant to decrease to half its initial value.**

In other words, half-life is the time required for one-half of the reaction to be completed. It is represented by the symbol $t_{1/2}$ or $t_{0.5}$.

Calculation of Half-life of a First order Reaction

The integrated rate equation (4) for a first order reaction can be stated as :

$$k = \frac{2.303}{t} \log \frac{[A]_0}{[A]}$$

where $[A]_0$ is initial concentration and [A] is concentration at any time t. Half-life, $t_{1/2}$, is time when initial concentration reduces to $\frac{1}{2}$ *i.e.*,

$$[A] = \frac{1}{2}[A]_0$$

Substituting values in the integrated rate equation, we have

$$k = \frac{2.303}{t_{1/2}} \log \frac{[A]_0}{1/2[A]_0} = \frac{2.303}{t_{1/2}} \log 2$$

or
$$t_{1/2} = \frac{2.303}{k} \log 2 = \frac{2.303}{k} \times 0.3010$$

or
$$t_{1/2} = \frac{0.693}{k}$$

It is clear from this relation that :

(1) half-life for a first order reaction is **independent of the initial concentration.**

(2) it is **inversely proportional to k,** the rate-constant.

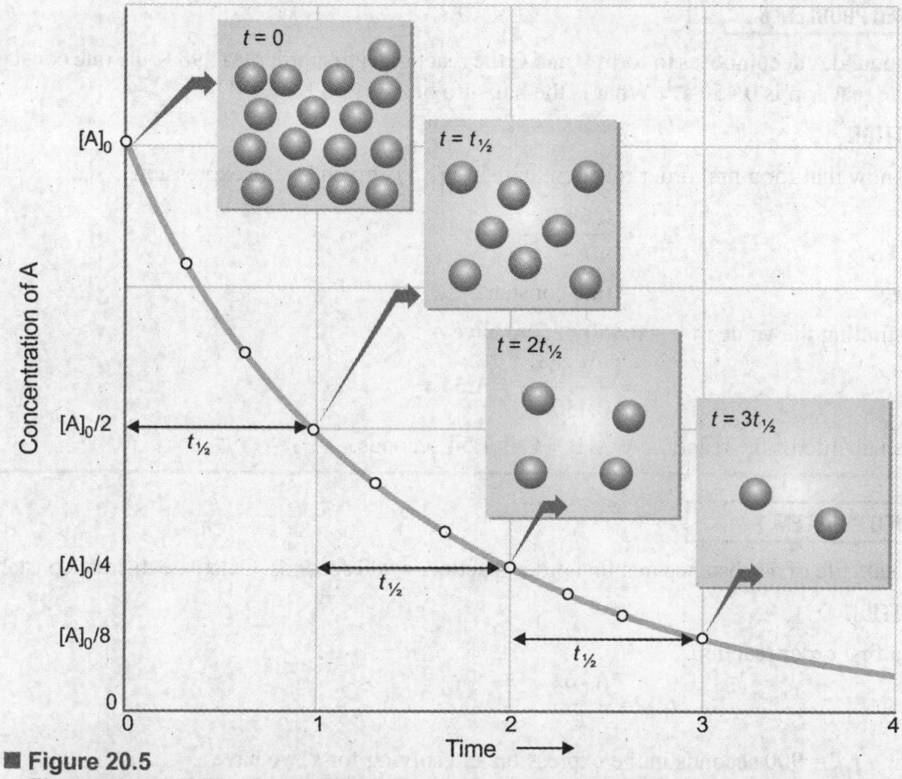

■ **Figure 20.5**
First-order half-life. Concentration of a reactant A as a function of time for a first-order reaction. The concentration falls from its initial value, $[A]_0$, to $[A]_0/2$ after one half-life, to $[A]_0/4$ after a second half-life, to $[A]_0/8$ after a third half-life, and so on. For a first-order reaction, each half-life represents an equal amount of time.

Calculation of Time for Completing any Fraction of the Reaction

As for half-change, we can calculate the time required for completion of any fraction of the reaction. For illustration, let us calculate the time in which two-third of the reaction is completed. First order integrated rate equation is

$$t = \frac{2.303}{k} \log \frac{[A]_0}{[A]}$$

Here, the initial concentration has $\frac{2}{3}$ reacted reducing it to $\frac{1}{3}$. Thus,

$$[A] = \frac{1}{3}[A]_0$$

Substituting values in the rate equation

$$t_{3/4} = \frac{2.303}{k} \log \frac{[A]_0}{1/3[A]_0} = \frac{2.303}{k} \log 3$$

$$= \frac{\mathbf{2.303}}{\boldsymbol{k}} \times \mathbf{0.4771}$$

SOLVED PROBLEM 6

Compound A decomposes to form B and C the reaction is first order. At 298 K the rate constant for the reaction is 0.450 s^{-1}. What is the half-life of A at 298 K?

SOLUTION

We know that for a first order reaction, half-life $t_{1/2}$, is given by the expression

$$t_{1/2} = \frac{0.693}{k}$$

where k = rate constant

Substituting the value of k = 0.450 s^{-1}, we have

$$t_{1/2} = \frac{0.693}{0.450\,s^{-1}} = \mathbf{1.54\ s}$$

Thus half-life of the reaction A → B + C is 1.54 seconds.

SOLVED PROBLEM 7

The half-life of a substance in a first order reaction is 900 seconds. Calculate the rate constant.

SOLUTION

For a first order reaction

$$t_{1/2} = \frac{0.693}{k}$$

Putting $t_{1/2}$ = 900 seconds in the expression and solving for k, we have

$$k = \frac{0.693}{t_{1/2}} = \frac{0.693}{900} = \mathbf{7.7 \times 10^{-4}\ seconds}$$

SOLVED PROBLEM 8

For the reaction

$$2N_2O_5 \longrightarrow 4NO_2 + O_2$$

the rate is directly proportional to $[N_2O_5]$. At 318 K, 90% of the N_2O_5 reacts in 3600 seconds. Find the value of the rate constant k.

SOLUTION

Since rate is ∝ $[N_2O_5]$ it is first order reaction. The integrated rate equation is

$$k = \frac{2.303}{t} \log \frac{[N_2O_5]_0}{[N_2O_5]}$$

When 90% of N_2O_5 has reacted, the initial concentration is reduced to $\frac{1}{10}$ That is,

$$[N_2O_5] = \frac{1}{10}[N_2O_5]_0$$

Substituting values in the rate equation,

$$k = \frac{2.303}{3600} \log \frac{[N_2O_5]_0}{\frac{1}{10}[N_2O_5]_0}$$

$$= \frac{2.303}{3600} \log 10 = \frac{2.303}{3600} \times 1$$

Thus $$k = \frac{2.303}{3600} = \mathbf{6.40 \times 10^{-4}\ s^{-1}}$$

SOLVED PROBLEM 9

The rate law for the decomposition of N_2O_5 (*l*) is : rate = $k\,[N_2O_5]$ where $k = 6.22 \times 10^{-4}\ s^{-1}$. Calculate half-life of N_2O_5 (*l*) and the number of seconds it will take for an initial concentration of N_2O_5 (*l*) of 0.100 M to drop to 0.0100 M.

SOLUTION

Calculation of half-life

$$t_{1/2} = \frac{0.693}{k} = \frac{0.693}{6.22 \times 10^{-4}\ \text{sec}^{-1}} = 1.11 \times 10^3\ \text{sec}$$

Calculation of time in seconds for drop of $[N_2O_5]$ from 0.100 M to 0.0100 M

From first order integrated rate equation,

$$t = \frac{2.303}{t} \log \frac{[N_2O_5]_0}{[N_2O_5]_t}$$

or $$t = \frac{2.303}{k} \log \frac{[N_2O_5]_0}{[N_2O_5]_t}$$

Substituting values

$$t = \frac{2.303}{6.22 \times 10^{-4}} \log \frac{0.100}{0.0100}$$

$$= \frac{2.303}{6.22 \times 10^{-4}} \times 1$$

$$= \mathbf{3.70 \times 10^3\ s}$$

SOLVED PROBLEM 10

For a certain first order reaction $t_{0.5}$ is 100 seconds. How long will it take for the reaction to be completed 75% ?

SOLUTION

Calculation of *k*

For a first order reaction

$$t_{1/2} = \frac{0.693}{k}$$

or $$100 = \frac{0.693}{k}$$

∴ $$k = \frac{0.693}{100} = 0.00693\ \text{sec}^{-1}$$

Calculation of time for 75% completion of reaction

The integrated rate equation for a first order reaction is

$$k = \frac{2.303}{t} \log \frac{[A]_0}{[A]}$$

or $$t = \frac{2.303}{k} \log \frac{[A]_0}{[A]}$$

When $\frac{3}{4}$ initial concentration has reacted, it is reduced to $\frac{1}{4}$

Substituting values in the rate equation

$$t_{3/4} = \frac{2.303}{0.00693} \log \frac{[A]_0}{\frac{1}{4}[A]_0}$$

$$= \frac{2.303}{0.00693} \log 4 = \mathbf{200\,sec}$$

SOLVED PROBLEM 11

A first order reaction is one-fifth completed in 2400 seconds. Calculate the time required for its 100% completion.

SOLUTION

Calculation of *k*

For a first order reaction

$$k = \frac{2.303}{t} \log \frac{[A]_0}{[A]}$$

After 2400 seconds, the initial concentration is reduced to $\frac{4}{5}$ That is,

$$[A] = \frac{4}{5}[A]_0$$

Substituting values in the equation above

$$k = \frac{2.303}{2400} \log \frac{[A]_0}{\frac{4}{5}[A]_0}$$

or $$k = \frac{2.303}{2400} \log 5 - \log 4 = 0.3348 \text{ seconds}$$

Calculation of time required for 100% completion

We know that for first order reaction

$$k = \frac{2.303}{t} \log \frac{[A]_0}{[A]}$$

$$t = \frac{2.303}{k} \log \frac{[A]_0}{[A]}$$

If reaction is 100% complete in, say, t_1 time, we have, [A] = 0. Thus,

$$t_1 = \frac{2.303}{0.3348} \log \frac{[A]_0}{0} = \infty$$

SOLVED PROBLEM 12

50% of a first order reaction is complete in 1380 seconds. Calculate the time required to complete 90% of the reaction.

SOLUTION

Calculation of *k*

$$t_{0.5} = \frac{0.693}{k}$$

or $$k = \frac{0.693}{t_{0.5}} = \frac{0.693}{600} = 1.807824 \text{ seconds}$$

Calculation of time for 90% completion of the reaction

For first order reaction, integrated rate equation is

$$k = \frac{2.303}{t} \log \frac{[A]_0}{[A]} \quad ...(1)$$

or $$t = \frac{2.303}{k} \log \frac{[A]_0}{[A]} \quad ...(2)$$

When 90% of the initial concentration has reacted, 10% of it is left. That is.

$$[A] = \frac{1}{10}[A]_0$$

Substituting values in equation (2)

$$t = \frac{2.303}{1.807824} \log \frac{[A]_0}{\frac{1}{10}[A]_0} = \frac{2.303}{1.807824} \log 10$$

$$= \frac{2.303}{1.807824} = \mathbf{4583.\overline{99} \text{ seconds}}$$

Half-life for a Second order Reaction

For the simple second order reaction 2A → Products, the integrated rate equation is

$$kt = \frac{1}{[A]} - \frac{1}{[A]_0}$$

where $[A]_0$ is the initial concentration and [A] is the concentration when time t has elapsed.

When one-half life has elapsed.

$$[A] = \frac{1}{2}[A]_0$$

and we have

$$k t_{1/2} = \frac{1}{\frac{1}{2}[A]_0} - \frac{1}{[A]_0}$$

or $$k t_{1/2} = \frac{2}{[A]_0} - \frac{1}{[A]_0}$$

Solving for $t_{1/2}$ we find that

$$t_{1/2} = \frac{1}{k[A]_0}$$

As in case of a first order reaction, half-life for a second order reaction is inversely proportional to rate constant k. While half-life of a first order reaction is independent of initial concentration, **half-life of a second order reaction depends on initial concentration.** This fact can be used to distinguish between a first order and a second order reaction.

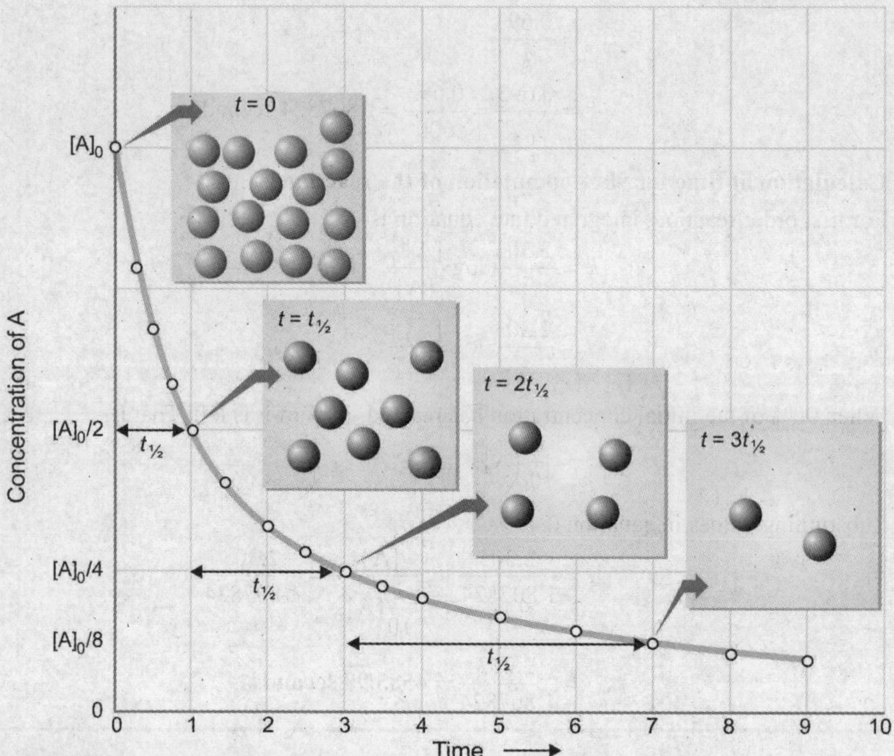

Figure 20.6
Second-order half-life. Concentration of a reactant A as a function of time for a second-order reaction. Note that each half-life is twice as long as the preceding one because $t_{1/2} = 1/k[A]_0$ and the concentration of A at the beginning of each successive half-life is smaller by a factor of 2.

HOW TO DETERMINE THE ORDER OF A REACTION

There are at least four different methods to determine the order of a reaction.

(1) Using integrated rate equations

The reaction under study is performed by taking different initial concentrations of the reactant (a) and noting the concentration ($a - x$) after regular time intervals (t). The experimental values of a, ($a - x$) and t are then substituted into the integrated rate equations for the first, second and third order reactions. **The rate equation which yields a constant value of k corresponds to the correct order of the reaction.** This method of ascertaining the order of a reaction is essentially a method of hit-and-trial but was the first to be employed. It is still used extensively to find the order of simple reactions.

(2) Graphical method

For reactions of the type A → products, we can determine the reaction order by seeing whether a graph of the data fits one of the integrated rate equations.

In case of First order

We have already derived the integrated rate equation for first order as

$$\ln \frac{a}{a-x} = kt$$

Simplifying, it becomes

$$\begin{array}{ccccc} \ln(a-x) & = & -kt & + & \ln a \\ \uparrow & & \uparrow & & \uparrow \\ y & = & mx & + & b \end{array}$$

Thus the two variables in the first order rate equation are :

$$\ln \frac{a}{a-x} \quad \text{and } t$$

Hence, if $\ln \frac{a}{a-x}$ is plotted against t and straight line results (Fig. 20.7), the corresponding reaction is of the first order. However, if a curve is obtained, the reaction is not first order.

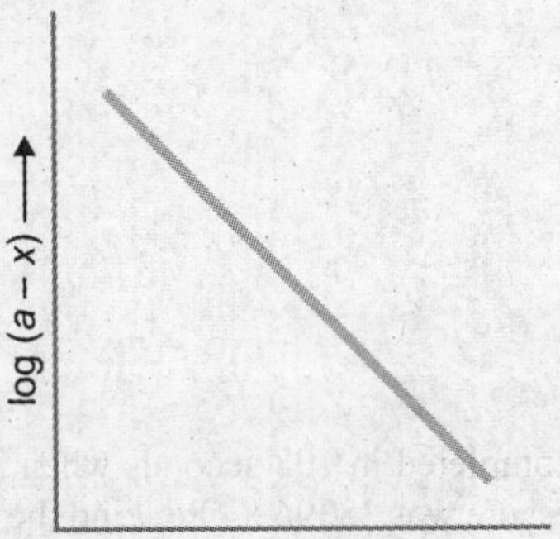

Figure 20.7
Plot of log(*a* – *x*) against *t* for a first order reaction.

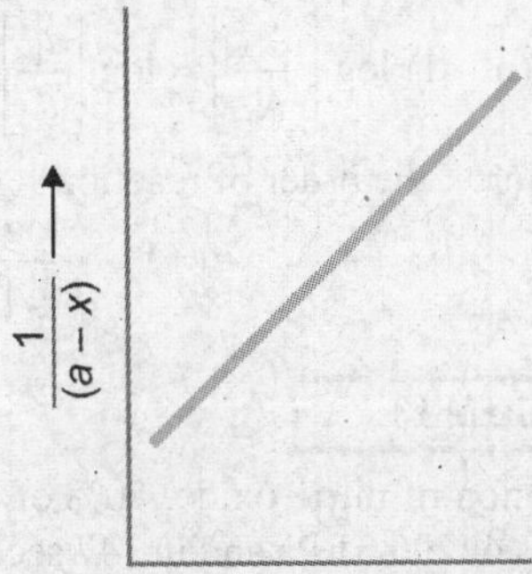

Figure 20.8
Plot of 1/(*a* – *x*) against *t* for a second order reaction.

In case of Second order

We have already shown that second order rate equation can be written as

$$\begin{array}{ccccc} \frac{1}{a-x} & = & kt & + & \frac{1}{a} \\ \uparrow & & \uparrow & & \uparrow \\ y & = & mx & + & b \end{array}$$

This is the equation of a straight line, $y = mx + b$. Here the two variables are

$$\frac{1}{a-x} \quad \text{and } t$$

Thus when $\frac{1}{a-x}$ is plotted against t and we get a straight line (Fig. 20.8), the reaction is second order. In case a curve is obtained, the reaction is not second order.

(3) Using half-life period

Two separate experiments are performed by taking different initial concentrations of a reactant. The progress of the reaction in each case is recorded by analysis. When the initial concentration is reduced to one-half, the time is noted. Let the initial concentrations in the two experiments be $[A_1]$ and $[A_2]$, while times for completion of half change are t_1 and t_2 respectively.

Calculation of order of reaction. We know that half-life period for a first order reaction is independent of the initial concentration, [A]. We also know :

$$\text{half-life} \propto \frac{1}{[A]} \qquad \text{for 2nd order reaction}$$

$$\text{half-life} \propto \frac{1}{[A]^2} \qquad \text{for 3rd order reaction}$$

$$\text{half-life} \propto \frac{1}{[A]^{n-1}} \qquad \text{for } n\text{th order reaction}$$

Substituting values of initial concentrations and half-life periods from the two experiments, we have

$$t_1 \propto \frac{1}{[A_1]^{n-1}} \qquad t_2 \propto \frac{1}{[A_2]^{n-1}}$$

and

$$\frac{t_2}{t_1} = \left[\frac{A_1}{A_2}\right]^{n-1}$$

$$(n-1)\log\left[\frac{A_2}{A_1}\right] = \log\left[\frac{t_1}{t_2}\right]$$

Solving for n, the order of reaction

$$n = 1 + \frac{\log [t_1/t_2]}{\log [A_2/A_1]}$$

SOLVED PROBLEM 13

In the reduction of nitric oxide, 50% of reaction was completed in 108 seconds when initial pressure was 44796.31 Pa and in 147 seconds initial pressure was 38396.84 Pa. Find the order of the reaction.

SOLUTION

We know that

$$\frac{t_2}{t_1} = \left[\frac{A_1}{A_2}\right]^{n-1} \qquad ...(1)$$

where t_1 and t_2 are half-life periods and $[A_1]$ and $[A_2]$ are the corresponding initial concentrations, while n is the order of the reaction.

Taking logs of the expression (1), we have

$$n = 1 + \frac{\log [t_2/t_1]}{\log [A_1/A_2]} \qquad ...(2)$$

Substituting values in expression (2),

$$n = 1 + \frac{\log 108/147}{\log 38396.84/44796.31}$$

$$= 1 + \frac{0.1339}{0.0669} = 1 + 2 = 3$$

Therefore, the reaction is of the **third order**.

(4) **The Differential method**

This method was suggested by van't Hoff and, therefore, it is also called **van't Hoff's differential method.** According to it, the rate of a reaction of the nth order is proportional to the nth power of concentration.

$$-\frac{dC}{dt} = k\,C^n$$

where C = concentration at any instant. In two experiments performed with different initial concentrations, we can write

$$-\frac{dC_1}{\text{dt}} = k\,C_1^n \quad ...(1)$$

$$-\frac{dC_2}{\text{dt}} = k\,C_2^n \quad ...(2)$$

Taking logs of the expression (1) and (2)

$$\log\left(-\frac{dC_1}{dt}\right) = \log k + n\,\log C_1 \quad ...(3)$$

$$\log\left(-\frac{dC_2}{dt}\right) = \log k + n\,\log C_2 \quad ...(4)$$

On substracting (4) from (3), we get

$$n = \frac{\log\left(-\frac{dC_1}{dt}\right) - \left(-\frac{dC_2}{dt}\right)}{\log C_1 - \log C_2} \quad ...(5)$$

To find n, $\left(-\frac{dC}{dt}\right)$ in the two experiments is determined by plotting concentrations against time (t). The slope, $\left(-\frac{dC}{dt}\right)$ at a given time interval is measured by drawing tangents. Using the values of slopes $\left(-\frac{dC_1}{dt}\right)$ and $\left(-\frac{dC_2}{dt}\right)$ in the equation (5), n can be calculated.

(5) Ostwald's Isolation method

This method is employed in determining the order of complicated reactions by 'isolating' one of the reactants so far as its influence on the rate of reaction is concerned. Suppose the reaction under consideration is :

$$A + B + C \longrightarrow \text{products}$$

The order of the reaction with respect to A, B and C is determined. For the determination of the order of reaction with respect to A, B and C are taken in a large excess so that their concentrations are not affected during the reaction. The order of the reaction is then determined by using any of the methods described earlier. Likewise, the order of the reaction with respect to B and C is determined. If n_A, n_B and n_C are the orders of the reaction with respect to A, B and C respectively, the order of the reaction n is given by the expression.

$$n = n_A + n_B + n_C$$

COLLISION THEORY OF REACTION RATES

According to this theory, **a chemical reaction takes place only by collisions between the reacting molecules.** But not all collisions are effective. Only a small fraction of the collisions produce a reaction. The two main conditions for a collision between the reacting molecules to be productive are :

(1) The colliding molecules must posses sufficient kinetic energy to cause a reaction.

(2) The reacting molecules must collide with proper orientation.

Now let us have a closer look at these two postulates of the collision theory.

(1) **The molecules must collide with sufficient kinetic energy**

Let us consider a reaction

$$A-A \;+\; B-B \longrightarrow 2A-B$$

A chemical reaction occurs by breaking bonds between the atoms of the reacting molecules and forming new bonds in the product molecules. The energy for the breaking of bonds comes from the kinetic energy possessed by the reacting molecules before the collision. Fig. 20.9 shows the energy of molecules A_2 and B_2 as the reaction $A_2 + B_2 \rightarrow 2AB$ progresses.

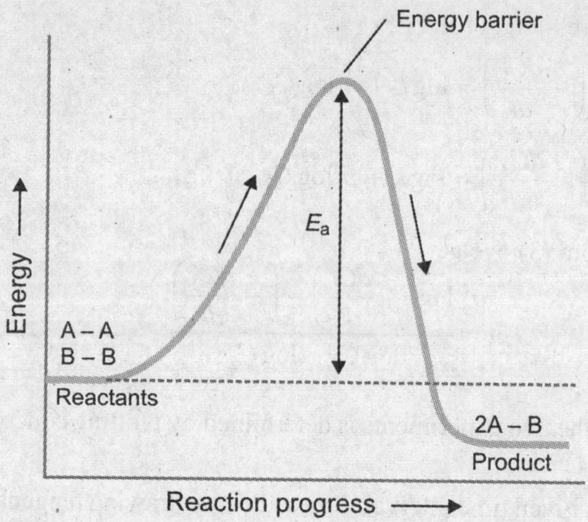

Figure 20.9
The energy of the colliding molecules as the reaction $A_2 + B_2 \rightarrow 2AB$ proceeds. The activation energy E_a provides the energy barrier.

The Fig. 20.9 also shows the **activation energy, E_a**, that is the minimum energy necessary to cause a reaction between the colliding molecules. Only the molecules that collide with a kinetic energy greater than E_a, are able to get over the barrier and react. The molecules colliding with kinetic energies less that E_a fail to surmount the *barrier*. The collisions between them are unproductive and the molecules simply bounce off one another.

(2) **The molecules must collide with correct orientation**

The reactant molecules must collide with favorable orientation (relative position). **The correct orientation is that which ensure direct contact between the atoms involved in the breaking and forming of bonds.** (Fig. 20.10)

From the above discussion it is clear that : **Only the molecules colliding with kinetic energy greater that E_a and with correct orientation can cause reaction.**

Collision Theory and Reaction Rate Expression

Taking into account the two postulates of the collision theory, the reaction rate for the elementary process.

$$A + B \longrightarrow C + D$$

is given by the expression

$$\text{rate} = f \times p \times z$$

where f = fraction of molecules which possess sufficient energy to react; p = probable fraction of collisions with effective orientations, and z = collision frequency.

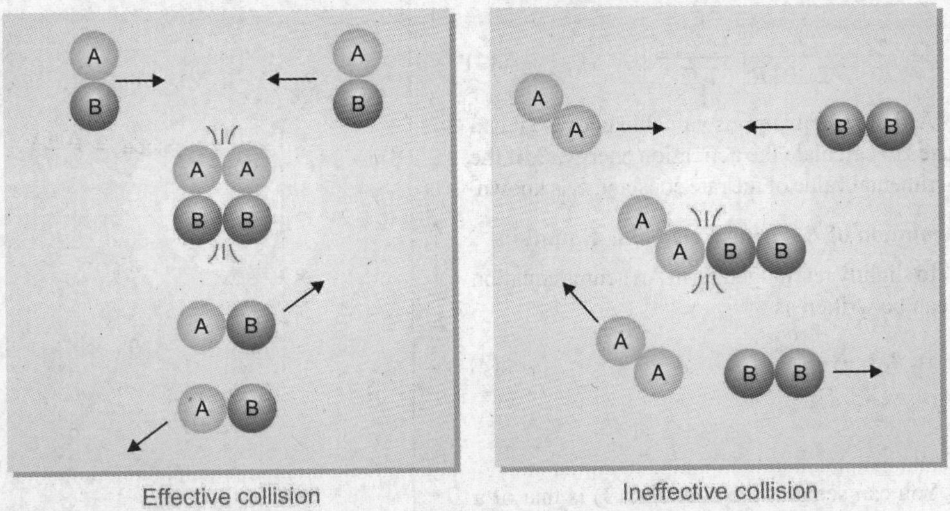

■ **Figure 20.10**
Orientations of reacting molecules A_2 and B_2 which lead to an effective and ineffective collision.

EFFECT OF INCREASE OF TEMPERATURE ON REACTION RATE

It has been found that generally an increase of temperature increases the rate of reaction. As a rule, an increase of temperature by 283 K doubles the reaction rate. Thus the ratio of rate constants of a reaction at two different temperatures differing by 10 degree is known as **Temperature Coefficient.**

$$\textit{i.e.,}\ \text{Temperature Coefficient} = \frac{k_{35^\circ}}{k_{25^\circ}} = \frac{k_{308}}{k_{298}}$$

$$= 2 \text{ to } 3$$

It has been found that **generally an increase of temperature increases the rate of reaction.** As a rule, an increase of temperature by 283 K doubles the reaction rate.

Temperature Dependence of Reaction Rate and Arrhenius Equation

We know that the kinetic energy of a gas is directly proportional to its temperature. Thus as the temperature of a system is increased, more and more molecules will acquire necessary energy greater than E_a to cause productive collisions. This increases the rate of the reaction.

In 1889, Arrhenius suggested as simple relationship between the rate constant, k, for a reaction and the temperature of the system.

$$k = A\, e^{-E_a/RT} \qquad \text{...(1)}$$

This is called the **Arrhenius equation** in which A is an experimentally determined quantity, E_a is the activation energy, R is the gas constant, and T is Kelvin temperature.

Taking natural logs of each side of the Arrhenius equation, it can be put in a more useful form :

$$\ln k = -\frac{E_a}{RT} + \ln A \qquad \text{...(2)}$$

$$\log k = \frac{-E_a}{2.303\,RT} + \log A \qquad \text{...(3)}$$

If k_1 and k_2 are the values of rate constants at temperatures T_1 and T_2 respectively, we can derive

$$\log \frac{k_2}{k_1} = \frac{E_a}{2.303\,R}\left[\frac{T_2 - T_1}{T_1\,T_2}\right] \qquad \text{...(4)}$$

Arrhenius equation is valuable because it can be used to calculate the activation energy, E_a if the experimental value of the rate constant, k, is known.

Calculation of E_a Using Arrhenius Equation

In slightly rearranged form Arrhenius equation (2) can be written as

$$\ln k = -\frac{E_a}{R}\left(\frac{1}{T}\right) + \ln A \qquad \text{...(3)}$$

$$\uparrow \qquad \uparrow \qquad \uparrow$$

$$y \quad = \quad mx \quad + \quad b$$

You can see that the equation (3) is that of a straight line, $y = mx + b$. The two variables in this equation are $\ln k$ and $1/T$.

Thus if we plot the natural logarithm of k against $1/T$, we get a straight line (Fig. 20.11). From the slope of the line, we can calculate the value of E_a.

$$\text{Slope} = -\frac{E_a}{R}$$

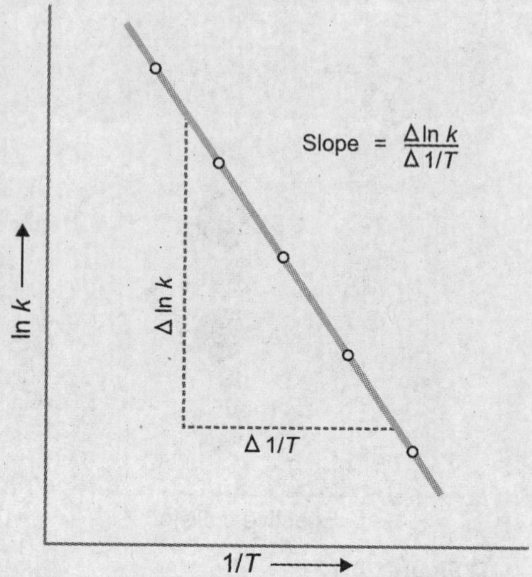

Figure 20.11
The plot of ln k versus $1/T$ gives a straight line. The slope of line $\Delta \ln k / \Delta 1/T$ gives E_a using the expression given above.

SOLVED PROBLEM 14

The values of the rate constant (k) for the reaction $2N_2O_5(g) \longrightarrow 4NO_2(g) + O_2(g)$ were determined at several temperatures. A plot of ln k *versus* $1/T$ gave a straight line of which the slope was found to be -1.2×10^4 K. What is the activation energy of the reaction?

SOLUTION

We know that

$$\text{Slope} = -\frac{E_a}{R}$$

$$E_a = -R \times (\text{slope})$$

Substituting the values, we have

$$E_a = (-8.314\ \text{JK}^{-1}\ \text{mol}^{-1})\,(-1.2 \times 10^4\ \text{K})$$
$$= 1.0 \times 10^5\ \text{J mol}^{-1}$$

Thus the activation energy for the reaction is **1.0×10^5 J mol^{-1}**

Calculation of E_a from the Values of k at Two Temperatures

The rate constant , k, is measured at two temperatures. E_a, is then calculated using the formula that can be derived as follows from equation (3) above.

At temperature T_1, where the rate constant is k_1,

$$\ln k_1 = -\frac{E_a}{RT_1} + \ln A \qquad \text{...(1)}$$

At temperature T_2, where the rate constant is k_2,

$$\ln k_2 = -\frac{E_a}{RT_2} + \ln A \qquad ...(2)$$

Subtracting the equation (1) from the equation (2), we have

$$\ln k_2 - \ln k_1 = \left(-\frac{E_a}{RT_2} + \ln A\right) - \left(-\frac{E_a}{RT_1} + \ln A\right)$$

$$= -\frac{E_a}{RT_2} + \frac{E_a}{RT_1}$$

or $$\ln\left(\frac{k_2}{k_1}\right) = \frac{E_a}{R}\left(\frac{1}{T_1} - \frac{1}{T_2}\right) \quad \text{or} \quad \log\frac{k_2}{k_1} = \frac{E_a}{2.303\,R}\left[\frac{T_2 - T_1}{T_1 T_2}\right]$$

Thus the values of k_1 and k_2 measured at T_1 and T_2 can be used to find E_a.

SOLVED PROBLEM 15

The gas-phase reaction between methane (CH_4) and diatomic sulfur (S_2) is given by the equation

$$CH_4\,(g) + 2S_2\,(g) \longrightarrow CS_2\,(g) + 2H_2S\,(g)$$

At 823 K the rate constant for this reaction is 1.1×10^{-3} m^3 mol^{-1} s. and at 898 K the rate constant is 6.4×10^{-3} mol^{-1} m^3 s. Calculate E_a for this reaction.

SOLUTION

Here $k_1 = 1.1 \times 10^{-3}$ m^3 mol^{-1} s $\qquad T_1 = 823$ K

$k_2 = 6.4 \times 10^{-3}$ m^3 mol^{-1} s $\qquad T_2 = 898$ K

Substituting the values in the equation

$$\ln\left(\frac{k_2}{k_1}\right) = \frac{E_a}{R}\left(\frac{1}{T_1} - \frac{1}{T_2}\right)$$

$$\ln\left(\frac{6.4\times10^{-3}}{1.1\times10^{-3}}\right) = \frac{E_a}{8.3145\,\text{J}\,\text{K}^{-1}\,\text{mol}^{-1}}\left(\frac{1}{823\,\text{K}} - \frac{1}{898\,\text{K}}\right)$$

Solving for E_a, gives

$$E_a = \frac{(8.3145\,\text{J}\,\text{K}^{-1}\,\text{mol}^{-1})\,\ln\left(\frac{6.4}{1.1}\right)}{\left(\frac{1}{823\,\text{K}} - \frac{1}{898\,\text{K}}\right)}$$

$$= \mathbf{1.4 \times 10^5\ J/mol}$$

SIMULTANEOUS REACTIONS

Sometimes there are some side reactions accompanying the main chemical reaction. Such reactions are known as **complex reactions** as these do not take place in a single step. In other words, such complex reactions proceed in a series of steps instead of a single step and the rate of overall reaction is in accordance with the stoichiometric equation for that reaction. Generally, following types of complications occur.

(1) **Consecutive reactions**

(2) **Parallel reactions**

(3) **Reversible or opposing reactions**

These are discussed as follows :

(1) Consecutive Reactions

The reactions in which the final product is formed through one or more intermediate steps are called **consecutive reactions.** These are also known as **sequential** reactions. In such reactions the product formed in one of the elementary reactions acts as the reactant for some other elementary reaction. Various step reactions can be written for the overall reaction as shown below :

	A	$\xrightarrow{k_1}$	B	$\xrightarrow{k_2}$	C
Initial conc.	$[A]_0$		0		0
Conc. after time t	[A]		[B]		[C]

In the above reaction the product C is formed from the reactant A through intermediate B. In this reaction each stage has its own different rate constants k_1 for the first step and k_2 for the second step. The net or overall rate of reaction depends upon the magnitude of these two rate constants. The initial concentration and concentration after time t are shown below each species in above reaction under consideration.

It is clear that

$$[A]_0 = [A] + [B] + [C]$$

The differential rate expressions are

$$\frac{-d[A]}{dT} = k_1[A]$$

$$\frac{d[B]}{dT} = k_1[A] - k_2[B]$$

and

$$\frac{d[C]}{dT} = k_2[B]$$

During the course of the reaction the concentration of A, B and C vary as shown in the Fig. 20.12

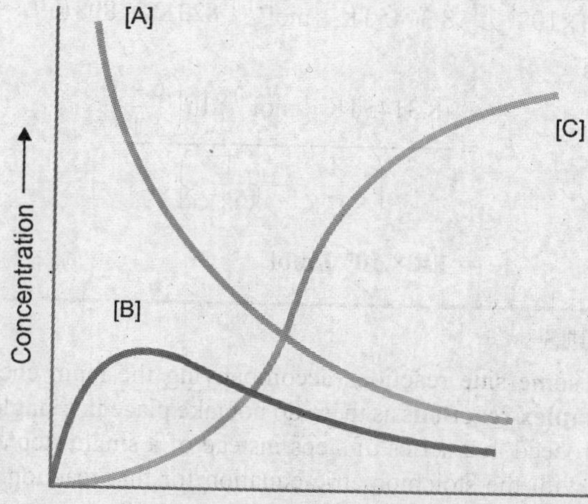

Figure 20.12
Variation of concentration of reactants and products in a consecutive reaction.

From the Fig 20.12 it is clear that the concentration of A decreases exponentially, the concentration of B first increases and then decreases and that of C increases (from zero) with time and finally attains the value equal to $[A]_0$ (initial concentration A) when all A has changed into the final product C.

Examples of First Order Consecutive Reactions

(*a*) Decomposition of dimethyl ether is gaseous phase

$$CH_3COCH_3 \xrightarrow{k_1} CH_4 + HCHO \xrightarrow{k_2} H_2 + CO$$

(*b*) Decomposition of Ethylene oxide

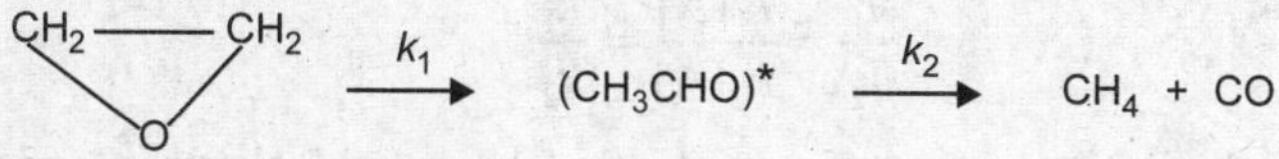

(*c*) Any radioactive decay of the type

$$^{218}_{84}Po \xrightarrow{-\alpha} {}^{214}_{82}Pb \xrightarrow{-\beta} {}^{214}_{83}Bi \xrightarrow{-\beta} {}^{214}_{84}Po$$

(2) **Parallel or Side Reactions**

In these reactions the reacting substance follows two or more paths to give two or more products. The preferential rate of such may be changed by varying the conditions like pressure, temperature or catalyst. The reaction in which the maximum yield of the products is obtained is called the main or major reaction while the other reaction (or reactions) are called side or parallel reactions. For example,

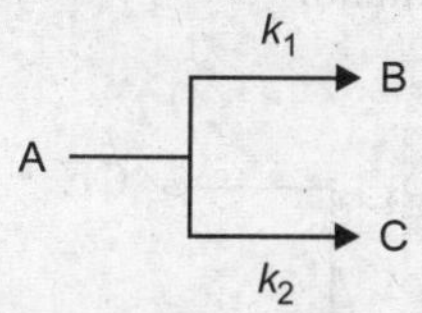

In the above reaction the reactant A gives two products B and C separately in two different reactions with rate constants k_1 and k_2 respectively. If $k_1 > k_2$ the reaction A $\longrightarrow$ B will be the major reaction and A $\longrightarrow$ C will be the side or parallel reaction. Let us assume that both these reactions are of first order and concentration of A is [A] at the time t. The differential rate expressions are

$$r_1 = \frac{-d[A]}{dT} = k_1[A] \qquad ...(i)$$

and

$$r_2 = \frac{-d[A]}{dT} = k_2[A] \qquad ...(ii)$$

The total rate of disappearance of A is given by

$$\frac{-d[A]}{dT} = r_1 + r_2 = k_1[A] + k_2[A]$$

$$= (k_1 + k_2)[A]$$

$$= k'[A] \qquad ...(iii)$$

where k' is the first order rate constant. It is equal to the sum of the two constants k_1 and k_2 of two side reactions.

Integrating equation (*iii*), we get

$$\int \frac{-d[A]}{dT} = \int k'[A]$$

applying the limits $[A_0]$ & $[A]_t$ and 0 & t, we have

$$\int_{[A]_0}^{[A]_t} \frac{-d[A]}{dt} = k' \int_0^t dt$$

$$\ln \frac{[A]_0}{[A]_t} = k't + (k_1 + k_2)t \qquad ...(iv)$$

where $[A]_0$ is the initial concentration of the reactant A and $[A]_t$ is concentration of A at time t.

The ratio of the rates of two side reactions is obtained by dividing equations (*i*) by (*ii*), we have

$$\frac{r_1}{r_2} = \frac{k_1[A]}{k_2[A]} = \frac{k_1}{k_2} \qquad ...(v)$$

From equations (*iv*) and (*v*) we can calculate the individual rate constant k_1 and k_2.

Examples of Parallel or Side Reactions

(*a*) Reaction of ethyl bromide with potassium hydroxide

$$CH_3CH_2Br + KOH \longrightarrow CH_2{=}CH_2 + KBr + H_2O$$

ethyl bromide → ethylene

$$CH_3CH_2Br + KOH \longrightarrow CH_3CH_2OH + KBr$$

ethyl bromide → ethyl alcohol

(*b*) Dehydration of 2-methyl–2–butanol

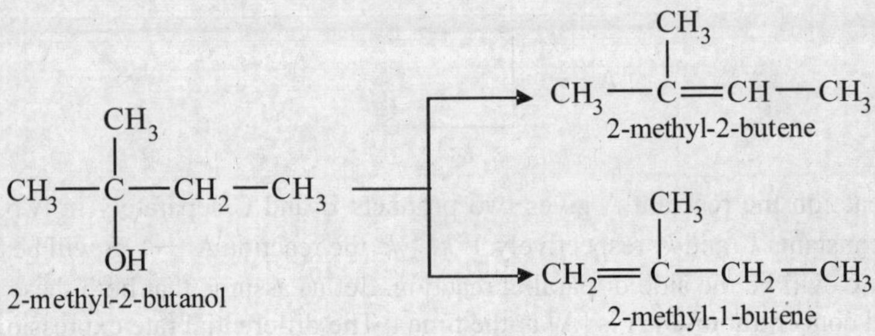

(*c*) Bromination of bromobenzene

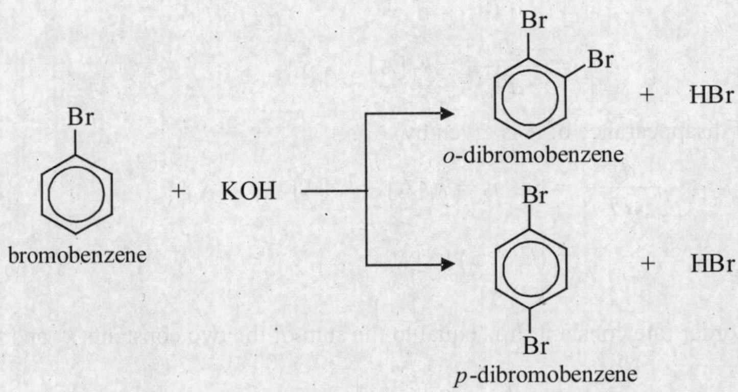

(3) **Reversible or Opposing Reactions**

In reversible or opposing reactions the products formed also react to give back the reactants. Initially, the rate of forward reaction is very large which decreases with passage of time and the rate of backward or reverse reaction is zero which increases with passage of item. A stage is reached when two rates become equal. This situation is called the chemical equilibrium. It is dynamic in nature *i.e.*, all the species are reaching at the rate at which they are being formed. A reaction of this type may be represented as

$$A \underset{k_b}{\overset{k_f}{\rightleftharpoons}} B$$

Initial concentration	$[A]_0$	0
Conc. after time t	[A]	[B]

where k_f and k_b are the rate constants of the forward and backward reactions respectively.

The overall rate of reaction is given by

Rate of Reaction = Rate of forward reaction − Rate of backward reaction

i.e.,
$$\frac{-d[A]}{dt} = \frac{d[B]}{dt} = k_f[A] - k_b[B] \qquad ...(i)$$

If $[A]_0$ is the initial concentration of A and x moles of it have reacted in time t

then $[A]_t = [A]_0 - x$

and $[B] = x$

Substituting these in equation (*i*), we get

$$\frac{dx}{dt} = k_f([A]_0 - x) - k_b x \qquad ...(ii)$$

At equilibrium
$$\frac{dx}{dt} = 0$$

Hence
$$k_f([A]_0 - x_{eq}) = k_b x_{eq} \qquad ...(iii)$$

where x_{eq} is the concentration of A that has reacted into B at equilibrium. From equation (*iii*) we have

$$k_b = k_f\left(\frac{[A]_0 - x_{eq}}{x_{eq}}\right)$$

Substituting the value of k_b in equation (*ii*), we get

$$\frac{dx}{dt} = k_f([A]_0 - x) - k_f\left(\frac{[A]_0 - x_{eq}}{x_{eq}}\right)x$$

Integrating this equation between the limits $t = 0, x = 0$ and $t = t, x = x_{eq}$, we have

$$\int_0^x \frac{dx}{x_{eq} - x} = k_f \frac{[A_0]}{x_{eq}} \int_0^t dt$$

$$= -\ln(x_{eq} - x) + \ln x_{eq} = k_f \frac{[A]_0 t}{x_{eq}}$$

or
$$\ln \frac{x_{eq}}{x_{eq} - x} = k_f \frac{[A]_0}{x_{eq}} t$$

From this equation we can find the value of k_f from the quantities $[A]_0$, x_{eq} and x at time t. All these quantities can be measured easily. From the value of k_f the value of k_b can be calculated by using the relation.

$$k_b = k_f\left(\frac{[A_0] - x_{eq}}{x_{eq}}\right)$$

Examples of Opposing Reactions

(*a*) Dissociation of hydrogen iodides

$$2HI \underset{k_b}{\overset{k_f}{\rightleftharpoons}} H_2 + I_2$$

(*b*) Isomerization of cyclopropane into propene

$$\underset{\text{cyclopropane}}{\triangle} \underset{k_b}{\overset{k_f}{\rightleftharpoons}} \underset{\text{propene}}{CH_3—CH{=}CH_2}$$

(*c*) Isomerization of ammonium cyanate into urea in aqueous solution.

$$\underset{\text{ammonium cyanate}}{NH_4CNO} \underset{k_b}{\overset{k_f}{\rightleftharpoons}} \underset{\text{urea}}{NH_2—\overset{\overset{O}{\|}}{C}—NH_2}$$

(*d*) Isomerization of alkyl ammonium cyanate into substituted urea in aqueous solution

$$\underset{\text{alkyl amm. cyanate}}{R^+NH_3CNO^-} \underset{k_b}{\overset{k_f}{\rightleftharpoons}} \underset{\text{alkylurea}}{NH_2—\overset{\overset{O}{\|}}{C}—NHR}$$

(*e*) Reaction between gaseous CO and NO_2

$$CO(g) + NO_2(g) \underset{k_b}{\overset{k_f}{\rightleftharpoons}} CO_2(g) + NO(g)$$

Limitations of the Collision Theory

The collision theory of reaction rates is logical and correct. However, it has been oversimplified and suffers from the following weaknesses.

(1) **The theory applies to simple gaseous reactions only.** It is also valid for solutions in which the reacting species exist as simple molecules.

(2) The values of rate constant calculated from the collision theory expression (Arrhenius equation) are in agreement with the experimental values only for simple bimolecular reactions. **For reactions involving complex molecules, the experimental rate constants are quite different from the calculated values.**

(3) There is no method for determining the steric effect (p) for a reaction whose rate constant has not been determined experimentally.

(4) In the collision theory it is supposed that only the kinetic energy of the colliding molecules contributes to the energy required for surmounting the energy barrier. There is no reason why the rotational and vibrational energies of molecules should be ignored.

(5) **The collision theory is silent on the cleavage and formation of bonds involved in the reaction.**

The various drawbacks in the simple collision theory do not appear in the modern transition-state theory.

TRANSITION STATE THEORY

The **transition state** or **activated complex theory** was developed by Henry Erying (1935). This theory is also called the **absolute rate theory** because with its help it is possible to get the absolute value of the rate constant. The transition state theory assume that simply a collision between the reactant molecules does not really causes a reaction. During the collision, **the reactant molecules form a transition state or activated complex which decomposes to give the products.**

Thus,

$$\underbrace{A + B—C}_{\text{Reactants}} \longrightarrow \underset{\substack{\text{Activated}\\\text{complex}}}{ABC^{\ddagger}} \longrightarrow \underbrace{A—B + C}_{\text{Products}}$$

The double dagger superscript (‡) is used to identify the activated complex.

The transition state theory may be summarized as follows :

(1) In a collision, the fast approaching reactant molecules (A and BC) slow down due to gradual repulsion between their electron clouds. In the process **the kinetic energy of the two molecules is converted into potential energy.**

(2) As the molecules come close, **the interpenetration of their electron clouds occurs which allows the rearrangement of valence electrons.**

(3) A partial bond is formed between the atoms A and B with corresponding weakening of B – C bond. This leads to formation of an **activated complex** or **transition state.** The activated complex is momentary and decomposes to give the products (A–B + C)

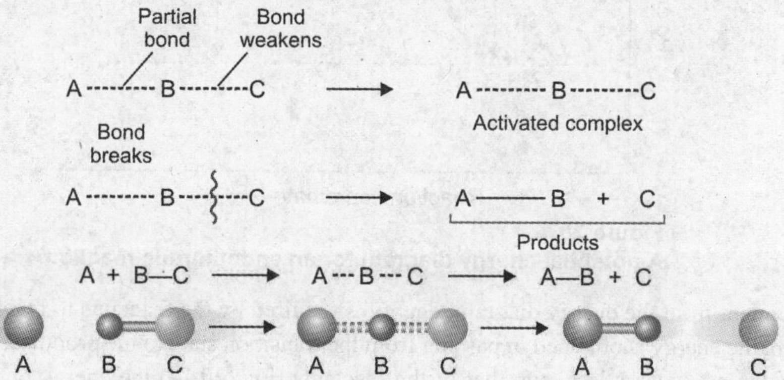

The activated complex theory may be illustrated by the **reaction energy diagram** (Fig. 20.13).

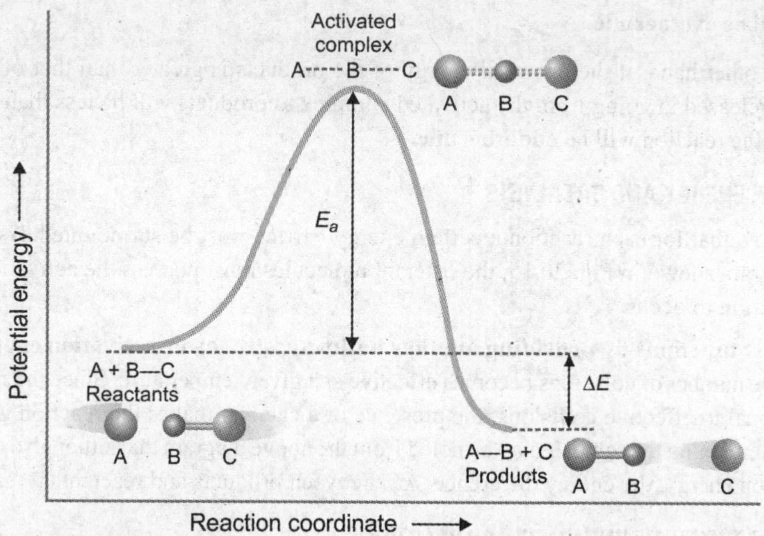

■ **Figure 20.13**
Change of potential energy during a collision between the reactant molecules for an exothermic reaction.

Here the potential energy of the system undergoing reaction is plotted against the **reaction coordinate** (the progress of the reaction). The difference in the potential energy between the reactants and the activated complex is the **activation energy,** E_a. The reactants must have this minimum energy to undergo the reaction through the transition state.

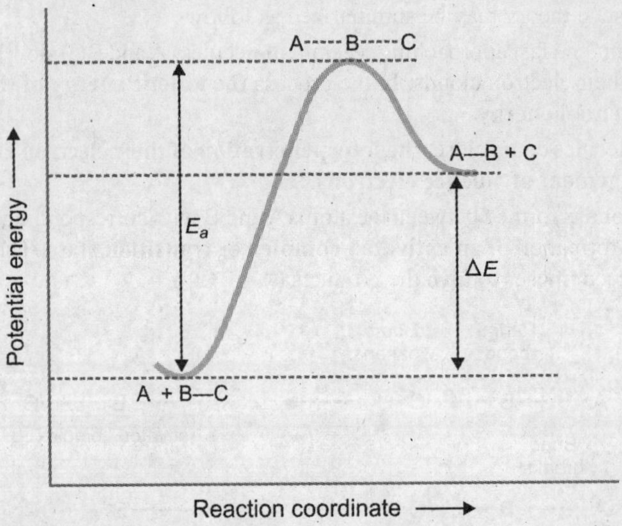

■ **Figure 20.14**
A potential energy diagram for an endothermic reaction.

As evident from the energy diagram, energy is required by the reactants to reach the transition state. Also, the energy is obtained in passing from the transition state to the products. If the potential energy of the products is less than that of the reactants (Fig. 20.14) the energy obtained in going from the activated complex to products will be more than the activation energy (E_a). Thus such a reaction will be **exothermic.**

On the other hand, if the potential energy of the products is greater than that of the reactants, the energy released in going from the activated complex to products will be less than the activation energy and the reaction will be **endothermic.**

ACTIVATION ENERGY AND CATALYSIS

We know that for each reaction a certain energy barrier must be surmounted. As shown in the energy diagram shown in Fig. 20.15, the reactant molecules must possess the activation energy, E_a, for the reaction to occur.

The catalyst functions by providing another pathway with lower activation energy, E_{cat}. Thus a much large number of collisions becomes effective at a given temperature. Since the rate of reaction is proportional to effective collisions, the presence of a catalyst makes the reaction go faster, other conditions remaining the same. It may be noted from the above diagram that although a catalyst lowers the activation energy, the energy difference, ΔE, between products and reactants remains the same.

LINDEMAN'S THEORY OF UNIMOLECULAR REACTIONS

A number of unimolecular reactions, for example,

$$2N_2O_5(g) \longrightarrow 4NO_2(g) + O_2(g)$$

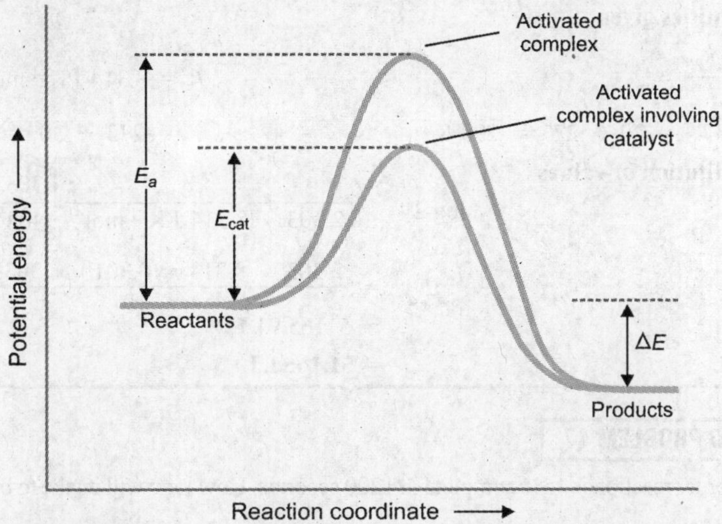

Figure 20.15
Energy diagram for a catalyzed and uncatalyzed reaction showing a lowering of activation energy by a catalyst.

are found to be of the first order. It is difficult to account for such reactions using the collision theory and the absolute reaction rate theory. If two molecules collide in order to provide necessary activation energy, a second order rate law should result. In 1922, Lindeman explained this anomaly by assuming that **there existed a time lag between activation and reaction of molecules.** During this time lag, the activated molecules could either react or be deactivated. Thus,

$$A + A \longrightarrow A + A^* \qquad \textit{activation} \;...(1)$$

$$A + A^* \longrightarrow A + A \qquad \textit{deactivation} \;...(2)$$

$$A^* \longrightarrow \text{products} \qquad \textit{reaction} \;...(3)$$

If the time lag is long, step (3) is slow, the reaction should follow first order kinetics. However, if A reacts as soon as formed, step (2) is the slow step, then the reaction should be second order.

The proof of Lindeman's theory is provided by studying the effect of change of pressure on the reaction. At high pressure the rate of deactivation will be dominant. On the other hand, at sufficiently low pressure all the activated molecules will react before they can be deactivated. Therefore the reaction kinetics should change from first order to second order with decreasing pressure. Several gases are known to exhibit this behavior.

ADDITIONAL SOLVED PROBLEMS

SOLVED PROBLEM 16

The rate constant for a reaction at 293 K is half of the rate constant at 303 K. Calculate the energy of activation of the reaction. [R = 8.314 J K^{-1} mol^{-1}]

SOLUTION

Formula used

$$\log \frac{k_2}{k_1} = \frac{E_a}{2.303 \times R}\left[\frac{T_2 - T_1}{T_1\, T_2}\right]$$

Quantities given

$$\frac{k_2}{k_1} = 2 \qquad R = 8.314 \text{ J K}^{-1} \text{ mol}^{-1}$$

$$T_1 = 20 + 273 = 293 \text{ K} \qquad T_2 = 273 + 30 = 303 \text{ K}$$

Substitution of values

$$\log 2 = \frac{E_a}{2.303 \times 8.314 \text{ J K}^{-1} \text{ mol}^{-1}} \left[\frac{303 \text{ K} - 293 \text{ K}}{303 \text{ K} \times 293 \text{ K}}\right]$$

or

$$E_a = \frac{2.303 \times 8.314 \times 0.3010 \times 303 \times 293}{10}$$

$$= 51165.91 \text{ J}$$

$$= \mathbf{51.165 \text{ kJ}}$$

SOLVED PROBLEM 17

A first order reaction is 15% complete in 1200 seconds. How long will it take to be 60% complete?

SOLUTION

(*a*) To calculate the value of rate constant

Formula used

$$k = \frac{2.303}{t} \log \frac{a}{a-x}$$

Quantities given

$a = 100 \qquad x = 15 \qquad a - x = 85 \qquad t = 1200$ seconds

Substitution of values

$$k = \frac{2.303}{1200} \log \frac{100}{85}$$

$$= \frac{2.303}{1200} \times 0.07058$$

$$= 0.162545/1200 \text{ s}^{-1}$$

$$= 1.35 \times 10^{-4} \text{ s}^{-1}$$

(*b*) To calculate the time when reaction is 60% complete.

Formula used

$$t = \frac{2.303}{k} \log \frac{a}{a-x}$$

Quantities given

$k = 0.4876 \text{ s}^{-1} \qquad a = 100 \qquad a - x = 40$

Substitution of values

$$t = \frac{2.303}{1.35 \times 10^{-4}} \log \frac{100}{40}$$

$$= \frac{2.303}{1.35 \times 10^{-4}} \times 0.39794001$$

$$= 0.6788 \times 10^{-4} = \mathbf{6788 \text{ s.}}$$

SOLVED PROBLEM 18

If a first order reaction is half completed in 100 seconds, calculate the time required for 99% of the reaction to be completed.

SOLUTION : (*a*) To calculate the value of rate constant

Formula used

$$k = \frac{0.693}{t_{1/2}}$$

Quantities given

$$t_{1/2} = 100 \text{ s}$$

Substitution of values

$$k = \frac{0.693}{100 \text{ sec}} = 0.00693 \text{ s}^{-1}$$

(*b*) To calculate the time for 99% of the reaction to be completed

Formula used

$$t = \frac{2.303}{k} \log \frac{a}{a-x}$$

Quantities given

$k = 0.00693 \text{ s}^{-1}$ $a = 100$ $x = 99$ $a - x = 1$

Substitution of values

$$t = \frac{2.303}{0.00693 \text{ sec}^{-1}} \log \frac{100}{1}$$

$$= 332.32 \times \log 100 \text{ s}$$

$$= 332.32 \times 2 \text{ s} = \mathbf{664.64 \text{ s}}$$

SOLVED PROBLEM 19

75% of a reaction of the first order was completed in 1920 seconds. When was it half completed ?

SOLUTION

(*a*) To calculate the value of rate constant

Formula used

$$k = \frac{2.303}{t} \log \frac{a}{a-x}$$

Quantities given

$t = 1920$ seconds $a = 100$ $x = 75$ $a - x = 25$

Substitution of values

$$k = \frac{2.303}{1920} \log \frac{100}{25}$$

$$= 0.001199 \log 4$$

$$= 0.001199 \times 0.6021$$

$$= \mathbf{7.22 \times 10^{-4} \text{ s}^{-1}}$$

(*b*) To calculate half life period

Formula used

$$t_{1/2} = \frac{0.693}{k}$$

Substitution of values

$$t_{1/2} = \frac{0.693}{7.22 \times 10^{-4} \text{ sec}^{-1}} = \mathbf{959.83 \text{ s}}$$

SOLVED PROBLEM 20

It was found that a solution of cane sugar in water hydrolyzed to a extent of 25% in 3600 seconds. Calculate the time that would be taken for the sugar to be 50% hydrolyzed assuming the reaction is of the first order.

SOLUTION : (*a*) To calculate the value of rate constant

Formula used

$$k = \frac{2.303}{t} \log \frac{a}{a-x}$$

Quantities given

$a = 100$ $x = 25$ $a - x = 75$ $t = 3600$ second

Substitution of values

$$k = \frac{2.303}{3600} \log \frac{100}{75}$$
$$= 0.0006397 \times 0.1249$$
$$= 7.99 \times 10^{-5}\ s^{-1}$$

(*b*) **To calculate the half time period**

Formula used

$$t_{½} = \frac{0.693}{k}$$

Substitution of value

$$t_{½} = \frac{0.693}{7.99 \times 10^{-5}}$$
$$= \mathbf{8673\ s}$$

SOLVED PROBLEM 21

The rate constant for the first order reaction at 318 K is twice that at 308 K. Find the energy of activation of the reaction.

SOLUTION

Formula used

$$\log \frac{k_2}{k_1} = \frac{E_a}{2.303 \times R}\left[\frac{T_2 - T_1}{T_1 T_2}\right]$$

Quantities given

$\frac{k_2}{k_1} = 2$ $R = 8.314\ J\ K^{-1}\ mol^{-1}$ $T_1 = 45 + 273 = 318$ K

$T_2 = 35 + 273 = 308$ K

Substitution of values

$$\log 2 = \frac{E_a}{2.303 \times 8.314\ J\ K^{-1}\ mol^{-1}}\left[\frac{318\ K - 308\ K}{318\ K \times 308\ K}\right]$$

or
$$= \frac{0.3010 \times 2.303 \times 8.314 \times 318 \times 308}{10}$$
$$= 56447.96\ J$$
$$= \mathbf{56.4979\ kJ}$$

SOLVED PROBLEM 22

For the first order reaction the half life period is 1800 seconds. What is the time taken for 75% of the completion of the reaction?

SOLUTION : (*a*) **To calculate the value of rate constant**

Formula used

$$k = \frac{0.693}{t_{½}}$$

Quantity given

$$t_{½} = 1800 \text{ seconds}$$

Substitution of value

$$k = \frac{0.693}{1800 \text{ sec.}} = 0.000385$$

(b) To calculate the time for 75% of completion of reaction

Formula used

$$k = \frac{2.303}{1800} \log \frac{a}{a-x}$$

Quantities given

$k = 0.000385$ s $\quad a = 100 \quad x = 75$

Substitution of values

$$t = \frac{2.303}{0.000385} \log \frac{100}{25}$$

$$= 5981.819 \times 0.6021$$

$$= 3601.41 \text{ s}$$

SOLVED PROBLEM 23

In the Arrhenius equation for a certain reaction, the value of A and E_a (activation energy) are 4×10^{13} s^{-1} and 98.6 kJ mol^{-1} respectively. If the reaction is of first order, at what temperature will its half life period be 600 seconds?

SOLUTION : (a) To calculate the value of rate constant

Formula used

$$k = \frac{0.693}{t_{1/2}}$$

Quantities given

$$t_{1/2} = 600 \text{ seconds}$$

Substitution of values

$$k = \frac{0.693}{600 \text{ sec}} = 1.155 \times 10^{-3} \text{ s}^{-1}$$

(b) To calculate the temperature

Formula used

$$k = Ae^{-Ea/RT}$$

or $$\log k = \frac{E_a}{2.303\, R\, T} + \log A$$

Quantities given

$k_2 = 1.55 \times 10^{-3}$ s^{-1} $\quad A = 4 \times 10^{13}$ s^{-1} $\quad E_a = 98.6$ kJ mol^{-1}

$R = 8.314 \times 10^{-3}$ kJ mol^{-1}

Substitution of values

$$\log 1.55 \times 10^{-3} = \log(4 \times 10^{13}) - \frac{98.6 \text{ kJ}}{2.303 \times 8.314 \times 10^{-3} \text{ kJ K}^{-1} \text{ mol}^{-1} \times T}$$

$$-2.9374 = 13.6021 - \frac{5.1496 \times 10^3}{T}$$

$$T = \frac{5.1496 \times 10^3}{13.6021 + 2.9374}$$

$$= \mathbf{311.35\ K}$$

SOLVED PROBLEM 24

The half life period of first order decomposition of nitramide is 7560 seconds at 288 K

$$NH_2NO_{2\,(aq)} \longrightarrow N_2O_{(g)} + H_2O_{(l)}$$

If 0.0062 kg of NH_2NO_2 is allowed to decompose, calculate

(a) time for NH_2NO_2 to decompose 99%; and

(b) volume of dry N_2O produced at this point measured at STP.

SOLUTION : (*a*) To calculate time for 99% decomposition

Formula used

$$k = \frac{2.303}{t} \log \frac{a}{a-x} \quad \text{or} \quad t = \frac{2.303}{k} \log \frac{a}{a-x}$$

Quantities given

$$a = 100 \qquad x = 99 \qquad a - x = 1 \qquad k = \frac{0.693}{t_{½}} = \frac{0.693}{7560 \text{ sec.}} = 9.17 \times 10^{-5}\ s^{-1}$$

Substitution of values

$$t = \frac{2.303}{9.17 \times 10^{-5}} \log \frac{100}{1}$$

$$= 25124.4 \text{ s} \times 2$$

$$= \mathbf{50248.8 \text{ second}}$$

(*b*) To calculate the volume of dry N_2O produced at STP

$$1 \text{ mole of } NH_2NO_2 = 1 \text{ mol of } N_2O$$

$$0.062 \text{ kg mol}^{-1} = 0.0224 \text{ m}^3 \text{ at STP}$$

$$0.062 \text{ kg} = \frac{0.0224 \times 0.062}{0.0062}$$

$$= 0.00224 \text{ m}^3$$

volume of N_2O produced at STP when % decomposition is 99%

$$= \frac{0.00224 \times 99}{100} \text{ m}^3$$

$$= \mathbf{0.0022176 \text{ m}^3}$$

EXAMINATION QUESTIONS

1. Define or explain the following terms :
   (*a*) Rate of reaction
   (*b*) Order of a reaction
   (*c*) Molecularity of a reaction
   (*d*) Rate constant
   (*e*) Half life of a reaction
   (*f*) Arrhenius equation
2. Derive mathematical expression for the rate constant of a reaction (A + B → Products) of the second order.
3. The first order rate constant for the decomposition of N_2O_2 of 273 K is $3.12 \times 10^{-4}\ s^{-1}$. If the energy of activation is 6200 joules per mole, calculate the rate constant at 298 K.

   Answer. 4.43×10^{-4} s

4. (*a*) Explain why the rate of a reaction cannot be measured by dividing the amount of reactants reacted by the time taken.

(*b*) What is activation energy? How is it determined?

5. (*a*) For the kinetics of bimolecular reactions briefly discuss the "Collision Theory".

(*b*) Derive an expression for the half life period of the following reaction:

$A \rightarrow B$, rate $\propto$ [A]

(*c*) With the help of two examples, show that the rate can be independent of initial concentration of the reactants. What is the order of such reactions?

(*d*) If the rate constant at one temperature alongwith the activation energy is given, how can the rate constant at any other temperature be determined?

6. Explain Arrhenius equation. Discuss Arrhenius concept of activation energy. Give graphical representation of activation energy diagram. (*Jiwaji BSc*)

7. (*a*) Explain, with examples, zero-order reaction. Write rate law expression for it.

(*b*) The reaction, A + B + C $\rightarrow$ Products, is found to obey the rate law

$$r = \frac{-d[A]}{dt} = k\,[A]^2\,[B]^{3/2}\,[C]^{-1/2}$$

What is the order of overall reaction? (*Purvanchal BSc*)

8. Derive an expression for rate constant of a bimolecular gaseous reaction on the basis of collision theory. (*Allahabad BSc*)

9. (*a*) Explain briefly the collision theory of reaction rates. What are its limitations and how far they are overcome by theory of absolute reaction rates?

(*b*) Write unit of rate constant for zero order reaction. (*Jamia Millia BSc*)

10. (*a*) A reactant R is converted into product by the following mechanism :

$$R \rightarrow x \rightarrow P$$

where both the steps are of first order. Find out how the concentration of x varies with time and give a physical interpretation of the resulting expression.

(*b*) A solution containing equal concentrations of ethyl acetate and NaOH is 25% saponified in 300 seconds. What will be the % saponification after 600 seconds.

Answer. 22.106 (*Vidyasagar BSc* (*H*))

11. (*a*) Explain various methods of determining the order of a reaction.

(*b*) Derive an expression for the dependence of rate constant of a reaction with temperature.

(*c*) Explain how the acid catalyzed hydrolysis of an ester is followed experimentally to show it to follow first order kinetics.

(*d*) What are zero order reactions? Give one example. (*Sri Venkateswara BSc*)

12. If 15% of a substance decomposes in first 600 seconds in a first order reaction, calculate how much of it would remain undecomposed after one hour?

Answer. 34.52% (*Vidyasagar BSc*)

13. Write short notes on :

(*a*) Activation energy (*b*) Energy barrier (*MD Rohtak BSc*)

14. The velocity constant for a certain reaction is 5.25×10^{-3} at 303 K and 11.1×10^{-3} at 314 K. Calculate the energy of activation of the reaction.

Answer. 53.849 kJ (*Mumbai BSc*)

15. The energy of activation for a reaction is 105 kJ mol^{-1}. Calculate the specific reaction rate for the reaction at 300 K, if the frequency factor is 2.5×10^{15} s^{-1}.

Answer. 1.3×10^{-3} s^{-1} (*Mumbai BSc*)

16. (*a*) Name different methods to determine the order of a reactions and describe the half-life method in detail.

(*b*) Give theory of absolute reaction rates. (*Jamia Millia BSc*)

17. If the half-life of a first order reaction in A is 900 seconds, how long it will take for [A] to reach 10 percent of the initial concentration?

Answer. 2990 s (*Arunachal BSc*)

18. The rate constant of a second order reaction is 5.70×10^{-5} dm^3 mol^{-1} s^{-1} at 298 K and 1.64×10^{-4} dm^3 mol^{-1} s^{-1} at 313 K. Calculate the activation energy of the reaction.(R = 8.314 JK^{-1} mol^{-1})

Answer. 54.645 kJ (*Mizoram BSc (H)*)

19. Explain the term collision frequency. On what factors does it depend? What is meant by collision diameter?

(*Guru Nanak Dev BSc*)

20. The rate constant for a first order reaction is 1.54×10^{-3} s^{-1}. Calculate its half-life period.

Answer. 450 s (*Allahabad BSc*)

21. How is the order of the reaction determined by Differential method and Half-life method?

(*Arunachal BSc*)

22. A second order reaction, in which the initial concentration of both the reactants is the same, is 25% completed in 600 seconds. How long will it take for the reaction to go to 75% completion?

Answer. 5400 s (*Jamia Millia BSc*)

23. Explain the following :

(*i*) If a reaction is pseudo order, the half life is always defined with respect to the species present in the smallest amount.

(*ii*) Under certain situation the rate of reaction and specific reaction rate become identical.

(*Delhi BSc*)

24. (*a*) Derive the Arrhenius equation and discuss how it is useful to calculate the energy of activation.

(*b*) Give one example each for pseudo unimolecular, second order, third order and zero order reactions.

(*Madurai BSc*)

25. Explain, with examples, the functions of catalytic promoters and poisons in chemical reactions.

(*Panjab BSc*)

26. The activation energy of a non-catalyzed reaction at 310 K is 83.68 kJ mol^{-1} and the activation energy for the same reaction catalyzed by enzymes is 25.10 kJ mol^{-1}. Calculate the ratio of the two rate constants.

Answer. 0.9975 (*Kalyani BSc*)

27. Write short notes on:

(*i*) Transition state theory of reaction rates

(*ii*) Parallel reactions

(*iii*) Opposing reactions (*Allahabad BSc*)

28. A ln k versus $1/T$ graph was plotted to calculate the activation energy of a reaction using Arrhenius equation for the effect of temperature on reaction rate. The slope of the straight line was found to be -2.55×10^4. Calculate the activation energy of the reaction.

Answer. 21.2007×10^4 J (*Guru Nanak Dev BSc*)

29. What do you understand by the term: 'rate determining step' of a complex reaction? What is steady–state hypothesis? (*Arunachal BSc*)

30. What would be the units of the rate constant in the following rate law :

rate = k [A][B]

if the concentration is expressed in mol/m^3 and time in seconds. (*Delhi BSc*)

31. Write down the rate and the differential rate expression for A + 2B → Products. (*Guru Nanak Dev BSc*)

32. Derive the rate expression for an opposing reaction in which the forward as well as reverse reactions are both first order. (*Kalyani BSc*)

33. (*a*) What are pseudo and true unimolecular reactions? Explain with examples.

(*b*) A first order reaction is 15% complete in 1200 seconds. What is the half-life period of the reaction?

Answer. (*b*) 5114 s (*Purvanchal BSc*)

34. (*a*) Derive an expression for the rate constant of a second order reaction, involving one reactant only. Also write how the equation you, derive, is employed in the graphical method to confirm second order.

(*b*) Describe the activated complex theory of bimolecular reactions. What are the advantages of this theory over collision theory? (*Guru Nanak Dev BSc*)

35. (*a*) Derive a second order rate equation for the reaction

$$2\text{ A} \rightarrow \text{Product.}$$

Explain that half-life period depends upon initial concentration of the reactants.

(*b*) Calculate the activation energy of a reaction whose rate constant at 300 K gets doubled for 283 K rise in temperature.

Answer. (*b*) 3561.84 J (*Agra BSc*)

36. Calculate the half-life period for the first order reaction whose rate constant is $1.052 \times 10^{-3}\ s^{-1}$.

Answer. 659 s (*Indore BSc*)

37. An acid solution of sucrose was hydrolyzed to the extent of 54% after 4020 seconds. Assuming the reaction to be of first order, calculate the time taken for 80% hydrolysis.

Answer. 8332.92 s (*Allahabad BSc*)

38. A second order reaction in which both the reactants have the same concentration is 25% complete in 10 minutes. How long will it take for the reaction to go to 90% completion?

Answer. 1618.71 s (*Purvanchal BSc*)

39. For a certain reaction, it takes 300 seconds for the initial concentration of 500 mol/m^3 to become 250 mol/m^3 and another 300 seconds to became 125 mol/m^3. (a) What is the order of the reaction ? (b) What is the rate constant of the reaction ?

Answer. (a) One (b) 8.16 s (*Mizoram BSc*)

40. If the half life of a first order in A is 120 seconds, how long will it take A to reach 25% of its initial concentration.

Answer. 240 s (*Delhi BSc*)

41. A reaction that is first order with respect to the reactant A has a rate constant of 360 s^{-1}. If we start $[A]$ = 500 mol/m^3 when would $[A]$ reach the value of 50 mol/m^3 ?

Answer. 6.397×10^{-3} s (*Mysore BSc*)

42. The rate of a particular reaction becomes three times when the temperature is increased from 298 K to 308 K. Calculate the energy of activation for the reaction.

Answer. 83.8 kJ mol^{-1} (*Agra BSc*)

43. Calculate the half life of a first order reaction where the specific rate constant is (a) 200 s^{-1} (b) 120 s^{-1}.

Answer. (a) 0.00346 s^{-1} ; (b) 0.005775 s^{-1} (*Patna BSc*)

44. (*a*) Derive an expression for the rate constant for 2nd order reaction assuming the initial concentration to be the same.

(*b*) 50% of a first order reaction is completed in 1380 seconds. Calculate the time required to complete 90% of the reaction.

Answer. (*b*) 4587.65 s (*Delhi BSc*)

45. (*a*) Derive an expression for rate constant (k) for the reaction A + B → Products on the basis of collision theory. Express the value of k in 10^{-6} m^3 mol^{-1} s^{-1}.

(*b*) The rate constants for a reaction at 300 K and 310 K are 4.5×10^{-5} s^{-2} and 9.0×10^{-5} s^{-2}. Evaluate the activation energy of the reaction.

Answer. 53.603 kJ *(Rajasthan BSc)*

46. Calculate the activation energy of a reaction whose reaction rate at 300 K gets doubled for 283 K rise in temperature (R = 8.314 J K^{-1} mol^{-1})

Answer. 28785.42 kJ *(Mysore BSc)*

47. (*a*) Define order of a reaction, molecularity of a reaction and half life period. Show that for first order reactions the half life period is independent of the initial concentration.

(*b*) The rate constant of a reaction is 1.2×10^{-3} s^{-1} at 303 K. Calculate the rate constant at 313 K, if the activation energy for the reaction is 44.12 kJ mol^{-1}.

Answer. 1.53×10^{-3} *(Guru Nanak Dev BSc)*

48. Derive the rate equation for the first order reaction and show that :

(*i*) Half-life is independent of initial concentration.

(*ii*) The rate constant is independent of concentration. *(Baroda BSc)*

MULTIPLE CHOICE QUESTIONS

1. Which of the following includes all the aims of kinetics?

(*i*) to measure the rate of a reaction

(*ii*) to be able to predict the rate of a reaction

(*iii*) to be able to establish the mechanism by which a reaction occurs

(*iv*) to be able to control a reaction

(*a*) (*i*), (*ii*) and (*iii*) (*b*) (*i*) and (*ii*)

(*c*) (*i*) and (*iii*) (*d*) (*i*), (*ii*), (*iii*) and (*iv*)

Answer. (*d*)

2. Reaction rates can change with

(*a*) temperature (*b*) the addition of a catalyst

(*c*) reactant concentrations (*d*) all of these

Answer. (*d*)

3. Reaction rates generally

(*a*) are constant throughout a reaction

(*b*) are smallest at the beginning and increase with time

(*c*) are greatest at the beginning of a reaction and decrease with time

(*d*) no such generalizations can be made

Answer. (*d*)

4. Consider the reaction in which nitric oxide is oxidized to nitrogen dioxide,
$2NO(g) + O_2(g) \rightarrow 2NO_2(g)$, for which the rate law is = $k\,[NO]^2\,[O_2]$. If this reaction takes place in a sealed vessel and the partial pressure of nitric oxide is doubled, what effect would this have on the rate of reaction?

(*a*) the reaction rate would triple (*b*) the reaction rate would double

(*c*) the reaction rate would quadruple (*d*) there would be no effect on the reaction rate

Answer. (*c*)

5. Which three factors affect the rate of a chemical reaction?
   (*a*) temperature, pressure and humidity
   (*b*) temperature, reactant concentration and catalyst
   (*c*) temperature, reactant concentration and pressure
   (*d*) temperature, product concentration and container volume
   Answer. (*b*)
6. For first-order reactions the rate constant, *k*, has the unit(s)
   (*a*) $l\ mol^{-1}$ (*b*) $time^{-1}$
   (*c*) $(mol/l)^{-1}\ time^{-1}$ (*d*) $time\ mol\ l^{-1}$
   Answer. (*b*)
7. What are the units of the rate constant for a reaction in solution that has an overall reaction order of two? (M is molarity, s is seconds.)
   (*a*) $M^{-1}\ s^{-1}$ (*b*) M^{-1}
   (*c*) s^{-1} (*d*) $M\ s^{-1}$
   Answer. (*a*)
8. The reaction A → B is a second-order process. When the initial concentration of A is 0.50 M, the half-life is 480 seconds. What is the half-life if the initial concentration of A is 0.10 M?
   (*a*) 96 s (*b*) 480 s
   (*c*) 2400 s (*d*) 960 s
   Answer. (*c*).
9. The quantity *k* in a rate law expression
   (*a*) is independent of concentration (*b*) is called the Arrhenius constant
   (*c*) is dimensionless (*d*) is independent of the temperature
   Answer. (*a*)
10. The first-order rate constant for the decomposition of N_2O_5 to NO_2 and O_2 at 343 K is $6.82 \times 10^{-3}\ s^{-1}$. Suppose we start with 0.300 mol of $N_2O_5(g)$ in a 0.500 L container. How many moles of N_2O_5 will remain after 90 seconds?
    (*a*) 0.081 mol (*b*) 0.555 mol
    (*c*) 0.325 mol (*d*) 0.162 mol
    Answer. (*d*)
11. As the temperature of a reaction is increased, the rate of the reaction increases because the
    (*a*) reactant molecules collide with greater energy
    (*b*) reactant molecules collide less frequently
    (*c*) reactant molecules collide less frequently and with greater energy
    (*d*) activation energy is lowered
    Answer. (*d*)
12. In a series of reactions, which is the rate-determining step?
    (*a*) the simplest reaction (*b*) the main reaction involving the major reactant
    (*c*) the slowest reaction (*d*) the fastest reaction
    Answer. (*c*)
13. Beaker A contains a 0.001 kg piece of zinc and beaker B contains 0.001 kg of powdered zinc. If $10^{-4}\ m^3$ of 0.1 M HCl is added to each of the beakers, how is the rate of reaction in beaker A compared to the rate of reaction in beaker B?
    (*a*) the rate in A is greater due to the smaller surface area of the zinc
    (*b*) the rate in A is greater due to the greater surface area of the zinc
    (*c*) the rate in B is greater due to the smaller surface area of the zinc
    (*d*) the rate in B is greater due to the greater surface area of the zinc
    Answer. (*d*)

14. The decomposition of ethylene oxide at 652 K

$(CH_2)_2O(g) \rightarrow CH_4(g) + CO(g)$

is first-order with $k = 0.12\ s^{-1}$. The activation energy of the reaction is 218 kJ/mol. Calculate the rate constant of the reaction at 525 K.

(*a*) $4.266 \times 10^{-5}\ s^{-1}$ (*b*) $12120\ s^{-1}$

(*c*) $5.22 \times 10^{-4}\ s^{-1}$ (*d*) $-852\ s^{-1}$

Answer. (*a*)

15. Here is a second order reaction A → P. If the initial concentration of A 0.0818 M goes down 30.0% in 289 seconds, what is the rate constant for the reaction?

(*a*) $0.0078\ l\ mol^{-1}\ s^{-1}$ (*b*) $1.7\ l\ mol^{-1}\ s^{-1}$

(*c*) $9.1\ l\ mol^{-1}\ s^{-1}$ (*d*) $16\ l\ mol^{-1}\ s^{-1}$

Answer. (*b*)

16. For a first–order reaction of the form A → P, $t_{½} = 3.24 \times 10^4$ seconds. If the concentration of A is 0.0013 M right now, what is the best estimate of what it was the day before yesterday?

(*a*) 0.0026 M (*b*) 0.0065 M

(*c*) 0.0052 M (*d*) 0.042 M

Answer. (*d*)

17. The rate law relates the rate of a chemical reaction to

(*a*) the concentrations of reactants (*b*) the temperature

(*c*) the activation energy (*d*) the reaction mechanism

Answer. (*a*)

18. Rate laws for chemical reactions are determined

(*a*) by examining the coefficients in the balanced chemical equation

(*b*) from the equilibrium constant

(*c*) from the rates of the forward and reverse reactions of the system at equilibrium

(*d*) by experiment

Answer. (*d*)

19. For a reaction, A → products, a graph of [A] versus time is found to be a straight line. What is the order of this reaction?

(*a*) zero order (*b*) first order

(*c*) second order (*d*) third order

Answer. (*a*)

20. To study the rate of a reaction, it is necessary to

(*a*) identify the reactants

(*b*) know the relative amounts of reactants used

(*c*) know the overall chemical equation for the reaction

(*d*) all of the above are necessary

Answer. (*d*)

21. Why do most chemical reaction rates increase rapidly as the temperature rises?

(*a*) the fraction of the molecules with kinetic energy greater than the activation energy increases rapidly with temperature

(*b*) the average kinetic energy increases as temperature rises

(*c*) the activation energy decreases as temperature rises

(*d*) more collisions take place between particles so that the reaction can occur

Answer. (*a*)

22. The rate constant for a reaction depends upon each of the following, Except:

(*a*) solvent for solutions (*b*) temperature

(*c*) concentration of reactants (*d*) nature of reactants

Answer. (*c*)

23. All of the following are true statements concerning reaction orders Except:

(*a*) the rate of a zero-order reaction is constant

(*b*) after three half-lives, a radioactive sample will have one-ninth of its original concentration

(*c*) the units for the rate constant for first order reactions are s^{-1}

(*d*) if doubling the concentration of a reactant doubles the rate of the reaction, then the reaction is first order in that reactant

Answer. (*b*)

24. The powers in the rate law are determined by

(*a*) the principle of detailed balance

(*b*) the physical states of the reactants and products

(*c*) experiment

(*d*) the coefficients in the balanced chemical reaction

Answer. (*c*)

25. Consider the reaction 3A $\rightarrow$ 2B.

The average rate of appearance of B is given by [B]/*t*. How is the average rate of appearance of B related to the average rate of disappearance of A?

(*a*) –2[A]/3*t* (*b*) [A]/*t*

(*c*) –[A]/*t* (*d*) –3[A]/2*t*

Answer. (*a*)

26. Which one of the following is incorrect for the reaction A $\rightarrow$ B?

(*a*) the half-life of a second-order reaction depends on the initial concentration

(*b*) the half-life is the time for one half of the reactant to be consumed

(*c*) the second-order rate constant can be found by plotting $1/[A]^2$ versus time, where [A] is the concentration of reactant

(*d*) the initial rates for a second-order reaction depend on the concentration of the reactant squared

Answer. (*c*)

27. Consider the reaction in which ammonia is synthesized from nitrogen and hydrogen gases :

$N_2(g) + 3 H_2(g) \rightarrow 2NH_3(g)$

How is the rate of formation of ammonia related to the rate of consumption of hydrogen?

(*a*) the rate of formation of ammonia is half the rate of consumption of hydrogen

(*b*) the rate of formation of ammonia is twice the rate of consumption of hydrogen

(*c*) the rate of formation of ammonia is equal to the rate of consumption of hydrogen

(*d*) the rate of formation of ammonia is two-thirds the rate of consumption of hydrogen

Answer. (*d*)

28. Which concentration plot is linear for a first-order equation? (A is one of the reactants).

(*a*) [A] versus time (*b*) square root of [A] versus time

(*c*) In[A] versus time (*d*) $[A]^2$ versus time

Answer. (*c*)

29. A reaction in which all reactants are in the same phase is called

(*a*) elementary (*b*) bimolecular

(*c*) homogeneous (*d*) heterogeneous

Answer. (*c*)

30. As temperature increases, the reaction rate .....

(*a*) decrease than increase (*b*) decreases

(*c*) increases (*d*) stays the same

Answer. (*c*)

31. For the reaction $2NO_2 + O_3 \rightarrow N_2O_5 + O_2$ the following observations are made :
Doubling the concentration of $[NO_2]$ doubles the rate, and doubling the concentration of $[O_3]$ doubles the rate. What is the rate law for the reaction?

(*a*) rate = $k\,[NO_2]$ (*b*) rate = $k\,[NO_2]^2\,[O_3]$

(*c*) rate = $k\,[NO_2]^2\,[O_3]^2$ (*d*) rate = $k\,[NO_2]\,[O_3]$

Answer. (*d*)

32. The half-life of a first-order process

(*a*) depends on the reactant concentration raised to the first power

(*b*) is inversely proportional to the square of the reactant concentration

(*c*) is inversely proportional to the reactant concentration

(*d*) is totally independent of the reactant concentration

Answer. (*d*)

33. The reaction, $2NO(g) \rightarrow N_2(g) + O_2(g)$, proceeds in a single elementary step. This reaction is thus

(*a*) the molecularity cannot be determined from the given information

(*b*) termolecular

(*c*) bimolecular

(*d*) unimolecular

Answer. (*c*)

34. Why is a minimum energy needed for an effective collision?

(*a*) energy is needed to break bonds

(*b*) energy is needed to orient the particles correctly

(*c*) a minimum energy is needed, so that the particles will collide many times per second

(*d*) enough energy is needed to give off heat in a reaction

Answer. (*a*)

35. Species that are formed in one step of reaction mechanism and used up in another step are called

(*a*) catalysts (*b*) intermediates

(*c*) inhibitors (*d*) activated complexes

Answer. (*b*)

36. For a certain reaction, the rate = $k\,[NO]^2\,[O_2]$, when the initial concentration of NO is tripled, the initial rate

(*a*) decreases by a factor of nine (*b*) increases by a factor of three

(*c*) increases by a factor of six (*d*) increases by a factor of nine

Answer. (*d*)

37. Which of the following statements associated with mechanisms of chemical reactions is incorrect?

(*a*) intermediates do not appear in the net chemical equation or overall rate law

(*b*) the first step in a mechanism always determines the rate of the reaction

(*c*) in elementary reactions, coefficients give the order with respect to reactants and products

(*d*) a plausible mechanism must account experimentally determined rate law

Answer. (*b*)

38. What happens in a steady state?

(*a*) product is being formed faster than reactants are regenerated

(*b*) heat is evolved

(*c*) the concentration of an intermediate is constant

(*d*) nothing is happening

Answer. (*c*)

39. What happens when molecules collide with less than the activation energy needed for the reaction?

(*a*) they stick together but do not react

(*b*) they react, but more slowly

(*c*) they react if the bonds are arranged in the correct orientation

(*d*) they do not react; they simply bounce off of each other

Answer. (*d*)

40. Why do fine iron wires burst into flame when lighted, while an iron frying pan does not?

(*a*) the finely divided iron has lower internal free energy than the frying pan

(*b*) the frying pan has much greater mass than the finely divided wires

(*c*) the frying pan is more dense than the wires

(*d*) the wires have greater surface area, enabling more iron particles to react more quickly

Answer. (*d*)

41. Which of the following terms describes a process in which two particles come together to form an activated complex?

(*a*) reaction mechanism (*b*) elementary process

(*c*) rate determining step (*d*) unimolecular

Answer. (*b*)

42. A + B + C $\rightarrow$ products is :

(*a*) unimolecular (*b*) trimolecular

(*c*) bimolecular (*d*) tetramolecular

Answer. (*b*)

43. Based on the collision model, the atoms at the top of the potential energy "hill" are called:

(*a*) top of the hill (*b*) activation energy

(*c*) transition state (*d*) steric factor

Answer. (*c*)

44. According to chemical kinetic theory, a reaction can occur

(*a*) if the reactants collide with the proper orientation

(*b*) if the reactants possess sufficient energy of collision

(*c*) if the reactants are able to form a correct transition state

(*d*) all of the above

Answer. (*d*)

45. What does termolecular refer to?

(*a*) 2 molecules colliding (*b*) 4 molecules colliding

(*c*) 3 molecules colliding (*d*) 1 molecule dissociating

Answer. (*c*)

46. What is the order of the absorption process for alcohol through the lining of the stomach and small intestine?

(*a*) zero order (*b*) first order

(*c*) second order (*d*) third order

Answer. (*b*)

47. The decomposition of nitrogen dioxide to nitrogen and oxygen is second-order with a rate constant $k = 12.5\ M^{-1}\ s^{-1}$. What is the half-life for the reaction if $[NO_2]_0 = 0.00260$ M?

(*a*) 0.0554 s (*b*) 30.8 s

(*c*) 385 s (*d*) 61.5 s

Answer. (*b*)

48. A first order reaction requires 2.32×10^7 s for the concentration of reactant to be reduced to 25.0% of its original value. What is the half life of the reaction?

(*a*) 1.16×10^7 s
(*b*) 5.8×10^7 s
(*c*) 2.32×10^7 s
(*d*) 4.64×10^7 s

Answer. (*a*)

49. In a reaction, the initial concentration of the reactant is 2.34 M. The equilibrium mixture contains 1.12 M of the reactant. Equilibrium was achieved in 523 seconds. What is the average rate of the reaction?

(*a*) 4.45×10^{-3} M/s
(*b*) 2.14×10^{-3} M/s
(*c*) 2.33×10^{-3} M/s
(*d*) 429 M/s

Answer. (*c*)

50. The half-life for a first order reaction is 8.73×10^{10} s. If the concentration after 3.49×10^{11} s is 0.0216 M, what was the initial concentration?

(*a*) 0.0690 M
(*b*) 0.345 M
(*c*) 0.173 M
(*d*) 1.000 M

Answer. (*b*)

51. For the hypothetical gas-phase reaction 3A(g) $\rightarrow$ 2B(g), the number of moles of A present at 1500 seconds is found to be 0.022. If the initial number of moles of A and B were 0.050 mol and 0.00 mol, respectively, how much B is present at 1500 seconds?

(*a*) 0.022 mol
(*b*) 0.056 mol
(*c*) 0.028 mol
(*d*) 0.019 mol

Answer. (*d*)

52. The number of undecayed nuclei in a sample of Bromine-87 decreased by a factor of 4 over a period of 112 seconds. What is the rate constant for the decay of Bromine-87?

(*a*) 56 s
(*b*) 6.93×10^{-1} s^{-1}
(*c*) 1.24×10^{-2} s^{-1}
(*d*) 6.19×10^{-3} s^{-1}

Answer. (*c*)

53. The half-life of radioactive sodium is 54000 seconds. How many hours would it take for a 0.064 kg sample to decay to one-eighth of its original concentration?

(*a*) 3
(*b*) 15
(*c*) 30
(*d*) 45

Answer. (*d*)

54. The decomposition of NOBr is second-order with respect to NOBr and second-order overall. If the initial concentration of NOBr is 0.102 M and the rate constant is 25 M^{-1} s^{-1}, what is [NOBr] after 60 seconds?

(*a*) 9.8 M
(*b*) 2.9×10^{-2} M
(*c*) 4.0×10^{-2} M
(*d*) 35 M

Answer. (*b*)

55. For a reaction of the type A + B + C going to product, the following observations are made: Doubling the concentration of A doubles the rate, doubling concentration of B has no effect on the rate, and tripling the concentration of C increases the rate by a factor of 9. What is the rate law for the reaction?

(*a*) rate = k [A]^2 [B] [C]^2
(*b*) rate = k [A] [C]^2
(*c*) rate = k [A]^2 [B] [C]
(*d*) rate = k [A] [B] [C]

Answer. (*b*)

56. If we have the reaction A(g) $\rightarrow$ 2B(g) and the number of moles of A is as follows,

time	0	300 s	300 s
moles A	0.100	0.085	0.070

what is the number of moles of B at 600 s?

(*a*) 0.030 mol
(*b*) 0.140 mol
(*c*) 0.060 mol
(*d*) 0.200 mol

Answer. (*c*)

57. From a plot or reasonable estimate, what is the instantaneous rate of change of A versus time at 600 s?

Time	0	2.50	5.00	7.50	10.0	12.5
[A]	0.500	0.389	0.303	0.236	0.184	0.143

(*a*) – 0.0209 (*b*) – 0.0164
(*c*) – 0.0316 (*d*) – 0.0184

Answer. (*d*)

58. A reaction requires 279 seconds in order to reach equilibrium. The concentration of one reactant changed from 2.13 M to 0.24 M. What is the average rate of the reaction?

(*a*) 7.63×10^{-3} M/s (*b*) 6.77×10^{-3} M/s
(*c*) 2.13 M/s (*d*) 148 M/s

Answer. (*b*)

59. In a first-order reaction, A → products, [A] = .0620 M initially and 0.520 M after 15.0 s. What is the half-life, $t_{1/2}$, of this reaction?

(*a*) 7.63×10^{-2} M/s^{-1} (*b*) 6.77×10^{-2} M/s^{-1}
(*c*) 2.13 M/s^{-1} (*d*) 148 M/s^{-1}

Answer. (*d*)

60. In a first-order reaction, A → product, [A] = 0.6200 M initially and 0.520. M after 15 s. What is the value of the rate constant?

(*a*) 9.12×10^{-2} s^{-1} (*b*) 5.09×10^{-2} s^{-1}
(*c*) 0.0117 s^{-1} (*d*) 0.154×10^{-1}

Answer. (*c*)

61. In a lab, each of the following factors will vary to affect reaction rate Except:

(*a*) catalyst used (*b*) concentration of reactants
(*c*) identity of reactants (*d*) oxygen availability

Answer. (*d*)

62. In a first-order reaction, A → products, $k = 0.0150$ s^{-1} If [A] is 0.4000 M initially, what will [A] be after 120 s ?

(*a*) 0.388 M (*b*) 1.92×10^{-3} M
(*c*) 0.487 M (*d*) 0.0661 M

Answer. (*d*)

63. In a second-order reaction the rate constant is 4.0×10^{-4} M^{-1} s^{-1}. What will be the concentration of reactant after 10 seconds if the initial concentration is 0.800 M?

(*a*) 1.49 M (*b*) 0.671 M
(*c*) 0.629 M (*d*) 0.300 M

Answer. (*b*)

64. In the reaction 2A + B → 2C + D, –Δ [A]/*t* is found to be 5.0 M s^{-1}. What is the rate of change of B?

(*a*) 2.5 M/s (*b*) 10 M/s
(*c*) 25 M/s (*d*) 5.0 M/s

Answer. (*a*)

65. The reaction A → P is a second-order process with $t_{1/2}$ = 23 seconds at an initial concentration of A = 0.500 M. What is the concentration of A after 60 seconds elapses?

(*a*) 0.263 M (*b*) 0.139 M
(*c*) 0.0822 M (*d*) 0.175 M

Answer. (*b*)

66. The thermal decomposition of $N_2O_5(g)$ to form $NO_2(g)$ and $O_2(g)$ is a first-order reaction. The rate constant for the reaction is $5.1 \times 10^{-4}\ s^{-1}$ at 318 K. What is the half-life of this process?

(*a*) 3.9×10^3 s (*b*) 2.0×10^3 s

(*c*) 1.0×10^{-3} s (*d*) 1.4×10^3 s

Answer. (*d*)

67. If the reaction, 2A + 3D → products, is first-order in A and second-order in D, then the rate law will have the form : rate =

(*a*) $k\,[A]\,[D]^2$ (*b*) $k\,[A]^2\,[D]$

(*c*) $k\,[A]^2\,[D]^2$ (*d*) $k\,[A]\,[D]$

Answer. (*a*)

68. If the reaction below is carried out in a closed, rigid container, how could you best follow the rate of the reaction?

$2SO_2(g) + O_2(g) \rightarrow 2SO_3(g)$

(*a*) monitor the density of the mixture (*b*) monitor the temperature

(*c*) monitor the pressure (*d*) monitor the change in concentration

Answer. (*c*)

69. Which of the following does not affect the rate of a chemical reaction?

(*a*) enthalpy of the reaction (*b*) concentration of reactants

(*c*) temperature (*d*) surface area

Answer. (*a*)

70. Which of the following rate laws is third-order overall?

(*a*) rate = $k\,[A]^3\,[B]^1$ (*b*) rate = $k\,[A]^3\,[B]^3$

(*c*) rate = $k\,[A]^5\,[B]^2$ (*d*) rate = $k\,[A]\,[B]^2$

Answer. (*d*)

71. For a certain reaction, a plot of In [A] versus *t* gives a straight line with a slope of $-1.46\ s^{-1}$. The order of the reaction in A is:

(*a*) 0 (*b*) 3

(*c*) 2 (*d*) 1

Answer. (*d*)

21 Catalysis

CONTENTS

Berzelius (1836) realized that there are substances which increase the rate of a reaction without themselves being consumed. He believed that the function of such a substance was to loosen the bonds which hold the atoms in the reacting molecules together. Thus he coined the term **Catalysis** (Greek *kata* = wholly, *lein* = to loosen).

There is no doubt that usually a catalyst accelerates a reaction as was originally thought by Berzelius. But a number of cases are now known where the catalyst definitely retards (slows down) the rate of reaction.

JÖNS JAKOB BERZELIUS

Jöns Jakob Berzelius (1779-1848), Swedish chemist, considered one of the founders of modern chemistry.

Berzelius was born near Linköping. While studying medicine at the University of Uppsala, he became interested in chemistry. After practicing medicine and lecturing, he became a professor of botany and pharmacy at Stockholm in 1807. From 1815 to 1832 he was professor of chemistry at the Caroline Medico-Chirurgical Institute in Stockholm. He became a member of the Stockholm Academy of Sciences in 1808 and in 1818 became its permanent secretary. For his contributions to science, Berzelius was made a baron in 1835 by Charles XIV John, king of Sweden and Norway.

Berzelius's research extended into every branch of chemistry and was extraordinary for its scope and accuracy. He discovered three chemical elements–cerium, selenium, and thorium–and was the first to isolate silicon, zirconium, and titanium. He introduced the term catalyst into chemistry and was the first to elaborate on the nature and importance of catalysis. He introduced the present system of chemical notation, in which each element is represented by one or two letters of the alphabet. In addition, Berzelius was primarily responsible for the theory of radicals, which states that a group of atoms, such as the sulfate group, can act as a single unit through a series of chemical reactions. He developed an elaborate electrochemical theory that correctly stated that chemical compounds are made up of negatively and positively charged components. All of his theoretical work was supported by elaborate experimental measurements. His greatest achievement was the measurement of atomic weights.

Statue of Berzelius in the center of Berzelii Park, Stockholm

A catalyst is defined as a substance which alters the rate of a chemical reaction, itself remaining chemically unchanged at the end of the reaction. The process is called Catalysis.

As evident from the above definition, a catalyst may increase or decrease the rate of a reaction.

A catalyst which enhances the rate of a reaction is called a **Positive catalyst** and the process **Positive catalysis** or simply **Catalysis.**

A catalyst which retards the rate of a reaction is called a **Negative catalyst** and the process **Negative catalysis.**

We will first proceed to discuss 'positive catalysis' or catalysis as it is commonly designated.

TYPES OF CATALYSIS

There are two main types of catalysis :

(*a*) Homogeneous catalysis

(*b*) Heterogeneous catalysis

Also, there is a third types of catalysis known as **Enzyme catalysis** which is largely of biological interest. This will be discussed separately at a later stage.

HOMOGENEOUS CATALYSIS

In homogeneous catalysis, **the catalyst is in the same phase as the reactants and is evenly distributed throughout.** This type of catalysis can occur in gas phase or the liquid (solution) phase.

Examples of Homogeneous Catalysis in Gas Phase

(*a*) Oxidation of sulfur dioxide (SO_2) to sulfur trioxide (SO_3) with nitric oxide (NO) as catalyst,

$$\underset{\text{gas}}{2SO_2} + \underset{\text{gas}}{O_2} + \underset{\text{gas}}{[NO]} \longrightarrow \underset{\text{gas}}{2SO_3} + [NO]$$

(*b*) Decomposition of acetaldehyde (CH_3CHO) with iodine (I_2) as catalyst,

$$\underset{\text{vapor}}{CH_3CHO} + \underset{\text{vapor}}{(I_2)} \longrightarrow \underset{\text{gas}}{CH_4} + \underset{\text{gas}}{CO}$$

Examples of Homogeneous Catalysis in Solution Phase

Many reactions in solutions are catalyzed by acids (H^+) and bases (OH^-).

(*a*) Hydrolysis of cane sugar in aqueous solution in the presence of mineral acid as catalyst,

$$\underset{\text{cane sugar}}{C_{12}H_{22}O_{11}} + H_2O \xrightarrow{H_2SO_4} \underset{\text{glucose}}{C_6H_{12}O_6} + \underset{\text{fructose}}{C_6H_{12}O_6} + [H_2SO_4]$$

(*b*) Hydrolysis of an ester in the presence of acid or alkali,

$$\underset{\text{ethyl acetate}}{CH_3COOC_2H_5} + H_2O \xrightarrow{H^+/OH^-} \underset{\text{acetic acid}}{CH_3COOH} + \underset{\text{ethanol}}{C_2H_5OH}$$

(*c*) Decomposition of hydrogen peroxide (H_2O_2) in the presence of iodide ion (I^-) as catalyst,

$$2H_2O_2 \xrightarrow{I^-} 2H_2O + O_2$$

HETEROGENEOUS CATALYSIS

The catalysis in which the catalyst is in a different physical phase from the reactants is termed Heterogeneous catalysis. The most important of such reactions are those in which the reactants are in the gas phase while the catalyst is a solid. The process is also called **Contact catalysis** since the reaction occurs by contact of reactants with the catalyst surface. In contact catalysis, usually the catalyst is a finely divided metal or a gauze. This form of catalysis has great industrial importance.

Examples of Heterogeneous Catalysis

Some examples of heterogeneous catalysis with reactants in the gas, liquid or the solid phase are listed below.

(1) **Heterogeneous catalysis with gaseous reactants (Contact catalysis)**

(*a*) Combination of sulfur dioxide (SO_2) and oxygen in the presence of finely divided platinum or vanadium pentoxide, V_2O_5, (*Contact Process for Sulfuric acid*).

$$\underset{\text{gas}}{2SO_2} + \underset{\text{gas}}{O_2} + \underset{\text{solid}}{[Pt]} \longrightarrow 2SO_3 + [Pt]$$

(*b*) Combination of nitrogen and hydrogen to form ammonia in the presence of finely divided iron, (*Haber Process for Ammonia*).

$$\underset{\text{gas}}{N_2} + \underset{\text{gas}}{3H_2} + \underset{\text{solid}}{[Fe]} \longrightarrow 2NH_3 + [Fe]$$

(*c*) Oxidation of ammonia to nitric oxide (NO) in the presence of platinum gauze (a stage in manufacture of Nitric acid).

$$\underset{\text{gas}}{4NH_3} + \underset{\text{gas}}{5O_2} + \underset{\text{solid}}{[Pt]} \longrightarrow 4NO + 6H_2O + [Pt]$$

(*d*) Hydrogenation reactions of unsaturated organic compounds are catalyzed by finely divided nickel.

$$\underset{\text{ethene(gas)}}{H_2C{=}CH_2} + \underset{\text{gas}}{H_2} + \underset{\text{solid}}{[Ni]} \longrightarrow \underset{\text{ethane(gas)}}{H_3C{-}CH_3} + [Ni]$$

Vegetable oils are triesters of glycerol with higher unsaturated acid (oleic acid). When hydrogen is passed through the *vegetable oils* in the presence of nickel, the carbon-carbon double bonds of the acid portions are hydrogenated to yield *solid fats* (*Vanaspati ghee*).

(2) **Heterogeneous catalysis with liquid reactants**

(*i*) The decomposition of aqueous solutions of hydrogen peroxide (H_2O_2) is catalyzed by manganese dioxide (MnO_2) or platinum in colloidal form,

$$\underset{\text{liquid}}{2H_2O_2} + \underset{\text{solid}}{[Pt]} \longrightarrow 2H_2O + O_2 + [Pt]$$

(*ii*) Benzene and ethanoyl chloride (CH_3COCl) react in the presence of anhydrous aluminium chloride to form phenyl methyl ketone ($C_6H_5COCH_3$),

$$\underset{\text{liquid}}{C_6H_6} + \underset{\text{liquid}}{CH_3COCl} + \underset{\text{solid}}{[AlCl_3]} \longrightarrow C_6H_5COCH_3 + HCl + [AlCl_3]$$

(3) **Heterogeneous catalysis with solid reactants**

The decomposition of potassium chlorate ($KClO_3$) is catalyzed by manganese dioxide (MnO_2).

$$\underset{\text{solid}}{2KClO_3} + \underset{\text{solid}}{[MnO_2]} \longrightarrow 2KCl + 3O_2 + [MnO_2]$$

CHARACTERISTICS OF CATALYTIC REACTIONS

Although there are different types of catalytic reactions, the following features or characteristics are common to most of them.

(1) **A catalyst remains unchanged in mass and chemical composition at the end of the reaction**

Qualitative and quantitative analysis show that a catalyst undergoes no change in mass of chemical nature. However, it may undergo a physical change. Thus granular manganese dioxide (MnO_2) used as a catalyst in the thermal decomposing of potassium chlorate is left as a fine powder at the end of the reaction.

(2) **A small quantity of catalyst is generally needed to produce almost unlimited reaction**

Sometimes a trace of a metal catalyst is required to affect very large amounts of reactants. For example, one ten-millionth of its mass of finely divided platinum is all that is needed to catalyze the decomposition of hydrogen peroxide.

On the other hand, there are catalysts which need to be present in relatively large amount to be effective. Thus in Friedel-Crafts reaction,

$$C_6H_6 + C_2H_5Cl \xrightarrow{AlCl_3} C_6H_5C_2H_5 + HCl$$

anhydrous aluminium chloride functions as a catalyst effectively when present to the extent of 30 per cent of the mass of benzene.

For the acid and alkaline hydrolysis of an ester,

$$\underset{\text{ester}}{RCOOR} + H_2O \xrightarrow{H^+ \text{ or } OH^-} RCOOH + R'OH$$

the rate of reaction is proportional to the concentration of the catalyst (H^+ or OH^-).

(3) A catalyst is more effective when finely divided

In heterogeneous catalysis, the solid catalyst is more effective when in a state of fine subdivision than it is used in bulk. Thus a lump of platinum will have much less catalytic activity than colloidal or platinized asbestos. Finely divided nickel is a better catalyst than lumps of solid nickel.

(4) A catalyst is specific in its action

While a particular catalyst works for one reaction, it will not necessarily work for another reaction. Different catalysts, moreover, can bring about completely different reactions for the same substance. For example, ethanol (C_2H_5OH) gives ethene (C_2H_4) when passed over hot aluminium oxide,

$$C_2H_5OH \xrightarrow{Al_2O_3} \underset{\text{ethene}}{CH_2{=}CH_2} + H_2O \qquad \textit{(Dehydration)}$$

but with hot copper it gives ethanal (CH_3CHO).

$$C_2H_5OH \xrightarrow{Cu} \underset{\text{ethanal}}{CH_3CHO} + H_2 \qquad \textit{(Dehydrogenation)}$$

(5) A catalyst cannot, in general, initiate a reaction

In most cases a catalyst speeds up a reaction already in progress and does not initiate (or start) the reaction. But there are certain reactions where the reactants do not combine for very long period (perhaps years). For example, a mixture of hydrogen and oxygen, which remains unchanged almost indefinitely at room temperature, can be brought to reaction by the catalyst *platinum black* in a few seconds.

$$H_2 + O_2 \xrightarrow{\text{room temp.}} \text{No reaction}$$

$$2H_2 + O_2 \xrightarrow{\text{Pt black}} 2H_2O$$

Thus it is now considered that the catalyst can initiate a reaction. According to this view, the reacting molecules (in the absence of catalyst) do not possess minimum kinetic energies for successful collisions. The molecules rebound from collision without reacting at all.

(6) A catalyst does not affect the final position of equilibrium, although it shortens the time required to establish the equilibrium

It implies that in a reversible reaction the catalyst accelerates the forward and the reverse reactions equally. Thus the ratio of the rates of two opposing reactions *i.e.*, the equilibrium constant, remains unchanged.

The effect of a catalyst on the time required for equilibrium to be established for the reaction

$$A + B \rightleftharpoons C + D$$

is illustrated in Fig. 21.1. To start with the concentrations of A and B are at the maximum and hence the rate of forward reaction is maximum. As the time passes the rate of the reaction decreases till the equilibrium is established. For the reverse reaction the initial concentrations of C and D are zero and the rate of reaction is lowest. As the time passes, the rate of reaction increases till the equilibrium is established. Similar curves of the rates of reactions with the catalyst show that the rates of the forward reaction and the reverse reaction are altered equally but the equilibrium is established in a much shorter time.

For example, in the **Haber Process for ammonia,**

$$N_2 + 3H_2 \underset{}{\overset{Fe}{\rightleftharpoons}} 2NH_3$$

the reaction is very slow. In the presence of the catalyst, the equilibrium is reached much sooner but the percentage yield remains unchanged. The iron catalyst shortens the time to attain equilibrium but cannot alter the percentage yield.

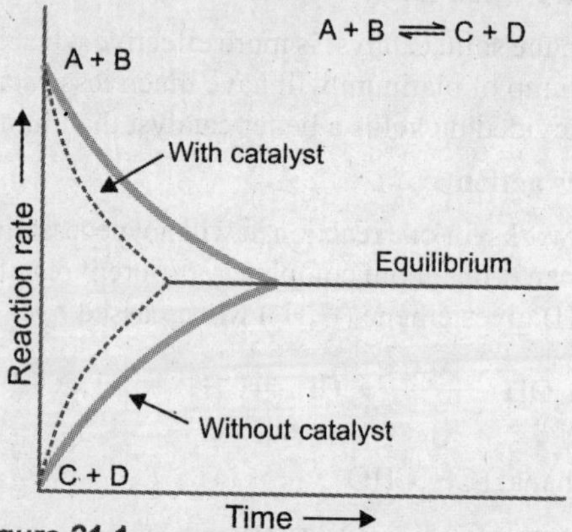

■ **Figure 21.1**
The effect of a catalyst on the time required for the equilibrium to be established.

Energy considerations also show that the final state of equilibrium cannot be changed by the catalyst. Suppose the catalyst accelerates the forward reaction more than the reverse reaction. This will shift the equilibrium point, which cannot happen without the supply of energy to the system. But a catalyst unchanged in mass and composition at the end of the reaction, cannot supply the required energy.

(7) **Change of temperature alters the rate of a catalytic reaction as it would do for the same reaction without a catalyst**

We have already studied the effect of temperature change on reversible reactions under Le Chatelier principle.

Some catalysts are, however, physically altered by a rise in temperature and hence their catalytic activity may be decreased. This is particularly true with colloidal solutions like that of platinum, since a rise in the temperature may cause their coagulation. In such a case the rate of reaction increases up to a certain point and then gradually decreases. **The rate of reaction is maximum at a particular temperature called the optimum temperature.**

PROMOTERS

The activity of a catalyst can often be increased by addition of a small quantity of a second material. This second substance is either not a catalyst itself for the reaction or it may be a feeble catalyst.

A substance which, though itself not a catalyst, promotes the activity of a catalyst is called a promoter.

Example of Promoters

Molybdenum (Mo) or aluminium oxide (Al_2O_3) promotes the activity of iron catalyst in the Haber synthesis for the manufacture of ammonia.

$$N_2 + 3H_2 \underset{+Mo}{\overset{Fe}{\rightleftharpoons}} 2NH_3$$

In some reactions, mixtures of catalysts are used to obtain the maximum catalytic efficiency.

For example, in the synthesis of methanol (CH_3OH) from carbon monoxide and hydrogen, a mixture of zinc and chromium oxide is used as a catalyst.

$$CO + 2H_2 \xrightarrow[Cr_2O_3]{ZnO} CH_3OH$$

Explanation of Promotion Action

The theory of promotion of a catalyst is not clearly understood. Presumably :

(1) **Change of Lattice Spacing.** The lattice spacing of the catalyst is changed thus enhancing the spaces between the catalyst particles. The absorbed molecules of the reactant (say H_2) are further weakened and cleaved. This makes the reaction go faster.

(2) **Increase of Peaks and Cracks.** The presence of the promoter increases the peaks and cracks on the catalyst surface. This increases the concentration of the reactant molecules and hence the rate of reaction.

The phenomenon of promotion is a common feature of heterogeneous catalysis.

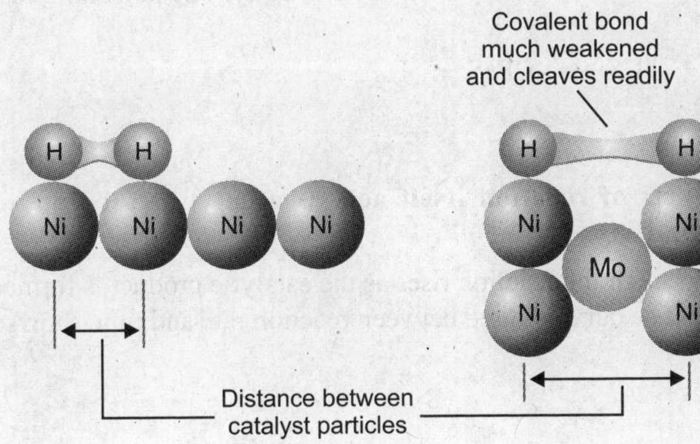

Figure 21.2
How the change of crystal lattice spacing of catalyst makes the reaction go faster?

CATALYTIC POISONING

Very often a heterogeneous catalyst is rendered ineffective by the presence of small amounts of impurities in the reactants.

A substance which destroys the activity of the catalyst to accelerate a reaction, is called a poison and the process is called Catalytic poisoning.

Examples of Catalytic Poisoning

(1) The platinum catalyst used in the oxidation of sulfur dioxide (Contact Process), is poisoned by arsenic oxide (As_2O_3)

$$SO_2 + O_2 \xrightarrow[As_2O_3]{Pt} 2SO_3$$

(2) The iron catalyst used in the synthesis of ammonia (Haber Process) is poisoned by H_2S.

$$N_2 + 3H_2 \xrightarrow[H_2S]{Fe} 2NH_3$$

(3) The platinum catalyst used in the oxidation of hydrogen is poisoned by carbon monoxide.

$$2H_2 + O_2 \xrightarrow[CO]{Pt} 2H_2O$$

Explanation of Catalytic Poisoning

The poison is adsorbed on the catalyst surface in preference to the reactants. Even a monomolecular layer renders the surface unavailable for further adsorption of the reactants. The poisoning by As_2O_3 or CO appears to be of this kind.

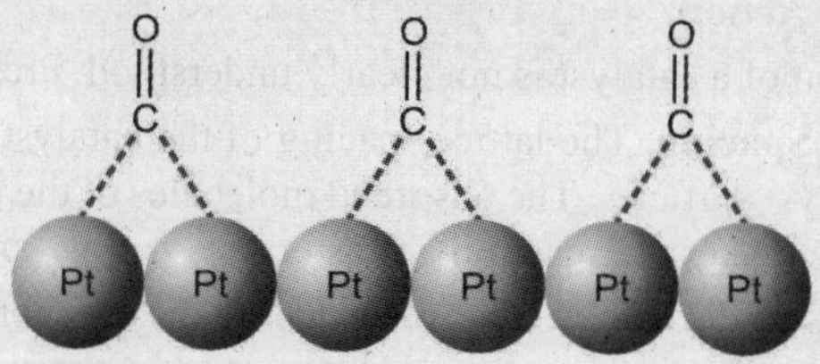

Figure 21.3
Poisoning of platinum catalyst by carbon monoxide.

(2) **The catalyst may combine chemically with the impurity.** The poisoning of iron catalyst by H_2S falls in this class.

$$Fe + H_2S \longrightarrow FeS + H_2$$

AUTOCATALYSIS

When one of the products of reaction itself acts as a catalyst for that reaction the phenomenon is called Autocatalysis.

In autocatalysis the initial rate of the reaction rises as the catalytic product is formed, instead of decreasing steadily (Fig. 21.4). The curve plotted between reaction rate and time shows a maximum when the reaction is complete.

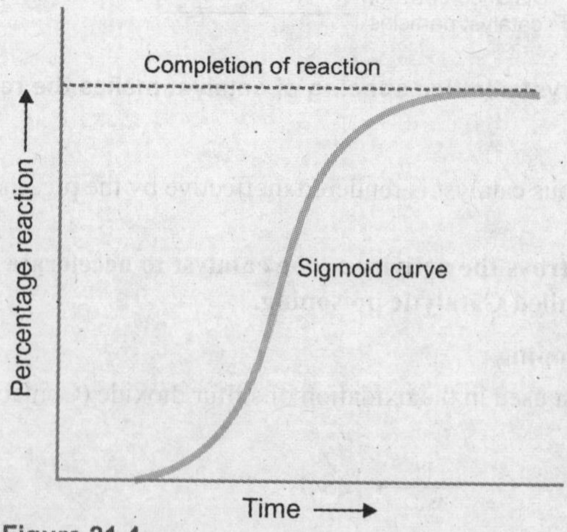

Figure 21.4
Curve showing the rise of rate of reaction with time.

Examples of Autocatalysis

(1) **Hydrolysis of an Ester.** The hydrolysis of ethyl acetate forms acetic acid (CH_3COOH) and ethanol. Of these products, acetic acid acts as a catalyst for the reaction.

$$CH_3COOC_2H_5 + H_2O \longrightarrow \underset{\text{catalyst}}{CH_3COOH} + C_2H_5OH$$

(2) **Oxidation of Oxalic acid.** When oxalic acid is oxidized by acidified potassium permanganate, manganous sulfate produced during the reaction acts as a catalyst for the reaction.

$$2KMnO_4 + 5H_2C_2O_4 + 3H_2SO_4 \longrightarrow \underset{\text{catalyst}}{2MnSO_4} + K_2SO_4 + 8H_2O + 10CO_2$$

(3) **Decomposition of Arsine.** The free arsenic produced by the decomposition of arsine (AsH_3) autocatalyses the reaction.

$$2AsH_3 \longrightarrow \underset{\text{catalyst}}{2As} + 3H_2$$

NEGATIVE CATALYSIS

When a catalyst reduces the rate of a reaction, it is called a Negative catalyst or Inhibitor. This phenomenon is called Negative catalysis or Inhibition. Negative catalysis is useful to slow down or stop altogether an unwanted reaction.

Examples of Negative Catalysis

(1) **Oxidation of Trichloromethane ($CHCl_3$)**

Trichloromethane (or chloroform) is used as anaesthetic. Upon oxidation by air it forms carbonyl chloride ($COCl_2$) which is a poisonous substance.

$$4CHCl_3 + 3O_2 \longrightarrow 4COCl_2 + 2H_2O + 2Cl_2$$

2 per cent of ethanol (C_2H_5OH) when added to chloroform acts as a negative catalyst and suppresses the formation of carbonyl chloride.

(2) **Decomposition of Hydrogen peroxide**

The decomposition of hydrogen peroxide,

$$2H_2O_2 \longrightarrow 2H_2O + O_2$$

is retarded by the presence of *dilute acids* or *glycerol.*

(3) **Tetraethyllead as Antiknock**

When tetraethyllead, $Pb(C_2H_5)_4$, is added to petrol, it retards the too repaid or explosive combustion of the fuel which is responsible for knocking of the engine.

Explanation of Negative Catalysis

The mechanism of negative catalysis could be different for different reactions.

(1) **By poisoning a catalyst.** A negative catalyst may function by poisoning a catalyst which already happens to be present in the reaction mixture. For example, the traces of alkali dissolved from the glass of the container, catalyze the decomposition of hydrogen peroxide (H_2O_2). But the addition of an acid would destroy the alkali catalyst and thus prevent decomposition.

(2) **By breaking a chain reaction.** In some cases negative catalysts are believed to operate by breaking the chain of reactions. For example, the combination of H_2 and Cl_2, which is a chain reaction, is negatively catalyzed by nitrogen trichloride (NCl_3).

$$Cl_2 \longrightarrow Cl^{\bullet} + Cl^{\bullet}$$

$$H_2 + C^{\bullet} \longrightarrow HCl + H^{\bullet}$$

$$H^{\bullet} + Cl_2 \longrightarrow HCl + H^{\bullet}$$

NCl_3 breaks the chain of reactions by absorbing the propagating species (Cl), and the reaction stops.

$$NCl_3 + Cl\cdot \longrightarrow \frac{1}{2}N_2 + 2Cl_2$$

ACTIVATION ENERGY AND CATALYSIS

According to the collision theory, a reaction occurs by the collisions between the reactant molecules (or ions). At ordinary temperature, the molecules do not possess enough energy and hence the collisions are not effective. However, when the temperature of the system is raised, the kinetic energy of the molecules increases. But the molecules do not react unless they attain a minimum amount of energy. **The minimum amount of energy required to cause a chemical reaction is known as the Activation Energy.** The activated molecules on collision first form an **Activated Complex.** As a result of breaking and forming of new bonds, the activated complex dissociates to yield product molecules.

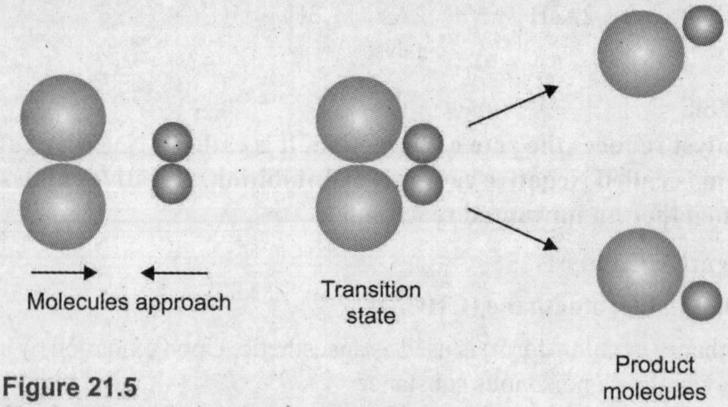

■ **Figure 21.5**
Mechanism of a molecular reaction.

A catalyst lowers the activation energy of the reaction by providing a new pathway (mechanism). Thus larger number of effective collisions occur in the presence of the catalyst than would occur at the same temperature without the presence of the catalyst. In this way the presence of the catalyst makes the reaction go faster, other conditions remaining the same.

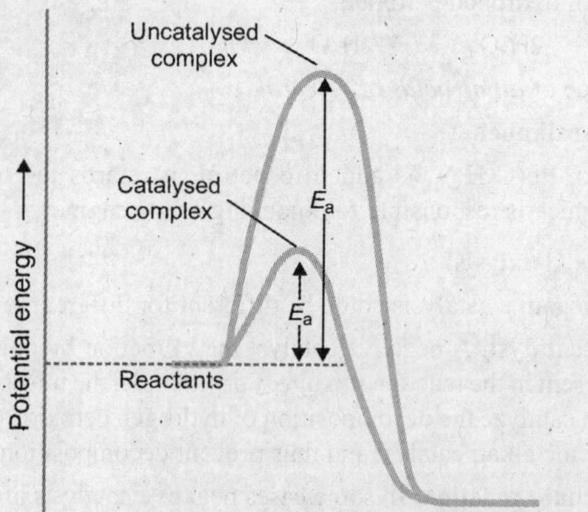

■ **Figure 21.6**
Energy diagram of a reaction with and without the catalyst, showing clearly the lowering of activation energy by the catalyst.

THEORIES OF CATALYSIS

There are two main theories of catalysis :

(1) Intermediate Compound Formation theory

(2) The Adsorption theory.

In general, the Intermediate Compound Formation theory applies to homogeneous catalytic reactions and the Adsorption theory applies to heterogeneous catalytic reactions.

The Intermediate Compound Formation Theory

As already discussed a catalyst functions by providing a new pathway of lower activation energy. In homogeneous catalysis, it does so by forming an intermediate compound with one of the reactants. The highly reactive intermediate compound then reacts with the second reactant to yield the product, releasing the catalyst. Let us illustrate it by taking the general reaction

$$A + B \xrightarrow{C} AB \qquad ...(1)$$

where C acts as catalyst. The reaction proceeds through the reactions :

$$A + C \longrightarrow \underset{\text{Intermediate}}{AC} \qquad ...(2)$$

$$AC + B \longrightarrow AB + C \qquad ...(3)$$

The activation energies of the reactions (2) and (3) are lower than that of the reaction (1). Hence the involvement of the catalyst in the formation of the intermediate compound and its subsequent decomposition, accelerates the rate of the reaction (1) which was originally very slow.

Example 1. Catalytic oxidation of sulfur dioxide (SO_2) in the presence of nitric oxide (NO) as catalyst. (Chamber Process of Sulfuric acid)

$$2SO_2 + O_2 \xrightarrow{NO} 2SO_3$$

Mechanism :

$$2NO + O_2 \longrightarrow 2NO_2 \text{ (Intermediate compound)}$$

$$NO_2 + SO_2 \longrightarrow SO_3 + NO$$

Example 2. Preparation of diethyl ether, $(C_2H_5)_2O$, from ethanol (C_2H_5OH) using sulfuric acid as catalyst.

$$C_2H_5OH + C_2H_5OH \xrightarrow{H_2SO_4} (C_2H_5)_2O + H_2O$$

Mechanism :

$$C_2H_5OH + H_2SO_4 \longrightarrow \underset{\substack{\text{Ethyl hydrogen sulfate}\\ \text{(Intermediate compound)}}}{C_2H_5HSO_4} + H_2O$$

$$C_2H_5HSO_4 + \underset{\substack{\text{(second}\\ \text{molecule)}}}{C_2H_5OH} \longrightarrow (C_2H_5)_2O + H_2SO_4$$

Example 3. Thermal decomposition of potassium chlorate ($KClO_3$) in the presence of manganese dioxide (MnO_2).

$$2KClO_3 \xrightarrow{MnO_2} 2KCl + 3O_2$$

Mechanism :

$$2KClO_3 + 6MnO_2 \longrightarrow \underset{\substack{\text{Intermediate}\\ \text{compound}}}{6MnO_3} + 2KCl$$

$$6MnO_3 \longrightarrow 6MnO_2 + 3O_2$$

Example 4. Formation of methylbenzene, $C_6H_5CH_3$ by reaction between benzene, C_6H_6, and methyl chloride, CH_3Cl, using aluminium chloride $AlCl_3$, as catalyst (Friedel-Crafts reaction),

$$C_6H_6 + CH_3Cl \xrightarrow{AlCl_3} C_6H_5CH_3 + HCl$$

Mechanism :

$$CH_3Cl + AlCl_3 \longrightarrow \underset{\text{Intermediate compound}}{[CH_3]^+ [AlCl_4]^-}$$

$$C_6H_6 + [CH_3]^+ [AlCl_4]^- \longrightarrow C_6H_5CH_3 + AlCl_3 + HCl$$

It may be noted that the actual isolation of intermediate compounds which would prove their existence is very difficult. As already stated, by their very nature they are unstable. In general, the intermediate compounds suggested as being formed are usually plausible rather than proved.

The Adsorption Theory

This theory explains the mechanism of a reaction between two gases catalyzed by a solid (*Heterogeneous or Contact Catalysis).* Here the catalyst functions by adsorption of the reacting molecules on its surface.

Generally speaking, four steps can be put forward for heterogeneous catalysis. For example, if the reaction is :

$$A(g) + B(g) \xrightarrow{\text{catalyst}} C(g) + D(g)$$

Step 1. Adsorption of Reactant molecules

The reactant molecules A and B strike the catalyst surface. They are held up at the surface by weak van der Waals forces (*Physical adsorption*) or by partial chemical bonds (*Chemisorption*).

Step 2. Formation of Activated complex

The particles of the reactants adjacent to one another join to form an *intermediate complex* (A – B). The activated complex is unstable. It has only fleeting existence.

Step 3. Decomposition of Activated complex

The activated complex breaks to form the products C and D. The separated particles of the products hold to the catalyst surface by partial chemical bonds.

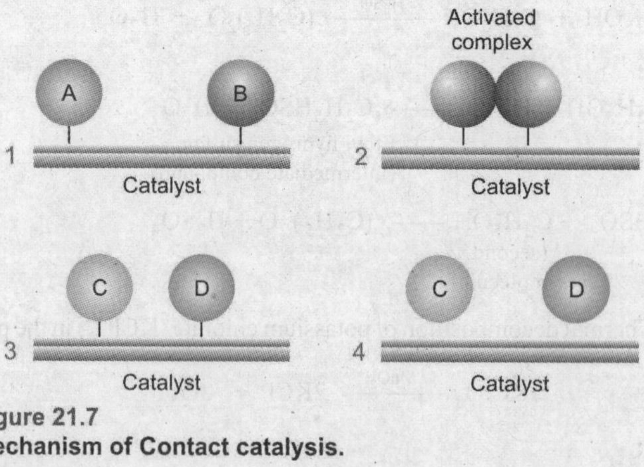

Figure 21.7
Mechanism of Contact catalysis.

Step 4. Desorption of Products

The particles of the products are *desorbed* or released from the surface. They are stable and can lead an independent existence.

The mechanism of contact catalysis may vary in details, depending on the nature of the reactants.

HYDROGENATION OF ETHENE (ETHYLENE) IN PRESENCE OF NICKEL

Ethene adds hydrogen in the presence of nickel as a catalyst to yield ethane.

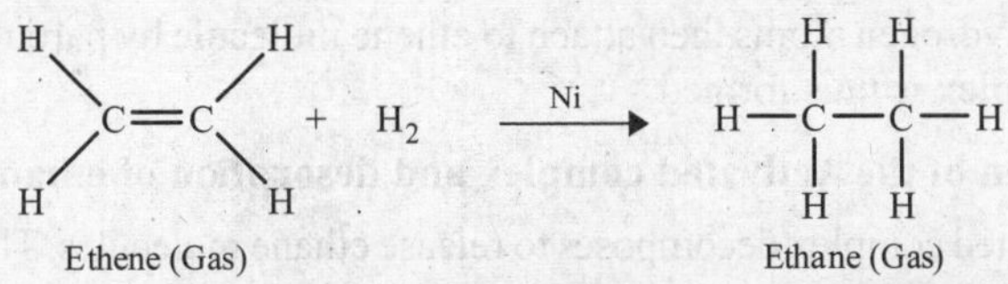

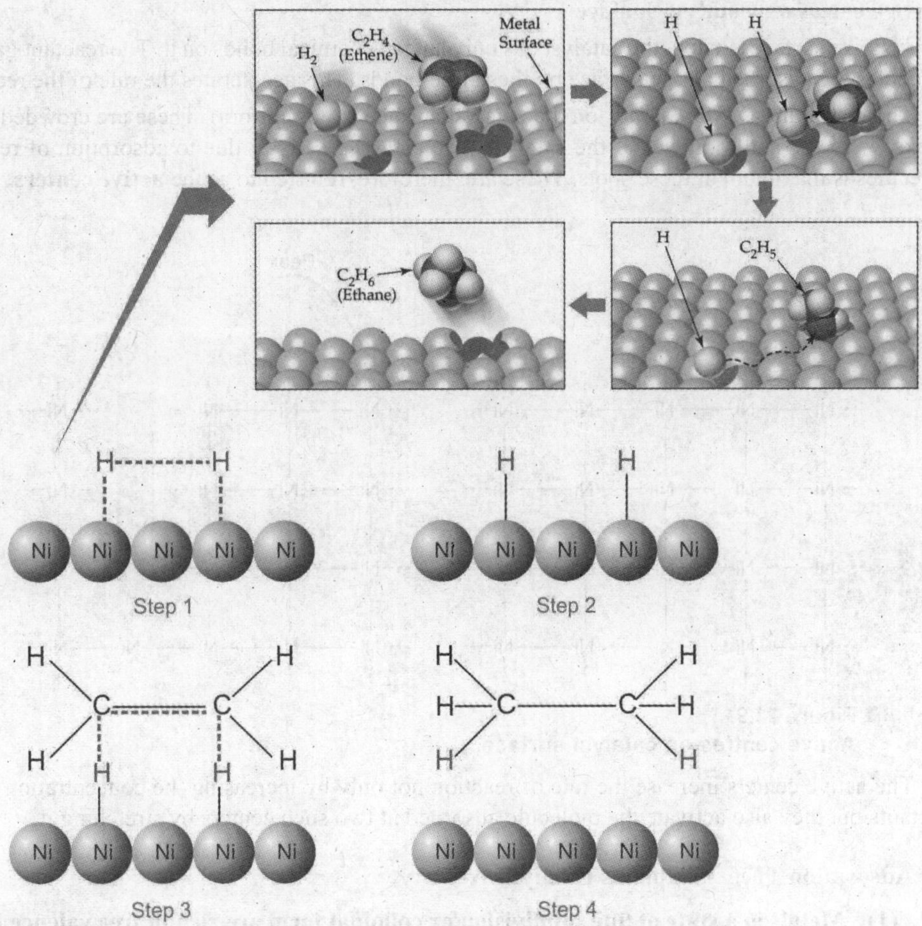

■ **Figure 21.8**
Mechanism of the hydrogenation of ethene on nickel surface.

The catalyst operates by the following steps.

Step 1. Adsorption of Hydrogen molecules

Hydrogen molecules are adsorbed on the nickel surface due to the residual valence bonds of the nickel atoms.

Step. 2 H–H Bonds are broken

The H–H bond is smaller (0.74×10^{-10} m) than Ni–Ni bond. Therefore, the H–H bond of the adsorbed hydrogen molecule is stretched and weakened. The weakened bond breaks, separating the hydrogen atoms. The separated hydrogen atoms are held to the nickel surface by chemical bonds.

Step 3. Formation of the Activated complex

The chemisorbed hydrogen atoms then attach to ethene molecule by partial chemical bonds. The unstable activated complex is thus formed.

Step 4. Decomposition of the Activated complex and desorption of ethane molecules

The unstable activated complex decomposes to release ethane molecules. The free catalyst surface is again available for further action.

Active Centers on Catalyst Surface

Just like surface tension, the catalyst has unbalanced chemical bonds on it. The reactant gaseous molecules are adsorbed on the surface by these free bonds. This accelerates the rate of the reaction.

The distribution of free bonds on the catalyst surface is not uniform. These are crowded at the 'peaks', 'cracks' and 'corners' of the catalyst. The catalytic activity due to adsorption of reacting molecules is maximum at these spots. These are, therefore, referred to as the **active centers.**

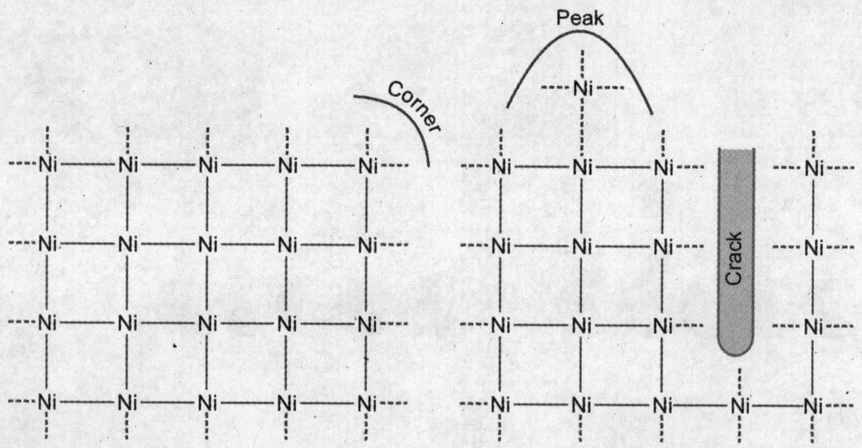

Figure 21.9
Active centres on catalyst surface.

The active centers increase the rate of reaction not only by increasing the concentration of the reactants but they also activate the molecule adsorbed at two such centers by stretching it.

The Adsorption Theory Explains Catalytic Activity

(1) **Metals in a state of fine subdivision or colloidal form are rich in free valence bonds and hence they are more efficient catalysts than the metal in lumps.**

(2) **Catalytic poisoning occurs because the so-called poison blocks the free valence bonds on its surface by preferential adsorption or by chemical combination.**

(3) **A promoter increases the valence bonds on the catalyst surface by changing the crystal lattice and thereby increasing the active centers.**

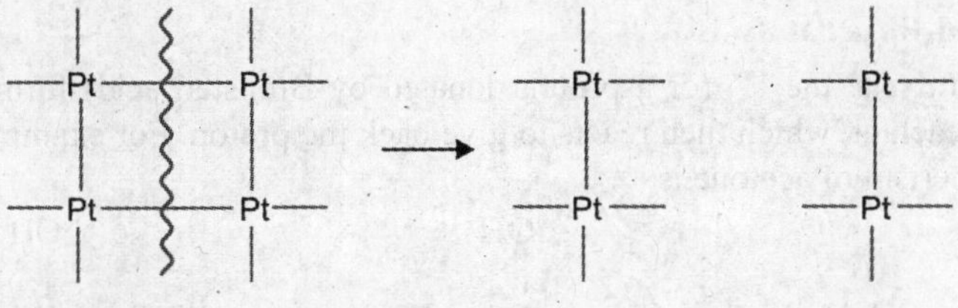

■ **Figure 21.10**
Subdivision of a catalyst makes it more efficient due to increase of free valence bonds.

ACID–BASE CATALYSIS

A number of homogeneous catalytic reactions are known which are catalyzed by acids or bases, or both acids and bases. These are often referred to as **Acid-Base catalysts.**

Arrhenius pointed out that acid catalysis was, in fact, brought about by H^+ ions supplied by strong acids, while base catalysis was caused by OH^- ions supplied by strong bases.

Examples of Acid-Base catalysis

(1) Inversion of Cane sugar :

$$\underset{\text{cane sugar}}{C_{12}H_{22}O_{11}} + H_2 \xrightarrow{H^+} \underset{\text{glucose}}{C_6H_{12}O_6} + \underset{\text{fructose}}{C_6H_{12}O_6}$$

(2) Keto-Enol tautomerism of Acetone :

$$\underset{\text{Keto form}}{CH_3-\overset{\overset{\displaystyle O}{\|}}{C}-CH_3} \xrightarrow{H^+} \underset{\text{Enol form}}{CH_3-\overset{\overset{\displaystyle OH}{|}}{C}=CH_2}$$

(3) Decomposition of Nitramide :

$$NH_2NO_2 \xrightarrow{H^+} N_2O + H_2O$$

(4) Hydrolysis of an Ester :

$$\underset{\text{ethyl acetate}}{CH_3COOC_2H_5} \xrightarrow[\text{or } OH^-]{H^+} \underset{\text{acetic acid}}{CH_3COOH} + \underset{\text{ethanol}}{C_2H_5OH}$$

General Acid-Base catalysis

More recently it has been found that :

(*a*) **Not only H^+ ions but all Bronsted bases (proton donors) cause acid catalysis.** Thus the general acid catalysts are :

H^+, undissociated acids (CH_3COOH), cations of weak bases (NH_4^+), and water (H_2O).

(*b*) **Not only OH^- ions but all Bronsted bases (proton acceptors) act as base catalyst.** Thus the general base catalysts are :

OH^-, undissociated bases, anions of weak acids (CH_3COO^-) and water (H_2O).

The catalysis brought about by general acids and bases is termed **General Acid-Base catalysis.** For elucidation, decomposition of nitramide is also catalyzed by acetate ions (CH_3COO^-).

$$NH_2NO_2 \xrightarrow{CH_3COO^-} N_2O + H_2O$$

Mechanism of Acid-Base catalysis

(*a*) In **acid catalysis,** the H^+ (or a proton donated by Bronsted acid) forms an intermediate complex with the reactant, which then reacts to give back the proton. For example, the mechanism of keto-enol tautomerism of acetone is :

$$\underset{\text{Acetone}}{CH_3-\overset{\overset{\displaystyle O}{\|}}{C}-CH_3} \xrightarrow{H^+} \underset{\text{Intermediate complex}}{CH_3-\overset{\overset{\displaystyle OH^+}{\|}}{C}-CH_3} \longrightarrow \underset{\text{Enol form}}{CH_3-\overset{\overset{\displaystyle OH}{|}}{C}=CH_2} + H^+$$

(*b*) In **base catalysis,** the OH^- ion (or any Bronsted base) accepts a proton from the reactant to form an *intermediate complex* which then reacts or decomposes to regenerate the OH^- (or Bronsted base). For example, the decomposition of nitramide by OH^- ions and CH_3COO^- ions may be explained as follows :

(*i*) **By OH^- ions**

$$NH_2NO_2 + OH^- \longrightarrow H_2O + \underset{\text{Intermediate complex}}{NHNO_2^-}$$

$$NHNO_2^- \longrightarrow N_2O + OH^-$$

(*ii*) **By CH_3COO^- ions**

$$NH_2NO_2 + CH_3COO^- \longrightarrow CH_3COOH + NHNO_2^-$$

$$NHNO_2^- \longrightarrow N_2O + OH^-$$

$$OH^- + CH_3COOH \longrightarrow H_2O + CH_3COO^-$$

ENZYME CATALYSIS

Numerous organic reactions are taking place in the body of animals and plants to maintain the life process. These reactions being slow remarkably catalyzed by the organic compounds known as **Enzymes.** All enzymes have been found to be complex protein molecules. Thus :

Enzymes are protein molecules which act as catalysts to speed up organic reactions in living cells. **The catalysis brought about by enzymes is known as Enzyme Catalysis.**

Each enzyme is produced in a particular living cell to catalyze a reaction occurring in that cell. Many enzymes have been identified and obtained in pure crystalline state from the cells to which they belong. However the first enzyme as prepared by synthesis in the laboratory in 1969.

Examples of Enzyme Catalysis

Some common examples of the biochemical reactions catalyzed by enzymes are :

(1) **Inversion of cane sugar** ($C_{12}H_{22}O_{11}$) by *Invertase present in yeast,*

$$C_{12}H_{22}O_{11} + H_2O \xrightarrow{\textit{invertase}} \underset{\text{glucose}}{C_6H_{12}O_6} + \underset{\text{fructose}}{C_6H_{12}O_6}$$

(2) **Conversion of glucose into ethanol** by *Zymase present in yeast,*

$$\underset{\text{glucose}}{C_6H_{12}O_6} \xrightarrow{\textit{zymase}} \underset{\text{ethanol}}{2C_2H_5OH} + 2CO_2$$

(3) **Hydrolysis of urea** ($H_2N—CO—NH_2$) by *Urease present in soya bean,*

$$H_2N-\overset{\overset{\displaystyle O}{\|}}{C}-NH_2 \; + \; H^+ \xrightarrow{urease} 2NH_3 \; + \; CO_2$$

MECHANISM OF ENZYME CATALYSIS

The long chains of the enzyme (protein) molecules are coiled on each other to make a rigid colloidal particle with cavities on its surface. These cavities which are of characteristic shape and abound in active groups (NH_2, COOH, SH, OH)] are termed **Active centers.** The molecules of substrate which have complementary shape, fit into these cavities just as key fits into a lock **(Lock-and-Key theory).** By virtue of the presence of active groups, the enzyme forms an activated complex with the substrate which at once decomposes to yield the products. Thus the substrate molecules enters the cavities, forms complex and reacts, and at once the products get out of the cavities.

Michaelis and Menten (1913) proposed the following mechanism for enzyme catalysis (Fig. 21.11).

$$E + S \rightleftharpoons \underset{\text{complex}}{ES} \longrightarrow P + E$$

where E = enzyme; S = substrate (reactant); ES = activated complex; P = products.

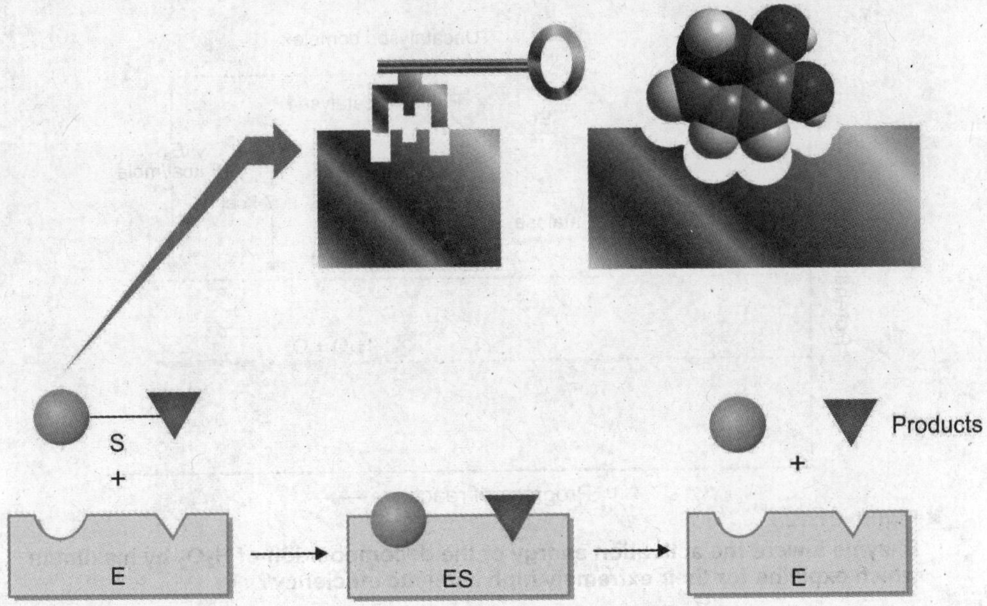

■ **Figure 21.11**
Illustration of the lock-and-key model of enzyme catalysis.

CHARACTERISTICS OF ENZYME CATALYSIS

In general, enzyme behave like inorganic heterogeneous catalysts. However, they are unique in their efficiency and high degree of specificity. Some more important features of enzyme catalysis are listed below.

(1) Enzymes are the most efficient catalysts known

The enzyme catalyzed reactions proceed at fantastic high rates in comparison to those catalyzed by inorganic substances. Thus one molecule of an enzyme may transform one million molecules of the substrate (reactant) per minute.

Like inorganic catalysts, enzymes function by taking an alternative pathway of lower activation energy. For example, the activation energy of the decomposition of hydrogen peroxide,

$$2H_2O_2 \longrightarrow 2H_2O + O_2$$

without a catalyst is 18 kcal/mol. With colloidal platinum (inorganic catalyst), the reaction can be accelerating by taking a pathway of lower activation energy of about 48.986 kJ. The enzyme *catalase* lowers the activation energy of the same reaction to less than 2 kcal/mol.

(2) Enzyme catalysis is marked by absolute specificity

An enzyme as a rule catalyses just one reaction with a particular substance. For example, *urease* (an enzyme derived from soya bean) catalyses the hydrolysis of urea and no other amide, not even methylurea.

$$\underset{\text{urea}}{H_2N{-}\overset{\overset{\large O}{\|}}{C}{-}NH_2} + H_2O \xrightarrow{\textit{urease}} 2NH_3 + H_2O$$

$$\underset{\text{methylurea}}{H_2N{-}\overset{\overset{\large O}{\|}}{C}{-}NHCH_3} + H_2O \xrightarrow{\textit{urease}} \text{No reaction}$$

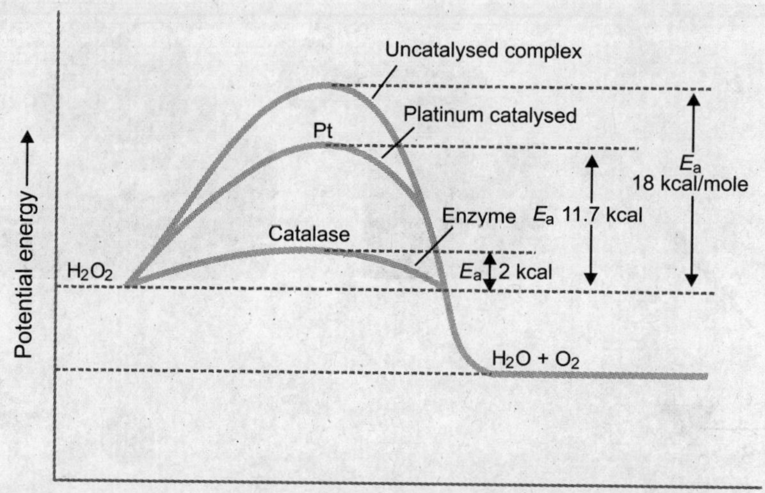

Figure 21.12

Enzyme lowers the activation energy of the decomposition of H_2O_2 by maximum which explains for their extremely high catalytic efficiency?

Enzyme catalyzed reactions are often marked by absolute specificity. Thus where a compound can exist in optically active isomers (identical in every respect except the space arrangement of groups), an enzyme which can act on one of the isomers is unable to act on the other. For example, the enzyme present in ordinary mould (*Penicillium glaucum*) when added to a (±)-mixture of tartaric acid, decomposes the (+)-form only, leaving the (–)-form behind.

(3) The rate of enzyme catalyzed reactions is maximum at the optimum temperature

The rate of an enzyme catalyzed reaction is increased with the rise of temperature but up to a certain point. Thereafter the enzyme is denatured as its protein structure is gradually destroyed. Thus the rate of reaction drops and eventually becomes zero when the enzyme is completely destroyed. The rate of an enzyme reaction with raising of temperature gives a bell-shaped curve. The temperature at which the reaction rate is maximum is called the **optimum temperature.**

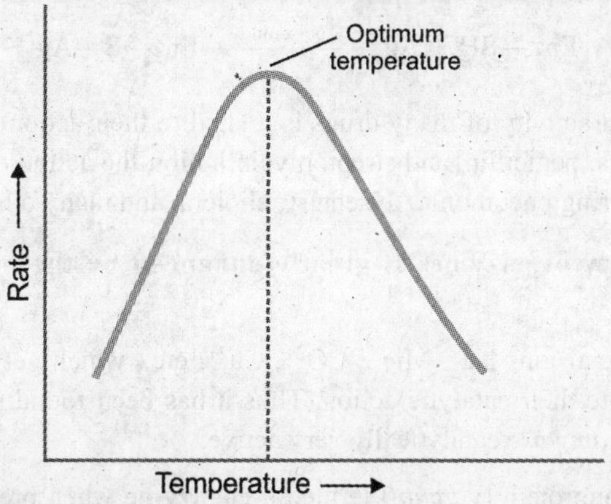

Figure 21.13
The variation of rate of an enzyme catalysed reaction with rise of temperature.

For example, the optimum temperatures, of enzyme reactions occurring in human body is 310 K (98.6°F). At much higher temperatures, all physiological reactions will cease due to loss of enzymatic activity. This is one reason why high body temperature (fever) is very dangerous.

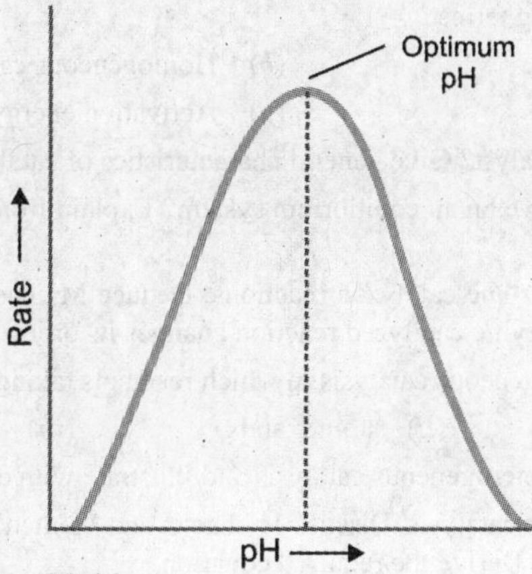

Figure 21.14
The variation of rate of an enzyme catalysed reaction with increase of pH.

(4) Rate of enzyme catalyzed reactions is maximum at the optimum pH

The rate of an enzyme catalyzed reaction varies with pH of the system. The rate passes through a maximum at a particular pH, known as the **optimum pH.** The enzyme activity is lower at other values of pH. Thus many enzymes of the body function best at pH of about 7.4, the pH of the blood and body fluids.

(5) Enzymes are markedly inhibited or poisoned

The catalytic activity of an enzyme is often reduced (*inhibited*) or completely destroyed (poisoned) by addition of other substances. These inhibitors or poisons interact with the active functional groups on the enzyme surface. For example, heavy metal ions (Ag^+, Hg^{2+}) react with the –SH groups of the enzyme and poison it.

$$\text{Enz—SH} + \text{Ag}^+ \xrightarrow{\text{poisoning}} \text{Enz—S—Ag} + \text{H}^+$$

The physiological activity of many drugs is related to their action as enzyme inhibitors in the body. Thus sulfa drugs, penicillin, and streptomycin inhibit the action of several bacteria and have proved effective in curing pneumonia, dysentery, cholera, and many other infectious diseases.

(6) Catalytic activity of enzymes is greatly enhanced by the presence of Activators or Coenzymes

Activators are metal ions Na^+, Mn^{2+}, CO^{2+}, Cu^{2+}, etc., which get weakly bonded to enzyme molecules and promote their catalytic action. Thus it has been found that the addition of sodium chloride (Na^+) makes *amylase* catalytically very active.

Often, a small nonprotein (*vitamin*) termed a **coenzyme** when present alongwith an enzyme, promotes the catalytic activity of the latter.

EXAMINATION QUESTIONS

1. Define or explain the following terms :
   (*a*) Catalyst (*b*) Homogeneous catalysis
   (*c*) Heterogeneous catalysis (*d*) Activation energy
2. What is meant by the term catalyst? Give general characteristics of catalytic reactions.
3. What effect does a catalyst have on an equilibrium system? Explain from the point of view of Kinetics and the Law of mass action.
4. Discuss the mechanism of enzyme catalyzed reactions. Deduce Michaelis Menten equation and give conditions under which an enzyme catalyzed reaction changes its order.
5. (*a*) Give examples of heterogeneous catalysis in which reactants taking part are
   (*i*) solid, (*ii*) liquid and (*iii*) gaseous state
   (*b*) Discuss the theories of heterogeneous catalysis and illustrate with examples.
6. Give characteristics of enzyme catalysis. Discuss Michaelis and Menten's enzyme mechanism in detail to express the rate of reaction. Derive the required equation.
7. (*a*) Discuss intermediate compound formation theory of catalysis.
   (*b*) Define the following :
   (*i*)Catalyst promoter (*ii*) Catalyst inhibitor (*iii*) Catalytic poison
   (*Kerala BSc*)
8. (*a*) Derive Michaelis-Menten equation for enzyme catalyzed reaction.
   (*b*) Explain the mechanism of heterogeneous catalysis by taking a suitable example.
   (*Madras BSc*)
9. Under what conditions an enzyme catalyzed reaction follows
   (*i*) Second order kinetics (*ii*) First order kinetics
   (*iii*) Zero order kinetics (*Nagpur BSc*)
10. Explain the important characteristics of enzyme catalysis. (*Assam BSc*)
11. What is heterogeneous catalysis? Explain giving one example. (*Guru Nanak Dev BSc*)
12. (*a*) By considering one example each of catalytic promoters and inhibitor, distinguish between the action of these promoters and inhibitors.

(*b*) What mechanism would you suggest for heterogeneous catalytic reactions? Clarify it with one example. (*MD Rohtak BSc*)

13. Derive an expression for the rate of an acid catalyzed reaction. (*Jamia Millia BSc*)

14. Explain the important characteristics of enzyme catalysis. (*Punjabi BSc*)

15. (*a*) Discuss the theories of heterogeneous catalysis.

(*b*) Give an account of enzyme catalysis. (*Purvanchal BSc*)

16. Write notes on :

(*i*) Catalyst and energy of activation

(*ii*) Specificity of catalytic action (*Mizoram BSc*)

17. (*a*) What is the function of catalytic promoters and inhibitors in chemical reactions?

(*b*) Explain the important characteristics of enzyme catalysis. (*Punjabi BSc*)

18. Explain the following with example:

(*i*) Negative catalyst (*ii*) Autocatalysis

(*iii*) Catalytic poison (*iv*) Promoters

(*v*) Specificity of a catalyst (*Purvanchal BSc*)

19. Derive an expression for the kinetics of enzyme catalysis of Michaelis equation. What will be the effect of very high substance concentrations? (*Arunachal BSc*)

20. Write the Michaelis-Menten equation for an enzyme-catalyzed reaction and define/explain all symbols used in it. (*Guru Nanak Dev BSc*)

21. Write down the mechanism of enzyme catalysis suggested by Michaelis and Menten. (*Meerut BSc*)

22. (*a*) Give an example each for enzyme catalysis and acid-base catalysis.

(*b*) Describe the theory of homogeneous and heterogeneous catalysis. (*Baroda BSc*)

23. Write notes on :

(*i*) Heterogeneous catalysis

(*ii*) Homogeneous catalysis (*Panjab BSc*)

24. Write notes on :

(*i*) Enzyme catalysis

(*ii*) Autocatalysis (*Delhi BSc*)

25. (*a*) What are characteristic of catalytic reactions?

(*b*) Explain with examples autocatalysis. (*Madras BSc*)

26. Explain how a catalyst increases the speed of a reaction. (*Sambalpur BSc*)

27. What is catalysis? Derive the Michaelis-Menten equation to explain enzyme catalysis. (*Bundelkhand BSc*)

28. Explain the difference between homogeneous and heterogeneous catalysis. Discuss the theories of catalysis. (*Kerala BSc, 2011*)

MULTIPLE CHOICE QUESTIONS

1. A catalyst

(*a*) does not react

(*b*) reacts and is produced in one of the later steps of a reaction

(*c*) reacts in an early step and is produced in a later step

(*d*) reacts but is not produced

Answer. (*c*)

2. A catalyst
(*a*) changes the free energy change of a reaction
(*b*) is included in the balanced chemical equation for a reaction
(*c*) changes the mechanisms for a reaction
(*d*) slows down the rate of a reaction
Answer. (*c*)

3. A catalyst can be described as a substance that:
(*a*) undergoes change to accelerate the rate of the reaction
(*b*) increases the kinetic energy of the reactants
(*c*) provides a path of lower activation energy for the reaction
(*d*) lowers the potential energy of the products with respect to the energy of the reactants
Answer. (*c*)

4. The minimum amount of energy needed to bring about a non-spontaneous change is the
(*a*) free energy (*b*) activation energy
(*c*) entropy (*d*) enthalpy
Answer. (*b*)

5. A catalyst will affect the rate of the forward reaction by changing the
(*a*) activation energy (*b*) heat of reaction
(*c*) heat of formation (*d*) potential energy of the products
Answer. (*a*)

6. Which of the following is most likely to increase the rate of a reaction?
(*a*) decreasing the temperature
(*b*) increasing the volume of the reaction vessel
(*c*) reducing the activation energy
(*d*) decreasing the concentration of the reactant in the reaction vessel
Answer. (*c*)

7. All of the following are true statements concerning catalysts except
(*a*) a catalyst will speed the rate-determining step
(*b*) a catalyst will be used up in a reaction
(*c*) a catalyst may induce steric strain in a molecule to make it react more readily
(*d*) a catalyst will lower the activation energy of a reaction
Answer. (*b*)

8. At equilibrium
(*a*) the forward reaction will continue
(*b*) a change in reaction conditions may shift the equilibrium
(*c*) the reverse reaction will not continue
(*d*) both (*a*) and (*b*)
Answer. (*d*)

9. Which statement is not correct regarding the function of a catalyst?
(*a*) it affects the rate of a chemical reaction
(*b*) it lowers the energy of the product, causing the reaction to be more exothermic
(*c*) it changes the mechanism of a reaction
(*d*) it lowers the activation energy
Answer. (*b*)

10. Which of the following will lower the activation energy for a reaction?
(*a*) increasing the concentrations of reactants
(*b*) raising the temperature of the reaction
(*c*) adding a suitable catalyst
(*d*) there is no way to lower the activation energy of a reaction
Answer. (*c*)

11. Which of the following reactions occurs at the fastest rate?
(*a*) one which is exothermic by 15 kcal/mol and has an activation energy of 20 kcal/mol
(*b*) one which is endothermic by 5 kcal/mol and has an activation energy of 30 kcal/mol
(*c*) one which is exothermic by 30 kcal/mol and has an activation energy of 15 kcal/mol
(*d*) one which is exothermic by 20 kcal/mol and has an activation energy of 20 kcal/mol
Answer. (*c*)

12. At which temperature does an aqueous solution of LiCl have the highest average kinetic energy?
(*a*) 373 K (*b*) 473 K
(*c*) 273 K (*d*) 373 K
Answer. (*b*)

13. The minimum amount of energy required to start a chemical reaction is called
(*a*) entropy (*b*) enthalpy
(*c*) free energy (*d*) activation energy
Answer. (*d*)

14. Which one of the following statements is incorrect?
(*a*) activation energies of simple reactions can be negative
(*b*) more molecules in a gas have energies above some threshold value as the temperature is increased
(*c*) the rate does not depend on the magnitude of the ΔE
(*d*) the lower the activation energy, the faster the rate if equal numbers of collisions are always involved
Answer. (*a*)

15. Which one of the following statements is incorrect?
(*a*) catalysts speed up a reaction by lowering the activation energy
(*b*) heterogeneous catalysts are often found on a solid in contact with gaseous reactants
(*c*) an automotive catalytic converter uses a catalyst to burn gasoline better for higher gas mileage
(*d*) enzymes are biological catalysts
Answer. (*c*)

16. The energy needed to overcome for a reaction to occur is called:
(*a*) activation energy (*b*) energy of vaporization
(*c*) specific heat (*d*) potential energy
Answer. (*a*)

17. When a catalyst is added to a reaction system,
(*a*) the reaction rate is changed (*b*) the equilibrium position of the reaction is shifted
(*c*) the ΔH of the reaction is changed (*d*) all of these
Answer. (*a*)

18. As a good rule of thumb, many reactions double in speed for a 283 K increase in T at room temperature. Taking the two temperatures to be 293 K and 303 K, the activation energy for such reactions would be:
(*a*) 148 kJ mol^{-1} (*b*) 0.35 kJ mol^{-1}
(*c*) 6.12 kJ mol^{-1} (*d*) 51 kJ mol^{-1}
Answer. (*d*)

19. A catalyst

(*a*) may be in same phase with the reactants or in a different phase
(*b*) may accelerate a reaction
(*c*) affects a reaction without being consumed in the process
(*d*) all of the above

Answer. (*d*)

20. For any chemical reaction at equilibrium, the rate of the forward reaction is

(*a*) less than the rate of the reverse reaction
(*b*) greater than the rate of the reverse reaction
(*c*) equal to the rate of the reverse reaction
(*d*) unrelated to the rate of the reverse reaction

Answer. (*c*)

21. A reaction requires the presence of a strip of metal in the reaction vessel, when the reactants are gases. This is an example of what kind of catalysis?

(*a*) homogeneous (*b*) heterogeneous
(*c*) equilibrium (*d*) thermodynamic

Answer. (*b*)

22. Which of the following steps occur in heterogeneous catalysis at a solid/gas surface?

(i) reactant molecules react on the surface
(ii) reactant molecules are adsorbed
(iii) reactant molecules collide in the gas phase
(iv) reactant molecules diffuse along the surface

(*a*) (i), (ii) and (iv) (*b*) (i) and (ii)
(*c*) (ii), (iii) and (iv) (*d*) (ii) and (iii)

Answer. (*a*)

23. For the reaction A → B, the activation energy is E_a = 125 kJ/mol and the heat of reaction, ΔH = 50 kJ/mol. What is the E_a for the reverse reaction in kJ/mol?

(*a*) –75 kJ/mol (*b*) 125 kJ/mol
(*c*) 175 kJ/mol (*d*) 7.5 kJ/mol

Answer. (*d*)

24. A reaction has an activation energy of 40.0 kJ/mol. At what temperature will the rate of this reaction be triple that at 300 K?

(*a*) 281 K (*b*) 322 K
(*c*) 1.89×10^{-4} K (*d*) 0.638 K

Answer. (*b*)

25. The optimum temperature is that

(*a*) when catalytic activity of the catalyst is minimum
(*b*) when catalytic activity of the catalyst is maximum
(*c*) when catalytic activity of the catalyst is zero
(*d*) when catalytic activity of the catalyst is negative

Answer. (*a*)

26. Enzymes are

(*a*) substances synthesized by chemists to decrease the reaction rate
(*b*) highly porous substances to activate acids and bases
(*c*) extremely poor in catalytic activity
(*d*) catalysts found in organisms

Answer. (*d*)

27. The name catalyst was given by

(*a*) Chadwick (*b*) J.J. Thomson

(*c*) Berzelius (*d*) Rutherford

Answer. (*c*)

28. Which one of the following statements is not true?

(*a*) enzymes require optimum temperature (*b*) enzymes required optimum pH

(*c*) enzymes increase activation energy (*d*) enzymes are highly specific in nature

Answer. (*c*)

29. A catalyst becomes ineffective by the catalytic poison because

(*a*) it combines chemically with one of the reactants

(*b*) it combines chemically with the catalyst

(*c*) it absorbs reactant molecules

(*d*) it is preferentially adsorbed on the catalyst

Answer. (*d*)

30. Which one of the following statements is correct in reversible reaction? A catalyst

(*a*) alters the equilibrium constant of the reaction

(*b*) decreases the rate of forward reaction

(*c*) increases the rate of forward reaction

(*d*) increases the rate of backward and forward reaction to equal extent

Answer. (*d*)

31. A catalyst is a substance which

(*a*) increases the equilibrium concentration of the products

(*b*) changes the equilibrium constant of the reaction

(*c*) supplies energy to the reactions

(*d*) shortens the time to reach the equilibrium

Answer. (*d*)

32. The enzyme which can catalyze the conversion of glucose to ethanol is

(*a*) maltase (*b*) diastase

(*c*) invertase (*d*) zymase

Answer. (*d*)

33. The decomposition of Nitroglycerine is an example of

(*a*) self catalytic reaction (*b*) enzyme catalyzed reaction

(*c*) autocatalytic reaction (*d*) induced catalysis

Answer. (*c*)

34. Enzymes are

(*a*) finely divided metals (*b*) transition metal ions

(*c*) auto-catalysts (*d*) proteins

Answer. (*d*)

35. Which of the following processes does not involve the use of a catalyst?

(*a*) Contact process (*b*) Ostwald process

(*c*) Lead chamber process (*d*) Thermite process

Answer. (*d*)

36. In Haber's process for the synthesis of ammonia, the use of finely divided iron (a catalyst) increases:

(*a*) the rate of formation of NH_3 as well as amount of ammonia formed

(*b*) the rate of formation of NH_3 without changing the amount of ammonia formed (*c*) the rate of formation of NH_3 but decreases the amount of ammonia formed

(*d*) the amount of ammonia formed without increasing the rate of formation

Answer. (*b*)

37. Which of the following statements is incorrect

(*a*) enzymes are catalysts (*b*) urease is an enzyme

(*c*) enzymes can catalyze any reaction (*d*) enzymes are in colloidal state

Answer. (*a*)

38. The efficiency of an enzyme in catalyzing a reaction is due to its capacity

(*a*) to change the shape of the substrate molecule

(*b*) to lower the activation energy of the reaction

(*c*) to form a strong enzyme substrate complex

(*d*) to decrease the bond energy of the substrate molecule

Answer. (*b*)

39. A promoter is a substance which

(*a*) lowers the kinetic energy of reactants (*b*) lowers the activation energy of reaction

(*c*) enhances the activity of the catalyst (*d*) enhances the concentration of the catalyst

Answer. (*c*)

40. Catalytic poisoning

(*a*) reduces the activity of the catalyst

(*b*) reduces the reaction rate of the reaction

(*c*) increases the activation energy of the reaction

(*d*) increases the temperature of the reaction

Answer. (*a*)

41. Arsenic oxide (AS_2O_3) acts in the Contact process as

(*a*) a catalyst (*b*) a promoter

(*c*) a poison (*d*) an enzyme

Answer. (*c*)

42. In the synthesis of ammonia by Haber's process, the substance which acts as catalytic poison is

(*a*) platinum (*b*) iron

(*c*) AS_2O_3 (*d*) H_2S

Answer. (*d*)

43. Tetraethyl lead, $Pb(C_2H_5)_4$, when added to petrol acts as

(*a*) a catalyst (*b*) a promoter

(*c*) an inhibitor (*d*) auto catalyst

Answer. (*c*)

44. An example of acid-base catalysis is

(*a*) inversion of cane sugar (*b*) keto-enol tautomerism

(*c*) decomposition of nitramide (*d*) all the above

Answer. (*d*)

45. Which is true?

(*a*) Metals in a state of fine sub-division are more efficient catalysts than metal in lumps

(*b*) Catalytic poisoning occurs due to blockage of free valence bonds on the surface

(*c*) A promoter increases the valence bonds on the catalyst surface by changing the crystal lattice

(*d*) all the above

Answer. (*d*)

46. In intermediate compound formation theory of catalysis, the intermediate complex formed has
    (*a*) lesser activation energy than uncatalyzed complex
    (*b*) more activation energy than uncatalyzed complex
    (*c*) lesser kinetic energy than the reactants
    (*d*) lesser kinetic energy than the products
    Answer. (*a*)
47. The intermediate compound formation theory generally applies to
    (*a*) homogeneous catalytic reactions (*b*) heterogeneous catalytic reactions
    (*c*) auto catalytic reactions (*d*) all of these
    Answer. (*a*)
48. Which of the following processes is an example of heterogeneous catalysis?
    (*a*) Contact process (*b*) Haber's process
    (*c*) Hydrogenation of vegetable oils (*d*) all of these
    Answer. (*d*)
49. The phenomenon of negative catalysis is also known as
    (*a*) auto catalysis (*b*) self catalysis
    (*c*) inhibition (*d*) enzyme catalysis
    Answer. (*c*)

22

Colloids

CONTENTS

Thomas Graham (1861) studied the ability of dissolved substances to diffuse into water across a permeable membrane. He observed that crystalline substances such as sugar, urea, and sodium chloride passed through the membrane, while others like glue, gelatin and gum arabic did not. The former he called **crystalloids** and the latter **colloids** (Greek, *kolla* = glue ; *eidos* = like). Graham thought that the difference in the behavior of 'crystalloids' and 'colloids' was due to the particle size. Later it was realized that **any substance, regardless of its nature, could be converted into a colloid by subdividing it into particles of colloidal size.**

WHAT ARE COLLOIDS?

In a **true solution** as sugar or salt in water, the solute particles are dispersed in the solvent as single molecules or ions. Thus the diameter of the dispersed particles ranges from 10^{-10} m to 10^{-9} m.

On the other hand, in a **suspension** as sand stirred into water, the dispersed particles are aggregates of millions of molecules. The diameter of these particles is of the order 2×10^{-7} m or more.

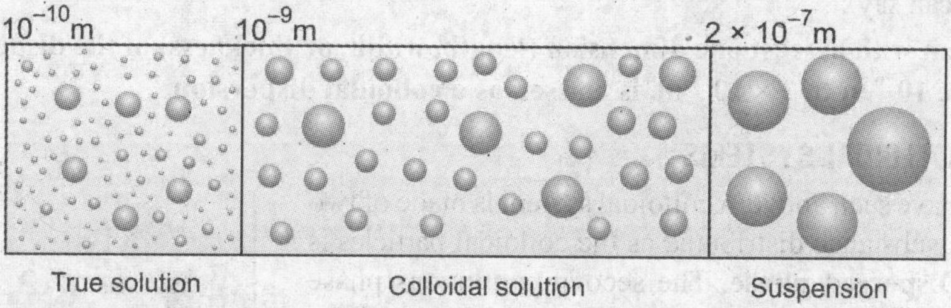

Figure 22.1
Particle size (indicated by diameter) range of true solution, colloidal dispersion, and suspension.

The colloidal solutions or colloidal dispersions are intermediate between true solutions and suspensions. In other words, the diameter of the dispersed particles in a colloidal dispersion is more than that of the solute particles in a true solution and smaller than that of a suspension.

When the diameter of the particles of a substance dispersed in a solvent ranges from about 10^{-9} m to 2×10^{-7} m, the system is termed a colloidal solution, colloidal dispersion, or simply a colloid.

The material with particle size in the colloidal range is said to be in the **colloidal state.**

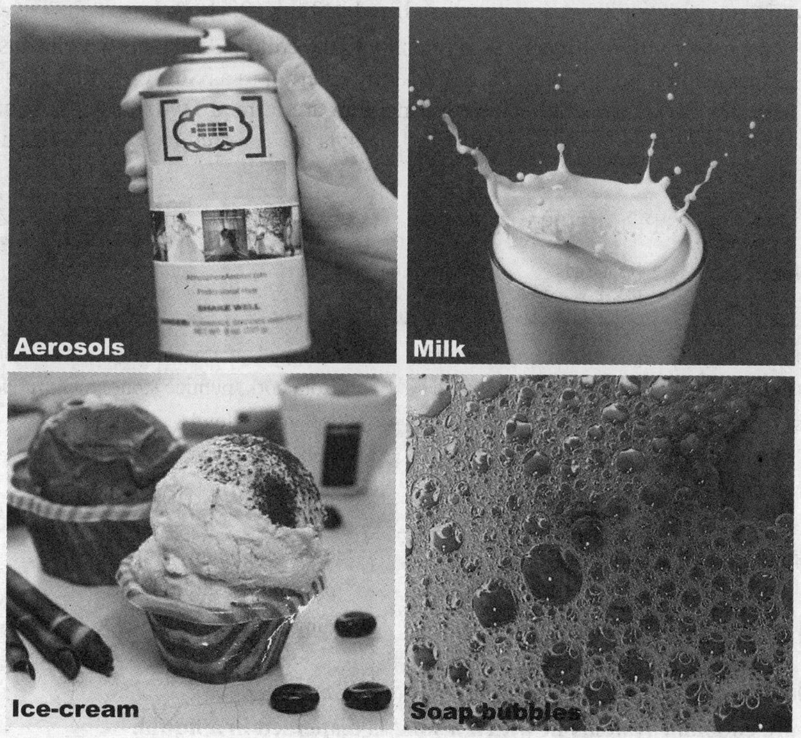

Figure 22.2
Common examples of colloids.

The colloidal particles are not necessarily corpuscular in shape. In fact, these may be rod-like, disc-like, thin films, or long filaments. For matter in the form of corpuscles, the diameter gives a measure of the particle size. However, in other cases one of the dimensions (length, width and thickness) has to be in the colloidal range for the material to be classed as colloidal. Thus in a broader context we can say :

A system with at least one dimension (length, width, or thickness) of the dispersed particles in the range 10^{-9} m to 2×10^{-7} m, is classed as a colloidal dispersion.

TYPES OF COLLOIDAL SYSTEMS

As we have seen above, a colloidal system is made of two phases. The substance distributed as the colloidal particles is called the **Dispersed phase.** The second continuous phase in which the colloidal particles are dispersed is called the **Dispersion medium.** For example, for a colloidal solution of copper in water, copper particles constitute the dispersed phase and water the dispersion medium.

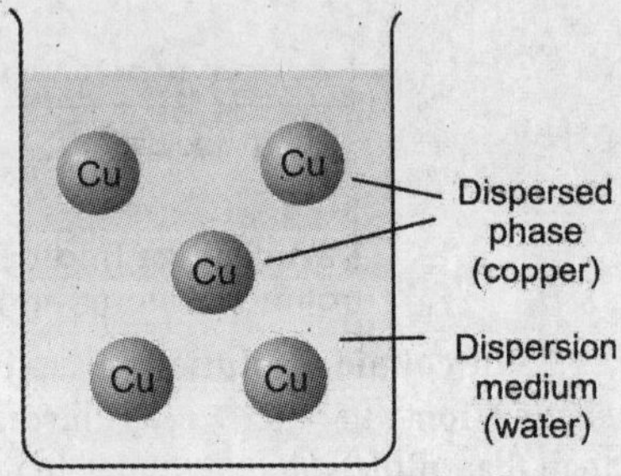

Figure 22.3
A colloidal system of copper in water

As stated above, a colloidal system is made of a dispersed phase and the dispersion medium. Because either the dispersed phase or the dispersion medium can be a gas, liquid or solid, there are eight types of colloidal systems possible. A colloidal dispersion of one gas in another is not possible since the two gases would give a homogeneous molecular mixture.

The various types of colloidal systems are listed in Table. 22.1.

In this chapter we will restrict our study mainly to the colloidal systems which consist of a solid substance dispersed in a liquid. These are frequently referred to as **Sols** or **Colloidal solution.** The colloidal solutions in water as the dispersion medium are termed **Hydrosols or Aquasols.** When the dispersions medium is alcohol or benzene, the sols are referred to as **Alcosols** and **Benzosols** respectively.

TABLE 22.1. TYPES OF COLLOIDAL SYSTEMS

Type Name	Dispersed Phase	Dispersion medium	Examples
Foam	gas	liquid	whipped cream, shaving cream, soda-water
Solid foam	gas	solid	froth cork, pumice stone, foam rubber
Aerosol	liquid	gas	fog, mist, clouds
Emulsion	liquid	liquid	milk, hair cream
Solid emulsion (gel)	liquid	solid	butter, cheese
Smoke	solid	gas	dust, soot in air
Sol	solid	liquid	paint, ink, colloidal gold
Solid sol	solid	solid	ruby glass (gold dispersed in glass), alloys.

LYOPHILIC AND LYOPHOBIC SOLS OR COLLOIDS

Sols are colloidal systems in which a solid is dispersed in a liquid.

These can be subdivided into two classes :

(*a*) Lyophilic sols (solvent-loving)
(*b*) Lyophobic sols (solvent-hating)

Lyophilic sols are those in which the dispersed phase exhibits a definite affinity for the medium or the solvent.

The examples of lyophilic sols are dispersions of starch, gum, and protein in water.

Lyophobic sols are those in which the dispersed phase has no attraction for the medium or the solvent.

The examples of lyophobic sols are dispersion of gold, iron (III) hydroxide and sulfur in water. The affinity or attraction of the sol particles for the medium, in a lyophilic sol, is due to hydrogen bonding with water. If the dispersed phase is a protein (as in egg) hydrogen bonding takes place between water molecules and the amino groups (–NH–, $-NH_2$) of the protein molecule. In a dispersion of starch in water, hydrogen bonding occurs between water molecules and the – OH groups of the starch molecule. **There are no similar forces of attraction when sulfur or gold is dispersed in water.**

CHARACTERISTICS OF LYOPHILIC AND LYOPHOBIC SOLS

Some features of lyophilic and lyophobic sols are listed below.

(1) Ease of preparation

Lyophilic sols can be obtained straightaway by mixing the material (starch, protein) with a suitable solvent. The giant molecules of the material are of colloidal size and these at once pass into the colloidal form on account of interaction with the solvent.

Lyophobic sols are not obtained by simply mixing the solid material with the solvent.

(2) Charge on particles

Particles of a hydrophilic sol may have little or no charge at all.

Particles of a hydrophobic sol carry positive or negative charge which gives them stability.

(3) Solvation

Hydrophilic sol particles are generally solvated. That is, they are surrounded by an adsorbed layer of the dispersion medium which does not permit them to come together and coagulate. Hydration of gelatin is an example.

There is no solvation of the hydrophobic sol particles for want of interaction with the medium.

(4) Viscosity

Lyophilic sols are viscous as the particle size increases due to solvation, and the proportion of free medium decreases. Warm solutions of the dispersed phase on cooling set to a gel *e.g.*, preparation of *table jelly*.

Viscosity of hydrophobic sol is almost the same as of the dispersion medium itself.

(5) Precipitation

Lyophilic sols are precipitated (or coagulated) only by high concentration of the electrolytes when the sol particles are dissolved.

Lyophobic sols are precipitated even by low concentration of electrolytes, the protective layer being absent.

(6) Reversibility

The dispersed phase of lyophilic sols when separated by coagulation or by evaporation of the medium, can be reconverted into the colloidal form just on mixing with the dispersion medium. Therefore this type of sols are designated as **Reversible sols.**

On the other hand, the lyophobic sols once precipitated cannot be reformed merely by mixing with dispersion medium. These are, therefore, called **Irreversible sols.**

(7) Tyndall effect

On account of relatively small particle size, lyophilic sols do not scatter light and show no Tyndall effect. Lyophobic sol particles are large enough to exhibit tyndall effect.

(8) Migration in electronic field

Lyophilic sol particles (proteins) migrate to anode or cathode, or not at all, when placed in electric field.

Lyophobic sol particles move either to anode or cathode, according as they carry negative or positive charge.

COMPARISON OF LYOPHILIC AND LYOPHOBIC SOLS

Lyophilic Sols	Lyophobic Sols
1. Prepared by direct mixing with dispersion medium.	**1.** Not prepared by direct mixing with the medium.
2. Little or no charge on particles.	**2.** Particles carry positive or negative charge.
3. Particles generally solvated.	**3.** No solvation of particles.
4. Viscosity higher than dispersion medium; set to a gel.	**4.** Viscosity almost the same as of medium; do not set to a gel.
5. Precipitated by high concentration of electrolytes.	**5.** Precipitated by low concentration of electrolytes.
6. Reversible.	**6.** Irrerversible.
7. Do not exhibit Tyndall effect.	**7.** Exhibit Tyndall effect.
8. Particles migrate to anode or cathode, or not at all.	**8.** Particles migrate to either anode or cathode.

PREPARATION OF SOLS

Lyophilic sols may be prepared by simply warming the solid with the liquid dispersion medium *e.g.*, starch with water. On the other hand, lyophobic sols have to be prepared by special methods. These methods fall into two categories :

(*a*) Dispersion Methods in which larger macro-sized particles are broken down to colloidal size.

(*b*) Aggregation Methods in which colloidal size particles are built up by aggregating single ions or molecules.

DISPERSION METHODS

In these methods, material in bulk is dispersed in another medium.

(1) Mechanical dispersion using Colloid mill

The solid along with the liquid dispersion medium is fed into a Colloid mill. The mill consists of two steel plates nearly touching each other and rotating in opposite directions with high speed. The solid particles are ground down to colloidal size and are then dispersed in the liquid to give the sol. 'Colloidal graphite' (a lubricant) and printing inks are made by this method.

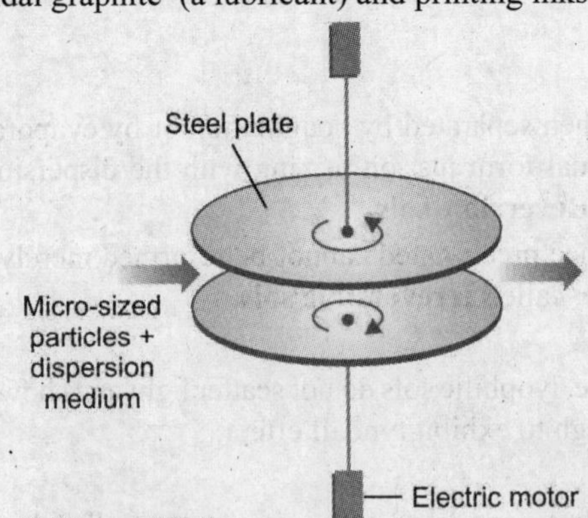

■ Figure 22.4
A Colloid mill.

Recently, mercury sol has been prepared by disintegrating a layer of mercury into sol particles in water by means of ultrasonic vibrations.

(2) Bredig's Arc Method

It is used for preparing hydrosols of metals *e.g.*, silver, gold and platinum. An arc is struck between the two metal electrodes held close together beneath *de-ionized* water. The water is kept cold by immersing the container in ice/water bath and a trace of alkali (KOH) is added. The intense heat of the spark across the electrodes vaporises some of the metal and the vapor condenses under water. Thus the atoms of the metal present in the vapor aggregate to form colloidal particles in water. Since the metal has been ultimately converted into sol particles (*via* metal vapor), this method has been treated as of dispersion.

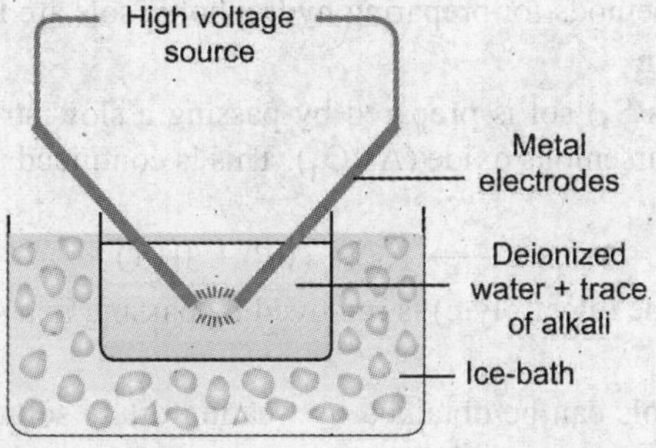

Figure 22.5
Bredig's Arc method.

Non-metal sols can be made by suspending coarse particles of the substance in the dispersion medium and striking an arc between iron electrodes.

(3) By Peptization

Some freshly precipitated ionic solids are dispersed into colloidal solution in water by the addition of small quantities of electrolytes, particularly those containing a common ion. The precipitate adsorbs the common ions and electrically charged particles then split from the precipitate as colloidal particles.

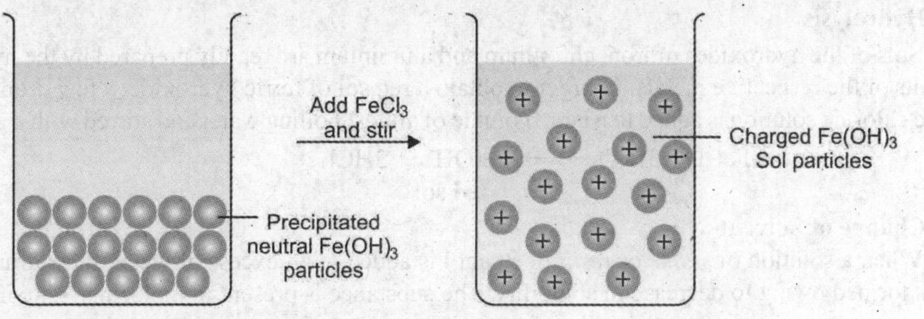

Figure 22.6
Sol of ferric hydroxide is obtained by stirring fresh precipitate of ferric hydroxide with a small amount of $FeCl_3$.

The dispersal of a precipitated material into colloidal solution by the action of an electrolyte in solution, is termed peptization. The electrolyte used is called a peptizing agent.

Peptization is the reverse of coagulation of a sol.

Examples of preparation of sols by peptization

(1) **Silver chloride,** Ag^+Cl^-, can be converted into a sol by adding hydrochloric acid (Cl^- being common ion.)

(2) **Ferric hydroxide,** $Fe(OH)_3$, yields a sol by adding ferric chloride (Fe^{3+} being common ion).

AGGREGATION METHODS

These methods consists of **chemical reactions or change of solvent** whereby the atoms or molecules of the dispersed phase appearing first, coalesce or aggregate to form colloidal particles. The conditions (temperature, concentration, etc.) used are such as permit the formation of sol particles but prevent the particles becoming too large and forming precipitate. The unwanted ions (spectator ions) present in the sol are removed by dialysis as these ions may eventually coagulate the sol.

The more important methods for preparing hydrophobic sols are listed below :

(1) Double Decomposition

An arsenic sulfide (As_2S_3) sol is prepared by passing a slow stream of hydrogen sulfide gas through a cold solution of arsenious oxide (As_2O_3). This is continued till the yellow color of the sol attains maximum intensity.

$$As_2O_3 + 3H_2S \longrightarrow As_2S_3\ (sol) + 3H_2O$$

Excess hydrogen sulfide (electrolyte) is removed by passing in a stream of hydrogen.

(2) Reduction

Silver sols and gold sols can be obtained by treating dilute solutions of silver nitrate or gold chloride with organic reducing agents like tannic acid or methanal (HCHO)

$$AgNO_3 + \text{tannic acid} \longrightarrow \text{Ag sol}$$

$$AuCl_3 + \text{tannic acid} \longrightarrow \text{Au sol}$$

(3) Oxidation

A sol of sulfur is produced by passing hydrogen sulfide into a solution of sulfur dioxide.

$$2H_2S + SO_2 \longrightarrow 2H_2O + S\downarrow$$

In qualitative analysis, sulfur sol is frequently encountered when H_2S is passed through the solution to precipitate group 2 metals if an oxidizing agent (chromate or ferric ions) happen to be present. It can be removed by boiling (to coagulate the sulfur) and filtering through two filter papers folded together.

(4) Hydrolysis

Sols of the hydroxides of iron, chromium and aluminium are readily prepared by the hydrolysis of salts of the respective metals. In order to obtain a red sol of ferric hydroxide, a few drops of 30% ferric chloride solution is added to a large volume of almost boiling water and stirred with a glass rod.

$$FeCl_3 + 3H_2O \longrightarrow \underset{\text{red sol}}{Fe(OH)_3} + 3HCl$$

(5) Change of Solvent

When a solution of *sulfur* or *resin* in ethanol is added to an excess of water, the sulfur or resin sol is formed owing to decrease in solubility. The substance is present in molecular state in ethanol but on transference to water, the molecules precipitate out to form colloidal particles.

PURIFICATION OF SOLS

In the methods of preparation stated above, the resulting sol frequently contains besides colloidal particles appreciable amounts of electrolytes. To obtain the pure sol, these electrolytes have to be removed. This purification of sols can be accomplished by three methods :

(*a*) Dialysis
(*b*) Electrodialysis
(*c*) Ultrafiltration

Dialysis

Animal membranes (bladder) or those made of parchment paper and cellophane sheet, have very fine pores. These pores permit ions (or small molecules) to pass through but not the large colloidal particles. When a sol containing dissolved ions (electrolyte) or molecules is placed in a bag of permeable membrane dipping in pure water, the ions diffuse through the membrane. By using a continuous flow of fresh water, the concentration of the electrolyte outside the membrane tends to be zero. Thus diffusion of the ions into pure water remains brisk all the time. In this way, practically all the electrolyte present in the sol can be removed easily.

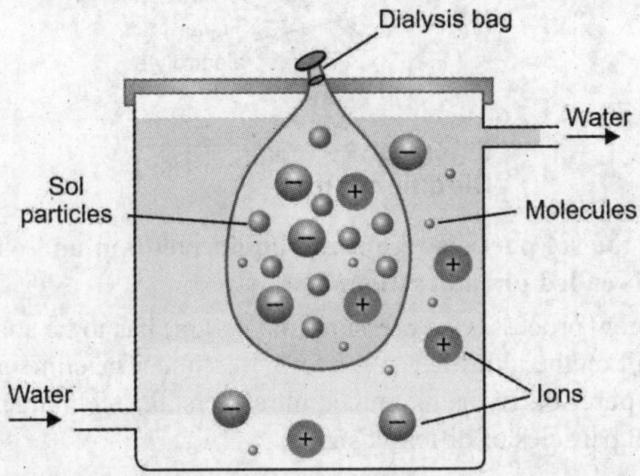

■ **Figure 22.7**
Dialysis of a sol containing ions and molecules.

The process of removing ions (or molecules) from a sol by diffusion through a permeable membrane is called Dialysis. The apparatus used for dialysis is called a **Dialyser.**

Example. A ferric hydroxide sol (red) made by the hydrolysis of ferric chloride will be mixed with some hydrochloric acid. If the impure sol is placed in the dialysis bag for some time, the outside water will give a white precipitate with silver nitrate. After a pretty long time, it will be found that almost the whole of hydrochloric acid has been removed and the pure red sol is left in the dialyser bag.

Electrodialysis

In this process, dialysis is carried under the influence of electric field (Fig. 22.8). Potential is applied between the metal screens supporting the membranes. This speeds up the migration of ions to the opposite electrode. Hence dialysis is greatly accelerated. Evidently **electrodialysis is not meant for nonelectrolyte impurities** like sugar and urea.

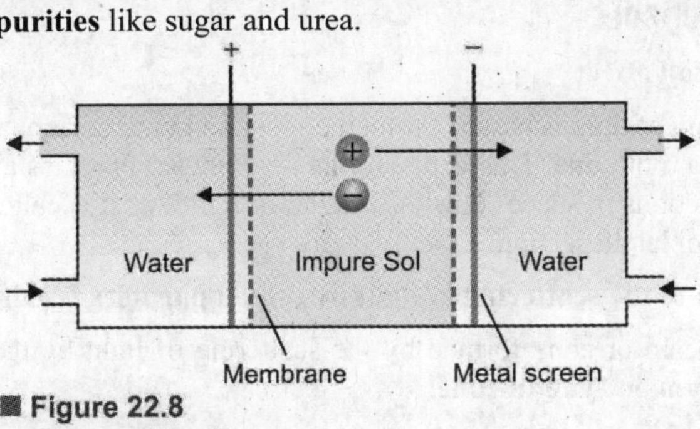

■ **Figure 22.8**
Electrodialysis.

Ultrafiltration

Sols pass through an ordinary filter paper, Its pores are too large to retain the colloidal particles. However, if the filter paper is impregnated with collodion or a regenerated cellulose such as *cellophane* or *visking*, the pore size is much reduced. Such a modified filter paper is called an **ultrafilter.**

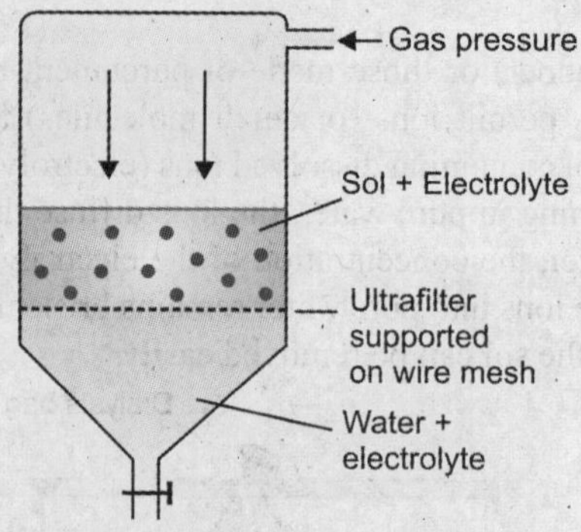

■ **Figure 22.9**
Ultrafiltration.

The separation of the sol particles from the liquid medium and electrolytes by filtration through an ultrafilter is called ultrafiltration.

Ultrafiltration is a slow process. Gas pressure (or suction) has to be applied to speed it up. The colloidal particles are left on the ultrafilter in the form of slime. The slime may be stirred into fresh medium to get back the pure sol. By using graded ultrafilters, the technique of ultrafiltration can be employed to separate sol particles of different sizes.

PROPERTIES OF SOLS–THEIR COLOR

The color of a hydrophobic sol depends on the wavelength of the light scattered by the dispersed particles. The wavelength of the scattered light again depends on the size and the nature of the particles. This is fully borne out from the following date in case of silver sols.

Color of Ag-Sol	Particle Diameter
Orange-yellow	6×10^{-8} m
Orange-red	9×10^{-8} m
Purple	13×10^{-8} m
Violet	15×10^{-8} m

The color changes produced by varying particles size have been observed in many other cases.

OPTICAL PROPERTIES OF SOLS

(1) Sols exhibit Tyndall effect

When a strong beam of light is passed through a sol and viewed at right angles, the path of light shows up as a hazy beam or cone. This is due to the fact that sol particles absorb light energy and then emit it in all directions in space. This 'scattering of light', as it is called, illuminates the path of the beam in the colloidal dispersion.

The phenomenon of the scattering of light by the sol particles is called Tyndall effect.

The illuminated beam or cone formed by the scattering of light by the sol particles is often referred as **Tyndall beam** or **Tyndall cone.**

The hazy illumination of the light beam from the film projector in a smoke-filled theatre or the light beams from the headlights of car on a dusty road, are familiar examples of the Tyndall effect. If the sol particles are large enough, the sol may even appear turbid in ordinary light as a result of Tyndall scattering.

True solutions do not show Tyndall effect. Since ions or solute molecules are too small to scatter light, the beam of light passing through a true solution is not visible when viewed from the side. Thus Tyndall effect can be used to distinguish a colloidal solution from a true solution.

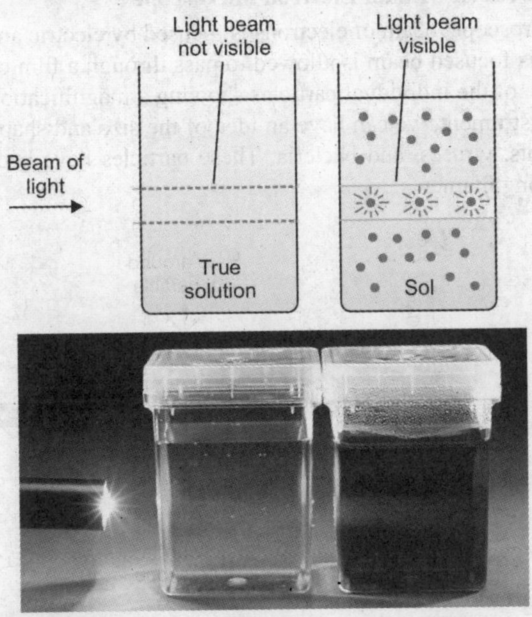

■ Figure 22.10
Tyndall effect.

■ Figure 22.11
Tyndall effect in nature.

(2) Ultramicroscope shows up the presence of individual particles

Sol particles cannot be seen with a microscope. Zsigmondy (1903) used the Tyndall phenomenon to set up an apparatus named as the **ultramicroscope.** An intense beam of light is focused on a sol contained in a glass vessel. The focus of light is then observed with a microscope at right angles to the beam. Individual sol particles appear as bright specks of light against a dark background (dispersion medium). It may be noted that under the ultramicroscope, the actual particles are not visible. It is the larger halos of scattered light around the particles that are visible. **Thus an ultramicroscope does not give any information regarding the shape and size of the sol particles.**

(3) Sol particles can be seen with an Electron microscope

In an electron microscope, beam of electrons is focused by electric and magnetic fields on to a photographic plate. This focused beam is allowed to pass through a film of sol particles. Thus it is possible to get a picture of the individual particles showing a magnification of the order of 10,000. With the help of this instrument, we can have an idea of the size and shape of several sol particles including paint pigments, viruses, and bacteria. These particles have been found to be spheroid, rod-like, disc-like, or long filaments.

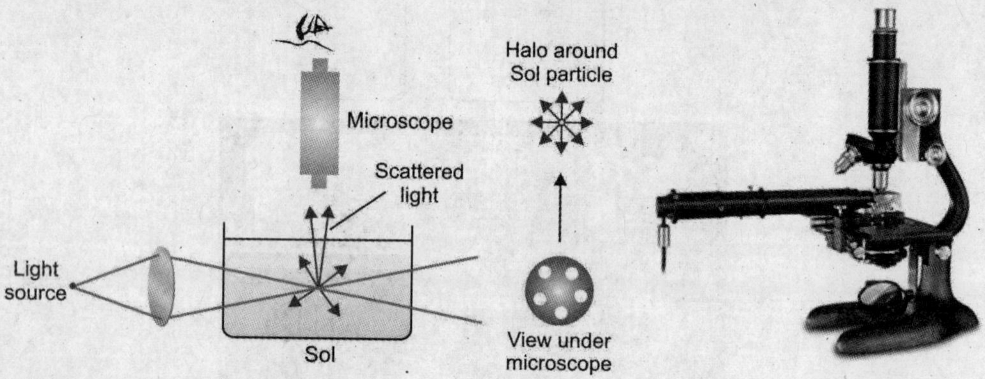

■ **Figure 22.12**
Principle of the Ultramicroscope.

KINETIC PROPERTIES OF SOLS

Brownian Movement

When a sol is examined with an ultramicroscope, the suspended particles are seen as shining specks of light. By following an individual particle it is observed that the particle is undergoing a constant rapid motion. It moves in a series of short straight-line paths in the medium, changing directions abruptly.

The continuous rapid zig-zag movement executed by a colloidal particle in the dispersion medium is called Brownian movement or motion.

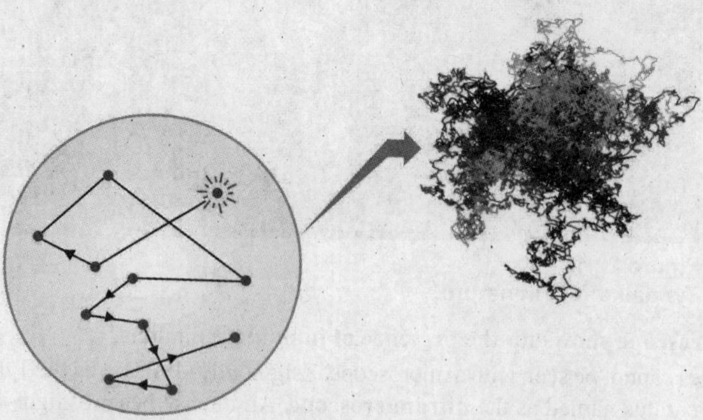

■ **Figure 22.13**
An illustration of Brownian movement.

This phenomenon is so named after Sir Robert Brown who discovered it in 1827.

Suspension and true solutions do not exhibit Brownian movement.

Explanation of Brownian movement. The explanation of Brownian movement was advanced by Albert Einstein around 1955 by mathematical considerations based on the kinetic molecular theory. According to him, at any instant a colloidal particle was being struck by several molecules of the dispersion medium. The movement of the particle was caused by unequal number of molecules of the medium striking it from opposite directions. When more molecules struck the particle on one side than on another, the direction of movement changed. Fig. 22.14 illustrates how a colloidal particle is knocked about in a zig-zag path by molecules of the dispersion medium.

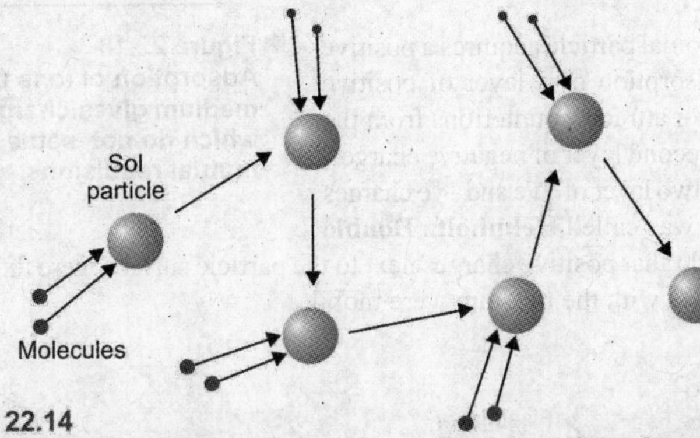

■ **Figure 22.14**
The bombardment on the sides of the colloidal particles by molecules of dispersion medium causes the random movement of the particle.

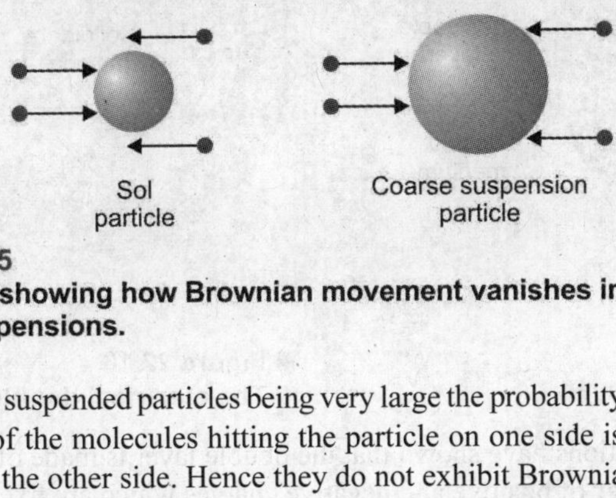

■ **Figure 22.15**
Illustration showing how Brownian movement vanishes in coarse suspensions.

In a suspension, the suspended particles being very large the probability of unequal bombardments diminishes. The force of the molecules hitting the particle on one side is cancelled by the force of collisions occurring on the other side. Hence they do not exhibit Brownian movement.

The phenomenon of Brownian movement is an excellent proof of the existence of molecules and their ceaseless motion in liquids. It also explains how the action of gravity, which would ordinarily cause the settling of colloidal particles, is counteracted. The constant pushing of the particles by the molecules of the dispersion medium has a stirring effect which does not permit the particles to settle.

ELECTRICAL PROPERTIES OF SOLS

(1) The sol particles carry an electric charge

The most important property of colloidal dispersions is that all the suspended particles posses either a positive or a negative charge. The mutual forces of repulsion between similarly charged particles prevent them from aggregating and settling under the action of gravity. This gives stability to the sol.

The sol particles acquire positive or negative charge by preferential adsorption of positive or negative ions from the dispersion medium. For example, a ferric hydroxide sol particles are positively charged because these adsorb Fe^{3+} ions from ferric chloride ($FeCl_3$) used in the preparation of the sol. Since the sol as a whole is neutral, the charge on the particle is counterbalanced by oppositely charged ions termed **counterions** (in this case Cl^-) furnished by the electrolyte in medium.

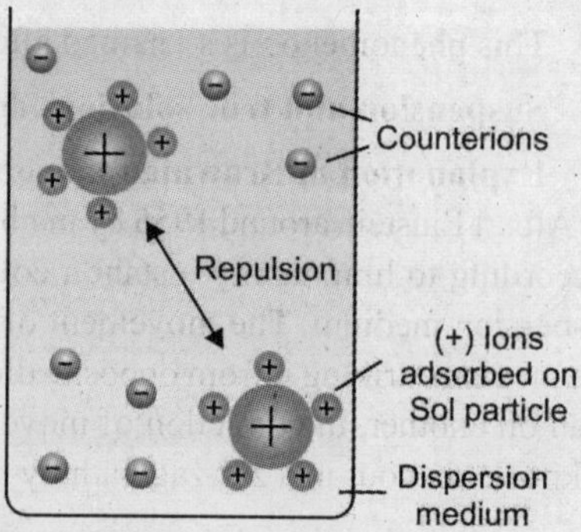

■ **Figure 22.16**
Adsorption of ions from dispersion medium gives charge to Sol particles which do not settle on account of mutual repulsions.

Electrical Double layer

The surface of colloidal particle acquires a positive charge by selective adsorption of a layer of positive ions around it. This layer attracts counterions from the medium which form a second layer of negative charges. The combination of the two layer of +ve and –ve charges around the sol particle was called **Helmholtz Double layer.** Helmholtz thought that positive charges next to the particle surface were fixed, while the layer of negative charges along with the medium were mobile.

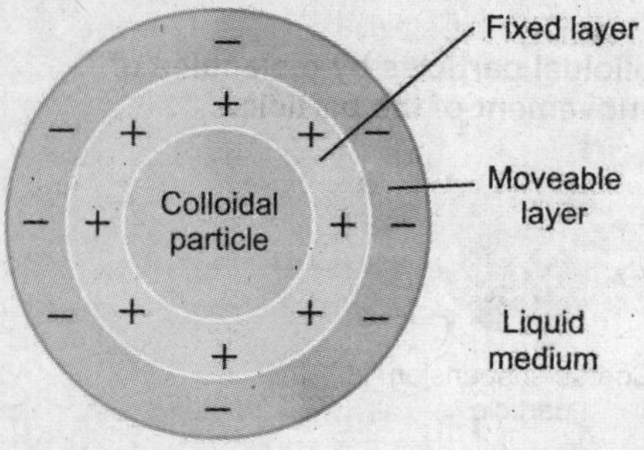

■ **Figure 22.17**
Helmholtz double layer.

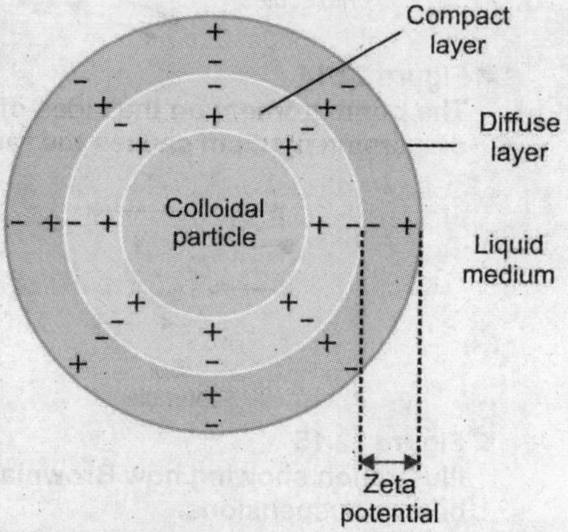

■ **Figure 22.18**
The electrical double layer (Stern).

More recent considerations have shown that the double layer is made of :

(*a*) a **Compact layer** of positive and negative charges which are fixed firmly on the particle surface.

(*b*) a **Diffuse layer** of counterions (negative ions) diffused into the medium containing positive ions.

The combination of the compact and diffuse layer is referred to as the **Stern Double layer** after the colloid chemist who first realized its significance. The diffuse layer is only loosely attached to the particle surface and moves in the opposite direction under an applied electric field. Because of the distribution of the charge around the particle, there is a difference in potential between the compact layer and the bulk of solution across the diffuse layer. This is called by **Electrokinetic** or **Zeta potential.** The presence of the double layer accounts for the electrical properties : (*a*) Cataphoresis; and (*b*) Electro-osmosis of colloids. It has been made possible to estimate the magnitude of the zeta potential with the help of these properties.

We have explained the theory of electrical double layer taking example of a positive sol. Our considerations could well be applied to a negative sol with the interchange of the disposition of positive and negative ions.

(2) Electrophoresis

If electric potential is applied across two platinum electrodes dipping in a hydrophilic sol, the dispersed particles move toward one or the other electrode.

The movement of sol particles under an applied electric potential is called electrophoresis or cataphoresis.

If the sol particles migrate toward the positive electrode, they carry a negative charge. On the other hand, if they move toward the negative electrode, they are positively charged. **Thus by noting the direction of movement of the sol particles, we can determine whether they carry a positive or negative charge.**

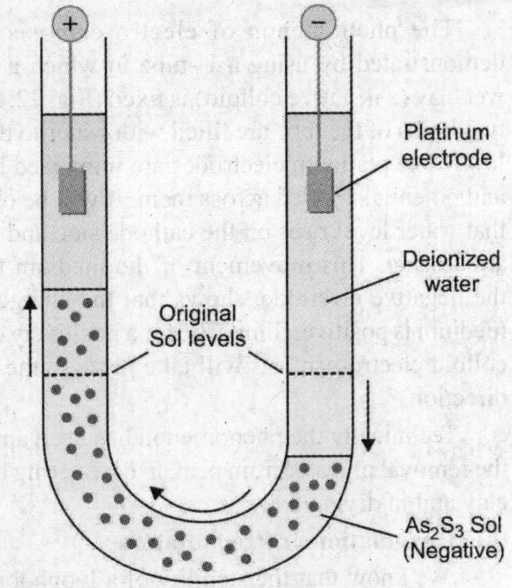

■ **Figure 22.19 Electrophoresis of a Sol.**

The phenomenon of electrophoresis can be demonstrated by placing a layer of As_2S_3 sol under two limbs of a U-tube. When a potential difference of about 100 volts is applied across the two platinum electrodes dipping in deionized water, it is observed that the level of the sol drops on the negative electrode side and rises on the positive electrode side (Fig. 22.19). This shows that As_2S_3 sol has migrated to the positive electrode, indicating that the particles are negatively charged. Similarly, a sol of ferric hydroxide will move to the negative electrode, showing that its particles carry positive charge.

Using water as the dispersion medium, the charge on the particles of some common sols determined by electrophoresis is given below.

Negatively Charged	Positively Charged
(1) Metals *e.g.*, Au, Ag, Pt	(1) Ferric hydroxide
(2) Arsenic sulfide	(2) Aluminium hydroxide
(3) Starch	(3) Basic dyes
(4) Clay	(4) Haemoglobin

Applications. Some important applications of electrophoresis are : (1) Removal of smoke from chimney gases; (2) Removal of suspended impurities; (3) Electro-plating of rubber on metal surfaces from latex (a sol); (4) painting of metal parts of cars from colloidal pigments.

(3) Electro-osmosis

A sol is electrically neutral. Therefore the dispersion medium carries an equal but opposite charge to that of the dispersed particles. Thus the medium will move in opposite direction to the dispersed phase under the influence of applied electric potential. When the dispersed phase is kept, stationary, the medium is actually found to move to the electrode of opposite sign that its own.

The movement of the dispersion medium under the influence of applied potential is known as electro-osmosis.

Electro-osmosis is a direct consequence of the existence of zeta potential between the sol particles and the medium. When the applied pressure exceeds the zeta potential, that *diffuse layer* moves and causes electro-osmosis.

The phenomenon of electro-osmosis can be demonstrated by using a U-tube in which a plug of wet clay (a negative colloid) is fixed (Fig. 22.19). The two limbs of the tube are filled with water to the same level. The platinum electrodes are immersed in water and potential applied across them. It will be observed that water level rises on the cathode side and falls on anode side. This movement of the medium towards the negative electrode, shows that the charge on the medium is positive. Similarly, for a positively charged colloid electro-osmosis will take place in the reverse direction.

Technically the phenomenon has been applied in the removal of water from peat, in dewatering of moist clay and in drying dye pastes.

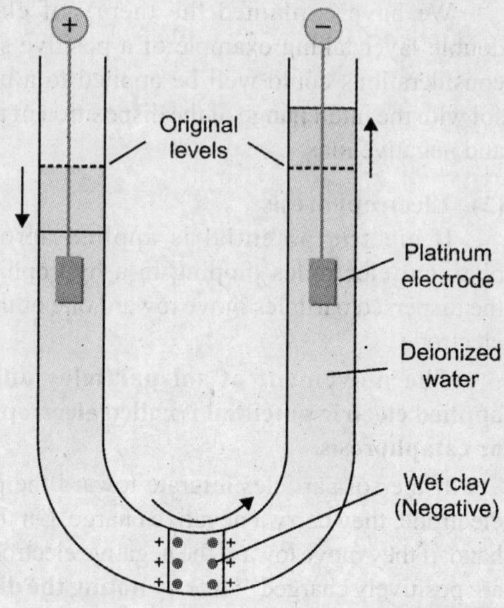

■ **Figure 22.20**
Illustration of Electro-osmosis.

(4) Coagulation or Precipitation

We know that the stability of a lyophobic sol is due to the adsorption of positive or negative ions by the dispersed particles. The repulsive forces between the charged particles do not allow them to settle. If, some how, the charge is removed, there is nothing to keep the particles apart from each other. They aggregate (or flocculate) and settle down under the action of gravity.

The flocculation and settling down of the discharged sol particles is called coagulation or precipitation of the sol.

How coagulation can be brought about?

The coagulation or precipitation of a given sol can be brought about in four ways :

(*a*) By addition of electrolytes
(*b*) By electrophoresis
(*c*) By mixing two oppositely charged sols
(*d*) By boiling

(*a*) **By addition of Electrolytes.** When excess of an electrolyte is added to a sol, the dispersed particles are precipitated. The electrolyte furnishes both positive and negative ions in the medium. The sol particles adsorb the oppositely charged ions and get discharged. The electrically neutral particles then aggregate and settle down as precipitate (Fig. 22.21).

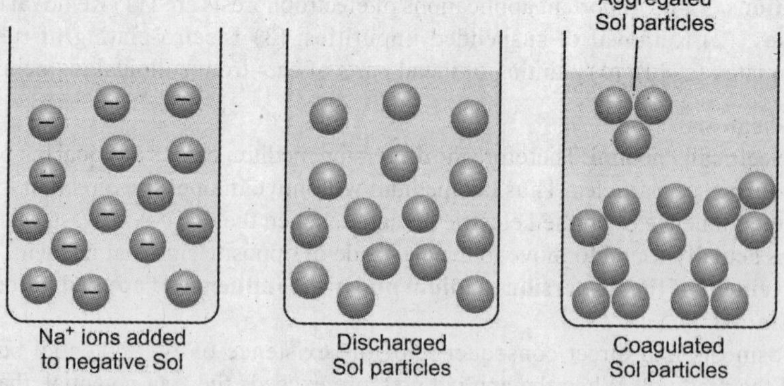

■ **Figure 22.21**
Coagulation of a Sol (illustration).

A negative ion (anion) causes the precipitation of a positively charged sol, and *vice versa.* The effectiveness of an anion or cation to precipitate a sol, will naturally depend on the magnitude of the charge or valence of the effective ion. From a study of the precipitating action of various electrolytes on particular sol, Hardy and Schulze gave a general rule.

Hardy-Schulze Rule states that **the precipitating effect of an ion on dispersed phase of opposite charge increases with the valence of the ion.**

The higher the valency of the effective ion, the greater is its precipitating power. Thus for precipitating an As_2S_3 sol (negative), the precipitating power of Al^{3+}, Ba^{2+}, Na^+ ions is in the order

$Al^{3+} > Ba^{2+} > Na^+$

Similarly, for precipitating $Fe(OH)_3$ sol (positive), the precipitating power of cations $[Fe(CN)_6]^{3-}$, SO_4^{2-}, Cl^- is in the order.

$[Fe(CN)_6]^{3-} > SO_4^{2-} > Cl^-$

The precipitation power of an electrolyte or ion is experimentally determined by finding the minimum concentration in millimoles per liter or mol/m^3 in SI unit required to cause the precipitation of a sol in 7200 seconds. This is called the **Flocculation value.** The smaller the flocculation value the higher the precipitating power of an ion.

TABLE 22.2. FLOCCULATION VALUES

$Fe(OH)_3$ Sol (+ve)		As_2S_3 Sol (–ve)	
Electrolyte	Concentration (mol / m^3)	Electrolyte	Concentration ($moles / m^3$)
NaCl, (Cl^-)	9.3	NaCl, (Na^+)	51
KCl, (Cl^-)	9.0	KCl, (K^+)	50
K_2SO_4, SO_4^{2-}	0.20	$BaCl_2$, (Ba^{2+})	0.69
$K_3Fe(CN)_6$, $[Fe(CN)_6]^{3-}$	0.096	$AlCl_3$, (Al^{3+})	0.093

It may be noted how rapidly the precipitation power increases with the increases of valence. The ratio for the mono-, di-, and trivalent anion or cation are approximately 1 : 40 : 90 for $Fe(OH)_3$ sol and 1 : 70 : 500 for the As_2S_3 sol.

(*b*) **By Electrophoresis.** In electrophoresis the charged sol particles migrate to the electrode of opposite sign. As they come in contact with the electrode, the particles are discharged and precipitated.

(*c*) **By mixing two oppositely charged sols.** The mutual coagulation of two sols of opposite charge can be effected by mixing them. The positive particles of one sol are attracted by the negative particles of the second sol. This is followed by mutual adsorption and precipitation of both the sols. Ferric hydroxide (+ve sol) and arsenious sulfide (–ve sol) form such a pair.

(*d*) **By boiling.** Sols such as sulfur and silver halides dispersed in water, may be coagulated by boiling. Increased collisions between the sol particles and water molecules remove the adsorbed electrolyte. This takes away the charge from the particles which settle down.

(5) Protective action of sols

Lyophobic sols are readily precipitated by small amounts of electrolytes. However these sols are often stabilized by the addition of lyophilic sols.

The property of lyophilic sols to prevent the precipitation of a lyophobic sol is called protection.

The lyophilic sol used to protect a lyophobic sol from precipitation is referred to as a **Protective colloid.**

Example. If a little gelatin (hydrophilic colloid) is added to a gold sol (hydrophobic sol), the latter is protected. The 'protected gold sol' is no longer precipitated on the addition of sodium chloride.

Explanation. The particles of the hydrophobic sol adsorb the particles of the lyophilic sol. Thus the lyophilic colloid forms a coating around the lyophobic sol particles. The hydrophobic colloid, therefore, behaves as a hydrophilic sol and is precipitated less easily by electrolytes.

Gold number

The lyophilic colloids differ widely in their powers of protection. The protective action of different colloids is measured in terms of the **'Gold number'** introduced by Zsigmondy. The gold number is defined as :**the number of milligrams of a hydrophilic colloid that will just prevent the precipitation of 10 ml of a gold sol on the addition of 1 ml of 10 per cent sodium chloride solution.**

The onset of precipitation of the gold sol is indicated by a color change from red to blue when the particle size just increases.

The gold numbers of hydrophilic colloids are given in Table 22.3. The smaller the gold number of a hydrophilic colloid, the greater is its protective power. Gelatin has a small gold number and is an effective protective colloid. Starch has a very high value, which shows that it is an ineffective protective colloid.

TABLE 22.3. GOLD NUMBERS OF SOME HYDROPHILIC COLLOIDS

Lyophilic colloid	Gold number
Gelatin	0.005 – 0.01
Egg albumen	0.08 – 0.10
Gum arabic	0.10 – 0.15
Potato starch	25

The use of protective colloids to stabilize colloidal systems is widespread. In the preparation of ice cream, gelatin is added to act as a protecting agent to the colloidal particles of ice. If the ice particles coagulate, the smooth texture of ice cream is lost. *Argyrol*, used in eye drops, is a sol of silver protected by organic material.

Origin of charge on sol particles

All the dispersed particles of a particular sol carry a positive or a negative charge. They acquire this charge by

(*a*) Adsorption of ions from the aqueous medium

(*b*) Ionization of surface groups

(1) By Adsorption of ions

In most cases **the charge on the sol particles originates by the selective adsorption of ions common to the particles from the dispersion medium.**

Examples. (*i*) Ferric hydroxide sol particles are positive because they adsorb the common ion Fe^{3+} from the aqueous medium.

$$Fe(OH)_3 + Fe^{3+} \longrightarrow \underset{\text{Positive ferric hydroxide sol particle}}{Fe(OH)_3/Fe^{3+}}$$

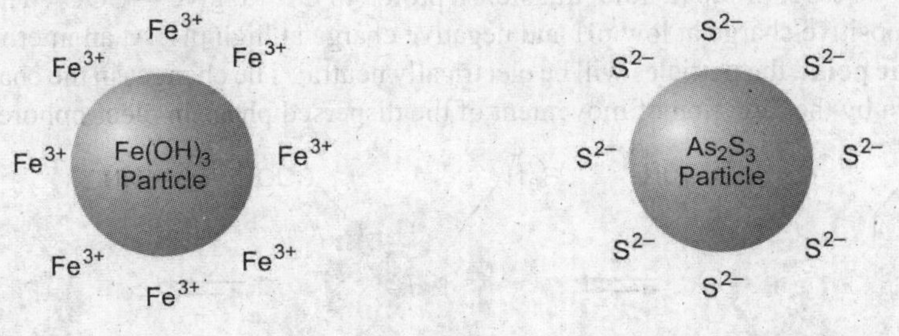

(a) (b)

■ **Figure 22.22**
(*a*) $Fe(OH)_3$, sol particles adsorb Fe^{3+} ions and become positive ;
(*b*) As_2S_3 sol particles adsorb S^{2-} ions and acquire negative charge.

(*ii*) Arsenic sulfide sol particles acquire a negative charge since they adsorb the common ion S^{2-} from the medium.

$$As_2S_3 + S^{2-} \longrightarrow As_2S_3/S^{2-}$$

Negative arsenic sulfide sol particle

It is not necessary that a particular sol particles always adsorb the same kind of ions. In fact, **the particles may adsorb the anions or cations whichever are in excess and acquire the corresponding charge.** For example, AgCl sol produced by the addition of $AgNO_3$ solution to sodium chloride solution, bears a positive charge if Ag^+ ions are in excess. On the other hand, if Cl^- ions are in excess, the AgCl sol particles acquire a negative charge.

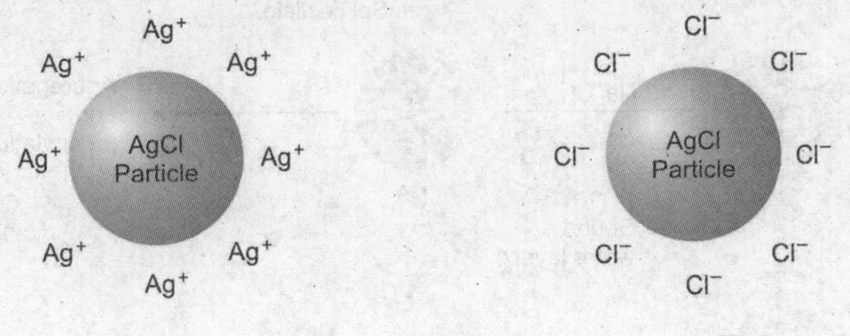

(a) (b)

■ **Figure 22.23**
(*a*) When Ag^+ ions are in excess, AgCl sol particle adsorbs these ions and becomes positive; (*b*) when Cl^- ions are in excess, the AgCl particle adsorbs these and acquires a negative charge.

(2) Ionization of Surface groups

(*a*) **Charge on Soaps and Detergent sols.** Soaps and detergent sol particles are aggregates of many molecules. The hydrocarbons tails of the molecules are directed to the center, while the groups $-COO^-Na^+$ (or $-OSO_3^-Na^+$) constitute the surface in contact with water. As a result of ionization of the surface groups, the particle surface is now made of the anionic heads $-COO^-$ (or $-OSO_3^-$). This makes the sol particle negative.

(*b*) **Charge on Protein sols.** Protein sol particles possess both acidic and basic functional groups. In aqueous solution at low pH, the $-NH_2$ group (basic) acquires a proton to give $-NH_3^+$ while at high pH the —COOH group (acidic) transfers a proton to OH^- to give —COO^-. Thus the protein sol particle has positive charge at low pH and negative charge at high pH. At an intermediate pH called the **isoelectric point,** the particles will be electrically neutral. The changes in the charge of the protein sol are shown by the direction of movement of the dispersed phase in electrophoresis.

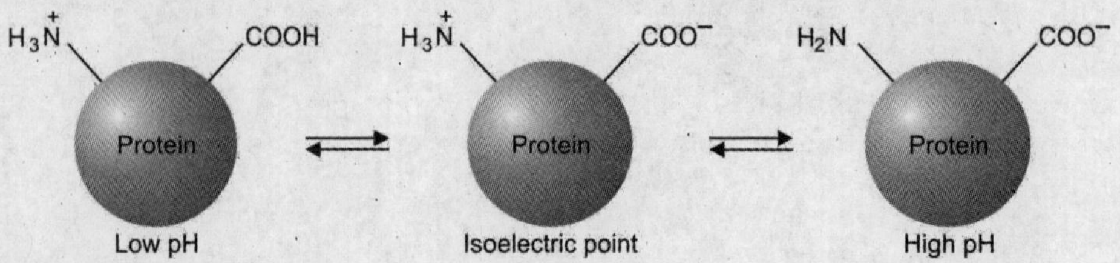

■ **Figure 22.24**
Charge on protein sol changes with pH.

STABILITY OF SOLS

A true colloidal solution is stable. Its particles do not ever coalesce and separate out. The stability of sols is mainly due to two factors :

(1) Presence of like charge on sol particles

The dispersed particles of a hydrophobic sol posses a like electrical charge (all positive or all negative) on their surface. Since like charges repel one another, the particles push away from one another and resist joining together. However, when an electrolyte is added to a hydrophobic sol, the particles are discharged and precipitated.

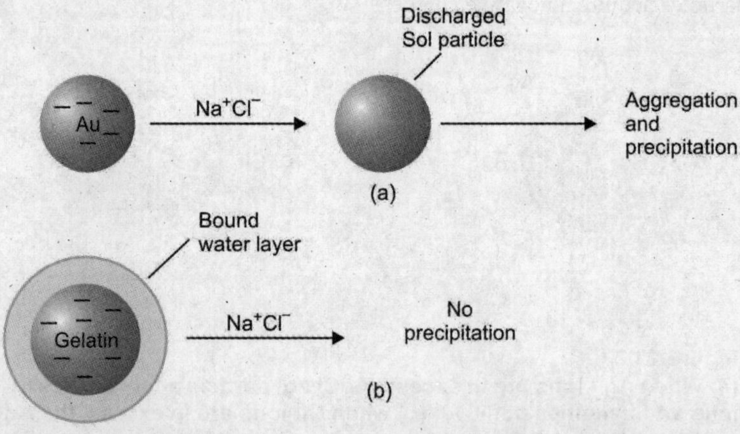

■ **Figure 22.25**
(*a*) A negatively charged gold particle is precipitated by Na^+ ions; (*b*) The water layer around gelatin particle does not allow Na^+ ions to penetrate and discharge the particle.

(2) Presence of Solvent layer around sol particle

The lyophilic sols are stable for two reasons. Their particles possess a charge and in addition have a layer of the solvent bound on the surface. For example, a sol particle of gelatin has a negative charge and a water layer envelopes it. When sodium chloride is added to colloidal solution of gelatin, its particles are not precipitated. The water layer around the gelatin particle does not allow the Na^+ ions to penetrate it and discharge the particle. The gelatin sol is not precipitated by addition of sodium chloride solution. Evidently, **lyophilic sols are more stable than lyophobic sols.**

ASSOCIATED COLLOIDS

The molecules of substances as soaps and artificial detergents are smaller than the colloidal particles. However in concentrated solutions these molecules form aggregates of colloidal size. **Substances whose molecules aggregate spontaneously in a given solvent to form particles of colloidal dimensions are called Associated or Association Colloids.**

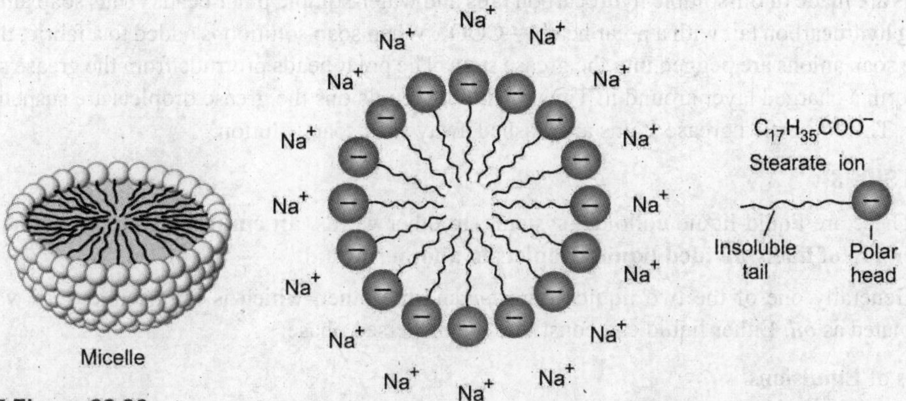

■ **Figure 22.26**
A soap micelle.

The colloidal aggregates of soap or detergent molecules formed in the solvent are referred to as **micelles.** Some examples of associated colloids are :

$C_{17}H_{35}—CO\bar{O}Na^+$
Sodium stearate

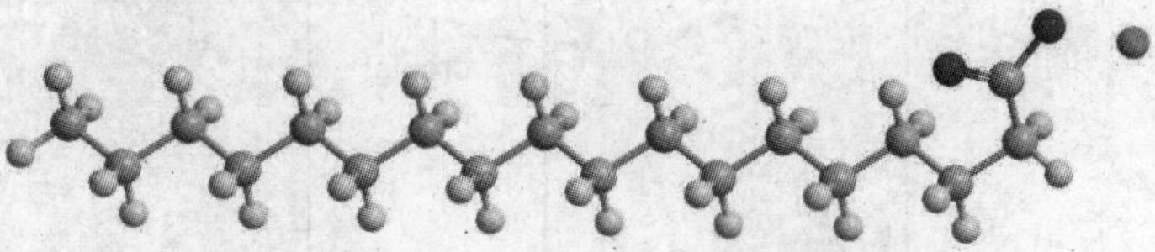

$C_{12}H_{25}—SO_2\bar{O}Na^+$
Sodium dodecyl sulphate

Explanation. Soap or detergent molecule ionizes in water to form an anion and sodium ion. Thus sodium stearate (a typical soap) furnishes stearate anion and sodium ion in aqueous solution.

$$\underset{\text{Sodium stearate}}{C_{17}H_{35}COO^- Na^+} \longrightarrow \underset{\text{Stearate ion}}{C_{17}H_{35}COO^-} + Na^+$$

As many as 70 stearate ions aggregate to form a micelle of colloidal size. The stearate ion has a long hydrocarbon chain (17 carbons) with a polar —COO⁻ group at one end. The zig zag hydrocarbon tail is shown by a wavy line and the polar head by a hollow circle. In the micelle formation, the tails being insoluble in water are directed toward the center, while the soluble polar heads are on the surface in contact with water (Fig. 22.26). The charge on the micelle due to the polar heads accounts for the stability of the particle.

Cleansing Action of Soaps and Detergents

The cleansing action of soap is due to

(1) Solubilization of grease into the micelle
(2) Emulsification of grease

Solubilization. In relatively strong solution the soap (or detergent) anions spontaneously form a micelle. The hydrocarbon tails are in the interior of the micelle and COO^- ions on the surface. The grease stain is thus absorbed into the interior of the micelle which behaves like liquid hydrocarbons. As the stain is detached from the fabric, the dirt particles sticking to the stain are also removed.

Emulsification. As already discussed the soap or detergent molecules are ionized in water. The anions are made of oil-soluble hydrocarbon tails and water-soluble polar heads. Thus soap anion has a long hydrocarbon tail with a polar head, —COO^-. When soap solution is added to a fabric, the tails of the soap anions are pegged into the grease stain. The polar heads protrude from the grease surface and form a charged layer around it. Thus by mutual repulsions the grease droplets are suspended in water. The emulsified grease stains are washed away with soap solution.

EMULSIONS

These are liquid-liquid colloidal systems. In other words, **an emulsion may be defined as a dispersion of finely divided liquid droplets in another liquid.**

Generally one of the two liquids is *water* and the other, which is immiscible with water, is designated as *oil.* Either liquid can constitute the dispersed phase.

Types of Emulsions

There are two types of emulsions.

(*a*) **Oil-in-Water type (O/W type) ;** (*b*) **Water-in-Oil type (W/O type)**

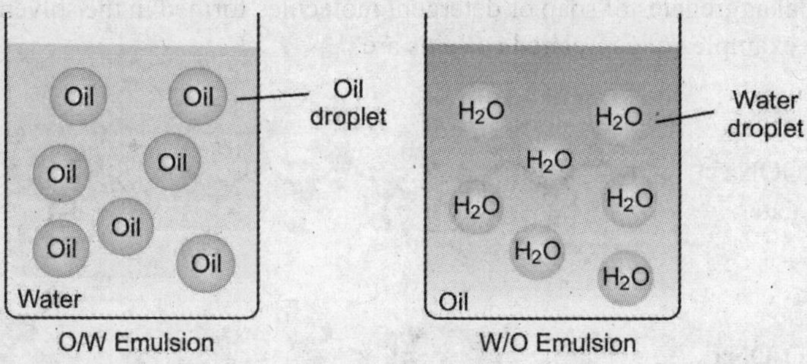

Figure 22.27
Two types of Emulsions.

Examples of Emulsions

(1) Milk is an emulsion of O/W type. Tiny droplets of liquid fat are dispersed in water.

(2) Stiff greases are emulsions of W/O type, water being dispersed in lubricating oil.

Preparation of Emulsions

The dispersal of a liquid in the form of an emulsion is called **emulsification.** This can be done by agitating a small proportion of one liquid with the bulk of the other. It is better accomplished by passing a mixture of the two liquid through a colloid mill known as **homogenizer.**

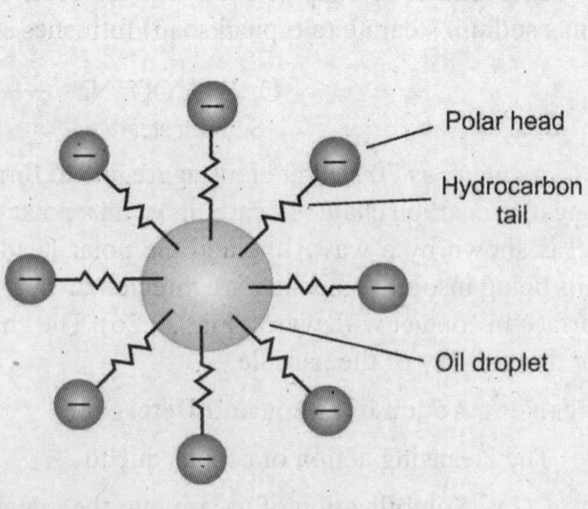

Figure 22.28
Role of emulsifier (Soap).

The emulsions obtained simply by shaking the two liquids are unstable. The droplets of the dispersed phase coalesce and form a separate layer. To have a stable emulsion, small amount of a third substance called the **Emulsifier** or **Emulsifying agent** is added during the preparation. This is usually a soap, synthetic detergent, or a hydrophilic colloid.

Role of Emulsifier

The emulsifier concentrates at the interface and reduces surface tension on the side of one liquid which rolls into droplets. Soap, for example, is made of a long hydrocarbon tail (oil soluble) with a polar head —COO^-Na^+ (water soluble). In O/W type emulsion the tail is pegged into the oil droplet, while the head extends into water. Thus the soap acts as go-between and the emulsified droplets are not allowed to coalesce.

Properties of Emulsions

(1) **Demulsification.** Emulsions can be broken or 'demulsified' to get the constituent liquids by heating, freezing, centrifuging, or by addition of appreciable amounts of electrolytes. They are also broken by destroying the emulsifying agent. For example, an oil-water emulsion stabilized by soap is broken by addition of a strong acid. The acid converts soap into insoluble free fatty acids.

(2) **Dilution.** Emulsions can be diluted with any amount of the dispersion medium. On the other hand the dispersed liquid when mixed with it will at once form a separate layer. This property of emulsions is used to detect the type of a given emulsion.

WHAT ARE GELS?

A gel is a jelly-like colloidal system in which a liquid is dispersed in a solid medium. For example, when a warm sol of gelatin is cooled, it sets to a semisolid mass which is a gel. The process of a gel formation is known as **Gelation.**

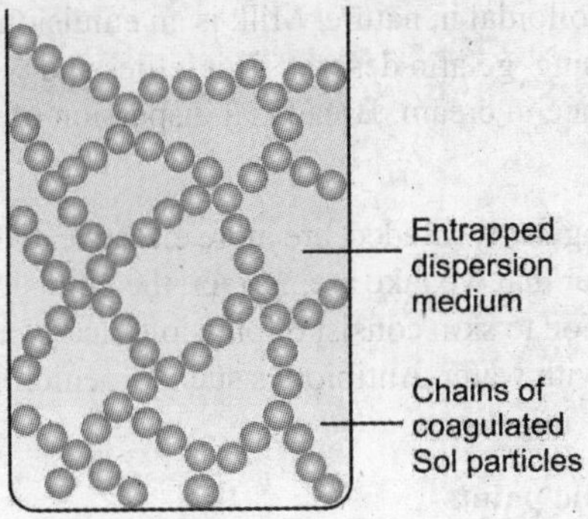

■ Figure 22.29
The network of the coagulated sol particles enclosing liquid, forming a gel.

Explanation. Gelation may be thought of as partial coagulation of a sol. The coagulating sol particles first unite to form long thread-like chains. These chains are then interlocked to form a solid framework. The liquid dispersion medium gets trapped in the cavities of this framework. The resulting semisolid porous mass has a gel structure. A sponge soaked in water is an illustration of gel structure.

Two types of Gels

Gels may be classified into two types :

(*a*) **Elastic gels** are those which posses the property of elasticity. They change their shape on applying force and return to original shape when the force is removed. Gelatin, starch and soaps are examples of substances which form elastic gels.

Elastic gels are obtained by cooling fairly concentrated lyophilic sols. The linkages between the molecules (particles) are due to electrical attraction and are not rigid.

(*b*) **Non-elastic gels** are those which are rigid *e.g.*, silica gel. These are prepared by appropriate chemical action. Thus silica gel is produced by adding concentrated hydrochloric acid to sodium silicate solution of the correct concentration. The resulting molecules of silicic acid polymerize to form silica gel. It has a network linked by covalent bonds which give a strong and rigid structure.

Properties of Gels

(1) **Hydration.** A completely dehydrated elastic gel can be regenerated by addition of water. But once a nonelastic gel is free from moisture, addition of water will not bring about gelation.

(2) **Swelling.** Partially dehydrate elastic gels imbibe water when immersed in the solvent. This causes increase in the volume of the gel and process is called **Swelling.**

(3) **Syneresis.** Many inorganic gels on standing undergo shrinkage which is accompanied by exudation of solvent. This process is termed **Syneresis.**

(4) **Thixotropy.** Some gels are semisolid when at rest but revert to liquid sol on agitation. This reversible sol-gel transformation is referred to as **Thixotropy.** Iron oxide and silver oxide gels exhibit this property. The modern thixotropic paints are also an example.

APPLICATIONS OF COLLOIDS

Colloids play an important role in our daily life and industry. A knowledge of colloid chemistry is essential to understand some of the various natural phenomena around us. Colloids make up some of our modern products. A few of the important applications of colloids are listed below.

(1) Foods

Many of our foods are colloidal in nature. Milk is an emulsion of butterfat in water protected by a protein, casein. Salad dressing, gelatin deserts, fruit jellies and whipped cream are other examples. Ice cream is a dispersion of ice in cream. Bread is a dispersion of air in baked dough.

(2) Medicines

Colloidal medicines being finely divided, are more effective and are easily absorbed in our system. Halibut-liver oil and cod-liver that we take are, in fact, the emulsions of the respective oils in water. Many ointments for application to skin consist of physiologically active components dissolved in oil and made into an emulsion with water. Antibiotics such as penicillin and streptomycin are produced in colloidal form suitable for injections.

(3) Non-drip or thixotropic paints

All paints are colloidal dispersions of solid pigments in a liquid medium. The modern *nondrip or thixotropic paints* also contain long-chain polymers. At rest, the chains of molecules are coiled and entrap much dispersion medium. Thus the paint is a semisolid gel structure. When shearing stress is applied with a paint brush, the coiled molecules straighten and the entrapped medium is released. As soon as the brush is removed, the liquid paint reverts to the semisolid form. This renders the paint 'non-drip'.

(4) Electrical precipitation of smoke

The smoke coming from industrial plants is a colloidal dispersion of solid particles (carbon, arsenic compounds, cement dust) in air. It is a nuisance and pollutes the atmosphere. Therefore, before allowing the smoke to escape into air, it is treated by **Cottrell Precipitator** (See Fig. 22.30). The smoke is let past a series of sharp points charged to a high potential (20,000 to 70,000 V). The points discharge high velocity electrons that ionize molecules in air. Smoke particles adsorb these positive ions and become charged. The charged particles are attracted to the oppositely charged electrodes and get precipitated. The gases that leave the *Cottrell precipitator* are thus free from smoke. In addition, valuable materials may be recovered from the precipitated smoke. For example, arsenic oxide is mainly recovered from the smelter smoke by this method.

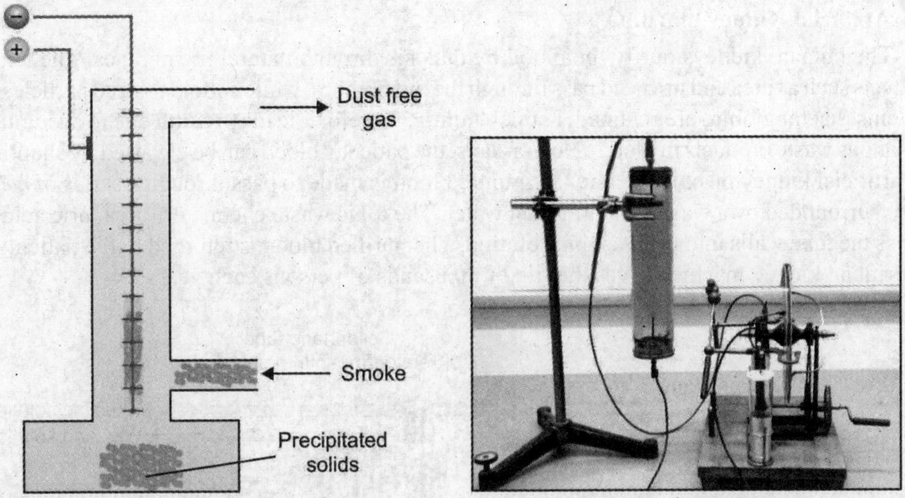

Figure 22.30
Cottrell Smoke precipitator.

(5) Clarification of Municipal water

The municipal water obtained from natural sources often contains colloidal particles. The process of coagulation is used to remove these. The sol particles carry a negative charge. When aluminium sulfate (*alum*) is added to water, a gelatinous precipitate of hydrated aluminium hydroxide (*floc*) is formed,

$$Al^{3+} + 3H_2O \longrightarrow Al(OH)_3 + 3H^+$$
$$Al(OH)_3+ + 4H_2O + H^+ \longrightarrow Al(OH)_3(H_2O)_4^+$$

The positively charged *floc* attracts to it negative sol particles which are coagulated. The *floc* along with the suspended matter comes down, leaving the water clear.

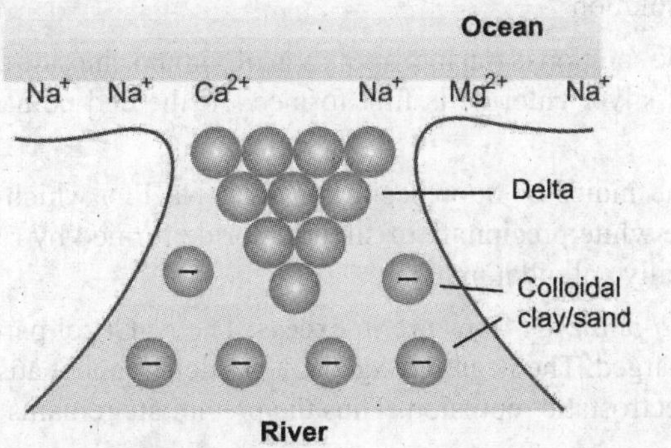

Figure 22.31
Formation of Delta.

(6) Formation of Delta

The river water contains colloidal particles of sand and clay which carry negative charge. The sea water, on the other hand, contains positive ions such as Na^+, Mg^{2+}, Ca^{2+}. As the river water meets sea water, these ions discharge the sand or clay particles which are precipitated as **delta.**

(7) Artificial Kidney machine

The human kidneys purify the blood by dialysis through natural membranes. The toxic waste products such as urea and uric acid pass through the membranes, while colloidal-sized particles of blood proteins (haemoglobin) are retained. Kidney failure, therefore, leads to death due to accumulation of poisonous waste products in blood . Now-a-days, the patient's blood can be cleansed by shunting it into an 'artificial kidney machine'. Here the impure blood is made to pass through a series of *cellophane tubes* surrounded by a washing solution in water. The toxic waste chemicals (urea, uric acid) diffuse across the tube walls into the washing solution. The purified blood is returned to the patient. The use of artificial kidney machine saves the life of thousands of persons each year.

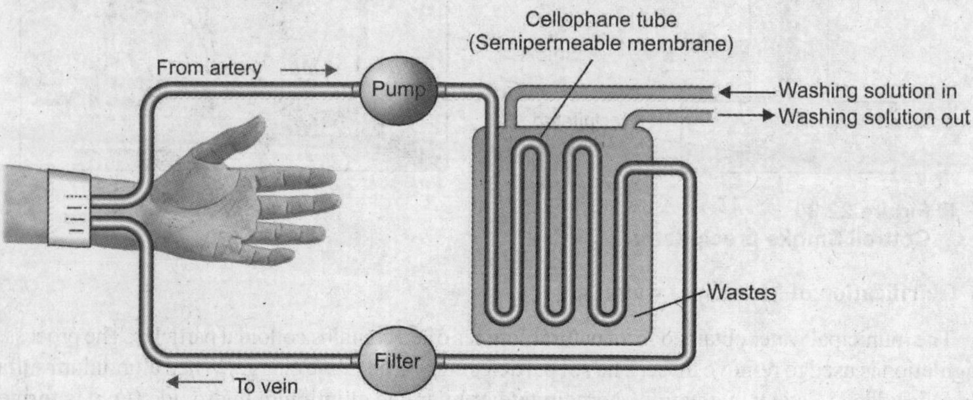

Figure 22.32
An artificial kidney machine for purification of blood by dialysis.

(8) Adsorption indicators

These indicators function by preferential adsorption of ions onto sol particles. Fluorescein (Na^+Fl) is an example of adsorption indicator which is used for the titration of sodium chloride solution against silver nitrate solution.

When silver nitrate solution is run into a solution of sodium chloride containing a little fluorescein, a white precipitate of silver chloride is first formed. At the end-point, the white precipitate turns sharply pink.

Explanation. The indicator fluorescein is a dye (Na^+Fl^-) which gives colored anion Fl^- in aqueous solution. The white precipitate of silver chloride formed by running $AgNO_3$ solution into NaCl solution is partially colloidal in nature.

(*a*) *Before the end-point,* Cl^- ions are in excess. The AgCl sol particles adsorb these ions and become negatively charged. The negative $AgCl/Cl^-$ particles cannot adsorb the colored fluorescein anions (Fl^-) due to electrostatic repulsion. Thus the precipitate remains white.

(*b*) *After the end-point,* Ag^+ ions become in excess. AgCl sol particles adsorb these and acquire positive charge. The positive $AgCl/Ag^+$ particles now attract the colored fluorescein anions (Fl^-) and turn rose-red.

Thus the end-point is marked by white precipitate changing to pink.

(9) Blue color of the sky

This is an application of Tyndall effect. The upper atmosphere contains colloidal dust or ice particles dispersed in air. As the sun rays enter the atmosphere (Fig. 22.33) these strike the colloidal particles. The particles absorb sunlight and scatter light of blue color ($46 \times 10^{-8} - 51 \times 10^{-8}$ m). The light that is incident at earth's surface is considerably reddened due to the removal of most of the blue light in the upper atmosphere.

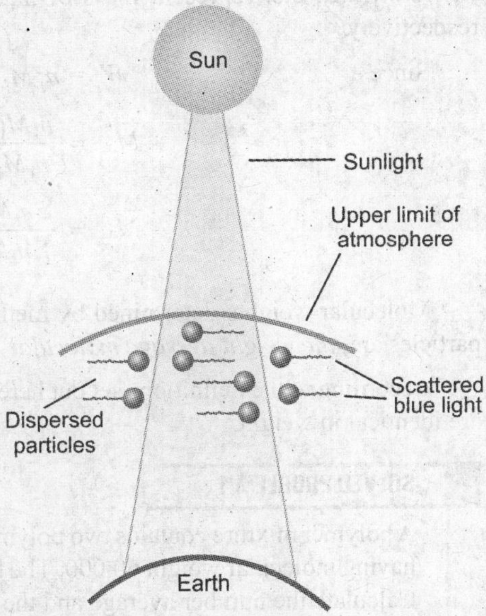

Figure 22.33
Blue color of sky (illustration).

WHAT ARE MACROMOLECULES?

Colloidal solutions are formed by aggregation of atoms or molecules to give particles of colloidal size. Yet there are **substances which are themselves composed of giant molecules and dissolve in a solvent to yield colloidal solutions directly.** These giant molecules are termed **macromolecules.** The dimensions of the macromolecules fall in a range between 10^{-9} and 10^{-6} m. Proteins (gelatin), synthetic polymers (plastics), synthetic rubber, cellulose and starch all possess macromolecules.

Solutions of macromolecules behave like reversible colloids or lyophilic sols. They show a weak *Tyndall effect* and possess high viscosity. **Macromolecules in solution do not carry an electric charge and do not show electrophoresis.**

Molecular weight of Macromolecules

The molecular weight is an important property of polymeric substances such as proteins, polymers (plastics, starch) and other macromolecules. **Generally, molecules of a protein or a polymer may not be of the same size.** Therefore all the experimental methods of molecular weight determination will give some kind of an average value. Two types of average molecular weights have been defined.

(1) **Number average molecular weight.** It is defined as :

$$\overline{M}_n = \frac{\text{Total weight, } W}{\text{Total number of particles}}$$

$$\overline{M}_n = \frac{n_1 M_1 + n_2 M_2 + ...}{n_1 + n_2 + ...}$$

$$\overline{M}_n = \frac{\Sigma n_i M_i}{\Sigma n_i}$$

$n_i M_i$ stands for the weight of macromolecules numbering n_i and having molecular weight M_i. The experimental methods based on properties which depend on the number of particles present *e.g., osmotic pressure, yield number average molecular weight.*

(2) **Weight average molecular weight.** It is defined as :

$$\overline{M}_w = \frac{m_1 M_1 + m_2 M_2 + ...}{m_1 + m_2 + ...}$$

$$= \frac{\Sigma m_i M_i}{\Sigma m_i}$$

where m_1, m_2, etc. represent mass of macromolecules having molecular weights M_1, M_2, etc. respectively.

Since $W_i = n_i M_i$

$$\overline{M}_v = \frac{n_1 M_1^2 + n_2 M_2^2 + ...}{n_1 M_1 + n_2 M_2 + ...}$$

$$\overline{M}_v = \frac{\Sigma n_i M_i^2}{\Sigma n_i M_i}$$

Molecular weights determined by methods based on properties dependent on the mass of the particles are *the weight average molecular weights.*

According to the definitions set out here, $\overline{M}_w > \overline{M}_n$. The two are equal only when all particles are identical in weight.

SOLVED PROBLEM 1

A polymer mixture contains two polymers, one having molecular weight 100,000 and the other having molecular weight 60,000. The two components are present in equimolar concentration. Calculate the number average and the weight average molecular weights.

SOLUTION

$$\overline{M}_n = \frac{n_1 M_1 + n_2 M_2}{n_1 + n_2}$$

$$= \frac{1 \times 100,000 + 1 \times 60,000}{1 + 1} = \frac{160,000}{2}$$

$$= 80,000$$

This is the number average molecular weight of the polymer mixture.

$$\overline{M}_w = \frac{\Sigma n_i M_i^2}{\Sigma n_i M_i}$$

$$= \frac{1 \times (100,000)^2 + 1 \times (60,000)^2}{1 \times 100,000 + 1 \times 60,000}$$

$$= \frac{100 \times 10^8 + 36 \times 10^8}{160,000} = \frac{136 \times 10^8}{16 \times 10^4}$$

$$= \frac{136}{16} \times 10^4$$

$$= 85,000$$

Thus the weight average molecular weight is **85,000**.

DETERMINATION OF MOLECULAR WEIGHTS OF MACROMOLECULES

There are a number of methods available for the determination of molecular weight of macromolecules. Here, we will discuss the more important ones.

(1) Osmotic Pressure Method

The van't Hoff equation for dilute solutions may be written as :

$$p = \frac{c}{M} RT$$

or
$$\frac{p}{c} = \frac{RT}{M}$$

where p = osmotic pressure, pascal ; c = concentration of solution kg/m^3; R = gas constant, 8.314 J mol^{-1} K^{-1}; T = kelvin temperature ; M = molecular weight of the solute (polymer).

In actual determination of molecular weight of a high polymer, osmotic pressure (p) of a series of small concentrations (c) is measured with the help of a special Osmometer shown in Fig. 22.35. The plot of p/c against c is a straight line (Fig. 22.34).

It is extrapolated to zero concentration. This gives RT/M as the intercept from which the molecular weight can be calculated .

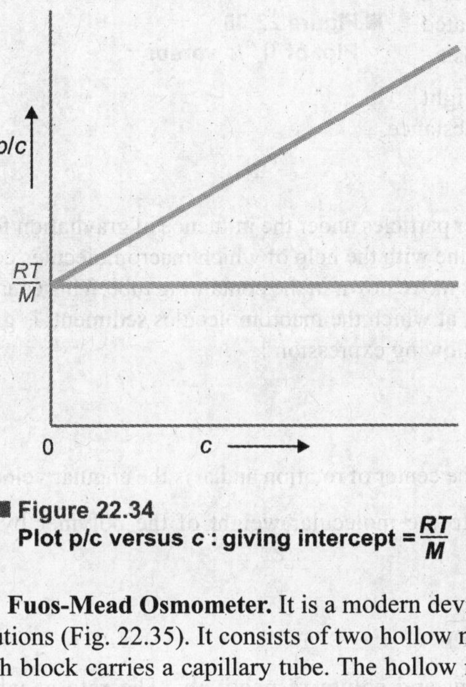

■ **Figure 22.34**
Plot p/c versus c : giving intercept = $\frac{RT}{M}$

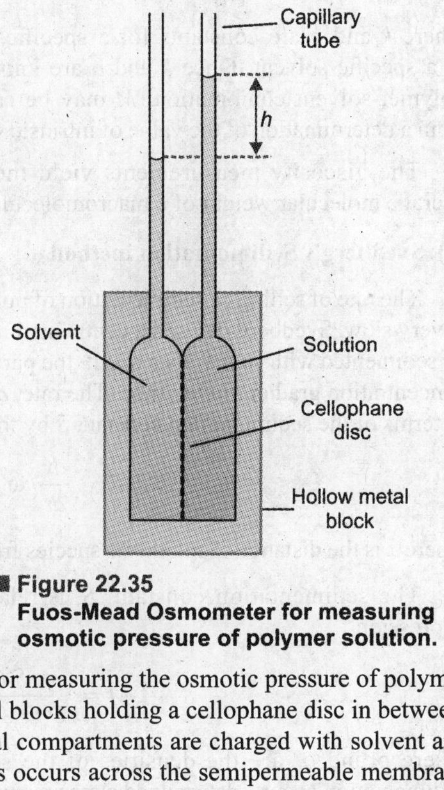

■ **Figure 22.35**
Fuos-Mead Osmometer for measuring osmotic pressure of polymer solution.

Fuos-Mead Osmometer. It is a modern device for measuring the osmotic pressure of polymer solutions (Fig. 22.35). It consists of two hollow metal blocks holding a cellophane disc in between. Each block carries a capillary tube. The hollow metal compartments are charged with solvent and solution through the side-tubes (not shown). Osmosis occurs across the semipermeable membrane (cellophane disc). The height of the solution in the capillary (h) is read off differentially to eliminate surface tension effect.

(2) Viscosity method

It is a very convenient method for determining the molecular weights of macromolecules in solution. The addition of macromolecules to a solvent increases its viscosity over that of pure solvent.

The **relative viscosity** of a solution of a polymer, denoted by η_r, is given by the expression

$$\eta_r = \frac{\eta}{\eta_0} \qquad ...(1)$$

where η is viscosity of solution and η_0 that of the solvent at the same temperature.

The **specific viscosity,** denoted by η_{sp}, is given by

$$\eta_{sp} = \eta_r - 1 \qquad ...(2)$$

In terms of (1) and (2), the intrinsic viscosity is defined as

$$[\eta] = \lim_{C \to 0} \left(\frac{\eta_{sp}}{c} \right)$$

where c is the concentration of the solute. The plot η_{sp}/c against η gives a straight line. Extrapolation to $c = 0$ yields the intrinsic viscosity.

It was shown by Staudinger that an empirical relationship exists between intrinsic viscosity $[\eta]$ and the molecular weight $[M]$ of the high polymer.

$$[\eta] = kM^a$$

where k and a are constants for a specific polymer in a specific solvent. Once k and a are known for a polymer-solvent combination, M may be calculated from a determination of the value of intrinsic viscosity.

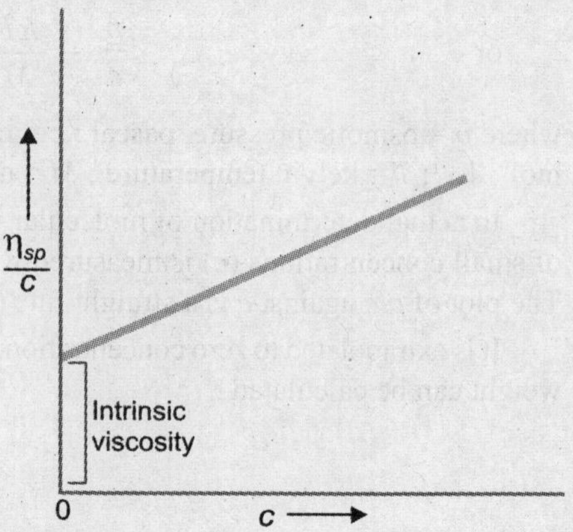

Figure 22.36
Plot of η_{sp}/c versus c.

The viscosity measurements yield the weight average molecular weight of a macromolecular substance.

(3) Svedberg's Sedimentation method

The rate of selling or sedimentation of polymer particles under the influence of gravitation force is very slow. Svedberg devised a centrifugal machine with the help of which macromolecules could be sedimented with speed. As a result, the particles move down in the containing tube. This causes a concentration gradient in the tube. The rate, dx/dt, at which the macromolecules sediment, is given in terms of the sedimentation constant S by the following expression :

$$S = \frac{dx}{dt} / \omega^2 x$$

where x is the distance of the solute species from the center of rotation and ω is the angular velocity.

The sedimentation constant, S, is related to the molecular weight of the polymer by the expression :

$$M = \frac{S\,RT}{D\,(1 - \rho/\rho')}$$

where ρ and ρ' are the densities of the solvent and solute respectively. The rate at which sedimentation occurs, determined experimentally, can thus be used for finding the molecular weight of macromolecules. Since the rate of sedimentation depends on the mass of the particle undergoing sedimentation, the molecular weight obtained by this method is the *weight average molecular weight,* $\overline{M}_w$.

(4) Sedimentation Equilibrium method

This method for determining molecular weight of a high polymer is quicker and convenient compared to method (3).

If a sol is whirled sufficiently long in an ultracentrifuge, a stage is reached at which the sol no longer settles. At this stage an equilibrium is reached between the centrifugal force and diffusion of the material in a direction opposite to the centrifugal force. If c_1 and c_2 be the concentrations of the particles at points x_1 and x_2 cm from the center of rotation, the molecular weight, M, of the high polymer is given by the relation

$$\ln \frac{c_2}{c_1} = \frac{M\omega^2(\rho - \rho')(x_2^2 + x_1^2)}{2RT\rho}$$

or
$$M = \frac{2RT\rho \ln c_2 / c_1}{\omega^2(\rho - \rho')(x_1^2 - x_2^2)}$$

By determining the concentrations c_1 and c_2 at the two levels x_1 and x_2 in the settling cell at sedimentation equilibrium, M can be calculated.

ADDITIONAL SOLVED PROBLEMS

SOLVED PROBLEM 2

Consider a polymer mixture composed of 5 molecules of molar mass 1 kg mol^{-1}, 5 molecules of molar mass 2 kg mol^{-1}, 5 molecules of molar mass 3 kg mol^{-1} and 5 molecules of molar mass 4 kg mol^{-1}. Calculate (i) the number average molecular weight and (ii) weight average molecular weight.

SOLUTION : **(*a*) To calculate the number average molecular weight**

Formula used

$$\overline{M_n} = \frac{\Sigma n_i M_i}{\Sigma n_i} = \frac{n_1 M_1 + n_2 M_2 + n_3 M_3 + n_4 M_4}{n_1 + n_2 + n_3 + n_4}$$

Quantities given

$M_1 = 1$ kg mol^{-1} $\quad M_2 = 2$ kg mol^{-1} $\quad M_3 = 3$ kg mol^{-1} $\quad M_4 = 4$ kg mol^{-1}

$n_1 = 5$ $\quad n_2 = 5$ $\quad n_3 = 5$ $\quad n_4 = 5$

Substitution of values

$$\overline{M_n} = \frac{1 \times 5 + 2 \times 5 + 3 \times 5 + 4 \times 5}{5 + 5 + 5 + 5} \text{ kg mol}^{-1}$$

$$= \frac{5 + 10 + 15 + 20}{20} \text{ kg mol}^{-1}$$

$$= \mathbf{2.5 \text{ kg mol}^{-1}}$$

(*b*) To calculate the weight average molecular weight

Formula used

$$\overline{M_w} = \frac{\Sigma m_i M_i}{\Sigma m_i} = \frac{w_1 M_1 + w_2 M_2 + w_3 M_3 + w_4 M_4}{m_1 + m_2 + m_3 + m_4}$$

Quantities given

$M_1 = 1$ kg mol^{-1} $\quad M_2 = 2$ kg mol^{-1} $\quad M_3 = 3$ kg mol^{-1} $\quad M_4 = 4$ kg mol^{-1}

$m_1 = 1$ $\quad m_2 = 2$ $\quad m_3 = 3$ $\quad m_4 = 4$

Substitution of values

$$\overline{M_w} = \frac{1 \times 1 + 2 \times 2 + 3 \times 3 + 4 \times 4}{1 + 2 + 3 + 4} \text{ kg mol}^{-1}$$

$$= \frac{30}{10} \text{ kg mol}^{-1}$$

$$= \mathbf{3 \text{ kg mol}^{-1}}$$

SOLVED PROBLEM 3

A sample of polymer contains 10, 20, 30 and 40 percent molecules of the polymer with molecular weights 10000, 12000, 14000 and 16000. Calculate the mole fraction of each type of polymer molecule. Also calculate the number and weight average molecular weights of the polymer sample.

SOLUTION : ***(a)* To calculate the mole fraction of each type of polymer**

Formula used

$$\textit{Mole fraction} \quad \overline{X_i} = \frac{\textit{Number of molecules of type i}}{\textit{Total number of molecules}}$$

Quantities given

No. of molecules of each type = 10, 20, 30 and 40

Total number of molecules = 100

Substitution of values

$$\text{Mole fraction of type } A = \frac{10}{100} = \mathbf{0.1}$$

$$\text{Mole fraction of type } B = \frac{20}{100} = \mathbf{0.2}$$

$$\text{Mole fraction of type } C = \frac{30}{100} = \mathbf{0.3}$$

$$\text{Mole fraction of type } D = \frac{40}{100} = \mathbf{0.4}$$

***(b)* To calculate the number average molecular weight**

Formula used

$$\overline{M_n} = \frac{\Sigma\, n_i M_i}{\Sigma\, n_i}$$

Quantities given

$n_1 = 0.1$ $\quad n_2 = 0.2$ $\quad n_3 = 0.3$ $\quad n_4 = 0.4$

$M_1 = 10000$ $\quad M_2 = 12000$ $\quad M_3 = 14000$ $\quad M_4 = 16000$

Substitution of values

$$\overline{M_n} = \frac{0.1 \times 10000 + 0.2 \times 12000 + 0.3 \times 14000 + 0.4 \times 16000}{0.1 + 0.2 + 0.3 + 0.4}$$

$$= 1000 + 2400 + 4200 + 6400$$

$$= \mathbf{14000}$$

***(c)* To calculate the weight average molecular weight.**

Formula used

$$\overline{M_w} = \frac{\Sigma\, X_i M_i^2}{\Sigma\, M_i}$$

$$= \frac{0.1 \times (10000)^2 + 0.2 \times (12000)^2 + 0.3 \times (14000)^2 + 0.4 \times (16000)^2}{14000}$$

$$= \frac{10^7 + 2.88 \times 10^7 + 5.88 \times 10^7 + 10.24 \times 10^7}{14000}$$

$$= \frac{20 \times 10^7}{14000}$$

$$= \mathbf{14286}$$

EXAMINATION QUESTIONS

1. Define or explain the following terms :
   (*a*) Catalyst
   (*b*) Homogeneous catalysis
   (*c*) Heterogeneous catalysis
   (*d*) Activation energy
   (*e*) Catalysis
   (*f*) Arrhenius equation
2. What are colloids? How are they classified? How would you prepare the colloidal solution of
   (*a*) gold
   (*b*) arsenious sulfide
   (*c*) sulfur
3. What is electrophoresis? How does this phenomenon provide information about the sign of charge on particles.
4. What are emulsions? How are they classified? How are they prepared?
5. (*a*) What are lyophillic colloids? Why are they called reversible colloids?
   (*b*) Why $MgCl_2$ is a better coagulant than KCl for As_2S_3 sol?
6. Discuss the origin of charge on colloidal particles. What is meant by electrical double layers? What is meant by Zeta potential?
7. (*a*) Explain the stability of colloids.
   (*b*) Write the difference between gels and emulsions.
   (*c*) Explain the Schulze-Hardy rule for coagulation.
8. State different methods of preparation of colloidal dispersions. Describe Bredig's arc method in detail.
9. (*a*) Explain Relative, Specific, and Intrinsic viscosities. How are they related to each other?
   (*b*) Describe the viscosity method for determining the molecular weight of a polymer.
10. (*a*) What do you understand by the number average and the weight average molecular weight of macromolecules?
    (*b*) Discuss in detail the sedimentation equilibrium method for determining the molar masses of polymers.
11. (*a*) Define number average and weight average molecular weights.
    (*b*) Describe osmometric method for the determination of molecular masses of polymers.
    (*c*) Write intrinsic viscosity-molecular weight relationship. Define the values of two constants appearing in the above relation when fractions of polymer of molecular weights 34000, 61000 and 130000 dissolved in an organic solvent gave the intrinsic viscosities 1.02, 1.60 and 2.75 respectively at 25°C.
12. (*a*) What are emulsions and what is emulsifying agent?
    (*b*) What are Micelles? Give examples.
13. (*a*) How is weight average and number average molecular weight ratio is used to predict the polydispersity in polymer systems?
    (*b*) "Even the 95% purity of chemicals is not sufficient for the synthesis of polymer chains containing more than 20 molecules". Justify the statement.

14. (*a*) Describe condensation polymerization and addition polymerization. Point out the type of polymerization process involved in the following :

(*i*) Polystyrene (*ii*) Nylon

(*iii*) Polyester (*iv*) Polyethylene

(*b*) Explain with examples the number average and weight average molecular weights of polymers. Which type of molecular weight is determined by the following techniques :

(*i*) Osmotic pressure method

(*ii*) Sedimentation equilibrium method (*Punjabi BSc*)

15. Discuss briefly the viscosity method for the determination of molecular weight of a polymer. (*Guru Nanak Dev BSc*)

16. For the determination of molecular weight of polymers how can the viscosity measurement be helpful? (*Panjab BSc*)

17. (*a*) What is the size range of colloidal particles? Distinguish between a molecular solution and a colloidal dispersion.

(*b*) Describe one method for the preparation of colloidal solutions. (*Mizoram BSc*)

18. Write a short note on 'origin of charge on colloidal particles'. (*Arunachal BSc*)

19. Give in details the methods of preparation of colloids. (*Sri Venkateswara BSc*)

20. Explain why :

(*a*) Alum is used in town water supply

(*b*) Alum is used in shaving

(*c*) Tyndall cone is formed when a beam of light is concentrated on a colloidal solution (*Agra BSc*)

21. Define gold number. The gold numbers of A, B, C and D are 0.005, 0.05, 0.5 and 5 respectively. Which of these has the greatest protective action. Explain your answer. (*Mysore BSc*)

22. Calculate the number average molecular weight of a sample of nylon having 1000 molecules of 5000 ; 2000 molecules of 6000 and 4000 molecules of 1000 molecular weight.

Answer. 3000 (*Sri Venkateswara BSc*)

23. Calculate the weight average molecular weight of a polymer containing equal number of particles with molecular weights 5000 and 10000.

Answer. 8333 (*Madurai BSc*)

24. Calculate weight average molecular weight of a sample containing equal number of particles with molecular weight 10,000 and 20,000.

Answer. 16667 (*Baroda BSc*)

25. Calculate number average molecular weight of a given sample of a polymer having 1000 molecules of 5000, 500 molecules of 4000 and 200 molecules of 3000 molecular weights.

Answer. 4470 (*Delhi BSc*)

26. (*a*) What is a protective colloid? How does a hydrophilic colloid stabilize a hydrophobic one? Give an account of Gold number in this context.

(*b*) What do you understand by the coagulation value of an electrolyte?

(*c*) What are gels? Give examples. (*Jammu BSc*)

27. (*a*) What is meant by peptization? Give a suitable example.

(*b*) Describe a method for purifying colloidal solutions.

(*c*) Discuss the function of a protective colloid. (*Agra BSc*)

28. (*a*) What is a colloid? Discuss the essential difference between lyophillic and lyophobic colloids.

(*b*) Write short notes on :

(*i*)Tyndall effect and Brownian movement

(*ii*)Electrophoresis and its applications (*Madras BSc*)

29. Give three points of difference between lyophobic and lyophillic colloids. (*Nagpur BSc*)

30. What are emulsions and gels? What distinguishes between elastic and non-elastic gels? (*Sri Venkateswara BSc*)

MULTIPLE CHOICE QUESTIONS

1. In true solutions, the diameter of the dispersed particles is in the range from
(a) 10^{-10} m to 10^{-9} m (b) 10^{-10} m to 10^{-8} m
(c) 10^{-8} m to 2×10^{-8} m (d) 2×10^{-8} m to 5×10^{-8} m
Answer. (a)

2. In a suspension the diameter of the dispersed particles is of the order
(a) 10^{-10} m (b) 10^{-8} m
(c) 10^{-7} m (d) 2×10^{-7} m
Answer. (d)

3. In a colloidal solution, the diameter of dispersed particle is in the range
(a) 10^{-10} m to 10^{-8} m (b) 10^{-10} m to 5×10^{-8} m
(c) 10^{-10} m to 10^{-7} m (d) 10^{-10} m to 2×10^{-7} m
Answer. (d)

4. A colloidal solution consists of
(a) a dispersed phase (b) a dispersion medium
(c) a dispersed phase in a dispersion medium (d) a dispersion medium in a dispersed phase
Answer. (c)

5. The sols in which the dispersed phase exhibits a definite affinity for the medium or the solvent is called
(a) lyophillic sols (b) lyophobic sols
(c) emulsions (d) hydrosols
Answer. (a)

6. The scattering of light by the dispersed phase is called
(a) Brownian movement (b) Tyndall effect
(c) adsorption (d) electrophoresis
Answer. (b)

7. In lyophobic sols, the dispersed phase has no ______ for the medium or solvent
(a) repulsion (b) attraction
(c) solvation (d) hydration
Answer. (b)

8. The lyophillic sols are
(a) reversible in nature
(b) irreversible in nature
(c) sometimes reversible sometimes nonreversible
(d) none of the above
Answer. (a)

9. The dispersal of a precipitated material into colloidal solution by the action of an electrolyte in solution is called
(a) coagulation (b) dialysis
(c) peptization (d) ultra-filtration
Answer. (c)

10. An arsenic sulfide sol (AS_2S_3) is prepared by the reaction

$$AS_2O_3 + H_2S \rightarrow AS_2S_3(\text{sol}) + 3H_2O$$

This method of preparing colloidal solution is
(a) reduction (b) oxidation
(c) hydrolysis (d) double decomposition
Answer. (d)

11. A sol of sulfur is produced by passing hydrogen sulfide into solution of sulfur dioxide

$$2H_2S + SO_2 \rightarrow 2H_2O + S$$

This method is known as

(*a*) reduction (*b*) oxidation
(*c*) hydrolysis (*d*) double decomposition

Answer. (*a*)

12. The process of removing ions (or molecules) from a sol by diffusion through a permeable membrane is called

(*a*) ultra-filtration (*b*) dialysis
(*c*) electrophoresis (*d*) osmosis

Answer. (*b*)

13. ______ do not show Tyndall effect

(*a*) true solution (*b*) colloidal solutions
(*c*) suspensions (*d*) none of these

Answer. (*a*)

14. The continuous rapid zig-zag movement executed by a colloidal particle in the dispersion medium is called

(*a*) Tyndall effect (*b*) Brownian movement
(*c*) electrophoresis (*d*) peptization

Answer. (*b*)

15. The explanation of Brownian movement was given by

(*a*) Robert Brown (*b*) Robert Boyle
(*c*) Albert Einstien (*d*) Tyndall

Answer. (*c*)

16. The movement of sol particles under an applied electric potential is called

(*a*) electrophoresis (*b*) electro-osmosis
(*c*) electrofiltration (*d*) none of these

Answer. (*a*)

17. The presence of the double layer in colloids accounts for

(*a*) kinetic properties (*b*) electrical properties
(*c*) optical properties (*d*) stability of colloids

Answer. (*b*)

18. If the sol particles in a given colloid moves towards the anode, they carry

(*a*) positive charge (*b*) negative charge
(*c*) no charge (*d*) sometimes positive sometimes negative charge

Answer. (*b*)

19. A sol of Ferric chloride moves to the negative electrode. The colloidal particles carry

(*a*) no charge (*b*) positive charge
(*c*) negative charge (*d*) none of these

Answer. (*b*)

20. The movement of the dispersion medium under the influence of applied potential is known as

(*a*) osmosis (*b*) diffusion
(*c*) electro-osmosis (*d*) electrophoresis

Answer. (*c*)

21. The flocculation and settling down of the discharged sol particles is called

(*a*) peptization (*b*) coagulation
(*c*) osmosis (*d*) diffusion

Answer. (*b*)

22. The precipitating effect of an ion in dispersed phase of opposite charge _______ with the valence of the ion

(*a*) decreases (*b*) increases

(*c*) no effect (*d*) none of these

Answer. (*b*)

23. The precipitating power of Al^{3+}, Na^+, Ba^{2+} is in the order

(*a*) $Na^+ > Ba^{2+} > Al^{3+}$ (*b*) $Ba^{2+} > Na^+ > Al^{3+}$

(*c*) $Al^{3+} > Na^+ > Ba^{2+}$ (*d*) $Al^{3+} > Ba^{2+} > Na^+$

Answer. (*d*)

24. The effectiveness of SO_4^{2-}, Cl^- and $[Fe(CN)_6]^{3-}$ ions for precipitating $Fe(OH)_3$ positive sol decreases in the order

(*a*) $[Fe(CN)_6]^{3-} > SO_4^{2-} > Cl^-$ (*b*) $Cl^- > SO_4^{2-} > [Fe(CN)_6]^{3-}$

(*c*) $SO_4^{2-} > Cl^- > [Fe(CN)_6]^{3-}$ (*d*) $[Fe(CN)_6]^{3-} > Cl^- > SO_4^{2-}$

Answer. (*a*)

25. The precipitating effect of an ion on dispersed phase increases with the valence of the precipitating ions. This rule is known as

(*a*) flocculation value rule (*b*) Hardy-Schulze rule

(*c*) Brownian rule (*d*) gold number rule

Answer. (*b*)

26. The charge on AS_2S_3 sol is due to

(*a*) absorption of H^+ ions (*b*) adsorption of H^+ ions

(*c*) adsorption of S^{2-} ions (*d*) absorption of S^{2-} ions

Answer. (*c*)

27. The number of milligrams of a hydrophilic colloid that will just prevent the precipitation of 1×10^{-5} m^3 of gold sol on addition of 1×10^{-6} m^3 of 10 per cent sodium chloride solution is called

(*a*) gold number (*b*) charge number

(*c*) absorption number (*d*) silver number

Answer. (*a*)

28. The _______ the gold number of a hydrophilic colloid, the greater is its protective power

(*a*) higher (*b*) lower

(*c*) constant (*d*) none of these

Answer. (*b*)

29. The charge on the sol particles is due to _______ of ions common to the particles from the dispersion medium

(*a*) selective adsorption (*b*) selective coagulation

(*c*) selective peptization (*d*) selective absorption

Answer. (*a*)

30. An emulsion is a colloidal solution of a _______ dispersed in another liquid

(*a*) solid (*b*) liquid

(*c*) gas (*d*) medium

Answer. (*b*)

31. A gel is a colloidal system in which a _______ is dispersed in a _______ medium.

(*a*) solid, liquid (*b*) liquid, solid

(*c*) liquid, liquid (*d*) liquid, gas

Answer. (*b*)

32. The human kidneys purify the blood by _______ through natural membranes

(*a*) osmosis (*b*) diffusion

(*c*) dialysis (*d*) emulsification

Answer. (*c*)

33. Milk is an example of

(*a*) sol (*b*) gel

(*c*) emulsion (*d*) true solution

Answer. (*c*)

34. The function of an emulsifier is to

(*a*) coagulate a colloidal solution (*b*) stabilize an emulsion

(*c*) stabilize a sol (*d*) electrify a colloidal solution

Answer. (*b*)

35. Ice cream is a dispersion of _______ in cream

(*a*) water (*b*) water vapor

(*c*) ice (*d*) cooling agent

Answer. (*c*)

36. The function of alum used for the purification of water is to

(*a*) coagulate the sol particles (*b*) disperse the sol particles

(*c*) emulsify the sol particles (*d*) absorb the sol particles

Answer. (*a*)

37. The macromolecules possess

(*a*) high viscosity (*b*) high molecular weights

(*c*) show a weak Tyndall effect (*d*) all of these

Answer. (*d*)

38. The blue color of the sky is due to

(*a*) Brownian movement (*b*) Tyndall effect

(*c*) the presence of macromolecules (*d*) electrophoresis

Answer. (*b*)

39. The weight average molecular mass of macromolecules is _______ than the number average molecular mass

(*a*) greater (*b*) lesser

(*c*) equal (*d*) none of these

Answer. (*a*)

40. Macromolecules in solutions do not show

(*a*) electrophoresis (*b*) osmosis

(*c*) Tyndall effect (*d*) Brownian movement

Answer. (*a*)

41. In osmosis pressure method for determination of molecular weight of macromolecules, a plot of p/c versus $\frac{RT}{M}$ is a _______

(*a*) straight line (*b*) curved line

(*c*) sine curve (*d*) none of these

Answer. (*a*)

42. Fog is an example of colloidal system of

(*a*) liquid dispersed in a liquid (*b*) solid dispersed in a solid

(*c*) gas dispersed in a liquid (*d*) liquid dispersed in a gas

Answer. (*d*)

43. Gold number gives

(*a*) the number of gold atoms present is one gram of gold

(*b*) the number of gold atoms required to coagulate one g of the colloidal solution

(*c*) the number of gold atoms present in one gram of gold alloy

(*d*) none of the above

Answer. (*d*)

44. Which of the following does not show Tyndall effect?

(*a*) colloidal solution (*b*) isotonic solution

(*c*) both of these (*d*) none of these

Answer. (*b*)

45. Which of the following has minimum flocculating power?

Pb^{4+}, Al^{3+}, Ba^{2+}, Na^{+}

(*a*) Na^{+} (*b*) Ba^{2+}

(*c*) Al^{3+} (*d*) Pb^{4+}

Answer. (*a*)

46. Smoke is an example of

(*a*) solid dispersed in solid (*b*) solid dispersed in liquid

(*c*) solid dispersed in gas (*d*) gas dispersed in solid

Answer. (*c*)

47. Which of the following is not a colloidal solution?

(*a*) brine solution (*b*) fog

(*c*) smoke (*d*) butter

Answer. (*a*)

48. Which of the following is the most effective in causing the coagulation of ferric hydroxide sol?

(*a*) NaCl (*b*) Na_2SO_4

(*c*) $K_2\ SO_4$ (*d*) $K_3\ [Fe\ (CN)_6]$

Answer. (*d*)

49. Peptization involves

(*a*) digestion of food (*b*) precipitation of colloidal solution

(*c*) hydrolysis of proteins (*d*) breaking of a precipitate to colloidal state

Answer. (*d*)

50. The cleansing action of soap is due to

(*a*) hydrolysis of salt present in soap (*b*) ionization of salt present in soap

(*c*) high molecular mass of soap (*d*) emulsification properties of soap

Answer. (*d*)

23

Adsorption

CONTENTS

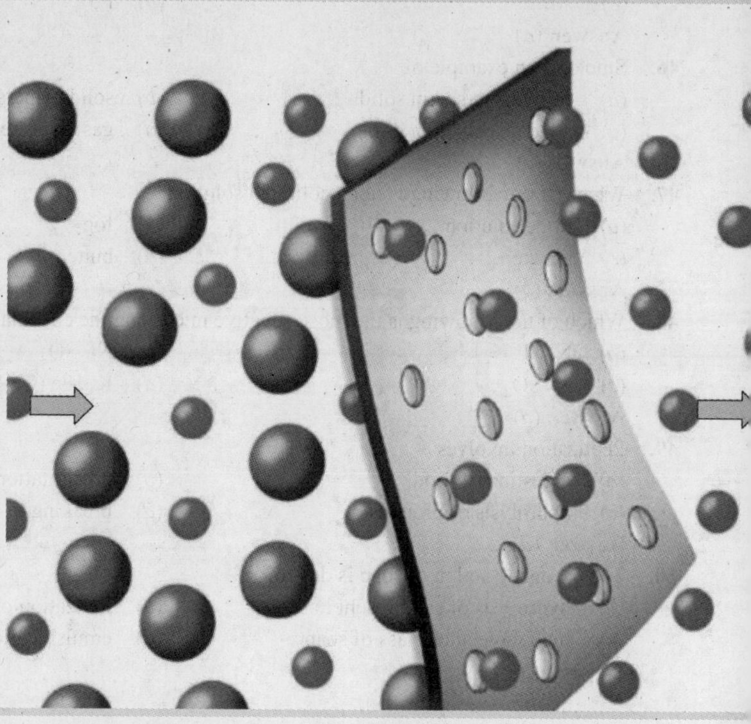

When a solid surface is exposed to a gas or a liquid, molecules from the gas or the solution phase accumulate or concentrate at the surface.

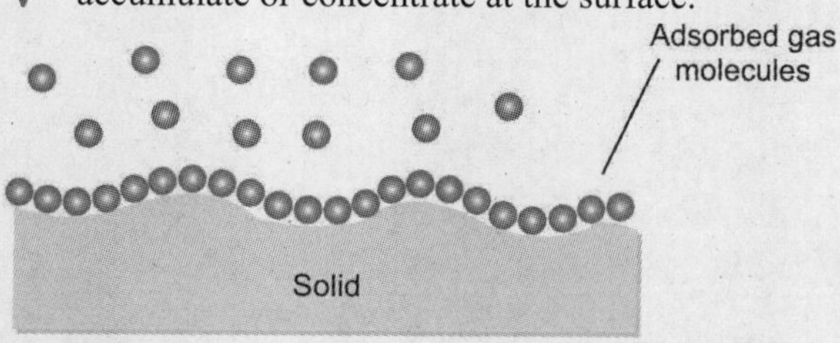

■ **Figure 23.1**
Adsorption of a gas at a solid surface.

The phenomenon of concentration of molecules of a gas or liquid at a solid surface is called adsorption.

The substance that deposits at the surface is called **Adsorbate** and the solid on whose surface the deposition occurs is called the **Adsorbent.**

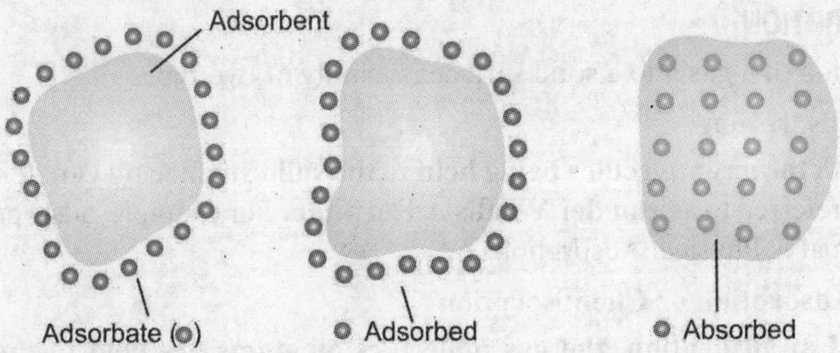

■ **Figure 23.2**
Adsorption *versus* absorption (illustration).

Examples of Adsorption

(1) Adsorption of a dye by a charcoal. If finely divided charcoal is stirred into a dilute solution of *methylene blue* (an organic dye), the depth of color of the solution decreases appreciably. The dye molecules have been adsorbed by charcoal particles.

(2) Adsorption of a gas by charcoal. If a gas (SO_2, Cl_2, NH_3) is treated with powdered charcoal in a closed vessel, the gas pressure is found to decrease. The gas molecules concentrate on charcoal surface and are said to be adsorbed.

Adsorption versus Absorption

The term 'adsorption' must be carefully distinguished from another like-sounding term 'absorption'. **While adsorption implies deposition at the surface only, absorption implies penetration into the body of the solid** (Fig. 23.2). For illustration a chalk crayon when dipped in ink adsorbs the later and on breaking it is found to be white from within. On the other hand, water is absorbed by a sponge and is distributed throughout the sponge uniformly.

Both adsorption and absorption often take place side by side. It is thus difficult to distinguish between the two processes experimentally. Mc Bain introduced the general term **Sorption** which includes both *the adsorption and absorption.*

Mechanism of Adsorption

Atoms or molecules of a solid surface behave like the surface molecules of a liquid. These are not surrounded by atoms or molecules of their kind. Therefore, they have unbalanced or residual attractive forces on the surface which can hold adsorbate particles.

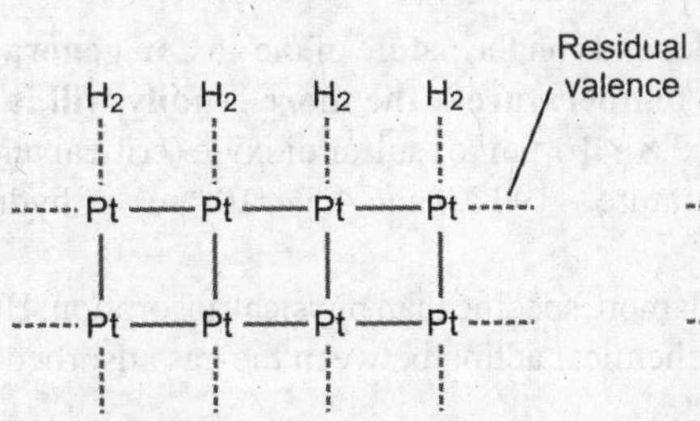

■ **Figure 23.3**
Hydrogen adsorbed molecularly on platinum.

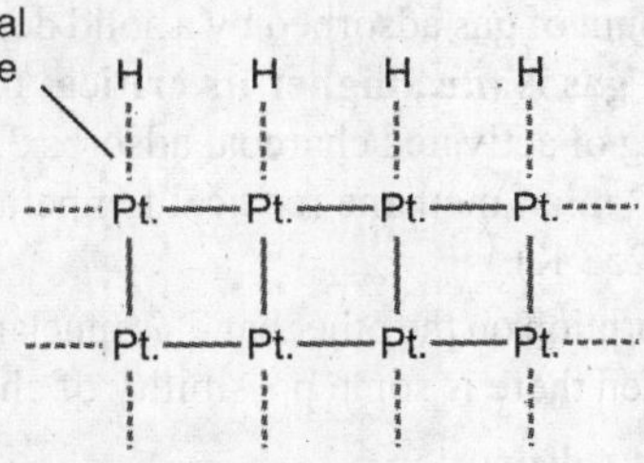

■ **Figure 23.4**
Hydrogen adsorbed atomically on platinum.

The adsorbed atoms or molecules can be held on the surface of a metal such as platinum (Pt) by physical van der Waal's force or chemical forces due to residual valence bonds. Thus the adsorption of hydrogen on platinum may take place in two ways (molecularly or atomically as shown above).

TYPES OF ADSORPTION

The adsorption of a gas into a solid surface is mainly of two types :

(*a*) Physical Adsorption

This is due to the gas molecules being held to the solid surface by van der Waal's attractive forces. It is also referred to as **van der Waal's Adsorption.** For example, adsorption of hydrogen or oxygen on charcoal is Physical Adsorption.

(*b*) Chemical Adsorption or Chemisorption

In this kind of adsorption, the gas molecules or atoms are held to the solid surface by chemical bonds. These bonds may be covalent or ionic in nature. For example, hydrogen is chemisorbed on nickel. Hydrogen molecules is first adsorbed by van der Waal's forces and then dissociates. The hydrogen atoms are thus chemisorbed on nickel.

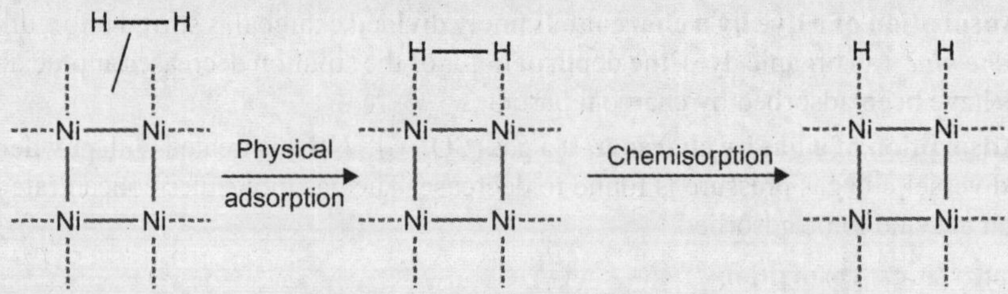

■ **Figure 23.5**
Hydrogen dissociates before it is chemisorbed on nickel.

Often adsorption is a combination of the two types of adsorption stated above.

ADSORPTION OF GASES BY SOLIDS

The adsorption of gases by solid adsorbents has certain characteristic feature. Physical adsorption and chemisorption are found to differ in many respects.

(1) Adsorption and Surface area

Adsorption being a surface phenomenon, the extent of adsorption depends on the surface area. **Increase in the surface area of the adsorbent, increases the total amount of the gas adsorbed.** Thus finely divided metals (nickel, platinum) and porous substances (charcoal, silica gel) provide a large surface area and are best solid adsorbents.

(2) Nature of Gas

The amount of gas adsorbed by a solid depends on the nature of the gas. **In general, more easily liquefiable a gas is (*i.e.*, higher its critical temperature), the more readily will it be adsorbed.** Thus 0.001 kg of activated charcoal adsorbs 3.8×10^{-4} m^3 of sulfur dioxide (critical temperature 430 K), 1.6×10^{-5} m^3 of methane (critical temperature – 190 K) and 4.5×10^{-6} m^3 of hydrogen (critical temperature 253 K).

Chemisorption on the other hand, is much more specific than physical adsorption. However, it will not occur when there is some possibility of chemical action between the gas adsorbed and the solid.

(3) Heats of Adsorption

Heat of adsorption is defined as the energy liberated when 1 gm mole of a gas is adsorbed on the solid surface. In physical adsorption, gas molecules concentrate on the solid surface. Thus it is similar to the condensation of a gas to liquid. Therefore, **adsorption like condensation is an exothermic process.** Since the attractions between gas molecules and solid surface are due to relatively weak van der Waal's forces, heats of adsorption are small (about 20.93 kJ mol^{-1}).

In chemisorption the attractive forces are due to the formation of true chemical bonds. Therefore, the heats of adsorption are large (83.74 to 418.68 kJ mol^{-1}).

(4) Reversible character

Physical adsorption is a reversible process. The gas adsorbed onto a solid can be removed (desorbed) under reverse conditions of temperature and pressure. Thus,

$$\text{Gas} \rightleftharpoons \text{Gas/Solid} + \text{Heat}$$

Chemisorption, on the contrary, is not reversible because a surface compound is formed.

(5) Effect of temperature

Physical adsorption occurs rapidly at low temperature and decreases with increasing temperature (Le Chatelier's Principle).

Chemisorption, like most chemical changes, generally increase with temperature. Thus a rise of temperature can often cause physical adsorption to change to chemisorption. Nitrogen, for example, is physically adsorbed on iron at 463 K but chemisorbed to form a nitride at 773 K.

(6) Effect of pressure

Since a dynamic equilibrium exists between the adsorbed gas and the gas in contact with the solid as stated in (4), Le Chatelier's Principle is applied. Actually it has been found that **increase of pressure leads to increase of adsorption and decrease of pressure causes desorption.**

(7) Thickness of Adsorbed layer of gas

From a study of the isotherms relating to the amount of gas adsorbed to the equilibrium pressure, Langmuir showed that at low pressure, the physically adsorbed gas forms only one molecular thick layer. However, above a certain pressure, multimolecular thick layer is formed.

COMPARISON OF PHYSICAL ADSORPTION AND CHEMISORPTION

Physical adsorption	Chemisorption
1. Caused by intermolecular van der Waal's forces.	**1.** Caused by chemical bond formation.
2. Depends on nature of gas. Easily liquefiable gases are adsorbed readily.	**2.** Much more specific than physical adsorption.
3. Heat of adsorption is small (about 20.93 kJ mol^{-1}).	**3.** Heat of adsorption is large (83.74–418.68 kJ mol^{-1}).
4. Reversible.	**4.** Irreversible.
5. Occurs rapidly at low temperature; decreases with increasing temperature.	**5.** Increases with increase of temperature.
6. Increase of pressure increases adsorption; decrease of pressure causes desorption.	**6.** Change of pressure has no such effects.
7. Forms multimolecular layers on adsorbent surface.	**7.** Forms unimolecular layer.

In chemisorption, the adsorbed layer of gas is one-molecule thick since chemical combination can take place with the adsorbent surface only directly.

ADSORPTION ISOTHERMS

The adsorption of a gas on a solid adsorbent in a closed vessel is a reversible process.

$$\text{Free Gas} \rightleftharpoons \text{Gas adsorbed on solid}$$

The amount of the gas adsorbed depends on equilibrium pressure (P) and temperature.

The relationship between the equilibrium pressure of a gas and its amount adsorbed on the solid adsorbent at any constant temperature is called an Adsorption isotherm. It may be given in the form of an equation or graphical curve.

Freundlich Adsorption Isotherm

Freundlich proposed an empirical relation in the form of a mathematical equation.

$$\frac{w}{m} = k\ P^{1/n}$$

where w is the mass of the gas adsorbed on a mass m of adsorbent at a pressure P; k and n are constants depending on the nature of the gas and the adsorbent and on temperature. This relation is generally represented in the form of a curve obtained by plotting the mass of the gas adsorbed per unit mass of adsorbent (w/m) against equilibrium pressure.

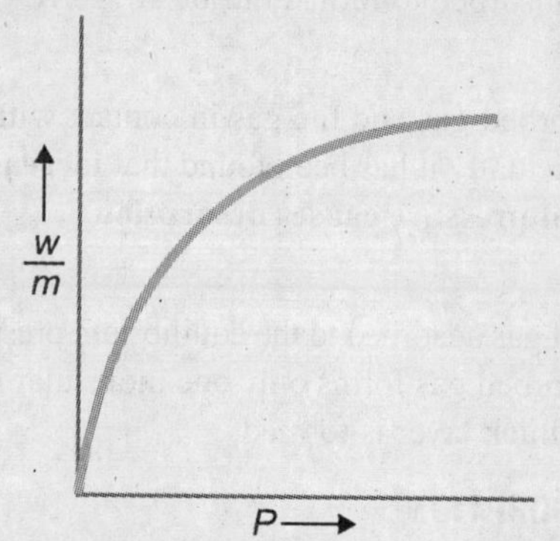

Figure 23.6
Freundlich isotherm; a plot of mass of adsorbed gas per unit mass of adsorbent.

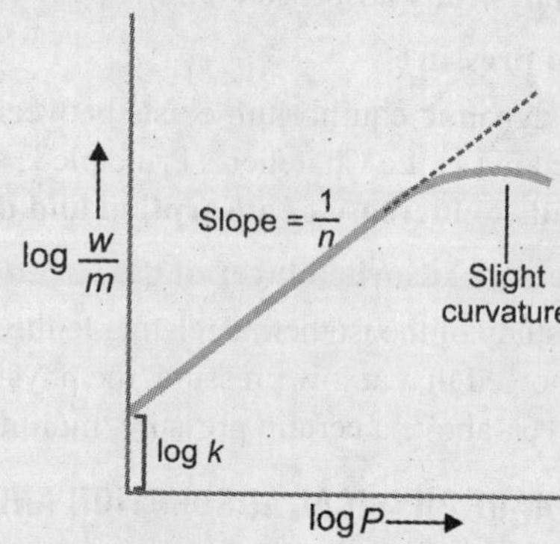

Figure 23.7
Plot of log w/m against log P shows slight curvature at higher pressures.

Freundlich isotherm is not applicable at high pressures. Taking logarithms on both sides of Freundlich equation, we have

$$\log \frac{w}{m} = \log k + \frac{1}{n} \log P$$

This is equation for a straight line. Thus a plot of log (w/m) against log P should be a straight line with slope $1/n$ and intercept log k. However, it is actually found that the plots were straight lines at low pressures, while at higher pressure they showed a slight curvature, especially at low temperature. This indicated that Freundlich equation is approximate and does not apply to adsorption of gases by solids at higher pressures.

LANGMUIR ADSORPTION ISOTHERM

Langmuir (1916) derived a simple adsorption isotherm based on theoretical considerations. It was named after him.

Assumptions

Langmuir made the following assumptions.

(1) The layer of the gas adsorbed on the solid adsorbent is one-molecule thick.

(2) The adsorbed layer is uniform all over the adsorbent.

(3) There is no interaction between the adjacent adsorbed molecules.

Derivation of Langmuir Isotherm

Langmuir considered that the gas molecules strike a solid surface and are thus adsorbed.

Some of these molecules then evaporate or are 'desorbed' fairly rapidly. A dynamic equilibrium is eventually established between the two opposing processes, adsorption and desorption.

If θ is the fraction of the total surface covered by the adsorbed molecules, the fraction of the naked area is (1 – θ). **The rate of desorption (R_d) is proportional to the covered surface θ. Therefore,**

$$R_d = k_d\,\theta$$

where k_d is the rate constant for the desorption process.

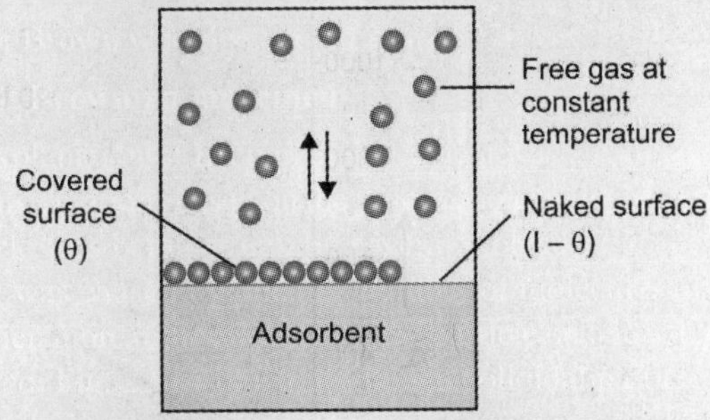

Figure 23.8
Dynamic equilibrium exists between free molecules and those adsorbed on the fraction of adsorbent surface.

The rate of adsorption (R_a) is proportional to the available naked surface (1 – θ) and the pressure (*P*) of the gas.

$$R_a = k_a\,(1-\theta)\,P$$

where k_a is rate constant for the adsorption process.

At equilibrium the rate of desorption is equal to the rate of adsorption. That is,

$$k_d\,\theta = k_a\,(1-\theta)\,P$$

or $$\theta = \frac{k_a P}{K_d + k_a P}$$

or $$\theta = \frac{(k_a / k_d)\,P}{1 + (K_a / k_d)\,P}$$

or $$\theta = \frac{K P}{1 + K P}$$

where *K* is the equilibrium constant and is referred to as the **adsorption coefficient.**

The amount of the gas adsorbed per gram of the adsorbent, *x*, is proportional to θ.

Hence, $$x \propto \frac{K P}{1 + K P}$$

or $$x = K' \frac{K P}{1 + K P} \qquad \text{...(1)}$$

where *K'* is a new constant. Equation (1) gives the relation between the amount of gas adsorbed to the pressure of the gas at constant temperature and is known as **Langmuir Adsorption isotherm.**

In order to test the Langmuir isotherm, equation (1) is rearranged so that

$$\frac{P}{x} = \frac{1}{K'} + \frac{P}{K''} \qquad \text{...(2)}$$

where *K"* constant = *K'*/*K*.

The equation (2) is similar to an equation for a straight line. Thus if *P*/*x* is plotted against *P*, we should get a straight line with slope 1/*K"* and the intercept 1/*K'*. It was found in most cases that the actual curves were straight lines. Thus Langmuir isotherm stood verified.

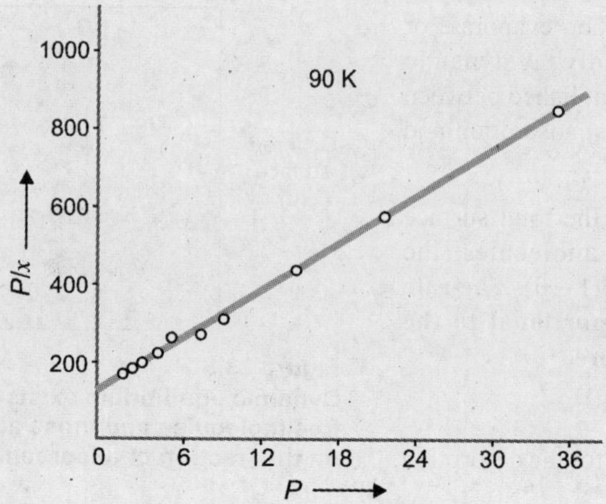

Figure 23.9
Verification of Langmuir isotherm for adsorption of N_2 on mica at 90 K.

Langmuir Isotherm holds at low pressures but fails at high pressures

As stated above, Langmuir Adsorption isotherm may be written as

$$\frac{P}{x} = \frac{1}{K'} + \frac{P}{K''}$$

If the pressure (P) is very low, the factor P/K'' may be ignored and the isotherm assumes the form

$$x = K'P \qquad \textit{(at low pressure)}$$

If the pressure (P) is very high, the factor $1/K'$ may be ignored and the isotherm becomes

$$x = K'' \qquad \textit{(at high pressure)}$$

Hence, at low pressures, the amount of gas adsorbed (x) is directly proportional to pressure (P). At high pressures the mass adsorbed reaches a constant value K'' when the adsorbent surface is completely covered with a unimolecular layer of the gas. At this stage adsorption is independent of pressure.

ADSORPTION OF SOLUTES FROM SOLUTIONS

Porous or finely divided solid substances can also adsorb dissolved substances from solution. Thus activated charcoal is used to remove colored impurities from solutions. Charcoal will also adsorb many dyestuffs. When a solution of acetic acid is shaken with activated charcoal, part of the acid is removed by adsorption and concentration of solution is decreased. Again, precipitates obtained in qualitative analysis often act as absorbents. For example, magnesium hydroxide when precipitated in the presence of the dye-stuff *magneson* forms a blue 'lake'.

Adsorption from solution generally follows the same principles as laid down for adsorption of gases by solids and is subject to the same factors. Thus,

(1) Some adsorbents specifically adsorb certain solutes more effectively than others.

(2) An increase of temperature decreases the extent of adsorption.

(3) An increase in surface area increases the extent of adsorption.

(4) Adsorption of solutes also involves the establishment of an equilibrium between the amount adsorbed and the concentration of the solute in solution.

The precise mechanism of adsorption from solution is not clear. However there is a limit to the adsorption by a given mass of adsorbent and hence possibly adsorption takes place unless a unimolecular layer is formed. **Freundlich Isotherm, using concentration instead of pressure is obeyed by adsorption from solution.**

That is,

$$\frac{w}{m} = k \times C^{\frac{1}{n}}$$

where w = mass of solute adsorbed on a mass m of adsorbent ; C = equilibrium concentration of the solution ; and k and n are constants. Taking logs of the above equation

$$\log \frac{w}{m} = \log k + \frac{1}{n} \log C$$

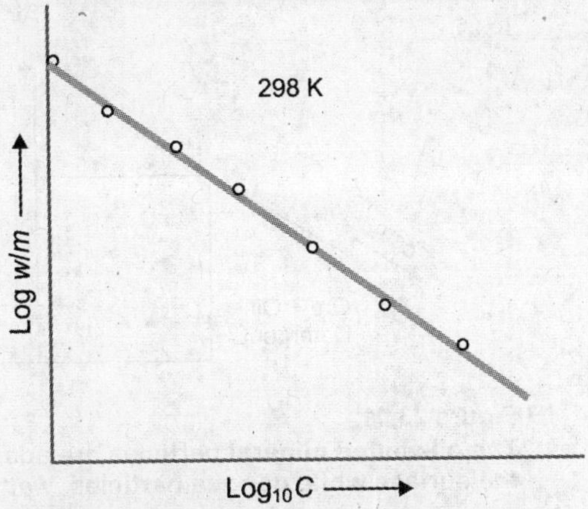

■ **Figure 23.10**
Verification of Freundlich equation by applying it to adsorption of acetic acid on charcoal at 298 K.

This implies that a plot of log w/m against log C should be a straight line. The validity of Freundlich isotherm has been tested by plotting the experimental values of log w/m versus log C determined for adsorption of acetic acid on charcoal at 298 K.

APPLICATIONS OF ADSORPTION

Adsorption finds numerous applications both in the laboratory and industry. Some of these are listed below.

(1) Production of high vacua

If a partially evacuated vessel is connected to a container of activated charcoal cooled with liquid air, the charcoal adsorbs all the gas molecules in the vessel. This results in a very high vacuum. This process is used in high vacuum equipments as Dewar flask for storage of liquid air or liquid hydrogen. Silica gel is also useful as an adsorbent in production of high vacua.

(2) Gas mask

All gas masks are devices containing an adsorbent (activated charcoal) or a series of adsorbents. These adsorbents remove poisonous gases by adsorption and thus purify the air for breathing.

(3) Heterogeneous catalysis

In heterogeneous catalysis, the molecules of reactants are adsorbed at the catalyst surface where they form an 'adsorption complex'. This decomposes to form the product molecules which then take off from the surface.

(4) Removal of coloring matter from solutions

Animal charcoal removes colors of solutions by adsorbing colored impurities. Thus in the manufacture of cane-sugar, the colored solution is clarified by treating with animal charcoal or activated charcoal.

(5) Froth Flotation process

The low grade sulfide ores (PbS, ZnS, Cu_2S) are free from silica and other earthy matter by *Froth Flotation Process.* The finely divided ore is mixed with oil (pine oil) and agitated with water containing a detergent (foaming agent). When air is bubbled into this mixture, the air bubbles are stabilized by the detergent. These adsorb mineral particles wetted with oil and rise to the surface. The earthy matter wetted by water settles down at the bottom.

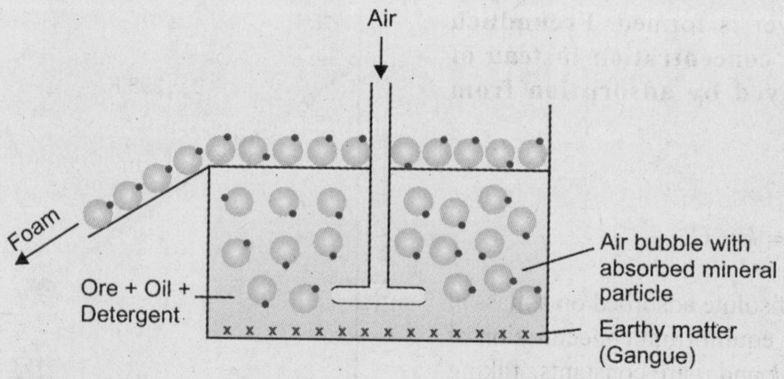

■ **Figure 23.11**
The oil-wetted mineral particles are adsorbed by stabilized air-bubbles which rise to the surface while gangue particles wetted by water settle down.

(6) Chromatographic analysis

Mixtures of small quantities of organic substances can be separated with the help of *Chromatography* which involves the principles of selective adsorption.

The mixture is dissolved in a suitable solvent (*hexane*) and poured through a tube containing the adsorbent (alumina). The component most readily adsorbed is removed in the upper part of the tube. The next most readily adsorbed component is removed next, and so on. Thus the material is separated into 'bands' in different parts of the tube. Now pure solvent is poured through the tube. Each component dissolved in the solvent comes down by turn and is collected in a separate receiver.

Mixtures of gases can be separated by selective adsorption of gases by liquids (*Gas chromatography*).

ION–EXCHANGE ADSORPTION

In recent years, many synthetic resins have been made which function as ion-exchangers. In effect, the resin has one ion adsorbed on it. The resin releases this ion and adsorb another like ion. The process is called **ion-exchange adsorption**. When cations are exchanged, the resin is known as **cation exchanger**. When anions are exchanged, it is referred to as **anion exchanger** .

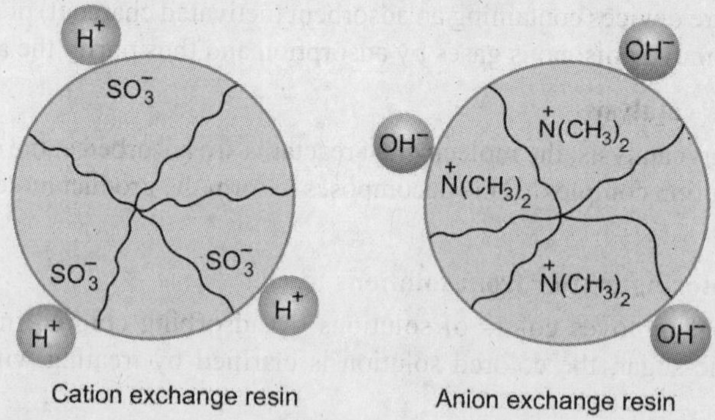

■ **Figure 23.12**
Macromolecules of ion-exchange resins.

Cationic exchange

The cationic exchangers are high polymers containing acidic groups such as sulfonic acid group, $–SO_3$, H. The resulting macro-anion has adsorbed H^+ ions. When solution of another cation (Na^+) is allowed to flow over it, H^+ ions are exchanged for Na^+ ions. This process in fact, consist of desorption of H^+ ions and adsorption of Na^+ ions by the resin.

$$\underset{\text{cationic resin}}{R^-H^+} + Na^+ \rightleftharpoons \underset{\text{sodium 'salt'}}{R^-Na^+} + H^+$$

Since the above cationic exchange is reversible, the sodium 'salt' upon treatment with an acid regenerates the original resin.

Anionic exchange

A resin containing a basic group such as quaternary ammonium hydroxide, $-N^+R_3O^-H$, will act as anion exchanger. It will, for example, exchange OH^- ion for Cl^-.

$$\underset{\text{anionic resin}}{R^+OH^-} + Cl^- \rightleftharpoons \underset{\text{resin 'chloride'}}{R^+Cl^-} + OH^-$$

The original anion exchange resin can be regenerated by treatment of the resin 'chloride' with a base (OH^- ions).

APPLICATIONS OF ION–EXCHANGE ADSORPTION

Ion-exchange adsorption has many useful applications in industry and medicine.

(1) Water softening

Hard water contains Ca^{2+} ions and Mg^{2+} ions. These form insoluble compound with soap and the later does not function as detergent. Hard water is softened by passing through a column packed with sodium cation-exchanger resin, R^-Na^+. The Ca^{2+} and Mg^{2+} ions in hard water are replaced by Na^+ ions.

$$2R^-Na^+ + Ca^{2+} \longrightarrow R_2^-Ca^{2+} + 2Na^+$$

(2) Deionization of water

Water of very high purity can be obtained by removing all dissolved salts. This is accomplished by using both a cation and anion exchange resin. The water free from all ions (cations and anions) is referred to as **Deionized or Demineralized water**.

The water is first passed through a column containing a cation-exchange resin, R^-H^+. Here any cations in water (say Na^+) are removed by exchange for H^+. The water is then passed through a second column packed with an anion-exchanger, R^+OH^-. Any anions (Cl^-) are removed by exchange of OH^- for Cl^-.

$$R^-H^+ + Na^+ \longrightarrow R^-Na^+ + H^+ \quad \textit{(First Column)}$$

$$R^+OH^- + Cl^- \longrightarrow R^+Cl^- + OH^- \quad \textit{(Second Column)}$$

The H^+ and OH^- ions thus produced react to form water.

$$H^+ + OH^- \longrightarrow H_2O$$

Thus the water coming out of the second column is entirely free from ions, whether cations or anions. The water is purer than distilled water and is called **Conductivity water**.

In another process, which is more common way, the tap water is passed into a column containing both types of resin (cation and anion exchanger). Here cations and anions are removed simultaneously.

Electrical demineralization of water

Ion-exchange resins supported on paper or fiber can be used as membranes through which only cations or anions will pass. Such membranes are used in electrical demineralization of water and they act as ionic sieves (Fig. 23.13). Upon application of the electric current, cations move through the cation-exchanger membrane to the negative electrode. The anions move in the opposite direction through the anion exchange membrane. Thus the water in the middle compartment is demineralized.

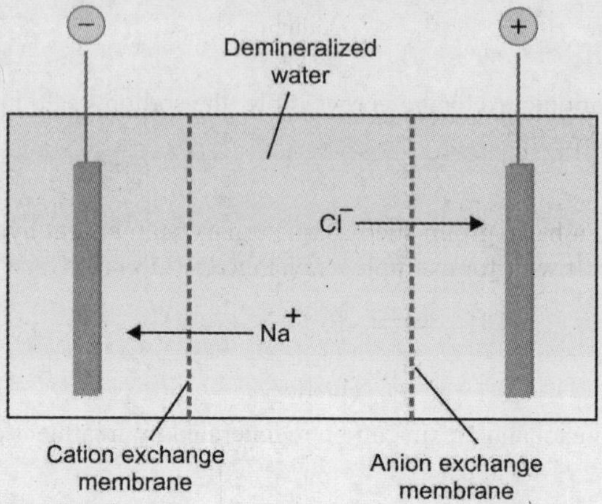

Figure 23.13
Electrical demineralization of saline water.

(3) Medical uses

Excess sodium salts can be removed from the body fluids by giving the patient a suitable ion-exchanger to eat. Weakly basic anion-exchangers are used to remove excess acid or 'acidity' in the stomach.

EXAMINATION QUESTIONS

1. Define or explain the following terms :
   (*a*) Adsorption
   (*b*) Physical Adsorption
   (*c*) Chemical Adsorption
   (*d*) Freundlich adsorption isotherm
   (*e*) Langmuir adsorption isotherm
2. What is adsorption? Define the terms 'adsorbent' and adsorbate' giving suitable examples. Describe the phenomenon of the adsorption of solids from a solution.
3. What is the effect of temperature on adsorption of gases on solids?
4. Write the assumptions of Langmuir adsorption isotherm and derive the equation pertaining to it.
5. (*a*) Write the main points of Langmuir's theory of adsorption.
   (*b*) Draw adsorption isobars for physical adsorption and chemical adsorption.
6. Discuss Langmuir theory of adsorption and derive expression for Langmuir monolayer adsorption isotherm.
7. Write what do you understand by the term adsorption. Give four points of differences between physical adsorption and chemical adsorption.
8. Derive Langmuir's adsorption isotherm stating the assumptions on which it is based. Show that for a moderate range of pressures it reduces to Freundlich adsorption isotherm.
9. Distinguish between physical adsorption and chemical adsorption. What are adsorption isobars? (*Madurai BSc*)
10. Distinguish between physical adsorption and chemical adsorption. (*Delhi BSc*)

11. Discuss Freundlich adsorption isotherm of a gas on a solid surface. How are the constants in this isotherm equation determined? How will you prove that Langmuir adsorption isotherm is superior to Freundlich adsorption isotherm? *(Agra BSc)*

12. Write down the Langmuir's adsorption isotherm. Also write two objections in Langmuir's theory. *(Mysore BSc)*

13. (*a*) In a particular experiment it is required to have a large amount of gas absorbed on the surface of a solid. Suggest a few factors which may be helpful alongwith suitable explanations.
    (*b*) What signs of ΔH and ΔS in the case of physical adsorption are expected? Justify your answer.
    (*c*) Write a note on chemisorption. *(Panjab BSc)*

14. What are the postulates of Langmuir adsorption isotherm? On the basis of these postulates, derive Langmuir equation. How it can be used to determine the surface area of an adsorbent? *(Baroda BSc)*

15. (*a*) What do you understand by the term adsorption? What are the factors which affect adsorption?
    (*b*) Discuss the effect of temperature and pressure on the adsorption of a gas on a solid surface. *(Lucknow BSc)*

16. Show that Freundlich isotherm is a special case of Langmuir isotherm. *(Jamia Millia BSc)*

17. What are the postulates of Langmuir theory of adsorption? Derive Langmuir adsorption equation in the form $y = \dfrac{ap}{1+bp}$. How is this equation verified? *(Jamia Millia BSc)*

18. (*a*) Distinguish between physisorption and chemisorption.
    (*b*) How is Langmuir adsorption isotherm related to Freundlich's isotherm? How are shapes of adsorption isotherms modified when multilayer adsorption takes place. *(Kalyani BSc)*

19. Deduce Gibbs adsorption equation thermodynamically. *(Allahabad BSc)*

20. (*a*) Give two applications of adsorption.
    (*b*) Explain the reason why a finally powdered substance is more effective adsorbent? *(Arunachal BSc)*

21. (*a*) How is chemisorption distinguished from physisorption on the basis of number of adsorbed layer?
    (*b*) How can Langmuir adsorption isotherm equation be used to explain the observation—the decomposition of PH_3 gas on tungsten metal surface is first order at low pressure and zeroth order at high pressure? *(Guru Nanak Dev BSc)*

22. (*a*) Derive Langmuir isotherm equation. How does this isotherm interpret the kinetics of unimolecular reactions catalysed by the solid surfaces?
    (*b*) Explain the term adsorption and write why it is caused.
    (*c*) Give an account of the ways by which physisorption and chemisorption are distinguished from each other. *(Guru Nanak Dev BSc)*

23. Draw the typical adsorption isotherms obtained in the case of unimolecular and multimolecular adsorption. *(Dibrugarh BSc)*

24. What are the characteristics of adsorption? Derive expression for Langmuir's adsorption isotherm. *(Delhi BSc)*

25. Derive Langmuir adsorption isotherm equation. Show under what conditions it becomes identical with Freundlich adsorption isotherm equation. *(Gulbarga BSc)*

26. (*a*) Distinguish between the terms absorption and adsorption.
    (*b*) What is an adsorption isotherm? Give applications of adsorption. *(Avadh BSc)*

27. 0.01 kg of oxygen is adsorbed on 0.0025 kg of metal powder at 273 K and 101325.00 Pa pressure. Calculate the volume of the gas adsorbed per gram of adsorbent.

    Answer. 2.80005 m^3 *(Madurai BSc)*

28. 0.004 kg of a gas (oxygen) is adsorbed on 0.0015 kg of metal powder at 300 K and 70927.5 Pa. Calculate the volume of the gas at STP adsorbed per kilogram of adsorbent.

Answer. 2.9303 m^3

(Punjabi BSc)

29. For an adsorbent – adsorbate system obeying the Langmuir adsorption isotherm, a = 0.48 Pa·and b = 0.16 Pa. At what pressure will 50 % of the surface be covered ?

Answer. 0.07 Pa

(Agra BSc)

30. 5 kg of a catalyst absorb 400 m^3 of N_2 at STP to form a monolayer. What is the surface area per gram if the area occupied by a molecule of N_2 is 16×10^{-10} m.

Answer. 344 $m^2\ g^{-1}$

(Madras BSc)

31. (*a*) Show diagrammatically the different types of adsorption isotherms obtained for adsorption of gases on solids.

(*b*) Discuss the behavior of Langmuir adsorption isotherm at very low and very high pressures.

(Kalyani BSc)

32. What is an adsorption isotherm? Deduce Langmuir's adsorption isotherm. *(Bundelkhand BSc)*

33. Distinguish between Adsorption and Absorption. Discuss the factors which affect the adsorption of a gas on a solid adsorbent. Discuss in brief the type of adsorption isotherms commonly observed for the adsorption of gases on a variety of adsorbents at different temperatures. *(Baroda BSc)*

34. 10^{-4} m^3 of 0.3 M acetic acid is shaken with 0.0008 kg of wood charcoal. The final concentration of the solution after adsorption is 0.125 M. Calculate the weight of acetic acid adsorbed per kilogram of carbon.

Answer. 0.00131 kg

(Allahabad BSc)

35. Derive Freundlich adsorption isotherm from the Gibbs adsorption isotherm applied to a gas.

(Arunachal BSc)

MULTIPLE CHOICE QUESTIONS

1. The phenomenon of concentrations of molecules of a gas or liquid at a solid surface is called

(*a*) absorption (*b*) adsorption

(*c*) catalysis (*d*) none of these

Answer. (*b*)

2. Adsorbate is that substance

(*a*) which concentrates on the surface

(*b*) where adsorption takes place

(*c*) which evaporates from the surface of metals

(*d*) none of these

Answer. (*a*)

3. The adsorption of gases on metal surfaces is called

(*a*) catalysis (*b*) occlusion

(*c*) adsorption (*d*) absorption

Answer. (*b*)

4. Increase in _______ of the adsorbent increases the total amount of the gas adsorbed

(*a*) density (*b*) volume

(*c*) surface area (*d*) surface tension

Answer. (*c*)

5. _______ the critical temperature of the gas, the more readily will it be adsorbed
(*a*) lower (*b*) higher
(*c*) intermediate (*d*) none of these
Answer. (*b*)

6. The process of adsorption is
(*a*) exothermic
(*b*) endothermic
(*c*) sometimes exothermic, sometimes endothermic
(*d*) none of the above
Answer. (*a*)

7. Physical adsorption is a _______ process
(*a*) reversible (*b*) irreversible
(*c*) exothermic (*d*) none of these
Answer. (*a*)

8. Physical adsorption occurs rapidly at _______ temperature
(*a*) low (*b*) high
(*c*) absolute zero (*d*) none of these
Answer. (*a*)

9. Physical adsorption generally _______ with increasing temperature
(*a*) decreases (*b*) increases
(*c*) sometimes decreases, sometime increases (*d*) none of these
Answer. (*a*)

10. Chemisorption generally _______ with temperature
(*a*) increases (*b*) decreases
(*c*) remains the same (*d*) none of these
Answer. (*a*)

11. Multi-molecular layers are formed in
(*a*) absorption (*b*) physical adsorption
(*c*) chemisorption (*d*) reversible adsorption
Answer. (*c*)

12. The relationship between equilibrium pressure of a gas and its amount adsorbed on the solid adsorbent at constant temperature is called
(*a*) chemisorption (*b*) adsorption isobars
(*c*) adsorption isotherms (*d*) none of these
Answer. (*c*)

13. Freundlich isotherms is not applicable at
(*a*) high pressure (*b*) low pressure
(*c*) 273 K (*d*) room temperature
Answer. (*a*)

14. At low pressures, the amount of the gas adsorbed is _______ proportional to the pressure
(*a*) directly (*b*) inversely
(*c*) sometimes directly, sometimes inversely (*d*) none of these
Answer. (*a*)

15. Which of the following is not an equation for Freundlich isotherm?
(*a*) $\frac{w}{m} = k \times C^{\frac{1}{n}}$ (*b*) $\frac{w}{m} = k \times P^{\frac{1}{n}}$

(c) $\log \frac{w}{m} = \log k + \frac{1}{n} \log P$

(d) $\log \frac{w}{m} = \log k - \frac{1}{n} \log P$

Answer. (*d*)

16. In gas masks the poisonous gases are removed by the adsorbent by the process of

(*a*) absorption
(*b*) adsorption
(*c*) catalysis
(*d*) none of these

Answer. (*b*)

17. Froth flotation process for the concentration of sulfide ore makes use of the process of

(*a*) adsorption
(*b*) heterogeneous catalysis
(*c*) absorption
(*d*) equilibrium

Answer. (*a*)

18. The water free from all ions (cations and anions) is referred to as

(*a*) heavy water
(*b*) concentrated water
(*c*) mineral water
(*d*) demineralized water

Answer. (*d*)

19. Heat of adsorption is defined as the energy liberated when _______ of a gas is adsorbed on the solid surface.

(*a*) 1 molecule
(*b*) 10^{-3} kg
(*c*) 10^{-3} kg mol
(*d*) 1 kg

Answer. (*c*)

20. In physical adsorption the gas molecules are held to the solid surface by

(*a*) hydrogen bond
(*b*) sigma bond
(*c*) pi bond
(*d*) van der Waal's forces

Answer. (*d*)

21. The adsorption of hydrogen on charcoal is

(*a*) physical adsorption
(*b*) chemical adsorption
(*c*) sorption
(*d*) none of these

Answer. (*a*)

22. The process of desorption increases with _______ of pressure

(*a*) decrease
(*b*) increase
(*c*) sometime increases, sometimes decreases
(*d*) none of these

Answer. (*a*)

23. Adsorption takes place with

(*a*) decrease in enthalpy of the system
(*b*) increase in enthalpy of the system
(*c*) no change in enthalpy of the system
(*d*) none of these

Answer. (*a*)

24. The heat of adsorption in physical adsorption lies in the range

(*a*) 1–10 kJ mol^{-1}
(*b*) 10 – 400 kJ mol^{-1}
(*c*) 40 – 100 kJ mol^{-1}
(*d*) 40 – 400 kJ mol^{-1}

Answer. (*a*)

25. In chromatographic analysis, the principle used is

(*a*) absorption
(*b*) adsorption
(*c*) distribution
(*d*) evaporation

Answer. (*b*)

26. Which of the following is not a characteristic of physical adsorption?

(*a*) adsorption is reversible (*b*) multi molecular layer is formed

(*c*) ΔH is of the order 400 kJ (*d*) occurs rapidly at low temperature

Answer. (*c*)

27. In gas masks, the poisonous gases are adsorbed by activated charcoal. The activated charcoal acts as

(*a*) adsorbate (*b*) adsorbent

(*c*) catalysis (*d*) adsorption agent

Answer. (*b*)

28. Which is incorrect statement?

(*a*) physical adsorption is irreversible in water

(*b*) physical adsorption involves multi-molecular layers

(*c*) the energy evolved is small

(*d*) physical adsorption is caused by van der Waal's forces

Answer. (*a*)

29. Which out of the following is incorrect?

(*a*) chemisorption is reversible in nature (*b*) physical adsorption is reversible in nature

(*c*) ΔH is small in physical adsorption (*d*) ΔH is large in chemical adsorption

Answer. (*a*)

30. The efficiency of adsorbent increases with increase in

(*a*) viscosity (*b*) surface tension

(*c*) surface area (*d*) number of ions

Answer. (*c*)

31. The ion-exchange resins are the compounds with

(*a*) high molecular masses (*b*) high surface tension

(*c*) low viscosities (*d*) high surface area

Answer. (*a*)

32. Hard water is made soft by passing it through a column packed with high polymer resin. This process makes use of

(*a*) cation-exchange (*b*) chromatographic analysis

(*c*) adsorption of –vely charged ions (*d*) heterogeneous catalysis

Answer. (*a*)

33. Which of the following is not an application of adsorption?

(*a*) gas masks (*b*) heterogeneous catalysis

(*c*) froth flotation process (*d*) softening of water by boiling

Answer. (*d*)

34. Langmuir while deriving adsorption isotherms did not make the following assumptions

(*a*) the layer of the gas adsorbed on the solid surface is one-molecule thick

(*b*) the adsorbed layer is uniform

(*c*) there is no attraction between the adjacent molecules

(*d*) the attraction between the adsorbent molecule is extremely large

Answer. (*d*)

35. Which of the following is incorrect?

(*a*) chemisorption is caused by bond formation

(*b*) chemisorption is specific in nature

(*c*) chemisorption is reversible

(*d*) chemisorption increases with increase in temperature

Answer. (*c*)

36. The process represented by the equation

$R'H^+ + Na^+ \rightarrow R'Na^+ + H^+$ is

(*a*) cation exchange (*b*) anion exchange

(*c*) resin exchange (*d*) chromatographic analysis

Answer. (*a*)

37. An anion exchange process is represented by

(*a*) $R'H^+ + Na^+ \rightleftharpoons R'Na^+ + H^+$ (*b*) $R'OH + Cl^- \rightleftharpoons R'Cl^- + OH^-$

(*c*) $2R\,Na^+ + Ca^{2+} \rightleftharpoons R_2Ca^{2+} + 2Na$ (*d*) none of these

Answer. (*b*)

38. In an adsorption process unimolecular layer is formed. It is

(*a*) physical adsorption (*b*) chemical adsorption

(*c*) ion-exchange (*d*) chromatographic analysis

Answer. (*b*)

39. The rate of desorption R_d is given by (where θ is the fraction of total surface covered by the adsorbed molecules)

(*a*) $R_d = k_d\,\theta$ (*b*) $R_d = k_d\theta^2$

(*c*) $R_d = k_d/\theta$ (*d*) $R_d = k_d/\theta^2$

Answer. (*a*)

40. Langmuir Isotherms holds at low pressure but fails at

(*a*) low temperature (*b*) high pressure

(*c*) intermediate pressure (*d*) none of these

Answer. (*b*)

24 Electrolysis and Electrical Conductance

CONTENTS

Water-soluble substances are distinguished as *electrolytes* or *nonelectrolytes*.

Electrolytes are electrovalent substances that form ions in solution which conduct an electric current. Sodium chloride, copper (II) sulfate and potassium nitrate are examples of electrolytes.

Nonelectrolytes, on the other hand, are covalent substances which furnish neutral molecules in solution. Their water-solutions do not conduct an electric current. Sugar, alcohol and glycerol are typical nonelectrolytes.

An electrolyte invariably undergoes chemical decomposition as a result of the passage of electric current through its solution.

The phenomenon of decomposition of an electrolyte by passing electric current through its solution is termed Electrolysis (*lyo* = breaking).

The process of electrolysis is carried in an apparatus called the **Electrolytic cell.** The cell contains water-solution of an electrolyte in which two metallic rods (electrodes) are dipped. These rods are connected to the two terminals of a battery (source of electricity). The electrode connected to the positive terminal of the battery attracts the negative ions (anions) and is called **anode.** The other electrode connected to the negative end of the battery attracts the positive ions (cations) and is called **cathode.**

MECHANISM OF ELECTROLYSIS

How the electrolysis actually takes place, is illustrated in Fig 24.1. The cations migrate to the cathode and form a neutral atom by accepting electrons from it. The anions migrate to the anode and yield a neutral particle by transfer of electrons to it. As a result of the loss of electrons by anions and gain of electrons by cations at their respective electrodes chemical reaction takes place.

Example. Let us consider the electrolysis of hydrochloric acid as an example. In solution, HCl is ionized,

$$HCl \longrightarrow H^+ + Cl^-$$

In the electrolytic cell Cl^- ions will move toward the anode and H^+ ions will move toward the cathode. At the electrodes, the following reactions will take place.

At cathode :

$$H^+ + e^- \longrightarrow H \qquad \textit{(Reduction)}$$

As you see, each hydrogen ion picks up an electron from the cathode to become a hydrogen atom. Pairs of hydrogen atoms then unite to form molecules of hydrogen gas, H_2.

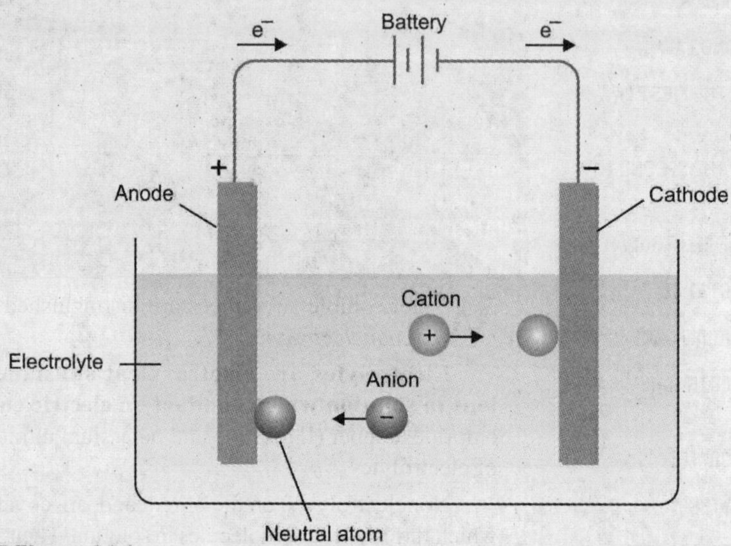

■ **Figure 24.1**
The mechanism of electrolysis.

At Anode :

$$Cl^- \longrightarrow Cl + e^- \qquad \textit{(Oxidation)}$$

After the chloride ion loses its electron to the anode, pair of chlorine atoms unite to form chlorine gas, Cl_2.

The net effect of the process is the decomposition of HCl into hydrogen and chlorine gases. The overall reaction is :

$$2HCl \longrightarrow H_2 + Cl_2 \qquad \textit{(Decomposition)}$$

ELECTRICAL UNITS

There are a few electrical units which we should understand before taking up the study of quantitative aspects of electrolysis. These are :

Coulomb

A coulomb is a **unit quantity of electricity.** It is the amount of electricity which will deposit 0.001118 gram of silver from a 15 per cent solution of silver nitrate in a coulometer.

Ampere

An ampere is a **unit rate of flow of electricity.** It is that current which will deposit 0.001118 gram of silver in one second. In other words, an ampere is a current of one coulomb per second.

Ohm

An ohm is a **unit of electrical resistance.** It is the resistance offered at 0ºC to a current by a column of mercury 106.3 cm long of about 1 sq mm cross-sectional area and weighing 14.4521 grams.

Volt

A volt is a **unit of electromotive force.** It is the difference in electrical potential required to send a current of one ampere through a resistance of one ohm.

FARADAY'S LAWS OF ELECTROLYSIS

Michael Faraday studied the quantitative aspect of electrolysis. He discovered that there exists a definite relationship between the amounts of products liberated at the electrodes and the quantity of electricity used in the process. In 1834, he formulated two laws which are known as **Faraday's Laws of Electrolysis.** These are :

First Law

The amount of a given product liberated at an electrode during electrolysis is directly proportional to the quantity of electricity which passes through the electrolyte solution.

Second Law

When the same quantity of electricity passes through solutions of different electrolytes, the amounts of the substances liberated at the electrodes are directly proportional to their chemical equivalents.

Definition of Electrochemical equivalent in light of First Law

If m is the mass of substance (in grams) deposited on electrode by passing Q coulombs of electricity, then

$$m \propto Q \qquad \textit{(First Law)}$$

We know that $Q = I \times t$

where I is the strength of current in amperes and t is the time in second for which the current has been passed.

Therefore, $m \propto I \times t$

or $m = Z \times I \times t$

where Z is the constant known as the Electrochemical equivalent of the substance (electrolyte).

If $I = 1$ ampere and $t = 1$ second, then

$$m = Z$$

Thus, **the electrochemical equivalent is the amount of a substance deposited by 1 ampere current passing for 1 second (*i.e.*, one coulomb).**

The Electrical unit Faraday

It has been found experimentally that the quantity of electricity required to liberate one gram-equivalent of a substance is 96,500 coulombs. This quantity of electricity is known as **Faraday** and is denoted by the symbol ***F***.

It is obvious that the quantity of electricity needed to deposit 1 mole of the substance is given by the expression.

$$\text{Quantity of electricity} = n \times F$$

where n is the valency of its ion. Thus the quantity of electricity required to discharge :

$$\text{one mole of } Ag^+ = 1 \times F = 1F$$
$$\text{one mole of } Cu^{2+} = 2 \times F = 2F$$
$$\text{one mole of } Al^{3+} = 3 \times F = 3F$$

We can represent the reactions on the cathode as :

$$Ag^+ + e = Ag$$
$$Cu^{2+} + 2e = Cu$$
$$Al^{3+} + 3e = Al$$

It is clear that the moles of electrons required to discharge one mole of ions Ag^+, Cu^{2+} and Al^{3+} is one, two and three respectively. Therefore it means that the quantity of electricity in one Faraday is one mole of electrons. Now we can say that.

1 Faraday = 96,500 coulombs = 1 Mole electrons

Importance of the First law of Electrolysis

With the help of the first law of electrolysis we are able to calculate :

(1) the value of electrochemical equivalents of different substances; and

(2) the masses of different substances produced by passing a known quantity of electricity through their solutions.

Verification of the Second law of Electrolysis

According to this law when the same quantity of electricity is passed through different electrolyte solutions, the masses of the substances deposited on the electrodes are proportional to their chemical equivalents.

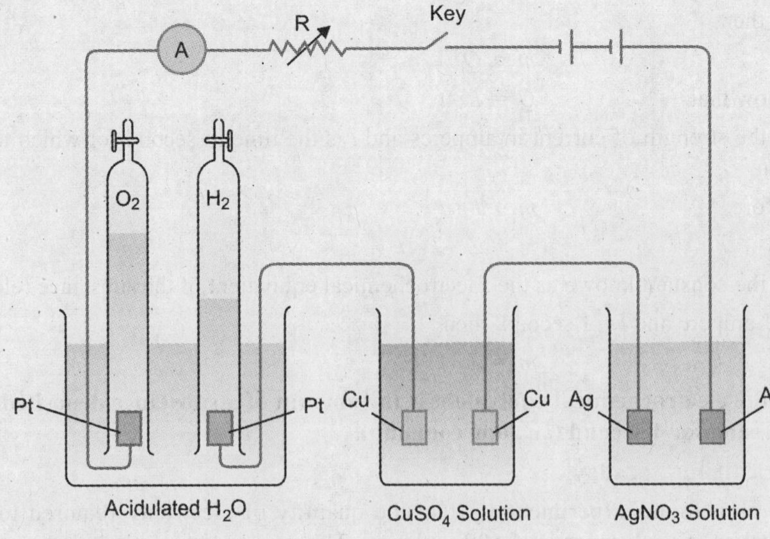

■ **Figure 24.2**
Illustrating Faraday's Second Law of Electrolysis.

To verify the law, let us take an arrangement of the type shown in Fig. 24.2. Pass the same quantity of electricity through the three **coulometers** (the term 'coulometer' is now in practice replaced by the older term 'voltameter') containing solution of dilute H_2SO_4, $CuSO_4$ and $AgNO_3$ respectively. These coulometers are fitted with platinum, copper and silver electrodes as shown in Fig. 24.2. The masses of hydrogen, copper and silver liberated/deposited at the respective cathodes are in the ratio of their equivalent weights. That is,

$$\frac{\text{mass of hyrogen liberated}}{\text{mass of copper deposited}} = \frac{\text{Eq. Wt. of hydrogen}}{\text{Eq. Wt. of copper}}$$

and $$\frac{\text{mass of copper deposited}}{\text{mass of silver deposited}} = \frac{\text{Eq. Wt. of copper}}{\text{Eq. Wt. of silver}}$$

From this experiment, we can calculate the mass of hydrogen, copper and silver liberated at their respective cathodes by one coulomb of electricity. We find these are always :

$$\text{Hydrogen} = 1.036 \text{ kg}$$
$$\text{Copper} = 3.292 \times 10^{-7} \text{ kg}$$
$$\text{Silver} = 1.118 \times \times 10^{-6} \text{ kg}$$

Since the equivalent weights of hydrogen, copper and silver are 1, 31.78 and 107.88 respectively, it follows that **the chemical equivalents are proportional to the chemical equivalents** (or equivalent weights).

Importance of the Second law of Electrolysis

The second law of electrolysis helps to calculate :

(1) the equivalent weights of metals
(2) the unit of electric charge
(3) the Avogadro's number

SOLVED PROBLEM 1

1.978×10^{-4} kg of copper is deposited by a current of 0.2 ampere in 3000 seconds. What is the electrochemical equivalent of copper?

SOLUTION

Here $t = 3000$ seconds ; $I = 0.2$ ampere. Quantity of electricity used is

$$Q = I \times t = 0.2 \times 3000 = 600 \text{ coulombs}$$

Amount of copper deposited by 600 coulombs $= 1.978 \times 10^{-4}$ kg

$$\text{Amount of copper deposited by 1 coulomb} = \frac{1.978 \times 10^{-4}}{600} = 3.296 \times 10^{-7} \text{kg}$$

$\therefore$ Electrochemical equivalent of copper $= 3.296 \times 10^{-7}$ kg

SOLVED PROBLEM 2

What current strength in amperes will be required to liberate 0.01 kg of iodine from potassium iodide solution in one hour (3600 seconds)?

SOLUTION

0.127 kg of iodine is liberated by = 96,500 coulomb

$$\therefore \quad \text{0.01 kg of iodine is liberated by} = \frac{96{,}500}{0.127} \times 0.01 \text{ coulomb}$$

Let current strength be $= I$

Time in seconds $= 1 \times 60 \times 60$

We know that the quantity of electricity, Q, used is given by the expression

$$Q = I \times \text{time in seconds}$$

$\therefore$ Current strength, $I = \dfrac{Q}{t} = \dfrac{96,500 \times 0.01}{0.127 \times 3600}$

$= \mathbf{2.11\ ampere}$

SOLVED PROBLEM 3

An electric current is passed through three cells in series containing respectively solution of copper sulfate, silver nitrate and potassium iodide. What weights of silver and iodine will be liberated while 1.25×10^{-3} kg of copper is being deposited?

SOLUTION $\dfrac{\text{Wt. of copper}}{\text{Wt. of Iodine}} = \dfrac{\text{Eqvt. wt. of copper}}{\text{Eqvt. wt. of Iodine}}$

or $\dfrac{1.25 \times 10^{-3}}{x} = \dfrac{31.7 \times 10^{-3}}{0.127}$

Hence $x = 5 \times 10^{-3}$ kg of iodine

Also, $\dfrac{\text{Wt. of copper}}{\text{Wt. of silver}} = \dfrac{1.25 \times 10^{-3}}{y} = \dfrac{\text{Eqvt. wt. of Cu } (= 31.7 \times 10^{-3}\text{ kg})}{\text{Eqvt. wt. of silver } (= 108 \times 10^{-3}\text{ kg})}$

$\therefore$ Wt. of silver $(y) = \dfrac{108 \times 10^{-3} \times 1.25 \times 10^{-3}}{31.7 \times 10^{-3}} = \mathbf{4.26 \times 10^{-3}\ kg\ \ or\ \ 4.26\ g}$

CONDUCTANCE OF ELECTROLYTES

We have seen that electrolyte solutions conduct electric currents through them by movement of the ions to the electrodes. The power of electrolytes to conduct electric currents is termed **conductivity** or **conductance.** Like metallic conductors, electrolytes obey Ohm's law. According to this law, the current I flowing through a metallic conductor is given by the relation.

$$I = \frac{E}{R}$$

where E is the potential difference at two ends (in volts); and R is the resistance measured in ohms (or Ω). The resistance R of a conductor is directly proportional to its length, l, and inversely proportional to the area of its cross-section, A. That is,

$$R \propto \frac{l}{A}$$

or $$R = \rho \times \frac{l}{A} \quad ...(1)$$

where ρ "rho" is a constant of proportionality and is called **resistivity** or **specific resistance.** Its value depends upon the material of the conductor. From (1) we can write

$$\rho = R \times \frac{A}{l}$$

If $l = 1$ cm and $A = 1$ sq cm, then

$$\rho = R$$

Thus it follows that the **Specific resistance** of a conductor is the resistance in ohms which one centimeter cube of it offers to the passage of electricity.

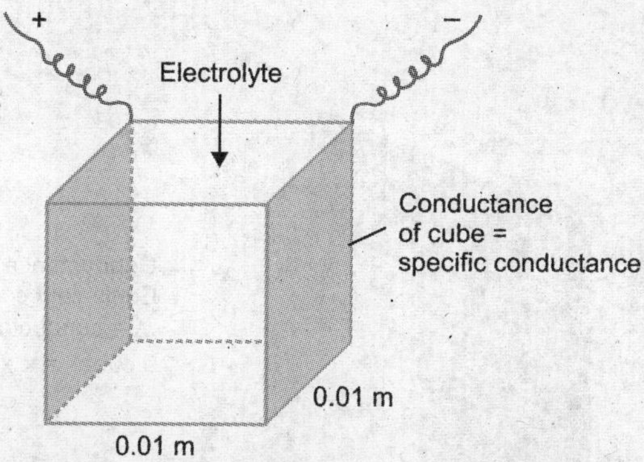

Figure 24.3
Diagrammatic illustration of definition of specific conductance.

Specific Conductance

It is evident that a substance which offers very little resistance to the flow of current allows more current to pass through it. Thus the power of a substance to conduct electricity or conductivity is the converse of resistance. The reciprocal of specific resistance is termed **Specific conductance** or **Specific conductivity.** It is defined as : **the conductance of one centimeter cube (cc) of a solution of an electrolyte.**

The specific conductance is denoted by the symbol κ (kappa). Thus,

$$\kappa = \frac{1}{\rho} = \frac{1}{R} \times \frac{l}{A}$$

Units of Specific conductance

Specific conductance is generally expressed in reciprocal ohms (r.o) or **mhos** or **ohm^{-1}**. Its unit can be derived as follows :

$$\kappa = \frac{1}{A} \times \frac{l}{R} = \frac{1}{\text{ohm}} \times \frac{\text{cm}}{\text{cm}^2}$$

$$= \mathbf{ohm^{-1}cm^{-1}}$$

The internationally recommended unit for ohm^{-1} (or mho) is **Siemens, S.** When S is used, the conductance is expressed as **S cm^{-1}**. It may be noted that Siemens is not a plural, the unit is named after Sir William Siemens–a noted electrical engineer.

The specific conductance increases with : (*i*) ionic concentration, and (*ii*) speeds of the ions concerned.

In measuring the specific conductance of the aqueous solution of an electrolyte, the volume of water in which a certain amount of the electrolyte is dissolved is always measured in cubic centimeters (cc) and this is known as **dilution.** If the volume of a solution is V_{cc}, the specific conductance of the solution is written as κ.

Equivalent Conductance

It is defined as **the conductance of an electrolyte obtained by dissolving one gram-equivalent of it in V cc of water.**

The equivalent conductance is denoted by Λ. It is equal to the product of the specific conductance, κ and the volume *V* in cc containing one gram-equivalent of the electrolyte at the dilution *V*.

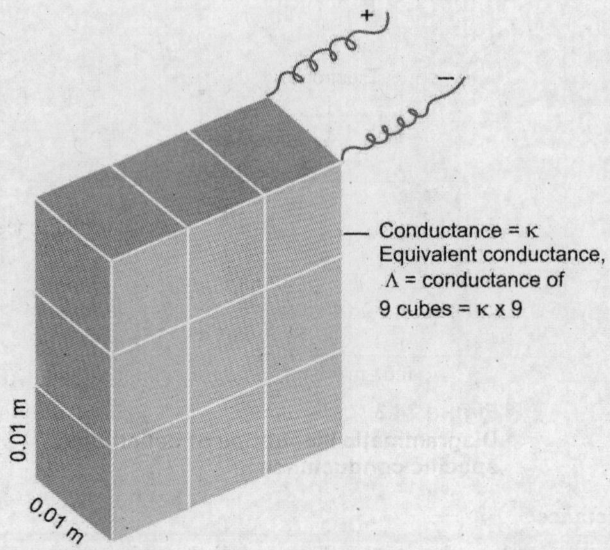

Figure 24.4
Solution of 1 g-eqvt. dissolved in 9 cc water between electrode plates 1 cm apart has L = κ x 9.

Thus,

$$\Lambda = \kappa \times V$$

This is illustrated in Fig. 24.4. A solution having one gram-equivalent of the electrolyte dissolved in, say, 9×10^{-6} m^3 water be placed between two electrodes 1×10^{-2} m apart. The solution could be considered as consisting of nine cubes, each of which has a conductance k (specific conductance). Thus the total conductance of the solution will be $9 \times \kappa$. Similarly, V m^3 of solution will form V cubes and the total conductance will be $\kappa \times V$.

In general, if an electrolyte solution contains N gram-equivalents in 1000 cc of the solution, the volume of the solution containing 1 gram-equivalent will be 1000/N. Thus,

$$\Lambda = \frac{\kappa \times 1000}{N}$$

Unit of Equivalent conductance

The unit of equivalent conductance may be deduced as follows :

$$\Lambda = \kappa \times V$$

$$= \frac{1}{R} \times \frac{l}{A} \times V$$

$$= \frac{1}{\text{ohm}} \times \frac{\text{m}}{\text{m}^2} \times \frac{\text{m}^3}{\text{eqvt}}$$

$$= \text{Siemens m}^2 \text{ eqvt}^{-1}$$

Variation of Equivalent conductance with Concentration (or Dilution)

The equivalent conductance of a solution does not vary linearly with concentration. The effect of concentration on equivalent conductance can be studied by plotting Λ values against the square root of the concentration. It has been found that variation of equivalent conductance with $\sqrt{C}$ depends upon the nature of electrolyte. Fig. 24.5 shows the behavior of strong and weak electrolytes with change of concentration.

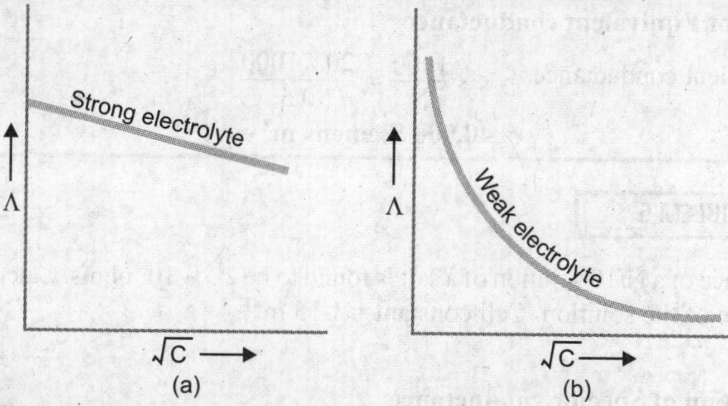

Figure 24.5
Variation of equivalent conductivity, L with $\sqrt{C}$:
(*a*) for strong electrolyte; (*b*) for weak electrolyte.

Strong electrolytes are completely ionized at all concentrations (or dilutions). The increase in equivalent conductance is not due to the increase in the number of current carrying species. This is, in fact, due to the decrease in forces of attraction between the ions of opposite charges with the decrease in concentration (or increase in dilution). At higher concentration, the forces of attraction between the opposite ions increase ($F \propto q_1 q_2 / r^2$). Consequently, it affects the speed of the ions with which they move towards oppositely charged electrodes. This phenomenon is called **ionic interference. As the solution becomes more and more dilute, the equivalent conductance increases, till it reaches a limitary value.** This value is known as equivalent conductance at infinite dilution (zero concentration) and is denoted by A.

Weak electrolytes have low ionic concentrations and hence interionic forces are negligible. Ionic speeds are not affected with decrease in concentration (or increase in dilution). The increase in equivalent conductance with increasing dilution is due to the increase in the number of current-carrier species. In other words, the degree of ionization (α) increases. Thus **increase in equivalent conductance (Λ) in case of a weak electrolyte is due to the increase in the number of ions.**

In case of a weak electrolyte Λ_∞ is the equivalent conductance when ionization is complete. So, the conductance ratio Λ/Λ_∞ is the degree of ionization. That is,

$$\alpha = \frac{\Lambda}{\Lambda_\infty}$$

SOLVED PROBLEM 4

0.5 Normal solution of a salt placed between two platinum electrodes, 0.2 m apart and of area of cross-section 0.0004 m^2 has a resistance of 25 ohms. Calculate the equivalent conductance of the solution.

SOLUTION

Calculation of specific conductance

$l = 0.2$ m $\quad A = 0.0004$ m^2 $\quad R = 25$ ohms

$$\text{Specific conductance } \kappa = \frac{1}{R} \times \frac{l}{A}$$

$$= \frac{1}{25} \times \frac{0.2}{0.0004}$$

$$= 20 \text{ siemens m}^{-1}$$

Calculation of Equivalent conductance

$$\text{Equivalent conductance} = \kappa \times \frac{1000}{N} = \frac{20 \times 1000}{0.5}$$

$$= \mathbf{40{,}000\ Siemens\ m^2\ eqvt^{-1}}$$

SOLVED PROBLEM 5

The resistance of a N/10 solution of a salt is found to be 2.5×10^3 ohms. Calculate the equivalent conductance of the solution. Cell constant = 1.15 m^{-1}.

SOLUTION

Calculation of Specific conductance

$$\text{Specific conductance, } \kappa = \frac{1}{R} \times \text{cell constant}$$

$$= \frac{1}{2.5 \times 10^3} \times 1.15$$

Calculation of Equivalent conductance

$$\text{Equivalent conductance} = \frac{\kappa \times 1000}{N}$$

$$= \frac{1.15 \times 1000}{2.5 \times 10^3 \times 0.1} = \frac{115}{25}$$

$$= \mathbf{4.60\ ohm^{-1}\ m^2\ eqvt^{-1}}$$

Molar Concentration

It is another quantity which helps in comparing the conductivities of electrolytes. It is defined as: **the conductance of all ions produced by one mole (one gram-molecular weight) of an electrolyte when dissolved in a certain volume V m^3.**

Molar conductance is denoted by μ. Its value is obtained by multiplying the specific conductance, κ, by the volume in cc containing one mole of the electrolyte.

Thus,

Molar conductance, μ = k × V where V is the volume of the solution in cc containing one mole of the electrolyte.

Units of Molar Concentration

Since
$$\kappa = \frac{1}{R} \times \frac{l}{A}$$

$$\mu = \frac{1}{R} \times \frac{l}{A} \times V$$

$$= \frac{1}{\text{ohm}} \times \frac{\text{m}}{\text{m}^2} \times \frac{\text{m}^3}{\text{mol}}$$

$$= \text{ohm}^{-1}\ \text{m}^2\ \text{mol}^{-1}$$

Calculation of Molar conductance

Molar conductance can be calculated by using the relation :

$$\mu = \frac{\kappa \times 1000}{M}$$

where M is the number of moles of the electrolyte present in 1000 cc of solution.

Upon dilution specific conductance decreases, while Equivalent conductance and Molar conductance increases.

It is important to note that specific conductance decreases with dilution. It is the conductance of one cc of the solution. Upon diluting the solution, the concentration of ions per cc decreases. Hence the specific conductance falls. On the other hand, the equivalent and molar conductance show an increase as these are the products of specific conductance and the volume of the solution containing one gram-equivalent or one mole of the electrolyte respectively. With dilution, the first factor decreases, while the other increases. The increase in the second factor is much more than the decrease in the first factor. The specific and molar conductance of NaCl solution at 18°C are shown in Table 24.1.

TABLE 24.1. SPECIFIC AND MOLAR CONDUCTANCE OF NaCl SOLUTION AT 291 K

Volume, V in cc containing 1 g mol	Volume , V m^3 1 kg mol	Specific conductance, $ohm^{-1}\ m^2\ eqvt^{-1}$	Molar conductance $ohm^{-1}\ m^2\ mol^{-1}$
1,000	0.001	0.0744×10^{-4}	74.4×10^{-4}
5,000	0.005	0.01760×10^{-4}	88.2×10^{-4}
20,000	0.020	0.0479×10^{-4}	95.9×10^{-4}
500,000	0.500	0.000213×10^{-4}	106.7×10^{-4}
1,000,000	1.000	0.0001078×10^{-4}	107.3×10^{-4}
2,000,000	2.000	0.0000542×10^{-4}	108.5×10^{-4}
5,000,000	5.000	0.0000218×10^{-4}	109.2×10^{-4}
10,000,000	10.000	$0.00001097 \times 10^{-4}$	109.7×10^{-4}

The equivalent conductance of some common electrolytes at 18°C is given in Table 24.2.

TABLE 24.2. EQUIVALENT CONDUCTANCE OF SOME COMMON ELECTROLYTES AT 291 K

Volume, V in m^3 containing 1 kg equivalent	Equivalent conductance Λ, $ohm^{-1}\ m^2\ eqvt^{-1}$				
	NaOH	KCl	HCl	CH_3COOH	CH_3COONa
0.001	160×10^{-4}	98.3×10^{-4}	301×10^{-4}	1.32×10^{-4}	41.2×10^{-4}
0.002	172×10^{-4}	120.4×10^{-4}	327×10^{-4}	2.01×10^{-4}	49.4×10^{-4}
0.010	183×10^{-4}	112.0×10^{-4}	351×10^{-4}	4.60×10^{-4}	61.1×10^{-4}
0.020	190×10^{-4}	$115\ 9 \times 10^{-4}$	360×10^{-4}	6.48×10^{-4}	64.2×10^{-4}
0.100	200×10^{-4}	122.4×10^{-4}	370×10^{-4}	14.3×10^{-4}	70.2×10^{-4}
0.200	203×10^{-4}	124.4×10^{-4}	372×10^{-4}	20.0×10^{-4}	72.4×10^{-4}
0.500	206×10^{-4}	126.3×10^{-4}	376×10^{-4}	30.2×10^{-4}	74.3×10^{-4}
	210×10^{-4}	127.3×10^{-4}	377×10^{-4}	41.0×10^{-4}	75.2×10^{-4}

Summary of Electrochemical Quantities

A summary of the electrochemical terms, their symbols and units in which they are expressed are listed in Table 24.3 for reference.

TABLE 24.3. THE ELECTROCHEMICAL QUANTITIES, THEIR SYMBOLS AND UNITS

Quantity	Symbol	Unit	SI Unit
Resistance	R	ohm or Ω	ohm
Resistivity or Specific resistance	ρ (rho)	ohm cm	ohm m
Conductance	$1/R$	ohm^{-1} or Siemens	Siemens
Specific conductance	κ (kappa)	$ohm^{-1}\ cm^{-1}$	$S\ m^{-1}$
Dilution	V	cc	m^3
Equivalent conductance	Λ	$ohm^{-1}\ cm^2\ eqvt^{-1}$	$Siemens\ m^2\ eqvt^{-1}$
Molar conductance	μ	$ohm^{-1}\ cm^2\ mol^{-1}$	$Siemens\ m^2\ mol^{-1}$

Variation of Conductance with Temperature

The conductance of a solution of an electrolyte generally increases with rise in temperature. It has been found by experiment that the conductance of a given solution increases by 2-3 per cent for one degree rise in temperature. For example, the conductances of 0.1 M KCl at two different temperatures are

$$1.12\ ohm^{-1}\ m^{-1} \text{ at } 291\ K$$

$$1.29\ ohm^{-1}\ m^{-1} \text{ at } 298\ K$$

The conductance of a given electrolyte depends on two factors :

(1) The number of ions present in unit volume of solution

(2) The speed at which ions move towards the electrodes

At a given temperature, the first factor remains the same for a particular electrolyte. Thus the increase in conductance with rise in temperature is due to the influence of factor (2). With rise in temperature the viscosity of the solvent (water) decreases which makes the ions to move freely toward the electrodes.

For weak electrolytes, the influence of temperature on conductance depends upon the value of ΔH accompanying the process of ionization. If the ionization is exothermic (–ΔH), the degree of ionization is less at higher temperature (Le Chatelier's principle) and conductance decreases. Conversely, if the ionization is endothermic (+ΔH), the degree of ionization is more at higher temperature and conductance increases.

STRONG AND WEAK ELECTROLYTES

Electrolytes may be divided into two classes :

(*a*) Strong electrolytes

(*b*) Weak electrolytes

Strong Electrolytes

A strong electrolyte is a substance that gives a solution in which almost all the molecules are ionized. The solution itself is called **a strong electrolytic solution.** Such solutions are good conductors of electricity and have a high value of equivalent conductance even at low concentrations. The strong electrolytes are :

(1) **The strong acids** *e.g.*, HCl, H_2SO_4, HNO_3, $HClO_4$, HBr and HI.

(2) **The strong bases** *e.g.*, NaOH, KOH, Ca $(OH)_2$, Mg $(OH)_2$, etc.

(3) **The salts.** Practically all salts (NaCl, KCl, etc) are strong electrolytes.

Weak Electrolytes

A weak electrolyte is a substance that gives a solution in which only a small proportion of the solute molecules are ionized. Such a solution is called **a weak electrolytic solution,** that has low value of equivalent conductance. The weak electrolytes are :

(1) **The weak acids :** All organic acids such as acetic acid, oxalic acid, sulfurous acid (H_2SO_3) are examples of weak electrolytes.

(2) **The weak bases :** Most organic bases *e.g.*, alkyl amines ($C_2H_5NH_2$) are weak electrolytes.

(3) **Salts.** A few salts such as mercury (II) chloride and lead (II) acetate are weak electrolytes.

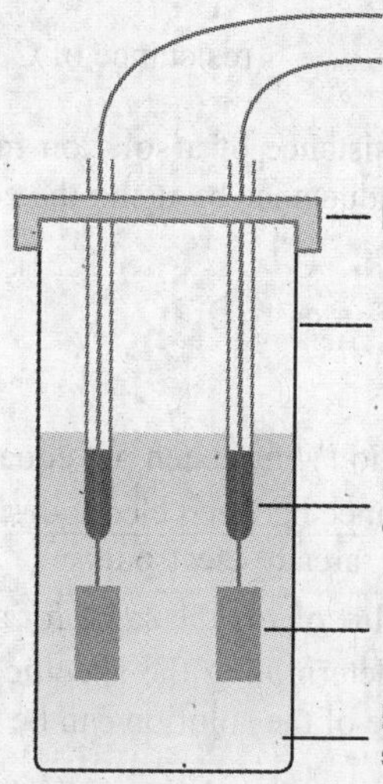

■ **Figure 24.6**
A Conductance cell.

Measurement of Electrolytic conductance

We know that conductance is the reciprocal of resistance. Therefore it can be determined by measuring the resistance of the electrolytic solution. This can be done in the laboratory with the help of a **Wheatstone bridge.**

The solution whose conductance is to be determined is placed in a special type of cell known as the **conductance cell.**

A simple type of conductance cell used in the laboratory is shown in Fig. 24.6. The electrodes fitted in the cell are made of platinum plates coated with platinum black. These are welded to platinum wires fused in two thin glass tubes. The contact with copper wires of the circuit is made by dipping them in mercury contained in the tubes.

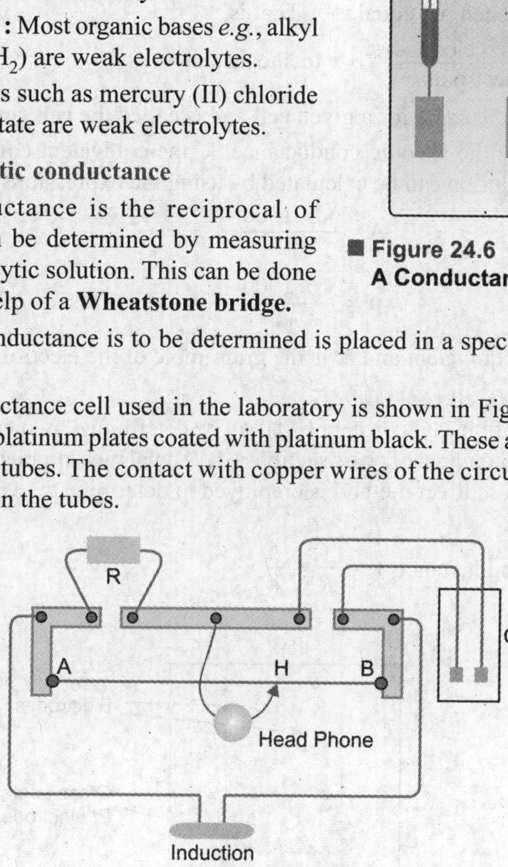

■ **Figure 24.7**
Apparatus for Conductance measurement.

The arrangement commonly used for the measurement of resistance of the conductance cell is shown in Fig. 24.7. It may be noted that a *head-phone* is used in place of a galvanometer. AB is a manganin wire tightly stretched over a meter rule graduated in millimeters. A sliding contact H (shown by arrow-head) moves along this wire. R is a resistance box. C is the conductance cell containing electrolytic solution. I is the induction coil from which alternating current is led as shown in the diagram. When the current is flowing, any resistance is unplugged in the resistance box R. The sliding contact H is moved until the sound in the head-phone is minimum. When this occurs, we have

$$\frac{\text{resistance of C}}{\text{resistance of R}} = \frac{\text{resistance BH}}{\text{resistance AH}} = \frac{\text{length BH}}{\text{length AH}}$$

$$\text{or} \qquad \text{resistance of C} = \frac{\text{length BH}}{\text{length AH}} \times \text{resistance R}$$

The resistance of a solution in the conductance cell as measured above can be converted to specific conductance by using the equation

$$\kappa = \frac{1}{R} \times \frac{l}{A} \qquad ...(1)$$

$$\text{or} \qquad \kappa = \frac{1}{R} \times x \qquad ...(2)$$

The ratio l/A has been put equal to x. That is,

$$\frac{\text{distance between electrodes}}{\text{area of electrode}} = x \text{ (cell constant)}$$

The value of x is the same for a given cell and is called the **cell constant.**

After determining the specific conductance, k, the equivalent conductance, Λ, and the molar conductance of the solution can be calculated by using the expressions.

$$\Lambda = \frac{\kappa \times 1000}{N}$$

$$\mu = \frac{\kappa \times 1000}{M}$$

where N is the gram-equivalent and M is the gram-mole of the electrolyte.

Determination of the Cell constant

The exact value of the cell constant (l/A) can be determined by measuring the distance between the electrodes (l) and their area of cross sections (A). Actual measurement of these dimensions is very difficult. Therefore an indirect method is employed to determine the value of cell constant.

We know that :

$$\text{specific conductance } \kappa = \frac{1}{R} \times \frac{l}{A}$$

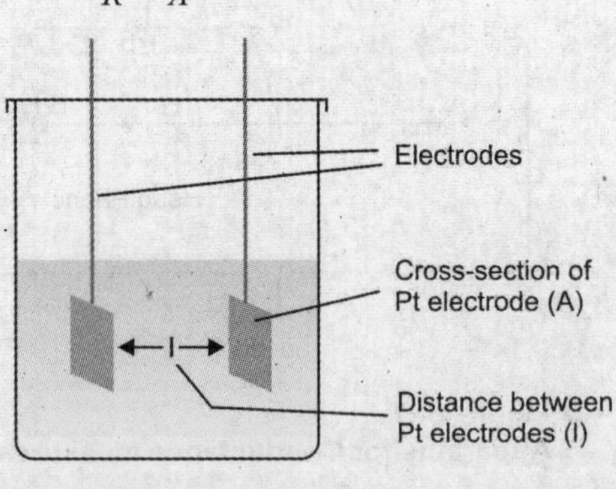

■ **Figure 24.8**
Dimensions of a Conductance cell.

$$\text{or} \qquad \kappa = \text{observed conductance} \times \text{cell constant}$$

$$\therefore \qquad \text{cell constant, } x = \frac{\text{specific conductance}}{\text{observed conductance}}$$

To determine the cell constant, a standard solution of KCl whose specific conductance at a given temperature is known, is used. Then a solution of KCl of the same strength is prepared and its conductance determined experimentally at the same temperature. Substituting the two values in the above expression, the cell constant can be calculated.

For example, according to Kohlrausch the specific conductance of N/50 solution at 25ºC is 0.002765 mho. Now, an N/50 solution of KCl is prepared by dissolving 0.372 g pure KCl in 250 cc 'extra-pure' water (conductance water) and its conductance determined at 25ºC. The cell constant is then calculated by substituting the observed conductance in the expression

$$\text{cell constant} = \frac{0.002765}{\text{observed conductance}}$$

SOLVED PROBLEM 6

The specific conductance of a N/50 solution of KCl at 298 K is 0.002765 Siemens. If the resistance of a cell containing this solution is 400 ohms, what is the cell constant?

SOLUTION cell constant, $x = \dfrac{0.002765}{\text{observed conductance}}$

$= 0.002765 \times \text{resistance}$

$= 0.002765 \times 400$

$= \mathbf{1.106}$

SOLVED PROBLEM 7

The resistance of decinormal solution of a salt occupying a volume between two platinum electrodes 1.80×10^{-2} m apart and 5.4×10^{-4} m^2 in area was found to be 32 ohms. Calculate the equivalent conductance of the solution.

SOLUTION

Here $l = 1.80 \times 10^{-2}$ m and $A = 5.4 \times 10^{-4}$ m^2

$\therefore$ cell constant $x = \dfrac{l}{A} = \dfrac{1.80\times10^{-2}}{5.4\times10^{-4}} = \dfrac{100}{3}$

observed conductance $= \dfrac{1}{32}$ mhos

Since the solution is $N/10, V = 10^{-2}$ m^2

Now, specific conductance $= x \times$ obs. conductance

$$\kappa = \frac{100}{3} \times \frac{1}{32} = \frac{100}{96} \text{ mhos}$$

or $\Lambda = \kappa \times V$

$= \dfrac{100}{96} \times 10^{-2}$

$= \mathbf{0.01042\ mhos\ m^2\ equiv^{-1}}$

SOLVED PROBLEM 8

A conductance cell on being filled with a 0.02 molar solution of KCl at 298 K showed a resistance of 165 ohms. The specific conductance of the KCl solution used is 0.277 mho m^{-1}. The same cell containing 0.01 molar NaCl solution gave an electrical resistance of 384 ohms. Calculate the specific and equivalent conductance of the NaCl solution.

SOLUTION cell constant $= \dfrac{\text{specific conductance}}{\text{observed conductance}}$

Given conductance of KCl = 0.277 mho m^{-1}

Observed conductance of KCl $= \frac{1}{165}$ mho

$$\therefore \quad \text{Cell constant} = \frac{0.277 \times 10^{-3}}{1/165} = 45.705\,\text{m}^{-1}$$

Specific conductance of NaCl = cell constant × obs. conductance

$$= 45.705 \times 1/384 = 1.1902 \times 10^{-1}\ \text{ohm}^{-1}\ \text{m}^{-1}$$

$$\therefore \quad \text{Eqvt conductance of NaCl} = \frac{\text{sp. conductance} \times 1000}{N}$$

$$= \frac{1.1902 \times 10^{-1} \times 0.01\,\text{m}^2}{0.01}$$

$$= \mathbf{0.11902\ ohm^{-1}\ m^2\ eqvt^{-1}} = \mathbf{0.11902\ Siemens\ m^2\ eqvt^{-1}}$$

ADDITIONAL SOLVED PROBLEMS

SOLVED PROBLEM 9

The resistance of N/10 solution is found to be 2.5×10^3 ohms. Calculate the equivalent conductance of the solution. Cell constant $= 1.15\ \text{m}^{-1}$

SOLUTION : ***(a)*****To calculate specific conductance**

Formula used

$$\kappa = \frac{l}{R} \times \textit{cell constant}$$

Quantities given

$R = 2.5 \times 10^3$ ohms *cell constant* $= 1.15\ \text{m}^{-1}$

Substitution of values

$$\kappa = \frac{1}{2.5 \times 10^3\ \text{ohm}} \times 1.15\ \text{m}^{-1}$$

$$= 4.60 \times 10^{-4}\ \text{ohm}^{-1}\ \text{m}^{-1}$$

(*b*) To calculate equivalent conductance

Formula used

$$\text{Equivalent conductance} = \frac{\kappa \times 1000}{N}$$

Quantities given

$\kappa = 4.60 \times 10^{-4}\ \text{ohm}^{-1}\ \text{m}^{-1}$ $N = 0.1\ \text{eqvt}^{-1}$

Substitution of values

$$\text{Equivalent conductance} = \frac{4.60 \times 10^{-4}\ \text{ohm}^{-1}\ \text{m}^{-1} \times 0.01\ \text{m}^3}{0.1\ \text{eqvt}^{-1}}$$

$$= \mathbf{0.000046\ ohm^{-1}\ m^2\ eqvt^{-1}}$$

$$= \mathbf{0.000046\ Siemens\ m^2\ eqvt^{-1}}$$

SOLVED PROBLEM 10

The specific conductance of N/5 KCl solution at 298 K is 0.002780 Siemens. The resistance of the cell containing this solution is 500 ohm. Calculate the cell constant.

SOLUTION

Formula used

$$\text{Specific conductance} = \frac{1}{R} \times \text{cell constant}$$

or

$$\text{Cell constant} = \text{Specific conductance} \times \text{Resistance}$$

Quantities given

Specific conductance = 0.002780 Siemens $\qquad R = 500$ ohm

Substitution of values

$$\text{Cell constant} = 0.002780 \text{ Siemens} \times 500 \text{ ohm}$$
$$= \mathbf{1.39}$$

SOLVED PROBLEM 11

The resistance of a 0.5 N solution of an electrolyte occupying a volume between two platinum electrodes which are 1.72×10^{-2} m apart and have an area of 4.5×10^{-4} sq m is 25 ohms. Calculate the equivalent conductance of the solution.

SOLUTION : (*a*) To calculate specific conductance

Formula used

$$\text{Specific conductance} = \frac{1}{R} \times \frac{l}{a}$$

Quantities given

$R = 25$ ohm $\qquad l = 1.72 \times 10^{-2}$ m $\qquad a = 4.5 \times 10^{-4}$ sq m

Substitution of values

$$\text{Specific conductance} = \frac{1}{25 \text{ ohm}} \times \frac{1.72 \times 10^{-2} \text{ m}}{4.5 \times 10^{-4} \text{ sq m}}$$
$$= 38.22 \text{ ohm}^{-1} \text{ m}^{-1}$$

(*b*) To calculate equivalent conductance

Formula used

$$\text{Equivalent conductance} = \frac{\textit{Specific conductance} \times 0.001 \text{ m}^3}{N}$$

Quantities given

Specific conductance = $38.22 \text{ ohm}^{-1} \text{ m}^{-1}$ $\qquad N = 0.5$ eqvt

Substitution of values

$$\text{Equivalent conductance} = \frac{38.22 \text{ ohm}^{-1} \text{ m}^{-1} \times 0.001 \text{ m}^3}{0.5 \text{ eqvt}}$$
$$= \mathbf{0.07644 \text{ ohm m}^2 \text{ eqvt}^{-1}}$$

SOLVED PROBLEM 12

Following results were obtained by conductance measurements of potassium sulfate using the cell with cell constant 0.2281. The observed conductance of potassium sulfate solution is 3×10^{-3} Siemens. The equivalent conductance of potassium sulfate solution is 140×10^{-4} mho m^2. Calculate the concentration of potassium sulfate solution..

SOLUTION : (*a*) To calculate the specific conductance

Formula used

$$\text{Cell constant} = \frac{\textit{Specific conductance}}{\textit{Observed conductance}}$$

Quantities given

Cell constant = 0.2281 *Observed conductance* = 3×10^{-3} Siemens

Substitution of values

$$0.2281 = \frac{\textit{Specific conductance}}{3 \times 10^{-3} \text{ Siemens}}$$

or
$$\text{Specific Conductance} = 3 \times 10^{-3} \text{ Siemens} \times 0.2281 = 0.6843 \times 10^{-3} \text{ Siemens}$$

(*b*) To calculate concentration of the potassium sulfate solution

Formula used

$$\text{Equivalent conductance} = \frac{\textit{Specific conductance} \times \textit{1000}}{N}$$

Quantities given

Specific conductance = 0.6843×10^{-3} Siemens Equivalent conductance = 140×10^{-4} mhos m^2

Substitution of values

$$140 \times 10^{-4} \text{ Siemens m}^2 \text{ eqvt}^{-1} = \frac{0.6843 \times 10^{-3} \text{ Siemens} \times 0.001 \text{ m}^3}{N}$$

or
$$N = \frac{0.6843 \times 10^{-3} \times 0.001 \text{ m}^3}{140 \times 10^{-4}} \text{ eqvt}$$

$$= \mathbf{4.89 \times 10^{-5} \text{ eqvt}}$$

SOLVED PROBLEM 13

0.1 N solution of sodium acetate was placed between two electrodes which were 0.72×10^{-2} m apart and each has a cross-section 2.25×10^{-4} m^2. The resistance of solution was 52.40 ohms. Find the specific and equivalent conductance.

SOLUTION : (*a*) To calculate the specific conductance

Formula used

$$\text{Specific conductance} = \frac{1}{R} \times \frac{l}{a}$$

Quantities given

R = 52.40 ohm $l = 0.72 \times 10^{-2}$ m $a = 2.25 \times 10^{-4}$ m^2

Substitution of values

$$\text{Specific conductance} = \frac{1}{52.40 \text{ ohm}} \times \frac{0.72 \times 10^{-2} \text{ m}}{2.25 \times 10^{-4} \text{ m}^2} = \mathbf{0.6107 \text{ ohm}^{-1} \text{ m}^{-1}}$$

(*b*) To calculate the equivalent conductance

Formula used

$$\text{Equivalent conductance} = \frac{\textit{Specific conductance} \times \textit{1000}}{N}$$

Quantities given

Specific conductance = 0.6107 ohm^{-1} m^{-1} N = 0.1 eqvt

Substitution of values

$$\text{Equivalent conductance} = \frac{0.6107 \text{ ohm}^{-1} \text{ m}^{-1} \times 0.001 \text{ m}^3}{0.1 \text{ eqvt}}$$

$$= \mathbf{0.006107 \text{ ohm}^{-1} \text{m}^2 \text{eqvt}^{-1}}$$

SOLVED PROBLEM 14

A conductance cell when filled with 0.05 solution of KCl records a resistance of 410.5 ohm at 298 K. When filled with calcium chloride solution (11g $CaCl_2$ in 500 ml) it records 990 ohms. If the specific conductance of 0.05M KCl solution is 0.189 mho m^{-1}, calculate (i) the cell constant, (ii) specific conductance (iii) molecular conductance of $CaCl_2$.

SOLUTION : ***(a)* To calculate the cell constant**

Formula used

$$\text{Cell constant} = \frac{\textit{Specific conductance}}{\textit{Observed conductance}}$$

Quantities given

Specific conductance = 0.189 mho m^{-1}

$$\text{Observed conductance} = \frac{1}{R} = \frac{1}{410.5 \text{ ohm}}$$

Substitution of values

$$\text{Cell constant} = \frac{0.189 \text{ mho m}^{-1}}{1 / 410.5 \text{ ohm}} = \mathbf{77.58\ m^{-1}}$$

***(b)* To calculate specific conductance of $CaCl_2$ Solution**

Formula used

$$\text{Specific conductance} = \frac{1}{R} \times \textit{cell constant}$$

Quantities given

R = 990 ohm *cell constant* = 77.58 m^{-1}

Substitution of values

$$\text{Specific conductance} = \frac{1}{990 \text{ ohm}} \times 77.58 \text{ m}^{-1}$$

$$= \mathbf{0.0784\ ohm^{-1}\ m^{-1}}$$

***(c)* To calculate molar conductance of $CaCl_2$**

Formula used

$$\text{Molar conductance, } \mu = \frac{\kappa \times 0.001 \text{ m}^3}{M}$$

Quantities given

κ = 0.0784 ohm^{-1} m^{-1} $M = \dfrac{22 \text{ g}}{111 \text{ g mol}^{-1}} = 0.198$ mol

Substitution of values

$$\mu = \frac{0.0784 \text{ ohm}^{-1} \text{ m}^{-1} \times 0.001 \text{ m}^3}{0.198 \text{ mol}}$$

$$= \mathbf{3.96 \times 10^{-4}\ m^2\ mol^{-1}}$$

SOLVED PROBLEM 15

The conductance of N/10 $AgNO_3$ solution taken in a cell with cell constant 95.55 m^{-1} is 0.0099 ohm^{-1}. Calculate (i) specific conductance (ii) equivalent conductance

SOLUTION : ***(a)* To calculate specific conductance**

Formula used

$$\kappa = \frac{1}{R} \times \text{Cell constant}$$

Quantities given

$\frac{1}{R} = 0.0099 \text{ ohm}^{-1}$ *cell constant* $= 95.55 \text{ m}^{-1}$

Substitution of values

$$\kappa = 0.0099 \text{ ohm}^{-1} \times 95.55 \text{ m}^{-1}$$
$$= \mathbf{0.946 \text{ ohm}^{-1} \text{ m}^{-1}}$$

(*b*) To calculate equivalent conductance

Formula used

$$\text{Equivalent conductance} = \frac{\textit{Sp. conductance} \times \textit{1000}}{N}$$

Quantities given

Sp. conductance = 0.946 $\text{ohm}^{-1}\text{m}^{-1}$ $N = 1/10 = 0.1$ eqvt

Substitution of values

$$\text{Equivalent conductance} = \frac{0.946 \text{ ohm}^{-1}\text{m}^{-1} \times 0.001 \text{ m}^3}{0.1 \text{ eqvt}}$$
$$= \mathbf{0.00946 \text{ Siemens m}^2\text{eqvt}^{-1}}$$

SOLVED PROBLEM 16

The specific conductance of N/50 KCl solution is 0.2765 $\text{ohm}^{-1}\text{m}^{-1}$ at 298 K. If the resistance of the solution contained in the cell is 1000 ohms, calculate the cell constant.

SOLUTION

Formula used

$$\text{Cell constant} = \frac{\textit{Specific conductance}}{\textit{Observed conductance}}$$

Quantities given

$$\text{Specific conductance} = 0.2765 \text{ ohm}^{-1} \text{ m}^{-1}$$
$$\text{Observed conductance} = \frac{1}{R} = \frac{1}{1000 \text{ ohm}} = 0.001 \text{ ohm}^{-1}$$

Substitution of values

$$\text{Cell constant} = \frac{0.2765 \text{ ohm}^{-1} \text{ m}^{-1}}{0.001 \text{ ohm}^{-1}}$$
$$= \mathbf{276.5 \text{ m}^{-1}}$$

SOLVED PROBLEM 17

What current strength in ampere will be required to deposit 0.1 kg of silver from $AgNO_3$ solution in 3600 seconds?

SOLUTION

Formula used

$$m = Z \times I \times t$$

Quantities given

$t = 3600$ s $m = 0.1$ kg $Z = \frac{108 \times 10^{-3} \text{ g}}{96500 \text{ C}}$

Substitution of values

$$0.1 = \frac{108 \times 10^{-3} \text{ kg}}{96500} \times I \times 3600 \text{ s}$$

or

$$I = \frac{0.1 \times 96500}{108 \times 10^{-3} \times 3600} \text{ ampere}$$
$$= \mathbf{24.82 \text{ ampere}}$$

EXAMINATION QUESTIONS

1. Define or explain the following terms :
   (*a*) Electrolysis (*b*) Coulomb
   (*c*) Faraday's laws of electrolysis (*d*) Specific conductance
   (*e*) Equivalent conductance (*f*) Molar conductance
   (*g*) Degree of Dissociation (*h*) Cell constant
2. (*a*) Define specific and molar conductance.
   (*b*) In a particular cell, 0.01 M solution of KCl gave a resistance of 15.0 ohms at 298 K while 0.01 M solutions of HCl gave a resistance of 51.4 ohm at the same temperature. If the specific conductance of 0.01 M KCl is 0.1409 S m^{-1} at 280 K, calculate the cell constant, specific conductance and equivalent conductance of the HCl solution.
   Answer. 211.2 m^{-1}; 4.10 ohm^{-1} m^{-1}; 41 $\times 10^4$ ohm^{-1} m^2 $eqvt^{-1}$
3. The specific conductance of N/5 KCl solution at 298 K is 0.002780 Siemens. The resistance of the cell containing this solution is 500 ohm. Calculate cell constant.
   Answer. 1.39
4. If equivalent conductance at infinite dilution of NaCl, HCl and CH_3COONa are 126.45, 426.16 and 91.0 respectively, find the equivalent conductance of acetic acid at infinite dilution.
   Answer. 390.71 Siemens
5. The equivalent conductance of ammonium chloride at infinite dilution is 149.7 Siemens; for sodium hydroxide it is 247.8 Siemens; and for sodium chloride is 126.45 Siemens at 298 K. Calculate the equivalent conductance for ammonium hydroxide in mho at infinite dilution at the same temperature.
   Answer. 271.05 Siemens
6. How is the specific conductance of an electrolyte solution determined? Describe the experimental method. The resistance of a N/100 solution of an electrolyte was found to be 210 ohm at 298 K. Calculate the equivalent conductance of the solution at 298 K. (Cell constant = 88 m^{-1})
   Answer. 0.0419 ohm^{-1} m^2 $eqvt^{-1}$
7. A conductance cell had a resistance of 165 ohms when filled with 0.02 molar KCl solution at 298 K. For such solution specific conductance is 0.00277 Siemens/m. The same cell filled with 0.01 molar NaCl had a resistance of 384 ohms. Calculate specific conductance and equivalent conductance of solution.
   Answer. 1.190×10^{-3} ohm^{-1} m^{-1}; 1.190×10^{-4} ohm^{-1} m^{-1} $eqvt^{-1}$
8. The conductance of N/10 $AgNO_3$ solution taken in a cell with cell constant 95.55 m^{-1} is 0.0099 ohm^{-1}. Calculate :
   (*a*) Specific conductance (*b*) Equivalent conductance
   Answer. (*a*) 0.946 mho m^{-1}; (*b*) 94.59×10^{-4} Siemens m^2 $eqvt^{-1}$
9. The resistance of a cell containing 0.02 M KCl solution at 298 K was found to be 175 ohms. The specific conductance of this solution is 27.7×10^{-2} ohm^{-1} m^{-1} at 298 K. An exactly 0.01 M solution of another substance in the same cell had a resistance of 579 ohms. Calculate the equivalent conductance of this substance.
   Answer. 83.72×10^{-4} ohm^{-1} m^2 $eqvt^{-1}$
10. Define conductance, specific conductance, equivalent conductance and molar conductance.
    (Himachal Pradesh BSc)
11. Define specific conductance and equivalent conductance. Derive the relationship between them.
    (Himachal Pradesh BSc)

12. Calculate equivalent conductivity from the following data :

(*i*) 0.1 N solution has a resistance of 2.5×10^2 ohms

(*ii*) Cell constant = 1.15×10^2 m^{-1}

Answer. 46×10^{-4} ohm^{-1} m^2 eqvt^{-1} *(Andhra BSc)*

13. (*a*) Write a short note on variation of equivalent conductance with dilution.

(*b*) Write down the methods for determination of equivalent conductance at infinite dilution for strong and weak electrolytes.

(*c*) How would you measure the conductivity of an aqueous salt solution? *(Aligarh BSc)*

14. Explain the difference between molecular and equivalent conductance of a solution. What are the units for the two terms. *(Delhi BSc)*

15. "On progressive dilution, specific conductance of an electrolyte decreases but molar conductance increases" discuss. *(Arunachal BSc)*

16. Define specific conductance, conductivity, equivalent conductivity and molar conductivity. What are the relations between them? What effect do they produce upon dilution? *(HS Gaur BSc)*

17. Describe the applications of conductance measurements. *(Allahabad BSc)*

18. Show that:

$$\text{Equivalent conductance} = \text{specific conductance} \times \frac{1000}{C}$$

where '*C*' is concentration in g eqvt. per liter. *(Allahabad BSc)*

19. 0.5 N solution of a salt surrounding two platinum electrodes 0.021 cm apart and 2.4×10^{-4} square meter in area was found to offer a resistance of 250 ohms. Calculate the equivalent conductance of the solution.

Answer. 7.0×10^{-6} ohm^{-1} m^2 eqvt^{-1} *(Arunachal BSc)*

20. Using the plot of molar conductance versus concentration of the solution, discuss how the molar conductance at infinite dilution of the solution can be obtained. *(Kalyani BSc)*

21. Define equivalent conductance and molar conductance. Establish the relation between specific conductance and equivalent conductance. *(Kalyani BSc)*

22. The molar conductance of a solution of aluminium chloride is found to be 130×10^{-4} ohm m^2 mol^{-1} at 298 K. What would be its equivalent conductance at the same temperature.

Answer. 43.33×10^{-4} ohm^{-1} m^2 eqvt^{-1} *(Sambalpur BSc)*

23. The conductivity of a solution containing 0.001 kg of anhydrous $BaCl_2$ in 200×10^{-6} m^3 of water has been found to be 0.58 ohm^{-1} m^{-1}. What is the

(*i*) Molar conductance and

(*ii*) Equivalent conductance of the solution? (Atomic mass of Ba = 137 and Cl = 35.5)

Answer. 241.26×10^{-4} ohm^{-1} m^2 mol^{-1}; 120.63×10^{-4} ohm^{-1} m^2 eqvt^{-1} *(Arunachal BSc)*

24. 0.5 N NaCl is placed between two electrodes 1.5×10^{-2} m apart and having an area of each 3×10^{-4} sq m offered a resistance of 25.0 ohms. Calculate the equivalent conductance.

Answer. 40×10^{-3} ohm^{-1} m^2 eqvt^{-1} *(Andhra BSc)*

25. (*a*) Define Molar, Equivalent and Specific conductance and discuss the effect of dilution.

(*b*) Resistance of 0.01 M aqueous solution of an electrolyte at room temperature is 420 ohm and cell constant is 0.84×10^2 m^{-1}. Calculate the Molar conductance.

Answer. (*b*) 2×10^4 ohm^{-1} m^2 mol^{-1} *(Madras BSc)*

26. When molten lithium chloride, LiCl, is electrolyzed, lithium metal is liberated at the cathode. How many grams of lithium are liberated when 500 coulomb of charge passes through the cell ?

Answer. 0.0360×10^{-3} *(Sambalpur BSc)*

27. Calculate the quantity of electricity that will be required to produce 355 kg chlorine gas by the electrolysis of a concentrated solution of NaCl. Also calculate the volume of hydrogen gas liberated at 300 K and 1 Pa pressure during the process.

Answer. 10 F ; 123.07 m^3 *(Andhra BSc)*

28. Specific conductance of 0.1 N solution of an electrolyte is 2 $ohm^{-1}m^{-1}$. Calculate its equivalent conductance.

Answer. 2×10^4 $ohm^{-1}m^2eqvt^{-1}$ *(Agra BSc)*

29. A solution of copper sulfate weighing 20 kg was electrolyzed using 0.02 Faraday of electricity. Calculate the weight of the resulting solution.

Answer. 9.205 kg *(Madurai BSc)*

30. Specific conductance of 0.02 N KCl at 298 K is 0.002768 ohm^{-1} and it has resistance of 500 ohms. An 0.25 N solution of another salt kept in the same cell was found to have resistance of 300 ohms at 298 K. Calculate the cell constant and equivalent conductance of the salt solution.

Answer. 1.384; 18.45 *(Vidyasagar BSc)*

31. A current 4.0 amperes is passed for 8 seconds between nickel electrodes in 500 m^3 of 2M solution of nickel nitrate. What will be the molarity of the solution at the end of electrolysis ?

Answer. 0.806 M *(Baroda BSc)*

32. The resistance of 0.1 N solution of a electrolyte in a cell of cell constant 84 m^{-1} is 60 ohm. Calculate the resistivity of the cell.

Answer. 7.1429×10^{-5} *(Shivaji BSc)*

33. The resistance of a 0.5 N solution of an electrolyte occupying a volume between two platinum electrodes which are 1.72×10^{-2} m apart and have an area of 4.5×10^{-4} sqm is 25 ohms. Calculate the equivalent conductance of the solution.

Answer. 3.058×10^3 *(Aligarh BSc)*

34. The resistance of an N/10 solution is found to be 2.5×10^3 ohms. Calculate the equivalent conductance of the solution. Cell constant = 1.15×10^2 m^{-1}.

Answer. 4.6×10^2 ohm^{-1} $eqvt^{-1}$ *(Anna BSc)*

35. State and explain Faraday's laws of Electrolysis. Following results were obtained by conductance measurements of potassium sulfate using the cell with cell constant 0.2281. The observed conductance of potassium sulfate solution is 3×10^{-3} Siemens. The equivalent conductance of potassium sulfate solution is 140×10^{-3} Siemens m^2. Calculate the concentration of potassium sulfate solution.

Answer. 4.888 N *(Baroda BSc)*

36. What is cell constant? How is it determined experimentally? The specific conductance of N/50 KCl solution is 0.2765 ohm^{-1} m^{-1} at 298 K. If the resistance of the solution contained in the cell is 100 ohms, calculate the cell constant.

Answer. 27.65 m^{-1} *(Madras BSc)*

37. (*a*) Define (*i*) Specific conductance (*ii*) Equivalent conductance (*iii*) Molecular conductance and (*iv*) Cell constant. Give their units.

(*b*) Find specific conductivity of the solution if 0.5 solution of a salt occupying a volume between two Pt electrodes 1.72×10^{-2} m apart and of area 4.5×10^{-4} sqm has a resistance of 15 ohms.

Answer. (*b*) 2.548 ohm^{-1} m^{-1} *(Himachal Pradesh BSc)*

MULTIPLE CHOICE QUESTIONS

1. One faraday of electricity is passed through an HCl electrolyte solution. Select the correct electrode result.

(*a*) 1×10^{-3} kg of chloride ions is deposited at the anode
(*b*) 1×10^{-3} kg of hydrogen ions is deposited at the cathode
(*c*) 5×10^{-3} kg of hydrogen ions are deposited at the anode
(*d*) 35×10^{-3} kg of chloride ions are deposited at the anode

Answer. (*b*)

2. One faraday will oxidize _____ mole(s) of Cu to Cu^{2+} ions.
(*a*) 0 (*b*) 1/2
(*c*) 1/4 (*d*) 1
Answer. (*b*)

3. How much time (in hours) is required to plate out 25 kg of gold metal from a solution of $Au(NO_3)_3$ when the current is 2.00 amperes and the electrode efficiency is only 65%?
(*a*) 9.36 hr (*b*) 2.88 hr
(*c*) 3.11 hr (*d*) 7.85 hr
Answer. (*d*)

4. How many faradays of charge are required to electroplate 127 kg of copper from a 2 M cuprous chloride solution?
(*a*) 1 (*b*) 2
(*c*) 4 (*d*) 6
Answer. (*d*)

5. Cu metal displaces $Ag^+(aq)$ from an aqueous solution. Which of the following is correct?
(*a*) Ag is a better reducing agent than Cu (*b*) Ag^+ is easier to reduce than Cu^{2+}
(*c*) Cu^{2+} is a better oxidizing agent than Ag^+ (*d*) Ag is easier to oxidize than Cu
Answer. (*b*)

6. Predict the products from the electrolysis of aqueous silver sulfate.
(*a*) silver metal and sulfur (*b*) hydrogen and oxygen
(*c*) hydrogen and sulfur (*d*) silver metal and oxygen
Answer. (*d*)

7. Which of these metals will not dissolve in hydrochloric acid under standard conditions?
(*a*) zinc (*b*) aluminium
(*c*) copper (*d*) magnesium
Answer. (*c*)

8. Which two of the following metals do not react with HCl(aq)?
Mg, Ag, Zn, Fe, Au
(*a*) Ag and Zn (*b*) Mg and Fe
(*c*) Ag and Au (*d*) Zn and Mg
Answer. (*c*)

9. What mass in kilograms of copper will be deposited from a solution of Cu^{2+} by a current of 2.50 A in 2 seconds?
(*a*) 23.7 (*b*) 0.187
(*c*) 1.65 (*d*) 5.93
Answer. (*d*)

10. How many ml of $H_2(g)$, measured at STP, are produced at a platinum cathode in the electrolysis of $H_2SO_4(aq)$ by 2.45 A of electric current in 5.00 seconds?
(*a*) 7.70×10^{-3} ml H_2 (*b*) 85.4 ml H_2
(*c*) 171 ml H_2 (*d*) 1.42 ml H_2
Answer. (*b*)

11. In the electrolysis of aqueous NaCl, how many cubemeter of $Cl_2(g)$ (at STP) are generated by a current of 7.50 A for a period of 100 seconds?
(*a*) 10.4 m^3 (*b*) 45000 m^3
(*c*) 5.22 m^3 (*d*) 0.466 m^3
Answer. (*c*)

12. What volume (in *liters*) of dry oxygen gas (at STP) could be generated at the anode by the electrolysis of water over a 6.0 hour period with a current of 1.5 amperes?
(*a*) 1.9 L (*b*) 2.1 L
(*c*) 2.5 L (*d*) 3.2 L
Answer. (*a*)

13. Among lead, iron, chromium, and copper, the least easily oxidized metal is
(*a*) chromium (*b*) copper
(*c*) iron (*d*) lead
Answer. (*b*)

14. Select the incorrect statement among the following :
(*a*) gold is the least active metal (*b*) iron will replace maganese in a compound
(*c*) potassium is most easily oxidized (*d*) silver is relatively unreactive
Answer. (*b*)

15. Select the incorrect statement about the chemical activity at electrodes during electrolysis.
(*a*) anions give up electrons (*b*) cations take up electrons
(*c*) oxidation occurs at the anode (*d*) proton transfer occurs in the reactions
Answer. (*d*)

16. Copper metal will replace silver ions in solution, resulting in the production of silver metal and copper ions. This indicates that
(*a*) silver has a higher oxidation potential than copper
(*b*) a combustion reaction is occurring
(*c*) copper has a higher oxidation potential than silver
(*d*) silver is much less soluble than copper
Answer. (*c*)

17. How many grams of copper would be produced by the reduction of Cu^{2+} if 3.0 amperes of current are passed through a copper (II) nitrate solution for one second?
(*a*) 18.20 (*b*) 3.56
(*c*) 31.80 (*d*) 63.50
Answer. (*b*)

18. How many grams of Ni can be electroplated from a solution of nickel chloride by four faradays of electricity?
(*a*) 29.3 (*b*) 58.7
(*c*) 117.4 (*d*) 176.1
Answer. (*c*)

19. How many grams of Cu could be produced from $CuSO_4$ by 0.5 faradays of charge?
(*a*) 15.9 (*b*) 63.5
(*c*) 127.0 (*d*) 31.75
Answer. (*a*)

20. How many grams of copper will be deposited from a solution of $CuSO_4$ by a current of 3 amperes in 2 second?
(*a*) 5 (*b*) 7
(*c*) 8 (*d*) 11
Answer. (*b*)

21. Equal volumes of the following solutions are electrolyzed, using inert platinum electrodes and a current of 1.00 A, until the solution concentration falls to one–half its initial value. Which solution will take the longest time?
(*a*) 0.50 M $Cu(NO_3)_2$(aq) (*b*) 0.30 M $Zn(NO_3)_2$
(*c*) 0.80 M $AgNO_3$(aq) (*d*) 0.25 M $Au(NO_3)_3$(aq)
Answer. (*a*)

22. An electrolytic cell is set up for the production of aluminum, which involves the reduction of Al^{3+} to Al. The external source passes a current of 11.2 A through the cell with an emf of 6.0 V. How long does it take for the cell to produce a (454 kg) of aluminum metal?

(*a*) 226 (*b*) 40
(*c*) 121 (*d*) 3.26×10^3

Answer. (*c*)

23. A current of 5.0 Amps is passed through molten magnesium chloride, $MgCl_2$, for 3.0 seconds. How many grams of magnesium, Mg(s), can be produced by this reduction?
Faraday constant : 1 F = 96,485 C/mol. molar atomic mass (Mg) = 24.305.

(*a*) 0.30 (*b*) 6.80
(*c*) 17.6 (*d*) 24.5

Answer. (*b*)

24. What mass of Cu(s) can be produced by electrolysis of Cu^{2+} using a current of 2.5 A over a period of 2.0 hr?

(*a*) 5.9 (*b*) 1.8×10^4
(*c*) 9.3×10^{-2} (*d*) 11.9

Answer. (*a*)

25. Chromium metal can be produced from an acidic solution of chromate, CrO_4^{2-}.

$$8\,H^+(aq) + CrO_4^{2-}(aq) + 6e^- \rightarrow Cr(s) + 4\,H_2O(\ell)$$

How many grams of chromium can be produced by the passage of 1 Amp for 10 hours?
Faraday constant : 1 F = 96,485 C/mol. molar atomic mass (Cr) = 51.996×10^{-3} kg/mol.

(*a*) 0.40×10^{-3} kg (*b*) 3.23×10^{-3} kg
(*c*) 19.4×10^{-3} kg (*d*) 40.8×10^{-3} kg

Answer. (*b*)

26. Specific conductance is the conductance of

(*a*) one centimeter cube of solution of an electrolyte
(*b*) one centimeter cube of a solid electrolyte
(*c*) one gram of the solution of an electrolyte
(*d*) one gram of the solid electrolyte

Answer. (*a*)

27. The units of specific conductance are

(*a*) ohm cm (*b*) ohm cm^{-1}
(*c*) ohm^{-1} cm (*d*) ohm^{-1} cm^{-1}

Answer. (*d*)

28. The equivalent conductance of a solution of an electrolyte

(*a*) increases with dilution (*b*) decreases with dilution
(*c*) does not vary with dilution (*d*) none of these

Answer. (*a*)

29. The SI units of equivalent conductance are

(*a*) Siemens m^2 eqvt (*b*) ohm^{-1} cm^{-1} $eqvt^{-1}$
(*c*) ohm^{-1} cm^2 $eqvt^{-1}$ (*d*) ohm^{-1} cm^{-2} $eqvt^{-1}$

Answer. (*c*)

30. The molar conductance of solution of an electrolyte is measured in

(*a*) ohm cm mol^{-1} (*b*) ohm^{-1} cm^{-1} mol^{-1}
(*c*) ohm cm^{-1} mol^{-1} (*d*) ohm^{-1} cm^2 mol^{-1}

Answer. (*d*)

31. With rise in temperature the conductance of a solution of an electrolyte generally

(*a*) decreases (*b*) increases
(*c*) remains constant (*d*) none of these

Answer. (*b*)

32. The cell constant can be obtained by

(*a*) dividing specific conductance by observed conductance

(*b*) dividing observed conductance by specific conductance

(*c*) multiplying specific conductance by observed conductance

(*d*) multiplying specific conductance by equivalent conductance

Answer. (*a*)

33. The cell constant is the ratio of

(*a*) distance between electrodes to area of electrode

(*b*) area of electrode to distance between electrodes

(*c*) specific conductance to area of electrode

(*d*) specific conductance to distance between the electrodes

Answer. (*d*)

34. When electricity is passed through acidulated water, 224 m^3 of hydrogen gas at STP is collected at the cathode in 965 s. The current passed in ampere is

(*a*) 0.5 (*b*) 1.0

(*c*) 1.5 (*d*) 2.0

Answer. (*d*)

35. What weight of copper will be deposited by passing 1 Faraday of electricity through cupric salt?

(*a*) 3.175 (*b*) 2.0

(*c*) 63.5 (*d*) 31.75

Answer. (*d*)

36. Which out of the following will decompose on passing electric current?

(*a*) glucose (*b*) urea

(*c*) silver nitrate (*d*) ethyl alcohol

Answer. (*c*)

37. An ion is reduced to the element when it absorbs 6×10^{19} electrons. The number of equivalents of the ion is

(*a*) 0.0001 (*b*) 0.001

(*c*) 0.01 (*d*) 0.1

Answer. (*a*)

38. On passing 0.1 Faraday of electricity through $AlCl_3$, the amount of aluminium metal deposited on the cathode is (at mass of Al = 27×10^{-3})

(*a*) 0.27×10^{-3} kg (*b*) 0.81×10^{-3} kg

(*c*) 0.9×10^{-3} kg (*d*) 13.5×10^{-3} kg

Answer. (*c*)

39. When 96500 coulombs of electricity is passed through an aqueous solution of Nickel Chloride (at mass of Ni = 58.5×10^{-3}) the weight of nickel metal deposited would be

(*a*) 5.85×10^{-3} kg (*b*) 29.25×10^{-3} kg

(*c*) 58.5×10^{-3} kg (*d*) 117×10^{-3} kg

Answer. (*b*)

40. The equivalent weight of a metal is given by

(*a*) $\frac{c \times t \times 96500}{m}$ (*b*) $\frac{96500 \times m}{c \times t}$

(*c*) $\frac{c \times m}{t \times 96500}$ (*d*) $\frac{c \times t}{m \times 96500}$

Answer. (*b*)

41. Two electrolytic cells, one containing Cuprous chloride and the other Cupric chloride, are connected in series. The ratio of ions deposited at cathodes in two cells when electricity is passed through the cells will be

(*a*) 1:1 (*b*) 1:2
(*c*) 2:1 (*d*) 1:3

Answer. (*c*)

42. On passing 96500 coulombs of electricity through a dilute solution of an acid, the volume of hydrogen collected at STP is

(*a*) 1120 ml (*b*) 11200 ml
(*c*) 2240 ml (*d*) 22400 ml

Answer. (*b*)

43. A solution of sodium sulfate in water is electrolyzed using platinum electrodes. The products at anode and cathode respectively are

(*a*) SO_2, O_2 (*b*) Na, O_2
(*c*) H_2, O_2 (*d*) O_2, H_2

Answer. (*d*)

44. The cathodic reaction in electrolysis of dilute sulfuric acid with platinum electrode is

(*a*) neutralization (*b*) oxidation
(*c*) reduction (*d*) oxidation & reduction

Answer. (*a*)

45. If the specific conductance and conductance of a solution are same, then the cell constant is equal to

(*a*) 0 (*b*) 0.5
(*c*) 1.0 (*d*) 10.0

Answer. (*c*)

46. On passing one faraday of electricity, one mole of metal is deposited from the solution of

(*a*) KCl (*b*) $BaCl_2$
(*c*) $AlCl_3$ (*d*) none of these

Answer. (*a*)

47. A certain current liberated 1.008×10^{-3} g of hydrogen in 2 seconds. How many grams of copper can be deposited by the same current flowing for the same time in $CuSO_4$ solution. (at mass of Cu = 63.5)

(*a*) 31.75×10^{-3} kg (*b*) 63.5×10^{-3} kg
(*c*) 127×10^{-3} kg (*d*) 15.875×10^{-3} kg

Answer. (*a*)

48. The specific conductance of a 0.01 M solution of KCl is 1.4×10^{-3} ohm^{-1} m^{-1} at 298 K. Its equivalent conductance is

(*a*) 0.14 (*b*) 1.4
(*c*) 14.0 (*d*) 140

Answer. (*d*)

49. The distance between two electrodes of a cell is 3.0 m and area of each electrode is 6.0 m. The cell constant is

(*a*) 2.0 (*b*) 1.0
(*c*) 0.5 (*d*) 18

Answer. (*c*)

50. The specific conductance of NaCl solution at 18°C is 0.0124 ohm^{-1} m^2 eqvt^{-1} and the resistance of the cell containing the solution at the same temperature is 50.0 ohm. The cell constant will be

(*a*) 0.62 (*b*) 0.31
(*c*) 0.124 (*d*) 0.000248

Answer. (*a*)

25 Theory of Electrolytic Dissociation

CONTENTS

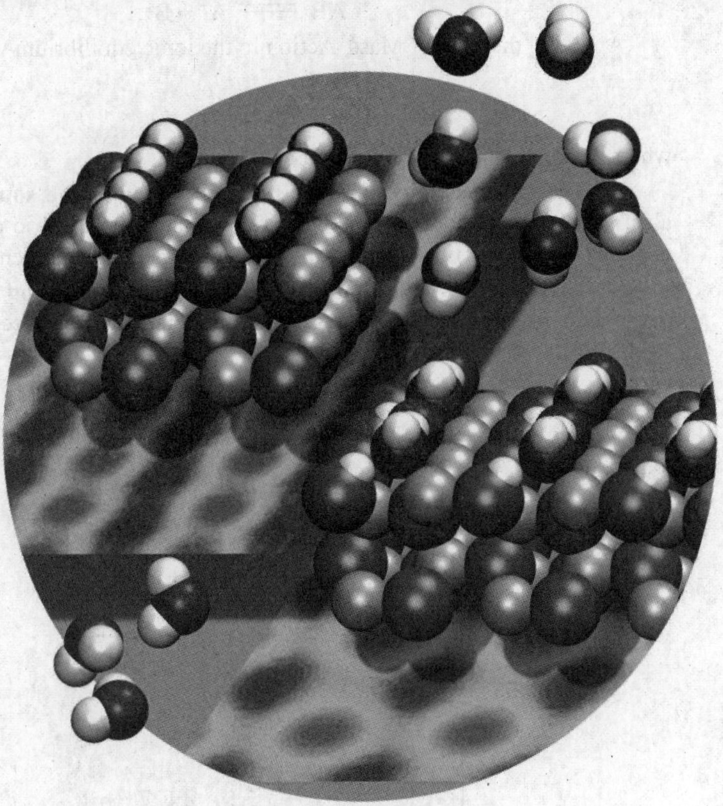

ARRHENIUS THEORY OF IONIZATION

Savante Arrhenius studied the conduction of current through water solutions of electrolytes. He came to believe that the conductivity of solutions was due to the presence of ions. In 1884, Arrhenius put forward his theory of ionization. **Arrhenius theory of ionization** may be stated as :

(1) When dissolved in water, **neutral electrolyte molecules are split up into two types of charged particles.** These particles were called ions and the process was termed **ionization.** The positively charged particles were called **cations** and those having negative charge were called **anions.**

In its modern form, the theory assumes that the ions are already present in the solid electrolyte and these are held together by electrostatic force. When placed in water, these neutral molecules *dissociate* to form separate anions and cations. Thus,

$$AB \longrightarrow A^+ + B^- \quad (\textit{Old view})$$

$$A^+B^- \longrightarrow A^+ + B^- \quad (\textit{Modern view})$$

For that reason, this theory may be referred to as the theory of electrolytic dissociations.

(2) The **ions present in solution constantly reunite to form neutral molecules.** Thus there is a state of equilibrium between the undissociated molecules and the ions.

$$AB \rightleftharpoons A^+ + B^-$$

Applying the Law of Mass Action to the ionic equilibrium we have,

$$\frac{[A^+][B^-]}{[AB]} = K$$

where K is called the **Dissociation constant.**

(3) The **charged ions are free to move through the solution to the oppositely charged electrode.** This movement of the ions constitutes the electric current through electrolytes. This explains the conductivity of electrolytes as well as the phenomenon of electrolysis.

(4) The **electrical conductivity of an electrolyte solution depends on the number of ions present in solution.** Thus the degree of dissociation of an electrolyte determines whether it is a strong electrolyte or a weak electrolyte.

MIGRATION OF IONS

We know that electrolytes dissociate in solution to form positive ions (cations) and negative ions (anions).

$$AgNO_3 \longrightarrow Ag^+ + NO_3^-$$

$$CuSO_4 \longrightarrow Cu^{2+} + SO_4^{2-}$$

$$H_2SO_4 \longrightarrow 2H^+ + SO_4^{2-}$$

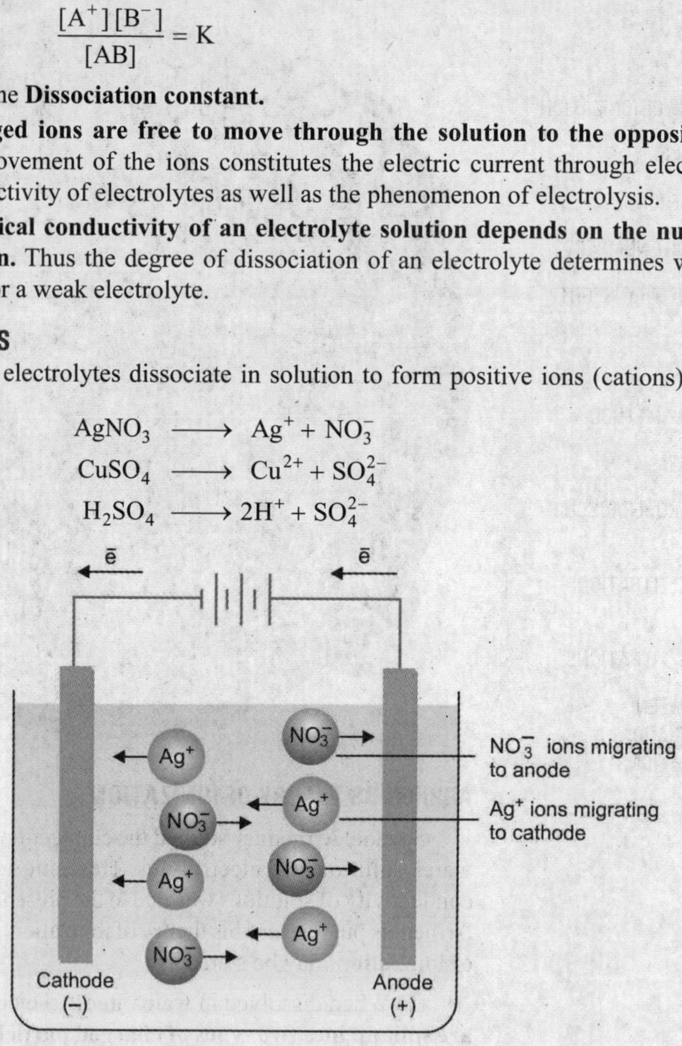

■ **Figure 25.1**
Migration of ions through electrolytic solution to opposite electrodes.

As the current is passed between the electrodes of the electrolytic cell, the ions migrate to the opposite electrodes. Thus in the electrolytic solution of $AgNO_3$, the cations (Ag^+) will move to the cathode and anions (NO_3^-) will move to the anode. Usually different ions move with different rates.

The migration of ions through the electrolytic solution can be demonstrated by the following experiments.

(1) Lodge's moving boundary experiment

The apparatus used consists of a U-tube which has a long horizontal portion. It is fitted with electrodes in the side limbs. The horizontal portion is filled with a jelly of agar-agar treated with a trace of alkali. This is then made red by addition of a few drops of phenolphthalein. When the jelly is set, dilute sulfuric acid is added in the anodic limb of the tube. Sodium sulfate solution is added in the cathodic limb. On passing the current, H^+ ions in the left limb solution eventually move into the agar-agar jelly. Their passage is marked by the gradual discharge of the red color due to the neutralization of the alkali by H^+ ions. The movement of the red boundary through the agar-agar jelly shows that H^+ ions migrate to the cathode limb.

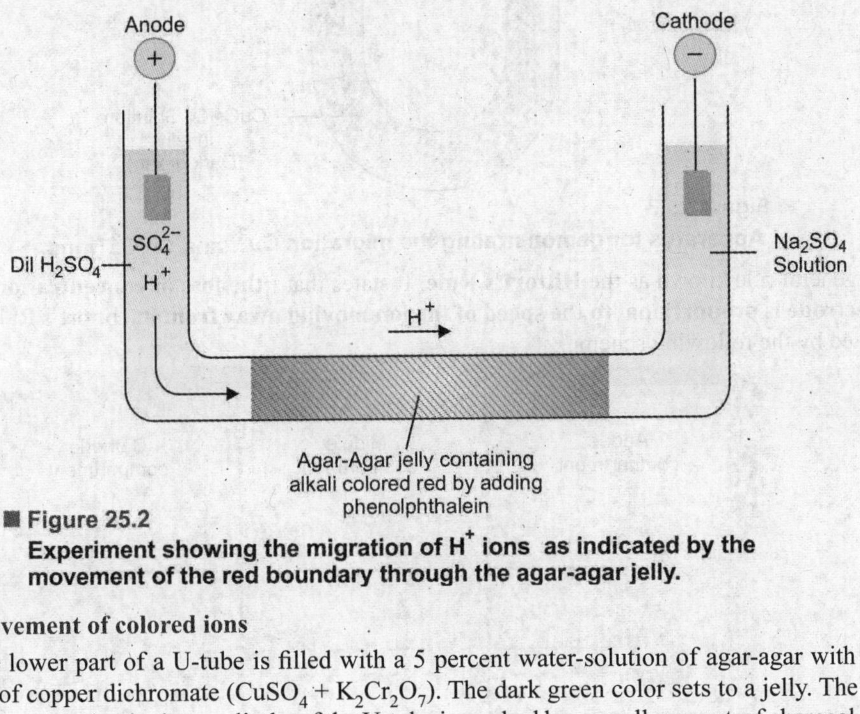

Figure 25.2
Experiment showing the migration of H^+ ions as indicated by the movement of the red boundary through the agar-agar jelly.

(2) Movement of colored ions

The lower part of a U-tube is filled with a 5 percent water-solution of agar-agar with a small amount of copper dichromate ($CuSO_4 + K_2Cr_2O_7$). The dark green color sets to a jelly. The surface of the green solution in the two limbs of the U-tube is marked by a small amount of charcoal. In both the limbs is then placed a layer of solution of potassium nitrate and agar-agar. This is also allowed to set. Over this second layer is placed some solution of potassium nitrate in pure water and the two electrodes are inserted in it. As the current is turned on, the blue color of Cu^{2+} ions rises into the jelly under the cathode. The reddish yellow dichromate ions ($Cr_2O_7^{2-}$) move up under the anode. After some time the two types of ions are seen rising with well-defined boundaries. The use of jelly in this experiment prevents the mixing of the solutions by diffusion.

RELATIVE SPEED OF IONS

We have already considered that ions move to the oppositely charged electrodes under the influence of the electric current. But the speeds of cations migrating towards the cathode and those of anions migrating towards the anode are not necessarily the same. However, the speed of a cation moving away from the anode will be proportional to the fall of concentration of these ions at the anode. Similarly, the speed of an anion moving away from the cathode will be proportional to the fall of concentration of anions around the cathode. Hittorf studied such changes experimentally and

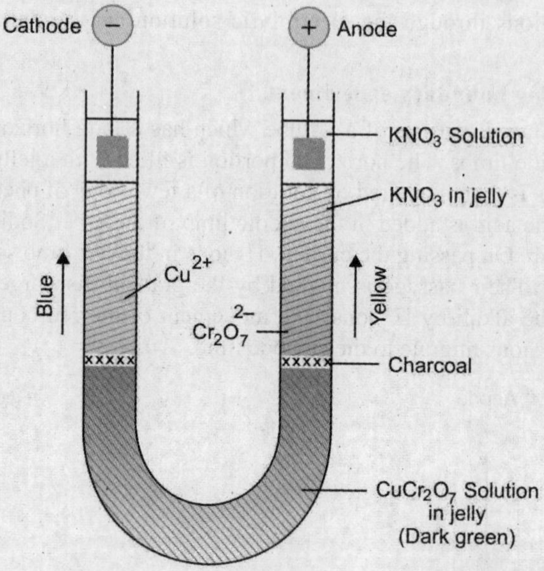

■ **Figure 25.3**
Apparatus for demonstrating the migration Cu^{2+} and $Cr_2O_7^{2-}$ ions.

gave a general rule known as the **Hittorf's Rule**. It states that : **the loss of concentration around any electrode is proportional to the speed of the ion moving away from it.** Hittorf's Rule may be illustrated by the following scheme.

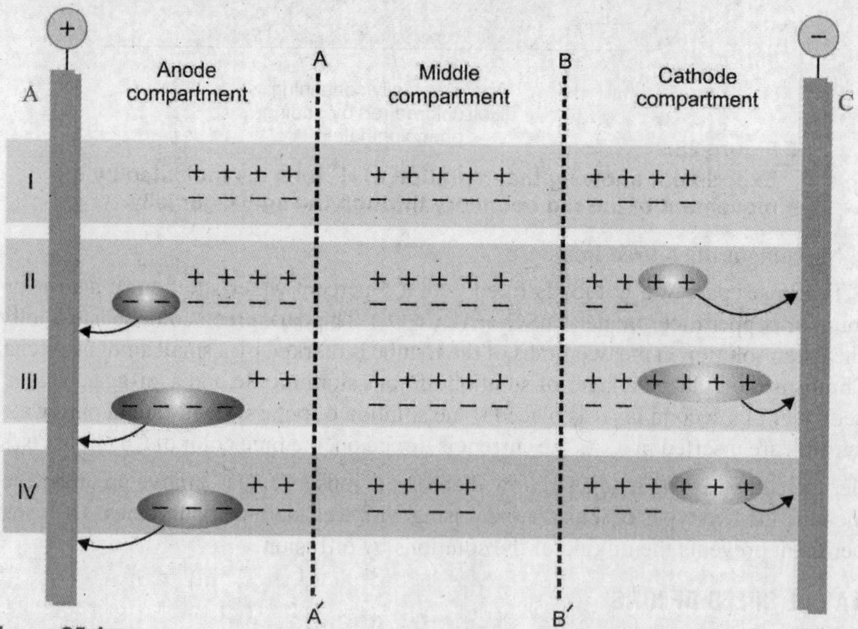

■ **Figure 25.4**
Scheme showing that the loss of concentration around any electrode is proportional to the speed of the ion moving away from it.

In Fig. 25.4, A is anode and C is cathode. AA′ and BB′ are two imaginary planes which divide the cell into three compartments, **the anode compartment, the middle compartment and the cathode compartment.** The sign (+) represents a cation while the sign (–) represents an anion.

Before electrolysis, let there be 13 ion-pairs in the cell. The number of ion-pairs in the two outer compartments is 4 each and there are 5 ion-pairs in the middle compartment (Position I). Now let us consider the following cases:

(*i*) Let the anions alone be capable of movement

When, say, two anions have moved towards the anode, we get the position as shown in II. The cations have not moved at all. But inspite of that the number of discharged anions and cations is the same *viz.*, 2. *The concentration in the anode compartment, however, has not altered while in the cathode compartment it has fallen by two ion-pairs.*

(*ii*) Let the anions and cations move at the same rate

When two ions of each type have crossed over towards the opposite electrodes, we get the condition as shown in III. The number of discharged anions and cations is the same *viz.*, 4. *The concentration of both the anode compartment and the cathode compartment has fallen to the same extent viz, by two ion-pairs.*

(*iii*) Let the cations move at twice the speed of the anions

In this case, when cations have moved to the cathode compartment, one anion passes into the anode compartment. The state of affairs is shown in IV. The total number of discharged anions and cations is again the same *viz.*, 3. Although the concentration in the cathode compartment has fallen by one ion-pair, the concentration in the anode compartment decreases by two ion-pairs.

It is evident from the above considerations that ions are always discharged in equivalent amounts on the opposite electrodes. It is really due to the difference in the speeds of anions and cations that the changes in concentration around electrodes are caused. We also conclude that the **loss in concentration around any electrode is proportional to the speed of the ion moving away from it.**

This is **Hittorf's Rule**. We can write the expression :

$$\frac{\text{Fall around anode}}{\text{Fall around cathode}} = \frac{\text{Speed of cation}}{\text{Speed of anion}} = \frac{v_+}{v_-}$$

where v_+ and v_- is the speed of cations and anions respectively.

In deriving the above relation, an important assumption has been made, namely, the discharged ions do not react with the material of the electrodes. However, in many cases they combine with the material of the electrodes rather than depositing on it. This results in an increase in concentration around such an electrode instead of a decrease.

Migration Speed of Ag^+ and NO_3^- ions

Let us study the electrolysis of a solution of silver nitrate in a cell of the type described above but using silver electrodes. We find that instead of a fall in concentration of silver nitrate around the anode, it increases. This is due to the fact that every NO_3^- ion that arrives at the anode dissolves from it one Ag^+ ion to form $AgNO_3$. If the electrodes were of platinum, the state of affairs would be as shown in Fig. 25.4 III when ions moved at the same rate. That is, there should have taken place a fall in concentration by 2 ion-pairs in each compartment. With silver electrodes, however, we have the condition shown in Fig. 25.5.

It is clear that the concentration around the cathode has fallen from 4 to 2 ion-pairs but around the anode it has increased from 4 to 6 ion-pairs. It is very easy to calculate the speed ratio of ions in this case as well by putting a silver coulometer in the circuit. This would give us the total current passed or increase in concentration around the anode as if no silver ions had migrated towards the cathode. The difference of the total increase and the actual increase around the anode would give the fall in concentration around the anode due to the migration of silver ions.

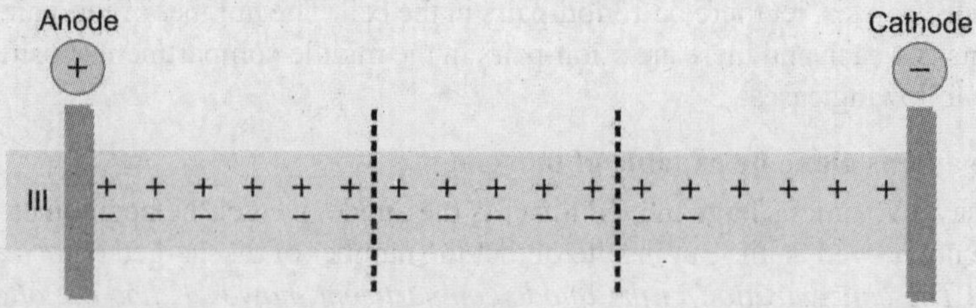

Figure 25.5
With silver electrodes, concentration around the cathode falls while it increases around the anode.

SOLVED PROBLEM 1

In an experiment, the increase in concentration of $AgNO_3$ around the silver anode was 5.6×10^{-6} kg of silver. 10.73×10^{-6} kg of silver were deposited by the same current in the silver coulometer placed in series. Find the speed ratio of Ag^+ and NO_3^- ions.

SOLUTION

Fall of concentration around cathode (= rise of concentration around anode) = 5.6×10^{-6} kg

If no Ag^+ ions had migrated from the anode, the increase in concentration around the anode would have been 10.73×10^{-6} kg silver. But the actual increase is 5.6×10^{-6} kg

$\therefore$ Fall around the anode due to migration of $Ag^+ = (10.73\times10^{-6} - 5.6\times10^{-6}\text{ kg}) = 5.13\times10^{-6}$ kg

$$\frac{\text{Speed of } Ag^+}{\text{Speed of } NO_3^-} = \frac{\text{Fall around anode}}{\text{Fall around cathode}} = \frac{5.13\times10^{-6}}{5.6\times10^{-6}} = \mathbf{0.916}$$

WHAT IS TRANSPORT NUMBER?

During electrolysis the current is carried by the anions and the cations. **The fraction of the total current carried by the cation or the anion is termed its Transport number or Hittorf's number.** If v_+ represents the speed of migration of the cation and v_- that of the anion,

$$\text{the transport number of cation} = \frac{v_+}{v_+ + v_-}$$

$$\text{the transport number of anion} = \frac{v_-}{v_+ + v_-}$$

The transport number of the cation is represented by t_+ and that of the anion by t_-.

Thus $$t_+ = \frac{v_+}{v_+ + v_-} \quad \text{and} \quad t_- = \frac{v_-}{v_+ + v_-}$$

or $$\frac{t_+}{t_-} = \frac{v_+}{v_-} \quad \text{and} \quad t_+ + t_- = 1$$

If the speed ratio v_+/v_- be denoted by r, we have,

$$r = \frac{t_+}{t_-} = \frac{t_+}{1 - t_+}$$

and $$t_- = \frac{1}{1 + r}$$

SOLVED PROBLEM 2

The speed ratio of silver and nitrate ions in a solution of silver nitrate electrolyzed between silver electrodes is 0.916. Find the transport number of the two ion.

SOLUTION

We know that $$t_{-} = \frac{1}{1+r}$$

where t_{-} is the transport number of the anion and r is the speed ratio of the anion and the cation.

$$\therefore \qquad t_{NO_3^-} = \frac{1}{1 + 0.916} = 0.521$$

and $$t_{Ag^+} = 1 - t_{NO_3^-} = 1 - 0.521 = \mathbf{0.479}$$

DETERMINATION OF TRANSPORT NUMBER

There are two methods for determination of the transport number of an ion :

(1) Hittorf's method

(2) Moving Boundary method

Hittorf's Method

This method of determining transport number is based on **Hittorf's Rule.** According to this rule, **the loss of concentration around any electrode is proportional to the speed of the ion moving away from that electrode.** The transport number of an ion is calculated from the changes in concentration found experimentally around the electrodes. The apparatus used in this method consists of two vertical glass tubes joined together through a U-tube in the middle. All the three tubes are provided with stopcocks at the bottom. The U-tube is also provided with stopcocks at the tops of the two limbs. By closing these stopcocks, the communication between the solutions in the cathode and anode limbs can be stopped. The silver anode is sealed in a glass-tube and the cathode is a piece of freshly silvered foil. The apparatus is filled with a solution of silver nitrate and a steady current of about 0.01 ampere is passed for two to three hours. It is an **important precaution** that the current is passed only for a short time so that too large a change in concentrating does not take place. The apparatus is connected with a silver or copper coulometer as shown in Fig. 25.6, which shows the circuit for the experiment.

When the current has been passed for about three hours, the stopcocks at the top of the U-tube are closed. The whole of the liquid in the anode compartment is carefully drained into a weighed flask and its weight determined. Its silver content is determined by titrating against a standard solution of potassium thiocyanate. The weight of silver deposited in the silver coulometer is also noted. If a copper coulometer is used in place of silver coulometer, the weight of silver equivalent to the copper deposited is calculated by multiplying it with 108/31.5. There should be no change in the concentration of the solution in the U-tube if the experiment has been successfully performed.

If the above experiment has been performed by using silver electrodes, in this case nitrate ions attack the silver anode. Consequently, there is an increase in concentration of Ag^+ ions rather than decrease. The same experiment can also be performed by using platinum electrodes to avoid the attack of anions on the anode.

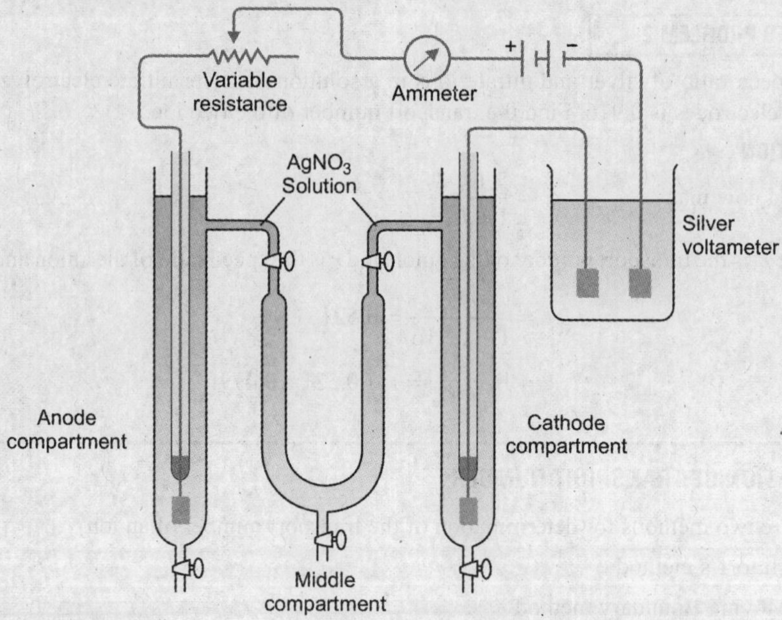

■ **Figure 25.6**
Hittorf's apparatus for determining Transport numbers.

Calculations

Two different cases may arise :

Case 1. When electrodes are unattackable (Pt electrodes are used).

After passing electric current :

Let the weight of anodic solution taken out $= a\ g$

weight of $AgNO_3$ present in it by titration $= b\ g$

weight of water $= (a - b)\ g$

Before passing electric current :

Let weight of $AgNO_3$ in $(a - b)$ g of water before passing electric current be $= c\ g$

$\therefore$ Fall in concentration

$$= (c - b)\ g \text{ of } AgNO_3 = \frac{(c-b)}{170}\ g \text{ eqvt of } AgNO_3$$

$$= \frac{(c-b)}{170}\ g \text{ eqvt of Ag} = d \text{ (say)}$$

Let the weight of silver deposited in silver coulometer be

$$= w_1 g$$

$$= \frac{w_1}{108} \text{ g eqvt of Ag}$$

$$= W \text{ (say) g eqvt of Ag}$$

$$\text{Transport number of } Ag^+ \left(t_{Ag^+}\right) = \frac{\text{Fall in concentration around anode in g eqvt}}{\text{Amt. of Ag deposited in g eqvt}} = \frac{d}{W}$$

$$\text{and Transport number of } NO_3^- \text{ ion } \left(t_{NO_3^-}\right) = 1 - \frac{d}{W}$$

Case 2. When electrodes are attackable (Ag electrodes are used).

Increase in conc. of anodic solution = $(b - c)$ g of $AgNO_3$

$$= \frac{(b-c)}{170} \times 108\ g \text{ of Ag}$$

$$= \frac{(b-c)}{170}\ g \text{ eqvt of Ag}$$

$$= e \text{ (say)}$$

If no Ag^+ ions had migrated from the anode, the increase in concentration of Ag^+ ions would have been equal to W.

$\therefore$ Fall in concentration due to migration of Ag^+ ion = $W - e$

Hence, transport number of Ag^+ ion $\left(t_{Ag^+}\right) = \dfrac{W-e}{W}$

and transport number of NO_3^- ion $\left(t_{NO_3^-}\right) = 1 - \dfrac{W-e}{W}$

SOLVED PROBLEM 3

A solution of silver nitrate containing 12.14×10^{-3} kg of silver in 50×10^{-6} m^3 of solution was electrolyzed between platinum electrodes. After electrolysis, 50×10^{-6} m^3 of the anode solution was found to contain 11.55×10^{-3} kg of silver, while 1.25×10^{-3} kg of metallic silver was deposited on the cathode. Calculate the transport number of Ag^+ and NO_3^- ions.

SOLUTION

Weight of Ag in 50×10^{-6} m^3 of the solution before electrolysis = 12.14×10^{-3} kg

Weight of Ag in 50×10^{-6} m^3 of the solution after electrolysis = 11.55×10^{-3} kg

$\therefore$ Fall in concentration of Ag = 12.14×10^{-3} kg – 11.55×10^{-3} kg

$$= 0.59 \times 10^{-3} \text{ kg}$$

$$= 0.0055 \text{ g eq}$$

Weight of Ag deposited in silver coulometer = 1.25×10^{-3} kg

$$= \frac{1.25\times10^{-3}}{108\times10^{-3}} \text{ kg eq}$$

$$= 0.0116$$

Hence, transport number of $Ag^+ \left(t_{Ag^+}\right) = \dfrac{\text{Fall in conc. around anode}}{\text{No. of g eqvt deposited in silver coulometer}}$

$$= \frac{0.0055\times10^{-3}}{0.0116\times10^{-3}}$$

$$= 0.474$$

and Transport number of NO_3^- ion $(t_{NO_3^-}) = 1 - 0.474$

$$= \mathbf{0.526}$$

SOLVED PROBLEM 4

In an electrolysis of copper sulfate between copper electrodes the total mass of copper deposited at the cathode was 0.153×10^{-3} kg and the masses of copper per unit volume of the anode liquid before and after electrolysis were 0.79×10^{-3} kg and 0.91×10^{-3} kg respectively. Calculate the transport number of the Cu^{2+} and SO_4^{2-} ions.

SOLUTION

$$\text{Wt. of copper in the anode liquid before electrolysis} = 0.79 \times 10^{-3} \text{ kg}$$

$$\text{Wt. of copper in the anode liquid after electrolysis} = 0.91 \times 10^{-3} \text{ kg}$$

$$\text{Increase in weight} = 0.91 \times 10^{-3} \text{ kg} - 0.79 \times 10^{-3} \text{ kg} = 0.12 \times 10^{-3} \text{ kg}$$

$$\text{Increase in weight of copper cathode in the coulometer} = 0.153 \times 10^{-3} \text{ kg}$$

This means that if no copper had migrated from the anode, increase in weight would have been 0.153×10^{-3} kg.

$$\text{But actual increase} = 0.12 \times 10^{-3} \text{ kg}$$

$$\text{Fall in concentration due to migration of } Cu^{2+} \text{ ions} = 0.153 \times 10^{-3} \text{ kg} - 0.12 \times 10^{-3} \text{ kg} = 0.033 \times 10^{-3} \text{ kg}$$

$$\therefore \quad \text{Transport number of } Cu^{2+} \text{ ion} = \frac{0.033 \times 10^{-3}}{0.153 \times 10^{-3}} = 0.216$$

$$\text{and Transport number of } SO_4^{2-} \text{ ion} = (1 - 0.216) = \mathbf{0.784}$$

Moving Boundary Method

The moving boundary method is based on the direct observation of migration of ions under the influence of applied potential. This method is very accurate and has been used in recent years for precision measurements.

The apparatus used consists of a long vertical tube fitted with two electrodes at the two ends (Fig. 25.7). The tube is filled with a solution of cadmium chloride ($CdCl_2$) at the lower end and hydrochloric acid at the upper end in a way that there is a sharp boundary between the two (due to difference in their refractive indices).

The platinum cathode dipped in HCl solution is inserted at the top and the anode (cadmium stick) is introduced at the bottom.

On passing electric current through the apparatus, hydrogen gas is evolved at the cathode and H^+ ions move toward the anode. The H^+ ions are replaced by Cd^{2+} ions and hence the boundary line moves in the upward direction. By noting the length through which the boundary moves and the quantity of electricity passed through the cell, the transport number of H^+ ion can be calculated.

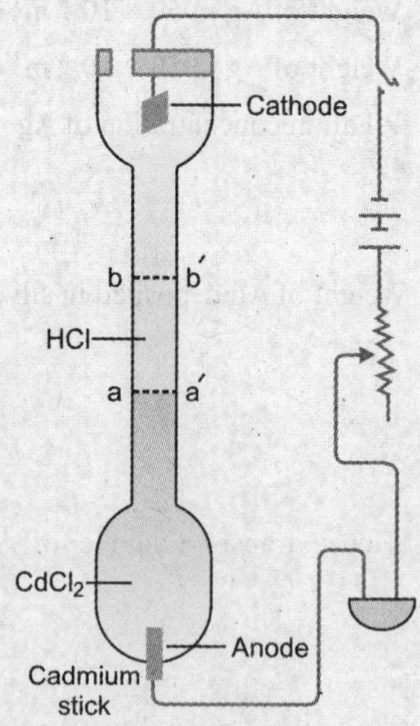

Figure 25.7
Determination of transport number by moving boundary method.

In general, if the transport number of a cation A^+ is to be determined, the electrolyte AX solution is taken in the upper part of the apparatus and a layer of another electrolyte BX having the common ion X^- is introduced in the lower part of the apparatus. The electrolyte BX is selected so that the velocity of B^+ ion is less than that of A^+ ion. In such a case, the situation is described in Fig. 25.8.

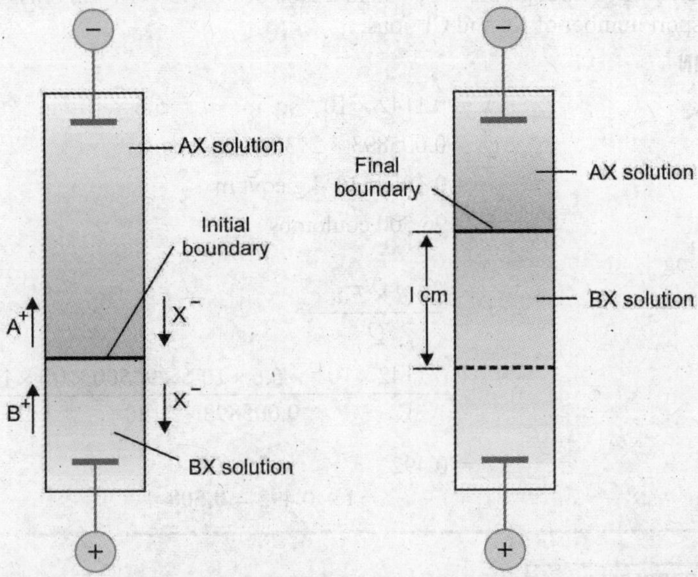

■ **Figure 25.8**
Moving boundary experiment for determining the transport number of A^+ ion.

Calculations

Let c be the original concentration of A^+ ions in gram-equivalents per ml of the solution.

If the distance through which boundary moves $= l$ cm, and the area of cross-section of the tube $= s$ sq cm.

Then the number of equivalents of A^+ moving upward $= s \times l \times c$

Let the number of Faradays of current passed $= n$

The fraction of current carried by A^+ ions $= n \times t_{A^+}$

Hence $$n \times t_{A^+} = s \times l \times c$$

and $$t_{A^+} = \frac{s \times l \times c}{n} \quad \text{... (1)}$$

where $n = Q/F$. Q being the quantity of current passed and F stands for Faraday (= 96, 500 coulombs).

The above expression can now be written as

$$t_{A^+} = \frac{s \times l \times F \times c}{Q} \quad \text{... (2)}$$

With the help of the expressions (1) and (2), the value of t_{A^+} (transport number of A^+), can be found from the experimental values of s, l, c and n.

SOLVED PROBLEM 5

In a moving boundary experiment with 0.100×10^3 KCl using 0.065×10^3 LiCl as indicator solution, a constant current of 0.005893 amp was passed for 2130 seconds. The boundary was observed to move through 5.6×10^{-2} m in a tube of 0.1142×10^{-4} m cross-section. Calculate the transport number of K^+ and Cl^- ions.

SOLUTION

Here, $s = 0.1142 \times 10^{-4}$ sq. m $\quad l = 5.6 \times 10^{-2}$ m

$Q = 0.005893 \times 2130$ coulombs

$c = 0.100 \times 10^3$ kg eqvt m^3

$F = 96{,}500$ coulombs

Applying

$$t_k = \frac{s \times l \times F \times c}{Q}$$

$$= \frac{0.1142 \times 10^{-4} \times 5.6 \times 10^{-2} \times 96500 \times 0.1 \times 10^3}{0.005893 \times 2130}$$

$$= 0.492$$

and $t_{cl^-} = 1 - t_{k^+} = 1 - 0.492 = \mathbf{0.508}$

SOLVED PROBLEM 6

Calculate the transport number of H^+ ion from the following data obtained by moving boundary method :

Concentration of HCl solution $= 0.10 \times 10^3$ N

Weight of silver deposited in the coulometer $= 0.12 \times 10^{-3}$ kg

Distance moved by the boundary $= 7.5 \times 10^{-2}$ m

Cross-section of the tube $= 1.25 \times 10^{-4}$ sq m

Equivalent weight of silver $= 108 \times 10^{-3}$ kg

SOLUTION

Here, $s = 1.25 \times 10^{-4}$ sq m $\quad l = 7.5 \times 10^{-2}$ m

$c = 0.1$ g eqvt/liter

$= 0.1 \times 10^{-3}$ kg eqvt/m^3

We know 108×10^{-3} kg silver $= 1$ Faraday

$$0.12 \times 10^{-3} \text{ kg silver} = \frac{0.12 \times 10^{-3}}{108 \times 10^{-3}} = 0.00111$$

Applying $\quad t_{H^+} = \dfrac{s \times l \times c}{n}$

We have $\quad t_{H^+} = \dfrac{1.25 \times 10^{-4} \times 7.5 \times 10^{-2} \times 0.1 \times 10^3}{0.0011}$

$$= 0.852$$

$\therefore$ The transport number of H^+ is **0.852**.

KOHLRAUSCH'S LAW

From a study of the equivalent conductances of different electrolytes at infinite dilution (λ_∞), Kohlrausch discovered that each ion contributes to the conductance of the solution. In 1875, he enunciated a generalization which is called the **Kohlrausch's Law.** It states that : **the equivalent conductance of an electrolyte at infinite dilution is equal to the sum of the equivalent conductances of the component ions.**

The law may be expressed mathematically as :

$$\lambda_\infty = \lambda_a + \lambda_c$$

where λ_a is the equivalent conductance of the anion and λ_c that of the cation.

For example, the equivalent conductance of NaCl at infinite dilution at 298 K is found to be 126.45. The equivalent conductance of Na^+ and Cl^- ion is 50.11 siemens and 76.34 siemens respectively. Thus,

$$\lambda_\infty(\text{NaCl}) = \lambda_{Cl^-} + \lambda_{Na^+}$$

or

$$126.45 = 50.11 + 76.34$$

This is in conformity with the Kohlrausch's Law.

Applications of Kohlrausch's Law

Each ion has the same constant ionic conductance at a fixed temperature, no matter of which electrolyte it forms a part. It is expressed in siemens m^2 and is directly proportional to the speeds of the ions.

$$\lambda_a \propto v_- \quad \text{or} \quad \lambda_a = k \times v_-$$

$$\lambda_c \propto v_+ \quad \text{or} \quad \lambda_c = k \times v_+$$

where k is the proportionality constant.

Also

$$\lambda_\infty \propto (\lambda_a + \lambda_c) = k(v_- + v_+)$$

$$\therefore \quad \frac{\lambda_a}{\lambda_\infty} = \frac{kv_-}{k(v_- + v_+)} = \frac{v_-}{v_- + v_+} = t_- \qquad ...(1)$$

and

$$\frac{\lambda_c}{\lambda_\infty} = \frac{kv_+}{k(v_- + v_+)} = \frac{v_+}{v_- + v_+} = t_+ = 1 - t_- \qquad ...(2)$$

Dividing (1) by (2), we have

$$\frac{\lambda_a}{\lambda_c} = \frac{t_-}{1 - t_-} \qquad ...(3)$$

With the help of the expression (3), we can determine ionic conductances from the experimental values of the transport number of the ions.

(1) Calculation of λ_∞ for Weak electrolytes

Weak electrolytes do not ionize to a sufficient extent in solution and are far from being completely ionized even at very great dilution. The practical determination of λ_∞ in such cases is, therefore, not possible. However, it can be calculated with the help of Kohlrausch's law.

$$\frac{\lambda_a}{\lambda_c} = \frac{t_-}{1 - t_-}$$

$$\lambda_a - t_-\lambda_a = t_-\lambda_c$$

$$\lambda_a = t_-(\lambda_a + \lambda_c)$$

$$= t_-\lambda_\infty$$

Thus the ionic conductance of an ion is obtained by multiplying the equivalent conductance at infinite dilution of any strong electrolyte containing that ion by its transport number.

In this manner, the ionic mobilities of the two ions present in the weak electrolyte can be calculated. Thus we can get the equivalent conductance of the electrolyte at infinite dilution by adding up these two values.

SOLVED PROBLEM 7

At 298 K the transport number of H^+ ion in HCl and CH_3COO^- ion in CH_3COONa are 0.81 and 0.47 respectively. The equivalent conductances at infinite dilution of HCl and CH_3COONa are 426 siemens m^2 $eqvt^{-1}$ and 91.0 siemens m $eqvt^{-1}$ respectively. Calculate the equivalent conductance of acetic acid at infinite dilution.

SOLUTION

$$\text{Conductance of } H^+ \text{ ion} = \lambda_\infty(\text{HCl}) \times t_{H^+}$$
$$= 426 \times 0.81 = 345.06$$

$$\text{Conductance of } CH_3COO^- \text{ ion} = \lambda_\infty(CH_3COOH) \times t_{CH_3COO^-}$$
$$= 91 \times 0.47 = 42.77$$

$\therefore$ Conductance of acetic acid at infinite dilution $= \lambda_a + \lambda_c$

$$= 42.77 + 345.06$$
$$= \mathbf{387.83\ siemens}$$

SOLVED PROBLEM 8

Calculate the equivalent conductance at 293 K of NH_4OH at infinite dilution. Given :

$$\lambda_\infty(NH_4Cl) = 130 \qquad \lambda_\infty(OH^-) = 174 \qquad \lambda_\infty(Cl^-) = 66$$

SOLUTION

Applying Kohlrausch's law,

$$\lambda_\infty(NH_4OH) = \lambda_\infty(NH_4Cl) + \lambda_\infty(OH^-) - \lambda_\infty(Cl^-)$$
$$= 130 + 174 - 66$$
$$= \mathbf{238\ Siemen^{-1}\ m^2\ eqvt^{-1}}$$

(2) Calculation of Absolute Ionic mobilities

The absolute ionic mobility of an ion is defined as **the velocity of an ion in centimeters per second under a potential gradient of one volt per meter.** (Potential gradient = applied EMF/distance between (the electrodes). It is expressed in m s^{-1}.

For example, let the velocity of the ion at infinite dilution be U m per second when the distance between the electrodes is 20×10^{-2} m and the voltage 100 V. Then the potential difference is $100/20 \times 10^{-2}$ *i.e.*, 5×10^2 volts per m and the ionic mobility is $U/5 \times 10^2$ m s^{-1}.

It has been found that the ionic conductance is directly proportional to the ionic mobility *i.e.*,

$$\lambda_a \propto U_a \quad \text{and} \quad \lambda_c \propto U_c$$

or

$$\lambda_a = k\,U_a \quad \text{and} \quad \lambda_c = k\,U_c$$

where k is the proportionality constant. Its value is equal to the charge on one gram-equivalent of the ion under the potential gradient of 1 volt per cm *i.e.*, k = 96,500 coulomb (1 Faraday). Therefore, the ionic mobility is obtained by dividing the ionic conductance by 96,500 coulombs.

$$U_\neq = \frac{\lambda_\neq}{96{,}500}$$

or

$$\lambda_\neq = FU_\neq$$

SOLVED PROBLEM 9

The molecular conductance at infinite dilution of KCl is 130.1. The Hittorf number, t_- for the chloride ion in very dilute solution is 0.505. Calculate the mobilities in m s^{-1} of potassium and chloride ions.

SOLUTION

$$\text{Ionic conductance of Cl}^- \text{ ion, } \lambda_a = \lambda_\infty \times t_-$$

$$\therefore \text{ Ionic mobility of Cl}^- \text{ ion, } U_a = \frac{130.1 \times 0.505}{96,500}$$

$$= \mathbf{6.81 \times 10^{-6}\ m\ s^{-1}}$$

$$\text{Ionic conductance of K}^+ \text{ ion, } \lambda_c = \lambda_\infty (1 - t_-)$$

$$= 130.1 \times 0.495$$

$$\therefore \text{ Ionic mobility of K}^+ \text{ ion, } U_c = \frac{130.1 \times 0.495}{96,500}$$

$$= \mathbf{6.67 \times 10^{-6}\ m\ s^{-1}}$$

(3) Calculation of the solubility of sparingly soluble salts

Substance like AgCl or $PbSO_4$ which are ordinarily called insoluble do possess a definite value of solubility in water. This can be determined from conductance measurements of their saturated solutions (*Chapter* 24). Since a very small amount of solute is present it must be completely dissociated into ions even in a saturated solution so that the equivalent conductance κV is equal to equivalent conductance at infinite dilution. This according to Kohlrausch's Law is the sum of the ionic mobilities.

i.e., $\kappa V = \lambda_\infty = \lambda_a + \lambda_c$

knowing κ and λ_∞, V can be found out which is the volume in 10^{-3} l containing 10^{-3} kg-eqvt of the electrolyte.

SOLVED PROBLEM 10

The specific conductance of saturated solution of silver chloride at 291 K is 1.24×10^{-6} siemens after subtracting that of water. The mobilities of Ag^+ and Cl^- ions at this temperature are 53.8 and 65.3 respectively. Calculate the solubility of silver chloride in kg per m^3.

SOLUTION

Here $\kappa = 1.24 \times 10^{-6}$ siemens

and $\lambda_\infty = \lambda Ag + \lambda Cl = 53.8 + 65.3 = 119.1$

Now $\lambda_\infty = \kappa \times V$

or $119.1 = 1.24 \times 10^{-6} \times V$

$$\therefore \quad V = \frac{119.1 \times 10^{-6}}{1.24 \times 10^{-6}} \text{ m}^3$$

i.e., $\frac{119.1 \times 10^{-6}}{1.24 \times 10^{-6}}$ m^3 contains AgCl = 1 g-eqvt = 143.5

$$\therefore \quad 10^{-3} \text{ m}^3 \text{ contain AgCl} = \frac{143.5 \times 1.24 \times 10^{-6}}{119.1 \times 10^{-6}} \times 10^{-3}$$

$$= \frac{143.5 \times 1.24}{119.1} \times 10^{-3}$$

$$= \mathbf{1.494 \times 10^{-6}\ kg}$$

(4) Calculation of the Degree of Dissociation or Conductance Ratio

The apparent degree of dissociation, α, of an electrolyte at the dilution *V* is given by $\alpha = \lambda_v/\lambda_\infty$, where λ_v is the equivalent conductance of the electrolyte at the dilution *V* and λ_∞ is its equivalent conductance at infinite dilution. This according to Kohlrausch's Law is the sum of λ_a and λ_c.

SOLVED PROBLEM 11

The conductance of silver ion at 291 K is 55.7 and of the nitrate ion 60.8. If the specific conductance of $AgNO_3$ in a decinormal solution at 291 K is 0.00947 siemens, what will be the percentage of dissociation of the salt at this concentration ?

SOLUTION

Specific conductivity (κ) in N/10 solution = 0.00947 siemens

Volume containing 1 kg equivalent = 0.1 m^3

∴ Eqvt conductance at this dilution = 0.00947 × 0.1

= 94.7 Siemens

Eqvt conductance at infinite dilution = 55.7 + 60.8 = 116.5

$$\alpha = \frac{94.7}{116.5} = 0.8128$$

Hence the salt is **81.28%** dissociated.

SOLVED PROBLEM 12

The equivalent conductance at 291 K of a normal solution of KCl is 98.2 and for infinite dilution at the same temperature 131. Calculate the degree of dissociation of KCl at this dilution.

SOLUTION

We know $\dfrac{\lambda_v}{\lambda_\infty} = a$

Here, $\lambda_v = 98.2, \quad \lambda_\infty = 131$

$$\therefore \quad \frac{\lambda_v}{\lambda_\infty} = \frac{98.2}{131} = 0.75$$

Hence the degree of dissociation of KCl in normal solution is **0.75**.

(5) Calculation of the Ionic product for Water

The observed specific conductance of the purest water at 298 K is 5.54×10^{-8} siemens. The conductance of one liter of water containing 1 gram eqvt of it would be :

$$\lambda_{H_2O} = 5.54 \times 10^{-8} \times 1000 = 5.54 \times 10^{-5} \text{ siemens}$$

At the same temperature the conductance of H^+ ions and OH^- ions are :

$$\lambda_{H^+} = 349.8 \text{ siemens}$$

$$\lambda_{OH^-} = 198.5 \text{ siemens}$$

According to Kohlrausch's Law

$$\lambda_{H_2O} = \lambda_{H^+} + \lambda_{OH^-}$$

$$= 349.8 + 198.5 = 548.3 \text{ siemens}$$

One molecule of water gives one H^+ ion and one OH^- ion

$$H_2O = H^+ + OH^-$$

Assuming that ionic concentration is proportional to conductance, we have

$$[H^+] = [OH^-] = \frac{5.5 \times 10^{-5}}{548.3} = 1.01 \times 10^{-7} \text{ g ion litre}^{-1}$$

The ionic product of water is then

$$K_w = [H^+][OH^-] = 1.02 \times 10^{-14} \text{ at } 298 \text{ K}$$

For most purposes, the value of K_w is taken to be 10^{-14}

CONDUCTOMETRIC TITRATIONS

Titrations in which conductance measurements are made use of in determining the end-point of acid-alkali reactions, some displacement reactions or precipitation reactions are called **Conductometric titrations.** In these titrations, advantage is taken of the fact that the **conductance of a solution at a constant temperature depends upon the number of ions present in it and their mobility.** For this purpose, the titrant is added from a burette into a measured volume of the solution to be titrated which is taken in a conductance cell and the conductance readings corresponding to the various additions are plotted against the volume of the titrant. In this way two linear curves are obtained, the point of intersection of which is the **end-point.** Several phenomena like hydrolysis of reactants or products or partial solubility of a precipitated product give rise, however to a curvature in the curves. The shapes of curves obtained in certain types of titration are discussed below :

(1) Titration of a Strong acid against a Strong base

Consider the reaction in which hydrochloric acid solution is titrated against a solution of sodium hydroxide. Take 20 ml of the acid solution in the conductance cell placed in a thermostat and determine its conductance. Now add 1 ml sodium hydroxide solution from the burette at a time. After each addition, determine the conductance of the solution after through mixing and plot the conductance of the solution against the volume of the alkali added. It will be observed that the points lie on two lines which are almost straight. The point of intersection of the interpolated lines will be the end point and the volume of alkali corresponding to this point is the volume of alkali required to neutralize 20 ml of the acid (Fig. 25.9). The reason for this is that before the addition of alkali, the conductance of the solution is due to presence of H^+ and Cl^- ions. Since hydrogen ions possess the greatest mobility of any ion, the greater part of the conductance is due to it. As alkali solution is added, the hydrogen ions are removed by combination with the hydroxyl ions forming feebly ionized water molecules and their place is taken up by comparatively slow moving Na^+ ions.

$$H^+ + Cl^- + Na^+ + OH^- \longrightarrow Na^+ + Cl^- + H_2O \qquad \textit{(feebly ionized)}$$

As a result of this, the conductance of the solution decreases and continues to fall with every subsequent addition of alkali till the end-point is reached. After the equivalence point, the further addition of sodium hydroxide solution results in an increase of conductance since the hydroxyl ions are no longer removed in the chemical reaction in the form of feebly ionized water. **The point of minimum conductance, therefore, coincides with the end-point of the titration.**

In order to get accurate results, the volume change during titration should be as little as possible. The titrant should, therefore, be about 10 times as strong as the acid solution in the conductance cell in order to keep the volume change small. If this is not so, a correction to the readings has to be applied *viz*,

$$\text{actual conductance} = \left(\frac{v + V}{V}\right) \times \text{observed conductance}$$

where v is the volume of the titrant and V is the original volume of the solution to be titrated.

(2) Titration of a Weak acid against a Strong alkali

When a weak acid like acetic acid is titrated against a strong alkali like sodium hydroxide, we get a curve of the type shown in Fig. 25.10. The initial conductance of the solution is low because

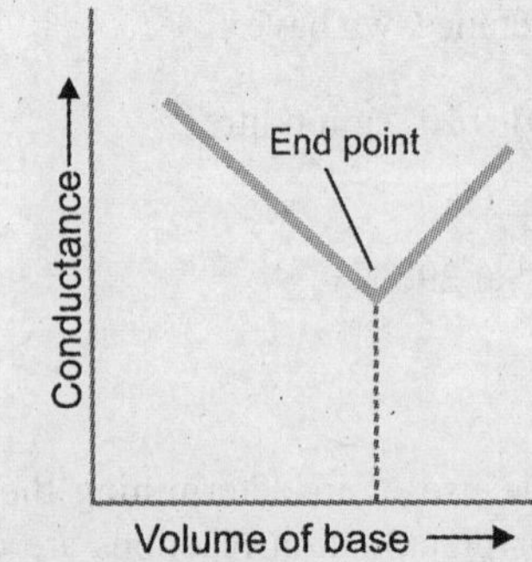

Figure 25.9
Conductometric titration curve for strong acid and strong base.

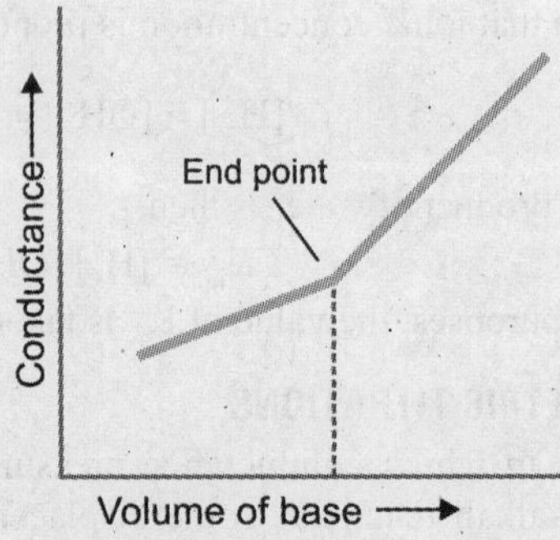

Figure 25.10
Curve for titration of a weak acid against a strong base.

of the poor dissociation of the weak acid. On adding alkali, highly ionized sodium acetate is formed. The acetate ions at first tend to suppress the ionization of acetic acid still further due to *Common Ion Effect* but after a while the conductance begins to increase because the conducting power of highly ionized salt exceeds that of the weak acid.

$$CH_3COOH + Na^+ + OH^- \longrightarrow CH_3COO^- + Na^+ + H_2O \quad \textit{(feebly ionized)}$$

Immediately after the end point, further addition of sodium hydroxide introduces the fast moving hydroxyl ions. Thus, the *conductance value shows a sharp increase*. The point of intersection of the two curves, gives the end-point.

$$H^+ + Cl^- + NH_4OH \longrightarrow NH_4^+ + Cl^- + H_2O \quad \textit{(feebly ionized)}$$

After the end-point has been reached, the addition of ammonium hydroxide will not cause any appreciable change in conductance value as it is a weak electrolyte and its conductance is very small compared with that of the acid or its salt. The shape of this part of the curve will, therefore, be as shown in the figure.

(3) Titration of a Strong acid against a Weak base

The curve obtained for the titration of a strong acid against a weak base is shown in Fig. 25.11. In this case, the conductance of the solution will first decrease due to the fixing up of the fast moving H^+ ions and their replacement by slow moving NH_4^+ ions.

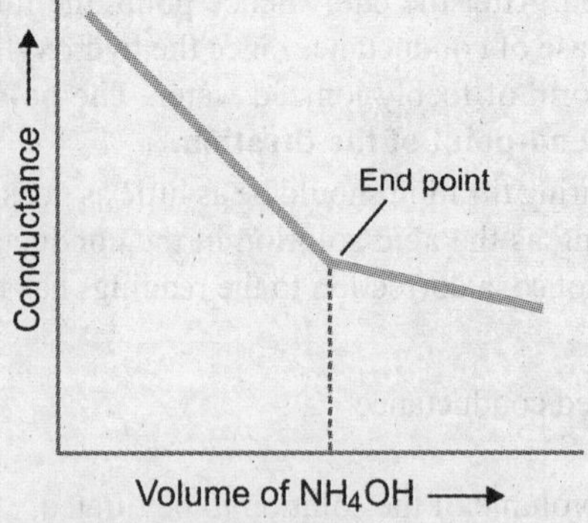

Figure 25.11
Curve for titration of a strong acid against a weak base.

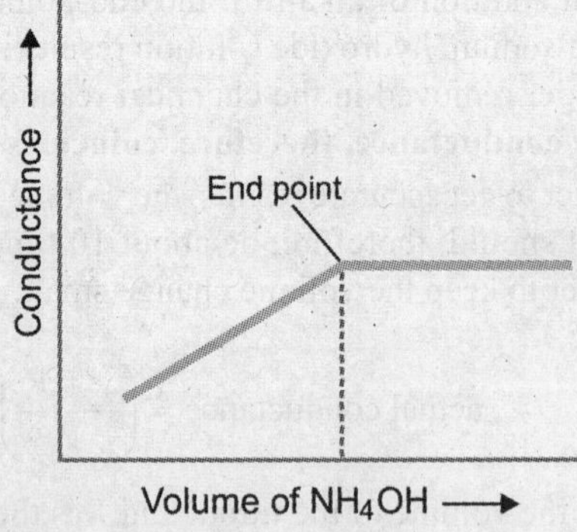

Figure 25.12
Curve for titration of acetic acid against ammonium hydroxide.

(4) Titration of a Weak acid against a Weak base

The conductometric method is particularly suitable as such titrations do not give a sharp end-point with indicators. Consider the titration of acetic acid with ammonium hydroxide. The complete titration curve is shown in Fig. 25.12. The initial conductance of the solution in this case is also low due to the poor dissociation of the weak acid. But it starts increasing as the salt CH_3COONH_4 is formed. After the equivalence point, the conductivity remains almost constant because the free base NH_4OH is a weak electrolyte. The end-point is quite sharp.

(5) Precipitation reactions

The end-point in precipitation reactions can be accurately determined by conductometric titration. The accuracy is further increased by working with fairly dilute solutions and addition of alcohol which reduces the solubility of the precipitate and prevents adsorption. In the titration of potassium chloride against silver nitrate, for example, the change in conductance on the addition of silver nitrate is not much since the mobility of the potassium ion and the silver ion is of the same order. Thus the curve is nearly horizontal.

$$Ag^+ + NO_3^- + K^+ + Cl^- \longrightarrow K^+ + NO_3^- + AgCl\ (ppt)$$

After the end-point, there is a sharp increase in conductance due to an increase in the number of free ions in solution.

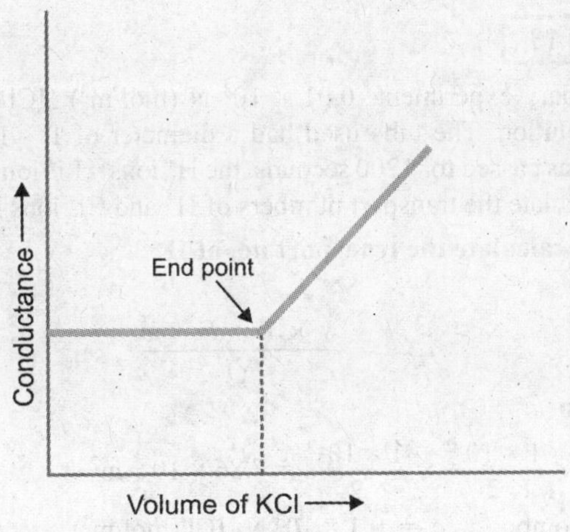

■ Figure 25.13
Titration of potassium chloride against silver nitrate.

Advantage of Conductometric Titrations

(1) Colored solutions where no indicator is found to work satisfactorily can be successfully titrated by this method.

(2) This method is useful for the titration of weak acids against weak bases which do not give a sharp change of color with indicators in ordinary volumetric analysis.

(3) More accurate results are obtained because the end -point is determined graphically.

Precautions. (*i*) It is necessary to keep the temperature constant throughout the experiment. (*ii*) In acid-alkali titrations, the titrant should be about 10 times stronger than the solution to be titrated so that the volume change is as little as possible.

DIFFERENCES BETWEEN CONDUCTOMETRIC AND VOLUMETRIC TITRATIONS

Conductometric Titrations	Volumetric Titrations
1. Conductance measurements are done to check end points.	1. Volume measurements are done to check end points.
2. Titrations can be carried out even with colored solution.	2. These titrations fail in colored solutions as suitable indicators are not available some times.
3. Accurate results are obtained.	3. Results are not so accurate.
4. End point is determined graphically.	4. End point is determined by change in color of indicator.
5. In case of polybasic acids conductometric titrations can be used.	5. In case of polybasic acids volumetric titrations do not give correct end points.
6. These are successful even in weak acids and strong bases.	6. These are not successful in weak acids and strong bases.

ADDITIONAL SOLVED PROBLEMS

SOLVED PROBLEM 13

In a moving boundary experiment 0.01×10^3 N (mol/m^3) HCl solution was floated on a lithium chloride solution. The tube used had a diameter of 1×10^{-2} m. When a current of 11×10^{-3} ampere was passed for 1200 seconds, the H^+ ions – Li^+ ions boundary moved through 13.9×10^{-2} m. Calculate the transport numbers of H^+ and Cl^- ions in HCl solution used.

SOLUTION : (*a*) To calculate the transport no. of H^+

Formula used

$$t_{H^+} = \frac{s \times l \times F \times c}{Q}$$

Quantities given

$s = \pi r^2 = \frac{22}{7} \times \frac{1 \times 10^{-2}}{2} \times \frac{1 \times 10^{-2}}{2} = 7.86 \times 10^{-5}\ \text{m}^2$ $l = 13.9 \times 10^2$ m

$F = 96500$ coulomb $c = 0.1 \times 10^3$ N HCl (mol/m^3) $Q = 11 \times 10^{-3} \times 1200$

Substitution of values

$$t_{H^+} = \frac{7.86 \times 10^{-5} \times 13.9 \times 10^{-2} \times 96500 \times 0.1 \times 10^3}{11 \times 10^{-3} \times 1200}$$

$$= \mathbf{0.7984}$$

(*b*) To calculate the transport no. of Cl^-.

Formula used

$$t_{Cl^-} = 1 - t_{H^+}$$

Quantities given

$$t_{H^+} = 0.7984$$

Substitution of values

$$t_{Cl^-} = 1 - 0.7984$$

$$= \mathbf{0.2016}$$

SOLVED PROBLEM 14

In a moving boundary experiment with 0.1×10^3 N (mol/m^3) KCl using 0.65×10^3 N (mol/m^3) LiCl as indicator solution, a constant current of 0.006 amp was passed for 1900 seconds and the boundary was observed to move through 5×10^{-2} m in a tube of 0.1142×10^{-4} m^2 cross-section. Calculate the transport number of K^+ and Cl^- ions.

SOLUTION : (*a*) To calculate the transport number of K^+ ions.

Formula used

$$t_{K^+} = \frac{s \times l \times F \times c}{Q}$$

Quantities given

$s = 0.1142 \times 10^{-4}$ m^2 $\quad l = 5 \times 10^{-2}$ m $\quad F = 96500$ coulomb

$c = 0.1 \times 10^3$ N (mol/m^3) $\quad Q = 0.006 \times 1900$ coulomb

Substitution of values

$$t_{K^+} = \frac{0.1142 \times 10^{-4} \times 5 \times 10^{-2} \times 96500 \times 0.1 \times 10^3}{0.006 \times 1900}$$

$$= \mathbf{0.4833}$$

(*b*) To calculate the transport number of Cl^- ions

Formula used

$$t_{Cl^-} = 1 - t_{K^+}$$

Quantities given

$$t_{K^+} = 0.4833$$

Substitution of values

$$t_{Cl^-} = 1 - 0.4833$$

$$= \mathbf{0.5167}$$

SOLVED PROBLEM 15

A dilute solution of silver nitrate was electrolyzed using platinum electrodes. After passing small current for 10800 seconds, a fall of 0.00064×10^{-3} kg equivalent of electrolyte occurred in the anionic solution. The weight of copper deposited was found to be 0.05088×10^{-3} kg. Calculate the transport number of silver ions.

SOLUTION

Formula used

$$\text{Transport No. of } Ag^+ \text{ ions} = \frac{\textit{Fall in concentration}}{\textit{Amt. of Cu deposited in g eqvt}}$$

Quantities given

Fall in Concentration $= 0.00064 \times 10^{-3}$ kg eqvt.

$$\textit{Amt. of Cu deposited in g eqvt} = \frac{0.05088 \times 10^{-3}}{31.5}$$

Substitution of values

$$\text{Transport No. of } Ag^+ \text{ ions} = \frac{0.00064 \times 10^{-3} \times 31.5}{0.05088 \times 10^{-3}}$$

$$= \mathbf{0.40}$$

SOLVED PROBLEM 16

A solution of $AgNO_3$ containing 0.5066×10^{-3} kg of the salt in 60×10^{-3} kg of solution was electrolyzed between silver electrodes. After electrolysis 50×10^{-3} kg of the anode solution found to contain 0.5186×10^{-3} kg of $AgNO_3$ when a current of one ampere was passed for 118 seconds. Calculate the transport numbers of Ag^+ and NO_3^- ions.

SOLUTION : (*a*) To calculate the transport number of Ag^+ ions.

After electrolysis

Weight of anodic solution taken = 50×10^{-3} kg

Weight of $AgNO_3$ present in it = 0.5186×10^{-3} kg

Weight of water = 49.4814×10^{-3} kg

Before Electrolysis

Weight of $AgNO_3$ solution = 60×10^{-3} kg

Weight of $AgNO_3$ present = 0.5066×10^{-3} kg

Weight of water = 59.4934×10^{-3} kg

Weight of $AgNO_3$ present in 49.4814g

$$\text{of water} = \frac{0.5066 \times 10^{-3}\text{ kg}}{59.4934 \times 10^{-3}\text{ kg}} \times 49.4814 \times 10^{-3}\text{ kg}$$

$$= 0.421 \times 10^{-3}\text{ kg}$$

Observed increase in the weight of $AgNO_3$ in anodic compartment due to electrolysis

$$= 0.5186 \times 10^{-3}\text{ kg} - 0.421 \times 10^{-3}\text{ kg}$$

$$= 0.0976 \times 10^{-3}\text{ kg}$$

$$= \frac{0.0976 \times 10^{-3}\text{ kg}}{170 \times 10^{-3}\text{ kg}}\text{ kg equiv of } AgNO_3$$

$$= 5.741 \times 10^{-4}\text{ kg}$$

Quantity of electricity passed $= I \times t$

$$= 1\text{ amp} \times 118\text{ s}$$

$$= 118\text{ coulombs}$$

Amount of $AgNO_3$ liberated when 118 coulombs of electricity are passed $= \frac{118}{96500}$

$$= 1.223 \times 10^{-3}$$

Loss around the anode due to migration of ions

$$= 1.223 \times 10^{-3} - 5.741 \times 10^{-4}$$

$$= 6.489 \times 10^{-4}$$

$$\text{Transport number of } Ag^+ \text{ ions} = \frac{6.489 \times 10^{-4}}{1.223 \times 10^{-3}\text{ g}}$$

$$= \mathbf{0.530}$$

(*b*) To calculate the transport number of NO_3^- ions

Transport number of NO_3^- ions = 1 – transport number of Ag^+

$$= 1 - 0.530$$

$$= \mathbf{0.470}$$

SOLVED PROBLEM 17

A 0.05×10^3 M NaOH solution offered a resistance of 31.6 ohm in a conductivity cell at 298 K. If the cell constant of the cell is 0.367×10^2 m^{-1} find out the molar conductivity of sodium hydroxide solution.

SOLUTION : **(*a*) To calculate the specific conductivity of NaOH solution**

Formula used

$$\text{Specific conductivity, } \kappa = \frac{\text{Cell constant}}{\text{Resistance}}$$

Quantities given

Cell constant $= 0.367 \times 10^2 \text{ m}^{-1}$ *Resistance* $= 31.6$ ohm

Substitution of values

$$\text{Specific conductivity, } \kappa = \frac{0.367 \times 10^2 \text{ m}^{-1}}{31.6 \text{ ohm}}$$

$$= 1.16 \text{ ohm}^{-1} \text{ m}^{-1}$$

(*b*) To calculate the molar conductivity of NaOH solution

Formula used

$$\Lambda_m = \frac{1000 \times \kappa}{c}$$

Quantities given

$\kappa = 1.16 \text{ ohm}^{-1} \text{ m}^{-1}$ $c = 0.05 \times 10^3$ M

Substitution of values

$$\Lambda_m = \frac{1000 \times 1.16 \text{ ohm}^{-1} \text{ m}^{-1}}{0.05 \times 10^3 \text{ ohm}^{-1} \text{ m}^{-1}}$$

$$= \mathbf{23.2 \text{ ohm}^{-1} \text{ m}^{-1}}$$

SOLVED PROBLEM 18

The resistance of 0.1×10^3 M (mol/m^3) KCl solution in a conductance cell is 300 ohm and specific conductance of 0.1 M KCl is 1.29 siemens m^{-1}. The resistance of a 0.1×10^3 M (mol/m^3) NaCl solution in the same cell is 380 ohm. Calculate the equivalent conductance of the 0.1 M NaCl solution.

SOLUTION : **(*a*) To calculate the cell constant**

Formula used

$$\kappa = \frac{1}{R} \times \text{cell constant}$$

Quantities given

$\kappa = 1.29 \text{ siemens m}^{-1}$ $R = 300$ ohm

Substitution of values

$$1.29 \text{ siemens m}^{-1} = \frac{1}{300 \text{ ohm}} \times \text{cell constant}$$

$$\text{cell constant} = 300 \text{ ohm} \times 1.29 \text{ siemens m}^{-1}$$

$$= 3.87 \text{ m}^{-1}$$

(*b*) To calculate the equivalent conductance of 0.1 M NaCl solution

Formula used

$$\Lambda = \kappa \times \frac{1000}{N}$$

Quantities given

$\kappa = \frac{1}{R} \times \textit{cell constant} = \frac{1}{380} \times 3.87 = 1.018 \text{ siemens m}^{-1}$ $N = 0.1 \times 10^3$

Substitution of values

$$\Lambda = \frac{1.018 \times 1000}{0.1 \times 10^3} \text{ ohm}^{-1} \text{ m}^2 \text{ eqvt}^{-1}$$

$$= \mathbf{0.1018 \text{ ohm}^{-1} \text{ eqvt}^{-1} \text{ m}^2}$$

SOLVED PROBLEM 19

A solution of silver nitrate containing 6.07×10^{-3} kg silver in 25×10^{-6} m^3 solution was electrolyzed between platinum electrodes. After electrolysis, 5.77×10^{-3} kg of silver was present in 25×10^{-6} m^3 of anode solution while 0.625×10^{-3} kg of metallic silver was deposited on the cathode. Calculate the transport number of Ag^+ and NO_3^- ions.

SOLUTION : (*a*) To calculate the transport no. of Ag^+ ions

Formula used

$$\text{Transport number of } Ag^+ \text{ ions} = \frac{\text{Fall in conc. around anode}}{\text{No. of g eqvt deposited in silver coulometer}}$$

Quantities given

$$\text{Fall in conc. around anode} = 6.07\times10^{-3} - 5.77\times10^{-3} = \frac{0.3\times10^{-3}}{108\times10^{-3}} = 0.002777 \text{ kg eqvt.}$$

$$\text{No. of eqvt. deposited in silver coulometer} = \frac{0.625\times10^{-3}}{108\times10^{-3}} = 0.005787 \text{ kg eqvt.}$$

Substitution of values

$$\text{Transport number of } Ag^+ \text{ ions} = \frac{0.002777}{0.005787}$$

$$= 0.4799$$

(*b*) To calculate the transport number of NO_3^- ions.

$$\text{Transport number of } NO_3^- \text{ ions} = 1 - \text{transport no. of } Ag^+ \text{ ions}$$

$$= 1 - 0.480$$

$$= \mathbf{0.520}$$

EXAMINATION QUESTIONS

1. Define or explain the following terms :
   (*a*) Arrhenius theory
   (*b*) Transport number
   (*c*) Hittorf's method
   (*d*) Moving boundary method
   (*e*) Kohlrausch's law
   (*f*) Degree of dissociation
2. (*a*) Explain ionic conductance, transport number of an ion, and absolute ionic velocity.
   (*b*) The absolute velocity of Ag^+ is 0.00057×10^{-2} m s^{-1} and of the NO_3^- is 0.00069×10^{-2} m s^{-1}. Assuming complete dissociation, calculate the specific conductivity of 0.01 M solution of silver nitrate.

   Answer. (*b*) 0.00126×10^{-2} mhos
3. Describe Hittorf's method for the experimental determination of the transference number. The speed ratio of silver and nitrate ions in $AgNO_3$ electrolyzed between silver electrodes was found to be 0.89. Calculate the transference number of silver and nitrate ions.

   Answer. $Ag^+ = 0.4708$; $NO_3^- = 0.5292$
4. (*a*) What do you understand by the transport number of an ion?
   (*b*) In a moving boundary experiment with 0.1×10^3 N (mol/m^3) KCl using 0.65×10^3 N (mol/m^3) LiCl as indicator solution, a constant current of 0.006 amp was passed for 1900 seconds and the boundary was observed to move through 5×10^{-2} m in a tube of 0.1142×10^{-4} m^2 cross-section. Calculate the transport number of K^+ and Cl^- ions.
   Answer. (*b*) $K^+ = 0.4833$; $Cl^- = 0.5167$

5. (*a*) What is meant by transport number of ions? Describe Hittorf's method for the determination of transport number of silver ions.

(*b*) A dilute solution of silver nitrate was electrolyzed using platinum electrodes. After passing small current for 10800 seconds, a fall of 0.00064×10^{-3} kg equivalent of electrolyte occurred in the anionic solution. The weight of copper deposited was found to be 0.05088×10^{-3} kg. Calculate the transport number of silver ions.

Answer. (*b*) 0.40

6. (*a*) Explain Kohlrausch's law of ionic mobility. How does it help in determining the equivalent conductivity of infinite dilution of weak electrolyte?

(*b*) Describe the Hittorf's method for determining the transport number of Ag^+ and NO_3^- ions in solution of silver nitrate.

7. The specific conductance of a saturated solution of AgCl is 1.55×10^{-4} ohm^{-1} m^{-1} at 298 K. The ion conductance of Ag^+ Cl^{-1} are 61.94 and 76.34 ohm^{-1} respectively. Calculate the solubility of AgCl in gram equivalent per liter at 298 K. Neglect the specific conductance of water.

Answer. 12.802 kg/m^3

8. In a moving boundary experiment with 0.1×10^3 N KCl solution, the boundary moved 4.94×10^{-2} m during 4020 seconds when a current of 5.21 A was used. The cross-sectional area of the tube was 0.230×10^{-4} m^2. Calculate the transport number of K^+ ions.

Answer. 0.523×10^{-3}

9. Write short notes on :

(*a*) Transport number

(*b*) Kohlrausch law of independent migration of ions (*Delhi BSc*)

10. (*a*) How is solubility of sparingly soluble salt is determined by conductance measurement?

(*b*) Discuss the application of Kohlrausch's law. (*Lucknow BSc*)

11. If the conductance at infinite dilution of NaCl, HCl and CH_3COONa are 125.45×10^{-8}, 426.1×10^{-8} and 91.0×10^{-8} m^2 eqvt^{-1} respectively, calculate equivalent conductance of acetic acid at infinite dilution.

Answer. 391.65×10^{-8} ohm^{-1} m^2 eqvt^{-1} (*Guru Nanak Dev BSc*)

12. Define Transport number. Describe briefly the principle of experimental determination of Transport number by Hittorf's method. (*Panjab BSc*)

13. (*a*) State the principle of conductometric titrations. Draw the titration curves for

(*i*)Weak acid with a strong base

(*ii*)Na_2SO_4 solution with $BaCl_2$ solution

(*b*) Calculate the ionization constant of water if the specific conductivity of water is 5.51×10^{-10} ohm^{-1} m^{-1} and ionic conductance of H^+ and OH^- ions at the same temperature are 349.8×10^{-4} and 196.5×10^{-4} respectively.

Answer. 1.017×10^{-8} (*Arunachal BSc*)

14. Derive the expression for the determination of Transport number by Hittorf's method when electrodes are not attacked? (*Arunachal BSc*)

15. Explain the factors on which the transport number depends. (*MD Rohtak BSc*)

16. Explain why the transport number of cadmium ion in cadmium iodide solution at high concentration may become zero or negative also. (*Arunachal BSc*)

17. Explain how is Kohlrausch's law helpful in determining the equivalent conductance of weak electrolyte at infinite dilution. (*Jammu BSc*)

18. Discuss the effect of concentration on transference number. (*Kalyani BSc*)

19. (*a*) Explain the term 'ionic mobility'. Describe Hittorf's theoretical device to show that although most of the ions differ largely in their mobilities, but their equivalent amounts are discharged on electrolysis at appropriate electrodes.

(*b*) In moving boundary experiment with 0.100 N KCl using 0.065 N LiCl as indicator solution, a constant current of 0.005893 amp was passed for 2130 seconds. The boundary was observed to move through 5.6×10^{-2} m in a tube of 0.1142×10^{-24}sq.m cross-section. Calculate the transport number of K^+ and Cl^- ions.

Answer. 0.492; 0.508 (*Rajasthan BSc*)

20. What is meant by limiting equivalent conductance of weak electrolyte? How do you determine it by the help of Kohlrausch's law? *(Sambalpur BSc)*

21. Explain the use of Kohlrausch's law to determine the solubility of a sparingly soluble salt. *(Nagpur BSc)*

22. (*a*) Define transport number.

(*b*) A solution of silver nitrate containing 0.51 gm of the salt in 60.40 gm of the solution was electrolyzed under similar electrodes. After electrolysis 64 gm of the anode solution was found to contain 0.56 gm of silver nitrate. A current of 0.04 amp was passed for 2950 seconds. Calculate the transference number of nitrate ion.

Answer. (*b*) 0.520 *(Delhi BSc)*

23. (*a*) How is Kohlrausch's law helpful in determining the solubility of a sparingly soluble salt in water?

(*b*) How can you find the transference number of an ion by Hittorf's method? *(Madurai BSc)*

24. A solution of HCl was electrolyzed in a cell between two platinum electrodes. The cathode compartment contained 0.354×10^{-3} kg of chloride ions before electrolysis and 0.32×10^{-3} kg after electrolysis. A silver coulometer in series had a deposit of 0.532×10^{-3} kg of silver after passing the same amount of current. Calculate the transport number of H^+ and Cl^- ions.

Answer. 0.936 ; 0.064 *(Rajasthan BSc)*

25. In the Hittorf cell using silver electrodes and $AgNO_3$ as an electrolyte, a certain amount of current was passing which deposited 9.886×10^{-7} kg equivalent of silver in the coulometer. The anode compartment had the composition 28.235×10^{-3} kg of H_2O and 0.099×10^{-3} kg of $AgNO_3$ before electrolysis and 28.435×10^{-3} kg of water and 0.1874×10^{-3} kg of $AgNO_3$ after electrolysis. Calculate the transport number of Ag^+ and NO_3^- ions.

Answer. 0.48 and 0.52 *(Kalyani BSc)*

26. In a moving boundary experiment with 1 N KCl solution using $CaCl_2$ as indicator, a current of 0.0115 A was passed for 1800 seconds and the boundary moved through a volume of 0.106×10^{-6} m^3. Calculate the transport number of K^+ ion.

Answer. 0.494 *(Vikram BSc)*

27. A conductivity cell whose cell constant is 2 m^{-1}, is filled with 0.1×10^3 M acetic acid solution. Its resistance is found to be 3765 siemens. Calculate the degree of dissociation of acetic acid.

Answer. 0.0136 *(Mysore BSc)*

28. In a moving boundary method, a current of 30×10^{-3} A was passed for 80 seconds. During this time the boundary of a solution of HCl containing 10 mol/m^3 moved 17×10^{-2} m towards the cathode. Calculate the transport number of H^+ and Cl^- ions. (The cross-sectional area of the glass tube is 1.0×10^{-5} m^2).

Answer. 0.68 ; 0.32 *(Agra BSc)*

29. What do you understand by transport number? Give one method for the determination of it. *(Anna BSc)*

30. How does Kohlrausch's law helps in determining the dissociation constants of a weak acid by conductance measurements? *(Shivaji BSc)*

31. What do you understand by the terms: transport number and ionic mobility of an ion? How are these related to each other? Write their units. *(Panjab BSc)*

32. How does Kohlrausch's law helps in determining the dissociation constants of a weak acid by conductance measurements? *(Madurai BSc)*

33. During the electrolysis of a solution of potassium chloride between platinum electrodes, 0.0137×10^{-3} kg of chloride was lost from the anodic compartment and 0.0857×10^{-3} kg of silver was deposited in a silver coulometer connected in series with the cell. Find at the transport number of K^+ and Cl^- ions.

Answer. 0.4854; 0.5146 *(Delhi BSc)*

34. Describe Hittorf's method for determining the transport number of Ag^+ and NO_3^- ions in a solution of $AgNO_3$ when the electrodes are made up of silver. *(Rajasthan BSc)*

MULTIPLE CHOICE QUESTIONS

1. According to Arrhenius theory an electrolyte when dissolved in water gives two types of
(*a*) charged particles (*b*) molecules
(*c*) ion pairs (*d*) fundamental particles
Answer. (*a*)

2. For an electrolyte AB, the ionization constant is given by the expression
(*a*) $K = \frac{[A][B]}{[AB]}$ (*b*) $K = \frac{[A^+]+[B^+]}{[AB]}$
(*c*) $K = \frac{[A^+]-[B^+]}{[AB]}$ (*d*) $K = \frac{[A^+][B^-]}{[AB]}$
Answer. (*d*)

3. The conductivity of an electrolyte is due to the
(*a*) presence of ions in the electrolyte (*b*) free movement of ions in the solution
(*c*) reunion of ions in the solution (*d*) release of heat energy due to ionization
Answer. (*b*)

4. The electrical conductivity of an electrolyte depends upon
(*a*) the number of molecules in the electrolyte
(*b*) the number of ions present in the electrolyte
(*c*) the number of ions present in the solution
(*d*) the number of molecules of the solvent
Answer. (*c*)

5. On passing electrical current through an electrolyte solution
(*a*) cations move towards anode
(*b*) anions move towards cathode
(*c*) cations move towards cathode and anions towards anode
(*d*) both cations and anions move in same direction
Answer. (*c*)

6. On passing electrical current through an electrolyte solution, the cations
(*a*) move towards cathode with the speed equal to that of anions towards anode
(*b*) move with faster speed than that of anions
(*c*) move with different speed as compared to that of anions
(*d*) move with slower speed than that of anions
Answer. (*c*)

7. The Hittorf's rule states that
(*a*) the loss of concentration around any electrode is proportional to the speed of the ions moving towards it
(*b*) the loss of concentration around any electrode is proportional to the speed of the ions moving away from it
(*c*) the loss of concentrations around both the electrodes is proportional to the sum of speed of cations and anions
(*d*) none of the above
Answer. (*b*)

8. The Hittorf's rule can be represented by the expression

(*a*) $\frac{\text{Fall around anode}}{\text{Rise around cathode}} = \frac{\text{speed of cation}}{\text{speed of anion}}$ (*b*) $\frac{\text{Fall around anode}}{\text{Fall around cathode}} = \frac{\text{speed of anion}}{\text{speed of cation}}$

(*c*) $\frac{\text{Fall around anode}}{\text{Fall around cathode}} = \frac{\text{speed of cation}}{\text{speed of anion}}$ (*d*) $\frac{\text{Rise around cathode}}{\text{Fall around anode}} = \frac{\text{speed of anion}}{\text{speed of cation}}$

Answer. (*c*)

9. The fraction of the total current carried by the cation or anion is termed as

(*a*) fractional number (*b*) speed number
(*c*) carrier number (*d*) transport number

Answer. (*d*)

10. If v_+ is the speed of cation and v_- that of anion, the transport number of cation is given by

(*a*) $\frac{v_+}{v_+ - v_-}$ (*b*) $\frac{v_+}{v_- - v_+}$

(*c*) $\frac{v_+}{v_+ + v_-}$ (*d*) $\frac{v_-}{v_+ + v_-}$

Answer. (*c*)

11. If v_+ is the speed of cation and v_- that of anion, then the speed ratio is given by

(*a*) $r = \frac{t_+}{t_-}$ (*b*) $r = \frac{t_+}{1 + t_+}$

(*c*) $r = \frac{t_-}{1 - t_+}$ (*d*) $r = \frac{t_+}{1 - t_-}$

Answer. (*d*)

12. The sum of the transport number of cation and anion is equal to

(*a*) 1 (*b*) 0
(*c*) 0.5 (*d*) ∞

Answer. (*a*)

13. In Hittorf method for determination of transport numbers we make use of a

(*a*) H-tube (*b*) V-tube
(*c*) U-tube (*d*) L-tube

Answer. (*c*)

14. While determining transport number by Hittorf method, _______ electrodes are used when electrodes are attackable

(*a*) silver (*b*) copper
(*c*) mercury (*d*) platinum

Answer. (*d*)

15. In the moving boundary method for the determination of transport number, the formula used is (where *n* is the number of faradays of current passed)

(*a*) $t_{A^+} = \frac{slc}{n}$ (*b*) $t_{A^+} = \frac{slc}{n^2}$

(*c*) $t_{A^+} = \frac{slc^2}{n}$ (*d*) $t_{A^+} = \frac{sl^2c}{n}$

Answer. (*a*)

16. One Faraday is equal to

(*a*) 96.500 coulombs (*b*) 9650 coulombs
(*c*) 965.00 coulombs (*d*) 96500 coulombs

Answer. (*d*)

17. The statement of Kohlrausch's law is

(*a*) the equivalent conductance of an electrolyte at infinite dilution is equal to the product of equivalent conductances of the component ions

(*b*) the equivalent conductance of an electrolyte at infinite dilution is equal to the difference of equivalent conductances of the component ions

(*c*) the equivalent conductance of an electrolyte at infinite dilution is equal to the sum of equivalent conductances of the component ions

(*d*) none of the above

Answer. (*c*)

18. Kohlrausch's law can be expressed as

(*a*) $\lambda_\alpha = \lambda_a - \lambda_c$ (*b*) $\lambda_\alpha = \lambda_c - \lambda_a$

(*c*) $\lambda_\alpha = \lambda_a + \lambda_c$ (*d*) $\lambda_\alpha = \lambda_a \times \lambda_c$

Answer. (*c*)

19. Kohlrausch's law can be used to determine

(*a*) λ_α for weak electrolytes (*b*) absolute ionic mobilities

(*c*) solubility of sparingly soluble salts (*d*) all of these

Answer. (*d*)

20. If λ_α is the equivalent conductance at infinite dilution and λ_v the equivalent conductance of the electrolyte at the dilution *v*, the degree of dissociation is given by

(*a*) $\alpha = \dfrac{\lambda_\alpha}{\lambda_v}$ (*b*) $\alpha = \dfrac{\lambda_v}{\lambda_\alpha}$

(*c*) $\alpha = \lambda_\alpha - \lambda_v$ (*d*) $\alpha = \lambda_v - \lambda_\alpha$

Answer. (*b*)

21. When a strong acid is titrated against a strong base the end point is the point of

(*a*) zero conductance (*b*) maximum conductance

(*c*) minimum conductance (*d*) none of these

Answer. (*c*)

22. During titration of a weak acid against a weak base, there is a sharp increase in ________ at the end point

(*a*) conductivity (*b*) equivalent conductance

(*c*) specific conductance (*d*) none of these

Answer. (*a*)

23. If the equivalent conductance of a certain solution of acetic acid is 0.003907 siemens m^2 eqvt^{-1}. If λ_α of CH_3COOH is 390.7×10^{-4}, the degree of dissociation of acetic acid is

(*a*) 0.1 (*b*) 0.2

(*c*) 0.5 (*d*) 0.75

Answer. (*a*)

24. The equivalent conductance at infinite dilution of NaCl, HCl and CH_3COONa at 298 K are 0.0126, 0.0426 and 0.0091 siemens m^2 respectively. The equivalent conductance of acetic acid at infinite dilution at 298 Kwill be

(*a*) 643.0×10^{-4} (*b*) 517.0×10^{-4}

(*c*) 217.0×10^{-4} (*d*) 391.0×10^{-4}

Answer. (*d*)

25. The equivalent conductance of decinormal solution of acetic acid at 298 K is 80 and at infinite dilution 400 siemens. The percentage dissociation of acetic acid is

(*a*) 10% (*b*) 20%

(*c*) 50% (*d*) 100%

Answer. (*b*)

26. If the transport number of K^+ is 0.492 in KCl solution. The transport number of Cl^- ion will be

(*a*) 0.984 (*b*) 0.492
(*c*) 0.508 (*d*) 0.016

Answer. (*c*)

27. In a binary electrolyte AB of certain concentration, the transport number of cation is 0.4. The transport number of the anion will be

(*a*) 0.2 (*b*) 0.4
(*c*) 0.6 (*d*) 0.8

Answer. (*c*)

28. The equivalent conductance at 219 K of a normal solution of KCl is 98.2 and at infinite dilution at the same temperature is 131. The degree of dissociation of KCl will be

(*a*) 131/98.2 (*b*) 98.2/131
(*c*) 98.2/131 + 98.2 (*d*) 131/131 + 98.2

Answer. (*b*)

29. If the degree of dissociation of weak electrolyte at a certain temperature is 0.1, the percentage of the molecules undissociated will be

(*a*) 10% (*b*) 20%
(*c*) 50% (*d*) 90%

Answer. (*d*)

30. For strong electrolytes, the degree of dissociation is

(*a*) nearly equal to one (*b*) nearly equal to zero
(*c*) nearly equal to infinity (*d*) nearly equal to 0.5

Answer. (*a*)

31. The equivalent conductance at infinite dilution of NH_4Cl, NaOH and NaCl at 291 K are respectively, 0.01298, 0.02274 and 0.01089 siemens m^2 g $eqvt^{-1}$. The equivalent conductance of NH_4OH at infinite dilution at 291 K will be

(*a*) 128.8×10^{-4} (*b*) 108.9×10^{-4}
(*c*) 227.4×10^{-4} (*d*) 238.3×10^{-4}

Answer. (*d*)

32. At infinite dilution the conductance of NH_4Br, KOH and KBr at 298 K are 151×10^{-7}, 271×10^{-7} and 151×10^{-7} siemens m^2 kg $eqvt^{-1}$ respectively. If the equivalent conductance of NH_4OH at 0.01 N concentration at 298 K is 27.1×10^{-7} ohm^{-1} kg $eqvt^{-1}$, the degree of dissociation of NH_4OH would be

(*a*) 0.1×10^{-7} (*b*) 0.01×10^{-7}
(*c*) 0.001×10^{-7} (*d*) 1.0×10^{-7}

Answer. (*a*)

33. λ_o for $Ba(OH_2)$, $BaCl_2$ and NH_4Cl are 228×10^{-7}, 120×10^{-7} and 129×10^{-7} siemens m^2 kg $eqvt^{-1}$. The value of λ_o for NH_4OH

(*a*) 108×10^{-7} siemens m^2 kg $eqvt^{-1}$ (*b*) 249×10^{-7} siemens m^2 kg $eqvt^{-1}$
(*c*) 348×10^{-7} siemens m^2 kg $eqvt^{-1}$ (*d*) 183×10^{-7} siemens m^2 kg $eqvt^{-1}$

Answer. (*d*)

34. The ionic mobility of an ion is given by the relation

(*a*) $\lambda_\alpha \div 96500$ (*b*) $\lambda_\alpha \times 96500$
(*c*) $\lambda_\alpha + 96500$ (*d*) $\lambda_\alpha - 96500$

Answer. (*a*)

35. If the velocity of an ion at infinite dilution be *u* m/s when the distance between the electrodes is 0.20 m and the voltage 100 V, the ionic mobility is

(*a*) $u/5 \times 10^{-2}$ m s^{-1} (*b*) $5\,u \times 10^{-2}$m s^{-1}
(*c*) $u + 5 \times 10^{-2}$ m s^{-1} (*d*) $u - 5 \times 10^{-2}$ m s^{-1}

Answer. (*a*)

36. The ionic product of water is

(*a*) 1×10^{-7} (*b*) 1×10^{-10}

(*c*) 1×10^{14} (*d*) 1×10^{-14}

Answer. (*d*)

37. The solubility, s, of a sparingly soluble salt is related to its equivalent conductance at infinite dilution by the relation (*K* is specific conductance)

(*a*) $s = \dfrac{K \times 1000}{\lambda_\alpha}$ (*b*) $s = \dfrac{c \times 1000}{\lambda_\alpha - \lambda}$

(*c*) $s = \dfrac{K \times 1000}{\lambda_\alpha - \lambda}$ (*d*) $s = \dfrac{c \times 1000}{\lambda_\alpha}$

Answer. (*a*)

38. For an electrolyte of the type AB (sparingly soluble in water), the solubility is related to its solubility product by the relation

(*a*) $K_{sp} = \sqrt{s}$ (*b*) $K_{sp} = s^2$

(*c*) $K_{sp} = 4s^2$ (*d*) $K_{sp} = 4s^3$

Answer. (*b*)

39. The transport number of an ion depends on the mobility of that ion. It also depends upon the

(*a*) mobility of the solvent

(*b*) mobility of the electrolyte

(*c*) mobility of the other ion with which it is associated

(*d*) none of the above

Answer. (*c*)

40. The ionic mobility of Li^+ is small as compared to that of K^+ ion. It is because

(*a*) Li^+ ion is very small

(*b*) Li^+ ion is heavily hydrated

(*c*) Li^+ ion has only one principal quantum number

(*d*) none of the above

Answer. (*b*)

41. The λ_o for K^+ ion is 73.52×10^{-5} siemens m^{-1} kg $eqvt^{-1}$. Its ionic mobility will be

(*a*) $73.52 \times 10^{-5} \div 96500$ (*b*) $73.52 \times 10^{-5} \times 96500$

(*c*) $73.52 \times 10^{-5} + 96500$ (*d*) $73.52 \times 10^{-5} - 96500$

Answer. (*a*)

26 Ionic Equilibria Solubility Product

CONTENTS

OSTWALD'S DILUTION LAW

According to the Arrhenius Theory of dissociation, an electrolyte dissociates into ions in water solutions. These ions are in a state of equilibrium with the undissociated molecules. This equilibrium is called the **Ionic equilibrium.** Ostwald noted that **the Law of Mass Action can be applied to the ionic equilibrium as in the case of chemical equilibria.**

Let us consider a binary electrolyte AB which dissociates in solution to form the ions A^+ and B^-.

$$AB \rightleftharpoons A^+ + B^-$$

Let C moles per m^3 be the concentration of the electrolyte and α (alpha) its degree of dissociation. The concentration terms at equilibrium may be written as :

$$[AB] = C(1-\alpha) \text{ mol/m}^3$$
$$[A^+] = C\alpha \text{ mol/m}^3$$
$$[B^-] = C\alpha \text{ mol/m}^3$$

Applying the Law of Mass Action :

Rate of dissociation $= k_1 \times C(1-\alpha)$

Rate of combination $= k_2 \times C\alpha \times C\alpha$

At equilibrium :

$$k_1 \times C(1-\alpha) = k_2 \times C\alpha \times C\alpha$$

or
$$\frac{C\alpha \times C\alpha}{C(1-\alpha)} = \frac{k_1}{k_2} = K_c$$

or
$$K_c = \frac{\alpha^2 C}{(1-\alpha)} \text{ mol m}^{-3} \quad \text{...(1)}$$

The equilibrium constant K_c is called the **Dissociation constant or Ionization constant.** It has a constant value at a constant temperature.

If one mole of an electrolyte be dissolved in V liter of the solution, then

$$C = \frac{1}{V}$$

V is known as the **Dilution** or the solution. Thus the expression (1) becomes

$$K_c = \frac{\alpha^2}{(1-\alpha)V} \quad \text{...(2)}$$

This expression which correlates the variation of the degree of dissociation of an electrolyte with dilution, is known as Ostwald's Dilution Law.

For Weak Electrolytes

For weak electrolytes, the value of α is very small as compared to 1, so that in most of the calculation we can take $1-\alpha \simeq 1$. T1hus the Ostwald's Dilution Law expression becomes

$$K_c = \frac{\alpha^2}{V}$$

It implies that **the degree of dissociation of a weak electrolyte is proportional to the square root of the dilution** *i.e.*,

$$\alpha \propto \sqrt{K_c V}$$

or
$$\alpha = K'\sqrt{V}$$

For Strong Electrolytes

For strong electrolytes, the value of α is large and it cannot be neglected in comparison with 1. Thus we have to use the original expression (2). That is,

$$K_c = \frac{\alpha^2}{(1-\alpha)V} \quad \text{or} \quad \alpha^2 = K_c V - \alpha K_c V$$

which gives a quadratic equation

$$\alpha^2 + \alpha K_c V - K_c V = 0$$

from this equation the value of α can be evaluated.

Experimental Verification of Ostwald's Law

The Ostwald's Dilution law can be verified if the values of α, the degree of dissociation, at different dilutions are known. The values of α are determined experimentally by using the relation.

$$\alpha = \frac{\lambda_v}{\lambda_\infty}$$

where λ_v and λ_∞ are the equivalent conductances at dilution V and infinite dilution respectively. Their values are found by conductance measurements and Kohlrausch's law. The value of α at various dilutions thus determined are inserted in the expression :

$$K_c = \frac{\alpha^2}{(1 - \alpha)\ V}$$

If the values of K_c come out to be constant, the Ostwald's law stands verified.

Limitation of Ostwald's Law

Ostwald's Dilution law **holds good only for weak electrolytes and fails completely when applied to strong electrolytes.** For strong electrolytes, which are highly ionized in solution, the value of the dissociation constant K, far from remaining constant, rapidly falls with dilution. In Table 26.1, the values of K at different dilutions illustrate the point.

Table 26.1. Dissociation of KCl at 291 K λ_∞ = 129.9

V	λ_v	$\alpha = \lambda_v / 129.9$	$K = \alpha^2 / (1 - \alpha)\ V$
1	98.27	0.565	2.350
2	102.41	0.7883	1.434
5	107.96	0.831	0.8154
20	115.75	0.891	0.3642
200	119.96	0.9234	0.2221
500	124.41	0.9577	0.1084

Factors that explain the failure of Ostwald's law in case of strong electrolytes

(1) The law is based on Arrhenius theory which assumes that only a fraction of the electrolyte is dissociated at ordinary dilutions and complete dissociation occurs only at infinite dilution. However, this is true for weak electrolytes. Strong electrolytes are almost completely ionized at all dilutions and λ_v/λ_∞ does not give the accurate value of α.

(2) The Ostwald's law is derived on the assumption that the Law of Mass Action holds for the ionic equilibria as well. But when the concentration of ions is very high, the presence of charges affects the equilibrium. Thus the Law of Mass Action in its simple form cannot be applied.

(3) The ions obtained by dissociation may get hydrated and may affect the concentration terms. Better results are obtained by using *activities* instead of concentrations.

THEORY OF STRONG ELECTROLYTES

A number of theories have been put forward by different workers in order to explain the high conductance of strong electrolytes. Southerland (1906) held the view that ions in solution were surrounded by a large number of ions of opposite charge. Due to the weakening of interionic forces, the ionic velocities were accelerated. This resulted in the increase of conductance of the electrolyte solution. However, not much notice was taken of Southerland's view.

Ghosh's Formula

In 1918 J.C. Ghosh revived the above theory. He assumed that though the electrolyte is completely ionized, all the ions are not free to move owing to the influence of electric charges and it is only the *mobile ions* which contribute to the conductance of the solution. The value α represents the "active" proportion of the electrolyte and can be determined by purely electrical data, the Law of Mass Action playing no part whatsoever. His formula

$$\sqrt[3]{V} \log \alpha = K$$

was applicable to univalent strong electrolytes.

Debye-Huckel Theory

In 1923 Debye and Huckel and in 1926 Onsagar put forward the **modern theory of strong electrolytes** in which account is taken of the electrostatic forces between the ions. Without going into its mathematical details, a brief outline of the main ideas of the theory is given below :

(1) **The strong electrolyte is completely ionized at all dilutions.** The present position as it has emerged from the study of Raman spectra, X-ray analysis of crystals, Distribution coefficients and vapor pressures is that there is a very small amount of unionized substance also present and therefore instead of saying 'completely ionized' we should say 'almost completely ionized'.

(2) **Since oppositely charged ions attract each other, it suggests that anions and cations are not uniformly distributed in the solution of an electrolyte but that the cations tend to be found in the vicinity of anions and *vice-versa*** (Fig. 26.1). Though the solution is on the whole neutral, there is in the vicinity of any given ion a predominance of ions of opposite charge which we call as **counter ions.** The ions are all the time on the move in all directions but on the average, more *counter ions* than like *ions* pass by any given ion. This spherical haze of opposite charge is called **ionic atmosphere.**

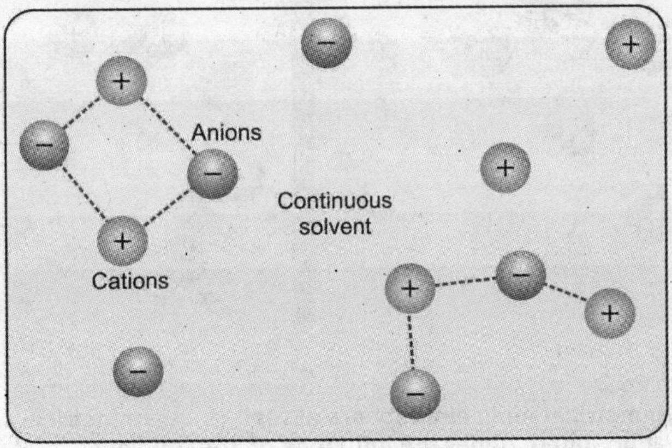

Figure 26.1
The ions of a particular charge are surrounded by more ions of the opposite charge and solvent molecules. Because of the large number of ions in concentrated solution,the ion activity is reduced due to hindered movement of the ions.

(3) **Decrease in equivalent conductance with increase in concentration is due to fall in mobilities of the ions due to greater inter-ionic effect and *vice-versa.***

(4) The ratio $\lambda_v / \lambda_\infty$ does not correctly give the degree of dissociation α for strong electrolytes but only the conductance or conductance coefficient f_c.

(5) In spite of almost complete ionization, λ_v is much less than λ_∞.

The observed deviations are due to the following reasons :

(1) **Asymmetry or Relaxation Effect.** Imagine a central negative ion. This is surrounded by a number of positively charged ions which form its 'ionic atmosphere.' This atmosphere is symmetrically

situated in the absence of any electrical field and the force of attraction exerted by the atmosphere on the central ion is uniform in all directions. When an electric field is applied, the negative ion moves towards the anode and the positive ionic atmosphere towards the cathode. This leaves a large number of positive ions behind it than there are in front of the negative ion with the result that the symmetry of the atmosphere about the central negative ion is destroyed (Fig. 26.2) and it becomes distorted.

Whereas initially the force of attraction exerted by the atmosphere on the central ion was uniform it becomes greater now behind the ion than in front. As a result of this the negative ion experiences a force which tends to drag it backwards and this slows down its movement in the forward direction. This behavior is known as **Asymmetry Effect.** It may be said that the negative ion which leaves the ionic atmosphere of positive ions behind to die away would build a new ionic atmosphere and the asymmetry would be corrected. If this process of building up and dying away were instantaneous, there would be no cause for asymmetry and the atmosphere would always be symmetrically placed about the ion. But, as it is, the formation of the new ionic atmosphere does not take place at the same rate at which the old one decays and the later lags behind or takes more time, known as **'relaxation time'.** During this interval, there is a preponderance of positive ions to the left of the central negative ion which is under move and these tend to drag it back. For this reason asymmetry effect is also known as **Relaxation Effect.** Moreover, the central ion moves into a place where there is an excess of ions of its own sign and this has also a retarding effect.

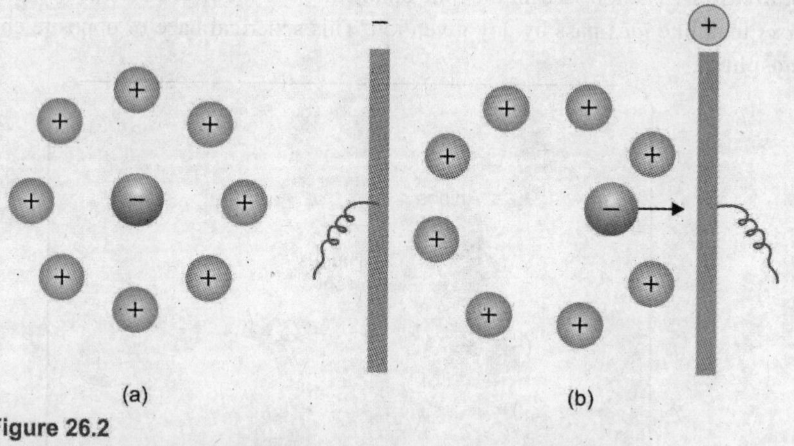

■ **Figure 26.2**
***(a)* Symmetrical ionic atmosphere at rest; *(b)* Asymmetrical ionic atmosphere under the influence of applied field.**

(2) **Electrophoretic Effect.** Another factor which acts as a drag and tends to retard the motion of an ion in solution is the tendency of the applied field to move the ionic atmosphere (to which solvent molecules are also attached) in a direction opposite to that in which the central ion associated with solvent molecules is moving. Thus the central negative ion moving towards the anode has to make its way through the ionic atmosphere with its associated solvent molecules which is moving in the opposite direction *i.e.*, towards the cathode. This causes a retarding influence on the movement of the ion the effect of which is equal to the increase in the viscous resistance of the solvent. By analogy to the resistance acting on the movement of a colloidal particle under an electrical field (See Chapter 22 on Colloids), this effect is called **Electrophoretic Effect.**

Both the above causes reduce the velocity of the ion and operate in solutions of strong electrolytes with the result that a value of equivalent conductance (λ_v) lower than the value at infinite dilution (λ_∞) is obtained. At infinite dilution since the electrical effects are practically absent, the two values tend to approach each other.

Debye-Huckel-Onsager Conductance Equation takes these causes into account and for a univalent electrolyte supposed to be completely dissociated is written in the form

$$\lambda_v = \lambda_\infty - (A + B\lambda_\infty)\sqrt{C}$$

where A and B are constants and c is the concentration in kg-equivalents per m³. These constants depend only on the nature of the solvent and the temperature and are given by the relationships :

$$A = \frac{82.4}{(DT)^{1/2}\,\eta} \quad \text{and} \quad B = \frac{8.20 \times 10^5}{(DT)^{3/2}}$$

where D and η are the dielectric constant and coefficient of viscosity of the *medium* respectively at the absolute temperature T. The constant A is a measure of the electrophoretic effect while B is that of the asymmetry effect. For *water* at 298 K with D = 78.5 and $\eta \times 10^3 = 8.95$, the value of A is 60.20 and that of B is 0.229. On substituting these values in the above equation, we have

$$\lambda_v = \lambda_\infty - (60.20 + 0.229\lambda_\infty)\sqrt{C}$$

It follows, therefore, that if the above equation is correct, a straight line of slope equal to (60.20 + 0.229 λ_∞) should be obtained by plotting observed equivalent conductance (λ_v) against the square root of the corresponding concentration *i.e.*, $\sqrt{C}$. The Onsager equation responds to this test favorably and for aqueous solutions of univalent electrolytes, it is closely obeyed at concentrations upto about 2×10^{-3} equivalents per m³ (Fig. 26.3). Slight deviations at higher concentrations are attributed to certain approximations assumed in deriving the equation.

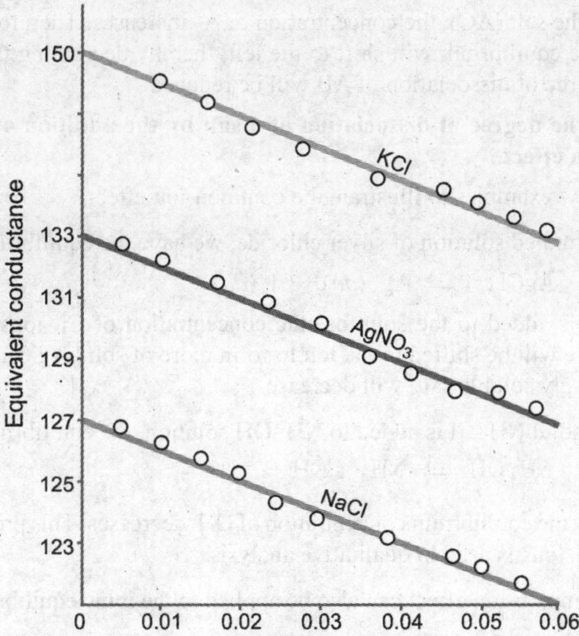

■ Figure 26.3
Testing of Onsager equation. The circles represent the observed value of l_n whereas the dotted line stands for the theoretical values of the slopes expected from the Onsager equation.

DEGREE OF DISSOCIATION

When a certain amount of electrolyte ($A^+ B^-$) is dissolved in water, a small fraction of it dissociates to form ions (A^+ and B^-). When the equilibrium has been reached between the undissociated and the free ions, we have

$$AB \rightleftharpoons A^+ + B^-$$

The fraction of the amount of the electrolyte in solution present as free ions is called the Degree of dissociation.

If the degree of dissociation is represented by x, we can write

$$x = \frac{\text{amount dissociated (mol/m}^3\text{)}}{\text{initial concentration (mol/m}^3\text{)}}$$

The value of x can be calculated by applying the Law of Mass Action to the ionic equilibrium stated above :

$$K = \frac{[A^+][B^-]}{[AB]}$$

If the value of the equilibrium constant, K, is given, the value of x can be calculated.

THE COMMON–ION EFFECT

When a soluble salt (say A^+C^-) is added to a solution of another salt (A^+B^-) containing a common ion (A^+), the dissociation of AB is suppressed.

$$AB \rightleftharpoons A^+ + B^-$$

By the addition of the salt (AC), the concentration of A^+ increases. Therefore, according to Le Chatelier's principle, the equilibrium will shift to the left, thereby decreasing the concentration of A^+ ions. Or that, the degree of dissociation of AB will be reduced.

The reduction of the degree of dissociation of a salt by the addition of a common-ion is called the Common-ion effect.

Let us consider a few examples to illustrate the common-ion effect.

Example 1. In a saturated solution of silver chloride, we have the equilibrium

$$AgCl\ (s) \rightleftharpoons Ag^+\ (aq) + Cl^-\ (aq)$$

When sodium chloride is added to the solution, the concentration of Cl^- ions will increase. The equilibrium shown above will be shifted to the left to form more of solid Ag Cl. Thus the solubility of AgCl, a typical sparingly soluble salt, will decrease.

Example 2. When solid NH_4 Cl is added to NH_4 OH solution, the equilibrium

$$NH_4OH \rightleftharpoons NH_4^+ + OH^-$$

shifts to the left. Thereby the equilibrium concentration of OH^- decreases. This procedure of reducing the concentration of OH^- ions is used in qualitative analysis.

Example 3. The common-ion effect can also be applied to the ionic equilibrium of a weak acid as HF.

$$HF \rightleftharpoons H^+ + F^-$$

NaF is added to the equilibrium mixture. The concentration of F^- (common ion) is increased. Thus the equilibrium shifts to the left. In other words, the degree of dissociation of HF decreases. It was found by experiment that the degree of dissociation of HF in 1M solution is 2.7, while the value reduces to 7.2×10^{-4} after the addition of 1M NaF.

SOLVED PROBLEM 1

Find the degree of dissociation of HF in 1M aqueous solution. The value of K for the ionic equilibrium $HF \rightleftharpoons H^+ + F^-$ is 7.2×10^{-4}.

SOLUTION

HF dissociates in water to form H^+ and F^- ions. On reaching the equilibrium we have

$$HF \rightleftharpoons H^+ + F^-$$

Thus one mole of HF taken initially dissociates to yield 1 mole of H^+ and 1 mole of F^-. If x be the degree of dissociation, the concentration terms at equilibrium are :

$$[HF] = 1 - x \text{ mol/m}^3$$

$$[F^-] = x \text{ mol/m}^3$$

$$[H^+] = x \text{ mol/m}^3$$

Substituting these values in the equilibrium expression, we have

$$K = 7.2 \times 10^{-4} = \frac{[H^+][F^-]}{[HF]} = \frac{(x)(x)}{1.00 - x}$$

If x is very small compared to 1, we can write :

$$7.2 \times 10^{-4} = \frac{x^2}{1.00}$$

$$\therefore \quad x^2 = (7.2 \times 10^{-4})(1.00) = 7.2 \times 10^{-4}$$

$$\text{or} \quad x = \sqrt{7.2 \times 10^{-4}} = 2.7 \times 10^{-2}$$

Thus the degree of dissociation of HF in 1M solution is **2.7×10^{-2}**

FACTORS WHICH INFLUENCE THE DEGREE OF DISSOCIATION

The *degree of dissociation* of an electrolyte in solution depends upon the following factors :

(1) Nature of Solute

The nature of solute is the chief factor which determines its degree of dissociation in solution. Strong acids and strong bases, and the salts obtained by their interaction are almost completely dissociated in solution. On the other hand, weak acids and weak bases and their salts are feebly dissociated.

(2) Nature of the solvent

The nature of the solvent affects dissociation to a marked degree. It weakens the electrostatic forces of attraction between the two ions and separates them. This effect of the solvent is measured by its '*dielectric constant*'. The **dielectric constant** of a solvent may be defined as its capacity to weaken the force of attraction between the electrical charges immersed in that solvent.

The dielectric constant of any solvent is evaluated considering that of vacuum as unity. It is 4.1 in case of ether, 25 in case of ethyl alcohol and 80 in case of water. The higher the value of the dielectric constant the greater is the dissociation of the electrolyte dissolved in it because the electrostatic forces vary inversely as the dielectric constant of the medium. Water, which has a high value of dielectric constant is, therefore, a strong dissociating solvent. The electrostatic forces of attraction between the ions are considerably weakened when electrolytes are dissolved in it and as a result, the ions begin to move freely and there is an increase in the conductance of the solution.

(3) Concentration

The extent of dissociation of an electrolyte is inversely proportional to the concentration of its solution. The less concentrated the solution, the greater will be the dissociation of the electrolyte. This is obviously due to the fact that in a dilute solution the ratio of solvent molecules to the solute molecules is large and the greater number of solvent molecules will separate more molecules of the solute into ions.

(4) Temperature

The dissociation of an electrolyte in solution also depends on temperature. The higher the temperature greater is the dissociation. At high temperature the increased molecular velocities overcome the forces of attraction between the ions and consequently the dissociation is great.

SOLUBILITY EQUILIBRIA AND THE SOLUBILITY PRODUCT

When an ionic solid substance dissolves in water, it dissociates to give separate cations and anions. As the concentration of the ions in solution increases, they collide and reform the solid phase. Ultimately, a dynamic equilibrium is established between the solid phase and the cations and anions in solution. For example, for a sparingly soluble salt, say Ag Cl, we can write the equilibrium equations as follows :

$$\underset{\text{solid}}{AgCl} \rightleftharpoons Ag^+ + Cl^-$$

At equilibrium the solute continues to dissolve at a rate that exactly matches the reverse process, the return of solute from the solution. Now the solution is said to be saturated.

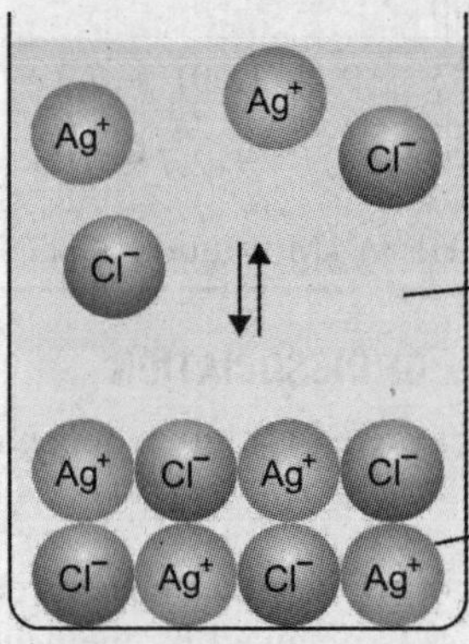

■ **Figure 26.4**
Equilibrium between an ionic solid and ions in a saturated solution.

A Saturated solution is a solution in which the dissolved and undissolved solute are in equilibrium.

A saturated solution represents the limit of a solute's ability to dissolve in a given solvent. This is a measure of the "solubility" of the solute.

The Solubility (*S*) of a substance in a solvent is the concentration in the saturated solution. Solubility of a solute may be represented in kg per 10^{-4} m^3 of solution. It can also be expressed in moles per m^3.

Molar Solubility is defined as the number of moles of the substance per one liter (l) (or m^3) of the solution.

The value of solubility of a substance depends on the solvent and the temperature.

Applying the Law of Mass Action to the above equilibrium for Ag Cl, we have

$$K = \frac{[Ag^+]\,[Cl^-]}{[Ag\,Cl]}$$

The amount of Ag Cl in contact with saturated solution does not change with time and the factor [Ag Cl] remains the same. Thus the equilibrium expression becomes

$$K_{sp} = [Ag^+]\,[Cl^-]$$

where $[Ag^+]$ and $[Cl^-]$ are expressed in mol/m^3. The equilibrium constant in the new context is called the **Solubility Product Constant** (or simply the **Solubility Product**) and is denoted by K_{sp}. The value of K_{sp} for a particular solubility equilibrium is constant at a given temperature. The product $[Ag^+]$ $[Cl^-]$ in the K_{sp} expression above is also called the **Ionic Product** or **Ion Product.**

TABLE 26.2. SOLUBILITY PRODUCT VALUES FOR SOME SPARINGLY SOLUBLE SALTS

Substance	Formula	K_{sp} at 298 K
Aluminium hydroxide	$Al(OH)_3$	$[Al^{3+}][OH^-]^3 = 1.6 \times 10^{-33}$
Barium fluoride	BaF_2	$[Ba^{2+}][F^-]^2 = 1.7 \times 10^{-6}$
Barium carbonate	$BaCO_3$	$[Ba^{2+}][CO_3^{2-}] = 8.1 \times 10^{-9}$
Calcium fluoride	CaF_2	$[Ca^{2+}][F^-]^2 = 4.0 \times 10^{-11}$
Copper (I) bromide	CuBr	$[Cu^{+1}][Br^-] = 4.2 \times 10^{-8}$
Copper (I) sulfide	Cu_2S	$[Cu^+]^2[S^{2-}] = 8.5 \times 10^{-45}$
Iron (III) hydroxide	$Fe(OH)_3$	$[Fe^{3+}][OH^-]^3 = 2.0 \times 10^{-39}$
Lead (II) bromide	$PbBr_2$	$[Pb^{2+}][Br^-]^2 = 7.9 \times 10^{-5}$
Lead (II) iodide	PbI_2	$[Pb^{2+}][I^-]^2 = 1.4 \times 10^{-8}$
Silver Chloride	AgCl	$[Ag^+][Cl^-] = 1.8 \times 10^{-10}$

The K_{sp} expression may be stated as : **the product of the concentration of ions (mol/m^3) in the saturated solution at a given temperature is constant.** This is sometimes called the **Solubility product principle.**

NUMERICAL PROBLEMS

While discussing numerical problems on solubility product principle, three cases may arise:

Case 1. Calculation of K_{sp} from Solubility

The solubility of a sparingly soluble salt as AgCl may be defined as **the moles of solid AgCl which dissolve per one meter (1.0 m^3) of solution to reach the equilibrium with the excess solid.**

Here, the equilibrium reaction is

$$AgCl\,(s) \rightleftharpoons Ag^+ + Cl^-$$

One mole of AgCl gives one mole of Ag^+ ion and one mole of Cl^- ion. If S mol/m^3 be the solubility of AgCl, the equilibrium concentrations of Ag^+ and Cl^- are :

$$[Ag^+] = S \text{ mol/m}^3$$

$$[Cl^-] = S \text{ mol/m}^3$$

Substituting the values in the K_{sp} expression,

$$\begin{aligned} K_{sp} &= [Ag^+]\,[Cl^-] \\ &= [S\ mol/m^3]\,[S\ mol/m^3] \\ &= S^2\ mol^2/m^6 \end{aligned}$$

Knowing the value of S, K_{sp} can be calculated.

The units for K_{sp} values are usually omitted.

SOLVED PROBLEM 2

The solubility of CuBr is found to be 2.0×10^{-4} mol/m³ at 298 K. Calculate K_{sp} value for CuBr.

SOLUTION

The solubility of CuBr is = 2.0×10^{-4} mol/m³. This means that the equilibrium concentration of CuBr = 2.0×10^{-4} mol/m³.

From the reaction

$$\underset{\text{solid}}{CuBr} \rightleftharpoons Cu^+ + Br^-$$

it is evident that 1 mole of CuBr yields one mole of Cu^+ and one mole of Br^-. Thus 2.0×10^{-4} mole of CuBr would yield 2.0×10^{-4} mol of Cu^+ and 2.0×10^{-4} mol of Br^-.

∴ Equilibrium concentration of $Cu^+ = 2.0 \times 10^{-4}\ mol\ m^{-3}$

Equilibrium concentration of $Br^- = 2.0 \times 10^{-4}\ mol\ m^{-3}$

Substituting these in the expression

$$\begin{aligned} K_{sp} &= [Cu^+]\,[Br^-] \\ &= [2.0 \times 10^{-4}\ mol\ m^{-3}]\,[2.0 \times 10^{-4}\ mol\ m^{-3}] \\ &= \mathbf{4.0 \times 10^{-8}\ mol^2\ m^{-6}} \end{aligned}$$

Case 2. Calculation of Solubility from K_{sp}

We have seen how the experimentally determined value of solubility of an ionic solid can be used to calculate its K_{sp} value. The reverse is also possible. We can calculate the solubility of an ionic compound if the K_{sp} value is given. Here, we assume that the solubility is S. Then the equilibrium concentrations of the ions into which the compound dissociates are found. The value of K_{sp} is found by substituting these in the K_{sp} expression.

SOLVED PROBLEM 3

Calculate the solubility of $NiCO_3$ in moles per m³ and kilograms per m³. The value of K_{sp} for $NiCO_3 = 1.4 \times 10^{-7}$.

SOLUTION

$$\begin{aligned} NiCO_3\,(S) &\rightleftharpoons Ni^{2+} + CO_3^{2-} \\ K_{sp} &= [Ni^{2+}]\,[CO_3^{2-}] \\ 1.4 \times 10^{-7} &= [Ni^{2+}]\,[CO_3^{2-}] \end{aligned}$$

In the above equilibrium reaction, one mole of $NiCO_3$ dissociates to produce one mole of Ni^{2+} and one mole of CO_3^{2-}. If S mol/m³ be the solubility of $NiCO_3$, the equilibrium concentration of Ni^{2+} is S mol/m³ and that of CO_3^{2-} also it is S mol/m³.

Substituting these values in the K_{sp} expression, we have

$$1.4 \times 10^{-7} = [S\ mol/m^3]\,[S\ mol/m^3]$$

or $S^2 = 1.4 \times 10^{-7}$

$$S = \sqrt{1.4 \times 10^{-7}} = 3.7 \times 10^{-4}\ \text{mol/m}^3$$

This means that the solubility of $NiCO_3$ is 3.7×10^{-4} mol/m³. Therefore, the solubility in kg/m³ is = $3.7 \times 10^{-4} \times 118.7 \times 10^{-3}$ = **44×10^{-6} kg/m³** where 118.7 is the molecular mass of $NiCO_3$.

SOLVED PROBLEM 4

Calculate the K_{sp} for Bismuth sulfide (Bi_2S_3), which has a solubility of 1.0×10^{-12} mol/m³ at 298 K.

SOLUTION

Here, the equilibrium reaction is

$$Bi_2S_3 \rightleftharpoons 2Bi^{3+} + 3S^{2-}$$

One mole of Bi_2S_3 gives two moles of Bi^{3+} ion and three moles of S^{2-} ion. Thus 1.0×10^{-12} mole Bi_2S_3 will give 2 (1.0×10^{-12} mol) of Bi^{3+} and 3 (1.0×10^{-12} mol) of S^{2-}. We can write the equilibrium concentrations of Bi^{3+} and S^{2-} as :

$$[Bi^{3+}] = 2.0 \times 10^{-12}\ \text{mol/m}^3$$
$$[S^2] = 3.0 \times 10^{-12}\ \text{mol/m}^3$$

Substituting these values in the K_{sp} expression

$$K_{sp} = [Bi^{3+}]^2 [S^{2-}]^3$$
$$= (2.0 \times 10^{-12})^2 (3.0 \times 10^{-12})^3$$
$$= 1.08 \times 10^{-60}$$

Thus the value of K_{sp} is **1.08×10^{-60}**.

SOLVED PROBLEM 5

The solubility of $BaSO_4$ is 2.33×10^2 kg/m³ at 293 K. Calculate the solubility product of $BaSO_4$ assuming that the salt is completely ionized.

SOLUTION

$$\underset{S\ \text{mol/m}^3}{BaSO_4} \rightleftharpoons \underset{S\ \text{mol/m}^3}{Ba^{2+}} + \underset{S\ \text{mol/m}^3}{SO_4^{2-}}$$

Where S represents the solubility of $BaSO_4$ at 293 K.

Amount of $BaSO_4$ in one ml = 2.3×10^2 kg

Amount of $BaSO_4$ in one l = $2.3 \times 10^2 \times 10^3$ kg

No. of moles of $BaSO_4$ per m³ = $\dfrac{2.3 \times 10^2 \times 10^{-3}}{233 \times 10^{-3}} = 1.0 \times 10^{-3}$

$$\therefore \quad K_{sp} = [Ba^{2+}][SO_4^{2-}]$$
$$= (1.0 \times 10^{-3})(1.0 \times 10^{-3})$$
$$= 1.0 \times 10^{-6}$$

Thus the solubility product of $BaSO_4$ at 293 K is **1.0×10^{-6}**.

Case 3. Whether precipitation will occur on mixing solutions

When two reacting solutions are mixed, calculate the concentration of each ion in the solution in which precipitation is produced. The ionic product, Q, is then calculated. We know that K_{sp} is the ionic product when the ions are in contact with the solid substance. Therefore, the precipitation will occur for any higher ion concentrations. In other words, the precipitation will take place if $Q > K_{sp}$.

If $Q < K_{sp}$, no precipitation will occur.

SOLVED PROBLEM 6

A 200 m³ of 1.3×10^{-3} M $AgNO_3$ is mixed with 100 m³ of 4.5×10^{-5} M Na_2S solution. Will precipitation occur ? ($K_{sp} = 1.6 \times 10^{-49}$)

SOLUTION

The reaction that would cause precipitation is

$$2Ag^+ + S^{2-} = Ag_2S\ (s)$$

The ion product is

$$Q = [Ag^+]^2\ [S^{2-}]$$

Let us now calculate the molar concentrations of Ag^+ and S^{2-} ions :

$$[Ag^+] = 1.3 \times 10^{-3}\ \text{M} \times \frac{200\,\text{m}^3}{(200 + 100)\,\text{m}^3} = 8.7 \times 10^{-4}\ \text{M}$$

$$[S^{2-}] = 4.5 \times 10^{-5}\ \text{M} \times \frac{100\,\text{m}^3}{300\,\text{m}^3} = 1.5 \times 10^{-5}\ \text{M}$$

$\therefore$ $Q = (8.7 \times 10^{-4})\ (1.5 \times 10^{-5}) = 13.05 \times 10^{-9}$

K_{sp} for the reaction is $= 1.6 \times 10^{-49}$ (given)

$\therefore$ $Q > K_{sp}$ and **precipitation will occur**.

SOLVED PROBLEM 7

50 m³ of 6.0×10^{-3} M $CaCl_2$ is mixed with 30 m³ of 0.04 M NaF_2. Will precipitation of CaF_2 occur ? (K_{sp} for $CaCl_2 = 4.0 \times 10^{-11}$)

SOLUTION

The reaction that can cause precipitation is

$$Ca^{2+} + 2F^- = CaF_2(s)$$

The concentration of each ion in solution is :

$$[Ca^{2+}] = 6.0 \times 10^{-3}\ \text{M} \times \frac{50.0\,\text{m}^3}{(50.0 + 30.0)\,\text{m}^3} = 3.75 \times 10^{-3}\ \text{M}$$

$$[F^-] = 0.040\,\text{M} \times \frac{30.0\,\text{m}^3}{80.0\,\text{m}^3} = 0.015\,\text{M}$$

$\therefore$ $Q = [Ca^{2+}]\ [F^-]^2 = (3.75 \times 10^{-3})\ (0.015)^2 = 8.4 \times 10^{-7}$

Since $Q > K_{sp}$, **precipitation will occur**.

Case 4. Calculation of solubility in presence of a common ion

Let us consider how to find the solubility of a solid when water contains an ion in common with the dissolving salt. Assuming that S mol/m³ is the solubility, the concentrations of ions are expressed in terms of S. It may be noted that the concentration of the common ion is added to the equilibrium concentration to get the final concentration. The values of the final concentrations are then substituted into the K_{sp} expression and the value of S calculated.

SOLVED PROBLEM 8

Calculate the solubility of silver chromate, Ag_2CrO_4, in a 0.100 M solution of $AgNO_3$. (K_{sp} for $Ag_2CrO_4 = 9.0 \times 10^{-12}$)

SOLUTION

Ag_2CrO_4 dissolves according to the equilibrium reaction

$$Ag_2\,CrO_4(s) \rightleftharpoons 2Ag^+ + CrO_4^{2-}$$

If S mol/m³ be the solubility, we can say that :

$$S\,\text{mol/m}^3\ Ag_2CrO_4 \longrightarrow 2S\,\text{mol/m}^3\ Ag^+ + S\,\text{mol/m}^3\ CrO_4^{2-}$$

∴ The equilibrium concentrations in terms of S are.

$$[Ag^+] = 0.100 \quad \text{(already present + 2 } S)$$

$$[CrO_4^{2-}] = S$$

Substituting these in the K_{sp} expression, we have

$$9.0 \times 10^{-12} = [Ag^+][CrO_4^{2-}] = (0.100 + 2S)^2$$

$$= (0.100)^2\,(S)$$

(Since S is very small compared to 0.100 M, the factor $2S^2$ is ignored)

Hence $$S = \frac{9.0 \times 10^{-12}}{(0.100)^2} = 9.0 \times 10^{-10}\ \text{mol/m}^3$$

Therefore, the solubility of Ag_2CrO_4 in 0.1 M $AgNO_3$ is S = **9.0×10^{-10} mol/m³**.

APPLICATION OF SOLUBILITY PRODUCT PRINCIPLE IN QUALITATIVE ANALYSIS

Precipitation reactions are used to identify cations present in an unknown mixture of salts.

The Ion product and K_{sp}. The solubility product (K_{sp}) of an insoluble substance is the product of the concentrations of its ions at equilibrium. However, **the ion product is the product of actual concentrations of ions** which may or may not be in equilibrium with the solid. For example,

$$AgCl(s) \rightleftharpoons Ag^+ + Cl^-$$

If S be the solubility of AgCl, we have

$$K_{sp} = [S\,\text{mol/m}^3\ Ag^+]\ [S\,\text{mol/m}^3\ Cl^-]$$

Suppose 0.25 mol/m³ excess of HCl is added to the solution. Then Q, the ion product, will be

$$Q = [S\,\text{mol/m}^3\ Ag^+]\ [(S + 0.25)\ \text{mol/m}^3\ Cl^-]$$

The increase in concentration of Cl^- will shift the equilibrium to the left to form a precipitate of AgCl.

Thus, if

$Q > K_{sp}$ the precipitation will occur

$Q = K_{sp}$ the solution is saturated

$Q < K_{sp}$ no precipitation will occur, more solid salt will dissolve

SELECTIVE PRECIPITATION

The scheme of Qualitative Analysis is based on the **Principle of Selective Precipitation.** Let us consider a mixture containing Ag^+, Cu^{2+} and Fe^{3+}.

Step 1 : Add dil HCl to the solution of the mixture. Ag^+ will precipitate as AgCl. It is removed by filtration.

Step 2 : Pass H_2S through the filtrate. Cu^{2+} is precipitated as CuS and removed by filtration.

Step 3 : Add NH_4OH or NaOH solution to filtrate. Fe^{3+} is precipitated as $Fe(OH)_3$.

In this way ions are precipitated from solution one by one. **The precipitation of cations from solution one at a time is called Selective precipitation.**

Separation of the Basic ions into Groups

Qualitative analysis of a mixture containing all the common cations involves first their separation into six major Groups based on solutions. Each group is then treated further to separate and identify the individual cations. The scheme of separation of cations into different Groups is listed in Fig.26.5.

Group I — Insoluble Chlorides

When excess dil HCl is added to a solution containing a mixture of common cations, only Ag^+, Pb^{2+} and Hg^{2+} will be precipitated as insoluble chlorides. The K_{sp} of these chlorides is very low and is easily exceeded. The K_{sp} value of other chlorides is high and they remain dissolved. For example, the K_{sp} of CuCl is 1.9×10^{-7} and, therefore, it will not be precipitated.

Group II — Sulfides insoluble in acid solution

After removing the insoluble chlorides in Group I, the solution is still acidic. When H_2S is passed through this solution, only the insoluble sulfides of Hg^{2+}, Cd^{2+}, Bi^{3+}, Cu^{2+}, and Sn^{4+} will precipitate. It is so as the concentration of S^{2-} ion is relatively low because of the higher concentration of H^+ (Common Ion effect).

$$H_2S\,(g) \rightleftharpoons 2H^+ + S^{2-} \qquad \text{(equilibrium shifted to left)}$$

Other cations Mn^{2+}, Zn^{2+}, Ni^{2+}, CO^{2+} with higher value of K_{sp} of the respective sulfides will remain dissolved because the ionic product is less than K_{sp}.

Group III — Insoluble Hydroxide in $NH_4OH + NH_4Cl$

To the filtrate from Group II is added NH_4Cl and then NH_4OH. NH_4OH is a weak base and the concentration of OH^- is low. NH_4Cl provides more NH_4^+ ions which shifts the equilibrium to the left (Common Ion effect).

$$NH_4OH \rightleftharpoons NH_4^+ + OH^-$$

Thus some OH^- ions are converted to undissociated NH_4OH. So that the concentration of OH^- ions becomes extremely low. But the ion product, Q, is greater than K_{sp} values for the hydroxides of Al^{3+}, Fe^{3+} and Cr^{3+} which precipitate out. On the other hand, the ion product is lower than the K_{sp} values for Zn $(OH)_2$ Mn $(OH)_2$, and $Mg(OH)_2$ which remain dissolved.

	$Zn(OH)_2$	$Mn(OH)_2$	$Mg(OH)_2$
K_{sp}	4.5×10^{-17}	4.6×10^{-14}	1.5×10^{-11}

Group IV — Insoluble Sulfides in basic solution

The filtrate from Group III contains OH^- ions and is basic. H_2S is passed through it. In the presence of OH^- ions, the H^+ ions produced from H_2S form unionized water.

$$H_2S \rightleftharpoons 2H^+ + S^{2-}$$

$$H^+ + OH^- \longrightarrow H_2O \text{ (nonionised)}$$

Thus the above equilibrium shifts to the right and the concentration of S^{2-} ions increases. This increases the actual ion product for the reaction

$$\underset{\text{metal sulphide}}{MS(s)} \rightleftharpoons M^{2+} + S^{2-}$$

which becomes greater than the K_{sp} value for the insoluble sulfides of Mn^{2+}, Zn^{2+}, Ni^{2+}, and CO^{2+}. Thus MnS, ZnS, NiS and CoS are therefore, precipitated.

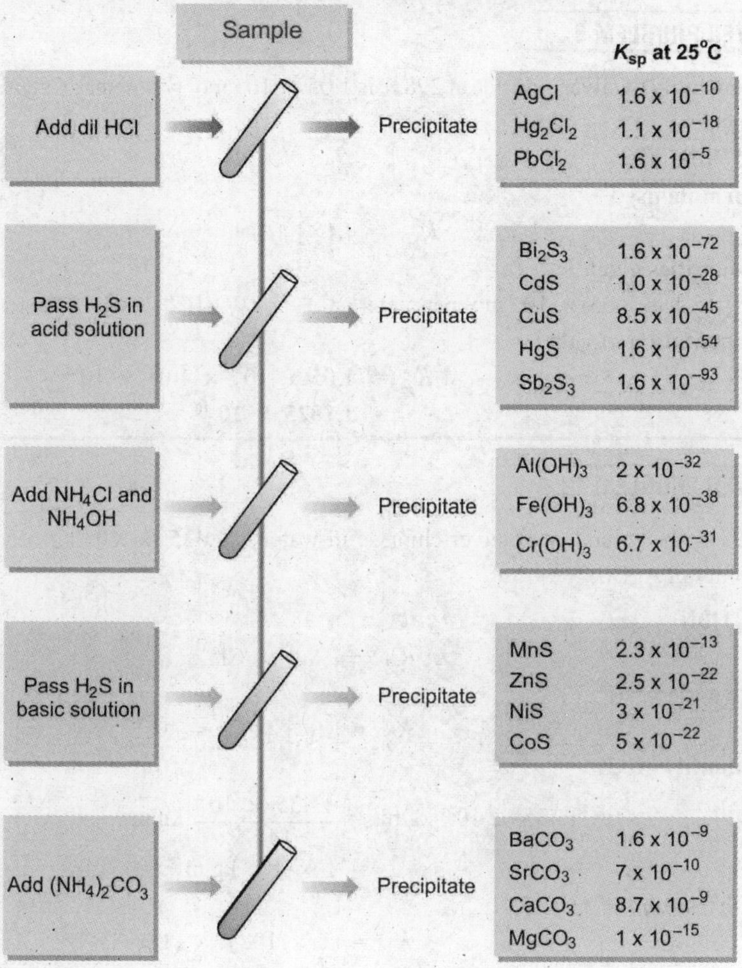

■ Figure 26.5
Precipitation into the qualitative analysis groups.

Group V — Insoluble Carbonates

To the filtrate from Group IV is added excess of sodium carbonate solution. The excess CO_3^{2-} ions in solution will drive the equilibrium to the left for $BaCO_3$, $SrCO_3$, $MgCO_3$ and $CaCO_3$. Thus,

$$BaCO_3 \rightleftharpoons Ba^{2+} + \underset{\text{(excess)}}{CO_3^{2-}}$$

This leads to the precipitation of the carbonates of Ba^{2+}, Sr^{2+}, Mg^{2+}, and Ca^{2+} to completion.

Group VI. The cations Na^+ and K^+, if present in the mixture, are identified by individually testing by the flame test.

ADDITIONAL SOLVED PROBLEMS

SOLVED PROBLEM 9

The solubility of silver chloride at 298 K is 1.05×10^{-5} mol per meter^3. Calculate the solubility product.

SOLUTION

Formula used

$$K_{sp} = [Ag^+]\ [Cl^-]$$

Quantities given

$[Ag^+] = 1.05 \times 10^{-5}$ mol per meter^3 $\quad$ $[Cl^-] = 1.05 \times 10^{-5}$ mol per meter^3

Substitution of values

$$K_{sp} = 1.05 \times 10^{-5} \times 1.05 \times 10^{-5}$$
$$= \mathbf{1.1025 \times 10^{-10}}$$

SOLVED PROBLEM 10

At 293 K the solubility of silver chloride in water is 1.435×10^{-3} kg per meter^3. Find the solubility of AgCl.

SOLUTION

$$AgCl \rightleftharpoons Ag^+ + Cl^-$$

Formula used

$$K_{sp} = [Ag^+]\ [Cl^-] = s^2$$

Quantity given

$$s = 1.435 \times 10^{-3}\ kg/m^3 = \frac{1.435 \times 10^{-3}}{143.5}\ kg\ m^{-3}$$
$$= 1 \times 10^{-5}\ kg\ m^{-3}$$

Substitution of values

$$K_{sp} = (1 \times 10^{-5}) \times (1 \times 10^{-5})$$
$$= \mathbf{1 \times 10^{-10}}$$

SOLVED PROBLEM 11

Calculate the pH value of a solution obtained by mixing 500 m^3 of 0.1 N CH_3COOH and 500 m^3 of 0.1 N CH_3COONa. K_a for acetic acid is 1.8×10^5.

SOLUTION :

Formula used

$$pH = pK_a + \log \frac{[Salt]}{[Acid]}$$

Quantities given

$$pK_a = -\log 1.8 \times 10^5 = 4.774$$

$[Salt] = \frac{0.1}{2} = 0.05\ N$ $\qquad [Acid] = \frac{0.1}{2} = 0.05\ N$

Substitution of values

$$pH = 4.774 + \log \frac{0.05}{0.05} = \mathbf{4.774}$$

SOLVED PROBLEM 12

Calculate the pH value of a solution whose hydrogen ion concentration is 0.006 kg ion/meter^3.

SOLUTION

Formula used

$$pH = -\log [H^+]$$

Quantity given

$$[H^+] = 0.006\ kg/m^{-3} = \frac{0.006}{2}\ mol\ lit^{-1} = 0.003\ mol\ m^{-3}$$

Substitution of values

$$pH = -\log (0.003)$$
$$= -\log (3 \times 10^{-3})$$
$$= (3 - 0.4771)$$
$$= \mathbf{2.5229}$$

SOLVED PROBLEM 13

What is the pH of the buffer composed of 0.1M solution of HCN in 0.1M KCN ? The dissociation constant of HCN is 0.01.

SOLUTION

Formula used

$$pH = pK_a + \log \frac{[Salt]}{[Acid]}$$

Quantities given

$pK_a = -\log K_a = -\log 0.02 = 1.6990$

$[Salt] = 0.1\ M$ $\qquad [Acid] = 0.1\ M$

Substitution of values

$$pH = 1.6990 + \log \frac{0.1}{0.1}$$
$$= \mathbf{1.6990}$$

SOLVED PROBLEM 14

The specific conductivity of a saturated solution of silver chloride is 1.33×10^{-6} mhos at 298 K. Given that the ionic conductivities for Ag^+ and Cl^- ions are 56.9 and 68.4 mhos respectively. Calculate the solubility and solubility product of silver chloride.

SOLUTION : (*a*) To calculate the solubility of AgCl.

Formula used

$$\lambda_\infty = \kappa \times V$$

Quantities given

$$\lambda_\infty = \Lambda_{Ag^+} + \Lambda_{Cl^-} = 56.9 + 68.4 = 125.3 \qquad \kappa = 1.33 \times 10^{-6}$$

Substitution of values

$$125.3 = 1.33 \times 10^{-6} \times V$$

or $$V = \frac{125.3}{1.33 \times 10^{-6}} = 94.2 \text{ m}^3$$

$$94.2 \text{ m}^3 \text{ contain AgCl} = 1 \text{ kg equivalent} = 143.5 \times 10^{-3} \text{ kg AgCl}$$

$$10^{-3} \text{ m}^3 \text{ contain AgCl} = \frac{143.5 \times 10^{-3}}{94.2}$$

$$= \mathbf{1.523 \times 10^{-3} \text{ kg m}^{-3}}$$

(*b*) To calculate the solubility product of AgCl

Formula used

$$K_{sp} = [Ag^+]\ [Cl^-]$$

Quantities given

$$[Ag^+] = [Cl^-] = \frac{1.523 \times 10^{-3} \text{ kg m}^{-3}}{143.5 \text{ kg m}^{-3}} = 1.06 \times 10^{-5} \text{ mol m}^{-3}$$

Substitution of values

$$K_{sp} = (1.06 \times 10^{-5})\ (1.06 \times 10^{-5})$$

$$= \mathbf{1.1236 \times 10^{-10}}$$

SOLVED PROBLEM 15

The solubility of a salt of type AB_2 with mol mass 78×10^{-3} kg in water is 1.6×10^{-5} kg m^{-3} at 293 K. Calculate its solubility product.

SOLUTION : The equilibrium is

$$AB_2 \rightleftharpoons A^{2+} + 2B^-$$

The solubility of the salt AB_2 in moles per liter

$$= \frac{1.6 \times 10^{-5} \text{ kg m}^{-3}}{78 \times 10^{-3} \text{ kg mol}^{-1}}$$

$$= 2.05 \times 10^{-4} \text{ mol m}^{-3}$$

Formula used

$$K_{sp} = [A^{2+}]\ [B^-]^2$$

Quantities given

$$[A^{2+}] = 2.05 \times 10^{-4} \text{ mol m}^{-3} \qquad [B^-] = 2 \times 2.05 \times 10^{-4} \text{ mol m}^{-3} = 4.1 \times 10^{-4} \text{ mol m}^{-3}$$

Substitution of values

$$K_{sp} = 2.05 \times 10^{-4} \times (4.1 \times 10^{-4})^2$$

$$= \mathbf{3.446 \times 10^{-11}}$$

EXAMINATION QUESTIONS

1. Define or explain the following terms :
   (*a*) Ostwald's dilution law
   (*b*) Degree of dissociation
   (*c*) Common ion effect
   (*d*) Solubility equilibria
   (*e*) Solubility product
2. (*a*) Define the terms 'Solubility' and 'Solubility product'. Explain the use of solubility product in qualitative analysis.
   (*b*) The solubility of $BaSO_4$ is 2.33×10^{-4} kg per m^3 at 293 K. Calculate the solubility product of $BaSO_4$ assuming that the dissolved salt is completely ionized.
   Answer. (*b*) 1.0×10^{-6}
3. (*a*) Give the relation between solubility and solubility product.
   (*b*) The solubility product of silver chromate in water is 2.0×10^{-12} at 298 K. Calculate the solubility at that temperature.
   Answer. (*b*) 7.935×10^{-5} mol/m^3
4. Define :
   (*a*) (*i*) Solubility; and (*ii*) Solubility product.
   (*b*) The solubility of silver chloride at 298 K is 1.05×10^{-5} mol per meter^3. Calculate the solubility product.
   Answer. 1.1025×10^{-10}
5. (*a*) Zinc sulfide is precipitated by hydrogen sulfide from ammoniacal solution. Explain.
   (*b*) The solubility of $BaSO_4$ at 291 K is 0.00233 kg per meter^3. Calculate the solubility product of $BaSO_4$. (Molecular weight of $BaSO_4$ is 233).
   Answer. 1×10^{-10}
6. (*a*) What is the Henderson equation?
   (*b*) Calculate the pH value of a solution obtained by mixing 500 m^3 of 0.1N CH_3COOH and 500 m^3 of 0.1N CH_3COONa. K_a for acetic acid is 1.8×10^5.
   Answer. 4.774
7. (*a*) What is meant by pH of a solution? A solution has a pH = 6. Is it acidic or alkaline?
   (*b*) Calculate the pH of (*i*) 10^{-8} N aqueous HCl solution; and (*ii*) 10^{-7} aqueous NaOH solution.
   Answer. (*a*) Acidic; (*b*) (*i*) 8; (*ii*) 7
8. State and explain Ostwald's Dilution law. Can this law be applied to the dissociation of HCl in aqueous solution?
9. What is the ionic strength of an electrolyte? Explain how mean activity coefficient of an electrolyte can be determined from the Debye-Huckel theory of activity coefficients? (*Delhi BSc*)
10. How does the stability constant of a complex ion in solution predict about the stability of complex towards its dissociation. (*Himachal Pradesh BSc*)
11. Explain briefly how does Debye-Huckel theory explain the increase in equivalent conductance with dilution in case of strong electrolyte. (*Punjabi BSc*)
12. With the help of conductivity how can you determine :
    (*i*) Solubility product of AgCl
    (*ii*) Degree of dissociation of a weak electrolyte
    (*iii*) Ionic product of water (*Panjab BSc*)

13. Briefly discuss the Debye-Huckel theory of activity coefficients of strong electrolytes. *(Guru Nanak Dev BSc)*

14. Define solubility and solubility product. Give its two applications. *(Jiwaji BSc)*

15. What is Ostwald's dilution law? How it is verified? What are its limitations? *(HS Gaur BSc)*

16. Explain any two applications of common ion effect. *(Venkateshwar BSc)*

17. Explain how conductance measurements can be used to determine the solubility of sparingly soluble salts. *(Mizoram BSc)*

18. The dissociation constant of a weak mono basic acid is 4×10^{-10}. Calculate the hydrogen ion concentration of 0.1 N solution of this acid.

Answer. 5.198 *(Jiwaji BSc)*

19. How do you determine the degree of ionization of an electrolyte by conductance method? *(Sambalpur BSc)*

20. (*a*) What is solubility product of an electrolyte? Explain giving three examples, the use of solubility product in qualitative analysis.

(*b*) At 293 K the solubility of silver chloride in water is 1.435×10^{-3} kg per meter^3. Find the solubility product of AgCl.

Answer. 1×10^{-10} *(Agra BSc)*

21. Calculate the solubility product of silver chromate Ag_2CrO_4 at 298 K if the concentration of Ag^+ ion 1.5×10^{-4} mol meter^3 is in a saturated solution of silver chromate at 298 K.

Answer. $K_{sp} = 1.6875 \times 10^{-12}$ *(Sri Venkateswara BSc)*

22. The solubility of lead sulfate in water 0.038 kg m^{-3} at 298 K. Calculate its solubility product at 298 K. (molar mass of $PbSO_4$ = 303).

Answer. 1.5725×10^{-8} *(HS Gaur BSc)*

23. When one liter of saturated solution of lead chloride, $PbCl_2$ is evaporated to dryness, the residue is found to weight 4.5 kg. Calculate the value of K_{sp} for $PbCl_2$.

Answer. 1.7×10^{-5} *(Jammu BSc)*

24. The solubility product of AgCl in water is 1.5×10^{-10}. Calculate its solubility in 0.01 M NaCl solution.

Answer. 1.5×10^{-8} mol m^{-3} *(Panjab BSc)*

25. Calculate the solubility product of AgCl if its solubility at 293 K is 1.435×10^{-5} g/meter^3.

Answer. 1×10^{-10} *(Madurai BSc)*

26. (*a*) State Ostwald's dilution law. How is it experimentally verified?

(*b*) Calculate pH value of a solution whose hydrogen ion concentration is 0.006 gm ion/liter.

Answer. (*b*) 2.5 *(Jivaji BSc)*

27. (*a*) State the principle of solubility product. How is the solubility of a salt is affected by the presence of a common ion?

(*b*) The solubility product of lead bromide is 8.0×10^{-5}. If the salt is 80% dissociated in the saturated solution, find the solubility of the salt.

Answer. 12.44 kg/m^3 *(Madurai BSc)*

28. What is the pH of the buffer composed of 0.1M solution of HCN in 0.1M KCN? The dissociation constant of HCN is 0.01 M.

Answer. 2 *(Delhi BSc)*

29. (*a*) How will you determine the solubility and solubility product of sparingly soluble salts through conductometric measurements?

(*b*) The specific conductivity of a saturated solution of silver chloride is 1.33×10^{-6} siemens at 298 K. Given that the ionic conductivities for Ag^+ and Cl^- ions are 56.9 and 68.4 siemens respectively. Calculate the solubility and solubility product of the silver chloride.

Answer. (*b*) 1.523×10^{-3} kg m^{-3}; 1.1264×10^{-10} *(Agra BSc)*

MULTIPLE CHOICE QUESTIONS

1. The Ostwald's dilution law when applied to a binary electrolyte gives the expression (α is degree of dissociation)

(*a*) $K_c = \frac{\alpha^2}{(1-\alpha)V}$ (*b*) $K_c = \frac{\alpha}{(1-\alpha)V}$

(*c*) $K_c = \frac{\alpha^2}{(1-\alpha)V^2}$ (*d*) $K_c = \frac{\alpha^2}{(1-\alpha)^2 V}$

Answer. (*a*)

2. Which of the following expression holds good for a weak electrolyte (α is the degree of dissociation)

(*a*) $K_c = \frac{\alpha}{V}$ (*b*) $K_c = \frac{\alpha^2}{V}$

(*c*) $K_c = \frac{\alpha}{V^2}$ (*d*) $K_c = \frac{\alpha^2}{V^2}$

Answer. (*b*)

3. If λ_∞ and λ_v are the equivalent conductances at infinite dilution and at *V* dilution, the degree of dissociation, α, is given by

(*a*) $\lambda = \frac{\lambda_\infty}{\lambda_v}$ (*b*) $\lambda = \frac{\lambda_\infty}{\lambda_v^2}$

(*c*) $\lambda = \frac{\lambda_v}{\lambda_\infty}$ (*d*) none of these

Answer. (*c*)

4. Which of the following statement is true?
   (*a*) Ostwald's dilution law holds good only for strong electrolytes and fails completely when applied to weak electrolytes
   (*b*) Ostwald's dilution law holds good only for weak electrolytes and fails completely when applied to strong electrolytes
   (*c*) Ostwald's dilution law holds good for both weak and strong electrolytes
   (*d*) Ostwald's dilution law does not hold good for both weak and strong electrolytes

Answer. (*b*)

5. The failure of Ostwald's dilution law in case of strong electrolytes is due to
   (*a*) strong electrolytes are almost completely dissociated at all dilutions and λ_v/λ_α does not give accurate value of α
   (*b*) the law of mass action in its simple form cannot be applied when the concentration of ions is very high
   (*c*) the ions get hydrolyzed and affect the concentration terms
   (*d*) all of the above

Answer. (*d*)

6. Which of the following postulates of Debye-Huckel theory is/are true?
   (*a*) the strong electrolyte is completely ionized at all dilutions
   (*b*) the oppositely charged ions are completely distributed in the solution but the cations tend to be found in the vicinity of anions and vice-versa
   (*c*) decrease in equivalent conductance with increase in concentration is due to fall in mobilities of ions due to inter-ionic effect
   (*d*) all of the above

Answer. (*d*)

7. The effect that tends to retard the mobilities of ions in solution is
(*a*) asymmetry effect
(*b*) relaxation effect
(*c*) electrophoretic effect
(*d*) all of these
Answer. (*d*)

8. The value of equilibrium constant K for the equilibrium $AB \rightleftharpoons A^+ + B^-$ is
(*a*) $K = \frac{[AB]}{[A^+][B^-]}$
(*b*) $K = \frac{[A^+][B^-]}{[AB]}$
(*c*) $K = \frac{[AB]^2}{[A^+][B^-]}$
(*d*) $K = \frac{[AB]}{[A^+]^2\,[B^-]^2}$
Answer. (*b*)

9. When a salt is added to a solution of another salt having a common ion, the degree of dissociation, α,
(*a*) increases
(*b*) remains the same
(*c*) decreases
(*d*) none of these
Answer. (*c*)

10. When NH_4Cl is added to a solution of NH_4OH, the equilibrium $NH_4OH \rightleftharpoons NH_4^+ + OH^-$ shifts towards
(*a*) left
(*b*) right
(*c*) remains unchanged
(*d*) none of these
Answer. (*a*)

11. Which of the following electrolytes when added to the equilibrium $AgCl(s) \rightleftharpoons Ag^+(aq) + Cl^-(aq)$ disturbs the equilibrium towards left thereby decreasing the solubility of AgCl.
(*a*) KBr
(*b*) NaCl
(*c*) $NaNO_3$
(*d*) HNO_3
Answer. (*b*)

12. The degree of dissociation of an electrolyte depends upon
(*a*) nature of solute
(*b*) nature of solvent
(*c*) concentration of solute
(*d*) all of these
Answer. (*d*)

13. A saturated solution is that in which
(*a*) the concentration of dissolved and undissolved solute is the same
(*b*) there exists an equilibrium between dissolved and undissolved solute
(*c*) the concentration of undissolved solute is almost negligible
(*d*) none of the above
Answer. (*b*)

14. The solubility (s) of a substance in a solvent is the concentration in ______ solution.
(*a*) unsaturated
(*b*) supersaturated
(*c*) saturated
(*d*) none of these
Answer. (*c*)

15. Molar solubility is the number of ______ of the substance per meter^3 of the solution.
(*a*) grams
(*b*) kilograms
(*c*) g-equivalents
(*d*) moles
Answer. (*d*)

16. The solubility product of AgCl is given by the expression
(*a*) $K_{sp} = [Ag^+]\,[Cl^-]$
(*b*) $K_{sp} = [Ag^+]^2\,[Cl^-]^2$
(*c*) $K_{sp} = [Ag^+]\,[Cl^-]^2$
(*d*) none of these
Answer. (*a*)

17. The product of concentration of ions (mol meter^{-3}) in the saturated solution at a given temperature is called

(*a*) ionic product (*b*) solubility product

(*c*) solubility (*d*) solubility constant

Answer. (*c*)

18. If '*s*' is the solubility of AgCl in water the solubility product K_{sp} is given by

(*a*) $K_{sp} = s$ (*b*) $K_{sp} = s^2$

(*c*) $K_{sp} = s^3$ (*d*) $K_{sp} = \sqrt{s}$

Answer. (*b*)

19. For CuBr, the solubility is 2.0×10^{-4} mol meter^3 at 298 K. The solubility product would be

(*a*) 1.0×10^{-4} mol^2 meter^{-2} (*b*) 1.414×10^{-2} mol^2 meter^{-2}

(*c*) 4.0×10^{-8} mol^2 meter^{-2} (*d*) none of these

Answer. (*c*)

20. If '*s*' is the solubility of CaF_2, the solubility product would be given by the expression

(*a*) $K_{sp} = s^2$ (*b*) $K_{sp} = s^3$

(*c*) $K_{sp} = 4s^3$ (*d*) $K_{sp} = 4s^2$

Answer. (*c*)

21. The precipitation of a salt takes place if its ionic product is

(*a*) equal to its solubility product (*b*) less than its solubility product

(*c*) greater than its solubility product (*d*) none of the above

Answer. (*c*)

22. When excess of dil HCl is added to a solution containing a mixture of common cations, only Ag^+, Pb^{2+} and Hg^+ will be precipitated. This is because of the reason that K_{sp} of these chlorides is

(*a*) very low (*b*) very high

(*c*) equal to zero (*d*) equal to infinity

Answer. (*a*)

23. ZnS is not precipitated when H_2S is passed through the filtrate of group I during salt analysis. This is so because

(*a*) the K_{sp} for ZnS is very high (*b*) the K_{sp} for ZnS is very low

(*c*) Zns evaporates (*d*) none of these

Answer. (*a*)

24. The 'solubility' of an electrolyte of type AB_2 is *s* (ignoring the units), the expressions for solubility product will be

(*a*) $K_{sp} = s^2$ (*b*) $K_{sp} = 4s^2$

(*c*) $K_{sp} = 4s^3$ (*d*) $K_{sp} = 4s^4$

Answer. (*c*)

25. The expression for the solubility product of $Al(OH)_3$ would be (*s* is its solubility)

(*a*) $K_{sp} = s \times 3s^2$ (*b*) $K_{sp} = s \times 9s^2$

(*c*) $K_{sp} = s \times 9\,s^3$ (*d*) $K_{sp} = s \times 4s^2$

Answer. (*c*)

26. A solution is unsaturated if its

(*a*) ionic product < K_{sp} (*b*) ionic product = K_{sp}

(*c*) ionic product > K_{sp} (*d*) none of these

Answer. (*a*)

27. For a supersaturated solution, the ionic product is

(*a*) greater than the solubility product (*b*) equal to its solubility product

(*c*) lesser than its solubility product (*d*) none of these

Answer. (*a*)

28. For an electrolyte of 1:2 type, the solubility is given by the expression

(*a*) $s = \sqrt{K_{sp}}$ (*b*) $s = \sqrt[3]{K_{sp}}$

(*c*) $s = \sqrt{\frac{K_{sp}}{4}}$ (*d*) $s = \sqrt[3]{\frac{K_{sp}}{4}}$

Answer. (*d*)

29. For $Fe(OH)_3$, the solubility '*s*' is related to its solubility product K_{sp} by the relation

(*a*) $s = \sqrt[3]{\frac{K_{sp}}{4}}$ (*b*) $s = \sqrt[3]{\frac{K_{sp}}{27}}$

(*c*) $s = \sqrt[4]{\frac{K_{sp}}{27}}$ (*d*) $s = \sqrt[4]{\frac{K_{sp}}{4}}$

Answer. (*c*)

30. The solubility product of $BaSO_4$ in water is 1.5×10^{-9}. Its solubility would be

(*a*) 3.87×10^{-5} mol meter^{-3} (*b*) 1.224×10^{-5} mol meter^{-3}

(*c*) 3.87×10^{-4} mol meter^{-3} (*d*) 12.24×10^{-5} mol meter^{-3}

Answer. (*a*)

31. The solubility of $Mg(OH)_2$ in pure water is 1.65×10^{-4} mol meter^{-3}. Its solubility product would be

(*a*) 1.796×10^{-11} (*b*) 17.96×10^{-11}

(*c*) 4.49×10^{-12} (*d*) 4.49×10^{-11}

Answer. (*a*)

32. When NH_4Cl is added to NH_4OH solution, the dissociation of NH_4OH is suppressed. It is due to

(*a*) NH_4Cl is more soluble than NH_4OH

(*b*) NH_4Cl is having one ion common to NH_4OH

(*c*) the K_{sp} of NH_4Cl is greater than that of NH_4OH

(*d*) NH_4Cl is easily oxidized

Answer. (*b*)

33. The solubility product of a salt of the type AB_3 is 4×10^{-12}. Its solubility in mol meter^{-3} would be

(*a*) 1×10^{-3} (*b*) 1×10^{-4}

(*c*) 1×10^{-6} (*d*) 1.587×10^{-4}

Answer. (*b*)

34. The solubility of a salt is '*s*' and the solubility product is $4s^3$. The ratio of cations to anions in the salt is

(*a*) 1:1 (*b*) 1:2

(*c*) 1:3 (*d*) 2:3

Answer. (*b*)

35. The solubility product of a salt is equal to the square of its solubility. The ratio of cations to anions in the salt is

(*a*) 1:2 (*b*) 1:3

(*c*) 2:3 (*d*) 1:1

Answer. (*d*)

36. The addition of HCl will not suppress the ionization of

(*a*) CH_3COOH (*b*) H_2SO_4

(*c*) H_2S (*d*) C_6H_5COOH

Answer. (*b*)

37. Pure NaCl gets precipitated when HCl gas is passed through the saturated solution of NaCl. It is because

(*a*) the ionic product of NaCl becomes more than its solubility product

(*b*) the ionic product of NaCl becomes lesser than its solubility product

(*c*) the ionic product of NaCl becomes equal to its solubility product

(*d*) the ionic product of NaCl becomes zero

Answer. (*a*)

38. Ostwald's dilution law is applicable to

(*a*) all electrolytes (*b*) strong electrolytes
(*c*) weak electrolyte (*d*) non-electrolytes

Answer. (*c*)

39. Which pair, out of the following, will show common ion effect?

(*a*) $HCl + NHO_3$ (*b*) $HCl + H_2S$
(*c*) $HCl + H_2SO_4$ (*d*) $HCl + NaCl$

Answer. (*b*)

40. Which of the following pairs of electrolytes will not show common ion effect?

(*a*) $NH_4Cl + NH_4OH$ (*b*) $HCl + H_2S$
(*c*) $HCl + CH_3COOH$ (*d*) $NaCl + NH_4Cl$

Answer. (*d*)

41. The solubility of an electrolyte of the type AB_3 is 0.25×10^{-4} mol meter^{-3}. Its solubility product will be

(*a*) 1.56×10^{-14} (*b*) 6.25×10^{-14}
(*c*) 6.25×10^{-10} (*d*) 1.56×10^{-12}

Answer. (*b*)

42. The solubility of a salt of the type A_3B_2 type in water is *s* moles meter^{-3}. Its solubility product will be

(*a*) $25\ s^2$ (*b*) $125\ s^3$
(*c*) $81\ s^4$ (*d*) $108\ s^5$

Answer. (*d*)

43. The solubility product of a salt AB_2 is 4.0×10^{-9}. Its solubility will be

(*a*) 2×10^{-3} mol meter^{-3} (*b*) 1×10^{-3} mol meter^{-3}
(*c*) 1.58×10^{-3} mol meter^{-3} (*d*) 1.33×10^{-3} mol meter^{-3}

Answer. (*b*)

44. For the equilibrium $PbCl_2 \rightleftharpoons Pb^{2+} + 2Cl^-$ the solubility product is

(*a*) $K_{sp} = [Pb^{2+}]^2\,[Cl^-]^2$ (*b*) $K_{sp} = [Pb^{2+}]\,[Cl^-]^2$
(*c*) $K_{sp} = [Pb^{2+}]^2\,[Cl^-]$ (*d*) $K_{sp} = [Pb^{2+}]\,[Cl^-]$

Answer. (*b*)

45. If the solubility of $Mg(OH)_3$ is $\sqrt{2}$, its solubility product will be2

(*a*) 8 (*b*) $4\sqrt{2}$
(*c*) $8\sqrt{2}$ (*d*) $9\sqrt{2}$

Answer. (*c*)

46. The solubility of Calcium fluoride is 1×10^{-4} mol meter^3. Its solubility product will be

(*a*) 2×10^{-8} (*b*) 4×10^{-8}
(*c*) 8×10^{-12} (*d*) 4×10^{-12}

Answer. (*d*)

47. Which pair out of the following will have similar expression for solubility product?

(*a*) $AgCl, PbCl_2$ (*b*) $PbCl_2, Hg_2Cl_2$
(*c*) $Mg(OH)_2, CaF_2$ (*d*) $MgCl_2, AlCl_3$

Answer. (*c*)

48. When HCl gas is passed through a saturated solution of NaCl, the solubility of NaCl

(*a*) will increase (*b*) will remain the same
(*c*) will decrease (*d*) will become zero

Answer. (*c*)

49. The solubility product of a salt A_2B with mol mass 100 is 4.0×10^{-9}. How many grams of it will dissolve in water to make one liter of saturated solution?

(*a*) 0.1 g (*b*) 0.01 g

(*c*) 0.001 g (*d*) 0.0001 g

Answer. (*a*)

50. The solubility product of salt of bivalent metal with chloride ions is 4×10^{-6}. Its solubility in mol meter^{-3} will be

(*a*) 10^{-1} (*b*) 10^{-2}

(*c*) 10^{-3} (*d*) 10^{-4}

Answer. (*b*)

27

Acids and Bases

CONTENTS

There are **three concepts of acids and bases** in current use. Each has its own peculiar advantages. The student should understand all the three concepts :

(*a*)Arrhenius concept

(*b*)Bronsted-Lowry concept

(*c*)Lewis concept

ARRHENIUS CONCEPT

Savante Arrhenius (1884) proposed his concept of acids and bases. According to this concept, **an acid is a compound that releases H^+ ions in water; and a base is a compound that releases OH^- ions in water.**

For example, HCl is an Arrhenius acid and NaOH is an Arrhenius base.

$$HCl(aq) \longrightarrow H^+(aq) + Cl^-(aq)$$
$$NaOH(aq) \longrightarrow OH^-(aq) + Na^+(aq)$$

Limitations of Arrhenius Concept

Arrhenius concept of acids and bases proved to be very useful in the study of chemical reactions. However, it has the following limitations :

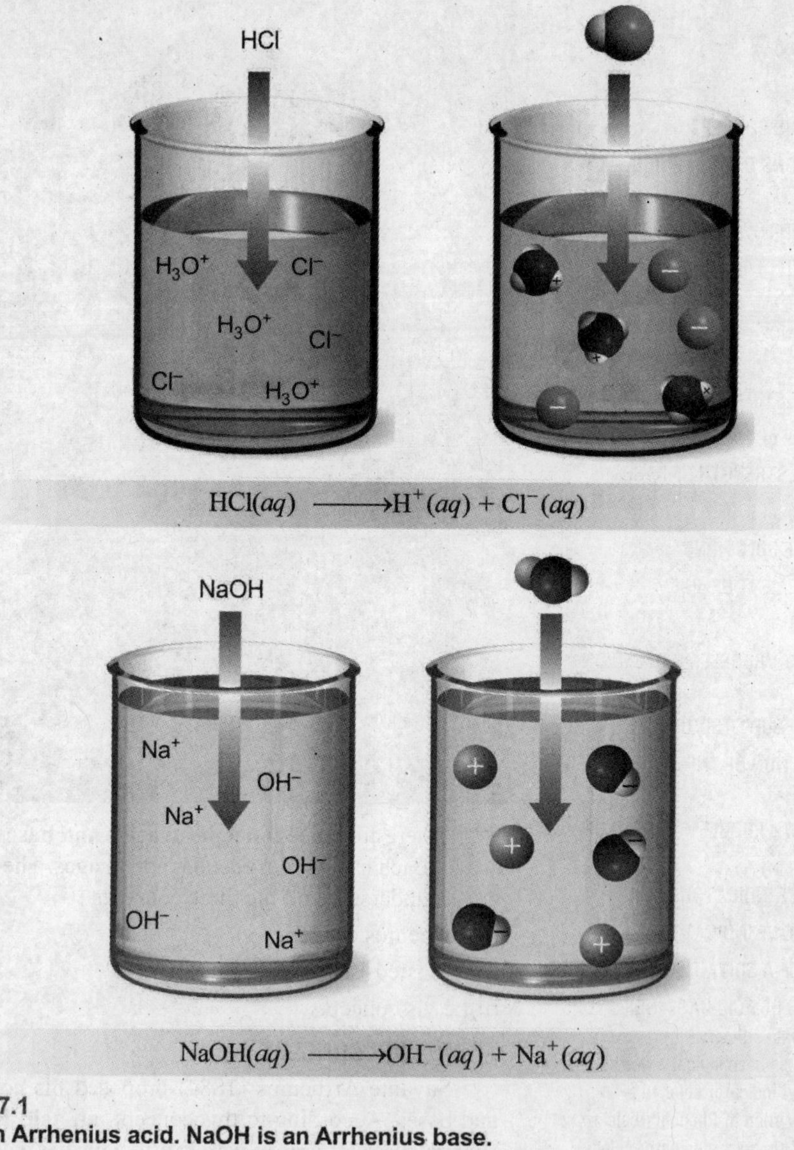

Figure 27.1
HCl is an Arrhenius acid. NaOH is an Arrhenius base.

(1) **Free H^+ and OH^- ions do not exist in water.** The H^+ and OH^- ions produced by acids and bases respectively do not exist in water in the free state. They are associated with water molecules to form complex ions through hydrogen bonding. Thus the H^+ ion forms a **hydronium ion** :

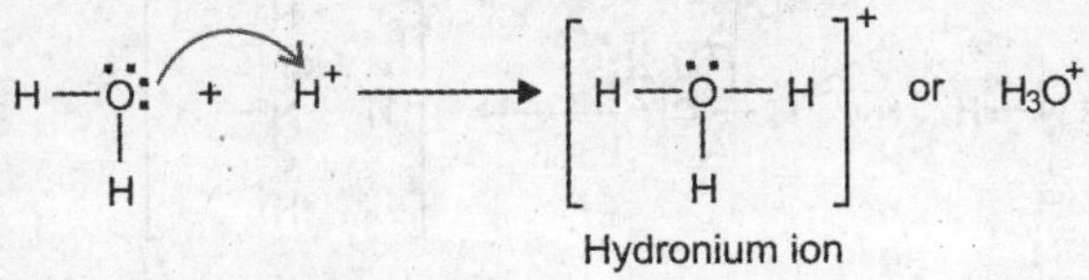

Similarly, OH^- ion forms the complex $H_3O_2^-$.

Although the hydrogen and hydroxyl ions are associated with water molecule, for simplicity we shall generally write them H^+ and OH^-.

(2) **Limited to water only.** Arrhenius defined acids and bases as compounds producing H^+ and OH^- ions in water only. But a truly general concept of acids and bases should be appropriate to other solvents as well.

(3) **Some bases do not contain OH^-.** Arrhenius base is one that produces OH^- ions in water. Yet there are compounds like ammonia (NH_3) and calcium oxide (CaO) that are bases but contain no OH^- ions in their original formulation.

Arrhenius models of acids and bases, no doubt, proved very helpful in interpreting their action. However on account of its limitations the Arrhenius concept needed to be modified.

BRONSTED–LOWRY CONCEPT

In 1923 J.N. Bronsted and J.M. Lowry independently proposed a broader concept of acids and bases. According to this theory,

an acid is any molecule or ion that can donate a proton (H^+)

a base is any molecule or ion that can accept a proton

For brevity we can say that an acid is a proton donor while a base is a proton acceptor.

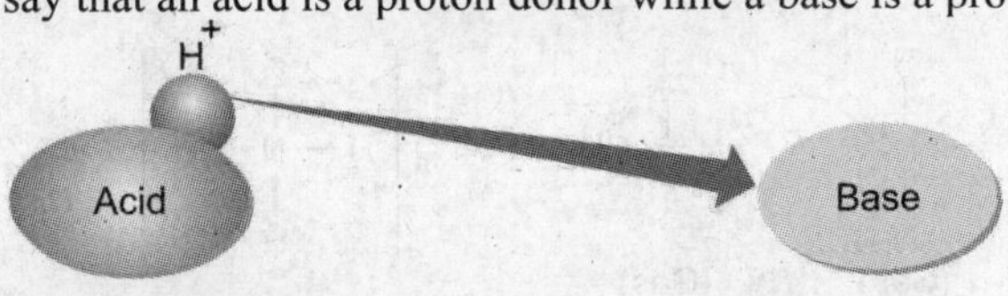

Figure 27.2
A Bronsted acid is a proton donor and a Bronsted base is a proton acceptor.

An acid qualifying Bronsted-Lowry concept is termed a **Bronsted-Lowry acid** or simply **Bronsted acid.**

A base qualifying Bronsted-Lowry concept is termed a **Bronsted-Lowry base** or simply **Bronsted base.**

Examples of Bronsted acids and bases

(1) **HCl gas and H_2O.** When dry HCl gas dissolves in water, each HCl molecule donates a proton to a water molecule to produce hydronium ion.

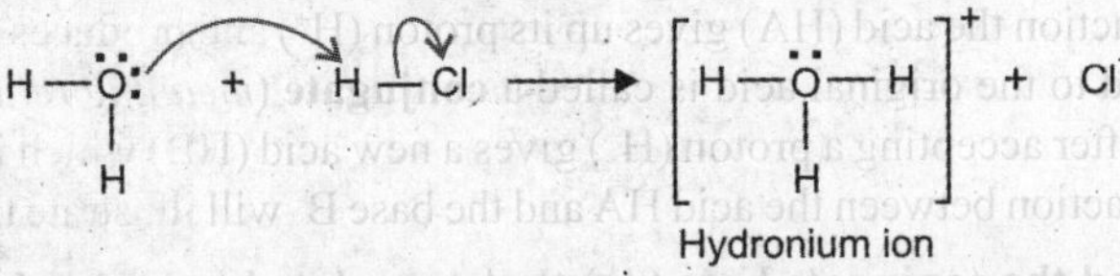

Thus HCl gas is a Bronsted acid and water that accepts a proton is a Bronsted base.

(2) **HCl and Ammonia, NH_3.** HCl gas reacts with ammonia (NH_3) to form solid NH_4Cl.

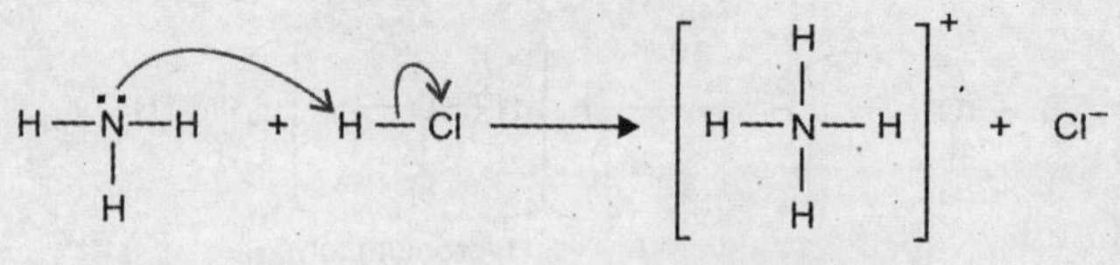

Ammonium ion

HCl is a proton donor and hence a Bronsted acid, while NH_3 is a proton acceptor and a Bronsted base.

(3) **Calcium oxide and H_2O.** When calcium oxide is dissolved in water, it is converted to calcium hydroxide. Here a water molecule donates a proton to oxide ion, O^{2-}, and is a Bronsted acid. The oxide ion accepts a proton and give $2OH^-$ ions, hence is a Bronsted base.

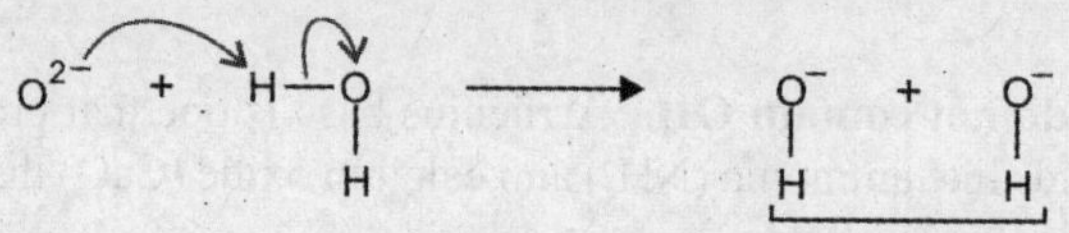

Two OH^- ions

Bronsted-Lowry concept is superior to Arrhenius concept

(1) **Much wider scope.** Arrhenius concept of acids and bases is restricted to the study of substances which can release H^+ or OH^- ions in water. Bronsted-Lowry concept embraces all molecules and ions that can donate a proton (acids) and those which can accept a proton (bases).

(2) **Not limited to aqueous solutions.** The Bronsted-Lowry model is not limited to aqueous solutions as is the case with Arrhenius model. It can be extended even to the gas phase. For example, gaseous ammonia (a Bronsted base) can react with hydrogen chloride gas (a Bronsted acid) to give ammonium chloride.

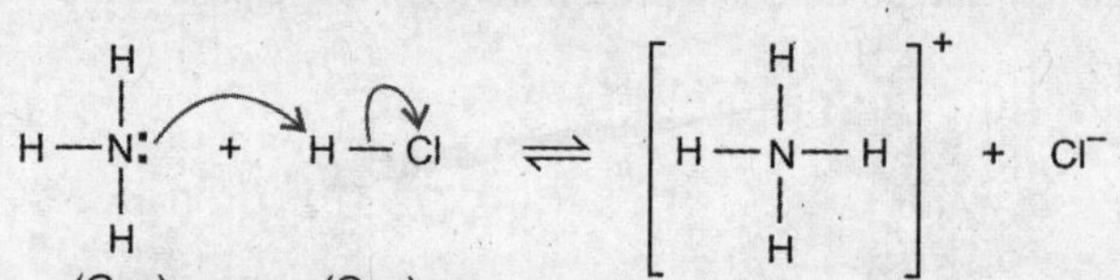

Here a proton is donated by HCl to NH_3 as shown above. Note that this is not considered as an acid-base reaction according to Arrhenius concept.

(3) **Release of OH^- not necessary to qualify as a base.** Arrhenius base is a substance that releases OH^- ions in water. On the other hand, Bronsted base is a substance that accepts a proton. Thus liquid ammonia (NH_3) does not produce OH^- ions in water but it is a recognized base. But according to Bronsted-Lowry model, it qualifies as a base since it can accept a proton to form NH_4^+ (an acid).

$$\underset{\text{base}}{NH_3} + H^+ \rightleftharpoons \underset{\text{acid}}{NH_4^+}$$

Conjugate Acid-Base pairs

In an acid-base reaction the acid (HA) gives up its proton (H^+) and produces a new base (A^-). The new base that is related to the original acid is called a **conjugate** (*meaning related*) **base.** Similarly the original base (B^-) after accepting a proton (H^+) gives a new acid (HB) which is called a **conjugate acid.** A hypothetical reaction between the acid HA and the base B^- will illustrate the above definitions.

The acid (HA) and the conjugate base (A^-) that are related to each other by donating and accepting a single proton, are said to constitute a conjugate Acid-Base pair.

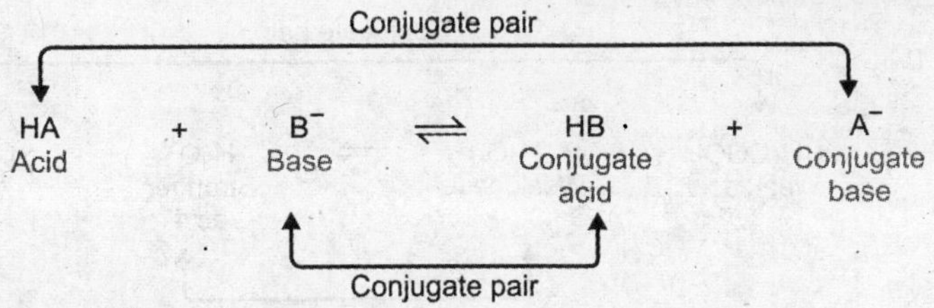

■ **Figure 27.3**
Illustration of conjugate pairs.

It may be noted that in any acid-base reaction, there are two conjugate acid-base pairs, Thus, in the above equation, the two conjugate pairs are : HA and A^-; and HB and B^-.

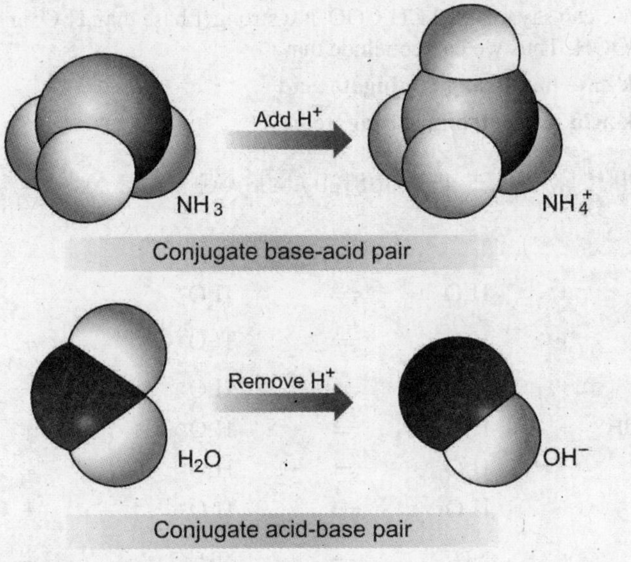

Examples of Conjugate Acid-base pairs

Let us consider the reaction between HCl (*aq*) and NH_3 (*aq*) which is illustrated below in Fig. 27.4.

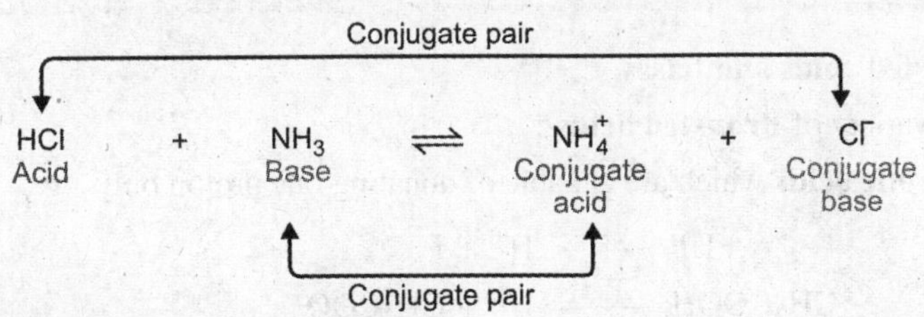

■ **Figure 27.4**
In every acid-base reaction involving H^+ transfer, there are two acid-base conjugate pairs.

In this case HCl (acid) and Cl^- (conjugate base) constitute one acid-base conjugate pair. Base NH_3 and the conjugate acid (NH_4^+) comprise the second acid-base conjugate pair.

Now let us consider the reaction between acetic acid and water to form the conjugate base CH_3COO^- and the conjugate acid H_3O^+.

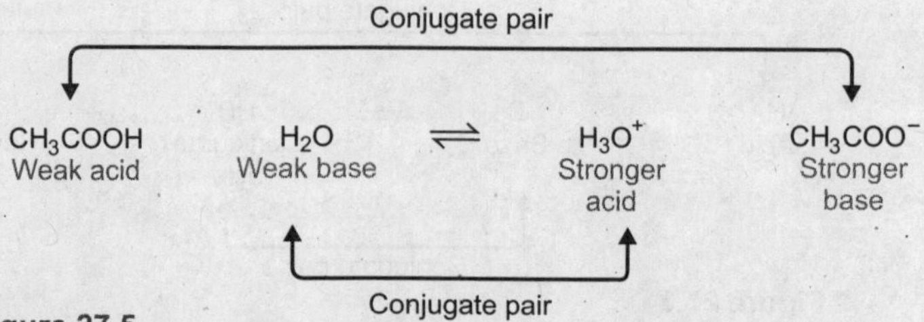

Figure 27.5
A weak acid (CH_3COOH) produces a strong conjugate base (CH_3COO^-); and a weaker base (H_2O) gives a stronger conjugate acid (H_3O^+).

We know that acetic acid is less than 1% ionized in water. Since the equilibrium is displaced toward the left, we can say that : (*i*) CH_3COO^- is a stronger base than H_2O; and (*ii*) H_3O^+ is a stronger acid than CH_3COOH. Thus we can conclude that :

(*a*) **a weak base has strong conjugate acid**

(*b*) **a weak acid has a strong conjugate base**

TABLE 27.1. SOME EXAMPLES OF BRONSTED ACIDS AND BASES

Acid		Base		Conjugate Acid		Conjugate Base
HCl	+	H_2O	$\rightleftharpoons$	H_3O^+	+	Cl^-
HNO_3	+	H_2O	$\rightleftharpoons$	H_3O^+	+	NO_3^-
HCO_3^-	+	H_2O	$\rightleftharpoons$	H_3O^+	+	CO_3^{2-}
CH_3COOH	+	H_2O	$\rightleftharpoons$	H_3O^+	+	CH_3COO^-
HCN	+	H_2O	$\rightleftharpoons$	H_3O^+	+	CN^-
H_2S	+	H_2O	$\rightleftharpoons$	H_3O^+	+	HS^-
H_2O	+	NH_3	$\rightleftharpoons$	NH_4^+	+	OH^-
H_2O	+	CO_3^{2-}	$\rightleftharpoons$	HCO_3^-	+	OH^-
H_2O	+	H_2O	$\rightleftharpoons$	H_3O^+	+	OH^-

Classes of Bronsted acids and bases

There are a variety of **Bronsted acids :**

(1) **Monoprotic acids** which are capable of donating one proton only *e.g.*,

$$HF \longrightarrow H^+ + F^-$$

$$CH_3COOH \longrightarrow H^+ + CH_3COO^-$$

(2) **Polyprotic acids** which are capable of donating two or more protons *e.g.*, H_2SO_4, H_3PO_4, carbonic acid (H_2CO_3), hydrosulfuric acid, etc.

$$H_2S \longrightarrow 2H^+ + S^{2-}$$

$$\begin{array}{c} COOH \\ | \\ COOH \end{array} \longrightarrow 2H^+ + \begin{array}{c} COO^- \\ | \\ COO^- \end{array}$$

oxalic acid → oxalate ion

Similarly, there are **Bronsted bases :**

(1) **Monoprotic bases** which can accept one proton.

$$HS^- + H^+ \longrightarrow H_2S$$
$$H_2O + H^+ \longrightarrow H_3O^+$$

(2) **Polyprotic bases** which can accept two or more protons *e.g.*, anions of diprotic and triprotic acids.

$$SO_4^{2-} + 2H^+ \longrightarrow H_2SO_4$$
$$PO_4^{3-} + 3H^+ \longrightarrow H_3PO_4$$

Amphiprotic substances

Molecules or ions that can behave both as Bronsted acid and base are called amphiprotic substances. For example, with HCl, water acts as a base in accepting a proton from the acid.

$$\underset{\text{acid}}{HCl} + \underset{\text{base}}{H_2O} \longrightarrow H_3O^+ + Cl^-$$

However, water is an acid while donating a proton to ammonia.

$$\underset{\text{base}}{NH_3} + \underset{\text{acid}}{H_2O} \longrightarrow NH_4^+ + OH^-$$

Listed below are some **amphiprotic ions :**

$$H_2S \xleftarrow{+H^+} HS^- \xrightarrow{-H^+} S^{2-}$$
$$H_2CO_3 \xleftarrow{+H^+} HCO_3^- \xrightarrow{-H^+} CO_3^{2-}$$
$$HPO_4^- \xleftarrow{+H^+} HPO_4^{2-} \xrightarrow{-H^+} PO_4^{3-}$$

Strength Bronsted acids and bases

The strength of a Bronsted acid depends upon its tendency to donate a proton. The strength of a Bronsted base depends on its ability to accept a proton. For example, HCl is nearly 100% ionized in water. Its reaction with water can be depicted by the equation :

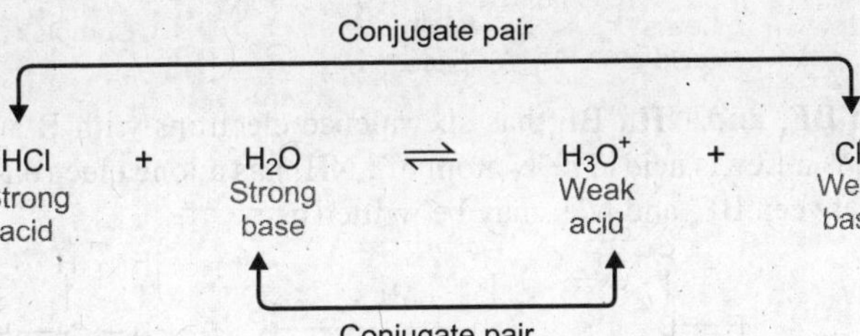

■ **Figure 27.6**
A strong acid HCl gives a weak conjugate base (Cl^-); and a strong base (H_2O) gives a weak conjugate acid (H_3O^+).

Since the reaction has proceeded almost completely to the right, it means that HCl has a strong tendency to lose a proton. Also, the base H_2O has a strong ability to accept a proton. The overall situation is that the acid and base on the left are each stronger than the conjugate acid and conjugate base on the right. That is why the equilibrium is displaced to the right. Thus we can conclude that :

a strong acid has a weak conjugate base
a strong base has a weak conjugate acid

LEWIS CONCEPT OF ACIDS AND BASES

In the early 1930s, G.N. Lewis proposed even a more general model of acids and bases. According to Lewis theory,

an acid is an electron-pair acceptor

a base is an electron-pair donor

Lewis pictured an acid and base as sharing the electron pair provided by the base. This creates a covalent bond (or coordinate bond) between the **Lewis acid** and the **Lewis base.** The resulting combination is called a **Complex**. If the Lewis acid be denoted by A and the Lewis base by B, then the fundamental equation of the Lewis theory can be written as :

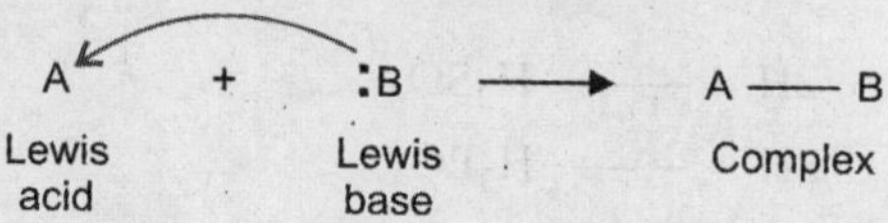

It may be noted that : (1) all cations or molecules short of an electron-pair act as Lewis acids; and (2) all anions or molecules having a lone electron-pair act as Lewis bases.

Examples of Lewis reactions

(1) **Between H^+ and NH_3.** Proton (H^+) is a Lewis acid as it can accept an electron-pair. Ammonia molecule (: NH_3) has an electron-pair which it can donate and is a Lewis base. Thus the Lewis reaction between H^+ and NH_3 can be written as :

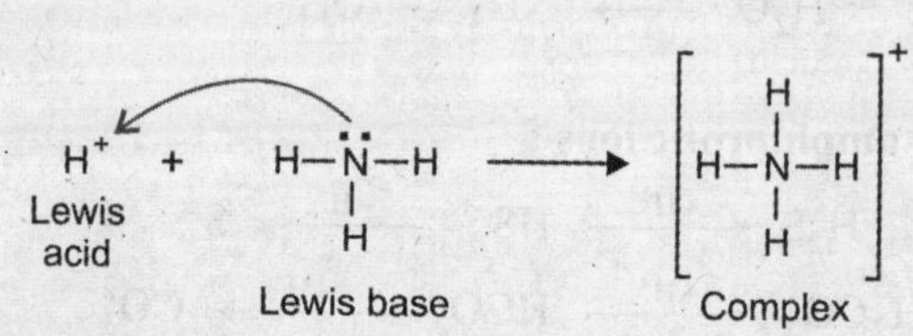

(2) **Between H^+ and OH^-.** A proton (H^+) is an electron-pair acceptor and, therefore, a Lewis acid. The OH^- is an electron-pair donor and hence a Lewis base. Thus Lewis reaction between H^+ and OH^- can be written as :

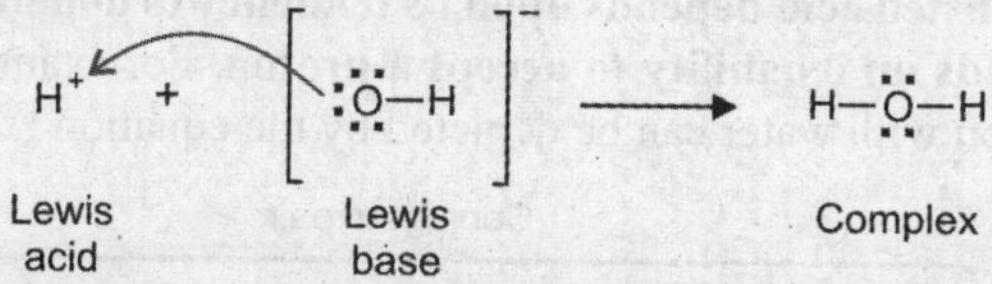

(3) **Between BF_3 and NH_3.** BF_3 has six valence electrons with B atom which can accept an electron-pair and is a Lewis acid. The N atom of : NH_3 has a lone electron-pair and is a Lewis base. Lewis reaction between BF_3 and NH_3 may be written as :

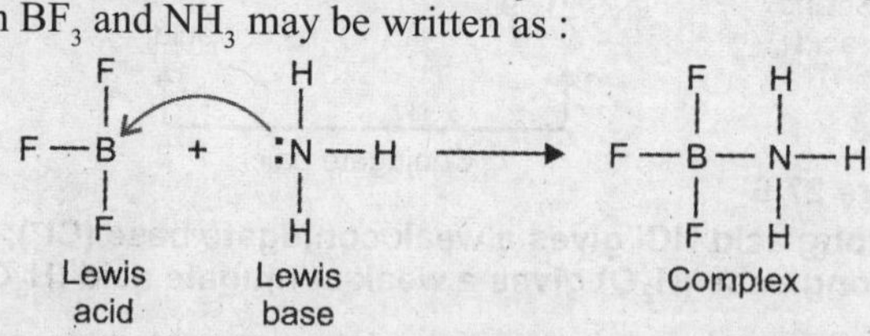

(4) **Hydration of Al^{3+}.** The hydration of a metal ion such as Al^{3+} is also a Lewis reaction.

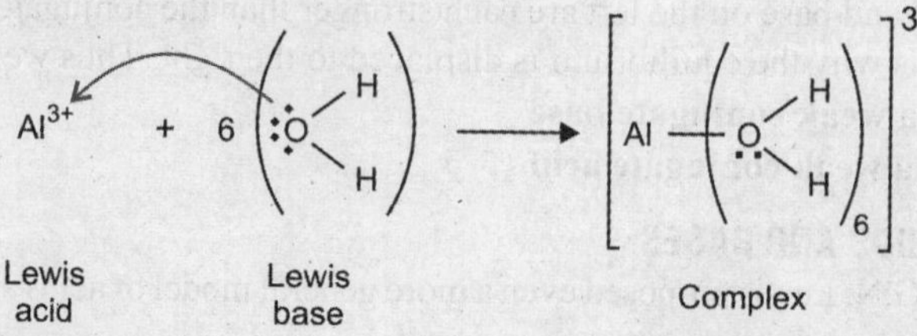

Superiority of Lewis model of acids and bases

The useful but limited model of Arrhenius was replaced by a more general model of Bronsted and Lowry. Even a more general model was proposed by Lewis. However, the Bronsted-Lowry model is now used in common practice.

TABLE 27.2. THREE MODELS FOR ACIDS AND BASES

Model	Definition of acid	Definition of base
Arrhenius (1884)	H^+ producer	OH^- producer
Bronsted-Lowry (1923)	H^+ donor	H^+ acceptor
Lewis (1939)	electron-pair acceptor	electron-pair donor

The advantages of the Lewis acid-base model are :

(1) All the Bronsted-Lowry acid base reactions are covered by the Lewis model. It is so because the transfer or gain of a proton is accompanied by the loss or donation of an electron-pair in both types of reactions.

(2) Many reactions which do not involve transfer of a proton *e.g.*,

$$BF_3 + NH_3 \longrightarrow BF_3 - NH_3$$

are also covered by the Lewis theory.

Neutralization in the Bronsted-Lowry theory

According to Bronsted-Lowry theory, all acid-base neutralization reactions in aqueous solution can be represented by a single chemical reaction :

$$H_3O^+ + OH^- \longrightarrow 2H_2O$$

This is so because the acid molecule always produces H_3O^+ by donating a proton to a H_2O molecule, and the base molecule always produces OH^- by accepting a proton from another H_2O molecule.

Water can act both as an acid and a base

Water is an amphoteric substance. It can behave either as an acid or a base. One molecule of water transfers a proton to another molecule. There results a hydronium ion (H_3O^+) and a hydroxyl ion (OH^-).

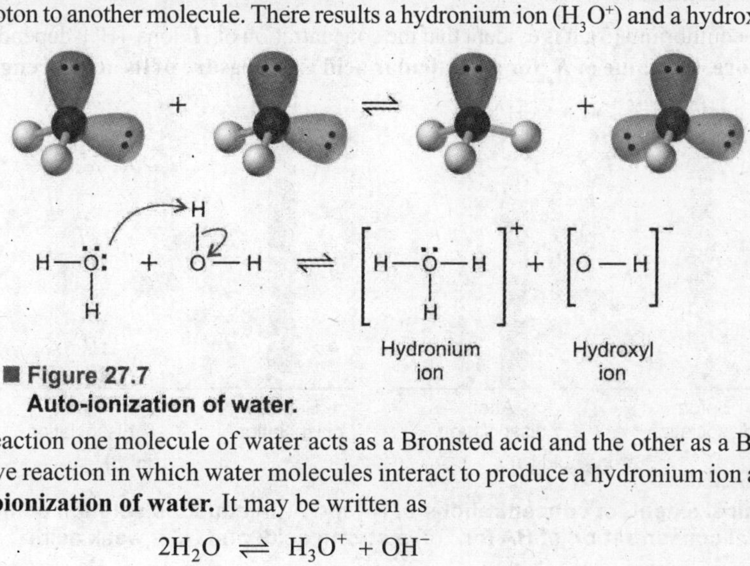

Figure 27.7
Auto-ionization of water.

In this reaction one molecule of water acts as a Bronsted acid and the other as a Bronsted base.

The above reaction in which water molecules interact to produce a hydronium ion and a OH^- ion is called **autoionization of water.** It may be written as

$$2H_2O \rightleftharpoons H_3O^+ + OH^-$$

That autoionization actually occurs was proved by Friedrich Kohlrausch (1840-1910). He found that even the purest water conducts electricity to a very small extent which was due to the generation of H_3O^+ and OH^- ions in water by ionization.

Water autoionization is most fundamental to our study of acids and bases. Obviously, H^+ ions are associated with water to give hydronium ions (H_3O^+). But for simplicity we shall generally write the dissociation equilibrium of water as

$$H_2O \rightleftharpoons H^+ + OH^-$$

RELATIVE STRENGTH OF ACIDS

The strength of an acid depends on its ability to transfer its proton (H^+) to a base to form its conjugate base. When a monoprotic acid (HA) dissolves in water, it transfers its proton to water (a Bronsted base) to form hydronium ion (H_3O^+) and a conjugate base.

$$HA + H_2O \rightleftharpoons H_3O^+ + \underset{\substack{\text{conjugate}\\ \text{base}}}{A^-} \qquad ...(1)$$

For simplifying our discussion, we take

$$H_3O^+ = H^+$$

Thus we can write the equilibrium reaction (1) as

$$HA + H_2O \rightleftharpoons H^+ + A^- \qquad ...(2)$$

This equation represents the dissociation of the acid HA into H^+ ion and A^- ion.

Applying the Law of Mass action to the acid dissociation equilibrium, we can write

$$K_a = \frac{[H^+][A^-]}{[HA]} \qquad ...(3)$$

where K_a is called the **acid dissociation constant.** In dilute solution of the acid (HA) we note that the concentration of liquid water remains essentially constant. Therefore, the terms included in the equilibrium expression (3).

The strength of an acid is defined as the concentration of H^+ ions in its aqueous solution at a given temperature.

From the equilibrium (3), it is evident that the concentration of H^+ ions, $[H^+]$, depends on the value of K_a. **Therefore, the value of K_a for a particular acid is a measure of its acid strength or acidity.**

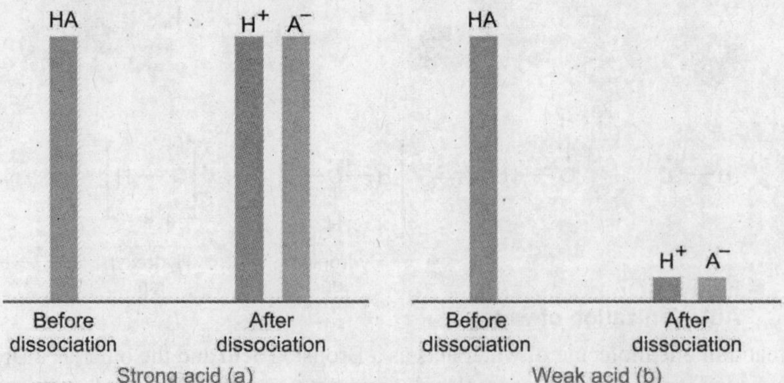

■ **Figure 27.8**
Graphical extent of concentrations of H^+ and A^- in aqueous solution compared to original concentration of HA for : (*a*) a strong acid; and (*b*) a weak acid.

TABLE 27.3. VALUES OF K_a FOR SOME COMMON MONOPROTIC ACIDS

Formula	Name	K_a
HSO_4^-	Hydrogen sulfate Ion	1.2×10^{-2}
$HClO_2$	Chlorous acid	1.2×10^{-2}
$ClCH_2COOH$	Monochloroacetic acid	1.35×10^{-3}
HF	Hydrofluoric acid	7.2×10^{-4}
HNO_2	Nitrous acid	4.0×10^{-4}
CH_3COOH	Acetic acid	1.8×10^{-5}
HOCl	Hypochlorous acid	3.5×10^{-8}
HCN	Hydrocyanic acid	6.2×10^{-10}
NH_4^+	Ammonium ion	5.6×10^{-10}
HOC_6H_5	Phenol	1.6×10^{-10}

Increasing acid strength (↑)

In any aqueous solution of a strong acid, practically all the original acid (HA) is dissociated and the value of K_a is large. On the other hand, a weak acid in aqueous solution is dissociated to a very small extent and the value of K_a is also small. Thus in general we can say that **the value of acid dissociation constant is large for a strong acid while it is small for a weak acid.**

The units of K_a are mol/l but are customarily omitted. The strong acids are not listed in the above table because practically all the acid is dissociated and its concentration at equilibrium cannot be measured accurately.

Calculation of Relative strength of Weak acids from K_a

We have seen that for an acid in aqueous solution we have, $HA \rightleftharpoons H^+ + A^-$

and
$$K_a = \frac{[H^+][A^-]}{[HA]}$$

Let C moles per liter be the concentration of the acid and α its degree of dissociation. Then,

$$[H^+] = C\alpha$$
$$[A^-] = C\alpha$$
$$[HA] = C(1-\alpha)$$

Substituting the values in the equilibrium expression we have

$$K_a = \frac{C\alpha \times C\alpha}{C(1-\alpha)}$$
$$= \frac{C\alpha^2}{(1-\alpha)}$$

For weak acids $1 - \alpha \simeq 1$. Therefore,

$$K_a = C\alpha^2$$

For two different acids, 1 and 2, let the degree of dissociation be α_1 and α_2; and the dissociation constants K_1 and K_2. Then,

for acid 1 $\qquad K_1 = C\alpha_1^2 \qquad$ ...(1)

for acid 2 $\qquad K_2 = C\alpha_2^2 \qquad$ ...(2)

Dividing equation (1) by (2), we get,

$$\frac{\alpha_1}{\alpha_2} = \sqrt{\frac{K_1}{K_2}}$$

Since $[H^+]$ is a measure of acid strength and it depends on the degree of dissociation α, we can write

$$\frac{\text{Strength of acid 1}}{\text{Strength of acid 2}} = \sqrt{\frac{K_1}{K_2}}$$

Evidently, **the ratio $\sqrt{K_1/K_2}$ would give us the relative strengths of the two acids.**

SOLVED PROBLEM 1

The dissociation constants of formic acid and acetic acid are 21.4×10^{-5} and 1.81×10^{-5} respectively. Find the relative strengths of the acids.

SOLUTION

$$\frac{\text{Strength of HCOOH}}{\text{Strength of } CH_3COOH} = \sqrt{\frac{K_{HCOOH}}{K_{CH_3COOH}}}$$

$$= \sqrt{\frac{21.4 \times 10^{-5}}{1.81 \times 10^{-5}}}$$

$$= 3.438$$

Thus formic acid is **3.438** times stronger than acetic acid.

SOLVED PROBLEM 2

Two hypothetical acids HA and HB have the dissociation constants 1×10^{-3} and 1×10^{-5} respectively in water at 298 K. Calculate the strength of HA with respect to HB.

SOLUTION

$$\frac{\text{Strength of HA}}{\text{Strength of HB}} = \sqrt{\frac{K_{HA}}{K_{HB}}}$$

$$= \sqrt{\frac{1 \times 10^{-3}}{1 \times 10^{-5}}}$$

$$= \mathbf{10}$$

Thus HA is ten times stronger than HB.

RELATIVE STRENGTH OF BASES

According to the Arrhenius concept, a base is a substance which produces OH^- ions in aqueous solution. The basic properties of such a substance are due to these hydroxyl ions. Let us consider a base BOH whose dissociation can be represented as

$$BOH \rightleftharpoons B^+ + OH^- \qquad ...(1)$$

Applying the Law of Mass action to the above equilibrium we can write the equilibrium expression as

$$K_b = \frac{[B^+]\,[OH^-]}{[BOH]} \qquad ...(2)$$

K_b is called the **base dissociation constant or base ionization constant.**

The strength of a base is defined as the concentration of OH^- ions in its aqueous solution at a given temperature.

From the equilibrium expression (2), it is evident that the concentration of OH^- ions, $[OH^-]$, depends on the value of K_b. Therefore, the value of K_b for a certain base is a measure of its base strength. In the aqueous solution of a strong base, practically all the original base is dissociated and the value of K_b is large. In the case of a weak base, it is dissociated in aqueous solution to a very small extent and the value of K_b is also small.

Calculation of K_b

The equilibrium expression for the dissociation of a base is

$$K_b = \frac{[B^+][OH^-]}{[BOH]}$$

If C is the molar concentration of the base and α its degree of dissociation, the equilibrium concentration of the various species are

$$[BOH] = C(1-\alpha) \text{ mol/l}$$
$$[B^+] = C\alpha \text{ mol/l}$$
$$[OH^-] = C\alpha \text{ mol/l}$$

Substituting the values in the equilibrium expression

$$K_b = \frac{C\alpha \times C\alpha}{C(1-\alpha)}$$
$$= \frac{\alpha^2 C}{(1-\alpha)}$$

Knowing the values of α and C, K_b can be calculated.

***K_b* for Bronsted bases**

Not all bases conform to the Arrhenius definition of a base. Ammonia (NH_3) and amines are the examples. Bronsted equilibrium for a base in aqueous solution may be written as :

$$\underset{\text{base}}{B} + \underset{\text{acid}}{H_2O} \rightleftharpoons \underset{\substack{\text{conjugate}\\\text{acid}}}{BH^+} + \underset{\substack{\text{conjugate}\\\text{base}}}{OH^-}$$

The equilibrium constant K_b for this general reaction is

$$K_b = \frac{[BH^+][OH^-]}{[B]}$$

The Bronsted equilibrium expression is virtually the same as that of the Arrhenius equilibrium. The concentration of OH^- ions determines the strength of a Bronsted base.

The values of K_b for some weak bases at 298 K are listed in Table 27.4.

Table 27.4. VALUES OF K_b FOR SOME COMMON WEAK BASES AT 298 K

Formula	Name	K_b
NH_3	Ammonia	1.8×10^{-5}
CH_3NH_2	Methylamine	4.38×10^{-4}
$C_2H_5NH_2$	Ethylamine	5.6×10^{-4}
$C_6H_5NH_2$	Aniline	3.8×10^{-10}
C_5H_5N	Pyridine	1.7×10^{-9}

MEASUREMENT OF pH

pH can be measured:

- by addition of a pH indicator into the solution under study. The indicator color varies depending on the pH of the solution. Using indicators, qualitative determinations can be made with universal indicators that have broad color variability over a wide pH range and quantitative determinations can be made using indicators that have strong color variability over a small pH range. Extremely precise measurements can be made over a wide pH range using indicators that have multiple equilibriums in conjunction with spectrophoto-metric methods to determine the relative abundance of each pH-dependent component that make up the color of solution.

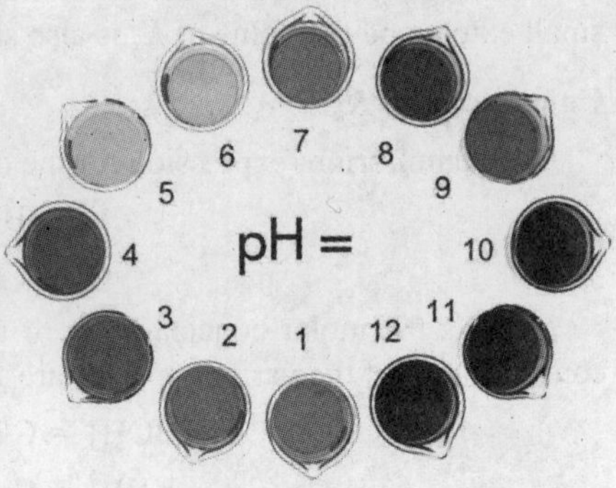

Color Key

- by using a pH meter together with pH-selective electrodes (pH glass electrode, hydrogen electrode, quinhydrone electrode, ion sensitive field effect transistor and others).
- by using pH paper, indicator paper that turns color corresponding to a pH on a color key. pH paper is usually small strips of paper (or a continuous tape that can be torn) that has been soaked in an indicator solution, and is used for approximations.

Substance	pH
Hydrochloric Acid 10M	-1.0
Battery acid	0.5
Gastric acid	1.5 – 2.0
Lemon juice	2.4
Cola	2.5
Vinegar	2.9
Orange or apple juice	3.5
Beer	4.5
Acid Rain	<5.0
Coffee	5.0
Tea or healthy skin	5.5

Substance	pH
Milk	6.5
Pure Water	7.0
Healthy human saliva	6.5 – 7.4
Blood	7.34 – 7.45
Seawater	7.7 – 8.3
Hand soap	9.0 – 10.0
Household ammonia	11.5
Bleach	12.5
Household lye	13.5
Caustic Soda	13.9

As the pH scale is logarithmic, it doesn't start at zero. Thus the most acidic of liquids encountered can have a pH of as low as **5**. The most alkaline typically has pH of **14**.

The *Hydrangea macrophylla* blossoms in pink or blue, depending on soil pH. In acidic soils, the flowers are blue; in alkaline soils, the flowers are pink.

THE pH OF SOLUTIONS

A knowledge of the concentration of hydrogen ions (more specifically hydronium ions) is of the greatest importance in chemistry. Hydrogen ion concentrations are typically quite small numbers. Therefore, chemists report the hydrogen ion concentration of a solution in terms of pH. It is defined as **the negative of the base-10 logarithm (log) of the H⁺ concentration.** Mathematically it may be expressed as

$$\mathbf{pH = -\log [H^+]}$$

where $[H^+]$ is the concentration of hydrogen ions in moles per liter.

Alternative and more useful forms of pH definition are :

$$\mathbf{pH = \log \frac{1}{[H^+]}}$$

and $$\mathbf{[H^+] = 10^{-pH}}$$

The pH concept is very convenient for expressing hydrogen ion concentration. It was introduced by Sorensen in 1909. It is now used as a general way of expressing other quantities also, for example,

(*a*) Concentration of OH^- ions in aqueous solution of a base is expressed as

$$p[OH^-] = -\log [OH^-]$$

(*b*) Equilibrium constant for water is written as

$$pK_w = -\log [K_w]$$

For any quantity *X*, we can write

$$pX = -\log X$$

The "**p**" in these expression means **" – log of the quantity"**

The measurement of pH

The pH of a given solution can be measured with the help of an apparatus called **pH meter.** This consists of a *voltameter* connected to two electrodes (Fig. 27.9).

(*a*) **a standard electrode** of known potential; and

(*b*) **a special electrode** (the probe) enclosed in a glass membrane that allows migration of H^+ ions. The glass case contains a reference solution of dilute HCl.

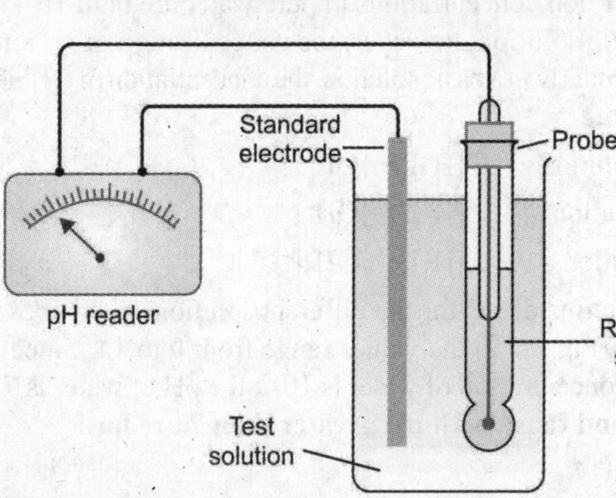

■ Figure 27.9
A pH meter.

The two electrodes are dipped in the solution to be tested. If this solution has a different pH from the solution in the probe, an electrical potential results. Thus the potential between the standard electrode and the glass electrode varies with the pH of the solution under test. This potential is recorded by an inbuilt potentiometer of the pH meter. The potentiometer reading is automatically converted electrically to a direct reading of the pH of the unknown solution. Knowing the pH of the solution its hydrogen ion concentration can be calculated.

pH Scale

In order to express the hydrogen ion concentration or acidity of a solution, a pH scale was evolved. The pH is defined as

$$pH = -\log [H^+] \quad \text{or} \quad [H^+] = 10^{-pH}$$

The hydrogen ion concentration of different acidic solutions were determined experimentally. These were converted to pH values using the above relations. Then these pH values were computed on a scale taking water as the reference substance. **The scale on which pH values are computed is called the pH scale.**

Water dissociates to H^+ and OH^- ions to a very small degree so that we have the equilibrium.

$$H_2O \rightleftharpoons H^+ + OH^- \qquad ...(1)$$

We can write the equilibrium expression as

$$K = \frac{[H^+][OH^-]}{[H_2O]} \qquad ...(2)$$

Since water is so little dissociated, the concentration of undissociated molecules, $[H_2O]$, is presumed to be constant. Therefore from expression (2) we can write

$$[H^+][OH^-] = K[H_2O] = K_w = \text{a constant.} \qquad ...(3)$$

where K_w is called the **water dissociation constant or the water ionization constant.**

When the concentrations of H^+ and OH^- ions in water are expressed in mole per liter, the value of K_w found experimentally is 1.0×10^{-14}. From the equation (1) it is obvious that one molecule of water dissociates to give one H^+ ion and one OH^- ion. This means that the concentration of H^+ and OH^- ions in pure water is equal. Using the expression (3), we have

$$[H^+][OH^-] = \sqrt{K_w} = \sqrt{1.0 \times 10^{-14}}$$

or $$[H^+] = [OH^-] = 1.0 \times 10^{-7} \text{ mol/l or } 1.0 \times 10^{-4} \text{ mol/m}^3$$

Thus the H^+ ion and OH^- ion concentrations in pure water are both 10^{-7} mol l^{-1} at 298 K and it is said to be neutral. In acidic solution, however, the concentration of H^+ ions must be greater than 10^{-7} mol l^{-1} or 10^{-4} mol/m^3. Similarly in a basic solution, the concentration of OH^- ions must be greater than 10^{-7} mol l^{-1}. Thus we can state :

neutral solution	$[H^+] = [OH^-]$
acidic solution	$[H^+] > [OH^-]$
basic solution	$[H^+] < [OH^-]$

Expressing the $[H^+]$ in terms of pH for the different solutions cited above, we get what we call the **pH scale.** On this scale (Fig. 27.9) the values range from 0 to 14. Since pH is defined as – log $[H^+]$ and the hydrogen ion concentration of water is 10^{-7}, the pH of water is 7. **All solutions having pH less than 7 are acidic and those with pH greater than 7 are basic.**

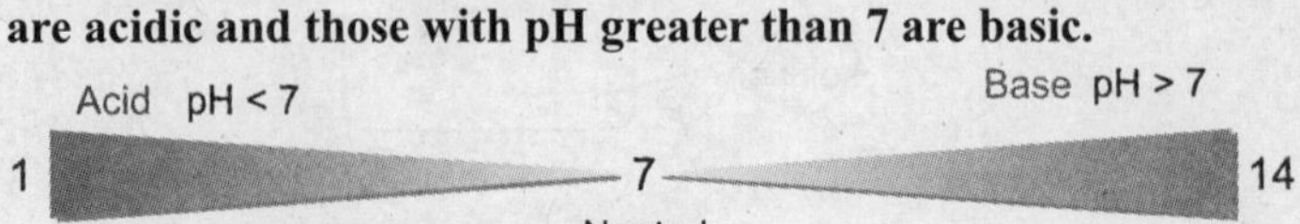

As shown by the pH scale, pH decreases with the increase of $[H^+]$. **The lower the pH, higher is the $[H^+]$ or acidity.**

NUMERICAL PROBLEMS BASED ON pH

To understand the pH concept fully, one must have a good exercise in the related numerical problems. Here we will review and discuss certain useful relations that the student is expected to know.

How to calculate $[H^+]$ and $[OH^-]$ from K_w. In any aqueous solution, the product of $[H^+]$ and $[OH^-]$ is always equal to K_w. This is so irrespective of the solute and relative concentrations of H^+ and OH^- ions. However, the value of K_w depends on temperature. At 298 K it is 1.0×10^{-14}. Thus,

$$[H^+]\ [OH^-] = 1.0 \times 10^{-14}$$

Each of $[H^+]$ and $[OH^-]$ in pure water at 298 K is 10^{-7}. The concentration of H^+ and OH^- ions are expressed in gram moles per liter. However 298 K the SI Unit of concentration is mol/m^3.

The concentration of H^+ and OH^- ions can be calculated from the expressions :

$$[H^+] = \frac{K_w}{[OH^-]}$$

$$[OH^-] = \frac{K_w}{[H^+]}$$

Relation between pH and pOH

We have already stated that pH concept can also be used to express small quantities as $[OH^-]$ and K_w. Thus we have

$$pOH = -\log [OH^-]$$

$$pK_w = -\log k_w$$

Let us consider the log form of the expression

$$K_w = [H^+]\ [OH^-]$$

That is, $\log K_w = \log [H^+] + \log [OH^-]$

or $-\log K_w = -\log [H^+] - \log [OH^-]$

Thus $pK_w = pH + pOH$

Since $K_w = 1.0 \times 10^{-14}$

$pK_w = -\log (1.0 \times 10^{-14}) = 14.00$

Hence, for any aqueous solution at 298 K, pH and pOH add up to 14.00. That is,

$$\mathbf{pH + pOH = 14.00}$$

In general, the pH problems may be of the following types :

TYPE 1. Interpreting the pH

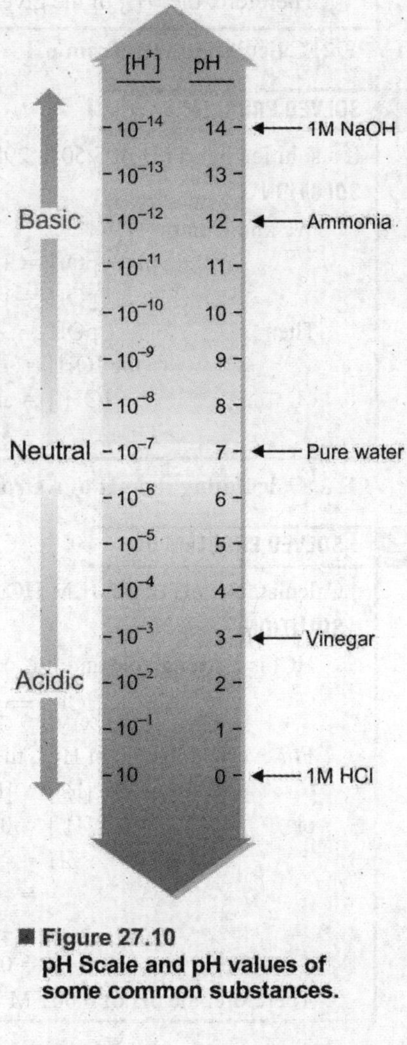

Figure 27.10
pH Scale and pH values of some common substances.

SOLVED PROBLEM 3

The hydrogen ion concentration of a fruit juice is 3.3×10^{-2} M. What is the pH of the juice ? Is it acidic or basic ?

SOLUTION

The definition of pH is

$$pH = -\log [H^+]$$

We are given $[H^+] = 3.3 \times 10^{-2}$

Substituting into the definition of pH, we get

$$pH = -\log (3.3 \times 10^{-2})$$
$$= -(-1.48)$$
$$= \mathbf{1.48}$$

Since the pH is less than 7.00 the **solution is acidic**.

SOLVED PROBLEM 4

If a solution has a pH of 7.41, determine its H^+ concentration.

SOLUTION

Since $pH = -\log [H^+]$

$-pH = \log [H^+]$

Now $[H^+]$ can be calculated by taking antilog of – pH :

$$[H^+] = \text{antilog } (-\text{pH})$$

In this case,

$$[H^+] = \text{antilog } (-7.41) = 10^{-7.41}$$

Therefore, the $[H^+]$ of the given solution is $\mathbf{3.9 \times 10^{-8}\ M}$.

TYPE 2. Calculating OH^- from pH

SOLVED PROBLEM 5

If a solution has a pH of 5.50 at 298 K, calculate its $[OH^-]$.

SOLUTION

We know that :

$$pH + pOH = 14.00$$
$$pOH = 14.00 - pH = 14.00 - 5.50 = 8.50$$

Then $pOH = -\log [OH^-] = 8.50$

$$\log [OH] = -8.50$$
$$[OH^-] = \text{antilog } (-8.50)$$
$$= \mathbf{3.2 \times 10^{-9}\ M}$$

TYPE 3. Calculating the pH of a strong acid

SOLVED PROBLEM 6

Calculate the pH of 0.001 M HCl.

SOLUTION

HCl is a strong acid and it is completely dissociated in aqueous solution.

$$\underset{0.001M}{HCl} \rightleftharpoons \underset{0.001M}{H^+} + Cl^-$$

For every molecule of HCl, there is one H^+. Therefore,

$$[H^+] = [HCl]$$

or $[H^+] = 0.001\ M$

$\therefore$ $pH = -\log (0.001)$

$$= -\log (1 \times 10^{-3})$$
$$= -\log 1 + 3 \log 10$$
$$= 0 + 3 = 3$$

Therefore, the pH of 0.001 M HCl is **3**.

SOLVED PROBLEM 7

The pH of a solution of HCl is 2. Find out the amount of acid present in a m^3 of the solution.

SOLUTION

$$pH = 2$$

$$-\log [H^+] = 2 \qquad \textit{(By definition)}$$

The dissociation of HCl takes place according to the equation :

$$HCl \longrightarrow H^+ + Cl^-$$

One molecule of HCl gives one ion of H^+. Therefore,

$$[H^+] = [HCl] = 10^{-2}\ M$$

$$\therefore \text{ Amount of HCl in one } m^3 = 10^{-2} \times \text{mol mass of HCl}$$

$$= 10^{-2} \times 36.5$$

$$= \mathbf{0.365\ gl^{-1}\ or\ 0.36\ kg/m^3}$$

TYPE 4. Calculating the pH of a Strong base

SOLVED PROBLEM 8

Determine the pH of 0.10 M NaOH solution.

SOLUTION

NaOH is a strong base and it is completely dissociated in aqueous solution.

$$\underset{0.10M}{NaOH} \rightleftharpoons Na^+ + \underset{0.10M}{OH^-}$$

Therefore the concentration of OH^- ions is equal to that of the undissociated NaOH.

$$[OH^-] = [NaOH] = 0.10\ M$$

$[H^+]$ can be calculated by applying the expression by substituting the value of $K_w = 1 \times 10^{-14}$ and of $[OH^-]$

$$[H^+] = \frac{K_w}{[OH^-]} = \frac{1 \times 10^{-14}}{0.10} = \frac{1 \times 10^{-14}}{10^{-1}} = 10^{-13}\ M$$

By definition

$$pH = -\log [H^+] = -\log^{-13}$$

$$= 13$$

Therefore the pH of 0.10 M NaOH is **13**.

SOLVED PROBLEM 9

Calculate the pH of a 0.020 M $Ba(OH)_2$ solution.

SOLUTION

Barium hydroxide dissociates according to the equation

$$Ba(OH)_2 \rightleftharpoons Ba^{2+} + 2OH^-$$

One molecule of barium hydroxide dissociates to give two OH^- ions. Therefore we have

$$[OH^-] = 0.040\ M$$

The pOH of the solution is, therefore,

$$pOH = -\log 0.040 = 1.40$$

We have that $pH + pOH = 14.00$

$$\therefore \quad pH = 14.00 - 1.40 = 12.60$$

The pH of 0.020 M $Ba(OH)_2$ solution is **12.60**.

TYPE 5. Calculating pH of a Weak acid

SOLVED PROBLEM 10

Calculate the pH of 0.1 M CH_3COOH. The dissociation constant of acetic acid is 1.8×10^{-5}.

SOLUTION

Acetic acid dissociates in aqueous solution as

$$CH_3COOH \rightleftharpoons H^+ + CH_3COO^-$$

The K_a for acetic acid is stated as

$$K_a = \frac{[H^+][CH_3COO^-]}{[CH_3COOH]} = 1.8 \times 10^{-5} \qquad ...(1)$$

One molecule of CH_3COOH dissociates to form one H^+ and one CH_3COO^-. Thus,

$$[H^+] = [CH_3COO^-]$$

If x be the concentration of H^+ ion at equilibrium, the various concentrations at the equilibrium may be written as

$$[CH_3COOH] = 0.10 - x$$
$$[H^+] = x$$
$$[CH_3COO^-] = x$$

Since the value of x is negligible relative to 0.10, $(0.10 - x)$ is approximately equal to 0.10. Substituting the values in (1), we have

$$K_a = \frac{x \times x}{0.10} = 1.8 \times 10^{-5}$$

Solving for x,

$$\frac{x^2}{0.10} = 1.8 \times 10^{-5}$$

or

$$x^2 = (1.0 \times 10^{-1})(1.8 \times 10^{-5}) = 1.8 \times 10^{-6}$$

$$x = \sqrt{1.8 \times 10^{-6}} = 1.34 \times 10^{-3} \text{ M}$$

Thus the hydrogen concentration $[H^+]$ of 0.1 M CH_3COOH is 1.34×10^{-3} M

$$\text{pH} = -\log[H^+] = -\log(1.34 \times 10^{-3})$$
$$= +3 - \log 1.34$$
$$= 2.87$$

Thus the pH of 0.1 M CH_3COOH solution is **2.87**.

SOLVED PROBLEM 11

Find out the pH of a 0.002 M acetic acid solution if it is 2.3% ionized at this dilution.

SOLUTION

The dissociation equation for acetic acid is

$$CH_3COOH \rightleftharpoons H^+ + CH_3COO^-$$

If the dissociation is 2.3%, it means that 2.3×0.002 molecules dissociated and generated 2.3×0.002 ions of H^+.

$$[H^+] = 0.002 \times 0.023$$

$$\therefore \quad = 0.000046$$

$$pH = -\log [H^+]$$
$$= -\log (0.000046) = -\log (4.6 \times 10^{-5})$$
$$= +5 - \log 4.6$$
$$= +5 - 0.6628$$
$$= \mathbf{4.3372}$$

TYPE 6. Calculating pH of a Weak base

SOLVED PROBLEM 12

Calculate the pH of 0.1 M NH_3 solution. The ionization constant, K_b, for NH_3 is 1.8×10^{-5}.

SOLUTION

The equilibrium reaction is

$$\underset{0.10-x}{NH_3} + H_2O \rightleftharpoons \underset{x}{NH_4^+} + \underset{x}{OH^-}$$

We assume $[OH^-] = x$ when equilibrium is reached. Thus we can write the equilibrium expression as

$$K_b = \frac{[NH_4^+][OH^-]}{[NH_3]} = \frac{x^2}{0.10 - x}$$

Since x is insignificant relative to 0.1, we can write

$$K_b = \frac{x^2}{0.10}$$

Solving for x,

$$x = \sqrt{0.10 \times K_b} \quad where\ K_b = 1.8 \times 10^{-5}$$

$$x = \sqrt{0.10 \times (1.8 \times 10^{-5})} = 1.34 \times 10^{-3}\ M$$

Since $x = [OH^-]$, we have

$$pOH = -\log (1.34 \times 10^{-3}) = 2.87$$

We know that

$$pH + pOH = 14.00$$

$$\therefore \quad pH = 14.00 - 2.87 = 11.13$$

Thus the pH of 0.1 M NH_3 solution is **11.1**.

TYPE 7. Calculating pH of aqueous Sulfuric acid

SOLVED PROBLEM 13

Calculate the pH of 0.100 M H_2SO_4 at 298 K. K_{a_1} is large and $K_{a_2} = 0.012$ M.

SOLUTION

Sulfuric acid ionizes in two steps,

$$H_2SO_4 \longrightarrow H^+ + HSO_4^- \qquad ...(1)$$

$$\underset{\text{weak acid}}{HSO_4^-} \rightleftharpoons H^+ + SO_4^{2-} \qquad ...(2)$$

The pH of aqueous solution of H_2SO_4 depends on the total hydrogen ion concentration with both ionization steps taken into account.

First Ionization Step

The first ionization is complete and it results in a hydrogen ion concentration of the acid, 0.100 M.

$\therefore$ pH due to step (1) is = 1.0

Second Ionization Step

The second ionization step adds to the hydrogen ion concentration slightly. Therefore, the overall pH of the solution will be slightly less than 1.

In step (2), HSO_4^- is treated as a weak acid. The initial concentration of HSO_4^- is 0.100 M. If x concentration is ionized at equilibrium, the concentrations at this point are :

$$[H^+] = 0.100\text{ M} + x$$
$$[SO_4^{2-}] = x$$
$$[HSO_4^-] = 0.100\text{ M} - x$$

Since K_{a_2} is not very small, x cannot be ignored. Thus, the ionization constant expression is

$$K_{a_2} = \frac{[SO_4^{2-}][H^+]}{[HSO_4^-]} = \frac{x \times (0.100\text{ M} + x)}{0.100\text{ M} - x}$$

Substituting K_{a_2} and rearranging we get

$$x^2 + (0.112\text{ M})\,x - 0.0012\text{ M}^2 = 0$$

Solving for x using the quadratic formula

$$x = \frac{-b \pm \sqrt{b^2 - 4ac}}{2a}$$
$$= \frac{-0.112 + \sqrt{(0.112)^2 + 4 \times 0.0012}}{2} = 0.010\text{ M}$$
$$= \frac{-0.112 - \sqrt{(0.112)^2 + 4 \times 0.0012}}{2} = -0.122\text{ M}$$

Since the $[H^+]$ at equilibrium is $0.100 - x$, these two solutions are

$$[H^+] = 0.110 \quad \text{or} \quad -0.022\text{ M}$$

The second solution makes no sense and can be discarded. Thus,

$$\text{pH} = -\log 0.110 = \mathbf{0.96}$$

This is slightly lower than 1.0 as expected.

WHAT IS A BUFFER SOLUTION ?

It is often necessary to maintain a certain pH of a solution in laboratory and industrial processes. This is achieved with the help of **buffer solutions, buffer systems or simply buffers.**

A buffer solution is one which maintains its pH fairly constant even upon the addition of small amounts of acid or base.

In other words, a buffer solution resists (or buffers) a change in its pH. That is, we can add a small amount of an acid or base to a buffer solution and the pH will change very little. Two common types of buffer solutions are :

(1) a weak acid together with a salt of the same acid with a strong base. These are called **Acid buffers** *e.g.*, $CH_3COOH + CH_3COONa$.

(2) a weak base and its salt with a strong acid. These are called **Basic buffers.** *e.g.*, $NH_4OH + NH_4Cl$.

Let us illustrate buffer action by taking example of a common buffer system consisting of solution of acetic acid and sodium acetate (CH_3COOH/CH_3COONa).

ACID RAIN

Acid Rain, form of air pollution in which airborne acids produced by electric utility plants and other sources fall to Earth in distant regions. The corrosive nature of acid rain causes widespread damage to the environment. The problem begins with the production of sulfur dioxide and nitrogen oxides from the burning of fossil fuels, such as coal, natural gas, and oil, and from certain kinds of manufacturing. Sulfur dioxide and nitrogen oxides react with water and other chemicals in the air to form sulfuric acid, nitric acid, and other pollutants. These acid pollutants reach high into the atmosphere, travel with the wind for hundreds of miles, and eventually return to the ground by way of rain, snow, or fog, and as invisible "dry" forms.

Damaged by acid rain

Damage from acid rain has been widespread in eastern North America and throughout Europe, and in Japan, China, and Southeast Asia. Acid rain leaches nutrients from soils, slows the growth of trees, and makes lakes uninhabitable for fish and other wildlife. In cities, acid pollutants corrode almost everything they touch, accelerating natural wear and tear on structures such as buildings and statues. Acids combine with other chemicals to form urban smog, which attacks the lungs, causing illness and premature deaths.

$$CH_3COOH \rightleftharpoons H^+ + CH_3COO^-$$

$$CH_3COONa \longrightarrow Na^+ + CH_3COO^-$$

since the salt is completely ionized, it provides the common ions CH_3COO^- in excess. The common ion effect suppresses the ionization of acetic acid. This reduces the concentration of H^+ ions which means that pH of the solution is raised. Thus, a 0.1 M acetic acid solution has a pH of 2.87 but a solution of 0.1 M acetic acid and 0.1 M sodium acetate has a pH of 4.74 (Fig. 27.11). Thus 4.74 is the pH of the buffer. On addition of 0.01 mol NaOH the pH changes from 4.74 to 4.83, while on the addition of 0.01 mol HCl the pH changes from 4.74 to 4.66. Obviously the buffer solution maintains fairly constant pH and the changes in pH could be described as marginal.

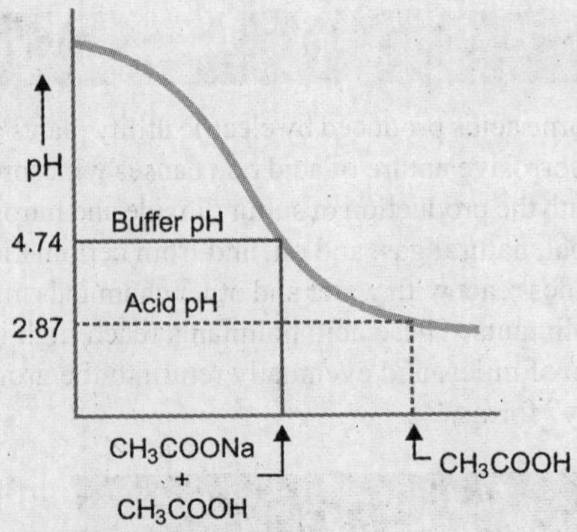

■ **Figure 27.11**
The buffer solution (CH_3COOH/CH_3COONa) has a higher pH than the acid itself.

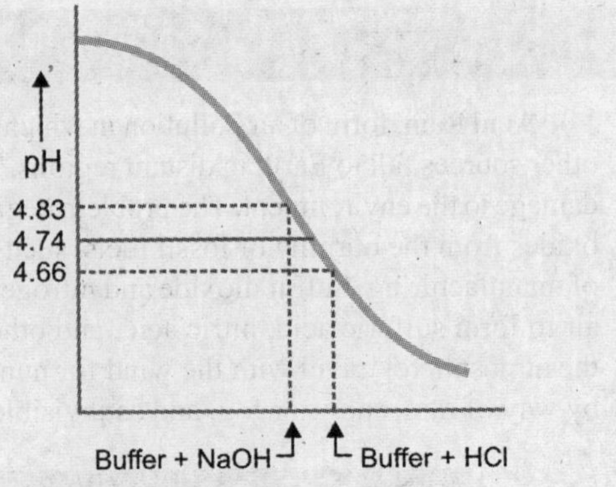

■ **Figure 27.12**
The pH of buffer changes only slightly upon addition of an acid (0.01 moles) or base (0.01 mole NaOH).

HOW A BUFFER OPERATES?

We have already stated that a buffer solution containing equimolar amounts (0.10 M) of acetic acid and sodium acetate has pH 4.74. Now we proceed to discuss how the addition of a small amount of HCl or NaOH to the buffer solution affects its pH.

The pH of the buffer is governed by the equilibrium

$$CH_3COOH \rightleftharpoons CH_3COO^- + H^+ \qquad ...(1)$$

The buffer solution has a large excess of CH_3COO^- ions produced by complete ionization of sodium acetate,

$$CH_3COONa \longrightarrow CH_3COO^- + Na^+ \qquad ...(2)$$

(1) **Addition of HCl.** Upon the addition of HCl, the increase of H^+ ions is counteracted by association with the excess of acetate ions to form unionized CH_3COOH. Thus the added H^+ ions are neutralized and the pH of the buffer solution remains virtually unchanged. However owing to the increased concentration of CH_3COOH, the equilibrium (1) shifts slightly to the right to increase H^+ ions. This explains the marginal increase of pH of the buffer solution on addition of HCl (Fig. 27.12)

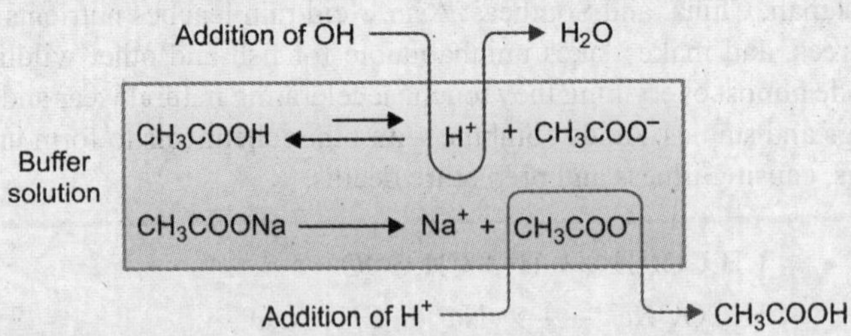

■ **Figure 27.13**
Mechanism of Buffer action of an acid buffer.

(2) **Addition of NaOH.** When NaOH is added to the buffer solution, the additional OH^- ions combine with H^+ ions of the buffer to form water molecules. As a result the equilibrium (1) shifts to the right to produce more and more H^+ ions till practically all the excess OH^- ions are neutralized and the original buffer pH restored. However, a new equilibrium system is set up in which [CH_3COOH]

is lower than it was in the original buffer. Consequently [H^+] is also slightly less and pH slightly higher than the buffer pH values (Fig. 27.13).

Operation of a Basic buffer as NH_4OH/NH_4Cl can also be explained on the same lines as of an acid buffer (Fig. 27.14) upon addition of HCl the H^+ ions combine with OH^- ions of the buffer to form water molecules. The equilibrium,

$$NH_4OH \rightleftharpoons NH_4^+ + OH^-$$

is shifted to the right till all the additional H^+ ions are neutralized and the original buffer pH restored. When NaOH is added to the buffer solution, OH^- ions associate with excess of NH_4^+ ions to form unassociated NH_4OH. Thus the pH of the buffer is maintained approximately constant.

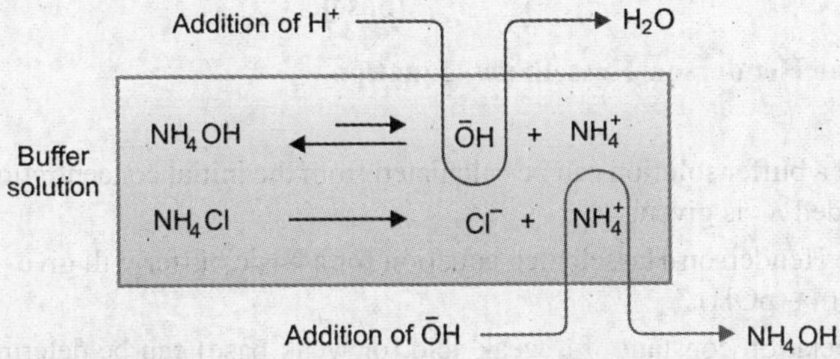

■ Figure 27.14
Mechanism of Buffer action of a basic buffer.

CALCULATION OF THE pH OF BUFFER SOLUTIONS

The pH of an acid buffer can be calculated from the dissociation constant, K_a, of the weak acid and the concentrations of the acid and the salt used.

The dissociation expression of the weak acid, HA, may be represented as

$$HA \rightleftharpoons H^+ + A^-$$

and $$K_a = \frac{[H^+][A^-]}{[HA]}$$

or $$[H^+] = K_a \times \frac{[HA]}{[A^-]} \quad ...(1)$$

The weak acid is only slightly dissociated and its dissociation is further depressed by the addition of the salt (Na^+A^-) which provides A^- ions (Common ion effect). As a result the equilibrium concentration of the unionized acid is nearly equal to the initial concentration of the acid. The equilibrium concentration [A^-] is presumed to be equal to the initial concentration of the salt added since it is completely dissociated. Thus we can write the equation (1) as

$$[H^+] = K_a \times \frac{[\text{acid}]}{[\text{salt}]} \quad ...(2)$$

where [acid] is the initial concentration of the added acid and [salt] that of the salt used.

Taking negative logs of both sides of the equation (2), we have

$$-\log[H^+] = -\log K_a - \log\frac{[\text{acid}]}{[\text{salt}]} \quad ...(3)$$

But $-\log[H^+] = \text{pH}$ and $\log K_a = pK_a$

Thus from (3) we have

$$pH = pK_a - \log\frac{[\text{acid}]}{[\text{salt}]} = pK_a + \log\frac{[\text{salt}]}{[\text{acid}]}$$

Hence $$\mathbf{pH = pK_a + \log\frac{[salt]}{[acid]}}$$

This relationship is called the **Henderson-Hasselbalch equation** or simply **Henderson equation.**

In a similar way, the Henderson-Hasselbalch equation for a basic buffer can be derived. This can be stated as :

$$\mathbf{pOH = pK_b + \log\frac{[salt]}{[base]}}$$

Significance of the Henderson-Hasselbalch equation

With its help :

(1) The pH of a buffer solution can be calculated from the initial concentrations of the weak acid and the salt provided K_a is given.

However, the Henderson-Hasselbalch equation for a basic buffer will give pOH and its pH can be calculated as (14 – pOH).

(2) The dissociation constant of a weak acid (or weak base) can be determined by measuring the pH of a buffer solution containing equimolar concentrations of the acid (or base) and the salt.

$$pH = pK_a + \log\frac{[\text{salt}]}{[\text{acid}]}$$

Since $$[\text{salt}] = [\text{acid}], \quad \log\frac{[\text{salt}]}{[\text{acid}]} = \log 1 = 0$$

$$\therefore \quad pK_a = pH$$

The measured pH, therefore, gives the value of pK_a of the weak acid.

Likewise we can find the pK_b of a weak base by determining the pH of equimolar basic buffer.

(3) A buffer solution of desired pH can be prepared by adjusting the concentrations of the salt and the acid added for the buffer.

It is noteworthy that buffer solution are most effective when the concentrations of the weak acid (or weak base) and the salt are about equal. This means that pH is close to the value of pK_a of the acid (or pK_b of the base).

NUMERICAL PROBLEMS BASED ON BUFFERS

The numerical problems pertaining to buffer solutions may be classified into the following types:

TYPE 1. Calculation of pH of Acid Buffers

SOLVED PROBLEM 14

Find the pH of a buffer solution containing 0.20 mol per liter CH_3COONa and 0.15 mol per liter CH_3COOH. K_a for acetic acid is 1.8×10^{-5}.

SOLUTION

$$K_a = 1.8 \times 10^{-5}$$

$$pK_a = -\log(1.8 \times 10^{-5}) = 4.7447$$

$$pH = pK_a + \log\frac{[\text{salt}]}{[\text{acid}]} \qquad \textit{(Henderson-Hasselbalch equation)}$$

$$= 4.7447 + \log \frac{0.20}{0.15}$$

$$= 4.7447 + \log \frac{4}{3}$$

$$= 4.7447 + 0.6021 - 0.4771$$

$$= \mathbf{4.8697}$$

ALTERNATIVE SOLUTION

The equilibrium equation for dissociation of acetic acid is

$$CH_3COOH \rightleftharpoons H^+ + CH_3COO^-$$

$$K_a = \frac{[H^+][CH_3COO^-]}{[CH_3COOH]}$$

Since CH_3COOH is feebly dissociated in the presence of CH_3COO^- ions (Common ion effect), $[CH_3COOH]$ is the same as of acid taken initially. Again $[CH_3COOH^-]$ is equal to the initial concentration of the salt CH_3COONa as it is completely dissociated. Thus we have

$$K_a = \frac{[H^+] \times 0.20}{0.15} = 1.8 \times 10^{-5}$$

or $$[H^+] = \frac{(1.8 \times 10^{-5}) \times 0.15}{0.20} = 1.35 \times 10^{-5}$$

$$pH = -\log (H^+)$$

$$= -\log (1.35 \times 10^{-5})$$

$$= 5 - 0.1303$$

$$= \mathbf{4.8697}$$

SOLVED PROBLEM 15

Estimate the pH at 298 K containing 0.10 M sodium acetate and 0.03 M acetic acid pK_a for $CH_3COOH = 4.57$.

SOLUTION

$$pH = pK_a + \log \frac{[\text{salt}]}{[\text{acid}]}$$ *(Henderson-Hasselbalch equation)*

$$= 4.57 + \log \frac{0.10}{0.03}$$

$$= 4.57 + 0.52 = \mathbf{5.09}$$

SOLVED PROBLEM 16

Calculate the pH of a buffer solution that is 0.250 M in formic acid, HCOOH, and 0.100 M in sodium formate, HCOONa. K_a for formic acid is 1.8×10^{-4}.

SOLUTION

$$K_a = 1.8 \times 10^{-4}$$

$$pK_a = -\log (1.8 \times 10^{-4})$$

$$= 4 - \log 1.8 = 3.7447$$

$$pH = pK_a + \log \frac{[\text{salt}]}{[\text{acid}]}$$ *(Henderson-Hasselbalch equation)*

$$= 3.7447 + \log \frac{[0.100]}{[0.250]}$$

$$= 3.7447 - 0.3979 = \mathbf{3.3468}$$

ALTERNATIVE SOLUTION

$$HCOOH \rightleftharpoons H^+ + HCOO^-$$

$$K_a = \frac{[H^+][HCOO^-]}{[HCOOH]}$$

or

$$[H^+] = \frac{K_a[HCOOH]}{[HCOO^-]}$$

Since HCOOH is weakly ionized in the presence of $HCOO^-$, the [HCOOH] at equilibrium is equal to its initial concentration. Again, HCOONa is fully ionized and hence the concentration of $HCOO^-$ ions is the same as that of the salt taken.

Substituting values in the above expression

$$[H^+] = \frac{1.8 \times 10^{-4} \times 0.250}{0.100} = 4.5 \times 10^{-4}\text{ M}$$

$$pH = -\log [H^+]$$
$$= -\log (4.5 \times 10^{-4})$$
$$= 4 - \log 4.5$$
$$= 4 - 0.6532$$
$$= 3.3468$$

$\therefore$ $$pH = \mathbf{3.3468}$$

SOLVED PROBLEM 17

The K_a of propionic acid is 1.34×10^{-5}. What is the pH of a solution containing 0.5 M propionic acid, C_2H_5COOH, and 0.5 sodium propionate, C_2H_5COONa. What happens to the pH of this solution when volume is doubled by the addition of water?

SOLUTION

The pH of solution can be calculated by substituting the given values in the Henderson-Hasselbalch equation.

$$pK_a = 1.34 \times 10^{-5}$$

$$pK_a = -\log K_a = -\log 1.34 \times 10^{-5} = 4.87$$

$$pH = pK_a + \log \frac{[\text{salt}]}{[\text{acid}]} \qquad \textit{(Henderson-Hasselbalch equation)}$$

$$= 4.87 + \log \frac{0.5}{0.5}$$

$$= 4.87 + 0$$

$$= \mathbf{4.87}$$

ALTERNATIVE SOLUTION

The equation for the dissociation of propionic acid is

$$C_2H_5COOH \rightleftharpoons C_2H_5COO^- + H^+$$

$$K_a = \frac{[C_2H_5COO^-][H^+]}{[C_2H_5COOH]} = \frac{0.5 \times [H^+]}{0.5} = [H^+]$$

$\therefore$ $$-\log [H^+] = -\log K_a \quad \text{or} \quad pH = pK_a$$

$$pH = \mathbf{4.87}$$

When the volume of the solution is doubled, there will be no change in pH of the solution because the ratio [salt] / [acid] remains the same.

TYPE 2. Calculation of pH of Basic or Alkaline Buffers

Henderson-Hasselbalch equation for basic buffers is first used to determine the value of pOH.

$$pOH = pK_b + \log\frac{[salt]}{[base]}$$

where K_b is the dissociation constant of the base.

We know that,

$$pH + pOH = 14.00$$
$$pH = 14.00 - pOH$$

The rest of the procedure followed is the same as for calculating the pH of acid buffers.

SOLVED PROBLEM 18

A buffer solution contains 0.015 mol of ammonium hydroxide and 0.025 mol of ammonium chloride. Calculate the pH value of the solution. Dissociation constant of NH_4OH at the room temperature is 1.80×10^{-5}.

SOLUTION

In this case,

$$pOH = pK_b + \log\frac{[salt]}{[base]}$$

But

$$pK_b = -\log K_b$$
$$= -\log(1.80 \times 10^{-5})$$
$$= 4.7447$$

∴

$$pOH = 4.7447 + \log\frac{0.025}{0.015}$$
$$= 4.7447 + 0.2218$$
$$= 4.9665$$

Since

$$pH + pOH = 14$$

∴

$$pH = 14 - 4.9665$$
$$= \mathbf{9.0335}$$

SOLVED PROBLEM 19

A buffer solution contains 0.25 M NH_3 and 0.40 M NH_4Cl. Calculate the pH of the solution. K_b for ammonia = 1.8×10^{-5}.

SOLUTION

$$NH_3 + H_2O \rightleftharpoons NH_4^+ + OH^-$$

$$K_b = \frac{[NH_4^+]\ [OH^-]}{[NH_3]}$$

or

$$[OH^-] = \frac{K_b \times [NH_3]}{[NH_4^+]}$$

$$= \frac{1.8 \times 10^{-5} \times 0.25}{0.40} = 1.125 \times 10^{-5}$$

$$pOH = -\log(1.125 \times 10^{-5}) = 4.945$$
$$pH = 14.00 - 4.945 = \mathbf{9.055}$$

This problem can also be solved with the help of Henderson-Hasselbalch equation.

TYPE 3. Calculation of pH of buffer solution after addition of HCl or NaOH

The pH of a buffered solution changes upon the addition of a strong acid or strong base. The changed pH can be calculated as follows:

(1) **When HCl is added.** Let the buffer contain a mol/m^3 of weak acid HA and b mol/m^3 of the salt $Na^+ A^-$. That is,

$$\underset{a}{HA} \rightleftharpoons H^+ + \underset{b}{A^-}$$

If x mol of HCl/m^3 is added, it forms x mol/m^3 of H^+ ions. These combine with (x mol) A^- ions and the equilibrium is pushed to the left to form x mol undissociated acid. In other words, the concentration of A^- ions is decreased by x mol/m^3 and that of HA is increased by x mol/m^3.

$$\underset{(a+x)}{HA} \rightleftharpoons H^+ + \underset{(b-x)}{A^-}$$

Thus we can write the equilibrium expression for the buffer solution with changed concentration as

$$K_a = \frac{[H^+][A^-]}{[HA]} = \frac{[H^+](b-x)}{(a+x)}$$

or

$$[H^+] = \frac{K_a(a+x)}{(b-x)}$$

Knowing the value of K_a, $[H^+]$ of the buffer can be calculated and

$$pH = -\log[H^+] = -\log\frac{K_a(a+x)}{(b-x)}$$

(2) **When NaOH is added.** If x mol/m^3 of NaOH is added, it produces x mol/m^3 OH^- ions. Thus combine with x mol/m^3 H^+ ions of the buffer to form water. The equilibrium

$$HA \rightleftharpoons H^+ + A^-$$

is thus shifted to the right till x mol/m^3 H^+ ions are produced that are needed to neutralize the added OH^- ions. The net result is that x mol/m^3 HA dissociates to yield x mol/m^3 A^- ions. Hence, the concentration of HA is decreased by x mol/m^3 and the concentration of A^- is increased by x mol/m^3.

$$\underset{(a-x)}{HA} \rightleftharpoons H^+ + \underset{(b+x)}{A^-}$$

We can now write the following expression for the new equilibrium with changed concentration as

$$K_a = \frac{[H^+](b+x)}{(a-x)}$$

or

$$[H^+] = \frac{K_a(a-x)}{(b+x)}$$

$$pH = -\log[H^+] = -\log\frac{K_a(a-x)}{(b+x)}$$

The pH of a buffer after the addition of HCl or NaOH can also be calculated with the help of Henderson-Haselbalch equation. It may be noted that **any volume changes caused in these additions are ignored.**

SOLVED PROBLEM 20

The pH of a buffer solution containing 0.5 mol/m^3 of CH_3COOH and 0.5 mol/m^3 CH_3COONa has been found to be 4.76. What will be the pH of this solution after 0.1 mol/m^3 HCl has been added to the buffer ? Assume that the volume is unchanged. $K_a = 1.75 \times 10^{-5}$.

SOLUTION

The concentrations of each of CH_3COOH and CH_3COO^- for the initial equilibrium

$$CH_3COOH \rightleftharpoons H^+ + CH_3COO^-$$

are 0.5 mol/m³. This is so because the acid is very slightly dissociated and CH_3COONa strongly so.

When 0.1 mol/m³ acid is added, 0.1 mol/m³ H^+ ions are produced. The equilibrium is pushed in the reverse direction to form 0.1 mol/m³ CH_3COOH. Thereby, the concentration of CH_3COOH is increased by 0.1 mol/m³ and that of CH_3COO^- ion is decreased by 0.1 mol/m³. Thus the concentration of the acid and salt when the new equilibrium is set up after the addition of HCl, are :

$$[CH_3COOH] = (0.5 + 0.1) = 0.6 \text{ mol/m}^3$$

$$[CH_3COO^-] = (0.5 - 0.1) = 0.4 \text{ mol/m}^3$$

The equilibrium expression for the equilibrium can, now, be written as

$$K_a = \frac{[H^+][CH_3COO^-]}{[CH_3COOH]} = \frac{[H^+] \times 0.4}{0.6}$$

or $$[H^+] = \frac{K_a \times 0.6}{0.4} = \frac{1.75 \times 10^{-5} \times 0.6}{0.4} = 2.63 \times 10^{-5}$$

∴ $$pH = -\log (2.63 \times 10^{-5}) = \mathbf{4.58}$$

SOLVED PROBLEM 21

A 1 m³ of solution containing 0.1 mol of CH_3COOH and 0.1 mol of CH_3COONa provides a buffer of pH 4.74. Calculate the pH of solution after the addition of 0.02 mol NaOH. $K_a = 1.8 \times 10^{-5}$.

SOLUTION

0.02 mol NaOH yields 0.02 mol of OH^- ions. These combine with 0.02 mol H^+ ions to form H_2O and the equilibrium

$$CH_3COOH \rightleftharpoons H^+ + CH_3COO^-$$

is shifted to the right (Le Chatelier's principle). Thereby 0.02 mol of CH_3COOH dissociates to form 0.02 mol H^+ needed to neutralize OH^- ions and also 0.02 mol of CH_3COO^- ions. Hence the concentration of CH_3COOH is decreased by 0.02 mol/m³ and that of CH_3COO^- ions is increased by 0.02 mol/m³.

$$\underset{(0.1-0.02)}{CH_3COOH} \rightleftharpoons H^+ + \underset{(0.1+0.02)}{CH_3COO^-}$$

Thus the concentrations in mol/m³ when the new equilibrium is set up are

$$[CH_3COOH] = 0.1 - 0.02 = 0.08$$

$$[CH_3COO^-] = 0.1 + 0.02 = 0.12$$

The equilibrium expression may now be written as :

$$K_a = \frac{[H^+][CH_3COO^-]}{[CH_3COOH]} = \frac{[H^+] \times 0.12}{0.08} = 1.8 \times 10^{-5}$$

or $$[H^+] = \frac{1.8 \times 10^{-5} \times 0.08}{0.12} = 1.2 \times 10^{-5}$$

$$pH = -\log [H^+] = -\log (1.2 \times 10^{-5}) = \mathbf{4.92}$$

TYPE 4. To obtain a buffer of given pH

Here we first find the $[H^+]$ corresponding to the required pH. Then we apply the equilibrium expression.

$$K_a = \frac{[H^+][A^-]}{[HA]}$$

or

$$[H^+] = K_a \times \frac{[HA]}{[A^-]}$$

Knowing the values of $[H^+]$, [HA] and K_a, we can calculate the ratio $[HA]/[A^-]$ or $[A^-]$ from the above expression.

SOLVED PROBLEM 22

A chemist needs a buffered solution of propanoic acid, CH_3CH_2COOH, and its salt, CH_3CH_2COONa. Calculate the ratio $[CH_3CH_2COOH]/[CH_3CH_2COONa]$ required to yield a pH of 4.30. K_a for propanoic acid is 1.3×10^{-5}.

SOLUTION

A pH of 4.30 corresponds to

$$[H^+] = 10^{-4.30} = \text{antilog}(-4.30) = 5.0 \times 10^{-5}\text{ M}$$

$$CH_3CH_2COOH \rightleftharpoons CH_3CH_2COO^- + H^+$$

$$K_a = \frac{[CH_3CH_2COO^-][H^+]}{[CH_3CH_2COOH]}$$

or

$$[H^+] = K_a \frac{[CH_3CH_2COOH]}{[CH_3CH_2COO^-]} = 5.0 \times 10^{-5}$$

$$\therefore \quad \frac{[CH_3CH_2COOH]}{[CH_3CH_2COO^-]} = \frac{5.0 \times 10^{-5}}{K_a} = \frac{5.0 \times 10^{-5}}{1.3 \times 10^{-5}} = \mathbf{3.8}$$

SOLVED PROBLEM 23

Calculate the concentration of sodium formate, HCOONa, that must be present in a 0.10 M solution of formic acid to produce a pH of 3.80. K_a for formic acid is 1.8×10^{-4}.

SOLUTION

A pH of 3.80 corresponds to

$$[H^+] = 10^{-3.80} = \text{antilog}(-3.80) = \text{antilog}(-4 + 0.20)$$
$$= 0.0001585$$
$$= 1.585 \times 10^{-4}$$

$$HCOOH \rightleftharpoons H^+ + HCOO^-$$

$$K_a = \frac{[H^+][HCOO^-]}{HCOOH}$$

$$[HCOO^-] = \frac{K_a[HCOOH]}{[H^+]}$$

Substituting the values

$$[HCOO^-] = \frac{1.8 \times 10^{-4} \times 0.10}{1.585 \times 10^{-4}}$$

$$= \mathbf{0.113\ M}$$

SOLVED PROBLEM 24

A chemistry student desires to prepare one m³ of a solution buffered at pH 9.00. How many kilograms of ammonium chloride have to be added to one m³ of 0.20 M NH_3 to make such a buffer. pK_b value of ammonia is 4.74 in the equation

$$NH_3 + H_2O \rightleftharpoons NH_4^+ + OH^-$$

SOLUTION

The equilibrium expression is

$$K_b = \frac{[NH_4^+][OH^-]}{[NH_3]}$$

Henderson-Hasselbalch equation may be used with convenience.

$$pOH = pK_b + \log\frac{[salt]}{[base]}$$

Since $\quad pH + pOH = 14$

$$pOH = 14 - 9 = 5$$

$\therefore \quad 5 = 4.74 + \log\frac{[salt]}{0.20}$

Thus $\quad \log\frac{[salt]}{0.20} = 5 - 4.74 = 0.26$

$$\frac{[salt]}{0.20} = 10^{0.26} = 1.8$$

$$[salt] = 1.8 \times 0.20 = \mathbf{0.36\ M}$$

Thus the concentration of NH_4Cl in the buffer must be 0.36 M for the buffer to hold a pH of 9.00. The amount of NH_4Cl per liter = 0.36 × 53.5 = **19.26 g or 19.26 × 10⁻³ kg**

ACID–BASE INDICATORS

In an acid-base titration the base solution can be added gradually from a burette into an acid solution contained in a receiver flask. When the amount of the base added equals the amount of the acid in the flask, the **equivalence point** or the **end-point** is reached. The end-point of a titration is shown by color change of an indicator previously added to the acid solution in the receiver flask.

An acid-base indicator is an organic dye that signals the end-point by a visual change in color.

Phenolphthalein and methyl orange are two common examples of acid-base indicators. Phenolphthalein is pink in base solution and colorless in acid solution. Thus when added to the acid solution in the receiver flask, it shows no color. As the added base is in slight excess, it becomes pink. Thus phenolphthalein signals the end-point by a color change from colorless to pink. Similarly methyl orange indicates the end-point by a color change from red (in acid) to yellow (in base).

pH range of indicators

Most indicators do not change color at a particular pH. They do so over a range of pH from two to three units. This is called the **pH range** which is different for various indicators.

TABLE 27.5. pH RANGES OF SOME ACID-BASE INDICATORS

Indicator	Color change (acid-base)		pH range
Methyl orange	red-orange	■ → ■	3.1 – 4.4
Methyl red	red-yellow	■ → □	4.4 – 6.0
Litmus	red-blue	■ → ■	5.0 –8.0
Bromothymol blue	yellow-blue	□ → ■	6.0 – 7.6
Phenolphthalein	colorless-pink	□ → ■	8.3 – 10.0

pH curves and Indicator range

During an acid-base titration the pH of the solution in the receiver flask changes with the addition of the titrant from the burette. A plot of pH against the volume of the solution being added is known as **pH curve** or **titration curve.** For illustration, the pH curve produced by titration of HCl solution with NaOH solution is shown in Fig. 27.15. As NaOH is added, the pH of the solution increases slowly at first, then rapidly in the vicinity of the equivalence point and again slowly. The equivalence point lies in the middle of the vertical portion of the curve (pH = 7). It must be clearly understood that equivalence point is the theoretical end-point of a titration. The end-point of a titration determined by a color change of the indicator in titration solution is the experimental estimate of the equivalence point.

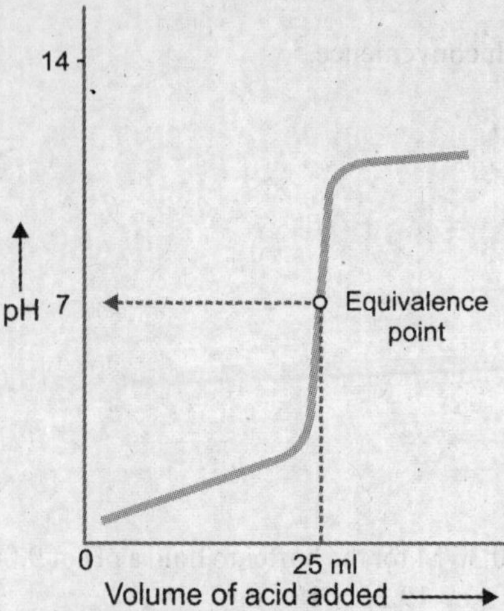

■ **Figure 27.15**
pH curve for titration of a strong base with strong acid.

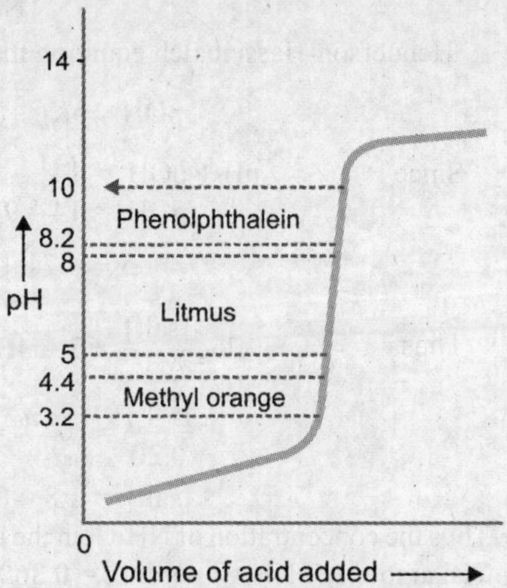

■ **Figure 27.16**
Indicator ranges for phenolphthalein, litmus and methyl orange.

The titration curve in Fig. 27.15. shows that it remains vertical around the equivalence point. From a study of this part of the curve, it is evident that the volume of m^3 used at the experimental end-point will be very nearly the same as for the equivalence point provided that : (*i*) the indicator used has a small pH range and; (*ii*) the range wholly falls on the vertical portion of the curve.

Thus, **a suitable indicator for a given titration may be defined as one which has as narrow a pH range as possible that lies entirely on the upright part of the titration curve.** For example, as shown in Fig. 27.16, phenolphthalein, litmus and methyl orange may be used as indicators for acid-base titrations.

CHOICE OF A SUITABLE INDICATOR

The choice of a suitable indicator for a particular acid-base titration depends on the nature of the acid and the base involved in the titration. We may have the titration of :

(*a*) a strong acid with a strong base
(*b*) a weak acid with a strong base
(*c*) a strong acid with a weak base
(*d*) a weak acid with weak base

Which indicator is suitable for a given titration, can be found by examining the titration curve of that titration. We have already discussed that a suitable indicator is one which has a small pH range that falls wholly on the upright portion of the titration curve.

All the pH curves given in Fig. 27.17 refer to addition of 0.1 M monoacid base to 25 ml or 25×10^{-6} of 0.1 M of monobasic acid. The equivalence point in all cases is at pH 7 when all the acid has been neutralized by the base to form a salt. If the titration is performed so that acid is added to the base, the pH curve is the mirror image of that shown.

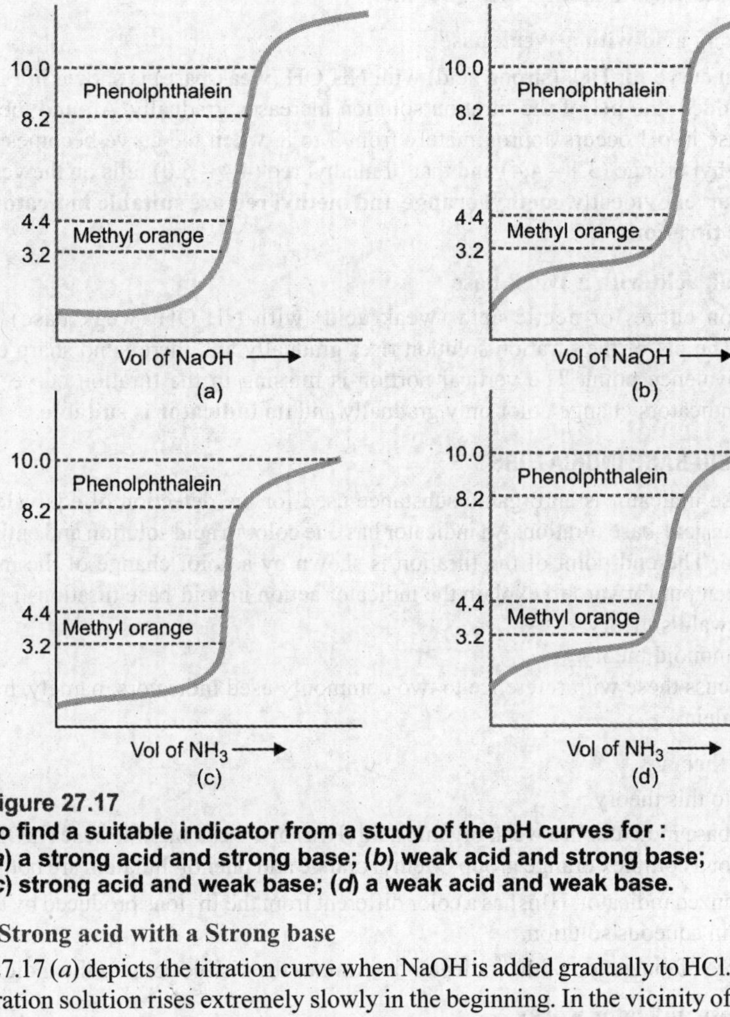

Figure 27.17
To find a suitable indicator from a study of the pH curves for : (*a*) a strong acid and strong base; (*b*) weak acid and strong base; (*c*) strong acid and weak base; (*d*) a weak acid and weak base.

Titrating a Strong acid with a Strong base

Figure 27.17 (*a*) depicts the titration curve when NaOH is added gradually to HCl. It shows that the pH of the titration solution rises extremely slowly in the beginning. In the vicinity of the equivalence point, the pH rises dramatically and the curve becomes vertical. Beyond this, the curve becomes almost flat that shows a slight rise of pH when only excess base is present in the titration solution.

The vertical portion of the curve extends from pH 3 to pH 7. The pH ranges of methyl orange (3.1– 4.4), and phenolphthalein (8.30 – 10.0) are fairly narrow and fall on the vertical curve. **Thus both methyl orange and phenolphthalein are suitable indicators** for strong acid/strong base titrations. Litmus with an exceptionally wide pH range (4.5 – 8.3) is seldom used. Its color does not change sharply from red to blue but goes through various shades of purple.

Titrating a Weak acid with a Strong base

Figure 27.17 (*b*) represents the titration curve when NaOH (strong base) is added to acetic acid (weak acid). The pH curve rises slowly in the beginning but near the equilibrium point, the pH changes abruptly from 6 to 11 and the curve becomes vertical. Beyond this the shape of the titration curve is similar to that for strong acid/strong base.

Phenolphthalein has pH range 8.3 – 10.0 that falls on the vertical part of the titration curve as marked in the figure. The pH range of methyl orange (3.1 – 4.4), on the other hand, does not fall on the vertical curve. Thus if methyl orange is used as indicator, the experimental end-point will be reached earlier than the equivalence point. Therefore for weak acid-strong base titration **Phenolphthalein is a suitable indicator, while methyl orange is not.**

Titrating a Strong acid with a Weak base

The titration curve for HCl (strong acid) with NH_4OH (weak base) is shown in Fig. 27. 17 (*c*). As NH_4OH is added, the pH of the titration solution increases gradually. Around the equivalence point, a sharp rise in pH occurs approximately from 3 to 8, when the curve becomes vertical. The pH range of methyl orange (3.1 – 4.4) and that of methyl red (4.4 – 6.0) falls on the vertical portion of the titration curve. Evidently, **methyl orange and methyl red are suitable indicators for strong acid/weak base titrations.**

Titrating a Weak acid with a Weak base

The titration curve for acetic acid (weak acid) with NH_4OH (weak base) is shown in Fig. 27.17 (*d*). The pH of the titration solution rises gradually and there is no sharp change in pH around the equivalence point. The vertical portion is missing in the titration curve. Under these conditions, all indicators change color only gradually and **no indicator is suitable.**

THEORIES OF ACID-BASE INDICATORS

An acid-base indicator is an organic substance used for the detection of equivalence point or neutral point in an acid-base titration. An indicator has one color in acid solution and entirely different in basic solution. The end-point of the titration is shown by a color change of the indicator. Two theories have been put forward to explain the indicator action in acid-base titrations :

(1) The Ostwald's theory

(2) The Quinonoid theory

We will discuss these with reference to two commonly used indicators, namely, methyl orange and phenolphthalein.

The Ostwald's theory

According to this theory :

(1) an acid-base indicator is a weak organic acid (HIn) or a weak organic base (InOH), where the letter In stands for a complex orange group. Methyl orange and phenolphthalein are both weak acids.

(2) the unionized indicator, HIn, has a color different from the In^- ions produced by the ionization of the indicator in aqueous solution.

(3) the degree of ionization of the indicator determines the visible color of the indicator solution.

How an acid-base indicator works

Let us explain the indicator action by taking example of methyl orange. Methyl orange is a weak acid and gives the following ionization equilibrium in solution.

$$\underset{\text{red}}{\text{HIn}} \rightleftharpoons H^+ + \underset{\text{yellow}}{In^-}$$

In accordance with the law of mass action,

$$K_{in} = \frac{[H^+][In^-]}{[HIn]} \qquad ...(1)$$

where K_{in} is the dissociation constant of the indicator and is called the **Indicator constant.**

The anion In^- is yellow and the nonionized form HIn^- is red. If an acid is added to the solution, the hydrogen ion concentration, $[H^+]$, in the equilibrium expression (1) increases. To maintain K_{in} constant, the equilibrium shifts to the left. Thereby the concentration of $[In^-]$ is reduced and the concentration of [HIn] increases so that the solution is red. On the other hand, upon addition of a base to the solution, H^+ ions are removed as H_2O by reacting with OH^- ions of the base. This shifts the equilibrium to the right, resulting in the increase of In^- ions that are yellow. Thus in acid solution the unionized HIn molecules predominate and the solution is pink, while in basic solution In^- ions are in excess and the solution is yellow.

Relation of Indicator color to pH

The indicator solution contains both the yellow In^- and the red HIn molecules. The actual color shade of the indicator depends on the ratio of concentration of In^- and HIn present in solution. From the equilibrium constant expression (1) we can write

$$[H^+] = K_{in}\frac{[In^-]}{[HIn]} \qquad ...(2)$$

If $[H^+]$ is large, the concentration of In^- ions is also large and the color is yellow. When $[H^+]$ is small, [HIn] is large and the solution is red. At the equivalence point, [In] = [HIn] and the color is orange (red + yellow). Obviously the **indicator color is controlled by hydrogen ion concentration or pH of the solution.**

Taking logarithms and using definition of pH and K_{in}, the expression (2) can be converted to the Henderson-Hasselbalch equation.

$$pH = pK_{in} + \log\frac{[In^-]}{[HIn]} \qquad ...(3)$$

At the equivalence point, $[In^-]$ = [HIn] and methyl orange in solution is orange. Then,

$$pH = pK_{in}$$

The numerical value of the indicator constant K_{in} for methyl orange is 3.6 and the pH of the orange solution is, therefore, about 4. As the values of K_{in} for the various indicators are different, they will have intermediate intense colors (**middle tint**) at different pH values. When a base is added to an acid solution in a titration, the color change of the indicator is gradual. It just becomes visible to the human eye when $[In^-]/[HIn] = \frac{1}{10}$ and pH calculated from equation (3) is 3.1. The color of the indicator continues to change till $[In^-]$ / [HIn] = 10 when pH is 4.4. The pH range between 3.1 (red) and 4.4 (yellow) is called the **color change interval** of methyl orange. The visible indicator color change takes place between these pH values.

Indicator action of Phenolphthalein

It can be explained as in case of methyl orange. It is a weak acid and exists as the following equilibrium in solution,

$$\underset{\text{colourless}}{HIn} \rightleftharpoons H^+ + \underset{\text{pink}}{In^-}$$

HIn molecules are colorless, while In^- ions are pink. Thus in acid solution, phenolphthalein is colorless and in basic solution it is pink. The value of K_{in} = 9.6 and the pH of the intermediate intense pink tint is also 9.6. The color change interval of phenolphthalein is 8.1 – 10.0.

Quinonoid theory of Indicator color change

The Ostwald's theory takes care of the quantitative aspect of indicator action adequately. The Quinonoid theory, on the other hand, tells us the cause of color change of an indicator in acid-base solutions. It lays down that :

(1) the unionized HIn molecule and the anion In^- are tautomeric forms of the indicator which is an organic dye.

(2) one tautomeric form possesses the quinonoid structural unit and is called the **quinonoid form.**

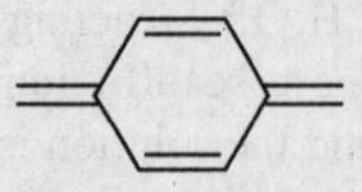

It has a deep color. The other form has a lesser coloring group, say, –N = N – and or simply benzene rings and is called the **benzenoid form.** This form has a light color or no color.

(3) the color change of the indicator occurs when one tautomeric form is transformed into the other due to change of pH of the solution.

Let us illustrate the Quinonoid theory by taking example of methyl orange and phenolphthalein.

Methyl Orange. The red quinonoid form of methyl orange exists in acid solution. It is converted to yellow benzenoid form when pH alters to the basic side.

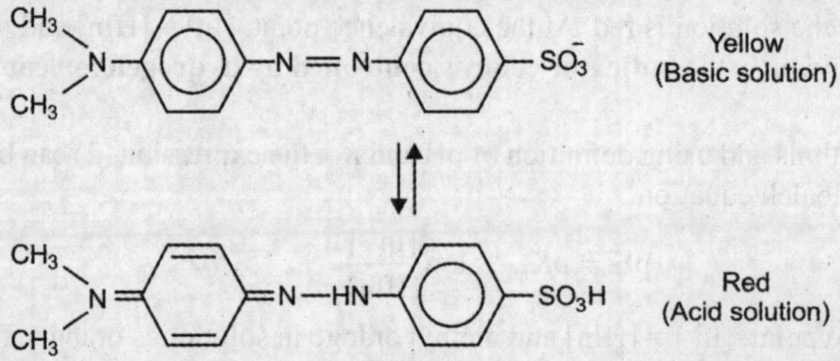

Phenolphthalein. Phenolphthalein exists in two tautomeric forms : (*i*) the benzenoid form which is yellow and present in basic solution; and (*ii*) the quinonoid form which is pink and present in acid solution.

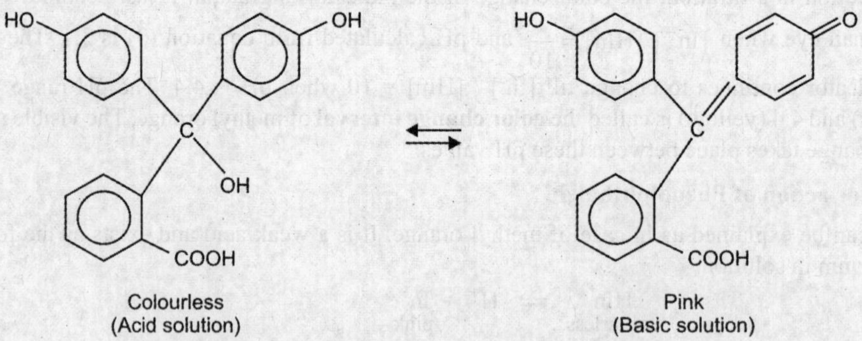

ADDITIONAL SOLVED PROBLEMS

SOLVED PROBLEM 25

Dissociation constant of propionic acid at 298 K is 1.34 × 10⁻⁵. Calculate the degree of dissociation and *pH* in 0.1 M acid.

SOLUTION : (*a*) To calculate the degree of dissociation of propionic acid

Formula used

$$K_a = \frac{\alpha^2 C}{1-\alpha} \text{ or } \alpha^2 C \qquad [\because 1-\alpha \approx 1]$$

Quantities given

$K_a = 1.34 \times 10^{-5}$ $C = 0.1$ M

Substitution of values

$$1.34 \times 10^{-5} = \alpha^2 \times 0.1$$

or $$\alpha^2 = \frac{1.34 \times 10^{-5}}{0.1}$$

or $$\alpha = \sqrt{1.34 \times 10^{-4}}$$

$$= 0.0115$$

(*b*) To calculate the pH of 0.1 M propionic acid

Formula used

$$pH = -\log [H^+]$$

Quantities given

$$[H^+] = 0.0115 \text{ M}$$

Substitution of value

$$pH = -\log (0.0115)$$
$$= \mathbf{1.939}$$

SOLVED PROBLEM 26

Calculate the *pH* of a 0.1 M acetic acid when it is half neutralized by NaOH solution ($K_a = 1.7 \times 10^{-5}$).

SOLUTION : Acetic acid dissociates as

$$CH_3COOH \rightleftharpoons CH_3COO^- + H^+$$

$$K_a = \frac{[CH_3COO^-]\,[H^+]}{[CH_3COOH]}$$

Let x be the concentration of H^+ ions at equilibrium, then

$$[H^+] = [CH_3COO^-] = x$$

$$[CH_3COOH] = 0.05 - x \cong 0.05$$

Substituting the values we get

$$K_a = \frac{x \times x}{0.05}$$

Quantity given

$$K_a = 1.7 \times 10^{-5}$$

Substitution of values

$$1.7 \times 10^{-5} = \frac{x^2}{0.05}$$

or $$x^2 = 8.5 \times 10^{-7}$$

or $$x = 9.219 \times 10^{-4}$$

Now $$pH = -\log [H^+]$$

$$= -\log (9.219 \times 10^{-4})$$

$$= \mathbf{3.0353}$$

SOLVED PROBLEM 27

Calculate the pH of the buffer solution containing 0.04 M NH_4Cl and 0.02 M NH_4OH. K_b for NH_4OH is 1.8×10^{-5}.

SOLUTION

Formula used

$$pOH = pK_b + \log \frac{[Salt]}{[Base]}$$

Quantities given

$pK_b = -\log\ 1.8 \times 10^{-5} = 4.7447$ $\quad [Salt] = 0.04$ M $\quad [Base] = 0.02$ M

Substitution of values

$$\begin{aligned} pOH &= 4.7447 + \log \frac{0.04}{0.02} \\ &= 4.7447 + \log 2 \\ &= 4.7447 + 0.3010 \\ &= 5.0457 \end{aligned}$$

Since

$$\begin{aligned} pH + pOH &= 14 \\ pH &= 14 - pOH \\ &= 14 - 5.0457 \\ &= \mathbf{8.9543} \end{aligned}$$

SOLVED PROBLEM 28

Calculate the pH of solution having hydrogen ion concentration 2.5×10^{-3} M.

SOLUTION

Formula used

$$pH = -\log [H^+]$$

Quantity given

$$[H^+] = 2.5 \times 10^{-3}\ \text{M}$$

Substitution of values

$$\begin{aligned} pH &= -\log (2.5 \times 10^{-3}) \\ &= -\log (0.0025) \\ &= -(-2.6020) \\ &= \mathbf{2.6020} \end{aligned}$$

SOLVED PROBLEM 29

Calculate the pH of 0.01 M NH_4OH solution. Dissociation constant of NH_4OH is 1.8×10^{-5}.

SOLUTION : (*a*) To calculate the concentration of OH^- ions.

$$\underset{0.01-\alpha}{NH_4OH} \rightleftharpoons \underset{\alpha}{NH_4^+} + \underset{\alpha}{OH^-}$$

Formula used

$$\begin{aligned} K_b &= \frac{[NH_4^+]\ [OH^-]}{[NH_4OH]} \\ &= \frac{\alpha^2}{0.01 - \alpha} \\ &= \frac{\alpha^2}{0.01} \qquad [\because 0.01 - \alpha \approx 0.01] \end{aligned}$$

or $$\alpha = \sqrt{0.01 \times K_b}$$

Substitution of values

$$\alpha = \sqrt{0.01 \times 1.8 \times 10^{-5}}$$

or $$\alpha = \sqrt{1.8 \times 10^{-7}}$$

$$[OH^-] = \alpha = 4.2426 \times 10^{-4}$$

(*b*) To calculate pOH value

Formula used

$$pOH = -\log [OH^-]$$

Quantity given

$$[OH^-] = 4.2426 \times 10^{-4}$$

Substitution of value

$$pH = 14 - pOH$$
$$= 14 - 3.3723$$
$$= \mathbf{10.6277}$$

SOLVED PROBLEM 30

Calculate the *pH* volume of a solution obtained by mixing 0.083 mol of acetic acid and 0.091 mol of sodium acetate and making the volume 500 ml. K_a for acetic acid is 1.75×10^{-5}.

SOLUTION

Formula used

$$pH = pK_a + \log \frac{[Salt]}{[Acid]}$$

Quantities given

$pK_{a,} = -\log 1.75 \times 10^{-5} = 4.7569$ $[Salt] = 2 \times 0.091 \text{ M} = 0.182$

$[Acid] = 2 \times 0.083 \text{ M} = 0.166$

Substitution of values

$$pH = 4.7569 + \log \frac{0.182}{0.166}$$
$$= 4.7569 + \log 1.0964$$
$$= 4.7569 + 0.0399$$
$$= \mathbf{4.7968}$$

SOLVED PROBLEM 31

What will be the H^+ ion concentration of solutions having (i) $pH = 5.0$ and (ii) $pH = 7.8$?

SOLUTION : (*a*) To calculate the $[H^+]$ of solution having pH = 5

Formula used

$$pH = -\log [H^+]$$

or $$[H^+] = \text{Antilog} (-pH)$$

Quantity given

$$pH = 5$$

Substitution of value

$$[H^+] = \text{Antilog } (-5)$$
$$= \mathbf{1 \times 10^{-5}\ M}$$

(*b*) To calculate the [H⁺] of solution having pH = 7.8

Formula used

$$pH = -\log [H^+]$$

or $$[H^+] = \text{Antilog } (-pH)$$

Quantity given

$$pH = 7.8$$

Substitution of value

$$[H^+] = \text{Antilog } -7.8$$
$$= \mathbf{1.584 \times 10^{-8}}$$

SOLVED PROBLEM 32

Calculate the amount of NH_3 and NH_4Cl required to prepare a buffer solution of *pH* 9.0 when total concentration of buffering reagent is 0.6 mol m^{-3}. (pK_b for NH_3 = 4.7) Also take log 2 = 0.3.

SOLUTION

Formula used

$$pOH = pK_b + \log \frac{[Salt]}{[Base]}$$

Quantities given

$pOH = 14 - pH = 14 - 9.0 = 5.0$ $\qquad pK_b = 4.7$

Substitution of values

$$5 = 4.7 + \log \frac{[Salt]}{[Base]}$$

or $$\log \frac{[Salt]}{[Base]} = 0.3$$
$$= \log 2$$

or $$\frac{[Salt]}{[Base]} = 2$$

Let the amount of NH_4Cl be x mol m^{-3}. Then

$$\frac{x}{0.6 - x} = 2$$

or $$x = 1.2 - 2x$$

or $$3x = 1.2$$
$$= 0.4 \text{ mol m}^{-3}$$

Thus the amount of NH_4Cl = **0.4 mol m^{-3}**

and the amount of NH_3 = 0.6 – 0.4 = **0.2 mol m^{-3}**

EXAMINATION QUESTIONS

1. Define or explain the following terms :
   (*a*) Arrhenius acids (*b*) Arrhenius bases
   (*c*) Bronsted acids (*d*) Bronsted bases
   (*e*) Lewis acids (*f*) Lewis bases
   (*g*) pH of solutions (*h*) Buffer solution
2. (*a*) What are strong and weak electrolytes?
   (*b*) Define pH of a solution. Calculate pH of 0.085 N monobasic acid which dissociates 25%.
   Answer. (*b*) 1.67
3. (*a*) Compare the Lewis theory of acids and bases with the Bronsted-Lowry concept.
   (*b*) What are conjugate acid-base pairs? Give two examples.
4. State and explain Lowry-Bronsted theory and Lewis theory of acids and bases. In what way Lewis acid differs from Bronsted acid?
5. A buffer solution is prepared by mixing 3 kg of acetic acid and 1.30 kg of sodium acetate (CH_3COONa $3H_2O$) and making the total volume to one m^3. Calculate the pH of the buffer ($K_a = 1.8 \times 10^{-5}$; H = 1; C = 12; O = 16; Na = 23)
   Answer. 5.4437
6. Do you expect the pH of pure water at 373 K to be less than 7, or more than 7. Explain your answer. *(Nagpur BSc)*
7. Why a solution of NaCl does not act as buffer? *(Himachal Pradesh BSc)*
8. How will you explain the buffer action of aqueous solution of ammonium acetate? *(Punjabi BSc)*
9. (*a*) What is a Buffer solution? Give examples.
   (*b*) Explain the buffer action of an acidic buffer. *(Madurai BSc)*
10. Explain how does a buffer solution work? *(Kathmandu BSc)*
11. What are acids and bases? Explain the different theories of acids and bases. *(Kathmandu BSc)*
12. Explain : The second dissociation constant of a dibasic acid is always greater than its first dissociation constant. *(Delhi BSc)*
13. (*a*) Define Lewis acids and Lewis bases. Give one example of each.
    (*b*) What do you understand by conjugate acids and bases in the following reactions :
    (*i*) $NH_3 + H_2O \rightleftharpoons NH_4^+ + OH^-$
    (*ii*) $HNO_3 + H_2O \rightleftharpoons H_3O^+ + NO_3^-$
    (*c*) Calculate the pH of a solution having hydrogen ion concentration 2.5×10^{-3} M.
    Answer. 2.6 *(Arunachal BSc)*
14. What are buffer solutions? Derive Henderson's equation. *(Purvanchal BSc)*
15. Explain with an example why pH of a buffer solution does not change significantly on small addition of acids or bases. *(Guru Nanak Dev BSc)*
16. Derive the following equations :
    (*i*) $\text{pH} = \text{p}K_a + \log \dfrac{[\text{Salt}]}{[\text{Acid}]}$
    (*ii*) $\text{pH} = \left[-\dfrac{1}{2}(\log K_w + \log K_a - \log C)\right]$ *(Panjab BSc)*
17. A solution is found to contain 0.63 kg of nitric acid per 100 m^3 of the solution. If the acid is completely dissociated, calculate the pH of the solution.
    Answer. 1 *(Delhi BSc)*

18. What is meant by buffer capacity? What is its value when 0.01 mol of NaOH is added to 9 buffer solution to change its pH value from 4.745 to 4.832?
Answer. 0.11494 (*Sambalpur BSc*)

19. Calculate the pH of 0.01 M solution of acetic acid. K_a for CH_3COOH is 1.8×10^{-5} at 298 K.
Answer. 3.3725 (*Delhi BSc*)

20. The buffer solution of pH = 5 is to be prepared by using acetic acid and sodium acetate. Calculate the ratio [Salt] : [Acid] that must be used. $K_a = 1.8 \times 10^{-5}$.
Answer. 0.5691 (*Goa BSc*)

21. (*a*) What do you understand by a buffer solution? Give examples of such solutions.
(*b*) Given the dissociation constant of acetic acid at 298 K as 1.8×10^{-5}, calculate the pH of a solution containing 0.185 kg mol acetic acid and 3.015 kg mol sodium acetate per m^3.
Answer. (*b*) 5.956 (*Dibrugarh BSc*)

22. What is a buffer solution? Calculate the pH of 0.001 M HCl and 0.01 M NaOH.
Answer. 3; 12 (*Agra BSc*)

23. An aqueous solution of aniline of concentration 0.24 M is prepared. What concentration of sodium hydroxide is needed in this solution so that anilinium ion concentration remains at 1×10^{-8} M ?
(K_a for $C_6H_5NH_3^+ = 2.4 \times 10^{-5}$)
Answer. 1×10^{-2} (*Punjabi BSc*)

24. Calculate the *pH* of the following solution :
(i) 0.0002 M HCl
(ii) 0.1 M NH_4OH ($K_b = 1.8 \times 10^{-5}$)
Answer. (i) 3.6990 ; (ii) 11.13 (*Madurai BSc*)

25. Calculate the *pH* of a solution made by mixing 50 m^3 of 0.01 M $Ba(OH)_2$ solution with 50 m^3 of water.
Answer. 12 (*Mysore BSc*)

26. What is the pH of 0.50 M aqueous NaCN solution ? pK_b of CN^- is 4.70`
Answer. 11.5 (*Baroda BSc*)

27. Calculate the hydrogen ion concentration in a sample of blood having *pH* = 7.4.
Answer. 3.98×10^{-8} M (*Goa BSc*)

28. (*a*) Discuss the Bronsted-Lowry concept and Lewis concept of acids and bases.
(*b*) Write three methods for determining the strength of acids. (*Bundelkhand BSc*)

29. Calculate the degree of hydrolysis of ammonium acetate. The dissociation constant of NH_4OH is 1.8×10^{-5} and that of acetic acid is 1.8×10^{-5} and $K_w = 1.0 \times 10^{-14}$.
Answer. 0.55×10^{-2} (*Rajasthan BSc*)

30. (*a*) Derive Henderson's equation to calculate the pH of a buffer solution.
(*b*) Calculate the pH of a 0.1M acetic acid when it is half neutralized by NaOH solution ($K_a = 1.7 \times 10^{-5}$)
Answer. (*b*) 4.7695 (*Arunachal BSc*)

31. Calculate the pH of the buffer solution containing 0.04 M NH_4Cl and 0.02 M NH_4OH. K_b for NH_4OH is 1.8×10^{-5}.
Answer. 8.9543 (*Purvanchal BSc*)

32. Calculate the pH of solution having hydrogen ion concentration 2.5×10^{-3} M.
Answer. 2.6020 (*Agra BSc*)

33. (*a*) Calculate the pH of 0.01 M NH_4OH solution. Dissociation constant of NH_4OH is 1.8×10^{-5}.
(*b*) Draw and explain strong acid-strong base pH neutralization curve. Why weak acid-weak base titration is difficult to carry out?
Answer. (*a*) 10.6275 (*Madras BSc*)

34. (*a*) Explain the buffer action of a mixture of ammonium hydroxide and ammonium chloride solution.
(*b*) The pH scale ranges from 0–14. Explain. (*Kalyani BSc*)

MULTIPLE CHOICE QUESTIONS

1. An acid is a compound that gives H^+ ions in water and a base is a compound that gives OH^- ions in water. This concept was given by
   (a) Arrhenius (b) Lewis
   (c) Bronsted (d) Lowry
   Answer. (a)
2. Which one of the following is the limitation of Arrhenius concept?
   (a) free H^+ and OH^- ions do not exist in water (b) this concept is limited to water only
   (c) some bases do not contain OH^- ions (d) all of these
   Answer. (d)
3. According to Bronsted-Lowry concept, an acid is a substance that
   (a) accepts a proton (b) releases a proton
   (c) accepts an electron pair (d) releases an electron pair
   Answer. (b)
4. A substance accepts a proton. According to Bronsted-Lowry concept it is
   (a) an acid (b) a base
   (c) a neutral substance (d) amphoteric
   Answer. (b)
5. In the following reaction
   $H_2O + HCl \rightarrow H_3O^+ + Cl^-$
   the species that acts as a Bronsted base is
   (a) H_2O (b) HCl
   (c) H_3O^+ (d) Cl^-
   Answer. (b)
6. When calcium oxide is dissolved in water, following reaction takes place
   $O^{2-} + H_2O \rightarrow 2OH^-$
   The Bronsted acid is
   (a) O^{2-} (b) H_2O
   (c) OH^- (d) none of these
   Answer. (b)
7. In the following hypothetical reaction
   $HA + B^- \rightleftharpoons HB + A^-$
   The conjugate base of the acid HA is
   (a) B^- (b) HA
   (c) A^- (d) none of these
   Answer. (c)
8. In the reaction between NH_3 and HCl
   $HCl + NH_3 \rightleftharpoons NH_4^+ + Cl^-$
   the conjugate acid of NH_3 is
   (a) HCl (b) NH_4^+
   (c) Cl^- (d) none of these
   Answer. (b)
9. A weak base has _______ conjugate acid and a weak acid has a _______ conjugate base
   (a) strong, strong (b) weak, strong
   (c) strong, weak (d) weak, weak
   Answer. (a)

10. Molecules or ions that can behave both as Bronsted acid and base are called

(*a*) monoprotic acids (*b*) polyprotic acids

(*c*) amphiprotic substances (*d*) polyprotic bases

Answer. (*c*)

11. HCO_3^- is an example of

(*a*) conjugate acid (*b*) conjugate base

(*c*) amphiprotic ion (*d*) amphoteric ion

Answer. (*c*)

12. The strength of a Bronsted acid depends upon its tendency to ______ a proton

(*a*) gain (*b*) donate

(*c*) react with (*d*) none of these

Answer. (*b*)

13. According to Lewis concept an acid is ______ acceptor

(*a*) proton (*b*) base

(*c*) electron (*d*) electron pair

Answer. (*d*)

14. All cations and molecules that are short of an electron pair act as ______

(*a*) Lewis acids (*b*) Lewis bases

(*c*) Bronsted acid (*d*) Bronsted bases

Answer. (*a*)

15. According to Lewis concept, a reaction between an acid and a base is the transfer of

(*a*) a proton from acid to base (*b*) OH^- ion from bases to acid

(*c*) electron pair from acid to base (*d*) electron pair from base to acid

Answer. (*d*)

16. The autoionization of water can be represented by

(*a*) $H_2O \rightleftharpoons H^+ + OH^-$ (*b*) $H_2O + H^+ \rightleftharpoons H_3O^+ + OH^-$

(*c*) $H_2O + H_2O \rightleftharpoons H_3O^+ + OH^-$ (*d*) $H_2O + HA \rightleftharpoons H_3O^+ + A^-$

Answer. (*c*)

17. The strength of an acid depends upon

(*a*) the concentration of H^+ ions in aqueous solution

(*b*) the concentration of electrons pairs in aqueous solution

(*c*) the tendency to release the electron pair

(*d*) the tendency to gain the electron pair

Answer. (*a*)

18. The dissociation constant of an acid HA is given by

(*a*) $K_a = \dfrac{[H^+][A^-]}{[HA]}$ (*b*) $K_a = \dfrac{[H^+]^2[A^-]^2}{[HA]}$

(*c*) $K_a = \dfrac{[H^+][A]^2}{[HA]}$ (*d*) $K_a = \dfrac{[H^+]^2[A]}{[HA]}$

Answer. (*a*)

19. For a weak acid with α as its degree of dissociation, the value of dissociation constant is given by (*C* is concentration of acid in mol/m³)

(*a*) $K_a = C\alpha$ (*b*) $K_a = C\alpha^2$

(*c*) $K_a = C^2\alpha$ (*d*) $K_a = C^2\alpha^2$

Answer. (*b*)

20. The strength of a base depends upon the

(*a*) concentration of H^+ ions in aqueous solution

(*b*) concentration of OH^- ions in aqueous solution

(*c*) concentration of electron pairs in aqueous solution

(*d*) concentration of the base

Answer. (*b*)

21. The dissociation constant K_b for a base BOH is given by the expression

(*a*) $K_b = \frac{[B^+][H^+]}{[BOH]}$ (*b*) $K_b = \frac{[B^+][OH^-]}{[BOH]}$

(*c*) $K_b = \frac{[B^+][OH^-]^2}{[BOH]}$ (*d*) $K_b = \frac{[B^+]^2[OH^-]}{[BOH]}$

Answer. (*b*)

22. Which is correct about pH?

(*a*) $pH = -\log[H^+]$ (*b*) $pH = \log \frac{1}{[H^+]}$

(*c*) $pH = 10^{-pH}$ (*d*) all of these

Answer. (*d*)

23. The concept of pH was introduced by

(*a*) Arrhenius (*b*) Bronsted

(*c*) Lewis (*d*) Sorensen

Answer. (*d*)

24. Which of the following is correct?

(*a*) pH – pOH = 14 (*b*) pH + pOH = 7

(*c*) pH – pOH = 0 (*d*) pH + pOH = 14

Answer. (*d*)

25. The pH of 0.001 M HCl is

(*a*) 0.001 (*b*) 10^{-3}

(*c*) 3 (*d*) –3

Answer. (*c*)

26. The pH of a solution of HCl is 1. The amount of acid present in one m^3 of the solution will be

(*a*) 3.65 kg/m^3 (*b*) 0.365 kg/m^3

(*c*) 36.5 kg/m^3 (*d*) 1.0 kg/m^3

Answer. (*a*)

27. The pH of 0.1 M NaOH solution will be

(*a*) 1 (*b*) 0.1

(*c*) 13 (*d*) 14

Answer. (*c*)

28. The pOH of 0.1 M KOH is

(*a*) 0.1 (*b*) 1.0

(*c*) –1.0 (*d*) 13

Answer. (*b*)

29. The fraction of total molecules which is ionized in a solution of an electrolyte is known as

(*a*) dissociation constant (*b*) mole fraction of the electrolyte

(*c*) degree of dissociation (*d*) electrolyte constant

Answer. (*c*)

30. Acetic acid is a weak electrolyte because

(*a*) its molecular mass is high (*b*) it exists as dimer
(*c*) it is highly unstable (*d*) it has low ionization

Answer. (*d*)

31. The strength of an acid depends on the

(*a*) number of H atoms present in its molecule (*b*) concentration of protons given by it on ionization
(*c*) concentration of water (*d*) concentration of acid

Answer. (*b*)

32. Which of the following does not act as Bronsted acid?

(*a*) CH_3COO^- (*b*) NH_4^+
(*c*) HCO_3^- (*d*) HSO_3^-

Answer. (*a*)

33. The pH of a solution is 3, the hydrogen ion concentration is mol m^{-3} is

(*a*) 1×10^{-1} (*b*) 2×10^{-1}
(*c*) 1×10^{-3} (*d*) 3×10^{-2}

Answer. (*c*)

34. The pOH and pH of 0.1 M aqueous HCl will be

(*a*) 13, 1 (*b*) 1, 13
(*c*) 14, 0 (*d*) 0, 14

Answer. (*a*)

35. The pH of a solution increases from 1 to 2. The concentration of H^+ ions

(*a*) decreases (*b*) increases
(*c*) remains the same (*d*) becomes zero

Answer. (*a*)

36. On the addition of a small amount of an acid or a base, the pH value of a buffer solution ________

(*a*) increases (*b*) decreases
(*c*) remains unaltered (*d*) none of these

Answer. (*c*)

37. The Henderson equation for an acidic buffer is

(*a*) $pH = pK_a + \log \frac{[salt]}{[acid]}$ (*b*) $pOH = pK_a + \log \frac{[salt]}{[acid]}$
(*c*) $pH = pK_a - \log \frac{[salt]}{[acid]}$ (*d*) $pH = pK_a - \log \frac{[acid]^2}{[acid]}$

Answer. (*a*)

38. For a basic buffer, the Henderson equation is

(*a*) $pH = pK_b + \log \frac{[salt]}{[base]}$ (*b*) $pOH = pK_b - \log \frac{[salt]}{[base]}$
(*c*) $pOH = pK_b + \log \frac{[salt]}{[base]}$ (*d*) $pH = pK_b + \log \frac{[salt]}{[base]}$

Answer. (*c*)

39. When a small quantity of HCl is added to a mixture of sodium acetate and acetic acid (equimolar), the pH value

(*a*) increases (*b*) decreases
(*c*) remains the same (*d*) increases abruptly

Answer. (*c*)

40. The following equilibrium exists in aqueous solution $CH_3COOH \rightleftharpoons H^+ + CH_3COO^-$. When dil HCl is added

(*a*) the equilibrium constant will increase
(*b*) acetate ion concentration decreases
(*c*) acetate ion concentration increases
(*d*) the equilibrium constant will decrease

Answer. (*b*)

41. The compound that is not a Lewis acid is

(*a*) BF_3
(*b*) $AlCl_3$
(*c*) $BeCl_2$
(*d*) $BaCl_2$

Answer. (*d*)

42. The ionic product of water will increase if

(*a*) temperature is reduced
(*b*) temperature is increased
(*c*) pressure is decreased
(*d*) pressure is increased

Answer. (*b*)

43. The solution A has pH equal to 2 and solution B has pH of 4. Which one of the following is true?

(*a*) solution B is 100 times more acidic than solution A
(*b*) solution B is twice more acidic than solution A
(*c*) solution A is twice more acidic than solution B
(*d*) solution A is 100 times more acidic than solution B

Answer. (*d*)

44. The pH of 10^{-10} M solution of NaOH is nearly

(*a*) –10
(*b*) 10
(*c*) 7
(*d*) zero

Answer. (*c*)

45. The pH of a solution is enhanced from 2 to 3. The concentration of H^+ in the new solution

(*a*) is three times the original solution
(*b*) is about 1.5 times the original solution
(*c*) decreases 10 times
(*d*) increases 10 times

Answer. (*c*)

46. Which of the following is not a buffer solution?

(*a*) $NH_4Cl + NH_4OH$
(*b*) $CH_3COONa + CH_3COOH$
(*c*) NaOH + HCl
(*d*) HCOONa + HCOOH

Answer. (*c*)

47. Which one of the following statement is correct?

(*a*) $pK_w = 14$ at all temperatures
(*b*) pK_w = pH at 273 K
(*c*) pK_w decreases with increase temperature
(*d*) pK_w increases with increase in temperature

Answer. (*d*)

48. All nucleophiles are

(*a*) Arrhenius acid
(*b*) Bronsted bases
(*c*) Lewis acids
(*d*) Lewis bases

Answer. (*d*)

49. The best indicator for titrating HCl with NH_4OH is

(*a*) litmus
(*b*) phenolphthalein
(*c*) methyl orange
(*d*) phenol red

Answer. (*c*)

28 Salt Hydrolysis

CONTENTS

WHAT IS HYDROLYSIS?

The salt of a weak acid, HA and a strong base dissolves in water to form the anion A^-. The A^- anion tends to react with water by drawing a proton (H^+) from its molecule to form the unionized molecule.

$$A^- + H-O-H \rightleftharpoons HA + \bar{O}-H \quad ...(1)$$

Similarly, the salt of a weak base, BOH, and a strong acid dissolves in water to form the cation B^+. The cation B^+ reacts with water by accepting OH^- ions from its molecule.

$$B^+ + H-O-H \rightleftharpoons BOH + H^+ \quad ...(2)$$

The reaction of an anion or cation with water accompanied by cleavage of O–H bond is called Hydrolysis.

The term hydrolysis is derived from *hydro*, meaning water, and *lysis*, meaning breaking. It may be noted that in **anionic hydrolysis** shown in (1) the solution becomes slightly basic (pH > 7) due to the generation of excess OH^- ions. In **cationic hydrolysis** shown in (2), there is excess of H^+ ions which makes the solution slightly acidic (pH < 7).

BRONSTED–LOWRY CONCEPT OF HYDROLYSIS

HA and A^- are conjugate acid-base pair

$$\underset{\text{weak acid}}{HA} + H_2O \rightleftharpoons H_3O^+ + \underset{\text{conjugate base}}{A^-}$$

Since HA is a weak acid (poor proton donor), its conjugate base, A^-, must be relatively strong (good proton acceptor). Owing to this fact, A^- ions tend to react with water by accepting proton from the later to form HA molecule (anionic hydrolysis),

$$A^- + H_2O \longrightarrow HA + OH^-$$

The presence of OH^- ions makes the solution basic.

Similarly, BOH and B^+ are a conjugate acid-base pair. Since BOH is a weak base, its conjugate acid, B^+, would be relatively strong. Thus B^+ would accept OH^- ions from water to form BOH molecules.

$$B^+ + H_2O \longrightarrow BOH + H^+$$

The presence of excess H^+ ions makes the solution acidic.

Why NaCl Solution is Neutral?

NaCl dissociates in water to give the anion Cl^-. HCl and Cl^- constitute an acid-base conjugate pair.

$$HCl + H_2O \rightleftharpoons H_3O^+ + \underset{\text{conjugate base}}{Cl^-}$$

Since HCl is a strong acid, Cl^- is very weak base. Cl^- is unable to accept a proton (H^+) from an acid, particularly water. That is why Cl^- does not hydrolyze. It cannot generate OH^- ions as follows:

$$Cl^- + H_2O \not\longrightarrow HCl + OH^-$$

The pH of sodium chloride solution remains unaffected.

EXAMPLES OF HYDROLYSIS

The different salts may be classified into the following types according to their hydrolytic behavior:

(1) Salts of Weak acids and Strong bases

(2) Salts of Weak bases and Strong acids

(3) Salts of Weak acids and Weak bases

We have already considered that a salt of strong acid and strong base *e.g.*, NaCl, does not show hydrolysis.

Salts of Weak acids and Strong bases

Sodium acetate, CH_3COONa, and sodium cyanide, NaCN, are examples of this type of salts.

Sodium acetate, CH_3COONa. This is a salt of the weak acid, CH_3COOH, and strong base, NaOH. It ionizes in aqueous solution to form the anion CH_3COO^-. Being the conjugate base of a weak acid, CH_3COOH, it is a relatively strong base. Thus CH_3COO^- accepts H^+ ion from water and undergoes hydrolysis.

$$CH_3COO^- + H_2O \longrightarrow CH_3COOH + OH^-$$

The resulting solution is slightly basic due to excess OH^- ions present.

Sodium Cyanide, NaCN. It is the salt of a weak acid, HCN, and a strong base, NaOH. It ionizes to form CN^- anions. Being conjugate base of a weak acid, CN^- is relatively strong base. Thus the anion CN^- accepts a H^+ ion from water and undergoes hydrolysis.

$$CN^- + H_2O \rightleftharpoons HCN + OH^-$$

The solution becomes basic due to the generation of OH^- ions.

Salts of Weak bases and Strong acids

Some salts of weak bases and strong acids undergo cationic hydrolysis and yield slightly acidic solutions.

Ammonium chloride is a typical example of this class of salts. It is the salt of a weak base. NH_4OH, and strong acid, HCl. It ionizes in aqueous solution to form the cation, NH_4^+.

$$NH_4OH \rightleftharpoons \underset{\text{conjugate acid}}{NH_4^+} + OH^-$$

NH_4^+ is a Bronsted conjugate acid of the weak base NH_4OH. Therefore, it is a relatively strong acid. It accepts OH^- ion from water (H_2O) and forms the unionized NH_4OH and H^+ ion.

$$NH_4^+ + H_2O \rightleftharpoons NH_4OH + H^+$$

The accumulation of H^+ ions in solution makes it acidic.

The other examples of this type of salts are ferric chloride, aluminium chloride, and copper sulfate.

Salts of Weak acids and Weak bases

The examples of this type of salts are ammonium acetate, ammonium cyanide and ammonium fluoride. Both the anion and the cation produced by ionization of the salt undergo hydrolysis. The resulting solution is neutral, basic or acidic depending on the relative hydrolysis of the anions and the cations.

Ammonium acetate, CH_3COONH_4. It is the salt of weak acid, CH_3COOH, and weak base, NH_4OH. In aqueous solution it ionizes to form the anion CH_3COO^- and the cation NH_4^+. Since the acid and the base are both weak, their conjugate base (CH_3COO^-) and conjugate acid (NH_4^+) are relatively strong. They accept H^+ and OH^- ions respectively from water and undergo considerable hydrolysis.

$$\underset{\text{conjugate base}}{CH_3COO^-} + H_2O \rightleftharpoons CH_3COOH + OH^- \quad ...(1)$$

$$\underset{\text{conjugate acid}}{NH_4^+} + H_2O \rightleftharpoons NH_4OH + H^+ \quad ...(2)$$

The overall hydrolysis may be represented as

$$CH_3COO^- + NH_4^+ + H_2O \rightleftharpoons CH_3COOH + NH_4OH$$

We have stated above that pH of the resulting solution will depend on the relative extent of anionic hydrolysis (1) and cationic hydrolysis (2). If both the ions react to the same extent (as shown for CH_3COONH_4), $[OH^-] = [H^+]$ and the solution is neutral. If the cation reacts to a greater extent, the solution is slightly acidic. If the anion is a little more reactive, the solution will be basic. Thus, a solution of CH_3COONH_4 is neutral, a solution of NH_4CN is slightly basic and a solution of NH_4F is slightly acidic.

QUANTITATIVE ASPECT OF HYDROLYSIS

Hydrolysis is a reversible reaction. The equilibrium constant derived by application of Law of Mass action to a hydrolysis (or hydrolytic) reaction is called the **Hydrolysis constant** or **Hydrolytic constant.** The hydrolysis constant is represented by K_h.

Now, we proceed to discuss the mathematics of hydrolysis of the various types of salts.

Salt of a Weak acid and Strong base

The general hydrolysis reaction of a salt of weak acid (HA) and strong acid can be written as

$$A^- + H_2O \rightleftharpoons HA + OH^-$$

This leads to the equilibrium constant expression

$$K_h = \frac{[HA]\,[OH^-]}{[A^-]\,[H_2O]}$$

The concentration of water, $[H_2O]$, is very large and is regarded as practically constant.

Thus the hydrolysis constant expression assumes the form

$$K_h = \frac{[HA]\,[OH^-]}{[A^-]} \qquad ...(1)$$

Relation between K_h, K_w and K_a

We know that the ionic product of water, K_w, is expressed as

$$K_w = [H^+]\,[OH^-] \qquad ...(2)$$

For the dissociation of a weak acid, HA,

$$HA \rightleftharpoons H^+ + A^-$$

the acid dissociation constant, K_a, is expressed as

$$K_a = \frac{[H^+]\,[A^-]}{[HA]} \qquad ...(3)$$

Dividing (2) by (3)

$$\frac{K_w}{K_a} = \frac{[OH^-]\,[HA]}{[A^-]} = K_h \qquad \text{[by equation (1)]}$$

or

$$\frac{K_w}{K_a} = K_h \qquad ...(4)$$

It is clear from (4) that the hydrolysis constant (K_h) of the salt varies inversely as the dissociation constant K_a of the weak acid. Therefore, **weaker the acid greater is the hydrolysis constant of the salt.**

Relation between Hydrolysis constant and Degree of hydrolysis

The degree of hydrolysis is the fraction of the salt which has undergone hydrolysis when equilibrium is established. It is generally represented by α.

Suppose we start with one mole of the salt dissolved in V m^3 of solution. Then the equilibrium concentrations are :

$$A^- + H_2O \rightleftharpoons HA + OH^-$$

$$\text{Equilibrium concentrations} \qquad \frac{1-\alpha}{V} \qquad\qquad \frac{\alpha}{V} \qquad \frac{\alpha}{V}$$

Hence the hydrolysis constant K_h is given by

$$K_h = \frac{[HA]\,[OH^-]}{[A^-]} = \frac{\alpha/V \times \alpha/V}{1 - \alpha/V} = \frac{\alpha^2}{(1-\alpha)V}$$

If α is small, (1 – α) may be taken as equal to one. Then,

$$K_h = \frac{\alpha^2}{V}$$

or

$$\alpha^2 = K_h V = \frac{K_w}{K_a} V$$

∴

$$\alpha = \sqrt{\frac{K_w V}{K_a}}$$

$$= \sqrt{\frac{K_w}{K_a C}}$$

where C is the initial concentration of the salt. Knowing the values of K_w, K_a and C, the degree of hydrolysis can be calculated.

Derivation of pH. The pH of an aqueous solution of weak acid and strong base can be derived as follows :

From earlier discussion, we know that

$$[OH^-] = \frac{\alpha}{V} = \alpha C$$

and

$$[H^+] = \frac{K_w}{[OH^-]}$$

∴

$$[H^+] = \frac{K_w}{\alpha C}$$

But

$$\alpha = \sqrt{\frac{K_w}{K_a C}}$$

∴

$$[H^+] = \frac{K_w}{C}\sqrt{\frac{K_a C}{K_w}} = \sqrt{\frac{K_w K_a}{C}}$$

Taking logarithms and reversing the sign throughout

$$-\log[H^+] = -\frac{1}{2}\log K_w - \frac{1}{2}\log K_a + \frac{1}{2}\log C$$

$$pH = \frac{1}{2}pK_w + \frac{1}{2}pK_a + \frac{1}{2}\log C$$

$$= 7 + \frac{1}{2}pK_a + \frac{1}{2}\log C$$

It is evident that pH of the solution will always be greater than 7. Thus **aqueous solution of salt of weak acid and strong base will be always alkaline.**

SOLVED PROBLEM 1

Calculate the hydrolysis constant and pH of 0.625 M solution of CH_3COONa. K_a 1.754 × 10^{-5}.

SOLUTION

Calculation of K_h

We know that for salt of a weak acid and strong base

$$K_h = \frac{K_w}{K_a}$$

Substituting the values of K_w and K_a

$$K_h = \frac{10^{-14}}{1.754 \times 10^{-5}} = \mathbf{5.701 \times 10^{-10}}$$

Calculation of pH

The hydrolysis equation is :

$$CH_3COO^- + H_2O \rightleftharpoons CH_3COOH + OH^-$$

Thus, $$K_h = \frac{[CH_3COOH]\,[OH^-]}{[CH_3COO^-]}$$

If x be the concentration of CH_3COOH and OH^- ions at equilibrium, the concentration of CH_3COO^- ions is $0.625 - x$. Substituting these values in the hydrolysis constant expression

$$K_h = \frac{x \times x}{0.625 - x} = 5.701 \times 10^{-10}$$

Assuming that x is negligible as compared to 0.625,

$$x^2 = 0.625 \times (5.701 \times 10^{-10}) = 3.56 \times 10^{-10}$$

$$x = \sqrt{3.56 \times 10^{-10}} = 1.89 \times 10^{-5}\ \text{M}$$

Hence $$[OH^-] = 1.89 \times 10^{-5}\ \text{M}$$

$$pOH = -\log [OH^-] = -\log (1.89 \times 10^{-5}) = 4.72$$

$$pH = 14.00 - pOH = 14.00 - 4.72$$

$$= \mathbf{9.28}$$

SOLVED PROBLEM 2

What is the pH of a 0.2 M solution of NaCN? K_a for HCN = 4.0 × 10^{-10}.

SOLUTION

Calculation of Hydrolysis Constant

NaCN is the salt of weak acid HCN and strong base. Therefore,

$$K_h = \frac{K_w}{K_a} = \frac{1 \times 10^{-14}}{4 \times 10^{-10}} = \mathbf{2.5 \times 10^{-5}}$$

Calculation of pH

From the hydrolysis reaction,

$$CN^- + H_2O \rightleftharpoons HCN + OH^-$$

$$K_h = \frac{[HCN]\,[OH^-]}{[CN^-]} = 2.5 \times 10^{-5}$$

Let x be the concentration of OH^- ions produced by hydrolysis. Therefore at equilibrium, we have

$$[HCN] = x$$
$$[OH^-] = x$$
$$[CN^-] = 0.2 - x = 0.2 \text{ because } x \text{ is negligible}$$

Substituting these values in the expression above,

$$K_h = \frac{x \times x}{0.2} = 2.5 \times 10^{-5}$$

$$x = \sqrt{2.5 \times 0.2 \times 10^{-5}}$$

Hence, $[OH^-] = x = 2.24 \times 10^{-3}$

$$pOH = -\log(2.24 \times 10^{-3}) = 2.65$$
$$pH = 14 - pOH = \mathbf{11.35}$$

Salts of Weak bases and Strong acids

The hydrolysis of a salt of a weak base BOH (*e.g*, NH_4OH) and a strong acid may be represented by the equation :

$$B^+ + H_2O \rightleftharpoons BOH + H^+$$

Hydrolysis Constant. Applying the Law of Mass Action to the above hydrolysis reaction, the hydrolysis constant, K_h, is given by

$$K_h = \frac{[H^+][BOH]}{[B^+][H_2O]}$$

Since $[H_2O]$ is very large, it is taken to be constant and the hydrolysis constant expression is reduced to

$$K_h = \frac{[H^+][BOH]}{[B^+]} \quad ...(1)$$

Relation between K_h, K_w and K_b. We know that the ionic product of water K_w is expressed as :

$$K_w = [H^+][OH^-] \quad ...(2)$$

For the dissociation of a weak base, BOH

$$BOH \rightleftharpoons B^+ + OH^-$$

The dissociation constant, K_b, can be expressed as :

$$K_b = \frac{[B^+][OH^-]}{[BOH]} \quad ...(3)$$

Dividing (2) by (3)

$$\frac{K_w}{K_b} = \frac{[H^+][BOH]}{[B^+]} = K_h$$

or $$\frac{K_w}{K_b} = K_h \quad ...(4)$$

Thus the hydrolysis constant, K_h, varies inversely as the dissociation constant, K_b, of the base. Therefore **weaker the base greater will be the hydrolysis constant of the salt.**

Relation between Hydrolysis constant and degree of hydrolysis. Suppose we start with one mole of the salt dissolved in V liters of solution. Then the concentrations when equilibrium is attained are :

$$B^+ + H_2O \rightleftharpoons BOH + H^+$$

Equilibrium concentrations $\quad \frac{1-\alpha}{V} \qquad \frac{\alpha}{V} \quad \frac{\alpha}{V}$

Applying the Law of Mass Action, the hydrolysis constant, K_h is given by the expression

$$K_h = \frac{[H^+][BOH]}{[B^+]} = \frac{\alpha/V \times \alpha/V}{(1-\alpha)/V} = \frac{\alpha^2}{(1-\alpha)V}$$

When α is small, $(1 - \alpha)$ may be considered as equal to one. Then we have

$$K_h = \frac{\alpha^2}{V}$$

or $\quad K_h \times V = \alpha^2$

or $\quad \alpha = \sqrt{K_h \times V}$

From (4) we have

$$K_h = \frac{K_w}{K_b}$$

$$\therefore \quad \alpha = \sqrt{\frac{K_w}{K_b} \times V} \qquad ...(5)$$

$$= \sqrt{\frac{K_w}{K_b \times C}} \qquad \left[\because C \propto \frac{1}{V}\right]$$

where C is the initial concentration of the salt.

Derivation of pH. From the above discussion it is clear that

$$[H^+] = \frac{\alpha}{V} = \alpha \times C$$

Substituting the value of α from equation (5), we have

$$[H^+] = \frac{1}{V}\sqrt{\frac{K_w \times V}{K_b}} = \sqrt{\frac{K_w}{K_b V}} = \sqrt{\frac{K_w \times C}{K_b}}$$

Taking logarithms and reversing the signs

$$-\log[H^+] = -\frac{1}{2}\log K_w - \frac{1}{2}\log C + \frac{1}{2}pK_b$$

or $\quad pH = 7 + \frac{1}{2}pK_b = \frac{1}{2}\log C$

In this case it is evident that pH will always be less than 7. Thus, **the solution of a salt of weak base and strong acid will always be acidic.**

SOLVED PROBLEM 3

Calculate the pH of a 0.20 M solution of ammonium chloride, $K_b = 1.8 \times 10^{-5}$.

SOLUTION

Calculation of K_h

NH_4Cl is salt of a weak base NH_4OH and strong acid HCl. Therefore,

$$K_h = \frac{K_w}{K_b} = \frac{1.0 \times 10^{-14}}{1.8 \times 10^{-5}} = \mathbf{5.56 \times 10^{-10}}$$

Calculation of pH

The hydrolysis of NH_4Cl can be represented as

$$\underset{0.20-x}{NH_4^+} + H_2O \rightleftharpoons \underset{x}{NH_4OH} + \underset{x}{H^+}$$

Let x be the concentration of NH_4OH and H^+ at equilibrium. The concentration of NH_4^+ will be $(0.20 - x)$.

Thus,

$$K_h = \frac{[NH_4OH][H^+]}{[NH_4^+]}$$

$$= \frac{x \times x}{0.20} \qquad (\because x \text{ is small})$$

or $$x^2 = K_h \times 0.20 = (5.56 \times 10^{-10}) \times 0.20$$

$\therefore$ $$x = \sqrt{1.11 \times 10^{-10}} = [H^+] = 1.053 \times 10^{-5} \text{ mol} l^{-1}$$

and $$pH = -\log [H^+] = -\log (1.053 \times 10^{-5}) = \mathbf{4.9775}$$

SOLVED PROBLEM 4

Calculate the pH value of 0.15 M solution of ammonium chloride if the dissociation constant for ammonia is 1.80×10^{-5}.

SOLUTION

This is an alternative solution to that given for Example 1.

Here we use the expression

$$pH = \frac{1}{2}pK_w - \frac{1}{2}pK_b - \frac{1}{2}\log C$$

Now $$pK_w = -\log K_w = -\log 1 \times 10^{-14} = 14$$

$$pK_b = -\log K_b = -\log (1.8 \times 10^{-5}) = 4.7444$$

and $$\log C = \log (0.15) = -0.8239$$

$\therefore$ $$pH = 7 - 2.3722 - (-0.4119)$$

$$= \mathbf{5.0397}$$

Salts of Weak acids and Weak bases

In this type of salt, both the anion of weak acid (X^-) and the cation of weak base (B^+) undergo hydrolysis simultaneously.

$$B^+ + X^- + H_2O \rightleftharpoons BOH + HX$$

Hydrolysis constant. Applying Law of Mass Action to the above hydrolysis reaction we have the hydrolysis constant, K_h.

$$K_h = \frac{[BOH][HX]}{[B^+][X^-][H_2O]}$$

$[H_2O]$ is very large and is taken to be constant. The hydrolysis constant expression, therefore, becomes

$$K_h = \frac{[BOH][HX]}{[B^+][X^-]} \qquad ...(1)$$

Relation between K_h, K_w, K_a and K_b. Applying Law of Mass Action to the ionization of weak acid, HX, weak base, BOH, and water, we can write

$$HX \rightleftharpoons H^+ + X^- \qquad K_a = \frac{[H^+][X^-]}{[HX]} \qquad ...(2)$$

$$BOH \rightleftharpoons B^+ + OH^- \qquad K_b = \frac{[OH^-][B^+]}{[BOH]} \qquad ...(3)$$

$$K_w = [H^+][OH^-] \qquad ...(4)$$

Dividing (4) by (3) and (2), we have

$$\frac{K_w}{K_a \times K_b} = \frac{[H^+][OH^-][HX][BOH]}{[X^-][H^+][B^+][OH^-]}$$

$$\therefore \qquad = \frac{[HX][BOH]}{[X^-][B^+]}$$

$$\frac{K_w}{K_a \times K_b} = K_h \qquad ...(5)$$

Relation between Hydrolysis constant and Degree of Hydrolysis. Let us start with 1 mole of the salt of a weak acid and weak base. If α is the degree of hydrolysis (fraction hydrolyzed), the equilibrium concentrations are :

$$\underset{\frac{1-\alpha}{V}}{B^+} + \underset{\frac{1-\alpha}{V}}{X^-} + H_2O \rightleftharpoons \underset{\frac{\alpha}{V}}{BOH} + \underset{\frac{\alpha}{V}}{HX}$$

Substituting these in the hydrolysis constant expression (1)

$$K_h = \frac{\alpha/V \times \alpha/V}{1-\alpha/V \times 1-\alpha/V}$$

or $$K_h = \frac{\alpha^2}{(1-\alpha)^2}$$

When α is small, (1 – α) may be taken as equal to one. Thus we have

$$K_h = \alpha^2$$

or $$\alpha = \sqrt{K_h}$$

From equation (5) $$K_h = K_w / K_a \times K_b$$

$$\therefore \qquad \alpha = \sqrt{\frac{K_w}{K_a \times K_b}} \qquad ...(6)$$

Derivation of pH. Hydrogen ion concentration of the solution of a salt of weak acid and weak base can be derived from the dissociation equilibrium of the weak acid, HX.

$$HX \rightleftharpoons H^+ + X^-$$

$$K_a = \frac{[H^+][X^-]}{[HX]}$$

$$\therefore \qquad [H^+] = \frac{K_a[HX]}{[X^-]}$$

From the hydrolysis reaction of the salt, we know that

$$[HX] = \frac{\alpha}{V} \quad \text{and} \quad [X^-] = \frac{1-\alpha}{V}$$

Substituting these values we have

$$[H^+] = \frac{K_a \times \frac{\alpha}{V}}{\frac{1-\alpha}{V}} = K_a \left(\frac{\alpha}{1-\alpha}\right)$$

Ignoring α as compared to unity,

$$[H^+] = K_a \times \alpha$$

Substituting the value of α from equation (6),

$$[H^+] = K_a \sqrt{\frac{K_w}{K_a\, K_b}} = \sqrt{\frac{K_w\, K_a}{K_b}}$$

Taking logarithms and reversing the sign throughout

$$-\log[H^+] = -\frac{1}{2}\log K_w - \frac{1}{2}K_a + \frac{1}{2}\log K_b$$

or $$\text{pH} = \frac{1}{2}\text{p}K_w + \frac{1}{2}\text{p}K_a - \frac{1}{2}\text{p}K_b$$

If $$\text{p}K_a = \text{p}K_b$$

i.e., the dissociation constant of the acid is equal to that of the base,

$$\text{pH} = \frac{1}{2}\text{p}K_w = 7$$

Thus the solution will be neutral despite the fact that hydrolysis has taken place. Since the dissociation constant of acetic acid is almost the same as that of ammonium hydroxide, the solution of ammonium acetate is neutral *i.e.*, it has a pH of 7.

If $\text{p}K_a > \text{p}K_b$ *i.e.*, the acid is relatively weaker than the base, the solution will be alkaline as pH is more than 7. If $\text{p}K_a < \text{p}K_b$ *i.e.*, the acid is relatively stronger, the solution will be acidic as pH will be less than 7.

SOLVED PROBLEM 5

Calculate the pH of a solution of ammonium acetate. Given that : $K_a = 1.75 \times 10^{-5}$, $K_b = 1.8 \times 10^{-5}$ and $K_w = 1.0 \times 10^{-14}$.

SOLUTION

We know that

$$\text{pH} = \frac{1}{2}\text{p}K_w + \frac{1}{2}\text{p}K_a - \frac{1}{2}\text{p}K_b$$

Now, $$\frac{1}{2}\text{p}K_w = -\frac{1}{2}\log(1.0 \times 10^{-14}) = 7$$

$$\frac{1}{2}\text{p}K_a = -\frac{1}{2}\log K_a = -\frac{1}{2}\log(1.75 \times 10^{-5})$$

$$= \frac{1}{2} \times 4.7570 = 2.3785$$

and $$\frac{1}{2}\log K_b = -\frac{1}{2}\log(1.8 \times 10^{-5})$$

$$= \frac{1}{2} \times 4.7447 = 2.3723$$

∴ $$\text{pH} = 7 + 2.3785 - 2.3723$$

$$= \mathbf{7.006}$$

DETERMINATION OF DEGREE OF HYDROLYSIS

The degree of hydrolysis of a salt can be determined by a number of methods. The more important ones are described below.

Dissociation Constant Method

The degree of hydrolysis, α, is related to the ionic product of water, K_w, and the dissociation constant of the weak acid, K_a, or of the base, K_b, from which the salt is obtained. The various relationship are listed below :

(*i*) For salt of a Weak acid and Strong base :

$$\alpha = \sqrt{\frac{K_w}{K_a \times C}}$$

(*ii*) For salt of a Weak base and Strong acid :

$$\alpha = \sqrt{\frac{K_w}{K_b \times C}}$$

(*iii*) For salt of a Weak acid and Weak base :

$$\alpha = \sqrt{\frac{K_w}{K_a \times K_b}}$$

Substituting the values of K_w, K_a, K_b and C, the initial concentration of the salt, in the appropriate expression, α can be calculated.

It may be noted that the degree of hydrolysis for the salt of a weak acid and weak base is independent of the concentration. However in this case, the value of α is not small and (1 – α) cannot be taken as equal to one. Therefore the relationship for calculating the degree of hydrolysis is considered in the form

$$\frac{\alpha^2}{(1-\alpha)^2} = \sqrt{\frac{K_w}{K_a \times K_b}}$$

This is by far the most accurate method for determining the degree hydrolysis of a salt and is used in all modern work.

SOLVED PROBLEM 6

What is the percentage hydrolysis of NaCN in N/80 solution when the dissociation constant for NaCN is 1.3×10^{-9} and $K_w = 1.0 \times 10^{-14}$.

SOLUTION

Since NaCN is the salt of a weak acid (HCN) and strong base (NaOH), the degree of hydrolysis, α is given by the expression

$$\alpha = \sqrt{\frac{K_w}{K_a \times C}}$$

$$= \frac{1.0 \times 10^{-14} \times 80}{1.3 \times 10^{-9}}$$

$$= \sqrt{6.16 \times 10^{-4}}$$

$$= 2.48 \times 10^{-2}$$

∴ Percentage hydrolysis of NaCN in N/80 solution is **2.48.**

SOLVED PROBLEM 7

Calculate the hydrolysis constant and degree of hydrolysis of NH_4Cl in 0.001 M solution. $K_b = 1.8 \times 10^{-5}$, $K_w = 1.0 \times 10^{-14}$.

SOLUTION

Since NH_4Cl is the salt of a weak base and a strong acid, the degree of hydrolysis, α, is given by the expression

$$\alpha = \sqrt{\frac{K_w}{K_b \times C}}$$

$$= \sqrt{\frac{1.0 \times 10^{-14}}{1.8 \times 10^{-5} \times 0.001}}$$

$$= \sqrt{5.5 \times 10^{-7}}$$

$$= 7.4 \times 10^{-4} = 0.00074$$

For the salt of a weak base and strong acid,

$$K_h = \frac{K_w}{K_b}$$

$$= \frac{10^{-14}}{1.8 \times 10^{-5}} = \mathbf{5.55 \times 10^{-10}}$$

SOLVED PROBLEM 8

Calculate the degree of hydrolysis of ammonium acetate, if the dissociation constant of ammonium hydroxide is 1.8×10^{-5}, that for acetic acid is 1.8×10^{-5} and the ionic product of water is 1.0×10^{-14}

SOLUTION

Ammonium acetate is the salt of a weak acid (CH_3COOH) and weak base (NH_4OH).

$$\therefore \quad K_h = \frac{K_w}{K_a \times K_b}$$

We know that

$$K_h = \frac{\alpha^2}{(1-\alpha)^2}$$

Thus, $$\frac{\alpha^2}{(1-\alpha)^2} = \frac{1.0 \times 10^{-14}}{1.8 \times 10^{-5} \times 1.8 \times 10^{-5}}$$

or $$\frac{\alpha}{1-\alpha} = \sqrt{\frac{1.0 \times 10^{-14}}{1.8 \times 10^{-5} \times 1.8 \times 10^{-5}}}$$

$$= \frac{1.0 \times 10^{-2}}{1.8} = \frac{1}{180}$$

Hence $$\alpha = 0.0055 = \mathbf{0.55 \times 10^{-2}}$$

From Conductance measurements

The degree of hydrolysis, α, of a salt can be determined by conductance measurements. Let us consider a solution containing the salt of a weak base and a strong acid. The hydrolysis reaction can be written as

$$\underset{(1-\alpha)}{B^+} + H_2O \rightleftharpoons \underset{\alpha}{BOH} + \underset{\alpha}{H^+}$$

If it be assumed that the base is so weak that it is not dissociated at all, it will contribute nothing to conductance of the solution. The equivalent conductance of the salt, therefore, consists of :

(*a*) that due to $(1 - \alpha)$ equivalents of the salt.

(*b*) that due to α equivalents of the acid produced by hydrolysis.

Thus, we can say that :

$$\wedge = (1-\alpha)\wedge_{salt} + \alpha\wedge_{acid}$$

$$\wedge - \wedge_{salt} = \alpha(\wedge_{acid} - \wedge_{salt})$$

or

$$\alpha = \frac{\wedge - \wedge_{salt}}{(\wedge_{acid} - \wedge_{salt})}$$

$\wedge$ is found by conductance measurements. $\wedge_{acid}$ is taken as the value for strong acid at infinite dilution. $\wedge_{acid}$ is determined by adding excess of weak base to the solution to suppress hydrolysis so that the resulting experimentally determined value of $\wedge$ can be taken as that of the unhydrolyzed salt.

SOLVED PROBLEM 9

The equivalent conductance of a solution of aniline hydrochloride, $C_6H_5NH_3Cl$, was found to be 0.0144 siemens m^2 $eqvt^{-1}$ at a certain dilution and at 298 K. In the presence of excess of aniline, the value of conductance is 0.01036 siemens m^2 $eqvt^{-1}$. If $\wedge$ for HCl at this temperature is 0.0383, calculate the degree of hydrolysis.

SOLUTION

$$\alpha = \frac{\wedge - \wedge_{salt}}{\wedge_{HCl} - \wedge_{salt}}$$

Substituting the values

$$\alpha = \frac{0.0144 - 0.01036}{0.0383 - 0.01036}$$

$$= \frac{0.00404}{0.02794} = \mathbf{0.1446}$$

ADDITIONAL SOLVED PROBLEMS

SOLVED PROBLEM 10

A 0.02 M solution of sodium acetate in water at 298 K has a hydrogen concentration of 3×10^{-9} M. What is the hydrolysis constant of the salt ?

SOLUTION : (*a*) To calculate the K_a of the acid

Formula used

$$[H^+] = \sqrt{\frac{K_w K_a}{C}} \quad \text{or} \quad [H^+]^2 = \frac{K_w K_a}{C}$$

Quantities given

$[H^+] = 3 \times 10^{-9}$ $\qquad$ $K_w = 1 \times 10^{-14}$ $\qquad$ $C = 0.02$

Substitution of values

$$(3 \times 10^{-9})^2 = \frac{10^{-14} \times K_a}{0.02}$$

or $$K_a = \frac{9 \times 10^{-18} \times 0.02}{10^{-14}} = 1.8 \times 10^{-5}$$

(*b*) To calculate the hydrolysis constant of sodium acetate.

Formula used

$$K_h = \frac{K_w}{K_a}$$

Quantities given

$K_w = 10^{-14}$ $\qquad K_a = 1.8 \times 10^{-5}$

Substitution of values

$$K_h = \frac{10^{-14}}{1.8 \times 10^{-5}}$$

$$= \mathbf{5.5 \times 10^{-10}}$$

SOLVED PROBLEM 11

The dissociation constant of acetic acid is 1.8×10^{-5} at 291 K. The ionic product of water is 10^{-14} at 291 K. What would be the degree of hydrolysis in a 0.012 N solution of sodium acetate?

SOLUTION

Formula used

$$\alpha = \sqrt{\frac{K_w}{K_a \times C}}$$

Quantities given

$K_w = 10^{-14}$ $\qquad K_a = 1.8 \times 10^{-5}$ $\qquad C = 0.012$

Substitution of values

$$\alpha = \sqrt{\frac{10^{-14}}{1.8 \times 10^{-5} \times 0.012}}$$

$$= \sqrt{4.623 \times 10^{-9}}$$

$$= \mathbf{6.799 \times 10^{-5}}$$

SOLVED PROBLEM 12

Calculate the percentage of hydrolysis of sodium acetate in 0.1 N solution at 291 K using the following data. It is to be assumed that the salt is completely dissociated.

$K_a = 1.8 \times 10^{-5}$; $K_w = 1.02 \times 10^{-14}$.

SOLUTION

Formula used

$$\alpha = \sqrt{\frac{K_w}{K_a \times C}}$$

Quantities given

$K_w = 1.02 \times 10^{-14}$ $\qquad K_a = 1.8 \times 10^{-5}$ $\qquad C = 0.1$

Substitution of values

$$\alpha = \sqrt{\frac{1.02 \times 10^{-14}}{1.8 \times 10^{-5} \times 0.1}}$$

$$= \mathbf{7.527 \times 10^{-5}}$$

SOLVED PROBLEM 13

Calculate the hydrolysis constant, degree of hydrolysis and *pH* value of 10^{-2} M NH_4Cl solution at 290 K. ($K_b = 1.8 \times 10^{-5}$ and $K_w = 1.0 \times 10^{-14}$).

SOLUTION : (*a*) To calculate the hydrolysis constant

Formula used

$$K_h = \frac{K_w}{K_b}$$

Quantities given

$K_w = 1.0 \times 10^{-14}$ $\qquad$ $K_b = 1.8 \times 10^{-5}$

Substitution of values

$$K_h = \frac{1.0 \times 10^{-14}}{1.8 \times 10^{-5}}$$

$$= \mathbf{5.55 \times 10^{-10}}$$

(*b*) To calculate the degree of hydrolysis

Formula used

$$\alpha = \sqrt{\frac{K_w}{K_b \times C}}$$

Quantities given

$K_w = 1 \times 10^{-14}$ $\qquad$ $K_b = 1.8 \times 10^{-5}$ $\qquad$ $C = 10^{-2}$

Substitution of values

$$\alpha = \sqrt{\frac{1 \times 10^{-14}}{1.8 \times 10^{-5} \times 10^{-2}}}$$

$$= \sqrt{5.555 \times 10^{-8}}$$

$$= \mathbf{2.357 \times 10^{-4}}$$

(*c*) To calculate the pH

$$pH = ½\, pK_w - ½\, pK_b + ½ \log C$$

Quantities given

$pK_w = 14$ $\qquad$ $pK_b = -\log 1.8 \times 10^{-5} = 4.7447$ $\qquad$ $C = 10^{-2}$

Substitution of values

$$pH = ½ \times 14 - ½\ (4.7447) + ½ \log 10^{-2}$$

$$= 7 - 2.37235 + ½\ (-2)$$

$$= \mathbf{3.62765}$$

SOLVED PROBLEM 14

Calculate the hydrolysis constant and degree of hydrolysis of 0.1M sodium acetate solution.
Given : $K_w = 1.0 \times 10^{-14}$ and $K_a = 1.75 \times 10^{-5}$

SOLUTION : (*a*) To calculate hydrolysis constant

Formula used

$$K_h = \frac{K_w}{K_a}$$

Quantities given

$K_w = 1.0 \times 10^{-14}$ $\qquad$ $K_a = 1.75 \times 10^{-5}$

Substitution of values

$$K_h = \frac{1.0 \times 10^{-14}}{1.75 \times 10^{-5}}$$

$$= 5.71 \times 10^{-10}$$

(*b*) To calculate the hydrolysis constant

Formula used

$$\alpha = \sqrt{\frac{K_h}{C}}$$

Quantities given

$K_h = 5.71 \times 10^{-10}$ $\qquad$ $C = 0.1$

Substitution of values

$$\alpha = \sqrt{\frac{5.71 \times 10^{-10}}{0.1}}$$

$$= \mathbf{7.55 \times 10^{-5}}$$

SOLVED PROBLEM 15

Calculate the *pH* and hydrolysis constant of 0.05 M NH_4Cl solution.

$K_{NH4OH} = 1.75 \times 10^{-5}$ and $K_w = 1 \times 10^{-14}$.

SOLUTION : (*a*) To calculate pH value

Formula used

$$pH = ½\ pK_w - ½\ pK_b - ½ \log C$$

Quantities given

$pK_w = -\log K_w = -\log 1 \times 10^{-14} = 14$ $\qquad$ $pK_b = -\log K_b = -\log 1.75$ $\qquad$ $C = 0.05M$

Substitution of values

$$pH = ½ \times 14 - ½ \times (4.7569) - ½ \log 0.05$$

$$= 7 - 2.37845 - (-1.3010)$$

$$= \mathbf{5.92255}$$

(*b*) To calculate the hydrolysis constant

Formula used

$$K_h = \frac{K_w}{K_b}$$

Quantities given

$K_w = 1 \times 10^{-14}$ $\qquad$ $K_b = 1.75 \times 10^{-5}$

Substitution of values

$$K_h = \frac{1 \times 10^{-14}}{1.75 \times 10^{-5}}$$

$$= \mathbf{5.714 \times 10^{-10}}$$

SOLVED PROBLEM 16

Calculate the degree of hydrolysis and *pH* of 0.02M solution of sodium acetate. Hydrolysis constant of sodium acetate is 5.6×10^{-10}.

SOLUTION : (*a*) To calculate the degree of hydrolysis

Formula used

$$K_a = \sqrt{\frac{K_h}{C}}$$

Quantities given

$K_h = 5.6 \times 10^{-10}$ $\qquad$ $C = 0.02\text{M}$

Substitution of values

$$\alpha = \sqrt{\frac{5.6 \times 10^{-10}}{0.02}}$$

$$= \sqrt{2.8 \times 10^{-8}}$$

$$= \mathbf{1.67 \times 10^{-4}}$$

(*b*) To calculate the pH of the solution

Formula used

$$pH = \tfrac{1}{2}\ pK_w + \tfrac{1}{2}\ pK_a + \tfrac{1}{2}\ \log C$$

Quantities given

$pK_w = 14$ $\qquad$ $pK_a = -\log K_a = -\log \dfrac{K_w}{K_h} = -\log \dfrac{10^{-14}}{5.6 \times 10^{-10}}$

C = 0.02 M $\qquad$ $= -\log 1.786 \times 10^{-5} = +4.7482$

Substitution of values

$$pH = \tfrac{1}{2} \times 14 + \tfrac{1}{2}\ (+4.7482) + \tfrac{1}{2}\ \log (0.02)$$

$$= 7 + 2.3741 + \tfrac{1}{2}\ (-1.6990)$$

$$= 9.3741 - 0.8495$$

$$= \mathbf{8.5246}$$

SOLVED PROBLEM 17

Calculate the *pH* of 0.50 M aqueous NaCN solution. pK_b of CN^- is 4.70.

SOLUTION

Formula used

$$pH = ½\ pK_w + ½\ pK_a + ½\ \log C$$

Quantities given

$pK_w = 14$ $\quad pK_a = 14 - pK_b = 14 - 4.70 = 9.30$ $\quad C = 0.50$ M

Substitution of values

$$pH = ½ \times 14 + ½ \times 9.30 + ½\ \log 0.50$$

$$= 7 + 4.65 + ½\ (-0.3010)$$

$$= 11.65 - 0.1505$$

$$= \mathbf{11.4995}$$

EXAMINATION QUESTIONS

1. Define or explain the following terms :
   (*a*) Anionic Hydrolysis (*b*) Cationic Hydrolysis
   (*c*) Hydrolysis constant (*d*) Degree of Hydrolysis
2. Sodium phenate is hydrolyzed to the extent of 0.03% in 0.1 M aqueous solution at 298 K. Calculate
   (*i*) The hydrolysis constant of the salt; and (*ii*) the ionic product of water at 298 K. The dissociation constant of phenol is 1.3×10^{-10} at 298 K.
   Answer. (*i*) 9×10^{-5}; (*ii*) 1.17×10^{-14}
3. (*a*) What is hydrolysis constant of salt? Why aqueous solution of sodium carbonate is alkaline? Derive an expression for the hydrolysis constant and pH of this solution.
   (*b*) Calculate the pH of a decinormal solution of ammonium chloride. ($pK_a = 5.7$ and $pK_w = 14$)
   Answer. 10.35
4. What is meant by the terms 'Degree of Hydrolysis' and 'Hydrolysis constant'? Deduce the relation between hydrolysis constant and the dissociation constant of the base in the case of the hydrolysis of a salt of a strong acid and a weak base.
5. Derive the relation between hydrolysis constant ionic product of water and dissociation constant of a strong acid and a weak base.
6. Ammonium hydroxide undergoes hydrolysis in aqueous solution. Give the equation for the hydrolysis constant and show that it is related to dissociation constant of ammonium hydroxide.
7. Calculate the hydrolysis constant, degree of hydrolysis and pH value of 10^{-2} M NH_4Cl solution at 298 K. ($K_b = 1.8 \times 10^{-5}$ and $K_w = 1.0 \times 10^{-14}$)
   Answer. 5.555×10^{-10}; 2.357×10^{-9}; 3.372
8. Deduce an expression for the degree of hydrolysis of a salt of a weak acid and a strong base.
9. What is hydrolysis? For a salt of weak acid and weak base, derive

$$K_h = \frac{K_w}{K_a\ K_b}$$

10. The hydrogen ion concentration of 0.02 M sodium acetate solution is found to be 3.0×10^{-9} M at 298K. Calculate the hydrolysis constant of this salt. $K_w = 1.0 \times 10^{-14}$.

Answer. 5.555×10^{-10}

11. Obtain expression for the hydrolysis constant and degree of hydrolysis for the salt of a strong acid and weak base. *(Panjab BSc)*

12. Derive
$$K_h = \frac{K_w}{K_a} \quad \text{or} \quad K_h = \frac{K_w}{K_b}$$
and write the equation for pH of this solution using K_w, K_a, K_b and c. *(Jiwaji BSc)*

13. What is meant by the terms 'Degree of hydrolysis' and 'Hydrolysis constant'? *(MD Rohtak BSc)*

14. Find out the expression for hydrolysis constant of a salt of strong acid and weak base in terms of K_w and K_h. *(Purvanchal BSc, 2002)*

15. Show that the degree of hydrolysis of ammonium acetate is independent of concentration. *(Kalyani BSc)*

16. (*a*) Explain why an aqueous solution of $CuSO_4$ is acidic and that of NaCl is neutral?
(*b*) Calculate the percentage of hydrolysis of sodium acetate in 0.1 N solution at 298 K using the following data. It is to be assumed that the salt is completely dissociated. $K_a = 1.8 \times 10^{-5}$; $K_w = 1.02 \times 10^{-14}$.

Answer. (*b*) 7.452×10^{-5} *(Mysore BSc)*

17. Calculate the *pH* at the equivalence point when a solution of 0.10 M acetic acid is titrated with a solution of 0.10 M NaOH. K_a for acetic acid is 1.9×10^{-5}.

Answer. 8.71 *(Baroda BSc)*

18. Calculate the percentage hydrolysis of sodium acetate in 0.1 N solution at 298 K, assuming the salt to be completely dissociated. (K_a for Acetic acid = 1.8×10^{-5})

Answer. 0.0075% *(Nagpur BSc)*

19. What happens to the *pH* of 500 m³ of solution that is 0.1 molar in sodium acetate and 0.1 molar in acetic acid when 10 m³ of 0.1 M NaOH is added?

Answer. *pH* will increase *(Dibrugarh BSc)*

20. A 0.02 M solution of sodium acetate in water at 298 K has a hydrogen ion concentration of 3×10^{-9} M. What is the hydrolysis constant of the salt?

Answer. 5.5×10^{-10} *(Agra BSc)*

21. (*a*) Define salt hydrolysis and degree of hydrolysis. Establish a relation between hydrolysis constant and dissociation constant of a salt of weak acid and weak base.
(*b*) Calculate the hydrolysis constant and degree of hydrolysis of 0.1 M sodium acetate solution. Given : $K_w = 1.0 \times 10^{-14}$, $K_a = 1.75 \times 10^{-6}$

Answer. 7.5×10^{-5} *(Sambalpur BSc)*

22. (*a*) What is hydrolysis? Derive an expression for the hydrolysis constant of a salt of a weak acid and a strong base in terms of dissociation constant of a weak acid and ionic product of water.
(*b*) Calculate the degree of hydrolysis of sodium acetate. Dissociation constant of acetic acid is 1.80×10^{-5}. Ionic product of water is 1×10^{-14}.

Answer. (*b*) 7.452×10^{-5} *(Madurai BSc)*

23. 20 m³ of 0.2 M NaOH solution be treated with 40 m³ of 0.2 M acetic acid solution to give 70 m³. Calculate the *pH* of the solution.

Answer. 4.5684 *(Shivaji BSc)*

24. The dissociation constant of acetic acid is 1.8×10^{-5} at 291 K. The ionic product of water is 10^{-14} at 291 K. What would be the degree of hydrolysis in a 0.012 N solution of sodium acetate?

Answer. 2.150×10^{-6} *(Kalyani BSc)*

MULTIPLE CHOICE QUESTIONS

1. The reaction of an anion or cation with water accompanied by cleavage of O–H bond is called
(*a*) neutralization (*b*) hydrolysis
(*c*) acidification (*d*) ionization
Answer. (*b*)

2. In anionic hydrolysis the pH of solution is
(*a*) greater than 7 (*b*) equal to 7
(*c*) less than 7 (*d*) less than zero
Answer. (*c*)

3. In cationic hydrolysis, the resulting solution is
(*a*) acidic (*b*) basic
(*c*) neutral (*d*) sometimes acidic, sometimes basic
Answer. (*b*)

4. A salt of weak acid and strong base on hydrolysis yields a solution which is
(*a*) slightly acidic (*b*) slightly basic
(*c*) neutral (*d*) highly acidic
Answer. (*b*)

5. NaCN on hydrolysis produces a solution which has
(*a*) pH > 7 (*b*) pH = 7
(*c*) pH < 7 (*d*) pH = 0
Answer. (*b*)

6. A salt of weak base and strong acid on hydrolysis gives a solution which has
(*a*) more H^+ ions than OH^- (*b*) more OH^- ions than H^+
(*c*) equal H^+ and OH^- ions (*d*) no H^+ ions
Answer. (*a*)

7. A salt of weak acid and weak base on hydrolysis gives a solution which is
(*a*) acidic (*b*) basic
(*c*) neutral (*d*) sometimes acidic, sometimes basic
Answer. (*d*)

8. The hydrolysis constant of a salt of weak acid and strong base is given by the expression
(*a*) $K_h = \frac{[HA][OH^-]}{[A^-]}$ (*b*) $K_h = \frac{[HA][H^+]}{[A^-]}$
(*c*) $K_h = \frac{[HA][A^-]}{[OH^-]}$ (*d*) $K_h = \frac{[HA][A^-]}{[OH^-]}$
Answer. (*a*)

9. The dissociation constant of weak acid, K_a and that of a base K_b are related to hydrolysis constant of the salt by the relation
(*a*) $K_w = K_a \times K_b$ (*b*) $K_a = K_w \times K_b$
(*c*) $K_b = K_w \times K_a$ (*d*) $K_w = K_a / K_b$
Answer. (*a*)

10. Weaker the acid, greater is the
(*a*) ionic product (*b*) dissociation constant
(*c*) hydrolysis constant (*d*) degree of ionization
Answer. (*c*)

11. The degree of hydrolysis of a salt is that fraction of it which undergoes _______ when equilibrium is established.

(*a*) dissociation (*b*) racemization
(*c*) saponification (*d*) hydrolysis

Answer. (*d*)

12. The aqueous solution of salt of a weak acid and strong base will always be

(*a*) acidic (*b*) alkaline
(*c*) neutral (*d*) none of these

Answer. (*b*)

13. The pH of an aqueous solution of weak acid and strong base is given by the relation

(*a*) $pH = 7 + ½\,pK_a + ½\log C$ (*b*) $pH = 7 + ½\,pK_a - ½\log C$
(*c*) $pH = 7 - ½\,pK_a + ½\log C$ (*d*) $pH = 7 - ½\,pK_a - ½\log C$

Answer. (*a*)

14. The hydrolysis constant of a salt of weak base and strong acid is given by

(*a*) $K_h = \dfrac{[H^+][BOH]}{[B^+]}$ (*b*) $K_h = \dfrac{[H^+][B^+]}{[BOH]}$

(*c*) $K_h = \dfrac{[H^+][OH^-]}{[B^+]}$ (*d*) $K_h = \dfrac{[H^+][B^+]}{[BOH]^2}$

Answer. (*a*)

15. The dissociation constant K_b, the hydrolysis constant K_h and ionic product K_w are related to each other by the relation

(*a*) $K_w / K_b = K_h$ (*b*) $K_w / K_h = K_b$
(*c*) $K_w = K_h \times K_b$ (*d*) all of these

Answer. (*d*)

16. The hydrolysis constant K_h of a weak base and a strong acid is _______ to the dissociation constant K_b, of the base

(*a*) directly proportional (*b*) inversely proportional
(*c*) equal (*d*) not equal

Answer. (*b*)

17. The degree of hydrolysis of a weak base and strong acid is given by the relation

(*a*) $\alpha = \sqrt{\dfrac{K_b}{K_h \times C}}$ (*b*) $\alpha = \dfrac{K_b}{K_h \times C}$

(*c*) $\alpha = \dfrac{K_b}{K_h \times C^2}$ (*d*) $\alpha = \dfrac{K_b}{K_h \times \sqrt{C}}$

Answer. (*a*)

18. The pH of a solution of a salt of weak base and strong acid is

(*a*) greater than 7 (*b*) less than 7
(*c*) equal to 7 (*d*) equal to zero

Answer. (*b*)

19. The pH of a solution of a salt of weak base and strong acid is given by the expression

(*a*) $pH = 7 + ½\,pK_b + ½\log C$ (*b*) $pH = 7 + ½\,pK_b - ½\log C$
(*c*) $pH = 7 - ½\,pK_b + ½\log C$ (*d*) $pH = 7 - ½\,pK_b - ½\log C$

Answer. (*b*)

20. Which is the correct relation for a salt of weak acid and a weak base?

(*a*) $K_h = \dfrac{K_w}{K_a \times K_b}$ (*b*) $K_h = \dfrac{K_w \times K_a}{K_b}$

(*c*) $K_h = \frac{K_w \times K_b}{K_a}$ (*d*) $K_h = K_w \times K_a \times K_b$

Answer. (*a*)

21. The degree of hydrolysis of a salt of weak acid and weak base is given by the expression

(*a*) $\alpha = \sqrt{\frac{K_w \times K_b}{K_a}}$ (*b*) $\alpha = \sqrt{\frac{K_w \times K_a}{K_b}}$

(*c*) $\alpha = \sqrt{\frac{K_w}{K_a \times K_b}}$ (*d*) $\alpha = \sqrt{K_w \times K_a \times K_b}$

Answer. (*c*)

22. The aqueous solution of a salt of weak acid and weak bases is

(*a*) always acidic (*b*) always basic

(*c*) always neutral (*d*) sometimes acidic, sometimes basic

Answer. (*d*)

23. Sodium acetate and sodium cyanide are the salts of weak acids and strong bases. Their aqueous solution will be

(*a*) acidic (*b*) basic

(*c*) neutral (*d*) sometimes acidic, sometimes basic

Answer. (*b*)

24. NH_4Cl, $AlCl_3$ and $FeCl_3$ are the salts of ______ acids and ______ bases.

(*a*) weak, weak (*b*) weak, strong

(*c*) strong, weak (*d*) strong, strong

Answer. (*b*)

25. NH_4F, NH_4CN and CH_3COONH_4 are the salts of ______ acid and ______ bases.

(*a*) strong, strong (*b*) strong, weak

(*c*) weak, strong (*d*) weak, weak

Answer. (*d*)

26. When a salt of strong acid and weak base is dissolved in water ______ occurs to give ______ solution

(*a*) ionization, basic (*b*) ionization, acidic

(*c*) hydrolysis, acidic (*d*) hydrolysis, basic

Answer. (*c*)

27. Hydrolysis is reverse of ______.

(*a*) ionization (*b*) neutralization

(*c*) acidification (*d*) saponification

Answer. (*b*)

28. Sodium sulfide (Na_2S) on hydrolysis give a solution which is

(*a*) neutral (*b*) acidic

(*c*) basic (*d*) may be acidic or basic

Answer. (*c*)

29. The heat of neutralization of all strong acids and strong bases is

(*a*) equal to zero (*b*) nearly the same

(*c*) not fixed (*d*) varies from acid to acid

Answer. (*b*)

30. Which salt out of the following will be hydrolyzed to give basic solution.

NaCN, NaCl, $NaNO_3$, NH_4Cl

(*a*) NaCN (*b*) NaCl

(*c*) $NaNO_3$ (*d*) NH_4Cl

Answer. (*a*)

31. Borax ($Na_2B_4O_7$) on hydrolysis produces a solution which is

(*a*) acidic (*b*) basic

(*c*) neutral (*d*) sometimes acidic, sometimes basic

Answer. (*b*)

32. The solution of Na_2CO_3 is alkaline due to

(*a*) hydrolysis of CO_3^{2-} ions (*b*) hydrolysis of Na^+ ions

(*c*) neutralization of CO_3^{2-} ions (*d*) neutralization of Na^+ ions

Answer. (*a*)

33. A weak base becomes weaker in the presence of its salt. This statement is

(*a*) false (*b*) true

(*c*) cannot be predicted (*d*) none of these

Answer. (*b*)

34. NH_4OH becomes weaker in the presence of NH_4Cl due to

(*a*) ionization (*b*) hydrolysis

(*c*) neutralization (*d*) common ion effect

Answer. (*d*)

35. The pH value of an aqueous solution is

(*a*) equal to 7 (*b*) equal to 0

(*c*) less than 7 (*d*) more than 7

Answer. (*d*)

36. The pH value of $AlCl_3$ solution is

(*a*) less than 7 (*b*) greater than 7

(*c*) equal to 7 (*d*) equal to 0

Answer. (*a*)

37. The aqueous solutions of KNO_3, $ZnCl_2$, and K_2CO_3 separately are

(*a*) neutral, acidic and basic respectively (*b*) acidic, neutral and basic respectively

(*c*) basic, neutral and acidic respectively (*d*) neutral, basic and acidic respectively

Answer. (*a*)

38. The aqueous solution of $ZnCl_2$ is acidic due to

(*a*) cation hydrolysis (*b*) anion hydrolysis

(*c*) hydrolysis of both cation & anion (*d*) ionization

Answer. (*a*)

39. A salt 's' is dissolved in pure water of pH = 7. The resulting solution is having pH > 7. The salt is made up of

(*a*) a strong acid and a weak base (*b*) a weak acid and a strong base

(*c*) a weak acid and a weak base (*d*) a strong acid and a weak base

Answer. (*b*)

40. An aqueous solution of ammonium carbonate is

(*a*) weakly acidic (*b*) weakly basic

(*c*) strongly acidic (*d*) neutral

Answer. (*d*)

41. Which one of the following aqueous solutions will have highest pH value?

(*a*) NaCl (*b*) KNO_3

(*c*) $ZnCl_2$ (*d*) Na_2CO_3

Answer. (*d*)

42. An aqueous solution of $CuSO_4.5H_2O$ turns blue litmus red. It is due to the

(*a*) presence of Cu^{2+} ions (*b*) presence of SO_4^{2-} ions

(*c*) hydrolysis of Cu^{2+} ions (*d*) hydrolysis of SO_4^{2-} ions

Answer. (*c*)

43. One or both the ions of a salt react with water to produce acidic, basic or neutral solution. This process is called

(*a*) neutralization (*b*) ionization

(*c*) saponification (*d*) hydrolysis

Answer. (*d*)

44. When a pinch of NaCN is added to pure water, the pH

(*a*) increases (*b*) decreases

(*c*) remains the same (*d*) none of these

Answer. (*a*)

45. A salt undergoes cationic hydrolysis in water. The pH of the resulting solution would be

(*a*) less than 7 (*b*) greater than 7

(*c*) equal to 7 (*d*) equal to 0

Answer. (*a*)

46. Which statement is correct?

(*a*) NH_4Cl gives alkaline solution (*b*) sodium acetate gives acidic solution in water

(*c*) $ZnCl_2$ gives basic solution in water (*d*) KNO_3 gives neutral solution in water

Answer. (*d*)

47. Which one of the following will not be hydrolyzed?

(*a*) KNO_3 (*b*) K_2CO_3

(*c*) KCN (*d*) CH_3COOK

Answer. (*a*)

48. The degree of hydrolysis of ammonium acetate

(*a*) depends upon its concentration

(*b*) does not depend upon its concentration

(*c*) directly proportional to the square of its concentration

(*d*) does not depend upon temperature

Answer. (*b*)

49. A solution of ammonium acetate is _______ and its pH value is _______.

(*a*) acidic, less than 7 (*b*) basic, more than 7

(*c*) neutral, less than 7 (*d*) basic, more than 14

Answer. (*c*).

29 Electromotive Force

CONTENTS

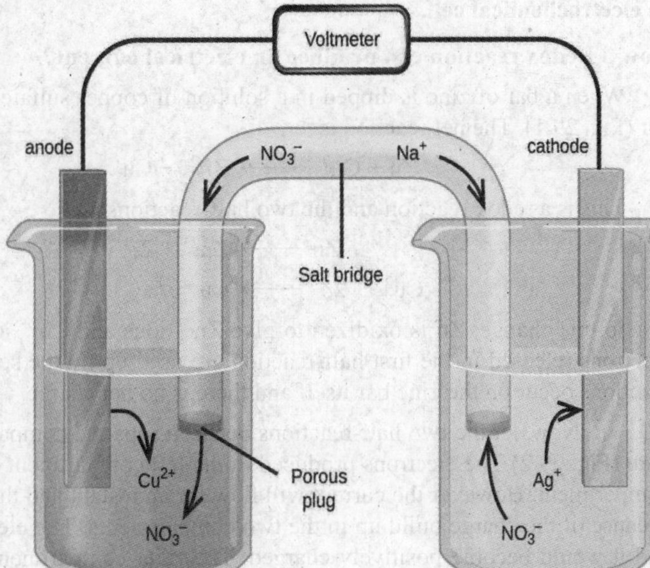

WHAT ARE HALF REACTIONS?

Let us consider the reaction

$$2\,Na + Cl_2 \longrightarrow 2Na^+ + 2Cl^-$$

It occurs by the transfer of electrons from Na to Cl. Na loses an electron and is said to be oxidized to Na^+ ion. At the same time, Cl gains an electron and is reduced to Cl^- ion. Such a reaction which is brought about by loss of electrons (oxidation) and gain of electrons (reduction) simultaneously, is called an **Oxidation-Reduction reaction or Redox reaction** in brief. It may be noted that in the overall redox reaction no free electrons are generated.

The redox reaction can be considered as made up of two reactions. For example, the redox reaction

$$2Na + Cl_2 \longrightarrow Na^+ + 2Cl^-$$

is composed of two half-reactions :

$$2Na \longrightarrow 2Na^+ + 2e^- \quad (oxidation)$$

$$Cl_2 + 2e^- \longrightarrow 2Cl^- \quad (reduction)$$

Each of the two reactions shows just its oxidation or just the reduction portion of the overall redox reaction. Being half components of the redox reaction, these reactions are called **Half-reactions.** The first half-reaction that proceeds by oxidation is often referred to as the **Oxidation half-reaction.** The second half-reaction that occurs by reduction, is referred to as the **Reduction half-reaction.** When the two half-reactions are added together, the sum is the net redox reaction.

We will see presently how redox reactions are a source of electric current in electrochemical cells.

ELECTROCHEMICAL CELLS

A device for producing an electrical current from a chemical reaction (redox reaction) is called an **electrochemical cell.**

How a Redox reaction can produce an electrical current?

When a bar of zinc is dipped in a solution of copper sulfate, copper metal is deposited on the bar (Fig. 29.1). The net reaction is

$$Zn + Cu^{2+} \longrightarrow Zn^{2+} + Cu$$

This is a redox reaction and the two half-reactions are :

$$Zn \longrightarrow Zn^{2+} + 2e^-$$

$$Cu^{2+} + 2e^- \longrightarrow Cu$$

In this change, Zn is oxidized to give Zn^{2+} ions and Cu^{2+} ions are reduced to Cu atoms. The electrons released in the first half-reaction are used up by the second half-reaction. Both the half-reactions occur on the zinc bar itself and there is no net charge.

Now, let the two half-reactions occur in separate compartments which are connected by a wire (Fig. 29.2) The electrons produced in the left compartment flow through the wire to the other compartment. However the current will flow for an instant and then stop. The current stops flowing because of the charge build up in the two compartments. The electrons leave the left compartment and it would become positively charged. The right compartment receives electrons and becomes negatively charged. Both these factors oppose the flow of electrons (electrical current) which eventually stops.

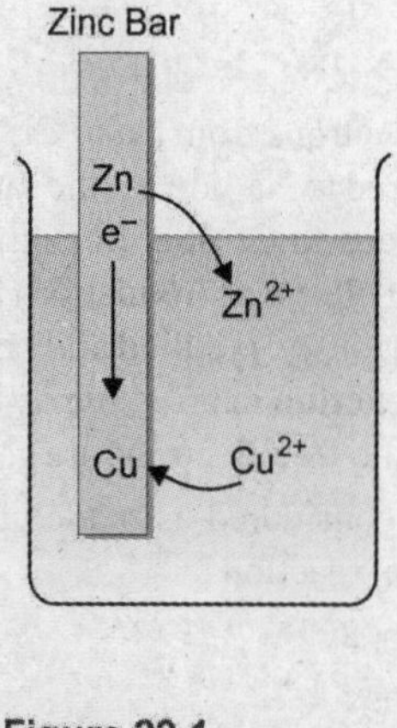

■ **Figure 29.1**
Redox reaction taking place on zinc bar itself.

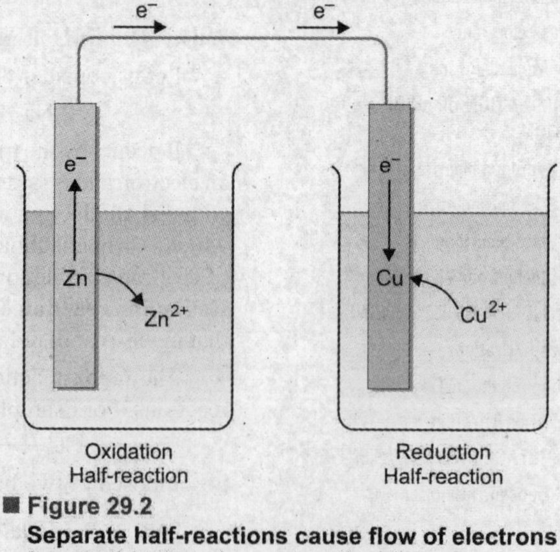

■ **Figure 29.2**
Separate half-reactions cause flow of electrons (current) in the wire connecting them.

This problem can be solved very simply. The solutions in the two compartments may be connected, say, by a **salt bridge.** The salt bridge is a U-tube filled with an electrolyte such as NaCl, KCl, or K_2SO_4. It provides a passage to ions from one compartment to the other compartment without extensive mixing of the two solutions. With this ion flow, the circuit is complete and electrons pass freely through the wire to keep the net charge zero in the two compartments.

Voltaic Cells

A **Voltaic cell,** also known as a **galvanic cell** is one in which electrical current is generated by a spontaneous redox reaction. A simple voltaic cell is shown in Fig. 29.3. Here the spontaneous reaction of zinc metal with an aqueous solution of copper sulfate is used.

$$Zn + Cu^{2+} \longrightarrow Zn^{2+} + Cu$$

A bar of zinc metal (anode) is placed in zinc sulfate solution in the left container. A bar of copper (cathode) is immersed in copper sulfate solution in the right container. The zinc and copper electrodes are joined by a copper wire. A salt bridge containing potassium sulfate solution interconnects the solutions in the anode compartment and the cathode compartment.

The oxidation half-reaction occurs in the anode compartment.

$$Zn \longrightarrow Zn^{2+} + 2e^-$$

The reduction half-reaction takes place in the cathode compartment.

$$Cu^{2+} + 2e^- \longrightarrow Cu$$

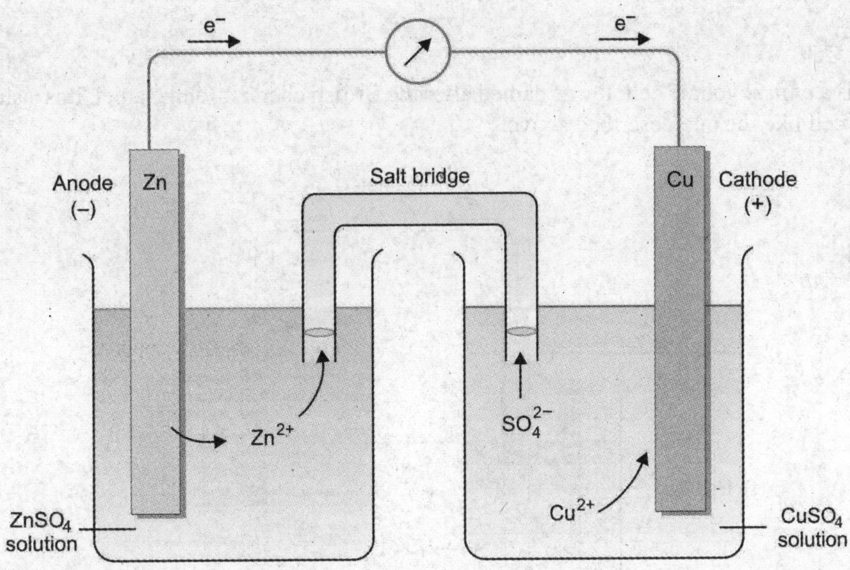

Figure 29.3
A simple voltaic (galvanic) cell.

When the cell is set up, electrons flow from zinc electrode through the wire to the copper cathode. As a result, zinc dissolves in the anode solution to form Zn^{2+} ions. The Cu^{2+} ions in the cathode half-cell pick up electrons and are converted to Cu atoms on the cathode. At the same time, SO_4^{2-} ions from the cathode half-cell migrate to the anode half-cell through the salt bridge. Likewise, Zn^{2+} ions from the anode half-cell move into the cathode half-cell. This flow of ions from one half-cell to the other completes the electrical circuit which ensure continuous supply of current. The cell will operate till either the zinc metal or copper ion is completely used up.

CELL TERMINOLOGY

Before taking up the study of the electrochemical cells, we should be familiar with a few common terms.

Current is the flow of electrons through a wire or any conductor.

Electrode is the material : a metallic rod/bar/strip which conducts electrons into and out of a solution.

Anode is the electrode at which oxidation occurs. It sends electrons into the outer circuit. It has negative charge and is shown as (–) in cell diagrams.

Cathode is the electrode at which electrons are received from the outer circuit. It has a positive charge and is shown as (+) in cell diagrams.

Electrolyte is the salt solutions in a cell.

Anode compartment is the compartment of the cell in which oxidation half-reaction occurs. It contains the anode.

Cathode compartment is the compartment of the cell in which reduction half-reaction occurs. It contains the cathode.

Half-cell. Each half of an electrochemical cell, where oxidation occurs and the half where reduction occurs, is called the half cell.

Daniel Cell

It is a typical voltaic cell. It was named after the British chemist John Daniel. It is a simple zinc-copper cell like the one described above.

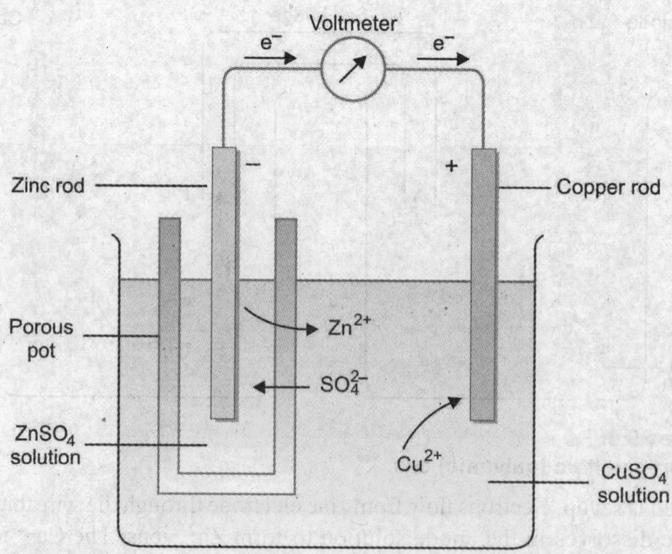

■ **Figure 29.4**
Daniel Cell.

In this cell the salt-bridge has been replaced by a porous pot. Daniel cell resembles the above voltaic cell in all details except that Zn^{2+} ions and SO_4^{2-} ions flow to the cathode and the anode respectively through the porous pot instead of through the salt-bridge. Inspite of this difference, the cell diagram remains the same.

Porous Pot

$$\underset{\text{Anode}}{Zn \mid Zn^{2+}} \parallel \underset{\text{Cathode}}{Cu^{2+} \mid Cu}$$

Cell reaction

The flow of electrons from one electrode to the other in an electrochemical cell is caused by the half-reactions taking place in the anode and cathode compartments. The net chemical change obtained by adding the two half-reactions is called the **cell reaction.** Thus, for a simple voltaic cell described above, we have

(*a*) Half-reactions :

$$Zn(s) \longrightarrow Zn^{2+}(aq) + 2e$$

$$Cu^{2+}(aq) + 2e^- \longrightarrow Cu(s)$$

(*b*) Cell reaction by adding up the half-reactions :

$$Zn(s) + Cu^{2+}(aq) \longrightarrow Zn^{2+}(aq) + Cu(s)$$

Cell potential or emf

In a Zn-Cu voltaic cell, electrons are released at the anode and it becomes negatively charged. The negative electrode pushes electrons through the external circuit by electrical repulsions. The copper electrode gets positive charge due to the discharge of Cu^{2+} ions on it. Thus electrons from the outer circuit are attracted into this electrode. The flow of current through the circuit is determined by the 'push', of electrons at the anode and 'attraction' of electrons at the cathode. These two forces constitute the 'driving force' or 'electrical pressure' that sends electrons through the circuit. This driving force is called the **electromotive force** (abbreviated **emf**) or **cell potential.** The emf of cell potential is measured in units of volts (V) and is also referred to as **cell voltage.**

Cell diagram or Representation of a Cell

A cell diagram is an abbreviated symbolic depiction of an electrochemical cell. For this purpose, we will consider that a cell consists of two half-cells. Each half-cell is again made of a metal electrode contact with metal ions in solution.

IUPAC Conventions. In 1953 IUPAC recommended the following conventions for writing cell diagrams. We will illustrate these with reference to Zinc-Copper cell.

(1) a **single vertical line** (|) represents a phase boundary between metal electrode and ion solution (electrolyte). Thus the two half-cells in a voltaic cell are indicated as

Phase Boundary

$$\underset{\text{Anode Half-Cell}}{Zn \mid Zn^{2+}} \qquad \underset{\text{Cathode Half-Cell}}{Cu^{2+} \mid Cu}$$

It may be noted that the metal electrode **in anode** half-cell is on the left, while **in cathode** half-cell it is on the right of the metal ion.

(2) A **double vertical line** (||) represents the salt bridge, porous partition or any other **means** of permitting ion flow while preventing the electrolyte from mixing.

(3) **Anode half-cell is written on the left** and **cathode half-cell on the right.**

(4) In the **complete cell diagram, the two half-cells are separated by a double vertical line (salt bridge) in between.** The zinc-copper cell can now be written as

$$\underset{\text{Anode Half-Cell}}{Zn \mid Zn^{2+}} \;\overset{\text{Salt Bridge}}{\parallel}\; \underset{\text{Cathode Half-Cell}}{Cu^{2+} \mid Cu}$$

(5) The symbol for an **inert electrode,** like the platinum electrode is often enclosed in a bracket. For example,

$$Mg \mid Mg^{2+} \parallel H^{+} \mid \underset{}{\overset{\text{Inert Electrode}}{H_2(Pt)}}$$

(6) The value of emf of a cell is written on the right of the cell diagram. Thus a zinc-copper cell has emf 1.1V and is represented as

$$Zn \mid ZnSO_4 \parallel CuSO_4 \mid Cu \qquad E = +1.1\ V$$

Convention regarding sign of emf value

The magnitude of the emf of a cell reflects the tendency of electrons to flow externally from one electrode to another. The electrons are transported through the cell solution by ions present and pass from the positive electrode (Cu in case of Daniel cell) to the negative electrode. This corresponds to a clockwise flow of electrons through the external circuit. Thus the emf of the cell is given the **+ve sign.** If the emf acts in the opposite direction through the cell circuit, it is quoted as **–ve value.** For example, Daniel cell has an emf of 1.1V and the copper electrode is positive. This can be expressed in two ways :

$$Zn \mid ZnSO_4 \parallel CuSO_4 \mid Cu \qquad E = +1.1\ V$$
$$Cu \mid CuSO_4 \parallel ZnSO_4 \mid Zn \qquad E = -1.1\ V$$

The negative sign indicates that the cell is not feasible in the given direction. The reaction will take place in the reverse direction.

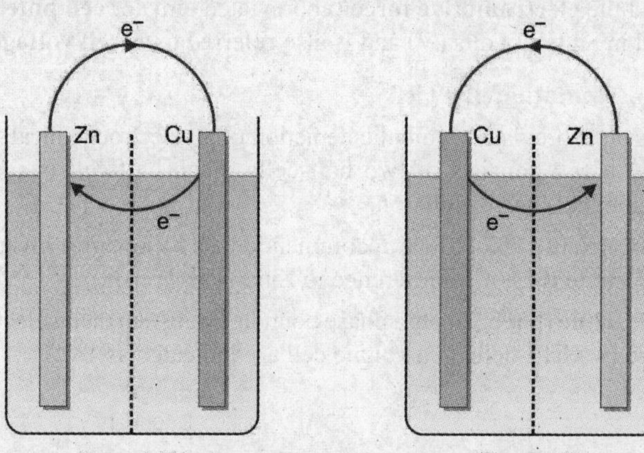

Figure 29.5
Illustration of emf sign in Daniel cell.

Calculating the emf of a cell

The emf of a cell can be calculated from the half-cell potentials of the two cells (anode and cathode) by using the following formula

$$E_{cell} = E_{cathode} - E_{anode}$$
$$= E_R - E_L \qquad \text{(IUPAC convention 3)}$$

where E_R and E_L are the reduction potentials of the right-hand and left-hand electrodes respectively. It may be noted that absolute values of these reduction potentials cannot be determined. These are found by connecting the half-cell with a standard hydrogen electrode whose reduction potential has been arbitrarily fixed as zero.

MEASUREMENT OF EMF OF A CELL

The emf of an unknown cell can be measured with the help of a **potentiometer** (Fig. 29.6). It consists of a wire AB which is about one meter long. The two ends of this wire are connected to a working battery W. A standard cell C_1 (*i.e.*, a cell of known emf) is connected to the end A. At the other end, the cell C_1 is connected to a galvanometer through a key K_1. The galvanometer is then joined to a sliding contact that moves on the wire AB. The cell C_2 whose emf is to be measured is similarly connected to the key K_2, the galvanometer and then the sliding contact. By using the key K_1, the cell C_1 is put into the circuit and the contact is moved to and fro along AB. When no current flows through the galvanometer, the point of contact X_1 is recorded. Then by using the key K_2, the cell C_2 is put into the circuit and the procedure is repeated to find the corresponding point X_2. The emf of the cell C_2 is calculated by using the following equation :

$$\frac{\text{emf of } C_2}{\text{emf of } C_1} = \frac{\text{distance } AX_2}{\text{distance } AX_1}$$

Cell emf's can also be measured by electronic voltmeters of the analog or digital type.

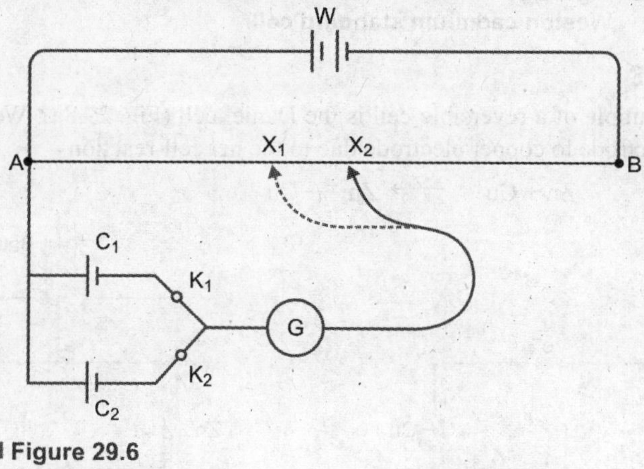

Figure 29.6
Measuring the emf of a cell with a potentiometer.

WESTON STANDARD CELL

A standard cell is one which provides a constant and accurately known emf. The Weston cadmium cell is most widely used as a standard cell. It is constructed in a H-shaped glass tube (Fig. 29.7). The positive electrode consists of mercury covered with a paste of solid mercurous sulfate (Hg_2SO_4) over which is placed a layer of cadmium sulfate crystals. The negative electrode is 12.5% cadmium amalgam, Cd(Hg), covered with cadmium sulfate crystals. The entire cell is filled with saturated cadmium sulfate solutions and sealed.

The cell reactions are :

at anode : $\quad Cd + SO_4^{2-} \rightleftharpoons CdSO_4 + 2e^-$

at cathode : $\quad Hg_2SO_4 + 2e^- \rightleftharpoons 2Hg + SO_4^{2-}$

Overall reaction : $\quad Cd + Hg_2SO_4 \rightleftharpoons CdSO_4 + 2Hg$

The emf of a cadmium standard cell is 1.0183 (V) at 293 K. It varies only slightly (0.0000406 volt/degree) with change of temperature. The emf remains constant for many years.

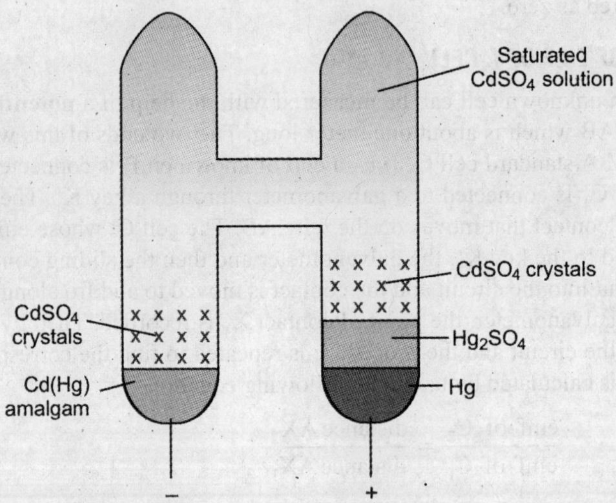

■ Figure 29.7
Weston cadmium standard cell.

REVERSIBLE CELLS

A familiar example of a reversible cell is the Daniel cell (Fig. 29.8a). We know that electrons flow from zinc electrode to copper electrode due to the net cell-reaction.

$$Zn + Cu^{2+} \longrightarrow Zn^{2+} + Cu$$

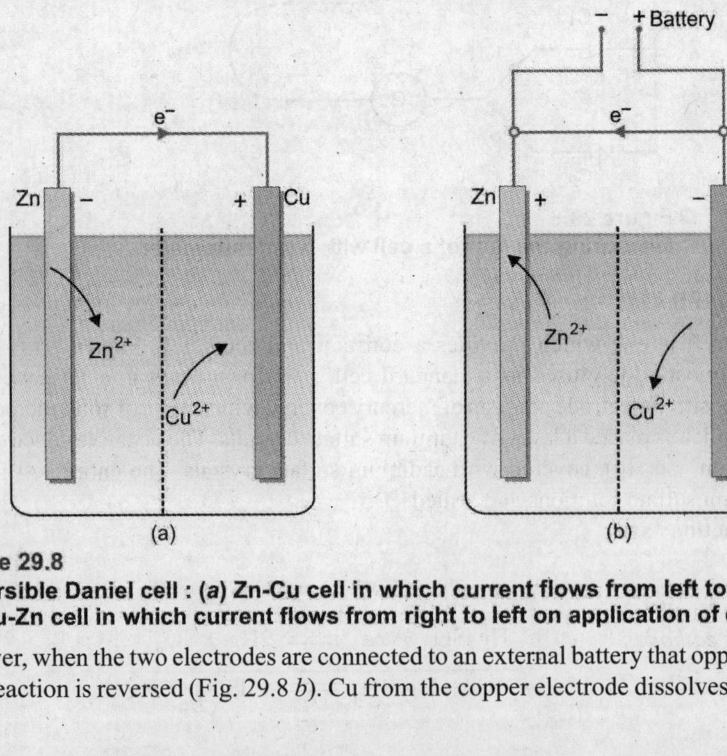

■ Figure 29.8
Reversible Daniel cell : (*a*) Zn-Cu cell in which current flows from left to right; (*b*) Cu-Zn cell in which current flows from right to left on application of external emf.

However, when the two electrodes are connected to an external battery that opposes the cell emf, the above reaction is reversed (Fig. 29.8 *b*). Cu from the copper electrode dissolves to form Cu^{2+} ion and Zn^{2+} ion is discharged on the zinc electrode to give Zn atom. The overall reaction taking place in the cell may be written as

$$Cu \longrightarrow Cu^{2+} + 2e^-$$
$$Zn^{2+} + 2e^- \longrightarrow Zn$$
$$\overline{Cu + Zn^{2+} \longrightarrow Cu^{2+} + Zn}$$

With the help of a potentiometer (Fig. 29.6), the cell emf is exactly balanced by the external emf. At the balance point, no current will flow through the circuit. Now let the external emf increase and then decrease by an infinitesimal amount. A minute current will flow first to the left and then to the right. This reversal of the cell current is accompanied by a corresponding change in the direction of the cell reaction. This type of reversible behavior is a feature of the reversible cells. **A reversible cell** may be defined as : **a cell that operates by reversal of the cell current and direction of cell reaction by infinitesimal change of emf on either side of the balance point.**

All voltaic cells are reversible cells.

RELATION BETWEEN EMF AND FREE ENERGY

When a cell produces a current, the current can be used to do work–to run a motor, for instance. Thermodynamic principles can be employed to derive a relation between electrical energy and the maximum amount of work, W_{max}, obtainable from the cell. The maximum amount of work obtainable from the cell is the product of charge flowing per mole and maximum potential difference, E, through which the charge is transferred.

$$W_{max} = -nFE \qquad ...(1)$$

where n is the number of moles of electrons transferred and is equal to the valence of the ion participating in the cell reaction. F stands for Faraday and is equal to 96,500 coulombs and E is the emf to the cell.

According to thermodynamics, the maximum work that can be derived from a chemical reaction is equal to the free energy (ΔG) for the reaction.

$$W_{max} = \Delta G \qquad ...(2)$$

Therefore from (1) and (2), we can write

$$\Delta G = -nFE \qquad ...(3)$$

According to Gibbs-Helmholtz equation, the decrease in free energy of a system at constant pressure is given by the expression

$$-\Delta G = -\Delta H - T\left(\frac{\partial(\Delta G)}{\partial T}\right)_p$$

where $-\Delta H$ is the decrease in heat content of the system or heat of the reaction at constant pressure.

Substituting the value of ΔG from (3)

$$nFE = -\Delta H - T\left(\frac{\partial(-nFE)}{\partial T}\right)_p$$

$$nFE = -\Delta H + nFT\left(\frac{\partial E}{\partial T}\right)_p$$

$$E = \frac{-\Delta H}{nF} + T\left(\frac{\partial E}{\partial T}\right)_p$$

The temperature coefficient of the emf of the cell ($\partial E/\partial T$), at constant pressure, therefore, determines whether the electrical energy would be equal to, more or less than the heat of reaction. Accordingly, three cases may arise:

Case 1. When the temperature coefficient is zero *i.e.*,

$$(\partial E/\partial T)_p = 0, \text{ then}$$

$$nFE = -\Delta H$$

Case 2. When the temperature coefficient is +ve *i.e.*,

$$(\partial E/\partial T)_p > 0, \text{ then}$$

$$nFE > -\Delta H$$

The additional energy will come either from the surroundings or the temperature of the cell would fall.

Case 3. When the temperature coefficient is negative *i.e.*,

$$(\partial E/\partial T)_p < 0, \text{ then}$$

$$nFE < -\Delta H$$

The excess energy will be given either to the surroundings or the temperature of the cell would rise.

If the heat of the reaction (or free energy change) and the temperature coefficient of a cell are known, we can calculate the emf, E, of the cell. For example, in case of a Daniel cell $(\partial E/\partial T)_p$ is nearly zero, $n = 2$ and $\Delta H = -50{,}100$ cals.

Since 1 Cal = 4.185 Volt-Coulomb,

$$E = \frac{-\Delta H}{nF} = \frac{-4.185 \times (-50{,}100)}{2 \times 96{,}500}$$

$$= 1.09 \text{ V}$$

Thus the emf of Daniel cell is 1.09 volts. Conversely, if the emf of a reversible cell and its temperature coefficient $(\partial E/\partial T)_p$ are known, ΔH (or ΔG) can be calculated. The heats of reaction calculated from emf measurements are nearly the same as derived from thermal measurements.

Single electrode potential

An electrochemical cell consists of two half-cells. With an open-circuit, the metal electrode in each half-cell transfers its ions into solution. Thus an individual electrode develops a potential with respect to the solution. The potential of a single electrode in a half-cell is called the **Single electrode potential.** Thus in a Daniel cell in which the electrodes are not connected externally, the anode Zn/Zn^{2+} develops a negative charge and the cathode Cu/Cu^{2+}, a positive charge. The amount of the charge produced on individual electrode determines its single electrode potential.

The single electrode potential of a half-cell depends on : (*a*) concentration of ions in solution; (*b*) tendency to form ions; and (*c*) temperature

Standard emf of a cell

The emf generated by an electrochemical cell is given by the symbol E. It can be measured with the help of a potentiometer. The value of emf varies with the concentration of the reactants and products in the cell solutions and the temperature of the cell. When the emf of a cell is determined under standard conditions, it is called **the standard emf.** The standard conditions are (*a*) 1 M solutions of reactants and products; and (*b*) temperature of 25°C. Thus standard emf may be defined as : **the emf of a cell with 1 M solutions of reactants and products in solution measured at 25°C.**

Standard emf of a cell is represented by the symbol **E°.** With gases 1 atm pressure is a standard condition instead of concentration.

For a simple Zn-Cu voltaic cell, the standard emf, E°, is 1.10 V. This means that the emf of the cell operated with [Cu^{2+}] and [Zn^{2+}] both at 1 M and 298 K is 1.10 V. That is,

$$Zn \mid Zn^{2+} (aq, 1\underline{M}) \parallel Cu^{2+} (aq, 1\underline{M}) \mid Cu \qquad E^\circ = +1.1\ V$$

Determination of emf of a half-cell

By a single electrode potential, we also mean the emf of an isolated half-cell or its half-reaction. The emf of a cell that is made of two half-cells can be determined by connecting them to a voltmeter. However, there is no way of measuring the emf of a single half-cell directly. A convenient procedure to do so is to combine the given half-cell with another standard half-cell. The emf of the newly constructed cell, E, is determined with a voltmeter. The emf of the unknown half-cell, E°, can then be calculated from the expression

$$E_{measured} = E_R - E_L$$

If the standard half-cell acts as anode, the equation becomes.

$$E_R = E_{measured} \qquad (\because E_L = 0)$$

On the other hand, if standard half-cell is cathode, the equation takes the form

$$E_L = -E_{measured} \qquad (\because E_R = 0)$$

The standard hydrogen half-cell or **Standard Hydrogen Electrode (SHE),** is selected for coupling with the unknown half-cell. It consists of a platinum electrode immersed in a 1 M solution of H^+ ions maintained at 298 K. Hydrogen gas at 101325 Pa enters the glass hood and bubbles over the platinum electrode. The hydrogen gas at the platinum electrode passes into solution, forming H^+ ions and electrons.

$$H_2 \longrightarrow 2H^+ + 2e^-$$

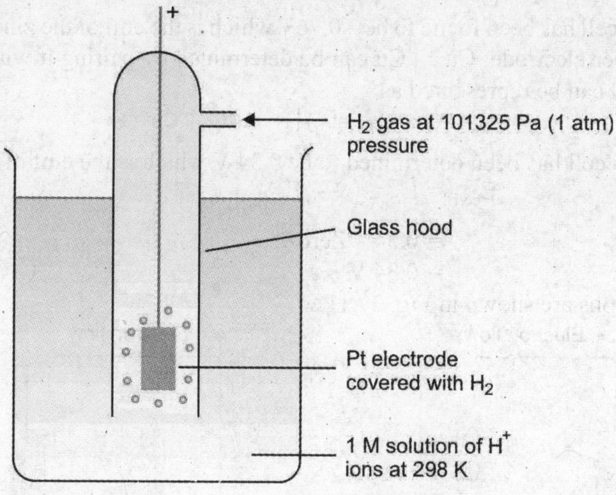

■ **Figure 29.9**
The standard hydrogen electrode.

The emf of the standard hydrogen electrode is arbitrarily assigned the value of zero volts. So, SHE can be used as a standard for other electrodes.

The half-cell whose potential is desired, is combined with the hydrogen electrode and the emf of the complete cell determined with a voltmeter. The emf of the cell is the emf of the half-cell.

$$E^\circ_{cell} = E^\circ_R - E^\circ_L$$

$$= 0 - 0.76 = -0.76\ V$$

For example, it is desired to determine the emf of the zinc electrode, $Zn \mid Zn^{2+}$. It is connected with the SHE as shown in Fig. 29.10. The complete electrochemical cell may be represented as :

$$Zn \mid Zn^{2+} \parallel H^{+} \mid H_2\ (1\ atm),\ Pt$$

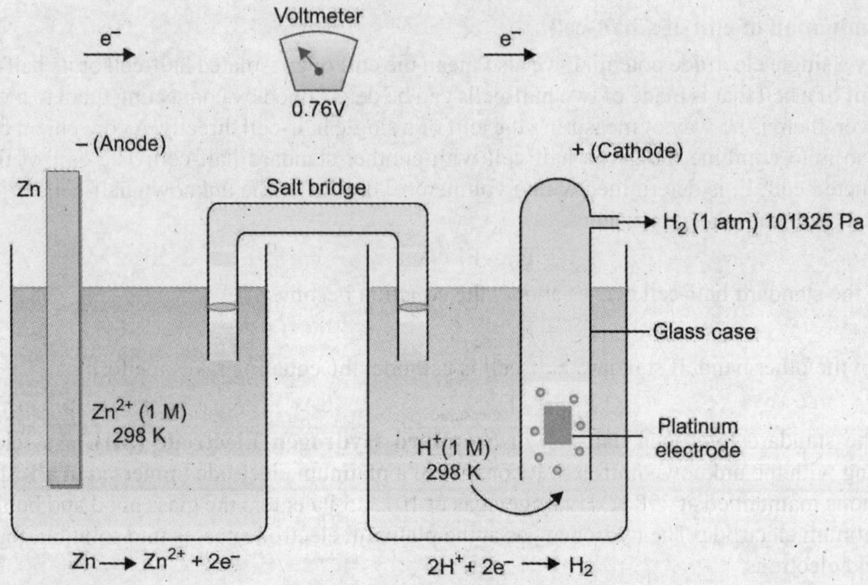

■ **Figure 29.10**
The zinc electrode (Zn, Zn^{2+}) coupled with hydrogen electrode.

The emf of the cell has been found to be – 0.76V which is the emf of the zinc half-cell. Similarly, the emf of the copper electrode, $Cu^{2+} \mid Cu$ can be determined by pairing it with the SHE when the electrochemical cell can be represented as :

$$Pt,\ 2H_2\ (1\ atm) \mid H^{+} \parallel Cu^{2+} \mid Cu$$

The emf of this cell has been determined to be 0.34 V which is the emf of the copper half-cell.

$$E^{\circ}_{cell} = E^{\circ}_{Cu/Cu^{2+}} - E^{\circ}_{SHE}$$
$$= 0.34 - \text{Zero}$$
$$= 0.34\ V$$

The two situations are shown in Fig. 29.11.

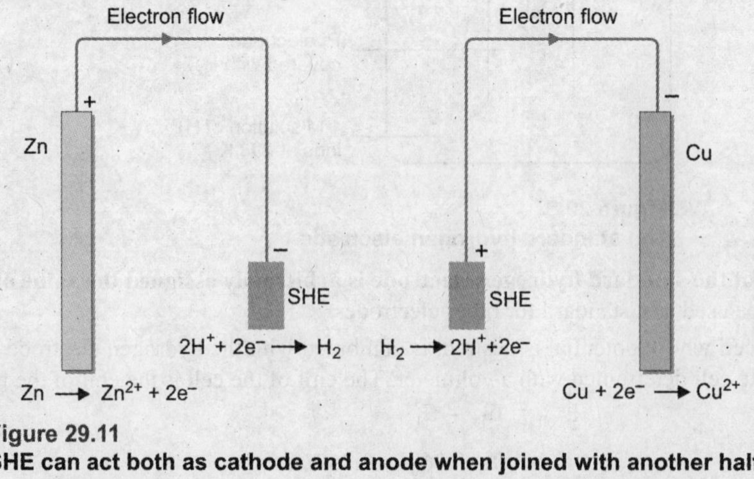

■ **Figure 29.11**
SHE can act both as cathode and anode when joined with another half-cell.

When it is placed on the right-hand side of the Zinc electrode, the hydrogen electrode reaction is

$$2H^+ + 2e^- \longrightarrow H_2$$

The electrons flow to the SHE and it acts as the cathode.

When the SHE is placed on the left hand side, the electrode reaction is

$$H_2 \longrightarrow 2H^+ + 2e^-$$

The electrons flow to the copper electrode and the hydrogen electrode as the anode. Evidently, **the SHE can act both as anode and cathode** and, therefore can be used to determine the emf of any other half-cell electrode (or single electrode).

IUPAC convention places the SHE on the left-hand side

In the procedure for determining the emf of a given half-cell, the standard hydrogen electrode can be placed on the left-hand or the right-hand. In the convention adopted by the IUPAC (International Union of Pure and Applied Chemistry), the SHE is always placed on the left-hand side of the half-cell under study. The electrons flow from left-to-right and the given half-cell electrode gains electrons (reduction). The observed emf of the combined electrochemical cell is then the emf of the half-cell on the right-hand. Such emf values of half-cells, or half reactions, are known as the **Standard reduction potentials or Standard potentials.** However, if the SHE be placed on the right-hand side of the given half-cell, the potential so obtained is called as the **Standard oxidation potential.** The latter potentials are the standard potentials with the sign reversed, the only difference being that cells have been turned around.

According to IUPAC convention, the standard reduction potentials alone are the standard potentials. The values of the standard potentials at 25°C (298 K) for some common Reduction Half-reactions are listed in Table **29.1.**

TABLE 29.1. STANDARD REDUCTION POTENTIALS AT 25°C (298 K)

Reduction Half-reaction	E° (V)
$F_2 + 2e^- \longrightarrow 2F^-$	2.87
$H_2O_2 + 2H^+ + 2e^- \longrightarrow 2H_2O$	1.78
$PbO_2 + SO_4^{2-} + 4H^+ + 2e^- \longrightarrow PbSO_4 + 2H_2O$	1.69
$Au^{3+} + 3e^- \longrightarrow Au$	1.50
$Cl_2 + 2e^- \longrightarrow 2Cl^-$	1.36
$O_2 + 4H^+ + 4e^- \longrightarrow 2H_2O$	1.23
$MnO_2 + 4H^+ + 2e^- \longrightarrow Mn^{2+} + 2H_2O$	1.21
$Br_2 + 2e^- \longrightarrow 2Br^-$	1.09
$Fe^{3+} + e^- \longrightarrow Fe^{2+}$	0.77
$MnO_4^- + e^- \longrightarrow MnO_4^{2-}$	0.56
$Cu^{2+} + 2e^- \longrightarrow Cu$	0.34
$\mathbf{2H^+ + 2e^- \longrightarrow H_2}$	**0.00**
$Sn^{2+} + 2e^- \longrightarrow Sn$	– 0.14
$Cd^{2+} + 2e^- \longrightarrow Cd$	– 0.40
$Zn^{2+} + 2e^- \longrightarrow Zn$	– 0.76
$H_2 + 2e^- \longrightarrow 2H^-$	– 2.23
$Mg^{2+} + 2e^- \longrightarrow Mg$	– 2.37
$Na^+ + e^- \longrightarrow Na$	– 2.71
$Li^+ + e^- \longrightarrow Li$	– 3.05

USING STANDARD POTENTIALS

In Table 29.1 the standard reduction potentials (E°) are arranged in the order of increasing potentials. The relative position of electrodes (M/M^+) in the table can be used to predict the reducing or oxidizing ability of an electrode.

The electrodes that are relatively positive indicate that reduction reaction involving addition of electrons,

$$M^+ + e^- \longrightarrow M$$

is possible. In case of relatively negative potential involving loss of electrons,

$$M \longrightarrow M^+ + e^-$$

is indicated. It also follows that the system with higher electrode potential will be reduced by the system with lower electrode potential.

Predicting the Oxidizing or Reducing Ability

Let us consider a series of elements Cu, H_2,Ni, Zn and their ions. These four elements could act as reducing agents. On the other hand, their ions Cu^{2+}, H^+, Ni^{2+} and Zn^{2+} can act as electron acceptors or oxidizing agents. If we list the respective half-reactions (or electrodes) in order of descending E° values, we will have placed the oxidizing agents in descending order of their ability to attract electrons.

Increasing Strength as Oxidising Agent ↑ | Increasing Strength as Reducing Agent ↓

$$Cu^{2+} + 2e^- \longrightarrow Cu \qquad E° = +0.34\ V$$

$$2H^+ + 2e^- \longrightarrow H_2 \qquad E° = 0.00$$

$$Ni^{2+} + 2e^- \longrightarrow Ni \qquad E° = -0.25$$

$$Zn^{2+} + 2e^- \longrightarrow Zn \qquad E° = -0.76$$

It is noteworthy that the value of E° becomes more negative down the series. This means that Cu^{2+} is the best oxidizing agent (most electron-attracting ion) of those in the list. That is, Cu^{2+} shows the greatest tendency to be reduced. Conversely, Zn^{2+} is the worst oxidizing agent, being the least electron-attracting ion. Of the elements Cu, H_2, Ni and Zn, Zn is the best reducing agent (best electron donor), since E° for the half-reaction

$$Zn \longrightarrow Zn^{2+} + 2e^- \qquad E° = +0.76\ V$$

has the most positive value. By the reasoning, Cu is the worst reducing agent.

The table of half reaction potentials above tells us that at standard conditions the following reactions occur spontaneously.

Increasing Oxidising Ability ↑

Cu^{2+} can oxidize H_2, Ni, Zn

H^+ can oxidize Ni, Zn

Ni^{2+} can oxidize Zn

Increasing Reducing Ability ↓

H_2 can reduce Cu^{2+}

Ni can reduce H^+ and Cu^{2+}

Zn can reduce Ni^{2+}, H^+ and Cu^{2+}

Some important points concerning the Table of Standard Reduction Potentials (Table 29.1) are:

(1) The more positive the value of E°, the better the oxidizing ability (the greater the tendency to be reduced) of the ion or compound, on moving upward in the Table.

(2) The more negative the value of E° the better the reducing ability of the ions, elements or compounds on moving downward in the Table.

(3) Under standard conditions, any substance in this Table will spontaneously oxidize any other substance lower than it in the Table.

Predicting cell emf

The standard emf, E°, of a cell is the standard reduction potential of right-hand electrode (cathode) minus the standard reduction potential of the left-hand electrode (anode). That is,

$$E^\circ_{cell} = E^\circ_{right} - E^\circ_{left}$$
$$= \text{Cathode potential} - \text{Anode potential}$$

Let us predict the emf of the cell

$$Zn(s) \mid Zn^{2+}(aq) \parallel Ag^{+}(aq) \mid Ag$$

by using the E° values from the Table

$$E^\circ_{cell} = E^\circ_R - E^\circ_L$$
$$= 0.80 - (-0.763)$$
$$= 0.80 + 0.763$$
$$= 1.563 \text{ V}$$

The answer is so clear from Fig. 29.12.

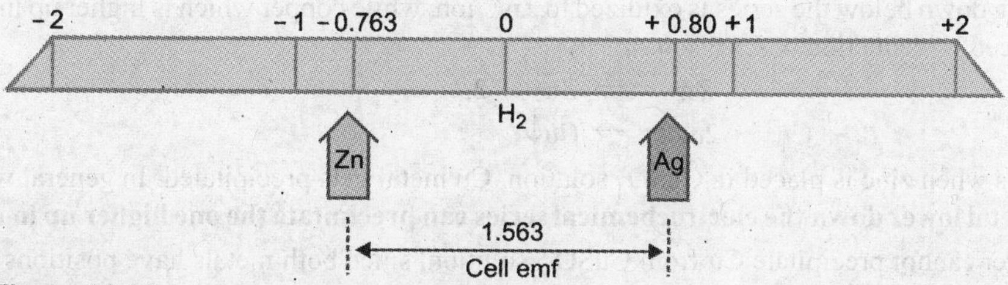

■ **Figure 29.12**
Diagrammatic representation of Cell emf.

Predicting Feasibility of Reaction

The feasibility of a redox reaction can be predicted with the help of the electrochemical series. The net emf of the reaction, E_{cell}, can be calculated from the expression

$$E^\circ_{cell} = E^\circ_{cathode} - E^\circ_{anode}$$

In general, if E°_{cell} = + ve, the reaction is feasible

E°_{cell} = – ve, the reaction is not feasible

SOLVED PROBLEM 1

Predict whether the reaction

$$2\,Ag(s) + Zn^{2+}(aq) \longrightarrow Ag^{+}(aq) + Zn(s)$$

is feasible or not. Consult the table for the E° values.

SOLUTION

The cell half reactions are

Anode : $2Ag(s) \longrightarrow 2Ag^{+}(aq) + 2e^{-}$ $\quad E^\circ = 0.80$ V

Cathode : $Zn^{2+}(aq) + 2e^{-} \longrightarrow Zn(s)$ $\quad E^\circ = -0.763$ V

$$E^\circ_{cell} = E^\circ_{cathode} - E^\circ_{anode}$$
$$\therefore \quad E^\circ_{cell} = -0.763 \text{ V} - 0.80$$
$$= -1.563$$

Since E°_{cell} is negative, the given reaction is **not feasible**.

SOLVED PROBLEM 2

Determine the feasibility of the reaction

$$2Al(s) + 2Sn^{4+}(aq) \longrightarrow 2Al^{3+} + 3Sn^{2+}(aq)$$

Consult the table for the E° values.

SOLUTION

The given reaction consists of the following half reactions

Anode : $2Al(s) \longrightarrow 2Al^{3+} + 6e^-$ $E° = -1.66$ V

Cathode : $3Sn^{4+} + 6e^- \longrightarrow 3Sn^{2+}$ $E° = +0.15$

$$E°_{cell} = 0.15 - (-1.66)$$
$$= 1.81 \text{ V}$$

Since $E°_{cell}$ is positive, the reaction is **feasible.**

Predicting whether a metal will displace another metal from its salt solution or not

As already shown, the metals near the bottom of the electrochemical series are strong reducing agents and are themselves oxidized to metal ions. On the contrary, the metals lying higher up in the series are strong oxidizing agents and their ions are readily reduced to the metal itself. For example, zinc lying down below the series is oxidized to Zn^{2+} ion, while copper which is higher up in the series is produced by reduction of Cu^{2+} ion.

$$Zn \longrightarrow Zn^{2+} + 2e^-$$
$$Cu^{2+} + 2e^- \longrightarrow Cu\downarrow$$

Thus when zinc is placed in $CuSO_4$ solution, Cu metal gets precipitated. In general we can say that **a metal lower down the electrochemical series can precipitate the one higher up in the series.**

Silver cannot precipitate Cu from $CuSO_4$, solution, since both metals have positions higher up in the series and are strong oxidizing agents.

Predicting whether a metal will displace hydrogen from a dilute acid solution

Any metal above hydrogen in the electrochemical series is a weaker reducing agent than hydrogen itself and cannot reduce H^+ to H_2. Any metal lying below hydrogen is a stronger reducing agent than hydrogen and will convert H^+ to H_2. This explains why Zn lying below hydrogen reacts with dil H_2SO_4 to liberate H_2, while Cu lying above hydrogen does not react.

$$Zn + H^+ \text{ (dil } H_2SO_4) \longrightarrow Zn^{2+} + H_2\uparrow$$
$$Cu + H^+ \text{ (dil } H_2SO_4) \not\longrightarrow Cu^{2+} + H_2$$

THE NERNST EQUATION

We know experimentally that the potential of a single electrode or half-cell varies with the concentration of ions in the cell. In 1889 Walter Nernst derived a mathematical relationship which enable us to calculate the half-cell potential, E, from the standard electrode potential, E°, and the temperature of the cell. This relation known as the **Nernst equation** can be stated as

$$E = E° - \frac{2.303\,RT}{nF}\log K \qquad ...(1)$$

where
- $E°$ = standard electrode potential
- R = gas constant
- T = Kelvin temperature
- n = number of electrons transferred in the half-reaction
- F = Faraday of electricity
- K = equilibrium constant for the half-cell reaction as in equilibrium law

Calculation of Half-cell potential

For an oxidation half-cell reaction when the metal electrode M gives M^{n+} ion,

$$M \longrightarrow M^{n+} + ne^-$$

the Nernst equation takes the form

$$E = E° - \frac{2.303\,RT}{nF}\log\frac{[M^{n+}]}{[M]} \qquad ...(2)$$

The concentration of solid metal [M] is equal to zero. Therefore, the Nernst equation can be written as

$$E = E° - \frac{2.303\,RT}{nF}\log[M^{n+}] \qquad ...(3)$$

Substituting the values of R, F and T at 298 K, the quantity 2.303 RT/F comes to be 0.0591. Thus the Nernst equation (3) can be written in its simplified form as

$$E = E° - \frac{0.0591}{n}\log[M^{n+}]$$

This is the equation for a half-cell in which oxidation occurs. In case it is a reduction reaction, the sign of E will have to be reversed.

SOLVED PROBLEM 3

What is the potential of a half-cell consisting of zinc electrode in 0.01M $ZnSO_4$ solution at 298 K, E° =0.763 V.

SOLUTION

The half-cell reaction is

$$Zn \longrightarrow Zn^{2+} + 2e^- \qquad (oxidation)$$

The Nernst equation for the oxidation half-cell reaction is

$$E = E° - \frac{0.0591}{n}\log[Zn^{2+}]$$

The number of electrons transferred $n = 2$ and E° =0.763 V.
Substituting these values in the Nernst equation we have

$$E = 0.763 - \frac{0.0591}{2}\log(0.01)$$

$$= 0.763 - \frac{0.0591}{2}(-2)$$

$$= 0.763 + 0.0591 = \mathbf{0.8221\ V}$$

Calculation of Cell potential

The Nernst equation is applicable to cell potentials as well. Thus,

$$E_{cell} = E°_{cell} - \frac{0.0591}{n}\log K$$

K is the equilibrium constant of the redox cell reaction.

SOLVED PROBLEM 4

Calculate the emf of the cell.

$$Zn \mid Zn^{2+}(0.001M) \parallel Ag^{+}(0.1M) \mid Ag$$

The standard potential of Ag/Ag^+ half-cell is + 0.80 V and Zn/Zn^{2+} is – 0.76 V.

SOLUTION

Step 1. Write the half-cell reactions of the anode and the cathode. Then add the anode and cathode half reactions to obtain the cell reaction and the value of $E°_{cell}$.

Cathode :	$2Ag^+ + 2e^- \longrightarrow 2Ag$	$E° = +0.80$ V
Anode :	$Zn \longrightarrow Zn^{2+} + 2e^-$	$E° = -0.76$ V
Cell :	$Zn + 2Ag^+ \rightleftharpoons Zn^{2+} + 2Ag$	$E° = 1.56$ V

Step 2. K for the cell reaction $= \dfrac{[Zn^{2+}]}{[Ag^+]^2}$

substitute the given values in the Nernst equation and solving for E_{cell}, we have

$$E_{cell} = E^{\circ}_{cell} - \frac{0.0591}{n}\log K$$

$$= 1.56 - \frac{0.0591}{2}\log\frac{[Zn^{2+}]}{[Ag^+]^2}$$

$$= 1.56 - \frac{0.0591}{2}\log\frac{[10^{-3}]}{[10^{-1}]^2}$$

$$= 1.56 - 0.02955\ (\log 10^{-1})$$

$$= 1.56 + 0.02955$$

$$= \mathbf{1.58955\ V}$$

Calculation of Equilibrium constant for the cell reaction

The Nernst equation for a cell is

$$E_{cell} = E°_{cell} - \frac{0.0591}{n}\log K$$

At equilibrium, the cell reaction is balanced and the potential is zero. The Nernst equation may, now, be written as

$$0 = E°_{cell} - \frac{0.0591}{n}\log K$$

or

$$\log K = \frac{nE°_{cell}}{0.0591}$$

SOLVED PROBLEM 5

Calculate the equilibrium constant for the reaction between silver nitrate and metallic zinc.

SOLUTION

Step 1. Write the equation for the reaction

$$2Ag^+ + Zn \rightleftharpoons Zn^{2+} + 2Ag \qquad E°cell = 1.56\ V$$

Step 2. Substitute values in the Nernst equation at equilibrium

$$\log K = \frac{nE°_{cell}}{0.0591}$$

$$0 = 1.56 - 0.03 \log K$$

$$-1.56 = -0.03 \log K$$

$$\log K = \frac{-1.56}{-0.03} = 52$$

$$\mathbf{K = 1 \times 10^{52}}$$

OTHER REFERENCE ELECTRODES

The standard hydrogen electrode (SHE) is not the most convenient standard electrode to use in the laboratory. The gas has to be carefully controlled and hydrogen gas can form explosive mixtures with air. Any other electrode system whose potential has been determined relative to the SHE can also be used as **Secondary standard electrode.**

(1) The Standard Silver-Silver Electrode

In this electrode system, silver wire (or strip) is covered with silver chloride (a highly insoluble substance). It is dipped in potassium chloride solution in which the concentration of Cl^- ion is 1 M This electrode can be represented as

$$Ag, AgCl \mid Cl^- (1\ M)$$

The half-cell reaction of the Ag-AgCl electrode is

$$AgCl + e^- \longrightarrow Ag + Cl^-$$

We can set up a cell involving this electrode and the hydrogen electrode

$$Pt, H_2 \mid H^+ (1\ M) \parallel Cl^- (1\ M) \mid AgCl, Ag$$

with a KCl salt bridge connecting the two solutions. The emf of the combined cell is found to be 0.2224 V. Thus the standard electrode potential for the silver-silver chloride electrode is 0.2224 V.

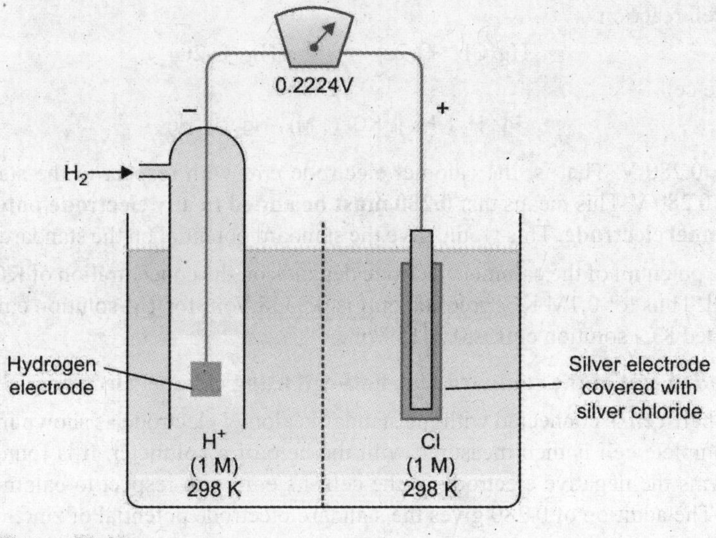

■ Figure 29.13
Schematic diagram of Ag-AgCl electrode coupled with SHE (salt bridge replaced by porous partition).

(2) The Calomel Electrode

It is the most commonly used secondary standard reference electrode. The standard calomel electrode, **SCE,** consists of a wide glass-tube with a narrow side-tube. It is set up as illustrated in Fig. 29.14. A platinum wire is dipping into liquid mercury covered with solid mercurous chloride (Hg_2Cl_2, calomel). The tube is filled with a 1 M solution of KCl (or saturated KCl solution). The side-tube containing KCl solution provides the salt bridge which connects the electrode to any other electrode.

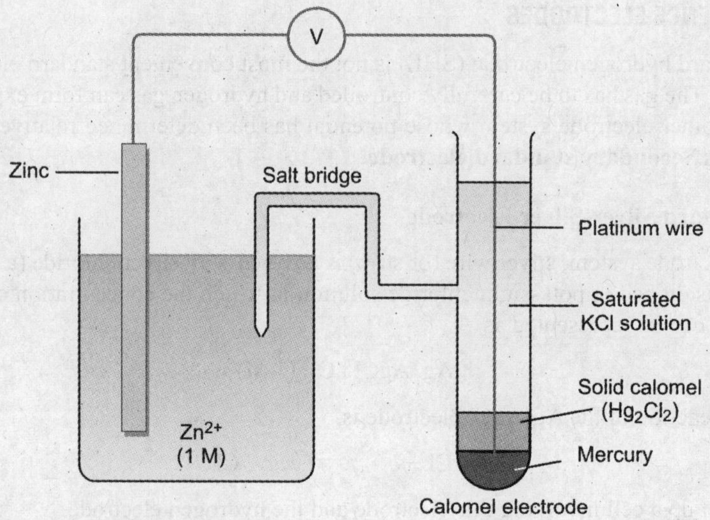

Figure 29.14
The calomel electrode coupled with zinc electrode for emf determine.

The calomel electrode is represented as

$$Cl^- \text{ (1M)} \mid Hg_2Cl_2, Hg$$

and the half-cell reaction is

$$Hg_2Cl_2 + 2e^- \longrightarrow 2Hg + 2Cl^-$$

The emf of the cell

$$Pt, H_2 \mid H^+ \parallel KCl \text{ (1 M)} \mid Hg_2Cl_2, Hg$$

at 298 K is + 0.280 V. That is, the calomel electrode emf with respect to the standard hydrogen electrode is + 0.280 V. This means that **0.280 must be added to any electrode potential measured against a calomel electrode.** This would give the standard potential on the standard hydrogen scale.

Note. The potential of the calomel electrode depends on the concentration of KCl solution taken in the half-cell. Thus for 0.1M KCl solution emf is 0.3338 Volt; for 1M solution emf is 0.2800 Volt; and for saturated KCl solution emf is 0.2415 Volt.

Determination of emf of the standard zinc half-cell using calomel electrode

The zinc half-cell is connected with the standard calomel electrode as shown in Fig. 29.14. The emf of the complete cell is then measured with the help of a voltmeter. It is found to be 1.040 V. Since zinc forms the negative electrode of the cell, its emf with respect to calomel electrode will be – 1.040 V. The addition of 0.280 gives the standard electrode potential of zinc.

$$(-1.040 + 0.280) = -0.76 \text{ V}$$

The dipping calomel electrode

For many purpose, a calomel electrode is manufactured as a self-contained unit. It is used by simply dipping the tip of the electrode in the solution of another electrode whose emf is to be determined. This type of electrode is often referred as the dipping calomel electrode (Fig. 29.15).

The mercury and calomel (Hg_2Cl_2) are held in a narrow central tube by a porous cotton wool plug. A platinum wire dips into the mercury. The KCl saturated solution contained in the electrode makes contact with the other electrode solution through a sintered glass disc. This disc functions as a 'salt bridge' between the two electrode solutions. Thus a galvanic cell is set up and its emf is measured with the help of a voltmeter.

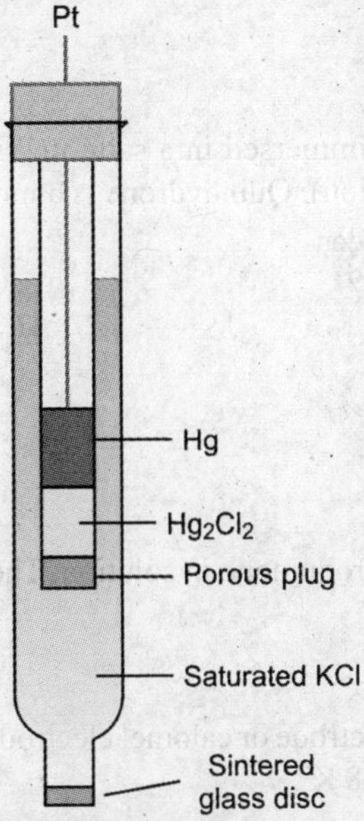

■ **Figure 29.15**
A dipping calomel electrode.

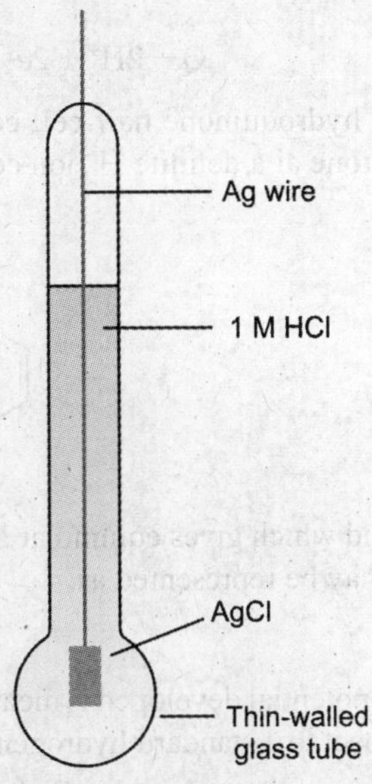

■ **Figure 29.16**
Glass electrode

The Glass Electrode

A commonly used secondary standard electrode is the so-called glass electrode. Its emf is determined by coupling with a standard calomel electrode (SCE). The glass electrode provides one of the easiest methods for measuring the pH of a given solution.

A simple type of glass electrode (Fig. 29.16) consists of a glass tube having a thin-walled bulb at the lower end. The bulb contains a 1M HCl solution. Sealed into the glass-tube is a silver wire coated with silver chloride at its lower end. The lower end of this silver wire dips into the hydrochloric acid, forming silver-silver chloride electrode. The glass electrode may be represented as

$$\text{Ag, AgCl} \mid 1\text{ M HCl} \,\vdots\, H^+ \text{ (Test Solution)}$$

When placed in a solution, the potential of the glass electrode depends on the H^+ ion concentration of the solution. The potential develops across the glass membrane as a result of a concentration difference of H^+ ions on the two sides of the membrane. This happens much in the same way as the emf of a concentration cell develops.

The potential of a glass electrode can be determined against a standard calomel electrode (SCE).

Quinhydrone Electrode

It is a widely used secondary standard electrode. It involves the redox reaction between quinone (Q) and hydroquinone (QH_2),

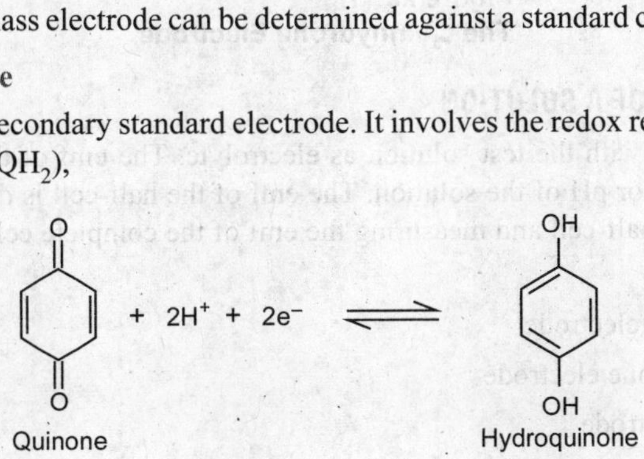

or, simply

$$Q + 2H^+ + 2e^- \rightleftharpoons QH_2$$

The hydroquinone half-cell consists of a platinum strip immersed in a saturated solution of quinhydrone at a definite H^+ ion concentration (buffered solution). Quinhydrone is a molecular

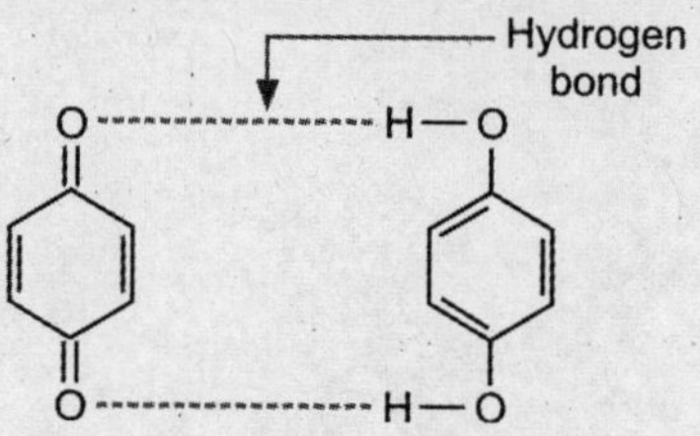

compound which gives equimolar amounts of quinone and hydroquinone in solution. The electrode system may be represented as

$$Pt \mid QH_2, Q, H^+$$

The potential developed is measured against a hydrogen electrode or calomel electrode. The emf with respect to a standard hydrogen electrode is 0.2875 V at 298 K.

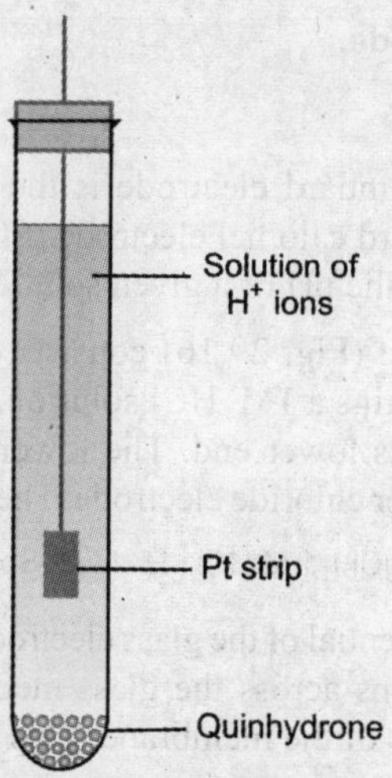

■ **Figure 29.17**
The Quinhydrone Electrode.

DETERMINATION OF pH OF A SOLUTION

A half-cell is set up with the test solution as electrolyte. The emf of the cell depends on the concentration of H^+ ions or pH of the solution. The emf of the half-cell is determined by coupling it with another standard half-cell and measuring the emf of the complete cell. The commonly used standard electrodes are :

(*a*) The hydrogen electrode

(*b*) The quinhydrone electrode

(*c*) The glass electrode

Using Hydrogen electrode

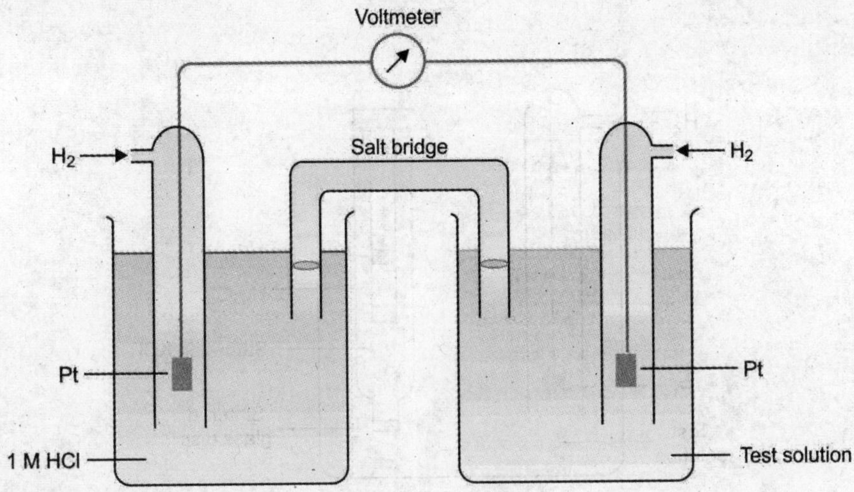

Figure 29.18
Determination of pH with Hydrogen electrode.

A standard hydrogen electrode is coupled with another hydrogen electrode which contains the solution of unknown pH (Fig. 29.18). In both half-cells hydrogen gas is used at 1 atm pressure or 101325.00 Pa and 298 K. The emf of the complete cell

$$Pt \mid H_2(1\ atm) \mid H^+(1M) \parallel H^+(Unknown) \mid H_2(1\ atm) \mid Pt$$

is recorded experimentally.

Calculations. The second electrode reaction is

$$H^+ + e^- \rightleftharpoons \tfrac{1}{2}H_2$$

The electrode potential of the second half-cell is given by the Nernst equation

$$E = E° + \frac{2.303\,RT}{nF}\log\frac{[H^+]}{H_2^{1/2}}$$

Since $H_2^{1/2} = 1$ and $E° = 0$, we have

$$E = \frac{2.303\,RT}{nF}\log[H^+] = \frac{2.303\,RT}{nF}\log[H^+]$$

Substituting the values of R, T and n (charge number of H^+) and F (Faraday constant), the expression becomes

$$E = 0.0591 \log [H^+]$$

or $$E = -0.0591\ pH \qquad ...(A)$$

The emf of the E_{cell}, is given as

$$E_{cell} = E_{right} - E_{left}$$

The left-hand electrode is the standard hydrogen electrode and its emf will be zero.

∴ $$E_{cell} = E_{right} = -(-0.0591)\ pH$$

Using A, $$E_{cell} = -(-0.0591 \times pH) = 0.0591 \times pH$$

or $$pH = \frac{E_{cell}}{0.059}$$

Knowing the observed value of emf of the cell, we can calculate pH of the test-solution.

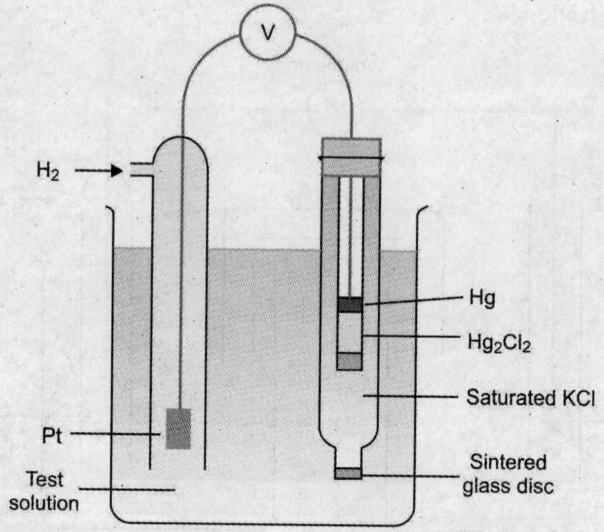

Figure 29.19
A calomel electrode coupled with unknown hydrogen electrode.

Using SCE instead of SHE

A hydrogen electrode containing solution of unknown pH is paired with a standard calomel electrode (Fig. 29.19). The complete cell may be represented as

$$\text{Pt, } H_2 \text{ (101325 Pa)} \mid H^+ \text{ (Unknown)} \parallel \text{KCl (Sat Solution)} \mid Hg_2Cl_2 \mid Hg$$

By noting the emf of the cell with the help of a voltmeter, the pH of the unknown solution can be calculated as follows :

The emf of the cell will be given by the expression

$$E_{cell} = E_{right} - E_{left}$$

$$= 0.2415 - (-0.0591 \times pH) \quad \text{...(from eq A)}$$

$$E_{cell} = 0.2415 + 0.0591 \times pH$$

or $$pH = \frac{E_{cell} - 0.2415}{0.0591}$$

Merits and demerits. The hydrogen electrode give absolute values of pH while other electrodes yield relative values. However, it is not convenient for routine measurement of pH because :

(1) It requires hydrogen gas and is difficult to set up and transport.
(2) It requires considerable volume of test solution.
(3) The test solution might 'poison' the surface of the platinum electrode.
(4) The potential of the electrode is already by changes in barometric pressure.

SOLVED PROBLEM 6

The emf of the following cell at 298 K is 0.445V.

$$\text{Pt, } H_2 \text{ (1 atm or 101325.00 Pa)} \mid H^+ \text{ (test solution)} \parallel \text{KCl (sat solution)} \mid Hg_2Cl_2 \mid Hg$$

Calculate the pH of the unknown solution, E_{cell}=0.2415.

SOLUTION

$$E_{cell} = 0.445 \text{ V (given)}$$

$$E°_{cell} = E°_{right} - E°_{left} = 0.2415 - 0 = 0.2415 \text{ V}$$

Using the relation

$$E_{cell} = E_{right} - E_{left}$$
$$0.445 = 0.2415 - (-0.0591 \times pH)$$
$$0.445 = 0.2415 + 0.0591 \times pH$$
$$\therefore \quad pH = \frac{0.445 - 0.2415}{0.0591} = \frac{0.2035}{0.0591}$$
$$= \mathbf{3.44}$$

Using Glass electrode

A glass electrode is immersed in the solution of unknown pH. It is coupled with a standard calomel electrode (SCE) as shown in Fig. 29.20. The emf of the complete cell

Ag , AgCl | 1M HCl | Glass || Solution of Unknown pH || SCE

can be determined experimentally.

Calculations

The potential of the glass electrode, E_G, at 298 K is given by equation

$$E_G = E°_G + \frac{2.303\,RT}{F} pH$$
$$= E°_G + 0.0591 \text{ pH} \qquad \text{...(A)}$$

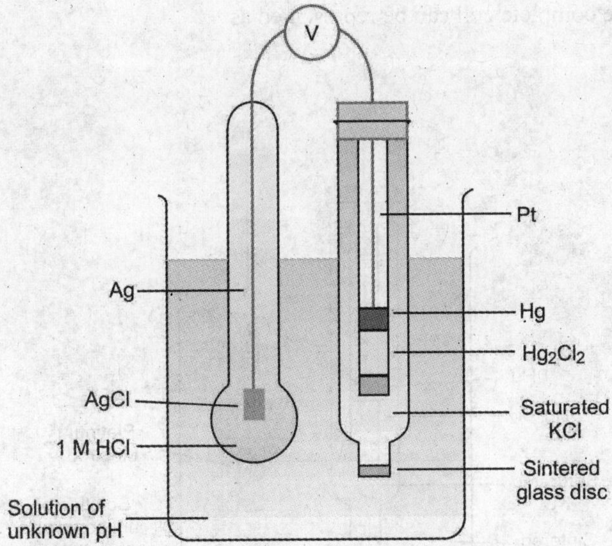

■ **Figure 29.20**
A glass electrode coupled with standard calomel electrode.

We know that

$$E_{cell} = E_R - E_L$$

i.e.,
$$E_{cell} = E_{SCE} - E_G$$

Substituting the value of E_G from (A)

$$E_{cell} = E_{SCE} - E°_G - 0.0591 \times pH$$

or
$$pH = \frac{E_{SCE} - E°_G - E_{cell}}{0.0591} \qquad \text{...(B)}$$

The value of the potential of calomel electrode is known while E_{cell} can be found experimentally. Therefore, we can find pH of a given solution if $E°_G$ is known. It can be determined by using a solution of known pH in the cell and measuring $E_{cell.}$ This value of $E°_G$ is constant for a particular glass electrode and can be used for any subsequent determinations of pH of unknown solutions with the help of equation (B).

The potential of the cell, E_{cell}, cannot be measured using ordinary potentiometer or voltmeter as the resistance of the glass membrane is very high and the current small. Therefore, an electronic voltmeter is required which reads pH directly.

Merits and demerits of Glass electrode

A glass electrode is universally used because

(1) It is simple to operate.

(2) It is not easily poisoned.

(3) Its activity is not affected by strong oxidizing and reducing agents.

(4) Since E°_G depends on a particular glass electrode used, it is not a universal constant and also changes with time. Hence a glass electrode only compares pH values while the hydrogen electrode measures pH absolutely.

Using Quinhydrone electrode

A platinum electrode is suspended in a solution whose pH is to be determined. The solution is saturated with quinhydrone compound. This half-cell is then combined with a standard calomel electrode (SCE). The complete cell can be represented as

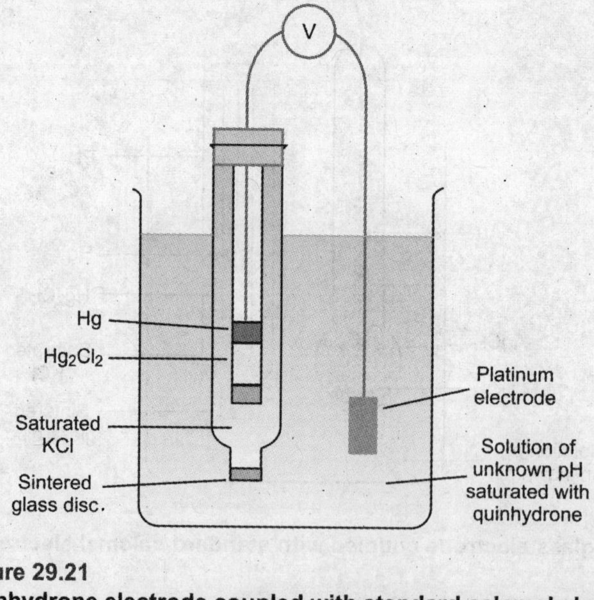

■ Figure 29.21
Quinhydrone electrode coupled with standard calomel electrode.

SCE ‖ Solution of Unknown pH Saturated With Quinhydrone ‖ Pt

The emf of the complete cell (E_{cell}) is determined with the help of a voltmeter.

Calculations. The reduction half-cell reaction of quinhydrone electrode is

$$Q + 2H^+ + 2e^- \rightleftharpoons QH_2$$

The potential, E_Q, of the quinhydrone electrode depends on the concentration of H^+ ions in solution. As worked out from Nernst equation, it is given by the relation

$$E_Q = E_Q^\circ - \frac{2.303\,RT}{F}\log[H^+]$$

or
$$E_Q = E_Q^\circ + \frac{2.303\,RT}{F}\text{pH} \quad ...(1)$$

The standard reduction potential, E_Q°, of the quinhydrone electrode is 0.6996. Thus from (1) we have

$$E_Q = 0.6996 + 0.0591 \times \text{pH} \quad ...(2)$$

We know that

$$E_{cell} = E_{right} - E_{left}$$

i.e.,
$$E_{cell} = E_Q - E_{SCE}$$

Substituting the value of emf of standard calomel electrode, E_{SCE}, which is 0.2415, and the value of E_Q from (2), we have

$$E_{cell} = 0.6996 + 0.0591 \times \text{pH} - 0.2415$$

Hence
$$0.0591\ \text{pH} = 0.4581 - E_{cell}$$

$$\text{pH} = \frac{0.4581 - E_{cell}}{0.0591}$$

Merits and Demerits. (1) Quinhydrone electrode is easily set up by simply immersing a platinum strip in the test solution.

(2) The pH values are very accurate even in the presence of oxidizing ions which interfere with the working of a hydrogen electrode.

(3) It does not give satisfactory results for solutions whose pH is more than 8.5 due to the ionization or oxidation of hydroquinone.

SOLVED PROBLEM 7

Find the pH of a solution placed in a hydroquinone half-cell which was coupled with standard calomel electrode. The emf of the combined cell was determined to be 0.123 V at 298 K.

$$E_{calomel} = 0.2415\text{ V};\ E_Q^\circ = 0.6996\text{ V}$$

SOLUTION

$$E_{cell} = E_Q - E_{SCA} \quad ...(1)$$

But we know that

$$E_Q = E_Q^\circ - \frac{2.303\,RT}{F}\text{pH}$$

$$= 0.6996 - 0.0591\ \text{pH}$$

Substituting the values of E_Q and E_{SCA} in equation (1), we have

$$0.123 = (0.6996 - 0.0591\ \text{pH}) - 0.2415 = 0.4581 - 0.0591\ \text{pH}$$

$$\text{pH} = \frac{0.4581 - 0.123}{0.0591}$$

$$\text{pH} = \frac{0.3351}{0.0591} = \mathbf{5.67}$$

POTENTIOMETRIC TITRATIONS

In a potentiometric titration, a suitable electrode immersed in the solution to be titrated acts as the 'indicator'. The **indicator electrode** is paired with a reference electrode and the two electrodes are connected to an electronic voltmeter. The emf of the indicator electrode changes gradually with the change of concentration of ions caused by the addition of titrant from the burette. The equivalence point is indicated by a sharp change in electrode potential.

Since the reference electrode potential has a constant value, any change in the indicator electrode potential is reflected by a similar change in the cell potential. Therefore, the equivalence point can be found by plotting a graph between the cell emf and the volume of titrant added from the burette. A sharp rise of the curve shows the equivalence point and the corresponding volume on the graph is the volume of the solution used for the titration.

The potentiometric titrations may be of three types :

(*a*) Acid-base titrations
(*b*) Oxidation-reduction titrations
(*c*) Precipitation titrations

Acid-base Titrations

The apparatus used for potentiometric acid-base titrations is shown in Fig. 29.22.

A hydrogen electrode or a glass electrode is immersed in solution of the acid whose strength is to be determined. The glass electrode is coupled with a standard calomel electrode. The cell thus formed is connected to the potentiometer or electronic voltmeter. When alkali is added, pH of the solution changes. The emf of the cell also changes with pH of the solution in accordance with the relation.

$$E = E° + 0.0591 \text{ pH}$$

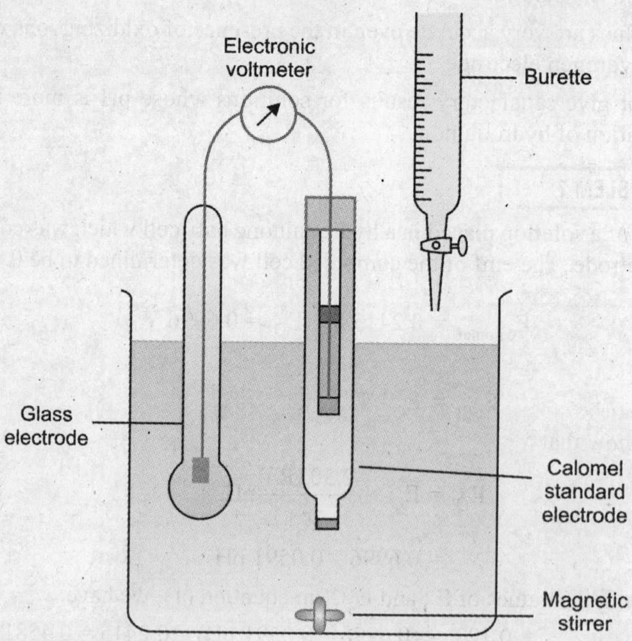

Figure 29.22
Apparatus for potentiometric acid-base titrations.

The standard alkali solution is then added from the burette in small volumes. After each addition, the emf of the cell is recorded. The emf is then plotted against the volume of alkali added. The shape of the curve for the titration of a strong acid against strong alkali (HCl versus NaOH) is shown in

Fig. 29.23 (*a*). The steepest portion of the curve indicates the equivalence point. However, when the solutions are very dilute, or weak acids or bases, are involved, the steepness of the curve is less marked and it is difficult to judge the end-point. In such a case, we plot the slope of the curve, ΔE/ΔV against the volume of alkali used. The maximum of the curve indicates the end-point.

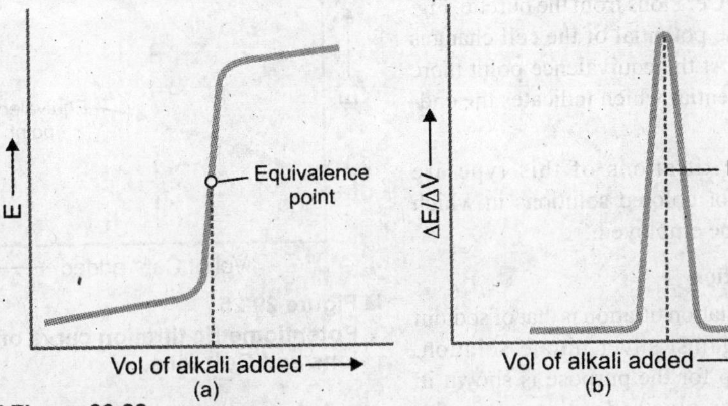

■ **Figure 29.23**
Potentiometric titration curve of an acid and a base.

Oxidation-reduction Titrations

The titration of ferrous ions (Fe^{2+}) with ceric ions (Ce^{4+}) is an example of oxidation-reduction (or redox) titration. Fe^{2+} ion is oxidized to Fe^{3+} ion, while Ce^{4+} is reduced to Ce^{3+} ion.

$$Fe^{2+} + Ce^{4+} \longrightarrow Fe^{3+} + Ce^{3+}$$

This titration can be carried in the apparatus shown in Fig. 29.24. The indicator electrode is a shiny platinum strip dipping in the solution of Fe^{2+} ions, and it is connected to a standard calomel electrode. The Ce^{4+} solution is added from the burette and the cell potential, E, recorded after each addition.

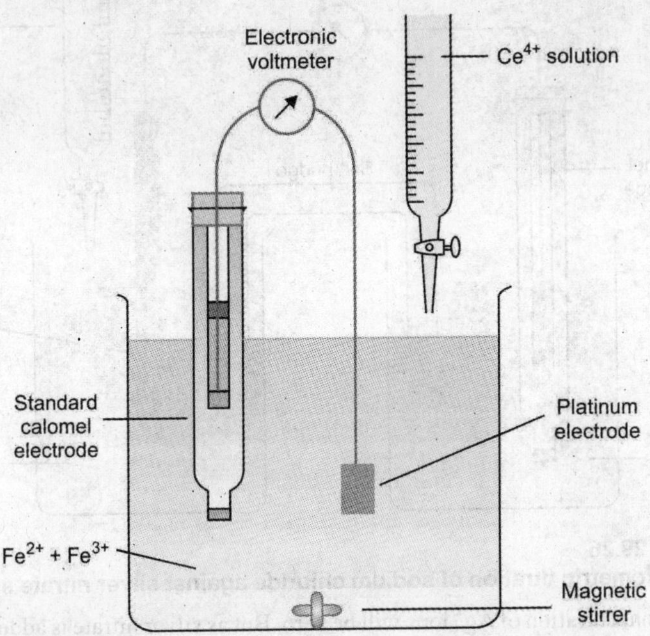

■ **Figure 29.24**
Apparatus for potentiometric titration of Fe^{2+} with Ce^{4+}.

The potential of the platinum electrode depends on the ratio $[Fe^{3+}]/[Fe^{2+}]$. The potential of the cell, E, also changes with the change of the ratio $[Fe^{3+}]/[Fe^{2+}]$. Therefore, the cell potential changes with the addition of Ce^{4+} ions from the burette. Fig. 29.25 shows how the potential of the cell changes during the titration. At the equivalence point there is a sharp rise of potential which indicates the end-point.

Potentiometric titrations of this type are particularly useful for colored solutions in which an indicator cannot be employed.

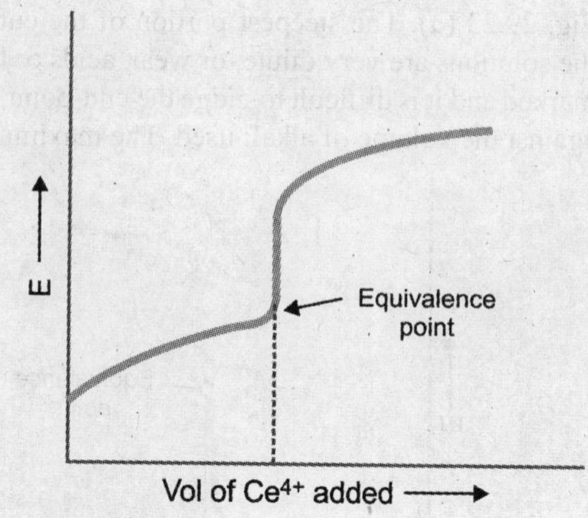

Figure 29.25
Potentiometric titration curve of Fe^{2+} ions and Ce^{4+} ions.

Precipitation Titration

A typical precipitation titration is that of sodium chloride solution against silver nitrate solution. The apparatus set up for the purpose is shown in Fig. 29.26. A silver electrode dipping in the unknown sodium chloride solution is coupled with a calomel electrode through a salt bridge. However, if the calomel electrode were in direct contact with a solution containing excess silver ions, chloride would seep through the sintered base and react to form an insoluble layer of silver chloride.

Any change in the cell potential is due to changes in concentration of Ag^+ ions around the silver electrode.

$$Ag^+ + e \longrightarrow Ag$$

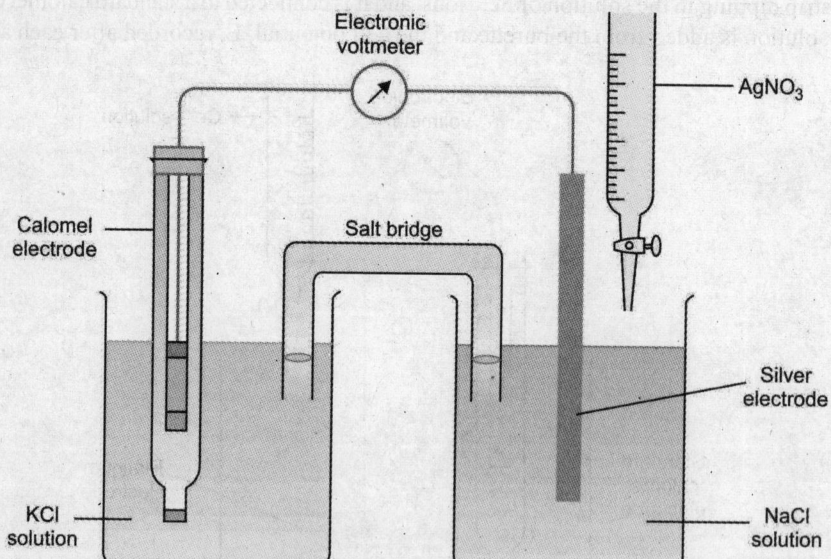

Figure 29.26
Potentiometric titration of sodium chloride against silver nitrate solution.

Initially the concentration of Ag^+ ions will be zero. But as silver nitrate is added from the burette, silver chloride is precipitated. Now the solution will contain a small concentration of Ag^+ ions formed by the slight dissociation of silver chloride. This concentration will increase slightly as Cl^- ions are removed in order to maintain the solubility product $K_{SP} = [Ag^+][Cl^-]$. After the equivalence point, the concentration of Ag^+ ions and, therefore the silver electrode potential will rise very sharply owing to the presence of excess of Ag^+ ions. The volume of $AgNO_3$ solution used to reach the equivalence point as shown in Fig. 29.27.

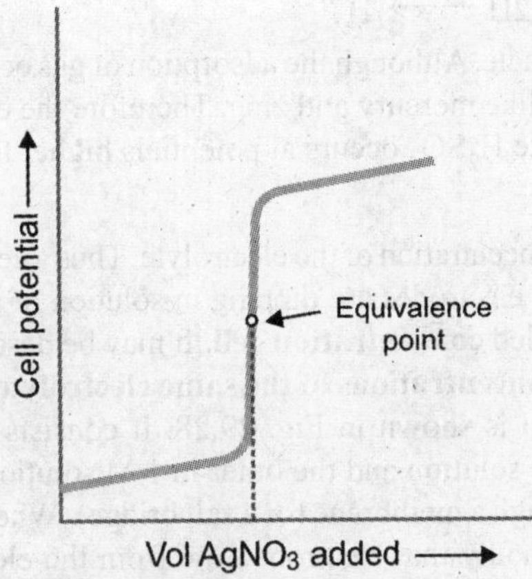

Figure 29.27
Potentiometric titration curve.

OVERVOLTAGE OR OVERPOTENTIAL

In an electrolysis cell the discharge of an ion on the cathode would occur at the standard half-reaction potential indicated in Table 29.1. Thus the H^+ ions will discharge at the cathode at $E° = 0.00$ V. It has been experimentally found to be nearly true using platinum black (*i.e.*, finely divided platinum) as the cathode. However with other metal electrodes, for example, mercury and zinc, the voltage needed for the discharge of H^+ ions (evolutions of H_2) is considerable higher than 0.00 V. The difference between the observed voltage (E) and the standard half-reaction voltage (E°) is called-the **hydrogen overpotential.** The values for hydrogen overpotentials on some metals are listed in Table 29.2.

TABLE 29.2. HYDROGEN OVERPOTENTIALS ON SOME METAL CATHODES IN DILUTE H_2SO_4

Cathode	Overpotentials
Mercury	0.78 V
Zinc	0.70 V
Copper	0.23 V
Silver	0.15 V
Platinum black	0.00 V

The hydrogen potentials are particularly well known. But an overpotential exists for any gas at any electrode and is defined as : **the additional potential, over and above the standard electrode potential, which is needed to secure the evolution of the gas.**

$$\text{Overpotential} = E - E°$$

How Overvoltage Occurs?

The overpotentials arise on account of the energy required for the adsorption of gas to occur on metal surface. The mechanism of, for example, of liberation of hydrogen gas by the discharge of H^+ ions is

Step 1. Discharge of H^+ ions to form H atoms :

$$H^+ + e^- \longrightarrow H$$

Step 2. Adsorption of H atoms on the cathode surface.

Step 3. Combination of adsorbed atoms to give hydrogen molecules :

$$2H \longrightarrow H_2$$

The step 2 may present obstacle. Although the adsorption of gas occurs readily on platinum black, it is very difficult on the metals like mercury and zinc. Therefore the evolution of hydrogen on these metals upon electrolysis of dilute H_2SO_4 occurs at potentials higher than E°.

Concentration Cells

Cell potentials depend on concentration of the electrolyte. Thus a cell can be constructed by pairing two half-cells in which identical electrodes are dipping in solution of different concentrations of the same electrolyte. Such a cell called **concentration cell.** It may be described as : **a cell in which emf arises as a result of different concentrations of the same electrolyte in the component half-cells.**

A typical concentration cell is shown in Fig. 29.28. It consists of two silver electrodes, one immersed in 0.1 M silver nitrate solution and the other in 1 M solution of the same electrolyte. The two solutions are in contact through a membrane (or a salt bridge). When the electrodes are connected by a wire, it is found experimentally that electrons flow from the electrode in more dilute (0.1M) solution to that in the more concentrated (1 M) solution.

Explanation. The concentration of Ag^+ ions in the left compartment is lower (0.1M) and in the right compartment it is higher (1M). There is a natural tendency to equalize the concentration of Ag^+ ions in the two compartments. This can be done if the electrons are transferred from the left compartment to the right compartment. This electron transfer will produce Ag^+ ions in the right compartment by the half-cell reactions :

$$Ag - e^- \longrightarrow Ag+ \qquad \textit{(left compartment)}$$

$$Ag^+ + e^- \longrightarrow Ag \qquad \textit{(right compartment)}$$

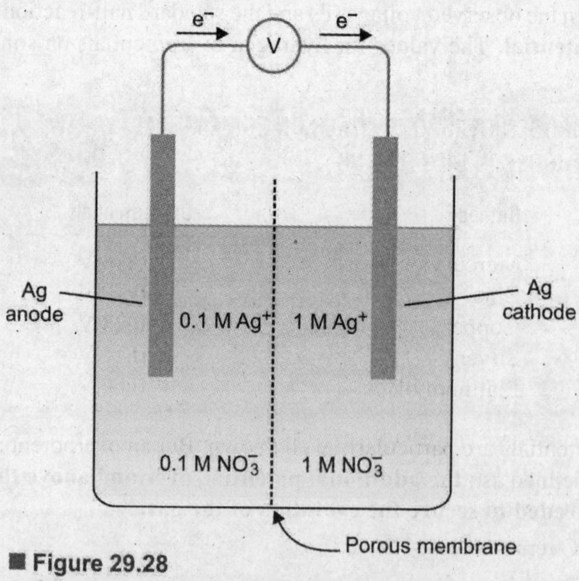

■ **Figure 29.28**
A typical concentration cell.

Thus in a bid to equalize concentration of Ag^+ ions in the two compartments the cell will develop emf (Le Chatelier's Principle) to cause the transfer of electrons. Eventually, the solutions in two compartments will have equal Ag^+ ion concentration and there will be no emf recorded.

Emf of Concentration Cell

Suppose the concentrations in the two half-cells are C_1 and C_2 at 298 K, C_2 being greater than C_1. Then emf, E, of the concentration cell will be given by the difference between the two electrode potentials. In terms of Nernst equation

$$E = \left(E_M + \frac{0.0591}{n}\log C_2\right) - \left(E_M + \frac{0.0591}{n}\log C_1\right)$$

$$= \frac{0.0591}{n}\log\left[\frac{C_2}{C_1}\right]$$

where E_M is the standard electrode potential of the metal M and n is the valence of the ions in contact with it. For example, the emf, E, of the concentration cell

$$Ag \mid Ag^+ (1\ M) \| (0.1M)\ Ag^+ \mid Ag$$

can be found by substituting the values in the above equation.

$$E = \frac{0.0591}{n}\log\frac{C_2}{C_1}$$

$$= 0.0591$$

Determination of Solubility of Sparingly soluble salts

The ionic concentration of a solution can be calculated from the emf of a concentration cell. In case of a sparingly soluble salt, the salt can be supposed to be completely ionized even in saturated solutions. Hence the ionic concentration is proportional to the solubility of the salt.

Suppose we want to find the solubility of silver chloride. This can be done by measuring the emf, E, of the cell.

$$Ag \mid N/100\ AgNO_3 \mid Glass \| Saturated\ AgCl \mid Ag$$

The emf of the cell at 298 K is given by the relation

$$E = \frac{0.0591}{n}\log\frac{C_2}{C_1}$$

Here, n, the valence of Ag+ ion is 1 and the concentration of Ag^+ in N/100 $AgNO_3$ solution is 0.01 kg/m^3.

Hence, $$E = \frac{0.0591}{1}\log\frac{0.01}{C_1}$$

From the above expression, the concentration of AgCl in kilogram ions of silver per metre^3 can be calculated. Multiplying this by 143.5, the molecular weight of silver chloride, we get the solubility of AgCl in kilograms per metre^3.

SOLVED PROBLEM 8

The emf of the cell

$$Ag \mid AgI\ in\ 0.045\ M\ KI \| 0.045\ M\ AgNO_3 \mid Ag$$

is 0.788 at 298 K. Calculate (*i*) the solubility product of AgI and (*ii*) the solubility of AgI in water at 298 K.

SOLUTION

Calculation of K_{SP}

At 298 K the concentration of Ag^+ in the cathodic half-cell is 0.045M and the concentration of I^- ion in anodic half-cell is 0.045 M KI.

Let the concentration of Ag^+ ion in the anodic half-cell due to solubility of AgI be C_1.

$$\therefore \qquad E = \frac{0.0591}{1} \log \frac{0.045}{C_1}$$

or

$$0.788 = \frac{0.0591}{1} \log \frac{0.045}{C_1}$$

or

$$\log \frac{0.045}{C_1} = \frac{0.788}{0.0591} = 13.33$$

or

$$\frac{0.045}{C_1} = 2.138 \times 10^{13}$$

$$C_1 = \frac{0.045}{2.138 \times 10^{13}} = \mathbf{2.105 \times 10^{-15}}$$

Calculation of Solubility

$$\text{Solubility of AgI} = \sqrt{K_{SP}}$$

$$= \sqrt{0.9472 \times 10^{-18}}$$

$$= 0.9732 \times 10^{-9} \text{ g mol liter}^{-1}$$

$$= 0.9732 \times 10^{-9} \times 143.5 \text{ g liter}^{-1}$$

$$= \mathbf{1.396 \times 10^{-6} \text{ g liter}^{-1} \text{ or } 1.396 \times 10^{-6} \text{ kg/m}^3}$$

Determination of Valence

The expression for the emf, *E*, of a concentration cell is

$$E = \frac{0.059}{n} \log_{10} \frac{C_2}{C_1} \qquad \text{...(A)}$$

where *n* is the valence of the metallic ion in solution, while C_1 and C_2 are the concentrations of the ions in the two half-cells. Knowing the experimental values of E, C_1 and C_2, *n* can be calculated.

For example, the valence of mercury in mercurous nitrate, $Hg_2(NO_2)_2$, can be determined by the **Concentration cell method.** The following cell is constructed

$$Hg \mid 0.05N\ Hg_2(NO_3)_2 \parallel 0.5N\ Hg_2(NO_3)_2 \mid Hg$$

and its emf found experimentally is 0.029 Volts.

Let C_1 be the concentration of mercurous ion in 0.05 N $Hg_2(NO_3)_2$ solution in the left half-cell and C_2 be the concentration of mercurous ion in 0.5N $Hg_2(NO_3)_2$ solution in the right half-cell.

Substituting the values in the expression (A), we have

$$0.029 = \frac{0.0591}{n} \log \frac{0.5}{0.05}$$

$$= \frac{0.0591}{n} \log 10 = \frac{0.0591}{n}$$

Therefore, $$n = \frac{0.0591}{0.029} = 2$$

ADDITIONAL SOLVED PROBLEMS

SOLVED PROBLEM 9

Calculate the emf of the following cell

$$Sn \mid Sn^{++},\ 0.75\ N \mid\mid Sn^{++},\ 0.5\ N \mid Sn$$

SOLUTION

Formula used

$$E_{cell} = \frac{0.0591}{n} \log \frac{C_2}{C_1}$$

Quantities given

$n = 2$ $\quad C_2 = 0.5$ N $\quad C_1 = 0.75$ N

Substitution of values

$$E_{cell} = \frac{0.0591}{2} \log \frac{0.5}{0.75}$$

$$= \frac{0.0591}{2} \log \frac{2}{3}$$

$$= \frac{0.0591}{2} \times (-0.1760)$$

$$= \mathbf{-0.0052}$$

SOLVED PROBLEM 10

The emf of the cell set up as under was found to be 0.1182 volt at 298 K. Calculate the solubility of AgCl.

$$Ag \mid \textit{Saturated Solution of AgCl} \mid\mid AgNO_3\ (0.001\ N) \mid Ag$$

SOLUTION

Formula used

$$E_{cell} = \frac{0.0591}{n} \log \frac{C_2}{C_1}$$

Quantities given

$E_{cell} = 0.1182$ V $\quad C_2 = 0.001$ $\quad n = 1$

Substitution of values

$$0.1182 = \frac{0.0591}{n} \log \frac{0.001}{C_1}$$

or $$\log \frac{0.001}{C_1} = \frac{0.1182}{0.0591} = 2$$

or $$\frac{0.001}{C_1} = \text{antilog } 2 = 100$$

or $$C_1 = \frac{0.001}{100}$$

$$= 1 \times 10^{-8} \text{ kg mol}^{-1}$$

∴ Solubility, $$C_1 = 1 \times 10^{-8} \times 143.5 \text{ kg m}^{-3}$$

$$= \mathbf{1.435 \times 10^{-6}\ kg/m^3}$$

SOLVED PROBLEM 11

The emf of the following cell,

$$Hg \mid Hg_2Cl_2, KCl\ (Sat) \parallel H^+ (a_1)\ Quinhydrone \mid Pt$$

was found to be 0.228 volt at 298 . Calculate the pH *of the solution.* (E_{cal} = *0.2415 Volt*)

SOLUTION

Formula used

$$\text{pH} = \frac{0.4581 - E_{cell}}{0.0591}$$

Quantities given

$$E_{cell} = 0.228 \text{ V}$$

Substitution of values

$$pH = \frac{0.4581 - 0.228}{0.0591}$$

$$= \mathbf{3.89}$$

SOLVED PROBLEM 12

The emf of the cell set up by using an acid solution, standard calomel electrode and quinhydrone electrode, was found to be 0.1595 volt at 298 K. Calculate the *pH* of the acid solution. (emf of standard calomel electrode = – 0.2440 V and emf quinhydrone electrode = 0.6990 V)

SOLUTION

Formula used

$$E_{cell} = \left[E_{Q+} - \frac{2.303\,RT}{n}\,pH \right] - E_{SCE}$$

Quantities given

E_{Q+} = 0.6990 V $\qquad E_{SCE}$ = – 0.2440

$\frac{2.303\,RT}{F}$ = 0.0591 $\qquad E_{cell}$ = 0.1595

Substitution of values

$$0.1595 = [\,0.6990 - 0.0591\,pH\,] - 0.2440$$

$$0.1595 = 0.6990 - 0.0591\,pH - 0.2440$$

$$0.1595 = 0.4550 - 0.0591\,pH$$

or $$pH = \frac{0.4550 - 0.1595}{0.0591}$$

$$= \mathbf{5}$$

SOLVED PROBLEM 13

Calculate the emf of the cell :

$$Ag_{(s)} \mid Ag^{+} (a = 0.1) \parallel Zn^{2+} (a = 0.1) \mid Zn_{(s)}$$

Given $E^{o}_{Ag/Ag+} = 0.799$ V $E^{o}_{Zn^{2+}/Zn} = 0.763$ V

Is the reaction spontaneous?

SOLUTION

Formula used

$$EMF\ of\ the\ cell = E_R - E_L$$

Quantities given

$E_R = -0.763$ V $E_L = 0.799$ V

Substitution of values

$$EMF\ of\ the\ cell = -0.763\ \text{V} - (0.799\ \text{V})$$
$$= \mathbf{-1.562\ V}$$

Since the emf of the cell is negative, the reaction is not spontaneous.

SOLVED PROBLEM 14

Calculate the free energy change of the following cell at 298 K.

$$Sn \mid Sn^{2+} (0.5N) \parallel Pb^{2+} (0.3N) \mid Pb$$

The standard emf of the cell is 0.14 V.

SOLUTION

Formula used

$$\Delta G = -nFE$$

Quantities given

$n = 2$ $F = 96500$ $E = 0.14$ V

Substitution of values

$$\Delta G = -2 \times 96500 \times 0.14\ \text{J}$$
$$= \mathbf{27020\ J}$$

SOLVED PROBLEM 15

When hydrogen electrode and normal calomel electrode are immersed in a solution at 298 K, a potential of 0.624 is obtained. Calculate the *pH* of the solution.

SOLUTION

Formula used

$$pH = \frac{E_{cell} - 0.2415}{0.0591}$$

Quantities given

$$E_{cell} = 0.624\ \text{V}$$

Substitution of values

$$\text{pH} = \frac{0.624 - 0.2415}{0.0591}$$
$$= \mathbf{6.47}$$

EXAMINATION QUESTIONS

1. Define or explain the following terms :
   (*a*) Redox reaction (*b*) Electrochemical cells
   (*c*) EMF (*d*) Free energy
   (*e*) Nernst equation (*f*) Concentration cells
2. What are concentration cells? Describe and discuss a concentration cell without transference. (*Nehu BSc*)
3. What is meant by Hydrogen Electrode? How would you make use to this electrode for determination of H^+ ion concentration in a solution? What are the advantages and disadvantages of this electrode? (*Agra BSc*)
4. Describe the construction and working of calomel electrode. (*Delhi BSc*)
5. (*a*) What is a reference electrode? Name two and discuss the working of any one reference electrode.
   (*b*) Write the cell reactions for the following electrochemical cells :

$$Cd \mid Cd^{2+} \parallel KCl \parallel Hg_2Cl_2(s) \mid Hg$$

$$Pt \mid H_2\ (1\ bar),\ HCl \parallel AgCl(s) \mid Ag$$

(*Madras BSc*)
6. Give an account of electrochemical series and its applications. (*Jiwaji BSc*)
7. What do you understand by the liquid junction potential? How does it arise? How is the liquid junction potential eliminated? (*Madurai BSc*)
8. (*a*) What is meant by standard electrode potential? Name and write the half reaction of a Calomel reference electrode.
   (*b*) Explain why metallic sodium reacts with H^+ to liberate H_2, whereas platinum metal is unreactive to H^+. (*Mizoram BSc*)
9. (*a*) What do you mean by electrode potential?
   (*b*) Describe the important applications of emf measurements. (*MD Rohtak BSc*)
10. (*a*) Discuss the principle of determination of pH of a solution with the help of a glass electrode.
    (*b*) From the cell reaction in a cell

$$Hg_2Cl_2(s) + H_2(101325.00\ Pa) \rightarrow 2Hg + 2H^+(a = 1)$$

the $E°$ of the cell at 298 K is 0.2676 volt and $\left(\frac{\partial E^o}{\partial T}\right)_P = -31.9 \times 10^{-4}$ volt deg^{-1}. Calculate $\Delta H°$ and $\Delta S°$.

Answer. 33.299 kJ; –6.1567 J K^{-1} (*Vidyasagar BSc*)
11. (*a*) Find out EMF of a concentration cell without transference.
    (*b*) Calculate the free energy change of the following cell at 298 K.

$$Sn \mid Sn^{2+}\ (0.5\ N) \parallel Pb^{2+}\ (0.3\ N) \mid Pb$$

The standard EMF of the cell is 0.14 V.

Answer. –25.7539 kJ (*Allahabad BSc*)
12. Derive an expression for ΔG, ΔH and ΔS in terms of emf of a cell and temperature coefficient of emf. (*Jamia Millia BSc*)
13. (*a*) How is equilibrium constant of the cell reaction calculated?
    (*b*) Calculate the EMF of the cell :

$$Zn \mid Zn^{2+}\ (0.001\ M) \parallel Ag^+\ (0.1\ M) \mid Ag$$

The standard potential of Ag | Ag^+ half cell is 0.080 V and Zn | Zn^{2+} is –0.76 volt.

Answer. 1.492 V (*Arunachal BSc*)

14. Calculate EMF of the following cell at 298 K.

$$Cd\ (Hg)\ (a = 0.1)\ |\ Cd\ SO_4\ (Soln)\ |\ Cd\ (Hg)\ (a = 0.01)$$

Answer. 0.0295 V *(Nagpur BSc)*

15. Calculate the standard emf and standard free energy change for the reaction :

$$\tfrac{1}{2}Cu(s) + \tfrac{1}{2}Cl_2(g) \rightleftharpoons \tfrac{1}{2}Cu^{2+} + Cl^- \quad \text{at 298 K}$$

Given $E^\circ_{Cl_2, Cl^-} = 1.36$ V; $E^\circ_{Cu^{+2}, Cu} = 0.34$V; F = 96500

Answer. 196.86 kJ *(Mizoram BSc (H))*

16. The equilibrium constant for the reaction :

$$Zn(s) + Sn^{2+} \rightleftharpoons Zn^{2+} + Sn(s)$$

is 1.122×10^{21} at 298 K. Calculate the standard potential of Zn | Zn^{2+} if $E^\circ_{Zn^{2}/Zn} = 0.140$ V

Answer. –0.4820 V *(Mumbai BSc)*

17. (*a*) Consider a cell in which the overall reaction is 2 $Cu^{2+} \rightarrow Cu^{2+}$ + Cu.
Calculate ΔG° for the reaction, if $E^\circ = + 0.184$ V at 298 K.

(*b*) Employing Nernst equation for the dependence of H^+ ion concentration on the potential of hydrogen electrode, show that

$$E_{H^+, H_2} = 0.0591 \text{ pH at 298 K}$$

Answer. –35.512 kJ *(Guru Nanak Dev BSc)*

18. Calculate the free energy change of the following cell at 298 K:

$$Sn\ |\ Sn^{2+}\ (0.5\ N)\ ||\ Pb^{2+}\ (0.3\ N)\ |\ Pb$$

The standard EMF of the cell is 0.14 volt.

Answer. – 27.020 kJ *(Allahabad BSc)*

19. Write the cell reaction for the cell

$$Zn\ |\ ZnSO_4(aq)\ ||\ CuSO_4(aq)\ |\ Cu$$

and calculate the equilibrium constant. Given $E^\circ_{Zn\,|\,Zn^{2+}} = 0.76$V; $E^\circ_{Cu\,|\,Cu^{2+}} = -0.37$ V at 298 K.

Answer. 5.7×10^{-39} *(Goa BSc)*

20. For the cell :

$$Zn\ |\ Zn^{2+}\ (c = 1.0\ M)\ ||\ Cu^{2+}\ (c = 1.0\ M)\ |\ Cu$$

the standard emf is 1.10 volts at 298 K . Write down the cell reaction and calculate standard Gibbs free energy change (ΔG).

Answer. –212.3 kJ *(Avadh BSc)*

21. (*a*) What are primary and secondary reference electrodes? Give one example of each.

(*b*) Describe saturated calomel electrode. Write the electrode reaction when the electrode acts as cathode. *(Arunachal BSc)*

22. (*a*) Discuss the effect of temperature on the EMF of an electrochemical cell.

(*b*) Discuss the use of EMF measurements to determine the hydrolysis constant of a salt. *(Nagpur BSc)*

23. (*a*) Derive Nernst equation showing effect of electrolyte concentration on the potential of an oxidation-reduction electrode.

(*b*) "A salt bridge can minimize the liquid junction potential" – Explain. For the potentiometric titration of $AgNO_3$ against KCl what type of salt bridge should be used and why? *(Kalyani BSc)*

24. (*a*) What is a reversible cell? Derive expression for the thermodynamic parameters from the EMF of such a cell.

(*b*) Write down the electrode and overall cell reaction for the cell :

$$Pt\ |\ H_2(101325.00\ Pa),\ H^+(a = 1)\ ||\ KCl(a = 1),\ AgCl(s)\ |\ Ag$$ *(Sambalpur BSc)*

25. What is a salt bridge? Explain its function in an electrochemical cell. *(Punjabi BSc)*

26. What are oxidation and reduction potentials? What is their use for predicting the feasibility of a reaction? *(Purvanchal BSc)*

27. (*a*) How are the emf of a cell and entropy of the reaction related?

(*b*) Explain the phenomenon of liquid junction potential. *(Guru Nanak Dev BSc)*

28. Calculate the free energy change of the following cell at 298 K :

$$Sn \mid Sn^{2+}(a = 0.6) \parallel Pb^{2+}(a = 0.3) \mid Pb$$

the standard emf of the cell is 0.014 volt.

Answer. 986.23 J *(Delhi BSc)*

29. (*a*) Calomel electrode is used as a reference electrode, give reasons. Describe its construction and working.

(*b*) What is meant by temperature coefficient of EMF? How is it related to ΔS?

(*c*) With the help of EMF measurements how is solubility and solubility product of a sparingly soluble salt determined? *(Dibrugarh BSc)*

30. (*a*) What is the basic difference between the voltage and emf of the cell? Explain.

(*b*) Define electrode potential. How is it different from liquid junction potential? How and why the later is eliminated?

(*c*) What will be the reaction in the cell which is made by combining standard Cd half cell with standard Cu half cell? Also calculate the emf of the cell. The standard oxidation potential of half cells are 0.403 and –0.337 volt.

Answer. (*c*) – 0.790V *(Madurai BSc)*

31. (*a*) What are concentration cells? How do they differ from chemical cells? Derive an expression of the emf of a concentration cell with transference reversible to cations.

(*b*) Describe with the help of diagram the working of a glass electrode and show how it can be used to determine the pH of a solution. *(Banaras BSc)*

32. Calculate the emf of a cell containing two hydrogen electrodes, the negative one is in contact with 10^{-6} M OH^- ions and the positive one is in contact with 0.05 M H^+.

Answer. 0.395 V *(Avadh BSc)*

33. Determine the equilibrium constant of the following reaction at 298 K

$$2Fe^{3+} + Sn \rightleftharpoons 2Fe^{2+} + Sn^{4+}$$

From the obtained value of equilibrium constant predict whether Sn^{2+} ion can reduce Fe^{3+} to Fe^{2+} quantitatively or not.

Answer. 1.4215×10^{31} *(Delhi BSc)*

34. Calculate the standard electrode potential of Ni^{2+}/Ni electrode, if the cell potential of the cell

$$Ni \mid Ni^{2+}\ (0.01\ M) \parallel Cu^{2+}\ (0.1\ M) \mid Cu$$

is 0.59 V.

Answer. – 0.2205 V *(Madras BSc)*

35. For a cell

$$Ag(s) \mid AgNO_3\ (0.01\ M) \parallel AgNO_3\ (1.0\ M) \mid Ag(s)$$

(a) Calculate the emf at 298 K

(b) Will the cell generate emf when two concentrations become equal.

Answer. 0.1184 V , No *(Baroda BSc)*

36. A cell contains two hydrogen electrodes. The negative electrode is in contact with solution of 10^{-6} M hydrogen ions. The emf of the cell is 0.118 volt at 298 K. Calculate the concentration of H^+ ion at positive electrode ?

Answer. 1×10^{-4} M *(Mumbai BSc)*

37. Calculate the emf of the cell

$$Cr \mid Cr^{3+}\ (0.1\ M) \parallel Fe^{2+}\ (0.01\ M) \mid Fe$$

Given $E^o_{Cr^{3+}/Cr} = 0.75$ V $\qquad E^o_{Fe^{2+}/Fe} = -0.45$ V

Answer. 0.2607 V *(Kanpur BSc)*

38. (*a*) Explain why emf of an electrochemical cell cannot be measured with the help of a voltmeter.

(*b*) Describe the applications of emf measurement in the determination of equilibrium constant of a cell reaction. *(Delhi BSc)*

39. (*a*) Distinguish between cell potential and electromotive force.

(*b*) Summarize the differences between galvanic and electrolytic cell

(*c*) Describe the use of emf measurements to determine the pH of aqueous solutions. Include in your answer a reference of both hydrogen electrode and glass electrode. *(Allahabad BSc)*

40. (*a*) Describe how hydrogen electrode is used for the measurement of pH of aqueous solutions.

(*b*) Give one example each of electrolytic concentration cells with and without transference. Also write the cell reactions and expressions for the emf of these cells. *(Kalayni BSc)*

41. Write short notes on :

(*a*) Calomel electrode (*b*) Standard hydrogen electrode

(*c*) Standard Cadmium cell (*d*) Amalgam electrode *(Agra BSc)*

42. (*a*) Write the type of the following reversible electrodes :

(*i*) Quinhydrone (*ii*) Saturated Calomel

(*iii*) Normal hydrogen electrode (*iv*) Silver-Silver electrode

(*b*) Write cell reaction and calculate E° for the cell :

$$Zn \mid Zn^{2+}\ (1\ M) \parallel Fe^{2+}\ (1\ M),\ Fe^{3+}\ (1\ M);\ Pt$$

Given $E°\ (Fe^{3+}, Fe^{2+}) = 0.77$ V, $E°\ (Zn^{2+}, Zn) = 0.76$ V

Answer. 0.1534 V *(Panjab BSc)*

43. Discuss in brief the determination of the following from emf measurements:

(*a*) pH of a solution (*b*) Dissociation constants of acids

(*c*) Solubility of sparingly soluble salt *(Guru Nanak Dev BSc)*

MULTIPLE CHOICE QUESTIONS

1. In the reaction

$$Zn(s) + 2HCl \rightarrow ZnCl_2(aq) + H_2(g)$$

(*a*) zinc is oxidized

(*b*) the oxidation number of chlorine remains unchanged

(*c*) the oxidation number of hydrogen changes from +1 to 0.

(*d*) all are correct

Answer. (*d*)

2. Electrolytic cells are electrochemical cells in which ______ reactions are forced to occur by the input of electrical energy.

(*a*) spontaneous (*b*) non-spontaneous

(*c*) exothermic (*d*) endothermic

Answer. (*b*)

3. In any electrochemical cell, the cathode is always ______.
   (*a*) a nonmetal
   (*b*) attached to a battery
   (*c*) the electrode at which some species gain electrons
   (*d*) the electrode at which some species lose electrons
   Answer. (*c*)
4. In an electrolytic cell, the charge on the electrode that gives electrons to the species in solution is ______; the chemical change that occurs at this electrode is called ______.
   (*a*) positive; oxidation (*b*) positive, reduction
   (*c*) negative, oxidation (*d*) negative, reduction
   Answer. (*d*)
5. In a galvanic cell the following reaction takes place:
   $$2H_2O \leftrightarrow O_2(g) + 4H^+ + 4e^-$$
   It occurs at the
   (*a*) cathode (*b*) anode
   (*c*) cathode and anode (*d*) none of these
   Answer. (*b*)
6. The site of oxidation in an electrochemical cell is
   (*a*) the anode (*b*) the cathode
   (*c*) the electrode (*d*) the salt bridge
   Answer. (*a*)
7. Which statement below is not true for the reaction?
   $$Fe^{3+} + e^- \rightarrow Fe^{2+}$$
   (*a*) Fe^{3+} is being reduced
   (*b*) the oxidation state of Fe has changed
   (*c*) Fe^{3+} could be referred to as an oxidizing agent in this reaction
   (*d*) both Fe^{3+} and Fe^{2+} are called anions
   Answer. (*d*)
8. $Zn \rightarrow Zn^{2+} + 2e^-$, $\varepsilon° = +0.76$ V, $Cr^{3+} + 3e^- \rightarrow Cr$, $\varepsilon° = -0.74$ V
   The anode in this cell is
   (*a*) Zn (*b*) Cr
   (*c*) Zn^{2+} (*d*) Cr^{3+}
   Answer. (*a*)
9. Which of the following statements is correct concerning the reaction:
   $$Fe^{2+} + 2H^+ + NO_3^- \rightarrow Fe^{3+} + NO_2 + H_2O$$
   (*a*) Fe^{3+} is oxidized and H^+ is reduced
   (*b*) Fe^{2+} is oxidized and nitrogen is reduced
   (*c*) Fe^{2+} and H^+ are oxidized
   (*d*) Oxygen is oxidized
   Answer. (*b*)
10. Which of the following is a half-reaction?
   (*a*) $Zn + Cu_2^+ \rightarrow Zn_2^+ + Cu$ (*b*) $H^+ + OH^- \rightarrow H_2O$
   (*c*) $Ag^+ + e^- \rightarrow Ag$ (*d*) $Ag^+ + Cl^- \rightarrow AgCl$
   Answer. (*c*)

11. Which of the following statements associated with electrochemical cells is incorrect?

(*a*) the function of a salt bridge in an electrochemical cell is to complete the circuit

(*b*) cell potential is the potential difference in a voltaic cell

(*c*) a Bronsted-Lowry acid-base reaction can be the basis of the net reaction in a chemical cell

(*d*) a half-reaction corresponds to one electrode in a voltaic cell

Answer. (*c*)

12. Write the cell diagram for the reaction below

$$Cl_2(g) + 2Ag(s) \rightarrow 2Ag^+(aq) + 2Cl^-(aq)$$

(*a*) $Ag \mid Ag^+(aq) \mid Cl_2(g), Cl^-(aq) \mid Pt$ (*b*) $Ag \mid Ag^+(aq), Cl^-(aq) \mid Cl_2(g) \mid Pt$

(*c*) $Pt, Cl_2(g) \mid Cl^-(aq) \parallel Ag^+(aq) \mid Ag$ (*d*) $Ag \mid Ag^+(aq) \parallel Cl_2(g), Cl^-(aq) \mid Pt$

Answer. (*d*)

13. Write the balanced equation for the voltaic cell made from Ag^+/Ag and Cu^{2+}/Cu and calculate $E°_{cell}$.

(*a*) $2\,Ag^+(aq) + Cu(s) \rightarrow 2\,Ag(s) + Cu_2^+$, $E°_{cell} = 1.260$ V

(*b*) $2\,Ag^+(aq) + Cu(s) \rightarrow Ag(s) + 2\,Cu_2^+$, $E°_{cell} = 1.140$ V

(*c*) $2\,Ag^+(aq) + Cu(s) \rightarrow 2\,Ag(s) + Cu^{2+}$, $E°_{cell} = 0.460$ V

(*d*) $2\,Ag^+(aq) + Cu(s) \rightarrow 2\,Ag(s) + Cu_2^+$, $E°_{cell} = 1.140$ V

Answer. (*c*)

14. Which of the following statements associated with batteries is incorrect?

(*a*) in a dry cell, the reaction $Zn \rightarrow Zn^{2+}$ continues to occur even when the battery is not being used

(*b*) secondary batteries are rechargeable

(*c*) the cell reaction in a primary battery is not reversible

(*d*) electrodes with greater surface area give a greater potential

Answer. (*d*)

15. When this redox reaction

$$PbO + [Co(NH_3)_6]^{3+} \rightarrow PbO_2 + [Co(NH_3)_6]^{2+}$$

is balanced in BASIC solution, there are

(*a*) reactants include $2OH^-$. Products include $2H_2O$

(*b*) reactants include $2OH^-$. Products include $2Co(NH_3)_6^{\ 2+}$

(*c*) reactants include $2H_2O$. Products include $2OH^-$

(*d*) products include $2OH^-$ and $1Co(NH_3)_6^{\ 2+}$

Answer. (*b*)

16. Given the nickel-cadmium battery reaction :

$$2NiOOH + Cd + 2H_2O \underset{\text{charge}}{\overset{\text{discharge}}{\rightleftharpoons}} 2Ni(OH)_2 + Cd(OH)_2$$

What occurs during discharge in the nickel-cadmium battery?

(*a*) Ni^{3+} is reduced to Ni^{2+} (*b*) Ni^{2+} is reduced to Ni^{3+}

(*c*) Ni^{3+} is oxidized to Ni^{2+} (*d*) Ni^{2+} is oxidized to Ni^{3+}

Answer. (*a*)

17. What is indicated when a chemical cell's voltage ($E°$) has dropped to zero?

(*a*) the concentration of the reactants has increased

(*b*) the concentration of the products has decreased

(*c*) the cell reaction has reached equilibrium

(*d*) the cell reaction has completely stopped

Answer. (*c*)

18. Given the redox reaction :

$$2Cr(s) + 3Cu^{2+}(aq) \rightarrow 3Cr^{3+}(aq) + 3Cu(s)$$

Which reaction occurs at the cathode in an electrochemical cell?

(*a*) reduction of $Cu^{2+}(aq)$ (*b*) reduction of Cu(s)

(*c*) oxidation of $Cr^{3+}(aq)$ (*d*) oxidation of Cr(s)

Answer. (*a*)

19. Which metal is used as a coating on steel to limit corrosion?

(*a*) Na (*b*) Ca

(*c*) K (*d*) Zn

Answer. (*d*)

20. Given the cell reaction:

$$2H_2O(\ell) \longrightarrow 2H_2(g) + O_2(g)$$

The cell is best described as

(*a*) an electrolytic cell in which an exothermic reaction occurs

(*b*) an electrolytic cell in which an endothermic reaction occurs

(*c*) a galvanic cell in which an exothermic reaction occurs

(*d*) a galvanic cell in which an endothermic reaction occurs

Answer. (*b*)

21. Given the reaction:

$$Pb(s) + Cu^{2+}(aq) \rightarrow \rightarrow Pb^{2+}(aq) + Cu(s)$$

What is the reducing agent?

(*a*) $Pb^{2+}(aq)$ (*b*) $Cu^{2+}(aq)$

(*c*) Pb(s) (*d*) Cu(s)

Answer. (*c*)

22. Given the reaction:

$$4HCl(aq) + MnO_2(s) \rightarrow MnCl_2(aq) + 2H_2O(\ell) + Cl_2(g)$$

The manganese is

(*a*) reduced and its oxidation number changes from +4 to +2

(*b*) reduced and its oxidation number changes from +2 to +4

(*c*) oxidized and its oxidation number changes from +4 to +2

(*d*) oxidized and its oxidation number changes from +2 to +4

Answer. (*a*)

23. What occurs when an atom is oxidized in a chemical reaction?

(*a*) a loss of electrons and a decrease in oxidation number

(*b*) a loss of electrons and an increase in oxidation number

(*c*) a gain of electrons and a decrease in oxidation number

(*d*) a gain of electrons and an increase in oxidation number

Answer. (*b*)

24. Standard cell potential is

(*a*) measured at a temperature of 298 K

(*b*) measured when ion concentrations of aqueous reactants are 1.00 M

(*c*) measured under the conditions of 101325.00 Pa for gaseous reactants

(*d*) all of the above

Answer. (*d*)

25. The standard reduction potentials in volts for Pb^{2+} and Ag^+ are –0.13 and +0.80, respectively. Calculate $E°$ in volts for a cell in which the overall reaction is

$$Pb + 2Ag^+ \rightarrow Pb^{2+} + 2Ag$$

(*a*) 1.73 (*b*) 0.67

(*c*) 0.93 (*d*) 1.47

Answer. (*c*)

26. Given the following information,

$$Fe^{3+}(aq) + H_2(g) \rightarrow 2H^+ + Fe^{2+}, E°_{cell} = 0.77$$

Determine $E°$ for the reaction :

$$e^- + Fe^{3+}(aq) \rightarrow Fe^{2+}(aq)$$

(*a*) 1.54 (*b*) 0.77

(*c*) 0.39 (*d*) – 0.77

Answer. (*b*)

27. Breathalyzers determine alcohol content via the redox reaction:

$$Cr_2O_7^{2-} + C_2H_5OH \rightarrow Cr^{3+} + C_2H_4O_2$$

Which substance is a reductant (reducing agent) and which is an oxidant (oxidizing agent)?

(*a*) C_2H_5OH, reductant; no oxidant (*b*) C_2H_5OH, reductant; $Cr_2O_7^{2-}$, oxidant

(*c*) C_2H_5OH, oxidant; Cr^{3+}, reductant (*d*) C_2H_5OH, reductant; Cr^{3+}, oxidant

Answer. (*b*)

28. Predict the products in the electrolysis of aqueous potassium bromide.

(*a*) hydrogen and bromine (*b*) potassium metal and oxygen

(*c*) oxygen and bromine (*d*) potassium metal and bromine

Answer. (*a*)

29. Which of the following statements associated with corrosion is incorrect?

(*a*) iron corrodes more readily than aluminium because iron is more active than aluminium

(*b*) cathodic protection prevents corrosion by using a sacrificial anode

(*c*) a corroding metal has both anodic and cathodic areas

(*d*) corrosion involves both oxidation and reduction

Answer. (*a*)

30. Which of the following shows a metal being oxidized?

(*a*) $2Na + 2H_2O \rightarrow 2NaOH + H_2$ (*b*) $Cu \rightarrow Cu^{2+} + 2e^-$

(*c*) $Cu^{2+} + 2e^+ \rightarrow Cu$ (*d*) Both (*a*) and (*b*)

Answer. (*d*)

31. It is possible to generate an electrical potential by inserting two strips of different metals into an acidic citrus fruit such as a lemon. Among other functions, the lemon serves as a salt bridge. Which shorthand notation would best describe a lemon into which has been inserted a strip of zinc and a strip of copper, with the two metal strips connected by a wire?

(*a*) $Zn(s) \mid Zn^{2+}(aq) \parallel O_2(g) \mid H_2O(I) \mid Cu(s)$ (*b*) $Zn(s) \mid Zn^{2+}(aq) \parallel H^+(aq) \mid H^2(g) \mid Cu(s)$

(*c*) $Cu(s) \mid Cu^{2+}(aq) \parallel O_2(g) \mid H_2O(I) \mid Zn(s)$ (*d*) $Zn(s) \mid Zn^{2+}(aq) \parallel Cu^{2+}(aq) \mid Cu(s)$

Answer. (*a*)

32. Write the net equation for the redox reaction that occurs in the voltaic cell with a $Cu(s) – Cu^{2+}(aq)$ electrode and a $Ag(s) – Ag^+(aq)$ electrode.

(*a*) $Cu^{2+}(aq) + 2Ag(s) \rightarrow Cu(s) + 2Ag^+(aq)$ (*b*) $Cu(s) + Cu^{2+}(aq) \rightarrow 2Ag^+(aq) + Ag(s)$

(*c*) $Cu(s) + 2Ag^+(aq) \rightarrow Cu^{2+}(aq) + 2Ag(s)$ (*d*) $2Ag^+(aq) + Ag(s) \rightarrow Cu(s) + Cu^{2+}(aq)$

Answer. (*c*)

33. Which of these statements about a galvanic cell are not true?

i. the cathode carries a positive sign

ii. the anions migrate toward the cathode

iii. the electrons are released through the anode

iv. reduction occurs at the anode

(*a*) i and iii (*b*) i and ii

(*c*) ii and iii (*d*) ii and iv

Answer. (*d*)

34. The half–reaction that occurs at the cathode during the electrolysis of molten sodium iodide is

(*a*) $2I^- \rightarrow I_2 + 2e^-$ (*b*) $I_2 + 2e^- \rightarrow 2I^-$

(*c*) $Na^+ + e^- \rightarrow Na$ (*d*) $Na \rightarrow Na^+ + e^-$

Answer. (*c*)

35. For a voltaic cell using Ag+ (1 M)/Ag and Cu^{2+}(1M)/Cu half cells, which of the following statements is false?

(*a*) electrons will flow through the external circuit from the copper electrode to the silver electrode

(*b*) reduction occurs at the silver electrode as the cell operates

(*c*) the mass of the copper electrode will decrease as the cell operates

(*d*) the concentration of Ag^+ will increase as the cell operates

Answer. (*d*)

36. In the lead storage battery

(*a*) a reversible reaction can occur to recharge the battery

(*b*) lead is oxidized to create a flow of electrons

(*c*) lead forms the cathode when it is being reduced

(*d*) all of the above

Answer. (*d*)

37. The advantages of fuel cells include

(*a*) they can be recharged by the addition of more material to be oxidized and/or reduced

(*b*) they can be made to produce little or no harmful pollutants

(*c*) they can be made to run very quietly

(*d*) all of the above

Answer. (*d*)

38. For the reaction

$$2In(s) + 6H^+(aq) \rightarrow 2In^{3+}(aq) + 3H_2(g),\ E° = +0.34\ V$$

Determine the value of $E°_{red}$ for the half-reaction

$$In^{3+}(aq) + 3e^- \rightarrow In(s)$$

(*a*) – 0.17 (*b*) – 0.34

(*c*) 0.17 (*d*) 0.34

Answer. (*b*)

39. From a consideration of the following two half-reactions

Half Reaction	*E° (Volts)*
$Mn^{2+} + 2e^- \rightarrow Mn$	–1.18
$AuCl_4^- + 3e^- \rightarrow Au + 4Cl^-$	1.00

What is the standard cell potential for the reaction

$$3Mn + 2AuCl_4^- \rightarrow 3Mn^{2+} + 2Au + 8Cl^-$$

(*a*) –2.18 V (*b*) –0.18 V

(*c*) 0.18 V (*d*) 2.18 V

Answer. (*d*)

40. Determine the values of $E°_{cell}$ and $\Delta G°$ for the reaction below.

$$O_2(g) + 4I^-(aq) + 4H^+(aq) \rightarrow 2H_2O(\ell) + 2I_2(s)$$

(*a*) $E°_{cell} = 0.159$ V and $\Delta G° = -2.68 \times 10^5$ J (*b*) $E°_{cell} = 0.694$ V and $\Delta G° = -2.68 \times 10^5$ J
(*c*) $E°_{cell} = 0.694$ V and $\Delta G° = +2.68 \times 10^5$ J (*d*) $E°_{cell} = 1.764$ V and $\Delta G° = -6.81 \times 10^5$ J

Answer. (*b*)

41. The standard reduction potentials of Cu^{2+} and Ag^+ in V are +0.34 and +0.80, respectively. Determine the value of E in volts for the following cell at 298 K

$$Cu \mid Cu^{2+} (1.00\ M) \parallel Ag^+ (0.0010\ M) \mid Ag$$

(*a*) 0.37 V (*b*) 0.55 V
(*c*) – 0.28 V (*d*) 0.28 V

Answer. (*d*)

42. What is the potential of a half cell consisting of a platinum wire dipped into a solution 0.01M in Sn^{2+} and 0.001M in Sn^{4+} at 298 K?

(*a*) $E°_{oxid.} + 0.059$ (*b*) $E°_{red.} - \frac{0.059}{2}$
(*c*) $E°_{red.} + \frac{0.059}{2}$ (*d*) $E°_{oxid.} = -0.059$

Answer. (*b*)

43. A galvanic cell can be represented as

$Pt(s) \mid Sn^{2+}$ (aq, 1 M), Sn^{4+} (aq, 1 M) $\parallel Fe^{2+}$ (aq, 1 M) Fe^{3+} (aq, 1 M) $\mid Pt(s)$

What reaction is occurring at the anode?

(*a*) $Pt \rightarrow Pt^{2+} + 2e^-$ (*b*) $Sn^{2+} \rightarrow Sn^{4+} + 2e^-$
(*c*) $Pt \rightarrow Sn^{2+} + 2e^-$ (*d*) $Fe^{2+} \rightarrow Fe^{3+} + 1e^-$

Answer. (*d*)

44. What is the cell voltage of $Zn \mid Zn^{2+}$ (0.1 M) $\parallel Ag^+$ (0.1 M) $\mid Ag$?

The standard reduction potential for $Zn^{2+} + 2e^- \rightarrow Zn$ is –0.76 V and for $Ag^+ + e^- \rightarrow Ag$ is +0.80 V.

(*a*) –0.76 V (*b*) +0.80 V
(*c*) +1.53 V (*d*) +1.59 V

Answer. (*c*)

45. Calculate the voltage (*E*) of a cell with $E° = 1.1$ volts, if the copper half-cell is at standard conditions but the zinc ion concentration is only .001 molar. Temperature is 298 K. The overall reaction is

$$Zn + Cu^{+2} \rightarrow Cu + Zn^{+2}$$

(*a*) 0.39 volt (*b*) 1.43 volt
(*c*) 6.19 volt (*d*) 1.19 volt

Answer. (*d*)

46. Suppose that an alkaline dry cell was manufactured using cadmium metal rather than zinc. What effect would this have on the cell emf?

(*a*) no change (*b*) the voltage would increase by 0.360 V
(*c*) the voltage would decrease by 0.360 V (*d*) it would not work

Answer. (*c*)

47. Given $Zn \rightarrow Zn^{+2} + 2e^-$ with $E° = +0.763$, calculate E for a Zn electrode in which $Zn^{+2} = 0.025$ M.

(*a*) 1.00 V (*b*) 0.621 V
(*c*) 0.810 V (*d*) 0.124 V

Answer. (*c*)

48. A strip of zinc is dipped in a solution of copper sulfate. Select the correct occurring half–reaction.

(*a*) $Co^{++} + 2e^- \rightarrow Co$, reduction (*b*) $Cu + 2e^- \rightarrow Cu^{++}$, reduction
(*c*) $Cu \rightarrow Cu^{++} + 2e^-$, reduction (*d*) $Zn \rightarrow Zn^{++} + 2e^-$, oxidation

Answer. (*d*)

49. A concentration cell is constructed by placing identical Zn electrodes in two Zn^{2+} solutions. If the concentrations of the two Zn^{2+} solutions are 0.10 M and 0.00010 M, respectively, what is the potential of the cell?

(*a*) + 0.763 V (*b*) + 0.089 V
(*c*) + 0.053 V (*d*) + 0.24 V

Answer. (*b*)

50. Calculate the potential (in volts) for the following voltaic cell at 298 K:

$$Cr/Cr^{3+}\ (0.10\ M)\ ||\ Cu^{2+}\ (0.0010\ M)\ /Cu$$

(*a*) 1.25 V (*b*) 1.33 V
(*c*) 1.41 V (*d*) 1.57 V

Answer. (*b*)

51. Calculate the cell potential for the voltaic cell that results when the following two half-cells are connected at 298 K:

(1) a platinum electrode inserted into a solution of 0.10 M Co^{3+} and 0.0010 M Co^{2+}
(2) a copper electrode inserted into a solution of 0.010 M Cu^{2+} ions

(*a*) 1.56 V (*b*) 1.30 V
(*c*) 1.48 V (*d*) 1.66 V

Answer. (*d*)

52. Calculate the potential (in volts) for the following voltaic cell at 298 K:

$$Ag/Ag^{+}\ (0.01\ M)\ ||\ MnO_4^-\ (0.1\ M);\ H^+\ (1\ M);\ Mn^{2+}\ (0.001\ M)/Pt$$

(*a*) +0.57 V (*b*) 0.71 V
(*c*) +0.85 V (*d*) +0.91 V

Answer. (*c*)

53. What is the equilibrium constant for the following at 298 K?

$$3Mn^{2+} + 2Cr \rightarrow 3Mn + 2Cr^{3+}$$

(*a*) 5.1×10^{44} (*b*) 1.3×10^{21}
(*c*) 2.5×10^{-45} (*d*) 8.0×10^{-23}

Answer. (*c*)

54. A voltaic cell has an $E°$ value of –1.00 V. The reaction _____.

(*a*) is spontaneous (*b*) has a positive $\Delta G°$
(*c*) has a negative $\Delta G°$ (*d*) has $K = 1$

Answer. (*b*)

55. Which of the following is FALSE regarding the salt bridge used in voltaic cells? The salt bridge _____.

(*a*) allows for the two half-cells to be kept separated
(*b*) maintains the electrical neutrality in each half cell
(*c*) allows mixing of the two electrode solutions
(*d*) is made of a medium through which ions can slowly pass

Answer. (*c*)

56. Which of the following can we use to measure pH?

(*a*) a glass electrode (*b*) a concentration cell
(*c*) a hydrogen electrode (*d*) all of these

Answer. (*d*)

57. Based on the following information, which will be the most effective oxidizing agent?

$Na^+ + e^- \rightarrow Na$ $E° = -2.71$

$O^2 + 4e^- + 2H_2O \rightarrow 4OH^-$ $E° = +0.40$

$Cl_2 + 2e^- \rightarrow 2Cl^-$ $E° = +1.36$

(*a*) Na (*b*) Na^+
(*c*) O_2 (*d*) Cl_2

Answer. (*d*)

58. What is $\Delta G°$ at 298 K for the reaction

$$Hg(\ell) + 2Fe^{3+}(aq) \rightarrow Hg^{2+}(aq) + 2Fe^{2+}(aq)?$$

(*a*) +314 kJ (*b*) –16 kJ
(*c*) –314 kJ (*d*) 16 kJ

Answer. (*d*)

59. The salt bridge in the electrochemical cell serves to

(*a*) increase the rate at which equilibrium is attained
(*b*) increase the voltage of the cell
(*c*) maintain electrical neutrality
(*d*) increase the oxidation/reduction rate

Answer. (*c*)

60. Complete and balance the following equation. (All stoichiometric coefficients must be integers)

$$MnO_4^-(aq) + Cl^-(aq) \rightarrow Mn^{2+}(aq) + Cl_2(g) \text{ (acidic solution)}$$

How many hydrogen ions are needed and on which side of the equation must they appear?

(*a*) 16, on the left (*b*) 8, on the left
(*c*) 16, on the right (*d*) 4, on the left

Answer. (*a*)

61. Complete and balance the following equation. (All stoichiometric coefficients must be integers)

$$HClO(aq) + Br_2(g) \rightarrow BrO_3^-(aq) + Cl_2(g) \text{ (acidic solution)}$$

How many hydrogen ions are needed and on what side of the equation must they appear?

(*a*) 0 hydrogen ions are needed (*b*) 10, on the right
(*c*) 12, on the right (*d*) 2, on the left

Answer. (*a*)

62. What is $\Delta G°$ at 298 K for the reaction:

$$2VO_2^+(aq) + 4H^+(aq) + Cd(s) \rightarrow 2VO^{2+}(aq) + 2H_2O(\ell) + Cd^{2+}(aq)?$$

(*a*) –271 kJ (*b*) 1.403 J
(*c*) –135 kJ (*d*) –115 kJ

Answer. (*a*)

63. From a consideration of the following two half-reactions at 298 K,

Half Reaction	*E° (Volts)*
$Pb^{2+}(aq) + 2e^- \rightarrow Pb(s)$	– 0.126
$Fe^{2+}(aq) + 2e^- \rightarrow Fe(s)$	– 0.447

What is the equilibrium constant for the following equation?

$Pb^{2+}(aq) + Fe(s) \rightarrow Pb(s) + Fe^{2+}(aq)$ $K_{eq} = e^{-\Delta G/RT}$

Faraday constant: 1 F = 96,485 C/mol. $R = 8.314$ J mol^{-1} K^{-1}

(*a*) 3.7×10^{-6} (*b*) 2.7×10^5
(*c*) 7.2×10^{10} (*d*) 2.4×10^{19}

Answer. (*c*)

64. From a consideration of the following two half-reactions

Half Reaction	*E°(Volts)*
$Cl_2(g) + 2e^- \rightarrow 2Cl^-(aq)$	1.36
$Br_2(\ell) + 2e^- \rightarrow 2Br^-(aq)$	1.07

what is the standard free energy change at 298 K for the following reaction?

$$Cl_2(g) + 2Br^-(aq) \rightarrow Br_2(\ell) + 2Cl^-(aq)$$

Faraday constant: 1 F = 96,485 C/mol. $R = 8.314$ J mol^{-1} K^{-1}

(*a*) –112 kJ/mol (*b*) –56 kJ/mol
(*c*) –28 kJ/mol (*d*) 470 kJ/mol

Answer. (*b*)

30 Photochemistry

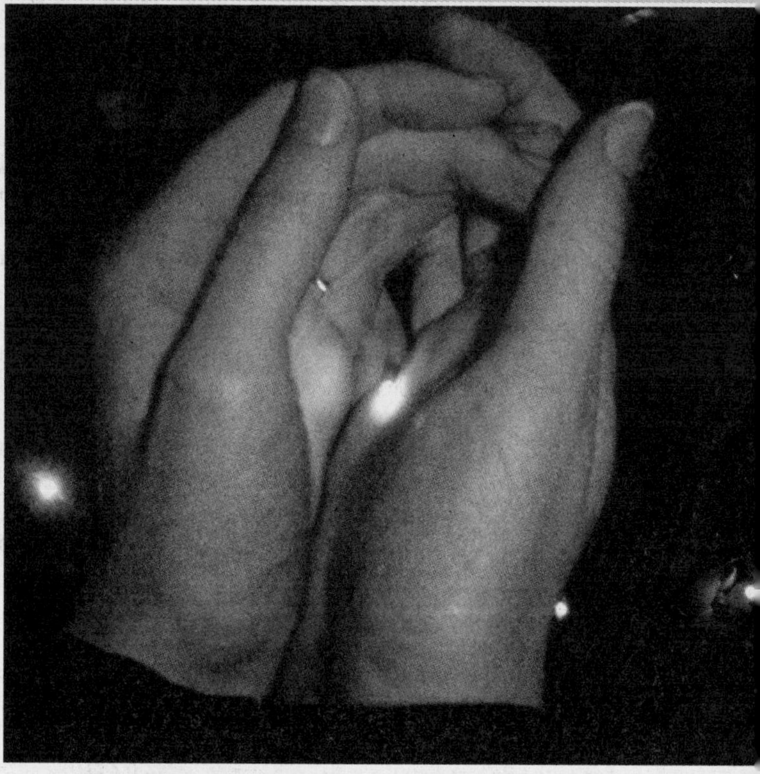

CONTENTS

PHOTOCHEMICAL REACTIONS

Ordinary reactions occur by absorption of heat energy from outside. The reacting molecules are energized and molecular collisions become effective. These bring about the reaction. The reactions which are caused by heat and in absence of light are called **thermal** or **dark reactions.**

On the other hand, some reactions proceed by absorption of light radiations. These belong to the visible and ultraviolet regions of the electromagnetic spectrum (2×10^{-7} to 8×10^{-7} m). The reactant molecules absorbs photons of light and get excited. These excited molecules then produce the reactions.

A reaction which takes place by absorption of the visible and ultraviolet radiations is called a photochemical reaction.

The branch of chemistry which deals with the study of photochemical reactions is called **photochemistry.**

Demonstration of a Photochemical reaction

A mixture of hydrogen and chlorine remains unchanged with lapse of time. But when exposed to light, the reaction occurs with a loud explosion.

$$H_2 + Cl_2 \xrightarrow{\text{dark}} \text{No reaction}$$

$$H_2 + Cl_2 \xrightarrow{\text{light}} 2HCl$$

A bottle is filled with equimolar amounts of hydrogen and chlorine (Fig. 30.1). It is tightly stoppered with a handball. When the lamp is turned on, a beam of light falls on the mixture through the bottom of the bottle. The reaction occurs with an explosion. The ball is expelled with high velocity so that it strikes the opposite wall of the lecture theatre.

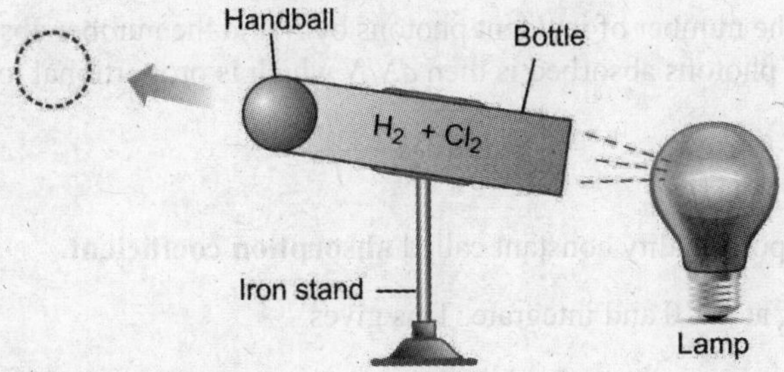

Figure 30.1
The 'HCl-cannon' experiment.

DIFFERENCE BETWEEN PHOTOCHEMICAL AND THERMOCHEMICAL REACTIONS

Photochemical Reactions	Thermochemical Reactions
1. These involve absorption of light radiations.	1. These reactions involve absorption or evolution of heat.
2. The presence of light is the primary requirement for reactions to take place.	2. These reactions can take place in dark as well as in light.
3. Temperature has a very little effect on the rate of photochemical reactions.	3. Temperature has a significant effect on the rate of a thermochemical reaction.
4. ΔG for photochemical spontaneous reactions may be +ve or –ve.	4. ΔG for a thermochemical reaction is always negative.
5. Photochemical activation is highly selective. The absorbed photon excites a particular atom or group of atoms which become site for the reaction.	5. Thermochemical activation is not selective in nature.

LIGHT ABSORPTION

When light is passed through a medium, a part of it is absorbed. It is this absorbed portion of light which causes photochemical reactions. Let a beam of monochromatic light pass through a thickness dx of the medium. The intensity of radiation reduces from I to $I\text{-}dI$.

The intensity of radiation can be defined as the number of photons that pass across a unit area in unit time.

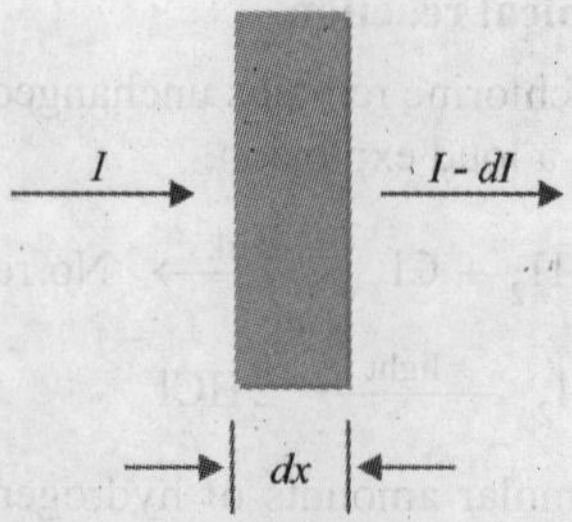

■ **Figure 30.2**
As a beam of intensity I passes through a medium of thickness dx, the intensity of the beam is reduced to I - dI.

Let us denote the number of incident photons by N and the number absorbed in thickness dx by dN. The fraction of photons absorbed is then dN/N which is proportional to thickness dx. That is,

$$\frac{dN}{N} = b\ dx = -\frac{dI}{I}$$

where b is proportionality constant called **absorption coefficient.**

Let us set $I = I_0$ at $x = 0$ and integrate. This gives

$$I = I_0\,(-bx)$$

or $$\ln\left(\frac{I}{I_0}\right) = -bx \qquad ...(1)$$

Lambert first derived equation (1) and it is known as **Lambert Law.** Beer extended this relation to solutions of compounds in transparent solvents. The equation (1) then takes the form (2).

$$\ln\left(\frac{I}{I_0}\right) = -\in C\,x \qquad ...(2)$$

where C = molar concentration; $\in$ is a constant characteristic of the solute called the *molar absorption coefficient.* The relation (2) is known as the **Lambert-Beer Law.** This law forms the basis of spectrophotometric methods of chemical analysis.

DETERMINATION OF ABSORBED INTENSITY

A photochemical reaction occurs by the absorption of photons of light by the molecules. Therefore, it is essential to determine the absorbed intensity of light for a study of the rate of reaction.

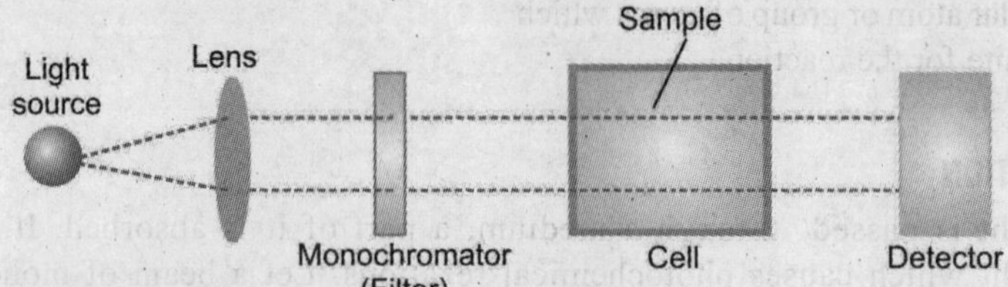

■ **Figure 30.3**
Schematic diagram of the apparatus used for measurement of light intensity.

An experimental arrangement for the purpose is illustrated in Fig. 30.3.

Light beam from a suitable source (tungsten filament or mercury vapor lamp) is rendered parallel by the lens. The beam then passes through a 'filter' or monochromator which yields light of one wavelength only. The monochromatic light enters the reaction cell made of quartz. The part of light that is not absorbed strikes the *detector*. Thus the intensity of light is measured first with the empty cell and then the cell filled with the reaction sample. The first reading gives the incident intensity, I_0, and the second gives the transmitted intensity, I. The difference, $I_0 - I = I_a$, is the absorbed intensity.

The detector generally used for the measurement of intensity of transmitted light is :

(*a*) a thermopile (*b*) photoelectric cell (*c*) a chemical actinometer.

Thermopile

It is made of a series of thermocouples in which unlike metals such as bismuth and silver are joined together. One end of the couple is blackened with lamp black and the other end is left as such. When the radiation strikes the black end it absorbs energy and is heated up. The temperature difference between the two ends causes a current to flow in the circuit as indicated by the galvanometer. **The current is proportional to intensity of radiation.** The thermopile is previously calibrated against a standard source of light.

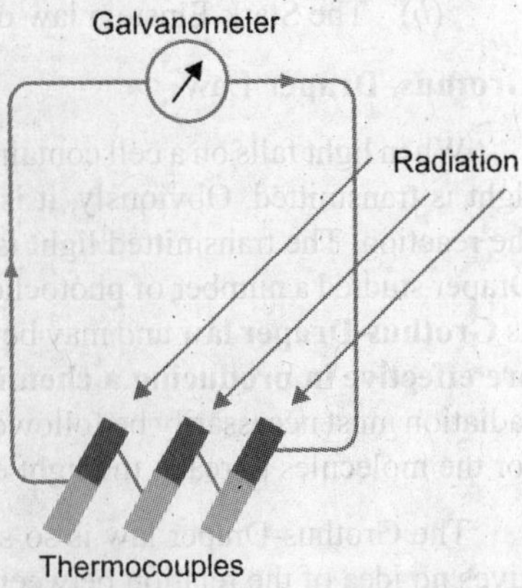

Figure 30.4
A thermopile.

Photoelectric Cell

A photoelectric cell (Fig. 30.5) can be conveniently used for measuring intensity of light. The light striking the active metal electrode (cesium, sodium or potassium) causes the emission of electrons. A current flows through the circuit which can be measured with an ammeter. The intensity of light is proportional to the current.

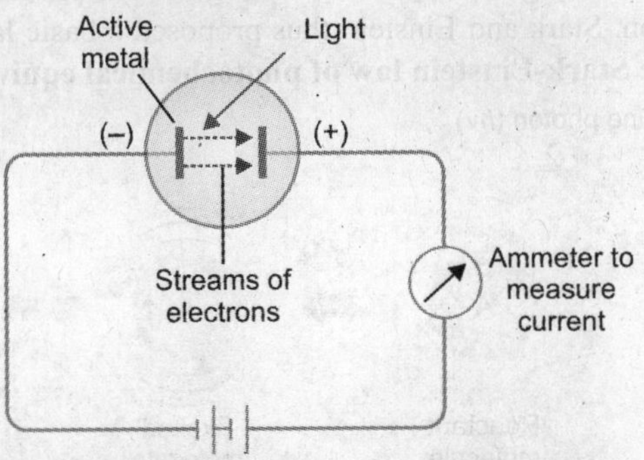

Figure 30.5
A photoelectric cell (illustration).

Chemical Actinometer

A chemical actinometer uses a chemical reaction whose rate can be determined easily. One such simple device is **Uranyl oxalate actinometer.** It contains 0.05 M oxalic acid and 0.01 M uranyl sulfate in water. When it is exposed to radiation, oxalic acid is decomposed to CO_2, CO and H_2O.

$$UO_2^{2+} + h\nu \longrightarrow (UO_2^{2+})^*$$

$$(UO_2^{2+})^* + \begin{array}{l}COOH\\|\\COOH\end{array} \longrightarrow UO_2^{2+} + CO_2 + CO + H_2O$$

oxalic acid

The concentration of oxalic acid that remains can be found by titration with standard $KMnO_4$ solution. The used up concentration of oxalic acid is a measure of the intensity of radiation.

LAWS OF PHOTOCHEMISTRY

There are two basic laws governing photochemical reactions :

(*a*) The Grothus-Draper law

(*b*) The Stark-Einstein law of Photochemical Equivalence

Grothus–Draper Law

When light falls on a cell containing a reaction mixture, some light is absorbed and the remaining light is transmitted. Obviously, it is the absorbed component of light that is capable of producing the reaction. The transmitted light is ineffective chemically. Early in the 19th century, Grothus and Draper studied a number of photochemical reactions and enunciated a generalization. This is known as **Grothus-Draper law** and may be stated as follows : **It is only the absorbed light radiations that are effective in producing a chemical reaction.** However, it does not mean that the absorption of radiation must necessarily be followed by a chemical reaction. When the conditions are not favorable for the molecules to react, the light energy remains unused. It may be re-emitted as heat or light.

The Grothus-Draper law is so simple and self-evident. But it is purely qualitative in nature. It gives no idea of the relation between the absorbed radiation and the molecules undergoing change.

Stark-Einstein Law of Photochemical Equivalence

Stark and Einstein (1905) studied the quantitative aspect of photochemical reactions by application of *Quantum theory of light.* They noted that each molecule taking part in the reaction absorbs only a single quantum or photon of light. The molecule that gains one photon-equivalent energy is activated and enters into reaction. Stark and Einstein thus proposed a basic law of photochemistry which is named after them. The **Stark-Einstein law of photochemical equivalence** may be stated as :

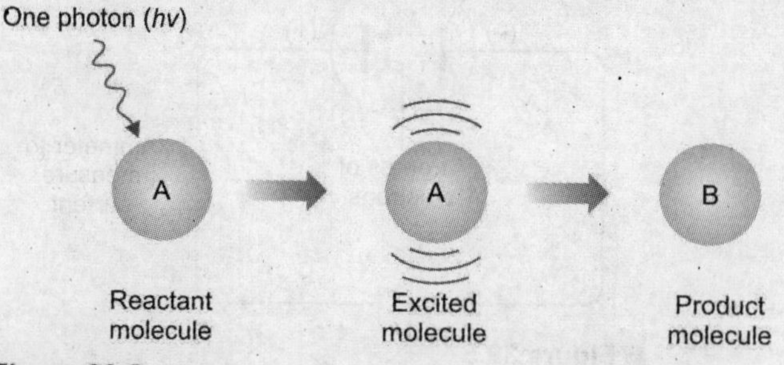

Figure 30.6
Illustration of Law of Photochemical equivalence; absorption of one photon decomposes one molecule.

In a photochemical reaction, each molecule of the reacting substance absorbs a single photon of radiation causing the reaction and is activated to form the products.

The law of photochemical equivalence is illustrated in Fig. 30.6 where a molecule 'A' absorbs a photon of radiation and gets activated. The activated molecule (A*) then decomposes to yield B. We could say the same thing in equation form as :

$$\begin{array}{l} A + h\nu \longrightarrow A^* \\ A^* \longrightarrow B \\ \hline \text{overall} \quad A + h\nu \longrightarrow B \end{array}$$

In practice, we use molar quantities. That is, one mole of A absorbs one mole of photons or one einstein of energy, *E*. The value of *E* can be calculated by using the expression given below:

$$E = \frac{2.859}{\lambda} \times 10^5 \text{ kcal mol}^{-1} \text{ or } 859 \times 4.187 \times 10^5 \text{ kJ mol}^{-1}$$

Primary and Secondary reactions

The overall photochemical reaction may consist of :

(*a*) a primary reaction

(*b*) secondary reactions

A primary reaction proceeds by absorption of radiation.

A secondary reaction is a thermal reaction which occurs subsequent to the primary reaction.

For example, the decomposition of HBr occurs as follows :

$$\begin{array}{lll} HBr + h\nu \longrightarrow H + Br & & \text{Primary reaction} \\ HBr + H \longrightarrow H_2 + Br & & \text{Secondary reaction} \\ Br + Br \longrightarrow Br_2 & & \text{Secondary reaction} \\ \hline 2HBr + h\nu \longrightarrow H_2 + Br_2 & & \text{Overall reaction} \end{array}$$

Evidently, the primary reaction only obeys the law of photochemical equivalence strictly. The secondary reactions have no concern with the law.

Quantum yield (or Quantum efficiency)

It has been shown that not always a photochemical reaction obeys the Einstein law. The number of molecules reacted or decomposed is often found to be markedly different from the number of quanta or photons of radiation absorbed in a given time.

The number of molecules reacted or formed per photon of light absorbed is termed Quantum yield. It is denoted by ϕ so that

$$\phi = \frac{\text{No. of molecules reacted or formed}}{\text{No. of photons absorbed}}$$

For a reaction that obeys strictly the Einstein law, one molecule decomposes per photon, the quantum yield $\phi = 1$. When two or more molecules are decomposed per photon, $\phi > 1$ and the reaction has a **high quantum yield.** If the number of molecules decomposed is less than one per photon, the reaction has a **low quantum yield.**

Cause of high quantum yield

When one photon decomposes or forms more than one molecule, the quantum yield $\phi > 1$ and is said to be high. The chief reasons for high quantum yield are :

(*a*) **Reactions subsequent to the Primary reaction.** One photon absorbed in a primary reaction dissociates one molecule of the reactant. But the excited atoms that result may start a subsequent secondary reaction in which a further molecule is decomposed

$$\begin{array}{lll} AB + h\nu \longrightarrow A + B & & \text{Primary} \\ AB + A \longrightarrow A_2 + B & & \text{Secondary} \end{array}$$

Obviously, one photon of radiation has decomposed two molecules, one in the primary reaction and one in the secondary reaction. Hence the quantum yield of the overall reaction is 2.

(*b*) **A reaction chain forms many molecules per photon.** When there are two or more reactants, a molecule of one of them absorbs a photon and dissociates (primary reaction). The excited atom that is produced starts a secondary reaction chain.

$$A_2 + h\nu \longrightarrow 2A \quad ...(1) \quad \text{Primary}$$
$$A + B_2 \longrightarrow AB + B \quad ...(2)$$
$$B + A_2 \longrightarrow AB + A \quad ...(3)$$
(2) and (3): Secondary Reaction chain

It is noteworthy that A consumed in (2) is regenerated in (3). This reaction chain continues to form two molecules each time. Thus the number of AB molecules formed in the overall reaction per photon is very large. Or that the quantum yield is extremely high.

Examples of high quantum yield

The above reasons of high quantum yield are illustrated by citing examples as below :

(*i*) **Decomposition of HI.** The decomposition of hydrogen iodide is brought about by the absorption of light of less than 4×10^{-7} m. In the primary reaction, a molecule of hydrogen iodide absorbs a photon and dissociates to produce H and I. This is followed by secondary steps as shown below :

$$HI + h\nu \longrightarrow H + I \quad ...(1) \quad \text{Primary}$$
$$H + HI \longrightarrow H_2 + I \quad ...(2)$$
$$I + I \longrightarrow I_2 \quad ...(3)$$
(2) and (3): Secondary

$$2HI + h\nu \longrightarrow H_2 + I_2 \quad \text{Overall reaction}$$

In the overall reaction, two molecules of hydrogen iodide are decomposed for one photon ($h\nu$) of light absorbed. Thus the quantum yield is 2.

(*ii*) **Hydrogen-Chlorine reaction.** This is a well known example of a **photochemical chain reaction.** A mixture of hydrogen and chlorine is exposed to light of wavelength less than 4×10^{-7} m. The hydrogen and chlorine react rapidly to form hydrogen chloride. In the primary step, a molecule of chlorine absorbs a photon and dissociates into two Cl atoms. This is followed by the secondary reactions stated below :

$$Cl_2 + h\nu \longrightarrow 2Cl \quad ...(1) \quad \text{Primary reaction}$$
$$Cl + H_2 \longrightarrow HCl + H \quad ...(2)$$
$$H + Cl_2 \longrightarrow HCl + Cl \quad ...(3)$$
(2) and (3): Secondary reactions

The Cl atom used in step (2) is regenerated in step (3). Thus the steps (2) and (3) constitute a self-propagating chain reaction. This produces two molecules of HCl in each cycle. Thus one photon of light absorbed in step (1) forms a large number of HCl molecules by repetition of the reaction sequence (2) and (3). The chain reaction terminates when the Cl atoms recombine at the walls of the vessel where they lose their excess energy.

$$2Cl \xrightarrow{\text{walls}} Cl_2$$

The number of HCl molecules formed for a photon of light is very high. The quantum yield of the reaction varies from 10^4 to 10^6.

Causes of low quantum yield

The chief reasons of low quantum yield are :

(*a*) **Deactivation of reacting molecules.** The excited molecules in the primary process may be deactivated before they get opportunity to react. This is caused by collisions with some inert molecules or by fluorescence.

$$A + h\nu \longrightarrow A^* \quad \textit{Activation}$$
$$A^* \longrightarrow A + h\nu' \quad \textit{Fluorescence}$$

(*b*) **Occurrence of reverse of primary reaction.** Here the primary reaction generally yields a polymer. The product then undergoes a thermal reaction giving back the reactant molecules.

$$2A \underset{\text{thermal}}{\overset{h\nu}{\rightleftharpoons}} A_2$$

The reverse thermal reaction proceeds till the equilibrium state is reached.

(*c*) **Recombination of dissociated fragments.** In a primary process the reactant molecules may dissociate to give smaller fragments. These fragments can recombine to give back the reactant.

$$(AB) + h\nu \longrightarrow A + B$$
$$A + B \longrightarrow (AB)$$

Thus the secondary reactions involving the fragments to form the product will not occur. This will greatly lower the yield.

The yield of particular photochemical reaction may be lower than expected for more than one reason cited above.

Examples of low quantum yield

The examples listed below will illustrate the above causes of low quantum yield:

(*i*) **Dimerization of Anthracene.** When anthracene, $C_{14}H_{10}$, dissolved in benzene is exposed to ultraviolet light, it is converted to dianthracene, $C_{28}H_{20}$.

$$2C_{14}H_{10} + h\nu \longrightarrow C_{28}H_{20}$$

Obviously, the quantum yield should be 2 but it is actually found to be 0.5. The low quantum yield is explained as the reaction is accompanied by fluorescence which deactivates the excited anthracene molecules. Furthermore, the above reaction is reversible.

$$2C_{14}H_{10} \underset{\text{thermal}}{\overset{h\nu}{\rightleftharpoons}} C_{28}H_{20}$$

The transformation of the product back to the reactant occurs till a state of equilibrium is reached. This further lowers the quantum yield.

(*ii*) **Combination of H_2 and Br_2.** When a mixture of hydrogen and bromine is exposed to light, hydrogen bromide is formed. The reaction occurs by the following possible steps.

$$Br_2 + h\nu \longrightarrow 2Br \quad ...(1) \quad \text{Primary reaction}$$
$$\left.\begin{aligned} Br + H_2 &\longrightarrow HBr + H \quad ...(2) \\ H + Br_2 &\longrightarrow HBr + Br \quad ...(3) \\ H + HBr &\longrightarrow H_2 + Br \quad ...(4) \\ Br + Br &\longrightarrow Br_2 \quad ...(5) \end{aligned}\right\} \text{Secondary reactions}$$

The reaction (2) is extremely slow. The reactions (3), (4) and (5), depend directly or indirectly on (2) and so are very slow. Therefore most of the Br atoms produced in the primary process recombine to give back Br_2 molecules. Thus the HBr molecules obtained per quantum is extremely small. The quantum yield of the reaction is found to be 0.01 at ordinary temperature.

CALCULATION OF QUANTUM YIELD

By definition, the quantum yield, ϕ, of a photochemical reaction is expressed as :

$$\phi = \frac{\text{Number of molecules decomposed or formed}}{\text{Number of photons of radiation energy absorbed}}$$

or
$$\phi = \frac{\text{Number of moles decomposed or formed}}{\text{Number of moles of radiation energy absorbed}}$$

Thus we can calculate quantum yield from :

(*a*) The amount of the reactant decomposed in a given time and

(*b*) The amount of radiation energy absorbed in the same time

The radiation energy is absorbed by a chemical system as photons. Therefore we should know the energy associated with a photon or a mole of photons.

The energy of photons; einstein

We know that the energy of a photon (or quantum), $\in$, is given by the equation.

$$\in = h\nu = \frac{hc}{\lambda} \qquad ...(1)$$

where h = Planck's constant (6.624×10^{-34} J s)

ν = frequency of radiation

λ = wavelength of radiation

c = velocity of light (3×10^8 m s^{-1})

If λ is given in m, the energy is expressed in J.

The energy, *E*, of an Avogadro number (*N*) of photons is referred to as one einstein. That is,

$$E = \frac{Nhc}{\lambda} \qquad ...(2)$$

Substituting the values of N (= 6.02×10^{23}), h and c, in (2), we have

$$E = \frac{1.196 \times 10^{1}}{\lambda} \text{ J mol}^{-1}$$

If λ is expressed in Å units (1Å = 10^{-8} cm),

$$E = \frac{1.196 \times 10^{9}}{\lambda} \text{ J mol}^{-1} \qquad ...(3)$$

Since 1 cal = 4.184×10^7 erg, energy in calories would be

$$E = \frac{1.196 \times 10^{9}}{\lambda \times 4.184 \text{ J}} \qquad ...(4)$$

$$= \frac{2.859}{\lambda} \times 10^{8} \text{ J mol}^{-1}$$

or
$$E = \frac{2.859}{\lambda} \times 10^{5} \text{ kJ mol}^{-1} \qquad ...(5)$$

It is evident from (3) that the numerical value of einstein varies inversely as the wavelength of radiation. **The higher the wavelength, the smaller will be the energy per einstein.**

SOLVED PROBLEM 1

Calculate the energy associated with (*a*) one photon; (*b*) one einstein of radiation of wavelength 8×10^{-7} m. $h = 6.63 \times 10^{-34}$ J-s; $c = 3 \times 10^8$ m s^{-1}.

SOLUTION

(*a*) Energy of a photon $= \dfrac{hc}{\lambda} = \dfrac{6.62 \times 10^{-34} \times 3 \times 10^{8}}{8 \times 10^{-7}}$

$$= 2.4825 \times 10^{-19} \text{ J}$$

(*b*) Energy per einstein $= \dfrac{Nhc}{\lambda} = \dfrac{6.02 \times 10^{23} \times 6.62 \times 10^{-34} \times 3 \times 10^{8}}{8 \times 10^{-7}}$

$$= \mathbf{1.4945 \times 10^3 \text{ J}}$$

SOLVED PROBLEM 2

When a substance *A* was exposed to light, 0.002 mol of it reacted in 1204 seconds. In the same time *A* absorbed 2.0×10^6 photons of light per second. Calculate the quantum yield of the reaction. (Avogadro number $N = 6.02 \times 10^{23}$)

SOLUTION

Number of molecules of *A* reacting $= 0.002 \times N = 0.002 \times 6.02 \times 10^{23}$

Number of photons absorbed per second $= 2.0 \times 10^6$

Number of photons absorbed in 1204 seconds $= 2.0 \times 10^6 \times 1204$

$$\text{Quantum yield } \phi = \frac{\text{No. of molecules reacted}}{\text{No. of photons absorbed}}$$

$$= \frac{0.002 \times 6.02 \times 10^{23}}{2.0 \times 10^6 \times 1204} = \mathbf{5.00 \times 10^{11}}$$

SOLVED PROBLEM 3

When irradiated with light of 5×10^{-7} m wavelength, 1×10^{-4} mol of a substance is decomposed. How many photons are absorbed during the reaction if its quantum efficiency is 10.00. (Avogadro number $N = 6.02 \times 10^{23}$)

SOLUTION

Quantum efficiency of the reaction = 10.00

No. of moles decomposed $= 1 \times 10^{-4}$

No. of molecules decomposed $= 1 \times 10^{-4} \times 6.02 \times 10^{23}$

$$= 6.02 \times 10^{19}$$

we know that,

$$\phi = \frac{\text{No. of molecules decomposed}}{\text{No. of photons absorbed}}$$

$$= \frac{6.02 \times 10^{19}}{\text{No. of photons absorbed}}$$

$$\text{No. of photons absorbed} = \frac{6.02 \times 10^{19}}{10} = \mathbf{6.02 \times 10^{18}}$$

SOLVED PROBLEM 4

When propionaldehyde is irradiated with light of $\lambda = 3.02 \times 10^{-7}$ m, it is decomposed to form carbon monoxide.

$$CH_3CH_2CHO + h\nu \longrightarrow CH_3CH_3 + CO$$

The quantum yield for the reaction is 0.54. The light energy absorbed is 1.5×10^3 J mol in a given time. Find the amount of carbon monoxide formed in moles in the same time.

SOLUTION

From expression (3), we have

$$\text{one einstein } (E) = \frac{1.196 \times 10^9}{\lambda} \text{ J mol}$$

$$\text{when } \lambda = 3.02 \times 10^{-7} \text{ m, one einstein} = \frac{1.196 \times 10^9}{3.02 \times 10^{-7}} \text{ J mol}$$

$$\text{or } 1.5 \times 10^3 \text{ J mol of energy} = \frac{15000 \times 3.02}{1.196 \times 10^{16}} = 3.78 \times 10^{-13} \text{ einstein}$$

$$\text{But} \quad \phi = \frac{\text{No. of moles of CO formed}}{\text{No. of einsteins absorbed}} = 0.54$$

Hence the amount of CO formed = $0.54 \times 3.78 \times 10^{-13}$ = **2.04×10^{-13} mol**

PHOTOSENSITIZED REACTIONS

In many photochemical reactions the reactant molecule does not absorb the radiation required for the reaction. Hence the reaction is not possible. In such cases the reaction may still occur if a foreign species such as mercury vapor is present. The mercury atom absorbs the incident radiation and subsequently transfers its energy to the reactant molecule which is activated. Thus the reaction occurs. A species which can both absorb and transfer radiant energy for activation of the reactant molecule, is called a **photosensitizer.** The reaction so caused is called a **photosensitized reaction.**

The role of mercury vapor is that of a go-between. The mercury atom absorbs the incident radiation and is excited. The excited atom collides with a reactant molecule (A) and transfer to it the excitation energy. This energy is enough to activate the molecule (A). The mercury atom returns to the original unactivated state.

$$Hg + h\nu \longrightarrow Hg^*$$

$$Hg^* + A \longrightarrow A^* + Hg$$

Examples of Photosensitized reactions

(*a*) **Reaction between H_2 and O_2.** This reaction is photosensitized by mercury vapor. The product is hydrogen peroxide, H_2O_2.

$$Hg + h\nu \longrightarrow Hg^* \qquad \text{Primary absorption}$$

$$Hg + H_2 \longrightarrow 2H + Hg \qquad \text{Energy transfer}$$

$$\left.\begin{array}{l} H + O_2 \longrightarrow HO_2 \\ HO_2 + HO_2 \longrightarrow H_2O_2 + O_2 \end{array}\right\} \text{Reaction}$$

Hydrogen peroxide may decompose to form water, H_2O.

(*b*) **Reaction between H_2 and CO.** Mercury vapor is used as photosensitizer. The product is formaldehyde, HCHO.

$$Hg + h\nu \longrightarrow Hg^* \qquad \text{Primary absorption}$$

$$Hg + H_2 \longrightarrow 2H + Hg \qquad \text{Energy transfer}$$

$$H + CO \longrightarrow HCO$$

$$\left.\begin{array}{l} HCO + H_2 \longrightarrow HCHO + H \\ 2HCO \longrightarrow HCHO + CO \end{array}\right\} \text{Reaction}$$

Some glyoxal, CHO-CHO, is also formed by dimerization of formyl radicals, HCO.

PHOTOPHYSICAL PROCESSES

If the absorbed radiation is not used to cause a chemical change, it is re-emitted as light of longer wavelength. The three such photophysical processes which can occur are :

(*a*) Fluorescence (*b*) Phosphorescence (*c*) Chemiluminescence

Fluorescence

Certain molecules (or atoms) when exposed to light radiation of short wavelength (high frequency), emit light of longer wavelength. The process is called **fluorescence** and the substance that exhibits fluorescence is called florescent substance. Florescence stops as soon as the incident radiation is cut off.

Examples. (*a*) a solution of quinine sulfate on exposure to visible light, exhibits blue fluorescence.

(*b*) a solution of chlorophyll in ether shows blood red fluorescence.

■ **Figure 30.7**
Fluorescent minerals, shown under ultraviolet light.

■ **Figure 30.8**
Tonic water is clear under normal light, but vividly fluorescent under ultraviolet light, due to the presence of the quinine used as a flavoring.

■ **Figure 30.9**
A solution of chlorophyll in ether solution shows blood red fluorescence.

Explanation. When a molecule absorbs high energy radiation, it is excited to higher energy states. Then it emits excess energy through several transitions to the ground state. Thus the excited molecule emits light of longer frequency. The color of fluorescence depends on the wavelength of light emitted.

Phosphorescence

When a substance absorbs radiation of high frequency and emits light even after the incident radiation is cut off, the process is called phosphorescence. The substance which shows phosphorescence is called **phosphorescent substance.**

Phosphorescence is chiefly caused by ultraviolet and visible light. It is generally shown by solids.

Examples. (*a*) Sulfates of calcium, barium and strontium exhibit phosphorescence.

(*b*) Fluorescein in boric acid shows phosphorescence in the blue region at 5.7×10^{-7} m wavelength.

Explanation. As in fluorescence, a molecule absorbs light radiation and gets excited. While returning to the ground state, it emits light energy of longer wavelength. In doing so the excited molecule passes from one series of electronic states to another and gets trapped. This shows the emission of light which persists even after the removal of light source. Thus **phosphorescence could be designated as delayed fluorescence.**

$$A + h\nu \longrightarrow A^* \xrightarrow{\text{slow}} h\nu'$$

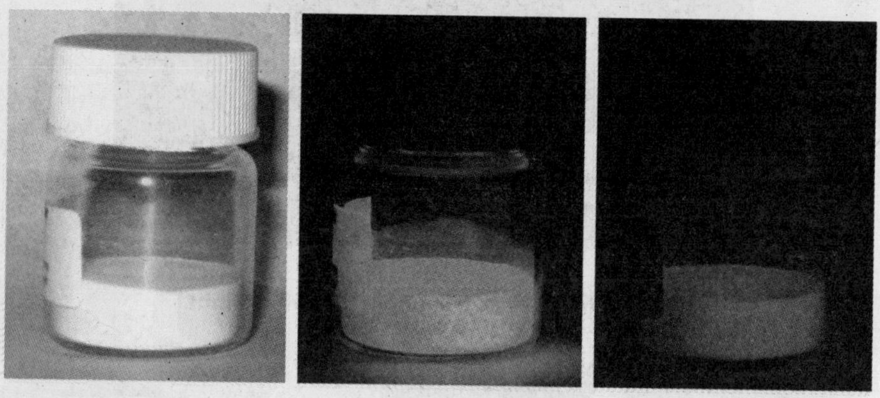

■ **Figure 30.10**
Phosphorescent powder under visible light, ultraviolet light, and total darkness

Chemiluminescence

Some chemical reactions are accompanied by the emission of visible light at ordinary temperature. **The emission of light as a result of chemical action is called chemiluminescence.** The reaction is referred to as **a chemiluminescent reaction.** Such a reaction is the reverse of a photochemical reaction which proceeds by absorption of light. The light emitted in a chemiluminescent reaction is also called **'cold light'** because it is produced at ordinary temperature.

Examples. (*a*) The glow of fireflies due to the aerial oxidation of *luciferin* (a protein) in the presence of enzyme *luciferase.*

(*b*) The oxidation of 5-*aminophthalic* cyclic hydrazide (*luminol*) by hydrogen peroxide in alkaline solution, producing bright green light.

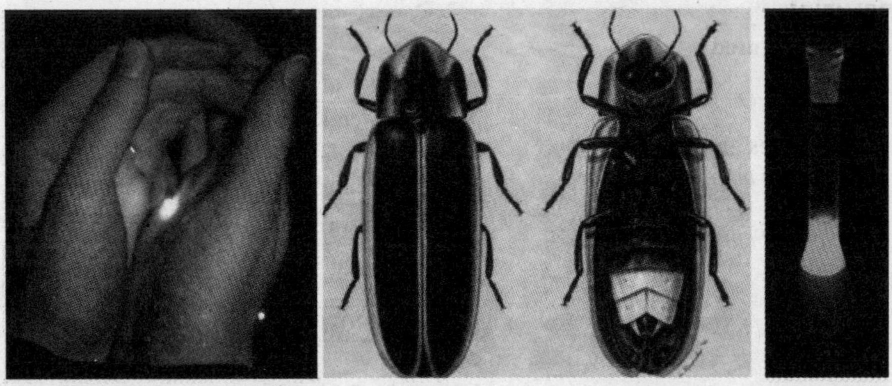

■ **Figure 30.11**
Chemiluminescence of fireflies and luminol.

Explanation. In a chemiluminescent reaction, the energy released in the reaction makes the product molecule electronically excited. The excited molecule then gives up its excess energy as visible light while reverting to ground state.

ADDITIONAL SOLVED PROBLEMS

SOLVED PROBLEM 5

Calculate the energy of the photon corresponding to wavelength 4.8×10^{-7} m.
($h = 6.624 \times 10^{-34}$ J s ; $c = 3 \times 10^8$ m s^{-1})

SOLUTION

Formula used

$$E = \frac{hc}{\lambda}$$

Quantities given

$h = 6.624 \times 10^{-34}$ J s $c = 3 \times 10^8$ m s^{-1} $\lambda = 4800$ Å $= 4.8 \times 10^{-7}$ m

Substitution of values

$$E = \frac{6.624 \times 10^{-34} \times 3 \times 10^8}{4.8 \times 10^{-7}} \text{ J}$$

$$= 4.14 \times 10^{-19} \text{ J}$$

SOLVED PROBLEM 6

For a photochemical reaction A $\longrightarrow$ B, 1.0×10^{-5} mol of B were formed on absorption of 6.0×10^7 J at 3.6×10^{-7} meter. Calculate the quantum efficiency of the reaction.
($N = 6.62 \times 10^{23}$; $h = 6.62 \times 10^{-34}$ J s)

SOLUTION

Formula used

$$\textit{Quantum efficiency} = \frac{\textit{No. of moles of B formed}}{\textit{No. of einsteins absorbed}}$$

Quantities given

$$\textit{No. of moles of B formed} = 1.00 \times 10^{-5}\text{ mol}$$

$$\textit{No. of einsteins absorbed} = \frac{\text{Energy absorbed}}{\text{Energy per einstein}}$$

$$= \frac{6.0 \times 10^{7}\text{ J}}{\dfrac{1.196 \times 10^{9}\text{ J}}{\lambda}} \quad \left[\because 1\text{ einstein} = \frac{1.196 \times 10^{9}\text{ J}}{\lambda}\right]$$

$$= \frac{6.0 \times 10^{9} \times 3.6 \times 10^{-7}}{1.196 \times 10^{9}} \quad [\because \lambda = 3.6 \times 10^{-7}\text{ m}]$$

$$= 1.806 \times 10^{-8}\text{ m}$$

Substitution of values

$$\text{Quantum efficiency, } \phi = \frac{1.0 \times 10^{-5}}{1.806 \times 10^{-8}}$$

$$= \mathbf{0.554 \times 10^{3}}$$

SOLVED PROBLEM 7

Calculate the values of frequency, quantum energy and einstein for 5×10^{-7} m radiation.

SOLUTION : (*a*) To calculate the value of frequency

Formula used

$$\upsilon = \frac{c}{\lambda}$$

Quantities given

$c = 3 \times 10^{8}\text{ m s}^{-1}$ $\quad\quad$ $\lambda = 5 \times 10^{-7}$ m

Substitution of values

$$\upsilon = \frac{3 \times 10^{8}}{5 \times 10^{-7}}$$

$$= \mathbf{6 \times 10^{14}}$$

(*b*) To calculate the quantum energy

Formula used

$$E = \frac{h\,c}{\lambda}$$

Quantities given

$h = 6.62 \times 10^{-34}$ J s $\quad\quad$ $c = 3 \times 10^{8}\text{ m s}^{-1}$ $\quad\quad$ $\lambda = 5 \times 10^{-7}$ m

Substitution of values

$$E = \frac{6.62 \times 10^{-34} \times 3 \times 10^{8}}{5 \times 10^{-7}}$$

$$= \mathbf{3.972 \times 10^{-19}\text{ J}}$$

(*c*) To calculate the value of one einstein

Formula used

$$\textit{Energy per einstein} = \textit{Avogadro's No.} \times \textit{Energy per photon}$$

Quantities given

Avogadro's No. $= 6.02 \times 10^{23}$ $\quad\quad$ *Energy per photon* $= 3.972 \times 10^{-19}$ J

Substitution of values

$$\textit{Energy per einstein} = 6.02 \times 10^{23} \times 3.972 \times 10^{-19}\ \text{J}$$

$$= 2.39 \times 10^{-18}\ \text{J mol}^{-1}$$

SOLVED PROBLEM 8

In a photochemical combination of H_2 and Cl_2 a quantum yield of 1×10^6 is obtained with a wavelength of 4.8×10^{-7} m. How many moles of HCl would be produced under these conditions per Joule of radiation energy absorbed ?

SOLUTION : (*a*) To calculate no. of einsteins absorbed per Joule of energy

Formula used

$$\textit{No. of einsteins} = \frac{\textit{Energy absorbed}}{\textit{Energy per einstein}}$$

Quantities given

$$\textit{Energy absorbed} = 1\ \text{cal} = 4.184 \times 10^7\ \text{erg}\ (4.184\ \text{J})$$

$$\textit{Energy per einstein} = \frac{1.196 \times 10^9}{\lambda}\ \text{J mol}^{-1}$$

$$\lambda = 4.8 \times 10^{-7}\ \text{m}$$

Substitution of values

$$\text{No. of einsteins} = \frac{4.184\ \text{J} \times 4.8 \times 10^{-7}\ \text{m}}{1.196 \times 10^9\ \text{J mol}^{-1}}$$

$$= 1.679 \times 10^{-5}\ \text{m}$$

(*b*) To calculate the no. of moles of HCl produced

Formula used

$$\phi = \frac{\textit{No. of moles of HCl produced}}{\textit{No. of einsteins absorbed}}$$

Quantities given

$\phi = 1 \times 10^6$ *No. of einsteins absorbed* $= 1.679 \times 10^{-5}$

Substitution of values

$$1 \times 10^6 = \frac{\text{No. of moles of HCl produced}}{1.679 \times 10^{-5}}$$

or No. of moles of HCl produced = **16.79 mol**

SOLVED PROBLEM 9

The quantum yield for reaction $2\,HI \longrightarrow H_2 + I_2$ is 2. Calculate the number of photons absorbed in an experiment in which 0.01 mol of HI are decomposed. ($N = 6.02 \times 10^{23}$)

SOLUTION

Formula used

$$\phi = \frac{\textit{No. of molecules reacted}}{\textit{No. of photons absorbed}}$$

Quantities given

$\phi = 2$ *No. of molecules reacted* $= 0.01 \times 6.02 \times 10^{23}$

Substitution of values

$$2 = \frac{0.01 \times 6.02 \times 10^{23}}{\textit{No. of photons absorbed}}$$

$$\textit{No. of photons absorbed} = \frac{0.01 \times 6.02 \times 10^{23}}{2}$$

$$= \mathbf{3.01 \times 10^{21}}$$

SOLVED PROBLEM 10

A certain system absorbs 8.81×10 J of radiation of the wavelength 2.5×10^{-7} m in a certain time. It is observed that 1.12×10^{-4} mol of the irradiated substance has reacted in the same time. What is the quantum efficiency of the process?

$N = 6.023 \times 10^{23}$ $\quad h = 6.625 \times 10^{-34}$ $\quad c = 2.998 \times 10^{8}$ m s^{-1}

SOLUTION : (*a*) To calculate the energy per einstein

Formula used

$$\textit{Energy per einstein} = \frac{N h c}{\lambda}$$

Quantity given

$N = 6.02 \times 10^{23}$ $\quad h = 6.62 \times 10^{-34}$ J s $\quad c = 2.998 \times 10^{8}$ m s^{-1} $\quad \lambda = 2.5 \times 10^{-7}$ m

Substitution of values

$$\text{Energy per einstein} = \frac{6.02 \times 10^{23} \times 6.62 \times 10^{-34} \times 2.998 \times 10^{8}}{2.5 \times 10^{-7}}$$

$$= 4.77 \times 10^{5} \text{ J}$$

(*b*) To calculate the quantum efficiency

Formula used

$$\phi = \frac{\textit{No. of molecules reacted}}{\textit{No. of einsteins absorbed}}$$

Quantities given

$$\textit{No. of molecules reacted} = 1.12 \times 10^{-4} \text{ mol}$$

$$\textit{No. of einsteins absorbed} = \frac{\textit{Energy absorbed}}{\textit{Energy per einstein}}$$

$$= \frac{8.81 \times 10 \text{ J}}{4.77 \times 10^{5} \text{ J}}$$

$$= 1.847 \times 10^{-4}$$

Substitution of values

$$\phi = \frac{1.12 \times 10^{-4}}{1.847 \times 10^{-4}}$$

$$= \mathbf{0.6064}$$

SOLVED PROBLEM 11

A certain system absorbs 3.0×10^{16} quantum of light per second on irradiation for 600 s. 0.002 mol of reactant was found to have reacted. Calculate the quantum efficiency of the process. ($N = 6.023 \times 10^{23}$)

SOLUTION

Formula used

$$\phi = \frac{\textit{No. of molecules reacted}}{\textit{No. of einsteins absorbed}}$$

Quantities given

$$\textit{No. of molecules reacted} = 0.002 \times 6.023 \times 10^{23}$$

$$\textit{No. of photons absorbed} = 3.0 \times 10^{16} \times 600$$

Substitution of values

$$\phi = \frac{0.002 \times 6.023 \times 10^{23}}{3.0 \times 10^{16} \times 600}$$

$$= \mathbf{66.92}$$

SOLVED PROBLEM 12

If one mole of a compound absorbs light of wavelength 2.537 × 10[−7] J, calculate the amount of energy absorbed in kilo joules per mole.

SOLUTION

Formula used

$$\textit{Energy absorbed} = \frac{N h c}{\lambda}$$

Quantities given

$N = 6.02 \times 10^{23}$ $\qquad h = 6.62 \times 10^{-34}$ J s

$c = 3.0 \times 10^{8}$ m s^{-1} $\qquad \lambda = 2.537 \times 10^{-7}$ J

Substitution of values

$$= \frac{6.02 \times 10^{23} \times 6.62 \times 10^{-34} \times 3 \times 10^{8}}{2.537 \times 10^{-7}}$$

$$= 4.7125 \times 10^{5} \text{ J}$$

$$= 4.7125 \times 10^{2} \text{ kJ}$$

$$= \mathbf{471.25\ kJ}$$

SOLVED PROBLEM 13

In the photochemical reaction between Cl_2 and H_2, 2.5×10^{-10} mol of chlorine have been converted to hydrochloric acid by irradiation with light of 4.5×10^{-7} m. What is the quantum efficiency of the reaction if light absorbed in the system is 3.0×10^{-15} J?

SOLUTION : (*a*) To calculate the no. of photons absorbed

Formula used

$$\textit{No. of photons absorbed} = \frac{\textit{Total energy absorbed}}{\textit{Energy of a photon}}$$

Quantities given

$$\textit{Total energy absorbed} = 3.0 \times 10^{-15}$$

$$\textit{Energy of a photon} = \frac{hc}{\lambda} = \frac{6.62 \times 10^{-34} \times 3 \times 10^{8}}{4.5 \times 10^{-7}}$$

$$= 4.41 \times 10^{-19} \text{ J}$$

Substitution of values

$$\text{No. of photons absorbed} = \frac{3.0 \times 10^{-15} \text{ J}}{4.41 \times 10^{-19} \text{ J}}$$

$$= 6.80 \times 10^{3}$$

(*b*) To calculate the quantum efficiency

Formula used

$$\textit{Quantum efficiency} = \frac{\textit{No. of molecules reacted}}{\textit{No. of photons absorbed}}$$

Quantities given

$$\textit{No. of photons absorbed} = 6.80 \times 10^{3}$$

No. of molecules of Cl_2 *reacted* $= 2.50 \times 10^{-10} \times 6.02 \times 10^{23}$

$= 1.505 \times 10^{14}$

Substitution of values

$$\text{Quantum efficiency} = \frac{1.505 \times 10^{14}}{6.80 \times 10^{3}} = 2.213 \times 10^{10}$$

EXAMINATION QUESTIONS

1. Define or explain the following terms :
   (*a*) Photochemical reaction
   (*b*) Lambert law
   (*c*) Lambert-Beer law
   (*d*) Grothus-Draper law
   (*e*) Stark Einstein law
   (*f*) Quantum yield
   (*g*) Quantum Efficiency
   (*h*) Energy of photons
   (*i*) Einstein

2. (*a*) What is meant by quantum energy and Einstein energy? State Einstein law of photochemical equivalence.
   (*b*) In the photochemical reaction B → C, 1.00×10^{-5} mol of C is formed as a result of the absorption of 6 J at 3.6×10^{-7} m. Calculate the Quantum yield.
   Answer. (*b*) 0.553×10^{6}

3. (*a*) How would you explain very high and very low quantum efficiencies of some photochemical reactions?
   (*b*) For the photochemical reaction A → B, 1.0×10^{-5} mol of B were formed on absorption of 6×10^{7} J at 3.6×10^{-7} m. Calculate the quantum efficiency of the reaction.
   ($N = 6.02 \times 10^{23}$; $h = 6.0 \times 10^{-34}$ J/s)
   (*c*) What is an actinometer? Describe how a uranyl oxalate actinometer may be used.
   Answer. (*b*) 90.92%

4. Calculate the values of frequency, quantum energy and einstein for 5×10^{-7} m radiation.
   $c = 3.0 \times 10^{8}$ m/s; $N = 6.02 \times 10^{23}$; $h = 6.62 \times 10^{-27}$ erg second.
   Answer. 6×10^{14}; 3.972×10^{-19} J; 2.39×10^{-18} J mol^{-1}

5. (*a*) Explain briefly fluorescence and chemiluminescence.
   (*b*) In a photochemical combination of H_2 and Cl_2 a quantum yield of 1×10^{6} is obtained with a wavelength of 4.8×10^{-7} m. How many moles of HCl would be produced under these conditions per Joules of radiation energy absorbed?
   Answer. (*b*) 16.79 mol

6. A beam of monochromatic light was passed through a 1.5 m long cell filled with a solution of concentration *c* and 12 percent of the incident intensity was absorbed. What must be the length of another cell which is filled with a solution of concentration 1.5 *c* and which absorbs 48 percent of incident intensity?
   Answer. 1.3863 m

7. Radiation of wavelength 2.5×10^{-7} m was passed through a cell containing 10^{-5} m^3 of a solution which was 0.05 molar in oxalic acid and 0.01 molar in uranyl sulfate. After absorption of 80 joules of radiation energy, the concentration of oxalic acid was reduced to 0.04 molar. Calculate the quantum yield for the photochemical decomposition of oxalic acid at the given wavelength.
   Answer. 59.8

8. (*a*) 10% incident light is transmitted after passing through 0.02 m thick glass. If glass is 0.01 m thick, then how much light is absorbed of the same wavelength ?

(*b*) The quantum efficiency for the hydrogen chlorine reaction is very high, why?

(*c*) Explain extinction coefficient and molar absorption coefficient.

Answer. 68.38%

9. (*a*) State Beer-Lambert law and mention its uses.

(*b*) Write a note on photodimerization of anthracene. (*Madurai BSc*)

10. (*a*) Explain - Phosphorescence.

(*b*) Define quantum yield. How can it be experimentally determined? (*Jamia Millia BSc*)

11. State Stark-Einstein law of photochemical equivalence. (*Guru Nanak Dev BSc*)

12. (*a*) Define quantum yield.

(*b*) What are the reasons for abnormal quantum yield.

(*c*) Explain fluorescence and phosphorescence. (*Andhra BSc*)

13. Calculate :

(*i*) Wave number and

(*ii*) Wave length of the radiation with frequency of 4×10^{14} s^{-1}.

(Given velocity of light = 3×10^8 m s^{-1})

Answer. Wavelength = 0.75×10^{-6} m; Wavenumber = 1.33×10^6 m^{-1} (*Nagpur BSc*)

14. (*a*) Derive the mathematical expression of Beer-Lambert Law.

(*b*) Write the mathematical expression for quantum efficiency.

(*c*) Write a short note on 'Chemiluminescence' (*Arunachal BSc*)

15. (*a*) What are photochemical reactions? Explain the photochemical decomposition of HI.

(*b*) Write a short note on Stark-Einstein Law of photochemical equivalence.

(*Purvanchal BSc*)

16. (*a*) What are the primary and secondary photochemical reactions?

(*b*) How do you apply Lambert-Beer Law to determine unknown concentrations in solution?

(*Sambalpur BSc*)

17. Discuss the following :

(*a*) Quantum yield (*b*) Fluorescence

(*c*) Chemiluminescence (*Arunachal BSc*)

18. The optical density of 0.001 M solution of A in a cell of 0.1×10^{-2} m path length is 0.162. Calculate its extinction coefficient.

Answer. 162×10^3 mol^{-1} m^{-1} (*Sambalpur BSc*)

19. A certain substance in a cell of length '*l*' absorbs 10% of incident light. What %age of light will be absorbed in cell which is five times as long?

Answer. 41% (*Delhi BSc*)

20. A certain system absorbs 3×10^8 quanta of light per second. On irradation for 400 s, 0.001 mol of the reactant was found to have reacted. Calculate the quantum yield for the process. (Avogadro's number = 6.023×10^{23} mol^{-1})

Answer. 50.19×10^8 (*Marathwada BSc*)

21. A certain system absorbs 2×10^{16} quanta of light per second. On irradiation for 600 sec, 0.001 mol of the reactant was found to have reacted. Calculate the quantum yield for the process.(Avogadro's number = 6.024×10^{23})

Answer. 50.2 (*Mumbai BSc*)

22. State and explain the law of photochemical equivalence and calculate the value of 1 einstein for light having $\lambda = 2 \times 10^{-7}$ m.

Answer. 597.79 kJ *(Meerut BSc)*

23. Calculate the energy of a photon corresponding to wave length 360 nm. Given :

velocity of light = 3×10^8 m s^{-1}; $h = 6.62 \times 10^{-34}$ J s^{-1}.

Answer. 5.52×10^{-19} J *(Delhi BSc)*

24. Calculate percentage of light transmitted through 0.005 m length of a liquid of absorption coefficient 2.5.

Answer. 28.65% *(Bundelkhand BSc)*

25. What percentage of light will be transmitted through two cells put together in the path of light, if their individual transmissions are 60% and 30%.

Answer. 18% *(Kalyani BSc)*

26. A 0.002 m thick plate of a material transmits 70% of the incident light. Calculate the percentage of light transmitted if the thickness of the plate is 0.0005 m.

Answer. 91.47 *(Mysore BSc)*

27. (*a*) Distinguish between fluorescence and phosphorescence. Draw energy level diagrams to explain these processes and discuss them briefly.

(*b*) Briefly discuss flash photolysis. *(Bundelkhand BSc)*

28. (*a*) Explain Lambert-Beer Law while discussing physical significance of extinction coefficient and molar extinction coefficient.

(*b*) What does the quantum efficiency or quantum yield signify? *(Madurai BSc)*

29. A certain system absorbs 8.81×10^5 J of radiation of the wavelength 2.540×10^{-7} m in a certain time. It is observed that 1.12×10^{-4} mol of the irradiated substance has reacted in the same time. What is the quantum efficiency of the process?

$N = 6.023 \times 10^{23}$; $h = 6.625 \times 10^{-34}$ J s; $c = 2.998 \times 10^8$ m/s.

Answer. 5.9842×10^5 *(Mumbai BSc)*

30. (*a*) Name and state the basic laws of photochemistry. Give the numerical value of one Einstein in different units in terms of wavelength in cm and Å.

(*b*) A system is irradiated for 1200 second and is found to absorb 4×10^{18} quantum per second. If the amount decomposed is 3×10^{-3} mol and N = 6.023×10^{23}, calculate the quantum efficiency of the reaction.

Answer. 376 *(Marathwada BSc)*

31. A 0.001 M aqueous solution of a certain substance absorbs 10% of the incident light in a Lambert-Beer law cell of path length 0.01 m. Calculate the concentration required for the absorption of the incident light.

Answer. 21.8 mol/m^3 *(Arunachal BSc)*

32. (*a*) State and explain the law of photochemistry. What do you mean by quantum yield? How is it determined experimentally?

(*b*) What do you mean by one Einstein of energy? How is it related to wavelength? *(Jiwaji BSc)*

MULTIPLE CHOICE QUESTIONS

1. A photochemical reaction takes place by the absorption of

(*a*) visible and ultraviolet radiations (*b*) Infrared radiations

(*c*) heat energy (*d*) none of these

Answer. (*a*)

2. Photochemistry deals with the study of
   (*a*) photons
   (*b*) photos
   (*c*) reactions which proceed with absorptions of UV light
   (*d*) reactions which proceed with absorption of IR light

Answer. (*c*)

3. The wavelength of ultraviolet and visible regions of electromagnetic spectrum is
   (*a*) less than 2×10^{-7} m (*b*) more than 8×10^{-7} m
   (*c*) 2×10^{-7} m to 8×10^{-7} m (*d*) none of these

Answer. (*c*)

4. Which of the following statements about the photochemical reactions is true?
   (*a*) the presence of light is the primary requirement for reactions to take place
   (*b*) temperature has a very little effect on the rate of photochemical reactions
   (*c*) ΔG for photochemical spontaneous reactions may +ve or –ve
   (*d*) all of the above

Answer. (*d*)

5. Photochemical activation is highly selective. This statement is
   (*a*) true (*b*) false
   (*c*) sometimes true (*d*) none of these

Answer. (*a*)

6. The number of photons that pass through a unit area in a unit time is called
   (*a*) amplitude of light (*b*) frequency of light
   (*c*) intensity of light (*d*) wavelength of light

Answer. (*c*)

7. The absorption coefficient is given by
   (*a*) $b = \frac{-dI}{I} \times \frac{1}{dx}$ (*b*) $b = \frac{-dI}{I} \times \frac{dN}{N}$
   (*c*) $b = \frac{dI}{I} \times \frac{1}{dx}$ (*d*) $b = \frac{dI}{I} \times \frac{dn}{N}$

Answer. (*a*)

8. The equation for the Lambert's law is
   (*a*) $ln\left(\frac{I_o}{I}\right) = -bx$ (*b*) $ln\left(\frac{I}{I_o}\right) = -bx$
   (*c*) $ln\left(\frac{I}{I_o}\right) = -\in Cx$ (*d*) $ln\left(\frac{I}{I_o}\right) = \in Cx$

Answer. (*b*)

9. "It is only the absorbed light radiations that are effective in producing a chemical reaction." This is the statement of
   (*a*) Lambert law (*b*) Lambert-Beer law
   (*c*) Grothus-Draper law (*d*) Stark-Einstein law

Answer. (*c*)

10. "In a photochemical reaction each molecule of the reacting substance absorbs a single photon of radiation causing the reaction and is activated to form the products." This is the statement of
   (*a*) Lambert-Beer's law (*b*) Grothus-Draper law
   (*c*) Stark-Einstein law (*d*) Lambert's law

Answer. (*c*)

11. One einstein of energy is

(*a*) $E = \frac{2.859}{\lambda} \times 10^5$ ca mol^{-1} (*b*) $E = \frac{2.859}{\lambda} \times 10^5$ kcal mol^{-1}

(*c*) $E = \frac{2.859}{\lambda} \times 10^5$ J mol^{-1} (*d*) $E = \frac{2.859}{\lambda} \times 10^5$ kJ mol^{-1}

Answer. (*b*)

12. Which of the following statements is true?

(*a*) it is the secondary reaction in which absorption of radiation takes place
(*b*) it is the primary reaction in which absorption of radiation takes place
(*c*) the absorption of radiation takes place in both – the primary and secondary reactions
(*d*) none of the above

Answer. (*b*)

13. The number of molecules reacted or formed per photon of light absorbed is called

(*a*) yield of the reaction (*b*) quantum efficiency
(*c*) quantum yield (*d*) quantum productivity

Answer. (*c*)

14. For a reaction that obeys Einstein law,

(*a*) $\phi = 1$ (*b*) $\phi > 1$
(*c*) $\phi < 1$ (*d*) $\phi = \alpha$

Answer. (*a*)

15. In some photochemical reactions low quantum yield is obtained. It is due to

(*a*) deactivation of reacting molecules (*b*) occurrence of reverse primary reaction
(*c*) recombination of dissociated fragments (*d*) all of these

Answer. (*d*)

16. The quantum yield, ϕ, of a photochemical reaction is expressed as

(*a*) $\phi = \frac{\text{no. of molecules decomposed or formed}}{\text{no. of photons of radiation energy absorbed}}$

(*b*) $\phi = \frac{\text{no. of molecules activated}}{\text{no. of photons of radiation energy absorbed}}$

(*c*) $\phi = \frac{\text{no. of molecules of reactants}}{\text{no. of photons of radiation energy absorbed}}$

(*d*) none of the above

Answer. (*a*)

17. The energy associated with a photon is given by the equation

(*a*) $\in = h \times \lambda$ (*b*) $\in = h \times \nu$
(*c*) $\in = h \times c$ (*d*) $\in = h \times c^2$

Answer. (*b*)

18. One einstein is the energy associated with

(*a*) one molecule (*b*) one photon
(*c*) Avogadro number of photons (*d*) Faraday number of photons

Answer. (*c*)

19. One einstein is given by (N is Avogadro's number)

(*a*) $E = \frac{Nhc^2}{\lambda}$ (*b*) $E = \frac{Nhc}{\lambda^2}$

(*c*) $E = \frac{Nhc}{\lambda}$ (*d*) $E = \frac{Nh}{c\lambda}$

Answer. (*c*)

20. The energy per einstein depends upon the wavelength of photon. The higher the wavelength, the _______ will be the energy per einstein.

(*a*) higher (*b*) smaller

(*c*) zero (*d*) infinity

Answer. (*b*)

21. The ratio of energy per einstein and that of a photon is

(*a*) equivalent number (*b*) Einstein number

(*b*) Lambert's number (*d*) Avogadro's number

Answer. (*d*)

22. A species which can both absorb and transfer radiant energy for activation of the reactant molecule is called

(*a*) radioactive substance (*b*) an ionizer

(*c*) a photochemical substance (*d*) a photosensitizer

Answer. (*d*)

23. The substances that when exposed to light radiations of short wavelength emit light of longer wavelength are called

(*a*) photosensitized substances (*b*) phosphorescent substances

(*c*) fluorescent substances (*d*) none of these

Answer. (*c*)

24. _______ stops as soon as the incident radiation is cut off

(*a*) fluorescence (*b*) phosphorescence

(*d*) chemiluminescence (*d*) none of these

Answer. (*a*)

25. Sulfates of calcium, barium and strontium exhibit

(*a*) chemiluminescence (*b*) fluorescence

(*c*) phosphorescence (*d*) none of these

Answer. (*c*)

26. The emission of light as a result of chemical action is called _______

(*a*) phosphorescence (*b*) fluorescence

(*c*) chemiluminescence (*d*) none of these

Answer. (*c*)

27. The light emitted in a chemiluminescent reaction is also called

(*a*) cold light (*b*) hot light

(*c*) bright light (*d*) none of these

Answer. (*a*)

28. The glow of fireflies is due to the aerial oxidation of luciferin. It is an example of

(*a*) fluorescence (*b*) phosphorescence

(*c*) chemiluminescence (*d*) none of these

Answer. (*c*)

29. A solution of quinine sulfate on exposure to visible light exhibits

(*a*) fluorescence (*b*) phosphorescence

(*c*) chemiluminescence (*d*) none of these

Answer. (*a*)

30. The reactions which are caused by heat and in absence of light are called

(*a*) photochemical reactions (*b*) catalytic reactions

(*c*) exothermic reactions (*d*) thermal or dark reactions

Answer. (*d*)

31. A glass of certain thickness is found to have a transmission of 70% of light. If the thickness of glass is reduced, the transmission of light would

(*a*) decrease (*b*) increase
(*c*) remains the same (*d*) reduce to zero

Answer. (*a*)

32. A substance in a cell length (l) absorbs 20% of incident light. If the cell length is changed to $5l$, the fraction of incident light that will be absorbed is

(*a*) also increased (*b*) decreased
(*c*) unchanged (*d*) none of these

Answer. (*a*)

33. "Only the fraction of incident light that is absorbed by the substance can bring about a chemical change". is

(*a*) First law of photochemistry (*b*) Second law of photochemistry
(*c*) Third law of photochemistry (*d*) none of these

Answer. (*a*)

34. The energy of an einstein of radiation of wavelength 4×10^{-7} m is _______ than that of radiation of 300 nm

(*a*) lesser (*b*) greater
(*c*) equal to (*d*) none of these

Answer. (*a*)

35. Photochemical decomposition of a substance is called

(*a*) thermal dissociation (*b*) thermolysis
(*c*) photolysis (*d*) none of the above

Answer. (*c*)

36. Out of the photons corresponding to light of wavelengths 2×10^{-7} m, 4×10^{-7} m, 6×10^{-7} m and 8×10^{-7} m, the one with highest energy will be

(*a*) photon of light with 2×10^{-7} m wavelength (*b*) photon of light with 4×10^{-7} m wavelength
(*c*) photon of light with 6×10^{-7} m wavelength (*d*) photon of light with 8×10^{-7} m wavelength

Answer. (*a*)

37. The substances which initiate a photochemical reaction but itself does not undergo any chemical change is called

(*a*) catalysis (*b*) fluorescent
(*c*) sensitizer (*d*) none of these

Answer. (*c*)

38. Organic dyes like eosin, chlorophyll, ultrarine etc. show _______ in the visible or UV region

(*a*) fluorescence (*b*) phosphorescence
(*c*) chemiluminescence (*d*) none of these

Answer. (*a*)

39. In photochemical reactions, the absorption of light takes place in

(*a*) primary processes only (*b*) secondary processes only
(*c*) either primary or secondary process (*d*) both primary and secondary processes

Answer. (*a*)

31 SI Units

CONTENTS

What about Instagram?

COMMON SYSTEMS OF MEASUREMENTS

There are two common systems of measurement.

(1) Metric System

This is a decimal system of weights and measures originally based on the meter as the unit of length and the kilogram as the unit of mass.

(2) SI System

The International system of units was adopted by the 11th General Conference of Weights and Measures in 1960. The **SI units are widely used and they have been fully accepted by the scientific community.**

In this book we have used SI units throughout.

The six SI base units are building blocks of the system (Table 31.1).

TABLE 31.1. SI BASE UNITS

Physical Quantity	Unit	Symbol
Length	meter	m
Mass	kilogram	kg
Time	second	s
Temperature	kelvin	K
Electric current	ampere	A
Number of particles	mole	mol

SI UNITS OF LENGTH

The SI unit of length is the meter (m). Fractions and multiples of SI units are named by adding appropriate prefixes. The commonly used metric length units are listed in Table 31.2.

TABLE 31.2. COMMON METRIC LENGTH UNITS

Unit	Symbol	Relation
meter	m	
kilometer	km	1 km = 10^3 m
decimeter	dm	1 dm = 10^{-1} m
centimeter	cm	1 cm = 10^{-2} m
millimeter	mm	1 mm = 10^{-3} m
micrometer	mm	1 μm = 10^{-6} m
nanometer	nm	1 nm = 10^{-9} m
picometer	pm	1 pm = 10^{-12} m
angstrom	Å	1Å = 10^{-8} cm = 10^{-10} m

Even though the unit **angstrom** (Å) is not part of the SI system, it is still used for distances between atoms. Currently, the interatomic distances are sometimes reported in units of nanometers (nm) or picometer (pm).

$$1 \text{ nm} = 10 \text{ Å}$$
$$1 \text{ pm} = 10^{-2} \text{ Å}$$
$$1 \text{ nm} = 10^{-3} \text{ pm}$$

It may be noted that **the metric symbols are not changed into plurals.** Thus five centimeters of length is written as

Correct	**Incorrect**		
5 cm	5 cm.	5 c.m.	5 cms

SI UNITS OF VOLUME

The derived SI unit of volume is

Cubic meter m^3

This is the volume of a cube that is 1 meter on each edge. The related units of volume which are also used are :

Cubic centimeter cm^3

Cubic decimeter dm^3

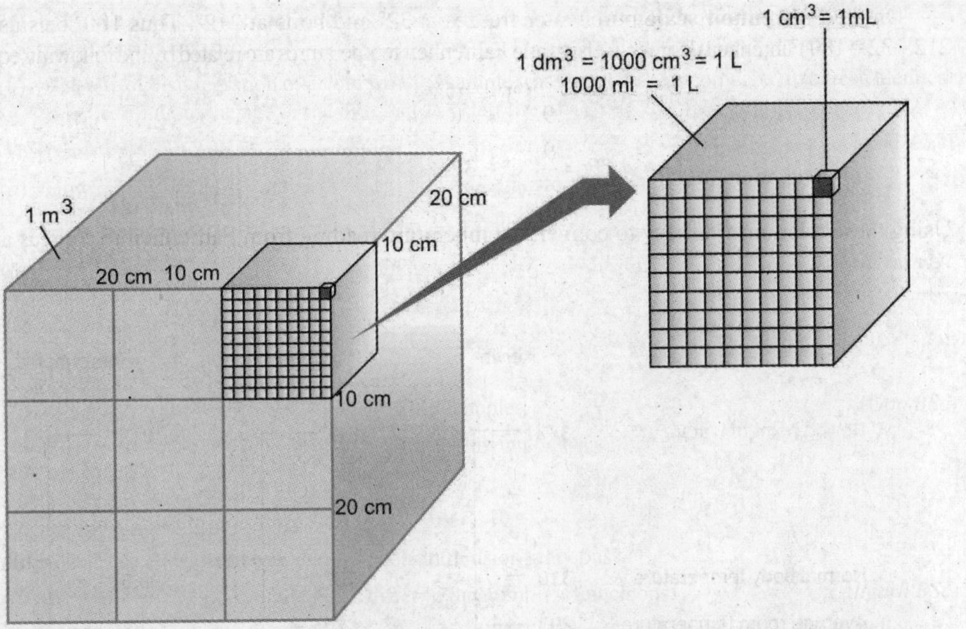

■ **Figure 31.1**
Relationship between length and volume.

Another common measure of volume is the litre (a non-SI unit) which is denoted by **L** (ℓ or l). **A liter is the volume occupied by a cube 10 cm on edge.** That is,

$$1 \text{ L} = (10 \text{ cm})^3 = 1000 \text{ cm}^3$$

Also $\quad 1 \text{ L} = 1000 \text{ mL}$

Therefore $\quad 1000 \text{ mL} = 1000 \text{ cm}^3$

$$1 \text{ mL} = 1 \text{ cm}^3 \text{ (cc)}$$

Hence **the volume units milliliter (mL) and cubic centimeter (cc) can be used interchangeably.** It may again be stated that metric symbols are not changed into plurals. Thus,

Correct	Incorrect
mL (or ml)	mLs (mls), m.l., ml.

SI UNIT OF TEMPERATURE

The series of markings on a thermometer which read temperature is called a **temperature scale.**

A temperature scale in which 0° is assigned to the freezing point of pure water and 100° to the boiling-point is known as the **Celsius scale.** The temperatures are expressed in **degrees Celsius** (°C). **Room temperature on the celsius scale is taken to be 25°C.** The celsius scale is not a part of the SI system. Since it is widely used in scientific literature, it is difficult to abandon it.

The SI system uses the Kelvin scale. A degree on the Kelvin scale has the same magnitude as a degree on the celsius scale but zero on the Kelvin scale equals – 273.15°C. Thus the temperature (0 K) is often referred to as the **absolute zero.** Celsius and Kelvin temperature are related as

$$\mathbf{K = {}^\circ C + 273.15, \quad {}^\circ C = K - 273.15}$$

It may be noted that the unit for temperature on the Kelvin scale is **K and not °K.** This notation has been approved by IUPAC and is now used by chemists all over the world. **Thus it may be noted that a degree sign (°) is not used with the Kelvin scale.**

On the **Fahrenheit scale** pure water freezes at 32° and boils at 212°. Thus 100° celsius equals 212 – 32 = 180 Fahrenheit degrees. Celsius and Fahrenheit temperatures are related by the following equations.

$$°C = \frac{5}{9}(°F - 32)$$

$$°F = \frac{9}{5}°C + 32$$

Using these relations it is easy to convert a temperature reading from Fahrenheit to Celsius and *vice versa.*

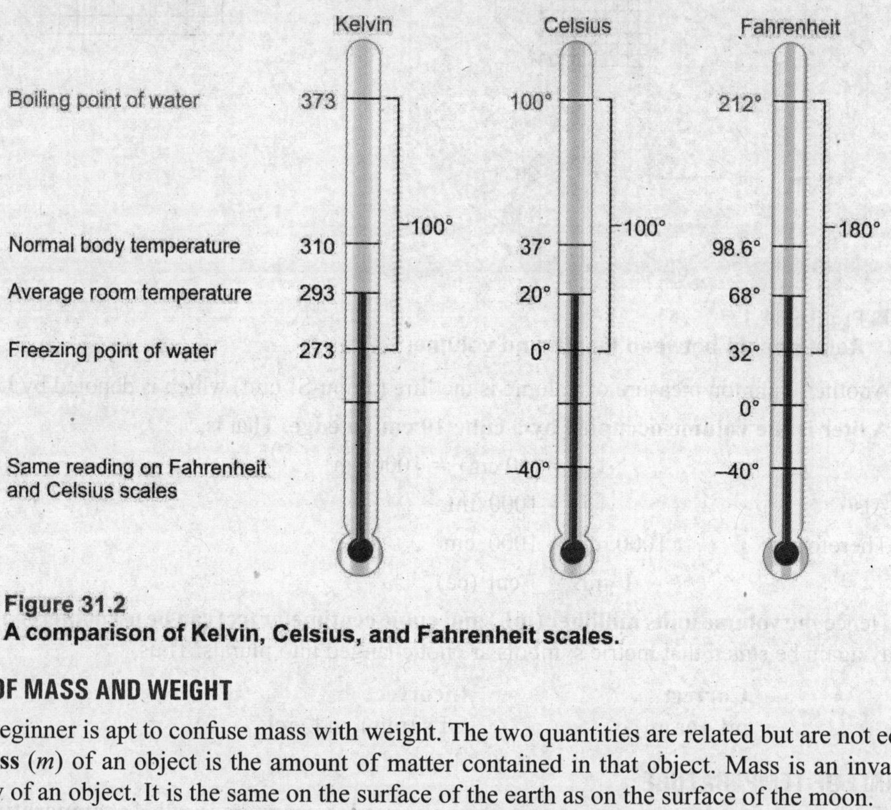

■ Figure 31.2
A comparison of Kelvin, Celsius, and Fahrenheit scales.

UNITS OF MASS AND WEIGHT

A beginner is apt to confuse mass with weight. The two quantities are related but are not equal. The **mass** (*m*) of an object is the amount of matter contained in that object. Mass is an invariant property of an object. It is the same on the surface of the earth as on the surface of the moon.

The **weight** (*w*), on the other hand, is force and not mass. It can be calculated by multiplying mass with the gravitational acceleration (*g*). That is,

$$w = m \times g$$

The gravitational pull on an object decreases as the object is moved farther from the center of the earth. Thus astronauts lose weight as they move higher and higher from earth. It follows, therefore, that **even though the weight of an object can vary at different places, its mass stays the same.**

Although mass and weight are not the same, the two terms are used interchangeably even by the scientific community. This is so because an object of a certain mass weigh with virtually the same anywhere on the earth. Known masses, for example, are measured by a process termed 'weighing' with a balance.

The basic unit of mass in the metric system (or SI system) is gram. The commonly used units based on the gram are listed in Table 31.3.

TABLE 31.3. COMMONLY USED METRIC WEIGHT AND MASS UNITS

Unit	Symbol	In terms of grams
gram	g	
Kilogram	kg	1 kg = 10^3 g
milligram	mg	1 mg = 10^{-3} g
microgram	μg	1 μg = 10^{-6} g

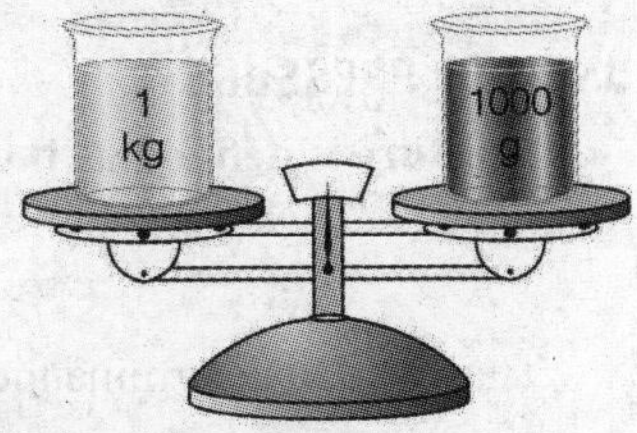

The British system of metric weights is also used by chemists in which

ounce (1 lb = 16 oz)	1 lb = 453.6 g
pound 1 lb	2.205 lb = 1kg
ton (1 ton = 2000 lb)	

It may be noted that metric units are not pluralised. Thus,

Correct	**Incorrect**
2 g	2gs, 2gms, 2g.m.

UNITS OF FORCE

Force (F) is defined as the product of mass (m) and acceleration (a).

$$\mathbf{F = m \times a}$$

Acceleration is the change in velocity (v) per unit time (t). Velocity is the change in distance (l) per unit time. Using SI base units, we can derive the unit for acceleration.

		UNIT
distance	l	m
velocity $\left(\frac{\text{distance}}{\text{time}}\right)$	v	$\frac{m}{s}$
acceleration $\left(\frac{\text{change in velocity}}{\text{time}}.\right)$	a	$\frac{m}{s^2}$

The derived SI unit for force, then, is kg ms^{-2}. The unit is called **newton** and has the symbol N. Thus,

$$\mathbf{1\ N = 1\ kg\ ms^{-2}}$$

UNITS OF WORK AND HEAT ENERGY

Work has been defined as the product of the force and the distance through which it operates

$$w = f \times d$$

Because force is expressed in newtons and distance in meters, **the SI units of work and energy is the newton-meter.** It is also called **Joule (J).**

$$\mathbf{1J = 1\ Nm}$$

Heat is energy that flows from one object to another because of a temperature difference between the objects. The quantity of heat transferred is best expressed in joules. But it is often given in **calories (cal).** One calorie is defined as exactly 4.184 joules. Thus,

$$\mathbf{1\ cal = 4.184\ J}$$

One calorie of energy will raise the temperature of 1 g of liquid water by 1°C. *The calorie is a non-SI unit, but like the joule it can be used for any form of energy.* The calorie written with a capital C is equal to one kilocalorie, 1000 calories. Thus,

$$\mathbf{1\ C = 1000\ cal}$$

UNITS OF PRESSURE

Pressure is defined as the force per unit area exerted on a surface. That is,

$$P = \frac{F}{A}$$

Thus we can determine the SI unit for pressure as :

	UNIT
Force F	kg ms^{-2} or N
Area A	m^2
Pressure $\frac{P}{A}$	kg m^{-1} s^{-2} or $\mathbf{Nm^{-2}}$

The SI unit Nm^{-2} is named **pascal** and given the symbol **Pa.**

Three other units which have been traditionally used are :

atmosphere, symbol **atm,** is defined as the pressure exerted by a column of mercury 760 mm in height at 0ºC.

torr, symbol **Torr,** is defined as the pressure exerted by a 1 mm column of mercury at 0ºC.

millimeter of mercury or mm Hg, which is the height in millimeters of mercury that the pressure can support.

The various units of pressure are related as

$$\mathbf{1\ atm = 760\ Torr = 76\ mm\ Hg = 1.013 \times 10^5\ Pa}$$

The three non-SI units *viz.*, Torr and mm Hg are still commonly used in current practice and it will take quite some time before the scientific community adopts the SI unit Pa.

UNITS OF DENSITY

One of the physical properties of a solid, a liquid, or a gas is its density (*d*). **Density is defined as mass per unit volume.** This may be expressed mathematically as

$$d = \frac{m}{V}$$

By using the base SI units and remembering that the unit for volume is m^3, we can derive the SI unit for density.

$$\mathbf{\frac{kg}{m^3}\ \text{or}\ kg\ m^{-3}}$$

The other units of density commonly used are

g cm^{-3} or g ml^{-1} for liquid or solid densities

g L^{-1} or g dm^{-3} for gas densities

The term **specific gravity is the ratio of the density of a substance to the density of a reference substance.** The reference substance for solids and liquids is usually water.

$$\text{sp gr} = \frac{\text{density of a substance}}{\text{density of reference substance}}$$

Specific gravity, being the ratio of two densities has no units.

32 Mathematical Concepts

CONTENTS

LOGARITHMIC FUNCTIONS

We know that $\log_2 8$ is the number to which 2 must be raised to get 8.

$$\therefore \qquad \log_2 8 = 3$$

In general, if $a^x = y$, $(a > 1)$, then we say that $\log_a y = x$. If $e^x = y$, then we say that the natural logarithm of y is x and we write $\log y = x$. In other words, if the base of a logarithm is not mentioned, then it is understood that the base is e. In fact, we cannot think of logarithm of a number without any base.

Two Important Results

Prove that (*i*) $\log_a 1 = 0$

(*ii*) $\log_a a = 1$

Proof : (*i*) $\because a^\circ = 1$; $\quad \therefore \log_a 1 = 0$

(*ii*) $\because a^1 = a$ $\quad \therefore \log_a a = 1$

SOLVED PROBLEM 1

Find value of $\log_{.5} 256$.

SOLUTION. Let $\log_{.5} 256 = x$

$$\therefore \quad \left(\frac{1}{2}\right)^x = 2^8$$

$$\therefore \quad 2^{-x} = 2^8$$

$$\Rightarrow \quad -x = 8, \text{ or } x = -8$$

$$\therefore \quad \log_{.5} 256 = -8$$

SOLVED PROBLEM 2

If $x = 2^{\frac{-1}{3}\log_2 64}$, find x

SOLUTION. Let $\log_2 64 = y$

$$\therefore \quad 2^y = 64$$

$$\Rightarrow \quad 2^y = 2^6$$

$$\Rightarrow \quad y = 6$$

$$\Rightarrow \quad \log_2 64 = 6$$

$$\therefore \quad x = 2^{\frac{-1}{3}\log_2 64}$$

$$\Rightarrow \quad x = 2^{\frac{-1}{3}(6)} = 2^{-2} = \frac{1}{4}$$

Fundamental Properties of Logarithms

(1) **Product Formula.** Logarithm of the product of two numbers to any base is equal to the sum of logarithms of the number to the same base.

i.e. $$\log_a (mn) = \log_a m + \log_a n$$

(2) **Quotient Formula.** Logarithm of the quotient of two numbers to any base is equal to the difference of logarithms of the numerator and denominator to the same base.

i.e. $$\log_a \left(\frac{m}{n}\right) = \log_a m - \log_a n$$

(3) **Power Formula.** Logarithm of a number raised to a power is equal to the index of the power multiplied by the logarithm of the number to the same base.

i.e. $$\log_a m^n = n \log_a m$$

(4) **Base Changing Formula**

$$\log_n m = \frac{\log_a m}{\log_a n}$$

Two Systems of Logarithms

(1) **Natural Logarithms.** Logarithms to the base e (= 2.7183 approximately) are called natural logarithms. They are used in all theoretical calculations.

(2) **Common Logarithms.** Logarithms to base 10 are called common logarithms. They are used in arithmetical calculations.

CHARACTERISTIC AND MANTISSA

The integral part of the logarithm of a number, after expressing the decimal part as positive, if not already so, is called the **characteristic** and the positive decimal part is called **mantissa.** The mantissa is always positive.

Two Rules to Find the Characteristics

Rule 1. The characteristic of the logarithm of a number greater than 1 is positive and 1 less than the number of digits in the integral part of the number.

Example. (*i*) If the number is 732, then the characteristic of the logarithm = 2 (= 3 – 1)

(*ii*) If the number is 7.8256, then the characteristic of the logarithm = 0 (= 1 –1)

Rule 2. The characteristic of the logarithm of a number of a positive number less than 1 is negative and numerically 1 more than the number of zeros immediately after the decimal point.

Example. (*i*) Consider the number .1205, which is positive but less than 1. There is no zero immediately after decimal point.

$\therefore$ Characteristic of logarithm $= -(0 + 1)$

$= -1 = \bar{1}$

(*ii*) Consider the number ·002007. This number is positive and less than 1. Also there are two zeros immediately after the decimal.

$\therefore$ Characteristic of logarithm $= -(2 + 1)$

$= -3 = \bar{3}$

Rule to Find Mantissa

We can explain this rule by an example. Suppose we wish to find mantissa of log 57.6932.

(*i*) Remove the decimal point from 57.6932 we get 576932. We take its first four significant figures only. Therefore, number is 5769. The first two figures from the left form 57, the third figure is 6 and the fourth is 9.

(*ii*) In the table of logarithms, we find 57 in the first column.

(*iii*) In the horizontal row beginning with 57 and under the column headed by 6, we find the number 7604 at the intersection. We note it down.

(*iv*) In continuation of this row and under the small column on the right headed by 9, we find the number 7 at the intersection.

(*v*) Adding 7 to 7604, prefixing the decimal point, the mantissa = ·7611.

ANTILOGARITHM

The number whose logarithm is *x*, is called the antilogarithm of *x* and is written as antilog *x*.

Example. $\because$ log 3 = .4771

$\therefore$ antilog (.4771) = 3

Rule to Find Antilog of a Number

We can find the number whose logarithm is 2.6078.

(*i*) The characteristic of the logarithm = 2. This is less than the number of digits in the integral part of the required number.

$\therefore$ Number of digits in the integral part of the required number = 2 + 1 = 3

(*ii*) Removing the integral part 2 from the given logarithm 2.6078, we get, .6078 from the table of antilogarithms.

$$\begin{aligned} \text{The number corresponding to } .607 &= 4046 \\ \text{Mean difference for } 8 &= 7 \\ \therefore \quad \text{Number corresponding to } \cdot 6078 &= 4053 \\ \therefore \quad \text{Required number} &= 405.3 \end{aligned}$$

SOLVED PROBLEM 3

Given log 2 = .30103, find the number of digits in 2^{64}.

SOLUTION. Let

$$\begin{aligned} x &= 2^{64} \\ \therefore \quad \log x &= \log 2^{64} = 64 \ \log 2 = 64\ (0.30103) \\ \therefore \quad \log x &= 19.26592 \\ \therefore \quad \text{Characteristic} &= 19 \end{aligned}$$

$\therefore$ Number of digits in x or $2^{64} = 19 + 1 = 20$

SOLVED PROBLEM 4

Find the fifth root of 8.012.

SOLUTION. Let

$$\begin{aligned} x &= (8.012)^{1/5} \\ \therefore \quad \log x &= \log\ (8.012)^{1/5} \\ &= \frac{1}{5} \log (8.012) = \frac{1}{5} (0.9037) \\ \therefore \quad \log x &= 0.1807 \\ \therefore \quad x &= 1.516 \end{aligned}$$

SOLVED PROBLEM 5

Evaluate $\sqrt{\dfrac{.0075 \times .014}{80.35}}$ using log tables.

SOLUTION. Let

$$x = \sqrt{\frac{.0075 \times .014}{80.35}}$$

$$\begin{aligned} \therefore \quad \log x &= \log\left[\frac{.0075 \times .014}{80.35}\right]^{\frac{1}{2}} \\ &= \frac{1}{2} \log\left[\frac{.0075 \times .014}{80.35}\right] \\ &= \frac{1}{2} [\log (.0075 \times .014) - \log 80.35] \\ &= \frac{1}{2} [\log .0075 + \log .014 - \log 80.35] \\ &= \frac{1}{2} [\bar{3} - .8751 + \ \bar{2} - .1461 - 1.9050] \\ &= \frac{1}{2} [-3 + .8751 - 2 + .1461 - 1.9050] \end{aligned}$$

$$= \frac{1}{2}\,[-6.9050 + 1.0212]$$

$$= \frac{1}{2}\,(-5.8838)$$

$$= -2.9419 = -2 - .9419$$

$$= (-2 - 1) + (1 - .9419) = -3 + .0581$$

$$\therefore \quad \log x = \bar{3}\,.0581$$

$$\therefore \quad x = .001143$$

EXPONENTIAL FUNCTIONS

Let $f(x) = e^x$ where $e = 2.7182818$......... The function e^x is also written as exponential (x) or, in short, as exp(x). For Instance, the exponential function $e^{-E/RT}$ which we shall frequently came across in the text, is written as exp ($-E/RT$). The algebraic operations with respect to exponential functions are

$$e^x \,.\, e^y = e^{x+y}$$

$$e^x \,/\, e^y = e^{x-y}$$

The exponential and logarithmic functions are related as $\frac{1}{e^x} = \ln x$ and $\frac{1}{10^x} = \log x$

Polynomial

A polynomial is a function such as

$$f(x) = a_0 + a_1\,x + a_2 x^2 + \ldots\ldots\ldots\ldots\ldots\ldots + a_n x^n$$

where $a_1, a_2, a_3,$ .... a_n are constants and exponent n, which is a positive integer, is called degree of polynomial.

Curve Sketching

The relationship between the x and y co-ordinates of points lying on a straight line is represented by a straight line graph.

DISPLACEMENT–TIME GRAPHS

The shortest distance between the initial and final positions of an object is called its displacement. When the displacement of a moving object is plotted against time, we obtain displacement– time graph. For plotting this graph, time is represented on x-axis and displacement on y-axis as shown in Fig. 32.1 :

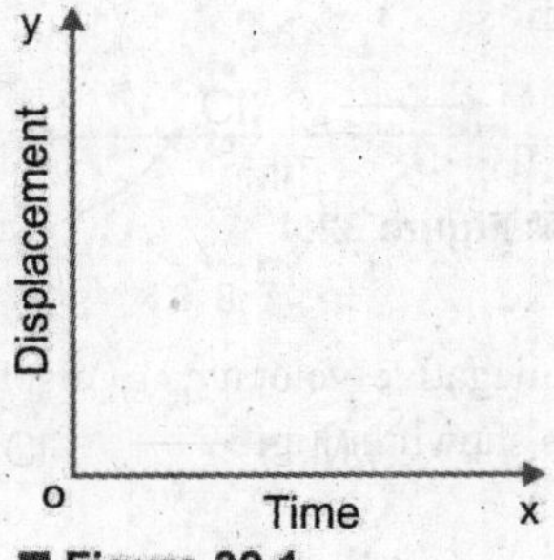

■ **Figure 32.1**

Types of Diaplacement–Time Graphs

(1) This graph shows uniform positive velocity *i.e.* displacement increases with time in this type as shown in Fig. 32.2 :

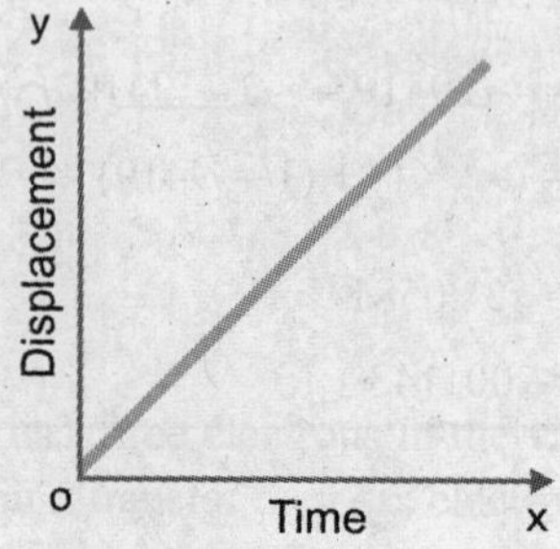

■ **Figure 32.2**

(2) This graph shows uniform negative velocity *i.e.* displacement decreases with time in this type as shown in Fig. 32.3 :

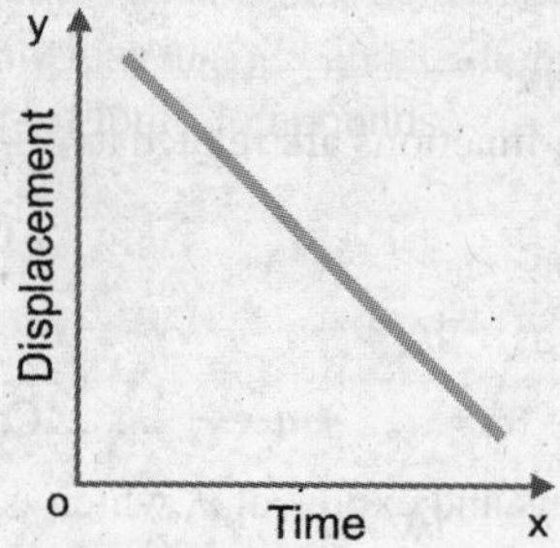

■ **Figure 32.3**

(3) This graph shows variable positive velocity *i.e.* displacement increases but variably as shown in Fig. 32.4 :

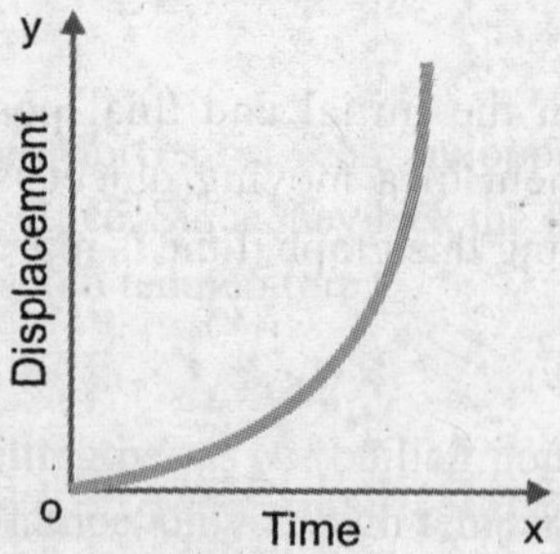

■ **Figure 32.4**

(4) This graph shows variable negative velocity *i.e.* displacement decreases with time in a variable manner *i.e.* not constantly as shown in Fig. 32.5

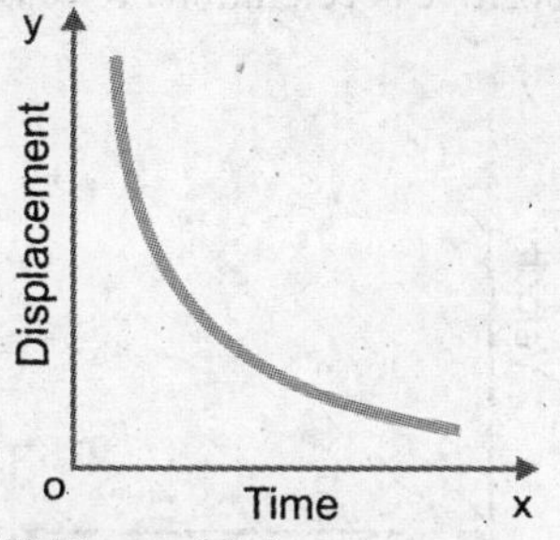

■ **Figure 32.5**

Notes :

(*i*) In displacement–time graph velocity $= \dfrac{\text{Change in displacement}}{\text{Change in time}}$

$$= \frac{\text{Final displacement} - \text{Initial displacement}}{\text{Final time} - \text{Initial Time}}$$

(*ii*) The graph as shown in Fig. 32.6 is impossible as displacement is changing without any change in time.

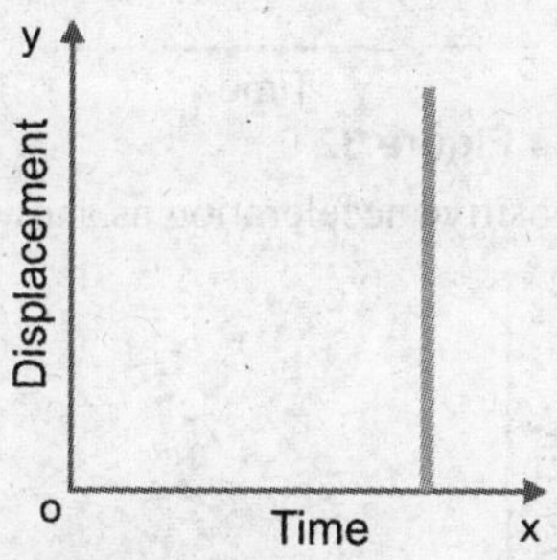

■ **Figure 32.6**

VELOCITY–TIME GRAPHS

When the velocity of an object is plotted against time, the graph so obtained is called velocity–time graph. For plotting this graph, the time is represented along x-axis and velocity along y-axis (Fig. 32.7).

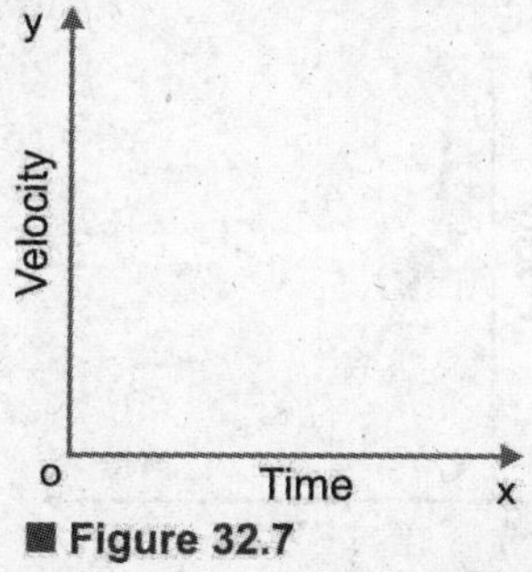

■ **Figure 32.7**

Types of Velocity–Time Graphs

(1) This graph shows uniform positive acceleration *i.e.* velocity increases with time in this type as shown in Fig. 32.8 :

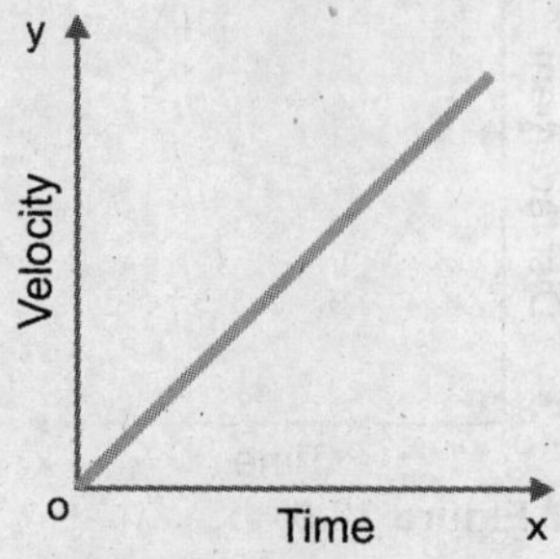

■ **Figure 32.8**

(2) This graph shows uniform negative acceleration *i.e.* velocity decreases with time in this type as shown below in Fig. 32.9 :

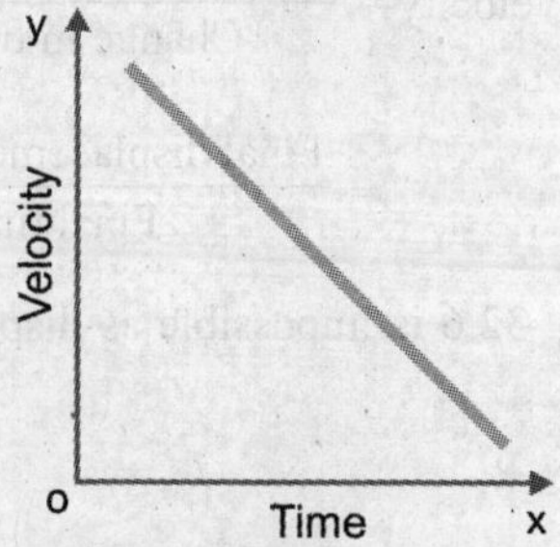

■ **Figure 32.9**

(3) This graph shows variable positive acceleration as shown below in Fig. 32.10 :

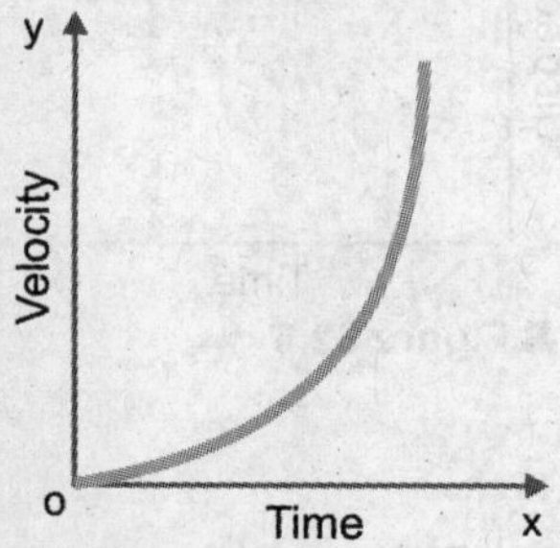

■ **Figure 32.10**

(4) This graph shows variable negative acceleration as shown below in Fig. 32.11 :

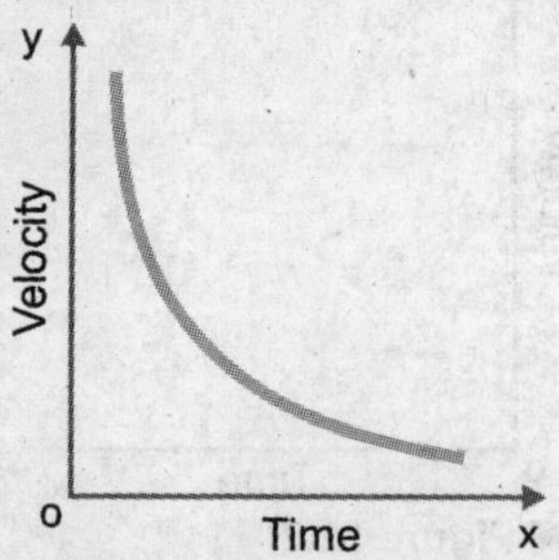

■ **Figure 32.11**

Notes :

(*i*) In velocity–time graphs,

$$\text{Acceleration} = \text{Rate of change of velocity}$$
$$= \frac{\text{Change in velocity}}{\text{Change in time}}$$
$$= \frac{\text{Final velocity} - \text{Initial velocity}}{\text{Final time} - \text{Initial time}}$$

(*ii*) For any time–interval, the area enclosed between the velocity–time graph and x-axis is equal to the distance travelled in that interval.

(*iii*) The graph as shown below in Fig. 32.12 is impossible state. It shows that velocity is changing without any change in time, which is not possible.

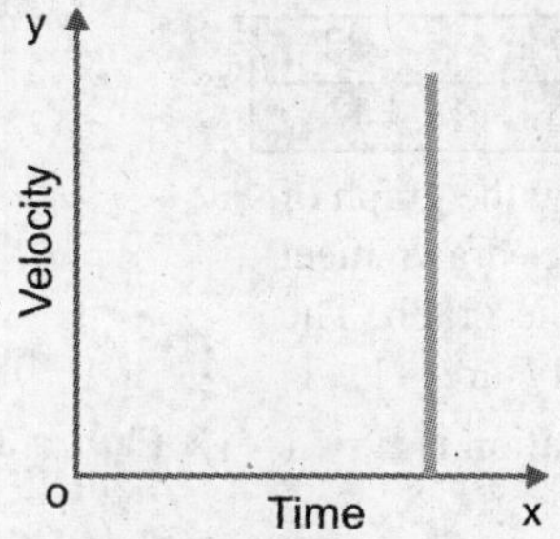

■ **Figure 32.12**

Graphs of Linear Equations

The graph of linear equation of the form $y = mx + c$ will always be a straight line.

It is very easy to draw the graph of linear equations as illustrated in following example.

SOLVED PROBLEM 6

Draw the graph of the equation $y = x + 3$

SOLUTION. We draw the following value table :

x	0	1	2
y	3	4	5

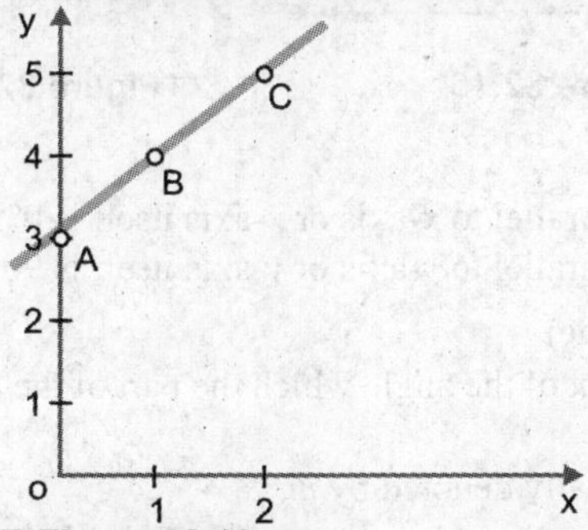

■ **Figure 32.13**
Graph of equation y = x + 3

We now plot the points A (0, 3), B (1, 4) and C (2, 5) on graph paper and join the points A, B and C and produce on either side. The line ABC is required graph of given equation.

SOLVED PROBLEM 7

Solve the given equations graphically,

$$2x - 3y = 1 \quad \text{and} \quad 3x - 4y = 1$$

SOLUTION. The given equations can be written as :

$$y = \frac{2}{3}x - \frac{1}{3} \quad \text{and} \quad y = \frac{3}{4}x - \frac{1}{4}$$

Table of values for $2x - 3y = 1$

x	2	5	8
y	1	3	5

Table of values for $3x - 4y = 1$

x	3	7	11
y	2	5	8

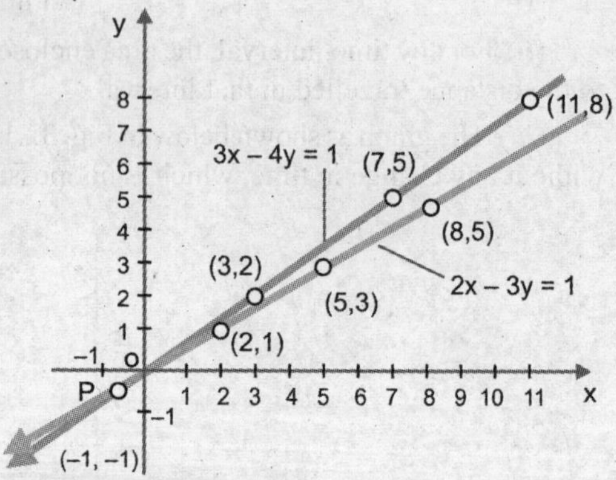

■ Figure 32.14

On the same graph paper draw the graph of each given equation. Both lines drawn meet at point P as is clear from the graph. The co-ordinates of common point P are $(-1, -1)$

Hence, solution of given equation is $x = -1$ and $y = -1$

SLOPE OF A LINE

(1) Inclination of a straight line

The angle which the line makes with positive direction of x-axis measured in anti-clockwise direction is called the inclination of the line. It is denoted by θ (Fig. 32.15).

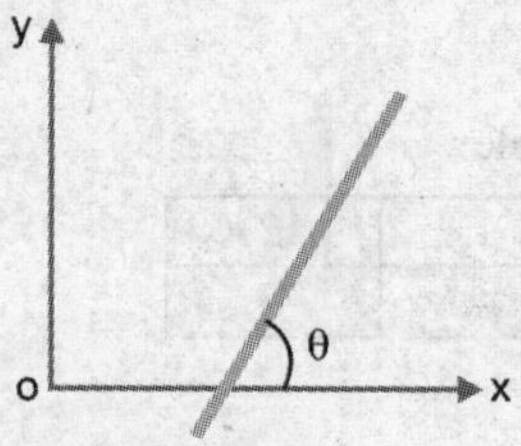

■ Figure 32.15

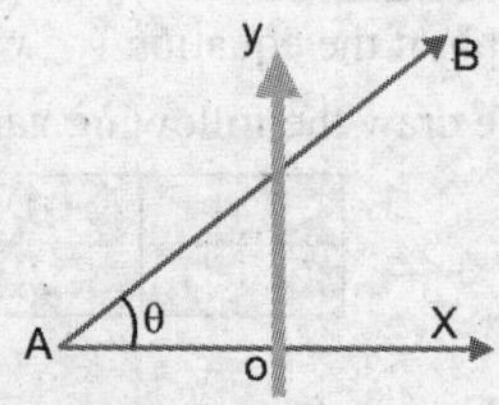

■ Figure 32.16

It should be noted that :

(*i*) inclination of line parallel to x-axis or x-axis itself is 0°.

(*ii*) inclination of line parallel to y-axis or y-axis itself is 90°.

(2) Slope (or Gradient of a line)

Slope of a line is the tangent of the angle which the part of the line above the x-axis makes with positive direction of x-axis.

The slope of a line is generally denoted by m.

$\therefore$ If θ is the angle which the line AB makes with x-axis, then slope is $\tan\theta$ (Fig. 32.16).

$$\therefore \qquad m = \tan\theta$$

Note : If $\theta = \frac{\pi}{2}$, then $m = \tan\frac{\pi}{2}$, which is not defined.

$\therefore$ Slope of a vertical line is not defined.

(3) Slope of a line passing through two fixed points

The slope of a line passing through two fixed points $A\,(x_1, y_1)$ and $B\,(x_2, y_2)$ is given by

$$\text{Slope} = \frac{y_2 - y_1}{x_2 - x_1} \quad i.e. \quad m = \frac{y_2 - y_1}{x_2 - x_1}$$

(4) Slope–Intercept form

$y = mx + b$ where m is the slope of the line and b is the intercept on the y–axis.

(5) Slope–Point form

The equation of a line with slope m and passing through a point (x_1, y_1) is

$$y - y_1 = m\,(x - x_1)$$

(6) Two–Point form

The equation of a line passing through two points (x_1, y_1) and (x_2, y_2) is

$$y - y_1 = \frac{y_2 - y_1}{x_2 - x_1}\,(x - x_1)\,;\quad x_2 \neq x_1$$

(7) The slope of the line $ax + by + c = 0$

The slope of line $ax + by + c = 0$ is

$$\text{is } \frac{-a}{b} = -\frac{\text{coeff. of } x}{\text{coeff. of } y}$$

(8) Slope of parallel lines

Two lines are parallel if and only if their slopes are equal.

(9) Slope of perpendicular lines

Two lines are perpendicular if and only if the product of their slope is –1.

Note. In cases, 8 and 9 the lines taken should be non-vertical.

(10) Intercept form $\dfrac{x}{a} + \dfrac{y}{b} = 1$

where x-intercept of the line is a and y-intercept is b.

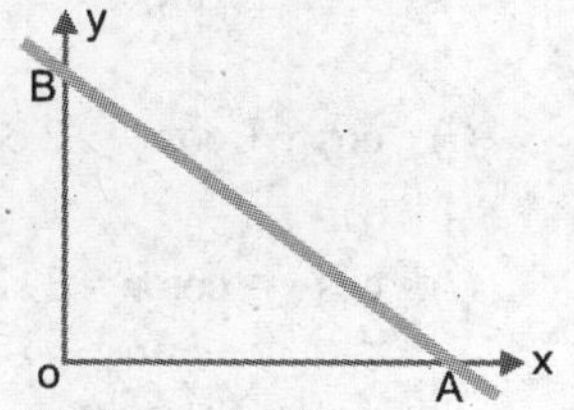

■ **Figure 32.17**

Notes :

Let a straight line AB, meet x-axis in A and y-axis in B. Then

(*i*) OA is called intercept of the line on x-axis.

(*ii*) OB is called the intercept of the line on y-axis and

(*iii*) The two together, taken in this order, are called the intercepts of the line on the axes.

SOLVED PROBLEM 8

Find the slope of the line whose inclination is 45°.

SOLUTION. Let m be the slope of line,

Then, $\qquad m = \tan 45° = 1$

SOLVED PROBLEM 9

The equation of a line is $2x - 2y - 5 = 0$. Find inclination and gradient of the line.

SOLUTION. The equation of line is $2x - 2y - 5 = 0$.

$$\therefore \qquad \text{gradient} = \frac{2}{-2} = 1$$

Let θ be inclination of the line, then

$$\tan\theta = m = 1$$

$$\Rightarrow \quad \theta = 45^\circ$$

SOLVED PROBLEM 10

A straight line passes through the points $P(4, -5)$ and $Q(6, 7)$. Find the slope of the line PQ.

SOLUTION. Given points are $P(4, -5)$ and $Q(6, 7)$

$$\therefore \text{The slope of the line } PQ = \frac{7-(-5)}{6-4} = \frac{12}{2} = 6$$

TRIGONOMETRIC FUNCTIONS

The trigonometric functions such as $f(x) = \sin x, \cos x, \tan x, \cot x, \sec x$ and $\operatorname{cosec} x$ are periodic. For example, $\sin x$ and $\cos x$ have the period $= 2\pi$ radian. Some fundamental relations among trigonometric functions are mentioned below :

(*a*) $\sin^2 x + \cos^2 x = 1;\ \sec^2 x = 1 + \tan^2 x,\ \operatorname{cosec}^2 x = 1 + \cot^2 x$

(*b*) $\sin(x \pm y) = \sin x \cos y \pm \cos x \sin y$

(*c*) $\cos(x \pm y) = \cos x \cos y \pm \sin x \sin y$

(*d*) $\tan(x \pm y) = (\tan x \pm \tan y) / (1 \pm \tan x \tan y)$

(*e*) $\cos 2x = \cos^2 x - \sin^2 x,\ \sin 2x = 2\sin x \cos x = \dfrac{2\tan x}{1+\tan^2 x}$

(*f*) $\sin^2 x = \dfrac{1}{2}(1 - \cos 2x);\ \cos^2 x = \dfrac{1}{2}(1 + \cos 2x)$

(*g*) $\sin 3x = 3\sin x - 4\sin^3 x;\ \cos 3x = 4\cos^3 x - 3\cos x$

(*h*) $\sin x + \sin y = 2\sin\dfrac{x+y}{2}\cos\dfrac{x-y}{2};\ \sin x - \sin y = 2\cos\dfrac{x+y}{2}\sin\dfrac{x-y}{2}$

(*i*) $\cos x + \cos y = 2\cos\dfrac{x+y}{2}\cos\dfrac{x-y}{2}$

(*j*) $\cos x - \cos y = 2\sin\dfrac{x+y}{2}\sin\dfrac{x-y}{2} = -2\sin\left(\dfrac{x+y}{2}\right)\sin\left(\dfrac{x-y}{2}\right)$

Inverse Trigonometric Functions

Inverse trigonometric functions such as $f(x) = \sin^{-1}x, \cos^{-1}x, \tan^{-1}x$ etc. are the inverse of corresponding trigonometric functions.

For example, if $\quad x = \sin y$

Then $\quad y = \sin^{-1} x$

DIFFERENTIATION

Derivative of a function

The derivative of a function $y = f(x)$ at a point x is defined as

$$f'(x) = \lim_{h\to 0} \frac{f(x+h) - f(x)}{h}$$

$$= \lim_{\Delta x\to 0} \frac{\Delta y}{\Delta x} = \frac{dy}{dx}$$

where $\Delta x = h$ and $\Delta y = f'(x+h) - f(x)$ are the increments in the variables x and y, respectively. The derivative of $f(x)$ is denoted by $\frac{dy}{dx}$ provided the limit exists, *i.e.*

$$\frac{dy}{dx} = \lim_{h \to 0} \frac{f(x+h) - f(x)}{h}$$

and may be interpreted as the rate of change of y w.r.t. x. The process of finding the derivative is called differentiation. The derivative of a function $f(x)$ at a given point represents a slope of the tangent drawn to the curve of $y = f(x)$ at a point where the function is defined. The derivative is also called differential coefficient.

SOLVED PROBLEM 11

If $y = x^2 + 6x + 8$, find $\frac{dy}{dx}$

SOLUTION

$$y + \Delta y = (x+h)^2 + 6(x+h) + 8$$

$$= x^2 + 2xh + h^2 + 6x + 6h + 8$$

$$\therefore \quad \Delta y = (2x+6)\,h + h^2$$

$$\therefore \quad \frac{\Delta y}{\Delta x} = \frac{(2x+6)\,h + h^2}{h} = 2x + 6 + h$$

$$\therefore \quad \frac{dy}{dx} = \lim_{\Delta x \to 0} \frac{\Delta y}{\Delta x} = \lim_{h \to 0} (2x + 6 + h) = 2x + 6$$

SOLVED PROBLEM 12

Differentiate $\sin x$ with respect to x from first principles.

SOLUTION

$$\frac{d}{dx}\sin x = \lim_{h \to 0} \frac{\sin(x+h) - \sin x}{h}$$

$$= \lim_{h \to 0} \frac{2\cos\left(x + \frac{h}{2}\right)\sin\frac{h}{2}}{h}$$

$$= \lim_{h \to 0} \cos\left(x + \frac{h}{2}\right) \frac{\sin\frac{h}{2}}{\frac{h}{2}} = \cos x \quad \left[\because \lim_{h \to 0} \frac{\sin\frac{h}{2}}{\frac{h}{2}} = 1\right]$$

Derivatives of Some simple functions

The following differentiation formulas should be memorized by the reader. We have assumed in these formulas that u and v are differentiable functions of x and c, n are arbitrary constants :

(1) $$\frac{d}{dx}(c) = 0$$

(2) $$\frac{d}{dx}(u+v) = \frac{du}{dx} + \frac{dv}{dx}$$

(3) The Constant Multiple Rule

$$\frac{d}{dx}(cu) = \frac{cdu}{dx}$$

(4) The Product Rule

$$\frac{d(uv)}{dx} = u\frac{dv}{dx} + v\frac{du}{dx}$$

(5) The Quotient Rule

$$\frac{d}{dx}(u/v) = \frac{v\dfrac{du}{dx} - u\dfrac{dv}{dx}}{v^2}, \quad v \neq 0$$

(6) The Power Rule for Positive and Negative Integers

$$\frac{d}{dx}x^{n} = x^{n-1}$$

(7) $$\frac{d}{dx}e^{x} = e^{x}$$

(8) $$\frac{d}{dx}a^{x} = a^{x} \log a$$

(9) $$\frac{d}{dx}(\log x) = \frac{1}{x}$$

(10) $$\frac{d}{dx}(\sin x) = \cos x$$

(11) $$\frac{d}{dx}(\cos x) = -\sin x$$

(12) $$\frac{d}{dx}(\tan x) = \sec^2 x$$

(13) $$\frac{d}{dx}(\cot x) = -\operatorname{cosec}^2 x$$

(14) $$\frac{d}{dx}(\sec x) = \sec x \, . \, \tan x$$

(15) $$\frac{d}{dx}(\operatorname{cosec} x) = -\operatorname{cosec} x \cot x$$

(16) $$\frac{d}{dx}(\sin^{-1} x) = \frac{1}{\sqrt{1-x^2}}$$

(17) $$\frac{d}{dx}(\cos^{-1} x) = \frac{-1}{\sqrt{1-x^2}}$$

(18) $$\frac{d}{dx}(\tan^{-1} x) = \frac{1}{1+x^2}$$

(19) $$\frac{d}{dx}(\cot^{-1} x) = \frac{-1}{1+x^2}$$

(20) $$\frac{d}{dx}(\sec^{-1} x) = \frac{1}{x\sqrt{x^2-1}}$$

(21) $$\frac{d}{dx}(\operatorname{cosec}^{-1} x) = \frac{-1}{x\sqrt{x^2-1}}$$

Chain Rule. According to this rule the derivative of the composite of two differentiable functions is the product of their derivatives evaluated at appropriate points.

Thus, $$\frac{dy}{dx} = \frac{dy}{du} \cdot \frac{du}{dx}$$

SOLVED PROBLEM 13

Find the differential coefficients of

$(i)\ x^{4/3} \quad (ii)\ \log(1+x) \quad (iii)\ a^{mx}$

SOLUTION. (*i*) Let $y = x^{4/3}$

$$\Rightarrow \frac{dy}{dx} = \frac{4}{3}x^{4/3-1} = \frac{4}{3}x^{1/3}$$

(*ii*) Let $y = \log(1+x)$

$$\Rightarrow \frac{dy}{dx} = \frac{1}{1+x}$$

(*iii*) Let $y = a^{mx}$

$$\Rightarrow \frac{dy}{dx} = a^{mx}\, m \log a$$

SOLVED PROBLEM 14

Find the differential coefficients of

$(i)\ 4^x \quad (ii)\ e^x \quad (iii)\ \log_{10}x$

SOLUTION. (*i*) Let $y = 4^x$

$$\Rightarrow \frac{dy}{dx} = 4^x \log 4$$

(*ii*) Let $y = e^x$

$$\Rightarrow \frac{dy}{dx} = e^x$$

(*iii*) Let $y = \log_{10} x$

$$\Rightarrow \frac{dy}{dx} = \frac{1}{x}\log_{10} e$$

PARTIAL DIFFERENTIATION

Partial Derivatives

Consider $z = f(x, y)$

Then $\underset{h\to 0}{\text{Lt}} \dfrac{f(a+h, b) - f(a, b)}{h}$ if it exists, is said to be partial derivatives of f w.r.t. x at (a, b) and is denoted by $\left(\dfrac{\partial z}{\partial x}\right)_{(a,b)}$ or $fx\,(a, b)$.

Again $\underset{k\to 0}{\text{Lt}} \dfrac{f(a, b+k) - f(a, b)}{k}$ if it exists, is said to be partial derivatives of f w.r.t. x at (a, b) and is denoted by $\left(\dfrac{\partial z}{\partial y}\right)_{(a,b)}$ or $fy\,(a, b)$.

Partial Differentiation of Higher Orders

Partial derivatives of first order can be formed as we formed those of f.

Therefore,

$$\frac{\partial}{\partial x}\left(\frac{\partial z}{\partial x}\right) = \frac{\partial^2 z}{\partial x^2} \quad ...(1)$$

$$\frac{\partial}{\partial y}\left(\frac{\partial z}{\partial x}\right) = \frac{\partial^2 z}{\partial y \partial x} \quad ...(2)$$

$$\frac{\partial}{\partial y}\left(\frac{\partial z}{\partial y}\right) = \frac{\partial^2 z}{\partial y^2} \quad ...(3)$$

(1), (2) and (3) can be denoted as f_{xx}, f_{yx} and f_{yy} respectively.

SOLVED PROBLEM 15

Differentiate $(3x^2 + 1)^2$ with respect to x.

SOLUTION. Let $y = (3x^2 + 1)^2 = u^2$

where $u = 3x^2 + 1$

Hence

$$\frac{dy}{dx} = \frac{dy}{du} \cdot \frac{du}{dx} = \frac{d(u^2)}{du} \frac{d}{dx}(3x^2 + 1)$$

$$= (2u)\ (6x)$$

$$= 2\,(3x^2 + 1)\,6x$$

$$= 36x^3 + 12x$$

SOLVED PROBLEM 16

Find the first order partial derivatives of

(*i*) $u = y^x$ (*ii*) $u = \log(x^2 + y^2)$

SOLUTION. (*i*)

$$u = y^x$$

$$\log u = x \log y$$

$$\Rightarrow \quad \frac{1}{u}\frac{\partial u}{\partial x} = \log y$$

$$\Rightarrow \quad \frac{\partial u}{\partial x} = u \log y = y^x \log y$$

Also

$$\log u = x \log y$$

$$\Rightarrow \quad \frac{1}{u}\frac{\partial u}{\partial y} = x\frac{1}{y}$$

$$\Rightarrow \quad \frac{\partial u}{\partial y} = u\frac{x}{y} = y^x\ \frac{x}{y} = xy^{x-1}$$

(*ii*)

$$u = \log(x^2 + y^2)$$

$$\Rightarrow \quad \frac{\partial u}{\partial x} = \frac{1}{x^2 + y^2}(2x) = \frac{2x}{x^2 + y^2}$$

Also

$$\frac{\partial u}{\partial y} = \frac{1}{x^2 + y^2}(2y) = \frac{2y}{x^2 + y^2}$$

MAXIMA AND MINIMA

Graphs of functions show maxima and/or minima and in some cases the functions are merely increasing or decreasing.

In Fig. 32.18, the function $y = f(x)$ has maximum values at the points a_1, a_2, a_3 .... and minimum values at the points b_1, b_2, b_3, ...

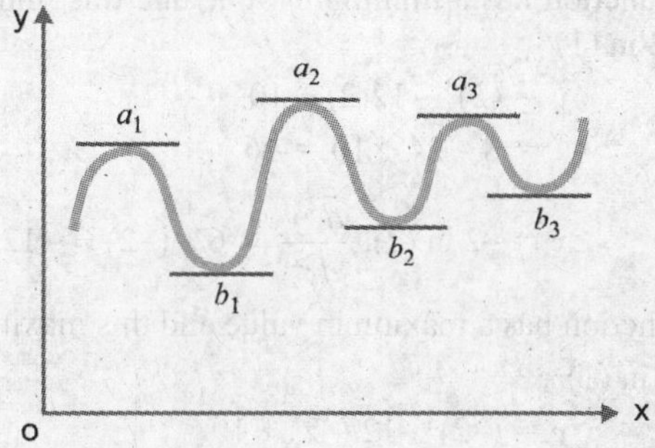

■ Figure 32.18

Working rules for finding maximum and minimum values of a function

Step 1. Put the given function = y

Step 2. Find $\frac{dy}{dx}$ and $\frac{d^2y}{dx^2}$

Step 3. Put $\frac{dy}{dx} = 0$ and solve this equation.

Step 4. Put $x = a$ in $\frac{d^2y}{dx^2}$. If the result is –ve, the function is maximum at $x = a$ and maximum $y = f(a)$.

If by putting $x = a$ in $\frac{d^2y}{dx^2}$, result is +ve, the function has minimum value at $x = a$ and minimum $y = f(a)$.

Similarly, test for other values of b, c, ... of x found in Step 4.

SOLVED PROBLEM 17

Find maximum and minimum values of $x^3 - 12x + 10$

SOLUTION. Let $$y = x^3 - 12x + 10 \quad ...(1)$$

$$\Rightarrow \quad \frac{dy}{dx} = 3x^2 - 12$$

For maximum and minimum values, $\frac{dy}{dx} = 0$

$$\Rightarrow \quad 3x^2 - 12 = 0$$
$$\Rightarrow \quad x^2 - 4 = 0$$
$$\Rightarrow \quad (x-2)(x+2) = 0$$
$$\Rightarrow \quad x = 2, -2$$

Also $\dfrac{d^2y}{dx^2} = 6x$ ...(2)

putting $x = 2$ in (2), $\dfrac{d^2y}{dx^2} = 6 \times 2 = 12 > 0$

$\therefore$ at $x = 2$, the function has a minimum value and this minimum value is obtained by putting $x = 2$ in (1)

$\therefore$ Minimum $y = (2)^3 - 12(2) + 10$
$= 8 - 24 + 10 = -6$

Again putting $x = -2$ in (2), $\dfrac{d^2y}{dx^2} = 6 \times (-2) = -12 < 0$

$\therefore$ at $x = -2$, the function has a maximum value and this maximum value is obtained by putting $x = -2$ in (1)

$\therefore$ Maximum $y = (-2)^3 - 12(-2) + 10$
$= -8 + 24 + 10$
$= 26$

SOLVED PROBLEM 18

Find the maximum and minimum values of $2x^3 - 9x^2 + 12x + 6$

SOLUTION. Let $y = 2x^3 - 9x^2 + 12x + 6$ ...(1)

$\Rightarrow \dfrac{dy}{dx} = 6x^2 - 18x + 12$

For maximum and minimum values, $\dfrac{dy}{dx} = 0$

$\Rightarrow 6x^2 - 18x + 12 = 0$

$\Rightarrow x^2 - 3x + 2 = 0$

$\Rightarrow (x - 1)(x - 2) = 0$

$\Rightarrow x = 1, 2$

Also, $\dfrac{d^2y}{dx^2} = 12x - 18$ ...(2)

Putting $x = 1$ in (2), $\dfrac{d^2y}{dx^2} = 12 - 18 = -6 < 0$

$\therefore$ at $x = 1$, the function has a maximum value and this maximum value is obtained by putting $x = 1$ in (1).

$\therefore$ Maximum $y = 2.1^3 - 9.1^2 + 12.1 + 6 = 11$

Again putting $x = 2$ in (2)

$\dfrac{d^2y}{dx^2} = 24 - 18 = 6$ which is +ve

$\therefore$ at $x = 2$, the function has a minimum value and this minimum value is obtained by putting $x = 2$ in (1).

Minimum $y = 2.2^3 - 9.2^2 + 12.2 + 6$
$= 10$

INTEGRATION

Integration is the process which is inverse of differentiation. In differentiation, a function is given and it is required to find its differential coefficient. But integration is its reverse process *i.e.*, given the differential coefficient of a function, it is required to find the function.

Thus if $\frac{d}{dx}[f(x)] = g(x)$, then $g(x)$ is called an integral of $g(x)$ w.r.t. x and is written as

$$\int g(x)\, dx = f(x)$$

The function $g(x)$ to be integrated is called **Integrand** and function sought is called **integral** or **primitive,** here $f(x)$. The symbol dx after the integrand $g(x)$ denotes that x is the independent variable and integration is done w.r.t. x. The process of finding the integral of a function is called **Integration.** For example,

Since $\frac{d}{dx}(x^4) = 4x^3$

$\therefore$ $\int 4x^3\, dx = x^4$

Constant of Integration

If $\frac{d}{dx}[f(x)] = g(x)$, then we know that $\frac{d}{dx}[f(x) + c]$ is also equal to $g(x)$, c being an arbitrary constant. It, therefore, follows that $\int g(x)\, dx = f(x) + c$, also and the arbitary c is called the constant of integration.

Some Important Results

(1) $\int x^n = \frac{x^{n+1}}{n+1}$ [where $n \neq 1$]

(2) (*i*) $\int 1\, dx = x$

(*ii*) $\int 0\, dx = c$ (constant)

(3) $\int \frac{1}{x}\, dx = \log x$

(4) $\int e^x\, dx = e^x$

(5) $\int a^x\, dx = \frac{a^x}{\log a}$

(6) $\int (ax+b)^n\, dx = \frac{(ax+b)^{n+1}}{(n+1)\, a}$

(7) $\int \frac{1}{ax+b}\, dx = \log(ax+b)$

(8) $\int \sin x\, dx = -\cos x$

(9) $\int \cos x\, dx = \sin x$

(10) $\int \sec^2 x\, dx = \tan x$

(11) $\int (\sec x \times \tan x)\, dx = \sec x$

(12) $\int \operatorname{cosec}^2 x\, dx = -\cot x$

(13) $\int \operatorname{cosec} x \cot x\, dx = -\operatorname{cosec} x$

SOLVED PROBLEM 19

Integrate the following functions w.r.t. x :

(*i*) x^7 (*ii*) $e^{-nx}\,dx$ (*iii*) a^{7x+8}

SOLUTION.(*i*) $$\int x^7\,dx = \frac{x^{7+1}}{7+1} = \frac{x^8}{8}$$

(*ii*) $$\int e^{-nx}\,dx = \frac{e^{-nx}}{-n}$$

(*iii*) $$\int a^{7x+8}\,dx = \frac{a^{7x+8}}{7\log a}$$

SOLVED PROBLEM 20

Evaluate the following integrals :

(*i*) $\displaystyle\int \frac{\sin^2 x}{1+\cos x}\,dx$ (*ii*) $\displaystyle\int \left[\frac{2\cos x}{3\sin^2 x} + 5x^2 - 6\right] dx$

SOLUTION. (*i*) $$\int \frac{\sin^2 x}{1+\cos x}\,dx = \int \frac{1-\cos^2 x}{1+\cos x}\,dx$$

$$= \int \frac{(1-\cos x)(1+\cos x)}{(1+\cos x)}\,dx$$

$$= \int (1-\cos x)\,dx$$

$$= x - \sin x + c$$

(*ii*) $$\int \left[\frac{2\cos x}{3\sin^2 x} + 5x^2 - 6\right] dx$$

$$= \int \left[\frac{2\cos x}{3\sin x}\operatorname{cosec} x + 5x^2 - 6\right] dx$$

$$= \int \left[\frac{2}{3}\cot x \operatorname{cosec} x + 5x^2 - 6\right] dx$$

$$= \frac{-2}{3}\operatorname{cosec} x + \frac{5}{3}x^3 - 6x + c$$

PERMUTATIONS AND COMBINATIONS

Factorial of an Integer

Let $n \in$ N. The combined product of first n natural number is called the factorial n. It is denoted by $\underline{|n}$.

$$\therefore \quad \underline{|n} = n(n-1)(n-2)\,\ldots\ldots\ldots\ldots\, 4.3.2.1$$

$$= 1.2.3.\,\ldots\ldots\ldots\, n$$

$$\underline{|1} = 1$$

$$\underline{|3} = 3 \times 2 \times 1 = 6$$

Factorial Zero

$$\underline{|0} = 1$$

or

$$0\,! = 1$$

Permutations

It is an arrangement that can be made by taking some or all of a number of given things.

Meaning of ^nP_r

It means the number of permutations of n different things taken r at a time.

Illustrations

Consider three letters a, b, c. The permutations of three letters taken two at a time are :

$$ab,\ bc,\ ca$$
$$ba,\ cb,\ ac$$

$\therefore$ The number of arrangements of three letters taken two at a time is 6 *i.e.* $^3P_2 = 6$

Note. ^nP_r is also written as $P(n, r)$

Combination

It is a group (or selection) that can be made by taking some or all of a number of a given things at a time.

Meaning of ^nC_r

^nC_r means the number of combinations of n different things taken r at a time.

Illustration

Consider three letters a, b, c the group of these 3 letters taken two at a time are ab, bc, ca.

As far as group is concerned ac or ca is the same group, as in a group we are concerned with the number of things contained, whereas in the case of arrangement we have to take into consideration the order of things.

Note. ^nC_r is also written as $C(n, r)$

Theorem

Let the number of distinct objects be n and $1 \leqslant r \leqslant n$. Then the number of all permutations of n objects taken r at a time is given by $\dfrac{|n}{|n-r}$.

SOLVED PROBLEM 21

Evaluate :

(*i*) $P(9, 5)$ (*ii*) $P(12, 0)$

SOLUTION. (*i*) We have
$$P(9, 5) = \frac{|9}{|9-5} = \frac{|9}{|4} = \frac{(9)(8)(7)(6)(5)|4}{|4} = 15120$$

(*ii*) We have
$$P(12, 0) = \frac{|12}{|12-0} = \frac{|12}{|12} = 1$$

SOLVED PROBLEM 22

Determine the number of different 5-letter words formed from the letters of the word 'EQUATION'.

SOLUTION. The given word is 'EQUATION'.

$\therefore$ number of letters = 8

Number of letters to be taken at a time = 5

$$\therefore \text{ Total number of words formed } = P(8, 5) = 8 \times 7 \times 6 \times 5 \times 4 = 6720$$

Notes. (1) The number of permutations of n dissimilar things taken r at a time is given by

$$^{n}P_{r} = \frac{n!}{(n-r)!}$$

(2) Number of permutations of n dissilmilar things, taken all at a time is $^{n}P_{n} = n!$

(3) Number of circular permutations of n different things taken all at a time is $(n-1)!$

(4) Number of permutations of n things, taken all at a time, when P_1, are alike of one kind, P_2 are alike of second kind, ... P_r are alike of rth kind is given by

$$^{n}C_{r} = \frac{n!}{(P_1!)(P_2!)\,..........\,(P_r!)}$$

Notes. (1) Number of combinations of n different things taken r at a time is given by

$$\frac{n!}{(n-r)!\,(r!)} = \frac{^{n}P_{r}}{r!}$$

(2) $\quad ^{n}C_{0} = 1, {}^{n}C_{n} = 1$

(3) $\quad ^{n}C_{p} = {}^{n}C_{q}$

$\Rightarrow \quad p + q = n \text{ or } p = q$

(4) $\quad ^{n}C_{r} = {}^{n}C_{n-r}$

(5) $\quad ^{n}C_{r-1} + {}^{n}C_{r} = {}^{n+1}C_{r}$

(6) Number of combinations of n different things, taken r at a time when p particular things always occur

$$= {}^{n-p}C_{r-p}$$

(7) Number of combinations of n different things taken r at a time when p particular things never occur is

$$= {}^{n-p}C_{r}$$

SOLVED PROBLEM 23

Verify that $C(8, 4) + C(8, 3) = C(9, 4)$

SOLUTION.

$$\text{L.H.S.} = C(8, 4) + C(8, 3) = {}^{8}C_{4} + {}^{8}C_{3} = \frac{8 \times 7 \times 6 \times 5}{1 \times 2 \times 3 \times 4} + \frac{8 \times 7 \times 6}{1 \times 2 \times 3} = 70 + 56 = 126$$

$$\text{R.H.S.} = C(9, 4) = {}^{9}C_{4} = \frac{9 \times 8 \times 7 \times 6}{1 \times 2 \times 3 \times 4} = 126$$

$$\therefore \quad \text{R.H.S.} = \text{L.H.S.}$$

SOLVED PROBLEM 24

Show that total number of ways in which six '+' and four '–' signs can be arranged in a line such that no two '–' signs occur together is 35.

SOLUTION. $x + x + x + x + x + x + x$

Since no two '–' signs occur together 4 '–' signs can be arranged in 7x marked places in $^6C_6$ ways.

$\therefore$ Required number of ways

$$= {}^7C_4 \times {}^6C_6 = \frac{7 \times 6 \times 5 \times 4}{1 \times 2 \times 3 \times 4} \times 1$$

$$= 35$$

PROBABILITY

Introduction

When we perform an operation again and again under the same conditions, than

(*i*) either the result is certain,

(*ii*) or the result is not unique but may be one of the several possibilities.

Suppose we toss a coin. Then we are not certain of head or tail. In this case, we talk of chance or probability which is taken to be quantitative measure of certainity.

Now we define certain terms which are used frequently:

Trial and Event. An experiment repeated under essentially identical conditions may not give unique result but may result in any one of the several possible outcomes. The experiment is called a Trial **(or random experiment)** and the outcomes are known as events or cases. For example, Tossing of a coin is **trial** and getting head or tail is an **event**.

Exhaustive Events. The total number of possible outcomes in any trial is known as exhaustive events. For example, in tossing a coin, there are two exhaustive cases, head and tail.

Favourable events or cases. The number of cases favourable to an event in a trial is the number of outcomes which ensure the happening of the event. For example, in tossing a die, the total number of cases favourable to the appearance of a multiple of 3 are two viz., 3 and 6.

Mutually Exclusive Events. Events are said to be mutually exclusive or incompatiable if the happening of any one of them rules out the happening of all others. For example, in tossing a coin the events head and tail are mutually exclusive.

Independent and Dependent Events. Two or more events are said to be independent if the happening or non-happening of any one does not depend (or is not affected) by the happening or non-happening of any other. Otherwise they are said to be dependent. For example, if a card is drawn from a pack of well shuffled cards and replaced before drawing. The second card, the result of second draw is independent of first draw. However, if the first card is not replaced, then the second draw is dependent on the first draw.

Probability. If a trial results in *n* exhaustive, mutually exclusive and equally likely cases and *m* of them are favourable to the happening of an event *E*, then probability of happening of *E* is given by :

$$P \text{ or } P(E) = \frac{\text{Favourable Number of cases}}{\text{Exhaustive Number of cases}}$$

SOLVED PROBLEM 25

A coin is tossed once. What are all possible outcomes? What is the probability of the coin coming up "Heads"?

SOLUTION. The coin can come up either "Heads" (H) or "tails" (T).

$\therefore \quad S = \{H, T\}$

$\therefore$ Total number of possible ways = 2

Number of favourable ways = 1

$\therefore \quad$ Required probability $= \dfrac{1}{2}$

SOLVED PROBLEM 26

In a single throw of two dice, what is the probability of obtaining a total greater than 10?

SOLUTION. Total number of outcomes = 6 × 6 = 36

Favourable outcomes are (5, 6), (6, 5), (6, 6)

$\therefore$ Number of favourable outcomes = 3

$\therefore \quad$ Required probability $= \dfrac{3}{36} = \dfrac{1}{12}$

EXAMINATION QUESTIONS

1. Find value of
   (*a*) $\log_{81} 243$ (*b*) $\log_{\frac{1}{3}} 243$
   (*c*) $\log_{.5} .04$ (*d*) $\log_{\sqrt{2}} 324$
   Answer. (*a*) $\dfrac{5}{4}$; (*b*) –5 ; (*c*) –2 ; (*d*) 4
2. If $\log_x 243 = 5$; find x
   Answer. 3
3. Given log 3 = .4771, find
   (*a*) the number of digits in 3^{62}.
   (*b*) the number of digits in 27^5.
   (*c*) the position of first significant figure in 3^{-65}.
   (*d*) the position of first significant figure in 3^{-32}.
   Answer. (*a*) 30; (*b*) 8; (*c*) 32nd; (*d*) 16th
4. (*a*) Find the seventh root of .03457.
   (*b*) Find the fifth root of .096.
   (*c*) Find the seventh root of .00001427.
   (*d*) Find the seventh root of .001.
   Answer. (*a*) .6183; (*b*) .6259; (*c*) .2035; (*d*) .3727

5. Find the values of :

(*a*) $\dfrac{0.0518 \times 4.68}{.0054 \times 25.5 \times 0.9}$ (*b*) $\dfrac{368.4361 \times .006143}{4384.612 \times 0.8391}$

(*c*) $\dfrac{(435)^3 \sqrt{.056}}{(380)^4}$

Answer. (*a*) 1.956; (*b*) .0006149; (*c*) .0009342

6. Evaluate :

(*a*) $\log_2 3$ (*b*) $\log_{43} 57$

Answer. (*a*) 1.59 approximately; (*b*) 1.07 approximately

7. Draw graph for the equation : $2x + 3y = 6$

8. Solve graphically the following equations :

$$x + y = 3, \quad 2x + 5y = 12$$

Answer. $x = 1,\ y = 2$

9. Find the slope of a line whose inclination is :

(*a*) 30° (*b*) 45°

Answer. (*a*) $\dfrac{1}{\sqrt{3}}$; (*b*) 1

10. Find the inclination of a line whose gradient is

(*a*) $\dfrac{1}{\sqrt{3}}$ (*b*) 1

(*c*) $\sqrt{3}$

Answer. (*a*) 30°; (*b*) 45°; (*c*) 60°

11. Find the slope of line $-5y + 1 = 0$.

Answer. 0

12. Find the differential coefficient of following functions :

(*a*) $x^{5/3}$ (*b*) $(3x + 5)^{7/3}$

(*c*) $\log_5 x$ (*d*) e^{3x}

(*e*) $\dfrac{3^x}{2^x}$

Answer. (*a*) $\dfrac{5}{3}x^{2/3}$; (*b*) $7\,(3x + 5)^{4/3}$; (*c*) $\dfrac{1}{x}\log_5 e$; (*d*) $3e^{3x}$; (*e*) $\dfrac{3^x}{2^x}\log\dfrac{3}{2}$

13. Find the maximum and minimum values of

(*a*) $x^4 - 14x^2 + 21x + 9$ (*b*) $x^3 + 2x^2 - 4x + 8$

Answer. (*a*) Maximum Value = 20 at $x = 1$

Minimum Value = 17 at $x = 2$

(*b*) Minimum Value = $\dfrac{-256}{27}$ at $x = \dfrac{2}{3}$

Maximum Value = 0 at $x = -2$

14. Integrate the following functions w.r.t. x.

(*a*) x^{10} (*b*) a^{5x+7}

(*c*) $\int \frac{1}{1+\sin x}\,dx$

(*d*) $\int (x^a + a^x + e^x a^x + \sin a)\,dx$

(*e*) $\int \sin x \sec^2 x\,dx$

Answer. (*a*) $\frac{x^{11}}{11}$; (*b*) $\frac{a^{5x+7}}{5\log a}$; (*c*) $\tan x - \sec x$; (*d*) $\frac{x^{a+1}}{a+1} + \frac{a^x}{\log a} + \frac{(ae)^x}{\log (ae)} + \sin a x$; (*e*) $\sec x$

15. Evaluate :

(*a*) $P(15,3)$

(*b*) $P(30,2)$

Answer. (*a*) 2730; (*b*) 870

16. How many different words containing all the letters of word 'TRIANGLE' can be formed?

Answer. 40320

17. Evaluate : $C(19,17) + C(19,18)$

Answer. 190

18. In a single throw of a pair of two dice, write the probability of getting a doublet of even numbers.

Answer. $\frac{1}{12}$

MULTIPLE CHOICE QUESTIONS

1. The value of log1 is

(*a*) 1
(*b*) 0
(*c*) 2
(*d*) 3

Answer. (*b*)

2. The value of log (*xy*) is

(*a*) 0
(*b*) 1
(*c*) $\log x + \log y$
(*d*) 2

Answer. (*c*)

3. $\log\left(\frac{x}{y}\right)$ is equal to

(*a*) $\log x - \log y$
(*b*) 0
(*c*) 1
(*d*) 3

Answer. (*a*)

4. What is the value of $\log x^y$?

(*a*) $\log x + \log y$
(*b*) $\log x - \log y$
(*c*) $y \log x$
(*d*) zero

Answer.(*c*)

5. The value of $\log_{81} 243$ is

(*a*) $\frac{6}{4}$
(*b*) $\frac{5}{4}$
(*c*) $\frac{7}{4}$
(*d*) $\frac{9}{4}$

Answer. (*b*)

6. The fifth root of .096 is

(*a*) .6253 (*b*) .6257

(*c*) .6259 (*d*) .6371

Answer. (*c*)

7. The following graph shows

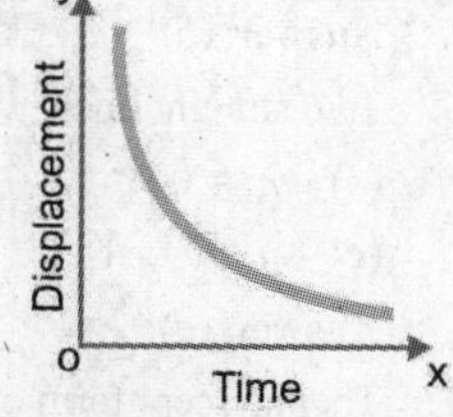

(*a*) uniform positive velocity

(*b*) uniform negative velocity

(*c*) variable positve velocity

(*d*) variable negative velocity

Answer. (*d*)

8. The following graph shows

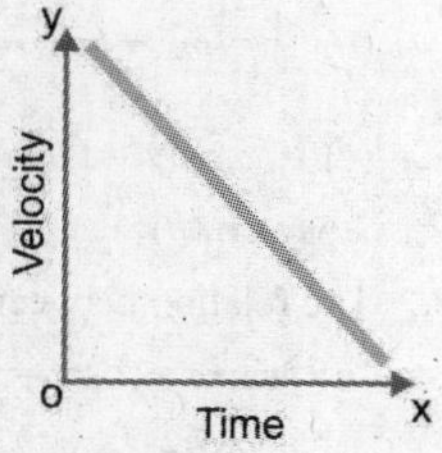

(*a*) uniform positive acceleration

(*b*) uniform negative acceleration

(*c*) variable positive acceleration

(*d*) variable negative acceleration

Answer. (*b*)

9. The following graph shows

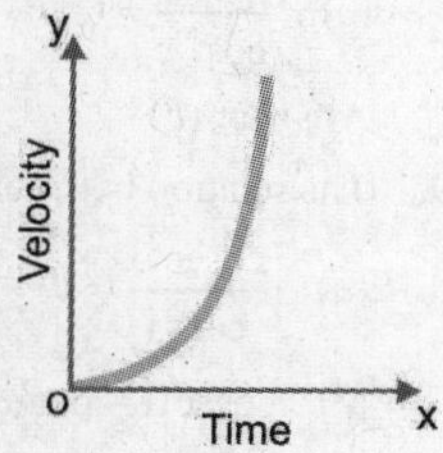

(*a*) uniform positive acceleration

(*b*) uniform negative acceleration

(*c*) variable positive acceleration

(*d*) variable negative acceleration

Answer. (*c*)

10. The graph of linear equation is always in the form of

(*a*) circle (*b*) sphere

(*c*) straight line (*d*) curve

Answer. (*c*)

11. Slope of a line is not defined if the line is

(*a*) parallel to x-axis (*b*) parallel to the line $x - y = 0$

(*c*) parallel to the line $x + y = 0$ (*d*) parallel to y-axis

Answer. (*d*)

12. Slope of any line parallel to x-axis is

(*a*) 1 (*b*) –1

(*c*) 0 (*d*) not defined

Answer. (*d*)

13. The equation $y - y_1 = m(x - x_1)$, $m \in R$ represents the line

(*a*) parallel to x-axis (*b*) parallel to y-axis

(*c*) parallel to the line $x - y = 0$ (*d*) parallel to the line $x + y = 0$

Answer. (*b*)

14. The equation $y - y_1 = m(x - x_1)$ for different values of m and (x_1, y_1) fixed, represents

(*a*) a family of parallel lines (*b*) a straight line

(*c*) a family of lines which are concurrent (*d*) a family of concurrent lines

Answer. (*d*)

15. The straight lines $y = m_1x$ and $y = m_2x$ are perpendicular to each other if

(*a*) $m_1 = 1/m_2$ (*b*) $m_1m_2 = -1$

(*c*) $m_1 = m_2$ (*d*) $m_1 = -m_2$

Answer. (*b*)

16. Two straight lines, whose gradients are m_1, m_2 respectively are parallel if

(*a*) $m_1 = 0$ (*b*) $m_2 = 0$

(*c*) $m_1m_2 = -1$ (*d*) $m_1 = m_2$

Answer. (*d*)

17. The intercept form of line is given by

(*a*) $y = mx + b$ (*b*) $\frac{x}{a} + \frac{y}{b} = 1$

(*c*) $2x + 3y = 1$ (*d*) $x + 2y = 2$

Answer. (*b*)

18. The relation between $\sin^2 x$ and $\cos^2 x$ is

(*a*) $\sin^2 x - \cos^2 x = 1$ (*b*) $\sin^2 x + \cos^2 x = 1$

(*c*) $\frac{\sin^2 x}{\cos^2 x} = 1$ (*d*) $\sin^2 x \times \cos^2 x = 1$

Answer. (*b*)

19. The relation between $\text{cosec}^2 x$ and $\cot^2 x$ is

(*a*) $\frac{\text{cosec}^2 x}{\cot^2 x} = 1$ (*b*) $\text{cosec}^2 x + \cot^2 x = 1$

(*c*) $\text{cosec}^2 x - \cot^2 x = 1$ (*d*) $\text{cosec}^2 x \times \cot^2 x = 1$

Answer. (*c*)

20. The value of $\sin (x + y)$ is

(*a*) $\sin x \cos y - \cos x \sin y$ (*b*) $\sin x \cos y + \cos x \sin y$

(*c*) $\sin x \sin y + \cos x \cos y$ (*d*) $\sin x \sin y - \cos x \cos y$

Answer. (*b*)

21. The largest value of $\sin \theta . \cos \theta$ is

(*a*) 1 (*b*) $\frac{1}{2}$

(*c*) $\frac{1}{\sqrt{2}}$ (*d*) $\frac{\sqrt{3}}{2}$

Answer. (*b*)

22. If $\sqrt{x} + \sqrt{y} = 1$, then $\frac{dy}{dx}$ at $\left(\frac{1}{4}, \frac{1}{4}\right)$ is

(*a*) $\frac{1}{2}$ (*b*) 1

(*c*) –1 (*d*) 2

Answer. (*c*)

23. If $x = a(t - \sin t)$, $y = a(1 + \cos t)$, then $\frac{dy}{dx}$ equals

(*a*) $-\tan \frac{t}{2}$ (*b*) $\cot \frac{t}{2}$

(*c*) $-\cot \frac{t}{2}$ (*d*) $\tan \frac{t}{2}$

Answer. (*c*)

24. $\frac{d}{dx}\left(\cos^{-1} x + \sin^{-1} x\right)$ is

(*a*) $\frac{\pi}{2}$ (*b*) 0

(*c*) $\frac{2}{\sqrt{1-x^2}}$ (*d*) none of these

Answer. (*b*)

25. If $x = a\cos^3 t,\ y = a\sin^3 t$, then $\frac{dy}{dx}$ is

(*a*) $\cos t$ (*b*) $\cot t$

(*c*) $\operatorname{cosec} t$ (*d*) $-\tan t$

Answer. (*d*)

26. If $y = t^2 + t - 1$, then $\frac{dy}{dx}$ is equal to

(*a*) $2t + 1$ (*b*) $t^2 + t - 1$

(*c*) 0 (*d*) not defined

Answer. (*d*)

27. If *x* be real, the minimum value of $x^2 - 8x + 17$ is

(*a*) -1 (*b*) 0

(*c*) 1 (*d*) 2

Answer. (*c*)

28. The maximum value of $\frac{\log x}{x}$ is

(*a*) 1 (*b*) $\frac{2}{e}$

(*c*) e (*d*) $\frac{1}{e}$

Answer. (*d*)

29. The function $f(x) = 2x^3 - 3x^2 - 12x + 4$ has

(*a*) two maxima (*b*) two minima

(*c*) one maxima and one minima (*d*) no maxima and minima

Answer. (*c*)

30. At $x = \frac{5\pi}{6}$, $f(x) = 2\sin 3x + 3\cos 3x$ is

(*a*) maximum (*b*) minimum

(*c*) zero (*d*) none of these

Answer. (*d*)

31. $\frac{d}{dx}\left(\int f(x)\,dx\right)$ is equal to

(*a*) $f'(x)$ (*b*) $\frac{(f(x))^2}{2}$

(*c*) $f(x)$ (*d*) none of these

Answer. (*c*)

32. $\int f(x)\, dx = f(x)$, then

(*a*) $f(x) = x$ (*b*) $f(x) =$ constant

(*c*) $f(x) = 0$ (*d*) $f(x) = e^x$

Answer. (*d*)

33. $\int \frac{dx}{\sqrt{4x^2 - 1}}$ is

(*a*) $\frac{1}{2}\tan^{-1} 2x$ (*b*) $\log\left(\sqrt{4x^2 - 1}\right)$

(*c*) $\sqrt{4x^2 - 1}$ (*d*) none of these

Answer. (*d*)

34. $\int x^2 e^{x^2}\, dx$ is equal to

(*a*) $x^2 (e^{x^2} - 1)$ (*b*) $\frac{1}{2}x^2 (e^{x^2} - 1)$

(*c*) $\frac{1}{2}e^{x^2} (x^2 - 1)$ (*d*) $\frac{1}{2}(e^{x^2} - 1)$

Answer. (*c*)

35. $\int (\tan x + \cot x)\, dx$ is equal to

(*a*) $\log (C \tan x)$ (*b*) $\log (\sin x + \cos x) + C$

(*c*) $\log x + C$ (*d*) none of these

Answer. (*a*)

36. $\int \frac{1}{x \log x}\, dx$ equals

(*a*) $\log x$ (*b*) $\log (\log x)$

(*c*) $\log (\log (\log x)$ (*d*) none of these

Answer. (*b*)

37. $\int \frac{\sin (\log x)}{x}\, dx$ equals

(*a*) $\cos (\log x)$ (*b*) $\sin (\log x)$

(*c*) $-\cos (\log x)$ (*d*) $-\sin (\log x)$

Answer. (*a*)

38. $\int \log x\, dx$ will be equal to

(*a*) $x \log\left(\frac{x}{e}\right)$ (*b*) $x \log x$

(*c*) $\frac{\log x}{x}$ (*d*) $x \log\left(\frac{e}{x}\right)$

Answer. (*a*)

39. If $C(n, 10) = C(n, 12)$ then n is equal to

(*a*) 2 (*b*) 10×12

(*c*) 22 (*d*) none of these

Answer. (*c*)

40. The number of diagonals of a hexagon is

(*a*) 3 (*b*) 6

(*c*) 9 (*d*) 12

Answer. (*c*)

41. The probability of a sure event is

(*a*) 1 (*b*) 2

(*c*) $\frac{1}{2}$ (*d*) unlimited

Answer. (*a*)

42. The probability of an impossible event is

(*a*) 1 (*b*) 2

(*c*) $\frac{1}{2}$ (*d*) 0

Answer. (*d*)

43. The probability of having at least one tail in 4 throws with a coin is

(*a*) $\frac{15}{16}$ (*b*) $\frac{1}{16}$

(*c*) $\frac{1}{4}$ (*d*) 1

Answer. (*a*)

44. A dice is thrown once. Then the probability of getting a number greater than three is

(*a*) $\frac{1}{2}$ (*b*) $\frac{2}{3}$

(*c*) 6 (*d*) 0

Answer. (*a*)

45. The probability that a leap year selected at random will contain 53 Sundays is

(*a*) $\frac{1}{7}$ (*b*) $\frac{2}{7}$

(*c*) $\frac{6}{7}$ (*d*) $\frac{6}{14}$

Answer. (*b*)

Physical Constants

Quantity	Symbol	Traditional units	SI units
Atomic mass unit ($\frac{1}{12}$th mass of ^{12}C atom)	amu	1.6606×10^{-2} g	1.6606×10^{-27} kg
Avogadro's number	N	6.022×10^{23} particles/mol	6.022×10^{23} particles/mol
Bohr radius	a_0	0.52918 Å	5.2918×10^{-13} m
Boltzmann constant	k	1.3807×10^{-16} erg/K	1.3807×10^{-23} J/K
Charge-to-mass ratio of electron	e/m	1.7588×10^{8} Coulomb/g	1.7588×10^{11} C/kg
Electron rest mass	m_e	9.1095×10^{-28} g	9.1095×10^{-31} kg 0.00054859 amu
Faraday constant	F	96,487 coulombs/mole^{-1}	96,487 J/V mol^{-1}
Gas constant	R	$0.08206 \dfrac{\text{L atm}}{\text{mol K}}$	$8.3145 \dfrac{\text{Pa dm}^3}{\text{mol K}}$
Gravitational acceleration	g	980.6 cm/s	9.906 m/s
Molar volume (STP)	V_m	22.414 L/mol	22.414×10^{-3} m^3/mol
Neutron rest mass	m_n	1.67495×10^{-24} g	1.67495×10^{-27} kg 1.008665 amu
Planck's constant	h	6.6262×10^{-27} erg sec	6.6262×10^{-27}
Proton rest mass	m_p	1.6726×10^{-27} erg sec	1.6726×10^{-27} kg 1.0077277 amu
Velocity of ligh (in vaccum)	c	2.9979×10^{10} cm/s 186,281 miles/s	2.9979×10^{8} m/s
Rydberg constant	R_z	3.289×101^{5} cycles/s 2.1799×10^{-11} erg	1.0974×10^{7} m^{-1} 2.1799×10^{-18} J

Conversion Factors

cm → in $\quad \frac{1\ \text{in}}{2.54\ \text{cm}}$

cm^3 → in^3 $\quad \frac{1\ \text{in}^3}{2.54\ \text{cm}^3}$

cm → in → ft

$\frac{1\ \text{in}}{2.54\ \text{cm}} \quad \frac{1\ \text{ft}}{12\ \text{in}}$

in → cm $\quad \frac{2.54\ \text{cm}}{1\ \text{in}}$

in^2 → cm^2 $\quad \frac{(2.54\ \text{cm})^2}{(1\ \text{in})^2}$

km^2 → m^2 $\quad \frac{(1000)^2}{(1\ \text{km})^2}$

km → mi $\quad \frac{0.6214\ \text{m}}{1\ \text{km}}$

km → mi → laps

$\frac{0.6214\ \text{mi}}{1\ \text{km}} \quad \frac{1\ \text{lap}}{0.250\ \text{mi}}$

m → mm $\quad \frac{1\ \text{mm}}{0.001\ \text{m}}$

ft → in → m

$\frac{12\ \text{in}}{1\text{ft}} \quad \frac{1\ \text{m}}{39.37\ \text{in}}$

in → cm $\quad \frac{2.54\ \text{cm}}{1\ \text{in}}$

dm^3 → m^3 → cm^3 → in^3

$\frac{(0.1\ \text{m})^3}{(1\ \text{dm})^3} \quad \frac{(1\ \text{cm})^3}{(0.01\ \text{m})^3} \quad \frac{(1\ \text{in})^3}{(2.54\ \text{cm})^3}$

Appendix B

Conversion Factors

L → qt → cu: $\frac{1.057\ \text{qt}}{1\ \text{L}}$, $\frac{4\ \text{cu}}{1\ \text{qt}}$

g → mL → L: $\frac{1\ \text{mL}}{0.84\ \text{g}}$, $\frac{1\ \text{L}}{1000\ \text{mL}}$

g → cm^3 → mL: $\frac{1\ \text{cm}^3}{1.32\ \text{g}}$, $\frac{1\ \text{mL}}{1\ \text{cm}^3}$

kg → g → cm^3: $\frac{1000\ \text{g}}{1\ \text{kg}}$, $\frac{1\ \text{cm}^3}{0.752}$

kg → g → cm^3 → mL → L: $\frac{1000\ \text{g}}{1\ \text{kg}}$, $\frac{1\ \text{cm}^3}{0.789\text{g}}$, $\frac{1\ \text{mL}}{1\text{cm}^3}$, $\frac{1\ \text{L}}{100\ \text{mL}}$

K → °C: $K = °C + 273$

°C → K: $K = °C + 273$

°F → °C: $°C = \frac{[°F - 32]}{1.8}$

°C → °F: $°C = \frac{[°F - 32]}{1.8}$

K → °C → °F: $K = °C + 273$, $°C = \frac{[°F - 32]}{1.8}$

m, V → d: $d = \frac{m}{v}$

Cal → cal → J: $\frac{1000\ \text{cal}}{1\ \text{cal}}$, $\frac{4.184\text{J}}{1\ \text{cal}}$

Dissociation constants of acids at 25ºC

Name	Formula	Ka_1	Ka_2	Ka_3
Acetic acid	CH_3COOH	1.8×10^{-5}		
Arsenic acid	H_3AsO_4	5.6×10^{-3}	1.0×10^{-7}	3.0×10^{-12}
Arsenious acid	H_3AsO_3	6.0×10^{-10}		
Benzoic acid	C_6H_5COOH	6.5×10^{-5}		
Boric acid	H_3BO_3	5.8×10^{-10}		
Carbonic acid	H_2CO_3	4.3×10^{-7}	5.6×10^{-11}	
Chloroacetic acid	$ClCH_2COOH$	1.4×10^{-3}		
Formic acid	$HCOOH$	1.8×10^{-4}		
Hydrocyanic acid	HCN	4.9×10^{-10}		
Hydrofluoric acid	HF	6.8×10^{-4}		
Hydrogen peroxide	H_2O_2	2.4×10^{-12}		
Hydrogen sulphate ion	HSO_4^-	1.2×10^{-2}		
Hydrogen sulphide	H_2S	5.7×10^{-8}		
Hypobromous acid	$HBrO$	2.0×10^{-9}		
Hypochlorous acid	$HClO$	3.0×10^{-8}		
Hypoiodous acid	HIO	2.0×10^{-11}		
Lactic acid	$CH_3(OH)COOH$	1.4×10^{-4}		
Malonic acid	$CH_3(OH)COOH$	1.4×10^{-4}		
Malonic acid	$CH_2(COOH)_2$	1.5×10^{-3}	2.0×10^{-6}	
Nitrous acid	HNO_2	4.5×10^{-4}		
Oxalic acid	$(COOH)_2$	5.9×10^{-2}	6.4×10^{-5}	
Phenol	C_6H_5OH	1.3×10^{-10}		
Phosphoric acid	H_3PO_4	7.5×10^{-3}	6.2×10^{-8}	4.2×10^{-13}
Propionic acid	CH_3CH_2COOH	1.3×10^{-5}		
Sulphuric acid	H_2SO_4	strong acid	1.2×10^{-2}	
Sulphurous acid	H_2SO_3	$1.7 \times 1^{0-2}$	6.4×10^{-8}	
Tartaric acid	$(CHOHCOOH)_2$	1.0×10^{-3}	4.6×10^{-5}	

Periodic Table

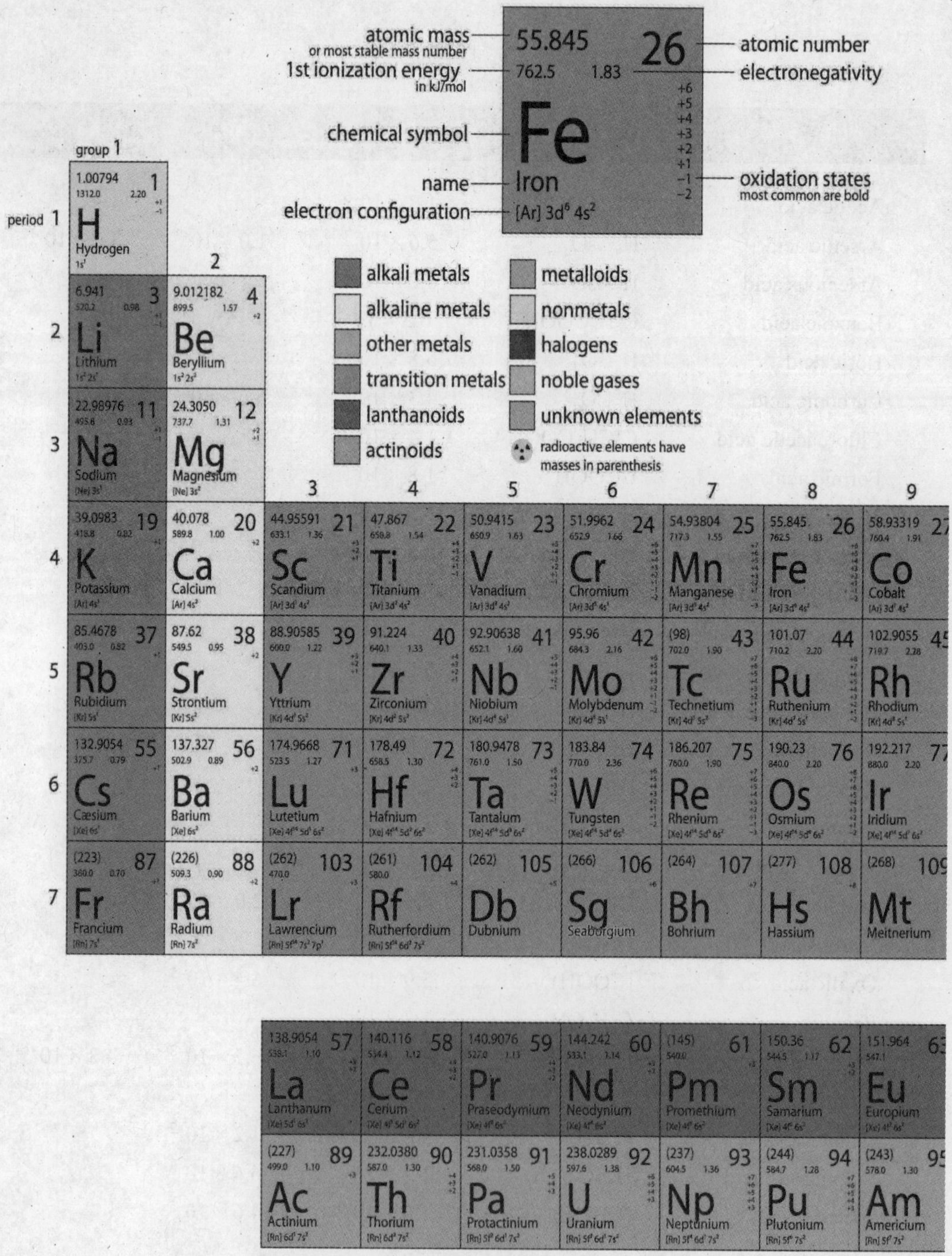

Periodic Table

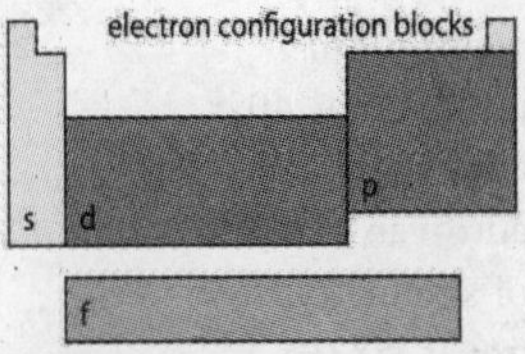

notes

- as of yet, elements 112-118 have no official name designated by the IUPAC.
- 1 kJ/mol ≈ 96.485 eV.
- all elements are implied to have an oxidation state of zero.

10	11	12	13	14	15	16	17	18
								4.002602 2 2372.3 **He** Helium $1s^2$
			10.811 5 800.6 2.04 **B** Boron $1s^2 2s^2 2p^1$	12.0107 6 1086.5 2.55 **C** Carbon $1s^2 2s^2 2p^2$	14.0067 7 1402.3 3.04 **N** Nitrogen $1s^2 2s^2 2p^3$	15.9994 8 1313.9 3.44 **O** Oxygen $1s^2 2s^2 2p^4$	18.998403 9 1681.0 3.98 **F** Fluorine $1s^2 2s^2 2p^5$	20.1797 10 2080.7 **Ne** Neon $1s^2 2s^2 2p^6$
			26.98153 13 577.5 1.61 **Al** Aluminium [Ne] $3s^2 3p^1$	28.0855 14 786.5 1.90 **Si** Silicon [Ne] $3s^2 3p^2$	30.97696 15 1011.8 2.19 **P** Phosphorus [Ne] $3s^2 3p^3$	32.065 16 999.6 2.58 **S** Sulfer [Ne] $3s^2 3p^4$	35.453 17 1251.2 3.16 **Cl** Chlorine [Ne] $3s^2 3p^5$	39.948 18 1520.6 **Ar** Argon [Ne] $3s^2 3p^6$
58.6934 28 737.1 1.88 **Ni** Nickel [Ar] $3d^8 4s^2$	63.546 29 745.5 1.90 **Cu** Copper [Ar] $3d^{10} 4s^1$	65.38 30 906.4 1.65 **Zn** Zinc [Ar] $3d^{10} 4s^2$	69.723 31 578.8 1.81 **Ga** Gallium [Ar] $3d^{10} 4s^2 4p^1$	72.64 32 762.0 2.01 **Ge** Germanium [Ar] $3d^{10} 4s^2 4p^2$	74.92160 33 947.0 2.18 **As** Arsenic [Ar] $3d^{10} 4s^2 4p^3$	78.96 34 941.0 2.55 **Se** Selenium [Ar] $3d^{10} 4s^2 4p^4$	79.904 35 1139.9 2.96 **Br** Bromine [Ar] $3d^{10} 4s^2 4p^5$	83.798 36 1350.8 3.00 **Kr** Krypton [Ar] $3d^{10} 4s^2 4p^6$
106.42 46 804.4 2.20 **Pd** Palladium [Kr] $4d^{10}$	107.8682 47 731.0 1.93 **Ag** Silver [Kr] $4d^{10} 5s^1$	112.441 48 867.8 1.69 **Cd** Cadmium [Kr] $4d^{10} 5s^2$	114.818 49 558.3 1.78 **In** Indium [Kr] $4d^{10} 5s^2 5p^1$	118.710 50 708.6 1.96 **Sn** Tin [Kr] $4d^{10} 5s^2 5p^2$	121.760 51 834.0 2.05 **Sb** Antimony [Kr] $4d^{10} 5s^2 5p^3$	127.60 52 869.3 2.10 **Te** Tellurium [Kr] $4d^{10} 5s^2 5p^4$	126.9044 53 1008.4 2.66 **I** Iodine [Kr] $4d^{10} 5s^2 5p^5$	131.293 54 1170.4 2.60 **Xe** Xenon [Kr] $4d^{10} 5s^2 5p^6$
195.084 78 870.0 2.28 **Pt** Platinum [Xe] $4f^{14} 5d^9 6s^1$	196.9665 79 890.1 2.54 **Au** Gold [Xe] $4f^{14} 5d^{10} 6s^1$	200.59 80 1007.1 2.00 **Hg** Mercury [Xe] $4f^{14} 5d^{10} 6s^2$	204.3833 81 589.4 1.62 **Tl** Thallium [Xe] $4f^{14} 5d^{10} 6s^2 6p^1$	207.2 82 715.6 2.33 **Pb** Lead [Xe] $4f^{14} 5d^{10} 6s^2 6p^2$	208.9804 83 703.0 2.02 **Bi** Bismuth [Xe] $4f^{14} 5d^{10} 6s^2 6p^3$	(210) 84 812.1 2.00 **Po** Polonium [Xe] $4f^{14} 5d^{10} 6s^2 6p^4$	(210) 85 890.0 2.20 **At** Astatine [Xe] $4f^{14} 5d^{10} 6s^2 6p^5$	(220) 86 1037.0 **Rn** Radon [Xe] $4f^{14} 5d^{10} 6s^2 6p^6$
(271) 110 **Ds** Darmstadium	(272) 111 **Rg** Roentgenium	(285) 112 **Cn** Copernicium	(284) 113 **Uut** Ununtrium	(289) 114 **Uuq** Ununquadium	(288) 115 **Uup** Ununpentium	(292) 116 **Uuh** Ununhexium	117 **Uus** Ununseptium	(294) 118 **Uuo** Ununoctium

157.25 64 593.4 1.20 **Gd** Gadolinium [Xe] $4f^7 5d^1 6s^2$	158.9253 65 565.8 **Tb** Terbium [Xe] $4f^9 6s^2$	162.500 66 573.0 1.22 **Dy** Dysprosium [Xe] $4f^{10} 6s^2$	164.9303 67 581.0 1.23 **Ho** Holmium [Xe] $4f^{11} 6s^2$	167.259 68 589.3 1.24 **Er** Erbium [Xe] $4f^{12} 6s^2$	168.9342 69 596.7 1.25 **Tm** Thulium [Xe] $4f^{13} 6s^2$	173.054 70 603.4 **Yb** Ytterbium [Xe] $4f^{14} 6s^2$
(247) 96 581.0 1.30 **Cm** Curium [Rn] $5f^7 6d^1 7s^2$	(247) 97 601.0 1.30 **Bk** Berkelium [Rn] $5f^9 7s^2$	(251) 98 608.0 1.30 **Cf** Californium [Rn] $5f^{10} 7s^2$	(252) 99 619.0 1.30 **Es** Einsteinium [Rn] $5f^{11} 6s^2$	(257) 100 627.0 1.30 **Fm** Fermium [Rn] $5f^{12} 7s^2$	(258) 101 635.0 1.30 **Md** Mendelevium [Rn] $5f^{13} 7s^2$	(259) 102 642.0 1.30 **No** Nobelium [Rn] $5f^{14} 7s^2$

Index